Canadian Special Publication of Fisheries and Aquatic Sciences 121

Climate Change and Northern Fish Populations

Edited by
R.J. Beamish

Department of Fisheries and Oceans
Pacific Biological Station
Nanaimo, British Columbia
Canada V9R 5K6

NATIONAL RESEARCH COUNCIL OF CANADA
Ottawa 1995

ISBN 0-660-15780-2
ISSN 0706-6481
NRCC No. 38663

Canadian Cataloguing in Publication Data

Symposium on Climate Change and Northern Fish Populations (1992: Victoria, B.C.)
Climate change and northern fish populations

(Canadian Special Publication of Fisheries and Aquatic Sciences, ISSN 0706-6481; no. 121)
Proceedings of the Symposium on Climate Change and Northern Fish Populations held in Victoria, British Columbia, October 19th to 24th 1992.
Includes abstracts in French.
ISBN 0-660-15780-2

1. Climatic changes - North Pacific Ocean - Congresses; 2. Fish populations - North Pacific Ocean - Congresses; 3. Climatic changes - North Atlantic Ocean - Congresses; 4. Fish poulations - North Atlantic Ocean - Congresses.
I. Beamish, R.J. (Richard James), 1942- ; II. National Research Council Canada. III. Title; IV. Series.

QC981.8.C5C54 1994   551.5'25   C94-980352-9

This publication is available from:

Subscription Office, Research Journals, National Research Council of Canada, Building M-55, Montreal Road, Ottawa, Ontario, Canada K1A 0R6

Enquiries: Tel.: 613-993-9084   FAX No.: 613-952-7656

Remittances should be made payable to the Receiver General for Canada, credit National Research Council of Canada.

NRC, CISTI, Monograph Publishing Program
Editor-in-chief Robert H. Haynes

Publishing Department, NRC
Director Aldyth Holmes
Head, Monograph Publishing Gerald J. Neville
Contract Editor Paul D. Payson

Printed on recycled paper.

[*The views expressed are those of the authors and not necessarily those of the Department of Fisheries and Oceans or the National Research Council of Canada.*]

Correct citation for this publication:

Beamish, R.J. [ed.]. 1995. Climate change and northern fish populations. Can. Spec. Publ. Fish. Aquat. Sci. 121: 739 p.

Contents

TOPIC 2. Effects of Climate Change on Fish Populations and Fisheries

POSTER SESSION

TOPIC 3. Preparing for Climate Change - Workshops

SYMPOSIUM SUMMARY

Abstract

Beamish, R.J. [ed.]. 1995. Climate change and northern fish populations. Can. Spec. Publ. Fish. Aquat. Sci. 121: 739 p.

These proceedings summarize some of the recent studies of the relationships among climate, the aquatic environment, and the dynamics of fish populations. The studies are mostly from the North Pacific Ocean, but there are reports of investigations from the North Atlantic Ocean and from fresh water. The various papers include numerous examples of the relationships between fish abundance trends and the environment.

Résumé

Beamish, R. J. [ed.]. 1995. Climate change and northern fish populations. Can. Spec. Publ. Fish. Aquat. Sci. 121: 739 p.

Il s'agit d'un sommaire de quelques récentes études sur les relations entre le climat, le milieu aquatique et la dynamique des populations de poissons. Les études ont surtout eu lieu dans le Pacifique Nord, bien que certaines aient été effectuées dans l'Atlantique Nord et dans des milieux d'eau douce. Les divers documents citent de nombreux exemples des relations entre les tendances de l'abondance du poisson et le milieu ambiant.

Sponsors

Canada Department of Fisheries and Oceans
Dairyland Foods
International Pacific Halibut Commission (IPHC)
Ministry of Agriculture, Fisheries and Food, Province of British Columbia
North Pacific Marine Science Organization (PICES)
Sun-Rype Products Ltd.

Coordinating Committee

V. Alexander (U.S.A.)
W. Aron (U.S.A.)
R.J. Beamish (Canada) - Chairman
J.S. Campbell (Canada)
M. Chadwick (Canada)
J.C. Davis (Canada)
J. Ito (Japan)
S. Karnicki (Poland)
G.D. Koshinsky (Canada)
P.H. LeBlond (Canada)
H. Loeng (Norway)
D.A. McCaughran (U.S.A.)
W.D. McKone (Canada)
N.P. Novikov (Russia)
S. Pennoyer (U.S.A.)
H. Regier (Canada)
B.J. Rothschild (U.S.A.)
M. Sinclair (Canada)
M.P. Sissenwine (U.S.A.)
J.H. Steele (U.S.A.)
S.A. Studenetsky (Russia)
Q. Tang (China)
A. Thompson (Canada)
W.S. Wooster (U.S.A.)

Steering Committee

(Department of Fisheries and Oceans, Pacific Region,
Biological Sciences Branch, Pacific Biological Station,
Nanaimo, B.C., Canada)

R. Beamish - Chairman
J. Boutillier
K. Francis
V. Haist
T. Hamer
M. Henderson
R. Mair
D. Noakes
I. Perry
A. Steele
A. Thompson
D. Welch

List of Contributors

Aglen, A., Institute of Marine Research, P.O. Box 1870 Nordnes, N-5024, Bergen, Norway
Anderson, D.M., Technology Policy Analysis Department, Pacific Northwest Laboratory, P.O. Box 999, Richland, Washington 99352, USA
Azumaya, T., Faculty of Fisheries, Hokkaido University, 3-1-1 Minato-cho, Hakodate 041, Japan
Armstrong, D.A., School of Fisheries, University of Washington, Seattle, Washington 98195, USA
Bakun, A., Pacific Fisheries Environmental Group, Southwest Fisheries Science Center, National Marine Fisheries Service, NOAA, P.O. Box 837, Monterey, California 93942, USA
Beamish, R.J., Department of Fisheries and Oceans, Biological Sciences Branch, Pacific Biological Station, Nanaimo, B.C. V9R 5K6
Belyaev, V.A., Pacific Research Institute of Fisheries and Oceanography (TINRO), 4 Shevchenko Alley, Vladivostok 690600, Russia
Bienfang, P.K., The Oceanic Institute, Makapuu Point, P.O. Box 25280, Honolulu, Hawaii 96825, USA
Bingham, C., School of Statistics, University of Minnesota, 352 Classroom-Office Building, 1994 Buford Avenue, St. Paul, Minnesota 55108, USA
Bograd, S.J., Joint Institute for the Study of the Atmosphere and Ocean, Seattle, Washington 98115, USA
Bouillon, D.R., ESSA - Environmental and Social Systems Analysts, 3rd Floor, 1765 West 8th Avenue, Vancouver, B.C. V6J 5C6
Brodeur, R.D., Alaska Fisheries Science Center, National Marine Fisheries Service, 7600 Sand Point Way N.E., Seattle, Washington 98115-0070, USA
Bulatov, O.A., Pacific Research Institute of Fisheries and Oceanography (TINRO), 4 Shevchenko Alley, Vladivostok 690600, Russia
Butler, V.L., Environmental Sciences Department, Pacific Northwest Laboratory, P.O. Box 999, Richland, Washington 99352, USA
Byrd, G.V., Jr., U.S. Fish and Wildlife Service, Alaska Maritime National Wildlife Refuge, Homer, Alaska 99603, USA
Carscadden, J., Department of Fisheries and Oceans, Northwest Atlantic Fisheries Centre, P.O. Box 5667, St. John's, Newfoundland A1C 5X1
Chatters, J.C., Environmental Sciences Department, Pacific Northwest Laboratory, P.O. Box 999, Richland, Washington 99352, USA
Cohen, Y., Department of Fisheries and Wildlife, University of Minnesota, 200 Hodson Hall, 1980 Folwell Avenue, St. Paul, Minnesota 55108, USA
Collie, J.S., Graduate School of Oceanography, University of Rhode Island, Narragansett, Rhode Island 02882-1197, USA
Conner, D.J., Tribal Fisheries Assessment Unit, Red Lake Band of Chippewa Indians, P.O. Box 279, Red Lake, Minnesota 56671, USA
Cooney, R.T., Institute of Marine Science, University of Alaska Fairbanks, Fairbanks, Alaska 99775, USA
Cunningham, P.K., Bureau of Fish Management, Wisconsin Department of Natural Resources, 101 South Webster Street, P.O. Box 7921, Madison, Wisconsin 53707-7921, USA
Cury, P., Pacific Fisheries Environmental Group, Southwest Fisheries Science Center, National Marine Fisheries Service, NOAA, P.O. Box 831, Monterey, California 93942, USA
Decker, M.B., Department of Ecology and Evolutionary Biology, University of California, Irvine, California 92717, USA
Dempson, J.B., Department of Fisheries and Oceans, Northwest Atlantic Fisheries Centre, P.O. Box 5667, St. John's, Newfoundland A1C 5X1
Deriso, R.B., Inter-American Tropical Tuna Commission, c/o Scripps Institute of Oceanography, La Jolla, California 92093, USA
Diewert, R.E., Department of Fisheries and Oceans, Biological Sciences Branch, 555 West Hastings Street, Vancouver, B.C. V6B 5G3
Doyle, M.J., Department of Fisheries and Wildlife, Oregon State University, Corvallis, Oregon 97331-3803, USA
Eaton, J.G., Environmental Research Laboratory-Duluth, U.S. Environmental Protection Agency, Duluth, Minnesota 55804, USA
Francis, R.C., Fisheries Research Institute WH-10, University of Washington, Seattle, Washington 98195, USA
Hanawa, K., Department of Astronomy and Geophysics, Faculty of Science, Tohoku Unviersity, Aoba-ku, Sendai 980, Japan
Hare, S.R., Fisheries Research Institute WH-10, University of Washington, Seattle, Washington 98195, USA
Hartmann, J., Institute for Lake Research, D-88085, Langenargen, Germany
Healey, M.C., Westwater Research Centre, University of British Columbia, Vancouver, B.C. V6T 1Z2
Helle, J.H., Auke Bay Laboratory, Alaska Fisheries Science Center, National Marine Fisheries Service, National Oceanic and Atmospheric Administration, 11305 Glacier Highway, Juneau, Alaska 99801-8626, USA
Henderson, M.A., Department of Fisheries and Oceans, Biological Sciences Branch, 555 West Hastings Street, Vancouver, B.C. V6B 5G3
Hinch, S.G., Westwater Research Centre, University of British Columbia, Vancouver, B.C. V6T 1Z2
Hoffman, M.S., Auke Bay Laboratory, Alaska Fisheries Science Center, National Marine Fisheries Service, National Oceanic and Atmospheric Administration, 11305 Glacier Highway, Juneau, Alaska 99801-8626, USA
Hollowed, A.B., National Marine Fisheries Service, Alaska Fisheries Science Center, BIN C15700, 7600 Sand Point Way N.E., Seattle, Washington 98115-0070, USA
Hondzo, M., Saint Anthony Falls Hydraulic Laboratory, University of Minnesota, Minneapolis, Minnesota 55414, USA
Huang, S., Second Institute of Oceanography, State Ocean Administration, Hangzhou, Zhejiang, 310012, People's Republic of China
Hunt, G.L., Jr., Department of Ecology and Evolutionary Biology, University of California, Irvine, California 92717, USA

Hurrell, J.W., National Center for Atmospheric Research, P.O. Box 3000, Boulder, Colorado 80307, USA

Husby, D.M., Pacific Fisheries Environmental Group, Southwest Fisheries Science Center, National Marine Fisheries Service, NOAA, P.O. Box 831, Monterey, California 93942, USA

Ingraham, W.J., Jr., National Marine Fisheries Service, Alaska Fisheries Science Center, 7600 Sand Point Way N.E., Seattle, Washington 98115, USA

Iseki, K., Fisheries Oceanography Division, Seikai National Fisheries Research Institute, 49 Kokubu-machi, Nagasaki 850, Japan

Ishida, Y., National Research Institute of Far Seas Fisheries, Fisheries Agency of Japan, 5-7-1, Orido, Shimizu, Shizuoka 424, Japan

Iversen, S.A., Institute of Marine Research, P.O. Box 1870 Nordnes, N-5024, Bergen, Norway

Kasai, A., Ocean Research Institute, University of Tokyo, 1-15-1, Minami-Dai, Nakanoku, Tokyo, Japan

Kawasaki, T., Tohoku University, Kugenuma-Fujigaya 1-10-6, Fujisawa 251, Japan

Kishi, M.J., Ocean Research Institute, University of Tokyo, 1-15-1, Minami-Dai, Nakanoku, Tokyo, Japan

Koizumi, N., Ocean Research Institute, University of Tokyo, Minami-Dai 1-15-1, Nakanoku, Tokyo 164, Japan

Krovnin, A.S., Russian Federal Research Institute of Fisheries and Oceanography (VNIRO), 17a, Verkhnyaya Krasnoselskaya, Moscow 107140, Russia

Kuznetsov, V.V., Russian Federal Research Institute of Fisheries and Oceanography (VNIRO) 17, Verkhnyaya Krasnoselskaya, Moscow 107140, Russia

Lappalainen, J., Finnish Game and Fisheries Research Institute, P.O. Box 202, 00151 Helsinki, Finland

Larkin, P.A., Fisheries Centre, University of British Columbia, Room 18, Hut B-3, 6248 Biological Sciences Road, Vancouver, B.C. V6T 1Z4

Lehtonen, H., Finnish Game and Fisheries Research Institute, P.O. Box 202, 00151 Helsinki, Finland

Levy, D.A., Levy Research Services Limited, 102-2221 Folkstone Way, West Vancouver, B.C. V7S 2Y6

Li, L., Institute of Oceanology, Academia Sinica, Qingdao 266071, People's Republic of China

Lin, C., Second Institute of Oceanography, State Ocean Administration, Hangzhou, Zhejiang, 310012, People's Republic of China

Lin, P., Institute for Environmental Studies, University of Toronto, Toronto, Ontario M5S 1A4

Mackas, D.L., Department of Fisheries and Oceans, Biological Sciences Branch, Institute of Ocean Sciences, P.O. Box 6000, Sidney, B.C. V8L 4B2

Macklin, S.A., Pacific Marine Environmental Laboratory, Seattle, Washington 98115, USA

Matsubara, T., Hokkaido National Fisheries Research Institute, Katsurakoi 116, Kushiro, Hokkaido 085, Japan

Matsumiya, Y., Ocean Research Institute, University of Tokyo, Minami-Dai 1-15-1, Nakanoku, Tokyo 164, Japan

McConnaughey, R.A., National Marine Fisheries Service, Alaska Fisheries Science Center, 7600 Sand Point Way N.E., Seattle, Washington 98115, USA

McCormick, J.H., Environmental Research Laboratory-Duluth, U.S. Environmental Protection Agency, Duluth, Minnesota 55804, USA

McFarlane, G.A., Department of Fisheries and Oceans, Biological Sciences Branch, Pacific Biological Station, Nanaimo, B.C. V9R 5K6

Megrey, B.A., National Marine Fisheries Service, Alaska Fisheries Science Center, 7600 Sand Point Way N.E., Seattle, Washington 98115, USA

Mendelssohn, R., Pacific Fisheries Environmental Group, Southwest Fisheries Science Center, National Marine Fisheries Service, NOAA, P.O. Box 8321, Monterey, California 93942, USA

Minns, C.K., Department of Fisheries and Oceans, Great Lakes Laboratory for Fisheries and Aquatic Sciences, Bayfield Institute, 867 Lakeshore Road, P.O. Box 5050, Burlington, Ontario L7R 4A5

Moore, J.E., Department of Fisheries and Oceans, Great Lakes Laboratory for Fisheries and Aquatic Sciences, Bayfield Institute, 867 Lakeshore Road, P.O. Box 5050, Burlington, Ontario L7R 4A5

Müter, F.-J., Institute of Marine Science, School of Fisheries and Ocean Sciences, University of Alaska Fairbanks, Fairbanks, Alaska 99775-1080, USA

Narayanan, S., Department of Fisheries and Oceans, Northwest Atlantic Fisheries Centre, P.O. Box 5667, St. John's, Newfoundland A1C 5X1

Neitzel, D.A., Environmental Sciences Department, Pacific Northwest Laboratory, P.O. Box 999, Richland, Washington 99352, USA

Niebauer, H.J., Institute for Marine Science, School of Fisheries and Ocean Sciences, University of Alaska Fairbanks, Fairbanks, Alaska 99775, USA

Norcross, B.L., Institute of Marine Science, School of Fisheries and Ocean Sciences, University of Alaska Fairbanks, Fairbanks, Alaska 99775-1080, USA

O'Connell, M.F., Department of Fisheries and Oceans, Northwest Atlantic Fisheries Centre, P.O. Box 5667, St. John's, Newfoundland A1C 5X1

Ogura, M., National Research Institute of Far Seas Fisheries, Fisheries Agency of Japan, 5-7-1, Orido, Shimizu, Shizuoka 424, Japan

Ohtani, K., Faculty of Fisheries, Hokkaido University, 3-1-1 Minato-cho, Hakodate 041, Japan

Olsen, J., Prince William Sound Aquaculture Corporation, P.O. Box 1110, Cordova, Alaska 99574, USA

Page, J.S., Department of Fisheries and Oceans, Centre for Ocean Climate Chemistry, Institute of Ocean Sciences, P.O. Box 6000, Sidney, B.C. V8L 4B2

Parker, K.S., National Oceanic and Atmospheric Administration, Pacific Marine Environmental laboratory, NOAA Building No. 3, 7600 Sand Point Way N.E., Seattle, Washington 98115, USA

Parrish, R.H., Pacific Fisheries Environmental Group, Southwest Fisheries Science Center, National Marine Fisheries Service, NOAA, P.O. Box 831, Monterey, California 93942, USA

Parsons, L.S., Department of Fisheries and Oceans, Biological Sciences Directorate, 200 Kent Street, Ottawa, Ontario K1A 0E6

Pereira, D.L., Minnesota Department of Natural Resources, 1200 Warner Road, St. Paul, Minnesota 55106, USA
Prinsenberg, S., Department of Fisheries and Oceans, Bedford Institute of Oceanography, P.O. Box 1006, Dartmouth, Nova Scotia B2Y 4A2
Quinn, T.J, II, Juneau Center, School of Fisheries and Ocean Sciences, University of Alaska Fairbanks, Juneau, Alaska 99801-8677, USA
Reddin, D.G., Department of Fisheries and Oceans, Northwest Atlantic Fisheries Centre, P.O. Box 5667, St. John's, Newfoundland A1C 5X1
Regier, H.A., Institute for Environmental Studies, University of Toronto, Toronto, Ontario M5S 1A4
Rice, J., Department of Fisheries and Oceans, Biological Sciences Branch, Pacific Biological Station, Nanaimo, B.C. V9R 5K6
Ricker, W.E., c/o Department of Fisheries and Oceans, Biological Sciences Branch, Pacific Biological Station, Nanaimo, B.C. V9R 5K6
Rodionov, S.N., National Center for Atmospheric Research, P.O. Box 3000, Boulder, Colorado 80307-3000, USA
Rothschild, B.J., Chesapeake Biological Laboratory, Center for Environmental and Estuarine Studies, P.O. Box 38, Solomons, Maryland 20688, USA
Roy, C., Pacific Fisheries Environmental Group, Southwest Fisheries Science Service, NOAA, P.O. Box 831, Monterey, California 93942, USA
Royer, T.C., Institute of Marine Science, School of Fisheries and Ocean Sciences, University of Alaska Fairbanks, Fairbanks, Alaska 99775-1080, USA
Rugen, W.C., National Marine Fisheries Service, Alaska Fisheries Science Center, 7600 Sand Point Way N.E., Seattle, Washington 98115, USA
Schumacher, J.D., Pacific Marine Environmental Laboratory, Seattle, Washington 98115, USA
Schweigert, J.F., Department of Fisheries and Oceans, Biological Sciences Branch, Pacific Biological Station, Nanaimo, B.C. V9R 5K6
Scott, M.J., Technology Policy Analysis Department, Pacific Northwest Laboratory, P.O. Box 999, Richland, Washington 99352, USA
Shackell, N., Department of Fisheries and Oceans, Northwest Atlantic Fisheries Centre, P.O. Box 5667, St. John's, Newfoundland A1C 5X1
Sharp, D., Alaska Department of Fish and Game, P.O. Box 669, Cordova, Alaska 99574, USA
Sharr, S., Alaska Department of Fish and Game, P.O. Box 669, Cordova, Alaska 997574, USA
Shatilina, T.A., Pacific Research Institute of Fisheries and Oceanography (TINRO), 4 Shevchenko Alley, Vladivostok 690600, Russia
Skagen, D.W., Institute of Marine Research, P.O. Box 1870 Nordnes, N-5024, Bergen, Norway
Smestad, O., Institute of Marine Research, P.O. Box 1870 Nordnes, N-5024, Bergen, Norway
Spangler, G.R., Department of Fisheries and Wildlife, University of Minnesota, 200 Hodson Hall, 1980 Folwell Avenue, St. Paul, Minnesota 55108, USA
Stabeno, P.J., Pacific Marine Environmental Laboratory, Seattle, Washington 98115, USA
Steele, J.H., Woods Hole Oceanographic Institution, Woods Hole, Massachusetts 02543, USA
Stefan, H.G., Saint Anthony Falls Hydraulic Laboratory, University of Minnesota, Minneapolis, Minnesota 55414, USA
Stockner, J.G., Department of Fisheries and Oceans, Biological Sciences Branch, 4160 Marine Drive, West Vancouver,
B.C. V7V 1N6
Studenetsky, S.A., All-Union Academy of Agricultural Sciences (VASKHNIL), 17, Verkhnaya Krasnoselskaya, Moscow,
B-140, 107140, Russia
Sugimoto, T., Ocean Research Institute, University of Tokyo, 1-15-1, Minami-Dai, Nakanoku, Tokyo, Japan
Sundby, S., Institute of Marine Research, P.O. Box 1870 Nordnes, N-5024, Bergen, Norway
Svendsen, E., Institute of Marine Research, P.O. Box 1870 Nordnes, N-5024, Bergen, Norway
Tang, Q., Yellow Sea Fisheries Research Institute, 19 Laiyang Road, Qingdao 266003, People's Republic of China
Trenberth, K.E., National Center for Atmospheric Research, P.O. Box 3000, Boulder, Colorado 80307, USA
Tyler, A.V., School of Fisheries and Ocean Sciences, University of Alaska Fairbanks, Fairbanks, Alaska 99775-7220, USA
Verkhunov, A.V., Russian Federal Research Institute of Fisheries and Oceanography (VNIRO), Moscow 107140, Russia
Wada, T., Hokkaido National Fisheries Research Institute, Katsurakoi 116, Kushiro, Hokkaido 085, Japan
Ware, D.M., Department of Fisheries and Oceans, Biological Sciences Branch, Pacific Biological Station, Nanaimo,
B.C. V9R 5K6
Welch, D.W., Department of Fisheries and Oceans, Biological Sciences Branch, Pacific Biological Station, Nanaimo,
B.C. V9R 5K6
Whitney, F.A., Department of Fisheries and Oceans, Centre for Ocean Climate Chemistry, Institute of Ocean Sciences,
P.O. Box 6000, Sidney, B.C. V8L 4B2
Willette, T.M., Alaska Department of Fish and Game, P.O. Box 669, Cordova, Alaska 99574, USA
Wong, C.S., Department of Fisheries and Oceans, Centre for Ocean Climate Chemistry, Institute of Ocean Sciences,
P.O. Box 6000, Sidney, B.C. V8L 4B2
Wooster, W.S., School of Marine Affairs HF-05, University of Washington, Seattle, Washington 98195, USA
Wyllie Echeverria, T., 17635 - 96th Place S.W., Vashon, Washington 98070, USA
Xia, S., Yellow Sea Fisheries Research Institute, Chinese Academy of Fishery Sciences, Qingdao 266003, People's Republic of China
Xu, B., Second Institute of Oceanography, State Ocean Administration, Hangzhou, Zhejiang, 310012, People's Republic of China

Yang, J., Institute of Oceanology, Academia Sinica, Qingdao 266071, People's Republic of China
York, A.E., National Marine Mammal Laboratory, Northwest and Alaska Fisheries Center, National Marine Fisheries Service, National Oceanic and Atmospheric Administration, 7600 Sand Point Way N.E., BIN C15700, Seattle, Washington 98115, USA
Zebdi, A., Department of Fisheries and Wildlife, Natural Resources Building, Michigan State University, East Lansing, Michigan 48824-1222, USA
Zeng, J., Department of Oceanography, University of British Columbia, 6270 University Boulevard, Vancouver, B.C. V6T 1Z4
Ziemann, D.A., The Oceanic Institute, Makapuu Point, P.O. Box 25280, Honolulu, Hawaii 96825, USA

List of Contributors — Poster Session

Anderson, C.S., Minnesota Department of Natural Resources, Section of Fisheries, 500 Lafayette Road, St. Paul, Minnesota 55155, USA
Bjørke, H., Institute of Marine Research, P.O. Box 1870 Nordnes, N-5024 Bergen, Norway
Bograd, S.J., National Oceanic and Atmospheric Administration, Pacific Marine Environmental Laboratory, NOAA Building No. 3, 7600 Sand Point Way N.E., Seattle, Washington 98115, USA
Bond, N.A., National Oceanic and Atmospheric Administration, Pacific Marine Environmental Laboratory, NOAA Building No. 3, 7600 Sand Point Way N.E., Seattle, Washington 98115, USA
Carlson, S.R., National Oceanic and Atmospheric Administration, National Marine Fisheries Service, Alaska Fisheries Science Center, Auke Bay Laboratory, 11305 Glacier Highway, Juneau, Alaska 99801-8626, USA
Chadwick, E.M.P., Department of Fisheries and Oceans, Gulf Fisheries Centre, P.O. Box 5030, Moncton, New Brunswick E1C 9B6
Green, R., 433A Midwood Crescent, Waterloo, Ontario N2L 3G1
Hermann, A.J., National Oceanic and Atmospheric Administration, Pacific Marine Environmental Laboratory, NOAA Building No. 3, 7600 Sand Point Way N.E., Seattle, Washington 98115, USA
Ignell, S.E., National Oceanic and Atmospheric Administration, National Marine Fisheries Service, Alaska Fisheries Science Center, Auke Bay Laboratory, 11305 Glacier Highway, Juneau, Alaska 99801-8626, USA
Kashiwai, M., Hokkaido National Fisheries Research Institute, 116 Katsurakoi, Kushiro, Hokkaido 085, Japan
Klyashtorin, L., Russian Institute for Fisheries and Oceanography (VNIRO), 107140 Moscow, Russia
Loeng, H., Institute of Marine Research, P.O. Box 1870 Nordnes, N-5024 Bergen, Norway
Minns, C.K., Department of Fisheries and Oceans, Great Lakes Laboratory for Fisheries and Aquatic Sciences, Bayfield Institute, 867 Lakeshore Road, P.O. Box 5050, Burlington, Ontario L7R 4A5
Moore, J.E., Department of Fisheries and Oceans, Great Lakes Laboratory for Fisheries and Aquatic Sciences, Bayfield Institute, 867 Lakeshore Road, P.O. Box 5050, Burlington, Ontario L7R 4A5
Ottersen, G., Institute of Marine Research, P.O. Box 1870 Nordnes, N-5024 Bergen, Norway
Pereira, D.L., Minnesota Department of Natural Resources, 1200 Warner Road, St. Paul, Minnesota 55106, USA
Radomski, P.J., Minnesota Department of Natural Resources, Section of Fisheries, 500 Lafayette Road, St. Paul, Minnesota 55155, USA
Randall, R.G., Department of Fisheries and Oceans, Great Lakes Laboratory for Fisheries and Aquatic Sciences, Bayfield Institute, 867 Lakeshore Road, P.O. Box 5050, Burlington, Ontario L7R 4A5
Rumbaugh, R.A., National Oceanic and Atmospheric Administration, National Marine Fisheries Service, Alaska Fisheries Science Center, Auke Bay Laboratory, 11305 Glacier Highway, Juneau, Alaska 99801-8626, USA
Sak, T., Minnesota Department of Natural Resources, Section of Fisheries, 500 Lafayette Road, St. Paul, Minnesota 55155, USA
Smirnov, B., Russian Institute for Fisheries and Oceanography (VNIRO), 107140 Moscow, Russia
Stabeno, P.J., National Oceanic and Atmospheric Administration, Pacific Marine Environmental Laboratory, NOAA Building No. 3, 7600 Sand Point Way N.E., Seattle, Washington 98115, USA
Stevens, A., Minnesota Department of Natural Resources, 1200 Warner Road, St. Paul, Minnesota 55106, USA

ACKNOWLEDGEMENTS

The organization of this Symposium and the publication of the Proceedings was a task made possible by the support and cooperation of a very large number of individuals. The financial support provided by our sponsors enabled many foreign scientists to attend the symposium, as well as facilitating a number of other services providing during the symposium. The efforts of both the Organizing and Steering committees were absolutely crucial to the success of the symposium, dealing with the myriad of details that are typical of large scientific meetings. Largely owing to the ingenuity and hard work of the Steering Committee, it proved to be a memorable event, both scientifically and socially.

Dr. Warren Wooster, Chairman of PICES, provided financial support, advice, and encouragement. The symposium actually was a result of his earlier efforts to bring fisheries scientists and ocean scientists together to study the impact of ocean changes on fish populations.

The difficult task of rewriting some of the foreign papers was shared by several people, including J. Westrheim and W.E. Ricker. The technical editing was carried out in excellent style by Y. Van Ruskenveld, West Coast Editorial Association, Victoria, B.C. This task was particularly challenging, due to the wide variety of styles common to the various organizations in participating countries. Ray Scarsbrook is acknowledged for his work in redrawing some figures.

However, in terms of the administration and coordination of the symposium, there is no question that Ann Thompson, the Symposium secretary, played a pivotal role. With every day came new challenges, all of which she faced with great aplomb and a cheerful smile. Whether it was dealing with visas, arrangements for meeting rooms, hotel reservations, or keeping track of manuscripts, she was indefatigable. Ann assisted in putting the proceedings together, and is largely responsible for getting the symposium proceedings published in a reasonable time.

The following persons assisted in a number of ways such as giving invited talks, acting as session chairpersons, organizing tours, reviewing manuscripts, and assisting participants, and their enthusiastic support helped to make the symposium successful.

Dr. V. Alexander; Dr. W. Aron; The Honourable Mr. B. Barlee; Dr. T. Beacham; Dr. J.S. Beckett; Mr. D. Bouillon; Mr. J. Boutillier; Dr. R. Brodeur; Mr. T. Butler; Mr. J. Caldwell; Dr. J.S. Campbell; Dr. D. Cayan; Dr. M. Chadwick; Dr. C. Clarke; Dr. L. Coady; Dr. R. Cooney; Dr. D. Cushing; Dr. P. Cury; Dr. J.C. Davis; Dr. K. Denman; Mr. L. Dickie; Dr. W.G. Doubleday; Ms. K. Francis; Dr. R. Francis; Dr. C. Groot; Dr. V. Haist; Ms. T. Hamer; Dr. S. Hare; Dr. P. Harrison; Dr. D. Hay; Dr. M. Healey; Dr. J. Helle; Dr. M. Henderson; Dr. A. Hollowed; Mr. M. Huntley; Dr. Y. Ishida; Dr. J. Ito; Dr. G. Jamieson; Dr. S. Karnicki; Dr. J. Kitchell; Dr. G.D. Koshinsky; Dr. Y. Kushnir; Dr. P.A. Larkin; Dr. P.H. LeBlond; Dr. H. Loeng; Dr. D. McCaughran; Mr. S. McKinnell; Dr. W.D. McKone; Dr. D. Mackas; Ms. R. Mair; Dr. D. Noakes; Dr. N.P. Novikov; Dr. S. Parsons; Dr. S. Pennoyer; Dr. I. Perry; Dr. R. Peterman; Dr. H. Regier; Dr. J. Rice; Dr. W.E. Ricker; Dr. B. Rothschild; Dr. J. Schweigert; Dr. M.P. Sissenwine; Mr. B. Skud; Dr. G. Stauffer; Ms. A. Steele; Dr. J.H. Steele; Dr. S.A. Studenetsky; Dr. S. Sundby; Dr. Q. Tang; Dr. K.A. Thomson; Ms. A. Thompson; Dr. A. Trites; Ms. D. Tubman; Dr. A. Tyler; Dr. D. Ware; Dr. D. Welch; Mr. J. Westrheim; Dr. W. Wooster; Dr. T. Wyllie Echeverria.

Introduction

The need to understand the relationship between climate and the dynamics of fish populations

R. J. Beamish

This conference and recent problems in key fisheries in the North Atlantic and North Pacific oceans emphasize the importance of understanding the effects of climate on fish population dynamics. This understanding is particularly important if we are to manage fisheries during a period of climate change. Global climate change may result in significant changes in the planet's ecosystems by the year 2050. Add to this a doubling of the planet's population and a possible 5 to 10 fold increase in economic levels by the year 2050 (World Commission on Environment and Development 1987), and it becomes clear how critical it is that we address the issue of how environmental factors affect fish populations. The issue is potentially so overwhelming, both in terms of technical complexity and in the time available for the research, that there may be a tendency to delay the appropriate research until the case for global climate change is proven, or even worse, until there is a crisis. Scientists estimate that it may be more than 10 years before man-made changes can be detected against the natural fluctuations in climate (Wigley et al. 1992), thus we should not expect a resolution to the debate over global warming in the near future. It is urgent that we study the effects of climate on our marine ecosystems now, because we must understand the impacts of natural environmental changes before we can adapt to and mitigate the changes resulting from man-made changes. It is clear that we have much to learn and that we are going to have to learn it quickly. One senior scientist believes that the implications for fisheries management are profound, calling for little less than a revolution in the approach to managing fish stocks (Mann 1993).

All species of fish have limits to their abundance. The limits may fluctuate, and in some cases the fluctuations may be large, but, in general, we do not see evidence of large random variation in abundances among the various commercially important species. The apparent stability in the relative abundance of many species is evidence that carrying capacities (defined here as all factors in an ecosystem that interact to regulate fish abundance) for species in a particular ecosystem also have some stability. However, there is evidence that shifts in carrying capacities occur (Beamish and Bouillon 1993; Hare and Francis, this volume; Wooster and Hollowed, this volume). If as a result of such a shift, abundance declines and reaches a new equilibrium, rebuilding of fish stocks to the previous higher level may not be possible without manipulating the ecosystem. Sustaining a fishery, therefore, may not be maintaining a catch (sometimes considered to be something approaching the largest ever recorded), but maintaining the optimal catch at the existing carrying capacity for a species. A key consideration, therefore, is the need to understand the biological basis for fish abundance changes. If the abundance change results from overfishing, fishing theory provides the procedures for determining how to rebuild stocks to make maximum use of the carrying capacity. However, if the decline in abundance is mainly a result of a natural change in carrying capacity, management actions would not be expected to rebuild abundance to the previous levels. The lower levels of catch at the lower carrying capacity could still support a fishery, and traditional stock assessment approaches could be used to establish catch quotas that would be lower than in the past. The optimal catch, in this case, would ensure that an adequate spawning level existed so that the population would increase when more favourable environmental conditions returned.

It is difficult to understand the reasons for abundance changes by only studying catch. Catch dynamics and population dynamics are not necessarily the same thing. A problem in fisheries management is that it has been easier to sample and measure catch than it has been to understand all the important factors affecting the dynamics of populations. Fortunately, catch trends can be a measure of the dynamics of a population, and the use of catch to manage fisheries appears to be useful in the short term. However, we can not determine how long "short term" is unless we understand more about how environmental factors affect fish populations. Abundance changes that occur for environmental reasons may not be apparent from catch data because it is recruitment that may be affected first. Major changes in recruitment are not usually evident until the affected recruits enter a fishery, and this can range from a few years to more than 10, depending on the species. The reasons for a change in recruitment would be apparent, however, if we were able to consider all the factors that affect the dynamics of populations or if we could identify the changes in the environment that would cause a shift in the marine survival of a species. There is no reason to believe that the parameters that cause the year-to-year variations are the same parameters that

R.J. Beamish. Department of Fisheries and Oceans, Biological Station, Nanaimo, BC V9R 5K6, Canada.

cause long-term shifts. The year-to-year changes in fish abundance can be unsettling to a manager, but these changes tend to oscillate around a mean. It is the long-term carrying-capacity shifts that can devastate local economies.

All resource managers are interested in the effects of short-term and long-term environmental changes on the dynamics of the stocks they manage, but in the absence of reliable relationships, managers are forced to use data and relationships that best measure the population dynamics process. At this stage in fisheries management, we understand very little about the interannual and interdecadal effects of the environment on population dynamics of fishes; we understand even less about how to separate natural and man-made environmental changes. If modern fisheries management started with Bill Ricker's (1954) paper on stock and recruitment, then we have had 40 years to refine our understanding of the dynamics of fisheries. If over the next 40–50 years we are to experience major changes in the environment, then we have little time to make progress. One method of accelerating learning is to present ideas to a group of peers. This can be particularly effective when a group of peers has expertise in a range of subjects.

In 1987, the International North Pacific Fisheries Commisson (INPFC) and International Recruitment Investigations in the Subarctic Pacific (IRIS) held a conference on the effects of ocean variability on recruitment (Beamish and McFarlane 1989). Two of the goals of this conference were to exchange scientific information that was available in Canada, the United States, Japan, Russia, and China, and to encourage fisheries scientists and ocean scientists to integrate their science. From the fisheries side, we were looking for ways that fisheries management might benefit from a better understanding of the impact of the marine environment on fish abundance trends. The conference proceedings added to the evidence that identified relationships between environmental factors and the population dynamics process of fishes, but the environmental factors were almost exclusively oceanographic parameters. There was only a suggestion that climate effects were important. Because of the concern about global climate change, the organizers believed that another conference that emphasized the relationships among climate, oceans, and fish was needed. This symposium on climate change and northern fish populations was organized to bring climatologists, oceanographers, and fisheries scientists together to consider how we can improve our management advice through a better understanding of the interrelationships among environmental factors, fishing effects, and fish population dynamics.

The organizers of this symposium were fortunate that the new Marine Science Organization for the North Pacific (PICES) was created by international treaty between Canada, China, Japan, and the United States in 1992. The purpose of PICES is to promote and coordinate marine scientific research to advance scientific knowledge of the North Pacific Ocean and its living resources. With this emphasis on oceans and fish, it was logical that PICES host this symposium. This symposium, therefore, was held at the first Annual Meeting of PICES, in Victoria, British Columbia, Canada, October 19–24, 1992.

The proceedings of this symposium are not the first on the topic, but they are one of the first. The intent of the symposium was not only to accumulate a collection of original scientific material, but also to record observations, explanations, and data series from marine and some freshwater ecosystems that link fish population dynamics to their environment. Because it was not a requirement to include new material, papers may contain a mixture of new and older data. Because of the limits of time and space, the papers in these proceedings contain more marine than freshwater information, and more information from the Pacific Ocean and the Bering Sea than from other oceans. However, the science in the papers will be relevant to other areas.

When we planned the conference, we expected relatively few papers. Our initial request for papers resulted in an overwhelming response. We received approximately 120 requests to present papers, about four times what we expected. We accepted as many papers as possible, requiring some compromises in the editing, that resulted in some inconsistencies in formats. Some of the papers that were not originally written in English have been rewritten, and although they have been checked by the authors, we are not absolutely certain that our rewritten version is precisely the same as the original version. All papers were reviewed, but not all differences between the reviewers and the authors were resolved. In some cases resolution was achieved by publishing a shortened version of the original paper, because we considered the views and data to be important. The papers in the these proceedings, therefore, contain new information, provide reviews of existing information, and provide an introduction to current research relating oceanography and climatology to fisheries management. These proceedings should provide the reader with useful background material, and hopefully stimulate the imaginations of researchers at a time when creativity is needed urgently.

References

Beamish, R.J., and D.R. Bouillon. 1993. Pacific salmon production trends in relation to climate. Can. J. Fish. Aquat. Sci. 50: 1002–1016.

Beamish, R.J., and G.A. McFarlane [ed.]. 1989. Effects of ocean variability on recruitment and an evaluation of parameters used in stock assessment models. Can. Spec. Publ. Fish. Aquat. Sci. 108: 379 p.

Mann, K.H. 1993. Physical oceanography, food chains, and fish stocks: a review. ICES J. Mar. Sci. 50: 105-119.

Ricker, W.E. 1954. Stock and recruitment. J. Fish. Res. Board Can. 11: 559-623.

Wigley, T.M. L., G.I. Pearman, and P.M. Kelly. 1992. Indices and indicators of climate change: issues of detection, validation and climate sensitivity, p. 85–96. *In* J.M. Mintzer [ed.] Confronting climate change: risks, implications, and responses. Cambridge University Press, Cambridge, U.K. 382 p.

World Commission on Environment and Development. 1987. Our common future. Oxford University Press, Oxford, U.K. 400 p.

Opening remarks

L. S. Parsons

Good morning ladies and gentlemen. On behalf of the Canadian Department of Fisheries and Ocean, it gives me great pleasure to welcome you to beautiful Victoria, and to this Symposium on Climate Change and Northern Fish Populations. It is fitting that the inaugural meeting of the North Pacific Marine Science Organization, PICES, is being held in conjunction with this Symposium. I hope these meetings will foster a spirit of scientific cooperation among disciplines, and contribute to the flow of ideas and information which must be a cornerstone of PICES.

The subject of this Symposium is of great importance to marine scientists around the world. PICES will likely become heavily involved in many of the topics discussed here. For too long, traditional fisheries models have ignored the influence of environmental factors on the living marine resources. Improving our knowledge of climate variability and global climate change is essential to further our understanding of the marine environment and the resources that live there.

Ocean science and fisheries science have made advances in our knowledge in relation to certain earth processes. The studies of ocean-atmosphere interaction, for example, have provided insight into the association of the EL NIÑO current, atmospheric pressure phenomena, and the processes that govern fish behaviour and distribution. Atmospheric models have been developed to predict climate change and scientists are actively developing ocean sub-models. The interaction of the ocean and the atmosphere cannot be ignored in any attempt to understand the global system and climate change. Similarly, the influence of the oceans is determined not only by large-scale horizontal and vertical water transport but also by its chemistry and biology. Such programs as WOCE, GLOBEC, and JGOFS are being implemented to improve our knowledge of these matters.

Traditional models of fish population dynamics are inadequate to explain changes in abundance. These changes are in some way linked to the physical and chemical processes in the ocean which affect year-class strength. Continued expansion of industrial activities is contributing to global climate change. This has added another dimension to the equation. Global change will affect entire ecosystems. We can expect changes that will affect the population dynamics of important marine stocks in the North Pacific. Management of fisheries will have to take these changes into consideration. We will need to understand how to incorporate these changes into the scientific advice we provide.

The North Pacific plays a critical role in global weather and climate processes. It is one of the major fish producing areas of the world. The region is unique in the number and abundance of species, which migrate enormous distances and intermix. This is a region of complex oceanographic and biological interaction. Unexpected impacts are sometimes felt from apparently unconnected causes. It is obvious that the North Pacific must be studied as an integral system.

For almost a hundred years, scientists have been preoccupied with the biological and environmental processes that contribute to changes in productivity. At the beginning of this century Hjort (1914) hypothesized that recruitment variability was related to the transport of eggs and larvae away from favourable nursery areas. Many researchers have subsequently worked on the relationship between parent stock size and recruitment, and environmental variability. Among these was Bill Ricker who in the 1950s developed sophisticated models of stock and recruitment relationships which had a major influence on the way we view fish population dynamics.

In recent years, various researchers have studied the relationship between recruitment and environmental conditions associated with ocean transport (Nelson et al. 1977; Bailey 1981; Sinclair et al. 1985). Several studies have demonstrated the influence of oceanographic and meteorological processes on recruitment of various Pacific coast stocks. For example, Francis et al. 1989 concluded Pacific hake recruitment appears to be independent of stock size and strongly correlated with environmental conditions occurring during early life history. They similarly linked the production of several important fisheries with large scale environmental dynamics.

Several lines of evidence point to a change in North Pacific productivity, particularly at higher trophic levels. Brodeur and Ware (1992) showed that the standing stock of zooplankton doubled in the Gulf of Alaska between the 1950s and 1980s. C. S. Wong from the Institute of Ocean Sciences has shown that nitrate levels increased substantially at about the same time. Odate (1986) also found similar changes off the coast of Japan over similar time scales. Changes in a wide range of physical oceanographic variables have occurred in the eastern Pacific. These may be associated with the intensification of the Aleutian Low that has occurred since the late 1970s. Evidence is accumulating that trends in the fish production in the North Pacific may be linked to trends in climate.

Beamish and Bouillon (in press) have recently examined the relationship between climate and salmon production in the North Pacific. They suggest that the trends in salmon production from 1925 to 1989 were not primarily the result of fishing effort, management actions or artificial rearing, but rather the result of environmental influences.

On the east coast of Canada, ocean environment is thought to have contributed to the increase in lobster abundance from historical lows in 1974 to levels in the 1980s not seen since the last century. This increase occurred even though fishing effort remained at about the same level as during the previous decline.

L.S. Parsons. Department of Fisheries and Oceans, Biological Sciences Directorate, 200 Kents Street, Ottawa, ON K1A 0E6, Canada.

Recently, northern cod on the east coast of Canada has undergone an abrupt and drastic decline in abundance. In 1991, extremely low water temperatures were associated with a substantial sudden decrease in abundance of the older fully recruited year-classes. Several other East Coast groundfish stocks are also in decline. It has been suggested that environmental influences are a major contributing factor.

On a broader front, during the 1980s, climate change and global warming became a major concern worldwide. If recent trends continue, it is estimated that CO_2 and other greenhouse gases in the atmosphere could double pre-industrial levels, possibly by the 2030s. This could lead to a rise in global mean temperatures greater than any in man's history.

Far reaching impacts on marine resources can be expected if ocean temperature changes and sea level rises occur as predicted. Climate is a powerful determinant of the distribution, abundance and species mix of fishes in coastal waters. One challenge now facing marine scientists is to distinguish the added effect of global warming from the natural variability inherent in the marine ecosystem, and determine how these changes affect the living resources in the oceans and those who depend upon them. The economic and social impact of such changes can be severe.

Brian Rothschild in his book "Dynamics of Marine Populations" succinctly described why scientists have been so preoccupied recently with ocean climate variability:

"Marine fish populations undergo dramatic fluctuations in abundance over time, a phenomenon that has mystified scientists and lay people for centuries. These fluctuations obviously have considerable economic and social impact and it is important to understand why they occur."

A recent example of this type of impact is the decline in northern cod on the Atlantic coast of Canada resulting in a moratorium on fishing for two years. This threw thousands of fishermen and plant workers out of work. The Canadian government has had to intervene and provide substantial funding over the next couple of years to aid fishermen during the downturn and to restructure this fishery.

We are only on the threshold of understanding the factors determining climate variability, and the impact of these factors on ocean resources. To achieve better understanding, the scientific community will have to undertake a massive exchange of information and ideas, and engage in extensive debate. Cooperation and collaboration are required at both national and international levels. Only through such collaboration can progress be made.

This Symposium will probably see the largest collection of scientific papers ever presented at one time on the topic of climate change and northern fish populations. We hope that it will make an important contribution towards improving our ability to understand climate change and its impact on managing fish populations. PICES is co-sponsoring this Symposium because this topic is of considerable interest to member countries. The results of the Symposium should assist the Scientific Committees of PICES in developing their agenda for the future. It will also undoubtedly set a precedent for the standards that PICES-sponsored symposia will aspire to in the future.

I hope that this Symposium will point the way to the research necessary for better understanding of marine climate processes and their effect on living marine resources.

Best wishes for a very successful Symposium.

References

Bailey, K. M. 1981. Larval transport and recruitment of Pacific hake *Merluccius productus*. Mar. Ecol. Prog. Ser. 6: 1-9.

Beamish, R. J., and D. R. Bouillon. 1993. Pacific salmon production trends in relation to climate. Can. J. Fish. Aquat. Sci. 50: 1002-1016.

Brodeur, R. D., and D. M. Ware. 1992. Long-term variability in zooplankton biomass in the subarctic Pacific ocean. Fish. Oceanogr. 1: 32-38.

Francis, R. C., S. A. Adlerstein, and A. Hollowed. 1989. Importance of environmental fluctuations in the management of Pacific hake (*Merluccius productus*), p. 51-56. *In* R. J. Beamish and G. A. McFarlane [ed.] Effects of ocean variability on recruitment and an evaluation of parameters used in stock assessment models. Can. Spec. Publ. Fish. Aquat. Sci. 108.

Hjort, J. 1914. Fluctuations in the great fisheries of northern Europe viewed in the light of biological research. Rapp. P.-v. Réun. Cons. int. Explor. Mer 20: 228 p.

Odate, K. 1986. Automatic data processing for estimation of abundance of zooplankton in the waters off northeastern Honshu, Japan, 1951-1976. Bull. Tohoku Reg. Fish. Res. Lab. 48: 31-47. (In Japanese)

Nelson, W. R., M. Ingham, and W. E. Schaaf. 1977. Larval transport and year class strength of Atlantic menhaden, *Brevoortia tyranus*. Fish. Bull. (U.S.) 75: 23-41.

Rothschild, B. J. 1986. Dynamics of marine fish populations, 251-269. Harvard University Press.

Sinclair, M., M. J. Tremblay, and P. Bernal. 1985c. El Niño events and variability in a Pacific mackerel (*Scomber japonicus*) survival index: Support for Hjort's second hypothesis. Can. J. Fish. Aquat. Sci. 42: 602-608.

Keynote Address

Climate Change and Community Structure

John H. Steele

Steele, J.H. 1995. Keynote Address: Climate Change and Community Structure, p. 5-9. *In* R.J. Beamish [ed.] Climate change and northern fish populations. Can. Spec. Publ. Fish. Aquat. Sci. 121.

Abstract: Decadal changes in the physical environment are likely to have impacts on marine communities rather than on individual fish stocks. The relevant spatial scales will be regional rather than global, or local. These space/time scales will require new management structures for the research between the existing national efforts and the emerging global programs.

Résumé: Les changements décennaux dans le milieu physique auront vraisemblablement des répercussions sur les biocénoses marines plutôt que sur les différentes populations de poissons. Les échelles spatiales pertinentes seront de niveau régional plutôt que global ou local. Elles exigeront de nouvelles structures de gestion de la recherche qui se situeront entre l'effort actuel à l'échelle nationale et les nouveaux programmes globaux qui se dessinent.

Introduction

The title of this paper is deliberately ambiguous. I shall use the term "community structure" in two different ways: to refer to the marine communities, especially fish stocks, which are potentially subject to the impacts of climate change; and then in the context of the communities of research scientists who are putting together new programs to clarify these ocean problems.

My thesis is that we are hoping to resolve new questions about the nature of physical/biological interactions at much larger space and longer time scales. This obviously requires research projects that comprehend these larger scales and are interinstitutional or international in their structure.

Our response to this challenge has been the emergence of "global" programs that undertake to display these questions at the largest possible scales. The reasons for this global vision are not always scientific. However, the field research is necessarily regional and often process oriented. It can appear that such regional projects demean or deny the global assumptions. I shall argue that, in most cases, this intermediate scale is most appropriate for the scientific problems. This is true for the study of fish populations in northern environments.

Oceanographic Processes

The title of this volume presupposes relatively strong relations between physical and ecological processes and is the explicit objective of GLOBEC (for Global Ocean Ecosystem Dynamics; see Table 1). This close correspondence is revealed in the Stommel diagram (Haury et al. 1978) for zooplankton, where different ecological processes have space/time scales comparable to the physical scales (Stommel 1963). Similarly, trophic interactions among phytoplankton, zooplankton, and fish can be superimposed on the same space/time scale (Steele 1978). One can generalize by saying that interactions between trophic levels are interactions between space and time scales. But, obviously, the individual processes must have characteristic temporal (and therefore spatial) scales. Thus zooplankton diurnal behavior is critical for phytoplankton population dynamics. Similarly, the feeding of larval fish depends on the growth rates and patchiness of cohorts of copepods. For these

Table 1. Goals of global oceanographic programs.

(1) World Ocean Circulation Experiment (WOCE)
To understand the general circulation of the global ocean well enough to be able to predict ocean response and feed back to long-term changes in the atmosphere.

(2) Global Ooean Flux Study (GOFS)
To understand the processes governing the production and fate of biogenic materials in the sea well enough to predict their influences on, and responses to, global scale perturbations.

(3) Global Ocean Ecosystem Dynamics (GLOBEC)
To understand ocean ecosystem dynamics and how they are influenced by physical processes so that the predictability of population fluctuations in a changing global climate can be assessed.

J.H. Steele. Woods Hole Oceanographic Institution, Woods Hole, Massachusetts 02543, USA.

reasons, the physical processes at critical time scales (days, months, years, decades) will have differing effects on plant and animal populations, depending on *their* time scales.

The most difficult term to define is "population," as it is used in the GLOBEC objective (Table 1) or in most fishery research. Sinclair (1988), in his study of herring populations, used the concept "of a self-sustaining component of a particular species." The idea of a "self-sustaining component" implies something about the dynamics. This is close to the "stock" concept used in fishery management, and here again, it is not clear how far this definition is a practical convenience for managers, or whether it provides insight into the natural regulation of the populations.

What is clear, however, is that we regard something much smaller than the biogeographical province of a species as the proper unit for study. The geographical extremes of a distribution may not be self-sustaining, but may depend on dispersion. At small scales, the barnacles on a rock are not a population, because of the open nature of the life cycle resulting from larval dispersion (Roughgarden et al. 1988).

This inherently meso-scale character of marine population dynamics focuses attention on the corresponding scales in the physics. There is the assumption that the general regularity or predictability in major ocean features, such as fronts or gyres, are utilized by these populations during their life cycles (Steele 1985). However, it is now recognized that these features are variable on longer time scales and can induce random walks that would lead to extinction without some density dependence. For fish population models, this prospect can be resolved by the black box mechanism of a stock-recruitment relation. All the normally used functions are compensatory (Rothschild 1988), and they are usually very poor fits to the data. It is unlikely that longer time series will resolve this problem (Solow and Steele 1990), and so some knowledge of the

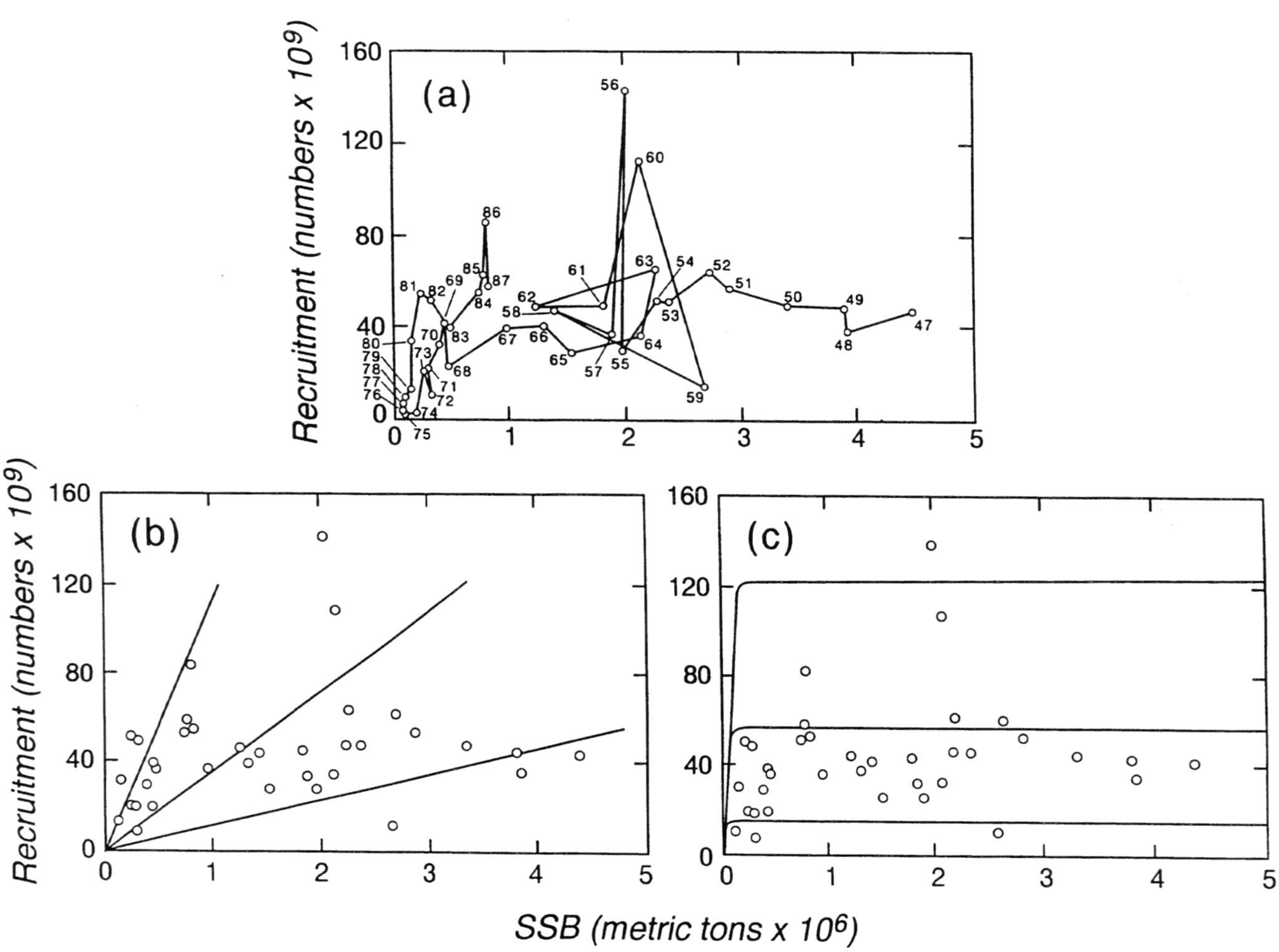

Fig. 1. Relations between recruitment and spawning stock biomass for North Sea herring: (a) consecutive years, 1947–87, (b) density-independent relations, and (c) density dependence.

processes is required. In particular, we would like to be able to separate the density-independent component due to physical variability from the ecological processes producing compensation (or density dependence, or negative feedback). Especially we want to do this at decadal time scales, where the population changes are most apparent and the management consequences most severe.

The problem can be exemplified by North Sea herring data (Fig. 1). Figures 1(b) and (c) illustrate density-independent and density-dependent recruitment per unit of stock. Neither seems particularly good by itself. A comparison with time series, Fig. 1(a), suggests there are shifts in population abundance between the periods 1950–76 and 1977–87, with dependence operating. A physical explanation of these marked longer-term changes has some basis for the North Sea (see Bailey and Steele 1992, for a review). If this is accepted as the source of density-independent variations, there is still a need to introduce ecological processes to provide dependence and explain the other components of variability on shorter time scales. These ecological mechanisms can be presented as a separate black box, such as the logistic relation, to safisfy the component illustrated in Fig. 1(c). If we seek realistic processes as explanations of this compensatory effect, is it adequate to confine the search to the one population? Or, as seems more likely, do we need to consider interactions with other species as prey, predators or competitors, especially at longer time scales.

Scales of Variance

The impact of variability at different time scales will be very dependent on the amplitude of the variance at these different scales. The major difference between the atmosphere and the ocean is that, up to decadal or longer time scales, the atmospheric variance approximates white noise (excluding predictable annual cycles). The ocean, on the other hand, exhibits red noise at all observed frequencies in the range of seconds to years (Wunsch 1981). A speculative extrapolation to geological time scales (Steele 1985, and references therein) implies that the redness extends to those scales at which the ocean and atmosphere are a single system.

Figure 2 combines these observations and speculations into one schematic representation of oceanic variance. This can be used to explain why marine pelagic organisms are more closely coupled to their physical environment than terrestrial ones (Steele 1991). The short-term variance in the sea is much less than in the atmosphere, so that the changes at larger spatial scales are more regular or reliable. Within the time scales from diurnal to annual, this regularity can be enhanced by particular features, fronts, eddies, and gyres (see Steele 1991 for a fuller discussion).

The converse of this argument is that variance in the sea as a function of time scale increases with longer periods, so that the price paid for utilizing the shorter-term regularity in the ocean is accommodating to the larger variance at decadal time scales and beyond. As Fig. 2 indicates, we have no direct information on the exact nature of the variance at decadal to centure scales. The general prediction is that the variance will increase monotonicaly toward the geological scales. The question is what will be the response of communities at these longer scales, which are greater than the lifetimes of their components

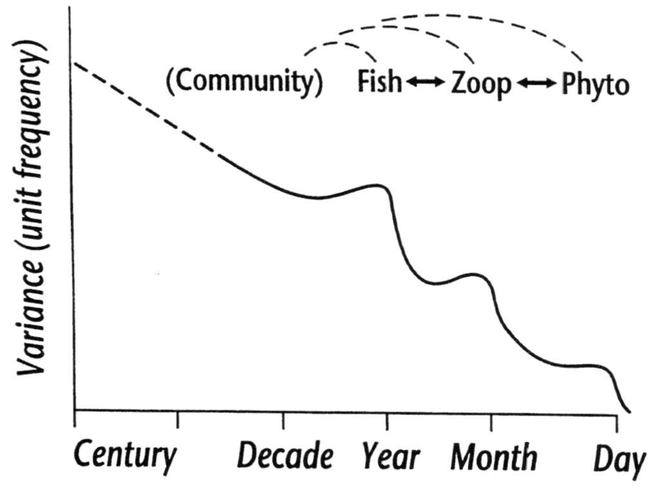

Fig. 2. An idealized representation, on log-log scale, of the ocean power spectra as displayed in temperature data. The time scales for individual trophic levels are indicated. The time scales for changes in community structure are suggested.

(excluding marine mammals). This is very different from terrestrial systems, where the basic plant components generally have lifetimes (above or below ground) on the order of decades to centuries or longer.

Steele and Henderson (1984) used a model of a single fish population with two equilibria to show how red noise forcing would cause switching between high and low abundance at time scales of the order of decades. Implicit in such a simple model is that other components of the ecosystem must have correspondingly large changes for the overall system to balance (Steele 1974). For the North Sea, Jones (1984) showed that the marked switches in demersal and pelagic fish stock abundances appeared to keep the energy budget about the same. Contemporaneous changes of several species at different trophic levels are a feature of other studies, such as the Russell cycle in the English Channel (Southward 1980).

In such calculations, the spatial scales are quite large and the time scale is decadal. The problem, as Cushing (1982) has pointed out, is to explain such changes in terms of competition between particular species in specific locations and at critical times such as metamorphosis. The relevant scales, however, are much less than global. If we extrapolate the approximate space/time relation, km = days (Steele 1978), defining the core of the physical/biological scale interactions, then a decade = 3.10^3 km. This, in turn, is equivalent to 30° latitude, greater than most "large marine ecosystems" (Sherman et al. 1990). However, there are also topographic constraints on many fish populations. The North Atlantic cod stocks (Fig. 3) are a good example of the different spatial scales. At decadal periods, we may no longer be looking at the dynamics of separate stocks, nor even of particular species. We cannot ignore the community dynamics within which each of these stocks is embedded.

The spatial scales in Fig. 3 are very much less than global, so that data such as globally averaged temperatures are not relevant. The physical dynamics of interest are, for example, circulation within the North Sea or the Gulf of Maine, or the effects of variations at the open boundaries. In turn, there are

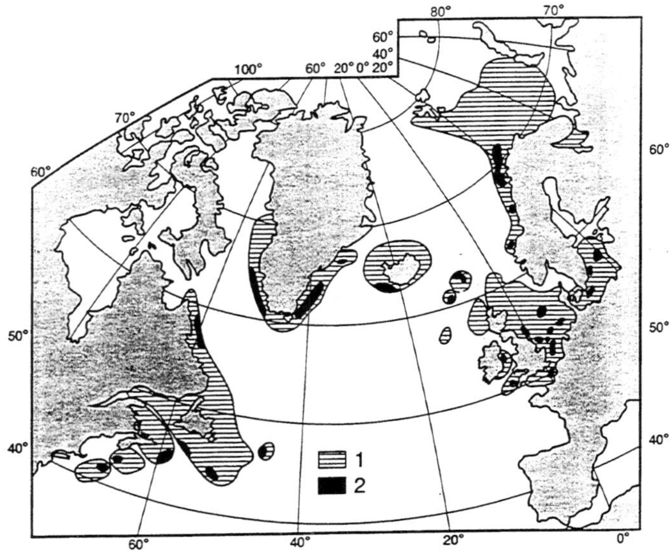

Fig. 3. The distribution of North Atlantic cod stocks showing (1) area of each stock and (2) location of spawning. (Redrawn from FAO 1981).

questions about the exchange of various zooplankton and larval fish communities across such boundaries. The problem is to couple ocean basin simulations to models at the appropirate meso-scales.

The topics that attract general attention may be global ocean climate change and fish population abundance. But the actual connections will be at physically smaller scales, local and regional; and for ecologically more complex communities. Our science is grappling with the definition of these interactions and with the new research required to study them. This, in turn, requires new ways of organizing that research so that the scope of the effort matches the scales of the problems. It is this aspect that I wish to review next.

Research Communities

The impetus for our global concerns is partly sociological—the long, hot summer of 1988 in the U.S.; partly technological—the new perspective provided by satellites. The involvement of marine physics is necessarily inderect. We have made valiant efforts to identify the oceans as the controlling element in climate change at decadal time scales. This has significant but limited success in the scientific community, where cloud physics is considered the critical factor.

For the more general public, the marine aspects are often biological, such as coral reef bleaching and sports fish abundance. These factors that are not necessarily economic and are local or regional in their immediate impact. The overall consequences may become worldwide as the result of cumulative examples, and may have economic effects through tourism as well as from alteration in harvesting of renewable resources. But, although the problems may be given a global gloss, an understanding of the processes involved will require study at scales intermemdiate between the local and the global, as described in the previous section.

The scientific questions at these scales result in problems of organization to design, implement, and fund appropriate research. The base we started from was national commitment to local or regional research programs for 1 to 3 years' duration. The new element, beginning about 1978, was the global perspective of surface temperature, topography, and color provided by satellites. These technical capabilities were the impetus for programs such as WOCE and GOFS (Table 1), which stressed the global context and emphasized prediction as their aim. These programs became part of IGBP (International Geosphere, Biosphere Program), which accentuates further the global participation in the research. Yet the most successful program, in terms of predictive capability, is TOGA (Tropical Ocean-Global Atmosphere), which focuses on El Niño events in the central Pacific and on 6-month to 1-year forecasts. Also, it is noteworthy that implementation of these global programs tends to be through the study of critical processes at much smaller scales, such as subduction of water masses, or spring blooms in the NE Atlantic.

The ecological components may appear to have a hard time competing in this global marketplace. There is no simple technological fix available by remote sensing. There is no general theory such as the Navier-Stokes equations in geophysical fluid dynamics. However, there is now considerable controversy over the links between anthropogenic perturbations and biogenic fluxes (Broecker 1991) so the global perspective may not be such a good marketing device after all. Anyway the most significant fluctuations are in the species composition of the higher trophic levels on the continental shelves of the world's oceans (Cushing 1982). The phyiscal nature of these environments, the life cycles of the fish stocks, the geographical separation between stocks, all require collaborative studies at regional scales across ocean basins.

The scientific assumption is that different cod stocks in the North Atlantic (Fig. 3), or sardine stocks in the North Pacific (Kawasaki et al. 1991), can be regarded as case histories exemplifying the same general patterns. The patterns, however, depend on some combination of physical variability and community dynamics. Thus the unit of study is the fish stock, but this requires understanding of its relation with other species in each location, as well as comparison of patterns at different locations.

The scientific organization required for this is more than can be put in place by a single laboratory or by a single country; but it is less than global. In fact, a global base for organizing the research would be inappropriate. This is not to say that the underlying questions are not worldwide in concept. But the realization of these concepts, the physical causes of variability in communities, will be regional and will require regional organizations.

The populations, or stocks, or particular fish species may be the focus of the effort, because they are longest-lived, because they are of commercial importance, and, especially, because we have longer runs of data than for any other marine parameter (except sea-surface temperature and sea level). But the work must also depend on a comparable range of studies in the plankton. Further, the physical links between open ocean and shelf seas appear to be the critical component for the variance at longer time scales.

The organization of this kind of research would be poorly served by incorporation with global programs that are explicitly disciplinary in character (WOCE and GOFS) and that also separate the shelf from the open ocean as regions of study. The need is for less geographically diffuse organizations with greater interaction among disciplinary components. There is a role for regional councils such as ICES (International Council for Exploration of the Sea), which operates in the North Atlantic; and the PICES (Pacific ICES), whose formation is the occasion for this meeting. Such organizations can provide an umbrella to shelter specific programs, such as the North Atlantic Cod and Climate Change and to develop new initiatives.

References

Bailey, R.S., and J.H. Steele. 1992. North Sea herring fluctuations, p. 213–230. *In* M.H. Glantz [ed.] Climate variability, climate change and fisheries. Cambridge University Press, Cambridge, UK.

Broecker, W.S. 1991. Keeping global change honest. Global Biogeochem. Cycles 5: 191–192.

Cushing, D.H. 1982. Climate and fisheries. Academic Press, New York, NY.

Food and Agriculture Organization. 1981. Atlas of the living resources of the sea. Rome.

Haury, L.R., J.A. McGowan, and P.H. Wiebe. 1978. Patterns and process in the time space scales of plankton distributions. p. 277–328. *In* J.H. Steele [ed.] Spatial pattern in plankton communities. Plenum Press, New York, NY.

Jones, R. 1984. Some observations on energy transfer through the North Sea and Georges Bank food webs. Rapp. P.-v. Reun. Cons. Int. Explor. Mer 183: 204–217.

Kawasaki, T., S. Tanaka, Y. Toba, and A. Taniguchi [ed.]. 1991. Long-term variability of pelagic fish populations. Pergamon Press, New York, NY. 402 p.

Rothschild, B.J. 1986, Dynamics of marine fish populations. Harvard University Press, Cambridge, MA. 277 p.

Roughgraden, J., S. Gaines, and H. Possingham. 1988. Recruitment dynamics in complex life cycles. Science. 241: 1460–1466.

Sherman, K., M.A. Lewis, and B.G. Gold [ed.]. 1990. Large marine ecosystems. Amer. Ass. Adv. Sci., Washington, D.C. 242 p.

Sinclair, M. 1988. Marine populations. University of Washington Press. 252 p.

Solow, A., and J.H. Steele. 1990. On sample size, statistical power and the detection of densigy dependences. J. Anim. Ecol. 59: 1073–1076.

Southward, A.J. 1980. The western English Channel—an inconstant ecosystem. Nature 285: 361–366.

Steele, J.H. 1974. The structure of marine ecosystems. Harvard University Press, Cambridge, MA. 128 p.

Steele, J.H. 1978. Some comments on plankton patches, p. 1–20. *In* J.H. Steele [ed.] Spatial pattern in plankton communities. Plenum, New York, NY.

Steele, J.H. 1985. Comparison of marine and terrestrial ecological systems. Nature 313: 355–358.

Steele, J.H. 1991. Can ecological theory cross the land-sea boundary? J. Theor. Biol. 153: 425–436.

Stommel, H. 1963. Varieties of oceanographic experience. Science 139: 572–576.

Wunsch, C. 1981. Low frequency variability of the sea, p. 342–374. *In* B.A. Warren and C. Wunsch [ed.] Evolution of physical oceanography. MIT Press, Cambridge, MA.

Banquet Address

P. A. Larkin

I must begin by saying how pleased I am to have the privilege of speaking to you this evening. This first annual meeting of PICES is a significant historical occasion, especially because it is about a century overdue. And there are a number of things I wish to say to mark the occasion.

First, I should tell you that I had decided about a year ago that I had given my last speech. I had noticed that I was starting to get a few grey hairs, and that I enjoyed a nap after meals much more than standing up and trying not to doze off along with the audience. But when Dick Beamish called to invite me to speak, he said in his usual diplomatic way, it would be good for the young turks to hear from an old fart. That got to me, and I told him I **would** speak, giving the young farts a message from an old turk.

The second thing on my mind is the importance of acknowledging the contributions that some of you present here this evening have made to the launching of this PICES enterprise. I recall that it has not been easy. Whenever some countries were ready, others were not, and the various countries seemed to take turns at being not ready. The disappointments along the way might have daunted ordinary mortals, but thanks to the perseverance of a few individuals with a vision, particularly Warren Wooster, Dick Beamish, John Davis, and a few others, PICES is now launched on what will surely be a highly successful scientific voyage of discovery.

But before I develop that theme I might just say a word or two about Warren's long association with international matters concerning oceanography. I first met Warren in Paris when he was part of a United States delegation to a meeting of the International Oceanographic Congress, and I was one of the two members of the Canadian delegation. One of the items of business before the Congress was approval of a plan for the study of the Kuroshio current. Warren had been told that Henry Stommel, a very knowledgeable theoretician had grave concerns that the proposed frequency and spacing of transect samples would not be sufficient to detect the periodicities of physical events in the current. Warren had a copy of a paper Stommel had written on the topic and to say the least it was quite mathematical. Warren was very well known to all attending, so well known that if he tried to present Stommel's arguments everyone would know that he did not know what he was talking about. So the suggestion was made that perhaps someone in the Canadian delegation could make the point. The senior member of the Canadian delegation was Harry Hachey who was also very known to all, more for his interests in French cuisine than applied mathematics. So they asked me to present the case because nobody knew how little I knew.

I memorized a little script that evening and the following morning when the Kuroshio item came up on the agenda, the chairman (another Canadian, Bill Cameron) asked "Are there any comments and questions, ah yes, Dr. Larkin from Canada". I made my little speech, there were a few sceptical comments, but in the best tradition of international diplomacy, Warren Wooster rose to suggest to the chair that perhaps the Congress could approve the proposed plan in principle, subject to an examination by a technical committee of the point that had been raised. The chairman then said, "If that is agreed, shall we break for coffee". And during the coffee break he suggested to me that perhaps I should go back to Canada before anyone found out how little I did know about the mathematics of sampling in the ocean!

In a multitude of ways, Warren Wooster has made major contributions to international oceanography for a period of more than 40 yr. If PICES was a university, Warren would be the first and most illustrious chancellor and recipient of honorary degrees for diplomacy, perseverance, and good humour (but not mathematics). Dick Beamish would not get a degree in mathematics either.

To return to my theme (I do have one), the establishment of PICES is indeed a momentous occasion. Alfred Needler, the former Deputy Minister of Fisheries for Canada often remarked that the truly lasting contribution of the International North Pacific Fisheries Commission would be the exchange and publication of scientific information. In his view, that was the cornerstone of every international fisheries treaty. Given the free exchange of scientific information and research findings, there is a base of understanding from which all parties can argue their national case. Before the INPFC, no one knew where salmon were distributed in the North Pacific, there was only a feeble appreciation of the oceanography of the region, the groundfish resources of the Bering Sea had yet to be tapped and were an unknown quantity, the dynamics of North Pacific ecosystems were not even vaguely known. But the intensive INPFC field studies, particularly in the 1960s, and their subsequent analysis since, have given us all a broad sketch of the nature of the North Pacific and adjacent seas.

Thus, those of my generation, the old turks, did all the hard stuff and it only remains for the young farts to fill in the details. Or so we thought.

As you know times change. This first PICES conference, fittingly enough, is addressed to the question of climate change and northern fish populations, a topic which is not mentioned in any of the INPFC publications. To some extent one might argue that climate change is something of a hypothetical proposition (the former Governor of the State of Washington is certainly sceptical) and if it develops as slowly as seems likely, few of you here today will be alive to see whether your predictions were accurate - which should be a great incentive to be daring. Certainly, oceanography seems poised to become more of a predictive than a descriptive science. Just as the meteorologists forecast the weather for a few days, oceanographers will be telling us what kind of ocean conditions to expect for a few months, or years. One can only hope they will do better than our local meteorologists who have lately had trouble even describing yesterday's weather, let alone predicting tomorrow's.

Climate is not the only thing changing nowadays. Since the halcyon days of the INPFC there have been major changes in the fisheries of the North Pacific. The wave of fishing activity that began after World War II and precipitated the INPFC was transformed by the institution of 200-mile zones. The United States developed major groundfish fisheries taking 1 million t a year. Japan and the United States launched major salmon enhancement activities. In 1989 almost 5 billion young salmon were released from hatcheries into the North Pacific.

Awareness grew of the complex of interactions that link the physical conditions of the ocean with the dynamics of marine ecosystems. Spectacular oscillations in the abundance of herring and sardines, the collapse of king crab fisheries, the increase in the survival of salmon, and the decline of fur seal and sea lion populations might all be traced at least in part to changes in oceanographic circulation as much as to exploitation, management, or enhancement activities.

Gaining the ability to predict long-term oceanographic changes will certainly be of major importance to the fisheries of the future. Many years ago oceanography was seen as a useful tool for finding where fish would congregate and could be caught. I particularly recall the many Japanese contributions to the literature and particularly those of Professor Uda.

But the technologies of echo sounding made that kind of oceanography largely superfluous for fishing. Why bother trying to deduce where the fish might be when you can scan the ocean and see them? The study of fine-scaled oceanographic structure now has its greatest application to the understanding of the role of patchiness in production and year-class strength and the anomalies of fish migration.

The prediction of long-term changes in ocean climate and circulation has much more far reaching implications for the strategics of commercial exploitation. Will Pacific salmon disappear from the southern extremities of this range? Will the same species of groundfish continue to dominate catches in the Bering Sea? Will the North Pacific sustain the high levels of productivity of the past decade? These sorts of questions are of major significance to commercial fisheries.

My remarks have to this point been largely focused on oceanography, which was chosen as the topic for this first annual meeting of PICES. But as we all know, there will be much more to PICES than just oceanography. Indeed a major issue to be faced will be just how wide a scope the organization will have.

The routine exchange of fisheries statistics is an essential prerequisite for knowledgeable fisheries management whether or not there are international conservation agreements. The fish do not pay any attention to the political boundaries of the oceans, nor should the science. Rather evidently PICES will have a large fisheries biology and management component. But will PICES also foster exchanges on fish-processing technologies? How about enhancement technologies, fish-hatchery techniques, trophic-fertilization techniques? And how about the economics of fisheries? The bioeconomics and sociobiology of rights-based fishing?

Then there is aquaculture. As some of you will know there has been much discussion in recent years about how world fish production is to grow from the present level of 100 million t to a level of 160 million t by the year 2025 when the world population is projected to be 8 billion people. Most experts are agreed that there are few prospects for increasing production from the capture fisheries of the world so the spotlight is on aquaculture which has been growing at a steady rate since the late 1970s. Indeed, half of the increase in total world fish production in the past 15 yr has come from aquaculture and the prospect is that virtually all of the future increase will come from aquaculture.

Many years ago, that great British oceanographer, Sir Alastair Hardy was convinced that the lanternfishes of the world's oceans would be the great fish food source of the future. He predicted that by 1984 they would yield 150 million t a year. Well so far, nobody seems to have found a way to make a go of lanternfish, even with tomato sauce. Krill have been a greater market success, but production seems unlikely to go beyond 1 million t for some years yet. So aquaculture is the big wave of the future.

Will PICES consider aquaculture, at least if it is in salt water? Will PICES have conferences on offshore engineering for large scale mariculture? How abut a conference on various means of fertilizing large scale oceanic impoundments for fish production?

And then there is pollution. Will PICES address the issues of oil spills in the North Pacific? How about the problems of estuarial pollution throughout the Pacific Rim? And how about disposals of plastics at sea?

Will PICES follow the same paths as its Atlantic inspiration ICES or will PICES take on a quite different character? Will PICES expand its vision to the Central and South Pacific? Will the activities of PICES inspire a truly international management of the resources of the North Pacific? Will PICES foster the training of young oceanographers and fisheries biologists at sea?

Now in the last 5 min I have asked altogether a total of 14 questions about the future of PICES. I would like the answers please by the end of the week. And while you are at it, I would like an explanation for how you get the acronym PICES out of North Pacific Science Organization. I like the acronym better than the name of the organization so I will ask one more questions. If it is not too late to change, why not a new name for the organization. The Pacific International Convention for Exploration of the Sea? That has a nice historic ring. Or how about the Pacific Institution for the Conservation of Entropy at Sea?

Whatever the name, I wish the new organization the greatest of success in the future. Just do like the old turks did and you too will work to help make the world a better place for fishing.

TOPIC 1.

Evidence for Changes in Climate and the Resulting Effects in Freshwater and Marine Environments

Decadal coupled atmosphere—ocean variations in the North Pacific Ocean

K.E. Trenberth and J.W. Hurrell

Trenberth, K.E., and J.W. Hurrell. 1995. Decadal coupled atmosphere–ocean variations in the North Pacific Ocean, p. 15–24. *In* R.J. Beamish [ed.] Climate change and northern fish populations. Can. Spec. Publ. Fish. Aquat. Sci. 121.

Abstract: Observational evidence is reviewed for a substantial decade-long change in the North Pacific atmosphere and ocean lasting from about 1976 to 1988. Significant changes in atmospheric circulation for the winter half year were associated with changes in surface wind stress and the Sverdrup transport in the North Pacific, storm tracks, surface ocean sensible and latent heat fluxes, and mixing in the ocean that affected nutrient availability. The circulation changes produced substantial increases in sea-surface temperatures along the west coast of North America and Alaska, decreases in sea-surface temperatures over the central North Pacific, and decreases in sea ice in the Bering Sea. Large changes in nutrients led to increases in phytoplankton and zooplankton, which, along with the altered ocean currents and temperatures, changed the migration patterns and increased the stock of many fish species. A hypothesis is proposed that stresses the role of tropical forcing but with important feedbacks and effects in the extratropics that emphasize decadal relative to interannual time scales.

Résumé : On passe en revue les données d'observation sur les bouleversements qui ont marqué l'atmosphère et l'eau du Pacifique Nord pendant la décennie de 1976 environ à 1988. Des changements importants survenus pendant le semestre d'hiver dans la circulation atmosphérique se sont ajoutés à d'autres changements à la force d'entraînement du vent de surface et au transfert de masse de Sverdrup dans le Pacifique Nord, aux trajectoires des tempêtes, aux flux thermiques sensibles et latents de surface et au brassage des eaux, changements qui ont modifié la disponibilité des nutriments. Les changements dans la circulation atmosphérique ont engendré des hausses substantielles de températures de la surface de la mer le long de la côte ouest de l'Amérique du Nord et de l'Alaska, mais des diminutions de ces mêmes températures dans la partie centrale du Pacifique Nord ainsi qu'une diminution des glaces marines dans la mer de Béring. Les importants changements sur le plan des nutriments ont quant à eux entraîné des hausses dans les populations de phytoplanctons et de zooplanctons qui, de concert avec les courants et températures océaniques modifiés, ont transformé les comportements migratoires et fait augmenter les stocks de nombreuses espèces de poissons. L'hypothèse proposée est que le forçage tropical aurait joué un rôle important dans ces phénomènes mais que d'importantes réactions et répercussions dans la zone extratropicale accentueraient l'importance des échelles de temps décennales par rapport aux interannuelles.

Introduction

It has often been stated that the atmosphere has a very short memory while the ocean has enormous thermal inertia allowing it to provide the important memory for climate variations. In fact, however, it is the collaborative interaction between the atmosphere, the ocean, and other parts of the climate system that gives rise to important climate variations. In this paper, we document recent important variations on decadal time scales involving both the atmosphere and the ocean in the North Pacific, and we explore links to other parts of the globe. The variations described emphasize the very large physical scales that are apparently set by the atmospheric planetary-scale waves. More local effects are also often important in the marine environment affecting fisheries, but the patterns described provide the large-scale context and therefore provide the link between what otherwise appear to be rather disparate changes in many species of fish throughout the North Pacific.

Recently, considerable evidence has emerged of a substantial decade-long change in the North Pacific atmosphere and ocean that began about 1976. Changes in atmospheric circulation throughout the troposphere at this time have been documented by Trenberth (1990; see Fig. 1) and Nitta and Yamada (1989) for the winter half year. Associated changes occurred in surface wind stress and, by inference, in the Sverdrup transport in the North Pacific Ocean (Trenberth 1991) which have been directly measured in the Gulf of Alaska (Royer 1989). Bakun (1990) has noted changes in the longshore wind

K.E. Trenberth and **J.W Hurrell.** National Center for Atmospheric Research, P. O. Box 3000, Boulder, CO 80307, USA. The National Center for Atmospheric Research is sponsored by the National Science Foundation.

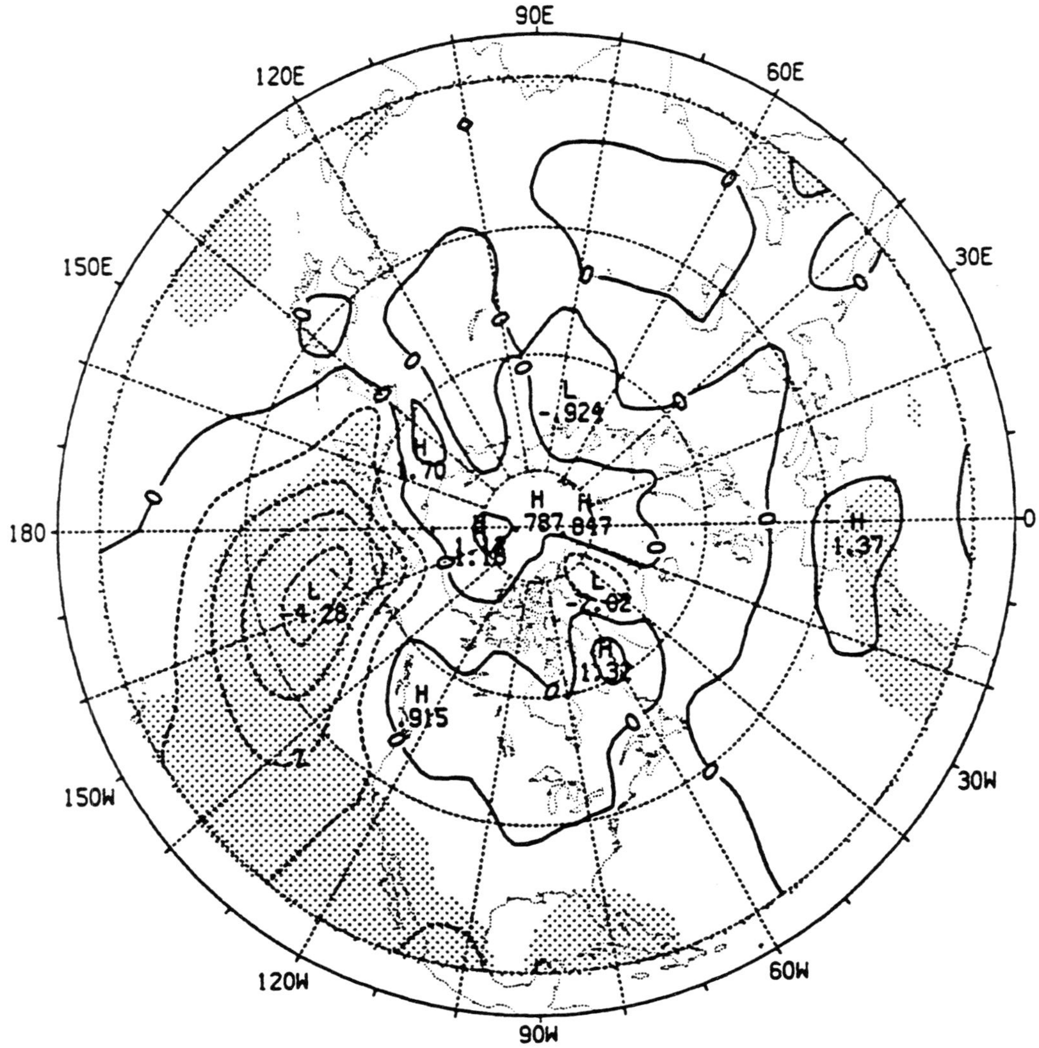

Fig. 1. The difference in mean sea level pressures for November through March 1977–88 versus 1924–76 (mbar). Stippling indicates statistical significance at 5%. From Trenberth (1990).

stress off California that are probably related. Changes in the monthly mean flow also imply changes in storm tracks and synoptic eddy activity (e.g., Lau 1988; Rogers and Raphael 1992) and in the surface ocean sensible and latent heat fluxes (Cayan 1992). The circulation changes were accompanied by changes in heat and moisture advection by the quasi-stationary flow (e.g., Rogers and Raphael 1992), so that there were substantial changes in the temperatures along the west coast of North America, in sea-surface temperatures (SSTs) over the North Pacific (Trenberth 1990), as well as in coastal rainfall and streamflow (Cayan and Peterson 1989), and in sea ice in the Bering Sea (Manak and Mysak 1987). Also accompanying the changes in the Pacific was a higher incidence of cold outbreaks across the plains of North America, ultimately leading to major freezes affecting the Florida citrus crop (Rogers and Rohli 1991).

In addition to changes in the physical environment, Venrick et al. (1987) observed associated large changes in the epipelagic ecosystem in the North Pacific, with increases in total chlorophyll in the water column and thus in phytoplankton. McFarlane and Beamish (1992) link these changes to substantial increases reported in the dominant zooplankton (calanoid copepods) after 1976. Mysak (1986) earlier noted that the changes in ocean currents and temperatures had altered the migration patterns of fish, in particular tuna (*Thunnus thynnus*) and salmon (*Oncorhynchus spp.*), in the northeast Pacific (see also Hamilton (1987)). Ebbesmeyer et al. (1991) found that the "step in Pacific climate" in 1976 had a profound effect on 40 environmental variables. They contrasted the periods 1968–75 with 1977–84 and noted that, in addition to the above parameters, climate-related changes were found in the behaviour of Canada geese (*Branta canadensis mofitti*), fish species other than salmon, and Dungeness crabs (*Cancer magister*) in the Northeast Pacific, and salinity, water temperature, and mollusk (*Macoma carlottensis*) abundance, in Puget Sound. McFarlane and Beamish (1992) proposed that the cooling of the central North Pacific Ocean increased nutrient supply to the surface waters, thereby improving primary productivity, to the benefit of many fish species. They further link the years of good recruitment to a more intense Aleutian Low in winter.

The changes in fish recruitment and catch following the mid-1970s, as described by Beamish and Bouillon (1993), Brodeur and Ware (1995), Polovina et al. (1994) and several papers in this volume, were profound. Large increases in productivity have been noted including substantial increases in sablefish (*Anoplopoma fimbria*) (McFarlane and Beamish 1992); roughly doubled catches and biomass of several varieties of salmon (notably pink (*O. gorbuscha*), coho (*O. kisutch*), chum (*O. keta*), and sockeye (*O. nerka*)); substantial increases in flying squid (*Ommastrephes bartrami*), walleye pollock (*Theragra chalcogramma*), northern anchovy (*Engraulas mordax*), halibut (*Hippoglossus stenolepis*) and Pacific hake (*Merluccius productus*), and, to a lesser extent, in herring (*Clupea harengus pallasi*); but with decreases in jack mackeral (*Trachurus symmetricus*), Alaskan king crab (*Prarlithodes camtschatica*), and northern fur seals (*Callorhinus ursinus*).

The change in 1976 is but one of several large changes that have occurred in the North Pacific. Namias (1959, 1963, 1969) has documented decadal-scale variations in the past, with coherent signals in the atmospheric circulation and in the SSTs, and with teleconnections downstream across North America (see also Douglas et al. (1982) and Namias et al. (1988)). The fairly sluggish response of the midlatitude ocean to changes in ocean forcing through surface momentum and heat fluxes effectively serves as a low-pass filter and emphasizes the longer time scales. More recent studies throw further light on these aspects and on the physical links between the atmosphere and the ocean in the North Pacific, and will be discussed below. New evidence has also emerged on the teleconnections downstream across North America, associated, in particular, with the so-called Pacific – North American (PNA) teleconnection pattern.

Possible causes of the changes were discussed by Trenberth (1990), who noted the close link between North Pacific changes on the decadal time scale with those in the tropical Pacific and Indian oceans, and the changes in frequency and intensity of El Niño versus La Niño events. This expanded on the similar link noted during El Niño events by Bjerknes (1969). These aspects will be pursued and quantified further below. In addition, we update the time series and carry out more comprehensive correlation analyses with surface temperature and SST analyses, and examine further the links with the Southern Oscillation and tropical Pacific Ocean. The report given here is part of a more complete study by Trenberth and Hurrell (1994).

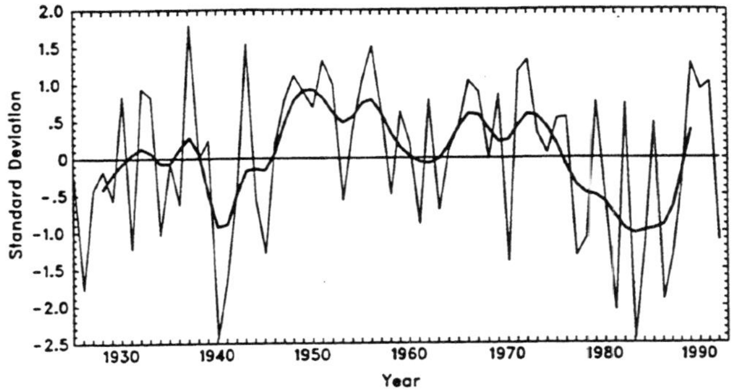

Fig. 2. Time series of the NP index (mean North Pacific sea-level pressures averaged over 30–65°N, 160°E to 140°W) for the months November through March beginning in 1925 and smoothed with the low-pass filter to bring out decadal and longer periods.

Table 1. The values of the NP index (area-averaged sea-level pressures over the North Pacific from 30–65°N, 140°E to 60°W) for the 5 winter months November through March, expressed as anomalies. The year shown is identified with January (i.e., the value for 1900 is for November 1899 through March 1900). Values are considered less reliable earlier this century, especially before 1924 because of the sparse data base (see Trenberth and Paolino 1980). The values are departures from the 92-yr mean of 1009.39 mb.

Year	00	01	02	03	04	05	06	07	08	09
1900	−3.67	0.61	0.09	4.12	3.43	−0.90	0.91	4.76	2.44	1.21
1910	2.50	3.40	0.05	0.03	−0.32	1.07	3.45	2.77	0.92	−0.82
1920	2.59	0.44	1.62	1.48	−0.94	−1.06	−4.24	−1.42	−0.87	−1.72
1930	1.55	−3.17	1.77	1.27	−2.72	−0.48	−1.88	3.53	−0.45	0.00
1940	−5.70	−4.03	−1.29	2.95	−1.59	−3.31	−0.06	1.29	1.95	1.53
1950	1.01	2.51	1.71	−1.79	0.43	1.83	2.92	0.92	−1.56	0.92
1960	−0.13	−2.52	1.30	−1.92	−0.31	0.50	1.89	1.57	−0.37	1.44
1970	−3.65	2.10	2.45	0.13	−0.32	0.64	0.71	−3.43	−2.89	1.17
1980	−1.70	−4.95	1.22	−5.92	−3.10	0.62	−4.59	−3.42	−1.25	2.36
1990	1.52	1.86	−2.93	−0.64	0.88					

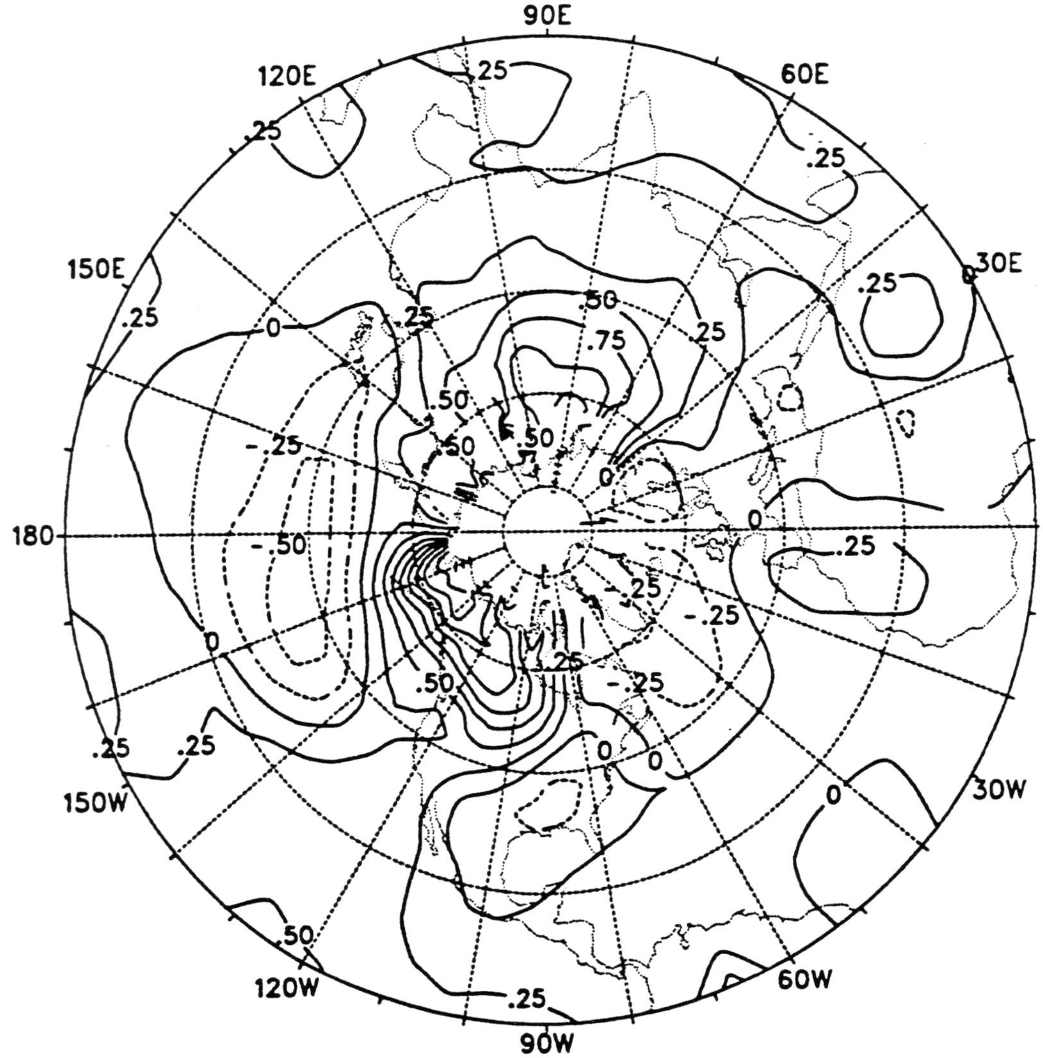

Fig. 3. Twelve-year (1977–88) average surface-temperature or SST anomalies as departures from the 1951–80 mean. Contours every 0.25°C. Shown are the anomalies for the 5 winter months (November to March). Broken lines are negative.

North Pacific Observed Large-scale Trends

In the North Pacific, a close association between SST anomalies and atmospheric circulation has been well recognized. Changes in surface temperatures arise from changes in temperature and moisture advection over the oceans by anomalous winds and through the associated changes in surface fluxes and vertical mixing within the ocean. Anomalous northerly winds over the ocean are typically not only cold but also dry, so that large increases in surface fluxes of both sensible and latent heat can be expected into the atmosphere, thereby cooling the ocean. Convection and mechanical mixing in the ocean can spread those influences through considerable depth and may also entrain water from below the thermocline, giving the anomalies a finite lifetime (e.g., Frankignoul 1985).

To explore the decadal-scale variations, a simple robust index has been developed using sea-level pressures. Following Trenberth (1990) and guided by Fig. 1, the index is the area-weighted mean sea level pressure over the region 30–65°N, 160°E to 140°W. We refer to this as the NP (North Pacific) index. This is slightly smaller than the area used in Trenberth (1990) but corresponds better to that involved in the decadal time-scale variations (see Fig. 1). Trenberth (1990) noted that the correlation of the NP index with the mid-tropospheric (500 mbar) PNA teleconnection pattern of Wallace and Gutzler (1981) is highly significant at −0.92 for 1947–87. An evaluation of sea level pressure charts (Trenberth and Paolino 1980) shows them to be most reliable after 1924.

Time scales of the NP index were explored by Trenberth and Hurrell (1994) who found that it is desirable to average over the 5 winter months, November through March to remove noise arising from high-frequency weather systems. Accordingly, Fig. 2 shows the NP time series for the 5-mo winter average from 1925, along with a low-pass curve designed to bring out the decadal time scales. The 1977–88 period was characterized by unusually low values of the NP index; in fact, the only previous time that comparable values occurred was during the much shorter interval from 1940 to 1941. There is,

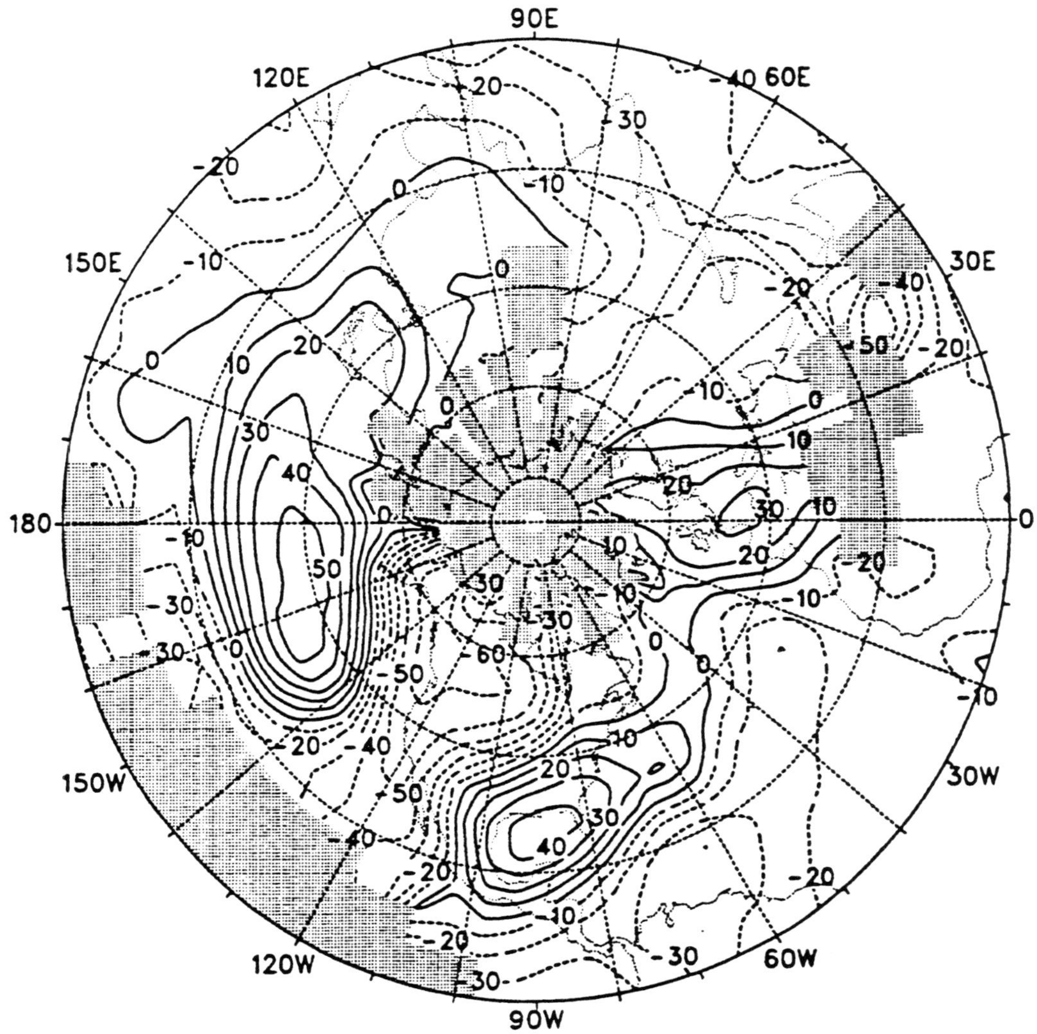

Fig. 4. Correlations between the NP index and surface temperatures for 1935–90. Dashed lines are negative. The 5% significance level is 0.27. Shaded areas indicate insufficient data.

nevertheless, considerable interannual variability present. Values of the NP index from 1900–94 are given in Table 1, but values earlier this century should be regarded as less reliable owing to the limited data available.

In the Aleutian Low for November through March from 1977 to 1988, pressures were lower by ~3 mbar averaged over the NP index area of the North Pacific. Lower pressures were present individually in all 5 winter months and are highly statistically significant (Fig. 1). No such change is present in other months of the year. The winter changes correspond to the centre of the low farther east and deeper on average by 4.3 mbar for the 5 winter months, and deeper by 7–9 mbar in January (Trenberth 1991; see also Fig. 1).

Climatological mean surface wind stress for 1980–86 revealed changes in the North Pacific from means based on ship data before 1977 that help confirm the reality of the sea-level pressure changes (Trenberth 1991), as do analyses with independent data sets (Nitta and Yamada 1989). Moreover, the associated changes in the curl of the wind stress and the corresponding Sverdrup transport (Trenberth 1991) over such a long period imply significant changes in North Pacific Ocean currents.

The corresponding changes in surface temperatures are shown in Fig. 3. The surface temperatures are taken from the updated Intergovernmental Panel on Climate Change data set (IPCC 1990, 1992), which consists of land-surface data from the University of East Anglia (Jones 1988) blended with SSTs from the United Kingdom Meteorological Office (Bottomley et al. 1990). Shown are the surface-temperature anomalies, expressed as departures from the 1951–80 means, for the 5 winter months (November–March) averaged over the 1977–88 period. The temperature anomalies are strongly regional and of both positive and negative signs. The 12-yr period features very large North Pacific basin temperature anomalies, with warming of over 1.5°C in Alaska and cooling of >0.5°C in the central North Pacific. The pattern in Fig. 3 over the North Pacific is very similar to the first empirical orthogonal function (EOF) of SSTs (Davis 1976; Lau and Nath

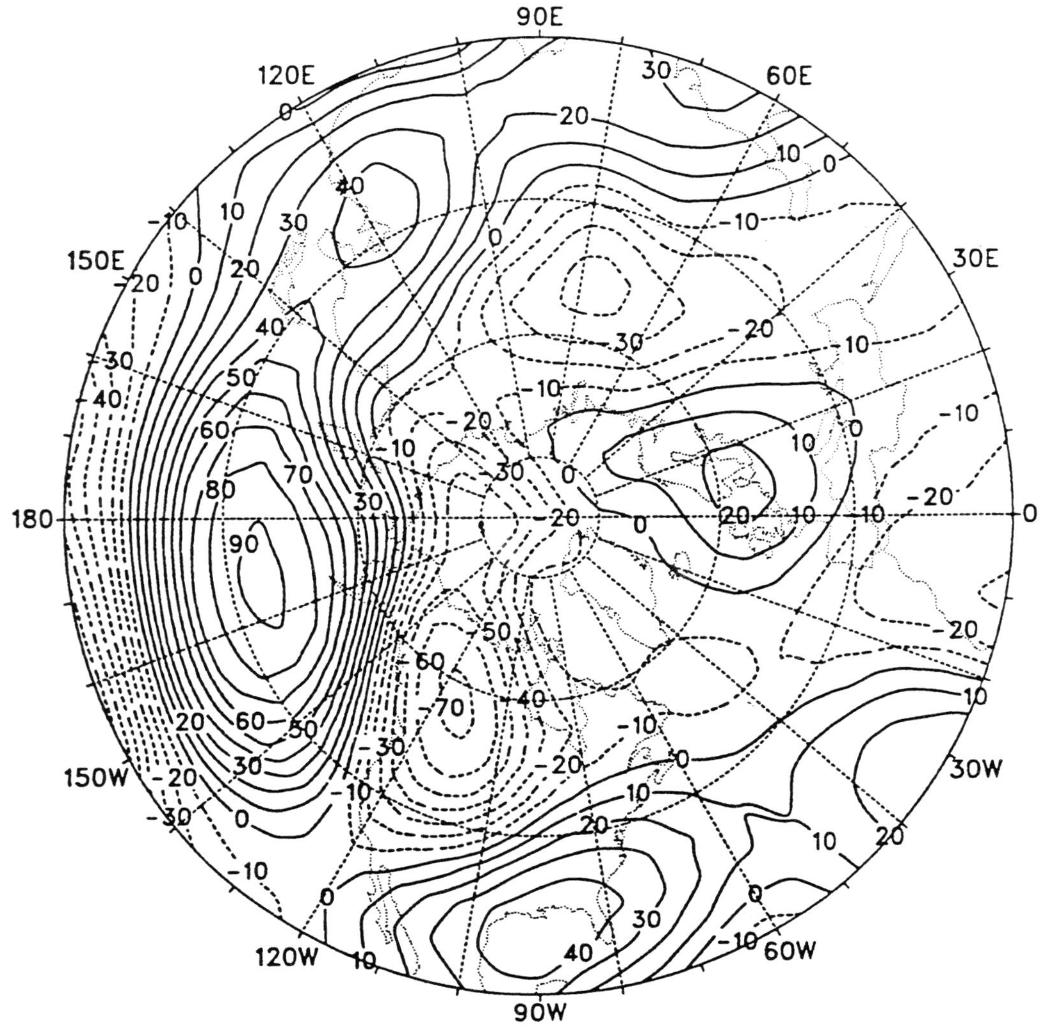

Fig. 5. Correlations of the November–March NP index with 500-mb heights for 1948–91. Broken lines are negative. Correlations over the 30–65°N, 140°E to 60°W, region account for 26% of the variance.

1990). Over the North Pacific, the SST anomalies are sufficiently persistent that there is little difference between annual and 5-mo mean anomalies. The winter pattern also reveals below-normal temperatures over the southeastern part of the United States, illustrating the PNA teleconnection. This is reflected in a higher than usual incidence of major freezes affecting the Florida citrus crop after 1977 (Rogers and Rohli 1991).

The very large regional Pacific temperature anomalies for 1977–88 in Fig. 3 provide the most compelling argument that the changes in sea-level pressure are real. The warming over Alaska and along the west coast of North America and the cooling in the central and western North Pacific would be expected with a stronger Aleutian Low because of thermal advection (Rogers and Raphael 1992), increased ocean mixing, and changes in the surface fluxes (Cayan 1992; Alexander 1992a, b). The increased southward gradient flow in the eastern North Pacific, revealed by the pressure pattern in Fig. 1, would bring warmer and moister air into Alaska and along the west coast of North America, while anomalous northerly winds would create colder than normal conditions in the central and western North Pacific. Lower SSTs result from large sensible and latent heat fluxes into the atmosphere combined with increased mixing in the ocean (Cayan 1992).

To reveal further the nature of the surface temperature changes associated with the NP index, Fig. 4 shows the correlations for the 5-mo average for November–March from 1935 to 1990. For each grid point, "seasonal" values were computed only when data existed for at least 3 of the 5 mo defining the season. Correlations between variables were not computed if the two variables had fewer than 75% of the total number of seasons in common. Across the North Pacific and North American regions, the pattern shows that the anomaly featured in Fig. 3 is consistent with the whole record: below-normal NP values are associated with below-normal temperatures over the North Pacific and southeastern United States and above-normal surface temperatures along the west coast, extending throughout Alaska and across most of Canada. We have also investigated these relationships as a function of various lags. To objectively decide how much variance is explained by the correlations across an area, we averaged the correlation coefficient squared for the region 140°E to 60°W, 30–65°N. The largest surface temperature variance explained by the NP index for this region occurs with NP leading by 1–2 mo (r^2 values with NP leading by 3, 2, 1, and 0 mo are 0.15, 0.19, 0.20, and 0.16, respectively). The pattern is similar to that at zero lag but the correlation coefficients increase most places by about 0.1 (to >0.6 over the North Pacific, and to <–0.7 over British Columbia).

The above results are consistent with those of Davis (1976). The link between SSTs in the North Pacific and the overlying atmospheric circulation has become well established. The main relationship seems to be one where the changes in atmospheric circulation are responsible for SST changes, as shown, for instance, by simultaneous and lag correlations (Davis 1976, 1978; Lanzante 1984; Wallace et al. 1990). Nevertheless, there is the strong expectation that extratropical SST anomalies also influence and may reinforce the atmospheric circulation (Kushnir and Lau 1992).

Figure 5 shows the correlations of the NP index with the 500-mbar height field. The PNA pattern emerges strongly, with all the PNA centres identified by Wallace and Gutzler (1981) appearing.

Causes of Changes in the North Pacific

Examination of the possible causes of the changes focuses attention on the association between large-scale coherent climate variations and changes in atmospheric waves. The stationary planetary waves in the atmosphere are forced by orography and patterns of diabatic heating arising from the distribution of land and sea, both in the extratropics and in the tropics (e.g., Frankignoul 1985).

When possible causes of changes are considered for the North Pacific, one prospect is in situ forcing through the influence on the circulation of extratropical SST anomalies in the North Pacific (Namias 1959, 1963). It has been difficult to substantiate such influences either statistically (Davis 1976, 1978) or with models (Ting 1991; Kushnir and Lau 1992). Recent modeling studies of SST anomalies in the northern hemisphere indicate that changes in storm tracks alter the eddy vorticity fluxes in the upper troposphere in such a way that they often reinforce and help maintain the circulation anomalies (Lau and Nath 1990; Ting 1991; Kushnir and Lau 1992).

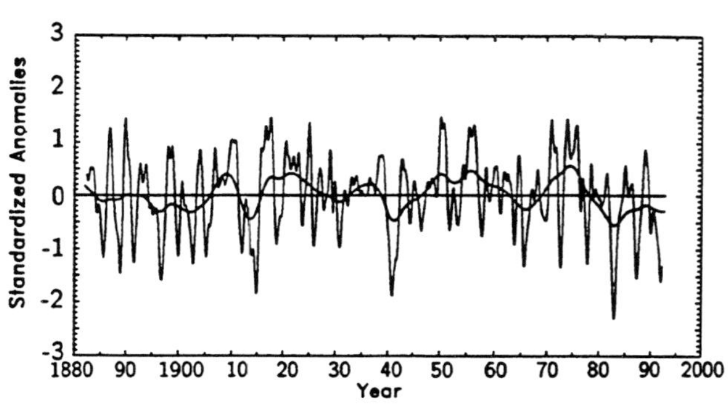

Fig. 6. Time series of the normalized monthly Southern Oscillation Index (minus Darwin sea level pressure anomalies), filtered with an 11-term low-pass filter designed to remove fluctuations less than 1 yr (Trenberth 1984) and with a low-pass spline smoothing filter that removes periods less than 10 yr.

While the changes in eddy transports from the altered weather systems are one major complication, another is that atmospheric heating effects may not be local. The sensible heat exchanged between ocean and atmosphere is realized locally, but the latent heat lost by the ocean through evaporation is realized only as an increase in moisture, and the actual atmospheric heating is not realized until precipitation occurs, often far downstream. This latter aspect depends on the prevailing weather at the time and varies with location according to the prevailing winds and background climatological flow. These nonlocal effects are therefore a sensitive function of position, and they add a large nondeterministic component to any forcing. This means that it is much more difficult to detect any systematic effects in both the real atmosphere and models. It also helps account for differences in results from many different model experiments, because inserted SST anomalies vary in location and intensity and the climates of models vary.

Another possible cause of changes in the North Pacific comes from changes in teleconnections. The best known examples of global impacts of local forcing are with changes in tropical SSTs such as the El Niño – Southern Oscillation (ENSO) phenomenon, whereby coupling between changes in the atmosphere and the underlying ocean in the tropical Pacific are linked by teleconnections to higher latitudes (Bjerknes 1969; Horel and Wallace 1981).

Links with the Tropical Pacific

The period of the deeper Aleutian Low regime extends from 1977 to 1988 when there were three El Niño (warm) events in the tropical Pacific but no compensating La Niño (cold) events, so that the tropical Pacific experienced above normal SSTs and a persistently negative Southern Oscillation Index (SOI) for that period (Fig. 6). Here we have defined an SOI using only the pressure at Darwin because it provides the longest, most reliable and homogeneous index available (Trenberth 1984; Trenberth and Shea 1987). It is expressed as minus Darwin in Fig. 6 to be compatible with the standard Tahiti minus Darwin index (see Trenberth 1984). A low-pass filter (smoothing spline fit) is also shown to bring out the periods greater than 10 yr.

Modeling studies (e.g., Blackmon et al. 1983; Alexander 1992b) confirm the causal link between SSTs in the tropics and the North Pacific circulation, with a deeper Aleutian Low set up as a teleconnection because of El Niño conditions. Alexander (1992a, b) further showed that the observed changes in North Pacific Ocean SSTs can be accounted for largely by atmospheric changes resulting from associated changes in surface fluxes, mixing of the upper layers of the ocean, and deepening of the mixed layer through entrainment. But the results obtained here are not simply due to the 1982–83 and 1986–87 El Niños; the Aleutian Low was also much deeper than normal in several other years and especially in the winter of 1980–81. Note, however, from Fig. 2 that the previous timewhen comparably low values occurred over the North Pacific was during the major 1939–42 El Niño event.

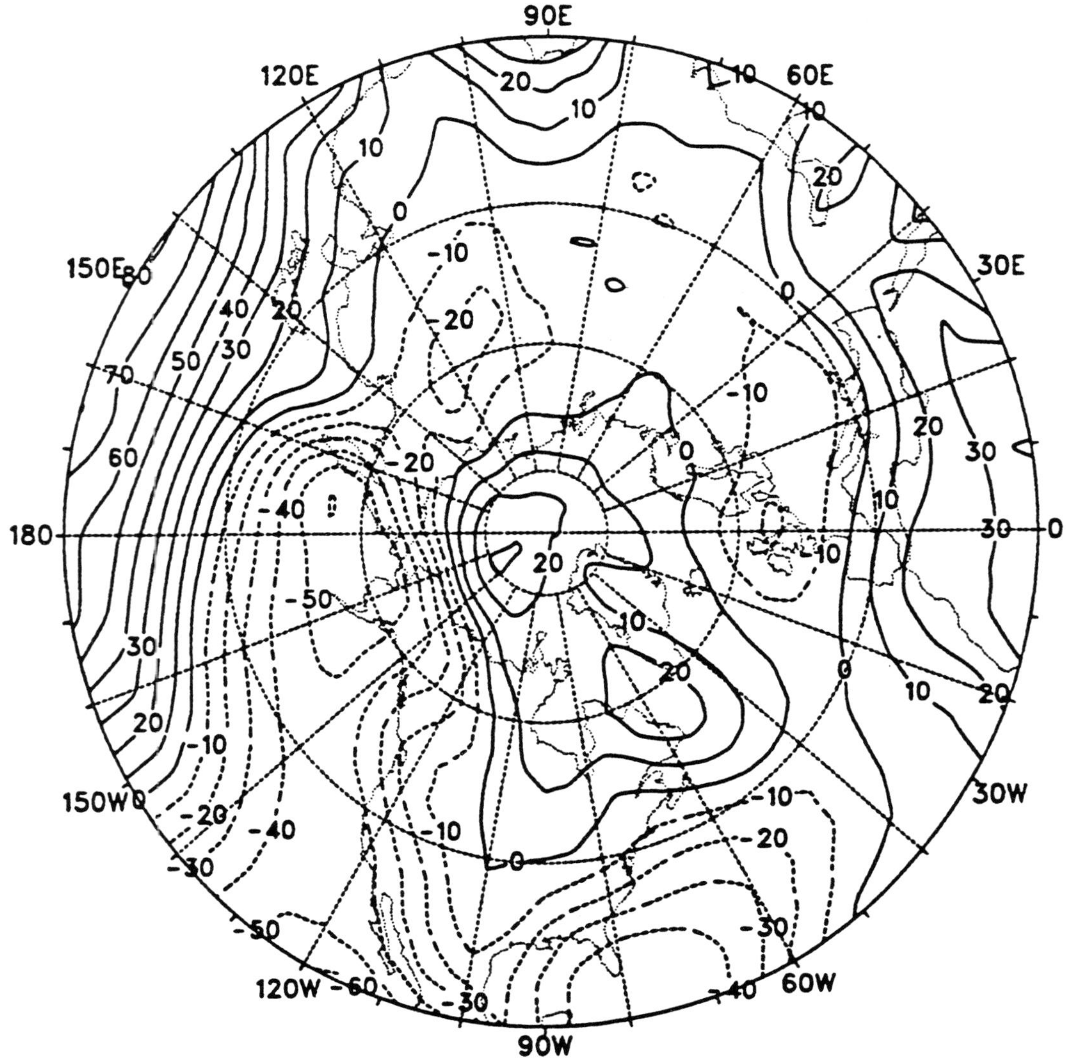

Fig. 7. Correlations of the 5-mo mean SSTs in the region 5°N to 10°S, 90°W to 180° for August through December with November through March sea-level pressures over the northern hemisphere for 1935–91 (i.e., the tropical SSTs lead by 3 mo). The 1% significance level is 0.34.

We have examined the correlation of the NP index with the more complete Tahiti minus Darwin SOI for the period 1935–91 using the 5 winter month average. Highest cross correlations of 0.53 occur at zero lag ±1 mo and are highly statistically significant. However, it is well established that the Southern Oscillation and the SST fields in the tropical Pacific evolve as El Niño events develop (Trenberth 1976; Rasmusson and Carpenter 1982; Trenberth and Shea 1987; Wright et al. 1988).

We have therefore examined in more detail the relationships between SSTs in the tropics and the NP index. Problems with data coverage are severe in the tropics before 1951. Results reveal (Trenberth and Hurrell 1994) that changes in SST throughout much of the tropical Pacific lead the NP index by about 3 mo, although the cross correlation is not sharply defined and values are only slightly smaller at zero lag. Nevertheless, these results emphasize the involvement of tropical SST variations in the atmospheric and surface-temperature variations over the North Pacific and North America. To summarize the results, we have compiled a tropical SST time series for the region 5°N to 10°S, 90°W to 180° (this region has SST anomalies of much the same sign during ENSO events and is extensive enough to include desirable off-equatorial data). Correlations of these tropical SSTs leading the November–March northern hemisphere sea level pressures by 3 mo show the link with the North Pacific and extension across North America (Fig. 7). Note the values of opposite sign over North America, which are very important as part of the overall pattern. The anomalous wind flow accompanying this pattern is indeed one where stronger southerlies along the west coast of North America accompany a negative SOI (i.e., El Niño conditions).

Discussion and Conclusions

The picture emerging from these empirical and modeling studies is not yet fully clear, but the evidence suggests the following hypothesis. In the tropics, coupled ocean–atmosphere interactions result in coupled modes of which ENSO is the most prominent. This results in large interannual variability in the Pacific sector with preferred time scales of

2–7 yr, but with small amplitude decadal variations. All these fluctuations have manifestations in higher latitudes through teleconnections within the atmosphere. In the North Pacific, ENSO variability is found in the PNA pattern (and the NP index) but is best seen when averages are taken over the 5 winter months, as the noise level associated with natural weather variability is high on monthly time scales. The deepened Aleutian Low in ENSO events results in a characteristic SST anomaly pattern that, on average, is enhanced through positive feedbacks from (1) effects of the extratropical SST anomaly itself, and (2) changes in momentum (and vorticity) fluxes associated with changes in high-frequency storm tracks (Kushnir and Lau 1992; Trenberth and Hurrell 1994). The same influences are present on long time scales; but, whereas surface fluxes and mixed-layer processes are dominant in changing SSTs on interannual time scales, changes in ocean currents must also become a factor on decadal time scales and would reinforce the SST changes. Moreover, the long time scale involved in changing the currents and the Sverdrup circulation adds further persistence to the extratropical system that, along with heat storage in the top 500 m of the ocean, evidently serves to emphasize the decadal over interannual time scales relative to the tropics.

Aspects of the above hypothesis have appeared in Namias (1959, 1963, 1969) but here we have emphasized much more the links with the tropics. A major but as yet unanswered question is whether either the intensity or frequency of ENSO events might change as a result of global warming. The long observational record (Fig. 6) reveals that the frequency and intensity of ENSO events have changed in the past (Trenberth and Shea 1987), with strong ENSO fluctuations from about 1880 to 1920. Aside from the major event from 1939 to 1942, stronger and more regular ENSO events only resumed in the 1950s. However, the low-pass curve in Fig. 6 indicates that the recent imbalance between the occurrence of warm versus cold events in the tropical Pacific is unprecedented.

Whether the unusual 1977–88 imbalance can be ascribed to a specific cause or is merely a part of natural variability is a very difficult question. The major change that occurred in March–April 1988, with a transition from El Niño to a very strong La Niña (Fig. 2), apparently ended the climate regime although the underlying ocean currents and heat storage must still be perturbed and the pattern could re-emerge. Indeed, the recent 1991–92 ENSO event was noted for exceptionally warm water along the west coast of both North and South America in early 1992.

Whatever the cause, the effects on fisheries in the North Pacific are profound. The stronger Aleutian Low and southward-shifted storm tracks evidently increase the mixing in the central North Pacific Ocean bringing more nutrients to the upper layers and enhancing productivity. However, these links involve many trophic levels and need to be explored in considerably more detail before they are fully understood.

Acknowledgements.

We specially thank Dennis Shea for preparing some of the figures. This research is partially sponsored by the Tropical Oceans Global Atmosphere Project Office under grant NA86AANRG0100.

References

Alexander, M. 1992a. Midlatitude atmosphere–ocean interaction during El Niño. I. The North Pacific Ocean. J. Clim. 5: 944–958.

Alexander, M. 1992b. Midlatitude atmosphere–ocean interaction during El Niño Pt. II. The Northern Hemisphere atmosphere. J. Clim. 5: 959–972.

Bakun, A. 1990. Global climate change and intensification of coastal ocean upwelling. Science (Wash., D.C.) 247: 198–201.

Beamish, R.J., and D.R. Boullon. 1993. Pacific salmon production trends in relation to climate. Can. J. Fish. Aquat. Sci. 50: 1002–1016.

Bjerknes, J. 1969. Atmospheric teleconnections from the equatorial Pacific. Mon. Weather. Rev. 97: 163–172.

Blackmon, M.L., J.E. Geisler, and E.J. Pitcher. 1983. A general circulation model study of January climate anomaly patterns associated with interannual variation of equatorial Pacific sea surface temperatures. J. Atmos. Sci. 40: 1410–1425.

Bottomley, M., C.K. Folland, J. Hsiung, R.E. Newell, and D.E. Parker. 1990. Global ocean surface temperature atlas. The UK Meteorological Office, London. 20 p. 313 plates.

Brodeur, R.D., and D.M. Ware. 1995. Interdecadal variability in distribution and catch rates of epipelagic nekton in the Northeast Pacific Ocean, p. 329–356. *In* R.J. Beamish [ed.] Climate change and northern fish populations. Can. Spec. Publ. Fish. Aquat. Sci. 121.

Cayan, D.R. 1992. Latent and sensible heat flux anomalies over the northern oceans: the connection to monthly atmospheric circulation. J. Clim. 5: 354–369.

Cayan, D.R., and D.H. Peterson. 1989. The influence of North Pacific atmospheric circulation on streamflow in the West. Geophys. Monogr. 55: 375–397.

Davis, R. 1976. Predictability of sea surface temperature and sea level pressure anomalies over the North Pacific Ocean. J. Phys. Oceanogr. 6: 249–266.

Davis, R. 1978. Predictability of sea-level pressure anomalies over the North Pacific Ocean. J. Phys. Oceanogr. 8: 233–246.

Douglas, A.V., D.R. Cayan and J. Namias. 1982. Large-scale changes in North Pacific and North American weather patterns in recent decades. Mon. Weather Rev. 112: 1851–1862.

Ebbesmeyer, C.C., D.R. Cayan, D.R. McLain, F.H. Nichols, D.H. Peterson, and K.T. Redmond. 1991. 1976 step in the Pacific climate: Forty environmental changes between 1968–1975 and 1977–1984, p. 120–141. *In* J.L. Betancourt and V.L. Sharp [ed.] Proceedings 7th Annual Pacific Climate (PACLIM) Workshop, April 1990. California Deptartment of Water Resources, Sacramento, Calif. Tech. Rep. 26. Interagency Ecological Studies Program.

Frankignoul, C. 1985. Sea surface temperature anomalies, planetary waves, and air–sea feedbacks in middle latitudes. Rev. Geophys. 8: 233–246.

Hamilton, K. 1987. Interannual environmental variation and North American fisheries. Bull. Am. Meteorol. Soc. 68: 1541–1548.

Horel, J.D., and J.M. Wallace. 1981. Planetary-scale atmospheric phenomena associated with the Southern Oscillation. Mon. Weather Rev. 109: 813–829.

IPCC. 1990. Scientific assessment of climate change. IPCC WG I, WMO, UNEP. J. T. Houghton, G. J. Jenkins and J.J. Ephraums [ed] Cambridge University Press, Cambridge, UK. 365 p.

IPCC. 1992. Climate Change 1992. J.T. Houghton, B.A. Callander, and S.K. Varney [ed] Cambridge University Press, Cambridge, UK. 200 p.

Jones, P.D. 1988. Hemispheric surface air temperature variations: recent trends and an update to 1987. J. Clim. 1: 654–660.

Kushnir, Y., and N.-C. Lau. 1992. The general circulation model response to a North Pacific SST anomaly: Dependence on time scale and pattern polarity. J. Clim. 5: 271–283.

Lanzante, J.R. 1984. A rotated eigenanalysis of the correlation between 700-mb heights and sea surface temperatures in the Pacific and Atlantic. Mon. Weather. Rev. 112: 2270–2280.

Lau, N.-C. 1988. Variability of the observed midlatitude storm tracks in relation to low frequency changes in the circulation pattern. J. Atmos. Sci. 45: 2718–2743.

Lau, N.-C., and M.J. Nath. 1990. A general circulation model study of the atmospheric response to extratropical SST anomalies observed in 1950-79. J. Clim. 3: 965–989.

Manak, D.K., and L.A. Mysak. 1987. Climatic atlas of arctic sea ice extent and anomalies, 1953–1984. Climate Research Group Rep. 87-8.

McFarlane, G.A., and R.J. Beamish. 1992. Climatic influence linking copepod production with strong year-classes in sablefish, *Anoplopoma fimbria*. Can. J. Fish. Aquat. Sci. 49: 743–753.

Mysak, L.A. 1986. El Niño interannual variability and fisheries in the northeast Pacific Ocean. Can. J. Fish. Aquat. Sci. 43 : 464–497.

Namias, J. 1959. Recent seasonal interactions between North Pacific waters and the overlying atmospheric circulation. J. Geophys. Res. 64: 631–646.

Namias, J. 1963. Large-scale air–sea interactions over the North Pacific from summer 1962 through the subsequent winter. J. Geophys. Res. 68: 6171–6186.

Namias, J. 1969. Seasonal interactions between the North Pacific Ocean and the atmosphere during the 1960s. Mon. Weather Rev. 97: 173–192.

Namias, J., X. Yuan, and D.R. Cayan. 1988. Persistence of North Pacific sea surface temperature and atmospheric flow patterns. J. Clim. 1: 682–703.

Nitta, T., and S. Yamada. 1989. Recent warming of tropical sea surface temperature and its relationship to the Northern Hemisphere circulation. J. Meteorol. Soc. Jpn. 67: 375–383.

Polovina, J.J., G.T. Mitchum, N.E. Graham, M.P. Craig, E.E. DeMartini, and E.N. Flint. 1994. Physical and biological consequences of a climate event in the central North Pacific. Fish. Oceanogr. 3:1, 15–21.

Rasmusson, E.M., and T.H. Carpenter. 1982. Variations in tropical sea surface temperature and surface wind fields associated with the Southern Oscillation/El Niño. Mon. Weather Rev. 110: 354–384.

Rogers, J.C., and M.N. Raphael. 1992. Meridional eddy sensible heat fluxes in the extremes of the Pacific/North American teleconnection pattern. J. Clim. 5: 127–139.

Rogers, J.C., and R.V. Rohli. 1991. Florida citrus freezes and polar anticyclones in the Great Plains. J. Clim. 4: 1103–1113.

Royer, T.C. 1989. Upper ocean temperature variability in the Northeast Pacific Ocean. Is it an indicator of global warming? J. Geophys. Res. 94: 18 175 – 18 183.

Ting, M. 1991. The stationary wave response to a midlatitude SST anomaly in an idealized GCM. J. Atmos. Sci. 48 : 1249–1275.

Trenberth, K.E. 1976. Spatial and temporal variations of the Southern Oscillation. Q. J. R. Meteorological Soc. 102: 639–653.

Trenberth, K.E. 1984. Signal versus noise in the Southern Oscillation. Mon. Weather Rev. 112: 326–332.

Trenberth, K.E. 1990. Recent observed interdecadal climate changes in the Northern Hemisphere. Bull. Am. Meteorological. Soc. 71: 988–993.

Trenberth, K.E. 1991. Recent climate changes in the Northern Hemisphere, p. 377–390. *In* M. Schlesinger, [ed.], Greenhouse-Gas-Induced Climate Change. A Critical Appraisal of Simulations and Observations, 8–12 May 1989, Amherst, Mass. DOE Workshop, Elsevier, New York.

Trenberth, K.E., and J.W. Hurrell. 1994. Decadal ocean-atmosphere variations in the Pacific. Clim. Dyn. 9: 303–319.

Trenberth, K.E., and D.A. Paolino. 1980. The Northern Hemisphere sea-level pressure data set: trends, errors, and discontinuities. Mon. Weather Rev. 108: 855–872.

Trenberth, K.E., and D.J. Shea. 1987. On the evolution of the Southern Oscillation. Mon. Weather Rev. 115: 3078–3096.

Venrick, E.L., J.A. McGowan, D.A. Cayan, and T.L. Hayward. 1987. Climate and chlorophyll *a*: long-term trends in the central North Pacific Ocean. Science (Wash., D.C.), 238: 70–72.

Wallace J.M., and D.S. Gutzler. 1981. Teleconnections in the geopotential height field during the Northern Hemisphere winter. Mon. Weather Rev. 109: 784–812.

Wallace J.M., C. Smith and Q. Jiang. 1990. Spatial patterns of atmosphere–ocean interaction in the northern winter. J. Clim. 3: 990–998.

Wright, P.B., J.M. Wallace, T.P. Mitchell, and C. Deser. 1988. Correlation structure of the El Niño Oscillation phenomenon. J. Clim. 1: 609–625.

Long-Term Variations in SST Fields of the North Pacific Ocean

Kimio Hanawa

Hanawa, K. 1995. Long-term variations in SST fields of the North Pacific Ocean, p. 25–36. *In* R.J. Beamish [ed.] Climate change and northern fish populations. Can. Spec. Publ. Fish Aquat. Sci. 121.

Abstract: The interannual to interdecadal variations of sea-surface temperature (SST) and the upper-ocean thermal conditions in the North Pacific Ocean are described, mostly based on recent studies by the author's group. The SST fields included a large interdecadal (>5 years) and El Niño – Southern Oscillation (3–4 years) time-scale signals. In the upper-ocean thermal condition, interdecadal time-scale signals were clearly detected even at 200 m, especially in the central Pacific at midlatitudes, i.e., the boundary region between the Subtropical and Subarctic Gyres. By comparing upper-ocean thermal conditions in the north-western Pacific of two pentads (1938–1942; 1978–1982), spin-up – spin-down of the Subtropical Gyre and the temperature change of the North Pacific Subtropical Mode Water were also detected.

Résumé : Sont décrites, basées surtout sur de récentes études effectuées par le groupe de l'auteur, les variations interannuelles à interdécennales des températures de la surface de la mer (TSM) et des conditions thermiques de la couche supérieure du Pacifique Nord. Les champs de la TSM comprenaient d'importants signaux d'échelles de temps interdécennaux (DC; > 5 ans) et attribuables à El Niño–Oscillation australe (3–4 ans). Dans les conditions thermiques de la couche supérieure, les signaux DC ont été clairement détectés même à 200 m, surtout aux latitudes moyennes du Pacifique central, c'est-à-dire dans la région frontalière entre les circulations subtropicale et subarctique. En comparant les conditions thermiques de la couche supérieure du nord-ouest du Pacifique sur deux quinquennats (1938–1942; 1978–1982), on a également décelé une accélération ou une décélération de la circulation subtropicale ainsi que le changement de température des eaux du mode subtropical du Pacifique Nord.

Introduction

It is important to know whether the recent global warming is a long-term trend due to increasing greenhouse gasses, or is merely due to natural fluctuations with decadal, or longer, time scales. In such studies, it is essential to clarify mutual interaction processes of atmosphere–ocean system at various time scales of climate changes. Also important is the need to extract the specific pattern of changes of oceanic and atmospheric conditions at various time scales, from practical viewpoints such as fisheries.

There are many studies that focused on specific time scales such as those for the El Niño – Southern Oscillation (ENSO) and the Quasi-Biennial Oscillation. Hanawa et al. (1988) discussed the relationship between the phase of ENSO and appearance of sea-surface temperature (SST) anomalies in the Northwest Pacific. By composite analysis, they showed the well-organized SST anomaly pattern in the ENSO-year winter (during the ENSO event) in the Northwest Pacific, and that in the next-year winter is almost the same pattern as the ENSO winter, but with sign reversed. Iwasaka et al. (1988) reported that the arrangement of SST anomalies was very similar to that of subdomains of the Northwest Pacific, derived by cluster analysis based on the similarity of temporal variability of SST anomalies. Hanawa (1991) provided a comprehensive review of this subject.

Recently, decadal variation in SST fields and atmospheric conditions over the north Pacific were reported by Kashiwabara (1987), Nitta and Yamada (1989), and Trenberth (1990). Nitta and Yamada (1989) showed that SST fields in the global ocean exhibited decadal time scale signals. They further noted SST-anomaly differences between two decades (1967–1976 and 1977–1986), in which the north-south oscillating pattern of the North Pacific appears. Trenberth (1990) showed the interdecadal atmospheric change in sea-level pressure over the North Pacific, and temperature fields in the Northern Hemisphere winter. He pointed out that the Aleutian Low of the 1976–1977 winter was relatively intense, and that this depressed condition has remained in recent years.

Nitta and Yamada (1989) and Trenberth (1990) suggested that the intensification of the Aleutian Low was associated with the Pacific – North American (PNA) teleconnection pattern of Wallace and Gutzler (1981), which has been excited by the high SST fields maintained for about 10 years in the mid- to eastern-equatorial Pacific, from the middle 1970s to the late 1980s.

In this report, first we will show the dominant spatial patterns of SST variations in the North Pacific, and those temporal evolutions, with special reference to a decadal time-scale fluctuation (Tanimoto et al. 1993). Second, we will show

K. Hanawa. Department of Astronomy and Geophysics, Faculty of Science, Tohoku University, Aoba-ku, Sendai 980-77, Japan.

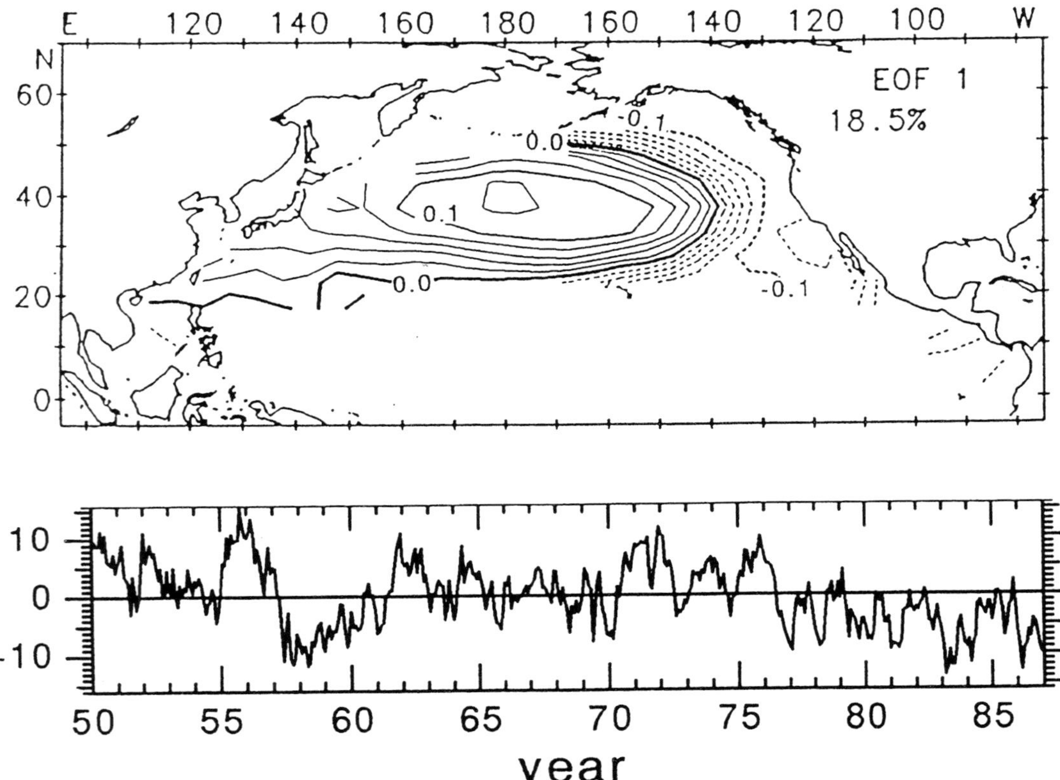

Fig. 1. Pattern of the leading EOF mode of unfiltered SST anomalies in the north Pacific (upper panel) and its time coefficient (lower panel). This mode accounts for 18.5% of the total variance. Solid (broken) lines represent positive (negative) values.

the change of upper-ocean thermal conditions during two recent decades corresponding to the abovementioned decadal time scale SST variations (Watanabe et al. 1993). Last, we will provide evidence of spin-up – spin-down of the North Pacific Subtropical Gyre, by comparing upper-ocean thermal conditions in the Northwest Pacific during two pentads (1938–1942 and 1978–1982) (Bingham et al. 1992). Here, spin-up and spin-down mean swift and slow rotation of gyre circulation, respectively; during the spun-up (spun-down) period, the Kuroshio and the North Equatorial Current as parts of the north Pacific Subtropical Gyre flow faster (slower) than normal.

Decadal Time-Scale SST Variations

Tanimoto et al. (1993) tried to detect the dominant patterns in SST fields at various time scales, by adopting cluster, spectral, and empirical orthogonal function (EOF) analyses of newly prepared SST data. As a supplement, they used 500-hPa geopotential-height data to find the large-scale atmospheric variations associated with SST variations.

In this section, their results are shown, with respect to cluster, spectral, and EOF analyses of unfiltered SST anomalies, EOF analyses of filtered SST anomalies, and decadal variability in the ocean and atmosphere.

Data Sources

An accurate, 37-year data set (1950–86) of SSTs was prepared from the Comprehensive Ocean Atmosphere Data Set (COADS), in monthly, 5° latitude by 5° longitude grids. Although COADS contains a huge number of historical SST data, mainly taken by merchant ships, the areas where time-series analyses could be performed were limited in the middle and high latitudes of the north Pacific. In the low latitudes, data were insufficient even after 1950. Therefore, low-latitude data were excluded in the time-series analysis, but included in the composite analysis. At each grid point, a 37-year mean climatological cycle was created, from which anomalies were calculated.

The 500-hPa geopotential-height data set was provided by the Japan Meteorological Agency.

Cluster, spectral, and EOF analyses of unfiltered SST anomalies

Figure 1 shows a pattern of the leading mode and time coefficient of an EOF analysis for the unfiltered SST anomalies, which can explain 18.5% of the total variance. This pattern is characterized by an elliptical monopole centered around 40°N lat and 180°, accompanied by anomalies with opposite sign along the North American coast. This result is very similar to the dominant EOFs obtained in previous studies (Weare et al. 1976; Iwasaka et al. 1987; Nitta and Yamada 1989). The spectrum of the time coefficient showed several peaks with almost the same energy levels (not shown here).

Figure 2*a* shows the four subdomains (I–IV) of the north Pacific extracted by cluster analysis based on the temporal similarity of SST anomalies. Regions and arrangement are exactly the same as those of Iwasaka et al. (1988), using a

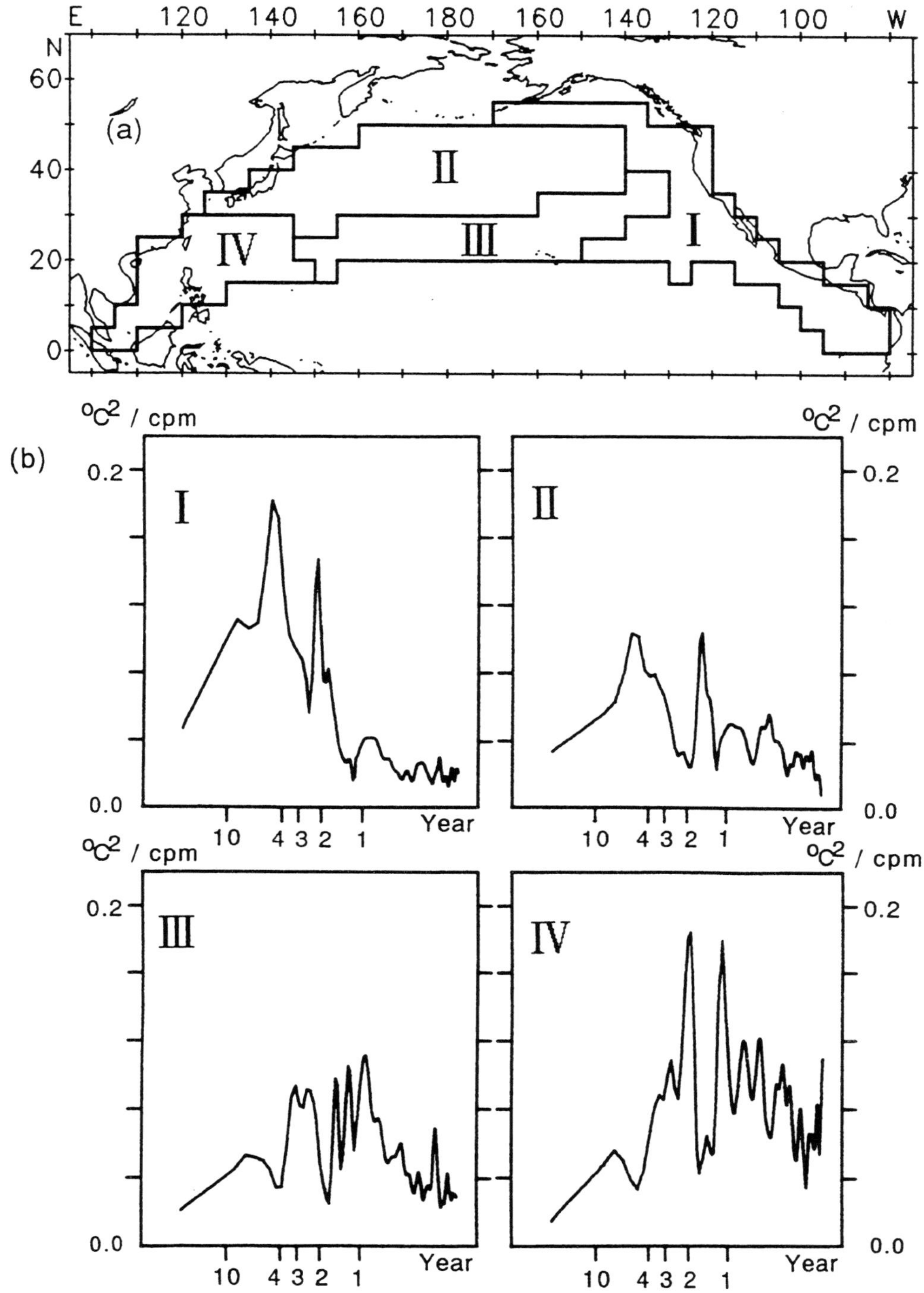

Fig. 2. (a) Sub-domains (I – IV) extracted by cluster analysis of unfiltered SST anomalies. (b) FFT spectra of time series of SST anomalies spatially averaged in each subdomain.

shorter data period. Figure 2*b* shows spectra for SST anomalies spatially averaged in each subdomain, which was extracted by cluster analysis. It is interesting that spectral forms differ among subdomains, and several peaks appear in each spectrum. These results strongly suggest that there are signals with various time scales in the North Pacific SST fields, and the dominant patterns at various time scales may have different centers of action, i.e., regions of maximum amplitude.

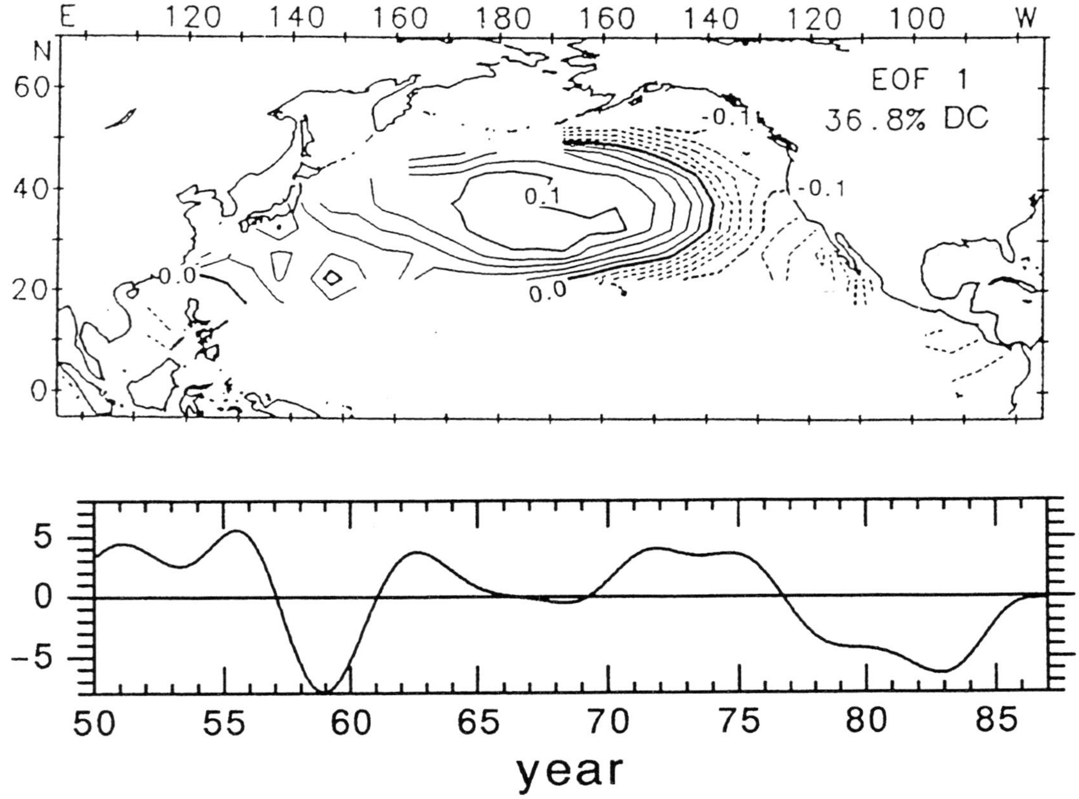

Fig. 3. Pattern of the leading EOF mode of filtered SST anomalies in the North Pacific, for the DC (>60 months) time scale (upper panel), and its time coefficient (lower panel). This mode can account for 36.8% of the total DC time scale variance.

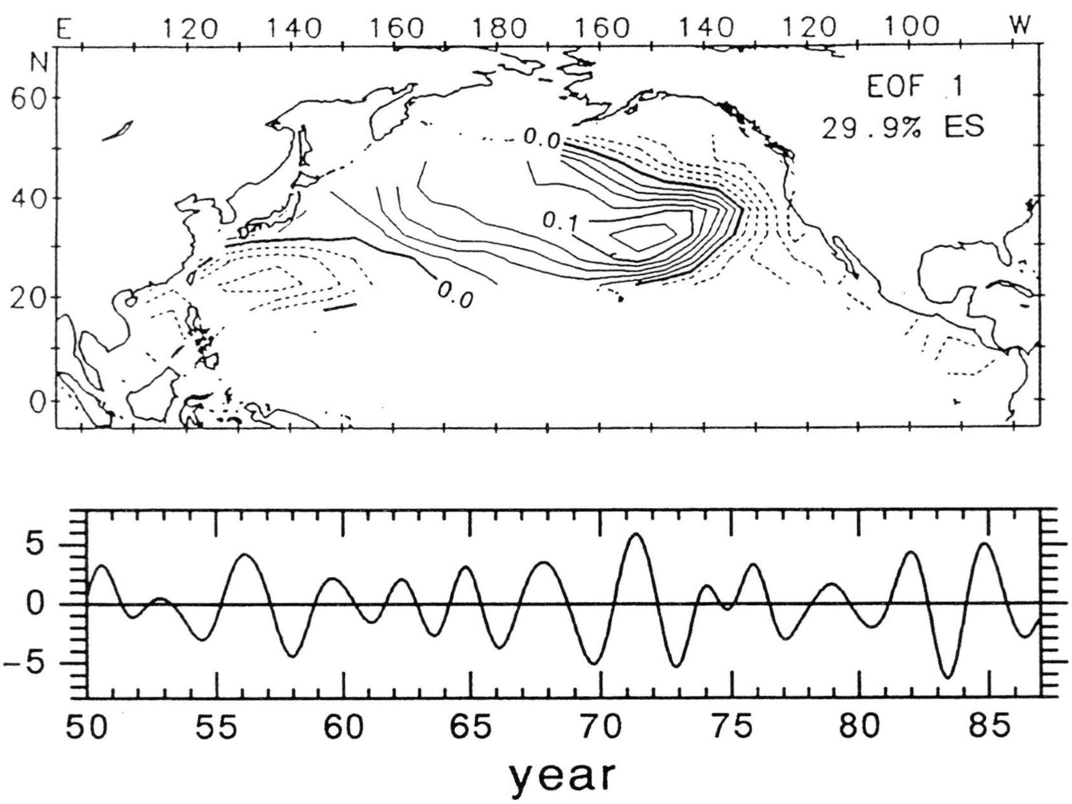

Fig. 4. Pattern of the leading EOF mode of filtered SST anomalies in the north Pacific, for the ES (24–60 months) time scale (upper panel), and its time coefficient (lower panel). This mode can account for 29.9% of the total ES time scale variance.

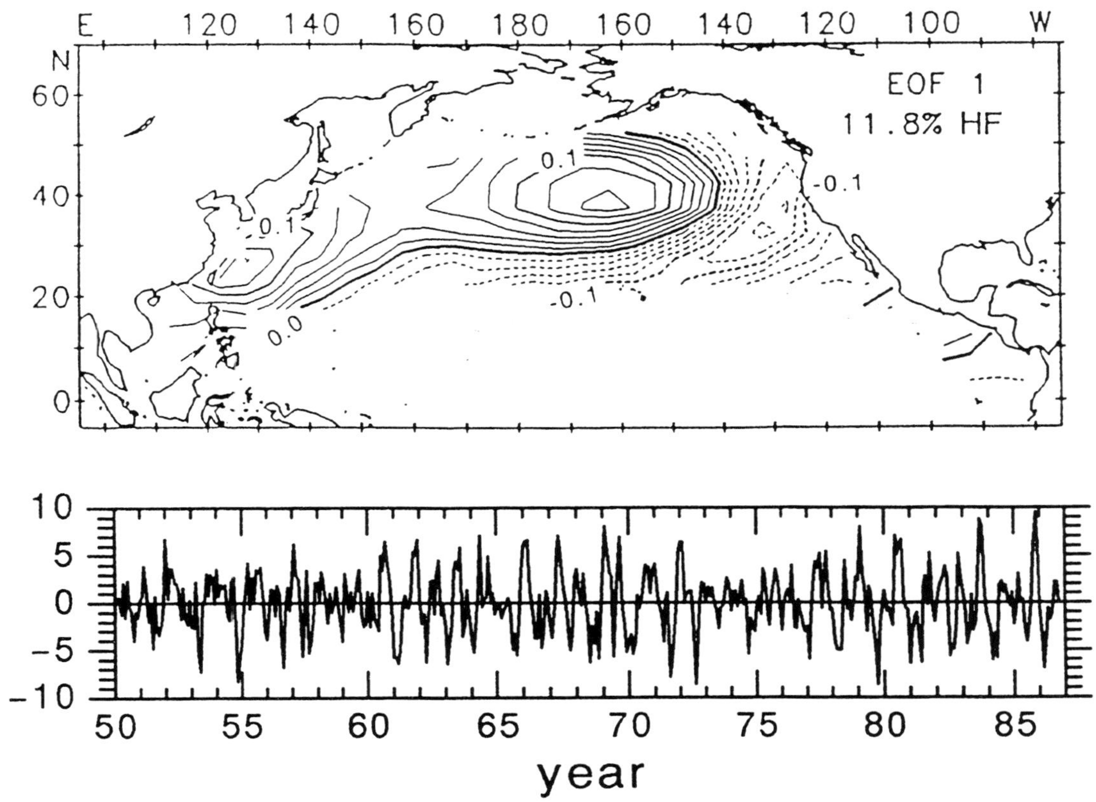

Fig. 5. Pattern of the leading EOF mode of filtered SST anomalies in the north Pacific, for the HF (<24 months) time scale (upper panel), and its time coefficient (lower panel). This mode can account for 11.8% of the total HF time scale variance.

EOF analyses for filtered SST anomalies

To extract characteristic patterns of SST fields and those time evolutions, based on the above results, Tanimoto et al. (1993) divided the SST signals into three time scales, and applied an EOF analysis. The three time scales were: (i) decadal (DC; >60 months); (ii) ENSO (ES; 24–60 months); and (iii) high frequency (HF; <24 months). Recursive-type digital filters were used to extract the time-scale signals.

Figures 3–5 show the leading EOF modes for the DC, ES, and HF time scales, respectively. The pattern for the DC time scale (Fig. 3), which can explain 36.8% of the total DC variance, strongly resembles the first EOF of the unfiltered SST anomalies shown in Fig. 1. The leading mode of the HF time scale (Fig. 5; 11.8% of the total HF variance) over the eastern half of the Pacific is very similar to that of the DC time scale, but with the center of action located eastward at around 40°N and 160°W. In the western half, however, there is another center of action with the same sign at around 25°N and 125°E. The first EOF pattern for the ES time scale (Fig. 4; 29.9% of the total ES variance) is somewhat different from those for DC and HF time scales. That is, the center of action in the mid-Pacific is shifted considerably southward, and the western Pacific is covered by anomalies with the opposite sign. Most of the significant positive values of the time coefficient correspond to the ENSO cold events (La Niña; 1956–1957, 1966–1967, 1970–1971, 1980–1981, and 1983–1984), and most of the negative values correspond to warm events (El Niño; 1957–1958, 1965–1966, 1968–1969, 1972–1973, and 1982–1983). This result confirms that the north Pacific SST anomalies with the ES time scale are strongly influenced by tropical atmosphere–ocean interactive phenomena.

Decadal Variability in the Ocean and Atmosphere

In the temporal evolution of the first EOF mode for the DC time scale (Fig. 3), we can recognize a quasi-steady anomaly period immediately followed by another persistent anomaly period with opposite sign. Therefore, based on the value of the time coefficient, the individual years treated were categorized into three periods: L_+ (18 years; 1950–1956, 1961–1964, and 1970–1976), L_- (13 years; 1957–1960 and 1977–1985) and L_0 (6 years; 1965–1969 and 1986).

Figure 6 shows the composite maps of the unfiltered SST anomalies averaged over the three periods. For L_+ and L_-, the extratropical SST anomalies are characterized by the pattern associated with the first EOF mode shown in Fig. 3. This means that SST fluctuations with the DC time scale are very significant even in unfiltered SST anomaly fields. On the other hand, anomalies with opposite sign appear over the tropics. That is, fluctuations of the equator to pole gradient of SSTs exist over the North Pacific.

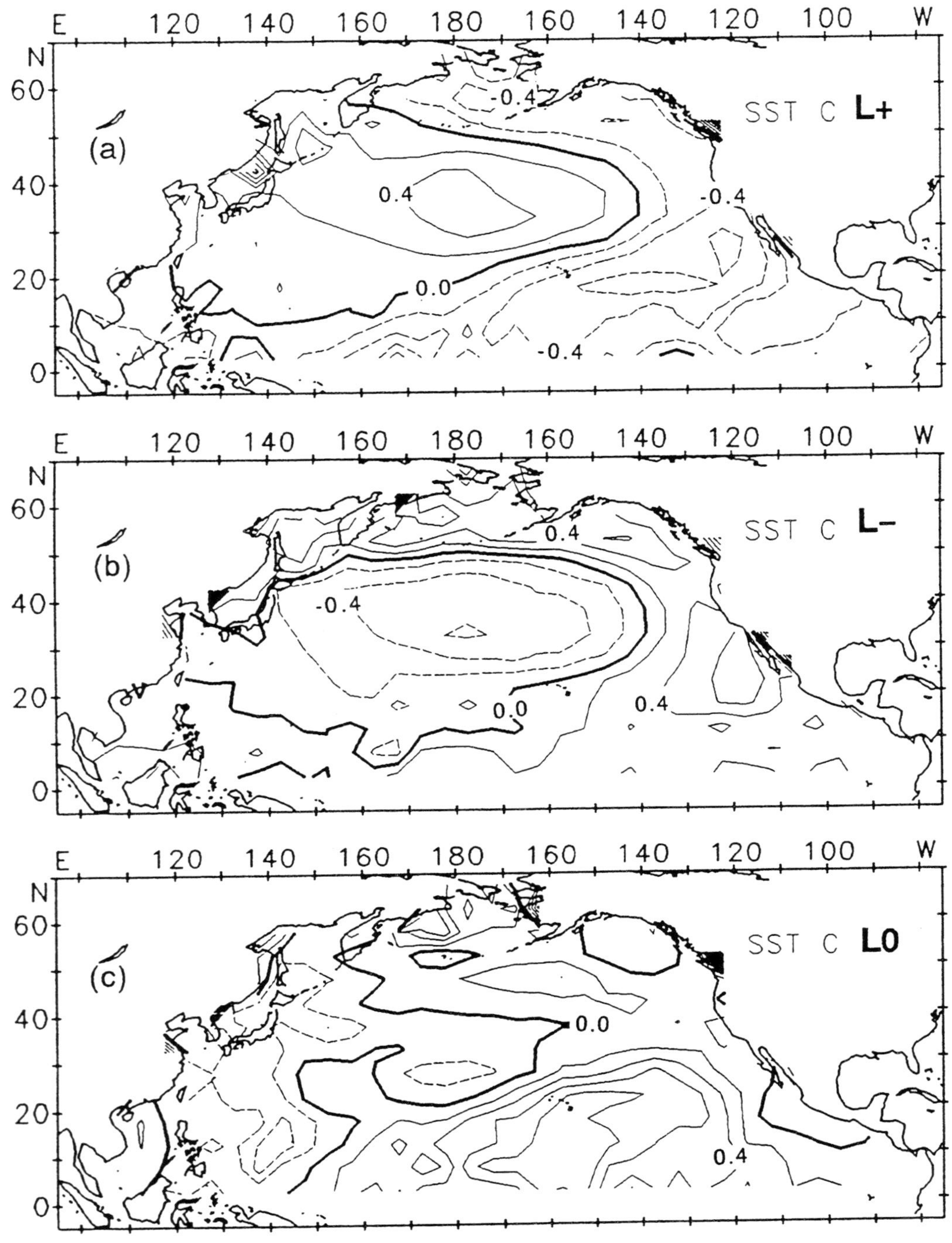

Fig. 6. Composite maps of unfiltered SST anomalies for three categories of decadal scale variation: (*a*) L_+; (*b*) L_-; and (*c*) L_0. Contour interval is 0.2°C. Solid (broken) lines represent positive (negative) values. (See text for category definitions.)

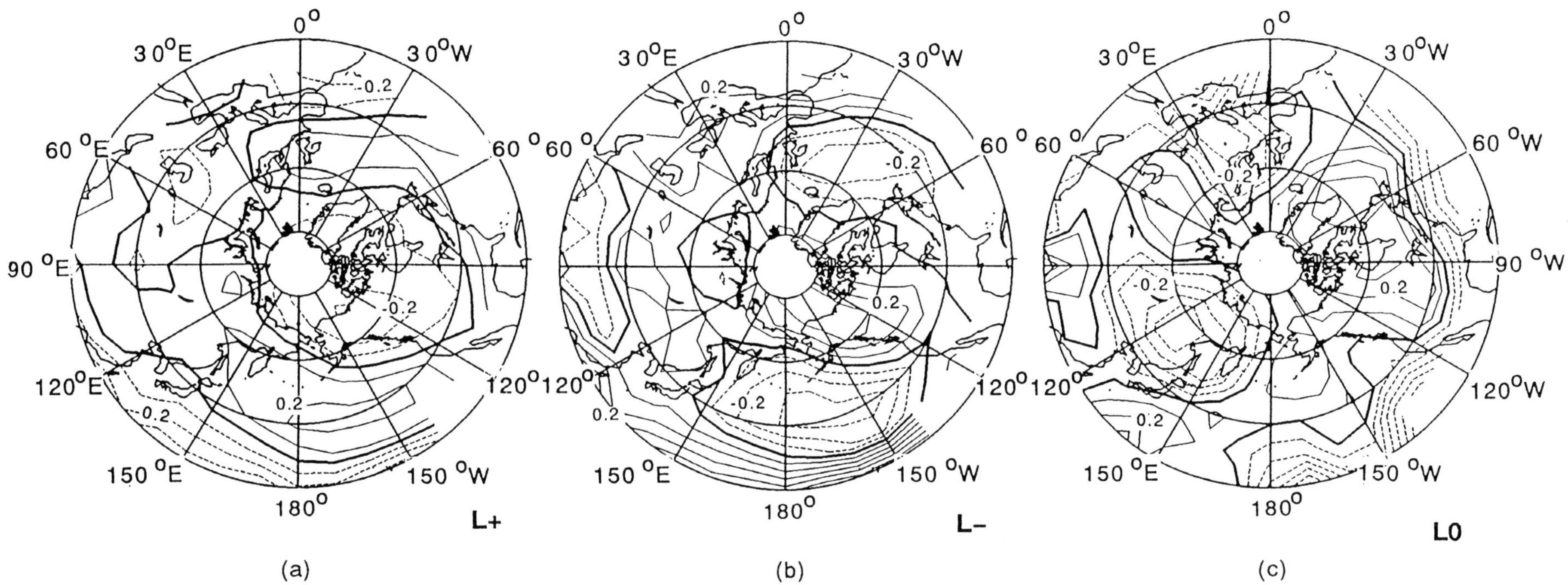

Fig. 7. Composite maps of winter (December–February) 500-hPa geopotential height anomaly fields, corresponding to DC time scale signals in SST anomaly fields, for three categories of decadal scale variation: *(a)* L_+; *(b)* L_-; and *(c)* L_0. Anomalies are normalized by the standard deviation at each grid point. Solid (broken) lines represent positive (negative) values. (See text for category definitions.)

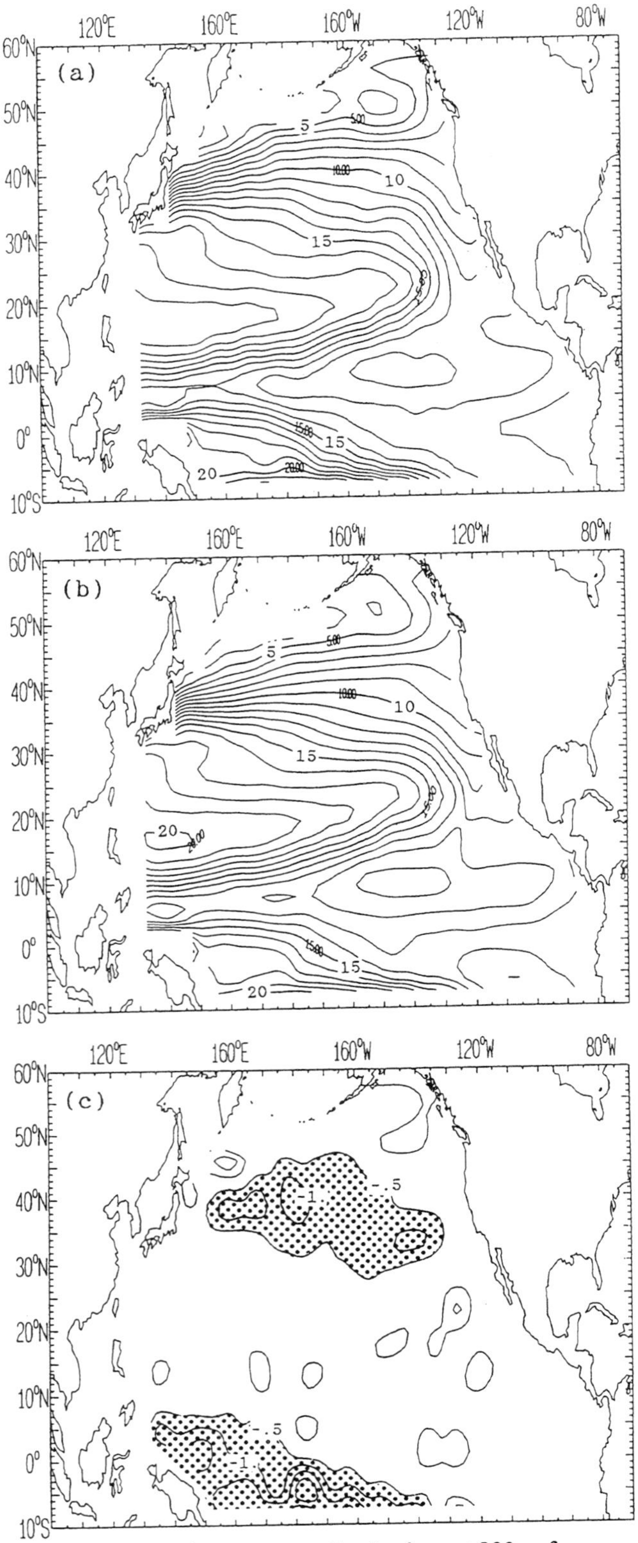

Fig. 8. Horizontal temperature distributions at 200 m for (*a*) period I (1964–1975) contour interval 1°C and (*b*) period II (1976–1988), contour interval 1°C and (*c*) contour interval 0.5°C temperature differences (period II – I) between the two pentads. (Panel c). Stippled regions denote temperature difference less than −0.5°C. Zero contours are not drawn in Fig. 8.

Tanimoto et al. (1993) prepared a composite map, similar to Fig. 6, of 500-hPa height anomalies, but only for the winter season, December–February (Fig. 7). For the L_- (L_+) period, PNA-like (anti-PNA-like) anomalies appear over the north Pacific and the North America. This means the jet stream over the Pacific tends to be extended further downstream than normal in the L_- period of PNA-like anomalies.

This composite analysis for SST and 500-hPa height anomalies suggests that persistent SST anomalies with a positive (negative) sign in the equatorial region excite a PNA-like (anti-PNA-like) pattern over the North Pacific and North America. As a result, intensified (weakened) westerlies cause negative (positive) SST anomalies in the central region of the mid-latitudes.

Unsolved Questions

This study demonstrated the existence of dominant DC-scale SST variations and an accompanying change in atmospheric conditions. What is the trigger for transition from one persistent period of DC-scale variations to another? What mechanism works in this process? These are unsolved important questions, and future studies are needed.

Upper-Ocean Thermal Conditions

Watanabe et al. (1993) investigated the depth to which the DC time scales, in SST fields, penetrated in the ocean. They used all subsurface temperature data archived in the National Oceanographic Data Center to date, and data compiled by the National Research Institute of Far Seas Fisheries. The total number of observations was 1.4×10^6. After a quality control of the data, they made monthly 5° latitude × 5° longitude grids for the Pacific.

According to the time coefficient of the first EOF mode for the DC time scale (Fig. 3), the mean temperature fields were computed for two periods: 1964–1975 (I) and 1976–1986 (II). Period I (period II) roughly corresponds to the L_+ (L_-) period described in the previous section. Patterns of SST differences detected in the data set (not shown) exhibited almost the same distribution of SST anomalies as that detected in the previous section (see Fig. 6). That is, negative SST anomalies appeared in the central Pacific at midlatitudes, and positive anomalies appeared along the North American coast and in the central to eastern equatorial region.

Figures 8*a* and 8*b* show the horizontal temperature distributions at 200 m in periods I and II, respectively. Figure 8*c* presents the temperature difference between the two (period II - period I). Negative temperature anomalies less than −0.5°C are clearly detected in the central North Pacific even at 200 m. In several regions, temperature anomalies exceeded −1.0°C. From the observation of several meridional cross-sections (not shown here), it was also found that this temperature decrease in period II occurred much deeper ca 600 m in the boundary region between the Subtropical and Subarctic Gyres in the central north Pacific.

In Fig. 8*c*, negative SST anomalies appeared in the western half of the equatorial region, and positive anomalies appeared in the eastern half. This is because the thermocline was raised

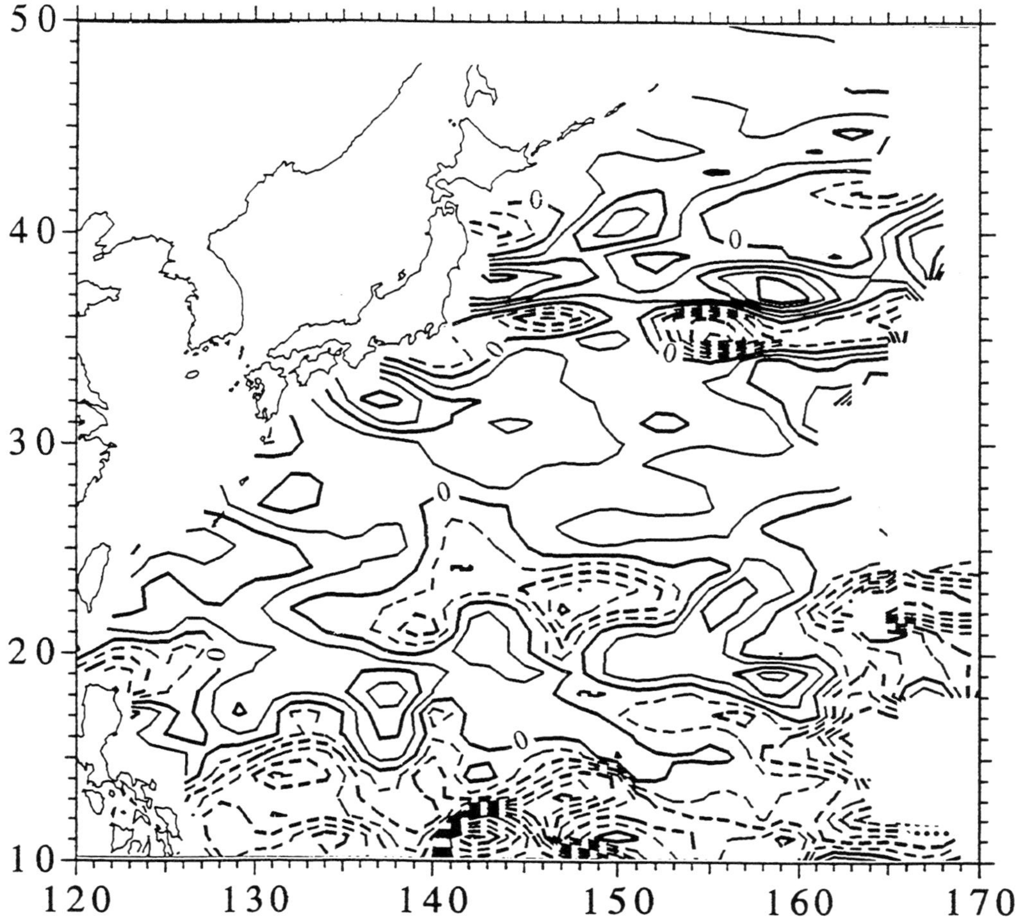

Fig. 9. Temperature difference at 200 m depth between two pentads: P78 (1978–1982) - P34 (1938–1942). Contour interval is 0.5°C. Stippled areas denote negative differences.

in the western region and lowered in the eastern region during period II. The change in thermocline depth was responsible for the change in SST fields through the upwelling process.

The above result is preliminary, and it is expected that more detailed analyses for this data set will provide useful information on long-term oceanic climate change.

Spin-up – spin-down: North Pacific Subtropical Gyre

Bingham et al. (1992) compared thermal conditions in two pentads, P34 (1938–1942) and P78 (1978–1982), in a study of long-term changes in the temperature structure of the upper Northwest Pacific. The P34 data were taken mostly by the Japanese Imperial Navy in a series of hydrographic surveys (Teramoto 1972). The P78 data were mostly from XBTs taken as part of the TRANSPAC program (White and Bernstein 1979). For each pentad, the data were interpolated to a set of standard depths, put through quality control procedures, and averaged for each 1° latitude by 1° longitude grid.

Temperature differences at 200 m between the two pentads (P78 – P34) are shown in Fig. 9. Temperature cross sections along 145°E for P34 and P78, respectively, are shown in Figs. 10*a* and 10*b*, and the differences (P78 – P34) in Fig. 10*c*. A large area of the Subtropical Gyre was warmer during P78, while the southern Subtropical Gyre (in the area of the North Equatorial Current) was warmer during P34. This suggests that the transports of the Kuroshio and North Equatorial currents were larger during P78. That is, the North Pacific Subtropical Gyre was spun-up in P78, compared with P34.

Properties of the North Pacific Subtropical Mode Water (NPSTMW) were also compared between the two pentads. The NPSTMW is important, because it is considered to be the ocean's memory of past atmospheric forcing. Because NPSTMW is characterized as a subsurface thermostad (Suga et al. 1989), it can be detected by a vertical temperature gradient. The low temperature gradient layer between the seasonal and main thermoclines corresponds to NPSTMW.

Meridional (longitudinal) cross sections of the vertical temperature gradient are shown in Figs. 11*a*–11*d*; Figs. 11*a* and 11*c* for P34, and Figs. 11*b* and 11*d* for P78. The core layer of NPSTMW was the layer with a temperature gradient less than 1.5°C/100 m. The NPSTMW was thicker, more uniform in temperature, and more confined geographically during P34 than during P78. A greater thickness was considered to be the result of stronger winter cooling during P34. Actually, the

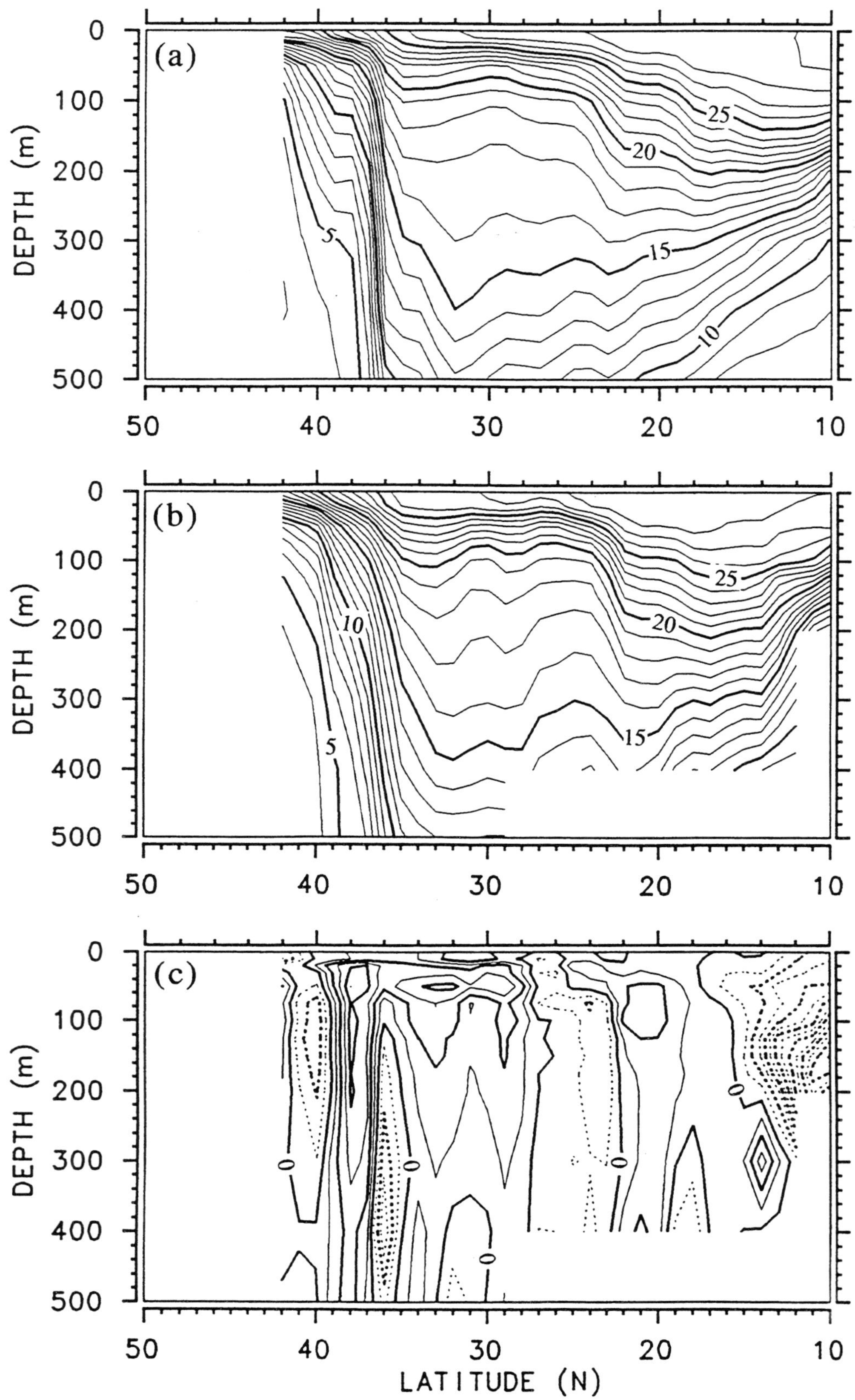

Fig. 10. Temperature cross sections along 145°C for pentads P34 and P78, and their temperature difference (P78 – P34). Contour interval is 0.5°C. Broken contours in Fig. 10c represent negative differences.

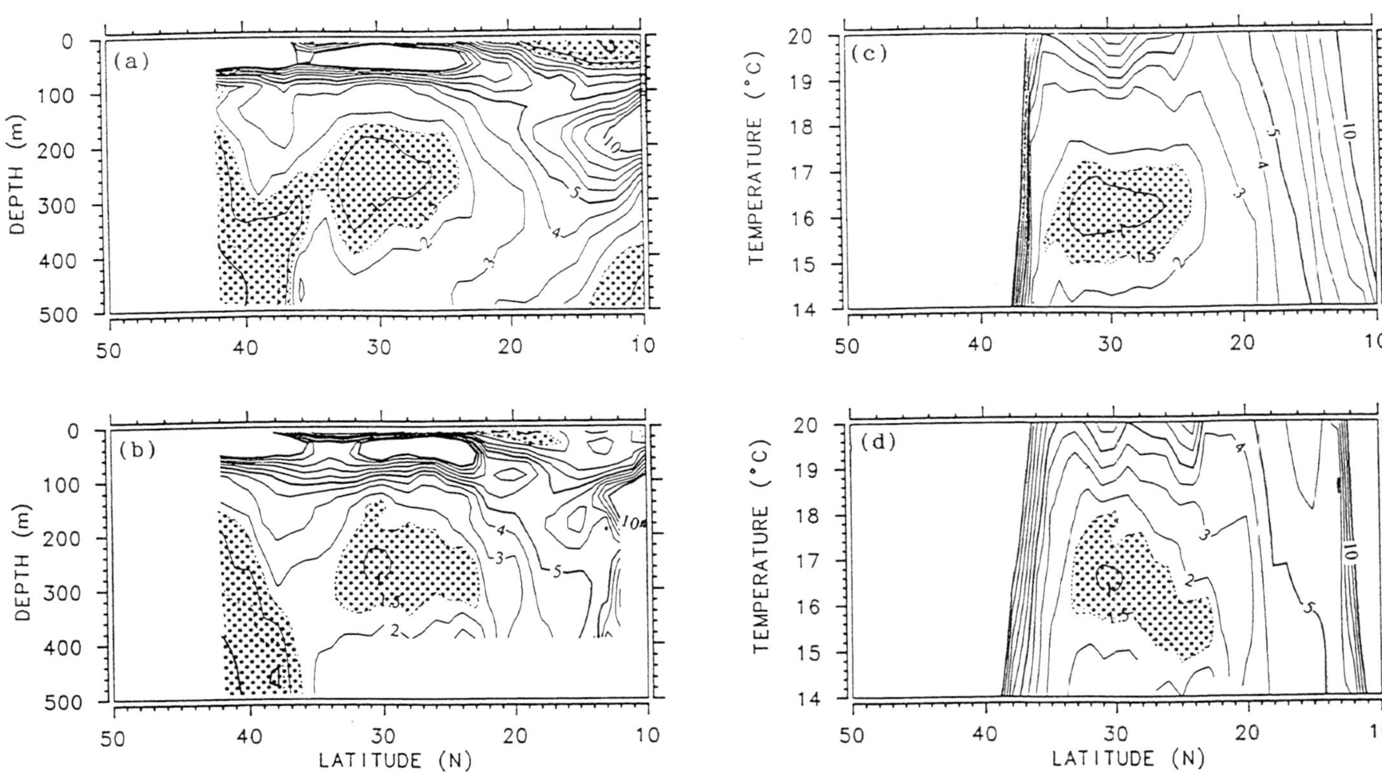

Fig. 11. Cross sections of vertical temperature gradients, by latitude, along 145°E with depth as the vertical coordinate for pentads P34 (*a*) and P78 (*b*); and temperature as the vertical coordinate for pentads P34 (*c*) and P78 (*d*). Contours are °C/100 m. Stippled areas (<1.5°C/100 m) represent the core layer of NPSTMW.

Monsoon Index was larger during P34 than during P78. The index is defined as the difference in sea-level pressure between Kushiro, Japan, and Irkutsk, Russia, and can represent the strength of winter East Asian Monsoon. Changes in the geographic extent of NPSTMW probably resulted from reduced advection by the Kuroshio Current system during P34. One of reasons for the reduced advection may have been the spin-down of the Subtropical Gyre, as mentioned above. The other may be the large meander of the Kuroshio during a larger portion of P34. The meander can cut off advection west of the Izu Ridge.

Summary and Discussion

This report was based on recent studies by the author's group (Bingham et al. 1992; Tanimoto et al. 1993; Watanabe et al. 1993). It described the interannual and interdecadal variations in SST fields and the upper-ocean thermal conditions of the north Pacific. Decadal time-scale signals were found in SST fields, and in the upper-ocean thermal conditions.

So far, the most effort has been expended on clarification of relatively short-term changes in ocean–atmosphere interactions, especially those associated with ENSO events. During the recent decade, our understanding of ENSO events and their impact on extratropical climate has been greatly deepened. However, as shown in this report, there are many time scales in climate variation. Studies on variation with decadal time scales are few, but beginning. The author believes our next target is this time-scale phenomenon. Because the various data sets concerning elements representing oceanic and atmospheric conditions, such as COADS, have been prepared, we now have a good starting point. Clarifications of this time-scale phenomenon will also provide useful information to fisheries scientists.

Acknowledgments

The author thanks Dr. R.J. Beamish and Dr. D. Ware, Canadian Department of Fisheries and Ocean, Pacific Biological Station, Nanaimo, B.C., for inviting him to the symposium, and for their support.

References

Bingham, F.M., T. Suga, and K. Hanawa. 1992. Comparison of upper ocean thermal conditions in the western North Pacific between two pentads: 1938–42 and 1978–82. J. Oceanogr. 48: 405–425.

Hanawa, K. 1991. Long-term variations in the physical environments in the Pacific Ocean, p. 19–28. *In* T. Kawasaki, S. Tanaka, Y. Toba, and A. Taniguchi [ed.] Long-term variability of pelagic fish populations and their environment. Pergamon Press, Oxford.

Hanawa, K., T. Watanabe, N. Iwasaka, T. Suga, and Y. Toba. 1988. Surface thermal condition in the western North Pacific during ENSO events. J. Meteorol. Soc. Jpn. 66: 445–456.

Iwasaka, N., K. Hanawa, and Y. Toba. 1987. Analysis of SST anomalies in the North Pacific and their relation to 500mb height anomalies over the Northern Hemisphere. J. Meteorol. Soc. Jpn. 65: 103–114.

Iwasaka, N., K. Hanawa, and Y. Toba. 1988. Partition of the North Pacific ocean based on similarity in temporal variations of the SST anomaly. J. Meteorol. Soc. Japan 66: 433–443.

Kashiwabara, T. 1987. On the recent winter cooling in the North Pacific. Tenki, 34: 777–781. (in Japanese)

Nitta, T. and S. Yamada. 1989. Recent warming of tropical sea surface temperature and its relationship to the Northern Hemisphere circulation. J. Meteorol. Soc. Jpn. 67: 375–383.

Suga, T., K. Hanawa, and Y. Toba, 1989: Subtropical mode water in the 137°E section. J. Phys. Oceanogr. 19: 1605–1618.

Tanimoto, Y., N. Iwasaka, K. Hanawa, and Y. Toba. 1993. Characteristic variations of sea surface temperature with multiple time scales in the North Pacific. J. Clim. 6: 1153–1160.

Teramoto, T. 1972. History of the Japanese observation program of the Kuroshio and adjacent regions, p. 3–52. *In* H. Stommel and K. Yoshida [ed.] Kuroshio, its physical aspects. University of Tokyo Press, Tokyo, Japan.

Trenberth, K.E. 1990. Recent observed interdecadal climate changes in the Northern Hemisphere. Bull. Am. Meteorol. Soc. 71: 988–993.

Wallace, J.M. and D.S. Gutzler. 1981. Teleconnections in the geopotential height field during the Northern Hemisphere winter. Mon. Weather Rev. 109: 784–812.

Watanabe, T., K. Mizuno, and K. Hanawa. 1993. Preliminary report on interannual variability of the North Pacific thermal conditions, p. 387–388. *In* Book of Abstracts of 1993 Spring Meeting of Japan Oceanographic Society, Tokyo, Japan. (in Japanese)

Weare, B.C., A.R. Navato, and R.E. Newell. 1976. Empirical orthogonal Function analysis of the Pacific Ocean surface temperatures. J. Phys. Oceanogr. 6: 671–678.

White, W.B. and R. Bernstein. 1979. Design of an oceanographic network in the mid-latitude North Pacific. J. Phys. Oceanogr. 9: 592–606.

The effects of climate on the year-class variations of certain freshwater fish species

Hannu Lehtonen and Jyrki Lappalainen

Lehtonen, H., and J. Lappalainen. 1995. The effects of climate on the year-class variations of certain freshwater fish species, p. 37–44. *In* R.J. Beamish [ed.] Climate change and northern fish populations. Can. Spec. Publ. Fish Aquat. Sci. 121.

Abstract: The variations in the year-class strength of three cool-water species, perch (*Perca fluviatilis*), pikeperch (*Stizostedion lucioperca*), and roach (*Rutilus rutilus*), and one cold-water species, whitefish (*Coregonus lavaretus*), were analyzed using the age distribution of catches among 22 populations in the Baltic Sea and one pikeperch population in Lake Lohjanjärvi, southern Finland. Strong year-classes were evident in all the species studied in 1972 and 1979; weak ones were observed in 1981. The year-classes of percid fishes (perch and pikeperch) correlated together as did different geographical sites. Perch and pikeperch year-class strength correlated statistically significantly with the air temperature in the first summer of life, while the year-class strength of roach and whitefish did not. The correlations between year-class strength and wind velocity, precipitation, and the severity of winter were mainly insignificant in the species studied.

Résumé : Les variations dans l'effectif des classes annuelles dans 22 populations de la mer Baltique de trois espèces vivant en eaux fraîches, la perche (*Perca fluviatilis*), le doré (*Stizostedion lucioperca*) et le gardon (*Rutilus rutilus*), et d'une espèce vivant en eaux froides, le corégone (*Coregonus lavaretus*), ainsi que dans une population de dorés du lac Lohjanjärvi, dans le sud de la Finlande, ont été analysées à l'aide de la distribution de l'âge des prises. La présence de classes annuelles fortes était évidente dans toutes les espèces étudiées en 1972 et en 1979; des classes annuelles faibles ont été observées en 1981. Les classes annuelles des percidés (perche et doré) étaient en corrélation les unes avec les autres, de la même façon que les différents lieux géographiques. Les effectifs des classes annuelles des perches et des dorés ont présenté une corrélation statistiquement significative avec la température de l'air au cours du premier été de vie, contrairement à ceux des gardons et des corégones. Dans l'ensemble, les corrélations entre l'effectif d'une classe annuelle et la vélocité du vent, les précipitations et la rigueur de l'hiver n'étaient pas statistiquement significatives dans le cas des espèces étudiées.

Introduction

The role of climate in the regulation of year-class strength is a question of major interest for scientists when estimating the effects of global climatic change on fish populations and fisheries. It has been shown that much of the variation in year-class strength in northern freshwater fish species can be attributed to temperatures during their early life history (Neuman 1976; Svärdson and Molin 1981; Shepherd et al. 1984; Kallemeyn 1987; Regier et al. 1990; Böhling et al. 1991; Buijse et al. 1992; Salojärvi 1992).

Observations of year-class strengths in populations living in varying environments suggest that many warm- and cool-water fish species have stronger and more stable year-classes in warm waters (Karås 1987; Sandström and Svensson 1990). The distribution range of a fish species is dependent on the ability of larvae to survive over the first winter, i.e., on the ability of young fish to attain a minimum amount of growth during the first growth season of life (Shuter and Post 1990).

H. Lehtonen and J. Lappalainen. Finnish Game and Fisheries Research Institute, P.O. Box 202, 00151 Helsinki, Finland.

Growth is highly dependent on temperature and other climatological processes and increased temperatures will increase the growth rate of most fish species, including larvae (Karås and Neuman 1981; Karås 1987; Trippel et al. 1991). Water temperature is also affected by winds, which may have a crucial importance in shallow waters by an open coastline, and hence in the formation of year-class strength (Böhling et al. 1991). On the other side, correlations between climate and year-class strength may be unpredictable owing to interspecific interactions. These include, for example, predation and competition.

Pikeperch, perch, roach, and whitefish are ecologically different subjects on which to test the effects of climate on year-class strength. Pikeperch, perch, and roach are spring spawners with a relatively high temperature preference. The whitefish is a fall spawner, spawning on gravel and stony substrates at the edge of the open sea. The larvae of sea-spawning whitefish live about 2 mo on clean, shallow, sandy shores at depths of less than a 0.5 m (Leskelä et al. 1991). Perch and roach spawn soon after the ice melts in April or May, while pikeperch normally spawn in June, at temperatures of 14–16°C (Erm 1981). Whitefish spawn, in the north, during the second half of October, and in the south, in early November. The eggs develop under the ice and usually hatch in the short period

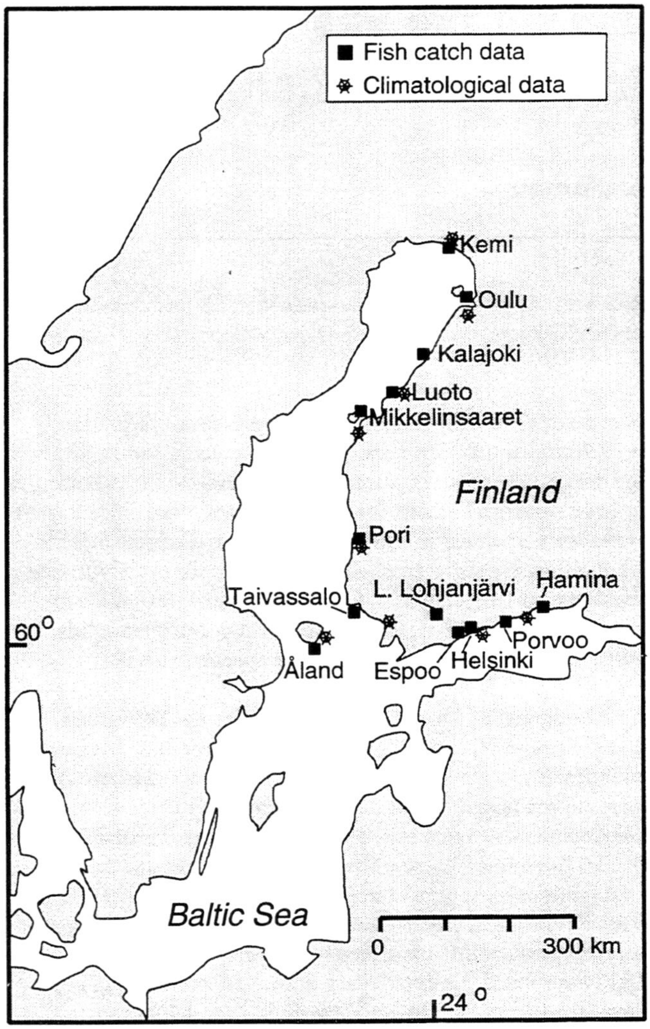

Fig. 1. The study sites.

when the ice melts. Because of these specific differences between the species studied, egg and early larval development occurs in stages over various times and temperatures.

The purpose of this study is to examine the dependence of year-class strength on climatological variables on the basis of fish data collected during the 1970s, 1980s, and 1990s and available meteorological records. The last decade was of particular interest for this study, because the climate in Finland was exceptionally variable.

Study Area

The study area covers the Baltic coastal waters between latitudes 60 and 66°N and Lake Lohjanjärvi in southern Finland (Fig. 1). The Baltic Sea is the largest brackish water basin in the world. In the northern part of the study area, the salinity of the surface water is 0.1–0.3% and in the southwestern coast of Finland 0.5–0.6% (Table 1). There is no tide in the Baltic and the water level depends mostly on winds.

In shallow inlets and bays, higher temperatures (20–25°C) are reached almost every year. The coastal waters are normally ice covered for 3–4 mo in southwestern Finland, and more than 6 mo in the northernmost parts of the Baltic Sea. The

Table 1. Environmental characteristics of the areas investigated.

Area	Character	Salinity (ppt)	Ice cover (d)
Kemi	Open coast	2–3	190
Oulu	Open coast	3	180
Kalajoki	Open coast	3	170
Luoto	Archipelago	3–4	160
Mikkelinsaaret	Archipelago	4	150
Pori	Archipelago	5	130
Taivassalo	Archipelago	6	120
Åland	Archipelago	6	100
Espoo	Archipelago	5	120
Helsinki	Archipelago	5	120
Porvoo	Archipelago	5	120
Hamina	Archipelago	4	130
L. Lohjanjärvi	Lake	0	140

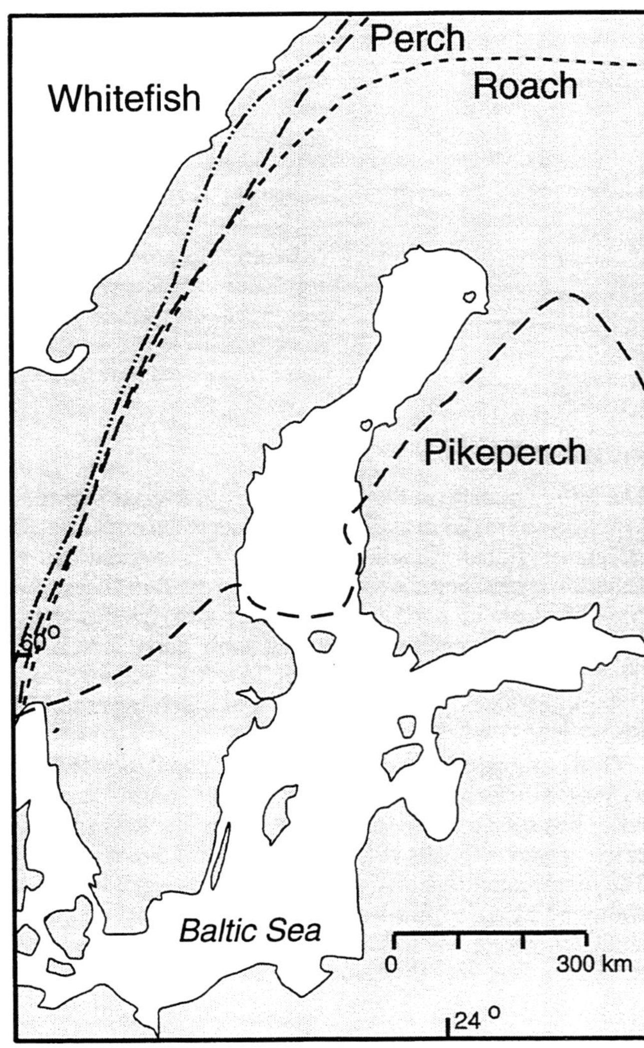

Fig. 2. The northern range of distribution for the species studied.

interannual variations are small in the northern Baltic, while ice-free winters occur outside the archipelago in southwestern Finland. The only lake in this study, Lake Lohjanjärvi, is ice covered an average of 140 d each year (Table 1).

The composition of the fish fauna varies among the areas. Cold-water species, such as salmonids and coregonids, are more common and warm- and cool-water species, such as cyprinids and percids, are less common in the northern parts of the study area. Pikeperch (*Stizostedion lucioperca*) does not occur in the northern part of the study area, whereas the other species, perch (*Perca fluviatilis*), roach (*Rutilus rutilus*) and whitefish (*Coregonus lavaretus*), all live in each of the study sites (Fig. 2).

Material and Methods

The relative strength of each year-class was analyzed from 22 freshwater fish populations representing three cool-water species (pikeperch, perch, and roach) and one cold-water species (sea-spawning whitefish). Material was collected from both commercial whitefish and pikeperch catches and from our own test fishing. Whitefish and pikeperch were sampled using gill nets; perch, using wire traps and multimeshed gill nets; and roach, using wire traps and angling. Altogether 42 789 fish were sampled, measured, and aged. Samples were taken mainly during the spawning period.

Pikeperch, roach, and whitefish were aged from scales and perch from opercular bones. The relative year-class strength was then analyzed retrospectively on the basis of yearly age distributions in catch samples. Years having small sample sizes were omitted. The year-class strength was calculated in almost all cases from the annual catch sample of 100–300 fish.

The indices of year-class strengths were calculated according to the method of Svärdson (1961), and adjusted following Neuman (1974). The prerequisite for this method is that all samples are caught using the same method in different years. The selectivity of the gear has no importance provided that it is the same throughout the study period. Calculations were restricted to fishes that were old enough to be recruited to the sampling gear and to ages that were reasonably well represented in the samples. The year-class strength were estimated stepwise beginning with the calculation of the age distribution percentage in the yearly samplings. Thereafter, the mean age distribution percentage for the whole period was established. In the next step, the different year-classes in different years were expressed as percentages of this mean distribution. Thus, if 5 yr old in 1985 (1980 year-class) constituted 20% of the 4- to 7-yr-old fish compared with 40% in the mean distribution in the actual area, they reached 50% of the average for this year. If the corresponding value was 70% in 1986 (6 yr), 60% in 1987 (7 yr), and 60% in 1988 (8 yr), the index of the strength of the year-class 1980 would be 60%.

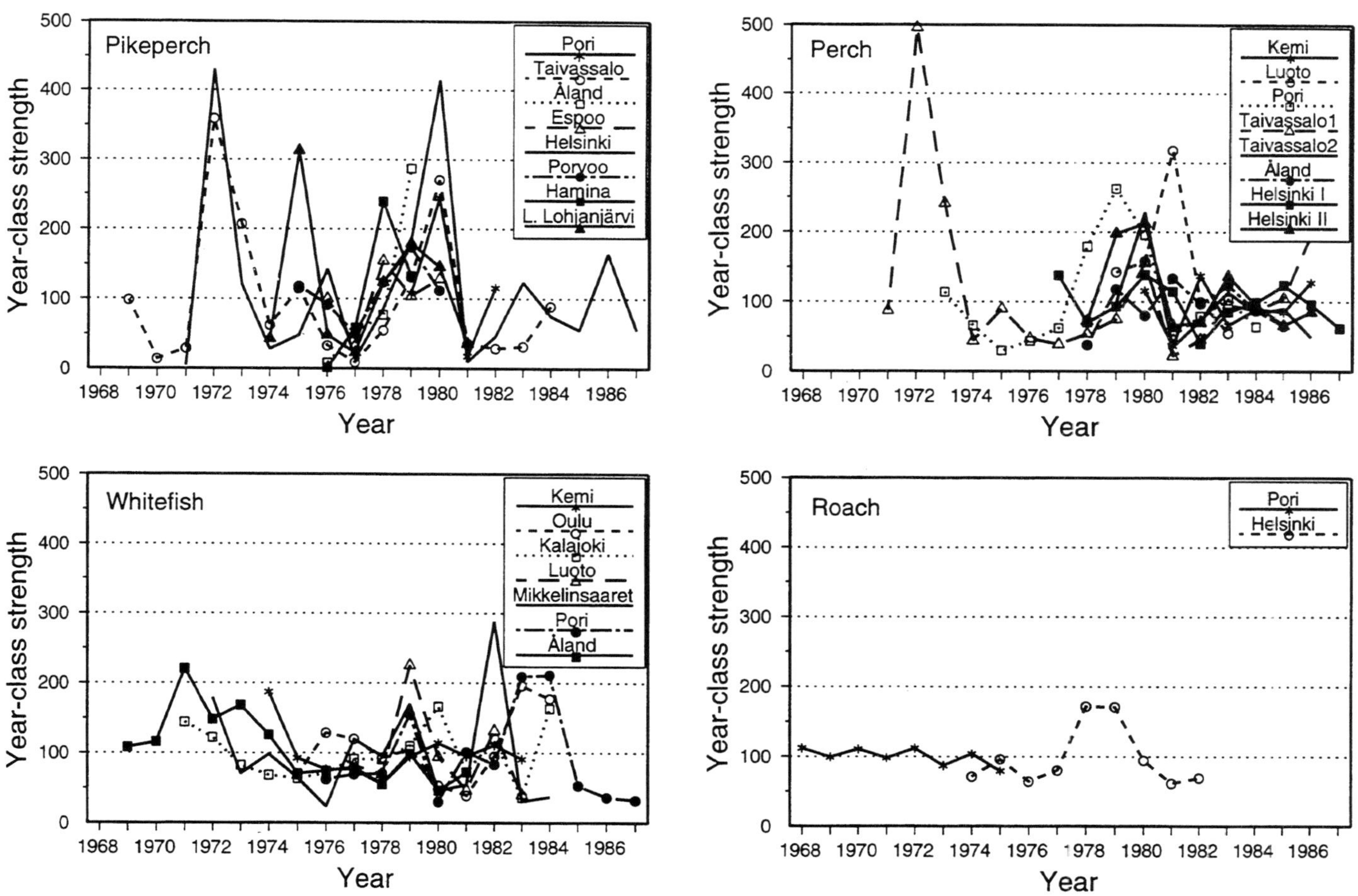

Fig. 3. The year-class variations in species studied.

Only year-class indices based on at least 2 yr were used. With this method, the average year-class strength in a population is 100. A year-class was considered strong when the index was over 130 and weak when it was less than 70.

The year-class variation and the climate data of different locations were analyzed with the Pearson correlation analysis (SAS Institute Inc. 1985). The year-class data were also grouped by species to study the effect of first summer temperatures on year-class strength. Because the distributions of grouped data were approximately lognormal, a logarithmic transformation (Sokal and Rohlf 1981) was applied.

The meteorological data were obtained from the Finnish Meteorological Institute. The variables analyzed were daily mean air temperature (degrees Celsius), wind velocity (metres per second), and precipitation (millimetres). The data on the formation, duration, and melting of the ice cover were obtained from the publications of the Institute of Marine Research (Palosuo 1971; Kalliosaari 1978, 1982; Kalliosaari and Seinä 1987; Seinä and Kalliosaari 1991). The ice data were correlated with the first winter of pikeperch, perch, and roach larvae. Variations in year-classes of whitefish were also correlated with the winter the eggs were developing.

Table 2. Strong and weak year-classes in the species studied. The years used for estimation of the year-class strength are in parentheses.

Species	Strong year-classes	Weak year-classes
Pikeperch (1969–1987)	1972, 1973, 1979, 1980, 1986	1977, 1981
Perch (1971–1986)	1972, 1973, 1979, 1980	1974, 1977, 1981
Whitefish (1971–1987)	1971, 1972, 1979, 1982, 1984	1975, 1981
Roach (1968–1982)	1978, 1979	1976, 1977, 1981

Results

Strong and weak year-classes

Year-classes of all species in 1972 and 1979 were generally strong (Fig. 3). Weak year-classes appeared in 1977 and 1981, except for whitefish in 1977 (Table 2). In percid fish (pikeperch and perch), strong year-classes appeared simultaneously. In Lake Lohjanjärvi, a strong year-class also appeared in 1975. The perch year-class in 1981 was strong in Luoto and Åland, although it was clearly below average in the other six sites. Variations in whitefish year-class strengths between different populations were much smaller than in percid fish. The

Table 3. Significant correlations between species and locations. N = number of year-classes.

Species and location		r	p	N
Within species				
Perch, Taivassalo[a]	vs. perch, Helsinki II	0.79	<0.05	8
Pikeperch, Taivassalo	vs. pikeperch, Åland	0.97	<0.05	4
Pikeperch, Taivassalo	vs. pikeperch, Helsinki	0.87	<0.001	13
Pikeperch, Åland	vs. pikeperch, Helsinki	0.98	<0.05	4
Whitefish, Oulu	vs. whitefish, Pori	0.74	<0.05	9
Whitefish, Pori	vs. whitefish, Åland	0.85	<0.05	6
Between species				
Roach, Helsinki	vs. perch, Pori	0.81	<0.01	9
Whitefish, Oulu	vs. pikeperch, Åland	−0.96	<0.05	4
Whitefish, Oulu	vs. pikeperch, Pori	0.98	<0.05	4
Whitefish, Kalajoki	vs. pikeperch, Åland	0.98	<0.05	4
Perch, Taivassalo[b]	vs. pikeperch, Taivassalo	0.88	<0.001	14
Perch, Taivassalo[b]	vs. pikeperch, Helsinki I	0.74	<0.01	16
Perch, Taivassalo[b]	vs. pikeperch, Åland	0.99	<0.01	4
Perch, Taivassalo[a]	vs. pikeperch, Taivassalo	0.88	<0.05	6
Perch, Taivassalo[a]	vs. pikeperch, Helsinki I	0.88	<0.01	8
Perch, Helsinki I[c]	vs. pikeperch, Helsinki II	0.84	<0.01	9
Perch, Helsinki I[c]	vs. pikeperch, Hamina	0.99	<0.01	4
Perch, Helsinki II[d]	vs. pikeperch, Taivassalo	0.82	<0.05	7

Taivassalo[a] = Fyke net.
Taivassalo[b] = Wire trap.
Helsinki I[c] = Bay of Vanhankaupunginlahti.
Helsinki II[d] = Bay of Vartiokylänlahti.

variation in the year-class strength of roach in Pori was small and no exceptionally strong or weak year-classes were detected during the period 1968–1975.

Correlations between species and locations

Areas situated nearby had the strongest correlations among year-classes. There was a statistically significant correlation between the pikeperch year-classes of Taivassalo and Helsinki ($p = 0.0001$), and Taivassalo and Åland ($p = 0.0292$) (Table 3). There were no significant correlations between the year-class variations of the pikeperch in Porvoo or Lake Lohjanjärvi and other populations studied. A statistically significant correlation among perch year-classes was found between Taivassalo and Helsinki II ($p = 0.0193$). In whitefish, statistically significant positive correlations were found between Pori and Oulu ($p = 0.0231$) and Pori and Åland ($p = 0.0311$).

The year-class strengths of different species studied were mainly positively correlated in different locations (Table 3). The strongest correlation between perch and pikeperch was in Taivassalo. The variations in roach year-classes differed clearly from the other species studied. The only correlation found was between roach in Helsinki and perch in Pori.

Correlations between environmental factors and year class strength

The correlations with ice cover, precipitation, and wind were, in general, insignificant, while many positive correlations with summer temperature during the first summer of life were obvious, especially in perch and pikeperch (Table 4). There were, however, some exceptions (pikeperch in Porvoo and Hamina). Roach year-classes were correlated with the

Table 4. Probabilities for correlations between air temperature data and year-class strength. Negative correlations are marked with (-). ns = Not significant.

Area	*N*	May	June	July	August	Sepember	October	May–September	May–October	June–August
Whitefish										
Kemi	10	ns	ns	ns	ns	<0.05	ns	<0.10	<0.05	
Oulu	10	ns	ns	ns	ns	ns	ns	ns	ns	
Kalajoki	14	ns	ns	ns	ns	ns	ns	ns	ns	
Luoto	7	ns	ns	ns	<0.05	ns	<0.01(-)	ns	ns	
Mikkelinsaaret	13	<0.10	ns	ns	<0.10	ns	ns	ns	ns	
Pori	12	0.05	ns	ns	ns	ns	ns	<0.10	<0.10	
Åland	13	ns	ns	ns	ns	ns	ns	ns	ns	
Pikeperch										
Pori	4	ns	ns	ns	ns	ns				<0.10
Taivassalo	16	ns	<0.05	<0.01	<0.10	ns				<0.001
Åland	4	ns	<0.10	<0.01	ns	ns				<0.05
Helsinki	17	ns	<0.10	<0.10	ns	ns				<0.01
Porvoo	6	ns	ns	ns	ns	ns				ns
Hamina	4	ns	ns	ns	<0.10	ns				ns
L. Lohjanjärvi	8	ns	ns	ns	<0.05	ns				<0.01
Perch										
Kemi	7	ns	ns	ns	ns	ns				ns
Luoto	5	ns	ns	ns	ns	ns				ns
Pori	12	ns	<0.05	ns	ns	ns				ns
Taivassalo[a]	16	ns	<0.05	<0.01	ns	ns				<0.001
Taivassalo[b]	8	ns	ns	ns	ns	<0.10				<0.01
Åland	8	ns	ns	ns	ns	ns				ns
Helsinki I[c]	11	ns	<0.10	ns	ns	ns				ns
Helsinki II[d]	9	ns	<0.10	ns	ns	ns				<0.05
Roach										
Pori	8	ns	<0.10	ns	ns	ns				ns
Helsinki	9	ns	<0.10	ns	ns	ns				ns

[a]=Fyke net.
[b]=Wire trap.
[c]=Bay of Vanhankaupunginlahti.
[d]=Bay of Vartiokylänlahti.

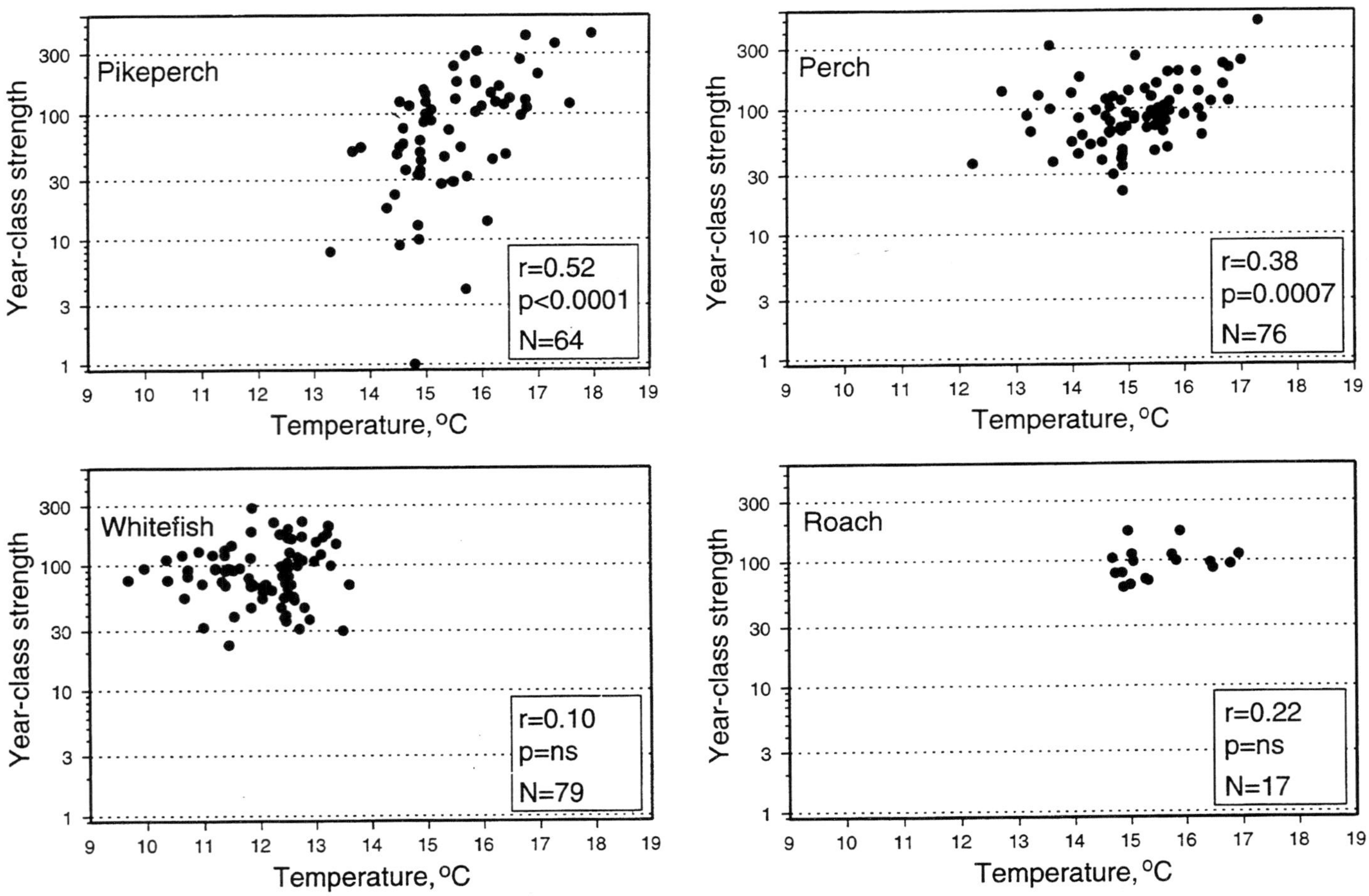

Fig. 4. The correlation between summer temperatures and year-class strengths. Temperature is the mean air temperature (°C) from June to August with the exception of May to September for whitefish.

temperature in June. Whitefish year-classes did not correlate with temperature, and there were dicrepancies in the correlations that did appear.

The grouped year-class data showed the importance of first summer temperatures on pikeperch and perch (Fig. 4). The same was not observed with roach or whitefish.

Discussion

The dependence of the year-class strength of perch and pikeperch on temperature during the first months of life was clearly detected in this study. Post-hatching temperature is likely the most important abiotic factor controlling the growth of fish larvae and is, therefore, of crucial importance to the survival of the larvae (McCauley and Kilgour 1990). Even a brief and sudden food shortage may cause starvation and possibly lead to losses in larval density (McLean et al. 1981; Willemsen 1983; Viljanen 1988; Johnson and Evans 1991). Perch and pikeperch are inactive and their growth ceases when the water temperature is below 10°C (Karås 1987, 1990).

In the Baltic Sea, sudden changes in temperature may occur because of upwellings. Temperature drops of more than 10°C on an open coast have been observed within 1 h. The importance of winds on the formation of perch year-class strengths has been observed when long-term southwesterly winds have dominated and caused differences between the western and eastern coasts of the Gulf of Bothnia (Böhling et al. 1991). The importance of winds on the year-class strength of roach was emphasized also by Ponton and Gerdeaux (1987).

The nursery areas of the fish populations studied are, however, in sheltered bays, inlets, estuaries, or archipelagoes. Their water temperature follow air temperature fairly consistently (McCombie 1959), provided that long-term winds do not disturb the situation. Therefore, the correlations between year-class strength and air temperature were positive in almost all cases. The nonsignificant relationships between wind velocity or precipitation and year-class variations may be a consequence of the fact that only monthly mean values were available. All sampling areas except whitefish areas in Kemi, Oulu, and Kalajoki were situated in sheltered areas, which reduces the importance of wind as well.

Although northern fish populations inhabit waters encompassing a wide range of temperatures, each species has a narrower range in which their survival and growth are maximized. Perch and pikeperch have similar environmental temperature gradients, and this may explain the strong covariation in their year-class strength. Roach and whitefish indicate a different pattern, which cannot be explained by the temperature alone. A deviating variation in roach year-classes might be caused by the predation of strong perch and pikeperch year-classes (Raat 1990). Althought hatching later, pikeperch grow more rapidly and roach yearlings are vulnerable to pikeperch predation already in their first autumn (Van Densen and Grimm 1988; Buijse and Houthuijzen 1992).

Whitefish year-classes were least dependent on temperature in the fish species studied. According to a review by Salojärvi (1992), the most important regulating mechanisms for European whitefish are the abundance of food and the number of possible competitors and predators. However, there are also indications that high spring and summer temperatures have positive effects on year-class strength (Hudd et al. 1988; Trippel et al. 1991).

The shortcomings involved in correlation analysis in this kind of study are obvious. The spurious correlations were minimized by presupposing that the year-class strengths should vary in a similar pattern over a larger geographical range (Shephard et al. 1984; Böhling et al. 1991). We also suggest that the correlations between the year-class strength and environmental factors were meaningless unless the same relation could also be shown in nearby populations within the species studied.

In conclusion, the present data showed that temperature alone affects the variation in the year-class strength of perch and pikeperch. Whitefish and roach year-classes are affected by other factors as well and their year-class strengths cannot be forecasted exclusively on the basis of water temperature. We cannot, however, exclude the importance of climate when forecasting for the year-classes of these species. Strong simultaneous year-classes occurred in all species studied in 1972 and 1979 and weak ones in 1981. The summers of 1977 and 1981 were the coldest while 1972 was the warmest during the study period. Therefore, more detailed research is still needed to shed light on the possible cause and effect relationship between temperature and year-class variations.

Acknowledgements

We thank Karl Sundman, Kari Hietanen, Irmeli Torvi, and Alpo Huhmarniemi for the age determinations and the Department of Sports and Recreation of the City of Helsinki for the roach data. Richard Hudd and Mika Rahikainen provided some of the perch material. The study was partially financed by the climate change programme (SILMU) of the Academy of Finland and the Finnish Foundation for Natural Resources. The English language was corrected by Kathleen Tipton.

References

Böhling, P., R. Hudd, H. Lehtonen, P. Karås, E. Neuman, and G. Thoresson. 1991. Variations in year-class strength of different perch (*Perca fluviatilis*) populations in the Baltic Sea with special reference to temperature and pollution. Can. J. Fish. Aquat. Sci. 48: 1181–1187.

Buijse, A.D., W.L.T. Van Densen, and L.A. Schaap. 1992. Year-class strength of Eurasian perch (*Perca fluviatilis* (L.)) and pikeperch (*Stizostedion lucioperca* (L.)) in relation to stock size, water temperature and wind in Lake Ijssel, the Netherlands, 1966–1989, p. 31–69. *In* Buijse, [ed.] A. D. Dynamics and exploitation of unstable percid populations. Landbouwuniversiteit, Wageningen, The Netherlands.

Buijse, A.D., and R.P. Houthuijzen. 1992. Piscivory, growth and size-selective mortality of age 0 pikeperch, *Stizostedion lucioperca* (L.). p. 71–87. *In* Buijse, A.D. [ed.] Dynamics and exploitation of unstable percid populations. Landbouwuniversiteit, Wageningen, The Netherlands.

Erm, V. 1981. Pikeperch. Tallinn Valgus. 127 p. (In Estonian)

Hudd, R., H. Lehtonen, and I. Kurttila. 1988. Growth and abundance of fry; factors which influence the year-class strength of whitefish (*Coregonus widegreni*) in the southern Bothnian bay (Baltic). Finn. Fish. Res. 9: 213–220.

Johnson, T.B., and D.O. Evans. 1991. Behaviour, energetics, and associated mortality of young-of-the-year white perch (*Morone americana*) and yellow perch (*Perca flavescens*) under simulated winter conditions. Can. J. Fish. Aquat. Sci. 48: 672–680.

Kallemeyn, L.W. 1987. Correlations of regulated lake levels and climatic factors with abundance of young-of-the-year walleye and yellow perch in four lakes in Voyageurs National Park. N. Am. J. Fish. Manage. 7: 513–521.

Kalliosaari, S. 1978. Ice winters 1971–75 along the Finnish coast. Finn. Mar. Res. 245: 1–63.

1982. Ice winters 1976–80 along the Finnish coast. Finn. Mar. Res. 249: 3–63.

Kalliosaari, S., and A. Seinä. 1987. Ice winters 1981–85 along the Finnish coast. Finn. Mar. Res. 254: 3–63.

Karås, P. 1987. Food consumption, growth and recruitment in perch (*Perca fluviatilis* L.). Acta Univ. Ups. 108: 1–19.

Karås, P. 1990. Seasonal changes in growth and standard metabolic rate of juvenile perch, *Perca fluviatilis* L. J. Fish Biol. 37: 913–920.

Karås, P., and E. Neuman. 1981. First-year growth of perch (*Perca fluviatilis* L.) and roach (*Rutilus rutilus* (L.)) in a heated Baltic bay. Rep. Inst. Freshwater Res. Drottningholm 59: 48–63.

Leskelä, A., R. Hudd, H. Lehtonen, A. Huhmarniemi, and O. Sandström. 1991. Habitats of whitefish (*Coregonus lavaretus* (L.) s.l.) larvae in the Gulf of Bothnia. Aqua Fenn. 21: 145–151.

McCauley, R.W., and M. Kilgour. 1990. Effect of air temperature on growth of largemouth bass in North America. Trans. Am. Fish. Soc. 119: 276–281.

McCombie, A.M. 1959. Some relations between air temperature and surface water temperatures of lakes. Limnol. Oceanogr. 4: 252–258.

McLean, J.A., B.J. Shuter, H.A. Regier, and J.C. MacLeod. 1981. Temperature and year-class strength of smallmouth bass. Rapp. P.-v. Réun. Cons. int. Explor. Mer 178: 30–40.

Neuman, E. 1974. The effects of temperature on the growth of perch (*Perca fluviatilis* L.) in some Baltic archipelagoes. Inf. Sötvattenslab. Drottningholm 6: 104 p.

Neuman, E. 1976. The growth and year-class strength of perch (*Perca fluviatilis* L.) in some Baltic archipelagoes, with special reference to temperature. Rep. Inst. Freshwater Res. Drottningholm 55: 51–70.

Palosuo, E. 1971. Ice winters 1966-70 along the Finnish coast. Merentutkimuslaitoksen Julk. 234: 1–63.

Ponton, D., and D. Gerdeaux. 1987. Roach (*Rutilus rutilus* (L.)) in Lake Geneva in 1983–85. Age structure, recruitment and growth. Bull. Fr. Piscic. 305: 43–53. (In French with an English summary)

Raat, A.J.P. 1990. Production, consumption and prey availability of northern pike (*Esox lucius*), pikeperch (*Stizostedion lucioperca*) and European catfish (*Silurus glanis*): a bioenergetics approach. Hydrobiologia, 200/201: 497–509.

Regier, H.A., J.A. Holmes, and D. Pauly. 1990. Influence of temperature changes on aquatic ecosystems: an interpretation of empirical data. Trans. Am. Fish. Soc. 119: 374–389.

Salojärvi, K. 1992. The role of compensatory processes in determining the yield from whitefish (*Coregonus lavaretus* L. s.l.) stocking in inland waters in northern Finland. Finn. Fish. Res. 13: 1–30.

Sandström O., and B. Svensson. 1991. Biological effects of cooling waters. Studies in Biotestbasin, Forsmark, 1984–1988. 55 p. Grafik Prab Stockholm, Sweden (In Swedish with an English summary)

SAS Institute Inc. 1985. Procedures guide for personal computers, version 6 edition. SAS Institute Inc., Cary, N.C. 373 p.

Seinä, A., and S. Kalliosaari. 1991. Ice winters 1986–1990 along the Finnish coast. Finn. Mar. Res. 259: 3–61.

Shepherd, J.G., J.G. Pope, and R.D. Cousens. 1984. Variations in fish stocks and hypotheses concerning their links with climate. Rapp. P.-v. réun. Cons. int. Explor. Mer 185: 255–267.

Shuter, B.J., and J.R. Post. 1990. Climate, population viability, and the zoogeography of temperate fishes. Trans. Am. Fish. Soc. 119: 314–336.

Sokal, R.R., and F.J. Rohlf. 1981. Biometry. 2nd ed. W.H. Freeman & Co., San Francisco. 859 p.

Svärdson, G. 1961. No effect of whitefish cultivation in Kalmarsund. Sven. Fisk. Tidskr. 70: 23–26. (In Swedish)

Svärdson, G., and G. Molin. 1981. The impact of eutrophication and climate on a warmwater fish community. Rep. Inst. Freshwater Res. Drottningholm 59: 142–151.

Trippel, E.A., R. Eckmann, and J. Hartmann. 1991. Potential effects of global warming on whitefish in Lake Constance, Germany. Ambio, 20: 226–231.

Van Densen, W.L.T., and M.P. Grimm. 1988. Possibilities for stock enhancement of pikeperch (*Stizostedion lucioperca*) in order to increase predation on planktivores. Limnologica, 19: 45–49.

Viljanen, M. 1988. Relations between egg and larval abundance, spawning stock and recruitment in vendace (*Coregonus albula* L.). Finn. Fish. Res. 9: 271–289.

Willemsen, J. 1983. Biology and management of pikeperch, p. 115–125. *In* Proceedings, 3rd British Freshwater Conference.

Atmospheric teleconnections and coherent fluctuations in recruitment to North Atlantic cod (*Gadus morhua*) stocks

S.N. Rodionov

Rodionov, S.N. 1995. Atmospheric teleconnections and coherent fluctuations in recruitment to North Atlantic cod (*Gadus morhua*) stocks, p. 45–55. *In* R.J. Beamish [ed.] Climate change and northern fish populations. Can. Spec. Publ. Fish Aquat. Sci. 121.

Abstract: The concept of "biological teleconnections" is considered, based on the analysis of year-classes of 10 cod (*Gadus morhua*) stocks in the northern North Atlantic and European Arctic seas. It is shown by means of factor analysis that a significant part of the total variance of recruitment to the cod stocks may be explained by two factors. The first factor reflects the west–east opposition in year-class strength and appears to be related to atmospheric teleconnections. In fact, all the analyzed teleconnection patterns (four at sea level and three at the 500-hPa level) carry a signature of out of phase fluctuations between sea-surface temperature – surface-air temperature anomalies in the northwest Atlantic and northern European seas. The second factor exhibits north–south opposition in year-class strength and is probably associated with the different reaction of the cod stocks during periods of general warming and cooling in the North Atlantic.

Résumé : On examine le concept de «télécorrespondances biologiques», en s'inspirant de l'analyse de classes annuelles de 10 stocks de morues de l'Atlantique (*Gadus morhua*) dans l'Atlantique Nord et dans les mers arctiques de l'Europe. On a constaté, grâce à l'analyse des facteurs, qu'une partie importante de la variance totale du recrutement des stocks de morues s'explique par deux facteurs. Le premier traduit l'opposition ouest-est dans l'effectif des classes annuelles et semble être lié aux télécorrespondances atmosphériques. En fait, toutes les télécorrespondances analysées (quatre au niveau de la mer et trois au niveau de 500 kPa) portent la signature de fluctuations déphasées entre les anomalies des températures de la surface de la mer et de l'air à la surface dans le nord-ouest de l'Atlantique et dans le nord des mers européennes. Le deuxième facteur montre l'opposition nord-sud dans l'effectif des classes annuelles et est probablement associé à la réaction différente des stocks de morues durant les périodes de réchauffement et de refroidissement généraux dans l'Atlantique Nord.

Introduction

In meteorology, the term "teleconnections" refers to patterns of significant simultaneous correlations between weather and climatic anomalies at widely separated "centers of action." The first evidence of global teleconnections emerged from analyses of sea-level pressure (SLP) and surface-air temperature (SAT) data in the early decades of the 20th century (Walker and Bliss 1932). A vast body of literature now exists that confirms the reality of teleconnections not only in SLP (Barnett 1985) and SAT (Girskaya 1981), but also in sea-surface temperature (SST) (Hsiung and Newell 1983), geopotential heights (Gambo and Kudo 1983), cloudiness (Dvoryaninov and Goloborod'ko 1989), precipitation (Lau and Sheu 1991), and other meteorological and oceanographic fields.

S.N. Rodionov. National Center for Atmospheric Research, P.O. Box 3000, Boulder, Colorado 80307-3000, USA. The National Center for Atmospheric Research is sponsored by the National Science Foundation.

Linkages between fluctuations in fish stocks that occur over relatively large distances in response to climatic processes was first studied in the pioneering work of Izhevskii (1961). Proceeding from the philosophical idea of the universal interrelation between natural processes on Earth, he suggested a new concept for fisheries oceanography based on existence of so-called "systems of nature." According to Izhevskii, each of these systems represents a complex of mutually dependent physical, chemical, and biological processes fluctuating as one phase within the system and out of phase in relation to the correspondent processes in neighbouring systems. On the basis of his subjective analysis of linear temporal correlation in station air-temperature data and the position of the oceanic polar front, he found two such systems in the North Atlantic sector, referred to as Greenland – North American and Atlantic (Fig. 1). To the east of the Atlantic system, he placed the Continental system that includes the Black, Caspian, and Aral Seas. In a later study, Izhevskii (1964) noted the possible existence of two more systems, which he named as East Siberian and Pacific, according to their geographical positions.

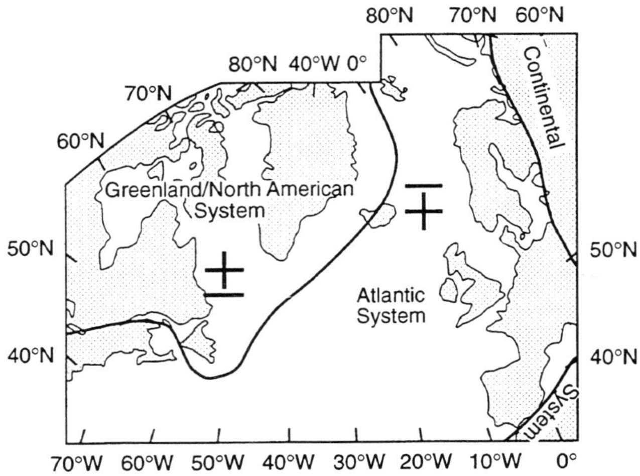

Fig. 1. Izhevskii's "systems of nature" in the North Atlantic sector. Plus and minus signs indicate out of phase fluctuations in neighbouring systems.

As a confirmation of out of phase biological processes in the Greenland – North American and Atlantic systems, Izhevskii showed that strong year-classes of the west of Greenland cod (*Gadus morhua*) often correspond to poor year-classes of cod spawning in the Norwegian Sea and vice versa. A similar negative correlation was found between Arcto-Norwegian and Newfoundland cod stocks. As for the Iceland cod, according to Izhevskii, its population dynamics are controlled by the processes in the Greenland – North American system. Therefore, it should fluctuate in phase with the Newfoundland and west of Greenland cod and out of phase with the Arcto-Norwegian cod.

In a recent paper, Elizarov et al. (1990) found some evidence confirming Izhevskii's hypothesis on out of phase fluctuations in the Greenland – North American and Atlantic systems. The opposition in cod fluctuations turned out to be most pronounced between cod stocks around the British Isles and Newfoundland stocks located near the southern boundary of the Greenland – North American system. Some indications of out of phase fluctuations were also noted for the North Sea and west of Scotland cod in relation to the west of Greenland cod.

There are also other viewpoints on the question of teleconnections in fish stocks. It has often been suggested that successful year-classes occur in the same year in widely separated areas (Martin and Kohler 1965; Cushing 1982; Kawasaki 1991). Templeman (1965) showed that there were periods of 2 or 3 successive years when successful year-classes of cod, haddock (*Melanogrammus aeglefinus*), and herring (*Clupea harengus*) were produced simultaneously off Greenland, Iceland, and in the Norwegian – Barents Sea area. There were also intervening years or groups of years when successful year-classes were relatively scarce.

In a later paper, Templeman (1972) pointed out that successful year-classes of cod in the Northeast Atlantic and European Arctic often occurred 1 yr after good year-classes of cod in the Sable Island to southern Grand Bank area and on Georges Bank, or 1 or 2 yr after a good year-class of cod on Browns Bank. Templeman explained this phenomenon by sequences of favourable hydrological and meteorological events developed from the southern Grand Bank to Iceland, to the Barents Sea. The time lag in this succession of good year-classes was matched with the time required for surface drift of temperature anomalies across the North Atlantic.

Garrod and Colebrook (1978) extended Templeman's approach considerably with 18 stocks (including gadoids) for 22 yr, from 1950 to 1972. They found it possible to distinguish two areas, the Arctic rim and the oceanic Atlantic, where the biological effects of environmental variation are different.

Unlike Garrod and Colebrook, Shepherd et al. (1984) found no strong correlation between fish stocks in distinctly separated geographical areas. They considered correlations between recruitment to 18 fish stocks in the Northeast Atlantic and European Arctic and concluded that the majority of significant correlations occurred between neighbouring stocks. They also determined that correlations between remote areas were barely more than could be expected by chance.

Dragesund (1971) also noted no teleconnections when he carried out a comparative analysis of year-class strength in self-contained stocks of abundant fish species (herring, cod, haddock, saithe (*Pollachius virens*), redfish (*Sebastes spp.*), and capelin (*Mallotus villosus*)) in the North Atlantic. He concluded that even in neighbouring regions no apparent covariation could be found.

As for the Pacific Ocean, Kawasaki (1991) demonstrated that three sardine (*Sardinops sagax*) populations, namely the Far Eastern sardine around Japan, the California sardine off North America, and the Chilean sardine off South America, have repeated "coincidental, enormous amplitude and long-period fluctuations." At the same time, the Pacific herring (*Clupea pallasi*) and the Atlantic herring have shown the fluctuation patterns similar to, but completely out of phase with, those of sardine (Kawasaki 1991).

It is also worth noting the apparent opposition in oceanographic conditions between the Northeast Atlantic and the Bering Sea. During the past several decades, there were three distinct climatic regimes in the Northeast Atlantic. A cold regime of the 1960s, particularly notable by the end of the decade, was followed by a warm regime of the 1970s, which in turn was replaced by a new cold regime since 1977 (Rodionov and Krovnin 1992). Similar regimes, but of opposite signs, occurred in the Bering Sea (Rodionov and Krovnin 1991). In both regions, changes in multiyear climatic regimes had a significant effect on fish populations. For example, during the early 1970s, when the climate was warm in the Northeast Atlantic, recruitment to the Arcto-Norwegian cod increased (Fig. 2). Particularly strong was the 1970 year-class, which, according to data from Dragesund (1971) and the International Council for the Exploration of the Seas (ICES) (1990), was the strongest year-class for the entire period since 1931, when the data became available. On the other hand, recruitment of walleye pollock (*Theragra chalcogramma*) in the eastern Bering Sea, the other gadoid stock, was relatively poor during the 1970–76 period (except the 1972 year-class) compared with that in the 1960s and late 1970s (Rodionov and Krovnin 1991). This example provides strong supporting evidence for the hypothesis that the driving force of "biological teleconnection" is climate operating on a global scale. The fact that the regions are well separated suggests that the linkages are not mediated through the ocean but rather have their origin in the atmosphere.

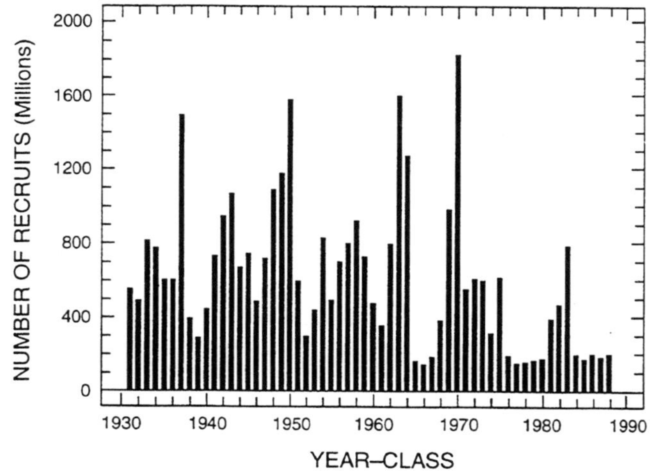

Fig. 2. Recruitment to the Arcto-Norwegian cod stock, 1931–1988.

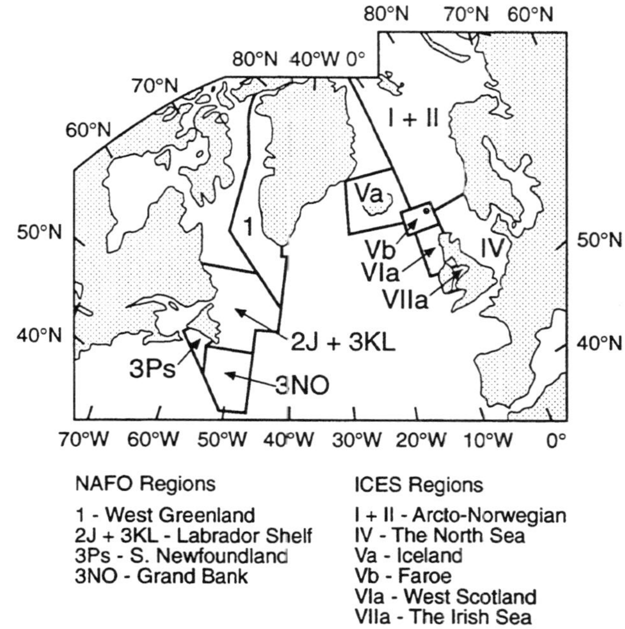

Fig. 3. Location of the fisheries regions (NAFO and ICES classification).

Unlike atmospheric teleconnections, whose theoretical and observational aspects have been extensively examined in numerous publications, the concept of biological teleconnections is only at the beginning stage of exploratory analysis. The objective of this study is to explore the existence of biological teleconnections through analysis of 10 cod stocks in the North Atlantic region, and compare them with known atmospheric teleconnection patterns.

Data

To determine the indices of atmospheric teleconnection patterns, monthly mean SLP and 700-hPa geopotential heights in grids 5° latitude by 10° longitude were used (available at the National Center for Atmospheric Research (NCAR)). The SLP indices were computed for the period from 1924 to 1990, and the 700-hPa height indices from 1948 to 1990. All the calculations were made only for winter months (December through February) because many teleconnection patterns are most pronounced in the cold season. When translating the atmospheric teleconnections to the SAT and the sea-surface temperature (SST) the data set prepared by P.D. Jones (University of East Anglia, Norwich, UK.) and also available at NCAR was used. This data set contains gridded (5° resolution) SAT (over land) and SST (over ocean) anomalies, based on the 1950–79 means; the period used was from 1948 to 1990.

Ten cod stocks throughout the North Atlantic were examined (Fig. 3). In the Northwest Atlantic (Northwest Atlantic Fisheries Organization (NAFO) area), these are: west of Greenland (or subarea 1, according to NAFO classification; 1924–88 year-classes), Labrador shelf (division 2J + 3KL; 1962–80), Grand Bank (division 3NO; 1959–88), and south of Newfoundland (division 3Ps; 1959–84). In the Northeast Atlantic and European Arctic (ICES area), these are: Arcto-Norwegian (subareas I and II, according to ICES classification; 1931–88), the North Sea (subarea IV; 1968–87), Iceland (division Va; 1924–84), Faroe (division Vb; 1961–87), west of Scotland (division VIa; 1968–87), and the Irish Sea (division VIIa; 1968–88). The majority of data on year-class strength were derived from the NAFO (1989) and ICES (1990) reports. In addition, data from Dragesund (1971) and Malmberg (1986) were taken into account. As a rule, year-class strength was assessed as number of recruits at age 3 years.

It is known that very large cross covariances can be generated between two uncorrelated processes as a result af the large autocovariances within the two processes (Jenkins and Watts 1968). To avoid overestimation of the correlation coefficients, a preliminary procedure of differencing of the time series was applied. This procedure serves as a high-pass filter, so that resultant time series of first differences (that is changes from year to year) contain no linear trends and no or significantly reduced autocorrelations (Box and Jenkins 1970). This is particularly important for short time series, when the possibility of spurious correlations is rather high. For example, the correlation coefficient between recruitment to Labrador

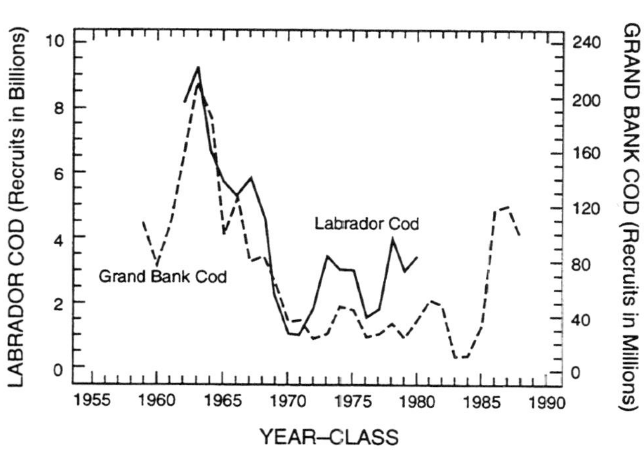

Fig. 4. Recruitment to the Labrador (solid line) and Grand Banks (broken line) cod stocks.

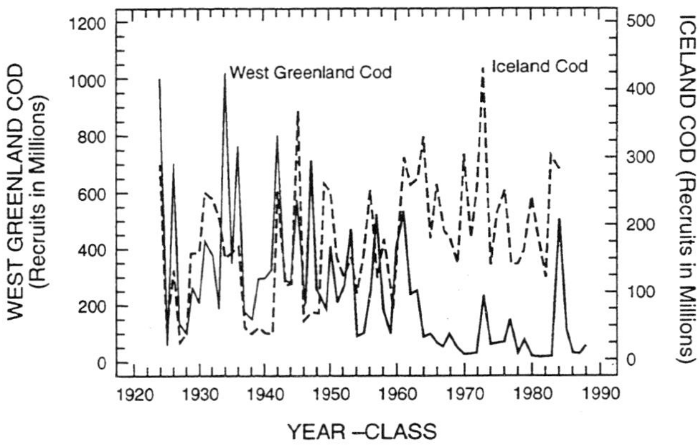

Fig. 5. Recruitment to the West of Greenland (solid line) and Iceland (broken line) cod stocks.

and Grand Bank cod stocks for the period from 1962 to 1980 is $r = 0.90$. To a large extent, this high correlation may be explained by a strong downward trend in both time series, which drastically reduces the amount of independent observations (Fig. 4). When the differential filter was applied, autocorrelation at a time lag of 1 yr (i.e., between neighbouring observations) decreased from 0.76 to 0.03 (for the Labrador cod) and from 0.80 to −0.26 (for the Grand Bank cod). As a result, the correlation coefficient reduced to $r = 0.40$.

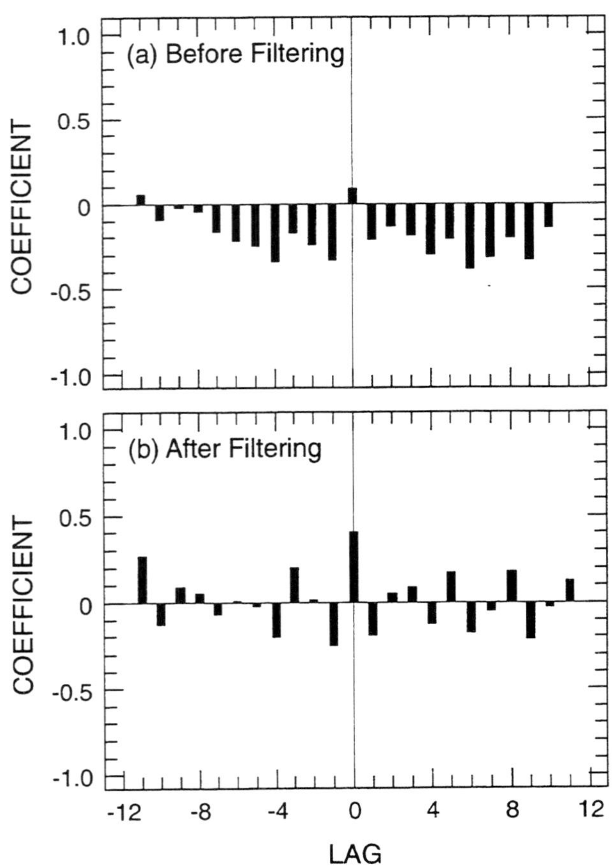

Fig. 6. Cross-correlation function between recruitment to the West of Greenland and Iceland cod stocks (*a*) before and (*b*) after filtering.

Another advantage of this filter is that it helps to mitigate the problem of heterogeneity of some time series of fishery statistics, which may be caused, for example, by overfishing. Thus, owing to overfishing, a stock can collapse and continue to fluctuate at a new, lower level, which is significantly different from that before the human's intervention. Data on such a stock becomes heterogeneous, and their analysis statistically incorrect. Differencing the time series underscores relative fluctuations in the stock, in that it shows how strong this year-class is relative to the previous one. Because the stock continues to react to environmental changes in the same way as it did before the collapse, producing a stronger year class in more favourable conditions and vice versa, the time series of the first differences is more homogeneous and better reflects changes in environmental conditions than the original time series.

As an example, let us consider fluctuations in recruitment to the west of Greenland and Iceland cod stocks (Fig. 5). Beginning with the 1960s, the year-class strength of the west of Greenland cod has been significantly reduced (except the 1982 year-class), that of the Icelandic cod was somewhat higher than in the preceding several decades. As a result of these opposite trends, the cross-correlation function between recruitment to the cod stocks is just slightly positive at zero time lag and negative for the other time lags within ±11-year interval (Fig. 6a). When differencing was applied, the cross-correlation function has become significantly (at 5% significance level) positive at zero time lag, randomly fluctuating as the lag (of both signs) increased on modulus (Fig. 6*b*).

Atmospheric teleconnection patterns

Research on atmospheric teleconnections has a longer history than on their biological counterparts. The principal teleconnection pattern, the North Atlantic Oscillation (NAO), was first identified by Walker (1924). The NAO is generally represented by a north–south dipole in the SLP with centres near Greenland and the central Atlantic west of the Azores. At present, it is a well-documented pattern both in pressure and temperature fields (e.g., Lamb and Peppler 1987). It has also been confirmed that the NAO is a dominant mode of interannual variability in winter atmospheric circulation (Kushnir and Wallace 1989). Recently, apart from the NAO, Rogers (1990) identified three other low-frequency SLP variability patterns in the Atlantic–European sector with centres of action over the eastern Atlantic (EATL), southern Europe and northern Atlantic (SENA), and Scandinavia and central North Atlantic (SCAN). Teleconnection indices for each pattern were computed as the equally weighted sums of normalized (by standard deviation) anomalies of SLP in the corresponding centres of action with signs given in Fig. 7*a*.

Teleconnection patterns at the midtropospheric level differ from those at the surface. Wallace and Gutzler (1981) identified two prominent structures in the 500-hPa height field in the Atlantic–European sector, which they labeled on the basis of their geographical location as western Atlantic (WA) and eastern Atlantic (EA) (Fig. 7*b*). The latter has the third centre near the Black Sea, which was not taken into account in this study because low-frequency perturbations in the height field become zonally elongated and their continental features weaken for the time scales longer than 30 d (Kushnir and

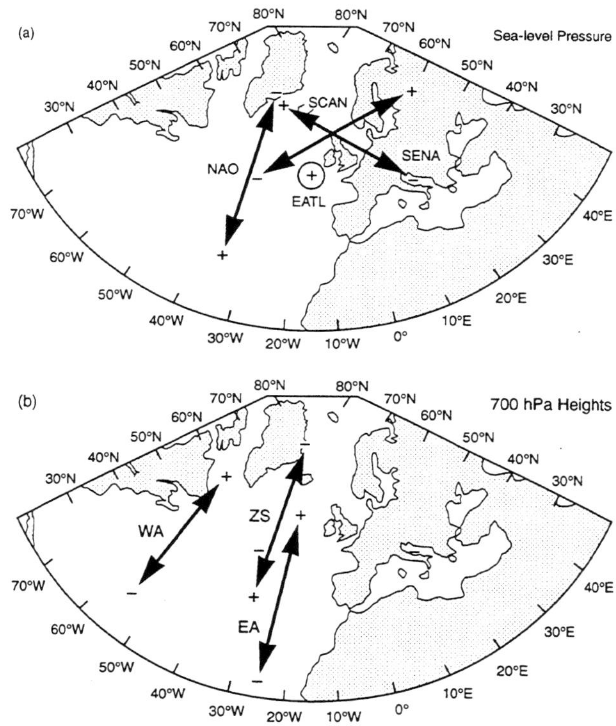

Fig. 7. Teleconnection centres of action in (*a*) sea level pressure and (*b*) 700-hPa geopotential heights. Arrows connect centers with strongly opposite fluctuations. Plus and minus signs indicate with what signs LSP and 700-hPa height anomalies were taken to calculate correspondent teleconnection indices. See text for abbreviations.

Wallace 1985). Later, Esbensen (1984) identified a zonally symmetric seesaw in the 700-hPa height field between polar and middle latitudes, the Atlantic segment of which resembles the NAO signal in the 500-hPa field as documented by Kushnir and Wallace (1989). This zonally symmetric (ZS) pattern explains a major fraction of the interannual variability over the North Atlantic (Esbensen 1984).

Analysis of correlation coefficients between teleconnection indices (Table 1) shows that these patterns are not independent. Thus, the NAO turns out to be highly correlated with the SENA pattern, with the correlation coefficient $r = -0.80$. Rogers (1990) showed that the winter-averaged time series of SENA and NAO indices are both associated with an opposition in mean winter temperature anomalies between Greenland and northern Europe. Actually, the SENA pattern closely resembles the winter (W1) and autumn (A2) SLP eigenvectors identified by Trenberth and Paolino (1981), who interpreted them as being the NAO. Another pattern, which contains elements in common with the NAO, is the WA; the correlation coefficient between the corresponding indices $r = -0.78$.

Atmospheric processes affect fish stocks, first of all, through changes in temperature conditions. Therefore, atmospheric teleconnection patterns were translated into SAT–SST anomaly patterns by computing composite maps for upper and lower terciles of teleconnection indices. The differences in corresponding composites for each atmospheric pattern are shown in Fig. 8. The most striking is the fact that all the composite maps demonstrate a more or less pronounced west–east opposition in SST–SAT field, which is characteristic for the classic NAO. This result confirms that the NAO is the principal pattern in the North Atlantic region and the other teleconnection patterns actually represent some distortions of it. This result agrees with Wallace and Smith (1990), who found that leading eigenvectors of SST appear to be related to the NAO in atmospheric circulation.

Patterns in year-class strength

The shortage, and less reliable character, of the data makes the identification of spatial patterns in recruitment to the cod stocks a more difficult task than that for the atmospheric patterns. In the present study, an attempt was made to resolve this by means of factor analysis, which extracts principal components for variables under study using a correlation matrix. It is similar to principal component analysis, except that it scales factor weights so that the sum of squares is equal to the associated eigenvalue of the correlation matrix and so is related to the total variance explained by that factor. Factor analysis and its modifications along with an analysis of one-point correlation maps represent two main approaches used in atmospheric teleconnection studies.

Table 2 is the correlation matrix of the time series of annual recruitment to 10 North Atlantic cod stocks (upper right part of table) and its first differences (lower left part of table). As the table shows, the number of significant correlation coefficients (at the 10% level) between first differences exceeds that between original data. The highest correlation coefficients turned out to be between the neighbouring stocks. Thus, the North Sea cod is highly correlated with the west of Scotland cod, which in turn is in good agreement with the Irish Sea cod. The other group with high correlation coefficients is composed of cod stocks located off the North American seaboard (Labrador, south Newfoundland, and Grand Bank). For all these stocks high year-classes occurred during the first half of the 1960s followed by a significant reduction in year-class strength in the 1970s. However, the correlation between these

Table 1. Correlation coefficients (expressed in percent) between atmospheric teleconnection indices, 1948–1990. See text for explanation of abbreviations.

	SENA	SCAN	EATL	ZS	WA	EA
NAO	−80*	−18	21	59*	−78	2
SENA		0	−18	−77*	62*	−12
SCAN			−20	−26	24	68*
EATL				54*	29	39
ZS					−37	9
WA						32

* Significant at the 0.01 level.

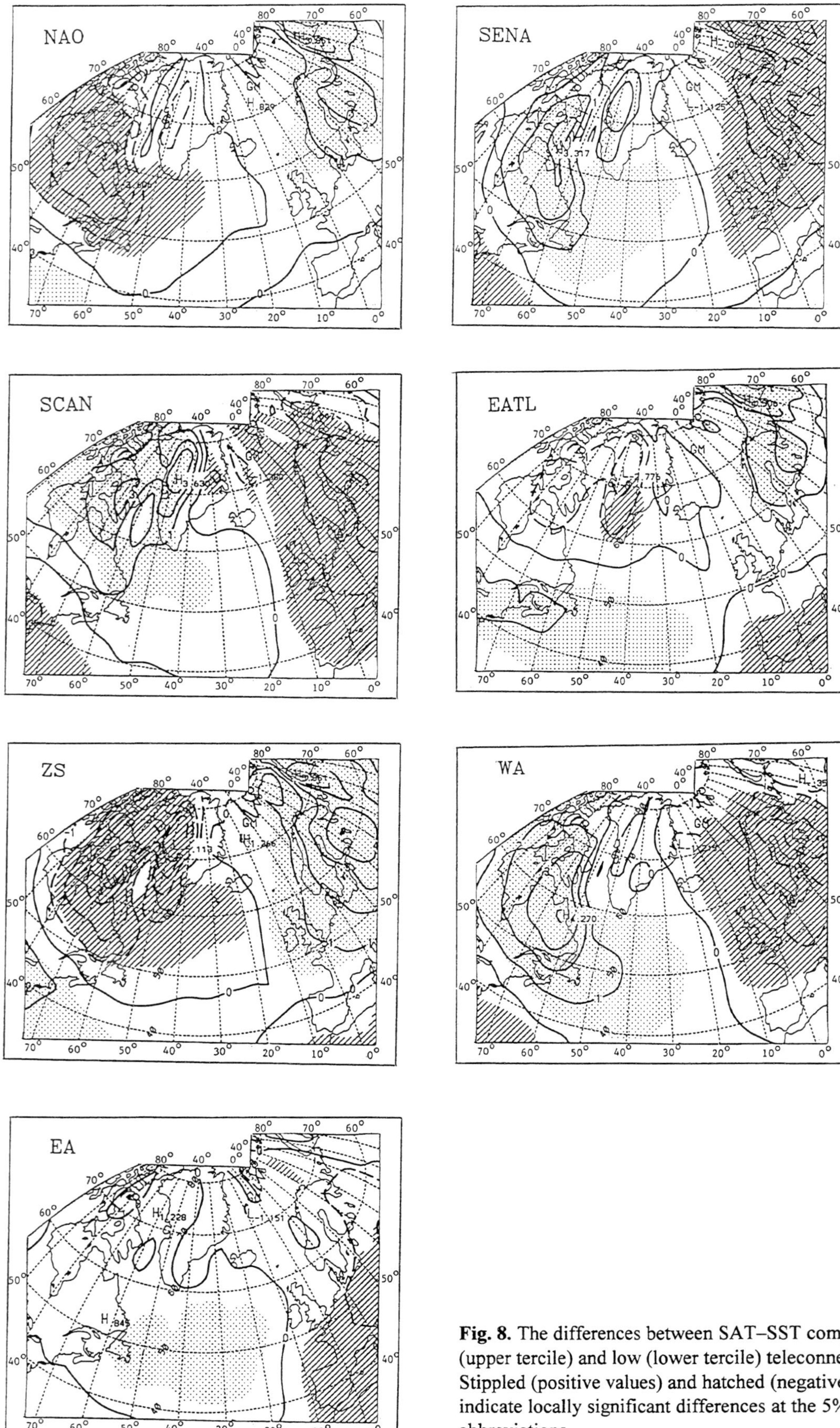

Fig. 8. The differences between SAT–SST composites for high (upper tercile) and low (lower tercile) teleconnection indices. Stippled (positive values) and hatched (negative values) areas indicate locally significant differences at the 5% level. See text for abbreviations.

cod stocks becomes much lower for the time series of the first differences; that is, when the downward trend in recruitment is removed. On the whole, the correlation between the group of stocks around the British Isles and that along the North American coast is negative.

As shown above, when the trend is removed from the time series of recruitment to the west of Greenland and Iceland cod stocks, the correlation coefficient between them becomes significant at 0.1% level. This result may serve as a confirmation of the hypothesis that the appearance of cod on the offshore banks west of Greenland in the 1920s was the result of a drift of eggs, larvae, and subsequently 0-group fish from the Icelandic spawning grounds when the climate became warmer (Cushing 1982).

Factor analysis applied to the correlation matrix of the first differences showed that the first two factors explain 49% of the total variance of the recruitment to the cod stocks. Factor weights (or eigenvectors) for the first factor (which explains 29% of the total variance) and the second factor (20% of total variance) are shown in Fig. 9*a* and Fig. 9*b*, respectively. Figure 9*a* demonstrates marked opposition in fluctuations of recruitment to cod stocks in the Northwest and Northeast Atlantic, or in terms of Izhevskii's "systems of nature," between the Greenland – North American and Atlantic systems. One example of this opposition is given in Fig. 10, where time series of recruitment to the North Sea and south of Newfoundland cod stocks are presented. The difference between the Izhevskii's scheme (Fig. 1), and Fig. 9*a* is that, in the latter case, the border separating the Greenland – North American and Atlantic

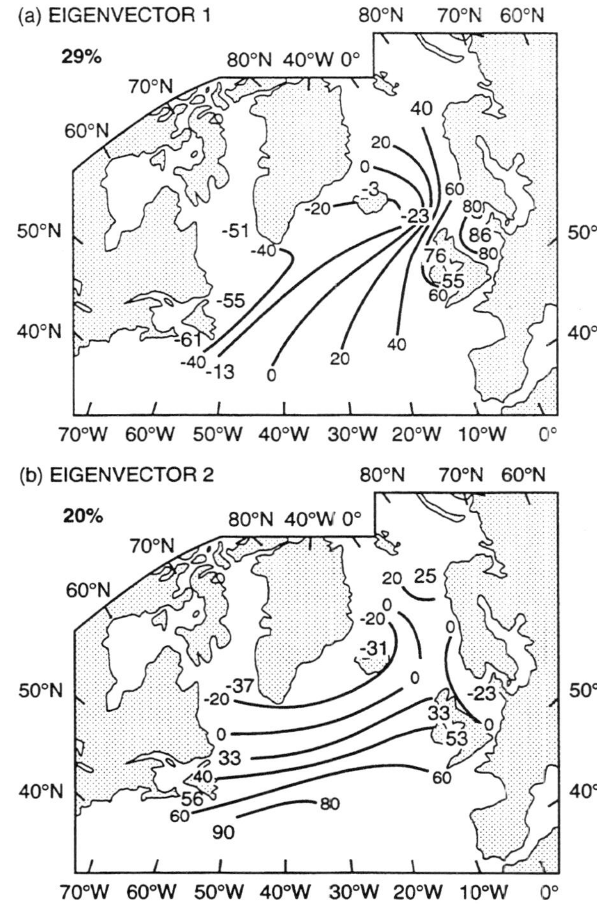

Fig. 9. The first (*a*) and second (*b*) eigenvectors of recruitment to North Atlantic cod stocks based on the whole available data set.

Table 2. Correlation coefficients (expressed in percent) between recruitment to cod stocks (above diagonal) and their first differences (below diagonal). Numbers of paired observations are given in parentheses.

Stocks	Arcto-Norwegian	North Sea	Iceland	Faroe	West Scotland	Irish Sea	West Greenland	Labrador	Grand Banks	South Newfoundland
Arcto-Norwegian		38 (20) †	14 (53)	0 (27)	−21 (20)	12 (21)	17 (58)	19 (19)	34 (30) †	10 (26)
North Sea	46 (19) †		−11 (17)	−14 (20)	32 (20)	18(20)	−32 (20)	−39 (13)	−28 (20)	−50 (17) †
Iceland	28 (51) †	11 (16)		9 (24)	−20(17)	−8 (17)	10 (60)	22 (19)	12 (25)	18 (25)
Faroe	2 (26)	−10 (19)	7 (23)		−12 (20)	−12 (20)	−14 (27)	−14 (19)	−16 (27)	29 (24)
West Scotland	22 (19)	51 (19) †	−5 (16)	−30 (19)		67 (20) †	−27 (20)	3 (13)	25 (20)	−9 (17)
Irish Sea	25 (20)	27 (19)	−13 (16)	−4 (19)	66 (19) †		−18 (21)	−10 (13)	26 (21)	−12 (17)
West Greenland	−10 (57)	−40 (19) †	42 (58) †	−9 (26)	−35 (19)	−21 (20)		61 (19) †	17 (30)	25 (26)
Labrador	7 (18)	−52 (12) †	13 (18)	−2 (18)	−21 (12)	−17 (12)	30 (18)		90 (19) †	70 (19) †
Grand Banks	29 (29)	−33 (19)	−4 (23)	−7 (26)	25 (19)	25 (20)	−24 (29)	40 (18) †		50 (26) †
South Newfoundland	−14 (25)	−53 (16) †	−21 (23)	26 (23)	−45 (16) †	3 (16)	7 (25)	24 (18)	53 (25) †	

† Significant at the 0.10 level.

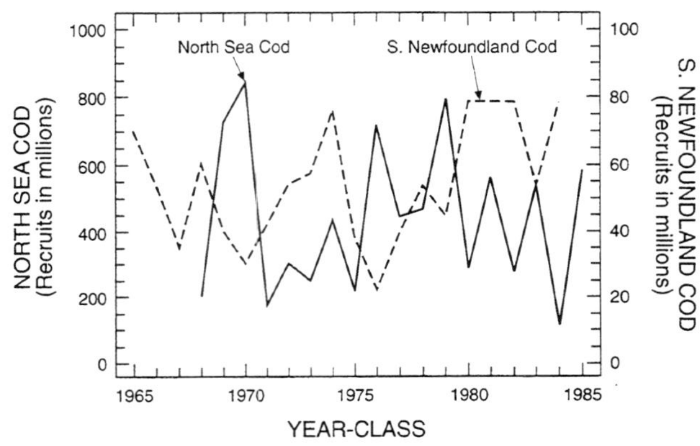

Fig. 10. Recruitment to the North Sea (solid line) and south of Newfoundland (broken line) cod stocks.

systems is shifted eastward, so that not only Iceland cod (as assumed by Izhevskii) but even Faroe cod belong to the Greenland – North American system.

The spatial distribution of the second factor weights (Fig. 9*b*) demonstrates the north–south opposition in year-class strength. The signs of this opposition, however, are different in the western and central parts of the region (where they are negative in the north and positive in the south) and its eastern part (positive for the Arcto-Norwegian cod and negative for the North Sea cod).

When the factor weights for the first two factors are plotted (Fig. 11), it becomes possible to classify the cod stocks in three groups on a regional basis: (A) northwestern Atlantic (Labrador, south Newfoundland, and Grand Bank); (B) central North Atlantic (west of Greenland, Iceland, and Faroe); and (C) northern European seas (Arcto-Norwegian, west of Scotland, the Irish Sea, and the North Sea). It should be noted, however, that the quality of such classification (a ratio of within-group to among-group variance) is not high or, in other words, the groups are not very compact.

This result is different from that of Garrod and Colebrook (1978), who applied principal components analysis to 17 stocks of herring and gadoids throughout the North Atlantic. They classified them into four groups (A through D), with two of them, A and C, containing the majority of stocks. Group A included Arcto-Norwegian and Gulf of St. Lawrence cod stocks, whereas group C included Faroe, Iceland, south of Newfoundland, and Nova Scotia cod stocks. According to Garrod and Colebrook, group A is composed of stocks inhabiting an environment most subject to the influence of the continental land mass, whereas group C predominantly contained stocks close to the axis of the Gulf Stream and the North Atlantic Drift or lying in narrow shelf waters more heavily influenced by oceanic conditions. It is interesting to note that the west of Greenland cod was in neither of these two groups and, together with east of Greenland cod, formed group B, which was still closer to group A.

The data set on recruitment to cod stocks used in this study consists of time series of different lengths; the data are overlapped only for 1968–1980. Factor analysis was repeated for this shorter data set, containing only synchronous observa-

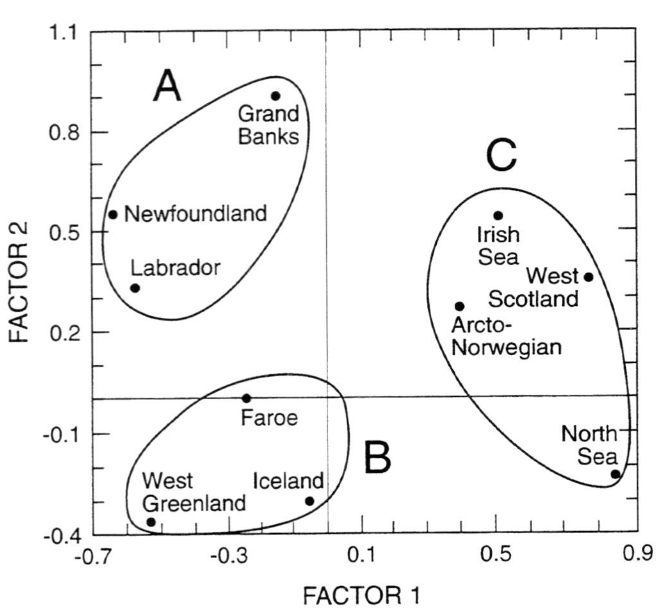

Fig. 11. Plots of the first two factor weights and classification of the cod stocks, based on the whole available data set.

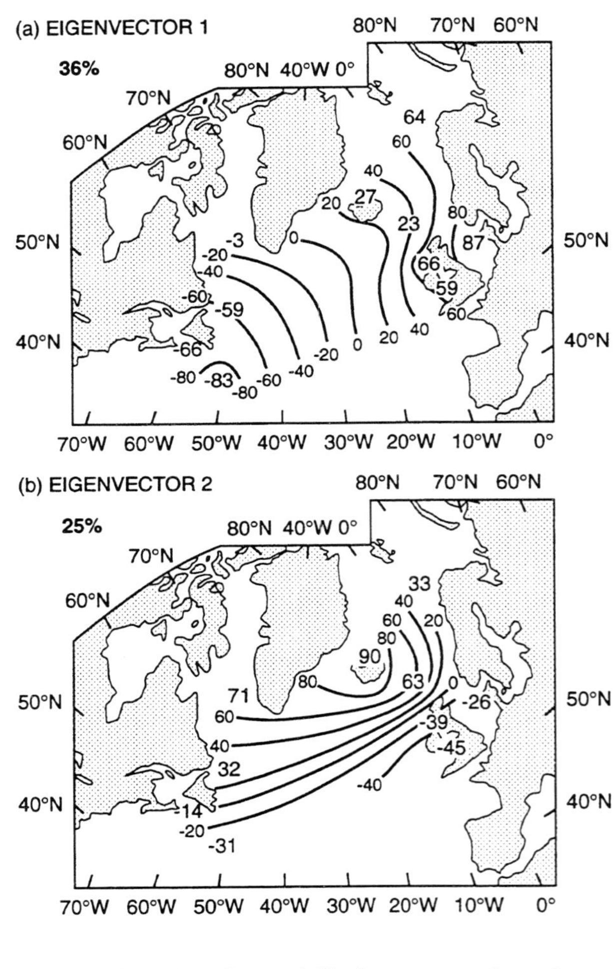

Fig. 12. The first (*a*) and second (*b*) eigenvectors of recruitment to North Atlantic cod stocks for the 1968–1980 period.

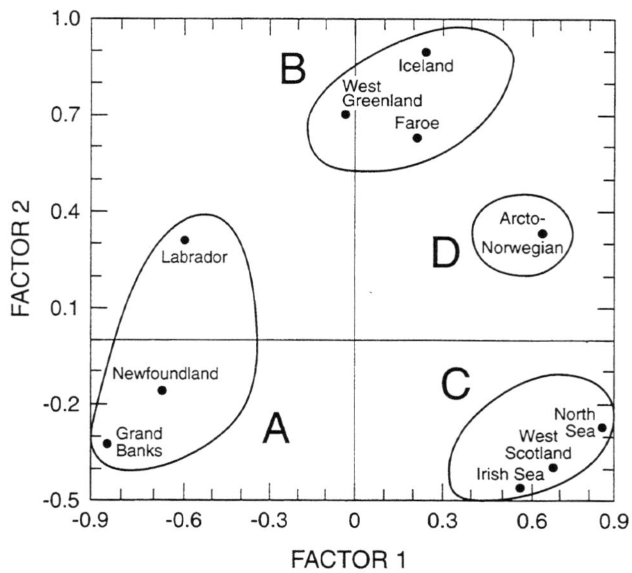

Fig. 13. Plots of the first two factor weights, and classification of the cod stocks for the 1968–1980 period.

tions, and the results for the first two factors, which now describe 61% of the total variance, are presented in Figs. 12 and 13. Comparing Fig. 12*a* with Fig. 9*a*, one can see that both exhibit west–east opposition, while in the latter case the 0 contour is shifted westward so that the Iceland and Faroe cod stocks belong to the Atlantic system (in Izhevskii's terms). The north–south opposition in Fig. 12*b* is even more pronounced than in Fig. 9*b* and includes all the region under study. As for the classification in Fig. 13, the Arcto-Norwegian cod stock is now located at about an equal "distance" between groups B and C and, hence, should be singled out.

Discussion

The similarity between the first eigenvector (Figs. 9*a* and 12*a*) and SST–SAT anomaly patterns (Fig. 8) suggests that the first factor, describing the largest part of the total variance of year to year changes in recruitment to the cod stocks, is probably related to atmospheric teleconnections and, above all, to the NAO. However, the majority of correlation coefficients between the atmospheric teleconnection indices and the time series of first differences in year-class strength turned out to be insignificant. Just a few of them exceeded a 10% significance level, and only two of them (between the WA index and the North Sea cod ($r = 0.54$), and between the SCAN index and the west of Greenland cod ($r = -0.25$)) exceeded the 5% level, the latter still being low in absolute value. Unfortunately, the available data on recruitment to the cod stocks allow construction of only a short time series of the first factor (1968–1980), which devalues any correlation analysis. Therefore, one cannot rule out that the resemblance between the first eigenvector and the SST–SAT anomaly pattern is just a random coincidence.

The pattern of cod year-class variability associated with the second eigenvector appears to reflect the different responses of cod stocks to temperature changes. It should be noted that while, for some cod stocks (e.g., the Arcto-Norwegian cod) the effect of sea temperature is quite well known and the results of numerous studies are more or less unambiguous (at least as to the sign of the relationship), this is not the case for other stocks (e.g., the Labrador cod), for which the results are less certain and even controversial. A brief review of the cod – sea temperature relationship for the cod stocks under study is given below.

Characterizing the thermal regime in the Barents Sea, Izhevskii (1961) broke down the range of sea temperatures on the Kola section to four categories (very cold, cold, warm, and very warm) and found that they matched very well with the corresponding categories (very poor, poor, rich, and very rich) of the year-class strength of the Arcto-Norwegian cod. Similar results have been obtained by Kislyakov (1961), Saetersdal and Loeng (1986), Elizarov and Borisov (1989), and many other researchers, so that the existence of a positive correlation between sea temperature and the recruitment to the Arcto-Norwegian cod is well documented.

Dickson et al. (1973) showed for the North Sea cod that the survival of juveniles is higher when the sea is cold. One good example supporting this result is the outburst of gadoids in the 1960s when the climate in the North Atlantic became colder (Cushing 1982).

Malmberg (1986) identified warm and cold periods in Icelandic waters on the order of several years, and showed that recruitment of 3-year-old cod was well above normal during warm periods and decreased during cold ones. Furthermore, during periods of unfavorable hydrological conditions the 0-group indices of cod were low, even in years of relatively high zooplankton densities.

Hermann (1953) and Elizarov (1963) found a positive correlation between recruitment to the west of Greenland cod and the thermal regime. As an indicator of thermal regime Hermann (1953) used the bottom temperature on Fyllas Bank, while Elizarov (1963) used the average temperature in the upper 200 m layer on the section from Labrador to the southern tip of Greenland. In a more recent study, Buch and Hansen (1986), having compared the cod catches off west of Greenland with air and sea temperature variations, found no simple relationship between them. Nevertheless, some interesting patterns were noted:

1) When the warming of the climate began, cod returned to west of Greenland waters.

2) When the cooling began around 1970, catches began to decrease nearly instantaneously.

3) When a second very strong cooling occurred in 1982–83, catches of cod declined to almost nil.

As for cod stocks on the Labrador Shelf and Newfoundland Banks, Elizarov (1963) noted that good year-classes coincided with rises in air and sea temperature, explaining this relationship as follows. In years with weak northerly winds, the Labrador Current is also weak; the transport of cold Arctic water becomes less intensive and the relatively warm, nutrient-rich waters of the Atlantic origin spread farther north.

On the other hand, in their study of cod in the southern Gulf of St. Lawrence, Paloheimo and Kohler (1968) found no noticeable correlation between annual mean SST and the number of recruits at age 6 for the 1943–59 year-classes. As well, Martin and Kohler (1965) found a negative correlation for the 1921–62 period between annual mean SST at St. Andrews, N.B., and cod landings from the Georges Bank 4 yr later. The existence of a negative correlation between sea temperature and recruitment to the Labrador cod was found by Borovkov (1978), who gave the following explanation of the relationship. A decrease in sea temperature is indicative of a strengthening of the Labrador Current, which leads to a more intensive transport of eggs and larvae to the south. Hence, it takes less time for them to reach areas with more favourable conditions that facilitate survival of juveniles and eventually result in a successful year-class. Alternatively, a weakening of the Labrador Current, accompanied by an increase in sea temperature, leads to the appearance of a poor year-class.

It is evident, even from this brief review, that the effects of environmental changes on cod stocks in the North Atlantic are different in different areas. A general rule describing these different effects was proposed by Templeman (1972). He points out that during periods of warming recruitment to cod stocks tends to increase at the northern end of the cod range and becomes scarce to absent at the southern end, whereas during periods of cooling cod year-classes become successful to the south and poor to the north. The north–south opposition in the second eigenvector (Fig. 9*b* and 12*b*) may be considered as evidence supporting Templeman's hypothesis.

While the mechanism linking atmospheric teleconnections with recruitment to cod stocks is not clear, it may be assumed that it encompasses atmospheric and oceanic processes of different time scales. First of all, it is important to bear in mind that SST is capable of responding to atmospheric forcing on a time scale of days, as evidenced by episodic cooling and deepening of the mixed layer that takes place in response to the passage of storms. Sometimes even one strong storm is enough to produce a poor year-class even if all other conditions are not critical. This means that, to a great extent, changes in recruitment represent a random process, which is hard to predict. The role of atmospheric teleconnections consists of the formation of different storm tracks (for the correspondence between teleconnection patterns and storm tracks see Rogers 1990), and hence, in a change in the probability of a storm passing over the spawning grounds. In turn, the frequency of any particular modes of teleconnection patterns depends on more general climatic conditions, such as the climatic warming in the North Atlantic during the first half of the century and the subsequent cooling.

In sum, the analysis of 10 cod stocks in the North Atlantic region has shown that a significant part (up to 60%) of the interannual variability in recruitment to the cod stocks may be described by two factors. The first factor reflects the west–east opposition in year-class strength and appears to be related to atmospheric teleconnections. In fact, all the analyzed teleconnection patterns (four at sea level and three at the 500-hPa level) after their translation to the temperature field, carry a signature of out of phase fluctuation between the SST–SAT anomalies in the northwest Atlantic and northern European seas. The second factor reflects north–south opposition in year-class strength and seems to be associated with the different responses of cod stocks during years of general warming and cooling in the North Atlantic. It is not clear, however, what the mechanism is that links atmospheric teleconnections and recruitment to cod stocks and why the latter should exhibit west–east and north–south variations. More in-depth ecological study is needed.

Acknowledgements

The author wishes to thank Steven L. Rhodes for his helpful comments and editorial assistance.

References

Barnett, T.P. 1985. Variations in near-global sea level pressure. J. Atmos. Sci. 42: 478–501.

Borovkov, V.A. 1978. Effects of hydrometeorological factors on year-class strength of the Labrador cod. Tr. PINRO, 40: 140–147 (In Russian)

Box, G.E.P., and G.M. Jenkins. 1970. Time series analysis, forecasting and control. 2nd printing. Holden-Day, San Francisco, Calif. 575 p.

Buch, E., and H.H. Hansen. 1986. Climate and cod fishery at West of Greenland, p. 345–364. International Symposium Long Term Changes in Marine Fish Populations, Vigo, Spain.

Cushing, D.H. 1982. Climate and fisheries. Academic Press, London. 373 p.

Dickson, R.R., J.G. Pope, and M.J. Holden. 1973. Environmental influences on the survival of North sea cod, p. 69-80. *In* J.H.S.Blaxter [ed.] The early life history of fish. Springer-Verlag, Heidelberg and New York.

Dragesund, O. 1971. Comparative analysis of year-class strength among fish stocks in the North Atlantic. Fiskeridir Skr. Ser. Havunders. 16: 49–64.

Dvoryaninov, G.S., and Goloborod'ko, S.V. 1989. Teleconnections in the variation of the cloud-amount field above the North Atlantic. Izv. Acad. Sci. USSR. Atmos. Oceanic Phys. Engl. Transl. 25: 766–775.

Elizarov, A.A. 1963. On oceanographic condition influencing year-class strength of commercially important species in the Northwest Atlantic. Oceanologiya 3: 1065–1078 (In Russian)

Elizarov, A.A., and V.M. Borisov. 1989. Long-term changes in environmental and biological conditions in the ecosystem of the Barents and Norwegian seas, p. 69–83. *In* Biologicheskie osnovy dinamiki chislennosti i prognozirovanija vilova ryb, VNIRO, Moscow. (In Russian)

Elizarov, A.A., B.N. Kotenev, and S.N. Rodionov. 1990. Izhevskii's systems of nature in the light of modern investigations. ICES, Copenhagen. C.M. C:5/Ref G+H. 21 p.

Esbensen, S.K. 1984. A comparison of intermonthly and interannual teleconnections in the 700 mb geopotential height field during the Northern Hemisphere winter. Mon. Weather Rev. 112: 2016–2032.

Gambo K., and K. Kudo. 1983. Three-dimensional teleconnections in the zonally assymmetric height field during the Northern Hemisphere winter. J. Meteorol. Soc. Jpn. 61: 36–50.

Garrod, D.J., and J.M. Colebrook. 1978. Biological effects of variability in the North Atlantic ocean. Rapp. P.-v. Réun. Cons. Int. Explor. Mer 173: 128–144.

Girskaya E.Ya. 1981. Temperature field and world weather. Tr. Gl. Geofiz. Obs. 458: 38–49 (In Russian)

Hermann, F. 1953. Influence of temperature on strength of cod year-classes. Ann. Biol. Copenhagen, 9: 1–31.

Hsiung, J., and R.E. Newell. 1983. The principal modes of variation of global sea surface temperature. Bull. Am. Meteorol. Soc. 64: 1957–1969.

ICES. 1990. Reports of the ICES Advisory Committee on Fishery Management, 1989. ICES, Copenhagen Coop. Res. Rep. 168: 361 p.

Izhevskii, G.K. 1961. Oceanographic basis of formation of the fisheries productivity of seas. Pishepromizdat, Moscow. 216 p. (In Russian)

Izhevskii, G.K. 1964. System basis of prediction of oceanographic conditions and reproduction success of commercial species of fish. VNIRO, Moscow. 165 p. (In Russian)

Jenkins, G.M., and D.G. Watts. 1969. Spectral analysis and its applications. 2nd printing. Holden-Day, San Francisco, Calif. 525 p.

Kawasaki, T. 1991. Effects of global climatic change on marine ecosystems and fisheries, p. 291–299. *In* J. Jager and H.L. Ferguson [ed.] Climate change: science, impacts and policy. Cambridge University Press, Cambridge, UK.

Kislyakov, A.G. 1961. Relationship between hydrological conditions and variations in abundance of cod. Tr. sovesh. Ikhtiol. Kom. Akad. Nauk. SSR 13: 260–264.

Kushnir, Y., and J.M. Wallace. 1989. Low-frequency variability in the Northern Hemisphere winter: geographical distribution, structure and time-scale dependence. Mon. Weather Rev. 46: 3122–3142.

Lamb, P.J. and R.A. Peppler. 1987. North Atlantic Oscillation: concept and an application. Bull. Amer. Meteorol. Soc. 68: 1218–1225.

Lau, K.-M., and P.J. Sheu. 1991. Teleconnections in global rainfall anomalies: seasonal to inter-decadal time scales, p. 227–256. *In* M.H. Glantz, R.W. Katz, and N. Nicholls [ed.] Teleconnections linking worldwide climate anomalies. Cambridge University Press, Cambridge, UK.

Malmberg, S.-A. 1986. Ecological impact of hydrographic conditions in Icelandic waters, p. 345–364. *In* International Symposium on Long Term Changes in Marine Fish Populations, Vigo, Spain.

Martin, W.R., and A.C. Kohler. 1965. Variation in recruitment of cod (Gadus morhua L.) in southern ICNAF waters, as related to environmental changes. ICNAF Spec. Publ. 6: 833–846.

NAFO. 1989. NAFO Scientific Council Report NAFO, Dartmouth, N.S. 151 p.

Paloheimo, J.E., and A.C. Kohler. 1968. Analysis of the southern Gulf of St. Lawrence cod population. J. Fish. Res. Bd. Can. 25: 555–578.

Rodionov, S.N., and A.S. Krovnin. 1991. Interannual Variability of Thermal Conditions in the Bering Sea, p. 49–55. *In* Proceedings Interntemational Symposium on the Bering Sea Fisheries, April 2–5, 1990, Khabarovsk, Russia. Alaska Fisheries Science Center, Seattle, Wash.

Rodionov, S.N., and A.S. Krovnin. 1992. The 1980's in the context of climate changes in the North Atlantic Region. ICES Mar. Sci. Symp. 195: 86–95.

Rogers, J.C. 1990. Patterns of low-frequency monthly sea level pressure variability (1899–1986) and associated wave cyclone frequencies. J. Clim. 3: 1364–1379.

Saetersdal, G., and H. Loeng. 1986. Ecological adaptation of reproduction in Arctic cod, p. 393–410. *In* A Workshop on Comparative Biology, Assessment, and Management of Gadoids from the North Pacific and Atlantic Oceans. Part 1, Seattle, Wash.

Shepherd, J.G., J.G. Pope, and R.D. Cousens. 1984. Variations in fish stocks and hypotheses concerning their links with climate. Rapp. P.-v. Réun. Cons. int. Explor. Mer, 185: 225–267.

Templeman, W. 1965. Relation of periods of successful yearclasses of haddock on the Grand Bank to periods of success of yearclasses for cod, haddock and herring in areas to the north and east. ICNEF Spec. Publ. 6 Environ. Symp. 6: 523–533.

Templeman, W. 1972. Year-class success in some North Atlantic stocks of cod and haddock. ICN Spec. Publ. 8: 223–241.

Trenberth, K.E., and D.A. Paolino, Jr. 1981. Characteristic patterns of variability of sea level pressure in the Northern Hemisphere. Mon. Weather Rev. 109: 1169–1189.

Walker, G.T. 1924. Correlation in seasonal variations of weather. IX. A further study of world weather. Mem. Indian Meteorol. Dep. 24: 275–332.

Walker, G.T., and E.W. Bliss. 1932. World weather V. Mem. R. Meteorol. Soc. 4: 53–84.

Wallace, J.M., and D.S. Gutzler. 1981. Teleconnections in the geopotential height field during the Northern Hemisphere winter. Mon. Weather Rev. 109: 784–812.

Wallace, J.M., and C. Smith. 1990. Spatial patterns of atmosphere–ocean interaction in the northern winter. J. Clim. 3: 990–998.

Predicted effects of global climate change on fishes in Minnesota lakes

Heinz G. Stefan, Midhat Hondzo, John G. Eaton, and J. Howard McCormick

Stefan, H.G., M. Hondzo, J.G. Eaton, and J.H. McCormick. 1995. Predicted effects of global climate change on fishes in Minnesota lakes, p. 57–72. *In* R.J. Beamish [ed.] Climate change and northern fish populations. Can. Spec. Publ. Fish Aquat. Sci. 121.

Abstract: According to global climate change models, Minnesota's annual mean air temperature will increase by approximately 4.0°C as the mean atmospheric CO_2 doubles. This is likely to have many environmental consequences e.g. on temperatures and dissolved oxygen concentrations in lakes. The interaction between climate, lake water quality, and fish populations has been investigated through model simulations for 27 classes of Minnesota lakes. Effects of projected climate change were calculated for lake temperatures and dissolved oxygen, including stratification dynamics. Benthic areas and water volumes having good growth promoting temperature and dissolved oxygen characteristics, before and after the projected climate change, were calculated. These areas and volumes are designated as good growth habitat area (GGHA) and good growth habitat volume (GGHV), respectively. Estimated values indicate that after the projected climate change, GGHA and GGHV will decrease for cold-water fish. In contrast, GGHA and GGHV will increase for cool-water and warmwater fish. Fish habitat will change by approximately the same percentage in terms of GGHA or GGHV. The reduction in GGHA and GGHV for cold-water fishes will be about twice as high for southern Minnesota lakes as it will be for northern lakes. The increases for cool-water and warmwater fishes will be two times greater for northern lakes than for southern lakes. The models and assessment techniques used here can serve as templates for analyzing projected climate change impacts on fishes of other regions.

Résumé : Selon les modèles globaux de changement climatique, la température moyenne annuelle de l'air au Minnesota augmente d'environ 4° C lorsque la teneur moyenne de CO_2 dans l'atmosphère double. Ce phénomène risque d'avoir de nombreuses conséquences environnementales, soit notamment une hausse des températures et des concentrations d'oxygène dissous dans les lacs. L'interaction entre le climat, la qualité de l'eau des lacs et les populations de poissons a fait l'objet de recherches par modélisation pour 27 catégories de lacs du Minnesota. L'incidence des changements climatiques prévus sur les températures des lacs et sur leurs concentrations en oxygène dissous, y compris sur leur dynamique de stratification, a été calculée. On a mesuré les zones benthiques et les volumes d'eau présentant des caractéristiques de températures et de concentrations en oxygène dissous qui favorisent la croissance, avant et après le changement climatique prévu. Ces zones et volumes sont désignés respectivement «secteur d'habitat de bonne croissance» (GGHA pour *good-growth habitat area*) et «volume d'habitat de bonne croissance» (GGHV pour *good-growth habitat volume*). Selon les résultats estimatifs, après le changement climatique prévu, les GGHA et les GGHV seront en déclin pour le poisson d'eau froide. En revanche, les deux seront en hausse pour le poisson d'eau fraîche et celui d'eau chaude. Les GGHA et GGHV de tous les habitats de poissons seront modifiés dans à peu près les mêmes proportions. La réduction des GGHA et des GGHV pour les poissons d'eau froide sera environ deux fois plus forte dans les lacs du sud du Minnesota qu'elle ne le sera dans ceux qui sont situés plus au nord. L'augmentation pour les poissons d'eau fraîche et d'eau chaude sera deux fois plus élevée dans les lacs du nord que dans ceux du sud. Les modèles et techniques d'évaluation utilisés ici peuvent servir de gabarits pour l'analyse des effets prévus du changement climatique sur les poissons d'autres régions.

Heinz G. Stefan, Midhat Hondzo. Saint Anthony Falls Hydraulic Laboratory, University of Minnesota, Minneapolis, MN 55414, USA

John G. Eaton, and J. Howard McCormick. Environmental Research Laboratory-Duluth., U.S. Environmental Protection Agency, Duluth, MN 55804, USA

Introduction

The University of Minnesota's St. Anthony Falls Hydraulic Laboratory and the United States Environmental Protection Agency (U.S. EPA) Environmental Research Laboratory-Duluth have cooperated to establish a methodology to estimate projected global climate change impacts on Minnesota lake and stream temperatures and dissolved oxygen concentrations. These projections are subsequently used to predict effects on indigenous fishery resources (Stefan et al. 1992*a*). The methodology uses some concepts previously outlined by Magnuson et al. (1979), Meisner et al. (1987), and Coutant (1990).

According to global climate change models, e.g., that developed by Columbia University's Goddard Institute for Space Studies (GISS), Minnesota's mean annual air temperature will increase by an average of approximately 4.0°C as the mean atmospheric CO_2 doubles (R. Jenne and D. Joseph, National Center for Atmospheric Research, personal communication). This is likely to have many environmental consequences, including changes in water temperatures and dissolved oxygen (DO) concentrations. These changes in lake and stream physical characteristics are in turn expected to have a profound effect on indigenous fish populations. These interactions between climate, lake water quality, and fish populations has been investigated through model simulations and presented here using Minnesota lakes as a case study. Results of this study were given by Stefan et al. (1992*a*, *b*). A summary is given herein, but background information from the previous reports will be necessary to fully appreciate the methodology and results.

The description of fish habitat includes lake benthic area and lake volume lying within the appropriate thermal and DO range for good growth of various species of fish. The use of habitat characteristics of lakes to estimate fish productivity or yield is not new (e.g., Moyle 1956; Ryder 1965, 1982; Schlesinger and Regier 1982). Christie and Regier (1988) related the estimated optimal temperature range for growth (optimal thermal niches) of four fish species to their yield as a function of lake benthic or water column (pelagic) habitat space having corresponding summer temperatures. Regression equations for sustained yield were developed by integrating daily pelagic or benthic optimal habitat space over the summer. Sustained yield of all four species in 21 lakes was significantly correlated with optimal thermal habitat area, and for two species was even more strongly correlated with benthic habitat volume. An attempt was made to relate habitat preference (benthic vs. pelagic) with the strength of the volume or area correlation. Unfortunately, detailed water temperature measurements in lakes are seldom available to extend the use of these concepts.

Numerical simulation models are useful for predicting thermal structure and dissolved oxygen distributions in lakes of different geometries and latitudes and for extrapolating to possible future climates. Such models calculate heat transfer from the atmosphere to the water and within the water column. Once water temperature profiles in lakes are computed, regressions between good growth habitat factors and sustained yield can be used to predict the effect of climate change on fish growth potential in lakes.

The first question we address herein is how good growth habitat area (GGHA) and good growth habitat volume (GGHV) vary among different lake classes and the three thermal guilds of fish (cold-water, cool-water, and warm water fishes) in Minnesota lakes. Both GGHA and GGHV are of interest because different species of fish have different degrees of dependence on benthic and pelagic food sources. Secondly, we will estimate the climate change impact on these two yield-determinant factors. Lastly, we will evaluate the impact on benthic and pelagic habitat available for fishes in 3002 Minnesota lakes.

Methods of analysis

Concepts

A one-dimensional lake water quality model (Riley and Stefan 1988) was adapted for this study (Stefan et al. 1993*a*; Hondzo and Stefan 1993*a*). This model has been successfully applied

Table 1. Physical parameters used to define 27 Minnesota lake classes.

Lake parameter	Descriptive term	Representative value used	Range	Cumulative frequency
Maximum depth (m)	Shallow	4.0	< 5.0	Lower 30%
	Medium	13.0	5.0–20.0	Central 60%
	Deep	24.0	> 20.0	Upper 10%
Surface area (km^2)	Small	0.2	< 0.4	Lower 30%
	Medium	1.7	0.4–5.0	Central 60%
	Large	10.0	> 5.0	Upper 10%
Secchi depth (m)	Eutrophic	1.2	< 1.8	Lower 20–50%
	Mesotrophic	2.5	1.8–4.5	Central 20–50%
	Oligotrophic	4.5	> 4.5	Upper 0–10%

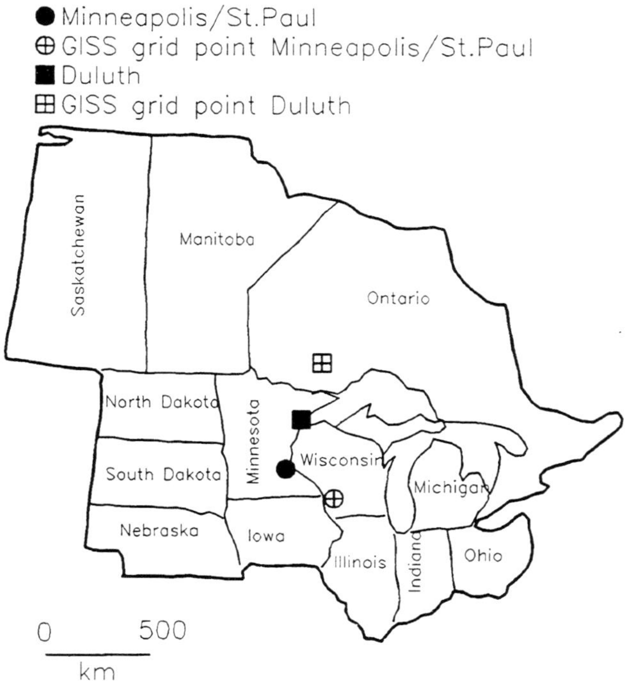

Fig. 1. Geographical location of the grid points closest to Minneapolis – St. Paul and Duluth, Minn., in the Columbia University Goddard Institute for Space Studies (GISS) climate model.

for several years to simulate hydrothermal and kinetic processes in individual north-central U.S. lakes and for a variety of meteorological conditions (Riley and Stefan 1988; Hondzo and Stefan 1991, 1993*a*). In the model, a lake is described by a system of horizontal layers, each of which is well mixed. The vertical temperature profile in the lake is computed from a balance between incoming heat from solar and longwave radiation and the outflow of heat through evaporation, convection and back radiation. The vertical DO profile in the lake is computed from a balance between oxygen sources (reaeration and photosynthesis) and the oxygen sinks (sedimentary oxygen demand, biochemical oxygen demand, and respiration) (Stefan et al. 1993*a*; Stefan and Fang 1993, 1994).

Simulations of water temperatures were made for 27 classes of lakes in southern and northern Minnesota (Hondzo and Stefan 1993*b*). These classes were based on surface area, maximum depth, and trophic state (Table 1). Representative area–depth relationships for three different lake classes (by surface area) were obtained from 122 lakes, which covered the entire range of distributions in a set of 3002 lakes. Simulations were made on a daily time step for past weather conditions (1955–1979) and for the 2 CO_2 (doubling of CO_2) GISS model climate scenario. The GISS grid points closest to stations having the best long-term meteorological records in northern (Duluth) and southern (Minneapolis – St. Paul) Minnesota are given in Fig. 1. For the 2 CO_2 GISS climate scenario, past weather conditions were modified by the monthly increments

Table 2. Expected changes in climate characteristics for Minneapolis – St. Paul and Duluth, Minn., following a doubling of atmospheric CO_2, using the Columbia University Goddard Institute for Space Studies (GISS) climate model. AT = air temperature, DT = dew point temperature, SR = solar radiation, and WS = wind speed.

Month	Minneapolis/St. Paul				Duluth				Difference (Duluth/Minneapolis/St. Paul)			
	AT (°C) diff[a]	DT (°C) diff	SR (–) ratio[b]	WS (–) ratio	AT (°C) diff	DT (°C) diff	SR (–) ratio	WS (–) ratio	AT (°C) diff	DT (°C) diff	SR (–) ratio	WS (–) ratio
March	5.20	6.54	0.98	0.47	4.80	7.24	1.04	0.82	–0.40	0.70	0.06	0.35
April	5.05	4.65	1.03	0.69	4.97	4.49	1.00	0.85	–0.08	–0.16	–0.03	0.16
May	2.63	3.71	1.00	0.67	1.54	1.96	1.04	0.57	–1.09	–1.75	0.04	–0.10
June	3.71	3.59	0.99	0.85	3.51	5.20	1.00	0.74	–0.20	1.61	0.01	–0.11
July	2.15	0.93	0.98	0.93	2.59	2.14	0.97	0.75	0.44	1.21	–0.01	–0.18
August	3.79	3.69	1.04	1.00	2.80	3.06	0.98	0.88	–0.99	–0.63	–0.06	–0.12
September	7.02	4.68	1.04	1.07	3.96	2.66	1.01	0.81	–3.06	–2.02	–0.03	–0.26
October	3.73	2.74	1.12	2.23	3.89	0.97	0.76	0.76	0.16	–1.77	–0.15	–1.47
November	6.14	5.71	1.03	5.00	5.93	5.51	0.95	1.06	–0.21	–0.20	–0.08	–3.94

[a]Difference = (2 CO_2 GISS – Past).

[b]Ratio = (2 CO_2 GISS)/Past.

or ratios of change. These increments were monthly additive or multiplicative values. Numerical values for the meteorological variable increments or ratios used in the simulations are given in Table 2. Output from the regional lake water temperature model (Stefan et al. 1993*a*; Hondzo and Stefan 1993*b*) was used in the regional DO model to predict oxygen concentrations in all 27 lake classes for the same time period (Stefan and Fang 1994). Resulting computations provided daily thermal and DO structures in a given lake class for estimated past and predicted future climatic conditions.

Temperature criteria for three fish guilds were developed from laboratory and field data as described by Eaton et al. (1995) and the guild designations for various species as suggested by Hokanson (1977). Table 3 gives temperature limits

Table 3. Thermal criteria for fish (guild means and ranges for species within a guild; Stefan et al. 1992*a*).

Guild	Lower good-growth temp. LGGT (°C)	Upper good-growth temp. UGGT (°C)	Upper lethal temperature LT (°C)	Optimum temperature OT (°C)
Cold water				
Mean	9.0	18.5	23.4	15.3
Range	(6.4–118)	(15.5–21.2)	(22.1–26.6)	(11.5–18.7)
Cool water				
Mean	16.3	28.2	30.4	25.1
Range	(13.2–18.2)	(27.7–28.8)	(28.0–32.3)	(24.0–25.7)
Warm water				
Mean	19.7	32.3	Undetermined[a]	29.2
Range	(17.7–22.5)	(31.4–34.7)		(27.0–32.0)

[a]Greater than 32.3°C (UGGT)

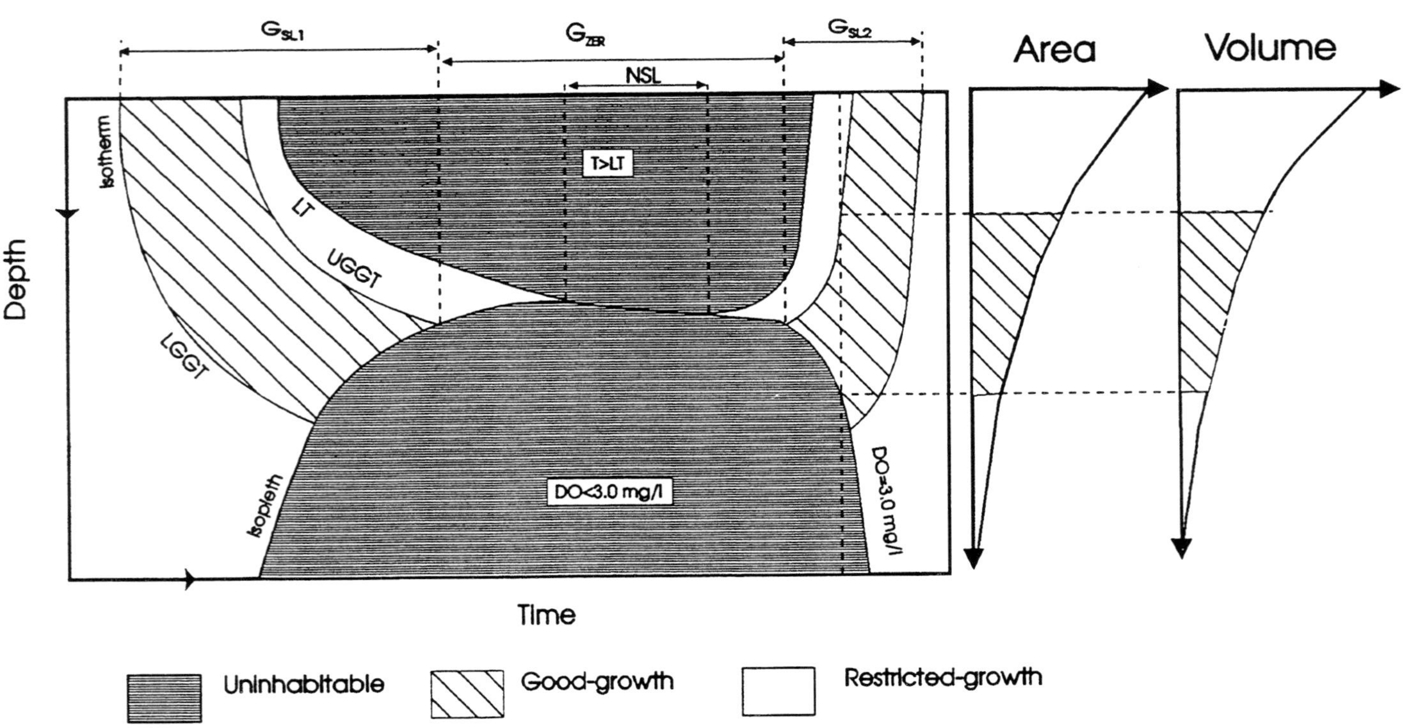

Fig. 2. Schematic of the distribution over time and depth of those isotherms and dissolved oxygen isopleths which are considered for the survival and growth of a fish species or thermal guild in a lake. LGGT=lower good growth temperature; UGGT=upper good growth temperature; LT=upper lethal temperature; NSL=nonsurvival period; G_{SL}=growth season length; G_{ZER}=nogrowth period; DO=dissolved oxygen.

for the three thermal guilds of fish which comprised a total of 28 species. Fish survival and fish growth temperature criteria were related to simulated daily water temperature and DO as shown schematically in Fig. 2: Three isotherms were plotted for each guild. They designate the lethal temperature threshold (LT), the upper good growth temperature limit (UGGT), and the lower good growth temperature limit (LGGT), respectively (Stefan et al. 1992*a*). In essence, the LT was determined from field data that relate maximum annual water temperature to geographic distribution (Eaton et al. 1995). The laboratory-derived LGGT consisted of the mean temperature between zero net growth and maximum growth, and the UGGT is the U.S. EPA upper temperature criterion for growth (optimum plus 1/3(ultimate incipient lethal temperature minus optimum temperature); U.S. EPA 1976). Guild values are the means of available LT, LGGT, and UGGT values for species in each guild. Where sufficient experimental data were available, growth values were obtained by mathematically fitting a curve to laboratory growth rate data.

Dissolved oxygen limits of 2.5 $mg \cdot L^{-1}$ for warmwater fish and 3.0 $mg \cdot L^{-1}$ for cool- and cold-water fish were based on the U.S. EPA Water Criteria document (1986) and discussions with G. Chapman (U.S. EPA Environmental Research Laboratory, Corvallis, Oreg.). The isopleth that designates the critical DO survival value is shown in Fig. 2. Between the lines in Fig. 2, three habitats can be identified:

(1) Uninhabitable space when the temperature is above, or the DO is below the survival or threshold limit for seven consecutive days.

(2) Good growth habitat if the temperature is between the upper and lower good growth limits and the DO is above the survival limit.

(3) Restricted-growth habitat if the temperature is above the upper good growth limit but below the upper survival limit; or if the temperature is below the lower good growth limit, and the DO will permit survival.

Validation of the predicted effects of these two parameters on fish populations was approached by matching actual fish observations with simulated suitable fish habitat in Minnesota lakes. Good agreement between fish observations and numerical simulations of fish habitat defined by water temperature and DO concentrations was found. In 759 out of 771 southern Minnesota lakes, predicted fish-guild presence was in agreement with surveys of fish observations (Stefan et al. 1993*b*). Assessment techniques employed herein are therefore deemed reasonable for predicting environmental conditions suitable for presence of various fishes in lakes.

Good growth habitat area

For a given lake class, lake depths (meters) are recorded over time (days) where and when water temperatures are (1) within the upper and lower good growth temperature limits, and (2) the DO is greater than the survival limit. Figure 3 is an illustrative example showing depth–time isotherms and DO isopleths for cold-water fish in an oligotrophic, medium-size (1.7 km^2), medium-depth (13 m), northern Minnesota lake for past climate (1955–1979), and for the future 2 CO_2 GISS climate scenario. Relationships of the same kind have been determined mathematically and graphically for all 27 lake classes in northern and southern Minnesota (Stefan et al. 1992*a*). Although useful, the good growth depth alone is not a reliable indicator of fish growth in a lake. Depth implicitly ignores lake morphometry (i.e., the area–depth–volume relationship) including littoral area within acceptable temperature and DO limits.

Good growth habitat depth can be used together with a lake's area–depth relationship to estimate good growth habitat area. For a given lake class, lake areas (square kilometres) are integrated over time (days) where and when (1) water temperatures are within the upper and lower good growth temperature limits, and (2) DO is equal to or greater than the survival limit. Integration begins when the lower good growth limit and DO survival limits are first exceeded. This is then the summation of possible good growth areas over a growing season. Figure 4 illustrates relative lake area habitats for cold-water fish at the beginning of specific months in oligotrophic, medium-size, medium-depth, northern Minnesota lakes for past climate (1955–1979) and for future 2 CO_2 GISS climate scenarios. A circle was chosen as a good visualization to represent the area (bottom) of a lake. The full circle represents total lake area and the center of the circle represents the point of maximum depth. The picture given for each month can be interpreted as a top view of the bottom contours (of a circular lake) separating uninhabitable, good growth, and restricted-growth areas.

Good growth habitat volume

For a given lake class, lake volumes (cubic meters) are integrated over time (days) where and when water temperatures are (1) within the upper and lower good growth temperature limits, and (2) DO is greater than the survival limit. Integration begins when the lower good growth limit and DO survival limit are first exceeded. The result is the summation of possible good growth volumes over a growing season. Figure 5 illustrates relative lake volume habitats for cold-water fishes by months in oligotrophic, medium-size, medium depth, northern Minnesota lakes. Past climate conditions (1966–79) are shown on the left and the future 2 CO_2 GISS climate scenario on the right. A square represents the total volume of a lake at the beginning of the noted months. The shape of a square was chosen as an abstract representation of lake volume.

Results

Projected climate change impact on good growth habitat areas

Individual lakes

Good growth bottom areas were integrated over time (square kilometres times days) for all 27 classes of southern and northern Minnesota lakes by Stefan et al. (1992*a*, *b*). The results expressed as percentages of the GGHA are given herein. Loss or gain of GGHA is expressed as a percentage also. The difference in GGHA for the past and future climate scenarios is defined as $GGHA_{GISS}$ minus $GGHA_{Past}$. Projected differences were normalized with respect to the $GGHA_{past}$ and multiplied by 100. The percentages of GGHA for all 27 classes of southern and northern Minnesota lakes are given

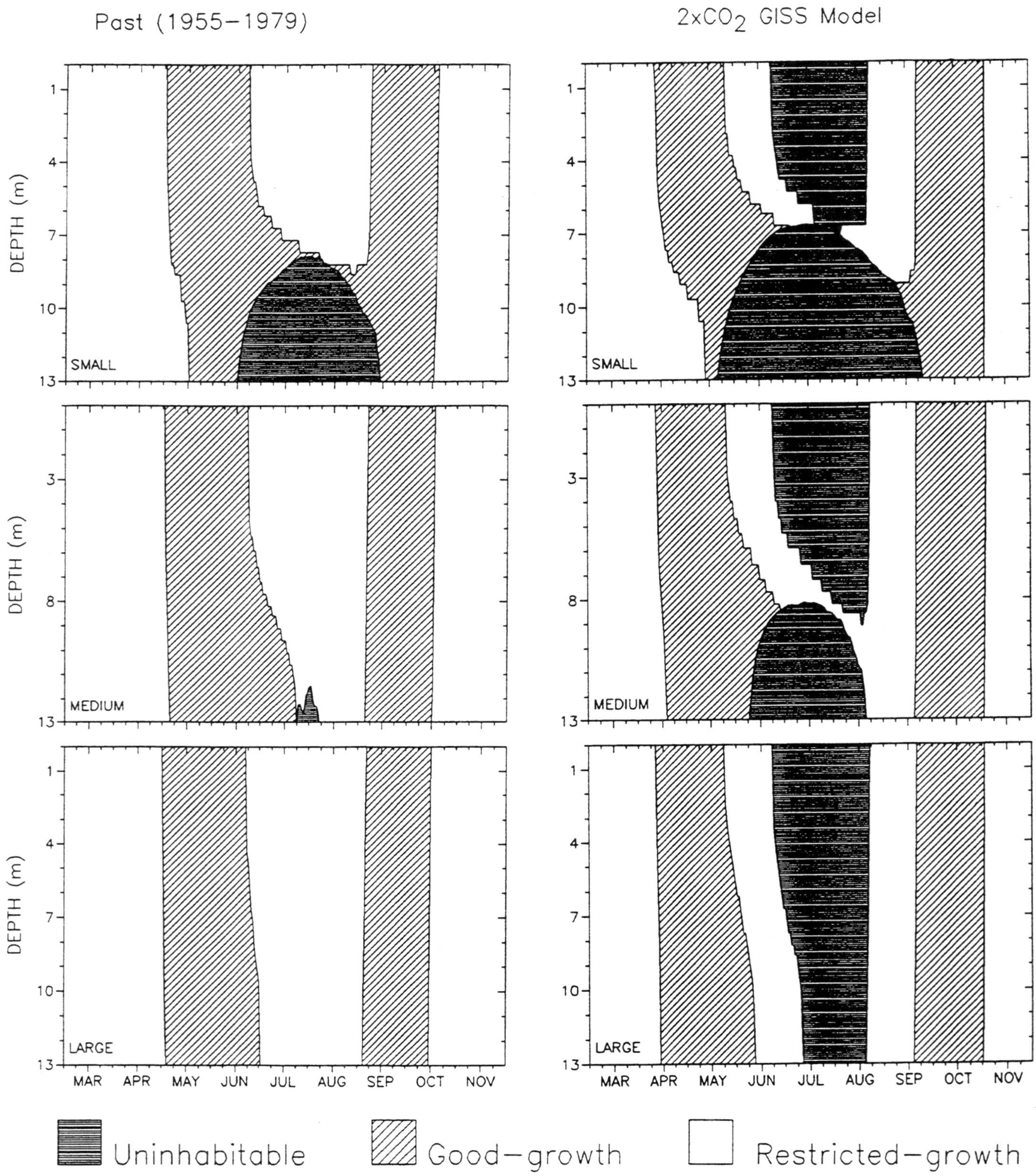

Fig. 3. Depth–time contours of cold-water fish habitats in oligotrophic medium-size (1.7 km^2), medium-depth (13 m), northern Minnesota lakes under past climate conditions (1955–1979) (i.e.,, average water temperature and DO conditions) and for the future 2 CO_2 GISS climate scenario.

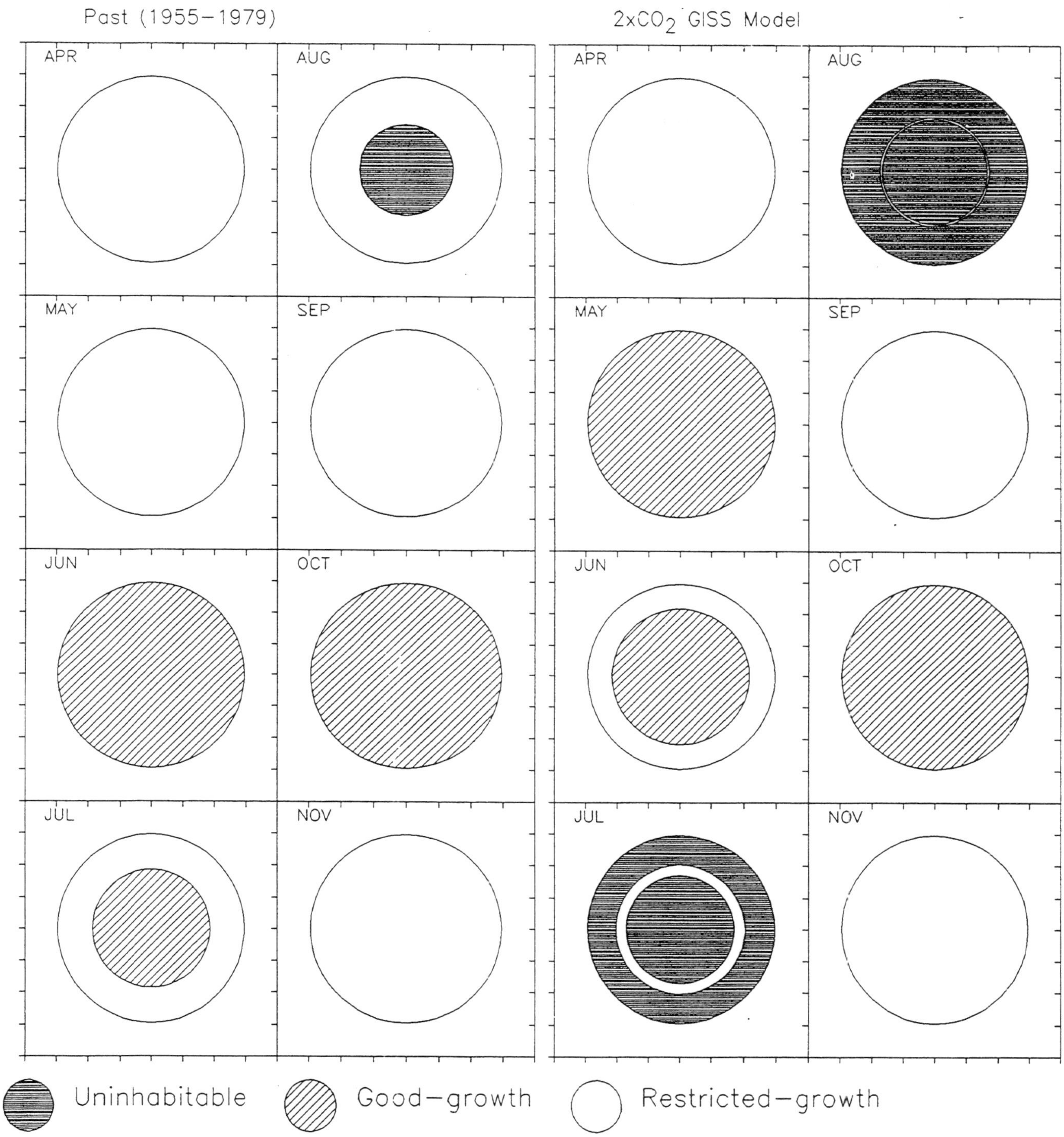

Fig. 4. Relative area of cold-water fish habitats, by month, in oligotrophic medium-size (1.7 km^2), medium-depth (13 m), northern Minnesota lakes under past climate conditions (1955–1979) and for the future 2 CO_2 GISS climate scenario.

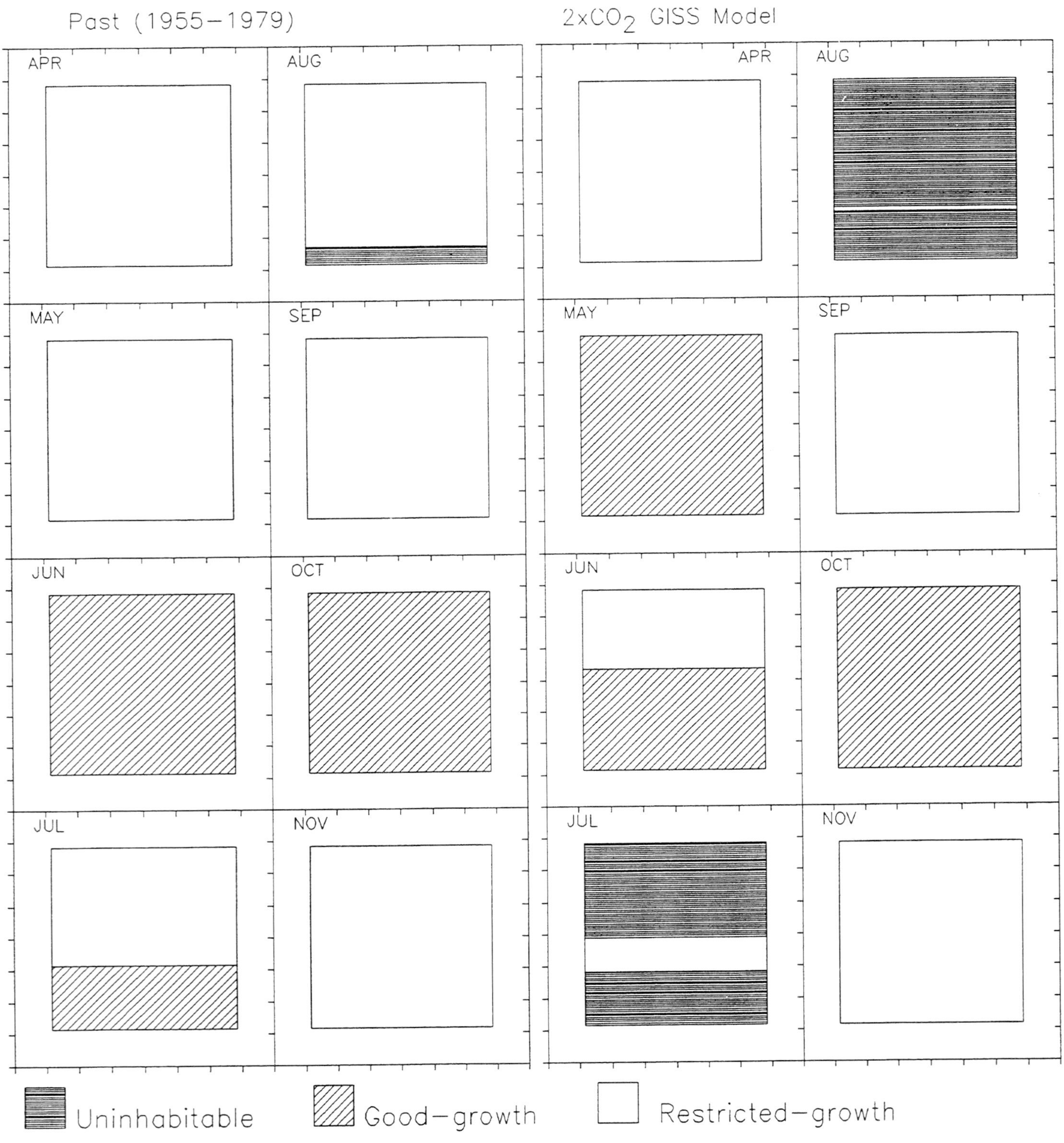

Fig. 5. Relative volume of cold-water fish habitats, by month, in oligotrophic medium-size (1.7 km^2), medium-depth (13 m), northern Minnesota lakes under past climate conditions (1955–1979) and for the future 2 CO_2 GISS climate scenario.

Table 4. Loss (–) or gain (+) of good growth bottom area in Minnesota lakes expressed as a percentage of past values [(GISS – Past)/Past] following a doubling of atmospheric CO_2. E = eutrophic, M = mesotrophic, O = oligotrophic, and U.C. = uninhabitable conditions

Lake characteristics			South (771 lakes)				North (2231 lakes)				Minnesota (3002 lakes)			
Maximum depth (m)	Surface area (km²)	Trophic status	No. of lakes	Cold	Cool	Warm	No. of lakes	Cold	Cool	Warm	No. of lakes	Cold	Cool	Warm
Shallow (4.0)	Small (0.2)	E	45	U.C.	8	27	184	−100	30	91	229	−100	25	72
		M	12	U.C.	13	29	187	−100	28	91	199	−100	27	86
		O	1	U.C.	13	38	2	−100	29	92	3	−100	23	66
	Medium (1.7)	E	170	U.C.	11	31	151	−100	20	81	321	−100	15	47
		M	24	U.C.	12	39	100	−100	29	91	124	−100	25	75
		O	2	U.C.	10	40	2	−100	27	82	4	−100	18	55
	Large (10.0)	E	19	U.C.	16	38	7	−100	27	89	26	−100	18	46
		M	2	U.C.	13	39	3	−100	28	91	5	−100	21	62
		O	0	U.C.	12	39	0	−100	28	92	0	0	0	0
Medium (13.0)	Small (0.2)	E	83	U.C.	−4	13	84	−19	7	68	167	−19	1	31
		M	70	U.C.	4	22	352	−100	20	65	422	−100	17	54
		O	6	U.C.	2	23	78	−100	28	69	84	−100	26	63
	Medium (1.7)	E	142	U.C.	0	15	102	−15	4	53	244	−15	1	25
		M	101	U.C.	6	21	532	−12	17	65	633	−12	15	54
		O	7	U.C.	6	26	76	−5	24	64	83	−5	22	58
	Large (10.0)	E	17	U.C.	12	25	14	−100	15	58	31	−100	13	36
		M	5	U.C.	14	29	45	−100	25	65	50	−100	23	59
		O	0	U.C.	13	35	1	−100	31	71	1	−100	31	71
Deep (24.0)	Small (0.2)	E	1	U.C.	−8	8	6	−100	11	66	7	−100	8	51
		M	7	U.C.	1	17	26	−13	23	71	33	−12	17	52
	O	0	U.C.	3	20	26	−5	35	83	26	−5	35	83	
	Medium (1.7)	E	10	U.C.	−8	6	3	−100	4	52	13	−100	−5	11
		M	33	U.C.	−2	13	104	−13	20	68	137	−13	13	46
		O	3	−27	1	18	71	−7	33	79	74	−8	31	74
	Large (10.0)	E	5	−100	−3	12	5	−10	14	66	10	−52	4	29
		M	6	−100	0	15	53	−5	20	77	59	−14	17	65
		O	0	−10	5	19	17	0	29	81	17	0	29	81

in Table 4. Uninhabitable conditions in a given lake class are indicated by U.C. Figure 4 gives examples of fish benthic habitat areas in oligotrophic northern Minnesota lakes. The results can be interpreted as follows.

A) *Cold-water fish:* encounter uninhabitable conditions in the past as well as under future climate scenarios in 22 of 27 lake classes in southern Minnesota lakes (Table 4). Good-growth conditions are present only in rare, oligotrophic, large-area, deep lakes. After climate change, if such conditions are still present, GGHA is reduced in all lake classes in southern Minnesota lakes.

For northern Minnesota, the simulations indicate habitable conditions in all classes of lakes under past climate conditions (Table 4). After climate change, shallow lakes, large lakes of medium depth, and small lakes of medium depth (except when they are eutrophic) will no longer support cold-water fish. These are essentially all well-mixed lakes. The GGHA will be entirely eliminated from these lakes. In all other lakes, GGHA will be reduced (Table 4).

B) *Cool-water fish:* have habitable conditions in all lake classes in southern Minnesota (Table 4). Climate change will reduce GGHA moderately in eutrophic, small-, medium-, and large-area deep lakes, and in eutrophic, small-area, medium-depth lakes (Table 4). Time-integrated percentages of differences in GGHA will be reduced by an average of 5% in these lakes. For other lake classes, GGHA will increase after climate change. The highest increase will be in well-mixed lakes. All shallow lakes, and large, medium-depth lakes will increase their GGHA by an average of 13%. This occurs because, with climate change, temperatures suitable for good growth will be reached 23–30 days earlier and will end 3–10 days later than before climate warming.

Cool-water fish can survive in all northern Minnesota lakes. Climate change will increase GGHA for all lake classes. Time-integrated GGHA will be increased by an average of 22%. This occurs because with climate change, the good growth season will be 29–41 days longer. In addition, increases in GGHA will be two times greater for northern than for southern Minnesota lakes. Good growth conditions currently prevail throughout the summer in all Minnesota lakes, and according to the predictions for northern Minnesota lakes, there will be no change after climate change.

C) *Warmwater fish:* currently have habitable conditions in all lake classes in northern and southern Minnesota lakes. After climate change, GGHA will increase in all Minnesota lakes. Time-integrated GGHA will increase on average by 25% in southern, and 75% in northern lakes (Table 4). This occurs because with climate change, the good growth season will be 42–52 days longer for northern Minnesota lakes. On the average, differences will be three times greater for northern than southern Minnesota lakes. Among the three fish guilds, the largest increases in GGHA were estimated for warmwater fishes in northern Minnesota lakes.

Regional impact

The regional impact of climate change was determined for each of the 27 lake classes. The difference for the past and future climate scenario ($GGHA_{GISS} - GGHA_{Past}$) is multiplied by the number of lakes in each class (Table 4), and normalized by the sum of differences ($\Sigma GGHA_{GISS} - GGHA_{Past}$) in southern or northern or all Minnesota lakes, respectively (Table 5). The percentages describe the change in GGHA in a particular lake class compared with the total change for all lake classes.

A) *Cold–water fish:* have uninhabitable conditions in all southern Minnesota lakes (Table 5). Although simulations indicate habitable conditions for deep oligotrophic lakes under past climates, the considerable length of the restricted growth– stress season has meant that cold-water fish have rarely been reported in these lakes. With climate change, GGHA will be eliminated from these lakes (Stefan et al. 1993*b*). Deep oligotrophic lakes are uncommon in southern Minnesota (1.8%), thus a reduction in GGHA is generally not considered significant.

Cold-water fish have suitable habitat conditions in all northern Minnesota lakes under average past climate. After climate change, habitable conditions will be eliminated in 16 of 27 lake classes (Stefan et al. 1992*b*) and GGHA will be eliminated in 50% of lakes in northern Minnesota. The highest reduction is expected in large and medium mesotrophic lakes of medium depth (Table 5).

B) *Cool-water fish:* have habitable conditions in all Minnesota lakes under average past and future projected climate conditions. With climate change, GGHA will be reduced in small, medium-depth eutrophic lakes, medium, deep eutrophic and mesotrophic lakes, and large, deep eutrophic lakes. These lakes comprise 17% of lakes in southern Minnesota. The highest increase in GGHA will occur in medium and large shallow eutrophic lakes, and in large eutrophic lakes of medium depth. Cool-water fish, according to the simulation results, have proportionally greater habitable areas in northern than in southern Minnesota lakes. With climate change, GGHA will increase in all lakes. The highest change will occur in mesotrophic lakes of medium size and medium depth.

C) *Warmwater fish:* find habitat in all southern and northern Minnesota lakes under average past and future projected climate conditions. Under projected climate changes, GGHA will increase in all southern and northern Minnesota lakes. The greatest increase in GGHA will occur in eutrophic, shallow-depth, medium and large lakes in southern Minnesota, and mesotrophic, medium-depth, medium and large lakes in northern Minnesota.

Projected impact of climate change on good growth habitat volumes

Individual lakes

The GGHV integration over times (cubic metres times days) between upper and lower good-growth limits in all 27 classes of southern and northern Minnesota lakes is given by Stefan et al. (1992*a*, *b*). Projected parameters, which quantify changes in GGHV, are defined in analogy to the GGHA. For the three thermal guilds of fish survival conditions and good growth season length will be similar numerically to those for GGHA. Projected percentages of differences in GGHV in individual Minnesota lakes due to climate change are given in Table 6. Figure 5 gives a graphical example of fish habitat volumes in oligotrophic northern Minnesota lakes.

A) *Cold-water fish:* habitat is not abundant in southern lakes and even the small GGHV will be diminished by about 14% with projected climate change (Table 6). Cold-water fish will then have only a remote chance of surviving and then only in deep oligotrophic lakes, of which there are only three in the sample set, all of medium size.

With climate change, GGHV will be reduced by an average of 8% in northern Minnesota lakes. An increase in the midsummer stress period of 0–65 days, depending on the lake class, will be associated with climate change.

B) *Cool-water fish:* will gain GGHV with climate change in all southern Minnesota lakes except in eutrophic, small, medium-depth and deep lakes. GGHV will increase by an average of 8% in the south and 25% in the north, after climate change.

C) *Warmwater fish:* will gain GGHV in all Minnesota lakes. GGHV will increase by an average of 27 and 76% in southern and northern Minnesota lakes, respectively.

Regional impact

Projected percentages of differences in GGHV for each of the 27 lake classes are given in Table 7. The percentages describe the change in GGHV in a particular lake class compared with the total change for all lake classes.

Table 5. Loss (–) or gain (+) of good growth bottom area in Minnesota lakes expressed as a percentage of total changes [(GISS – Past)/Σ(GISS – Past)] following a doubling of atmospheric CO_2. E = eutrophic, M = mesotrophic, O = oligotrophic, and U.C. = uninhabitable conditions.

Lake characteristics			South (771 lakes)			North (2231 lakes)			Minnesota (3002 lakes)		
Maximum depth (m)	Surface area (km^2)	Trophic status	Cold	Cool	Warm	Cold	Cool	Warm	Cold	Cool	Warm
Shallow	Small	E	U.C.	1	1	2	2	2	2	1	1
(4.0)	(0.2)	M	U.C.	0	0	2	2	2	2	1	1
		O	U.C.	0	0	0	0	0	0	0	0
	Medium	E	U.C.	30	28	16	8	10	15	12	14
	(1.7)	M	U.C.	5	5	11	8	8	10	7	7
		O	U.C.	0	0	0	0	0	0	0	0
	Large	E	U.C.	30	24	5	3	3	5	8	8
	(10.0)	M	U.C.	3	3	2	1	1	2	2	2
		O	U.C.	0	0	0	0	0	0	0	0
Medium	Small	E	U.C.	0	0	0	0	0	0	0	0
(13.0)	(0.2)	M	U.C.	0	1	4	1	1	4	1	1
		O	U.C.	0	0	1	1	0	1	0	0
	Medium	E	U.C.	0	7	2	1	3	2	1	4
	(1.7)	M	U.C.	8	9	7	23	22	7	18	19
		O	U.C.	1	1	0	5	4	0	4	3
	Large	E	U.C.	18	13	10	3	4	9	6	6
	(10.0)	M	U.C.	7	5	31	15	15	29	16	13
		O	U.C.	0	0	1	1	0	1	0	0
Deep	Small	E	U.C.	0	0	0	0	0	0	0	0
(24.0)	(0.2)	M	U.C.	0	0	0	0	0	0	0	0
		O	U.C.	0	0	0	0	0	0	0	0
	Medium	E	U.C.	–1	0	0	0	0	0	0	0
	(1.7)	M	U.C.	0	1	1	3	2	1	2	2
		O	1	0	0	1	4	2	1	3	2
	Large	E	43	–1	1	0	1	1	3	0	1
	(10.0)	M	55	0	2	2	13	13	6	11	11
		O	0	0	0	0	7	5	0	6	4

A) *Cold-water fish:* encounter uninhabitable conditions in most southern Minnesota lakes. With climate change, these habitat conditions will worsen. The GGHV will be eliminated in 16 of 27 lake classes in Minnesota. These are shallow lakes (all sizes), and medium-depth large and small lakes (Table 7).

B) *Cool-water fish:* will gain GGHV with climate change in all southern Minnesota lakes except eutrophic small and medium, deep and medium-depth lakes. The greatest change is expected in medium-depth large lakes. Climate change will increase GGHV in all northern Minnesota lakes. The greatest GGHV change is expected in mesotrophic, medium, medium-depth, and large, deep lakes.

C) *Warmwater fish:* will gain GGHV with climate change in all Minnesota lakes. The highest increase in GGHV is expected in large, medium-depth and deep lakes in southern Minnesota, and in mesotrophic, medium-depth, medium size lakes in northern Minnesota.

Table 6. Loss (–) or gain (+) of good growth volumes in Minnesota lakes expressed as percentages of past values [(GISS – Past)/Past] following a doubling of atmospheric CO_2. E = eutrophic, M = mesotrophic, O = oligotrophic, and U.C. = uninhabitable conditions.

Lake characteristics			South (771 lakes)				North (2231 lakes)				Minnesota (3002 lakes)			
Maximum depth (m)	Surface area (km²)	Trophic status	No. of lakes	Cold	Cool	Warm	No. of lakes	Cold	Cool	Warm	No. of lakes	Cold	Cool	Warm
Shallow (4.0)	Small (0.2)	E	45	U.C.	11	31	184	−100	30	92	229	−100	25	73
		M	12	U.C.	11	39	187	−100	29	92	199	−100	28	86
		O	1	U.C.	13	39	2	−100	29	90	3	−100	23	66
	Medium (1.7)	E	170	U.C.	13	35	151	−100	25	88	321	−100	18	52
		M	24	U.C.	12	40	100	−100	29	91	124	−100	25	75
		O	2	U.C.	10	40	2	−100	28	81	4	−100	18	55
	Large (10.0)	E	19	U.C.	16	37	7	−100	28	91	26	−100	18	46
		M	2	U.C.	13	38	3	−100	28	92	5	−100	21	62
		O	0	U.C.	12	38	0	−100	28	91	0	0	0	0
Medium (13.0)	Small (0.2)	E	83	U.C.	0	17	84	−41	11	70	167	−41	5	35
		M	70	U.C.	7	24	352	−100	24	66	422	−100	20	55
		O	6	U.C.	5	26	78	−100	30	69	84	−100	28	64
	Medium (1.7)	E	142	U.C.	5	18	102	−29	11	57	244	−29	7	29
		M	101	U.C.	10	25	532	−14	23	66	633	−14	20	55
		O	7	U.C.	8	28	76	1	28	66	83	1	26	61
	Large (10.0)	E	17	U.C.	14	28	14	−100	21	63	31	−100	17	39
		M	5	U.C.	15	31	45	−100	28	67	50	−100	26	61
		O	0	U.C.	12	34	1	−100	33	70	1	−100	33	70
Deep (24.0)	Small (0.2)	E	1	U.C.	−6	10	6	−100	13	66	7	−100	10	52
		M	7	U.C.	3	19	26	−46	24	71	33	−46	19	53
		O	0	U.C.	6	21	26	−38	35	83	26	−38	35	83
	Medium (1.7)	E	10	U.C.	−4	9	3	−100	9	55	13	−100	−2	14
		M	33	U.C.	2	15	104	−49	23	70	137	−49	16	48
		O	3	−19	6	20	71	−40	34	80	74	−39	33	75
	Large (10.0)	E	5	−100	1	15	5	−37	18	68	10	−67	9	32
		M	6	−100	4	18	53	−30	24	78	59	−36	21	66
		O	0	−8	10	21	17	−20	32	80	17	−20	32	80

Summary of total regional impact

A summary of daily GGHA and GGHV values integrated over a growing season is given in Table 8 for 771 southern and 2231 northern Minnesota lakes. Although GGHA and GGHV do not have the same dimensions, it is evident that both parameters have the same trends. With climate change, GGHA or GGHV will be reduced for cold-water fish and increased for cool- and warmwater fish. Relative changes due to climate effects expressed as percentages of the values before climate change are given in Table 9. Cold-, cool-, and warmwater fish habitats will experience changes in GGHA and GGHV. There are substantial differences between northern and southern Minnesota lakes. The percent loss for cold-water fish in the south will be about twice that in the north. The percent gain for cool-water and warmwater fish will be three times larger in the north than in the south. It should be emphasized that this analysis is not capable of predicting the individual species that will occur in a given category of lakes after climate change, but only the habitat available for representatives of warm-, cool, and cold-water guilds.

Table 7. Loss (–) or gain (+) of good growth volume in Minnesota lakes expressed as a percentage of total changes [(GISS – Past)/Σ(GISS – Past)] following a doubling of atmospheric CO_2. E = eutrophic, M = mesotrophic, O = oligotrophic, and U.C. = uninhabitable conditions.

Lake characteristics			South (771 lakes)			North (2231 lakes)			Minnesota (3002 lakes)		
Maximum depth (m)	Surface area (km²)	Trophic status	Cold	Cool	Warm	Cold	Cool	Warm	Cold	Cool	Warm
Shallow (4.0)	Small (0.2)	E	U.C.	0	0	1	0	0	1	0	0
		M	U.C.	0	0	1	0	0	1	0	0
		O	U.C.	0	0	0	0	0	0	0	0
	Medium (1.7)	E	U.C.	14	14	5	2	3	4	4	5
		M	U.C.	12	2	3	2	2	3	2	2
		O	U.C.	0	0	0	0	0	0	0	0
	Large (10.0)	E	U.C.	10	9	1	1	1	1	2	2
		M	U.C.	1	1	0	0	0	0	0	0
		O	U.C.	0	0	0	0	0	0	0	0
Medium (13.0)	Small (0.2)	E	U.C.	0	1	0	0	0	0	0	0
		M	U.C.	1	1	4	2	2	4	2	2
		O	U.C.	0	0	1	1	0	1	0	0
	Medium (1.7)	E	U.C.	12	15	3	2	3	3	3	5
		M	U.C.	19	16	8	22	23	7	22	22
		O	U.C.	1	1	0	4	4	0	4	3
	Large (10.0)	E	U.C.	25	17	8	3	3	7	6	6
		M	U.C.	8	6	24	14	12	22	13	11
		O	U.C.	0	0	1	0	0	0	0	0
Deep (24.0)	Small (0.2)	E	U.C.	0	0	0	0	0	0	0	0
		M	U.C.	0	0	0	0	0	0	0	0
		O	U.C.	0	0	0	0	0	0	0	0
	Medium (1.7)	E	U.C.	–1	1	1	0	0	1	0	0
		M	U.C.	1	4	10	6	5	9	5	5
		O	1	1	1	6	7	5	6	6	4
	Large (10.0)	E	44	1	4	2	2	2	6	1	2
		M	55	5	6	18	22	22	21	20	20
		O	0	0	0	4	10	8	3	9	7

The consequence of complete elimination of suitable habitat, where it was once present, i.e., no production of that species or guild of fish, is one interpretation of the results of this study. But what are the consequences of reductions or increases in GGHA or GGHV of various magnitudes? Here we generalize and assume that increases in GGHA or GGHV will have a beneficial effect on fish production. The value of GGHA, at least theoretically, as an indicator of potentially available benthic feeding area has real potential as a measure of fish production, although sediment composition, food availability, and other factors, are also important but are not yet factored into the calculations. In a practical sense, it can be assumed that if GGHA increases, GGHV will also increase but this interpretation is also dependent on whether the fish community involved is benthic or pelagic. Christie and Regier (1988) assert that for walleye (*Stizostedion vitreum*) and northern pike (*Esox lucius*) GGHA is a better independent variable for sustained yield prediction than is GGHV. In contrast,

Table 8. Total good growth areas (billion m²·d) and volumes (billion m³·d) integrated over time for Minnesota lakes. Past indicates long–term average (1955–79), and 2 CO_2 indicates a doubling of atmospheric CO_2 following the GISS climate model.

Climate scenario	Cold–water fish: Total bottom area	Cold–water fish: Total volume	Cool–water fish: Total bottom area	Cool–water fish: Total volume	Warmwater fish: Total bottom area	Warmwater fish: Total volume
South (771 lakes)						
Past	11.0	86.0	165.0	631.0	127.0	481.0
2 CO_2	0.4	4.0	179.0	684.0	161.0	597.0
Difference	–10.6	–82.0	14.0	53.0	34.0	115.0
North (2231 lakes)						
Past	365.0	1 872.0	313.0	1 627.0	164.0	869.0
2 CO_2	217.0	1 102.0	378.0	2 028.0	282.0	1489.0
Difference	–148.0	–769.0	65.0	401.0	118.0	620.0
Minnesota (3002 lakes)						
Past	376.0	1 958.0	477.0	2 258.0	291.0	1 351.0
2 CO_2	218.0	1 106.0	557.0	2 713.0	442.0	2 086.0
Difference	–158.0	–851.0	80.0	454.0	151.0	735.0

Table 9. Projected changes in good growth areas and good growth volumes expressed as percentages of past values [(GISS – Past)/Past] following a doubling of atmospheric CO_2.

Region	Cold–water fish: Total bottom area	Cold–water fish: Total volume	Cool–water fish: Total bottom area	Cool–water fish: Total volume	Warm–water fish: Total bottom area	Warm–water fish: Total volume
South (771 lakes)	–96	–96	8	8	27	24
North (2231 lakes)	–41	–41	21	25	72	71
Minnesota (3002 lakes)	–42	–43	17	20	52	54

GGHV was considered better than GGHA for lake trout (*Christivomer namaycush*) and lake whitefish (*Coregonus clupeaformis*). They also demonstrated that linear relationships between fish productivity and GGHA and GGHV existed for the four species. However, GGHA and GGHV were defined by a 4°C temperature band near the optimum temperature for growth (Magnuson et al. 1979) rather than the good growth temperature ranges of 9.0–18.5°C, 16.3–28.2°C, and 19.7–32.3°C given in Table 3 and used for cold-, cool-, and warmwater fish guilds, respectively, in this study. The wider ranges used in our study were intended to indicate total quantitative changes in growth with climate change, i.e., to predict with a high degree of confidence the proportion of a total resource affected.

Using the relationships between sustained yield and GGHA or GGHV developed by Christie and Regier (1988), we have used estimated lake GGHA and GGHV (when temperatures and DO are within the good growth limits) as independent variables in estimating the total sustained yield of each of four commercially important fish species, i.e., lake trout, lake whitefish, walleye, and northern pike. Our GGHA and GGHV values were adjusted by the relative temperature ranges 4°/9.5° and 4°/11.9° for the cold- and cool-water species, respectively, before use in the Christie and Regier equations. The rough estimates of the total sustained yield (kilograms per year) before and after climate change for Minnesota lakes are given in Table 10. Lake trout and lake whitefish production in Minnesota lakes are projected to be reduced with climate

Table 10. Rough estimate of total sustained fish yield (kg·y^{-1}) for Minnesota lakes. Past indicates long–term average (1955–79), and 2 CO_2 indicates a doubling of atmospheric CO_2 following the GISS climate model. Fish yield values are based on simulated habitat areas and volumes as independent variables.

	Lake trout		Lake whitefish		Walleye		Northern pike	
Climate scenario	Area	Volume	Area	Volume	Area	Volume	Area	Volume
South (771 lakes)								
Past	2 897	6 647	86 977	91 693	243 852	56 621	125 704	45 855
2 CO_2	19	554	11 907	28 577	261 756	59 093	131 035	46 902
Change (%)	−99	−92	−86	−69	7	4	4	2
North (2231 lakes)								
Past	593 850	80 590	711 123	295 599	425 636	93 540	174 245	59 781
2 CO_2	269 408	52 466	520 529	241 689	501 572	105 125	191 846	63 585
Change (%)	−55	−35	−27	−18	18	12	10	6
Minnesota (3002 lakes)								
Past	621 265	83 576	723 906	300 688	614 082	111 284	216 012	65 527
2 CO_2	271 297	52 620	521 967	242 022	702 763	122 656	233 786	68 983
Change (%)	−56	−37	−28	−20	14	10	8	5

change by 37 and 20%, respectively, if the projected doubling of atmospheric CO_2 takes place. Walleye and northern pike production in Minnesota lakes are projected to increase by 14 and 8%, respectively.

Conclusions

Lake depth, lake size, and trophic status are the three lake parameters that were used as independent variables in this study. Their effects on simulated survival and good growth of fish in Minnesota lakes can be summarized as follows:

1) Lake depth. The losses in cold-water fish good growth potential will be larger in shallow lakes than in medium-depth and deep lakes. The gains in the cool-water fish good growth potential will be larger in well-mixed lakes, i.e., shallow-depth, and large, medium-depth lakes. The gains in warmwater fish good growth potential will be similar for all lake classes.

2) Lake size. The losses in cold-water fish good growth potential will be larger in small and large lakes than in medium lakes. The gains in the cool- and warm-water fish good growth parameters will be larger in small lakes and similar for medium and large lakes.

3) Trophic status. The effect of lake trophic status on fish in shallow lakes is small. Good growth potential is nearly the same in oligotrophic and eutrophic shallow lakes, both of which tend to be well mixed. In deep lakes, which have a seasonal stratification and are dimictic, oligotrophy is usually associated with higher growth potential. This trend continues after climate change. The largest losses in good growth volume due to climate change appear to occur in eutrophic lakes (for cold-water fish); the largest gains (for cool- and warm-water fishes) are projected to occur in oligotrophic lakes.

4) Cold-water fish are projected to lose GGHA and GGHV by the same percentage if the projected climate change occurs. Cold-water fish, now rare in southern Minnesota lakes, will virtually disappear and will experience a habitat reduction of 41% in northern Minnesota lakes.

5) Cool-water fish and warmwater fish will gain GGHA and GGHV. The increase will be two or three times higher for northern Minnesota lakes than for southern Minnesota lakes.

Acknowledgements

We are grateful to the U.S. Environmental Protection Agency, Office of Program Planning and Evaluation, Washington, D.C., and Environmental Research Laboratory, Duluth, Minn., for support of this study. B.E. Goodno and K.E.F. Hokanson provided lake data bases and fish–temperature data bases, respectively, without which this work could not have been done. We are also grateful to the reviewers of this paper for their suggestions.

References

Christie, C.G., and H.A. Regier. 1988. Measurements of optimal thermal habitat and their relationship to yields for four commercial fish species. Can. J. Fish. Aquat. Sci., 45: 301–314.

Coutant, C.C. 1990. Temperature-oxygen habitat for freshwater and coastal striped bass in a changing climate. Trans. Am. Fish. Soc. 119(2): 240–253.

Eaton, J.G., J.H. McCormick, B.E. Goodno, D.G. O'Brien, H.G. Stefan, and M. Hondzo. 1995. A field information based system for estimating fish temperature requirements. Fisheries 20(4). (In press)

Hokanson, K.E.F. 1977. Temperature requirements of some percids and adaptations to the seasonal temperature cycle. J. Fish. Res. Board Can. 34(10): 1524–1550.

Hondzo, M., and H.G. Stefan. 1991. Three case studies of lake temperature and stratification response to warmer climate. Water Resour. Res. AGU. 27(8): 1837–2846.

Hondzo, M., and H.G. Stefan. 1993*a*. Lake water temperature simulation model. J. Hydraul. Eng. 119(11): 1251–1273.

Hondzo, M., and H.G. Stefan. 1993*b*. Regional water temperature characteristics of lakes subjected to climate change, Clim. Change, 24: 187–211.

Magnuson, J.J., L.B. Crowder, and P.A. Medvick. 1979. Temperature as an ecological resource. Am. Zool. 19: 331–343.

Meisner, J.D., J.L. Goodier, H.A. Regier, B.J. Shuter, and W.J. Christie. 1987. An assessment of the effects of climate warming on Great Lakes Basin fishes. J. G. Lakes Res. 13: 340–352.

Moyle, J.B. 1956. Relationships between the chemistry of Minnesota surface waters and wildlife management. J. Wild. Manage. 20: 303–320.

Riley, M.J., and H.G. Stefan. 1988. A dynamic lake water quality simulation model. Ecol. Modelling 43: 155–182.

Ryder, R.A. 1965. A method for estimating the potential fish production of north-temperate lakes. Trans. Am. Fish. Soc. 94: 214–218.

Ryder, R.A. 1982. The morphoedaphic index—use, abuse and fundamental concepts. Trans. Am. Fish. Soc. 111: 154–164.

Schlesinger, D.A., and H.A. Regier. 1982. Climatic and morphoedaphic indices of fish yields from natural lake. Trans. Am. Fish. Soc. 111: 141–150.

Stefan, H.G., and X. Fang. 1994. Dissolved oxygen model for regional lake analysis. Ecol. Modelling 71: 37–68.

Stefan, H.G. and X. Fang. 1994. Model simulations of dissolved oxygen characteristics of Minnesota lakes: past and future. Environ. Management 18(1): 73–92.

Stefan, H.G, M. Hondzo, B. Sinokrot, X. Fang, J.G. Eaton, B.E. Goodno, K.E.F. Hokanson, J.H. McCormick, D.G. O'Brien, and J.A. Wisniewski. 1992*a*. A methodology to estimate global climate change impacts on lake and stream environmental conditions and fishery resources with application to Minnesota. 2nd ed. St. Anthony Falls Hydraulic Laboratory, University of Minnesota, Minneapolis. Proj. Rep. 323: 222 p.

Stefan, H.G., M. Hondzo, J.G. Eaton, and J.H. McCormick. 1992b. Predicted effects of global climate change on fishes of Minnesota lakes. St. Anthony Falls Hydraulic Laboratory, University of Minnesota, Minneapolis. Proj. Rep. 323: 45 p.

Stefan, H.G., M. Hondzo, and X. Fang. 1993*a*. Lake water quality modeling for projected future climate scenarios. J. Environ. Qual. 22(3): 417–431.

Stefan, H.G., M. Hondzo, J.G. Eaton, and J.H. McCormick. 1993*b*. Comparisons of actual fish observations with simulated suitable fish habitat in Minnesota lakes. St. Anthony Falls Hydraulic Laboratory, University of Minnesota, Minneapolis. Proj. Rep. 347: 30 p.

U.S. EPA. 1976. Quality criteria for water. Washington D.C., 256 p.

U.S. EPA. 1986. Ambient aquatic life criteria for dissolved oxygen. Washington, D.C. Publ. 440/5-86-003.

Climate change and variability of the Bering Sea current system

Andrew V. Verkhunov

Verkhunov, A.V. 1995. Climate change and variability of the Bering Sea Current system, p. 73–80. *In* R.J. Beamish [ed.] Climate change and northern fish populations. Can. Spec. Publ. Fish Aquat. Sci. 121.

Abstract: During 1990–92, oceanographic data were collected from four quasi-synoptic cruises in the Aleutian and Kamchatka basins of the Bering Sea. Current patterns, derived from dynamical topography maps, exhibited explicit seasonal and interannual variations. Seasonal variations were attributed to wind-stress curl and the heat budget of the basins. Large-scale circulation was most intensive in srping. while mesoscale eddy activity seemed to be higher in late autumn. The most relaxed flow conditions occurred at the end of summer. The flow pattern was modified by an anomalously weak inflow of the Alaskan Stream in August 1991, which occurred during the transfer period from warm- to cold-year regimes. Interannual variations may be governed by variations in the Alaskan Stream and the Western Subarctic Gyre. Stability in the Bering Sea Current system is maximal during cold years, and instability is likely to increase at the end of warm-year periods. Substantial changes observed in thermohaline and nutrient fields probably promote ecosystem ramifications that should be considered by resource specialists.

Résumé : Durant la période 1990–1992, des données océanographiques ont été recueillies à la suite de quatre croisières quasi-synoptiques dans les bassins des Aléoutiennes et de Kamchatka de la mer de Béring. La physionomie des courants marins, dérivée de cartes topographiques dynamiques, montrait des variations saisonnières et interannuelles explicites. Les premières ont été attribuées au vecteur rotationnel de la tension du vent et au budget calorifique des bassins. La circulation à grande échelle a été plus intensive au printemps, tandis que l'activité des tourbillons à moyenne échelle semblait supérieure à la fin de l'automne. La période la plus calme a eu lieu à la fin de l'été. Le profil de la circulation a été modifié par un afflux anormalement faible du courant de l'Alaska en août 1991, soit pendant la période de transition entre le régime des années chaudes et celui des années froides. Les variations interannuelles peuvent être déclenchées par les variations du courant de l'Alaska et de la circulation subarctique de l'ouest. Le système de courants de la mer de Béring est à son plus stable durant les années froides, et son instabilité augmente vraisemblablement à la fin des années chaudes. Les modifications importantes observées dans la circulation thermohaline et dans les nutriments favorisent probablement la présence de ramifications des écosystèmes qui devraient être étudiées par les spécialistes en ressources.

Introduction

The Bering Sea current system forming the Bering Sea Gyre is the northernmost part of the Subarctic Pacific Region (Favourite et al. 1976). Its main source is the Alaskan Stream, an extension of the Alaska Current of the Alaska Gyre (Gulf of Alaska). Basically, inflow occurs through eastern Near Strait (~170°E) and Amchitka Pass (~180°). Although some of the dilute, warm Alaskan Stream water flows through the eastern and central Aleutian passes, the net transport there is negligible (Arsenyev 1967; Favourite 1974). The Bering Slope Current (BSC) flows northwestward along the eastern side of the deep Bering Sea, and its source is the inflow in the eastern and central passes (Kinder et al. 1975). Its branch turning southwestward along the continental slope of Siberia is augmented by the northward flow from Near Strait, and forms the Kamchatka Current, just east of Shirshov Ridge (Hughes et al. 1974). Waters of the Kamchatka Current flowing out of the Bering Sea through Kamchatka Strait give rise to the East Kamchatka Current, western component of the Western Subarctic Gyre (Favourite et al. 1976). The contributions of Dobrovolski and Arsenyev (1959, 1961), Arsenyev (1967), Ohtani (1973), Hughes et al. (1974), Favourite et al. (1976), Sayles et al. (1979), and Khen (1989) complete our understanding of the mean circulation in the Bering Sea.

However, little is yet known about seasonal and interannual variations in the Bering Sea circulation. Arsenyev (1967) and Favorite (1974) computed strong seasonal changes in transport via Aleutian passes as a result of annual wind stress cycle, but found no major effects on its current pattern. Furthermore, Favorite (1974) and Reed et al. (1980) reported poor agreement between windstress and observed transport. Natarov (1991) described observed substantial circulation simplification as a consequence of winter intensification. This was in agreement with the date and conclusions of Reid (1973).

A.V. Verkhunov. Russian Federal Research Institute of Fisheries and Oceanography, Moscow 107 140, Russia. VNIRO

Synoptic surveys necessary for near-real-time determination of circulation patterns in the Bering Sea were impossible until the development of quick and precise hydrocasting equipment in 1970s and 1980s. In spite of persistent interest of hydrographical data to the fisheries, few publications were found to discuss the problem.

This paper presents evidence of major changes in the deep Bering Sea current system, based on quasi-synoptic oceanographic data collected during 1990–92. The possible character of these changes is discussed with respect to season and interannual variability.

Materials and methods

During 1990–92, VNIRO (Russian Federal Research Institute of Fisheries and Oceanography, Moscow) conducted three hydrographic research cruises: (i) April–May 1990, aboard R/V **Mlechny Put;** (ii) October–November 1990, along with Pacific Research Institute of Fisheries and Oceanography, Vladivostok (TINRO), aboard R/V **Professor Soldatov** (Verkhunov and Tkachenko 1992); and (iii) June 1992, along with TINRO and the Pacific Institute of Oceanology, Vladivostok, in the Rusian Extended Economic Zone, aboard the R/V **Akademik Nesmeyanov**. In August 1991, there was a joint United States – Russia cruise aboard the U.S. R/V **Miller Freeman** that covered the entire Bering Basin, and during which satellite-tracked drifters and long-term moorings were deployed (Reed et al. 1992*a*). Because the main features of the Bering Sea circulation with respect to depth are assumed to remain constant (Arsenyev 1967; Sayles et al. 1979; Verkhunov and Tkachenko 1992), this discussion is primarily limited to the surface waters.

On Russian vessels, data were collected with a Neil Brown CTD, calibrated at the shoreside laboratory, and corrected at sea. Accuracy is believed to be about 0.002°C, 0.01 practical salinity units (psu), and 2 dbar (Verkhunov and Tkachenko 1992). On the U.S. vessel, a Seabird SBE-9 CTD was used (Reed et al. 1992*a*). Geographical names used were taken from Sayles et al. (1979).

Results

Circulation

In April–May 1990 circulation fitted the classical pattern (Fig. 1A). The Kamchatka Basin cyclonic gyre, centered near 56°N and 168°E, was well defined. In the North Aleutian

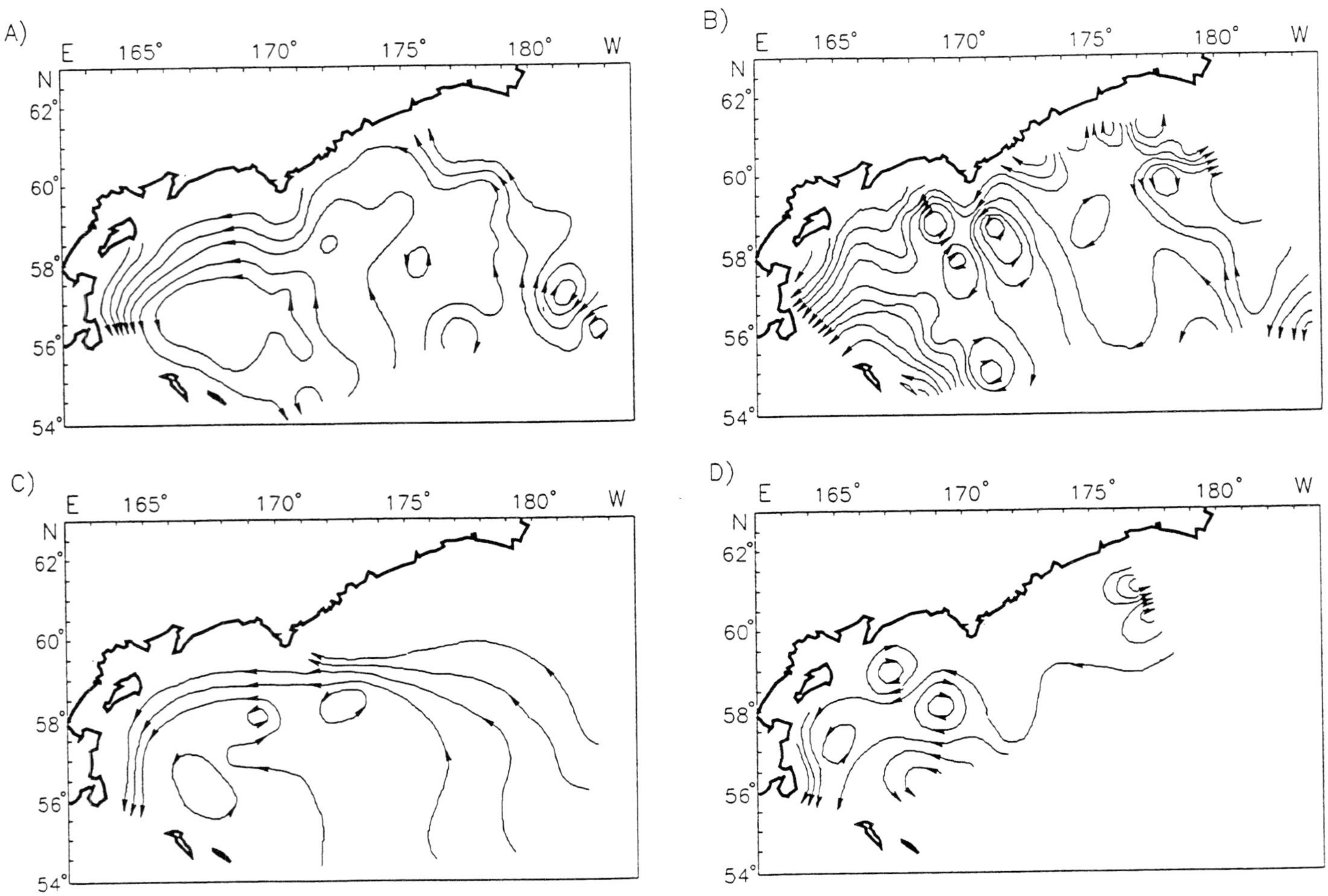

Fig. 1. Current patterns in the Bering Sea Basin during the 1990–92 hydrographic surveys: (A) April–May 1990; (B) October–November 1990; (c) August 1991; and (D) June 1992.

Basin, cyclonic circulation also occurred. The Kamchatka Current began as an intense flow near 175°E long. The BSC seemed to provide some of its source. Regions where flow entered in the central and the eastern Aleutian passes were not covered by these surveys, but according to geostrophic calculations, the flow through Near Strait provided perhaps 7 Sv, and the BSC carried a 4 Sv. The Kamchatka Current transport, near its exit through Kamchatka Strait, was calculated to be 11 Sv (Verkhunov and Tkachenko 1992). To meet such a pattern, the flow through Near Strait had to enter the Bering Sea in its eastern part, and leave in the western part.

In October–November 1990, a strong flow of about 11 Sv entered Kamchatka Strait from western Near Strait, while the Kamchatka Current total transport near Kamchatka Strait was only 6–7 Sv (Fig. 1B). There was also increased complexity and absence of explicit large scale cyclonic gyres. At the same time, some Pacific waters flowed directly into the Kamchatka Basin (Verkhunov and Tkachenko 1992).

In August 1991, circulation was generally slower (Fig. 1C). Weak inflow of 5 Sv through Near Strait resulted in a weak Kamchatka Current transport (6–7 Sv) (Reed et al. 1992*b*).

In June 1992, the western Bering survey showed a typical transport of 11 Sv for the Kamchatka Current. At the same time the Kamchatka Current pattern was disturbed. According to the dynamic calculations and direct current measurements, there were mutiple streams of opposit directions that could be related to eddies (Fig. 1D).

Water properties

Sea-surface temperature (SST) and salinity usually change seasonally owing to annual cycles of solar radiation and ice melt. Distribution of these properties reflects changes in circulation as a result of advection. The SST, surface salinity, vertical profiles of temperature and salinity, water exchange, and surface nutrient distribution are discussed below.

Surface temperature

The SSt pattern in April–May 1990 (Fig. 2A) exhibited a typical winter distribution, as described by Khen (1989). The southeast Aleutian Basin had the warmest temperatures (~3.5°C) and the coldest (−0.5°C) were observed in the northwest Kamchatka Basin, near the Karaginsky Shelf. In October–November 1990, the striking feature was pronounced eddylike distribution in the vicinity of the continental slope, especially in the Aleutian Basin (Fig. 2B). The frontlike isotherm compression over the Shirshov Ridge and in the south Kamchatka Basin appears to follow the discontinuity in the spacing of the survey stations.

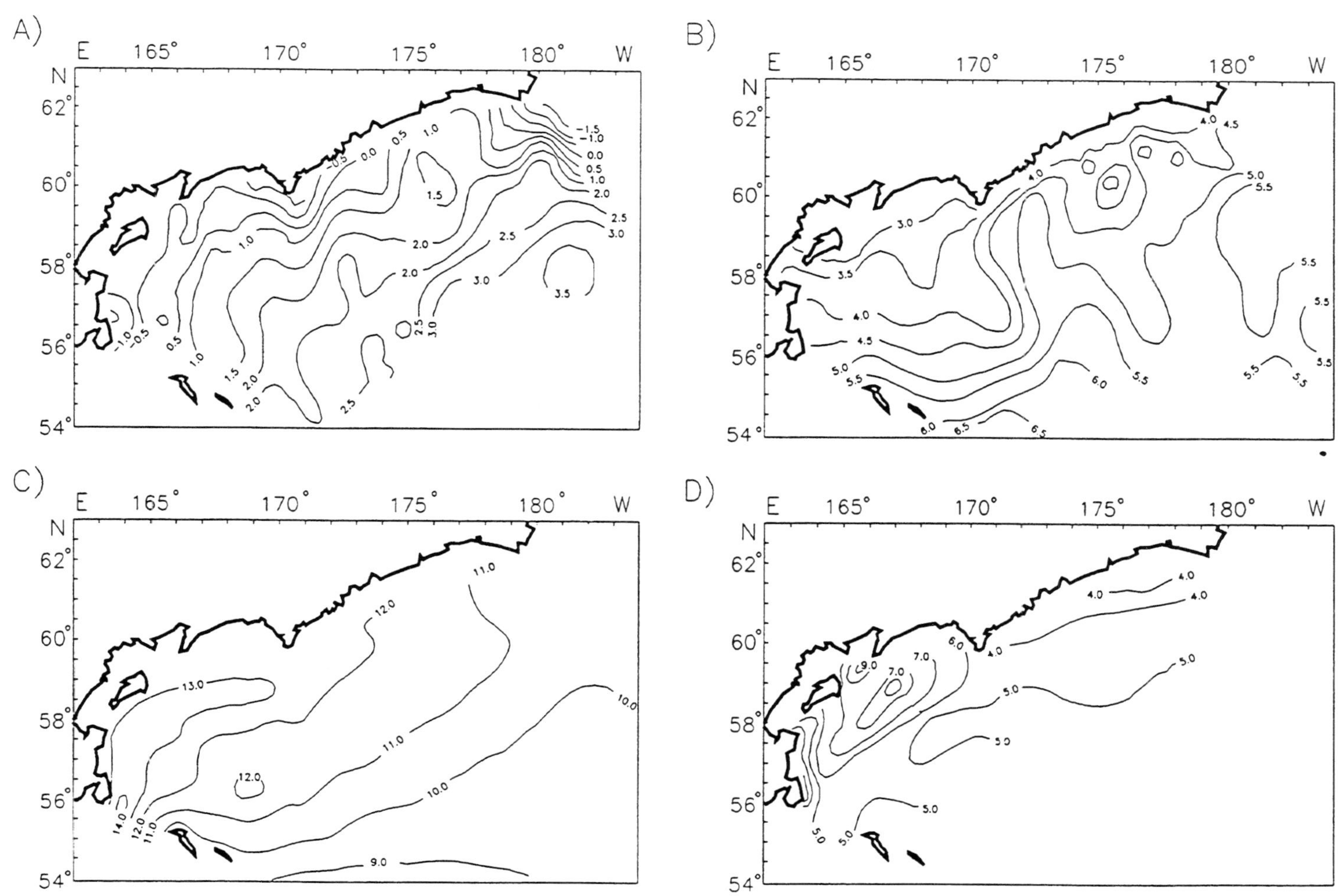

Fig. 2. Sea surface temperatures in the Bering Sea Basin during the 1990–92 hydrographic surveys: (A) April–May 1990; (B) October–November 1990; (c) August 1991; and (D) June 1992.

The transects of the August 1991 and June 1992 cruises were too remote from each other to delineate mesoscale features. In August 1991, isotherms were parallel to shore, and SSTs increased shoreward (Fig. 2C). The situation, not typical, exhibited the relaxed dynamic conditions in the Basin. In June 1992, such a pattern was characteristic of the Aleutian Basin (Fig. 2D), but in the Kamchatk Basin the Kamchatka Current was a well defined, relatively cold stream.

Surface salinity

Distribution of the near slope front was the strongest and most seaward just after the ice-melting period (Figs. 3A and 3D). In summer and autumn there was also a strong shelf frontal zone along the north Koryak Coast, which indicated the existence of the dilute Anadyr Current (Verkhunov 1993) (Fig. 3B–3D). Eddylike structures noted for October–November 1990 in the SST distribution also existed in the salinity field (Fig. 3A). Mean salinity values in autumn 1990 in the deep Bering Sea were less than the long-term monthly mean values of Khen (1989). Values for August 1991 and June 1992 were slightly less than in 1990. The cause would be a sequence of either extreme icy winters, or slowed water transport to the Pacific (especially for August 1991).

Vertical profiles

Temperature and salinity profiles at selected stations (closest to one another, among cruises) in the Kamchatka and Aleutian basins are shown in Fig. 4. They indicate that temporal changes in 1990–92 reached the 250–300 dbar level. The SSTs, depth of upper mixed layer, and relative intermediate layer temperature varied according to the seasonal cycle of solar radiation and convection development. The strongest interannual changes were associated with the depth of the winter ventilation layer. The curves for June 1992 show that year to have the deepest position of the lower boundary of this layer, and at the same time, less salinity in the main thermocline. It proves that during the winter of 1992 the water was anomalously cold and less saline.

Water exchange

The variations in water exchange with the North Pacific seem to be very important. In spite of the fact that only the August 1991 cruise covered all major straits, water temperature at the intermediate temperature maximum depth was a good indicator of the process (Fig. 5). Temperatures higher than 3.8°C on the 26.90 Sigma-*t* level indicated an inflow of Pacific water (Fig. 5). The only source of relatively warm waters at that level

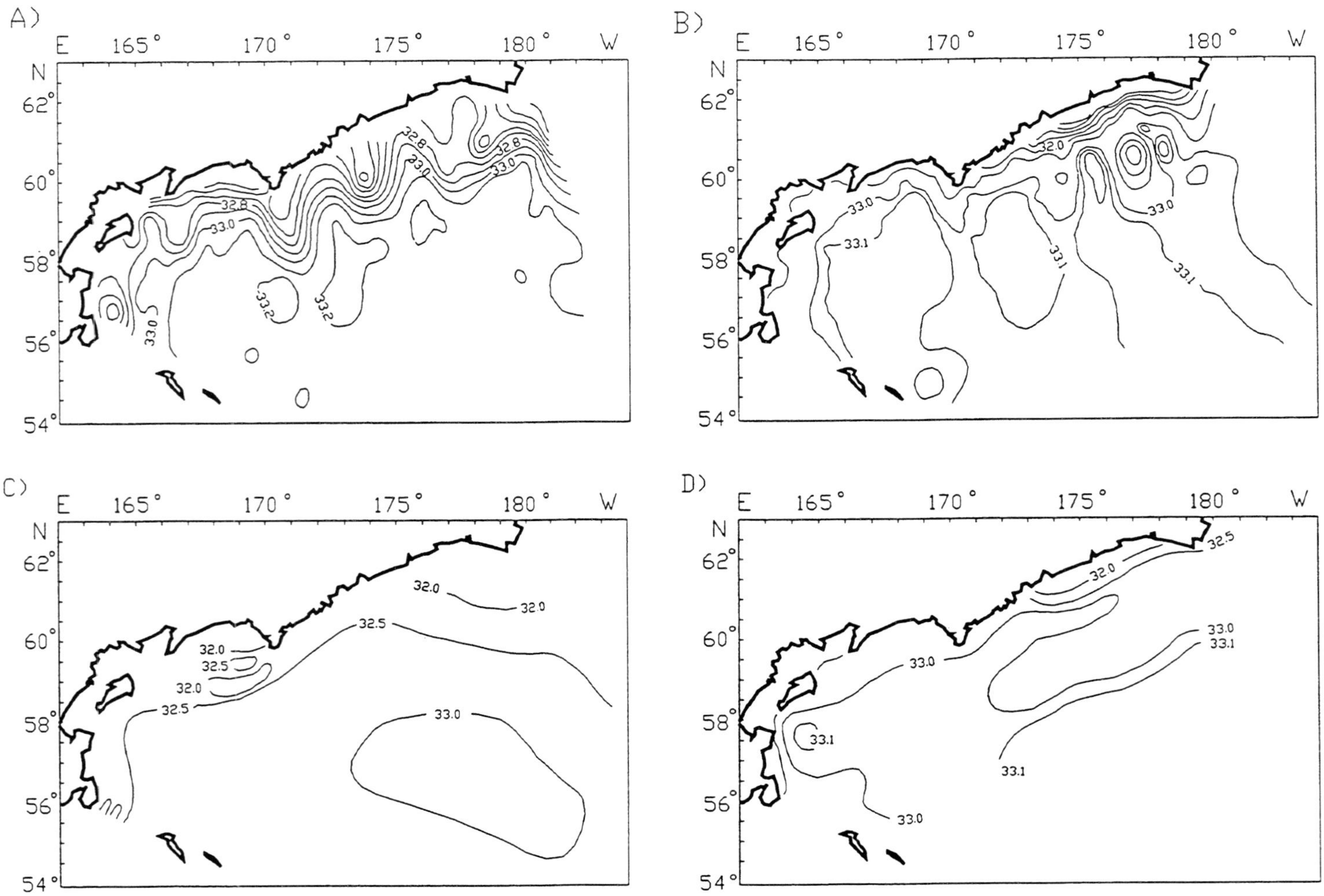

Fig. 3. Surface salinities in the Bering Sea Basin during the 1990–92 hydrographic surveys: (A) April–May 1990; (B) October–November 1990; (c) August 1991; and (D) June 1992.

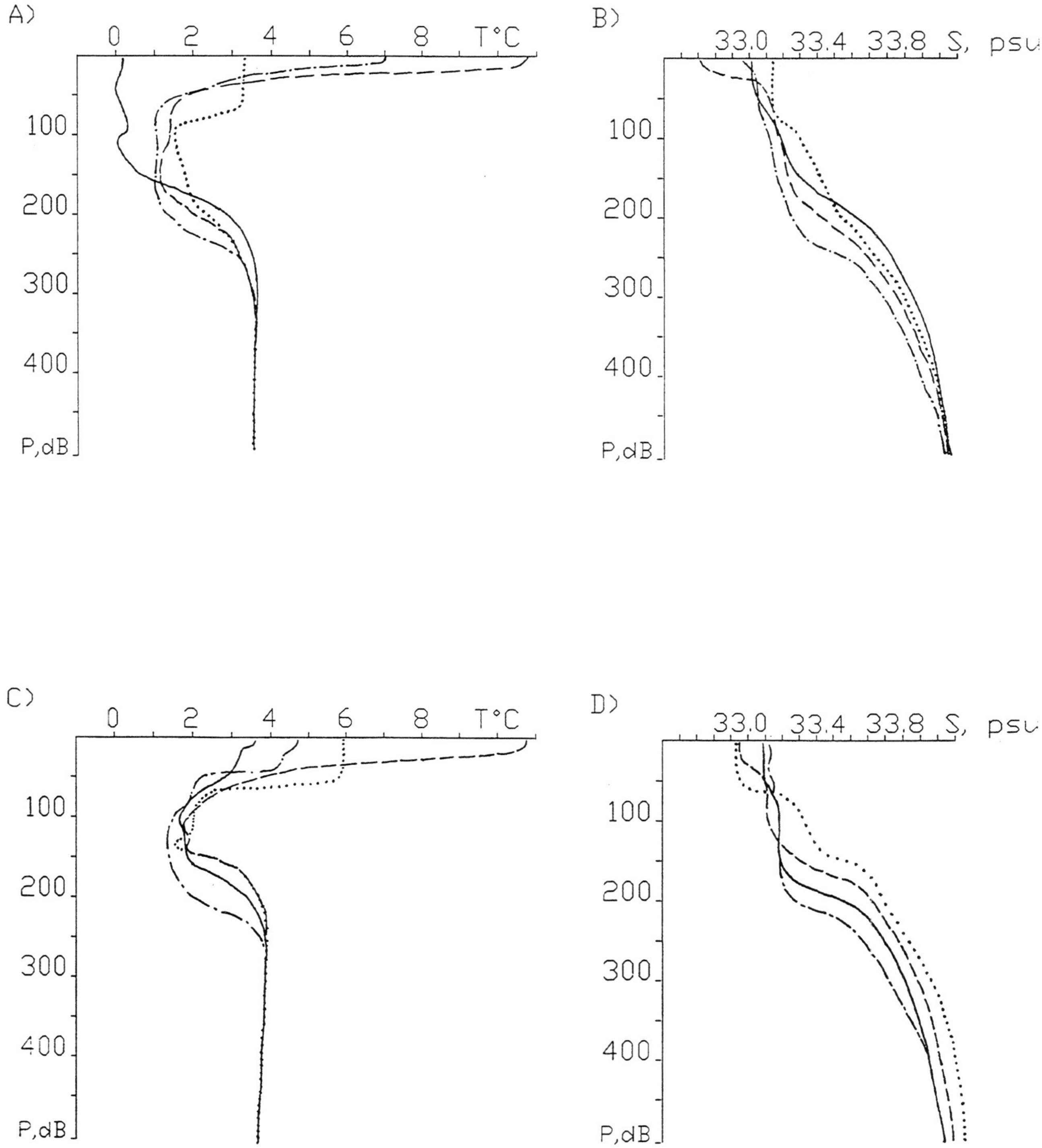

Fig. 4. Temperature (A, C) and salinity (B, D) profiles at selected stations in the Kamchatka (A, B) and the Aleutian (C, D) Basins. Solid line, Station 144, April 1990 (58°N, 168°E); Sotted line, Station 157, October–November, 1990 (58°00, 168°00); Broken line, Station 68, August 1991 (59°08, 68°31); Dash-dot line, Station 2113, June 1992 (58°44, 167°38).

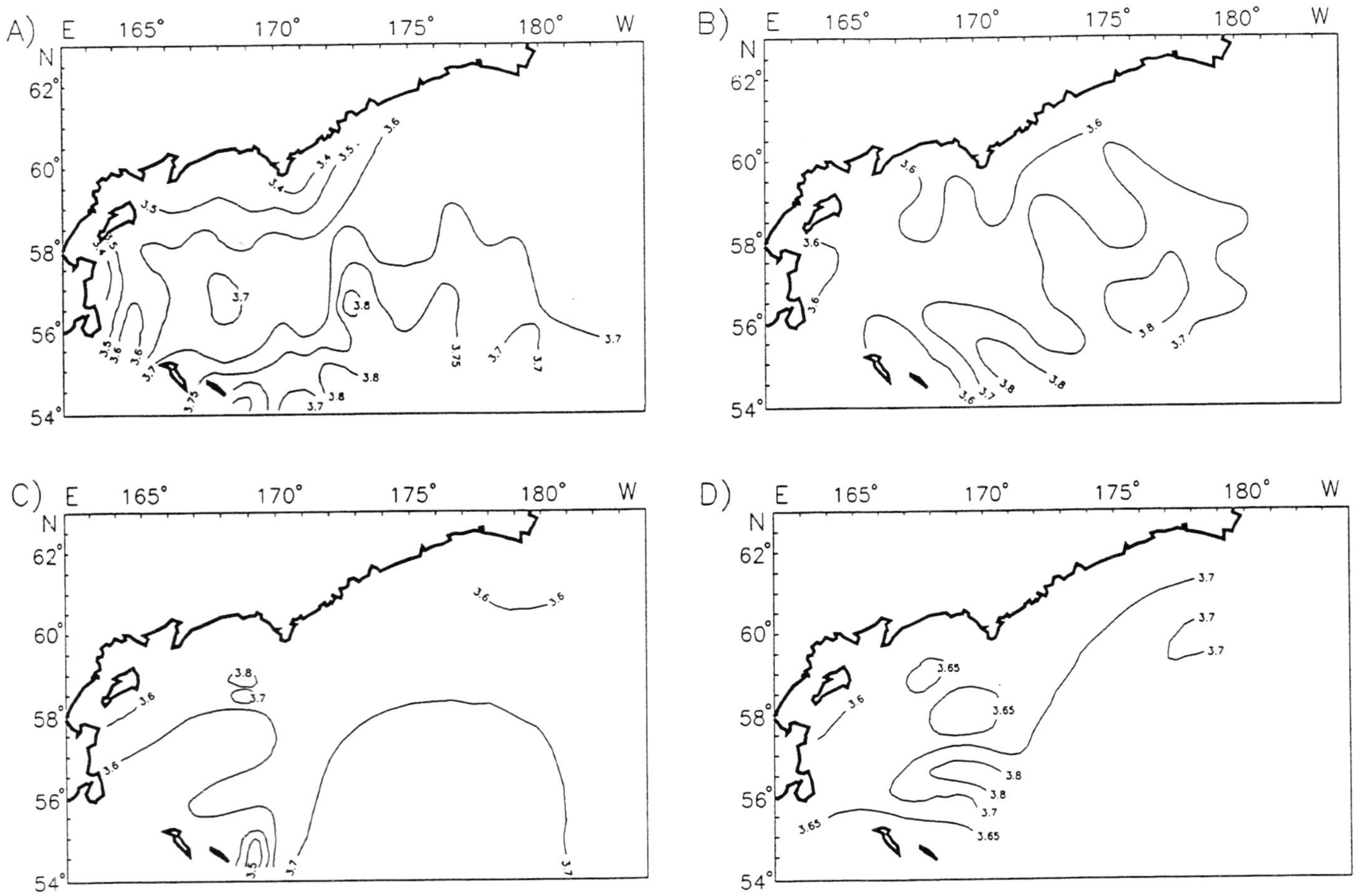

Fig. 5. Temperatures on the 26.90 Sigma-*t* surface in the Bering Sea Basin during the 1990–92 hydrographic surveys: (A) April–May 1990; (B) October–November 1990; (c) August 1991; and (D) June 1992.

in the Bering Sea is the Alaskan Stream (Arsenyev 1967; Favorite 1974). Thus, one can see an influx of warm water along the western Shirshov Ridge in April 1990 (Fig. 5A); its sharp turn to the southwest in June 1992 (Fig. 5B); its cessation in August 1991 (Fig. 5C) and both western and northern propagation in October–November 1990 (Fig. 5D). These not only confirm the circulation patterns described above, but lead to the conclusion that the Alaskan Stream did not enter the Bering Sea in August 1991 (Reed et al. 1992*b*). The latter was proved by the movements of satellite-tracked buoys (Stabeno and Reed 1992). Resumption of the Alaskan Stream entry into the Bering Sea in 1992 was accompanied by direct Pacific inflow to the Kamchatka Basin, as in November 1990, when the anomaly began (Fig. 5B and 5D).

Surface nutrient distribution

Simultaneous distributions of phosphate, silicate, and nitrate in the Bering Sea exhibited similar patterns in particular years during the study period (Mordasova et al. 1991; Sapodznikov et al. 1993); thus, only that of phosphate will be discussed here as an example. Background values for phosphate distribution during April–May and October–November 1990 were about the same in both the Aleutian and Kamchatka basins, and decreased from 2.0 mg P/L in spring to 1.6–1.8 mg P/L in autumn as a result of warm-season bioconsumption. In August 1991, concentrations in the Kamchatka Basin were extremely small (~0.4 mg P/L), compared with 1.5 mg P/L in the Aleutian Basin. Distribution in June 1992 differed from that in 1990 and 1991. A hydrochemical surface front was observed along the southern boundary of the survey, where seaward values were 1.6–1.8 mg P/L, and shoreward values were ~1 mg P/L in the Kamchatka Basin and 0.2–0.6 mg P/L in the Aleutian Basin. Thus surface nutrient concentrations in the summers of 1991 and 1992 were slightly less in both basins, with remarkable minima in the Kamchatka Basin in August 1991, and in the northern Aleutian Basin in 1992. These phenomena may be the result of the cessation of Alaskan Stream inflow observed in 1991 (Stabeno and Reed 1992), for they were not typical. At the same time, major changes in the current system pattern from spring to autumn 1990 were not revealed in the surface distribution of nutrients.

Discussion

It is reasonable to expect seasonal-scale manifestations as a result of the strong variability in regional meteorology. According to Gill and Niiler (1973), there are three main factors responsible for a large-scale seasonal variability in the ocean: (1) response to changes in atmospheric pressure; (2) barotropic response to changes in the wind stress; and, (3) steric changes in sea level, mostly produced by expansion and contraction of the water column above the seasonal thermocline, owing to changing fluxes of heat and water across the surface.

Seasonal changes in solar radiation are not of primary importance here, because substantial cloudiness occurs year round (Karpova 1963). Response to atmospheric pressure changes in considered to be dynamically negligible (Gill and Niiler 1973). Thus, barotropic wind stress response has to be a main factor in annual circulation changes in the Bering Sea. Arsenyev (1967) estimated that winter intensification of the sea currents varied by a factor of two. Overland et al. (1992) used the Navy, layered-ocean, nonlinear model with monthly wind stress curl values, and demonstrated a well-defined annual cycle. Seasonal features of the model were similar to our results. All components of the Bering Sea large-scale circulation, besides inflow through Near Strait, has pronounced transport maxima at the end of winter, and minima at the end of summer. The Kamchatka Current typically has transports of 7–8 Sv in summer, and 12–14 Sv in winter. Simulated buoys deployed at a common start point on the northern flank of the Alaskan Stream in April enter the sea via the eastern side of Near Strait and traverse the eastern portion of the Kamchatka Basin. Those released in July enter the sea via the western side of Near Strait and do not reach the eastern Basin. Those released in September do not enter the Basin.

Another factor of net flow intensification would be a seasonal transformation of the thermohaline structure in the Bering Basin. Comparison between seasonal structures of the Kamchatka Current shows different baroclinic slopes. Besides wind-stress effects, winter convection and intermixing of cold shelf waters with basin water masses in the interaction zone deepens the seasonal thermocline and sharpens the baroclinic slope of the boundary current. Warming of the current during summer leads to the opposite result. In addition to winter convection, we think that eddies were responsible for vertical transport of energy and properties through the seasonal thermocline. Satellite altimetry (Solomon and Ahlnas 1978) and drift-buoy (Kinder et al. 1975, 1980; Kinder and Coachman 1977; Reed and Stabeno 1990) data provided evidence that mesoscale eddies were possibly permanent features in the structure of the Bering Sea boundary currents, such as the Kamchatka Current, and the BSC. Intensifications of eddy activity was most likely during transfer periods in the boundary current structure (Kamenkovich et al. 1986).

The Bering Sea climate is characterized by quasi 11 year cycles of warm and cold regimes (Khen 1989, 1991). Cold years imply severe winter conditions with strong zonal atmospheric transport. Anomalously warm years are preceded by meridian air mass transport with frequent cyclones. In cold years, winter cooling on the shelf before ice cover, and convection near the ice edge, have to be the strongest. Cold intermediate waters preserved under seasonal thermocline in such years create favorable conditions for pronounced baroclinic flow year round. In contrast, the winter ventilation layer is shallower in warm years, and energy input for synoptic disturbances from atmospheric cyclones is higher.

Fluctuation of the Alaskan Stream is assumed to be one of the main factors in the variations of the Bering Sea Current system (Stabeno and Reed 1992; Reed et al. 1992*b*). The Alaskan Stream tended to exhibit interannual variations rather than seasonal ones (Reed 1984; Cummins 1989; Stabeno and Reed 1992). Our surface nutrient data showed mostly interannual change as well. Major variations in nutrient distributions were observed between autumn 1990 and summer 1991 at exactly the period when the pronounced anomaly in the Alaskan Stream flow into the Bering Sea occurred (Stabeno and Reed 1992).

Oceanographic data collected in 1990–1992 showed explicit variations in the Bering Sea Current system, both seasonal and interannual. The real nature of these changes is not clear yet, but so far we can conclude, from data analysis and other studies, that seasonal wind stress curl and heat budget on the margins of the basin are responsible for seasonal variations. Variations in the Alaskan Stream and Western Subarctic Gyre systems may govern the interannual variations. Finally, we observe some common effects. Thus, during cold years, the most stable conditions in the Bering Sea Current system are suspected to occur. Otherwise, at the end of warm-year periods, instability is likely to increase. Substantial changes in thermohaline and nutrient fields would promote possible ecosystem ramifications that should be considered by resource specialists.

Acknowledgements

We appreciate the support of B.N. Kotenev and V.V. Sapodznikov (VNIRO, Moscow). We thank R.K. Reed (PMEL, Seattle, Wash.) for valuable advice and the latest transcripts from PMEL scientists. Y.Y. Tkachenko (VNIRO) permanently assisted us in making observations at sea and discussions of the results. We used hydrochemical data processed by T.E. Whitledge (University of Texas, Austin) and hydrochemical staff of Marine Ecology Laboratory (VNIRO).

References

Arsenyev, V.S. 1967. Currents and water masses of the Bering Sea. Izd. Nauka, Moskva. (In Russian with English summary) (Transl. from Russian by 1968, National Marine Fisheries Service, Northwest Fisheries Center, Seattle, Wash.). 135 p.

Cummins, P.F. 1989. A quasi-geostrophic circulation model of the northeast Pacific. II. Effects of topographies and seasonal forcing. J. Phys. Oceanogr. 19: 1649–1668.

Dobrovolsky, V.D., and V.S. Arsenyev. 1959. On the question of the Bering Sea currents. Izv. Akad. Nauk. SSSR., 3: 3–9. (In Russian)

1961. The Bering sea hydrological description. T. Inst. Okeanol. im. P.P. Shirshova Akad. Nauk. SSSR., 38: 64–97. (In Russian)

Favorite, F. 1974. Flow into the Bering sea through Aleutian island passes, p. 3–37. *In* D.W. Hood and E.J. Kelly [ed.] Oceanography of the Bering Sea. Institute of Marine Sciences, University of Alaska, Fairbanks, Alaska.

Favorite, F., A.J. Dodimead, and K. Nasu. 1976. Oceanography of the Subarctic Pacific region, 1960–1971. Int. North Pacific Fish. Comm., Bull. 33: 187 p.

Gill, A.E., and P.P. Niiler. 1973. The theory of the seasonal variability in the ocean. Deep Sea Res. 20(2): 141–177.

Hughes, F.W., L.K. Coachman, and K. Aagard. 1974. Circulation, transport, and water change in the Western Bering sea, p. 59–98. *In* D.W. Hood and E.J. Kelly [ed.] Oceanography of the Bering Sea. Institute of Marine Sciences, University of Alaska, Fairbanks, Alaska.

Kamenkovich, V.M., M.N. Koshlyakov, and A.S. Monin. 1986. Synoptic eddies in the Ocean. D. Reidel, Boston. 264 p.

Karpova, L.A. 1963. The main features of the Bering sea climate, p. 97–110. *In* Sovietskie Rybokhozyaistvennye Issledovaniya v Severnoi Chasti Tihkogo Okeana. Vypusk Pervyi. Tr. VNIRO 28. – Izv. TINRO 50. (In Russian)

Kinder, T.H., L.K. Coachman, and J.A. Galt. 1975. The Bering Slope Current system. J. Phys. Oceanogr. 5(2): 231–244.

Kinder, T.H., and L.K. Coachman. 1977. Observation of a bathimetrically trapped current ring. J. Phys. Oceanogr. 7: 946–952.

Kinder, T.H., J.D. Schumacher, and D.V. Hansen. 1980. Observation of the baroclinic eddy: an example of mesoscale variability in the Bering Sea. J. Phys. Oceanogr. 10: 1228–1245.

Khen, G.V. 1989. Oceanographic conditions and the Bering Sea biological productivity, P. 79–89. *In* Proceedings, International Symposium on the Biology and Management of Walleye Pollock, Nov. 1988, Anchorage, Alaska, University of Alaska, Anchorage, Alaska.

Khen, G.V. 1991. To the anomalous warming of the Bering and Okhotsk Seas in 1980, p. 65–72. *In* D.E. Gershanovich [ed.] Monitoring usloviy sredy v raionah mopcko rybnogo promysla v 1989–1990. Collection of papers. Izd. VNIRO, Moskva. (In Russian with English summary)

Mordasova, N.V., A.K. Gruzevich, and V.L. Zubarevich. 1991. To pecularities of hydrochemical conditions in the Far East seas (winter–summer 1990), p. 73–88. *In* D.E. Gershanovich [ed.] Monitoring usloviy sredy v raionah mopckogo rybnogo promysla v 1989–1990. Collection of papers. Izd. VNIRO, Moskva. (In Russian with English summary)

Natarov, V.V. 1991. On the Bering sea water masses and currents, p. 111–132. *In* Sovietskie Rybokhozyaistvennye Issledovaniya v Severnoi Chasti Tihkogo Okeana. Vypusk pervyi. R. VNIRO 28. – Izv. TINRO 50. (In Russian)

Ohtani, K. 1973. Oceanographic structure in the Bering Sea. Mem. Fac. Fish. Hokkaido Univ. 21(1): 64–106.

Overland, J.E., M.C. Spillane, H.E. Hurlburt, and A.J. Wallcraft. 1992. A numerical study of the circulation of the Bering Sea Basin and exchange with the North Pacific Ocean. J. Phys. Oceanogr. In press.

Reed, R.K. 1984. Flow of the Alaskan Stream and its variations. Deep Sea Res. 31: 369–386.

Reed, R.K., and P.J. Stabeno. 1990. Flow trajectories in the Bering Sea: evidence for chaos. Geophys. Res. Lett. 17: 2141–2144.

Reed, R.K., R.D. Muench, and J.D. Schumacher. 1980. On baroclinic transport of the Alaskan Stream near Kodiak Island. Deep Sea Res. 27: 509–523.

Reed, R.K., A.V. Verkhunov, G.V. Khen, E.D. Cokelet, J.E. Overland, and T.E. Whitledge. 1992*a*. Recent U.S.–U.S.S.R. cruise in the Bering Sea. EOS Trans. Am. Geophys. Union, 73(16): 184.

Reed, R.K., G.V. Khen, and A.V. Verkhunov. 1992*b*. Water properties and flow over the Deep Bering Sea, Summer 1991. Deep Sea Res. In press.

Reid, J.L. 1973. Northwest Pacific Ocean waters in winter. Johns Hopkins Oceanogr. Stud. 5: 96 p.

Sapodznikov, V.V., N.V. Ardzanova, A.K. Gruzevich, V.L. Zubarevich, M.V. Karpushin, and D.O. Tolmachev. 1993. Hydrochemistry of the Western Bering Sea. *In* Ecology of the Bering Sea. Izd. VNIRO, Moskva. In press. (In Russian)

Sayles, M.A., K. Aagaard, and L.K. Coachman. 1979. Oceanographic atlas of the Bering Sea basin. University of Washington Press, Seattle and London. 158 p.

Solomon, H., and K. Ahlnas. 1978. Eddies in the Kamchatka current. Deep Sea Res. 25(4): 403–410.

Stabeno, P.J., and R.K. Reed. 1992. A major circulation anomaly in the western Bering Sea. Geophys. Res. Lett. In press.

Verkhunov, A.V. 1993. Thermohaline structure of the western Bering Sea shelf fronts. Oceanologiya. In press.

Verkhunov, A.V., and Y.Y. Tkachenko. 1992. Recent observations of variability in the western Bering Sea current system. J. Geophys. Res. 97: 14 369–14 376.

October–November, 1990 (58°00', 168°00'); Broken line, Station 68, August 1991 (59°08', 68°31'); Dash-dot line, Station 2113, June 1992 (58°44', 167°38').

Decadal-scale variations in the eastern subarctic Pacific. I. Winter ocean conditions

Warren S. Wooster and Anne Babcock Hollowed

Wooster, W.S., and A.B. Hollowed. 1995. Decadal-scale variations in the eastern subarctic Pacific. I. Winter ocean conditions, p. 81–85. *In* R.J. Beamish [ed.] Climate change and northern fish populations. Can. Spec. Publ. Fish. Aquat. Sci. 121.

Abstract: Sea-surface temperature patterns in the eastern North Pacific (ENP) Ocean exhibit an alternation between warm and cool eras with an average period of about 17 yr. Offshore, beyond about 140°W, opposite conditions occur. Decadal variations in the eastern Bering Sea are similar to, but not identical with, those inshore farther south. Near-surface ocean conditions in the ENP are linked to behaviour of the Aleutian Low, which is intense and east of its average position during warm eras. The initiation of such eras and the occurrence of anomalous warm years are related to certain El Niño – Southern Oscillation (ENSO) events that have a strong extratropical effect. Control of the timing and duration of decadal eras in the ENP appears to lie in the dynamics of the Southern Oscillation.

Résumé : L'évolution des températures de la surface de la mer dans l'est du Pacifique Nord montre une alternance entre les périodes chaudes et les périodes fraîches, qui ont une durée moyenne de 17 ans. Au large, au-delà d'environ 140° O, des conditions contraires règnent. Les variations décennales dans l'est de la mer Béring sont semblables, mais non identiques, à celles que l'on retrouve près des côtes plus au sud. Les conditions superficielles que l'on retrouve dans l'est du Pacifique Nord sont liées au comportement de la dépression des Aléoutiennes, qui est intense et se retrouve à l'est de sa position moyenne durant les périodes chaudes. Le début de ces périodes et la présence d'années anormalement chaudes s'expliquent par certaines activités d'El Nino – Oscillation australe qui ont un fort effet extratropical. La régulation de la synchronisation et de la durée des périodes décennales dans l'est du Pacifique Nord semble être liée à la dynamique de l'Oscillation australe.

Introduction

The abundance of stocks of fish and shellfish is affected by changes in environmental conditions in the upper few hundred metres of the ocean. While the links are not fully understood, there is enough evidence of biological response to environmental changes to warrant investigation of the relationship. Our approach has been to examine and compare the interannual and longer period patterns of variability in fish stock recruitment or abundance and in near-surface ocean conditions, seeking similarities in timing and intensity.

In a previous study, Hollowed and Wooster (1992) compared winter environmental conditions and year-class strength for a set of northeast Pacific groundfish stocks. We found a distinctive pattern of environmental change associated with the synchronous occurrence of strong recruitment during favourable environmental conditions. In the present paper, we examine in more detail the pattern of environmental change in the eastern North Pacific, associated changes in the Bering Sea, and the possible causes of the observed patterns. An accompanying paper (Hollowed and Wooster 1995), looks at the response of northeast Pacific fish stocks.

Warren S. Wooster. School of Marine Affairs HF-05, University of Washington, Seattle, WA 98195, USA.
Anne Babcock Hollowed. National Marine Fisheries Service, Alaska Fisheries Science Center, 7600 Sand Point Way NE, Seattle, WA 98115, USA.

Pattern of ocean variability

In earlier papers, we analyzed near-surface winter ocean conditions in the eastern North Pacific. Sources of data are described in Hollowed and Wooster (1992), and the pattern of variability is described there and in Wooster and Hollowed (1991). Our basic finding was that winter conditions appear to alternate between warm and cool eras, each lasting 6–12 yr (average 9.3 yr).

Temperature data from 1946 to 1988 were reanalyzed using local means and standard deviations from the 1970–1981 averages. The alternating pattern of warm and cool conditions is evident in Fig. 1 and is summarized in Table 1. Tentative identification of eras before 1932 was based on data from Roden (1989) and Ebbesmeyer et al (1989); for conditions between 1932 and 1945, see Hollowed and Wooster (1992).

When analyzed in this way, the average time from the start of one warm era to the next is about 17 yr. In the period 1946–1984, cool, mixed, and warm eras last 5–8 yr (average 6.5 yr). This can be compared with the findings of Royer (1989), who observed low frequency temperature fluctuations with a period of 20–30 yr in the northern Gulf of Alaska. In a

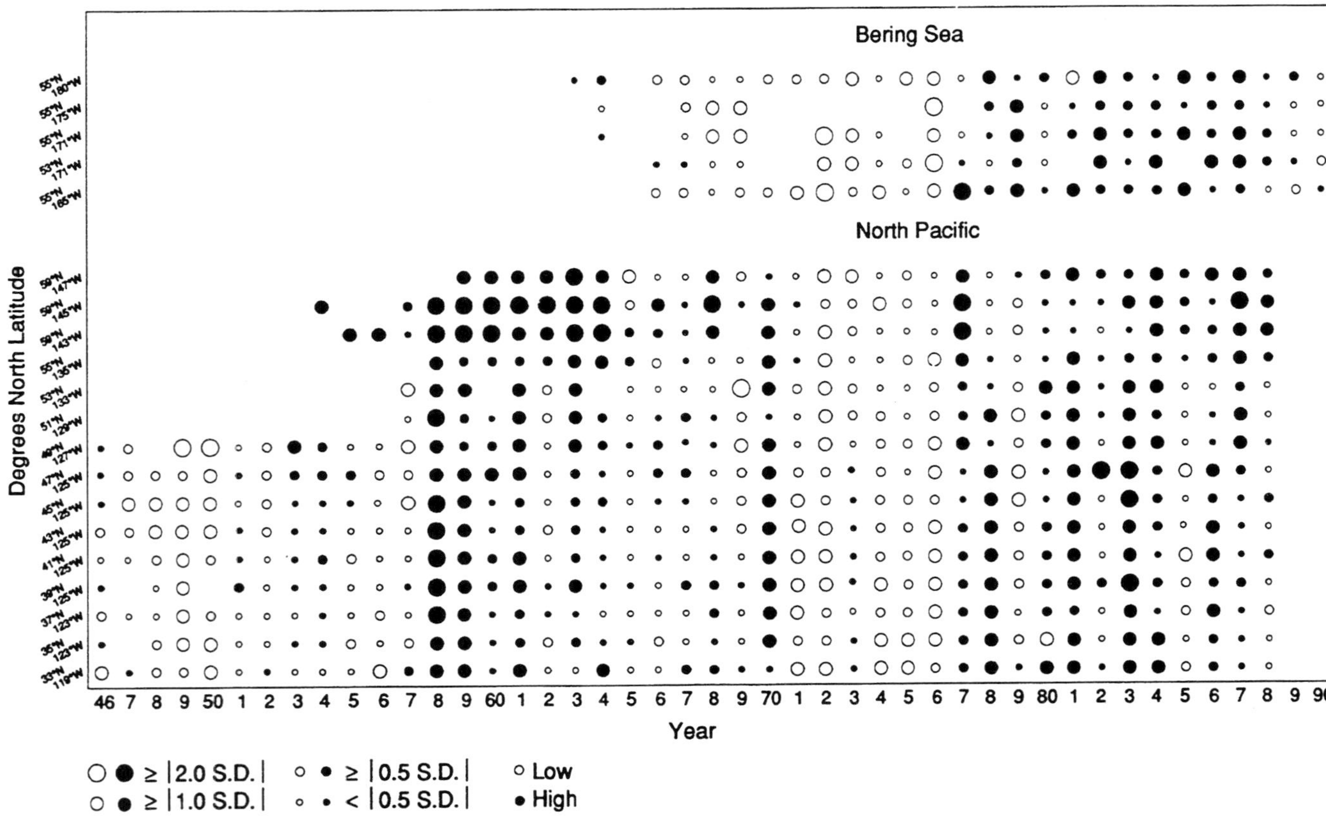

Fig. 1. Eastern North Pacific winter (January–March) SST anomalies (from a 1970–1981 base) in 2° rectangles between 32° and 60°N, 1946–1990. Values in the upper panel are from the Bering Sea.

Table 1. Summary of alternating pattern of warm and cool conditions since 1908. M and S refer to Medium and Strong Ninõ events.

Years	Character	El Niño event
1908–1913	Warm	1907 (M)
1914–1924	Cool	
1925–1931	Warm	1925 (S)
1932–1939	Cool	
1940–1945	Warm	1939 (M)
1946–1952	Cool	
1953–1957	Mixed	
1958–1964	Warm	1957 (S)
1965–1970	Mixed (70 warm)	
1971–1976	Cool	
1977–1984	Warm (79 cool)	1976 (M)
1985–????	Mixed	

later note, Royer (1993) examined a long time series of air temperatures at Sitka and concluded that the period of fluctuations was 18.6 yr, identical with that of the lunar nodal tide.

A hint of the possible origin of the decadal pattern (see later) is given by the observation that initiation of warm eras and isolated warm years (e.g., 1970) appears to be related in each instance to identifiable El Niño events (Quinn et al. 1987), often with a lag of 1 yr (see Table 1).

Contrary to the conclusion of Royer (1989) that the decadal fluctuations existed only northward from 55°N, the pattern of warm and cool eras extends south to at least 30°N. Farther offshore, beyond about 140°W, opposite conditions prevail. McLain (1984) designated the conditions as WC (warm offshore, cool inshore) and CW (cool offshore, warm inshore). The CW conditions are common in our warm eras and WC in our cool eras. This onshore–offshore distinction was illustrated by Namias (1972) in his correlation of sea-surface temperature (SST) anomalies at 40°N, 130°W with those elsewhere in the North Pacific. This was seen as evidence of quasi-stationary planetary waves with horizontal dimensions of about 7400 km from crest to crest.

Our data for three rectangles in the Bering Sea (upper panel of Fig. 1) also show decadal variation but with the latest warm era beginning in 1978 rather than 1977. This is also indicated by Niebauer's (1988) SST and Pribilof air temperatures. Another indicator of Bering Sea conditions is winter ice cover, in which Niebauer (1988) found a marked decrease after 1977. This is also evident in data for the maximum southern extent of ice along 169°W (T. Wyllie Echeverria, 17 635 96th Place SW, Vashon, Wash. 98070, personal communication) which reached 56°N in 1976, 57°N (the average for 1972–1989) in 1977, and 58°N in 1978 (see also Mysak and Manak 1989).

Our data are inadequate to confirm the timing of earlier eras in the Bering Sea, other than to suggest that the cool era from about 1965 also occurred there, although the 1970 anomaly was not evident (Fig. 1). Niebauer (1988), on the other hand, showed an intermediate warm era from 1965 or 1966 to 1970, preceded by a cool era extending back to 1960 or 1961, during the warm era south of the Aleutian Islands. While Bering Sea eras are not necessarily synchronous with those farther south, they do seem to reflect major shifts, such as that in the late 1970s.

Origin of decadal pattern

The SST is locally affected, directly by insolation and by fluxes of latent and sensible heat, and indirectly by wind stirring and convective overturn. However, these processes by themselves are unlikely to account for variations of the large time and space scales observed, which seem to call for more profound changes in the surface layer and its circulation as it responds to atmospheric forcing. Relevant processes include changes in the intensity and direction of flow as they relate to the gradients of properties of interest and the relocation of frontal and transitional regions. For example, in the eastern North Pacific, a change in the bifurcation of the west wind drift as it approaches the coast can significantly affect coastal ocean conditions.

Changes in atmospheric and oceanic circulation appear to be coupled so that the Alaska Gyre is strong while the California Current is weak and vice versa. Such coupling was proposed by Chelton and Davis (1982) and was discussed in our earlier paper (Hollowed and Wooster 1992). We postulated two categories of conditions:

Type	Era	Aleutian Low	Alaska	California	ENSO
A	cool	weak, west	weak	strong	no
B	warm	strong, east	strong	weak	yes

Alaska refers to circulation in the Alaska Gyre, California, to flow in the California Current and intensity of coastal upwelling; and ENSO, to the occurrence of El Niño – Southern Oscillation events with extratropical impact. Type A eras are cool because northerly winds inhibit northward flow along the Alaskan coast and strengthen upwelling farther south. Type B eras are warm because southerly winds associated with the eastern limb of the Aleutian Low inhibit upwelling and advect warm and moist air into the region. Offshore, northerlies on the west side of the low bring cold Arctic air to the central Pacific (Trenberth 1990).

This link between near-surface ocean conditions in the eastern North Pacific and behaviour of the Aleutian Low was suggested by Emery and Hamilton (1985). It can be tracked by a variety of indices that measure the position and intensity of the low. One of the more convenient, the Central North Pacific index, CNP, is the average of sea-level pressure anomalies over the region 35–55°N, 170°E–150°W (Cayan and Peterson 1989). In a series extending from 1899 to 1986, Cayan and Peterson found 25 strong (i.e., unusually low values of CNP) and 25 weak winters. The warm eras of our analysis coincide with strong CNP winters and with the incidence of ENSO events, while cool eras coincide with weak CNP winters.

Another set of indices (Seckel 1988) gives more detailed information on the winter Aleutian Low. Averages of measures of the intensity and location of the low during warm and cool eras are shown in Table 2.

During warm eras, the low moved 752–1011 km east of its normal position, and the pressure gradient east of the centre approximately doubled. Anomalous years also fit this pattern. For example, during 1970, a warm year in an otherwise mixed era, the low was more than 1400 km east of its average position, and the pressure gradient to the east was 2.5 times average. Conversely, during the anomalously cool year 1979 in an otherwise warm era, the low returned to its average longitude and the pressure gradient was only 27% greater than average. The mixed eras showed intermediate values of the Seckel indices.

The same pattern is evident in the winter values of the Bakun index at 48° N, another estimator of the Aleutian Low, derived from large scale geostrophic wind stress (Bakun 1973). As Table 2 shows, the index has high values during cool eras (and 27 in 1979), low during warm eras (and –38 in 1970), and is close to average in mixed eras.

Table 2. Averages of measures of the intensity and location of the Aleutian Low during warm and cool eras. The duration and character of the eras are explained in the text. The pressure gradient is the gradient east of the Aleutian Low, in mbar · 1000 km^{-1}. Distance is the distance in kilometres between the longitude of the lowest pressure at 50°N and longitude 130°W. Anomaly refers to the anomaly at 48°N of the average winter (December–February) Bakun index from the long-term average (1947–1985) value of –82 $t \cdot s^{-1} \cdot 100\ m^{-1}$ of coastline (Bakun 1973; Mason and Bakun 1986).

Era Years	Character	Pressure gradient	Distance (km)	Anomaly
1947–1972	Average	2.59	4242	
1946–1952	Cool	2.50	4360	24
1953–1957	Mixed	2.98	3959	0
1958–1964	Warm	4.85	3231	–13
1965–1970	Mixed	2.98	4065	– 5
1971–1976	Cool	2.67	3830	23
1977–1984	Warm	5.25	3490	–22

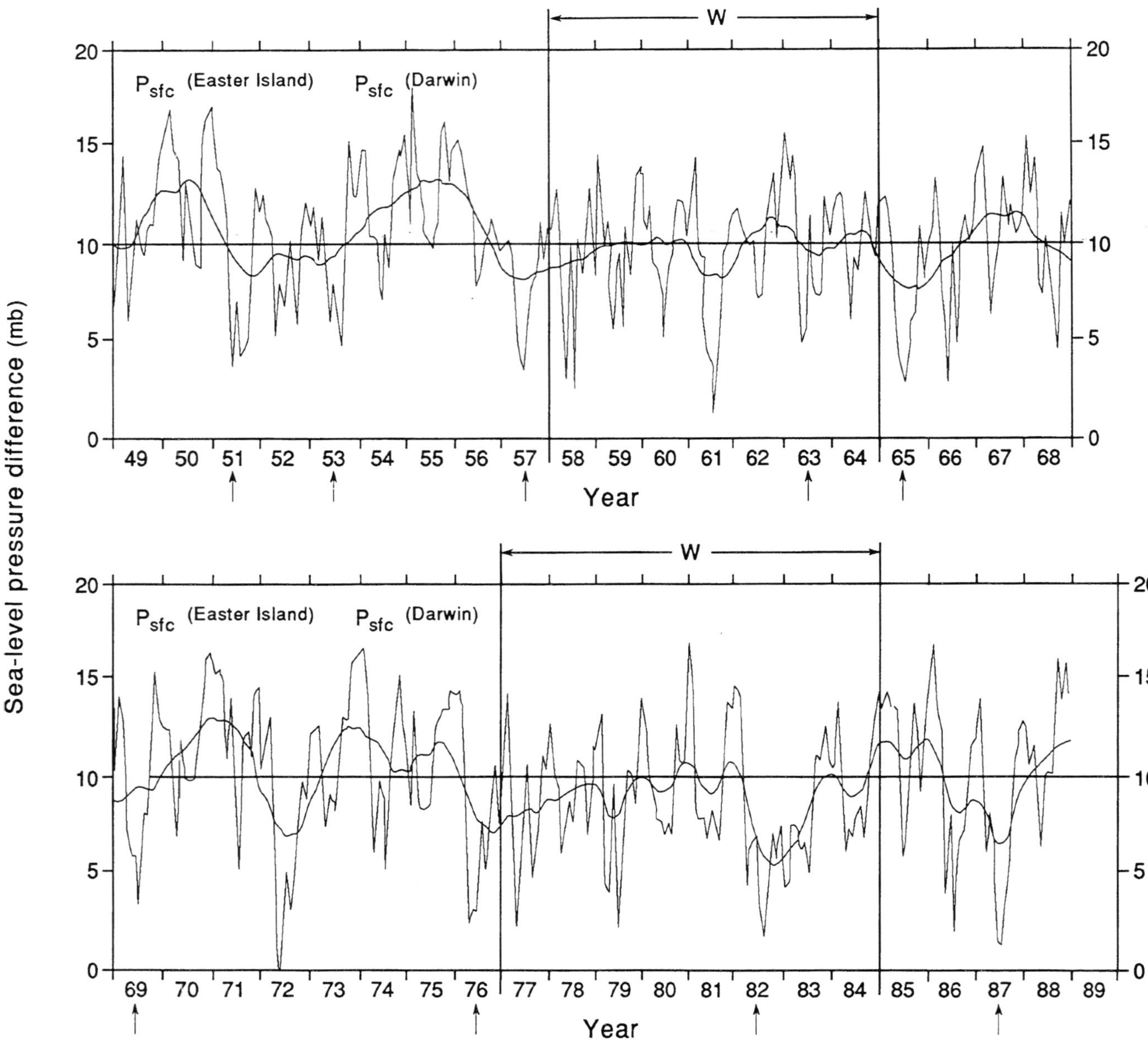

Fig. 2. Southern Oscillation index, the surface pressure difference between Easter Island and Darwin in mbar (fine line, monthly mean; thick line, 12-mo running mean). Arrows indicate El Niño events (Peixoto and Oort 1992). Warm eras in 1958–1964 and 1977–1984 are indicated (see text).

As noted earlier, the beginning of warm eras, the occurrence of anomalous warm years, and the occurrence of intensified and eastward-lying Aleutian Lows are associated with identifiable ENSO events. Those with greatest extratropical effects are associated with anomalous warming in the western tropical Pacific (Hamilton 1988). These links suggest that timing of the observed decadal pattern of variation in the eastern North Pacific might be governed by teleconnections with the Southern Oscillation (SO).

That possibility is supported by the subtle but significant difference in the character of the SO during cool and warm eras (Fig. 2). The average sea-level pressure difference between Darwin and Easter Island is 10.3 mbar (Wyrtki 1982). The SO anomalies decline sharply to negative values during ENSO events. As Trenberth (1990) has noted, the period since 1977 has been remarkable in its paucity of positive anomalies, or La Ninas. This seems also to have been the case in the previous warm era, 1958–1964.

Thus control of the timing and duration of decadal eras in the eastern North Pacific should be sought in the dynamics of the SO and their effects on atmospheric pressure distribution and circulation in the northern hemisphere.

Acknowledgements

We wish to thank Jim Shumacher, Richard Methot, Jim Ingraham, and Steven Bograd for reviewing this paper. We also appreciate the work of Ingrid Spies, who assisted in compiling the sea surface temperature data.

References

Bakun, A. 1973. Coastal upwelling indices, west coast of North America, 1946–71. U.S. Department of Commerce, Washington, D.C. NOAA Tech. Rep. NMFS SSRF-671: 103 p.

Cayan, D.R., and D.H. Peterson. 1989. The influence of North Pacific atmospheric circulation on streamflow in the West, p.375–397. *In* D.H.Peterson [ed.] Aspects of Climate Variability in the Pacific and the western Americas. American Geophysical Union, Geophys. Monogr. 55.

Chelton, D.B., and R.E. Davis. 1982. Monthly sea level variability along the western coast of North America. J. Phys. Oceanogr. 12: 757–784.

Ebbesmeyer, C.C., C.A. Coomes, G.A. Cannon, and D.E.Bretschneider. 1989. Linkage of ocean and fjord dynamics at decadal period, p. 399–417. *In* D.H.Peterson [ed.] Aspects of Climate Variability in the Pacific and the Western Americas. American Geophysical Union, Geophys. Monogr. 55.

Emery, W.J., and K.Hamilton. 1985. Atmospheric forcing of interannual variability in the northeast Pacific Ocean: connections with El Niño. J. Geophys. Res. **90**: 857–868.

Hamilton, K. 1988. A detailed examination of the extratropical response to tropical El Niño – Southern Oscillation events. J. Climatol. 8: 67–86.

Hollowed, A.B., and W.S. Wooster. 1993. Variability of winter ocean conditions and strong year classes of northeast Pacific groundfish. ICES Mar. Sci. Symp. 195: 433–444.

Hollowed, A.B., and W.S. Wooster. 1993. Decadal-scale variations in the eastern subarctic Pacific. II. Response of northeast Pacific fish stocks, p. 373–385. *In* R.J. Beamish [ed.] Climate change and northern fish populations. Can. Spec. Publ. Fish. Aquat. Sci. 121.

Mason, J.E., and A.Bakun. 1986. Upwelling index update, U.S. West Coast, 33N–48N latitude. U.S. Department of Commerce, Washington, D.C. NOAA Tech. Memo. NMFS-SWFC-67: 81 p.

McLain, D.R. 1984. Coastal ocean warming in the northeast Pacific, 1976-83, p.61–86. *In* W.G.Pearcy [ed.]. The Influence of ocean conditions on the production of salmonids in the North Pacific. Oregon State University, Corvallis, Oreg. ORESU-W-83-001.

Mysak, L.A., and D.K. Manak. 1989. Arctic sea-ice extent and anomalies, 1953–1984. Atmos. Ocean, 27: 376–405.

Namias, J. 1972. Space scales of sea-surface temperature patterns and their causes. Fish. Bull. U.S. 70: 611–617.

Niebauer, H.J. 1988. Effects of El Niño – Southern Oscillation and North Pacific weather patterns on interannual variability in the subarctic Bering Sea. J. Geophys. Res. 93: 5051–5068.

Peixoto, J.P., and A.H. Oort. 1992. Physics of climate. American Institute of Physics, New York, N.Y. 520 p.

Quinn, W.H., V.T. Neal and S.E. Antunez de Mayolo. 1987. El Niño occurrences over the past four and a half centuries. J. Geophys. Res. 92: 14 449 – 14 461.

Roden, G.R. 1989. Analysis and interpretation of long-term climatic variability along the west coast of North America, p. 93–111. *In* D.H.Peterson [ed.] Aspects of Climate variability in the Pacific and the western Americas. American Geophysical Union, Washington, D.C. Geophys. Monogr. 55.

Royer, T.C. 1989. Upper ocean temperature variability in the northeast Pacific Ocean: is it an indicator of global warming? J. Geophys. Res. 94: 18 175 – 18 183

Royer, T.C. 1993. High-latitude oceanic variability associated with the 18.6 year nodal tide. J. Geophys. Res. 98: 4639–4644.

Seckel, G.R. 1988. Indices for mid-latitude North Pacific winter wind systems; an exploratory investigation. Geojournal, 16: 97–111.

Trenberth, K.E. 1990. Recent observed interdecadal climate changes in the Northern Hemisphere. Bull. Am. Meteorol. Soc. 71: 988–993.

Wooster, W.S., and A.B. Hollowed. 1991. Decadal scale changes in the northeast Pacific Ocean. Northwest Environ. J. 7: 361–363.

Wyrtki, K. 1982. The Southern Oscillation, ocean–atmosphere interaction and El Niño. Mar. Technol. Soc.J. 16: 3–10.

Influence of interannual changes in ocean conditions on the abundance of walleye pollock (*Theragra chalcogramma*) in the eastern Bering Sea

Kiyotaka Ohtani and Tomonori Azumaya

Ohtani, K., and T. Azumaya. 1995. Influence of interannual changes in ocean conditions on the abundance of walleye pollock (*Theragra chalcogramma*) in the eastern Bering Sea, p. 87–95. *In* R.J. Beamish [ed.] Climate change and northern fish populations. Can. Spec. Publ. Fish. Aquat. Sci. 121.

Abstract: Interannual changes in sea conditions on the continental shelf in the eastern Bering Sea were investigated on the basis of long-term hydrographic data from 1963 to 1990. Winter sea conditions were estimated from summer data in relation to air temperature, salinity, wind, and from the amount of drift ice in winter. There was a cold period from 1971 to 1976 and a warm period from 1978 to 1981. Extremely cold conditions resulted from lower air temperatures and a large amount of drift ice during the previous winter, both brought by prevailing northerly winds. Warm conditions resulted from southerly winds. No appreciable relation between fecundity of walleye pollock (*Theragra chalcogramma*) and the number of young fish at age 2 was found. However, survival rations of walleye pollock at age 2 relate to the minimum water temperatures in the cold bottom water when the fish are age 1. The interannual change in the abundance of walleye pollock depends on fecundity and winter conditions at age 1.

Résumé : Dans le but de recueillir des données hydrographiques à long terme, on a étudié entre 1963 et 1990 les changements interannuels de l'état de la mer sur le plateau continental de l'est de la mer de Béring. Les conditions hivernales de la mer sur les plans de la température de l'air, de la salinité, des vents et de la quantité de glaces en dérive ont été estimées à partir des données estivales. On distingue une période froide de 1971 à 1976 et une période chaude de 1978 à 1981. Des conditions de froid extrême ont eu pour résultat d'abaisser les températures de l'air et de favoriser la présence d'une grande quantité de glaces en dérive durant l'hiver précédent, le tout apporté par des vents dominants du nord. Les conditions de chaleur étaient attribuables aux vents du sud. Aucune relation significative entre la fécondité de la morue du Pacifique occidental (*Theragra chalcogramma*) et le nombre de jeunes morues de 2 ans n'a été constatée. Toutefois, les coefficients de survie des jeunes morues de 2 ans sont liés aux températures minimales de l'eau dans les eaux froides profondes lorsque les individus sont âgés d'un an. Le changement interannuel dans l'abondance de la morue du Pacifique occidental dépend de la fécondité de l'espèce et des conditions hivernales que doivent affronter les individus d'un an.

Introduction

The southern region of the wide continental shelf in the eastern Bering Sea is known as a good fishing ground for the benthonic fish. Huge catches of yellowfin sole (*Pleuronectes asper*), walleye pollock (*Theragra chalcogramma*), Alaska king crab (*Paralithodes camtschaticus*) and others have been caught by commercial fisheries from this region throughout several decades.

Some benthonic species migrate together onto the continental shelf region from the deep ocean to spawn during early spring. The larvae and young fish are affected by sea conditions such as temperature, salinity, and the amount of food available through their initial growth stage. However, these conditions are set by the weather conditions and the development of drift ice throughout the winter. Some kinds of young fish live in the continental shelf region through their first winter to avoid cold conditions. This region acts as a nursery for the northern fish populations in the Bering Sea.

According to summer hydrographic observations, cold bottom waters are preserved under a sharp thermocline along 50–100 m isobaths from south of St. Lawrence Island to Bristol Bay. These cold bottom waters are inferred to be the remnants of late winter cooling and drift ice cover (Ohtani 1969, 1973).

The purposes of this article are to describe annual variations in the properties of bottom waters, to estimate sea conditions during the early life history of benthic fish from the summer data, and to discuss how interannual changes in winter cooling affect the abundance of walleye pollock.

K. Ohtani and T. Azumaya. Faculty of Fisheries, Hokkaido University, 3-1-1 Minato-cho, Hakodate 041, Japan.

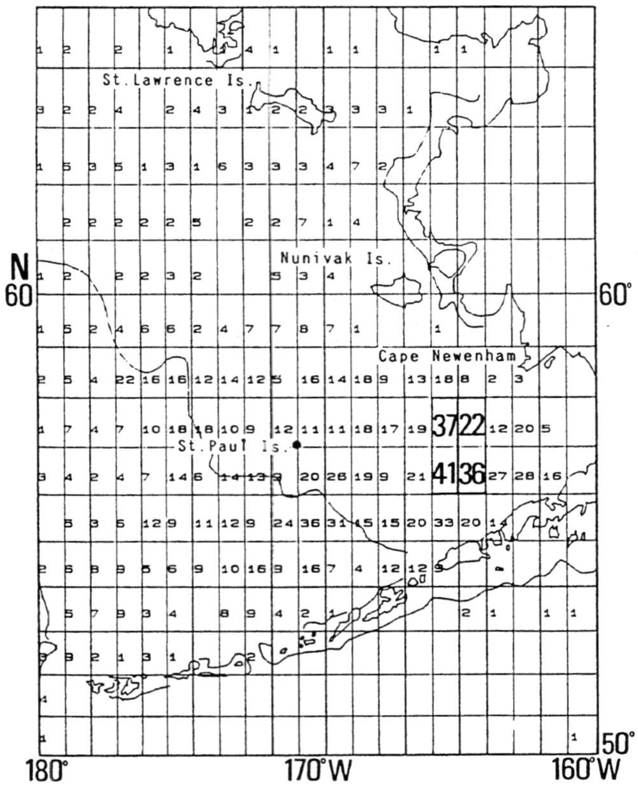

Fig. 1. Number of stations in each quadrangle. Data inside the quadrangle were used for interannual analysis.

Observations and statistics

Between 1963 and 1990, hydrographic observations were made at 1526 stations during summer voyages of the T.S. Oshoromaru by the research staff, officers, and graduate students of fisheries of the Faculty of Fisheries, Hokkaido University. Until 1983, observations were made with Nansen bottles carrying reversing thermometers; since 1984, these observations were made using Neil Brown CTD system (Faculty of Fisheries, Hokkaido University, 1964–1991). Distribution of the number of stations inside 1° × 1° quadrangles are shown in Fig. 1. Averaged distributions of temperature and salinity in the layer of minimum temperature are shown in Fig. 2.

Cold waters spread over the middle of the continental shelf from southwest of St. Lawrence Island toward the southeast. This cold area is bounded by the warmer basin water coming from the Alaskan Stream through the passes near the Alaska peninsula along the shelf break (Schumacher et al. 1982) and by the warmer shallow waters along the Alaskan coast. Although the coastal waters cool to near freezing in winter (Kinder and Schumacher 1981*a*), the summer heating effect could easily be transmitted to the bottom water by strong tidal mixing in a shallow region. However, the cold bottom waters are preserved by a sharp thermocline and a halocline between the surface and bottom layers. Vertical heat flux between these layers was estimated to be small by Koto and Fujii (1958), Reed (1978), and Kinder et al. (1978). Minimum temperatures at the stations in the area covered by the dense drift ice or in an area where ice is forming have a value near the melting point (−1.5°C) or the freezing point (−1.7°C) until midsummer.

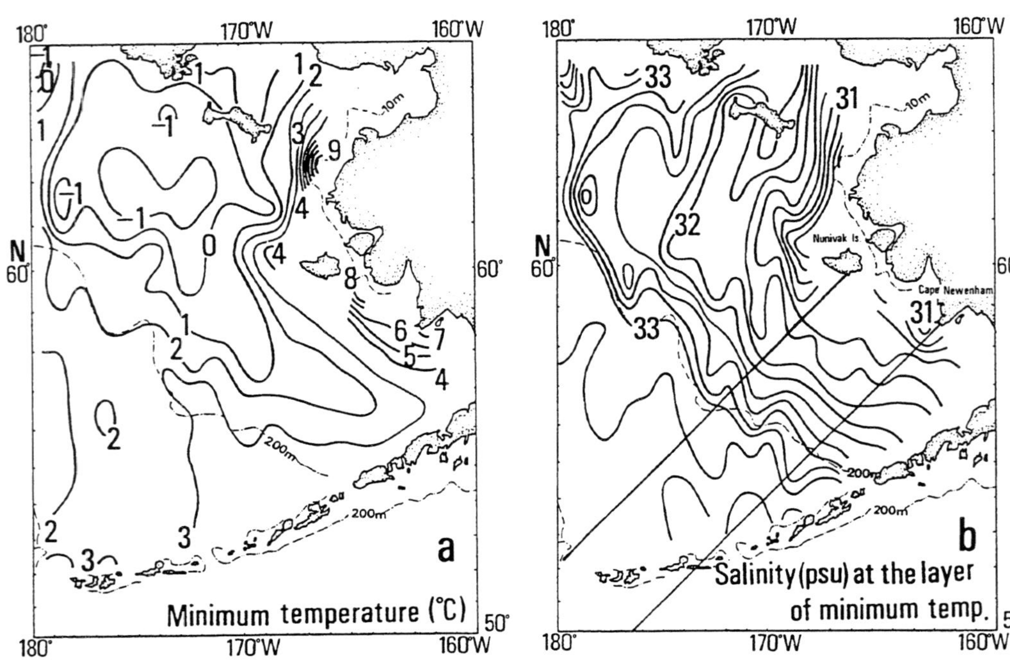

Fig. 2. Averaged distributions of temperature (*a*) and salinity (*b*) in the layer of minimum temperature. Thin lines indicate the direction to measure the distance of the 32-psu isohaline from the Alaskan coast.

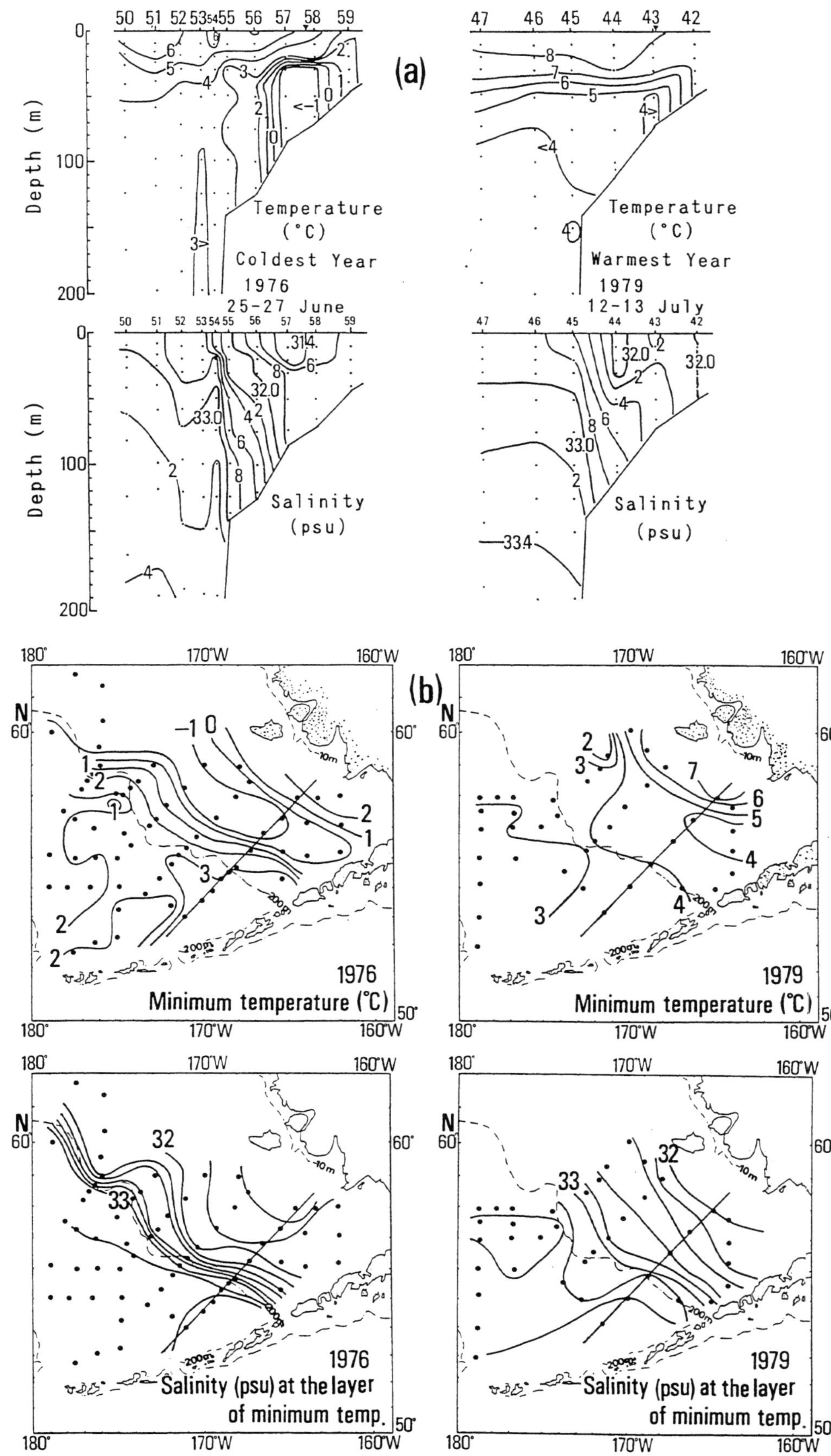

Fig. 3. (*a*) Vertical structures of temperature and salinity in the coldest year (1976) and in the warmest year (1979) during 1963–1990. (*b*) Horizontal distributions of temperature and salinity at the layer of minimum temperature.

Salinity distribution decreases from the shelf break to the Alaskan coast, so the isohalines run almost parallel with the isobaths. Temperature and salinity fronts enclosing the cold bottom water are consequently formed along the shelf break as a boundary of different water masses and along the offshore of the Alaskan coast as a tidal front, respectively (Ohtani 1969; Schumacher et al. 1979).

Thus the water masses on the continental shelf, the oceanic water, and the fresh water coming from the Alaskan coast are in the process of being mixed by the tidal current and the effects of winter cooling. There is no measurable constant flow in the midarea (Kinder and Schumacher 1981*b*), so the cold bottom waters remain along the central part of the shelf beneath the thermocline until autumn.

Although both Figs. 2*a* and 2*b* may be biased in these northern areas because of sparse observations, similar patterns indicated in the southern areas were obtained in every year except for 1979, the warmest year.

Annual sea conditions

Annual measurements were made between late June and mid-July. The number of stations and their locations are different each year, but the southeasternmost isotherms around the cold bottom water are indicated in every year (T. Azumaya and K. Ohtani, unpublished data).

Minimum temperatures of the basin water along the shelf break are always higher than those of the cold bottom water. Differences between them increase in cold years and decrease in warm years, so a temperature front along the shelf break is strengthened in cold years. This attributed to a difference in the thickness of the winter convection layer between the shelf break and the shelf.

A salinity front coincident with the temperature front clearly forms along the shelf break. Although the isohalines of 33.0 practical salinity units (psu) always run along the shelf break with little fluctuation, the 32-psu isohalines move off the Alaskan coast some distance, and run along the southern edge of the cold bottom in cold years. On the other hand, the temperature front along the shelf break is relatively weakend in the warm years, so the cold bottom waters are restricted within narrow limits in the northern area. A salinity front exists every year along the shelf break; however, the horizontal gradient of the salinity front is weakend, and the 32-psu isohalines shift toward the Alaskan coast in warm years.

Typical temperature and salinity structures in cold years and in warm ones are shown in Fig. 3. In 1976, extremely cold bottom water (near freezing) was found on the continental shelf under the thermocline accompanied by the halocline. The shelf-break front was composed of a large salinity front and a sharp inner temperature front. The salinity of the cold bottom water was less than 31.8 psu. The tidal front was indicated as a temperature front without a salinity front.

In the warm year 1979, there was no inner temperature front at the shelf break, but there was relatively warm bottom water. Saltier waters entered the shallow area.

Interannual changes of sea conditions

To represent the development and diminution of the bottom cold waters, annual minimum temperatures were used inside the quadrangle shown in Fig. 1 where continuous data are available. Interannual changes in deviations from the mean value of minimum temperatures in the area are shown in Fig. 4*a*. The mean value of minimum temperatures in the quadrangle for 28 yr is 2.02°C. Until 1968, the minimum temperature was around the mean value. This was followed by a cold period from 1971 to 1976, a warm period from 1978 to 1981, and a period of small fluctuations since 1982. The coldest year (1976) and the warmest year (1979) are indicated during the period. It is noticeable that in every year at the end of an El Niño event, the temperature is higher than in the year at the beginning of the event.

Yearly displacement of 32-psu isohalines is indicated by the mean values of distances from Nunivak Island and Cape Newenham in a southwestward direction along the lines shown in Fig. 2*b*. Deviations in distance from the mean value in each year are shown in Fig. 4*b*. A comparison of Figs. 4*a* and 4*b* shows that the 32-psu isohalines are close to the shelf break in cold years and farther from it in warm years. The correlation between distance and temperature is obtained by a negative linear equation, as shown in Fig. 5. This result suggests that drift ice caused an unusual decline in the temperature of the cold bottom waters.

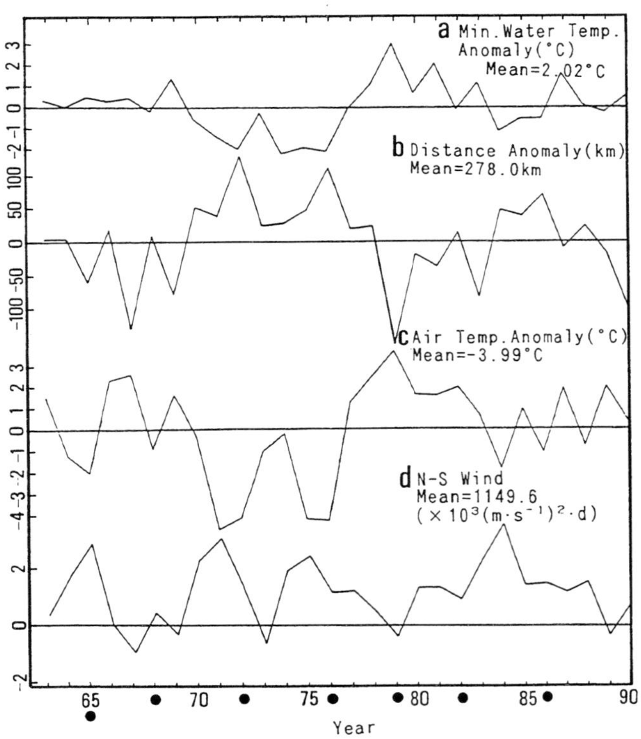

Fig. 4. Interannual values of minimum water temperature at the bottom in the quadrangle (*a*), distance anomalies of the 32-psu isohalines in the southwestward direction from Nunivak Island and Cape Newenham (*b*), anomalies in air temperatures for 3 mo (January–March) averaged at St. Paul Island (*c*), and the north–south component of integrated wind power in $(m \cdot s^{-1})^2 \cdot d$ (January–March positive: north component) (*d*). Black circles under the lower frame indicate the beginning of an El Niño – Southern Oscillation event.

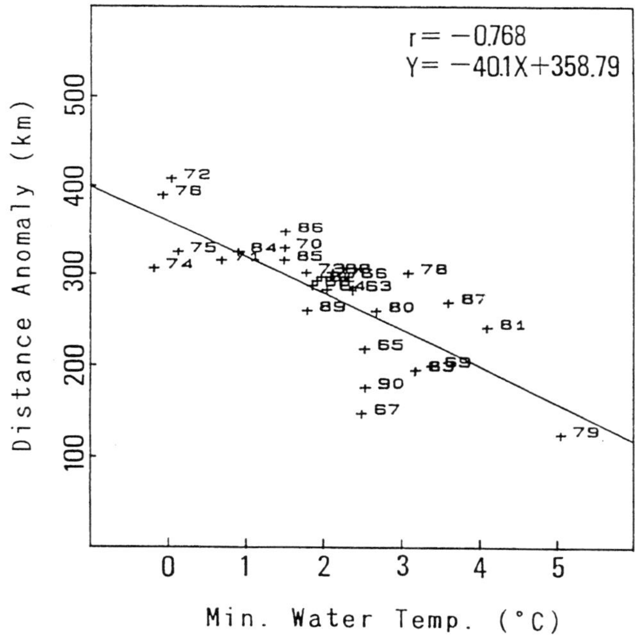

Fig. 5. Relationship between minimum temperatures in the quadrangle and the mean distance of 32-psu isohalines from Nunivak Island and Cape Newenham.

Effect of meteorological conditions on cold bottom waters

Winter cooling of shelf waters is basically caused by cold air temperature, as shown in Fig. 4*c* by the 3-mo mean air temperature during January to March at St. Paul Island. A positive correlation of minimum water temperatures and 3-mo mean air temperatures is shown in Fig. 6 as indicated by Coachman and Charnell (1979). However, this correlation has relatively large deviations (e.g., 1974).

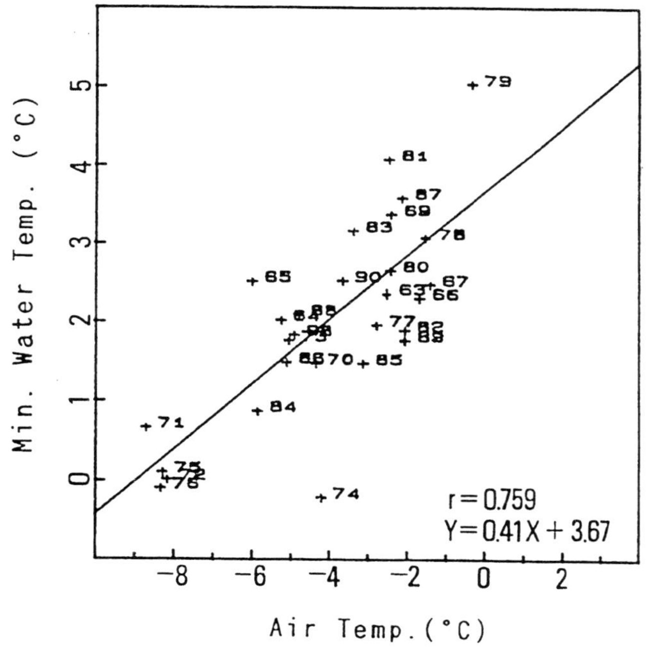

Fig. 6. Relationship between the 3-mo mean air temperatures at St. Paul Island and the minimum water temperatures in the quadrangle.

Next we consider the effect of drift ice. Niebauer (1981), McNutt (1981), and Cavalieri and Parkinson (1987) reported that the southern limits of drift ice covered this area in cold years. A decrease in temperature of 1°C in a water column 80 times the depth of drift ice can absorb the latent heat of drift ice. This is accompanied by a dilution of salinity in the upper layer during the thawing process (Ohtani 1989). Fig. 7 gives the regression of the minimum water temperatures over the salinity in the quadrangle. All points are dispersed but each group of points separated into warmer and colder than the mean air temperature; Fig. 7 shows a linear relation. The regression equation for the cold-year group has a higher correlation coefficient and a lower regression line with a larger gradient than that of the warm-year group. Although the 1974 value is not consistent with the linear relation in air and water temperatures shown in Fig. 6, it fits the cold-year relation shown in Fig. 7. The results shown in Figs. 5–7 indicate that the cooling effect of drift ice prevails in cold years.

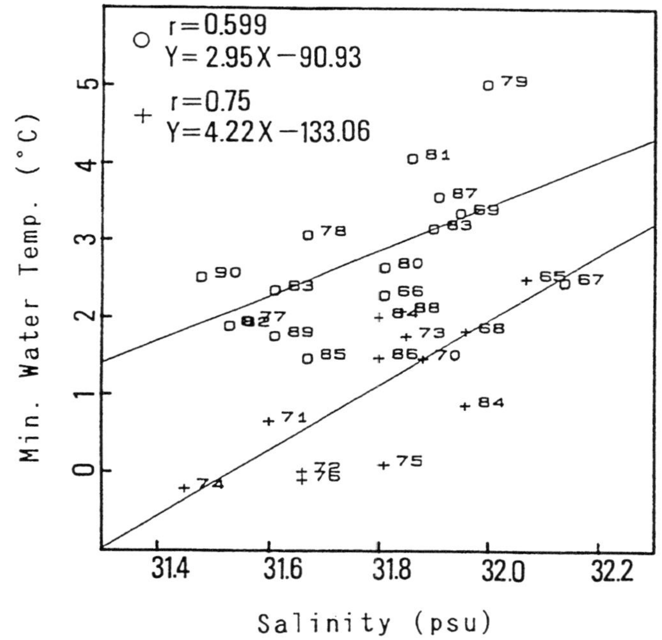

Fig. 7. Relationship between minimum water temperatures and salinity in the quadrangle. Open circles are warm years and crosses are cold years.

Because the cold air mass and the southerly extent of drift ice are brought by northerly winds, the relations between integrated wind power and the mean air temperature were investigated. The mean air temperature fitted for the sum total of north wind components gives $r = 0.71$ (Fig. 8). However, the wind components correlate slightly to minimum water temperatures, salinities, and the distances of isohalines ($r < 0.5$). A polynomial regression for minimum water temperature is given by the independent variables of mean air temperatures (degrees Celsius), salinities (psu) at minimum water temperatures, and north–south wind components (metres per second squared times days) as shown in Fig. 9. Although regression gives $r = 0.84$, it seems that most of the wind effects are included in the temperature and salinity variables.

These results prove that the summer data for the bottom waters are proportional to the winter sea conditions.

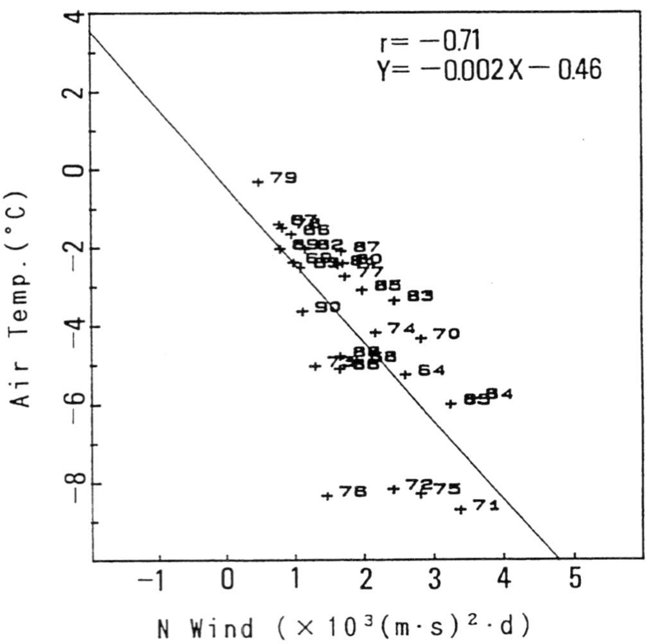

Fig. 8. Relationship between integrated wind power and mean air temperatures for the north component.

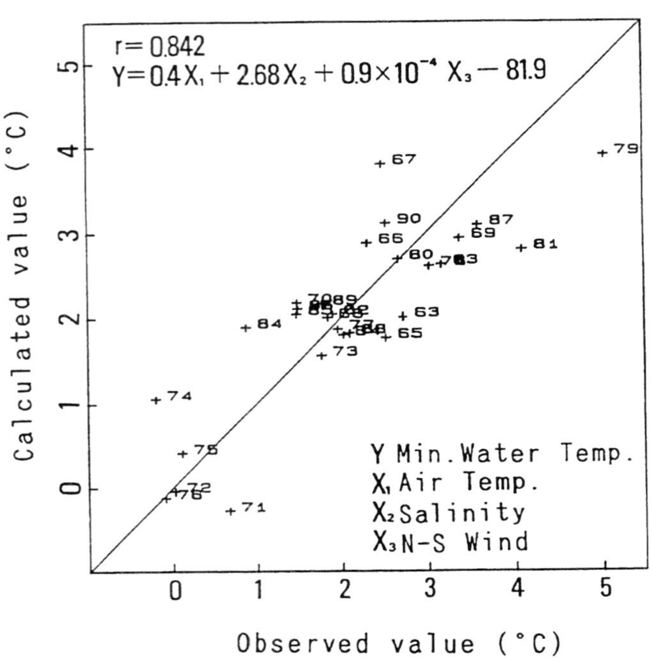

Fig. 9. Regression of minimum water temperatures and the variables of mean air temperatures, salinities, and north–south wind components.

Influence of winter sea conditions on the abundance of walleye pollock

Estimated numbers of walleye pollock at each age class in the eastern Bering Sea have been reported by means of cohort analysis (Wespestad and Traynor 1990). Because pollock are first taken by the commercial fisheries at age 2, natural mortality is a principal reason for the variability of recruitment until age 2. The annual numbers of age-2 pollock are shown in Fig. 10*b*. The numbers of spawners and fecundity, based on Sakurai (1982); cited in Kendall and Nakatani (1992), are shown in Fig. 10*a*. Annual numbers of age-2 fish and fecundity are of the same order, respectively, whereas there is a large difference of 10^5 in abundance between fecundity and age 2. The survival rations (SR_2) shown in Fig. 10*c* are the numbers of walleye pollock at age 2 (N_2) divided by their mothers' fecundity (F). Extraordinarily large values of survival rations for 1964–1967 can be eliminated because the numbers of spawners are too small in spite of the virginal stock, and they are at the beginning of the cohort analysis.

Except for the 1964–1967, correlations between survival ratios and minimum water temperatures at the year of birth (T_0) and at age 1 (T_1) are shown in Fig. 11*a* and 11*b* respectively. A correlation is found in Fig. 11*b* rather than in Fig. 11*a*. The regression equation is

$$(1)\ SR_2 = (0.278T_1 + 0.378) \times 10^{-5}$$

This result implies that the numbers of age-2 pollock depend on the mothers' fecundity, and the first order of variations of abundance of age-2 pollock should occur during the age-1 winter. By the following eq. 2, the number of age-2 pollock (N_2) can be predicted from their mothers' fecundity and the minimum temperature of the next winter:

$$(2)\ N_2 = F\,(0.278T_1 + 0.378) \times 10^{-5}$$

Discussion and Conclusion

It has been determined that the summer data are proportional to winter sea conditions. However, the results are probably biased because of the lack of continuous information throughout the cold season, such as the amount of sea ice and its distribution, and sea temperatures; moreover, summer observations are sparse. A more detailed analysis of the relationship between meteorological factors and sea conditions might reveal more precise interannual changes in the sea conditions.

Some questions arise from the survival ratios analysis.

(1) Why does the abundance of walleye pollock at age 2 change within a same order whereas a fifth order decrease in abundance occurs from egg to age 2?

(2) Why does the abundance of age-2 pollock depend on their mothers' fecundity in spite of the large decrease in abundance from egg to age 2?

(3) Why do survival ratios at age 2 correlate to minimum water temperatures at age 1?

We propose hypothese to answer these questions. The population of walleye pollock has a longer spawning period than the period during which foods suitable for juveniles are produced. In a spawning ground, variation with time in the number of eggs spawned makes a curve with a flat top rather

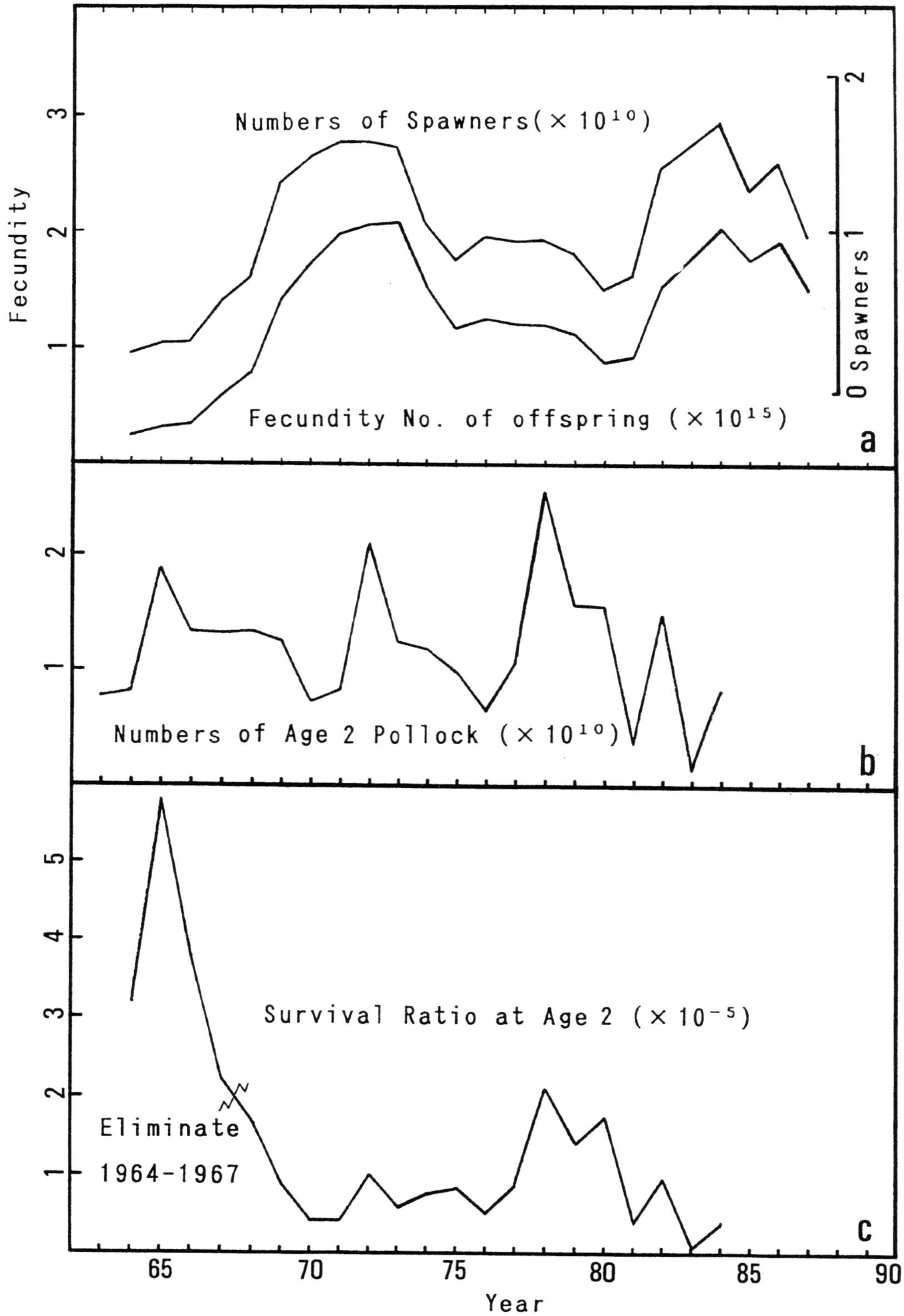

Fig. 10. Statistics of walleye pollock in the eastern Bering Sea. (*a*) Numbers of spawners and fecundity. (*b*) Number of age-2 pollock. (*c*) Survival rate at age 2.

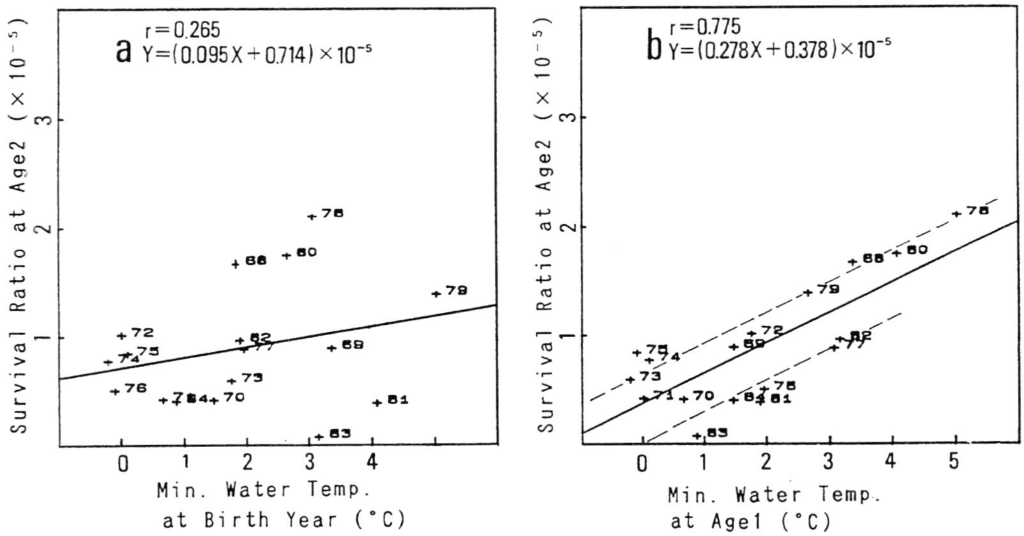

Fig. 11. Relationship between survival ratios at age 2 and minimum water temperature at the year of birth (*a*) and the minimum temperature at age 1 (*b*). In (*b*), broken regression line for the upper group is $SR_u = F\,(0.282T_1 + 0.635) \times 10^{-5}$, $r = 0.969$; and for the lower group is $SR_l = F\,(0.282T_l - 0.007) \times 10^{-5}$, $r = 0.898$.

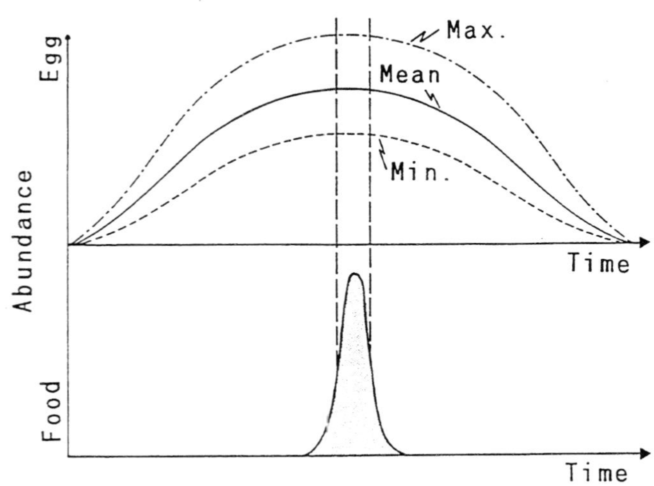

Fig. 12. Conceptual illustration of the spawning strategy of walleye pollock with respect to the availability of food for young pollock in a spawning ground.

than a steep normal distribution (Sakurai 1982; cited in Kendall and Nakatani 1992). The height of the curves is proportional to the abundance of yearly spawners as shown in Fig. 12. On the other hand, food production, in the form of the spring bloom, finishes within a short period, thus creating a curve with a steep peak. Therefore, those juveniles survive that are able to take advantage of the period of food production, while most of the others do not survive. Although the initial decrease is large, survival is relatively stable and depends on the abundance of spawners. Despite the great losses, walleye pollock seem to have a successful spawning strategy.

Young pollock remain on the shelf through their first winter. An area that acts as a nursery spreads over the whole southern region in warm winters; however, drift ice limits it to a narrow area in cold winters. Thus, the density of young pollock as prey increases in cold years; so predators easily catch young pollock at age 1. Thus the survival ratio of age-2 pollock correlates to the minimum water temperature at age 1.

These hypotheses suggest interpretations for the relationship between the lives of marine fish and their surroundings. More detailed and precise investigations should be continued cooperatively to study the reasons for interannual changes in marine resources.

Acknowledgements

We especially thank all of the research staff, and former Captain T. Fujii, Captain K. Masuda, the officers, and the crews of the T.S. Oshoromaru who spent many summers at sea to collect the oceanographic data. We very much appreciate the referees for their kind criticism and helpful suggestions to our manuscript. We also thank K. Sato and Y. Hamada of our laboratory for their help in preparing the manuscript.

References

Cavalieri, D.J., and C.L. Parkinson. 1987. On the relationship between atmospheric circulation and the fluctuations in the sea ice extents of the Bering and Okhotsk Seas. J. Geophys. Res. 92: 7141–7162.

Coachman, L.K., and R.L. Charnell. 1979. On lateral water mass interaction—case study, Bristol Bay, Alaska. J. Phys. Oceanogr. 9: 278–297.

Faculty of Fisheries, Hokkaido University. 1964–1991. Data record of oceanographic observations and exploratory fishing. Faculty of Fisheries, Hokkaido University, Hakodate, Japan. Publ. 7–34.

Kendall, A.W.J., and T. Nakatani. 1992. Comparisons of early-life history characteristics of walleye pollock *Theragra chalcogramma* in Shelikof Strait, Gulf of Alaska, and Funka Bay, Hokkaido, Japan. Fish. Bull. 90: 129–138.

Kinder, T.H., and J.D. Schumacher. 1981*a*. Hydrographic structure over the continental shelf of the southeastern Bering Sea, p. 31–52. *In* D.W. Hood, and J.A. Calder [ed.] The eastern Bering Sea shelf: oceanography and resources. University of Washington Press, Seattle, Wash. 625 p.

Kinder, T.H., and J.D. Schumacher. 1981*b*. Circulation over the continental shelf of the southeastern Bering Sea, p. 53–75. *In* D.W. Hood, and J.A. Calder [ed.] The eastern Bering Sea shelf: oceanography and resources. University of Washington Press, Seattle, Wash. 625 p.

Kinder, T.H., J.D. Schumacher, R.B. Tripp, and J.C. Haslett. 1978. The evolution of the hydrographic structure over the continental shelf near Bristol Bay, Alaska, during summer 1976. Department of Oceanography, University of Washington, Seattle, Wash. Tech. Rep. M 78–16.

Koto, H., and T. Fujii. 1958. Structure of the waters in the Bering Sea and the Aleutian region. Bull. Fac. Fish. Hokkaido Univ. 9: 149–170.

McNutt, S.L. 1981. Remote sensing analysis of ice growth and distribution in the eastern Bering Sea, p. 141–165. *In* D.W. Hood and J.A. Calder [ed.] The eastern Bering Sea shelf: oceanography and resources. University of Washington Press, Seattle, Wash. 625 p.

Niebauer, H.J. 1981. Recent fluctuations in sea ice distribution in the eastern Bering Sea, p. 133–140. *In* D.W. Hood and J.A. Calder [ed.] The eastern Bering Sea shelf: oceanography and resources. University of Washington Press, Seattle, Wash. 625 p.

Ohtani, K. 1969. On the oceanographic structure and the ice formation on the continental shelf in the eastern Bering Sea. Bull. Fac. Fish. Hokkaido Univ. 20: 94–117. (In Japanese with English abstract)

Ohtani, K. 1973. Oceanographic structure in the Bering Sea. Memr. Fac. Fish. Hokkaido Univ. 21: 65–106.

Ohtani, K. 1989. The role of the Sea of Okhotsk on the formation of the Oyashio water. Umi to Sora, 65: 63–83. (In Japanese with English abstract)

Reed, R.K. 1978. The heat budget of a region in the eastern Bering Sea, summer 1976. J. Geophys. 83: 3635–3645.

Sakurai, T. 1982. Reproductive ecology of walleye pollock *Theragra chalcogramma* (Pallas). Ph.D. thesis, Hokkaido University, Hakodate, Hokkaido, Japan. 178 p. (transl. from Japanese by Alaska Fisheries Science Centre, NMFS, NOAA, Seattle, Wash. 98115-0070)

Schumacher, J.D., T.H. Kinder, D.J. Pashinski, and R.L. Charnell. 1979. A structural front over the continental shelf of the eastern Bering Sea. J. Phys. Oceanogr. 9: 79–87.

Schumacher, J.D., C.A. Pearson, and J.E. Overland. 1982. On the exchange of water between the Gulf of Alaska and the Bering Sea through Unimak Pass. J. Geophys. Res. 87: 5785–5795.

Wespestad, V.G., and J.J. Traynor. 1990. Walleye pollock in condition of groundfish resources in the Bering Sea – Aleutian Islands region as assessed in 1988. NOAA Tech. Memo. NMFS FINWC-178. 224 p.

Effects of climate change on resource populations in the Yellow Sea ecosystem

Qisheng Tang

Tang, Q. 1995. The effects of climate change on resource populations in the Yellow Sea ecosystem, p. 97–105. *In* R.J. Beamish [ed.] Climate change and northern fish populations. Can. Spec. Publ. Fish Aquat. Sci. 121.

Abstract: Both physical and biological factors may perturb the biomass yields of resource populations in the Yellow Sea ecosystem. Natural factors may have had an important effect on long-term changes in dominant species, and changes in the abundance of the species of various ecotypes are correlated with climate change. It seems to be accepted that pelagic and shellfish species are sensitive to climate change, such as Pacific herring (*Clupea pallasi*) and prawn (*Penaeus orientalis*). The major environmental factors affecting fluctuations in recruitment appear to be rainfall, river runoff, daylight, wind, temperature, and salinity.

Résumé : Des facteurs physiques autant que biologiques peuvent perturber le rendement en biomasse des populations ressources de l'écosystème de la Mer Jaune. Des facteurs naturels peuvent avoir eu un effet important sur les changements à long terme dans les espèces dominantes, et certains changements dans l'abondance de l'espèce des divers écotypes sont en corrélation avec le changement climatique. Il semble accepté que les espèces pélagiques, notamment le hareng du Pacifique (*Clupea pallasi*) et les crustacés, dont certaines espèces de crevettes (*Penaeus orientalis*), sont sensibles au changement climatique. Les principaux facteurs environnementaux qui font fluctuer le recrutement semblent être la pluviosité, l'écoulement fluvial, la lumière du jour, les vents, les températures et la salinité.

Introduction

The Yellow Sea is one of the most intensively exploited areas in the world. Commercial utilization of the living resources in the ecosystem dates back several centuries. With a remarkable increase in fishing effort and its expansion to the entire Yellow Sea, nearly all the major stocks were fully fished by the mid-1960s. From the early 1960s through the 1980s, the resource populations declined in biomass and species shifted noticeably in dominance. Changes in the quantity and quality of biomass yields of the resource populations are attributed principally to human predation, especially in demersal species. They cannot, however, be explained merely by fishing pressure. Under the same fishing pressure, biomass yields of some species appear to be fairly stable, and changes in abundance are correlated with environmental variability. Therefore, the purpose of this paper is to evaluate the effects of climate change on the resource populations in the Yellow Sea ecosystem.

Following the introduction, a brief description of the natural setting of the ecosystem is given, and the relationship between changes in major resources populations and environment are reported. Discussions of the effects of climate on resource population dynamics and recommendations for future research are offered in the final section.

Q. Tang. Yellow Sea Fisheries Research Institute, 19 Laiyang Road, Qingdao, 26003, Peoples Republic of China

The natural setting

Physical environment

The Yellow Sea is the semienclosed body of water bounded by the Chinese mainland to the west, the Korean peninsula to the east, and a line running from just north of the mouth of the Changjiang (Yangtze) River to Cheju Island. It is about 1000 km long and 700 km wide with a mean depth of 44 m. The central part of the sea is traditionally called the Yellow Sea Depression with depths in the range of about 70–80 m and a maximum depth of 140 m. It is the major overwintering ground for most fish and invertebrates in the Yellow Sea ecosystem. The sea annually receives more than 1.6×10^9 t of sediments, mostly from the west Yellow Sea coast and the Changjiang River, which has formed a large delta. Turbidity and sediment type appear to be the major parameters that affect the distribution of planktonic and benthic organisms in the coastal waters of the sea.

Wind over the Yellow Sea has distinct monsoon characteristics and determines the variations in the sea's water properties. In winter, cold and dry winds are usually from the north-northwest every 3–8 d, and gale-strength winds usually prevail after the passage of a cold front. In summer, winds are from the south to south-southeast, and gale or stronger winds are associated with the passing of a typhoon, which occurs with a frequency of slightly less than twice per year.

The Yellow Sea Cold Water Mass is one of the most outstanding and important components in the hydrography of the sea. It is a water mass characterized by low temperature and higher salinity in the central part of its deep bottom layers, and is believed to be the remnants of local water left over from the

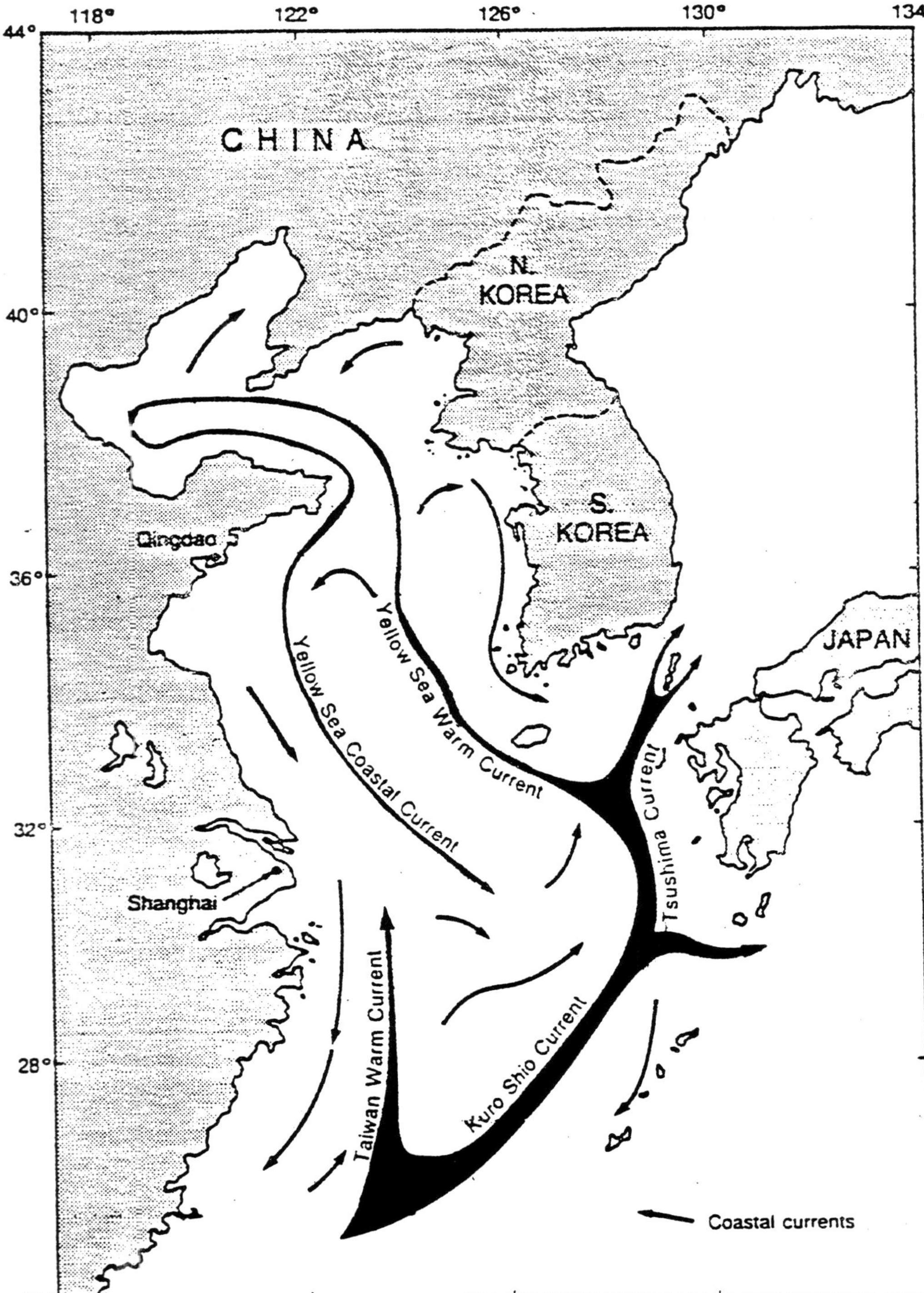

Fig. 1. Schematic diagram of the major winter current system in the Yellow Sea.

previous winter (Ho et al. 1959; Guan 1963; Zhang and He 1989). Stratification is strongest in summer, with a vertical temperature gradient greater than 10° · C m^{-1}. Weng et al. (1989) have reported that the distribution range of the cold water mass has distinct variations over a long period of time and its "relative volume" in a stronger year is 2.2 times that in a weaker year.

The seasonal mean circulation in the Yellow Sea is a basin-wide cyclonic gyre formed by the Yellow Sea Coastal Current and the Yellow Sea Warm Current. The Yellow Sea Warm Current, which is a branch of the Tsushima Warm Current from the Kuroshio region in the East China Sea, carries water of relatively high salinity (>33‰) and high temperature (>21°C) to the north along 124°E and then to the northwest. This current

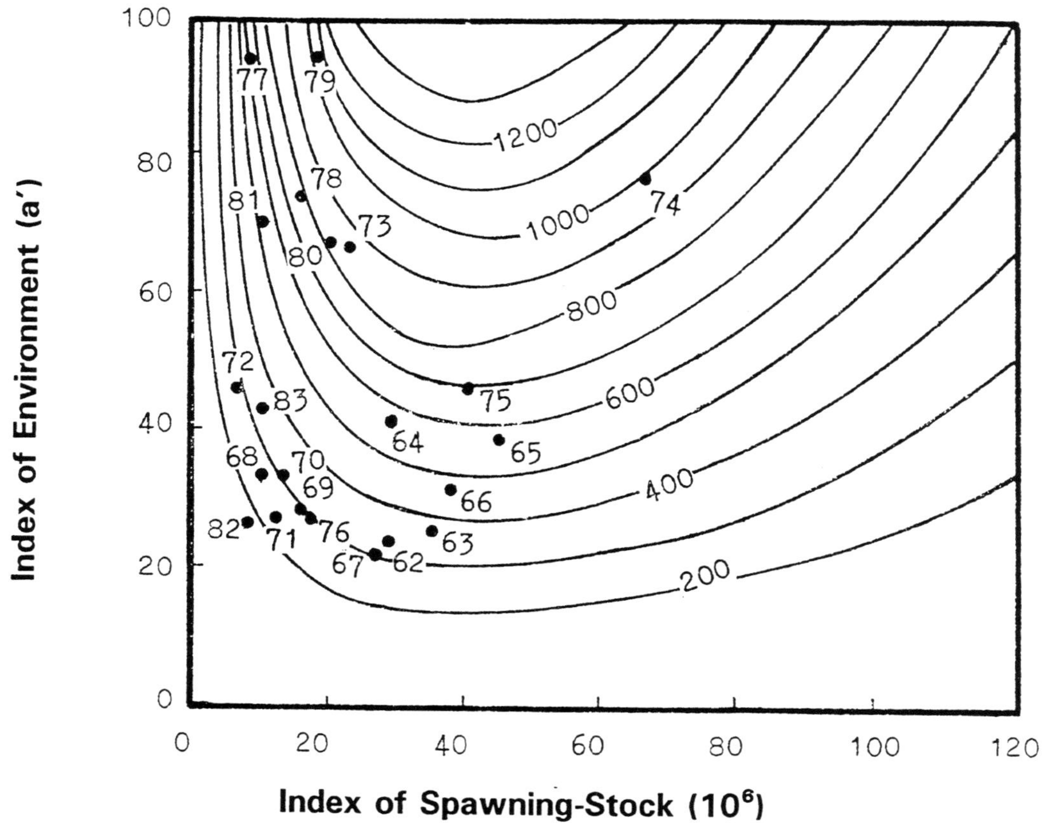

Fig. 2. Calculated recruitment isopleths for prawns in the Bohai Sea, based on observed indices (dots) of environment, spawning-stock abundance, and recruitment, 1962–83. The *a* is an index of environmental conditions selected by step-wise regression. Based on the data of the potential environmental factors affecting recruitment in 1962–83, river runoff, rainfall, daylight, and salinity were selected for inclusion in the index of environmental conditions (from Tang et al. 1989).

with its extension through most of the year and the Coastal Current flowing southward along the coast, plays an important role in exchanging the waters in the semienclosed ecosystem (Fig. 1).

Ecological characteristics

The Yellow Sea lies in the warm temperate zone, and the fauna are recognized as a sub-East Asia province of the North Pacific temperate zone (Cheng 1959; Liu 1963; Dong 1978; Zhao et al. 1990). The biotic communities of the sea are complex in species composition, spatial distribution, and community structure,and marked seasonal variations are the main characteristics of all components of the biotic communities.

As compared with other shelf regions in the northwest Pacific, the Yellow Sea ecosystem has a relatively low primary production, about 60 g C $\cdot$ m^{-2} $\cdot$ yr^{-1}, and the production of phytoplankton in the sea is estimated to be 0.52 10^9 t $\cdot$ km^{-2} $\cdot$ yr^{-1}. Over the past 30 yr, the phytoplankton biomass has been relatively stable, but from 1959 to 1986 the zooplankton declined in annual biomass by about 50%, a trend similar to that in the East China Sea (Tang 1989; Chen et al. 1990).

Possibly because of the complicated oceanographic conditions of the sea, the resource populations are composed of species of various ecotypes. Warm temperate species are the major component of the biomass, accounting for about 60% of the total biomass of the resource populations. Warmwater species and cold temperate species account for about 15 and 25%, respectively; demersal and semidemersal species account for about 58%; and pelagic species for about 42%. Because most of the species inhabit the Yellow Sea year round, the resource populations in the fauna have formed an independent community. The diversity and abundance of the community are comparatively lower than those in the East China Sea and South China Sea. The Shannon–Wiener diversity index (H,) and the Simpson ecological dominance index (C) of the resource populations were determined to be 2.3 and 0.34, respectively (Tang 1990).

Changes in resource populations

Changes in demersal species

Although we believe that the cause of decline in abundance for the most important commercial demersal species in the Yellow Sea, such as small yellow croaker *(Pseudosciaena polyactis)*, hairtail *(Trichiurus haumela)*, large yellow croaker *(Pseudosciaena crocea)*, flatfish (Pleuronectidae), Pacific cod (*Gadus macrocephalus*), searobin *(Lepidotrigla microptera)*, red seabream *(Pagrosomus major)*, *Miichthys miiuy*, and *Nibea albiflora*, is extreme overexploitation, including growth overfishing and recruitment overfishing (Xia 1978; Liu 1979; Chikuni 1985; Tang 1989; Zhao et al. 1990), fluctuations in abundance

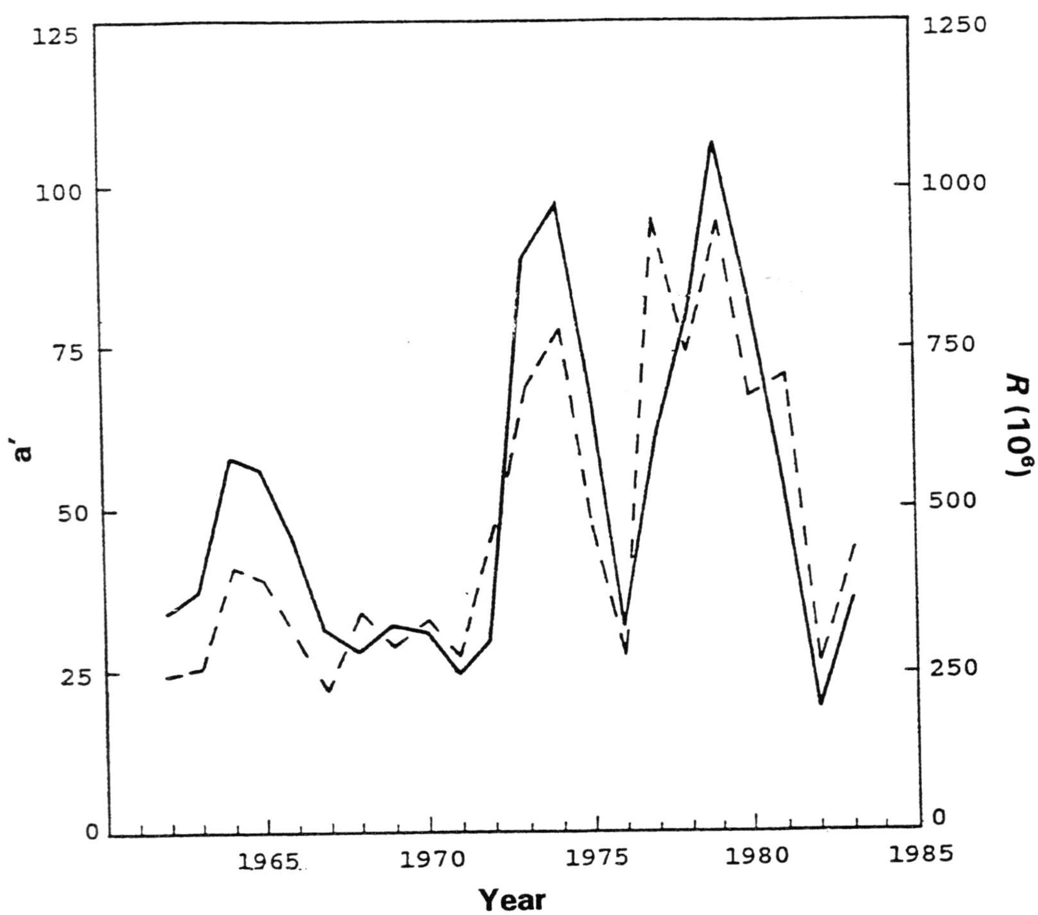

Fig. 3. Indices of environmental conditions (a = solid line) and recruitment (*R* = broken line) for prawns in the Bohai Sea, 1962–83 (from Tang et al. 1989).

for some demersal species may be affected by both natural and anthropogenic factors. The population dynamics of the fleshy prawn *(Penaeus orientalis)* provides a good example.

The prawn is a valuable crustacean that is widely distributed in the Bohai Sea and Yellow Sea. Its main spawning grounds lie near the estuaries along the coast of the Bohai Sea. Its life span is about 1 yr. Spawning occurs in late spring, after which the prawns grow rapidly and reach commercial size in September. When water temperatures begin to drop significantly in late autumn, the prawns migrate out of the Bohai Sea to overwinter in the Yellow Sea depression. The biomass yields of the fleshy prawn vary from year to year, and annual catch ranged from 7000 to 50 000 t during 1953–90. We found that both the environment and spawning-stock size were related to fluctuations in recruitment, and that the relative importance of the two factors varied from year to year. Examples of the three possible combinations of dominant factors have been noted in the 1960–83 records of the prawn: environment, spawning-stock size, and environment plus spawning-stock size (Fig. 2). For example, 1979 has a high *a* value (*a* is an index of environmental conditions selected by stepwise regression), and 1976 has a low *a* value; in both years, the spawning stock was the same small size, but the 1979 year-class produced three times the recruitment of the 1976 year-class because of the favourable environment. Environmental effects were well delineated for many years, and favourable environmental conditions plus a suitable large spawning stock resulted in a large recruitment, as in 1973–74 and 1978–80 (Fig. 3). The major environmental factors affecting recruitment may be river runoff, rainfall, daylight, and salinity (Tang et al. 1989). It is of interest to note that Y. Gao and Y. Liu (unpublished data, Marine Fisheries Research Institute of Shandong, 40 Yaojin Road, Yiantai, 264000, P.R. China) have reported a correlation between El Niño and biomass yields of prawn: prawn catch is reduced in El Niño years.

Changes in pelagic species

Pacific herring *(Clupea harengus pallasi)*, chub mackerel *(Scomber japonicus)*, Spanish mackerel *(Scomberomorus maculatus)*, and silver pomfret *(Stromaleoides argenteus)* are the major and larger sized pelagic species in the Yellow Sea. The annual catch from 1953 to 1988 fluctuated wildly, ranging from 30 000 to 300 000 t annually. The causes of fluctuations in the abundance of these pelagic species are more complicated than those for demersal species. There may be two patterns of population dynamics (Fig. 4). Spanish mackerel and silver pomfret stocks appear to be relatively steady, and both abundance and catch have tended to increase steadily since the

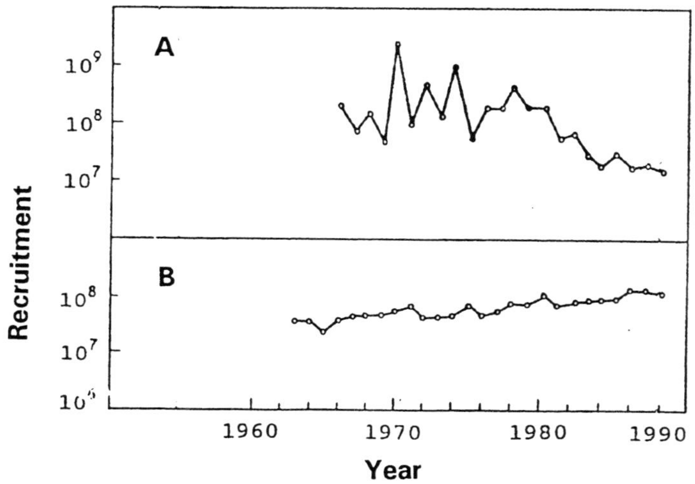

Fig. 4. Recruitment of Pacific herring (A) and Spanish mackerel (B) in the Yellow Sea.

species began to be utilized in the 1960s. The reason for this is not clear. Perhaps it is due to an unusual combination of natural and anthropogenic conditions; the changes are particularly large for Pacific herring and chub mackerel stocks. Detailed information on the population dynamics of these species follows.

Pacific herring in the Yellow Sea has a long history of exploitation, and the fishery is full of drama. This is demonstrated by the existence of villages and localities named for their association with it. In this century, the commercial fishery has experienced two peaks (in about 1900 and 1938), followed by a period of little or no catch. In 1967, a large number of 1-yr-old herring began appearing in the bottom trawl catch because of the recovery of the stocks. Since then, the catch has increased rapidly to a peak of 200 000 t in 1972, because the very strong 1970 year class reached a maximum historic abundance of 26.8 10^8 (age 1). Since 1982, the stock apparently declined because of the occurrence of a series of weak year-classes. The catch has declined to below 1000 t in 1989–90. Obviously, herring in the Yellow Sea have been characterized by strong fluctuations in recruitment, which directly affect the fishable stock. There is no strong relationship between spawning stock and recruitment (Tang 1981, 1987). Environmental effects on herring recruitment in the Yellow Sea were investigated by cluster analysis, based on the data of recruitment and environmental conditions of the spawning ground in 1968–1977. Results of these analyses indicated that rainfall, wind, and daylight appear to be the major factors affecting fluctuations in recruitment (Fig. 5), and long-term changes in abundance may be correlated with the 36-yr cycle of dryness–wetness oscillations in eastern China (Fig. 6). Li (1992) reported that Southern Oscillation (SO) affected the variation in dryness and wetness in eastern China during the last 400 yr. When the SO index is low, eastern China (e.g., Yellow River Valley) is likely to be dry.

Trends in the catch of chub mackerel are similar to those of Pacific herring. It has been generally accepted that environmental conditions may have had an important effect on the long-term changes in the abundance of chub mackerel. Ye (1985) suggested that the catch variations have a distinct 18-yr cycle.

Paste shrimp *(Acetes chinensis)* is a typical pelagic shrimp with a lifespan of 10–11 mo. It is abundantly distributed in the Bohai Sea and supports an important coastal fishery in northern China. The annual catches ranged from 16000 to 120000 t during 1954–90. Good catch years occurred in 1954–58, 1973–78, 1985, and 1990 with catches in the range of 80 000 – 120 000 t, and poor catch years occurred in 1961–68, 1981, and 1987 with catches in the range of 16 000 – 50 000 t. This shrimp prefers a neritic environment influenced by considerable discharge of freshwater from the land. It commonly lives near the estuaries of Liaodong Bay, Laizhou Bay, and Bohai Bay in

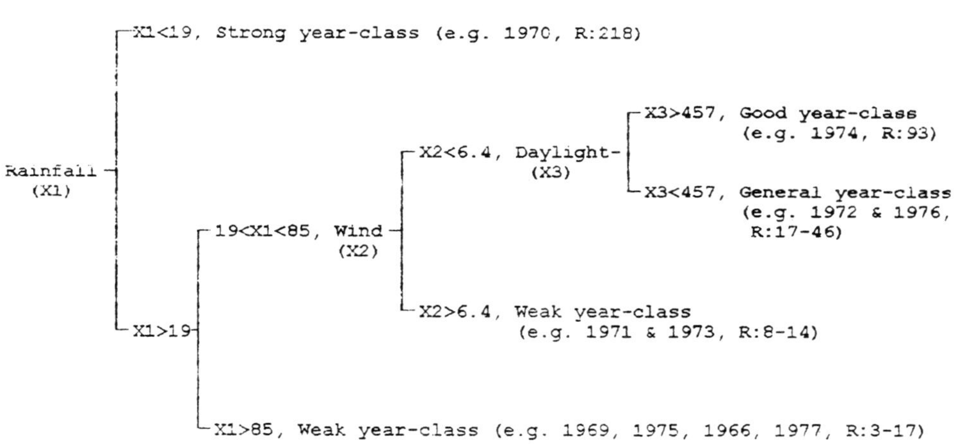

Fig. 5. Diagram of relationship between index of recruitment (R) and environmental factors (*X1*, *X2*, and *X3*), using cluster analysis (from Tang 1981).

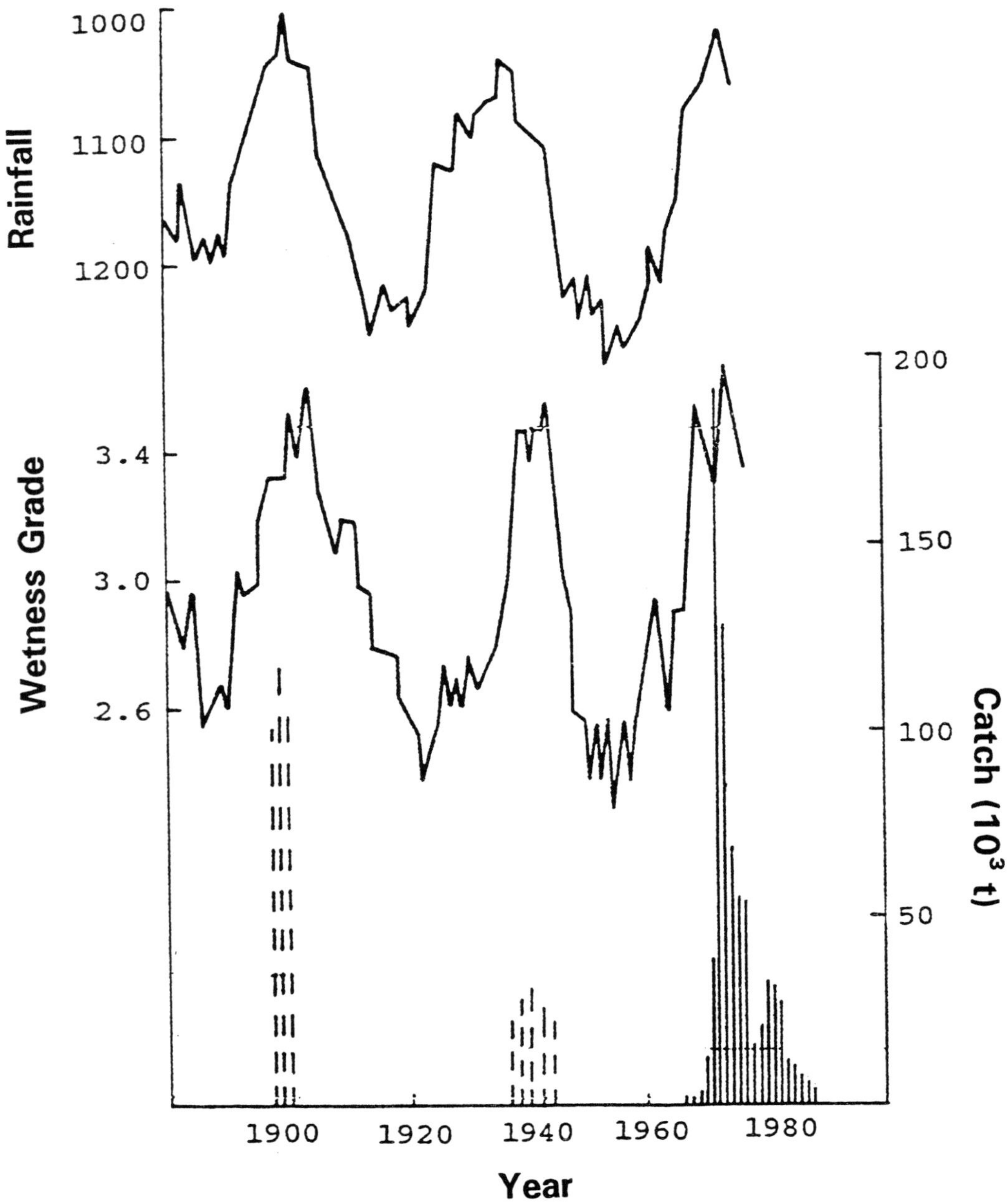

Fig. 6. Relationship between fluctuations in herring abundance in the Yellow Sea and the 36-yr cycle of wetness oscillations in eastern China (adapted from Tang 1981).

the Bohai Sea. Fluctuations in catch seem to have been caused mainly by natural factors; first by fluctuations in recruitment resulting from variations in early survival rate, and second by changes in availability associated with particular oceanographic conditions and shoaling behaviour (Chikuni 1985). The major environmental factors affecting abundance are considered to be rainfall, river runoff, and temperature (Meng and Zhou 1982; Wu 1991).

Species shifts in dominance

The living resources in the Yellow Sea ecosystem are multi-species in nature. About 20 major species account for 92% of the total biomass of the resource populations, and about 80 species account for the other 8%. From the 1950s through the 1980s, dramatic shifts in species dominance occurred. The dominant species in the 1950s and early 1960s were small yellow croaker and hairtail, while Pacific herring and chub mackerel became dominant in the 1970s. Some small-size, fast-growing, short-lived, and low-value species such as Japanese anchovy *(Engraulis japonicus)* and half-fin anchovy *(Setipinna gilberti)* increased markedly in abundance in the 1980s, and have taken a prominent position in the ecosystem resources.

As shown in Fig. 7, the major resource populations in 1958–59 were small yellow croaker, flatfish, Pacific cod, hairtail, skates *(Raja* and *Dasystis)*, searobin, and angler *(Lophins litulon)*, accounting for 71% of the total biomass. Of these, 11% were planktophagic species, 46% benthophagic species, and 43% ichthyophagic species. In 1985–86, the

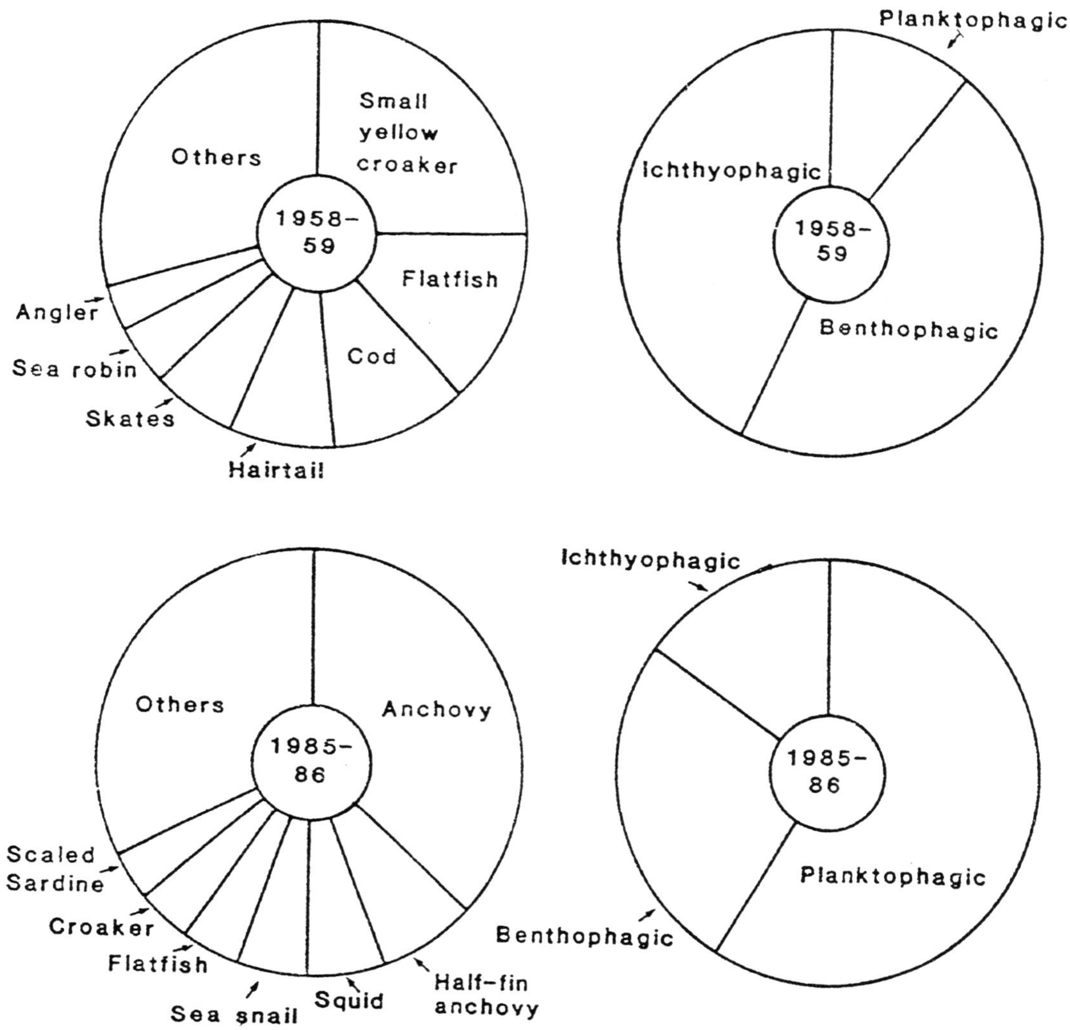

Fig. 7. Proportion of major species and various feeding habits in the biomass yields, based on bottom trawl surveys of the Yellow Sea by research vessel in 1958–59 and 1985–86.

major resource populations were Japanese anchovy, half-fin anchovy, Japanese squid *(Loligo japonica)*, seasnail *(Liparis tanakae)*, flatfish, small yellow croaker, and scaled sardine *(Harengula zunasi)*, accounting for 67% of the total biomass. Of these, 59% were planktophagic species, 26% benthophagic species, and 16% ichthyophagic species. As a result, some commercially important, major demersal species of large size and higher trophic level were replaced by less valuable pelagic species of smaller size and lower trophic level.

There may be two types of species shifts in the ecosystem resources: systematic replacement and ecological replacement. Systematic replacement occurs when one dominant species declines in abundance or is depleted by overexploitation and another competitive species uses surplus food and vacant space to increase its abundance. Ecological replacement occurs when minor changes in the natural environment effect stock abundance, especially pelagic species. The data, based on catch, indicate that warm and temperate species tend to increase in abundance during warm years (e.g., half-fin anchovy and cuttlefish *(Sepiella maindroni)* increased in the 1980s, while boreal species (e.g., Pacific herring) tend to increase during cold years, such as the 1970s (Fig. 8). Thus, natural factors may have an important effect on long-term changes in dominant species of various ecotypes.

Conclusion

Both physical and biological factors may affect biomass yields of resource populations in the Yellow Sea ecosystem. In general, the changes in quantity and quality of the biomass yield are attributed principally to human predation. But this is not the case for all of species. Based on this study, fluctuations in recruitment of some pelagic and shellfish species, and species shifts in dominance are correlated with environmental variability (e.g., river runoff, rainfall, daylight, wind, salinity, and temperature). It seems to be accepted that climate change may have important effects on resource populations in the sea, especially on pelagic species and shellfish, resulting in long-term changes. In the long term, the effects of the two perturbations on resource populations may be mingled, so that the causes of variability are extremely difficult to isolate.

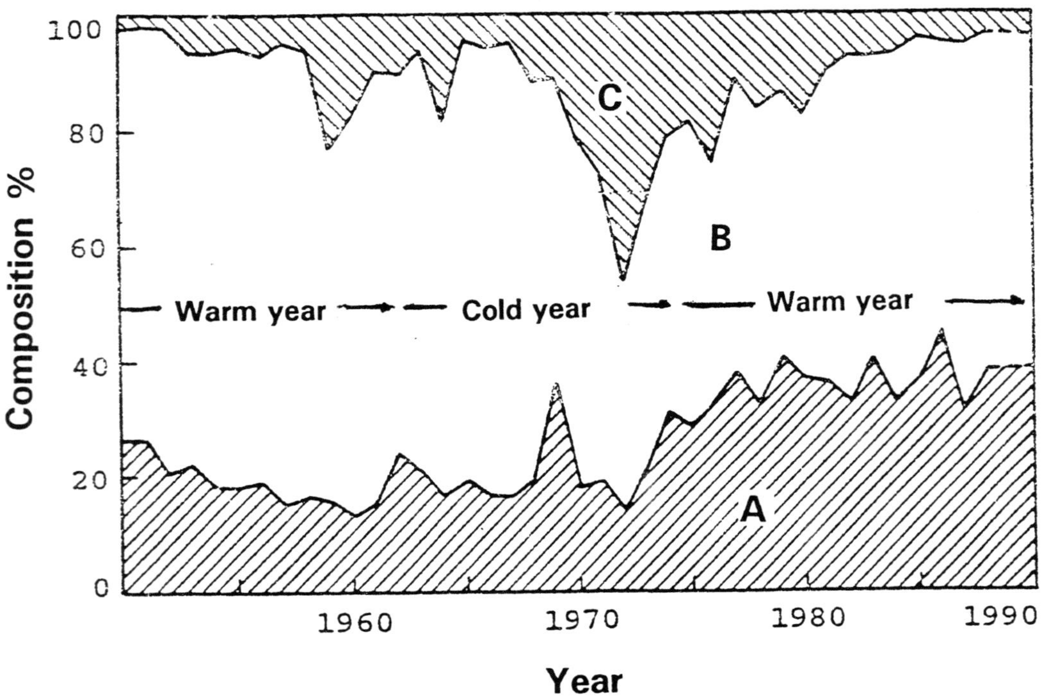

Fig. 8. Trends in relative abundance of various ecotypes comprising the annual catch in the Yellow Sea, and long-term changes in environmental conditions. (A) Warmwater species; (B) temperate-water species; (C) boreal species.

The above facts indicate that an ecosystem-perspective study should be considered. One approach is to study the Yellow Sea ecosystem dynamics through assessment and monitoring. The primary emphasis of research should be focused on population-dynamics processes and ocean climate variability, especially for those processes, such as secondary production and recruitment, which are critical to the maintenance of production of economically important resource populations.

References

Cheng, C. 1959. Notes on the economic fish fauna of the Yellow Sea and the East China Sea. Oceanol. Limnol. Sin. 2(1): 53–60. (In Chinese).

Chikuni, S. 1985. The fish resources of the northwest Pacific. FAO Fish. Tech. Pap. 266.

Dong, J. 1978. The geographical distribution of cephalopods in Chinese waters. Oceanol. Limnol. Sin. 9(1): 108–116. (In Chinese with English abstract).

Chen, G., X. Gu, H. Gao, and G. Yang (ed). 1990. Marine Fishery Environment of China. Zhejiang Science and Technological Press, Hangzhou, China. (In Chinese).

Guan, B. 1963. A preliminary study of the temperature variations and the characteristics of the circulation of the cold water mass of the Yellow Sea. Oceanol. Limnol. Sin. 5(4): 255–284. (In Chinese).

Ho, C., Y. Wang, Z. Lei, and S. Xu. 1959. A preliminary study of the formation of the Yellow Sea cold water mass and its properties. Oceanol. Liminol. Sin. 2: 11–15. (In Chinese).

Li, Y. 1992. A preliminary analysis on the relationship between Southern Oscillation and dryness/wetness in the eastern China during the last 400 years. Mar. Sci. 82(4): 37–40. (In Chinese).

Liu, J. 1963. Zoogeographical studies on the micrurus crustacean fauna of the Yellow Sea and East China Sea. Oceanol. Limnol. Sin. 5(3): 230–244. (In Chinese).

Liu, X. 1979. Status of fishery resources in the Bohai and Yellow Sea. Mar. Fish. Res. (Qingdao), Paper 26: 1–17. (In Chinese).

Meng, T., and B. Zhou. 1982. The application of A.I.D. method to study quantitative forecast of paste shrimp (*Acetes chinensis*) in the Bohai Sea. Acta Oceanol. Sin. 4(4): 490–496. (In Chinese with English abstract).

Tang, Q. 1981. A preliminary study on the cause of fluctuations in year class size of Pacific herring in the Yellow Sea. Trans. Oceanol. Limnol. 2: 37–45. (In Chinese with English abstract).

Tang, Q. 1987. Estimation of fishing mortality and abundance of Pacific herring in the Yellow Sea by cohort analysis (VPA). Acta Oceanol. Sin. 6(1): 132–141.

Tang, Q. 1989. Changes in the biomass of the Yellow Sea ecosystem, p. 7–35. *In* K. Sherman an L. M. Alexaner [ed.] Biomass and geography of large marine ecosystem. AAAS Sel. Symp. 111.

Tang, Q. 1990. Ecological dominance and diversity of fishery resources in the Yellow Sea. Mar. Sci. 2(1): 57–67.

Tang, Q., J. Deng, and J. Zhu. 1989. A family of Ricker SRR curves of the prawn under different environment conditions and its enhancement potential in the Bohai Sea. Can. Spec. Publ. Fish. Aquat. Sci. 108: 335–339.

Weng, X., Y. Zhang., C. Wang, and Q. Zhang. 1989. The variation characteristics of the Yellow Sea cold water mass. J. Ocean Univ. Qingdao, 19(1–2): 119–131.

Wu, J. 1991. Paste shrimp, p. 590–612. *In* J. Deng and C. Zhao [ed.] Marine fishery biology. Agriculture Publishing House, Beijing. 686 p. (In Chinese).

Xia, S. 1978. An analysis of changes in fisheries resources of the Bohai Sea, Yellow Sea, and East China Sea. Mar. Fish. Res. (Qingdao), Paper 25: 1–13. (In Chinese).

Ye, M. 1985. A preliminary analysis of the population dynamics of pelagic fishes in the Yellow Sea and Bohai Sea. Mar. Bull. 4(1): 63–67. (In Chinese).

Zhang, Y., and X. He. 1989. The annual variation and its forecasting of the intensity of cold water mass of the northwest Yellow Sea in spring. J. Ocean Univ. Qingdao, 19 (1–2): 173–283.

Zhao, C., X. Liu, B. Zhen, and Y. Yui (ed). 1990. Marine fishery resources of China. Zhejiang Science and Technology Press, Hangzhou, China. 178 p. (In Chinese).

Analysis of trends in primary productivity and chlorophyll-*a* over two decades at Ocean Station P (50°N, 145°W) in the Subarctic Northeast Pacific Ocean

C.S. Wong, F.A. Whitney, K. Iseki[1], J.S. Page, and Jiye Zeng

Wong, C.S., F.A. Whitney, K. Iseki, J.S. Page, and J. Zeng. 1995. Analysis of trends in primary production and chlorophyll-*a* over two decades at Ocean Station P (50°N, 145°W) in the subarctic Northeast Pacific Ocean, p. 107–117. *In* R.J. Beamish [ed.] Climate change and northern fish populations. Can. J. Fish. Aquat. Sci. 121.

Abstract: Measurements of in situ primary production were made using a clean sampling technique for ^{14}C uptake and incubation of seawater in polycarbonate bottles suspended from a free-drifting buoy at Ocean Station P (50°N, 145°W) for the period 1984–90. The rates of primary production for the seasons spring (March–May), summer (June–August), autumn (September–November), and winter (December–February) were 415, 466, 366 and 283 mg C · m^{-2} · d^{-1} respectively. The annual production was estimated to be 140 g C · m^{-2} · y^{-1}. The subarctic waters in the Northeast Pacific Ocean are more productive than previously thought by a factor of two, even though chlorophyll-*a* concentrations have remained near 0.4 mg · m^{-3} over a 27-yr period. Earlier inference of low production was based on a technique in which samples were possibly contaminated by metals and/or organics from several potential sources. Causes of high production in subarctic waters and possible enhancement by climatic factors are examined.

Résumé : On a mesuré la production primaire sur place en utilisant une technique d'échantillonnage propre permettant l'absorption et l'incubation de ^{14}C d'eau de mer dans des bouteilles de polycarbonate suspendues à une bouée dérivant librement à la station océanique P (50° N, 145° O) pendant la période 1984–1990. Les taux de production primaire pour les saisons du printemps (mars–mai), de l'été (juin–août), de l'automne (septembre–novembre), et de l'hiver (décembre–février) ont été respectivement de 415, 466, 366 et 283 mg C · m^{-2} · j^{-1}. La production annuelle a été évaluée à 140 g C · m^{-2} · an^{-1}. Les eaux subarctiques du nord-est du Pacifique sont deux fois plus productives qu'on ne le croyait, même si les concentrations de chlorophylle-*a* sont restées à près de 0,4 mg · m^{-3} sur une période de 27 ans. Les hypothèses antérieures de faible production étaient basées sur une technique faisant en sorte que les échantillons étaient peut-être contaminés par des métaux et (ou) par des produits organiques provenant de plusieurs sources potentielles. Les causes de la production élevée dans les eaux subarctiques et de son amélioration possible grâce à des facteurs climatiques sont examinées.

C.S. Wong, F.A. Whitney, K. Iseki[1], J.S. Page. Centre for Ocean Climate Chemistry, Institute of Ocean Sciences, P.O. Box 6000, Sidney, BC V8L 4B2, Canada.

[1] Present address: Fisheries Oceanography Division, Seikai National Fisheries Research Institute, 49 Kokubu-machi, Nagasaki, 850 Japan.

J. Zeng. Department of Oceanography, University of British Columbia, 6270 University Boulevard, Vancouver, BC V6T 1Z4, Canada.

Introduction

Ocean Station P at 50°N, 145°W in the Northeast Pacific Ocean is a location where there has been a sustained effort in measuring biological productivity for more than three decades. Early results estimated primary production to be 60 g $C \cdot m^{-2} \cdot yr^{-1}$ (McAllister 1972). This relatively low level of production meshed well with an earlier concept of low phytoplankton growth, which never produced blooms nor depleted macronutrients (Anderson et al. 1969; Parsons and Lalli 1988). However, Welschmeyer et al. (1991) using a clean sampling technique obtained a much higher production estimate of 230 g $C \cdot m^{-2} \cdot yr^{-1}$ (631 mg $C \cdot m^{-2} \cdot d^{-1}$), although they lacked winter data. Falkowski and Wilson (1992) also estimated primary production for broad areas of the North Pacific using Secchi disc data, and suggested a production rate of 167 g $C \cdot m^{-2} \cdot yr^{-1}$ in subarctic areas such as Station P.

Interannual variations in zooplankton biomass were observed in time-series measurements at Ocean Station P (Wickett 1967; Longhurst et al. 1972; Parslow 1981; Frost 1983; Miller et al. 1991; Brodeur and Ware 1992). Recently, Brodeur and Ware observed a doubling of zooplankton biomass between the 1950s and 1980s in the Alaska Gyre, and attribute this increase to an intensification of winter winds. They hypothesized that either an increase in grazing efficiency or in primary production was supporting the larger zooplankton stock.

We compared our ^{14}C primary production measurements that were made during 1984–90 with historical data at Station P for evidence of a similar increase in production and possible causes of changes due to climatic events in the Northeast Pacific Ocean. We also examined the validity of the data set collected prior to the introduction of clean sampling techniques (Carpenter and Lively 1980; Fitzwater et al. 1982).

Materials and methods

Our recent measurements of primary production were made by an in situ ^{14}C incubation technique using light and dark bottles suspended from a surface float (a single floatation buoy fitted with a radio beacon and flasher to assist in recovery). The ^{14}C uptake rates were determined by clean sampling techniques (Carpenter and Lively 1980; Fitzwater et al. 1982) with precautions against metal and organic contamination, which may drastically reduce uptake rates. Seawater samples were collected before sunrise from depths of 2, 5, 10, 20, 30, 50, 70, and 100 m using acid-cleaned 5-L Niskin PVC samplers with Teflon coated steel springs or 10-L Go-Flow samplers (for all sampling since 1987), and transferred into acid-cleaned 250-mL polycarbonate bottles. The seawater in the bottle was spiked with 5 or 12.5 µCi of a $NaH^{14}CO_3$. This radioactive solution was prepared by diluting $NaH^{14}CO_3$ (New England Nuclear) with a solution of Na_2CO_3 in deionized water (pH 9) and stored in acid-cleaned Teflon or polyethylene bottles, according to the precautions described in Fitzwater et al. (1982). To check on nonbiological uptake of ^{14}C, a zero-time control bottle was occasionally prepared and was found to be similar to dark bottle uptake. An array of incubation bottles consisted of one dark and two light bottles suspended in a clear Plexiglas tube on light nylon rope at the sampling depths.

The array was set to drift free from the ship to avoid shadows. The bottles were incubated in situ from dawn to local noon (i.e., half the daylight period). Immediately after recovery of the array, samples were stored in the dark, and sample filtering was begun through either a HA Millipore (0.45-µm pore for all samples since 1987) or Nuclepore polycarbonate filter (0.4-µm pore) using vacuum filtration at a pressure of 25 mm Hg (1 mm Hg = 133.322 Pa). All samples were filtered within 2–3 h of collection. The filters were stored in glass scintillation vials in the ship's freezer at −20°C. The radioactivity retained on the filters was determined on shore by adding 10 mL of Aquasol to each vial and counting β emissions with a Beckman LSC-100 or LS-3133 liquid scintillation counter. For estimates of ^{14}C primary production, dark bottle activity was subtracted from the average counts of two light bottles (standard deviation of light bottles = ±0.05µg $C \cdot m^{-3}$ over the incubation period, or about 1% of uptake rates in surface samples). The carbon fixation rates for the morning incubation (milligrams C per cubic meter) were extrapolated to daily estimates (milligrams C per cubic meter per day) in two ways: (1) for all data from 1987–90, by multiplying by a ratio equal to the daily irradiance divided by the irradiance received during the incubation period, or (2) for data previous to 1987, by calculating an hourly rate for the incubation period, and computing a daily rate dependent on the number of hours of daylight. The difference between these two estimates is less than 10% (standard deviation = 8.8%, for 10 comparisons in 1987–88). No correction was made for nighttime respiration loss, which was probably less than 15% of the production (preliminary time series results from Station P show that ^{14}C uptake is comparable over daylight and 24 h periods, suggesting that respiration losses are negligible). Primary production was calculated assuming the total CO_2 content in seawater to be 2.1 mM (25.2 mg $C \cdot L^{-1}$) and using the equations of Strickland and Parsons (1972).

Uptake rates were integrated from the surface to 80 m, a depth below which significant primary production was not recorded. Discrete rates were assumed to be representative of a layer equal to half the depth between adjacent samples both above and below. For example, the 5 m sample would be integrated between 3.5 and 7.5 m, and the 50 m sample spans 30–60 m. Between 80 and 95% of the carbon uptake occurred in the upper 40 m at all times of the year.

Differences in historical procedures include water sampling by Van Dorn bottles (with internal rubber tubing, which is known to be toxic to phytoplankton; Price et al. 1986; Williams and Robertson 1989), ^{14}C solution storage in glass ampoules, use of glass BOD bottles for incubations, formalin fixation of samples before filtration, acid fuming of filters in early work, and β counting by low efficiency Geiger counters (McAllister et al. 1960). Much of the primary production data from 1959–77 (e.g., Stephens 1977) shows large variability (by a factor of 10 or more) in the upper 20 m at Station P, whereas in situ rates collected for this study rarely vary by than a few percent between 0 and 20 m.

Between 1983–90, chlorophyll-*a* was determined by filtering at vacuum pressure of 25 mm Hg, a known volume of sea water through a Whatman GF/F glassfibre filter to retain phytoplankton cells, which were then extracted with 90% acetone for fluorometric determination as described

Table 1. ^{14}C primary production at Ocean Station P (50°N, 145°W).

Date	Primary production 0–80 m (mg C · m^{-2} · d^{-1})	Date	Primary production 0–80 m (mg C · m^{-2} · d^{-1})
Measurements by the Centre for Ocean Climate Chemistry			
May 03/84	262	Aug. 17/85	864
May 12/84	343	Aug. 18/85	526
May 17/84	609	Apr. 22/86	299
Aug. 04/84	341	July 07/86	411
Aug. 09/84	290	Apr. 04/87	356
Aug. 15/84	443	Jul. 18/87	437
Aug. 20/84	674	Sept. 29/87	573
Nov. 13/84	230	May 9/88	378
Apr. 21/84	1022	Jul 04/88	394
Apr. 22/85	620	July 06/88	311
Apr. 23/85	465	Feb. 19/89	226
May 05/85	328	Feb. 20/89	340
May 06/85	441	May 07/89	422
May 07/85	276	Oct. 13/89	296
May 08/85	214	May 16/90	443
Published results using clean procedures.			
June 78	494 [a]	—	—
May 84	415 [b]	Sept.–Oct. 87	664 [b]
Aug. 84	444 [b]	May 88	546 [b]
June 87	868 [b]	Aug. 88	578 [b]

[a]Booth et al. 1988.
[b]Miller et al. 1991 from SUPER cruises 1–6.

in Strickland and Parsons (1972) and Parsons et al. (1984). Samples were either stored frozen for analysis on land, or analyzed at sea. K. Stephens (Pacific Biological Station, Nanaimo, B.C., V9R 5K6, Canada, personal communication) supplied historical chlorophyll data (1964–77). Much of this data is summarized in Anderson et al. (1977) and Parsons and Lalli (1988).

Tabata and Peart (1985) and S. Tabata (Institute of Ocean Sciences, Sidney, B.C., V8L 4B2, Canada, personal communication) supplied sea-surface temperature and salinity data.

Results and discussion

Table 1 displays our primary productivity data for 1984–90. The historical data collected during the period 1969–76 (Stephens 1977) are shown in Fig. 1. The composite seasonal production rates are shown in Fig. 2. Also shown for comparison, are values obtained by the University of Washington using a similar clean sampling technique near Station P during the same period (Booth et al. 1988), values from the SUPER program in 1984–88 (Frost 1987; Welschmeyer et al. 1991; Miller et al. 1991), and a historical data set of 1960–66 (McAllister 1972; Parsons and Lalli 1988). The 1960–66 average of 162 mg C · m^{-2} · d^{-1} was 25% higher than the 1969–76 average of 121 mg C · m^{-2} · d^{-1} (Fig. 1), but were considered to be consistent due to the high variability of those two historical data sets. The monthly average of chlorophyll-*a* at Ocean Station P was compiled for 1964–90 as in Fig. 3; the annual average for 1964–90 is shown in Fig. 4.

Discrepancies between historical and recent data: methodology problems

The apparent difference between the ^{14}C production rates obtained in 1984–90 and that in 1960–76 is similar to the decadal trend of chlorophyll-*a* distribution in the central North Pacific Ocean (Venrick et al. 1987) between the 1980s and 1970s, and in the zooplankton biomass distribution between 1980–89 and 1956–62 (Brodeur and Ware 1992). Both show a doubling on a decadal time scale. Unfortunately, this apparent decadal difference in production may be masked by the uncertainties of the method used for earlier measurements in the 1950s and 1960s. The standard method described in Strickland and Parsons (1972) was subject to contamination present in stock reagents and sampling bottles. Carpenter and Lively (1980) showed that the ^{14}C uptake measurements in coastal waters using the conventional chromic acid cleaning and without special precautions to avoid contamination in reagent stock solutions and containers could be lowered by 25–35%,

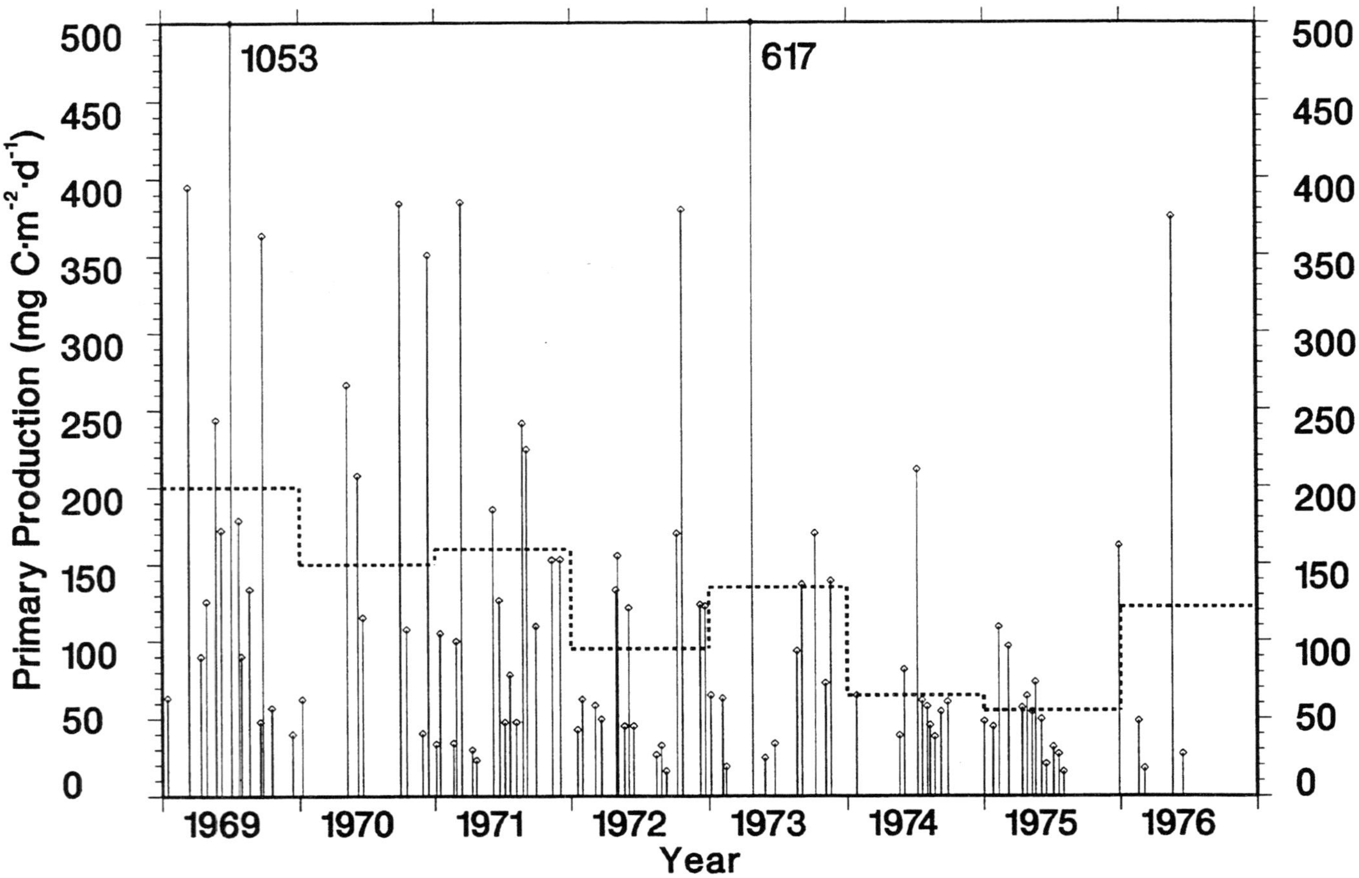

Fig. 1. Historical ^{14}C primary production at Ocean Station P (50°N, 145°W), 1969–76 (from Stephens 1977). Yearly averages are shown as a dashed line.

compared with techniques using clean reagents and acid-cleaned polycarbonate bottles. Fitzwater et al. (1982) found a 60% decrease in the ^{14}C uptake rate for open ocean waters. The variations in filtration and ^{14}C counting techniques alone were shown to have an interlaboratory variation of 20–25% (Richardson 1991). The combined effects of both metal contamination and filtration could produce an uncertainty of 100%.

Clean sampling protocol might be another factor. Fitzwater et al. (1982) showed that, for open ocean waters in the North Pacific Gyre, the values obtained using the "clean" approach were two to three times those using the "standard" technique. The use of nylon or Kevlar ropes instead of metal hydrowires and suspension from a drifting buoy away from the ship, as done in this paper and the SUPER cruises, further eliminated shipboard contaminants that might suppress production. For the 1956–76 data set, Van Dorn plastic bottles with rubber hoses were used, and distilled water was prepared using a metal still. These were additional factors that could cause the discrepancy. It is difficult to reconstruct the past technique with the specific equipment and reagents used at that time for comparison with the present technique. As pointed out by Welschmeyer et al. (1991), it is still inconclusive to attribute the high values to clean sampling protocols. Their highest values at Station P of >1300 mg C · m^{-2} · d^{-1} were even higher than the 1050 mg C · m^{-2} · d^{-1} (Martin et al. 1987) following very strict ultraclean protocols to eliminate trace element contamination such as iron. However, iron enrichment via contamination is unlikely to cause an increase in ^{14}C uptake over 6 h incubations because the effects of iron enhancement are only observed after 24 h (N. Welschmeyer, Moss Landing Marine Laboratories, Moss Landing, CA 95039, U.S.A., personal communication; P.W. Boyd, C.S. Wong, R.H. Goldblatt, R. Chretien, K.J. Orians, D. Varela. P.J. Harrison, and F. Whitney, unpublished data).

The historical data in Fig. 1 were collected at a much higher sampling frequency than newer data using the clean technique. This raises the question whether recent data are indicating natural changes in primary production. The spatial variability of surface chlorophyll-*a* within a 50 × 50 km grid centered on Station P obtained by Mackas (Frost 1991) showed a patchiness with a variability of a factor of two within a 30h survey. Our data in Table 1 showed a drastic reduction in production rates during April 21–23, 1985, from 1022 to 465 mg C · m^{-2} · d^{-1}, a factor of two within 48 h. The natural change, detected by the same technique, within a short time span was of the same order of magnitude as the difference in rates suggested in the literature for the two techniques. The sample population of the new data set was too small compared with the historical data set, which also showed the same natural changes within short

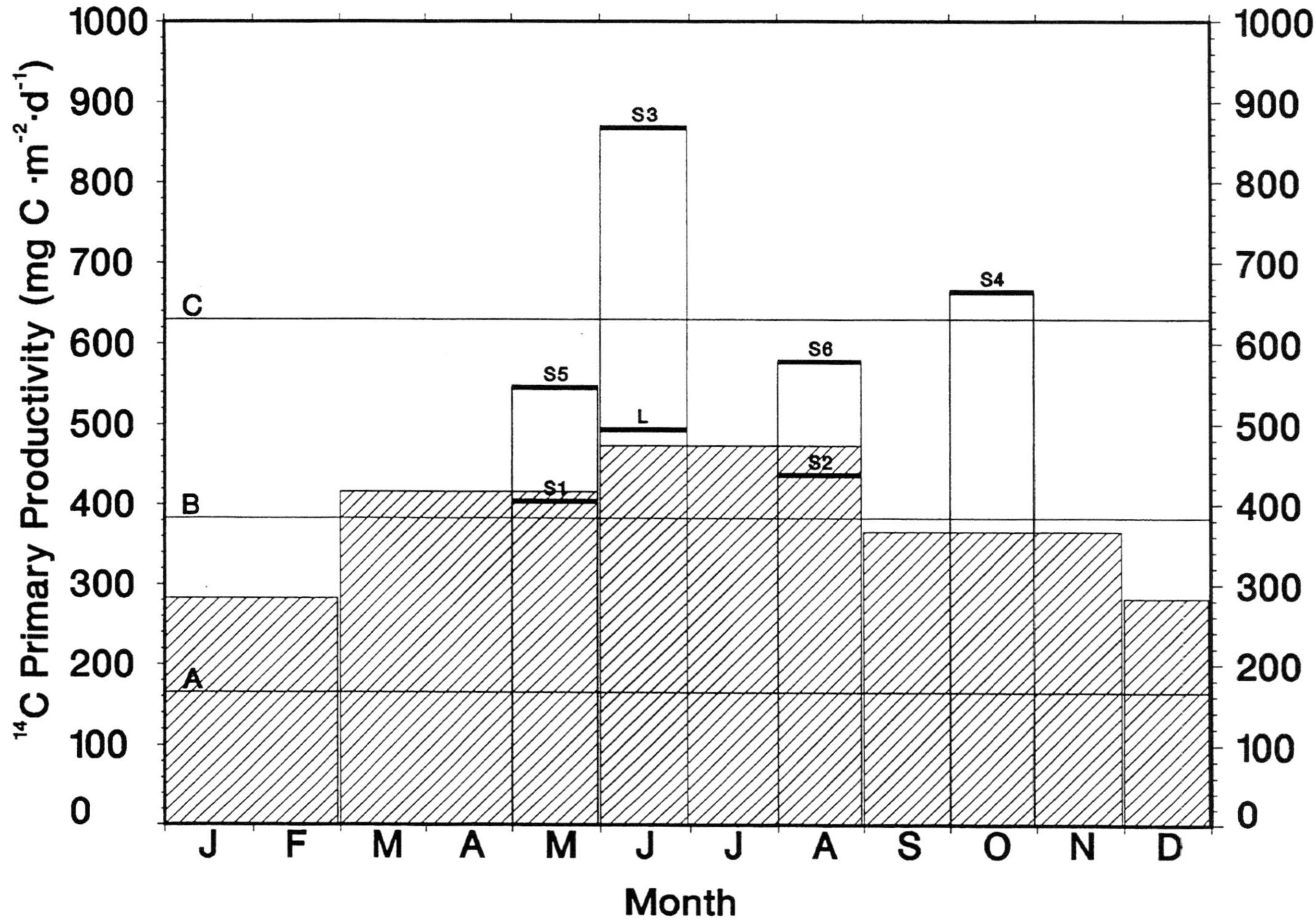

Fig. 2. ^{14}C primary production at Ocean Station P (50°N, 145°W) for the seasons, spring (March–May), summer (June–August), autumn (September–November), and winter (December–February) using the composite data set of 1984–90, collected by the Centre for Ocean Climate Chemistry. Hatched area = seasonal average as above; A = annual primary production of McAllister (1972); B = annual primary production of Wong et al. (this paper); C = annual primary production of Welschmeyer et al. (1991), based on the SUPER cruises, shown as S1, S2, S3, S4, S5 and S6; L = data collected by Lorenzen in Booth et al. (1988).

periods. Thus, the historical data could not be discounted completely without a detailed scrutiny of the methods in 1960–76 and in 1980s to define the difference quantitatively. Moreover, most launching of the free-floating productivity arrays took place during better weather and oceanic conditions, which coincide with periods of calm and above average-irradiation. The historical data set did serve a useful role in providing a complete picture of seasonal changes in primary production in the Northeast Pacific Ocean. More experimental work on methodology comparison is needed to help resolve the controversy.

Recent measurements of ^{14}C uptake rates

The new data set in the 1980s was obtained mainly in the spring and summer months and thus was insufficient in seasonal and interannual coverage to produce a composite annual cycle quantitatively or to allow us to infer interannual changes reliably. Our mean integrated values (0–80 m) of ^{14}C primary production at Station P for the four seasons spring (March–May), summer (June–August), autumn (September–November), and winter (December–February) as compiled from the 1984–90 data set, were 415 (8), 466 (5), 366 (3), and 283 (2) mg C · m^{-2} · d^{-1} (number of observations) respectively, as plotted in Fig. 2. Our average values and the SUPER data set of 1987 and 1988 (Welschmeyer et al. 1991; Miller et al. 1991) are both higher than historical data of McAllister et al. (1960) and McAllister (1972). The average of measurements taken in May 1988, June 1978 and 1987, August 1988, and October 1987 was 630 mg C · m^{-2} · d^{-1} (Welschmeyer et al. 1991).

Our data set included the only winter measurements using the clean technique and more values for the other seasons. The average annual production at Ocean Station P was estimated to be 383 mg C · m^{-2} · d^{-1}, or about 140 g C · m^{-2} · yr^{-1}. This estimate is more representative than the SUPER production rate, which was based mainly on late spring and summer data, thus biased on the high side. Our annual production rate was

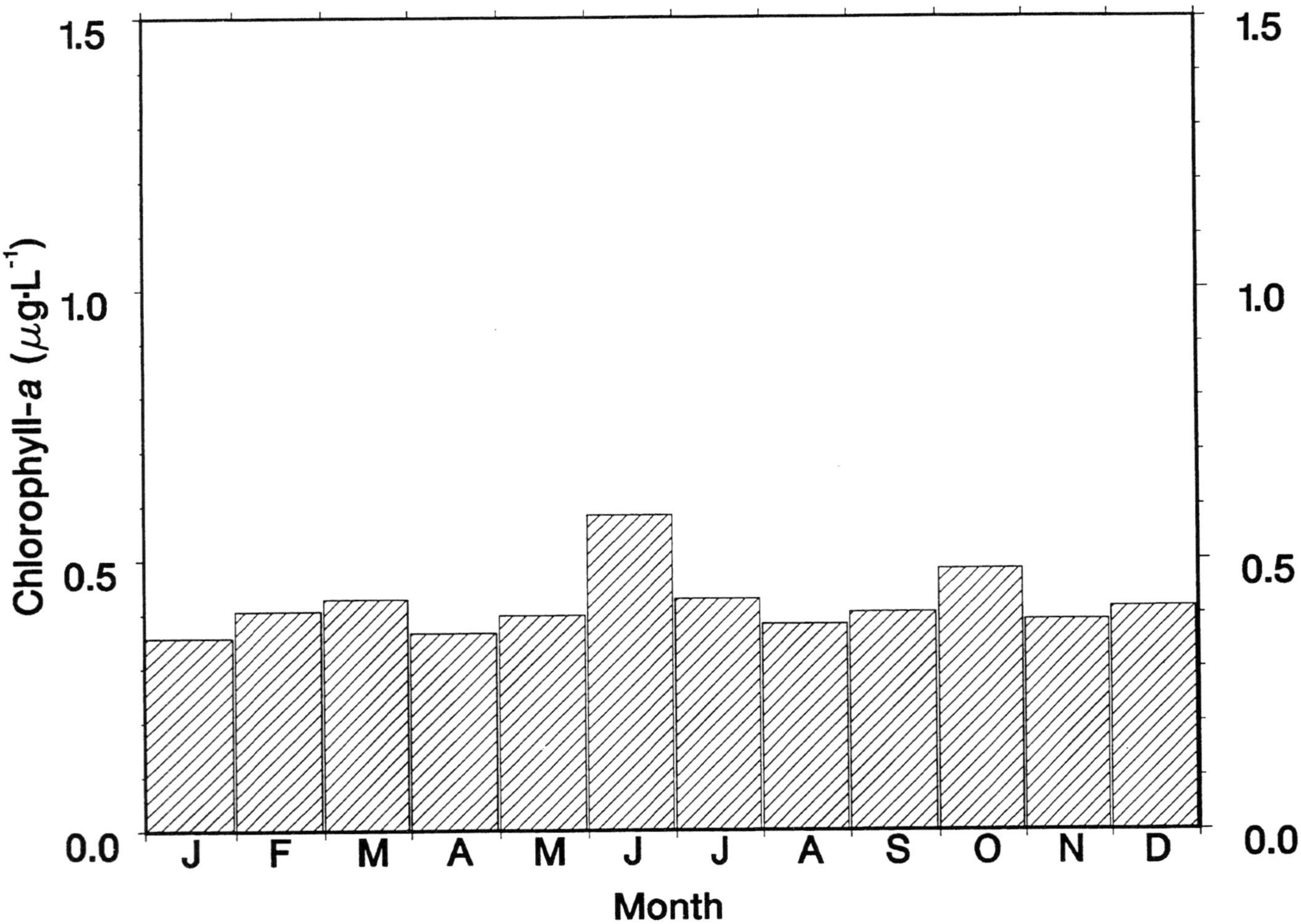

Fig. 3. Chlorophyll-*a* at Ocean Station P (50°N, 145°W) 1964–91, as monthly average.

more than double the 60 g C · m^{-2} · yr^{-1} of McAllister (1972) derived from the 1960–66 data set (Parsons and Lalli 1988) and in good agreement with the estimate made by Falkowski and Wilson (1992) of 167 g C · m^{-2} · yr^{-1}. The subarctic waters are as productive as equatorial Pacific waters at 176 g C · m^{-2} · yr^{-1} (Chavez and Barber 1987; Eppley 1989) and at 75–150 g C · m^{-2} · yr^{-1} based on changes in new nitrate production between El Niño and non-El Niño years (Wong et al. 1993).

Our annual primary production rate is consistent with estimates based on new production deduced from changes in dissolved nitrate and from exported particulate nitrogen at Ocean Station P. The seasonal new nitrate at Station P is about 7 µmo · L^{-1} (Parsons and Lalli 1988), based on seasonal change from the winter surface maximum of 13 µmo · L^{-1} to a late summer minimum of 6 µmol · L^{-1} (Anderson et al. 1969). First, we estimated the primary production to be 133 g C · m^{-2} · yr^{-1} by using a carbon to nitrogen ratio of 6.6 in a euphotic zone of 80 m, and by employing the relationship of Eppley and Peterson (1979; primary production^2 = 400× new production). Second, we estimated it to be 120 g C · m^{-2} · yr^{-1} using the formula of Betzer et al. (1984) as reviewed by Hargrave (1985), i.e., $F = AD^aP^b$, where F is the carbon flux or new production expressed as carbon, D is the depth, P is the total production, and A, a, and b are constants. Although the approaches using nitrate depletion and particle flux are affected by uncertainties, e.g., in the first case, assumption of the depth of the euphotic zone, and advective and diffusive processes (Miller et al. 1991), their tentative agreement pointed to a much higher production rate in subarctic waters.

Causes of higher production rates in subarctic waters in recent years

A doubling of production rates between the period 1960–76 and the 1980s was not evident in other indicators of productivity changes in Northeast Pacific waters. Primary production inferred from new production derived from seasonal new nitrate at Station P, extracted from long time series measurements, did show enhancement during El Niño years (C.S. Wong, unpublished data). The deviations of the annual values from the mean of 1966–90 average seasonal new nitrate could be as high as 70% of the mean, implying a possible change of up to 100 g C · m^{-2} · yr^{-1} in primary production.

The period 1959–78 was considered to be cold, followed by a warm period of 1979–90, as indicated by the anomaly of annual sea-surface temperature (Fig. 5). The primary production rates inferred from seasonal nitrate (C.S. Wong, unpublished data) were 145 and 128 g C · m^{-2} · yr^{-1} for the cold and warm periods respectively. The monthly average of

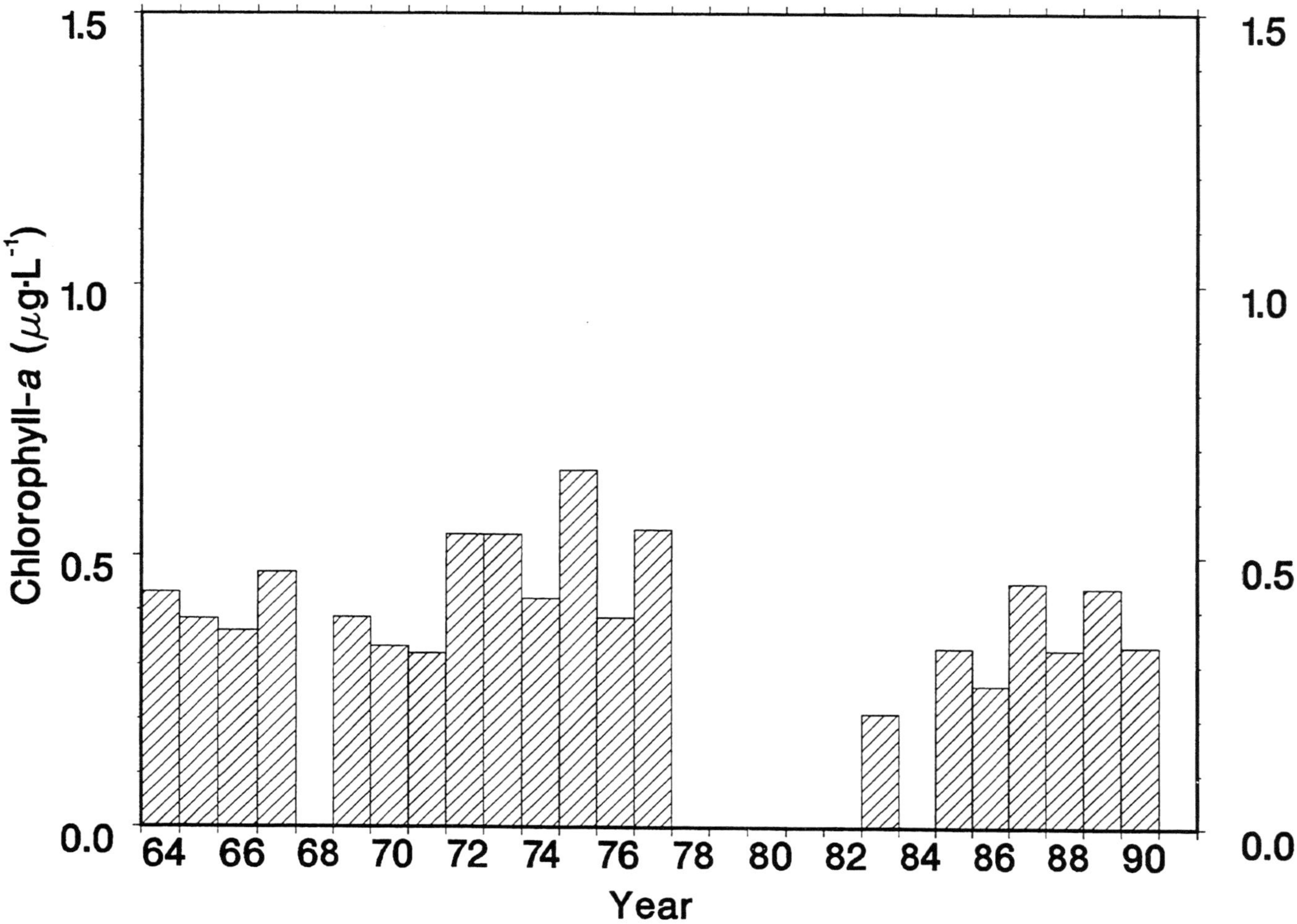

Fig. 4. Chlorophyll-*a* at Ocean Station P (50°N, 145°W), 1964–91, as annual average.

chlorophyll-*a* changed little throughout the year (Fig. 3). For the period 1964–92 (Fig. 4), chlorophyll-*a* showed a relatively steady annual level of about 0.4 µg · L^{-1} in the surface waters at Station P, with little change interannually, thus indicating low but constant phytoplankton biomass in the surface waters. However, increased nitrate levels reflect greater vertical mixing, which may selectively increase the population of diatoms and larger algal species through an increased supply of growth limiting nutrients (such as iron; P.W. Boyd, C.S. Wong, R.H. Goldblatt, R. Chretien, K.J. Orians, D. Varela. P.J. Harrison, and F. Whitney, unpublished data) Larger algal species are hypothesized to be controlled by the efficient grazing of larger zooplankton (Welschmeyer et al. 1991). This would lead to increased exported production without a bloom and without changes in the low level of chlorophyll-*a* in the upper ocean.

New production in subarctic waters is 25% of the total production based on an averaged annual *f*-ratio of 0.25 at Station P by D. Valera (Department of Oceanography, University of British Columbia, Vancouver, B.C., V6T 1Z4, Canada, personal communication) Her *f*-ratio includes both urea and ammonia in the regenerated production, and is about double the *f*-ratios for central ocean and coastal N.E. Pacific of Martin et al. (1987) quoted in Eppley (1989), if a factor of 0.85 is applied to their ratios based on ammonia only. If the change in new production does not invoke a proportional change in regenerated production, the doubling of zooplankton biomass consequently does not require a doubling of algal biomass for both macro- and micro-phytoplankton. Parslow (1981) examined the interannual variation in phytoplankton biomass indicated by chlorophyll-*a* and primary production rate at Station P during 1964–76. No correlation between years of high chlorophyll-*a* and high zooplankton biomass was found. A model of grazing control of phytoplankton stock for the subarctic Pacific Ocean by mesozooplankton could produce a scenario of doubling of zooplankton biomass without a corresponding shift in the phytoplankton biomass (B.W. Frost, School of Oceanography, University of Washington, Seattle, WA, 98195, U.S.A., personal communication) Miller et al. (1984) suggested that the enhanced productivity in spring and summer at Station P, as shown in the 40% depletion of the winter level of nitrate in the mixed layer, would lead to an increase in copepod biomass but not in phytoplankton.

Primary production may be affected by climatic events; however, the mechanisms are obscure. The total upward flux of nitrate, both advective and diffusive, was assessed to be 70% of the total nitrate change or almost double the seasonal new nitrate determined from nitrate profiles (Miller et al.

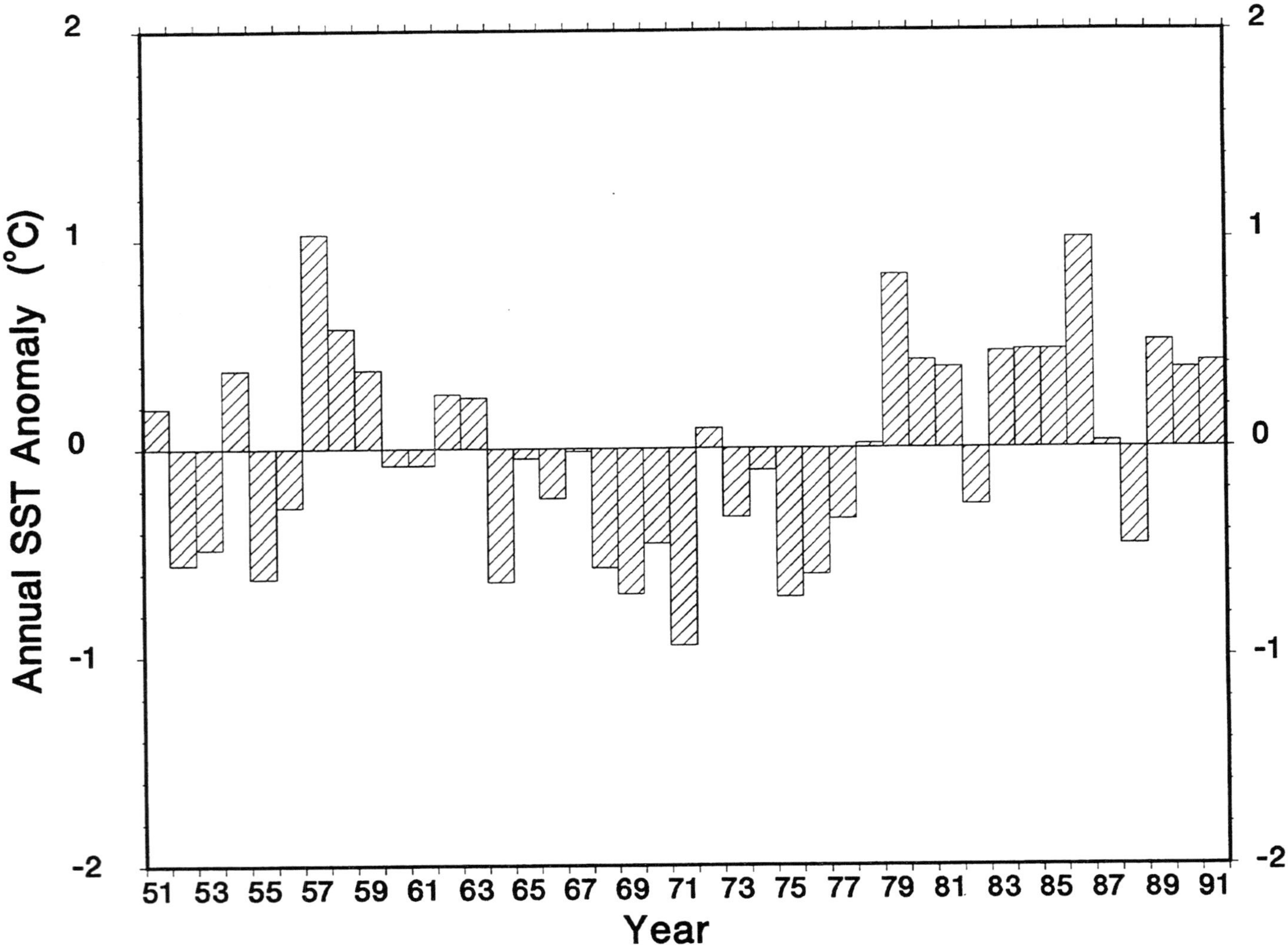

Fig. 5. Anomaly of annual sea-surface temperature (SST) at Ocean Station P (50°N, 145°W) as deviations from the 1950–89 mean.

1991). The advective flux was affected by Ekman pumping, i.e., it was sensitive to the climatic factor of wind stress. Mysak (1985) suggested an intensification of the Aleutian Low in the Gulf of Alaska during the winter of the years with El Niño – Southern Oscillation (ENSO) events resulted in similar intensification of the eddies. Salinity was suggested as an indicator of the extent of vertical mixing (Wickett 1967; Frost 1983). The anomaly of annual sea-surface salinity at Ocean Station P as deviations from the 1956–91 mean was plotted in Fig. 6. The period 1980–88 was one of low salinity (Tabata 1991), inferred to be a period of less vertical mixing and thus lower vertical supply of nitrate. If the period 1984–90, with a higher ^{14}C production rate, was indeed one with a lesser supply of deep nitrate, one would expect surface waters in the colder period of 1960–76, with higher salinity, to be replenished with more deep nutrients and hence with slightly higher production. In contrast, the observed ^{14}C production rates showed the exact opposite trend for these periods.

Nitrate is not the limiting nutrient in the subarctic waters. Iron deficiency was shown to be the controlling factor (Martin and Fitzwater 1988; P.W. Boyd, C.S. Wong, R.H. Goldblatt, R. Chretien, K.J. Orians, D. Varela. P.J. Harrison, and F. Whitney, unpublished data) At Station P, dissolved iron has a vertical distribution similar to nitrate (Bruland et al. 1991). Vertical mixing supplies both nitrate and iron from deeper waters below 100 m. However, the supply of Fe(II) is below the requirement to fully utilize nitrate, thus leaving an excess of this nutrient in the subarctic waters of the Northeast Pacific. The dominant source of new iron in the open ocean is from the atmosphere (Duce and Tindale 1991). Episodic meteorological events over China often lead to large dust storms in the late spring and early summer. Dust and aerosols from the Chinese deserts and loess plateaus are transported to the Northeast Pacific in about 5–10 days. The injection of Fe(III) and Fe(II) into the ocean from these episodic events would enhance productivity. Fe(II) was found to be significant in remote aerosols with $56 \pm 32\%$ of the total Fe being Fe(II). Because of high acidity in aerosols, solubilization of Fe(III) to Fe(II) by photoreduction occurs during long range transport (Zhuang et al. 1992). Upon deposition into the euphotic zone in the ocean, Fe(II) being more soluble than Fe(III), should be more readily available for use as a micronutrient by phytoplankton. Then, decadal changes in the amount and frequency of dust events could be an important factor to look for in understanding ocean

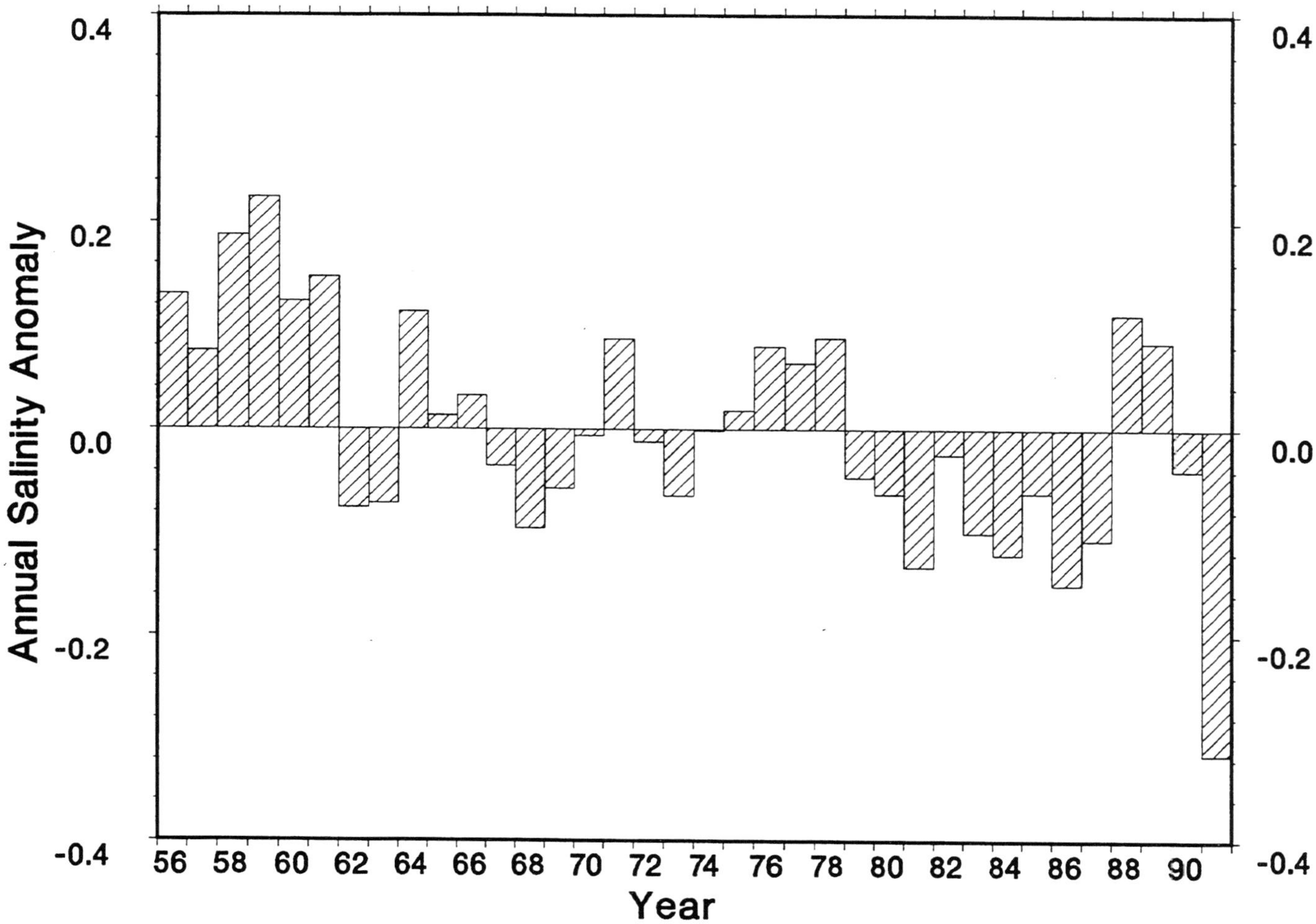

Fig. 6. Anomaly of annual sea surface salinity at Ocean Station P (50°N, 145°W) as deviations from the 1956–91 mean.

productivity changes in nutrient rich but iron poor waters. The warm period of the 1980s had noticeably more frequent El Niño events (IPCC 1990), more intensive Aleutian low pressure systems and stronger atmospheric circulation over the northern North Pacific Ocean. The atmospheric iron input into this area could thus enhance the productivity over decadal periods.

We conclude that subarctic primary production should be much higher than historical estimates as a result of more reliable data from improved techniques, although natural changes such as higher input of aerosol iron must be considered.

Conclusions

1) In situ ^{14}C primary production at Station P (50°N, 145°W) using a clean sampling technique and free-floating suspension of incubated polycarbonate bottles was a factor of two higher than values obtained by the conventional method of the 1950s and 1960s.
2) For the limited data set of 1984–90, production rates based on the clean sampling technique were 415 (8), 466 (5), 366 (3), and 283 (1) mg C · m^{-2} · d^{-1} (no. of observations) for the seasons spring (March–May), summer (June–August), autumn (September–November) and winter (December– February), respectively.
3) The annual production rate at Station P in subarctic Northeast Pacific waters was 140 g C · m^{-2} · yr^{-1}, as high as that of equatorial waters (Chavez and Barber 1987) and double the estimate of McAllister (1972).
4) Time-series of chlorophyll-*a* at Station P between 1964 and 1991 showed little interannual variation in the annual average from the low level of 0.4 μg · L^{-1} .
5) The production in subarctic waters at Station P could be enhanced by climatic factors, but this is likely not significant enough to account for the decadal doubling of zooplankton biomass described by Brodeur and Ware (1991).

Acknowledgements

We thank Paul Harrison, Philip Boyd and other colleagues for constructive review of the manuscript. This work was part of the Canadian JGOFS Program, and was supported by the Federal Energy R&D Program (Project Number 48105: Fossil Fuel CO_2 Emissions and Oceanic Uptake) and A-Base of the Federal Department of Fisheries and Oceans, Canada.

References

Anderson, G.C., T.R. Parsons, and K. Stephens. 1969. Nitrate distribution in the subarctic northeast Pacific Ocean. Deep Sea Res. 16: 329–334.

Anderson, G.C., R.K. Lam, B.C. Booth, and J.M. Glass. 1977. A description and numerical analysis of the factors affecting the processes of production in the Gulf of Alaska. Department of Oceanography, University of Washington, Seattle, Wash. Spec. Rep. 76: 231 p.

Betzer, P.R., W.J. Showers, C.D. Winn, G.R. DiTullio, and P.M. Kroopnick. 1984. Primary productivity and particle fluxes on a transect of the equator at 153° in the Pacific Ocean. Deep Sea Res. 31: 1–12.

Booth, B.C., J. Lewin, and C.J. Lorenzen. 1988. Spring and summer growth rates of subarctic Pacific phytoplankton assemblages determined from carbon uptake and cell volumes estimated using epifluorescence microscopy. Mar. Biol. 98: 287–298.

Brodeur, R.D., and D.M. Ware. 1992. Long-term variability in zooplankton biomass in the subarctic Pacific Ocean. Fish. Oceanogr. 1: 32–38.

Bruland, K.W., J.R. Donat, and D.A. Hutchins. 1991. Interactive interfluences of bioactive trace metals on biological production in oceanic waters. Limnol. Oceanogr. 36: 1555–1577.

Carpenter, E.J., and J.S. Lively. 1980. Review of estimates of algae growth using ^{14}C tracer technique, p. 161–178. *In* P.G. Falkowski [ed.] Primary productivity in the sea. Plenum Press, New York.

Chavez, F.P., and R.T. Barber. 1987. An estimate of new production in the equatorial Pacific. Deep Sea Res. 34: 1229–1243.

Duce, R.A., and N.W. Tindale. 1991. Atmospheric transport of iron and its deposition in the ocean. Limnol. Oceanogr. 36: 1715–1726.

Eppley, R.W. 1989. New production: History, methods, problems, p. 85–97. *In* W.H. Berger, V.S. Smetacek and G. Wefer [ed.] Productivity of the ocean: Present and past. John Wiley & Sons, New York.

Eppley, R.W., and B.J. Peterson. 1979. Particulate organic matter flux and planktonic new production in the deep ocean. Nature (London) 282: 677–680.

Falkowski, P.G., and C. Wilson. 1992. Phytoplankton productivity in the North Pacific ocean since 1900 and implications for absorption of anthropogenic CO_2. Nature (London) 358: 741–743.

Fitzwater, S.E., G.A. Knauer, and J.H. Martin. 1982. Metal contamination and its effect on primary production measurements. Limnol. Oceanogr. 277: 544–551.

Frost, B.W. 1983. Interannual variation of zooplankton standing stock in the open Gulf of Alaska, p. 146–157. *In* W.S. Wooster [ed.] From year to year, interannual variability of the environment and fisheries of the Gulf of Alaska and the eastern Bering Sea. University of Washington, Seattle, Wash. Wash. Sea Grant Program. WSG-WO-83-33,

Frost, B.W. 1987. Grazing control of phytoplankton stock in the open subarctic Pacific Ocean: a model assessing the role of mesozooplankton, particularly the large calanoid copepods, *Neocalanus* spp. Mar. Ecol. Prog. Ser. 39: 49–68.

Hargrave, B.T. 1985. Particle sedimentation in the ocean. Ecol. Model. 30: 229–2446.

Intergovernmental Panel on Climate Change.1990. Climate change, the IPCC scientific assessment. J.T. Houghton, G.J. Jenkins, and J.J. Ephraums [ed.] Cambridge Univerity Press, Cambridge, U.K.

Longhurst, A.R., N. Colebrook, J. Gullard, R. LeBrasseur, C. Lorenzen, and P. Smith. 1972. The instability of ocean populations. New Sci. 54: 500–502.

Martin, J.H., and S.E. Fitzwater. 1988. Iron deficiency limits plankton growth in the northeast Pacific Subarctic. Nature (London) 331: 341–343.

Martin, J. H., G. A. Knauer, D. M. Carl, and W. W. Broenkow. 1987. VERTEX: carbon cycling in the northeast Pacific. Deep Sea Res. 34: 267–285.

McAllister, C.D. 1972. Estimates of the transfer of primary production to secondary production at Ocean Station P, p. 575– 580. *In* A.Y. Takenouti [ed.] Biological oceanography of the northern North Pacific Ocean. Idemitsu Shoten, Tokyo, Japan.

McAllister, C.D., T.R. Parsons and J.D.H. Strickland. 1960. Primary production at Ocean Station P in the northeast Pacific Ocean. J. Cons. Cons. int. Explor. Mer, 25: 240–259.

Miller, C.B., B.W. Frost, H.P. Batchelder, M.J. Clemons, and R.E. Conwat. 1984. Life history of large, grazing copepods in a subarctic ocean gyre: *Neocalanus plumchrus*, *Neocalanus cristatus*, and *Eucalanus bungii* in the Northeast Pacific. Prog. Oceanogr. 13: 201–243.

Miller, C.B., B.W. Frost, P.A. Wheeler, M.R. Landry, N. Welschmeyer, and T.M. Powell. 1991. Ecological dynamics in the subarctic Pacific, a possible iron-limited ecosystem. Limnol. Oceanogr. 36: 1600–1615.

Mysak, L.A. 1985. On the interannual variablity of eddies in the northeast Pacific Ocean, p. 977–106. *In* W.S. Wooster and D.L. Fluharty [ed.] El Niño North. El Niño effects in the eastern subarctic Pacific Ocean. University of Washington, Seattle, Wash. Washington Sea Grant Program. WSG-WO 85-3.

Parslow J.S. 1981. Phytoplankton–zooplankton interactions: data analysis and modelling (with particular reference to Ocean Station P 50°N, 145°W and Controlled Ecosystem Enclosure Experiments). Ph.D. thesis, University of British Columbia, Vancouver, B.C. 400 p.

Parsons, T.R., and C.M. Lalli. 1988. Comparative oceanic ecology of the plankton communities of the subarctic Atlantic and Pacific Oceans. Oceanogr. Mar. Biol. Annu. Rev. 26: 317–359.

Parsons, T.R., Y. Maita, and C.M. Lalli. 1984. A manual of chemical and biological methods for seawater analysis. Pergamon Press, Oxford. 173 p.

Price, N.M., P.J. Harrison, M.R. Landry, F. Azam, and K.J.F. Hall. 1986. Toxic effects of latex and Tygon tubing on marine phytoplankton, zooplankton and bacteria. Mar. Ecol. Prog. Ser. 34: 41–49.

Richardson, K. 1991. Comparison of ^{14}C primary production determinations made by different laboratories. Mar. Ecol. Prog. Ser. 72: 189–201.

Stephens, K. 1977. Primary productivity data from weatherships occupying Ocean Station "P" 1969 to 1975. Fish. Mar. Serv. Data Rep. 38: 88 p.

Strickland, J.D.H., and T.R. Parsons. 1972. A practical handbook of seawater analysis. 2nd ed. Bull. Fish. Res. Bd. Can. 1677: 310 p.

Tabata, S. 1991. Annual and interannual variablity of baroclinic transports across Line P in the northeast Pacific Ocean. Deep Sea Res. 338: S221–S245.

Tabata, S., and J.L. Peart. 1985. Statistics of Oceanographic data based on hydrographic/STD casts made at Ocean Station P during August 1956 through June 1981. Can. Data Rep. Hydrogr. Ocean Sci. 31: 133 p.

Venrick, E.L., J.A. McGowan, D.R. Cayan, and T.L. Hayward. 1987. Climate and chlorophyll-*a*: long-term trends in the Central North Pacific Ocean. Science (Washington, D.C.) 238: 70–72.

Welschmeyer, N., R. Goericke, S. Strom, and W. Peterson. 1991. Phytoplankton growth and herbivory in the subarctic Pacific: a chemotaxonomic analysis. Limnol. Oceanogr. 36: 1631–1649.

Wickett, W.P. 1967. Ekman transport and zooplankton concentration in the North Pacific Ocean. J. Fish. Res. Bd. Can. 24: 581–594.

Williams, P.J. leB. and J.I. Robertson. 1989. A serious inhibition problem from a Niskin sampler during plankton productivity studies. Limnol. Oceanogr. 34: 1300–1305.

Wong, C.S., Y.-H. Chan, J.S. Page, G.E. Smith, and R.D. Bellegay. 1993. Changes in equatorial CO_2 flux and new production estimated from CO_2 and nutrient levels in Pacific surface waters during the 1986/87 El Niño. Tellus, 45B: 64–79.

Zhuang, G., R.A. Duce, and P.R. Brown. 1992. Link between iron and sulfur cycles suggested by detection of Fe(II) in remote marine aerosols. Nature (London) 355: 537–539.

Do cyclic temperatures cause cyclic fisheries?

Franz-Josef Müter, Brenda L. Norcross, and Thomas C. Royer

Müter, F-J., B.L. Norcross and T.C. Royer. 1995. Do cyclic temperatures cause cyclic fisheries? p. 119–129. *In* R.J. Beamish [ed.] Climate change and northern fish populations. Can. Spec. Publ. Fish. Aquat. Sci. 121.

Abstract: There is a low frequency (approximately 25 yr) cyclic fluctuation in the sea-surface temperature (SST) in the northeast quadrant (northeast of 52.5°N, 177.5°W) of the northern North Pacific. A similar cyclic fluctuation was not found in the other quadrants. Temperature followed a decreasing trend in those areas. Although there appear to be 14-yr cyclic patterns in landings of red king crab (*Paralithodes camtschatica*) and walleye pollock (*Theragra chalcogramma*) in the Gulf of Alaska and Bering Sea, those patterns were not significantly correlated to the cyclic fluctuations in water temperature. King crab and pollock cycling in antiphase may result from fisheries rather than biological interactions. Significant correlations between water temperature and red king crab indicated decreased catches of crab 10 yr after a low SST anomaly and 3–4 yr after a high one. Pollock correlated significantly with water temperatures around time of catch. Large catches were associated with warm temperatures in the Gulf of Alaska and with cold temperatures in the Bering Sea, suggesting effects of temperature on the distribution and availability of pollock. No evidence was found for an effect of temperature on year-class strength of walleye pollock. Longer time series of fisheries-independent data are needed to test hypothetical relationships between temperature and fish abundance more accurately.

Résumé : On distingue une fluctuation cyclique de faible fréquence (environ 25 ans) dans la température de la surface de la mer (TSM) du quadrant nord-est (nord-est de 52,5° N, 177,5° O) du nord du Pacifique Nord. Il n'existe pas de fluctuation cyclique similaire dans les autres quadrants. La température a suivi une tendance à la baisse dans ces régions. Bien qu'il semble y avoir une fluctuation cyclique de 14 ans dans les débarquements de crabe royal (*Paralithodes camtschatica*) et de morue du Pacifique occidental (*Theragra chalcogramma*) dans le golfe d'Alaska et dans la mer de Béring, ces tendances n'étaient aucunement en corrélation significative avec les fluctuations cycliques de la température de l'eau. Il est possible que le déroulement en opposition de phase du cycle du crabe royal et de la morue soit attribuable à la pêche plutôt qu'à des interactions biologiques. Les corrélations significatives entre la température de l'eau et l'abondance du crabe royal montraient une diminution des prises de crabes dix ans après une baisse de la TSM et 3–4 ans après une hausse de la TSM. Les prises de morue étaient en corrélation importante avec les températures de l'eau durant la période de capture. Les prises importantes étaient associées à des températures chaudes dans le golfe d'Alaska et à des températures froides dans la mer de Béring, ce qui donne à croire que la température avait un effet sur la répartition et sur la disponibilité de la morue. Aucun signe n'a été constaté quant à l'effet de la température sur l'effectif des classes annuelles de morue du Pacifique occidental. Il faudrait disposer de séries chronologiques plus longues de données ne provenant pas de la pêche pour vérifier de façon plus précise les relations hypothétiques qui semblent exister entre la température de l'eau et l'abondance du poisson.

F.-J. Müter, B.L. Norcross, and T.C. Royer. Institute of Marine Science, School of Fisheries and Ocean Sciences, University of Alaska Fairbanks, Fairbanks, AK 99775-1080, U.S.A.

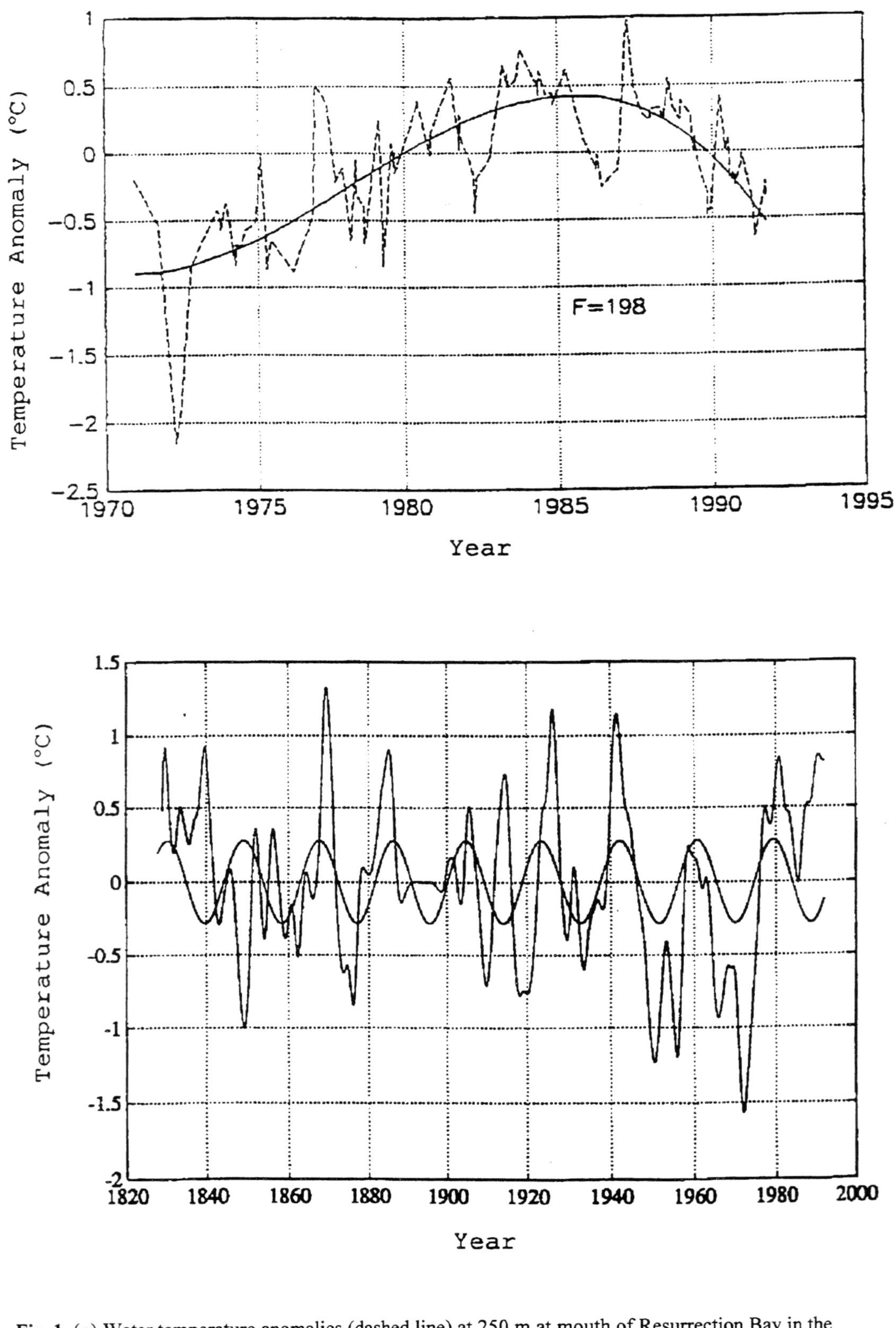

Fig. 1. (*a*) Water temperature anomalies (dashed line) at 250 m at mouth of Resurrection Bay in the Gulf of Alaska fitted with a third order polynomial (solid line) (from Royer 1992). (*b*) Low pass filtered Sitka air temperature anomalies. The smooth curve is the fitted 18.6-yr nodal tide signal for the entire record (from Royer 1992).

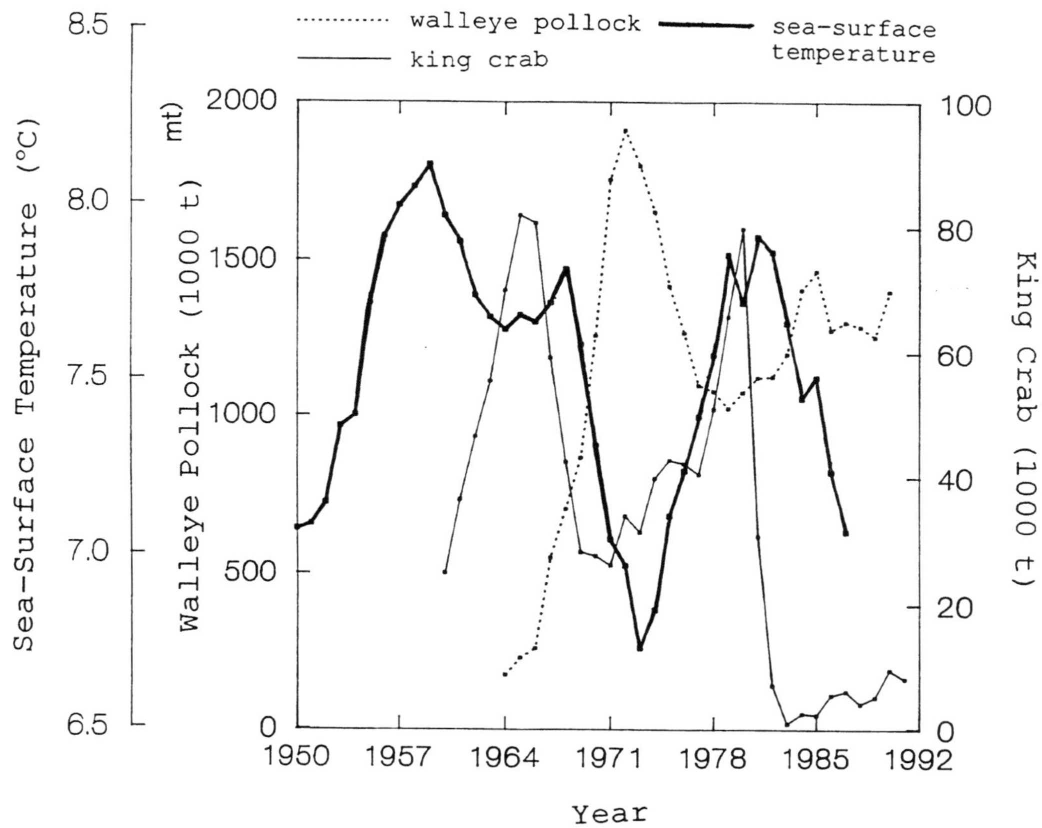

Fig. 2. Annual sea-surface temperatures (heavy line) and landings data (thousands of metric tons) of red king crab (solid line) and walleye pollock (broken line) in the northeast quadrant (Gulf of Alaska and eastern Bering Sea) of the North Pacific Ocean.

Introduction

Speculation that global warming may change ocean temperatures led us to look for relationships between water temperature and fish catch patterns. The upper 250 m of the Northeast Pacific Ocean exhibit temperature fluctuations with an amplitude greater than 1°C on a decadal time scale (Royer 1989). These fluctuations have been observed throughout the water column at a station near Seward, Alaska, in Resurrection Bay in the northern Gulf of Alaska (Fig. 1a). The amplitude of the ocean temperature signal increases with latitude and is highest at 55°N and poleward (Royer 1989). This decadal fluctuation is not confined to the Gulf of Alaska region but extends into the eastern Bering Sea (Royer 1989). It is well correlated and in phase with air temperatures at coastal stations along the Gulf of Alaska (Beegle 1986).

The Sitka, Alaska, air temperature record (1828–present; Fig. 1b) shows a long period (15–25 yr) cycle with an amplitude of about 0.34°C that coincides with fluctuations in ocean temperatures over the last 40 yr (Royer 1989, 1993). This suggests that both the ocean and atmosphere are acting in unison on those time scales, resulting in similar long-term cyclic changes in temperature for the upper waters of the North Pacific Ocean.

Cyclic changes in water temperature and other oceanographic conditions can directly and indirectly affect marine fishes and shellfishes. Fish are able to perceive temperature changes of less than 0.1°C (Bull 1952), which is well within the range of the observed temperature changes. Temperature directly modifies metabolism and behaviour, and all fish have specific temperature tolerances and preferences that may change with development (Laevastu and Hayes 1981). Temperature changes thus affect distribution, feeding, growth, reproduction, and survival of fish. Indirectly, temperature affects feeding, growth, and survival by affecting the abundance of food and predators.

Temperature anomalies and inter- and intra-specific interactions, such as predation and cannibalism, can cause natural fluctuations in the abundance of fish (Laevastu and Marasco 1982). Effects of temperature on year-class strength have been documented in many cases (Saetersdal and Loeng 1987; Ellertsen et al. 1989; Thompson and Hilden 1989). Furthermore, many commercially exploited fish stocks for which long term records exist show aperiodic fluctuations. Examples are several Atlantic herring (*Clupea harengus harengus*) and cod (*Gadus morhua*) stocks (Cushing 1982), North Sea plaice (*Pleuronectes platessa*) (Cushing 1982), North Sea sole (*Solea solea*) (De Veen 1978), Pacific sardine (*Sardinops sagax*) (Smith 1978), and Dungeness crab (*Cancer magister*) off California (Armstrong 1983). Low-frequency fluctuations in some fish stocks may be linked to periodic temperature fluctuations through the multiyear effects of negative or positive temperature anomalies.

In the Northeast Pacific, like elsewhere, many commercially exploited fish populations show large fluctuations in abundance. In this paper, we focus on red king crab (*Paralithodes*

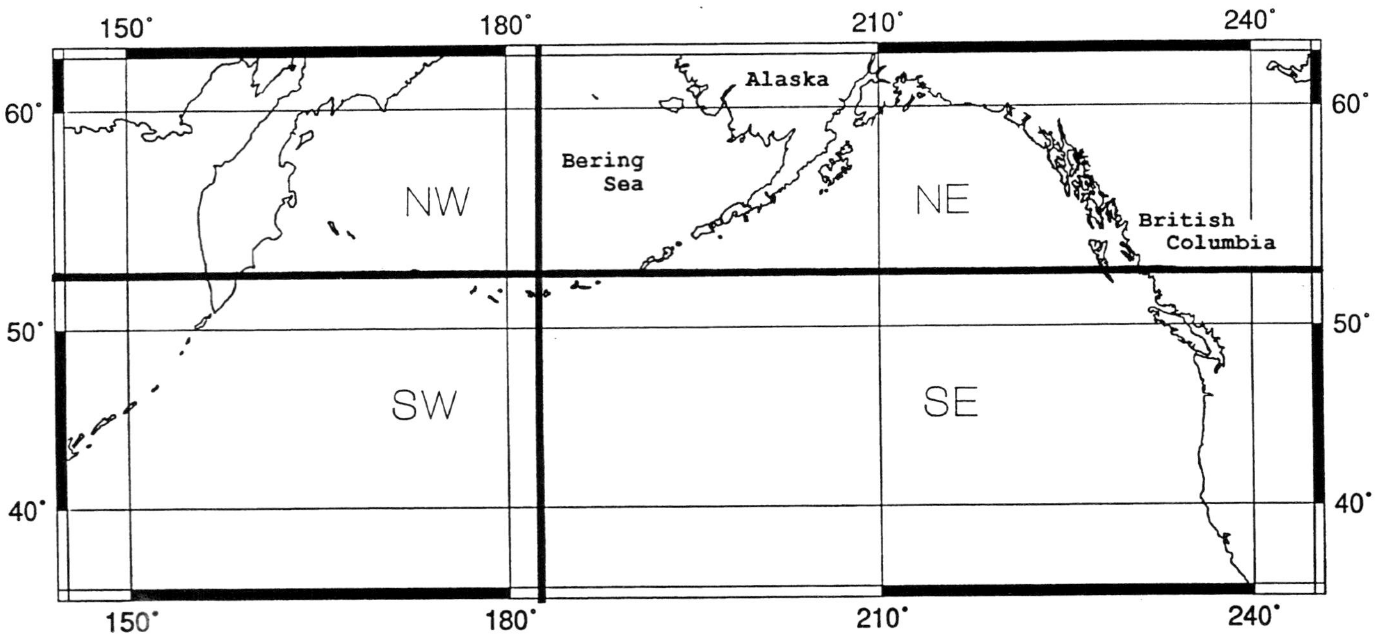

Fig. 3. North Pacific Ocean divided into quadrants at 52.5°N and 177.5°W. The southern limit is 40°N and the northern limit is 60°N. The coastlines are the east and west limits.

Table 1. Table of fisheries data sets used in this study.

Data set	Description	Geographical area	Years	Source
1	Walleye pollock, U.S. and foreign commercial catch (WPGAK)	Gulf of Alaska	1971–90	Alton et al. 1987 Low 1991
2	Walleye Pollock, U.S. and foreign commercial catch (WPEBS)	Eastern Bering Sea	1964–90	Bakkala et al. 1987 Low 1991
3	Walleye Pollock, U.S. and foreign commercial cath (WPNE)	Gulf of Alaska + Eastern Bering Sea (1 and 2 combined)	1964–90	Alton et al. 1987 Bakkala et al. 1987 Low 1991
4	Age 4 Walleye Pollock, estimated abundance (WP4)	Gulf of Alaska	1964–91	Hollowed and Megrey 1991
5	Red King Crab, U.S. commercial catch (KCKD)	Kodiak registration area 'K'	1960–91	Westward Region Shellfish Staff, Alaska Department of Fish and Game
6	Red King Crab, U.S. and foreign commercial catch (KCEBS)	Eastern Bering Sea	1960–91	Otto 1989 Westward Region Shellfish Staff, Alaska Department of Fish and Game
7	Red King Crab, total commercial catch (KCNE)	Eastern Bering Sea, Kodiak area and Alaska Peninsula	1960–91	Otto 1989 Westward Region Shellfish Staff, Alaska Department of Fish and Game
8	Red King Crab catch per unit effort (crabs per pot, U.S fishery) (CPUEKD)	Kodiak area	1965–83	Westward Region Shellfish Staff, Alaska Department of Fish and Game
9	Red King Crab catch per unit effort (crabs per pot, U.S. fishery) (CPUEBB)	Bristol Bay	1966–91	Westward Region Shellfish Staff, Alaska Department of Fish and Game

camtschatica) and walleye pollock (*Theragra chalcogramma*) stocks, which appear to demonstrate cyclic changes in catch data within past decades (Fig. 2). The periods of the cycles are approximately 14 yr for both king crab and pollock, but they are out of phase. Pollock maxima coincide with a decline in crab, and crab maxima coincide with a minimum catch of pollock. These observed fluctuations in catch occurred simultaneously with the long-term cyclic changes in water temperature that have been observed in the Northeast Pacific (Fig. 2). Thus, we investigate relationships between water temperature and catch and/or abundance data for these two species. We hypothesize that fluctuations in the abundance of these two major commercial species in the Northeast Pacific are associated with temperature fluctuations in the water column.

Materials and methods

To investigate temperature fluctuations in different parts of the northern North Pacific we used the sea-surface temperature (SST) fields from 1947 through 1990. The temperature data we used were Pacific SSTs from ship observations, analyzed by the Climate Research Group at Scripps Institution of Oceanography ("Namias" data series; Chelton 1984) . These Sea-surface temperatures data are averaged into 5° squares monthly. SSTs have been shown to be highly correlated with water-column temperatures (Royer 1989); therefore, SST data can be used as proxies for water-column temperatures. We limited our analysis to SSTs between 40°N, the axis of the North Pacific Current, and 60°N, the northern limit of the SST observations. This region was subdivided into four quadrants (northeast, southeast, northwest, and southwest) along 177.5°W and 52.5°N for analysis (Fig. 3). The boundaries were chosen such that the northeast quadrant corresponds approximately to the area for which we obtained fisheries data. Monthly SST anomalies were used for autocorrelations and best linear fits. Yearly averages of the monthly anomalies were used for cross correlations with fisheries data.

We used a variety of data sets to investigate fluctuations in the abundance of red king crab and walleye pollock. Table 1 lists all data sets used, the geographical area for each data set, the length of the record, and the source. Most data sets consist of reported commercial catches from U.S. and foreign fisheries. In an attempt to eliminate the effects of the fishery, we examined the catch per unit effort (CPUE) for red king crab from Bristol Bay and the CPUE for red king crab from the Kodiak area (data sets 8 and 9 in Table 1). No CPUE data for pollock were available. We did investigate year-class strength by examining fluctuations in estimated abundance of age-4 walleye pollock from model results of stock assessment in the Gulf of Alaska (Hollowed and Megrey 1991).

Catch data from the commercial fisheries should always be used with caution, because catches do not necessarily reflect abundances. However, there are fisheries-independent data that indicate that the peaks and subsequent sharp declines evident in the catch data for red king crab and walleye pollock reflect true peaks in abundances of these species (Bakkala et al. 1987; Low 1991; Westward Region Shellfish Staff, Alaska Department of Fish and Game 1992). Since long-term data sets of fisheries-independent data are not available, we used catch data as an indicator of overall abundance integrated over many year classes.

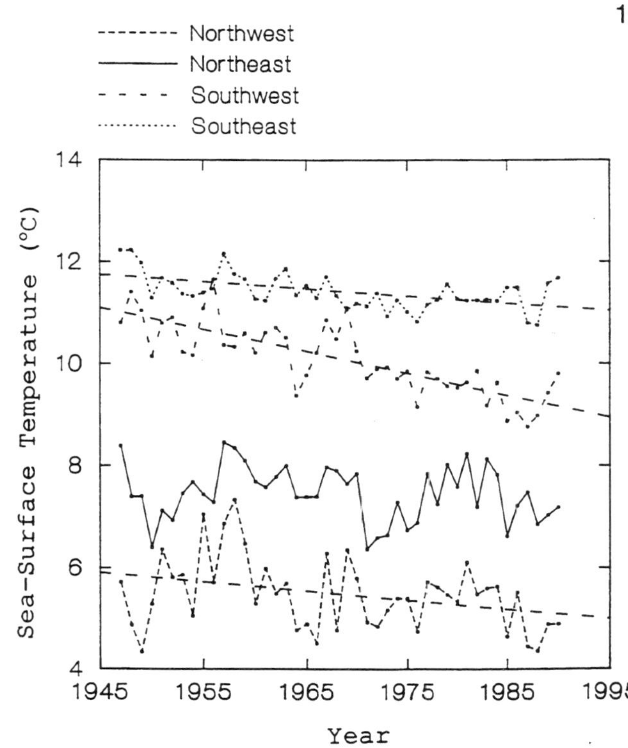

Fig. 4. Annual sea-surface temperatures for the quadrants of the North Pacific Ocean with linear fits for the southeast, southwest, and northwest quadrants.

Autocorrelations for all data sets and cross correlations between catch–abundance data and SST were done using a software program for time series analysis by SYSTAT (Wilkinson 1990). Commercial catches of red king crab and pollock and crab CPUE in the eastern Bering Sea were cross-correlated with SST anomalies in the southeastern Bering Sea around 55°N, 165°W (SSTEBS in Table 2). Similarly, commercial catches of red king crab and pollock, crab CPUE, and age-4 pollock from the Gulf of Alaska or Kodiak were cross-correlated with SST anomalies around 55°N, 150°W (SSTKD in Table 2), just southeast of Kodiak Island. The total crab and pollock data combined for the Gulf of Alaska and eastern Bering Sea were cross-correlated with SST from the northeast quadrant averaged over the northern Gulf of Alaska north of 52.5° N and the eastern Bering Sea east of 177.5°W (SSTNE in Table 2).

We correlated SST anomalies with the raw data, as well as with differenced crab and pollock data. Differencing replaces each value in a time series with the difference between that value and the previous value, thus removing linear trends in the data. Because the original data show strong trends and thus dependence (high autocorrelations at small lags), the effective degrees of freedom are reduced considerably. Thus, conservative significance levels (SL) were assigned using Student's *t*-statistic ($t=r/\mathrm{SE}$), standard errors (SE) calculated according to Box and Jenkins (1976), and an effective N_{eff} calculated according to Chelton (1984) as

$$N_{\mathrm{eff}} = \frac{N}{\Sigma r_{(xx)}(k) r_{(yy)}(k)}$$

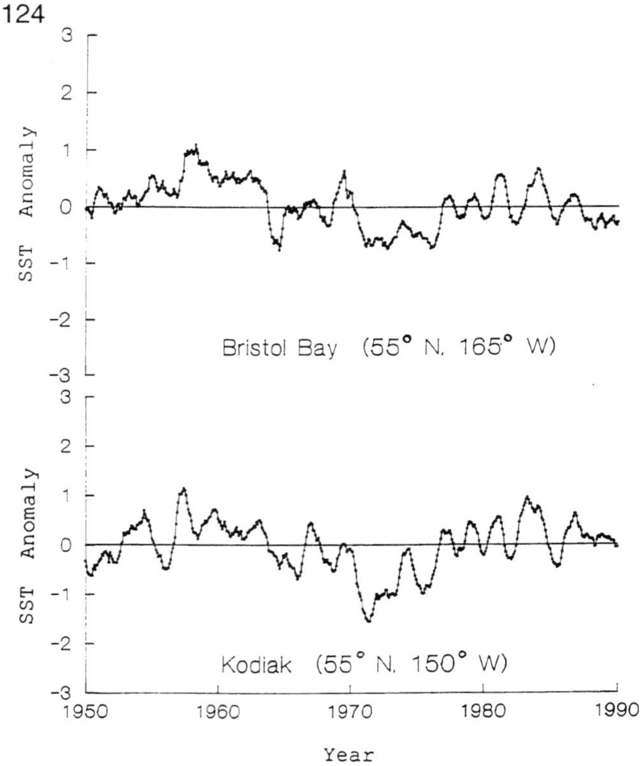

Fig. 5. Monthly sea-surface temperature anomalies smoothed over 13 mo for representative locations of Bristol Bay, eastern Bering Sea, and near Kodiak Island, Gulf of Alaska.

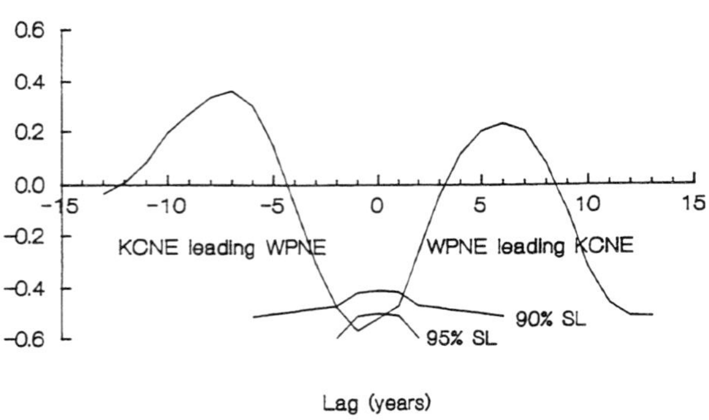

Fig. 6. Cross correlation between red king crab (KCNE) and walleye pollock (WPNE) in the northeast quadrant of the North Pacific Ocean. Significance levels are indicated.

where $r_{(xx)}(k)$ and $r_{(yy)}(k)$ are the autocorrelations of time series $x(t)$ and $y(t)$ at lag k, and N is the number of observations common to both time series.

Results

Temperatures

Annual averages for SSTs for the southeast, southwest, northeast, and northwest quadrants of the North Pacific over 44 yr (Fig. 4) illustrate that the amplitude of fluctuations was higher in the northern quadrants. The northeast quadrant showed a pattern that is consistent with that of Royer (1989), with SST at a maximum in 1957, decreasing into the early 1970s, and increasing again to another maximum in the early 1980s. The northwest quadrant SST was at a maximum in 1958, but with no pronounced increase after the early 1970s. Neither southern quadrant contained the low-frequency signal found in the northeast quadrant. To investigate trends in the data, linear regressions were fitted to the SSTs. The F-values were 71 (southwest), 16 (southeast), 5.1 (northwest), and 1.3 (northeast). The first three F-values were significant at 95% SL. The linear regressions suggest a decrease in SST of 0.043 ± 0.010°C/yr in the southwest, 0.014 ± 0.008°C/yr in the southeast, and 0.018 ± 0.016°C/yr in the northwest (Fig. 4). Thus, the northeast quadrant was the only area that showed no decreasing trend in SST over the past 40 yr. Furthermore it was the only area where a very low frequency fluctuation was evident in the record.

An autocorrelation for the differenced SST in the Northeast Pacific showed a high negative correlation ($r = -0.404$, $p = 0.01$) for a lag of 1 yr, indicating that temperatures tend to decrease in the year following an increase or increase in a year following a decrease. The short period (2 yr) pattern is apparent in the SST anomalies of Kodiak and Bristol Bay (Fig. 5). These fluctuations in SST have been observed throughout the northern North Pacific and were shown to have a period of about 2.5 yr (Salmon 1992).

Red king crab and walleye pollock

We cross-correlated catches of red king crab and walleye pollock from the eastern Bering Sea and the Gulf of Alaska between 1964 and 1990 (Fig. 6). There is a high negative correlation between walleye pollock catches and king crab catches in the same year and 1 yr earlier (95% SL). A linear regression between red king crab catches and pollock catches in the same year confirmed this inverse relationship ($r = 0.61$, $F = 14.84$, $p = 0.001$).

Red king crab

While the commercial catch of total northeastern king crab appeared to have cyclic behaviour with a period of 14 yr (Fig. 2), autocorrelations were not significant at the 95% level for any lags greater than 2. However, they were negative for lags between 6 and 11 and showed a peak at a lag of 14 yr for the 32-yr data set ($r = 0.21$, $0.5 < p < 0.6$). Cross correlations (Table 2) indicated that red king crab catches in the Northeast Pacific significantly increased 10–12 yr after a high SST anomaly (and decreased after a low SST anomaly). The negative correlation between the SST anomaly in Bristol Bay and CPUE of red king crab 3 yr later (Table 2) indicated a drop in crab catches 3 yr after a high SST or an increase 3 yr after a low SST. The cross correlation function also indicated a positive correlation at a lag of 10 yr (SST leading king crab CPUE), but it was not significant because of the low effective N at high lags ($r = 0.39$, $N_{eff} = 6$). King crab CPUE and SST anomaly around Kodiak showed no significant correlations for positive lags (SST leading) (Table 2).

Walleye pollock

Like red king crab, total commercial catch of walleye pollock showed a 14-yr cycle (Fig. 2), but the autocorrelation was not significant at the 95% level after a lag of 1 yr. However, it showed a strong negative correlation at a lag of 7 yr ($r = -0.489$, $p < 0.2$). Again, we correlated water temperature with fisheries data to test our hypothesis that cyclic temperatures cause cyclic fisheries.

Table 2. Table of significant correlation coefficients of catch and abundance data with SST. For abbreviations see Table 1. Positive lags are for SST leading. *, 95% significance level ($p < 0.05$); † 90% significance level ($p < 0.1$). Correlations at 90% significance are not shown if they occurred adjacent to correlations that are 95% significant.

	SSTKD			SSTEBS			SSTNE		
	Lag	*r*	t	Lag	*r*	t	Lag	*r*	t
WPGAK	0	0.54	2.39*						
	1	0.55	2.40*						
WPEBS				−1	−0.48	2.43*			
				−7	0.53	2.36†			
WPNE							0	0.49	2.54*
							1	−0.53	2.70*
							−7	0.64	2.87*
WP4	−1	0.52	2.64*						
KCKD	−6	0.53	2.26*						
KCEBS				4	−0.37	1.94†			
KCNE							−3	0.45	2.38*
							12	0.39	1.96†
KCNE differenced							−4	0.55	2.78*
							10	0.44	1.98†
CPUEKD	All ns								
CPUEBB				3	−0.54	2.52*			

In the Gulf of Alaska, the SST anomaly was positively correlated with pollock catches in the same year and the following year (Table 2). That is, when the SST was high, catches of pollock were high and when SST was low catches of pollock were low. A linear regression of pollock catch in the Gulf of Alaska on SST near Kodiak (55°N, 150°W) during the same year confirmed this significant relationship ($r = 0.536$, $F = 7.26$, $p = 0.015$). The differenced data showed no significant correlations.

In contrast to the direct relationship between pollock and temperature in the Gulf of Alaska, landings of walleye pollock in the eastern Bering Sea were negatively correlated with SST anomalies in the following year (Table 2), and at the 90% SL for the same year and 2 yr later. Thus, catches were low when SST was high and catches were high when SST was low. There were no significant correlations for catch following temperature. A linear regression of pollock catch in the eastern Bering Sea on SST the same year indicated a slightly negative relationship ($r = -0.33$, $F = 3.03$, $p = 0.09$). The relationship became highly significant when the three years of low catches during the development of the fishery (1964–66) were excluded from analysis ($r = -0.57$, $F = 10.33$, $p = 0.004$). Again, we found no significant correlations for the differenced data.

We expected to find evidence for an effect of temperature on year-class strength at hatching or during the first year. However, abundance of age-4 pollock in the Gulf of Alaska was not significantly correlated with SST anomalies 3 and 4 yr earlier ($r < 0.08$). There was a significant correlation at a lag of −1 yr (Table 2). Four-year-old pollock appear to positively lead the SST by 1 yr. That is consistent with pollock catch in the Gulf of Alaska, which contains many 5-yr-old fish, being in phase with the temperature.

Discussion

Temperatures

Sea-surface temperatures in the Northeast Pacific Ocean fluctuate at very low frequencies and are currently declining. In our Northeast Pacific quadrant, temperatures decreased and increased by approximately 2°C over a period of about 25 yr (Fig. 4 and 5). Temperatures in the other quadrants of the northern North Pacific show a significantly decreasing trend since at least 1947 (Fig. 4).

McLain (1984) suggested that there has been a large-scale warming of coastal SSTs in the Northeast Pacific since 1976. He compared this warming trend with similar periods of unusually warm coastal water in the past and suggested shifts in the wind field and El Niño effects as possible causes. Chelton (1984) examined Pacific SSTs from the equator northward using the same "Namias" time series that was used in this study. The warming trend in the Northeast Pacific since 1976 is clearly evident in his northeast quadrant, which includes average SSTs north of 42.5° N and east of 172.5° W, approximately the same area as our northeast and south east quadrants. He found no similar warming trend for other parts of the North Pacific. Royer (1989), addressing individual 5° × 5° areas within the northeast quadrant, suggested that the warming trend is part of a very low frequency fluctuation that is evident in air temperature records from several near-coastal stations in Alaska. The 18.6-yr nodal tide cycle accounts for 30% of the variation of the low frequency signal (Royer 1993). There are some differences in the results of Chelton (1984) and Royer (1989) because Royer (1989) addresses individual 5° × 5° locations, whereas Chelton (1984) (and this paper) average the SSTs over large regions. The conclusions do not differ, with the exception of the containment of higher frequency components in this study, that Royer (1989) and Chelton (1984) filtered out with a 25-mon filter. The present study suggests that the observed temperature cycle for SSTs in offshore areas may be limited to the eastern Bering Sea and the northern Gulf of Alaska. Thus, a large part of the temperature variations may be due to changes in local heating, cooling, or advection rather than large-scale phenomena like El Niño or rising CO_2 levels in the atmosphere.

Cyclic temperatures and cyclic fisheries

There is widespread belief that cyclic changes or global trends in ocean temperatures will strongly affect fishes and fisheries (Cushing 1975, 1982; Francis 1990; Regier et al. 1990). Over the last 30 yr while temperatures in the Northeast Pacific fluctuated widely as part of a very low frequency cycle, walleye pollock catches and king crab catches were well correlated and out of phase (Fig. 6) but did not vary on the same time scale as temperatures.

The peaks in catches and CPUE of red king crab (Fig. 2) are separated by 14 yr (1965–66, 1980). Likewise, peaks in catches of walleye pollock (Fig. 2) and abundance estimates from cohort analysis are separated by 14 yr (Bakkala et al. 1987; Low 1991). Though the 14-yr cyclic pattern in each species is apparent in the catch data, it is not significant in the autocorrelations at a 14-yr lag because of the short record lengths. In contrast, peaks in the most recent temperature cycle are separated by approximately 25 yr (Fig. 1, 2, 4, and 5). Thus, abundances of crab and pollock do not seem to fluctuate in phase with water-column temperatures. The period of the most recent temperature cycle is almost twice the period of fluctuations in king crab and pollock abundance over the past 30 yr. The fluctuations in red king crab and pollock fisheries appear to coincide with parts of the temperature cycle (Fig. 2), but they are not in synchrony with either the 2°C 25-yr cycle that is indicated in the northeast SSTs nor the 18.6-yr cycle which is superimposed on the temperature fluctuations and accounts for up to 30% of the low frequency variance in temperatures (Royer 1993).

Though the periodicity is different, we found some interesting correlations between temperature and catch data that might help to explain some of the observed changes in the abundance of crab and pollock. King crab abundance seems to increase 9–11 yr after a peak in SST. Crabs hatched during warm years will enter the fishery approximately 8–10 yr after the peak in SST. Because we found a positive correlation between SST and differenced catches of crab 10 yr later (Table 2), and a similar pattern, though not significant ($p > 0.1$), in the raw data, it is possible that warm periods result in strong year-classes, and/or cool periods have an adverse effect on year-class strength. Temperature has been shown to have a strong effect on eggs, time of hatching, and larval development in red king crab (Shirley and Shirley 1989). Because larvae must feed within the first 4 d after hatching (Paul and Paul 1980), the timing of hatching may be critical for larval survival. Cold temperatures may delay hatching sufficiently to result in a mismatch between the larvae and their food. Temperature also has a linear effect on the growth of juvenile king crab (Stevens and Munk 1990), with warm temperatures leading to an increased growth rate (Stevens 1990). Rodin (1990) indicated that mortality of king crab larvae on the West Kamchatka Shelf is higher in cold years, and year-classes hatched in cold years are much smaller. If this is also true in the Northeast Pacific, it could provide a mechanism for the observed relationship between SST and king crab abundance.

Sea-surface temperature is inversely related to CPUE of Bristol Bay red king crab 3 yr later (Table 2). This could indicate an adverse effect of warm temperatures on king crab that can significantly affect the abundance of crab. Warm temperatures may be a stressor for adult crab and could lead to increased susceptibility to parasites and disease (Post 1983). Increased mortality due to epizootic disease has been proposed as a possible cause for the sharp decline in the king crab population after 1983 (Otto 1990), which followed a substantial increase in water temperature. Because king crab in the eastern Bering Sea and the Gulf of Alaska constitute independent stocks (Wooster 1990), the simultaneous precipitous decline of both stocks in the early 1980s indicates that there may be a single mechanism affecting all stocks in the Northeast Pacific. Our results suggest that the combination of anomalously cold temperatures in the early 1970s and anomalously warm temperatures in the early 1980s contributed to the population crash of king crab after 1983.

Bulatov (1989) has suggested that temperature is also a strong determinant of year-class strength for walleye pollock in the eastern Bering Sea. Thus, we expected to find a significant correlation between year-class strength in the Gulf of Alaska, represented by age-4 pollock abundances, and temperatures 3–4 yr earlier. However, the correlation was near 0. This, and the absence of a significant correlation between any of the differenced pollock data sets and temperature, may indicate the lack of a causal relationship between temperature and abundance of walleye pollock. There appears to be little influence of water-column temperature on early life history success of pollock in the Gulf of Alaska. Likewise, in the Bering Sea, an anomalously cold (1972) and an anomalously warm (1978) year produced the two strongest year-classes on record (Bakkala et al. 1987). Thus, temperatures cannot be used to predict successful year-classes. Other environmental or biological factors must be more important in determining survival. Results from trophic studies (Francis and Bailey 1983) and ecosystem modeling (Laevastu and Larkins 1981) have suggested that cannibalism may be an extremely important factor in the regulation of pollock abundance.

There is a direct correlation between high catches of pollock and high SST in the Gulf of Alaska in the same year (Table 2). However, only one peak in catches of pollock in the Gulf of Alaska (1984–85) coincides with a warming period. Thus, despite its significance, this may be a spurious correlation. In the Bering Sea, the relationship is opposite, with high catches of pollock associated with low temperatures. National Marine Fisheries Service (NMFS) surveys in the eastern Bering Sea indicate that the distribution of pollock on the shelf differs in cold and warm years (Bakkala et al. 1987). Cold water temperatures tend to concentrate pollock on the outer shelf, while they intrude into inner-shelf waters in warm years. They are more concentrated in cold years, affecting both catch and abundance estimates. Thus the observed relationships may in part be due to the fact that the catch data used for analysis is not a good indicator of abundance and reflect in part the availability of pollock to the fishery.

The fact that walleye pollock and king crab seem to cycle in antiphase raises the possibility that a biological mechanism causes the observed fluctuations. Possible cause and effect relationships are competition and predation. Differences in habitat, feeding mode, and mobility make it unlikely that king crab and walleye pollock compete for the same food items, except possibly in the larval stage. Diets of adult crab and pollock do not overlap. Crab primarily consume mollusks, echinoderms, crustaceans, polychaetes, and scavenged fish (Jewett and Feder 1982). Pollock feed on crustaceans, euphausiids, fish, and juvenile pollock (Dwyer et al. 1987). While other fishes like Pacific cod (*Gadus macrocephalus*) and Pacific halibut (*Hippoglossus stenolepis*) can eat substantial amounts of adult king crab, walleye pollock stomachs have never been observed to contain remains of king crab nor do king crab stomachs contain identifiable pollock (M. Holmberg, Institute of Marine Sciences, University of Alaska Fairbanks, Fairbanks, AK 99775, unpublished data). Thus it seems that both predation and competition can be excluded as possible mechanisms for the observed relationship.

What then can cause the antiphase cycles of pollock and crab? Rather than a biological relationship, this may be due to a shift to other fisheries after crab catches plummet. As crab was declining in the late 1960s, pollock became a target species in the eastern Bering Sea and expanded rapidly to a major fisheries (Bakkala et al. 1987). Likewise, coincident with the crash of the crab fishery in the early 1980s was the discovery of the Shelikof Strait stock of pollock. Some fishing boats converted from crabbers to trawlers despite the expenses and an initial loss in revenue (Wooster 1990). Thus, there was an influence of the crab fishery on the pollock fishery that had no biological cause. Alternatively, the relationship could be caused by an unknown factor that influences king crab and walleye pollock on the same time scales but with opposite effects on abundance.

In summary, there is a very significant cyclic pattern with a period of approximately 25 yr that is evident in the water temperatures of the Gulf of Alaska and Bering Sea north of 52.5°N and west of 177.5°W. The water temperatures in this area decreased from the late 1950s to the early 1970s and then increased through the mid-1980s. They are currently decreasing. This cyclic pattern is only apparent in the northeast quadrant of the North Pacific, but not in the southeast, southwest, or northwest quadrants, which have been experiencing a decreasing trend since the early 1950s.

Coincident with the 25-yr temperature cycle in the Northeast Pacific are large fluctuations in catches of red king crab and walleye pollock with peaks approximately 14 yr apart. However, while the water temperature data can be related to air temperature and proxied back to 1828 to verify a cyclic pattern, the fisheries data do not constitute a long enough time series to reveal a significant cyclic pattern and have no alternative source of data to be substituted as long term proxy.

The significant cross correlation between temperature and king crab 10 yr later supports evidence of other researchers that year-class strength of red king crab is influenced by water temperature. Our results show no evidence for an effect of temperature on year-class strength of walleye pollock and the relationship between pollock and temperature at time of catch may be due to effects on the distribution of the fish. However, while we can speculate on displacement of fish species and/or food or nutrients with temperature changes we can neither prove nor disprove such relationships here because the data are insufficient. To verify a cyclic fluctuation requires a time series containing about five cycles. For crab and pollock with speculated 14-yr cycles, that means data for at least 70 yr. This demonstrates the need for long time series of fisheries-independent data coincident with physical data.

Acknowledgements

We thank Dr. Mark Johnson and Maggie Castellini for reviewing an early draft of the manuscript. Funding for this study was provided by the State of Alaska. Dr. Royer was supported by NOAA Grant NA90AA-D-AC805.

References

Armstrong, D.A. 1983. Cyclic crab abundance and relationship to environmental causes. p. 102–110. *In* W.S. Wooster [ed.]. From year to year. Interannual variability of the environment and Fisheries of the Gulf of Alaska and the eastern Bering Sea. University of Washington, Seattle, Wash. Wash. Sea Grant Program Publ. WSG-WO 83-3.

Bakkala, R.G., V.G. Wespestad, and L. Low. 1987. Historical trends in abundance and current condition of walleye pollock in the Eastern Bering Sea. Fish. Res. 5: 199–215.

Beegle, C.J. 1986. Time and space scales of some oceanic and atmospheric parameters in the Gulf of Alaska. Master's thesis, University of Alaska, Fairbanks, Alaska. 99 p.

Box, G.E.P., and G.M. Jenkins. 1976. Time series analysis: Forecasting and control. Holden-Day, Inc., Oakland, Calif. p. 376–377.

Bulatov, O. A. 1989. The role of environmental factors in fluctuations of stocks of walleye pollock (*Theragra chalcogramma*) in the Eastern Bering Sea. *In* Effects of ocean variability on recruitment and an evaluation of parameters used in stock assessment models. Can. Spec. Publ. Fish. Aquat. Sci. 108: 353–357.

Bull, H.O. 1952. An evaluation of our knowledge of fish behaviour in relation to hydrography. Rapp. ICES, 131: 8–23.

Chelton, D.B. 1984. Short term climatic variability in the Northeast Pacific Ocean, p. 87–99. *In* W.G. Pearcy [ed.] The Influence of Ocean Conditions on the Production of Salmonids in the North Pacific, a workshop. Oregon State University, Corvallis, Oreg. Sea Grant Publ. ORESU-W-83-001.

Cushing, D.H. 1975. Marine ecology and fisheries. Cambridge University Press, Cambridge, UK. 278 p.

Cushing, D.H. 1982. Climate and fisheries. Academic Press, London. 373 p.

De Veen, J.F. 1978. Changes in North Sea sole stocks (*Solea solea*, L.). Rapp. P. -V. Réun. Cons. int. Explor. Mer, 172: 124–136.

Dwyer, D.A., K.M. Bailey, and P.A. Livingston. 1987. Feeding habits and daily ration of walleye pollock (*Theragra chalcogramma*) in the eastern Bering Sea, with special reference to cannibalism. Can. J. Fish. Aquat. Sci. 44: 1972–1984.

Ellertsen, B., P. Fossum, P. Solemdal, and S. Sundby. 1989. Relation between temperature and survival of eggs and first-feeding larvae of northeast Arctic cod (*Gadus morhua* L.), p. 209–219. *In* The Early Life History of Fish. Third ICES Symposium, 3–5 Oct. 1988, Bergen, Norway. Rapp. P. -V. Réun. Ciem. 191: 209–219.

Francis, R.C. 1990. Climate change and marine fisheries. Fisheries, 15: 7–9.

Francis, R.C. and K.M. Bailey. 1983. Factors affecting recruitments of selected gadoids in the Northeast Pacific and eastern Bering Sea, p. 35–60. *In* W.S. Wooster [ed.]. From year to year. Interannual variation of the environment and Fisheries of the Gulf of Alaska and the eastern Bering Sea. University of Washington, Seattle, Wash. Washington Sea Grant Program Publ. WSG-WO 83-3.

Hollowed, A.B., and B.A. Megrey. 1991. Stock assessment and fishery evaluation report for the 1992 Gulf of Alaska groundfish fishery, Table 16. North Pacific Fishery Management Council, Anchorage, Alaska.

Jewett, S.C., and H.M. Feder. 1982. Food and feeding habits of the king crab *Paralithodes camtschatica* near Kodiak Island, Alaska. Mar. Biol. 66(3): 243–250.

Laevastu, T., and M.L. Hayes. 1981. Fisheries oceanography and ecology. Fishing News Books, Farnham, England. 199 p.

Laevastu, T., and H.A. Larkins. 1981. Marine fisheries ecosystem: its quantitative evaluation and management. Fishing News Books, Ltd., Farnham, England. 162 p.

Laevastu, T., and R. Marasco. 1982. Fluctuations of fish stocks and the consequences of the fluctuations to fishery and its management. U.S. Department of Commerce, Washington, D.C. NOAA Tech. Memo. NMFS F/NWC 27. 53 p.

Low, L. [ed.] 1991. Status of living marine resources off Alaska as assessed in 1991. U.S. Department of Commerce, Washington, D.C. NOAA Tech. Memo. NMFS F/NWC-211.

McLain D.R. 1984. Coastal ocean warming in the Northeast Pacific, 1976–83, p. 87–89. *In* W.G. Pearcy [ed.] The Influence of Ocean Conditions on the Production of Salmonids in the North Pacific, a Workshop. Oregon State University, Corvallis, Oreg. Sea Grant Publ. ORESU-W-83-001.

Otto, B.S. 1990. An overview of Eastern Bering Sea king and tanner crab fisheries, p. 9–26. *In* Proceedings, International Symposium on King and Tanner Crabs, 28–30 Nov. 1989, Anchorage, Alaska. University of Alaska, Fairbanks, Alaska. Alaska Sea Grant Program Rep. 90-04.

Paul, A.J., and J.M. Paul. 1980. The effect of early starvation on later feeding success of king crab zoea. J. Exp. Mar. Biol. Ecol. 44: 247–251.

Post, G.W. 1983. Text book of fish health. TFH publications, Neptune City, N.J. 256 p.

Regier, H.A., J.J. Magnuson, and C.C. Coutant. 1990. Introduction to Proceedings: Symposium on Effects of Climate Change on Fish. Trans. Am. Fish. Soc. 119(2): 173–174.

Rodin, V.E. 1990. Population biology of the king crab *Paralithodes camtschatica* Tilesius in the North Pacific Ocean, p. 133–144. *In* Proceedings, International Symposium on King and Tanner Crabs, 28–30 Nov. 1989, Anchorage, Alaska. University of Alaska, Fairbanks, Alaska. Alaska Sea Grant Program Rep. 90-04.

Royer, T.C. 1989. Upper ocean temperature variability in the Northeast Pacific Ocean: is it an indicator of global warming? J. Geophys. Res. 94: 18 175 – 18 183.

Royer, T.C. 1993. High latitude oceanic variability associated with the 18.6 year nodal tide. J. Geophys. Res. 98: 4639–4644.

Saetersdal, G., and H. Loeng. 1987. Ecological adaptation of reproduction in northeast Arctic cod. Fish. Res. 5: 253–270.

Salmon, D. 1992. On interannual variability and climate change in the North Pacific. Ph.D. thesis, University of Alaska Fairbanks, Fairbanks, Alaska. 219 p.

Shirley, S.M., and T.C. Shirley. 1989. Interannual variability in density, timing and survival of Alaskan red king crab larvae. Mar. Ecol. Prog. Ser. 54: 51–59.

Smith, P.E. 1978. Biological effects of ocean variability: time and space scales of biological response. Rapp. P. -V. Réun. Cons. int. Explor. Mer, 173: 117–127.

Stevens, B.G. 1990. Temperature-dependent growth of juvenile red king crab (*Paralithodes camtschatica*) and its effects on size-at-age and subsequent recruitment in the eastern Bering Sea. Can. J. Fish. Aquat. Sci. 47: 1307–1317.

Stevens, B.G., and J.E. Munk. 1990. A temperature-dependent growth model for juvenile red king crab, *Paralithodes camtschatica*, in Kodiak, Alaska. *In* Proceedings, International Symposium on King and

Tanner Crabs, 28–30 Nov. 1989, Anchorage, Alaska. University of Alaska, Fairbanks, Alaska. Alaska Sea Grant Program Rep. 90-04: 293–304.

Thompson, A.B., and M. Hilden. 1989. Relationship between March sea temperature and year-class strength for selected North Sea fish stocks, p. 437. *In* The Early Life History of Fish. Third ICES symposium, 3–5 Oct. 1988, Bergen, Norway. Rapp. P. -V. Réun. Ciem. 191.

Wilkinson, L. 1990. SYSTAT: the system for statistics. SYSTAT, Inc., Evanston, Ill.

Wooster, W.S. 1990. King crab deposed, p. 13–17. *In* M.H. Glantz, and L.E. Feingold [ed.] Climate variability, climate change and fisheries. National Center for Atmospheric Research, Boulder, Colo.

Westward Region Shellfish Staff, Alaska Department of Fish and Game. 1992. Annual management report for the shellfish fisheries of the westward region, 1991. Westward Region Shellfish Staff, Alaska Department of Fish and Game, Juneau, Alaska. Regional Inf. Rep. 4K92-9.

Sea-ice conditions and the distribution of walleye pollock (*Theragra chalcogramma*) on the Bering and Chukchi Sea Shelf

Tina Wyllie Echeverria

Wyllie Echeverria, T. 1995. Sea-ice conditions and the distribution of walleye pollock (*Theragra chalcogramma*) on the Bering and Chukchi Sea Shelf, p. 131–136. *In* R.J. Beamish [ed.] Climate change and northern fish populations. Can. Spec. Publ. Fish. Aquat. Sci. 121.

Abstract: The Bering Sea Shelf is a large shallow shelf that is seasonally ice covered and supports large commercial fisheries. Sea ice occurs on the shelf for a period of up to 8 months. Its maximum southern extent varies as much as 30% interannually. The pattern of ice cover depends primarily on wind speed and direction, and secondarily on thermodynamics. Sea ice profoundly affects the underlying pelagic and benthic habitats. Pelagic species from the North Pacific dominate the outer shelf domain during the open-water season. This area is the object of multi-national fishing efforts focused on harvesting historically abundant resources of walleye pollock (*Theragra chalcogramma*). A hypothesis is presented that species utilization of the shelf will vary with ice cover. Benthic and pelagic species will respond differently to an open water outer shelf. I compare abiotic and biotic variability and demonstrate pollock can occupy areas further north during years of reduced ice cover.

Résumé : Le plateau de la mer de Béring est une grande plate-forme peu profonde couverte de glace pendant l'hiver et où se pratique un importante pêche commerciale. Il est gelé pendant une période pouvant aller jusqu'à 8 mois. Son extension vers le sud varie jusqu'à 30 % d'une année à l'autre. La configuration de la couverture de glace dépend surtout de la vitesse et de la direction des vents, et accessoirement de la thermodynamique. La présence de glace a d'importantes répercussions sur les habitats pélagiques et benthiques. Les espèces pélagiques du Pacifique Nord dominent la partie au large du plateau durant la saison où les eaux sont libres de glace. Ce secteur fait l'objet de pêches multinationales axées sur la récolte de ressources historiquement abondantes de morues du Pacifique occidental (*Theragra chalcogramma*). Une hypothèse est présentée voulant que l'utilisation du plateau par cette espèce varie selon la couverture de glace. Les espèces benthiques et pélagiques réagissent différemment à un plateau extérieur en eaux libres. Je compare la variabilité non biotique et biotique et démontre que la morue peut occuper des zones plus au nord durant les années où la couverture de glace est réduite.

Introduction

This work is an inquiry into the relationship of seasonal sea-ice cover to the population of walleye pollock (*Theragra chalcogramma*) inhabiting the Bering and Chukchi Sea Shelf. This area is inhabited by both arctic and subarctic species. Seasonal and interannual variability in ice cover is compared with the distribution of pollock populations. Historical data sets documenting interannual variability over the shelf and surveys documenting the distribution of walleye pollock are used.

As management units, large marine ecosystems commonly are multinational in their boundaries and use. The coastal zone fisheries on the Bering Sea Shelf are managed by Russia and the United States. The fishery is dominated by groundfish, with pollock and flatfish the principal catch. The distribution of pollock changes seasonally, with migrations off the shelf during the ice season (Fadeyev 1988). Spring heralds the time when the population moves back to the shelf to spawn (Hinckley 1987).

Sea ice covers most of the shelf each year. This seasonal ice expands, retreats, and is maintained during winter months. Its extent correlates with north–south winds (Overland and Pease 1982; Mysak and Manak 1989) and, to a lesser extent, sea-surface and air temperatures (Skjoldal and Rey 1989). Seasonal change in sea-ice cover affects the salinity, temperature, and stability of the underlying water masses. A warm period has been identified for the North Pacific and the Bering Sea beginning in 1976 (Niebauer 1988; Trenberth 1990; Salmon 1992), evidenced by a reduction in the areal ice cover. Northward blowing winds, generated by an intensified Aleutian Low pressure system, push the ice northward, resulting in reduced ice cover during warm periods. Sea-ice extent is at a maximum during March and at a minimum during September. Figure 1 shows the area of the shelf seasonally covered by ice in 1990. These conditions extend into the Chukchi Sea as well, crossing the arctic–subarctic front (Mysak and Manak 1989).

T. Wyllie Echeverria. School of Fisheries and Ocean Sciences, Institute of Marine Sciences, University of Alaska Fairbanks, Fairbanks, AK 99775, U.S.A. Current address: NOAA/PMEL, 7600 Sand Point Way NE., Seattle, WA 98115-0070, USA.

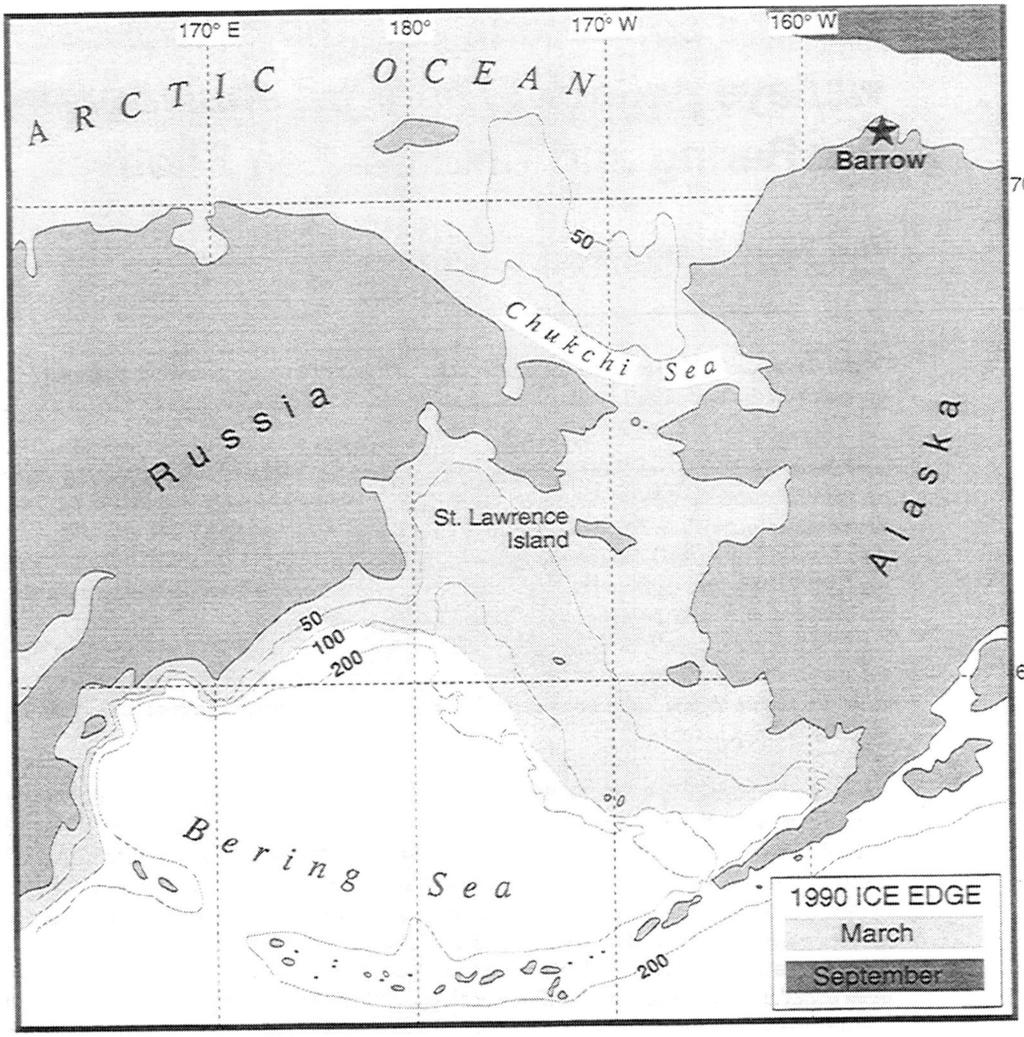

Fig. 1. Areal extent of sea ice during maximum extent in March and minimum extent in September 1990.

Movements of gadid populations in the Atlantic have been related to such physical variables as salinity, temperature, and current movements (see Cushing 1982). The range of pollock, a North Pacific gadid, reaches the arctic–subarctic border and could be expected to respond to a changing environment there.

Patterns, cycles, and regime changes in the physical conditions on the shelf may be useful indicators of environmental changes that affect fish populations. Although pollock may not be directly affected by sea ice, populations may change in concert with environmental changes. Investigation of long-term, periodic changes, rather than interannual variability, may be necessary to detect population changes.

Methods

The species composition of the ichthyofauna in the northeastern Chukchi Sea was investigated from August through September 1989–92. Various midwater larval fish nets (BONGO, Isaac's-Kidd, BEAM, Frame) and bottom trawls (Otter) were used in a survey funded by Minerals Management Service (Wyllie Echeverria et al. 1995; Smith et al. 1995). Surveys were also conducted aboard the Japanese fisheries training vessel the R/V Oshoro Maru during July 1990 and 1991 using nets similar to those in the above survey (Faculty of Fisheries 1991). These surveys afford the opportunity to compare interannual variations in the ichthyofauna of the region.

The Alaska Science Center of the National Marine Fisheries Service (NMFS-ASC) conducts annual surveys of the fisheries resources in the southeastern Bering Sea. Abundance and distribution estimates have been made annually since 1971 (Walters 1983), and have been extended over the northern Bering Sea shelf triannually since 1976. This survey covers the area of the shelf inside U.S. boundaries where walleye pollock are abundant. The NMFS survey data is used to illustrate the known areas of distribution for walleye pollock.

Sea-ice conditions have been evaluated in several ways. Mysak and Manak (1989) provide a comprehensive study of the Arctic Ocean and its marginal seas. They evaluate sea-ice extent and anomalies for the period 1953–84 on the Bering and Chukchi seas using data gathered by the Navy – National Oceanographic and Atmospheric Administration Joint Ice Center (Navy – NOAA JIC) and the U.S. Naval Oceanographic Office. By removing the climatological annual cycle

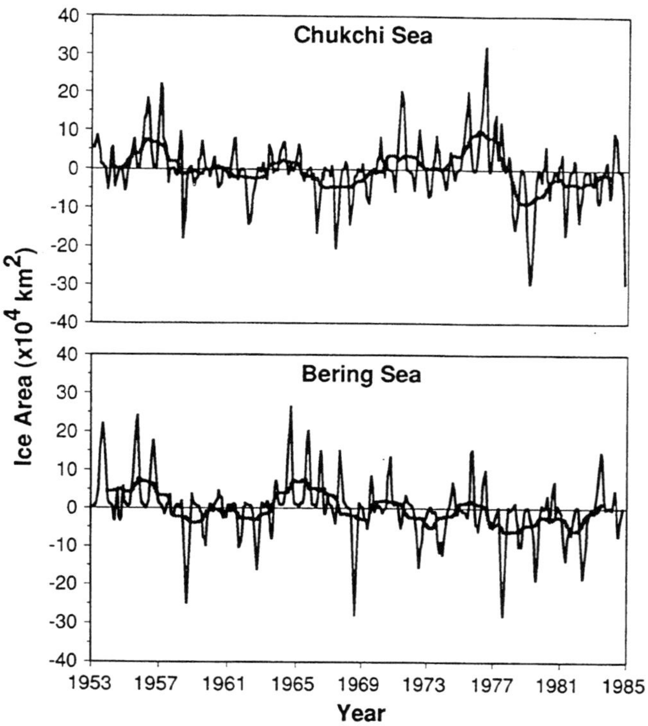

Fig. 2. Smoothed anomalies of areal sea-ice extent for the Chukchi and Bering seas (Mysak and Manak 1989).

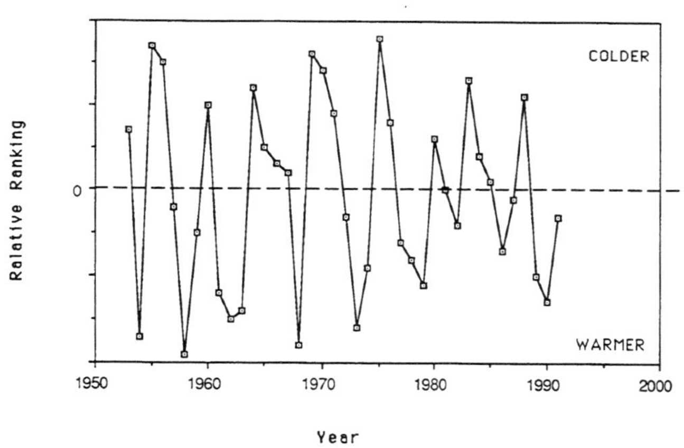

Fig. 3. Sea-ice severity index for the north coast of Alaska. Years are ranked around a mean (0), more severe ice years have a positive sign. A 4–6 yr period of heavy and light ice years are noted (Barnett 1980). Note a reduction in the amplitude of heavy–light ice cover is since 1976.

from the monthly values of areal ice extent, interannual fluctuation in ice cover can be studied (Fig. 2) (Mysak and Manak 1989). The fluctuations in their data set for the Bering and Chukchi seas are related to the North Pacific Pressure Index (NEPPI). Barnett (1980) developed a sea-ice severity index using six parameters dealing with sea-ice extent and concentration along the northern shore of Alaska. This index is used by shipping vessels and updated by the Navy – NOAA JIC annually. It provides a relative ranking of summer ice conditions near Barrow, Alaska for the period from 1953 to the present (Fig. 3).

Table 1. Catches of walleye pollock during the open water seasons of 1989–1991.

Year	Life history stage	Gear	Number
1989	Age 0	IKMT	0
	Age 1	Otter trawl	1
1990	Age 0	Beam, Frame, Bongo	51
		IKMT, Bongo	7
	Age 1+	Otter trawl	179
	Adults	Otter trawl	14
1991	Age 0	Beam	1
	Age 1+		0

Results

The ichthyoplankton surveys to the Chukchi Sea in 1989–92 revealed the presence of walleye pollock (Table 1). Pollock were most abundant, and the surveyed area was most extensive in 1990 (Fig. 4). The most abundant life history stages were young-of-the-year fish, and age-1 fish.

Adult pollock were sampled by otter trawl from three stations in the Chukchi Sea. The four males and nine females were 65–81 cm fork length, and 7–12 yr old, and in a nonreproductive stage. All pollock stomachs contained food items, primarily Arctic cod (*Boreogadus saida*), sculpins (*Cottidae*), eelpouts (*Zoarcidae*), benthic worm (*Echiuris* spp., *Glycerus* spp.), and shrimp (*Argis* spp., *Pandalus* spp., *Crangon* spp.) (Fig. 5).

Discussion

The position of the Aleutian Low (of which NEPPI is a measure) has been used to predict the percent ice cover. The NEPPI provides a time series of physical parameters so that long-term patterns can be inferred. An intensification and a westward displacement of the Aleutian Low are associated with years of light ice cover. In cold years (1973–77), with associated winds blowing to the south, sea ice is pushed over the entire Chukchi and Bering seas south to the shelf break (latitude 56°N). In some warm years (1978–90), ice only reached latitude 59°N. In the Bering and Chukchi seas, a 4 to 6-yr period of heavy to light ice concentrations is evident (Fig. 3) (Barnett 1980; Mysak and Manak 1989). This pattern persists within the larger-scale periods of warm and cold (Fig. 2).

Pollock are distributed in the eastern and western North Pacific Ocean northward to Bering Strait (Andriyashev 1954). Pollock were not reported north of Bering Strait before 1976; however, sampling during that period was sparse (Table 2). Since that time, several surveys to that region landed pollock. This apparently expanded distribution has coincided with a

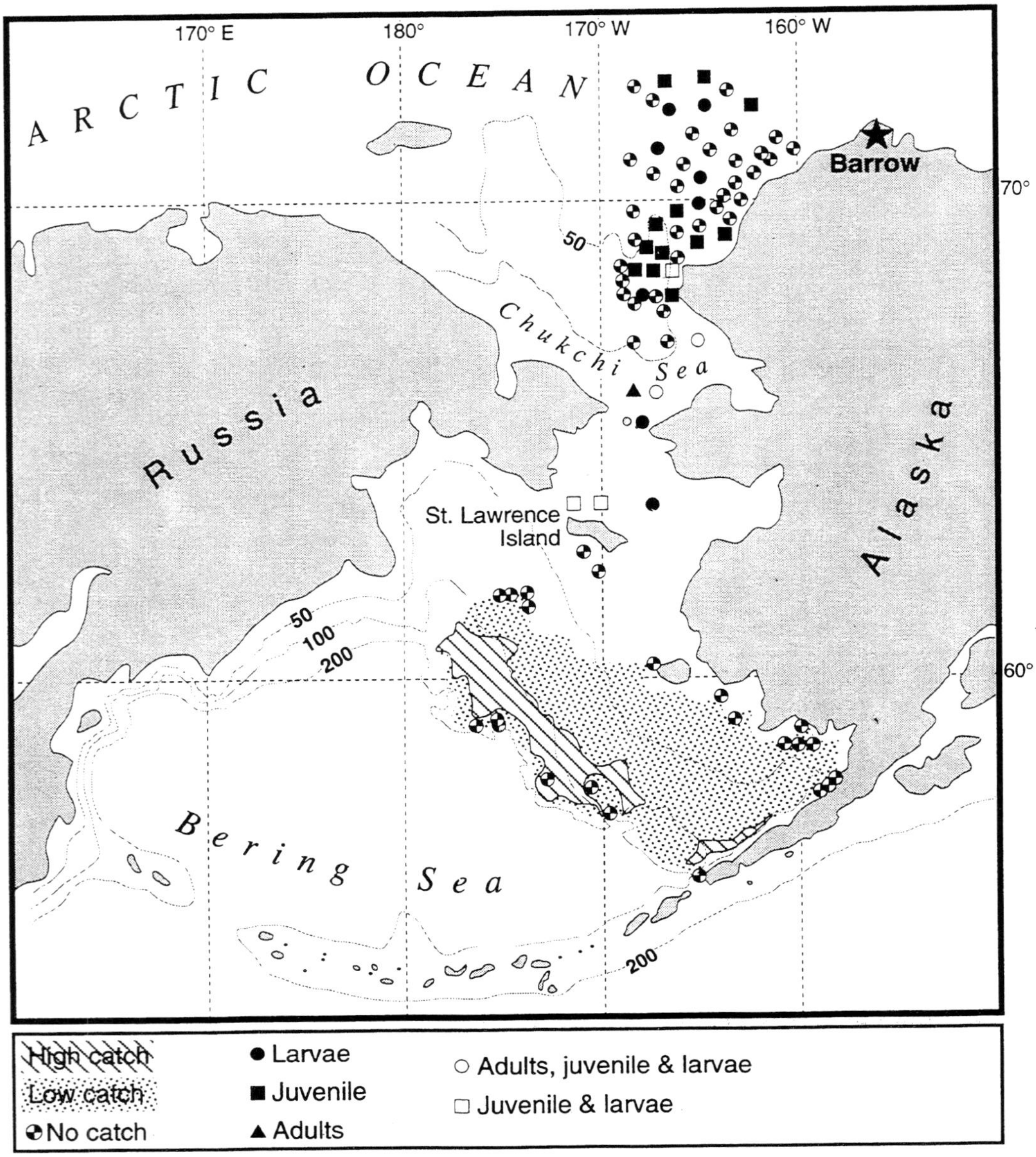

Fig. 4. Distribution of larval, age-1, and adult pollock caught during various surveys in 1990.

Table 2. The occurrence of walleye pollock in ichthyofaunal surveys to the Chukchi Sea.

Presence	Year	Area	Life history stage	Reference
No	pre-1953	Chukchi	Adult	Walters 1955
No	1959	SE Chukchi	Larvae, adult	Alverson and Wilimovsky 1966
No	1970	NE Chukchi	Larvae, adult	Quast 1972
Yes	1976	NE Chukchi	Age-1+	Wolotira et al. 1977
Yes	1989	SE Chukchi	Larvae, Age 1+	Wyllie Echeverria et al. 1995
Yes	1990	SE Chukchi	Larvae, Age-1+	Wyllie Echeverria et al. 1995
Yes	1990	SE Chukchi	Larvae, Age-1+	Fig. 4
			Adult	Faculty of Fisheries 1991
Yes	1991	NE Chukchi	Age-1+	Wyllie Echeverria et al. 1995

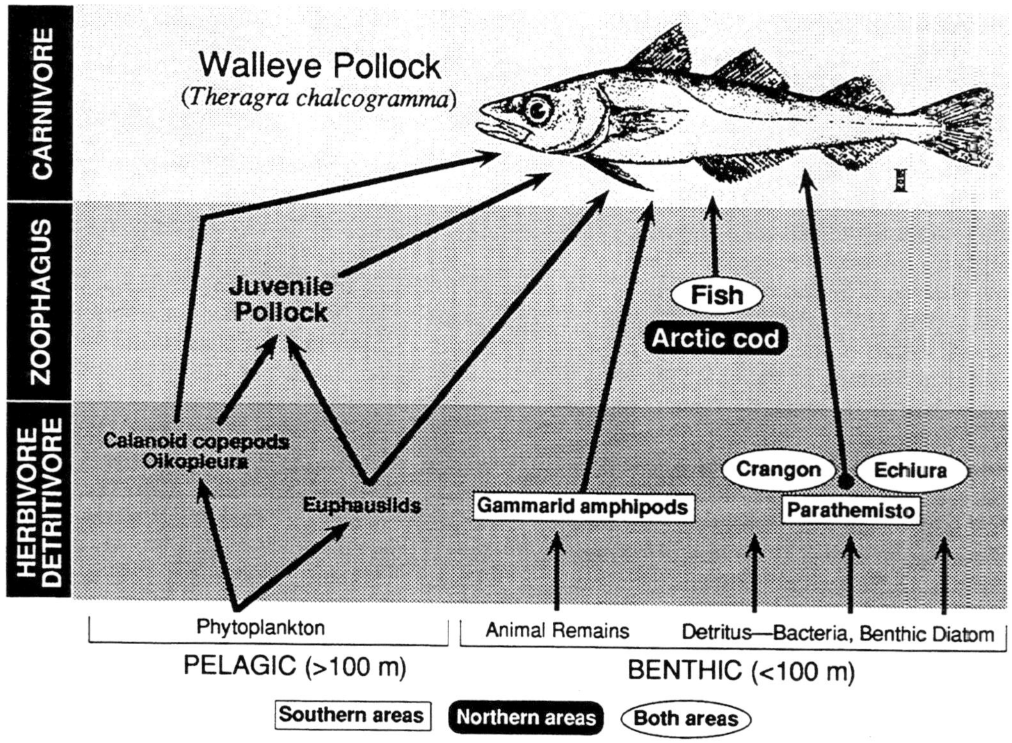

Fig. 5. Food web of adult pollock in the southeastern Bering Sea compared with the central Chukchi Sea. Fish in benthic habitats consume similar prey.

period of reduced ice cover (since 1976) and a year-class of extremely high abundance in 1978 (Wespestad and Traynor 1988).

Age-1 pollock were first reported north of Bering Strait in 1976 (Wolotira et al. 1977). A few larval pollock were observed in the Chukchi Sea in August 1988 indicating that eggs and larvae survive in waters northward of their reported habitat and a northern spawning population may exist (Wyllie Echeverria and McRoy 1992).

The presence of adult walleye pollock, suggests that a portion of a population could be moving northward seasonally and possibly spawning, or alternatively, that a small resident population could exist. During periods of light ice cover, pollock are present north of Bering Strait, where they actively feed and possibly reproduce. These specimens may be a segment of the main population that has roamed outside the boundaries of the species optimal habitat. These vagrants may not be an integral segment of the population (Sinclair 1987).

Strickland and Sibley (1984) predicted a possible northward expansion of walleye pollock habitat under a global warming scenario. Changes in the environment such as reduced ice cover, increased water temperature and reduced stratification of the water column could lead to habitat expansion, increased food supply, increased spawning area, and increased cannibalism for pollock. Young pollock populations are subjected to heavy cannibalism (Livingston 1988), and Francis and Bailey (1983) hypothesized that ice cover may provide some shelter from adult pollock. The current study, with the documentation of a Chukchi Sea component of the Bering Sea pollock population, provides insight into the capabilities of pollock to occupy new territory.

Summary

1) A warm regime dominated the Bering and Chukchi Sea Shelf environment from 1976 until at least 1988.
2) Reduced northward transport dominated the shelf environ ment from 1974 to at least 1985.
3) Pollock were not reported north of Bering Strait before 1971, and since then have been caught during 1976, 1989, 1990, and 1991.
4) Pollock feed on similar prey in the shallow benthic habitat of the southeastern Bering Sea and the Chukchi Sea.
5) The wider distribution may indicate the potential of pollock to change areal patterns or it may reflect the presence of vagrants (Sinclair 1987), which would be lost as part of the main population.

Acknowledgements

Thanks are extended to the captain and crew of the Oshoro Maru, the Ocean Hope 3, and the Surveyor. Tsuneo Nishiyama, Willard Barber, and Sandy Wyllie Echeverria assisted in the conception and scope of this investigation. Patricia Livingston and Gary Walters of the Alaska Science Center, National Marine Fisheries Service, provided unpublished data. Warren Wooster and Richard Strickland provided valuable reviews. This work was partially funded by Minerals Management

Service, U.S. Department of the Interior, Anchorage, Alaska, contract No. 14-35-0001-3-559, and National Science Foundation under Grant No. DPP8405286.

References

Alverson, D.L., and N.J. Wilimovsky. 1966. Fishery investigation of the southeastern Chukchi Sea, p. 843–860. *In* J.J. Wilimovsky [ed.] Environment of the Cape Thompson Region, Alaska. U.S. Atomic Energy Commission, Washington, D.C.

Andriyashev, A.P. 1954. Fishes of the northern seas of the U.S.S.R. Zoological Institute of the U.S.S.R. Academy of Sciences No. 53. (Transl. by Israel Program Sci. Transl., 1964. Available from U.S. Dep. Commer., Natl. Tech. Inf. Serv., Springfield, VA. Publ. OTS 63-11160.) 617 p.

Barnett, D.G. 1980. A long-range forecasting method for the north coast of Alaska, p. 360–373. *In* R.S. Pritchard [ed.] Proceeding of the Arctic Ice Dynamics Joint Experiment, International Commission on Snow and Ice Symposium. University of Washington Press, Seattle, Wash.

Cushing, D.H. 1982. Climate and fisheries. Academic Press, New York. 373 p.

Faculty of Fisheries. 1991. Data record of oceanographic observations and exploratory fishing No. 34, p. 161–189. Faculty of Fisheries, Hokkaido University, Hakodate, Hokkaido, Japan.

Fadeyev, N.S. 1988. Spatial and temporal variability of the eastern Bering Sea walleye pollock size composition in relation to its migrations, p. 479–508. *In* Proceedings International Symposium on the Biology and Management of Walleye Pollock. University of Alaska, Fairbanks, Alaska. Alaska Sea Grant Rep. 89-1.

Francis, R.C., and K.M. Bailey. 1983. Factors affecting recruitment of selected gadoids in the northeast Pacific and east Bering Sea, p. 35–60. *In* W. Wooster [ed.] From year to year: Interannual variability of the environment and fisheries of the Gulf of Alaska and the eastern Bering Sea. Washington Sea Grant, University of Washington Press, Seattle, Wash.

Hinckley, S. 1987. The reproductive biology of walleye pollock, *Theragra chalcogramma*, in the Bering Sea, with reference to spawning stock structure. Fish. Bull. (U.S.) 85: 481–498.

Livingston, P.A. 1988. Interannual trends in walleye pollock, *Theragra chalcogramma*, cannibalism in the eastern Bering Sea, p. 275–296. *In* Proceedings International Symposium on the Biology and Management of Walleye Pollock. University of Alaska, Fairbanks, Alaska. Alaska Sea Grant Rep. 89-1.

Mysak, L.A., and Manak, D.K. 1989. Arctic sea-ice extent and anomalies, 1953–1984. Atmos.-Ocean, 27: 376–405.

Niebauer, H.J. 1988. Effects of El Nino–Southern Oscillation and North Pacific weather patterns on interannual variability in the subarctic Bering Sea. J. Geophys. Res. **93**(C5): 5051–5068.

Overland J.E., and C.H. Pease. 1982. Cyclone climatology of the Bering sea and its relation to sea ice extent. Mon. Weather Rev. 110: 5–13.

Quast, J. 1972. Preliminary report on the fish collected on WEBSEC-70, p. 203–206. *In* WEBSEC-70 An ecological survey in the eastern Chukchi Sea, September–October 1970. U.S. Coast Guard Oceanogr. Rep. 50.

Salmon, D.K. 1992. On interannual variability and climate change in the North Pacific. Ph.D. thesis, University of Alaska, Fairbanks, Alaska. 219 p.

Sinclair, M. 1987. Marine populations: an essay on population regulation and speciation. Washington Sea Grant. University of Washington Press, Seattle, Wash. 252 p.

Skjoldal, H.R., and F. Rey. 1989. Pelagic production and variability of the Barents Sea ecosystem, p. 241–286. *In* K. Sherman and L.M. Alexander [ed.] Biomass yields and geography of large marine ecosystems. AAAS Selected Symp. 111.

Smith, R.L., W.E. Barber, M. Vallarino, J. Gillispie, and A. Ritchie 1995. Biology of *Gymnocanthus tricuspis*, the Arctic staghorn sculpin, from the northeastern Chukchi Sea. J. Am. Fish. Soc. (In press)

Strickland, R.M., and T. Sibley. 1984. Projected effects of CO_2- induced climate change on the Alaska walleye pollock (*Theragra chalcogramma*) fishery in the eastern Bering Sea and Gulf of Alaska. University of Washington, Seattle,Wash. FRI-UW-8411. 112 p.

Trenberth, K.E. 1990. Recent observed interdecadal climate changes in the northern hemisphere. Bull. Am. Meteorol. Soc. **71**: 988–993.

Walters, V. 1955. Fishes of Arctic America and Siberia. Bull. Am. Mus. Nat. History 106(5): 257–371.

Walters, G.E. 1983. An atlas of demersal fish and invertebrate community structure in the Eastern Bering Sea: Part 2, 1971–77. U.S. Department of Commerce, Washington, D.C. NOAA Tech. Memo. NMFS F/NWC-40. 152 p.

Wespestad, V.G., and J.J. Traynor. 1988. Walleye pollock, p. 11–32. *In* R.G. Bakkala [ed.] Condition of groundfish resources of the eastern Bering Sea and Aleutian Islands region in 1987. U.S. Department of Commerce, Washington, D.C. NOAA Tech. Memo. NMFS F/NWC-139.

Wolotira, R.J. Jr., T.M. Sample, and M. Morin, Jr. 1977. Demersal fish and shellfish resources of Norton Sound, the southeastern Chukchi Sea, and adjacent waters in the baseline year 1976. NAFC Processed Report, U.S. Department of Commerce, NOAA-NMFS, Seattle, Wash. 292 p.

Wyllie Echeverria, T., and C.P. McRoy. 1992. Larval fish distribution, p. 195–197. *In* P.S. Nagel [ed.] Results of the joint U.S.–U.S.S.R. Bering and Chukchi Seas expedition (BEWRPAC), summer 1988. U.S. Fish and Wildlife Service, Washington, D.C.

Wyllie Echeverria, T., S. Wyllie Echeverria, and W.E. Barber. 1995. Young-of-the-year fishes in the northeastern Chukchi Sea. J. Am. Fish. Soc. (In press)

Factors limiting the distributions of Ontario's freshwater fishes: the role of climate and other variables, and the potential impacts of climate change

Charles K. Minns and James E. Moore

Minns, C.K., and J.E. Moore. 1995. Factors limiting the distributions of Ontario's freshwater fishes: the role of climate and other variables, and the potential impacts of climate change, p. 137–160. *In* R.J. Beamish [ed.] Climate change and northern fish populations. Can. Spec. Publ. Fish. Aquat. Sci. 121.

Abstract: According to the predictions of various global climate models, if atmospheric CO_2 doubles, Canada's freshwater resources will be directly, and adversely, affected by increasing temperature and changing patterns of precipitation, runoff, and terrestrial vegetation. Freshwater fish species have distributions and patterns of abundance that are constrained by regional factors such as climate, physiography, and their biogeographic history, as well as by local factors such as ecosystem dimensions and biotic interactions. A fish distribution data base of species presence–absence by tertiary watershed in Ontario was analyzed to assess the relative importance of constraining factors, both climatic and geographic. The variables examined included mean annual air temperature, distances from the Great Lakes and the sea, and elevations above the Great Lakes and the sea. Total species richness by tertiary watershed was strongly related to mean annual air temperature. The presence–absence patterns of 61 species were subjected to stepwise logistic regression analysis. In 33 species, models including a temperature effect were obtained. Using a geographic information system, the models were used to predict the potential spatial extent and degree of impacts on freshwater fish species pools.

Résumé : Selon les prévisions obtenues au moyen de divers modèles climatiques globaux, si la teneur en CO_2 de l'atmosphère double, les hausses de température et la modification des profils de précipitations, d'écoulement des eaux et de croissance de la végétation terrestre auront une influence directe et néfaste sur les ressources en eau douce du Canada. Les espèces de poissons d'eau douce ont des répartitions et des courbes d'abondance limitées par des facteurs régionaux comme le climat, la physiographie et l'histoire biogéographique, ainsi que par des facteurs locaux comme les dimensions de l'écosystème et les interactions biotiques. Dans le but d'analyser l'importance relative des facteurs contraignants tant climatiques que géographiques, on a analysé une base de données sur la répartition du poisson en fonction de la présence ou de l'absence d'une espèce donnée par bassin hydrographique tertiaire en Ontario. Les variables examinées comprenaient la température annuelle moyenne de l'air, les distances depuis les Grands Lacs et la mer, ainsi que les altitudes par rapport aux Grands Lacs et à la mer. La richesse totale en espèces par bassin tertiaire était fortement liée à la température moyenne annuelle de l'air. La structure présence-absence de 61 espèces a fait l'objet d'une analyse de régression séquentielle logistique. Chez 33 espèces, on a obtenu des modèles, y compris un effet de la température. Grâce à un système d'information géographique, les modèles ont servi à prédire l'ampleur et le niveau potentiels sur le plan spatial des effets sur les bassins d'espèces d'eau douce.

Introduction

In 1990, the Intergovernmental Panel on Climate Change (IPCC 1990) offered its consensus judgements that (1) the Earth's mean surface air temperature has increased by 0.3–0.6°C since the end of the 19th century; (2) temperature is likely to increase a further 3°C before the end of the 21st century, given a business as usual scenario for emissions of greenhouse gases; and (3) the Earth's ecosystems will be greatly altered by the extent and rapidity of the changes. Alternative, more optimistic emission scenarios, as yet largely unimplemented, will only lower the rate of increase and not reverse the trend. Climate change is more than just warming. The suite of changes in precipitation, runoff, wind, soil moisture, sea level, snow cover, etc., as well as temperature, will induce substantial changes in all aspects of life on Earth.

C.K. Minns and J.E. Moore. Great Lakes Laboratory for Fisheries and Aquatic Sciences, Department of Fisheries and Oceans, Bayfield Institute, 867 Lakeshore Road, P.O. Box 5050, Burlington, Ontario L7R 4A5, Canada.

These dire predictions of our future are largely based on the output of global climate models (GCMs) and are unfortunately clouded by uncertainty in the structures and parameters of the many GCMs now operating and in the historical records of climate in the recent past. However, the size, rapidity, and spatial scope of the climate change predicted are too great to be ignored.

In this context, the custodians of renewable resources such as air, soils, forests, waters, and fisheries have two compelling reasons to assess the potential impacts of climate change. The first reason is to provide further evidence to strengthen support for tighter emission controls to limit, or even reverse, climate change. The second reason, given limited success with the first, is to prepare for the new array of management issues arising in altered ecosystems. As human beings continue to degrade and modify the world's ecosystems, we should assume that induced climate change will also continue. The impacts of that climate change on fresh waters and their fish assemblages are of particular concern to humans because we depend on fresh waters for our survival.

Canadians are the custodians of a disproportionately large share of the world's accessible fresh waters in lakes and rivers. Having a large share brings Canada added responsibility and an opportunity to show leadership in the assessment of the potential impacts and the formulation of sustainable management actions. A comparison of world, North American, and Canadian resources shows the extent of Canada's share (Table 1) and shows that a country with 0.5% of world population controls 7% of runoff, 43% of significant Ramsar wetlands, 50% of lake and river surface area, and 15% of the surface volume. Canada's per capita renewable water supply of 109 000 $m^3 \cdot ayr^{-1}$ is 14 times the world average.

Of a global estimate of 8411 species of freshwater fish, 621 (7.4%) are reported to be found in North America (Nelson 1984). In 1971, Scott and Crossman (1973) estimated there were 177 (2.1% of the total) freshwater species in Canada. In Canada, native, commercial, and recreational freshwater fisheries yielded 9.30, 54.46, and 115.00×10^3 t, respectively, in 1985 (Pearse 1988). The market value of the commercial fishery was $140 million while sportfishing expenditures were estimated at $2487 million. Besides providing the bases for significant commercial, native, and recreational fisheries, all these fish species are the "miner's canaries" signalling the health and safety of Canada's freshwater ecosystems. In assessing the status of Canada's freshwater fisheries, first Beamish et al. (1986) and then Pearse (1988) indicated that the resources were generally in decline because of habitat loss and/or overexploitation despite the fact that the Fisheries Act, first enacted in 1857, provides the federal government with a powerful legislative basis for protecting fish and fish habitat. Pearse (1988) quoted Ken Loftus, a past director of fisheries in Ontario, describing the deteriorating condition of freshwater fishery resources in Canada as a "creeping crisis." Gradual incremental changes to the climate will exacerbate this crisis.

Freshwater fish are at greater risk than marine ones. Freshwater fish are more constrained in responding to climate change and are more exposed to multiple human stresses. While marine fish distributions are primarily determined by temperature regimes, freshwater ones are determined by temperature, salinity, and topographic constraints (Nelson 1984). While marine fish largely adapt to climate changes through rapid redistribution, freshwater fish are slower to adapt, increasing the likelihood of ecosystem disruptions. In

Table 1. World, North American, and Canadian land, human population, and freshwater resources.

Measures	World	North America		Canada	
		Absolute	%	Absolute	%
Land area (km3, x103)	130691	21317	16.3	9161	7.0
Human population (1990) ($\times 10^6$)	5300	426	8.0	27	0.5
Water consumption ($km^3 \cdot yr^{-1}$)	4130	-	-	-	-
Runoff					
Total ($km^3 \cdot yr^{-1}$)	40673	6945	17.1	2901	7.1
Per capita (10^3 $m^3 \cdot yr^{-1}, \times 10^3$)	77	16	211.4	109	1422.2
Wetlands					
Ramsar (ha, $\times 10^6$)	30	15	50.1	13	42.8
Total (ha, $\times 10^6$)	-	-	127.0	-	-
Freshwater lakes & rivers					
Area (km^2)	1236000 ?	1060671	70.0 ?	755180	50.0 ?
Volume (km^3)	125000	37750 ?	30.2 ?	18750 ?	15.0
Freshwater fish species	8411	621	7.4	177	2.1
Freshwater fish harvest					
Commercial ($t \cdot yr^{-1}, \times 10^3$)	12641	310	2.5	55	0.4
Recreation		-	-	115	-
Native		-	-	9	-

Sources: UNEP 1991; Leeden 1990; Law 1992, Geraghty et al. 1973; Healey and Wallace 1987; Scott and Crossman 1973; Nelson 1984; Pearse 1988; Beamish et al. 1986; Darnay 1992.

marine ecosystems, the greater disruptions may be in the patterns of human exploitation of marine resources. We should not confuse ecosystem and human concerns.

Human impacts on fish habitats have usually occurred earlier in freshwater ecosystems and the size of the impact is often greater on a local, human scale. These impacts include habitat destruction and modification, overfishing, exotic introductions, pollution of all kinds (eutrophication, acidification, contamination, thermal pollution), and now climate change. Severe eutrophication of lakes preceded regional marine effects (e.g., the North Sea) throughout the world. Human dependence on fresh water for drinking, irrigation, and industrial uses has ensured that settlement and development have encroached on and degraded freshwater ecosystems. Fresh waters will be among the earliest ecosystems to show signs of climate change, and changes in their fish assemblages will provide the strongest signals.

Tonn (1990) has suggested a simple cascading model of freshwater fish species pools and spatial- and temporal-scale filters linking global and continental species pools to the assemblages found in particular lakes and rivers. This model provides a suitable framework for assessing the potential impact of natural and human-made changes on a series of spatial scales. A global pool of species passes through a continental filter to yield a continental pool. The continental filter includes processes operating on global spatial and geological time scales, such as plate tectonics and glaciation. The continental pool then passes through a regional filter where major soil, landscape, and climatic differences, and dispersal barriers, such as watershed boundaries, are important features. Then, the regional pool is filtered again to yield a lake-type pool, which is in turn filtered to give the final local pool found in a single lake or drainage basin. The lake-type filter has features reflecting the distribution, abundance, and variability of limnological resources, while the local filter has features associated with lake morphometry and biotic interactions. Tonn (1990) reviewed much of the research on the nature of these filters and their impact on North American freshwater fish species pools. Climate-related changes can be expected to be most apparent in altered regional patterns of fish species distributions.

Earlier, Coutant (1981) identified three main areas of concern regarding the impact of climate change on fresh waters: (1) diminished water supply and related water quality aspects; (2) changed balance of photosynthesis and decomposition; and (3) changes to species distributions. The first two are more likely to be detected at the lake-type and local scales in Tonn's (1990) scheme, while the changed distributions will be most evident regionally. Studies of the hydrological implications of climate change are still in the rudimentary stages (Gleick 1989; Chang et al. 1992), inhibiting efforts to assess the fisheries impacts. Schindler et al. (1990) have shown evidence of a changed thermal regime in a small lake in northwestern Ontario over 20 yr. They found phytoplankton biomass and diversity increased but no consistent pattern in primary production measurements. There is little other evidence for overall changes in ecosystem production processes. At present, temperature effects on the distribution of freshwater fish are some of the most accessible for projecting the potential climate change impacts.

The analysis of regional filters has received relatively less attention than others. McAllister et al. (1986) presented analyses of North American freshwater fish distribution on a grid, showing that species richness was related to a number of climatic indices, particularly warmth and aridity. Other factors, such as wind, number of fog-free days, and number of days with fog and/or snow, were also significant but appeared to be surrogates for proximity to the sea. They also showed that range size was related to fish size. Legendre and Legendre (1984) performed a regional analysis of the postglacial spread of freshwater fishes in Quebec. Their analysis of regional occurrence patterns did not include any explicit quantitative consideration of the climatic, topographic, geological, and salinity factors, although they drew on these factors to interpret the results of spatial clustering. Minns (1989) analyzed the patterns of freshwater fish species in Ontario's inland lakes, showing the regional clustering of species occurrence patterns and developing a regression model of richness with a mixture of regional (tertiary watershed species richness, elevation) and lake-type or local (area, depth, pH) factors. In the Great Lakes region, Meisner et al. (1987) have provided an overview assessment of the potential impact of climate change, noting many freshwater fish species found in Ontario whose northern range limits are thermally bound and species whose ranges might increase or decrease given climate warming. Mandrak (1989) identified the potential for 27 of 58 species to invade the Great Lakes either from the Mississippi or the Atlantic coastal basins.

The object of this study is to quantitatively assess the climatic and topographic factors determining the regional distributions of freshwater fish species in Ontario and to develop predictive models. Where temperature is a significant determinant, the resultant models will be used to assess the potential regional impact of climate change.

Materials and methods

Ontario fish species distribution

The Ontario fish species distribution patterns analyzed here were derived from an exhaustive compilation of distribution records assembled by the Fisheries Branch of the Ontario Ministry of Natural Resources (OMNR) (G.E. Gale, Fisheries Branch, OMNR, 99 Wellesley Street West, Toronto, Ont. M7A 1W3, personal communication). This compilation brought together fish observation records from the ministry's lake inventory, stream inventory, stream assessment, and areas of concern programs, and from the archives of the Royal Ontario Museum, Toronto, Ont., and the National Museum of Natural History, Ottawa, Ont.

The Ontario fish distribution data base contained 117 725 records, consisting of site name, latitude, longitude, quaternary watershed code, capture date, OMNR fish species code, and a data source code. The data base was summarized by tertiary watershed to give a binary presence–absence matrix. All records for hybrids and occurrences above the species level in the data base were ignored. There were over 10 000 unique site–date combinations recording one or more fish species occurrences.

The division of areas into a hierarchy of primary through quaternary watershed units is documented in the Reference Index Hydrometric Map Supplement (Water Survey of

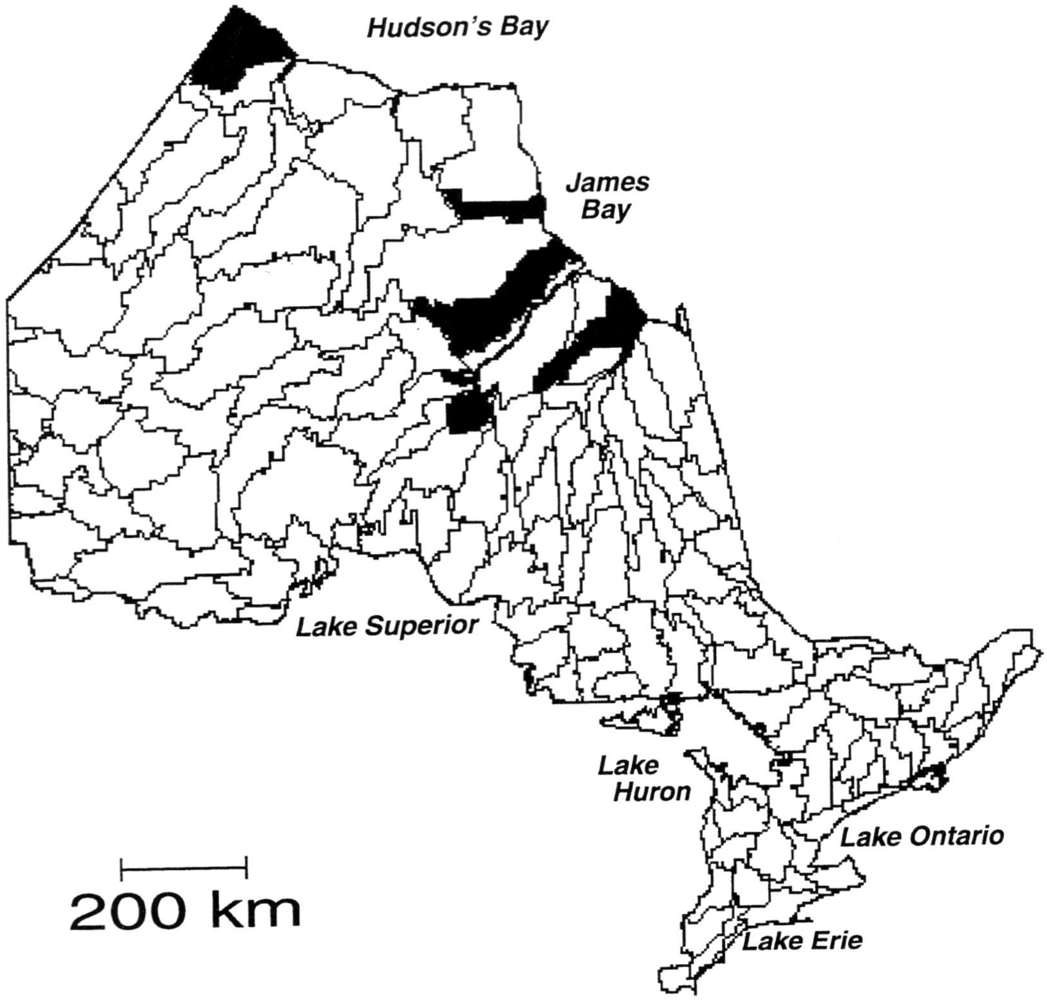

Fig. 1. A map of Ontario showing the tertiary watershed boundaries. The black area indicate watersheds in which no fish species occurrences are recorded in the OMNR database.

Canada 1977) and in the Ontario compilation of lake counts and measures by watershed (Cox 1978). The watershed unit system is based on the delineation of major river systems at the higher levels and on the division of drainage areas into similar-size units at lower levels. There are 137 whole or partial tertiary watershed units with fish distribution data within Ontario (Fig. 1).

Climate and topographical variables

Measures by tertiary watershed of mean annual, January, and July air temperatures for 1951–80, elevations above sea level for all watersheds and above the nearest Great Lake for all St. Lawrence drainage watersheds, and distances from the sea and the Great Lakes were obtained using a national-scale data base assembled in a geographic information system (Intera-Tydac's SPANS GIS).

Mean air temperatures and increases predicted with climate change

To obtain mean annual, January, and July temperatures, we used a summary data set for 1495 long-term weather stations across Canada, assembled and provided by the Atmospheric Environment Service (AES) (Climate Data Services Branch, AES, Environment Canada, Downsview, Ont.). Using the linear contouring feature of our GIS, we generated a map with 1°C contour intervals. Then we overlaid the temperature maps with a tertiary watershed map and computed area mean temperatures for each watershed. Because areas between contours were assigned to a 1°C class, the watershed areal means were computed using class values and then converted to temperature by assigning the value of the midpoint of the interval to integer classes. This effectively limits the estimate of watershed mean temperatures to a rectangular confidence interval of ±0.5°C. Over the broad range of temperatures encountered regionally this uncertainty is unlikely to compromise the analysis of temperature-related distribution patterns by tertiary watershed unit.

For climate change scenarios, we selected the predictions of the second-generation, high-resolution global circulation model developed at the Canadian Climate Centre (CCC-GCM) (AES 1990). According to IPCC (1990), the CCC-GCM predicts temperature and precipitation increases in the midrange of the many GCM predictions available. The second-generation model uses a 3.75° by 3.75° grid, has more detailed treatment of diurnal and annual cycles, and provides improved representation of oceanic transports and ice, and of terrestrial hydrological processes. The CCC-GCM predicts a global surface air temperature increase of 3.5°C, 3.8% increases in precipitation and evaporation, and a 6.6% loss of soil moisture. We used the differences between averaged outputs of $1\times CO_2$ and $2\times CO_2$ simulation runs to derive a predicted annual mean air temperature increase by grid polygon. To obtain mean temperatures after climate change by tertiary watershed unit, we assigned the AES weather stations to their GCM polygons by latitude and longitude, added the predicted temperature increase to the observed 1951–80 mean, and then repeated the contouring and area-averaging steps performed for the current climate.

Topographic variables

For the distance measures, we used the GIS to compute a geometric centroid position for each tertiary watershed as a reference point. We derived two sets of line vectors describing the marine and Canadian coastal shorelines from the GIS data base. We then used a simple computer program, previously designed to compute fetch values, to find the minimum distances between the centroids of each tertiary watershed and the line vectors, giving distances from the sea (DSEA) and the Great Lakes (DGL). We defined an additional distance variable (DWAT) as the minimum of the other two distances. To derive the elevations above sea level (ELSEA), we used an overlay of tertiary watersheds on an elevation map prepared from data for North America provided on a 5 grid by the National Geophysical Data Center (National Geophysical Data Center, NOAA, 325 Broadway, Boulder, Col., U.S.A) to compute mean elevations. To derive the elevations above the nearest Great Lake for St. Lawrence watersheds (ELWAT), we obtained a long-term mean elevation above sea level for each Great Lake (Table 2), assigned groups of tertiary watersheds to appropriate lakes, and subtracted the lake elevations from the elevations above sea level.

A listing of values obtained by tertiary watershed is given in Appendix Table A.

Statistical analyses and procedures

GLM logistic regression

Nicholls (1989) outlined a general strategy for modelling species distributions on a regional scale using generalized linear models (GLM), particularly stepwise logistic regression. The species presence–absence records (Y_i = 1 or 0) are modelled against a set of independent variables (X_1, X_2, ...) using the following equations for the *i*th species:

(1) $Y_i = \exp(U_i) / [1 + \exp(U_i)]$

(2) $U_i = B_0 + B_1X_{1i} + B_2X_{2i} + ...$

The fitting process is governed by a maximum likelihood (L) equation measuring deviance reduction (DR) as follows:

(3) $L = -2\Sigma[Y_i\log_e(Y_{est}) + (1 - Y_i)\log_e(1 - Y_{est})]$

which applies for ungrouped binary data (McCullagh and Nelder 1983). Austin et al. (1984) and Braak and Looman (1986) have described the use of logistic regression in direct gradient analysis and provided an interpretation of the coefficients obtained. In our study, we were particularly interested in temperature-related distribution patterns. There are four non-random patterns of interest (Fig. 2): pattern A shows a northern low-temperature distribution boundary; B, a southern high-temperature limit; C, a distribution with both limits; and for completeness, D, an unlikely distribution with an interval break in the distribution along the temperature gradient. Patterns A and B are obtained if temperature enters eq. 2 as a linear term. Patterns C and D are obtained if, in addition, temperature also enters eq. 2 as a squared, or quadratic, term (Braak and Looman 1986). For this reason, both linear and quadratic forms of temperature were assessed in the logistic modelling.

The fitting process proceeds in a stepwise manner: first B_0 is fitted (representing the mean proportion (p) of presences in the data set) and then B_1, B_2, etc. Each successive deviance reduction is approximately distributed as a χ^2 with one degree of freedom (df). The DR for a two-variable model is defined as:

(4) $DR_{X^1X^2}$ = Minimum (L_{X^1}, L_{X^2}) – $L_{X^1X^2}$

The maximum deviance is a function of the number (N) of binary observations ($-2N\log_e(0.5)$) and is obtained by assuming an equal random probability of occurrence (0.5). In this study,

Table 2. Long-term annual mean water elevation values, metres above the International Great Lakes Datum (IGLD), used to adjust the elevations above sea level (ELSEA) for tertiary watersheds associated with the Great Lakes gauging stations to produce elevations above major water bodies (ELWAT).

Tertiary watershed site	Station location (Ontario)	Elevation above IGLD (m)	Period of record	Secondary watersheds adjusted
2HM	Lake Ontario, Kingston	74.64	1963–87	2H, 2J, 2K, 2L, 2M
2HA	Lake Erie, Port Colborne	173.93	1913–87	2G
2GH	Lake St.Clair, Belle River	174.93	1961–87	
2FE	Lake Huron, Goderich	176.27	1910–87	2C, 2D, 2E, 2F
2BD	Lake Superior, Michipicoten	183.11	1922–87	2A, 2B

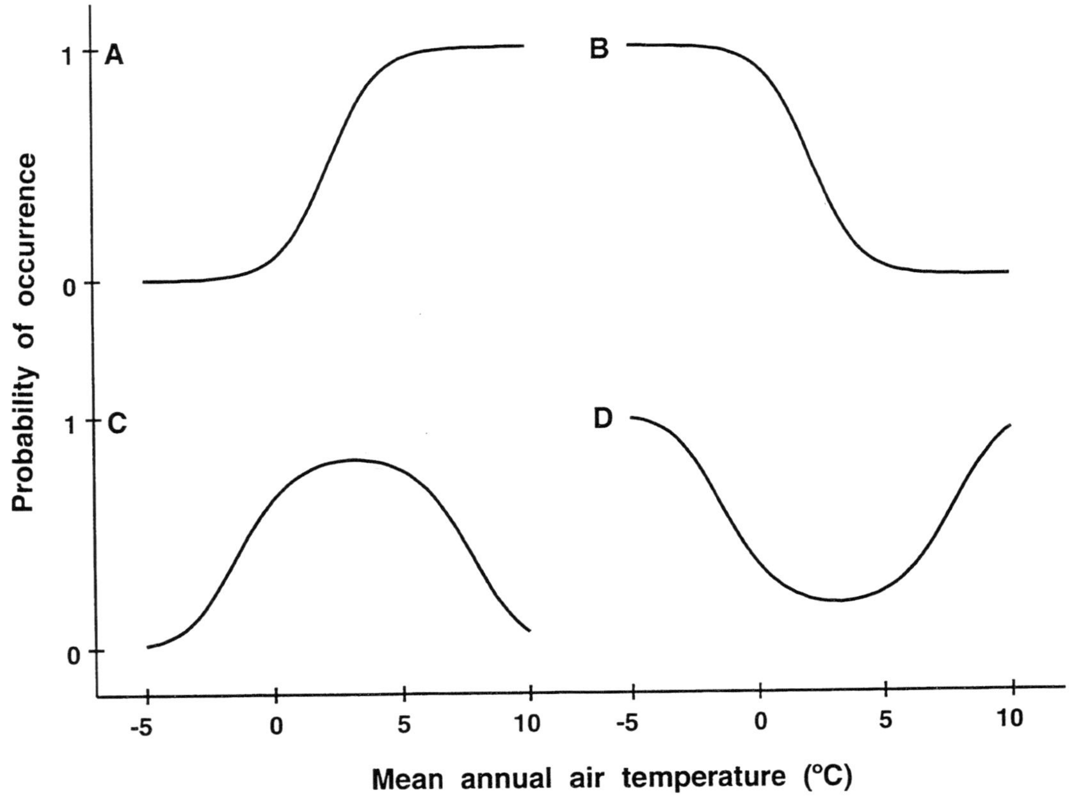

Fig. 2. Potential, structured patterns of occurrence in relation to temperature obtainable using logistic regression modelling: A) low temperature boundary, B) high temperature boundary, C) low and high temperature boundaries, and D) an intermediate break in the distribution.

given that presence–absence data for 137 tertiary watersheds were used, the maximum deviance was 189.92. When the intercept B_0 is fitted, the total deviance is equal to $-2[pN\log_e(p) + (1-p)N\log_e(1-p)]$; $p = \exp(B_0)/(1+\exp(B_0))$. The fit of succeeding variables added is measured against this reference likelihood, L_p, established by fitting B_0.

A posteriori K-testing

Titus et al. (1984) described the use of Cohen's *K*-statistic to test the results of discriminant analyses and subsequent predicted classifications. The *K*-statistic provides a chance-corrected assessment of the agreement between observed and predicted group memberships. Kappa and its standard error (SE) are calculated as:

(5) $K = (p_{Obs} - p_{Exp})/(1 - p_{Exp})$

(6) $SE_K = \sqrt{[p_{Exp}/(N(1 - p_{Exp}))]}$

where p_{Obs} is the proportion of N samples where observed = predicted, and p_{Exp} is the proportion expected by chance alone.

A test of significance (Z = K/SE) is approximately a random normal deviate. Here we used *K*-testing to screen the logistic models *a posteriori* for predictive strength and rejected those with probabilities less than 0.01. In applying the logistic models, we consistently used the predicted 0.5 occurrence level as the boundary between presence and absence.

Data handling and statistical procedures

Organization, manipulations, and simple statistical analyses of data were performed using RS/1 (BBN Software Products Corp. 1986). The K-statistics and testing were performed using a custom-written RS/1 procedure. More complex statistical analyses, such as logistic regression and principle components analysis, were performed using SYSTAT (Wilkinson 1990). Logistic regressions were performed using the NONLINEAR module of SYSTAT with the following MODEL and LOSS equations:

(7) MODEL $Y_{Obs} = \exp(B_0 + B_1X_1 + B_2X_2 + \ldots)/(1 + \exp(B_0 + B_1X_1 + B_2X_2 \ldots)$

(8) $\text{LOSS} = -2[Y_{Obs}\log(Y_{Est}) + (1 - Y_{Obs})\log(1 - Y_{Est})]$

The significance of statistical results was judged routinely with a probability of 0.01. This more stringent criterion was chosen to compensate in part for the large number of tests being performed. In most instances, results were significant at $P = 0.001$ or less. Higher levels of significance are required if the resulting models are to be used for predictive purposes.

Table 3. Summary statistics for nine independent variables for logistic regression modelling fish species' distribution patterns and regional fish species richness by tertiary watershed in Ontario.

Acronym	Description	Mean	Median	SD	Max	Min
TANN	Mean annual air temperature, 1951–80 (°C)	2.3	1.9	3.36	8.9	−4.5
TJAN	Mean January air temperature, 1951–80 (°C)	−15.7	−16.7	5.70	−4.9	−25.5
TJUL	Mean July air temperature, 1951–80 (°C)	18.0	18.0	1.77	22.0	13.8
$T2CO_2$	Mean annual air temperature at $2{\times}CO_2$ (predicted, CCC-GCM) (°C)	7.0	6.6	3.47	14.4	0.4
DGL	Distance to the Great Lakes (km)	226.9	170.7	203.58	738.1	7.8
DSEA	Distance to the sea (km)	491.5	499.0	237.02	1000.4	21.4
DWAT	Minimum (DGL, DSEA) (km)	161.2	118.6	143.84	595.9	7.8
ELSEA	Elevation above sea level (m)	278.5	291.0	96.26	453.6	51.4
ELWAT	Elevation above water (m)	212.4	204.3	119.97	453.6	−3.6
NSPP	No. of fish species recorded	42.5	37.0	21.17	106.0	1.0

Results

Independent variables

Mean annual air temperatures 1951–80 (TANN) varied from −4.5 to 8.9°C with a mean of 2.3°C (Table 3). January temperatures (TJAN) ranged from −25 to −5°C while July temperatures (TJUL) ranged from 14 to 22°C. After adding the temperature increases predicted by the CCC-GCM, mean annual air temperatures (T2CO2) ranged from 0.4 to 14.4°C with increases ranging from 4.5 to 5.5°C. Mean January and July air temperatures were highly correlated with mean annual temperature, $r^2 = 0.967$ and 0.861 (significant at $P = 0.001$ on 136 df). Linear regression models yielded the following equations (coefficient ± SE):

(9) $TJAN = (-19.51 \pm 0.11) + (1.68 \pm 0.03)TANN$

(10) $TJUL = (16.86 \pm 0.07) + (0.49 \pm 0.02)TANN$

Using these equations, TANN values of −5 and 5°C correspond to TJAN values of −27.9 and −11.1°C, and to TJUL values of 14.4 and 19.3°C, respectively. Given the close agreement among temperature variables, only the variable mean annual air temperature (TANN) was used in subsequent analyses.

Distances from the Great Lakes shoreline (DGL) or the sea (DSEA) to the centroids of tertiary watersheds ranged from 8 to 740 and 21 to 1000 km, respectively, while the mininum of the distances (DWAT) ranged from 8 to 600 km (Table 3). Elevations above sea level (ELSEA) ranged from 51 to 454 m. The minimum elevation above water (ELWAT) was −4 m, a negative reflecting the inclusion of watersheds draining into the St. Lawrence below the level of Lake Ontario. Correlations among the topographic variables and with TANN and $TANN^2$ were variable (Table 4). TANN was significantly correlated with all variables but ELSEA. Only four of the correlations were nonsignificant ($P = 0.01$) (DSE vs. DWAT, DGL vs. ELSEA, DSEA vs. ELWAT, and TANN vs. ELSEA). Both DGL and ELWAT were negatively correlated with TANN, $TANN^2$, and DSEA. ELSEA was negatively correlated with $TANN^2$ and DGL. Correlations among the independent variables create the potential for confounding effects in model development. One possible alternative strategy of using principal components analysis to extract a series of uncorrelated factor scores was rejected as the interpretation of temperature-related effects would have become difficult.

Table 4. Pearson correlations among the climatic and topographic variables and with total freshwater fish species richness by tertiary watershed in Ontario N=137, $r \geq 0.218$ is significant at $P = 0.01$. See table 3 for definition of variables.

Variable	$TANN^2$	DSEA	DGL	DWAT	ELSEA	ELWAT	NSPP
TANN	0.802	0.718	−0.830	−0.513	0.002	−0.423	0.828
$TANN^2$		0.557	−0.514	−0.556	−0.399	−0.652	0.768
DSEA			−0.491	−0.013	0.338	−0.033	0.656
DGL				0.660	−0.146	0.359	−0.696
DWAT					0.371	0.717	−0.503
ELSEA						0.766	−0.050
ELWAT							−0.445

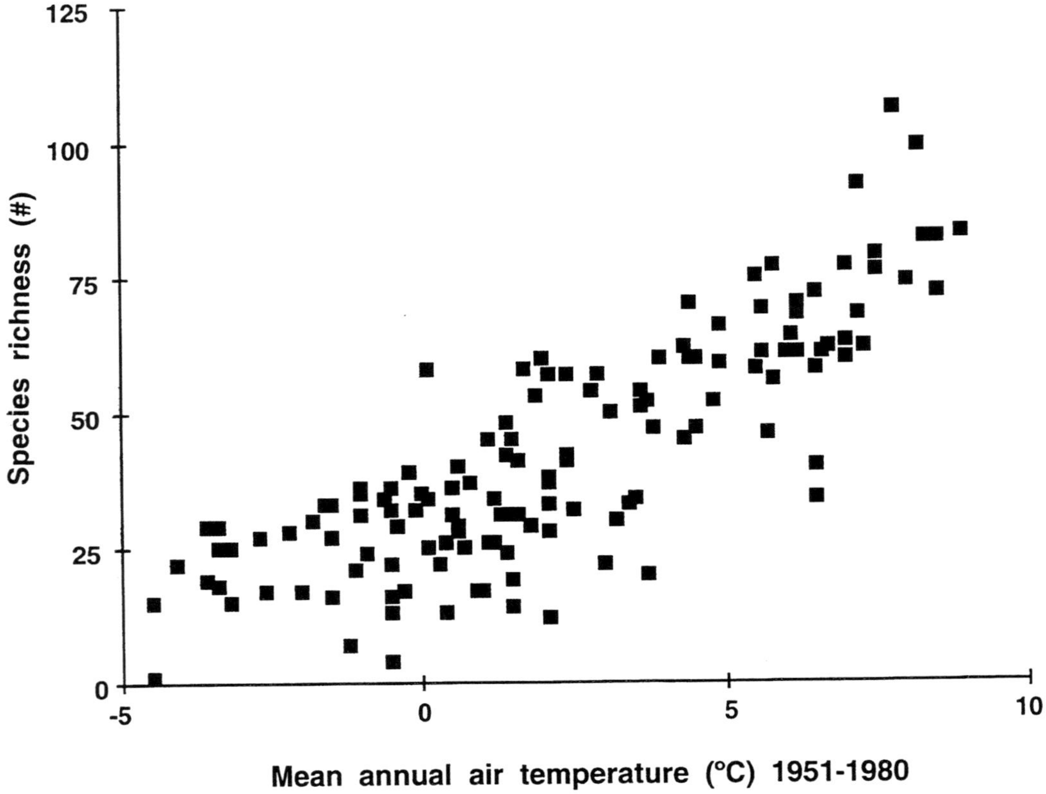

Fig. 3. Regional species richness versus mean annual air temperature (°C), 1951–1980, by tertiary watershed in Ontario.

Tertiary watershed species richness

In the 137 tertiary watershed units in Ontario, there were records of 150 separate species. This is greater than the 132 reported by Scott and Crossman (1973). The difference reflects the earlier incompleteness of the Ontario species records. Species richness (NSPP) by tertiary watershed varied from 1 to 106. NSPP was significantly correlated ($P = 0.01$) with several of the independent variables (Table 4), positively with TANN, TANN^2, and DSEA, and negatively with DGL, DWAT, and ELWAT. The greatest correlations were with TANN and TANN^2. A graph of the data indicated a slight upward curvilinearity in the relationship between species richness and TANN (Fig. 3). In a stepwise multiple regression, TANN and TANN^2 both entered significantly ($P = 0.001$), giving the following model with an $r^2 = 0.717$:

(11) $NSPP = (28.95 \pm 1.26) + (3.75 \pm 0.49)TANN + (0.31 \pm 0.08)TANN^2$

The next variable to enter the regression would have been DGL with a negative coefficient, but the fit was marginal, significant at $P = 0.05$ but not at $P = 0.01$. None of the other topographic variables was included in the regression model. This is evidence of the importance of temperature in determining overall regional species richness.

Species' presence–absence patterns

For the analysis of single species' presence–absence patterns, we initially selected 61 species with occurrences between 25 and 125 among 137 watersheds (Table 5). Species with higher or lower occurrences were excluded from the analysis logistic regression. Significant deviance reductions are difficult to obtain once the mean proportional frequency of occurrence has been used to estimate the intercept B_0. While this selection process retained a wide-ranging representation of the Ontario ichthyofauna, many familiar species with very extensive distributions in Ontario, such as yellow perch (*Perca flavescens*), white sucker (*Catostomus commersoni*), and northern pike (*Esox lucius*), are excluded. Subcontinental regions larger than Ontario must be examined to determine the distributional constraints on Ontario's ubiquitous species. According to Legendre and Legendre's (1984) listing, the majority of the 61 species selected were stenohaline. Euryhaline species included alewife (*Alosa pseudoharengus*), brook trout (*Salvelinus fontinalis*), rainbow trout (*Onchorhynchus mykiss*), brown trout (*Salmo trutta*), rainbow smelt (*Osmerus mordax*), American eel (*Anguilla rostrata*), and ninespine stickleback (*Pungitius pungitius*). Several species have been introduced or have invaded the region recently, including alewife, rainbow and brown trout, rainbow smelt, and carp (*Cyprinus carpio*), which is stenohaline. The ranges of species, such as smallmouth bass (*Micropterus dolomieu*), largemouth bass (*Micropterus salmoides*), and brook trout, have been greatly altered by deliberate stocking, while many of the smaller species have likely had their distributions extended by unconscious transfers in bait buckets.

To begin, we computed the intercepts B_0 and the associated deviance reduction for all 61 species and fitted each of the independent variables separately to a logistic regression (Table 6), except for TANN^2, which was fitted after both the

Table 5. List of OMNR fish species codes, common and Latin names, and the frequency of occurrence: 137 Ontario tertiary watersheds for the 61 fishes selected for further distribution analysis out of 150 present.

OMNR code	Common name	Latin name	Frequency	OMNR code	Common name	Latin name	Frequency
S031	Lake sturgeon	*Acipenser fulvescens*	48	S209	Fathead minnow	*Pimephales promelas*	117
S041	Longnose gar	*Lepisosteus osseus*	25	S210	Blacknose dace	*Rhinichthys atratulus*	56
S061	Alewife	*Alosa pseudoharengus*	27	S211	Longnose dace	*Rhinichthys cataractae*	100
S076	Rainbow trout	*Onchorhynchus mykiss*	73	S212	Creek chub	*Semotilus atromaculatus*	80
S078	Brown trout	*Salmo trutta*	32	S213	Fallfish	*Semotilus corporalis*	35
S080	Brook trout	*Salvelinus fontinalis*	99	S214	Pearl dace	*Semotilus margarita*	111
S081	Lake trout	*Salvelinus namaycush*	83	S232	Yellow bullhead	*Ictalurus natalis*	32
S091	Lake whitefish	*Coregonus clupeaformis*	113	S233	Brown bullhead	*Ictalurus nebulosus*	67
S093	Lake herring	*Coregonus artedii*	114	S234	Channel catfish	*Ictalurus punctatus*	32
S102	Round whitefish	*Prosopium cylindraceum*	30	S236	Tadpole madtom	*Noturus gyrinus*	25
S121	Rainbow smelt	*Osmerus mordax*	48	S251	American eel	*Anguilla rostrata*	26
S132	Muskellunge	*Esox masquinongy*	43	S261	Banded killifish	*Fundulus diaphanus*	37
S141	Central mudminnow	*Umbra limi*	62	S271	Burbot	*Lota lota*	107
S162	Longnose sucker	*Catostomus catostomus*	97	S281	Brook stickleback	*Culaea inconstans*	125
S168	Silver redhorse	*Moxostoma anisurum*	52	S283	Ninespine stickleback	*Pungitius pungitius*	93
S171	Shorthead redhorse	*Moxostoma macrolepidotum*	87	S291	Trout-perch	*Percopsis omiscomaycus*	112
S182	Northern redbelly dace	*Phoxinus eos*	100	S311	Rock bass	*Ambloplites rupestris*	79
S183	Finescale dace	*Phoxinus neogaeus*	101	S313	Pumpkinseed	*Lepomis gibbosus*	66
S185	Lake chub	*Couesius plumbeus*	94	S314	Bluegill	*Lepomis macrochirus*	51
S186	Carp	*Cyprinus carpio*	43	S316	Smallmouth bass	*Micropterus dolomieu*	77
S189	Brassy minnow	*Hybognathus hankinsoni*	47	S317	Largemouth bass	*Micropterus salmoides*	53
S194	Golden shiner	*Notemigonus crysoleucas*	93	S319	Black crappie	*Pomoxis nigromaculatus*	36
S196	Emerald shiner	*Notropis atherinoides*	97	S332	Sauger	*Stizostedion canadense*	49
S198	Common shiner	*Notropis cornutus*	97	S334	Walleye	*Stizostedion vitreum vitreum*	125
S199	Blackchin shiner	*Notropis heterodon*	56	S338	Iowa darter	*Etheostoma exile*	122
S200	Blacknose shiner	*Notropis heterolepis*	117	S341	Johnny darter	*Etheostoma nigrum*	121
S202	Rosyface shiner	*Notropis rubellus*	34	S342	Logperch	*Percina caprodes*	110
S203	Spotfin shiner	*Notropis spilopterus*	29	S381	Mottled sculpin	*Cottus bairdi*	113
S204	Sand shiner	*Notropis stramineus*	28	S382	Slimy sculpin	*Cottus cognatus*	96
S206	Mimic shiner	*Notropis volucellus*	80	S383	Spoonhead sculpin	*Cottus ricei*	33
S208	Bluntnose minnow	*Pimephales notatus*	81				

intercept and TANN. Then, we computed the next stepwise addition to the logistic model by fitting the intercept, TANN, and separately each of the five topographical variables and assessing the deviance reductions (Table 7). The species were divided into four groups depending on the significance and magnitude of the deviance reductions. Group 1 contained 17 species and was separated from the other three groups because the deviance reduction for TANN was not significant at $P = 0.01$. Apart from those species in this group with significant quadratic temperature fits, analyses of group-1 species were not pursued. While there may be other significant models not related to temperature, further analysis was not warranted given the climate focus of the study.

A sampling of the species' presence–absence data in relation to TANN shows the variety of patterns observed (Fig. 4). Two species were selected from each of the four groups identified to show the patterns for the highest and lowest deviance reductions obtained with TANN, except for lake trout (*Salvelinus namaycush*) in group 1, which had the highest quadratic temperature DR. The temperature-related separation of presences and absences into a type A (Fig. 2) pattern is clear for common shiner (*Notropis cornutus*), central mudminnow (*Umbra limi*), and smallmouth bass. The pattern for ninespine stickleback, though fuzzy, represents a high-temperature southern boundary, type B (Fig. 2). Lake trout is the only one of the species with significant quadratic temperature fits that recognizably shows a type C (Fig. 2) pattern.

In group 2, TANN was significant but one or more of the other variables had a greater DR. Group 2 contained 15 species. Within this group, a subset of seven species was identified where both TANN and one of the topographical variables entered the stepwise model at $P = 0.01$. In two species, common shiner and Iowa darter (*Etheostoma exile*), TANN entered the model significantly after the most significant topographical variable, DGL and ELSEA, respectively. For five species (alewife, brown trout, lake whitefish (*Coregonus clupeaformis*), blacknose shiner (*Notropis heterolepis*), and mimic shiner (*Notropis volucellus*)), the topographical variable that singly produced the greatest DR did not produce the greatest DR when taken in combination with TANN.

Table 6. Deviance reductions obtained by stepwise fitting the mean and then one of the temperature (TANN, TANN + TANN^2), distance (DGL, DSEA, DWAT), and elevation (ELSEA, ELWAT) variables in logistic regressions for 61 fish species organized in four groups. Deviance reduction is χ^2 with 1 df, values greater than 6.64 are significant at $P = 0.01$.

OMNR Code	Mean	TANN	TANN^2	DGL	DSEA	DWAT	ELSEA	ELWAT
Mean Annual Air Temperature (TANN) not significant								
S031	12.46	0.10	0.04	0.19	0.16	0.08	0.09	0.03
S081	6.19	3.38	26.35	9.60	5.41	0.00	20.93	5.16
S093	65.93	4.42	10.65	0.04	3.00	1.05	9.25	10.49
S102	45.90	3.85	2.83	10.33	3.26	11.95	2.83	1.96
S162	24.45	3.53	0.30	0.36	4.35	3.37	2.68	0.97
S171	10.12	3.99	0.65	6.74	5.19	0.03	5.22	0.91
S185	19.45	2.07	0.95	1.54	5.18	5.94	2.89	0.24
S211	30.09	1.88	0.85	4.94	3.89	7.88	0.87	1.01
S214	56.78	4.28	3.99	7.90	2.49	2.76	15.37	1.97
S271	45.90	0.88	9.87	0.99	0.09	1.72	15.83	10.35
S291	59.74	0.07	5.16	2.60	0.20	0.09	5.75	3.16
S332	11.26	0.06	0.06	0.30	5.77	6.53	3.13	3.24
S334	108.56	0.99	7.54	5.80	1.22	1.29	13.16	6.20
S341	91.15	4.04	0.00	0.61	9.24	3.33	3.93	2.69
S381	62.78	0.47	0.66	2.78	0.50	0.04	0.36	0.59
S382	22.72	1.50	0.09	1.91	5.35	0.39	3.49	0.05
S383	38.65	0.15	0.23	0.00	0.18	0.00	2.97	0.57
TANN significant but other variables have greater χ^2								
S061	53.93	46.18	0.39	66.49	21.10	59.15	13.01	55.20
S076	0.59	72.36	11.18	77.78	16.94	49.15	1.22	15.22
S078	40.99	52.84	4.08	63.32	28.83	52.32	0.38	23.37
S080	28.14	24.96	1.01	43.60	0.02	56.72	0.00	11.66
S091	62.78	13.88	7.38	2.19	10.69	5.28	8.05	15.63
S121	12.46	49.75	7.93	76.10	21.13	53.12	1.67	37.47
S168	8.03	21.42	0.06	13.36	24.69	0.90	0.59	0.59
S182	30.09	43.77	5.54	59.52	9.04	29.02	5.46	1.87
S183	32.12	9.38	11.30	17.50	0.47	12.06	5.11	0.01
S196	24.45	7.26	1.04	6.58	18.74	0.01	1.42	0.28
S198	24.45	76.88	0.02	82.51	17.54	27.44	6.61	1.49
S200	76.03	33.05	2.22	29.40	34.00	0.06	21.81	1.53
S206	3.88	27.86	0.21	18.26	33.73	0.05	6.47	0.28
S281	108.56	12.02	2.19	19.31	1.99	11.28	2.31	0.58
S338	95.27	18.39	2.62	15.42	15.93	0.50	20.78	4.12
TANN significant and greatest χ^2, but other variables enter significantly after								
S041	59.74	48.40	1.53	39.85	15.16	31.44	13.26	45.07
S141	1.24	124.16	0.55	63.13	80.88	30.07	0.91	26.26
S204	51.16	59.16	0.15	48.09	22.71	38.54	11.70	49.62
S208	4.59	86.39	0.64	53.81	64.14	14.12	3.28	6.31
S213	34.22	27.17	3.38	17.84	1.19	12.82	3.63	5.69
S236	59.74	53.65	0.34	24.11	41.77	14.82	5.59	19.36
S251	56.78	51.71	3.84	34.44	12.79	25.00	11.71	24.74
S283	17.92	20.83	5.03	4.18	10.06	2.79	7.59	13.06
S313	0.18	112.10	0.04	43.45	77.14	12.89	0.01	12.69

Table 6 (*concluded*)

OMNR Code	Mean	TANN	$TANN^2$	DGL	DSEA	DWAT	ELSEA	ELWAT
TANN significant and greatest χ^2, and no other variables enter significantly								
S132	19.45	38.58	10.38	17.25	24.65	3.62	0.00	3.01
S186	19.45	82.29	0.10	69.66	48.70	50.69	3.24	37.02
S189	13.73	84.68	25.23	64.11	27.20	41.97	0.67	22.24
S194	17.92	66.30	1.45	46.52	18.79	13.66	5.44	1.40
S199	4.59	57.63	10.44	37.97	27.73	13.61	0.02	6.41
S202	36.39	69.83	6.85	44.16	24.57	30.54	4.49	30.62
S203	48.49	71.74	13.03	31.74	35.74	29.80	15.35	43.49
S209	76.03	33.40	0.08	23.03	29.26	0.94	13.73	0.47
S210	4.59	54.13	0.44	43.55	39.63	30.05	0.12	17.46
S212	3.88	80.18	12.62	55.75	40.68	33.95	0.77	12.19
S232	40.99	105.59	2.84	56.58	36.54	45.06	9.79	38.19
S233	0.07	109.56	1.18	62.87	40.90	28.89	0.81	22.59
S234	40.99	62.80	1.28	35.09	22.92	22.41	6.79	25.95
S261	30.09	88.38	9.02	54.99	27.29	39.88	9.33	36.02
S311	3.23	95.77	0.73	53.64	55.24	16.34	2.93	6.75
S314	9.04	98.37	1.50	61.27	41.66	36.28	0.86	22.07
S316	2.11	121.68	0.09	90.51	61.05	38.33	1.92	15.06
S317	7.08	109.82	2.14	70.34	42.44	43.46	2.09	28.41
S319	32.12	54.98	0.04	35.61	34.89	20.85	7.22	26.29
S342	53.93	9.45	0.01	9.34	5.97	0.51	3.26	0.61

However, another topographical variable fitted with TANN produced a combined DR greater than the single most significant variable. The topographic variables for the five species were, in order, ELWAT, DWAT, DGL, ELSEA, and DWAT. In the remaining eight species, either TANN did not enter the model significantly or the combination of TANN with any topographic variable produced a DR smaller than that obtained with one of the topographic variables singly.

In group 3 with nine species, TANN produced the greatest DR (Table 6) but one of the five topographical variables then entered the model with a significant DR in the next step (Table 7). In six species (central mudminnow, bluntnose minnow (*Pimephales notatus*), fallfish (*Semotilus corporalis*), tadpole madtom (*Noturus gyrinus*), American eel, and pumpkinseed (*Lepomis gibbosus*)), the topographic variable entering the model was DSEA. For longnose gar (*Lepisosteus osseus*) and sand shiner (*Notropis stramineus*), the variable was ELWAT and for ninespine stickleback it was DGL.

Finally, in group 4 with 20 species, TANN produced the greatest DR (Table 6) and then none of the topographical variables entered significantly in the next fitting step (Table 7).

A posteriori *K* Testing

In the logistic regression modelling analysis step, we found 36 species (7 from group 2, 9 from group 3, and 20 from group 4) with significant deviance reductions on fitting a linear temperature term. Testing of the predictive performance of the 36 models resulted in the number of useful models being reduced to 33.

Before pursuing the models obtained using TANN only, we further analyzed those species where we obtained a significant deviance reduction when $TANN^2$ was added after the intercept and TANN had been fitted. The results suggested that most of the quadratic temperature fits were spurious (Table 8). Goodness of fit analyses using K showed that in three of four species from group 1, two of four from group 2, and none of seven in group 4, the quadratic models were poor predictors of presence–absence. Further, in the group-4 species, agreement between observed and predicted occurrences was only slightly better than in the linear model for one species, virtually the same for four species, and worse for two (rosyface shiner (*Notropis rubellus*) and spotfin shiner (*Notropis spilopterus*)). We used the coefficients in the quadratic temperature models to compute upper and lower temperature thresholds (T_{50a} and T_{50b}) for the 15 species. In many instances, one or both of the T_{50} thresholds lay close to the extremities of the data set, and visual inspection of the data usually revealed one or few outliers for presence or absence. In six of the group-4 cases, the T_{50a} value lay within 2°C of the T_{50} obtained with the linear model. In two species, spotfin shiner and creek chub (*Semotilus atromaculatus*), the B_2 coefficient was positive, indicating a model that predicts a break in the distribution in relation to TANN (Fig. 2D).

Of the 15 species with significant quadratic DR, the most convincing example of a quadratic model was obtained with, lake trout (Fig. 4). That the predictive model depends on air temperature is inconsistent for a coldwater species occupying the hypolimnia of deep oligotrophic lakes. The temperature-related pattern of lake trout occurrence may coincide with the discontinuous regional distribution of deep glacial lakes in Ontario or may be related to temperature-related limitations on shallow-water spawning, which is a feature of lake trout life history in Ontario's inland lakes.

We thus obtained for 36 of the original 61 species significant one- or two-variable logistic regressions that included TANN as one of the variables (seven species from group 2, nine from

Table 7. Deviance reductions obtained for 61 species after fitting combinations of TANN and each of the topographical variables in a logistic regression ($B_0 + B_1$TANN + B_2Topographic variable).

OMNR Code	Group	Deviance reduction for TANN plus				
		DGL	DSEA	DWAT	ELSEA	ELWAT
S031	1	1.50	0.00	0.27	0.09	0.12
S081	1	1.87	0.06	1.12	3.72	9.74
S093	1	8.26	0.12	0.04	4.00	0.36
S102	1	1.23	0.35	0.00	3.79	0.27
S162	1	16.67	0.32	12.37	2.52	0.02
S171	1	0.10	0.27	1.82	3.83	3.78
S185	1	21.52	0.06	11.47	2.00	0.03
S211	1	0.83	0.01	0.00	0.77	0.27
S214	1	0.34	0.02	0.64	3.08	5.36
S271	1	11.09	2.02	0.08	0.99	0.31
S291	1	4.31	0.01	0.23	0.03	1.20
S332	1	0.14	8.03	1.63	0.06	0.32
S334	1	4.67	0.08	3.08	0.35	4.48
S341	1	2.82	0.11	9.95	2.94	6.04
S381	1	1.81	0.07	0.03	0.32	1.17
S382	1	0.02	0.43	0.00	1.36	0.09
S383	1	0.46	0.01	0.05	0.15	0.01
S061	2	1.63	0.00	7.15	6.04	*11.42*
S076	2	5.92	5.66	10.61	1.97	0.55
S078	2	3.51	0.84	*11.85*	4.46	0.33
S080	2	1.08	30.23	4.39	0.92	5.25
S091	2	*8.76*	0.91	0.00	6.18	3.70
S121	2	0.00	0.05	11.69	0.17	10.20
S168	2	0.05	2.89	4.20	1.64	3.41
S182	2	0.18	5.31	10.74	1.45	0.04
S183	2	0.77	4.57	2.75	4.12	1.10
S196	2	0.32	0.41	1.83	1.03	0.24
S198	2	*7.44*	6.97	7.05	0.35	0.46
S200	2	0.95	5.41	6.66	*11.38*	5.50
S206	2	0.02	3.07	*9.01*	7.02	3.30
S281	2	0.57	2.43	6.33	0.23	0.14

Table 7. *(Concluded)*

OMNR Code	Group	Deviance reduction for TANN plus				
		DGL	DSEA	DWAT	ELSEA	ELWAT
S338	2	0.20	1.74	6.41	*10.90*	9.88
S041	3	2.18	2.25	1.69	5.22	*11.25*
S141	3	2.00	*20.49*	2.46	0.64	0.07
S204	3	2.63	0.34	2.21	2.90	*10.94*
S208	3	0.01	*7.67*	0.81	2.31	0.75
S213	3	0.00	*21.36*	0.29	1.51	0.48
S236	3	2.87	*7.39*	3.25	0.00	0.44
S251	3	0.13	*6.52*	0.03	3.52	0.30
S283	3	*9.82*	0.00	1.42	6.65	2.56
S313	3	11.51	*14.30*	13.21	0.70	4.41
S132	4	2.40	1.24	6.02	1.66	5.51
S186	4	4.30	5.23	3.23	0.16	1.67
S189	4	1.63	1.40	0.87	3.16	0.64
S194	4	0.14	3.05	0.16	1.32	1.50
S199	4	0.01	0.11	0.81	1.68	3.33
S202	4	0.01	0.80	0.00	0.16	0.32
S203	4	4.02	1.18	0.29	4.40	3.67
S209	4	0.11	3.62	0.91	4.55	2.51
S210	4	1.39	4.24	2.51	0.24	0.56
S212	4	0.30	0.48	3.87	0.01	0.44
S232	4	0.17	0.10	0.18	1.15	0.66
S233	4	0.25	0.03	0.31	3.31	0.14
S234	4	0.38	0.60	0.66	0.34	0.05
S261	4	0.02	2.43	0.00	0.64	0.11
S311	4	0.52	2.68	0.92	2.57	1.46
S314	4	0.00	0.44	0.06	2.30	1.72
S317	4	0.08	0.17	0.01	0.12	0.33
S316	4	3.44	2.09	1.69	0.92	0.14
S319	4	0.02	2.70	0.08	1.99	1.15
S342	4	0.74	0.12	0.52	2.02	3.62

group 3, and 20 from group 4). We then tested the predictive power of those regression models using K-tests (Tables 9 and 10). Of the species in groups 2 and 3 with significant TANN models, only that for lake whitefish produced a nonsignificant K-test ($P = 0.01$). The percent agreement between observed and predicted (p_{Obs}) ranged from 0.76 to 0.92. Because of the inclusion of a topographic variable in these models, a single threshold temperature for presence–absence could not be calculated. Instead, we noted the minimum temperature (T_{Min}) for which presence was predicted in the data set and the maximum temperature (T_{Max}) for which absence was predicted. The absolute species' ranges spanned by these temperatures varied from 1.1 to 13.4. Two species, lake whitefish and ninespine stickleback, had values of 13.4, the range of temperatures in the data set. Both species have model coefficients (B_1 negative in Table 9) indicating a high-temperature, southern distribution limit. Lake whitefish had already been excluded because of its nonsignificant K-test. Ninespine stickleback was retained for subsequent analyses. Excluding those two species, the range was relatively narrow, 1.1–4.0, with a mean of 2.3 for 14 species with low-temperature, northern distribution limits.

We used the logistic equations, involving only TANN, obtained for the group-4 species (Table 6) to test the predictive success, using the K-statistic (Table 10). The percent agreement between observed and predicted presence–absence patterns (p_{Obs}) varied from 76 to 96% and in 18 of 20 cases the prediction success was significantly greater ($p < 0.01$) than that expected by chance alone. The presence–absence threshold mean annual air temperatures ranged from −5.1 to 5.7. In all group-4 species, the raw data and the logistic model fitted represented minimum temperature distribution boundaries.

Meisner et al. (1987) listed 32 Ontario species with northern low-temperature distributions bounded by July temperatures between 16.7 and 21°C, of which 12 overlapped with the 61 species selected for analysis here. Among those 12, seven were

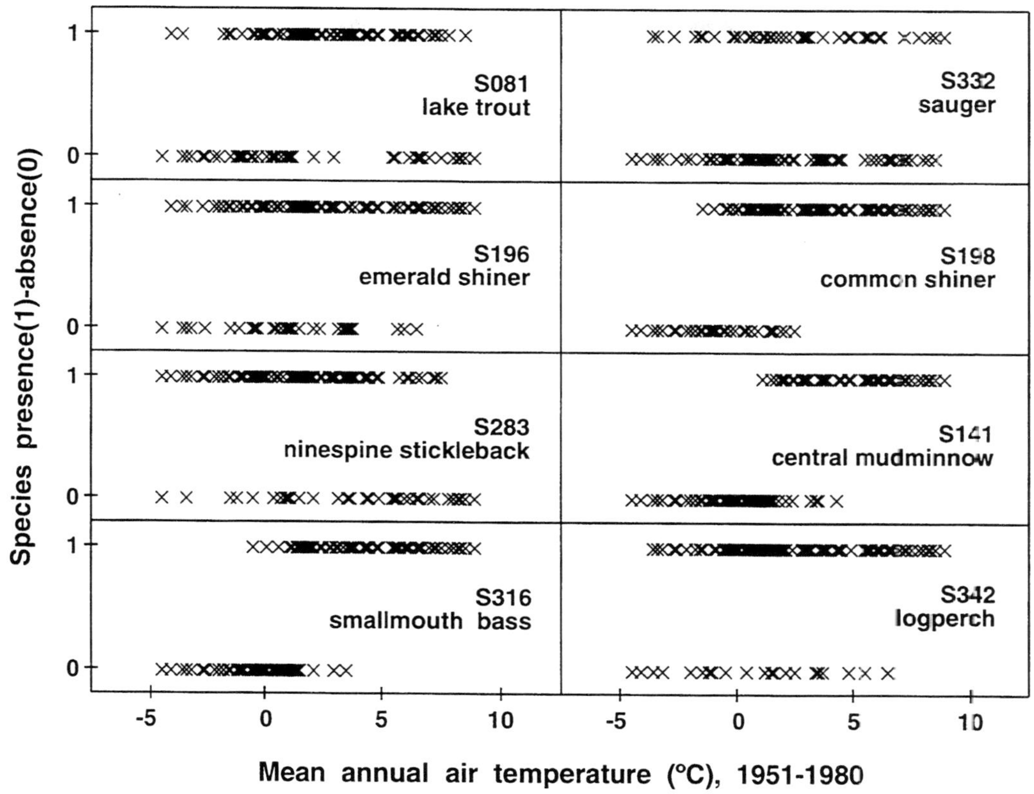

Fig. 4. Plots of observed presence–absence patterns versus mean annual air temperature (°C), 1951–80 (TANN), by tertiary watershed for 2 species from each of the 4 groups identified (Table 6): group 1 = S081 (lake trout), S332 (sauger); group 2 = S196 (emerald shiner), S198 (common shiner); group 3 = S283 (ninespine stickleback), S141 (central mudminnow); and group 4 = S316 (smallmouth bass), S342 (logperch).

found in group 4, one in group 3, three in group 2, and one in group 1, giving good agreement. All 12 were drawn from the groups with 16.7 and 18.0°C thresholds. The modelled temperature thresholds (Tables 9 and 10) ranged from −3.6 to 1.3°C for the five species in the 16.7°C group (blacknose shiner, mimic shiner, fathead minnow (*Pimephales promelas*), rock bass (*Ambloplites rupestris*), and Iowa darter) while the range was higher, from 0.4 to 5.3°C, for the 18.0°C group (rosyface shiner, bluntnose minnow, creek chub, brown bullhead (*Ictalurus nebulosus*), and smallmouth and largemouth bass).

Potential climate change impacts

The temperature-driven regression model of total tertiary watershed species richness (Equation 11) provided the first basis for predicting impacts. Given the range of temperature increases, 4.5–5.5°C, predicted for Ontario watersheds in the CCC-GCM, the model predicted richness increases ranging from 11.7 to 60.4 species with a mean of 31.8 for the 137 tertiary watersheds in the data set. The higher richness increases occurred in the south and close to the Great Lakes and should include new species, such as those identified by Mandrak (1989) in the Mississippi and Atlantic drainage basins. The lower values occurred in the north and east and should include many of the species already present in Ontario and identified in the species-by-species analyses as having temperature-related distribution limits.

The predicted regional species richness increases have large implications for lake-type and local species pool sizes. Minns (1989) showed that tertiary watershed species richness influenced local, lake-level, species richness. The slope in a linear regression model was 0.103. This means that effectively for every 10-species increase in tertiary watershed richness, species richness per lake will be predicted to increase by one. Given the regional predictions above, this implies richness increase of one to six species per lake, with a mean of three. This could be expected to have considerable impact, given that fish species richness per lake is typically three to six throughout much of Ontario.

Within the context of the overall species richness changes projected above, potential changes in the distributions of species already present in Ontario are nested. In the southern, more species-rich areas, richness increases will have to come from species, such as those identified by Mandrak (1989), from outside the region. Species present in Ontario and having temperature-limited distributions will be important in the more northerly, colder parts of the region. To address that aspect, we used those 33 species presence–absence models associated with significant temperature parameters and a significant K-test to assess the impact of climate change (Table 11). Taking the 33 species' distribution models, we predicted presence–absence for both current and future mean annual air temperatures. The absolute difference between

Table 8. The presence–absence threshold temperatures (T_{50a}, T_{50b}) obtained with a quadratic logistic temperature equation ($B_0 + B_1$TANN + B_2TANN^2), and the results of a posteriori K-testing of goodness-of-fit for the 15 species in groups 1, 2, and 4 (Table 6) with significant deviance reductions. p_{obs} = proportional agreement of observed and predicted species occurrences; p_{exp} = proportion expected by chance; Z = random normal deviate; P = probability of obtaining the Z value by chance.

OMNR Code	Group	T_{50a}	T_{50b}	P_{Obs}	P_{Exp}	K	SE	Z	P
S081	1	−0.9	7.2	0.693	0.553	0.31	0.095	3.30	0.0010
S093	1	−5.0	7.9	0.847	0.803	0.22	0.173	1.29	0.1990
S271	1	−4.3	8.0	0.788	0.752	0.15	0.149	0.98	0.3290
S334	1	−4.6	10.1	0.912	0.912	0.00	0.276	0.00	0.9990
S076	2	1.3	10.2	0.818	0.506	0.63	0.086	7.30	0.0001
S091	2	−6.3	7.4	0.818	0.782	0.16	0.162	1.00	0.3160
S121	2	3.0	9.6	0.766	0.532	0.50	0.091	5.51	0.0001
S183	2	−1.9	9.6	0.752	0.689	0.20	0.127	1.59	0.1110
S132	4	3.2	8.3	0.766	0.553	0.48	0.095	5.02	0.0001
S189	4	3.1	8.5	0.883	0.536	0.75	0.092	8.16	0.0001
S199	4	2.3	9.6	0.803	0.510	0.60	0.087	6.86	0.0001
S202	4	4.4	9.5	0.832	0.608	0.57	0.107	5.36	0.0001
S203	4	5.9	−5.7	0.898	0.675	0.69	0.123	5.57	0.0001
S212	4	1.7	−5.1	0.861	0.503	0.72	0.086	8.39	0.0001
S261	4	4.2	9.4	0.912	0.586	0.79	0.102	7.77	0.0001

Table 9. The coefficients of the logistic regressions ($B_0 + B_1$TANN $+ B_2$Topographic variable), the predicted minimum presence and maximum absence threshold temperatures (T_{Min}, T_{Max}), and the results of the a posteriori K-testing of goodness-of-fit for 16 fish species in groups 2 and 3 (Table 6) where TANN and a topographic variable entered significantly in the stepwise fitting procedure. ASE = asymptotic standard error; p_{Obs} = proportional agreement of observed and predicted species occurrences; p_{Exp} = proportion expected by chance; SE = standard error; Z = random normal deviate; P = probability of obtaining the Z value by chance.

OMNR Code	B_0	ASE	B_1	ASE	B_2	ASE	Variable	T_{Min}	T_{Max}	P_{Obs}	P_{Exp}	K	SE	Z	P
Group 2 species with second significant variable															
S061	0.116	0.627	0.305	0.077	−0.0178	0.0035	ELWAT	4.9	7.2	0.898	0.710	0.65	0.134	4.84	0.0001
S078	−0.967	0.625	0.344	0.091	−0.0172	0.0050	DWAT	3.9	6.5	0.869	0.634	0.64	0.112	5.70	0.0001
S091	5.139	1.050	−0.636	0.149	−0.0075	0.0024	DGL	−4.5	8.9	0.818	0.792	0.12	0.167	0.75	0.4560
S198	2.667	0.645	0.515	0.156	−0.0080	0.0018	DGL	−0.9	1.5	0.898	0.593	0.75	0.103	7.27	0.0001
S200	−1.406	0.621	0.396	0.077	0.0113	0.0025	ELSEA	−3.4	−0.4	0.905	0.766	0.59	0.155	3.84	0.0001
S206	−1.378	0.491	0.437	0.090	0.0048	0.0018	DWAT	−1.0	2.5	0.766	0.521	0.51	0.089	5.74	0.0001
S338	−1.149	0.561	0.273	0.066	0.0125	0.0025	ELSEA	−3.6	−1.1	0.920	0.834	0.52	0.191	2.71	0.0068
Group 3 species															
S041	−0.968	0.718	0.377	0.091	−0.0132	0.0032	ELWAT	4.9	7.2	0.854	0.720	0.48	0.137	3.49	0.0005
S141	−9.847	1.384	1.494	0.199	0.0108	0.0018	DSEA	1.4	3.2	0.920	0.502	0.84	0.086	9.77	0.0001
S204	−1.090	0.743	0.456	0.097	−0.0132	0.0032	ELWAT	4.9	7.2	0.905	0.696	0.69	0.129	5.31	0.0001
S208	−2.595	0.593	0.663	0.129	0.0040	0.0013	DSEA	0.4	1.6	0.854	0.519	0.70	0.089	7.84	0.0001
S213	0.471	0.619	0.951	0.185	−0.0093	0.0023	DSEA	4.9	8.9	0.832	0.687	0.46	0.127	3.65	0.0003
S236	−7.248	1.126	0.552	0.106	0.0056	0.0016	DSEA	6.2	7.3	0.854	0.729	0.46	0.140	3.28	0.0010
S251	−2.352	0.760	1.072	0.183	−0.0064	0.0022	DSEA	5.6	7.0	0.883	0.702	0.61	0.131	4.65	0.0001
S283	3.619	0.849	−0.573	0.132	−0.0060	0.0021	DGL	−4.5	8.9	0.759	0.582	0.42	0.101	4.20	0.0001
S313	−5.969	0.924	1.007	0.149	0.0069	0.0015	DSEA	1.4	2.9	0.905	0.500	0.81	0.086	9.47	0.0001

Table 10. The coefficients of the logistic regressions ($B_0 + B_1$TANN), the computed absence–presence threshold temperature (T_{50}), and the results of the a posteriori K-testing of goodness-of-fit for the 20 fish species in group 4 (Table 6) where TANN was the only variable to enter in the stepwise fitting procedure. ASE = asymptotic standard error; P_{obs} = proportional agreement of observed and predicted species occurrences; p_{exp} = proportion expected by chance; SE = standard error; Z = random normal deviate; P = probability of obtaining the Z value by chance.

OMNR	Logistic regression coefficients					Goodness of fit testing					
code	B_0	ASE	B_1	ASE	T_{50}	P_{Obs}	P_{Exp}	K	SE	Z	P
S132	−2.00	0.34	0.41	0.08	4.92	0.76	0.59	0.41	0.103	3.94	0.0001
S186	−3.36	0.45	0.77	0.10	4.36	0.88	0.57	0.73	0.099	7.30	0.0001
S189	−3.06	0.42	0.77	0.10	3.98	0.88	0.56	0.74	0.096	7.66	0.0001
S194	−0.13	0.23	0.69	0.11	0.18	0.82	0.57	0.59	0.099	5.93	0.0001
S199	−1.72	0.32	0.51	0.08	3.36	0.80	0.52	0.57	0.089	6.41	0.0001
S202	−3.91	0.52	0.74	0.10	5.27	0.85	0.63	0.61	0.111	5.50	0.0001
S203	−4.97	0.66	0.87	0.12	5.72	0.90	0.67	0.69	0.121	5.75	0.0001
S209	1.46	0.23	0.57	0.10	−2.54	0.87	0.79	0.37	0.167	2.21	0.0272
S210	−1.65	0.31	0.49	0.08	3.39	0.78	0.52	0.54	0.089	6.07	0.0001
S212	−0.94	0.26	0.75	0.12	1.25	0.83	0.52	0.65	0.088	7.42	0.0001
S232	−7.86	0.97	1.48	0.18	5.31	0.96	0.63	0.88	0.112	7.83	0.0001
S233	−2.34	0.36	1.06	0.15	2.20	0.87	0.50	0.74	0.086	8.60	0.0001
S234	−3.83	0.52	0.69	0.10	5.53	0.84	0.64	0.55	0.114	4.82	0.0001
S261	−4.49	0.57	0.92	0.12	4.87	0.91	0.61	0.76	0.107	7.10	0.0001
S311	−1.26	0.28	0.95	0.14	1.33	0.80	0.51	0.60	0.088	6.79	0.0001
S314	−3.19	0.43	0.90	0.12	3.56	0.89	0.53	0.76	0.092	8.35	0.0001
S316	−2.12	0.36	1.46	0.21	1.45	0.90	0.51	0.79	0.087	9.13	0.0001
S317	−3.47	0.45	1.04	0.13	3.33	0.88	0.53	0.74	0.090	8.17	0.0001
S319	−3.01	0.43	0.57	0.09	5.27	0.82	0.62	0.54	0.109	4.95	0.0001
S342	1.07	0.23	0.21	0.07	−5.06	0.80	0.80	0.00	0.172	0.00	0.9990

observed and predicted species richness, out of 33, gave a measure of the prediction uncertainty. Absolute differences ranged from 0 to 10 with a mean of 3.2. Most differences were less than 5. In contrast, the mean predicted richness values rose from 13.3 to 24.1 between current (1951–80) and predicted $2xCO_2$ temperatures. The increment between predictions average 10.8 with a range of 0–24. The predicted increase in species richness was more than three times the uncertainty differences. As with uncertainty for the temperature increases themselves, much of the temperature-induced change in species richness will have to occur before increased richness is distinguishable from variation.

Overall, the pattern of species-richness changes for these 33 species can be captured by summing the predicted occurrences versus mean annual air temperature (Fig. 5). We assumed that group-4 species were either present (1) or absent (0) depending on T_{50} (Table 10), while for group-2 and -3 species we assumed absence below T_{Min}, presence above T_{Max}, and a linear increase from 0 to 1 between T_{Min} and T_{Max} (Table 9). Most of the transitions occurred between mean annual air temperatures of 1 and 6°C. Temperature increases of 4.5–5.5°C will effectively shift the curve to the left by onethird of the x-axis. The main area predicted to have the largest levels of invasions from the 33 species lies in a band across northern Ontario roughly level with the top edge of Lake Superior (Fig. 6). Lesser changes are predicted north and south of that band. North of the band, the temperature increase would not be sufficient to allow all species to survive. South of the band, many of the 33 species are already present to varying extents and there is less scope for large richness increases.

The predicted changes in fish species distributions could have significance on a species by species basis. For instance, consider smallmouth bass, a highly prized game fish, and carp, a disastrous introduction that has contributed to the deterioration of fish assemblages and fish habitat in the Great Lakes region. Both have distributions related to mean annual air temperature and are predicted to have greatly increased potential ranges after climate change (Fig. 7). The pattern obtained for smallmouth bass is similar to that obtained by Shuter and Post (1990) using a bioenergetic–population dynamics approach. The threshold temperature for smallmouth bass, 1.45°C, was close to the point where mean recruitment was predicted to increase and variability of recruitment to decrease. Expansion of smallmouth bass and carp distributions will increase the overlap with the productive portions of the distributions of walleye (*Stizostedion vitreum vitreum*), northern pike, lake trout, and lake whitefish (Minns and Moore 1992) across northern central Ontario.

Table 11. Frequency distributions and summary statistics of observed and predicted species richness, of the 33 species with significant temperature-related distribution models, and differences and increases in species richness, given the current 1951–80 temperature regime and a future one, after 2× CO_2 (CCC–GCM II), for 137 tertiary watersheds in Ontario.

Variable	Current species richness of 33			After 2×CO_2 temperature increase	
	Observed	Predicted	Difference	Predicted	Increase
Frequency class					
0–4	30	41	102	4	32
5–8	31	17	27	5	17
9–12	12	25	8	11	24
13–16	10	4	0	6	38
17–20	7	15	0	19	21
21–24	19	2	0	6	5
25–28	18	11	0	36	0
29–33	10	22	0	50	0
Statistics					
Mean	13.6	13.3	3.2	24.1	10.8
SD	9.9	10.6	2.6	8.3	6.2
Minimum	0	1	0	3	0
Maximum	31	32	10	32	24

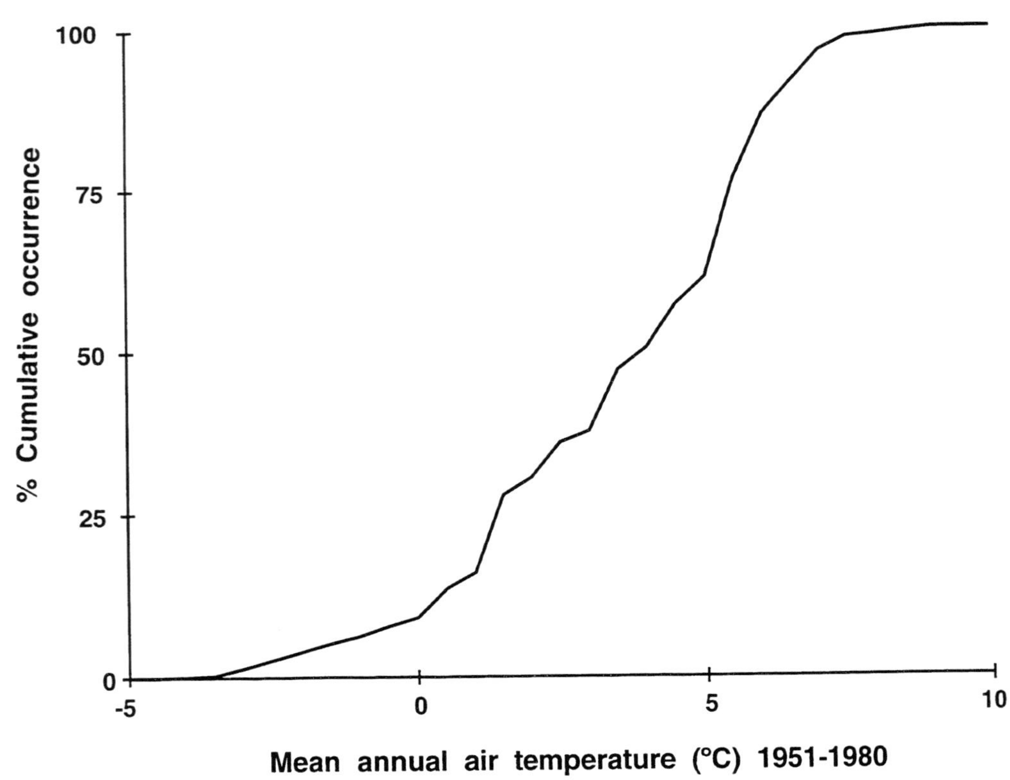

Fig. 5. Cumulative sum of predicted presence-absence occurrences for 33 species showing temperature-linked distribution patterns versus mean annual air temperatures (°C) from −5 to 10° by 0.5°C intervals.

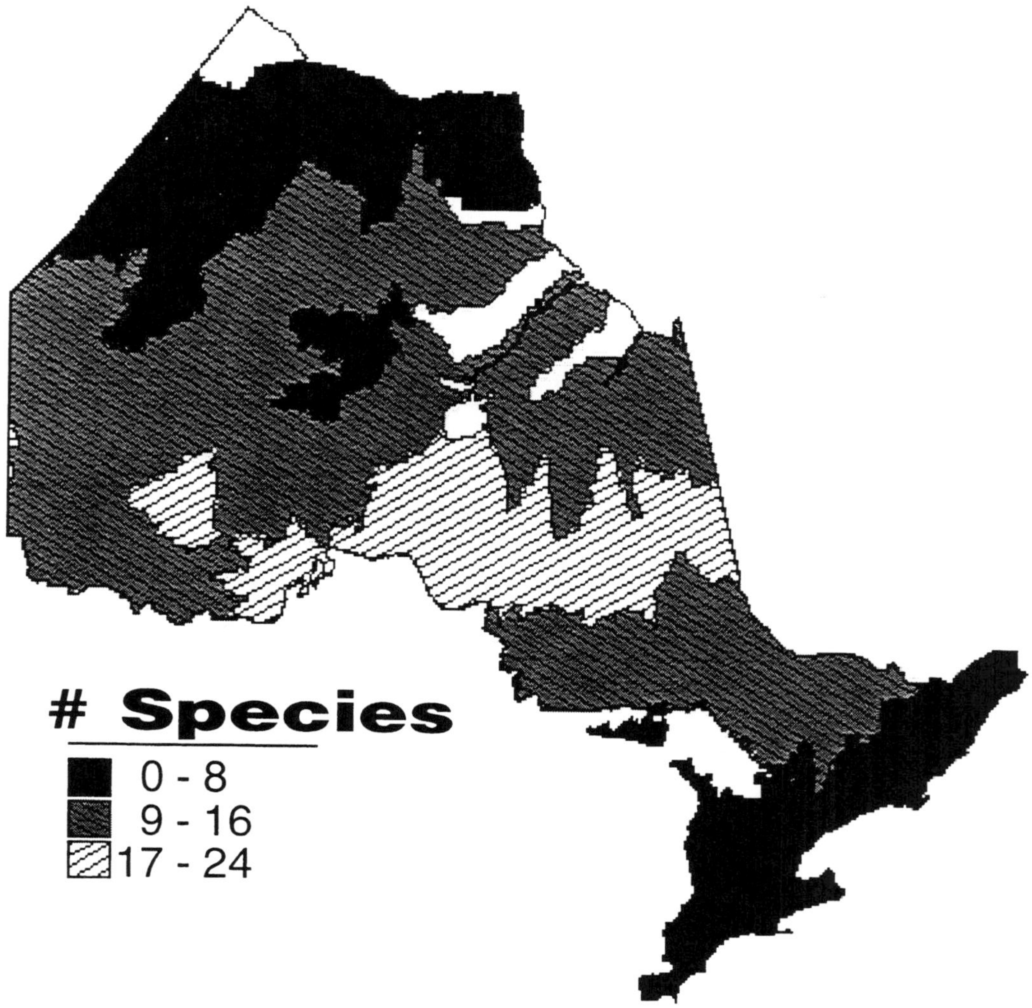

Fig. 6. A map of Ontario showing those tertiary watersheds where 0–8,9–16, and 17–24 of the 33 freshwater fish species with temperature-determined distribution boundaries are predicted to be able to invade following climate warming of 4.5–5.5°C.

Discussion

The discussion of these results addresses four main areas. The assumptions and data limitations are clarified so that the reader can determine the strength of the results and predictions. The significance and implications of the results and predictions are placed in a broader context. The potential scope for future additions and extensions to the approach used are suggested. Finally, the reader is offered summary conclusions and recommendations.

Assumptions and limitations

The fish species distribution data, the selection of climatic and topographic variables, and the modelling methodology could have an impact on the results obtained in this study. Each of these three aspects involves many assumptions and limitations. We chose to use binary presence–absence data by tertiary watershed unit to provide an initial, coarse-scale analysis of patterns. A coarse binary regional data set should be least influenced by the ecosystem-type and local factors that determine site-by-site occurrences. The OMNR inventory data could be summarized to give the proportion of sites where each species occurs by watershed or the raw data could be analyzed by site. The quaternary watershed units could be used to provide a finer scale resolution. These alternatives pose difficulties as they result in a confounding of species' pool filters at different spatial scales. Both Minns (1989) and Tonn (1990) have given examples of the interplay of local- and regional-scale processes for lake fish assemblages. Using proportional occurrence values offers an intermediate solution. However, the proportions may be affected by regional variation in the relative supply of lake types that favour a particular species or group of species. Combining lake and river sites may cause some blurring of distributional boundaries. While lakes may offer fewer thermal restrictions, fish distributions are still ultimately determined for most species by river and stream dispersal. Lake distributions should be enveloped by lotic ones.

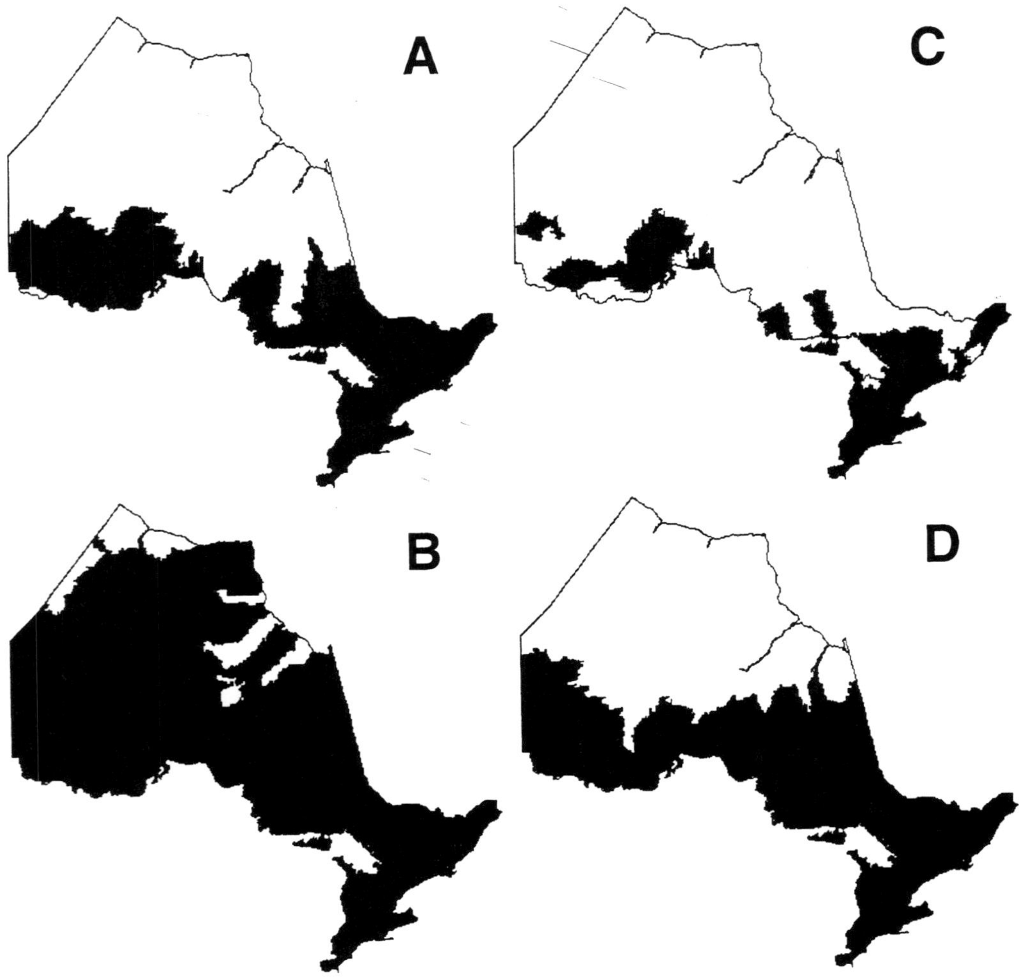

Fig. 7. Observed current and model-predicted future potential distributions of smallmouth bass (A and B respectively) and common carp (C and D) by tertiary watershed units in Ontario.

Tertiary watershed units represent fairly large areas of the province with considerable size variation (Fig. 1) and require a substantial pooling of fish occurrences. However, the watershed units are determined by the drainage networks of rivers, the principal routes for the dispersal of freshwater fish.

None of these occurrence data types are free from the effects of incomplete surveys or coverage. Minns (1989) showed that the likelihood of encountering species was related to their size. In Ontario, the intensity of fish inventory decreases northward, and the observed total richness values are probably underestimates. On balance, the coarse presence–absence data are the least affected by the survey limitations and are most likely to reflect the impact of regional filtering processes envisioned in Tonn's (1990) cascade model of species pools.

The independent quantitative variables used in this study were selected to mimic the factors that have shaped the development of regional fish species pools. Air temperature is a primary indicator of climate, and the correlations among annual, January, and July temperatures indicated that a single variable was sufficient. McAllister et al. (1986) found an aridity variable was significant in predicting North American species richness. Hydrological factors, such as annual and seasonal runoff and precipitation levels, do have regional patterns different from those of the variables we used (OMNR 1984) but were not considered in this study. The topographic variables were selected to imitate the processes and factors influencing the movement of species throughout the region since the recession of the last glaciers (cf. Hocutt and Wiley 1986). Distances from the Great Lakes and from the sea indicate the main origins for species moving into the province. The elevations above the sea and above the Great Lakes indicated the topographic barriers to fish that exacerbate the straightforward effects of distance and time. In the fitted models, we interpreted the inclusion of a topographic variable in a species' model as an indication of an additional constraint on the observed distribution. Alternatively, the inclusion may indirectly reflect some aspect of the biogeographic history and origins of the species.

There is a broader issue of whether we can expect to develop numeric explanatory models for the distributions of all species. In more than half the fitted models (18 of 33), temperature was the only variable fitted, but in 28 of 61 species, no temperature-related model was obtained. Certainly a wider range of independent variables may help us predict the patterns. However, as we pass from the regional to continental ranges, we begin to encroach on evolutionary processes, such as speciation and species exclusions, which will be more difficult to represent numerically. One way to broaden the scope of distribution modelling might be to consider ecomorphometric and life-history characteristics on a comparative basis among groups of species, e.g., taxon groups or ecological guilds. Minns (1989) had some success with explaining patterns of species richness and encounter probabilities in Ontario lakes by including a fish-size factor. Mahon (1984) and Portt et al. (1988) have begun to build the necessary data bases and understanding for application to distribution prediction.

Temperature change was the only impact of climate change considered here. Fresh waters are intimately bound into the hydrological cycles and changes in runoff will affect lakes and rivers. In addition, climate change will change the terrestrial landscapes surrounding fresh waters and feeding them with nutrients and organic matter. Rizzo and Wiken (1992) used an ecological map of Canada and predictions of a GCM to project landscape changes. In Ontario, much of the boreal forest would be replaced by grasslands. Thus, examining the temperature-related boundaries of fish distributions can only give a small glimpse of the impacts of climate change on fresh waters.

The incompleteness of the climate considerations is compounded by the uncertainties associated with the GCMs (IPCC 1990). Despite being noted as a high-resolution, second-generation model, the CCC-GCM still has a very coarse spatial resolution in relation to the species distribution patterns being analyzed. We partially compensate for the mismatch of resolutions by adding GCM-predicted changes on a grid basis to observed conditions in the reference period 1951–80 on the finer but uneven patterns of climate stations.

The logistic modelling approach used here has been used successfully by Nicholls (1989) with terrestrial plants in Australia, by Minns (1990) with native fish distributions in New Zealand, and by Osborne and Tigar (1992) with birds in Lesotho, Africa. The application here was consistent with other direct gradient analyses (Austin et al. 1984; Braak and Looman 1986). However, as we found here using *K*-statistics, the significance of deviance reductions cannot be the sole guide for selecting predictive models. Generally, significant K-tests were obtained when the deviance reduction obtained with the fitted model was about five times the threshold χ^2 -value needed for significance at $P = 0.01$. This result is consistent with the rule of thumb for reliable prediction with linear regression models, that the regression *F*-value should be 5–10 times the value required for significance.

Significance and implications

The magnitudes of the predicted changes in overall fish species richness and of the pattern changes forecast for 33 species already established in Ontario are staggering. The combination of the northward expansion of the species identified here and in Meisner et al. (1987) and the invasions projected by Mandrak (1989) will profoundly alter the fish assemblages of Ontario's lakes and rivers. The sigmoid pattern (Fig. 5) of species richness response to temperature is similar to that seen for pH in eastern Canadian lakes (Minns et al. 1990), except that, while richness decreases with acidification, it increases with warming.

Invasions of many species farther onto the Canadian shield with its many oligotrophic waters will bring these species into the central, most productive portions of the ranges of important sport and commercial species such as lake whitefish, walleye, northern pike, and lake trout (Minns and Moore 1992). The implications of such invasions, although clearly a disruption of natural conditions, are unclear. Welcomme (1992) presented a global overview of the consequences of 291 introductions of freshwater species, mostly fish, in 148 countries. He reported that most introductions, 63%, were viewed as benign, with 17% having negative impacts, 15% having positive impacts, and 5% having a mixture. In the Great Lakes region, many of the introductions, or invasions, are regarded negatively. Species such as carp, rainbow smelt, alewife, sea lamprey (*Petromyzon marinus*), and more recently the European percid, ruffe (*Gymnocephalus cernua*), are seen to have had negative impacts on native fish communities. Imposing extensive species changes in addition to the overfishing and habitat degradation pressures cited in the Introduction may exacerbate the perturbations. The numbers of species implied in predicted increases, 12–60 per tertiary watershed, almost guarantee that some invasions, like that of carp in the Great Lakes, will have negative impacts.

The observed positive relationship between species richness and temperature has been shown for other taxa (France 1991, 1992), such as amphipods and crayfish in fresh waters. A climatic explanation of latitudinal variation in species richness has become dominant (France 1991) and is consistent with the fundamental temperature links of ecological energy processes (Regier et al. 1990). Of course, the importance of other factors related to geological history and accidents of topography cannot be ignored (Tonn 1990; France 1992). Part of the explanation of the north–south species richness gradient may be gradients of nutrient and major ion concentrations in waters, which relate to the underlying geology (OMNR 1984). Rich sedimentary soils and rich waters in the south give way to igneous rocks and dilute waters to the north, and in the far north, to the vast wetlands of the Hudsons Bay and James Bay lowlands that overlie sedimentary deposits. The contribution of topographic variables to the distribution models of several species was consistent with the expected role of the geographic origins of and constraints on portions of the species pools.

There are few process-oriented studies of fish distributions, such as that of Shuter and Post (1990) for smallmouth bass and yellow perch, which can be compared with the empirical results obtained here. For individual fish species, the patterns of change may depend on the width of their temperature-related distribution band. France (1991) has reviewed the hypotheses concerning the responses of distributions to changing climate and presented an empirical model pointing to a narrowing of range caused by increased partitioning of thermal habitat as temperature increases. This factor was not addressed here because few, if any, of the species analyzed had complete ranges defined, but it is clearly an important factor. The narrowing of ranges could be complicated by a consideration of the time needed for species to expand their ranges. Both the modelling done here and that suggested by France (1991) involve steady-state predictions and take no account of the time involved for redistribution. Because the present species distributions occurred during the 7000 – 14 000 yr since the last glaciation (Hocutt and Wiley 1986), it is unreasonable to think that major redistributions can occur within the 21st century. The narrowing of range and the inability to move rapidly could place a strain on many species. In the shortterm, reductions in species richness might occur before overall richness can compensate in the long term.

Additions and extensions

The geographic scope of distribution modelling and climate change assessment can be expanded by extending the spatial boundaries of the data base. A freshwater fish distribution data base already exists for the tertiary watersheds east of Ontario (Minns and Moore 1992), although it lacks the intensity and rigour of surveys conducted in Ontario. Analysis of a larger region should increase the number of fish-species distribution boundaries, including high- and low-temperature ones. Within Ontario, the more detailed records should be pursued on a finer scale for lesser numbers of species with the addition of lake-type and local filter processes and factors. Minns (1990) was able to relate the occurrences of endemic freshwater fish in New Zealand to a variety of land-use and geological factors. Expanding the array of variables used to explain fish distributions should allow predictive models to be developed for more species. As noted above, the search for the links between distribution patterns and ecological characteristics of species should be expanded. Ecomorphometrics may be one fruitful avenue in the absence of more complete life-history assessments of more species. As a reading of Scott and Crossman (1973) reveals, most knowledge of Canadian freshwater fish species is anecdotal, inconsistent, and incomplete.

In broader ecological terms, the framework of Tonn's (1990) cascade model of species pools should be extended. Analyses of filters in different spatial contexts require different but complementary indicators and measures. The presence–absence prediction modelling of overall distributions must be formally linked to considerations of the population dynamics (Shuter and Post 1990) and of production and yields (Minns and Moore 1992). The eventual goal should be an examination of the distribution and productivity effects of climate change on freshwater fish resources and their habitats on a national scale.

Conclusions and recommendations

Recent invasions and/or introductions of zebra mussels (*Driessena polymorpha*), sea lamprey, carp, and other species into the Great Lakes region are seen by many to be calamitous and greatly regretted. The future prospect of extensive regional modifications of freshwater species distributions as a result of climate change should be greeted with alarm. Even the simplest analyses, such as those shown here and elsewhere, indicate that climate change could present humans with unprecedented changes to the freshwater ecosystems they share with fish and other species. In the growing institutional commitment to, and scientific literature on, the problem of climate change, the emphasis has largely been on reducing the uncertainty in predictions of future climatic conditions worldwide. Given the consensus opinion (IPCC 1990) that a significant amount of climate change is certain, more emphasis should be placed on assessing the forms and intensity of impacts on the world's ecosystems. In this, and most other ecosystem problems (eutrophication, acidification, habitat loss, introduction, overfishing, etc.), fresh waters and their biota, most noticeably fish, are affected soonest and to the greatest degree. Assessments of renewable resource impacts have a greater potential to mobilize the will to remedy the problems and, if all else fails, will ensure that we are better prepared for the changes. Because the IPCC (1990) is advising everyone that some climate change will be unavoidable, devoting more attention to natural resource impacts, especially in fresh waters, is only prudent.

As Canada is steward of a disproportionate share of the world's fresh waters, it is appropriate that Canada first set an example by shouldering its stewardship responsibility and second assume a leadership role in encouraging other nations to conserve their freshwater resources. Fresh waters are an indispensable part of human life and activity, and the health of those freshwater ecosystems is visible in the well-being of the fish assemblages they support.

Acknowledgements

First we thank Robin Green, a co-op student, for performing much of the preparatory work on the GIS, which eased the workload on this project. Dr. Bob Randall provided important feedback and encouragement during the development of the analytical approach. The provision by OMNR of a copy of the fish species distribution data base for this study is gratefully acknowledged.

References

AES. 1990. Application of the Canadian Climate Centre general circulation model output for regional climate impact assessment, version 1, April 1990. Canadian Climate Centre, Atmospheric Environment Service, Environment Canada, Downsview, Ont. 18 p.

Austin, M.P., R.B. Cunningham, and P.M. Fleming. 1984. New approaches to direct gradient analysis using environmental scalars and statistical curve-fitting procedures. Vegetatio, 55: 11–27.

BBN Software Products Corp. 1986. RS/1 version 12.1 for IBM personal computer and compatibles. Cambridge, Mass.

Beamish, F.W.H., P.J. Healey, and D. Griggs. 1986. Freshwater fisheries in Canada. Report on Phase I of a National Examination. Canadian Wildlife Federation, Ottawa, Ont. 155 p.

Chang, L.H., C.T. Hunsaker, and J.D. Draves. 1992. Recent research on effects of climate change on water resources. Water Resour. Bull. 28: 273–286.

Coutant, C.C. 1981. Foreseeable effects of CO_2-induced climatic change: freshwater concerns. Environ. Conserv. 8: 285–297.

Cox, E.T. 1978. Counts and measures of Ontario lakes. Ontario Ministry of Natural Resources MS Rep. 114 p.

Darnay, A.J. 1992. Statistical record of the environment. Gale Research Inc., Detroit, Mich. 855 p.

France, R.L. 1991. Empirical methodology for predicting changes in species range extension and richness associated with climate warming. Int. J. Biometeorol. 34: 211–216.

France, R.L. 1992. The North American latitudinal gradient in species richness and geographical range of freshwater crayfish and amphipods. Am. Nat. 139: 342–354.

Geraghty, J.J., D.W. Miller, F. van der Leeden, and F.L. Troise. 1973. Water atlas of the United States. Water Information Centre, Port Washington, N.Y.

Gleick, P.H. 1989. Climate change, hydrology, and water resources. Rev. Geophys. 27: 329–344.

Healey, M.C., and R.R. Wallace. 1987. Canadian aquatic resources. Can. Bull. Fish. Aquat. Sci. 215: 533 p.

Hocutt, C.H., and E.O. Wiley. 1986. The zoogeography of North American freshwater fishes. Wiley, New York. 866 p.

IPCC. 1990. Climate change: the IPCC scientific assessment. Intergovernmental Panel on Climate Change. J.T. Houghton, G.J. Jenkins, and J.J. Ephrauems [ed.] Cambridge University Press, Cambridge, UK. 365 p.

Law, B. 1992. Corpus almanac and Canadian sourcebook. 27th ed. Southam Business Communications, Don Mills, Ont. p. 3–9.

Legendre, P., and V. Legendre. 1984. Postglacial dispersal of freshwater fishes in the Quebec peninsula. Can. J. Fish. Aquat. Sci. 41: 1781–1802.

Mahon, R. 1984. Divergent structures of fish taxocenes of north temperate streams. Can. J. Fish. Aquat. Sci. 41: 330–350.

Mandrak, N.E. 1989. Potential invasion of the Great Lakes by fish species associated with climatic warming. J. Great Lakes Res. 15(2): 306–316.

McAllister, D.E., S.P. Platania, F.W. Schueler, M.E. Baldwin, and D.S. Lee. 1986. Ichthyological patterns on a geographical grid, p. 17–51. *In* C.H. Hocutt and E.O. Wiley [ed.] The zoogeography of North American freshwater fishes. Wiley, New York.

McCullagh, P., and J.A. Nelder. 1983. Generalized linear models. Chapman and Hall, London. 261 p.

Meisner, J.D., J.L. Goodier, H.A. Regier, B.J. Shuter, and W.J. Christie. 1987. An assessment of the effects of climate warming on Great Lakes basin fishes. J. Great Lakes Res. 13: 340–352.

Minns, C.K. 1989. Factors affecting fish species richness in Ontario lakes. Trans. Am. Fish. Soc. 118: 533–545.

Minns, C.K. 1990. Patterns of distribution and association of freshwater fish in New Zealand. N. Z. J. Mar. Freshwater Res. 24: 31–44.

Minns, C.K., and J.E. Moore. 1992. Predicting the impact of climate change on the spatial pattern of freshwater fish yield capability in eastern Canadian lakes. Clim. Change, 22: 327–346.

Minns, C.K., J.E. Moore, D.W. Schindler, and M.L. Jones. 1990. Assessing the potential extent of damage to inland lakes in eastern Canada due to acidic deposition. III. Predicted impacts on species richness in seven groups of aquatic biota. Can. J. Fish. Aquat. Sci. 47: 821–830.

Nelson, J.S. 1984. Fishes of the world. 2nd ed. John Wiley & Sons, New York. 523 p.

Nicholls, A.O. 1989. How to make biological surveys go further with Generalised Linear Models. Biol. Conserv. 50: 51–75.

OMNR. 1984. Water quantity resources of Ontario. Ontario Ministry of Natural Resources, Toronto, Ont. 72 p.

Osborne, P.E., and B.J. Tigar. 1992. Interpreting bird atlas data using logistic models: an example from Lesotho, Southern Africa. J. Appl. Ecol. 29: 55–62.

Pearse, P.H. 1988. Rising to the challenge: a new policy for Canada's freshwater fisheries. Canadian Wildlife Federation, Ottawa, Ont. 180 p.

Portt, C.B., C.K. Minns, and S.W. King. 1988. Morphological and ecological characteristics of fishes common in Ontario lakes. Can. Manuscr. Rep. Fish. Aquat. Sci. 1991: 37 p.

Regier, H.A., J.A. Holmes, and D. Pauly. 1990. Influence of temperature changes on aquatic ecosystems: an interpretation of empirical data. Trans. Am. Fish. Soc. 119: 374–389.

Rizzo, B., and E. Wiken. 1992. Assessing the sensitivity of Canada's ecosystems to climatic change. Clim. Change, 22: 37–55.

Schindler, D.W., K.G. Beaty, E.J. Fee, D.R. Cruikshank, E.R. DeBruyn, D.L. Findlay, G.A. Linsey, J.A. Shearer, M.P. Stainton, and M.A. Turner. 1990. Effects of climatic warming on lakes of the central boreal forest. Science (Washington, D.C.), 250: 967–970.

Scott, W.B., and E.S. Crossman. 1973. Freshwater fishes of Canada. Bull. Fish. Res. B. Can. 184: 966 p.

Shuter, B.J., and J.R. Post. 1990. Climate, population viability, and the zoogeography of temperate fishes. Trans. Am. Fish. Soc. 119: 314–336.

ter Braak, C.J.F., and C.W.N. Looman. 1986. Weighted averaging, logistic regression and the Gaussian response model. Vegetatio, 65: 3–11.

Titus, K., J.A. Mosher, and B.K. Williams. 1984. Chance-corrected classification for use in discriminant analysis: ecological applications. Am. Midl. Nat. 111: 1–7.

Tonn, W.M. 1990. Climate change and fish communities: a conceptual framework. Trans. Am. Fish. Soc. 119: 337–352.

UNEP. 1991. United Nation Environment Programme. Environmental Data Report 3rd ed. Basil Blackwell, Oxford. 408 p.

van der Leeden, F. 1990. The water encyclopedia. Lewis Publishers, Chelsea, Mich. 808 p.

Water Survey of Canada. 1977. Reference index, hydrometric map supplement, Canada 1977. Inland Water Directorate, Water Survey of Canada, Environment Canada, Ottawa, Ont.

Welcomme, R.J. 1992. A history of international introductions of inland aquatic species. ICES Mar. Sci. Sympo. 194: 3–14.

Wilkinson, L. 1990. SYSTAT: the system for statistics. SYSTAT, Inc. Evanston, Ill.

Appendix

Listing of geometric latitude and longitude centroids, independent climate and topographic variables, and total freshwater species richness for 137 tertiary watershed units in Ontario. TWS = Tertiary watershed code; Latitude = Geometric latitude centroid; Longitude = Geometric longitude centroid; January = Mean January air temperature (°C) 1951–80; July = Mean July air temperature 1951–80; Annual = Mean annual air temperature 1951–80; 2×CO_2 = Mean annual air temperature after 2×CO_2; DGL = Distance from the Great Lakes; DSEA = Distance from the sea (km); ELSEA = Elevation above sea level (m); ELWAT = Elevation above the sea or Great Lakes (m); NSPP = Freshwater fish species richness.

TWS	Latitude (N)	Longitude (W)	January	July	Annual	2×CO_2	DGL	DSEA	ELSEA	ELWAT	NSPP
02AA	48.1519	−89.7821	−15.5	18.2	2.5	7.0	20.4	685.0	261.8	78.7	32
02AB	48.7173	−89.7266	−16.4	17.5	2.0	6.5	52.2	640.8	402.0	218.8	60
02AC	48.9174	−88.7542	−16.5	17.3	1.7	6.2	21.6	571.5	288.6	105.5	58
02AD	49.8080	−88.6235	−19.3	16.7	0.1	4.6	97.3	504.8	351.2	168.1	58
02AE	49.1093	−87.8001	−16.5	17.1	1.6	6.1	24.5	505.3	288.6	105.5	31
02BA	49.0495	−86.8375	−16.7	17.2	1.4	5.9	29.9	461.0	249.0	65.9	42
02BB	49.1663	−86.0058	−17.2	16.9	1.3	5.8	55.3	412.5	311.3	128.2	31
02BC	48.5325	−85.5046	−15.7	17.5	2.1	6.6	56.5	448.5	376.8	193.7	38
02BD	48.1434	−84.4920	−15.0	17.5	2.4	6.9	36.9	429.8	369.6	186.4	42
02BE	47.5369	−84.1981	−13.7	17.5	3.4	7.9	39.9	473.8	390.7	207.6	33
02BF	46.9938	−84.1527	−12.3	18.1	3.9	8.4	21.1	525.8	376.0	192.9	60
02CA	46.5862	−83.8052	−11.5	18.5	4.4	8.9	26.3	556.2	311.8	135.6	70
02CB	47.1498	−83.0404	−13.0	17.9	3.5	8.0	105.5	476.0	453.5	277.2	34
02CC	46.5996	−83.0838	−11.7	18.5	4.3	8.8	47.1	535.7	369.6	193.3	45
02CD	46.3610	−82.7161	−11.4	18.5	4.5	9.0	17.8	547.0	276.6	100.3	47
02CE	46.8173	−82.0535	−13.1	18.4	3.6	8.1	77.6	479.2	394.8	218.6	54
02CF	46.5710	−81.3305	−13.0	18.5	3.8	8.3	57.1	486.2	326.5	150.3	47
02CG	45.7907	−82.2431	−10.3	18.5	4.9	9.4	8.1	592.0	204.1	27.9	66
02DA	47.0865	−81.0564	−14.3	18.5	3.2	7.7	118.6	424.8	400.2	224.0	30
02DB	46.3711	−80.8582	−13.0	18.5	3.7	8.2	39.7	499.0	252.0	75.7	20
02DC	46.8833	−80.2103	−14.5	18.4	3.1	7.6	108.8	433.4	322.3	146.0	50
02DD	46.1580	−80.0321	−12.8	18.5	3.7	8.2	63.2	512.8	250.5	74.3	52
02EA	45.6025	−79.8936	−11.4	18.5	4.5	9.0	28.0	574.4	272.7	96.4	60
02EB	45.2552	−79.2607	−11.1	18.5	4.4	8.9	58.5	612.2	327.0	150.7	60
02EC	44.5146	−79.2732	−9.2	18.9	5.6	10.5	43.5	695.9	262.0	85.7	69
02ED	44.3912	−79.9033	−9.2	18.7	5.5	10.8	16.5	711.0	288.3	112.0	75
02FA	44.8074	−81.2439	−7.5	18.5	5.8	10.9	8.9	677.8	216.5	40.3	77
02FB	44.4973	−80.6733	−8.0	18.5	5.5	11.0	13.5	705.0	336.9	160.6	58
02FC	44.1720	−80.9486	−7.4	19.0	6.2	11.7	46.9	744.6	357.4	181.1	70
02FD	44.0587	−81.5694	−6.5	19.5	6.7	12.2	16.1	766.2	249.8	73.5	62
02FE	43.7523	−81.2082	−7.1	19.5	6.6	12.1	57.4	795.1	344.2	168.0	61
02FF	43.2460	−81.7008	−6.3	20.4	7.5	13.0	10.6	859.1	246.0	69.7	79
02GA	43.6352	−80.4700	−7.9	19.6	6.2	11.7	64.2	800.2	378.3	204.3	61
02GB	43.1122	−80.1908	−5.8	21.0	8.0	13.5	31.7	857.8	223.4	49.4	74
02GC	42.8223	−80.6028	−5.8	20.5	7.8	13.3	26.5	893.5	196.2	22.3	106
02GD	43.2121	−81.0424	−6.9	20.1	7.0	12.5	60.1	853.8	295.2	121.3	60
02GE	42.5548	−81.8790	−5.7	21.2	8.3	13.8	17.8	939.3	200.6	26.7	82
02GF	42.4741	−81.7300	−5.5	21.5	8.5	14.0	7.8	946.1	170.3	−3.6	72
02GG	42.7658	−82.0565	−5.6	21.1	8.2	13.7	35.1	918.7	199.1	25.2	99
02GH	42.1565	−82.7822	−5.0	22.0	8.9	14.4	16.2	1000.4	186.5	12.6	83
02HA	43.0581	−79.4324	−4.9	21.5	8.5	14.0	15.6	825.0	181.8	107.2	82
02HB	43.5329	−79.9263	−6.7	20.6	7.2	12.7	25.4	808.6	244.6	170.0	92
02HC	43.8246	−79.4746	−7.1	20.3	7.0	12.5	25.1	774.3	210.3	135.6	77
02HD	43.9857	−78.4056	−7.5	19.9	6.5	12.0	55.2	695.2	158.7	84.0	58
02HE	44.0070	−77.2575	−7.7	21.4	7.5	12.2	8.4	615.6	90.6	16.0	76
02HF	44.9362	−78.5936	−10.7	18.5	4.8	9.3	92.6	650.8	353.2	278.6	52
02HG	44.1953	−78.8747	−8.1	19.5	6.5	12.0	45.0	716.1	284.3	209.7	34
02HH	44.5523	−78.3982	−9.3	19.3	5.7	10.4	83.9	662.7	294.5	219.8	46

Appendix (*continued*)

TWS	Latitude (N)	Longitude (W)	January	July	Annual	$2\times CO_2$	DGL	DSEA	ELSEA	ELWAT	NSPP
02HJ	44.2639	−78.2362	−8.2	19.5	6.5	11.5	52.8	666.9	238.1	163.5	40
02HK	44.5249	−77.8008	−9.0	19.7	6.0	10.6	51.3	621.6	246.6	171.9	61
02HL	44.5684	−77.3718	−8.8	20.5	6.6	11.1	45.0	588.8	222.1	147.5	61
02HM	44.4164	−76.9084	−8.2	21.2	7.2	11.7	25.7	565.9	154.7	80.0	68
02JC	47.9382	−79.8843	−16.6	17.5	1.6	6.1	227.4	313.2	283.0	208.4	41
02JD	47.5766	−80.5572	−15.7	17.6	2.4	6.9	176.3	362.0	392.6	318.0	41
02JE	46.8413	−78.9906	−14.2	18.0	2.9	7.4	170.7	434.5	321.4	246.8	57
02KA	46.1858	−77.8401	−13.3	18.0	3.6	8.1	197.2	519.1	277.7	203.1	51
02KB	45.8941	−78.2617	−12.8	18.0	3.6	8.1	153.2	546.0	373.1	298.4	54
02KC	45.6577	−77.0500	−11.8	19.7	4.9	9.4	163.7	513.6	209.8	135.2	59
02KD	45.3469	−77.9666	−11.7	18.5	4.3	8.8	143.3	594.1	413.7	339.1	62
02KE	45.2405	−76.9045	−10.5	20.4	5.8	10.3	116.0	520.6	284.0	209.4	56
02KF	45.1096	−76.5009	−10.0	20.5	6.1	10.6	91.6	498.0	198.5	123.8	64
02LA	44.9293	−76.0396	−9.6	20.5	6.2	10.7	58.0	475.5	130.4	55.8	68
02LB	45.2971	−75.1359	−10.7	20.4	5.6	10.1	114.7	392.2	82.8	8.2	61
02MA	44.5001	−76.2787	−8.5	21.2	7.3	11.8	21.1	518.1	128.5	53.9	62
02MB	44.6070	−75.7276	−8.8	20.8	7.0	11.5	23.5	474.6	101.2	26.6	63
02MC	45.1660	−74.6070	−10.0	20.5	6.5	11.0	132.1	364.5	72.4	−2.2	72
04AC	53.5466	−93.4689	−23.8	16.7	−2.0	2.5	623.7	450.1	305.6	305.6	17
04AD	54.0526	−92.6869	−25.0	15.7	−3.4	1.1	639.8	376.3	274.2	274.2	25
04AE	55.0387	−91.3069	−25.5	15.1	−4.1	0.4	699.4	241.0	175.2	175.2	22
04BA	55.6915	−88.8634	−25.5	14.3	−4.5	0.9	738.1	77.8	104.9	104.9	1
04CA	52.7903	−93.5296	−22.5	17.3	−1.0	3.5	561.8	508.7	342.8	342.8	36
04CB	52.7620	−91.5470	−22.8	16.9	−1.6	3.0	480.1	427.4	352.4	352.4	33
04CC	54.5332	−89.5986	−24.7	15.3	−3.6	1.8	618.3	195.7	179.6	179.6	29
04CD	54.5109	−90.9954	−24.8	15.6	−3.4	1.5	638.5	259.6	211.6	211.6	29
04CE	54.1622	−89.1385	−24.5	15.5	−3.2	2.3	573.7	215.9	200.1	200.1	25
04DA	52.6030	−89.4655	−22.9	16.7	−1.8	3.3	409.4	384.1	307.6	307.6	30
04DB	53.5550	−88.0759	−23.7	15.9	−2.7	2.7	502.9	248.6	203.0	203.0	27
04DC	54.4970	−86.2417	−23.8	14.6	−3.4	2.1	619.2	92.6	112.8	112.8	18
04DD	55.4821	−86.8244	−24.7	13.8	−4.5	1.0	718.8	38.4	95.2	95.2	15
04EA	53.4380	−85.6623	−22.8	15.3	−2.6	2.8	517.3	187.8	138.9	138.9	17
04EC	54.2246	−83.1151	−22.8	13.9	−3.2	2.3	644.3	45.5	51.4	51.4	15
04ED	54.7434	−84.3727	−23.5	13.9	−3.6	1.9	674.1	55.6	107.9	107.9	19
04FA	51.8496	−89.9656	−22.0	17.4	−1.0	3.5	341.6	472.7	361.8	361.8	31
04FB	52.1288	−87.1849	−22.5	16.5	−1.5	3.0	352.3	327.8	241.4	241.4	27
04FC	52.7631	−84.4907	−22.2	15.4	−2.2	3.1	462.8	148.8	113.4	113.4	28
04GA	51.1335	−91.2606	−20.8	17.5	−0.2	4.3	323.0	577.6	409.5	409.5	39
04GB	50.4132	−89.5177	−20.6	16.9	−0.5	4.0	186.7	529.6	408.6	408.6	36
04GC	51.4343	−88.8676	−21.8	17.0	−1.0	3.5	275.0	451.6	339.4	339.4	35
04GD	51.3807	−87.4972	−21.6	16.5	−1.5	3.0	267.3	360.9	283.2	283.2	33
04GE	51.0998	−86.6864	−21.5	16.5	−0.9	3.6	253.9	317.2	257.1	257.1	24
04GF	51.5912	−85.4712	−21.6	16.5	−1.2	3.3	320.8	220.2	163.0	163.0	7
04HA	51.4178	−83.1566	−21.0	15.9	−1.1	3.4	376.1	91.9	100.9	100.9	21
04HB	51.7570	−81.6170	−20.8	15.3	−1.5	3.0	476.8	40.9	61.6	61.6	16
04JA	49.0439	−84.3295	−17.3	17.2	1.2	5.7	130.9	346.6	364.0	364.0	26
04JB	50.3417	−83.8015	−19.8	16.5	−0.5	4.0	259.2	217.0	199.3	199.3	22
04JC	49.5231	−84.7512	−18.4	16.6	0.5	5.0	149.8	329.1	310.7	310.7	31
04JD	49.8517	−85.9834	−19.3	16.5	0.0	4.5	125.4	353.7	301.9	301.9	35
04JE	50.4711	−85.8015	−20.5	16.5	−0.5	4.0	195.2	298.0	237.4	237.4	16
04JF	50.5979	−86.5821	−20.7	16.5	−0.5	4.0	199.8	335.5	281.2	281.2	32
04LA	47.8961	−81.5663	−15.7	17.6	2.1	6.6	202.3	353.9	350.8	350.8	37
04LB	48.9487	−81.7336	−17.8	17.0	1.1	5.6	258.3	255.0	291.0	291.0	26
04LC	48.0777	−82.4594	−15.6	17.5	2.1	6.6	180.5	363.7	397.0	397.0	33
04LD	49.0646	−82.2250	−17.8	17.1	1.0	5.5	232.8	254.8	301.7	301.7	17
04LE	48.1838	−83.1477	−15.5	17.5	2.1	6.6	131.3	372.6	427.7	427.7	28

Appendix (*concluded*)

TWS	Latitude (N)	Longitude (W)	January	July	Annual	$2\times CO_2$	DGL	DSEA	ELSEA	ELWAT	NSPP
04LF	49.2536	−82.5867	−18.1	16.7	0.7	5.2	224.0	248.3	300.0	300.0	25
04LG	50.5760	−81.8727	−20.0	16.0	−0.4	4.1	366.1	101.8	129.5	129.5	29
04LH	48.4908	−83.6115	−16.1	17.5	1.8	6.3	112.6	360.2	420.1	420.1	29
04LJ	49.2364	−83.5387	−18.0	16.8	0.9	5.4	175.9	291.1	326.6	326.6	17
04LK	49.9315	−83.4069	−19.3	16.5	0.1	4.6	247.6	231.5	272.4	272.4	25
04LL	49.6866	−82.7896	−18.9	16.6	0.4	4.9	248.3	217.8	294.1	294.1	26
04LM	50.5411	−82.4730	−19.7	16.4	−0.5	4.0	338.8	138.8	121.7	121.7	4
04MA	48.7061	−79.5389	−17.7	16.8	0.8	5.3	316.7	225.6	306.2	306.2	37
04MB	48.4339	−80.4256	−17.5	17.5	1.5	6.0	272.4	267.0	315.7	315.7	31
04MC	49.0493	−80.7552	−18.4	16.7	0.6	5.1	328.8	212.2	293.1	293.1	28
04MD	48.5462	−81.0714	−17.4	17.3	1.2	5.7	277.5	272.6	315.0	315.0	34
04ME	49.8224	−81.2366	−19.0	16.4	0.1	4.6	339.5	152.0	244.6	244.6	34
04MF	50.2943	−80.7760	−19.5	16.0	−0.3	4.2	397.8	95.7	236.9	236.9	17
04NB	49.4825	−79.3836	−18.5	16.4	0.3	4.8	402.4	139.0	301.4	301.4	22
04NC	50.5568	−79.4688	−19.7	15.6	−0.6	3.9	489.1	21.4	211.6	211.6	34
05PA	48.4188	−91.2656	−16.1	18.6	2.8	7.3	130.4	751.2	449.7	449.7	54
05PB	48.9705	−92.0662	−17.5	18.5	2.1	6.6	210.3	766.2	410.9	410.9	57
05PC	48.7075	−93.9885	−16.8	19.5	3.0	7.6	333.1	902.8	347.5	347.5	22
05PD	49.3476	−94.1848	−18.0	19.4	2.4	7.2	367.6	845.6	347.6	347.6	57
05PE	49.9455	−94.6377	−18.5	19.5	1.9	6.7	425.4	804.3	337.4	337.4	53
05PF	50.2624	−95.0925	−19.1	18.7	1.5	7.0	470.8	792.7	329.4	329.4	14
05PG	49.7653	−95.0321	−18.5	19.5	2.1	7.6	442.0	836.1	330.1	330.1	12
05PJ	50.5977	−94.8552	−19.5	18.5	1.5	6.3	474.9	752.7	371.8	371.8	19
05QA	49.7905	−91.2883	−18.9	18.1	1.1	5.6	215.6	673.0	453.6	453.6	45
05QB	50.5604	−92.2210	−19.9	18.3	0.6	5.1	322.6	662.5	397.9	397.9	40
05QC	51.0963	−93.4799	−20.4	18.4	0.5	5.0	427.4	653.0	396.4	396.4	36
05QD	49.7902	−93.0639	−18.8	18.5	1.5	6.0	320.4	765.3	406.3	406.3	45
05QE	50.4114	−94.0039	−19.4	18.5	1.4	5.9	412.6	736.6	372.8	372.8	48
05RA	50.7631	−95.0487	−19.8	18.5	1.4	6.4	495.8	746.0	357.3	357.3	24
05RB	51.1452	−94.7022	−20.4	18.5	0.6	5.1	497.9	697.7	376.6	376.6	29
05RC	51.8367	−93.8038	−21.3	17.7	−0.1	4.4	500.0	599.5	383.8	383.8	32
05RD	51.6223	−95.0092	−21.1	18.4	0.4	4.9	546.0	670.9	355.6	355.6	13
05RE	52.3791	−94.9247	−21.8	17.5	−0.5	4.0	595.9	607.1	344.1	344.1	13

The El Niño of 1983 as reflected in the ichthyoplankton off Washington, Oregon, and northern California

Miriam J. Doyle

Doyle, M.J. 1995. The El Niño of 1983 as reflected in the ichthyoplankton off Washington, Oregon, and northern California, p. 161–180. *In* R.J. Beamish [ed.] Climate change and northern fish populations. Can. Spec. Publ. Fish. Aquat. Sci. 121

Abstract: Anomalies observed in the ichthyoplankton off the U.S. northwestern coast during 1983 included temporal shifts in peak abundance of eggs and larvae among certain species, reduced abundance of eggs and larvae among other species, the occurrence of rare southern species, and changes in distribution patterns. These anomalies are attributed to changes in spawning patterns and advection of eggs and larvae in response to the physical oceanographic anomalies characterizing the 1983 El Niño event in this region. Such changes, which resulted from the physical forcing on the ecosystem, seem to have been mediated mainly through water temperature and transport anomalies. A high level of stability in the spawning and early life history patterns of fish species in this region is implied by the fact that only a small portion of the dominant species seem to have been affected by the strong El Niño event of 1983. As well, in 1984 and 1985, abundance and distribution patterns in the ichthyoplankton had returned to the apparently normal patterns observed in 1980, 1981, and 1982.

Résumé : Les anomalies observées dans l'ichtyoplancton de la côte nord-ouest américaine en 1983 comprenaient des changements temporaires dans l'abondance de pointe des oeufs et des larves de certaines espèces, une abondance réduite d'oeufs et de larves chez d'autres espèces, la présence de rares espèces venant du sud et des changements dans la répartition des populations. Ces anomalies sont attribuées à des modifications dans les comportements de frai et à l'advection d'oeufs et de larves en réponse aux anomalies océanographiques physiques qui ont caractérisé l'événement El Niño de 1983 dans cette région. De tels changements, attribuables au forçage physique de l'écosystème, semblent avoir été indirectement causés par la température de l'eau et par les anomalies de transport. Le fait que seule une faible fraction des espèces dominantes semblent avoir été touchées par le puissant épisode El Niño de 1983 démontre un niveau élevé de stabilité des caractéristiques du frai et des premiers stades de vie de diverses espèces de poissons dans cette région. En outre, en 1984 et en 1985, les tendances de l'abondance et de la répartition de l'ichtyoplancton sont revenues aux valeurs apparemment normales observées en 1980, 1981 et 1982.

Introduction

An El Niño – Southern Oscillation event is usually manifest off the U.S. west coast in the following anomalous oceanographic conditions: positive sea-temperature anomalies, increased sea levels, continued poleward transport of water because of unusually persistent southerly winds, diminished coastal upwelling, and shoreward transport of oceanic water (Lynn 1983; McGowan 1984; Reed 1984; Huyer and Smith 1985; Fahrbach et al. 1991). Biological consequences of an El Niño in the Northeast Pacific Ocean can include northern range extensions and shifts in population distributions of many species of marine organisms, and a reduction in primary production that causes proportional reductions in the growth and reproductive success of zooplankton, fish, birds, and marine mammals (Barber and Chavez 1983; Bailey and Incze 1985; Miller et al. 1985; Smith 1985; Pearcy and Schoener 1987; Arntz et al. 1991; Simpson 1992). All El Niño events are different, however, and considerable variation exists in their effects on the oceanography and biology of the Northeast Pacific. The 1983 tropical El Niño resulted in exceptionally strong physical and biological anomalies throughout the Pacific Ocean, including U.S. coastal waters (Cane 1983; Reed 1984; Schoener and Fluharty 1985; Pearcy and Schoener 1987; Fahrbach et al. 1991; Arntz et al. 1991).

Although El Niño events can have drastic ecological and economical consequences for fisheries, this is documented primarily for subtropical species and little is understood about the effects of El Niño on subarctic fish (Bailey and Incze

M.J. Doyle. Department of Fisheries and Wildlife, Oregon State University, Corvallis, OR 97331-3803, U.S.A. Present address: NOAA, National Marine Fisheries Service, Alaska Fisheries Science Center, 7600 Sand Point Way NE, Seattle, WA 98115-0070, U.S.A.

1985). Little is known in particular about the effects of El Niño on spawning patterns and the early life history of fish in the temperate zone of the Pacific. Off the U.S. northwestern coast, only one study to date has focused on the influence of an El Niño event on the ichthyoplankton. Brodeur et al. (1985) investigated patterns in distribution and abundance of ichthyoplankton in the coastal upwelling zone off Newport, Oreg., during the 1983 El Niño and compared these with patterns documented for this region during the spring and summer of previous years. The observed unusual larval fish catches in 1983 were correlated with the anomalous oceanographic conditions of that year suggesting that El Niño events alter the distribution and extent of spawning effort, as well as the transport and survival of eggs and larvae, among fish populations off the northern U.S. west coast.

Ichthyoplankton sampling over a much broader area off Washington, Oregon, and northern California was carried out from 1980 to 1987 and incorporated the 1983 El Niño (Doyle 1992*a*). These data are examined here for the purpose of identifying differences or anomalies in the distribution and abundance patterns of ichthyoplankton species during 1983 with those in other years during the 1980s. The results are interpreted with respect to the anomalous oceanographic conditions observed off the U.S. west coast during 1983. They provide further insight into the effects of El Niño events on the spawning patterns and early life history of fish in this region.

Methods

Sampling program

Ichthyoplankton surveys conducted off the U.S. northwest coast during the 1980s involved the U.S. Northwest and Alaska Fisheries Center in Seattle, Wash., and the U.S.S.R. Pacific Research Institute in Vladivostok, Russia. From 1980 to 1987, 10 cruises were carried out; during each one a maximum of 125 stations was occupied (Table 1). Seasonal coverage was limited: six of the 10 cruises took place in spring (March to early June); one in summer (August 1980); one in winter (January 1987); and two in autumn (October–November 1981 and November–December 1983). The station grid covered an area of approximately 249 000 km^2 off Washington, Oregon, and northern California (Fig. 1).

Table 1. Timing of west coast ichthyoplankton surveys, with number of cruises undertaken and number of stations sampled.

Year	No. of cruises	No. of stations	Months sampled
1980	2	216	April, May, August
1981	2	248	May, October, November
1982	1	124	May
1983	2	237	April, May, November
1984	1	124	March, April
1985	1	124	April, May
1987	1	88	January
	10	1161	

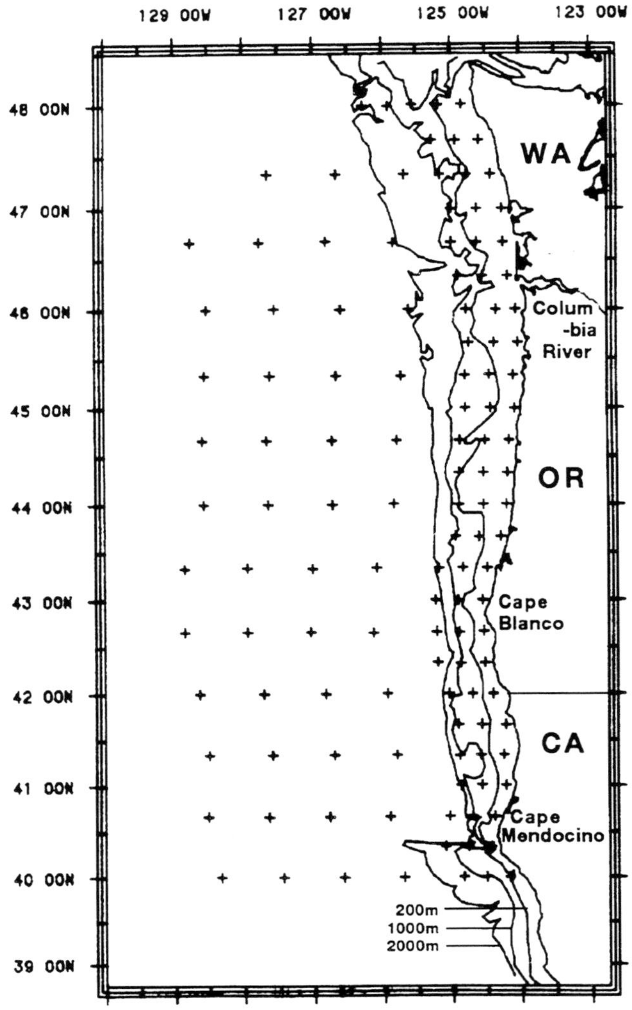

Fig. 1. Survey area showing grid of sampling stations and bathymetry.

During each cruise, plankton sampling for fish eggs and larvae and hydrographic casts for determination of temperature and salinity were carried out at all stations. Water samples were collected with Niskin bottles at nominal depths of 0, 5, 10, 15, 20, 25, 30, 35, 50, 75, 100, 200, 250, and 300 m as depth permitted. Paired neuston tows were conducted at each station using Sameoto samplers (Sameoto and Jaroszynski 1969) with 0.3 m high by 0.5 m wide frames and 0.505 mm mesh netting. The samplers were towed at a speed of 2.0 knots for 10 minutes and sampled the upper approximately 15 cm of the water column. Following standard procedures, oblique tows to 200 m depth, or 5 m from the bottom in water shallower than 200 m, were carried out at each station using 60 cm frame bongo samplers fitted with 0.505 mm mesh nets (Smith and Richardson 1977). Flow meters in the mouths of the neuston and bongo samplers were used to determine the volume of water filtered by each net.

The plankton samples were preserved in a 5% buffered formalin solution. One of each of the paired neuston and bongo samples was retained by the Americans and the other by the

Soviets. The American plankton samples were processed by the Polish Plankton Sorting Center in Szczecin, Poland, and subsequently by the Northwest and Alaska Fisheries Center in Seattle. Fish eggs and larvae were removed from the samples, identified to the lowest taxonomic level possible, counted, and measured. Counts of fish eggs and larvae were converted to number per 10 m^2 of sea-surface area for the bongo samples and number per 1000 m^3 for the neuston samples.

Data analysis

The multivariate analytical technique of numerical classification (Clifford and Stephenson 1975) was used to investigate multispecies spatial patterns in the ichthyoplankton data. Four data sets were analyzed in this manner and included data from the bongo collections only. The first two consisted of mean levels of abundance of species of eggs and larvae, respectively, at positions throughout the sampling grid for all six spring cruises combined. The second two consisted of abundance of species of eggs and larvae for the spring cruise in 1983. These data sets were chosen so that multispecies patterns in distribution and abundance of ichthyoplankton could be compared for general spring conditions and spring conditions during an El Niño year. Only the dominant ichthyoplankton taxa that contributed significantly to the overall spatial patterns were included in this analysis. Taxa occurring in less than 5% of the samples for a particular data set were removed. In addition the data were log transformed prior to analysis.

Normal and inverse classifications were carried out on the data sets; that is, both the species and the stations were classified into groups. The first step in the numerical classification procedure calculated correlation coefficients for each pair of species or stations in a data set. In this instance the Bray–Curtis dissimilarity coefficient (Bray and Curtis 1957) was used. An agglomerative, hierarchical sorting strategy was then used to produce dendrograms depicting clusters of stations and species. The "flexible sorting" strategy was used and a recommended value of −0.25 was chosen as the clustering intensity coefficient (Lance and Williams 1967; Boesch 1977).

To aid in identification of groups from the dendrograms, the original data sets (species abundance by stations) were rearranged in two-way tables according to the order that species and stations appeared in the dendrograms. In this manner, it was possible to see how a chosen group of stations was characterized by the occurrence or definitive range of abundance of a particular species or group of species. After the final species and station groups were chosen, the two-way species abundance by station tables were reduced by calculating the mean abundance of each species within the different species groups, for each station group. The station groups were then plotted on maps of the sampling area to aid in the identification of geographically distinct ichthyoplankton assemblages. Only the two-way tables and station-group plots are presented in the results.

Results

Oceanographic anomalies

The 1983 physical anomalies of elevated sea temperatures and suppressed coastal upwelling are reflected in the temperature, salinity, and upwelling data collected during the west coast ichthyoplankton surveys.

Temperature data collected during April and May of 1983 (the El Niño year) are compared with the data collected during the same months of 1980 and 1985 (taken to be normal years). The typical horizontal pattern of water temperature in the

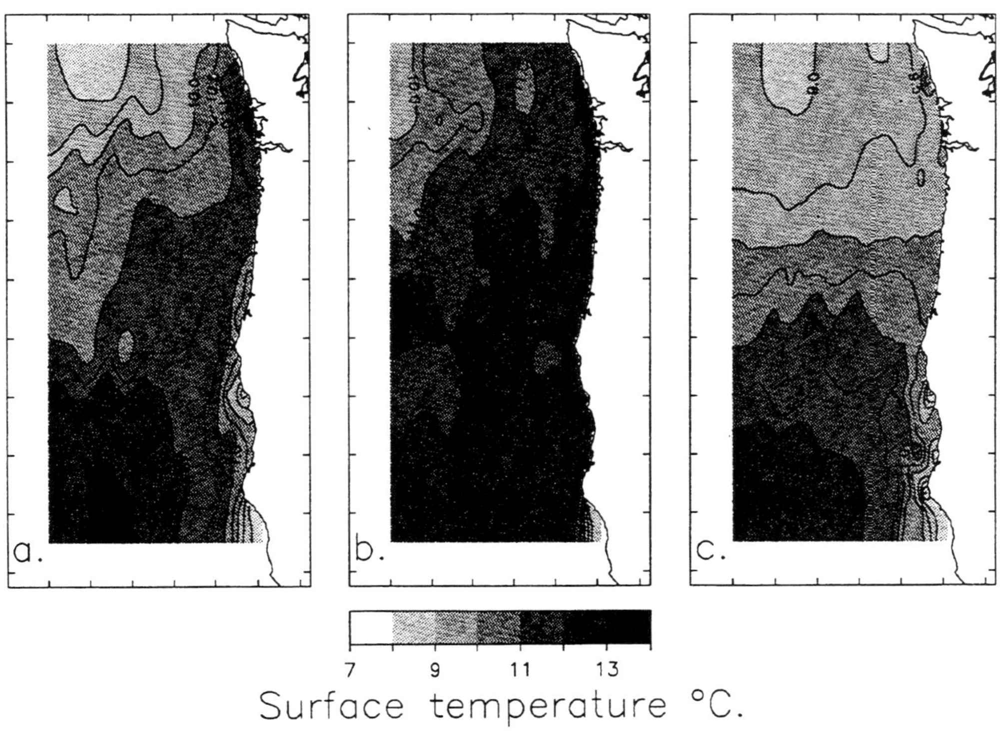

Fig. 2. Surface temperature during (*a*) April–May 1980, (*b*) April–May 1983, and (*c*) April–May 1985.

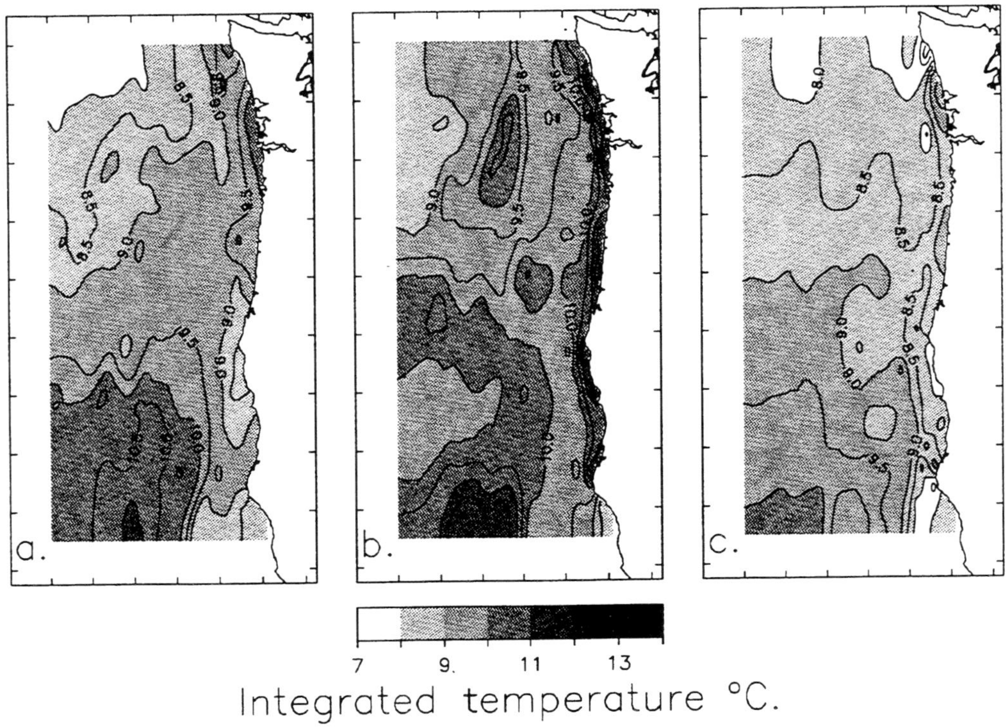

Fig. 3. Temperature, integrated over the water column to a maximum depth of 200 m, during (*a*) April–May 1980, (*b*) April–May 1983, and (*c*) April–May 1985.

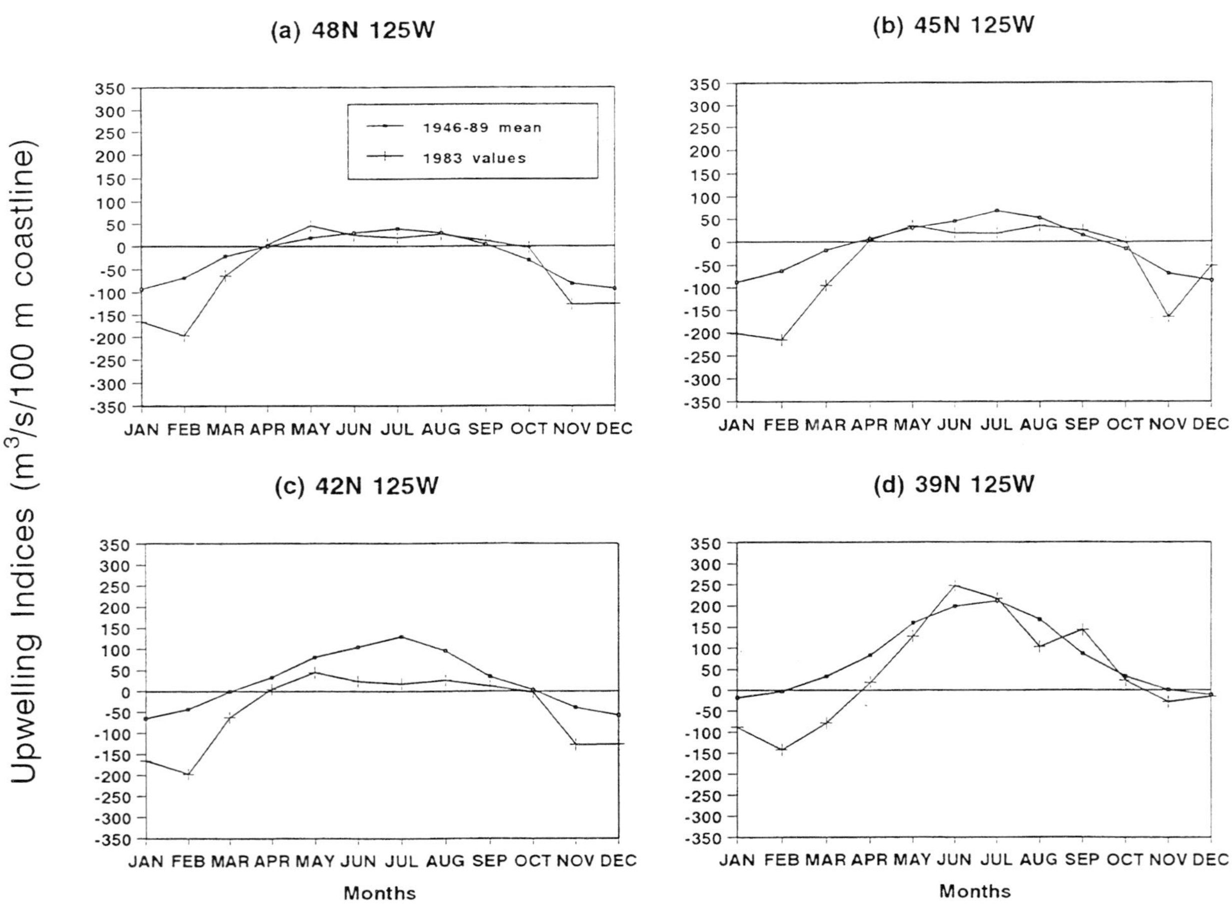

Fig. 4. Monthly upwelling indices (Bakun 1973) for 1983 at four locations along the 125°C meridian plotted with mean values for 1946 to 1989. (*a*) northern Washington. (*b*) northern Oregon. (*c*) Oregon–California border. (*d*) northern California.

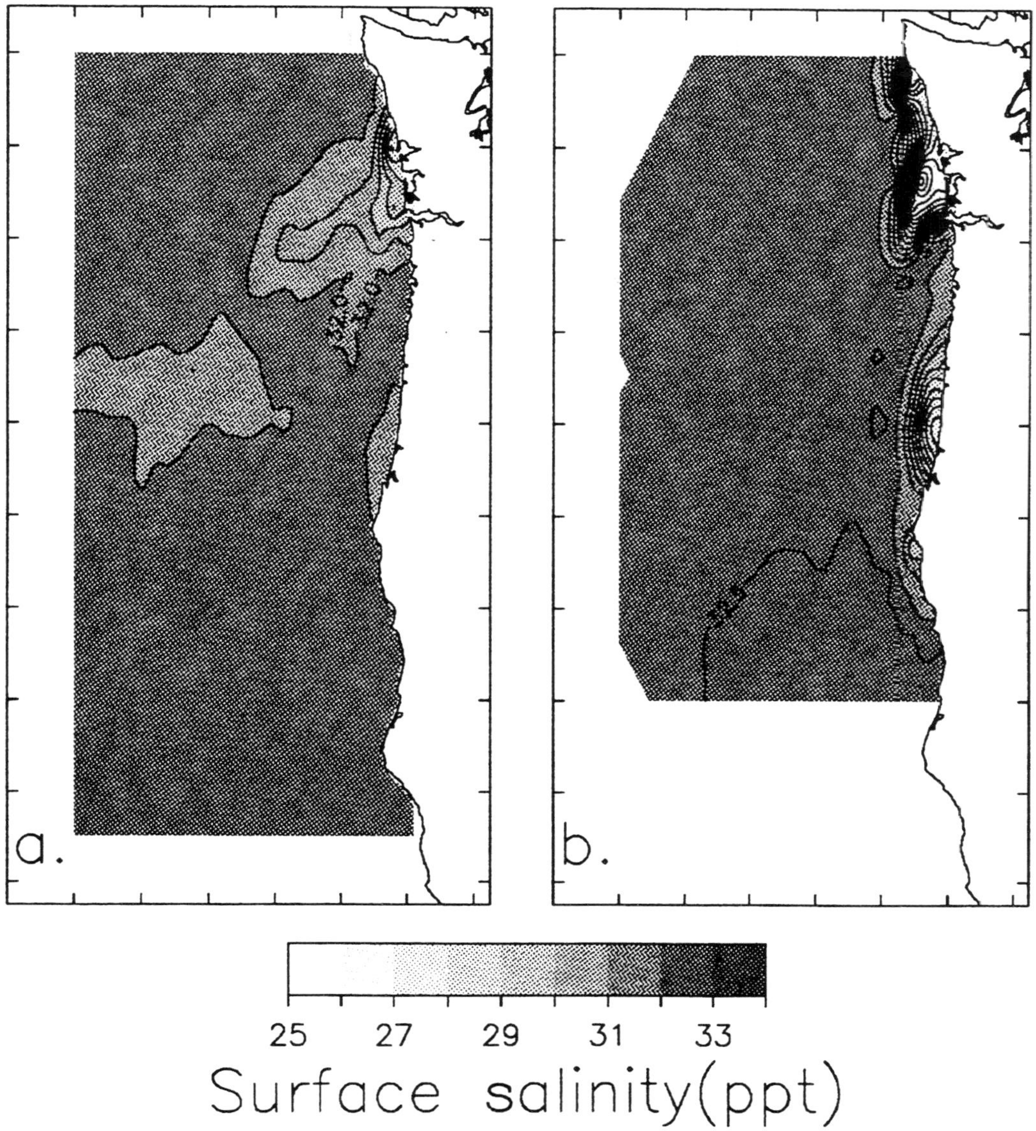

Fig. 5. Surface salinity during (*a*) October–November 1981 and (*b*) November 1983.

sampling area during spring is characterized by an increase in temperature from north to south throughout the oceanic zone and a decrease in a southerly direction along the coast (Figs. 2*a*, 2*c*, 3*a*, and 3*c*). For both surface temperature and temperature integrated for the water column, the presence of colder water along the southern Oregon and northern California coasts is detectable and reflects the commencement of upwelling along this section of the coast early in spring. Off Washington and northern Oregon, upwelling does not usually become established until early summer and so during April–May, water in the coastal zone remains relatively warm. During April–May 1983, water temperatures were higher throughout the sampling area than in 1980 and 1985, and upwelling of cold water along the coast was not apparent except for a very small region south of Cape Mendocino off northern California (Figs. 2*b* and 3*b*). In spring 1983, the 12°C surface isotherm extended much farther north (Fig. 2*b*) than in the springs of 1980 (Fig. 2*a*) and 1985 (Fig. 2*c*), and temperatures in the coastal zone were primarily 12–13°C.

The 1983 El Niño event is reflected in the annual pattern of upwelling indices for that year (Fig. 4). Off Washington and Oregon, index values were depressed during the summer (Figs. 4*a* and 4*b*), and at 42°N off the Washington–Oregon border they were below the mean during all months of the year (Fig. 4*c*). Off northern California, values were significantly below normal during the first half of the year but similar to mean values from June through December. For this section of the coast, the commencement of upwelling seemed to be later than usual, whereas farther north, upwelling was weak and diminished from spring through autumn.

A comparison of horizontal patterns of surface salinity in the autumns of 1981 and 1983 also indicates diminished upwelling in the coastal zone during the El Niño year. During the summer months, when offshore Ekman transport, which drives the coastal upwelling, is at a peak, the low-salinity water of the Columbia River plume extends offshore covering a large area of the oceanic region off Washington and Oregon (Fiedler and Laurs 1990). By autumn, the plume is less extensive

Table 2. List of dominant ichthyoplankton taxa and their association as eggs or larvae with the different sampling gears.
*Taxa for which anomalous patterns in distribution or abundance were observed during 1983.

Family	Genus or species	Neuston		Bongo	
		Eggs	Larvae	Eggs	Larvae
Engraulidae	*Engraulis mordax**	X	X	X	X
Osmeridae	Unidentified				X
Bathylagidae	Unidentified			X	
	Bathylagus spp.			X	
	*Bathylagus ochotensis**			X	X
	Bathylagus pacificus				X
Chauliodontidae	*Chauliodus macouni*	X		X	X
Melanostomidae	*Tactostoma macropus*	X			
Myctophidae	Unidentified			X	X
	*Diaphus theta**				X
	Protomyctophum crockeri				X
	Protomyctophum thompsoni				X
	Tarletonbeania crenularis		X		X
	*Stenobrachius leucopsarus**				X
Scomberesocidae	*Cololabis saira*		X		
Trachipteridae	*Trachipterus altivelis**	X		X	
Scorpaenidae	Sebastes spp.		X		X
	Sebastolobus spp.				X
Anoplopomatidae	*Anoplopoma fimbria*		X		
Hexagrammidae	*Ophiodon elongatus*		X		
	Hexagrammos decagrammus		X		
	Hexagrammos lagocephalus		X		
Cottidae	*Hemilepidotus hemilepidotus*		X		
	*Hemilepidotus spinosus**		X		
	Scorpaenichthys marmoratus		X		
Bathymasteridae	*Ronquilus jordani*		X		
Cryptacanthodidae	*Cryptacanthodes aleutensis*		X		
Icosteidae	*Icosteus aenigmaticus*	X		X	
Ammodytidae	*Ammodytes hexapterus*		X		
Centroplophidae	*Icichthys lockingtoni**	X		X	
Paralichthyidae	Icitharichthys spp.*	X		X	
	Citharichthys sordidus				X
	Citharichthys stigmaeus				X
Pleuronectidae	Unidentified	X		X	
	Errex zachirus	X		X	X
	Pleuronectes isolepis	X			X
	Eopsetta exilis	X		X	X
	Microstomus pacificus	X		X	
	Pleuronectes vetulus	X		X	X
	Psettichthys melanostictus	X		X	X

offshore but still apparent as a remnant of the strong offshore transport during summer (Fig. 5*a*). In contrast, during autumn 1983, the Columbia River plume remained close to the coast (Fig. 5*b*) reflecting the diminished offshore transport and upwelling during the summer of that year.

Anomalous patterns in abundance and distribution of dominant ichthyoplankton species

The dominant taxa taken during the west coast ichthyoplankton surveys comprised a total of 32 species in 19 families along with four categories identified to genus level only: *Bathylagus* spp., *Sebastes* spp., *Sebastolobus* spp., and *Citharichthys* spp. (Table 2 from Doyle 1992*a*). In addition, many eggs belonging to the families Bathylagidae, Myctophidae, Paralichthyidae, and Pleuronectidae were not identified beyond the family level, as was the case with osmerid larvae. These taxa together accounted for greater than 80% of total abundance of eggs and larvae in the neuston and bongo samples. The association of the dominant taxa of eggs and larvae with the different sampling gears indicates the existence of a unique assemblage of larvae in the neuston (Doyle 1992*b*). These larvae were absent or scarce in the bongo samples and include the taxa *Cololabis saira*, *Anoplopoma fimbria*, the hexagrammids, the cottids, and *Cryptacanthodes aleutensis*.

Table 3. Variation in abundance among sampling periods for dominant species that displayed anomalous abundance or anomalous timing of peak abundance during 1983 relative to other years. Mean abundance of eggs or larvae for each sampling period is expressed as no.·10 mu^{-2} for the bongo samples and no.·1000 mu^{-3} for the neuston samples. Sampling periods arranged in seasonal order. – = Species absent from samples; * = species occurred in <5% of the samples; x = no eggs identified from neuston samples collected during May 1982.

Species	Stage	March–April 1984	April–May 1985	April–May 1980	April–May 1983	May 1982	May 1981	August 1980	October–November 1981	November 1983	January 1987
Bongo samples											
Engraulis mordax	Eggs	–	–	22.84	31.66	–	–	1.78	–	*	–
	Larvae	–	–	–	9.81	–	–	138.97	0.52	–	–
Osmeridae	Larvae	2.67	26.20	8.72	*	3.76	*	*	–	–	12.71
Bathylagus ochotensis	Larvae	40.15	13.92	13.24	28.00	11.92	9.10	0.20	*	*	8.58
Citharichthys spp.	Eggs	2.86	*	32.53	47.73	68.07	37.08	23.25	174.67	4.84	51.10
	Larvae	*	–	*	2.55	*	0.81	0.51	17.70	3.40	1.84
Neuston Samples											
Engraulis mordax	Eggs	–	–	86.65	156.23	×	1.14	46.51	–	*	–
	Larvae	–	–	*	11.87	–	–	51.60	34.27	–	*
Hemilepidotus spinosus	Larvae	75.07	83.70	52.59	13.63	11.19	–	–	–	*	160.54
Citharichthys spp.	Eggs	16.38	9.45	183.07	670.76	×	167.62	567.02	1024.53	27.65	1138.38

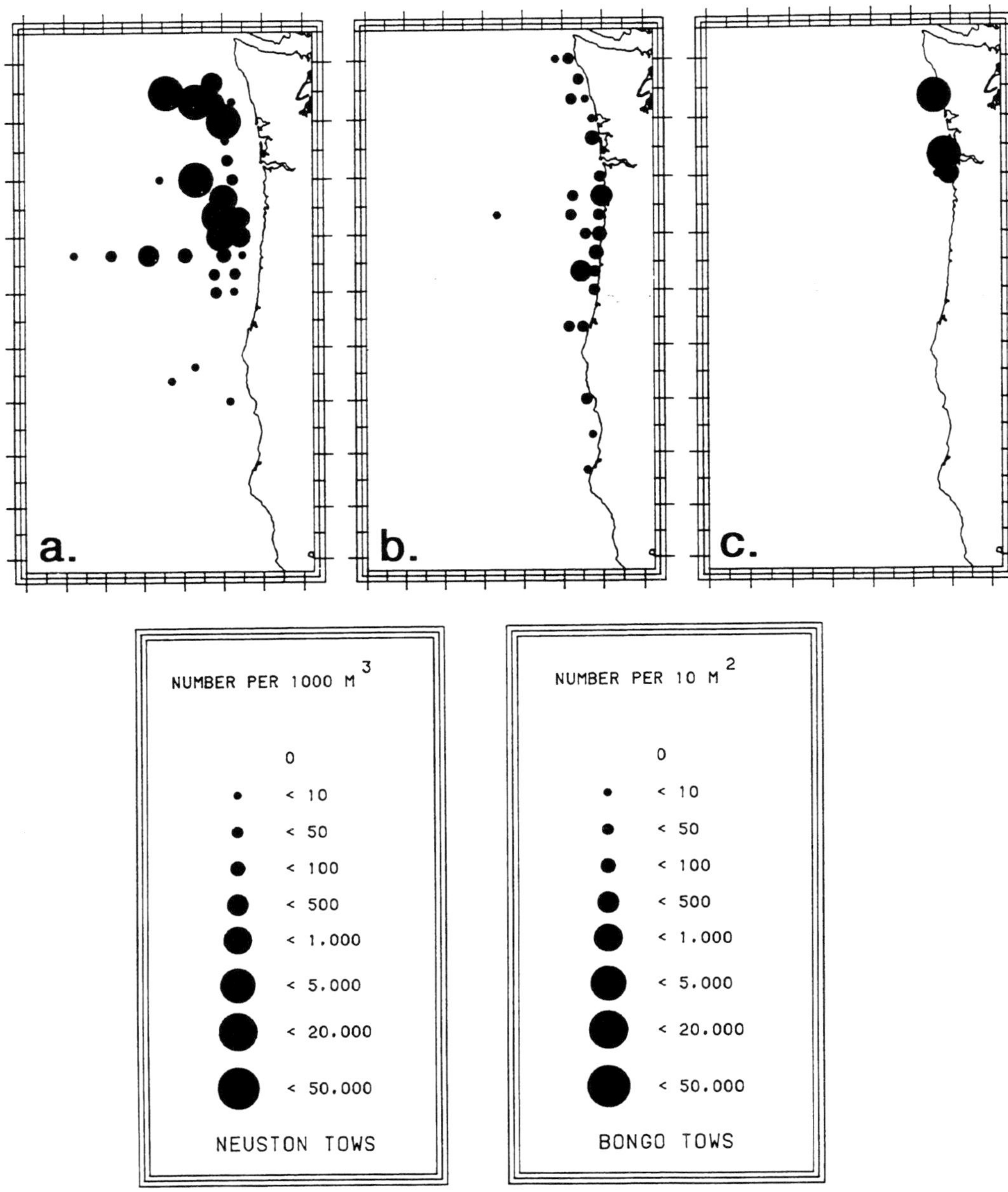

Fig. 6. Distribution patterns for *Engraulis mordax*. (*a*) Larvae from bongo samples taken during August 1980. (*b*) Larvae from bongo samples taken during April–May 1983. (*c*) Eggs from bongo samples taken during April–May 1983.

Anomalous patterns in abundance or distribution were observed during 1983 for several of the dominant taxa.

Engraulis mordax (northern anchovy)

The main spawning season for the northern subpopulation of *E. mordax* off the U.S. west coast is summer to early fall, and the pelagic eggs and larvae are usually found in association with the warm near-surface waters of the Columbia River plume (Laroche and Richardson 1980; Richardson 1981; Doyle 1992*a*). During the present study, peak densities of anchovy larvae were recorded during the one summer cruise of August 1980 (Table 3). The distribution of larvae on this occasion reflected the extensive offshore orientation of the Columbia River plume at this time of year (Fig. 6*a*).

During spring months, anchovy eggs and larvae are usually scarce in the ichthyoplankton off the northwest U.S. coast. Interannual variations in this pattern are observed, however, and occasionally spawning seems to commence as early as March or April (Richardson et al. 1980; Brodeur et al. 1985). During April–May of 1980 and 1983, anchovy eggs were unusually abundant in the sampling area, suggesting an early commencement of spawning in both these years. Such early spawning has been associated previously with elevated water temperatures (Brodeur et al. 1985). Water temperatures during spring 1980 were slightly higher than normal and in 1983 substantially higher than usual because of the El Niño event that year (Doyle 1992*a*).

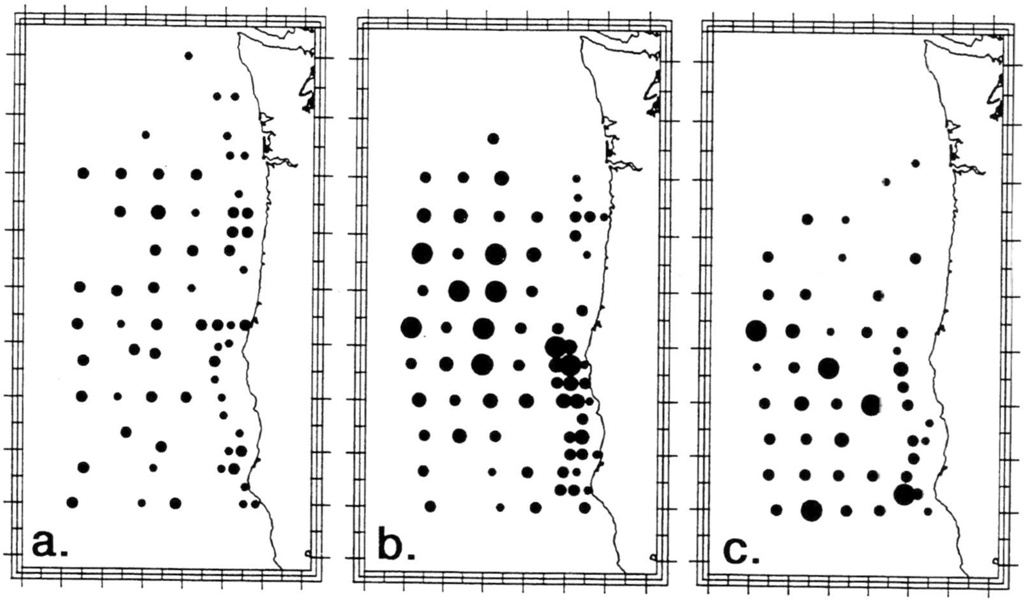

Fig. 7. Distribution patterns for *Bathylagus ochotensis* larvae from bongo samples taken during (*a*) April–May 1980, (*b*) April–May 1983, and (*c*) April–May 1985. See Fig. 6 for key to numbers of larvae.

In April–May 1983, the abundance of anchovy eggs was significantly higher than in April–May 1980, and in addition, larvae were relatively abundant, unlike any of the other spring cruises (Table 2). Spawning during spring 1983 seems, therefore, to have been more extensive than in spring 1980. The distribution of anchovy eggs and larvae during April–May 1983 (Figs. 6*b* and 6*c*) indicates that spawning was associated with the unusually warm coastal waters off Washington and Oregon (Figs. 2 and 3). The coastal distribution of the eggs and larvae on this occasion contrasts with the more usual offshore pattern observed when spawning is at a peak during summer months (Fig. 6*a*). A comparison of the records of anchovy larvae for the autumns of 1981 and 1983 (Table 3), seems to indicate that spawning finished earlier than usual in 1983, as larvae were scarce on this occasion.

Osmeridae (smelts)

Smelts are small, shallow-water schooling fish that spawn demersal eggs on beaches and in estuaries mainly during late winter and early spring. Osmerid larvae were common in ichthyoplankton samples taken during the winter and spring months of the surveys in the 1980s; peak densities were recorded during April–May 1985 (Table 3). During April–May 1983, osmerid larvae were unusually scarce. Brodeur et al. (1985) made the same observation for Oregon coastal waters during 1983. The unusually warm conditions in the coastal zone during spring 1983 may have brought about a change in the spawning patterns or early life history of osmerids that year.

Bathylagus ochotensis (eared blacksmelt)

The deep sea smelt, *B. ochotensis*, is common in the epi- and meso-pelagic zones throughout the Northeast Pacific. Spawning occurs throughout the year but with a significant peak during late winter and early spring off the U.S. northwest coast when eggs and larvae are common in the plankton (Doyle 1992*a*). Bathylagid eggs are difficult to identify to species and because of their small size are not sampled efficiently by standard bongo gear. Larval data, therefore, provides a better indication of seasonal and interannual variation in abundance.

During the west coast surveys, peak densities of *B. ochotensis* larvae were recorded during March to early April 1984 (Table 3). For the years 1980 to 1985 levels of abundance during the months April and May were moderate. In April–May 1983, however, larvae were significantly more abundant than for these months of the other years (Table 3, Fig. 7). The increased abundance may reflect enhanced spawning intensity or larval survival associated with the El Niño warming event of that year. There were also increased occurrences of larvae in coastal waters during spring 1983, particularly off southern Oregon and northern California between Cape Blanco and Cape Mendocino (Fig. 7). This may be related to the anomalous onshore transport which characterized the 1983 El Niño.

Diaphus theta (California headlightfish)

Diaphus theta is the second most abundant myctophid in the west coast survey area. Eggs and larvae are epi- and meso-pelagic, and spawning seems to take place throughout the year but with a significant peak during spring (Doyle 1992*a*).

The anomalous pattern for this species during 1983 is the increased occurrence of larvae close to the coast from Cape Blanco southwards (Fig. 8*b*). This shelf area is usually free of larvae (Figs. 8*a* and 8*c*) because of the association of *D. theta* with the oceanic zone and the normal offshore Ekman transport prevalent at this time of year. The anomalous onshore transport of warm, offshore water that prevailed through the spring and summer months of 1983 probably resulted in the advection of *D. theta* larvae shoreward along this section of the coast.

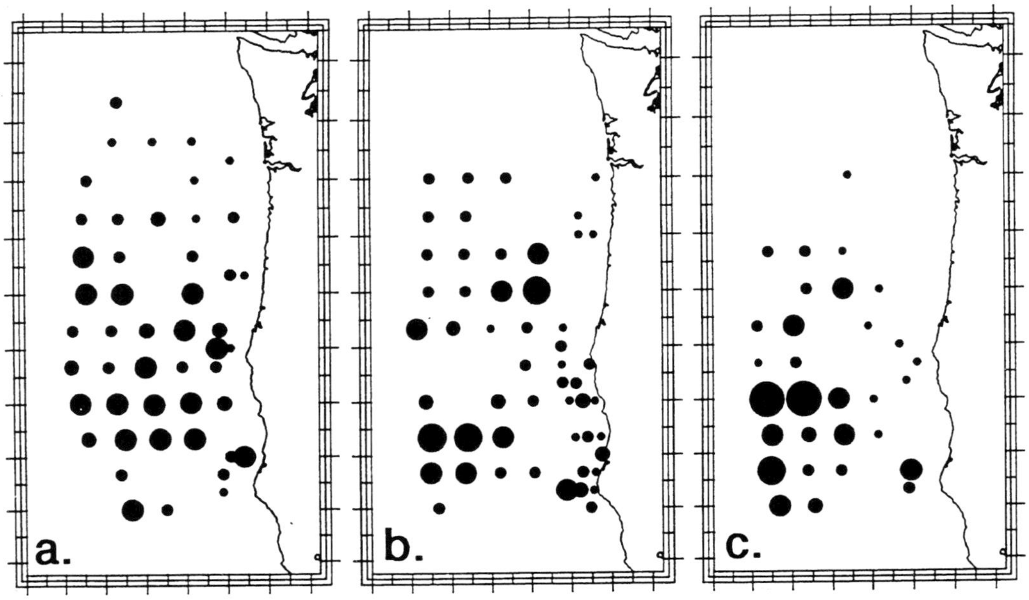

Fig. 8. Distribution patterns for *Diaphus theta* larvae from bongo samples taken during (*a*) May 1981, (*b*) April–May 1983, and (*c*) April–May 1985. See Fig. 6 for key to numbers of larvae.

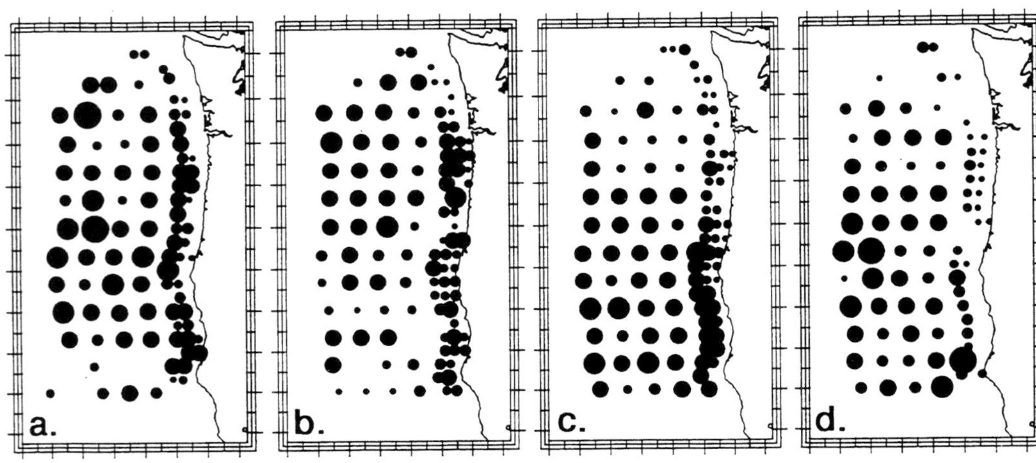

Fig. 9. Distribution patterns for *Stenobrachius leucopsarus* larvae from bongo samples taken during (*a*) May 1981, (*b*) April–May 1983, (*c*) March–April 1984, and (*d*) April–May 1985. See Fig. 6 for key to numbers of larvae.

Stenobrachius leucopsarus (northern lampfish)
This epi-, meso- and bathy-pelagic species is the most abundant myctophid in the northeast Pacific (Smoker and Pearcy 1970). It was also by far the most abundant species of larvae in the west coast ichthyoplankton collections (Doyle 1992*a*). Eggs and larvae are pelagic and present in the plankton throughout the year with a major peak in abundance occurring during spring.

As for *D. theta*, the incidence of occurrence of *S. leucopsarus* larvae in coastal waters was much higher during spring 1983 in comparison with the spring of other years (Fig. 9). The usual pattern of distribution is characterized by the absence or scarcity of larvae close to the coast (Figs. 9*a*, 9*c*, and 9*d*). During April–May 1983, however, larvae were common and relatively abundant all along the Oregon and northern California shelf (Fig. 9*b*). Again, this pattern can be related to the anomalous onshore transport that prevailed during 1983 as a result of the major El Niño event.

Trachipterus altivelis (king-of-the-salmon)
In the eastern temperate Pacific, this species is common in the epi- and meso-pelagic zones (Matarese et al. 1989). Eggs and larvae are pelagic, and in the survey area, spawning seems to occur throughout the year with no major peak (Doyle 1992*a*). Eggs were extremely abundant in the neuston samples and moderately abundant in the bongo collections. Larvae of *T. altivelis* were absent from the neuston and scarce in the bongo samples.

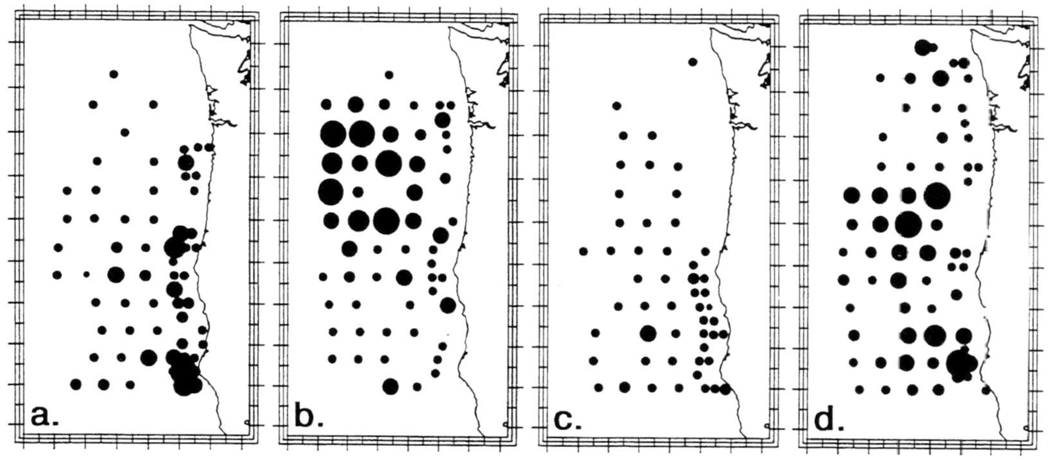

Fig. 10. Distribution patterns for *Trachipterus altivelis* eggs from neuston samples taken during (*a*) October–November 1981, (*b*) April–May 1983, (*c*) November 1983, and (*d*) April–May 1985. See Fig. 6 for key to numbers of larvae.

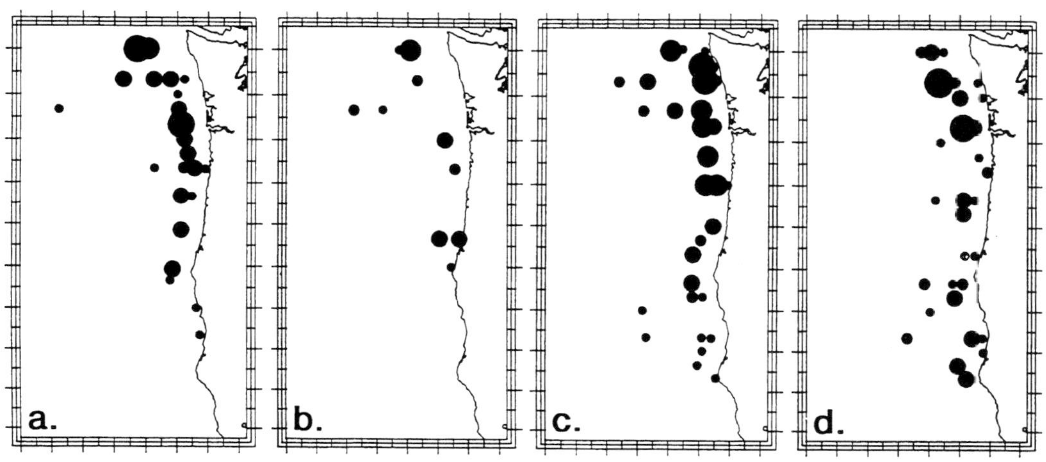

Fig. 11. Distribution patterns for *Hemilepidotus spinosus* larvae from neuston samples taken during (*a*) April–May 1980, (*b*) April–May 1983, (*c*) March–April 1984, and (*d*) April–May 1985. See Fig. 6 for key to numbers of larvae.

The 1983 anomaly for this species consisted of a more northerly distribution of eggs in the neuston than usual (Fig. 10). The usual pattern of egg distribution for this species is characterized by the occurrence of peak densities in the southern section of the sampling area, particularly in slope and deep water off southern Oregon and northern California (Figs. 10*a*, 10*c*, and 10*d*). This suggests a positive association between spawning and water temperature, as temperature increases in a southerly direction in the deep water zone. During April–May 1983, peak densities of eggs were centred in deep water off northern Oregon (Fig. 10*b*), implying a northerly extension in peak spawning activity. This apparent northerly extension in spawning may be related to the anomalous high water temperatures throughout the sampling area during spring 1983.

Hemilepidotus spinsosus (brown Irish lord)
The cottid *H. spinosus* occurs in shallow water from the intertidal zone down to approximately 100 m depth in the Northeast Pacific. Spawning occurs primarily from December to February or March, and a northerly progression in spawning occurs along the coast (Doyle 1992*a*). Larvae of *H. spinosus* seem to be obligate members of the neuston (Doyle 1992*b*).

Densities of *H. spinosus* larvae were anomalously low during April–May 1983 (Table 3, Fig. 11). Perhaps there was a reduction in spawning activity during spring 1983 because of the occurrence of higher temperatures in the coastal zone. It is also possible that peak spawning may have taken place earlier than usual that year in response to anomalously high temperatures during late winter. The apparent northerly progression in spawning for this species along the west coast supports such a positive association between spawning activity and temperature.

Icichthys lockingtoni (medusafish)
The medusafish occurs in coastal and offshore waters throughout the entire North Pacific. It lives commensally in the bell of large jellyfish, but large medusafish are thought to live a solitary existence (Fitch and Lavenberg 1968). Eggs and

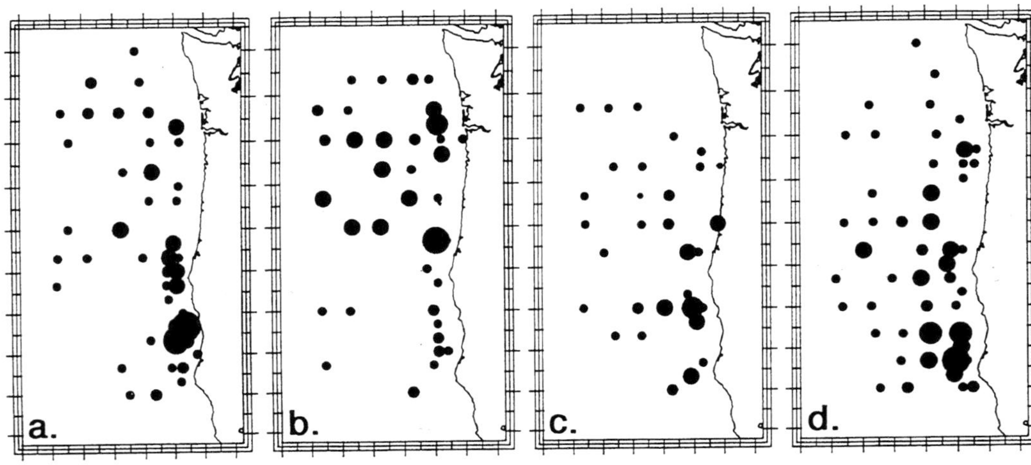

Fig. 12. Distribution patterns for *Icichthys lockingtoni* eggs from neuston samples taken during (*a*) May 1981, (*b*) April–May 1983, (*c*) March–April 1984, and (*d*) April–May 1985. See Fig. 6 for key to numbers of larvae.

larvae are pelagic but little else is known about the biology or early life history of this species. Spawning occurs throughout the year with a possible extended peak from mid-spring through late summer (Doyle 1992*a*). Medusafish larvae were scarce in the neuston and bongo samples. However, eggs were abundant in both, particularly in the neuston. It seems that the eggs are positively buoyant and accumulate in the surface layer.

Eggs were, on average, widely distributed throughout the sampling area. Abundance tended to be higher in the southern sector, however, and peak densities were generally along the shelf edge and slope region off southern Oregon and northern California (Figs. 12*a*, 12*c*, and 12*d*). A more northerly distribution occurred during the spring of 1983 (Fig. 12*b*). This difference again suggests an association between spawning and water temperature.

Citharichthys spp. (sanddabs)

Members of the left-eyed flatfish family Paralichthyidae live demersally in shelf and coastal waters of the North Pacific. Two species occur in the study area: *Citharichthys sordidus* (the Pacific sanddab) and *C. stigmaeus* (the speckled sanddab) (Matarese et al. 1989). Both species have pelagic eggs and larvae. They can be found in the plankton of coastal waters during all seasons although spawning intensity seems to be strongly seasonal (Doyle 1992*a*). Poor sampling coverage during summer and autumn, however, precludes an accurate description of the annual spawning pattern. Nevertheless, the data suggest a peak spawning period of August to January for these species in the sampling area. At present, *Citharichthys* spp. eggs cannot be identified with certainty to species (Matarese et al. 1989). Eggs were considerably more abundant in the west coast samples than larvae, especially in the neuston, where they seem to accumulate because of positive buoyancy.

Interannual variation in egg and larval abundance suggests that 1983 was an unusual year with an increased intensity of spawning during spring and diminished spawning in autumn (Doyle 1992*a*). Eggs and larvae were significantly more abundant in April–May 1983 than in the same period in other years and were considerably less abundant in autumn 1983 than in autumn 1981 (Table 3). It seems that, as for anchovy, there was a temporal shift in spawning of the sanddabs during 1983 and that this earlier spawning may be related to the anomalously high temperatures that prevailed off the U.S. west coast during the spring of that El Niño year.

Unusual occurrences of rare species

Several species of ichthyoplankton that are normally absent or very scarce off Washington and Oregon occurred anomalously in the sampling area during 1983 and 1984 (Table 4). All these fish species are abundant off California and their eggs and larvae are prominent in California Cooperative Oceanic and Fisheries Investigations (CalCOFI) ichthyoplankton collections (Loeb et al. 1983; Moser et al. 1987; Stevens et al. 1987, 1988, 1990).

Apart from a single occurrence in May 1982, eggs of *Nansenia crassa* (family Argentinidae) occurred in the west coast samples only during November 1983 and March–April 1984. Similarly, there was a single occurrence of larvae of the bathylagid *Leuroglossus stilbius* in May 1982, and five specimens were taken in March–April 1984. This bathylagid characterizes a northern complex of the larval fish assemblages in the CalCOFI survey area (Moser et al. 1987). The myctophid *Lampanyctus ritteri* belongs to a southern complex among the larval fish assemblages off California (Moser et al. 1987). Low numbers of these larvae were taken during most of the west coast cruises (Table 4). However, occurrences during 1983 and 1984 were considerably higher than for the other years.

Pacific whiting or hake (*Merluccius productus*) is one of the most abundant marine fish in the California Current system. It spawns during the winter off Baja California. During spring, juveniles and adults migrate northwards, as far as British Columbia, before returning to their winter spawning grounds (Bailey et al. 1982). During the springs of 1983 and 1984, *M. productus* eggs and larvae occurred anomalously in the west coast ichthyoplankton samples (Table 4). There were no records of hake eggs or larvae during other years of the west coast surveys or in other investigations off Oregon. It is known that all age groups of Pacific hake migrated farther north in 1983 than in other years (Arntz et al. 1991) and it is postulated that the traditional spawning location of hake shifted north in 1983 and 1984 (Hollowed 1992).

Table 4. Variation in occurrence among sampling periods for rare species that displayed anomalous occurrence during 1983 or 1984 relative to other years. Numbers given are the total number of specimens of eggs or larvae caught during each sampling period. Sampling periods arranged in seasonal order. – = Species absent from samples.

Species	Stage	March–April 1984	April–May 1985	April–May 1980	April–May 1983	May 1982	May 1981	August 1980	October–November 1981	November 1983	January 1987
Bongo samples											
Nansenia crassa	Eggs	4	–	–	–	1	–	–	–	6	–
Leuroglossus stilbius	Larvae	5	–	–	–	1	–	–	–	–	–
Lampanyctus ritteri	Larvae	46	2	5	41	4	26	6	2	6	–
Merluccius productus	Eggs	46	–	–	23	–	–	–	–	4	–
	Larvae	36	–	–	1	–	–	–	–	–	–
Neuston Samples											
Nansenia crassa	Eggs	6	–	–	–	–	–	–	–	1	–

Table 5. Two-way coincidence table resulting from numerical classification of egg abundance data from bongo samples taken during spring cruises (data from six cruises combined). Mean abundance in station groups expressed as no.·10 m^{-2}.

Species groups		Station groups 1	2	3	4
1	Bathylagidae (unidentified)	16.1	7.3	19.2	0.1
	Trachipterus altivelis	8.6	2.9	5.0	0.9
	Bathylagus spp.	6.3	3.2	6.5	0.1
	Icichthys lockingtoni	3.1	4.4	3.9	0.2
	Chauliodus macouni	5.3	0.7	2.0	0.0
	Bathylagus ochotensis	2.7	2.4	0.8	0.0
	Myctophidae (unidentified)	228.7	34.5	0.5	0.1
	Icosteus aenigmaticus	2.6	1.9	9.9	0.7
2	*Errex zachirus*	0.0	5.6	0.5	10.9
	Eopsetta exilis	0.2	21.9	0.3	41.4
	Microstomus pacificus	0.1	9.7	0.9	2.6
	Pleuronectidae (unidentified)	0.0	0.1	0.1	33.0
	Paralichthyidae (unidentified)	0.1	3.8	0.6	79.0
3	*Pleuronectes vetulus*	0.0	0.0	0.0	0.9
	Citharichthys spp.	0.0	0.2	0.1	4.3
	Engraulis mordax	0.1	0.1	0.0	28.9
	Total fish eggs (dominant taxa)	273.9	97.7	50.3	203.1

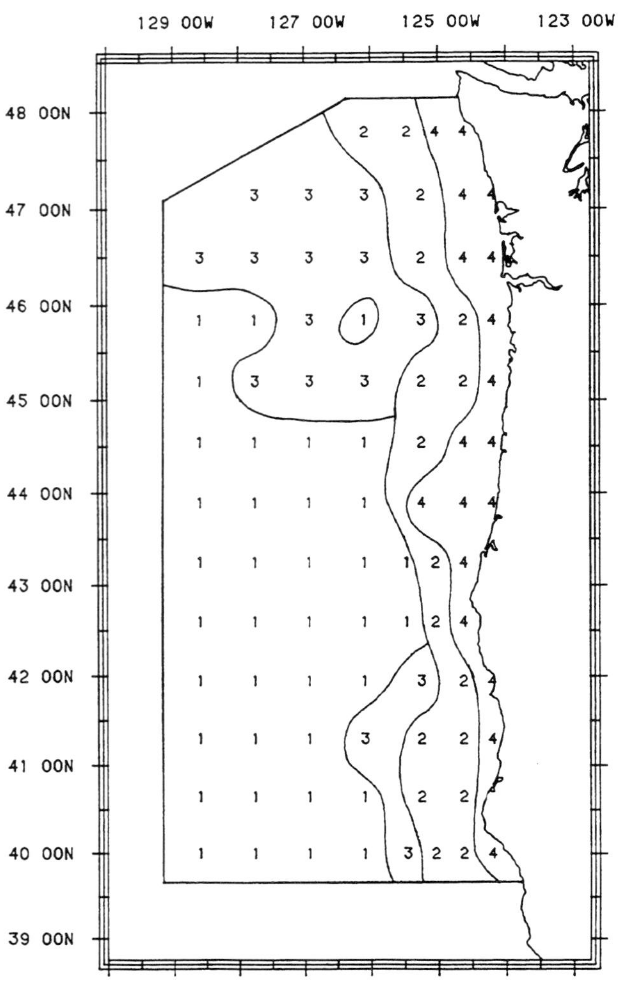

Fig. 13. Distribution of station groups for dominant taxa of eggs in bongo samples, spring cruises.

The unusual occurrences of all the above species off the U.S. northwest coast during 1983 and 1984 may be related to the anomalously warm water and enhanced poleward transport characterizing the 1983 El Niño event.

Anomalous patterns in distribution of ichthyoplankton assemblages

Several subsurface ichthyoplankton assemblages have been identified in the survey area, and their distribution and seasonal variation have been described (Doyle 1992*a*; Doyle et al. 1993). In general, the boundaries to these assemblages are fluid. Unique coastal and oceanic assemblages are present during all seasons of the year, but abundance of the constituent species vary seasonally. Northern and southern components to the assemblages, distinguished on the basis of species abundance, may manifest during the coldest months of the year. A slope–transitional assemblage, associated with the continental slope region, can be identified during spring and summer. It is defined by species for which the adult habitat is the outer continental shelf and slope but also can contain species from the coastal and oceanic assemblages. The Columbia River plume assemblage occurs only during summer when the warm waters of the plume extend offshore, and it is defined essentially by the occurrence of peak densities of anchovy larvae.

The west coast ichthyoplankton assemblages, which have been identified by numerical classification of abundance and distribution data, are compared here for general spring conditions and the spring of the El Niño year, 1983. Eggs and larvae are treated separately.

Table 6. Two-way coincidence table resulting from numerical classification of egg abundance data from bongo samples taken during April/May 1983. Mean abundance in station groups expressed as no.·10 m^{-2}.

Species groups		Station groups				
		1	2	3	4	5
1*a*	*Chauliodus macouni*	0.0	0.0	0.0	3.7	5.2
	Trachipterus altivelis	2.7	0.0	1.3	11.5	8.6
	Myctophidae (unidentified)	0.0	7.3	0.0	170.3	0.0
1*b*	*Merluccius productus*	3.0	1.3	2.2	0.3	0.5
	Bathylagus ochotensis	0.0	0.8	0.0	2.3	0.0
1*c*	Bathylagidae (unidentified)	1.1	5.9	1.7	13.1	1.8
	Bathylagus spp.	0.2	10.0	0.0	4.4	1.7
	Icichthys lockingtoni	0.6	8.4	0.5	1.4	0.0
	Icosteus aenigmaticus	1.6	9.2	0.5	1.2	0.5
2*a*	*Errex zachirus*	18.5	0.5	5.7	0.3	0.0
	Eopsetta exilis	78.8	0.5	17.7	0.0	0.0
	Microstomus pacificus	20.8	4.1	2.4	1.5	1.4
	Paralichthyidae (unidentified)	7.1	0.4	193.2	0.3	0.0
2*b*	Pleuronectidae (unidentified)	0.5	0.0	16.5	0.3	0.0
	Engraulis mordax	0.3	0.0	140.0	0.0	0.0
	Citharichthys spp.	0.0	0.0	10.3	0.0	0.0
	Total fish eggs	135.3	48.8	396.3	226.1	21.4

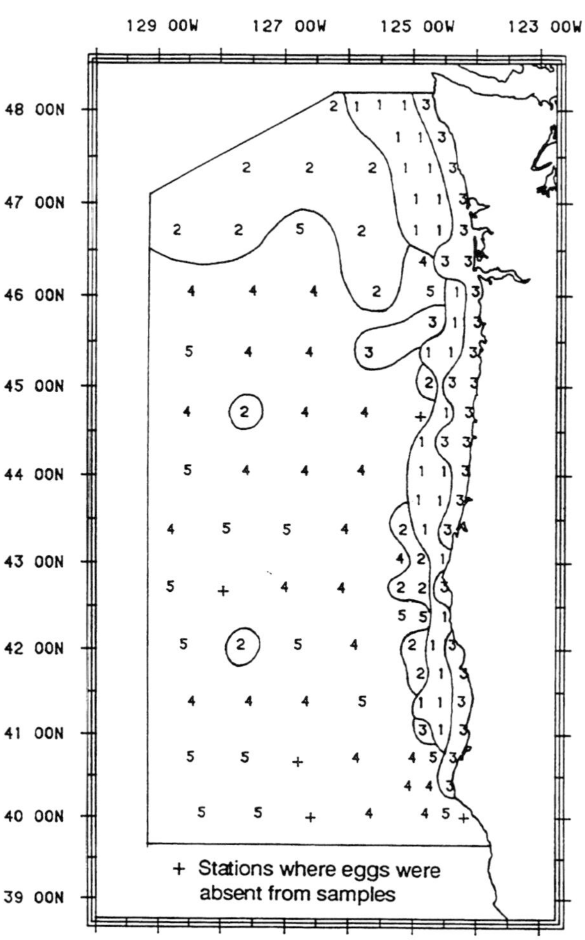

Fig. 14. Distribution of station groups for dominant taxa of eggs in bongo samples, April–May 1983.

The nature and distribution of the station groups, identified for the dominant taxa of eggs in bongo samples from all spring cruises, reflects the occurrence of the three main ichthyoplankton assemblages in the survey area during usual spring conditions (Table 5, Fig. 13). Station group 4 represents the coastal assemblage and contained the highest numbers of all pleuronectid taxa (except for *Microstomus pacificus*), the paralichthyids (including the two *Citharichthys* species), and *E. mordax*. Along the slope and adjacent deep water, station group 2 represents the slope–transitional assemblage. It is characterized by peak densities of *M. pacificus* and low to moderate densities of the other flatfish taxa and most of the mesopelagic taxa. The oceanic assemblage is described by station groups 1 and 3 in the oceanic zone, which contained the mesopelagic taxa almost exclusively. Peak densities of myctophids, *Trachipterus altivelis* and *Chauliodus macouni* occurred in station group 1 to the south of this zone, whereas *Icosteus aenigmaticus* was most abundant in station group 3 to the north. The bathylagids and *Icichthys lockingtoni* were relatively evenly distributed among these two station groups.

The distribution of station groups for eggs in the bongo samples taken during the spring of the 1983 El Niño year (Table 6, Fig. 14) again reflects the occurrence of the three main ichthyoplankton assemblages. Differences are apparent, however, when compared with the general spring patterns. For instance, the paralichthyids (*Citharichthys* spp.) and *E. mordax*, which characterize the coastal assemblage (station group 3), were considerably more abundant during 1983. In

Table 7. Two-way coincidence table resulting from numerical classification of larval abundance data from bongo samples taken during spring cruises (data from six cruises combined). Mean abundance in station groups expressed as no.·10 m^{-2}.

Species groups		Station groups 1	2	3	4
1	*Chauliodus macouni*	2.8	1.7	0.4	0.0
	Nansenia candida	4.4	1.5	0.2	0.0
	Lampanyctus spp.	3.8	1.2	0.2	0.0
	Lampanyctus ritteri	4.8	1.0	0.2	0.0
	Lestidiops ringens	4.2	0.3	0.1	0.0
	Bathylagus pacificus	1.3	3.8	2.7	0.1
	Protomyctophum thompsoni	1.7	2.4	1.2	0.1
	Sebastolobus spp.	2.5	2.8	1.0	0.4
	Myctophidae (unidentified)	7.2	1.4	0.9	0.3
	Trachipterus altivelis	0.6	0.1	0.0	0.0
2	*Tarletonbeania crenularis*	22.8	11.9	3.6	0.1
	Bathylagus ochotensis	51.0	28.8	4.2	0.2
	Diaphus theta	118.8	20.0	1.9	0.3
	Protomyctophum crockeri	10.0	4.0	1.1	0.1
	Stenobrachius leucopsarus	214.2	144.9	87.6	6.4
	Sebastes spp.	4.8	15.8	31.4	9.6
3	*Pleuronectes vetulus*	0.0	0.0	0.7	11.8
	Engraulis mordax	0.0	0.2	1.1	6.2
	Osmeridae (unidentified)	0.0	0.0	0.3	35.4
	Citharichthys spp.	0.0	0.1	0.6	0.9
	Errex zachirus	0.4	0.4	1.8	1.5
	Eopsetta exilis	0.5	3.6	6.0	5.2
	Total fish larvae (dominant taxa)	455.8	245.9	147.2	78.6

addition, the distribution of the coastal (station group 3) and slope (station group 1) assemblages were restricted to a narrower band along the shelf and slope, and the oceanic assemblage (station groups 2, 4, and 5) extended closer to shore than usual (Fig. 14). The southern component of the oceanic assemblage, represented by station groups 4 and 5 and containing the highest densities of myctophids, and the species *Trachipterus altivelis* and *Chauliodus macouni*, also extended farther to the north during 1983.

In general, for larvae in the bongo samples during spring, the various assemblages of species displayed a distinct north–south trend as well as an inshore–offshore trend (Table 7, Fig. 15). The coastal assemblage, represented by station group 4 and characterized by the occurrence of anchovy, osmerid, pleuronectid, and *Citharichthys* larvae, was most prominent off Washington and northern Oregon and relatively scarce off southern Oregon and northern California. Station group 3 representing the slope–transitional assemblage (characterized by highest densities of *Sebastes* spp. and the deepwater pleuronectids, and low densities of the mesopelagic taxa) extended over a wide area of the slope and adjacent oceanic zone off Washington and northern Oregon but was confined essentially to a narrow band of shelf and slope off southern Oregon and northern California. In the oceanic zone, the southern component of the oceanic assemblage was represented by station group 1 occurring in the deepest water off southern Oregon and northern California. Larval fish abundance was at a maximum in this group with peak densities of most mesopelagic taxa, particularly myctophids. The other component of the oceanic assemblage consisted of station group 2, which occurred most extensively off northern Oregon. Species composition here was the same as for the southern component, but larval abundance was lower overall mainly because of the reduction in numbers of the dominant myctophids.

Quite a different pattern in the distribution of species assemblages was apparent for larvae in the bongo samples during spring 1983 (Table 8, Fig. 16). The coastal and slope–transitional assemblages, represented by station groups 3 and 2, respectively, were restricted mainly to a narrow band along the shelf. No northern and southern components were apparent in the oceanic assemblage (described by station groups 1, 4, 5, and 6), which occurred thoughout most of the sampling area and extended much closer to the coast than usual.

The usual north–south trends in the assemblages, which have been observed during "normal" spring conditions in the survey area, reflect the north-south trends in water temperature and coastal upwelling that are prevalent at that time of year (Doyle 1992*a*; Doyle et al. 1993). For instance, the reduced abundance of species belonging to the coastal assemblage off southern Oregon and northern California can be related to the

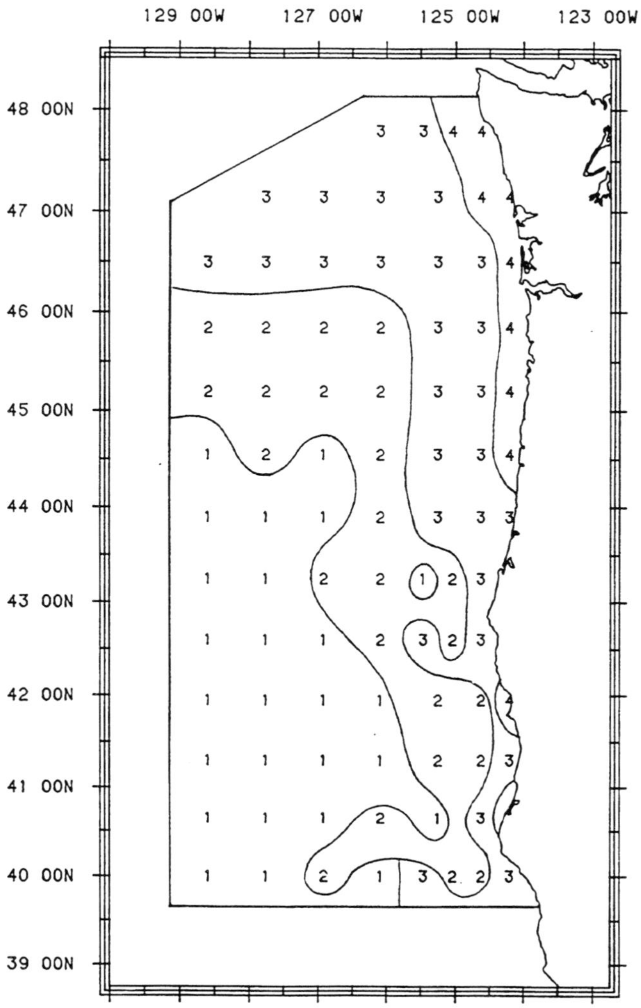

Fig. 15. Distribution of station groups for dominant taxa of larvae in bongo samples, spring cruises.

early commencement of offshore transport of surface water, and upwelling of cold oceanic water here, compared with farther north (Figs. 2*a*, 2*c*, 3*a*, 3*c*, and 4). Spawning, and the abundance of eggs and larvae of these species, are likely to be reduced in this region because of the occurrence of extensive drift offshore, away from nursery grounds and adult habitats.

Concomitantly, the north–south trend in abundance of mesopelagic fish eggs and larvae, which gives rise to northern and southern components in the oceanic assemblage, seems to be related to temperature. The greater abundance of eggs and larvae of many of the mesopelagic species, particularly myctophids, observed in the southern part of the oceanic zone suggests that spawning is associated with the warmer water in this region (Figs. 2*a*, 2*c*, 3*c*, and 4).

During spring 1983, temperatures along the coast were uniformly high, and offshore transport and coastal upwelling were diminished (Figs. 2*c*, 3*c*, and 4). The lack of any north–south trend in the coastal assemblage of ichthyoplankton on this occasion seems to reflect this feature of oceanographic uniformity along the coast. Similarly, the relatively even distribution of mesopelagic taxa in the oceanic assemblage of ichthyoplankton during spring 1983 may be related to the extension of anomalously warm water throughout the oceanic zone (Figs. 2*c* and 3*c*). The onshore extension of the oceanic assemblage of ichthyoplankton during March–April 1983 can be related to the anomalous shoreward transport of water that prevailed that year, particularly in the southern portion of the survey area.

Discussion

The anomalous patterns observed in the ichthyoplankton off the U.S. northwest coast during 1983 are attributed to the following changes in spawning and early life-history patterns among an assortment of fish species:

1) An early commencement and completion of spawning among certain summer spawners such as *Engraulis mordax* and *Citharichthys* spp.
2) Diminished spawning or temporal shift in spawning among some coastal winter–spring spawners such as Osmeridae and *Hemilepidotus spinosus*.
3) A northerly shift in the main spawning area among certain mesopelagic taxa such as *Bathylagus ochotensis*, *Trachipterus altivelis*, and *Icichthys lockingtoni*.
4) An extensive northerly shift in spawning range among several "southern" species that do not usually spawn in this region, such as *Nansenia crassa*, *Leuroglossus stilbius*, and *Merluccius productus*.
5) A shoreward extension of the oceanic assemblage of ichthyoplankton and concomitant retention of the coastal and slope–transitional assemblages closer to the coast.

These biological anomalies can be correlated with the physical oceanographic anomalies that characterized the 1983 El Niño. The changes in the ichthyoplankton that resulted from the physical forcing on the ecosystem seem to have been mediated mainly through water temperature and transport anomalies.

Increased temperatures throughout the survey area during 1983 are considered to have caused both the temporal shift in spawning of certain species and the northern extension in the spawning range or peak spawning area among other species. In the CalCOFI survey area, Moser et al. (1987) observed that the northern complex of larval fish species declined and their southern distributional limits contracted northward during the 1957 El Niño, while species of the southern complex of fish larvae showed a relative increase and a concomitant northward expansion of their northern distributional limits. They associate these northerly shifts in larval distributions with a northward expansion of spawning as a result of the warming trend that began in 1957.

It is also possible that the occurrence of spawning adults so far north and the presence of their eggs and larvae may have resulted from the anomalous northward transport of water that occurred in the California Coastal Current region during the 1983 El Niño. Miller et al. (1985) documented the anomalous occurrence of southern species in the zooplankton off Oregon during spring and summer 1983 and attribute these occurrences to the increased poleward advection of water.

Even though sea temperatures, sea level, and water transport patterns seem to have returned to normal in U.S. coastal waters during 1984, in general the biological effects of the strong El Niño event lingered, and were pronounced in 1984 as well as

Table 8. Two-way coincidence table resulting from numerical classification of larval abundance data from bongo samples taken during April–May 1983. Mean abundance in station groups expressed as no.·10 m^{-2}.

Species groups		Station groups 1	2	3	4	5	6
1a	*Diaphus theta*	2.1	0.4	2.5	110.1	54.0	58.3
	Tarletonbeania crenularis	3.5	3.3	0.0	32.8	12.1	7.6
	Stenobrachius leucopsarus	70.9	90.4	8.4	203.0	135.1	42.3
	Bathylagus ochotensis	16.5	3.0	1.3	64.6	47.5	41.5
1b	*Errex zachirus*	0.0	11.3	4.2	0.0	8.5	0.0
	Eopsetta exilis	2.8	19.9	6.5	1.5	12.6	0.0
	Sebastes spp.	4.5	27.6	18.9	4.0	45.0	0.0
2	*Bathylagus pacificus*	1.8	2.3	0.6	0.9	3.0	0.0
	Protomyctophum thompsoni	1.5	0.0	0.0	3.5	0.9	0.0
	Microstomus pacificus	0.6	3.9	0.7	1.1	0.9	0.0
	Icichthys lockingtoni	0.0	0.0	0.0	0.9	3.4	1.3
	Sebastolobus spp.	0.0	2.3	1.3	4.8	10.1	9.9
	Lampanyctus spp.	0.3	0.5	0.0	2.0	0.4	0.0
3	*Chauliodus macouni*	0.6	0.0	0.0	6.2	0.0	2.5
	Lampanyctus ritteri	0.0	0.0	0.0	7.8	3.3	6.9
	Protomyctophum crockeri	1.3	0.9	0.0	5.7	2.5	16.6
	Lestidiops ringens	0.0	0.0	0.0	3.1	0.4	9.1
	Myctophidae (unidentified)	2.4	0.0	0.2	3.0	0.0	0.0
	Melamphaeidae (unidentified)	0.6	0.0	0.0	2.5	0.0	0.0
	Nansenia candida	0.3	0.4	0.3	3.1	0.8	0.0
4	*Pleuronectes vetulus*	0.6	0.9	2.8	0.0	0.4	0.0
	Psettichthys melanostictus	0.0	0.0	7.6	0.0	0.0	0.0
	Citharichthys spp.	0.3	2.4	9.7	0.0	1.7	0.0
	Engraulis mordax	0.0	25.7	32.4	0.3	1.7	0.0
	Microgadus proximus	0.0	0.0	3.4	0.0	0.0	0.0
	Pleuronectes isolepis	0.0	0.3	6.5	0.0	0.0	0.0
	Cyclopteridae (unidentified)	0.9	3.8	2.0	0.0	0.4	0.0
	Total Fish Larvae	116.3	207.1	128.8	470.1	353.6	201.0

1983 (Huyer and Smith 1985; Arntz et al. 1991). The anomalous occurrences of the rare "southern" species of fish eggs and larvae off the U.S. northwest coast during 1984 may be part of this lingering effect.

The shoreward extension of the oceanic assemblage of ichthyoplankton and the retention of the coastal and slope–transitional assemblages closer to the coast during 1983 can be related directly to the anomalous shoreward Ekman transport prevalent that year along the coast, particularly off southern Oregon and northern California. Diminished production in the plankton of U.S. west coast waters during the 1983 El Niño may also have influenced the abundance of fish larvae in this region by reducing the food organisms available to them.

It is difficult to draw any conclusions regarding the effects of the 1983 El Niño on survival and recruitment among fish species off the U.S. northwestern coast. No subsequent population abundance data are available for those species for which anomalies in early life-history patterns were observed during 1983. No data are available either on the effects of the El Niño anomalies on such factors as larval growth rates, condition, and survival, or growth and survival at the juvenile stage. Along with changes in spawning patterns and egg and larval distribution patterns, changes in any or all of these factors as a result of an El Niño could have potentially drastic effects on recruitment levels (Bailey and Incze 1985). Brodeur et al. (1985) speculate that although northern anchovy larvae had unusually broad spatial and temporal distributions off Oregon in 1983, prospects for the eventual survival of the 1983 year-class of northern anchovy as well as other spring- and summer-spawning fish species in this region are uncertain because of the low levels of plankton production associated with the much-reduced upwelling.

It is interesting to note that patterns in distribution or abundance of eggs and larvae during 1983 were anomalous for only a small portion of the dominant fish species occurring off the U.S. northwest coast. This suggests a high level of stability among the spawning and early life-history patterns of fish species in the survey area. Further evidence for such stability is the apparent return to "normal" patterns in 1984 and 1985. The ichthyoplankton patterns observed during the springs of

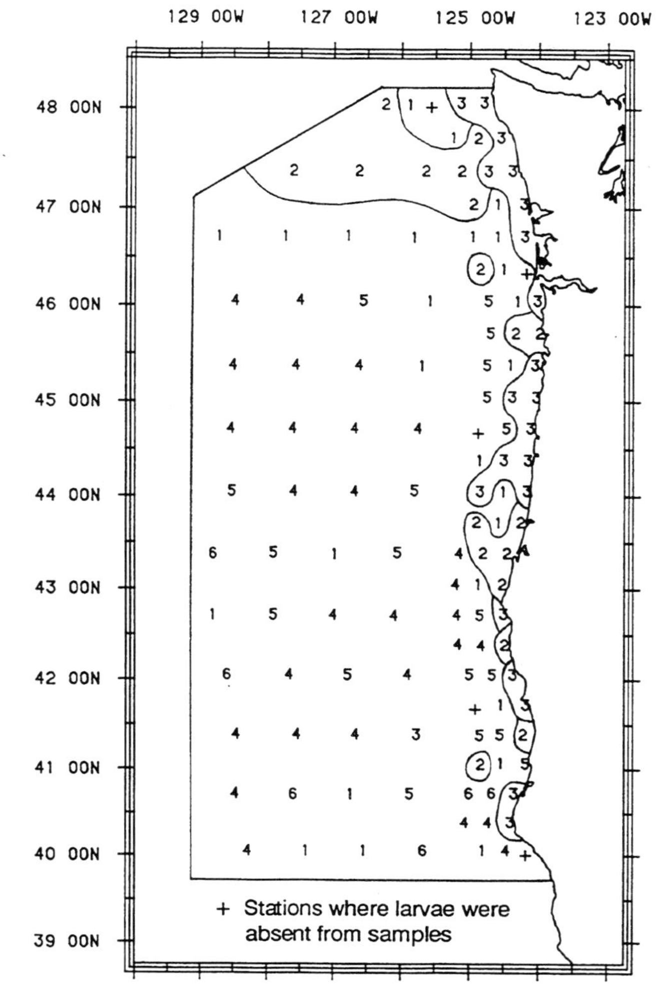

Fig. 16. Distribution of station groups for dominant taxa of larvae in bongo samples, April–May 1983.

1984 and 1985 were comparable with those observed during the springs of 1980, 1981, and 1982 and seemed within the normal range of interannual variation in such patterns. Similarly, Moser et al. (1987) noted the conservative quality among larval fish assemblages in the southern domain of the California Current region despite some changes in distribution patterns brought about by the 1957 El Niño.

References

Arntz, W., W.G. Pearcy, and F. Trillmich. 1991. Biological consequences of the 1982–83 El Niño in the eastern Pacific, p. 22–42. *In* F. Trillmich and K.A. Ono [ed.] Pinnipeds and El Niño – Responses to environmental stress. Springer-Verlag, Berlin.

Bailey, K.M., R.C. Francis, and P.R. Stevens. 1982. The life history and fishery of Pacific whiting, *Merluccius productus*. Calif. Coop. Oceanic. Fish. Invest. Rep. 23: 81–98.

Bailey, K.M., and L.S. Incze. 1985. El Niño and the early life history and recruitment of fishes in temperate marine waters, p. 143–165. *In* W.S. Wooster and D.L. Fluharty [ed.] El Niño North. Washington Sea Grant Program, University of Washington Press, Seattle, Wash.

Bakun, A. 1973. Coastal upwelling indices, west coast of North America, 1946–71. NOAA Tech. Rep. NMFS, SSRF-671. 103 p.

Barber, R.T., and F.P. Chavez. 1983. Biological consequences of El Niño. Science (Wash., D.C.), 222: 1203–1210.

Boesch, D.F. 1977. Application of numerical classification in ecological investigations of water pollution. U.S. Environmental Protection Agency, Washington, D.C. Ecol. Res. Ser. EPA-600/3-77-033: 115 p.

Bray, J.R., and J.T. Curtis. 1957. An ordination of the upland forest communities of southern Wisconsin. Ecol. Monogr. 27: 325–349.

Brodeur, R.D., D.M. Gadomski, W.G. Pearcy, H.P. Batchelder, and C.B. Miller. 1985. Abundance and distribution of ichthyoplankton in the upwelling zone off Oregon during anomalous El Niño conditions. Estuanine Coastal Shelf Sci. 21: 365–378.

Cane, M.A. 1983. Oceanographic events during El Niño. Science (Wash., D.C.), 222: 1189–1195.

Clifford, H.T., and W. Stephenson. 1975. An introduction to numerical classification. Academic Press, New York. 229 p.

Doyle, M.J. 1992*a*. Patterns in distribution and abundance of ichthyoplankton off Washington, Oregon, and northern California (1980 to 1987). NOAA, NMFS, AFSC Proc. Rep. 92-14: 344 p.

Doyle, M.J. 1992*b*. Neustonic ichthyoplankton in the northern region of the California Current ecosystem. Calif. Coop. Oceanic. Fish. Invest. Rep. 33: 141–161.

Doyle, M.J., W.W. Morse, and A.W. Kendall Jr. 1993. A comparison of larval fish assemblages in the temperate zone of the northeast Pacific and northwest Atlantic. Bull. Mar. Sci. 53(2): 588–644.

Fahrbach, E., F. Trillmich, and W. Arntz. 1991. The time sequence and magnitude of physical effects of El Niño in the eastern Pacific, p. 8–21. *In* F.Trillmich and K.A. Ono [eds] Pinnipeds and El Niño – Responses to environmental stress. Springer-Verlag, Berlin.

Fiedler, P.C., and R.M. Laurs. 1990. Variability of the Columbia River plume observed in visible and infrared satellite imagery. Int. J. Remote Sensing 11(6): 999–1010.

Fitch, J.E. and R.J. Lavenberg. 1968. Deep-water fishes of California. University of California Press, Berkeley, Calif. 155 p.

Hollowed, A.B. 1992. Spatial and temporal distributions of Pacific Hake, *Merluccius productus*, larvae and estimates of survival during early life stages. Calif. Coop. Ocean. Fish. Invest. Rep. 33: 100–123.

Huyer, A., and R.L. Smith. 1985. The signature of El Niño off Oregon, 1982–1983. J. Geophys. Res. 90: 7133–7142.

Lance, G.N., and W.T. Williams. 1967. A general theory of classificatory sorting strategies. 1. Hierarchical systems. Comput. J. 9: 373–380.

Laroche, J.L., and S.L. Richardson. 1980. Reproduction of northern anchovy, *Engraulis mordax*, off Oregon and Washington. Fish. Bull. U.S. 78: 603–618.

Loeb, V.J., P.E. Smith, and H.G. Moser. 1983. Geographical and seasonal patterns of larval fish species structure in the California Current area, 1975. Calif. Coop. Oceanic. Fish. Invest. Rep. 28: 97–127.

Lynn, R.J. 1983. The 1982–83 warm episode in the California Current. Geophys. Res. Lett. 10: 1093–1095.

Matarese, A.C., A.W. Kendall Jr., D.M. Blood, and B.M. Vinter. 1989. Laboratory guide to early life history stages of northeast Pacific fishes. NOAA Tech. Rep. NMFS-80. 625 p.

McGowan, J.A. 1984. The California El Niño, 1983. Oceanus, 27: 48–51.

Miller, C.B., H.P. Batchelder, R.D. Brodeur, and W.G. Pearcy. 1985. Response of the zooplankton and ichthyoplankton off Oregon to the El Niño event of 1983, p. 185–187. *In* W.S. Wooster and D.L. Fluharty [eds.] El Niño North. Washington Sea Grant Program, University of Washington Press, Seattle, Wash.

Moser, H.G., P.E. Smith, and L.E. Eber. 1987. Larval fish assemblages in the California Current region, 1954–1960, a period of dynamic environmental change. Calif. Coop. Oceanic Fish. Invest. Rep. 28: 97–127.

Pearcy, W.G., and A. Schoener. 1987. Changes in the marine biota coincident with the 1982 to 1983 El Niño in the northeastern subarctic Pacific ocean. J. Geophys. Res. 92C: 14 417 – 14 428.

Reed, R.K. 1984. Oceanographic observations off the Pacific Northwest following the 1982 El Niño event. Mar. Fish. Rev. 46: 7–12.

Richardson, S.L. 1981. Spawning biomass and early life of northern anchovy, *Engraulis mordax*, in the northern subpopulation off Oregon and Washington. Fish. Bull. U.S. 78: 855–876.

Richardson, S.L., J.L. Laroche, and M.D. Richardson. 1980. Larval fish assemblages and associations in the Northeast Pacific ocean along the Oregon coast, winter-spring 1972–1975. Est. Coast. Mar. Sci. 11: 671–699.

Sameoto, D.D., and L.O. Jaroszynski. 1969. Otter surface trawl: a new neuston net. J. Fish. Res. Board. Can. 26: 2240–2244.

Schoener, A., and D.L. Fluharty. 1985. Biological anomalies off Washington in 1982–1983 and other major Niño periods, p. 211–225. *In* W.S. Wooster and D.L. Fluharty [ed.] El Niño North. Washington Sea Grant Program, University of Washington Press, Seattle, Wash.

Simpson, J.J. 1992. Response of the Southern California Current system to the mid-latitude North Pacific coastal warming events of 1982–1983 and 1940–1941. Fish. Oceanogr. 1: 57–79.

Smith, P.E. 1985. A case history of an anti-El Niño to El Niño transition on plankton and nekton distribution and abundances, p. 121–142. *In* W.S. Wooster and D.L. Fluharty [ed.] El Niño north. Washington Sea Grant Program, University of Washington Press, Seattle, Wash.

Smith, P.E., and S.L. Richardson. 1977. Standard techniques for pelagic fish egg and larval surveys. FAO Fish. Tech. Pap. 175. 100 p.

Smoker, W., and W.G. Pearcy. 1970. Growth and reproduction of the lanternfish *Stenobrachius leucopsarus*. J. Fish. Res. Board Can. 27: 1265–1275.

Stevens, E.G., R.L. Charter, H.G. Moser, and M.S. Busby. 1987. Ichthyoplankton and station data for California Cooperative Oceanic Fisheries Investigations survey cruises in 1959. NOAA Tech. Memo. NMFS-SWFC-87: 273 p.

Stevens, E.G., R.L.Charter, H.G. Moser, and L.R. Zins. 1988. Ichthyoplankton and station data for California Cooperative Oceanic Fisheries Investigations survey cruises in 1969. NOAA Tech. Memo. NMFS-SWFC-100: 265 p.

Stevens, E.G., R.L. Charter, H.G. Moser, and C. Meyer. 1990. Ichthyoplankton and station data for California Cooperative Oceanic Fisheries Investigations survey cruises in 1984. NOAA Tech. Memo. NMFS-SWFC-141: 157 p.

A comparative study of climatic changes in the North Pacific and North Atlantic and their relation to the abundance of fish stocks

A.S. Krovnin

Krovnin, A.S. 1995. A comparative study of climatic changes in the North Pacific and North Atlantic and their relation to the abundance of fish stocks, p. 181–198. *In* R.J. Beamish [ed.] Climate change and northern fish populations. Can. Spec. Publ. Fish Aquat. Sci. 121.

Abstract: Spatial and temporal features of mean winter sea-surface temperature (SST) anomaly variations in the North Atlantic and North Pacific during the last several decades were studied. In each ocean, several large-scale subdomains with the coherent SST anomaly fluctuations were defined on the basis of Ward's method (1963), a hierarchical clustering method. The SST anomaly variations in the northeastern and southwestern North Atlantic are opposite to those in the northwestern and southeastern North Atlantic. The spatial structure of the SST anomaly fluctuations in the North Pacific is characterized by two independent patterns: changes in the SST anomalies in the eastern and central North Pacific, and those in the northwestern and southwestern North Pacific, are out of phase. Strong relationships between the SST anomaly variations in each subdomain and the well-known teleconnection patterns in the atmosphere were identified. Examination of the temporal variability of the SST anomalies within each subdomain in the North Pacific and North Atlantic allowed the definition of several periods with essentially different anomaly variations. Some evidence for climatic effects on variations in the abundance of commercial fish stocks were also obtained.

Résumé : Les caractéristiques spatiales et temporelles des variations des anomalies de la moyenne hivernale de la température superficielle de la mer (SST) dans l'Atlantique Nord et le Pacifique Nord pendant les dernières décennies ont été étudiées. Dans chaque océan, plusieurs sous-domaines à grande échelle présentant des fluctuations cohérentes des anomalies de la SST ont été définis à partir de la méthode de Ward (1963), méthode de groupement hiérarchique. Les variations des anomalies de la SST dans le nord-est et le sud-ouest de l'Atlantique Nord sont à l'opposé de celles du nord-ouest et du sud-est de l'Atlantique Nord. La structure spatiale des fluctuations des anomalies dans le Pacifique Nord se caractérise par deux régimes indépendants : les changements des anomalies qui se produisent dans l'est et le centre du Pacifique Nord sont déphasés par rapport à ceux qui sont observés dans le nord-ouest et le sud-ouest du Pacifique Nord. Nous avons repéré de fortes relations entre les variations des anomalies dans chaque sous-domaine et les schémas bien connus de téléconnection dans l'atmosphère. L'examen de la variabilité temporelle des anomalies de la SST dans chaque sous-domaine du Pacifique Nord et de l'Atlantique Nord a permis de définir plusieurs périodes de variations profondément différentes des anomalies. Nous avons aussi recueilli certaines preuves des effets des variations climatiques sur l'abondance des stocks commerciaux de poisson.

Introduction

Surface air temperature averaged globally or hemispherically is often used as an indicator of climate variability (e.g., Vinnikov 1986). However, regional and global climatic anomalies may differ considerably and even reveal opposit tendencies. To interpret local climatic tendencies correctly, one must consider them in the context of global climatic processes. To do that, one must have some idea of the spatial and temporal structures of the climate system.

A.S. Krovnin. Russian Federal Research Institute of Fisheries and Oceanography (VNIRO)17a, Verkhnyaya Krasnoselskaya, Moscow, 107140, Russia.

To study the spatial structure of the global climate system so-called teleconnections are used. These are simultaneous correlations between geopotential heights on a given pressure surface at widely separated points on the earth. Wallace and Gutzler (1981) described the five main teleconnection patterns in the middle troposphere of the northern hemisphere during the winter season. These atmospheric patterns are regional in scale, and their recurrent character results in the existence of multiple quasi-stationary regimes in the atmosphere.

Because of the atmosphere–ocean interaction, one could expect that anomalies in oceanographic characteristics, and in particular, sea surface temperature (SST), are closely related to large-scale anomalies in atmospheric circulation. This supposition is confirmed by the results of Namias (1972), who found strong spatial coherence between SST anomalies and the overlying mid-tropospheric height anomalies in the North Pacific. Such coherent anomalies, positive or negative, may cover as much as one third to one half of the North Pacific. Hence, the spatial scale of extensive regions with characteristic fluctuations in SST is comparable with that in the atmosphere. Identification of these regions may give some information about the mechanisms governing large-scale SST fluctuations.

The teleconnections in the atmosphere and ocean may give clues for understanding the relationships between various biological and fishery characteristics in distant regions of the world's oceans. Many examples of such relationships are documented by Cushing (1982).

The main objectives of this paper are to: (1) study the spatial and temporal structures of the SST anomaly variations in the North Atlantic and North Pacific; (2) analyze the relationships between SST anomaly fluctuations and atmospheric teleconnections; (3) compare climatic changes in the North Atlantic and North Pacific during the last decades; and (4) investigate possible relationships between variations in the abundance of some commercial fish stocks in the North Atlantic and North Pacific.

Data

As a basis for the study, monthly mean values of SST in the North Atlantic (from 20 to 65°N) and North Pacific (from 20 to 55°N), and geopotential heights on the 700-hPa surface (20–90°N, 0–355°E) at grid points of 5° latitude by 5° longitude for the period 1957–91 were used. Data on SST are available from the Russian Hydrometeorological Center, and those on geopotential heights are available from the U.S. National Center for Atmospheric Research. Additionally, for computation of indices of the North Atlantic Oscillation (NAO) and North Pacific Oscillation (NPO), monthly means of surface air temperature at certain points on a 5° by 10° grid were taken from the Russian Hydrometeorological Centre.

To obtain some evidence for relationships between fluctuations in the abundance of fishes, data on recruitment in some North Atlantic fish stocks were used. These are northeastern Arctic haddock (*Melanogrammus aeglefinus*) (1963–88), cod (*Gadus morhua*) (1950–88), saithe (*Pollachius virens*) (1962–88), North Sea haddock (1969–88), cod (1969–88), saithe (1960–88), plaice (*Pleuronectes platessa*) (1958–88), and whiting (*Micromesistius poutassou*) (1969–88); west of Scotland herring (*Clupea harengus*) (1970–88), haddock (1969–88), cod (1969–88), saithe (1968–88), whiting (1968–88); west of Ireland herring (1970–78); Irish Sea cod (1968–88); Gulf of Maine herring (1950–78); Fortune Bay herring (1964–78); southern Gulf of St. Lawrence spring- and fall-spawning herring (1958–71;and cod in Northwest Atlantic Fishery Organization (NAFO) Divisions 2J+3KL (1962–80), 3NO (1959–1985), and 3Ps (1959–84). The years in parentheses are the periods for which data are available. All data on recruitment in the Northeast Atlantic fish stocks are taken from the International Council for the Exploration of the Sea (ICES) (1990). Data on recruitment in cod stocks in NAFO Divisions 3NO and 3Ps are available from NAFO (1988, 1989), and those in NAFO Division 2J+3KL are from Baird and Bishop (1986). Data on Gulf of Maine herring are taken from Anthony and Fogarty (1985), Fortune Bay herring from Winters et al. (1985), southern Gulf of St. Lawrence herring from Winters (1976), and Georges Bank haddock from Cohen (1986). For the North Pacific, only data on the relative abundance of pollack (*Theragra chalcogramma*) stock in the eastern Bering Sea for the period of 1963–83 are used (Hen 1988).

Methods

For each point on a 5° by 5° grid, mean winter SST anomalies (January–April) and mean winter geopotential height anomalies (December–February) were calculated. The selection of the months from January to April for calculation of mean winter SST anomalies is based on a preliminary analysis of mean long-term annual changes averaged over each ocean. It was determined that the period from January through April is the coldest in both oceans.

The North Atlantic and North Pacific were partitioned into several regions, with coherent SST anomaly fluctuations within each region, by using the hierarchical clustering method, known as Ward's method (Ward 1963). This method is based on the supposition that groups produced by a clustering procedure should have maximum internal homogeneity. Ward proposed a technique that was specifically developed to minimize the pooled within-group sums of squares (defined as the sum of the squared distances from each point to its cluster centre) at each level. In other words, the two groups to be combined are those whose fusion produces the least increase in the within-group sum of squares. The algorithm for Ward's method used in this paper can be summarized as follows:

1) The squared Eucllidian distances between SST anomalies at *i*th and *j*th grid points are calculated:

$$\mathrm{d}_{ij}^2 = \sum_{l=1}^{P}\sum_{n=1}^{P}(T_{il} - T_{jl})(T_{in} - T_{jn})r_{ln}$$

where T_{il}, T_{jl}, T_{in}, and T_{jn} are the SST anomalies at grid points i and j in lth and nth years, respectively; r_{ln} is the correlation coefficient between fields of the SST anomalies in lth and nth years; and P is the number of observations at each grid point ($P=35$). In this step, each grid point is considered to be a single-point cluster.

2) The smallest element d_{ij}^2 remaining in matrix D^2 is determined.
3) Fusion of i and j into a single group, k, is carried out.
4) The new squared distances d_{km}^2 (where m represents each of the remaining points or groups) are calculated:

$$d_{km}^2 = \{(N_m + N_i)\, d_{im}^2 + (N_m + N_j)\, d_{jm}^2 - N_m\, d_{ij}^2\}\ (N_m + N_k)$$

where N_i and N_j are the numbers of individuals in groups i and j; $N_k = N_i + N_j$; and N_m is the number of individuals in group m. These distances replace d_{im}^2 and d_{jm}^2 in matrix D^2.

5) Step (2) is repeated for $(N - 2)$ cycles, where N is the number of grid points.

Before applying Ward's method, the linear trend was removed from the time series of SST anomalies at each grid point in both oceans using the spline-function (Vager and Serkov 1987), and residuals at each grid point were normalized by their standard deviation.

No hierarchical clustering method has hard criteria to determine the final number of clusters. The pattern correlation between the centre of particular cluster (in our case this is the

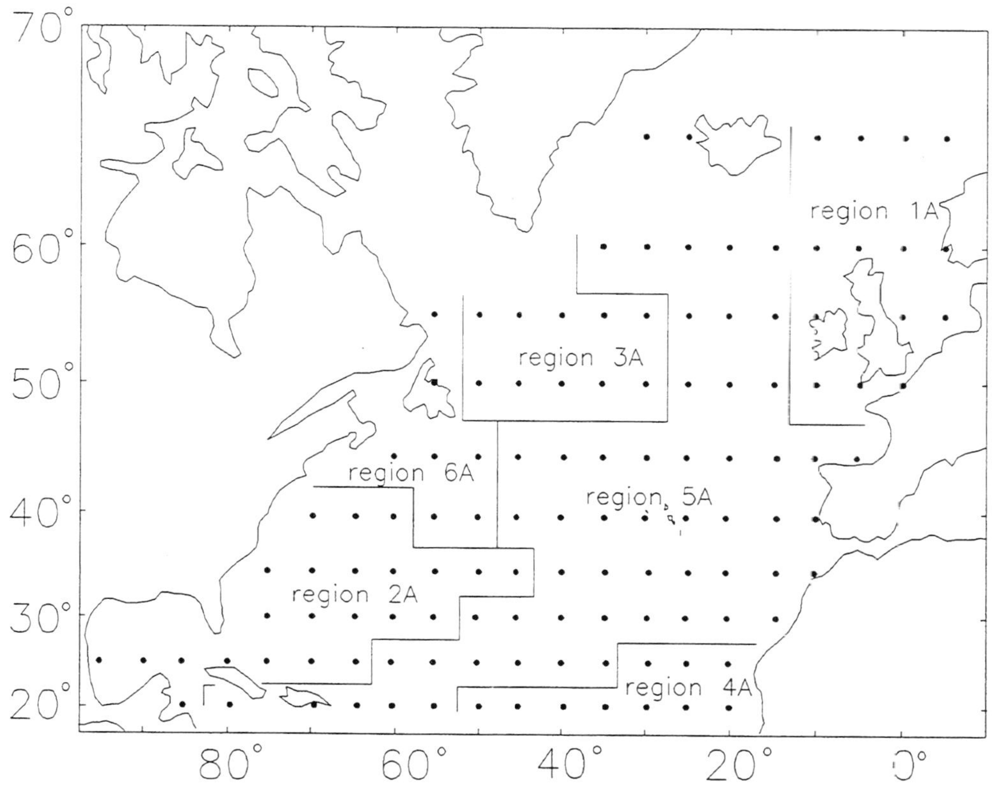

Fig. 1. Results of the cluster analysis for the SST anomaly field in the North Atlantic. Dots show the position of the 5° latitude by 5° longitude grid for the SST data set.

area-averaged time series of SST anomalies) and any element in the cluster (the time series of SST anomalies at grid points forming this cluster) should exceed a threshold, r_{th}. The threshold value of 0.53 used in this study corresponds to $1.5r_s$ (r_s is the 95% significance level for the 35-yr data and is equal to 0.35).

The NAO and NPO indices were constructed as differences between normalized winter (December through February) mean surface air temperature anomalies at grid points 60°N, 10°E, 70°N, 50°W for the NAO and 55°N, 110°W, 55°N, 170°W for the NPO, respectively. These two pairs of grid points are the closest to the stations used by van Loon and Rogers (1978) in their study of the Greenland–Scandinavia air temperature seesaw and by Rogers (1981) in his study of the NPO.

Results

The North Atlantic

Results of the cluster analysis show that the North Atlantic can be divided into six major regions with respect to the similarity in the SST anomaly variations: the northeastern part of the North Atlantic including the southern Norwegian Sea and the North Sea (region 1A), the southwestern part (region 2A), the northwestern part (region 3A), the southeastern part (region 4A), the central part (region 5A), and the small region around Newfoundland (region 6A) (Fig. 1).

Figures 2A–2F demonstrate the distribution of correlation coefficients between the time series of area-averaged SST anomalies for each region and SST anomalies at every grid point in the North Atlantic. These time series were calculated as the spatial average of the non-normalized local time series over the region specified by the corresponding cluster. As shown in Fig. 2, the resulting clusters satisfy the above criterion for determination of their final number. In addition, Fig. 2A to 2D show that the SST anomaly fluctuations in regions 1A and 2A are opposite to those in regions 3A and 4A.

Results shown in Fig. 2 are summarized in Table 1. As could be expected, there are statistically significant cross-correlation coefficients between the area-averaged SST anomalies for regions 1A through 4A.

Table 1. Simultaneous cross-correlation coefficients between the area-averaged SST anomalies for regions 1A to 6A in the North Atlantic. An asterisk indicates a coeffecient exceeding the 95% significance level.

	2A	3A	4A	5A	6A
region 1A	0.54*	−0.46*	−0.28	0.23	0.17
region 2A		−0.64*	−0.63 *	−0.01	0.23
region 3A			0.55 *	0.30	−0.14
region 4A				0.40 *	−0.17
region 5A					0.23
region 6A					

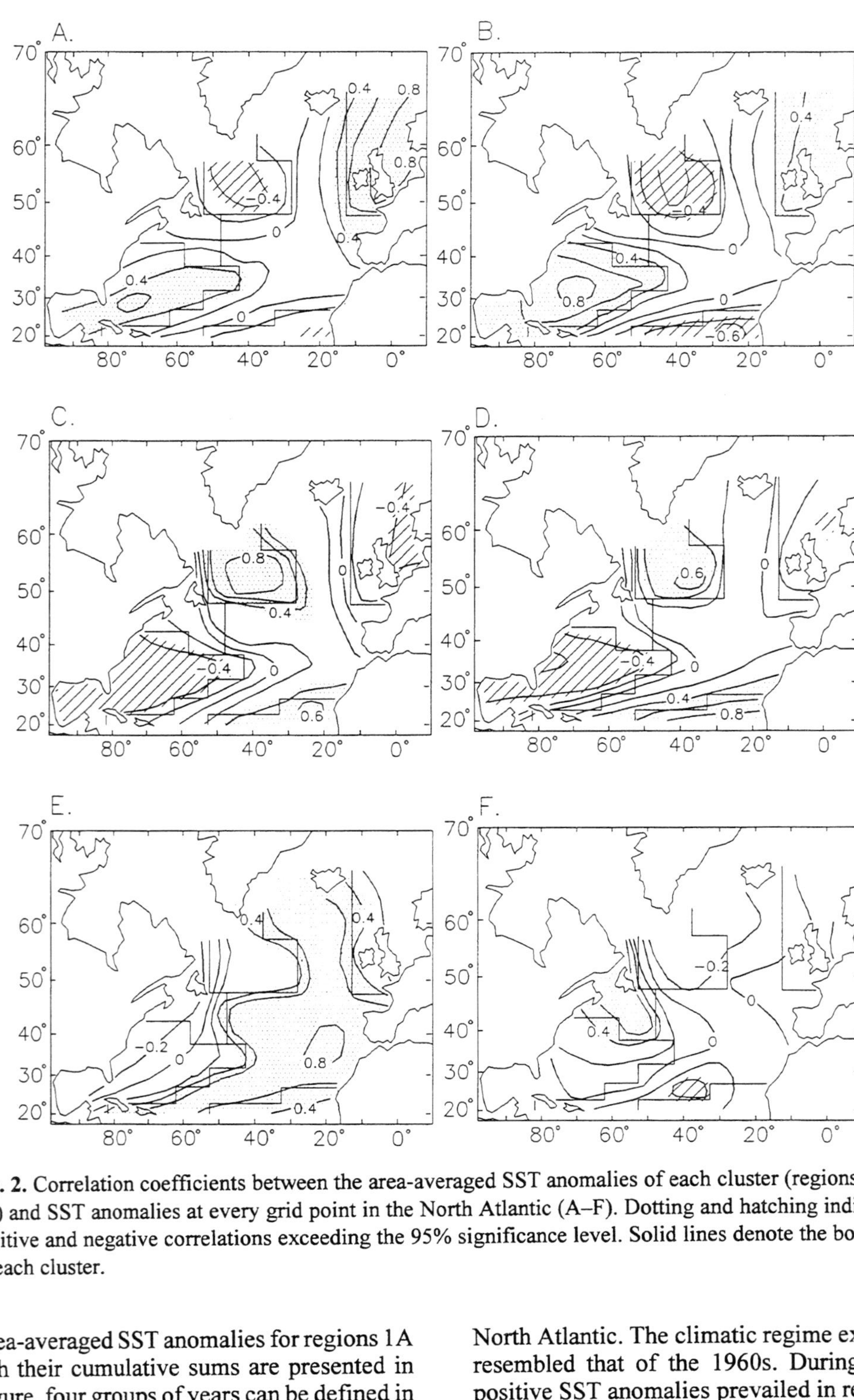

Fig. 2. Correlation coefficients between the area-averaged SST anomalies of each cluster (regions 1A to 6A) and SST anomalies at every grid point in the North Atlantic (A–F). Dotting and hatching indicate positive and negative correlations exceeding the 95% significance level. Solid lines denote the borders of each cluster.

Time series of area-averaged SST anomalies for regions 1A to 6A together with their cumulative sums are presented in Fig. 3. From this figure, four groups of years can be defined in regions 1A to 5A with esentially different regimes of the SST anomaly variations (1962–70, 1971–76, 1977–82, and 1983–91). It is rather difficult to distinguish the same regimes in SST anomaly fluctuations in region 6A. The period 1962–70 was characterized by cooling in regions 1A and 2A, and associated warming in regions 3A, 4A, and 5A. In the early 1970s, the cold regime in the Northeast Atlantic and the southwestern North Atlantic was replaced by a warm one, while SST began to decrease in the northwestern, central, and southeastern North Atlantic. The climatic regime existing during 1977–82 resembled that of the 1960s. During the 1983–91 period, positive SST anomalies prevailed in region 2A, and negative ones in region 4A. At the same time, pronounced warming in region 1A started only in 1986. It should also be noted that warming in the central North Atlantic started about 1977 and continued during 1983–91, while SST anomaly variations in Region 3A did not reveal apparent cooling or warming tendency during that time.

The same regimes are revealed in variations in the NAO index (Fig. 4). Correlation coefficients between the NAO index and area-averaged SST anomalies for regions 1A to 6A

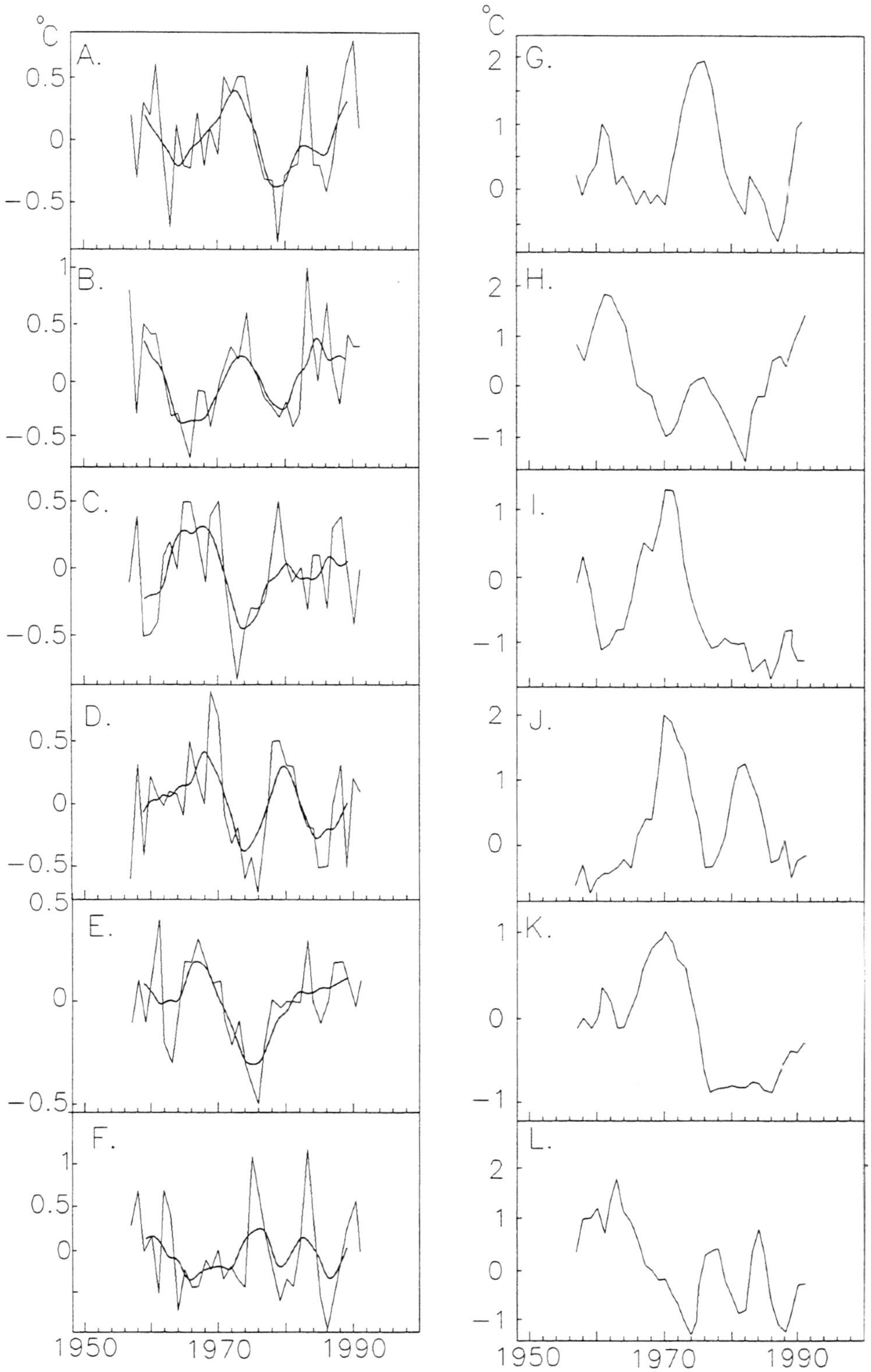

Fig. 3. Variations in the area-averaged SST anomalies for regions 1A to 6A in the North Atlantic (A–F) and their cumulative sums (G–L). Heavy curved lines represent 5-yr moving averages.

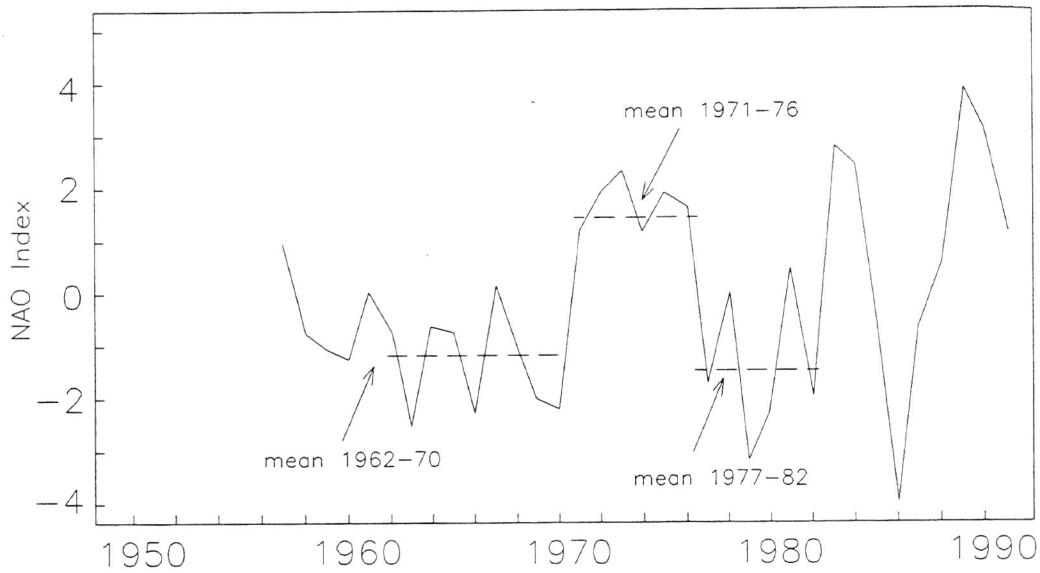

Fig. 4. Index of the North Atlantic Oscillation (NAO).

are 0.70, 0.46, −0.40, −0.40, −0.05, and 0.47, respectively. All correlation coefficients except for Region 5A exceed the 95% significance level. Hence, it can be supposed that, during the winter, the SST anomaly variations in regions 1A to 4A are determined in large extent by the features of the NAO pattern. The SST anomaly variations in region 6A also appear to be affected by the NAO pattern.

Figures 5A to 5F show the pattern of correlations for regions 1A to 6A between the area-averaged SST anomalies and the anomalies in geopotential heights on the 700-hPa surface. As shown in Fig. 5A, the SST anomaly fluctuations in region 1A during the winter season may be connected with the eastern Atlantic (EA) teleconnection pattern of Wallace and Gutzler (1981), while those in regions 2A and 3A may be forced by the western Atlantic (WA) pattern and, to some extent, by the EA pattern (Fig. 5B and 5C). At the same time, the distribution of correlation coefficients between the area-averaged SST anomalies in region 4A and the geopotential-height anomalies exhibits features of both the eastern Atlantic and Pacific – North American (PNA) patterns (Fig. 5D). Moreover, as shown in Fig. 5E, the SST anomaly variations in the central North Atlantic (region 5A) appear to be connected with the PNA pattern, while there is no significant relationship with the North Atlantic teleconnection patterns. Figure 5F demonstrates the absence of association between the SST anomaly fluctuations in region 6A and the known teleconnection patterns in the atmosphere.

Thus, there appears to be consistent patterns of association between the SST anomaly variations in all regions in the North Atlantic (except for region 6A) and the winter atmospheric teleconnections. It is interesting that, unlike those in regions 1A to 4A, the SST anomaly fluctuations in the central North Atlantic are not connected with the North Atlantic teleconnection patterns but instead reveal an apparent association with the PNA pattern, which is the dominant feature of the North Pacific climate system. As for region 6A, its spatial position allows one to suppose that changes in SST anomalies in this region are determined mainly by the interaction between the cold waters of the Labrador Current and the warm waters of the Gulf Stream. A major freshwater outflow from the St. Lawrence River and Great Lakes may also strongly influence this region.

The North Pacific

The results of the cluster analysis for the North Pacific are presented in Fig. 6. As shown in this figure, the North Pacific can be divided into five major regions: the eastern part of the North Pacific (region 1P), the central part (region 2P), the northwestern part (region 3P), the southwestern part (region 4P), and the southern part (region 5P).

Figures 7A to 7E show distributions of the correlation coefficients among the area-averaged SST anomalies for each region and the time series of the SST anomalies in all grid points in the North Pacific. There is an apparent out-of-phase relationship in the SST anomaly variations between the eastern (region 1P) and central (region 2P) North Pacific, and between the northwestern (region 3P) and southwestern (region 4P) North Pacific.

Cross-correlation coefficients between the area-averaged time series of the SST anomalies for each region are given in Table 2. As it could be expected from Fig. 7, the cross-correlation coefficient between the area-averaged SST anomalies in Regions 1P and 2P, and that for Regions 3P and 4P are negative and exceed the 95% significance level.

Time series of the area-averaged SST anomalies for regions 1P to 5P together with their cumulative sums are presented in Fig. 8. As shown in Fig. 8A and 8F, three regimes in the SST anomaly variations can be defined in regin 1P. They are 1962–70, 1971–76, and 1977–91. The periods 1962–70 and 1971–76 were characterized by cooling in the eastern

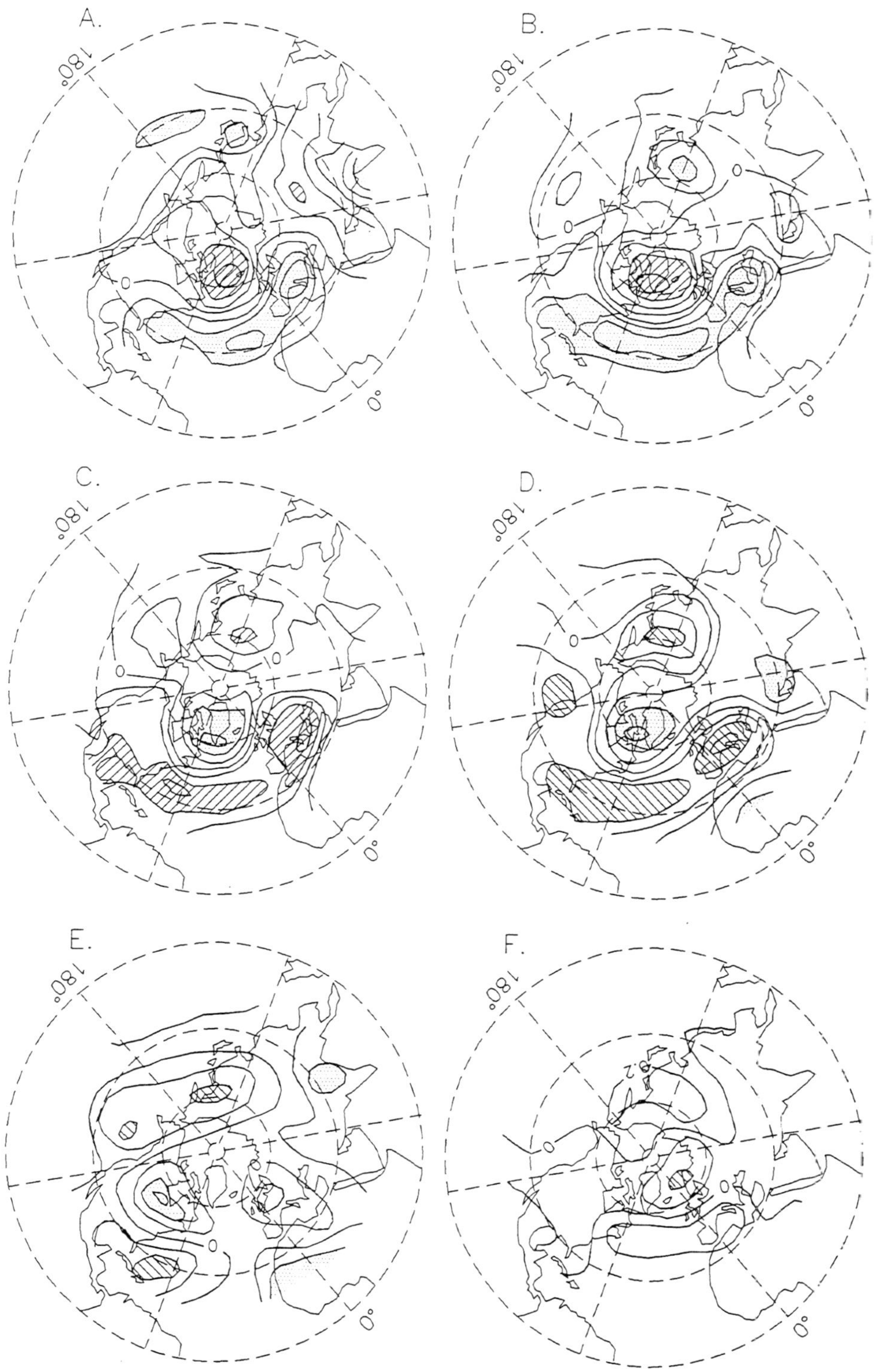

Fig. 5. Distributions of correlation coefficients between the area-averaged SST anomalies for regions 1A to 6A and the 700 hPa winter mean geopotential height anomalies. The dotted areas indicate the correlation coefficients that are above 0.4, and the hatched areas indicate those below –0.4. The contour interval is 0.2.

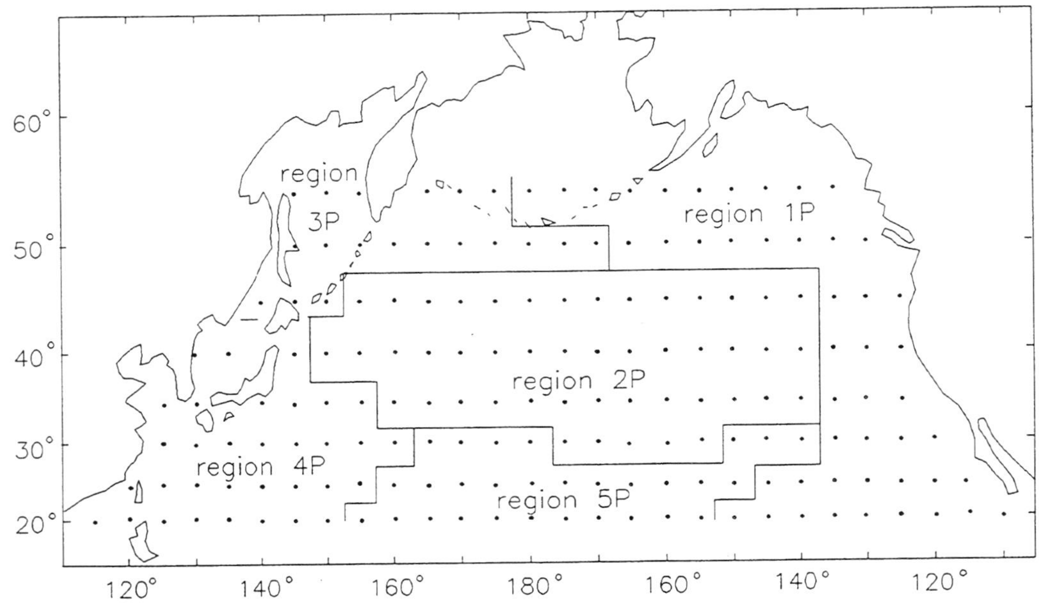

Fig. 6. Results of the cluster analysis for the SST anomaly field in the North Pacific. Dots show the position of the 5° latitude by 5° longitude grid for the SST data set.

North Pacific, but this cooling was most pronounced after 1970. Beginning in 1977–78, the cold regime in this region was replaced by a warm one, although a tendency toward slight cooling has appeared since 1986. The same time periods can be seen in the SST anomaly fluctuations in regions 2P (Fig. 8B and 8G). The warm regime that existed in 1962–76 was followed by a cold one, which continued from 1977 to 1987. But, unlike region 1P, there were no such sharp changes in the character of the SST anomaly variations in region 2P from 1962–70 to 1971–76. Moreover, beginning in 1988, a marked tendency toward warming has been observed in this region.

The warming during the 1960s and the first half of the 1970s was also evident in regions 4P and 5P (Fig. 8D and 8I; 8E and 8J). This warming was followed by cooling, which continued until 1985. At the same time, the SST anomaly variations in region 3P did not clearly show regimes similar to those distinguished for other regions in the North Pacific (Fig. 8C and 8H).

Table 2. Simultaneous cross-correlation coefficients between the area-averaged SST anomalies for Regions 1P to 5P in the North Pacific. An asterisk indicates a coefficient exceeding the 95% significance level.

	2P	3P	4P	5P
region 1P	−0.38 *	0.17	0.19	−0.02
region 2P		0.26	0.15	0.26
region 3P			−0.42 *	−0.07
region 4P				0.31
region 5P				

The similar regimes are also revealed in variations of the NPO index (Fig. 9). The correlation coefficients between the NPO index and the time series of area-averaged SST anomalies for regions 1P to 5P are −0.02, −0.02, −0.08, 0.37, and 0.06, respectively. Thus, only the correlation coefficient for region 4P exceeds the 95% significance level.

Geographical distribution of the correlation coefficients between the area-averaged SST anomalies in the North Pacific and geopotential-height anomalies are presented in Fig. 10A to 10E. The SST anomaly fluctuations in regions 1P, 2P, and 5P appear to be associated with the PNA pattern, while those in Regions 3P and 4P are associated with the western Pacific (WP) pattern of Wallace and Gutzler (1981).

Comparison of climatic changes in the North Atlantic and North Pacific

Some of the results of studying spatial and temporal structures of the SST anomaly variations in the North Atlantic and North Pacific may serve as a basis for comparing climatic changes in both oceans during the last several decades. As described above, several periods with different regimes of the SST anomaly fluctuations were identified in both oceans. They were 1962–70, 1971–76, 1977–82, and 1983–91.

The first three groups are distinguished rather clearly in almost all regions in the North Atlantic and North Pacific. But this is not so for the period from 1983 to 1991. For example, warming in the eastern North Pacific and central North Atlantic that started in 1977–78 continued during 1983–91. The marked warming in the Northeast Atlantic and central North Pacific started in 1986–87, while signs of an SST decrease appeared in the southwestern North Atlantic at that time. Based on a detailed examination of the SST anomaly variations during 1983–91, this period can be divided into two subperiods: 1983–86 and 1987–91.

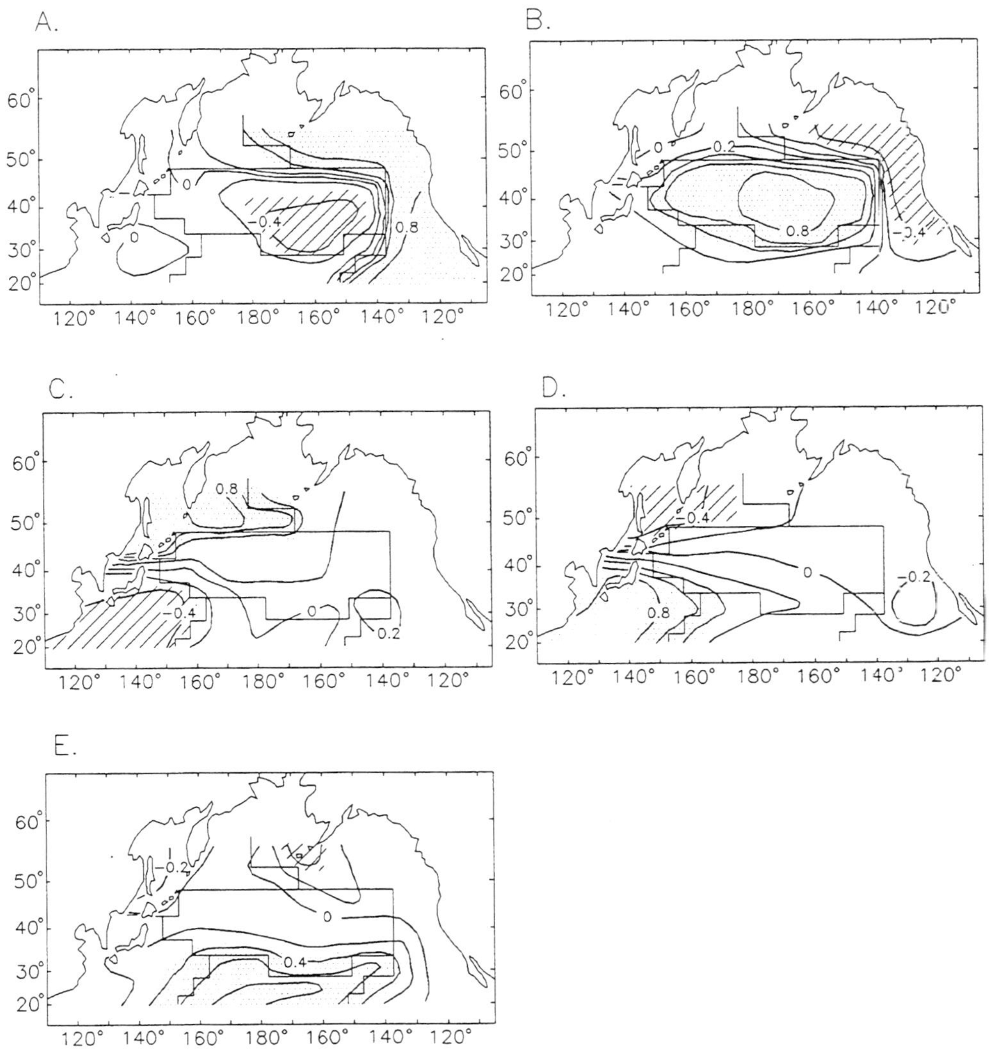

Fig. 7. Correlation coefficients between the area-averaged SST anomalies of each cluster (regions 1P to 5P) and SST anomalies at every grid point in the North Pacific (A–E). Dotting and hatching indicate positive and negative correlations exceeding the 95% significance level. Solid lines denote the borders of each cluster.

Figure 11A shows the differences between mean winter geopotential-height anomalies on the 700-hPa surface averaged over the periods 1971–76 and 1962–70. The period 1971–76 was characterized by an increase in geopotential heights over almost the whole North Pacific. Together with a slight decrease in geopotential heights over western North America, this led to more frequent northerly air flows off the west coast of North America and corresponding cooling in the eastern North Pacific. It also led to enhanced southerlies and warming in the central North Pacific, although this warming was not pronounced (Fig. 11B). A marked warming was evident in the Okhotsk Sea (region 3P).

Changes in atmospheric circulation over the North Atlantic between the above two periods were more dramatic. A significant decrease in geopotential heights over and east of Greenland and their increase in the middle latitudes of the North Atlantic resulted in strengthening of the zonal atmospheric circulation during 1971–76. As a result, an apparent warming occurred in the northeastern and southwestern parts of the North Atlantic, while marked tendencies toward cooling were observed in other regions of the ocean.

Opposite changes in atmospheric circulation and in the character of SST anomaly variations occurred between the periods 1977–82 and 1971–76 compared with those between 1971–76 and 1962–70 (Fig. 12). A decrease in geopotential heights over the middle latitudes of the North Pacific and North Atlantic and their increase over Canada and Greenland resulted in a significant lessening of zonal atmospheric circulation over both oceans during 1977–82. These atmospheric changes apparently led to the pronounced cooling in the central North Pacific, and the northeastern and southwestern North Atlantic, and the warming in the eastern North Pacific

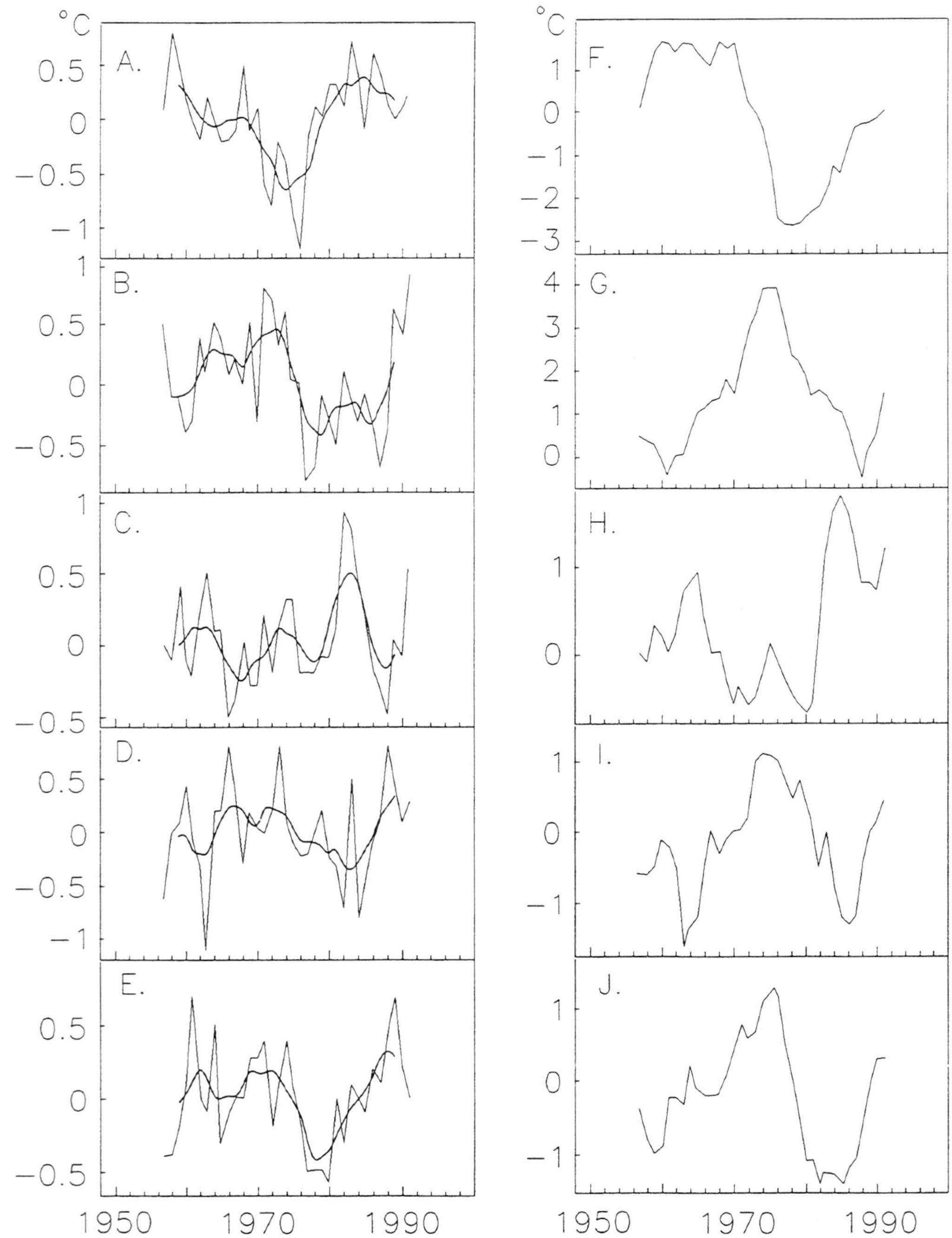

Fig. 8. Variations in the area-averaged SST anomalies for regions 1P–5P in the North Pacific (A–E) and their cumulative sums (F–J). Heavy curved lines represent 5-yr moving averages.

and the rest of the North Atlantic. A marked cooling was also evident in the southwestern North Pacific, while the SST anomalies increased slightly in northwestern Pacific.

Differences between geopotential height anomalies averaged over the periods 1983–86 and 1977–82 are illustrated by Fig. 13A. As shown in this Figure, signs of an intensification in zonal atmospheric circulation were observed in both oceans during 1983–86 but were more pronounced in the North Atlantic. Sea surface temperatures did not change significantly in most of the North Pacific between the periods 1983–86 and 1977–86 (except for the rather marked warming in the northeastern and southeastern North Pacific), while a pronounced warming was observed in region 2A in the North Atlantic (Fig. 13B).

Drastic changes occurred between 1987–91 and 1983–86 (Fig. 14). As shown in Fig. 14A, geopotential heights over most of the North Pacific increased significantly, while they decreased over most of the North Atlantic. Enhanced northerlies

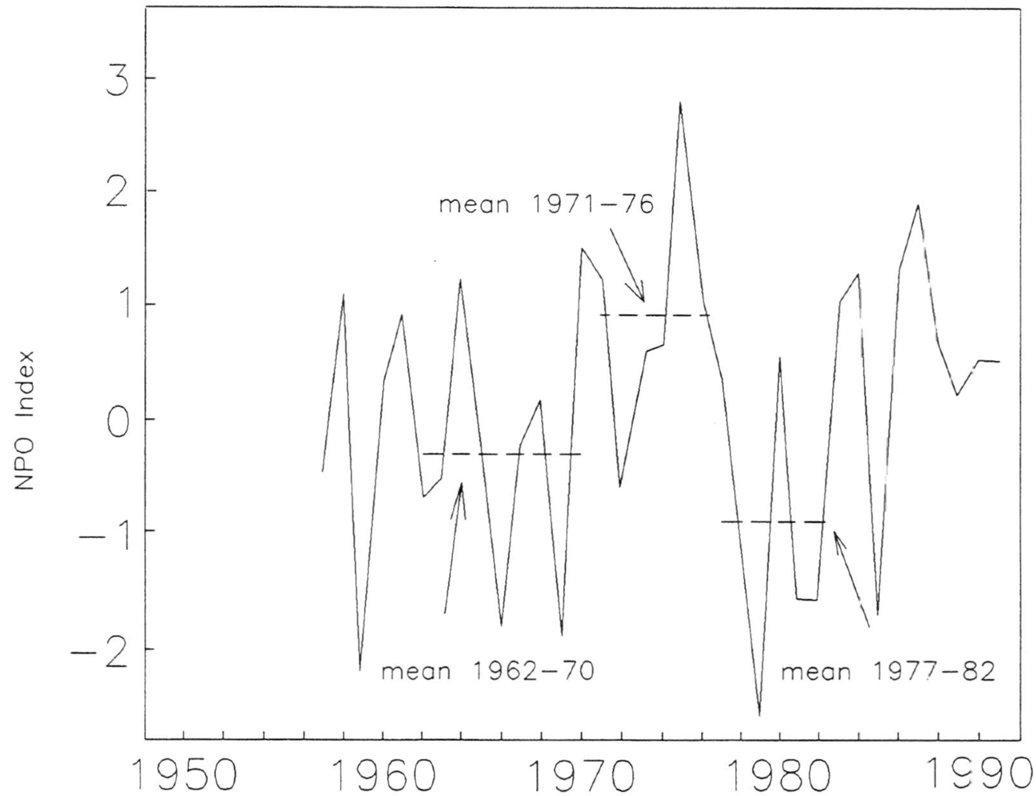

Fig. 9. Index of the North Pacific Oscillation (NPO).

along the western seaboard of North America resulted in a decrease in SST in the eastern North Pacific during 1987–91 (Fig. 14B). More frequent southerly air flows favoured a warming in the central and southwestern parts of the ocean. At the same time, pronounced cooling was evident in the Okhotsk Sea. But if the situation observed in the atmosphere over the North Pacific during 1987–91 was similar to that during 1971–76, the character of the atmospheric circulation over the North Atlantic was rather unusual. Apparently frequent outbreaks of cold continental air resulted in pronounced cooling in the western North Atlantic. Differences between the SST anomalies averaged over 1987–91 and 1983–86 exceeded minus 2°C. At the same time advection of warm air from the south led to a rise in SST in the northeastern Atlantic.

Thus, there is a great deal of similarity in the changes in atmospheric circulation and the character of the SST anomaly variations in the North Pacific and North Atlantic. The sharpest changes (i.e., transitions from one climatic regime to another) occurred in the same years in both the atmosphere and the ocean.

Table 3 shows the cross-correlation coefficients between the time series of area-averaged SST anomalies in the North Pacific and those in the North Atlantic. Most of the coefficients do no exceed the 95% significance level. A strong positive correlation was found between the SST anomaly fluctuations in the eastern North Pacific (region 1P) and those in the central North Atlantic (region 5A). This result is not unexpected, considering that both regions reveal an apparent association of the same sign with the PNA pattern. Significant positive correlations were obtained also between the SST anomaly variations in region 1A and those in regions 2P, 3P and 5P.

Relationships between abundance of some commercial fish stocks

It has often been suggested that strong year-classes in commercial fish stocks appear in the same year in distant regions of the ocean. For example, Templeman (1965) examined 10 stocks of cod, haddock, and herring in the North Atlantic between 1902 and 1962 and showed that in some years (e.g., 1949, 1950, and 1956) good year-classes are common to six or more stocks.

We calculated the cross-correlation coefficients between recruitments in the 22 North Atlantic commercial fish stocks. In all, 253 coefficients were calculated. Most of correlations wre statistically insignificant. In only 17 cases were the coefficients higher that 95% significance level (Table 4).

Strong linear correlations between recruitments occur in the North Sea and the waters surrounding the British Isles. Habitat areas of all these stocks lie within region 1A, which has been defined on the basis of the spatial strucutre of the SST anomaly fluctuations in the North Atlantic. Hence, the above relationships may be considered as some evidence for climatic effects on variations in recruitment. Indeed, the apparent negative correlations between North Sea haddock and North Sea plaice

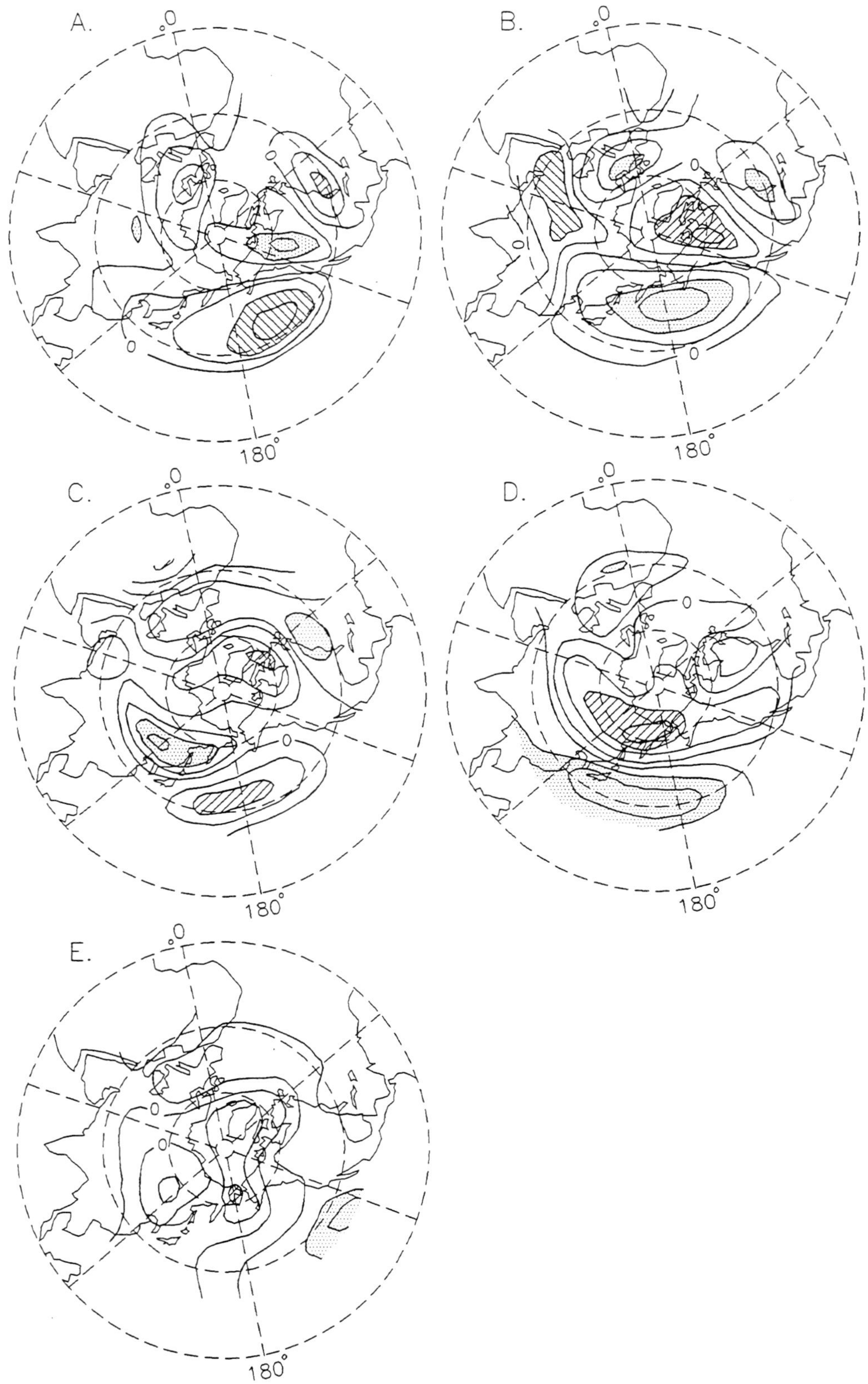

Fig. 10. Distributions of correlation coefficients between the area-averaged SST anomalies for Regions 1P to 5P and the 700 hPa winter mean geopotential height anomalies. The dotted areas indicate the correlation coefficients that are above 0.4, and the hatched areas indicate those below –0.04. The contour interval is 0.2.

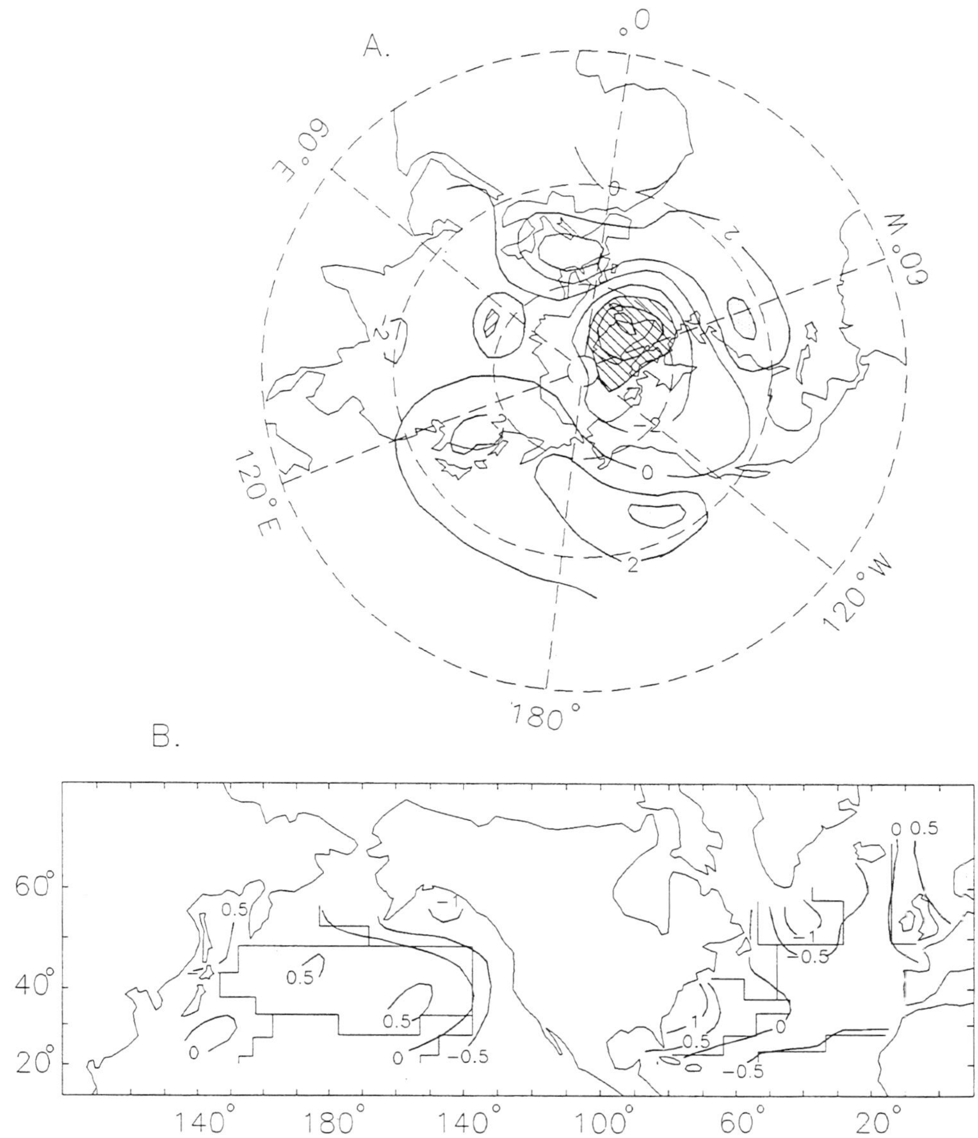

Fig. 11. (*a*) Changes in 700-hPa height anomalies and (*b*) SST anomalies in the North Pacific and North Atlantic. Differences are mean winter values averaged over 1971–76 minus mean winter values averaged over 1961–70. The dotted areas in Fig. 11*a* show differences that are above 4 dm and the hatched areas show differences that are below –4 dm.

($r = -0.45$), and between west of Scotland herring and North Sea saithe ($r = -0.56$) may be taken as additional evidence for climatic effect, because North Sea haddock and saithe stocks are directly related to water temperature, while North Sea plaice and west of Scotland herring stocks are inversely related to water temperature (Dementyeva 1976; Shepherd et al. 1984).

Rather strong linear correlations between variations in recruitment are also obtained for region 6A.

Fig. 15 demonstrates two examples of variations in the abundance of commercial fish stocks. As shown in this figure, recruitment of the North Sea haddock stock was at a higher level during 1970–74 than during 1975–82. These periods coincided roughly with the warm and cold regimes in the Northeast Atlantic (region 1A) during the first and second half of the 1970s, respectively. Figure 15B, taken from Rodionov and Krovnin (1991), shows pronounced variations in the relative abundance of the pollack stock of the eastern Bering Sea. These variations correspond strongly to climatic regimes in both the North Pacific and North Atlantic.

The above examples may serve as good indications of climatic effects on specific commercial fish populations. It is important that drastic changes in the state of a population occur during the transition from one climatic regime to another.

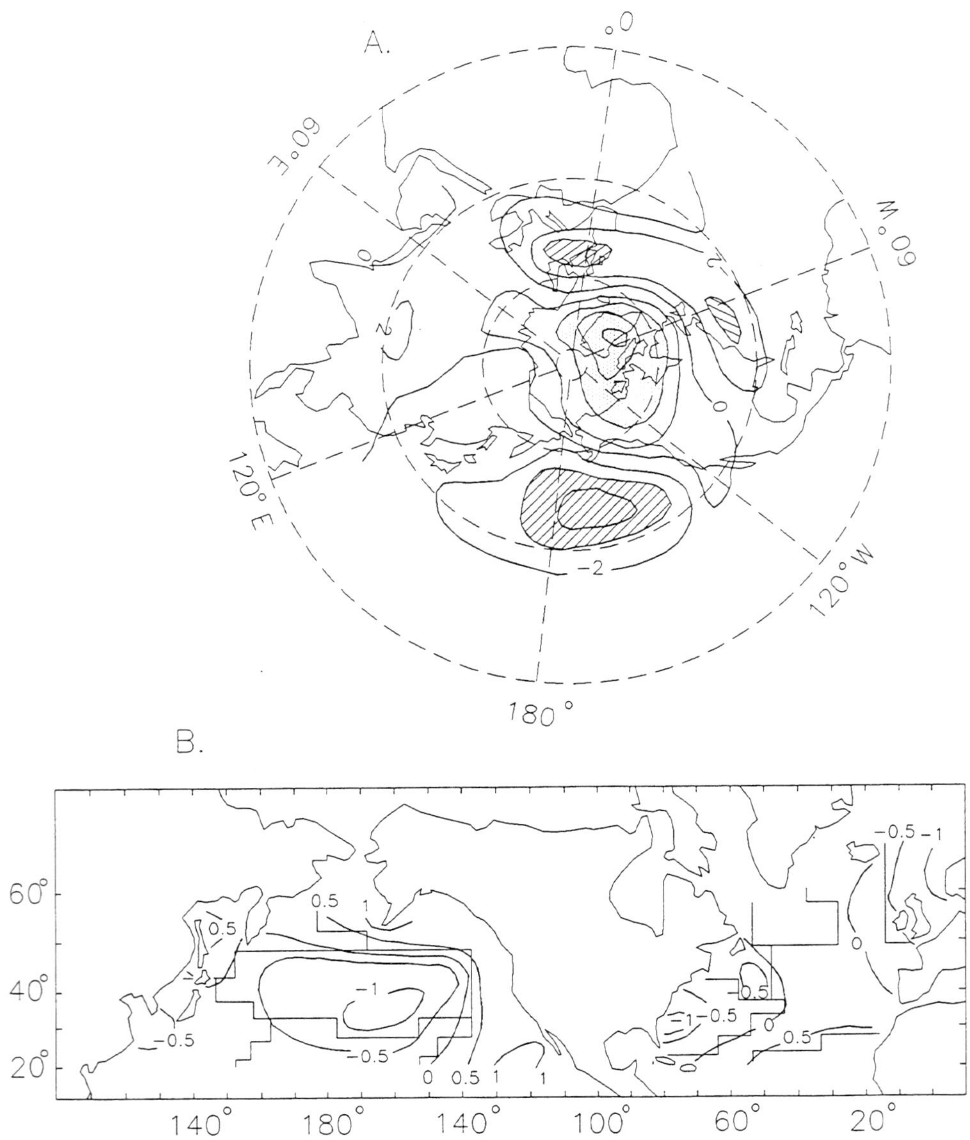

Fig. 12. (*a*) Changes in 700-hPa height anomalies and (*b*) SST anomalies in the North Pacific and North Atlantic. Differences are mean winter values averaged over 1977–82 minus mean winter values averaged over 1971–76. The dotted areas in Fig. 12*a* show differences that are above 4 dm and the hatched areas show differences that are below –4 dm.

Discussion

Dickson and Namias (1976) demonstrated that the climate of the North Atlantic was determined in large extent by the presence or absence of the high-pressure anomaly cell over Greenland. Build-up of this anomaly is highly dependent on the features of a strong baroclinic zone in the atmosphere along the east coast of North America. During warm winters in the southeastern United States, wave cyclones form and move northeastward across the North Atlantic, developing rapidly into large-scale occluded systems and eventually, in the statistical aggregrate, forming the Icelandic Low.

During cold winters the enhanced baroclinicity along the eastern seaboard of North America results in cyclone development well to the southwest of the normal zone of maximum storm activity; thus favouring an increase in pressure over Greenland.

As shown by Namias (1970), the establishment of cold winter regime in the southeastern United States is associated with a long-wave pattern over the northern hemisphere. It corresponds to an upper ridge over northwestern North America and a trough over the southeastern part of the continent. Such distribution leds to frequent northly outbreaks in eastern North America.

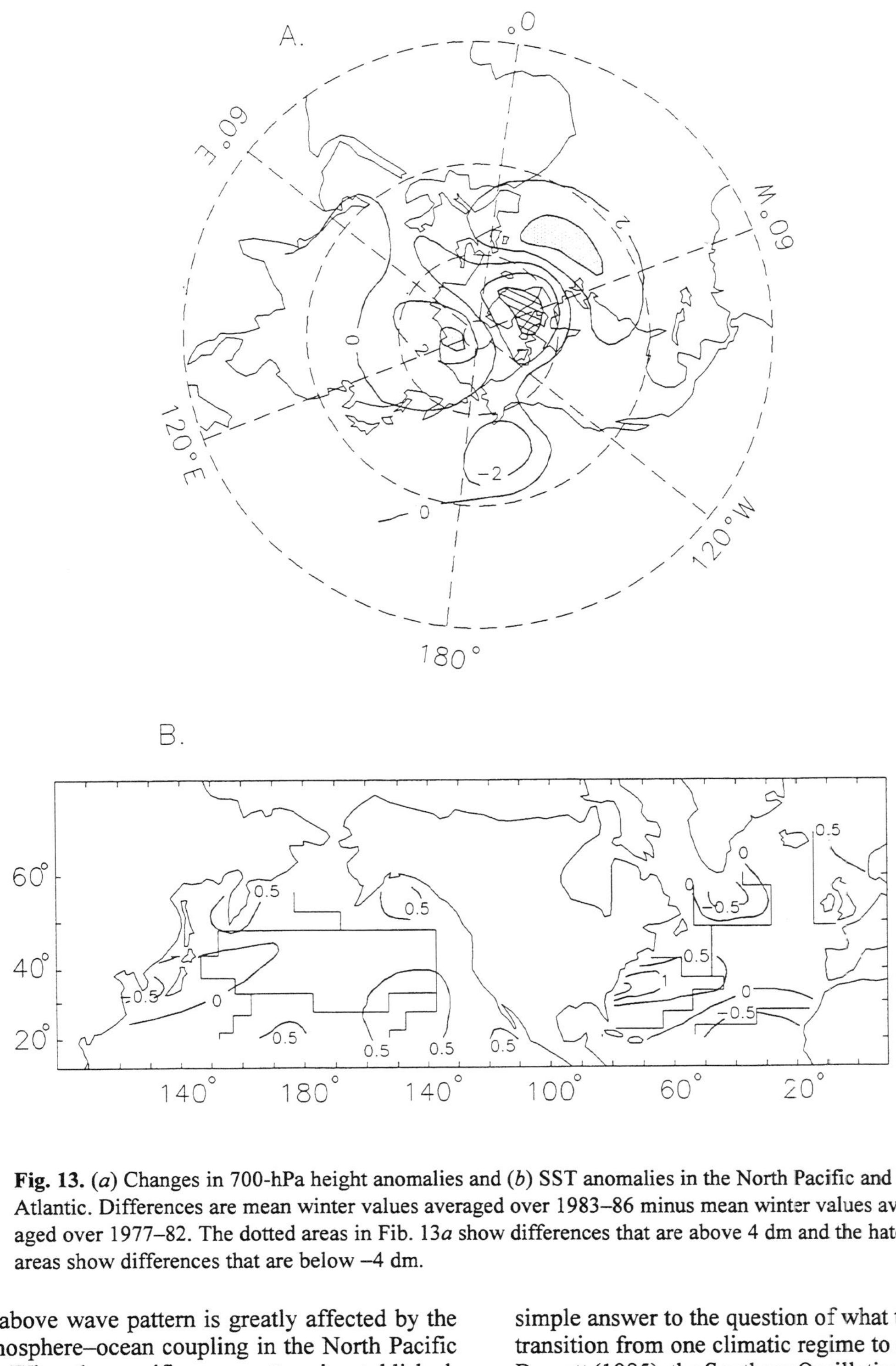

Fig. 13. (*a*) Changes in 700-hPa height anomalies and (*b*) SST anomalies in the North Pacific and North Atlantic. Differences are mean winter values averaged over 1983–86 minus mean winter values averaged over 1977–82. The dotted areas in Fib. 13*a* show differences that are above 4 dm and the hatched areas show differences that are below −4 dm.

In turn, the above wave pattern is greatly affected by the large-scale atmosphere–ocean coupling in the North Pacific (Namias 1970). When the specific wave pattern is established, the thermal conditions of the underlying surface may be responsible, through feedback mechanisms at certain key locations, for encouraging the prolongations or repeated occurrence of the wave pattern as a whole, thus turning a short-term change into a longer term climatic regimes (Dickson and Namias 1976).

The scheme described here provides a good explanation for the results of our comparison of climatic changes in the North Pacific and North Atlantic. However, it is impossible to give a simple answer to the question of what the reasons are for the transition from one climatic regime to another. As shown by Barnett (1985), the Southern Oscillation, monsoon variations, and well-known teleconnection patterns are all part of a single, global-scale signal in the sea-level pressure (SLP) field. He pointed out further that snow albedo feedback over Asia and its coupling to the SLP signal might be crucial to an explanation of the observations on climatic fluctuations, particularly the alternating plus and minus signs of the anomaly field. He also pointed out that the role of SSTs as an initiator of drastic climate changes was not as obvious as one might expect.

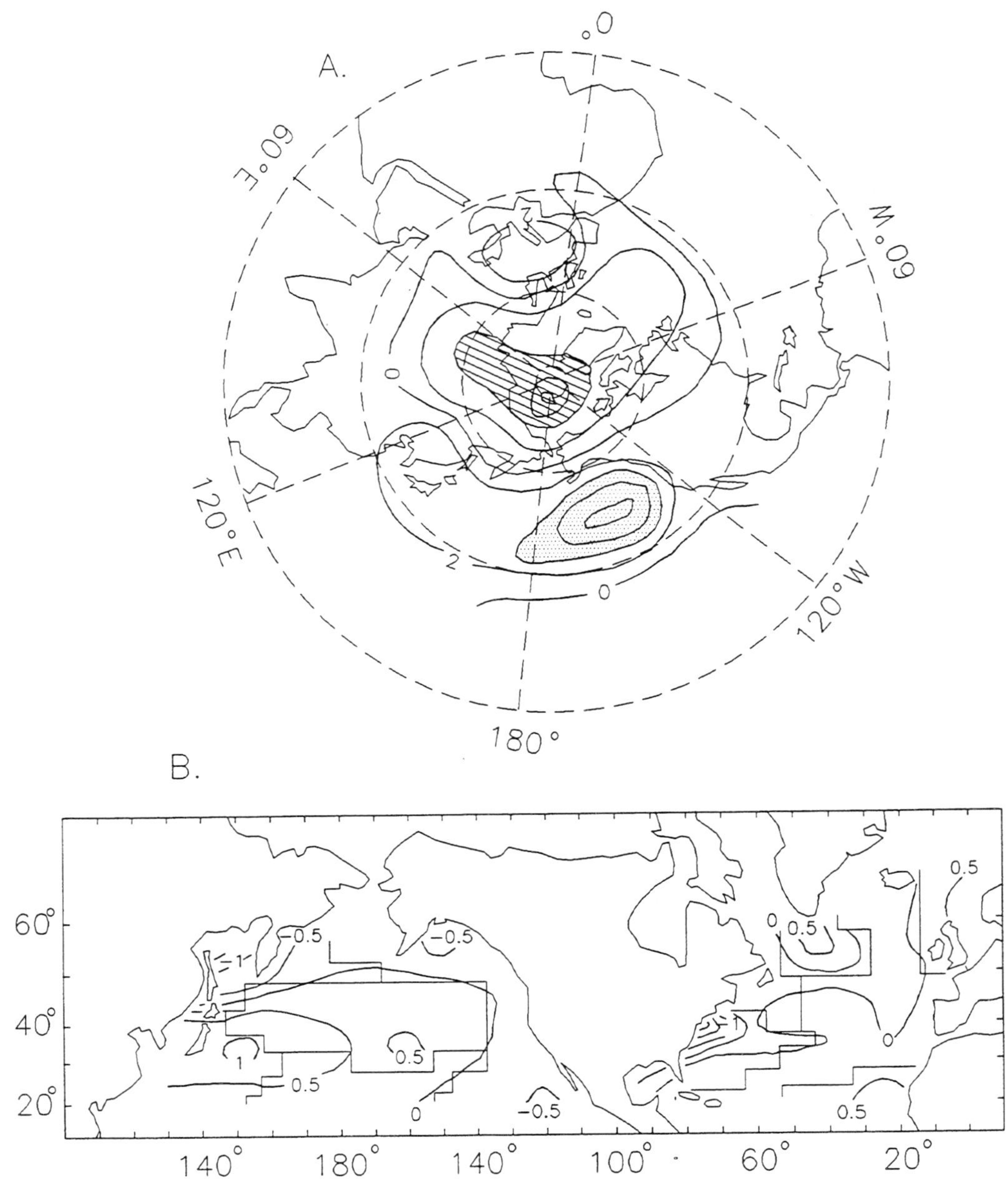

Fig. 14. (*a*) Changes in 700-hPa height anomalies and (*b*) SST anomalies in the North Pacific and North Atlantic. Differences are mean winter values averaged over 1987–91 minus mean winter values averaged over 1983–86. The dotted areas in Fig. 14*a* show differences that are above 4 dm and the hatched areas show differences that are blow –4 dm.

Table 3. Simultaneous cross-correlation coefficients between the area-averaged SST anomalies for each region in the North Atlantic and those in the North Pacific. An asterisk indicates the coefficient exceeding the 95% significance level.

	1A	2A	3A	4A	5A	6A
Region 1P	–0.12	0.19	0.12	0.24	0.59 *	–0.01
Region 2P	0.44 *	0.12	0.01	–0.20	–0.11	0.04
Region 3P	0.06	0.31	–0.23	–0.37 *	–0.21	0.39 *
Region 4P	0.46 *	0.02	0.00	0.22	0.39 *	–0.18
Region 5P	0.61 *	0.23	–0.06	–0.05	0.31	–0.13

Summary

A study of the spatial structure of the SST anomaly variations in the North Atlantic shows this structure can be divided into six subdomains. The SST anomaly fluctuations in the northeastern and southwestern North Atlantic (regions 1A and 2A, respectively) are opposite to those in the northwestern and southeastern North Atlantic (regions 3A and 4A, respectively).

The SST anomaly variations in the North Pacific are characterized by the existence of two independent spatial patterns. These are significant out-of-phase relationships between the SST anomaly variations in the eastern (region 1P) and central (region 2P) North Pacific, and those in the northwestern (region 3P) and southwestern (region 4P) North Pacific.

Table 4. Significant correlations at the 95% level between recruitments of some North Atlantic commercial fish stocks. Gulf spring herring = southern Gulf of St. Lawrence spring-spawning herring; Gulf fall herring = southern Gulf of St. Lawrence fall-spawning herring.

Fish stock		Correlation coefficient	Sample size
North Sea haddock	North Sea plaice	−0.49	19
North Sea haddock	West of Scotland haddock	0.82	20
North Sea saithe	West of Scotland saithe	0.49	20
North Sea saithe	West of Scotland herring	−0.56	19
North Sea whiting	West of Scotland whiting	0.71	20
West of Scotland haddock	West of Scotland cod	0.59	19
West of Scotland haddock	Irish Sea cod	0.53	20
Northeastern Arctic haddock	Northeastern Arctic cod	0.55	26
Gulf of Maine herring	Georges Bank haddock	0.41	29
Fortune Bay herring	Gulf spring herring	0.76	8
Cod in NAFO Division 3NO	Gulf fall herring	0.53	13
Cod in NAFO Division 3NO	Cod in NAFO Division 3Ps	0.72	26
Cod in NAFO Division 3NO	Georges Bank haddock	0.63	24
Cod in NAFO Division 2J+3KL	Cod in NAFO Division 3NO	0.68	19
Cod in NAFO Division 2J+3KL	Cod in NAFO Division 3Ps	0.57	19
Cod in NAFO Division 2J+3KL	West of Ireland herring	−0.57	11
Georges Bank haddock	North Sea plaice	0.51	25

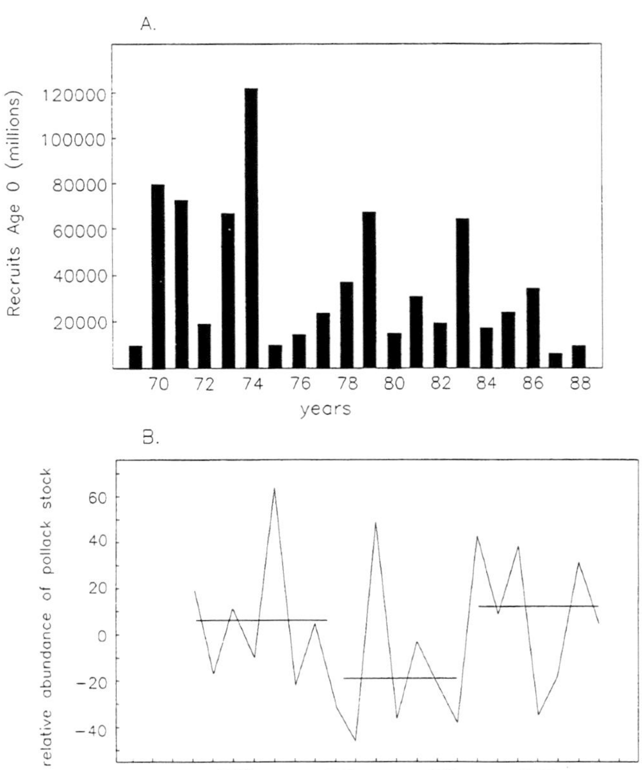

Fig. 15. (*a*) Variations in recruitment of North Sea haddock stock (ICES 1990) and (*b*) relative abundance of pollock stock in the eastern Bering Sea. (Taken from Rodionov and Krovnin 1991.)

The SST anomaly fluctuations in all regions of both the North Atlantic and North Pacific (except for region 6A) are associated with the well-known teleconnection patterns in the atmosphere.

In both oceans, several periods were identified in atmospheric circulation and the character of SST anomaly fluctuations. These periods are 1962–70, 1971–76, 1977–82, 1983–86, and 1987–91. Transition from one regime to another occurs in the same year in both the atmosphere and the ocean. This feature may be explained in terms of the existence of a single, global-scale signal in the SLP field that includes the Southern Oscillation, monsoon variations, and well-known teleconnection patterns (Barnett 1985).

Some evidence exists for the effects of climate on commercial fish stocks. In some cases, the above-mentioned regimes are clearly reflected in variations of abundance of fish populations.

References

Anthony, V.C., and M.J. Fogarty. 1985. Environmental effects on recruitment, growth, and vulnerability of Atlantic herring (*Clupea harengus harengus*) in the Gulf of Main region. Can. J. FIsh. Aquat. Sci. 42(Suppl. 1): 158–173.

Baird, J.W., and C.A. Bishop. 1986. Assessment of the cod stock in NAFO Divisions 2J+3KL. NAFO SCR Doc. 86/67. Ser. N1163: 65 p.

Barnett, T.P. 1985. Variations in near-global sea level pressure. J. Atmos. Sci. 42: 478–501.

Cohen, E.B. 1986. Some factors affecting the recruitment of Georges Bank cod and haddock III. p. 899–916. *In* M. Alton [comp.] Proceedings of a Workshop on Comparative Biology, Assessment, and Management of Gadoids from the North Pacific and Atlantic Oceans. Northwest and Alaska Fisheries Center, Seattle, WA.

Cushing, D.H. 1982. Climate and fisheries. Academic Press, London. 373 p.

Dementyeva, T.F. 1976. Biological basement of fishery forecasts. Pishchevaya promyshlennost, Moscow. 237 p. (In Russian)

Dickson, R.R., and J. Namias. 1976. North American influence on the circulation and climate of the North Atlantic sector. Mon. Weather Rev. 104: 1255–1264.

Hen, G.V. 1988. Seasonal and interannual variability of the Bering Sea waters and its impact on distribution and abundance of hydrobionts. Ph.D. thesis, Vladiviostok, Russia. 160 p. (In Russian)

ICES. 1990. Reports of the Advisory Committee on Fishery Management. Coop. Res. Rep. 168(1): 361 p.

Namias, J. 1970. Climatic anomaly over the United States during the 1960's. Science (Wash., D.C.) 170: 741–743.

1972. Space scales of sea-surface temperature patterns and their causes. Fish. Bull. U.S. 70: 611–617.

NAFO. 1988. Scientific council reports. Dartmouth, N.S. 149 p.

NAFO. 1989. Scientific council reports. Dartmouth, N.S. 180 p.

Rodionov, S.N., and A.S. Krovnin. 1991. Influence of thermal conditions on abundance of pollack stodk in the eastern Bering Sea. Ryb. Khoz. 3: 21–26. (In Russian)

Royers, J.C. 1981. The North Pacific oscillation. J. Climatol. 1: 39–57.

Shepherd, J.G., J.G. Pope, and R.D. Cousens. 1984. Variations in fish stocks and hypotheses concerning their links with climate. Rapp. P. V. Réun. Cons. int. Explor. Mer 185: 255–267.

Templeman, W. 1965. Relation of periods of successful yearclasses of haddock on the Grand Bank to periods of success of yearclasses for cod, haddock and herring in areas to the north and east. ICNAF Spec. Publ. 6: 523–533.

Vager, B.G., and N.K. Serkov. 1987. Using of splines in meterology and hydrology. Gidrometeoizdat, Leningrad. 160 p. (In Russian)

van Loon, H., and J.C. Rogers. 1978. The seesaw in winter temperatures between Greenland and northern Europe. I. General description. Mon. Weather Rev. 106: 296–310.

Vinnikov, K.Y. 1986. Sensitivity of climate. Gidrometeoizdat, Leningrad. 224 p. (In Russian)

Wallace, J.M., and D.S. Gutzler. 1981. Teleconnections in the geopotential height field during the Northern Hemisphere winter. Mon. Weather Rev. 109: 784–912.

Ward, J.H., Jr. 1963. Hierarchical grouping to optimize an objective function. J. Am. Stat. Assoc. 58: 236–244.

Winters, G.H. 1976. Recruitment mechanisms of southern Gulf of St. Lawrence Atlantic herring (*Clupea harengus harengus*). J. Fish. Res. Board Can. 33: 1751–1763.

Winters, G.H., E.L. Dalley, and J.A. Moores. 1985. Fortuity disguised as fisheries management: the case history of Fortune Bay herring. Can. J. Fish. Aquat. Sci. 42(Suppl. 1): 263–274.

TOPIC 2
Effects of climate change on fish populations and fisheries

Fishstock fluctuations as indicators of multidecadal fluctuations in the biological productivity of the ocean

B.J. Rothschild

Rothschild, B.J. 1995. Fishstock fluctuations as indicators of multidecadal fluctuations in the biological productivity of the ocean, p. 201–209. *In* R.J. Beamish Climate change and northern fish populations. Can. Spec. Publ. Fish. Aquat. Sci. 121.

Abstract: Multidecadal abundance fluctuations have evidently been characteristic of fish population for centuries. Before the onset of intensive fishing, the fluctuations must have been induced by natural causes. The fluctuations have persisted into the industrial era. During the industrial era, some components of the fluctuations are related to fishing mortality. However, other components seem independent of fishing mortality. The fishing-independent components are related either to multidecadal fluctuations in primary production or to changes in the trophodynamic pathways by which primary production is transformed to fish biomass. This implies that the historical record of fish abundance might be used to make inferences on past changes in productivity. Conversely, the contemporaneous state (increasing or decreasing) of biological production, if known, might be used to determine whether the basic productivity of a fish stock is increasing or decreasing, information that might be critical to advancing the effectiveness of fishery management.

Résumé : Depuis des siècles, les fluctuations dans l'abondance multidécennale caractérisent de façon évidente les populations de poissons. Avant l'apparition de la pêche intensive, ces fluctuations étaient attribuables à des causes naturelles. Elles ont persisté durant l'ère industrielle. Au cours de cette ère, certains éléments de ces fluctuations sont liés à la mortalité par pêche, bien que d'autres en soient indépendants. Les éléments indépendants de la pêche sont liés soit aux fluctuations multidécennales dans la production primaire soit à des changements dans le mécanisme trophodynamique par lequel la production primaire est transformée en biomasse de poisson. Cela suppose que le dossier historique de l'abondance du poisson pourrait être utilisé pour que l'on tire des conclusions sur les changements passés en matière de productivité. Inversement, l'état contemporain de la production biologique (en croissance ou en décroissance), s'il est connu, pourrait être utilisé pour que l'on détermine jusqu'à quel point la productivité de base d'un stock de poissons augmente ou diminue, renseignement susceptible de s'avérer capital pour l'accroissement de l'efficacité de la gestion des pêcheries.

Introduction

This paper reviews some ideas relevant to the notion that multidecadal fluctuations in fish-stock abundance are driven by changes in biological productivity. (For this paper, "biological productivity" is the complex between the magnitude of primary production and the trophodynamic-pathway structure that converts the primary production into fish biomass. Obviously, changes in either the magnitude of primary production or in the trophodynamic-pathway structure can result in changes in fish production.) The existence of coupled fluctuations between fish-stock abundance and biology is important for two reasons. First, coupled fluctuations imply that the historical record of multidecadal changes in fish-population abundance indicates past changes in biological productivity. Second, coupled fluctuations suggest that evidence of changes in biological production can be used to signal the potential for future increases or decreases in fish stocks. If this were true, significant improvements in fish-management performance would be realized. In addition, the establishment of cause-and-effect linkages between biological productivity and fish-stock fluctuations might be useful in elucidating the critical question of why some stocks depleted by fishing do not increase in abundance in spite of substantial reduction or termination of fishing mortality.

Clues to linkages between multidecadal fluctuations in fish stocks and biological productivity are found in evidence that connects changes in fish-stock abundance with multidecadal changes in plankton standing stocks over broad reaches of the ocean (assuming that changes in standing stocks are associated with changes in biological productivity). The suggestion of multidecadal fluctuations in productivity raises questions concerning (1) how the fluctuations are excited; (2) whether

B.J. Rothschild. Chesapeake Biological Laboratory, Center for Environmental and Estuarine Studies, P.O. Box 38, Solomons, MD 20688, USA.

the assumption that the ocean is in a biological steady state, typically used in the development of some global change models, is supportable (see Falkowski and Wilson 1992); and (3) the fact that, if the ocean is not in a biological steady state, estimates of long-term average or climatological indices of overall biological productivity used in global change models and based on only a few decades of observation are biased.

The first part of this paper considers the existence of large, fishery-independent, multidecadal fluctuations in fish-stock abundance that must, by definition, result from changes in biological productivity. The second part of the paper considers how these fluctuations might be related to parallel changes in primary and secondary productivity and physical forcing.

Fluctuations in fish-stock abundance

Records of fish-stock fluctuations date back several centuries and represent both preindustrial and industrial periods (see Cushing 1988*a*). Records for the preindustrial period are based on a relatively few long time series and a few studies of fish-scale depositions in sediments. The industrial period, which began about 1900, is characterized by power propulsion; mechanical means of fishing-gear deployment; and relatively high fishing mortality.

The preindustrial period

During the preindustrial period, presumed relatively low levels of fishing mortality, by modern standards, suggest that landing weight is a reasonable index of stock abundance or availability. This implies that the catch records for the preindustrial period are closely related to fish-stock abundance.

As an example, Uda (1952) (Fig. 1) summarized the fluctuations in the abundance of major fish stocks off Japan. He documented the abundance of skipjack tuna, *Katsuwonus pelamis*, from about 1825 to 1875; the Far Eastern sardine, *Sardinops melanostictus*, from 1675 to 1750 and again from 1800 to 1875; and the Hokkaido herring, *Clupea pallasi*, from 1750 to 1775 and again from 1800 to 1850. As another example, Ljungman (1882) (Table 1) recorded periods of abundance and scarcity of Bohuslan herring, *Clupea harengus*, dating back to the 16th century. He showed periods of abundance and absence of roughly 50 yr (see also Cushing 1982).

Another type of preindustrial record is based on the examination of fish scales in sediments (reviewed by Shackleton 1987). For example, Soutar and Isaacs (1969, 1974) demonstrated large multidecadal fluctuations, dating back several centuries, in sardine, *Sardinops caerulea*, anchovy, *Engraulis mordax*, and hake, *Merluccius productus*, in varved sediments at a few study sites off California (see also Baumgartner et al. (1992) for a critical review and extension of the sardine–anchovy record). De Vries and Pearcy (1982) also considered the record of fish-scale deposition off Peru. These records all imply the existence of multidecadal variability with fluctuations as cycles or quasi cycles lasting as long as perhaps 80 yr.

When interpreting sets of observations from the preindustrial period, it is important to take into account issues related to "availability," which can be defined in this case as the fraction of a fish population available to be measured by some index such as relative abundance of scales (availability is usually defined as the fraction of a population exposed to

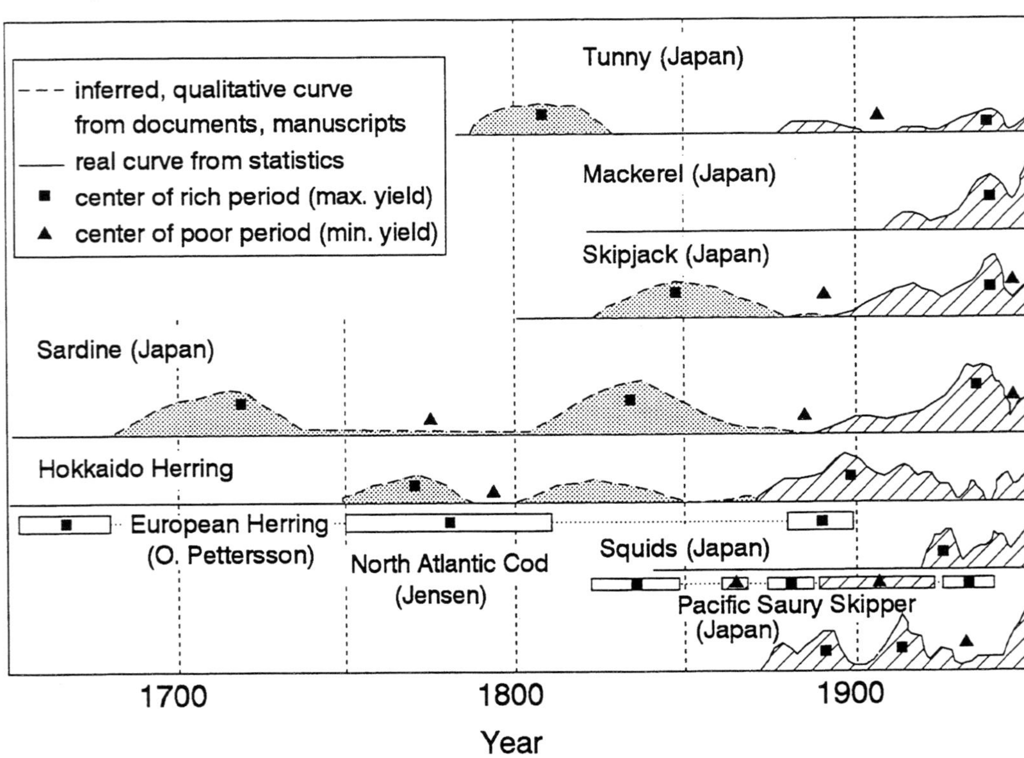

Fig. 1. Long-period variations of important fisheries (Uda 1952). Uda's figure suggests long-term quasi-cyclic variability in fish stocks, which is evident, in some cases, over the last two centuries.

Table 1. Alternation of Swedish and Norwegian fisheries according to Ljungman and Boeck. (From Cushing 1982).

Ljungman's periods of Bohuslan fisheries	Ljungman's comments	Boeck's dates of Norwegian fisheries
915–970	—	
971–1026	Good fisheries at least during	
1027–1082	the reign of Olav the Saint	
1083–1138	Commercial growth of Konungahalla;	
1139–1194	probably good fisheries	
1195–1250	Colonization of islands; convents. Probably good fisheries	
1251–1306	No good fisheries at beginnings and middle but probably good towards the end	
1307–1362	Particularly good fisheries at least during the last thirty years of the century	
1363–1418	No good fisheries	
1419–1474	Good fisheries at least during midcentury	
1475–1530	No good fisheries	1500–68 low abundance
1531–1586	Particularly good fisheries 1556–87	
1587–1642	No good fisheries	1600–48
1643–1698	Good fisheries at least between 1660 and 1680	
1699–1754	No especially good fisheries until 1747–81	1700–84 variable
1755–1810	Rich fisheries 1748–1808	
1811–1866	No good fishery	1818–70
1867–1922[a]	Rich fishery began in 1877 (and continued to 1906 or 1922)	1896 (or 1910) – 67

[a] Recorded since Ljungman's publication

fishing mortality). For example, inferred fluctuations in the abundance of the anchovy–sardine–hake complex reported by Soutar and Isaacs are based on samples taken at a localized area off California. Yet it is known from modern observations that when the sardine was at a high level of abundance and actively fished in the 1940s and 1950s, it ranged as far north as British Columbia, but when it was at a low level of abundance the population(s) tended to be localized off California. As another example, when the Far Eastern sardine was at a low level of abundance in the 1960s, it could be found only in waters adjacent to Japan, but when it was at a high level of abundance in the 1970s, its range extended over large areas of the western North Pacific Ocean. Range extensions are commonly associated with changes in stock biomass and suggest that localized changes in abundance often represent much larger changes in stock biomass and hence, by assumption, productivity. It is important to add that the evidence on long-term variability of several species, as inferred from the scale record, does not seem to reflect a correlation among the abundances of the species. This means that the species do not "replace" one another as is sometimes postulated. Rather, it seems that one or more species responds to changes in productivity, rather than to the abundances of each other.

Thus, even though the record is sparse, it is reasonable to infer that fish stocks have undergone multidecadal fluctuations in abundance over the centuries.

The industrial period

A large number of time series of fish-stock abundance record changes in stock and fishing effort during the industrial period. These time series reflect in more detail the existence of multidecadal fluctuations in fish-stock abundance.

Because fishing during the industrial period can be relatively intense, fluctuations in stock abundance have generally been thought to be closely coupled to the magnitude of fishing mortality. However, time-series of stock abundance and fishing mortality suggest that some components of stock-abundance variability is driven by changes in biological productivity, rather than by changes in fishing intensity. This is reflected by the fact that there are multiyear stanzas during the industrial period when stocks increase or decrease independent of any changes in fishing mortality. On one hand, the changes in stock abundance could relate to adjustments to equilibrium or to changes in the size or age-specific fishing-mortality vector. On the other hand, there is a reasonable likelihood that the increases or decreases are driven by changes in biological productivity.

To demonstrate this idea, contrast the temporal changes in the stock abundance per unit fishing mortality of the Icelandic summer-spawning herring and Norwegian spring spawning herring (Jakobsson 1985). Both stocks exhibited a dramatic decline in abundance between 1950 and 1970. The apparent causes of the decline appear to differ between the two stocks: the decline in abundance of Norwegian herring is closely

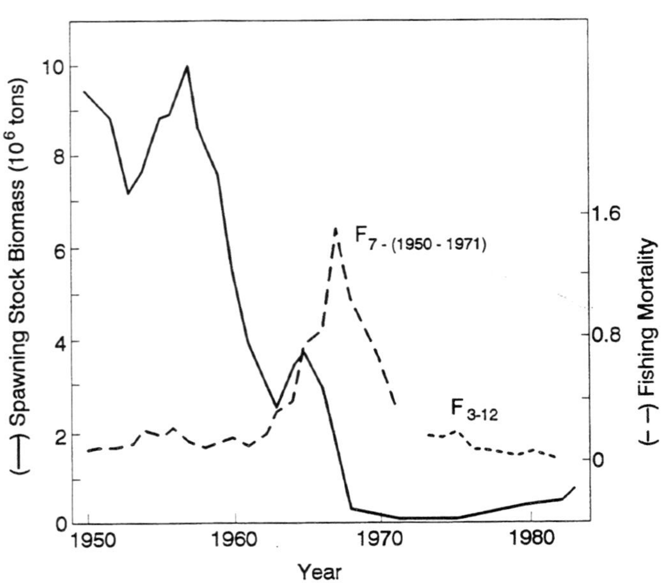

Fig. 2. Spawning stock biomass (Icelandic summer spawner) and fishing mortality rate (F) for 3-yr-old and older herring, 1950–1982. (based on Jakobsson (1985, Fig. 16); see also Rothschild (1986, Fig. 4.5*a*). Note that the stock begins to decline in the late 1950s before the substantial increase in fishing mortality in the mid-1960s.

coupled with increases in fishing mortality, while the decline in the abundance of Icelandic stock began well before there were any increases in fishing mortality (Rothschild 1986). Cushing (1988*b*) showed that the decline in the Icelandic stock may have been maintained by the relatively cool fresh water that characterized the North Atlantic, particularly from 1965 to 1971. The data can be taken to imply that the decline in the Icelandic herring was driven by changes in biological productivity, while that of the Norwegian herring was due to fishing (Fig. 2).

Examples of changes in biological productivity

Despite the caveats, it seems reasonable to infer that fishing records of the preindustrial era and inferences based on the deposition of fish scales in sediments reflect the existence of multidecadal fluctuations in fish-stock abundance over the centuries. Likewise, it is reasonable to conclude that some components of modern fish-stock abundance fluctuations are independent of fishing mortality. This section contains three examples suggesting linkages between the biological productivity of fish and that of plankton: the North Pacific, the Northeast Atlantic Ocean and the North Sea, and the Peru Current.

The North Pacific Ocean

Perhaps the most complete example of changes in multidecadal basin-scale productivity is associated with the North Pacific Ocean. The changes may have been signaled by the observation of Venrick et al. (1987) that the standing stock of summer chlorophyll north of Hawaii had doubled from 1968 to 1985. Brodeur and Ware (1992) showed that there was at least a doubling of large net zooplankton in the central subarctic Pacific Ocean (Fig. 3). In another paper, Brodeur and Ware (1995) showed that the biomass of nekton (14 species including salmonids, nonsalmonids, and cephlapods) had doubled between the late 1950s and 1980s over a large area of the Gulf of Alaska. During the same time period, in the eastern North Pacific, the Far Eastern sardine population "exploded" from a stock that yielded virtually no landings in 1972 to landings of nearly 4 million tonnes in 1985. In 1972, the Far Eastern sardine was restricted to the waters in the vicinity of the Japanese archipelago, but by 1985 its range had expanded to large reaches of the western North Pacific Ocean.

Thus the decades of the 1970s and 1980s showed evidence of broad increases in biological productivity in various areas of the North Pacific. It is difficult to say whether the increases in productivity were related to one another, and it is true, that while the stocks cited above increased, others decreased. Yet there is a distinct overall impression that the North Pacific Ocean underwent a remarkable increase in the productivity of many stocks from about 1970 to about 1990.

The Northeast Atlantic Ocean and the North Sea

Many authors have shown, based on continuous plankton recorder (CPR) records, that standing stocks of the planktonic flora and fauna and by implication, the biological production of the Northeast Atlantic Ocean and the North Sea have been declining for three decades: the 1950s, the 1960s, and the 1970s (e.g., Aebisher et al. 1990). It appears that these indices tended to increase in the 1980s.

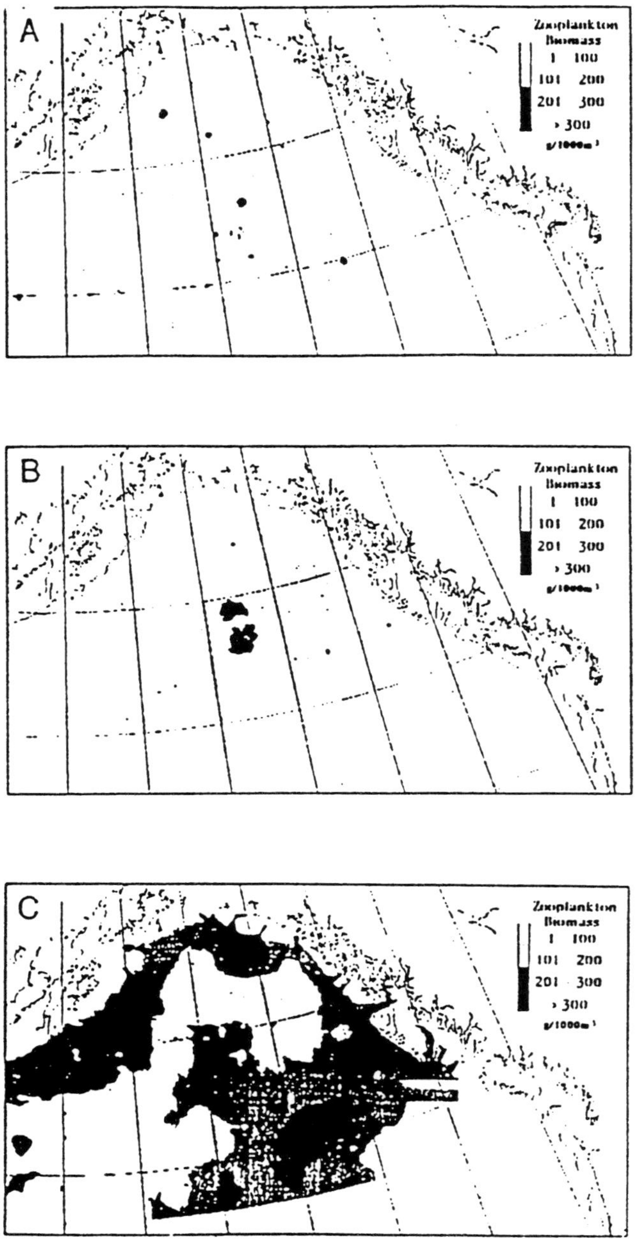

Fig. 3. Changes in zooplankton biomass in the North Pacific Ocean. Brodeur and Ware (1992) indicate that the standing stock of net zooplankton approximately doubled over two decades beginning in 1970. It is interesting to note that this implies a doubling in the downwelling of fecal pellets or export of fecal pellet related carbon to the deep ocean. (From Brodeur and Ware 1992).

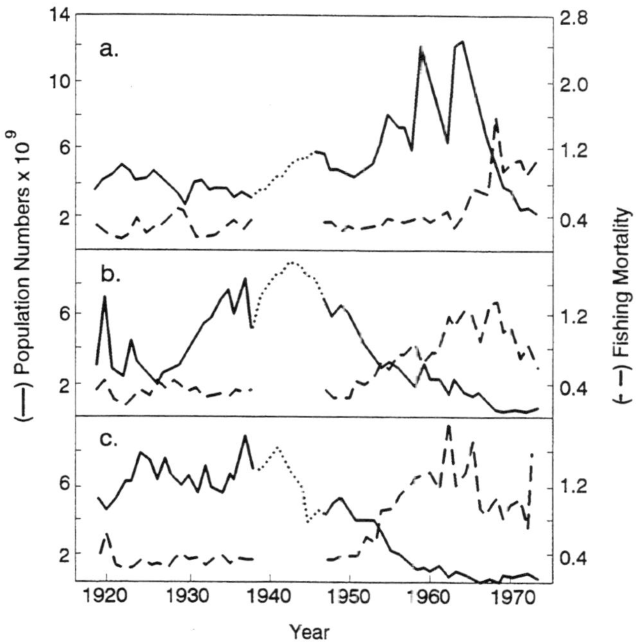

Fig. 4. Changes in the population size and fishing mortality of herring in the northern North Sea (division IVa), in the central North Sea (division IVb), and in the southern North Sea (division IVc). There was no fishing in the 1940s because of World War II. (From Burd (1978, Fig. 139). Note fluctuations in stock independent of fishing mortality. The implication is that the large fishery-independent changes in population result from changes in biological productivity rather then from fishing (see also Rothschild 1986).

Data for the North Sea are more detailed, and it is particularly interesting to contrast the changes in the North Sea herring stocks with the CPR data. For the purpose of herring fishery statistics, the North Sea has been divided into northern, central, and southern divisions (see Burd 1978). In all divisions, there was no fishing mortality during World War II (D.H. Cushing, personal communication, has pointed to the existence of some fishing on southern stock during the war.) In the northern division, fishing mortality was relatively low and virtually constant from the 1920s until the mid-1960s. In the central and southern divisions, fishing mortality was relatively low and virtually constant until the 1950s. Under low and constant fishing mortality and in the absence of fishing during World War II the stocks in the north remained relatively constant until 1950 and then increased sharply. The stocks in the southern division increased from the mid-1920s to the mid-1940s and then declined continually. While in the southern division, the stocks that were at a relatively high level of abundance declined subsequent to 1940. It is important to note that during the period there was a sharp increase in the abundance of gadoid fishes in the North Sea. An inspection of Burd's figure (Fig. 4) suggests that some fluctuations of herring stocks in the North Sea tended to be independent of fishing mortality. It is particularly interesting to compare the increase in the northern stock with the gadoid outburst that occurred at roughly the same time.

This can be seen from the fact that declines in stock, for example, occurred during periods of relatively low and constant fishing mortality (of course, fishing mortality increased after the stocks declined, as might be expected with relatively fixed capital and a declining stock). In addition, the declines in the herring stocks appear to be coupled with declines of two standard deviations in phytoplankton, zooplankton, and kittiwake, *Rissa tridactyla*, indices over a three-decade period (Aebisher et al. 1990) (Fig. 5).

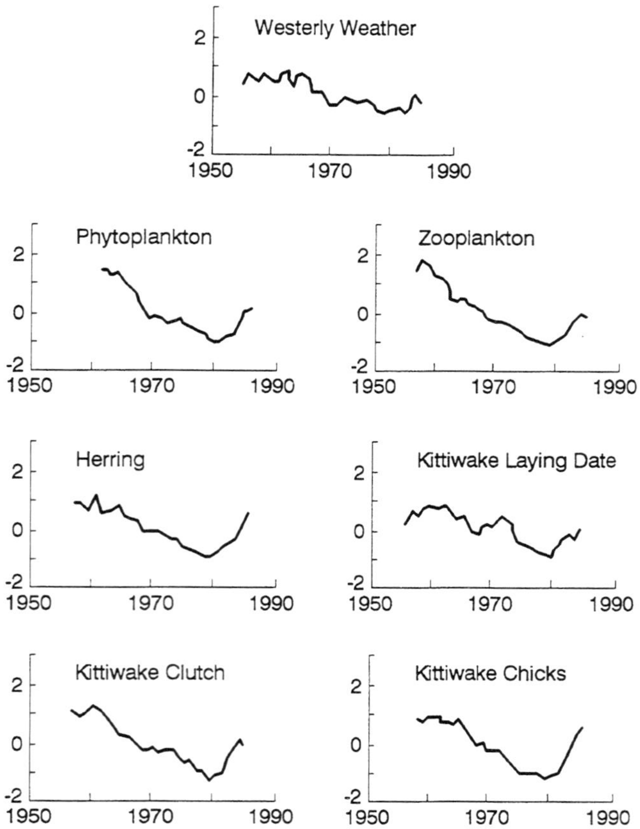

Fig. 5. Standardized (zero mean, unit variance) time series and 5-yr running means for abundances of phytoplankton, zooplankton, and herring, for kittiwake laying date, clutch size, and chick production, and for frequency of westerly weather, from 1955 to 1987. Ordinates in standardized units. In absolute terms, 5-yr running means for frequency of westerly weather ranged from 60 to 82 d·yr^{-1}, corresponding to a maximum decrease of 27%. (From Aebisher et al. 1990).

My personal examination of CPR data on a few key species of zooplankton in the North Sea reflects that these declines may be as high as a factor of about 10 for the *Paracalanus–Pseudocalanus* group and about 5 for *Calanus*. The declines have been attributed to various causes. Cushing (1992) assessing the changes in the Downs Stock linked declines with the herring stock–recruitment relationship. Aebisher et al. (1990) considered the decline in indices of westerly weather to be important, while Dickson et al. (1988) stressed the increase in northerly, relatively intense winds as a cause. In particular, they suggested that the intensity of winds delayed the onset of stratification and this reduced the standing stocks of phytoplankton and consequently zooplankton.

The Peruvian Current

The biomass of the Peruvian anchovetta, *Engraulis ringens*, was about 15 million tonnes in 1972 when coincident with the onset of an El Niño, it collapsed to about 4 million tonnes. Many thought that the collapse was related to the 1972 El Niño while others acknowledged the effect of the El Niño, but thought fishing was implicated. Some 20 yr later, the anchovetta has not recovered to its former level of abundance

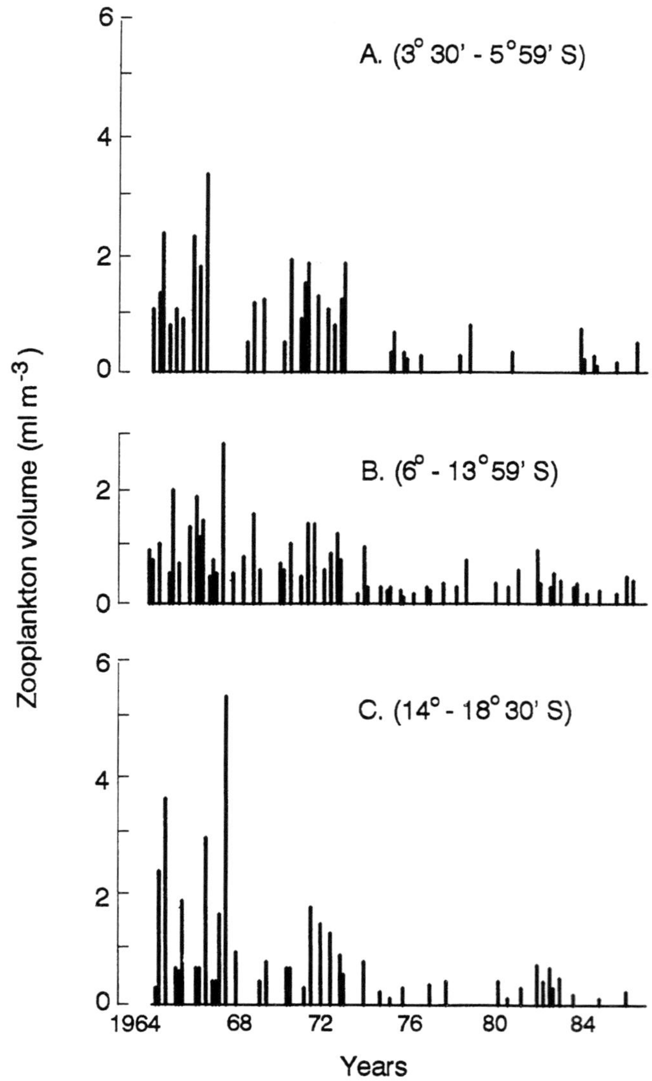

Fig. 6. Trend of zooplankton abundance off Peru, 1964–1987 (from Carrasco and Lozano 1989). Note that there were substantial declines in zooplankton starting in about 1972, the year that the anchovetta population collapsed. This was also the year of a major El Niño. Despite subsequent El Niño events and intervening periods neither the zooplankton nor the anchovetta have recovered to their former level of abundance.

(however, in approximately 1980, the sardine that inhabited waters to the south of the anchovetta area sharply increased in abundance).

Examination of zooplankton data suggests that the anchovetta collapse may not have been related to either the El Niño or to fishing, but rather to some long-term change in biological productivity. In the early 1970s, the stocks of zooplankton in the Peru–Chile region declined at the same time as the anchovetta (Fig. 6). The abundance of zooplankton is obviously not directly influenced by fishing, and the fact that El Niños occur every several years suggests that the zooplankton and anchovetta decline is not related to the 1972 El Niño event. Even though it can be argued that the eventual increase in the sardine population somehow keeps the zooplankton stocks at a low level, it nevertheless appears that a significant change in biological production, which occurred in the early 1970s, has evidently maintained itself, resulting in significantly depressed zooplankton and anchovetta populations.

Trophodynamic pathways

The above observations suggest that fish stocks undergo multidecadal fluctuations in abundance and that these changes are correlated with parallel changes in biological production. Several authors have linked these changes in production with variations in the wind field (Trenberth 1990; Cushing and Dickson 1976; Cury and Roy 1987), thereby linking large spatial-scale physical events with smaller-scale physical events (because surface wind velocity contributes to mixed-layer turbulent flow) (e.g., Rothschild 1989; Rothschild and Osborn 1988; see also Sundby and Fossum 1990; Kiørboe 1993) and changes in biological productivity.

However, the precise modalities by which changes in air–sea interactions drive changes in fish-stock production require considerable research. There are basically two intertwined issues. The first relates to the availability of evidence of any long-term change in primary production. The second relates to the efficiency by which primary production is transformed into fish biomass. The issues are intertwined because increases or decreases in primary production could result from changes in pathway efficiency because a substantial part of primary production is fueled by excretion from heterotrophs.

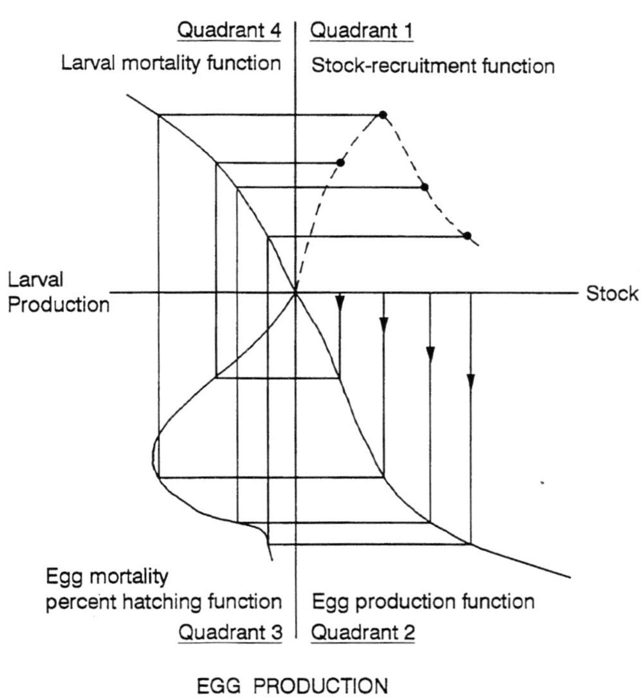

Fig. 7. A Paulik diagram showing the generation of a dome-shaped recruitment-stock function with a dome-shaped function in quadrant 3. The diagram is not drawn to scale. (From Rothschild 1986). The diagram implies that different trophodynamic pathways apply to each quadrant. A large year-class implies a highly efficient transfer in each quadrant.

The general problem of trophodynamic pathways is a fundamental component of biological oceanography and beyond the scope of this paper. However, in connection with the special case of fish production, there are three issues. These relate to how closely the productivity of various fish stocks is coupled to primary production; how various life-history stages track primary production; and the availability of primary production to fish.

The first pathway issue concerns the fact that the abundance of only some fish stocks can be expected to closely track changes in primary production. This is because some fish stocks are closely coupled trophodynamically to primary production, while others are not. For example, all fish larvae are planktivourous, but in some species, juveniles and adults are planktivourous as well. Yet in other species, the juveniles and adults feed on nekton or benthos. Species with planktivourous adults would be expected to track primary production more closely than fish that are piscivourous or eat large organisms as adults. The idea is that, during multiyear periods of relatively high plankton production, the abundance of planktivourous adult populations appears to be linked linearly with plankton abundance. On the other hand, during periods of low plankton abundance or, in the case of species in which the adults are not planktivourous, nonlinear population dynamics dominate, giving the impression that fish abundance and plankton abundance are uncorrelated, when in fact they are linked.

The second pathway issue is somewhat related to the first and concerns the dynamics of larval fish. These dynamics are generally thought to influence variability in recruitment. The dynamics of larval fish depend on their trophodynamic status. Yet the trophodynamic status of larval fish is different than the trophodynamic status of juvenile and mature fish. This can be seen in the Paulik diagram that shows the different trophodynamic pathways (Fig. 7). For production in a fish stock to increase, it is necessary for all pathways (i.e., all quadrants in the Paulik diagram) to be "turned on." The simultaneous increase in flow through all pathways on a multiyear scale would require a consistent change in production.

The third pathway issue concerns the extent to which primary production is fully available to fish. In many systems, because of the large production accounted for by organisms less than 100 μm in size, the production of fish may only be a small fraction of the total. On the other hand, it may be that variability in stocks or production in plankton may drive the variability in the fish stocks.

Applications

It is reasonable to infer that preindustrial and industrial-era multidecadal fluctuations in fish-stock abundance reflect changes in productivity or trophic pathways. Changes in the productivity of fish stocks seem evident and at times decoupled from fishing mortality as it increased in intensity during the industrial era.

There are two applications of these observations: (1) fluctuations in fish-stock abundance are important indicators of basin-scale long-term changes in productivity and are therefore of interest in determining the oceans' role in global change, and (2) the existence of long time-scale stock fluctuations (relative to fishery-management decisions) suggests that these fluctuations can be at least used for prediction in fishery management.

Global change issues
Global change is taken to mean a change or a departure from the statistical long-term average (or higher moments) of the complex of variables that represent the physics of air–sea interactions. (The interpretation of whether or not a global change has occurred depends on the time scales under consideration.)

Generally speaking, evidence can be found that changes in the air–sea interaction complex occur on century and decadal time scales. Some of these changes have obvious connections with the abundance and availability of fish. The obvious connections generally relate to temperature changes, which evidently are linked to spatial changes or to changes in abundance. For example, as the ocean warms and cools, the positions of isotherms move north and south along the coasts. The fish populations and other fauna tend to track the isotherms. Thus, in some cases, from a fixed point on the coast, the apparent abundance seems to increase or decrease, even though the mean stock abundance in relatively constant. In terms of abundance changes, the year-class strength of cod in northern Norway seems to be controlled by temperature since, when the temperature is relatively cold, good year-classes do not occur, but when the temperature is warm, poor and good year-classes occur.

Some of the changes in the air–sea interaction complex are more subtle than those related to temperature, that their effects are still poorly understood. They are changes in the wind field. These changes, some of which have been demonstrated to be persistent oscillations, undoubtedly affect production either through increased input of nutrients or in a modification of the pathways in production. Yet the overall details of these changes are not understood.

If we assume that the multidecadal fluctuations in fish stocks that are trophodynamically close to plankton are driven by major changes in biological productivity as *implied* in the North Pacific and North Atlantic Oceans, then we can see that these changes in fish-stock abundance can be indicative of fluctuations in global-change parameters. For example, the changes in chlorophyll standing stock in the North Atlantic and in a restricted area of the North Pacific Ocean imply that primary production may have increased in the North Pacific and decreased in the North Atlantic during the same period of time. The possible increases in primary production in the North Pacific and decreases in the North Atlantic each reflect a positive-feedback property in the carbon pump. For example, the inferred increases in productivity in the North Pacific resulted in a doubling in large zooplankton, thus increasing the production of fecal pellets and hence roughly doubling the downwelling of particulate carbon into the deep ocean. In addition to changing the interactions of the traditional food chain involving the large zooplankton, there is reason to infer that the small-particle food chain was also affected, although it is difficult to speculate on the magnitude and directions of the effect.

These observations tend to contradict the suggestion by Falkowski and Wilson (1992) that the ocean is in a biological steady state (an assertion frequently made in global-change models). The steady-state assumption, however, could be supported if the fluctuations described above were merely deviations from a long-term average rather than departures from a stationary state. On the other hand, if the fluctuations are periodic or quasi periodic, then one would need to be concerned about the parameter estimates in biological steady state models, in the sense that these estimates were not derived from steady state conditions, but rather from peaks or troughs in the fluctuations.

Fisheries management
For fisheries management, linkages between productivity and fish stock are of particular practical importance because it is useful to know whether the system is in a state of relatively high productivity or one of relatively low productivity or when the productivity is changing. This is because the theory of fishing, as we use it, essentially only applies to constant productivity. The shifts from high to low productivity or vice versa can be seen in plots of recruitment–stock relationships, which frequently exhibit stanzas of relatively high and low recruitment for the same fixed stock size. It is interesting, as well, that these shifts do not occur gradually, but rather in 1 yr or in only a few years' time. Understanding these periods of high and low productivity would do much to reduce the great degree of uncertainty in recruitment–stock relationships.

Acknowledgements

This work was supported by grants from U.S.AID through the Fisheries Stock Assessment–Collaborative Research Support Program.

References

Aebisher, N.J., J.C. Coulson, and J.M. Colebrook. 1990. Parallel long-term trends across four marine trophic levels and weather. Nature (Lond.) 347(6295): 753–755.

Baumgartner T.R., A. Soutar, and V. Ferriera-Bartrina. 1992. Reconstruction of the history of Pacific sardine and northern anchovy populations over the past two millennia from sediments of the Santa Barbara Basin, California. Calif. Coop. Oceanic Fish. Invest. Rep. 33: 24–40.

Brodeur, R.D., and D.M. Ware. 1992. Long-term variability in zooplankton biomass in the subarctic Pacific Ocean. Fish. Ocean 1(1): 32–38.

Brodeur, R.D., and D.M. Ware. 1995. Interdecadal variability in distribution and catch rates of epipelagic nekton in the northeast Pacific Ocean, p. 329–356. *In* R.J. Beamish [ed.] Climate change and northern fish populations. Can. Spec. Publ. Fish. Aquat. Sci. 121.

Burd, A.C. 1978. Long term changes in sea herring stocks. Rapp. P.V. Réun. Cons. int. Explor. Mer 172: 137–153.

Carrasco, S., and O. Lozano, 1989. Seasonal and long-term variations of zooplankton volumes in the Peruvian Sea. 1964–1987, p. 82–85. *In* D. Pauly, P. Muck, J. Mendo, and I. Tsukayama. [ed.] The Peruvian Upwelling Ecosystem: Dynamics and Interactions, Callao, Peru. Aug. 24–28, 1987 ICLARM Conf. Proc. 18: 483 p.

Cury, P., and C. Roy. 1987. Upwelling et peche des especes, pelagiques cotieres de Cote-d'Ivoire: une approache globale. Oceanol. Acta 10(3): 347–357.

Cushing, D.H. 1982. Climate and fisheries. Academic Press, London, UK.

Cushing, D.H. 1988*a*. Provident sea. Cambridge University Press. Cambridge, UK.

Cushing, D.H. 1988*b*. The northerly wind, p. 235–244. *In* B.J. Rothschild [ed.] Toward a theory on biological physical interactions in the world ocean. Kluwer Academic Publishers, Dordrecht, The Netherlands.

Cushing, D.H. 1992. A short history of the Downs stock of herring. ICES J. Mar. Sci. 49: 437–443.

Cushing, D.H., and R.R. Dickson. 1976. The biological response in the sea to climatic changes. Adv. Mar. Biol. 14: 1–22.

De Vries, T.J., and W.G. Pearcy. 1982. Fish debris in sediments of the upwelling zone off central Peru: a late Quaternary record. Deep-Sea Res. 28(1): 87–109.

Dickson, R.R., P.M. Kelly, J.M. Colebrook, W.S. Wooster, and D.H. Cushing. 1988. Winds and production in the eastern Atlantic. J. Plankton Res. 10: 151–169.

Falkowski, P.G., and C. Wilson. 1992. Phytoplankton productivity in the Pacific Ocean since 1900 and implications for absorption of anthropogenic CO_2. Nature (Lond.) 358: 741–743.

Jakobsson, J. 1985. Monitoring and management of the east Atlantic herring stocks. Can. J. Fish. Aquat. Sci. 42: 207–221.

Kiørboe, T. 1993. Turbulence, phytoplankton cell size and the structure of pelagic food webs. Adv. Mar. Biol. 29: 1–72.

Ljungman, A. 1882. Contribution towards solving the question of the secular periodicity of the great herring fisheries. U.S. Comm. Fish Fish. 7(7): 497–503.

Rothschild, B.J. 1986. Dynamics of marine fish populations. Harvard University Press, Cambridge, Mass.

Rothschild, B.J. 1989. On the causes for variability of fish populations—the linkage between large and small scales. *In* T. Kawasaki, S. Tanada, Y. Toba, and A. Taniguchi [ed.] Long-Term Variability of Pelagic Fish Populations and Their Environment. Proceedings of the International Symposium 1989, Sendai, Japan, 14–18 November 1989: 367–376.

Rothschild, B.J., and T.R. Osborn. 1988. Small-scale turbulence and plankton contact rates. J. Plankton Res. 10(3): 465–474.

Shackleton, L.Y. 1987. A comparative study of fossil fish scales from three upwelling regions. *In* A.I.L. Payne, G.J.A. Gulland, and K.H. Brink [ed.] S. Afr. J. Mar. Sci. 5: 79–84.

Soutar, A., and J.D. Issacs. 1969. History of fish populations inferred from fish scales in anaerobic sediments off California. Calif. Coop. Oceanic Fish. Invest. Rep. 13: 63–70.

Soutar, A., and J.D. Issacs. 1974. Abundance of pelagic fish during the 19th and 20th centuries as recorded in anaerobic sediment off the Californias. Fish. Bull. U.S. 72(2): 257–273.

Sundby, S., and P. Fossum. 1990. Feeding conditions of Arcto-Norwegian cod larvae compared at the Rothschild–Osborn theory on small scale turbulence plankton contact rates. J. Plankton Res. 12: 1153–1162.

Trenberth, K.E. 1990. Recent observed interdecadal climate changes in the northern Hemisphere. Bull. Am. Meteorol. Soc. 71: 988.

Uda, M. 1952. On the relation between the variation of the important fisheries conditions and the oceanographical conditions in the adjacent waters of Japan I. J. Tokyo Univ. Fish. 38(3): 364–389.

Venrick, E.L., J.A. McGowan, D.R. Cayan, and T.L. Hayward. 1987. Climate and chlorophyll a: long-term trends in the Central Pacific Ocean. Science, (Wash., D.C.) 238: 70–72.

Use of Arrhenius models to describe temperature dependence of organismal rates in fish

Ping Lin and Henry A. Regier

Lin, P., and H.A. Regier. 1995. Use of Arrhenius models to describe temperature dependence of organismal rates in fish, p. 211–225. *In* R.J. Beamish [ed.] Climate change and northern fish populations. Can. Spec. Publ. Fish. Aquat. Sci. 121.

Abstract: There is a copious and diverse literature on the role that habitat temperature plays in the physiology and ecology of ectotherms like fish. Generally and implicitly, rate processes of fish are taken to be more responsive to temperature changes than are anatomical or structural biotic features. Various forms of a general Arrhenius model are increasingly being used for descriptive empirical generalization, as well as analytical explication, of biochemical and biological processes affected by temperature. Appropriate versions of the Arrhenius model were fitted to data on several organismal rate processes for the cold-water sockeye salmon (*Oncorhynchus nerka*) and warmwater largemouth bass (*Micropterus salmoides*). According to several admittedly weak tests concerning the relevance of the model the results are promising. The tests involve statistical goodness of fit, appropriate temperature of maximal rate, appropriate breadth of thermal niche, general consistency of the magnitude of some coefficients with estimates from data in which temperature did not vary, and a reasonable general pattern among the model's coefficients.

Résumé : Il existe une documentation abondante et diversifiée sur le rôle que la température des habitats joue dans la physiologie et dans l'écologie des espèces ectothermes comme le poisson. De façon générale et implicite, chez le poisson, les processus faisant intervenir la vitesse sont jugés plus sensibles aux changements de température que les caractéristiques biotiques anatomiques ou structurales. Diverses formes d'un modèle général d'Arrhenius sont de plus en plus utilisées pour que l'on obtienne une généralisation empirique descriptive ainsi qu'une explication analytique des processus biochimiques et biologiques affectés par la température. Les versions pertinentes du modèle d'Arrhenius ont été appliquées aux données sur plusieurs processus faisant intervenir la vitesse chez le saumon sockeye (*Oncorhynchus nerka*) d'eau froide et chez l'achigan à grande bouche (*Micropterus salmoides*) d'eau chaude. D'après plusieurs critères dont on reconnaît la faiblesse pour ce qui est de la pertinence du modèle, les résultats sont prometteurs. Les critères comportent la qualité de l'ajustement, la température de vitesse maximale appropriée, la taille appropriée de la niche thermique, la compatibilité générale de l'ampleur de certains coefficients avec les estimations provenant de données dans lesquelles la température ne varie pas, ainsi qu'une configuration générale raisonnable des coefficients produits par le modèle.

Introduction

Climate changes, on a variety of time scales, generally involve temperature changes. The expected global climate warming will influence the habitat temperature of fish populations, for example, by changing surface heat flux (e.g., warming surface temperature in summer), altering the depth profile of the temperature structure, and changing the seasonal temperature cycle and dissolved oxygen concentrations of the water body (Blumberg and Di Toro 1990; Coutant 1990; McCormick 1990). Habitat temperatures directly influence fish physiology and behaviour (Fry 1947, 1971; Hokanson 1977; Magnuson et al. 1979; Weatherley 1990), and abundance and production of fish populations (Christie and Regier 1988).

Body size has long been recognized as an important variable reflecting many aspects of the physiology, ecology, and behaviour of living organisms (Peters 1983; Calder 1984; Dickie et al. 1987). Especially for fish, size-based scaling of physiological and ecological processes has been well reviewed for larval, juvenile, and adult stages (Brett and Groves 1979; Mittelbach 1981; Werner and Gilliam 1984; Miller et al. 1988); an exponential function of mass is commonly used for scaling purposes.

With climate change, warming rates of habitat may be rapid (0.3°C per decade), and genetic change may be too slow as an adaptive mechanism (Sugden 1992). A key issue concerning the influence of climate change on fish relates to

P. Lin and H.A. Regier. Institute for Environmental Studies, University of Toronto, Toronto, ON M5S 1A4, Canada.

the temperature effects on various physiological and ecological rate processes, especially growth, because the changes in rate processes may contribute to large-scale events with fish populations, such as geographic shifts and local extinction.

We have been searching for empirical generalizations, as well as analytical explanations, for the temperature dependence of such rates for different fish species. Cold-water sockeye salmon (*Oncorhynchus nerka*) and warmwater largemouth bass (*Micropterus salmoides*) were chosen to indicate what is known and what may be conjectured quantitatively about temperature dependence at the organismal level of organization as relevant to population phenomena. A brief survey of relevant temperature phenomena for organisms of these two species follows.

Temperature response of wild fish populations

Temperature is one of the most important factors affecting the life history characteristics of fish (Fry 1971; Cossins and Bowler 1987). Fish generally select the temperature at which they carry out physiological functions with relatively high efficiency, and compete and cooperate effectively with other species (Brett 1971; Fry 1971; Crawshaw 1977; Beitinger and Fitzpatrick 1979; Magnuson et al. 1979). Freshwater fish species from temperate North America roughly fall into three broad groups (or guilds) for which preferred temperatures have been estimated: cold-water fish, with thermal niches centred around 15°C; cool-water fish, with thermal niches centred around 24°C; and warmwater fish, with thermal niches centred around 28°C (Hokanson 1977). The spawning, metabolism, growth, development, and overall behaviour of fish respond strongly to the thermal regimes of diel, seasonal, and annual scales (Brett 1971; Kitchell et al. 1977; Rice and Cochran 1984; Cossins and Bowler 1987; Shuter and Post 1990; Weatherley 1990; Burgner 1991). Because of their

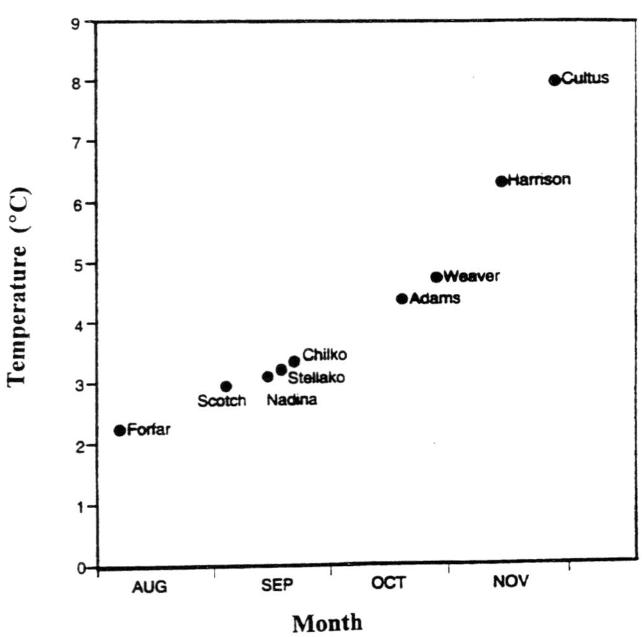

Fig. 1. Spawning times and mean incubation temperature of nine Fraser River sockeye stocks in British Columbia. (From Brannon 1987, Fig. 1)

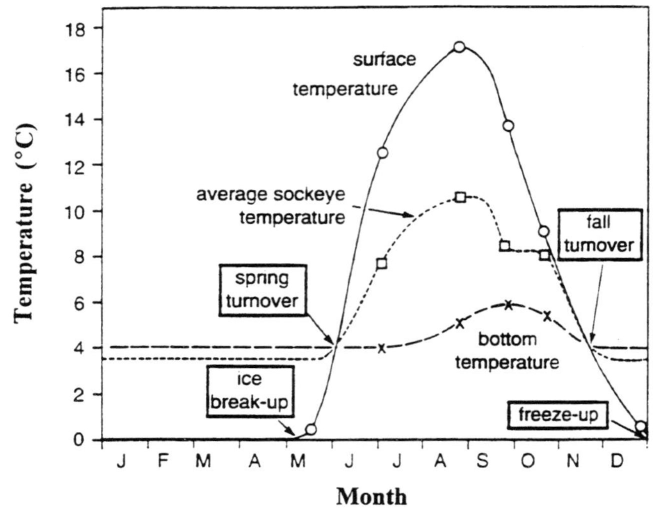

Fig. 2. Computed average temperatures experienced by juvenile sockeye during vertical migration, compared with surface and bottom temperature in Babine Lake, British Columbia, 1967. (From Brett 1971, Fig. 7)

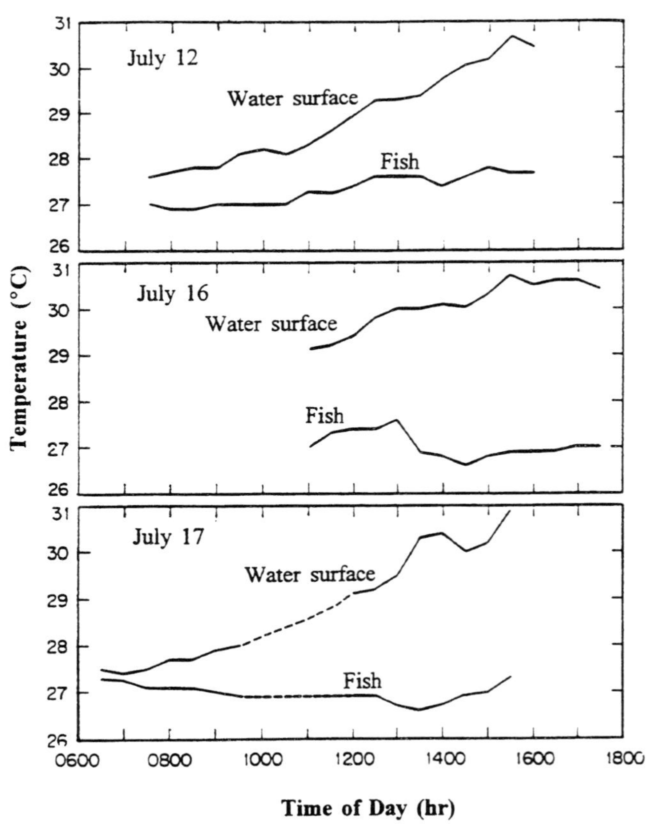

Fig. 3. Temperature selected by an adult largemouth bass and water-surface temperatures taken at half-hour intervals for 3 days in July 1974, K-25 pond, Oak Ridge, Tenn. (From Coutant 1975, Fig. 8)

sensitivity to temperature, fish species generally select habitat of a temperature within a fundamental thermal niche defined as ±2.0°C of their median preferred temperature (Magnuson et al. 1979). Small changes in temperature can have important bioenergetic consequences to an organism's survival, growth, or reproductive success (Kitchell et al. 1977; Crowder and Magnuson 1982). Temperature performance breadth is relevant to discussions of habitat occupancy and competition (Huey and Stevenson 1974; Magnuson et al. 1979).

For sockeye salmon, as expected, the rate of development of the embryos is faster at higher incubation temperatures. Differences among stocks in timing of spawning is related directly to the temperature regime of the spawning site (Fig. 1) (Brannon 1987). This flexibility among stocks in timing of spawning is an adaptation that may help to optimize the timing of dispersal of fry in spring into their feeding habitat, particularly to take advantage of the seasonal peak abundance of zooplankton of appropriate size (Brannon 1987). Fry that emerge earlier or later than the optimum may suffer greater mortality (Godin 1982). The trend of summer and autumn diel vertical migrations and feeding behaviour of underyearling sockeye is well illustrated by studies at Babine Lake, British Columbia (Fig. 2) (Brett 1971). The energetics analysis derives from studies of voluntary food intake, rate of digestion, and return of appetite, which are each distinctively temperature dependent (Brett 1971). Within a sockeye-producing lake, growth of juvenile sockeye may be directly related to temperature and inversely related to fish population density (Burgner 1991).

For warmwater largemouth bass, observations in the wild indicate that temperature is a key factor. Largemouth bass appear actively to select, if available, those temperatures at which they perform best (Fig. 3) (Coutant 1975). The general pattern of their vertical migrations is probably related to patchy availability of temperature and food resources (Diana 1984). Their growth and abundance may be directly related to temperature and prey density (Strawn 1961; Buynak et al. 1991).

Overall, the cold-water sockeye salmon and warmwater largemouth bass show different temperature preferences and adaptation to environmental seasonality as reflected in their life history cycle. The conceptual life cycle (Fig. 4) of cold-water and warmwater species living in temperate North America, which is modified from Shuter and Meisner (1992), is a helpful interpretation for the species discussed here. The cold-water sockeye salmon and warmwater largemouth bass show a similar pattern of response to a range of habitat temperature.

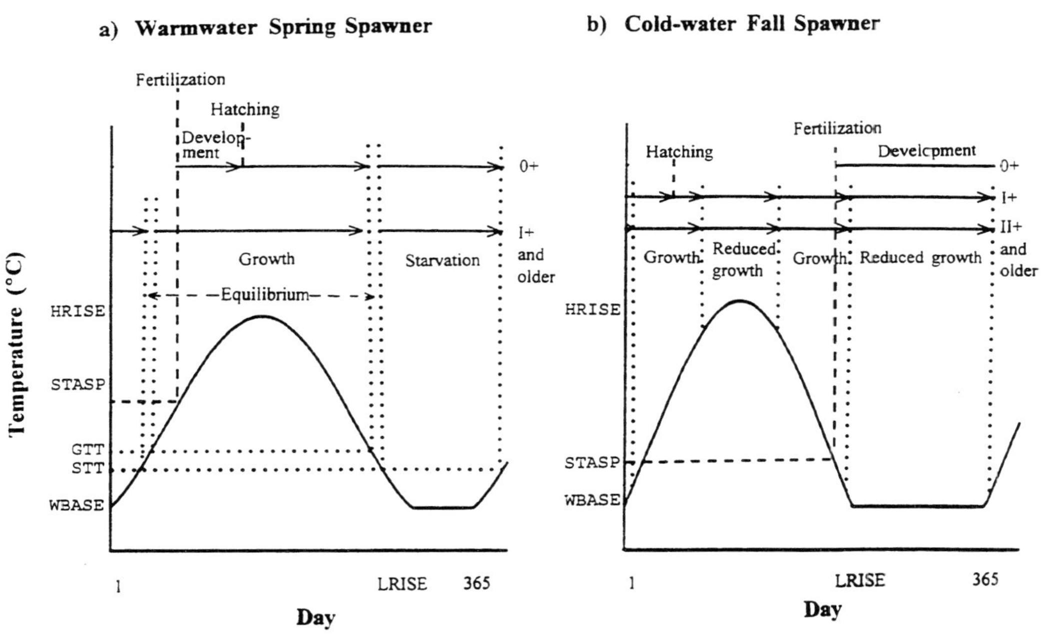

Fig. 4. Generalized annual growth and reproduction periods for typical warmwater and cold-water fish living in central North America. Life cycle events are superimposed on the annual cycle typical of surface water temperatures in North American lakes and rivers (—). Please note that the left and right edges of the figures overlap. LRISE is the number of days in a year that the temperature is greater than the winter base temperature (WBASE, taken as 4°C here for convenience); HRISE is the maximum daily temperature for the year. All fertilizations for the year occur when the temperature first reaches start spawning temperature (STASP) in spring (a) or autumn (b). Young-of-the-year and older growth occurs when the temperature is greater than growth threshold temperature (GTT), and starvation occurs when the temperature is less than starvation threshold temperature (STT), both with respect only to Fig. 4a because cold-water species can continue to grow at 4°C. (Modified from Shuter and Meisner 1992, Fig. 8)

Temperature dependence of particular rate processes

In the laboratory, various types of responses of fish species to temperature have been investigated. Typical examples of these kinds of studies are physiological effects related to acclimation temperatures and particular upper and lower lethal temperatures, as well as behavioral responses such as preference and avoidance reactions (Fry 1947, 1971). Regression relationships, among temperatures at which different kinds of rate processes are maximal or optimal within certain species, have been estimated across a range of species adapted to different temperatures (Jobling 1981). A mathematical expression that can incorporate all these concepts can be expressed as follows:

(1) ER or BT = *f*(TT, ATS, BTS, SE; MA; BE, RP; LH)

where ER is the relative rate at approximate equilibrium, d*R*/*R*d*t*, or BT is the behavioral threshold; TT is the test temperature; ATS is the acclimation temperature to which the organism has adapted over a short interval such as several days with all other factors constant; ATM is the acclimation temperature to which the organism has adapted over a medium-term period of some weeks; SE is the season of the year to which long-term evolutionary adaptation has occurred and within which medium-term (refer to ATM) and the short-term (refer to ATS) adaptations occur; MA is the mass (or "weight") of the biotic system under study, e.g., individual fish; BE is the biotic entity, e.g., organism of a particular species; RP is the kind of rate process; and LH is the life history stage, because patterns between life-history stages may differ beyond the difference associated with mass (MA) and seasonal (SE) considerations.

The independent variables on the right side of the function have been ordered empirically in four sets by semicolons, with the four sets taken to be relatively independent of each other in a causal sense, at least as far as we know.

For fish, many particular characterizations of this multidimensional surface have been estimated and published. In effect, the relationship of the equilibrium relative rate, ER, to one or two of the independent variables (e.g., test temperature and mass of the organism), with all other variables held constant, can be described using a variety of quantitative regression models and methods. For example, Brett (1971) assessed the essential energy relations of sockeye salmon; he found that growth and metabolism are related to body size, temperature, and feeding intensity. When such fundamental necessities as food are available in excess of needs, temperature may be the principal factor determining growth rates and total annual growth (Coutant 1975). Such a conclusion has been commonly accepted in quantitative research on fish (Kitchell et al. 1977; Rice and Cochran 1984; Dickie et al. 1987; Shuter and Post 1990). This paper illustrates an attempt to unify particular understandings of the effects of temperature on physiological responses of fish, using cold-water sockeye salmon and warmwater largemouth bass as examples.

Based on a general sense of the various thermal phenomena associated with ectothermic processes, we observe two kinds of temperature response: monotonic change between two extremes, at both of which the function falls to zero almost discontinuously, and with a maximum near the upper extreme; and a curved or humped function with the maximum well

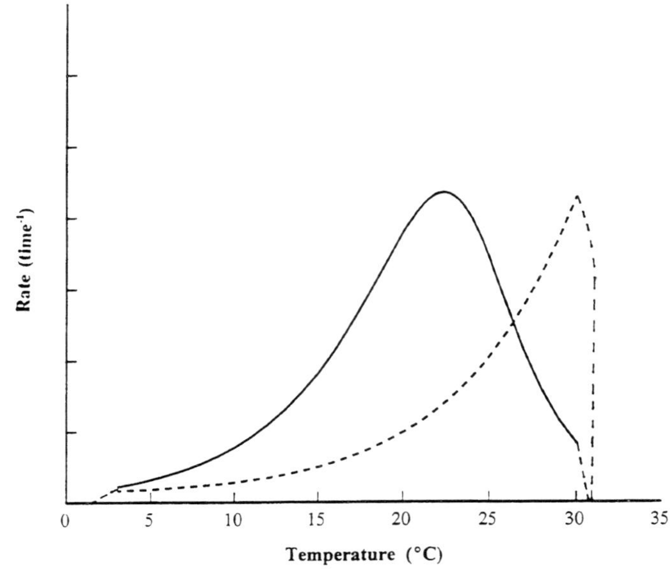

Fig. 5. Typical relationships between physiological rates of ectotherms and temperature: monotonic response, which is approximately exponential except near the extremes (---), and humped symmetrical or humped asymmetrical curves (—).

below the upper extreme and with gradual descents from the maximum to zero at the extremes, with some of them humped approximately symmetrically and others asymmetrically (Fig. 5). The rate of egg development and standard metabolism usually are of the first type, and active metabolism, growth, and critical swimming speed are of the second type.

A variety of quantitative formulae has been used to describe the temperature responses of ectotherms (Fry 1947; Cossins and Bowler 1986). We prefer the Arrhenius models because: some forms of thermodynamic Arrhenius models have been widely accepted in biochemistry, cell biology, physiology, and ecology; the models represent complex phenomena by means of a simple but effective mathematical structure; and there may be opportunity for generalization in the theory of open systems as reflected in the coefficients of the Arrhenius models (Sharpe and DeMichele 1977; Regier et al. 1990). Here a selected set of Arrhenius models is used to describe and compare the temperature dependence of various rate processes of sockeye salmon and largemouth bass.

The simple or simplified Arrhenius model was proposed by Arrhenius in 1898 (Arrhenius 1915) to describe the effects of temperature on various biochemical rate phenomena and was developed as a more general thermodynamic model. Regier et al. (1990) provide a rationale for its use in aquatic ecology with monotonic functions (see Fig. 5). The model is

(2) $$k = e^{A-B/T}$$

where k is the rate constant with units of time^{-1}, T is the Kelvin temperature, and A and B are constants to be fitted from data. Please note that the issues of whether T (with $T = 0$ at -273.16°C) in this expression should be replaced with another T (with $T = 0$ at an empirically determined temperature usually between -10 and 10°C) is addressed briefly in the discussion section below.

The compound Arrhenius model, applied to rate processes in plants by Johnson and Thornley (1984), has been explored here for application to fish. It may provide a useful tool to describe the rate processes with a humped function (i.e., with an optimum temperature well within the usual temperature range of a fish species, see Fig. 5):

$$(3) \qquad K' = \frac{e^{A-B/T}}{1 + e^{C-D/T}}$$

where *A, B, C, D* are constants to be fitted from the data. Possible thermodynamic meanings of all the parameters in Arrhenius models have been suggested by Sharp and DeMichele (1977) and Johnson and Thornley (1984).

Here we apply both the simple and compound Arrhenius models to some temperature-dependent rate processes of sockeye salmon and largemouth bass. For some physiological processes that show a monotonic response with respect to temperature except near the ends of the range, such as rate of egg development and standard metabolism, the simple Arrhenius model is chosen to fit the data excluding extreme points at both ends. For those that show a humped symmetrical or humped asymmetrical response over temperature, the compound Arrhenius model provides a reasonable description much of the full range of temperature, except perhaps at the low temperature extreme, see Discussion. Here only a few for Arrhenius curves are considered.

The standard metabolic rate (*SMR*, milligrams O_2 per gram per hour) is a function of body mass, W (milligrams), and acclimation temperature, T (Kelvin) (Robinson et al. 1983). A modified model expanded from the simple Arrhenius relationship and excluding the parts of the functional response at extremely low and extremely high temperatures is

$$(4) \qquad \text{SMR} = W^m e^{A-B/T}$$

here m is a constant ≈ -0.2 (Dickie et al. 1987; Zotin 1990).

Active metabolism relates to an excited, mobile, or working animal (Zotin 1990). When fish move in water, the resistance of the medium must be overcome. It has been found empirically that the relationship between oxygen consumption and the speed of swimming is approximately exponential at constant temperature (Brett 1964, 1965; Webb 1971):

$$(5) \qquad \text{AMR} = \text{SMR}e^{\alpha V}$$

where *AMR* is active metabolic rate (milligrams O_2 per gram per hour), α is a constant and V is the swimming speed (centimetres per second). As the effect of temperature and inhibition of high temperature are taken into consideration, the relevant active metabolic rate at critical swimming speed may be better described by the compound Arrhenius model (Holmes and Lin 1994):

$$(6) \qquad \text{AMR} = \frac{W^m e^{A-B/T}}{1 + e^{C-D/T}} e^{\alpha V'}$$

where critical swimming speed V' is determined empirically from experiments.

The rate of development between laying and hatching of the egg, as a simple Arrhenius function, is equal to the inverse of the length of time of development, H (day^{-1}) and mass of an individual egg (Pauly and Pullin 1988; Regier et al. 1990):

$$(7) \qquad H = W^m e^{A-B/T}$$

Here length of time of development is taken as the time to achieve a 50% hatch of a batch of eggs. If the masses of eggs in a batch are very similar, W^m may be taken to be a constant and becomes part of the A exponent.

Fish growth is a complex process. When food is available in excess of need, G (day^{-1}) may be directly controlled by the body size and temperature as

$$(8) \qquad G = \frac{W^m (e^{A-B/T})}{(1 + e^{C-D/T})}$$

When food is limited, the Winberg (1956) growth model provides an alternative approach.

Growth can be considered as the difference of input and output of appropriate free energy or the difference between anabolic and catabolic processes (Winberg 1956; Paloheimo and Dickie 1965):

$$(9) \qquad G = k_1 I - k_2 W^m$$

where k_1 and k_2 are temperature-related rate constants of a compound Arrhenius model and a simple Arrhenius type respectively. The input energy, I, is a function of the body size and the food ration (Paloheimo and Dickie 1965):

$$(10) \qquad I = pW^n$$

where n is constant, and p is the food ration.

Combining Equations 9 and 10, a growth model modified from the Winberg growth model with body size and food ration included and relevant to the normal growth temperature range is

$$(11) \qquad G = \frac{W^n (e^{A1-B1/T})}{(1 + e^{C1-D1/T})} - W^m e^{A2-B2/T}$$

Here we have used the first term for anabolism and the second term for catabolism.

The critical swimming speed or maximum sustained swimming speed (V', body lengths per second or centimetres per second) is a function of temperature and body size (Gibson and Fry 1954; Beamish 1970), which may have the following form:

$$(12) \qquad V' = \frac{W^m (e^{A-B/T})}{(1 + e^{C-D/T})}$$

This expression may have only descriptive significance.

Parameters for all the above models were estimated from appropriate data from the literature by linear regression using PROC REG and nonlinear least-squares regression using PROC NLIN (SAS Institute Inc. 1985) as appropriate (Table 1).

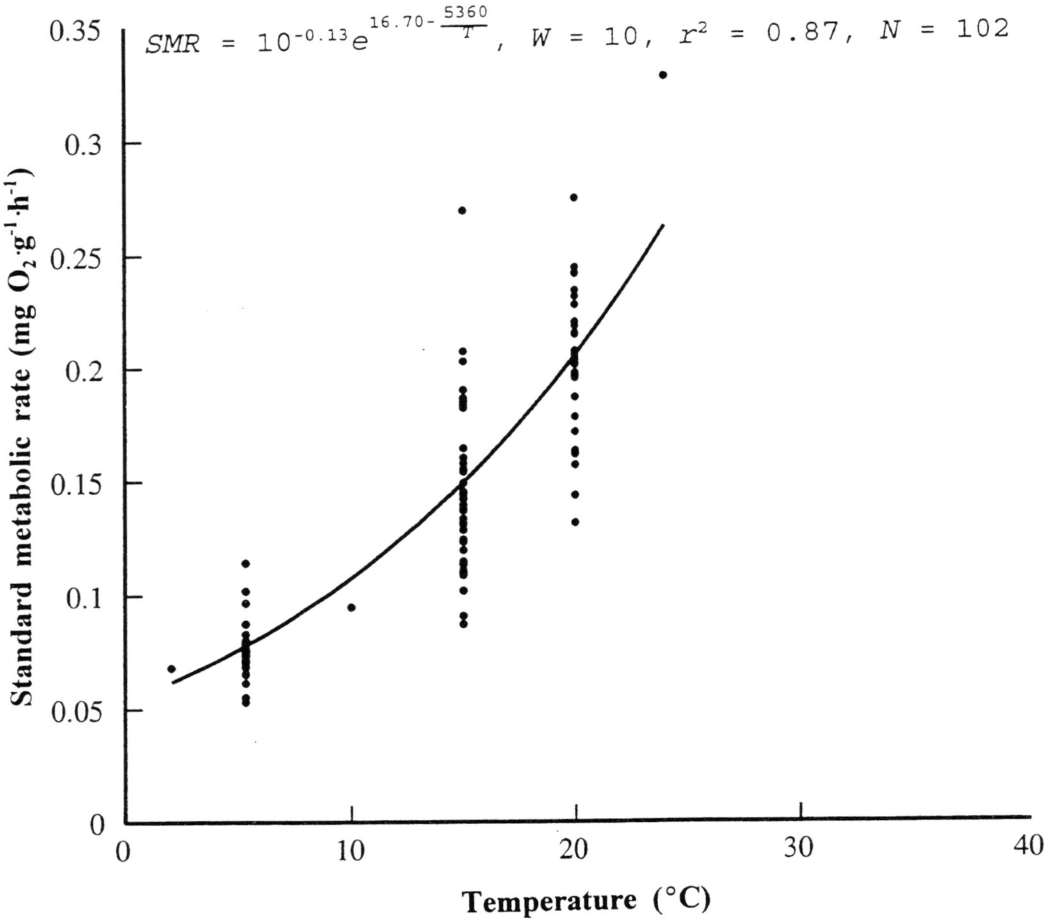

Fig. 6. Arrhenius plot of the standard metabolic rate (SMR, mg $O_2 \cdot g^{-1} \cdot h^{-1}$) as a function of temperature (°C) for sockeye salmon; data at lower and higher extremes are not indicated here. Equation 4 from the text was used for this figure.

Table 1: Comparison of physiological processes of cold-water sockeye salmon and warmwater largemouth bass to Kelvin temperature (T, K) and body size (W, weight in g), parameters of the models defined by Equations 4, 6, 7, 8, and 12. NA = not applicable. See text for description of physiological processes and parameters.

Physiological process	Parameters						r^2	N	Temperature range (°C)	Weight range (g)	Source[a]
	α ($\cdot 10^{-3}$)	m ($\cdot 10^{2}$)	A	B ($\cdot 10^{-3}$)	C ($\cdot 10^{-1}$)	D ($\cdot 10^{-3}$)					
Sockeye salmon (*Oncorhynchus nerka*)											
SMR (mg $O_2 \cdot g^{-1} \cdot h^{-1}$)	NA	−13.43	16.70	5.36	NA	NA	0.87	101	5–24	3–2162	4
AMR (mg $O_2 \cdot g^{-1} \cdot h^{-1}$)	0.124	−3.59	30.69	8.67	4.28	12.32	0.90	11	2–24	3–1432	4
H (d^{-1})	NA	−20.09[b]	26.31	8.67	NA	NA	0.97	64	1–16	0.058–0.155	1
G (d^{-1})	NA	−12.00	45.80	14.15	6.35	47.24	0.97	6	1–24	1–16	3,6
V' ($cm \cdot s^{-1}$)	NA	16.77	25.89	6.28	7.26	21.21	0.92	51	2–26	1–1962	4,5
Largemouth bass (*Micropterus salmoides*)											
SMR (mg $O_2 \cdot g^{-1} \cdot h^{-1}$)	NA	−42.59	12.70	4.01	NA	NA	0.54	17	5–35	8–98	8
AMR (mg $O_2 \cdot g^{-1} \cdot h^{-1}$)	14.00	−9.36	13.34	4.31	2.54	7.99	0.98	13	10–34	83–236	2
H (d^{-1})	NA	−20.09[b]	29.53	9.06	NA	NA	0.96	11	10–32	$1.7–2.6 \cdot 10^{-3}$	9
G (d^{-1})	NA	−39.00	42.80	13.76	14.09	42.45	0.78	48	15–35	0.01–0.15	7
V' ($cm \cdot s^{-1}$)	NA	34.02	14.32	3.50	4.79	14.72	0.99	6	10–34	118–164	2

[a] Source: 1. Beacham and Murray 1990. 2. Beamish 1970. 3. Biette and Geen 1979. 4. Brett 1964, 1965; Brett and Glass 1973. 5. Brett 1967. 6. Brett et al. 1969. 7. Coutant and DeAngelis 1983. 8. Fry and Brett 1974. 9. Heidinger 1975; 10. Kramer and Smith 1962.

[b] The egg weight exponent is not estimated here but is based on the estimate from a multispecies comparison (Pauly and Pullin 1988).

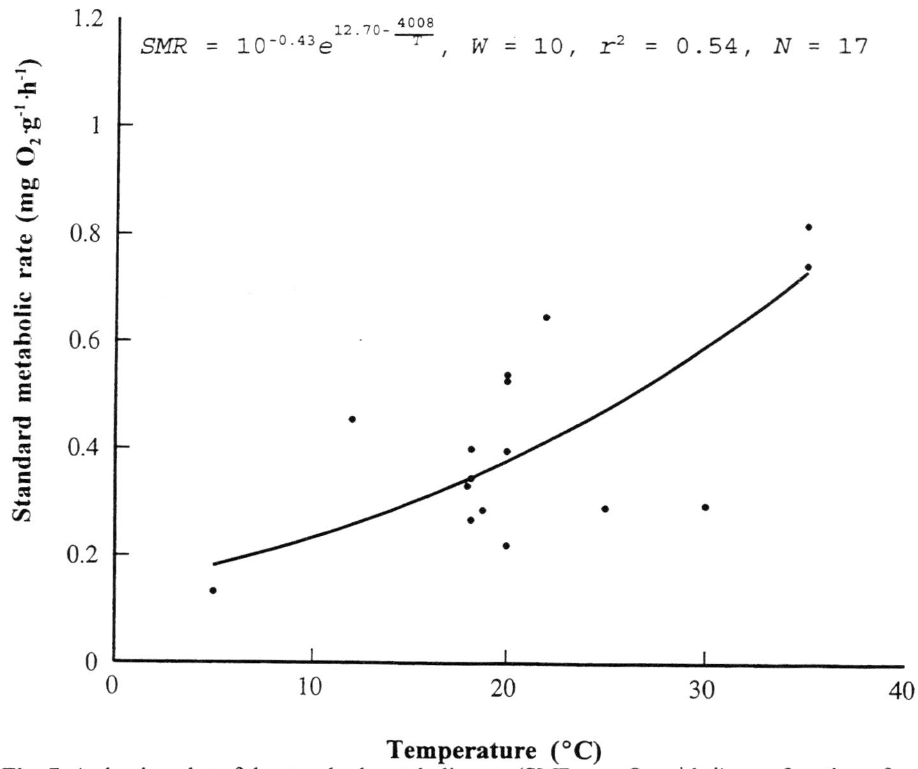

Fig. 7. Arrhenius plot of the standard metabolic rate (SMR, mg $O_2 \cdot g^{-1} \cdot h^{-1}$) as a function of temperature (°C) for largemouth bass; data at lower and higher extremes are not indicated here. Equation 4 from the text was used for this figure.

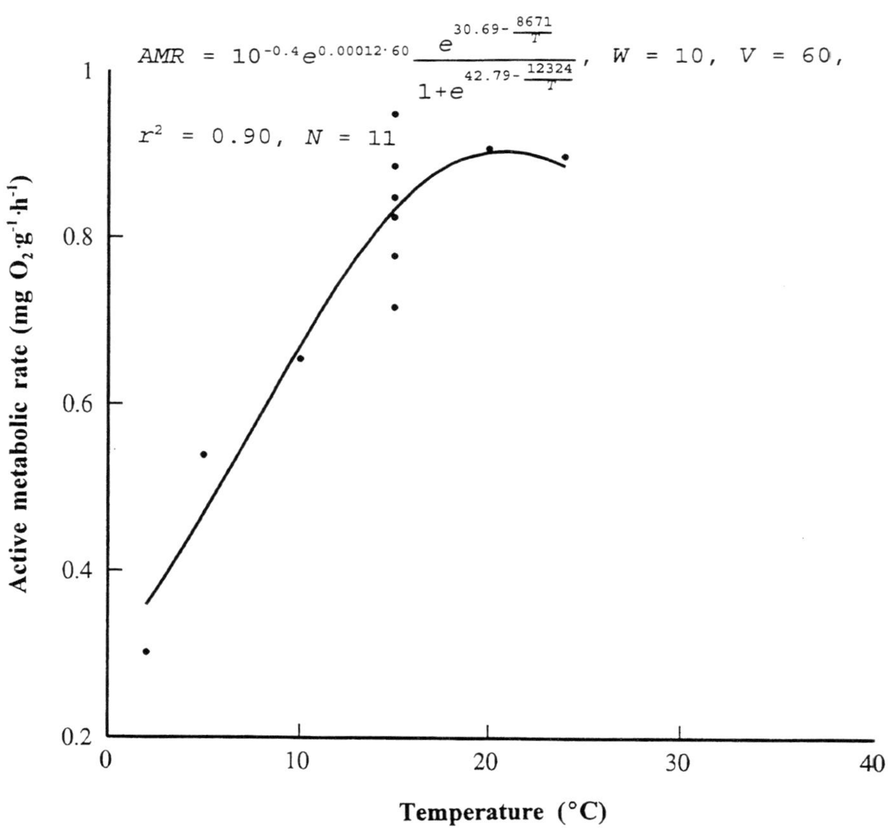

Fig. 8. Arrhenius plot of the active metabolic rate (AMR, mg $O_2 \cdot g^{-1} \cdot h^{-1}$) as a function of temperature (°C) for sockeye salmon; data at lower and higher extremes are not indicated here. Equation 6 from the text was used for this figure.

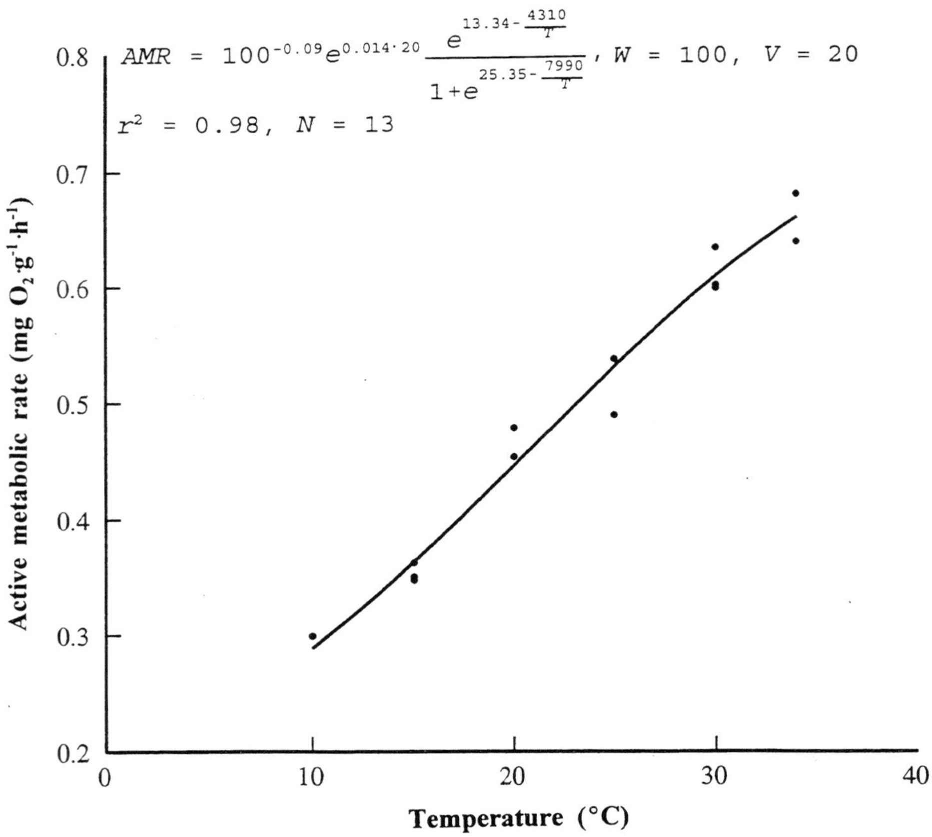

Fig. 9. Arrhenius plot of the active metabolic rate (AMR, mg $O_2 \cdot g^{-1} \cdot h^{-1}$) as a function of temperature (°C) for largemouth bass; data at lower and higher extremes are not indicated here. Equation 6 from the text was used for this figure.

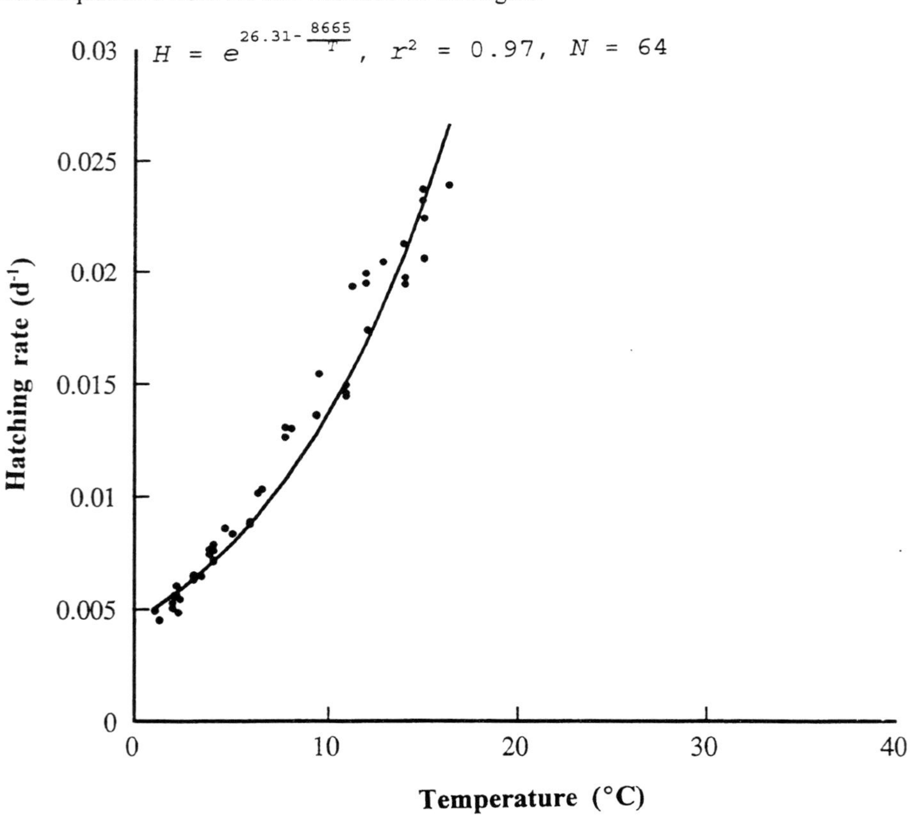

Fig. 10. Arrhenius plot of the hatching rate (d^{-1}) (inverse of days from fertilization to 50% hatch of eggs) as a function of temperature (°C) for sockeye salmon; data at lower and higher extremes are not indicated here. Equation 7 from the text was used for this figure.

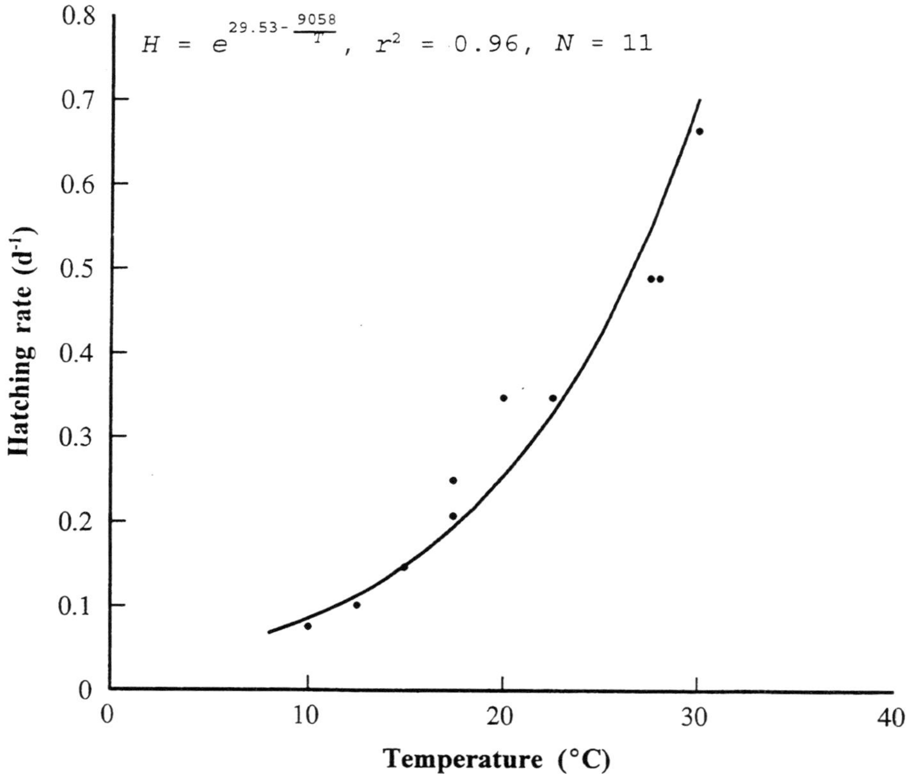

Fig. 11. Arrhenius plot of the hatching rate (d^{-1}) (inverse of days from fertilization to 50% hatch of eggs) as a function of temperature (°C) for largemouth bass; data at lower and higher extremes are not indicated here. Equation 7 from the text was used for this figure.

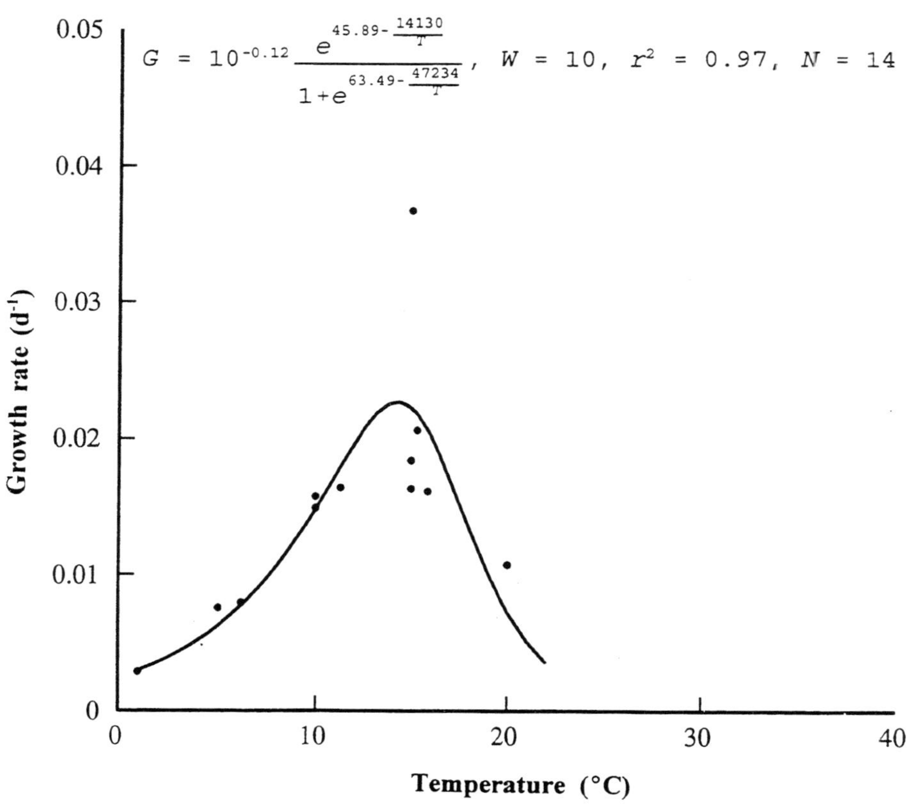

Fig. 12. Arrhenius plot of the growth rate (G, d^{-1}) as a function of temperature (°C) for sockeye salmon. Equation 8 from the text was used for this figure.

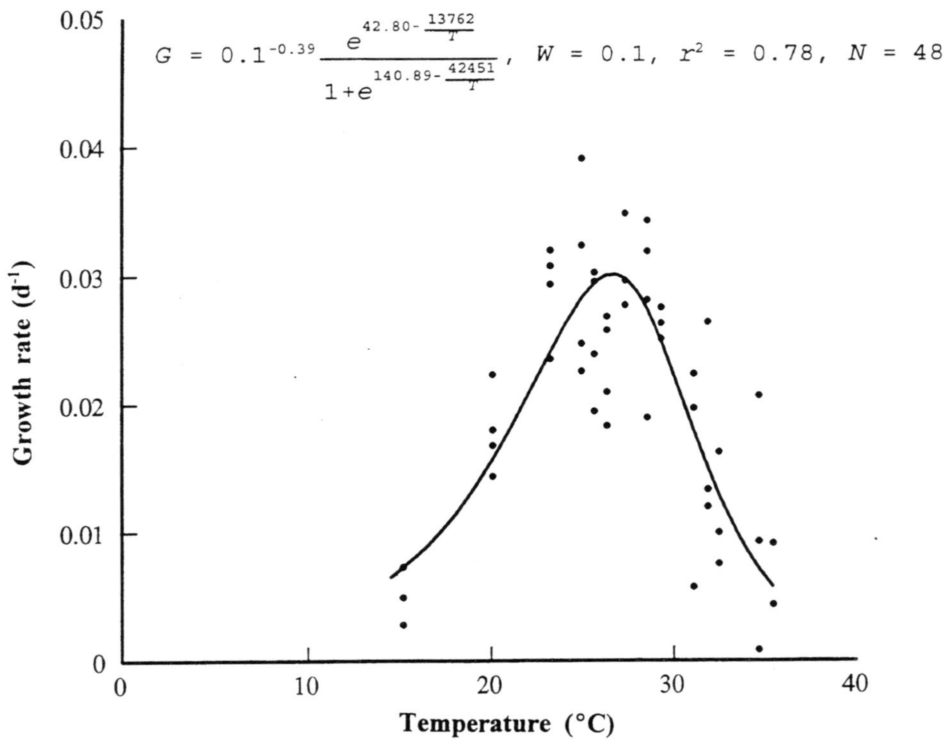

Fig. 13. Arrhenius plot of the growth rate (G, d^{-1}) as a function of temperature (°C) for largemouth bass. Equation 8 from the text was used for this figure.

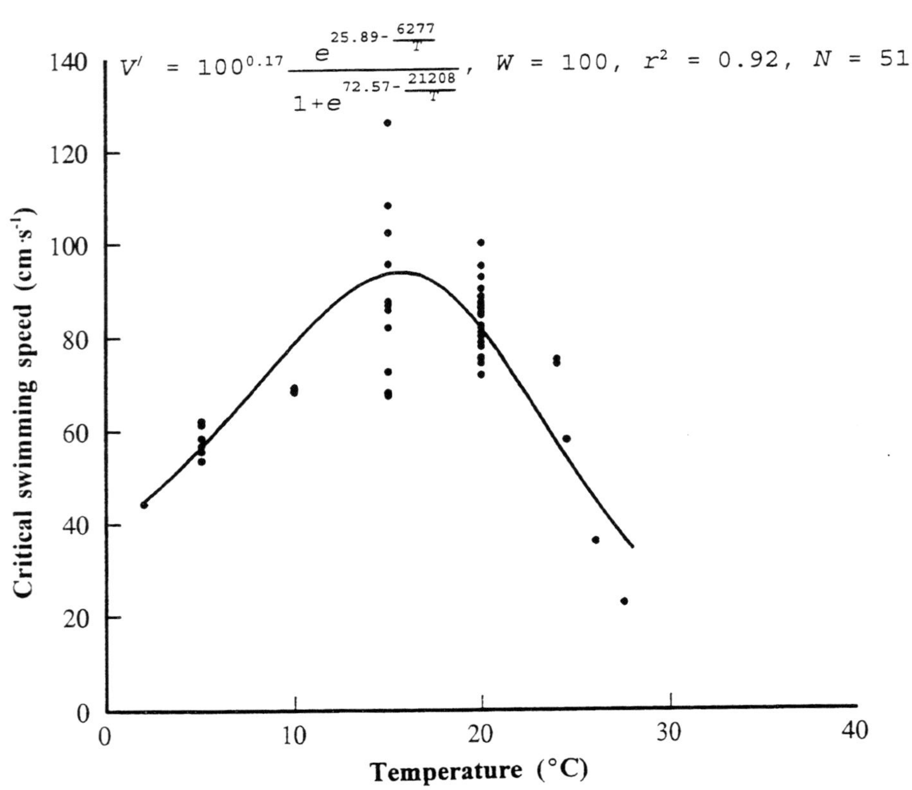

Fig. 14. Arrhenius plot of the critical swimming speed (CV, $cm \cdot s^{-1}$) as a function of temperature (°C) for sockeye salmon. Equation 12 from the text was used for this figure.

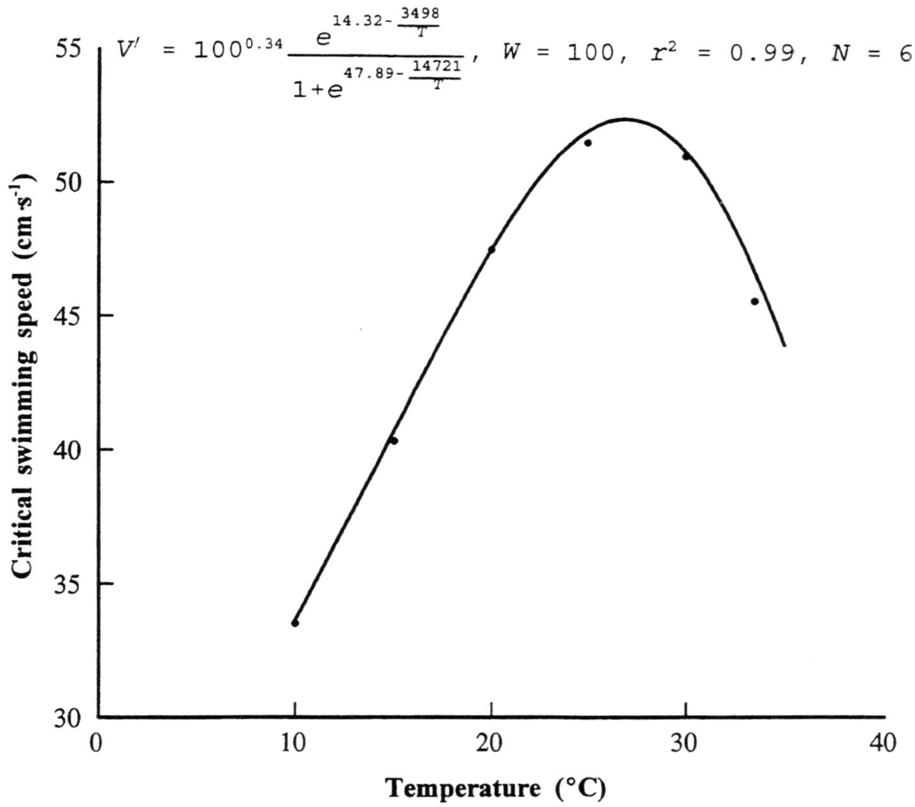

Fig. 15. Arrhenius plot of the critical swimming speed (CV, $cm{\cdot}s^{-1}$) as a function of temperature (°C) for largemouth bass. Equation 12 from the text was used for this figure.

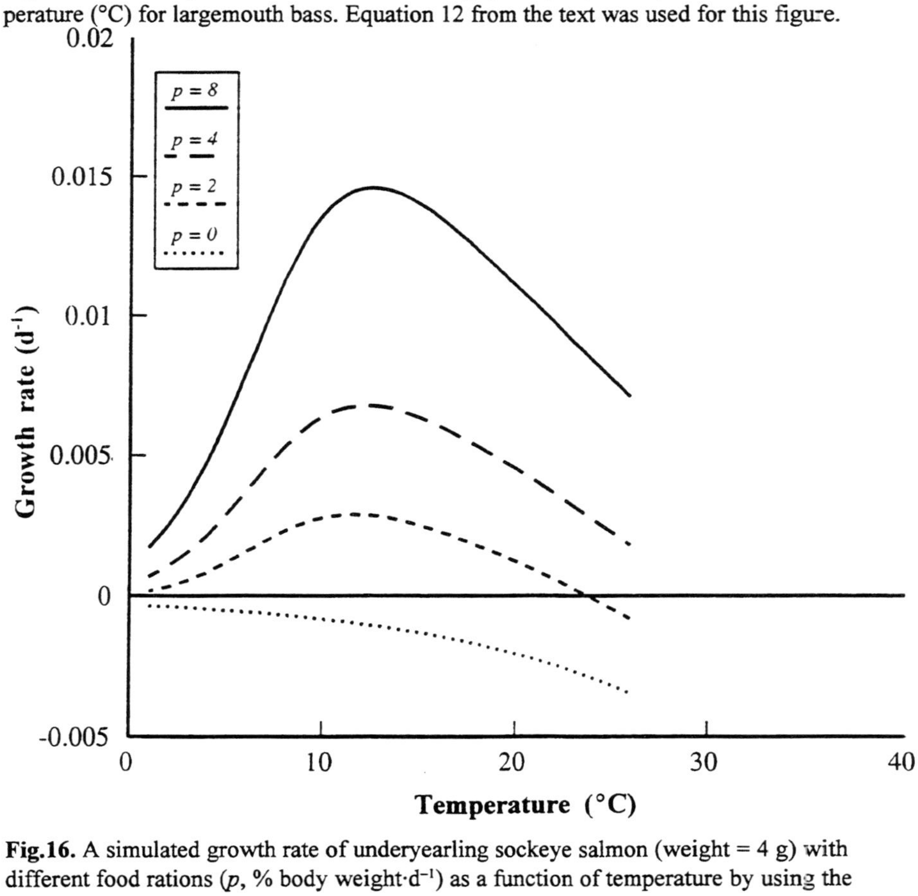

Fig.16. A simulated growth rate of underyearling sockeye salmon (weight = 4 g) with different food rations (p, % body weight·d^{-1}) as a function of temperature by using the modified Winberg growth equation (Equation 11):

$$G = pW^{-0.12}\,(e^{92.02-27320/T}) - W^{-0.33}e^{21.31-7521/T}\,(1 + e^{109.33-30679/T})$$

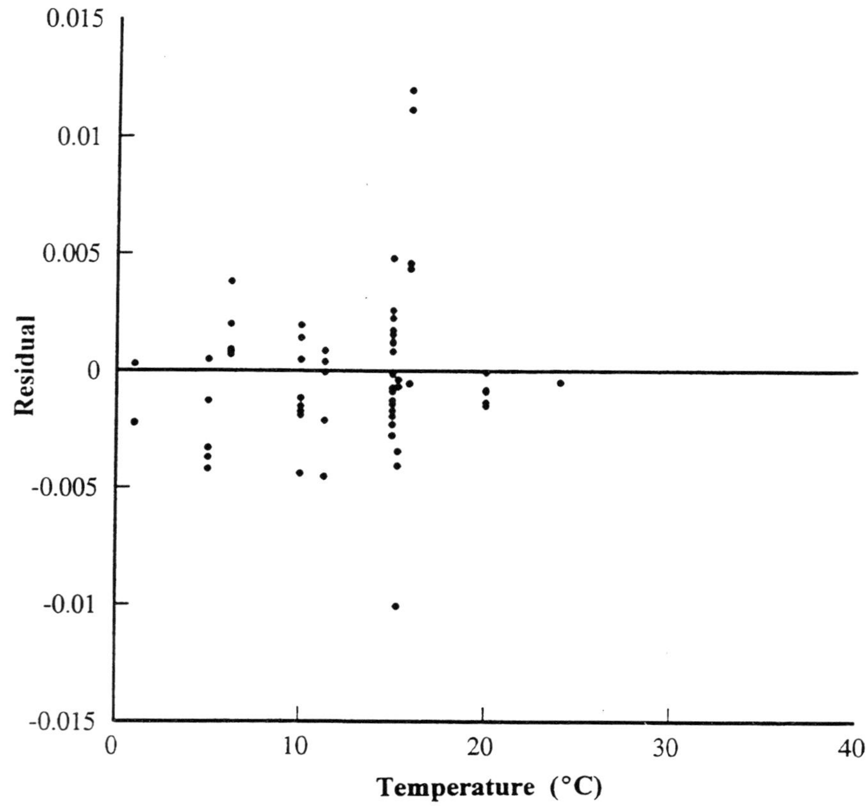

Fig. 17. The residuals from the modified Winberg growth model (Equation 11) fitted to laboratory observations (Brett et al. 1969; Brett and Glass 1973; Biette and Geen 1980) as a function of temperature (°C).

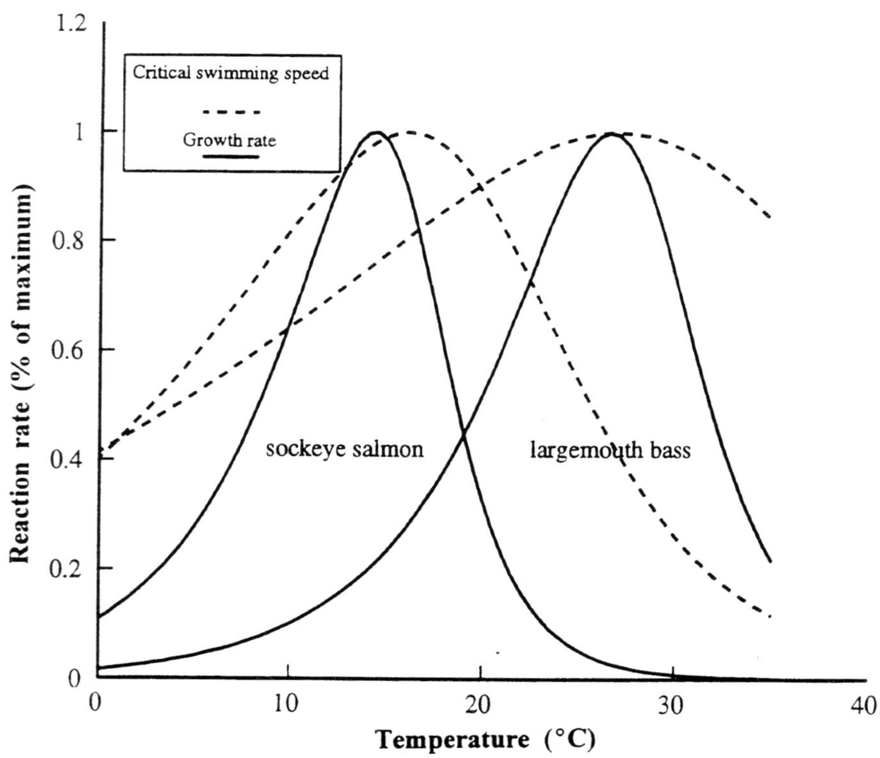

Fig. 18. Comparison of standardized temperature effects for cold-water sockeye salmon and warmwater largemouth bass with respect to growth rate (—) and critical swimming ability (---) based on the standard reaction rate (% maximum).

The temperature responses of different organismal rate processes of sockeye salmon and largemouth bass, with data points corrected for arbitrarily selected combinations of body mass or swimming speed, are shown in Figs. 6–15.

The modified Winberg growth model (Equation 11 above) is applied to laboratory data on growth phenomena under the influence of temperature and food ration (Brett et al. 1969; Brett and Glass 1973; Biette and Geen 1980), and its parameters are estimated as: $n = -0.12$, $m = -0.33$, $A_1 = 92.02$, $A_2 = 27320$, $C_1 = 109.33$, $D_1 = 30679$, $A_2 = 21.31$, and $B_2 = 7521$ ($r^2 = 0.86$, $N = 61$). A simulated growth rate with different food rations (*p*, percent body weight per day) as a function of temperature is provided in Fig. 16. The significance of the model to the data can be tested by the Shapiro–Wilk statistic ($P < 0.0001$) of residual values for the null hypothesis that the residual values are a random sample from a normal distribution (SAS Institute Inc. 1985); note the high r^2 and the plot of residual values presented in Fig. 17.

For the mathematical relationships described above, parameters estimated from laboratory data may not exactly equal those estimated from field observations because of the effects of other factors, but the general patterns should be similar. Figure 18 shows standardized temperature effects for cold-water sockeye salmon and warmwater largemouth bass with respect to growth rate and critical swimming speed.

Specifically, parameter *A* may be population related and all other parameters in the Arrhenius models may be species specific. The power function of body size may change with different life stages. The values for the parameters *n* and *m* estimated here (Table 1) are generally consistent with those in the literature (Peters 1983).

Discussion

In this study, we apply appropriate versions of the Arrhenius model, which relate to catabolism, anabolism and activity, respectively, to fit five different organismal rates of two fish species. Both goodness of fit of the proposed Arrhenius models to data sets and the plots of predicted and observed values verify that Arrhenius models can be used to describe the physiological phenomena of aquatic organisms. For those phenomena that have an optimum, the fitted models yield appropriate temperatures for maximal rates. The breadth of thermal niche is approximately equal to the breadth of temperature range in which growth of fish is nearly maximal (Fig. 18).

By comparing the coefficients for the two species in Table 1, a similar pattern can be observed among the coefficients of different rates. Other species (data not shown) also show similar patterns. As the reaction process becomes more complex, the coefficients *A, B, C, D* tend to be larger. We are exploring possible thermodynamic explanations. Whether the expression for critical swimming speed (Equation 12) and the relevant coefficients in Table 1 have thermodynamic meaning in a definitive sense is doubtful; this may be an example of a purely empirical descriptive fit.

Field studies of sockeye salmon and largemouth bass provide results that are broadly consistent with the temperature response derived from the laboratory (Brett 1971; Rice and Cochran 1984). Some general effects of climate change may be assessed with the help of these general temperature relationships (e.g., equations 4, 6, 7, 8, 11, and 12). The effects on a local fish population would be uncertain because the relevant rates would also be influenced by such factors as food availability, and interspecific and intraspecific competition. The result may be useful if we focus on large-scale statistical phenomena, such as the approximate limit of a geographic range.

Generally, body size plays a significant role in physiological and ecological processes, such as fish mortality, abundance of fish populations, and an organism's fitness. This fitness may depend on either the time the organism requires to reach a given size or the size when it reproduces (Dickie et al. 1987; Shuter and Post 1990; Regier et al. 1990).

Algae, aquatic insects, and zooplankton, the principal components of primary and secondary production in aquatic ecosystems, are all strongly affected by temperature (Vannote and Sweeney 1980; Lonsdale and Levinton 1985). Temperature affects growth, metabolism, reproduction, and the distribution of these organisms. Arrhenius models have been used with phytoplankton, insects, and plants (e.g., Johnson and Lewin 1946; Goldman and Carpenter 1974; Sharpe and DeMichele 1977; Johnson and Thornley 1984).

The modified Winberg growth model is a simplified bioenergetic model (see also Kitchell et al. 1977), which incorporates catabolism and anabolism. The fitting of the experimental data with its appropriate compound Arrhenius models shows that it may be possible to describe the effects of temperature and food ration in this way.

The Arrhenius models appear to have some explanatory capabilities with respect to physiological responses to temperature change. This approach to fish phenomena is still largely empirical and descriptive. We have noticed that the fittings or predictions at the extreme points (minimum and maximum survival temperature) are not fully satisfactory; the thermodynamic implications of the coefficients of the Arrhenius models for fish species still do not make definitive theoretical sense. A modification of the Arrhenius models, in which the Kelvin temperature scale is replaced by a scale with zero point usually within 10 degrees of Celsius zero, may result in better fits to data, especially near the lower temperature extremes (Kavanau 1950; Cossins and Bowler 1987). In general, the zero-point temperature for physiological processes falls between −10°C and 10°C. Whether generalizations on the zero point are inferrable as a function of the variables in Equations 1 and 2 above is now being explored by us. For example, the zero points for the sockeye salmon relationship appear to fall below 0°C, while those for largemouth bass fall above 0°C. Meanwhile, the relationships reported in the present paper, with zero taken at Kelvin zero, appear to provide useful general descriptions for the range of temperatures away from the extremes for each type of rate.

Acknowledgements

This study was supported by funding from the Natural Sciences and Engineering Research Council to H.A. Regier and an Open Fellowship of the University of Toronto to P. Lin. Advice from K.K. Ing, J.A. Holmes, J. Kay, and B.J. Shuter is appreciated. We thank L.M. Dickie and another reviewer for their helpful comments.

References

Arrhenius, S. 1915. Quantitative laws in biological chemistry. Bell, London. 164 p.

Beacham, T.D., and C.B. Murray. 1990. Temperature, egg size, and development of embryos and alevin of five species of Pacific salmon: a comparative analysis. Trans. Am. Fish. Soc. 119: 927–945.

Beamish, F.W.H. 1970. Oxygen consumption of largemouth bass, *Micropterus salmoides*, in relation of swimming speed and temperature. Can. J. Zool. 48: 1221–1228.

Beitinger, T.L., and L.C. Fitzpatrick. 1979. Physiological and ecological correlates of preferred temperature in fish. Am. Zool. 19: 319–329.

Biette, R.M., and G.H. Geen. 1980. Growth of underyearling sockeye salmon (*Oncorhynchus nerka*) under constant and cyclic temperatures in relation to live zooplankton ration size. Can. J. Fish. Aquat. Sci. 37: 203–210.

Blumberg, A.F., and D.M. Di Toro. 1990. Effects of climate warming on dissolved oxygen concentrations in Lake Erie. Trans. Am. Fish. Soc. 119: 210–223.

Brannon, E.L. 1987. Mechanisms stabilizing salmonid fry emergence timing, p.120–124. *In* H.D. Smith, L. Margolis, and C.C. Wood, [ed.]. Mechanisms of migration in fishes. NATO Conf. Ser. IV Mar. Sci. 14.

Brett, J.R. 1964. The respiratory metabolism and swimming performance of young sockeye salmon. J. Fish. Res. Board Can. 21: 1183–1226.

Brett, J.R. 1965. The relation of size to rate of oxygen consumption and sustained swimming speed of sockeye salmon (*Oncorhynchus nerka*). J. Fish. Res. Board Can. 22: 1491–1501.

Brett, J.R. 1967. Swimming performance of sockeye salmon (*Oncorhynchus nerka*) in relation to fatigue time and temperature. J. Fish. Res. Board Can. 24: 1731–1741.

Brett, J.R. 1971. Energetic responses of salmon to temperature. A study of some thermal relations in the physiology and freshwater ecology of sockeye salmon (*Oncorhynchus nerka*). Zoologist 11: 99–113.

Brett, J.R., and N.R. Glass. 1973. Metabolic rates and critical swimming speeds of sockeye salmon (*Oncorhynchus nerka*). J. Fish. Res. Board Can. 30: 379–387.

Brett, J.R., and T.D. Groves. 1979. Physiological energetics, p. 279–352. *In* W.S. Hoar and D.J. Randall [ed.] Fish physiology. Vol. 9. Academic Press, New York, NY.

Brett, J.R., J.E. Shelbourn, and C.T. Shoop. 1969. Growth rate and body composition of fingerling sockeye salmon, *Oncorhynchus nerka*, in relation to temperature and ration size. J. Fish. Res. Board. Can. 26: 2363–2394.

Burgner, R.L. 1991. Life history of sockeye salmon, p. 1–118. *In* C. Groot and L. Margolis [ed.] Pacific salmon life histories. UBC Press, Vancouver, B.C.

Buynak, G.L., W.N. McLemore, and B. Mitchell. 1991. Changes in largemouth bass populations at Kentucky and Barkley lakes: environmental or regulatory responses? North Am. J. Fish. Manage. 11: 285–292.

Calder, W.A., III. 1984. Size, function and life history. Harvard University Press, Cambridge, MA. 431 p.

Christie, G.C., and H.A. Regier. 1988. Measures of optimal thermal habitat and their relationship to yields for four commercial fish species. Can. J. Fish. Aquat. Sci. 45: 301–314.

Cossins, A.R., and K. Bowler. 1987. Temperature biology of animals. Chapman & Hall, London, UK. 339 p.

Coutant, C.C. 1975. Temperature selection by fish—a factor in power plant impact assessments, p. 575–597. *In* Environmental effects of cooling systems at nuclear power plants. International Atomic Energy Agency, Vienna, Austria.

Coutant, C.C. 1990. Temperature-oxygen habitat for freshwater and coastal striped bass in a changing climate. Trans. Am. Fish. Soc. 119: 240–253.

Coutant, C.C., and D.L. DeAngelis. 1983. Comparative temperature-dependent growth rates of largemouth and smallmouth bass fry. Trans. Am. Fish. Soc. 112: 416–423.

Crawshaw, L.I. 1977. Physiological and behavioral reaction of fishes to temperature change. J. Fish. Res. Board Can. 34: 730–734.

Crowder, L.B., and J.J. Magnuson. 1982. Thermal habitat shifts by fishes at the thermocline in Lake Michigan. Can. J. Fish. Aquat. Sci. 39: 1046–1050.

Diana, J.S. 1984. The growth of largemouth bass, *Micropterus salmoides* (Lacepede), under constant and fluctuating temperatures. J. Fish. Biol. 24: 165–172.

Dickie, L.M., S.R. Kerr, and P.R. Boudreau. 1987. Size-dependent processes underlying regularities in ecosystem structure. Ecol. Monogr. 57: 233–250.

Fry, F.E.J. 1947. Effects of the environment on animal activity. Univ. Toronto Stud. Biol. Ser. 55. Publ. Ontario Res. Lab. 68: p. 5–62.

Fry, F.E.J. 1971. The effect of environmental factors on the physiology of fish, p. 1–89. *In* W.S. Hoar and D.J. Randall [ed.] Fish Physiology. Vol. 6. Academic Press, New York, NY.

Fry, F.E.J., and J.R. Brett. 1974. Oxygen consumption: vertebrates, p. 1624–1630. *In* P.L. Altman and D.S. Dittmer [ed.] Biology data book. 2nd ed. Vol. 3. Federation of American Societies for Experimental Biology, Bethesda, MD.

Gibson, E.S., and F.E.J. Fry. 1954. The performance of the lake trout, *Salvelinus namaycush*, at various levels of temperature and oxygen pressure. Can. J. Zool. 32: 252–260.

Godin, J.G.J. 1982. Migrations of salmonid fishes during early life history phase: daily and annual timing, p. 22–50. *In* E.L. Brannon and E.O. Salo [ed.] Proceedings of the Salmon and Trout Migratory Behavior Symposium June 3–5, 1981, Univ. of Washington. School of Fisheries, University of Washington, Seattle, WA.

Goldman, J.C., and E.J. Carpenter. 1974. A kinetic approach to the effect of temperature on algal growth. Limnol. Oceanogr. 19: 756–766.

Heidinger, R.C. 1975. Life history and biology of the largemouth bass, p. 11–20. *In* H. Clepper [ed.] Black bass biology and management. Sport Fishing Institute, Washington, DC.

Hokanson, K.E.F. 1977. Temperature requirements of some percids and seasonal adaptations to the seasonal temperature cycle. J. Fish. Res. Board Can. 34: 1524–1550.

Holmes, J.A., and P. Lin. 1994. The thermal niche of larval sea lamprey. Can. J. Fish. Aquat. Sci. 51: 253–262.

Huey, R.B., and R.D. Stevenson. 1979. Integrating thermal physiology and ecology of ectotherms: a discussion of approaches. Am. Zool. 19: 357–366.

Jobling, M. 1981. Temperature tolerance and the final preferendum—rapid methods for the assessment of optimum growth temperatures. J. Fish Biol. 19: 439–455.

Johnson, F.H., and I. Lewin. 1946. The growth rate of E. *coli* in relation to temperature, quinine and coenzyme. J. Cell. Comp. Physiol. 28: 47–75.

Johnson, I.R., and J.H.M. Thornley. 1984. Temperature dependence of plant and crop processes. Ann. Bot. (Lond.) 55: 1–24.

Kavanau, J.L. 1950. Enzyme kinetics and rate of biological processes. J. Gen. Physiol. 34: 193–209.

Kitchell, J.F., D.J. Stewart, and D. Weininger. 1977. Applications of a bioenergetics model to yellow perch (*Perca flavescens*) and walleye (*Stizostedion vitreum*). J. Fish. Res. Board Can. 34: 1922–1940.

Kramer, R.H., and L.L. Smith, Jr. 1962. Formation of year classes in largemouth bass. Trans. Am. Fish. Soc. 91: 29–41.

Lonsdale, D.J., and J.S. Levinton. 1985. Latitudinal differentiation in copepod growth: an adaptation to temperature. Ecology 66: 1397–1407.

Magnuson, J.J., L.B. Crowder, and P.A. Medvick. 1979. Temperature as an ecological resource. Am. Zool. 19: 331–343.

McCormick, M.J. 1990. Potential changes in thermal structure and cycle of Lake Michigan due to global warming. Trans. Am. Fish. Soc. 119: 183–194.

Miller, T.J., L.B. Crowder, J.A. Rice, and E.A. Marshall. 1988. Larval size and recruitment mechanisms in fishes: toward a conceptual framework. Can. J. Fish. Aquat. Sci. 45: 1657–1670.

Mittelbach, G.G. 1981. Foraging efficiency and body size: a study of optimal diet and habitat use by bluegills. Ecology 62: 1370–1386.

Paloheimo, J.E., and L.M. Dickie. 1965. Food and growth of fishes. I. A growth curve derived from experimental data. J. Fish. Res. Board Can. 22: 521–542.

Pauly, D., and R.S.V. Pullin. 1988. Hatching time in spherical, pelagic, marine fish eggs in response to temperature and egg size. Environ. Biol. Fish. 22: 261–271.

Peters, R.H. 1983. The ecological implications of body size. Cambridge University Press, Cambridge, UK. 329 p.

Regier, H.A., J.A. Holmes, and D. Pauly. 1990. Influence of temperature changes on aquatic ecosystems: an interpretation of empirical data. Trans. Am. Fish. Soc. 119: 374–389.

Rice, J.A., and P.A. Cochran. 1984. Independent evaluation of bioenergetics model for largemouth bass. Ecology 65: 632–739.

Robinson, W.R., R.H. Peters, and J. Zimmermann. 1983. The effects of body size and temperature on metabolic rate of organisms. Can. J. Zool. 61: 281–288.

SAS Institute Inc. 1985. SAS user's guide: statistics, version 5 edition. Cary, NC. 957 p.

Sharpe, P.J.H., and D.W. DeMichele. 1977. Reaction kinetics of poikilotherm development. J. Theor. Biol. 64: 649–670.

Shuter, B.J., and J.D. Meisner. 1992. Tools for assessing the impact of climate change of freshwater fish populations. GeoJournal 28: 7–20.

Shuter, B.J., and J.R. Post. 1990. Climate, population variability, and the zoogeography of temperate fishes. Trans. Am. Fish. Soc. 119: 314–336.

Strawn, K. 1961. Growth of largemouth bass fry at various temperatures. Trans. Am. Fish. Soc. 90: 334–335.

Sugden, A.M. 1992. Using biotic interactions to forecast the consequences of global climate change. Trends Ecol. Evol. (TREE) 7(2): 35–36.

Vannote, R.L., and B.W. Sweeney. 1980. Geographic analysis of thermal equilibria: a conceptual model for evaluating the effect of natural and modified thermal regimes on aquatic insect communities. Am. Nat. 115: 667–695.

Weatherley, A.H., 1990. Approach to understanding fish growth. Trans. Am. Fish. Soc. 119: 662–672.

Webb, P.W. 1971. The swimming energetics of trout. II. Oxygen consumption and swimming efficiency. J. Exp. Biol. 55: 521–540.

Werner, E.E., and J.F. Gilliam. 1984. The ontogenetic niche and species interactions in size-structured populations. Annu. Rev. Ecol. Syst. 15: 393–425.

Winberg, G.G. 1956. Rate of metabolism and food requirements of fishes. Fish. Res. Board Can. Transl. Ser. 194: 253 p.

Zotin, A.I. 1990. Thermodynamic basis of biological processes. Walter de Gruyter, Berlin, Germany. 293 p.

An exploratory analysis of associations between biotic and abiotic factors and year-class strength of Gulf of Alaska walleye pollock (*Theragra chalcogramma*)

Bernard A. Megrey, Steven J. Bograd, William C. Rugen, Anne B. Hollowed, Phyllis J. Stabeno, S. Allen Macklin, James D. Schumacher, and W. James Ingraham, Jr.

Megrey, B.A., S.J. Bograd, W.C. Rugen, A.B. Hollowed, P.J. Stabeno, S.A. Macklin, J.D. Schumacher, and W.J. Ingraham, Jr. 1995. An exploratory analysis of associations between biotic and abiotic factors and year-class strength of Gulf of Alaska walleye pollock (*Theragra chalcogramma*), p. 227–243. *In* R.J. Beamish [ed.] Climate change and northern fish populations. Can. Spec. Publ. Fish. Aquat. Sci. 121.

Abstract: This analysis examines the relative importance of various physical and biological factors on year-class strength of walleye pollock (*Theragra chalcogramma*) in Shelikof Strait, Alaska. Biological data for the analysis consists of a recruitment time series covering the period 1962–89 and egg and larval abundance time series for 1981–89. Physical data describe atmospheric and oceanic features of the region including estimates of precipitation, turbulent wind mixing, volume transport, and sea-surface temperature. Principal component analysis showed that 37% of the variance in the physical data was captured by the first two modes. The second mode, attributed to large-scale atmospheric circulation, separated conditions that contribute to high and low recruitment. Univariate and multivariate statistical techniques suggested that age-0 and age-1 abundance estimates as well as age-2 recruitment estimates are related to precipitation, an index of atmospheric sea-level pressure gradient, and local wind mixing. From the results of these analyses, several hypotheses are proposed: spring wind mixing in Shelikof Strait affects larval survival; large-scale atmospheric circulation affects processes leading to recruitment; baroclinicity may generate mechanisms conducive to larval survival; and effects of the physical environment on predation and behaviour may affect juvenile survival.

Résumé : Cette analyse porte sur l'importance relative de divers facteurs physiques et biologiques pour l'effectif de classes annuelles de morues du Pacifique occidental (*Theragra chalcogramma*) dans le détroit de Shelikof, en Alaska. Les données biologiques de l'analyse comprennent une série chronologique de recrutement portant sur la période 1962–1989 et une autre sur l'abondance des oeufs et des larves pour la période 1981–1989. Les données physiques décrivent les caractéristiques atmosphériques et océaniques de la région, y compris des estimations des précipitations, du mélange turbulent dû aux vents, du transport de volumes, et de la température de la surface de la mer. L'analyse des principaux éléments a montré que les deux premiers modes accaparaient 37 % de la variance des données physiques. Le deuxième mode, attribué à la circulation atmosphérique à grande échelle, séparait les conditions qui contribuent à la force et à la faiblesse du recrutement. Les techniques statistiques à une et à plusieurs variables donnaient à croire que les estimations de l'abondance à l'âge 0 et 1 ainsi que celles du recrutement à l'âge 2 étaient liées aux précipitations, à un indice du gradient de la pression atmosphérique au niveau de la mer et à un mélange local dû aux vents. À partir des résultats de ces analyses, plusieurs hypothèses sont proposées : le mélange dû aux vents le de printemps dans le détroit de Shelikof influe sur la survie des larves; la circulation atmosphérique à grande échelle modifie les processus liés au recrutement; la baroclinicité est susceptible de générer des mécanismes favorisant la survie des larves; et enfin, les effets du milieu physique ambiant sur la prédation et sur le comportement sont susceptibles d'avoir un effet sur la survie des poissons juvéniles.

B.A. Megrey, W.C. Rugen, A.B. Hollowed, and W.J. Ingraham, Jr. National Marine Fisheries Service, Alaska Fisheries Science Center, Seattle, WA 98115, USA.

S.J. Bograd. Joint Institute for the Study of the Atmosphere and Ocean, Seattle, WA 98115, USA.

P.J. Stabeno, S.A. Macklin, J.D. Schumacher. Pacific Marine Environmental Laboratory, Seattle, WA 98115, USA.

Introduction

Walleye pollock (*Theragra chalcogramma*) is an important commercial species in the North Pacific Ocean and Bering Sea. Since the mid-80s, pollock catches have averaged between 6 and 7 million t annually (FAO 1990). In the decade of the 1980s, pollock was the single most important species by weight in the world fish catch (Bakkala et al. 1987). In the Exclusive Economic Zone of the United States, over a third of the total fish harvest is pollock.

Since the discovery in 1980 of a large spawning population of walleye pollock in Shelikof Strait (Fig. 1) Gulf of Alaska, extensive research has been conducted on the life history of the population (Kendall et al. 1987; Reed et al. 1989; Schumacher and Kendall 1991). The Fisheries–Oceanography Coordinated Investigations (FOCI) research program was initiated in 1984 to investigate processes that influence recruitment of pollock in Shelikof Strait.

Research conducted by investigators studying the Gulf of Alaska pollock population has established a large information base from which to begin an analysis of environmental associations with recruitment. This body of work involves investigations on every major pollock life stage and includes studies that enumerate abundance trends (Kim and Gunderson 1989; Megrey 1990; Picquelle and Megrey 1991,1993;Hinckley et al. 1991); examine biological processes affecting life stages (Kendall et al. 1987; Yamashita and Bailey 1989; Yoklavich and Bailey 1989, 1990; Kendall and Kim 1989; Hinckley et al. 1990; Kim and Bang 1990; Schumacher and Kendall 1991; Brodeur et al. 1991; Canino et al. 1991; Hinckley et al. 1991); map horizontal, vertical, and temporal distributions (Kendall et al. 1987; Kim and Kendall 1989; Kim and Gunderson 1989; Kendall and Picquelle 1990; Rugen 1990; Kim and Nunnallee 1991); describe the oceanic and atmospheric physical environment (Reed et al. 1987; Reed and Schumacher 1989*a*, *b*; Ortner et al. 1990; Schumacher et al. 1990; Bond and Macklin 1993; Macklin et al. 1993); and develop modeling techniques to examine recruitment-process hypotheses (Stabeno and Megrey 1991).

Many investigators have attempted to examine the causes of recruitment variation in fishes and to relate these to biotic and abiotic regulators (Nelson et al. 1977; Leggett 1977; Chadwick et al. 1977; Bailey 1981; Leggett et al. 1984; Parker 1989*b*). Hollowed (1992) lists no less than 47 studies for stocks along the Northeast Pacific coast alone. Some of these studies were not able to establish strong environmental links to recruitment, while others successfully underscored the importance of abiotic regulators of year-class strength in fishes.

A fundamental problem in trying to examine changes in recruitment resulting from variations in the physical environment stems from the fact that the abiotic environment is dominated by nonlinear dynamics and varies over wide temporal and spatial scales (Wooster and Bailey 1989). As pointed out by Walters and Ludwig (1981), the problem is compounded further by inaccuracies in measuring both the environmental variability and the biotic response. Additional complications arise because the magnitude of the mortality and differences in the importance of various sources of mortality change as individuals age through different life stages.

The mixed success in establishing a clear connection between variations in the abiotic environment and recruitment has resulted in two schools of thought regarding the utility of environment–recruitment studies. Walters and Collie (1988) criticized correlative environment–recruitment studies as futile because of biases, measurement error, and the near

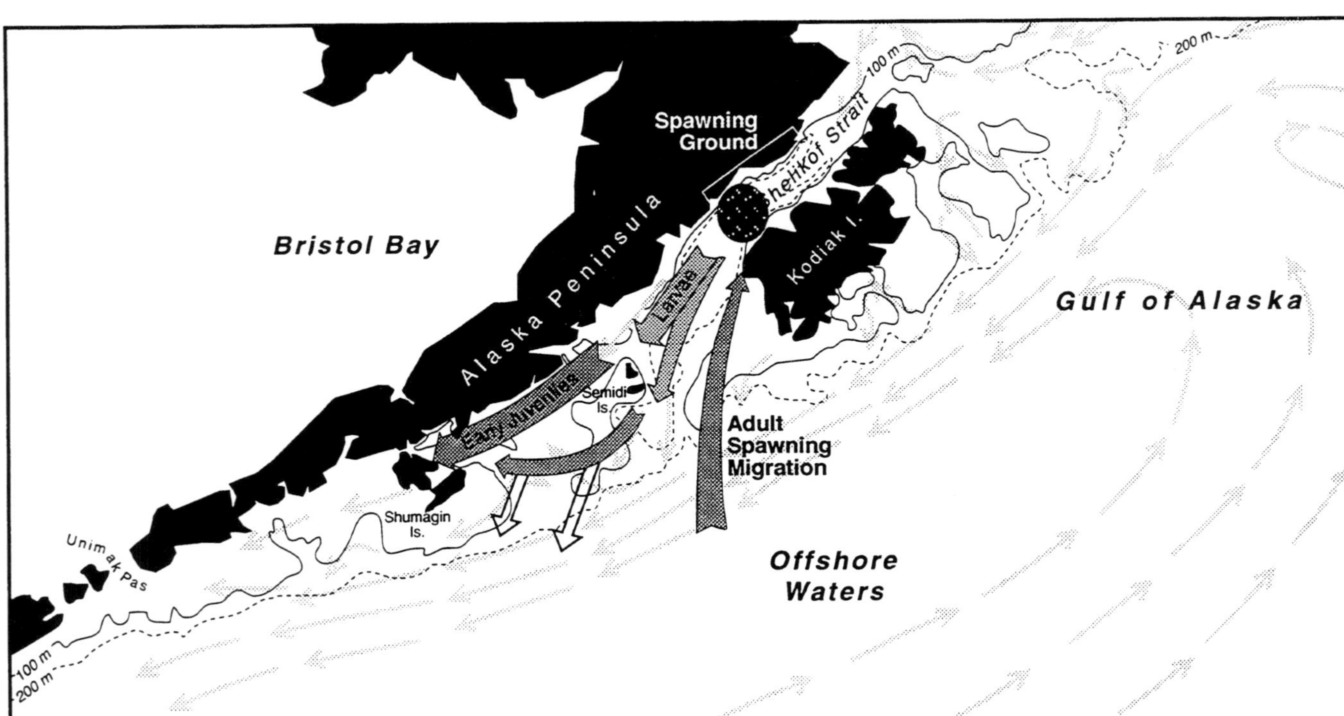

Fig. 1. Gulf of Alaska and the geographic location of different walleye pollock life stages.

certainty of spurious correlations. Others are optimistic. Kope and Botsford (1990) claim that correlative studies provide information on patterns that lead to the formulation of testable hypotheses.

Recently, Tyler (1992) and Hollowed (1992) have advocated the continued use of correlative studies as long as the analysis is based on a sound conceptual framework and judicious use of statistical methods. This approach can reduce the likelihood of spurious correlations. The analysis we describe follows the three-stage approach recommended by Tyler (1992) and Hollowed (1992). Our research blueprint begins with an examination of temporal and spatial distributions of various life stages of walleye pollock, then relates these to postulated key events in the life history that may influence survival. This important first step reduces the environmental variables to a manageable subset and identifies when they might affect survival. The second step involves examining spatial and temporal characteristics of biological distributions and relating them to potential environmental influences. The third step compares time series of selected environmental variables with the time series of recruitment.

Our aim in this analysis was to provide insight into the potential mechanisms underlying recruitment variability by using statistical models to investigate associations. Because this is the first work to deal with a statistical description of the recruitment processes in Shelikof Strait pollock, it is, by nature, exploratory. We were not interested in predicting or forecasting recruitment, although, from a fisheries management perspective, this is a reasonable long-term goal of such research. The work presented here concentrates on physical factors.

Methods

Recruitment time series extension

The central piece of information required for this analysis was a recruitment time series of suitable length for Gulf of Alaska pollock. Prior to this study, the recruitment series had only 17 data points (1973–89). We considered the length of this data set marginally useful. We extended this series to include the 1962–72 segment, giving us a total of 28 annual data points covering the period 1962–89.

The recruitment time series extension was accomplished by combining catch biomass and length-frequency data from 1964 to 1975. During this period, biological information such as length and sex were collected, but age structures were not. These data were combined with age and length data collected from the fishery after passage of the Magnuson Fishery and Conservation Management Act of 1975. Annual estimates of age composition were derived within a stock assessment model (Methot 1989, 1990) using a length-to-age conversion. This technique is similar to the iterative age–length key procedure of Kimura and Chikuni (1987), but allowed for simultaneous fit to all length distributions. Age-composition information from the 1976 commercial fishery data and from research surveys during 1973–75 were used to develop the age–length key. The reconstructed 1964–91 catch-at-age data segment, submitted to the stock assessment model, provided annual estimates of age-2 recruitment. Hollowed et al. (1991) supply details regarding analytical models and technical aspects of the data processing steps required to complete this analysis.

Identification of appropriate space and time scales

Past and ongoing research on pollock has provided detailed information on the geographic region, important biological processes, timing and location of life history stages, and physical setting. Figure 1 presents a schematic of the current understanding of the life history of pollock in this geographic setting. Walleye pollock spawn mainly in the deep-sea valley between March and mid-April (Kim and Nunnallee1991). Dense patches of eggs, found deep in the water column (Kendall and Kim 1989; Kendall and Picquelle 1990), hatch after about 2 wk, depending on temperature (Blood et al. 1994). Larvae rise into the mixed layer, and by late April, they can often be found in large identifiable patches along the Alaska Peninsula (Incze et al. 1990; Rugen 1990; Schumacher and Kendall 1991). Between April and June, larvae are advected southwestward with the Alaska Coastal Current. A portion of these may be advected off the shelf (Kim and Kendall 1989). By late June and July, the juveniles are distributed along the coast of the Alaska Peninsula near the Shumagin Islands (Hinckley et al. 1991). By the end of their first summer, some of the juveniles are found in the embayments along the Alaska Peninsula (Bailey and Spring 1992). In the late winter, adult pollock migrate from summer feeding grounds back to Shelikof Strait to spawn.

From this information, we concluded that the geographic region of interest spanned the area around Kodiak Island, Shelikof Strait, and the Alaska Peninsula west to the Shumagin Islands (Fig. 1). Thus, the geographic domain of our study area consisted of those areas of the Gulf of Alaska contained in the region defined by 55 and 59°N latitude and 152 and 160°W longitude, seaward from the Alaska Peninsula to the shelf break (2000-m depth contour), and from the troposphere to the bottom of the ocean.

We chose to examine environmental factors only during the early life historystages. Evidence that year-class strength is set by age-0 is indicated by the close correlation of age-0 pollock abundance to estimates of age-2 recruits (Schumacher and Kendall 1991; Bailey and Spring 1992). Thus, we are assuming that year-class strength is set by the fall of the year in which the fish were spawned. With this established, we restricted the examination of environmental data to the period January through August.

Formulation of hypotheses

To formulate recruitment hypotheses, we applied the events method of relating geography to various life stages (Tyler 1992). Using previous research findings, we determined the geographic location of a given life stage, the time of the year that the life stage occupied the location, and a suite of potential environmental factors that might influence survival. Potential mechanisms are described below.

Wind

Evidence exists that wind mixing affects larval survival in Shelikof Strait. Bailey and Macklin (1993) have shown that increased survival of first-feeding larvae can correspond with periods of calm winds. Low to moderate winds (mixing) deepen the mixed-layer and may enhance primary–secondary production and enrich the food chain (Lasker 1975, 1978), Thus, reducing starvation. At higher levels of turbulence,

Table 1. Description of environmental data series. Variables and locations in boldface were used in the regression and principal component analyses.

Data series	*Description*
Sea-surface temperature	Monthly COADS[a] average sea-surface temperature (°C) at three locations (**55°N, 159°W; 57°N, 155°W**; 59°N, 153°W)
Air temperature	Monthly average air temperature (°C) at Kodiak, Alaska (57° 47' N, 152° 24' W)
Air temperature	Monthly COADS average air temperature (°C) at three locations (**55°N, 159°W; 57°N, 155°W**; 59°N, 153°W)
Sea level pressure	Monthly COADS sea level pressure (mb) at three locations (55°N, 159°W; 57°N, 155°W; 59°N, 153°W)
Wind speed	Monthly COADS average wind speed ($m \cdot s^{-1}$) at three locations (55°N, 159°W; 57°N, 155°W; 59°N, 153°W)
Percent ice cover	Percent ice cover in Cook Inlet
Date of last ice	Julian date of last ice cover in Cook Inlet
Maximum ice extent	Julian date of maximum ice cover in Cook Inlet
OSCURS[b] model output	OSCURS model trajectories. A large-scale index of Gulf of Alaska circulation. Number of simulated drifters out of six released on Feb 1 at 55°N between 152°W and 137°W that cross 154°W by April 1.
Rainfall	Monthly average precipitation (dm) at Kodiak, Alaska (**57°47N, 152°24W**)
NEPPI[c]	Monthly average sea-level pressure index (ΔP mbar). Calculated as the difference in sea-level pressure between points over the north-central Pacific and near Reno, Nevada.
Freshwater runoff	Index of integrated mean winter Gulf of Alaska coastal freshwater discharge ($m^3 \cdot s^{-1}$) anomaly
Wind direction	Monthly NMC[d] grid-point average wind direction (degree) at two locations (55°N, 160°W; 57°N, 156°W)
Wind speed	Monthly NMC grid-point average wind speed ($m \cdot s^{-1}$) at two locations (55°N, 160°W; 57°N, 156°W)
Wind stress	Monthly NMC grid-point average wind stress ($N \cdot m^{-2}$) at two locations (55°N, 160°W; 57°N, 156°W)
Wind mixing	Monthly NMC grid-point wind mixing averages ($W \cdot m^{-2}$) at two locations (**55°N, 160°W; 57°N, 156°W**)

[a]COADS = Comprehensive Ocean–Atmosphere Data Set.
[b]OSCURS = Ocean Surface Current Simulations.
[c]NEPPI = Northeast Pacific Pressure Index.
[d]NMC = National Meteorological Center.

feeding could also be enhanced by the increased encounter rate of passively drifting larvae and their food (Rothschild and Osborne 1988). Conversely, stronger winds could result in overmixing, which may hinder larval survival by dissipating food aggregations (Cury andRoy 1989), and behavioural studies (Olla and Davis 1990*a*) indicate that pollock larvae may avoid wind-induced turbulence.

In this region, winds interact with the baroclinic structure to cause large fluctuations (>2 x 10^6 $m^3 \cdot s^{-1}$) in the transport of the Alaska Coastal Current (Schumacher et al. 1990). At times, local mountain-gap winds are associated with perturbations of the Alaska Coastal Current in Shelikof Strait (Vastano et al. 1992).

Freshwater input

The freshwater flux from precipitation and melting ice or snow increases the density gradient between water types in Shelikof Strait. This, in turn, supports baroclinic instability (Mysak et al. 1981; Schumacher and Kendall 1991) which can result in the formation of mesoscale eddies. These eddies have been observed to contain high concentrations of pollock larvae (Incze et al. 1989; Ortner et al. 1990; Vastano et al. 1992; Schumacher et al. 1993). The consequence of high larval concentration remains equivocal. If eddies concentrate food particles along with larvae, then larval survival should be enhanced, but eddies could also make larvae easier for predators to locate. Larvae have been retained on the continental shelf for long periods because of the effects of eddies, as opposed to being transported into the less productive Alaska Stream.

Transport

Both runoff and winds affect interannual variation in the strength of the Alaska Stream and the Alaska Coastal Current, which in turn directly affects the strength and character of the flow in and through Shelikof Strait. Weak transport keeps eggs and larvae in Shelikof Strait, which should improve chances for survival. Strong transport reduces time in Shelikof Strait and increases the likelihood that the animals will be transported off the continental shelf, where presumably they starve (Hinckley et al. 1991). The Northeast Pacific Pressure Index (NEPPI, described in the Appendix) of Emery and Hamilton (1985) is correlated with water transport through Shelikof Strait (Roach and Schumacher 1991).

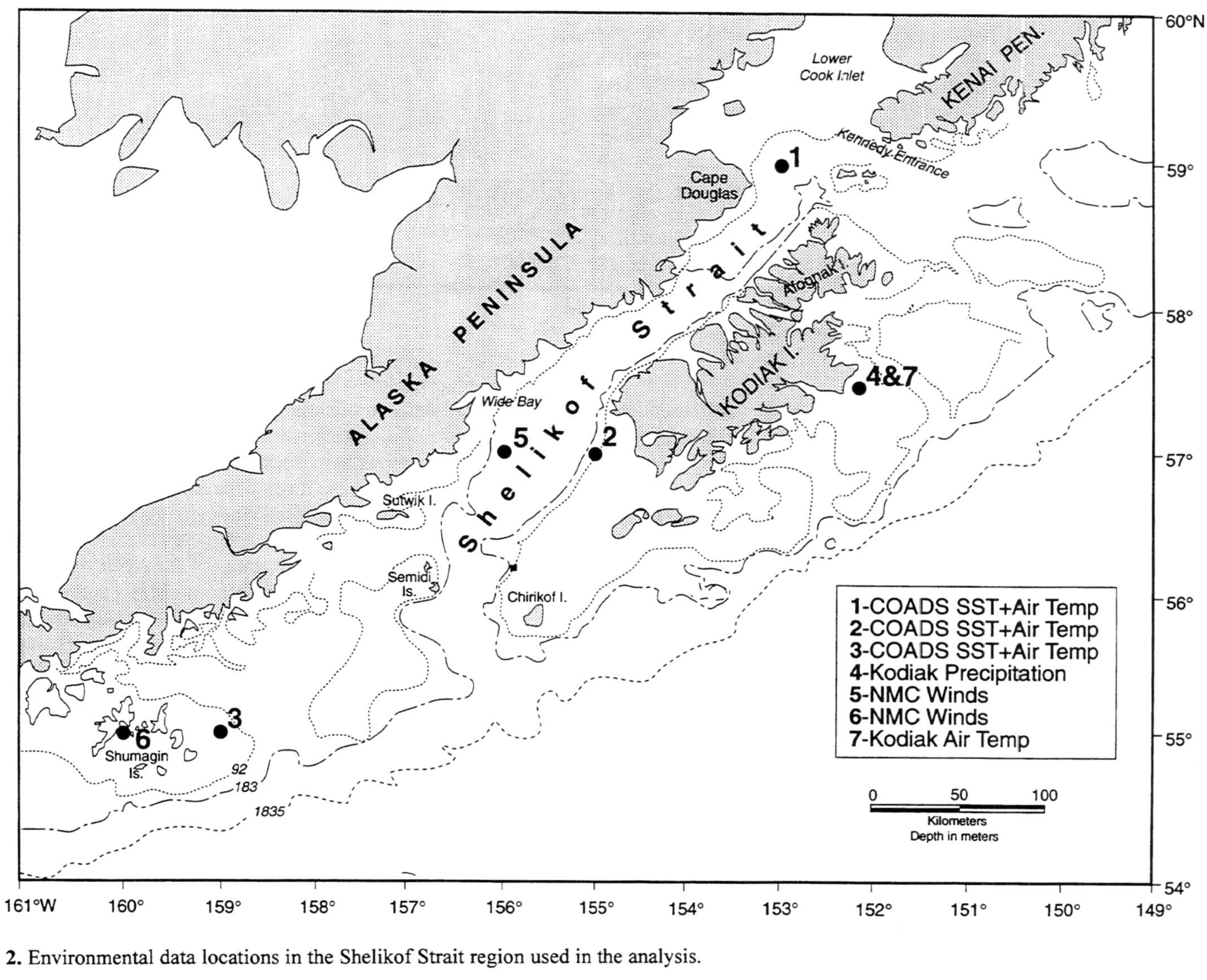

Fig. 2. Environmental data locations in the Shelikof Strait region used in the analysis.

Table 2. Alphabetical listing of data series variable names used in regression and principal component analyses.

Variable	Description
AGE2	Number of age-2 recruits (billions)
AGE0	Age-0 juvenile abundance index (billions)
AGE1	Age-1 juvenile abundance index (billions)
C4AIRJA	July–August COADS[a] air temperature (°C) at 55°N, 159°W
C4AIRMJ	May–June average COADS air temperature (°C) at 55°N, 159°W
C4SSTJA	July–August COADS sea-surface temperature (°C) at 55°N, 159°W
C4SSTMJ	May–June average COADS sea-surface temperature (°C) at 55°N, 159°W
C5SSTJF	January–February average COADS sea-surface temperature (°C) at 57°N, 155°W
C5AIRMA	March–April average COADS air temperature (°C) at 57°N, 155°W
C5SSTMA	March–April average COADS sea-surface temperature (°C) at 57°N, 155°W
lnAGE0	Natural log AGE0
lnAGE1	Natural log AGE1
lnAGE2	Natural log AGE2
KPTJA	July–August average Kodiak precipitation
KPTJF	January–February average Kodiak precipitation
KPTMA	March–April average Kodiak precipitation
KPTMJ	May–June average Kodiak precipitation
NEPAPR	Average April NEPPI[b] index
NEPAUG	Average August NEPPI index
NEPFEB	Average February NEPPI index
NEPJAN	Average January NEPPI index
NEPJUN	Average June NEPPI index
NEPMAR	Average March NEPPI index
NEPMAY	Average May NEPPI index
RUNOFF	Freshwater discharge index
WIN1MAY	May average wind mixing at 55°N, 160°W
WIN1JUN	June average wind mixing at 55°N, 160°W
WIN2MAR	March average wind mixing at 57°N, 156°W
WIN2MAY	May average wind mixing at 57°N, 156°W

COADS[a] = Comprehensive Ocean–Atmosphere Data Set.
NEPPI[b] = Northeast Pacific Pressure Index.

Temperature
Sea temperature directly affects all biological rates; growth rate is one of the most important for fish larvae. Temperature influences the egg stage mainly through its effect on development time: eggs hatch sooner at warmer temperatures (Blood et al. 1994). Faster growth of larvae may reduce the time they are vulnerable to size-selective predation. Air temperature affects melting of the snow pack.

Selection from time series
Using these hypotheses of the linkage between abiotic and biotic variables in the Gulf of Alaska, we selected a suite of biological and potentially influential environmental time series. A summary of the available environmental data series is listed in Table 1, and a description is provided in the Appendix. We examined Kodiak air temperature, lower Cook Inlet ice cover, and Comprehensive Ocean–Atmosphere Data Set (COADS) sea-level pressure data, but either the time series were too short or they were redundant with other data sources.

The initial data set was very large and often environmental data were available on time scales as fine as hourly. We used subjective and objective schemes to reduce the dimensionality of the environmental data set.

Using our knowledge of biological and physical time and space scales, we selected several geographic locations within the study domain for which to consider environmental data (Fig. 2). The northern edge of Shelikof Strait (59°N,153°W) was selected because it represents upstream flow conditions. The exit region of Shelikof Strait (57°N,155°W) was selected because it is an area where the earliest larval life stage is typically located. The area of the Shumagin Islands (55°N,159°W) was selected because it is an area where late-larval and young-of-the-year juvenile life stages are typically located. Information from areas outside the Shelikof Strait region was also considered. For example, data from lower Cook Inlet was considered because freshwater input from this area affects the flow through Shelikof Strait. Also, an indicator of circulation in the Gulf of Alaska was selected.

Subjective data screening was initially performed to eliminate redundant variables, data of poor quality (means calculated from few observations), and data series with incomplete coverage over the 1962–89 period. Dimensionality was further reduced through an objective correlation analysis of the abiotic time series. For example, if a time series showed high correlations between adjacent monthly averages, we concluded that six 2-mo averages would describe the data as well as 12 individual monthly averages and have the added advantage of eliminating the monthly serial correlation. We calculated bimonthly averages for other environmental variables as well.

We reduced the dimensionality further by matching life-stage location with the timing of appropriate environmental data series. For example, we know that larvae can be found in the exit region of Shelikof Strait only in late spring. We judged it unnecessary to consider environmental information from this location after the period when the cohort was present. If a variable had an influence, its effect would have manifested itself either before or during the time when the cohort occupied the location. Thus, we used January–February and March–April averages for the exit region of Shelikof Strait and May–June and July–August averages for the Shumagin Islands region. An alphabetical listing of the data series and variable names used in the statistical analyses and through the following dicussion are given in Table 2.

Application of univariate and multivariate statistical techniques

Univariate and multivariate statistical techniques were applied to the assembled data series to identify potentially important mechanisms affecting recruitment. Principal component analysis was applied to the suite of environmental variables in order to evaluate variance contributions of the various factors. The correlation matrix was factored using listwise deletion of missing values and no rotation.

As advocated by Tyler (1992), standard step wise multivariate regression analysis was used because it ranks the environmental variables according to the degree that they account for variance in recruitment. A multiple generalized linear model routine was used to perform forward step wise multiple regression using the environmental variables as independent variables and recruitment, larval, and juvenile abundance as the dependent variables. The probability of entry into the model was set to 0.05; the probability of exit was set to 0.15; and the minimum tolerance for a variable to enter the model was set to 0.5. All statistical procedures were conducted on a personal microcomputer using the SYSTAT statistical software analysis package (Wilkinson 1990).

Results

Recruitment time series

The 1972, 1975–79, 1984, and 1988 year-classes were above average (Fig. 3); the 1975–79 year-classes stand out. Whether the strong year-classes of the late 1970s are the result of a shift in the ecosystem is not known. However, the time period coincides with observations made by Hollowed and Wooster (1992) and Trenberth and Hurrell (1995), who report that atmospheric and oceanic conditions during the late 1970s were different from those that occurred during the early 1980s.

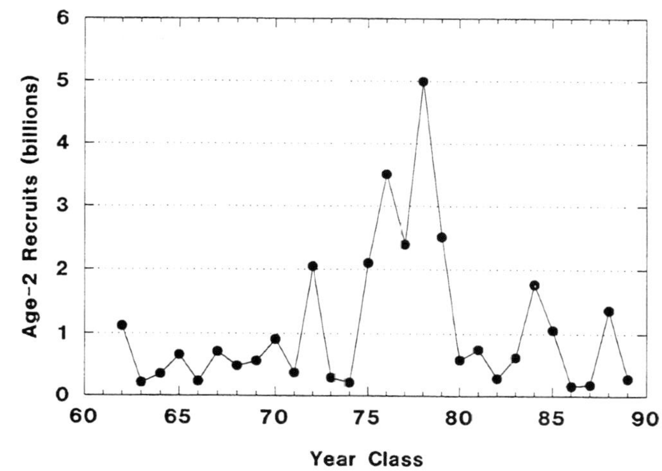

Fig. 3. Age-2 recruitment time series for Gulf of Alaska walleye pollock plotted against year-class, 1962–89.

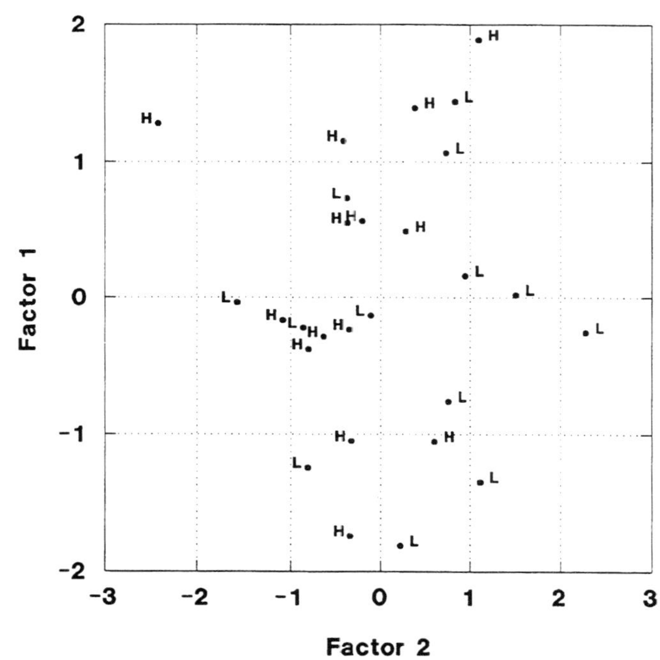

Fig. 4. High (H) and low (L) walleye pollock age-2 recruitment plotted against the first two principal component scores. The first principal component is related to mixing events and the second describes some aspect of large-scale atmospheric circulation.

Principal component analysis

Results from the multivariate principal component analyses showed that a total of 37% of the variance was captured by the first (25%) and second (12%) principal component modes. We plotted the recruitment data against the factor scores (Fig. 4), which shows the relationship between years of high (H) and low (L) recruitment and the first two principal component factor scores. High (low) recruitment values are those that fall above (below) median recruitment (0.64 billion). The second factor seems to have captured conditions that contribute to years of high and low recruitment. Ten out of 14 high recruitment years (71%) occurred when factor 2 scores are low, while eight out of 13 (62%) low recruitment years occurred when the factor 2 scores are high. High and low recruitment years were evenly distributed along the factor 1 axis. The same recruitment data labeled by year (Fig. 5) show that most annual recruitment points from the 1960s and up to the mid-70s occurred when factor 1 was low, while all of the 1980s (except for 1982, an El Niño year) and the late 1970 recruitment points

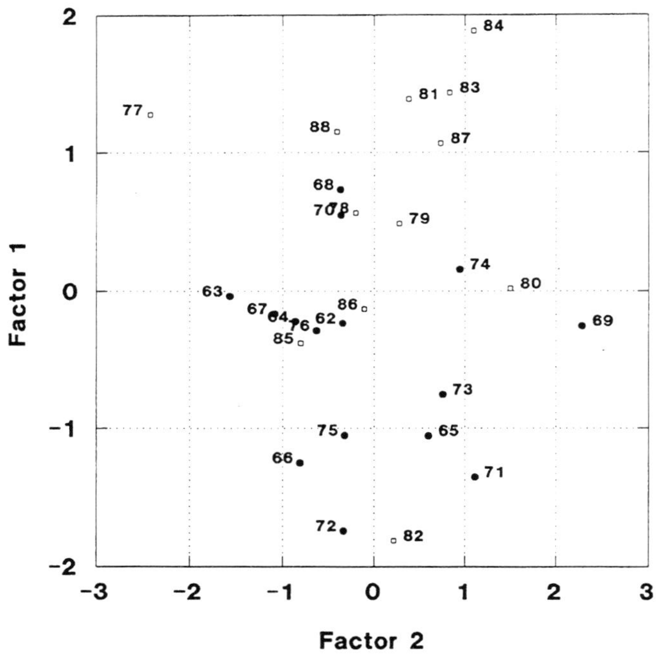

Fig. 5. Age-2 recruitment plotted by year-class against the first two principal component scores. The first principal component is related to mixing events and the second describes some aspect of large-scale atmospheric circulation.

occurred when factor 1 scores were high. The change in the late 1970s roughly corresponds to the observed regime shift in the physical environment reported by Trenberth and Hurrell (1995) and Wooster and Hollowed (1993).

Figure 6 shows environmental variables that had factor-loading scores (>|0.5|) for the first two principal components. The first component is positively related to air (AIR) and sea-surface temperature (SST) at two locations (C4, the Shumagin Islands region (55°N, 159°W) and C5, the exit region of Shelikof (57°N, 156°W)) during May–June (MJ), July–August (JA), January–February (JF), and March–April (MA); to Kodiak precipitation in January–February (KPTJF); to freshwater runoff (RUNOFF); and spring NEPPI values (NEPFEB, NEPMAR). The first component was negatively related to wind mixing at two locations during March, May, and June. This suggests that the first mode is related to mixing events. The second component is positively related to NEPPI (March through May) values (NEPMAR, NEPAPR, NEPMAY) and May Kodiak precipitation (KPTMA). The second component was negatively related to January and August NEPPI (NEPJAN, NEPAUG) and July–August Kodiak precipitation (KPTJA). This suggests that the second mode describes some aspect of the large-scale atmospheric circulation.

Regression analysis

Several multiple linear regression models were evaluated using the suite of environmental information to examine relationships with the biological variables of recruitment and age-0 and age-1 abundance. While correlative techniques have limited use in determining causal relationships, they have proved useful in a data exploration such as this.

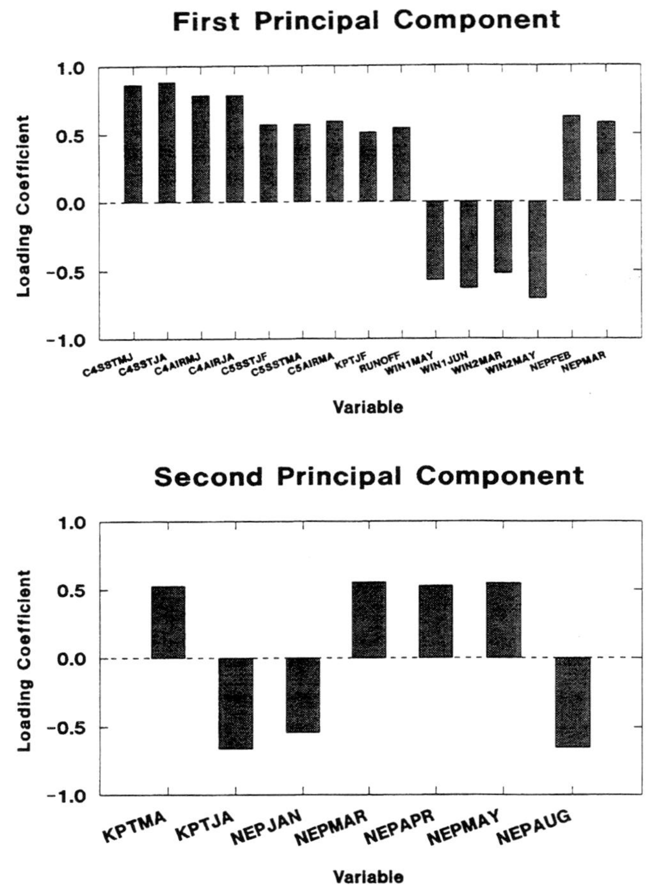

Fig. 6. Principal component loading coefficients for variables that contributed to the first two principal components. Variable abbreviations are explained in Table 2. The first principal component is related to mixing events and the second describes some aspect of large-scale atmospheric circulation.

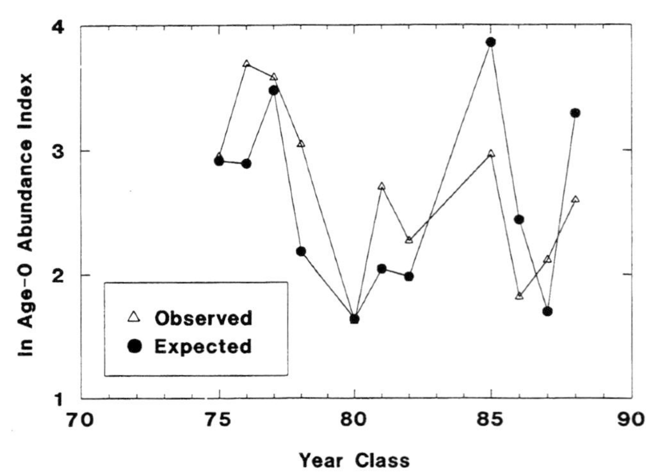

Fig. 7. Observed age-0 abundance index compared to values expected from the linear multiple regression model ln(AGE0) = f(KPTJA, NEPAUG). Variable abbreviations are explained in Table 2.

Table 3. Results of linear regression models relating the age-0 abundance index to environmental variables.

(A) Regression

Dependent variable	*N*	R^2	SEE[a]	Independent variable[b]	Regression coefficient	St	*t*	*p*
AGE0	11	0.877	7.606	KPTJA	15.135	1.789	8.459	<0.001
ln(AGE0)	11	0.955	0.646	KPTJA	1.771	0.173	10.213	<0.001
				NEPAUG	−0.214	0.067	−3.177	0.011

(B) ANOVA

Dependent variable and source	SS	df	MS	*F*	*p*
AGE0					
Regression	4140.198	1	4140.198	71.558	<0.001
Residual	578.581	10	57.858		
Total	4718.779	11			
ln(AGE0)					
Regression	79.159	2	39.580	94.976	<0.001
Residual	3.751	9	0.417		
Total	82.910				

[a]SEE = standard error of the estimate.
[b]KPTJA = July–August average Kodiak precipitation; NEPAUG = average August NEPPI index.

Models relating age-0 pollock abundance to environmental variables described a large percentage of the variation (Table 3). A model relating age-0 pollock abundance (AGE0) to July–August Kodiak precipitation (KPTJA) accounted for 88% of the variation. The best model for age-0 pollock related the natural log of age-0 abundance ln(AGE0) with July–August Kodiak precipitation (KPTJA) and August NEPPI (NEPAUG). This model explained 96% of the variation in age-0 abundance (Table 3). The sign for the precipitation variable was positive, and the sign for the NEPPI variable was negative. A plot of observed age-0 abundance versus linear model predictions (Fig. 7) shows good agreement despite several trend reversals and a range in abundance that spans two orders of magnitude.

Models relating age-1 pollock abundance to environmental variables occurring during the birth year also described a large percentage of the variation (Table 4). The best model for age-1 pollock related the natural log of age-1 abundance ln(AGE1) with May–June Kodiak precipitation (KPTMJ) and June NEPPI (NEPJUN). This model explained 96% of the variation in age-1 abundance. Similar to the age-0 model, the sign of the coefficients was positive for the precipitation variable and negative for the NEPPI variable. There were fewer data points for age-1 but again the linear model tracked observed abundance very well (Fig. 8).

The variances explained by the age-0 and age-1 linear models improved when the larval and juvenile data were put on the log scale. The high R^2 values from the age-0 and age-1 data remain suspect because of the low number of data points. Variables with very similar temporal scales consistently proved to be of high importance. For example, the model for age-1 juveniles included May–June Kodiak precipitation and June NEPPI variables, and the age-0 model included July–August Kodiak precipitation and August NEPPI variables.

The relationship yielding the highest explained variation in age-2 recruitment (AGE2), January–February Kodiak precipitation (KPTJF), and June NEPPI (NEPJUN) is summarized in Table 5. These results indicate that January–February Kodiak precipitation and June NEPPI values accounted for 65% of the observed variation in recruitment. The sign of the regression coefficient for KPTJF was positive, while that for NEPJUN was negative (Table 5). This is consistent with the hypothesis that freshwater input has a positive influence on year-class strength, while large positive NEPPI values, has a negative influence. Univariate analysis showed that KPTJF and NEPJUN were not equivalent in their contribution to explained variation in recruitment (KPTJF, r^2 = 0.580; NEPJUN, r^2 = 0.042). When the natural log of recruitment is used, as advised by Hennemuth et al. (1980), the best model was described by June NEPPI (NEPJUN) and March wind mixing (WIN2MAR) at the exit of Shelikof Strait (57°N, 156°W) (Table 5). These variables accounted for 35% of the observed variation in recruitment. The sign of the regression coefficients for NEPJUN and WIN2MAR were negative.

The June timing of the NEPPI variable indicates that a wintertime NEPPI signature (high positive values) in June is detrimental for larval survival. The actual mechanisms may be related to any of the NEPPI characteristics described in the Appendix.

Table 4. Results of linear regression models relating the age-1 abundance index to environmental variables.

(A) Regression

Dependent variable	N	R^2	SEE[a]	Independent variable[b]	Regression coefficient	St	t	p
ln(AGE1)	8	0.960	0.301	KPTMJ	0.826	0.069	11.948	<0.001
				NEPJUN	−0.149	0.027	−5.522	<0.001

(B) ANOVA

Source	SS	df	MS	F	p
Regression	13.103	2	6.552	72.280	<0.001
Residual	0.544	6	0.091		
Total	13.642	8			

[a]SEE = standard error of the estimate.
[b]KPTMJ = May–June average Kodiak precipitation; NEPJUN = average June NEPPI index.

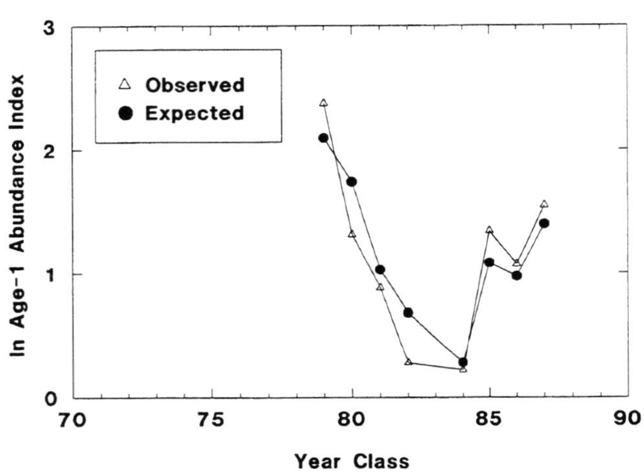

Fig. 8. Observed age-1 abundance index compared to values expected from the linear multiple regression model ln(AGE1) = f(KPTMJ, NEPJUN). Variable abbreviations are explained in Table 2.

A comparison of the observed recruitment time series and that predicted by the precipitation–NEPPI linear model (Fig. 9) shows good agreement, but the simple linear model fails to track the large 1972 and the 1975, 1976, 1978, and 1979 year-classes. These results indicate that there are important factors, unaccounted for in the linear model, that cause very strong year-classes.

Discussion

Results from the univariate and multivariate analyses suggested that freshwater input and wind play a major role in determining recruitment success. The influence of these factors was consistently demonstrated over the larval, juvenile, and adult life stages. Freshwater input during the months of January and February was important in describing age-2 recruitment. Rainfall during the winter months, which is stored as ice, can contribute large quantities of freshwater to the Gulf of Alaska later in the year. From these results we can draw several hypotheses: (1) spring wind mixing in Shelikof Strait affects larval survival; (2) large-scale atmospheric circulation directly affects the Shelikof region; (3) baroclinicity may generate mechanisms conducive to larval survival; and (4) effects of the physical environmental on predation and behaviour may affect juvenile survival.

The specific mechanisms that describe how the processes mentioned above affect recruitment success are more difficult to identify mainly because of complications associated with separating factors that predispose an animal to die (i.e., starvation) from those that are directly responsible for killing it (i.e., predation). Although the same linkages between the environment and recruitment operate at different time scales, the relative importance of any one influence may differ depending on whether additional mechanisms come into play. Furthermore the link's manifestation on recruitment may lag its occurrence.

There are several potential sources of error in estimates of recruitment derived from age-structured stock assessment models. Two of these problems are particularly relevant here. First, spurious trends in recruitment could be generated from incorrect parameter estimates in the age-structured analysis. Sims (1984), Bradford and Peterman (1989), and Megrey (1989) noted that incorrect selection of terminal fishing mortalities or natural mortalities produced spurious trends in recruitment patterns. Sampson (1988) showed the relative errors in successive cohort size estimates are not necessarily stable and were sensitive to errors in catch-at-age estimates and estimates of natural mortality. Lapointe and Peterman (1991) also observed that spurious correlation could result from errors in the natural mortality rate used in virtual

Table 5. Results of linear regression models relating the age-2 abundance index to environmental variables.

(A) Regression

Dependent variable	N	R^2	SEE[a]	Independent variable[b]	Regression coefficient	St	t	p
AGE2	28	0.651	0.967	KPTJF	0.687	0.102	6.731	<0.001
				NEPJUN	−0.129	0.056	−2.297	0.030
ln(AGE2)	28	0.353	0.853	NEPJUN	−0.104	0.050	−2.097	0.046
				WIN2MAR	−10.707	3.625	−2.953	0.007

(B) ANOVA

Dependent variable and source	SS	df	MS	F	p
AGE2					
Regression	45.317	2	22.658	24.233	<0.001
Residual	24.310	26	0.935		
Total	69.627	28			
ln(AGE2)					
Regression	10.322	2	5.161	7.089	0.003
Residual	18.928	26	0.728		
Total	29.250	28			

[a]SEE = standard error of the estimate.
[b]

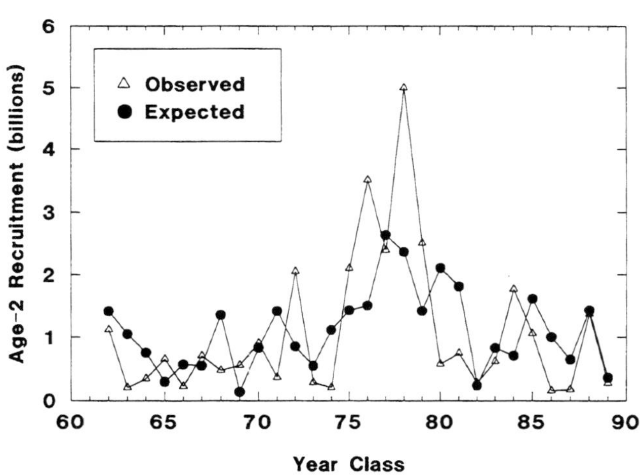

Fig. 9. Observed age-2 recruitment compared to values expected from the linear multiple regression model AGE2 = f(KPTJF, NEPJUN). Variable abbreviations are explained in Table 2.

population analysis. The second source of error is due to incorrect aging of the fish. This effect could decrease variability in recruitment and cause autocorrelations in recruitment, as is sometimes observed in recruitment time series (Sissenwine 1984). Kimura and Lyons (1991) examined among-reader bias and variability in the age determination process. Their study shows percent agreement for pollock otoliths falls below 50% at age 7. The effects of aging error may be less critical in this analysis because estimates of aging error were incorporated directly into the assessment procedure used to estimate the recruitment time series.

While these problems may lead to serious misconceptions in the time series of recruitment, the problems are mediated somewhat because we used the stock synthesis model, which is a newer stock assessment model that relies on the separability assumption. Even using newer models, Megrey et al. (1990) show that different results can be obtained from assessment models because of inconsistencies in the data sources. We assume that catch-at-age data were fairly accurate and that estimated trends in recruitment were reasonable representations of the historical patterns of recruitment. The additional information supplied by the longer recruitment time series gave us an opportunity to examine historical patterns in recruitment variability and long-term yield potential. It also gave us more data with which to examine relationships between year-class strength and environmental and biological factors, and an improved ability to evaluate the roles of density-dependent and density-independent effects on year-class success.

Despite these problems, we offer some speculations. Other research (Stabeno et al. 1995) has shown that strong winds and average rainfall enhance the formation and maintenance of an eddy in Shelikof Strait. Both cyclonic and anticyclonic eddies have been observed in the Shelikof Strait region several times during the past 5 yr (Vastano et al. 1992; Schumacher et al. 1993). The creation of eddies is sensitive to the timing of precipitation and storms. Perhaps strong wind followed by extended calm periods plus addition of freshwater create conditions favourable for larvae as a result of eddy formation.

Another important finding is that significant (yet unknown) mechanisms may be occurring during the summer months. Two pieces of evidence lead to this conclusion. The first is that age-2 recruitment predicted from the linear model seemed to track observed year-class abundance. However, the 1981 predicted value was well above the observed value. We know from other studies (Reed et al. 1989) that mortality for the 1981 year-class was unusually high. Thus, the linear regression model, which predicted an above-average year class in 1981 based solely on environmental data, erred in the correct direction. This suggests that a secondary forcing acting in late summer of 1981 was responsible for the weak year-class. Similar evidence is available from the linear age-0 and ln(age-1) models, which tracked the observed population abundances fairly accurately. The differences in absolute numbers indicate that other mechanisms may affect the total abundance but not the trend. The ln(age-0) and ln(age-1) models include a summer and/or fall variable. Moreover, the variables in any one model have a high degree of temporal overlap.

These clues imply that mechanisms late in the summer or fall also contribute to year-class variations. Pollock during this time of the year are no longer planktonic; they are capable of swimming against water movements, and they exhibit complex behaviour patterns (Olla and Davis 1990*a*, *b*, 1992; Baird et al. 1991; Sogard and Olla 1993). Pollock have also been observed to exhibit behaviour in response to features of the physical environment (Bailey 1989). Behavioural response of fishes to the environment has been observed to influence predatory species. For example, Haney (1991) noted that differences in water mass characteristics in the Bering Sea may influence the use of shallow, nearshore areas by large alcids. He observed that warmer, fresher water in nearshore areas of the Bering Sea tended to contain high concentrations of schooling fish, thereby retaining or attracting piscivorous seabirds such as common murres (*Uria aalge*), thick-billed murres (*Uria lomvia*), tufted puffins (*Fratercula cirrhata*), and horned puffins (*Fratercula corniculata*) to shallow areas near islands. If this occurs in the Gulf of Alaska and is related to freshwater input, it could affect predation depending on the response of juvenile pollock to this water type. These findings suggest that predation may be an important factor operating in the second half of the year. Consequently, effects of the physical environment on predators and behavioural interactions between predators and larval and juvenile pollock may affect pollock survival. Complete specification of the pollock environment, therefore, must include time series of prey and predator abundances.

The current study was successful in that we were able to identify variables or processes that appear to affect recruitment. Future work will concentrate on (1) use of nonlinear analyses such as those of Schnute and McKinnel (1984) and Cury et al. (1995); (2) development of models that factor in potential biological controls of recruitment such as the influence of spawner biomass or predators; (3) application of formal time series models and transfer function applications; and (4) determination of the degree to which better understanding of these processes can be used to assist fisheries management. Even though this analysis was by design exploratory in nature and limited in scope to simple linear analysis of only environmental effects on recruitment, our findings provide a useful basis for future studies on interannual variability and climate change in the region.

Acknowledgements

This research was funded by NOAA's Fisheries–Oceanography Coordinated Investigations and is FOCI Contribution FOCI-0179, NOAA/Pacific Marine Environmental Laboratory Contribution 1434, and Joint Institute for the Study of the Atmosphere and Ocean Contribution 254.

References

Bailey, K.M. 1981. Larval transport and recruitment of Pacific hake *Merluccius productus*. Mar. Ecol. Prog. Ser. 6: 1–9.

Bailey, K.M. 1989. Interaction between the vertical distribution of juvenile walleye pollock *Theragra chalcogramma* in the eastern Bering Sea, and cannibalism. Mar. Ecol. Prog. Ser. 53: 205–213.

Bailey, K.M., and S.A. Macklin. 1993. Analysis of patterns in larval walleye pollock (*Theragra chalcogramma*) survival and wind mixing events in Shelikof Strait, Gulf of Alaska. Mar. Ecol. Prog. Ser. 113:1–12.

Bailey, K.M., and S. Spring. 1992. Comparison of larval, age-0 juvenile and age-3 recruit abundance indices of walleye pollock *Theragra chalcogramma* in the western Gulf of Alaska. ICES J. Mar. Sci. 49: 297–304.

Baird, T.A., C.H. Ryer, and B.L. Olla. 1991. Social enhancement of foraging on an ephemeral food source in juvenile walleye pollock (*Theragra chalcogramma*). Environ. Biol. Fish. 31: 307–311.

Bakkala, R.G., V.G. Wespestad, and L.L. Low. 1987. Historical trends in abundance and current condition of walleye pollock in the eastern Bering Sea. Fish. Res. 5: 199–216.

Blood, D.B, A.C. Matarese, and M.M. Yoklavich. 1994. Embryonic development of walleye pollock *Theragra chalcogramma* from Shelikof Strait, Gulf of Alaska. Fish. Bull. U.S. 92: 207–222.

Bond, N.A., and S.A. Macklin. 1993. Research aircraft observations of offshore directed flow near Wide Bay of the Alaska Peninsula. Mon. Weather Rev. 121: 150–161.

Bradford, M.J., and R.M. Peterman. 1989. Incorrect parameter values used in virtual population analysis (VPA) generate spurious time trends in reconstructed abundances, p. 87–99. *In* R.J. Beamish and G.A. McFarlane [ed.] Effects of ocean variability on recruitment and an evaluation of parameters used in stock assessment models, Can. Spec. Publ. Fish. Aquat. Sci. 108.

Brodeur, R.D., K.M. Bailey, and S. Kim. 1991. Cannibalism on eggs by walleye pollock (*Theragra chalcogramma*) in Shelikof Strait, Gulf of Alaska. Mar. Ecol. Prog. Ser. 71: 207–218.

Canino, M.F., K.M. Bailey, and L.S. Incze. 1991. Temporal and geographic differences in feeding and nutritional condition of walleye pollock larvae, *Theragra chalcogramma*, in Shelikof Strait, Gulf of Alaska. Mar. Ecol. Prog. Ser. 79: 27–36.

Cayan, D.R., D.R. McLain, and W.D. Nichols. 1991. Monthly climate time series data for the Pacific Ocean and western America's. U.S. Geological Survey, Washington, D.C. Rep. OFR 91-92.

Chadwick, H.K., D.E. Stevens, and L.W. Miller. 1977. Some factors regulating the striped bass population in the Sacramento – San Joaquin estuary, California, p. 18–35. *In* W. Van Winckle [ed.] Assessing the effects of power plant-induced mortality on fish populations. Pergamon Press, New York. 388 p.

Cury, P., and C. Roy. 1989. Optimal environmental window and pelagic fish recruitment in upwelling areas. Can. J. Fish. Aquat. Sci. 46: 670–680.

Cury, P., C. Roy, R. Mendelssohn, A. Bakun, D.M. Husby, and R.H. Parrish. 1994. "Moderate" is "better": exploring nonlinear climatic effects on the California anchovy (*Engraulis mordax*), p. 417–424. *In* R.J. Beamish [ed.] Climate change and northern fish populations. 121.

Dunn, J.R., and W.C. Rugen. 1989. A catalog of Northwest and Alaska Fisheries Science Center ichthyoplankton cruises 1965–88. NWAFC Proc. Rep. 89-04, Northwest and Alaska Fisheries Center, National Marine Fisheries Service, NOAA, Seattle, WA, 98115-0070. 66 p.

Emery, W.J., and Hamilton. 1985. Atmospheric forcing of interannual variability in the northeast Pacific Ocean: connections with El Niño. J. Geophys. Res. 90: 857–868.

FAO. 1990. Yearbook of fishery statistics. Rome. 15 p.

Haney, J.C. 1991. Influence of pyncocline topography and water-column structure on marine distributions of alcids (Aves: Alcidae) in Anadyr Strait, Northern Bering Sea, Alaska. Mar. Biol. 110: 419–435.

Hennemuth, R.C., J.E. Palmer, and B.E. Brown. 1980. A statistical description of recruitment in eighteen selected fish stocks. J. Northwest Atl. Fish. Sci. 1: 101–111.

Hinckley. S., K.M. Bailey, S.J. Picquelle, P.J. Stabeno, and J.D. Schumacher. 1991. Transport, distribution, and abundance of larval and juvenile walleyepollock (*Theragra chalcogramma*) in the western Gulf of Alaska in 1987. Can. J. Fish. Aquat. Sci. 48: 91–98.

Hinckley. S., K.M. Bailey, S.J. Picquelle, P.J. Stabeno, and J.D. Schumacher. 1990. Realistic estimation of mortality rates of larval walleye pollock in the Gulf of Alaska as determined from field sampling. EOS, Trans. Am. Geophys. Union 71: 93

Hollowed, A.B. 1992. Spatial and temporal distributions of Pacific hake, *Merluccius productus*, larvae and estimates of survival during early life stages. Calif. Coop. Oceanic Fish. Invest. Rep. 33: 100–123.

Hollowed, A.B., B.A. Megrey, P. Munro, and W.C. Karp. 1991. Walleye pollock. *In* Stock assessment and fishery evaluation report for the 1992 Gulf of Alaska groundfish fishery. North Pacific Fisheries Management Council., Anchorage, AK. p. 1-1 to 1-92.

Hollowed, A.B., and W.S. Wooster. 1992. Variability of winter ocean conditions and strong year classes of Northeast Pacific groundfish. ICES Mar. Sci. Symp. 195: 433–444.

Huang, N.E. 1979. On the surface drift currents in the ocean. J. Fluid Mech. 91: 191–208.

Hubert, W.E., and T. Laevastu. 1965. Synoptic analysis and forecasting of surface currents. U.S. Navy Fleet Numerical Weather Facility, Monterey, CA. Tech. Note 9: 47 p.

Incze, L.S., A.W. Kendall, Jr., J.D. Schumacher, and R.K. Reed. 1989. Interactions of a mesoscale patch of larval fish (*Theragra chalcogramma*) with the Alaska Coastal Current. Cont. Shelf Res. 9: 269–284.

Incze, L.S., P. Ortner, and J.D. Schumacher. 1990. Microzooplankton, vertical mixing and advection in a larval fish patch. J. Plankton Res. 12: 365–379.

Ingraham, W.J., Jr., and R.K. Miyahara. 1988. Ocean surface current simulations in the North Pacific Ocean and Bering Sea (OSCURS — numerical model). NOAA Tech. Memo. NMFS F/NWC-130: 155 pp.

Ingraham, W.J., Jr., and R.K. Miyahara. 1989. Tuning of OSCURS numerical model to ocean surface current measurements in the Gulf of Alaska. NOAA Tech. Memo. NMFS F/NWC-168: 67 pp.

Ingraham, W.J., Jr., R.K. Reed, J.D. Schumacher, and S.A. Macklin. 1991. Interannual variability of circulation in the Gulf of Alaska in relation to water properties and fisheries resources. EOS, Trans. Am. Geophys. Union 72: 257–264.

Kendall, A.W., Jr., M.E. Clarke, M.M. Yoklavich, and G.W. Boehlert. 1987. Distribution, feeding, and growth of larval walleye pollock, *Theragra chalcogramma*, in Shelikof Strait, Gulf of Alaska. Fish. Bull. U.S. 85: 499–521.

Kendall, A.W., Jr., and S. Kim. 1989. Buoyancy of walleye pollock (*Theragra chalcogramma*) eggs in relation to water properties and movement in Shelikof Strait, Gulf of Alaska, p. 169–180. *In* R.J. Beamish and G.A. McFarlane [eds.] Effects of ocean variability on recruitment and an evaluation of parameters used in stock assessment models, Can. Spec. Publ. Fish. Aquat. Sci. 108.

Kendall, A.W., Jr., and S.J. Picquelle. 1990. Egg and larval distributions of walleye pollock *Theragra chalcogramma* in Shelikof Strait, Gulf of Alaska. Fish. Bull. U.S. 88: 133–154.

Kim, S., and B. Bang. 1990. Oceanic dispersion of larval fish and its implication for mortality estimates: case study of walleye pollock larvae in Shelikof Strait, Alaska. Fish. Bull. U.S. 88: 303–311.

Kim, S., and D.R. Gunderson. 1989. Cohort dynamics of walleye pollock in Shelikof Strait, Gulf of Alaska, during the egg and larval period. Trans. Am. Fish. Soc. 118: 264–273.

Kim, S., and A.W. Kendall, Jr. 1989. Distribution and transport of larval walleye pollock (*Theragra chalcogramma*) in Shelikof Strait, Gulf of Alaska, in relation to water movement. Rapp. P.-v. Réun. Cons. int. Explor. Mer 191: 127–136.

Kim, S., and E.P. Nunnallee. 1991. Distribution of walleye pollock, *Theragra chalcogramma*, spawning in Shelikof Strait, Gulf of Alaska, based on acoustic and ichthyoplankton surveys 1981, 1984 and 1985. Bull. Korean Fish. Soc. 23: 425–438.

Kimura, D.K., and S. Chikuni. 1987. Mixtures of empirical distributions: An iterative application of the age–length key. Biometrics. 43: 23–35.

Kimura, D.K., and J.J. Lyons. 1991. Between-reader bias and variability in the age-determination process. Fish. Bull. U.S. 89: 53–60.

Klein, P. 1980. A simulation of the effects of air–sea transfer variability on the structure of the marine upper layers. J. Phys. Oceanogr. 10: 1824–1841.

Kope, R., and L.W. Botsford. 1990. Determination of factors affecting recruitment of chinook salmon *Oncorhynchus tshawytscha* in central California. Fish. Bull. U.S. 88: 257–269.

Lapointe, M.F., and R.M. Peterman. 1991. Spurious correlations between fish recruitment and environmental factors due to errors in the natural mortality rate used in virtual population analysis (VPA). ICES J. Mar. Sci. 48: 219–228.

Larson, S.E. 1975. A 26-year time series of monthly mean winds over the oceans. Part 1. A statistical verification of computed surface winds over the North Pacific and North Atlantic. U.S. Navy Environmental Prediction Research Facility, Monterey, CA. ENVPREDRSCWFAC Tech. Pap. No. 8-75. 89 p.

Larson, S., and T. Laevastu. 1972. Numerical analysis of ocean surface currents, p. 55–74. *In* Studi in onore di Giuseppina Aliverti. Instituto Universitario Navale di Napoli, Napoli, Italy.

Lasker, R. 1975. Field criteria for survival of anchovy larvae: the relation between inshore chlorophyll maximum layers and successful first feeding. Fish. Bull. U.S. 73: 453–462.

Lasker, R. 1978. The relation between oceanographic conditions and larval anchovy food in the California Current: identification of factors contributing to recruitment failure. Rapp. P.-v. Réun. Cons. int. Explor. Mer 173: 212–230.

Leggett, W.C. 1977. Density dependence, density independence, and recruitment of American shad (*Alosa spaidissima)* population of the Connecticut River, p. 3–17. *In* W. Van Winckle [ed.] Assessing the effects of power plant-induced mortality on fish populations. Pergamon Press, New York, NY. 388 p.

Leggett, W.C., K.T. Frank, and J.E. Carscadden. 1984. Meteorological and hydrographic regulation of year-class strength of capelin (*Mallotus villosus*). Can. J. Fish. Aquat. Sci. 41: 1193–1201.

Macklin, S.A., P.J. Stabeno, and J.D. Schumacher. 1993. A comparison of gradient and observed over-the-water winds along a mountainous coast. J. Geophys. Res. 98: 16 555 – 16 569.

Megrey, B.A. 1989. Review and comparison of age-structured stock assessment models from theoretical and applied points of view, p. 8–48. *In* E.F. Edwards and B.A. Megrey [ed.] Mathematical analysis of fish stock dynamics. Am. Fish. Soc. Symp. 6.

Megrey, B.A. 1990. Population dynamics and management of walleye pollock stocks in the Gulf of Alaska, 1976–86. Fish. Res. 11: 321–354.

Megrey, B.A., A.B. Hollowed, and R.D. Methot. 1990. Integrated analysis of Gulf of Alaska walleye pollock catch-at-age and research survey data using two different stock assessment procedures, p. 279–302. *In* Loh-Lee Low [ed.] Proceedings of the Symposium on Application of Stock Assessment Techniques to Gadids. INPFC Bull. 50. 10/31–11/1, Seattle, WA.

Methot, R.D. 1989. Synthetic estimates of historical abundance and mortality for northern anchovy, p. 61–78. *In* E.F. Edwards and B.A. Megrey [ed.] Mathematical analysis of fish stock dynamics. Am. Fish. Soc. Symp. 6.

Methot, R.D. 1990. Synthesis model: an adaptable framework for analysis of diverse stock assessment data, p. 259–278. *In* Loh-Lee Low [ed.] Proceedings of the Symposium on Application of Stock Assessment Techniques to Gadids. INPFC Bull. 50. 10/31–11/1, Seattle, WA.

Mysak, L., R.D. Muench, and J.D. Schumacher. 1981. Baroclinic instability in a downstream varying channel: Shelikof Strait, Alaska. J. Phys. Oceanogr. 11: 950–969.

Niebauer, H.J. 1988. Effects of El Niño – Southern Oscillation and North Pacific weather patterns on interannual variability in the subarctic Bering Sea. J. Geophys. Res. 93: 5051–5068.

Nelson, W.R., M.C. Ingraham, and W.E. Schaff. 1977. Larval transport and year-class strength of Atlantic menhaden (*Brevoortia tyrannus)*. Fish. Bull. U.S. 75: 23–41.

Olla, B.L., and M.W. Davis. 1990a. Effects of physical factors on the vertical distribution of larval walleye pollock (*Theragra chalcogramma*) under controlled laboratory conditions. Mar. Ecol. Prog. Ser. 63: 105–112.

Olla, B.L., and M.W. Davis. 1990*b*. Behavioral responses of juvenile walleye pollock (*Theragra chalcogramma*) to light, thermoclines and food: possible role in vertical distribution. J. Exp. Mar. Biol. Ecol. 135: 59–68.

Olla, B.L., and M.W. Davis. 1992. Phototactic responses of unfed walleye pollock, *Theragra chalcogramma* larvae: comparisons with other measures of condition. Environ. Biol. Fish. 35: 105–108.

Ortner, P.B., D.R. Nieman, and L.S. Incze. 1990. Biological characterization of a coastal mesoscale eddy. EOS, Trans. Am. Geophys. Union 71: 177.

Parker, K.S. 1989a. Influence of oceanographic and meteorological processes on the recruitment of Pacific halibut, *Hippoglossus stenolepis,* in the Gulf of Alaska. Ph.D. dissertation, University of Washington, Seattle, WA. 154 p.

Parker, K.S. 1989*b*. Influence of oceanographic and meteorological processes on the recruitment of Pacific halibut, *Hippoglossus stenolepis*, in the Gulf of Alaska, p. 221–237. *In* R.J. Beamish and G.A. McFarlane [ed.] Effects of ocean variability on recruitment and an evaluation of parameters used in stock assessment models. Can. Spec. Publ. Fish. Aquat. Sci. 108.

Picquelle, S.J., and B.A. Megrey. 1991. A method of estimating spawning biomass from egg surveys with an application to walleye pollock, *Theragra chalcogramma*, in the Shelikof Strait, Alaska. AFSC Processed Report 91-21. NOAA/NMFS, Seattle, WA, 98115. 79 p.

Picquelle, S.J., and B.A. Megrey. 1993. A preliminary spawning biomass estimate of walleye pollock, *Theragra chalcogramma*, in the Shelikof Strait, Alaska, based on the annual egg production method, p. 230–252. *In* J.R. Hunter and N.C.H. Lo [ed.]Proceedings from the Symposium on the Advances in Estimating the Biomass of Marine Fish Stocks Using Ichthyoplankton. Proceedings of the 15th Annual Early Life History Section of the American Fisheries Society. Bull. Mar. Sci. 53(2).

Reed, R.K. 1980. Direct measurements of recirculation in the Alaskan Stream. J. Phys. Oceanogr. 10: 976–978.

Reed. R.K., L.S. Incze, and J.D. Schumacher. 1989. Estimation of the effects of flow on dispersion of larval pollock, *Theragra chalcogramma*, in Shelikof Strait, Alaska, p. 239–246. *In* R.J. Beamish and G.A. McFarlane [ed.] Effects of ocean variability on recruitment and an evaluation of parameters used in stock assessment models. Can. Spec. Pub. Fish. Aquat. Sci. 108.

Reed, R.K., and J.D. Schumacher. 1989a. Transport and physical properties in central Shelikof Strait, Alaska. Cont. Shelf Res. 9: 261–268.

Reed, R.K., and J.D. Schumacher. 1989*b*. Some mesoscale features of flow in Shelikof Strait, Alaska. J. Geophys. Res. 94: 12 603 – 12 606.

Reed, R.K., J.D. Schumacher, and L.S. Incze. 1987. Circulation in Shelikof Strait, Alaska. J. Phys. Oceanogr. 17: 1546–1554.

Richardson, S.L. 1981. Spawning biomass and early life of northern anchovy, *Engraulis mordax*, in the northern subpopulation off Oregon and Washington. Fish. Bull. U.S. 78: 855–876.

Roach, A.T., and J.D. Schumacher. 1991. Observations of seasonal and interseasonal variability in Shelikof Strait, Alaska, p. 3304–3317. *In* Proceedings of the Seventh Symposium on Coastal and Ocean Management, Coastal Zone 91. ASCE/Long Beach, CA. July 8–12, 1991.

Rothschild, B.J., and T.R. Osborne. 1988. Small-scale turbulence and plankton contact rates. J. Plankton Res. 10: 465–474.

Royer, T.C. 1982. Coastal fresh water discharge in the Northeast Pacific. J. Geophys. Res. 87: 2017–2021.

Rugen, W.C. 1990. Spatial and temporal distribution of larval fish in the western Gulf of Alaska, with emphasis on the period of peak abundance of walleye pollock (*Theragrachalcogramma*) larvae. NWAFC Processed Report 90-01, NOAA/NMFS, Seattle, WA, 98115. 162 p.

Sampson, D.B. 1988. The stability of virtual population analysis cohort size estimates. J. Cons. Cons. int. Explor. Mer 44: 135–142.

Schnute, J., and S. McKinnell. 1984. A biologically meaningful approach to response surface analysis. Can J. Fish. Aquat. Sci. 41: 936–953.

Schumacher, J.D., and A.W. Kendall, Jr. 1991. Some interactions between young walleye pollock and their environment in the western Gulf of Alaska. Calif. Coop. Oceanic Fish. Invest. Rep. 32: 22–40.

Schumacher, J.D., P.J. Stabeno, and S.J. Bograd. 1993. Characteristics of an eddy over a continental shelf: Shelikof Strait, Alaska. J. Geophys. Res. 98: 8395–8404.

Schumacher, J.D., P.J. Stabeno, and A.T. Roach. 1990. Volume transport in the Alaska Coastal Current. Cont. Shelf Res. 12: 1071–1083.

Sims, S.E. 1984. An analysis of the effects of errors in the natural mortality rate on stock-size estimates using virtual population analysis (cohort analysis). J. Cons. Cons. int. Explor. Mer 41: 149–153.

Sissenwine, M.P. 1984. Why do fish populations vary, p. 59–94. *In* R.M. May [ed.] Exploitation of marine communities. Springer-Verlag, New York, NY.

Sogard, S.M., and B.L. Olla. 1993. The influence of predator presence on utilization of artificial seagrass habitats by juvenile walleye pollock, *Theragra chalcogramma*. Environ. Biol. Fish. 37: 57–65.

Spring, S., and K. Bailey. 1991. Distribution and abundance of juvenile pollock from historical shrimp trawl surveys in the western Gulf of Alaska. AFSC Processed Report 91-18. NOAA/NMFS, Seattle, WA, 98115 66 p.

Stabeno, P.J., A.J. Hermann, N.A. Bond, and S.J. Bograd. 1995. Modeling the impact of climate variability on the advection of larval walleye pollock (*Theragra chalcogramma*) in the Gulf of Alaska, p. 719–727. *In* R.J. Beamish [ed.] Climate change and northern fish populations. Can. Spec. Publ. Fish. Aquat. Sci. 121.

Stabeno, P.J., and B.A. Megrey. 1991. Bio-physical models of the recruitment process in Gulf of Alaska walleye pollock. ICES CM 1991/Mini: 4.

Trenberth, K.E., and J.W. Hurrell. 1995. Decadal coupled atmosphere–ocean variations in the North Pacific Ocean. p. 15–24. *In* R.J. Beamish [ed.] Climate change and northern fish populations. Can. Spec. Publ. Fish. Aquat. Sci. 121.

Tyler, A.V. 1992. A context for recruitment correlations: why marine fisheries biologists should still look for them. Fish. Oceanogr. 1: 97–107.

Vastano, A.C., Schumacher, J.D., and L.S. INCZE. 1992. Observations and analysis of fishery processes: larval pollock at Shelikof Strait, Alaska. Fish. Oceanogr. 1: 20–31.

Walters, C.J., and J S. Collie. 1988. Is research on environmental factors useful for fisheries management? Can. J. Fish. Aquat. Sci. 45: 1848–1854.

Walters, C.J., and D. Ludwig. 1981. Effects of measurement errors on the assessment of stock–recruitment relationships. Can. J. Fish. Aquat. Sci. 38: 704–710.

Weber, J.E. 1983. Steady wind- and wave-induced currents in the open ocean. J. Phys. Oceanogr. 13: 524–530.

Wilkinson, L. 1990. SYSTAT: the system for statistics. SYSTAT Inc., Evanston, IL.

Witting, R. 1909. Zur Kenntnie den vom Winde Erzeigtem Oberflächenstromes. Ann. Hydrogr. Marit. Met. 73: 193 p.

Woodruff, S.D., R.J. Slutz, R.L. Jenne, and P.M. Steurer. 1987. Acomprehensive ocean–atmosphere data set. Bull. Am. Meteorol. Soc. 68: 1239–1250.

Wooster, W.S., and K.M. Bailey. 1989. Recruitment of marine fishes revisited, p. 153–159. *In* R.J. Beamish and G.A. McFarlane [ed.] Effects of ocean variability on recruitment and an evaluation of parameters used in stock assessment models. Can. Spec. Publ. Fish. Aquat. Sci. 108.

Wooster, W.S., and A.B. Hollowed. 1993. Decadal scale variations in the subarctic Pacific. I. Winter ocean conditions, p. 81–85. *In* R.J. Beamish [ed.] Climate change and northern fish populations. Can. Spec. Publ. Fish. Aquat. Sci. 121.

Yamashita, Y., and K.M. Bailey. 1989. A laboratory study of the bioenergetics of larval walleye pollock *Theragra chalcogramma*. Fish. Bull. U.S. 87: 525–536.

Yoklavich, M.M., and K.M. Bailey. 1989. Growth of larval and juvenile walleye pollock from Shelikof Strait, Gulf of Alaska, as determined from daily increments in otoliths, p. 241–251. *In* Proceedings of the International Symposium on the Biology and Management of Walleye Pollock, Nov. 14–16, 1988, Anchorage, AK, University of Alaska, Fairbanks, AK, Sea Grant Rep. 89-1.

Yoklavich, M.M., and K.M. Bailey. 1990. Hatching period, growth, and survival of young walleye pollock *Theragra chalcogramma* as determined from otolith analysis. Mar. Ecol Prog. Ser. 64: 13–22.

Appendix

Biological and environmental time series used

Egg and larval abundance indices

Egg and larval index values were calculated from data collected on Alaska Fisheries Science Center (AFSC) research cruises. Because cruises were conducted for a variety of purposes, station patterns and the number of stations sampled are not consistent among cruises (Dunn and Rugen 1989). To deal with these inconsistencies, index values were calculated for regions and times that are of historic importance to eggs and larvae and were sampled during most years. This was done to avoid extrapolating data to areas where no information was available.

Because mortality can have a large effect during an index interval (1 mo in this case), abundance values were standardized to the midpoint of the index period. The egg mortality rate (z = 0.186/day) was taken from Picquelle and Megrey (1991,1993) and larval mortality rate (z = 0.110/day) was taken from Yoklavich and Bailey (1990). The abundances for eggs and larvae in the index region were calculated by year using the Sette and Ahlstrom method (Richardson 1981). Abundances were calculated by determining the mean catch per 10 m^2 for each station and then weighting by the polygonal area of the station. The grand mean of all stations within the index region was then multiplied by the total area of the index region. Because early sampling for eggs was done to an insufficient depth to cover the whole vertical range of distribution, egg abundances were corrected for tows sampled to less than 250-m depth using the method described in Kendall and Kim (1989).

Age-0 and age-1 juvenile abundance

Indices of juvenile pollock were derived from data generated from shrimp and juvenile pollock surveys conducted in the Shelikof Strait region between 1975 and 1988. Spring and Bailey (1991) provide detailed descriptions of gear used to collect samples, geographic and temporal coverage of the surveys, gear-dependent mortality corrections, and all assumptions and data processing steps. Juvenile abundance values used in this study were corrected for gear and mortality effects. Data for age-0 pollock were available from 1975–88. Age-1 juvenile abundance indices were available from 1980–88. One year was subtracted from the age-1 indices to correctly associate a juvenile year-class abundance estimate with the correct calendar year of an environmental variable. Thus, the age-1 juvenile index ranged from 1979–87.

COADS SST, air temperature, and sea-level pressure

To account for the variability in fluid temperature affecting the ocean's mixed layer, we selected air temperature, sea- surface temperature (SST), and sea-level pressure from the monthly summaries trimmed group of the Comprehensive Ocean–Atmosphere Data Set (COADS) (Woodruff et al. 1987). These data are monthly averages of marine observations subjected to quality control to remove outliers and binned into 2° latitude by 2° longitude boxes. Our subset comprised monthly averaged data from three locations: 55°N,159°W; 57°N,155°W; and 59°N,153°W, over the period January 1962 through August 1989.

OSCURS Model Output

OSCURS (Ocean Surface Current Simulations) is an empirical ocean-wide model covering the subarctic Pacific region with a one-quarter mesh FNOC (U.S. Navy Fleet Numerical Oceanography Center) grid (about 90 km). Expanding the studies of Hubert and Laevastu (1965) and Larson and Laevastu (1972), the model combines long-term mean surface geostrophic currents (Ingraham and Miyahara 1988) with wind-generated surface mixed layer currents to form a resultant current vector at each grid point. Wind speed and direction are derived from daily FNOC sea-level pressure data to provide daily continuity from 1946 to 1990 following the methods of Larson (1975). Wind-induced ocean currents are then calculated from the empirical functions of the wind (Witting 1909; Huang 1979; Weber 1983). The Gulf of Alaska portion of the model was first tuned so that the model trajectories calculated for the period 21 September–31 December 1978 (about 3 mo) matched the trajectory of a satellite-tracked drifter (drogued at 20-m) from Reed (1980) for the same dates and starting locations (Ingraham and Miyahara 1989).

OSCURS data consist of an annual index (1946–91) of the tendency for surface currents to flow southwestward out of the Gulf of Alaska, as derived from model trajectory patterns. The numerical value of the index used in this study is defined as the number of trajectories out of six starting on 1 February that are along 55° N between 137° and 152° W and move west of 154° W by the end of April. This is the same index used by Ingraham et al. (1991) to show that large-scale interannual changes in surface currents in the Gulf of Alaska during February–April (1976–89) were connected to changes in water properties below sill depth in Shelikof Strait. The data for this study covers the period 1962–89, although the data are extended annually back to 1946.

Kodiak rainfall

Estimates of monthly precipitation for the years 1962–84 were obtained from U.S. Geological Survey 1991–92 report (Cayan et al. 1991). The station at the airport on Kodiak Island, Alaska, was used. These precipitation measures were updated for 1985–89 from the EarthInfo CD ROM (U.S. Geological Survey, Boulder, Colo.). The data series used in this study spans the period 1962–89.

NEPPI

Much of the variability in the physical environment of the Gulf of Alaska is due to large-scale atmospheric phenomena (Schumacher and Kendall 1991). The Aleutian Low dominates the variability of the atmospheric circulation over the Gulf of Alaska and plays a crucial role in the hydrological cycle (Neibauer 1988). NEPPI (Northeast Pacific Pressure Index (Emery and Hamilton 1985)), is a scalar index of the large-scale, sea-level pressure gradient across the northeast Pacific Ocean from (40°N,120°W) near Reno, Nev., to (50°N,170°W) south of Amukta Pass in the Aleutian Islands. NEPPI, which varies with the intensity of the atmospheric circulation and the track of storms over the Northeast Pacific Ocean, provides a measure of the strength, frequency, and location of the Aleutian Low. NEPPI correlates strongly with northeastern Pacific SST and adjusted coastal sea levels (Emery and Hamilton 1985), with coastal volume transport in the Shelikof Strait region (Roach and Schumacher 1991), and with Gulf of Alaska circulation (Ingraham et al. 1991). We computed a monthly mean NEPPI from twice daily, gridded (381 km at 60°N) sea-level pressures produced by the U.S. National Meteorological Center (NMC), and archived and averaged by the Department of Atmospheric Sciences, University of Washington, Seattle, Wash. Linear interpolation of gridded pressures yielded sea-level pressures at (40°N, 120°W) and (50°N,170°W).

Freshwater runoff

The Alaska Coastal Current has a strong buoyancy-driven component created by freshwater runoff originating from ice melt, precipitation, and coastal river discharge. Buoyancy-driven coastal flows are typical of subarctic regions where changes in salinity rather than temperature generate density gradients. Royer (1982) modeled the monthly freshwater runoff as a line source around the Gulf of Alaska by estimating the effects of insolation, local precipitation minus evaporation, and air temperature around the northern Gulf of Alaska. This series was modified (Parker 1989*a*, *b*) to include freshwater additions from Cook Inlet and sources along Shelikof Strait. Values used in this study are indices that represent deviations from the mean winter (November–April) estimates of integrated coastal freshwater discharge into the Gulf of Alaska as provided by Parker (1989*a*, *b*).

Wind mixing

Over-the-water winds transfer mechanical energy between the atmosphere and the ocean. The main effects of wind on early life stages of larvae are transport by wind-driven currents and deepening of the mixed layer. Wind stress, proportional to the square of the wind speed, causes the former, and wind mixing,proportional to the cube of the wind, causes the latter. The relative importance of these energy-transfer mechanisms depends on wind and current directions and the structure of the oceanic mixed layer (Klein 1980). To account for the maximum effect of wind on early life stages of pollock larvae, we picked mixing, the highest mode. A study of survival of first-feeding larvae with respect to wind mixing in Shelikof Strait (Bailey and Macklin 1993) supports this choice. Twice daily, winds were computed from the gridded NMC sea-level pressures by applying a geotriptic wind model tuned to the Shelikof Strait region (Macklin et al. 1993). The winds were cubed and monthly averages determined for the study period. We retained monthly averages to avoid the possibility of missing the influence of small-scale wind events, which are episodic in nature. One NMC time series was produced for wind mixing near the exit of Shelikof Strait (57°N,156°W) and one for the Shumagin Islands (55°N,160°W). An independent estimate of wind mixing was determined from COADS wind data. The COADS wind data comprised monthly averaged wind speed observations from three locations: 55°N,159°W; 57°N,155°W; and 59°N,153°W, over the period January 1962 through August 1989.

Size decline and older age at maturity of two chum salmon (*Oncorhynchus keta*) stocks in western North America, 1972–92

John H. Helle and Margaret S. Hoffman

Helle, J.H., and M.S. Hoffman. 1995. Size decline and older age at maturity of two chum salmon (*Oncorhynchus keta*) stocks in western North America, 1972–92, p. 245–260. *In* R.J. Beamish [ed.] Climate change and northern fish populations. Can. Spec. Publ. Fish Aquat. Sci. 121.

Abstract: Two populations of chum salmon (*Oncorhynchus keta*) in western North America were monitored for long-term changes in size and age at maturity. One population was from Fish Creek, a tributary of the Salmon River at the head of Portland Canal near Hyder, Alaska, where age and size samples were collected from 1972 to 1992. Chum salmon from this stream are known for their large size. The second population was from the Quilcene National Fish Hatchery on Hood Canal near Quilcene, Wash., where age and size samples were collected from 1973 to 1992. Both populations showed a significant decline in mean length at maturity of all age groups, starting about 1980. The carcass weight difference estimated from mean lengths of age-4 males for both populations between 1976 and 1991 was about 3 kg, a 46% reduction. The mean age at maturity for both populations increased as growth decreased. Decreased size could be caused by changes in oceanographic conditions, increased population density, or both. Regardless of the reason, the smaller size could result in lower survival of chum salmon stocks.

Résumé : On a surveillé les changements à long terme dans la taille et dans l'âge de maturité de deux populations de saumon kéta (*Oncorhynchus keta*) de l'ouest de l'Amérique du Nord. Une de ces populations provenait du ruisseau Fish, tributaire de la rivière Salmon à la tête du canal Portland près de Hyder, en Alaska, où les échantillons pertinents pour l'âge et pour la taille ont été recueillis de 1972 à 1992. Le saumon kéta de ce cours d'eau est reconnu pour sa taille imposante. La deuxième population venait de l'écloserie nationale de Quilcene, sur le canal Hood, près de Quilcene, dans l'État de Washington, où les échantillons utilisés pour déterminer l'âge et la taille ont été recueillis de 1973 à 1992. Les deux populations accusaient un important déclin de la longueur moyenne à maturité chez tous les groupes d'âges, à partir d'environ 1980. La différence de poids de carcasse estimée d'après la longueur moyenne des mâles d'âge 4 dans les deux populations entre 1976 et 1991 atteignait environ 3 kg. L'âge moyen à maturité des deux populations augmentait à mesure que la croissance diminuait. La diminution de taille pourrait être attribuable à des changements dans les conditions océanographiques, à la densité accrue de la population, ou aux deux. Quelle qu'en soit la raison, la réduction de la taille des stocks de saumons kéta risque d'abaisser leur taux de survie.

Introduction

Long-term records of salmon (*Oncorhynchus* spp.) size at maturity should reflect environmental changes experienced by the salmon at sea. For chum salmon (*O. keta*), both early marine growth and growth during their last season at sea have been shown to be correlated to sea-surface temperatures and other marine weather parameters (Helle 1979). Although changes in mean size at maturity of a stock probably reflect variation in physical oceanic conditions, density-dependent factors (e.g., intra- and inter-specific food competition) at sea may also influence size at maturity (Davidson and Vaughan 1941; Peterman 1978). If so, long-term records of size at maturity could also indicate the ocean's capacity to provide food and space for each salmon species. During the 1980s, chum salmon population levels may have approached the ocean's carrying capacity for this species (see Pearcy 1992).

Ideally, long-term observations of size and age at maturity should come from a single stock (spawning population). Records from mixed-stock fisheries may contain unquantifiable variability in size and age due to stock differences (Beacham 1984; Beacham and Murray 1985, 1987). Some size and age data on chum salmon from individual stocks are available, but most time series are of short duration (see Salo 1991, Tables 7–9). Helle (1979) provided data on growth and age and size at maturity of chum salmon from Olsen Creek in Prince William Sound, Alaska, over a 20-yr period (1959–78). Unfortunately, the Olsen Creek data base could not be continued through the 1980s, when many salmon populations in Alaska reached record levels.

J.H. Helle and M. S. Hoffman. Auke Bay Laboratory, Alaska Fisheries Science Center, National Marine Fisheries Service, National Oceanic and Atmospheric Administration, 11305 Glacier Highway, Juneau, AK 99801-8626, USA

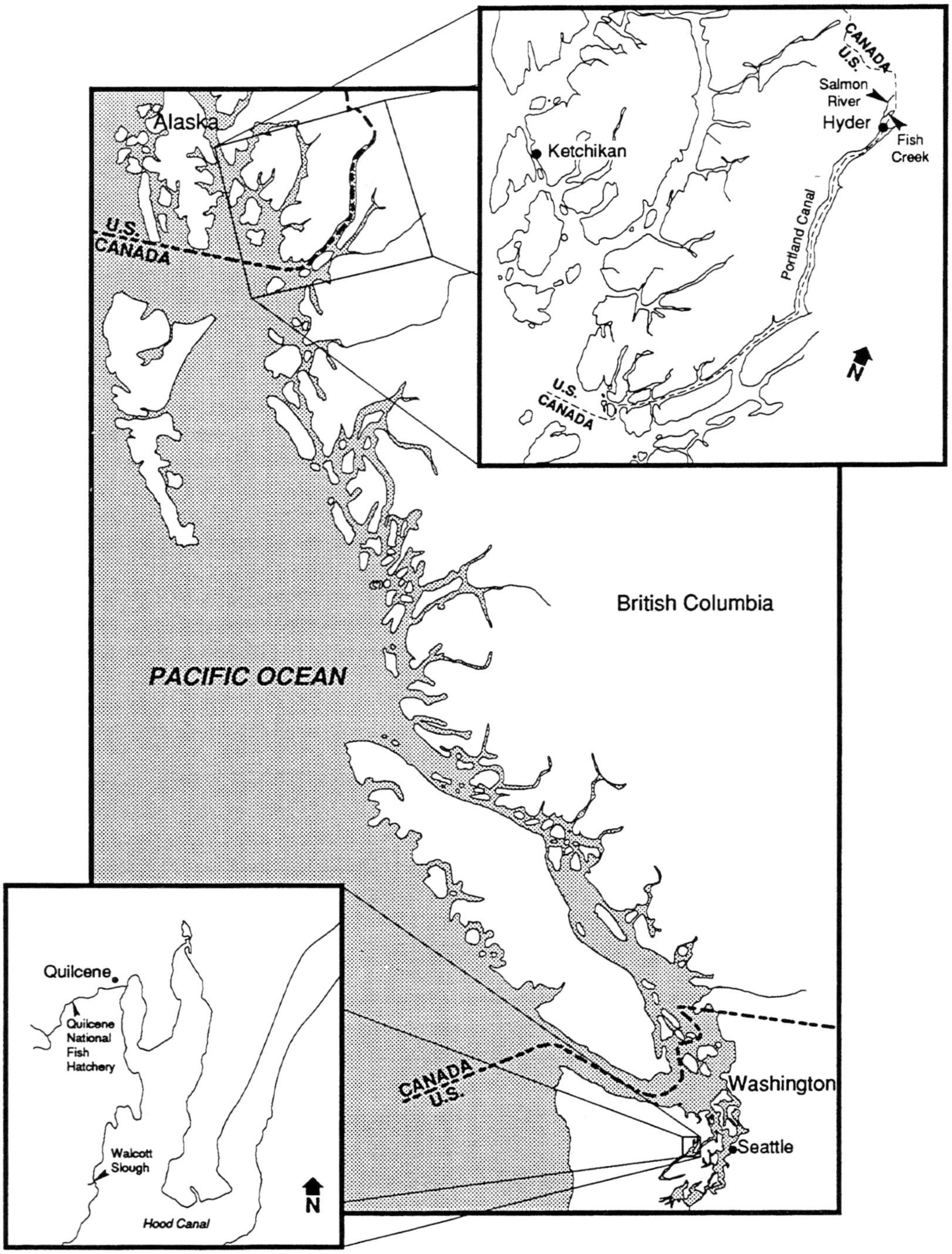

Fig. 1. Location of Fish Creek at the head of Portland Canal, near Hyder, Alaska, and Quilcene National Fish Hatchery and Walcott Slough, near Quilcene, Wash.

Long-term studies were initiated in 1972 by the Auke Bay Laboratory at Fish Creek near Hyder, Alaska, and at the Quilcene National Fish Hatchery in 1973 near Quilcene, Wash., to compare growth and age and size at maturity of two chum salmon stocks separated by distance and spawning time. In this paper, we report findings from 1972–1992 at Fish Creek and 1973–1992 at Quilcene National Fish Hatchery.

Study area

Fish Creek

Fish Creek is a tributary of the Salmon River, which flows into Portland Canal near Hyder, Alaska (Fig. 1). Portland Canal is a long (~100 km), narrow fjord that separates Alaska from British Columbia. Fish Creek, known for producing large chum salmon (Helle 1984), is accessible to salmon for about 5 km and flows into the east side of the Salmon River about 5 km upstream from Hyder. Stream locations where most chum salmon spawn are characterized by upwelling groundwater, which is cooler than the stream surface water in summer and warmer in winter (J.H. Helle, unpublished data).

Quilcene National Fish Hatchery

Quilcene National Fish Hatchery is about 5 km upstream from the mouth of the Quilcene River, which flows into Hood Canal in Puget Sound near Quilcene, Wash. (Fig. 1). This U.S. Fish and Wildlife Service hatchery produces chum salmon, chinook salmon (*O. tshawytscha*), and coho salmon (*O. kisutch*).

Methods

Fish Creek

The number of spawning chum salmon was estimated several times during each spawning season by observers who walked along the stream bank and counted the fish. Alaska Department of Fish and Game (ADFG) biologists from Ketchikan also counted the fish periodically during each season from fixed-wing aircraft flying <150 m above the stream.

Length measurements and scales were taken from spawned dead chum salmon. Peak spawning is usually in middle to late August. Carcasses were collected at approximately the same time (late August) each year, along the stream where most chum salmon spawn, between 1 km upstream and 2 km downstream of the Fish Creek bridge, which is about 6 km north of Hyder. Working downstream from the bridge, all carcasses were removed from the stream until sample size reached about 100 of each sex. In 1991, this sample size was unattainable even though the whole stream was searched for carcasses that year (Table 1).

Length was measured (to the nearest 1 mm) from the middle of the eye to the end of the hypural plate (MEHP) using a caliper rule with the fish flat on the ground. Common measurements such as fork length were not used because the caudal fin in spawned salmon (especially in females) is usually worn, and the jaws of mature male chum salmon are elongated. All lengths in all years were measured by one person (senior author) for continuity. Carcass weight was measured using a spring scale to the nearest 0.1 kg.

Table 1. Number of chum salmon by date and sex, sampled at Fish Creek and Quilcene National Fish Hatchery, 1972–92. Dash indicates no data.

	Fish Creek			Quilcene		
Year	Date	Males (no.)	Females (no.)	Date	Males (no.)	Females (no.)
1972	17 Aug.	90	100	—	—	—
1973	16 Aug.	108	88	10 Dec.	78	69
1974	22 Aug.	127	105	9 Dec.	110	124
1975	20 Aug.	100	99	10 Dec.	129	79
1976	24 Aug.	163	181	8 Dec.	150	109
1977	24 Aug.	99	103	9 Dec.	108	90
1978	29 Aug.	118	109	11 Dec.	118	103
1979	29 Aug.	131	133	10 Dec.	80	46
1980	7 Sept.	112	117	15, 17 Dec.	53	152
1981	6 Aug.	103	127	8, 10 Dec.	71	134
1982	24 Aug.	150	150	6 Dec.	106	98
1983	9, 10 Aug.	94	103	19 Dec.	132	154
1984	23 Aug.	147	176	10 Dec.	139	139
1985	27 Aug.	167	130	10 Dec.	124	82
1986	14, 15 Aug.	134	145	10 Dec.	168	196
1987	27 Aug.	159	153	10 Dec.	75	150
1988	29 Aug.	206	190	5 Dec.	152	154
1989	26 Aug.	152	160	4 Dec.	88	51
1990	19 Aug.	115	128	5 Dec.	184	89
1991	23–29 Aug.	47	45	16 Dec.	30	25
1992	25 Aug.	158	125	7 Dec.	99	36

Age was determined from carcass scales. Scales from carcasses of Pacific salmon that make long journeys up rivers to spawn are usually resorbed on the edge and marine annuli may be missing (Clutter and Whitesel 1956). Resorption is usually not excessive on scales from carcasses of chum salmon from coastal streams, and these scales can be used to age the fish accurately (Oakley 1966; Helle 1979). One scale was removed from the left side of each fish approximately four rows above the lateral line at a point below the posterior insertion of the dorsal fin. Each scale was cleaned between the fingers and immediately placed on a gummed card, 40 scales per card. A plastic impression was made of the gummed cards in the laboratory (Arnold 1951). Age was determined from a projected image of the scale magnified 80 diameters. All age determinations were made by one person (senior author) for continuity without reference to length or weight. Age determinations have been verified periodically by several biologists experienced in ageing chum salmon scales. Age designation followed the Gilbert and Rich (1927) method with the subscript for freshwater residence deleted because chum salmon juveniles migrate to sea shortly after emergence from stream gravel. Therefore, the parents of a 4-yr-old mature chum salmon in 1992 spawned in 1988. Statistics summarizing the length data by sex and age were obtained by a computer program (DiCostanzo 1965).

Quilcene National Fish Hatchery

Total number of fish that returned to the hatchery and were artificially spawned were counted by the hatchery staff each year. Before the late 1970s, chum fry from Quilcene National Fish Hatchery were released in Walcott Slough, a small spring-fed stream in the intertidal area near Brinnon, about 16 km south of the hatchery (Fig. 1). Adults returned to Walcott Slough, where they were captured in a trap and spawned; the gametes were reared at the hatchery. In the early 1980s, fry were released both at the hatchery into the Quilcene River and at Walcott Slough. Fry were not released at Walcott Slough after 1985 (L. Telles, Quilcene National Fish Hatchery, Quilcene, WA 98376, personal communication).

Length measurements and scales were taken from fish that were sacrificed for gametes for the hatchery. Fish were sampled and measured, and scales were collected as for Fish Creek. Again, all lengths (except in 1973) were measured by one person (senior author). In all years, samples were taken in early December, near the peak of the run. Weights taken from chum salmon at Quilcene hatchery are not directly comparable with those at Fish Creek, where samples were of carcasses (fish that had spawned and died naturally in the stream). Fish were not weighed every year.

Results

Abundance of spawners

Chum salmon returning to Fish Creek pass through commercial fisheries in southern southeast Alaska and northern British Columbia, as do many other chum salmon stocks in the region. Catches of chum salmon generally increased in Alaskan fisheries during the 1980s (Fig. 2). Most fisheries in this boundary area that capture chum salmon are targeting the more abundant pink salmon (*O. gorbuscha*) or sockeye salmon (*O. nerka*). The number of Fish Creek chum salmon caught in these fisheries is unknown.

Yearly peak counts of chum salmon that escaped through the fisheries and returned to Fish Creek were highly variable, ranging 13.5 fold: 1187 fish in 1991 and 16 080 in 1987. Average peak live counts of chum salmon in Fish Creek were about 4200 during 1972–84 and about 7800 during 1985–92. Because new chum salmon are entering the stream and spawned fish are dying throughout the season, determination of the total number of fish that spawned (escapement) requires specific studies. Helle (1970) demonstrated that the yearly peak count of pink salmon spawners at Olsen Creek in Prince William Sound can be multiplied by about three to estimate actual escapement.

Chum salmon that escaped fisheries and returned each year to Walcott Slough averaged 17 600 during 1973–78. The returns declined sharply after 1978 as the result of a planned reduction in chum salmon returning to Walcott Slough. During 1979–89, the average yearly return was 3200. The average yearly return direct to Quilcene National Fish Hatchery was 1300 chum salmon during 1983–91.

Age of spawners

Age composition of chum salmon returning to Fish Creek was highly variable during the study (Fig. 3). Age-4 and -5 fish were usually more abundant than age-3 fish. Of particular interest is the increase in age-5 fish after 1985 and the appearance of some age-6 fish in later years of the study (Fig. 3).

Age composition of the chum salmon returning to Quilcene National Fish Hatchery is shown in Fig. 4. Age-4 and -3 fish were usually more abundant than age-5 fish in this area. Although age composition was highly variable, the percentage of age-4 and -5 fish has increased since 1985 while the percentage of younger fish (age 3) has decreased (Fig. 4).

Size of spawners

Mean length of age-3, -4, and -5 males and females at Fish Creek has decreased over time (Fig. 5). The mean length of age-3 and -4 males and females at Quilcene National Fish Hatchery also shows a clear trend of decreasing size (Fig. 6).

The most abundant age group for both sexes during the study at both Fish Creek and Quilcene was usually age 4. Mean length of male and female age-4 fish and 95% confidence intervals for Fish Creek (Fig. 7) and Quilcene National Fish Hatchery (Fig. 8) trend downward. The lack of overlap of the 95% confidence intervals of the mean lengths (Figs. 7 and 8) between returns in the early 1970s and in the 1990s clearly illustrates the significance of the size declines.

Based on length–weight relationships for age-4 male chum salmon from Fish Creek and Quilcene National Fish Hatchery, the weight difference estimated from mean lengths of age-4 males at both Fish Creek and Quilcene National Fish Hatchery between 1976 (large mean length) and 1991 (smallest mean length) was about 3 kg, a 46% decline. The weights for Fish Creek are from spawned-out carcasses and the weights for Quilcene chum are from freshly killed, mature fish, so they are not directly comparable.

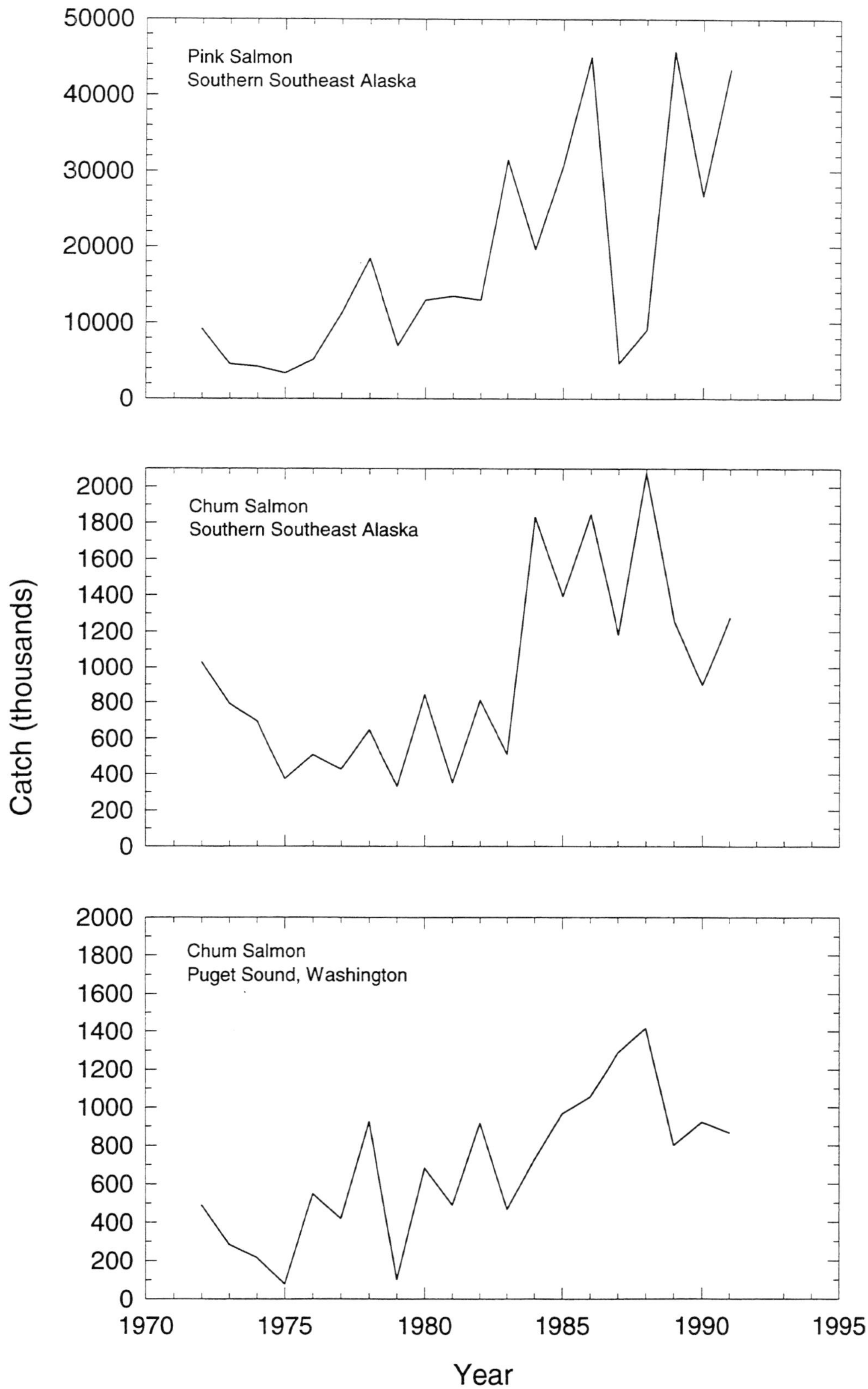

Fig. 2. Commercial catch of pink salmon and chum salmon in southern southeast Alaska and chum salmon in Puget Sound, Washington, 1972–91. Data are from Alaska Department of Fish and Game and Washington Department of Fisheries.

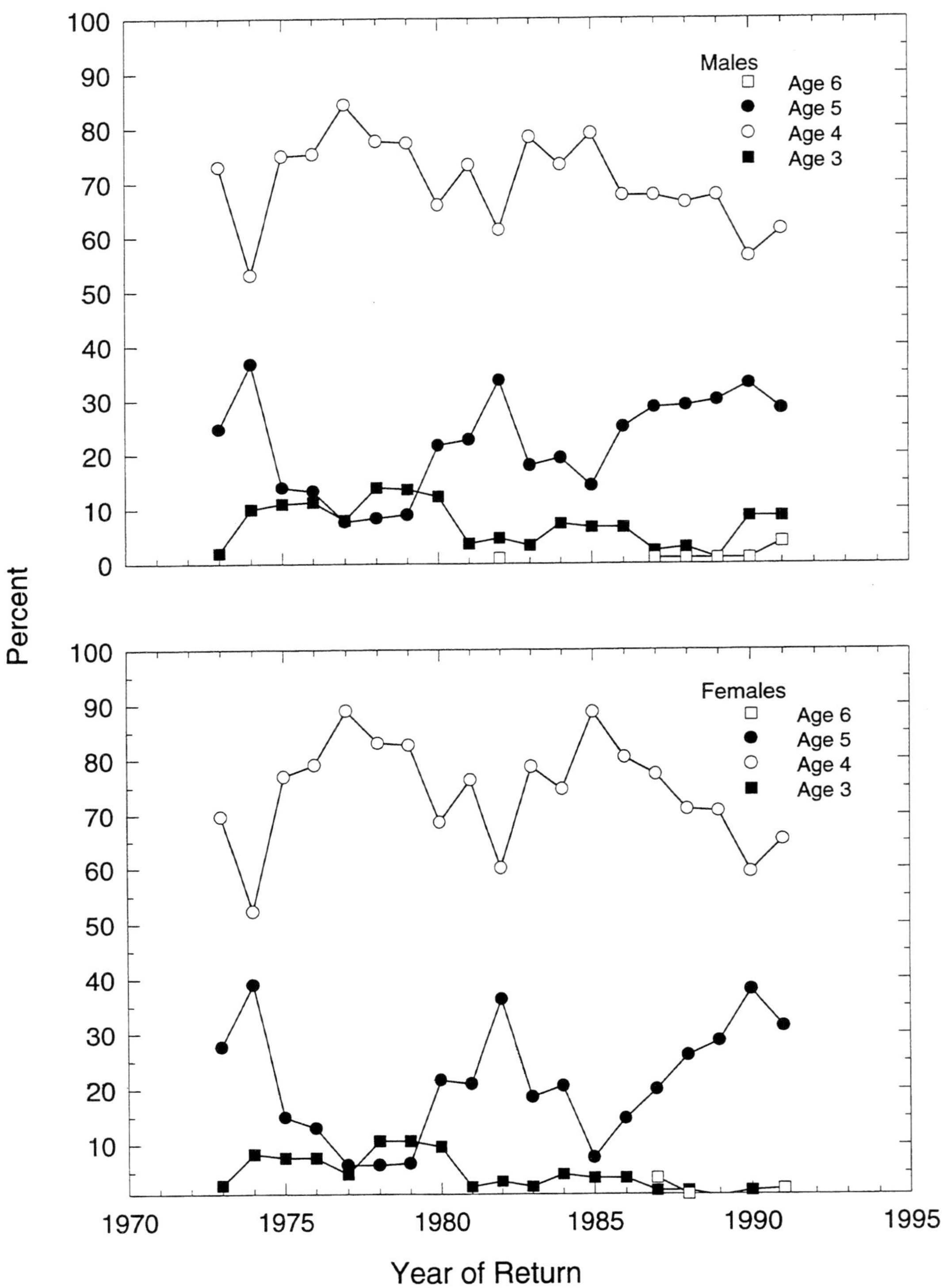

Fig. 3. Age composition of age-3, -4, -5, and -6 male and female chum salmon that returned to Fish Creek. Ages 3, 4, and 5 smoothed by moving average of three years, 1972–92.

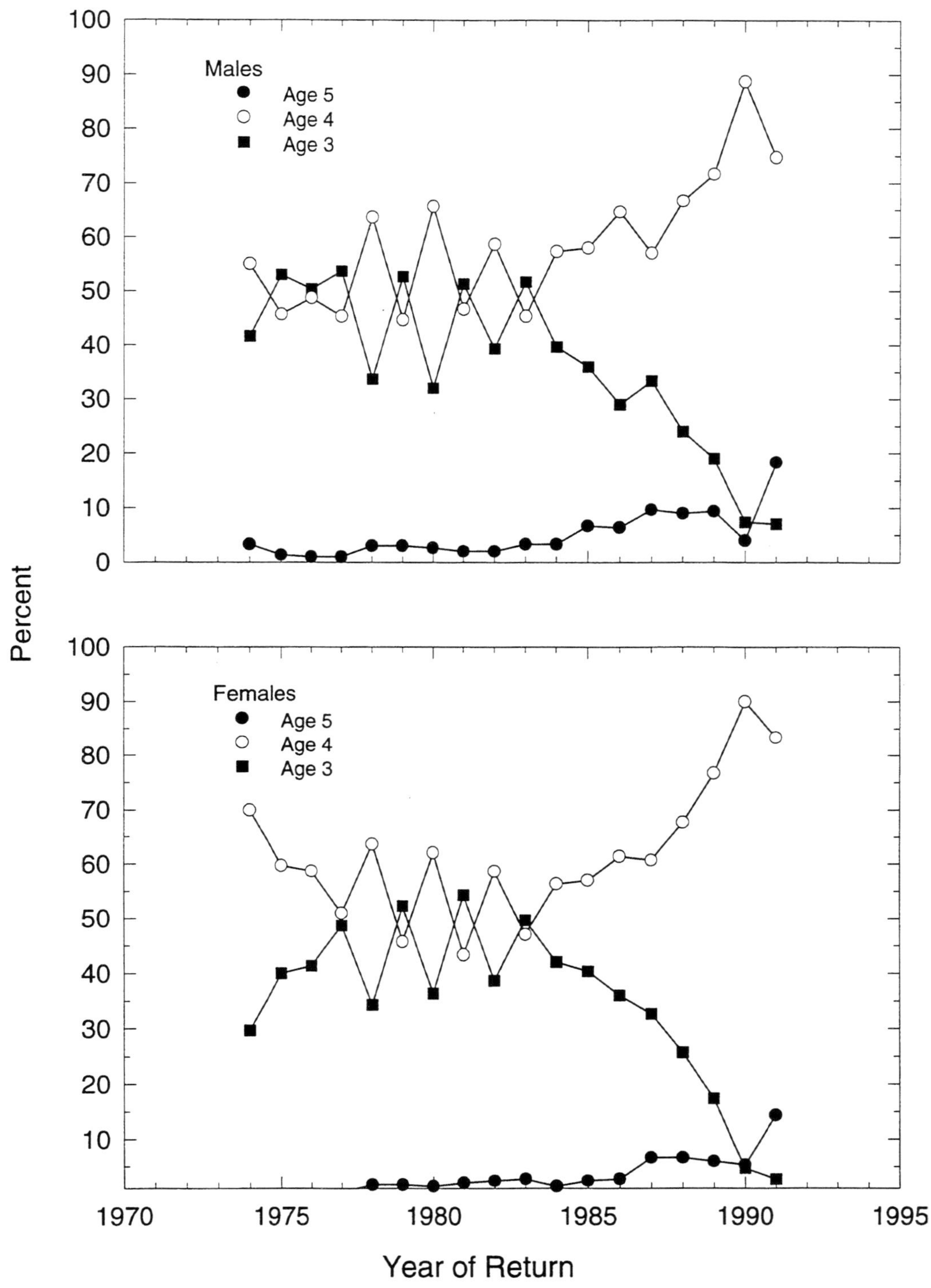

Fig. 4. Age composition of age-3, -4, and -5 male and female chum salmon that returned to Quilcene National Fish Hatchery. Data smoothed by moving average of three years, 1973–92.

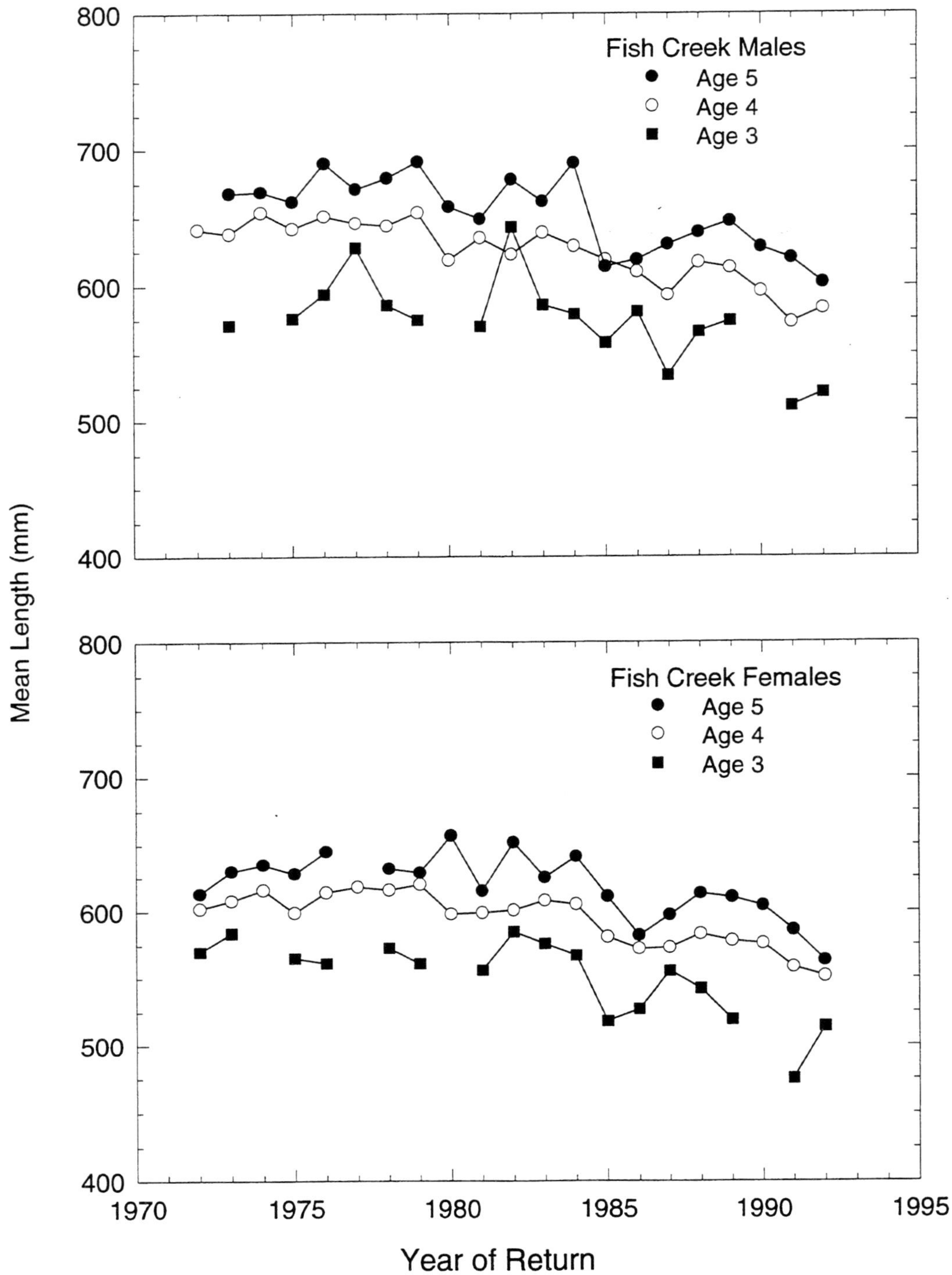

Fig. 5. Mean length of age-3, -4, and -5 male and female chum salmon spawners at Fish Creek, 1972–92.

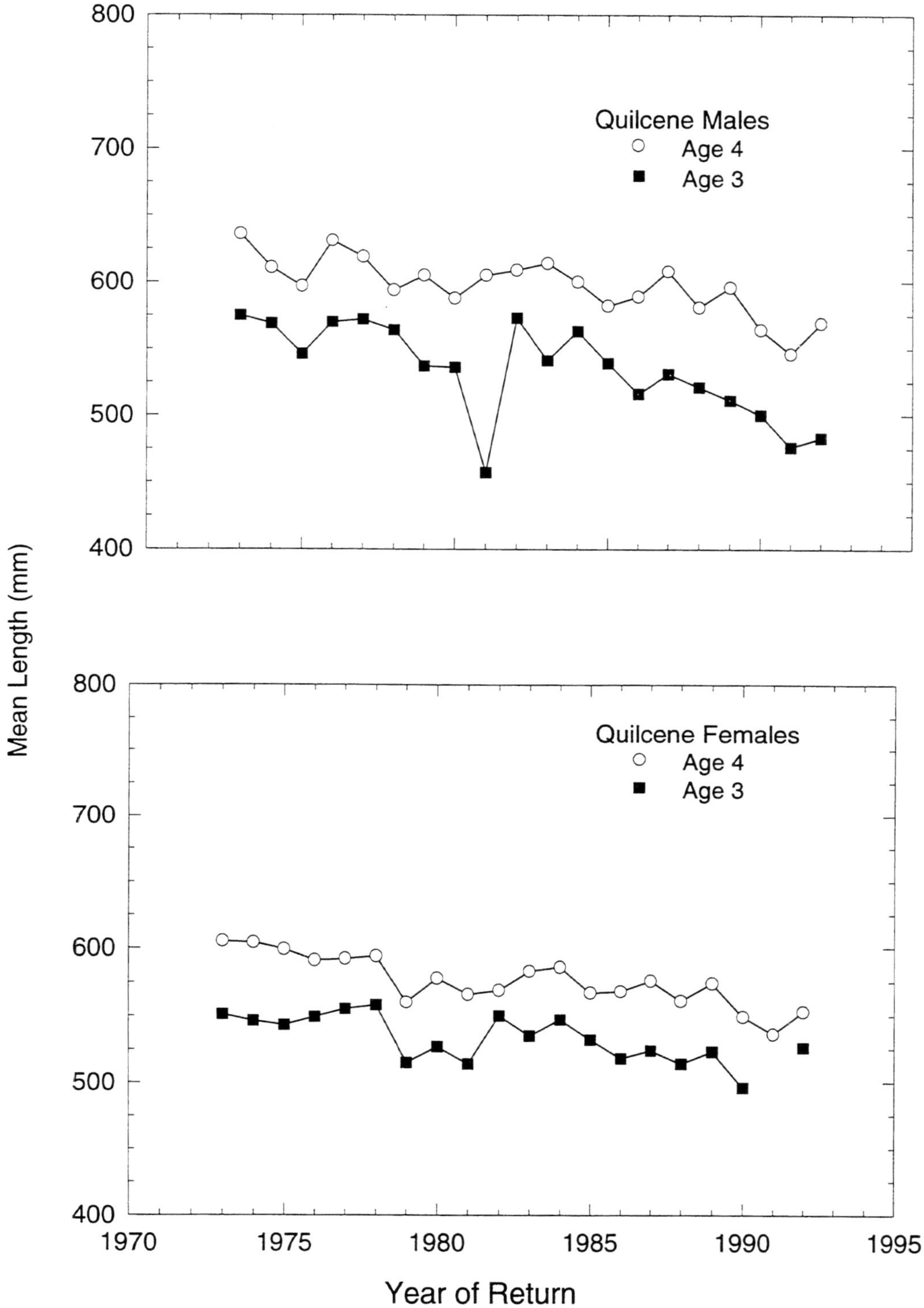

Fig. 6. Mean length of age-3 and -4 male and female chum salmon spawners at Quilcene National Fish Hatchery, 1973–92.

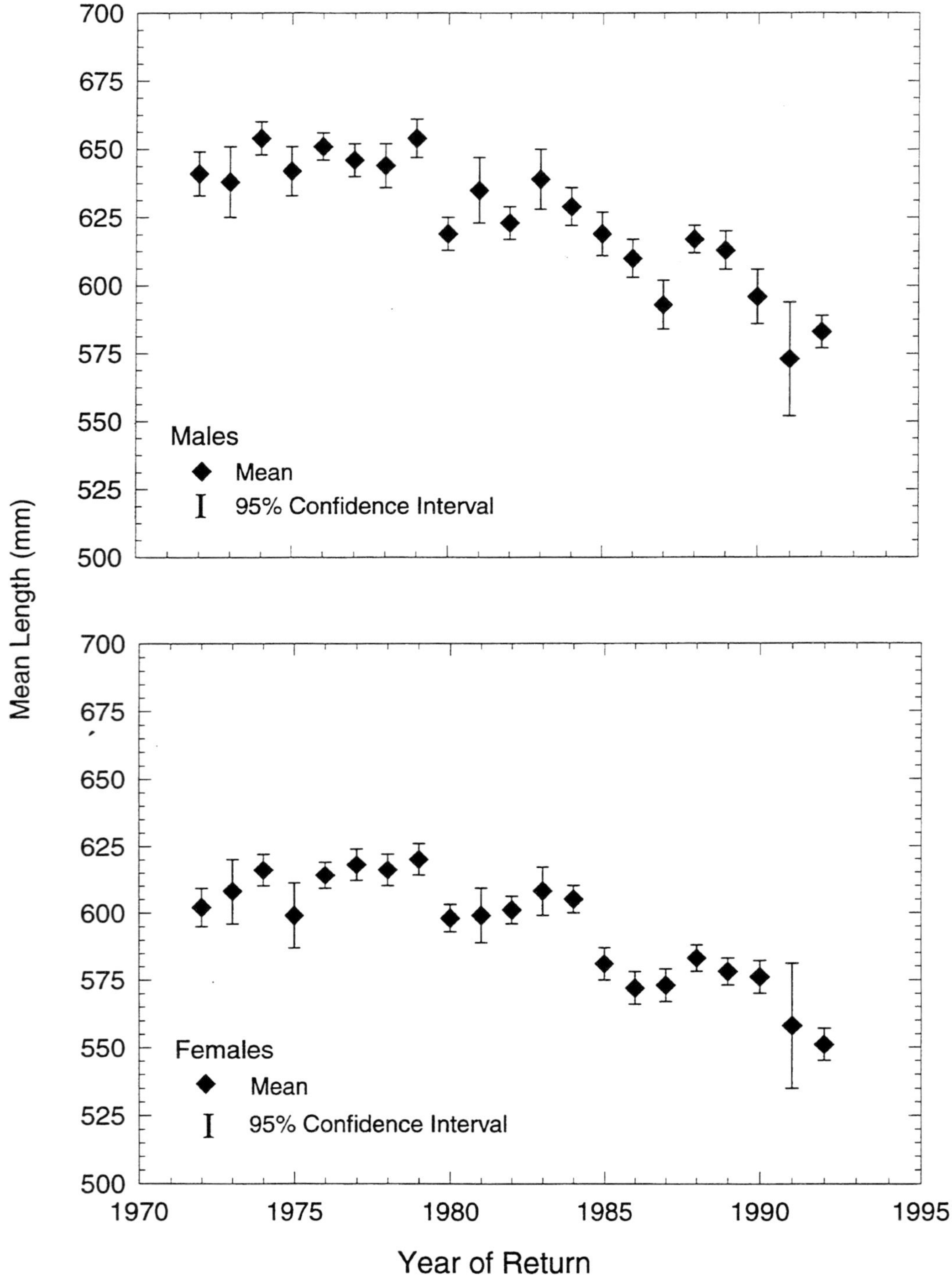

Fig. 7. Mean length and 95% confidence interval of age-4 male and female chum salmon spawners at Fish Creek, 1972–92.

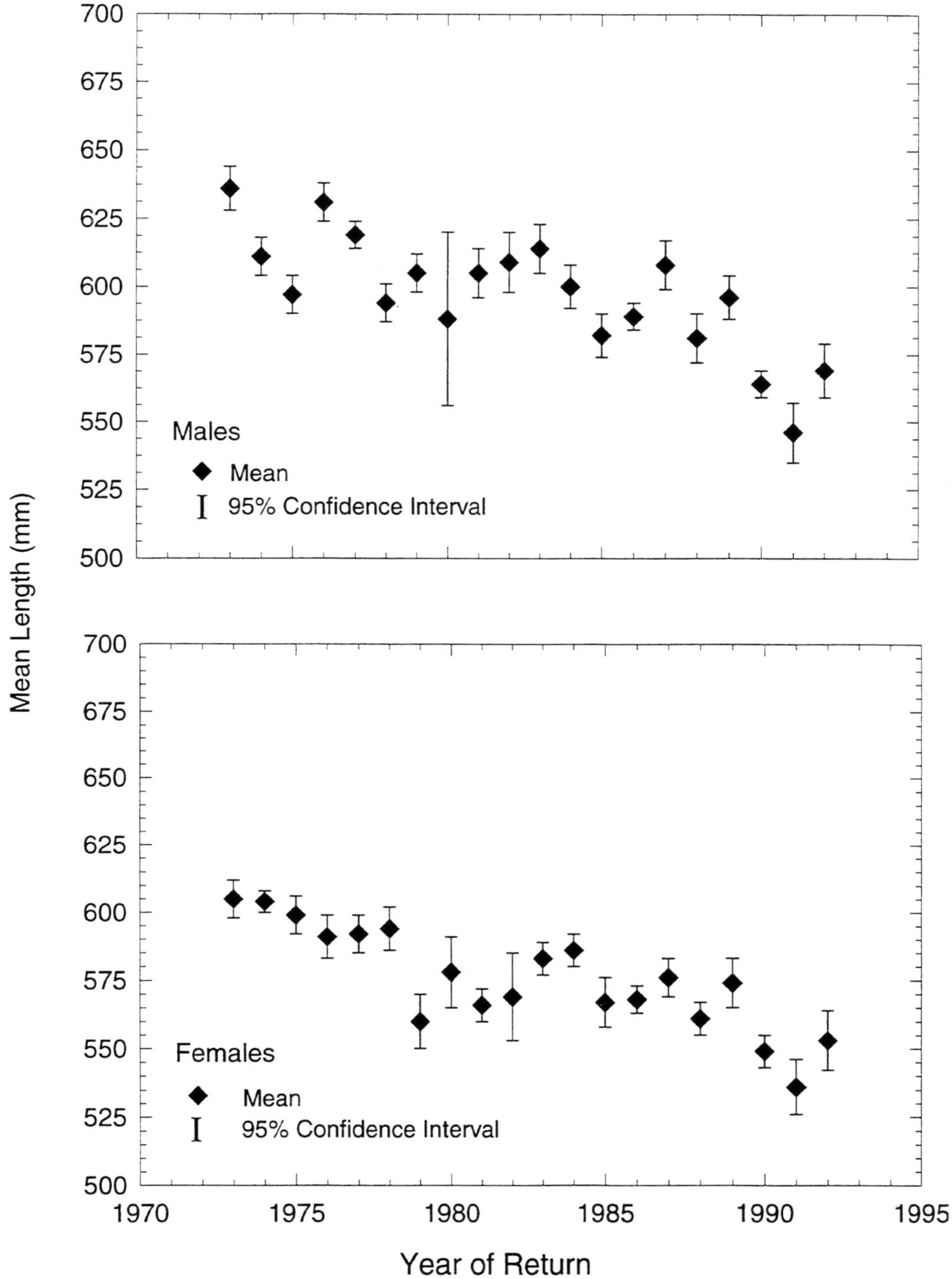

Fig. 8. Mean length and 95% confidence interval of age-4 male and female chum salmon spawners at Quilcene National Fish Hatchery, 1973–92.

Discussion

Abundance of spawners

The exploitation rate of chum salmon returning to Fish Creek in fisheries of southern southeast Alaska and northern British Columbia is not presently known. In the mid-1980s, however, the ADFG initiated a study to determine exploitation rates by marking juvenile chum in the snout with microwire tags and recovering adults in commercial fisheries, and as they entered the stream through a weir. Results from this study should be available in 1994 (J. Koerner, ADFG, 2030 Sea Level Drive, Ketchikan, AK 99901, personal communication). Because chum salmon are also captured incidentally in fisheries in northern British Columbia as well as in southeast Alaska for the more abundant pink salmon (Fig. 2), the escapement index (peak count) of chum salmon in Fish Creek may not reflect the total number of fish returning to the stream in any one year.

Chum salmon returning to Quilcene National Fish Hatchery probably pass through fisheries in southern British Columbia as well as Puget Sound and Hood Canal in Washington. The chum salmon fisheries in Washington, like those in southeast Alaska, showed a large increase in catches during the 1980s (Fig. 2). Returns to Quilcene National Fish Hatchery would probably also have shown large increases in numbers in the 1980s except for the planned reduction at Walcott Slough. Management agencies asked for the reduction because the late timing of the returns coincided with runs of several wild stocks and resulted in harvest conflicts (R. Boomer, U.S. Fish and Wildlife Service, 2625 Parkmont Lane, Olympia, WA 98507, personal communication). Therefore, Quilcene National Fish Hatchery stopped planting fry at Walcott Slough in 1985 and started releasing fry from the hatchery directly into the Quilcene River in 1981. Artificial propagation of chum salmon started at Walcott Slough in 1911.

Age of spawners

Age composition of maturing chum salmon returning to a stream changes during the season. Older fish return earlier than younger fish (Helle 1979). Therefore, fish need to be sampled throughout the entire run to obtain a reliable estimate of the age composition for the season. Our major objective for this study was to compare size changes over a 20-yr period, so we decided it was important to sample approximately the same time each year, close to the peak of the run (Table 1). Age composition changed during our study, and because we sampled at approximately the same time in most years, the changes should represent interannual changes. Mean age at maturity increased at both Fish Creek and Quilcene National Fish Hatchery (Fig. 9). Furthermore, because size at maturity declines, less growth could mean slower maturity. Helle (1979) indicated that mean age at maturity of chum salmon was negatively related to growth in the second year in the ocean.

Size decline in other species of salmon

Changes in average size and age of Pacific salmon have been reported for five species in North America (Ricker 1981). Ricker (1980a) described changes in size and age of chum

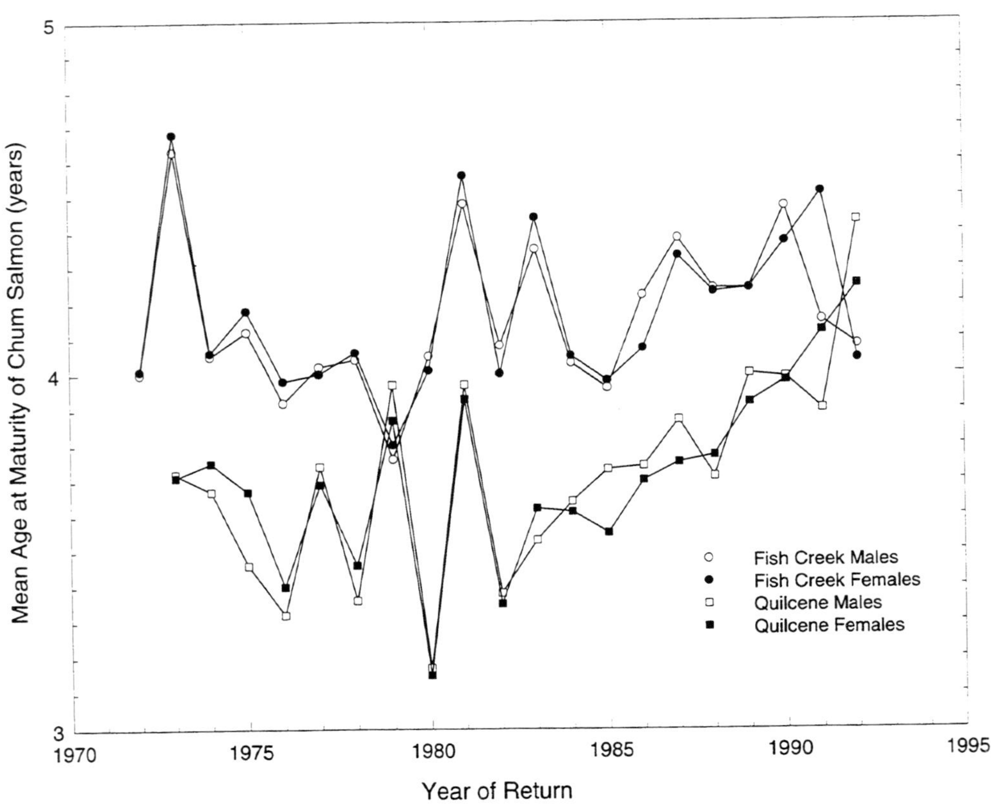

Fig. 9. Mean age at maturity of male and female chum salmon spawners at Fish Creek and Quilcene National Fish Hatchery.

salmon and decreases in size and age of chinook salmon (Ricker 1980b). Ricker et al. (1978) described size decreases in pink salmon; and Ricker and Wickett (1980), in coho salmon (*O. kisutch*). Ricker (1981) discussed environmental and possible genetic causes for these declines. Fagen (1988) discussed long-term declines in the size of sport-caught chinook salmon in Alaska. Both Ricker (1980b) and Fagen (1988) discussed the potential for gear-caused decreases in size at maturity of chinook salmon. This species of salmon is captured as immature as well as maturing fish in troll fisheries and some sport fisheries, so larger fish could be selected against. However, Bigler (B. Bigler, Wards Cove Packing Co., P.O. Box C-5030, Seattle, WA 98105, personal communication) has documented a size decline in chinook salmon caught in western Alaska (Yukon River, Kuskokwim River, and Nushagak Bay) from the mid-1960s through 1992. Commercial trolling is not allowed in these areas, and these fish were maturing when captured; however, the fish were captured mainly by gill nets, which can be selective for size. Mean size of sport-caught sockeye salmon in the Russian River also decreased between 1975 and 1985 (Nelson et al. 1986).

Most studies mentioned above describe size changes of salmon captured in mixed-stock fisheries, and in most cases the ages of the fish were not determined. Regardless, most of the studies agree that the salmon size is declining. The most definitive data on size and age changes in salmon populations would be from long-term studies on individual fish stocks.

Causes of size decline

Our study was relatively long term (21- and 20-yr span) and based on two individual stocks of chum salmon, about 1100 km apart. Chum salmon at Fish Creek are wild and Quilcene Hatchery chum are artificially propagated (since 1911). The similar declines in size at maturity shown by both stocks indicate that the declines result from a common cause in the marine environment and not from fish-culture practices or stock differences due to geographic distance (Figs. 5–8).

Recent reports of declining size at maturity and increasing age at maturity of Asian chum salmon stocks also indicate that some factor or factors in the marine environment are causing similar size and age changes in both Asian and North American chum salmon stocks. Kaeriyama (1989, 1995), Kaeriyama and Urawa (1992), Ishida et al. (1993), and Hayashizaki and Hitoshi (1995) discuss the recent size and age changes relative to the very large releases of juvenile chum salmon from artificial propagation facilities in Japan. These authors show that chum salmon growth decreased in the ocean and suggest that decreases in size and increases in age are due to population density-dependent factors in the ocean. Very large releases of chum salmon (and other Pacific salmon) have been made in the last two decades from artificial propagation facilities in Russia and North America as well as Japan (see Pearcy 1992). The decline in size at maturity of chum salmon from Fish Creek and Quilcene National Fish Hatchery coincides with the dramatic increase in numbers of Asian chum salmon (Fig. 10).

Fig. 10. Weight of world (North American and Asian) and North American catch of chum salmon, and mean length of age-4 chum salmon spawners from Fish Creek and Quilcene National Fish Hatchery. Data for 1978, 1981, and 1990 are from FAO (1979, 1983, 1992).

Peterman (1978, 1985) and Peterman and Wong (1984) provided evidence for and discussed the extent of density-dependent factors on growth and survival of salmon at sea. Peterman (1991) also recommended that density-dependent factors in the ocean be included in the design of large-scale artificial propagation facilities. Ishida et al. (1993) discussed the possibility that the increasing abundance of Japanese chum salmon has affected the survival rate of Russian chum salmon in the ocean through reduced growth rates. The evidence that density-dependent factors are responsible for decreases in size and increases in age of chum salmon and other Pacific salmon is convincing; however, changes in climatic and oceanographic conditions also have to be considered.

Climatic and oceanographic changes in the North Pacific Ocean since the early 1970s are well documented. McLain and Favorite (1976) documented effects on marine resources of the anomalously cold winters of 1971–75, followed by the coastal warming of the northeastern Pacific Ocean in 1976–83 (McLain 1984). Many authors have documented the significant climatic and oceanographic changes that occurred during the late 1970s and the 1980s (e.g., Favorite and Ingraham (1976), Tabata (1984, 1989), Freeland (1990), and Pearcy (1992)). These warm conditions were highly favourable for salmon survival, and enormous numbers of salmon were harvested in the 1980s in the North Pacific Ocean, especially in Alaska. Catches of sockeye in Bristol Bay, Alaska, and pink salmon in southern southeast Alaska exceeded the previous record catches of the 1930s (Eggers 1989; Rigby et al. 1991; Pearcy 1992). Mean catch of pink and sockeye salmon in southern southeast Alaska during the 1980s was similar to catches in the 1930s but exceeded the mean catches during the 1920s and 1940s–1970s (Fig. 11). However, mean chum salmon catch in southern southeast Alaska in the 1980s improved over catches in the 1960s and 1970s, but was lower than catches from the 1920s to 1950s (Fig. 11).

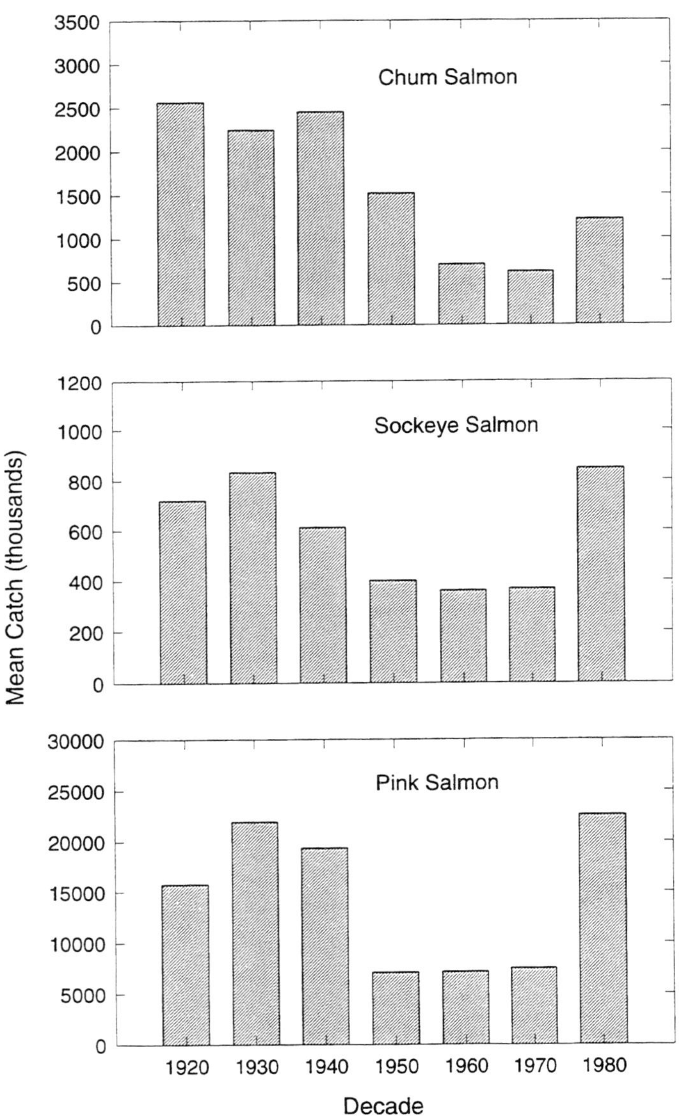

Fig. 11. Commercial harvest of chum, sockeye, and pink salmon in southern southeast Alaska by decade, 1920–80. Data are from Alaska Department of Fish and Game.

Interaction of North American and Asian chum salmon

Distribution of both North American and Asian chum salmon on the high seas has been documented by Manzer et al. (1965), Neave et al. (1976), Myers et al. (1990), and Salo (1991). North American and Asian chum salmon ranges in the ocean overlap considerably. Competition for food resources is mainly intraspecific because chum salmon utilize food items on a different trophic branch from the food consumed by the other Pacific salmon (Welch and Parsons 1993). Environmental conditions and density-dependent factors could affect North American and Asian chum salmon stocks similarly when they are present in the same area.

The arguments that environmental change or density-dependent factors are responsible for the decline in size at maturity and increase in age at maturity are convincing. Either factor alone or an interaction of the two could affect size and age at maturity. Also, intraspecific competition for food on the high seas between Asian and North American stocks of chum salmon is possible. Regardless, the trend for smaller size at maturity of chum salmon on both sides of the North Pacific Ocean could be a harbinger of a major change in chum salmon abundance. Helle (1989) reported that at Olsen Creek, Alaska, chum salmon survival to maturity is strongly related to mean size (length) at maturity of parents. If this relationship applies to most wild chum salmon stocks and artificially cultured stocks in both North America and Asia and the mean size at maturity does not start to increase significantly in the next few years, survival rates could decline.

Acknowledgements

Many people assisted in collecting samples and counting spawners, and we appreciate their efforts. We especially acknowledge those who assisted several years or more: Donald Siedelman (retired) and Steven Hoffman, ADFG, Sport Fish Division, Ketchikan; Dennis Blankenbeckler (retired), Donald House, Jerold Koerner, and Kathleen Wendt, ADFG, Commercial Fisheries Division, Ketchikan; James Dangel, ADFG, Commercial Fisheries Division, Sitka; Glen Contreras, U.S. Forest Service, Ketchikan; Monroe Dodson, Russell Ferg, Larry Telles, Ron Wong, and Ken Sexton, Quilcene National Fish Hatchery; Dan Fender, Don Cole (deceased),

Mike Paiya, and Charles Hamstreet, U.S. Fish and Wildlife Service, Olympia, WA; Steven Schroder, University of Washington (now with Washington Department of Fisheries); Howard Sears (dec.) and Jerome Pella (who also reviewed the manuscript), Auke Bay Laboratory. Data on chum salmon returns to the U.S. Fish and Wildlife Service's Quilcene National Fish Hatchery facilities were provided by Larry Telles.

References

Arnold, E.L., Jr. 1951. An impression method for preparing fish scales for age and growth analysis. Prog. Fish-Cult. 13(1): 11–16.

Beacham, T.D. 1984. Age and morphology of chum salmon in southern British columbia. Trans. Am. Fish. Soc. 113: 727–736.

Beacham, T.D., and C.B. Murray. 1985. Variation in length and body depth of pink salmon (*Oncorhynchus gorbuscha*) and chum salmon (*O. keta*) in southern British Columbia. Can. J. Fish. Aquat. Sci. 42: 312–319.

Beacham, T.D., and C.B. Murray. 1987. Adaptive variation in body size, age, morphology, egg size, and developmental biology of chum salmon (*Oncorhynchus keta*) in British Columbia. Can. J. Fish. Aquat. Sci. 44: 244–261.

Clutter, R.I., and L.E. Whitesel. 1956. Collection and interpretation of sockeye salmon scales. Int. Pac. Salmon Fish. Comm. Bull. 9: 159 p.

Davidson, F.A., and E. Vaughan. 1941. Relation of population size to marine growth and time of spawning migration in pink salmon (*Oncorhynchus gorbuscha*) of southeastern Alaska. J. Mar. Res. 4: 231–246.

DiCostanzo, C. 1965. Length–frequency distributions of red salmon by sex, age group and years of ocean life. Fortran IV Program FRD 297. University of Washington, College of Fisheries, Fisheries Research Institute, Seattle, WA.

Eggers, D.M. 1989. Alaska commercial salmon catches, 1878–1988. Alaska Dep. Fish Game Reg. Inf. Rep. 5J89-04: 69 p.

Fagen, R. 1988. Long-term trends in maximum size of sport-caught chinook salmon (*Oncorhynchus tshawytscha*): a data-analytic approach to weights of first-prize fish in four southeastern Alaska salmon derbies. Fish. Res. 6: 125–134.

Favorite, F., and W.J. Ingraham, Jr. 1976. Sunspot activity and oceanic conditions in the northern North Pacific Ocean. J. Oceanogr. Soc. Japan 32: 107–115.

Food and Agricultural Organization of the United Nations (FAO). 1979. Yearbook of fishery statistics: catches and landings, 1978. Vol. 46. Rome.

FAO. 1983. Yearbook of fishery statistics: catches and landings, 1981. Vol. 52. Rome.

FAO. 1992. Yearbook of fishery statistics: catches and landings, 1990. Vol. 70. Rome.

Freeland, H.J. 1990. Sea surface temperatures along the coast of British Columbia: regional evidence of a warming trend. Can. J. Fish. Aquat. Sci. 47: 346–350.

Gilbert, C.H., and W.H. Rich. 1927. Second experiment in tagging salmon in the Alaska Peninsula fisheries reservation, summer of 1923. Fish. Bull. U.S. 42: 27–75.

Hayashizaki, K.-i., and I. Hitoshi. 1995. Size decrease of chum salmon, *Oncorhynchus keta*, in Tohoku districts, Japan. *In* B.E. Riddell and C.C. Woods [eds.] Proceedings of International Symposium on Biological Interactions of Enhanced and Wild Salmonids. Can. J. Fish. Aquat. Sci. 52(Suppl.). (In press)

Helle, J.H. 1970. Biological characteristics of intertidal and fresh-water spawning pink salmon at Olsen Creek, Prince William Sound, Alaska, 1962–63. U.S. Fish Wildl. Serv. Spec. Sci. Rep. Fish. 602. 19 p.

Helle, J.H. 1979. Influence of marine environment on age and size at maturity, growth, and abundance of chum salmon, *Oncorhynchus keta* (Walbaum), from Olsen Creek, Prince William Sound, Alaska. Ph.D. thesis, Oregon State University, Corvallis, OR. 118 p.

Helle, J.H. 1984. Age and size at maturity of some populations of chum salmon in North America, p. 126–143. *In* P.A. Moiseev [ed.] Proceedings of the Pacific Salmon Biology Conference, 3–13 Oct. 1978, Yuzhno-Sakhalinsk, USSR. Pacific Sciences Institute of Fisheries and Oceanography (TINRO), 4 Shevchenko Alley, 690600, Vladivostok, USSR.

Helle, J.H. 1989. Relation between size-at-maturity and survival of progeny in chum salmon, *Oncorhynchus keta* (Walbaum). J. Fish Biol. 35(Suppl. A): 99–107.

Ishida, Y., S. Ito, M. Kaeriyama, S. McKinnell, and K. Nagasawa. 1993. Recent changes in age and size of chum salmon (*Oncorhynchus keta*) in the North Pacific Ocean and possible causes. Can. J. Fish. Aquat. Sci. 50: 290–295.

Kaeriyama, M. 1989. Aspects of salmon ranching in Japan. Physiol. Ecol. Jpn. Spec. Vol. 1: 625–638.

Kaeriyama, M. 1995. Production trends of salmon enhancement in Japan. *In* B.E. Riddell and C.C. Woods [eds.] Proceedings of International Symposium on Biological Interactions of Enhanced and Wild Salmonids. Can. J. Fish. Aquat. Sci. 52(Suppl.). (In press)

Kaeriyama, M., and S. Urawa. 1992. Future research by the Hokkaido Salmon Hatchery for the proper maintenance of Japanese salmonid stocks, p. 57–62. *In* Y. Ishida, K. Nagasawa, D.W. Welch, K.S. Myers, and A.P. Shershnev [eds.] Proceedings of the International Workshop on Future Salmon Research in the North Pacific Ocean, Shimizu, November 1991. National Research Institute of Far Seas Fisheries, Shimizu, Japan.

Manzer, J.I., I. Ishida, A.E. Peterson, and M.G. Hanavan. 1965. Salmon of the North Pacific Ocean. V. Offshore distribution of salmon. Int. North Pac. Fish. Comm. Bull. 15.

McLain, D.R. 1984. Coastal ocean warming in the northeast Pacific, 1976–83, p. 61–86. *In* W.G. Pearcy [ed.] The Influence of Ocean Conditions on the Production of Salmonids in the North Pacific. Proceedings of a Workshop. ORESU-W-83-001. Oregon State University Sea Grant Program, Corvallis, OR.

McLain, D.R., and F. Favorite. 1976. Anomalously cold winters in the southeastern Bering Sea, 1971–75. Mar. Sci. Comm. 2: 299–334.

Myers, K.W., R.V. Walker, S. Fowler, and M.L. Dahlberg. 1990. Known ocean ranges of stocks of Pacific salmon and steelhead as shown by tagging experiments, 1956–1989. INPFC Doc. FRI-UW-9009. Fisheries Research Institute, University of Washington, Seattle, WA.

Neave, F., T. Yonemori, and R.G. Bakkala. 1976. Distribution and origin of chum salmon in offshore waters of the North Pacific Ocean. Int. North Pac. Fish. Comm. Bull. 35. 79 p.

Nelson, D.C., D.E. Athons, and J.A. Carlon. 1986. Russian River sockeye salmon, p. 90–149. *In* Cook Inlet salmon studies. Federal Aid in Fish Restoration and Anadromous Fish Studies. Vol. 27. Alaska Dep. Fish and Game, Division of Commercial Fisheries, P.O. Box 25526, Juneau, AK 99802.

Oakley, A.L. 1966. A summary of information concerning chum salmon in Tillamook Bay. Res. Briefs Fish Comm. Oreg. 12(1): 1–17.

Pearcy, W.G. 1992. Ocean ecology of North Pacific salmonids. Washington Sea Grant Program, University of Washington Press, Seattle, WA. 179 p.

Peterman, R.M. 1978. Testing for density-dependent marine survival in Pacific salmonids. Can. J. Fish. Aquat. Sci. 35: 1434–1450.

Peterman, R.M. 1985. Patterns of interannual variation in age at maturity of sockeye salmon (*Oncorhynchus nerka*) in Alaska and British Columbia. Can. J. Fish. Aquat. Sci. 42: 1595–1607.

Peterman, R.M. 1991. Density-dependent marine processes in North Pacific salmonids: lessons for experimental design of large-scale manipulations of fish stocks. ICES Mar. Sci. Symp. 192: 69–77.

Peterman, R.M., and F.Y.C. Wong. 1984. Cross correlations between reconstructed ocean abundances of Bristol Bay and British Columbia sockeye salmon (*Oncorhynchus nerka*). Can. J. Fish. Aquat. Sci. 41: 1814–1824.

Ricker, W.E. 1980a. Changes in the age and size of chum salmon (*Oncorhynchus keta*). Can. Tech. Rep. Fish. Aquat. Sci. 930.

Ricker, W.E. 1980b. Causes of the decrease in age and size of chinook salmon (*Oncorhynchus tshawytscha*). Can. Tech. Rep. Fish. Aquat. Sci. 944.

Ricker, W.E. 1981. Changes in the average size and average age of Pacific salmon. Can. J. Fish. Aquat. Sci. 38: 1636–1656.

Ricker, W.E. 1982. Size and age of British Columbia sockeye salmon (*Oncorhynchus nerka*) in relation to environmental factors and the fishery. Can. Tech. Rep. Fish. Aquat. Sci. 1115.

Ricker, W.E., H.T. Bilton, and K.V. Aro. 1978. Causes of the decrease in size of pink salmon (*Oncorhynchus gorbuscha*). Can. Tech. Rep. Fish. Aquat. Sci. 820.

Ricker, W.E., and W.P. Wickett. 1980. Causes of the decrease in size of coho salmon (*Oncorhynchus kisutch*). Can. Tech. Rep. Fish. Aquat. Sci. 971.

Rigby, P., J. McConnaughey, and H. Savikko [eds.]. 1991. Alaska commercial salmon catches, 1878–1991. Regional Inf. Rep. 5J91-16. Alaska Dep. Fish and Game, Division of Commercial Fisheries, P.O. Box 25526, Juneau, AK 99802.

Salo, E.O. 1991. Life history of chum salmon (*Oncorhynchus keta*). *In* C. Groot and L. Margolis [eds.] Pacific salmon life histories. University of British Columbia Press, Vancouver, BC.

Tabata, S. 1984. Oceanographic factors influencing the distribution, migration, and survival of salmonids in the Northeast Pacific Ocean—a review, p. 128–160. *In* W.G. Pearcy [ed.] The Influence of Ocean Conditions on the Production of Salmonids in the North Pacific. Proceedings of a workshop. ORESU-W-83-001. Oregon State University Sea Grant Program, Corvallis, OR.

Tabata, S. 1989. Trends and long-term variability of ocean properties at Ocean Station P in the northeast Pacific Ocean. Geophys. Monogr. 55: 113–132.

Welch, D.W., and T.R. Parsons. 1993. $\delta^{13}C$–$\delta^{15}N$ values as indicators of trophic position and competitive overlap for Pacific salmon (*Oncorhynchus* spp.). Fish. Oceanogr. 2(1): 11–23.

The difficulty of tracing the effects of climate change on the fishes of Lake Constance

Jürgen Hartmann

Hartmann, J. 1995. The difficulty of tracing the effects of climate change on the fishes of Lake Constance, p. 261–270. *In* R.J. Beamish [ed.] Climate change and northern fish populations. Can. Spec. Publ. Fish Aquat. Sci. 121.

Abstract: Eight fish–weather regressions for the whitefish (*Coregonus lavaretus*), perch (*Perca fluviatilis*), and roach (*Rutilus rutilus*) of Lake Constance (Bodensee) were compared to determine how they reflected the effects of climate change. Three tentative generalizations were deduced. First, the weather parameters of spring and autumn are the most relevant. Second, wind velocity covaries with biological parameters as well as temperature. Third, the empirical regression patterns can be understood as fragments of theoretical S-shaped curves. I concluded that, at least at Lake Constance, the identification of the effects of climate change on fish biology will be difficult for several reasons: the relevance of specific seasons, discontinuous biological response, interactions with four other trends (including 11-year cycles), uncertainty about the role of specific meteorological factors, and year-to-year variability.

Résumé : Huit analyses de régression poisson-conditions météorologiques concernant le corégone (*Coregonus lavaretus*), la perche commune (*Perca fluviatilis*) et le gardon (*Rutilus rutilus*) du lac de Constance (Bodensee) ont été comparées pour que l'on détermine comment elles exprimaient les effets d'un changement climatique. Trois conslusions générales provisoires ont été tirées. Premièrement, les paramètres météorologiques du printemps et de l'automne sont des plus pertinents. Deuxièmement, la vélocité des vents varie de concert avec les paramètres biologiques et la température. Troisièmement, les courbes de régression empiriques peut être interprétées comme étant des fragments de courbes théoriques en S. J'en conclus que, du moins dans le lac de Constance, il sera difficile de déterminer les effets du changement climatique sur la biologie des poissons pour plusieurs raisons : la pertinence des saisons particulières, le caractère discontinu de la réaction biologique, les interactions avec quatre autres tendances (y compris les cycles sur 11 ans), l'incertitude sur le rôle de facteurs météorologiques précis et la variabilité d'une année à l'autre.

Introduction

This paper compares weather–fish regressions of Lake Constance to determine how fish biology may be affected by climate change. It aims at generalizations about the most influential season, the most influential proxy of climate, and the pattern of the climate–fish regression. The results are discussed with respect to the potential for identifying the effects of climate change on fish biology. The paper does not attempt description and discussion of each specific regression nor quantitative prognosis from the specific regressions.

Numerous weather–fish correlations have already been described (Shephard et al. 1984), and their use and misuse have often been discussed (Walters and Collie 1988). Whereas at most places single or very few time series on fish biology are available, at Lake Constance the data offer the chance for a comparative study. Restriction to such a comparison reduces the well-known danger: to be misguided by single not clear-cut or spurious correlations.

Contrary to North America (Regier et al. 1990*b*), in Europe very little attention has been paid to effects of climate change on fish biology, especially on freshwater fish. Only year-class strength of whitefish (*Coregonus lavaretus* L.) of Lake Constance has already been studied in the context of global warming (Trippel et al. 1991). But that study differs with respect to intention (quantitative prediction), method (less sophisticated correction of bias of the dependent parameter, temperature in April chosen as the independent parameter), and the result (exponential increase of year-class strength expected) from the one presented here. Lake Constance (Bodensee–Obersee) is a large (500 km in area; 95 m mean depth) European sub-alpine lake, bounded by Austria, Switzerland, and Germany. Descriptive data of the at-present mesotrophic lake were tabulated by Hartmann and Nümann

J. Hartmann. Institute for Lake Research, D-88085 Langenargen, Germany.

Table 1. Sunspot activity and cyclism in biological, limnological, and meteorological parameters of Lake Constance. +, maximum; –, minimum; (), extremly low or high value at the end of the time series (5-yr running means).

Parameter	Cycles
Air temperature in autumn (Sept.–Nov.)	– + – + – + – +
Annual near-bottom minimum of oxygen	+ – +++
Annual precipitation[a]	+ – + – + – + – + – (+)
Water level[a]	+ – + – + – + – + – +
Sunspot activity[b]	– + — + — + – + – + — + – +
Beginning of spawning time of whitefish	– ++ – ++ – + – + – ++ – + –
Growth of age-0+ whitefish	(–) + – + – ++
Scale of calendar years	I ---------- I ---------- I ---------- I ---------- I ---------- I ---------- I ---------- I ------ (1920, 1940, 1960, 1980)

[a]Luft et al. 1990.
[b]Hartmann 1984.

(1977). The actual state of the lake is described annually by the Internationale Gewässerschutzkomission (e.g., Müller 1993). After cultural eutrophication (starting in the 1950s), the lake has been undergoing oligotrophication since the early 1980s.The most important fish are whitefish and perch (*Perca fluviatilis*) (Hartmann and Nümann 1977).

Material and methods

Meteorological data

Wind velocity and water-surface temperature were measured daily at Constance by the Meteorological Station (Silvanerweg 6, D-78464 Konstanz) and the Landratsamt (Benediktinerplatz 1, D-78467 Konstanz). Because the wind station moved twice within the town, the wind data before August 1957 were tentatively corrected (means compared) by a factor of 1.3; those from August 1957 to mid-September 1972 by a factor of 0.9. While temperature (partly published by Anonymous 1965) and wind velocity were measured at the western edge (outlet) of the main basin, data on precipitation were available from a more central station at Langenargen (H. Quoss, Institut für Seenforschung, Untere Seestraße, 81 D-88085 Langenargen). Cycles (Table 1) of air temperature and annual near-bottom minimum of oxygen were studied with already seasonally pooled data from the Agrarmeteorologische Messstation at Romanshorn (Deucherstrasse 8 CH-8590 Romanshorn) and Müller (1993). The meteorological data are available from the above sources or from the author.

Table 2. Year-class strength of whitefish and perch in Lake Constance.

	Year-class strength ($\times 10^6$)			Year-class strength ($\times 10^6$)	
Year	Whitefish	Perch	Year	Whitefish	Perch
1954	1.0		1972	0.6	2.2
1955	2.2		1973	0.7	17.1
1956	1.9		1974	7.0	2.0
1957	2.8		1975	0.5	7.6
1958	3.6		1976	2.0	23.4
1959	1.2		1977	2.0	1.4
1960	1.8		1978	0.9	0.8
1961	2.1		1979	1.8	19.7
1962	0.7		1980	2.7	1.7
1963	2.5		1981	1.8	4.3
1964	0.3		1982	5.9	47.5
1965	0.4		1983	0.3	0.7
1966	1.0		1984	0.4	0.4
1967	1.6		1985	0.6	11.3
1968	2.7		1986	4.4	11.5
1969	0.5		1987	4.9	7.0
1970	0.4	6.2	1988		28.5
1971	3.4	13.7			

Table 3. Growth of whitefish and perch in recent years. These data complenent the data of Hartmann and Knöpfler (1986).

	Growth			
Year	0+ whitefish cm	0+ perch cm	1+ perch cm	2+ perch cm
1983	13.45	9.0	7.45	4.51
1984	12.60	7.5	6.1	2.52
1985	15.95	7.7	9.05	4.58
1986		7.25	7.85	4.30
1987			7.1	4.61
1988				5.32
1989				4.73
1990				3.68
1991				4.60

Biological data

Most biological data are given in the Tables 2 and 3. Year-class strength of whitefish was calculated by virtual population analysis (VPA). It had to be corrected for the fact that the nominal catch (mean nominal catch = 350 000 kg) represents a decreasing portion of the real landings (from about 82% in 1954 to about 52% in 1984 and later) (personal communications with local fishery wardens and fishermen; e.g., A. Göppinger, Andreas-Bruggerstrasse 13 D-88085 Langenargen). For the year 1954, for example, the unreported landings were estimated by the following calculation: 18/82 × 350 000 kg. Additionally, the numbers were multiplied by 1.2, taking into account that gutted fish are weighed by the fishermen, unlike in VPA. Year-class strength of perch (mean nominal catch = 400 000 kg) was calculated accordingly, assuming that the unreported portion increased from 53% (1972) to 64% (1984 and later). These procedures of bias correction differ from earlier ones (e.g., Trippel et al. 1991), e.g., by using correction factors for the landings of each calendar year i.e., by using more than one correction factor for each year-class. Experimental fishing took place at least once per month, and commercial catch is reported monthly. Experimental and commercial fishing used the same type of gear. Whitefish was caught by 38 – 40 mm gill nets before 1965 and with 44-mm nets later. Perch was caught by 28-mm nets and/ or 32-mm nets in 1990–91, and with 32-mm nets earlier.

Growth in length (Hartmann and Knöpfler 1986; Table 3) of age-0+ and age-1+ yr fish was calculated from scales (whitefish) or opercula (perch) by a simplified method (Hartmann 1984) that uses 20 females and 20 males (usually age-2+ yr) per year-class, caught nearest to the date when half the year-class is fished out. Growth of perch in the third year was calculated from females caught after their third growing season (September–May), neglecting bias (Hartmann 1992 a) by size-selective fishing.

The beginning of spawning time (Hartmann 1984, 1991) is calculated as the deviation (days) from the 5-yr running mean of the date when 50% of the female whitefish are ready for spawning (stage of ripeness: 6) or have already spawned.

The egg sizes and relative fecundity of whitefish are given by Hartmann and Quoss (1993). Growth of roach (*Rutilus rutilus* L.) is tabulated by Hartmann (1993).

Regressions and statistics

The regressions were kept as simple as possible to reduce the well-known danger of ending with spurious correlations (caused by data transformation and excessive trial and error). Therefore, Figs. 1–8 and the main discussion refer to the

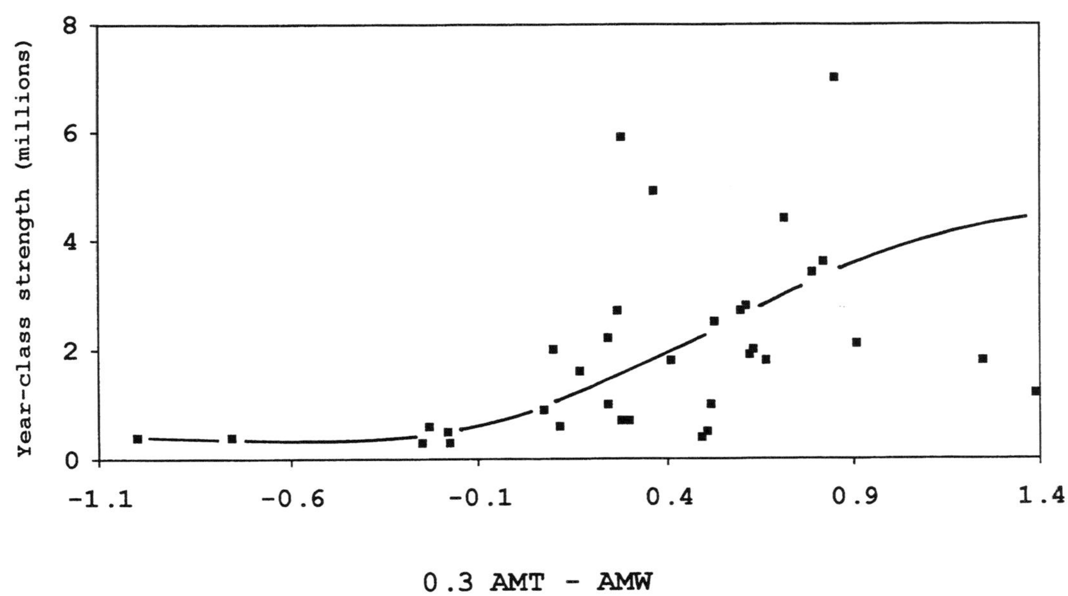

Fig. 1. Year-class strength of whitefish of Lake Constance and weather in spring of the first year. YCS, year-class strength; AMT, annual mean temperature (°C) at water surface from April to May; AMW, annual mean wind velocity (m/s) from April to May. AMT weighed by factor 0.3 according to the result of multiple linear regression. AMT – AMW, temperature minus wind.

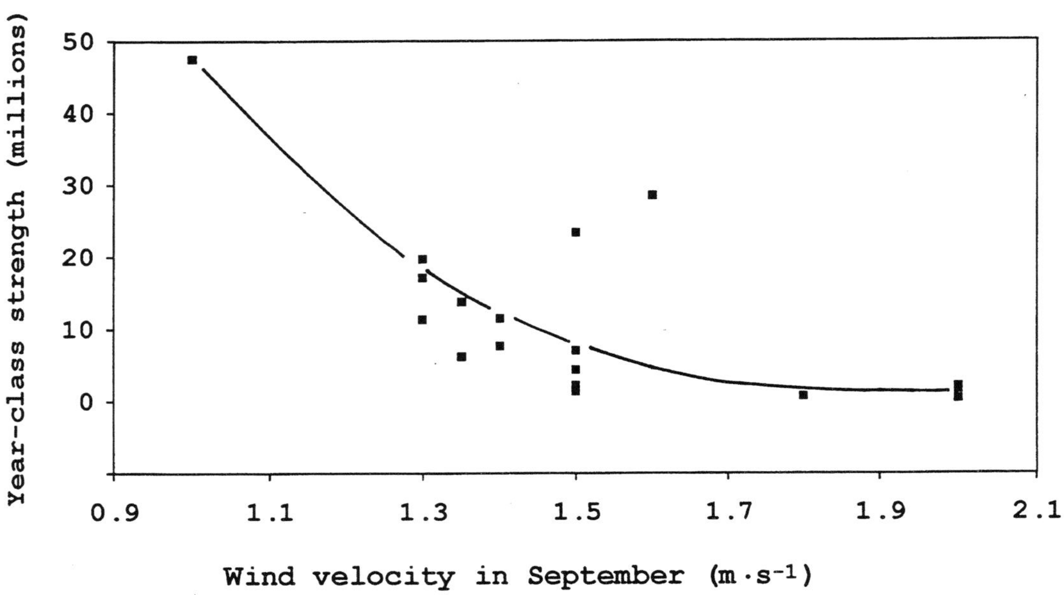

Fig. 2. Year-class strength of perch in Lake Constance and annual mean wind velocity in September of the first year.

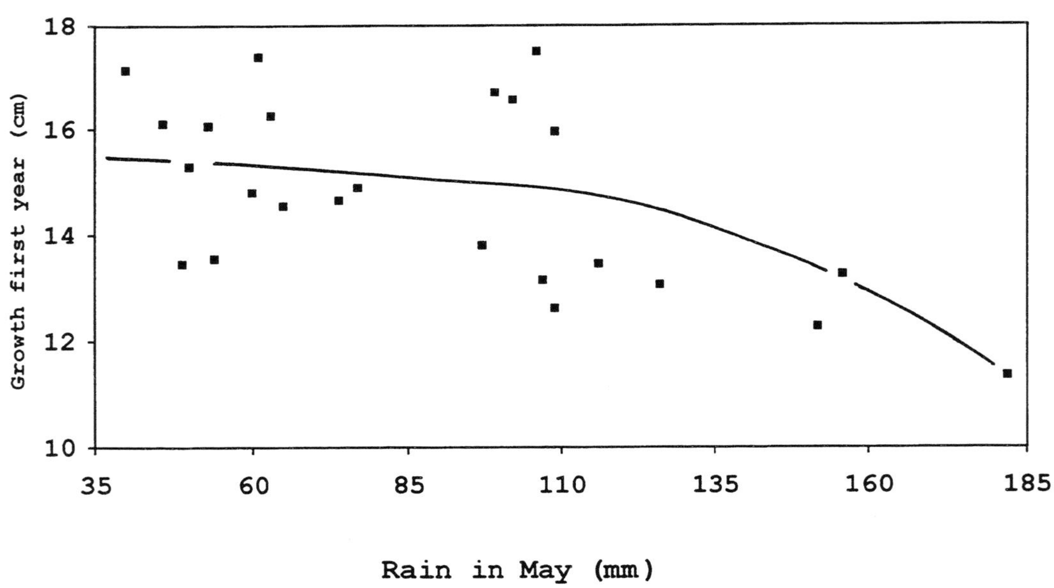

Fig. 3. Growth of age-0+ whitefish in Lake Constance and the sum of precipitation in May.

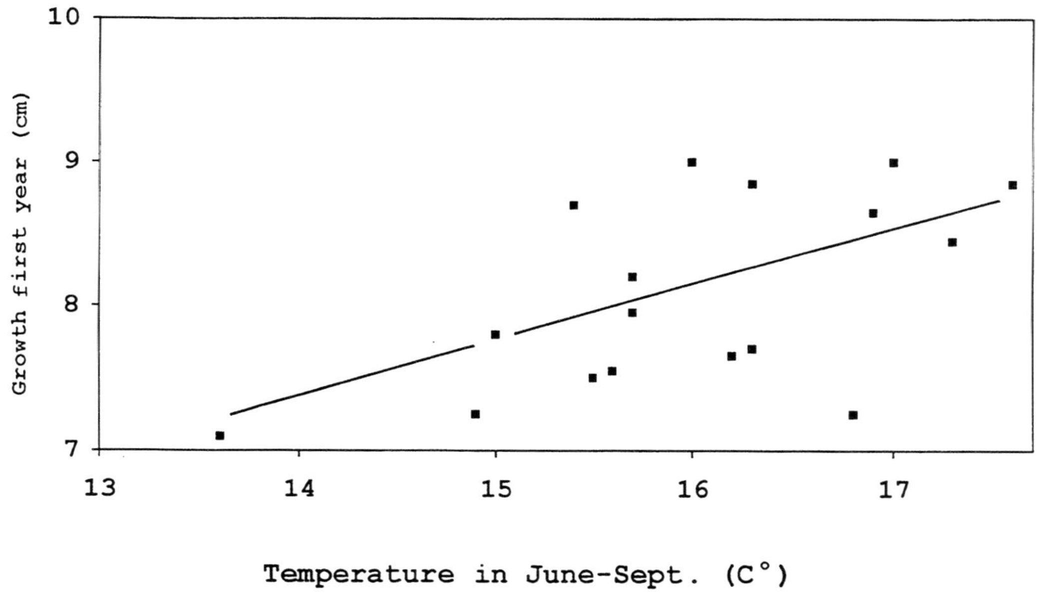

Fig. 4. Growth of age-0+ perch in Lake Constance and mean temperature at water surface from June to September.

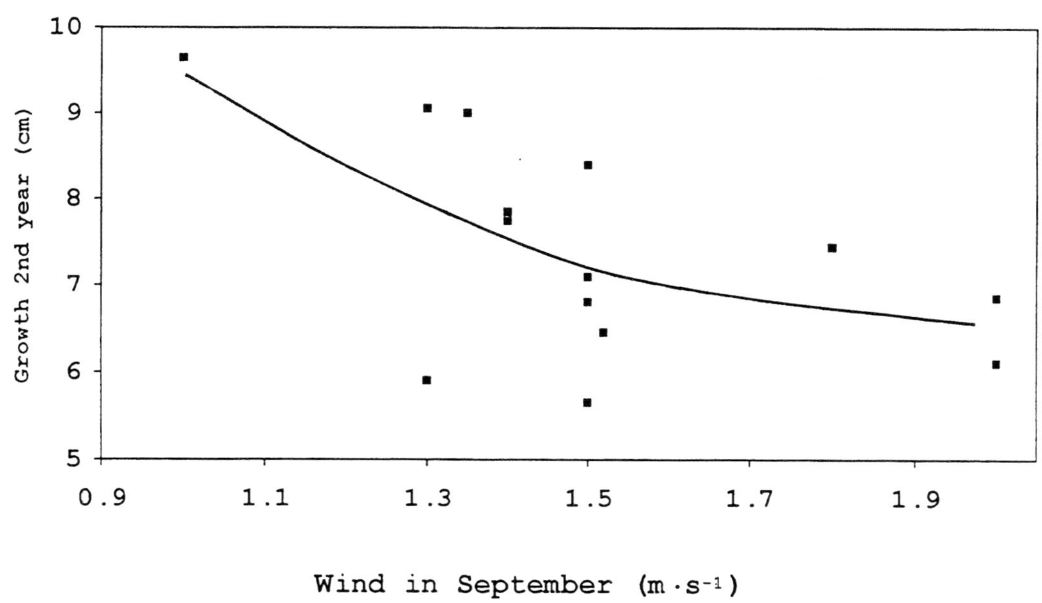

Fig. 5. Growth of age-1+ perch in Lake Constance and mean wind velocity in September.

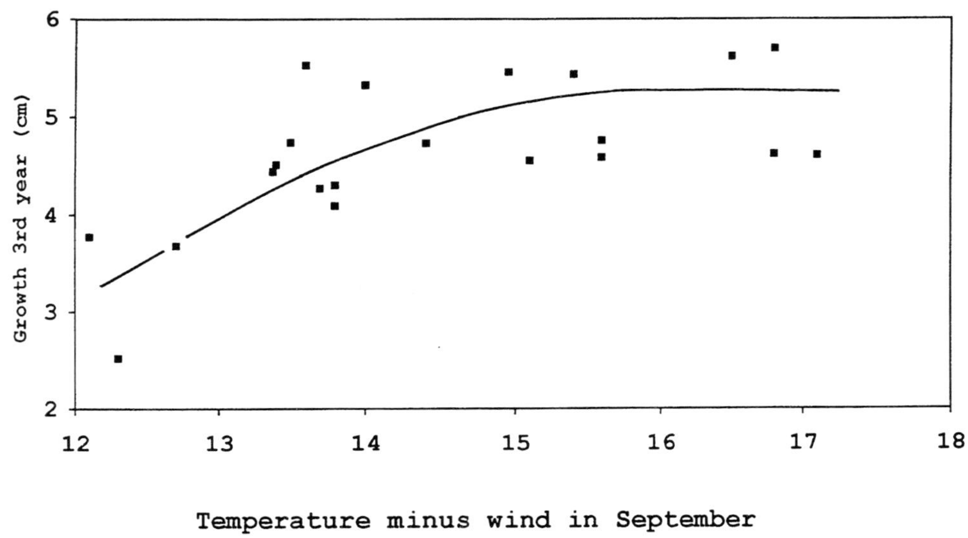

Fig. 6. Growth of age-2+ perch in Lake Constance and weather in September.

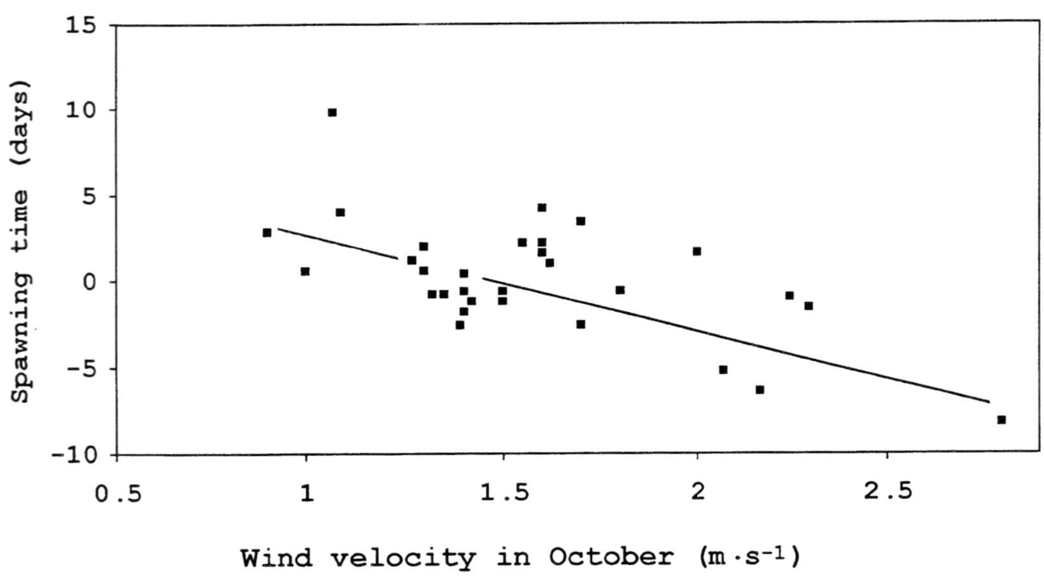

Fig. 7. Annual beginning of whitefish spawning in Lake Constance (December) and mean wind velocity in October.

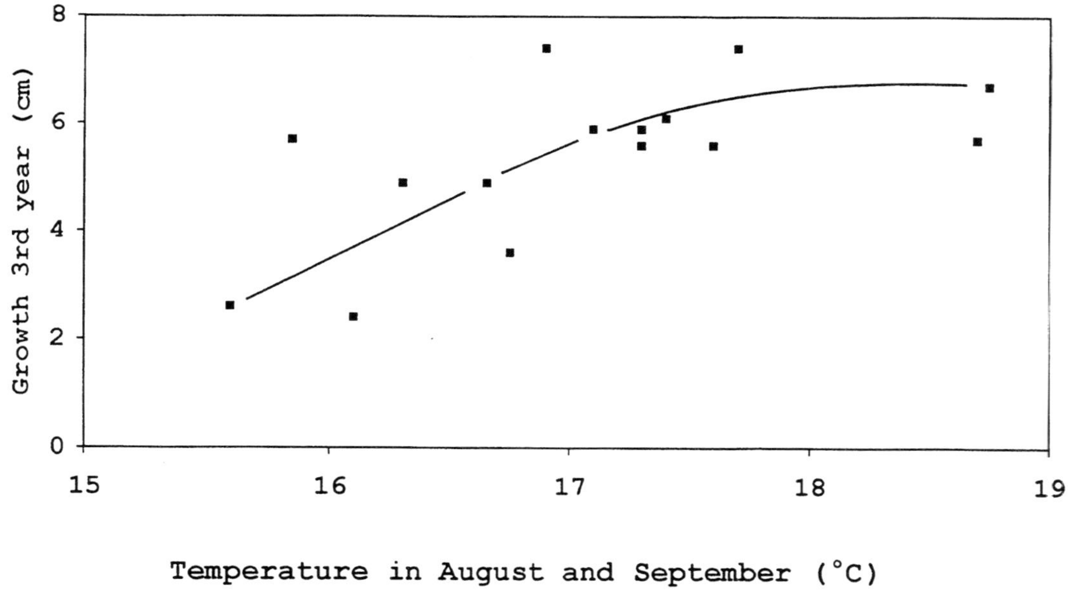

Fig. 8. Growth of age-2+ roach in Lake Constance and mean temperature of water surface in August and September.

Table 4. Seasonal distribution of meteorological parameters that explain variation in fish-biologial dependent variables of Lake Constance. Numbers = percentage of variation explained by mean surface temperature (T), mean wind velocity (W), or sum of precipitation (R) in linear regressions; +, additional percentage of variation explained by second independent variable.

		Months					Formula
Dependent variable	n	April (A)	May (M)	June–Sept. or Aug.–Sept. (JS) or (AS)	Sept. (S)	Oct. (O)	
Year-class strength of perch (PYCS)	19				W:	61	ln(PYCS) = 7.8 – 4.1 SW
Year-class strength of whitefish (CYCS)	34	T: 30 W: + 13					$100 \cdot CYCS^{-1} = 134 - 37AMT + 130AMW$
Beginning of spawning season of whitefish (CSP)	31					W: 39	CSP = 8 – 5OW
Growth of age-0+ whitefish (CG1)	24		R: 37				ln(CG1) = 2.9 – 0.002MR
Growth of age-0+ perch (PG1)	17			T: 34			PG1 = 1.8 + 0.4JS
Growth of age-1+ perch (PG2)	17				W: 35		PG2 = 9.1 – 4 ln(SW)
Growth of age-2+ perch (PG3)	21				W: 45		PG3 = 7.6 – 0.2SW
Growth of age-2+ roach (RG3)	15			T: 38			RG3 = –11.8 + TAS

nonlinearized data. Curves were fitted by eye. (Fitting by computer would have led, in spite of much more sophisticated formulas, to poorer fit and no additional information in our context of overall tendency of the climate–fish regression.)

The correlation coefficients in Table 4 refer to linear regressions. The r^2 in multiple regressions means adapted (corrected) r^2 (as suggested by text books). The r^2 and its improvement (in multiple regressions) by adding meteorological (independent) parameters is tested at the 5% level. Where trends were apparent in time series (spawning time, fecundity), deviations from 5-yr running means were used for the calculations.

Results

From eight weather–fish regressions, the following information is derived with respect to trackability of effects of climatic change on fish: seasons rather than annual means are relevant, several meteorological parameters are influential, the specific curves are not homogenously shaped, and much variability remains unexplained.

In all eight linear regressions shown in Table 4, meteorological parameters of spring and/or autumn (April–May; September–October) contributed to explain the variation of the fish parameters year-class strength, growth, and spawning time. In seven of the eight regressions meteorological data for May or September did so. Only in one case (the growth of perch in the first year) did summer temperature improve the model. No correlations were found between weather (temperature, wind, rain) and the whitefish parameters fecundity, egg size, and age-1+ growth.

Of seasonal or monthly mean surface-water temperature, wind velocity, and sum of precipitation, wind most often (5× wind, 3× temperature, 1× rain) fitted in the linear regressions (Table 5). But comparison (t-test) of the means (30 13; 27 21; 10 12%) of the correlation coefficients of linear equations with each of the three meteorological independent variables (Table 5) reveals that, on average, the biological dependent variables were as sensitive to temperature change as they were to wind velocity. Rain appears to be less influential.

When nonlinearized by data transformation, five of the eight correlations (Figs. 1–8) appear not to be homogenous. Flat curve phases appear along with rather steep slopes.

Variability unexplained by the specific models is high. This results in uncertainty about the definite shape of the specific regressions. For example, Fig. 4 can be interpreted as a linear or as a logistic regression.

Discussion

I do not intend to discuss the potential mechanisms (Hokanson 1977a; Shepherd et al. 1984) of the (formal) weather–fish relationships described here, in much detail. My main point is that the effects of climate change on the fish of Lake Constance may be difficult to trace for several reasons: the different relevance of the seasons, discontinuous biological response, interaction with additional trends, uncertainty about the role of specific meteorological factors, and year-to-year variability.

Temperature in spring may influence growth rate and thus year-class success by determining start of growing season (Trippel et al. 1991). The most important fact may be that fish larvae hunt prey more successfully in warmer water (Braum and Quoss 1981). Accordingly, perch larvae die from starvation when temperature remains below 15°C (Hokanson 1977b). Wind in autumn may work through the date of autumnal destruction of weeds. In summer, weeds are the hiding place of the young-of-the-year of perch and roach, but not of whitefish. This may partly explain the apparent difference between the species in respect to the most influential season. Rain, in this context, is probably just a conglomerate of diverse meteorological parameters. Weather in October influences spawning time of the whitefish probably by determining the date of downward migration of the species in October (Hartmann et al. 1991). This migration may open sexual synchronization of the fishes before main spawning within few days (in December).

Because trends of homologous meteorological time series of the seasons are not identical (Anonymous 1993; Irsch 1993; Trippel et al. 1991), one has to know which season is most relevant in order to identify (and predict) general or specific

Table 5. Percentages of variation in the fish biology of Lake Constance explained by temperature, wind velocity, and precipitation in linear regressions. Some n values are lower than those in Table 4; independent variables are not transformed.

Dependent variable	Fish species	n	Months	Temperature	Wind	Precipitation
Year-class strength	Whitefish	24	Apr. and May	46	29	17
Year-class strength	Perch	19	Sept.	38	61	3
Beginning of spawning time	Whitefish	27	Oct.	33	31	8
Age-0+ growth	Whitefish	24	May	10	2	37
Age-1+ growth	Perch	17	June–Sept.	34	8	4
Age-1+ growth	Perch	17	Sept.	9	32	1
Age-2+ growth	Perch	21	Sept.	34	45	7
Age-2+ growth	Roach	15	Aug. and Sept.	38	3	5

effects of climate change. One reason why the weather of the transitional seasons (spring and autumn) appears to be distinctly relevant for the fish dynamics of Lake Constance may be its ability to prolong or shorten the main growing and feeding season (Trippel et al. 1991; Hokanson 1977b; Holtby 1988). The literature contains more evidence for the relevance of weather in spring than for the importance of autumn weather (Clady 1976; Lawler 1965; Kipling and Frost 1970).

While especially in laboratory experiments, biological response to directed environmental change may often follow an optimum curve (Lin and Regier 1995), the figures here, derived from field data, can better be understood as fragments of negative or positive (logistic) S-shaped curves. It seems reasonable to assume that fish do not react continuously and monotonically to climatic change (Beverton et al. 1984; Ursin 1982). Reaction limits (Falconer 1989) followed by a phase of equilibrium (Zeeman 1976), or at least increased compensation (Regier et al. 1990*a*), can be expected. The literature contains very little information on discontinuous responses of fish to meteorological change (McCauley and Kilgour 1990; Ursin 1982), but the literature may be biased. Nonmonotonic curves are less easily identified than, for example, linear ones. Their calculation is more sophisticated and presupposes a higher number of more evenly distributed data pairs, including some extreme values. As Ursin (1982) points out, the literature is also biased against an interest in biological stability (no change). Data sets that show poor weather–fish correlation or others that suggest an exponential relationship may in fact cover phases (fragments) of S-curves. If so, the S-curve hypothesis is of practical relevance. Before predicting effects of climate change on fishes by extrapolation of a monotonic curve, a logistic correlation with limited trend should be considered. The exponential curve on year-class success of the whitefish of Lake Constance (Trippel et al. 1991) is one example. The S-curve hypothesis will be testable, when more data are available. Note that the largest data set exhibits the clearest logistic correlation (Fig. 1). But on the other hand, the longer a time series is, the more endangered is its methodological homogenity. The wind data used here in context with the year-class strength of the whitefish of Lake Constance are a good example.

Uncertainty with respect to the shape of the specific curves is neither surprising nor irritating in our context. As Cury et al. (1995) point out, low rather than high correlations are expected between climate variability and fish population responses. The observation of a high degree of uncertainty supports the view that prediction of effects of climatic change on fishes may be difficult to trace.

At Lake Constance, global warming is not the only factor that can be expected to contribute to future trends in the fish community. Size-selective gillnetting (Hartmann 1992*b*) may continue as well as the present oligotrophication (Müller 1993). Also a decrease in the present high fishing pressure might occur, and a meteorological and biological 11-year cycle seems to exist (Table 2). Possible interactions (Fig. 9) of such diverse trends render identification of climate-change effects on fish biology difficult.

Several meteorological factors contribute to climate (weather) and climate change. Their specific role in climate change (Palutikov et al. 1984; cited in George et al. 1990) and altering

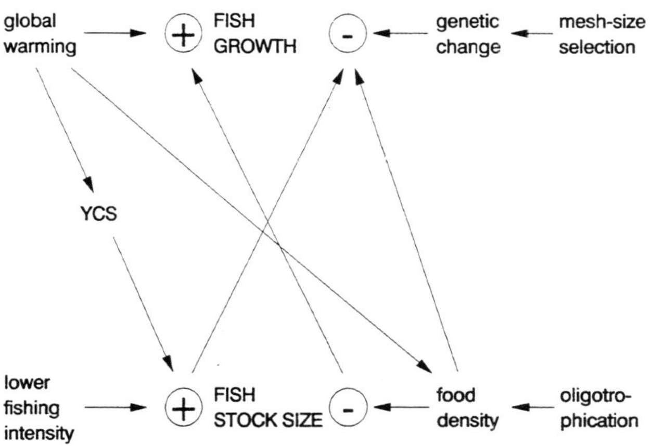

Fig. 9. Potential long-term effects of stressors on the fish population of Lake Constance.

fish biology may depend on the specific situation. In Europe, long-term trends of of seasonal temperature and precipitation differ between regions (Anonymous 1993). Surprises may derive from interferences and shifting relevance in the causal chain (DeAngelis and Cushman 1990). Finally, biological (Trippel et al. 1989) and meteorological short-term variations complicate detection of trends and correlations. The problems in tracing biological effects of climate change discussed here are probably not confined to the case of the fish of Lake Constance. To increase the number of data pairs (n), I intend to standardize the data of the eight sets and than to treat them as belonging to one ore two sets only.

References

Anonymous. 1965. Deutsches Gewässerkundliches Jahrbuch Sonderheft Land Baden-Württemberg Abflussjahr 1963. Landesstelle Gewässerkunde Karlsruhe [ed.] Braun, Karlsruhe. 111 p.

Anonymous. 1993. In Deutschland gibt es im Sommer immer weniger Niederschläge. BMFT-J. June: 12 p.

Beverton, R.J.H., J.G. Cooke, J.B. Csirke, R.W. Doyle, G. Hempel, S.J. Holt, A.D. MacCall, D.J. Policansky, J. Roughgarden, J.G. Shephard, M.P. Sissenwine, and P.H. Wiebe. 1984. Dynamics of single species. Dahlem Workshop Rep. 32: 13–58.

Braum, E., and H. Quoss. 1981. Beobachtungen über die Entwicklung des Blaufelchen (*Coregonus lavaretus wartmanni*) im Bodensee-Obersee. Schweiz. Z. Hydrol. 43: 114–125.

Clady, M.D. 1976. Influence of temperature and wind on the survival of early stages of yellow perch, *Perca flavescens*. J. Fish. Res. Board Can. 39: 1887–1893.

Cury, P., C. Roy, R. Mendelssohn, A. Bakun, D.M. Husby, and R.H. Parrish.1995. Moderate is better: exploring nonlinear climatic effect on Californian anchovy (*Engraulis mordax*), p. 000–000, p. 417–424. *In* R.J. Beamish [ed.] Climate change and northern fish populations. Can. Spec. Publ. Fish. Aquat. Sci. 121.

DeAngelis, D.L., and R.M. Cushman. 1990. Potential application of models in forecasting the effects of climate changes on fisheries. Trans. Am. Fish. Soc. 119: 224–239.

Falconer, D.S. 1989. Introduction to quantitative genetics. 3rd ed. Longman Scientific and Technical, New York, N.Y. 438 p.

George, D.G., D.P. Hewitt, J.W. Lund, and W.J.P. Smyly. 1990. The relative effects of enrichment and climate change on the long-term dynamics of *Daphnia* in Esthwaite Water. Freshwater Biol. 23: 55–70.

Hartmann, J. 1984. 11-year cycle of spawning time and growth of the whitefish (*Coregonus lavaretus*) of Lake Constance. Schweiz. Z. Hydrol. 46: 163–170.

Hartmann, J. 1991. Richtiger Zeitpunkt der Laichfischerei auf Blaufelchen (*Coregonus lavaretus*) im Bodensee? Österr. Fisch. 44: 229–235.

Hartmann, J. 1992*a*. Scheinbares Ausgleichswachstum, ein Kiemennetzeffekt bei Barsch (*Perca fluviatilis*) und Felchen (*Coregonus lavaretus*) des Bodensees. Österr. Fisch. 45: 232–235.

Hartmann, J. 1992*b*. Rückgang von Kondition und Wachstum bei Felchen (*Coregonus lavaretus*) durch selektive Kiemennetzbefischung? Fischökologie 6: 21–30.

Hartmann, J. 1993. Entwicklung der Cypriniden-(Weissfisch-) Bestände im Bodensee. Österr. Fisch. 46: 201–207.

Hartmann, J., and G. Knöpfler. 1986. Barschwachstum im Bodensee. Österr. Fisch. 39: 45–51.

Hartmann, J., G. Knöpfler, H. Löffler, and H. Quoss. 1991. Tages- und jahresperiodische Vertikalwanderung der Fische im Bodensee. Fischökologie 4: 67–74.

Hartmann, J., and W. Nümann. 1977. Percids of Lake Constance, a lake undergoing eutrophication. J. Fish. Res. Board Can. 34: 1670–1677.

Hartmann, J., and H. Quoss. 1993. Fecundity of whitefish (*Coregonus lavaretus* L.) during the eu- and oligotrophication of Lake Constance. Fish Biol. 43: 81–87.

Hokanson, K.E.F. 1977*a*. Temperature requirements of some percids and adaptations to the seasonal temperature cycle. J. Fish. Res. Board Can. 34: 1524–1550.

Hokanson, K.E.F. 1977*b*. Optimum culture requirements of the early life phases of the yellow perch. *In* R.W. Soderberg [ed.] Perch Fingerling Production for Aquaculture. Proceedings of a Conference, Madison, Wis. Dec. 12, 1977. University of Wisconsin, Madison, Wis. Sea Grant Coll. Program Advisory Rep. 421: 24–60.

Holtby, L.B. 1988. Effects of logging on stream temperatures in Carnation Creek, British Columbia, and associated impacts on the coho salmon (*Oncorhynchus kisutch*). Can. J. Fish. Aquat. Sci. 45: 502–515.

Irsch, W. 1993. Treibhauseffekt läßt Vögel sesshaft werden. Schwäbische Ztg. 128 of June 7.

Kipling, C., and W.E. Frost. 1970. Study of the mortality, population numbers, year class strengths, production and food-consumption of pike, *Esox lucius* L., in Windermere from 1944 to 1962. J. Anim. Ecol. 39: 115– 157.

Lawler, G.H. 1965. Fluctuations in the success of year-classes of whitefish populations with special reference to Lake Erie. J. Fish. Res. Board Can. 22: 1197–1227.

Lin, P., and H.A. Regier. 1995. Use of Arrhenius models to describe temperature dependence of organismal rates in fish, p. 211–225. *In* R.J. Beamish [ed.] Climate change on northern fish populations. Can. Spec. Publ. Fish. Aquat. Sci. 121.

Luft, G., G. van den Eertwegh, and H. Vieser. 1990. Veränderung der Bodensee-Wasserstände von 1887 bis 1987. Ministerium Umwelt Baden-Württemberg [ed.] Handb. Hydrol. Baden-Württemberg 6.2: 77 p.

McCauley, R.W., and D.M. Kilgour. 1990. Effect of air temperature on growth of largemouth bass in North America. Trans. Am. Fish. Soc. 119: 276–281.

Müller, H. 1993. Limnologischer Zustand des Bodensees. Jber. Int. Gewässerschutzkommiss. Bodensee 19: 11 p.

Regier, H.A., J.A. Holmes, and D. Pauly. 1990*a*. Influence of temperature changes on aquatic ecosystems: an interpretation of empirical data. Trans. Am. Fish. Soc. 119: 374–389.

Regier, H.A., J.J. Magnuson, and C.C. Coutant. 1990*b*. Introduction to proceedings: symposium on effects of climate change on fish. Trans. Am. Fish. Soc. 119: 173–175.

Shepherd, J.G., J.G. Pope, and R.D. Cousins. 1984. Variations in fish stocks and hypotheses concerning their links with climate. Rapp. P.-v. Reun. Cons. int. Explor. Mer 185: 255–267.

Trippel, E.A., R. Eckmann, and J. Hartmann. 1991. Potential effects of global warming on whitefish in Lake Constance, Germany. Ambio 20: 226–231.

Ursin, E. 1982. Stability and variability in the marine ecosystem. Dana 2: 51–67.

Walters, C.J., and J.S. Collie. 1988. Is research on environmental factors useful to fisheries management? Can. J. Fish. Aquat. Sci. 45: 1848–1854.

Zeeman, E.C. 1976. Catastrophe theory. Sci. Am. 234: 65–83.

Potential influence of North Pacific sea-surface temperatures on increased production of chum salmon (*Oncorhynchus keta*) from Japan

Yukimasa Ishida, David W. Welch, and Miki Ogura

Ishida, Y., D.W. Welch, and M. Ogura. 1995. Potential influence of North Pacific sea-surface temperatures on increased production of chum salmon (*Oncorhynchus keta*) from Japan, p. 271–275. *In* R.J. Beamish [ed.] Climate change and northern fish populations. Can. Spec. Publ. Fish. Aquat. Sci. 121

Abstract: We examined the potential influence of changes in sea-surface temperature (SST) in the North Pacific Ocean on Japanese chum salmon (*Oncorhynchus keta*) adult return rates and growth rates. SST near the Kuril Islands in July and SST in the central North Pacific in winter and spring showed significant decreasing trends from 1947 to 1988. Average return rate after the mid-1960s was higher than that occurring before the mid-1960s, when most juveniles were not fed prior to release. Return rate was negatively correlated with SST near the Kuril Islands and in the central North Pacific after the mid-1960s, but not before. Growth was positively correlated with spring SST in the central North Pacific. The present results and previously reported negative correlations between growth and fish density suggest that chum salmon production is enhanced in Japan by hatchery technology but that yields have been reduced by declining growth rates caused by decreasing SST and increasing fish density in the central North Pacific.

Résumé : Nous avons examiné l'influence éventuelle des changements dans la température de la surface de la mer (TSM) dans le Pacifique Nord sur les taux de retour et de croissance du saumon kéta (*Oncorynchus keta*) adulte du Japon. De 1947 à 1988, la TSM près des îles Kouriles en juillet et dans le centre du Pacifique Nord en hiver et au printemps accusait de fortes tendances à la baisse. Après le milieu des années 1960, le taux moyen de retour a été plus élevé que pendant la période précédente, où la plupart des juvéniles n'étaient pas alimentés avant leur libération. Après le milieu des années 1960, mais pas avant, le taux de retour était en corrélation négative avec la TSM près des îles Kouriles et dans le centre du Pacifique Nord. La croissance était en corrélation positive avec la TSM du printemps dans le centre du Pacifique Nord. Ces résultats et les corrélations négatives précédemment déclarées entre la croissance et la densité du poisson portent à croire que la production de saumon kéta au Japon est favorisée par la technologie des écloseries mais que les rendements ont été réduits en raison des taux de croissance en déclin causés par la diminution de la TSM et par l'augmentation de la densité du poisson dans le centre du Pacifique Nord.

Introduction

Sea-surface temperature (SST) changes are believed to affect the abundance, return timing, migration route, and growth of Pacific salmon (*Oncorhynchus* spp.) (Rogers 1984; Mysak 1986; Ricker 1981; Helle 1979; Nickelson 1986). In the eastern North Pacific, SST steadily decreased from the 1950s to the 1970s, followed by a warming trend beginning in the mid-1970s (Chelton 1984). Parallel with these changes in SST, the abundance of Pacific salmon in western and central Alaska declined and reached a minimum in the early 1970s, then began to increase in the mid-1970s. These changes in abundance are thought to be related to changes in temperature, particularly during winter months while salmon are at sea (Rogers 1984). In the western North Pacific, a similar warming trend after the mid-1970s is not evident (Chelton 1984), but chum salmon (*O. keta*) production in Japan began to increase significantly compared with chum salmon stocks in Russia and North America. Coincident with this increase in abundance, the size of returning adults decreased, possibly due to a combination of density-dependent effects and changes in ocean conditions (Kaeriyama 1989; Ishida et al. 1993).

The objectives of this study are to describe the SST changes in those regions of the North Pacific occupied by chum salmon from Japan and to examine the potential influence of changes in SST on two factors influencing chum salmon productivity: adult survival (return rate) and growth.

Y. Ishida and **M. Ogura**. National Research Institute of Far Seas Fisheries, Fisheries Agency of Japan, 5-7-1, Orido, Shimizu, Shizuoka 424 Japan.
D.W. Welch. Pacific Biological Station, Department of Fisheries and Oceans, Nanaimo, BC V9R 5K6 Canada.

Materials and methods

SST data were provided by the Climate Research Group (A-024), Scripps Institution of Oceanography, University of California. The data from 1947 to 1988 were averaged into 5° squares between latitudes 20°N and 60°N and longitudes 130°E and 110°W. SST were examined mainly near the Kuril Islands (45°N, 150°E) in July, and in the central North Pacific (40°–45°N, 170°E–170°W) in winter (January–March) and spring (April–June). It is thought that chum salmon juveniles from Japan are distributed near the Kuril Islands in summer and overwinter in the central North Pacific (Irie 1990; Salo 1991; Ueno et al. 1992).

The number of juveniles released and the number of adult returns were analyzed for brood years from 1955 to 1987 (Committee of the Centennial History of Salmon Enhancement in Hokkaido 1988; Association of Salmon Enhancement in Honshu 1988). The return rates were calculated as the ratio of adult returns to the number of juveniles released 4 yr earlier. Return rates were analyzed separately before and after 1966, because, before 1966, most juveniles were not fed prior to release (Kobayashi 1980). The percentages of juveniles being fed prior to release in Hokkaido and average body weights of age-4 chum salmon returning to the Ishikari River, Hokkaido, from 1953 to 1988 were also used in this study. Wild chum salmon stocks are negligible in Japan.

Results

Survival

SST near the Kuril Islands in July showed large year-to-year variation but decreased between 1947 and 1988 ($r = -0.36$, $P < 0.05$) (Fig. 1). Average SST in the 1980s was 10.2°C, 1.5°C lower than the average SST (11.7°C) between 1947 and 1979. SST in the central North Pacific in winter and spring also showed a decreasing trend ($r = -0.61$, $P < 0.01$; $r = -0.53$, $P < 0.01$, respectively). Average winter and spring SST in the central North Pacific were 8.2°C and 9.1°C in the 1980s, 0.8 and 0.9°C lower than the average SST (9.0 and 10.0°C, respectively) observed before 1980.

The number of juvenile chum salmon released from Japanese hatcheries averaged about 500 million juveniles through the 1960s, then increased sharply to about 2 billion juveniles by 1980 (Fig. 2). Juvenile releases have remained stable at

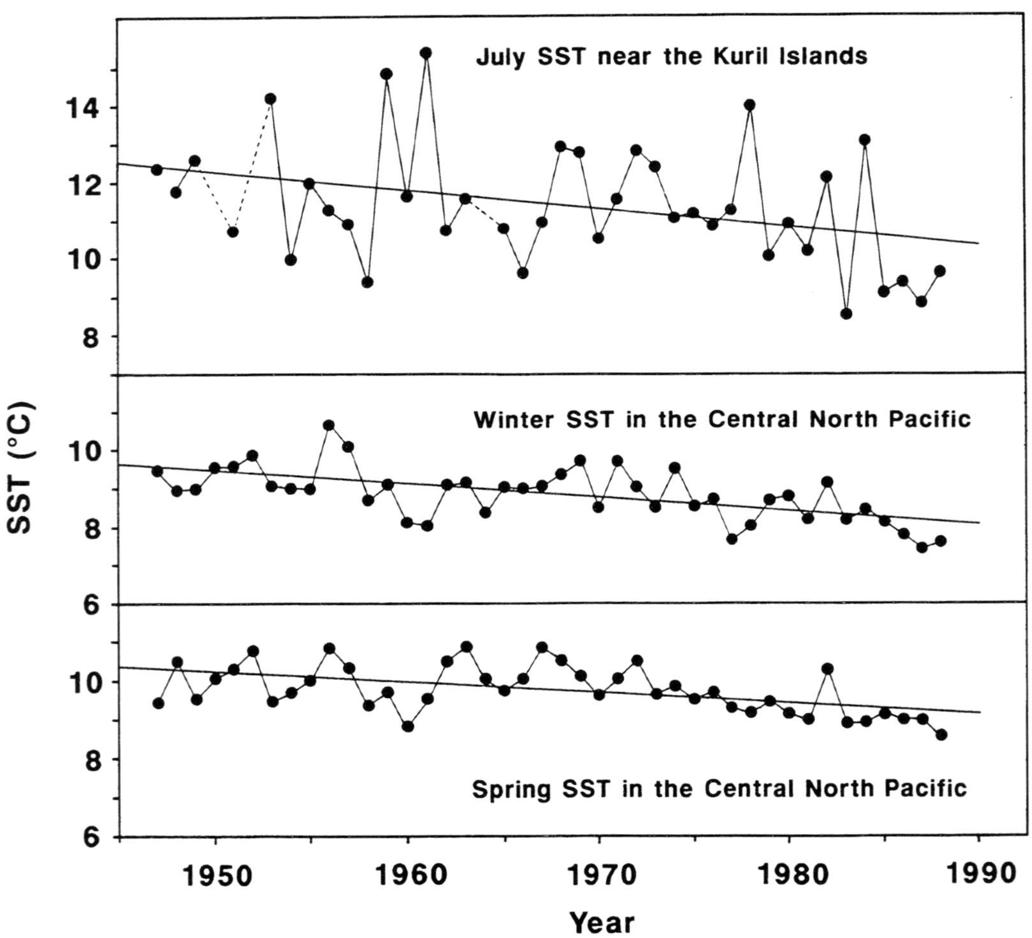

Fig. 1. Changes in sea-surface temperature (SST) in July near the Kuril Islands, and in the winter (January–March) and spring (April–June) in the central North Pacific Ocean (40°–45°N, 170°E–170°W). Broken lines indicate periods with missing data.

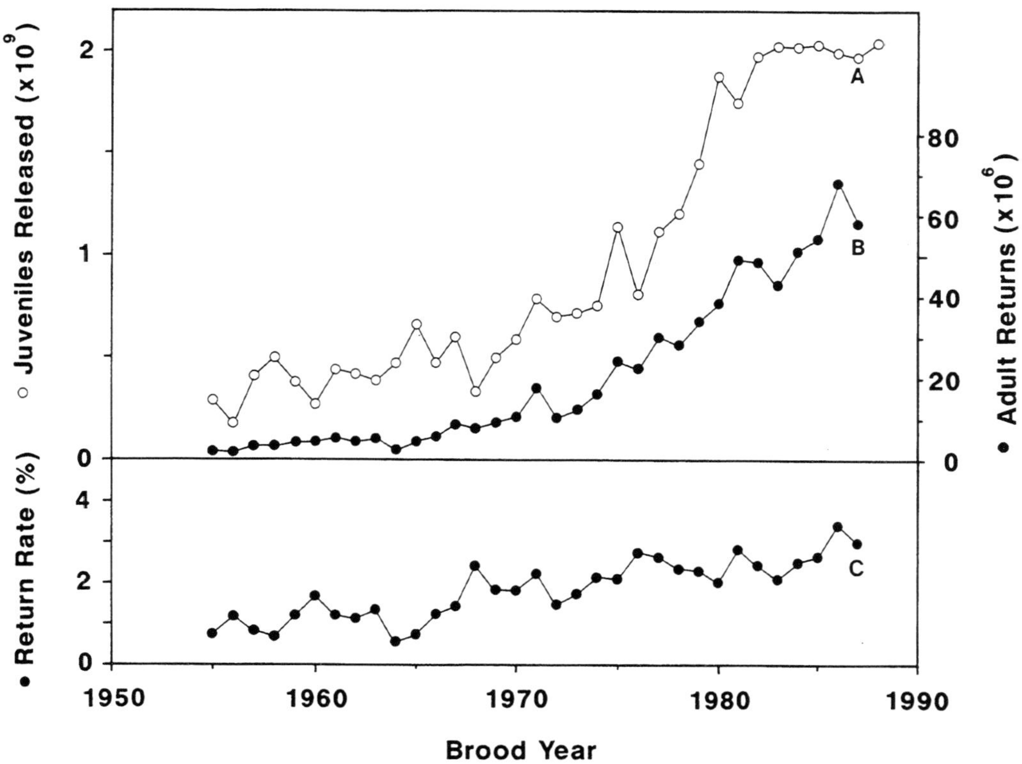

Fig. 2. Changes in number of juveniles released (A), number of adults returning (B), and the return rate (C) for Japanese chum salmon.

roughly 2 billion since that time. Adult returns also increased in response to the increased number of juveniles released from hatcheries, reaching more than 50 million fish by the mid-1980s. Survival from juvenile to adult after 1966 averaged 2.3%, significantly higher than the 1.0% average before 1966, when most juveniles released from hatcheries were not fed.

No significant correlation was found between return rate and July SST near the Kuril Islands either before or after 1966. Significant negative correlations were observed between return rate and both winter and spring SST in the central North Pacific Ocean after 1966, but no correlation was found before 1966 (Fig. 3). A significant correlation was found between return rate and the percentage of juveniles being fed after 1966, but not before (Fig. 4).

Growth

Significant positive correlations were also found between mean body weight of age-4 chum salmon returning in the fall as adults to the Ishikari River and SST occurring in the central North Pacific in the preceding three springs after 1966 (Fig. 5). Before 1966, relationships were not significant, at least in part due to the limited data available, but similar positive relationships were found.

Discussion

The present study shows that marine survival of chum salmon from Japan was negatively correlated with SST after 1966. However, the higher survival rate after 1966 is largely the result of changing hatchery practices, with juveniles being fed prior to release, and releases scheduled to coincide with appropriate conditions in coastal waters (Kobayashi 1980; Kaeriyama 1989).

For wild salmon stocks, several papers have previously suggested that ocean temperatures affect either the abundance or survival of Pacific salmon (Rogers 1984; Mysak 1986; Helle 1979). For Alaskan sockeye salmon (*O. nerka*), the increases in abundance were associated with increasing SST, particularly during the winter months (Rogers 1984). It was speculated that sockeye salmon were distributed farther south during cold years and may become more concentrated and hence more vulnerable to predators. In Russia, Birman (1959) found that abundance of chum salmon was higher than average when ocean water near the coast was relatively warm, but the inverse relation was found for pink salmon (*O. gorbuscha*). He suggested that temperature changes influenced survival of chum and pink salmon during their first winter at sea. He also pointed out that changes in temperature were unlikely to directly influence survival, but rather that temperature changes probably reflected other biological changes; for example, the influence of predators. The biological mechanisms underlying the different response to ocean conditions between different species are still unexplained, but predation might be a major factor controling the survival of salmon on the high sea.

Our study also indicates that year-to-year changes in SST appear to be associated with changes in growth of Japanese chum salmon. Ricker (1981) showed that the decreasing size of mature salmon returning to British Columbia generally did not exhibit significant correlations with available ocean

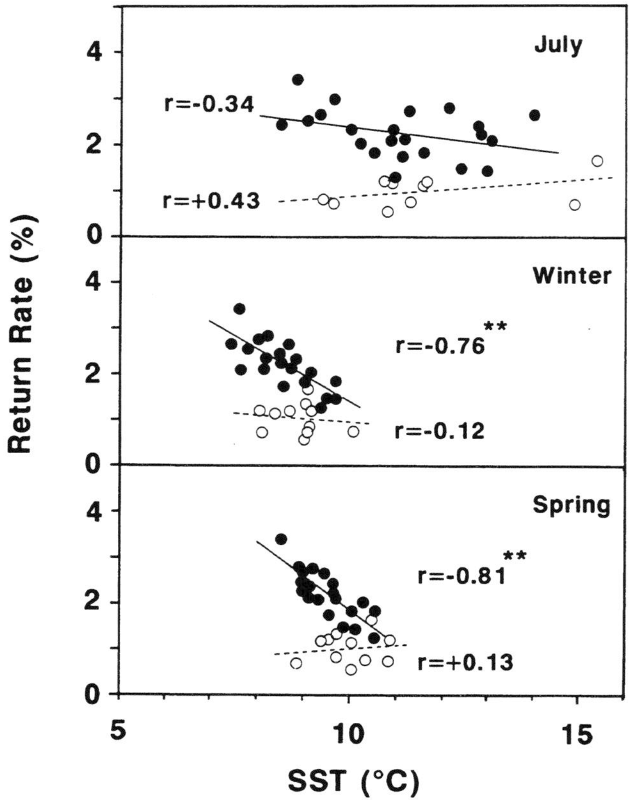

Fig. 3. Relationship between SST and return rates of Japanese chum salmon (*$P < 0.05$; **$P < 0.01$). Open circles indicate data for release years prior to 1966 and solid circles indicate data for release years 1966 and later.

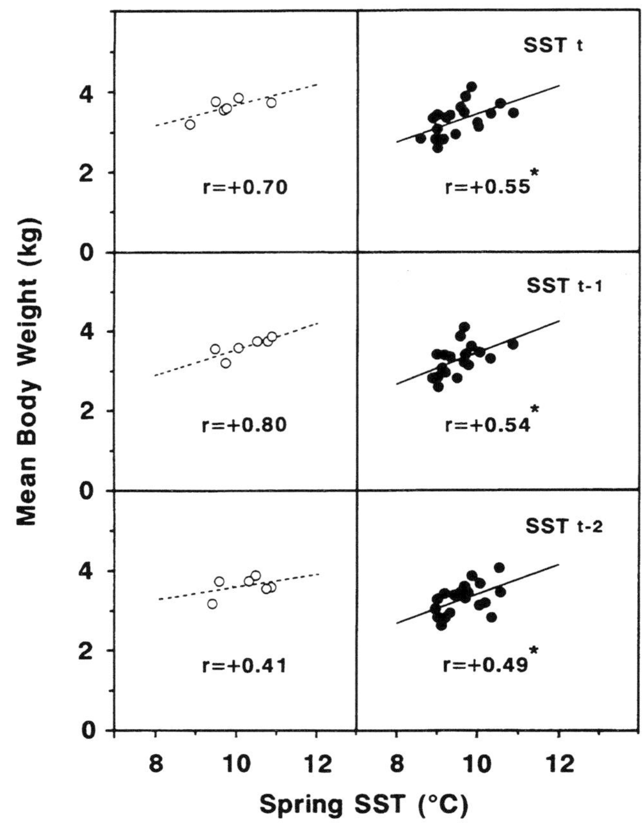

Fig. 5. Relationship between spring SST in the central North Pacific Ocean and mean body weight of age-4 chum salmon returning to the Ishikari River. Open circles indicate data based on juveniles released before 1966 and solid circles indicate data for juveniles released in 1966 and subsequent years.

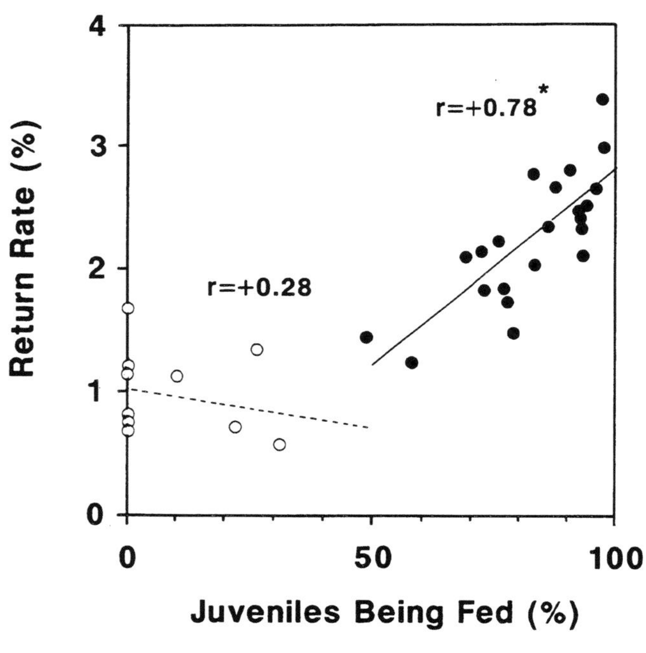

Fig. 4. Relationship between the percentage of juveniles being fed in Hokkaido and return rate of chum salmon in Japan. Open circles indicate data before 1966 and solid circles indicate data for 1966 and later years.

temperature or salinity series, although he did not rule out an effect of ocean conditions on the decreasing size of salmon. For sockeye salmon in Bristol Bay, Alaska, spring air temperature at Bristol Bay had a significant but small effect on changes in mean body weight (Rogers 1980).

For Japanese and Russian chum salmon, density dependence has been suggested as one of the possible causes for the recent changes in age and size (Kaeriyama 1989; Ishida et al. 1993). However, the positive relationship between spring SST and mean body weight of age-4 chum salmon in the Ishikari River before 1966 also indicates that ocean conditions contributed to the changes in chum salmon growth, because the abundance of Japanese chum salmon was still low and density-dependent growth was probably negligible before 1966. In contrast, zooplankton biomass and pelagic fish and squid abundance in the Gulf of Alaska has increased in recent years (Brodeur and Ware 1992). Mean biomass in summer tended to fluctuate at about 10-year intervals with low levels until the early 1960s and after 1980, and with high levels during the middle 1960s and late 1970s in the Oyashio waters of the western North Pacific (Kotani and Odate 1991). However, little information is available for the central North Pacific, especially with regard to year-to-year variation in zooplankton biomass (Kawamura 1988). Future studies should monitor ocean conditions not only for changes in physical conditions such as SST but also for changes in biological conditions. Such studies should include both primary production and the trophic levels forming the forage base for salmon in the central North Pacific Ocean.

Acknowledgments

We thank Dan Cayan, Emelia Bainto, and Larry Riddle, Climate Research Group A-024, Scripps Institution of Oceanography, University of California, for providing the SST data, and Takashi Sasaki and Kazuya Nagasawa, National Research Institute of Far Seas Fisheries, and Masahide Kaeriyama, Hokkaido Salmon Hatchery, for reviewing the manuscript.

References

Association of Salmon Enhancement in Honshu. 1988. Report of salmon enhancement in Honshu. Statistics from 1972 to 1981. Tokyo, Japan. 196 p.

Birman, I.B. 1959. More about the influence of the Kuroshio on the dynamics of abundance of salmon. Vopr. Ikhtiol. 13: 16–18. (Transl. from Russian by Fish. Res. Board Can. Transl. Ser. 269)

Brodeur, R.D., and D.M. Ware. 1992. Long-term variability in zooplankton biomass in the subarctic Pacific Ocean. Fish. Oceanogr. 1(1): 32–38.

Chelton, D.B. 1984. Short-term climatic variability in the northeast Pacific Ocean, p. 87–99. *In* W.G. Pearcy [ed.] The influence of ocean conditions on the production of salmonids in the North Pacific. Oregon State University, Corvallis, OR Sea Grant Coll. Program ORESU-W-83-001.

Committee of the Centennial History of Salmon Enhancement in Hokkaido. 1988. Centennial history of salmon enhancement in Hokkaido. Statistics. Sapporo, Japan. 432 p.

Helle, J.H. 1979. Influence of marine environment on age and size at maturity, growth, and abundance of chum salmon, *Oncorhynchus keta* (Walbaum), from Olsen Creek, Prince William Sound, Alaska. Ph.D. Thesis, Oregon State University, Corvallis, OR. 129 p.

Irie, T. 1990. Ecological studies on the migration of juvenile chum salmon, *Oncorhynchus keta*, during early ocean life. Bull. Seikai. Nat. Fish. Res. Inst. 68: 1–142.

Ishida, Y., S. Ito, M. Kaeriyama, S.M. McKinnell, and K. Nagasawa. 1993. Recent changes in age and size of chum salmon (*Oncorhynchus keta*) in the North Pacific Ocean and possible causes. Can. J. Fish. Aquat. Sci. 50: 290–295.

Kaeriyama, M. 1989. Aspects of salmon ranching in Japan. Physiol. Ecol. Spec. 1: 625–638.

Kawamura, A. 1989. Characteristics of the zooplankton biomass distribution in the standard Norpac net catches in the North Pacific region. Bull. Plankton Soc. 35(2): 175–177.

Kobayashi, T. 1980. Salmon propagation in Japan, p. 91–107. *In* J.E. Thorpe [ed.] Salmon ranching. Academic Press, New York, N.Y.

Kotani, Y., and K. Odate. 1991. Variations of zooplankton biomass in the Oyashio waters. Bull. Hokkaido Natl. Fish. Res. Inst. 55: 185–195.

Mysak, L.A. 1986. El Niño, year-to-year variability and fisheries in the northeast Pacific Ocean. Can. J. Fish. Aquat. Sci. 43: 464–497.

Nickelson, T.E. 1986. Influences of upwelling, ocean temperature, and smolt abundance on marine survival of coho salmon (*Oncorhynchus kisutch*) in the Oregon Production Area. Can. J. Fish. Aquat. Sci. 43: 527–535.

Ricker, W.E. 1981. Changes in the average size and average age of Pacific salmon. Can. J. Fish. Aquat. Sci. 38: 1636–1656.

Rogers, D.E. 1980. Density-dependent growth of Bristol Bay sockeye salmon, p. 267–283. *In* W.J. McNeil and D.C. Himsworth [ed.] Salmonid ecosystems of the North Pacific. Oregon State University Press, Corvallis, OR.

Rogers, D.E. 1984. Trends in abundance of northeastern Pacific stocks of salmon, p. 100–127. *In* W.G. Pearcy [ed.] The influence of ocean conditions on the production of salmonids in the North Pacific. Oregon State University, Corvallis, OR Sea Grant Coll. Program ORESU-W-83-001.

Salo, E.O. 1991. Life history of chum salmon (*Oncorhynchus keta*), p. 231–309. *In* C. Groot and L. Margolis [ed.] Pacific salmon life histories. University of British Columbia Press, Vancouver, B.C.

Ueno, Y., J. Seki, I. Shimizu, and A. P. Shershnev. 1992. Large juvenile chum salmon *Oncorhynchus keta* collected in coastal waters of Iturup Island. Nippon Suisan Gakkaishi 58(8): 1393–1397.

Effect of climate on herring (*Clupea pallasi*) population dynamics in the Northeast Pacific Ocean

Abdelkrim Zebdi[1] and Jeremy S. Collie[2]

Zebdi, A., and J.S. Collie. 1995. Effect of climate on herring (*Clupea pallasi*) population dynamics in the Northeast Pacific Ocean, p. 277–290. *In* R.J. Beamish [ed.] Climate change and northern fish populations. Can. Spec. Publ. Fish. Aquat. Sci. 121.

Abstract: The abrupt rise and fall of the world herring fisheries has motivated many research projects in fisheries oceanography. This study begins with an analysis of the stock of Pacific herring (*Clupea pallasi*) in Sitka Sound, Alaska. Time series of meteorological and hydrographic variables are used to classify the physical environment of Sitka Sound according to cold, dry and warm, wet years. Recruitment of Sitka herring was higher following warm, wet winters. Sea-surface temperature (SST) anomalies explained about 40% of the variation in herring recruitment; upwelling accounted for an additional 18%. Spawning occurred earlier in years with positive SST anomalies. An SST-dependent Ricker model was fitted to the data to distinguish the relative importance of environmental and density-dependent factors on herring recruitment. Herring growth was density dependent and not temperature dependent. Finally, we compared recruitment data from herring stocks in Alaska, British Columbia, and Washington state with the geographic pattern of SST anomalies. Patterns of recruitment synchrony were found to correspond to the hydrographic domains in the Northeast Pacific, as delineated with SST records.

Résumé : L'expansion et le déclin abrupts des pêches mondiales de hareng ont suscité le lancement de nombreux projets de recherche en océanographie des pêches. La présente étude commence par une analyse du stock de harengs du Pacifique (*Clupea pallasi*) dans le détroit de Sitka, en Alaska. Des séries chronologiques de variables météorologiques et hydrographiques sont utilisées pour classer le milieu ambiant physique du détroit de Sitka selon les années froides et sèches et les années chaudes et humides. Le recrutement du hareng de Sitka était plus élevé après les hivers chauds et humides. Les anomalies de la température à la surface de la mer (TSM) expliquaient environ 40 % de la variation du recrutement du hareng; les remontées d'eau en expliquaient une autre tranche de 18 %. Le frai avait lieu plus tôt les années où les anomalies de la TSM étaient positives. Un modèle de Ricker tributaire de la TSM a été appliqué aux données pour que l'on puisse distinguer l'importance relative des facteurs environnementaux et de densité sur le recrutement du hareng. La croissance du hareng dépendait de la densité et non de la température. Enfin, nous avons comparé les données sur le recrutement des stocks de harengs en Alaska, en Colombie-Britannique et dans l'État de Washington par rapport à la configuration géographique des anomalies de la TSM. Nous avons constaté que les tendances de la synchronie du recrutement correspondaient aux domaines hydrographiques du nord-est du Pacifique, dont les limites étaient calculées dans les données sur la TSM.

A. Zebdi and **J.S. Collie.** Juneau Center for Fisheries and Ocean Sciences, University of Alaska Fairbanks, 11120 Glacier Highway, Juneau, AK 99801, USA.

[1]Present address: Department of Fisheries and Wildlife, Natural Resources Building, Michigan State University, East Lansing, MI 48824-1222, USA.

[2]Present address: Graduate School of Oceanography, University of Rhode Island, Narragansett, RI 02882-1197, USA.

Introduction

Large fluctuations in fish abundance are documented in the catch histories of nearly all fisheries. Paleosedimentary records show that fish stocks fluctuated widely even before commercial fishing developed (Soutar and Isaacs 1974). Three processes determine fish population abundance: recruitment, growth, and mortality. These are influenced by environmental factors both localized (storms, currents) and widespread (El Niño) (Parrish et al. 1981; Norcross and Shaw 1984; Hamilton and Mysak 1986; Mysak 1986; Rose and Leggett 1988; Wespestad 1991). Fish are removed by fishing and natural mortality, which includes starvation and predation. Recruitment, the only process of abundance renewal, is influenced by the number of spawners and environmental conditions (Ricker 1975).

The clupeoids have an enormous catch variability: ratios of best: worst years range between 2: 1 and 25: 1 among different stocks (Cushing 1982). Such variations may be due to environmental factors acting on recruitment, overfishing, or both. Process-oriented and correlative studies have related environmental changes to fluctuations of fish stocks, especially pelagic ones such as sardines (*Sardinops* sp.), anchovies (*Engraulis mordax*) (Lasker 1978), mackerel (*Scomber japonicus*) (Prager and Hoenig 1989), saury (*Cololabis saira*), and Pacific whiting (*Merluccius productus*) (Bailey and Francis 1985). These studies suggest that year-to-year variation can be partly attributed to changes in oceanic variables generated by atmospheric fluctuations. Oceanic and atmospheric changes explain only a fraction of the interannual variability in fish population abundance, and are more likely to explain long-term trends and low-frequency cycles.

Herring populations have highly variable recruitment rate (e.g., Anthony and Fogarty 1985), which cannot be explained by stock–recruitment relationships alone. A relationship between recruitment and oceanographic–meteorological factors seems likely, but not obvious. Cushing (1982), in his research into the effects of climate on fisheries, related the alternation between Norwegian and Bohuslån herring (*Clupea harengus*) to warm and cold periods in the North Atlantic. He interpreted temporal patterns in catches or abundance, shared among geographically separated populations, as the effects of large-scale environmental forcing. For instance, periods of high and low catches of the Norwegian herring corresponded roughly to those of the Japanese sardine (*Sardinops melanostictus*) and the Californian sardine (*Sardinops sagax*).

In the Northeast Pacific, patterns of strong and weak year-classes are shared by different species in the same geographic region (Hollowed et al. 1987) and by adjacent stocks of the same species (e.g., herring; Ware and McFarlane 1989). Recruitment in widely separated stocks (e.g., California and

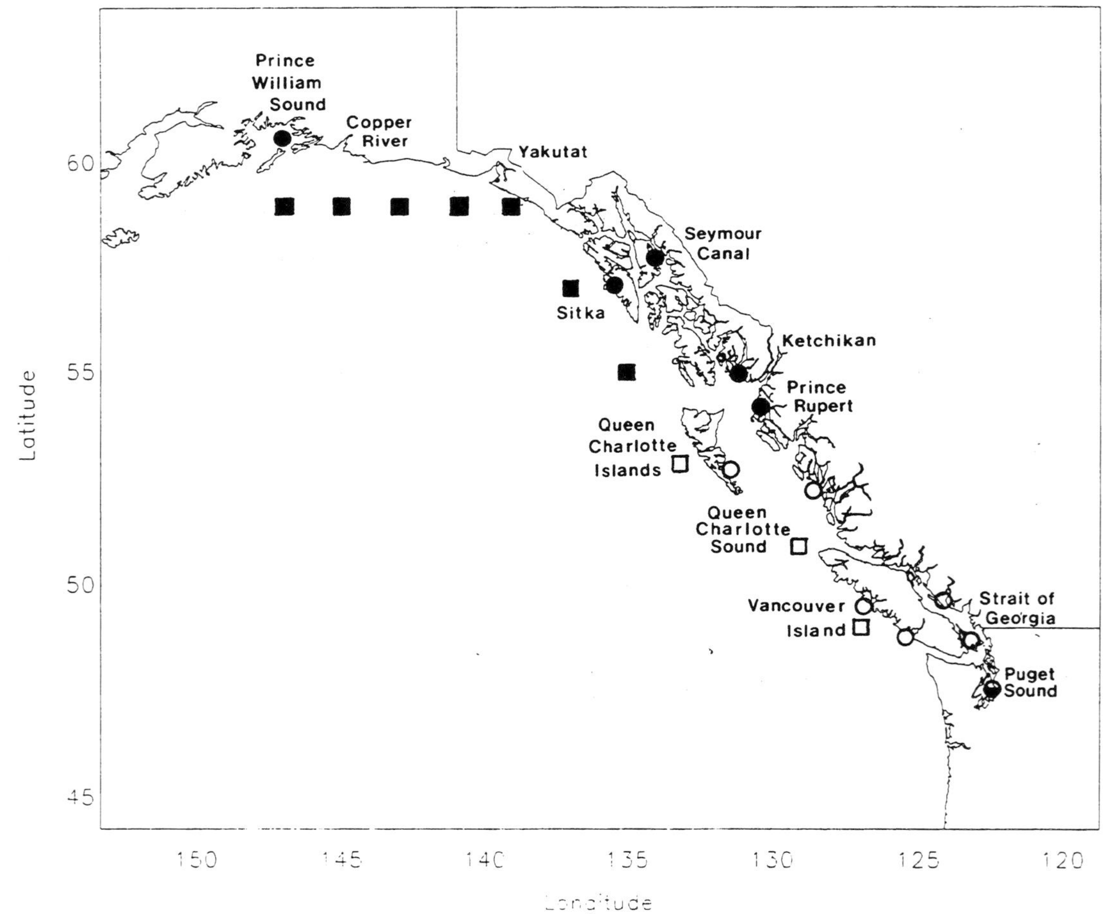

Fig. 1. The Northeast Pacific ocean with the locations of the different herring stocks (circles) and the SST stations (squares). The shading patterns indicate cluster membership, in the dendograms of Fig. 8 and 9.

Bering Sea) tends to be negatively correlated. More recent work seeks patterns of environmental forcing of the same spatial scale as the recruitment patterns. Hollowed and Wooster (1992) rationalized the large body of oceanographic and meteorological data from the Northeast Pacific by classifying years by type-A and type-B winters. Type-A winters are characterized by weak circulation of the Alaska Gyre, with strong upwelling farther south, low temperatures, and low sea level. Type-B winters have essentially the opposite conditions: an intense Aleutian low-pressure cell, strong circulation in the Gulf of Alaska, high sea-surface temperature (SST), and high sea level. Type-A and type-B conditions alternate on a decadal time scale (Hollowed and Wooster 1992).

This study focuses on physical variables and herring data from Sitka Sound, Alaska. Part I examines short- and long-term changes in the marine environment and relates local variability measured in Sitka to larger-scale patterns documented for the Gulf of Alaska. In Part II, interannual variability in recruitment of Sitka Sound herring is related to physical changes in the marine environment by considering different avenues through which physical changes may govern herring population abundance: maturation, time of spawning, larval survival, and growth. In Part III, recruitment time series from herring stocks in Alaska, British Columbia, and Washington are examined to determine the spatial extent of recruitment patterns.

Part I: Physical environment

Sitka Sound, the primary study area, is located in southeastern Alaska (Fig. 1). Sitka is unique in that one of the largest herring fisheries in the Northeast Pacific (5600 t average landings in 1980s) occurs near the oldest continuously operating magnetic observatory in North America. Some of the measurements, such as precipitation, began in 1842 when the Russians still occupied the area. However, gaps of up to 20 yr exist in the data.

The physical variables used in this study are: air temperature (AT), precipitation (PPT), sea-level height (SLH), SST, sea-surface salinity (SSS), coastal freshwater discharge (FWD; Royer 1982), Northeast Pacific pressure index (NEPPI; Emery and Hamilton 1985; Niebauer 1988), and coastal upwelling index (UPW; Bakun 1973). The data sets and sources are described more completely in Zebdi (1991).

To determine the most important driving variable in the system and to search for cycles relevant to herring fluctuations, time-series analysis of the physical variables was performed. We extracted seasonal cycles from the data sets and removed obvious trends such as that in SLH. The resultant time-series anomalies, or deviations from the overall mean, were analyzed for cyclic patterns and irregular fluctuations.

Trends and cyclic patterns

Correct interpretation of trends and cycles in the time series depends on the time span under consideration. A segment of a cyclic event can give the impression of a trend, and without specific detrending techniques, such as fitting a Gompertz or logistic curve, any attempt to interpret the so-called trend could lead to unwarranted conclusions. The major and most obvious trend we encountered in the time series anomalies considered here is the decrease in SLH. Sea level appears to decrease because of isostatic rebound of the land (Hicks and Shofnos 1965). At Sitka, the rate of rebound was estimated, by linear regression, to be 2.12 $mm \cdot yr^{-1}$ and was removed from the SLH anomalies. A trend of increasing SST began in the early 1970s (Fig. 2). It could be just part of the overall cyclic pattern seen in these data (Royer 1989), representing merely the commencement of a new cycle. The previous warming cycle, as it appears in the graph, began in the early 1950s and ended in the early 1970s.

We calculated autocorrelation functions of the yearly means of the physical variables to identify any long-term (very low frequency) cycles in the data. The autocorrelation coefficients

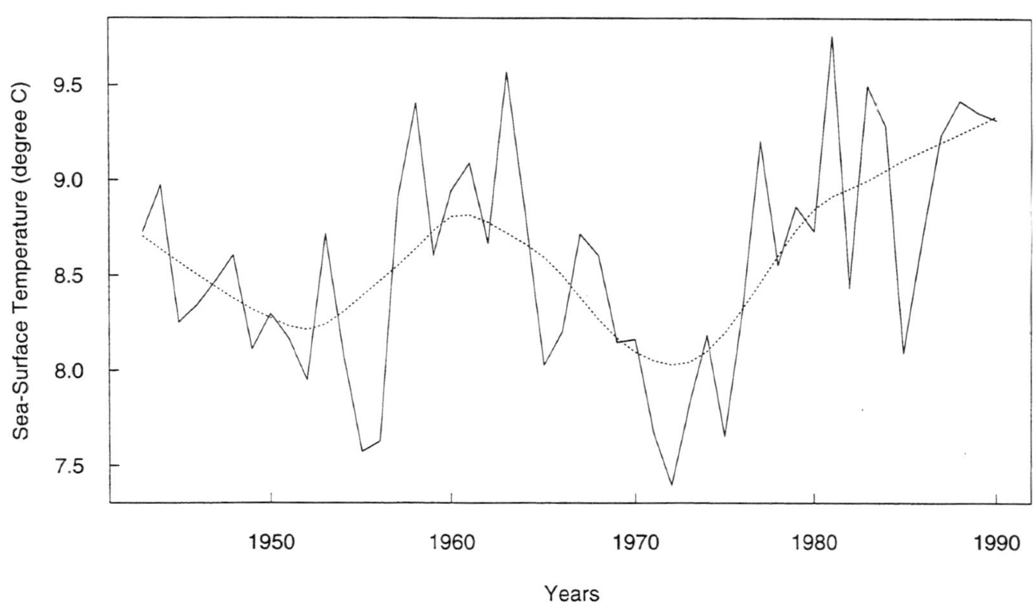

Fig. 2. Yearly means of sea-surface temperature in Sitka Sound (solid line) and smoothing line using LOWESS (broken line) with f = 0.25 (tension parameter).

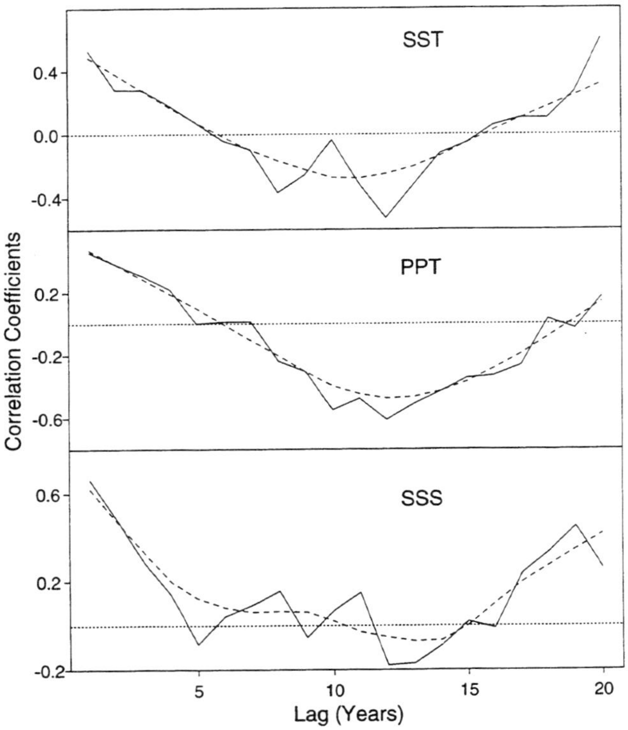

Fig. 3. Autocorrelograms of SST, PPT, and SSS (solid lines) and smoothed line using LOWESS (dashed line) with f = 0.4 (tension parameter). See text for explanation of abbreviations.

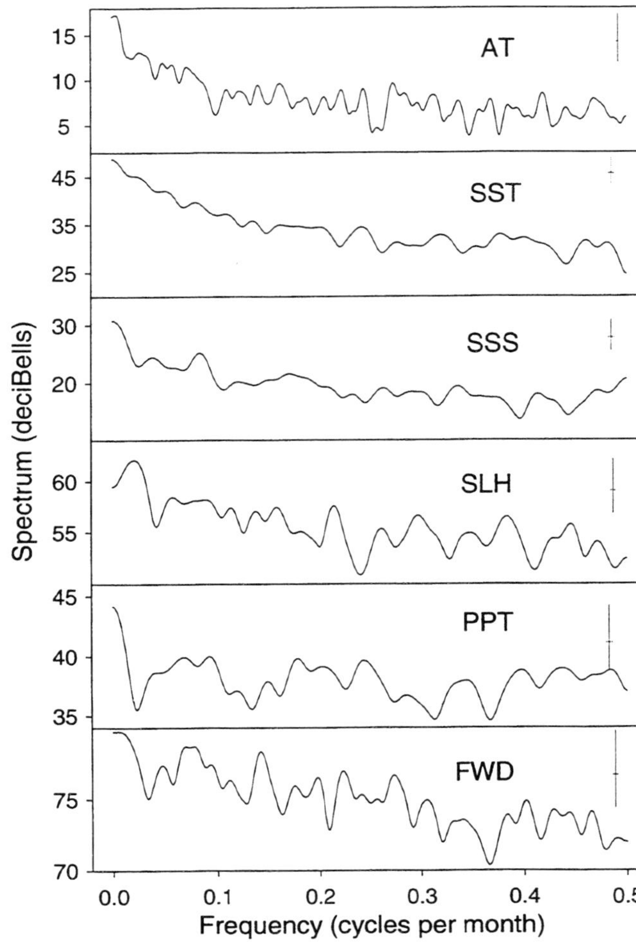

Fig. 4. Smoothed spectra for AT, SST, SSS, SLH, PPT and FWD. The graphs were obtained by applying a running mean filter over the raw periodograms to reduce variability in the spectrum estimates. The units are deciBells ($10\log_{10}(S(f))$, where S is the smoothed estimate and f is the frequency). The 95% confidence interval (vertical) and band width (horizontal) are represented graphically at the right hand side of the graphs. See text for explanation of abbreviations.

for lags of 1–20 yr were plotted as correlograms (Fig. 3). Smoothing was performed with locally weighted regression and smoothing scatter plots (LOWESS; Cleveland 1979). The most remarkable result is the same U-shaped form of the smoothed correlogram for AT, PPT, FWD, SSS, and SST, suggesting cycles of about 20 yr. Upwelling and NEPPI (not shown) have noticeably different patterns.

We used spectral analysis to identify and to quantify cycles that could have been missed in the autocorrelation part of the study. Running averages smoothed the variability in the sample spectra (Fig. 4) in which the band width and the spectrum confidence interval are given. Band width is a measure of the amount of smoothing. The degrees of freedom equal two when there is no smoothing and increase with the amount of smoothing. (The trade-off when using smoothing is between the variability of the estimates and the bias introduced by increasing the band width.) Unsmoothed spectra were examined first, then smoothing was added to see how much confidence could be given to the results.

The raw spectra of the physical variables showed some common peaks at relatively low frequencies, but these peaks disappeared with smoothing. For low-frequency cycles of interest in this study, no cycles can be detected with confidence. There are hints of power at very low frequency (e.g., AT, Fig. 4) that could correspond to the 21- to 22-yr cycle seen in the raw periodogram; however, this requires longer time series to be isolated with certitude. The 22-yr Hale cycle of solar activities and the 11-yr cycle of flare appearances on the sun's surface, recognized as sunspots, are well documented in astronomy and seem to be reflected by similar cycles in SST, FWD, SLH, and PPT. However, this reflection of solar cycles in oceanic and climatic conditions does not seem to be agreed upon in the literature because of low statistical significance limitations (Kerr 1988; Guiot 1988; Royer 1989).

Global warming is a highly debated and still controversial phenomenon. There is evidence for a warming trend in SST and AT (Fig. 2), but this trend cannot be indisputably considered as a signal of global warming, at least not for Sitka. It only suggests that there is a reversal of the long cooling period, begun in 1958, as part of the very low-frequency temperature oscillation. The trend is not a monotonic temperature increase associated with the increase of CO_2 and other greenhouse gases in the atmosphere, although a global warming signal might also be present in the data. Royer (1989) predicted that the recent warming trend would reverse, bringing below-normal water and air temperatures to the northwest Pacific over the next 5–15 yr.

Table 1. Correlation matrix for the physical variables and, in parentheses, the lag (in months) at which the maximum correlation occurred. The significance levels from an F-test are represented by (**) for $p < 0.01$ and by (*) for $p < 0.05$. The column headings are the lead variables (e.g., the maximum correlation between PPT and SST occurred when PPT led SST by 2 mo). See text for explanation of abbreviations.

	AT	FWD	NEP	PPT	SLH	SSS	SST
FWD	0.42 (0) **	1.00 (0)					
NEP	0.63 (0) **	0.40 (0) **	1.00 (0)				
PPT	0.19 (0) **	0.69 (0) **	0.25 (0) **	1.00 (0)			
SLH	0.12 (9) *	0.13 (0) **	0.12 (0) **	−0.10 (−2) *	1.00 (0)		
SSS	−0.14 (0) **	−0.32 (0) **	−0.13 (0) **	−0.45 (0) **	0.12 (−11) **	1.00 (0)	
SST	0.66 (0) **	0.27 (2) **	0.32 (0) **	0.18 (2) **	0.05 (3)	0.15 (0)	1.00 (0)
UPW	−0.10 (0) *	−0.12 (0) *	−0.21 (0) **	0.14 (1) **	−0.22 (0) **	−0.11 (0) *	−0.11 (−7) *

Oceanic and atmospheric forcing in the Northeast Pacific Lagged pairwise cross-correlations for the monthly anomalies provided a general perception of the coupling of the physical variables and the major environmental driving forces. Most of the highest correlation coefficients in the pairwise cross-correlations are obtained at zero lag, as shown in the correlation matrix (Table 1). Many of the statistically significant r values obtained are due to colinearity. The most noteworthy pairs are summarized in Fig. 5.

The signs of the correlation coefficients are consistent with Hollowed and Wooster's (1992) classification of type-A (cold, dry) and type-B (warm, wet) winters, from a southeastern Alaskan perspective (Table 2). Type-B conditions are characterized by a high value of the NEPPI index (intensified Aleutian Low), high temperatures (AT, SST), high rainfall and FWD, and correspondingly low SSS. The Aleutian Low intensifies winds from the south along the Alaska panhandle, resulting in coastal convergence of surface waters (negative upwelling) and increased SLH.

Cross-correlation analysis shows that AT leads SST, and that temperature anomalies on time scales of 1–12 mo are atmospherically driven (Davis 1976). The process of ocean–atmosphere heat transfer is very complicated, and whether the ocean heats the air or the air heats the ocean depends on the temporal and spatial scales considered. In the Gulf of Alaska, temperature anomalies to depths of 200 m are in phase and often larger at depth (Royer 1989). Apparently the temperature anomalies are propagated by horizontal convection, not vertically from the surface. In coastal waters and in sheltered areas, where the

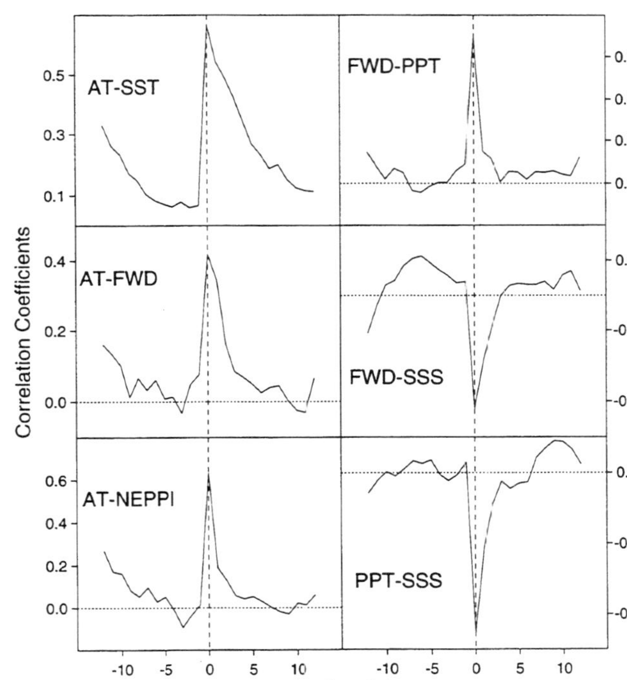

Fig. 5. Pairwise lagged correlograms for the most important pairs of physical variables. The first variable mentioned is the leading variable in the positive lag section of the graph (e.g., AT–SST: AT leads SST from lag 0 to 12 mo). See text for explanation of abbreviations.

continental effect is strongly felt, AT becomes the driving variable for the other oceanographic conditions. Perhaps this is why Freeland (1989) could not perceive the El Niño Southern Oscillation (ENSO) signal in the SST along the coast of British Columbia. In the Bering Sea, atmospheric forcing is the driving force and the only acceptable explanation for the ENSO signal (Niebauer 1988).

Two main scenarios of heat transfer emerge. In the first, ocean currents carry heat from the tropics and travel north, warming up the air and creating pressure anomalies that drive the weather patterns, and thus all the oceanic variables (Leetmaa 1989). The warm currents cool upon meeting cold regions at the northernmost latitudes, then travel back at depth toward the equator. The ENSO signal in this case would be transmitted north through the transformation of the equatorial Kelvin wave (which causes warming off the coast of Peru) into two coastal branches, one of which travels north with warm waters. The signal could easily be missed in coastal waters because of the strong continental effect; this was the case for British Columbia coastal records studied by Freeland (1989).

The second scenario is based on atmospheric forcing, whereby changes at the equator cause pressure anomalies that are teleconnected to high-latitude atmospheric systems that determine the origin of the air masses. Air masses come either from the north-central Pacific (warm and full of moisture) or from the north (cold and dry) depending on the intensity and position of the Aleutian Low and its counterpart, the North American High. These air masses determine the climatological and oceanographic conditions in the Gulf of Alaska. These two apparently contradictory scenarios of ocean–atmosphere heat transfer can be reconciled by considering them as part of a feedback cycle occurring at longer time scales over the entire Pacific Ocean (Trenberth 1994).

Part II: ecology and population dynamics of Pacific herring in Sitka Sound

Pacific herring recruit to the sac roe fishery at age 3. The stages of the recruitment process include maturation of the reproductive organs of the spawners, spawning, incubation, hatching, larval growth, metamorphosis, growth and survival in the nursery areas, and finally migration to the adult feeding grounds (Pitcher and Hart 1982). In addition to the direct effect of the number of eggs spawned, many factors have been related to herring recruitment variability. Hypotheses about recruitment variability often assume that survival during one of the stages is critical to recruitment.

Harsh overwintering conditions can weaken the spawners, thus reducing egg production either in number or size, and subsequently reducing recruitment. Size-specific fecundity was found to be correlated with water temperature during the prespawning months (Hay 1985; Winters and Wheeler 1987; Tanasichuk and Ware 1987). Eggs are also lost from predation and storms (Haegele et al. 1981). Changes in physical conditions at the shoreline during the incubation period could thus effect recruitment.

Survival during the larval stage is thought to be a key determinant of recruitment variability. Cushing (1975) hypothesized that larval survival depends on a temporal match between hatching of fish larvae and the plankton bloom. The presence of food at a certain level of the water column is required for successful first feeding of fish larvae (Lasker 1985); storms or wind-driven upwellings that disrupt the concentration of food in the water column can cause starvation. Sinclair (1988) transformed the transport–retention hypothesis (Parrish et al. 1981) into the member–vagrant hypothesis, giving it a general theoretical form with evolutionary implications. According to this hypothesis, drifting larvae are transported to favorable nursery areas where survival is maximized. Annual variations in physical circulation patterns result in different survival rates and hence different recruitment levels. There is evidence that size-selective predation on fish larvae at their early stages could be important for some populations. The growth rate of larvae, which can be affected by food availability and temperature, could determine survival and hence recruitment success (Anderson 1988).

Recruitment to the Sitka Sound herring stock was estimated with a modified version of GENCAGE (GENeralized Catch-at-AGE analysis (Funk and Sandone 1990)). Input to the program includes an assumed survival rate, initial estimates of exploitation rates, initial estimates of vulnerability for each age, initial population size estimates, tables of average weights-at-age and catch-at-age for each year, and auxiliary biomass estimates from spawn deposition surveys. The configuration of the model has catch occurring instantaneously at the start of each index year. GENCAGE uses a population model to reconstruct the numbers-at-age for each year by minimizing the difference

Table 2. Anomalies of physical variables during a Type-B winter and the relationship with herring recruitment.

Variable	Anomaly	Correlation with recruitment
Northeast Pacific Pressure Index	+	+
Air temperature	+	+
Sea-surface temperature	+	+
Precipitation	+	+
Freshwater discharge	+	+
Sea-surface salinity	−	−
Upwelling	−	−
Sea level height	+	+

between predicted and observed catches-at-age and between predicted and observed biomass. The analysis produces estimates of exploitation rates for each year, vulnerability of each age, parameters of the partial recruitment function, and the initial cohort sizes or recruitment estimates.

Estimated age-3 recruitment of the years 1971–90 was log transformed to obtain an index to compare with the physical variables. A lagged cross-correlation function was calculated over a 3-yr time span: the year during which the larvae had hatched, the year before, and the year after. Correlation coefficients were calculated with the statistical package S-plus (Becker et al. 1988) and plotted as correlograms of the physical variables versus recruitment (Fig. 6). The lagging did not reduce the number of pairs because the data for physical variables encompassed the lags considered. This means that r values with an absolute value greater than 0.44 are statistically significant at the 0.05 probability level. The physical variables: NEPPI, AT, SST, PPT, FWD, and SLH were generally positively correlated with recruitment, although the 0.05 significance level was hardly reached. During the year of spawning, SSS and UPW were generally negatively correlated with recruitment. The general signs of the correlations between recruitment and the physical variables were consistent with the signs of correlations among the physical variables (Table 2). To a large extent, changes in the physical variables are synchronous and herring recruitment has a consistent response to these fluctuations.

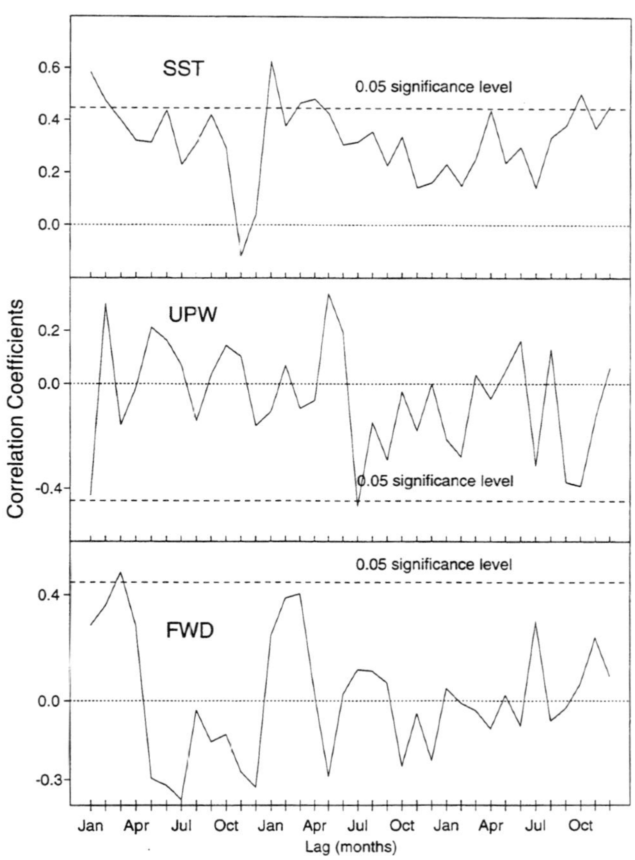

Fig. 6. Correlograms of SST, UPW, and FWD versus the recruitment index lagged over a period of 3 yr: the year before spawning occurs, the spawning year (April), and the year after. See text for explanation of abbreviations.

Stepwise regression analysis (Wilkinson 1990), with the recruitment index as dependent variable and the physical variables as independent variables, was used to find the "best" subset of explanatory variables. For each of the physical variables, the month with the highest correlation to recruitment was selected for the multiple regression. The procedure was performed forward and backward with different tolerance and α-levels. The foremost variable to be included in the regression model was January SST (Table 3). The only other variable to be entered in the model was July upwelling, resulting in a modest increase in r^2 and decrease of the p value (Table 3).

The regression of the recruitment index on SST was then analyzed with a set of regression diagnostics to examine the validity of the regression and to demonstrate that the model chosen was reasonable for the data at hand. The residuals of the least-squares regression, their standardized and studentized versions (Kleinbaum et al. 1988), as well as the residuals from the least median of squares regression (Rousseeuw 1984) and the robust regression (Hampel et al. 1986) were inspected. In all cases the regression assumptions were not violated. The variance of the residuals was reasonably constant over the range of SST (homoscedastic). There was little difference in the regression coefficients estimated by robust regression, least median of squares (LMS) regression, and least-squares regression.

To verify that the cyclic nature of the recruitment index did not affect the results by making the response values not statistically independent, a dummy variable (Z) was included in the model (Kleinbaum et al. 1988). For the years with the highest recruitment index (the 1972, 1976, 1980, and 1984 year-classes) Z equaled one; otherwise, Z equaled zero. With the inclusion of the dummy variable in the multiple regression, the coefficient of SST did not change considerably nor did its p value increase (Table 3). The dummy variable explained the peak values of the recruitment index (years for which Z equaled one), and therefore caused r^2 to increase from 0.40 with SST alone to 0.69 with the dummy variable. Finally, regression of SST on Z showed that SST was not correlated with the dummy variable ($r = 0.1$). The cycles in the recruitment index were not the major source of correlation with SST, even if part of the variability in those peaks was explained by Z, which is an unknown variable. The regression model with SST accurately reflected a true relationship in the data studied, and explained 40% of the recruitment variability.

Having established the correlation between herring recruitment and SST, we sought to validate the relationship by investigating longer time scales, potential mechanisms, and recruitment patterns in neighboring herring stocks. Recruitment for the period 1928–64 was estimated by cohort analysis of data from the herring reduction fishery in southeastern Alaska (V. Wespestad, NMFS, Seattle, WA, personal communication). The reduction fishery exploited a mixture of herring stocks including the Sitka Sound stock; therefore, the early and more recent data are not directly comparable. Sitka AT data were used because the SST data extend back only to 1943. AT is a valid proxy for SST because the two were highly correlated during years that they overlapped (Table 1). For the 1928–64 time period recruitment was positively correlated with AT at lags corresponding to 1 mo prior to spawning to 3 mo after the year-class was spawned. The maximum correlation of 0.4 was significant at the 0.05 probability level. The early recruitment data thus confirmed the more recent relationship.

Table 3. Results of the regressions between log recruitment and selected physical variables. df = degrees of freedom; *t* = Student's *t*-test.

	Residual SE	Multiple R^2	N	F	df	p	Regression Coefficient	SE	t	p
Recruitment on SST	1.3191	0.4017	20	12.086	1,18	0.0027				
Intercept							3.6996	0.2951	12.5384	<0.001
SST							0.0105	0.0030	3.4765	0.0027
Recruitment on SST and upwelling (UPW)	1.1348	0.5818	20	11.823	2,17	0.0006				
Intercept							3.6903	0.2560	14.0966	<0.0001
UPW							−0.08932	0.0330	−2.7054	0.0150
SST							0.0100	0.0026	3.8413	0.0013
Recruitment on SST and a dummy variable	0.9707	0.694	20	19.2784	2,17	<0.0001				
Intercept							3.2577	0.2433	13.3925	<0.0001
Dummy							2.1977	0.5454	4.0298	<0.0001
SST							0.0114	0.0022	5.1024	<0.0001
Environmentally dependent Ricker model	1.428	0.441	17	5.531	2,14	0.017				
Intercept							5.383	0.659	8.173	<0.001
Eggs							−1.103	0.379	2.912	0.011
SST							0.014	0.004	3.059	0.008

The spawning date of Pacific herring stocks occurs progressively later in direct relation to latitude (Hay 1985). For a given stock, such as Sitka, the onset of spawning varies linearly with winter temperature; cold temperatures delay spawning. In Sitka Sound, spawning occurs around the beginning of April, and the highest correlation between monthly SST and date of first spawn was obtained for March. The stage of the tidal cycle did not explain any of the residual variability in spawning date. The temperature–spawning date relationship is useful for predicting when the roe fishery will occur. Time of spawning could affect subsequent recruitment, but there is no obvious reason why spawning earlier in warm years would enhance survival.

An environmentally dependent Ricker model of the form:

$$\ln(R/E) = a - bE + c\text{SST} \quad (1)$$

was fitted to investigate the combined effects of egg production (E) and SST anomalies on recruitment (R). Parameters a, b, and c were all significant as estimated by linear regression (Fig. 7, Table 3). Inclusion of egg production decreased the residual sum of squares marginally from the model with SST only. However, the multiple regression model revealed compensation in herring recruitment that was not apparent without accounting for the joint effect of SST (Collie 1991).

Data on the mean length of age-3 herring (L_3) from 1971–90 were used to test for SST and density-dependent growth. We found a weak negative correlation ($r = -0.41$) between L_3 and the SST anomaly during the previous year. There was a stronger negative correlation ($r = -0.61$) between L_3 and total herring biomass the previous year. With both SST and biomass included in a multiple regression, the biomass effect was significant ($p = 0.03$), and the SST effect was insignificant ($p = 0.88$). Biomass and SST are positively correlated ($r = 0.64$), confounding the influence on L_3. We conclude that SST affects L_3 indirectly through the effect on recruitment, subsequently on biomass, and density-dependent growth.

Part III: spatial pattern of herring recruitment

Our final analysis to validate the relationship between herring recruitment and SST was to compare the spatial and temporal patterns of both variables. It has previously been noted that recruitment of the seven B.C. herring stocks is positively correlated (Ware and McFarlane 1989), and that the strength of the correlation is less between the northern and southern groups of stocks. Ware and McFarlane related this pattern to the hydrographic domains; the northern B.C. stocks are in the transitional zone and the southern stocks in the coastal upwelling domain. We extended the analysis of spatial recruitment patterns by including additional herring stocks. The hypothesis we tested is that the spatial pattern of recruitment variation matches the hydrographic domains as indicated by the spatial pattern of SST.

From north to south (Fig. 1), the stocks included in our analysis are Bering Sea (Zheng et al. 1993), Prince William Sound, Seymour Canal, Kah Shakes-Ketchikan (Fritz Funk, Alaska Department of Fish and Game, Juneau, AK, personal communication), Sitka Sound (Collie 1991), seven B.C. stocks (Haist and Schweigert 1992), and Puget Sound (Bargeman 1988). The recruitment data are from year-classes 1948–89,

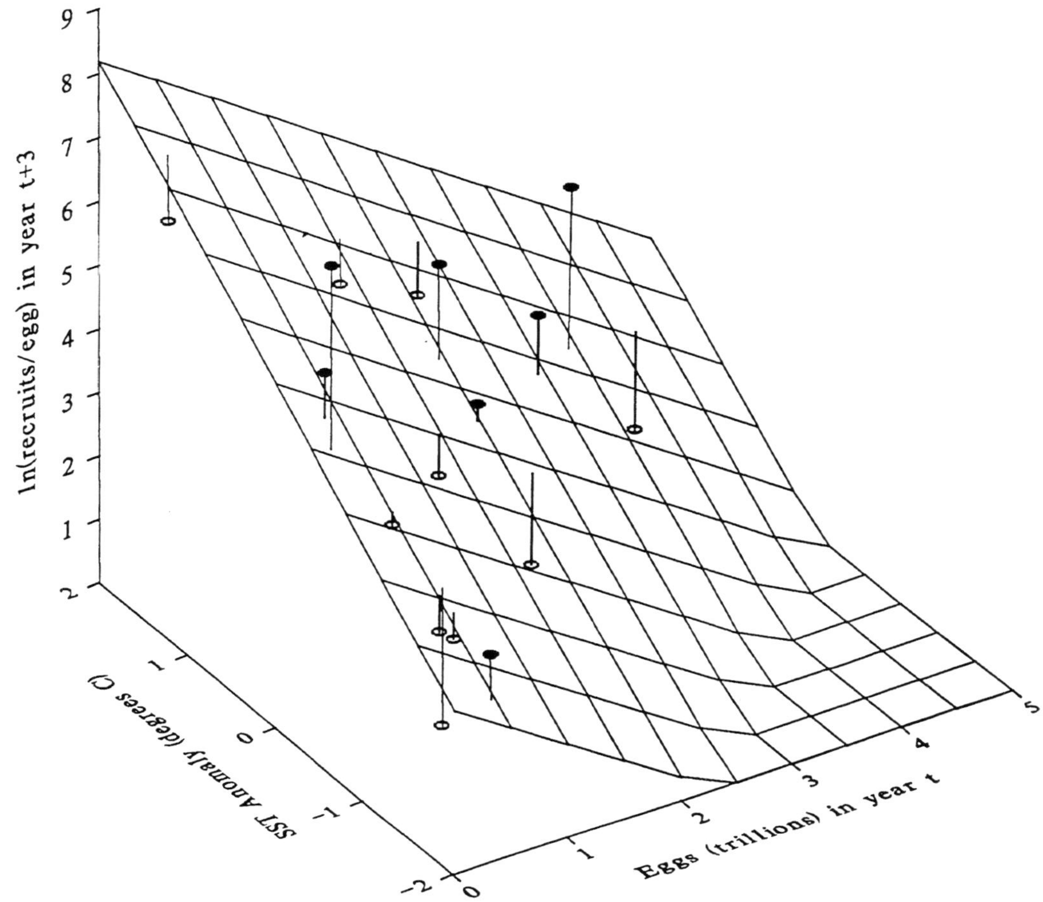

Fig. 7. Environmentally dependent Ricker model. The plane is the regression fit of equation 1 to the data (circles). Vertical lines measure the distance between the data point and the predicted value on the plane. Solid circles indicate positive residuals, open circles negative residuals. Eggs are measured in trillions and the resulting age-3 recruits in millions.

with many early years missing for the Alaska and Puget Sound stocks. The data were log transformed to normalize the distributions. Otherwise, synchronous strong year-classes would strongly weight the correlation coefficients. Cluster analysis was applied to the matrix of log recruitment values with group average sorting (Wilkinson 1990). The 13 herring stocks cluster more or less geographically (Fig. 8), even if the clustering algorithm reverses the order of some of the branches. The four cluster groups are southeastern Alaska plus Prince Rupert, Bering Sea, the rest of the B.C. stocks, and Puget Sound. Recruitment of the northern and southern clusters of stocks is negatively correlated and the split between the two occurs at approximately 54°N.

Monthly SST data from 2° × 2° quadrants bordering the Northeast Pacific coast were obtained from the Comprehensive Ocean–Atmosphere Data Set (COADS; Woodruff et al. 1987). Data from 1949–88 were selected for analysis as there were relatively few missing values during this period. Grand mean temperatures were calculated for each month of the year, weighted by the number of observations in each month or year cell. From each observation the grand mean for that month was subtracted to obtain SST anomalies with the seasonal cycle removed. Owing to the larger size of the data matrix (10 quadrants × 480 anomalies), the correlation matrix was calculated first for input to the group-average clustering.

As might be expected, SST anomalies along the Northeast Pacific coast are positively correlated and cluster almost geographically (Fig. 9). As hypothesized, the SST stations also clustered into two main groups with the division again occurring at 54°N. Thus, the spatial pattern of herring recruitment matches that of SST. The division at 54°N corresponds to the position where the subarctic boundary intersects the North American coast (Fulton and LeBrasseur 1985), at least in warm years. The subarctic boundary divides the transitional zone in the south from the coastal downwelling domain in southeastern Alaska.

Discussion

The interaction of fish populations with their marine environment has been substantiated through numerous works (Cushing 1982; Rothschild 1986; Sinclair 1988). Recruitment has received particular attention because of its crucial importance to fisheries as the sole process for renewal of harvestable fish. However, none of the hypotheses pertaining to recruitment

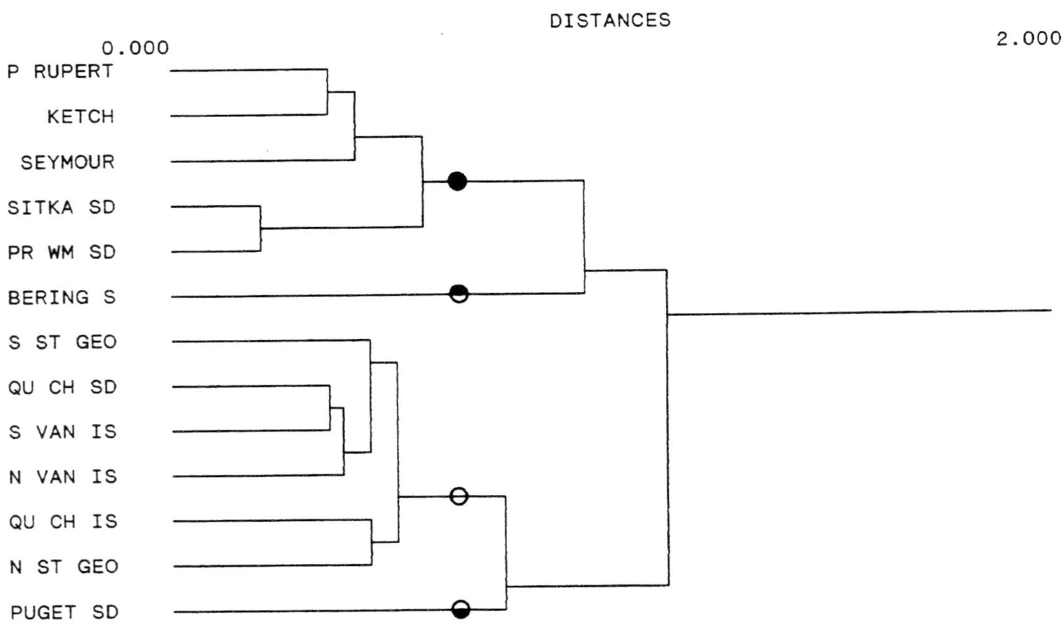

Fig. 8. Cluster diagram of log-transformed herring recruitment data from different stocks along the Northeast Pacific coast From top to bottom: Prince Rupert (BC), Ketchikan (Kah Shakes AK), Seymour Canal (AK), Sitka Sound (AK), Prince William Sound (AK), Bering Sea, Southern Strait of Georgia (BC), Queen Charlotte Sound (Central Coast BC), Southwest Vancouver Island (BC), Northwest Vancouver Island (BC), Queen Charlotte Islands (BC), Northern Strait of Georgia (BC), and Puget Sound (WA). Group-average sorting was used and the distance measure is $1 - r$, where r is the correlation coefficient.

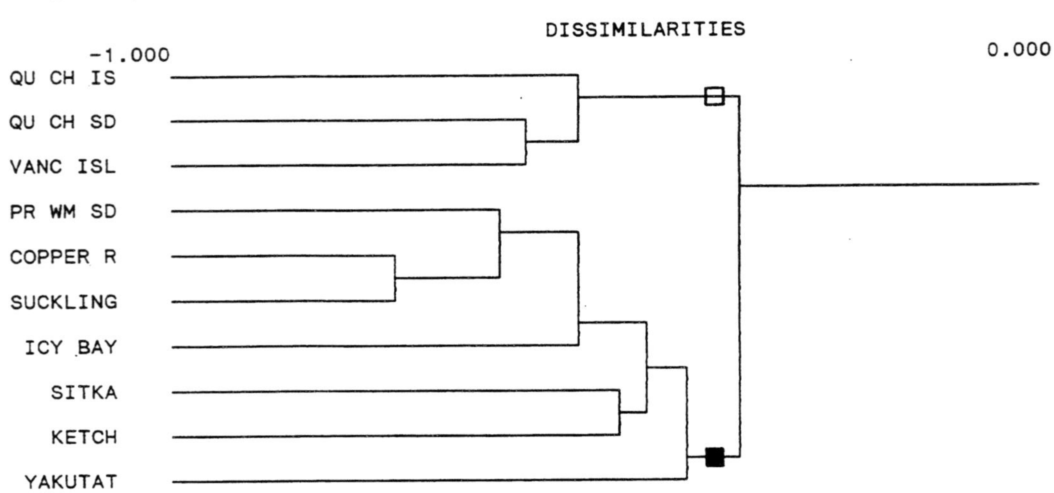

Fig. 9. Cluster diagram of SST anomalies from $2° \times 2°$ quadrants along the coast of the northeastern Pacific Ocean. The stations were given the name of a prominent position on the coast. From top to bottom: Queen Charlotte Islands, Queen Charlotte Sound, Vancouver Island, Prince William Sound, Copper River, Cape Suckling, Icy Bay, Sitka, Ketchikan, and Yukatat. Group-average sorting was used and the distance measure is $-r$.

fluctuations can be convincingly rejected, and the issue of what factors control fish survival prior to recruitment remains largely unresolved (Anderson 1988).

In our study of the possible effects of physical variables on herring recruitment, SST emerged as the physical parameter most highly correlated with recruitment variation. The winter, prespawning temperature explained 40% of the variation in log recruitment for the herring stock of Sitka Sound; the July upwelling index explained an additional 18% (Table 3). Overwintering temperature was similarly found to be the main correlate of recruitment variability for herring in the northwest Atlantic (Winters and Wheeler 1987). The variability in herring recruitment is probably due to a combination of events rather than a single factor. The 4-yr recruitment pattern that started with the 1976 year-class was not explained by SST or the other physical variables we considered. The inclusion of a dummy variable to account for the cycle, increased the r^2 of the multiple regression from 0.4 to almost 0.7.

The correlation coefficient between recruitment and SST was highest at a lag corresponding to the month of January prior to spawning, suggesting that the mechanism whereby temperature acts on recruitment may be related to the maturation processes. As poikilotherms, herring have a metabolism that depends directly on the temperature in the environment. If the temperature increases (decreases), the metabolic processes increase (decrease). During the overwintering period, a rise in temperature will influence the rate of development of sexual products and therefore will influence the arrival of a stock on the spawning ground and the time of spawning and hatching of eggs. If cold temperatures precede spawning, the result is late spawning and a shift of the whole cycle of egg development through metamorphosis to later in the season (Favorite et al. 1977). Timing is the key point in this explanation, but how late or early hatching affects survival remains speculative.

The match–mismatch hypothesis (Cushing 1975) could explain how warmer temperatures and maturation early in the season can be beneficial to herring survival. Herring with ripe gonadic products can be maintained under experimental conditions without spawning by lowering the water temperature for a relatively long period (up to 40 d and longer) (Hay 1985). This shows that full maturation is not the only factor that determines the readiness for spawning and that environmental cues (water temperature, tides (Hay 1990)) must be present for spawning to occur. If all the ecological conditions exist but maturation is late, spawning will be late, and the late-hatched larvae will not be able to find conditions favorable to their survival. Spawning can wait if maturation is complete, but environmental conditions will not wait if maturation is late.

A second explanation of recruitment variability is based not on the timing of the plankton bloom, but on the absolute level of productivity. Coincident with increasing temperatures from 1972–90 has been an apparent increase in zooplankton biomass in the Gulf of Alaska (Brodeur and Ware 1992). Feeding conditions for prerecruit herring therefore may have improved during this period.

The environmentally dependent Ricker model is useful for partitioning recruitment variability due to environmental causes from that due to density dependence. Similar recruitment models have been fitted to Atlantic herring stocks (Winters and Wheeler 1987) and Bering Sea herring (Wespestad 1991). Herring growth appears to depend more on population abundance than on SST. Back calculated age-0 growth of Bering Sea herring was positively related to year-class size and unrelated to SST (Wespestad 1991).

The cluster analysis of herring recruitment and SST data from a range of locations on the Northeast Pacific coast indicated that the appropriate spatial scale of environmental influences on recruitment is the oceanic domain. Hydrographic conditions experienced at coastal stations result from weather systems in the Gulf of Alaska and perhaps farther into the central North Pacific. Even though the processes affecting herring recruitment may be local, the environmental signal is propagated on a larger scale. The central North Pacific (CNP) winter atmospheric pressure index (Cayan and Peterson 1989) correlates negatively with NEPPI, such that a negative CNP anomaly corresponds with a positive SST anomaly in Sitka. Of 12 strong herring year-classes in southeastern Alaska during 1928–84, 7 were spawned in years with a negative CNP, 4 in years with 0 CNP and only 1 occurred in a year with a positive CNP index. Thus, variations in Pacific herring recruitment in southeastern Alaska are coherent with atmospheric pressure anomalies over the central North Pacific Ocean.

Although SST anomalies are coherent along the coast of the Northeast Pacific, a discontinuity occurs at approximately 54°N. The North Pacific Current reaches the North American coast at about this latitude and bifurcates, defining the coastal hydrographic domains. The northern branch forms the Alaska Current in the coastal downwelling domain and the southern branch becomes the California Current in the coastal upwelling domain (Ware and McFarlane 1989). Between the upwelling and downwelling domains is a transition zone, because the position of the subarctic boundary varies from year to year. The subarctic boundary blocks the northward flow of warm water and thus explains the cluster pattern in SST anomalies (Fig. 9).

Recruitment of Pacific herring is synchronous within oceanic domains but not between domains (Fig. 8). Many of the herring stocks exhibit a 4-yr periodicity in recruitment, but the northern and southern stocks are out of phase by 1 yr. In southeastern Alaska, the dominant year-classes have been 1976, 1980, 1984, and 1988; in British Columbia, strong cohorts followed 1 yr later in 1977, 1981, 1985, and 1989.

The recruitment response to SST anomalies also differs between oceanographic domains. The recruitment correlations were all positive within the four main clusters of Fig. 8 and mostly negative among clusters. Whereas recruitment of Sitka Sound herring was positively correlated with SST, recruitment of herring in the Strait of Georgia and on the west coast of Vancouver Island was negatively correlated with SST (Schweigert 1994). This reversal of the SST effect with latitude is consistent with the idea that stocks near the southern limit of their range should be negatively affected by temperature increases while stocks near the northern extent of the range respond positively. A negative correlation was found between herring recruitment and the estimated food ration consumed by Pacific hake (*Merluccius productus*) in the Canadian zone in summer (Ware and McFarlane 1989). The northward migration of hake is temperature dependent, possibly resulting in higher predation on prerecruits in warm years.

Our interpretation of the spatial patterns in herring recruitment differs somewhat from that of Ware and McFarlane (1989). An obvious difference is that we log transformed the recruitment data, which equalizes the influence of large and small year-classes on the correlation coefficient. The Prince Rupert or north coast stock has a higher similarity with the southeastern Alaska stocks than with the B.C. stocks. The remaining B.C. stocks clustered together without the geographic break illustrated by Ware and McFarlane (1989). Thus, we were able to discriminate herring recruitment patterns between the Coastal Downwelling and Transition Zones, but not between the Transition Zone and Coastal Upwelling Zone.

Herring populations are submitted to natural variability expressed in different aspects of the physical and biological worlds. Environmental factors that affect year-class success of herring probably range from single short-term events, such as a storm or freshet that affects the survival of a cohort in an isolated inlet, to large-scale events that affect the productivity and circulation of large areas for 1 yr or more. All these events can be related by one path or another to more subtle variations in the climate. Whether local or large scale, the ultimate reason for the environmental variability is that the ocean climate is highly unstable within limits. The fundamental unpredictability of atmospheric and climatological changes precludes any longer-term predictions of the biological response to climate change. Fluctuations in herring abundance that have occurred on time scales of 20–100 yr may be nonlinearly related to changes in physical variables such as ocean temperature. With the advent of chaotic modeling, an "orderly lack of periodicity" (Gleick 1987) may be expected.

Acknowledgements

We thank Fritz Funk, Vivian Haist, Jake Schweigert, Vidar Wespestad, and Jie Zheng for providing herring recruitment data; Anne Hollowed provided the COADS SST data. Reviews by Vivian Haist and an anonymous editor helped us improve the manuscript. This publication is the result of research sponsored by Alaska Sea Grant with funds from the National Oceanic and Atmospheric Administration, Office of Sea Grants, Department of Commerce, under grant no. NA90AA-D-SG066 (project no. R/07-B) and from the University of Alaska Fairbanks with funds appropriated by the state. The U.S. Government is authorized to produce and distribute reprints for government purposes, notwithstanding any copyright notation that may appear hereon.

References

Anderson, T. 1988. A review of size dependent survival during pre-recruit stages of fishes in relation to recruitment. J. Northwest Atl. Fish. Sci. 8: 55-66.

Anthony, V.C., and M.J. Fogarty. 1985. Environmental effects on recruitment, growth and vulnerability of Atlantic herring (*Clupea harengus harengus*) in the Gulf of Maine region. Can. J. Fish. Aquat. Sci. 42(Suppl. 1): 158–173.

Bailey, K.M., and R.C. Francis. 1985. Recruitment of Pacific Whiting, *Merluccius productus*, and the ocean environment. Mar. Fish. Rev. 47(2): 8–15.

Bakun, A. 1973. Coastal upwelling indices, west coast of North America, 1946–71. NOAA Tech. Rep. NMFS SSRF-671. NOAA, Seattle, WA.

Bargeman, G. 1988. Recruitment in the north Puget Sound herring stock. Unpublished manuscript, Washington Department of Fish and Game, Seattle, WA. 6 p.

Becker, R.A., J.M. Chambers, and A.R. Wilks. 1988. The new S language. A programming environment for data analysis and graphics. Wadsworth & Brooks/Cole Advanced Books & Software, Pacific Grove, CA.

Brodeur, R.D., and D.M. Ware. 1992. Long-term variability in zooplankton biomass in the subarctic Pacific Ocean. Fish. Oceanogr. 1: 32–38.

Cayan, D.R., and D.H. Peterson. 1989. The influence of North Pacific atmospheric circulation on streamflow in the west, p. 375–398. *In* D.H. Peterson [ed.] Aspects of climate variability in the Pacific and western Americas. Geophys. Monogr. 55.

Cleveland, W.S. 1979. Robust locally weighted regression and smoothing scatterplots. J. Am. Stat. Assoc. 74: 829–836.

Collie, J.S. 1991. Herring population dynamics and management in Sitka Sound, Alaska, p. 463–476. *In* Proceedings of the International Herring Symposium. Alaska Sea Grant Coll. Program Rep. 91-01.

Cushing, D.H. 1975. Marine ecology and fisheries. Cambridge University Press, Cambridge and New York.

Cushing, D.H. 1982. Climate and fisheries. Academic Press, London, New York. 373 p.

Davis, R.E. 1976. Predictability of sea-surface temperature and sea level pressure anomalies over the North Pacific Oceanogr. J. Phys. Ocean. 6: 249–266.

Emery, W.J., and K. Hamilton. 1985. Atmospheric forcing of interannual variability in the Northeast Pacific Ocean: connections with El Niño. J. Geophys. Res. 90: 857–868.

Favorite, F., T. Laevastu, and R.R. Straty. 1977. Oceanography of the northeastern Pacific Ocean and eastern Bering Sea, and relations to various living marine resources. Northwest and Alaska Fisheries Center, NMFS, NOAA, Seattle, WA.

Freeland, H.J. 1989. Sea-surface temperatures along the coast of British Columbia: regional evidence for a warming trend. Can. J. Fish. Aquat. Sci. 47: 346–350.

Fulton, J.D., and R.J. LeBrasseur. 1985. Interannual shifting of the subarctic boundary and some biotic effects on juvenile salmonids, p. 237–252. *In* W.S. Wooster and D.L. Fluharty [ed.] El Niño north. Wash. Sea Grant Publ. WSG-W085-3.

Funk, F.C., and G.J. Sandone. 1990. Catch–age analysis, growth, and natural mortality of Prince William Sound herring 1973–1988. Fish. Res. Bull. 90-01, Alaska Department of Fish and Game, Juneau, AK. 45 p.

Gleick, J. 1987. Chaos: making a new science. Penguin, London, UK.

Guiot, J. 1988. Climate of the Hudson Bay region and effects of solar activity, volcanic dust and lunar nodal tidal signal, p. 377–395. *In* C.G. Castagnoli [ed.] Solar–

terrestrial relationships and the earth's environment in the last millennia. North-Holland, Amsterdam, The Netherlands.

Haegele, C.W., R.D. Humphreys, and A.S. Hourston. 1981. Distribution of eggs by depth and vegetation type in Pacific herring (*Clupea harengus pallasi*) spawnings in southern British Columbia. Can. J. Fish. Aquat. Sci. 38: 381–386.

Haist, V., and J.F. Schweigert. 1992. Stock assessments for British Columbia herring in 1991 and forecasts of the potential catch in 1992. Can. MS. Rep. Fish. Aquat. Sci. 2146: 82 p.

Hamilton, K., and L.A. Mysak. 1986. Possible effects of the Sitka eddy on sockeye (*Oncorhynchus nerka*) and pink salmon (*Oncorhynchus gorbuscha*) migration off southeastern Alaska. Can. J. Fish. Aquat. Sci. 43: 498–504.

Hampel, F.R., E.M. Ronchetti, P.J. Rousseeuw, and W.A. Stahel. 1986. Robust statistics: the approach based on influence functions. Wiley, New York, NY.

Hay, D.E. 1985. Reproductive biology of Pacific herring (*Clupea harengus pallasi*). Can. J. Fish. Aquat. Sci. 42(Suppl. 1): 111–126.

Hay, D.E. 1990. Tidal influence on spawning time of Pacific herring (*Clupea harengus pallasi*). Can. J. Fish. Aquat. Sci. 47: 2390–2401.

Hicks, S.D., and W. Shofnos. 1965. The determination of land emergence from sea level observations in southeastern Alaska. J. Geophys. Res. 70: 3315–3320.

Hollowed, A.B., K.M. Bailey, and W.S. Wooster. 1987. Patterns of recruitment of marine fishes in the Northeast Pacific Ocean. Biol. Oceanogr. 5: 99–131.

Hollowed, A.B., and W.S. Wooster. 1992. Variability of winter ocean conditions and strong year classes of Northeast Pacific groundfish. *In* R.R. Dickson, P. Maelkki, G. Radach, R. Saetre, and M.P. Sissenwine [ed.] Hydrobiological variability in the ICES area, 1980–89. ICES Mar. Sci. Symp. 195: 433–444.

Kerr, R.A. 1988. Sunspot–weather link holding up. Science (Wash., D.C.) 242: 1124–1125.

Kleinbaum, D.G., L.L. Kupper, and K.E. Muller, 1988. Applied regression analysis and other multivariable methods. PWS-KENT Publishing C., Boston, MA.

Lasker, R. 1978. The relation between oceanographic conditions and larval anchovy food in the California current: identification of factors contributing to recruitment failure. Rapp. P.-v. Réun. Cons. int. Explor. Mer 173: 212–230.

Lasker, R. 1985. What limits clupeoid production? Can. J. Fish. Aquat. Sci. 42(Suppl.): 31–38.

Leetmaa, A. 1989. The interplay of El Niño and La Niña. Oceanus 32(2): 30–34.

Mysak, A.L. 1986. El Niño, interannual variability and fisheries in the Northeast Pacific Ocean. Can. J. Fish. Aquat. Sci. 43: 464–497.

Niebauer, H.J. 1988. Effects of El Niño–southern oscillation and north Pacific weather patterns on interannual variability in the subarctic Bering Sea. J. Geophys. Res. 93: 5051–5068.

Norcross, B.L., and R.F. Shaw. 1984. Oceanic and estuarine transport of fish eggs and larvae: a review. Trans. Am. Fish. Soc. 113: 153–165.

Parrish, R.H., C.S. Nelson, and A. Bakun. 1981. Transport mechanisms and reproductive success of fishes in the California Current. Biol. Oceanogr. 1: 175–203.

Pitcher, J.T., and P.J.B. Hart. 1982. Fisheries ecology. Avi Pubishing Co., Westport, CT. 414 p.

Prager, M.H., and J.M. Hoenig. 1989. Superposed epoch analysis: a randomization test of environmental effects on recruitment with application to chub mackerel. Trans. Am. Fish. Soc. 118: 608–618.

Ricker, W.E. 1975. Computation and interpretation of biological statistics of fish populations. Bull. Fish. Res. Board Can. 191: 382 p.

Rose, G.A., and W.C. Leggett. 1988. Atmosphere–ocean coupling and Atlantic cod migrations: effects of wind-forced variations in sea temperatures and currents on nearshore distributions and catch rates of *Gadus morhua*. Can. J. Fish. Aquat. Sci. 45: 1234–1243.

Rothschild, B.J. 1986. Dynamics of marine fish populations. Harvard University Press, Cambridge, MA. 277 p.

Rousseeuw, P.J. 1984. Least median of squares regression. J. Am. Stat. Assoc. 79: 871–880.

Royer, T.C. 1982. Coastal fresh water discharge in the northeast Pacific. J. Geophys. Res. 87: 2017–2021.

Royer, T.C. 1986. Temperature fluctuations in the Northeast Pacific from 1954 to 1985 in response to El Niño/southern oscillations and longer period forcing. Int. North Pac. Fish. Comm. Bull. 47: 203–208.

Royer, T.C. 1989. Upper ocean temperature variability in the Northeast Pacific Ocean: is it an indicator of global warming? J. Geophys. Res. 94: 18 175 – 18 183.

Schweigert, J.F. 1994. Environmental effects on long-term population dynamics and recruitment to Pacific herring (*Clupea pallasi*)populations in British Columbia, p. 569–583. *In* R.J. Beamish [ed.] Climate change and northern fish populations. Can. Spec. Publ. Fish Aquat. Sci. 121.

Sinclair, M. 1988. Marine populations: an essay on population regulation and speciation. University of Washington Press. Seattle, Wash. 252 p.

Soutar, A., and J.D. Isaacs. 1974. Abundance of pelagic fish during the 19th and 20th centuries as recorded in anaerobic sediment off the Californias. U.S. Natl. Mar. Fish. Serv. Fish. Bull. 72: 257–273.

Tanasichuk, R.W., and D.M. Ware. 1987. Influence of interannual variations in winter sea temperature on fecundity and egg size in Pacific herring (*Clupea harengus pallasi*). Can. J. Fish. Aquat. Sci. 44: 1485–1495.

Trenberth, K.E. 1994. Decadal climate variations in the North Pacific, p. 15–24. *In* R.J. Beamish [ed.] Climate change and northern fish populations. Can. Spec. Publ. Fish Aquat. Sci. 121.

Ware, D.M., and G.A. McFarlane. 1989. Fisheries production domains in the Northeast Pacific Ocean, p. 359–379. *In* R.J. Beamish and G.A. McFarlane [ed.]

Effects of ocean variability on recruitment and an evaluation of parameters used in stock assessment models. Can. Spec. Publ. Fish. Aquat. Sci. 108.

Wespestad, V.G. 1991. Pacific herring population dynamics, early life history, and recruitment variation relative to eastern Bering Sea oceanographic factors. Ph.D. thesis, University of Washington, Seattle, WA. 237 p.

Wilkinson, L., 1990. SYSTAT: the system for statistics. SYSTAT, Inc.Evanston, IL.

Winters, G.H., and J.P. Wheeler. 1987. Recruitment dynamics of spring-spawning herring in the northwest Atlantic. Can. J. Fish. Aquat. Sci. 44: 882–900.

Woodruff, S.D., R.J. Slutz, R.L. Jenne, and P.M. Steurer. 1987. A comprehensive ocean–atmosphere data set. Bull. Am. Meteorol. Soc. 68: 1239–1250.

Zebdi, A. 1991. Identification of environmental causes of recruitment variation in the herring stock of Sitka Sound, Alaska. M.Sc. thesis, University of Alaska, Fairbanks, AK. 132 p.

Zheng, J., F. Funk, G.H. Kruse, and R. Fagen. 1993. Threshold management strategies for Pacific herring in Alaska. Proceedings of the International Symposium on Management Strategies for Exploited Fish Populations, 21–24 Oct. 1992, Anchorage, Alaska.

Potential effects of global climate change on Dungeness crab (*Cancer magister*) populations in the northeastern Pacific Ocean

Robert A. McConnaughey and David A. Armstrong

McConnaughey, R.A., and D.A. Armstrong. 1995. Potential effects of global climate change on Dungeness crab (*Cancer magister*) populations in the northeastern Pacific Ocean, p. 291–306. *In* R.J. Beamish [ed.] Climate change and northern fish populations. Can. Spec. Publ. Fish. Aquat. Sci. 121.

Abstract: Predictions of the effects of global climate change on marine populations depend on a clear understanding of the link between population variability and the environment. This paper summarizes recent studies concerned with the mechanistic basis for large variations in Dungeness crab (*Cancer magister*) populations along the U.S. west coast. Trawl survey data, time-lagged commercial landings data and calculated Ekman and geostrophic transport vectors are the bases for a recruitment model incorporating restrictive juvenile habitat requirements and interannual variability in alongshore transport. Our analyses suggest that larvae in near-surface waters are advected considerable (and variable) distances alongshore, while being maintained close to shore and close to suitable juvenile habitat. This model contrasts with the prevailing paradigm for the species, which suggests that zoeae undergo progressive seaward transport after hatching followed by cross-shelf return of megalopae before settlement. Given the expected changes in temperature, salinity, and pH for the northeastern Pacific Ocean and the functional relationship between larval transport and year-class strength, we conclude that indirect effects of greenhouse warming on coastal circulation will be most significant to *C. magister* populations.

Résumé : Les prévisions quant aux effets du changement climatique global sur les populations marines dépendent d'une bonne compréhension du lien qui existe entre la variabilité d'une population et l'environnement. Le présent document résume de récentes études ayant porté sur la base mécaniste des importantes variations constatées dans les populations de crabes dormeurs (*Cancer magister*) le long de la côte ouest américaine. Les données recueillies par les chalutiers, les données différées sur les débarquements commerciaux ainsi que les vecteurs calculés d'Ekman et de déplacement géostrophique constituent les bases d'un modèle de recrutement incorporant les besoins restrictifs d'habitat des crabes juvéniles et la variabilité interannuelle du déplacement le long de la côte. Nos analyses portent à croire que les larves des eaux superficielles sont soumises au phénomène d'advection à des distances considérables (et variables) le long de la côte, tout en étant maintenues près de la côte et près des habitats appropriés pour les crabes juvéniles. Ce modèle contraste avec le paradigme dominant pour l'espèce, selon lequel les zoés subissent, après leur éclosion, un déplacement progressif vers la mer suivi d'un retour au plateau lorsqu'elles ont atteint le stade mégalope avant de se fixer. Étant donné les changements prévus dans la température, dans la salinité et dans le pH du nord-est du Pacifique, et la relation fonctionnelle qui existe entre le déplacement des larves et l'effectif des classes annuelles, nous concluons que les effets indirects du réchauffement dû à l'effet de serre sur la circulation côtière seront très importants pour les populations de *C. magister*.

Introduction

Factors that influence the distribution and abundance of meroplanktonic larvae ultimately affect the size of adult populations. In particular, recruitment levels to marine populations may vary considerably, at least in part, reflecting the vagaries of coastal circulation (Johnson and Hess 1990; Schumacher and Kendall 1991; McConnaughey et al. 1992, 1993). In response to this uncertainty, various life history strategies have evolved, including aspects of reproductive biology that are designed to optimize survival of the larval stages (Cushing 1975; Sinclair 1988) and maximize exposure to preferred substrate types at settlement (Caddy 1986). Life history characteristics that are common among species may be indicative of responses that are particularly adaptive. For example, many fish and invertebrate species along the U.S. west coast

R.A. McConnaughey. Alaska Fisheries Science Center, National Marine Fisheries Service, 7600 Sand Point Way N.E., Seattle, WA 98115, USA.

D.A. Armstrong. School of Fisheries, University of Washington, Seattle, WA 98195, USA.

spawn during the winter months (Parrish et al. 1981), suggesting that environmental conditions during this period are consistent with high larval survival.

Time-averaged life history responses, however, are not infallible as safeguards against year-class failure given the existence of environmental variability at a number of temporal scales. Indeed, the responsiveness of marine populations to environmental shifts over a range of periodicities may be inferred from a number of studies (interannual: Caddy and Gulland 1983; decadal: Hollowed et al. 1987; Baumgartner et al. 1992; geologic: van Geen et al. 1992). For example, Ware and Thomson (1991) presented a strong circumstantial argument that pelagic fish stocks in the southern California Bight tended to rise and fall in phase with a long-period (~40–60 yr) oscillation in primary and secondary production that is related to wind-forced coastal upwelling. Similarly, significant biological effects have been attributed to ephemeral physical manifestations of El Niño–Southern Oscillation (ENSO) events along the Pacific Coast (Wooster and Fluharty 1985). However, systematic and sustained changes in the environment (e.g., global greenhouse warming), which either alter the mean state or increase its variability, are likely to be of far greater consequence to marine populations. Changes of this type could have profound effects on populations of species that have evolved life history strategies keyed to a different marine regime.

Attempts to predict effects associated with global climate change must first identify extant links between populations and the environment before projecting a response to an undefined future state. Although many researchers propose that the globally averaged equilibrium surface temperature will increase from 1.5 to 4.5°C because of greenhouse warming (Leovy and Sarachik 1991), the effect on coastal circulation is unclear (cf. Bakun 1990 with Hsieh and Boer 1992). One obvious effect on the marine environment would be changes in sea-surface temperature. However, unless these changes are of sufficient magnitude to induce physiological stress, then indirect effects (such as changes in coastal circulation and primary production) will probably be more important to species with meroplanktonic larvae. Consequently, recruitment studies that describe functional links with the environment are an important component of interdisciplinary efforts to estimate the impacts of global climate change on marine ecosystems.

This paper summarizes recent studies concerned with the mechanistic basis for large variations in Dungeness crab (*Cancer magister*) populations along the U.S. west coast (McConnaughey et al. 1992, 1993). Because aspects of the *C. magister* life history are consistent with those of other commercially or ecologically important species, its recruitment mechanism may be representative of a larger group of coastal upwelling species and may thus provide insight beyond the present context. We begin our synthesis by summarizing relevant aspects of the *C. magister* life history and features of the physical oceanography along the west coasts of the United States and Canada. We then qualitatively compare alongshore and cross-shelf circulation during the pelagic larval phase with juvenile crab abundance along the Washington coast during 1983–88 to develop hypotheses for statistical analysis with a 40-yr (1951–90) time series of commercial landings data. A conceptual recruitment model is described, projected changes in the marine environment as a result of global warming are summarized, and the implications of these changes to the *C. magister* recruitment process are discussed.

The *C. magister* life history

Cancer magister supports important fisheries along the west coasts of the United States and Canada where populations are characterized by large interannual fluctuations in abundance (Fig. 1). The complex life history is typical of marine invertebrates with meroplanktonic larvae and a broadcast reproductive strategy. Eggs hatch in nearshore waters and larval ontogeny consists of five zoeal stages and a megalopa. Time to complete development increases from south to north, with approximately 4–5 mo (~January–May) required in northern waters (Lough 1976; Jamieson and Armstrong 1991). Larvae migrate vertically in the water column and generally occupy near-surface layers (Lough 1976; Reilly 1983; Jamieson et al. 1989; Hobbs and Botsford 1992). At the end of the pelagic larval phase, megalopae settle on sandy bottoms in coastal waters less than 50 m deep and in estuaries, with considerable variation in the magnitude of settlement from year to year (McConnaughey et al. 1992). Commercial exploitation begins approximately 4 yr after settlement (Botsford

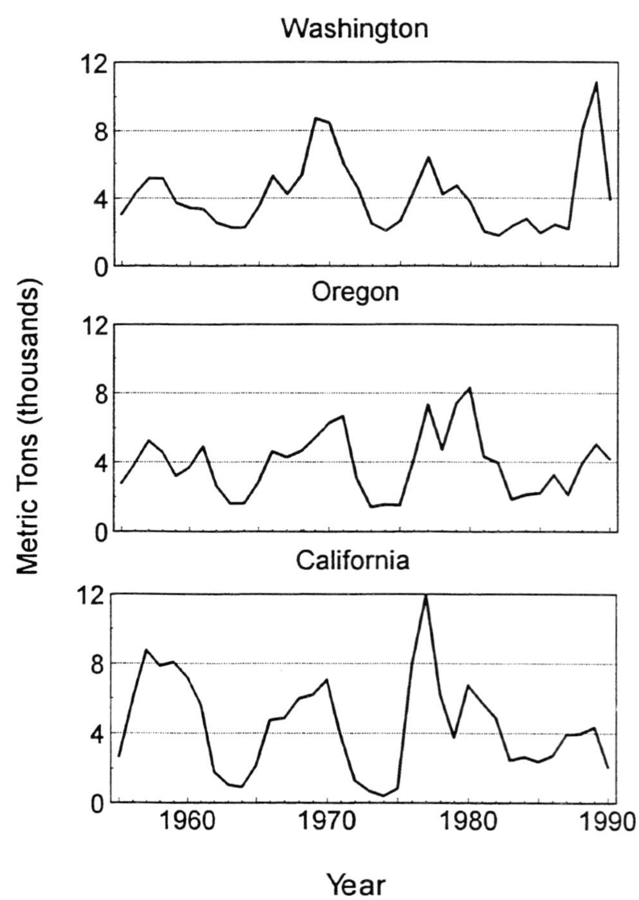

Fig. 1. Commercial fishery landings of *C. magister* along the U.S. west coast during 1955–90.

1984) and harvests also vary considerably (Fig. 1). A similar pattern of variation exists in central California, Oregon, and Washington, suggesting that either discrete populations are responding similarly to a large-scale environmental factor or that recruits to the populations are drawn from a common larval pool.

The functional basis for *C. magister* variability has received considerable attention, particularly since the unexpected collapse of the San Francisco Bay fishery in the early 1960s. Historically, this was an important fishery, which had produced crab for market continuously since the mid-1800s (Dahlstrom and Wild 1983). The explanations proposed by researchers have largely been correlative in nature and have generally focused on a single biotic or abiotic factor (reviewed by Hankin 1985; Botsford 1986). None of these, however, has found widespread acceptance, and many have been summarily rejected because of inappropriate time lags or the lack of a reasonable mechanism.

Relevant physical oceanography

The coastal waters of Washington state are part of the California Current system, an eastern boundary current (upwelling) system (Hickey 1979, 1989) (Fig. 2). The major components of this system are: (1) the California Current, which occurs well offshore and is characterized by a diffuse equatorward flow during all seasons; (2) the California Undercurrent, the jet-like counterflow beneath the California Current along the continental slope; and (3) the Davidson Current, which flows northward over the shelf during winter and is replaced during spring–summer by a coastal jet flowing equatorward. The continental shelf, delimited by the 200-m isobath, is broadest off northern Washington (~70 km) and narrows to 15 km or less to the south.

Nearshore currents are highly correlated with seasonal weather patterns and associated marine winds. Winter (~September–March) winds are largely from the southwest and associated with regular storm activity while summer (~April–September) winds are typically from the northwest. There are, however, important latitudinal differences in both the magnitude and the amplitude of the seasonal signal (Strub et al. 1987a). The mean cross-shelf circulation during winter (summer) is landward (seaward) in the surface layers and seaward (landward) at depth, with downwelling (upwelling) near the coast. Following the spring transition (March–April; Strub et al. 1987b), a rapid and large-scale change from the winter to the summer regime, there are periods of southward and landward flow along the Washington coast. This condition is the result of current forcing by remote winds (via coastal trapped waves) wherein alongshore currents lead the local wind stress by 1 or 2 mo (Hickey 1989).

Gyral circulation prevails near the Strait of Juan de Fuca, and the circulation off Vancouver Island is largely driven by the flux of low salinity water from the strait; both are generally unresponsive to seasonal changes in the wind regime (Fig. 3) (Thomson et al. 1989). The Vancouver Island Coastal Current is characterized by persistent poleward flow in areas landward of the 100 m isobath. This flow is continuous with circulation to the north and extends past the shelf break during winter. The Shelf-Break Current is centred over the 200-m isobath and reverses seasonally in response to prevailing wind patterns.

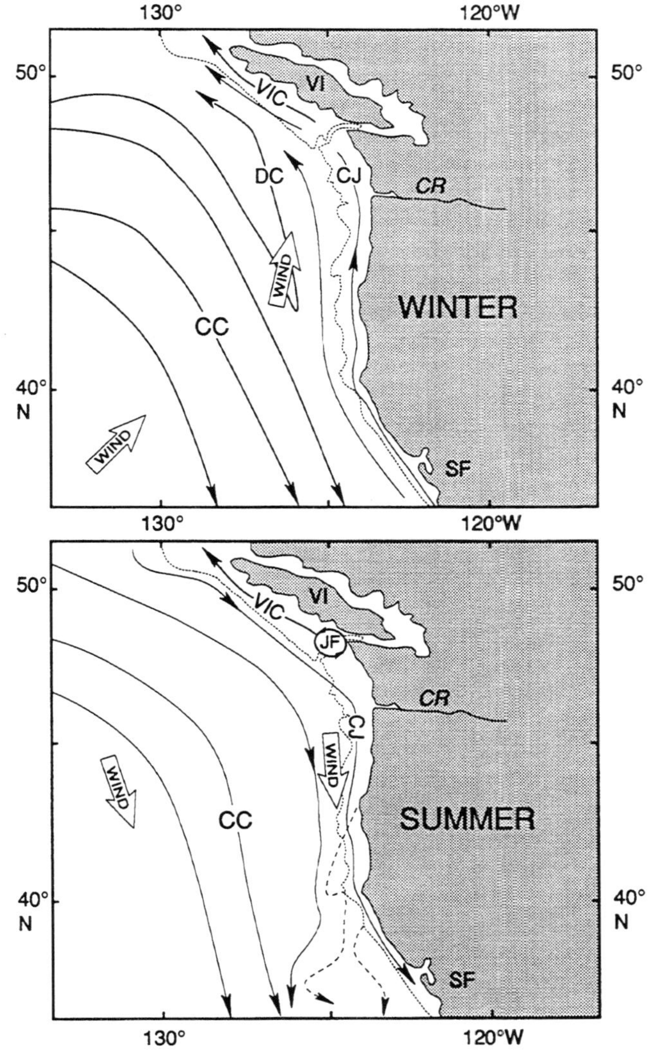

Fig. 2. Schematic of circulation in the California Current system during winter and summer. The two seasons are separated by a transitional period during spring (the "spring transition"), which generally occurs during March–April. The direction of prevailing winds and seasonal mean currents are indicated with arrows and the 200-m isobath is represented with a dotted line along the continental margin. Labeled in the winter diagram are the California Current (CC), the Davidson Current (DC), the coastal jet along the continental U.S. coast (CJ) and the Vancouver Island coastal current (VIC). In the summer diagram, JF identifies the Juan de Fuca Gyre while a time- and space-dependent meandering coastal jet south of Cape Blanco, Oregon, is represented with a dashed line. The CJ off Vancouver Island is commonly referred to as the Shelf-Break Current.

Superimposed on the seasonal patterns are event-scale fluctuations in wind stress that may temporarily reverse the flow of nearshore currents. These are high-frequency events with periodicity on the order of days, and associated current fluctuations are typically as large as or larger than the seasonal means. Historically, the periods of maximum variance in wind stress and associated currents off Washington and Oregon are

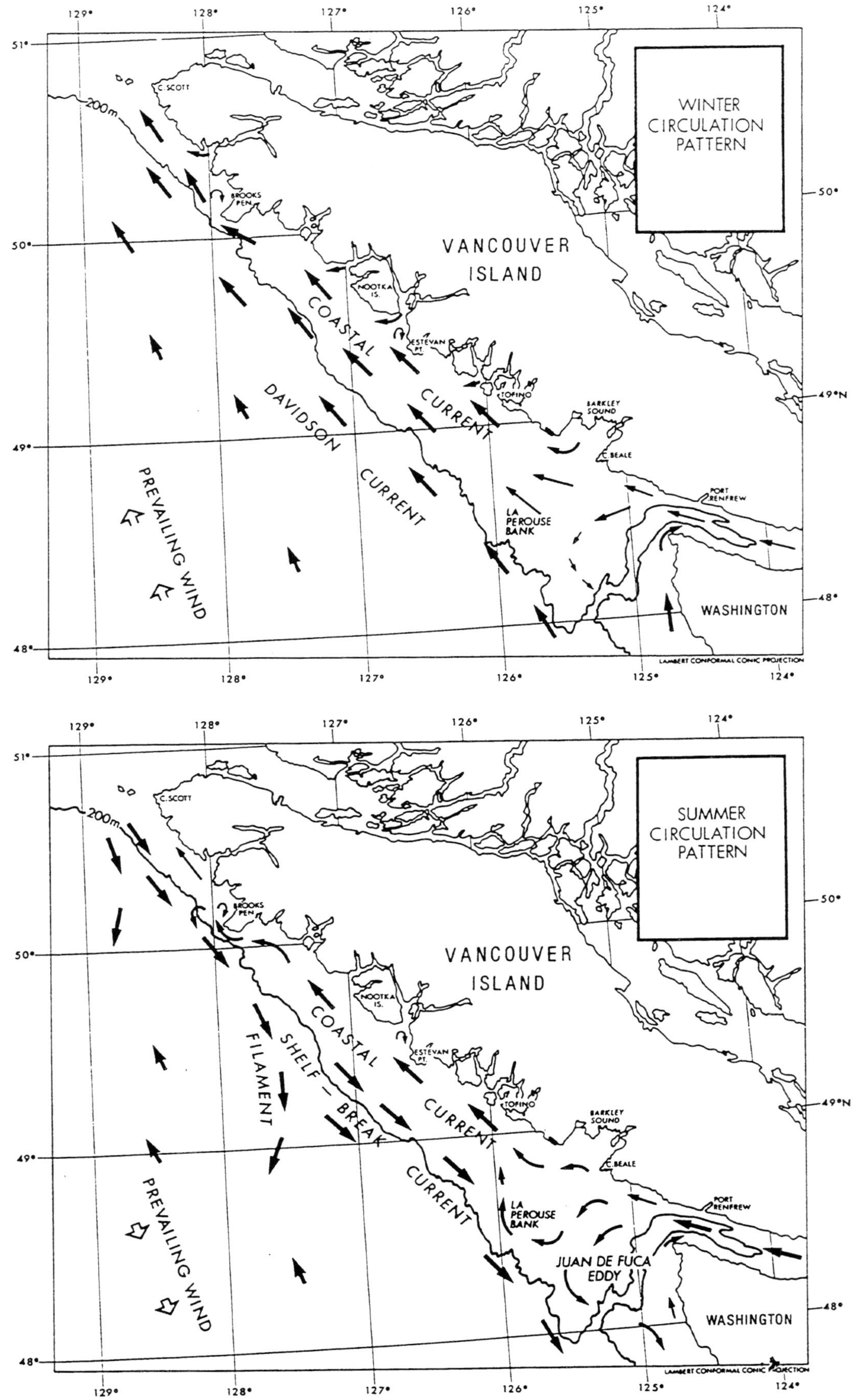

Fig. 3. Seasonal features of the coastal circulation along Vancouver Island, British Columbia. Note that alongshore flow in the Coastal Current does not reverse direction seasonally (adapted from Thomson et al. 1989).

December–January for the period of northward flow and May for southward flow. These are critical months in the ontogeny of *C. magister* larvae, corresponding to hatching of eggs and settlement of megalopae to the benthos.

Data and methods

Physical data

Geostrophic transport

Geostrophic transport (GT) functionally dominates alongshore flow. Off the U.S. west coast, alongshore wind stress can drive a relative (geostrophic) current that flows along the shelf. These currents represent the balance between pressure gradient forces associated with sea level slope and the Coriolis force. During winter, prevailing winds from the southwest cause a large buildup of water along the coast, resulting in strong poleward geostrophic flow. After the spring transition, more moderate winds from the northwest cause a landward sloping surface and weaker southward geostrophic flow.

Hourly staff-zero tide heights were obtained from the National Oceanic and Atmospheric Administration (NOAA), National Ocean Service Tidal Datum Section (S. Lyles, Rockville, MD 20852, personal communication) for a location near the mouth of Willapa Bay (46°42'N, 124°20'W; station no. 9440910). Adjusted sea levels were calculated from these heights and used to estimate geostrophic currents along the coast of Washington during January–May of 1983–1988 (McConnaughey et al. 1992).

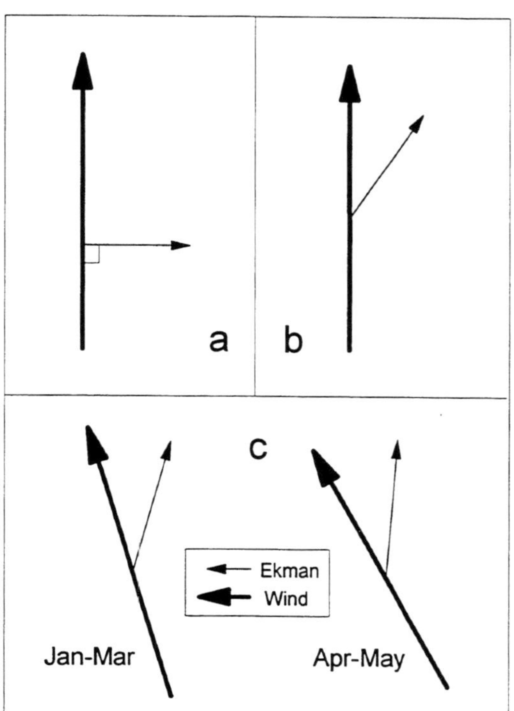

Fig. 4. Rotational corrections applied to **(a)** the original (Bakun) Ekman transport vectors in order to account for **(b)** the near-surface orientation of *C. magister* larvae (Ekman deflection limited to 45°) and **(c)** seasonal aspects of wind veerage near the Washington coast (calculated winds rotated counterclockwise by 26° (winter) and 39° (spring)).

Ekman transport

Ekman transport (ET) describes wind-driven circulation as a function of the wind stress (Ekman 1905; Stacey et al. 1986). It is of primary importance to cross-shelf circulation and also contributes to alongshore transport. In the absence of continuous and direct measurements, calculated winds derived from surface atmospheric pressure gradients are the primary source of long-term oceanic wind data (Thomson 1983). Thus, they are routinely used in oceanographic and fisheries investigations as indicators of wind-driven currents. They are, however, best adapted for open ocean conditions; the simple boundary corrections used (Bakun 1973) are not appropriate in nearshore areas because of wind veerage, a phenomenon that tends to align winds with the alongshore axis (Hsueh and Romea 1983; Thomson 1983; Halliwell and Allen 1984, 1987).

Calculated wind vectors, as well as northward and eastward components of associated ET were obtained from the National Marine Fisheries Service (NMFS) Pacific Fisheries Environmental Group (D. Husby, Monterey, CA 93942, personal communication) for a representative location along the southern Washington coast (46°55'N, 124°20'W). Monthly mean values were obtained for the period 1947–66 and 6-h means were available thereafter. These were seasonally corrected for near-surface coastal waters (Fig. 4) using values reported by Halliwell and Allen (1984, 1987). The ET deflection was set to 45°, to reflect the progressive rotation of wind-driven flow with depth (the familiar "Ekman spiral") and the reported near-surface orientation of *C. magister* zoeae and megalopae (Lough 1976; Jacoby 1982; Reilly 1983; Jamieson and Phillips 1988; Jamieson et al. 1989). This angle represents a time-averaged value for vertically migrating zoeae and megalopae that spend equal periods of time in neustonic waters, moving at low angles to the wind (Thomson 1981; Stacey et al. 1986), and at depth moving normal to the wind. (This angular deflection is comparable with one used by Hobbs et al. (1992).) Because suitable tide-height data were not available before 1973, the alongshore component of ET was used for comparison with the longer time series of commercial landings data. Although ET underestimates the magnitude of alongshore transport, it is highly correlated with GT and thus is a good relative index during both winter and summer. Uncorrected values are reported here although a linear regression suggests that GT is approximately five times greater than alongshore ET (GT = 4.8 ET – 1226 during winter, $P = 0.02$; GT = 4.6 ET – 162 during summer, $P = 0.01$). An Eulerian approach has been used to characterize coastal currents and larval advection. This assumes that transport values at a fixed point adequately reflect conditions experienced by larvae during transport away from this location. Spatial coherence scales for our study area are large and support the contention (>500 km alongshore and ~200 km cross shelf; Hickey 1989)

Biological data

Two data sets have been used to estimate *C. magister* year-class strength along the Washington coast: (1) direct estimates of early juvenile abundance based on trawl surveys during 1983–88, and (2) a 40-yr time series of commercial landings for the coastal Washington fishery (1951–90). Both data sets index cohort abundance because it is widely believed that

C. magister year-class strength is established during the larval phase. Although mortality during the juvenile stage can be considerable (Gunderson et al. 1990; Jamieson and Armstrong 1991; Fernandez et al. 1993), relative year-class strength generally is not affected (McConnaughey 1991). Furthermore, commercial landings will generally reflect trends in (adult) abundance because of the intensity of the fishery (Cleaver 1949; Methot 1986), and a number of other studies have used this approach (reviews by Hankin 1985 and Botsford 1986). There are, however, important differences in lag times for the oceanographic data. Ideally, time lags should be short so as to minimize confounding factors during the intervening period. Analysis of the trawl survey data used relatively short lags (a few months) between the oceanographic and biological indices; however, the time series itself was relatively short at 6 yr. The commercial landings time series, on the other hand, required a substantially longer lag (4 yr) but was of considerably greater scope. Protocols for data collection and the analyses performed are briefly described below with additional details provided elsewhere (McConnaughey 1991; McConnaughey et al. 1992, 1993).

Coastal Washington juvenile trawl survey

During May–September of 1983–88, monthly trawl surveys were conducted along the Washington coast and in two major coastal estuaries, Grays Harbor and Willapa Bay. Overall, 2218 individual samples of juvenile crab were collected using a modified 3-m plumb staff beam trawl designed for juvenile crab and flatfish (Gunderson and Ellis 1986). Sampling in the nearshore occurred at discrete depths along transects oriented normal to the coastline. Depths ranging from 5–59 m were sampled, with stratification defined by the bathymetry. The two estuaries were sampled using a geographically stratified random design. Crab in the samples were counted, measured inside the tenth antero-lateral spines and the sex was determined. Monthly estimates of recently settled (0+) crab populations in the nearshore and estuaries were calculated using area-swept algorithms maintained by the NMFS Alaska Fisheries Science Center (through arrangement with the director of the Resource Assessment and Conservation Engineering Division, G. Stauffer). Modes in monthly length–frequency plots were used to define the 0+ age-class (Fig. 5; T. Wainwright, NMFS Northwest Fisheries Science Center, Seattle, WA 98112, unpublished data), and the combined population estimates for June–September were averaged to produce an annual index of juvenile abundance.

Commercial fishery landings for coastal Washington

The coastal Washington *C. magister* fishery occurs primarily in nearshore waters between Point Grenville and the Columbia River; approximately 10% of the annual catch is taken in estuaries (Barry 1985). Commercial landings for this fishery are estimated from fish-receiving tickets recorded by buyer–processors; annual totals for 1951–90 were assembled for this study. A preliminary analysis of these data indicated a significant autocorrelation at a lag of 1 yr, suggesting that recruitment to the fishery was not "knife-edged." To avoid bias in the statistical analysis (Bakun 1985), the annual landings data were stratified for analysis relative to the mean catch for the 40-yr period (Fig. 6). We also considered it highly unlikely that relatively small differences in crab abundance could be explained on the basis of a single variable, larval advection. Moreover, this approach is consistent with our efforts to identify the crab recruitment mechanism, rather than develop a predictive model.

Analysis

Annual indices of *C. magister* year-class strength from the trawl surveys were qualitatively compared with total alongshore (Ekman plus geostrophic) and cross-shelf (Ekman) transport during the corresponding larval periods. Hypotheses from this exploratory analysis were then evaluated statistically

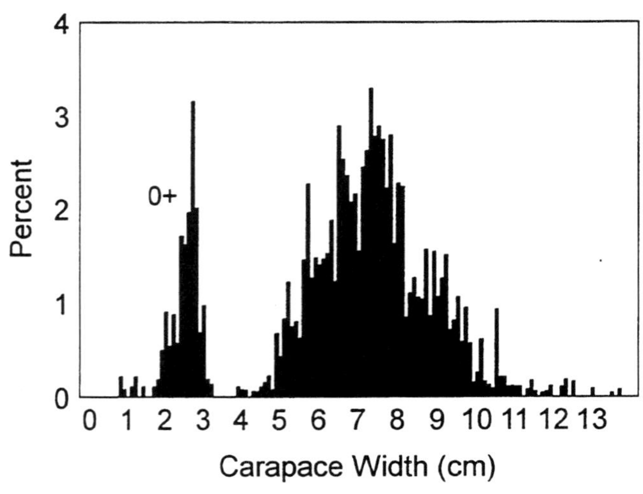

Fig. 5. Representative length–frequency distribution of carapace widths for *C. magister* from the coastal Washington trawl surveys. The limits of the distinct mode at left were used each month to identify young-of-the-year (0) for calculations of abundance using area-swept methodology.

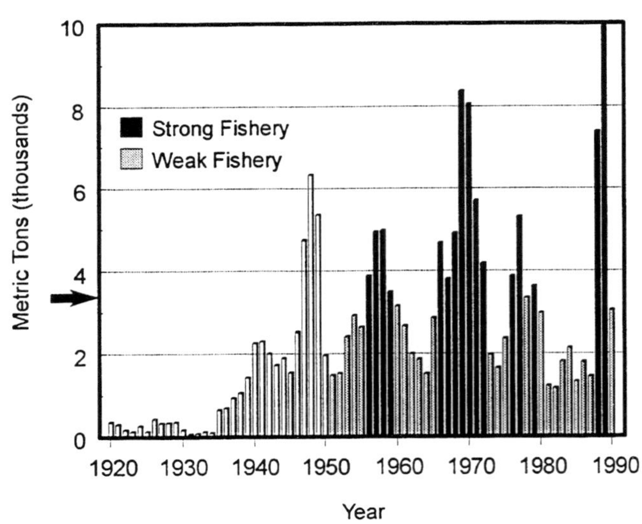

Fig. 6. Commercial fishery landings of *C. magister* for coastal Washington, 1920–90. Landings for 1951–90 were stratified relative to the long-term mean for the period (3454 t, indicated by arrow on the ordinate axis).

with the longer time series of commercial catch data being used as a proxy for year-class strength. One-tailed parametric *t*-tests compared mean alongshore and mean cross-shelf transport (1947–86) for the above-average and below-average year-classes identified from the commercial landings data (1951–90). Landings were lagged 4 yr to reflect mean time from settlement to capture by the fishery (Cleaver 1949; Botsford 1984). A simple time lag was used because a reliable model of crab growth is not available (Wainwright and Armstrong 1993). Although this approach only addresses mean conditions, it is conservative because only the statistical power of the tests was affected.

Results

Development of recruitment hypotheses

Juvenile crab abundance along the Washington coast, as determined by the trawl surveys, varied approximately 40-fold between 1983 and 1988 (Table 1). The composite annual index, based on the June–September average for the three areas, ranged from 12.7 (1986) to 500.6 (1985) million. Although densities in the nearshore were generally less than those in the estuaries, overall abundance was generally greater, reflecting the larger size of the nearshore survey area.

There was also considerable variation in the coastal circulation during the 1983–88 *C. magister* larval periods. Total alongshore transport during January–May (Ekman plus geostrophic) was characteristically northward, ranging from 1450 (1988) to 5050 km (1983). Conditions during 1985 were exceptional in that net alongshore transport was 155 km southward (Table 2). There was a distinct seasonal signal in that strong northward transport predominated during January–March while more moderate southward transport occurred during April and May (McConnaughey et al. 1993). Storm events were particularly important during January–February, with these months accounting for approximately 90% of the total alongshore transport during winter

Table 1. Millions of young-of-the-year (0+) *C. magister* along the southwestern Washington coast. Trawl survey estimates for June–September were averaged to provide an annual index of year-class strength. Values for Willapa Bay during 1983–84 were obtained using the monthly estimates of 0+ crab in the nearshore and Grays Harbor survey areas (R^2_{adj}=0.86; P<0.001; McConnaughey et al. 1992).

		0+ Abundance (millions)			
Year	Month	Grays Harbor	Willapa Bay	Nearshore	Total
1983	June	1.16	1.78	69.40	72.34
	July	4.60	4.82	8.79	18.21
	Aug.	6.38	4.72	8.37	19.47
	Sept.	7.05	3.69	3.15	13.90
	Mean				31.98
1984	June	6.18	9.04	391.73	406.95
	July	9.92	10.17	474.41	494.50
	Aug.	10.85	7.88	78.24	96.97
	Sept.	32.53	16.25	122.13	170.92
	Mean				292.34
1985	June	0.24	0.18	77.39	77.80
	July	2.00	0.97	1258.17	1261.14
	Aug.	5.89	2.58	375.96	384.43
	Sept.	5.60	2.28	271.08	278.95
	Mean				500.58
1986	June	0.04	0.05	0.40	0.50
	July	0.18	0.14	7.57	7.89
	Aug.	0.36	0.26	37.52	38.14
	Sept.	0.58	0.36	3.36	4.30
	Mean				12.71
1987	June	1.09	1.27	18.36	20.73
	July	9.38	5.16	83.91	98.45
	Aug.	4.43	3.47	41.15	49.05
	Sept.	9.07	6.44	339.80	355.31
	Mean				130.89
1988	June	0.54	3.60	14.44	18.58
	July	4.24	6.21	13.95	24.41
	Aug.	6.19	4.40	162.56	173.15
	Sept.	3.73	5.65	75.70	85.09
	Mean				75.31

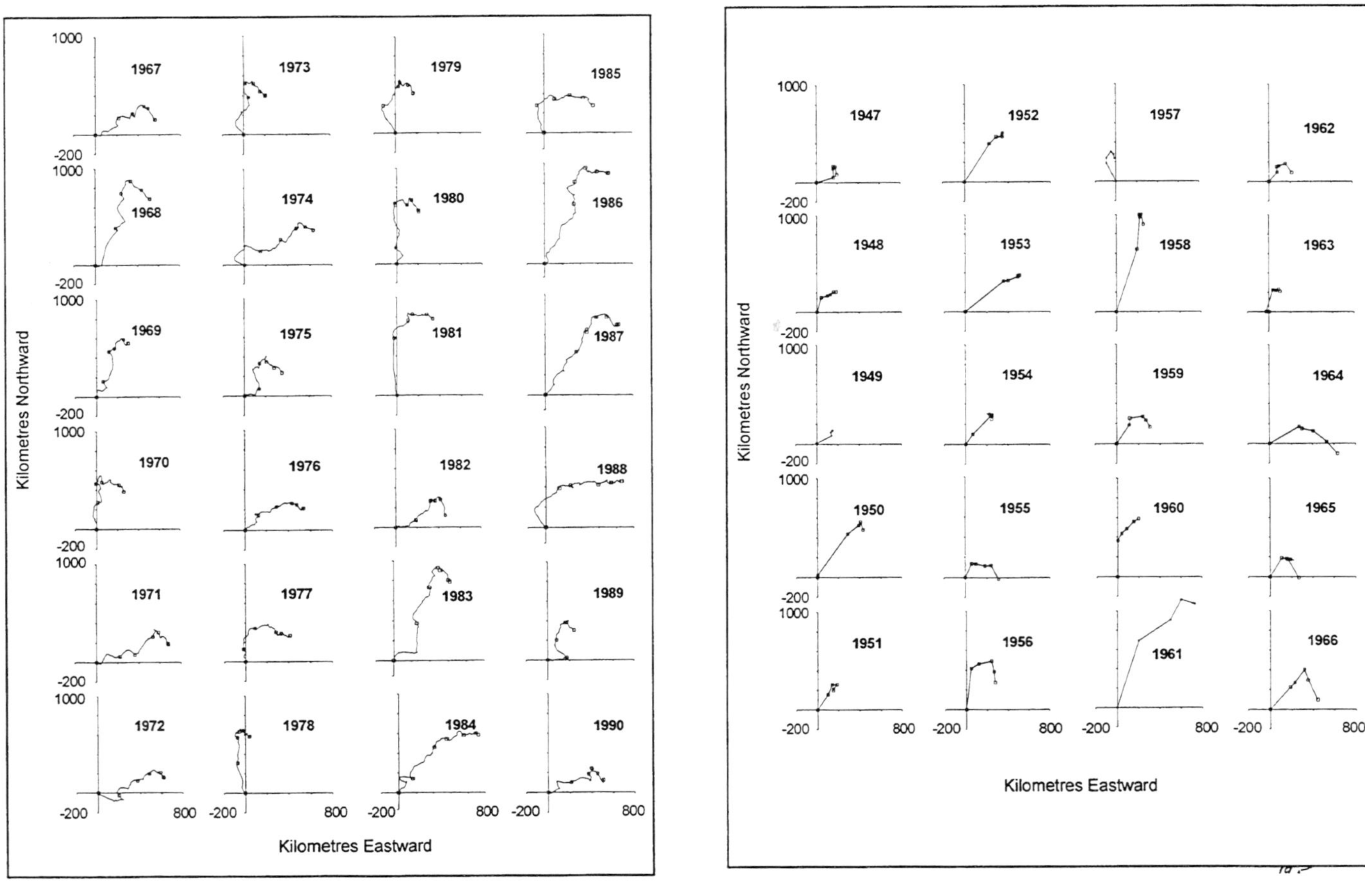

Fig. 7. Progressive vector diagrams of Ekman transport during January–May of 1947–90. The Cartesian planes are in standard geographic orientation (i.e., north to top) and small squares indicate 30-d intervals. The years 1947–86 correspond to commercial landings during 1951–90, and the plots reflect wind-driven transport of larvae in near-surface waters. The years 1987–90 are included as additional evidence of persistent landward and variable northward transport during the pelagic larval phase of *C. magister*.

(McConnaughey et al. 1992). The 1988 larval period was notable in that 74% of the combined alongshore transport occurred during the month of January. Net cross-shelf transport was consistently landward during the 6 yr examined and ranged from 275 (1985) to 460 km (1984). Although there were brief periods of seaward transport near the surface, these were unusual occurrences and were rapidly offset by ensuing landward transport (Fig. 7; 1983–88, as well as all other years). In addition to an obvious difference in the magnitudes of the alongshore and cross-shelf components, there was also an important distinction in variability. Overall, the alongshore component for 1983–88 was considerably more variable (coefficient of variation (CV) = 73%) than the cross-shelf component (CV = 21%).

Comparison of coastal transport during the pelagic larval phase with estimates of *C. magister* settlement magnitude suggested the following hypotheses (McConnaughey et al. 1992): (1) crab settlement (as measured by the 0+ index of abundance) is inversely related to the magnitude of (northward) alongshore transport during January–May, and (2) crab settlement is directly related to the amount of cross-shelf transport (Fig. 8). In particular, the lowest levels of crab settlement were associated with the strongest alongshore transport (1983, 1986) while substantially greater numbers of 0+ crab occurred when alongshore transport during the larval period was relatively weak (1984, 1985). The suggested relationship between crab settlement and cross-shelf transport was not as strong, although reduced landward transport during 1983 and 1986 was associated with lower crab abundance, while greater landward transport and higher levels of crab settlement occurred during 1984, 1987, and 1988.

Evaluation of recruitment hypotheses

Statistical analysis of the 40-yr time series supported the hypothesis that alongshore transport during the pelagic larval phase is an important factor influencing *C. magister* year-class strength along the Washington coast (McConnaughey et al. 1993). Average alongshore transport during the January–May larval period was significantly greater for below-average year-classes (434 km) than for above-average year-classes (241 km; $P < 0.01$; Fig. 9). (Recall that the ET index underestimates alongshore transport by a factor of approximately five, since GT is not accounted for; hence, the absolute difference between the means is much greater.) During 1947–86, the alongshore component of the circulation (CV = 77%) was once again more variable than the cross-shelf (CV = 54%) compo-

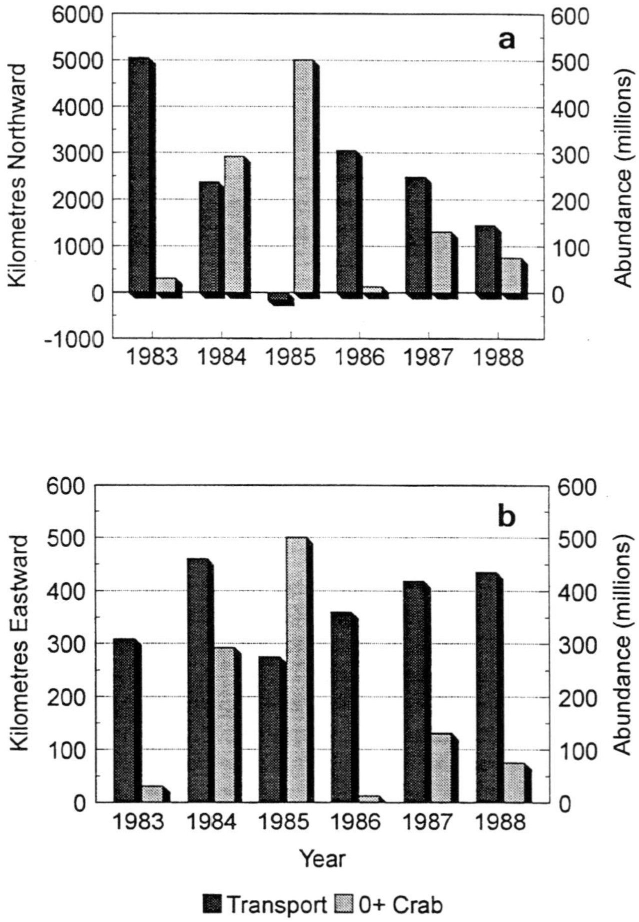

Fig. 8. Comparison of January–May totals of **(a)** alongshore transport (geostrophic and Ekman combined) and **(b)** cross-shelf (Ekman) transport with an annual index of young-of-the-year (0+) *C. magister* abundance along the Washington coast based on trawl survey data.

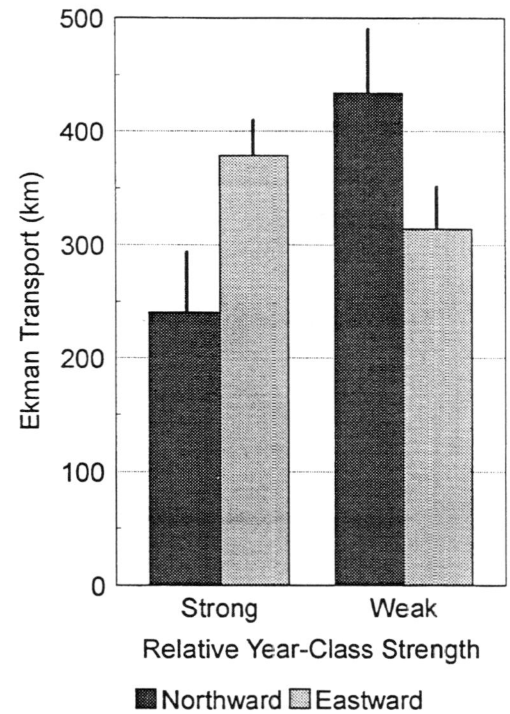

Fig. 9. Mean northward and eastward Ekman transport during January–May for above-average and below-average *C. magister* year-classes. Estimates of year class strength were based on commercial fishery landings in Washington (1951–90) and were lagged 4 yr relative to the transport data (1947–86) to reflect mean time required for recruitment to the fishery after settlement of the megalopae. Vertical lines above bars indicate one standard error for the means.

nent. The analysis of the cross-shelf data, although not statistically significant ($P = 0.14$), suggested that crab settlement was directly related to total landward transport during January–May (Fig. 9). Mean landward transport for the above-average year-classes (379 km) was approximately 20% greater than for the below-average year-classes (315 km). Net cross-shelf transport was landward in all years except one (1957; 14 km seaward), sustained periods of seaward transport were not apparent during any of the years examined (only 23 of 200 mo had net seaward transport), and a net landward position was established in virtually all cases by the end of February (Fig. 7). This suggests a more parsimonious interpretation of these data; namely, landward transport is not a factor that limits *C. magister* recruitment.

Discussion

Patterns of distribution and abundance of marine species often reflect environmental variability, and the scientific literature is replete with studies documenting specific examples. However, a large fraction of these case studies are purely correlative in nature and the underlying mechanism is often unclear. Included in this group are a number of studies examining *C. magister* variability in the northeastern Pacific. For example, Peterson (1973) reported a relationship between upwelling along the U.S. west coast during summer and commercial harvests of *C. magister* 0.5–1.5 yr later. The suggestion that this time lag reflects slow energy transfer from the water column to the benthos, affecting survival of benthic subadults, however, is not widely supported. Similarly, Botsford et al. (1982) discounted salmon predation on *C. magister* megalopae because, although salmon predation can be significant and the two time series were cyclic with the same period, the time lag was once again inappropriate. Cannibalism of juveniles by adults (Botsford and Wickham 1978), nemertean worm predation (Wickham 1979), and epibiotic fouling of egg masses (Fisher and Wickham 1976) have also been rejected as causative factors because of inappropriate time lags, noncyclic behaviour, and/or improbable functional relationships.

Although the prevailing paradigm for *C. magister* recruitment is openly problematic, it has nevertheless guided collective thinking about the species' dynamics for over a decade. Based on larval sampling off central California, Reilly (1983) hypothesized seaward drift of early-stage zoeae after hatching in nearshore waters followed by landward movement of megalopae from considerable distances offshore before settlement.

Table 2. Cumulative alongshore and cross-shelf transport for coastal Washington during the January–May pelagic larval phase of *C. magister* 1983–88. Alongshore transport is the combined total of the alongshore component of Ekman transport and geostrophic flow, while the cross-shelf estimates are the cross-shelf component of Ekman transport.

	Alongshore transport (km north)			Cross-shelf transport (km east)
Year	Geostrophic	Ekman	Total	Ekman
1983	4267	783	5050	309
1984	1790	582	2372	460
1985	–431	276	–155	275
1986	2153	899	3052	360
1987	1787	698	2485	418
1988	998	452	1450	435

However, the presumed cross-shelf transport is directly opposed to the prevailing movement of near-surface waters (Fig. 7) (Hickey 1979, 1989) and the dominant (alongshore) component of the coastal circulation is not adequately addressed (Table 2). The seasonal discrepancy in cross-shelf transport is particularly evident in central California (Wild et al. 1983), where surface waters typically move landward during November–February (early zoeal stages) and seaward during March–August (megalopae). It has been argued that late-stage zoeae reported at substantial distances from shore (i.e., outer shelf and beyond; Lough 1976; Reilly 1983; Hobbs et al. 1992) represent a relatively minor component of the larval population along the U.S. west coast and have received undue emphasis considering the limited scope of nearshore sampling (McConnaughey et al. 1992). Lough (1976), on the other hand, recognized that prevailing southwesterly wind and storm patterns during larval development would tend to retain larvae nearshore by virtue of landward Ekman transport, as verified in this paper (also see discussion by Jamieson et al. (1989) regarding Vancouver Island populations).

Mesoscale alongshore recruitment of *C. magister*

The restrictive juvenile habitat requirements, combined with the dynamics of coastal circulation and the protracted pelagic larval phase, suggest that interannual variability in advection of coastal larvae could be a significant factor affecting *C. magister* year-class strength. In contrast with the prevailing paradigm of larval advection and recruitment discussed above, our analyses suggest that *C. magister* larvae are retained near shore after hatching and may undergo considerable alongshore transport before settlement, with significant interannual variability in the magnitude of this transport. Other analyses indicate that successful recruitment of juveniles depends on encounters between megalopae and suitable substrate at time of settlement. Sandy substrates are the preferred habitat for juveniles (and adults) and these are common only in coastal estuaries and in nearshore areas less than 50 m deep. Finer materials such as silt and clay, common at greater depths, and coarser materials such as rock and gravel, which occur nearer the coast, are apparently unsuitable (McConnaughey et al. 1992). Given the habitat requirements and considerable substrate heterogeneity along the west coasts of Canada and the United States (Bornhold and Barrie 1991; McConnaughey et al. 1992), not all areas support *C. magister* juveniles and the "landing strip" for megalopae entrained in the coastal circulation is thus spatially constrained.

Despite strong dispersive tendencies, meroplanktonic larvae are fundamentally responsible for repopulating established areas. After hatching in the nearshore waters of Washington during winter, *C. magister* larvae are retained nearshore and close to suitable crab habitat by the persistent landward flow of near-surface waters (Fig. 7). Initially, the planktonic larvae undergo extended northward transport, which is offset to varying degrees by southward transport after the spring transition. Overall, the level of recruitment is inversely related to the extent of northward transport during January–May (Figs. 8 and 9) (McConnaughey et al. 1992, 1993). This process suggests that Washington populations receive a significant fraction of recruits from upstream sources (i.e., California, Oregon) and may lose large numbers of larvae to more northern populations. Similarly, recruitment from remote upstream sources has been proposed for ommastrephid (*Illex* spp; Rathjen 1981; Goldman and McGowan 1991), phyllosome (*Panulirus* spp.; Richards and Goulet 1977; Lee et al. 1992) and zoea (*Pandalus jordani*; Hannah 1993) larvae. Because the alongshore flow over the Vancouver Island continental shelf is northward and does not reverse seasonally in areas landward of the 100-m isobath (Fig. 3), larvae that are advected into this area may be effectively lost to the California–Oregon–Washington coastal system. Southward transport after the spring transition may provide for recruitment to southern populations and may be particularly important to northern populations during years characterized by strong northward transport during January–February (e.g., 1988). Under these circumstances, late-hatching larvae may be preferentially recruited because of proportionately less northward transport during their ontogeny

A few strong storms may have a disproportionate effect on recruitment to coastal populations because of correspondingly strong northward flow (e.g., 1988). Storm-related depletion of the larval pool may also result from deepening of the mixed layer and concomitant prey density reductions below a critical minimum level (Lasker 1975, 1978; Incze et al. 1987). Furthermore, storm activity may enhance diffusion of larvae away from suitable crab habitat at depths less than 50 m (i.e., ≤10–15 km from the Washington coast and even closer to the coasts of Oregon and California). This mechanism could explain low numbers of zoeae and megalopae reported at considerable distances from shore (Reilly 1983; Hobbs et al 1992). Similar to our findings, Ishimaru (1936) reported an inverse relationship between the total number of stormy days during the planktonic development of red king crab (*Paralithodes camtschaticus*) larvae and subsequent commercial landings of that year-class.

This characterization of *C. magister* recruitment focuses on general patterns of the coastal circulation (Figs. 2 and 3) by combining two of the major north–south regions along the U.S. west coast (Vancouver Island to Cape Blanco and Cape Blanco to Point Conception (Strub et al. 1987a)). Because of this, certain mesoscale features such as squirts, jets, and filaments, as well as alongshore differences in wind stress, intensity of coastal upwelling, coastal morphology, freshwater inflow, and alongshore advection, have not been addressed. In addition, the importance of variability in larval behaviour that influences vertical position in the water column has not been considered. These factors introduce a level of complexity that exceeds current computational and theoretical limits (Botsford et al. 1994). In the context of global climate change effects, these simplifying assumptions may introduce important variability into the scenarios described below and should eventually be considered.

Climate change and the coastal ecosystem

Existing general circulation models (GCMs) present divergent predictions about the effects of global warming and generally do not include an oceanic component (Glantz and Feingold 1990). Notwithstanding, a large-scale increase in the atmospheric equilibrium temperature is expected to have profound effects on the marine environment, given the fundamental coupling of the two systems. Many climate models predict a globally averaged equilibrium surface temperature increase of 1.5–4.4°C, in response to elevated levels of atmospheric CO_2 (Leovy and Sarachik 1991). In addition to changes in water temperature, other environmental factors that would likely be affected by global warming also influence the distribution and abundance of marine organisms (Sibley and Strickland 1985), including: (1) salinity, (2) pH, (3) vertical stability of the water column, and (4) horizontal currents. A corresponding increase in water temperatures is expected, particularly near the surface and at high latitudes (Sibley and Strickland 1985), with the greatest change expected during winter rather than summer (Leovy and Sarachik 1991). The GCMs also predict general increases in precipitation at high latitudes, particularly during winter (Leovy and Sarachik 1991). This could depress coastal salinities in the northeastern Pacific, which are characteristically low because of relatively high precipitation and land drainage. A general decrease in ocean pH of approximately 0.3 is also predicted, based on a doubling of atmospheric CO_2 (Sibley and Strickland 1985). Overall, however, unless changes in temperature, salinity and pH are of sufficient magnitude to induce physiological stress (e.g., Kondzela and Shirley 1993), indirect effects (i.e., changes in coastal circulation, primary production, and loss of habitat) will probably be most important to *C. magister* and other coastal upwelling species with meroplanktonic larvae.

Given that the general pattern of ocean circulation is controlled by the temperature gradient between the equator and the poles, preferential warming at high latitudes could reduce the magnitude of winds and ocean currents (Francis and Sibley 1991). Greater vertical stability of the water column is thus likely, except in areas influenced by upwelling and significant freshwater influx. However, the present generation of GCMs generally has inadequate spatial resolution (on the order of several degrees) to predict changes in coastal currents and deductive reasoning has therefore been the basis for current expectations. As a result, predicted responses are heavily dependent on the underlying assumptions and differences of opinion have occurred. For example, Bakun (1990) suggested that intensified alongshore winds and greater upwelling during summer will result from disproportionate heating of the land mass relative to the ocean. However, Hsieh and Boer (1992) concluded that the mid-latitude continents do not all follow the Bakun (1990) scenario, and the overall conclusions from the two models also diverge with respect to net change in biological productivity.

Climate change and *C. magister* recruitment

Although potential abiotic effects of global warming on the marine environment have been proposed, the corresponding biotic effects are not always clear because of complex and often opposing feedback mechanisms (Sibley and Strickland 1985). This is particularly true when the implications for a single species in a specific physical system are considered. The inherent variability of *C. magister* populations (Figs. 1 and 8) complicates the matter further. Unless a dramatic decline in abundance or shift in geographic range occurs, it may be difficult to detect a response to global warming in these populations. Notwithstanding, several potential effects of climate change are proposed below, based on anticipated changes in the marine environment and the particulars of the *C. magister* life history. In all likelihood, it will be the sensitivities of the most sensitive life history stage, the pelagic larval phase in this case (Reilly 1983), that will determine the net effect on the species.

Effects related to elevated water temperature

Water temperature increases associated with global warming will likely cause a poleward shift in species ranges, as suggested by anomalous occurrences of southern species during periodic ENSO events (Schoener and Fluharty 1985). Range shifts accompanying a more sustained increase in water temperature could result in increased competition between benthic stages of *C. magister* and previously disjunct populations of congeners. For example, the yellow rock crab (*C. anthonyi*), which presently ranges from Humboldt Bay into southern Baja California, attains carapace widths exceeding 178 mm and, like *C. magister*, prefers open sandy or soft bottom habitat (Parker 1992). Indeed, aggressive interaction and competition among sympatric congeners has been suggested for cancroid crabs along the southern California coast (Winn 1985), which, in that case, display considerably less overlap in habitat utilization than do *C. magister* and *C. anthonyi*. In addition to latitudinal range shifts and encroachments, high bottom temperatures in well-mixed areas could cause benthic juveniles and adults to redistribute in deeper waters. This could potentially expose them to a new suite of predators and competitors while, at the same time, increasing operational costs for commercial fisheries and reducing access for sport fisheries.

Direct effects of global warming on benthic species such as *C. magister* will probably be less profound than for holoplanktonic or pelagic species since, in most cases, temperature changes are expected to be greatest near the surface. However, Wild (1983) has demonstrated that survival and development rates of *C. magister* eggs and larvae are temperature sensitive and the collapse of the San Francisco Bay fishery (~1960), located near the current southern limit for the species, may be attributable to higher water temperatures associated with a northward and eastward shift of the California Current system (Huang 1972). On the other hand, elevated temperatures that are within the thermal tolerances of the species can support more rapid development rates, assuming adequate food resources are available. This would be advantageous to larvae because cumulative alongshore transport distances and advective losses due to export from the coastal system would be reduced, as would the period of exposure to planktivorous predators (e.g., Thomas 1985). However, spawning might occur earlier (Wild 1983) and advective losses could thereby increase because of an extended period of northward transport before the spring transition. Also, synchrony that has evolved between hatching and the availability of prey could be upset and delays in first-feeding could be unfavourable (Paul and Paul 1980; Anger et al. 1981).

Melting of polar ice packs is expected to contribute to a rise in sea level and resulting habitat losses may be important to benthic species with restrictive habitat requirements. Intertidal habitat in drowned river valleys such as Grays Harbor estuary provide important habitat for juvenile *C. magister* (Dumbauld et al. 1993; Fernandez et al. 1993), and many of these areas may be lost, at least temporarily, because low elevation land to replace them is scarce. It has been suggested that innundation of penaeid shrimp nursery areas in Gulf of Mexico estuaries would ultimately reduce juvenile recruitment and commercial fishery yields (Turner 1977; Browder et al. 1989). Similar reductions could occur in *C. magister* populations with important estuarine components.

Effects related to reduced coastal salinities

Direct effects related to changes in salinities are likely to be minor in comparison with those associated with temperature change. Indirect effects on larval advection, however, could be significant. The most dramatic salinity reductions would be restricted to the continental margin, reflecting increased inputs of freshwater from land drainage. Along the Washington coast, for example, the Columbia River discharges nearly 80% of the total drainage between San Francisco and the Strait of Juan de Fuca. Seasonally, the river's plume bends northward along the Washington coast during winter and southward during summer (Hickey 1979). Recent evidence suggests that, although alongshore flow typically follows isobaths, the plume causes seaward deflections during winter, such that crab larvae being advected northward along the coast may be diverted seaward, effectively bypassing stretches of the southern Washington coast (McConnaughey et al. 1993). Any magnification of this effect due to increased discharges could affect the distribution and abundance of *C. magister* recruits. These changes may also be evident in coastal estuaries where mean salinity reductions could reduce the range of suitable habitat for young-of-the-year crab in spring–summer and older juveniles year round. In addition, increased buoyancy flux from the Strait of Juan de Fuca would tend to amplify coastal currents along Vancouver Island (A. Hermann, NOAA Pacific Marine Environmental Laboratory, Seattle, WA 98115, personal communication), further reducing the likelihood of repatriating larvae exported into this area from the California–Oregon–Washington coastal system.

Effects related to lower pH

Changes in pH (generally estimated at −0.3) are not expected to have major effects on most marine organisms (Knutzen 1981). However, increased dissolution of calcium carbonate shells could make individuals more susceptible to injury and to invasions by pathogens (F. Morado, NMFS Alaska Fisheries Science Center, Seattle, WA 98115, personal communication).

Effects related to greater vertical stability or reduced turbulence

In many locations, increased surface heating and freshwater input will result in greater vertical stability (stratification) and reduced turbulence in the water column. This condition would likely confer energetic advantages on developing *C. magister* larvae as a result of heightened primary production at middle to high latitudes and reduced energetic costs associated with maintaining a position in the water column. Reduced wind mixing would result in a shallower mixed layer. This would effectively concentrate prey items and may improve larval feeding success and survival (Lasker 1975, 1978; Incze et al. 1987). However, upwelling and downwelling in areas of appropriate winds and coastline orientation may preclude this benefit in two ways: (1) these features may actually disperse prey, and (2) evidence suggests that upwelling may be responsible for a shift to smaller and generally less productive phytoplankton species (Greve and Parsons 1977; Sibley and Strickland 1985).

Effects related to changes in horizontal currents

Any changes in current patterns over the continental shelf during the pelagic larval phase of *C. magister* could have profound effects on the distribution and abundance of adult populations. Changes in storm tracks and the relative positions of centres of low and high pressure could cause large changes in wind fields and associated temporal and spatial aspects of coastal currents (Sibley and Strickland 1985; Bakun 1990; Hsieh and Boer 1992). Currently, cross-shelf (landward) flow along the U.S. west coast during winter (Fig. 7) is conducive to retention of larvae close to suitable juvenile and adult habitat, and *C. magister* recruitment varies directly with the magnitude of net alongshore transport (McConnaughey et al. 1992, 1993). Changes in the timing, direction, and magnitude of coastal winds related to greenhouse warming could disrupt the time-averaged balance between seasonal currents (Figs. 2 and 3), reproductive timing, and larval dispersal, ultimately affecting recruitment and commercial fishery production.

Various scenarios of change in the dynamics of *C. magister* larvae are imaginable. All other factors being equal, enhanced northward transport would likely reduce average recruitment to coastal Washington populations, as well as those to the south. This would be the result of greater advective losses of larvae beyond the Strait of Juan de Fuca (i.e., Vancouver Island and southeast Alaska) and, if successfully recruited there, would be a mechanism for the poleward shift in species range described above. It has been suggested that a warmer ocean could increase ENSO intensity and frequency (Zebiak and Cane 1991). Low crab settlement (McConnaughey et al. 1992) (Fig. 8) associated with elevated northward flow during ENSO events (Huyer and Smith 1985; U.S. GLOBEC 1992) may result. Reduced landward transport (or increased seaward transport), either as a result of an earlier spring transition or large-scale changes in the winter wind regime, could also have a negative effect on juvenile recruitment by transporting larvae outside the coastal landing strip defined by juvenile habitat requirements (as Bailey (1981) suggested for Pacific hake, *Merluccius productus*). However, these changes may also increase nutrient levels and primary production, thus promoting higher recruitment. Conversely, a reduction in average net northward transport, either due to weaker northward and/or stronger southward transport (whether in an absolute or relative sense) would be consistent with greater retention of potential recruits in California–Oregon–Washington waters.

Admittedly, the effects of global warming on marine ecosystems and their constituent populations are not clear and even the direction of change is equivocal. Despite this uncertainty, the potential magnitude of ecological and economic consequences of wholesale changes in the distributions and abundances of marine species indicates that continued research is prudent. The most efficient course is to identify representative species for major ecosystem types (e.g., coastal upwelling) and describe the mechanism and the nature of links between the biological and physical processes. Recent work with *C. magister* demonstrates the potential importance of larval advection to recruitment in coastal upwelling systems, and comparative studies with other species are underway. Results of this type should prove useful as an heuristic framework for ongoing modeling efforts (Botsford et al. 1994) and future large-scale field sampling programs (U.S. GLOBEC 1992).

Acknowledgements

The research summarized here was supported by an institutional grant from the Washington Sea Grant (NA86AA-D-SG044 Project R/F-68) and the U.S. Army Corps of Engineers (DACW67-85-C-0033). R.M. is particularly grateful for support provided by the Roy Jensen Memorial Fellowship. In addition, we are indebted to the many people who helped with the collection of data in the field and to our colleagues for spirited discussions. The contributions of Dr. B. Hickey, University of Washington School of Oceanography, were particularly significant. Contribution number 888, University of Washington, School of Fisheries, Seattle, WA 98195.

References

Anger, K., R.R. Dawirs, V. Anger, J.W. Goy, and J.D. Costlow. 1981. Starvation resistance in first stage zoeae of brachyuran crabs in relation to temperature. J. Crustacean Biol. 1: 518–525.

Bailey, K.M. 1981. Larval transport and recruitment of Pacific hake, *Merluccius producutus*. Mar. Ecol. Prog. Ser. 6: 1–9.

Bakun, A. 1973. Coastal upwelling indices, west coast of North America, 1967–1973. National Marine Fisheries Service, Seattle, WA. NOAA Tech. Rep. NMFS-SSRF-671: 103 pp.

Bakun, A. 1985. Comparative studies and the recruitment problem: searching for generalizations. Calif. Coop. Oceanic Fish. Invest. Rep. 26: 30–40.

Bakun, A. 1990. Global climate change and intensification of coastal ocean upwelling. Science (Wash., D.C.) 247: 198–201.

Barry, S. 1985. Overview of the Washington coastal Dungeness crab fishery, p. 33–36. *In* B.R. Melteff [ed.] Proceedings of the Symposium on Dungeness Crab Biology and Management. Alaska Sea Grant Prog. Rep. 85-3, University of Alaska, Anchorage, AK.

Baumgartner, T.R., A. Soutar, and V. Ferreira-Bartrina. 1992. Reconstruction of the history of Pacific sardine and northern anchovy populations over the past two millenia from sediments of the Santa Barbara Basin, California. Calif. Coop. Oceanic Fish. Invest. Rep. 33: 24–40.

Bornhold, B.D., and J.V. Barrie. 1991. Surficial sediments on the western Canadian continental shelf. Cont. Shelf Res. 11: 685–699.

Botsford, L.W. 1984. Effect of individual growth rates on expected behavior of the northern California Dungeness crab (*Cancer magister*) fishery. Can. J. Fish. Aquat. Sci. 41: 99–107.

Botsford, L.W. 1986. Population dynamics of the Dungeness crab (*Cancer magister*), p. 140–153. *In* G.S. Jamieson and N. Bourne [ed.] North Pacific Workshop on Stock Assessment and Management of Invertebrates. Can. Spec. Publ. Fish. Aquat. Sci. 92.

Botsford, L.W., R.D. Methot, and J.E. Wilen. 1982. Cyclic covariation in the California king salmon (*Oncorhynchus tschawytscha*), silver salmon (*O. kisutch*) and Dungeness crab (*Cancer magister*) fisheries. Fish. Bull. U.S. 80: 791–801.

Botsford, L.W., C.L. Moloney, A. Hastings, J.L. Largier, T.M. Powell, K Higgins, and J.F. Quinn. 1994. The influence of spatially and temporally varying oceanographic conditions on meroplanktonic meta populations. Deep-Sea Res. Part II, Topical studies in oceanography 41: 107–145.

Botsford, L.W., and D.E. Wickham. 1978. Behavior of age-specific, density dependent models and the northern California Dungeness crab (*Cancer magister*) fishery. J. Fish. Res. Board Can. 35: 833–843.

Browder, J.A., L.N. May, Jr.,A. Rosenthal, J. Gosselink, and R.H. Baumann. 1989. Modeling future trends in wetland loss and brown shrimp production in Louisiana using thermatic mapper imagery. Remote Sens. Environ. 28: 45–59.

Caddy, J.F. 1986. Modeling stock-recruitment processes in Crustacea: some practical and theoretical perspectives. Can. J. Fish. Aquat. Sci. 43: 2330–2344.

Caddy, J.F., and J.A. Gulland. 1983. Historical patterns of fish stocks. Mar. Policy 7: 267–276.

Cleaver, F.C. 1949. Preliminary results of the coastal crab (*Cancer magister*) investigation. Wash. Dep. Fish. Biol. Rep. 49A: 47–82.

Cushing, D.H. 1975. Marine ecology and fisheries. Cambridge University Press, London, UK. 278 p.

Dahlstrom, W.A., and P.W. Wild. 1983. A history of Dungeness crab fisheries in California, p. 7–24. *In* P.W. Wild and R.N. Tasto [ed.] Life history, environment and mariculture studies of the Dungeness crab, *Cancer magister*, with emphasis on the central California fishery resource. Calif. Dep. Fish Game Fish Bull. 172.

Dumbauld, B.R., D.A. Armstrong, and T.R. McDonald. 1993. Use of oyster shell to enhance intertidal habitat and mitigate loss of Dungeness crab (*Cancer magister*) caused by dredging. Can. J. Fish. Aquat. Sci. 50: 381–390.

Ekman, V.W. 1905. On the influence of the Earth's rotation on ocean currents. Ark. Mat. Astron. Fys. 2(11): 1–52.

Fernandez, M., O. Iribarne, and D.A. Armstrong. 1993. Habitat selection by young-of-the-year Dungeness crab, *Cancer magister,* and predation risk in intertidal habitat. Mar. Ecol. Prog. Ser. 92: 171–177.

Fisher, W.S., and D.E. Wickham. 1976. Mortalities and epibiotic fouling of eggs from wild populations of the Dungeness crab, *Cancer magister*. Fish. Bull. U.S. 74: 201–207.

Francis, R.C., and T.H. Sibley. 1991. Climate change and fisheries: what are the real issues? Northwest Envir. J. 7: 295–307.

Glantz, M.H., and L.E. Feingold. 1990. Introduction, p. 1–11. *In* M.H. Glantz and L.E. Feingold [ed.] Climate variability, climate change and fisheries. Environmental and Societal Impacts Group, National Center for Atmospheric Research, Boulder, CO.

Goldman, D.A., and M.F. McGowan. 1991. Distribution and abundance of omnastrephid squid paralarvae off the Florida Keys in August 1989. Bull. Mar. Sci. 49: 614–622.

Greve, W., and T.R. Parsons. 1977. Photosynthesis and fish production: hypothetical effects of climatic change and pollution. Helgol. Wiss. Meeresunters. 30: 666–672.

Gunderson, D.R., D.A. Armstrong, Y. Shi, and R.A. McConnaughey. 1990. Patterns of estuarine use by juvenile English sole (*Parophrys vetulus*) and Dungeness crab (*Cancer magister*). Estuaries 13: 59–71.

Gunderson, D.R., and I.E. Ellis. 1986. Development of a plumb staff beam trawl for sampling demersal fauna. Fish. Res. (Amst.) 4: 35–41.

Halliwell, G.R., Jr., and J.S. Allen. 1984. Large-scale sea level response to atmospheric forcing along the west coast of North America, summer 1973. J. Phys. Oceanogr. 14: 864–886.

Halliwell, G.R., Jr., and J.S. Allen.1987. The large-scale coastal wind field along the west coast of North America, 1981–1982. J. Geophys. Res. 92C: 1861–1884.

Hankin, D.G. 1985. Proposed explanations for fluctuations in abundance of Dungeness crabs: a review and critique, p. 305–326. *In* B.R. Melteff [ed.] Proceedings of the symposium on Dungeness Crab Biology and Management. Alaska Sea Grant Prog. Rep. 85-3. University of Alaska, Anchorage, AK.

Hannah, R.W. 1993. The influence of environmental variation and spawning stock levels on recruitment of ocean shrimp (*Pandalus jordani*). Can. J. Fish. Aquat. Sci. 50: 612–622.

Hickey, B.M. 1979. The California Current system—hypotheses and facts. Prog. Oceanogr. 8: 191–279.

1989. Patterns and processes of circulation over the Washington continental shelf and slope, p. 41–115. *In* M.R. Landry and B.M. Hickey [ed.] Coastal oceanography of Washington and Oregon. Elsevier, Amsterdam.

Hobbs, R.C., and L.W. Botsford. 1992. Diel vertical migration and timing of metamorphosis of larvae of the Dungeness crab *Cancer magister*. Mar. Biol. 112: 417–428.

Hobbs, R.C., L.W. Botsford, and A. Thomas. 1992. Influence of hydrographic conditions and wind forcing on the distribution and abundance of Dungeness crab, *Cancer magister*. Can. J. Fish. Aquat. Sci. 49: 1379–1388.

Hollowed, A.B., K.M. Bailey, and W.S. Wooster. 1987. Patterns in recruitment of marine fishes in the northeast Pacific Ocean. Biol. Oceanogr. 5: 99–131.

Hsieh, W.W., and G.J. Boer. 1992. Global climate change and ocean upwelling. Fish. Oceanogr. 1: 333–338.

Hsueh, Y., and R.D. Romea. 1983. A comparison of observed and calculated wintertime surface winds over the East China Sea. J. Geophys. Res. 88C: 9588–9594.

Huang, J.C.K. 1972. Recent decadal variation in the California Current system. J. Phys. Oceanogr. 2: 382–390.

Huyer, A., and R.L. Smith. 1985. The signature of El Niño off Oregon, 1982–1983. J. Geophys. Res. 90C: 7133–7142.

Incze, L.S., D.A. Armstrong, and S.L. Smith. 1987. Abundance of larval tanner crabs (*Chionoecetes* sp.) in relation to adult female and regional oceanography of the southeastern Bering Sea. Can. J. Fish. Aquat. Sci. 44: 1143–1156.

Ishimaru, J. 1936. Data for forecasting catch of king crab in Nemuro area, Japan. Hokkaido Suisan Shikenjo Jigyo Jumpo 332: 294–295.

Jacoby, C.A. 1982. Behavioral responses of the larvae of *Cancer magister* Dana (1852) to light, pressure and gravity. Mar. Behav. Physiol. 8: 267–283.

Jamieson, G.S., and D.A. Armstrong. 1991. Spatial and temporal recruitment patterns of Dungeness crab in the northeast Pacific. Mem. Queensl. Mus. 31: 365–381.

Jamieson, G.S., and A.C. Phillips. 1988. Occurrence of *Cancer* crab (*C. magister* and *C. oregonensis*) megalopae off the west coast of Vancouver Island, British Columbia. Fish. Bull. U.S. 86: 525–542.

Jamieson, G.S., A.C. Phillips, and W.S. Huggett. 1989. Effects of ocean variability on the abundance of Dungeness crab (*Cancer magister*) larvae, p. 305–325. *In* R.J. Beamish and G.A. McFarlane [ed.] Effects of ocean variability on recruitment and an evaluation of parameters used in stock assessment models. Can. Spec. Publ. Fish. Aquat. Sci. 108.

Johnson, D.F., and K.W. Hess. 1990. Numerical simulations of blue crab larval dispersal and recruitment. Bull. Mar. Sci. 46: 195–213.

Knutzen, J. 1981. Effects of decreased pH on marine organisms. Mar. Pollut. Bull. 12: 25–29.

Kondzela, C.M., and T.M. Shirley. 1993. Survival, feeding and growth of juvenile Dungeness crabs from southeastern Alaska reared at different temperatures. J. Crustacean Biol. 13: 25–35.

Lasker, R. 1975. Field criteria for survival of anchovy larvae: the relation between inshore chlorophyll maximum layers and successful first feeding. Fish. Bull. U.S. 73: 453–462.

Lasker, R. 1978. The relation between oceanographic conditions and larval anchovy food in the California Current: identification of factors leading to recruitment failure. Rapp. P.-v. Réun. Cons. int. Explor. Mer 173: 212–277.

Lee, T.N., C. Rooth, E. Williams, M. McGowan, A.F. Szmant, and M.E. Clarke. 1992. Influence of Florida Current, gyres and wind-driven circulation on transport of larvae and recruitment in the Florida Keys coral reefs. Cont. Shelf Res. 12: 971–1002.

Leovy, C., and E.S. Sarachik. 1991. Predicting climate change for the Pacific Northwest. Northwest Envir. J. 7: 169–201.

Lough, R.G. 1976. Larval dynamics of the Dungeness crab, *Cancer magister*, off the central Oregon coast, 1970–71. Fish. Bull. U.S. 74: 353–376.

McConnaughey, R.A. 1991. Factors affecting Dungeness crab (*Cancer magister*) year-class strength along the Washington coast. Ph.D. dissertation, University of Washington., Seattle, WA. 162 p.

McConnaughey, R.A., D.A. Armstrong, B.M. Hickey, and D.R. Gunderson. 1992. Juvenile Dungeness crab (*Cancer magister*) recruitment variability and oceanic transport during the pelagic larval phase. Can. J. Fish. Aquat. Sci. 49: 2028–2044.

McConnaughey, R.A., D.A. Armstrong, B.M. Hickey, and D.R. Gunderson. 1993. Interannual variability in coastal Washington Dungeness crab (*Cancer magister*) populations: larval advection and the coastal landing strip. Fish. Oceanogr. 3: 22–38.

Methot, R.D., Jr. 1986. Management of Dungeness crab fisheries, p. 326–334. *In* G.S. Jamieson and N. Bourne [ed.] North Pacific Workshop on Stock Assessment and Management of Invertebrates. Can. Spec. Publ. Fish. Aquat. Sci. 92.

Parker, D. 1992. Rock crabs, p. 18–20. *In* W.S. Leets, C.M. Dewees, and C.W. Haugen [ed.] California's living marine resources and their utilization. Calif. Sea Grant Exten. Publ. UCSGEP-92-12.

Parrish, R.H., C.S. Nelson, and A. Bakun. 1981. Transport mechanisms and reproductive success of fishes in the California Current. Biol. Oceanogr. 1: 175–203.

Paul, A.J., and J.M. Paul. 1980. The effect of early starvation on later feeding success of king crab zoeae. J. Exp. Mar. Biol. Ecol. 44: 247–251.

Peterson, W.T. 1973. Upwelling indices and annual catches of Dungeness crab, *Cancer magister*, along the west coast of the United States. Fish. Bull. U.S. 71: 902–910.

Rathjen, W.F. 1981. Exploratory squid catches along the continental slope of the eastern United States. J. Shellfish Res. 1: 153–159.

Reilly, P.N. 1983. Dynamics of Dungeness crab, *Cancer magister*, larvae off central and northern California, p. 57–84. *In* P.W. Wild and R.N. Tasto [ed.] Life history, environment and mariculture studies of the Dungeness crab, *Cancer magister*, with emphasis on the central California fishery resource. Calif. Dep. Fish Game Fish Bull. 172.

Richards, W.J., and J.R. Goulet, Jr. 1977. An operational surface drift model used for studying larval lobster recruitment and dispersal. FAO Fish. Rep. 200: 363–374.

Schoener, A., and D. Fluharty. 1985. Biological anomalies off Washington in 1982–83 and other major Niño periods, p. 211–225. *In* D.L. Fluharty and W.S. Wooster [ed.] El Niño north: Niño effects in the eastern subarctic Pacific Ocean. Wash. Sea Grant Prog. Rep. WSG-WO-85-3. University of Washington, Seattle, WA.

Schumacher, J.D., and A.W. Kendall. 1991. Some interactions between young walleye pollock and their environment in the western Gulf of Alaska. Calif. Coop. Oceanic Fish. Invest. Rep. 32: 1–19.

Sibley, T.H., and R.M. Strickland. 1985. Fisheries: some relationships to climate change and marine environmental factors, p. 211–225. *In* M.R. White [ed]. Characterization of information requirements for studies of CO_2 effects. Publ. DOE/ER-0236. U.S. Department of Energy, Washington, D.C.

Sinclair, M. 1988. Marine populations: an essay on population regulation and speciation. University of Washington Press, Seattle, WA. 252 p.

Stacey, M.W., S. Pond, and P.H. LeBlond. 1986. A wind-forced Ekman spiral as a good statistical fit to low-frequency currents in a coastal strait. Science (Wash., D.C.) 233: 470–472.

Strub, P.T., J.S. Allen, A. Huyer, R.L. Smith, and R.C. Beardsley. 1987a. Seasonal cycles of currents, temperatures, winds and sea level over the northeast Pacific continental shelf: 35°N to 48°N. J. Geophys. Res. 92C: 1507–1526.

Strub, P.T., J.S. Allen, A. Huyer, and R.L. Smith. 1987b. Large-scale structure of the spring transition in the coastal ocean off western North America. J. Geophys. Res. 92C: 1527–1544.

Thomas, D.H. 1985. A possible link between coho (silver) salmon enhancement and a decline in central California Dungeness crab abundance. Fish. Bull. U.S. 83: 682–691.

Thomson, R.E. 1981. Oceanography of the British Columbia coast. Can. Spec. Publ. Fish. Aquat. Sci. 56. 291 p.

Thomson, R.E. 1983. A comparison between computed and measured oceanic winds near the British Columbia coast. J. Geophys. Res. 88C: 2675–2683.

Thomson, R.E., B.M. Hickey, and P.H. LeBlond. 1989. The Vancouver Island coastal current: fisheries conduit or barrier?, p. 265–296. *In* R.J. Beamish and G.A. McFarlane [ed.] Effects of ocean variability on recruitment and an evaluation of parameters used in stock assessment models. Can. Spec. Publ. Fish. Aquat. Sci. 108.

Turner, R.E. 1977. Intertidal vegetation and commercial yields of penaeid shrimp. Trans. Am. Fish. Soc. 106: 411–416.

U.S. GLOBEC. 1992. Eastern boundary current program: report on climate change and the California Current ecosystem. Rep. 7. University of California, Davis, CA. 99 p.

van Geen, A., S.N. Luoma, C.C. Fuller, R. Anima, H.E. Clifton, and S. Trumbore. 1992. Evidence from Cd/Ca ratios in foraminifera for greater upwelling off California 4,000 years ago. Nature (Lond.) 358: 54–56.

Wainwright,T.C., and D.A. Armstrong. 1993. Growth patterns in the Dungeness crab (*Cancer magister*): synthesis of data and comparison of models. J. Crustacean Biol. 13: 36–50.

Ware D.M., and R.E. Thomson. 1991. Link between long-term variability in upwelling and fish production in the northeast Pacific Ocean. Can. J. Fish. Aquat. Sci. 48: 2296–2306.

Wickham, D.E. 1979. Predation by the nemertean *Carcinonemertes errans* on eggs of the Dungeness crab *Cancer magister*. Mar. Biol. (Berl.) 55: 45–53.

Wild, P.W. 1983. The influence of seawater temperature on spawning, egg development and hatching success of the Dungeness crab, *Cancer magister*, p. 197–214. *In* P.W. Wild and R.N. Tasto [ed.] Life history, environment and mariculture studies of the Dungeness crab, *Cancer magister*, with emphasis on the central California fishery resource. Calif. Dep. Fish Game Fish Bull. 172.

Wild, P.W., P.M.W. Law, and D.R. McLain. 1983. Variations in ocean climate and the Dungeness crab fishery in California, p. 175–188. *In* P.W. Wild and R.N. Tasto [ed.] Life history, environment and mariculture studies of the Dungeness crab, *Cancer magister*, with emphasis on the central California fishery resource. Calif. Dep. Fish Game Fish Bull. 172.

Winn, R.N. 1985. Comparative ecology of three cancrid crab species (*Cancer anthonyi*, *C. antennarius* and *C. productus*) in marine subtidal habitats in southern California. Ph.D. dissertation, University of Southern California, Los Angeles, CA. 235 p.

Wooster, W., and D.L. Fluharty. [ed.]. 1985. El Niño north: Niño effects in the eastern subarctic Pacific Ocean. Wash. Sea Grant Prog. Rep. WSG-WO-85-3. University of Washington, Seattle, WA. 312 p.

Zebiak S.E., and M.A. Cane. 1991. Natural climate variability in a coupled model, p. 457–469. *In* M.E. Schlesinger [ed.] Greenhouse-gas-induced climatic change: a critical appraisal of simulations and observations. Elsevier, New York.

Long-term variations in the oceanic environment of the East China Sea and their influence on fisheries resources

Chuanlan Lin, Bingrong Xu, and Shushen Huang

Lin, C., B. Xu, and S. Huang. 1995. Long-term variations in the oceanic environment of the East China Sea and their influence on fisheries resources, p. 307–315. *In* R.J. Beamish [ed.] Climate change and northern fish populations. Can. Spec. Publ. Fish. Aquat. Sci. 121.

Abstract: Temporal (1970–89) variations in oceanic elements affecting the East China Sea (ECS), their relationship to the catch of important fish species, and the extent of their fishing grounds were investigated. Local environment elements were: mean water temperature of the Kuroshio Current (*T*av); strength of the Kuroshio Front (KF); strength of the Taiwan Warm Current Front (TWCF); Changjiang River discharge (CRD) and the area of its dilute water mass (CDW); sea-surface temperature (SST) and sea-surface salinity (SSS) in the western East China Sea; and area of the cold water mass (CWM) in northern ECS. External environmental elements were the El Niño and La Niña events in the tropical Pacific Ocean. Spectral analysis of six environmental elements (*T*av, KF, TWCF, SST, CRD, CWM) revealed 7- and/or 2.5-yr periods of oscillation, and both periods were present in catches of all species and hairtail (*Trichiurus haumela*). Correlation and cross-spectrum analysis indicated significant coherence values and correlations, at 2.5- and 7-yr periods, among all species, and hairtail, catches with four environmental elements (*T*av, CRD, SST, SSS). Correlations suggested that Changjiang River discharge influenced abundance, or at least availability, of commercial species. El Niño (La Niña) events were associated with: negative (positive) anomalies of *T*av; weaker (stronger) KF; stronger (weaker) TWCF; larger (smaller) CDW mass; negative (positive) anomalies of SST; large (smaller) CWM; low (usually high) CPUEs; and contracted (expanded) fishing grounds.

Résumé : Les variations temporelles (1970–1989) dans les éléments océaniques touchant la mer de Chine orientale (MCO), leurs relations avec la capture d'importantes espèces de poissons et la superficie des pêcheries où se trouvent ces espèces ont fait l'objet d'une recherche. Les éléments de l'environnement local était les suivants : force du front du Kuroshio (KF); force du front du courant chaud de Taiwan (TWCF); décharge du fleuve Changjang (CRD) et zone de sa masse d'eau diluée (CDW); température de la surface de la mer (TSM) et salinité de la surface de la mer (SSM) dans l'ouest de la mer de Chine orientale; et zone de masse d'eau froide (CWM) dans le nord de la MCO. Les éléments environnementaux externes étaient les phénomènes El Niño et La Niña dans la région tropicale du Pacifique. L'analyse spectrale de six éléments environnementaux (Tav, KF, TWCF, TSM, CRD et CWM) a révélé l'existence de périodes d'oscillation de 7 ans et (ou) de 2,5 ans, et les deux périodes étaient manifestes dans les prises de toutes les espèces de poissons et de trichiures (*Trichiurus haumela*). L'analyse des corrélations et du spectre complet a révélé la présence de corrélations et de valeurs de cohérence importante, à des périodes de 2,5 et de 7 ans, entre les prises de toutes les espèces et de trichiures, et quatre éléments environnementaux (Tav, CRD, TSM et SSM). Les corrélations permettaient de conclure que l'élément CRD avait une influence sur l'abondance, ou du moins sur la disponibilité, des espèces commerciales. Les phénomènes El Nino (La Nina) ont été associés aux : anomalies négatives (positives) de l'élément Tav; plus grande faiblesse (plus grande force) de l'élément KF; plus grande force (plus grande faiblesse) de l'élément TWCF; plus grosse (plus petite) masse de l'élément CDW; anomalies négatives (positives) de l'élément TSM; plus grosse (plus petite) CWM; faible nombre (habituellement fort nombre) de prises par unité d'effort; et contraction (expansion) des pêcheries.

C. Lin, B. Xu, and **S. Huang.** Second Institute of Oceanography, State Ocean Administration, Hangzhou, Zhejiang, 310012, People's Republic of China.

Introduction

The East China Sea (ECS) is a middle-latitude continental shelf area. China, Japan, Korea, and other countries conduct year-round fisheries there. Since 1970, the traditional fisheries have exhibited substantial fluctuations in catch. Principal species were hairtail (*Trichiurus haumela*), yellow croakers (*Pseudosciaena polyactis* and *P. crocea*), Chinese herring (*Ilisha elongata*), and white Chinese croaker (*Argyrosomus argentatus*). Since 1983, their catches have all declined. Since the late 1970s, some previously unexploited species increased in abundance. For example, by the early 1980s, the population of the black scraper (*Navodon septentrionalis*) exceeded that of the hairtail. Also, some small pelagic fishery resources, such as the common Japanese mackerel (*Pneumatophorus japonicus*) and the Spanish mackerel (*Scomberomorus niponius*), suddenly increased in 1983 and 1984.

Studies have been undertaken on the interannual variations in the ECS oceanic environment (Guan 1979, 1983; Le 1983; Lin 1986; Wang and Weng 1988), as well as variations in fish populations and fishing grounds (Zhu 1987; Ma and Xu 1988; Hiroshi 1990). These results indicated some relationship of fish and the distribution of their fishing grounds. However, research on long-term relationships has been limited. This report describes the long-term variations elements of the oceanic environment in the ECS, and their relationship with the catch of important fish species and the distribution of fishing grounds.

Materials and methods

ECS hydrographic conditions are governed by four elements: inflowing Kuroshio Warm Current from the southeast and its branch, the Taiwan Warm Current (TWC); Changjiang River discharge (CRD) and its dilute offshore water mass (CDW); and the demersal Cold Water Mass (CWM) in northern ECS (Fig. 1).

Hydrographic data were taken from research vessel cruises (State Oceanic Administration of China), marine station records, and data published by the Japanese Meteorological Agency. Hydrographic sectional observations were collected seasonally, four times per year. Marine station records were collected monthly.

The following nine time series were utilized: (1) water temperature of the Kuroshio Current in the ECS; (2) strength of the Kuroshio; (3) strength of the Taiwan Warm Current Front (TWCF); (4) the Changjiang River discharge and the offshore dilute water area; (5) sea-surface temperature at Shengshan marine observation station in the western ECS; (6) sea-surface salinity at Shengshan marine observation station

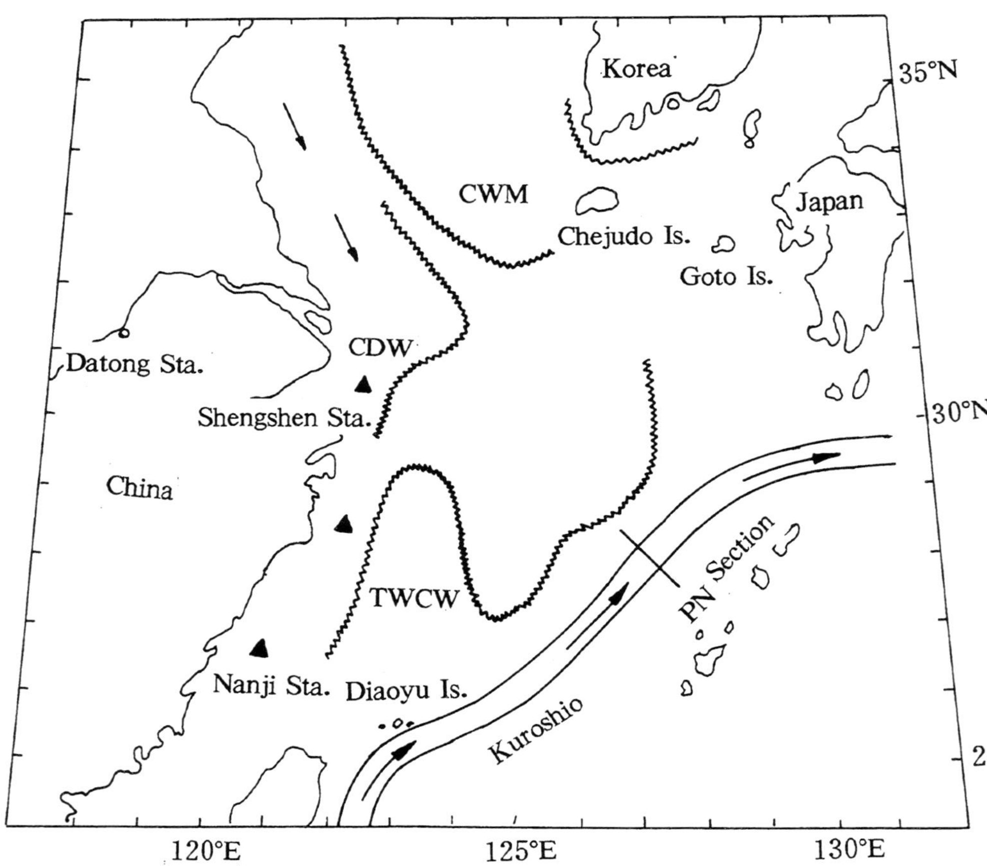

Fig. 1. Schematic diagram of the East China Sea showing the location of the Kuroshio Current, its front (KF), and its hydrographic section (PN), Taiwan Current (TWC) and its front (TWCF), Changjiang Dilute Water (CDW), and the Cold Water Mass (CWD). ▲ is position of marine stations.

in the western ECS; (7) area of the cold water mass in northern ECS; (8) catches of important commercial fish species; and (9) extent of the fishing grounds. Time series was 1970–1989, and seasonal divisions were December–February (winter), March–May (spring), June–August (summer), and September–November (autumn). Hydrographic and fisheries data were comparatively abundant during this period.

Mean water temperature of the Kuroshio Current (*T*av Lin et al. 1993a, b) were averages of temperature from the surface to 400 m along the PN section (Fig. 1). The Kuroshio Front (KF) is marked by a sharp horizontal gradient of temperature and salinity between the Kuroshio and the shelf water. When the KF intensifies, the gradient steepens. The strength of the KF is measured here as change in degrees Celsius per mile (1 mile = 1.6 km). TheTWC branches off the Kuroshio northeast of Taiwan Island. It moves northward offshore of Fujian and Zhejiang, and usually appears south of the mouth of the Changjiang River. TWC water is characterized by higher temperature and salinity. Its western side lies adjacent to the low-salinity ECS Coastal Current. There is a sharp salinity front between the two currents. The strength of the TWC is defined by the area surrounded by the 20°C isotherm of the bottom layer.

Offshore the CRD mixes with the surrounding sea water, and forms a strong dilute water mass known as the Changjiang diluted water (CDW). In winter, the CDW flows southward in a narrow band alongshore. In summer, the river is flooding and the surface plume flows northeastward towards Chejudo Island, almost covering the northwest ECS. The expansion, tuning, and mixing of the CDW may influence the size and shape of the fishing grounds. The outer boundary selected for the CDW was the 31 partial salinity units (PSU) isohaline. The observed area of CDW is directly proportional to the mean monthly discharge of the Changjiang River. Changjiang River discharge data were based on observations monthly at the Datong Hydrography Station. Sea-surface temperatures (SSTs) and sea-surface salinity (SSS) were taken from monthly observations at Shengshan marine observation station in western ECS (Fig. 1). The cold water mass (CWM) may exist year round, according to some studies, and fish may not be concentrated there. Its long-term variation is of great importance for this discussion. The strength of the CWM is measured by the area of bottom water less than 15°C, south of 32°N. Commercial fishery catch statistics were issued by the Huanghai Fisheries Institute and some marine fisheries companies of China.

Spectral analysis (Kanasewich 1981; Huang and Li 1984) was employed to investigate the oscillations in the various time series. Moving averages of seasonal values and normalized were calculated before the analysis. The expression of a time series of continuous power spectral density (S_L) is

$$(1)\quad S_L = \frac{1}{m}\,[r(0) + 2\sum_{\tau=1}^{m-1} r(\tau)\cos\frac{\pi L}{m}\tau + r(m)\cos L\pi]$$

$$(\tau = 0, 1, 2, ..., m)$$

where m is greatest lag number, L is wave number (L = 0, 1, ..., m), $r(\tau)$ is autocorrelation coefficient, its expression is

$$(2)\quad r(\tau) = \frac{1}{n-\tau}\sum_{t=1}^{n-\tau}\left(\frac{X_t - \bar{X}}{S}\right)\left(\frac{X_{t+\tau} - \bar{X}}{S}\right)$$

$$t = 1, 2, ..., n$$

where n is sample number of time series X_t and $\bar{X}$ and S are mean value and standard deviation of X_t, respectively.

Cross-spectrum analysis was used to determine the correlation and phase difference between two time series of various frequencies. Coherence square analysis was employed to determine the significance of correlation between two time series in same period. For two time series X_{t1} and X_{t2}, the expression of cross spectrum $S_{12}(L)$ is

$$(3)\quad S_{12}^2(L) = P_{12}(L) - iQ_{12}(L)$$

in Equation 3 the real part $P_{12}(L)$ is the cross covariance spectrum, the imaginary part $Q_{12}(L)$ is normal spectrum, their calculus follow Huang and Li (1984). The expression of the coherence square $R_{12}^2(L)$ is

$$(4)\quad R_{12}^2(L) = \frac{P_{12}^2(L) + Q_{12}^2(L)}{P_{11}(L)\,P_{22}(L)}$$

where P_{11} and P_{22} are continuous power spectral density of the time series X_{t1} and X_{t2}, respectively.

Significance for spectral density and for coherence square were carried out using red noise hypothesis spectrum and *f*-test, respectively.

Results

Long-term variations of the environment

All of the environmental elements examined exhibited fluctuations which were generally coincident with El Niño – La Niña events (Figs. 2 and 3). Mean seasonal water temperatures (*T*av) of the Kuroshio exhibited regular interannual oscillations during 1972–90, with large positive and negative anomalies, each approximately 2–3 yr in duration (Fig. 2A). Among-station anomalies were all positive during La Niña events (1973, 1975, 1984–85, and 1988; Wong 1990), and all negative during the El Niño events (1972, 1976, 1979, 1982–83, and 1986–87; Cane 1983; Luther 1983; Wong 1990). For the remaining environmental elements, El Niño (La Niña) events corresponded with weaker (stronger) KFs (Fig. 2B), stronger (weaker) CWM (Fig. 2F), stronger (weaker) TWC water (Fig. 2D), stronger (weaker) CDW mass (Fig. 2C and 2E), and negative (positive) anomalies of SSTs from western ECS.

Spectral shapes of the environmental elements were similar to each other (Figs. 4A–4F). Most had a significant peak at 2.5 yr, which suggests that this oscillation exists universally among the environmental elements. A 7-yr period was also noted for some elements.

Long-term variations of fisheries resources

Annual catches, during 1971–85, of commercially important species has a varied history (Fig. 5). Principal species, based on catch, were hairtail, black scraper, and Japanese mackerel.

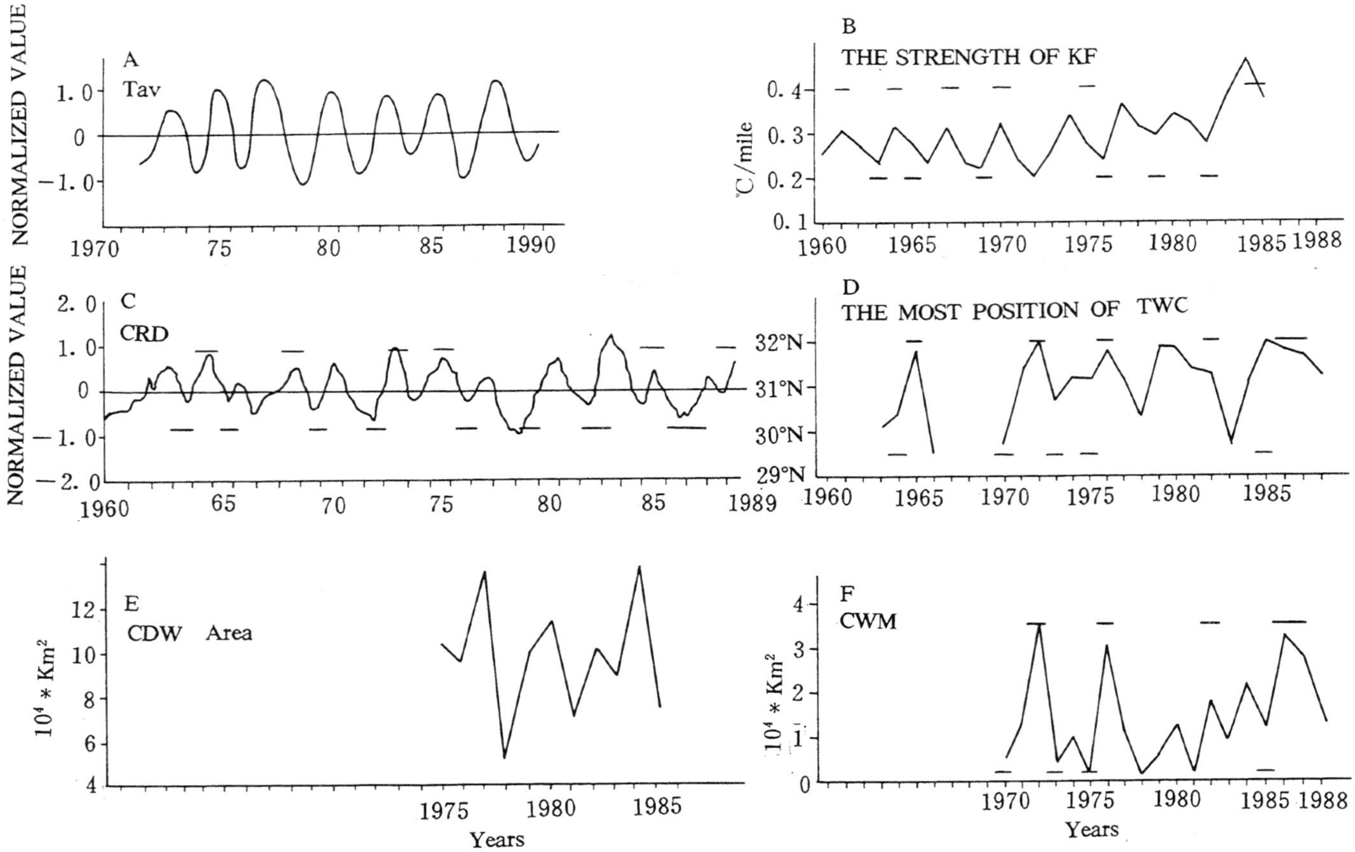

Fig. 2. Time series of six local environmental elements in the East China Sea, 1970–90: (A) mean seasonal temperature anomalies for the Kuroshio Current (*T*av); (B) strength of Kuroshio Front (°C change/mile); (C) Changjiang River Discharge anomalies; (D) most northern position (N lat) of Taiwan Warm Current; (E) area (10^4 km^2) Changjiang Dilute Water; and (F) area (10^4 km^2) of cold water mass. Heavy lines = El Niño events; light lines = La Niña events.

Hairtail catch rose irregularly from ca 30 000 t in 1971 to a peak of ca 120 000 t in 1983, then declined to ca 60 000 t in 1985 (Fig. 5A). Black scraper catch rose irregularly from nil in 1971 to an initial peak of ca 200 000 t in 1982; declined to ca 80 000 t in 1983; rose irregularly to a peak of ca 180 000 t in 1984; declined slightly to ca 150 000 t in 1985; rose again to new peak of ca 210 000 t in 1987; and declined to ca 180 000 t in 1988 (Fig. 5B). Japanese mackerel catch was less than 10 000 t/yr during 1971–80; rose sharply to a peak of ca 48 000 t in 1984; then declined to ca 8000 t in 1985 (Fig. 5C). Quasi-biennial fluctuations were apparent in most of the catch histories.

Shifts in importance of species were more apparent on a percent basis (Fig. 6B). Hairtail and yellow croakers predominated in the catches during the early 1970s (ca 74% in 1974, but were superseded in 1978 by the black scraper and, temporarily, by the Japanese mackerel. Hairtail proportion declined from ca 72% in 1973 to ca 30% in 1985. Yellow croaker catch was virtually nil after 1983.

Interannual fluctuations in all-species catch per unit effort (CPUE) exhibited a 2- to 4-yr period (Fig. 6A), similar to that exhibited by the environmental elements. Relatively high values appeared in 1971, 1974, 1978, 1982, and 1984, the times of La Niña events, except 1982, when fishing effort was very high. Relatively low values appeared in 1972, 1976, 1979, and 1983, the times of El Niño events.

Distribution of fishing grounds also exhibited variations that appeared to be related to environmental elements for the hairtail (CRD, SST, SSS), black scraper (*T*av), and Japanese mackerel (TWC, CRD). The hairtail fishing ground in western ECS was displaced southward (south of 31°N) and westward (west of 124°E) when the CRD and SST had negative anomalies and SSS had positive anomalies (El Niño events) (Fig. 7). Furthermore, the centre (high CPUE) of the fishing ground was smaller. During the CRD and SST had positive anomalies and SSS had negative anomalies (La Niña events), the fishing ground, and its centre, was more extensive. Its northern position was north of 32° or 34°N, and extended eastward to 125°E.

The black scraper fishing ground was extensive, e.g., winter–spring 1982, 1984, and 1987, when *T*av anomalies were negative (Fig. 8). The centre seemed to lie between Goto Island and nearby Diaoyu Island. Conversely, when the *T*av

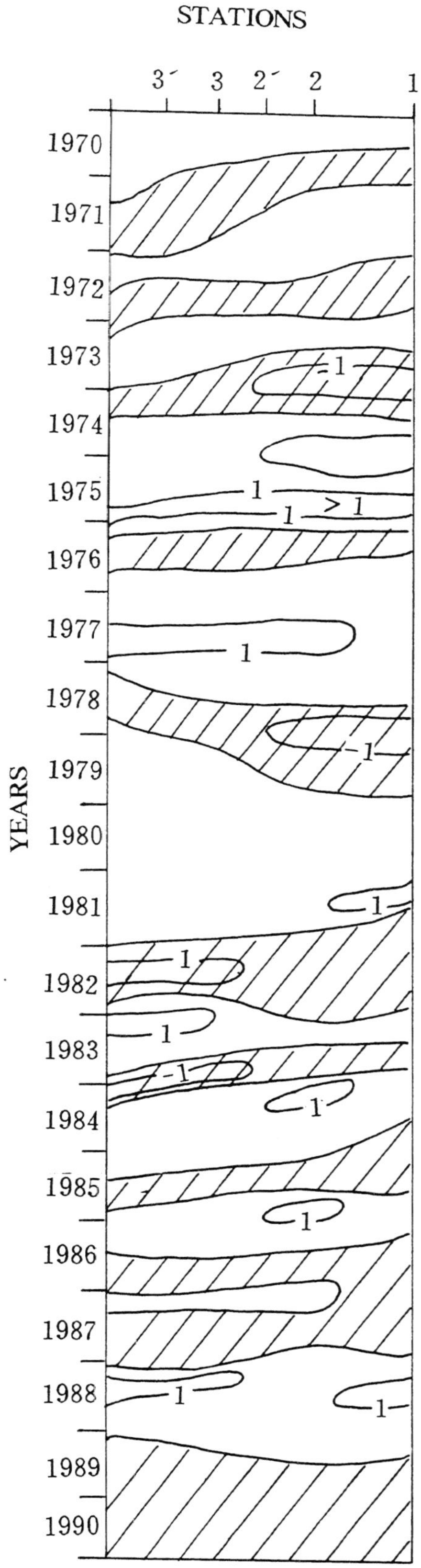

Fig. 3. Seasonal interstation temperature anomalies on the PN section of the Kuroshio Current, 1970–90. Striped area = negative anomalies; clear area = positive anomalies.

anomalies were positive, e.g., in winter–spring 1979, 1983, 1986, and 1988, the "edge" fishing ground could not form, and the central fishing ground was smaller.

The Japanese mackerel fishing ground and centre were extensive in 1983 and 1984, but reduced in 1985 (Fig. 9). The expansion may have been caused by environmental changes, viz., the extended CRD positive anomaly in the early 1980s, and the intensification, in 1984, of the oceanic front between the TWC and the ECS Coastal Current. The former element causes an increase in low-salinity plankton in the Changjiang River estuary and adjacent area, which is favourable to pelagic species. The latter creates a good fishing-ground environment.

Periodic oscillations, environmental elements and catch

Coherence-square analysis yielded significant coherence values (except the total catch per 2.5-years per SST cell) and correlations, for 2.5- and 7-yr periods, between all-species catch, and hairtail catch, and four environment elements (*T*av, CRD, SST, SSS) (Table 1). Correlations with all-species and hairtail catch were negative for *T*av and SSS, and positive for CRD and SST.

Summary and discussion

Long-term variations in the oceanic environment and its effect on fisheries resources of the ECS were investigated for the period 1971–1989. Local environmental elements were: *T*av (0–400 m) at PN section; strength of the KF; strength and location of the TWCF; CRD anomaly; area of the offshore CDW; SST and SSS anomalies at Shengshan marine observation station in western ECS; and area of the demersal CWM in northern ECS. External environmental elements were the El Niño and La Niña events in the tropical Pacific Ocean.

Spectral analysis of six environmental elements (*T*av, KF, TWC, CRD, SST, CWM) indicated that most exhibited a 2.5-yr period, and some a 7-yr period, of oscillation. All-species and hairtail catches exhibited both oscillation periods. Correlation and cross-spectrum analysis indicated significant coherence values and correlations, at 2.5- and 7-yr periods, between all-species and hairtail catches, and four environmental elements (*T*av, CRD, SST, SSS) consistent with most oceanic environment change. Similar variation patterns appeared with both the change in fishing grounds and oceanic environment. Correlations suggested that CRD and CDW influenced abundance or at least availability of commercial species.

El Niño – La Niña events were related to variations in environmental elements, fisheries and extent of fishing grounds. For environmental elements, El Niño (La Niña) events were associated with: negative (positive) anomalies of Kuroshio Current temperature; weaker (stronger) KF; stronger (weaker) TWCF; larger (smaller) CDW mass; negative (positive) anomalies of SST and positive (negative) anomalies of SSS from western ECS; larger (smaller) CWM; low (high) CPUEs; and contracted (expanded) fishing grounds. High CPUE did not occur in 1982 owing to excessive fishing effort.

The black scraper fishing ground could not form completely when *T*av anomalies were positive, but was extensive when *T*av anomalies were negative. These reactions were not necessarily related to the El Niño – La Niña events. The Japanese mackerel fishing ground was extensive in 1983 and 1984, but reduced in 1985. The expansion may have been caused by the strong 1982–83 El Niño event, which contributed to the

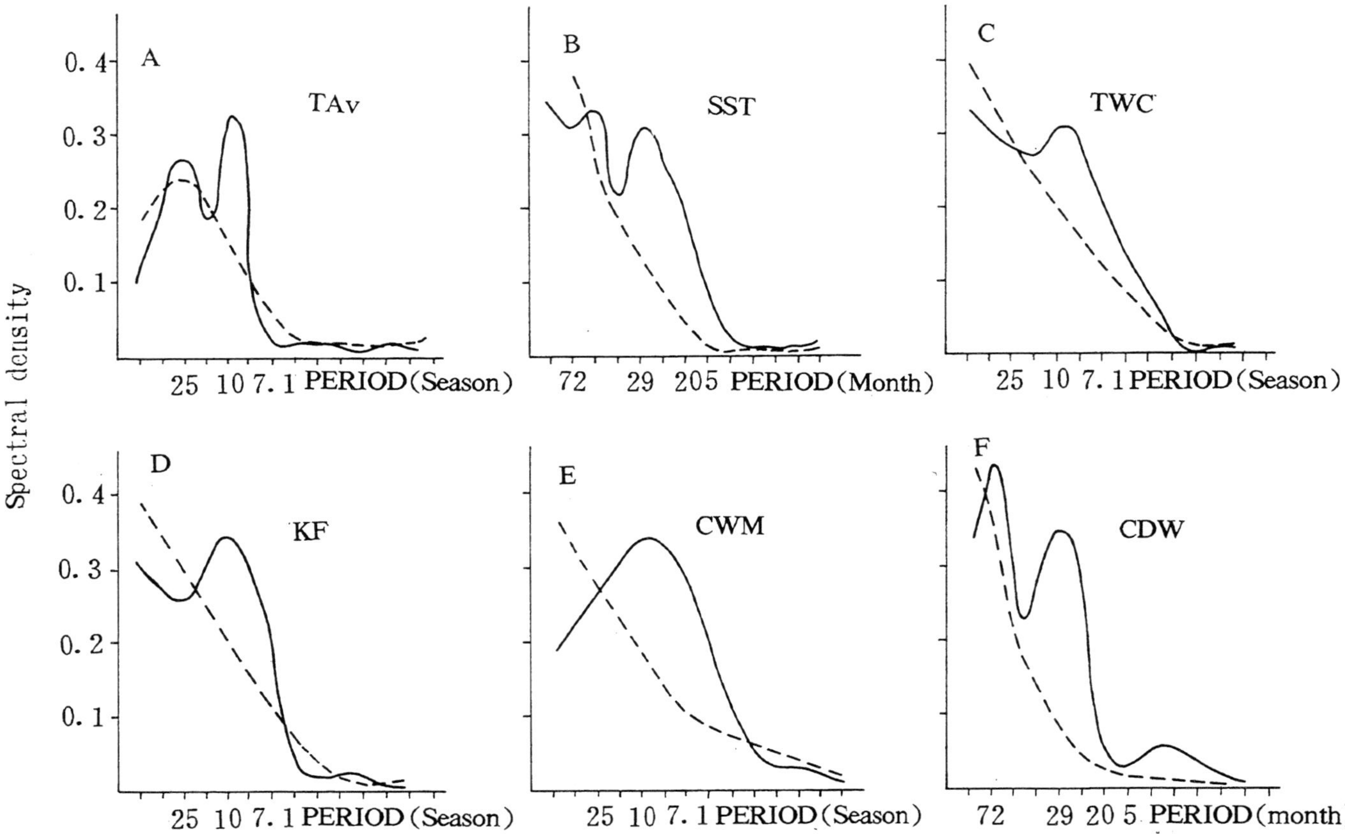

Fig. 4. Spectral densities of six local environmental elements in the East China Sea: *T*av, KF, TWC, CRD, SST, and CWM. Broken line is the red noise spectrum.

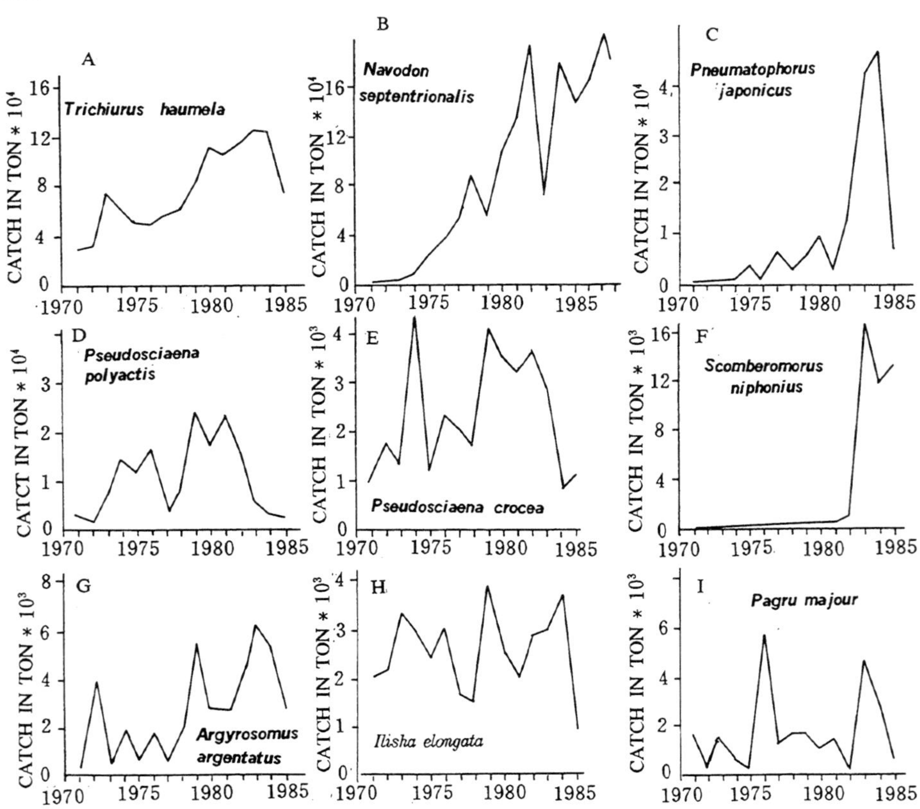

Fig. 5. Annual catches (t) of important commercial fishes in the East China Sea, 1971–1985.

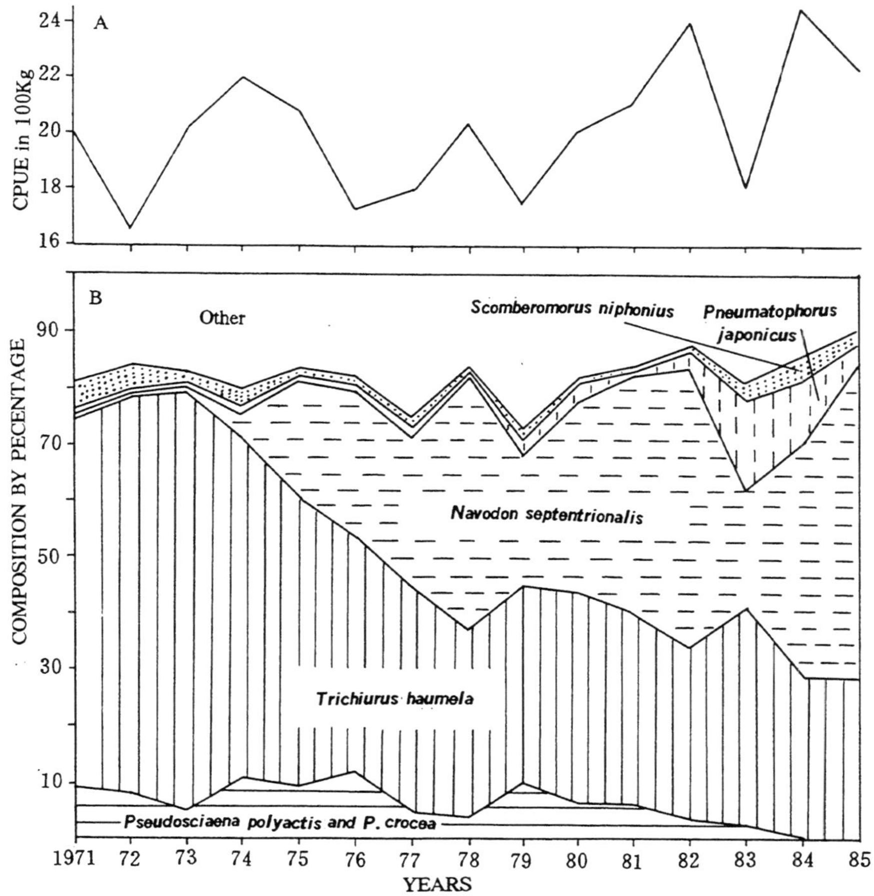

Fig. 6. Catch composition (%) of important fish species in the East China Sea, 1971–1985.

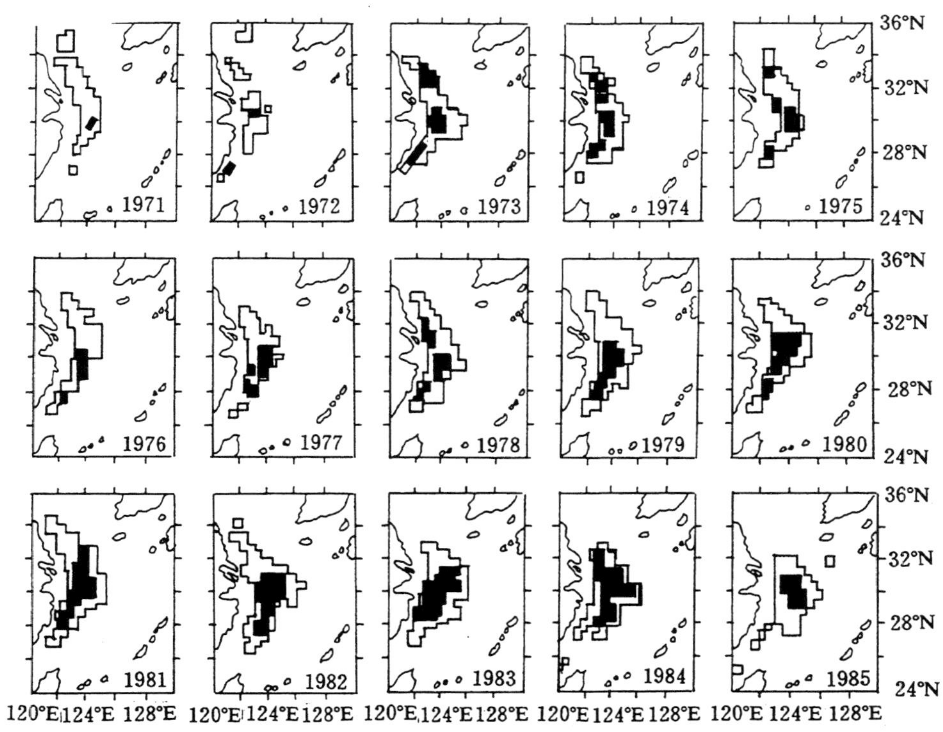

Fig. 7. Hairtail fishing grounds in the East China Sea, 1971–1985. (Shaded area = centres of high CPUE).

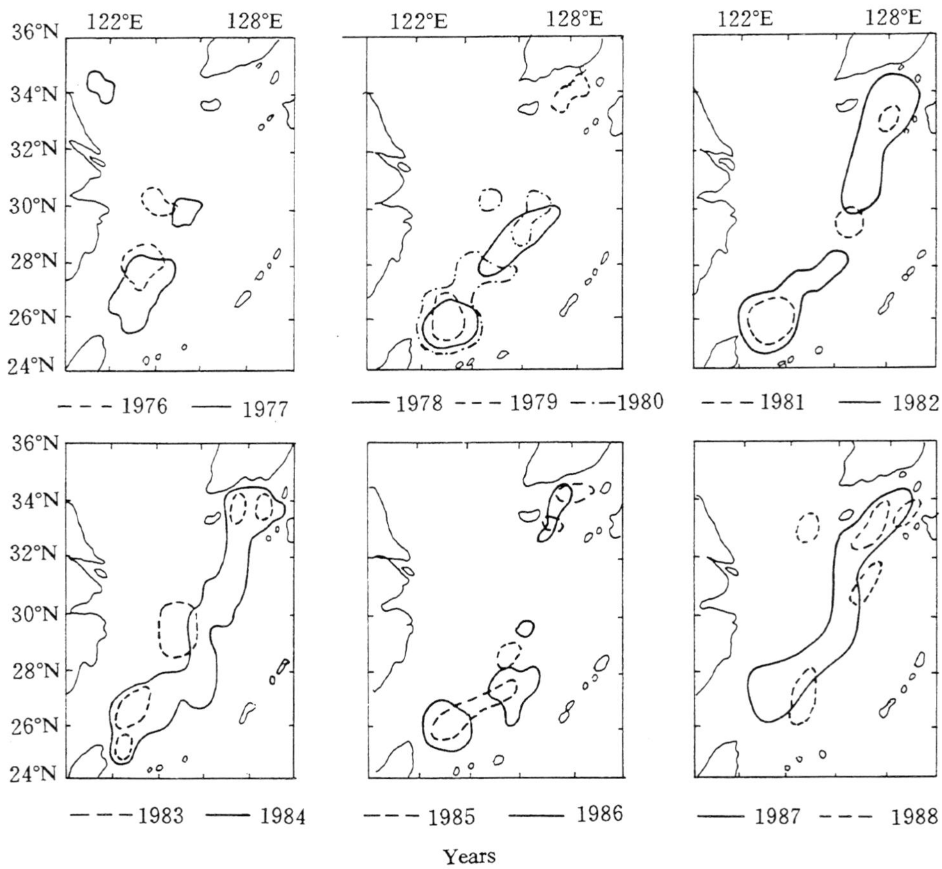

Fig. 8. Black scraper fishing grounds in the East China Sea, 1976–1988. (Broken lines enclose centres of high CPUE).

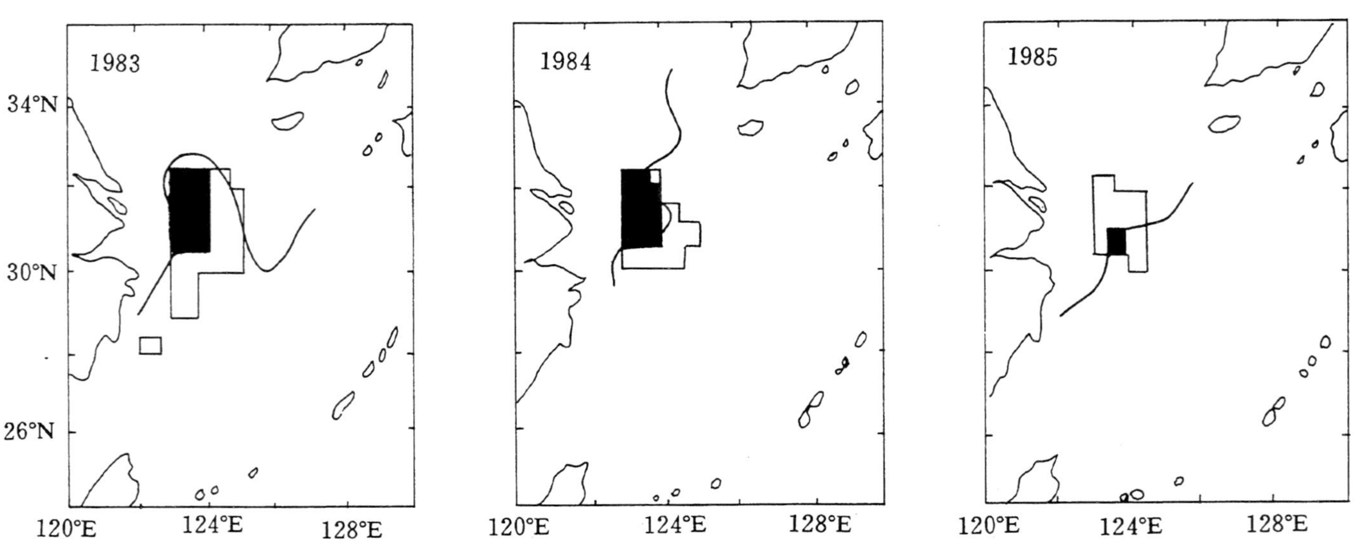

Fig. 9. Japanese mackerel fishing ground in the East China Sea, 1983–1985. (Shaded area = centres of high CPUE. Line indicates location of the Taiwan Warm Current Front).

Table 1. Coherence Square Analysis for all-species and hairtail catches, their 2.5- and 7-yr period oscillations, and four local environmental elements.

Catch	Period (yr)	Environmental Elements[a]			
		*T*av	CRD	SST	SSS
Total	2.5	0.56	0.68	0.30[b]	0.67
	7	0.40	0.78	0.57	0.73
Hairtail	2.5	0.51	0.88	0.48	0.69
	7	0.57	0.65	0.62	0.75
Correlation coefficients					
All-species catch		−0.64	0.68	0.62	−0.70
Hairtail catch		−0.72	0.61	0.58	−0.63

[a] CRD = Changjiang River Discharge; *T*av = mean temperature of the Kuroshio at the PN cross section (0–400 m); SST = sea-surface temperature; SSS = sea-surface salinity at Shengshan marine station in the western East China Sea.

[b] Not significant; all other coherence square and correlation coefficients are significant at the 1% level.

increased strength of the TWC Front, and the extended CRD large positive anomaly in 1983 (the second year of El Niño event).

In the East China Sea, both local and external environmental factors were found to affect the fishery resources, with respect to abundance of species and distribution of fishing grounds. Further research is needed on the relationship between fisheries and the oceanic environment.

Acknowledgments

This work was supported by the Zhejiang Provincial Nature Science Foundation of China.

References

Cane, M.A. 1983. Oceanographic events during El Niño. Science (Wash., D.C.) 222(4629): 1189–1195.

Guan, B. 1979. Some results of the study of variation of the Kuroshio in the East China Sea. Oceanol. Limnol. Sin. 10(4): 297–306. (In Chinese)

Guan, B. 1983. A sketch of the current structures and eddy characteristics in the East China Sea, p. 56–79. *In* Proceeding of the International Symposium on Sedimentation on the Continental Shelf, with Special Reference to the East China Sea, Hangzhou, April 12–16, 1983. Ed. by Acta Oceanologica Sinica, China Ocean Press, Beijing.

Hiroshi, M. 1990. Utilization of the demersal fish resources by Japanese fishing vessels in the East China and Yellow seas, p. 244–252. *In* East China Sea Symposium, Beijing, China, Nov. 2–3, 1990. Ed. and pub. by Research Institute for Ocean Development in Continental Shelf, Tokyo, Japan. (In Japanese)

Huang, J., and H. Li. 1984. The spectral analysis in meteorology, p. 27–141. China Meteorol. Press, Beijing. (In Chinese)

Kanasewich, E.R. 1981. Time sequence analysis in geophysics. p. 118–141. University of Alberta Press, Edmonton, AB.

Le, K. 1983. Some results from the study of the path of the Changjiang Dilute Water, p. 375–381. *In* Proceedings of International Symposium on Sedimentation on the Continental Shelf, with Special Reference to the East China Sea, Hangzhou, April 12–16, 1983, Ed. by Acta Oceanologica Sinica, China Ocean Press, Beijing.

Lin, C. 1986. Oceanographic characteristics of the Kuroshio Front in the East China Sea and the relationship to fisheries. Donghai Mar. Sci. 4(2): 7–15. (In Chinese)

Lin, C., B. Xu, W. Zheng, and S. Huang. 1993a. Interannual variability of water temperature of the Kuroshio in the East China Sea and fisheries of *Navodon septentrionalis*, p. 204–215, *In* Proceedings of the Symposium on the physical and chemical oceanography of the China seas, Eds. by Su Jilan, Wen-Sen Chuang, R. Ya Hsueh, China Ocean Press, Beijing, China.

Lin, C., B. Xu, W. Zheng, and S. Huang. 1993b. Interannual variation of temperature and salinity in relating to fishery condition of the Zhejiang offshore waters. J. Fish. China 17(2): 85–93.

Luther, D.S. 1983. Zonal winds in the central equatorial Pacific and El Niño. Science (Wash., D.C.) 222 (4621): 327–330.

Ma, Y., and J. Xu. 1989. Catch change pattern of hairtail in the East China Sea. Mar. Sci. 5: 37–39. (In Chinese)

Wang, C., and X. Weng. 1988. The variation of the Taiwan Warm Current in relation to marine conditions of the tropical Pacific. Mar. Sci. 1: 1–3. (In Chinese)

Wong, S. 1990. Review of the ENSO in the 80's. China Meteorol. Press 71(10): 1–12. (In Chinese)

Zhu, D. 1987. The relationship between fishery environment of the central section of the sea off Zhejiang and the fishing season of hairtail. J. Fish. China 11(3): 195–303. (In Chinese)

The Relationship of several environmental indices to the survival of juvenile male northern fur seals (*Callorhinus uarsinus*) from the Pribilof Islands

Anne E. York

York, A.E. 1995. The relationship of several environmental indices to the survival of juvenile male northern fur seals (*Callorhinus ursinus*) from the Pribilof Islands, p. 317–327. *In* R.J. Beamish [ed.] Climate change and northern fish populations. Can. Spec. Publ. Fish. Aquat. Sci. 121.

Abstract: This paper focuses on the relationship between the survival of northern fur seals (*Callorhinus ursinus*) and environmental time series (mainly air temperatures (ATs) and sea surface temperatures (SSTs)) collected over the species' range in the eastern North Pacific Ocean and Bering Sea. These relationships were studied by examining the cross correlation between the survival time series and the trend component of the environmental time series estimated from an seasonal trend decomposition based on loess (STL); standard errors were estimated using a parametric bootstrap procedure. Estimates of survival on land after removing the density dependent effects were positively correlated ($P < 0.01$) with ATs on St. Paul Island at the time of birth and negatively correlated ($P < 0.001$) with ATs at three Bering Sea stations 2–3 yr before the cohort was born. Survival at sea is significantly correlated with ATs at several Bering Sea stations ($P < 0.04$) at the time of birth and with near shore monthly mean SSTs and ATs in the Gulf of Alaska for several months during the year the cohort was born and 4 yr before the cohort was born ($P < 0.01$). Some hypotheses related to thermal requirements for northern fur seals or to climate caused changes in the food supply available to juveniles and/or pregnant or lactating females are suggested for explaining these correlations.

Résumé : Le présent document porte sur la relation qui existe entre la survie de l'otarie à fourrure de l'Alaska (*Callorhinus ursinus*) et les séries chronologiques environnementales (surtout les températures de l'air [TA] et les températures de la surface de la mer [TSM]) recueillies dans l'aire de répartition de cette espèce dans l'est du Pacifique Nord et dans la mer de Béring. Ces relations ont été étudiées du point de vue de la corrélation entre les séries chronologiques sur la survie et la tendance des séries chronologiques environnementales estimées à partir d'une décomposition des tendances saisonnières basée sur le loess; les erreurs-types ont été estimées au moyen d'une méthode bootstrap paramétriques. Les estimations de la survie sur terre après élimination des effets tributaires de la densité ont donné une corrélation positive ($P < 0,01$) avec les TA sur l'île Saint-Paul au moment de la naissance, et une corrélation négative ($P < 0,001$) avec les TA à trois stations de la mer de Béring 2–3 ans avant la naissance de la cohorte. La survie en mer est en forte corrélation avec les TA à plusieurs stations de la mer de Béring ($P < 0,04$) au moment de la naissance ainsi qu'avec les TSM et TA mensuelles moyennes près de la côte dans le golfe d'Alaska pendant plusieurs mois au cours de l'année de naissance de la cohorte et pendant 4 ans avant la naissance de la cohorte ($P < 0,01$). Certaines hypothèses liées aux besoins thermiques des otaries ou aux changements causés par le climat dans les disponibilités alimentaires des juvéniles et (ou) des femelles gestantes ou allaitantes sont proposées pour expliquer ces corrélations.

A.E. York. National Marine Mammal Laboratory, Northwest and Alaska Fisheries Center, National Marine Fisheries Service, National Oceanic and Atmospheric Administration, 7600 Sand Point Way N.E., Bin C15700, Seattle, WA 98115, USA

Introduction

The basic biology and natural history of the northern fur seal (*Callorhinus ursinus*) is available from several sources (e.g., Bartholomew and Hoel 1953; Peterson 1968; York 1987). Northern fur seals spend most of the year at sea and are distributed over a large area in the North Pacific Ocean (Fig. 1); during the summer, they go ashore to breed in large colonies called rookeries (Fig 1). Fur seals are not uniformly distributed throughout their range, but tend to accumulate in biologically productive areas, such as regions of upwelling along the coast of North America. Adult males arrive on the breeding islands from mid-May through early June and establish territories, which they maintain through early August. Pregnant females arrive from mid-June until early August; they give birth within 1 or 2 d to a single pup and mate within the following week. On the Pribilof Islands, the peak of births takes place in early July and the peak of breeding in mid-July. The postpartum female then begins a series of feeding–nursing cycles, lasting several days until her pup is weaned.

Pups are weaned during October and November and then begin their pelagic life. Mortality among juvenile fur seals is variable and often high. At least 50%, and sometimes more than 90%, die before the age of 2 yr (Lander 1979; York 1985). Mortality during the first 2 yr occurs in one of two stages: (1) 0-4 mo, when the animals are on land, or (2) age 4 mo – 2 yr, when they are at sea. Typically, 5–20% of pups die during their first 4 mo, while 50–80% of the remainder die during the following 20 mo (Lander 1979; York 1985). Mortality on land during the first 4 mo is correlated with population size and is considered density dependent (York 1985). The density effect, estimated via a logistic regression of the survival rate on land on the logarithm of numbers of pups born, accounts for about 30% of the variation in on-land survival. The time at sea between age 4 mo and 2 yr is a period of high mortality, but our knowledge of how pups acclimate to the pelagic environment and what causes their deaths is limited.

From 1958 through 1974, scientists from the United States and Canada conducted research cruises during which they collected northern fur seals. The locations of collection of all northern fur seals younger than 2 yr are shown in Fig. 2A and those older than 2 yr in Fig. 2B. The distribution of young animals is much more restricted than that of older animals; large numbers of younger animals were sampled off the coasts of Washington and Vancouver Island (Fig 2A). Kajimura (1979) summarized the available data on the distribution of young animals based on pelagic collections and stranding reports of tagged animals and suggested that the distribution and migration routes of young fur seals were strongly affected by ocean currents and prevailing wind regimes.

These same studies give us our best information on the diet of the northern fur seal. Fur seals feed on a variety of prey; 53 species of fish and 10 species of squid have been found in fur seal stomachs (Kajimura 1984). Kajimura (1984) describes the fur seal as an opportunistic feeder, preying on the most available species throughout its range. However, Perez and Bigg (1986) characterize its diet as about 60% small schooling fish, 23% other fish, and 17% squid, while Sinclair (1988) notes that over its range, the fur seals' prey is of similar size (generally ranging from about 10–20 cm.) and behaviour (large portions of the prey are schooling fish or squid).

Some work has been done on environmentally induced mortality on northern fur seals. Scheffer (1950a) reported large numbers of dead pups washed up on the Oregon and Washington coasts after a very stormy period in January and February 1950. Trites (1984) suggested that storm regimes and

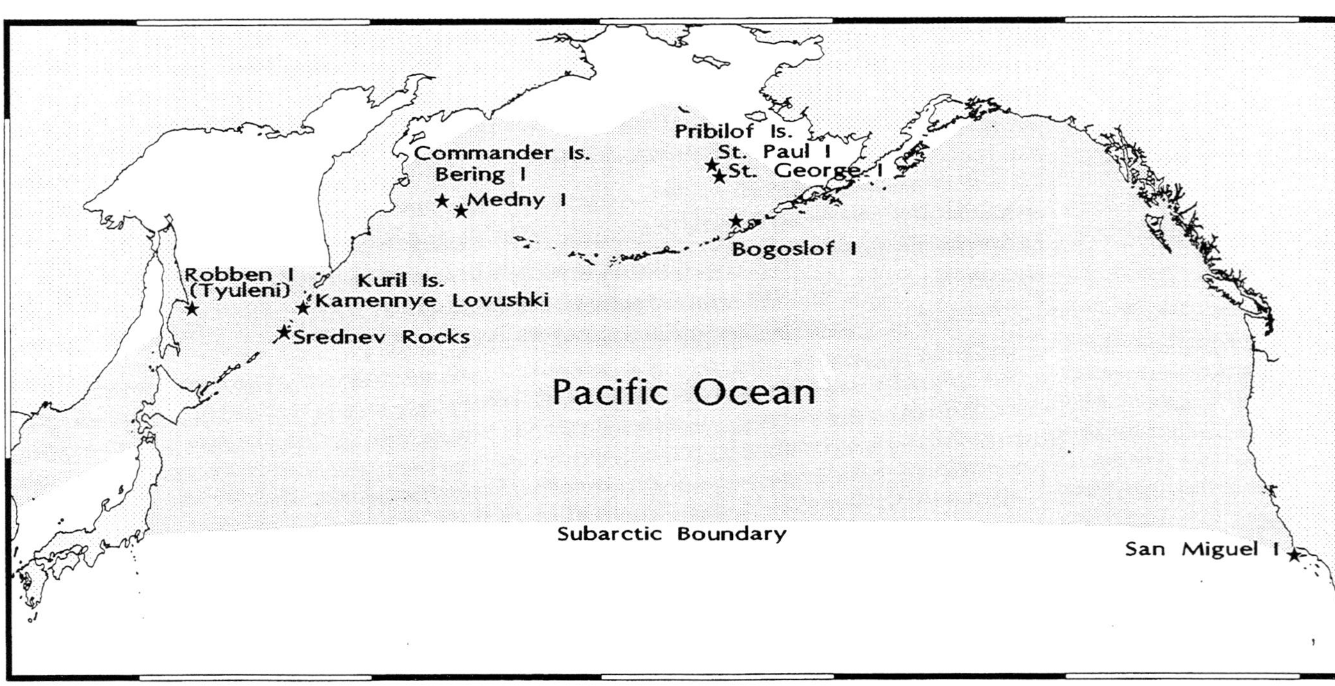

Fig. 1. General oceanic distribution of the northern fur seal, showing the breeding islands (after York 1987).

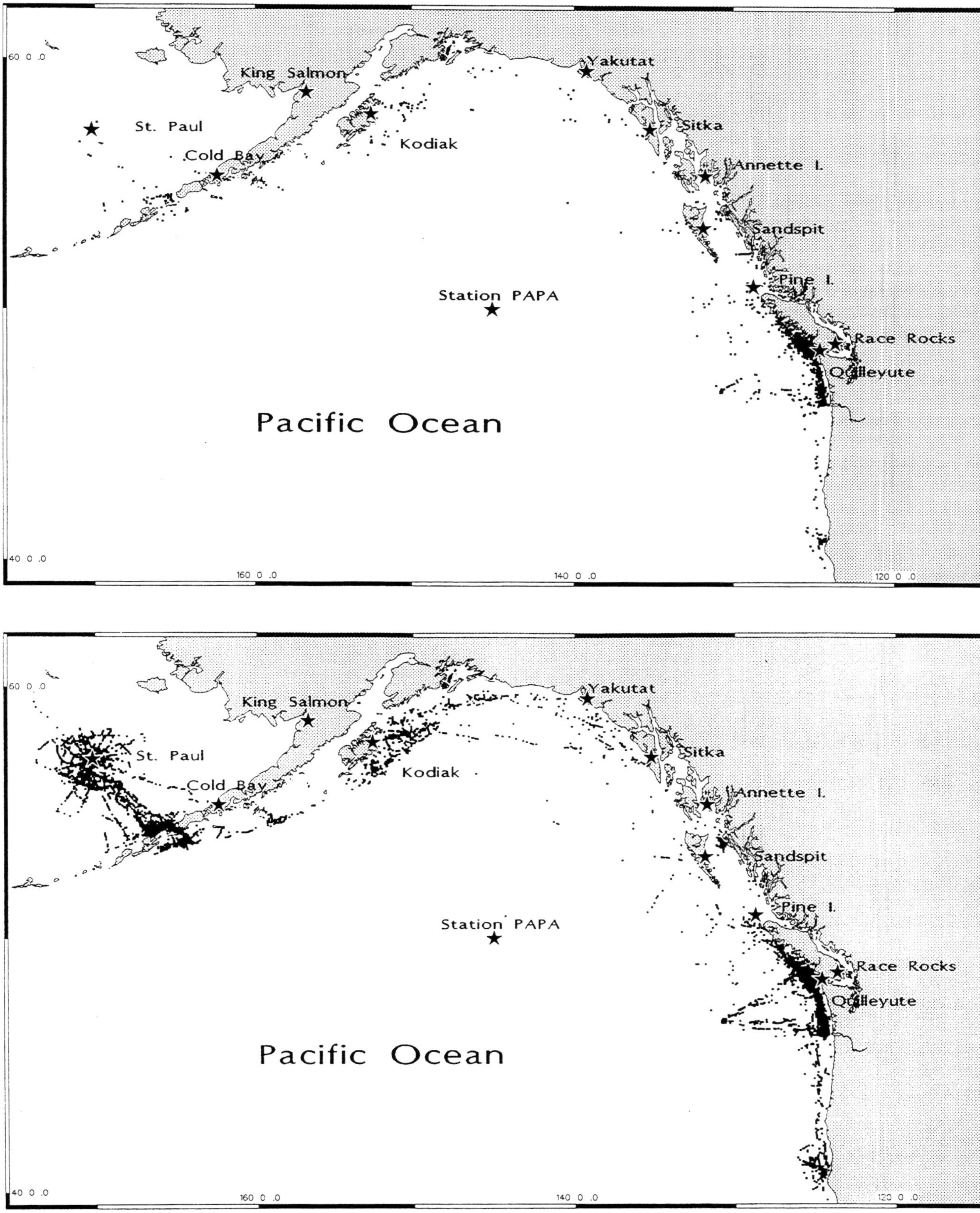

Fig. 2. (A) Locations of collection of northern fur seals younger than 2 yr in the combined U.S. and Canadian pelagic collections (1958–74). (B) Locations of collection of northern fur seals, 2 yr and older, in the combined U.S. and Canadian pelagic collections (1958–74).

cold weather in the mid-1970s in the North Pacific Ocean adversely influenced the survival of young fur seal pups and Trites (1990b) investigated the relationship between survival on land and air temperatures (ATs) on St. Paul Island. York (1991) investigated the relationship between early survival of the 1950–79 cohorts from the Pribilof Islands and sea-surface temperatures (SSTs) at Pine Island, British Columbia, and modeled survival as a function of SST in April of the birth year and March, 4 yr before the cohort was born. Others have compared attendance patterns of female northern fur seals and the growth and survival of pups during normal years and during the El Niño event of 1982–83. DeLong and Antonelis (1991) have shown that northern fur seals breeding on San Miguel Island, California, were adversely affected by the 1982–83 El Niño: females foraged for longer periods than normal, and weights of pups at age 3 mo and survival of pups during their first 3 mo were significantly less than other years. On the other hand, Gentry (1991) found no such effects for northern fur seals breeding on St. George Island (part of the Pribilof Islands) in Alaska; this is consistent with Niebauer's (1988) analysis that the 1982–83 El Niño had only limited effect in the Bering Sea. Both the Gentry and the DeLong and Antonelis studies were conducted during the summer breeding seasons.

The purpose of this paper is to investigate the relationship between the on-land and at-sea components of early survival of northern fur seals and several time series of ATs and SSTs within the range of the fur seal in the eastern North Pacific Ocean and Bering Sea (Table 1). In addition, I also investigate the relationship between fur seal survival and two indices of climate that have been cited as being related to the climate in the North Pacific Ocean. These indices are the sea-level pressures (SLP) at Darwin, Australia, and the Northeast Pacific Pressure Index (NEPPI), the difference in pressures between points near Reno, NV, (40°N, 120°W) and 50°N, 170°W in the Gulf of Alaska (Emery and Hamilton 1985). The SLPs at Darwin are highly correlated with the El Niño – southern oscillation (ENSO) (e.g., Trenberth 1976; Deser and Wallace 1987) and NEPPI is correlated with the year-class strength of walleye pollock (*Theragra chalcogramma*) in Shelikof Strait (Megry et al. 1995).

Methods and data

Northern fur seal survival data

A commercial harvest of sub-adult male fur seals (mostly ages 2–5 yr) was conducted on the Pribilof Islands from 1918 to 1984, and a small subsistence harvest has been conducted by Pribilof natives since 1985. Harvests were always conducted during late June through early August, and biological samples from the commercial and subsistence harvests have been collected for the determination of life history parameters since 1950. Scheffer (1950b) discovered that the age of fur seals could be determined by counting the number of growth rings in dentine layers of teeth, and a 20% subsample of animals from the harvest was routinely aged after 1950. Estimates of numbers of pups born and estimates of mortality on land from counts of dead pups are available for most years since 1950 (York and Kozloff 1987). Age-specific harvest rates of 2-, 3-, 4-, and 5-yr-old male fur seals for the 1950–79 cohorts (except 1971) were determined by dividing the number of males killed in each age class by the estimated number of male pups alive in September of their birth year. Lander's (1975) method, based on harvest rates, was used to estimate survival of juvenile males from age 4 mo to 2 yr for most cohorts born during 1950 through 1979; a modification of Lander's estimates of survival (Trites 1990a) are very similar and highly correlated with Lander's estimates ($r^2 = 0.98$) and would probably have given virtually identical results as provided here.

Estimates of survival of fur seals from birth to age 4 mo are available for those years when counts of dead pups and estimates of pups born were available; this includes all years from 1950–1979, except 1971.

Table 1. Location and duration of several environmental time series collected in the range of the northern fur seal or that may relate to the climate in the fur seal's range. SST = sea-surface temperature; AT = air temperature; SLP = sea-level pressure.

Location	Latitude	Longitude	Parameter	Range of years	Data Source
St. Paul, AK	57.2°N	170.2°W	AT	1946–1987	WeatherDisc Associates, Inc. 1990
Cold Bay, AK	55.2°N	170.2°W	AT	1949–1989	WeatherDisc Associates, Inc. 1990
King Salmon, AK	58.7°N	156.7°W	AT	1942–1987	WeatherDisc Associates, Inc. 1990
Annette Island, AK	55.0°N	131.6°W	AT	1946–1987	WeatherDisc Associates, Inc. 1990
Kodiak, AK	57.8°N	152.5°W	AT	1949–1989	Federal Climate Complex, 1990
Yakutat, AK	59.5°N	139.7°W	AT	1941–1987	WeatherDisc Associates, Inc. 1990
Sandspit, BC	53.3°N	131.8°W	AT	1948–1987	WeatherDisc Associates, Inc. 1990
Tattoosh, WA	48.0°N	124.6°W	AT	1898–1966	WeatherDisc Associates, Inc. 1990
Quilleyute, WA	48.4°N	124.7°W	AT	1966–1987	WeatherDisc Associates, Inc. 1990
Station PAPA, BC	50.0°N	145.0°W	SST	1950–1982	S. Tabata, Fisheries and Oceans, Canada
Race Rocks, BC	48.3°N	123.5°W	SST	1941–1982	S. Tabata, Fisheries and Oceans, Canada
Pine Island, BC	50.9°N	128.7°W	SST	1937–1982	S. Tabata, Fisheries and Oceans, Canada
Darwin, Australia	12.4°S	130.8°E	SLP	1880–1987	WeatherDisc Associates, Inc. 1990
Reno, NV	40.0°N	120.0°W	Pressure	1900–1987	A. Macklin, PMEL.
Gulf of Alaska	50.0°N	170.0°W	difference		

Environmental data and indices

I was interested in relating the survival of fur seals to monthly environmental parameters taken over a large part of their range and to other indices that might be related to climate within their range. To investigate the existence of lag effects (i.e., the relationship of survival to environmental parameters 4 yr or more before the cohort was born and up to 4 or 5 yr after the cohort was born), it was necessary to find monthly time series that extended from about 5 yr before to 5 yr after the birth of the cohort over the range of available fur seal data, i.e., from about 1945 to 1983. The availability of two CD-ROMs, WeatherDisc Associates, Inc. (1990) and Federal Climate Complex (1990), containing monthly summaries of station weather data made this task easy and most of the time series used in the analyses came from these sources (Table 1). Three time series of SSTs from British Columbia (S. Tabata, Canadian Fisheries and Oceans, Sydney, B.C., personal communication) were available (Table 1). The time series of NEPPI values, based on the work of Emery and Hamilton (1985), were provided by A. Macklin (PMEL, NOAA, Seattle, WA, 98115).

Analytical methods

The relationships between the time series of early survival of cohorts of northern fur seals and the environmental time series were studied in two phases. First each environmental time series was decomposed using the seasonal-trend decomposition method (STL), a filtering procedure based on loess (Cleveland et al. 1990). With this method, the series is reduced to a sum of seasonal, trend(s), and, remainder components. Then the relationship between fur seal survival and the environmental time series is investigated by constructing lagged cross correlations (Chatfield 1975) of the survival (or in the case of survival on land, the residual after accounting for the density-dependent effect) and the trend component of the two series. (The cross correlation between juvenile survival and SST lagged *m* months from January of the cohort year, say r_m is the simple correlation between the time series of survival rates and the SSTs *m* months after January of the cohort year. Thus, r_0 is the correlation between juvenile survival and the trend component of the environmental time series of interest for January of the cohort year; r_{-1} is the correlation between juvenile survival and the trend component for December of the year before the cohort year, etc.)

The cross correlations were calculated for *m* in a range of –60 to 60 mo from January in the cohort year. There were no significant (P = 0.05) autocorrelations (up to a lag 10 yr) in the time series of early oceanic survivals, and therefore, no correction for integral time scale (cf. Kundu and Allen 1976) was necessary. The break point between statistical significance and nonsignificance (P = 0.05) was estimated using Fisher's *Z*-transformation (Kendall and Stuart 1977). The level of correlation, which is significantly different from 0, changes for those time series not available over the entire range from 1945 to 1982 via a reduction in sample size.

The STL procedure is flexible and one can specify the amount of variation in the trend and seasonal component by changing the band width of the filters; for example, it is possible to determine if the seasonal component of the series changes over time. Because the purpose of this paper was to see if trends in environmental conditions were associated with trends in fur seal survival, the seasonal component was fitted to be periodic after the trend was filtered out. Thus, the main decision in the modeling process was choosing the band width of the filter for the trend component. Following the recommendation of Cleveland et al. (1990), and from an inspection of the cross correlation of the survival and the trend components over a range of band widths, there was very little difference in results for band widths ranging from 17 to 35 mo and a band width of 25 mo was used.

Confidence intervals for the cross correlations were estimated following a parametric bootstrap simulation method of Cleveland et al. (1983, 1990). This method requires that the residuals of the decomposed time series be independent Gaussian random variables. If the remainder terms described above failed the Shapiro–Wilk test of normality (Shapiro and Wilk 1965; Royston 1982) or if the Durbin–Watson test indicated significant autocorrelation (i.e., the remainders were not independent, normal, random variables), they were further reduced with a higher degree autoregressive model until both conditions were met by the residuals of the autoregressive model. Then, a synthetic remainder component was constructed by sampling from a normal distribution with variance equal to the variance of the residuals from the autoregressive process and computing the estimated remainder via the autoregressive model. The simulated environmental series was then calculated as the sum of this synthetic remainder plus the trend(s) and seasonal components from the original fit provided by STL. Start-up effects were eliminated by carrying out the calculation for 100 mo longer than the particular time series; then, the first 100 values were dropped from the synthetic series. STL was then applied to this simulated series to obtain new trend and seasonal components; then, the cross correlation of that trend component and the fur seal survival series was constructed. In the case of the Bering Sea series, the cross correlation was constructed with the on-land portion of survival adjusted for density dependence, as well. This computationally intensive procedure was repeated 200 times using the statistical package S-Plus[1] (Becker et al. 1988) on selected series.

Results

Early oceanic survival and environmental indices

The lagged cross correlations of early ocean survival and the trend components of the environmental time series for those stations outside the Bering Sea show two main patterns (Fig. 3). All, except those from the northern Gulf of Alaska, have two wide windows of significant positive correlation: the first, lasting 7–12 mo, is centred about the fall or winter months of the year before the cohort was born; this period begins with the month of conception and extends through the month of implantation and very early fetal growth of the cohort. The second period, lasting 8–12 mo, is centred during the spring, 4 or 5 yr before the cohort was born. Those from the northern Gulf of Alaska (Kodiak and Yakutat) do not have any large windows of significant positive correlation, but have a 12-mo

[1] Use of product names does not imply endorsement by the National Marine Fisheries Service.

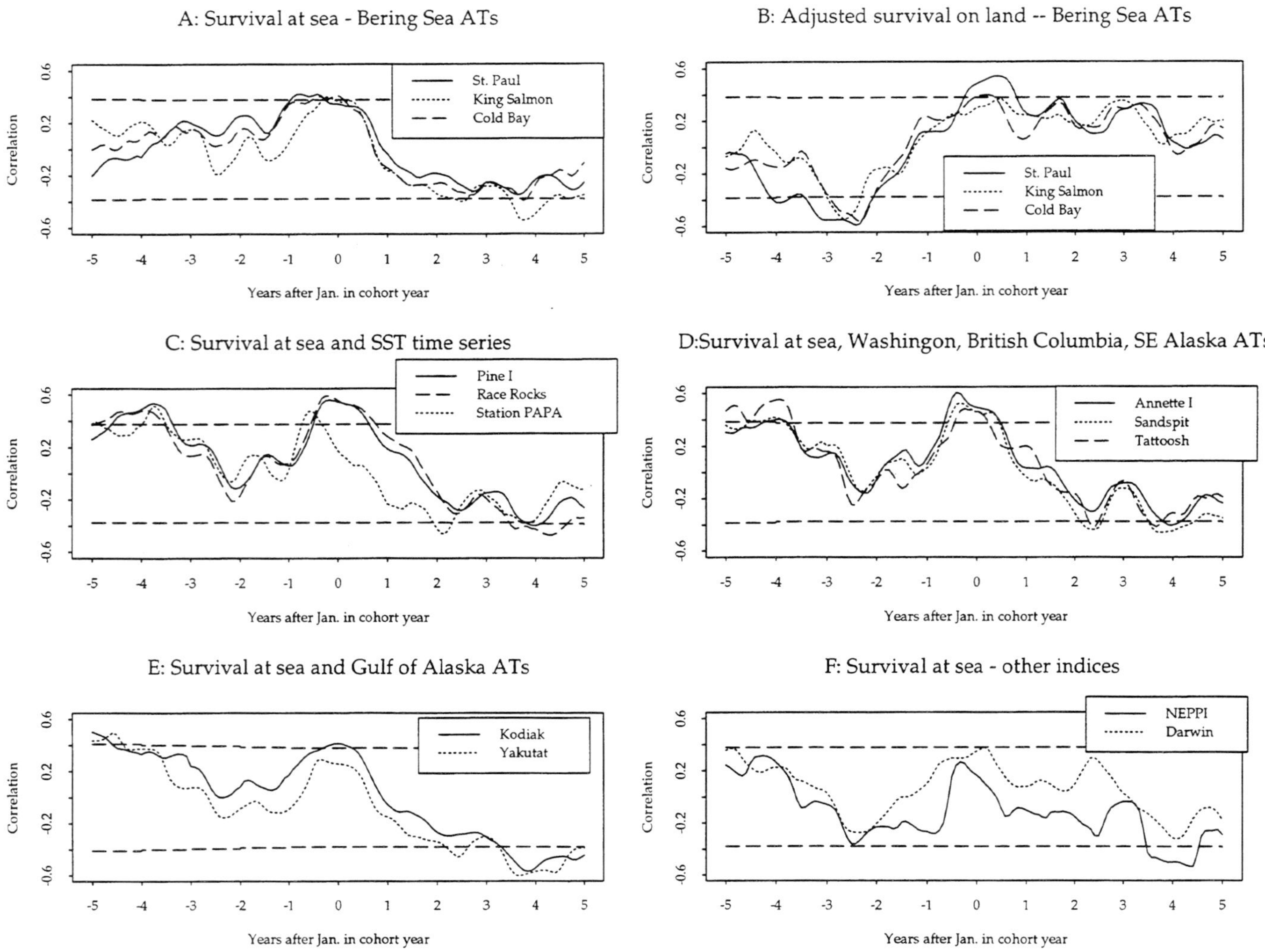

Fig. 3. Cross correlations of the trend component of several environmental time series decomposed using STL (see text for details) and the survival of juvenile male fur seals from the Pribilof Islands. (A) Air temperatures from St. Paul, King Salmon, and Cold Bay and survival on land (after adjusting for density dependence). (B) ATs from St. Paul, King Salmon, and Cold Bay and early oceanic survival. (C) SSTs from Pine Island, Race Rocks, and Station PAPA British Columbia, and early oceanic survival. (D) ATs from Annette Island, Alaska, Sand Spit, British Columbia, and Tatoosh, Washington and oceanic survival. (E) ATs from Kodiak and Yakutat, Alaska, and oceanic survival. (F) SLPs (Darwin, Australia), Northeast Pacific Pressure Index (NEPPI), and oceanic survival.

window of significant negative correlation centred in the winter, 4 yr after the cohort was born. The Bering Sea series, on the other hand, do not have a window of significant positive correlation with survival at sea for 4 yr before the cohort was born.

The lagged cross correlations of early ocean survival and the trend components of the Bering Sea ATs (Fig. 3A) are positive and significantly greater than 0 ($P = 0.05$) in a 4 to 7 mo window centred about January or February of the year of birth; they do not have a window of significant positive correlation ($P = 0.05$) with survival at sea for 4 yr before the cohort was born.

The lagged cross correlations of early ocean survival and the trend components of SSTs from Pine Island and Race Rocks (Fig. 3C) show significant ($P = 0.05$) positive correlations in a 13 to 17-mo window centred about February or March of the year of birth; significant ($P = 0.05$) positive correlations are also apparent in a 13 to 16-mo window centred about January, 4 yr before the cohort is born and a few significantly negative correlations ($P = 0.05$), 3 or 4 yr after the cohort is born (Fig. 3C).

Bootstrap simulations were computationally intensive and were only run for the cross correlations of survival at sea with the St. Paul, Pine Island, and Race Rocks temperature series. The bootstrap simulations preserved the centers of the windows of positive correlation, but in some cases, reduced their extent (Table 2). The windows of statistically significant ($P = 0.05$) negative correlation at age 4 yr (Fig. 2C) were not preserved in the simulation (Table 2).

Table 2. Comparison of cross correlation curves for the fitted model and simulation results (200 replicates). Months after January of the cohort year in which the cross correlation of the trend component of the decomposed series and the given survival series were statistically significant ($P = 0.05$) with a comparison with 200 replicated simulations of the same process.

	Months of significant correlations with fitted model		Months for which at least 97.5% of the simulations were significant	
Time series	Positive	Negative	Positive	Negative
Correlation with survival at sea				
St. Paul, AK	−11 to − 4	None	−10 to −2 0 to 3	None
Pine Island, B.C.	−55 to −41 − 5 to 7	46 to 49	−51 to −43 −4 to 5	None
Race Rocks, B.C.	−59 to −43 − 6 to 9	42 to 56	−49 to −48 − 3 to 2	None
Correlation with survival on land after adjusting for density dependence				
St. Paul, AK	− 3 to 9 −31 to −25	−39 to −35	−2 to 7 9 to 10	−38 to −27

Survival on land, adjusted for density dependence, and Bering Sea temperature

The pattern of correlation of survival on land (after accounting for density dependence) and the trend component of the Bering Sea ATs was significantly ($P = 0.05$) greater than 0 only for the St. Paul data in a window extending from October the year before birth until September of the birth year (Fig 3B). Bootstrap simulations were only run for the cross correlations with the St. Paul series and the bootstrap estimate of the window of statistically significant cross correlations ($P = 0.05$) was similar to that of the raw cross correlations (Table 2).

A window of significant negative correlation ($P = 0.05$) occurred all three series for most months during the winter and spring, 2 yr before the cohort was born. For the St. Paul series, the corresponding bootstrap window extended from November, 3 yr before the cohort was born until October, 2 yr before the cohort was born.

Discussion

Lander's (1979) estimates of mortality at sea are highly correlated ($r^2 = 0.98$; $P < 0.01$) with estimates of total harvest rates of cohorts of subadult male fur seals calculated by summing the age-specific harvest rates. Estimates of survival, based on Lander's method, were produced for cohorts that were at least age 5 yr before the commercial harvest ended in 1984. These estimates provide a rough index of the relative survival of the various cohorts and have been useful for modeling the population dynamics of the St. Paul Island and Robben Island populations (e.g., York and Hartley 1981; Frisman et al. 1982; Trites 1984). Similar estimates (Trites 1989) have been successfully used to model the fur seal population (Trites 1990a).

Two main hypotheses come to mind to explain the patterns of cross correlation of temperatures and survival of fur seals in Fig. 3. The first is simple and related to the direct effects of temperature and thermal stress on young animals. In this case, the timing of the correlations will have zero or at least, a very short lag, e.g., the air or water must be too cold while the animals are in that cold air or water. The second hypothesis involves indirect effects and relates an increase in survival of fur seals to changes in temperatures through an increase or decrease in abundance in some important prey of the fur seal. In this case, the timing of the correlation should be such that survival lags behind the temperature change. Because increased food supplies for pups or lactating or pregnant females might act to increase the survival of the pup, this form of this hypothesis has many possibilities.

The first hypothesis relates the biological relationship between temperatures and survival of fur seals through the thermal effects on young animals. If metabolic rate were higher in colder weather or colder water, then exposure could result in significantly higher physiological stress or higher total food requirements over time. This, in turn, could lead to lower survival rates during the colder years if prey were not sufficiently abundant or distributed in a manner easily accessible to the seals. In the case of survival on land, pups could be compromised if their mothers were not able to obtain enough prey to produce sufficient nourishment. Data in Kooyman et al. (1976) suggest that metabolic rate of northern fur seals increases in colder water; other experiments (G. Worthy, Texas A & M University, Galveston, TX, personal communication) show that the metabolic rate of California sea lions increases in colder water. However, most of the patterns in Fig. 3 are not consistent with this explanation. Except for Fig. 3B (the relationship between the Bering Sea temperature series and the non-density -dependent component of survival on land), there is no significant positive correlation between temperatures and survival when the animals are alive. On the other hand, the pattern of statistically significant correlations between the selected Bering Sea ATs (Fig. 3B) and survival on land (after removing the effect of density dependence), could be related to thermal effects. The significant correlations during those months occur for the St. Paul series only. They extend from the October before the cohort is born through October of the birth year and include the perinatal period up to weaning. Large numbers of dead pups often appear after heavy storms and cold temperatures and the correlation between St. Paul temperatures and survival during the breeding season is consistent with those observations. The lack of any significant correlation with the other two series suggests that very local conditions on St. Paul Island might relate to the thermal stress on individuals and be important for the year-class's early survival. This conclusion is somewhat different from Trites (1990b) wherein no significant relationships between survival on land and St. Paul temperatures were found; this difference is a consequence of examining the survival after adjusting it for density dependence.

Several studies in Trillmich and Ono (1991) show that food availability directly affects survival of young seals. Because the lack of food diminishes survival, high availability of preferred prey should then enhance survival. If the patterns in Fig. 3 are related to food resources for young fur seals or for their pregnant or lactating mothers, we must identify responses in the food chain caused by events 1–5 yr before the animals are weaned. An important step in explaining the timing of the bands of statistically significant correlations between temperatures and survival would be an analysis of the life history of the prey of the fur seal to generate hypotheses concerning the likely effects of a change in SST on food availability for young fur seals. However, given the fur seal's wide-ranging diet and migration, and the lack of data on the abundance and life history of its prey, especially of fishes that are not commercially important, we can only speculate on this relationship. Fisheries are complicated, and although a single environmental variable cannot be used to predict the year-class strengths of stocks of fish with much accuracy, environmental variables are significant factors and affect the size of a stock (cf. Favorite and McLain 1973; Sutcliffe et al. 1977).

The negative correlation between survival on land (adjusted for density dependence) and the selected Bering Sea AT trends during the winter and spring 2 yr before the cohort is born (months −27 through −18) is common to all three series and suggests a more widespread effect than merely on St. Paul Island (as above). Possibilities include a negative effect of cold on an organism that causes mortality in fur seals (e.g., hookworm) or a link through the food chain. Quinn and Niebauer (1995) have shown that the recruitment to age 2 yr of walleye pollock in the Bering Sea is negatively correlated with air temperatures in the Bering Sea 2 yr before the cohort is hatched; cool temperatures 2 yr before hatching are associated with larger than average recruitment at age 2. Because fur seals prefer to eat small age 0 or 1 pollock (Sinclair 1988), this negative correlation in Fig. 3B could be related to larger than normal quantities of age 0 or 1 pollock available for late-term pregnant or lactating females. For pregnant females, abundant prey could increase the size of the fetus, leading to heavier pups, and possibly to higher survival of the cohort (Baker and Fowler 1992). Easily available prey should lessen the duration of feeding trips that females must make, and thus increase the cohort's survival. Baker (1991) reported from an examination of nursing layers in teeth of subadult males from the commercial harvest, the mean number of nursing layers were 11 and no survivors were found with fewer than 8 layers.

The cross correlations of survival at sea of young animals and the Bering Sea ATs were significantly positive in a 13-mo window extending from February the year before birth through March of the birth year. Such a correlation could be related to response in the food chain, in which warmer temperatures the year before the cohort is born increases the availability of prey for young fur seals as they leave the Bering Sea in November or December. However, because pups leaving the Pribilof Islands in mid-November and journeying south spend on average only 10.5 days in the Bering Sea (Ragen et al. 1993) before they reach the Gulf of Alaska, the Bering Sea food sources would appear to be more important to females nursing pups then they are to pups shortly after weaning.

York (1991) speculated on the possible role Pacific herring (*Clupea pallasi*) could play in the survival of northern fur seals, and related some details on herring life history and their relationship to temperature change and how the correlations in Fig. 3 might be related to the availability of herring or fish with similar life histories. The stations, whose data have the strongest cross correlations with early oceanic survival, (Figs. 3C and D) surround the wintering ground for the British Columbia and Washington stocks of herring (Kajimura 1984). Pups appear on the British Columbia and Washington coasts in January (Kajimura 1984).

Herring is an important prey species for fur seals (Kajimura 1984; Perez and Bigg 1986); it ranks first or second in importance (based on percentage of volume in stomachs) for most months in British Columbia, the Gulf of Alaska, and Washington. Herring is used to feed captive fur seals and may be close to "ideal" for young animals: it is small, has high fat content, swims in large schools, and it would be readily available to young fur seals, especially during spawning. The distribution of herring spawning grounds from Auke Bay to Monterey Bay (cf. Haegele and Schweigert 1985) is similar to the distribution of young northern fur seals (those less than 2 yr) collected by scientists from the United States and Canada during 1958 through 1974 (Fig. 2A). In contrast to the distribution of all other fur seals taken during the same research cruises (Fig. 2B), few animals younger than 2 yr were found near the Aleutian chain or in the Bering Sea. In general, the younger animals were found closer to shore than the older animals; many were found near the mouths of bays, although this may be partially a result of sampling effort. Fur seals, with the exception of young animals or groups of animals following a spawning fish population, do not usually move into restricted bays (Kajimura 1984). Kajimura (1984) notes that the area off the Strait of Juan de Fuca and Hecate Strait are principal feeding grounds for juvenile and immature herring. Most young fur seals in pelagic collections were found in areas near spawning or feeding grounds of herring.

Herring is also an important commercial species, and there are data on year-class strengths for some stocks. Macy et al. (1978), Schoener and Fluharty (1985), and Bailey and Incze (1985) have reported that year-classes of most stocks of herring in Washington, British Columbia, and Alaska have been strong in years following El Niños, while those in California have been very weak. Favorite and McLain (1973) showed that strong cohorts of herring near the Queen Charlotte Islands followed years with anomalously high SST. Haist and Stocker (1985) showed that increases in SSTs during the summer are associated with higher growth rates for juvenile Pacific herring. An increase in SST the summer before the birth of cohort could therefore increase the availability of herring for young fur seals during the following winter and spring. Furthermore, a strong year-class of herring could give rise to another strong year-class *n* years following, when a large portion of the cohort reproduces at age *n*: a "baby-boom" effect. Herring are multiple spawners and are sexually mature in Alaska and British Columbia between the ages of 3 and 5 yr. It is commonly believed among herring fishermen in Alaska that large runs occur about every fifth year (Macy et al. 1978).

Other prey of fur seals also seem to respond to occurrences of El Niños. Schoener and Fluharty (1985) reported that the abundance of Pacific squid (*Loligo opalesens*) off Washington increased sharply following El Niño events in 1940–41, 1957–58, and 1982–83, and that the Pacific mackerel (*Scomber japonicus*) increased in abundance during the 1982–83 El Niño.

Thus, there is evidence that the availability of herring and, perhaps, squid and mackerel as food for fur seals is affected by changes in SST. How this relates to other prey of the fur seal is unknown. If other small schooling fish reacted similarly to temperature changes, it would be reasonable to think that an increase in abundance of these few species following a warming in SST could lead to an increase in early survival of fur seals, even though the fur seal is probably capable of switching to a different prey if its "preferred" prey were not available.

The effect of an increase in SST could be even more general. Frost (1983) showed that levels of zooplankton productivity in the Gulf of Alaska increased during the 1958–59 El Niño, whereas the abundance of zooplankton off California was reduced (Bernal and McGowan 1981). Bailey and Incze (1985) suggest that the warm water and increased productivity in the north were beneficial to species of fish in the north.

The lack of significant positive correlation with the SLPs from Darwin and NEPPI suggest that the fur seal survival is more related to changes in SSTs or ATs in areas that they occupy rather than to any general ENSO signal or relationship with the location of the Aleutian Low.

This work is speculative and shows a correlative relationship, not a causative one, between temperature and early oceanic survival of fur seals. It is not known how changes in temperature affect the assemblages of prey available to the fur seal over its extensive range in the North Pacific Ocean. The continued cooperation of mammalogists and fisheries biologists may enable us to understand how changes in environmental parameters, such as SSTs, and other factors, such as fishing effort, affect the availability of the prey species and how that is related to vital parameters of fur seals.

Acknowledgements

I thank A. Macklin, PMEL, Seattle, WA, for providing the Northeast Pacific Pressure Index data and S. Tabata, Fisheries and Oceans, Sydney, B.C. for the SST data from Pine Island, Race Rocks, and Station PAPA. Several co-workers at the National Marine Mammal Laboratory contributed to the formation of some of the ideas in this paper; among these are: George Antonelis, Jr., Elizabeth H. Sinclair, Richard Merrick, Harriet Huber, and Robert DeLong. I thank Rod Towell for making the maps and Jeffery Breiwick, Douglas DeMaster, Andrew Trites, Thomas Loughlin and Bryan Walker for helpful and careful reviews of the manuscript.

References

Bailey, K.M., and L.S. Incze. 1985. El Niño and the early life history and recruitment of fishes in temperate marine waters, p. 143–165. *In* W.S. Wooster and D.L. Fluharty [ed.] El Niño north. Washington Sea Grant Program, University of Washington, Seattle, WA.

Baker, J.D. 1991. Trends in female northern fur seal, *Callorhinus ursinus*, feeding cycles indicated by nursing lines in juvenile male teeth. M.S. thesis, University of Washington, Seattle, WA. 51 p.

Baker, J.D. and C.W. Fowler. 1992. Pup weight and survival of northern fur seals, *Callorhinus ursinus*. J. Zool. (Lond.) 227: 231–238.

Bartholomew, G.A., and P.G. Hoel. 1953. Reproductive behavior of the Alaska fur seal, *Callorhinus ursinus*. J. Mammal. 34: 417–436.

Becker, R.A., J.M. Chambers, and A.R. Wilks. 1988. The new S language, a programming environment for data analysis and graphics. Wadsworth and Brooks/Cole Advanced Books and Software, Pacific Grove, CA. 702 p.

Bernal, P.A., and J.A. McGowan. 1981. Advection and upwelling in the California current. p. 381–399. *In* F.A. Richards [ed.] Coastal and estuarine science 1. American Geophysical Union, Wash. D.C.

Chatfield, C. 1975. The analysis of time series: theory and practice. Chapman and Hall, London, UK. 263 p.

Cleveland, R.B., W.S. Cleveland, J.E. McRae, and I. Terpenning. 1990. STL: a seasonal-trend decomposition procedure based on loess. J. Off. Stat. 6: 3–73.

Cleveland, W.S., A.E. Freeny, and T.E. Graedel. 1983. The seasonal component of atmospheric CO_2: information from new approaches to the decomposition of seasonal time series. J. Geophy. Res. 88: 10 934 – 10 946.

DeLong, R.L., and G.A. Antonelis. 1991. Impact of the 1982–1983 El Niño on the northern fur seal population of San Miguel Island, California, p. 75–83. *In* F. Trillmich and K. Ono [ed.] Pinnipeds and El Niño: Responses to environmental stress. Springer Verlag, Heidelberg.

Deser, C., and J.M. Wallace. 1987. El Niño events and their relation to the Southern Oscillation: 1925–1986. J. Geophys. Res. 92: 14 189 – 14 196.

Emery, W.J., and K. Hamilton. 1985. Atmospheric forcing of interannual variability in the northeast Pacific Ocean: connections with El Niño. J. Geophys. Res. 90: 857–868.

Favorite, F., and D.R. McLain. 1973. Coherence in transpacific movements of positive and negative anomalies of sea surface temperature, 1953–60. Nature (Lond.) 244: 139–143.

Federal Climate Complex. 1990. International station meteorological climate summary CD-ROM, version 1.0. National Climatic Data Center, Asheville, NC.

Frisman, E. Ya., E.I. Skaletskaya, and A.E. Kuzin. 1982. A mathematical model of the population dynamics of a local northern fur seal herd. Ecol. Modell. 16: 151–172.

Frost, B.W. 1983. Interannual variation of zooplankton standing stock in the open Gulf of Alaska, p. 146–157. *In* W.S. Wooster [ed.] From year to year: Interannual variability of the environment and fisheries of the Gulf of Alaska and the Eastern Bering Sea. Washington Sea Grant Program, University of Washington, Seattle, WA.

Gentry, R.L. 1991. El Niño effects on adult northern fur seals at the Pribilof Islands, p. 84–93. *In* F. Trillmich and K. Ono [eds.] Pinnipeds and El Niño: Responses to environmental stress. Springer Verlag, Heidelberg.

Haegele, C.W., and J.F. Schweigert. 1985. Distribution and characteristics of herring spawning grounds and description of spawning behavior. Can. J. Fish. Aquat. Sci. 42 (Suppl.1): 39–55.

Haist, V., and M. Stocker. 1985. Growth and maturation of Pacific herring (*Clupea harengus pallasi*) in the Strait of Georgia. Can. J. Fish. Aquat. Sci. 42(Suppl. 1): 138–146.

Kajimura, H. 1979. Fur seal pup/yearling distribution in the eastern North Pacific, p. 9–50. *In* H.R. Kajimura, R.H. Lander, M.A. Perez, A.E. York, and M.A. Bigg [ed.] Preliminary analysis of pelagic fur seal data collected by the United States and Canada during 1958–74. Report submitted to the 22nd annual meeting of the Standing Scientific Committee, North Pacific Fur Seal Commission, 2–6 April 1979, Washington, D.C., USA.

Kajimura, H. 1984. Opportunistic feeding of the northern fur seal, *Callorhinus ursinus*, in the eastern North Pacific Ocean and eastern Bering Sea. U.S. Department of Commerce, Seattle, WA. NOAA Tech. Rep. NMFS SSRF-779: 49 p.

Kendall, M. and A. Stuart. 1977. The advanced theory of statistics. 4th ed. Vol. 1. Distribution theory. MacMillan, New York, NY. 490 p.

Kooyman, G.L., R.L. Gentry, and W.B. McAlister. 1976. Physiological impact of oil on pinnipeds. NWFC Processed Report, Northwest and Alaska Fisheries Center, National Marine Fisheries Service, NOAA, 2725 Montlake Boulevard E., Seattle, WA. 98112. 23 p.

Kundu, P.K., and J.S. Allen. 1976. Some three-dimensional characteristics of low-frequency current fluctuations near the Oregon coast. J. Phys. Oceanogr. 6: 181–199.

Lander, R.H. 1975. Method of determining natural mortality in the northern fur seal (*Callorhinus ursinus*) from known pups and kill by age and sex. J. Fish. Res. Board Can. 32: 2447–2452.

Lander, R.H. 1979. Role of land and ocean mortality in yield of male Alaskan fur seal, *Callorhinus ursinus*. Fish. Bull. U.S. 77: 311–314.

Macy, P.T., J.M. Wall, N.D. Lampsakis, and J.E. Mason. 1978. Resources of the non-salmonid pelagic fishes of the Gulf of Alaska and eastern Bering Sea, Part I. NWFC Processed Report, Northwest and Alaska Fisheries Center, National Marine Fisheries Service, NOAA, 2725 Montlake Boulevard E., Seattle, WA. 355 p.

Megrey, B.A., S.J. Bograd, W.C. Rugen, A.B. Hollowed, P.J. Stabeno, S.A. Macklin, J.D. Schumacher, and W.J. Ingraham, Jr. 1995. An exploratory analysis of associations between biotic and abiotic factors and year-class strength of Gulf of Alaska walleye pollock, p. 227–243. *In* R.J. Beamish [ed.] Climate change and northern fish populations. Can. Spec. Publ. Fish Aquat. Sci. 121.

Niebauer, H.J. 1988. Effects of El Niño – Southern Oscillation and North Pacific weather patterns on interannual variability in the Subarctic Bering Sea. J. Geophys. Res. 93(C5): 5051–5068.

Perez, M.A., and M.A. Bigg. 1986. Diet of northern fur seals, *Callorhinus ursinus*, off western North America. Fish. Bull. U.S. 84: 957–971.

Peterson, R.S. 1968. Social behavior in pinnipeds with particular reference to the northern fur seal, p. 3–53. *In* R.J. Harrison, R.C. Hubbard, R.S. Peterson, C.E. Rice, and R.J. Schusterman [ed.] The behavior and physiology of pinnipeds. Appleton-Century-Crofts, New York, NY.

Quinn, T.J. II, and H.J. Niebauer. 1995. Relation of eastern Bering Sea walleye pollock (*Theragra chalcogramma*) recruitment to environmental and oceanographic variables, p. 497–507. *In* R.J. Beamish [ed.] Climate change and northern fish populations. Can. Spec. Publ. Fish Aquat. Sci. 121.

Ragen, T.J. and G.A. Antonelis, and M. Kyota. 1993. The early phase of northern fur seal pup migration from St. Paul Island, AK, p. 88. *In* 10th Biennial Conference on the Biology of Marine Mammals, 11–15 Nov. 1993, Galveston, TX, USA.

Royston, J.P. 1982. The *W* test for normality. Appl. Stat. 31: 176–180.

Scheffer, V.B. 1950a. Winter injury to young fur seals on the northwest coast. Calif. Fish. Game 36: 378–379.

Scheffer, V.B. 1950b. Growth layers on the teeth of Pinnipedia as an indication of age. Science (Wash., D.C.) 112: 309–311.

Schoener, A., and D.L. Fluharty. 1985. Biological anomalies off Washington in 1982–83 and other major Niño periods, p. 211–225. *In* W.S. Wooster and D.L. Fluharty [ed.] El Niño north. Washington Sea Grant Program, University of Washington, Seattle, WA.

Shapiro, S.S. and M.B. Wilk. 1965. An analysis of variance test for normalilty (complete samples). Biometrika 52: 591–611.

Sinclair, E.H. 1988. Feeding habits of northern fur seals (*Callorhinus ursinus*) in the eastern Bering Sea. M.S. thesis, Oregon State University, Corvallis, OR. 93 p.

Sutcliffe, W.H., Jr., K. Drinkwater, and B.S. Muir. 1977. Correlations of fish catch and environmental factors in the Gulf of Maine. J. Fish. Res. Board Can. 34: 19–30.

Trenberth, K.E. 1976. Spatial and temporal variations of the Southern Oscillation. Q. J. R. Meterol. Soc. 102: 639–653.

Trillmich, F., and K. Ono [ed.] 1991. Pinnipeds and El Niño: responses to environmental stress. Springer Verlag, Heidelberg, Germany. 293 p.

Trites, A.W. 1984. Population dynamics of the Pribilof Islands North Pacific fur seal (*Callorhinus ursinus*). M. S. thesis. University of British Columbia, Vancouver, B.C., 123 p.

Trites, A.W. 1989. Estimating the juvenile survival rate of male northern fur seals (*Callorhinus ursinus*). Can. J. Fish. Aquat. Sci. 46: 1428–1436.

Trites, A.W. 1990a. The northern fur seal: biological relationships, ecological patterns and population management. Ph.D. thesis, University of British Columbia, Vancouver, B.C.

Trites, A.W. 1990b. Thermal budgets and climate spaces: the impact of weather on the survival of Galapagos (*Arctocephalus galapagoensis* Heller) and northern fur seal pups (*Callorhinus ursinus* L.). Funct. Ecol. 4: 753–768.

WeatherDisc Associates, Inc. 1990. World WeatherDisc (CD-ROM). WeatherDisc Associates, Inc., 4584 NE 89th Street, Seattle WA 98115.

York, A.E. 1985. Juvenile survival of fur seals, p. 34–45. *In* P. Kozloff [ed.] Fur seal investigations, 1982. NOAA Tech. Memo. NMFS-F/NWC-71.

York, A.E. 1987. Northern fur seal, *Callorhinus ursinus*: eastern Pacific population (Pribilof Islands and San Miguel Island), p. 9–21. *In* J.P. Croxall and R. Gentry [ed.] Status, Biology, and Ecology of Fur Seals: Proceedings of an international symposium and workshop. Cambridge, England, 23–27 April 1984. NOAA Tech. Rep. NMFS 51.

York, A.E. 1991. Sea surface temperatures and their relationship to the survival of juvenile male fur seals from the Pribilof Islands, p. 96–106. *In* F. Trillmich and K. Ono [ed.] Pinnipeds and El Niño: Responses to environmental stress. Springer Verlag, Heidelberg Germany.

York, A.E., and J.R. Hartley. 1981. Pup production following harvest of female northern fur seals. Can. J. Fish. Aquat. Sci. 38: 84–90.

York, A.E., and P. Kozloff. 1987. On the estimation of numbers of northern fur seal, *Callorhinus ursinus*, pups born on St. Paul Island, 1980–86. Fish. Bull. U.S 85: 367–375.

Interdecadal variability in distribution and catch rates of epipelagic nekton in the Northeast Pacific Ocean

Richard D. Brodeur and Daniel M. Ware

Brodeur, R.D., and D. M. Ware. 1995. Interdecadal variability in distribution and catch rates of epipelagic nekton in the Northeast Pacific Ocean, p. 329–356. *In* R.J. Beamish [ed.] Climate change and northern fish populations. Can. Spec. Publ. Fish. Aquat. Sci. 121.

Abstract: Catch data were analyzed from pelagic nekton research surveys over a wide area of the Gulf of Alaska during the late 1950s and the 1980s. Using a geographical information system for spatial analysis of standardized catch rates in surface gillnets, we show that the distribution patterns of 14 species of Pacific salmon (*Oncorhynchus* spp.), nonsalmonid fishes, and cephalopods have changed dramatically between these two periods. The distribution of most subarctic species was compressed to the northwest portion of the gulf whereas the transition species were found closer to the west coast of North America in the 1980s than in the earlier period. Most species examined showed marked increases in mean catch rate from the Central Subarctic Domain between these two decades. Jack mackerel (*Trachurus symmetricus*), the only species we examined that is predominantly a resident of the Coastal Upwelling Domain, was the only species to show a decline between the two periods. Estimated salmon biomass was nearly twice as high during the 1980s as it was in the late 1950s. Some possible links between these large-scale changes and the oceanic carrying capacity of the Northeast Pacific Ocean are considered.

Résumé : À la fin des années 1950 et pendant les années 1980, les données sur le necton pélagique capturé dans une vaste région du golfe d'Alaska dans le cadre de recherches et d'enquêtes ont été analysées. Grâce à un système d'information géographique permettant l'analyse spatiale des taux de capture normalisés dans les filets maillants de surface, nous montrons que la répartition de 14 espèces de saumons du Pacifique (espèce *Oncorhynchus*), de non-salmonidés et de céphalopodes a changé de façon spectaculaire entre les deux périodes. La distribution des espèces plus subarctiques a été comprimée dans la partie nord-ouest du golfe tandis que les espèces de transition étaient retrouvées plus près de la côte ouest de l'Amérique du Nord dans les années 1980 que pendant la période antérieure. La plupart des espèces examinées accusaient une hausse marquée du taux de prise moyen dans le domaine subarctique central entre les deux décennies. *Trachurus symmetricus (jack mackerel)*, unique espèce examinée qui réside principalement dans le domaine de la remontée côtière, était le seul à accuser un déclin entre les deux périodes. La biomasse de saumon estimative était presque deux fois plus élevée au cours des années 1980 qu'elle ne l'était à la fin des années 1950. Certains liens possibles entre ces changements à grande échelle et la capacité d'accueil du nord-est du Pacifique sont examinés.

Introduction

Understanding long-term changes in marine production systems is an important goal of fisheries research that is clearly relevant to fisheries management. Much progress toward this goal has been made for coastal ecosystems where traditional and continuous fisheries have existed for decades or more (Clark and Brown 1977; Mahon and Smith 1989). As the coastal fisheries become fully subscribed, the fishery may expand to offshore waters where often little is known about the standing stocks and production of the major species. The subarctic Pacific ecosystem has witnessed such an expansion with the initiation of a high seas driftnet fishery for Pacific salmon (*Oncorhynchus* spp.) and, more recently, squid.

The Central Subarctic Domain of the North Pacific is characterized by cool and relatively low-salinity water (Dodimead et al. 1963). Cyclonic circulation around the Gulf of Alaska generates upwelling in the centre of the gyre, bringing nutrient-laden water to the surface. Following stratification of the water column in spring, primary productivity increases rapidly but phytoplankton standing stocks apparently are held in check either by a large biomass of efficient zooplanktonic grazers or by limitation by some essential micronutrient (Miller

R.D. Brodeur. Alaska Fisheries Science Center, National Marine Fisheries Service, 7600 Sand Point Way NE, Seattle, WA 98115-0070, USA

D.M. Ware. Department of Fisheries and Oceans, Pacific Biological Station, Nanaimo, BC V9R 5K6, Canada

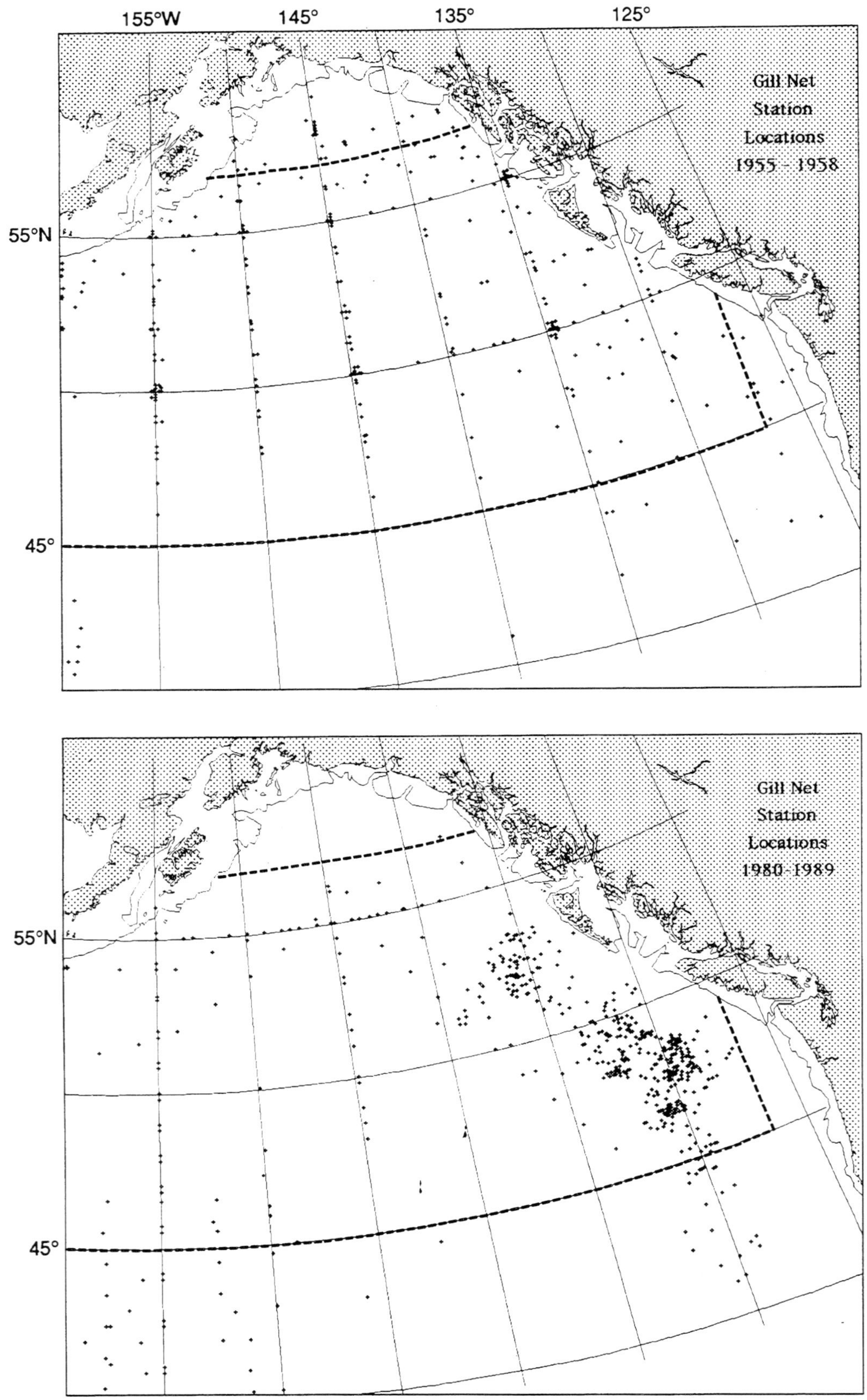

Fig. 1. Location of research gill net collections during 1955–58 and 1980–89. The dots indicate locations where the nets were set. The area over which catch rates were interpolated was bounded by the dashed lines and the 200-m isobath (inshore).

et al. 1991). Pacific salmon are the dominant top-level carnivores leading to a relatively simple pelagic food web for an oceanic ecosystem (Brodeur 1988; Pearcy 1991). This fairly productive region also supports substantial numbers of migratory epipelagic fishes, cephalopods, marine birds, and mammals, especially during the summer months (Sanger 1972; Taniguchi 1981; Kubodera 1986; Calkins 1987; DeGange and Sanger 1987; Yatsu et al. 1993).

Through the first half of this century, little was known about the species composition and distribution of the subarctic nektonic fauna. Since that time, the pelagic waters of the Gulf of Alaska have been sampled intensely during two different decades. The first occurred during the late 1950s when Canadian and U.S. researchers used variable-mesh gillnets to examine the distribution and biology of Pacific salmon; although a substantial amount of ancillary information was also gathered on nonsalmonid fishes and squid (Neave and Hanavan 1960; Larkins 1964; Fiscus and Mercer 1982). The most recent period of intensive sampling was during the 1980s, when Japanese and Canadian research vessels conducted surveys of the salmon and squid resources of the Gulf of Alaska. In the intervening period between these surveys, the physical and biological environment in the Gulf of Alaska changed substantially (Royer 1989; Tabata 1989; Brodeur and Ware 1992), and the salmonid resources were both augmented by enhancement and heavily depleted by the high-seas salmon driftnet fishery (Rogers 1987; Pearcy 1992). Having found a large-scale response by the zooplankton to the physical changes (Brodeur and Ware 1992), we were interested in seeing if there was any evidence of a similar response in the upper trophic levels of the Subarctic and Transition Domain ecosystems between these two periods.

In this paper, we provide estimates of the summer distribution patterns and catch rates of 14 pelagic nektonic species that dominate the northeast Pacific ecosystem. Because high-seas fisheries do not exist for most of the species we examined, our distributions are based on pelagic gill net research surveys conducted by several nations in the open waters of the Gulf of Alaska. We also derive estimates of the abundance and biomass of the five salmon species for each time period. Rather than attempting to convert gill net catch per unit effort (CPUE) to abundance or biomass directly, we used back-calculated abundances of salmon species during years for which we have reliable estimates to interpret the abundances of the remaining years.

Methods

Study area

The study area includes all waters beyond the continental shelf region between 45 and 57°N and between 127 and 160°W, and covers roughly 4.08' 10^6 km^2 (Fig. 1). This area includes much of the subarctic and the northern part of the Transition Zones of the Gulf of Alaska (Dodimead et al. 1963). The western boundary is somewhat arbitrary as there is no clearly defined physical feature delineating the study area from the western

Table 1. Nektonic species considered in this study and the domain of their primary zoogeographic affinity in the North Pacific.

Common name	Scientific name	Domain
Salmonid fishes		
Sockeye salmon	*Oncorhynchus nerka*	Subarctic
Pink salmon	*O. gorbuscha*	Subarctic
Chum salmon	*O. keta*	Subarctic
Coho salmon	*O. kisutch*	Subarctic
Chinook salmon	*O. tshawytscha*	Subarctic
Steelhead	*O. mykiss*	Subarctic
Nonsalmonid fishes		
Salmon shark	*Lamna ditropis*	Subarctic
Blue shark	*Prionace glauca*	Transitional
Pacific pomfret	*Brama japonica*	Transitional
Jack mackerel	*Trachurus symmetricus*	Coastal Upwelling/ Transitional
Albacore	*Thunnus alalunga*	Transitional
Cephalopods		
Neon flying squid	*Ommastrephes bartrami*	Transitional
Nail squid	*Onychoteuthis borealijaponicus*	Transitional
Eight-arm squid	*Gonatopsis borealis*	Subarctic

North Pacific. The boundary chosen represents the westward extent of most North American salmon stocks (exclusive of the Bering Sea stocks (French et al. 1975; Myers et al. 1990). In terms of biological production, the study area encompasses most of the Subarctic Domain, and part of the Coastal Downwelling Domain of Ware and McFarlane (1989) and the eastern portion of the Transition Domain of Pearcy (1991). We consider our analysis to be appropriate only for the surface layer normally sampled by gillnets.

Species considered

The species considered to be the dominant representatives of the nektonic megafauna exclusive of birds and mammals (Neave and Hanavan 1960; Fiscus and Mercer 1982; Kubodera 1986; Brodeur 1988; Ware and McFarlane 1989; Pearcy 1991) are listed in Table 1. One dominant member of the pelagic community, Pacific saury (*Cololabis saira*) was excluded from the analysis because studies have shown that this species, because of its attenuated body shape, is not adequately sampled by the gill net mesh sizes commonly used for the other species of interest (Wada and Kitakata 1982). Although some Pacific saury were caught during most cruises, their CPUEs were far below those made by quantitative small-mesh purse seines (Macy et al. 1978). Other species, such as Pacific herring (*Clupea pallasi*), spiny dogfish (*Squalus acanthias*), Atka mackerel (*Pleurogrammus monopterygius*), and Pacific mackerel (*Scomber japonicus*), are also occasionally found in substantial numbers at particular times and locations in high-seas gill net collections, but these species are more commonly associated with the Coastal Upwelling and Downwelling Domains (Ware and McFarlane 1989) and were excluded from the analysis. Species with more tropical affinities, such as skipjack (*Katsuwonus pelamis*) and bluefin (*Thunnus thynnus*) tuna, and yellowtail (*Seriola lalandi*), which migrate into the Transition Domain during the late summer (Mishima 1981; Shimazaki 1986; Pearcy 1991), were also excluded.

Catch rate estimates

To estimate the relative catch rates (CPUEs) of the species considered, catch rates in surface gillnets were calculated from Canadian, Japanese, and U.S. high-seas research cruises conducted during the last 35 yrs. The gear used was a standard gill net suspended by float lines from the surface. Although some subsurface gillnets were set during some cruises, these were not included in our analysis. From 1955 to 1958, multifilament nylon nets were used, but in the 1980s only monofilament nets were used. We have assumed that the fishing efficiencies of both nets were equal. In most cruises, the mesh sizes varied

Table 2. Summary data from U.S. and Canadian high-seas gill net surveys.

Vessel	Year	Dates	Mesh size (mm)	No. of sets	Reference
John N. Cobb	1955	6/25 – 9/13	83–152	46	Powell and Peterson 1957
Mitkof	1955	9/21 – 10/5	57–133	8	Powell and Peterson 1957
Key West	1956	5/23 – 7/20	62–133	19	Neave and Manzer 1957
Challenger	1956	6/3 – 9/10	62–133	25	Neave and Manzer 1957
John N. Cobb	1956	7/17 – 8/9	63–133	27	Hanavan and Tanonaka 1959
Celtic	1956	7/13 – 8/19	62–133	45	Hanavan and Tanonaka 1959
Paragon	1956	7/12 – 8/8	62–133	23	Hanavan and Tanonaka 1959
Key West	1957	5/13 – 8/22	62–133	26	Manzer and Neave 1958
Fort Ross	1957	5/19 – 8/28	62–133	35	Manzer and Neave 1958
Attu	1957	7/14 – 9/8	62–133	8	unpublished archival
Key West	1958	5/11 – 8/24	62–228	19	Manzer and Neave 1959
Fort Ross	1958	5/17 – 8/18	62–228	17	Manzer and Neave 1959
Hugh Smith	1958	7/29 – 8/3	62–133	6	McGary and Graham 1960
Total				304	

with every 50-m section (tan or shackle) of netting to catch a wide size range of nekton. The range of mesh sizes used (29–228 mm) was considered adequate to quantitatively sample the late juvenile and adult stages of all the species considered. The general procedure was to set the gear at sunset and retrieve it at dawn. A preliminary analysis of the catch rates of gillnets fished only during daylight hours showed that these were much lower than night catches, so the day sets were excluded.

We were interested in comparing decadal differences in typical summer catch rates; therefore, we grouped our data into two distinct time periods, regardless of month or year of collection. Three main data sources were used:

1) Canadian and U.S. high-seas salmon cruises. Collections are available from 13 summer cruises from 1955 through 1958 (Table 2). Although adult salmon were the target species, mesh sizes from 57 to 228 mm were used to collect all age groups on the high seas. Most of the tans were between 63 and 133 mm. Data were taken from International North Pacific Fisheries Commission (INPFC) documents, unpublished cruise reports, U.S. Fish and Wildlife Special Scientific Reports, and original log sheets archived at the Canadian Department of Fisheries and Oceans (DFO) Station in Nanaimo, B.C. Collections were generally made in a grid pattern of 1° squares. Altogether, 304 sets are available from these cruises.
2) Canadian–Japanese cooperative experimental squid and salmon surveys. Collections were made during 15 cruises aboard Canadian research vessels and Canadian and Japanese commercial gillnetters during the summers of 1980–89 (Table 3). The length and depth of net varied somewhat depending on the year and the vessel deploying it. The mesh sizes varied between 115 and 121 mm. Sampling was either along transects or concentrated within certain geographic areas. A total of 430 sets are available from these surveys. Data were taken from INPFC Documents, Canadian Industry and Technical Reports, and from original data sheets available at the DFO Station in Nanaimo.
3) Japanese research vessel data. Collections were made in the eastern North Pacific by the Hokkaido University Research Vessel *Oshoro maru* during the summers of 1980–85 and 1987–89 (Table 4). Collections were generally made at 1 or 2° intervals along latitudinal or longitudinal parallels and often were repeated in successive years. The length of the

Table 3. Summary data from Canadian-Japanese cooperative squid and salmon surveys.

Vessel	Year	Dates	Mesh size (mm)	No. of sets	Reference
Tomi maru	1980	8/15 – 8/29	121	12	Bernard 1981
Tenyu maru	1980	7/30 – 8/16	121	15	Bernard 1981
Tomi maru	1983	7/19 – 8/31	118–121	41	Sloan 1984
Simstar	1983	7/19 – 8/29	121	25	Robinson and Jamieson 1984
Tomi maru	1985	7/5 – 9/6	121	56	Jamieson and Heritage 1987
Tomi maru	1986	6/24 – 9/5	121	70	Jamieson and Heritage 1987
Ocean Pearl	1986	7/27 – 9/12	121	35	Jamieson and Heritage 1987
La Porsche	1986	8/9 – 9/14	121	28	Jamieson and Heritage 1987
W.E. Ricker	1986	8/26 – 9/11	115	11	Bernard 1986
Tomi maru	1987	6/15 – 8/24	115–121	66	Jamieson and Heritage 1988
Ocean Pearl	1987	6/13 – 8/5	118–121	49	Jamieson and Heritage 1988
W.E. Ricker	1987	6/3 – 6/26	115	6	LeBrasseur et al. 1987
W.E. Ricker	1988	6/12 – 7/2	115	6	LeBrasseur et al. 1988
Arctic Harvester	1989	7/17 – 8/15	115	10	McKinnell et al. 1989
Total			430		

gill nets ranged from 4.2 to 6.6 km; the nets fished to a depth of 8 m. Mesh size varied from 29 to 130 mm stretched mesh. Data were also available from a 1989 survey by the Fisheries Agency of Japan Research Vessel *Shoyo maru*, which used a net 2.5 km long and 8.5 m deep with mesh ranging from 48 to 157 mm. Altogether, 111 sets were available from these vessels within the study area. Data are taken from Hokkaido University data reports and INPFC documents.

The collection locations for the 1955–58 and 1980–89 periods are shown in Fig. 1. Most of the collections were made during July and August, and the monthly distribution of sampling effort was similar for both periods (Fig. 2). The actual catches per set of all species were recorded. The catches were expanded based on drop-off rates available from several cruises. The catches for all cruises were standardized to common CPUEs (catch·km^{-1}·12 h^{-1} fished) to correct for differences in net lengths and soak times among the different collections.

Distribution maps of CPUEs were drawn for each species using a microcomputer-based geographic information system (COMPUGRID, Geo-Spatial Systems Ltd.). CPUE values for all 7.2 km × 7.2 km pixels (sum = 78 824) within the Gulf of Alaska east of 160°W and west of 127°W longitude and between 45 and 57°N and seaward of the continental shelf (200-m isobath) were interpolated from the CPUE of the seven nearest stations using a euclidean distance ratio, which assigns greater weighting to the stations nearest the interpolated pixel. Catch rates from collections outside the sampling area were also used to interpolate the pixels near the edge of study area (Fig. 1). Mean estimates and 95% confidence intervals of CPUE for each species were calculated for both time periods.

Because we were interested in examining interdecadal changes in abundance and biomass of several dominant species, we converted gill net CPUEs to absolute abundance estimates. We also restricted our abundance estimation to the five salmon species because most of the mesh used in the gillnets may not be totally nonselective for several of the larger species caught (albacore (*Thunnus alalugna*) and shark (*Lamna ditropis* and *Prionace glauca*) (e.g., Nakano and Shimazaki 1989; Bartoo and Holts 1993). Rather than directly estimating abundance on an individual set basis (cf. Ishida 1986), we based our conversion on total run sizes of three salmon species for which long abundance time series are available. Rogers (1987) provided estimates of the numbers of pink *(O. gorbuscha)* and coho (*O. kisutch*) salmon returning to various areas of origin for a 35-yr period. We used only these species because they have only one year-class present in the open ocean during the summer (Takagi et al. 1981; Faculty of Fisheries 1981). Peterman and Wong (1984) also estimated abundances of the British Columbia and Bristol Bay sockeye salmon (*O. nerka*) fraction, which resides in the Gulf of Alaska during summer. We used their estimates and adjusted them for the remaining North American stocks not included in their analysis (Ware and McFarlane 1989).

We assumed a 15% natural mortality for all species between the period of residence on the high seas and the time they return to the river systems (Ricker 1976). We then regressed the yearly abundance estimates for pink, coho, and sockeye salmon mentioned above on the mean standardized catch rate for the same years from our analysis, with the constraint that the regression pass through the origin (zero catch = zero abundance). The overall catch rates for each

Table 4. Summary data from Japanese high-seas research surveys.

Vessel	Year	Dates	Mesh size (mm)	No. of sets	Reference
Oshoro maru	1980	7/10 – 7/15	29–130	6	Faculty of Fisheries 1981
Oshoro maru	1981	7/11 – 7/21	29–130	8	Faculty of Fisheries 1982
Oshoro maru	1982	7/16 – 7/23	29–130	8	Faculty of Fisheries 1983
Oshoro maru	1983	6/30 – 7/6	29–130	7	Faculty of Fisheries 1984
Oshoro maru	1984	7/16 – 7/30	29–130	8	Faculty of Fisheries 1985
Oshoro maru	1985	7/3 – 7/19	29–130	16	Faculty of Fisheries 1986
Oshoro maru	1987	7/14 – 7/24	29–130	11	Faculty of Fisheries 1988
Oshoro maru	1988	7/6 – 7/20	29–130	15	Faculty of Fisheries 1989
Oshoro maru	1989	7/1 – 7/17	29–130	17	Faculty of Fisheries 1990
Shoyo maru	1989	7/29 – 8/15	48–157	15	Kawasaki et al. 1989
Total				111	

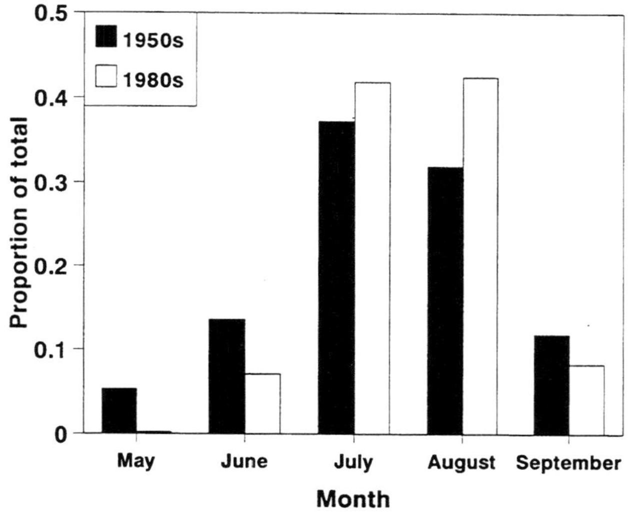

Fig. 2. Monthly distribution of sampling effort for the two time periods.

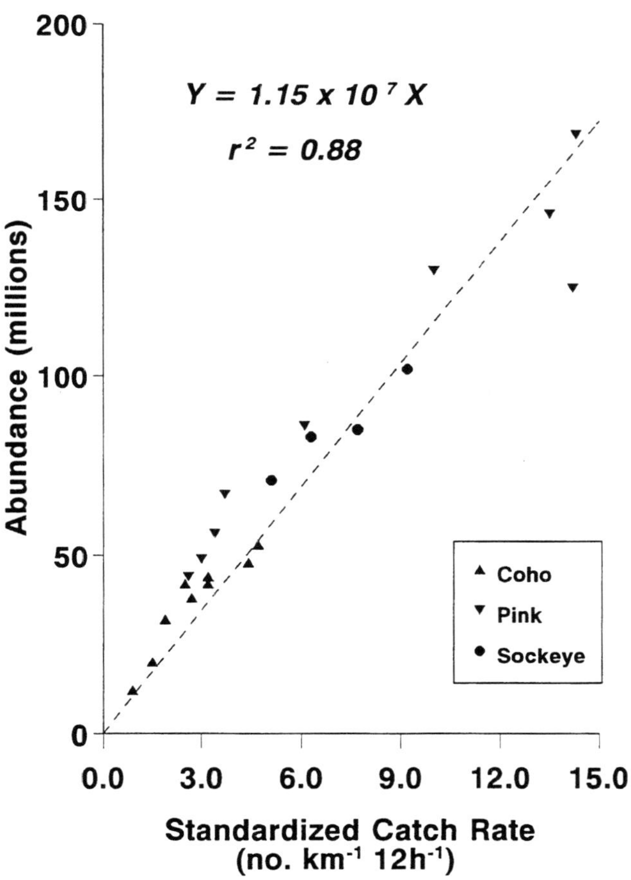

Fig. 3. Relationship between catch rate and estimated total summer abundance of salmon species in the Gulf of Alaska. The data points represent individual years or abundance and catch rate information by species. Abundance data are adjusted from the values given in Rogers (1987) for pink and coho salmon and from those in Peterman and Wong (1984) for sockeye salmon. The broken line represents the best linear fit to the data constrained through the origin. The regression equation is also given in the figure.

salmon species during both periods were converted to abundances using this relationship (Fig. 3). Finally, our abundance estimates were converted to biomass estimates based on the mean body weights of each species from each cruise when available or from mean lengths converted to weight from published length–weight relationships, when only lengths were measured.

Results

Distribution patterns

All five Pacific salmon species were distributed primarily in the subarctic region during both time periods, but the distributions were compressed toward the northern and western side of the Gulf of Alaska during the 1980s (Figs. 4–8). Although a substantial amount of fishing effort was directed south of 47°N (Fig. 1), there were relatively few salmon caught in this region during the 1980s. Sockeye, chum (*O. keta*), and pink salmon catch rates increased greatly in the northern and western gulf (Figs. 4–6). Coho and chinook salmon (*O. tshawytscha*) showed the highest increase in catch rates between 50 and 55°N in the western gulf (Figs. 7 and 8). The remaining salmonid species, steelhead (*O. mykiss*), also showed a pattern of widespread occurrence, but had low catch rates during the 1950s, with a greater concentration of catches in the northwest portion of the gulf during the 1980s (Fig. 9).

Blue shark (*Prionace glauca*) were distributed mostly in the Transition Domain during both periods (Fig. 10) and occurred in high concentrations in the southwest section and along the west coast of North America up to the northern tip of Vancouver Island (51°N). Salmon shark (*Lamna ditropis*) were distributed over a substantial portion of the gulf during both time periods, but shifted from the northern and western gulf to the southeast (Fig. 11).

Among the other pelagic fishes, jack mackerel (*Trachurus symmetricus*) were confined mainly to the eastern half of the gulf and had the highest catch rates in the Transition Domain (Fig. 12). Pacific pomfret (*Brama japonica*) occurred throughout much of the gulf in both periods (Fig. 13) but were patchily distributed and tended to be farther south in the 1980s. Albacore were caught mainly in the Transition Zone and had the most southerly distribution of the species examined (Fig. 14).

Neon flying squid (*Ommastrephes bartrami*) had a relatively low and uniform distribution over much of the southern half of the study area during the 1955–58 period, with the exception of the coastal areas (Fig. 15), whereas in the 1980s, it occurred in heavy concentrations near the continental margin. During the 1950s, nail squid (*Onychoteuthis borealijaponicus*) were not very abundant relative to the 1980s and were distributed in a small area of the Northeast Pacific (Fig. 16). Eight-arm squid (*Gonatopsis borealis*) showed a more boreal distribution than the other two cephalopod species and, similar to the salmonid species, was found mainly in the northwestern part of the gulf during the 1980s (Fig. 17).

Catch-rates estimates

During 1955–58, the mean catch rate of Pacific pomfret was almost twice that of the next most commonly caught species, chum salmon (Fig. 18). Jack mackerel also had a high catch rate during this time period. Cephalopods made up a relatively small proportion of the total nekton caught during this period. With the exception of jack mackerel, all species

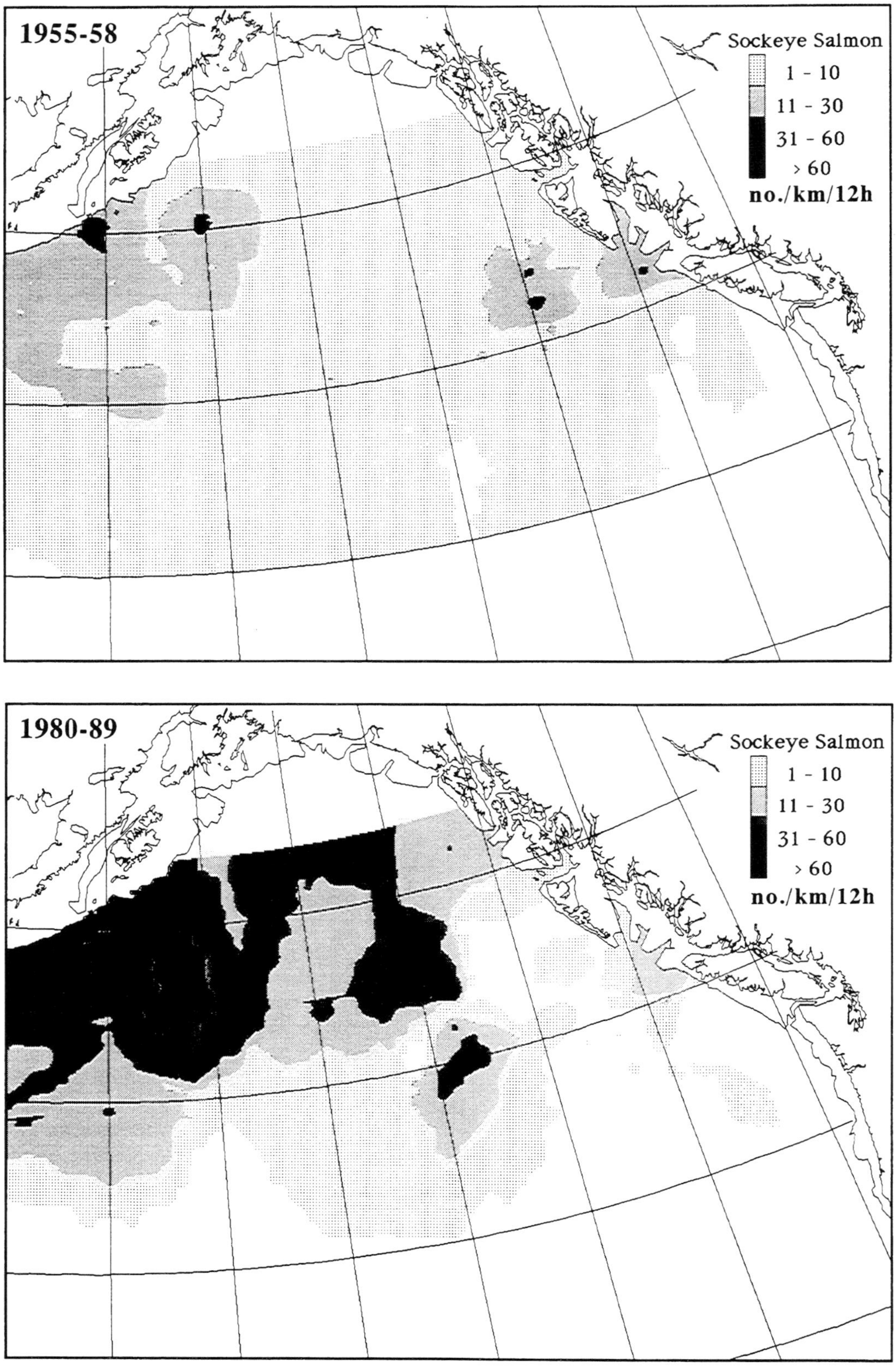

Fig. 4. Distribution of sockeye salmon catch rates (no.·km^{-1}·12 h^{-1}) for 1955–58 and 1980–89.

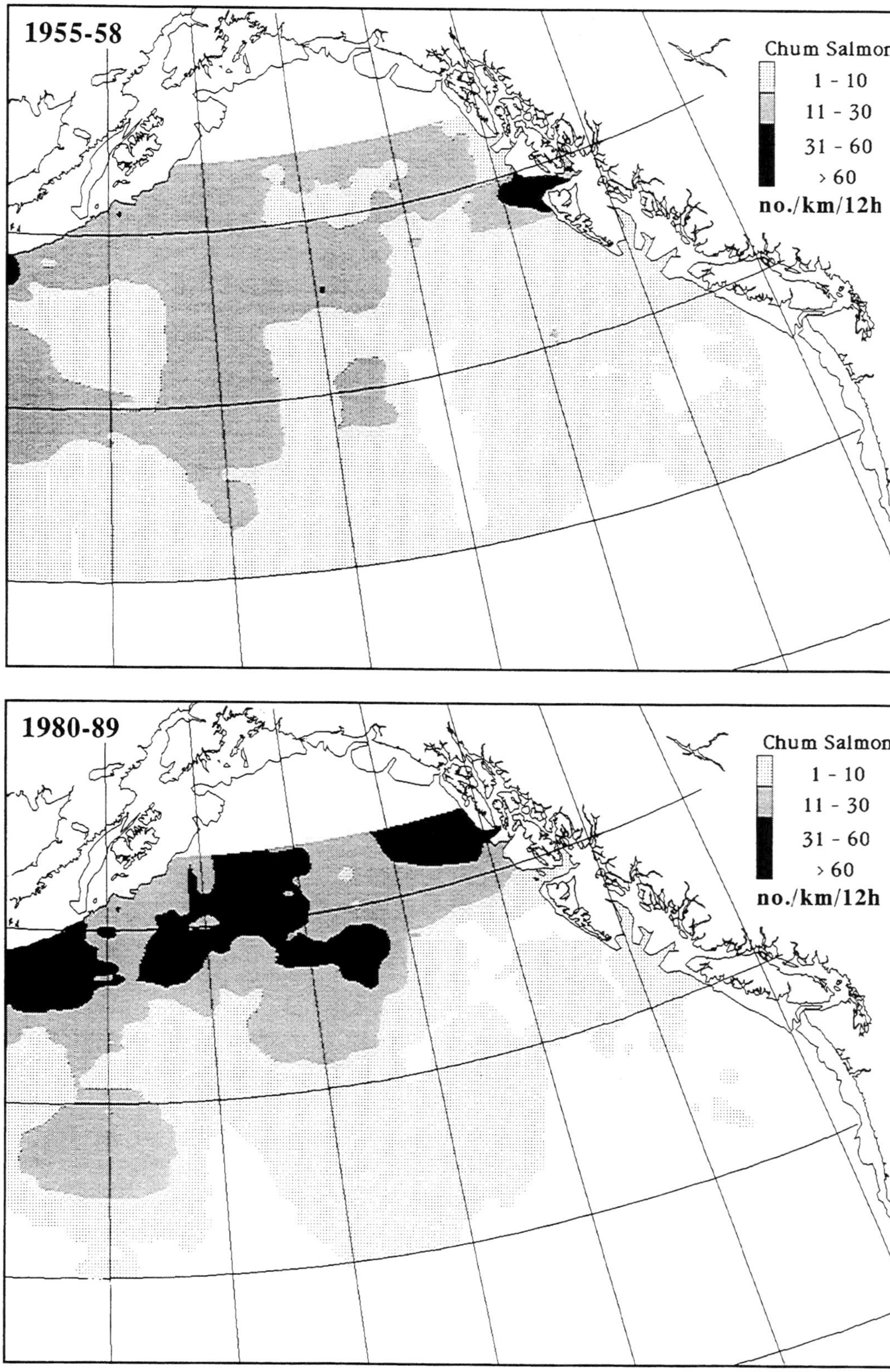

Fig. 5. Distribution of chum salmon catch rates (no.·km^{-1}·12 h^{-1}) for 1955–58 and 1980–89.

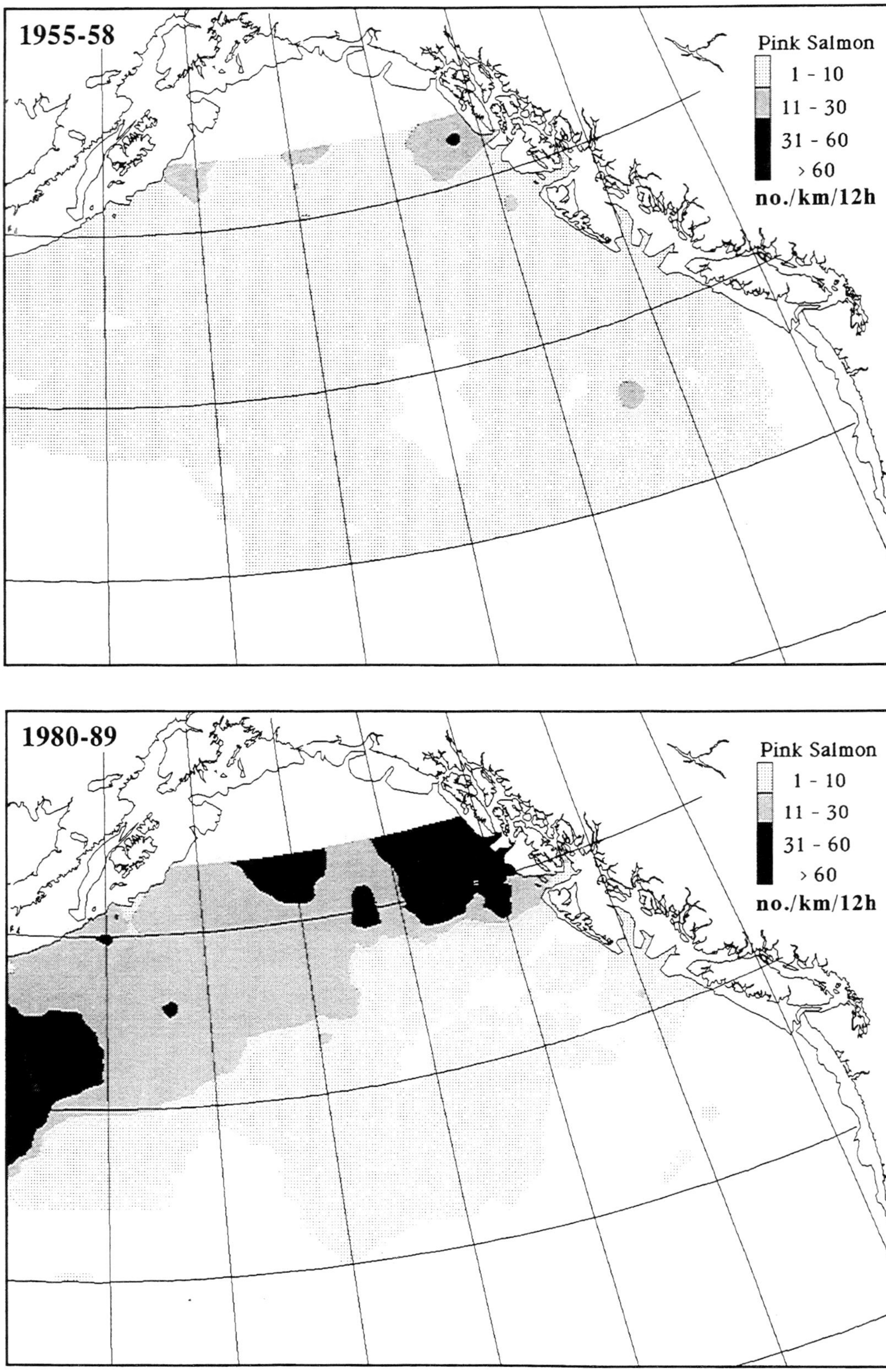

Fig. 6. Distribution of pink salmon catch rates (no.·km^{-1}·12 h^{-1}) for 1955–58 and 1980–89.

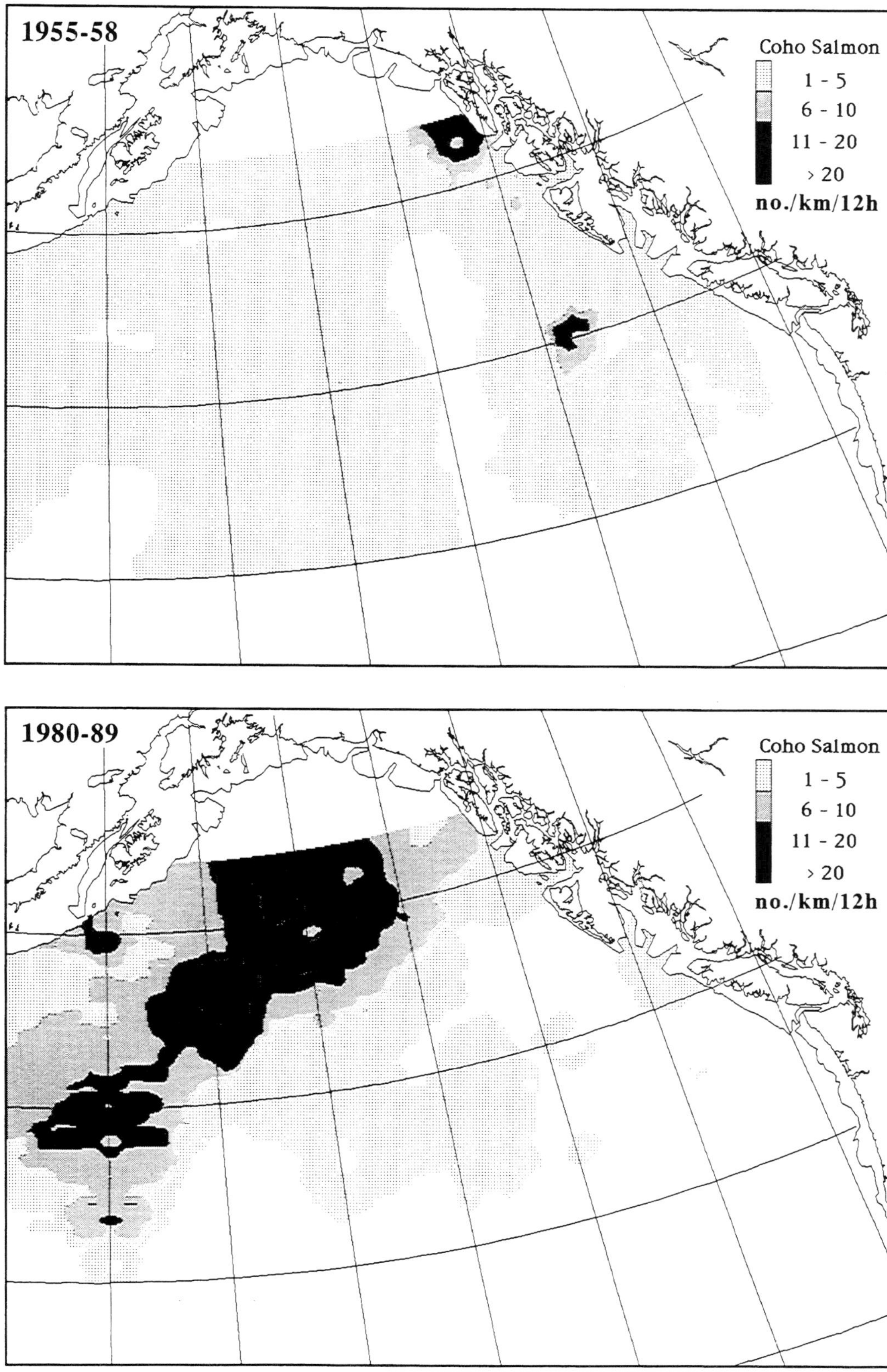

Fig. 7. Distribution of coho salmon catch rates (no.·km^{-1}·12 h^{-1}) for 1955–58 and 1980–89.

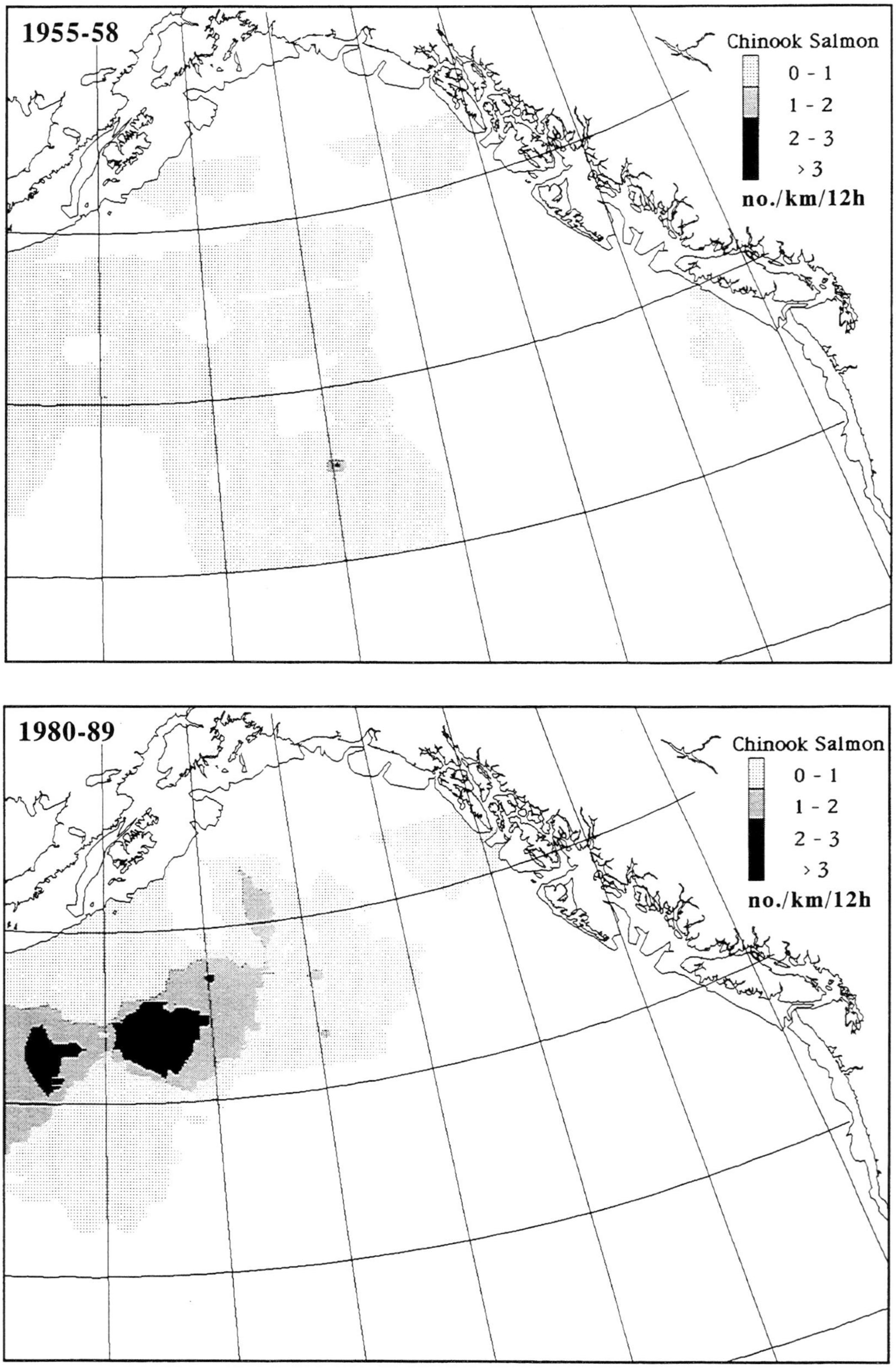

Fig. 8. Distribution of chinook salmon catch rates (no.·km^{-1}·12 h^{-1}) for 1955–58 and 1980–89.

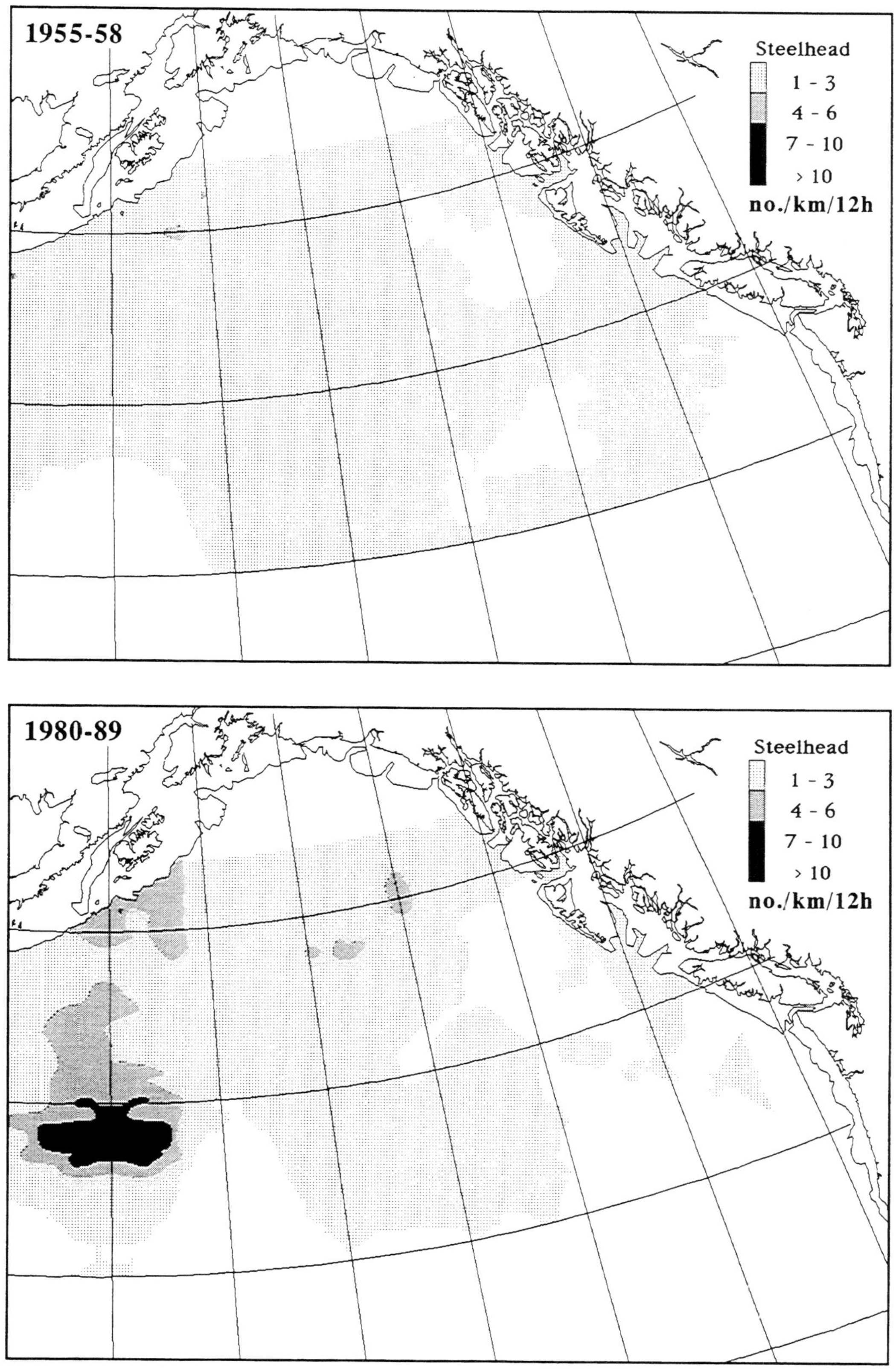

Fig. 9. Distribution of steelhead catch rates (no.·km^{-1}·12 h^{-1}) for 1955–58 and 1980–89.

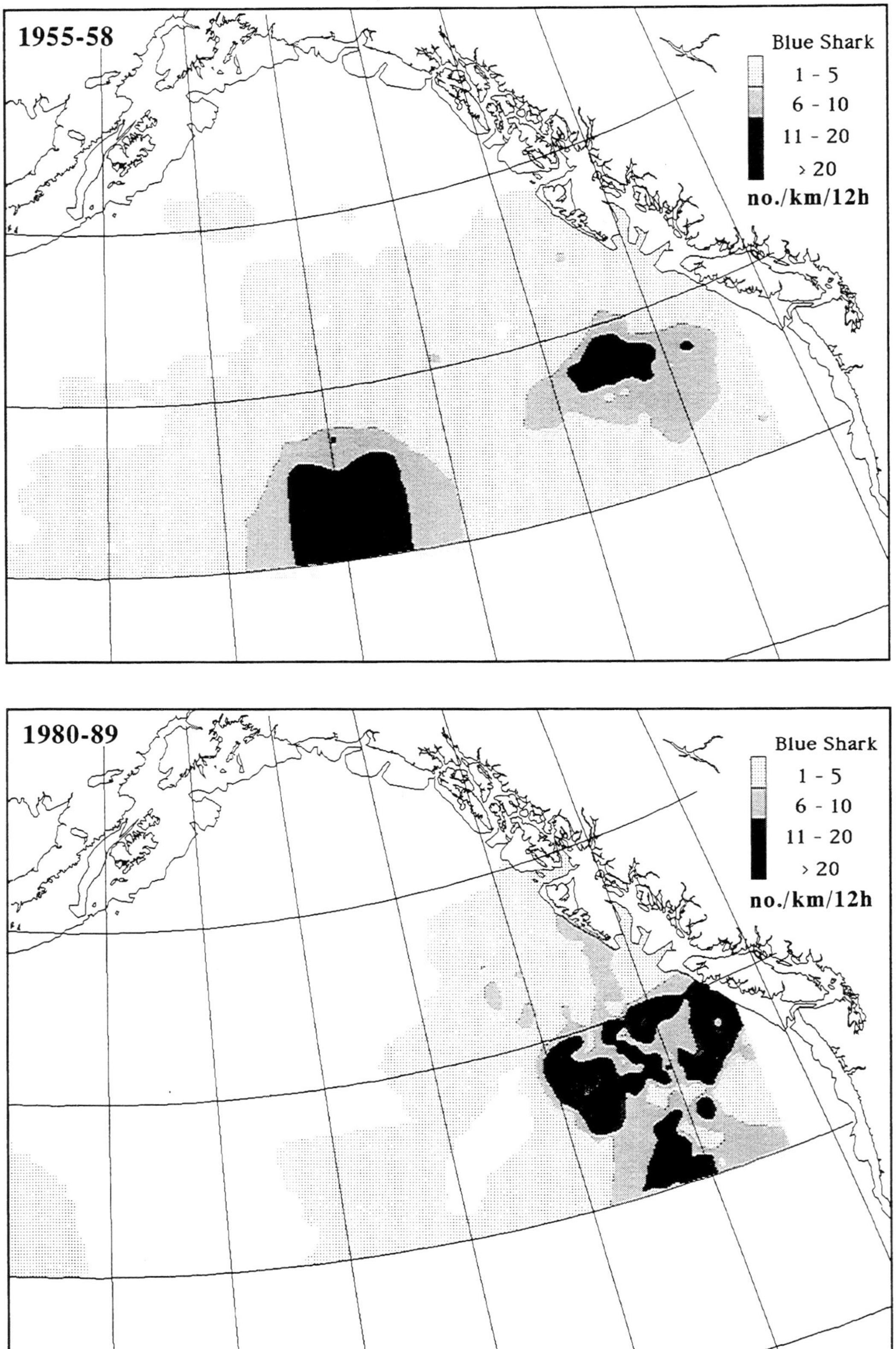

Fig. 10. Distribution of blue shark catch rates (no.·km^{-1}·12 h^{-1}) for 1955–58 and 1980–89.

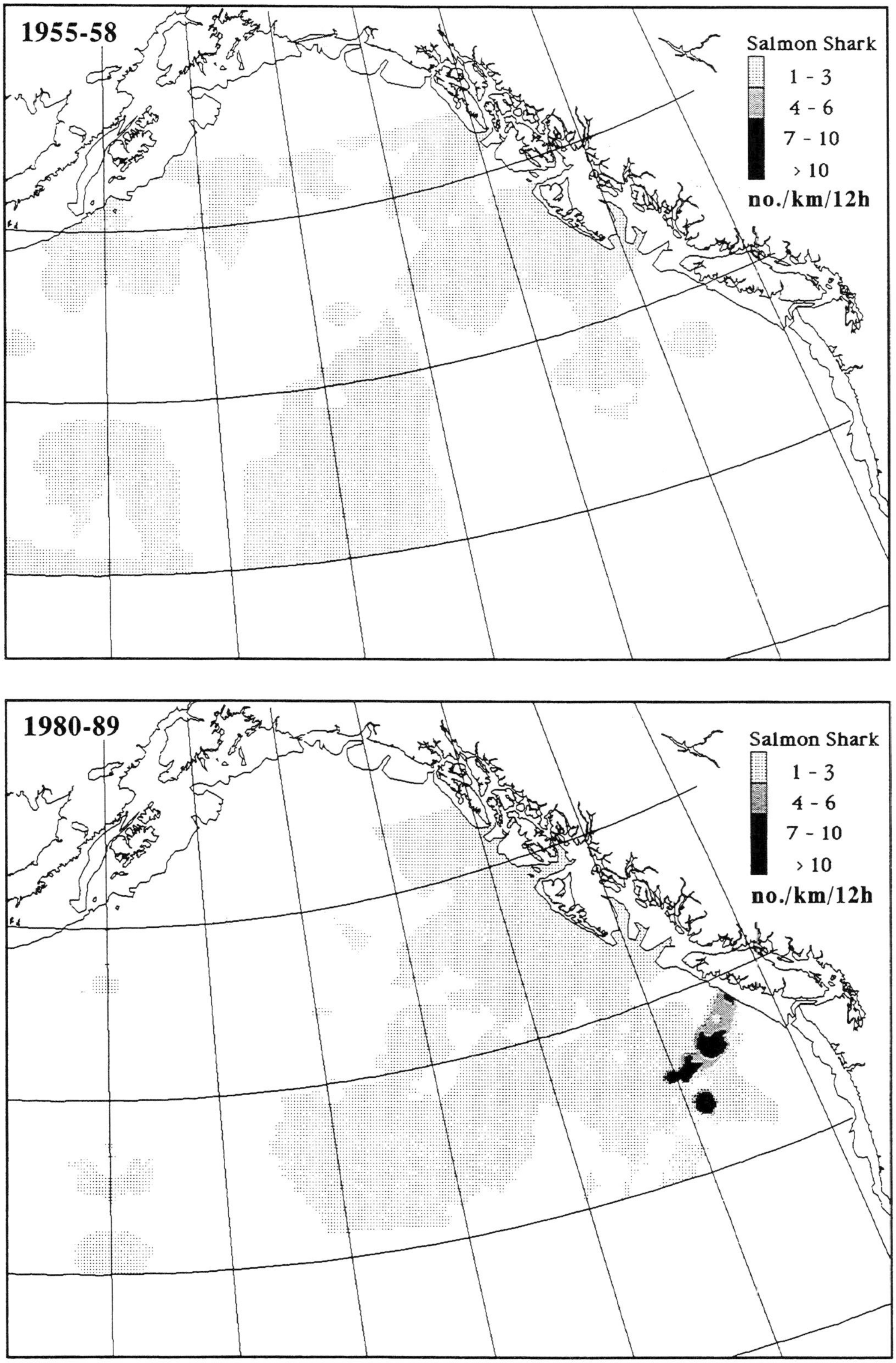

Fig. 11. Distribution of salmon shark catch rates (no.·km^{-1}·12 h^{-1}) for 1955–58 and 1980–89.

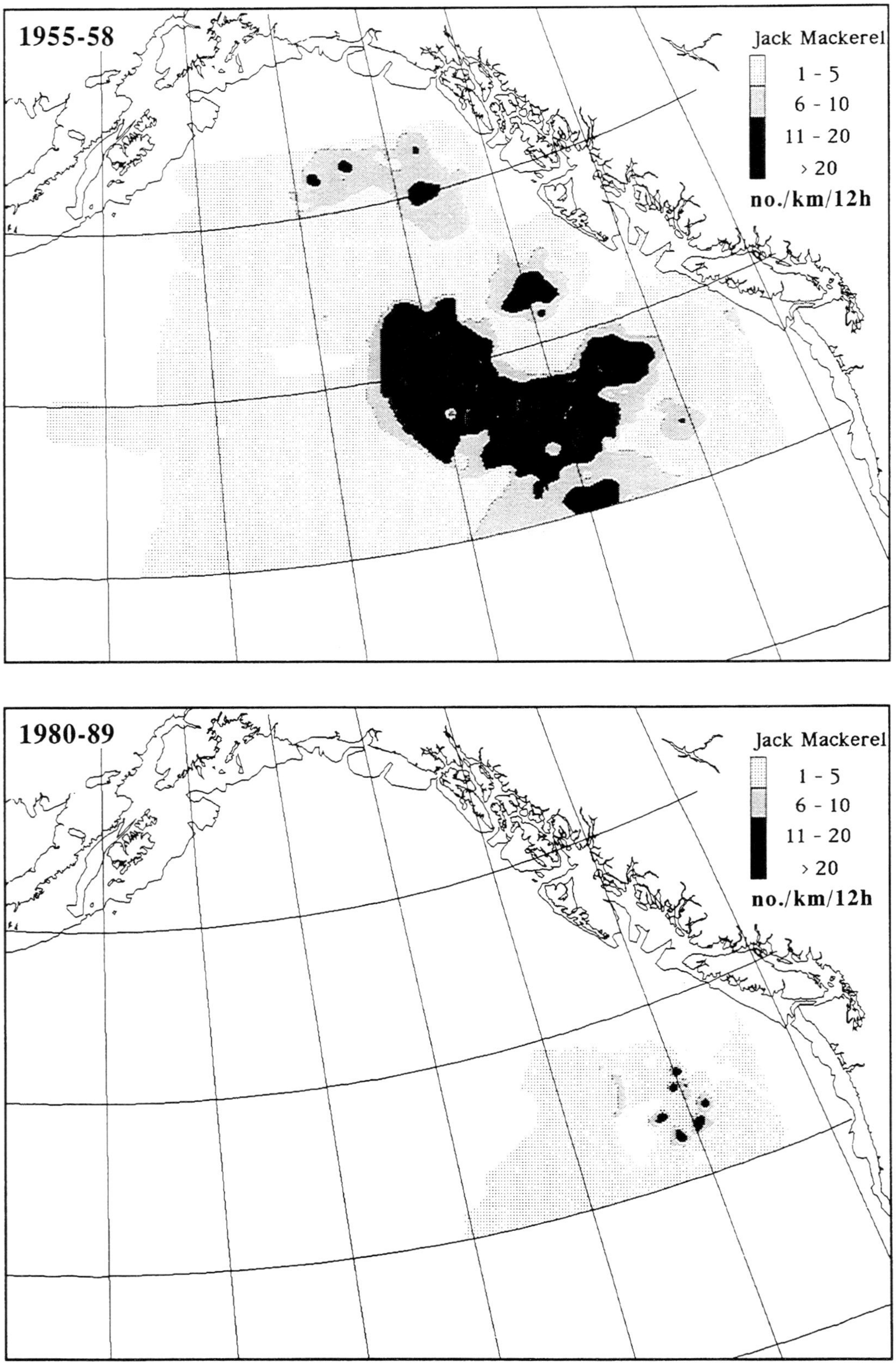

Fig. 12. Distribution of jack mackerel catch rates (no.·km^{-1}·12 h^{-1}) for 1955–58 and 1980–89.

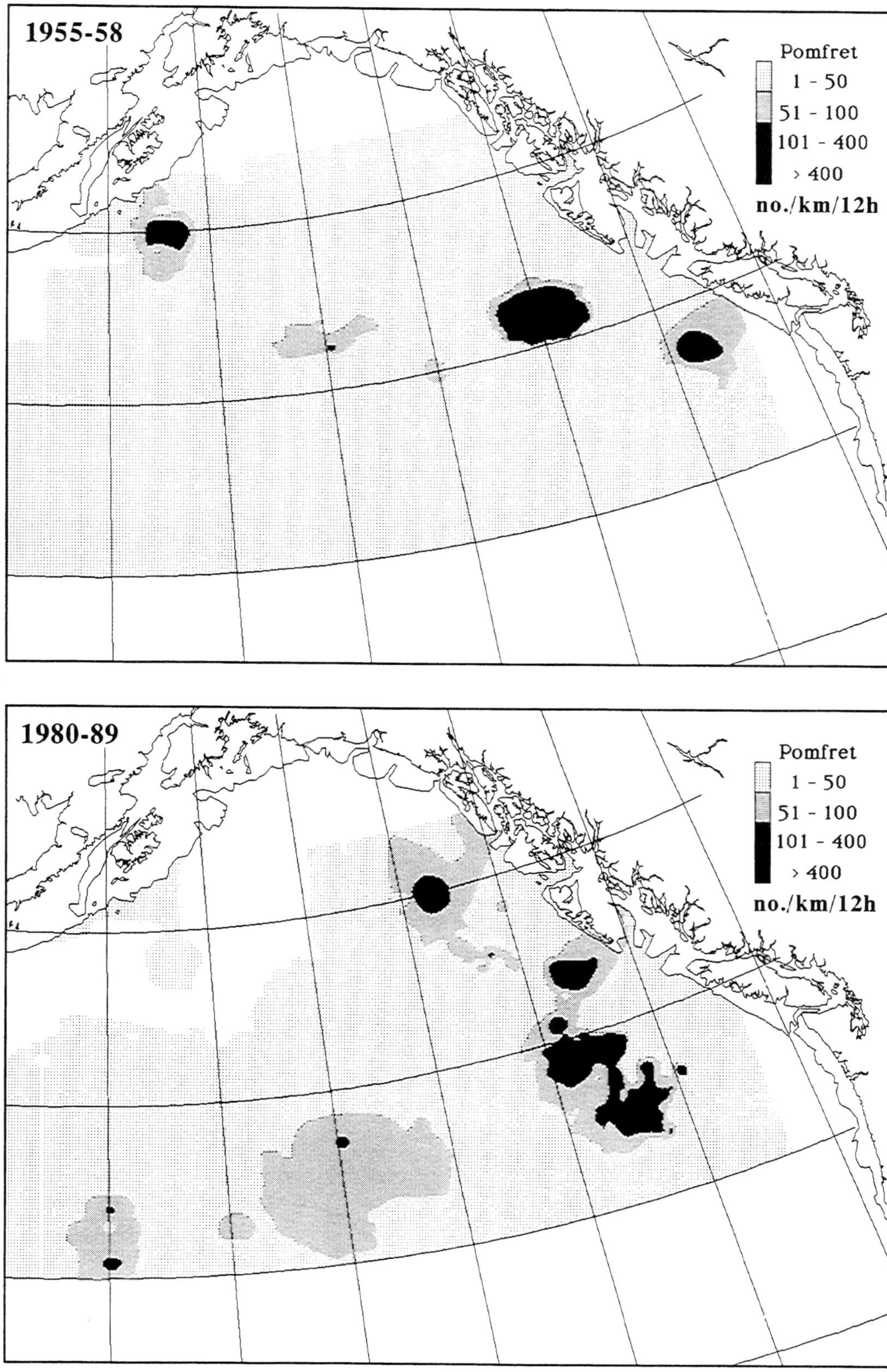

Fig. 13. Distribution of Pacific pomfret catch rates (no.$\cdot$km$^{-1}\cdot$12 h^{-1}) for 1955–58 and 1980–89.

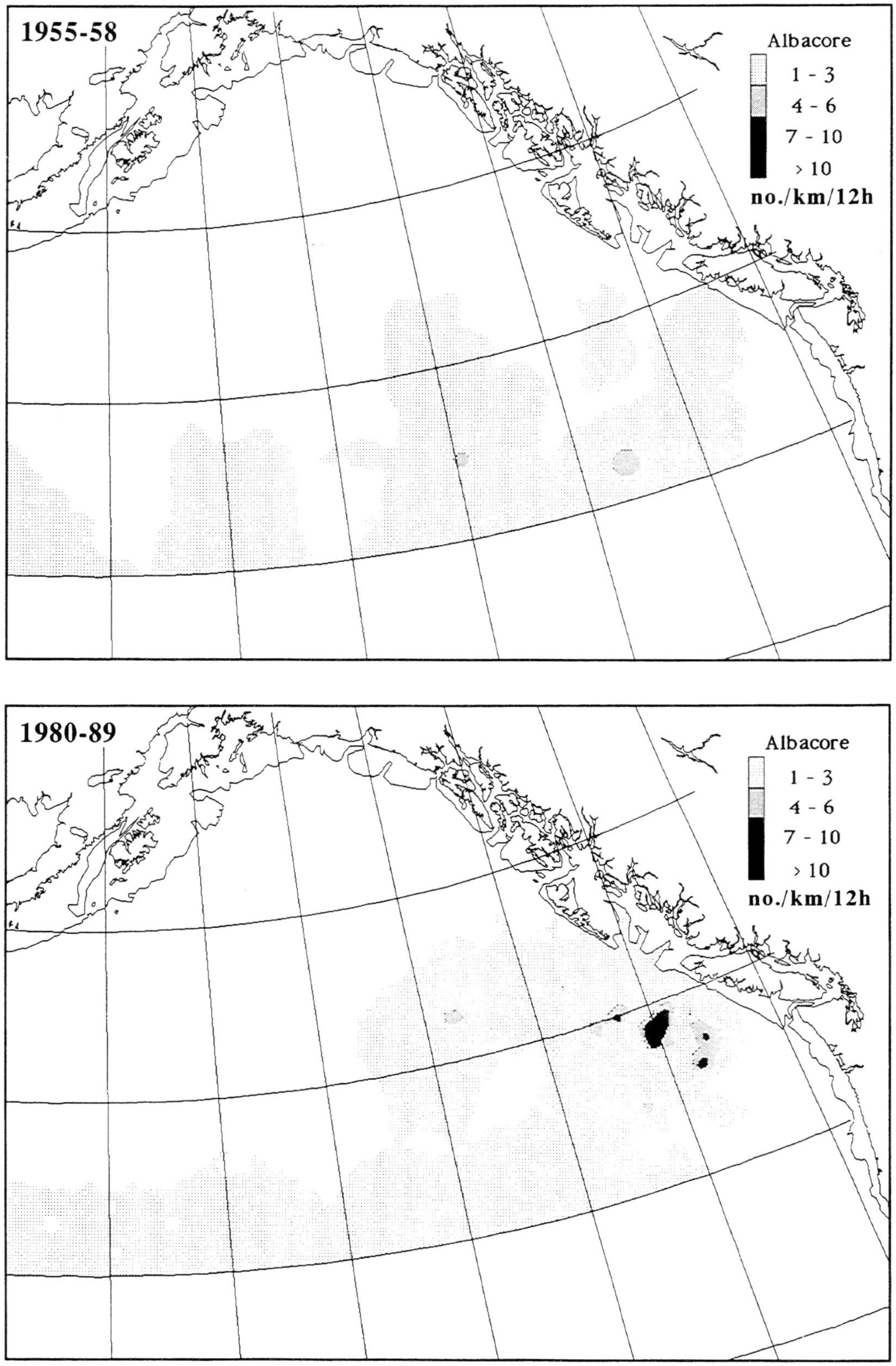

Fig. 14. Distribution of albacore catch rates (no.$\cdot$km$^{-1}\cdot$12 h^{-1}) for 1955–58 and 1980–89.

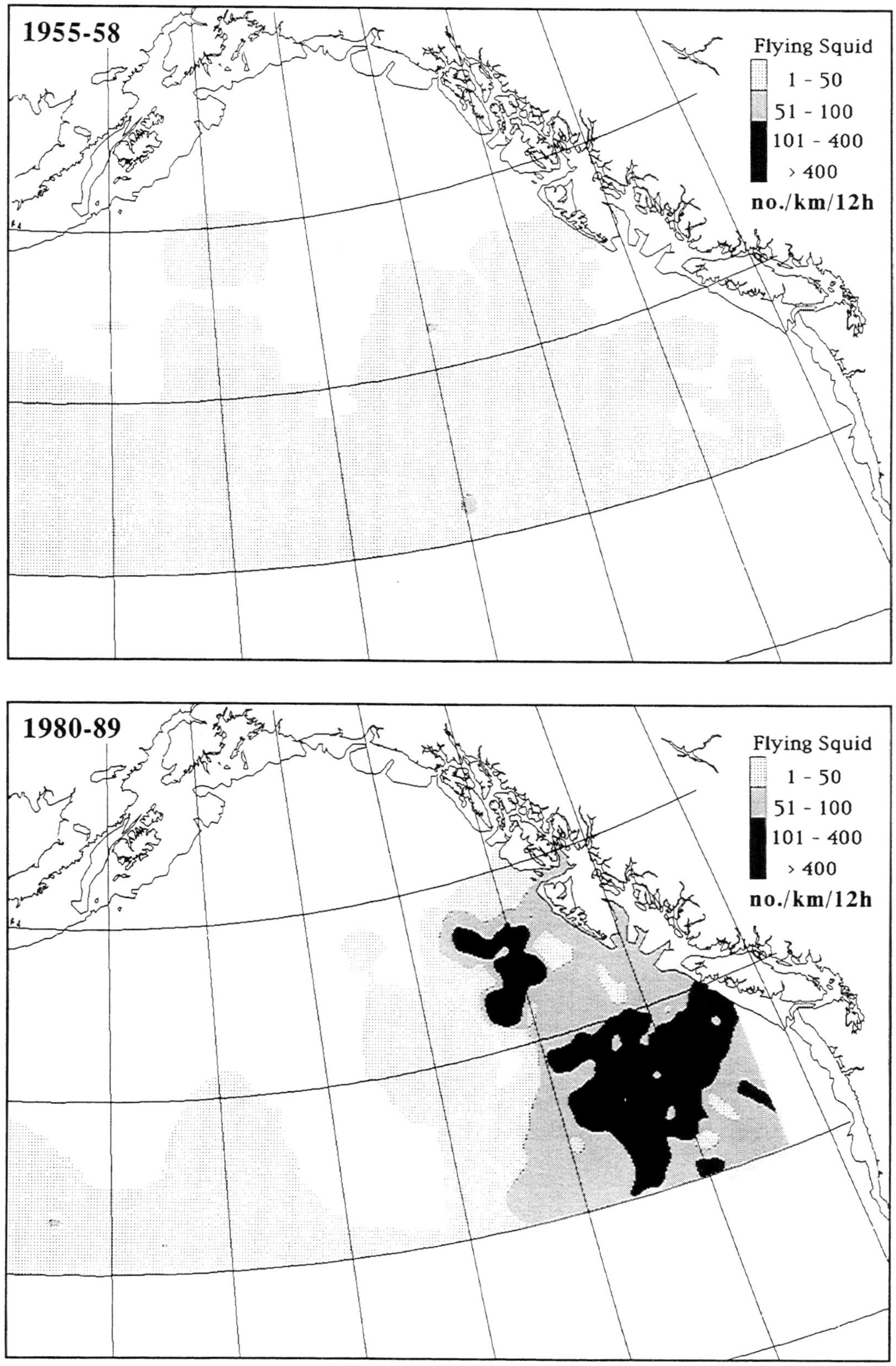

Fig. 15. Distribution of eon flying squid catch rates (no.·km^{-1}·12 h^{-1}) for 1955–58 and 1980–89.

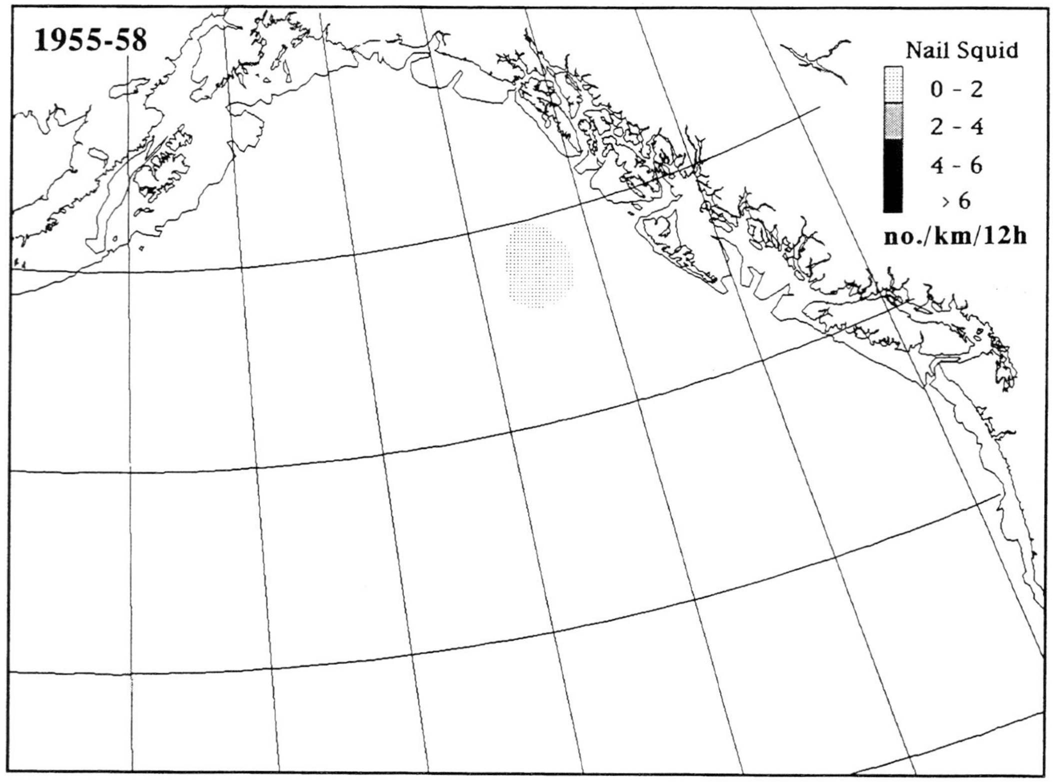

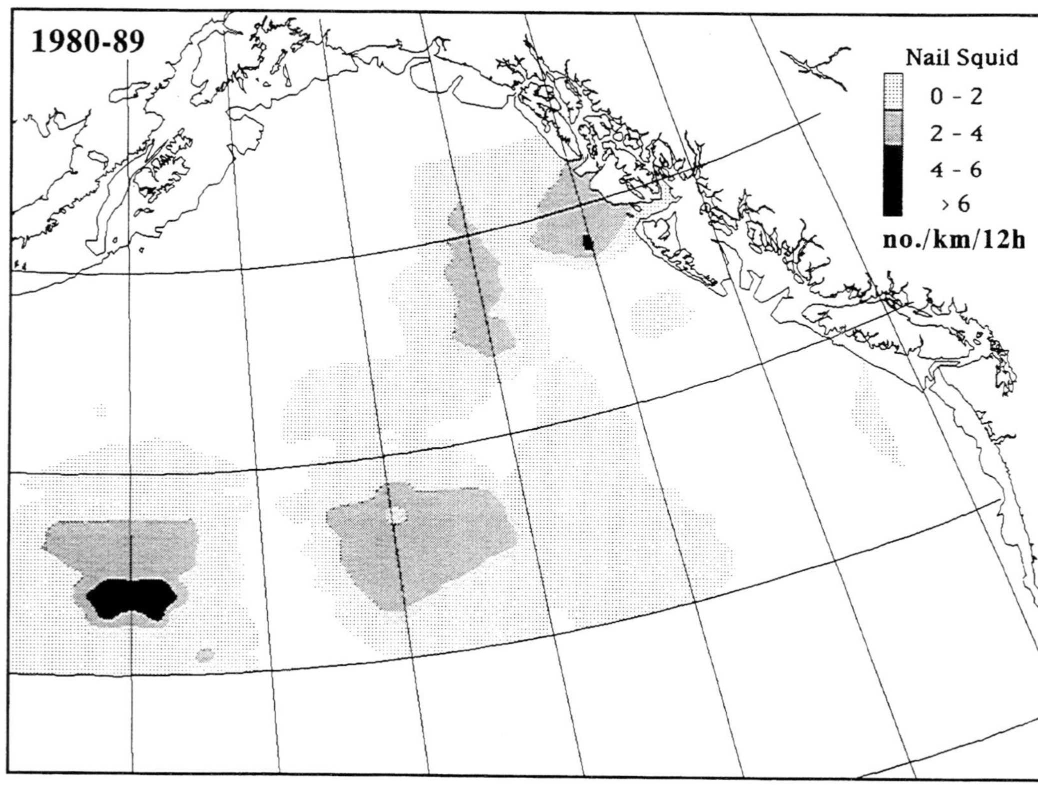

Fig. 16. Distribution of nail squid catch rates (no.·km^{-1}·12 h^{-1}) for 1955–58 and 1980–89.

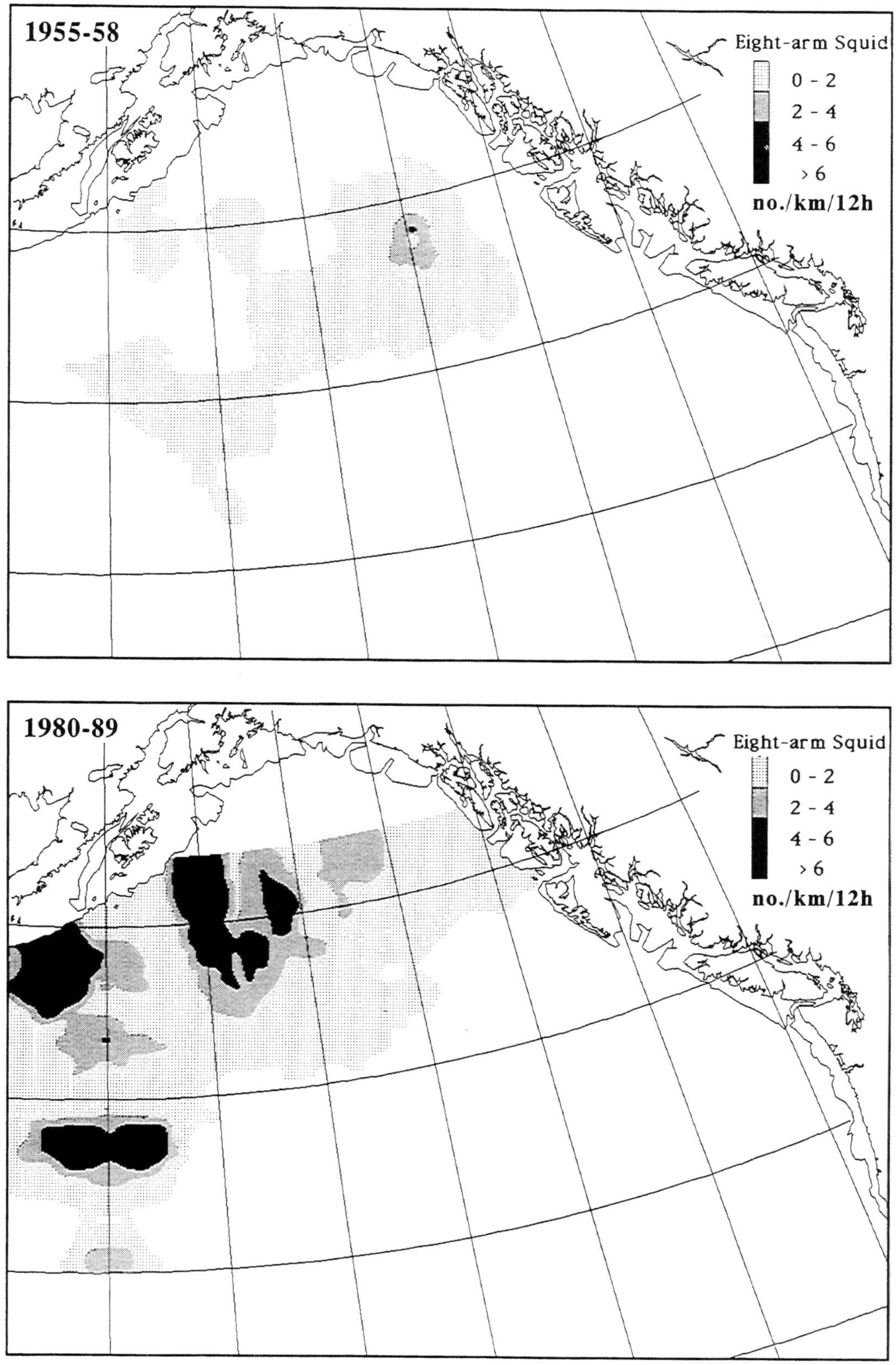

Fig. 17. Distribution of eight-arm squid catch rates (no.·km^{-1}·12 h^{-1}) for 1955–58 and 1980–89.

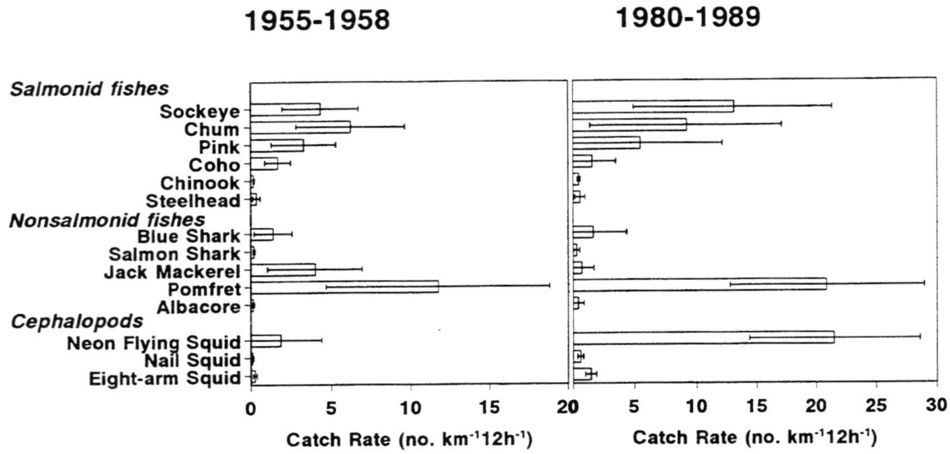

Fig. 18. Comparison of mean catch rates (wide horizontal bars) ±95% confidence intervals (error bars) for the dominant nektonic species for both time periods.

showed higher mean catch rates during 1980–89 (Fig. 18). Neon flying squid had the highest mean catch rate during this later period. Pacific pomfret and sockeye salmon were also dominant species during this period, and the catch rate for both these species approximately doubled between these two time periods. Jack mackerel had catch rates that were almost an order of magnitude lower than the late 1950s.

Salmon abundance and biomass

For all salmon species combined, the total mean estimated abundance almost doubled from 176 million to 345 million between the time periods, but because of a decrease in estimated overall mean size, there was only a 1.84-fold increase in biomass (0.36–0.67 million tonnes) between the periods (Tables 5 and 6). Sockeye salmon showed the largest increase in biomass among the individual species (3.1-fold) while chum and coho salmon showed the least (1.2-fold).

Discussion

The distribution patterns of the various nekton species presented here are not necessarily representative of what would be found in a synoptic survey of the Gulf of Alaska, but rather are composite distributions representing all the summer months and several years during each survey period. Consequently, there are several sources of potential bias that could cause some of the differences observed between the decades.

First, interannual variations in abundance and distribution of North Pacific epipelagic nekton are likely to be substantial, particularly during anomalously strong El Niño events (Radovich 1961; Wooster and Fluharty 1985). Although strong El Niños occurred during both sampling periods (1957–58 and 1982–83), the proportion of the total period represented by these anomalous conditions was greater in the earlier period, which could affect the reconstructed distribution patterns. Interannual differences in the distribution of mesoscale features such as the Sitka Eddy (Mysak 1985) or the large-scale circulation such as the Subarctic Gyre (Royer and Emery 1987) could also alter the distribution of pelagic nekton.

Second, the geographical distribution of sampling varied between periods (Fig. 1), with the 1950s effort concentrated in the northern and central gulf, and the 1980s effort more concentrated in the eastern and southwestern gulf. Large areas of the gulf were not sampled during each period, so species predominantly found in these areas are likely underrepresented in our analysis.

Third, seasonal trends are ignored in our analysis, some of which could affect our results. Although the distribution of sampling effort was fairly similar between the two periods (Fig. 2), the median sampling date was almost 2 wk earlier during 1955–58, which could have led to an underrepresentation of the Transition Domain species that migrate into the Gulf of Alaska in late summer (Neave and Hanavan 1960; Kubodera 1986; Shimazaki 1986). Inspection of Figs. 4–10 indicates that the summer boundary between the subarctic and Transition Domain species falls roughly along a diagonal stretching from 55°N, 135°W to 45°N, 160°W. This same diagonal roughly corresponds to the average position of the 14°C surface temperature isotherm in August (Dodimead et al. 1963). Consequently, our reconstructed distribution patterns of the nekton biomass in the Northeast Pacific reflect the midsummer pattern when the Subarctic Domain is seasonally at its minimal extent.

Fourth, we have assumed that there were no differences in gear efficiency between decades. Washington (1975) conducted an extensive comparison of the size, age, sex, and species composition of salmon collected with comparative settings of multifilament and monofilament nets. Monofilament nets were more efficient in capturing larger salmon and multifilament nets were more efficient in capturing the smaller individuals for the same mesh size because of the greater stretch capabilities of the monofilament nets. Although monofilament nets were more efficient at catching salmon than multifilament nets during the summer, the differences were generally not significant.

Finally, the cephalopod distributions during the early cruises of the 1950s reflect our best estimates but likely contain some errors. Although we consulted the original catch sheets when

Table 5. Summary of mean catch rates, abundance, and biomass estimates for the five Pacific salmon species during the period 1955–58. The 95% confidence interval values are given in parentheses below each mean.

Salmon species	Mean catch rate (no. · km^{-1} · 12 h^{-1})	Abundance (millions)	Mean weight (kg)	Biomass (thousands of tonnes)
Sockeye	4.33 (1.94–6.72)	49.9 (22.4–77.4)	2.17	108.29 (48.52–168.06)
Chum	6.23 (2.80–9.66)	71.8 (32.3–111.3)	2.05	147.19 (66.15–228.22)
Pink	3.27 (1.26–5.28)	37.7 (14.5–60.9)	1.58	59.54 (22.94–96.14)
Coho	1.29 (0.49–2.09)	14.9 (5.6–24.1)	2.70	40.14 (15.25–65.03)
Chinook	0.19 (0.15–0.23)	2.2 (1.7–2.7)	3.80	8.32 (6.57–10.07)
Total		176.4 (76.5–276.4)		363.48 (159.43–567.53)

Table 6. Summary of mean catch rates, abundance, and biomass estimates for the five Pacific salmon species during the period 1980–89. The 95% confidence interval values are given in parentheses below each mean.

Salmon species	Mean catch rate (no. · km^{-1} · 12 h^{-1})	Abundance (millions)	Mean weight (kg)	Biomass (thousands of tonnes)
Sockeye	13.14 (4.93–21.35)	151.4 (56.8–246.1)	2.22	336.18 (126.13–546.23)
Chum	9.25 (1.40–17.10)	106.6 (16.1–197.1)	1.63	173.76 (26.3–321.23)
Pink	5.50 (2.18–8.82)	63.4 (25.1–101.6)	1.46	92.45 (36.68–148.41)
Coho	1.59 (1.02–2.16)	18.3 (11.8–24.9)	2.69	49.29 (31.62–66.96)
Chinook	0.47 (0.34–0.60)	5.4 (3.9–6.9)	3.60	19.50 (14.11–24.89)
Total		345.0 (114.1–577.6)		671.28 (234.84–1107.72)

available, the data for squid were incomplete and the identifications were often tentative. From 1957 on, squid specimens were either preserved in formalin or frozen at sea and returned for identification in the laboratory (Fiscus and Mercer 1982).

Despite these possible sources of differences between the two periods, there are several reasons to believe that many of the differences in our reconstructed distributions and catch rates are real. Between the two periods we considered, some dramatic changes occurred in oceanographic conditions in the North Pacific Ocean. Indeed, a major shift in the physical regime took place around 1976–77 (summarized by Kerr (1992) and Trenberth and Hurrell (1995)), and it appears to have had substantial ramifications for nektonic production (Beamish and Bouillon 1993; Beamish 1994). Hare and Francis (1994), using intervention time-series analysis of salmon catch data, have identified the late 1970s as a transition period between low and high salmon production regimes. This increase in production in the mid-1970s appears to be pervasive, as it occurs in a number of stocks from both Asian and North American origin (Beamish and Bouillon 1993), which would indicate that it is probably a result of changes in marine survival rather than an increase in freshwater production. A combination of warmer water temperatures (Royer 1989; Tabata 1989) and increased zooplankton production (Brodeur and Ware 1992) could have fueled this substantial increase in nektonic production.

Because of the sampling biases shown to exist for pelagic gill-nets relative to other gear types, we consider our abundance and biomass estimates of the salmonids to be somewhat provisional. Hartt (1975) reviewed many of the limitations of gillnets and factors that affect their catching efficiency including mesh size, depth of gear, diel period fished, gear saturation, water transparency, predators on gilled fishes, and age composition and behaviour of individual fish. Our catch rate–abundance regression (Fig. 3) assumes that all salmonid species are equally susceptible to the gillnets. Differences in size, depth of swimming, and schooling behaviour weaken this assumption.

Fortunately for salmon, there are independent corroborative estimates of species abundance or biomass that bridge the two periods we considered. Long-term statistics on salmon catch and escapement by area exist from which estimates of yearly abundance and production can be estimated. Rogers (1987) has painstakingly compiled these records for all Pacific salmon species from 1950 to 1984. We have summed his annual production estimates for all species and all areas (exclusive of the Bering Sea stocks), which we present in Fig. 19. The combined total production (in tonnes) of salmon increased from an average of 335.5×10^3 tonnes) in the period 1955–58 to 530.8×10^3 tonnes for the 5-yr beginning in 1980, approximately a 1.6-fold increase. By comparison, our estimates for the same periods were 363.5×10^3 tonnes and 671.3×10^3 tonnes respectively, a 1.8-fold increase. The annual production numbers compiled by Rogers (1987) are expected to be lower than the actual biomass occurring in the gulf as they represent returns for the calendar year and do not account for either natural mortality on returning to coastal waters, or the fact that several species (particularly the abundant sockeye and

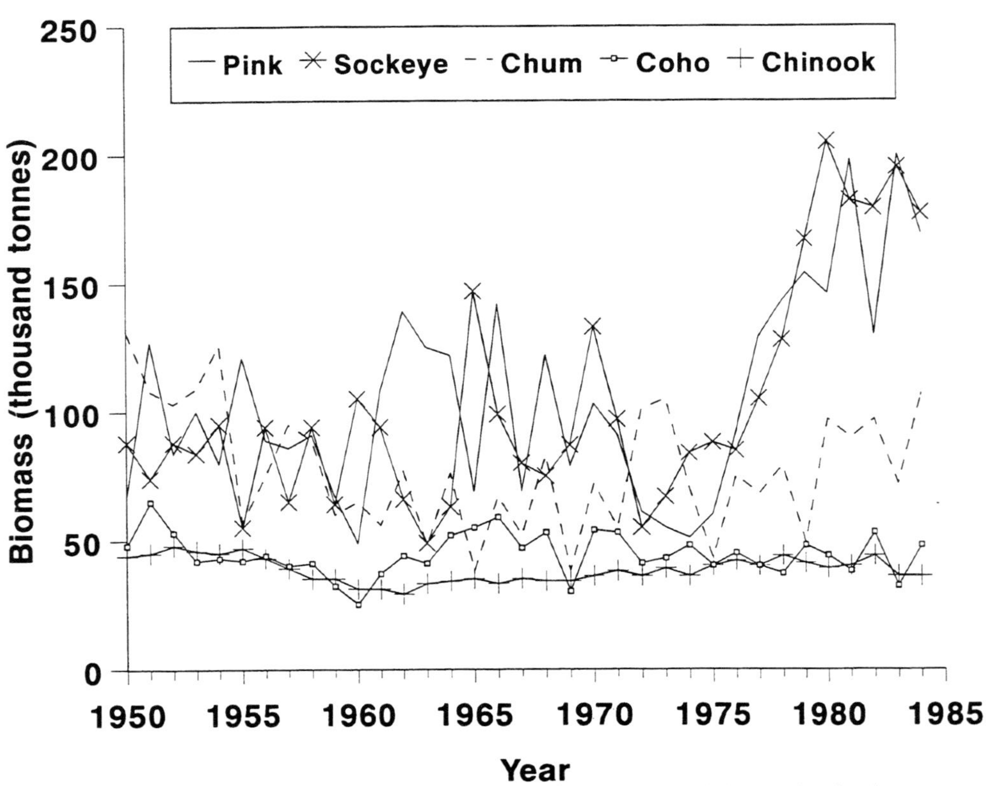

Fig. 19. Time series of total salmon production (catch plus escapement) by species for the years 1950–84 for stocks that are present only in the Gulf of Alaska. Data are from Rogers (1987).

chum salmon) have multiple year-classes present in the ocean at any time. Moreover, salmon production in Alaska continued to climb in the mid-1980s (see Pearcy 1992, Fig. 6.7), so the production for the early part of the decade probably underestimates the total production for that decade. We also note that our estimates of the combined biomass of salmon between 1955 and 1958 are 25% lower, but are probably more accurate than the preliminary figures worked up by Ware and McFarlane (1989) from a less extensive data base.

It is interesting that the catch rates of all nektonic species (particularly pomfret and neon flying squid) increased with the exception of jack mackerel, which showed a striking decline from an estimated mean CPUE of 3.99 in 1955–58 to 0.74 in the 1980s. This fivefold drop in catch may reflect a change in abundance or distribution, or a combination of these two factors. In contrast to the other species examined, jack mackerel straddles the Coastal Upwelling and Transitional Domains, spawning over an extensive area from Baja California to Washington. Consequently it is a very difficult species to assess. MacCall and Prager (1988) found that jack mackerel larvae were two to three times more abundant off California in the late 1950s than in the early 1980s. There has also been a general downward trend in southern California landings of jack mackerel since the early 1960s (Mason 1991). Both lines of evidence suggest that the biomass of this species may have decreased. However, because only a portion of the stock migrates north of 45°N in the Northeast Pacific in summer to feed, the decline in the catch rate of this stock north of 45°N in summer may also reflect an interdecadal decrease in the northward migration rate.

In conclusion, our analysis suggests that there has been roughly a twofold increase in salmonid biomass in the eastern subarctic Pacific since the 1950s. In an earlier paper, we found evidence of a large-scale doubling of the summer zooplankton biomass in the same region between the periods 1956–62, and 1980 –89 (Brodeur and Ware 1992). The biggest increases in zooplankton biomass occurred in areas adjacent to the shelf along the northern periphery of the Gulf of Alaska. Thus, the zooplankton increase would most benefit the subarctic salmonid species (sockeye, pink, chum, and steelhead salmon) rather than coho salmon, some stocks of which may reside entirely within the Coastal Upwelling Domain (Pearcy 1992). In particular, we hypothesized that more food would be advected onshore and thus be available to juvenile salmon during the summer months, resulting in better overall early ocean survival. It is highly likely that the increase in salmonid hatchery production beginning in the 1970s (Pearcy 1992; Beamish and Bouillon 1993) contributed substantially to our observed increases in salmon catch rates in the later period and perhaps occurred at a most opportune time considering the changes in coastal productivity. However, the substantial increases in adult salmon abundance, coupled with an apparent decrease in oceanic distributional ranges resulting from warmer temperatures, could have led to the decreasing mean weight-at-age observed for many salmon species (Welch 1992; Ishida et al. 1993).

Our interpretation of these findings is that there has been a general increase in productivity in the eastern subarctic Pacific, so that the doubling of the zooplankton biomass between the 1950s and the 1980s now supports twice as large a biomass of nekton, which is trophodynamically dependent on the plankton either directly or indirectly through some intermediate prey species. If this interpretation is correct, then it raises several important questions as to what the biological mechanism is behind the increase in plankton and large nekton biomass; what the response is of the intermediate trophic level (macrozooplankton and micronekton); and what likely triggers a shift to a lower productivity regime. The answers to these questions are important because when the next regime shift occurs, it is likely to have a significant negative impact on salmon production and once again lower the carrying capacity of the North Pacific Ocean for all epipelagic nekton.

Acknowledgements

We thank Glen Langford of Geo-Spatial Systems Ltd. for assistance with the GIS analysis. Dr. W.G. Pearcy, Dr. A.D. MacCall, and an anonymous reviewer provided helpful comments on the manuscript. Finally, we acknowledge and thank the many scientists and crew members from Japan, the United States, and Canada who spent many long hours at sea sampling and recording the vast amount of data that went into our analyses.

References

Bartoo, N., and D. Holts. 1993. Estimated drift gillnet selectivity for albacore *Thunnus alalunga*. Fish. Bull. U.S. 92: 371–378.

Beamish, R.J. 1994. Climate change and exceptional fish production off the west coast of North America. Can. J. Fish. Aquat. Sci. 50: 2270–2291.

Beamish, R.J. and D.R. Bouillon. 1993. Pacific salmon production trends in relation to climate. Can. J. Fish. Aquat. Sci. 50: 1002–1016.

Bernard, F.R. 1981. Canadian west coast flying squid experimental fishery. Can. Ind. Rep. Fish. Aquat. Sci. 122: 1–23.

Bernard, F.R. 1986. Data record: flying squid drift-net and oceanographic cruise by *W.E. Ricker*, August–September 1986. Submitted to the Annual Meeting of the International North Pacific Fisheries Commission. Department of Fisheries and Oceans, Fisheries Research Branch, Pacific Biological Station, Nanaimo, B.C., Canada, V9R 5K6.

Brodeur, R.D. 1988. Zoogeography and trophic ecology of the dominant epipelagic fishes in the northern North Pacific Bull. Ocean Res. Inst. 26(II): 1–27.

Brodeur, R.D., and D.M. Ware. 1992. Long-term variability in zooplankton biomass in the subarctic Pacific Ocean. Fish. Oceanogr. 1: 32–38.

Calkins, D.G. 1987. Marine mammals, p. 527–557. *In* D.W. Hood and S.T. Zimmerman [ed.] The Gulf of Alaska: physical environment and biological resources. Alaska Office Ocean Assessment Division, NOAA, Washington, DC.

Clark, S.H., and B.E. Brown. 1977. Changes in biomass of finfishes and squids from the Gulf of Maine to Cape Hatteras 1963–74, as determined by research vessel survey data. Fish. Bull. U.S. 75: 1–24.

DeGange, A.R., and G.A. Sanger. 1987. Marine birds, p. 479–524. *In* D.W. Hood and S.T. Zimmerman [ed.] The Gulf of Alaska: physical environment and biological resources. Alaska Office Ocean Assessment Division, NOAA, Washington, DC.

Dodimead, A.J., F. Favorite, and T. Hirano. 1963. Salmon of the North Pacific Ocean—Part II. Review of oceanography of the Subarctic Pacific region. Bull. Int. North Pac. Fish. Comm. 13: 195 p.

Faculty of Fisheries, Hokkaido University. 1981. Data record of oceanographic observations and exploratory fishing No. 24. Hakodate, Hokkaido, Japan. 299 p.

Faculty of Fisheries, Hokkaido University. 1982. Data record of oceanographic observations and exploratory fishing No. 25. Hakodate, Hokkaido, Japan. 351 p.

Faculty of Fisheries, Hokkaido University. 1983. Data record of oceanographic observations and exploratory fishing No. 26. Hakodate, Hokkaido, Japan. 364 p.

Faculty of Fisheries, Hokkaido University. 1984. Data record of oceanographic observations and exploratory fishing No. 27. Hakodate, Hokkaido, Japan. 375 p.

Faculty of Fisheries, Hokkaido University. 1985. Data record of oceanographic observations and exploratory fishing No. 28. Hakodate, Hokkaido, Japan. 390 p.

Faculty of Fisheries, Hokkaido University. 1986. Data record of oceanographic observations and exploratory fishing No. 29. Hakodate, Hokkaido, Japan. 348 p.

Faculty of Fisheries, Hokkaido University. 1988. Data record of oceanographic observations and exploratory fishing No. 31. Hakodate, Hokkaido, Japan. 396 p.

Faculty of Fisheries, Hokkaido University. 1989. Data record of oceanographic observations and exploratory fishing No. 32. Hakodate, Hokkaido, Japan. 433 p.

Faculty of Fisheries, Hokkaido University. 1990. Data record of oceanographic observations and exploratory fishing No. 33. Hakodate, Hokkaido, Japan. 374 p.

Fiscus, C.W., and R.W. Mercer. 1982. Squids taken in surface gillnets in the North Pacific Ocean by the Pacific Salmon Investigations Program, 1955–72. NOAA Tech. Memo. NMFS F/NWC-28: 32 p.

French, R.R., R.G. Bakkala, and D.F. Sutherland. 1975. Ocean distribution of Pacific salmon, *Oncorhynchus* spp., and steelhead trout, *Salmo gairdnerii*, as shown by tagging experiments. NOAA Tech. Rep. NMFS SSRF-689: 89 p.

Hanavan, M.G., and G. K. Tanonaka. 1959. Experimental fishing to determine distribution of salmon in the North Pacific Ocean and Bering Sea, 1956. U.S. Fish and Wildlife Service, Washington, DC. Spec. Sci. Rep. Fish. 302: 22 p.

Hare, S.R. and R.C. Francis. 1995. Climate change and salmon production in the Northeast Pacific Ocean, p. 357–372. *In* R.J. Beamish [ed.] Climate change and northern fish populations. Can. Spec. Publ. Fish Aquat. Sci. 121.

Hartt, A.C. 1975. Problems in sampling Pacific salmon at sea. Int. North Paci. Fish. Comm. Bull. 32: 165–231.

Ishida, Y. 1986. Estimation of catch efficiency and chum salmon abundance by parallel gillnet experiment. Bull. Jpn. Soc. Sci. Fish. 52: 239–248.

Ishida, Y., S. Ito, M. Kaeriyama, S. McKinnell, and K. Nagasawa. 1993. Recent changes in age and size of chum salmon (*Oncorhynchus keta*) in the North Pacific Ocean and possible causes. Can. J. Fish. Aquat. Sci. 50: 290–295.

Jamieson, G.S., and G.D. Heritage. 1987. Experimental flying squid fishing off British Columbia, 1985 and 1986. Can. Ind. Rep. Fish. Aquat. Sci. 179: 1–103.

1988. Experimental flying squid fishing off British Columbia, 1987. Can. Ind. Rep. Fish. Aquat. Sci. 186: 1–79.

Kawasaki, S., J. Ito, and K. Yamanaka. 1989. Cruise report of flying squid survey by the *Shoyo maru* in 1989. Submitted to the Annual Meeting of the International North Pacific Fisheries Commission. Fisheries Agency of Japan, Far Seas Fisheries Research Laboratory, Shimizu, Japan. 5 p.

Kerr, R.A. 1992. Unmasking a shifty climate system. Science (Wash., D.C.) 255: 1508–1510.

Kubodera, T. 1986. Relationships between abundance of epipelagic squids and oceanographical–biological environments in the surface waters of the Subarctic Pacific in summer. Bull. Int. North Pac. Fish. Comm. 47: 215–238.

Larkins, H.A. 1964. Some epipelagic fishes of the North Pacific Ocean, Bering Sea, and Gulf of Alaska. Trans. Am. Fish. Soc. 93: 286–290.

LeBrasseur, R.J., N.B. Hargreaves, and T. Gjernes. 1988. Canadian mid-ocean salmon studies, *W.E. Ricker*, 1 June – 12 July 1988. Submitted to the Annual Meeting of the International North Pacific Fisheries Commission. Department of Fisheries and Oceans, Fisheries Research Branch, Pacific Biological Station, Nanaimo, B.C., Canada, V9R 5K6. 19p.

LeBrasseur, R.J., B. Riddell, and T. Gjernes. 1987. Ocean studies in the Pacific Subarctic boundary area. Submitted to the Annual Meeting of the International North Pacific Fisheries Commission. Department of Fisheries and Oceans, Fisheries Research Branch, Pacific Biological Station, Nanaimo, B.C., Canada, V9R 5K6. 16 p.

MacCall, A.D., and M.H. Prager. 1988. Historical changes in abundance of six fish species off southern California, based on the CalCOFI egg and larval samples. Calif. Coop. Oceanic Fish Invest. Rep. 29: 91–101.

Macy, P.T., J.M. Wall, N.D. Lampsakis, and J.E. Mason. 1978. Resources of non-salmonid pelagic fishes of the Gulf of Alaska and eastern Bering Sea. Parts 1 and 2. Final Rep. OCSEAP, Bureau of Land Management, U.S. Department of the Interior, Washington, D.C. 714 p.

Mahon, R., and R.W. Smith. 1989. Demersal fish assemblages on the Scotian Shelf, Northwest Atlantic: spatial distribution and persistence. Can. J. Fish. Aquat. Sci. 46(Suppl. 1): 134–152.

Manzer, J.I., and F. Neave. 1958. Data record of Canadian exploratory fishing for salmon in the Northeast Pacific in 1957. Fish. Res. Board Can. MS. Rep. Ser. (Biol.) 659: 303 p.

Manzer, J.I., and F. Neave. 1959. Data record of Canadian Exploratory fishing for salmon in the Northeast Pacific in 1958. Fish. Res. Board Can. MS. Rep. Ser. (Biol.) 682: 223 p.

Mason, J.E. 1991. Variations in the catch of jack mackerel in the southern California purse seine fishery. Calif. Coop. Oceanic Fish. Invest. Rep. 32: 143–151.

McGary, J.W., and J.J. Graham. 1960. Biological and oceanographic observations in the Central North Pacific, July–September 1958. U.S. Fish and Wildlife Service, Washington, DC. Spec. Sci. Rept. Fish. 358: 107 p.

McKinnell, S.M., T. Gjernes, W. Shaw, and S. Whiteaker. 1989. Canadian North Pacific pelagic study, *Arctic Harvester*, July 12– August 22, 1989. Submitted to the Annual Meeting of the International North Pacific Fisheries Commission. Department of Fisheries and Oceans, Fisheries Research Branch, Pacific Biological Station, Nanaimo, B.C., Canada, V9R 5K6. 13 p.

Miller, C.B., B.W. Frost, P.A. Wheeler, M.R. Landry, N. Welschmeyer, and T.M. Powell. 1991. Ecological dynamics in the Subarctic Pacific, a possible iron-limited ecosystem. Limnol. Oceanogr. 36: 1600–1615.

Mishima, S. 1981. On the passing over of subtropic fishes to the Subarctic region in the summer season, p. 61–71. *In* S. Mishima [ed.] Res. Inst. North Pac. Fish. Hokkaido Univ. Spec. Vol.

Myers, K.W., R.V. Walker, S. Fowler, and M.L. Dahlberg. 1990. Known ocean ranges of stocks of Pacific salmon and steelhead as shown by tagging experiments, 1956–1989. Fishery Research Institute of Washington, Seattle, WA. FRI-UW-9009. 57 p.

Mysak, L.A. 1985. On the international variability of eddies in the Northeast Pacific Ocean. *In* W.S. Wooster and D.L. Fluharty [ed.] El Niño north: Niño effects in the eastern subarctic Pacific Ocean. Wash. Sea Grant Publ. University of Washington, Seattle, WA. p. 97–106.

Nakano, H., and K. Shimazaki. 1989. Gill net mesh selectivity for the blue shark (*Prionace glauca* L.). Bull. Fac. Fish. Hokkaido Univ. 40: 22–29.

Neave, F., and H.G. Hanavan. 1960. Seasonal distribution of some epipelagic fishes in the Gulf of Alaska region. J. Fish. Res. Board Can. 17: 221–233.

Neave, F., and J.I. Manzer. 1957. Data record of Canadian Exploratory fishing for salmon in the Northeast Pacific in 1957. Fish. Res. Board Can. MS. Rep. Ser. (Biol.) 659: 303 p.

Pearcy, W.G. 1991. Biology of the Transition Region, p. 39–55. *In* J. Wetherall [ed.] Biology, oceanography, and fisheries of the North Pacific transition zone and Subarctic frontal zone. NOAA Tech. Rep. NMFS 105.

Pearcy, W.G. 1992. Ocean ecology of North Pacific salmonids. University of Washington Press, Seattle, WA. 179 p.

Peterman, R.M., and F.Y.L. Wong. 1984. Cross-correlations between reconstructed ocean abundances of Bristol Bay and British Columbia sockeye salmon (*Oncorhynchus nerka*). Can. J. Fish. Aquat. Sci. 41: 1814–1824.

Powell, D.E., and A.E. Peterson. 1957. Experimental fishing to determine distribution of salmon in the North Pacific Ocean, 1955. U.S. Fish Wildlife Service, Washington, DC. Spec. Sci. Rep. Fish. 205 p.

Radovich, J. 1961. Relationships of some marine organisms of the North Pacific to water temperatures, particularly during 1957 through 59. Calif. Dep. Fish Game Fish. Bull. 112: 1–62.

Ricker, W.E. 1976. Review of the rate of growth and mortality of Pacific salmon in salt water, and noncatch mortality caused by fishing. J. Fish. Res. Board Can. 33: 1483–1524.

Robinson, S.M.C., and G.S. Jamieson. 1984. Report on a Canadian commericial fishery for flying squid using drifting gillnets off the coast of British Columbia. Can. Ind. Rep. Fish. Aquat. Sci. 150: 1–25.

Rogers, D.E. 1987. Pacific salmon, p. 461–476. *In* D.W. Hood and S.T. Zimmerman [ed.] The Gulf of Alaska: physical environment and biological resources. Alaska Office Ocean Assessment Division, NOAA, Washington, D.C.

Royer, T.C. 1989. Upper ocean temperature variability in the Northeast Pacific Ocean: is it an indicator of global warming? J. Geophys. Res. 94: 18 175 – 18 183.

Royer, T.C. and W.J. Emery. 1987. Circulation in the Gulf of Alaska, 1981. Deep-Sea Res. 34: 1361–1377.

Sanger, G.A. 1972. Preliminary standing stock and biomass estimates of seabirds in the Subarctic Pacific regions, p. 589–611. *In* A.Y. Takenouti [ed.], Biological oceanography of the northern North Pacific Ocean. Idemitsu-Shoten, Tokyo.

Shimazaki, K. 1986. Distribution of the pelagic fish community around the Subarctic Boundary in the North Pacific Ocean. Bull. Int. North Pac. Fish. Comm. 47: 247–264.

Sloan, N.A. 1984. Canadian–Japanese experimental fishery for oceanic squid off British Columbia, summer 1983. Can. Ind. Rep. Fish. Aquat. Sci. 152: 1–42.

Tabata, S. 1989. Trends in long-term variability of ocean properties at Ocean Station P in the northeast Pacific Ocean, p. 113–132. *In* D.H. Peterson [ed.] Aspects of climate variability in the Pacific and the Western Americas. Geophys. Monogr. 55.

Takagi, K., K.V. Aro, A.C. Hartt, and M.B. Dell. 1981. Distribution and origin of pink salmon (*Oncorhynchus gorbuscha*) in the offshore waters of the North Pacific Ocean. Bull. Int. North Pac. Fish. Comm. 40: 195 p.

Taniguchi, A. 1981. Plankton productivities in the Pacific Subarctic boundary zone; food conditions of the migrating pelagic fishes, p. 23–35. *In* S. Mishima [ed.] Res. Inst. North Pac. Fish Hokkaido Univ. Spec. Vol.

Trenberth, K.E., and J.W. Hurrell. 1995. Decadal coupled atmosphere–ocean variations in the North Pacific Ocean, p. 15–24. *In* R.J. Beamish [ed.] Climate change and northern fish populations. Can. Spec. Publ. Fish Aquat. Sci. 121.

Wada, T., and M. Kitakata. 1982. Sampling research with mid-water drift gillnet and behaviour of saury in the daytime. Bull. Hokkaido Reg. Fish. Res. Lab. 47: 11–22.

Ware, D.M., and G.A. McFarlane. 1989. Fisheries production domains in the Northeast Pacific Ocean, p. 359–379. *In* R.J. Beamish and G.A. McFarlane [ed.], Effects of ocean variability on recruitment and an evaluation of parameters used in stock assessment models. Can. Spec. Publ. Fish. Aquat. Sci. 108.

Washington, P. 1975. Comparison of the catches of salmon in multifilament and monofilament gillnets fished simultaneously on the high seas. Bull. Int. North Pac. Fish. Comm. 32: 42–70.

Welch, D.W. 1992. Possible future high seas salmon research by the Pacific Biological Station, p. 37–42. *In* Y. Ishida, K. Nagasawa, D.W. Welch, K.W. Myers, and A.P. Shershnev [ed.] Proceedings of the International Workshop on Future Salmon Research in the North Pacific Ocean. Nat. Res. Inst. Far Seas Res. Spec. Publ. 20.

Wooster W.S., and D.L. Fluharty [ed.] 1985. El Niño North: Niño effects in the eastern Subarctic Pacific Ocean. Wash. Sea Grant Publ., University of Washington, Seattle, WA. 312 p.

Yatsu, A., H. Shimada, and M. Murata. 1993. Distributions of epipelagic fishes, squids, marine mammals, seabirds and sea turtles in the central North Pacific. Bull. Int. North Pac. Fish. Comm. 53(II): 111–146.

Climate change and salmon production in the Northeast Pacific Ocean

S.R. Hare[1] and R.C. Francis

Hare, S.R. and R.C. Francis. 1995. Climate change and salmon production in the northeast Pacific Ocean, p. 357–372. *In* R.J. Beamish [ed.] Climate Change and Northern Fish Populations. Can. Spec. Publ. Fish. Aquat. Sci. 121.

Abstract: Alaskan salmon stocks have exhibited enormous fluctuations in production during the 20th century. In this paper, we investigate our hypothesis that large-scale salmon-production variability is driven by climatic processes in the Northeast Pacific Ocean. Using a time-series analytical technique known as intervention analysis, we demonstrate that Alaskan salmonids alternate between high and low production regimes. The transition from a high(low) regime to a low(high) regime is called an intervention. To test for interventions, we first fitted the salmon time series to univariate autoregressive integrated moving average (ARIMA) models. On the basis of tentatively identified climatic regime shifts, potential interventions were then identified and incorporated into the models, and the resulting fit was compared with the non-intervention models. A highly significant positive step intervention in the late 1970s and a significant negative step intervention in the late 1940s were identified in the four major Alaska salmon stocks analyzed. We review the evidence for synchronous climatic regime shifts in the late 1940s and late 1970s that coincide with the shifts in salmon production. Potential mechanisms linking North Pacific climatic processes to salmon production are identified.

Résumé : La production des stocks de saumons de l'Alaska a manifesté d'énormes fructuations au cours du XXe siècle. Dans le présent document, nous tentons de vérifier notre hypothèse selon laquelle la variabilité de la production du saumon à grande échelle dépend des processus climatiques dans le nord-est du Pacifique. Grâce à une technique analytique basée sur les séries chronologiques et connue sous l'appellation «analyse d'intervention», nous démontrons que les salmonidés de l'Alaska passent alternativement de régimes de production élevés à des régimes faibles. La transition entre un régime élevé (faible) et un régime faible (élevé) est appelée une intervention. Pour vérifier le nombre d'interventions, nous avons appliqué des modèles ARMMI (modèles autorégressifs à moyennes mobiles intégrées) à une variable à des séries chronologiques sur le saumon. En nous appuyant sur les changements de régimes climatiques que nous avons provisoirement identifiés, nous avons repéré les interventions potentielles que nous avons incorporées aux modèles, puis avons comparé les résultats aux modèles de non-intervention. Dans les quatre grands stocks de saumons de l'Alaska analysés, nous avons recensé une intervention (échelon positif) hautement significative au milieu des années 1970 et une intervention (échelon négatif) significative à la fin des années 1950. Nous examinons les données sur les changements synchrones de régimes climatiques au début des années 1950 et au milieu des années 1970 qui coïncident avec les variations de la production de saumon. Nous mettons au jour les mécanismes éventuels qui relient les processus climatiques du Pacifique Nord à la production de saumon.

Introduction

In the mid-1970s, ocean conditions in the North Pacific Ocean underwent a dramatic and abrupt change (Graham 1994). Coincident with the physical regime shift, Alaskan salmonids entered an era of greatly increased production that has persisted into the 1990s (Fig. 1). Throughout their long (over 100 yr) commercial exploitation history, several of the Alaskan salmon species have demonstrated "red noise" variability, wherein periods of high (low) production tend to persist for a lengthy period before abruptly reversing to the opposite state. For example, in the 1930s and early 1940s, salmon landings were high, followed by an era of low catches from the late 1940s to late 1970s. As Alaskan landings increased in the late 1970s, several North American west coast stocks, notably Oregon coho salmon (*Oncorhynchus kisutch*; Pearcy 1992), went into a prolonged period of decline.

Much early research on variability in salmon survival (and therefore production and catch) focused on the freshwater stage of their life cycle, attempting to link survival to conditions in their spawning and rearing habitat. The period spent at sea was regarded as relatively unimportant. There is now a growing awareness of the importance of the marine environment in determining salmon production (e.g., Pearcy 1984; Beamish and McFarlane 1989).

S.R. Hare and R.C. Francis. Fisheries Research Institute, WH-10, University of Washington, Seattle, WA 98195, USA.

[1]Present Address: International Pacific Halibut Commission, P.O. Box 95009, Seattle, WA 98145-2009, USA.

Variability in marine survival of salmon is poorly understood (Mathews 1984). Numerous studies have attempted to correlate survival with environmental factors, though few have proven useful in predicting salmon abundance or assisting in management decision making (Pearcy 1992). Part of the difficulty in elucidating the driving factors of survival is that the relationship between the environment and survival is clouded by many factors. Biotic (e.g., intra- and inter-specific competition, prey availability, predation) and abiotic (environmental variables, habitat) factors not only exhibit complex relationships with survival (non-linear, threshold) but are themselves often highly correlated.

Despite these drawbacks, the importance of attempting to understand the causes of variable survival should not be underestimated (Francis and Sibley 1991). In particular, understanding large-scale and long-term variability would benefit both fishery managers and fishermen (Shepherd et al. 1984).

Large marine ecosystems fluctuate in response to physical forcings that occur over a number of time intervals. There appears to be a nested hierarchy of interacting processes occurring on different time scales that are relevant to their dynamics, ranging from relatively discrete processes that occur over times on the order of 1 yr or less (e.g., the 1970 North Pacific winter atmospheric circulation pattern (Hollowed and Wooster 1992)), to processes that persist over long time periods and fluctuate at the inter-century level (Baumgartner et al. 1992). What we are most interested in identifying in this analysis are regimes that define points in time, separated by intervals on the order of decades, where major jumps or shifts in the level of abundance occur in large marine ecosystems. Therefore, in examining the interannual dynamics of various biological components of large marine ecosystems, what we see are responses to these nested hierarchies of interacting processes occurring at different time scales and working synergistically to create pattern. In this analysis, it is pattern at the regime level that we are trying to interpret.

We hypothesize that regional variability in salmon production is driven by large-scale climate change, reflected in North Pacific atmospheric–oceanic regime shifts. Under this hypothesis, salmon populations exhibit two characteristics: relatively stable production while a particular regime persists, followed by a rapid transition to a new production level in response to the physical regime shift. If large-scale salmon production is closely related to North Pacific climate processes, we should find coherent shifts in mean production levels across both species and area.

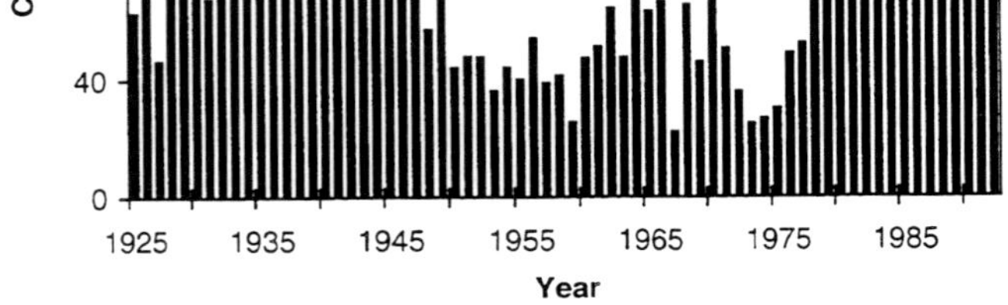

Fig. 1. Trend in total Alaskan salmon catch, 1925–1992.

In addition to the late 1970s regime shift, we surmise that an earlier shift, opposite in character, occurred in the late 1940s. Based on evidence summarized in the Discussion, we tentatively identify the regime shifts as taking place in the winters of 1946–47 and 1976–77. Our hypothesis suggests that two shifts in Alaskan salmon production should be detectable: a decrease in the late 1940s and an increase in the late 1970s.

To test this hypothesis, we proceed by statistically analyzing the historical production dynamics of four major Alaskan salmon stocks: western and central sockeye salmon (*Oncorhynchus nerka*), southeast and central pink salmon (*Oncorhynchus gorbuscha*). While many researchers have noted the aforementioned swings in production (e.g., Beamish and Bouillon 1995), there remained the possibility that the changes might be either random processes or nonsignificant, in a statistical sense. Owing to the high serial correlation (lack of independence between successive observations), the *t*-test for equality of means cannot be used to test for production shifts. We utilize a time-series technique known as intervention analysis to identify the significance, magnitude, and form of structural shifts (interventions) in the four time series. We identify and test the timing of the interventions by matching the onset of the physical regimes with the life history of the different species of salmon. Intervention analysis is a relatively recent statistical technique recommended as a method for detecting and quantifying non-random change in an unreplicated experiment (Carpenter 1990).

To test for interventions, we fitted univariate time-series models of the Box-Jenkins (1976) autoregressive integrated moving average (ARIMA) class. These ARIMA models provide a baseline fit to the correlation structure exhibited by the time series. Interventions are subsequently identified by analyzing model residuals. Model parameters are re-estimated incorporating the intervention(s), and the models compared on the basis of several criteria. After identifying the timing and nature of the interventions, we then review the evidence for synchronous large-scale physical regime shifts in the North Pacific.

Time-series modeling and intervention analysis

The use of time-series analysis to model fish population dynamics has increased in recent years. Most of the theoretical development and initial application has taken place in the econometric and business forecasting literature. Recognition of the potential applicability to ecological problems appears to have begun with Moran (1949).

There are five classes of commonly applied time-series models (Jenkins 1979). The simplest, and most widely known, comprise the so-called Box-Jenkins ARIMA univariate models. Simple ARIMA models utilize only the history of the time series to "explain" its observed variability. The second class comprises the transfer-function noise (TFN) models, which relate an output-series variability to both its own history and that of one or more explanatory variables. A third class, related to TFN models, comprises intervention models which incorporate the effects of unusual events, natural or human-made, to modify ARIMA models. The other two classes comprise multivariate models. Multivariate stochastic models permit feedback among several time series and are often referred to as vector ARIMA models. The final class includes explanatory variables giving a multiple input-multiple output mode and are sometimes referred to as multivariate transfer-function models.

In addition to these time-series models, there has been a parallel development of frequency-domain models, principally in the engineering literature. In the frequency-domain models, processes are modeled as combinations of cosine waves. While theoretically translatable to time-domain models, there have been few applications in ecology. More recently, state-space models have generated a great deal of attention. In state-space, or more generally, structural modeling, a time series is decomposed into linear, seasonal, and irregular components (Harvey 1989). The central feature of structural models is the use of the Kalman filter (Kalman 1960; Kalman and Bucy 1961) for parameter estimation and forecasting. The principal difference between traditional time-series and structural models is the manner in which the error component is modeled. Though neither method has emerged as clearly superior, structural models are likely to receive increased attention.

The first published use of time-series modeling in the fisheries literature was Dunn and Murphy (1976) and Murphy and Dunn (1977), who used univariate and transfer-function models to forecast fish catch in an Arkansas reservoir. Univariate and/or transfer-function models have been used to model the population dynamics of American lobster (*Homarus americanus*; Boudreault et al. 1977, Fogarty 1988a, Campbell et al. 1991), rock lobster (*Jasus edwardsii*; Saila et al. 1980), skipjack tuna (*Katsuwonus pelamis*; Mendelssohn 1981), yellowtail flounder (*Limanda ferruginea*: Kirkley et al. 1982), menhaden (*Brevoortia patronus*; Jensen 1985), haddock (*Melanogrammus aeglefinus*; Pennington 1985), Alaskan salmon (Quinn and Marshall 1989; Noakes et al. 1987), winter flounder (*Pseudopleuronectes americanus*; Jeffries et al. 1989), blue whiting (*Micromesistius poutassou*; Calderon-Aguilera 1991), pilchard (*Sardina pilchardus*; Stergiou 1989), and striped bass (*Morone saxatilis*; Tsai and Chai 1992). Intervention analysis has been applied to Dungeness crab (*Cancer magister*; Noakes 1986), geoduck clams (*Panope abrupta*; Noakes and Campbell 1992), power plant impact on yellow perch (*Perca flavescens*) and alewife (*Alosa pseudoharengus*; Madenjian et al. 1986), and to forecast invertebrate yield (Fogarty 1988b). Vector ARIMA models have been applied to Great Lakes pelagic species (Cohen and Stone 1987; Stone and Cohen 1990) and multivariate transfer-function models were used by Mendelssohn and Cury (1987, 1989) to explore catch per unit of effort in Ivory Coast pelagic fisheries.

In this paper, we use intervention models to determine if North Pacific regime shifts are reflected in Alaska salmonid time series. We provide a brief outline of the technique and explanation of time-series terminology and notation. Those seeking a more theoretical description should consult one of the numerous texts available including the seminal works on ARIMA model formulation (Box and Jenkins 1976) and intervention analysis (Box and Tiao 1975).

Notation

ARIMA and intervention models have several different representations. We employ the following notation:

$$1)\ \phi_p(B)\Phi_P(B^S)\nabla^d\nabla_s^D(Y_t^{(\lambda)}) = \theta_0 + \theta_q(B)\Theta_Q(B^S)a_t + \Sigma_{j=1}^{k}\frac{\omega_j(B)B^{bj}}{\delta_j(B)}I_{jt}^T$$

$Y_t^{(\lambda)}$ is the discrete time series, which may be transformed to stabilize the variance using the Box-Cox (1964) power transformation. The most common transformations are square root (λ=0.5), natural logarithm (λ=0.0), and inverse (λ=–1.0). No transformation is equivalent to a lambda value of 1.0. If required, a power transformation must be done as the first step in time-series modeling.

∇^d is an "integrating factor" (the "I" in ARIMA), better defined as a differencing operation to induce stationarity in the mean of a series. The number of differences taken (which can be at various lags) is indicated by *d*. If required, differencing is the second step in ARIMA modeling.

∇_s^D is a seasonal integrating factor(s) where *s* is the lag at which the *Dth* seasonal difference is taken. While seasonal models are generally applied to weekly, monthly, quarterly, etc. data, they may also be applied to non-seasonal data that exhibit seasonal (i.e., periodic) behavior.

θ_0 plays different roles depending on the value of *d* (order of differencing). For $d = 0$, θ_0 is equal to the estimated mean of the transformed input series multiplied by the sum of the autoregressive components and moved to the right-hand side of the equality. For $d \geq 1$, θ_0 is called the deterministic trend and is often omitted unless clearly called for (Wei 1990, p. 72).

a_t is a random error component assumed to be normally independently distributed with mean 0 and constant variance σ^2_a.

B is the backshift operator. By convention it is a special notation used to simplify the representation of lagged values: $By_t = y_{t-1}$, $B^s y_t = y_{t-s}$. Note also the following definition: $\nabla = 1 - B$, thus differencing is often represented by: $\nabla y_t = (1 - B)y_t$.

$\phi_p(B)$ is the autoregressive polynomial of the form $(1 - \phi_1 B - \phi_2 B^2 - ... - \phi_p B^p)$. The term "autoregressive" is in reference to how the value of *y* is being regressed on its own past values plus a random shock, thus relating the present value of a process to a linear combination of its past values. An autoregressive process can be written as $y_t = \phi_1 y_{t-1} + \phi_2 y_{t-2} + ... + \phi_P y_{t-P} + a_t$. An autoregressive process of order p is abbreviated AR(p), and lower orders than p need not be non-zero.

$\Phi_p(B^S)$ is the multiplicative seasonal autoregressive polynomial of the same form as the non-seasonal polynomial. Multiple seasonal autoregressive components may be included in the model, each of seasonality *S*. The subscript *P* identifies the presence of a seasonal component, and all coefficients other than that of the seasonal lag are set equal to 0.

$\theta_q(B)$ is the moving average polynomial of the form $(1 - \theta_1 B - \theta_2 B^2 - ... - \theta_q B^q)$. The moving average term models the persistence of random effects over time and can be written as $y_t = a_t + \theta_1 a_{t-1} + \theta_2 a_{t-2} + ... + \theta_p a_{t-p}$. A moving average process of order q is abbreviated MA(q), and lower orders than q need not be non-zero.

$\Theta_Q(B^S)$ is the multiplicative seasonal moving average polynomial of the same form as the non-seasonal polynomial. Multiple seasonal moving average components may be included in the model, each of seasonality S. The subscript Q identifies the presence of a seasonal component, and all coefficients other than that of the seasonal lag are set equal to 0.

I_{jt}^T represents the jth intervention and is analogous to a dummy variable in regression. Interventions can be either step ($I = 1$ for $t \geq T$, $I = 0$ otherwise) or pulse ($I = 1$ for $t = T$, $I = 0$ otherwise) functions. A step intervention indicates a permanent shift in the mean of a series, while a pulse indicates a one-time shock. There are several different system responses to step and impulse interventions, such as an abrupt permanent step, a step decay, and impulse decay.

$\omega_j(B)$ is a polynomial of the form $(\omega_0 - \omega_1 B - \omega_2 B^2 - ... - \omega_s B^s)$ representing the initial impact of the intervention.

$\delta_j(B)$ is a polynomial of the form $(1 - \delta_1 B - \delta_2 B^2 - ... - \delta_r B^r)$ representing the long-term impact of the intervention.

B^{b_j} models the delay in response associated with a particular intervention.

Nonseasonal ARIMA models use the notation (p, d, q) to compactly represent autoregressive, difference, and moving average orders. Seasonal models are expressed as $(p, d, q) \times (P, D, Q)_S$, with each seasonal component separately represented. Thus, a (1, 0, 5) model indicates the presence of additive lag 1 AR and lag 5 MA terms with smaller lag MA terms possibly present. A $(1, 0, 0) \times (0, 0, 1)_5$ model also has lag 1 AR and lag 5 MA terms, but the parameters are multiplicative rather than additive.

Model development

Univariate time-series model building, in the methodology of Box and Jenkins (1976), proceeds in the following fashion:

1) Model identification. In this step, tentative models are identified. Determination of the need for power transformation (for variance stabilization) and differencing (to render the series stationary in the mean) are first evaluated. Plots of the autocorrelation and partial autocorrelation functions (ACF and PACF respectively) of the possibly transformed series are examined to assist in determining the order of the AR and MA components (Box and Jenkins 1976). Several other identification tools are also available, such as the extended sample autocorrelation function (ESACF; Tsay and Tiao 1984), generalized partial autocorrelation coefficient (GPAC; Woodward and Gray 1981) and the prediction variance horizon (PVH; Parzen 1981).
2) Parameter estimation. Following selection of a potential model(s), estimates of the parameters are calculated. Access to time-series software is almost essential as ARIMA model parameters must be fitted using a nonlinear estimation routine (though the models themselves are usually linear). Maximum likelihood procedures, usually based on the Cholesky decomposition or the Kalman filter, have been developed as an alternative to the early methods of least squares and approximate likelihood utilized by Box and Jenkins (1976). Standard errors are also computed, and parameters judged to not be significantly different from zero can be dropped. The remaining parameters are then re-estimated.
3) Model diagnostic checking. With a tentative model selected and parameters estimated, the adequacy of the model must be assessed to determine if model assumptions are met. One basic assumption is that the residuals a_t form a white-noise series. A common test is the portmanteau test of Box and Pierce (1970), which uses the residual ACF to test the joint null hypothesis that all serial correlations are equal to zero. It is also common in time-series analysis that several models may be adequate in the sense that the model residuals are reduced to white noise. Several model selection criteria have been developed to assist in model selection. In this analysis, we compared competing models using five criteria: mean absolute error (MAE), which measures the average one-step-ahead prediction error; the unbiased residual variance σ^2_a, equal to the residual sum of squares divided by degrees of freedom; the coefficient of determination r^2, which is the amount of variance "explained" by the model; Akaike's Information Criterion (AIC; Akaike 1974); and Schwarz's Bayesian Criterion (SBC; Schwarz 1978). The AIC and SBC are performance statistics that balance statistical fit with model parsimony. The SBC utilizes a larger penalty function than the AIC, thus often suggesting a model with fewer parameters. Formulas for the model diagnostic and selection criteria are contained in the appendix.

Intervention detection and estimation

In intervention analysis, the correlation structure is initially assumed to be unaffected by the interventions that are modeled as deterministic functions of time. Once the best ARIMA model has been selected, the three-step modeling sequence is repeated to identify and test the significance of interventions.

The original intervention methodology developed by Box and Tiao (1975) permitted estimation of intervention effects when the timing of the interventions was known a priori. To handle the situation where the number and timing of potential interventions are unknown, Chang and Tiao (1983) proposed an iterative detection technique using a likelihood ratio test. Interventions are identified in a stepwise fashion beginning with the residuals from the univariate model. Following detection and estimation of an intervention, model parameters are estimated and the resultant intervention model compared with the univariate model using the criteria cited above. The new model residuals can then be re-analyzed for evidence of other interventions.

A good general review of intervention models is contained in Wei (1990), while Noakes (1986) discusses the applicability of intervention analysis to fisheries problems.

There are two types of interventions, pulse and step. The first represents a discrete system shock; the second a permanent change in the mean level of a process. In this analysis, we model step interventions that result in permanent shifts in the mean level of salmon production. Step interventions can be modeled as abrupt (i.e., a one time-step jump) or delayed (e.g., ramp, impulse decay) processes. It should be noted that testing for different types of interventions increases the probability of identifying a spurious intervention. However, our use of the AIC and SBC performance statistics should minimize this risk. Two software packages, AUTOBOX (Automatic Forecasting Systems, Inc. 1992), and SPSS Trends (SPSS, Inc. 1993), were used for all analyses.

Data

The salmon landings data used in this study were principally taken from an Alaska Department of Fish and Game (ADFG 1991) annual report. Data for 1992 were taken from Pacific Fishing (1994). We selected the four major regional groups of stocks: western Alaska sockeye salmon, central Alaska sockeye and pink salmon, and southeast Alaska pink salmon. Landings data for these regional stocks are more likely to reflect actual production than other Alaskan salmon stocks, as they have been the most intensively exploited stocks because of their high abundances and value. These four regional stocks accounted for over 80% of total Alaskan salmon catches (by number) for the period 1925–1992. To more accurately reflect salmon production by area (Fig. 2), we corrected the Alaskan landings for interceptions using data provided in Shepard et al. (1985), Harris (1989) and the Pacific Salmon Commission (1991). Details of the adjustments are provided in Francis and Hare (1994).

Catch data for these regional stocks are available from as early as the 1870s. We have restricted our analysis to 1925–1992 which we consider to be the period of full exploitation. If there is a "fishing up" effect in the early part of the record, the time-series analysis would be affected by this form of nonstationarity. Our time series span 68 years which is fully adequate for a proper time-series analysis (Newton 1988).

Results

Western Alaska sockeye

The western Alaska sockeye data required a square-root transformation to stabilize the variance. Differencing was not required. Examination of the ACF and PACF indicates rather complex dynamics in this time series, substantially different from the three other salmon time series (Fig. 3) Lags 1, 4, and 5 in the ACF and lags 1, 4, and 6 in the PACF were significant. A variety of models were fitted and compared. Initial identification indicated three candidate univariate models: (6, 0, 0), (1, 0, 5), and the seasonal model $(1, 0, 0) \times (1, 0, 0)_5$. Diagnostics indicated residual serial correlation at lag 3 for the seasonal model, thus a moving average term was added and the resultant $(1, 0, 0) \times (1, 0, 0)_5 \times (0, 0, 1)_3$ model compared with the nonseasonal models. On the basis of the diagnostic statistics, the (6, 0, 0) model was judged to be the most parsimonious at representing the catch dynamics. Within this model, the lag 2, 3, and 4 autoregressive terms were statistically insignificant and, therefore, dropped from the final model. Residual analysis indicated that all serial correlation had been accounted for by the model. The final fitted model parameter estimates and standard errors for the univariate and subsequent intervention models are given in Table 1. Model diagnostics for the univariate and intervention models are given in Table 2.

Based on the physical regime shifts that we tentatively identify occurring in the winters of 1946–47 and 1976–77 (Francis and Hare 1994), we hypothesize that interventions in the western Alaska sockeye salmon time series should be detected around 1949–50 and 1979–80. Sockeye salmon from this region spend 1 or 2 years rearing in freshwater before migrating to sea where they are first exposed (and, probably, most vulnerable) to oceanic conditions. Bristol Bay sockeye salmon, which comprise most of the western Alaska sockeye salmon, generally spend two years at sea, thus the year classes that entered the ocean in 1977 would be caught in 1979.

We fitted two intervention models, the first incorporating a 1979 step, the second also incorporating a 1949 step. For the one-intervention model, the 1979 step was highly significant ($p < 0.01$), and in the two-intervention model, both interventions were highly significant ($p < 0.01$). In both cases, the best statistical fit was provided by simple step (i.e., no delay) interventions. Both models substantially outperformed the nonintervention model. The coefficient of determination, r^2, improved from 0.459 to 0.575 with the 1979 intervention and further increased to 0.623 with inclusion of the 1949 intervention (all model diagnostics reflect model fit in the transformed metric; thus for western Alaska sockeye salmon, the statistics result from model fitting in square root space). Both the AIC and SBC decreased substantially with the addition of each intervention.

The 2 intervention model differed slightly from the two other models in its ARIMA components. The lag 1 AR term, which had decreased in significance from the no intervention to the one-intervention model, dropped out of the model and a lag 3 AR term was added. The AR(5) coefficient was positive and highly significant in all three models, likely reflecting the pseudo-regular 5 year cycle (Eggers and Rogers 1987). The decrease in significance of the AR(1) term with incorporation of interventions was a feature of the model building sequence for each of the salmon time series. One explanation for this result is that a time series that alternates between different levels (or regimes) will have the statistical appearance of a low frequency series with high apparent autocorrelation. Removing the "regime effect" from the time series often accounts for most of the low frequency (i.e., lag 1) autocorrelation.

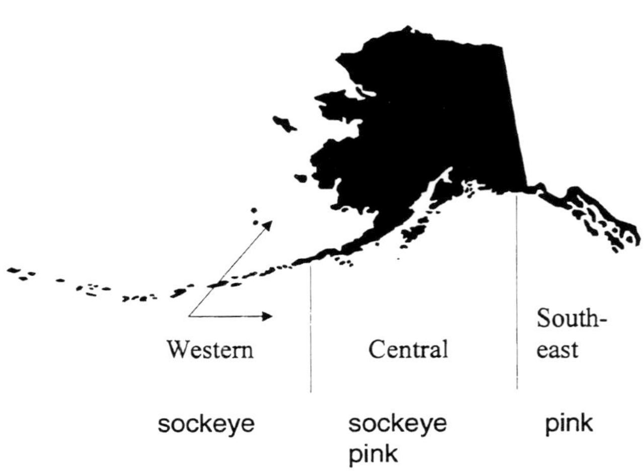

Fig. 2. ADFG statistical areas and regional salmon stocks used in this study.

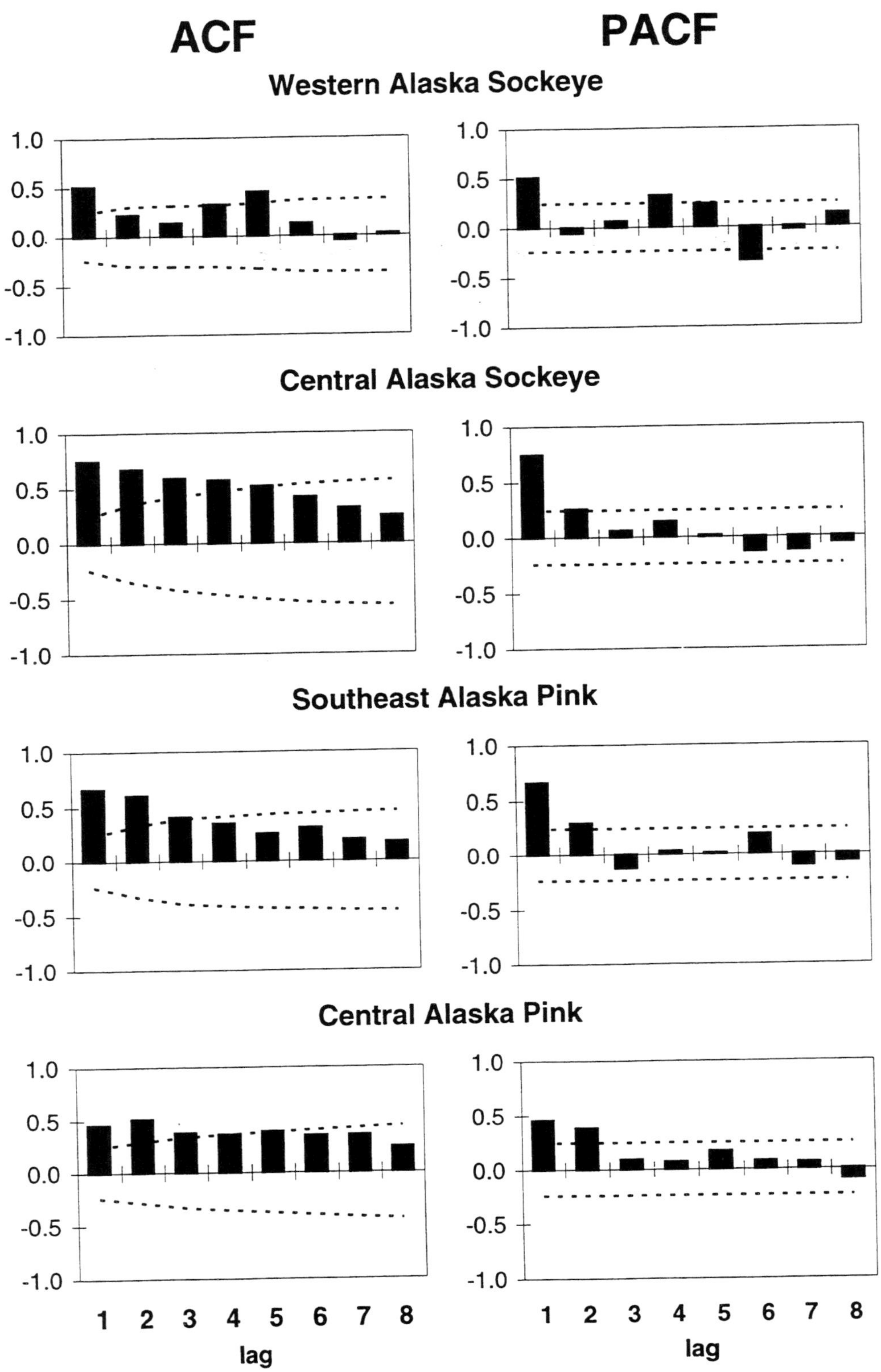

Fig. 3. Plots of the autocorrelation (ACF) and partial autocorrelation (PACF) functions for the four salmon time series. The ACF and PACF are computed for the appropriately differenced and transformed time series. The horizontal lines represent ± 2 standard errors of the sample lag autocorrelation estimates.

Table 1. Univariate and intervention ARIMA models with parameter estimates and associated standard errors developed for western Alaska sockeye salmon. Standard errors are given in parentheses below the equations.

Model	Parameter estimates and standard errors
Univariate	$(1 - 0.538B - 0.505B^5 + 0.369B^6)\sqrt{Y_t} = 1.209a_t$ (0.107) (0.111) (0.122) (0.107)
One intervention (1979)	$(1 - 0.299B - 0.499B^5 + 0.253B^6)\sqrt{Y_t} = 1.468 + a_t + 2.036I_t^{1979}$ (0.121) (0.109) (0.131) (0.105) (0.415)
Two interventions (1949, 1979)	$(1 + 0.305B^3 - 0.377B^5 + 0.225B^6)\sqrt{Y_t} = 4.206 + a_t - 0.754I_t^{1949} + 2.192I_t^{1979}$ (0.121) (0.114) (0.117) (0.161) (0.188) (0.223)

Table 2. Summary statistics for univariate and intervention ARIMA models developed for western Alaska sockeye salmon. MAE = mean absolute error of fitted values, σ^2_a = unbiased residual variance, r^2 = coefficient of determination, AIC = Akaike's Information Criterion, SBC = Schwarz's Bayesian Criterion, and Q = portmanteau residual autocorrelation test (up to lag 20) and associated p-value. All statistics are calculated in the transformed metric.

Model	MAE	σ^2_a	r^2	AIC	SBC	Q	p value
Univariate	0.741	0.836	0.459	186.6	195.5	15.17	0.767
One intervention	0.632	0.667	0.575	172.0	183.1	13.64	0.848
Two interventions	0.603	0.607	0.623	166.4	179.7	17.43	0.625

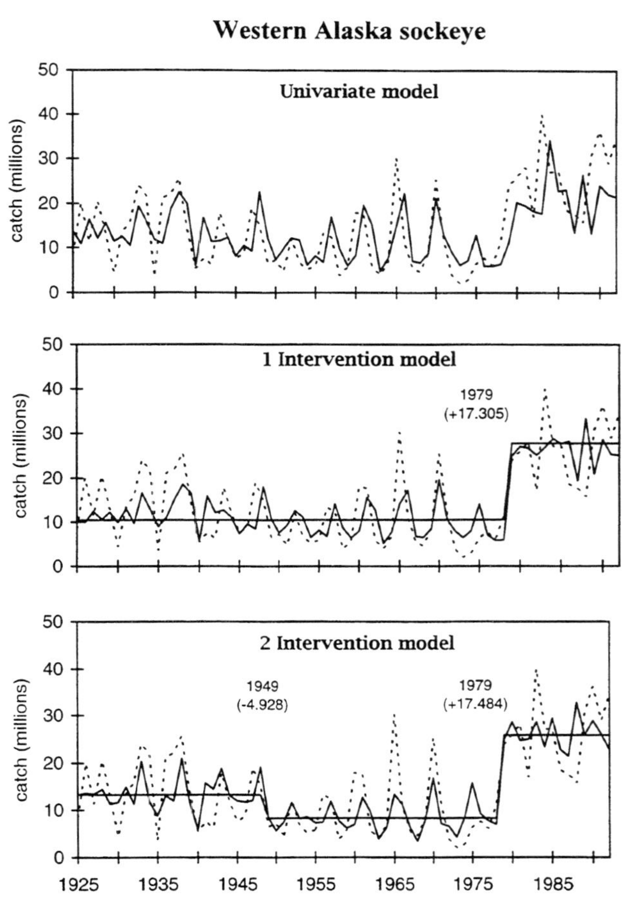

Fig. 4. Plots of model fits for ARIMA and intervention models developed for western Alaska sockeye salmon time series, 1925–1992. Landings data are indicated by dashed lines, fitted values by thick lines. Estimated means before and after interventions are shown by straight lines. Timing of the step interventions and resultant change in mean are also shown.

Resultant model fits and pre- and post-intervention means for the three models are illustrated in Fig. 4. For the one intervention (1979) model, estimates of the pre- and post-intervention means were 10.443 and 27.748 million, respectively, resulting in an estimated step intervention of 17.305 million. In the two-intervention model, the 1949 step was estimated at −4.928 million and the 1979 step at 17.484 million. The three means were estimated at: 13.287 (1925–1948), 8.359 (1949–1978), and 25.843 million (1979–1992).

Central Alaska sockeye

The central Alaska sockeye salmon time series dynamics were much less complex than those of the western Alaska sockeye salmon. The ACF and PACF for the natural logarithm transformed series (Fig. 3) indicated either a (2, 0, 0) or a (1, 0, 1) model. Model diagnostics indicated a better fit for a (2, 0, 0) model. The univariate model fit was the best among the four salmon time series (r^2=0.644). Model residuals showed no residual autocorrelation. Parameter estimates for the univariate and intervention models are given in Table 3, and model statistics in Table 4.

Table 3. Univariate and intervention ARIMA models with parameter estimates and associated standard errors developed for central Alaska sockeye salmon. Standard errors are given in paretheses below the equations.

Model	Parameter estimates and standard errors
Univariate	$(1 - \underset{(0.117)}{0.568}B - \underset{(0.121)}{0.316}B^2)\ln Y_t = \underset{(0.034)}{0.216} + a_t$
One intervention (1980)	$(1 - \underset{(0.102)}{0.572}B)\ln Y_t = \underset{(0.040)}{0.655} + a_t + \underset{(0.188)}{0.917}I_t^{1980}$
Two interventions (1950, 1980)	$(1 - \underset{(0.120)}{0.310}B)\ln Y_t = \underset{(0.058)}{1.197} + a_t - \underset{(0.112)}{0.409}I_t^{1950} + \underset{(0.135)}{1.145}I_t^{1980}$

Table 4. Summary statistics for univariate and intervention ARIMA models developed for central Alaska sockeye salmon. MAE = mean absolute error of fitted values, σ^2_a = unbiased residual variance, r^2 = coefficient of determination, AIC = Akaike's Information Criterion, SBC = Schwarz's Bayesian Criterion, and Q = portmanteau residual autocorrelation test (up to lag 20) and associated p-value. All statistics are calculated in the transformed metric.

Model	MAE	σ^2_a	r^2	AIC	SBC	Q	p value
Univariate	0.255	0.101	0.644	41.3	47.9	15.14	0.768
One intervention	0.234	0.094	0.672	35.7	42.4	14.13	0.824
Two interventions	0.213	0.087	0.704	31.1	40.1	9.92	0.970

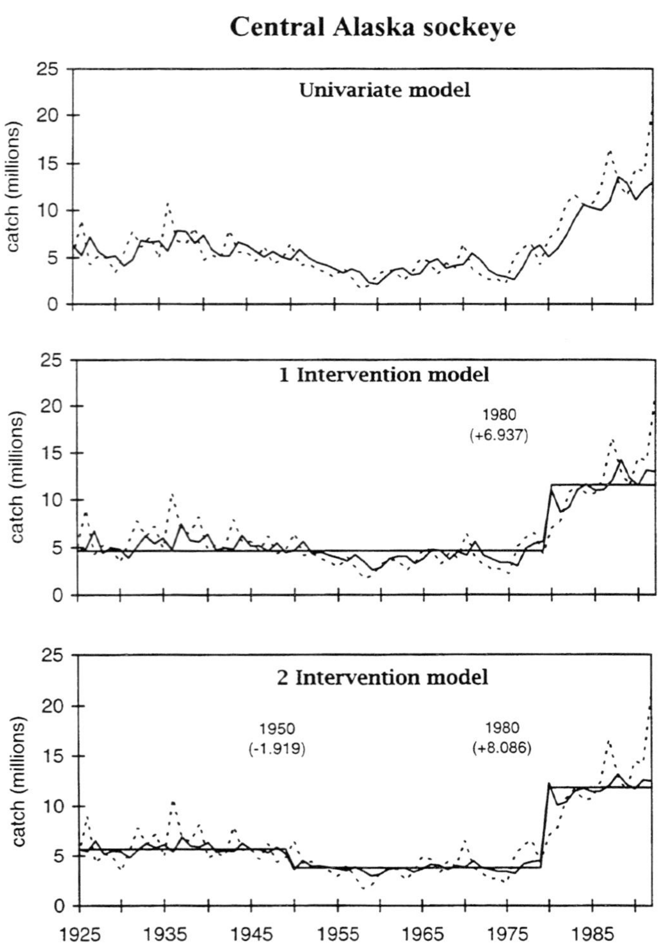

Fig. 5. Plots of model fits for ARIMA and intervention models developed for central Alaska sockeye salmon time series, 1925–1992. Landings data are indicated by dashed lines, fitted values by thick lines. Estimated means before and after interventions are shown by straight lines. Timing of the step interventions and resultant change in mean are also shown.

A large fraction of the central Alaska sockeye salmon (e.g., Kenai River, Chignik Lake runs) spend 3 years in the ocean before returning to spawn (Cross et al. 1983). In keeping with our hypothesis that the climate effect occurs during the first year of marine life, we tested for interventions in 1950 and 1980 for the central Alaska sockeye salmon time series. In the one-intervention (1980) model, the step intervention was highly significant ($p < 0.01$) and led to an improvement in all diagnostic statistics. The two-intervention model provided an equally large improvement as both interventions (1950, 1980) were highly significant. The lag 2 AR term, present in the no-intervention model, dropped out in each of the subsequent models. In addition, for reasons noted earlier, the magnitude of the AR 1 term also decreased with the incorporation of interventions.

The effective change in mean catch for the one intervention model (1980) was 6.937 million (Fig. 5). The estimated mean for the 1980–1992 period was 11.555 million, compared to an estimated mean of 4.618 million prior to the intervention effect. For the two-intervention model, the interventions were estimated to have decreased mean catch by 1.919 million

Table 5. Univariate and intervention ARIMA models with parameter estimates and associated standard errors developed for southeast Alaska pink salmon. Standard errors are given in paretheses below the equations.

Model	Parameter estimates and standard errors
Univariate	$(1 - 0.277B - 0.410B^2)\ln Y_t = 0.906 + a_t$
	(0.112) (0.115) (0.073)
One intervention (1978)	$(1 - 0.495B^2)\ln Y_t = 1.377 + a_t + 0.593I_t^{1978}$
	(0.108) (0.084) (0.310)
Two interventions (1948,1978)	$\ln Y_t = 3.284 + a_t - 1.032I_t^{1948} + 1.005I_t^{1978}$
	(0.121) (0.160) (0.183)

Table 6. Summary statistics for univariate and intervention ARIMA models developed for southeast Alaska pink salmon. MAE = mean absolute error of fitted values, σ^2_a = unbiased residual variance, r^2 = coefficient of determination, AIC = Akaike's Information Criterion, SBC = Schwarz's Bayesian Criterion, and Q = portmanteau residual autocorrelation test (up to lag 20) and associated p-value. All statistics are calculated in the transformed metric.

Model	MAE	σ^2_a	r^2	AIC	SBC	Q	p value
Univariate	0.484	0.397	0.348	133.7	140.3	14.43	0.808
One intervention	0.515	0.413	0.317	136.3	143.0	22.18	0.331
Two interventions	0.452	0.334	0.446	121.4	128.1	18.02	0.586

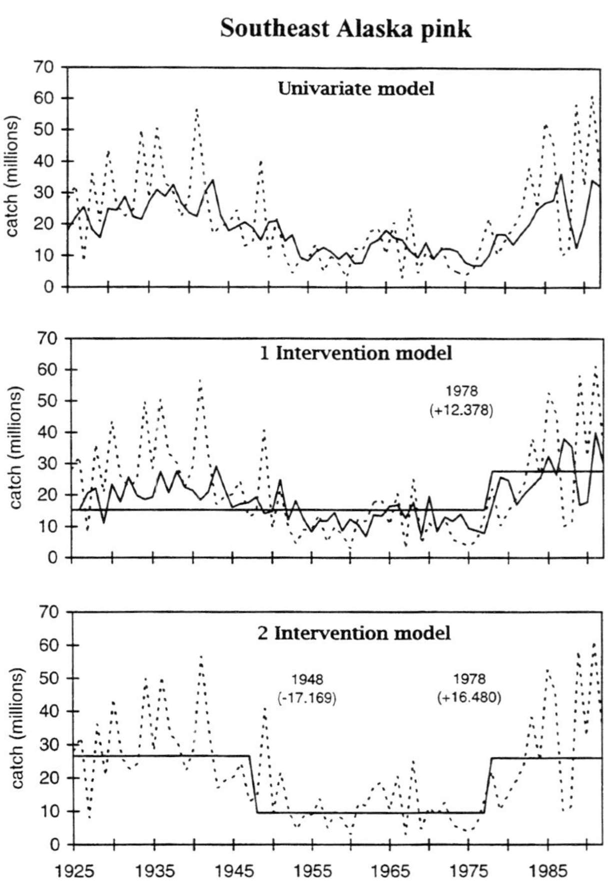

Fig. 6. Plots of model fits for ARIMA and intervention models developed for southeast Alaska pink salmon time series, 1925–1992. Landings data are indicated by dashed lines, fitted values by thick lines. Estimated means before and after interventions are shown by straight lines. Timing of the step interventions and resultant change in mean are also shown.

(from 5.665 to 3.746 million) between the 1925–1949 and 1950–1979 periods, and then increased mean catch by 8.086 million (to 11.832 million) for the 1980–1992 period.

Southeast Alaska pink

The southeast Alaska pink data required a natural logarithm transformation to stabilize the variance. The resultant ACF and PACF resembled central Alaska sockeye, indicating similar dynamics. The same two initial models, (2, 0, 0) and (1, 0, 1), were tested. The (2 ,0, 0) was eventually selected, the same model as for the central Alaska sockeye series. Model fit, however, was the poorest among the time series, as indicated by the r^2 value (0.348). Univariate and intervention model parameter estimates are listed in Table 5, and model statistics in Table 6.

Pink salmon migrate to the ocean in the spring following the year they were spawned and return the following year. Therefore, we tested for interventions in 1948 and 1978. In the one-intervention model, the 1978 intervention was highly significant, but the AR 1 term dropped out as its p-value increased above 0.05 (to 0.09). The one-intervention model

Table 7. Univariate and intervention ARIMA models with parameter estimates and associated standard errors developed for central Alaska pink salmon. Standard errors are given in paretheses below the equations.

Model	Parameter estimates and standard errors
Univariate	$(1 - 0.482B)\sqrt{Y_t} = 2.238 + (1 + 0.566B^2)a_t$
	(0.110) (0.178) (0.117)
One intervention (1978)	$(1 - 0.252B)\sqrt{Y_t} = 2.893 + (1 + 0.362B^2a_t + 2.089I_t^{1978}$
	(0.128) (0.163) (0.135) (0.433)
Two interventions (1948, 1978)	$\sqrt{Y_t} = 4.377 + (1 + 0.241B^2a_t - 0.946I_t^{1948} + 2.675I_t^{1978}$
	(0.219) (0.122) (0.289) (0.327)

Table 8. Summary statistics for univariate and intervention ARIMA models developed for central Alaska pink salmon. MAE = mean absolute error of fitted values, σ^2_a = unbiased residual variance, r^2 = coefficient of determination, AIC = Akaike's Information Criterion, SBC = Schwarz's Bayesian Criterion, and Q = portmanteau residual autocorrelation test (up to lag 20) and associated p-value. All statistics are calculated in the transformed metric.

Model	MAE	σ^2_a	r^2	AIC	SBC	Q	p value
Univariate	0.726	0.915	0.583	191.2	197.9	9.63	0.974
One intervention	0.628	0.797	0.653	181.9	190.8	10.81	0.951
Two interventions	0.608	0.745	0.672	177.1	1846.0	20.03	0.456

actually had a slightly worse fit than the no intervention model. Had the AR 1 term been retained, however, most diagnostics would have favored the one-intervention model. In the two-intervention model, both interventions (negative in 1948, positive in 1978) were also highly significant ($p < 0.01$). Interestingly, though, no ARIMA terms were significant after inclusion of the two interventions. The interpretation of this result is that Southeast Alaska pink salmon production (as indicated by catch) varies randomly about the various regime levels of production. Nearly half (r^2=0.446) of the total variation in Southeast Alaska pink salmon catch was accounted for by the two interventions.

The mean change in catch under the one-intervention model was 12.378 million, from a level of 15.280 million for the 1925–1977 period to a level of27.658 million for the 1978–1992 period (Fig. 6). Estimated average catch under the two-intervention model decreased by 17.169 million (from 26.678 to 9.509) from the 1925–1947 period to the 1948–1977 period and then increased by 16.480 (to 25.989) million during the 1978–1992 period.

Central Alaska pink

The central Alaska pink time series required a square-root transformation to stabilize the variance. Both the ACF and PACF of the transformed series show significant correlation at lags 1 and 2, indicating a mixed ARMA process. The best model we found was a (1,0,2) model with no MA(1) term. Parameter estimates and model statistics for the univariate and intervention models are listed in Tables 7 and 8, respectively.

In the one-intervention model, the highly significant step intervention identified in 1978 resulted in a mean level increase of 21.216 million, from 14.829 to 36.045 million (Fig. 7). The two-intervention model resulted in a further improvement of the model fit. Under this model, the mean level of production was 19.156 million during 1925–1947, then dropped by 7.383 million to a level of 11.773 million for the 1948–1977 period, then increased by 25.509 million to reach the modern catch level of 37.282 million.

Incorporation of the interventions reduced both the AR(1) and MA(2) parameters substantially as the "regime effect" accounted for an increasingly large part of the serial correlation. The AR(1) term was highly significant ($p < 0.01$) in the no-intervention model, remained barely significant ($p \sim 0.05$) in the one-intervention model, and was not retained in the two-intervention model, resulting in a (0, 0, 2) model. The MA(2) term reduced in magnitude from −0.566 (no-intervention model) to −0.241 (two-intervention model).

Discussion

Over the past seven decades, Alaskan salmon populations appear to have alternated between high and low production regimes. We propose that Alaskan salmon are responding to changes in North Pacific climate regimes. Under this hypothesis, each salmon population exhibits a unique smaller-scale variability about some mean level of production during a climatic regime. The transition from one regime to another occurs relatively rapidly, resulting in a shift in the mean production level of Alaskan salmon populations.

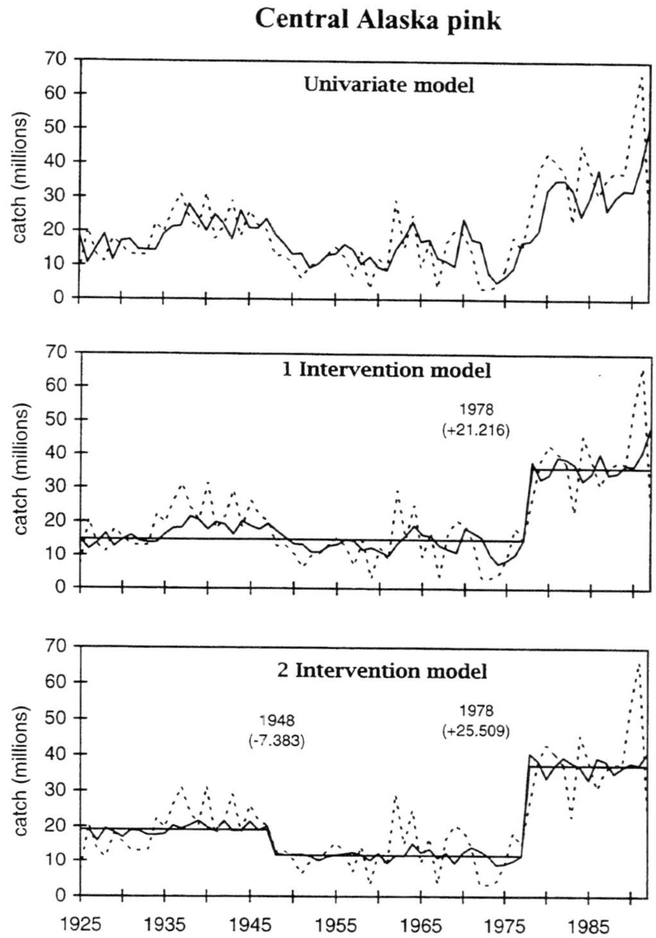

Fig. 7. Plots of model fits for ARIMA and intervention models developed for central Alaska pink salmon time series, 1925–1992. Landings data are indicated by dashed lines, fitted values by thick lines. Estimated means before and after interventions are shown by straight lines. Timing of the step interventions and resultant change in mean are also shown.

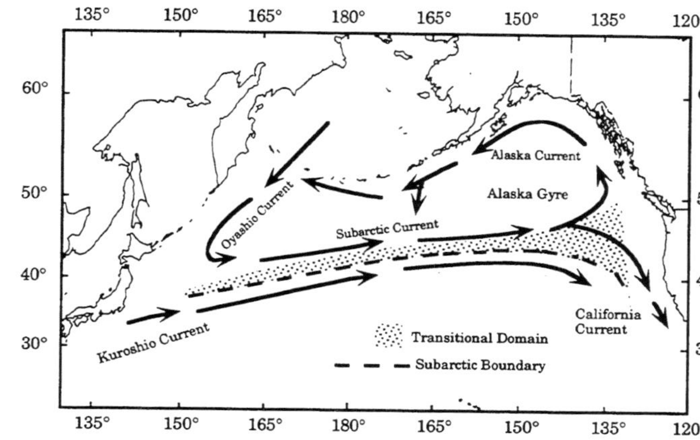

Fig. 8. Summary of major oceanographic features of the North Pacific.

In support of this hypothesis, we have demonstrated nearly synchronous production shifts in four regional Alaskan salmon stocks. These stocks include two different species from three widely separated geographic regions. Using the technique of intervention analysis, we identified three production regimes defined by two major production shifts, one in the late 1940s, the other in the late 1970s.

Alaskan pink and sockeye salmon spend the majority of their marine life cycle in the Central Subarctic Domain (CSD; Ware and McFarlane 1989) which encompasses the Gulf of Alaska (Fig. 8). The principal feature within the CSD is the Alaska Gyre, with an area of active upwelling at its core. The southern boundary of the CSD is defined by the Subarctic Current, whose latitudinal location varies yearly (Roden 1991, Ward 1993). During the seaward and return migrations, pink and sockeye salmon pass through the Coastal Downwelling Domain, a region extending from Queen Charlotte Sound to Prince William Sound dominated by the Alaska Current.

Any attempt to link physical processes in the marine environment to Alaskan salmon production must involve oceanographic conditions within these two regions. We now examine the two production-regime shifts in greater detail, summarize the change in production, and consider the evidence for concurrent climate-regime shifts. We then discuss potential mechanisms linking the physics and biology.

Late 1970s Shift

The increase in salmon production was highly significant in all four time series. In the two-intervention models, the smallest t-value (based on roughly 63 degrees of freedom) of the four late 1970s step intervention variables was 5.492 ($p < 0.0001$, southeast pink salmon). Both pink salmon time series showed a significant jump in 1978 to a higher production level. Because of the strength of the change in production, the timing of the intervention could also have been placed in 1977 or 1979, but model diagnostics indicated the best fit occurred in 1978. Additionally, we chose to test for a 1978 effect because, according to our hypothesis, the returning 1976 brood year class, first to be exposed to the new oceanic regime, should be the first to show a regime effect. A similar argument, based on the sockeye salmon life history, should lead to a 1979 or 1980 intervention for the two sockeye salmon time series, depending on whether the returning fish spent two or three years in the ocean. For the western Alaska sockeye, a 1979 intervention was statistically more significant than a 1980 intervention. The reverse was true for central Alaska sockeye.

Each of the four production groups is faced with a unique set of environmental conditions between their freshwater rearing habitat and entry into the marine feeding and migration grounds. The three geographic regions each contain numerous salmon-bearing rivers. Localized factors will, therefore, lead to some amount of unique variability added to the effect of the climatic regime on the population as a whole. This is reflected in the differing ARIMA structures among the four time series as well as the remaining unexplained variance. It is clear, however, that the four stocks entered an era of increased production in the late 1970s and have remained at that level in the 1990s. Combining the four series, we estimate that the increased production resulted in an annual mean catch increase of greater than 69 million salmon. This translates to a threefold difference in production between the previous regime of the late 40s–late 70s and the present regime beginning in the late 70s.

Evidence for the timing and strength of the late 1970s regime shift has been documented in numerous environmental and biological variables (Ebbesmeyer et al. 1991). The most obvious physical manifestations of the late 1970s shift include a strengthening and eastward shift of the Aleutian Low (Trenberth 1990) and warming of the surface waters in the Gulf of Alaska (Royer 1989). Defining the event as the onset of a new regime rather than a temporary system shock reflects the persistence of the new state variables. Most evidence pinpoints the winter of 1976–77 as the critical transition period. The shift appears to have been forced by an increasingly vigorous winter circulation over the North Pacific (Graham 1994), leading to more severe and frequent winter storms (Seymour et al. 1984), decreases in mid-Pacific sea-surface temperatures (SSTs), and basin-wide decreases in sea-level pressure (Trenberth 1990). The large-scale increase in central Pacific chlorophyll (and thus phytoplankton) during the 1970s has been attributed to persistence of warm SSTs in the summer months (Venrick et al. 1987). The increase in Alaskan air and sea-surface temperatures probably derived from warm air advected from the south by a strengthened Aleutian Low.

Hollowed and Wooster (1992) have hypothesized that the North Pacific alternates between two environmental states, with one transition occurring in 1977. The cool period prior to the transition, what they call a type A regime, is characterized by a weak winter Aleutian Low, enhanced westerly winds in the eastern Pacific, decreased advection into the Alaska Current, and negative coastal SST anomalies. A warm era (type B regime) is characterized by a strong winter Aleutian Low displaced to the east, enhanced southwesterly winds in the eastern Pacific, increased advection into the Alaska Current, and positive coastal SST anomalies.

The mechanisms driving the late 1970s regime shift are the subject of much intensive research. Several hypothesized mechanisms have suggested links between this regime shift in the North Pacific and an abrupt climate shift in the tropical Pacific, which occurred in the late 1970s. Kashiwabara (1987) and Nitta and Yamada (1989) have hypothesized that changes in the tropical Pacific forced the change in North Pacific winter circulation patterns. Trenberth (1990) noted that, in the period between 1976 and 1988, there were three warming El Niño events, but no cooling La Niña events. Graham (1994) holds that the El Niño–La Niña cycle continued but the background state was set to a different state. Miller et al. (1994) were able to reproduce the 1976–77 shift with a general circulation model driven by heat flux input, suggesting that the atmosphere (as opposed to an ocean-atmosphere feedback loop) was the primary force. On the basis of observational analyses, Trenberth and Hurrell (1994) attribute North Pacific atmosphere-ocean variability to both local (atmospheric) and remote (tropical oceanic) processes with mid-latitude feedback serving to emphasize decadal scale variability.

Late 1940s Shift

The negative production shifts identified in the late 1940s were all significant, but of lesser magnitude than those of the late 1970s. The *t*-values for the step interventions in the two-intervention models ranged from 6.45 ($p < 0.0001$, southeast pink salmon) to 3.27 ($p < 0.01$, central pink salmon). The timing of the interventions we tested were selected in the same manner as for the late 1970s shift. Assuming a climate shift in the winter of 1946–47, the appropriate years to test are 1948 (both pink time series), 1949 (western Alaska sockeye), and 1950 (central Alaska sockeye). We estimate the combined drop in catch following the late 1940s intervention at approximately 30 million salmon annually, a decrease of nearly 50%.

Evidence for an late 1940s regime shift is less confirming than for the late 1970s. To some extent, this may be due to the relative lack of data in comparison with that available for the later event. Also, if the salmon data are indicative of the physical data, the shift in physical variables is expected to be smaller and, therefore, more difficult to detect.

Francis and Hare (1994) found a statistically significant negative step in 1947 in Trenberth and Hurrell's (1994) North Pacific Index, a measure of winter atmospheric variability. Several researchers (Dzerdzeevskii 1962, Kutzbach 1970, Kalnicky 1974, Brinkmann 1981) have noted sharp changes in upper level atmospheric circulation patterns occurring in the late 1940s to early 1950s. Balling and Lawson (1982) and Granger (1984) showed that rainfall patterns over the southwestern United States changed in the early 1950s. Rogers (1984) presented average winter air temperatures for Kodiak and Bristol Bay from 1920–1983. With only a few exceptions, coastal Alaskan air temperatures remained anomalously low between the 1946–47 and the 1976–77 winters. Surface-temperature trends in the northern hemisphere were shown by Jones (1988) to be in a cool period between the late 1940s and late 1970s. The frequency and intensity of El Niño–Southern Oscillation events have undergone several changes in the past century (Trenberth 1990; Trenberth and Shea 1987) with strong events between 1880 and 1920, and 1950 and the present, and weak events between 1920 and 1950. Trenberth (1990) also noted the preponderance of cold (La Niña) tropical events during the 1950–1977 period compared with the present (1977–1993) imbalance marked by a greater number of warm (El Niño) events.

Several data sets that we examined dated back only to the late 1940s. While not capable of demonstrating a shift in the late 1940s, they do indicate a similarity of conditions for the 1947–1976 period. Between 1949 and 1976, Emery and Hamilton (1985) classified 22 of 28 North Pacific sea-level pressure patterns as either weak or near normal. Hollowed and Wooster (1992) identified 24 of 31 winter atmospheric circulation patterns between 1946 and 1976 as type A regimes (cool periods).

Potential Mechanisms

Establishing the mechanism whereby salmon production is driven by large-scale climate processes can only proceed by speculation at present. We alluded earlier to the general inability of most studies to establish predictable relationships between environmental variables and salmon survival and production that stand the test of time. Quinn and Marshall (1989), for example, found that inclusion of air and water temperature and freshwater discharge provided limited improvement to their time-series models of southeast Alaska salmon variability.

At least two speculative mechanisms have been advanced to help explain the late 1970s rise in Alaskan salmon production. Rogers (1984) proposed that the increase in catch derived from increased marine survival of migrating salmon in their last winter at sea. Anomalously warm surface temperatures in the Gulf of Alaska altered both the migration paths and timing of returning salmon thus lessening their vulnerability to predators (principally marine mammals). Additional evidence for this hypothesis may be provided by the 1970s and 1980s decline in northern fur seal (*Callorhinus ursinus*) and Steller's sea lion (*Eumetopias jubatus*) (Merrick et al. 1987; York 1987).

The second mechanism relates improved feeding conditions in the Alaska Current and Alaska Gyre to increased salmon production. Brodeur and Ware (1992) documented a twofold increase in zooplankton biomass between the 1950s and 1980s in the subarctic Pacific Ocean. They suggest that the primary beneficiaries of the elevated zooplankton biomass are juvenile salmon that migrate around the coastal margin of the CSD foraging on zooplankton advected to the oceanic shelf. Transport of zooplankton-rich waters derives from increased flow into the Alaska Current from the Subarctic Current (Pearcy 1992). Chelton (1984) has proposed that transport into the California and Alaska Currents fluctuates out of phase. This scenario suggests that the observed decrease in west coast salmon production may be due to poor feeding conditions resulting from decreased advection of subarctic water into the California Current (Pearcy 1992). Francis and Sibley (1991) illustrated opposite trends in production between Gulf of Alaska pink salmon and west coast coho salmon. The nature of the transitions from high (low) to low (high) production in both stocks suggests a single cause.

Perhaps the most interesting feature of the salmon regimes we have identified is the nature of the level of persistence exhibited by the different stocks. Hollowed and Wooster (1992) found synchronous recruitment patterns in several groundfish species corresponding to switches between type A and type B regimes. Strong year-classes apparently derived from the onset of type B regimes. Subsequent year-classes, however, were much smaller. This appears to be quite different from the situation we have documented for Alaskan salmon. In addition, the average duration of type A and B regimes was 7–10 yr, whereas we have identified much longer period regimes based on Alaskan salmon dynamics. This suggests that different components of the North Pacific large marine ecosystem respond to forcing factors of different scales.

Little is known about what causes low-frequency shifts in the structure and dynamics of large marine ecosystems. Margalef (1986) challenges us to develop a new paradigm in this regard. He suggests that infrequent and discontinuous changes in external (physical) energy are the most important factors affecting fluctuations in the biological production of these systems. These inputs, which he refers to as "kicks," disrupt established ecological relationships within an ecosystem.

Dr. John Steele (Woods Hole Oceanographic Institution, Woods Hole, MA 02543, personal communication) puts it another way. He feels that, in the ocean, the variances of biological processes that respond to both physical and biological forcings are inversely proportional to their frequencies. If the variance of a process is forced beyond certain bounds or tolerances, that part of the system "snaps," such as when an earthquake occurs, forcing repercussions throughout the ecosystem. As in the case of an earthquake, many system variables that "snap" at the time of the earthquake demonstrate no aberrant behaviors prior to the earthquake itself. So perhaps it is with large marine ecosystems.

Acknowledgments

This research was funded by Washington Sea Grant. We are indebted to Ray Hilborn, Jim Ianelli, Don Percival, Michael Ward and two anonymous reviewers for critically reviewing the manuscript. We also acknowledge Steven Riser, Warren Wooster, and Anne Hollowed for discussions that helped develop many of the ideas that appear in the paper.

References

Akaike, H. 1974. A new look at statistical model identification. IEEE Trans. Autom. Control AC-19: 716–723.

Alaska Department of Fish and Game. 1991. Alaska commercial salmon catches, 1878–1991. Reg. Info. Rep. 5J91-16. Division of Commercial Fish, Juneau, AK. 88 p.

Automatic Forecasting Systems, Inc. 1992. AUTOBOX. P.O. Box 563, Hatboro, PA 19040, USA.

Balling, R.C., Jr., and M.P. Lawson. 1982. Twentieth century changes in winter climatic conditions. Climatic Change 4: 57–69.

Baumgartner, T.R., A. Soutar, and V. Ferreira-Bartrina. 1992. Reconstruction of the history of Pacific sardine and northern anchovy populations over the past two millenia from sediments of the Santa Barbara Basin, California. Calif. Coop. Oceanic Fish. Invest. Rep. 33: 24–40.

Beamish, R.J., and D.R. Bouillon. 1995. Marine fish production trends off the Pacific Coast of Canada and the United States, p. 585–591. *In* R.J. Beamish [ed.] Climate change and northern fish populations. Can. Spec. Publ. Fish. Aquat. Sci. 121.

Beamish, R.J., and G.A. McFarlane [ed.] 1989. Effects of ocean variability on recruitment and an evaluation of parameters used in stock assessment models. Can. Spec. Publ. Fish Aquat. Sci. 108.

Boudreault, F.-R., J.-N. Dupont, and C. Sylvain. 1977. Modèles linéaires de prédiction des débarquements de homard aux Iles-de-la-Madeleine (Golfe du Saint-Laurent). J. Fish. Res. Board Can. 34: 379–383.

Box, G.E.P., and D.R. Cox. 1964. An analysis of transformations. J. R. Stat. Soc. Ser. B 26: 211–243.

Box, G.E.P., and G.M. Jenkins. 1976. Time series analysis: forecasting and control, Revised edition Holden-Day, San Francisco, CA. 575 p.

Box, G.E.P., and D.A. Pierce. 1970. Distribution of residual autocorrelations in autoregressive-integrated moving average time series models. J. Am. Stat. Assoc. 65: 1509–1526.

Box, G.E.P., and G.C. Tiao. 1975. Intervention analysis with applications to economic and environmental problems. J. Am. Stat. Assoc. 70: 70–79.

Brinkmann, W.A.R. 1981. Sea level pressure patterns over eastern North America, 1899–1976. Mon. Weather. Rev. 109: 1305–1317.

Brodeur, R.D., and D.M. Ware. 1992. Interannual and interdecadal changes in zooplankton biomass in the subarctic Pacific Ocean. Fish. Oceanogr. 1: 32–38.

Calderon-Aguilera, L.E. 1991. An autoregressive model of the temperature-growth relationship for the western Mediterranean blue whiting *Micromesistius poutassou*. Ecol. Modeling 56: 47–61.

Campbell, A., D.J. Noakes, and R.W. Elner. 1991. Temperature and lobster, *Homarus americanus*, yield relationships. Can. J. Fish. Aquat. Sci. 48: 2073–2082.

Carpenter, S.R. 1990. Large-scale perturbations: opportunities for innovation. Ecology 71: 2038–2043.

Chang, I., and G.C. Tiao. 1983. Estimation of time series parameters in the presence of outliers. Tech. Rep. 8. Statistics Research Center, University of Chicago,Chicago, IL.

Chelton, D.B. 1984. Short-term climate variability in the northeast Pacific Ocean, p. 87–99. *In* W.G. Pearcy [ed.] The influence of ocean conditions on the production of salmonids in the north Pacific. Oregon State Univ. Sea Grant, Corvallis, OR.

Cohen, Y., and J.N. Stone. 1987. Multivariate time series analysis of the Canadian fisheries system in Lake Superior. Can. J. Fish. Aquat. Sci. 44(Suppl. 2): 171–181.

Cross, B.A., D.R. Bernard, and S.L. Marshall. 1983. Returns-per-spawner ratios for sockeye salmon in upper Cook Inlet, Alaska. Alaska Dept. Fish Game Inform. Leafl. 221: 82 p.

Dunn, J.E., and C.O. Murphy. 1976. A study of the feasibility of ecosystems modelling of Beaver Reservoir. Univ. Arkansas Stat. Lab. Tech. Rep. 1. 87 p.

Dzerdzeevskii, B. 1962. Fluctuations of climate and general circulation of the atmosphere in extra-tropical latitudes of the Northern Hemisphere and some problems of dynamic climatology. Tellus 14: 328–336.

Ebbesmeyer, C.C., D.R. Cayan, D.R. Milan, F.H. Nichols, D.H. Peterson, and K.T. Redmond. 1991. 1976 step in the Pacific climate: forty environmental changes between 1968–1975 and 1977–1984, p.129–141. *In* J.L. Betancourt and V.L. Sharp [ed.] Proceedings of the Seventh Annual Pacific Climate (PACLIM) Workshop, April 1990, Asilomar, CA. California Department of Water Resources Interagency Ecological Studies Program Tech. Rep. 26.

Eggers, D.M., and D.E. Rogers. 1987. The cycle of runs of sockeye salmon (*Oncorhynchus nerka*) to the Kvichak River, Bristol Bay, Alaska: cyclic dominance or depensatory fishing?, p. 343–366. *In* H.D. Smith, L. Margolis, and C.C. Wood [ed.] Sockeye salmon (*Oncorhynchus nerka*) population biology and future management. Can. Spec. Publ. Fish. Aquat. Sci. 96.

Emery, W.J., and K. Hamilton. 1985. Atmospheric forcing of interannual variability in the northeast Pacific Ocean: connections with El Niño. J. Geophys. Res. 90: 857–868.

Fogarty, M.J. 1988a. Time series models of the Maine lobster fishery: the effect of temperature. Can. J. Fish. Aquat. Sci. 45: 1145–1153.

Fogarty, M.J. 1988b. Forecasting yield and abundance of exploited invertebrates. *In* J.F. Caddy [ed.] The scientific basis of shellfish management. John Wiley & Sons, New York and London.

Francis, R.C., and S.R. Hare. 1994. Decadal-scale regime shifts in the large marine ecosystems of the North-east Pacific: a case for historical science. Fish. Oceanogr. 3: 279–291.

Francis, R.C., and T.H. Sibley. 1991. Climate change and fisheries: what are the real issues? Northwest Environ. J. 7: 295–307.

Graham, N.E. 1994. Decadal-scale climate variability in the 1970s and 1980s: observations and model results. Clim. Dyn. 10: 135–159.

Granger, O.E. 1984. Twentieth-century climate anomaly patterns over the southwestern United States. Phys. Geogr. 5: 164–185.

Harris, C. 1989. The effects of international treaty changes on Japan's high seas salmon fisheries, with emphasis on their catches of North American sockeye salmon, 1972–1984. Ph.D. Dissertation, Univ. Washington, 231 p.

Harvey, A.C. 1989. Forecasting, structural time series models and the Kalman filter. Cambridge University Press. Cambridge. 554 p.

Hollowed, A.B., and W.S. Wooster. 1992. Variability of winter ocean conditions and strong year classes of Northeast Pacific groundfish. ICES mar. Sci. Symp. 195: 433–444.

Jeffries, P., A. Keller, and S. Hale. 1989. Predicting winter flounder (*Pseudopleuronectes americanus*) catches by time series analysis. Can. J. Fish. Aquat. Sci. 46: 650–659.

Jenkins, G.M. 1979. Practical experiences with modelling and forecasting time series. Gwilym Jenkins and Partners (Overseas) Ltd., Jersey, UK.

Jensen, A.L. 1985. Time series analysis and forecasting of Atlantic menhaden catch and CPUE. N. Am. J. Fish. Manage. 5: 78–85.

Jones, P.D. 1988. Hemispheric surface air temperature variations: recent trends and an update to 1987. J. Climate 1: 654–660.

Kalman, R.E. 1960. A new approach to linear filtering and prediction problems. ASME J. Basic. Eng. 82: 34–45.

Kalman, R.E., and R.S. Bucy. 1961. New results in linear filtering and prediction theory. ASME J. Basic Eng. 83: 95–107.

Kalnicky, R.A. 1974. Climatic change since 1950. Ann. Assoc. Am. Geogr. 64: 100–112.

Kashiwabara, T. 1987. On the recent winter cooling in the north Pacific. Tenki 34: 777–781. (In Japanese)

Kirkley, J.E., M. Pennington, and B.E. Brown. 1982. A short-term forecasting approach for analyzing the effects of harvesting quotas: application to the Georges Bank yellowtail flounder (*Limanda ferruginea*) fishery. J. Cons. Cons. int. Explor. Mer 40: 173–175.

Kutzbach, J.E. 1970. Large-scale features of monthly mean northern hemisphere anomaly maps of sea-level pressure. Mon. Weather. Rev. 98: 708–712.

Madenjian, C.P., D.J. Jude, and F.J. Tesar. 1986. Intervention analysis of power plant impact on fish populations. Can. J. Fish. Aquat. Sci. 43: 819–829.

Margalef, R. 1986. Reset successions and suspected chaos in models of marine populations, p. 321–343. *In* T. Wyatt, and M.G. Larrañeta [ed.] International Symposium on Long Term Changes in Marine Fish Populations, 18–21 Nov. 1986, Vigo, Spain, 1986.

Mathews, S.B.1984. Variability of marine survival of Pacific salmonids: a review, p. 167–182. *In* W.G. Pearcy [ed.] The influence of ocean conditions on the production of salmonids in the north Pacific. Sea Grant Publ., Oregon State University. Corvallis, OR.

Mendelssohn, R. 1981. Using Box-Jenkins models to forecast fishery dynamics: identification, estimation, and checking. Fish. Bull. U.S. 78: 887–896.

Mendelssohn, R., and P. Cury. 1987. Fluctuations of a fortnightly abundance index of the Ivorian coastal pelagic species and associated environmental conditions. Can. J. Fish. Aquat. Sci. 44: 408–421.

Mendelssohn, R., and P. Cury. 1989. Temporal and spatial dynamics of a coastal pelagic species, *Sardinella madarensis* off the Ivory Coast. Can. J. Fish. Aquat. Sci. 46: 1686–1697.

Merrick, R.L., T.R. Loughlin, and D.G. Calkins. 1987. Decline in abundance of the northern sea lion, *Eumetopias jubata*, in Alaska 1956–86. Fish Bull. 85: 351–365.

Miller, A.J., D.R. Cayan, T.P. Barnett, N.E. Graham, and J.M. Oberhuber. 1994. Interdecadal variability of the Pacific Ocean: model response to observed heat flux and wind stress anomalies. Clim. Dyn. 9: 287–302.

Moran, P.A.P. 1949. The statistical analysis of sunspot and lynx cycles. J. Anim. Ecol. 18: 115–116.

Murphy, C., and J.E. Dunn. 1977. Use of transfer function models to forecast sport fish harvest in Beaver Reservoir. Univ. Arkansas Stat. Lab. Tech. Rep. 6. 150 p.

Newton, H. J. 1988. TIMESLAB: a time series analysis laboratory. Wadsworth & Brooks/Cole Publishing Co., Pacific Grove, CA. 623 p.

Nitta, T., and S. Yamada. 1989. Recent warming of tropical sea surface temperature and its relationship to the northern hemisphere circulation. J. Meteorol. Soc. Jpn 67: 375–383.

Noakes, D. 1986. Quantifying changes in British Columbia Dungeness crab (*Cancer magister*) landings using intervention analysis. Can. J. Fish. Aquat. Sci. 43: 634–639.

Noakes, D.J., and A. Campbell. 1992. Use of geoduck clams to indicate changes in the marine environment of Ladysmith Harbour, British Columbia. Environmetrics 3(1): 81–97.

Noakes, D., D.W. Welch, and M. Stocker. 1987. A time series approach to stock-recruitment analysis: transfer function noise modelling. Nat. Resour. Modell. 2: 213–233.

Pacific Fishing. 1994. 1994 Yearbook. Vol XV, No. 3.

Pacific Salmon Commission. 1991. Second report on the parties' estimates of salmon interceptions. Rep. JIC (91)-1. Joint Interception Committee, Pacific Salmon Commission, Vancouver, B.C.

Parzen, E. 1981. Time series model identification and prediction variance horizon, p. 415–447. *In* D.F. Findley [ed.] Proceedings of the Second Applied Time Series Symposium, March 3–5, 1980, Tulsa OK. Academic Press, NY.

Pearcy, W.G. [ed.] 1984. The influence of ocean conditions on the production of salmonids in the north Pacific. Sea Grant Publ., Oregon State University, Corvallis, OR.

Pearcy, W.G. 1992. Ocean ecology of north Pacific salmonids. University of Washington Press, Seattle, WA. 179 p.

Pennington, M. 1985. Estimating the relative abundance of fish from a series of trawl surveys. Biometrics 41: 197–202.

Quinn, T.J., II, and R.P. Marshall. 1989. Time series analysis: quantifying variability and correlation in SE Alaska salmon catches and environmental data, p. 67–80. *In* R.J. Beamish and G.A. McFarlane [ed.] Effects of ocean variability on recruitment and an evaluation of parameters used in stock assessment models. Can. Spec. Publ. Fish. Aquat. Sci. 108.

Roden, G.I. 1991. Subarctic-subtropical transition zone of the north Pacific: large-scale aspects and mesoscale structure, p. 1–38. *In* J. Wetherall [ed.] Biology, oceanography, and fisheries of the north Pacific transition zone and subarctic frontal zone. NOAA Tech. Rep. NMFS 105.

Rogers, D.E. 1984. Trends in abundance of northeastern Pacific stocks of salmon, p. 100–127. *In* W.G. Pearcy [ed.] The influence of ocean conditions on the production of salmonids in the North Pacific. Oregon State University Press, Corvallis, OR.

Royer, T.C. 1989. Upper ocean temperature variability in the northeast Pacific Ocean: is it an indicator of global warming? J. Geophys. Res. 94: 18 175 – 18 183.

Saila, S.B., M. Wigbout, and R.J. Lermit. 1980. Comparison of some time series models for the analysis of fisheries data. J. Cons. Cons. int. Explor. Mer 39: 44–52.

Schwarz, G. 1978. Estimating the dimension of a model. Ann. Stat. 6: 461–464.

Seymour, R.J., R.R. Strange, D.R. Cayan, and R.A. Nathan. 1984. Influence of El Niños on California's wave climate, *In* Proceedings 19th International Conference on Coastal Engineering, American Society of Coastal Engineering, Houston, TX.

Shepard, M.P., C.D. Shepard, and A.W. Argue. 1985. Historic statistics of salmon production around the Pacific rim. Can. MS Rep. Fish. Aquat. Sci. 1819. 297 p.

Shepherd, J.G., J.G. Pope, and R.D. Cousens. 1984. Variations in fish stocks and hypotheses concerning their links to climate. Rapp. P.-v. Réun. Cons. int. Explor. Mer 36: 15–26.

SPSS, Inc. 1993. SPSS for Windows: Trends. 444 N. Michigan Ave., Chicago, IL. 60611

Stergiou, K.I. 1989. Modelling and forecasting the fishery for pilchard (*Sardina pilchardus*) in Greek waters using ARIMA time-series models. J. Cons. Cons. int. Explor. Mer 46: 16–23.
Stone, J.N., and Y. Cohen. 1990. Changes in species interactions of the Lake Superior fisheries system after the control of sea lamprey as indicated by time series models. Can. J. Fish. Aquat. Sci. 47: 251–261.
Trenberth, K.E. 1990. Recent observed interdecadal climate changes in the northern hemisphere. Bull. Am. Meteorol. Soc. 71: 988–993.
Trenberth, K.E., and J.W. Hurrell. 1994. Decadal atmosphere-ocean variations in the Pacific. Clim. Dyn. 9: 303–319.
Trenberth, K.E., and D.J. Shea. 1987. On the evolution of the Southern Oscillation. Mon. Weather. Rev. 115: 3078–3096.
Tsai, C.-F, and A.-L. Chai. 1992. Short-term forecasting of the striped bass (*Morone saxatilis*) commercial harvest in the Maryland portion of Chesapeake Bay. Fish. Res. 15: 67–82.
Tsay, R.S., and G.C. Tiao. 1984. Consistent estimates of autoregressive parameters and extended sample autocorrelation function for stationary and nonstationary ARMA models. J. Am. Stat. Assoc. 79: 84–96.
Venrick, E.L., J.A. McGowan, D.R. Cayan, and T.L. Hayward. 1987. Climate and chlorophyll a: long-term trends in the central north Pacific Ocean. Science (Wash., D.C.) 238: 70–72.
Ward, M.B. 1993. The physical dynamics of the North Pacific Ocean and its relation to salmon (*Oncorhynchus spp.*) production. M.S. Thesis, Univ. Washington. 109 p.
Ware, D.M., and G.A. McFarlane. 1989. Fisheries production domains in the northeast Pacific Ocean, p.359–379. *In* R.J. Beamish and G.A. McFarlane [ed.] Effects of ocean variability on recruitment and an evaluation of parameters used in stock assessment models. Can. Spec. Publ. Fish. Aquat. Sci. 108.
Wei, W.W.S. 1990. Time series analysis, univariate and multivariate methods. Addison-Wesley, Redwood City, CA. 478 p.
Woodward, W., and H.L. Gray. 1981. On the relationship between the S array and the Box-Jenkins method of model identification. J. Am. Stat. Assoc. 76: 579–587.
York, A.E. 1987. Northern fur seal, *Callorhinus ursinus*, eastern Pacific population (Pribilof Islands, Alaska, and San Miguel Island, California), p. 9–21. *In* J.P. Croxall and R.L. Gentry [ed.] Status, Biology and Ecology of Fur Seals. Proceedings of an International Symposium and Workshop, 23–27 April, 1984, Cambridge, UK, NOAA Tech. Rep. NMFS 51.

Appendix

The following time series model diagnostic and selection criteria were used.

Box–Pierce portmanteau test

The joint null hypothesis H_0: $\rho_1 = \rho_2 = \ldots = \rho_K = 0$ is tested with the statistic

$$Q = n(n+2)\sum_{k=1}^{k} \frac{1}{n-k}\,\hat{\rho}^2(k) \tag{A1}$$

The hypothesis of white noise is rejected if $Q > \chi^2_{\alpha,K-m}$, where K is the number of residuals calculated from the model and m is the number of estimated parameters.

Mean absolute error (MAE)

$$MAE = \frac{1}{K}\sum_{k=1}^{k} |a_k| \tag{A2}$$

Unbiased residual variance σ_a^2

$$\sigma_a^2 = \frac{RSS}{K-m} \tag{A3}$$

where RSS is the residual sum of squares and m is the number of estimated model parameters.

Coefficient of determination r^2

$$r^2 = 1 - \frac{RSS}{\sum (z - \bar{z})^2} \tag{A4}$$

where z represents the (possibly) transformed and differenced observed values.

Akaike's Information Criterion (AIC)

$$AIC = K \ln \tilde{\sigma}_a^2 + 2m \text{ where } \tilde{\sigma}^2_a = \frac{RSS}{K} \tag{A5}$$

where RSS is the residual sum of squares, K is the number of residuals, m is the number of estimated parameters, and σ_a^2 is the biased residual variance.

Schwarz Bayesian Criterion (SBC)

$$\text{SBC} = K \ln \tilde{\sigma}_a^2 + m \ln K \tag{A6}$$

where the parameters have the same interpretation as for the AIC.

Decadal-scale variations in the eastern subarctic Pacific: II. Response of Northeast Pacific fish stocks

Anne Babcock Hollowed and Warren S. Wooster

Hollowed, A.B. and W.S. Wooster. 1995. Decadal-scale variations in the eastern subarctic Pacific: II. Response of Northeast Pacific fish stocks, p. 373–385. *In* R.J. Beamish [ed.] Climate change and northern fish populations. Can. J. Fish. Aquat. Sci. 121.

Abstract: Patterns of variation in recruitment of marine fish stocks in the eastern subarctic Pacific Ocean were compared with winter environmental conditions for the period 1945–89. Strong year-classes were more common among stocks off the U.S. west coast than farther north in the Gulf of Alaska and Bering Sea. After 1977, the frequency of strong year-classes among several stocks off the U.S. west coast decreased. Among groundfish stocks, 4 yr of synchronous strong year-classes were identified: 1961, 1970, 1977, and 1984. The identity of most stocks participating in this synchrony differed with time. Many stocks had strong year-classes during warm years that were associated with El Niño – Southern Oscillation events, periods of intense Aleutian Lows, enhanced circulation in the Gulf of Alaska, and weakened coastal upwelling farther south. Time series of recruitment were examined for changes in the frequency of strong year-classes or average production during different temperature eras (warm or cool). This analysis showed that 43% of the stocks had more frequent strong year-classes and 65% of the stocks had higher average recruitment during a particular type of ocean tempearture era (warm or cool).

Résumé : Les tendances dans les variations du recrutement des stocks de poissons marins dans le secteur subarctique est du Pacifique ont été comparées aux conditions environnementales d'hiver pour la période 1945–1989. Les classes annuelles fortes étaient plus communes parmi les stocks au large de la côte ouest américaine que plus au nord, dans le golfe d'Alaska et dans la mer de Béring. Après 1977, la fréquence des classes annuelles fortes de plusieurs stocks au large de la côte ouest américaine a diminué. Parmi les poissons de fond, 4 années de classes annuelles fortes synchrones ont été relevées : 1961, 1970, 1977 et 1984. L'identité de la plupart des stocks qui contribuaient à cette synchronie variait avec le temps. De nombreux stocks avaient des classes annuelles fortes durant les années chaudes qui étaient associées aux événements El Niño – Oscillation australe, à des périodes d'intenses dépressions des Aléoutiennes, à une circulation améliorée dans le golfe d'Alaska et à un affaiblissement de la remontée côtière plus au sud. Les séries chronologiques sur le recrutement ont été examinées pour que l'on détermine s'il y avait eu des changements dans la fréquence des classes annuelles fortes ou de la production moyenne durant les quatre ères différentes de températures (chaudes ou fraîches). Cette analyse a montré que 43 % des stocks avaient eu des classes annuelles fortes plus fréquentes et que 65 % avaient eu un recrutement moyen plus élevé durant un type particulier d'ère de température océanique (chaude ou fraîche).

Introduction

Several studies have shown recruitment of many marine fish populations appears to be influenced by ocean temperature conditions. Temperature change may not be the proximal cause of shifts in recruitment success, but it often serves as a proxy for changes in circulation. Much of the evidence linking long-term patterns of recruitment and ocean temperature conditions comes from analyses in the North Atlantic and Barents Sea. For example, the rise of gadoid stocks in the North Sea in the 1960s occurred during a period of ocean cooling (Cushing 1982). Farther north, in the Barents Sea, Saetersdal and Loeng (1987) found that strong year-classes of cod generally occurred during the beginning of warm ocean conditions.

In the eastern North Pacific, several studies show temperature, or processes for which temperature is a proxy, may play an important role in determining year-class success (For a summary see Hollowed 1992; Bakun and Parrish 1980). We have shown that among 15 groundfish stocks, more than half had strong year-classes in 1961, 1970, and 1977 (Hollowed and Wooster 1992). Each of these 3 yr occurred during

A.B. Hollowed. National Marine Fisheries Service, Alaska Fisheries Science Center, Bin C15700, 7600 Sand Point Way NE, Seattle, WA 98115-0070, USA.

W.S. Wooster. School of Marine Affairs HF-05, University of Washington, Seattle, WA 98195, USA.

wintertime warm eras which tended to last for 6–12 yr and alternated with cool eras of about the same duration. The warm eras were commonly associated with El Niño – Southern Oscillation (ENSO) events and were characterized by intense Aleutian Lows, strong ocean circulation in the Gulf of Alaska, and weak upwelling farther south (Wooster and Hollowed 1995).

The most recent shift from cool to warm ocean conditions occurred in the late 1970s and is of particular interest because of the evidence that both pelagic and demersal fish stocks as well as other species were influenced by the shift. Catch statistics on high-latitude "northern" and low-latitude "southern" salmon populations show different trends. Prior to 1977, coho salmon populations (*Oncorhynchus kisutch*) in the south prospered while pink salmon populations (*Oncorhynchus gorbuscha*) in the north remained at relatively low levels. This pattern reversed after 1976 (Francis and Sibley 1991).

Pacific sardine (*Sardinops sagax*) and Pacific mackerel (*Scomber japonicus*) declined rapidly in the 1940s and remained low until the late 1970s. In the early 1980s, Pacific mackerel showed a strong recovery (MacCall et al. 1985). The California stock of Pacific sardine showed a modest increase during the 1980s (Barnes et al. 1992; Lluch-Belda et al. 1992). On a larger scale, there is evidence that in a similar period sardine populations have increased throughout the Pacific (Kawasaki 1991). Also in the early 1980s, the abundance of Pacific cod (*Gadus macrocephalus*) and walleye pollock (*Theragra chalcogramma*) in the Bering Sea increased dramatically while crab populations declined (Stevens et al. 1992; Thompson and Zenger 1991; Wespestad and Dawson 1991). Similarly, in the Gulf of Alaska, pandalid shrimp populations declined and walleye pollock increased during the 1980s (Albers and Anderson 1985; Hollowed et al. 1991).

These recent changes in fish and shellfish populations may also have influenced marine mammal populations. Between 1970 and 1990, the abundance of pinnipeds found off the coasts of Washington, Oregon, and California, (northern elephant seal (*Mirounga angustirostris*) and California sea lion (*Zalophus californianus*)), increased dramatically while northern populations (northern sea lions (*Eumetopias jubatus*), harbor seal (*Phoca vitulina*), and northern fur seal (*Callorhinus ursinus*)) declined (Lowry 1990; MacCall 1986; Merrick et al. 1987).

Such observations suggest that on decadal or longer time scales, specific environmental conditions favor some species thereby leading to shifts in the mix of species present in a given region. In this paper we explore the possibility that environmental forcing influences marine fish populations through changes in recruitment success. We examined recruitment of individual fish stocks in a region to determine the frequency of strong recruitment and the average production of recruits during specfiic environmental conditions.

Table 1. Summary of quantitative data sources.

Stock	Scientific name	Spawning location	Source
Walleye pollock	*Theragra chalcogramma*	BS	Wespestad and Dawson 1991
Yellowfin sole	*Pleuronectes asper*	BS	Wilderbuer 1992
Pacific ocean perch	*Sebastes alutus*	AI	Ianelli and Ito 1992
Atka mackerel	*Pleurogrammus monopterygius*	AI	Lowe 1992
Walleye pollock	*Theragra chalcogramma*	GOA	Hollowed et al. 1991
Pacific halibut	*Hippoglossus stenolepis*	CAN–GOA	P. Sullivan (personal communication)[a]
Sablefish	*Anoplopoma fimbria*	GOA	M. Sigler (personal communication)[b]
Pacific ocean perch	*Sebastes alutus*	GOA	Heifetz and Clausen 1992
Rock sole	*Lepidopsetta bilineata*	CAN–HS	Fargo 1989
Pacific cod	*Gadus macrocephalus*	CAN–3cd	Tyler and Foucher 1990
Pacific cod	*Gadus macrocephalus*	CAN–5cd	Tyler and Foucher 1990
Pacific ocean perch	*Sebastes alutus*	WA–OR	Ianelli et al. 1992
English sole	*Parophrys vetulus*	WA–OR	Golden et al. 1986
Bocaccio Rockfish	*Sebastes paucispinis*	CA–WA	Bence and Rogers 1992
Pacific hake	*Merluccius productus*	CA	Dorn et al. 1991
Widow rockfish	*Sebastes entomelas*	CA–WA	Hightower and Lenarz 1990
Dover sole	*Microstomus pacificus*	WA–OR	Turnock and Methot 1992
Pacific herring	*Clupea pallasi*	BS	Wespestad 1991
Pacific herring	*Clupea pallasi*	GOA	Collie 1991
Pacific herring	*Clupea pallasi*	CAN–CEN	Haist and Schweigert 1991
Pacific herring	*Clupea pallasi*	CAN–PR	Haist and Schweigert 1991
Pacific herring	*Clupea pallasi*	CAN–QC	Haist and Schweigert 1991
Pacific herring	*Clupea pallasi*	CAN–VIN	Haist and Schweigert 1991
Northern anchovy	*Engraulis mordax*	CA	Methot 1989

Abbreviations: AI = Aleutian Islands, BS = Bering Sea, CA = California; WA–OR = Washington–Oregon; CAN = Canada; CAN–HS = Hecate Strait; CAN–CEN = Central Canada; CAN–PR = Prince Rupert, Canada; CAN–WVI = North Vancouver Island, Canada; CAN–QC = Queen Charlotte Islands, Canada; GOA = Gulf of Alaska.
[a]International Pacific Halibut Commission, Seattle, WA.
[b]Alaska Fisheries Science Center, Auke Bay Laboratory, Auke Bay, AK.

Data sources

Recruitment time series were obtained for six Pacific herring (*Clupea pallasi*) stocks, northern anchovy (*Engraulis mordax*), and 16 groundfish stocks (Table 1). These stocks were located in four regions: Bering Sea, Gulf of Alaska, Canadian west coast, and U.S. west coast. In the North Pacific, these regions correspond to the downwelling (Gulf of Alaska), transitional (off southeastern Alaska and British Columbia) and upwelling (off the U.S. west coast) domains defined by Ware and McFarlane (1989).

Temperature eras defined by Wooster and Hollowed (1994) were used in this analysis. Using Comprehensive Ocean Atmosphere Data Set (COADS) sea-surface temperature data from 2 × 2° blocks located along the coast of the eastern the north Pacific, we found three types of ocean conditions (warm, cool, and mixed) in six eras over 19 years: 1946–52, cool; 1953–57, mixed; 1958–64, warm; 1965–70, mixed (1970 warm); 1971–76, cool; 1977–84, warm (1979 cool).

Methods

The available recruitment time series differed in length (Table 2). If stocks exhibited more frequent strong year classes during a time period associated with a temperature era, one would expect that the strongest variability to occur during a period that includes both a warm and a cool era. In order to facilitate between stock comparisons, a common period (1972–81) was used for calculating averages and standard deviations. This period bracketed the shift from cool to warm conditions and maximized the number of stocks available for comparison. Because most stocks exhibited occasional instances of extremely large year-classes, frequency histograms of the distribution of recruitment were based on log-transformed data.

The importance of selecting a common time periods for comparison of recruitment patterns is demonstrated using three stocks: Pacific hake (*Merluccius productus*), Pacific halibut (*Hippoglossus stenolepis*), and Pacific cod 3c, d (*Gadus macrocephalus*). These three stocks exhibited different recruitment patterns. The coefficients of variation for Pacific hake, Pacific cod, and Pacific halibut were 1.497, 0.611, and 0.366 respectively. Three time periods were considered: 1972–81 (cool – warm), 1964–76 (mixed – cool), and the long-term mean. In all three stocks the standard scores based on the 1972–81 base period were closer to the long-term mean for the above average year-classes (Figs. 1a–c).

Year-classes were classified into five categories based on the difference in standard scores of their log-transformed recruitment from the mean. The five categories were (1) ≤mean; (2) ≥mean; (3) ≥0.5 above mean; (4) ≥1.0 above

Table 2. Summary of strong year-classes by region and time period.

Stock	Area	Years	All Years		Before 1977		Since 1977	
			No. of years	%3–5[a]	No. of years	%3–5	No. of years	%3–5
Pollock	BS	1964–87	24	29	13	31	11	27
Yellowfin sole	BS	1964–83	20	25	13	31	7	14
Pacific ocean perch	AI	1950-89	39	26	27	26	12	25
Atka mackerel	AI	1972–89	18	39	5	20	13	46
Pollock	GOA	1962–89	28	14	15	7	13	23
Halibut	GOA	1950–81	34	32	32	31	4	40
Sablefish	GOA	1969–87	19	21	8	13	11	27
Rock sole	CAN	1950–81	21	66	21	78	5	0
Cod 3c,d	CAN	1954–86	33	42	23	57	10	10
Cod 5c,d	CAN	1959–86	28	68	18	67	10	40
English sole	WC[b]	1964–80	17	65	13	85	4	0
Bocaccio	WC	1970–81	12	50	7	71	5	20
Hake	WC	1956–87	32	31	21	33	11	27
Widow	WC	1970–81	12	50	7	28	5	80
Dover sole A[c]	WC	1961–87	27	19	16	25	11	9
Dover sole B[c]	WC	1966–87	22	59	11	63	11	54
Northern anchovy	WC	1964–89	26	23	13	23	13	23
Herring	BS	1957–85	29	28	20	25	9	33
Herring	GOA	1968–86	19	32	9	11	10	50
Her/CEN	CAN	1948–88	41	37	29	41	12	25
Her/PR	CAN	1948–88	41	41	29	38	12	50
Her/QC	CAN	1948–88	41	20	29	17	12	25
Her/VIN	CAN	1948–88	41	44	29	52	12	25

NOTE: See Table 1 for abbreviations.
[a]Percentage of year-classes in categories 3–5. See Methods for category designations.
[b]West Coast.
[c] Dover sole A = Columbia stock; Dover sole B = Eureka stock.

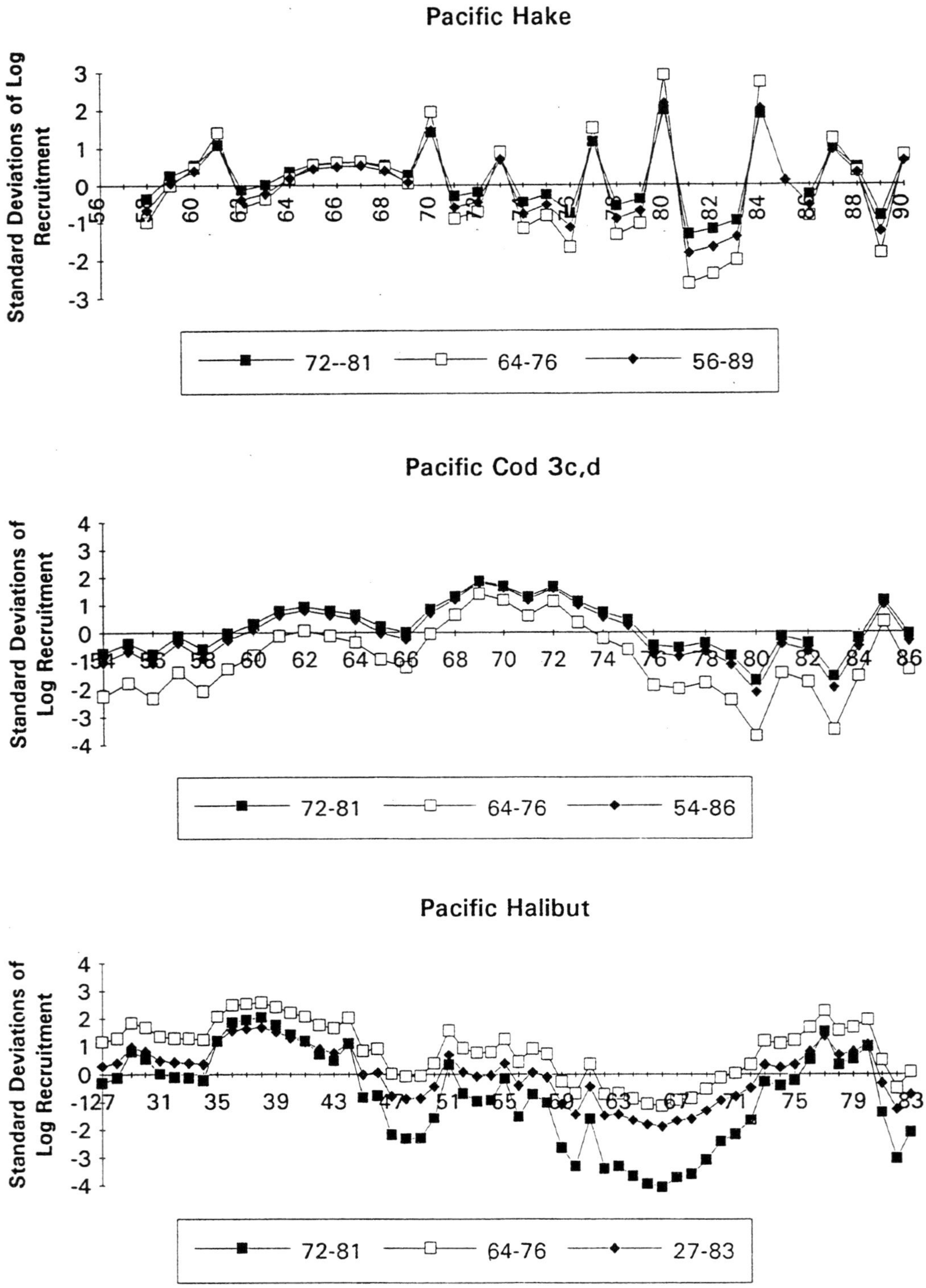

Fig. 1. Standard scores [(Obs. – $AVG_{(base\ period)}$)/$STD_{(base\ period)}$] of log transformed recruitment based on three different base time periods 1972–81, 1964–76, and the entire time series.

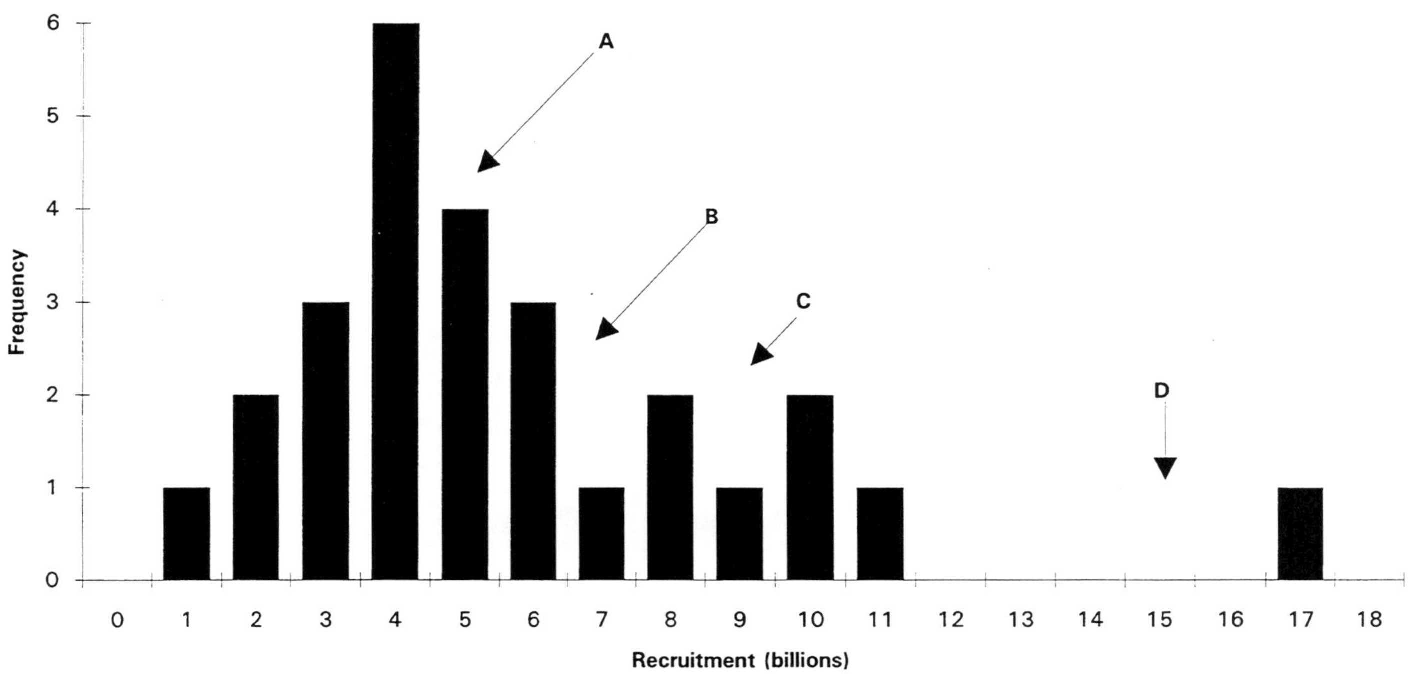

Fig. 2. Example of classification scheme used to describe year-class strength. The frequency histogram of Bering Sea pollock recruitment represents the number of observations that fell within the recruitment bins. The break points of the classification are identified as: (A) the antilog of the log transformed mean, (B) the antilog of 0.5 standard deviations (SD) above the log transformed mean, (C) the antilog of 1 SD above the log transformed mean, and (D) the antilog of 2 SD above the log transformed mean.

mean; (5) ≥2.0 above mean. Strong year-classes were considered to be those in categories 3–5. This approach is illustrated with Bering Sea walleye pollock data (Fig. 2).

We considered alternative classification schemes that incorporated spawner biomass, such as monitoring recruits per spawner or residuals from a functional spawner recruit relationship. These two techniques required strong assumptions regarding the role of spawners in determining future recruitment success and were not considered further.

Recruitment data categorized using the five point classification scheme were used to re-examine the occurrence of years of synchronous strong year-classes of groundfish according to the following criteria: (1) at least half of the stocks in a given year had recruitment in category 3 or above; and (2) of those year-classes, at least half had recruitment in category 4 or 5. These criteria were less satisfactory when data from few stocks was available (i.e., before 1960 and after 1987) than during the intervening years.

The temperature preference of stocks was determined by counting the occurrence of strong year-classes during four temperature eras: (1) 1958–64 (warm); (2) 1965–69 (mixed); (3) 1971–76 (cool); (4) 1977–84 (warm). For this analysis the isolated cool (1979) and warm (1970) years were included in temperature eras (3) and (4), respectively.

Temperature may influence production of marine fish stocks by increasing the frequency of occurrence of strong year-classes, or by increasing the magnitude of strong year-classes, or both. We examined the probability of observing strong year-classes in different temperature eras and the average production of strong year-classes during different temperature eras. Average production was measured as the average of the standard scores based on 1972–81 base period observed during a given temperature condition (warm, cool, or mixed). Temperature preferences were defined as periods when differences in the probability of strong year-classes differed by at least 0.10 between eras or when the average standard score from the base mean differed by at least 0.25 between eras.

Results

Frequency of strong year-classes

Recruitment success relative to average for the years 1972–81 differed significantly among stocks and regions (Fig. 3). A large number of very strong year-classes were observed in the Gulf of Alaska and west coast Pacific ocean perch (*Sebastes alutus*) stocks in the 1950s and 1960s (Fig. 3). Both of these stocks declined rapidly in the late 1960s and did not recover (Ianelli et al. 1992; Heifetz and Clausen 1992). Historical recruitment of these two stocks was substantially higher than that observed during the 1972–81 time period. This may be a case where the stock biomass was so low during the base time period that spawner biomass may have influenced recruitment. Because our goal was to evaluate a climate-driven response, stocks that exhibit a strong stock biomass driven

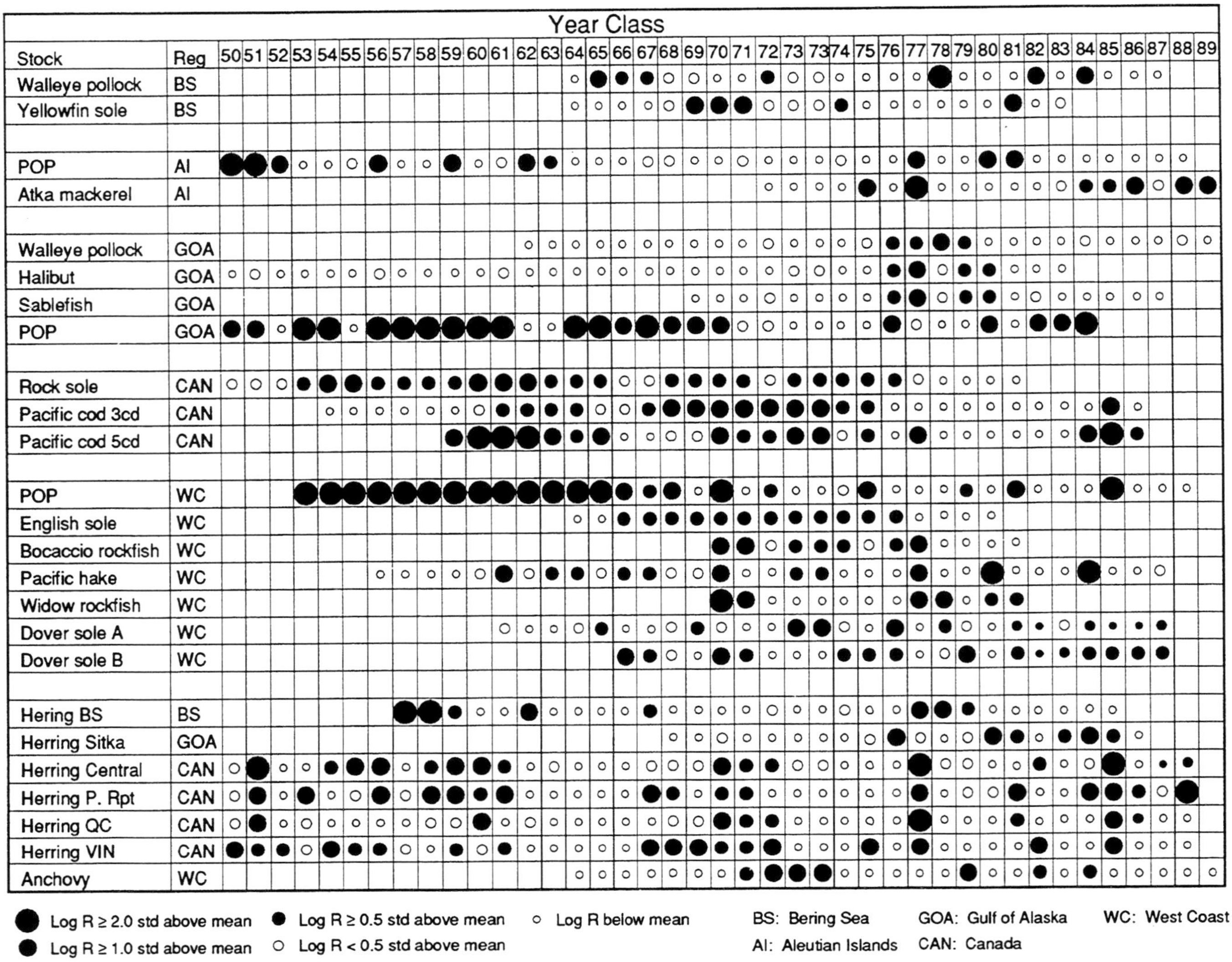

Fig. 3. Summary of year-class strength for Northeast Pacific marine fishes.

response should not be included in our analysis. Therefore, the Gulf of Alaska and west coast Pacific ocean perch stocks were not included in further analyses.

In the northernmost regions (Bering Sea, Aleutian Islands, and Gulf of Alaska), strong year-classes were infrequent, less than 30% of the time, for most stocks (Table 2). Most Canadian and U.S. west coast stocks, on the other hand, had relatively frequent strong year-classes with the notable exceptions of Pacific cod (3c,d), Pacific hake, and the Columbia River Dover sole (*Microstomus pacificus*) stock (Dover sole A). The most abundant stocks, Pacific hake, the Bering Sea stock of Pacific herring, walleye pollock, and northern anchovy, seldom had strong year-classes.

None of the herring stocks exhibited frequent strong year-classes. However, strong year-classes were somewhat less frequent in the northernmost stocks (Bering Sea, Gulf of Alaska, and Queen Charlotte) than they were farther south.

There was a noteworthy change in the frequency of strong year-classes after 1977. Of the seven stocks that exhibited strong year-classes 50% or more of the time before 1977, only the Eureka stock of Dover sole (Dover sole B) did so after 1977. On the other hand, two stocks (widow rockfish (*Sebastes entomelas*) and herring in the Gulf of Alaska and off Prince Rupert) had frequent strong year-classes only after 1977.

Synchronous strong year-classes

During the period when sufficient stocks were available for the comparison (1959–87), 7 years of synchronous strong year-classes among groundfish were identified: 1959, 1961, 1962, 1970, 1971, 1977, and 1984 (Table 3). This supports the earlier findings (Hollowed and Wooster 1992) and extends these key years to include 1984.

The years of synchronous strong year-classes did not always contain the same groundfish stocks. However, two stocks, Pacific cod 5c,d and Pacific hake, had year-classes more than 1 SD above the base average in each of 4 key years (1961, 1970, 1977, 1984) (Fig. 3). Records for widow rockfish and bocaccio rockfish (*Sebastes paucispinis*) did not include all key years; however, these stocks had year-classes more than 1 SD above the base average in both 1970 and 1977 (Fig. 3). Pacific cod 3c,d had strong year-classes only in the key years before 1977, others only in 1977 and/or 1984 (e.g.,

Bering Sea walleye pollock, Aleutian Islands Pacific ocean perch, Gulf of Alaska halibut, and Gulf of Alaska sablefish (*Anoplopoma fimbria*)) (Fig. 3).

The recruitment success of herring stocks showed some similarities to that of groundfish stocks. All Canadian herring stocks had strong year-classes in 1970, 1971, 1977, and 1985 (Fig. 3). Bering Sea herring year-classes were more than 1 SD above the base average in 1957, 1958, 1962, 1977, and 1978 (Fig. 3).

Ocean temperature preferences

Synchronous strong year-classes usually occurred during warm ocean conditions, often in connection with an ENSO event. The first period (1959–62) occurred during the warm era that followed the major 1957–58 event. The year 1961 was warmer than 1960 or 1962, and while no event for that year was included in the catalog of Quinn et al. (1987), the Southern Oscillation Index exhibited a major negative anomaly (Peixoto and Oort 1992, Fig. 16.9). During the second period (1970–71), 1970 was unusually warm and followed the weak 1969 event (Quinn et al. 1978). A warm year occurred in 1977 and followed the moderate 1976 event. The latest period (1984–85) followed the very strong event of 1982–83. Ocean conditions were warm throughout the region in 1984 and north of 54°N in 1985.

The association of synchronous strong year-classes with ENSO events can perhaps be extended back to 1951 when seven of the eight stocks available for analysis exhibited strong year-classes (Fig. 3). The Quinn et al. (1987) catalog shows a "near-moderate" event in that year.

Table 3. Summary of synchronous strong year-classes of groundfish stocks excluding Pacific ocean perch stocks from the Gulf of Alaska and West Coast. Categories 3–5 are as defined in Methods.

Year	No. of stocks	No. in categories 3–5	Percent in categories 3–5	No. in categories 4 and 5	Percentage 4 and 5's of 3–5's
1959	6	3	50	2	67
1960	6	2	33	2	100
1961	7	4	57	3	75
1962	8	4	50	3	75
1963	8	5	63	1	20
1964	11	4	36	0	0
1965	11	4	36	2	50
1966	12	4	33	1	25
1967	12	5	42	0	0
1968	12	3	25	1	33
1969	13	5	38	2	40
1970	15	9	60	7	78
1971	15	8	53	4	50
1972	16	4	25	1	25
1973	16	7	44	3	43
1974	16	6	3	0	0
1975	16	6	38	1	17
1976	1	8	50	1	13
1977	16	9	56		89
1978	16	4	25	3	75
1979	16	3	19	1	33
1980	16	5	31	2	40
1981	15	4	25	2	50
1982	11	1	9	1	100
1983	1	0	0	0	0
1984	10	5	50	3	60
1985	10	4	40	2	50
1986	10	3	30	1	33
1987	8	1	13	0	0
Mean			36		

Table 4. Probability of strong year-classes in different temperature eras. Values in brackets indicate the total number of years in the era. Values in columns indicate the total number of strong year-classes (categories 3–5) observed during the era; values in parentheses indicate the number of year-classes in categories 4 or 5.

Stock	Region	1958–64[7] Warm	1965–69[5] Mixed	1971–76+79[7] Cool	1977–84+70[8] Warm	temperature preference
Pollock	BS	—	3(1) 0.60	1(0) 0.14	3(3) 0.38	WM (not C)
Yellowfin sole	BS	—	1(1) 0.20	2(1) 0.29	2(2) 0.29[a]	?
Pacific ocean perch	AI	3(2) 0.43	0(0) 0.00	0(0) 0.00	3(3) 0.38	W
Atka mackerel	AI	—	—	1(1) 0.17[b]	2(1) 0.25	?
Pollock	GOA	—	0(0) 0.00	2(0) 0.29	3(1) 0.38	?
Pacific halibut	GOA	1(0) 0.14	0(0) 0.00	1(0) 0.14	2(1) 0.29[c]	?
Sablefish	GOA	—	—	2(0) 0.29	3(1) 0.38	?
Rock sole	CAN	7(3) 1.00	3(0) 0.60	5(0) 0.71	1(0) 0.20[d]	?
Pacific cod 3c,d	CAN	4(0) 0.57	3(2) 0.60	5(3) 0.71	1(1) 0.13	?
Pacific cod 5c,d	CAN	6(5) 1.00[e]	1(1) 0.20	4(1) 0.57	(3) 0.38	?
English sole	WC	—	4(0) 0.80	6(0) 0.86	—	?
Bocaccio rockfish	WC	—	—	4(1) 0.57	2(2) 0.25	C
Pacific hake	WC	3(1) 0.43	2(0) 0.40	1(0) 0.14	4(4) 0.50	WM (not C)
Widow rockfish	WC	—	—	1(1) 0.14	5(3) 1.00[f]	W
Dover sole A	WC	0(0) 0.00[g]	2(0) 0.40	2(2) 0.29	1(0) 0.13	CM (not W)
Dover sole B	WC	—	2(1) 0.40[h]	5(1) 0.71	3(2) 0.39	C
Herring	BS	3(2) 0.43	1(0) 0.20	1(0) 0.14	2(2) 0.25	?
Herring	GOA	—	—	1(1) 0.14	4(2) 0.50	W
Herring Central	CAN	4(3) 0.57	0(0) 0.00	3(1) 0.43	3(2) 0.38	?
Herring PR	CAN	4(3) 0.57	2(1) 0.40	1(0) 0.14	4(3) 0.50	W (not C)
Herring QC	CAN	1(1) 0.14	0(0) 0.00	3(1) 0.43	3(2) 0.38	?
Herring VIN	CAN	2(0) 0.29	3(3) 0.60	3(2) 0.43	3(2) 0.38	?
Northern anchovy	WC	—	0(0) 0.00	4(3) 0.57	(1) 0.38	C

NOTE: See Tables 1 and 2 for codes. Temperature codes: W = warm; WM = warm or mixed; C = cool; CM = cool or mixed; ? = no preference.
[a]Based on 7 years of data.
[b]Based on 6 years of data.
[c]Based on 7 years of data.
[d]Based on 5 years of data.
[d]Based on 5 years of data.
[e]Based on 6 years of data.
[f]Based on 5 years of data.
[f]Based on 4 years of data.
[g]Based on 4 years of data.
[h]Based on 4 years of data.

The temperature preferences[1] of individual stocks are shown in Tables 4 and 5. Of 23 fish stocks, 10 stocks (43%) showed clear temperature preferences based on the frequency of strong year-classes and the average production during an era. Fifteen of the 23 stocks (65%) exhibited temperature preferences based on the average production during an era. Clear temperature preferences were observed for several species that achieved large historical biomass levels: Bering Sea walleye pollock, Pacific ocean perch AI, and Pacific hake, and northern anchovy.

Over one half (9 of 16) of the groundfish stocks did not show clear differences in the probability of strong year-classes during temperature eras (Table 4). Among the remaining seven stocks, recruitment of four stocks was significantly stronger during warm or warm–mixed conditions (two stocks did particularly poorly during cool periods). Three stocks appeared to prefer cool or cool–mixed eras; these were Bocaccio rockfish, and Dover sole A, and B.

Most (11 of 16) of the groundfish stocks exhibited temperature preferences based on the average production during warm or cool temperature eras (Table 5). The four stocks that did not reveal a clear temperature preference were located in the Gulf of Alaska or Canadian regions. Six stocks showed a warm or warm–mixed temperature preference and six stocks showed a cool or cool–mixed preference.

Herring stocks from the Gulf of Alaska, and Prince Rupert preferred warm conditions and did poorly under cool (Tables 4 and 5). The average recruitment of the Vancouver Island herring stock was higher during cool ocean condtions (Table 5). Other Canadian herring stocks showed no clear preferences.

[1] The term preference is used here to indicate the probability of a strong year-class, or the average recruitment is higher during a particular temperature era when compared to alternative temperature conditions.

Table 5. Average standard score from mean recruitment (1972–1981) during different temperature eras. Values in brackets indicate the total number of years in the era.

Stock	Region	1958–64[7] Warm	1965–69[5] Mixed	1971–76+79[7] Cool	1977–84+70[8] Warm	Preference
Pollock	BS	—	0.36	−0.35	0.20	WM (not C)
Yellowfin sole	BS	—	−0.21	0.39	−0.05[a]	C (not W)
Pacific ocean perch	AI	0.46	−0.45	−0.55	0.16	W
Atka mackerel	AI	—	—	−0.20[b]	0.11	W
Pollock	GOA	—	−0.92	−0.24	−0.26	(not M)
Pacific halibut	GOA	−2.26	−2.86	−0.54	−0.68[c]	?
Sablefish	GOA	—	—	−0.33	−0.09	?
Rock sole	CAN	1.71	−0.55	−0.20	−1.98[d]	?
Pacific cod 3c,d	CAN	0.19	0.89	0.56	−0.29	CM
Pacific cod 5c,d	CAN	2.95[e]	−0.40	−0.14	−0.31	?
English sole	WC	—	0.81	0.56	—	M
Bocaccio rockfish	WC	—	—	0.34	−0.31	C
Pacific hake	WC	0.02	−0.08	−0.38	0.65	WM (not C)
Widow rockfish	WC	—	—	−0.15	3.31[f]	W
Dover sole A	WC	−0.44[g]	0.03	0.10	−0.39	CM (not C)
Dover sole B	WC	—	0.35[h]	0.44	−0.15	CM (not W)
Herring	BS	0.291	−1.083	−0.498	−0.757	?
Herring	GOA	—	—	−0.417	0.495	W
Herring central	CAN	0.632	−0.388	0.059	0.226	?
Herring PR	CAN	0.209	−0.115	−0.430	0.649	W
Herring QC	CAN	−0.333	−0.634	−0.076	0.063	?
Herring VIN	CAN	0.038	0.447	0.282	0.014	CM
Northern anchovy	WC	—	−0.868	0.347	0.025	C

NOTE: See Tables 1 and 2 for codes.
[a]Based on 7 years of data.
[b]Based on 6 years of data.
[c]Based on 7 years of data.
[d]Based on 5 years of data.
[d]Based on 5 years of data.
[e]Based on 6 years of data.
[f]Based on 5 years of data.
[f]Based on 4 years of data.
[g]Based on 4 years of data.
[h]Based on 4 years of data.
Temperature Codes: W = warm; WM = warm or mixed; C = cool; CM = cool or mixed; M = mixed; ? = no preference

Twelve stocks had time series long enough to evaluate stock responses to two warm periods. Similar shifts in the frequency of occurrence of strong year-classes in response to warm ocean conditions were observed in only five of the stocks (42%) (Table 4). Changes in the average productivity observed during warm ocean conditions were similar in seven of these 12 stocks (58%).

Discussion

In this study, we compared the patterns of variation in recruitment of marine fish stocks of the northeast Pacific Ocean with patterns in winter environmental conditions. Our findings, can be summarized as follows:

1) Strong year-classes are more frequent among fish stocks on the U.S. west coast than farther north.
2) The frequency of occurrence of strong year-classes appears to change over long periods.
3) Groundfish stocks participating in years of synchronous strong year-classes differ with time.
4) Years of synchronous strong year-classes and of strong year-classes in many individual stocks tend to occur during warm years and eras that are associated with ENSO events and periods of intense Aleutian Lows.
5) The probability of a strong year-class occurring and the average production changes in some stocks during climatic eras defined by winter sea surface temperature.

The difference in frequency of strong year-classes among northern stocks and among those farther south may reflect regional differences in predictability or stability of the environmental factors necessary for survival of prerecruits. Thus, the coincidence of the variety of conditions required to produce strong year-classes may be a rare event in the Bering Sea and the Gulf of Alaska as compared with the coastal upwelling regime farther south where the species present may be better adapted for the kind of environmental variability that they encounter.

Changes in the frequency of occurrence of strong year-classes may be influenced by the particular time period selected to standardize the recruitment times series. Most stocks that exhibited frequent strong year-classes, did so prior to 1977, but not after 1977. This suggests that the mean recruitment and the variability of recruitment shifted after 1977.

Several years of synchronous strong year-classes were identified in this analysis. Many of the recruitment times series were too short to permit careful examination of the changes in stocks participating in the key years of strong recruitment. At least two stocks, and perhaps as many as five appear to have had strong year-classes in most or all of the key years, while some others did so only in 1977 or later. Other stocks only prospered in one or two of the key years. These findings indicate that while warm years may be more likely to produce strong year-classes in marine fish stocks, these years do not consistently produce strong year-classes in marine fishes. Thus ocean conditions associated with warm temperature may be a necessary precondition for a strong year-class but they are not a sufficient condition necessary for production of strong year-classes.

Five (33%) of the stocks that exhibited temperature preferences based on average production did not show temperature preferences based on the probability of having a strong year-class. This may indicate that, for these stocks, ocean temperature did not increase the probability of several key events necessary to produce a strong year-class occurring, but it enhanced strong year-classes when they occurred.

Ten stocks (43%) exhibited temperature preferences based on both the probability of strong year-classes occurring, and the average production during the era. If this pattern persisted through time, long-term oscillations in the abundance of these stocks would be expected. This finding suggests that the abundance trends of some Northeast Pacific fish populations may be partially controlled by oscillations in ocean conditions. Ware (1992) made a similar conclusion in an analysis of Canadian herring stocks.

As mentioned in our introductory statements, several marine species exhibited marked shifts in abundance that coincided with the shift from warm to cool ocean conditions. One might ask whether these shifts are associated with decadal-scale changes in ocean conditions or whether these shifts represented processes that oscillate on a longer time period (i.e., regime shifts). Twelve of the recruitment time series examined in our study were long enough to evaluate two warm periods. Only five of 12 stocks (42%) showed common changes in the probability of occurrence of strong year-classes and seven of 12 (58%) showed similar changes in the average production during warm winter ocean conditions. Thus, definitive statements regarding the response of marine fish stocks to decadal-scale variations in winter ocean conditions may be premature.

Why do warm conditions favor synchrony in strong year-classes among groundfish stocks as well as strong year-classes in many individual stocks? These conditions are commonly associated with intensified winter Aleutian Lows, strong ocean circulation in the Gulf of Alaska, and weakened upwelling along the U.S. west coast. This "type-B circulation" (Hollowed and Wooster 1992; Wooster and Hollowed 1995) usually arises in association with ENSO events in the tropical Pacific, although some events have little effect in the subarctic Pacific Ocean, and the intensity of extratropical effects is poorly correlated with that of the tropical events.

Possible influences of these conditions on the success of recruitment include advective processes, direct effects on growth and physiology, and shifts in production. For species in the southern portion of the coastal upwelling domain, offshore transport of larvae spawned during the winter should be limited (Bailey 1981; Parrish et al. 1981; Johnson et al. 1986; Power 1986). Alongshore transport or persistent landward tranport of near-surface waters may be critical to species that spawn off the Oregon and Washington coast (McConnaughey et al. 1992). Enhanced onshore transport of eggs and larvae in the Gulf of Alaska may also be important. Ingraham et al. (1991) noted that large-scale circulation in the Alaska Gyre could influence bottom water properties, which may influence the spawning location as well as the transport of pollock eggs in Shelikof Strait. Parker (1989) found that strong wintertime wind conditions favored the production of large year-classes of Pacific halibut. In the transitional domain, wind-induced transport appears to play an important role on year-class success of Pacific cod (Tyler and Crawford 1991).

Ocean temperature itself may play an important role in year-class success. The role of temperature on rates of development, mortality, and survival of larval and post-larval fish has been examined by Pepin (1991) and Morse (1989). Pepin suggests that increased growth due to elevated temperatures may result in increased mortality because of the high ingestion rates required. However, warm years with elevated prey densities could result in increased survival. The importance of matching peak periods of larval abundance with the availability of suitable prey has been shown by Cushing (1969) and Lasker (1975, 1978).

There is evidence of enhanced production in the Alaska Gyre at lower trophic levels during periods of warm ocean conditions. McFarlane and Beamish (1992) noted that in the most recent warm period 1976–80 the mean concentration of copepods at Ocean Station P during March – May increased markedly. Brodeur and Ware (1992) found a positive correlation between the intensity of winter winds and subsequent summer zooplankton biomass in the subarctic gyre. In their view, more frequent winter storm tracks associated with type-B conditions may increase turbulence and increase the depth of the mixed layer in the Alaska Gyre, enabling zooplankton to graze phytoplankton more efficiently. Fifty percent of the species that spawn along the west coast of the United States showed cool ocean temperature preferences. In the California Current upwelling system, cool prefering species (such as northern anchovy) may benefit from enhanced equatorward advection of nutrients and zooplankton (Roesler and Chelton 1987) and enhanced nutrients from upwelling.

Warm conditions in the north may influence the amount of suitable habitat available to marine fish stocks (MacCall 1990; Iles and Sinclair 1982). Several species in the Northeast Pacific show shifts in distribution during warm ocean conditions associated with El Niño (Pearcy and Schoener 1987). If warm ocean conditions persist, the spawning distribution may expand.

In examining the effect of environmental conditions on the success of recruitment in important fish stocks, we do not propose that the effects of fishing on stocks have a negligible effect. But we do think that the rational management of fisheries requires understanding of all the factors that influence the abundance of fish stocks. While environmental factors cannot be controlled, we aspire to achieve an eventual understanding of the long-term influence of the environment that will permit a practical evaluation of their consequences for fish production and management.

Acknowledgements

We thank Kevin Bailey, Bernard Megrey, Richard Methot and two reviewers for their helpful comments.

References

Albers, W.D., and P.J. Anderson. 1985. Diet of Pacific cod, *Gadus macrocephalus*, and predation on the northern pink shrimp, *Pendalus borealis*, in Pavlof Bay, Alaska. Fish. Bull.U.S. 83(4): 601–610.

Bailey, K.M.. 1981. Larval transport and recruitment of Pacific hake, *Merluccius productus*. Mar. Ecol. Prog. Ser. 6: 1–9.

Barnes, J.T., L.D. Jacobson, A.D. MacCall, and P. Wolf. 1992. Recent population trends and abundance estimates for the Pacific sardine (*Sardinops sagax*). Calif. Coop. Oceanic Fish. Invest. Rep. 33: 60–75.

Bence, J.R. and J.B. Rogers. 1992. Status of Bocaccio in the Conception/Monterey/Eureka INPFC areas in 1992 and recommendations for management in 1993. *In* Appendices to the status of the Pacific coast groundfish fishery through 1992 and recommended acceptable biological catches for 1993. Pacific Fishery Management Council, Metro Center, Suite 420, 2000 SW First Avenue, Portland OR 97201.

Brodeur, R.D., and D.M. Ware. 1992. Long-term variability in zooplankton biomass in the subarctic Pacific Ocean. Fish. Oceanogr. 1: 32–38.

Collie, J.S. 1991. Herring population dynamics and management in Sitka Sound, Alaska, p. 463–476. *In* Proceedings of the International Herring Symposium. Lowell Wakefield Fisheries Symposium, Anchorage, AK. Alaska Sea Grant Rep. 91-01. University of Alaska, Fairbanks, AK.

Cushing, D.H. 1969. The fluctuation of year-classes and the regulation of fisheries. Fiskeridir. Skr. Ser. Havunders. 15: 368–379.

Cushing, D.H. 1982. Climate and fisheries. Academic Press, London. 373 p.

Dorn, M., R. Methot, E.P. Nunnallee, and M. Wilkins. 1991. Status of the coastal Pacific whiting resource in 1990. NOAA Tech. Memo. NMFS F/NWC 204: 97 p.

Fargo, J. 1989. Flatfish, p. 93–118. *In* J. Fargo and A.V. Tyler [ed.] Groundfish stock assessments off the west coast of Canada in 1988 and recommended total allowable catches for 1989. Can. Tech. Rep. Fish. Aquat. Sci. 1646.

Fargo, J. 1990. Flatfish, p. 95–125. *In* A.V. Tyler and J. Fargo [ed.] Groundfish stock assessments off the west coast of Canada in 1984 and recommended total allowable catches for 1985. Can. Tech. Rep. Fish. Aquat. Sci. 1723.

Francis, R.C., and T.H. Sibley. 1991. Climate change and fisheries: what are the real issues?. Northwest Environ. J. 7: 295–308.

Frank, K.T., and W.C. Leggett. 1985. Reciprocal oscillations in densities of larval fish and potential predators: a reflection of present or past predation? Can. J. Fish. Aquat. Sci. 42: 1841–1849.

Golden, J.T., R.L. Demory, E.K. Pikitch, and W.H. Barss. 1986. Results of virtual population analysis of English sole in the INPFC Columbia–Vancouver areas final report. Commer. Fish Res. Dev.Act Proj. 1-180-R2: NMFS, NOAA, U.S. Department of Commerce, Seattle, WA. 32 p.

Haist, V. and J.F. Schweigert. 1991. Stock assessments for British Columbia herring in 1990 and forcasts of the potential catch in 1991. Can. MS. Rep. Fish. Aquat. Sci. 2119.

Heifetz, J. and D.M. Clausen. 1992. Slope rockfish, Section 5, p. 1–37. *In* Stock assessment and fishery evaluation report for the 1993 Gulf of Alaska groundfish fishery. Gulf of Alaska Groundfish Plan Team, North Pacific Fishery Management Council, P.O. Box 103136, Anchorage, AK 99510.

Hightower, J.E., and W.H. Lenarz. 1990. Status of the widow rockfish fishery in 1990, Appendix F, p. 1–48. *In* Status of the Pacific coast groundfish fishery through 1990 and recommended acceptable biological catches for 1991. Pacific Fisheries Management Council, Portland, OR. Pacific Fishery Management Council, Metro Center, Suite 420, 2000 SW First Avenue, Portland OR 97201.

Hollowed, A.B. 1992. Spatial and temporal distributions of Pacific hake, *Merluccius productus*, larvae and estimates of survival during early life stages. Calif. Coop. Oceanic. Fish. Invest. Rep. 33: 100–123.

Hollowed, A.B., B.A. Megrey, P. Munro, and W. Karp. 1991. Walleye pollock, Section 1, p. 1–92. *In* Stock assessment and fishery evaluation report for the 1992 Gulf of Alaska groundfish fishery. Gulf of Alaska Groundfish Plan Team, North Pacific Fishery Management Council, P.O. Box 103136, Anchorage, AK 99510.

Hollowed, A.B., and W.S. Wooster. 1992. Variability of winter ocean conditions and strong year classes of Northeast Pacific groundfish. ICES Mar. Sci. Symp. 195: 433–444.

Ianelli, J.N., and D.H. Ito. 1992. Pacific ocean perch, Section 9, p. 2–36. *In* Stock assessment and fishery evaluation document for groundfish resources in the Bering Sea/Aleutian Islands region as projected for 1993. North Pacific Fisheries Management Council, P.O. Box 103136, Anchorage, AK 99510.

Ianelli, J.N., D.H. Ito, and M.E. Wilkins. 1992. Status and future prospects for the Pacific ocean perch resource in waters off Washington and Oregon as assessed in 1992, Appendix C, p. 1–45. *In* Appendices to the status of the Pacific Coast groundfish fishery through 1992 and recommended acceptable biological catches for 1993. Pacific Fishery Management Council, Metro Center, Suite 420, 2000 SW First Avenue, Portland, OR 97201.

Iles, T.D., and M. Sinclair. 1982. Atlantic herring: stock discreteness and abundance. Science (Wash., D.C.) 215: 627–633.

Ingraham, W.J., Jr., R.K. Reed, J.D. Schumacher, and S.A. Macklin. 1991. Circulation variability in the Gulf of Alaska. EOS Trans. Am. Geophys. Union 72(24): 257–264.

Johnson, D.F., L.W. Botsford, R.D. Methot, Jr., and T.C. Wainwright. 1986. Wind stress and cycles in Dungeness crab (*Cancer magister*) catch off California, Oregon, and Washington. Can. J. Fish. Aquat. Sci. 43: 838–845.

Kawasaki, T. 1991. Long-term variability in the pelagic fish populations, p. 47–60. *In* T. Kawasaki, S. Tanaka, Y. Toba, and A. Taniguchi [ed.] Long-term variability of pelagic fish populations and their environment. Pergamon Press, New York.

Lasker, R. 1975. Field criteria for survival of anchovy larvae: the relation between inshore chlorophyll maximum layers and successful first feeding. Fish. Bull. U.S. 73: 453–462.

Lasker, R. 1978. The relation between oceanographic conditions and larval anchovy food in the California Current: identification of factors contributing to recruitment failure. Rapp. P.-v. Réun. Cons. int. Explor. Mer 173: 212–230.

Lluch-Belda, D., S. Hernandez-Vazquez, D.B. Lluch-Cota, C.A. Salinas-Zavala, and R.A. Swartzlose. 1992. The recovery of the California sardine as related to global change. Calif. Coop. Oceanic Fish. Invest. Repts. 33: 50–59.

Lowe, S.A. 1992. Atka mackerel, Section 11, p. 1–46. *In* Stock assessment and fishery evaluation document for groundfish resources in the Bering Sea/Aleutian Islands region as projected for 1993. North Pacific Fisheries Management Council, P.O. Box 103136, Anchorage, AK 99510.

Lowry, L.F. 1990. Alaska's seals and sea lions: where are they going. Alaska's Wild. 1990 (Sept.–Oct.): 14–21.

MacCall, A.D. 1986. Changes in the biomass of the California Current ecosystem, p. 33–54. *In* K. Sherman and L.M. Alexander [ed.] Variability and Management of large marine ecosystems. Westview Press, Boulder, CO.

MacCall, A. 1990. Dynamic geography of marine fish populations. Books in Recruitment Fishery Oceanography, Wash. Sea Grant Program. University of Washington, Seattle, WA. 153 p.

MacCall, A.D., R.A. Klingbeil, and R.D. Methot. 1985. Recent increased abundance and potential productivity of Pacific mackerel (*Scomber japonicus*). Calif. Coop. Oceanic Fish. Invest. Rep. 26: 119–129.

McConnaughey, R.A., D.A. Armstrong, B.M. Hickey, and D.R. Gunderson. 1992. Juvenile Dungeness crab (*Cancer magister*) recruitment variability and oceanic transport during the pelagic larval phase. Can. J. Fish. Aquat. Sci. 49: 2028–2044.

McFarlane, G.A. and R.J. Beamish. 1992. Climatic influence linking copepod production with strong year-classes in sablefish, *Anoplopoma fimbria*. Can. J. Fish. Aquat. Sci. 49: 743–753.

Merrick, R.L., T.R. Loughlin, and D.G. Calkins. 1987. Decline in abundance of the northern sea lion, *Eumetopias jubatus*, in Alaska, 1956–1986. Fish. Bull. U.S. 85(2): 351–365.

Methot. R.D. 1989. Synthetic estimates of historical abundance and mortality for northern anchovy, p. 66–82. *In* E.F. Edwards and B.A. Megrey [ed.] Mathematical analysis of fish stock dynamics: reviews, evaluations, and current applications. Am. Fish. Soc. Symp. Ser. 6.

Morse, W.W. 1989. Catchability, growth, and mortality of larval fishes. Fish. Bull. U.S. 87: 417–446.

Parker, K.S. 1989. Influence of oceanographic and meteorological processes on the recruitment of Pacific halibut, *Hippoglossus stenolepis*, in the Gulf of Alaska, p. 221–237. *In* R.J. Beamish and G.A. McFarlane [ed.] Effects of ocean variability on recruitment and an evaluation of parameters used in stock assessment models. Can. Spec. Publ. Fish. Aquat. Sci. 108.

Parrish, R.H., C.S. Nelson, and A. Bakun. 1981. Transport mechanisms and reproductive success of fishes in the California Current. Biol. Oceanogr. 1(2): 175–203.

Pearcy, W.G., and A. Schoener. 1987. Changes in the marine biota coincident with the 1982–1983 El Niño in the northeastern subarctic Pacific ocean. J. Geophys. Res. 92(C13): 14 417 – 14 428.

Peixoto, J.P., and A.H. Oort. 1992. Physics of climate. American Institute of Physics, New York, NY. 520 p.

Pepin, P. 1991. Effect of temperature and size on development, mortality, and survival rates of the pelagic early life history stages of marine fish. Can. J. Fish. Aquat. Sci. 48: 503–518.

Power, J.H. 1986. A model of the drift of northern anchovy, *Engraulis mordax*, larvae in the California Current. Fish. Bull. U.S. 84(3): 585–603.

Quinn, W.H., V.T. Neal, and S. Antuneq de Mayolo. 1987. El Niño occurrences over the past four and a half centuries. J. Geophys. Res. 92: 14 449 – 14 461.

Quinn, W.H., D.O. Zopf, K.S. Short, and T.W. Kuo Yang. 1978. Historical trends and statistics of the Southern Oscillation, El Niño, and Indonesia droughts. Fish. Bull. U.S. 76: 663–678.

Roesler, C.S., and D.B. Chelton. 1987. Zooplankton variability in the California Current, 1951–1982. Calif. Coop. Oceanic Fish Invest. Rep. 28: 59–96.

Saetersdal, G., and H. Loeng. 1987. Ecological adaptation of reproduction in Northeast Arctic cod. Fish. Res. 5(1–2): 271–278.

Stevens, B.G., J.H. Bowerman, R.A. MacIntosh, and J.A. Haaga. 1992. Report to industry on the 1992 eastern Bering Sea carb survey. Processed Rep. 92-12. Alaska Fishery Science Centre, NOAA, Kodiak Laboratory, P.O. Box 1638, Kodiak, AK 99615. 57 p.

Thompson, and Zenger, 1991. Pacific cod, Section 2, p. 1–28. *In* Stock assessment and fishery evaluation document for groundfish resources in the Bering Sea/ Aleutian Islands region as projected for 1992. North Pacific Fisheries Management Council, P.O. Box 103136, Anchorage, AK 99510.

Turnock, B., and R. Methot 1992. Status of the Dover sole in 1991, Appendix E, p. 1–55. *In* Appendices to the status of the Pacific Coast groundfish fishery through 1992 and recommended acceptable biological catches for 1993. Pacific Fishery Management Council, Metro Center, Suite 420, 2000 SW First Avenue, Portland, OR 97201.

Tyler, A.V., and W.R. Crawford. 1991. Modeling of recruitment patterns in Pacific cod (*Gadus macrocephalus*) in Hecate Strait, British Columbia. Can. J. Fish. Aquat. Sci. 48: 2240–2249.

Tyler, A.V., and R.P. Foucher. 1990. Pacific cod, p. 59–73. *In* J. Fargo and A.V. Tyler [ed.] Groundfish stock assessments off the west coast of Canada in 1988 and recommended total allowable catches for 1989. Canadian Tech. Rep. Fish. Aquat. Sci. 1732.

Ware, D.M. 1992. Climate, predators and prey: behavior of a linked oscillating system, p. 279–291. *In* T. Kawasaki, S. Tanaka, Y. Toba, and A. Taniguchi [ed.] Long-term Variability of Pelagic Fish Populations and their Environment. Pergamon Press, New York.

Ware, D.M., and G.A. McFarlane. 1989. Fisheries production domains in the northeast Pacific Ocean, p. 359–379. *In* R.J. Beamish and G.A. McFarlane [ed.] Effects of ocean variability on recruitment and an evaluation of parameters used in stock assessment models. Can. Spec. Publ. Fish. Aquat. Sci. 108.

Wespestad, V.G. 1991. Pacific herring population dynamics, early life history, and recruitment variation relative to eastern Bering Sea oceanographic factors. Ph.D. dissertation, University of Washington, Seattle, WA. 237 p.

Wespestad, V.G., and P. Dawson. 1991. Walleye pollock, Section 1, p. 1–27. *In* Stock assessment and fishery evaluation document for groundfish resources in the Bering Sea/Aleutian Islands region as projected for 1992. North Pacific Fisheries Management Council, P. O. Box 103136, Anchorage, AK 99510.

Wilderbeuer, T.K. 1992. Yellowfin sole, Section 3, p. 1–25. *In* Stock assessment and fishery evaluation document for groundfish resources in the Bering Sea/Aleutian Islands region as projected for 1993. North Pacific Fisheries Management Council, P. O. Box 103136, Anchorage, AK 99510.

Wooster, W.S., and A.B. Hollowed. 1994. Decadal scale variations in the eastern subarctic Pacific: I. Winter ocean conditions, p. 81–85. *In* R.J. Beamish [ed.] Climate change and northern fish populations. Can. Spec. Publ. Fish Aquat. Sci. 121.

Influence of environment on stock fluctuations of Japanese sardine, *Sardinops melanostictus*

Tokio Wada, Takahiro Matsubara, Yoshiharu Matsumiya, and Noriyuki Koizumi

Wada, T., T. Matsubara, Y. Matsumiya, and N. Koizumi. 1995. Influence of environment on stock fluctuations of Japanese sardine, *Sardinops melanostictus*, p. 387–394. *In* R.J. Beamish [ed.] Climate change and northern fish populations. Can. Spec. Publ. Fish. Aquat. Sci. 121.

Abstract: The Japanese sardine (*Sardinops melanostictus*) stock reached a peak in the mid-1980s. However, the stock has been rapidly decreasing since the latter half of the 1980s. Using physical and biological time series spanning this period, we examined the influence of parameters of growth and mature condition, stock density, and environment on the reproductive success of Japanese sardine. Multiple regression analysis for biological indices suggests that sardine growth is density dependent and affected by sea-surface temperature (SST). Multiple regression analysis for reproductive indices shows that the reproduction rate is weakly density dependent and year-class strength is negatively correlated to SST in summer and autumn. However, there are no significant relationships between the reproductive indices and growth and maturation in the feeding ground. Results of a sardine rearing experiment indicate that gonad development is inhibited at 18–21°C. Therefore, changes in SST in the spawning grounds and on migration route would have some influences on the reproductive success of Japanese sardine.

Résumé : Le stock de sardines du Japon (*Sardinops melanostictus*) a atteint un sommet au milieu des années 1980. Toutefois, il diminue rapidement depuis la deuxième moitié de cette décennie. Grâce à des séries chronologiques physiques et biologiques couvrant cette période, nous avons examiné l'influence des paramètres de la croissance et de l'état mature, de la densité des stocks et de l'environnement sur la reproduction de la sardine du Japon. L'analyse de régression multiple des indices biologiques porte à croire que la croissance de la sardine dépend de la densité et est influencée par la température de la surface de la mer (TSM). L'analyse de régression multiple des indices de reproduction montre que le taux de reproduction dépend faiblement de la densité et que l'effectif des classes annuelles est en corrélation négative avec la TSM en été et en automne. Toutefois, il n'y a aucune relation statistiquement significative entre les indices de reproduction et la croissance et la maturation dans l'aire d'alimentation. Les résultats d'une expérience d'élevage de sardines indiquent que le développement des gonades est inhibé à 18–21°C. Par conséquent, les changements de TSM dans les aires de frai et sur la route migratoire exerceraient certains effets sur la reproduction de la sardine du Japon.

Introduction

Japanese sardine, *Sardinops melanostictus*, exhibits a repeating long-cycle fluctuation in its stock abundance. The catch of the sardine by Japan was less than 150 000 t/yr in the 1960s and reached 4.5 million t in the peak year of 1988, corresponding to the increase in the stock abundance. Since the latter half of the 1980s, the stock has been rapidly decreasing, and the catch was reduced to less than 3 million t in 1991.

T. Wada and **T. Matsubara.** Hokkaido National Fisheries Research Institute, Katsurakoi 116, Kushiro, Hokkaido 085, Japan.
Y. Matsumiya and **N. Koizumi.** Ocean Research Institute, University of Tokyo, Minami-Dai 1-15-1, Nakano-Ku, Tokyo 164, Japan.

In considering stock fluctuations of Japanese sardine, many authors have pointed out the influence of long-term and global changes in oceanographic conditions. Tomosada (1988) analyzed the time series of sea-surface temperatures (SST) in the coastal waters around Japan and found that the big catch periods of sardine corresponded to periods of warm temperatures in spawning and nursery grounds and cool temperatures in feeding grounds. Kawasaki (1991) found a positive correlation between the sardine catch and air temperature and suggested the influence of global long-term climate change.

In the Pacific Ocean, Japanese sardine spawns in the coastal waters from Honshu to Kyushu in winter and spring, and migrates into the waters from Hokkaido to the Kuril Islands for feeding in summer and autumn. Remarkable changes in distribution and individual growth have been observed in the feeding grounds, and are explained as density-dependent phenomena (Wada 1988; Wada and Kashiwai 1991).

In this paper, first, we examine the influence of sardine stock density and physical environment on the biological indices related to growth and maturation in the feeding grounds. Then, we analyze the relationships between reproductive success of the sardine and the biological indices, stock density, and physical environment by multiple regression analysis.

Effects of stock density and environment on biological indices

Biological indices

The biological indices were calculated using biological measurement data of Japanese sardine caught by purse seine in the coastal waters southeast of Hokkaido, in the Dohtoh region (Fig. 1), from 1978 to 1990. We used the data collected in October as the typical values of biological condition at the end of feeding period. In this paper, the condition factor given by the relation between body length and weight, the fatweight ratio, and the gonadosomatic index were used as biological indices of somatic condition, fat storage, and mature condition for spawning, respectively. Body length itself was also adopted as an index. Every index excluding body length was calculated for every individual using the following equations:

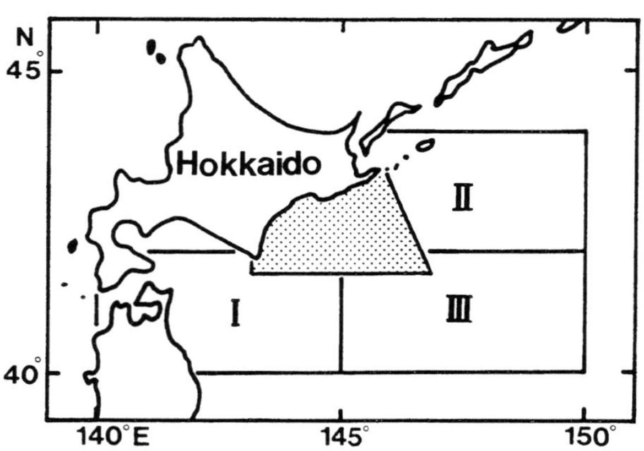

Fig. 1. Fishing grounds of the Japanese sardine in the Dohtoh region (dotted area), and three regions adjoining Dohtoh, used for the sea surface temperature (SST) anomaly data.

(1) Condition factor = body weight (grams) × body length (millimetres)$^{-3}$ × 10^6,

(2) Fat/weight ratio = fat weight around alimentary canal (grams) × body weight (grams)$^{-1}$ × 10^2,

(3) Gonadosomatic index = gonad weight (grams) × body weight (grams)$^{-1}$ × 10^2.

The gonadosomatic index was separately calculated by sex. Every index was averaged by year and age.

Table 1 shows the time series of the biological indices. The body length, condition factor, and fat/weight ratio were lower in the 1980s when stock abundance was at a high level and showed rapid recovery in recent years. Gonadosomatic indices of male and female did not show any clear changes with stock biomass and year.

Stock biomass

We used the stock biomass in the Dohtoh region as a density factor. The stock biomass was calculated as the product of the stock number by age in the Dohtoh region from 1976 to 1990 estimated by Wada (1991) and the average body weight at age for every year. Table 2 gives the time series of the stock biomass with the reproductive indices mentioned in latter part. Changes in the stock abundance in the Dohtoh region seem to correspond well to changes in overall sardine stock abundance in the Pacific Ocean (Wada 1988).

SST Anomalies

SST anomalies in the three regions adjoining Dohtoh (I, II, and III in Fig. 1) were used as environmental factors. SST anomalies in each region were obtained by reading anomaly areas divided into the first half (January–June) and the second half (July–December) of the respective years on the monthly SST anomaly charts (1970–90) issued by Hakodate Marine Meteorological Observatory, Japan Meteorological Agency. The information from the area measured was then converted to the half-yearly average of SST anomalies. The half-yearly SST anomalies had significant positive correlation among the regions, but no significant correlation between the anomalies of the first and second half year in every region. So the average value of the half-yearly SST anomalies in three regions was used in this paper.

Figure 2 shows the time series of the half-yearly SST anomalies. SST anomalies were mostly negative through the latter half of the 1970s and 1980s. But the anomalies abruptly turned to increase in 1987.

Multiple regression analysis

The relationships between the biological indices, stock biomass, and SST anomalies were examined using multiple regression analysis. The biological indices arranged by year (1978–1990) and age (1–4 yr old) were used as the dependent

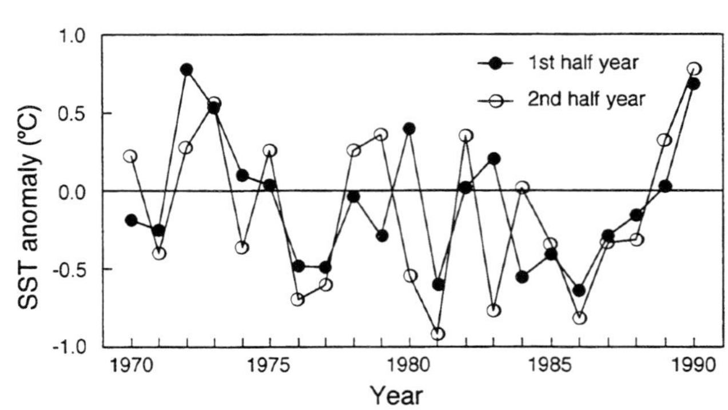

Fig. 2. Time series of SST anomalies (°C) in the first half and second half year in the Dohtoh and adjoining regions from 1970 to 1990.

Table 1. Time series of biological indices of Japanese sardine by age in the Dohtoh region in October from 1978 to 1990. The biological indices show the values of body length (BL; mm); condition factor (CF; $g \cdot mm^{-3} \times 10^6$); fat/weight ratio (FWR; %); gonadosomatic index of male (GSI-M; %); gonadosomatic index of female (GSI-F; %). Standard deviations are given in parentheses.

Age	Year	BL	CF	FWR	GSI-M	GSI-F
1	1978	177.8 (7.1)	14.27 (0.89)	5.19 (1.54)	0.080 (0.034)	0.247 (0.075)
	1979	171.3 (7.0)	12.42 (0.82)	4.60 (2.61)	0.119 (0.100)	0.309 (0.170)
	1980	—	—	—	—	—
	1981	147.6 (6.7)	12.06 (1.76)	4.67 (2.12)	0.048 (0.025)	0.144 (0.084)
	1982	155.2 (5.8)	11.80 (0.65)	3.55 (1.22)	0.076 (0.069)	0.196 (0.094)
	1983	159.8 (10.9)	11.61 (1.31)	—	0.091 (0.039)	0.256 (0.218)
	1984	152.5 (7.8)	12.40 (0.74)	3.22 (1.23)	0.051 (0.017)	0.264 (0.214)
	1985	161.1 (7.2)	11.68 (0.91)	—	—	—
	1986	159.6 (6.0)	11.67 (0.75)	5.00 (1.61)	0.026 (0.012)	—
	1987	159.0 (5.8)	11.41 (0.72)	3.44 (1.40)	0.058 (0.025)	0.235 (0.078)
	1988	159.7 (5.2)	11.74 (0.73)	2.86 (1.17)	0.071 (0.017)	0.218 (0.125)
	1989	—	—	—	—	—
	1990	175.4 (4.2)	12.96 (0.70)	3.42 (1.50)	0.079 (0.022)	0.349 (0.081)
2	1978	189.7 (11.9)	14.29 (1.09)	5.32 (1.59)	0.121 (0.048)	0.342 (0.131)
	1979	194.3 (8.4)	13.45 (0.82)	5.34 (1.97)	0.135 (0.062)	0.359 (0.170)
	1980	193.5 (12.0)	13.11 (0.01)	4.60 (1.00)	0.169*	0.435*
	1981	182.0 (17.0)	13.50 (1.82)	6.65 (0.68)	0.123 (0.009)	—
	1982	163.9 (8.2)	12.29 (0.95)	4.54 (1.66)	0.084 (0.035)	0.320 (0.095)
	1983	172.6 (6.4)	12.16 (0.61)	—	0.113 (0.032)	0.392 (0.081)
	1984	174.6 (5.6)	12.61 (1.20)	3.85 (0.84)	0.105 (0.015)	0.575*
	1985	176.8 (9.0)	12.96 (0.91)	—	—	—
	1986	182.2 (4.1)	12.88 (0.71)	4.82 (1.60)	0.122 (0.007)	0.429*
	1987	171.9 (7.6)	11.66 (0.81)	3.46 (1.34)	0.081 (0.034)	0.341 (0.142)
	1988	174.2 (6.3)	12.04 (0.66)	3.20 (1.22)	0.084 (0.026)	0.456 (0.095)
	1989	196.7 (2.5)	13.36 (1.02)	4.71 (1.16)	0.148 (0.086)	0.330 (0.131)
	1990	191.3 (7.2)	13.28 (0.32)	4.69 (0.82)	0.122*	0.340 (0.079)
3	1978	210.6 (9.5)	15.32 (1.27)	6.85 (2.50)	0.174 (0.048)	0.445 (0.134)
	1979	208.7 (8.2)	13.79 (0.79)	5.72 (1.74)	0.143 (0.072)	0.451 (0.129)
	1980	208.3 (9.4)	13.37 (0.77)	4.47 (1.58)	0.204 (0.052)	0.514 (0.119)
	1981	193.3 (4.5)	12.51 (1.26)	5.18 (2.35)	0.143 (0.024)	0.538*
	1982	179.3 (12.6)	12.81 (1.02)	4.64 (1.09)	0.117 (0.039)	0.290*
	1983	179.1 (7.4)	12.22 (0.84)	—	0.128 (0.042)	0.485 (0.157)
	1984	181.0 (7.6)	13.14 (1.35)	4.20 (1.68)	0.151 (0.054)	0.598 (0.519)
	1985	183.6 (12.6)	13.63 (0.91)	—	—	—
	1986	185.8 (6.0)	13.58 (1.31)	4.59 (1.49)	0.126 (0.049)	0.401 (0.035)
	1987	181.8 (5.3)	11.93 (0.70)	3.72 (1.33)	0.122 (0.033)	0.437 (0.058)
	1988	185.9 (5.1)	12.02 (0.85)	3.14 (1.28)	0.146 (0.039)	0.404 (0.069)
	1989	193.0 (5.8)	13.71 (0.87)	4.95 (1.48)	0.141 (0.058)	0.444 (0.099)
	1990	194.3 (5.1)	13.92 (0.91)	4.72 (1.60)	0.140 (0.035)	0.489 (0.110)
4	1978	217.0 (8.2)	15.62 (0.35)	7.13 (1.29)	0.162*	0.475 (0.009)
	1979	214.4 (9.0)	13.73 (0.79)	5.55 (1.86)	0.157 (0.080)	0.469 (0.194)
	1980	217.9 (6.4)	13.37 (0.72)	3.92 (1.56)	0.216 (0.066)	0.564 (0.111)
	1981	—	—	—	—	—
	1982	—	—	—	—	—
	1983	198.7 (8.8)	11.93 (0.96)	—	—	0.639 (0.106)
	1984	185.7 (6.1)	13.66 (1.10)	3.69 (1.02)	0.149*	0.605 (0.067)
	1985	189.3 (7.5)	13.29 (0.96)	—	—	—
	1986	192.5 (10.9)	13.31 (1.10)	4.57 (1.68)	0.139 (0.041)	0.408 (0.073)
	1987	192.6 (6.7)	12.00 (0.88)	2.94 (0.74)	0.154 (0.040)	0.500 (0.105)
	1988	195.4 (7.7)	12.73 (1.08)	3.86 (1.42)	0.162 (0.046)	0.554 (0.088)
	1989	197.2 (6.0)	13.89 (1.02)	4.60 (1.78)	0.158 (0.077)	0.483 (0.077)
	1990	198.4 (6.9)	13.99 (0.97)	5.19 (1.80)	0.164 (0.047)	0.514 (0.103)

* Number of specimens examined is one.

Table 2. Time series of stock biomass, number of parents spawning, number of mature progenies produced, reproduction rate, and year-class strength (available stock number at 1 yr old in Dohtoh region) of Japanese sardine.

Year or Year-class	Stock biomass ($\times10^3$ t)	Number of parents ($\times10^8$)	Number of progenies ($\times10^8$)	Reproduction rate	Year-class strength ($\times10^8$)
1975					29.2
1976	652				52.7
1977	1405	23.8	121.2	5.09	42.8
1978	1181	74.1	111.3	1.50	51.9
1979	2210	55.0	45.7	0.83	0.9
1980	1337	127.1	92.6	0.73	141.4
1981	2189	60.8	151.3	2.49	83.8
1982	2727	65.7	90.5	1.38	25.5
1983	2562	71.8	112.1	1.56	163.2
1984	3600	51.4	103.7	2.02	20.7
1985	2275	134.6	111.1	0.83	90.8
1986	2346	83.0	153.6	1.85	137.9
1987	3261	83.6	97.2	1.16	56.4
1988	3057	132.6	5.3	0.04	1.0
1989	1824	152.9			12.2
1990	1916	97.5			
1991		97.2			

variables of this analysis. The combination of stock biomass and SST anomaly for the current year and the previous year was used as a set of independent variables.

The multiple regression analyses were done according to the stepwise forward regression method based on Akaike's information criterion (AIC). In the case when some data for the dependent variables were not available, the analyses were done by eliminating the data of the independent variables corresponding to the dependent variables.

Table 3 shows the multiple regression analysis of every biological index by age.

Body length

The results of the multiple regression equations were significant at all ages. In the variables of the multiple regression equations, the stock biomass was used for all ages. The partial regression coefficients of the biomass in the current and previous years were negative. The SST anomaly was not significant as the variable of the regression equations except for age 1.

Condition factor

The results of the multiple regression equations were significant at all ages. In the variables of the regression equations, the stock biomass in the current year was used for all ages, and the partial regression coefficients were negative. The SST anomaly was used for all ages, and the partial regression coefficients were negative in the first half of the current year, and positive in the second half of the current year.

Fat/weight ratio

The results of the multiple regression equations were significant at all ages except age 4. Stock biomass was used as the variable in the regression equations, and the partial correlation coefficients were negative in both the current year and the previous year. The SST anomaly was also used, and the partial correlation coefficients were negative, especially in the first half of the current year. At age 3, the SST anomaly was significantly positive in the second half of the current year.

Gonadosomatic index of males

The results of the multiple regression equations were significant for all ages except age 1. The combination of variables used differed with age. Stock biomass in the current year was negatively correlated with the index at ages 2 and 3. The SST anomaly in the first half of the current year was positively correlated with the index at age 4.

Gonadosomatic index of females

The results of the multiple regression equations were significant at all ages except age 4. The SST anomaly was only used as the variable in the equations at age 2–4. The correlation between the index and the SST anomaly was mostly positive in the first half of the previous year. However, in other half-yearly periods, the signs of the partial correlation coefficients differed with age.

Effects of stock abundance, environment, and biological condition on reproduction

Reproductive indices

Reproduction rate and year-class strength were used as reproductive indices. Reproduction rate in a given year (r_t) was defined as the ratio of the number of mature progenies through the life (0–5 yr old) of a year-class (P_{rt}) to the number of parents spawning (P_{at}). The reproduction rates from 1977 to 1988 were calculated from Wada (1991) by using following equations:

Table 3. Multiple regression analyses for biological indices of Japanese sardine; body length (BL), condition factor (CF), fat/weight ratio (FWR), gonadosomatic index of fmale (GSI-M), and gonadosomatic index of female (GSI-F). Numbers indicate the standard partial regression coefficients of the independent variables with their significance. Significance with asterisk of the multiple regression equations, coefficients of determination (R^2), and degrees of freedom (df) are shown. *, significant at 5% level; **, significant at 1% level.

Index	Age	Stock weight		SST anomaly				Equation	R^2	df
		Previous year	Current year	Previous year		Current year				
				1st half	2nd half	1st half	2nd half			
BL	1		−0.4182	−0.2617	0.3577		0.4917*	*	0.826	4,6
	2		−0.7705**	0.3075				**	0.661	2,10
	3	−0.3735	−0.6522**					**	0.739	2,10
	4	−0.4502*	−0.5844*					**	0.744	2,8
CF	1		−0.7235*	0.3159	−0.3790		0.4845*	*	0.796	4,6
	2		−1.0100**	0.3153**		−0.5616**	0.4240**	**	0.968	4,8
	3		−0.8411**			−0.4831*	0.5946**	**	0.814	3,9
	4		−0.9914**	0.3180	−0.4435	−0.3799	0.4065	*	0.900	5,5
FWR	1		−0.9833**			−0.6764**		**	0.937	2,6
	2	−0.4150	−0.6940*	0.4089	−0.2552	−0.2544		*	0.903	5,5
	3	−0.3421**	−0.8924**		−0.2051*	−0.5941**	−0.5416**	**	0.988	5,5
	4	−0.4447	−0.4930						0.639	2,6
GSI–M	1				0.5004		0.4841		0.537	2,7
	2		−0.7800**	0.3426				**	0.698	2,9
	3		−0.5923*					*	0.351	1,10
	4					0.9327*	−0.5625	*	0.678	2,6
GSI–F	1		0.2486		0.6524*		0.6797*	*	0.841	3,5
	2			0.8875**	−0.5813*		−0.6009*	*	0.718	3,7
	3			0.7575**				**	0.574	1,10
	4			0.9258*	−0.7555	−0.9061*	−0.7780*		0.772	4,5

(4) $$r_t = P_{rt} / P_{at}$$

(5) $$P_{at} = \sum_{j=0}^{5} [(N_{t-1,j} - C_{t-1,j})p_{t-1,j}],$$

(6) $$P_{rt} = \sum_{j=0}^{5} [(N_{t+j,j} - C_{t+j,j})p_{t+j,j}].$$

Where $N_{t,j}$ is the number of available stock at age j in t year; $C_{t,j}$ is the number of catch at age j in t year; and $p_{t,j}$ is a ratio of the mature sardines at age j in year t. In the calculation of P_{at}, the differences in growth rate among year-classes were taken into consideration. It was assumed that the sardines over 169 mm in body length would join the next spawning before the 1980 and after the 1987 year-classes, and the sardines over 179 mm would spawn in the next winter to spring in the year-classes from 1980 to 1987. For the calculation of the reproduction rate from 1986 to 1988, we estimated the available stock abundance by age from 1991 to 1993 based on the abundance by age in 1990 supposing a constant annual survival rate (1.0). The year-class strength was defined as the available stock numbers at 1 yr old. The values for the 1975 to 1989 year-classes were prepared based on Wada's (1991) estimates.

Table 2 gives the number of parents spawning, the number of mature progenies produced, and the reproductive indices obtained. Reproduction rates were highest in 1977 and fluctuated, ranging from 0.7 to 2.5 progenies per parent during 1978 to 1987. However, reproduction suddenly collapsed in 1988 and the rate dropped to 0.04. Year-class strength fluctuated greatly year by year. In the 1980s, there were some good year-classes, such as 1980, 1983, and 1986; however, the year-classes in 1988 and 1989 were successively poor. A good year-class was not always related to a high reproduction rate, and there was no positive correlation between the two indices.

Multiple regression analysis

The relations between reproductive indices and stock biomass, SST anomalies, and biological indices were examined by multiple regression analysis according to the stepwise method based on AIC. The reproduction rate (1977–88) and year-class strength (1975–89) were used as the dependent variables. Stock weight, SST anomalies, combination of stock biomass and SST anomaly, and biological indices of 3 yr olds were used as the independent variables of the analysis. The multiple regression analysis was done by the same method described earlier.

In the case where the combination of stock biomass and SST anomaly was used as the independent variables, the multiple regression equation on the reproduction rate was significant. The stock biomass in the previous year and SST anomalies for the second half of the previous year and the current year were used as the variables. The partial correlation coefficients of these variables were all negative (Table 4). In the analysis with

Table 4. Multiple regression analyses for the reproduction rate of Japanese sardine. See Table 2 for explanation of abbreviations and asterisks.

Index	Stock weight		SST anomaly				Equation	R^2	df
	Previous year	Current year	Previous year		Current year				
			1st half	2nd half	1st half	2nd half			
(A) Regression with the combination of stock weight as the independent variable.									
Reproduction rate	−0.6132*			−0.3611		−0.4477	*	0.685	3,8

(B) Regression with the biological indices as the independent variable.

Index	Biological indices					Equation	R^2	df
	BL	CF	FWR	GSI-M	GSI-F			
Reproduction rate				0.5600			0.314	1,6

Table 5. Multiple regression analysis for year-class strength (available stock number at 1 yr old in Dohtoh region) of Japanese sardine. See Table 1 for explanation of abbreviations and asterisks.

Index	SST anomaly								Equation	R^2	df
	2 yr before		1 yr before		Current year		1 yr after				
	1st half	2nd half	1st half	2nd half	1st half	2nd half	1st half	2nd half			
Year-class strength					0.3898*	−0.7081**	−0.2654		**	0.720	3,11

the biological indices as the independent variables, the gonadosomatic index of male was used only as an independent variable, but the regression equation was not significant.

In the analysis of year-class strength, the combination of SST anomalies from 2 yr before to 1 yr after only provided a significant regression equation. The anomalies of the first and second half of the current year and the first half of the previous year were used as the variables, and the second half of the current year was significant. The sign of the partial correlation coefficient was negative (Table 5).

Discussion

Multiple regression analysis in this paper revealed that the biological indices of Japanese sardine are connected with the stock biomass (a density factor) and SST anomaly (an environmental factor). The indices, especially growth indices, such as condition factor and fat/weight ratio, have negative relationships with the stock biomass. Therefore, the indices have a density-dependent tendency. On the other hand, we could not find any significant relationship between the reproductive indices and the biological indices. This suggests that variability of the sardine recruitment can not be explained by density-dependent changes in the biological indices.

Tsuruta (1985) showed that, in a rearing experiment, nutritional conditions of female sardines at the beginning of sexual maturation have a major effect on the reproductive effort during the subsequent spawning season. In addition, Matsuura et al. (1991) found that female Japanese sardines with atretic ovaries showed lower values of the condition factor than females with no atresia in the ovaries. On the other hand, Murphy (1966) pointed out that the California sardine (*Sardinops sagax*) stock collapsed with two successive poor year-classes in 1949 and 1950. Kurita (1957) also showed the same phenomenon with the rapid stock decline of Japanese sardine in the 1940s. The sardine stock recovered rapidly during the 1970s. Matsuda et al. (1992) pointed out that the average reproduction rate of this period was significantly higher than those of other periods.

Density-dependent changes in the nutritional conditions of spawners would affect the interannual variability of the sardine recruitment. But this mechanism, in other words, a negative feedback system, would not be sufficient to explain the successive recruitment failure mentioned above. Therefore, we should consider any positive feedback system that allows the successive recruitment success or failure.

The multiple regression analysis on the reproductive indices is a little problematic because there are some strong correlations among the biological indices as the independent variables. We should create independent variables by principal component analysis on the biological indices in the future studies to get more clear relationships.

SST must be an important factor for sardine reproduction because it has strong influence on the biological indices and year-class strength. Kodama (1992) pointed out that stock abundance of 1-yr-old sardines and SST off the northern Honshu showed a negative relationship. Kuroda (1991) and Naganuma (1992) pointed out that the higher temperature in the spawning grounds and adjacent nursery grounds allowed good survival in the early life stages. However, Bulter (1991) mentioned that the relationship between larval California sardine mortality rate and sea water temperature in the California Current was not clear. On the other hand, Cury and Roy (1989) found the existence of favourable wind speed for the recruitment of sardines in Ekman-type upwelling regions, such as those off California. Ware and Thomson (1991) supported this and suggested that stock fluctuations of pelagic fishes in the area are consistent with a long-term (40–60 yr) fluctuation in primary and secondary production, which corresponds to a long-period oscillation of wind-induced upwelling.

SST anomalies in Dohtoh and adjacent waters were mostly negative in the 1980s (Fig. 2). On the other hand, the accumulated value of SST anomalies in the waters off southern Kyushu, which was the major spawning ground of sardine in the high stock abundance period in the 1980s, showed an increasing trend beginning in 1980 (Tomosada 1988; Kuroda 1991). The coincident opposite trends in SST anomalies between the spawning grounds and feeding grounds may be a reflection of synchronized spin-up (or spin-down) of the subarctic and subtropical gyres by a strong (or weak) winter monsoon over the North Pacific Ocean (Hanawa 1991). The successive serious decline in recruitment since 1988 corresponds with the abrupt increase of the SST anomalies. However, there is no hypothesis that can explain consistently the relationship between these phenomena. In the waters around Japan, which is a western boundary current area, we should examine the long-term changes in the biological production system as a background of the stock fluctuation of Japanese sardine.

A recent sardine rearing experiment suggests the direct physiological effect of temperature on the development of the gonads. Matsubara (1991 and unpublished data) reared adult sardines caught in the Dohtoh region in five different water temperatures from August to January and found that their gonad development was normal in 12 and 15°C. However, from 18 to 24°C, the development of gonads was inhibited and many atretic oocytes were found in the ovaries (Fig. 3).

Water temperature during the spawning season is about 17–21°C in southern Kyushu and 15–19°C in the waters off the Pacific coast of Honshu and Shikoku (Kuroda 1991). The upper limits of the temperatures in these areas fall within the range of temperatures at which gonad development is inhibited. The changes in SSTs in the spawning grounds and on the migration route from the feeding grounds cause the shift of location and timing of spawning and, in turn, would have strong effects on the survival on the early life stages.

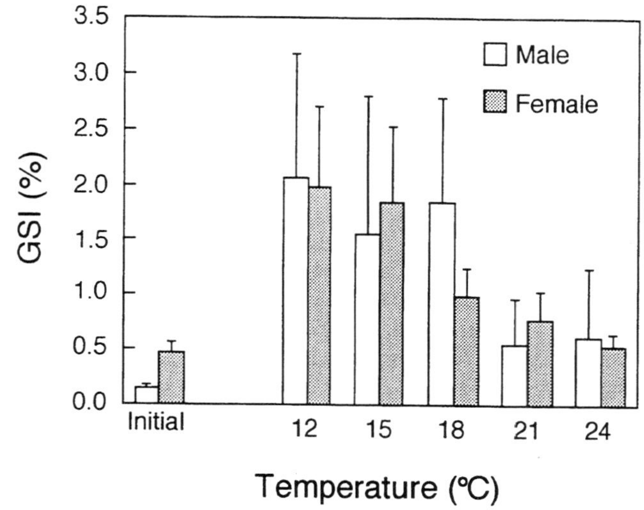

Fig. 3. Changes in gonadosomatic index (GSI, %) of the Japanese sardine under five different rearing temperatures (Matsubara unpublished data). Average GSI and standard deviation are in the initial condition and at the end of the rearing experiment (August 1991 – January 1992).

Acknowledgements

We are indebted to Dr. M. Kashiwai of the Hokkaido National Fisheries Research Institute and Dr. D. M. Ware of the Pacific Biological Station, Canada, for their careful readings of the manuscript and valuable comments. We extend our thanks to Mr. S. Honda of the Hokkaido National Fisheries Research Institute and Mr. Y. Mihara of the Hokkaido Prefectural Kushiro Fisheries Experimental Station for their kind assistance in data collection and analysis.

References

Bulter, J.L. 1991. Mortality and recruitment of Pacific sardine, *Sardinops sagax caerulea*, larvae in the California Current. Can. J. Fish. Aquat. Sci. 48: 1713–1723.

Cury, P., and C. Roy. 1989. Optimal environmental window and pelagic fish recruitment success in upwelling areas. Can. J. Fish. Aquat. Sci. 46: 670–680.

Hanawa, K. 1991. Long-term variations of the atmospheric circulation over the North Pacific and the Oyashio. Bull. Hokkaido Natl. Fish. Res. Inst. 55: 125–139.

Kawasaki, K. 1991. Long-term variability in the pelagic fish populations, p. 47–60. *In* T. Kawasaki, S. Tanaka, and Y. Toba [ed.] Long-term variability of pelagic fish populations and their environment. Pergamon Press, London, UK.

Kodama, J. 1992. The fluctuation of sea condition and fish condition in the waters off the coast of Kinkazan, especially the relation to long period fluctuation in the stock abundance of Japanese sardine. Bull. Jpn. Soc. Fish. Oceanogr. 56: 151–154.

Kurita, S. 1957. Causes affecting the size of sardine stock in the waters off Japan and adjacent regions. Bull. Tokai Reg. Fish. Res. Lab. 18: 1–14.

Kuroda, K. 1991. Studies on the recruitment process focusing on the early life history of the Japanese sardine, *Sardinops melanostictus* (Schlegel). Bull. Natl. Res. Inst. Fish. Sci. 3: 25–278.

Matsubara, T. 1991. Japanese sardine, p. 113–124. *In* K. Hirose [ed.] Reproductive rhythm in marine fish. Kouseisha Kouseikaku, Tokyo, Japan.

Matsuda, Y., T. Kishida, H. Shimizu, and T. Kidachi. 1992. Fluctuation of reproduction and stock management of the Japanese sardine (*Sardinops melanostictus*). Nippon Suisan Gakkaishi 58: 1861–1866.

Matsuura, S., M. Matsuyama, T. Otsuka, K. Takeshita, T. Mitani, and N. Ogawa. 1991. Histological maturity scales, maturation and spawning in Kyushu-subpopulation sardine, *Sardinops melanostictus*, p. 221–228. *In* T. Kawasaki, S. Tanaka, and Y. Toba [ed.] Long-term variability of pelagic fish populations and their environment. Pergamon Press, London, UK.

Murphy, G.I. 1966. Population biology of the Pacific sardine (*Sardinops caerulea*). Proc. Calif. Acad. Sci. 34: 1–84.

Naganuma, K. 1992. Variations in the ocean environments related to the variations in the fisheries resources. Bull. Jpn. Soc. Fish. Oceanogr. 56: 168–172.

Tomosada, A. 1988. Long term variation of sardine catch and temperature. Bull. Tokai Reg. Fish. Res. Lab. 126: 1–9.

Tsuruta, Y. 1985. Reproductive biology of Japanese sardine and anchovy in relation to their population dynamics. Bull. Jpn. Soc. Fish. Oceanogr. 51: 51–54.

Wada, T. 1988. Population dynamics on Japanese sardine, *Sardinops melanostictus*, caught by the domestic purse seine fishery in the waters off the coast of southeastern Hokkaido. Bull. Hokkaido Natl. Fish. Res. Inst. 52: 1–138.

Wada, T. 1991. The stock abundance of Japanese sardine. Bull. Jpn. Soc. Fish. Oceanogr. 55: 398–402.

Wada, T., and K. Kashiwai. 1991. Changes in growth and feeding ground with fluctuation in stock abundance, p. 181–190. *In* T. Kawasaki, S. Tanaka, and Y. Toba [ed.] Long-term variability of pelagic fish populations and their environment. Pergamon Press, London, UK.

Ware, D.M., and R.E. Thomson. 1991. Link between long-term variability in upwelling and fish production in the North Pacific ocean. Can. J. Fish. Aquat. Sci. 48: 2296–2306.

Influence of climatic changes on succession processes in the community of neritic pelagic fishes in the Kuroshio–Oyashio zone

V.V. Kuznetsov

Kuznetsov, V.V. 1995. Influence of climatic changes on succession processes in the community of neritic pelagic fishes in the Kuroshio–Oyashio zone, p. 395–404. *In* R.J. Beamish [ed.] Climate change and northern fish populations. Can. Spec. Publ. Fish Aquat. Sci. 121.

Abstract: In the community of neritic pelagic fishes in the Kuroshio–Oyashio zone, one can observe the interaction of allogenic and autogenic succession processes. Autogenic succession take place as a result of ecological interaction of the "competition–coexistence" type among the different populations. In this interaction, changes in species ratios and production processes are observed. The predominance of one or another species at different succession stages depends on their life cycle and ecology. In the initial stage of succession, *r*-strategists such as anchovy (*Engraulis japonicus*) and saury (*Cololabis saira*), have some advantages. Climax of the community is achieved when sarding (*Sardinops sagax melanosticta*) becomes dominant. At this stage, the community utilizes the resources of the biotope to the greatest extent and achieves the greatest total biomass. There is evidence that the community departs from this stage as a result of massive and prolonged pressure from unfavourable climatic and oceanographic factors. There is some indication that such an abnormal state of the environment resulting in a dop in pelagic fish abundance can be associated with the solar activity cycle.

Résumé : Dans l'ichtyofaune pélagique néritique de la zone Kuroshio-Oyashio, on peut observer l'interaction entre les processus allogénique et autogénique. La succession autogénique a lieu à la suite de l'interaction écologique du type «concurrence-coexistence» parmi les différentes populations. Dans cette interaction, on observe des changements dans les régimes alimentaires des différentes espèces et dans les processus de production. La prédominance d'une espèce ou de l'autre à différents stades de succession dépend de leur cycle de vie et de l'écologie. Dans le stade initial de succession, les stratèges de la régression, comme les anchois (*Engraulis japonicus*) et les scombrésoces (*Cololabis saira*), jouissent de certains avantages. Le climax de l'ichtyofaune est réalisé lorsque la sardine (*Sardinops sagax melanosticta*) devient dominante. À ce stade, la biocénose utilise au maximum les ressources du biotope et produit la plus grande quantité de biomasse totale. Des signes montrent que la biocénose s'éloigne de ce stade à la suite d'une pression massive et prolongée découlant de facteurs climatiques et océanographiques défavorables. Certaines indications portent à croire qu'un tel état anormal de l'environnement se traduisant par une baisse de l'abondance du poisson pélagique peut être associé au cycle de l'activité solaire.

Introduction

The community of warmwater neritic pelagic fishes that inhabits Northwest Pacific waters in the contact zone of the warm Kuroshio and the cold Oyashio currents has great signigicance for the world fisheries, as it supports an annual catch of up to 7.7×10^6 t (FAO 1991). This community consists of sardine (*Sardinops sagax melanosticta*), Japanese mackerel (*Scomber japonicus*), saury (*Cololabis saira*), anchovy (*Engraulis japonicus*), and Japanese horse mackerel (*Trachurus japonicus*), although the boundaries of the community in terms of its composition are relative.

The abundance of all these species fluctuates greatly and voluminous literature is devoted to these fluctuations. The investigations are stimulated by the great practical significance of the populations, along with the great instability of the fishery resource. The main orientation of the investigations has been the identification of oceanographic and climatic factors that determine the fluctuations in abundance (Kaganovsky 1945, 1960; Schmidt 1945, 1948; Uda 1952, 1957; Ito 1961; Nakai 1962; Bogdanov 1965; Novikov 1973, 1978; Shuntov 1978; Shuntov and Vasil'kov 1981, 1982; Belyaev and Kenya 1987; Novikov and Svirsky 1987), while great attention during the last three decades has been given to determining the cyclicity in these fluctuations, which is considered to be a reflection of a corresponding cyclicity in oceanographic and climatic factors. Within the past 20 yr, a connection has been identified between the cyclicity in Asian fish abundance and some changes in the atmospheric processes in the Pacific sector of the Northern Hemisphere (Davydov 1972, 1975). However, in spite of achievements in studying problems of multiannual cyclicity, attempts at long-

V.V. Kuznetsov. Russian Federal Research Institute of Fisheries and Oceanography, 17, V. Krasnoselskaya, Moscow, 107140, Russia.

term predictions of abundance fluctuations are usually unsuccessful (Shuntov 1978), which undoubtedly reflects an insufficient understanding of the processes.

In most publications, the abundance dynamics of the different neritic pelagic species of the Kuroshio–Oyashio zone are considered separately, the state of abundance of each species in considerable degree being determined as a derivative of the extent of its suitability for the climatic and oceanographic conditions. Abundance rations between species are also considered to be determined by the oceanographic conditions (Novikov and Svirsky 1987). In several papers, the importance of interspecies relations is also indicated (Uda 1957; Novikov 1967, 1973; Novikov and Svirsky 1987), and as Shuntov (1986) states, the problem of long-term fluctuations in fish abundance must be resolved with consideration of the biocenological background. However, in fact, the theoretical base for applying synecological data to the long-term predition of variations in commercial fish abundance does not exist.

The effect of successional processes on abundance variations of neritic pelagic fishes

In our view, failures in lont-term predictions of fish abundance changes are determined to a great degree by considering the biotic component of the ecosystem to be functionally dependent on the abiotic one. The autogenic processes in biological communities are not considered. However, the character of the community reaction on abiotic influence may in considerable degree be determined by the state of community. The methodology for the long-term prediction of large-scale abundance changes in these fishes ought to be based on a consieration of the development of biotic and abiotic ecosystem components. It is necessary to estimate the present ecological and climatological situations and their expected alterations.

In accordance with these principles, commercial catch statistics for neritic pelagic fishes in the Kuroshio–Oyashio zone, as well as the literature on the ecology of these fishes and environmental conditions, were analyzed (Kuznetsov and Kuznetsova 1988; Kuznetsov 1989*a*, *b*). The results of these investigations give evidence of the applicability to the neritic pelagic community of the ecological succession concept, which has been well elaborated on land communities. Applicability of this concept to aquatic communities was limited to shallow inshore areas near the coastline where the substrate character and its alterations by the vital functions of organisms are essential (Odum 1986).

The community of fish studied inhabits an unstable biotope, judging by its physical characteristics, at the point of junction of two currents having different vectors: subarctic and subtropical water masses. This instability results in the sporadic development of considerable anomalies, when a decline in the abundance and biomass of the major community components occurs. Such an environmental anomaly was recorded in 1940–44 (Schmidt 1945, 1948).

The considerable climatic and oceanological anomaly is an allogenic factor in relation to the community. Its powerful and intensive development might lead the community of warmwater neritic pelagic fishes to the initial point of ecological succession. This succession occurs here as a result of an interaction of the "competition–coexistence" type on the population level. The other important factor that affects ecological succession (the changing of the physical environment by the community; Odum 1986) is virtually absent here, which undoubtedly limits the scale of the ongoing succession processes. This limitation is in particular apparent in the absence of a classical series of succeeding communities. However, within the limits of the same community, most essential changes in the different species abundance as well as in production processes occur. Variations in the latter take place in the same direction as in the successional series of communities.

The best opportunities to increase the abundance at the initial state of ecological succession are available to those species characterized by the *r*-strategy (Odum 1986) and by the highest intrinsic population growth rate. These characteristics, according to the classification suggested by Kawasaki (1980), Kawasaki and Hashimoto (1983), and Kawasaki and Kumagai (1984) are typical of the species that belong to the 1A subtype. They are marked by a short lifespan, and the use of energy obtained form outside mainly for reproduction. In the community examined, the subtype 1A species are anchovy and saury. Their high intrinsic population growth rates allow these species to be the first at the initial stage in the competition for the common resource, given the low density of plankton-eating fish populations.

The pattern of feeding of these fishes corresponds to the pioneering role: they consume zooplankton, which has a high calorie value, is easily assimilated, and is preferred by many species. These species, especially anchovy, make up large juvenile concentrations and aggregations of adult individuals in the coastal waters of Japan. Apparently the large numbers of these fishes make the reproduction process of other species of the community more difficult. Juvenile anchovy concentrations in the inshore area are so great that in Japan this fish is subjected to directed fishing (Chikuni 1985).

Yet these species never reach the biomass of sardine or mackerel because of a number of limiting factors. These include the small body size of anchovy, which restricts its potential to inhabit vast water areas, and special features of feeding recorded in saury (Kun 1949; Novikov 1967; Odate 1977; Chikuni 1985). They cannot utilize available great plankton resource to a considerable extent. The development of highly abundant populations of small fish provide an extensive food base for fish-eating predators, which improves the feeding conditions for mackerel. When most accessible and preferred food resources are utilized in feeding areas, sardine begins to acquire greater advantages because it has a wide spectrum of feeding, it consumes in considerable extent phytoplankton, and it inhabits the water area quite widely.

Sardine and mackerel (*K*-strategists, subtype 1B Kawasaki) have a certain potential for increasing their abundance and biomass in conditions of relatively high population abundances of *r*-strategists owing to utilization of additional food resources.

The *K*-strategists live longer, and spend a greater part of energy obtained from outside to support the existence of adult individuals. Populations of fish in this group can accumulate parent stock in generation series, and are capable of reproducing abundant progeny under favourable conditions.

The biological characteristics of sardine allow it to be the dominant species despite the high abundance of all the other representatives of warmwater community. After sardine becomes the dominant species, it exceeds other species of the community considerably in biomass and the biomass of the other fishes decreases to some extent. There are grounds to believe that this is the climax state of community, which is a relatively stable stage. The stability of a superpopulation of sardine as the dominant component of the community is determined by a number of factors: the diversity and high biomass of its food sources, its resistance to short-term unfavourable climatic and oceanographic impacts in connection with its relatively long life history, and the extensiveness of the feeding and spawning area occupied. This stability increases even more because of the extent of a large phenotypic and apparently genotypic versatility (for example, stunted and rapidly growing individuals), and because of a decline in the energy expended on reproduction by the excess parent stock (slower maturation rate, lower fecundity). An upsurge in the abundance of sardine is a manifestation of the biological progress of its population. There are no mechanisms underlying that state that would provide a sharp decline in the abundance. No density-dependent factors in relation to sardine are known (e.g., a change in the substrate, or in the reproduction of parasites), which would tend to reduce its abundance to a very low level.

The succession process takes place under the modifying impact of both abiotic and biotic factors. The rate of development of this process, given large variability in physical environment, is determined in large measure by physical factors. Under the influence of allogenic factors the process can be reduced to an ititial point from any stage. It can slo be accelerated, decelerated, or averted at the expense of fulfilling the conditions that are favourable for the *K*- or *r*-strategists. Another important allogenous factor is intensive fishing. It harvests mostly *K*-strategists and therefore has a tendency to slow and avert the process of autogenous succession. However, until now, natural factors have prevailed in their effect on this particular community. Hence, the actual process is the result of a complex interaction between allogenous and autogenous succession.

The process of autogenous succession of a community is accompanied by an increase in the total biomass of its components, although powerful climatic and oceanographic anomalies may cause a biomass decrease at any stage. At the climax stage, the community utilizes the resources of its biotope most fully and reaches maximum total biomass. It is quite possible that the succession process within the pelagic fish community accompanies certain changes in the status of the plankton community, while there may be a correspondence between the climax status of the pelagic fish community and the state of the plankton community. Some sources indicate that when there is a low abundance of sardine in the Sea of Japan, an increase in the amount of predator plankton is observed (Dulepova et al. 1990), which may be of principal importance for ecosystem restructuring (Kawai and Isibasi 1986).

The climax stage dominated by sardine can last for an indefinitely long period of time. Departure from this relatively stable status becomes most probable as a result of a powerful impact of unfavourable climatic and oceanographic factors, which reduces the total biomass of community members and initiates a new cycle of competition for the common resource. The maturity state of the community by itself does not raise the probability of the onset of low catches, yet it determines the magnitude of the expected drop in the volume of catch.

If roughly assessed, a drop in the abundance of sardine can be regarded as the end of one succession cycle and the beginning of the next one. The lack of a close relationship between the cycle pattern of physical factors and abundance variations in neritic pelagic fishes, as described earlier, becomes understandable if we remember that these changes are governed by the results of a complex interaction of processes of entirely different origin.

Catch variation analysis for sardine and other neritic pelagic warmwater species in the 20th century shows that this concept is applicable. There are grounds to believe that the upsurge in the abundance of sardine in the 1920s and 1930s disappeared because of an unfavourable combination of factors, with temperature being the leading factor. An earlier assumption (Nakai 1962; Shuntov and Vasil'kov 1981, 1982) that the major cause of the decline in sardine was the process of kuroshio meandering that began in the mid-1930s and strengthened after 1938 was not confirmed during the present increase in sardine stock which was not influenced in any noticeable way by Kuroshio meandering.

In the Pacific, negative summer temperature anomalies were recorded east of Hokkaido in 1934–36 and 1941, as described by Arakawa (1975). Abnormally cold conditions began in the area of Sionomisaki (south of Honshu) in 1934–35, and reached their peak in 1937–41 (Uda 1952). The situation again became normal only in 1945. In a vast water area of the Sea of Japan, abnormally cold conditions were observed in 1940–44. Hence, cooling occurred in the ranges of both juveniles and adult fish. The development of these anomalies brought about a deterioration in the feeding conditions for sardine, and cut down the area used by the species, which led to a decrease in growth rate and an increase in natural mortality. The result was a lowering of the reproductive potential of the species. The loss of northern areas from the population's range might have contributed to confining the spawning to more southerly areas. Under these conditions, the formation of a Kuroshio meander could have aggravated the chain of unfavourable events in the status of the habitat, which resulted in a drop in the abundance of sardine.

The process of restoration of the abundance of the community's major populations in ensuing years occurred in keeping with the above biological characteristics of the coexisting species. Abundance and catches of *r*-strategists (saury and anchovy) increased (Fig. 1). Analysis of catches of Japanese horse mackerel indicates that the high abundance of this species coincided in time with the period of prevalence of anchovy and saury. Judging by the nature of their life histories, one might expect that the variations in the abundance of horse mackerel would be similar to those of mackerel and sardine. It should be pointed out, however, that its main spawning grounds are located south of those of other neritic pelagic fishes (Ermakov 1986); therefore, it has no strained competitive relations with the latter species in spawning and in early feeding areas. At the same time, when anchovy and saury dominate in the absence of abundant sardine and mackerel populations, horse mackerel finds a free ecological niche in the area, and its abundance goes up. The further process of the development of the community led to the increase in abun-

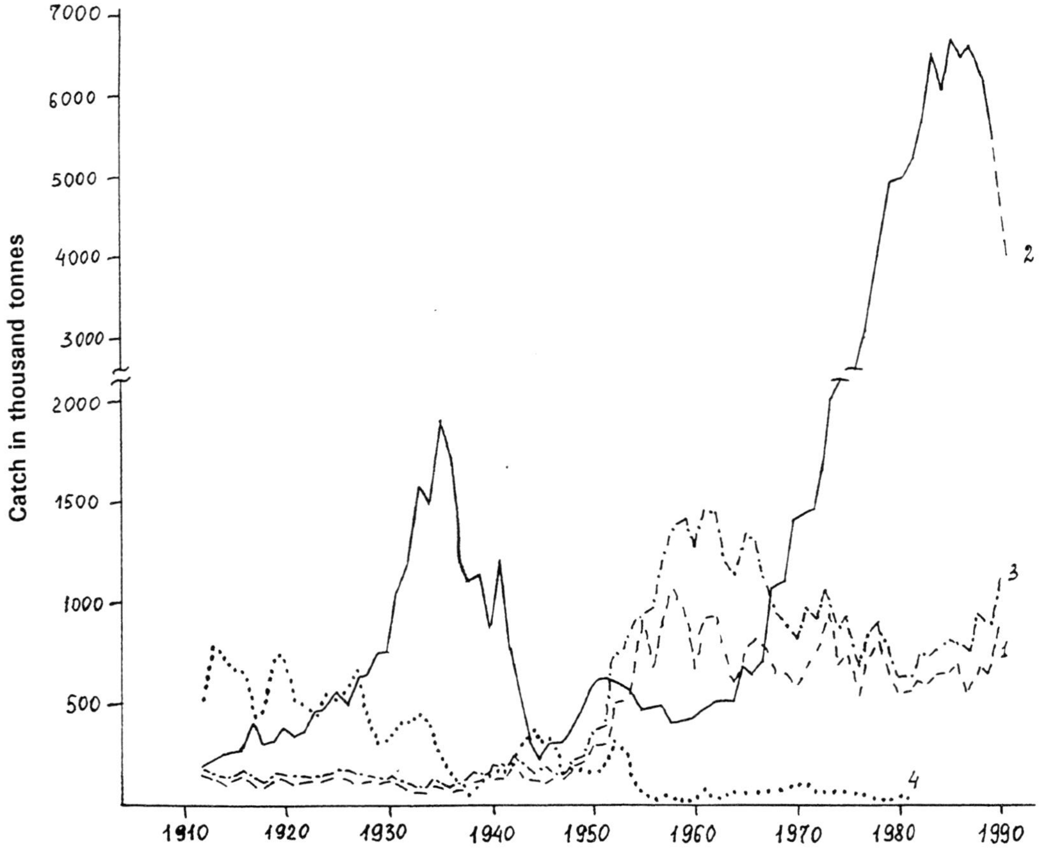

Fig. 1. Commercial catch dynamics of neritic pelagic fishes in the Kuroshio zone by groups: (1) saury and anchovy; (2) Japanese mackerel and sardine; (3) saury, anchovy, and horse-mackerel; (4) herring.

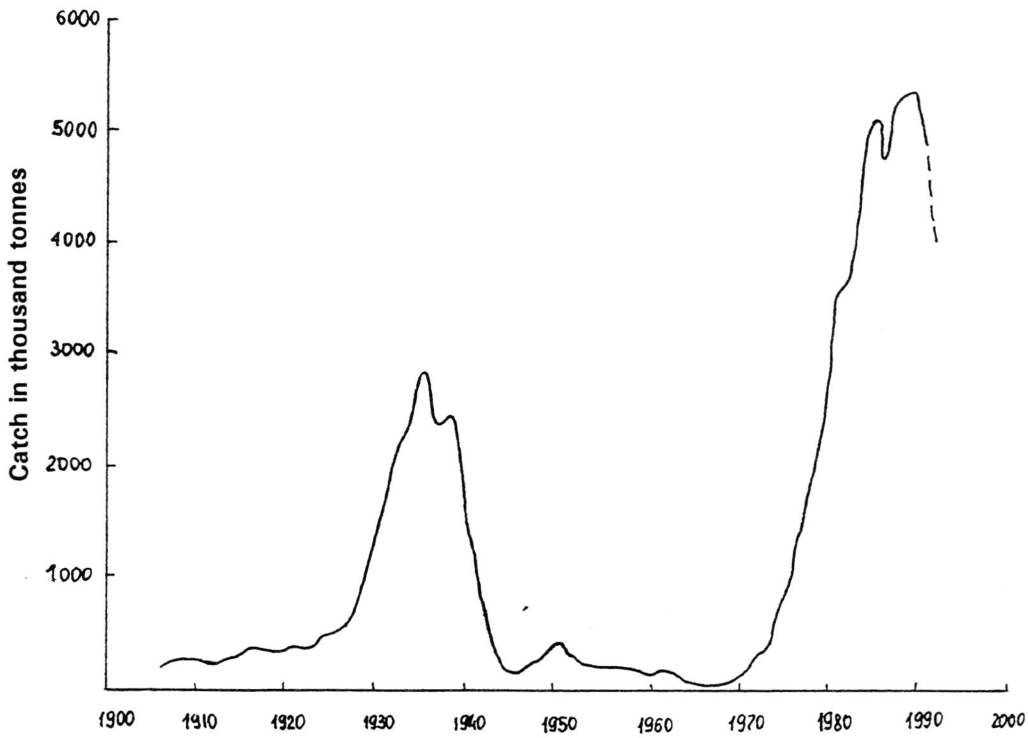

Fig. 2. Commercial catch dynamics of the Asian sardine.

Table 1. Chronology of succession cycles in the warmwater fish community based on large-scale abundance dynamics in sardine.

Abundance	Period	Conditional number of cycle duration 1	2	3	4	5	Average duration of periods (years)	Range of fluctuations (% of the average)
Low	Initial	?	1590	1725	1840	1942	—	—
	Final	1560	1690	1790	1928	1975	—	—
	Years	?	100	65	88	34	84	42
High	Initial	1560	1690	1790	1929	1976	—	—
	Final	1590	1725	1840	1941	?	—	—
	Years	30	35	50	13	>15	33	112
Full cycle	Initial	?	1590	1725	1840	1942	—	—
	Final	1590	1725	1840	1941	?	—	—
	Years	?	135	115	101	>49	117	29

dance of the main *K*-strategists in the area: mackerel and sardine. Here the longer cycle representative of this group (mackerel) was somewhat ahead of sardine, which had dominated in the last turn (Fig. 2).

Analysis of variations in sardine abundance

Some data are available for the periods of high sardine abundance observed in the 16–19th centuries (Uda 1952; Ito 1961). We put together a chronological table of succession cycles (Table 1), based on the ideas described of the regularities of such cycles in the warmwater fish community using data from Ito (1961) and other statistics available for the 20th century. The main feature and characteristics of these cycles are the large-scale fluctuations in the abundance of sardine. In calculating the averages for the table, we did not use the first and last cycle data since the total duration of these cycles is unknown to us. As Ito's basic data are approximate, and since there may only be a conditional boundary between high and low abundance, all the figures in the table are approximate.

The table shows that the average duration of a succession cycle was 117 years. The periods of high abundance of sardine varied between 13 and 50 yr. The duration of these periods is the most variable factor. The least variable factor is the overall duration of the cycle. Thus it turns out that, if the low-abundance phase is extended, the high-abundance phase corresponds to a short stretch of time, and vice versa. It should be pointed out, however, that the data series of observations does not allow us to speak of a definite trend.

The conditions that used to lead to declines in sardine abundance in the past are important as a matter of principle. If the biological properties of the community's climax stage of high sardine abundance allow this stage to exist for an indefinitely long period, the causes of downward trends in abundance ought to be sought in the effect of allogenous factors. Analysis of materials on the upsurge of abundance in the 1920s and 1930s showed the importance of the temperature factor. As for the remote past, there are individual historical records of climate anomalies; however, they are insufficient for the purpose of our study. Thus, we attempted to apply dendrometric data for the examination of climatic conditions in the 16–19th centuries. Digital data on annual increments in the multiannual cypress *Chamaecyparis_obtusa* of central Japan were tabulated by Arakawa (1975).

We thought it reasonable to address the analysis of wood rings to clarify the circumstances concurrent with the falls in sardine abundance for the following reasons: (1) in our view, termination of the periods of high abundance of sardine is associated with considerable changes in climate conditions, which might be reflected in the growth of the tree even under a great complexity of climatic impact; (2) the effect of long-term variability in the cypress growth rate on the results of analysis noted by Arakawa may become much lower when we take relatively limited spans of time; (3) in some cases, dendrometric data can be verified and supplemented with climate data obtained from other sources. Cypress annual growth data were levelled off by the 5-yr weighted running means, and are given in Fig. 3.

The abundant sardine period in the 16th century began during a period of rapid cypress growth. While the sardine abundance upsurge was going on, cypress growth declined gradually. Beginning about 1587, a very profound depression in cypress growth rate occured that lasted for about 20yr. In about 1590 there was a drop in sardine abundance. The coincidence recorded leads one to assume that the drastic deterioration of climate conditions beginning from 1587 resulted in a slowing of the cypress growth rate, and later in a decline in sardine abundance as well. The cypress growth pattern in the first half of the 17th century indicates that the climatic conditions had improved significantly, although this did not cause an increase in sardine abundance.

In the late 17th century, a peak in sardine abundance began during a high level of cypress ring growth. The end of the period of high sardine abundance falls at the time of a cypress-growth depression. It should be noted, however, that soon after the beginning of the sardine peak, there was also a sharp decline in the growth of cypress, yet the conditions that had caused it did not affect sardine abundance. It is not easy to determine the reason for such an accute deceleration of growth in an individual tree in the 1690s. One cannot expect full conformity between such different phenomena as tree growth

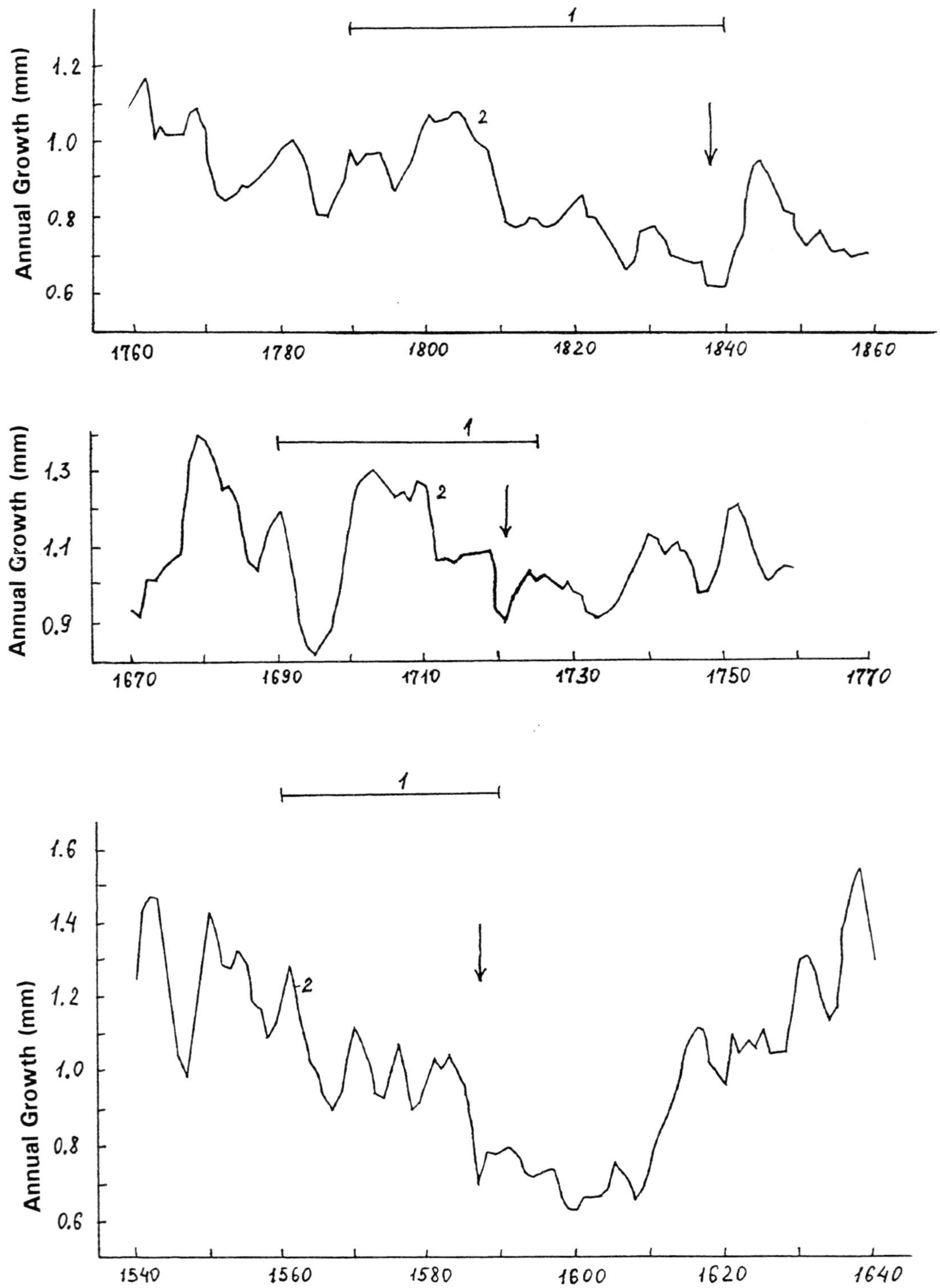

Fig. 3. Periods of high sardine abundance in the 16–19th centuries and growth dynamics of cypress in central Japan: (1) periods of sardine high abundance; (2) dynamics of cypress growth.

and sardine abundance. Tree growth can be affected by the reproduction of vermin, drought, and other causes that may not affect the abundance of sardine.

The rise in sardine abundance in the early 1790s began during a high level of cypress growth. The drop at about 1840 occurs during a period of deep depression in cypress growth. Additional information on that particular upsurge is available that provides some explanation of the circumstances concommittant with its eventual decline. Between 1835 and 1840 negative deflections of global average temperatures from normal levels were recorded; during 5 yr those deflections were quite notable. In 1833–34 and 1836–39, there were exceptionally high prices for rice in Kyoto connected with a poor crop caused by severe cold weather (Arakawa 1975). Prices were particularly high in 1837. A poor harvest brought famine in northern Japan, and the deaths of hundreds of thousands of people, which caused a noticeable drop in population. The exceptionally abnormal weather conditions of those times are associated with the large eruption of the Kosegina volcano in Nicaragua on January 20, 1835, which lowered direct solar radiation, and with northeastern winds ("yamase") that blew continously during the 3 summer months, affecting the rice and fruit harvests.

The data considered confirm our previous assumption that the climax stage of the succession process most likely terminates as a result of the development of a powerful and long-term climatic anomaly, and that conditions are important not only during spawning but also during feeding (summer). Resulting from this analysis of dendrometric data, the following principal conclusions can be made:

1) Sardine abundance increases occur during periods of good growth of cypress. The beginning of high-abundance periods does not coincide with the start of good cypress growth. It is likely that climatic conditions favourable to sardine are formed long before the beginning of the rise in sardine abundance. By the time sardine dominates, climatic conditions may begin to deteriorate, which reduces the duration of the period of high abundance.
2) Drops in the abundance of sardine agree with the periods of slow cypress growth. The abundance of sardines decreases with some delay in relation to the start of a decrease in cypress growth, which is explained by the great inertia in the process of fish abundance variations versus tree growth.
3) Some signs of a gradual deceleration in cypress growth are recorded when sardine abundance decreases, which may be a reflection of the gradual change in the climatic processes affecting the fish and the trees.

Potential forecasting of the status of the neritic pelagic fish community

Projection of future community dynamics can be made well in advance only on a probability basis. The timing of the termination of the climax stage in the development of the sardine-dominated community can be predicted by determining when the probability of formation of significant climatic and oceanographic anomalies would increase. Of special interest would be a collation of large-scale variations in sardine abundance and the century-long course of solar activity variations (Wolf numbers), with which the pattern of atmospheric transfers is connected. It is believed that the rise in solar activity causes longitudinal atmospheric transfers, while the fall in solar activity contributes to zonal transfers (Girs 1977). In longitudinal processes, the interlatitudinal transfer of heat increases, which makes those processes conducive to the development of climate anomalies that are not favourable to warmwater species. In recent times, very interesting data were obtained that pointed to a significant relationship between solar activity and the value of the solar constant (Willson et al. 1981, 1986; Willson and Hudson 1988, 1991) (i.e., solar activity affects the amount of energy that the sun radiates). Here, long-term climate fluctuations happen to be well-correlated with solar activity fluctuations (Chistyakov 1986). Solar activity data were taken from Loginov (1979), Russian (USSR) Academy of Sciences (1979–92), and Kopetsky (1991).

The results of that collation (Fig. 4) indicate that three peaks in sardine abundance, of the four known ones (disregarding the present one), ended in the period when the long-term rise in solar activity began. Two ended approximately during the development of the first maximum in the rise, and one ended somewhat earlier. One upsurge in sardine abundance (16th century) ended when solar acitvity was in a sharp decline. It is of special interest that during that period a very close relationship between the growth of cypress and the multiannual course of solar activity was observed. It is possible that the decline in solar activity in the second half of the 16th century was accompanied by a considerable deterioration in climate conditions in areas inhabited by sardine. This is supported by the fact of considerable general cooling of Earth's climate in the 15th and 16th centuries that occurred together with the onset of the Little Ice Age in the 16th and 17th centuries, which is associated with the Maunder minimum of solar activity (Willson and Hudson 1988). This case illustrates the ambiguity of the effect of solar activity variations on the fish community. This ambiguity is governed by the comprehensiveness of the climatic processes that occur. On the one hand, when solar activity decreases, the probability of the development of long-term

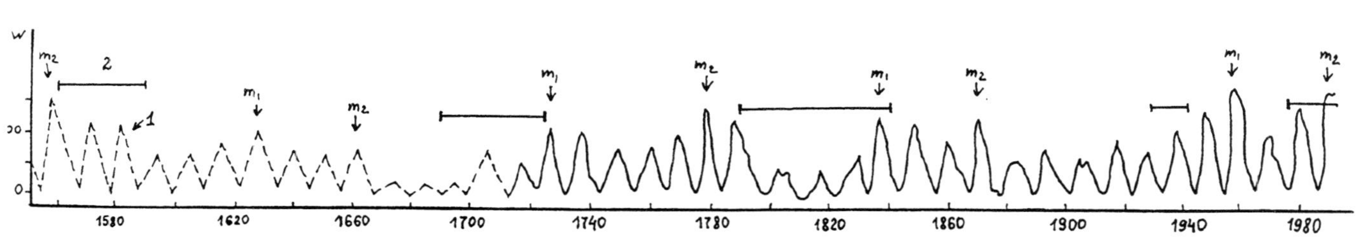

Fig. 4. Periods of high sardine abundance on the background of the long-term dynamics of solar activity: (1) chronological graphic of solar activity (broken part of line corresponds to reconstructed approximate data); (2) period of high sardine abundance. m_1 and m_2 show two maximums in every century rise of solar activity.

anomalies related to atmospheric transfers decreases too. On the other hand, the amount of solar energy received by the Earth diminishes, which in the long term leads to considerable cooling of the climate, even as far as the onset of the Little Ice Age.

As for the timing of the emergence of the peaks in sardine abundance, two of the four began soon after the passage of the second maximum in the long-term rise in activity; that is, at the early initial stage of the recession process on the century scale. The other two peaks began in the period of low activity on century scale, which means that there is no clearly expressed relationship between the start of an abundance upsurge and the status of solar activity. This fully agrees with the concept of autogenous succession cycles that occur in the community. On the whole, one might consider that the probability of a recession in sardine abundance increases greatly during the periods of the rises in solar activity in the century cycles. The fit of periods of low solar activity with sardine abundance was previously described by Shuntov and Vasil'kov (1982).

The growth in sardine abundance during the upsurge previous to the most recent one was accompanied by a pronounced process of warming in the 1920s and 1930s. The recession point coincided with the change in climate trend toward cooling, although temperature deviations from the norm were, on the whole, positive for the Northern Hemisphere during that period (Budyko 1977).

As for the present abundance peak, until recently there were grounds to believe that it would exist mostly under a low level of solar activity. A low activity level was expected until 2016 or even 2091 (Chistyakov 1983; Gregg 1984; Fairbridge and Shirley 1987). Such a situation would probably favour a flourishing population of sardines. But the actual solar activity has risen greatly since 1988, and its level in 1989, 1990, 1991, and early 1992 was very high. The present 11-yr cycle N22 is matched with the powerful second maximum of the long-term rise in solar activity. Population abundance and commercial catches of sardine should decrease rapidly given the status of this climate-forming factor. It looks as if the present succession cycle in the development of community is ending.

The special feature of the present peak in sardine abundance is that it occured during a relatively short-term decline in solar activity between the two maxima of the long-term rise. Thus, it appears to be "doomed" to be relatively transient. This is no less true of the previous upsurge, which began just before the start of the long-term rise in century-long cycle in solar activity.

In connection with the appearance of an "extraordinary" upsurge, there were two peaks in sardine abundance in this century, whereas previously such peaks appeared roughly once in a century. It is possible that the present global warming has played some role there.

The end of the succession cycle in the community's development will inevitably result in a fall in its overall production, and a decrease in catches. A community status where saury, anchovy, and horse mackerel are abundant appears to be probable in future.

As far as long-term prospects are concerned, it should be pointed out that at present the 115-yr cycle of solar activity that took the place of the 95-yr cycle in about 1905 is ending. These two cycles jointly make up a 210-yr cycle of solar activity. As the multicentury pattern shows, the century minimum at the point of junction of those cycles was "wide", i.e., the period of low level in solar activity between two rises was long (Chistyakov 1986). The long declines in solar activity known to have occured in the past (the Maunder minimum, 1645–1715; the Shperer, 1416–1534; and the Wolf, 1232–1342) also coincided with the end of 95-yr cycles. Now we are witnessing an acute downward trend in the115-yr cycle, which will be replaced by a 95-yr cycle. From the principle of mirror symmetry of century-long cycles, it is expected that the forthcoming decrease in activity will not last long. Its rise is expected to occur after 2020. From the positive correlation between solar radiation, together with the climate dependent on it, and solar activity, as well as in view of the human-induced climate warming, one might assume that future climate conditions will favour warmwater fishes. Between the 1990s and the 2030s, sardine will probably have an opportunity to increase its abundance greatly once again.

The existence of ties between the status of the community studied and solar activity, as anticipated, as well as the existence of certain possibilities for forecasting long-term solar cycles, provide some basis for projecting the future status of the community. Nevertheless, this matter needs deep study and a retrospective analysis. At present, virtually all those major phenomena on which such prospective forecasts could be based (processes in the sun, solar-earth relations and climate, and effects of climate variations on the fish community) have not been well studied, although the results obtained are encouraging.

By analogy with the case of sardine, it can be assumed that the present high level of biomass of the other mass pelagic species of the Pacific, pollock (*Theragra chalcogramma*), can also be related to the climax status of the community to which it belongs. The loss of stability in climate conditions that was stimulated by the recent rise in solar activity can affect populations of this and other species.

If we are to conserve abundant populations of pelagic fishes of high value, which exclusively provide high fish production (usually *K*-strategists), their commercial harvest should not be brought to estimated maximum sustainable yield (MSY) level, given the environmental instability caused by climatic variations. A considerable reserve must be left. Intensive fishing is good for the rejuvenation of fish populations and decreases their diversity in terms of age, size, and other characteristics, which cannot but reduce their stability when exposed to short-term adverse environmental effect (bring them closer to the characteristics of *r*-strategists). However, such a fishing strategy will not be productive if the abnormal environmental status is a long-term one. As we know, under low fishing intensity in previous centuries, highly abundant valuable populations suffered recessions repeatedly under the impact of natural factors. An alternative strategy would be maximum commercial utilization of populations that are doomed to a long depression inflicted by significant climate changes. However, given the contemporary potential for forecasting future climate changes, and the response of populations and communities to these changes, there is no reliable basis for making a choice between high-intensity and low-intensity fishing strategies.

References

Arakawa, H. 1975. Climate changes. Gidrometeoizdat, Leningrad. 107 p. (In Russian)

Belyaev, V.A., and Kenya V.S. 1987. Abundance and reproduction condition of far-eastern sardine in the northwest Pacific, p. 225–237. *In* Biological resources of the high seas. Nauka, Moscow. (In Russian)

Bogdanov, G.A. 1965. Some aspects of abundance dynamics of most important commercial sardine species. Sci. Inf. VNIRO 9: 3–11. (In Russian)

Budyko, M.I. 1977. Global ecology. Mysl', Moscow, 328 p. (In Russian)

Chikuni, S. 1985. The fish resources of the northwest Pacific. FAO, Rome. Fish. Tech. Pap. 226: 190 p.

Chistyakov, V.F. 1983. Prediction of solar activity till year 2030. Sol. Data 1: 97–100. (In Russian)

Chistyakov, V.F. 1986. One hundred and two hundred ten years cycles of solar activity. Sol. Data 6: 88–95. (In Russian)

Davydov, I.V. 1972. On oceanological condition of appearance of some year classes of herring in the northwest Bering Sea. Izv. TINRO 82: 281–307.

Davydov, I.V. 1975. Some features of atmospheric circulation upon the northwest Pacific and Their connection with water regime of Kamchatka's shelf. Izv. TINRO 97: 157–171. (In Russian)

Duplepova, E.P., Lapshina, V.I., and Blagoderov, A.I. 1990. Functioning elements of the pelagic system of the Sea of Japan, p. 61–70. *In* Results of ecosystem investigations of biological resources of the Far East seas. TINRO, Vladivostok, Russia. (In Russian)

Ermakov, Yu.R. 1986. Horse-mackerels, p. 247–258. *In* Biological resources of the Pacific ocean. Nauka, Moscow. (In Russian)

Fairbridge, R.W., and Shirley, J.H. 1987. Prolonged minima and the 179-yr cycle of the solar inertia motion. Sol. Phys. 10: 191–220.

FAO. 1991. Yearbook of fishery statistics 68. Rome. 518 p.

Girs, A.A. 1977. Large-scale interaction of atmosphere with ocean and its influence on reconstruction of atmospheric circulation types and climate oscillation. Tr. AANII 339: 160–169. (In Russian)

Gregg, D.P. 1984. A nonlinear solar cycle model with potential for forecasting on a decadal time scale. Sol. Phys. 90: 185–194.

Ito, S. 1961. Fishery biology of sardine, *Sardinops melanosticta* (Temm. et Schleg.), in the waters around Japan. Bull. Jpn. Reg. Fish. Res. Lab. 1: 227 p.

Kaganovsky, A.G. 1945. The problem of far eastern sardine. Rybn. Promyschl. USSR 3: 8–13. (In Russian)

Kaganovsky, A.G. 1960. Contemporary status of sardine stock. Tr. Soveshch. Ikhtiol. Kom. Acad. Nauk. USSR 10: 149–150. (In Russian)

Kawasaki, T. 1980. Fundamental relations among the selections of life history in the marine teleosts. Bull. Jpn. Soc. Sci. Fish. 46(3): 367–377.

Kawasaki, T., Hashimoto, H. 1983. Selections of life histories and its adaptive significance in a snailfish *Liparis tanakai* from Sendai Bay. Bull. Jpn. Soc. Sci. Fish. 49(3): 367–377.

Kawasaki, T., and Kumagai, A. 1984. Food habits of the far eastern sardine and their implication in the fluctuation pattern of the sardine stocks. Bull. Jpn. Soc. Sci. Fish. 50(10): 1657–1663.

Kopetsky, M. 1991. Height of maxima of 11-year cycles from 1500 to 1750. Sol. Data 12: 69–73. (In Russian)

Kun, M.S. 1949. Some materials on saury feeding. Izv. TINRO 31: 196–197. (In Russian)

Kuznetsov, V.V. 1989*a*. Variations in abundance of neritic pelagic fishes of the kuroshio zone and successional processes in ichthyocoenosis. Sci. Pap. Higher School Biol. Sci. 4: 25–31. (In Russian)

Kuznetsov, V.V. 1989*b*. Retrospective analysis of large-scale variations in West Pacific sardine abundnace, p. 84–97. *In* Biological background of abundance dynamics and fish catch prediction. VNIRO, Moscow. (In Russian)

Kuznetsov, V.V., and Kuznetsova, E.H. 1988. On ecological interrelations and abundance dynamics of warm water pelagic fishes in the zone of Kuroshio influence. Rybn. Khoz. 4: 51–54. (In Russian)

Loginov, V.F. [ed.] 1979. Catalogue of indices of solar and geomagnetical activity. All-Union Research Institute of Hydrometeorological Information, World Data Centre, Obninsk, Russia. 200 p. (In Russian)

Nakai, Z. 1962. Studies relevant to mechanisms underlying the fluctuation in the catch of the Japanese sardine, *Sardinops melanosticta* (Temm. et Scleg.). Japn. J. Ichthyol. 19(1–6): 1–113.

Novikov, Yu.V. 1967. Main features of biology and stock status of Pacific saury. Izv. TINRO 56: 3–50. (In Russian).

Novikov, Yu.V. 1973. Some specific features in studying the reproduction ecology of saury and other fish species of Kuroshio community. Izv. TINRO 87: 155–162. (In Russian)

Novikov, Yu.V., and Svirsky, V.G. 1987. Long-period abundance changes of basic commercial fishes in the North-West Pacific, p. 189–215. *In* Biological resources in the offshore ocean. Nauka, Moscow.

Odate, K. 1977. On the feeding habits of the Pacific saury, *Cololabis saira* (Brevoort). Bull. Tohoku Reg. Fish. Res. Lab. 38: 75–88.

Odum, E.P. 1986. Basic ecology. Vol. 2. Mir, Moscow. 376 p. (In Russian)

Russian (USSR) Academy of Sciences 1979–92. Solar data, monthly bulletin. Sain Petersburg.

Schmidt, P.Yu. 1945. Problem of far eastern sardine. Rybn. Promyschl. USSR 1: 3–13. (In Russian)

Schmidt, P.Yu. 1948. Fishes of the Pacific Ocean. Pischepromizdat. 124 p. (In Russian)

Shuntov, V.P. 1978. Fishes of Kuroshio Current, p. 29–44. *In* Ocean and man. Vladivostok. (In Russian)

Shuntov, V.P. 1986. The state of studies of many year cyclic changes in abundance of Far Eat sea fishes. Mar. Biol. 3: 3–14. (In Russian)

Shuntov, V.P., and Vasil'kov, V.P. 1981. Abundance dynamics of Far East and California sardine in the XX century. Vopr. Ikhtiol. 21(6): 963–975. (In Russian)

Shuntov, V.P., and Vasil'kov, V.P. 1982. Epochs of atmosperic circulation and circlicity of Far East and Kalifornia sardine abundance dynamics. Vopr. Ikhtiol. 22(2): 187–199. (In Russian)

Uda, M. 1952. On the relation between the variation of the important fisheries condition and the oceanographical condition in the adjacent waters of Japan. I. Tokyo Univ. Fish. 38: 363–389.

Uda, M. 1957. A consideration of the long years trend of the fisheries fluctuation in relation to sea conditions. Bull. Jpn. Soc. Sci. Fish. 23(7–8): 368–372.

Willson, R.C. 1981. Observation of solar irradiance variability. Science (Wash., D.C.) 211(4483): 700–702.

Willson, R.C. 1986. Long-term downward trend in total solar irradiance. Science (Wash., D.C.) 234(4780): 1114–1117.

Willson, R.C., and Hudson, H.S. 1988. Solar luminosity variations in solar cycle 21. Nature (Lond.) 332(6167): 810–812.

Willson, R.C., and Hudson, H.S. 1991. The sun's luminosity over a complete solar cycle. Nature (Lond.) 351(6321): 42–44.

Wind climate and foraging of larval and juvenile Arcto-Norwegian cod (*Gadus morhua*)

Svein Sundby

Sundby, S. 1995. Wind climate and foraging of larval and juvenile Arcto-Norwegian cod (*Gadus morhua*), p. 405–415. *In* R.J. Beamish [ed.] Climate change and northern fish populations. Can. Spec. Publ. Fish. Aquat. Sci. 121.

Abstract: This study demonstrates the effect of turbulence-induced contact rate after the first-feeding stage, as cod (*Gadus morhua*) larvae grow, increase their swimming speed, and change their diet to larger prey of lower concentration. Norwegian and Russian data are used to assess the change in diet of cod larvae as they grow and to model the time-dependent prey concentrations and swimming speed of cod. Time series of plankton surveys and wind data are applied to explore the interannual variations and the relative importance of prey concentration and turbulence to the overall contact rate between cod and their prey. The simulations show that the turbulence-induced contact rate decreases exponentially as cod grow, but significantly influences feeding until the stage of 2-mo-old juveniles. Interannual variations of wind speed substantially influence the overall contact rate between cod and prey. The year-class strengths of the Arcto-Norwegian cod are largely determined within the first months after hatching. Temperature has earlier been shown to be one of the factors influencing growth and survival during this stage. This study indicates that wind-induced turbulence is another important factor.

Résumé : Cette étude montre l'effet du coefficient de contact provoqué par la turbulence après le stade de la première alimentation, lorsque les larves de morues (*Gadus morhua L.*) croissent, augmentent leur vitesse de nage et modifient leur régime alimentaire pour s'attaquer à des proies plus grosses en concentration plus faible. On utilise les données norvégiennes et russes pour évaluer le changement de régime alimentaire des larves de morues en croissance et pour modéliser la vitesse de nage de la morue et les concentrations de proies selon le temps. On applique les séries chronologiques provenant des études sur le plancton et les données sur les vents pour examiner les variations interannuelles et l'importance relative de la concentration des proies et de la turbulence sur le coefficient de contact global entre la morue et ses proies. Les simulations montrent que le taux des contacts provoqués par la turbulence décroît de façon exponentielle à mesure que la morue grandit, mais influe de façon importante sur l'alimentation des morues juvéniles jusqu'au stade de deux mois. Les variations interannuelles de la vitesse du vent influent substantiellement sur le coefficient de contact global entre la morue et les proies. Les effectifs des classes annuelles de la morue arctico-norvégienne sont en grande partie déterminés au cours des premiers mois après l'éclosion. On a déjà montré que la température était un des facteurs qui influaient sur la croissance et sur la survie durant ce stade. L'étude montre que la turbulence provoquée par le vent constitue un autre facteur important.

Introduction

The influence of climate change on marine ecosystems is composed of the influences of changes in temperature, light, and motion. Many aspects of temperature have been focused on throughout the history of research on marine ecosystems, whereas the effects of motion, for many reasons, have been less studied. Even though the influence of mixing processes on plankton distribution must be considered as relatively well understood, we have just recently become aware of the effect of turbulence on contact rates between individual planktonic particles, theoretically outlined by Rothschild and Osborn (1988).

S. Sundby. Institute of Marine Research, P.O. Box 1870 Nordnes, N-5024 Bergen, Norway.

Based on field data from the coast of northern Norway, Sundby and Fossum (1990) calculated encounter rates between first-feeding cod (*Gadus morhua*) larvae and their prey and compared them to the theory of Rothschild and Osborn (1988); their results agreed well with the theory. Both indicated an enhancement of the encounter rate by a factor of two to three when the wind speed increased from 2 to 6 $m \cdot s^{-1}$. Recent field investigations on cod larvae gave additional support to the theory at higher wind speeds (Sundby et al. 1994). They found that the encounter rate increased by a factor of five as the wind speed increased from 2 to 10 $m \cdot s^{-1}$. Their data also indicated an even stronger effect on encounters in areas of strong tidal mixing. MacKenzie and Leggett (1991) used a function where the rate of turbulent energy dissipation was depth dependent through the mixed layer. They found that the contact rate between fish larvae and their prey could increase by a factor of 10 from nonturbulent conditions to turbulent conditions corresponding to wind speeds of 15 m·s^{-1}. However, turbulence may also influence several other processes, which may have some opposing effects on feeding and growth. Davies et al. (1991) made model simulations of the combined effects of turbulence on plankton encounter rates and on assumed prey patchiness. They pointed out that turbulence may have opposing effects for some particular distributions. Studies on grazing copepods also indicate that turbulence may have multiple effects on the feeding of zooplankton (Alcaraz et al. 1988; Saiz and Alcaraz 1992).

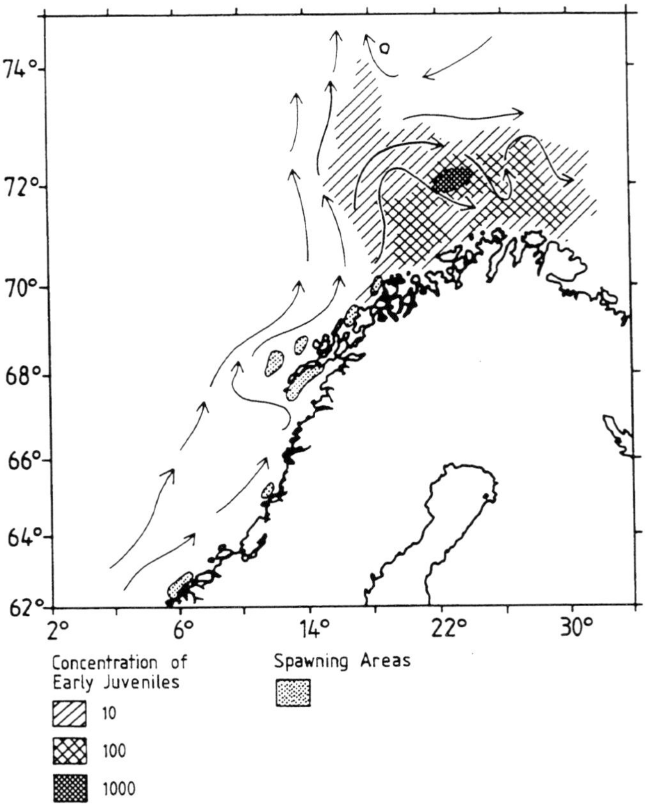

Fig. 1. Spawning locations (dotted areas) and distribution of early juveniles 2 mo after hatching (hatched areas) for the Arcto-Norwegian cod. Arrows indicate general circulation features.

The investigations referenced above all considered feeding by relatively small and slow-swimming organisms, from grazing copepods up to first-feeding fish larvae. However, larger-scale turbulence may also have the potential to enhance contact rates for larger organisms. In the present paper, I explore the effect of turbulence on Arcto-Norwegian cod larvae as they grow, increase their mobility, and change their feeding to larger prey of lower concentration. The upper limit of the influence of "small-scale" turbulence on encounter rate is modeled. During the modeling period, the cod larvae drift in the Norwegian Coastal Current from their spawning areas off the coast of northern Norway toward the northeast, and are gradually mixed into the Norwegian Atlantic Current. Figure 1 gives an overview of spawning locations and the general current pattern that takes the larvae from the spawning locations to their distribution at the early juvenile stage, 2 mo after hatching.

Materials and methods

To simulate contact rates between the cod larvae and their prey as the larvae grow, it is necessary to determine a set of parameters: the time-dependent speeds of predator and prey, the turbulent velocities of the pelagic layer derived from wind velocities, and prey concentrations.

Predator cruising speed

The speed of first-feeding larvae was taken from Solberg and Tilseth (1984), who studied swimming behaviour, together with growth, oxygen uptake, energy content, and feeding success in a small aquarium (volume of 3.5 L). The activity was at a maximum at the end of the yolk-sac stage. The average cruising speed then was 0.17 cm·s^{-1}, corresponding to 0.4 body lengths·s^{-1}. The speed of the larger cod larvae was estimated from Skiftesvik and Huse (1987) and Skiftesvik (1994). They studied cod larvae, kept in 3-5°C (similar to temperatures in the field) in larger aquaria (volume of 70 L), from the first-feeding stage to 70 d after hatching. The position of the larvae was recorded every second. They were actively swimming 50% of the time. The 5-min average cruising speed was lower than the speed recorded by Solberg and Tilseth (1984). The speed, measured in body lengths·s^{-1}, increased as the larvae grew, from 0.15 at first-feeding to 0.25 for 30-d-old larvae. In the present simulations, I used an average of the papers cited above, i.e., 0.33 body lengths·s^{-1}.

Length data for the first-feeding larvae in the field were taken from Ellertsen et al. (1989). The lengths of 18- to 38-d-old larvae were derived from dry weights in a study by Blom et al. (1991), who reared cod in large enclosures. The lengths of the early juveniles, 60–80 d old, were taken from a data base of the Norwegian post-larvae surveys conducted during 1977–91. The data are partly displayed in Bjørke and Sundby (1987). The length data were fitted to a Laird–Gompertz growth function (Bolz and Lough 1988):

$$L(t) = 4.6\exp\{5.9[1 - \exp(-0.0050t)]\} \quad (1)$$

where L is larval length in millimeters, and t is time in days (Fig. 2). Equation 1 is multiplied with the average cruising speed (in body lengths per second), and the larval cruising speed as a function of time, $v(t)$ in centimeters per second, becomes:

$$v(t) = 0.152\exp\{5.9[1 - \exp(-0.0050t)]\} \quad (2)$$

Prey cruising speed

Nauplii of *Calanus finmarchicus* are the main prey for first-feeding cod larvae (Ellertsen et al. 1989). The average length is about 250 μm. As the larvae grow they switch to larger stages of *C. finmarchicus* (Sysoeva and Degtereva 1965). The main diet of early juvenile cod (70 d) consists of copepodite stages IV and V (Helle 1993). The average length of these two stages is 2.2 mm. The average cruising speed was assessed as 0.5 body lengths·s^{-1} by Sundby and Fossum (1990). They derived the value from data on *Euphasia pacifica* by Torres and Childress (1983). From this, the cruising speed of nauplii is assessed as 0.01 cm·s^{-1}, increasing to 0.1 cm·s^{-1} for copepodite stages IV and V.

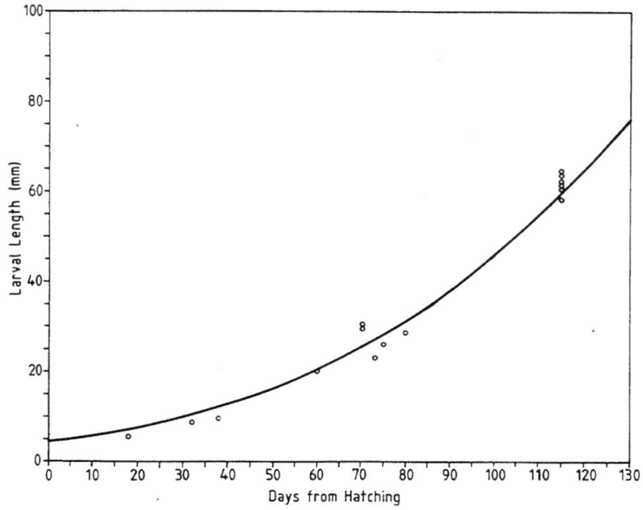

Fig. 2. Larval–juvenile lengths of cod and regression line with respect to Laird–Gompertz growth function. The data are taken from the international 0-group surveys, the early juvenile surveys and from experiments on cod rearing in large enclosures.

Turbulent velocities

The relevant scales of turbulent velocities for the present simulations are all larger than the Kolmogorov scale (see paragraph about prey concentrations below), and therefore they are within the range of scales where the turbulent kinetic energy spectrum is proportional to $k^{-5/3}$, where k is the wave number. Under these conditions the mean-square turbulent velocity, w^2, can be expressed (according to Rothschild and Osborn 1988):

$$w^2 = 3.62(\varepsilon r)^{2/3} \tag{3}$$

where ε is the rate of turbulent kinetic energy dissipation and r is the separation distance.

Cod larvae and early juveniles are confined to the upper mixed layer (above 40 m depth) with the peak concentration at about 10–15 m depth, and it is assumed that the relation between the rate of turbulent dissipation energy and wind speed established by Oakey and Elliott (1982) applies to the habitat of the cod larvae:

$$\varepsilon = (U/91)^3 \tag{4}$$

where U is the wind velocity in meters per second.

Prey concentrations and separation distance

Nauplii and juvenile copepod stages of *C. finmarchicus* are the main prey organisms for larval and early juvenile Arcto-Norwegian cod. The larvae start feeding on the nauplii (Wiborg 1948; Ellertsen et al. 1984). At this stage, more than 90% of the food is nauplii of *C. finmarchicus*. The hatching of cod larvae is normally more or less matched to the production of nauplii, depending on sea temperature, which strongly influences the onset of *Calanus* spawning (Ellertsen et al. 1989). As the cod larvae grow, they switch to the larger stages (Sysoeva and Degtereva 1965). At the beginning of July, the main prey is stage IV and V with cephalothorax lengths of 1.9 and 2.4 mm, respectively. *C. finmarchicus* is the predominant component in the diet not because of selection of the prey, but rather due to the predominant production of *C. finmarchicus* among zooplankton of suitable size for capture (Helle 1993). Pavshtiks and Timokhina (1972) estimated the total production of zooplankton in the Norwegian Sea in the period 1959–69. From their data it can be calculated that *C. finmarchicus* made up from 53–89% of the production of the mesozooplankton of suitable size. Lie (1961) found in the northeastern Norwegian Sea that *C. finmarchicus* was the dominating copepod in June and that more than 50% by biomass of the *C. finmarchicus* population consisted of the copepodite stages IV and V. During the early juvenile survey in July, the copepodite stages IV and V dominated the population of *C. finmarchicus* in the sea and in the stomach of the cod (Helle 1993).

Table 1 lists data from the literature on surveyed concentrations of prey organisms for larval and early juvenile cod. However, only two, Ellertsen et al. (1989) (data set A) and Helle (1993) (data set I), referred to zooplankton counts where cod was actually sampled and the gut content analysed. The other data in Table 1 (which show lower prey concentrations)

Table 1. Survey of characteristic concentrations (No.·L^{-1}) of prey organisms for Arcto-Norwegian cod along their drift route from first feeding (beginning of May) to early juvenile (mid-July) stages. * = averages of sections. The others are values from single stations. Reference E is *O*cean *W*eather *S*tation M(ike) in the Norwegian Sea 66°N, 2°E.

Ref. in Fig. 3	Region	Time	Prey type	Prey concentration (No.·l^{-1})	Reference
A	Lofoten	Early May	Copepod nauplii	1–50	Ellertsen et al. (1989)
B	NE Norwegian Sea	April–May	*C. finmarchicus* nauplii	0.4–3.5 *	Degtereva (1966)
C	NE Norwegian Sea	Early May	*C. finmarchicus* cop.	1 *	Timokhina (1962)
D	NE Norwegian Sea	Early June	*C. finmarchicus* cop.	0.3 *	Timokhina (1962)
E	OWS Mike	June	Copepods	2.5–5	Østvedt (1955)
F	NE Norwegian Sea	June–July	*C. finmarchicus* cop.	0.2–1.1 *	Sysoeva and Degtereva (1965)
G	NE Norwegian Sea	June–July	*C. finmarchicus* cop.	0.2–2.0 *	Degtereva (1966)
H	Norwegian Sea	Mid July	*Calanus*	0.15–0.6	Pavshtiks (1964)
I	NE Norwegian Sea	Mid July	*C. finmarchicus* cop.	0.1–5.0	Helle (1993)

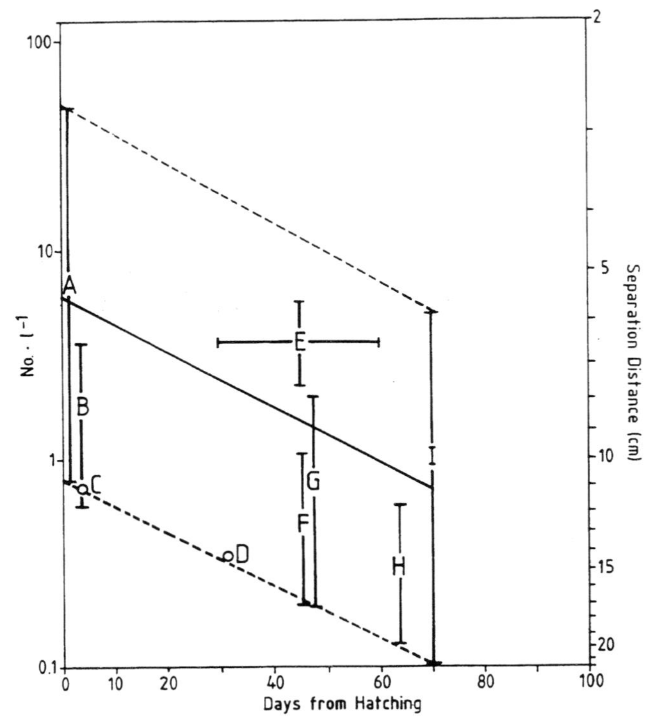

Fig. 3. Prey concentration for Arcto-Norwegian cod from first-feeding larvae to 70-d-old juveniles. Details of the data sets are in Table 1.

are averages of sections of a larger area partly outside the distribution of cod. The data are plotted in Fig. 3, which shows a general decrease in the prey concentration by about one order of magnitude as cod grow from first-feeding larvae (feeding on the naupliar stages) to the early juvenile stage (about 70 d). It is assumed that the data indicate an exponential decay of the prey concentration, which is also reasonable to assume from general considerations about mortality, growth, and dispersion of the zooplankton. The average concentrations of data set A for first-feeding larvae and data set I for early juveniles give the relation:

$$N(t) = N_0 \cdot e^{-0.03t} \tag{5}$$

where $N(t)$ is the prey concentration at time t (in days after first feeding) and N_0 is the initial prey concentration at first-feeding. This equation expresses how prey concentration changes as the cod pass from first-feeding to early juvenile stages. For the average concentration: $N_0 = 6 \cdot L^{-1}$; for the low concentration, $N_0 = 0.8 \cdot L^{-1}$; and for the high concentration, $N_0 = 50 \cdot L^{-1}$. The three functions are plotted in Fig. 3. It can be assumed that the separation distance, r, experienced by the cod larvae is approximately equal to $N^{-1/3}$.

Contact rate between cod larvae and their prey

The above data developed for the time-dependent swimming speed of cod larvae and their prey, and for turbulence at relevant scales is applied to the model of plankton contact rates by Rothschild and Osborn (1988) to explore the effect of wind-induced turbulence on the feeding of Arcto-Norwegian cod as they grow through the larval stages to become juveniles. The velocity component of the contact rate is

$$A(t) = \frac{u(t)^2 + 3v(t)^2 + 4w(t)^2}{3[v(t)^2 + w(t)^2]^{1/2}} \tag{6}$$

where $u(t)$ is the prey speed, $v(t)$ is the predator speed, and $w(t)$ is the root-mean-square turbulent velocity.

Interannual variations in prey availability

To investigate the effect of wind-induced turbulence on the interannual variations in the overall contact rate, Z, between cod and their prey, time series of wind and zooplankton data were compared to analyse the relative variation in the density (concentration) and velocity components of the contact rate:

$$Z = AN \tag{7}$$

where A is the velocity component and N is the density component of the contact rate.

Density component

Zooplankton data were taken from Russian zooplankton surveys conducted each year since 1959 along the drift route of larval and juvenile cod (Nesterova 1990). The aim of these surveys has been to monitor the food situation for the cod larvae. Consequently, the surveys cover the northeastern part of the Norwegian Sea and the southwestern part of the Barents Sea. There are two cruises per year: one in April–May, during the early larval period, and one in June–July, when the larvae are about 2 mo old. The zooplankton data are presented as mean values of biomass in milligrams per cubic meter for the upper 50 m. Because the data are available as biomass and not by numbers, it has to be assumed that the biomass is proportional to the number of suitable prey, and that the data display relative values for the density component.

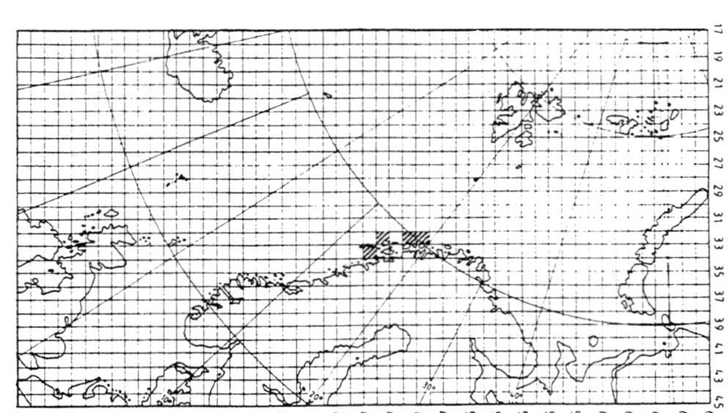

Fig. 4. Map of the grid net of the hindcast data base from the Norwegian Meteorological Institute (Eide et al. 1985). The wind data used in the present study are from the four hatched grid points.

Velocity component

Wind data were taken from the hindcast data base of the Norwegian Meteorological Institute (Eide et al. 1985). The data base consists of geostrophic surface winds for every 6 h in a grid net of 75 × 75 km. Data from four grid points off the coast of northern Norway, shown in Fig. 4, were used to calculate the monthly means of cubed wind speed for May and June each year. These data were then used in Equations 3, 4 and 5 to calculate the turbulent velocities on the relevant scales. Together with Equation 2, A_i is derived from Equation 6.

Results

Equations 2–5 are applied in Equation 6 to simulate the effect of wind-induced mixing on the contact rate between the cod and the copepod nauplii–copepodites, as the cod grows from larva to juvenile. Figure 5 shows the relative increase in contact rate from first-feeding larvae to early juveniles for three different wind speeds: 5, 10, and 15 m·s^{-1}, and for three different prey concentrations (low, average, and high). The three prey concentrations correspond to the low, average, and high concentration of Equation 5 (also plotted in Fig. 3). The lines converge toward 1 when the turbulent velocity becomes much less than the swimming speed of cod, and consequently, no longer contributes to the contact rate between the cod and its prey. The turbulence-induced contact rate is at its maximum for the slow-moving, first-feeding cod larvae. As Fig. 5 shows, the contact rate increases by an order of eight for the situation of low prey concentration under a wind speed of 15 m·s^{-1} compared with nonturbulent conditions, while it increases by a factor of five for high prey concentrations under the same wind speed. Hence, the turbulence-induced contact rate contributes to reducing the effects of variable prey concentrations, since the ratio between the maximum and minimum prey concentration, N_{high}/N_{low}, is 63, while the change in concentration experienced by the cod larvae (i.e., the change in the number of encounters) is less: $Z_{high}/Z_{low} = 39$.

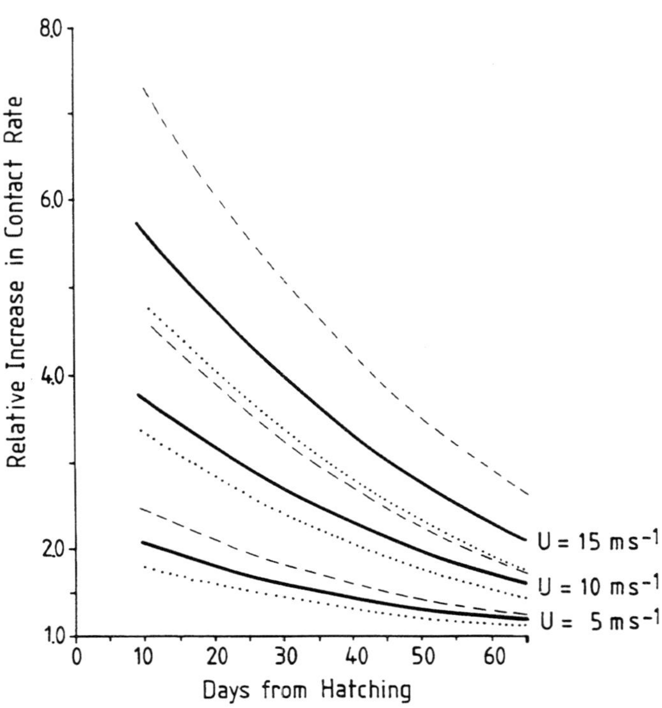

Fig. 5. The relative increase (compared to nonturbulent conditions) in the velocity component of the contact rate between Arcto-Norwegian cod and its prey from first-feeding stage to 65 d after hatching, for three wind speeds 5, 10, and 15 m·s^{-1}, and for three prey concentrations according to Equation (5). Hatched lines: $N_{low} = 0.8e^{-0.03\,t}$. Thick lines: $N_{ave.} = 6e^{-0.03\,t}$. Dotted lines: $N_{high} = 50e^{-0.03\,t}$.

As the cod larvae grow and increase their swimming speed, the influence of the turbulence-induced contact rate decreases. However, the simulations show that for strong mixing events the enhancement of contact rate is still considerable for 2-mo-old juveniles. At wind speeds of 15 m·s^{-1} and average prey concentrations, the contact rate is 2.5 times higher than it would be in nonturbulent conditions. The decreasing prey concentration as a function of time contributes to maintain the influence of turbulence as larger and larger scales of turbulence come into effect.

Figure 6a displays the interannual variations of zooplankton biomass (milligrams per cubic meter for the upper 50 m) in April–May for the period 1959–90 (Nesterova 1990). Even though the data do not show the actual zooplankton concentration where the cod larvae are found, and show biomass rather than numbers, it is assumed that they give a reasonably good picture of the typical interannual variations in prey abundance. Figure 6b shows the calculated interannual variations of the May mean of the velocity component of the contact rate. The data are derived from the wind data of the Norwegian hindcast data base (Eide et al. 1985). Figure 6c displays the product of the values in Figs. 6a and 6b (Z in Equation 7), and gives an indication of the mean May interannual variation in contact rate that first-feeding larvae experience. Figures 7a, 7b, and 7c show data similar to that in Figs. 6a, 6b, and 6c. The zooplankton data here (Fig. 7a), also based on Russian zooplankton surveys, are from June–July. The average mean monthly velocity component of the contact rate (Fig. 7b) is based on wind data from June, and Fig. 7c shows the product of the data in Figs. 7a and 7b. Hence, Fig. 7c gives an indication of the mean June interannual variation in contact rate for 2-mo-old juveniles.

Table 2 shows the May mean values, standard deviation, and range for the zooplankton data (Fig. 6a), the derived velocity component of the contact rate (Fig. 6b), and the overall contact rate, which is the product of these two parameters (Fig. 6c). Table 3 shows similar data for June (Figs. 7a, 7b, and 7c). According to Table 2, the standardized range and standard deviation of the interannual variation of the turbulence-induced contact rate are 0.55 and 0.14, respectively, while the corresponding values for the interannual variations of the density component are 2.26 and 0.52. This implies that the turbulence-induced contact rate contributes to about 20% in the overall variation of the contact rate as an average for May. The interannual variation of the velocity component of the contact rate decreases, as expected, from May to June as the influence of the speed of the juveniles increases. Hence, the standard deviation drops from 0.14 for May (Table 2) to 0.11 in June (Table 3). However, because the interannual variation of zooplankton biomass drops off even more, from 0.52 to 0.29, the relative contribution from the turbulence-induced

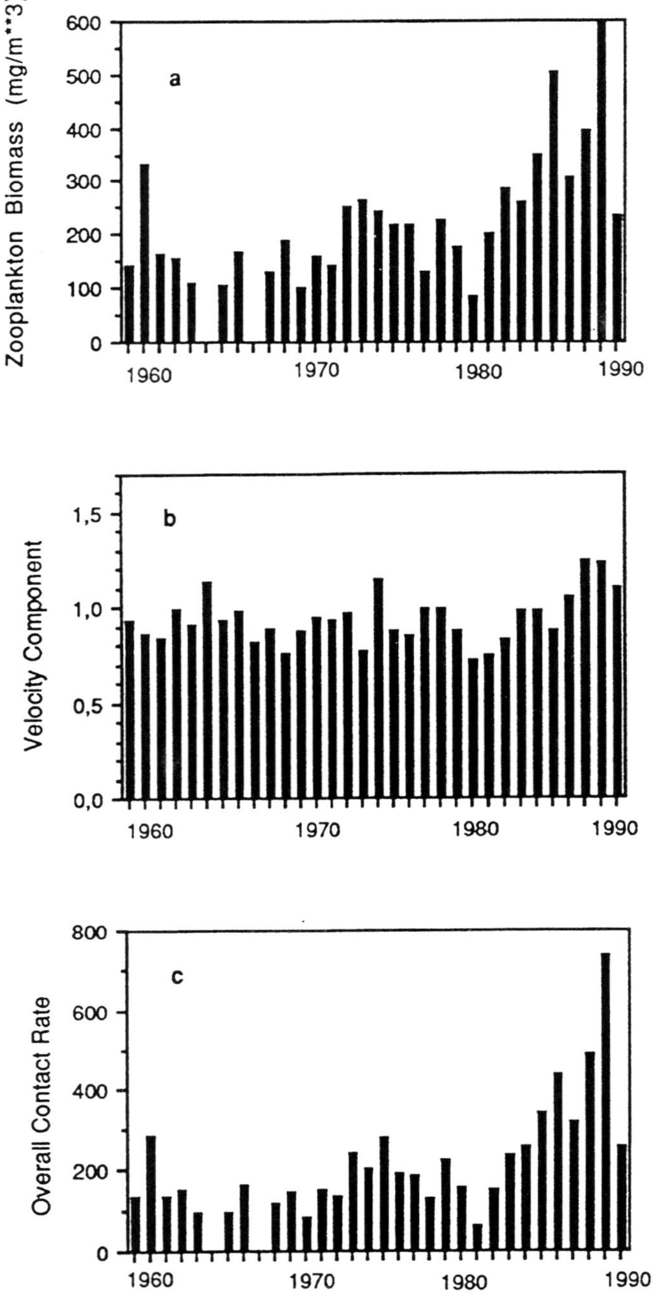

Fig. 6. (*a*) Mean zooplankton biomass (mg·m^{-3} of the upper 50 m) in the NE Norwegian Sea in April–May for the years 1959–90. After Russian zooplankton surveys (Nesterova 1990). (*b*) May monthly means of the velocity component of the contact rate between first-feeding cod larvae and their prey. (*c*) Monthly means of contact rate experienced by first-feeding cod larvae.

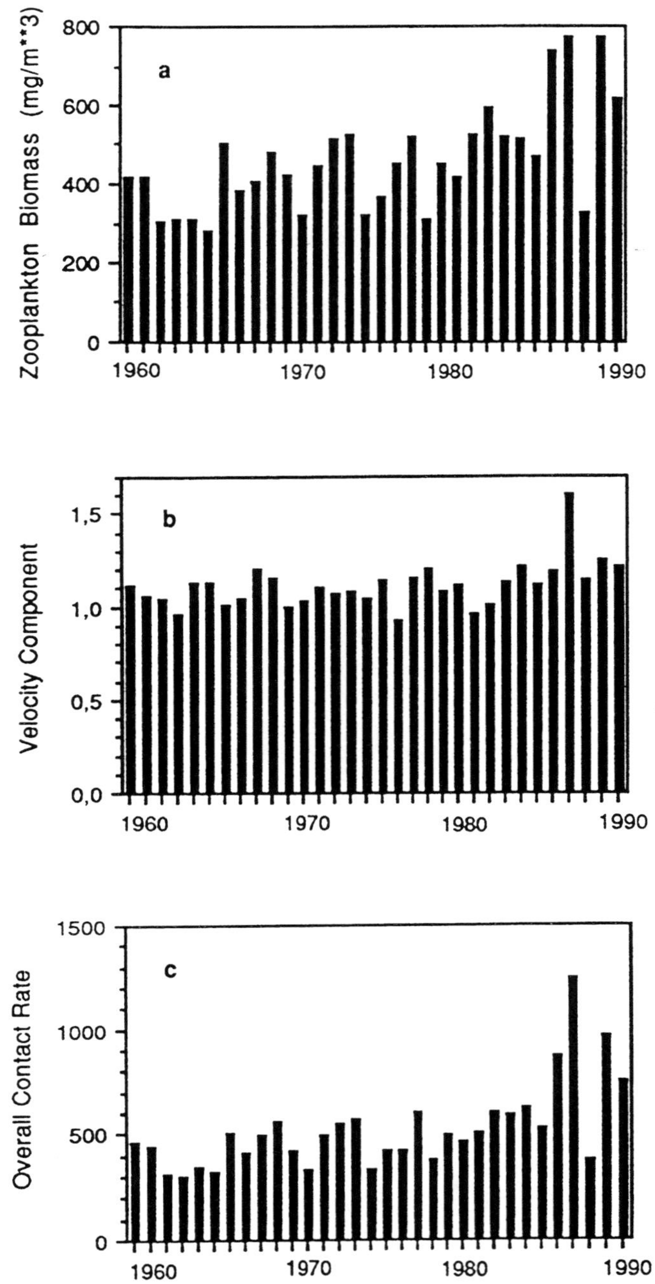

Fig. 7. (*a*) Mean zooplankton biomass (mg·m^{-3} of the upper 50 m) in the Northeastern Norwegian Sea and the Southwestern Barents Sea in June–July for the years 1959–90. After Russian zooplankton surveys (Nesterova 1990). (*b*) June monthly means of the velocity component of the contact rate between first-feeding cod larvae and their prey. (*c*) Monthly means of contact rate experienced by juvenile cod (2 mo).

Table 2. May mean monthly values, standard deviations and ranges for *N*, the zooplankton biomass (proxy for the density component of the contact rate); *A*, the velocity component of the contact rate; and *Z*, the overall contact rate (the product of *A* and *N*) for the period 1959–90. The zooplankton data are from Russian surveys (Nesterova 1990).

	Mean	SD	Standardized SD	Range	Standardized Range
Zooplankton biomass (*N*)	227	118	0.52	514	2.26
Velocity component of contact rate (*A*)	0.9374	0.13	0.14	0.511	0.55
Overall component of contact rate (*Z*)	220	140	0.64	674	3.07

Table 3. June mean monthly values, standard deviations and ranges for *N*, the zooplankton biomass (proxy for the density component of the contact rate); *A*, the velocity component of the contact rate; and *Z*, the overall contact rate (the product of *A* and *N*) for the period 1959–90. The zooplankton data are from Russian surveys (Nesterova 1990).

	Mean	SD	Standardized SD	Range	Standardized Range
Zooplankton biomass (*N*)	459	132	0.29	492	1.07
Velocity component of contact rate (*A*)	1.113	0.1211	0.11	0.686	0.62
Overall component of contact rate (*Z*)	519	201	0.39	938	1.81

velocity component of the contact rate to the overall interannual variation of the contact rate increases to about 30% in June. In general, the time series show no significant correlation between the zooplankton biomass and the velocity component of the contact rate. However, there is a long-term increase in both parameters after 1980, which gives a particularly strong increase in the overall contact rate during the last decade.

While Figs. 6 and 7 show interannual variations on monthly means, day-to-day variations are much larger. Figure 8 shows how the cubed wind speed varies continuously through May for a typical calm year (1981) and a windy year (1989) at the Lofoten first-feeding areas. The windy periods of 2–3 d in 1989 indicate the typical duration of cyclones passing through the region.

How sensitive are the simulations, displayed in Fig. 5, to changes in the assessed swimming speeds of the predator (0.33 body lengths·s^{-1}) and the prey (0.5 body lengths·s^{-1})? To investigate this, the relative change in contact rate was calculated for a 50% reduction and a doubling in swimming speeds. For the prey speed, the contact rates were practically insensitive to these changes, i.e., less than 1% for all wind

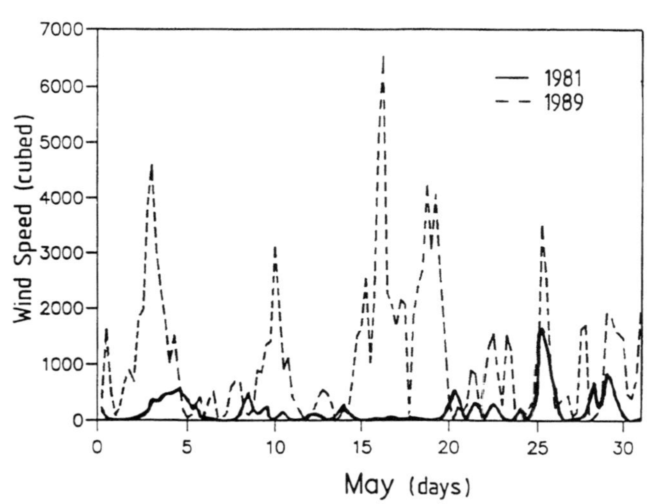

Fig. 8. Cubed wind speeds, at 6-h intervals, in areas of first-feeding cod larvae near Lofoten, northern Norway, during May in a calm (1981) and a windy (1989) year.

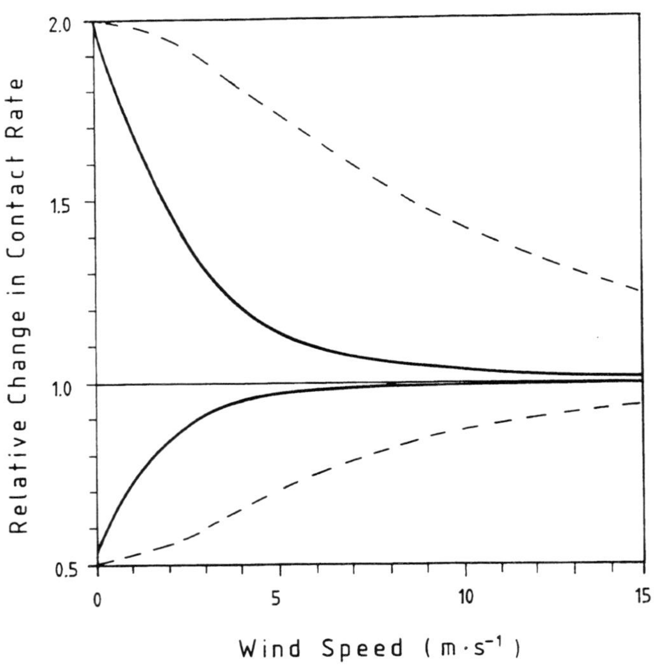

Fig. 9. Influence of changes in cod larval swimming speed on the contact rate for increasing wind speeds. Solid curves for first-feeding larvae, dotted curves for 70-d-old juveniles. The two curves below 1 show a 50% reduction in the assessed swimming speed of 0.33 body lengths·s^{-1}. The two curves above 1 show doubling in the assessed swimming speed.

speeds and all stages. However, the simulations were sensitive to such changes in the speed of the cod larvae–juveniles. Figure 9 shows the relative change in contact rates as wind speed increases from 0 to 15 m·s^{-1} for the first-feeding larvae (solid curve) and for 70-d-old juveniles (broken curve). The sensitivity to changes in swimming speed for first-feeding cod decreases rapidly as the wind speed (and, consequently the wind-induced turbulence) increases. At a wind speed of 4 m·s^{-1} the contact rate is only 20% higher for a doubling of the swimming speed and 5% lower for a 50% reduction. However, for 70-d-old juveniles the swimming speed contributes much more to the overall contact rate, and consequently the simulation are more sensitive to changes in swimming speed. At a wind speed of 15 m·s^{-1} the contact rate is still 25% higher for a doubling of the swimming speed and 6% lower for a 50% reduction.

Discussion

The simulations depend on a set of empirical and theoretical relations that need to be evaluated. The data on swimming speeds applied in Equation 2 rely on measurements in aquaria where the physical conditions, particularly motion, are far from natural. The average cruising speed measured by Solberg and Tilseth (1984) was higher than those measured by Skiftesvik and Huse (1987) and Skiftesvik (1994): 0.4 compared with about 0.2 body lengths·s^{-1}. Skiftesvik and Huse (1987) suggest that the higher speeds measured by Solberg and Tilseth (1984) are due to higher stress in small aquaria (e.g., wall effects). All experiments, however, had a very high prey concentration, from one to two orders of magnitude larger than in nature, and it should be expected that this would reduce the need to use a high swimming speed to encounter prey in the aquaria. On the other hand, both the absence of natural turbulence (which in nature increases the contact rate substantially), and the very high concentration of cod larvae (two to three orders of magnitude higher than in nature, which should be expected to increase the stress) would increase the swimming speed in the aquaria compared with that in nature. The average swimming speed of 0.33 body lengths·s^{-1} applied in the present simulation is in good agreement with experiments on herring by Blaxter and Staines (1971), who made their experiments in a 200-L tank.

Compared with other species, cod is a slow larva. Blaxter (1986) showed cruising speeds of several species; some of them were well above 2 body lengths·s^{-1}. It seems from Blaxter's data that the high-speed larvae are mainly those of species living in warm water. Lower metabolic rates in cold water may partly explain the lower swimming speeds. In addition, kinematic viscosity is higher in cold water, and this influences the smallest larvae. For example, the kinematic viscosity increases by 44% when the temperature drops from 15°C to 2°C. Therefore, it is to be expected that planktonic speeds in cold waters are lower, and the influence of turbulence to enhance planktonic contact rates should be higher.

The sensitivity analysis in Fig. 9 shows that the overall contact rate for first-feeding larvae is influenced by the larval swimming speed below wind speeds of about 4 m·s^{-1}. At higher wind speeds turbulence dominates the velocity component of the contact rate, and the simulations become insensitive to possible errors in larval swimming speed. However, the 70-d-old juveniles have attained such a high mobility that deviations from the assessed swimming speed will influence the simulations.

The swimming speed of the naupliar prey was assessed to be 0.5 body lengths·s^{-1} by Sundby and Fossum (1990). It is here assumed that this holds also for the larger stages of *C. finmarchicus*. However, the contribution of the prey swimming speed to the contact rate (in Equation 6) is negligible compared to the cod larvae speed and the turbulence. This the reason why the sensitivity analysis gave practically no changes in contact rate for a doubling in prey swimming speed. Consequently, even very large errors in the prey speed do not influence the simulations.

The decrease in prey concentration after first-feeding, shown in Fig. 3 and Equation 5, is less than one order of magnitude during the first 70 d. The decrease is surprisingly small compared with the decrease in the cod larvae concentration, which drops about three orders of magnitude during the same period. However, the differences in spawning period, and location and rate of stage development may explain a large part of this difference. Cod spawn within a time period of 2 mo at a limited number of locations along the coast (Fig. 1), and the concentrations of larvae will be exposed to strong advection and diffusion along the drift route to the Barents Sea. *Calanus finmarchicus* spawns over a longer period of time and over vast areas at the coast and in the Norwegian Sea. Consequently, horizontal diffusion does not substantially decrease the copepod concentration, so the drop in concentration must be largely due to mortality. Even though the instantaneous mortality rate is generally higher at lower trophic levels, the time for the development of stage V copepodites is limited to about 34 d at 8°C (Corkett et al. 1986). Besides, the duration of the the copepodite stages IV and V is considerably longer than the naupliar stages.

The change in diet of cod larvae as they grow was described in detail by Sysoeva and Degtereva (1965), and Helle (1993) has quantitatively studied the complete diet for early juveniles (70 d) and compared it with the available zooplankton in the sea. This confirmed the results of Sysoeva and Degtereva (1965). At this stage, the juveniles prey mainly on *C. finmarchicus* stages IV and V, primarily because that is the most available prey item and it has a suitable size. The diet becomes more complicated after this stage. The cod gradually switch to larger and more rapidly moving prey, such as the early stages of krill (Sysoeva and Degtereva 1965) and, in some years when their distributions overlap, capelin larvae (Helle 1993).

A large contribution to the overall uncertainty of the simulation is the estimate of the turbulent kinetic energy dissipation, ε. The relation established by Oakey and Elliott (1982) (Equation 4) was based on measurements of the upper mixed layer on the Scotian Shelf, a region where tidal mixing strongly contributes to the production of turbulent energy and where convection also contributes to mixing. This also occurs at the shelf off the coast of northern Norway. The estimates of eddy diffusivity based on vertical distribution of cod eggs in Lofoten (Sundby 1983) indicate a large contribution from tidal mixing when wind mixing was low. Sundby established a relation between the eddy diffusivity, K (square meters per second), of the upper mixed layer and the wind speed, U (meters per second):

$$K = 76.1 \times 10^{-4} + 2.26 \times 10^{-4} U^2 \tag{8}$$

Using the relation by Denman and Gargett (1983):

$$K = 0.25\varepsilon N^{-2} \tag{9}$$

where N is the Brunt–Vaisälä frequency (seconds^{-1}), it can be shown that the turbulent energy dissipation relation established by Oakey and Elliott (1982) for the Scotian Shelf corresponds well to the turbulent conditions off the coast of northern Norway. The larvae are, however, vertically distributed down to about 40 m, which implies that the turbulent energy dissipation experienced for the lower part of the distribution is lower than that used in the present simulation. MacKenzie and Leggett (1991) assembled a number of measurements on turbulent energy dissipation rate and fitted the data to a depth-dependent model for the turbulent energy dissipation. Their version for shelf regions is

$$\log \varepsilon(z) = -5.104\, z^{0.079} + 0.748 \log U^3 + 0.945 \tag{10}$$

where z is the depth.

Recalculating the velocity component of the contact rate using Equation 10 gives values that are about 20% lower. Although their model is a mix of different measurements and is partly influenced by wind mixing, tidal mixing, and convection, it seems to give a crude expression of the turbulent energy dissipation rate in the pelagic layer.

The discussion above displays some of the uncertainties in the data input to the simulations. Some of them may underestimate the turbulence-induced contact rate, some of them may overestimate it. The simulations are quite robust to errors in prey swimming speed and errors in predator swimming speed during first-feeding stage. The large uncertainties are linked to the predator swimming speed during the juvenile stages and to the depth-dependent rate of turbulent kinetic energy dissipation. More detailed studies on behaviour of juveniles are needed, and more understanding of the various processes causing turbulence of the mixed layer is needed (e.g., Haury et al. 1990; Yamazaki and Kamykowski 1991). However, these uncertainties cannot falsify the main results shown in Fig. 5: the feeding of cod larvae is strongly influenced by the turbulence-induced contact rate during first feeding, and even though the effect of turbulence is rapidly decreasing after first feeding, it cannot be neglected until the juveniles reach the age of about 2 mo.

It must be noted that the simulations showing the interannual variations in contact rate due to wind variations (Figs. 6 and 7) are based on turbulent kinetic energy dissipation estimates averaged out over 1 mo. The idea that recruitment processes may substantially influence the survival of larvae on a shorter time scale than 1 mo cannot be rejected. Figure 8 shows a period of 2 wk practically without wind in mid-May 1981. In a normal year, almost 50% of the recruits are at the first-feeding stage during this period. It is to be expected that the lack of wind would at least reduce the growth rate of the larvae.

On the other hand, a lower contact rate due to lower wind speeds would also reduce the potential predation on cod larvae from slow-moving predators. Our knowledge about the organisms preying on Arcto-Norwegian cod larvae is poor. Typical concentrations of cod larvae are 0.1–10·m^{-3}, which implies that the relevant scale in this process should be about one order of magnitude larger (r = 0.5–2.0 m) than for the interaction between cod larvae and copepods. Consequently, predators of much larger speeds than those of the cod larvae will also benefit from the process. We know that adult herring may prey on cod eggs (Melle 1985), but even though the turbulence scales are larger, the turbulence-induced contact rate can here be neglected because of the high speed of the herring. Besides, in this case, high turbulence distributes the eggs throughout the water column, reducing high concentrations at the surface and, hence, reducing the feeding efficiency of the herring. It has, however, been reported that small ctenophores and medusae may prey on cod eggs (Melle 1985), and these organisms would clearly benefit from turbulence-induced contact with the cod larvae.

With the strong efficiency of turbulence in providing contact between cod larvae and their prey, one may question the potential benefit for cod larvae in their first-feeding stage spending much energy to cruise for prey. Definitely they need their own movement to escape predators, ambush prey, and migrate vertically. But horizontal cruising for food is a process that they first need as larger organisms. Hence, letting the turbulence do the work, planktonic organisms can save substantial amounts of energy. As pointed out earlier (MacKenzie et al. 1990; Sundby and Fossum 1990), estimates of critical prey densities for fish larvae based on laboratory experiments (e.g., Solberg and Tilseth 1984) should be reconsidered to account for the difference in turbulence under laboratory and field conditions. The present study shows that not only does the interannual variation in prey concentration determine interannual variations in encounter rates, but the velocity component determined through the interannual variations in wind conditions influences the encounter rates.

Acknowledgements

I wish to thank PINRO, Murmansk, for making their data on zooplankton surveys available for the simulations, and one of the anonymous referees for helpful comments and advice.

References

Alcaraz, M., E. Saiz, C. Marrase, and D. Vaque. 1988. Effects of turbulence on the development of phytoplankton biomass and copepod populations in marine microcosms. Mar. Ecol. Prog. Ser. 49: 117–125.

Bjørke, H., and S. Sundby. 1987. Abundance indices and distribution of postlarvae and 0-group cod, p. 127–144. *In* H. Loeng [ed.] The effect of oceanographic conditions on distribution and population dynamics of commercial fish stocks in the Barents Sea. Proceedings of the Third Soviet–Norwegian Symposium, Murmansk, Russia, 26–28 May 1986. Institute of Marine Research, Bergen, Norway.

Blaxter, J.H.S. 1986. Development of sense organs and behaviour of teleost larvae with special reference to feeding and predator avoidance. Trans. Am. Fish. Soc. 115: 98–114.

Blaxter, J.H.S., and M.E. Staines. 1971. Food searching potential in marine fish larvae, p. 467–485. *In* D.J. Crisp [ed.] Proceedings of Fourth European Marine Biology Symposium 14–20 September 1969. Cambridge University Press, Cambridge, UK.

Blom, G., H. Otterå, T. Svåsand, T. Kristiansen, and B. Serigstad. 1991. The relationship between feeding conditions and production of cod fry (*Gadus morhua* L.) in a semi-enclosed marine ecosystem in western Norway, illustrated by use of a consumption model. ICES Mar. Sci. Symp. 192: 176–189.

Bolz, G.R., and G.R. Lough. 1988. Growth through the first six months of Atlantic cod, *Gadus morhua*, and haddock, *Melanogrammus aeglefinus*, based on daily otolith increments. Fish. Bull. U.S. 86(2): 223–236.

Corkett, C.J., I.A. McLaren, and J.-M. Sevigny. 1986. The rearing of marine calanoid copepods *Calanus finmarchicus* (Gunnerus), *C. glacialis* (Jaschnov) and *C. hyperboreus* (Krøyer) with comment on the equiproportional rule. Syllogeus 58: 539–546.

Davies, C.S., G.R. Flierl, P.H. Wiebe, and P.J.S. Franks. 1991. Micropatchiness, turbulence and recruitment in plankton. J. Mar. Res. 49: 109–151.

Degtereva, A.A. 1966. Zooplankton in the south-west of the Barents and the north-east of the Norwegian seas in 1959–61. Mater. Polyarn. Nauchno-Issled. Inst. Morsk Rybn. Khoz. Okeanogr. 7: 70–83. (In Russian)

Denman, K.L., and A.E. Gargett. 1983. Time and space scales of vertical mixing and advection of phytoplankton in the upper ocean. Limnol. Oceanogr. 28(5): 801–815.

Eide, L.I., M. Reistad, and J. Guddal. 1985. Database av beregnede vind og bølgeparametre for Nordsjøen, Norskehavet og Barentshavet, hver 6. time for årene 1955–81. (Data base of calculated wind and wave parameters for the North Sea, Norwegian Sea and Barents Sea, for every 6 hours during the period 1955–81). Det norske meteorologiske institutt, Oslo. 38 p. (In Norwegian)

Ellertsen, B., P. Fossum, P. Solemdal, S. Sundby, and S. Tilseth. 1984. A case study on the distribution of cod larvae and availability of prey organisms in relation to physical processes in Lofoten, p. 453–478. *In* E. Dahl, D. Danielssen, E. Moksness, and P. Solemdal [ed.] The propagation of cod. Flødevigen rapportser 1.

Ellertsen, B., P. Fossum, P. Solemdal, and S. Sundby. 1989. Relations between temperature and survival of eggs and first-feeding larvae of the north-east Arctic cod (*Gadus morhua* L.). Rapp. P.-v. Réun. Cons. int. Explor. Mer 191: 209–219.

Haury, L.R., H. Yamazaki, and E.C. Itsweire. 1990. Effects of turbulent shear flow on zooplankton distribution. Deep-Sea Res. 37(3): 447–461.

Helle, K. 1993. Distribution and abundance of early juvenile Arcto-Norwegian cod (*Gadus morhua* L.) in relation to water mass properties and prey abundance. Candidatus Scientiarum thesis, University of Bergen, Bergen, Norway. 116 p.

Lie, U. 1961. Zooplankton in relation to herring in the Norwegian Sea. Rep. Norw. Fish. Mar. Invest. 13(1): 1–66.

MacKenzie, B.R., and W.C. Leggett. 1991. Quantifying the contribution of small-scale turbulence to the encounter rates between larval fish and their zooplankton prey: effects of wind and tides. Mar. Ecol. Prog. Ser. 73: 149–160.

MacKenzie, B.R., W.C. Leggett, and R.H. Peters. 1990. Estimating larval fish ingestion rates: can laboratory derived values be reliably extrapolated to the wild? Mar. Ecol. Prog. Ser. 67: 209–225.

Melle, W. 1985. Predasjon på torskens egg og larver i Lofoten. (Predation on cod eggs and larvae in Lofoten.) Candidatus Scientiarum thesis, University of Bergen, Bergen, Norway. 142 p. (In Norwegian)

Nesterova, V.N. 1990. Plankton biomass along the drift route of cod larvae. PINRO, Murmansk. 64 p. (In Russian)

Oakey, N., and J. Elliott. 1982. Dissipation within the surface mixed layer. J. Phys. Oceanogr. 12: 171–185.

Østvedt, O.J. 1955. Zooplankton investigations from weather ship M in the Norwegian Sea, 1948–49. Hvalrådets Skr. 40: 1–93.

Pavshtiks, E.A. 1964. Long-term observations on the plankton distribution distribution during the period of spring–summer feeding herring in the Norwegian and the Greenland Seas (1951–1962). Tr. Polyarn. Nauchno-Issled. Inst. Morsk Rybn. Khoz. Okeanogr. 16: 151–164. (In Russian)

Pavshtiks, E.A., and A.F. Timokhina. 1972. History of investigations on plankton in the Norwegian Sea and the main results of Soviet investigations. Proc. R. Soc. Lond. Biol. Sci. 73: 267–278.

Rothschild, B.R., and T.R. Osborn. 1988. Small-scale turbulence and plankton contact rates. J. Plankton Res. 10: 465–474.

Saiz, E., and M. Alcaraz. 1992. Free-swimming behaviour of *Arcatia clausi* (Copepoda: Calanoida) under turbulent water movement. Mar. Ecol. Prog. Ser. 80: 229–236.

Skiftesvik, A.B. 1994. Impact of the physical environment on the behaviour of cod larvae. ICES mar. Sci. Symp. 198: 646–653.

Skiftesvik, A.B., and I. Huse. 1987. Behaviour studies of cod larvae, *Gadus Morhua* L. Sarsia 72: 367–368.

Solberg, T., and S. Tilseth. 1984. Growth, energy consumption and prey density requirements in first feeding larvae of cod (*Gadus morhua* L.), p. 145–166. *In* E. Dahl, D. Danielssen, E. Moksness, and P. Solemdal [ed.] The propagation of cod. Flødevigen rapportser 1.

Sundby, S. 1983. A one-dimensional model for the vertical distribution of pelagic fish eggs in the mixed layer. Deep-Sea Res. 30(6A): 645–661.

Sundby, S., and P. Fossum. 1990. Feeding conditions of Arcto-Norwegian cod larvae compared to the Rothschild–Osborn theory on small-scale turbulence and plankton contact rates. J. Plankton Res. 12(6): 1153–1162.

Sundby, S., B. Ellertsen, and P. Fossum. 1994. Encounter rates between first-feeding cod larvae and their prey during moderate to strong turbulent mixing. ICES mar. Sci. Symp. 198: 393–405.

Sysoeva, T.K., and A.A. Degtereva. 1965. The relation between the feeding of cod larvae and pelagic fry and the distribution and abundance of their principal food organisms. Int. Comm. Northwest Atl. Fish. Spec. Publ. 6: 411–416.

Timokhina, A.F. 1962. Seasonal changes in the age structure of abundance and in biomass of *Calanus finmarchicus* in Norwegian Sea. Tr. Polyarn. Nauchno-Issled. Inst. Morsk Rybn. Khoz. Okeanogr. 14: 223–234. (In Russian)

Torres, J.J., and J.J. Childress. 1983. Relationship of oxygen consumption to swimming speed in *Euphasia pacifica* 1. Effects of temperature and pressure. Mar. Biol. 74: 79–86.

Wiborg, K.F. 1948. Investigations on cod larvae in the coastal waters of northern Norway. Fiskeridir. Skr. Ser. Havunders. 9(3): 1–27.

Yamazaki, H., and D. Kamykowski. 1991. The vertical trajectories of motile phytoplankton in a wind-mixed water column. Deep-Sea Res. 38(2): 219–241.

Moderate is better: exploring nonlinear climatic effects on the Californian northern anchovy (*Engraulis mordax*)

Philippe Cury, Claude Roy, Roy Mendelssohn, Andrew Bakun, David M. Husby, and Richard H. Parrish

Cury, P., C. Roy, R. Mendelssohn, A. Bakun, D.M. Husby, and R.H. Parrish 1995. Moderate is better: exploring nonlinear climatic effects on the Californian northern anchovy (*Engraulis mordax*), p. 417–424. *In* R.J. Beamish [ed.] Climate change and northern fish populations. Can. Spec. Publ. Fish. Aquat. Sci. 121.

Abstract: The effect of environmental variability on marine population dynamics remains a challenging problem for fisheries science. The impact of four decades of environmental change on larval abundance of the northern anchovy (*Engraulis mordax*) off California is analyzed using recently developed nonlinear statistical techniques that extend linear multiple regression analysis. The three techniques used are: alternating conditional expectation, additivity and variance stabilization, and generalized additive interactive modelling. The dependencies of larval abundance on upwelling intensity and total anchovy biomass are nonlinear and account for 84% of the interannual variability; a linear regression analysis explained 37% of the variance. The higher larval abundances are produced with medium to large adult biomasses coupled with moderate upwelling intensity. These empirical results suggest that marine populations will respond to climatic trends, but will do so in a nonlinear manner. These nonlinear satistical algorithms, which are not yet widely used in ecology, provide a promising method for exploring the form of the relationships between environmental changes and populations responses.

Résumé : L'effet de la variabilité environnementale sur la dynamique des populations marines demeure un problème non résolu pour la science des pêches. Les conséquences de quatre décennies de changements environnementaux sur l'abondance des larves de l'anchois du Pacifique (*Engraulis mordax*) au large de la Californie sont analysées au moyen de méthodes statistiques non linéaires récemment mises au point qui complètent l'analyse de régression multiple linéaire. Les trois techniques utilisées sont : alternance des espérances mathématiques conditionnelles, additivité et stabilisation de la variance, et modélisation additive interactive généralisée. La dépendance de l'abondance des larves à l'égard de l'intensité des remontées d'eau et de la biomasse totale d'anchois est non linéaire et représente 84 % de la variabilité interannuelle. Une analyse de régression linéaire explique 37 % de la variance. Les abondances plus élevées de larves sont produites par des biomasses adultes moyennes à grandes jumelées à une intensité modérée des remontées d'eau. Les résultats empiriques portent à croire que les populations marines réagissent aux tendances climatiques mais de façon non linéaire. Ces algorithmes statistiques non linéaires, qui ne sont pas encore largement utilisés en écologie, représentent une méthode prometteuse d'analyse des formes de relations entre les changements environnementaux et les réactions des populations animales.

P. Cury and **C. Roy**. ORSTOM (Institut Français de Recherche Scientifique pour le Développement en Coopération), Département Terre–Océan–Atmosphère, 213 Rue La Fayette 75480 Paris Cedex 10, France.

R. Mendelssohn, **A. Bakun**, **D.M. Husby**, and **R.H. Parrish**. Pacific Fisheries Environmental Group, Southwest Fisheries Science Center, National Marine Fisheries Service, NOAA, P.O. Box 831, Monterey, CA 93942, USA.

Introduction

Precipitous declines or sudden recoveries have been observed in the world's largest fisheries of clupeoids. These boom and bust phenomena are found in the sardine and anchovy fisheries of the Benguela, the California, the Canary, and the Peru Current ecosystems (Sharp and Csirke 1983; Pauly and Tsukayama 1987). This instability appears to be due to changes in recruitment success (i.e., the addition of young fishes to the fishable population). Significant variations in recruitment are likely to arise both from excessive biomass depletion due to intensive fishing pressure and from alteration in the mortality rates of the embryonic and larval stages due to particular sensitivity to environmental change (Lasker 1981, 1985; Smith 1985; Beverton 1990). However, recent studies suggest that the final recruitment success is not necessarily linked to early larval abundance (Peterman et al. 1988) and that predation plays a major role in prerecruit mortality (Sissenwine 1984).

Environmental and ecological time series are now available for several decades. If some strong relationships do exist between environmental fluctuations and fish population dynamics, ecologists are now in a good position to identify emergent patterns. The California Cooperative Oceanic Fisheries Investigations (CalCOFI) data represent one of the most impressive marine data sets available, covering a multidecadal time period for the entire southern California ecosystem (Smith and Moser 1988). Since 1951, larvae from the central population of the northern anchovy (*Engraulis mordax*) have been sampled over the entire range of its distribution. Environmental data have been collected independently and during the sampling cruises.

Analysis of the relationships between two or more fundamental variables have most commonly employed linear statistical methods (Tyler 1992) or used an a priori transformation, such as a logarithmic transformation (Parrish and MacCall 1978; Crecco et al. 1986). However, a scattergram that reveals no linear relationship does not necessarily mean the absence of a tight link (Kareiva 1990; R. Mendelssohn, unpublished data). While nonlinearities are not new to ecologists, rigourous statistical methods to deal with them are. Recently developed nonlinear statistical methods provide useful tools to explore the relationship between different sets of variables (Breiman and Friedman 1985; Buja et al. 1989; Hastie and Tibshirani 1990; Gifi 1990). In this paper, we apply these nonlinear statistical techniques to the California Current anchovy and attempt to reconcile the empirical results to previous theoretical and field studies.

Data

Annual mean abundance of northern anchovy larvae in the principal reproductive area (32–34°N, 117.5–119°W) was calculated by averaging all sampled stations. From 1951 to 1990, 32 years were intensively sampled; a total of 3257 stations are represented. Mean annual larval number per square metre for the period 1951–75 are from Smith and Eppley (1982). P.E. Smith (Southwest Fisheries Center, National Marine Fisheries Service, La Jolla, CA 92038) provided us with the data from 1975 to 1990. Larvae are mostly found during the first two quarters; however, for certain years larvae were abundant during the third or fourth quarter (1954, 1955, 1956, 1966, 1967, 1972, 1989). Moreover, in some years, larvae were more abundant during the third and fourth quarters (1952, 1953). This led us to use annual mean abundance of northern anchovy larvae. No data were collected in 1970, 1971, 1973, and 1976. Poor data are available during 1974, 1977, 1982, and 1983 where only one quarter among four was sampled. For these reasons those 8 years were excluded from the analysis.

Anchovy biomass during this period was variable. By combining available fisheries data, sonar surveys, and egg production information, MacCall and Methot (1983), Methot (1989), and Jacobson and Lo (1991) have produced estimates of total anchovy biomass. We use these estimates in our analysis.

Because of our lack of understanding of the casual mechanisms that may be responsible for the observed variations in larval number, several environmental time series were selected for our analysis. Temperature has a strong effect on larval growth and consequently on larval stage duration (Blaxter and Hunter 1982; Houde 1989). To test for the relevance of low-frequency temperature variability, the temperature collected at each sampling station at 10 m depth was averaged over the year. The consistent presence of massive populations of small pelagic clupeoid fishes in coastal upwelling regions around the world suggests a strong link of the population dynamics of these fishes to the upwelling process. Coastal upwelling is a dominant oceanographic process off the California coast (Reid et al. 1958). Northerly winds cause coastal water to move offshore, and the deficit at the coast is compensated for by an upward flow of cold and nutrient-rich subsurface water. An estimate of the upwelling intensity is given by the offshore component of the Ekman transport (Bakun 1973), which is the wind-driven flow integrated over the ocean layer affected by the wind. The relative constancy in direction of the wind blowing along the California coast leads to a tight relationship between indices of upwelling and those of wind mixing that is roughly proportional to the cube of the wind speed (Elsberry and Garwood 1978). The upwelling index tracks variability in the main features of the upwelling process: the offshore flow of coastal water, the coastal enrichment by the upward nutrient input, and the mixing of the surface layers by the wind.

Variations in the upwelling process in the northern anchovy reproductive area from 1951 to 1990 are represented by an annual upwelling index derived from monthly values at 33°N, 119°W. This index is calculated using geostrophic winds derived from atmospheric pressure fields by the U.S. Navy Fleet Numerical Oceanography Center. Bakun (1973) details the procedure used to derive geostrophic winds from atmospheric pressure analyses and to estimate upwelling indices. As mentioned in Bakun (1973) these indices should be used with caution when considering long-term variability in this region. During the 1950s and 1960s the origin of the pressure fields charts changed (Bakun 1973). The procedure used to produce the pressure fields has also evolved from hand analysis to computer analysis. The bias introduced by these changes remains unknown. Discrepancies between the calculated and measured winds at the latitude of the Southern California Bight have also been noted (Bakun 1973, 1975; Halliwell and Allen 1987) and mean annual deviations range between 0 and 26%.

Methods

Present knowledge of the relationships between the anchovy larval number and relevant population and environmental variables is not sufficient to know a priori their correct form. Therefore, it is desirable to have methods that use the data (i.e., nonparametric) to estimate the appropriate functional form. Iterative algorithms have recently been developed that extend linear multiple regression analysis to generalized additive models (Hastie and Tibshirani 1990). These models are additive in empirically estimated transformations of the data. These procedures, which are not yet widely used in ecology, provide a method for exploring the form of the relationships between the response and the predictor variables when the forms of these relationships are unknown. Empirical, objective identification of possible nonlinear relationships are also made possible.

The usual linear mutiple regression model for predicting a response (or dependent) variable Y from p predictor variables X_i, where $i = 1, ..., p$, and for n observations, where $j = 1, ..., n$, is given by

$$Y(j) = \sum_{i=1}^{p} b_i X_i(j) + e(j)$$

where the $e(j)$ are independent, identically distributed, zero mean random variables, and a mean term is implicitly defined by putting X_i, for example, equal to one. The nonlinear multiple regression model (or generalized additive model is given by

$$S(Y(j)) = \sum_{i=1}^{p} b_i T_i(X_i(j)) + w(j)$$

where the functions $S(Y)$ and $T_i(X)$ are unknown and must be nonparametrically estimated from the data. The $w(j)$ are independent, identically distributed, zero mean random variables, usually assumed to be gaussian but not restricted to that assumption.

There are several approaches to estimating the latter equation under differing assumptions about the optimization criteron and about the transformations, including alternating conditional expectations (ACE) (Breiman and Friedman 1985), additivity and variance stabilization (AVAS) (Tibshirani 1988), and generalized additive interactive modelling (GAIM) (Hastie and Tibshirani 1990). The basic approach of each algorithm is iterative, where at each iteration, given the present estimates of $S()$ and $T_i()$, $i \neq k$ (where $i, k = 1,1 - p$), the partial residual:

$$S(Y(j)) - \sum_{j \neq k}^{p} b_i T_i(X_i(j))$$

is calculated, and the transformation $T_k()$ is calculated as the conditional expectation of the partial residual given X_k.This conditional expectation is estimated from the data using some form of scatterplot smoother. As the b_i are simply scaling factors, ACE and AVAS include the b_i in the function $T_i()$, while GAIM estimates the b_i's to perform analysis of deviance tests on the parameters, similar to the more usual analysis of variance tests in linear regression analysis.

The algorithms converge to optimal solutions for a given criterion (such as the maximum correlation between the transformed dependent variable and the transformed predictor variables in the ACE algorithm). Each algorithm has its own smoother and convergence criterion. There is a fundamental trade-off between bias and variance governed by the smoothing parameter. A very small value of the smoothing parameter produces a nearly exact fit but with high bias. A high value of the smoothing parameter will give a very smooth fit that is unbiased but with a high variance.

There are a number of properties of the ACE procedure that can lead to some anomalies (Breiman and Firedman 1985; Hastie and Tibshirani 1990). For a single predictor, ACE is symmetric in X and Y, whereas one expects a regression procedure to treat Y differently from X. ACE is not equivariant under monotone transformations of the predictors, exhibits strange behaviour in low-correlations settings, and does not reproduce model transformations. Moreover, disjoint clusters are collapsed by ACE, and the crossing of eigenvalues can cause discontinous behaviour.

AVAS is a modification of ACE designed for regression problems and uses an asymptotic variance-stabilizing transformation (the $S(Y(j))$ is assumed to be strictly monotone). While AVAS does not share many of ACE's anomalies, there is much less theoretical support for the technique than there is for ACE. In particular, global convergence of AVAS has not been established. Both algorithms use locally weighted nonlinear smoothers. GAIM uses cubic smoothing splines (which are linear smoothers) in the implementation that is available. GAIM also produces pointwise error bands that give better confidence in the observed nonlinearities that are found, but is restricted to transformations of the X variables only.

Thus, results from any one of these algorithms must be considered with caution. As in using any statistical method, interpretation of the results must consider potential biases due to the assumptions underlying the techniques. One way to deal with the problem is to examine results using the three different techniques for the same data set. When each of the algorithms produces similar results there is more confidence in the findings. Consistency of the results with present ecological knowledge also reinforces confidence in their correctness.

Results

The relationships between larval number and temperature, the upwelling index, and total anchovy biomass were initially explored by plotting the raw data (Fig. 1). No significant linear relationship was found between anchovy larval numbers and the temperature index as the data appear to be randomly scattered (Fig. 1A). When larval numbers are plotted versus the upwelling index (Fig. 1B), low larval numbers are observed for both weak and strong upwelling intensities. High larval numbers are to be found for moderate upwelling index values: about 1.5 $m^3 \cdot s^{-1}$ per metre of coastline. Larval number and the total anchovy biomass appear to be positively correlated; however, the data appear scattered for high biomass values (Fig. 1C).

A linear regression analysis using biomass and upwelling as predictor variables explained 37% of the variance in larval number, and the effect of upwelling on larval number was not

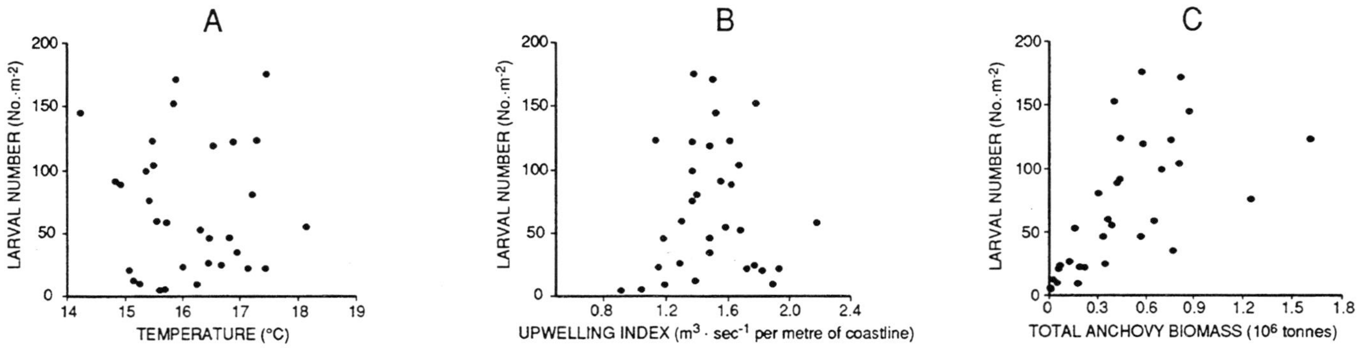

Fig. 1. Mean annual northern anchovy larval number for 1951–90 versus: (A) mean temperature at 10 m depth; (B) mean annual upwelling index; and (C) total anchovy biomass.

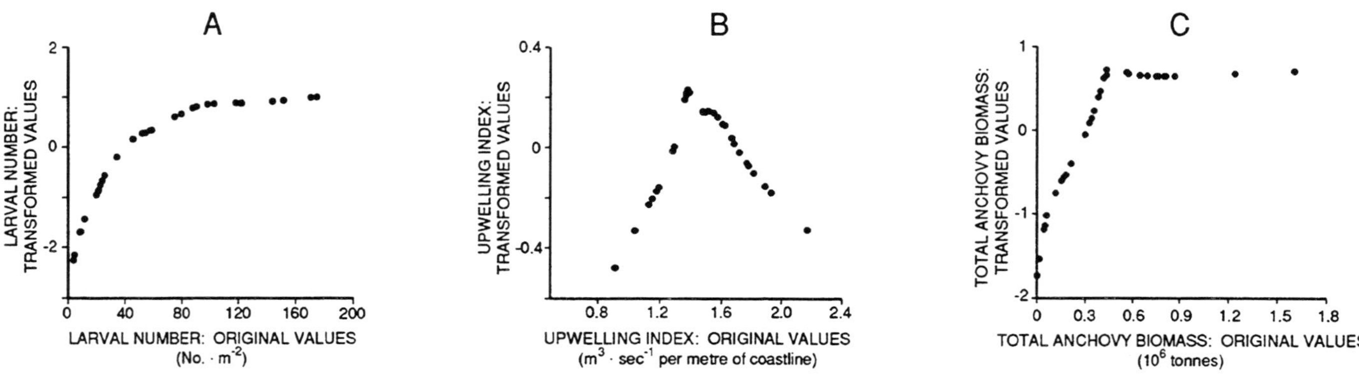

Fig. 2. Optimal empirical transformations from the ACE algorithm using larval number as the dependent variable and upwelling index and total anchovy biomass as the predictor variables. The shapes of the transformations are found by plotting the empirically transformed values of a variable versus their original values. The plots are for 1951–90. (A) Mean annual anchovy larval number (this variable is standardized in the ACE algorithm). (B) Mean annual upwelling index. (C) Total anchovy biomass.

significant. Using the ACE algorithm for the same variables explained 84% of the variability in larval number. The shape of the optimal empirical transformations of both the dependent and the predictor variables from this model are presented in Fig. 2. As the transformations $S()$ and $T_i()$ are conditional expectations empirically estimated by scatterplot smoothers, the iterative algorithms upon convergence return a new, transformed value for each observed value of Y or X_i. The form of the transformation is obtained by plotting the new, transformed value of Y or X_i versus the original, observed value. The estimated transformation of larval number is close to logarithmic in shape (Fig. 2A). The upwelling transformation is nearly dome-shaped with a maximum value around 1.5 $m^3 \cdot s^{-1}$ per metre of coastline (Fig. 2B). The transformation of the total biomass is approximately linear up to a value of about 0.5 Mt, but no real increase occurs above this value (Fig. 2C). When either the upwelling or the total anchovy biomass is separately considered in the analysis, each explains, respectively, 24% and 61% of the larval number variance. In each case, the shape of the transformations (figures not presented) is similar to that obtained when both variables are incorporated into the analysis. In the same way, considering separately biomass below and above 0.5 Mt gives similar transformations for the upwelling index. Identical results were obtained using AVAS (the plots are not presented here because they are similar in shape and the amount of explained variance is the same). Figure 3 presents the optimal empirical transformations from the GAIM algorithm using the logarithm of larval number as the response variable to make the results compatible with the previous results obtained with ACE (the GAIM algorithm only transforms the X variables). The r^2 is equal to 74%. The estimated fit for the transformed anchovy biomass values (Fig. 3A) is close to the transformation obtained using the ACE algorithm (Fig. 2C). The GAIM algorithm produces a smoother dome-shaped relationship for upwelling, in part due to the degrees of freedom used in the smoothing algorithm (Fig. 3B).

Estimating a generalized additive model for anchovy larval number, adult biomass, and upwelling can be viewed as an attempt, under additive restrictions, to empirically transform the three-dimensional surface formed from these observations so that the relationships between them become nearly linear. An interpolated estimate of the three-dimensional surface of the original data, using distance-weighted least squares

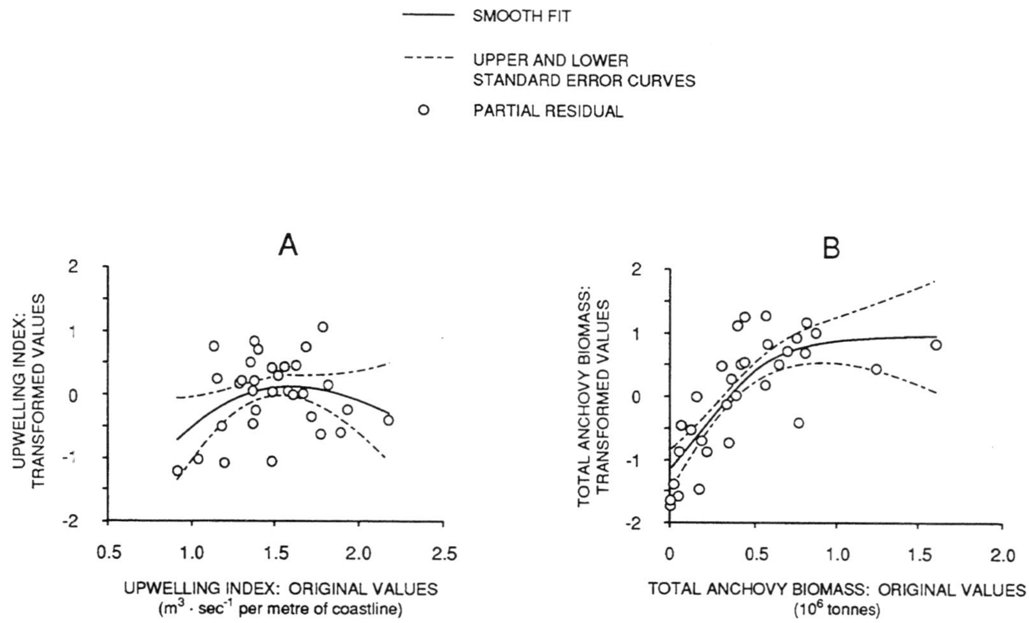

Fig. 3. Smooth fit, pointwise 2× standard error curves and partial residuals for the upwelling index and the total anchovy biomass using the GAIM algorithm. The logarithm of larval number was used as the dependent variable. The shapes of the transformations are found by plotting the empirically transformed values of a variable versus their original values. The plots are for 1951–90. (A) Mean annual upwelling index. (B) Total anchovy biomass.

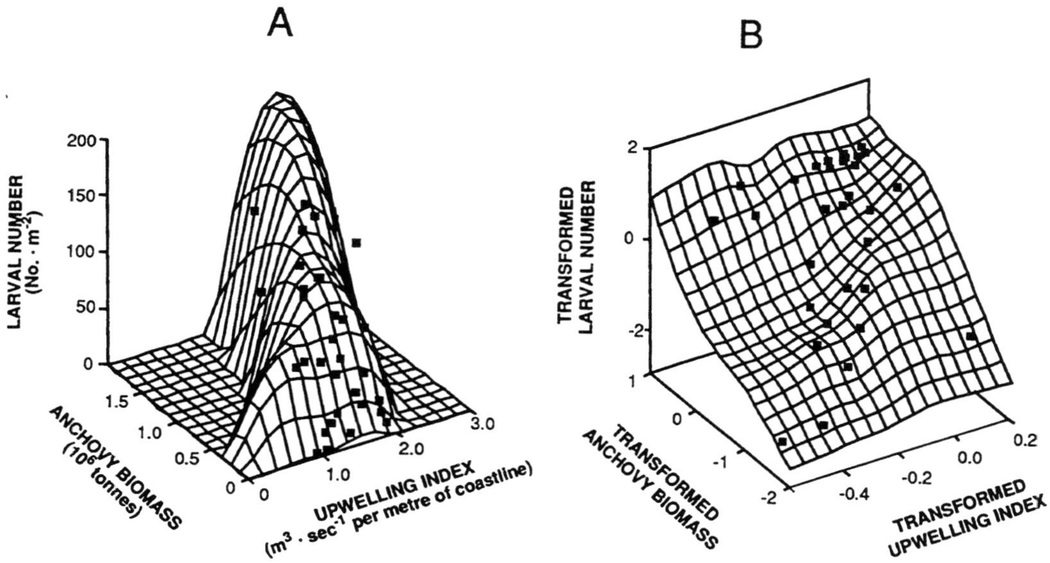

Fig. 4. Interpolated three-dimensional surfaces (using distance-weighted least squares) and the location of the actual values (squares). (A) Original data (larval number versus upwelling index and total anchovy biomass). This interpolated three dimensional surface shows similar nonlinear relationships between larval number and upwelling index, and also larval number and anchovy biomass as obtained from the ACE algorithm. (B) Empirically transformed data obtained from the ACE algorithm. The algorithm has successfully transformed the three variables so that the relationships between them is nearly linear.

(Fig. 4A), shows the dome-like relationship between larval number and upwelling, as well as the asymptotic relationship between larval number and adult biomass. The interpolated estimate of the three-dimensional surface for the transformed valules of the observations is nearly linear in all directions, which shows that the data have been successfully transformed so that the relationships are linear (Fig. 4B).

Discussion

Our results suggest that the larval anchovy number in the California Current has a dome-shaped relationship with upwelling and an asymptotic relationship with total biomass. Recent theories and hypotheses provide an interpretation of these results. In the present case, the abundance of larvae appears to be more related to the physical processes associated with upwelling than to direct effects of temperature. Indeed, this apparent absence of a relationship between larval number and temperature conforms to recent suggestions that variations in surface temperature are likely to have minor net effects on early life history survival (Pepin 1991). The graphical results suggest that larval number should be modeled on a log scale, indicating that the effects of the predictor variables are on a proportional rather than on an absolute basis. Density dependence, often invoked in fish population dynamics, appears to be nonlinear with a strong asymptotic effect on larval abundance at high biomass levels.

Various studies have identified both negative and positive effects of environmental factors on production and survival of larvae. Food production (Cushing 1990; Wroblewski and Richman 1987) and small-scale turbulence that increases the encounter rate between food particles and larvae (Rothschild and Osborn 1988; MacKenzie and Leggett 1991) are thought to be beneficial to larval survival. Food production may also benefit adult nutrition and therefore larval production (Smith and Eppley 1982). In contrast, intense wind-driven turbulent mixing that disaggregates patches of larval food appears to be detrimental (Lasker 1975; Peterman and Bradford 1987). It is also possible that destruction of food patches may alter adult nutrition. In addition, offshore transport that displaces larvae from favorable coastal areas appears to be a detrimental factor (Parrish et al. 1981).

For the northern anchovy, the relationship between larval number and upwelling intensity is dome shaped, suggesting that upwelling can be either beneficial or detrimental, depending on its intensity. Abundance of surviving larvae increases from low to moderate upwelling intensity, possibly because of the beneficial effect of increased food production and contact rates. It decreases for strong upwelling, possibly due to offshore transport and wind-driven turbulent mixing. Optimal conditions for a high number of surviving larvae correspond therefore to moderate upwelling intensity. These results are in striking conformity to other studies that have indicated that larval survival and recruitment success are regulated by a combination of different factors, each of which may depend in some way on upwelling intensity, rather than a single key factor (Therriault and Platt 1981; Husby and Nelson 1982; Wroblewski et al. 1989; Cury and Roy 1989; Roy et al. 1992).

To maintain, over decades, the homogeneity of environmental and ecological time series is clearly a difficult task. Analytical and methodological changes have certainly affected the consistency of the time series we use in the present analysis. As noted in the introduction, changes in the procedure used to produce the pressure fields have occurred since 1946: three different techniques have been used to estimate the total biomass of the northern anchovy and the samplilng frequency of the larval number has also changed through time. How it may affect our present analysis remains an open question.

Drastic fish population changes that take place within decades in many ecosystems are a social and economic challenge for fisheries (Glantz and Thompson 1981; Cury and Roy 1991) and of crucial interest for fish population modeling (Steele and Henderson 1984). Through modern electronic computation, new statistical methods permit exploration of nonlinear relationships between environmental changes and population responses. In ecology, these recently developed tools (Efron and Tibshirani 1991) are providing new insights and are refreshing our views of population dynamics (Mendelssohn and Cury 1987; Cury and Roy 1989; Swartzman et al. 1992). However, detecting nonlinear environmental effects on population dynamics remains a challenge for several reasons:

1) Several decades of data are sometimes available at different time scales but usually summarized to annual values, thereby only giving a few data points for the analysis. This is an important limitation because much more data are needed when exploring nonlinear relationships.
2) Collecting environmental and ecological data in an ecosystem during several decades is a long and exacting task. Regular financial support is difficult to find for these long-term projects.
3) The homogeneity of the time series over several decades is often questionable; analytical and methodological changes may affect their quality.
4) Our lack of understanding of causal mechanisms leads to an empirical choice between several variables; this choice may or may not be relevant to the population dynamics.
5) Low rather than high correlations are expected between climate variability and fish population responses, making definite relationships hard to establish.

Thus, caution should be added to optimism when exploring nonlinearities in fish population dynamics.

Over the past several decades, the stable global marine fish catch may have tended to mask the fact that the clupeoid contribution, which previously represented about one half of the marine fish catch in the upwelling areas, has now declined to only one third (Smith 1985). Recently it was suggested that the major coastal upwelling systems of the world have been increasing in upwelling intensity as greenhouse gases have accumulated in the earth's atmosphere (Bakun 1990). Our results suggest that the future of pelagic fish stocks will depend both on exploitation rates by fisheries and on climatic changes. Population fluctuations are not related in a linear manner to the density dependent and independent factors. Moderate upwelling intensity is better for larval survival. Thus, even if climatic changes are gradual and monotonic, the response of marine resource populations should not be expected to have the same well-behaved character. In that case, trend extrapolation may yield misleading predictions.

Acknowledgements

This is contribution No. 6 to the Climate and Eastern Ocean Systems (CEOS) project, a NOAA–ORSTOM–ICLARM cooperative project supported by the NOAA Climate and Global Change Marine Ecosystem Response Project and by the Terre–Océan–Atmosphère Department of ORSTOM. We thank P.E. Smith, who gave data and advice, and D.H. Cushing, L. Hutchings, L. Legendre, A. MacCall, B.R. MacKenzie, B. Peterson, R.M. Peterman, and F. Schwing, who provided valuable comments. A. Nichols drew the figures.

References

Bakun, A. 1973. Coastal upwelling indices, west coast of North America 1946–71. NOAA Tech. Rep. NMFS SSRF-671: 103 p.

Bakun, A. 1975. Daily and weekly upwelling indices, west coast of North America 1967–73. NOAA Tech. Rep. NMFS SSRF-693: 114 p.

Bakun, A. 1990. Global climate change and intensification of coastal ocean upwelling. Science (Wash., D.C.) 247: 198–201.

Beverton, R.J.H. 1990. Small marine pelagic fish and the threat of fishing: are they endangered? J. Fish Biol. 37: 5–16.

Blaxter, J.H.S., and J.R. Hunter. 1982. The biology of the clupeoid fishes. Adv. Mar. Biol. 20: 1–223.

Breiman, L., and J.H. Friedman. 1985. Estimating optimal transformations for mutiple regression and correlation. J. Am. Stat. Assoc. 80(391): 580–619.

Buja, A., T. Hastie, and R. Tibshirani. 1989. Linear smoothers and additive models (with discussion) Ann. Stat. 17: 453–555.

Crecco, V., T. Savoy, and W. Whitworth. 1986. Effects of density-dependent and climatic factors on American shad, *Alosa sapidissima*, recruitment: a predictive approach. Can. J. Fish. Aquat. Sci. 43: 457–463.

Cury, P., and C. Roy. 1989. Optimal environmental window and pelagic fish recruitment success in upwelling areas. Can J. Fish. Aquat. Sci. 46: 670–680.

Cury, P., and C. Roy [ed.]. 1991. Pêcheries ouest-africaines: variabilité, instabilité et changement. ORSTOM, Paris, France. 525 p.

Cushing, D.H. 1990. Plankton production and year-class strength in fish populations: an update of the match/ mismatch hypothesis. Adv. Mar. Biol. 26: 249–293.

Efron, B., and R. Tibshirani. 1991. Statistical data analysis in the computer age. Science (Wash. D.C.) 253: 390–395.

Elsberry, R.L., and R.W. Garwood. 1978. Sea surface temperature anomaly generation in relation to atmospheric storms. Bull. Am. Meteorol. Soc. 59(7): 786–789.

Gifi, A. 1990. Nonlinear multivariate analysis. John Wiley & Sons, Chichester, UK. 579 p.

Glantz, M.H., and J.D. Thompson [ed.]. 1981. Resource management and environmental uncertainty: lessons from coastal upwelling fisheries. Wiley-Interscience, London, UK. Publ. 11: 491 p.

Halliwell, G.R., Jr., and J.S. Allen. 1987. The large-scale coastal wind field along the west coast of North America 1981–1982. J. Geophys. Res. 92: 1861–1884.

Hastie, T., and R. Tibshirani. 1990. Generalized additive models. Chapman & Hall, London. 335 p.

House, E.D. 1989. Comparative growth, mortality, and energetics of marine fish larvae: temperature and implied latitudinal effects. Fish. Bull. U.S. 87: 471–496.

Husby, D.M., and C.S. Nelson. 1982. Turbulence and vertical stability in the California current. Calif. Coop. Oceanic Fish. Invest. Rep. 23: 113–129.

Jacobson, L.D., and N.C. Lo. 1991. Spawning biomass of the northern anchovy in 1991. NMFS, SWFC Admin. Rep. LJ 91-19. 53 p.

Kareiva, P.M. 1990. Stability from variability. Nature (Lond.) 344: 111–112.

Lasker R. 1975. Field criteria for survival of anchovy larvae: the relation between onshore chlorophyll maximum layers and successful first feeding. Fish. Bull. U.S. 73: 453–462

Lasker, R. [ed.]. 1981. Marine fish larvae. Morphology, ecology, and relation to fisheries. University of Washington Press, Seattle, WA. 131 p.

Lasker, R. 1985. What limits clupeoid production? Can. J. Fish. Aquat. Sci. 42: 31–38.

MacCall, A.D., and R.D. Methot. 1983. The historical spawning biomass estimates and population model in the 1983 anchovy fishery management plan. NMFS, SWFC Admin. Rep. LJ 83-17. 53 p.

MacKenzie, B.R., and W.C. Leggett. 1991. Quantifying the contribution of small-scale turbulence to the encounter rates between larval fish and their zooplankton prey: effects of wind and tide. Mar. Ecol. Prog. Ser. 73: 149–160.

Mendelssohn, R., and P. Cury. 1987. Fluctuations of a fortnightly abundance index of the Ivoirian coastal pelagic species and associated environmental conditions. Can. J. Fish. Aquat. Sci. 44: 408–428.

Methot, R.D. 1989. Synthetic estimates of historical abundance and mortality for northern anchovy. Am. Fish. Soc. Symp. 6: 66–82.

Parrish, R.H., C.S. Nelson, and A. Bakun. 1981. Transport mechanisms and reproductive success of fishes in the California current. Biol. Oceanogr. 1(2): 175–203.

Parrish, R.H., and A.D. MacCall. 1978. Climatic variation and exploitation in the Pacific mackerel fishery. Fish. Bull. U.S. 167: 1–110.

Pauly, D., and I. Tsukayama. 1987. The Peruvian anchoveta and its upwelling ecosystem: three decades of change. ICLARM Stud. Rev. 15: 351 p.

Pepin, P. 1991. Effect of temperature and size on development, mortality, and survival rates of the pelagic early life history stages of marine fish. Can. J. FIsh. Aquat. Sci. 48: 503–518.

Peterman, M.R., and M.J. Bradford. 1987. Wind speed and mortality rate of a marine fish, the northern anchovy (*Engraulis mordax*). Science (Wash., D.C.) 235: 354–356.

Peterman, M.R., M.J. Bradford, N.C. Lo, and R.D. Methot. 1988. Contribution of early life stages to interannual variability in recruitment of northern anchoby (*Engraulis mordax*). Can. J. Fish. Aquat. Sci. 45: 8–16.

Reid, J.L., Jr., G.I. Roden, and J.G. Wyllie. 1958. Studies of the California current system. Calif. Coop. Oceanic Fish. Invest. Prog. Rep. 23: 219–226.

Rothschild, B.J., and T.R. Osborn. 1988. The effects of turbulence on planktonic contact rates. J. Plankton Res. 10(3): 465–474.

Roy, C., P. Cury, and S. Kifani. 1992. Pelagic fish recruitment success and reproductive strategy in upwelling areas: environmental compromises. *In* A.I.L. Payne, K.H. Brink, K.H. Mann, and R. Hilborn [ed.] Benguela trophic functioning. S. Afr. J. Mar. Sci. 12: 135–146.

Sharp, G.D., and J. Csirke [ed.]. 1983. Proceedings of the expert consultation to examine changes in abundance and species composition of neritic fish resources. FAO Fish. Rep. 291: 1224 p.

Sissenwine, M.P. 1984. Why do fish populations vary?, p. 59–94. *In* R.M. May [ed.] Exploitation of marine communities. Dahlem Konferenzen. Springer-Verlag, Berlin, Germany. Life Sci. Res. Rep. 32.

Smith, P.E. 1985. Year-class strength and survival of 0-group clupeoids. Can. J. Fish. Aquat. Sci. 42: 69–82.

Smith, P.E., and R.W. Eppley. 1982. Primary production and the anchovy population in the southern California Bight: comparison of time series. Limnol. Oceanogr. 27(1): 1–17.

Smith, P.E., and H.B. Moser. 1988. CalCOFI time series: an overview of fishes. Calif. Coop. Oceanic Fish. Invest. Rep. 29: 66–78.

Steele, J.H., and E.W. Henderson. 1984. Modeling long-term fluctuation in fish stocks. Science (Wash., D.C.) 224: 985–987.

Swartzman, G., C. Huang, and S. Kaluzny. 1992. Spatial analysis of bering sea groundfish survey data using generalized additive models. Can. J. Fish. Aquat. Sci. 49: 1366–1378.

Therriault, J.C., and T. Platt. 1981. Environmental control of phytoplankton patchiness. Can. J. Fish. Aquat. Sci. 38: 638–641.

Tibshirani, R. 1988. Estimating transformations for regression via additivity and variance stabilization. J. Am. Stat. Assoc. 83(402): 394–405.

Tyler, A.V. 1992. A context for recruitment correlations: why marine fisheries biologists should still look for them. Fish. Oceanogr. 1(1): 97–107.

Wroblewski, J.S., J.G. Richman, and G.L. Mellor. 1989. Optimal wind conditions for the survival of larval northern anchovy, *Engraulis mordax*: a modeling investigation. Fish. Bull. U.S. 87: 387–398.

Wroblewski, J.S., and J.G. Richman. 1987. The non-linear response of plankton to wind mixing events—implications for survy of larval northern anchovy. J. Plankton Res. 9(1): 103–123.

The relationships among sea-surface temperature, the abundance of juvenile walleye pollock (*Theragra chalcogramma*), and the reproductive performance and diets of seabirds at the Pribilof Islands, southeastern Bering Sea

Mary Beth Decker, George L. Hunt, Jr., and G. Vernon Byrd, Jr.

Decker, M.B., G.L. Hunt, Jr., and G.V. Byrd, Jr. 1995. The relationships among sea-surface temperature, the abundance of juvenile walleye pollock (*Theragra chalcogramma*), and the reproductive performance and diets of seabirds at the Pribilof Islands, southeastern Bering Sea, p. 425–437. *In* R.J. Beamish [ed.] Climate change and northern fish populations. Can. Spec. Publ. Fish. Aquat. Sci. 121.

Abstract: Marine birds nesting on the Pribilof Islands, in the southeastern Bering Sea, exhibited decreases in reproductive performance (black-legged kittiwakes, *Rissa tridactyla*, and red-legged kittiwakes, *R. brevirostris*) and changes in food habits (black-legged kittiwakes and thick-billed murres, *Uria lomvia*) beginning about 1978. These changes coincided with the appearance of the unusually large 1978 year-class of walleye pollock (*Theragra chalcogramma*) and the beginning of a period of above-average sea-surface temperatures that lasted until 1984. Seabird reproductive performance and diets did not return to pre-1979 values after 1984, suggesting that the changes in the marine ecosystem to which the birds responded were of longer duration than the periods of warm and cool surface temperatures observed between 1975 and 1990.

Résumé : Les oiseaux marins qui nichent sur les îles Pribilof, dans le sud-est de la mer de Béring, accusent une diminution de leur reproduction effective (mouettes tridactyles, *Rissa tridactyla*, et mouettes à pattes rouges, *R. brevirostris*) ainsi que des changements dans leurs habitudes alimentaires (mouettes tridactyles et marmettes de Brünnich, *Uria lomvia*) depuis 1978 environ. Ces changements ont coïncidé avec l'apparition d'une classe annuelle (1978) inhabituellement nombreuse de morues du Pacifique occidental (*Theragra chalcogramma*) et avec le début d'une période de températures de la surface de la mer plus élevées que la moyenne, période qui a duré jusqu'en 1984. Après 1984, la reproduction effective et le régime alimentaire des oiseaux de mer ne sont pas revenus à leurs valeurs d'avant 1979, ce qui tend à démontrer que les changements dans l'écosystème marin auxquels ont réagi les oiseaux ont duré plus longuement que les périodes de températures de surface chaudes et fraîches observées entre 1975 et 1990.

Introduction

Between 1975 and 1990, the reproductive performance of marine birds breeding at the Pribilof Islands fluctuated widely (Byrd 1989; Schulmeister 1991; Springer 1992). During this same period, sea-surface temperatures fluctuated, as did juvenile walleye pollock (*Theragra chalcogramma*) populations (V.G. Wespestad, National Marine Fisheries Service, Seattle, WA, personal communication). Although fish populations are affected by oceanographic changes and commercial exploitation (Rothschild 1983), fluctuations in stock biomass are often difficult to monitor directly (Cushing 1988). Indication of variations in the availability of small fish and other prey stocks are provided by marine predators such as seabirds (Furness 1982; Nettleship et al. 1984; Springer et al. 1986; Cairns 1987; Croxall et al. 1988). Seabirds reflect the distribution and abundance of prey through their foraging ecology (Hunt 1990; Hunt and Harrison 1990; Hunt et al. 1990) and their reproductive biology (Hunt and Butler 1980; Anderson et al. 1982). Dietary changes of seabirds also reflect availability of prey and have been related to the collapse of commercial fisheries stocks (Montevecchi et al. 1988; Nettleship 1991). Thus, seabirds may indicate fluctuations in fish populations brought on by environmental perturbations, such as climate change or commercial harvesting.

M.B. Decker and G.L. Hunt, Jr. 1 Department of Ecology and Evolutionary Biology, University of California, Irvine, CA 92717, USA.

G.V. Byrd, Jr. U. S. Fish and Wildlife Service, Alaska Maritime National Wildlife Refuge, Homer, AK 99603, USA.

In this paper, we explore how variations in seabird reproductive performance and diets may have been influenced by fluctuations in sea-surface temperature and pollock abundance. We hypothesize that changes in the temperature regime of the southeastern Bering Sea resulted in changes in the availability or abundance of key prey required for successful seabird reproduction. Our analyses extend those of Springer (1992) in that we include information on seabird diets, and we examine separately each of three species of seabird on the two islands. By using annual variations in seabird diets and reproductive performance on the Pribilofs, we seek evidence for interdecadal changes in the nearby marine environment.

Changes in sea-surface temperature may affect growth, survival, or habitat choice of prey species used by seabirds. Specifically, we tested the hypothesis that changes in the physical state of the northern Pacific and Bering Sea, thought to have occurred around 1978, were reflected in the diets and reproduction of seabirds. Data gathered during PROBES (Processes and Resources of the Bering Sea Shelf) showed that interannual changes in sea-surface temperature could have influenced the development and standing stocks of copepods (Smith and Vidal 1986) and thus, presumably, the growth and distribution of young walleye pollock (Walsh and McRoy 1986), an important prey of seabirds nesting on the Pribilofs. We use data on seabird reproductive performance and diet as indications of changes in the availability of prey types to the birds. We assume that these changes in prey availability reflect regional changes within the marine ecosystem.

Methods

Study site

The Pribilof Islands are located in the southeastern Bering Sea near the continental shelf edge (Fig. 1). The waters surrounding the islands consist of discrete domains divided by oceanographic fronts (Kinder and Schumacher 1981). St. Paul Island is located in the middle domain and St. George Island,

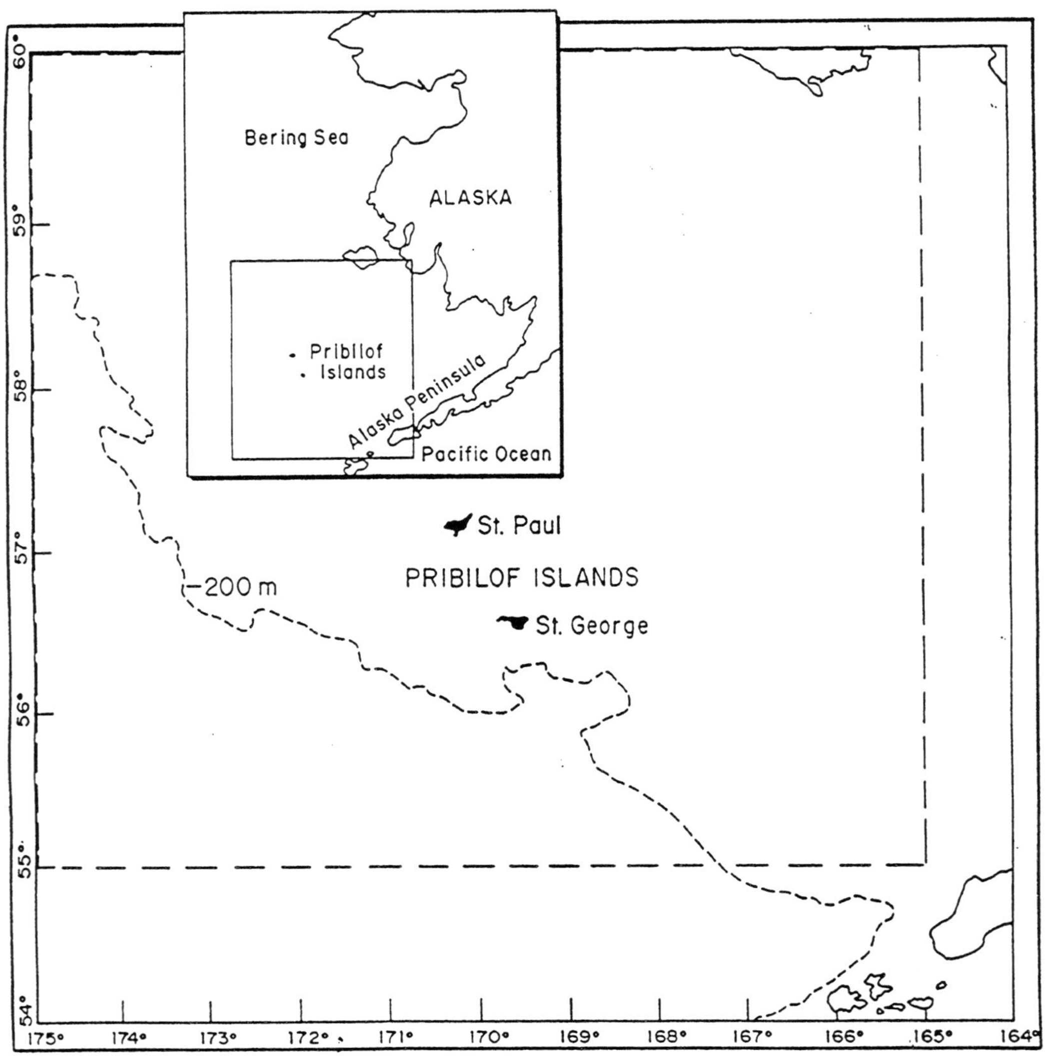

Fig. 1. Map of Pribilof Islands, Alaska. Hatched square indicates area in which sea-surface temperatures were recorded.

which is closer to the shelf edge, lies in the outer domain. The middle shelf is dominated by a benthic food web, whereas the outer shelf is dominated by pelagic species (Iverson et al. 1979; Schneider and Hunt 1982). Large populations of black-legged kittiwakes (*Rissa tridactyla*), red-legged kittiwakes (*R. brevirostris*) and thick-billed murres (*Uria lomvia*) nest on the Pribilof Islands (Hickey and Craighead 1977).

Analyses

Sea-surface temperature data were obtained from monthly mean values available for a 5° by 10° grid surrounding the Pribilof Islands from 1950 through 1991 (D. Cayan, Scripps Institution of Oceanography, La Jolla, CA, unpublished data) (Fig. 1). This area encompasses a major portion of the feeding area of Pribilof seabirds during the breeding season (Hunt et al. 1982). We calculated mean annual temperatures beginning in October of the year previous to the breeding season ending in September to reflect sea-surface temperatures for the 12 months before and during the seabird breeding season. Yearly averages of sea-surface temperature are presented as deviations from the long-term (1950–91) mean.

For this analysis, seabird reproductive performance and diet data from each of the three bird species on each island were considered to be independent samples of the marine environment surrounding the Pribilof Islands. Each of the three seabirds has a unique method of sampling the environment (Hunt et al. 1982). Thick-billed murres are diving birds that forage at depths up to 200 m (Croll et al. 1992) and are found predominantly in near-shore habitats. Black-legged kittiwakes are diurnal surface feeders that forage primarily in the middle domain of the continental shelf. Red-legged kittiwakes are also surface-feeders but they feed nocturnally in the outer-shelf and shelf-break regions. Schneider and Hunt (1984) found differences in diet between conspecific seabirds on the two islands and suggested that these results were related to the different oceanographic regimes surrounding the islands. Likewise, Hunt et al. (1986) showed that reproductive performance of birds nesting on the two islands differed. Therefore, because of these differences in mode and locale of foraging, we considered each bird species by island by year an independent sample of the prey available in the waters within 100 km of the Pribilof Islands, and of the birds' responses to changes in prey availability.

Reproductive performance data for black-legged kittiwakes, red-legged kittiwakes, and thick-billed murres were obtained from St. Paul and St. George islands in most years from 1975–90 (Byrd 1989; Schulmeister 1991). The single reproductive variable chosen to make comparisons with annual sea-surface temperature and diet composition was overall productivity (i.e., the number of chicks fledged / nest started). To test if reproductive success was simply correlated with sea-surface temperature, we performed Pearson's correlations for the period 1975–90. In addition, to test the hypothesis that changes in the ocean state that occurred around 1978 are reflected in seabird reproductive success, we conducted Mann-Whitney U tests to investigate differences in reproductive success between the periods 1975–78 and 1979–90. We used a criterion of $p = 0.10$ to reject the null hypothesis.

Seabird diet samples were obtained from the same three seabird species at the Pribilof Islands from 1975 to 1978 and 1984–88 (Hunt et al. 1982; Bradstreet 1985; Dragoo 1991; G.L. Hunt Jr., unpublished data). Diet samples were collected at the colony throughout the chick-rearing period from July through August. Three types of food samples were collected from seabirds on the Pribilof Islands: stomachs from shot adults, regurgitations from kittiwake chicks, and food loads brought to chicks in the bills of adult murres. Kittiwakes regurgitate meals to their chicks, and adult stomach contents are similar to chick regurgitations. However, adult murres bring large, single items to their chicks and consume for themselves different types of prey than those fed to chicks. Therefore, we present murre adult and chick diets separately.

For this analysis, we assume that each species sampled prey around the Pribilof Islands independently of other species and that individual samples obtained at the colony reflect independent foraging events. We base this assumption of independence on the long intervals between collection (shooting) of flying birds returning to the colony, the small number of nests from which samples were obtained on any one day, and the general scarcity of foraging flocks observed at sea (Hunt et al. 1982, G.L. Hunt Jr., unpublished observations). Diet data are presented in percent occurrence, which is the percentage of samples in which a prey item occurs, and percent volume, calculated by dividing the sum of the volumes of a particular prey item from all the samples by the total volume of all samples combined. For all of our analyses, we would have preferred to use percent volumes as the indication of the availability of prey, but percent volumes were not available for the 1985 and 1986 diet samples. Amphipods in the 1984 diets were not identified to lower taxa.

We tested for differences in the volumetric use of each prey species by seabirds between the two periods, 1975–78 and 1984 plus 1987-1988 using Mann–Whitney statistics. We used $p = 0.10$ as our criterion for rejecting the null hypothesis.

Table 1. Abundance of age-2 walleye pollock present on the eastern Bering Sea shelf (V.G. Wespestad, National Marine Fisheries Service, Seattle, WA, personal communication, May 1993).

Year	Abundance
1974	17 276 196 479
1975	11 656 877 299
1976	9 694 659 306
1977	9 604 853 528
1978	7 636 766 363
1979	11 272 767 252
1980	31 779 880 584
1981	12 188 817 965
1982	12 438 437 994
1983	5 780 141 392
1984	17 006 804 394
1985	4 680 695 666
1986	16 936 638 769
1987	11 073 137 364
1988	3 748 682 345
1989	1 193 048 352
1990	2 760 673 232
1991	6 951 349 211

Samples sizes were adequate to perform this test on black-legged kittiwake diets only. As a measure of dietary change in the seabird community as a whole, we tested whether the occurrence (presence or absence) of a prey item in the diets of all three bird species combined changed between the two time periods by using chi-square contingency tests. This procedure was completed separately for each major prey item.

Data on the numbers of age-2 walleye pollock present on the eastern Bering Sea shelf were used to create an estimate of the abundance of age-1 pollock, the size-class of pollock used by the seabirds during the early portion of the breeding season (V.G. Wespestad, personal communication) (Table 1). To obtain an estimate of age-1 pollock, we lagged age-2 numbers by 1 year. We assume that mortality from age 1 to age 2 has no significant trend over the period examined. Correlations of annual reproductive success with percent volume of pollock in the diet and with age-1 pollock numbers were calculated for each bird species.

Results

Sea-surface temperature

Low-frequency fluctuations in sea-surface temperature of over 1°C have been found in the northeastern Pacific and the Bering Sea (Royer 1989) (Fig. 2). Annual sea-surface temperature anomalies calculated for the bird reproductive year show a similar variation (Fig. 3). A cool period occurred during the 1970s, with 1978 being the last of this set of anomalously cool years. Sea-surface temperatures were warmer than average from 1979–84. Another cool period began in 1985.

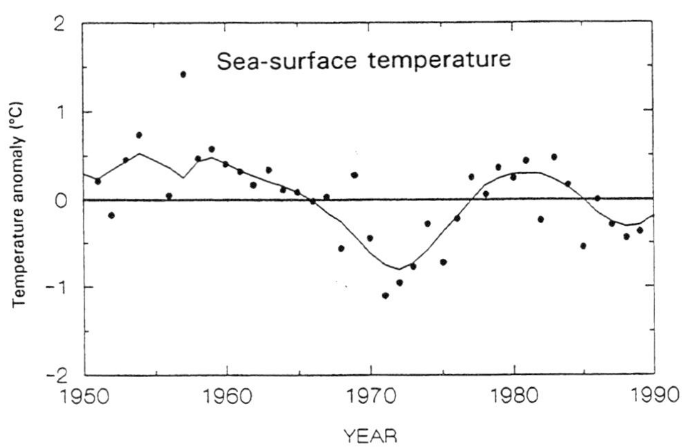

Fig. 2. Annual sea-surface temperature anomaly for the area around the Pribilof Islands from 1950–90.

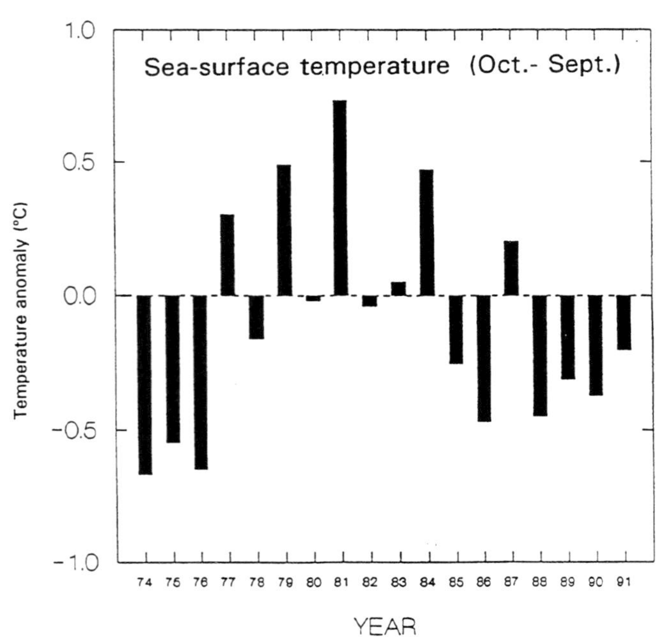

Fig. 3. Annual sea-surface temperature anomaly calculated from the months October to September to reflect temperatures before and during the seabird breeding season.

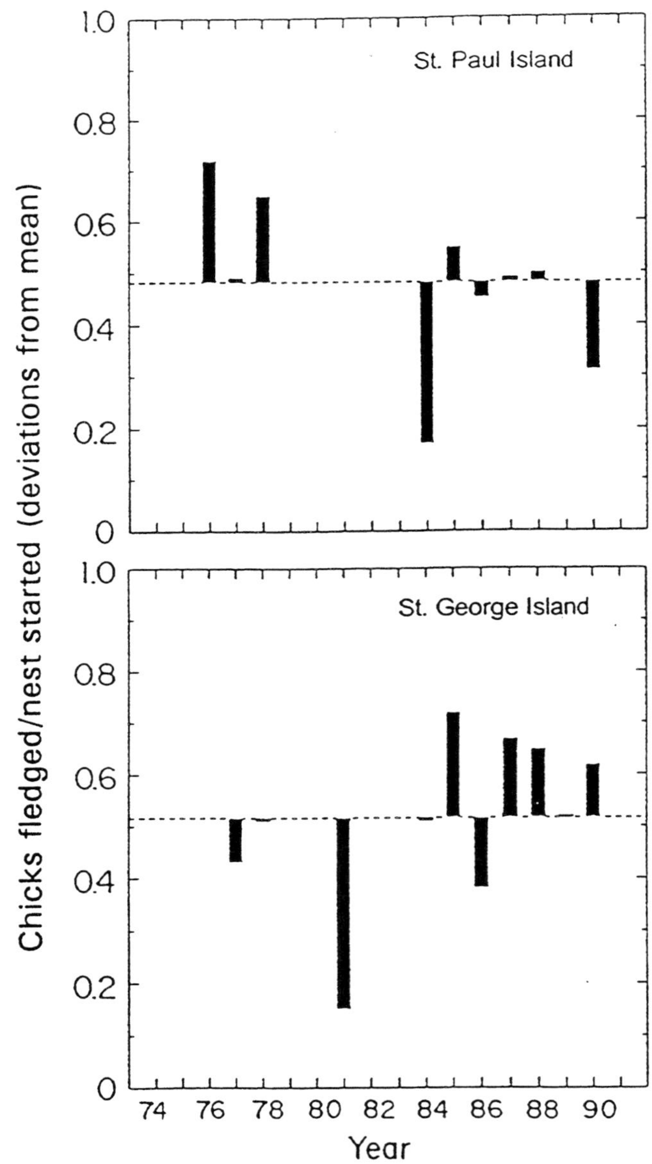

Fig. 4. Reproductive performance of thick-billed murres on St. Paul (upper) and St. George (lower) islands from 1975–90. The values for the number of chicks fledged / nest started are plotted as deviations from the long-term mean (St. Paul = 0.478; St. George = 0.538).

Reproductive performance of marine birds

Reproductive performance was negatively correlated with annual sea-surface temperature for subsurface-foraging thick-billed murres when data from populations on both islands were combined (r=–0.51, n=19, p=0.025) (Fig. 4). The trends were similar, but statistically significant on only one island when birds on each island were considered separately (St. Paul Island, r=–0.53, p=0.14; St. George Island, r=–0.55, p=0.10). Although reproductive performance of thick-billed murres on St. Paul Island appeared to have declined after 1978, reproductive performance on St. George Island was strongest in the cool period starting in 1985. The reproductive performance of thick-billed murres showed no significant difference between the two periods 1975–78 and 1979–90 when data for both islands were combined (Mann–Whitney U=42, p=0.516).

The reproductive performance of both surface-foraging black-legged and red-legged kittiwakes showed no significant correlations with sea-surface temperature when considered separately by island or when data from the two islands were combined. Both species of kittiwakes had poorer reproductive performance after 1978 (data from both islands combined: black-legged kittiwakes Mann–Whitney U=116, p=0.011; red-legged kittiwakes U=110.5, p=0.025) (Fig. 5 and 6). For the most part, during the warm period from 1979–84,

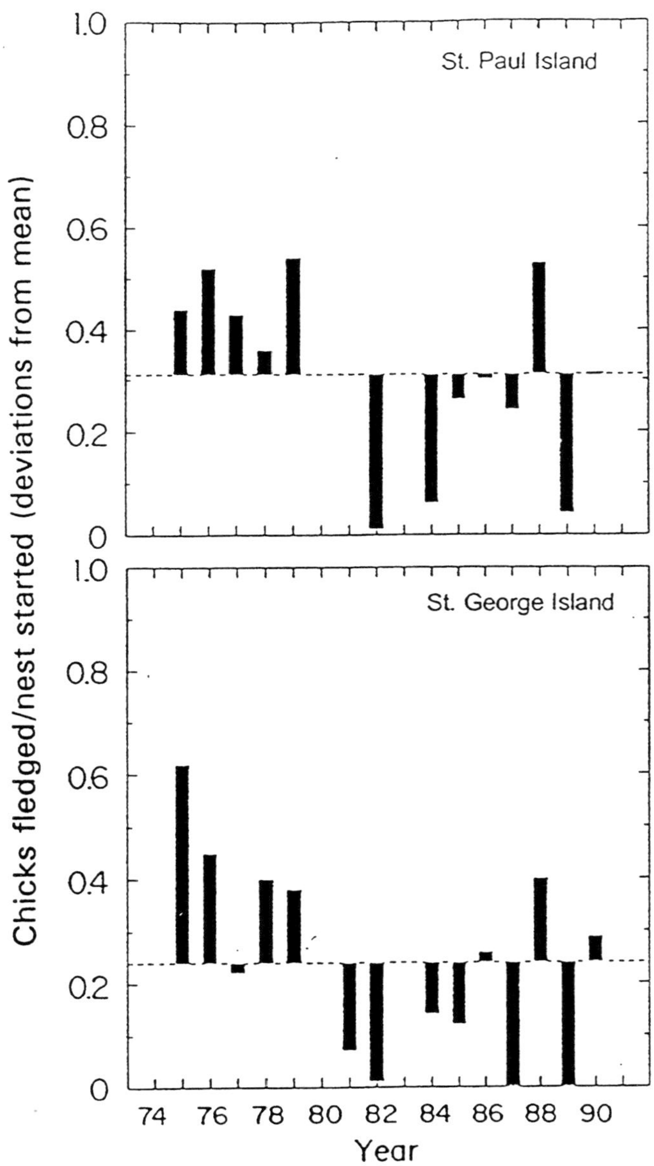

Fig. 5. Reproductive performance of black-legged kittiwakes on St. Paul (upper) and St. George (lower) islands from 1975–90. The values for the number of chicks fledged/nest started are plotted as deviations from the long-term mean (St Paul = 0.306, St. George = 0.261).

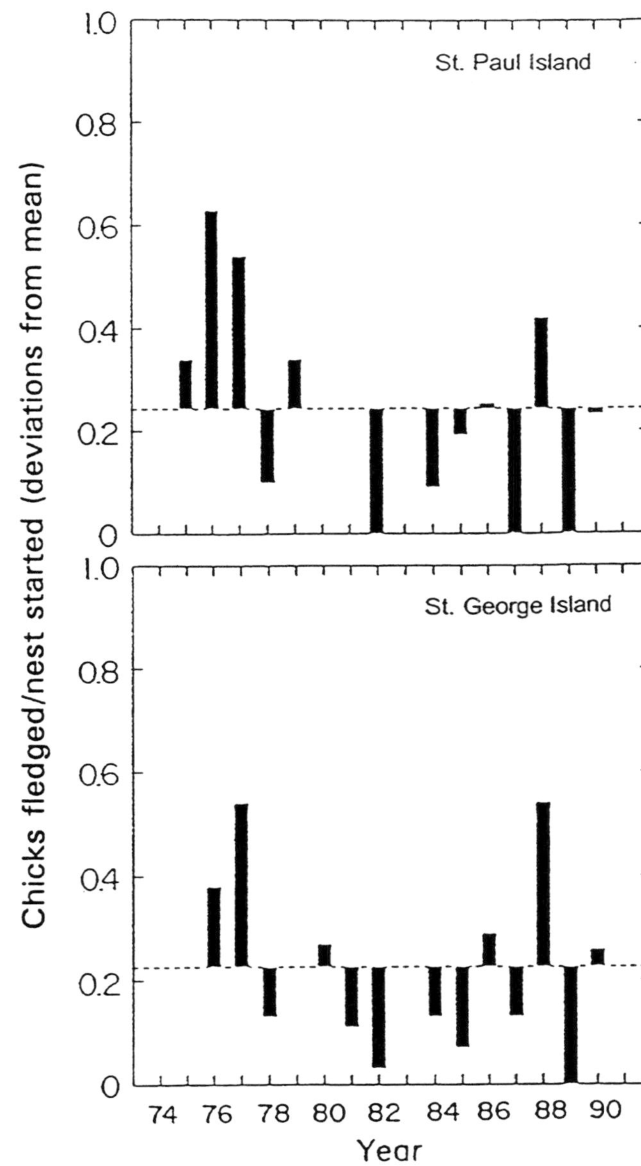

Fig. 6. Reproductive performance of red-legged kittiwakes on St. Paul (upper) and St. George (lower) islands from 1975–90. The values for the number of chicks fledged / nest started are plotted as deviations from the long-term mean (St. Paul = 0.243; St. George = 0.254).

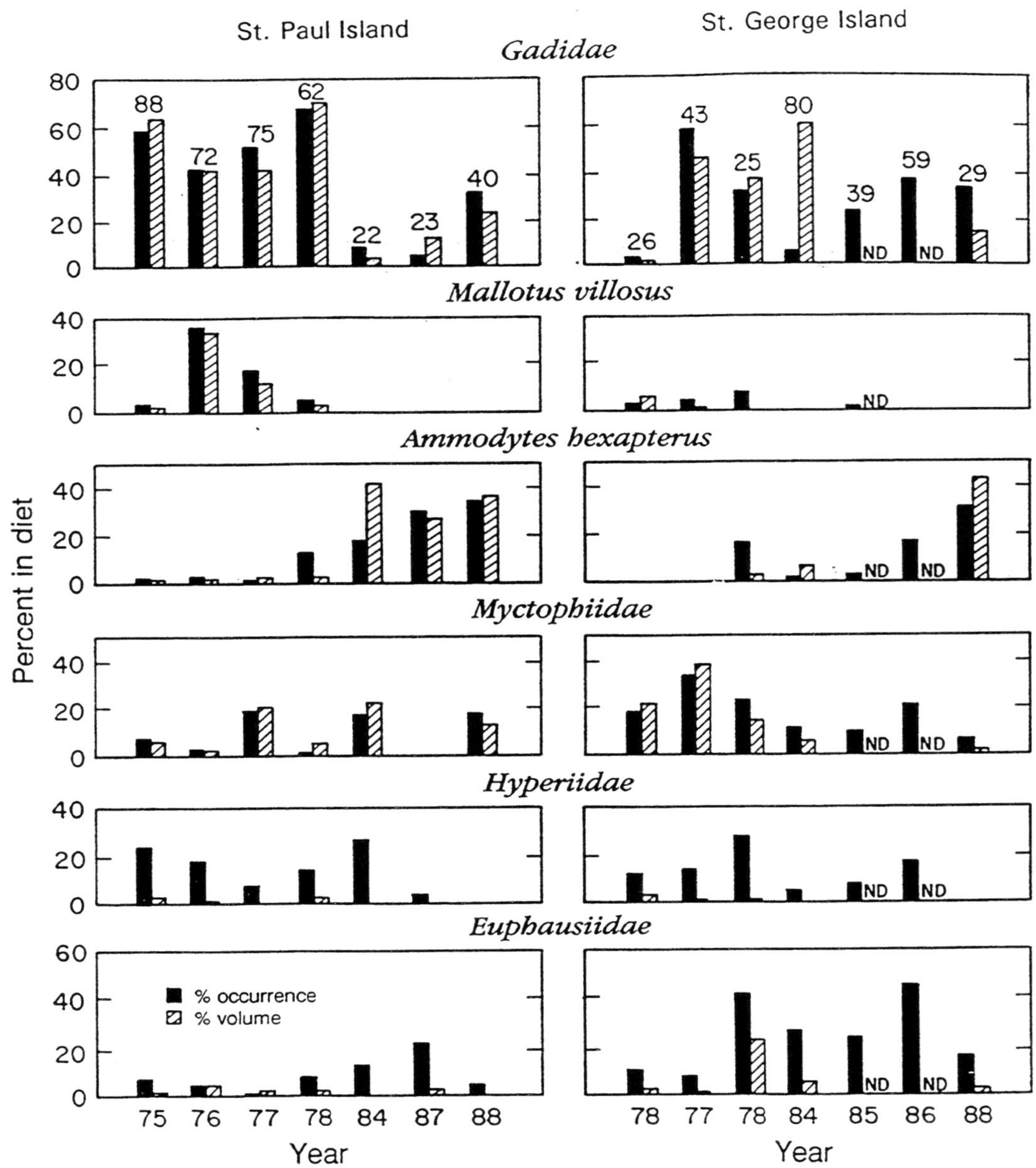

Fig. 7 Percent occurrence and percent volume of the major prey items found in black-legged kittiwakes from St. Paul and St. George islands. Numbers above the bars are the number of birds sampled. ND = no data.

reproductive performance of both species of kittiwakes was below the long-term average and remained below the average during the cool period beginning in 1985, with the exception of 1988. Thus, subsurface and surface-foraging birds showed different responses to changes in sea-surface temperature.

We examined an alternative hypothesis, that the variations in seabird reproductive performance correlated with the abundance of age-1 pollock in the Bering Sea, which in the 1975–79 period was the predominant age-class of pollock used by these birds early in the breeding season (Hunt et al. 1982). Correlations of annual reproductive performance and the number of age-1 pollock were not significant for any of the three seabirds over the period encompassed by this study.

Marine bird diets

Diets of the three seabirds are presented in Figs. 7 to 10 and indicate that each species of seabird showed similar changes in their diets over the time period of our study. The diets of surface-foraging black-legged kittiwakes changed between the periods before and after 1978 (Fig. 7). Capelin (*Mallotus villosus*) were present in the diets of birds on both islands during the cool period from 1975–78, but after 1978, capelin were no longer used (St. Paul, U=12.0, p=0.03; St. George, U=12.0, p=0.03). On St. Paul Island, hyperiid amphipods were less important in the diet of black-legged kittiwakes in the 1980s (U=12.0, p=0.03), as was the case on St. George Island (U=6.0, p=0.08). Black-legged kittiwake use of sand-

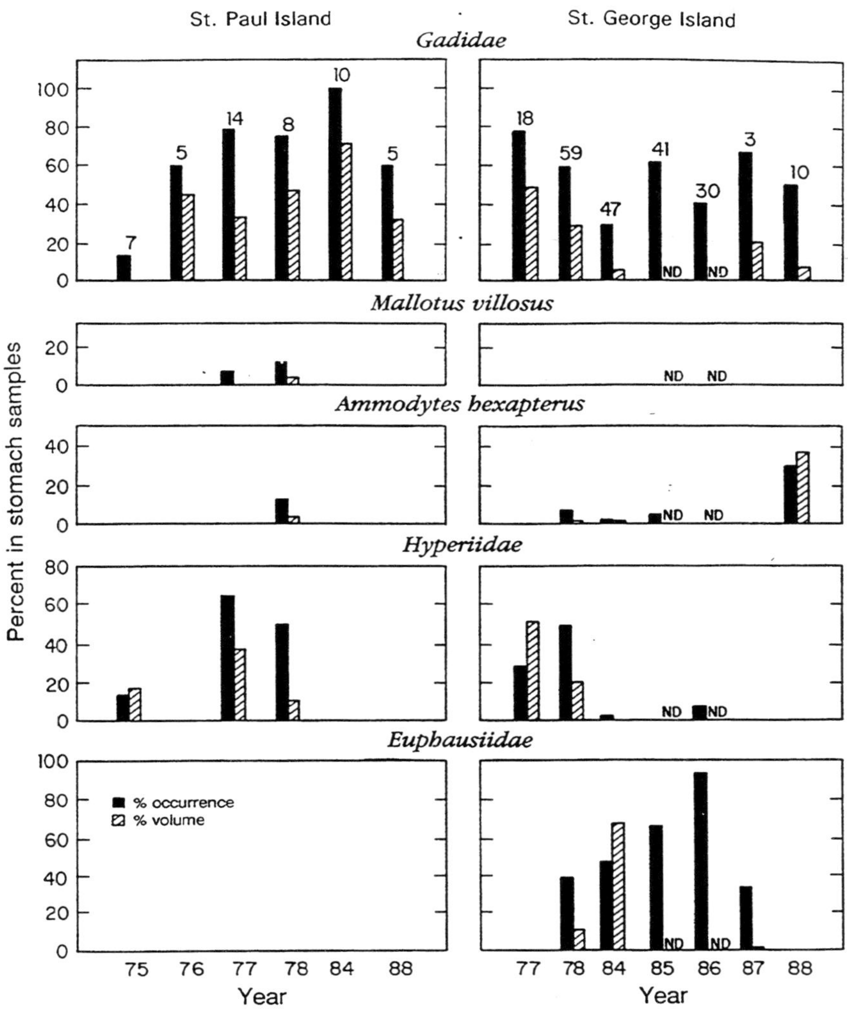

Fig. 8. Percent occurrence and percent volume of the major prey items found in the stomachs of adult thick-billed murres from St. Paul and St. George islands. Numbers above the bars are the number of birds sampled. ND = no data.

lance (*Ammodytes hexapterus*) increased after 1978 on St. Paul Island (U=12.0, p=0.03) but not on St. George Island (U=2.0, p=0.15). The use of gadids by black-legged kittiwakes declined after 1978 on St. Paul Island (U=12.0, p=0.03) but not on St. George Island (U=2.0, p=0.56).

Although inadequate sample sizes caused us to refrain from the use of statistical tests on the thick-billed murre diet data, changes similar to those found for black-legged kittiwakes were apparent (Figs. 8 and 9). In the early period, capelin and hyperiid amphipods were present in the diets of both adults and chicks, whereas in the latter period, capelin were no longer found in the diets and hyperiids appear to have been used less frequently by adults and chicks. Sandlance also appeared to increase in the diets of adult and young murres in the 1980s, especially on St. George Island. After 1978, cephalopods became important in the chick diets, particularly on St. George Island, and the use of gadids for chicks and adult murres appears to have declined on St. George Island but not on St. Paul Island.

Red-legged kittiwakes are surface foragers that specialize on myctophiids, which they capture at night near the shelf edge (Hunt at al. 1982). The use of myctophiids appeared to remain at approximately the same level throughout the study period. However, capelin, present as a trace in the diet of red-legged kittiwakes prior to 1978 at St. George Island, were not used in later years (Fig. 10). Overall, compared with the other seabird species, red-legged kittiwakes showed very little change in their diet over the course of this study.

When we combined data from the three species of seabirds on the two islands, we found three patterns in the use of foods. We found a significant drop in the presence of capelin in the diets after 1978, (Table 2) (X^2=13.61, $p < 0.001$). There was

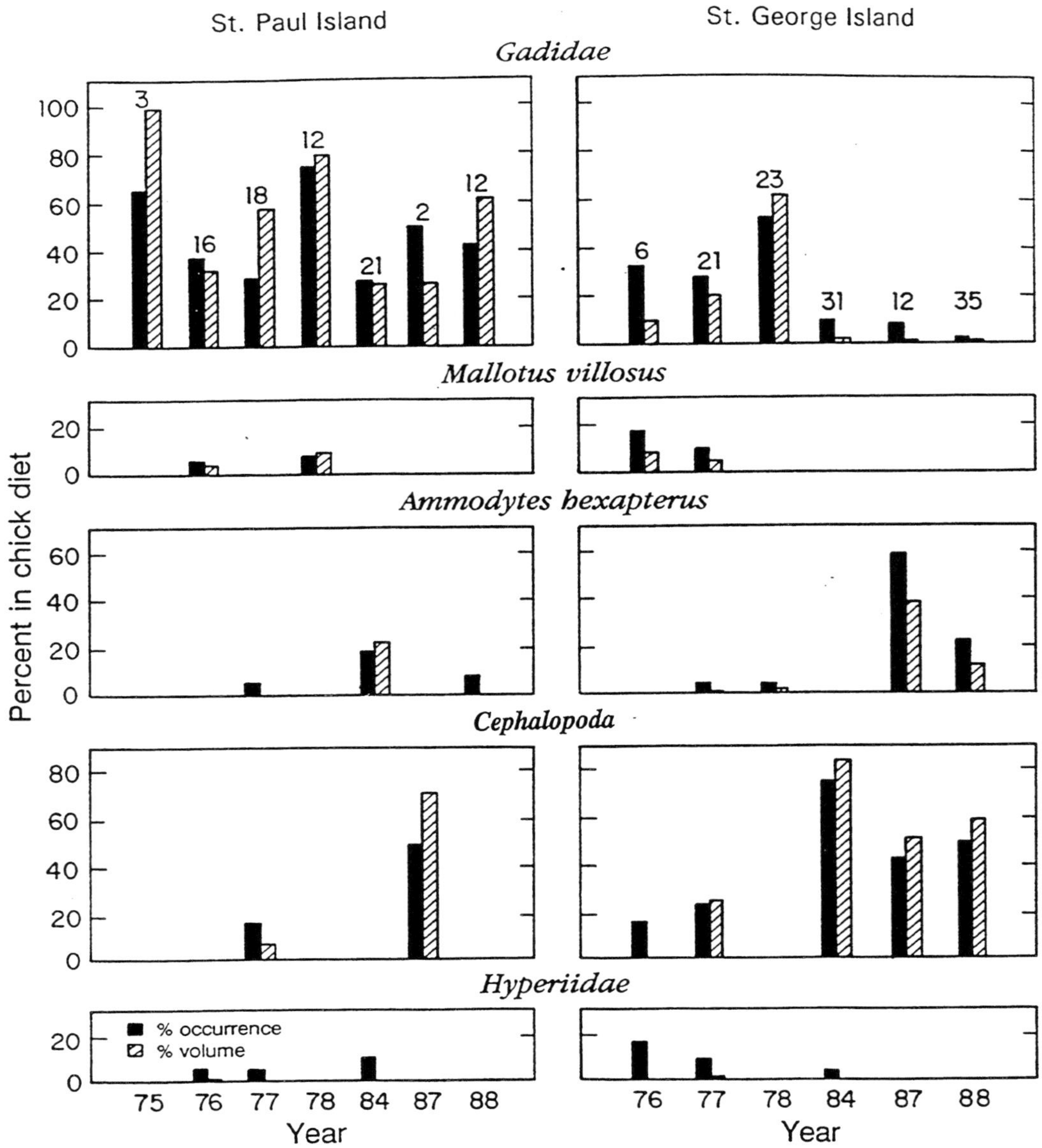

Fig. 9. Percent occurrence and percent volume of the major prey items found in regurgitations of thick-billed murre chicks from St. Paul and St. George islands. Numbers above the bars are the number of birds sampled. ND = no data.

Table 2. Number of years in which capelin was present or absent in seabird diets during two time periods; 1975–78 and 1984–88 ($x^2 = 13.61, p < 0.001$).

	1975–78	1984–88
Present	15	1
Absent	13	21

Table 3. Number of years in which hyperiid amphipods were present in seabird diets during two time periods; 1975–78 and 1984–88 ($x^2 = 4.28, p < 0.038$).

	1975–78	1984–88
Present	23	14
Absent	5	11

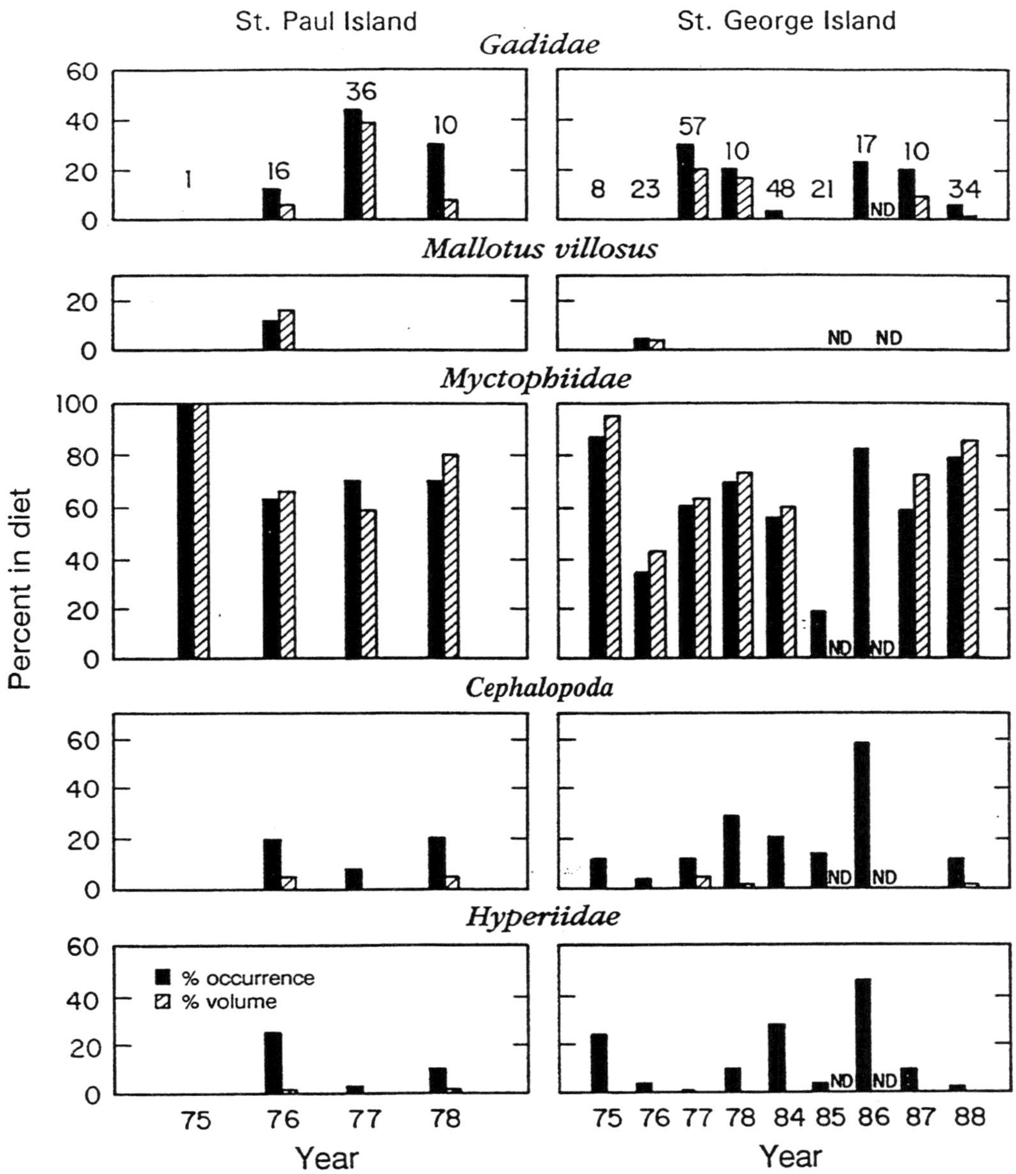

Fig. 10. Percent occurrence and percent volume of the major prey items found in red-legged kittiwakes from St. Paul and St. George islands. Numbers above the bars are the number of birds sampled. ND = no data.

also a significant decline in the use of hyperiids after 1978 (Table 3) (X^2=4.28, p=0.04). Although sandlance appeared was not statistically significant (X^2=1.17, p=0.3). The use ofto increase in occurrence in the diets after 1978, this increase gadids, although important in the diets of seabirds, showed no consistent or significant change through this period.

The percentage of pollock in the diets was not correlated with reproductive success, with the exception of a significant negative correlation for black-legged kittiwakes on St. George Island (r=–0.89, p=0.05). For black-legged kittiwakes on St. Paul Island, the percentage of pollock in the diets was significantly correlated with the index of age-1 pollock abundance (kittiwakes, r=0.836, p=0.02) The percentage of pollock in the diet of thick-billed murre chicks on St. George Island was also positively correlated with age-1 pollock numbers (r=0.77, p=0.075). In no other case was correlation of pollock use and number of age-1 pollock in the eastern Bering Sea statistically significant.

Discussion

Lloyd (1985) and Springer (1991, 1992) drew attention to a negative correlation between sea-surface temperature and reproductive performance of seabirds on the Pribilof Islands.

This pattern is opposite to that found at the Bluff colony in northern Norton Sound (Murphy et al. 1991; Springer 1991). We did not find a correlation between sea-surface temperature and seabird reproduction at the Pribilof Islands because reproductive performance failed to recover during the cool period beginning in 1985.

Cold water temperatures in the eastern Bering Sea may enhance the availability of young pollock to seabirds at the Pribilof Islands. Young pollock in the laboratory avoid cold water when a thermocline is present (Olla and Davis 1990). The presence of unusually cold bottom water on the eastern Bering Sea shelf could result in both the restriction of young fish to the upper water column for a longer period and possibly to the warmer waters of the outer shelf domain (Chen 1983; Ohtani and Azumaya 1995). If young pollock become concentrated in the waters of the outer shelf domain, they would then be more easily taken by predators (Ohtani and Azumaya 1995), including seabirds. Springer et al. (1986), Springer and Byrd (1989), and Springer (1992) have emphasized the potential importance of pollock recruitment for kittiwake (and other seabird) reproduction in the Pribilof Islands. In addition to recruitment, other more proximate factors may have a strong influence on prey availability in the Bering Sea. The concentration of small pollock in the surface waters near the islands, bottom temperatures on the shelf, thermocline strength, or predation pressure by large fish (Olla and Davis 1990; Sogard and Olla 1994; Ohtani and Azumaya 1995) all may contribute to the availability of pollock to Pribilof seabirds.

Variations in water temperature may also affect the distribution and abundance of capelin and of the hyperiid amphipod, *Parathemisto libellula*. *Parathemisto libellula* is often associated with cold or ice-covered waters (Lonne and Gulliksen 1991), and populations of this species may have decreased in the southeastern Bering Sea after the mild winters following 1978, when little ice was present (Niebauer 1980, 1983). Similarly, capelin have a restricted range of preferred temperatures for spawning (Pahlke 1985) and may have retreated to cooler, northern waters during the warming of the southeastern Bering Sea. An alternative explanation for the decreased use of capelin by surface-foraging kittiwakes is that capelin moved deeper in the water column (Methven and Piatt 1991) and avoided beach spawning (Pitt 1958) when water temperatures became warm. If so, capelin may have been present but unavailable to the kittiwakes, similar to the situation described by Baird (1990) near Kodiak Island. However, at Kodiak, deep-diving tufted puffins (*Fratercula cirrhata*), continued to capture capelin, which was not true for murres at the Pribilof Islands, which ceased to use capelin after 1978

Table 4. Abundance of capelin present near the Pribilof Islands (G. Walter, National Marine Fisheries Service, Seattle, WA. personal communication).

Year	Abundance
1975	52285349
1979	399460
1982	0
1983	763042
1984	564360
1985	33413
1986	51562
1987	0
1988	0
1989	0
1990	516515
1991	87678
1992	0

Fritz (cited in Springer 1992) suggested that there has been a decline in capelin populations in the Bering Sea since the mid-1970s. Records of capelin incidentally caught in groundfish surveys indicate that capelin abundance declined from the area around the Pribilof Islands after 1975 (Table 4) (G. Walters, National Marine Fisheries Service, Seattle, WA, personal communication). This decrease could be related to changes in sea temperatures, to competition with pollock for copepod prey (Springer 1992), or some other unknown factor. Because of the coincidence of ocean warming and the 1978 year-class of pollock, data are insufficient to evaluate the relative importance of these hypotheses. What is obvious is that capelin did not return to their former availability when sea-surface temperatures cooled in the mid-1980s.

Straty and Haight (1979), and Springer (1992) suggested an alternative hypothesis: that the observed changes in seabird reproduction and diets may have been caused by depletion of the birds' prey due to increased predation by adult pollock. Springer (1992) found that kittiwake productivity was negatively correlated with total pollock biomass and suggested that large numbers of adult pollock may have a negative effect on birds by depressing the abundance or availability of young pollock or other important prey (see also Dwyer et al. 1987). Little evidence is available to judge the veracity of these hypotheses. However, the apparent beginning to the decline of capelin prior to the arrival of the 1978 year-class of age-2 pollock in 1980 and the failure of capelin to rebound in numbers once the pollock biomass began to decline in 1986 does not argue for the hypothesis that predation by pollock is the principal factor controlling capelin near the Pribilofs. Likewise, *Parathemisto* apparently began to decrease in the birds' diets prior to 1979, when pollock would have been plentiful competitors for this food source. Similarly, we did not find a decline in the use of sandlance, which according to Springer (1992) should suffer from competition and possibly predation from pollock. In summary, although the hypothesis that pollock control the populations of alternative prey is attractive, most fragments of evidence available do not support the hypothesis.

Walleye pollock has been identified as a key prey species for marine birds and mammals in the southeastern Bering Sea (Hunt et al. 1982; Frost and Lowry 1986; Springer and Byrd 1989). Springer and Byrd (1989) found positive correlations between seabird reproductive performance and the abundance of age-1 pollock over the period 1975–82. In our study, we did not find correlations of abundance of age-1 pollock with annual reproductive performance of birds, but we examined a longer time series from 1975 to 1990 than did Springer and Byrd (1989). If pollock were the preferred prey species for marine birds, one might expect changes in the percentage of pollock in the diet (an index of availability of pollock) to affect reproductive performance of birds. However, the percentage of pollock in the diets was not correlated with reproductive success, with the exception of a negative relationship for black-legged kittiwakes on St. George Island. These findings suggest that the abundance of or rate of occurrence of pollock in the birds' diets do not drive the variation in reproductive performance of marine birds at the Pribilof Islands. Nevertheless, correlations between age-1 pollock abundance, as measured by fisheries surveys of the eastern Bering Sea shelf (V.G. Wespestad, National Marine Fisheries Service, Seattle, WA, personal communication) and the amount of pollock used by black-legged kittiwakes on St. Paul Island suggests that the diet of this seabird species may be reflecting the abundance or availability of age-1 pollock near the Pribilof Islands.

Examination of changes in seabird reproductive ecology with respect to the foraging locations of the various seabird species shows that both species of surface-foraging kittiwakes had reduced reproductive performance between the two time periods, whereas subsurface-foraging murres did not. Red-legged kittiwakes, which forage mostly at the shelf edge (Hunt et al. 1982; Schneider and Hunt 1984), showed little or no fundamental change in diet. In contrast, black-legged kittiwakes, which forage mostly in the middle domain, and thick-billed murres, which forage mostly near the islands, showed significant changes in diet composition. We interpret these patterns as indicating that the greatest changes in prey availability occurred on the shelf (primarily in the middle domain) and that changes occurred throughout the water column. The reduction in the availability of preferred prey in the water column was least felt by the deep-diving murres, which were able to switch to alternative prey such as cephalopods.

We conclude from these observations of marine bird ecology at the Pribilof Islands that there was a shift in the southeastern Bering Sea marine ecosystem that commenced about 1978. Increases in sea-surface temperature and total pollock abundance were contemporaneous with the shifts in seabird diets and reproductive performance. However, the relationship between prey availability, as judged by seabird usage, and either sea-surface temperature or pollock abundance is not simple. When temperatures became cooler after 1984, seabird reproductive performance and diets did not return to the pre-1978 values. Our results suggest that the Bering Sea ecosystem may not have returned to initial conditions after the change in physical state that occurred around 1978, and that although sea-surface temperature oscillates cyclically, the composition of the marine ecosystem may not simply fluctuate on the same cycle as the physical processes of the ocean. A model that assumed that biological entities would track physical conditions in the ocean would provide results at variance with the system as the seabirds reveal it.

Acknowledgements

We thank the many people who helped collect data on the Pribilof Islands. We also thank G. Updegraff and D. Newton for their help in developing software to access and analyze National Oceanographic Data Center data files. P. Hunt, J. Rice and P. Turner provided useful comments on an early draft of the manuscript. The research was funded in part by contracts under the Outer Continental Shelf Environmental Assessment Program to G. L. Hunt, Jr. and by National Science Foundation Grant DPP8521178 to GLH. M. B. Decker was supported by a NASA Graduate Student Fellowship in Global Change Research.

References

Anderson, D.W., F. Gress, and K.F. Mais. 1982. Brown pelicans: influence of food supply on reproduction. Oikos 39: 23–31.

Baird, P.H. 1990. Influence of abiotic factors and prey distribution on diet and reproductive success of three seabird species in Alaska. Ornis Scand. 21: 224–235.

Bradstreet, M.S.W. 1985. Feeding studies, p. 257–306. *In* S.R. Johnson [ed.] Population estimation, productivity, and food habits of nesting seabirds at Cape Pierce and the Pribilof Islands, Bering Sea, Alaska. Report by LGL Ecological Research Associates, Inc., to Minerals Management Service, Anchorage, AK.

Byrd, G.V. 1989. Seabirds in the Pribilof Islands, Alaska: trends and monitoring methods. M.Sc. Thesis, University of Idaho, Moscow, ID. 96 pp.

Cairns, D.K. 1987. Seabirds as indicators of marine food supplies. Biol. Oceanogr. 5: 261–271.

Chen, L.-F. 1983. The effects of water temperature on the seasonal distribution and growth of walleye pollock, *Theragra chalcogramma* (Pallas), in the southeastern Bering Sea. M.Sc. thesis, University of Alaska, Fairbanks, AK. 92 p.

Croll, D.A., A.J. Gaston, A.E. Burger, and D. Konnoff. 1992. Foraging behavior and physiological adaptation for diving in thick-billed murres. Ecology 73: 344–356.

Croxall, J.P., T.S. McCann, P.A. Prince, and P. Rothery. 1988. Reproductive performance of seabirds and seals at South Georgia and Signy Island, South Orkney Islands, 1976–1987: implications for southern ocean monitoring studies, p. 261–285. *In* D. Sahrhage [ed.]. Antarctic Ocean and Resources Variability. Springer-Verlag, New York, NY.

Cushing, D.H. 1988. The provident sea. Cambridge University Press, Cambridge, UK.

Dragoo, D.E. 1991. Food habits and productivity of kittiwakes and murres at St. George Island, Alaska. M.Sc. thesis, University of Alaska, Fairbanks, AK. 104 p.

Dwyer, D.A., K.M. Bailey, and P.A. Livingston. 1987. Feeding habits and daily ration of walleye pollock (*Theragra chalcogramma*) in the eastern Bering Sea, with special reference to cannibalism. Can. J. Fish. Aquat. Sci. 44: 1972–1984.

Frost, K.J., and L.F. Lowry. 1986. Sizes of walleye pollock, *Theragra chalcogramma*, consumed by marine mammals in the Bering Sea. Fish. Bull. U.S. 84: 192–197.

Furness, R.W. 1982. Competition between fisheries and seabird communities. Adv. Mar. Biol. 20: 225–307.

Hickey, J.J. and F.L. Craighead. 1977. A census of seabirds on the Pribilof Islands. Environmental Assessment of the Alaskan Continental Shelf. Annual reports of principle investigators. National Oceanic Atmospheric Administration, Boulder, CO. Environ. Res. Lab. 2: 96–195.

Hunt, G.L., Jr. 1990. The pelagic distribution of marine birds in a heterogeneous environment. Polar Res. 8: 43–54.

Hunt, G.L., Jr., and J.L. Butler. 1980. Reproductive ecology of western gulls and Xantus' murrelets with respect to food resources in the southern California Bight. Calif. Coop. Oceanic Fish. Invest. Rep. 21: p. 62–67.

Hunt, G.L., Jr., Z. Epply, B. Burgeson, and R. Squibb. 1982. Reproductive ecology, foods and foraging areas of seabirds nesting on the Pribilof Islands, 1975–1979. Environmental Assessment of the Alaskan Continental Shelf. Final reports of principle investigators. National Oceanic Atmospheric Administration, Boulder, CO. Environ. Res. Lab. Biol. Stud. 12: 1–258.

Hunt, G.L., Jr., Z. Epply, and D.C. Schneider. 1986. Reproductive performance of seabirds: the importance of population and colony size. Auk 103: 306–317.

Hunt, G.L., Jr., and N.M. Harrison. 1990. Foraging habitat and prey taken by least auklets at King Island, Alaska. Mar. Ecol. Prog. Ser. 65: 141–150.

Hunt, G.L., Jr., N.M. Harrison, and R.T. Cooney. 1990. The influence of hydrographic structure and prey abundance on foraging of least auklets. Stud. Avian Biol. 14: 7–22.

Iverson, R.L., L.K. Coachman, R.T. Cooney, T.S. English, J.J. Goering, G.L. Hunt, Jr., M.C. Macauley, C.P. McRoy, W.S. Reeburg, and T.E. Whitledge. 1979. Ecological significance of fronts in the southeastern Bering Sea, p. 437–466. *In* R. J. Livingston [ed.] Ecological processes in coastal marine systems. Plenum Press, New York, NY.

Kinder, T.H., and J.D. Schumacher. 1981. Hydrographic structure over the continental shelf on the southeastern Bering Sea, p. 31–51. *In* D.W. Hood and J.A. Calder [ed.] The eastern Bering Sea shelf: its oceanography and resources. University of Washington Press, Seattle, WA.

Lloyd, D.S. 1985. Breeding performance of kittiwakes and murres in relation to oceanographic and meteorologic conditions across the shelf of the southeastern Bering Sea. M.Sc. thesis, University of Alaska, Fairbanks, AK. 106 p.

Lonne, O.J., and B. Gulliksen. 1991. Source, density and composition of sympagic fauna in the Barents Sea, p. 289–294. *In* E. Sakshaug, C.C.E. Hopkins, and N.A. Oritsland. [ed.] Proceedings of the Pro Mare Symposium on Polar Ecology, Trondheim, Norway, 12–16 May 1990. Polar Res. 10(1).

Methven D.A., and J.F. Piatt. 1991. Seasonal abundance and vertical distribution of capelin (*Mallotus villosus*) in relation to water temperature at a coastal site off eastern Newfoundland. ICES J. Mar. Sci. 48: 187–193.

Montevecchi, W.A., V.L. Birt, and D.K. Cairns. 1988. Dietary changes of seabirds associated with local fisheries failures. Biol. Oceanogr. 5: 153–161.

Murphy, E.C., A.M. Springer, and D.G. Roseneau. 1991. High annual variability in reproductive success of kittiwakes (*Rissa tridactyla*) at a colony in western Alaska.J. Anim. Ecol. 60: 515–534.

Nettleship, D.N. 1991. The diet of Atlantic puffin chicks in Newfoundland before and after the initiation of an international capelin fishery, 1967–1984. Acta Congr. Int. Ornithol. 20: 2266–2271.

Nettleship, D.N., G.A. Sanger, and P.F. Springer [eds.]. 1984. Marine birds: their feeding ecology and commercial fisheries relationships. Proc. Pacific Seabird Group Symposium, Seattle, Washington, 6–8 Jan. 1982. 220 p. Spec. Publ. Canadian Wildlife Service, Ottawa, Ont.

Niebauer, H.J. 1980. Sea ice and temperature variability in the eastern Bering Sea and the relation to atmospheric fluctuations. J. Geophys. Res. 85: 7507–7515.

Niebauer, H.J. 1983. Multi-year sea ice variability in the eastern Bering Sea: an update. J. Geophys. Res. 88: 2733–2742.

Ohtani, K., and T. Azumaya. 1994. Inter-annual changes of the sea conditions on the continental shelf and its influence on abundance of walleye pollock in the eastern Bering Sea, p. 87–95. *In* R.J. Beamish [ed.] Climate change and northern fish populations. Can. Spec. Publ. Fish. Aquat. Sci. 121.

Olla, B.L., and M.W. Davis. 1990. Effects of physical factors on the vertical distribution of larval walleye pollock (*Theragra chalcogramma*) under controlled laboratory conditions. Mar. Ecol. Prog. Ser. 63: 105–112.

Pahlke, K.A. 1985. Preliminary studies of capelin (*Mallotus villosus*) in Alaskan waters. Alaska Dep. Fish Game Inf. Leafl. 250: 64 p.

Pitt, T.K. 1958. Distribution, spawning and racial studies of the capelin, *Mallotus villosus* (Muller), in the offshore Newfoundland area. J. Fish. Res. Board Can. 15: 275–293.

Rothschild, B.J. 1983. Global fisheries. Springer-Verlag, Heidelberg.

Royer, T.C. 1989. Upper ocean temperature variability in the northeast Pacific Ocean: is it an indicator of global warming? J. Geophys. Res. 94: 18 175 – 18 183.

Schneider, D., and G.L. Hunt, Jr. 1982. Carbon flux to seabirds in waters with different mixing regimes in the southeastern Bering Sea. Mar. Biol. 67: 337–344.

Schneider, D.C., and G.L. Hunt, Jr. 1984. A comparison of seabird diets and foraging distribution around the Pribilof Islands, Alaska, p. 86–95. *In* D.N. Nettleship, G.A. Sanger, and P.F. Springer [ed.] Marine birds: their feeding ecology and commercial fisheries relationships. Proc. of the Pacific Seabird Group Symposium, Seattle Washington, 6–8 Jan. 1982. Spec. Publ. Canadian Wildlife Service, Ottawa, Ont.

Schulmeister, S.D. 1991. Results of productivity monitoring of black-legged kittiwakes, red-legged kittiwakes, and thick-billed murres at St. George Island, Alaska, in 1990. Rep. U. S. Fish and Wildlife Service, Homer, AK. 26 p.

Smith, S.L., and J.Vidal. 1986. Variations in the distribution, abundance, and development of copepods in the southeastern Bering Sea in 1980 and 1981. Cont. Shelf Res. 5: 215–239.

Sogard, S.M., and B.L. Olla. (1993). Effects of light, thermoclines and predator presence on vertical distribution and behavioral interactions of juvenile walleye pollock, *Theragra chalcogramma*. J. Exp. Mar. Biol. Ecol. 167: 179–195.

Springer, A.M. 1991. Seabird distribution as related to food webs and the environment: examples form the North Pacific Ocean, p. 39–48. *In* W.A. Montevecchi and A.J. Gaston [ed.] Studies of high-latitude seabirds. 1. Behavioral, energetics and oceanographic aspects of seabird feeding ecology. Occas. Pap. 68. Canadian Wildlife Service, Ottawa, Ont.

Springer, A.M. 1992. A review: walleye pollock in the North Pacific—how much difference do they really make? Fish. Oceanogr. 1: 80–96.

Springer, A.M., and G.V. Byrd. 1989. Seabird dependence on walleye pollock in the southeastern Bering Sea, p. 667–677. *In* Proceedings of the International Symposium Biological Management of Walleye Pollock, Anchorage, AK, University of Alaska, Fairbanks, AK. Nov. 1988, Alaska Sea Grant Rep. N. 89–1.

Springer, A.M., D.G. Roseneau, D.S. Lloyd, C.P. McRoy, and E. Murphy. 1986. Seabird responses to fluctuating prey availability in the eastern Bering Sea. Mar. Ecol. Prog. Ser. 32: 1–12.

Straty, A.R., and R.E. Haight. 1979. Interactions among marine birds and commercial fish in the eastern Bering Sea, p. 201–219. *In* Bartonek, J.C., and D.N. Nettleship [ed.] Conservation of marine birds of northern North America. U.S. Department of the Interior, Fish and Wildlife Service, Washington, D.C. Wildl. Res. Rep. 11.

Walsh, J.J., and C.P. McRoy. 1986. Ecosystem analysis in the southeastern Bering Sea. Cont. Shelf Res. 5: 259–288.

Climate change and ocean energetics of Fraser River sockeye (*Oncorhynchus nerka*)

Scott G. Hinch, Michael C. Healey, Ron E. Diewert, and Michael A. Henderson

Hinch, S.G., M.C. Healey, R.E. Diewert, and M.A. Henderson. 1995. Climate change and ocean energetics of Fraser River sockeye (*Oncorhynchus nerka*), p. 439–445. *In* R.J. Beamish [ed.] Climate change and northern fish populations.Can. Spec. Publ. Can. Fish. Aquat. Sci. 121.

Abstract: We constructed a spatially explicit bioenergetics model to predict final ocean weight of Fraser River sockeye salmon (*Oncorhynchus nerka*), and to estimate the effects of climate warming on oceanic sockeye growth. The bioenergetics model successfully predicted average final ocean weight of the early Stuart stock under recent sea-surface temperature (SST) regimes. To assess the potential impact of climate warming on final ocean weight, we imposed a series of SST increases on the model. We found that predicted final ocean weights declined steadily as SSTs warmed above current values. An empirical correlation ($r = -0.433$, $P = 0.024$, n = 27 yr) between average final ocean weight of early Stuart sockeye and average SST at Station P (50°N, 145°W) during the month prior to migrating upriver also suggests that warm SSTs are associated with poorer growth. Global climate models predict that if atmospheric CO_2 doubles, Northeast Pacific Ocean SSTs will increase by about 3.5°C above present values. The bioenergetics model suggests that this increase will result in a 14% reduction in average final ocean weight. As a consequence, Fraser River sockeye would have fewer and smaller eggs, and they may have insufficient energy reserves to complete their river migration and spawn.

Résumé : Nous avons créé un modèle bioénergétique spatialement explicite pour prédire le poids final dans l'océan du saumon sockeye (*Oncorhynchus nerka*) du fleuve Fraser, ainsi que pour estimer les effets du réchauffement climatique sur la croissance du sockeye océanique. Le modèle bioénergétique a réussi à prédire le poids final océanique moyen du stock de la rivière Stuart à montée hâtive sous les récents régimes de température de surface de la mer (TSM). Pour évaluer les répercussions éventuelles du réchauffement climatique sur le poids final dans l'océan, nous avons appliqué au modèle une série d'augmentations de la TSM. Nous avons constaté que les poids finals prévus dans l'océan déclinaient de façon régulière à mesure que les TSM montaient au-dessus des valeurs courantes. Une corrélation empirique ($r = -0,433$, $P = 0,024$, n = 27 ans) entre le poids final moyen dans l'océan des sockeyes de la rivière Stuart à montée hâtive et la TSM moyenne à la station P (50° N, 145° O) durant le mois précédant la remontée migratoire donne également à croire que les TSM chaudes sont associées à une croissance plus faible. Les modèles climatiques globaux permettent de prévoir que si la teneur en CO_2 de l'atmosphère double, les TSM du nord-est du Pacifique monteront d'environ 3,5° C au-dessus des valeurs actuelles. Le modèle bioénergétique donne à croire que cette hausse se traduira par une réduction de 14 % du poids final moyen dans l'océan. En conséquence, le sockeye du fleuve Fraser aura moins d'oeufs et des oeufs plus petits, et risque de manquer de réserves énergétiques pour terminer sa migration et son frai dans le fleuve.

Introduction

Over the next 50–100 yr, CO_2 concentration in the atmosphere is predicted to double and there is strong scientific consensus that climate will change as a result (Houghton et al. 1990). Global climate models have been developed to help understand changes in climate systems and they predict that a doubling of atmospheric CO_2 will be associated with a 2–4°C increase in air temperature over the Northeast Pacific Ocean (Boer et al. 1992). Although regional predictions made by global climate models may not be as reliable as global predictions because of the models' coarse spatial resolution and highly simplified representations of heat transport within oceans (McBean 1990), they present compelling scenarios that could have serious consequences for oceanic fishes.

Of all the fish species in the Northeast Pacific, salmon (*Oncorhynchus* spp.) are likely to be the first affected by increases in water temperature because they spend the majority of their lives in surface waters and the ocean surface

S.G. Hinch and **M.C. Healey.** Westwater Research Centre, University of British Columbia, Vancouver, B.C. V6T 1Z2, Canada, Department of Oceanography, University of British Columbia, Vancouver, B.C. V6T 1Z4, Canada.

R.E. Diewert and **M.A. Henderson.** Canadian Department of Fisheries and Oceans, Biological Sciences Branch, 555 West Hastings Street, Vancouver, B.C. V6B 5G3, Canada.

will probably warm more quickly than deeper waters (McBean et al. 1991). Among the salmon species, sockeye salmon (*O. nerka*) may be most affected by changes in SSTs because of its long ocean migrations (Hartt 1966). Models have been developed to examine and predict the influence of SSTs on the distribution, abundance, production and migratory behaviour of sockeye salmon (Burgner 1980; Rogers 1984; Tabata 1984; Chelton 1984; Mysak 1986; Blackbourn 1987; Quinn and terHart 1987; Hsieh et al. 1991; Beamish and Bouillon 1993). Only a few, however, have explored the effects of temperature variation on oceanic growth rates of sockeye (Rogers 1980, 1984; Eggers et al. 1984). Because body size is an important determinant of migratory success and fecundity in sockeye (IPSFC 1959, 1980; Healey 1987), the relationship between temperature and growth rate should be examined in the context of a 3–4°C increase in SST in the Northeast Pacific Ocean.

We constructed an energetics model to explore the rate of energy accumulation (e.g., growth) that occurred each month during the oceanic phase of life for an "average" Fraser River sockeye salmon. Fraser River sockeye should be highly susceptible to the effects of climate warming because they are at the southern range of distribution of sockeye in North America. The validity of model predictions was assessed by comparing empirical measurements of mature adult river entry weight to model estimates of river entry weight using historical and current water temperature data. To examine the potential impact of climate change on sockeye growth, we imposed a series of temperature changes on the model. Although the model was run using weight and timing data on only one sockeye stock, the early Stuart, it should be generally applicable to other Fraser River sockeye stocks.

Methods

Model of oceanic sockeye distribution

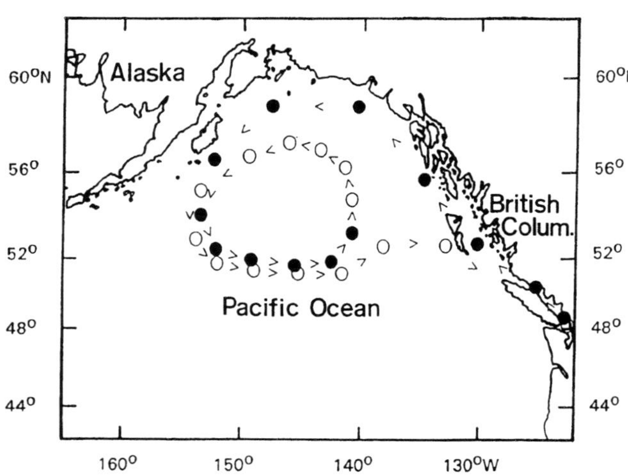

Fig. 1. Map of the Northeast Pacific Ocean showing average monthly locations of early Stuart sockeye (adapted from Brett (1983)). Closed and open circles represent the first and second year of ocean residency respectively. Arrowheads depict the monthly direction of movement. The first month of ocean life starts on May 25 at the mouth of the Fraser River (closed circle at 48°N); all other circles represent the 15th of each month.

After rearing for 1 yr in the Stuart Lake system, early Stuart sockeye smolts migrate to the ocean via the Fraser River. Once in the ocean, they migrate northward along the British Columbia and Alaska coasts and spend the next 2 yrs making two counterclockwise circuits in a region of the Northeast Pacific Ocean bounded by 50–58°N and 135–155°W (Fig. 1) (Brett 1983). A distribution model was developed by Brett (1983) to illustrate the monthly sea positions of an "average" British Columbia sockeye. Brett reported that monthly shifts in sockeye position were supported by mark–recapture and catch per unit effort studies (Neave 1964; Hartt 1966; French and McAlister 1970; French et al. 1976; Hartt and Dell 1978). Brett's model dealt with fish leaving and returning to the Skeena River, which is located on the north coast of British Columbia. We used general information on migration rates of Fraser River juveniles and mature adults (Groot and Cooke 1987; Healey 1989) to suggest monthly coastal positions for early Stuart sockeye that migrate along the south coast of British Columbia (Fig. 1). SST data from 1945 to 1987 that corresponded to these monthly positions were obtained from different sources (Fulton et al. 1968; Brett 1983; Slutz et al. 1985; Woodruff et al. 1987). We obtained more recent SST data from updates of the Comprehensive Ocean-Atmosphere Data Set (see Woodruff et al. 1987 for details).

Description of bioenergetic model

The bioenergetic model was based on the thermodynamic principle that food consumed (energy input) equals the amount of energy placed into growth, plus that used for metabolism, plus that lost through excretion and egestion. We obtained information for each of these processes from the literature, with the exception of growth, which was estimated using the model. Brett (1983) determined size-specific rates of food consumption (measured in units of energy) for sockeye which he presumed to correspond to the monthly locations of sockeye in his distribution model. We incorporated these values into our energetics model. Brett (1983) reported that empirical measurements of monthly growth were very similar to predicted monthly growth calculated from a general salmonid growth model that assumed maximum daily ration. Thus, growth rate in the ocean approximated maximum capacity, and consequently, Brett (1983) argued that feeding rate also approximated maximum daily intake. For the purposes of our model, we assumed that sockeye could obtain maximum ration so that ration was only constrained by temperature. Brett and Groves (1979) determined that over a temperature range that encompassed our data (2.5–15°C), a 1°C increase in temperature increased maximum ration for sockeye between 5 and 8%. We used an intermediate value of 6.5% as a factor by which food consumption changed corresponding to a 1°C change in temperature.

The weight- and temperature-dependent equations and parameters describing standard metabolic costs and costs of swimming were taken from Beauchamp et al. (1989). They based their equations on a model of optimum sockeye swimming speed in which a hydrodynamics approach was linked with optimal foraging theory. These equations are not, however, based on assessments of metabolic rates in nature so they do not consider energy expenditures associated with attack–avoid–escape patterns or that of direction-oriented migration. These additional costs are probably substantive for sockeye that face a gauntlet of predators and competitors in the ocean and that migrate over 12 000 km (Brett 1983). Brett (1983) showed that feeding metabolic rate for sockeye approximates one half of their active (constant swimming) metabolic rate. To assess total swimming metabolism, we thus doubled the influence of swimming speed in the equations that assessed temperature- and weight-dependent optimum swimming metabolism.

Apparent heat increment (also called apparent specific dynamic action), the metabolic cost associated with chewing and swallowing food and with postabsorptive processes, was set to equal 17% of the difference between energy consumed and energy egested (Stewart et al. 1983; Beauchamp et al. 1989). Apparent heat increment may not remain constant over the life of a sockeye (Beamish and Trippel 1990), but is probably fairly constant during the ocean phase of life when the diet is composed primarily of a single food, euphausiids (Brett 1983). An increase in omnivorous feeding will, however, increase apparent heat increment and could represent an additional energetic cost. Monthly specific levels of excretion–egestion were set to equal 30% of the monthly energy consumed (Brett 1983).

The temperatures of each monthly sea location and the average fish weight at the first site (i.e., the mouth of the Fraser River) were inputted to the model. Growth (in energy units) was calculated at each month by subtracting the amount of energy used for metabolism and excretion–egestion from the amount of energy consumed that month. Weight (in energy units) at month_{t+1} was determined by adding the estimated growth between month_t and month_{t+1} to the weight at month_t. We used equations in Brett (1983) to convert fish weights to energy equivalents.

The model operated under the following assumptions. (1) Monthly positions as presented in Fig. 1 were fixed. (2) Sockeye existed within the upper mixed layer of the ocean; thus, SSTs were representative of temperatures that the fish encountered. (3) Maximum daily ration was attainable. (4) Energetic costs associated with directed migration (e.g., energy used to move between monthly locations) and with biotic interactions were approximately equal to the costs arising from feeding. We will discuss the validity of these assumptions and the ramifications of violating them later in the manuscript.

Running the model

To test the performance of the model, we compared empirical measurements of river entry weights (e.g., final oceanic size) to river entry weights predicted by the model. The Fraser River gillnet fishery collected early Stuart sockeye from the lower Fraser River over a 4-wk period (June 15 to July 15) from 1967 to 1992 (Fig. 2). During the fishery, there is a complete lack of movement of sockeye in the Fraser River likely as a result of the fisheries' intensity, suggesting that gillnetting was not size selective in its removal of fish (J. Woodey, Pacific Salmon Commission, 600-1155 Robson Street, Vancouver, B.C. V6E 1B5, personal communication). In two separate runs of the model, we used site-specific temperature data that were based on among-year averages of data from 1960 to 1976 (1960s conditions), and on among-year averages from 1977 to 1992 (1980s conditions). This dichotomy was selected because a large-scale change occurred to northeast Pacific SSTs starting in 1976 (Namias et al. 1988). From 1976 to 1988, the Northeast Pacific Ocean (Fig. 1) experienced an increase in average annual SSTs of about 1°C at coastal sites and an increase of approximately 0.5°C at the high-sea sites (Hourston 1992). We did not have SST data from 1989 to 1992 that corresponded to the monthly positions illustrated in Fig. 1. However, an examination of unpublished monthly temperature data (provided by H. Freeland, Institute of Ocean Sciences, Sydney, B.C. V8L 4B2) collected from Amphitrite Point (49°N, 128.5°W) located on the west coast of Vancouver Island and from Station P (50°N, 145°W) located in the Northeast Pacific Ocean confirmed that the temperature patterns reported by Hourston (1992) were maintained from 1988 to 1992 at a coastal and a high-sea site.

To assess the effects of different temperature scenarios on river entry weights, we ran the model with temperatures that ranged from 3°C below average 1960s values to 5°C above 1960s values. This temperature range was chosen because, in particular years since 1945 in the Northeast Pacific Ocean, SSTs have been 2–3°C cooler than 1960s values (Slutz et al. 1985; Woodruff 1987), and global climate models predict a 2–4°C temperature increase from present temperatures over the next 100 yr (McBean et al. 1991).

In all model runs, the initial weight of sockeye juveniles at the Fraser River mouth was 5 g, an average value that is representative of smolt size for most Fraser River sockeye stocks (Henderson and Cass 1991). (However, results of preliminary model runs using different initial weights suggest that initial weight has only a small bearing on final ocean weight; final weights differed by less than 1% with initial weights ranging from 2.5 to 10 g.) The date of arrival at the mouth of the Fraser River was set to May 25, which corresponds to the peak arrival time for early Stuart sockeye (J. Woodey, Pacific Salmon Commission, 600-1155 Robson Street, Vancouver, B.C. V6E 1B5, personal communication). The model was allowed to run for 27 consecutive mo of ocean life, the amount of time it takes for average early Stuart sockeye to leave the mouth of the Fraser River as juveniles and return as mature adults. We concluded each model run on July 15, which is approximately 1 wk after the peak number of mature adults arrive at the Fraser River mouth (J. Woodey, Pacific Salmon Commission, 600-1155 Robson Street, Vancouver, B.C. V6E 1B5, personal communication) and is a date that corresponds with our final date of sampling for early Stuart sockeye in the lower Fraser River (Fig. 2).

We used output from the Canadian Climate Centre's global climate model (Boer et al. 1992), which is a comprehensive atmospheric model coupled to a simplified mixed layer ocean model, to generate SST scenarios under a doubling of atmospheric CO_2. The Canadian Climate Centre modelled the annual cycle of sea-surface temperature and sea-level pressure based on a 20-year climatology of the two times CO_2 scenario (i.e., double the present amount of atmospheric carbon dioxide) and the one times CO_2 "control" scenario.

Empirical relationship between SST and weight

Using daily sea-surface temperature records from Station P, average May SSTs were determined for years that corresponded with the river entry weight data. Among-year variability in

Station P SST is probably indicative of SST variability experienced by adult early Stuart sockeye that are migrating in the highseas. To assess a possible relationship between sockeye growth rate and SST, we correlated annual sockeye weights at return with annual May SSTs. We chose May of the final year of ocean life over other months for this analysis because output from the bioenergetics model shows that absolute growth is greatest during this time period so SST may have its greatest influence on growth during this month, and because the model suggests that SST has its greatest impact on final ocean size during the last spring in the ocean. The latter will be discussed below in more detail.

Results and discussion

Bioenergetics model

When temperature conditions characteristic of the 1960s were imposed on the model, the predicted weight at river entry was 2.67 kg, a value that was only 1.5% smaller than the overall average river weight (2.71 kg) calculated during the 10-yr period, 1967–76 (Fig. 2). When 1980s temperatures were imposed on the model, the weight at river entry was predicted to be 2.53 kg. This is slightly less than the model estimate for the 1960s and was only 5.9% smaller than the overall average river weight (2.68 kg) calculated from 1977 to 1992. This value was well within the range of average empirical river entry weights calculated from 1977 to 1992 (Fig. 2). The empirical data and model output suggest that the warmer time period (the 1980s) was associated with lighter sockeye, although the differences were small. We also concluded that the model should be useful for exploring the potential effects of alternative temperature scenarios on sockeye weights because it produced realistic estimates of river entry weights when recent temperature scenarios were employed.

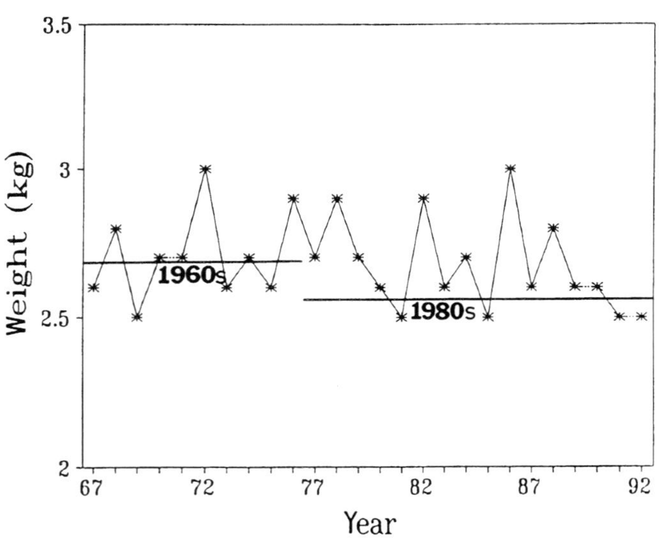

Fig. 2. Average annual weight (1967–92) of early Stuart sockeye caught in the lower Fraser River from June 15 to July 15 by the Fraser River gillnet fishery (from the Canadian Department of Fisheries and Oceans' catch statistic data base). Within-year sample size ranged from a low of 386 in 1968 to a high of 2.8′ 10^5 in 1977. The solid lines represent predicted river entry weight based on our model for 1960s (left line) and 1980s (right line) SSTs.

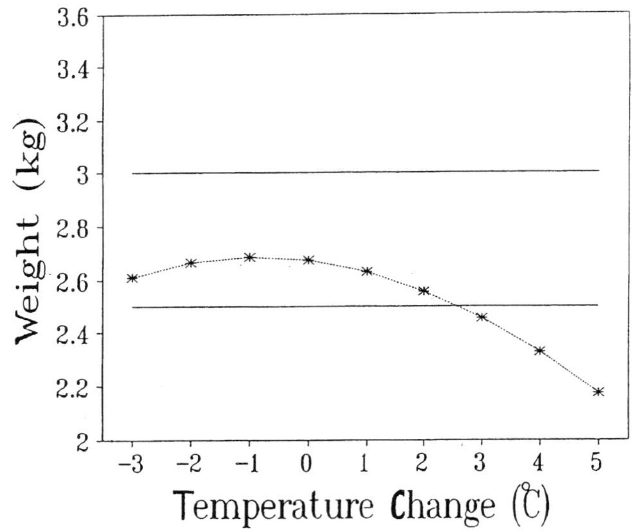

Fig. 3. Modeled relationship between weights of adult early Stuart sockeye when they enter the mouth of the Fraser River and SST change in the northeast Pacific Ocean. SST changes are relative to 1960s SSTs. A 1°C SST increase approximates the SST regime observed during the 1980s. Parallel lines represent the 26-yr minimum and maximum average annual river entry weight of adult early Stuart sockeye salmon (from Fig. 2).

We ran the model under oceanic temperatures 3°C below to 5°C above those observed during the 1960s (Fig. 3). For temperatures up to 3°C colder than the 1960s values, all predicted weights fell near the middle of the range of observed river entry weights during the 1960s. At 3°C below 1960s temperatures, average river entry weight was only 2.5% less than the predicted 1960s value. Maximum predicted river entry weight occurred when ocean temperatures were 1°C cooler than 1960s temperatures (approximately 2°C cooler than 1980s temperatures); however, this weight was only 0.4% greater than the predicted weight under 1960s temperatures. Thus, SSTs cooler than 1960s values appear to have little effect on growth and final adult size.

When SSTs were increased above 1960s values, the model predicted a steady decline in average river entry weight (Fig. 3). A 1°C increase (which approximated 1980s temperatures) decreased predicted river entry weight by 1.6% and a 2°C increase decreased predicted river entry weight by 4.4%. This latter value approached the smallest empirically observed, average river entry weight measured over the past 26 yr. SST increases of more than 2.5°C from 1960s values resulted in predicted river entry weights that were smaller than any empirical measure of average river entry weight. Therefore, temperature increases above 1960s values had more profound effects on river entry weight than temperature decreases. Because energetic costs of excretion–egestion were constant with respect to temperature, the energetic costs of metabolism apparently increased faster than the energy gained through

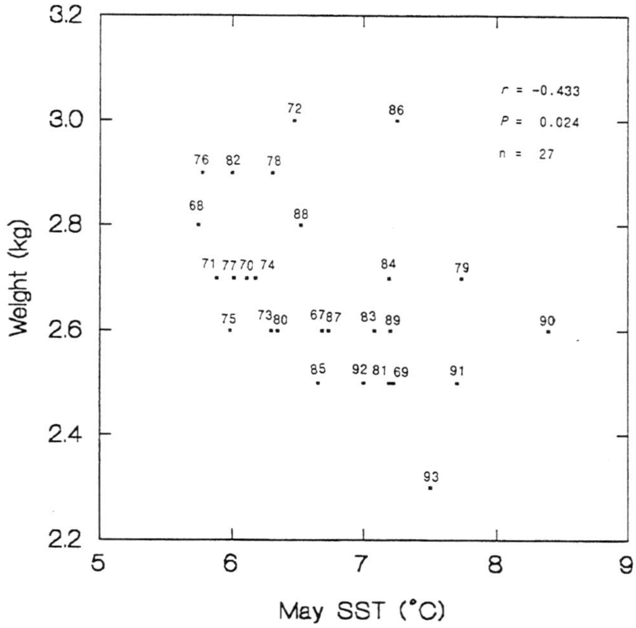

Fig. 4. The relationship from 1967 to 1993 between average river entry weight of early Stuart sockeye and May SST at Station P in the Northeast Pacific Ocean. Years are indicated on top of data points. Fish were captured and weights assessed in July, 2 mo after SST was determined. The Pearson correlation and associated probability are presented.

increased consumption rates in warmer waters. Thus, the balance between energy input and energy output that existed in cool temperatures broke down when temperatures increased modestly.

There appears to be a link between SSTs (1967 to 1993) and river entry weight with warmer years being correlated with lighter river entry weights (Fig. 4). Results of our bioenergetics modelling corroborates this by suggesting that once SSTs are greater than 1960s values, relatively warm temperatures will cause a reduction in growth rate.

The Canadian Climate Centre global climate model predicts that a doubling of atmospheric CO_2 will result in a 2–4°C increase in SSTs above present values in the Northeast Pacific (Boer et al. 1992). SSTs are predicted to increase by about 3.5°C at the sites that correspond with monthly sockeye locations in Fig. 1. If we assume that the past 30 years reflect "present" climate conditions, 3.5°C increase from a midpoint between 1960s and 1980s temperatures on Fig. 3 translates into a predicted river entry weight of 2.35 kg (Fig. 3). This value is 14% lighter than predicted weights under "present" climate conditions and 6.4% lighter than the smallest observed average river entry weight for early Stuart sockeye over the 26-yr period from 1967 to 1992 (Fig. 2).

Robustness of model predictions

How might violations of our bioenergetic model assumptions alter predictions about river entry weights of sockeye? The first assumption is that monthly positions in Fig. 1 are fixed. It has recently been demonstrated that the southern boundary for Pacific Ocean sockeye in the spring is determined by the position of the 8.9°C isothermal (D. Welch, Department of Fisheries and Oceans, Nanaimo, B.C. V9R 5K6, personal communication). Brett's (1983) distribution model may thus underestimate the southern extent of Fraser River sockeye in the northeast Pacific Ocean, so we needed to assess how a more southern distribution would affect predicted river entry weight. To do this, we initialized the model with the 1980 scenario SSTs that were modified by increasing the November to March average monthly SSTs to 8°C from their original 5–6°C. From a spatial perspective, these SST changes mean that we stretched the southwestern monthly locations in Fig. 1 farther south. Under this scenario, predicted river entry weight was 2.56 kg. Although this weight was 4.4% lighter than the 26-yr average, it was still within the 26-yr range of mean river entry weights. We examined several other ocean distributions and found that their influence on predicted river entry weight were also relatively small (unpublished data); therefore, it is unlikely that violating assumption 1 would seriously affect model predictions. However, we found that changes in the locations (i.e., SSTs) of the last 2 mo of ocean life could alter predicted river entry weights enough to make them fall out of the empirical range. This suggests that variation in river entry weight is largely caused during the last few months of ocean life. Thus, variation in SSTs may primarily influence river entry weights during the homeward ocean migration phase. We are currently conducting research to assess migration and growth patterns in Fraser River sockeye during their final few months in the ocean.

The second assumption is that SSTs are representative of temperatures that sockeye encounter. If sockeye can vertically migrate to depths below the upper mixed layer, then the SSTs we used would be overestimates. The magnitude of influence of colder SSTs on predicted river entry weights depends on monthly location and sockeye weight. For instance, during the final months of ocean migration, a large fish will accumulate more weight if it experiences slightly cooler than normal SSTs. However, there is no evidence to suggest that sockeye in the high seas migrate below the upper mixed layer.

The third assumption is that maximum temperature-dependent ration was attainable. Brett (1983) suggested maximum ration was attainable during the 1960s. Food availability (e.g., zooplankton biomass) in the Northeast Pacific Ocean has more than doubled since then (Brodeur and Ware 1992) so it is possible that sockeye were able to obtain a maximum temperature-dependent ration during the 1980s. The recent doubling of zooplankton biomass in the Northeast Pacific is correlated with changes in nutrient transport, not changes in temperature (Brodeur and Ware 1992), so it is difficult to speculate on how patterns of zooplankton production will change in response to future increases in SSTs. Because sockeye obtain a maximum ration in our model, the only way climate change can affect ration is by inhibiting food intake. If food availability was decreased and maximum ration was not attainable, perhaps as a result of increased inter- or intra-specific competition for food or because of a decline in food production, then our present estimates of weight loss are conservative. Sockeye could become lighter than we have suggested as temperatures warm.

The fourth assumption is that energetic costs of directed migration and biotic interactions approximate the energetic costs of foraging. Because the predicted final ocean weights from the bioenergetics models fell within the empirical long-term average final ocean weight, our estimates of total swimming metabolic expenditure may be reasonable. However, in the future if the strength of Northeast Pacific ocean currents is affected by doubling atmospheric CO_2 as some have suggested (McBean 1991; Hsieh and Boer 1992), then migration costs could be affected. For instance, intensified coastal currents would mean that less energy would be needed by juveniles to migrate with the surface currents in a northwest direction to the high seas, but more energy may be needed by maturing adults for the final months of homing migration, which is against the surface coastal currents in a southeast direction. The converse situation may arise if coastal currents weaken. It is possible that, regardless of the magnitude of change in currents, the energetic savings of migration by one ocean life-history stage may trade-off against energetic costs of migration by another stage. Predator–competitor regimes in the northeast Pacific Ocean may be altered by climate changes (McBean et al. 1991) which could affect the amount of energy that sockeye expend with biotic interactions. To resolve these complex issues, further research into the present and predicted future energetic costs associated with directed swimming and biotic interactions are needed.

Conclusions

Our study suggests that as ocean temperatures warm in response to increasing concentrations of atmospheric CO_2, Fraser River sockeye will become lighter. Lighter sockeye will have fewer and smaller eggs, and may therefore have lower reproductive value (Healey 1987). In addition, smaller sockeye run the risk of not having enough energy reserves to complete their river migration and spawning. For instance, early Stuart sockeye arrive at the spawning grounds with 90–95% of their fat reserves depleted (IPSFC 1959, 1980) and it seems plausible that smaller early Stuart sockeye may not be able to meet the energetic costs of river migration. Sockeye may be able to compensate for slower growth by increasing their ocean residency from 2 to 3 yr. However, by increasing their life cycle by 1 yr they put themselves at greater risk to predation and disease, and also lower their reproductive potential by increasing their generation time. As another strategy to increase their river entry weights, sockeye could compress their oceanic distribution into more northern areas thereby reducing the direct effects of warm SSTs. But this could cause a reduction in average growth rate as competitive interactions would probably increase in response to higher salmon densities in the northern portions of the northeast Pacific Ocean.

Acknowledgements

We are grateful to Keith Thomson and Roy Hourston for their comments and suggestions throughout this project. Insights provided by David Welch and Jim Woodey are appreciated. We thank Howard Freeland for providing unpublished SST data. Comments from two anonymous reviewers helped us to clarify several aspects of this paper. S.G.H. was supported by a postdoctoral fellowship from the Natural Sciences and Engineering Research Council of Canada.

References

Beauchamp, D.A., D.J. Stewart, and G.L. Thomas. 1989. Corroboration of a bioenergetics model for sockeye salmon. Trans. Am. Fish. Soc. 118: 597–607.

Beamish, F.W.H., and E.A. Trippel. 1990. Heat increment: a static or dynamic dimension in bioenergetic models? Trans. Am. Fish. Soc. 119: 649–661.

Beamish, R.J., and D.R. Bouillon. 1993. Pacific salmon production trends in relation to climate. Can. J. Fish. Aquat. Sci. 50: 1002–1016.

Blackbourn, D.J. 1987. Sea surface temperature and pre-season prediction of return timing in Fraser River sockeye salmon (*Oncorhynchus nerka*), p. 296–306. *In* H.D. Smith, L. Margolis, and C.C. Wood [ed.] Sockeye salmon (*Oncorhynchus nerka*) population biology and future management. Can. Spec. Publ. Fish. Aquat. Sci. 96.

Boer, G.J., N.A. McFarlane, and M. Lazare. 1992. Greenhouse gas-induced climate change simulated with the CCC second-generation general circulation model. J. Clim. 5: 1045–1077.

Brett, J.R. 1983. Life energetics of sockeye salmon, *Oncorhynchus nerka*. *In* W.P. Aspey and S.I. Lustik [ed.] Behavioural energetics: the cost of survival in vertebrates. Ohio State University Press, Columbus, OH.

Brett, J.R., and T.D.D. Groves. 1979. Physiological energetics. *In* W.S. Hoar, D.J. Randall, and J.R. Brett [ed.] Fish physiology. Vol. 8. Academic Press, London.

Brodeur, R.D. and D.M. Ware. 1992. Long-term variability in zooplankton biomass in the subarctic Pacific Ocean. Fish. Oceanogr. 1: 32–38.

Burgner, R.L. 1980. Some features of ocean migrations and timing of Pacific salmon, p. 153–164. *In* W.J. McNeil and D.C. Hinsworth [ed.] Salmonid ecosystems of the North Pacific. Oregon State University Press, Corvallis, OR.

Chelton, D.B. 1984. Short-term climatic variability in the northeast Pacific Ocean, p. 87–89. *In* W.G. Pearcy [ed.] The influence of ocean conditions on the production of salmonids in the North Pacific. Oregon State University Press, Corvallis, OR.

Eggers, D.M., C.P. Meacham, and D.C. Huttunen. 1984. Population dynamics of Bristol Bay sockeye salmon, 1956–1983, p. 200–205. *In* W.G. Pearcy [ed.] The influence of ocean conditions on the production of salmonids in the North Pacific. Oregon State University Press, Corvallis, OR.

French, R., and W.B. McAlister. 1970. Winter distribution of salmon in relation to currents and water masses in the northeastern Pacific Ocean and migration of sockeye salmon. Trans. Am. Fish. Soc. 99: 649–663.

French, R., H. Bilton, M. Osaka, and A.C. Hartt. 1976. Distribution and origin of sockeye salmon (*Oncorhynchus nerka*) in offshore waters of the north Pacific Ocean. INPFC Bull. 34.

Fulton, J.D., O.D. Kennedy, K. Stephens, and J. Skelding. 1968. Physical, chemical and biological data in Strait of Georgia, 1967. MS. Rep. Fish. Res. Board Can. 968.

Groot, C., and K. Cooke. 1987. Are the migrations of juvenile and adult Fraser River sockeye salmon (*Oncorhynchus nerka*) in near-shore waters related?, p. 53–60. *In* H.D. Smith, L. Margolis, and C.C. Wood [ed.] Sockeye salmon (*Oncorhynchus nerka*) population biology and future management. Can. Spec. Publ. Fish. Aquat. Sci. 96.

Hartt, A.C. 1966. Migrations of salmon in the North Pacific Ocean and Bering Sea as determined by survey and tagging, 1959–1960. INPFC Bull. 19.

Hartt, A.C., and M.B. Dell. 1978. Early oceanic migrations and growth of juvenile Pacific salmon and steelhead trout. INPFC Bull. 46.

Healey, M.C. 1987. The adaptive significance of age and size at maturity in female sockeye salmon (*Oncorhynchus nerka*), p. 110–117. *In* H.D. Smith, L. Margolis and C.C. Wood [ed.] Sockeye salmon (*Oncorhynchus nerka*) population biology and future management. Can. Spec. Publ. Fish. Aquat. Sci. 96.

Healey, M.C. 1989. Modelling the coastal migration of juvenile salmon through Hecate Strait, British Columbia. Can. Tech. Rep. Fish. Aquat. Sci. No. 1700.

Henderson, M.A., and A.J. Cass 1991. Effect of smolt size on smolt to adult survival for Chilko Lake sockeye salmon (*Oncorhynchus nerka*). Can. J. Fish. Aquat. Sci. 48: 988–994.

Houghton, J.T., G.J. Jenkins, and J.J. Ephraums. 1990. Climate change: the IPCC scientific assessment. Cambridge University Press, New York, NY.

Hourston, R.A. 1992. Variability of North Pacific Ocean surface sensible and latent heat fluxes. M.Sc. Thesis, University of British Columbia, Vancouver, B.C.

Hsieh, W.W., and G.J. Boer. 1992. Global climate change and ocean upwelling. Fish. Oceanogr. 1: 333–338.

Hsieh, W.H., W.G. Lee, and L.A. Mysak. 1991. Using a numerical model of the northeast Pacific Ocean to study the interannual variability of the Fraser River sockeye salmon (*Oncorhynchus nerka*). Can. J. Fish. Aquat. Sci. 48: 623–630.

International Pacific Salmon Fisheries Commission (IPSFC). 1959. The energy expenditures of Fraser River sockeye salmon during the spawning migration to Chilko and Stuart Lakes. IPSFC Prog. Rep. 6: 80 p.

International Pacific Salmon Fisheries Commission (IPSFC). 1980. Energy sources and expenditures in Fraser River sockeye salmon during their spawning migration. IPSFC Bull. 22: 51 p.

McBean, G.A. 1990. Possible impacts of climate change on the North Pacific Ocean and its marine resources. Trans. R. Soc. Can. 1(1): 435–452.

McBean, G.A., O. Slaymaker, T. Northcote, P. LeBlond, and T.S. Parsons. 1991. Review of models for climate change and impacts on hydrology, coastal currents and fisheries in British Columbia. Rep. 91-11, Canadian Climate Centre, Climate Information Branch, Atmospheric Environment Service, Downsview, Ont.

Mysak, L.A. 1986. El Niño, interannual variability and fisheries in the Northeast Pacific Ocean. Can. J. Fish. Aquat. Sci. 43: 464–497.

Namias, J., X. Yuan, and D.R. Cayan. 1988. Persistence of North Pacific sea surface temperature and atmospheric flow patterns. J. Clim. 1: 682–703.

Neave, F. 1964. Ocean migrations of Pacific salmon. J. Fish. Res. Board Can. 21: 1227–1244.

Quinn, T.P., and B.A. terHart. 1987. Movements of adult sockeye salmon (*Oncorhynchus nerka*) in British Columbia coastal waters in relation to temperature and salinity stratification: ultrasonic telemetry results, p. 61–77. *In* H.D. Smith, L. Margolis, and C.C. Wood [ed.] Sockeye salmon (*Oncorhynchus nerka*) population biology and future management. Can. Spec. Publ. Fish. Aquat. Sci. 96.

Rogers, D.E. 1980. Density-dependent growth of Bristol Bay sockeye salmon, p. 267–283. *In* W.J. McNeil and D.C. Hinsworth [ed.] Salmonid ecosystems of the North Pacific. Oregon State University Press, Corvallis, OR.

Rogers, D.E. 1984. Trends in abundance of northeastern Pacific stocks of salmon, p. 100–127. *In* W.G. Pearcy [ed.] The influence of ocean conditions on the production of salmonids in the North Pacific. Oregon State University Press, Corvallis, OR.

Slutz, R.J., S.J. Lubker, J.D. Hiscox, S.D. Woodruff, R.L. Jenne, D.H. Joseph, P.M. Steurer, and J.D. Elms. 1985. Comprehensive ocean–atmospheric dataset: Release I. NOAA Environmental Research Laboratories, Climate Research Program, Boulder CO. 286 p.

Stewart, D.J., D. Weininger, D.V. Rottiers, and T.A. Edsall. 1983. An energetics model for lake trout, *Salvelinus namaycush*: application to the Lake Michigan population. Can. J. Fish. Aquat. Sci. 40: 681–698.

Tabata, S. 1984. Oceanographic factors influencing the distribution, migration, and survival of salmonids in the northeast Pacific Ocean—a review, p. 128–160. *In* W.G. Pearcy [ed.] The influence of ocean conditions on the production of salmonids in the North Pacific. Oregon State University Press, Corvallis, OR.

Woodruff, S.D., R.J. Slutz, R.L. Leane, and P.M. Steurer. 1987. A comprehensive ocean–atmosphere dataset. Bull. Am. Meterol. Soc. 68: 1239–1250.

High-latitude climate forcing and tidal mixing by the 18.6-year lunar nodal cycle and low-frequency recruitment trends in Pacific halibut (*Hippoglossus stenolepis*)

Kenneth S. Parker, Thomas C. Royer, and Richard B. Deriso

Parker, K.S., T.C. Royer, and R.B. Deriso. 1995. High-latitude climate forcing and tidal mixing by the 18.6-yr lunar nodal cycle and low-frequency recruitment trends in Pacific halibut *(Hippoglossus stenolepis)*, p. 447–459. *In* R.J. Beamish [ed.] Climate change and northern fish populations. Can. Spec. Publ. Fish. Aquat. Sci. 121.

Abstract: Upper layer ocean temperatures in the northern North Pacific Ocean have increased by more than 2°C in the decade starting with 1972, and have declined since 1986. Proxy time series indicate this to be part of a low-frequency fluctuation in synchrony with atmospheric behavior. The 18.6-yr lunar synodic declination cycle is suggested as the cause for this variability through the systematic tidal modulation of the mean ocean circulation. Enhanced or declining production trends over a nearly 60-yr recruitment abundance record for Pacific halibut (*Hippoglossus stenolepis*) in the Gulf of Alaska have been found to correspond directly with this lunar nodal tide climate forcing. The number of age-8 recruits to a carefully controlled commercial fishery for the period 1927–83 displays a sinusoidal cycle with an 18.7-yr period clearly identified by spectral analysis. Sixty percent of the variance in the recruitment series is accounted for by the lunar nodal cycle. Furthermore, most of the recently observed ocean and air temperature increases in the North Pacific are associated with this decadal-frequency tidal forcing, rather than with global warming. Nutrient enhancement and productivity dynamics related to systematic long-period oscillations in the tidally-modulated mixing patterns and horizontal advection processes are discussed as induced mechanisms for possible causal relationships.

Résumé : Les températures des couches supérieures d'eau du nord du Pacifique Nord se sont accrues de plus de 2° C au cours de la décennie qui a commencé en 1972, mais décroissent depuis 1986. Des séries chronologiques établies par approximation montrent que ce phénomène fait partie d'une fluctuation de faible fréquence en synchronie avec le comportement de l'atmosphère. Le cycle de déclinaison synodique de la lune de 18,6 ans, à l'origine de la modulation systématique des marées de la circulation océanique moyenne, est présenté comme la cause possible de cette variabilité. On a constaté que les tendances à la hausse ou à la baisse de la production du flétan du Pacifique (*Hippoglossus stenolepis*) dans le golfe d'Alaska, d'après un dossier d'abondance du recrutement s'étendant sur presque 60 ans, correspondaient directement à ce forçage lunaire nodal du régime des marées. Le nombre de recrues d'âge 8 dans une pêcherie commerciale soigneusement contrôlée pour la période 1927–1983 affiche un cycle sinusoïdal avec une période de 18,7 ans clairement identifiée par analyse spectrale. Le cycle nodal lunaire représente 60 % de la variance de la série de recrutement. Par ailleurs, la plupart des augmentations de la température de l'océan et de l'air observées récemment dans le Pacifique Nord sont associées à ce forçage tidal décennal plutôt qu'au réchauffement global. L'amélioration des nutriments et la dynamique de productivité rattachées à des oscillations systématiques sur de longues périodes dans les configurations de mélanges modulées par les marées et dans les processus d'advection horizontaux sont discutés en tant que mécanismes induits qui pourraient servir à établir des relations causales.

K.S. Parker. NOAA Pacific Marine Environmental Laboratory, Seattle, WA 98115, USA.

T.C. Royer. Institute of Marine Science, University of Alaska, Fairbanks, AK 99775-1080, USA.

R.B. Deriso. Inter-American Tropical Tuna Commission, c/o Scripps Institute of Oceanography, La Jolla, CA 92093, USA.

Introduction

Within the growing community of fishery oceanographers, the tide appears to be turning from a more narrowly focused attention on the interannual scale of climatic variability to that of the interdecadal (e.g., Trenberth and Hurrell 1995; Wooster and Hollowed 1995). As an interesting and significant "disciplinary regime shift", this is complementary to the interest now being generated in "climatic regime shifts", such as the cooling trend in the North Pacific that ended in 1976. As such, suggestions of mechanisms inducing these lower frequency fluctuations may become increasingly helpful.

One intriguing possibility that may account for such longer period forcing is the lunar gravitational influence associated with the long-period oceanic nodal (synodic) tide cycle. The revolution of the lunar ascending node is represented as a precessing intersection of the solar ecliptic (apparent motion of the sun about the earth) and the projection of the moon's orbit about the earth. This so-called "regression of the nodes" entails a westward motion of about 19° per yr, which is sufficient to carry it entirely around a great astronomical circle (Fig. 1). It is also known as the lunar declination tide, referring to the inclination of the lunar orbit relative to the earth's equator (celestial equator) varying between 18.3° and 28.6° over a period of 18.61 yr. This motion also induces a significant modulation of the diurnal and semidiurnal lunar tides, which in turn produce variations in the strength of tidal currents to a degree that is dominant over other long-period cycles, such as the 8.85-yr lunar perigee cycle (Godin 1972) for which effects on sea-surface temperature (SST) have also been suggested.

Maksimov and Smirnov (1965, 1967), Maksimov and Sleptsov-Shevlevich (1970), Currie (1984, 1987), Currie and O'Brien (1988), and others have identified such oscillations in a variety of environmental signals, while proposing mechanisms associated with the nodal tide to produce them. Interestingly, a number of oceanographic and atmospheric variables appear to fluctuate in "lock–step" with the 18.6-yr cycle, including advective and mixing processes that are important to the biota. In fact, as the nature of the whole system is seen to vary at this frequency, we are presented with the kind of regime change that the biology can respond and adapt to. Several of these effects are noted in this paper, together with a discussion of the most probable mechanisms relating to the nodal tide effect on the Gulf of Alaska in terms of its coastal influence on biological productivity, which may ultimately be considered as one of several environmental determinants of Pacific halibut (*Hippoglossus stenolepis*) recruitment.

Pacific halibut recruitment variability (in biomass) has been approximately threefold over the history of the managed fishery, with a pattern that includes occasional extreme year-classes and long-period trends. Probable causes include changes in spawning biomass (generally discounted by previous analyses) and the variable survival of pre-recruits in response to environmental change. In this paper, we attempt to examine the apparent low-frequency, interdecadal oscillations of this

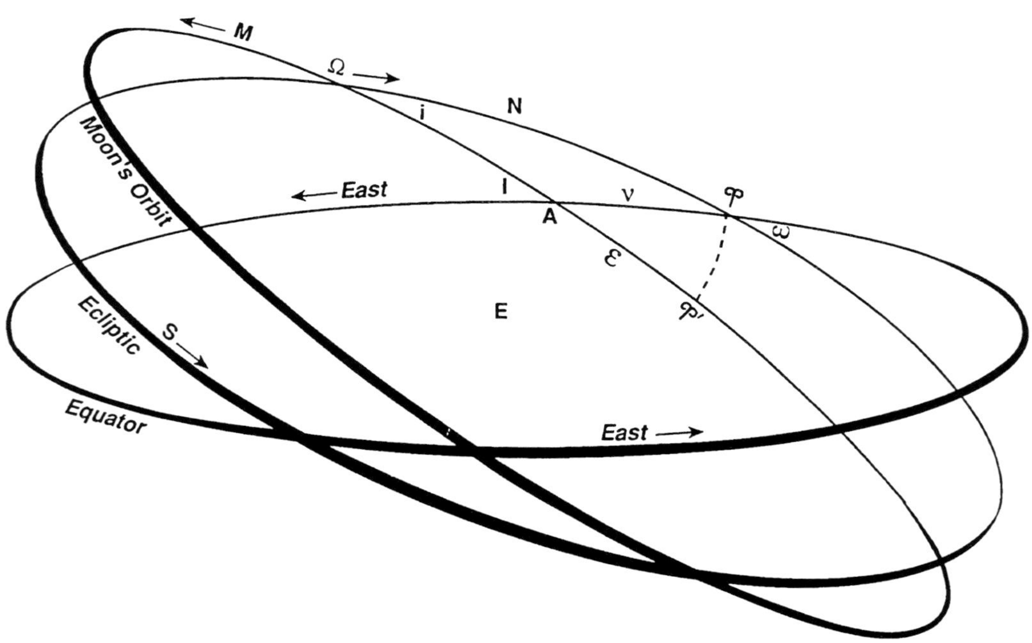

Fig. 1. The relationship between intersecting planes of the earth's (celestial) equator, the solar ecliptic, and the moon's orbit (about the earth) with the celestial sphere. Noted intersections are: the vernal equinox, ♈, the ascending lunar node, **Ω**, and the lunar intersection, ***A***.

historical pattern in relation to the possible influence of the lunar declinational cycle. While this match was previously discovered independently by K.S. Parker and R.B. Deriso (unpublished data), it went unreported until quite recently when Royer (1993) demonstrated corroborating physical evidence.

Data and Analysis

The reproductive strategy of the highly migratory Pacific halibut displays a pattern strongly dependent upon major features of onshelf flux and coastal transport processes. Pelagic larvae originating well offshore in winter must be carried in the prevailing coastal circulation to eventually complete development in shallow nursery areas during the critical spring productivity period. Mechanisms controlling shelf–slope exchange and the access of early life stages to such areas are thus important, particularly as mediated by the interannual fluctuations in the strength of the Alaska Coastal Current (ACC) (Royer 1981). Parker (1989b) and Parker and Schumacher (1984) previously demonstrated the primary importance of the winter wind-driven cross-shelf flux of slope water to interannual halibut larval success and consequent recruitment abundance. A more detailed presentation of life-history dynamics and the relevant environmental forcing on them can be found elsewhere (Parker 1988, 1989a).

Ultimately, halibut larvae may traverse 1000 km of coastal ocean while experiencing a wide spectrum of depth-dependent current velocities. This extensive, highly protracted pelagic phase from slope to shore is completed within 5–7 mo after hatching in the deep offshore waters. An energetic coastal circulation transports post-larval halibut alongshore and onshelf in a prevailing cyclonic downwelling pattern.

A time series of recruitment biomass for the Pacific halibut population in the Gulf of Alaska from 1927–83 (lagged back to year of spawning) has been created, incorporating many adjustments and spanning the full period of the fishery managed by the International Pacific Halibut Commission (IPHC). For the purposes of interannual, as well as interdecadal, comparisons in the present context, recruitment success is assumed to be set largely during the year of spawning, with survival levels remaining relatively intact through the 7–8 years preceding reproductive maturity and subsequent entry into the commercial fishery. Halibut catch statistics are widely recognized to be among the longest and most reliable abundance records of any fishery in the world, given the highly regulatory nature and relatively consistent fishing practices observed between only two cooperating nations for nearly 70 years.

As recruitment production may be estimated in terms of biomass or in numbers of fish, a consideration of the relative difference between the two methods is in order. The former is selected for use here because biomass captures the net effect of changes in both survival and growth. As there have indeed been substantial changes in both growth and survival for young halibut over the record, simple fish numbers do not reveal as accurately the true variations in reproductive value. Individual reproductive fitness seeks to maximize something closer to biomass rather than to number of progeny, i.e., in evolutionary terms biomass is the better estimator.

Furthermore, biomass is more reliably estimated than numbers of fish for the following reasons. First, the adjusted biomass of 8 yr olds (see below) takes into account incidental bycatch, which is measured only in biomass, and conversion of this factor to fish numbers is more crude. The subsequent adjustment of such incidental catches to equivalent 8 yr olds introduces even further error. Second, directed catches of halibut are measured to greater accuracy in terms of biomass (as expected, halibut CPUE estimates closely parallel biomass), as official weights are provided by the fish processors with subsamples of size composition taken by biologists from the catch. These values are then used to estimate the average weight of fish in the catch, which in turn is used to obtain estimates of numbers of fish in the catch. Because errors in estimates of fish numbers are negatively correlated with estimates of average weight of the catch, biomass estimates serve to remove some of these errors. We therefore conclude from stock assessment considerations, as well as from evolutionary theory, that production biomass represents the preferable natural currency to deal in, particularly for groundfish.

The catch-at-age methodology known as CAGEAN (catch-at-age analysis; Deriso et al. 1985) served as the basis for the abundance estimation procedure employed here. Absolute year-class strength was derived from relationships between commercial catch and various population parameters by age, weight, gear, and year. Unadjusted estimates of abundance were produced by a coast-wide Gulf of Alaska CAGEAN run, with catch-at-age (ages 8–20, 1935–91) and effort data for all IPHC regulatory areas combined, except the Bering Sea, used in the analysis (Myhre et al. 1977; A. Parma, IPHC, Seattle, WA, personal communication). Four groups of catchability corrections (1935–54, 1955–73, 1974–82, 1983–91) and three groups of gear selectivities (1935–54, 1955–73, 1974–91) were assumed. Recruitment values estimated by CAGEAN were also adjusted to incorporate losses from both bycatch and the capture of fish younger than age 8.

An abbreviated account of this procedure follows, with appropriate references included for further detail. After Quinn et al. (1983), Quinn and McGregor (1983), and Clark (1992), modified relationships for fish length–weight predictions derived from otolith weights were used to compute average weight and weight-at-age in the catch for CAGEAN, including a 10% adjustment since 1978 (Quinn et al. 1985). Regional catchability coefficients were applied to CPUE estimates, after Quinn et al. (1985), Deriso (1986), and Deriso and Neal (1988). To account for apparent shifts in halibut catchability, fishing gear selectivity factors were introduced to adjust for changes from J- to circle-hooks (Quinn et al. 1985) and fixed to snap gear (Deriso and Price 1987; Sullivan 1991), with CPUE estimates adjusted to reflect these new hook-type and spacing standards. A comparative examination of the effects that these modifications have made on biomass estimates is presented in recent IPHC stock assessment reports, including a useful discussion of weight modifications made in relation to

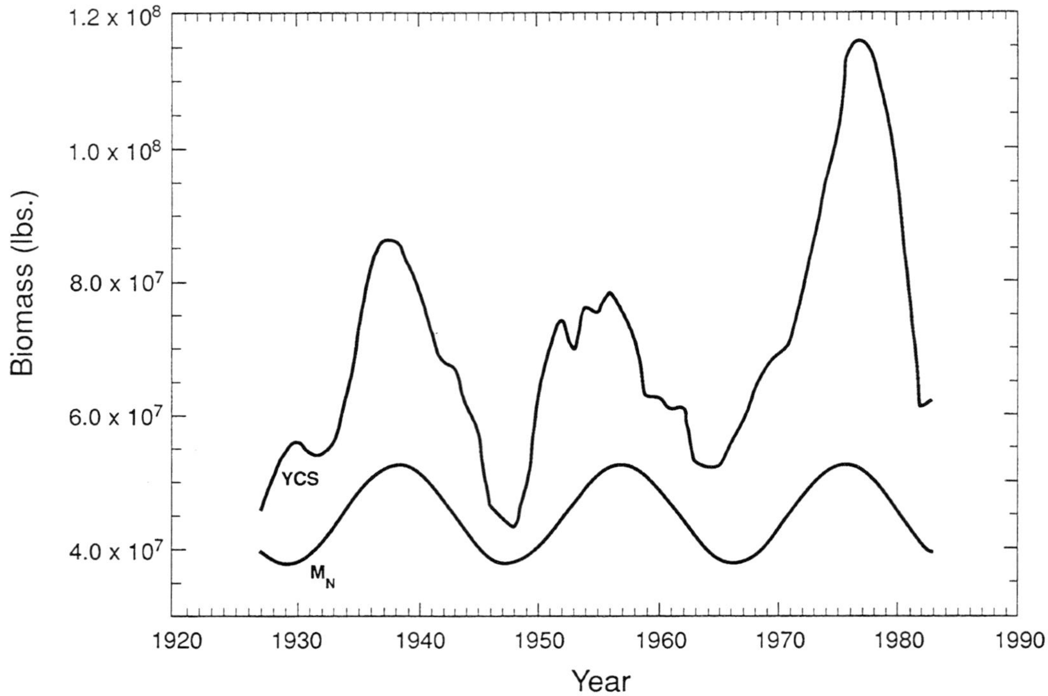

Fig. 2. The Pacific halibut recruitment biomass record, 1927–83 (YCS is year-class strength), with the "best fit" lagged 18.6-yr lunar nodal tide signal (M_N) superimposed. The abundance series has been corrected for regional catchabilities, gear selectivities, weights-at-age, and incidental bycatch, and smoothed with a 3-yr moving average.

catch in numbers versus stock biomass (such as an apparently sensitive 25% catchability correction that differentially affects the weight factors; e.g., IPHC 1991; A. Parma, IPHC, Seattle, WA, personal communication).

Finally, recruitment strength has been re-estimated by adjusting the estimates of 8-yr-old abundance obtained from CAGEAN to account for incidental losses and for fish younger than age 8 in the commercial catch. A revised procedure used to calculate recruitment losses from this bycatch is explained in detail elsewhere (IPHC 1990). Fish caught in the setline fishery at ages younger than 8 were converted into recruits by simply computing the number that would have survived to that age had they not been caught (with natural mortality set at 0.2). This adjustment turned out to be relatively minor.

Because of the usual vagaries of sampling a wild population from commercial catches, the representation of different age-classes over time is better estimated through the use of a moving average. In this case, a 3-years moving average was incorporated to produce a slightly smoothed abundance series, while allowing for more accurate estimation of each age-class by sampling all cohorts over 3 successive years. Some smoothing is desirable to reduce some of the variance in yearly estimates and thus permit a better indication of trend and recruitment over time. The resulting time series of recruitment biomass used for the present analyses is presented in Fig. 2.

Evidence of a marked cyclic component in the final time series is revealed from even a cursory glance. While this oscillatory nature is also seen when recruitment is given in numbers of fish (Parker 1989b), the empirical relationship between recruitment and spawning stock does differ somewhat. This is due to an increasing long-term trend in weight-at-age, as observed in the commercial catch (cf. above discussion; "Stock-recruitment analysis" by A.M. Parma and P.J. Sullivan in IPHC 1990). While recruitment biomass appears to have oscillated around a constant level, an overall declining trend is apparent when recruitment is measured in numbers of fish. In general, the highest recruitment levels have been found to correspond to the lowest stock sizes.

The sinusoidal 18.6-yr lunar nodal tide cycle is also depicted against the halibut recruitment series (Fig. 2). It represents the global node factor (f) according to the equilibrium tide (see Discussion) for the middle of each year, as expressed here in the nodal variations of the M_2 (principal semidiurnal) lunar tide component, 1927–83 (Schureman 1941; see Discussion). The signal was fitted by a simple least squares residual technique with a 2.5-yr phase shift, as determined by the cross-correlation matrix (Fig. 3). The correlation (r) value is 0.77 with 59% (r^2) of the variance being accounted for by the lunar declination series. As the correlation curve of this phase lag relation is relatively flat, it is relatively insensitive to the slight phase variance of approximately ±1 yr.

A simple FFT was then produced to characterize the frequency domain of the abundance record, and to help determine whether co-periodicity exists between both the biotic and abiotic time series (Fig. 4). A marked spectral peak is apparent at 0.05357 cycles · yr^{-1}, or 18.7 years. The series was first detrended and zero-extended, then a 3-point (yr) moving average was applied to produce the smoothed periodogram.

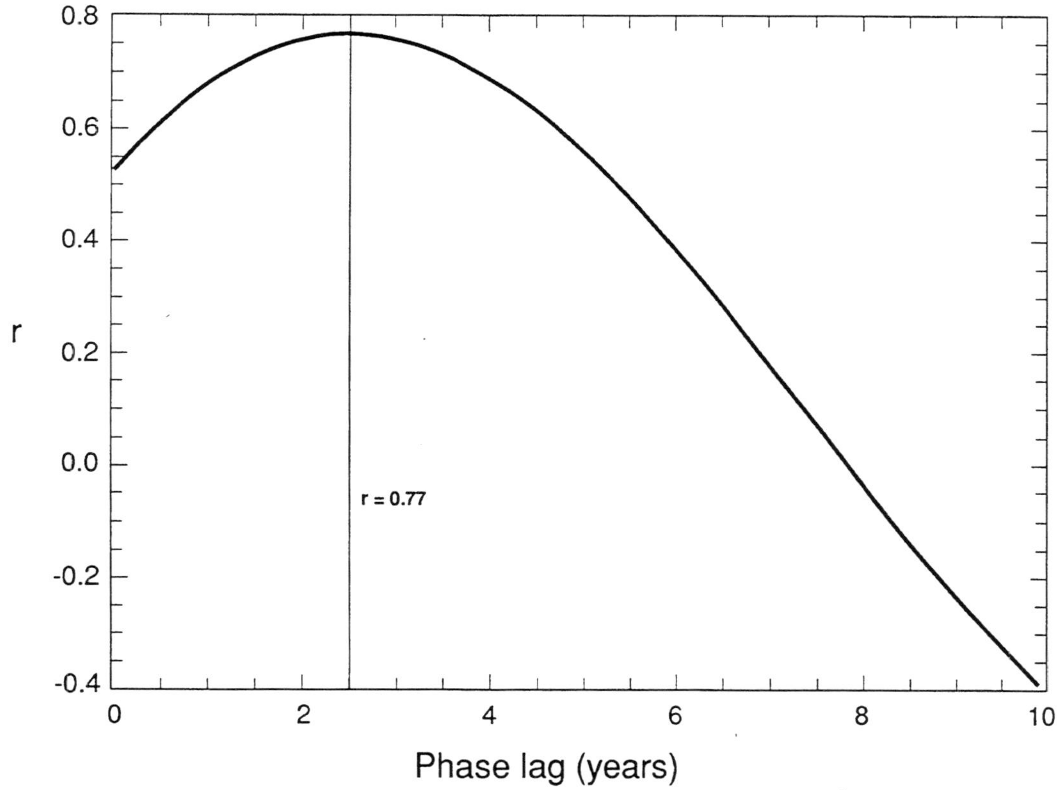

Fig. 3. The correlation function for the phase-lag relation with the Pacific halibut recruitment time series, and the nodal tide signal, for the period 1927–83. (r = 0.77, explained variance = 59%).

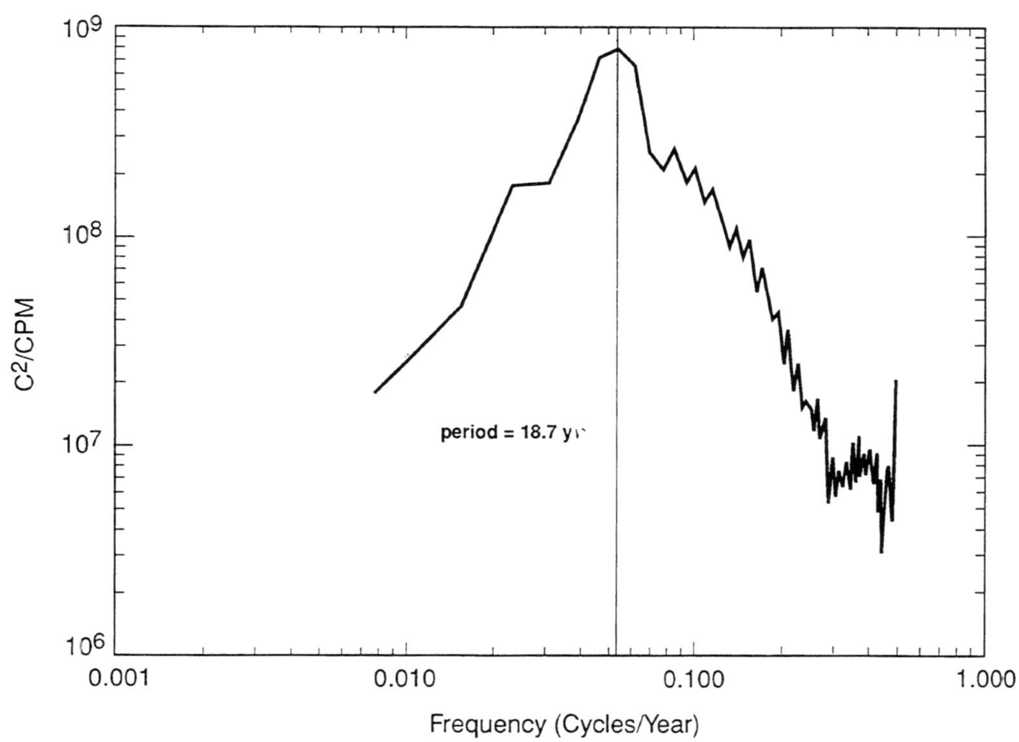

Fig. 4. The low-pass-filtered power spectra (FFT) of the Pacific halibut recruitment time series from 1927–83, detrended and zero-extended. The periodogram is depicted using a 3-point moving average with 10% off each end by a split cosine bell taper.

Lastly, the problem of possible leakage was addressed, whereby a large amplitude peak at a particular frequency can lead to higher than expected spectral estimates at neighboring frequencies. This factor was reduced by implementing a split cosine bell taper over 10% of the values at each end of the series.

In an attempt to corroborate this result and identify as precisely as possible the periodicity of the halibut abundance series, the more sophisticated Maximum Entropy Spectral Analysis (MESA) technique was also performed. Based on fitting a time series as an ARIMA (Autoregressive Integrated Moving Average) process, the spectrum is modeled as a transfer function of the autoregression (AR) series. The series is modeled as this AR portion plus a white-noise process, assuming a nondeterministic biological series may be described by low-order time domain models. By so doing, all but the residual variance is effectively filtered out, as the AR series serves as an optimal differencing filter. Simple mean subtraction or linear/polynomial detrending of the series (as in the FFT process) not only attenuates its low frequencies to an unknown extent, but also increases its sensitivity to outliers. The AR/MESA technique reduces this effect by localizing signal errors associated with outliers (only neighboring points are affected).

The appropriateness of invoking the MESA methodology is further underscored in this particular case where the FFT reveals a large line component (spike) in the spectrum, i.e., a spectral peak that is large relative to the background noise, thereby representing a regular signal that varies little in phase. In addition, because of the limited time period that spans but a few complete cycles, the ARIMA-based technique renders a more accurate spectral representation of the time series.

As with other methods based on fitting ARIMA models, the power spectrum *S(f)* of an AR process is

$$(1) \quad S(f) = \frac{\sigma_\varepsilon^2}{|1 - \alpha_1 \exp(-2\pi \iota f) - \cdots - \alpha_p \exp(-2\pi \iota p f)|^2}$$

where f is the frequency, σ^2e is the variance of the innovation process (white noise) $\varepsilon_{t'}$, and the filter coefficients $\alpha_1,\ldots,\alpha_p$ are a prediction error filter of order p. The denominator is in fact the power transfer function of the ARIMA filter as a function of frequency f.

The filter coefficients were obtained by invoking an AR estimation with Burg's (1967) algorithm, which is based on estimating the *k*th partial correlation coefficient by minimizing the sum of forward and backward prediction errors:

$$(2) \quad SS(a_{k,k}) = \sum_{t=k+1}^{n} \left\{ \begin{array}{l} [x_t - a_{1,k}x_{t-1} - \cdots - a_{k,k}x_{t-k}]^2 + \\ [x_{t-k} - a_{1,k}x_{t-k+1} - \cdots - a_{k,k}x_t]^2 \end{array} \right\}$$

Given all the coefficients for the order k–1 model, this is a function only of $a_{k,k}$. Equation 2 essentially measures how well the order k model predicts forwards and backwards and

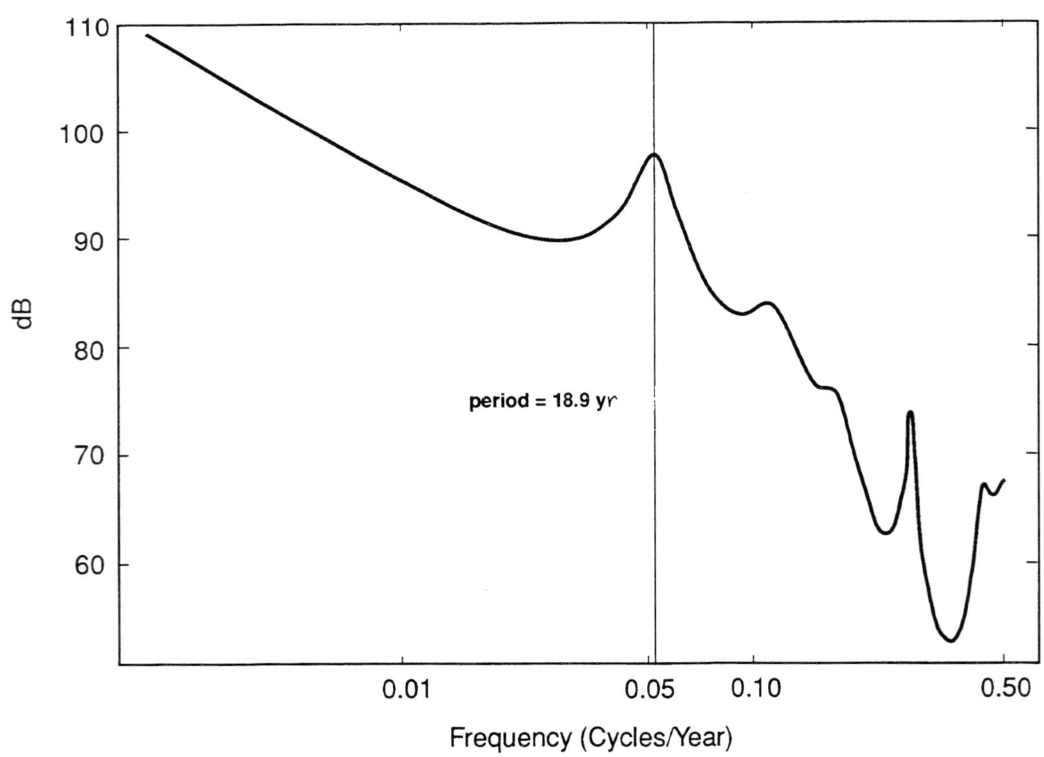

Fig. 5. The low-pass-filtered power spectra (MESA) of the Pacific halibut recruitment time series from 1927–83, detrended and zero-extended. The periodogram is depicted using Burg's algorithm with Maximum Entropy Spectral Analysis (MESA).

is optimal in the sense of maximizing a measure of entropy Burg (1967). It may be noted that order selection in this algorithm is implicitly based on the Akaike's Information Criterion (AIC) as a means of selecting the AR process order that best balances the reduction of estimated error variance with the number of parameters being fitted (Akaike 1974). For this case of an order k model, the AIC can be written as

$$(3)\quad \mathrm{AIC}(k) = n \log(\hat{\sigma}^2_{\varepsilon,k}) + 2k.$$

With the series modeled as an AR process, the value of k that minimizes $AIC(k)$ is an estimate of the specific order of the autoregression (13 in this case).

From the ratio of variances between the innovation process and the recruitment biomass (σ^2_{bm}) series:

$$(4)\quad \frac{\sigma^2_\varepsilon}{\sigma^2_{bm}} = \frac{12726626}{315689626} = 0.04$$

we see that 96% of the variance in the latter is removed by the application of this AR model, and the residuals after fitting were computed separately. The result of the MESA procedure using Burg's algorithm therefore corroborates that of the earlier FFT (Fig. 5). Once again, a strong — and even more reliable — spectral peak occurs at a period of 0.05294 cycles · yr^{-1}, or at 18.9 yr. Examination of any higher frequency spectral signals in the recruitment series is beyond the scope of this paper (e.g., the spike near 3 yr, the power of which is three orders of magnitude less that at 18.9 yr and may therefore be regarded as insignificant).

Discussion

Evidence appears to be mounting of biotic fluctuations on decadal time scales in marine ecosystems, beyond simply the interannual variability that has previously dominated the literature (e.g., Cushing 1982; Kawasaki et al. 1991; Kawasaki and Omori 1995). The risk-management implications (optimization of escapement with passive–active controls) in harvesting cyclic stocks when confronted with alternative recruitment hypotheses have previously been well addressed by Parma and Deriso (1990, for Pacific halibut) and Walters (1981, 1984, 1986) and will not be explored further here. While caution is advisable where various mechanisms confound environmental and stock-size effects in cyclic populations (e.g., Armstrong and Shelton 1988), the importance of possible decadal-scale oscillations is once again underscored in the present context by the strong relationship of explained variance in the Pacific halibut abundance series over nearly 60 yr, or three complete cycles. Indeed, the remarkable fit in this case of the halibut recruitment time series with a sinusoidal function (Fig. 2) exhibits a cyclic trend that appears temptingly consistent with some kind of extrinsic environmental forcing.

This low-frequency cycling (either periodic or quasi-periodic) has also been expressed in the long-term productivity fluctuations of a variety of other high-latitude fish stocks, e.g., northwest and northeast (Norwegian and Faeroes) Atlantic cod (*Gadus morhua*), Barents Sea cod (*Boreogadus saida*), North Atlantic (Scottish and Clyde) herring (*Clupea harengus*), Atlantic haddock (*Melanogrammus aeglefinus*), Atlantic and Cornish sardine (*Sardina pilchardus*), European and Celtic

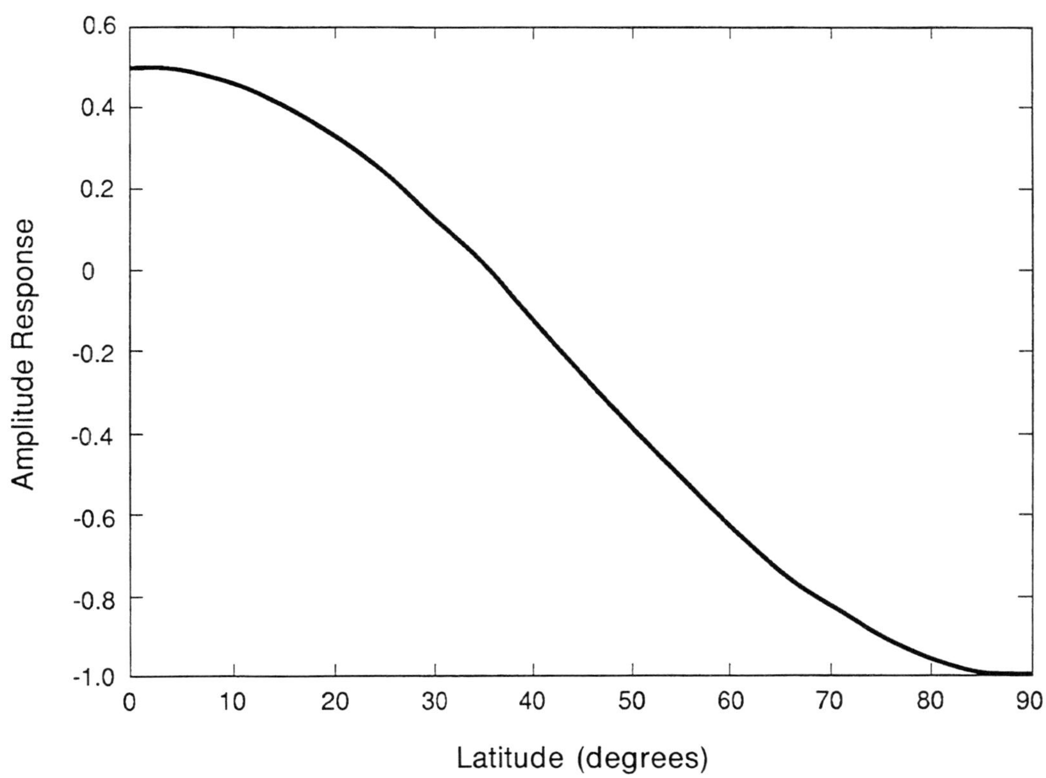

Fig. 6. The 18.6-yr equilibrium tidal amplitude response function for G_0 factors (from Doodson 1921).

Sea hake (*Merluccius merluccius*), American mackerel (*Scomber scombrus*), American striped bass (*Morone saxatilis*), American alewife (*Alosa pseudoharengus*), Gulf of Maine and Bay of Fundy scallops (*Argopecten irradians*), North Pacific pink (*Oncorhynchus gorbuscha*), chum (*Oncorhynchus keta*), sockeye (*Oncorhynchus nerka*), and coho (*Oncorhynchus kisutch*) salmon (Izhevskii 1964; Maksimov and Smirnov 1965, 1967; Southward et al. 1975; Caddy 1979; Van Winckle et al. 1979; Wyatt 1984; Johnson et al. 1986; Koslow et al. 1987; Wyatt and Vázquez 1988, cf. refs. therein; Hamre 1991; Francis and Sibley 1991; Cooney et al. 1995). Most of these recruitment records in fact demonstrate the existence of strong spectral peaks near the 19-yr synodic period.

Such a signal also appears prominently in a wide variety of climatological series, as well as terrestrial biological series, in both the northern and southern hemispheres, many of which are cited in Wyatt and Vázquez 1988 (cf. Currie 1983, 1984, 1991, 1992, 1993; Currie and O'Brien 1992; Peterson 1988). Some of these records are of sufficient length (i.e., tree rings) for the lunar nodal wavetrains to manifest 180· phase shifts at century-scale intervals. Most recently, the signal was also evident in a 130-yr global temperature series using the Comprehensive Ocean – Atmosphere Data Set (COADS) derived, combined terrestrial and marine data (Keeling and Whorf 1994).

The operative environmental mechanisms associated with the lunar declination cycle, however, remain difficult to clearly ascertain. Nevertheless a discussion of the leading hypotheses is in order, particularly from the perspective of a high-latitude forcing function that is most relevant here. The amplitude of the 18.6-yr equilibrium tide response is enhanced at higher latitudes, such as those of the northern Gulf of Alaska (Fig. 6) where the bulk of the halibut population has been historically distributed. In demonstrating the variance with latitude of nodal tide signal strength on both the Seward and Sitka ocean temperature records since 1970 (as well as on associated Sitka air temperature since 1828), Royer (1989, 1993) has noted the stronger high-latitude expression (e.g., Fig. 7). As a corollary to this argument, one would not expect as clear a representation of this relation in the more temperate and warm water populations of the low and mid-latitudes. Perhaps the evidence for synchronicity of productivity with the synodic oscillation may ultimately be found to increase with latitude.

In general, the means by which the lunar nodal forcing may influence biological relationships would be through the generation of systematic changes in horizontal circulation patterns and vertical mixing. Associated decadal-scale changes in recruitment potential are thus hypothesized that would track the zonally and meridionally specific timing of this cycle. It is in fact the meridional gradients of mean sea level in the North Pacific resulting from the nodal tide that are most important. The potential of this tide-generating force has been addressed quantitatively by Maksimov and Smirnov (1965) and has been recently re-emphasized by Trupin and Wahr's (1990) global sea level analyses that clearly identify the 18.6-yr signal as the largest component for periods greater than 1 yr. They make the case that ocean gradient force variations arise as a result of long-period sea level variations generated by the same long-period tidal circulatory oscillation along meridians (barotropic circulation changes).

The essential argument at work here is that current velocities in the system are intensified when the sea level slope from the equator toward the pole increases, and are weakened when the pole-to-equator slope increases. The resulting poleward velocity-gradient forcing would therefore also induce a degree of acceleration in the same direction, which by continuity might then lead to divergent processes and associated

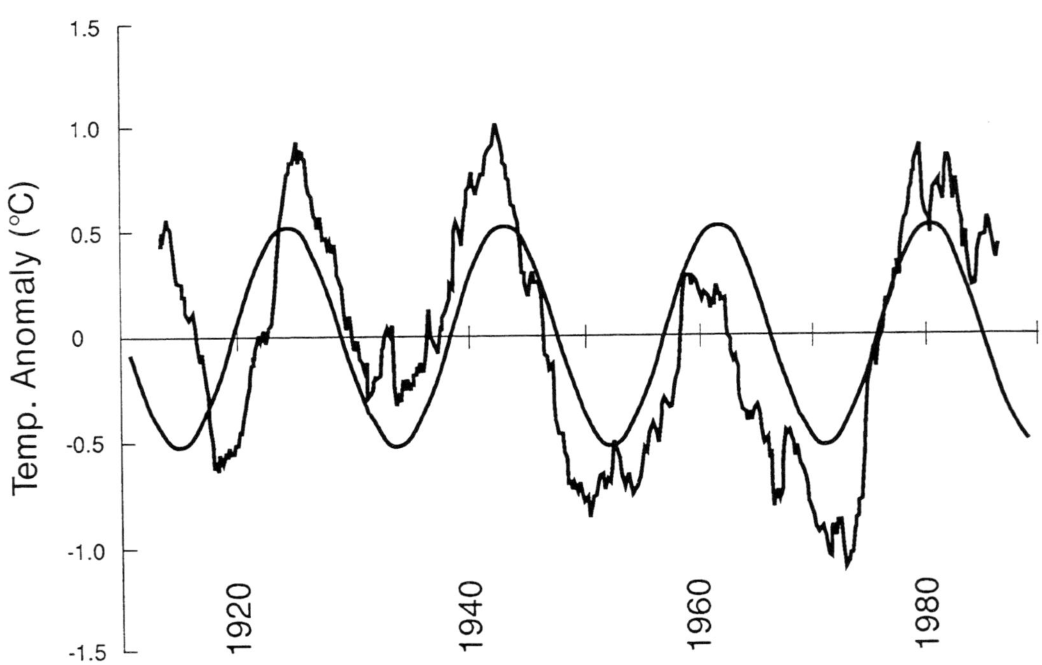

Fig. 7. Anomalies of the 5-yr-averaged Sitka, Alaska, air temperature time series, low-pass filtered, with the "best fit" 18.6-yr lunar tide signal superimposed.

upwelling. With such enhancement of upward vertical motion, it is also reasonable to assume a rise in productivity over the shallow coastal shelf regime due to higher ambient nutrient levels being entrained from deep waters in the central Gulf.

At the same time, such variations in the intensity of relevant currents should also affect their ambient temperatures. The poleward velocity component induced by a latitudinally-dependent lunar declinational tide would set up a regional thermal gradient that could represent a tracer of tidal excursions, with significant impact on the barotropic meridional heat flux. Indeed, Maksimov and Smirnov (1965) go on to demonstrate such direct velocity–temperature relations in the Gulf Stream between the Faeroes and the Shetland Islands, with sea temperatures rising or falling in accordance with the absolute values of a long-period tide-based index of Atlantic meridional circulation.

As meridional gradients serve to modify water and heat transport by the currents, similar long-period variations in the atmospheric circulation patterns also arise. Frequency analyses (Maksimov and Smirnov 1965, 1967) have revealed an approximately 19-yr component in long-period variations of the basic characteristics of atmospheric circulation and of the English climate, a finding that has been well corroborated around the globe by many more recent climatological analyses (e.g., Currie 1984, 1987, 1991, 1993; Currie and O'Brien 1988; Guiot 1988; Holdsworth et al. 1989; Feldman 1992). In noting the distribution of lunar nodal tide signals along northwestern North America as reported by these authors, Royer (1993) asserted consistency with the North Pacific as its source. Thus, upper layer temperatures are likely influencing low-frequency climate changes over North America, as well as in other high-latitude marine ecosystems. This corroborates Maksimov and Smirnov's (1965, 1967) arguments, while the out-of-phase (tidally forced) coupling of the Icelandic and Aleutian Lows' strengths may be reflecting a response to the nodal tide effect on ocean temperatures, in addition to the atmospheric seesaw linkage (Van Loon and Rogers 1978).

Maksimov and Smirnov (1965, 1967) also explore quantitatively the importance of such long-period atmospheric circulation changes to climate change and even forecasting, noting from observations how increased Gulf Stream intensity is associated with a distinct westward displacement ("east–west oscillation") of the Icelandic Low. The broad significance of such pressure-gradient patterns in the North Atlantic is emphasized through five consecutive 19-yr cycles, while providing empirical evidence of the nearly complete and cyclical cessation of meridional transport in North Atlantic air masses. Ultimately, the formation of tidal rhythms follows in the long-period variations of all indices of general atmospheric circulation in high latitudes. A similar phenomenon might in fact be postulated for the Aleutian Low in the North Pacific.

Loder and Garrett (1978) presented a model arguing for vertical mixing and SST variation in coastal seas induced by the modulation of semidiurnal and diurnal lunar tide forces over the 18.6-yr nodal cycle, and found evidence for such a mechanism at some North American locations. The relevance to the present context would, of course, be in the inshore nursery areas over the shallower shelf where halibut larvae must ultimately be transported, metamorphose, and succeed. Loder and Garrett argue that the average strength of tidal currents would affect the degree of vertical mixing, and that long-term systematic variations in oceanographic parameters are distinct from any claim for the effect of occasional tidal extrema. They find remarkable the coincidence of the phase of the fitted nodal cycle with that of the tidal variations, while speculating on the effect of SST-induced changes in local air temperatures and larger scale weather patterns. Moreover, they point out how a comprehensive study of air temperature in eastern North America (Mock and Hibler 1976) displays a 20-yr spectral peak in January while revealing minima very close to the times of maximum M_2 forcing over five full cycles. While uncertain of its ultimate significance, they believe that the empirical evidence for SST nodal cycling and its phase coincidence, together with the "firm physical grounds for expecting a reduction of SST at times of increased tidal currents, all suggest that the problem be given serious attention".

That such long-period temperature fluctuations, with a discernible ±1 °C range, may have important consequences for the regional subarctic biota has often been argued, although the exact stage of influence remains obscure. To be sure, the early life viability of many fish species in the Gulf of Alaska is temperature limited (e.g., Rogers 1986), and such thermal sensitivity extends to many interdependent biological populations critical to the regional ecosystem. It is then reasonable to speculate on the consequences to larval halibut of the availability and substitutions of key prey species whose ranges may shift with subtle thermal-regime changes. Bulatov (1989) found interannual variability in pollock year-classes in the Gulf to be similar to that in ambient water temperatures, and Cooney et al. (1995) also found such results with pink salmon and SST regime cycling in Prince William Sound at the head of the Gulf. Western Alaska (eastern Bering Sea) sockeye and chum salmon production (since 1925) also appear to track that of the pink, while Pacific Northwest coho abundance follows this same trend inversely (Francis and Sibley 1991; R.C. Francis unpublished data; R.T. Cooney, University of Alaska, Fairbanks,AK, personal communication). With the time scales of such changes being significant relative to the life history of Pacific halibut, as well as many other species, these dynamics can have a significant impact on the distribution and survival of populations. Furthermore, the specific use of the M_2 nodal series herein, taken with the work of Loder and Garrett (1978), also points to the importance of the additional biological response to the changing semidiurnal and diurnal tide components over the cycle, particularly at high latitudes.

A principal mechanism here for enhanced tidal mixing has also been proposed by Hibiya and LeBlond (1993) who recently modeled the modulation of mixing intensity of the diurnal and semidiurnal tides associated with the fortnightly spring-neap tidal cycle. Increasing mixing (spring tides) is seen to generate separate landward and seaward propagating internal waves in response to horizontal density gradients induced by the localized mixing. These waves then interact with the ambient steady shear flow to create strong upwelling over the mixing zone, with intensified seaward currents superimposed on the background estuarine flow. Observations of enhanced deep-water intrusion into coastal basins during the strong portion of the tidal cycle are reproduced by the model, thus lending strength to the argument for crucial tidal mixing effects on both the strength and timing of deep-water replacement. In essence, increased tidal mixing at this time would thereby replenish nutrients at a higher rate, together with an associated increase in cross-shelf advection. This exchange process, while varying systematically over the 18.6-yr cycle of diurnal and semi-diurnal tidal components, would then be highly relevant to the transport and dispersal of halibut larvae rising in the water column as they approach their essential shallow bay, estuarine destinations.

The degree of tidally modulated water column mixing and associated mixed layer depths represents an important mechanism to explain how these oscillations are mediated through the chain of events associated with primary and secondary productivity in the marine ecosystem. Again, this parameter should vary systematically to alter the air–sea heat exchange processes, and there may be a positive feedback involved due to the dependence of eddy diffusivity on density stratification (Wyatt and Vázquez 1988). Systematic changes in tidal forcing and local horizontal circulation patterns would also effect variability in larval dispersion patterns, possibly influencing early life history and prey base competition and survival modes. This would include horizontal advective water movement associated with the tidal modulation of shelf currents and their net poleward displacement.

Also of interest in this context, Ware (1991) reported an 18 to 20-yr peak in the North Pacific herring size-at-age time series, thereby indicating a possible relationship between adult success and this ambient productivity scenario. During times of enhanced and hence more vigorous tidal mixing along the shelf, prey base productivity may support elevated growth rates in this zooplankton-feeding population. Conversely, prey species substitutions may ensue during periods of reduced levels.

A process of *systematic amplification* may therefore be operating when the nodal tide influence reaches an as-yet-unknown threshold, raising the background level in support of total production potential in the critical surface layers where ichthyoplankton must survive, according to the mechanisms postulated above. Thinking in terms of productivity enhancement or attenuation then, one could envision oscillating patterns of positive feedback or resonance (or interference) with the varying lunar nodal signal as it is overlaid on other higher frequency conditions and/or environmental noise. Increased stock abundance in fisheries would then be linked with tidal events during the larval phase. This argument predicts the existence of the low-frequency tidal signal in both climatic data and fish populations, as well as in terrestrial populations where interannual changes in growth are dependent on those in temperature and rainfall (e.g., tree rings; Currie 1991). Most of the stocks noted above, in which the signal prevails, are exploited in sufficiently "open" ocean regimes to expect an arrival time of the nodal tide in accordance with theory (Wyatt and Vázquez 1988).

Another possible tidally-linked mechanism that can result in lateral advection of larvae drifting in the upper water column involves the effect of shear in an oscillatory (tidal) current on horizontal diffusion, known as the "shear effect" (Bowden 1963; Okubo 1967, 1968). This approach considers the contributions of velocity deviations from their depth means to the magnitude of the horizontal dispersion coefficient, which in turn varies inversely with the magnitude of the vertical eddy diffusion. A decrease in vertical diffusion due to, for instance, a stable density gradient, would thereby induce an increase in the horizontal dispersion. While this has been employed largely to describe salt diffusion in estuarine settings, it could also apply to the transport of larvae in the present context, particularly if they are vertically migrating in a mode adapted to consistent (and systematically fluctuating) lateral variations in the diurnal–semidiurnal tidal currents that vary with depth, e.g., about the spring pycnocline (H. Mofjeld, NOAA, PMEL, Seattle, WA, personal communication). In essence then, we speculate that vertical diffusion would interact with the horizontal oscillatory (M_2) currents to effect a substantial net horizontal dispersion of larvae. This process would then also fluctuate systematically over the 18.6-yr tidal cycle, as the diffusivity and shelf–slope tidal currents are modulated by the synodic oscillation.

Lastly, equilibrium tide theory, along with the corresponding locality-specific phase and amplitude for the nodal tide series employed here, presents a problematic element, but one that might eventually lead to some predictive capability for the phenomenon. The question of whether long-period astronomical tides such as this will even obey equilibrium law remains open. This is partially because of the steady tidal motions (i.e., tides of the "second class", which preclude the equilibrium tide assumption) which are unrelated to the tide-generating forces (Lamb 1932), and partially because of land-mass presence on the earth.

The unfortunate problem remains that to date no empirically-based global nodal tide model has been produced to account for the critical phasing and amplitude dynamics of this tide, which certainly vary both meridionally and zonally around the globe. Spherical harmonic decomposition analysis is needed to determine the precise degree of lunar forcing, as based on the moon's orbital declination angle. But no empirical–versus–theoretical modeling analysis is as yet available with which to determine the accurate global distribution and magnitude for a particular region of concern, and of course this remains outside the scope of this paper.

While various comparisons of nodal tide-reflecting sea level observations exist for many locations around the world (e.g., Lisitzin 1957; Rossiter 1962), an orderly pattern of the regional distribution of the nodal tide oscillation in sea level does not emerge. Lisitzin (1957) does note from her analysis of 27 stations from 1880–1954 that the differences between the North Atlantic west coast and the Gulf of Mexico are generally similar to those between the North Sea and the Baltic, while a retardation of the phase lag from north to south is also in evidence along the North American west coast. In demonstrating how enigmatic this accurate phase-lag determination is, the difference between lags for Ketchikan and San Francisco is considerable, and hardly to be expected from their latitudinal difference. For the most part, however, lag differences for stations near the same latitude on opposite coasts of an ocean are of similar magnitude to their difference in latitude.

With regard to the empirical fitting procedure employed here, the ultimate question of identifying an accurate phase lag for the northern Gulf of Alaska must therefore be left to future analyses. This frustrating but minor discrepancy notwithstanding, the results of the equilibrium model used herein fairly represent an appropriate first-order approximation, upon which an imposed 1 to 3-yr lag one way or the other can scarcely be criticized as significant.

Conclusions

The production and growth of marine fisheries is determined by a seemingly indecipherable complex of interactions among physical, chemical and biological processes, and these interactions are replete with positive and negative feedback mechanisms that undoubtedly compound the difficulty of identifying them. In particular, the amplitude and phase of available forcing for bottom and water-column mixing, on a wide spectrum of spatial and temporal scales, will determine the concentration of essential nutrient inputs that support standing stocks at lower trophic levels. From the considerations explored here, it is clear that the complex physical interactions between fluctuating tidal currents, mixing, and temperature changes make it very difficult to sort out, let alone quantify, the impacts of any specific process.

Nevertheless, the tidally-modulated reapportionment of nutrients, and/or the plankton that depends on them, is certainly mediated in response to an equally wide spectrum of climatic (including astronomic) factors. The total production capacity of the high-latitude North Pacific marine ecosystem will therefore be significantly influenced by both high- and low-frequency perturbations, which taken together, can produce the kind of energetic "kicks" to the system (Margalef 1986) that are reflected in the appearance of regime changes. The background of ambient conditions against which these fluctuations are observed might indeed be elucidated by a careful examination of such long-period oscillations as the lunar nodal tide, with particular attention to their systematic influence on the strength and intensity of diurnal–semidiurnal tidal mixing and upwelling processes in shallow shelf and embayment areas. Increased tidal mixing would thereby replenish nutrients at a higher rate, together with an associated increase in advection across the shelf. This exchange process, while varying systematically over the 18.6-yr diurnal–semidiurnal cycle, would then be highly relevant to the transport and dispersal of halibut larvae rising in the water column as they approach their critical shallow bay destinations.

It is apparent from the evidence presented or reviewed here that gradient forces associated with the lunar nodal tide, in response to adjustments of the mean sea level, would induce substantive changes in the intensity of shelf currents and their associated transport processes. This in turn induces a long-period variation in the amount of heat transported by these currents, which can lead to displacements of the major atmospheric pressure systems. Ultimately then, such tidal forcing leads indirectly to the development of tidal variations in all indices of general atmospheric circulation and climate at high latitudes. Hence the class of long-period variations in tidal forces are, as proposed by Maksimov and Smirnov (1965, 1967), capable of some climate-producing effect.

A significant portion of the 60-yr recruitment variation in Pacific halibut—along with many other northern fish stocks studied—is associated with the 18.6-yr lunar nodal cycle. Many air temperature, precipitation, and terrestrial biotic cycles also reflect this same periodicity, possibly in response to the ocean cycles. As Royer (1989, 1993) has shown, important information on the regional climate variations may lie in high-latitude subsurface waters such as the North Pacific. The lunar nodal cycle remains a critical element to be considered in circulation models of the ocean and atmosphere that address decadal time scales, as well as in investigations of biological productivity in the coastal regime. Indeed, given that their magnitudes are comparable with many of the global change forecasts, these ocean–atmosphere variations warrant general recognition in studies of high-latitude global climate change.

Acknowledgments

I would like to express my sincere appreciation to H. Mofjeld for enlightening discussions on the nodal tide, and P. Turet for his helpful discussions and substantial contribution to the statistical analysis. A. Parma provided generous assistance with the halibut time series and its methodology, and C. Keeling was helpful in discussing the tidal mechanism. Two anonymous reviewers also provided useful comments on the manuscript. This work was supported in part by the NOAA Climate and Global Change Program under grant No. NA90AA-D-AC805.

References

Akaike, H. 1974. A new look at the statistical model identification. IEEE Trans. Automa. Control 19: 716–723.

Armstrong, M.J., and P.A. Shelton. 1988. Bias in estimation of stock-recruit function parameters caused by nonrandom environmental variability. Can. J. Fish. Aquat. Sci. 45: 554–557.

Bowden, K.F. 1963. Horizontal mixing in the sea due to a shearing current. J. Fluid Mech. 21: 83–95.

Bulatov, O.A. 1989. Some data on mortality of walleye pollock *(Theragra chalcogramma)* in the early stages of ontogenesis. University of Alaska, Fairbanks, AK. Alaska Sea Grant Rep. 89-1: 185–196.

Burg, J.P. 1967. Maximum Entropy Spectral Analysis (MESA). Paper presented at 37th Ann. Intl. S.E.G. Mtg., Oklahoma, OK.

Caddy, J.F. 1979. Long-term trends and evidence for production cycles in the Bay of Fundy scallop fishery. Rapp. P.-v. Réun. Cons. int. Explor. Mer 175: 97–108.

Clark, W. 1992. Validation of the IPHC length–weight relationship for halibut. IPHC report of assessment and research activities, 1992. IPHC, Seattle, WA.

Cooney, R.T., T.M. Willette, S. Sharr, D. Sharp, and J. Olsen. 1995. The effect of climate on North Pacific pink salmon (*Oncorhynchus gorbuscha*) production: examining some details of a natural experiment, p. 475–482. *In* R.J. Beamish [ed.] Climate change and northern fish populations. Can. Spec. Publ. Fish. Aquat. Sci. 121.

Currie, R.G. 1983. Detection of 18.6-year nodal-induced drought in the Patagonian Andes. Geophys. Res. Lett. 10: 1089–1092.

Currie, R.G. 1984. On bistable phasing of 18.6-year nodal-induced flood in India. Geophys. Res. Lett. 11: 50–53.

Currie, R.G. 1987. Examples and implications of 18.6- and 11-year terms in world weather records, p. 378–403. *In* M.R. Rampino, J.E. Sanders, W.S. Newman, and L.K. Konigsson [ed.] Climate, history, periodicity and predictability. Van Nostrand Reinhold Co., New York.

Currie, R.G. 1991. Deterministic signals in tree-rings from Tasmania, New Zealand and South Africa. Ann. Geophys. 9: 71–81.

Currie, R.G. 1992. Deterministic signals in tree-rings from Europe. Ann. Geophys. 10: 241–253.

Currie, R.G.1993. Luni-solar 18.6- and 10-11-year signals in the USA air temperature records. J. Climatol. 13: 31–50.

Currie, R.G., and D.P. O'Brien. 1988. Periodic 18.6-year and cyclic 10 to 11-year signals in northeastern United States precipitation data. J. Climatol. 8: 255–281.

Currie, R.G., and D.P. O'Brien. 1992. Deterministic signals in USA precipitation records: Part II. Int. J. Climatol. 12: 281–304.

Cushing, D. 1982. Climate and fisheries. Academic Press, New York, NY. 373 p.

Deriso, R.B. 1986. 1985 adjustments to CPUE data. IPHC population assessment, 1985. IPHC, Seattle, WA.

Deriso, R.B., and P.R. Neal. 1988. Section 5. Comparison of catchability coefficients between regulatory areas. IPHC stock assessment document. III. Population assessment, 1987. IPHC, Seattle, WA.

Deriso, R.B., and R. Price. 1987. Derivation of CPUE estimates for regions 1 and 2. IPHC stock assessment document. III. Population assessment, 1986. IPHC, Seattle, WA.

Deriso, R.B., T.J. Quinn II, and P.R. Neal. 1985. Catch-age analysis with auxiliary information. Can. J. Fish. Aquat. Sci. 42: 815–824.

Doodson, A.T. 1921. The harmonic development of the tide-generating potential. Proc. R. Soc. Lond. Ser. A 100: 305–329.

Feldman, T.S. 1992. Climate and history in the late 18th and 19th centuries. EOS Trans. Am. Geophys. Union 73: 1–4.

Francis, R.C., and T.H. Sibley. 1991. Climate change and fisheries: what are the real issues? Northwest Environ. J. 7: 295–307.

Godin, G. 1972. The analysis of tides. University of Toronto Press, Toronto, Ont.

Guiot, J. 1988. Climate of the Hudson Bay region and effects of solar activity, volcanic dust and lunar nodal tidal signal, p. 377–395. *In* C.G. Castagnoli [ed.] Solar–terrestrial relationships and the Earth environment in the last millennia. North Holland, Amsterdam, The Netherlands.

Hamre, J. 1991. Interrelation between environmental changes and fluctuating fish populations in the Barents Sea, p. 259–270. *In* T. Kawasaki, S. Tanaka, Y. Toba, and A. Taniguchi [ed.] Long-term variability of pelagic fish populations and their environment. Pergamon Press, Tokyo, Japan.

Hibaya, T., and P.H. LeBlond. 1993. The control of fjord circulation by fortnightly modulation of tidal mixing. J. Phys. Oceanogr. 23: 2042–2052.

Holdsworth, G., H.R. Krouse, M. Nosal, M.J. Spencer, and P.A. Mayewski. 1989. Analysis of a 290-year net accumulation time series from Mt. Logan, Yukon, p. 71–79. *In* Snow cover and glacial variations. Proceedings, Baltimore Symposium, Baltimore, MD, May 1989. IAHS Publ. 183.

IPHC. 1990. Report of assessment and research activities, 1990. Seattle, WA.

IPHC. 1991. Report of assessment and research activities, 1991. Seattle, WA.

Izhevskii, G.K. 1964. Forecasting of oceanological conditions and the reproduction of commercial fish. All Union Sci. Res. Inst. Mar. Fish. Oceanogr. (Translated from Russian by the Israel Progr. Sci. Transl.). 95 p.

Johnson, D.F., L.W. Botsford, R.D. Methot, Jr., and T. Wainwright. 1986. Wind stress and cycles in Dungeness crab *(Cancer magister)* catch off California, Oregon, and Washington. Can. J. Fish. Aquat. Sci. 43: 838–845.

Kawasaki, T., and M. Omori. 1995. The impacts of climate change on Japanese fisheries, p. 523–528. *In* R.J. Beamish [ed.] Climate change and northern fish populations. Can. Spec. Publ. Fish. Aquat. Sci. 121.

Kawasaki, T., S. Tanaka, Y. Toba, and A. Taniguchi. [ed.]. 1991. Long-term variability of pelagic fish populations and their environment. Pergamon Press, Tokyo, Japan. 402 p.

Keeling, C.D., and T.P. Whorf. 1994. Decadal oscillation in global temperature and atmospheric carbon dioxide. *In* Natural climate variability on decade-to-century time scales. National Research Council, National Academy Press, Washington, DC.

Koslow, J.A., K.R. Thompson, and W. Silvert. 1987. Recruitment to Northwest Atlantic cod *(Gadus morhua)* and haddock *(Melanogrammus aeglefinus)* stocks: influence of stock size and climate. Can. J. Fish. Aquat. Sci. 44: 26–39.

Lamb, H. 1932. Hydrodynamics. Cambridge University Press, Cambridge, UK.

Lisitzin, E.J. 1957. The tidal cycle of 18.6 years in the oceans. J. Cons. Cons. Int. Explor. Mer 22: 147–151.

Loder, J.W., and C. Garrett. 1978. The 18.6-year cycle of sea surface temperature in shallow seas due to tidal mixing. J. Geophys. Res. 83: 1967–1970.

Maksimov, I.V., and N.P. Smirnov. 1965. A contribution to the study of the causes of long-period variations in the activity of the Gulf Stream. Oceanology 5: 15–24.

Maksimov, I.V., and N.P. Smirnov. 1967. A long-term circumpolar tide and its significance for the circulation of the ocean and atmosphere. Oceanology 7: 173–178.

Maksimov, N.V., and B.A. Sleptsov-Shevlevich. 1970. Long-term changes in the tide-generating force of the moon and the iciness of the Arctic Seas. Proc. Knipovich Polar Sci.-Res. Planning Inst. Mar. Fish. Oceanogr. (PINRO) 27: 22–40.

Margalef, R. 1986. Reset successions and suspected chaos in models of marine populations, p. 321–343. *In* Proceedings of the International Symposium Long Term Changes in Marine Fish Populations, Vigo, Spain.

Mock, S.J., and W.D. Hibler III. 1976. The 20-year oscillation in eastern North American temperature records. Nature (Lond.) 261: 484–486.

Myhre, R.J., G.J. Peltonen, G. St.-Pierre, B.E. Skud, and R.E. Walden. 1977. The Pacific halibut fishery: Catch, effort and CPUE, 1929–1975. IPHC Tech. Rep. 14: 94 p.

Okubo, A. 1967. The effect of shear in an oscillatory current on horizontal diffusion from an instantaneous source. Int. J. Oceanol. Limnol. 1: 194–204.

1968. Some remarks on the importance of the shear effect on horizontal diffusion. J. Oceanogr. Soc. J. 24: 60–69.

Parker, K.S. 1988. Pacific halibut, *Hippoglossus stenolepis,* in the Gulf of Alaska, p. 94–111. *In* N.J. Wilimovsky, L.S. Incze, and S.J. Westrheim [ed.] Species synopses: Life histories of selected fish and shellfish of the Northeast Pacific and Bering Sea. Washington Sea Grant Publ. 88-2. Fisheries Research Institute, University of Washington, Seattle, WA.

Parker, K.S. 1989a. Influence of oceanographic and meteorological processes on the recruitment of Pacific halibut, *Hippoglossus stenolepis,* in the Gulf of Alaska. Ph.D. dissertation, University of Washington, Seattle, WA. 154 p.

Parker, K.S. 1989b. Influence of oceanographic and meteorological processes on the recruitment of Pacific halibut, *Hippoglossus stenolepis,* in the Gulf of Alaska, p. 221–237. *In* R.J. Beamish and G.A. McFarlane [ed.] Effects of ocean variability on recruitment and an evaluation of parameters used in stock assessment models. Can. Spec. Publ. Fish. Aquat. Sci. 108.

Parker, K.S., and J.D. Schumacher. 1984. Interannual variations in Pacific halibut abundance and the Kenai Current in the Gulf of Alaska. EOS Trans. Am. Geophys. Union 64: 1080.

Parma, A.M., and R.B. Deriso. 1990. Experimental harvesting of cyclic stocks in the face of alternative recruitment hypotheses. Can. J. Fish. Aquat. Sci. 47: 595–610.

Parma, A.M., and P.J. Sullivan. 1990. Section 2. Stock-recruitment analysis. IPHC stock assessment document II: trends in the fishery, 1989. IPHC, Seattle, WA.

Peterson, R.G. 1988. Comparisons of sea level and bottom pressure measurements at Drake Passage. J. Geophys. Res. 93: 12 439 – 12 448.

Quinn, T.J., II, E.A. Best, L. Bjisterveld, and I.R. McGregor. 1983. Sampling Pacific halibut landings for age composition: history, evaluation, and estimation. IPHC Sci. Rep. 68: 56 p.

Quinn, T.J., II, R.B. Deriso, and S.H. Hoag. 1985. Methods of population assessment of Pacific halibut. IPHC Sci. Rep. 72: 52 p.

Quinn, T.J., II, and I.R. McGregor. 1983. Accuracy of fish weight predicted from otolith measurements. IPHC stock assessment document. I. Trends in the fishery, 1982. IPHC, Seattle, WA.

Rogers, D.E. 1986. *In* The Gulf of Alaska, p. 461. Alaska Office OAD, NOAA, Anchorage, AK.

Rossiter, J.R. 1962. Long-term variations in sea-level, p. 590–610. *In* M.N. Hill [ed.] The sea. Vol 1. Interscience Publishers, John Wiley & Sons, New York, NY. 864 p.

Royer, T.C. 1981. Baroclinic transport in the Gulf of Alaska. Part II. A fresh water driven coastal current. J. Mar. Res. 39: 251–256.

Royer, T.C. 1989. Upper ocean temperature variability in the Northeast Pacific Ocean: is it an indicator of global warming? J. Geophys. Res. 94: 18 175 – 18 183.

Royer, T.C. 1993. High-latitude oceanic variability associated with the 18.6-year nodal tide. J. Geophys. Res. 98: 4639–4644.

Schureman, P. 1941. Manual of harmonic analysis and prediction of tides. Coast Geodetic Survey Spec. Publ. 98. U.S. Department of Commerce, U.S. Government Printing Office, Washington, DC. 317 p.

Southward, A.J., E.I. Butler, and L. Pennycuik. 1975. Recent cyclic changes in climate and in the abundance of marine life. Nature (Lond.) 253: 714–717.

Sullivan, P.J. 1991. Population assessment, 1990. IPHC report of commission activities, 1991. IPHC, Seattle, WA.

Trenberth, K.E., and J.W. Hurrell. 1995. Decadal coupled atmosphere–ocean variations in the North Pacific Ocean, p. 15–24. *In* R.J. Beamish [ed.] Climate change and northern fish populations. Can. Spec. Publ. Fish. Aquat. Sci. 121.

Trupin, A., and J. Wahr. 1990. Spectroscopic analysis of global tide gauge sea level data. Geophys. J. Int. 100: 441–453.

Van Loon, H., and J. Rogers. 1978. The seesaw in winter temperatures between Greenland and northern Europe. Part I. General description. Mon. Weather Rev. 106: 296–310.

Van Winckle, W. , B.L. Kirk, and N.B.W. Rust. 1979. Periodicities in Atlantic coast striped bass (*Morone saxatilis*) commercial fisheries data. J. Fish. Res. Board Can. 36: 54–62.

Walters, C.J. 1981. Optimum escapements in the face of alternative recruitment hypotheses. Can. J. Fish. Aquat. Sci. 38: 678–689.

Walter, C.J. 1984. Managing fisheries under biological uncertainty, p. 263–274. *In* R.M. May [ed.] Exploitation of marine communities. Springer-Verlag Dahlem Konferenzen Series, Berlin, Germany.

Walter, C.J. 1986. Adaptive management of renewable resources. McMillan Publishing Co., New York, NY. 374 p.

Ware, D.M. 1991. Climate, predators and prey: behavior of a linked oscillating system, p. 279–291. *In* T. Kawasaki, S. Tanaka, Y. Toba, and A. Taniguchi [ed.] Long-term variability of pelagic fish populations and their environment. Pergamon Press, Tokyo, Japan.

Wooster, W.S., and A.B. Hollowed. 1995. Decadal-scale variations in the eastern subarctic Pacific: I. Winter ocean conditions, p. 81–85. *In* R.J. Beamith [ed.] Climate change and northern fish populations. Can. Spec. Publ. Fish. Aquat. Sci. 121.

Wyatt, T. 1984. Periodic fluctuations in marine fish populations. Environ. Educ. Inf. 3: 137–162.

Wyatt, T., and A. Vázquez. 1988. The interaction of lunar and solar time in relation to fish abundance, p.181–195. *In* T. Wyatt and M.G. Larrañeta [ed.] Long term changes in marine fish populations. Ediciones Real, Bayona, Spain.

Marine climate off Newfoundland and its influence on salmon (*Salmo salar*) and capelin (*Mallotus villosus*)

S. Narayanan, J. Carscadden, J.B. Dempson, M.F.O'Connell, S. Prinsenberg, D.G. Reddin, and N. Shackell[1]

Narayanan, S., J. Carscadden, J.B. Dempson, M.F. O'Connell, S. Prinsenberg, D.G. Reddin, and N. Shackell. 1995. Marine climate off Newfoundland and its influence on Atlantic salmon (*Salmo salar*) and capelin (*Mallotus villosus*), p. 461–474. *In* R.J. Beamish [ed.] Climate change and northern fish populations. Can. Spec. Publ. Fish. Aquat. Sci. 121.

Abstract: The climatic variability affecting ice cover and oceanographic conditions during 1980–92 is examined to compare conditions during anomalous years and to derive indices for environmental influences on Atlantic salmon (*Salmo salar*) and capelin (*Mallotus villosus*) on the continental shelf off Newfoundland and Labrador. The area of the cold intermediate layer (CIL), the mean thickness of the surface layer above the CIL, the day of the year when sea-surface temperature reached 4°C at Station 27, and the thickness of the surface layer at station 27 were the indices considered. Significant correlations were found between salmon migration time and the day of the year when SST reached 4°C at Station 27, between indices of salmon abundance and CIL area on the shelf, and between capelin spawning time and the upper layer temperatures in the spring. Furthermore, there is some evidence to suggest that salmon may modify its distribution in the inshore environment in response to oceanographic conditions. Capelin may also modify their distribution, moving in and out of the survey block in response to climatic variability.

Résumé : On examine la variabilité climatique qui a eu une influence sur la couverture de glace et sur les conditions océanographiques durant la période 1980–1992 pour comparer les conditions durant les années anormales et pour trouver des indices d'influences environnementales sur le saumon (*Salmo salar*) et le capelan (*Mallotus villosus*) du plateau continental de Terre-Neuve et du Labrador. Les indices considérés ont été la superficie de la couche intermédiaire froide (CIF), l'épaisseur moyenne de la couche de surface au-dessus de la CIF, le jour de l'année où la température de la surface de la mer (TSM) a atteint 4° C à la Station 27 et l'épaisseur de la couche superficielle à la Station 27. Des corrélations significatives ont été trouvées entre la période de migration du saumon et le jour de l'année où la TSM a atteint 4° C à la Station 27, entre les indices de l'abondance du saumon et la superficie de la CIF au-dessus du plateau continental, et entre la période de frai du capelan et les températures de la couche supérieure au printemps. En outre, certaines données portent à croire que le saumon peut modifier sa distribution dans le milieu côtier en réponse aux conditions océanographiques. Le capelan peut également modifier son aire de répartition de sorte qu'il peut entrer dans le secteur de recherche et en sortir en réponse à la variabilité climatique.

Introduction

In contrast to the general global trend of above-normal air temperatures, northeastern Canada, particularly Labrador and Newfoundland coastal regions, experienced below-normal air temperatures during most of the 1980s and 1990s (Drinkwater et al. 1992). An examination of data from individual years, however, shows that the 1980s started with positive anomalies before changing to negative anomalies in 1983. The cold period continued until 1992 with only a minor warming in 1987–88. The ice-cover and oceanographic conditions on the continental shelf off Newfoundland and Labrador also exhibited similar trends. The ice cover was more extensive and persisted for longer periods while water temperatures were below normal at all depths (Drinkwater et al. 1992; Narayanan et al. 1992).

Concurrent with these environmental changes, many fish stocks in the Northwest Atlantic are declining (Atkinson 1993). Furthermore, significant variability in the distribution and availability of various fish species have occurred as well. This unexpected coincidence of the decline in fish abundances with climatic perturbations provided the impetus for studies of environmental influences on biological processes. The

S. Narayanan, J. Carscadden, J.B. Dempson, M.F. O'Connell, D.G. Reddin, and N. Shackell. Department of Fisheries and Oceans, Northwest Atlantic Fisheries Centre, P.O. Box 5667, St. John's, NF A1C 5X1, Canada.

S. Prinsenberg. Department of Fisheries and Oceans, Bedford Institute of Oceanography, P.O. Box 1006, Dartmouth, NS B2Y 4A2, Canada.

[1] Authorship after the first author is assigned alphabetically.

purposes of this paper are to examine aspects of marine climate, and the interaction between the ocean environment and biological characteristics of two commercial species, Atlantic salmon (*Salmo salar*) and capelin (*Mallotus villosus*). Both species are pelagic and migratory, winter offshore, and move inshore for spawning during the spring and early summer months.

We confine ourselves to the area of the continental shelf from Hamilton Bank to Grand Bank (Fig. 1) because of the extensive spatial and temporal data coverage of the environment and these two fish species. We focus on aspects of marine environment pertinent to testing three hypotheses: (1) the timing of the onset and the rate of surface warming have a significant influence on the migration and spawning time of the two species; (2) the volume of the subzero water mass on the shelf influences the abundance of the two species; and (3) the temperature structure of the shelf (capelin) and adjacent waters (salmon) affects fish distribution. Because ice-cover affects the oceanographic conditions of the shelf, we review the aspects of ice dynamics that affect the temperature field over the shelf, and the onset and rate of surface warming. We examine in detail conditions that prevailed during selected anomalous years. We then focus on the oceanographic climate using water temperature as an indicator and examine the seasonal and interannual variability of the selected indices. Finally we attempt to relate how the climatic perturbations might have affected the spawning time, migration, abundance, and distribution of salmon and capelin.

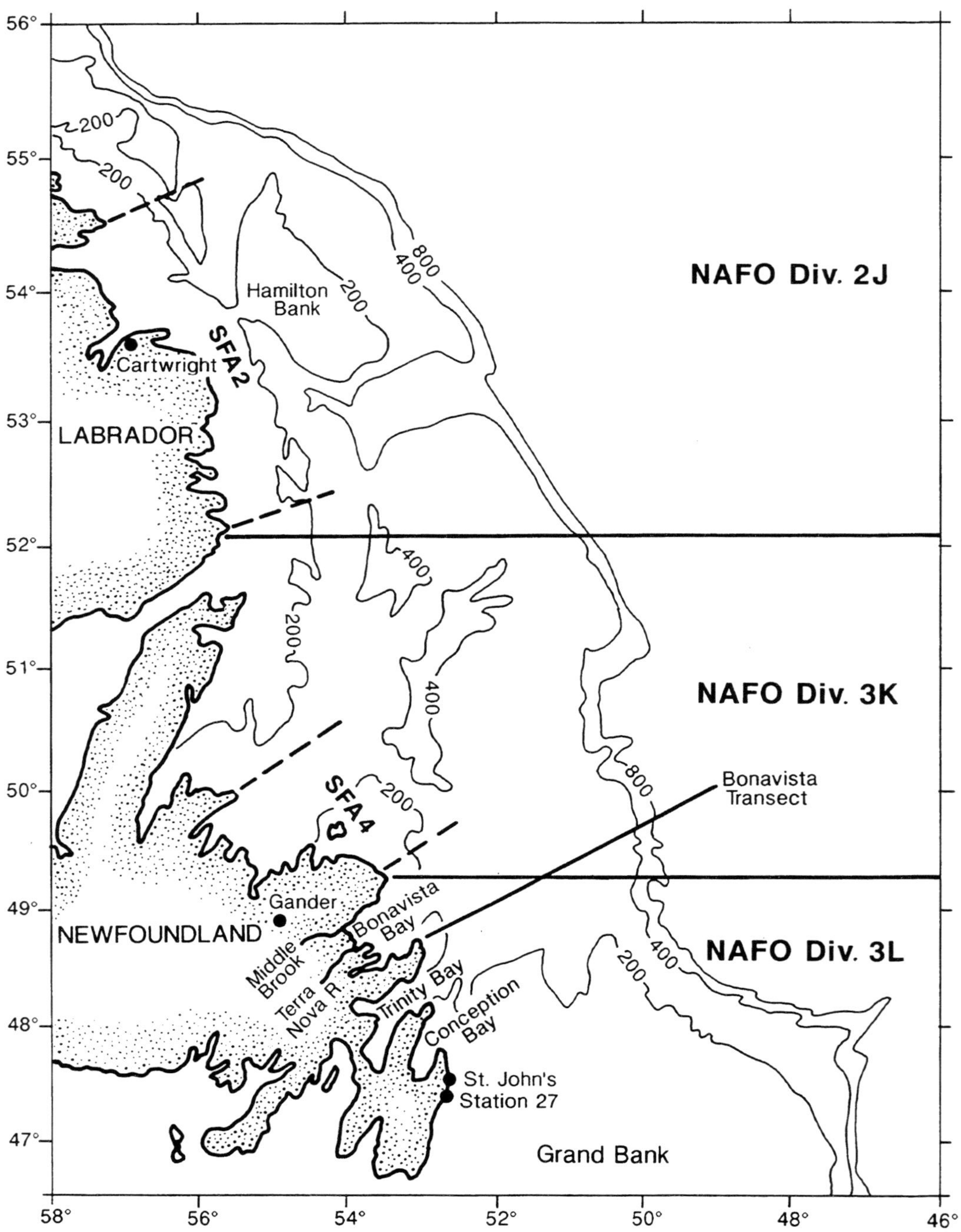

Fig. 1. Area map showing the NAFO Divisions, Salmon Fishing Areas, and oceanographic stations.

Ice dynamics and the atmosphere

Ice dynamics affect the oceanographic conditions on the shelf in several ways. Ice cover affects the air–sea interaction by insulating the ocean and by increasing the dynamic coupling between wind-stress and the ocean. Because an ice-covered area is dynamically rougher than an ice-free area, the increase in wind-stress coupling of an ice-covered area relative to an ice-free area causes a secondary circulation that includes vertical upwelling at the ice edge and a shoreward surface circulation. In addition, ice cover reduces the surface mixing by dampening the wind-generated surface waves.

Ice cover also affects ocean water properties directly. Salt rejection during ice growth enhances vertical mixing through buoyancy flux and thus has a significant effect on the upper layers (growth of 1.5 m of ice increases the salinity of a 40-m surface layer by 1 psu). When ice is continually advected away from an area, the ice production and the salinity in the surface layer are even larger. Mixed layers with constant salinity values and temperatures near the freezing point could extend to depths up to 200 m along the Labrador Shelf in midwinter. Because this water is transported southwards to the Newfoundland Shelf, greater ice production in the north leads to more cold water in the south at the start of the ice-melt cycle.

During the ice melt, fresh meltwater reduces the upper-layer salinities, thus stabilizing the water column. Furthermore, the proportion of incident solar radiation available for heating the oceanic surface layers is reduced by ice-cover reflection and by ice melt (the heat required to melt the 1.5 m thick ice cover would raise the temperature of a 60-m ocean layer by 2.0°C). A delay in ice retreat of 1 mo (for example) would eliminate the approximate 150 $W \cdot m^{-2}$ atmospheric heat flux into the water, which is equivalent to the heat required to raise the temperature of a 60 m ocean layer by 1.5°C (1 $W \cdot m^{-2} = 2.06\ cal \cdot cm^{-2} \cdot d^{-1}$).

Thus an increase in ice-cover extent in both space and time reduces the atmospheric heat flux into the ocean, which further delays the ice retreat from the area and further reduces the amount of heat left to warm the surface waters to their expected summer values. Consequently, variability in ice conditions has a significant effect on the oceanographic conditions of the shelf. In the following sections we examine the ice-cover data from 1963 to 1992 to identify anomalous years and to compare the two most recent cold years with each other and with a warm year. Based on estimated relationships between ice-cover characteristics and meteorological variables, we discuss the effects the prevailing winds and air temperatures during these years might have had on ice growth, ice drift, and ice-cover duration.

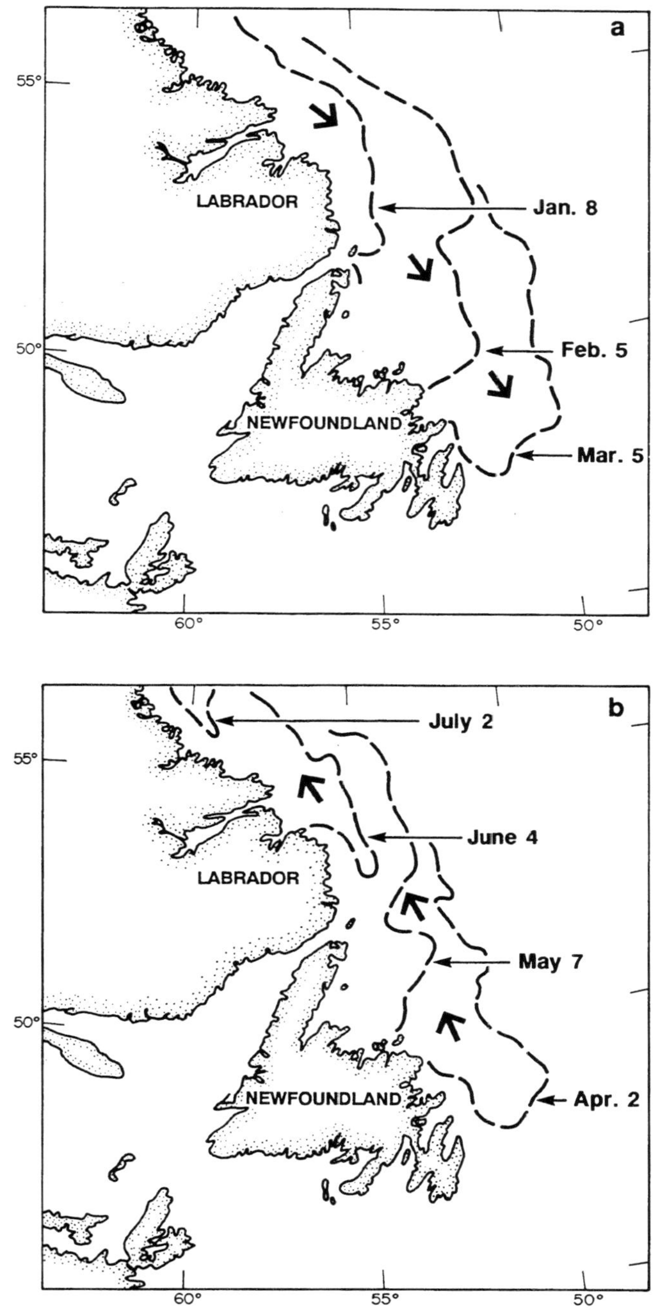

Fig. 2. Median ice cover extent for selected months (ice centre charts). (*a*) Ice advance. (*b*) Ice retreat.

Average ice conditions

The seasonal pack ice over the Newfoundland Shelf consists of locally grown ice, which starts to appear in late December after several weeks of below-zero air temperatures, and thicker and older ice originating from northern regions and transported into the area by the end of February by winds and currents. The volume flux associated with the advected ice determines the severity of the ice season and the southernmost extent of the ice, because this thick (greater than 1.5 m) and heavily rafted pack ice requires the longest time to decay. As Fig. 2 shows, the southern ice edge generally advances southwards until mid-March before retreating north.

Ice-cover severity on the shelf is influenced by air temperature and wind conditions. Cold arctic air not only causes ice growth along the Labrador and Newfoundland coasts, it is also a major contributor to the southward drift of pack ice along the coast. Peterson and Prinsenberg (1990) estimated that the ice drifts at 2.5% of the surface (10 m) wind speed in a direction 35° to the right of the wind. Using 25-year (1955–1980) average winds from Cartwright and Gander, we calculated the wind-induced mean ice-drift for the winter and spring months for Labrador and Newfoundland (Fig. 3).

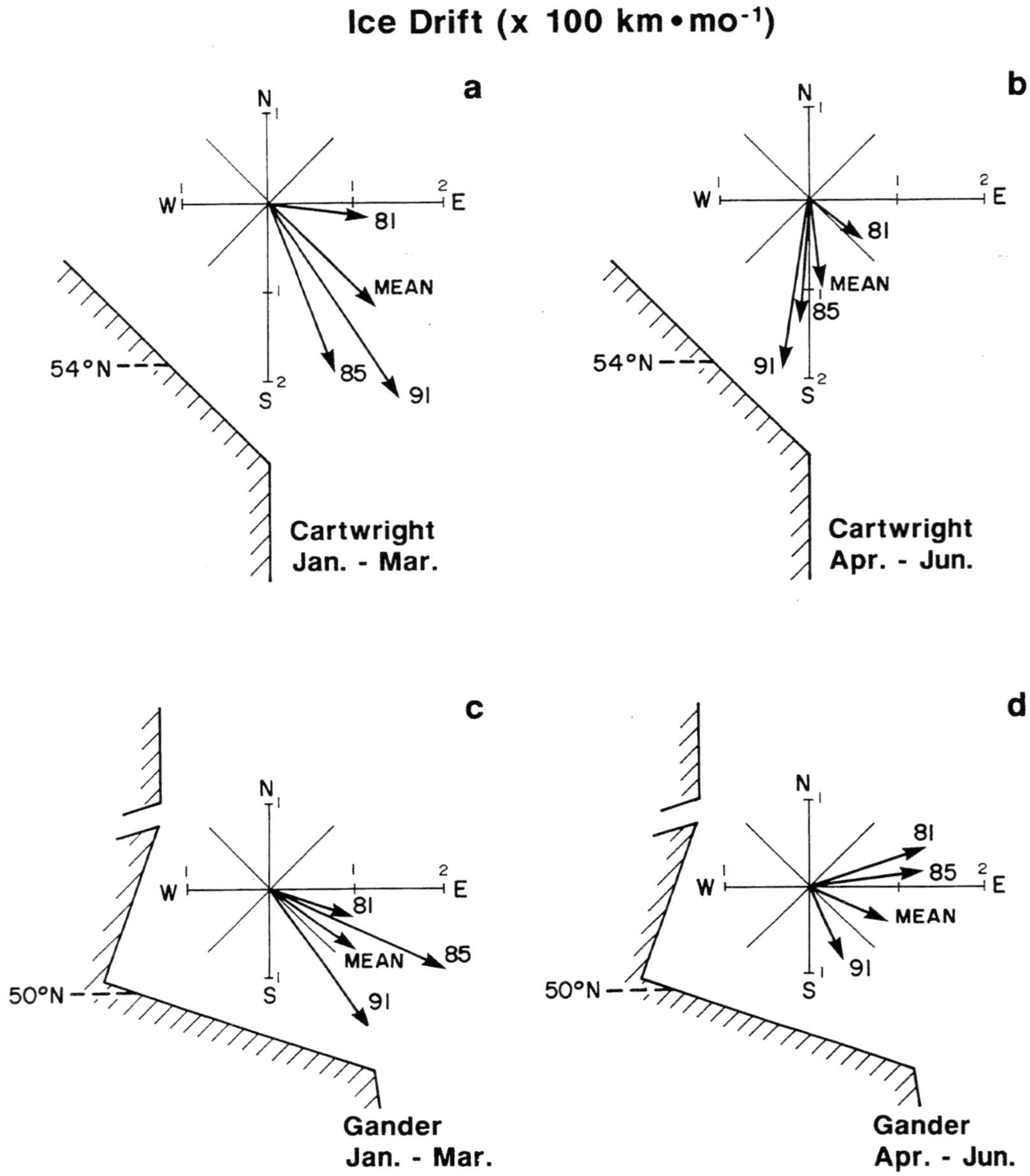

Fig. 3. Mean ice-drift vectors for 1981, 1985, and 1991. (*a*) Cartwright, January–March, (*b*) Cartwright, April–June, (*c*) Gander, January–March, (*d*) Gander, April–June.

Interannual variability in air temperature and ice cover

The ice-cover extent has noticeable interannual variability (Fig. 4) with the heavy-ice years coming in groups separated by light-ice periods. During the last 30 yr, there were three periods when ice-cover extent was above normal: 1972–1975, 1984–1986, and 1989 onwards. This variability in ice-cover extent and similar in-phase variability in air temperature (Petrie et al. 1992) introduce a similar variability in the ocean environment, which in turn affect the fish habitat. Before we examine how salmon and capelin respond to cold and warm conditions, we examine the conditions during the two most recent heavy-ice periods (1991 and 1985) and compare them with those during a light-ice period (1981).

1981 (warm, light-ice year)

Air temperature anomalies for 1980–81 (data from Cartwright is shown in Fig. 5 as an example) show that, except for December, warmer than normal temperatures persisted along the Canadian east coast. Based on a freezing degree-day model (Prinsenberg and Peterson 1992) the warmer conditions reduced ice growth by 19%. Furthermore, winter winds off the southern Labrador coast had a much larger southwest component, in addition to the prevailing northwest component (Fig. 3). The southwest wind component pushed the heavier pack ice offshore where it melted in the warmer surface water of the Labrador Sea. Although this ice was replaced by locally grown ice, this ice was thinner. The net result is a reduction in the ice volume transported to the Newfoundland Shelf. In

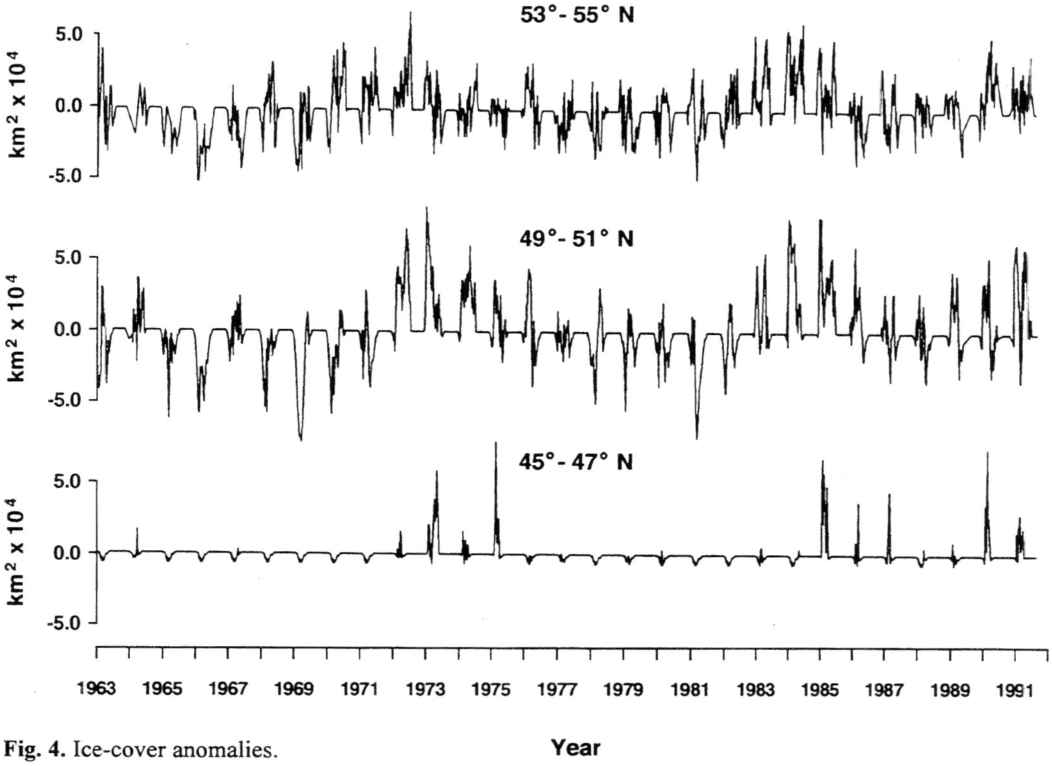

Fig. 4. Ice-cover anomalies.

addition, winds off Newfoundland were weaker and directed more offshore than normal during this period, causing less of the ice to move south and a larger ice loss because of the offshore wind component (Fig. 3).

During April, May, and June, when the southern ice edge retreats northward, air temperatures for southern Labrador and Newfoundland coastal regions were above normal (Fig. 5). The warmer temperatures enhanced the ice melt of the already thinner than normal ice. The larger offshore wind component during this period also increased the loss of ice from the region.

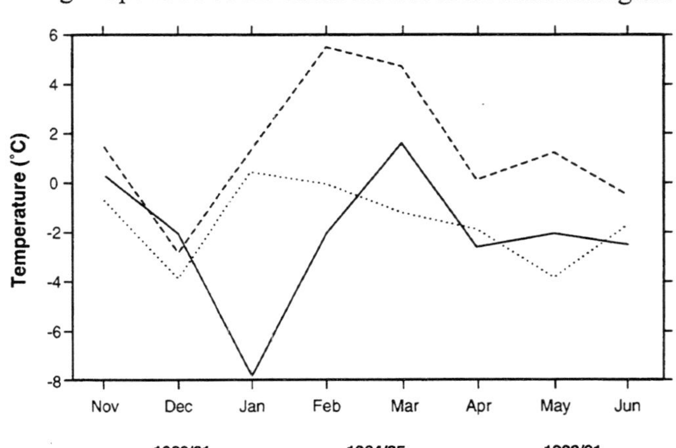

Fig. 5. Air temperature anomalies at Cartwright in 1981, 1985, and 1991.

Thus, compared with normal, pack ice in 1981 arrived later and left earlier, and the thickness and volume of ice melted on the southern Labrador and Newfoundland shelves were less.

1985 (cold, heavy-ice year)

The second heaviest ice year was the winter of 1984–85. Except for a warm spell in the north during January (Fig. 5), air temperatures were below normal for the entire winter and spring over Labrador and Newfoundland, which increased ice growth locally and decreased ice melt rates at the ice edge.

Winds throughout the ice-edge advance period were stronger than normal. Along the Labrador coast, the estimated ice drift was 25% larger than normal and had a smaller offshore component, reducing ice loss through melting in the warm Labrador Sea (Fig. 3). Over the Newfoundland Shelf, the winds during the winter months were almost 100% stronger than normal, pushing ice faster southeastwards. However, the offshore component on the Newfoundland Shelf over the total period was also stronger, indicating an increase in ice loss to offshore waters. The winds in the spring continued to show the same trend as in the winter in both areas: stronger in both regions, with a reduced offshore component on Labrador Shelf and a higher than normal offshore component on Newfoundland Shelf.

Thus in 1985, the ice volume was larger and pack ice was thicker than normal, and more ice drifted onto Newfoundland Shelf. Consequently, the ice cover appeared earlier, was larger in extent, and, in spite of the offshore ice-drift component, retreated later.

1991 (cold, Heavy-ice year)

Owing to the combined effect of stronger and more frequent onshore winds and below-normal air temperatures (Fig. 3 and 5) 1991 was the worst ice year in 30 yr. Below-normal air temperatures were first observed in the fall of 1990 in Baffin Bay, but extended over the entire Canadian east coast for the rest of the winter. The freezing degree-day model estimates that the temperature component of the ice growth was approximately 10% above normal for the January–March period; the wind component of the ice flux was also well above normal (Fig. 3). In March, a major storm with onshore winds caused severe rafting and increased the volume of ice transported onto the Newfoundland Shelf. The wind-induced drifts in 1991 were larger than normal by 37% for the winter off Newfoundland, and 55 and 90% larger for the winter and spring periods off southern Labrador (Narayanan et al. 1992).

The below-normal air temperatures and less-frequent offshore winds extended the ice season by delaying ice melt and reducing the ice loss by offshore ice flux. The net effect was that the pack ice arrived several weeks ahead of normal and retreated more slowly, extending the ice season by over 1 mo. Furthermore, it is believed that most of the ice transported onto the Newfoundland Shelf melted on the shelf itself thus reducing the proportion of solar radiation available for ocean surface layers.

A comparison of 1991 conditions with those in 1985 shows that air temperatures in 1991 were considerably lower than those in 1985, and winds were stronger and had a smaller offshore component. Furthermore, the differences in ice-drift patterns in March (onshore in 1991 and offshore in 1985) had a significant effect on the persistence and volume of ice on the Newfoundland Shelf. Consequently, 1991 was a heavier ice year than 1985.

Ocean Climate

A dominant oceanographic feature in the area is the temperature front located at the shelf break, separating the colder and fresher shelf waters from the warmer and saltier Labrador Sea waters. During the summer months, a subsurface layer of cold water exists inshore of this front; this layer is commonly referred to as the cold intermediate layer (CIL) whose boundaries are defined by the 0°C isotherm. During the winter months, the CIL extends to the surface and to the deeper layers as a result of surface cooling and storm mixing. In the fall, it shrinks to a minimum because of the combined effects of heat diffusion from the surface and advection (Fig. 6).

The summer CIL area along a transect across the shelf is an index of the combined effect of the severity of the winter and the rate of surface warming in the spring. Petrie et al. (1988) have shown that CIL areas along cross-shelf transects (Bonavista (Fig. 1), 47°N, southern Labrador, and White Bay) are significantly correlated with the monthly anomalies of ice-cover extent south of 55°N. This indicates that the low-frequency variability in the CIL area has a very large spatial scale and that CIL area is a robust index representing the variability in the areal extent of the ice cover. In this paper, we use the time series of CIL area along the Bonavista transect (Fig. 7*a*) collected during each summer (July–August).

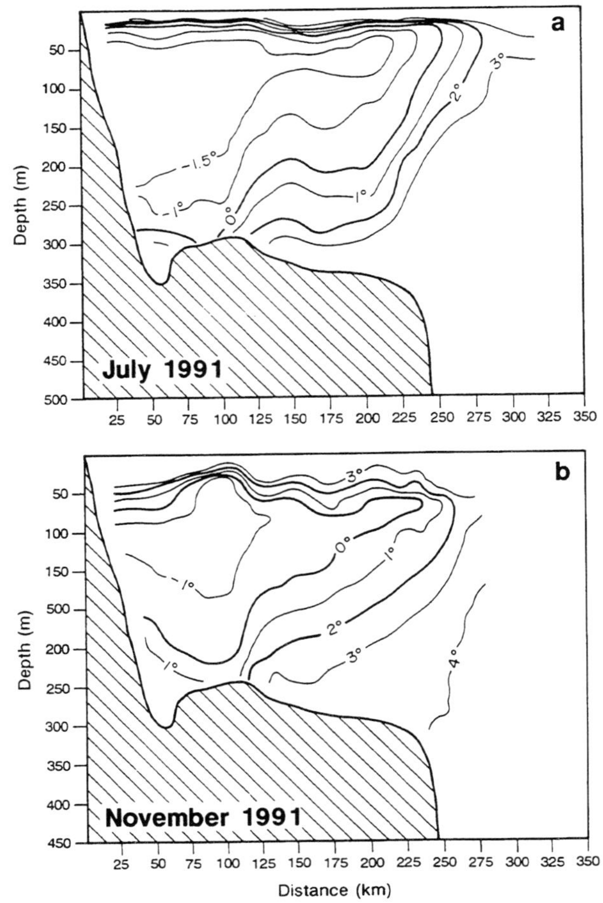

Fig. 6. Temperature sections along Bonavista transect in (*a*) summer and (*b*) fall.

Variability in CIL area may be due in part to increased onshore flux of warmer slope water along the bottom (Labrador Sea forcing) or to surface atmospheric forcing. To assess how much of the CIL variability is due to surface forcing, a mean surface layer (SL) thickness was estimated as the area above the CIL divided by the maximum width of the CIL (Fig. 7*a*).

The only long time series available for examining the interannual variability in the onset and rate of surface warming is the temperature profiles from Station 27 (47°32.8 N, 52°35.2 W; Fig. 1), a long-term monitoring station off St. John's. Petrie et al. (1991) have shown that the vertical structures of the annual and semiannual cycles at Station 27 represent the general characteristics at other locations on the shelf, and that temperature amplitude decreases with latitude but its phase remains approximately uniform over the region. Furthermore, the Station 27 temperature anomalies are significantly correlated with ice cover and CIL area.

The day of the year when sea-surface temperature (SST) reaches 4°C at Station 27 was selected as the index of the onset of surface warming (Fig. 7b). The 4°C was chosen because of

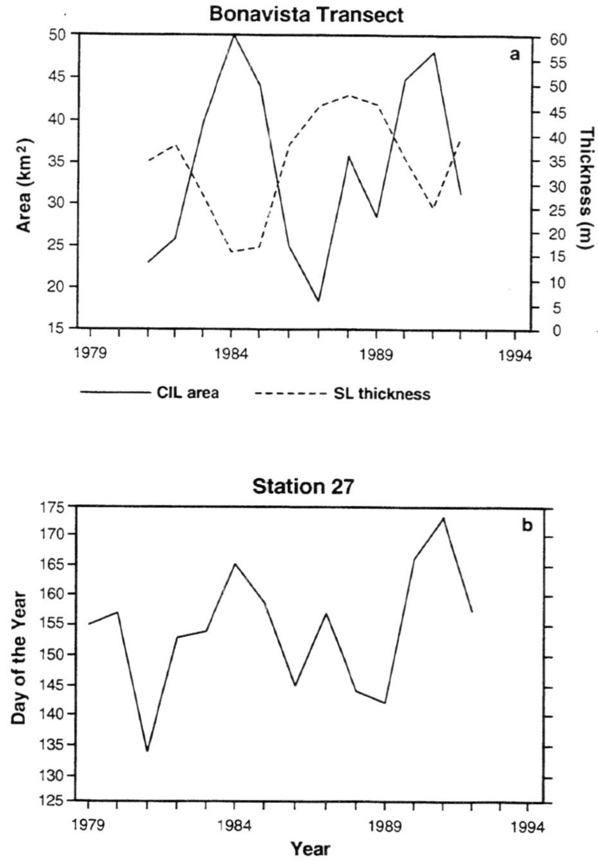

Fig. 7. (*a*) CIL area and mean SL thickness along Bonavista transect for 1979 to 1992. (*b*) Day of the year when SST reached 4°C at Station 27.

the known preference of salmon for areas with temperature 4–10°C (Reddin 198; Saunders 1986; Reddin and Friedland 1993). To compare the differences in the rate of surface warming between years, the thickness of the surface layer above the CIL was also calculated from each profile.

Interannual variability in water temperature

As in the case of ice cover, the temperature time series shows significant interannual variability. Petrie et al. (1992) found that the interannual variability in temperature exists at all depths, and for depths greater than 100 m, is exceeded or matched only by the annual component.

Both the transect data and the Station 27 data clearly indicate the two cold cycles in the 1980–92 period separated by the short warm cycle in 1986–87 (Fig. 7). In terms of CIL area, 1984 ranked first overall for the Bonavista transect, and 1991 ranked first (second overall) during the second negative anomaly cycle. Thus, for the first cycle, the heavy-ice year followed the CIL area maximum, whereas for the second, the two were concurrent. The CIL area in 1985 was also well above normal, and almost equal to that in 1990. In comparison, the 1981 CIL area was less than half of those in the two cold years. The SL thickness more or less followed the same trend: 1984 had the lowest value, followed by 1985 and 1991, and 1981 had a high value. Some of these differences in SL thickness are due to shifts in the dates of the survey period (the occupation date has a range of about 20 d) and hence does not totally reflect the interannual variability in surface warming. However, the SL thickness was found to be significantly correlated (inversely) with CIL area with an $r^2 = 0.52$ and $p < 0.01$.

The major difference between the two cold years, as highlighted in Fig. 7*b*, is the delayed surface warming. In 1991, the day of the year when SST reached 4°C was 19 d later than the 1979–92 average (154th day of the year), and 39 d later than in 1981; in 1984, the delay was only 11 d from the average.

The rate of surface warming and the maximum thickness of the surface layer (Fig. 8) also show significant differences among years. In 1981 at Station 27, the entire water column was at subzero temperatures for only a short period, from the end of January to the first week of April, when surface warming began. After the initial warming phase, when the thickness reached approximately 40 m, the rate of warming was slow until the last week of July. From then on, the thickness of the surface layer at Station 27 increased at a rate of 0.7 $m \cdot d^{-1}$, and reached a maximum value of about 125 m. In 1991, on the other hand, surface warming began only by the end of May and the rate was very slow (about half of the 1981 rate) until the beginning of September; the maximum SL thickness was only about 115 m. The second heaviest ice year, 1985, was similar to 1991 in that the surface warming, which began in early May, continued at a slow rate after the initial period until August; after that the rate increased to 0.6 $m \cdot d^{-1}$. In contrast, 1984 surface warming was delayed as in 1991, but once the warming began, it continued at a fast rate of 0.6 $m.d^{-1}$ until the maximum SL thickness of slightly over 130 m was reached.

The phase of the annual harmonic computed by Petrie et al. (1991) increased linearly with depth at an approximate rate of 1 $m.d^{-1}$. However, significant variability in the time of onset of surface warming, rate of warming, and maximum thickness of SL could occur in response to variability in ice cover and in the meteorological conditions.

As discussed in the previous section, the more ice production in the north the more cold water there is in the south at the start of the ice-melt cycle, and the same is true when more ice melts on the shelf. In 1991, both the ice production and the volume of ice melted on the shelf were considerably more than normal. Consequently, the southward extent of the subzero temperature water mass on the Grand Bank was considerably more in 1991 (Narayanan et al. 1992). An analysis of variance of average water temperature at 50 m (averaged over a longitudinal band on the shelf excluding the slope region), with year and latitude as factors, shows that both factors, and to a much lesser degree their interaction, were significant. More importantly, the year 1991 was significantly different from all other years (post-hoc contrast using interaction mean square as error term: $F = 30.543$, $p < 0.001$, df = 1,35) (N. Shackell, unpublished data).

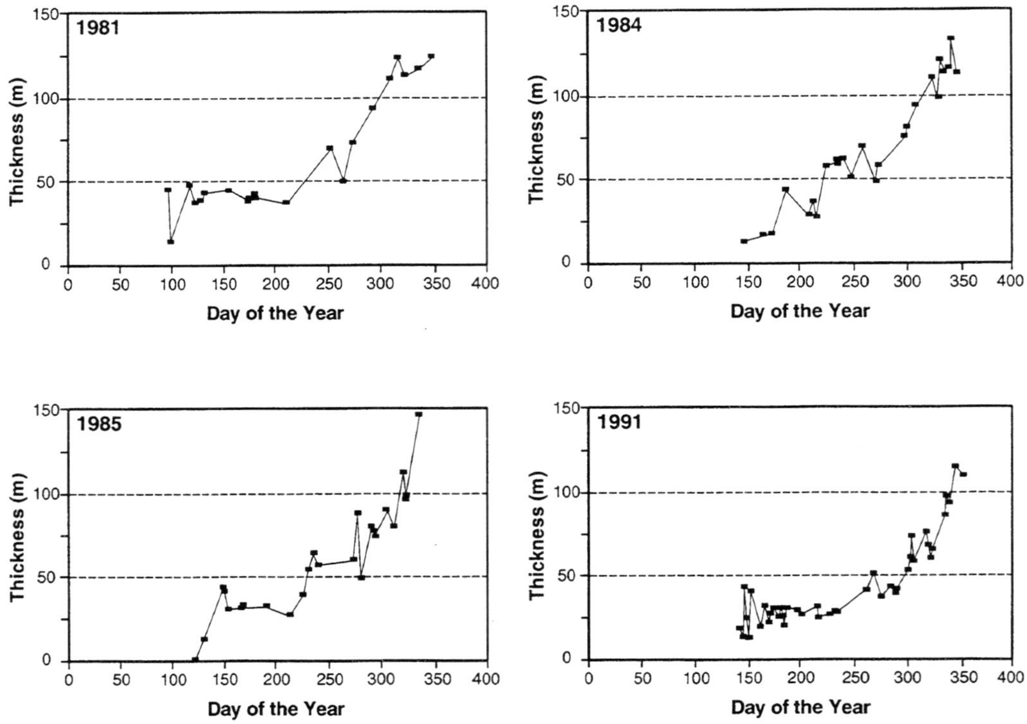

Fig. 8. SL thickness at Station 27 for 1981, 1984, 1985, and 1991.

Effect of climatic perturbations on salmon

The Atlantic salmon is an anadromous species, with freshwater residence of up to 8 yr (Randall and Schofield 1987) and marine life up to 4 yr (Saunders and Schom 1986). While feeding at sea, it lives in the upper layers of the Northwest Atlantic from the southern edge of the Grand Bank to the Davis Strait and the west coast of Greenland (Mills 1989). In spring, maturing one-sea-winter (1SW) and multiple-sea-winter (MSW) salmon return to their natural streams to spawn and are frequently caught in coastal fisheries off Newfoundland, Labrador, and Greenland. Salmon is an important recreational species of high economic value and a highly prized food item for humans.

Declines in Atlantic salmon stocks in the early 1980s led Canada to implement a 5-yr management plan beginning in 1984 (O'Connell et al. 1992*a*). In spite of this plan, directed at conserving MSW salmon stocks, catches of Atlantic salmon have continued to decline in Canada, as well as throughout most of the species' range in the north Atlantic (Working Group on North Atlantic Salmon 1992). Because these declines have been greater than could be accounted for from known sources, i.e., management changes to reduce commercial exploitation, and because they are spread widely over many stocks in many countries, the declines have been attributed to some unknown effect at sea, perhaps environmental in origin (Ritter 1989; Working Group on North Atlantic Salmon 1992; Friedland and Reddin 1993; Reddin and Friedland 1993). This lack of an explanation has been viewed with great concern by salmon resource interests and demonstrates the need for a better understanding of what is happening to Atlantic salmon in the sea (Nickson 1991).

SST is known to influence salmon abundance and distribution as estimated from research cruise and salmon catch rates (Reddin and Friedland 1993). In spite of known relationships between salmon and the marine environment, ocean climate has not been incorporated as a variable in salmon assessments nor has it been used as a covariate to predict salmon returns. This section examines information pertaining to the influence of marine climate on the migration timing and abundance of salmon from stocks spawning in rivers along the east coast of Newfoundland and Labrador.

Data used for these analyses are recreational and commercial catch data of small and large salmon, and counts of returning adult salmon from fishways or counting fences (O'Connell et al. 1992*b*). Commercial fishery statistics classify salmon as small (~2 kg) or large (~4 kg) on the basis of landed weight converted to whole weight. Typically, small salmon are 1SW in age (Allan and Ritter 1977), some are immature; large salmon are a mixture of virgin MSW, 1SW, and previous spawners (D. G. Reddin, unpublished data). In the recreational fishery and at counting facilities, salmon were classified as small (<63 cm) or large (>63 cm) according to length.

Catch information by week or day of the year was used to calculate Mundy's migration statistic, which is simply the mean date weighted to the catch in each time unit (Mundy 1982). Analyses were concentrated on Salmon Fishing Areas (SFAs) 2 and 4 (Fig. 1) because they both have a high percentage of local-origin salmon in their catches and because they contribute the largest portion of the catch over all SFAs. Note that both the mean arrival time and index of abundance are generally correlated among SFAs (large salmon had 16 out of 28 SFA comparisons with $p < 0.05$, and small salmon had 18 out of 28 with $p < 0.05$ for the mean arrival time, SFAs 1–8).

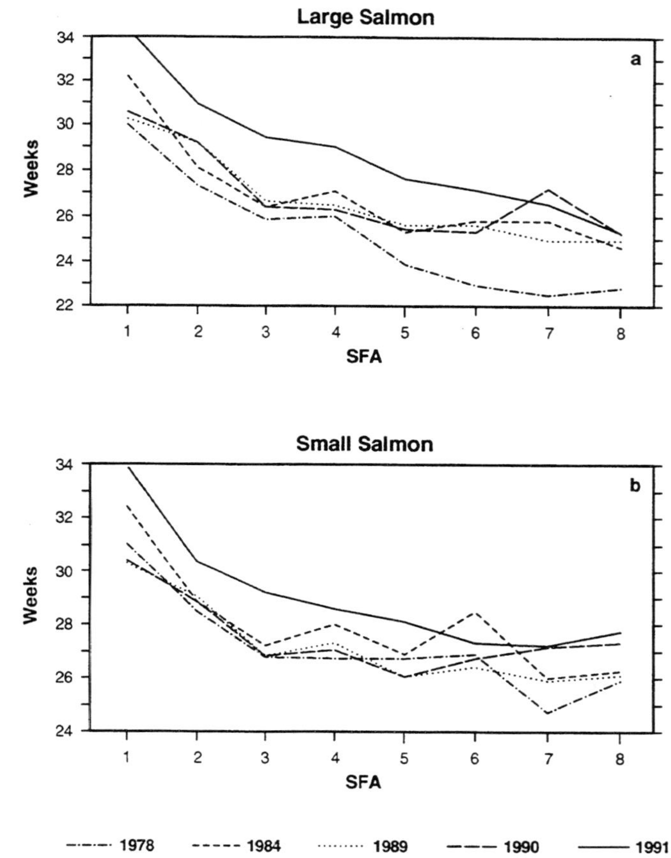

Fig. 9. Arrival times of (*a*) large and (*b*) small salmon at SFAs 1–8 from commercial fishery for selected years.

Salmon migration timing

The results presented here are based on a comparison of:

1) Mean arrival times in standard weeks (Working Group on North Atlantic Salmon 1992) calculated from weekly catches (by weight) of small and large salmon in the Newfoundland–Labrador commercial fishery in SFAs 2 and 4 (Fig. 1) with the day of the year when SST = 4°C at Station 27; and
2) Mean arrival times (Mundy 1982) calculated from daily counts of small salmon at enumeration facilities at Middle and Terra Nova rivers (Fig. 1) with the day of the year when SST = 4°C at Station 27.

Information on arrival time calculated from commercial catches (by weight) of small and large salmon for SFAs 1–8 for both small and large salmon indicates that salmon tend to arrive along the coast earlier in southern areas than in the north. Furthermore, it appears that the arrival time in any given SFA

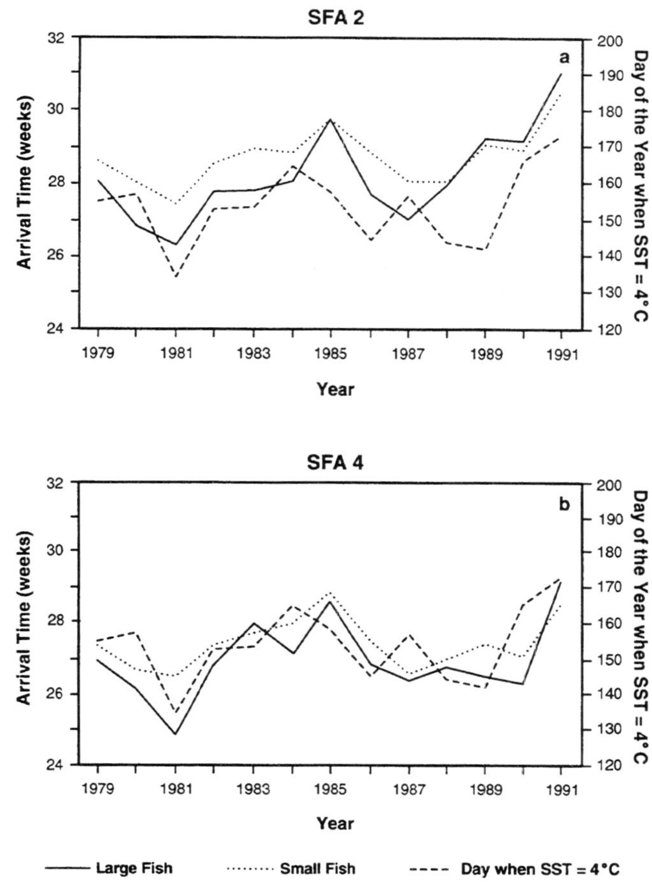

Fig. 10. Relationship between salmon arrival time at (*a*) SFA 2 and (*b*) SFA 4 from commercial fishery and surface warming for 1979–91.

varies among years (earlier in warmer years and later in cold years), the latest for almost all regions being in 1991. For both small and large salmon, catches in 1991 are generally the latest (Fig. 9) and lowest recorded.

The relationships between arrival time calculated from commercial catches for both small and large salmon in SFAs 2 and 4 (Fig. 10) are correlated with the day of the year when

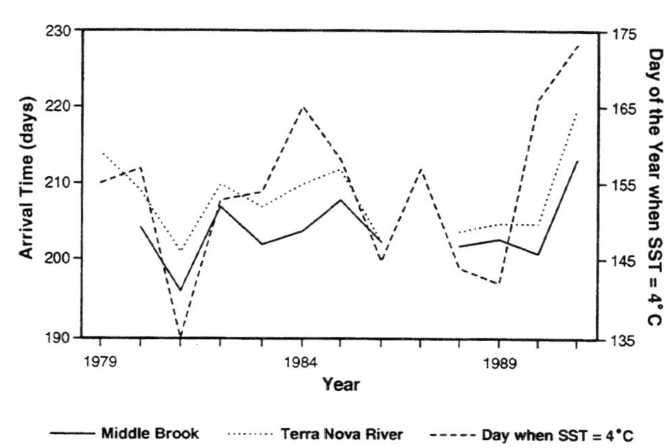

Fig. 11. Salmon arrival time at river fences in Newfoundland versus surface warming.

SST = 4°C at Station 27 (SFA 2, small: $r^2 = 0.39$, $p < 0.05$; SFA 2, large: $r^2 = 0.33$, $p < 0.05$; SFA 4, small: $r^2 = 0.27$, $p < 0.10$; SFA 4, large: $r^2 = 0.37$, $p < 0.05$). This suggests that salmon arrival time and distribution along the east coast of Newfoundland and Labrador is at least partially influenced by ocean climate. In years with warmer coastal sea temperatures salmon arrived earlier.

The relationships between arrival time calculated from counts at fishways on Middle Brook and Terra Nova River and the day of the year when SST at Station 27 first exceeded 4°C are also significantly correlated (Fig. 11). The model explained 37% of the variance in run timing for Middle Brook ($p < 0.05$) and 58% ($p < 0.01$) for Terra Nova River. This supports the findings from the commercial fishery regarding the climatic influence on salmon arrival timing in Newfoundland rivers.

Salmon abundance and distribution

The oceanographic influence on salmon abundance was examined using an index of abundance, calculated as the sum of the catch of small and large salmon in SFAs 2 and 4 during the first 2 wk (SFA 4) and 3 wk (SFA 2) of fishing in the commercial fishery standardized to a June 5 starting date. Catches in the first 2 and 3 wk of fishing were used for a standard because of changes that have occurred in the way in which the commercial and recreational fisheries for salmon are managed in Newfoundland and Labrador (O'Connell et al. 1992*a*). This index of abundance was then compared with CIL area along the Bonavista transect in July.

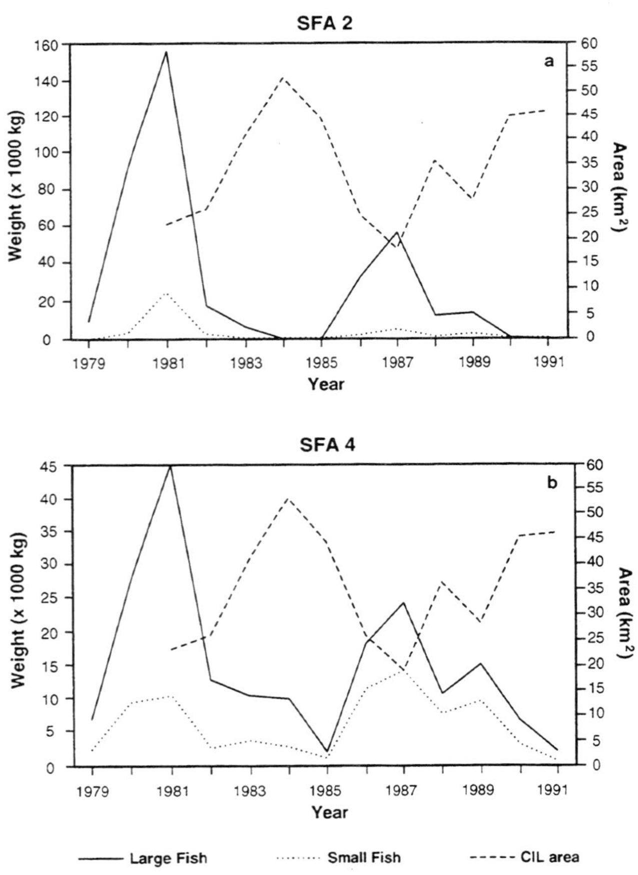

Fig. 12. Relationship between commercial salmon catches and CIL area at (*a*) SFA 2 and (*b*) SFA 4.

For commercial catches of large salmon in SFA 2, abundance was significantly related to CIL area ($r^2 = 0.40$, $p < 0.05$), but no relationship existed for small salmon ($r^2 = 0.22$, $p > 0.05$). In SFA 4, both small and large salmon abundance were correlated to CIL area (small: $r^2 = 0.60$, $p < 0.01$; large: $r^2 = 0.48$, $p < 0.05$). These correlations to test the hypothesis that marine climate influences the abundance of returning salmon indicated significant negative relationships in three out of four cases tested (Fig. 12). Correlations between the catches and the day of the year when SST reached 4°C were lower, but still significant, for the large salmon in both SFA 2 and 4 and for small salmon in SFA 4. This suggests that in years when the CIL area was larger and marine climate was colder, there were fewer salmon returning to coastal waters.

There are regional differences in the catches as well. Fig. 12 shows that, during cold years, salmon catches were almost nil in SFA 2, whereas in SFA 4, the decline was less dramatic. This may be due to a redistribution of salmon in response to adverse oceanographic conditions inshore.

Effect of climatic perturbations on capelin

Capelin is a small pelagic, schooling species circumpolar in distribution in the northern hemisphere. It is important as a prey species throughout its distribution and as a commercial species in the Barents Sea, Iceland, and the northwest Atlantic. Capelin stocks in the Northwest Atlantic have been managed throughout the 1980s using a conservative management strategy, specifically at an exploitation rate that does not exceed 10% of the projected biomass. In fact, the demand for capelin has been far less than the total allowable catch, and consequently, the landings between 1982 and 1989 averaged only 4.3% of the estimated biomass. Thus, it is probably safe to conclude that fishing has had little impact on the population dynamics of the capelin stocks in the Northwest Atlantic and that environmental perturbations have been the primary determinant in stock fluctuations.

While there have been extensive studies on the environmental factors affecting spawning and the survival of eggs and larvae (e.g., Frank and Leggett 1981*a*, *b*; Frank and Carscadden 1989), climatic influences on the timing of spawning or on population abundance (e.g., Leggett et al. 1984) had not been examined. The unusually late arrival of capelin on the beaches of Newfoundland in 1991 and the difficulty in finding capelin offshore during the spring and fall surveys point to the possibility that the oceanographic conditions may have a significant influence on capelin distribution and migration.

Timing of capelin spawning

Most spawning on the Newfoundland and Labrador beaches occurs during June and July. As in the case of salmon, there is a general trend of later spawning towards the north, and in a given area, there is some interannual variability in the timing. However, the temporal and spatial coverage of the capelin arrival on the beaches is considerably less compared to salmon so that only limited analysis is possible. Two sets of data are available that demonstrate the late spawning in 1991: peak spawning times determined by monitoring egg deposition on 15 study beaches in Conception Bay beginning in 1987 (Nakashima and Slaney 1992) and the peak spawning time as the day of the year when 11 500 kg of cumulative catch was reached, estimated from the detailed research log books of nine capelin trap-net fisherman from Conception and Trinity Bays since 1983. The latter measure was highly correlated with a shorter time series of direct observations of capelin spawning on a single spawning beach (Carscadden et al. 1992).

Nakashima and Slaney (1992) documented that spawning varied by only 7 d from 1987 to 1990, but in 1991 spawning was 30 d later than in 1987. Carscadden et al. (1992) found that spawning in 1991 was about 29 d later than the 1983–90 average and about 22 d later than the latest (1985) in the same time period. Furthermore, spawning in 1984 and 1985 occurred more or less on the same calendar date, whereas in 1990 (also a cold year), spawning was 7 d earlier than 1985.

Carscadden et al. (1992) have shown that spawning time is negatively correlated with water temperatures in the top 20 m during May (i.e., immediately prior to spawning and when gonads are maturing rapidly) and with the mean length in the mature population. When the temperature and mean length were combined in a multiple regression, 87% of the variation in spawning time was explained.

Observations on the inshore spawners during 1992 indicate that spawning in these two bays was once again delayed by 1 mo compared with the average; the spawning fish were smaller in size as well. This smaller overall mean length (150 mm compared with 174 mm 1981–90 average) was the result of a higher than normal proportion of age 2 fish in the spawning stock as well as smaller mean lengths at age (Nakashima and Carscadden 1993). The latter suggests that growth was reduced and this was probably the result of the unusually cold water temperatures, especially during 1991.

It is interesting to note the dramatic decline in the mean length of the spawning capelin in 1991 (mean length = 163 mm) and 1992 (mean length = 150 mm) compared with 1985 (mean length = 168 mm) and even more compared with 1984 (mean length = 177 mm), years when CIL area and ice cover were abnormally high. Note that 1984 was after a warm period, whereas 1985 followed one cold year, and 1991 and 1992 followed several cold years. The cumulative effect of several cold years appears to have a significant effect on growth and on the timing of spawning. Thus, just because CIL areas in any two summers or ice-cover areas during the winters are comparable, one cannot automatically conclude that the timing of spawning or size distribution during these 2 yr will also be the same.

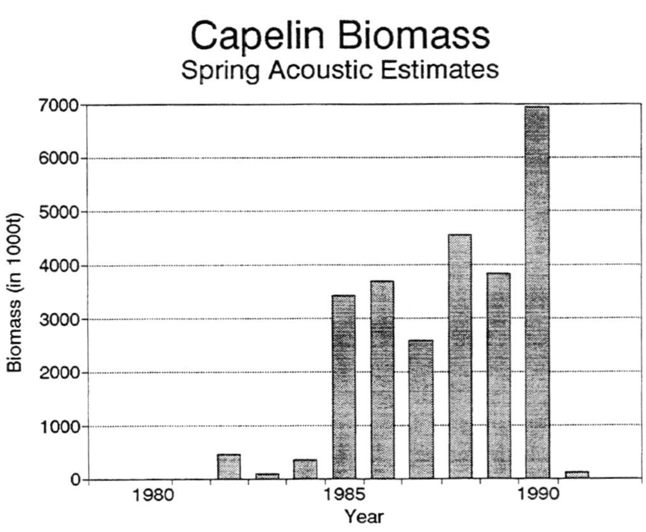

Fig. 13. Capelin biomass estimates from spring acoustic survey.

Capelin abundance and distribution

The abundance estimate of capelin is based on spring acoustic surveys (April–May) on the northern Grand Bank (Northwest Atlantic Fisheries Organization (NAFO) division 3L). Inshore trap catch rates in division 3L were found to be positively correlated with: (1) the mature biomass estimated from the spring surveys in the same year, and (2) the biomass projected to the spawning season based on the previous year's spring survey estimates (Carscadden 1992).

To assess whether the variabilities in the trap catch rates and the spring biomass estimates were caused by the environment, these two variables were regressed with the summer CIL area and SL thickness along the Bonavista transect and with the onset of surface warming (day of the year when SST reached 4°C). None of the correlations was significant.

The biomass estimates from the spring surveys, however, show a dramatic decline after the spring of 1990 (Fig. 13). Because the spring survey coverage in 1991 was restricted by ice, the survey was repeated during late June and early July to ensure that reduced geographical coverage was not the primary cause for the very low estimates; however, the estimate from this later survey was similar to the May estimate. The inshore trap catch rates and egg abundances in 1991, on the other hand were not unusually low (though the spawning was delayed), suggesting that the spring 1991 offshore acoustic surveys had failed to detect the mature fish. Note that, in 1991, the entire shelf area of division 3L at 50 m water depth had temperatures below zero (Narayanan et al. 1992). If capelin perceived the 1991 temperatures as suboptimal throughout the survey block, then a radical shift in spatial distribution could have occurred.

Factors affecting capelin distribution in the Northwest Atlantic during the offshore portion of their life cycle have received little attention (Schneider and Methven 1988). Shackell et al. (1994) recently examined the distribution of immature and mature capelin in division 3L, using data from hydroacoustic research surveys conducted in May from 1985 to 1992. From 1985 to 1990, immature capelin were generally distributed in the northern part of the survey area while mature capelin were distributed in the south. The authors showed that distribution of mature capelin was associated with temperature on relatively larger spatial (90 000 km^2) and temporal (among years) scales, but not on small spatial (2000 km^2) and temporal (within years) scales. Basically, in 1991 and 1992, there were no mature capelin in the southern part of the survey block. Given that mature capelin have been observed to migrate from north into the southern part of the survey block in the spring, the authors suggest that pre-spawning migration is related to state of maturity and hence migration may be delayed in cold years. Annual surveys conducted at the same time each year over a limited area may be out of phase with the biological migration phase, thus biomass may be underestimated in cold years if spring migration from the north of division 3L into the survey area is delayed.

Discussion

A closer look at the meteorological and oceanographic conditions that prevailed during the last several years has revealed significant differences between anomalous years. We have shown that summer CIL area, a robust index frequently used to represent climatic variability, does not necessarily exhibit similar differences between anomalous years, but indices such as the day of the year when SST reaches 4°C and rate of surface warming, seem to be more appropriate for representing these differences.

The extent to which the biological processes are influenced by the environment is determined by magnitude of interannual variability, as well as by the relative phases of biological and oceanographic processes. This is especially true for biological processes that occur during a period when oceanographic conditions are changing rapidly, for example, during early spring or late fall. We have shown that, the timing and rate of surface warming capture some of the interannual variability in the conditions during the spring to summer transition period. Thus they may be suitable indices to explain some of the variability in the migration time for the two species considered here, whereas the summer CIL characteristics may be more appropriate for explaining biomass variability.

Results presented in this paper indicate that the arrival time of Atlantic salmon in southern Labrador and along the northeast coast of Newfoundland may be influenced by the timing of the onset of surface warming of the ocean. Similarly, the timing of salmon spawning migrations to two rivers situated along the northeast Newfoundland coast was associated with ocean climate data. The climatic influence was also apparent in comparisons of salmon abundance: cold years were typically associated with years when salmon abundance was low. Thus, migrating salmon appear to have a preference for certain oceanographic conditions, probably for certain temperature ranges, and can shift the migration period and migration route, if necessary. This result is consistent with the conclusions in Reddin and Friedland (1993). Based on data from research vessel cruises since 1965, they have shown that offshore catch rates of salmon are significantly correlated with SST, and that salmon appear to have a habitat preference for 4–10°C water in the offshore environment. With respect to inshore areas, results of earlier tagging studies have indicated that salmon may modify their inshore migration routes in response to variable oceanographic conditions (Reddin and Shearer 1987).

In contrast with the above results, Chadwick and Claytor (1989, 1990) concluded that salmon-run timing was constant and was not related to environmental factors. While their results contradict those presented in this paper, it should be noted that their results pertain to a different geographic area, the Gulf of St. Lawrence. Because these salmon will be exposed not only to the conditions in the Northwest Atlantic but also to the marine climate in the Gulf of St. Lawrence, the links between their run time and oceanography may be different. Moreover, Dempson (1992) recently reported that the latest arrival time of salmon to a south coast Newfoundland river occurred in 1991. This was also a year in which returns to the river were the lowest recorded over a 6-yr period. Perhaps it is only in those years where extreme anomalies occur, such as in 1991, that fisheries in the Gulf of St. Lawrence respond in such negative ways.

Timing of capelin spawning on the beaches of eastern Newfoundland was also found to be influenced by oceanographic conditions in the spring (Carscadden et al. 1992). However, neither the inshore catch rates nor the offshore biomass estimated from the spring acoustic survey had any significant relationship to any of the oceanographic indices considered here. Furthermore, the 1991 spring biomass estimate was abnormally low, whereas the inshore catch rate was not significantly reduced, which suggests a change in capelin distribution in response to the abnormal conditions in 1991.

Changes in distribution associated with temperature have been well documented in the Barents Sea capelin, which have distinct migration patterns corresponding to warm and cold periods. It appears that distribution changes in response to large-scale temperature changes can even be delayed up to 1–2 yr (Loeng 1989). If northern Grand Bank capelin alter their spatial distribution in response to temperature changes, then this may be: (1) in pursuit of temperature–dependent prey, as is hypothesized for Barents Sea capelin; (2) to optimize growth when food is not a factor; or (3) due to a complex suite of factors. Nevertheless, dramatic changes in capelin biology as evidenced by an unexpected decline in offshore abundance, decline in mean size and delayed spawning, all coincident with abnormally cold water and extensive ice cover, suggest of an environmental effect with temperature as the dominant mechanism.

The fact remains that the 1991 abundance estimates for salmon and capelin were the lowest on record. In the absence of legal sea fisheries for salmon in 1992, preliminary indications are that salmon abundance in Newfoundland is at best comparable with that in 1991. The May 1992 acoustic estimate of capelin biomass was also low (Miller 1992). While direct evidence for cause and effect related to abundance and environment may be somewhat lacking, it is difficult to dismiss the broad scale pattern of low salmon returns to many Newfoundland rivers, and apparent low abundance of capelin and some groundfish stocks, coincident with some of the harshest environmental conditions to affect these waters over the past 30 yr.

Acknowledgments

The authors wish to acknowledge the anonymous reviewers for their many valuable suggestions and comments, and I. Peterson for assistance in producing the ice diagrams for this paper.

References

Allan, I.R.H., and J.A. Ritter. 1977. Salmonid terminology. J. Cons. Cons. int. Explor. Mer, 37: 293–299.

Atkinson, D.B. 1993. Some observations on the biomass and abundance of fish captured during stratified random bottom trawl surveys in NAFO divisions 2J3KL, Fall 1982–1991. NAFO SCR Doc. 93/29: 18 p.

Carscadden, J.E. 1992. Capelin in NAFO division 2J3KL. CAFSAC Res. Doc. 92/18: 14 p.

Carscadden, J.E., K.T. Frank, and B.S. Nakashima. 1992. Why was inshore capelin (*Mallotus villosus*) spawning delayed during 1991. CAFSAC Res. Doc. 92/61: 12 p.

Chadwick, E.M.P., and R.R. Claytor. 1989. Run timing of pelagic fishes in Gulf of St. Lawrence: area and species effects. J. Fish Biol. 35(Suppl. A): 215–223.

Chadwick, E.M.P., and R.R. Claytor. 1990. Predictability in a small commercial Atlantic salmon fishery in western Newfoundland. Fish. Res. 10: 15–27.

Dempson, J.B. 1992. Status of the Atlantic salmon population of Conne River, Newfoundland, in 1991. CAFSAC Res. Doc. 92/4: 28 p.

Drinkwater, K.F., B. Petrie, and S. Narayanan. 1992. Overview of environmental conditions in the Northwest Atlantic. NAFO SCR Doc 92/73: 36 p.

Frank, K.T., and J.E. Carscadden. 1989. Factors affecting recruitment variability of capelin (*Mallotus villosus*) in the Northwest Atlantic. J. Cons. Cons. int. Explor. Mer 45: 146–164.

Frank, K.T., and W.C. Leggett. 1981*a*. Wind regulation of emergence times and early larval survival in capelin (*Mallotus villosus*). Can J. Fish. Aquat. Sci. 38: 215–223.

Frank, K.T., and W.C. Leggett. 1981*b*. Prediction of egg development and mortality rates in capelin (*Mallotus villosus*) from meteorological, hydrographic, and biological factors. Can. J. Fish Aquat. Sci. 38: 1327–1338.

Friedland, K.F,. and D.G. Reddin. 1993. Marine survival in Atlantic salmon from indices of postmolt growth and sea temperature, p. 119–138. *In* D. Mills [ed.] Salmon in the sea. Blackwell Scientific Publications Ltd., London, UK.

Leggett, W.C., K.T. Frank, and J.E. Carscadden. 1984. Meteorological and hydrographic regulation of year-class strength in capelin (*Mallotus villosus*). Can. J. Fish. Aquat. Sci. 41: 1193–1201.

Loeng, H. 1989. The influence of temperature on some fish population parameters in the Barents Sea. J. Northwest Atl. Sci. 9: 103–113.

Miller, D.S. 1992. Results from acoustic survey for capelin (*Mallotus villosus*) in NAFO division 3L in 1992. NAFO SCR Doc. 92/52, Ser. 2110: 4 p.

Mills, D.H. 1989. Ecology and management of Atlantic salmon. Chapman and Hall Ltd., London, UK. 351 p.

Mundy, P.R. 1982. Computation of migratory timing statistics for adult chinook salmon in the Yukon River, Alaska, and their relevance to fisheries management. N. Am. J. Fish. Manage. 2: 359–370.

Nakashima, B.S. and J. Carscadden. 1993. The inshore capelin fishery in NAFO division 3K in 1992 and a comparison of mean lengths in NAFO divisions 3K and 3L. DFO Atl. Res. Doc. 93/8: 26 p.

Nakashima, B.S., and B.W. Slaney. 1992. Capelin (*Mallotus villosus*) egg deposition on fifteen spawning beaches in Conception Bay, Newfoundland in 1987–91. NAFO SCR Doc. 92/2, Ser. N2035: 8 p.

Narayanan, S., S. Prinsenberg, and E.B. Colbourne. 1992. Overview of environmental conditions in 2J3KL NAFO subareas in 1991. NAFO SCR Doc. 92/6: 30 p.

Nickson, Sir D. 1991. Progress Report, December 1991. Atlantic Salmon Trust. Pitlochry, Perthshire PH16 5JQ, Scotland. 49 p.

O'Connell, M.F., Dempson, J.B., and Reddin, D.G. 1992*a*. Evaluation of the impacts of major management changes in the Atlantic salmon (*Salmo salar* L.) fisheries of Newfoundland and Labrador, Canada, 1984–88. ICES J. Mar. Sci. 49: 69–87.

O'Connell, M.F., Dempson, J.B., Porter, T.R., Reddin, D.G., Ash, E.G.M., and N.M. Cochrane. 1992*b*. Status of Atlantic salmon (*Salmo salar* L.) stocks of the Newfoundland Region, 1991. CAFSAC Res. Doc. 92/22: 56 p.

Peterson, I., and S. Prinsenberg. 1990. Sea ice fluctuations in the western Labrador Sea (1963–1988). Can. Tech. Rep. Hydrogr. Ocean Sci. 123: 130 p.

Petrie, B., S. Akenhead, J. Lazier, and J. Loder. 1988. The cold intermediate layer on the Labrador and northeast Newfoundland Shelves, 1978–86. NAFO Sci. Counc. Stud. 12: 57–69.

Petrie, B., J.W. Loder, S. Akenhead, and J. Lazier. 1991. Temperature and salinity variability on the eastern Newfoundland shelf: the annual harmonic. Atmos.-Ocean, 29: 14–36.

Petrie, B., J.W. Loder, and J. Lazier. 1992. Temperature and salinity variability on the eastern Newfoundland shelf: the residual field. Atmos.-Ocean, 30: 120–139.

Prinsenberg, S., and I. Peterson. 1992. Sea ice properties off Labrador and Newfoundland during LIMEX 89. Atmos.-Ocean, 30: 207–222.

Randall, R.G., and E. J. Schofield. 1987. Status of Atlantic salmon in the Miramichi River, 1986. CAFSAC Res. Doc. 87/5: 10 p.

Reddin, D.G. 1988. Ocean life of Atlantic salmon (*Salmo salar*) in the Northwest Atlantic, p. 438–511. *In* D. Mills and D. Piggins [eds.] Atlantic salmon: Planning for the Future, Proceedings of the Third International Atlantic salmon symposium, Biarritz, France, Oct. 21–23, 1986. Croom Helm, London, England.

Reddin, D.G., and K.F. Friedland. 1993. Marine environmental factors influencing the movement and survival of Atlantic salmon, p. 79–103. *In* D. Mills [ed.] Salmon in the sea. Blackwell Scientific Publications Ltd., London, UK.

Reddin, D.G., and W.M. Shearer. 1987. Sea-surface temperature and distribution of Atlantic salmon in the Northwest Atlantic Ocean, p. 262–275. *In* M.J. Dadswell et al. [eds.] Common strategies of anadromous and catadromous fishes. Am. Fish. Soc. Symp. 1.

Ritter, J.A. 1989. Marine migration and natural mortality of North American Atlantic salmon (*Salmo salar* L.). Can. MS Rep. Fish. Aquat. Sci. 2041: 136 p.

Saunders, R. L. 1986. The thermal biology of Atlantic salmon: influence of temperature on salmon culture with particular reference to constraints imposed by low temperature. Institute of Freshwater Research, Drottningholm, The Netherlands. Rep. 63: 68–81.

Saunders, R.L., and C.B. Schom. 1986. Importance of the variation in life history parameter of Atlantic salmon (*Salmo salar*). Can. J. Fish. Aquat. Sci. 42: 615–618.

Schneider, D.C., and D.A. Methven. 1988. Response of capelin to wind-induced thermal events in the southern Labrador Current. J. Mar. Res. 46: 105–118.

Shackell, N.L., J.E. Carscadden, and D.S. Miller. 1994. Migration of pre-spawning capelin (*Mallotus villosus*) as related to temperature on the northern Grand Bank, Newfoundland. ICES J. of Mar. Sci. In press.

Working Group on North Atlantic Salmon. 1992. Report of the Working Group on North Atlantic Salmon. Cons. int. Explor. Mer. C.M. 1992/Assess 15: 155 p.

The effect of climate on North Pacific pink salmon (*Oncorhynchus gorbuscha*) production: examining some details of a natural experiment

R. Ted Cooney, T. Mark Willette, Sam Sharr, Dan Sharp, and Jeff Olsen

Cooney, R.T., T.M. Willette, S. Sharr, D. Sharp, and J. Olsen. 1995. The effect of climate on North Pacific pink salmon (*Oncorhynchus gorbuscha*) production: examining some details of a natural experiment, p. 475–482. *In* R.J. Beamish [ed.] Climate change and northern fish populations. Can. Spec. Publ. Fish. Aquat. Sci. 121.

Abstract: Positive correlations between pink salmon (*Oncorhynchus gorbuscha*) production and sea-surface and air temperatures in the Gulf of Alaska demonstrate linkages with long-term, low-frequency oceanographic and meteorological variability. A comparison of a warm spring (1990) and a cold spring (1991) in Prince William Sound, Alaska, suggests that one way these influences may manifest is by modifying the duration of the annual fry emigration into coastal nursery areas. Differences in air and sea-surface temperatures approximately 2°C apparently resulted in an extension of the pink salmon emigration by 38 d in the cold year. Evidence is presented that between-year differences in cloud cover contributed to the variability in thermal histories measured in the spring of 1990 and 1991. An interpretation linking the duration of the wild-fry emigration to levels of adult production is proposed on the basis of experimental results from an ocean-ranched pink salmon program in the region.

Résumé : Les corrélations positives entre la production de saumon rose (*Oncorhynchus gorbuscha*) et les températures de la surface de la mer et de l'air dans le golfe d'Alaska montrent l'existence de liens avec la variabilité océanographique et météorologique à long terme et à faible fréquence. Une comparaison d'un printemps chaud (1990) et d'un printemps froid (1991) dans la baie du Prince Guillaume, en Alaska, porte à croire qu'une façon dont ces influences peuvent se manifester est par la modification de la durée de l'émigration annuelle des alevins dans les zones d'alevinage côtières. Des différences d'environ 2° C dans les températures de l'air et de la surface de la mer se sont apparemment traduites par une prolongation de 38 jours de la durée de l'émigration du saumon rose au cours de l'année froide. Selon les données présentées, les différences d'une année à l'autre dans la couverture nuageuse ont contribué à la variabilité des conditions thermiques mesurées au printemps de 1990 et de 1991. Une interprétation rattachant la durée de l'émigration des alevins sauvages aux niveaux de la production des adultes est proposée à la suite des résultats expérimentaux d'un programme d'élevage en enclos marin du saumon rose dans la région.

Introduction

Francis and Sibley (1991) illustrate close correspondence between cycles in air and sea-surface temperatures and pink salmon (*Oncorhynchus gorbuscha*) production in the Gulf of Alaska over a 61-yr period. Royer (1989; T.C. Royer, personal communication) demonstrates the presence of long-term, low-frequency temperature cycles in the North Pacific. An hypothesis is advanced relating these temperature differences to very small changes in the north–south exchange of ocean waters forced by the lunar nodal tidal cycle of 18.6 yr. Historic fluctuations in pink salmon production in the Gulf of Alaska appear to track this 18- to 20-yr cycle in ocean temperature.

There are probably several ways that pink salmon production responds to low-frequency temperature fluctuations in the Gulf of Alaska (Francis and Sibley 1991). These range from reactions to ecosystem changes in lower trophic level production to the direct effects of temperature on metabolism, growth, and survival during all life stages, eggs to adults. Brodeur and Ware (1992) describe interannual and longer period changes

R. T. Cooney. Institute of Marine Science, University of Alaska Fairbanks, Fairbanks, Alaska 99775, USA.

T. M. Willette, S. Sharr, and **Dan Sharp.** Alaska Department of Fish and Game, P.O. Box 669, Cordova, Alaska 99574, USA.

J. Olsen. Prince William Sound Aquaculture Corporation, P.O. Box 1110, Cordova, Alaska 99574, USA.

in macrozooplankton standing stock in the subarctic Pacific in relation to the intensity of the winter winds. Salmon (1992) demonstrates associations between wind-stress curl and subsequent warming and cooling cycles in the North Pacific. These findings disclose that the Gulf of Alaska changes environmental–ecological conditions for salmon growth and survival on decadal time scales.

A growing body of literature supports the idea that run strength for juvenile Pacific salmon entering the ocean as fry (pink and chum (*O. keta*)) is set during early marine residence by losses to predators (Bax 1983; Hargreaves and LeBrasseur 1985; Hartt 1980; Healey 1982; Parker 1964, 1965, 1968, 1971). Growth conditions influencing the length of time that juveniles remain in the smallest, most vulnerable stages are implicated as important factors. Temperature and food are generally considered the variables defining the growth environment for juveniles (Healey 1980, 1991; Mortensen 1983; Urquhart 1979; Walters et al. 1978). Matching the emigration from natal streams and rivers with appropriate forage is a prerequisite for optimal growth. Presumably, local stocks have evolved reproductive patterns for maximizing the probability that a large percentage of emigrating fry will experience preferred growth conditions in coastal nurseries (Beacham and Murray 1987; Olson and McNeil 1967). Fry generally enter coastal waters in the northern Gulf of Alaska over a period of about 60 d that begins in late March or early April (Bailey 1969; Taylor 1988).

In 1990 and 1991, fry weirs were placed on six pink salmon index streams in southwestern Prince William Sound to enumerate the annual fry emigration. Studies of coastal ocean climate in the region, initiated by the Alaska Sea Grant College Program (Cooperative Fisheries and Oceanographic Studies), were also in place at this time. We combine these observations to describe the apparent modification of the wild pink salmon emigration by temperature. Fortuitously, 1990 was a very warm spring while 1991 was colder than average. This natural experiment provided a means to investigate the emigration process with respect to questions about climate change and pink salmon production. Experimental results from a large pink salmon ocean-ranching program in Prince William Sound are used to interpret the production consequences of early and late wild-fry emergence.

Methods

A comprehensive program of oceanographic and meteorological monitoring is in place in Prince William Sound as part of a cooperative study of marine conditions influencing the

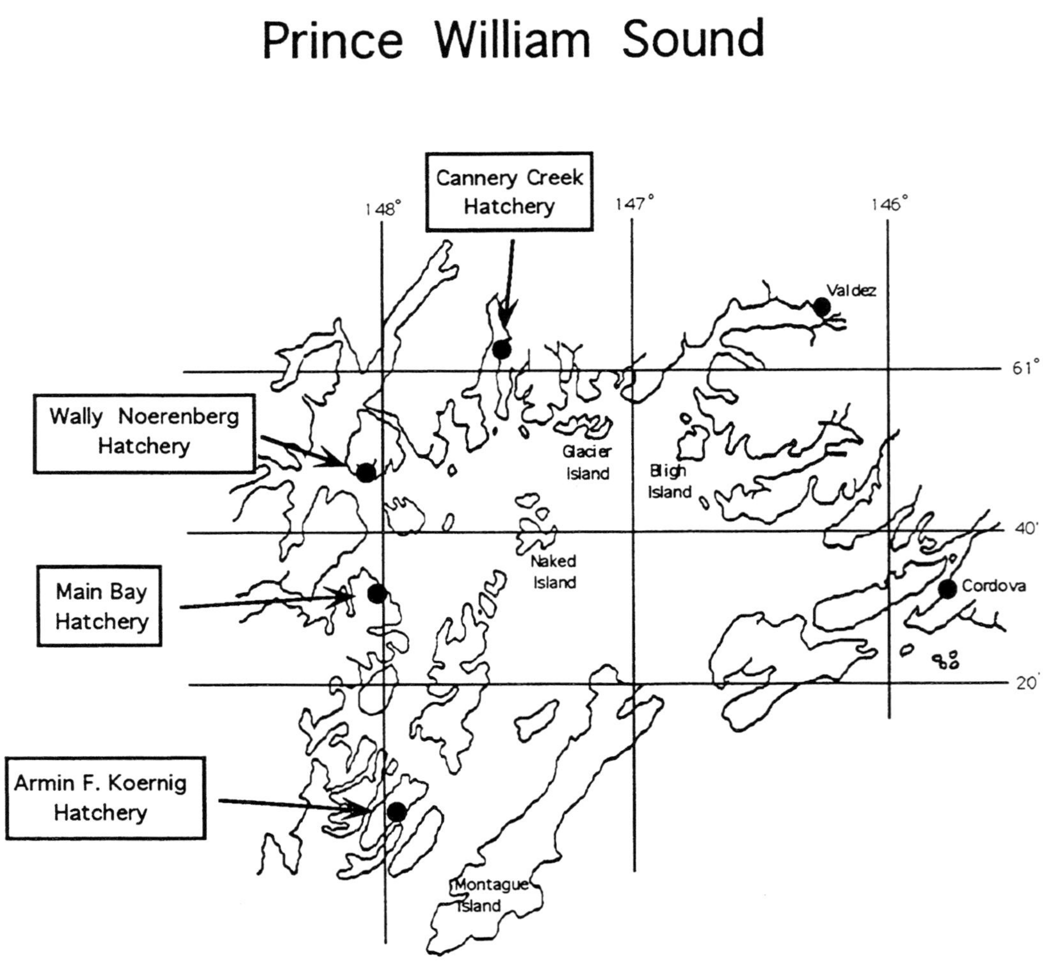

Fig. 1. The location of salmon hatcheries participating in the meteorological, ocean temperature, and plankton watch program in Prince William Sound.

growth of juvenile pink and chum salmon in the region (Cooney and Willette 1991). Measures of air and surface water temperature and estimates of cloud cover (in percent) are recorded daily at four salmon hatcheries (Fig. 1). A long-term sea surface temperature time series is also available for Prince William Sound and the northern Gulf of Alaska from the National Climate Center. Monthly mean temperatures from this data base for a region centered at 60°N 145°W, and for larger portions of the North Pacific, were provided by the Climate Research Group at Scripps Institution of Oceanography.

Salmon hatcheries in Prince William Sound also participate in a plankton watch during the winter, spring, and early summer months of each year. Plankton is sampled twice weekly using 0.5-m nets (0.250-mm Nitex) towed vertically from 20 m to the surface at locations near each hatchery. Settled volumes are recorded to the nearest milliliter and the percentage of the volume contributed by phytoplankton (green and brown) and zooplankton (pink) is estimated by hatchery personnel. The most comprehensive record of fry forage has been compiled at the Armin F. Koernig (AFK) Hatchery in southwestern Prince William Sound.

Pink salmon fry emigrating from six natal streams were enumerated by the Alaska Department of Fish and Game (ADFG) using large winged fyke nets and floating collecting boxes tended daily from early April to mid-July in 1990 and 1991 (Sharp 1991). Numbers of fry captured daily in the weirs were recorded, and for purposes of our evaluation, these totals were summed across all streams for each of the two years.

The Prince William Sound Aquaculture Corporation (PWSAC) and ADFG maintain a coded-wire fry-tagging program for a portion of all juvenile pink salmon released from hatcheries. The information from tag recoveries is used as a tool for in-season management of harvests, but also provides a means for determining the marine survivals of ocean-ranched pink salmon. Beginning with brood year 1985, PWSAC began to evaluate the importance of releasing pink salmon fry in relation to the timing of a consistent early season zooplankton bloom. In one treatment, short-term reared fry (fed approximately 10 d) were released into the early stages of the developing zooplankton bloom, while in another treatment, fry were held longer (fed more than 10 d) and released after the bloom was in decline. Different tag codes allow an evaluation of the marine survivals resulting from these two release strategies for brood years 1985, 1987, 1988, and 1989.

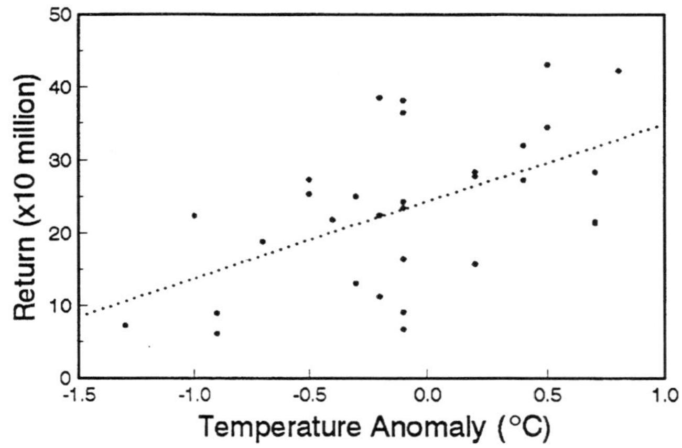

Fig. 2. A regression of pink salmon returns in the northern Gulf of Alaska over a 30-yr period on average sea surface temperature anomalies (May–June) for the northeastern Pacific Ocean (45–55°N, 135–165°W); $R^2 = 0.29$

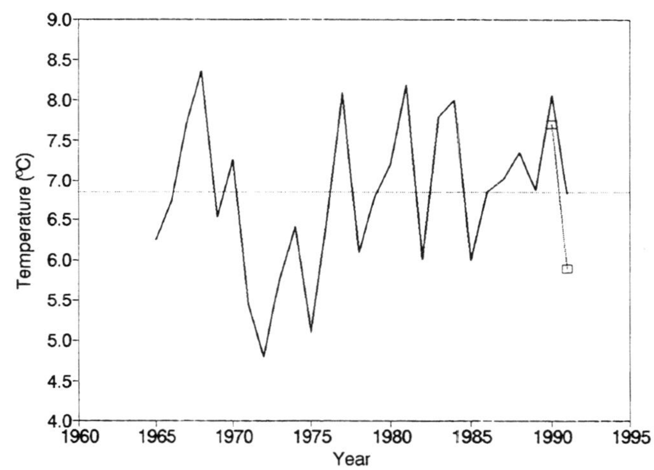

Fig. 3. Interannual differences in April–June sea surface temperatures (National Climate Center; 60°N, 145°W) in the northern Gulf of Alaska from 1965 to 1991. Measurements from hatcheries in 1990 and 1991 are depicted separately as open squares.

Results

A linear regression of the returns of pink salmon for Prince William Sound, lower Cook Inlet, and Kodiak on sea-surface temperatures for the northeastern Pacific Ocean demonstrates that about 30% of the interannual variability in production can be accounted for by upper-ocean temperatures measured in the spring and summer of the juvenile year (Fig. 2). A time series of average sea-surface temperatures for the months of April through June 1965–91 depicts year-to-year and longer period variability around a mean of 6.8°C for a 5° × 5° region of the northern Gulf that includes Prince William Sound (Fig. 3). Temperatures measured by hatchery personnel document 1990 as warmer than average and 1991 as colder than average. In both years, the lowest daily ocean temperatures occurred in March and averaged about 4°C. The seasonal warming trend in 1990 began in early to mid-April. This same trend was delayed about 3 wk in 1991 (Fig. 4). During the spring of both years, air and surface water temperatures were highly correlated.

One difference between these years was the extent of cloud cover. The warming trend in the warm year (1990) began during a period of cloud-free days in early April and appeared to be maintained by intermittent periods of clear weather throughout the spring (Fig. 5). In contrast, a long period of mostly cloudy weather beginning in April and extending into early May apparently delayed the start of seasonal warming in 1991. This correspondence between cloud-free days and episodes of warming and cooling in both years implicates insolation as important in determining spring air and water temperatures in the region.

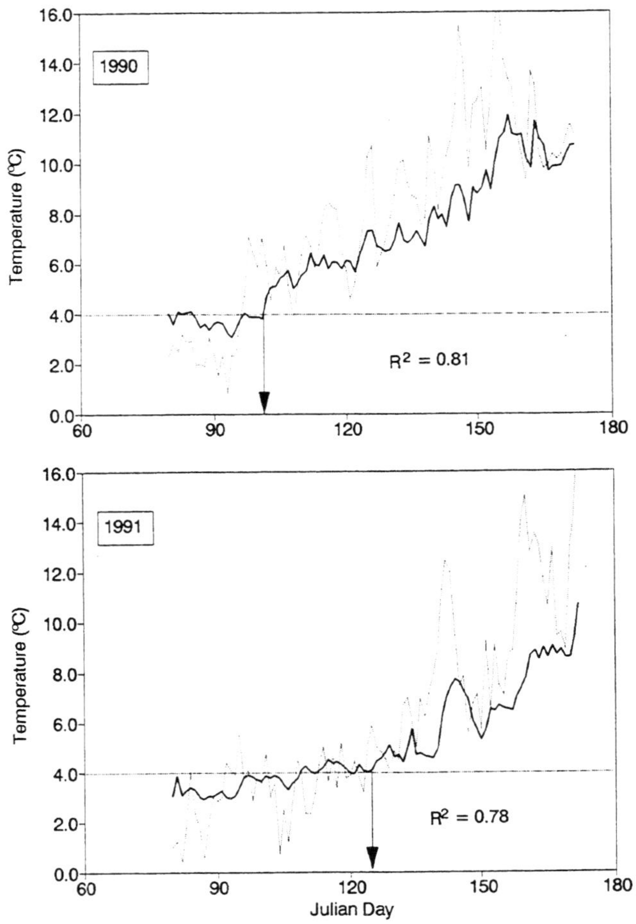

Fig. 4. Spring air (light line) and surface water (dark line) temperatures measured in Prince William Sound (average of all hatcheries) and correlation coefficients for water on air temperature. The initiation of the surface water warming cycle is indicated (arrows).

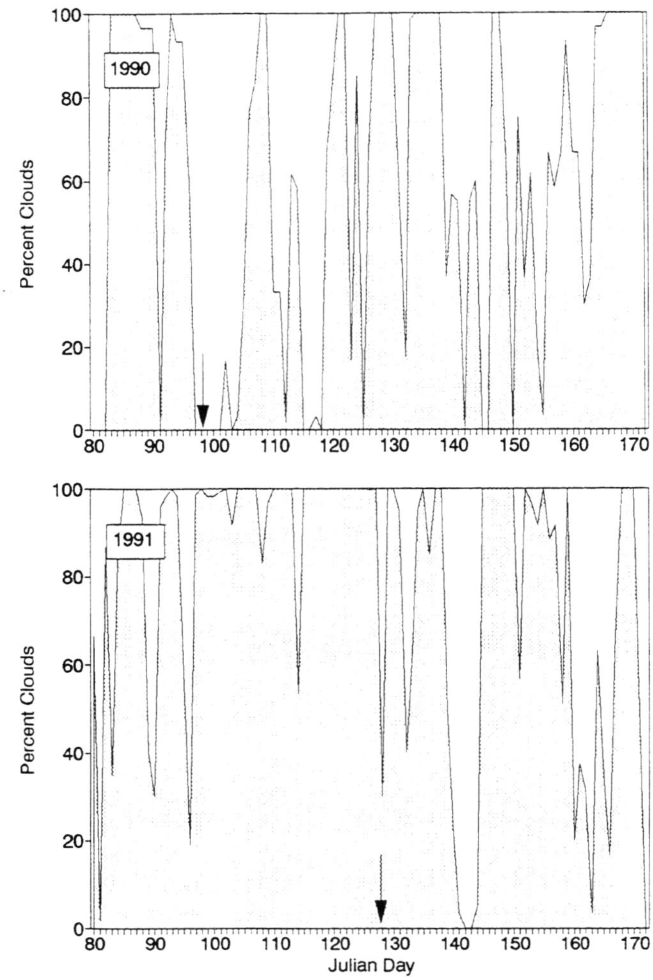

Fig. 5. Spring cloud cover in Prince William Sound (average of all hatcheries) for the years 1990 and 1991. The initiation of the surface water warming cycle is indicated (arrows).

Table 1. A comparison of oceanographic, meteorological, and pink salmon fry emigration results for a warm and cold spring in Prince William Sound; 21 March – 21 June.

	Year	
Variable	1990 (Warm)	1991 (Cold)
Mean air temperature (°C)	7.38	5.32
Mean water temperature (°C)	6.49	4.77
Cloud cover ≥90% (d)	30	58
Duration of emigration (d)	53	91
Fry out by 15 May (%)	76	56
Total emigration	2.10×10^6	1.97×10^6

The pink salmon fry emigration from six index streams began in mid-April both years, but was extended by more than a month in 1991, the cold year (Fig. 6). Approximately 76% of the emigration was completed by May 15 in 1990, whereas only 56% of the fry had emigrated by this same date in 1991. The total fry emergence was remarkably similar for both years, but lasted 38 d longer in 1991 (Table 1).

Records of zooplankton settled volumes, averaged over 11 yr at the AFK Hatchery, illustrate a general pattern of increasing standing stock by mid-April that peaks in May and declines by early June (Fig. 7). When this pattern is compared with the timing and duration of the pink salmon emigration for 1990 and 1991, it is apparent that under warm conditions a larger fraction of fry leaving natal streams match up with the peak of the zooplankton bloom than is the case in the cold year (Fig. 8).

The marine rearing strategy of releasing early fed pink salmon fry into the developing zooplankton bloom each year enhances the survival of those juveniles relative to fry released after being held and fed until late May and early June (Table 2).

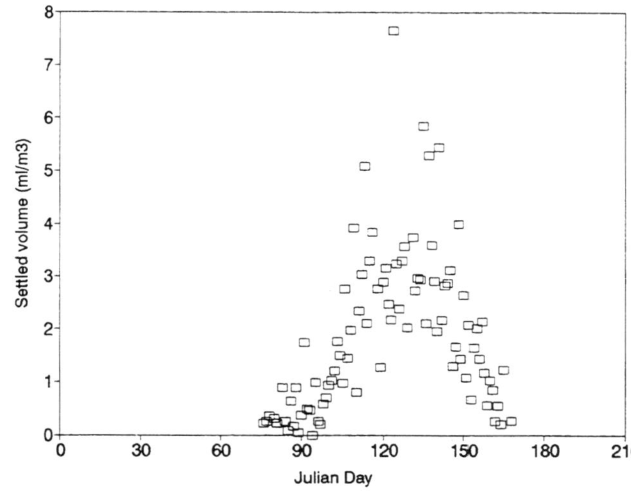

Fig. 7. Average daily zooplankton settled volumes measured at the AFK Hatchery in Prince William Sound for the period 1981 through 1991 (each square represents the mean of all observations for that Julian date).

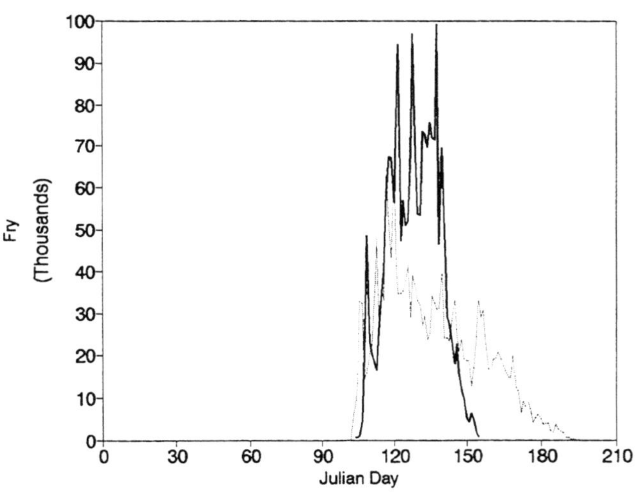

Fig. 6. The duration and magnitude of the pink salmon emigration from six natal streams in Prince William Sound for the years 1990 (dark line) and 1991 (light line).

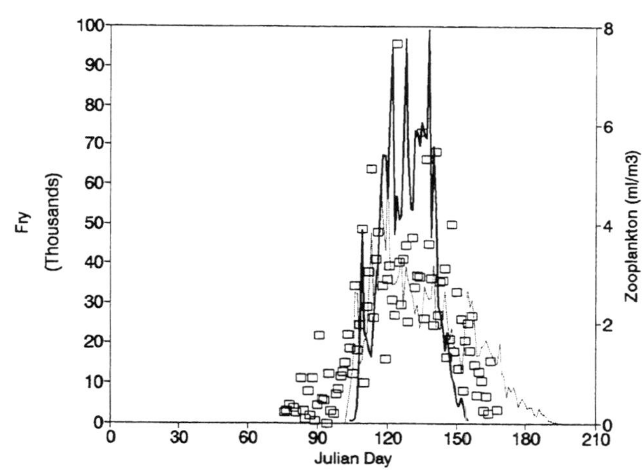

Fig. 8. A comparison between the timing of the fry emigration in 1990 (dark line) and 1991 (light line) and the timing and magnitude of spring zooplankton stocks (open squares) averaged for the years 1981 through 1991.

Table 2. The percent marine survival for pink salmon fry released under different conditions of pen rearing in relation to the timing of an early season zooplankton bloom.

Strategy	Mean weight (g)	Brood Year 1985 (%)	1987 (%)	1988 (%)	1989 (%)
Early release[a]	0.26	6.0	6.8	8.6	6.4
Late release[b]	0.32	3.5	1.1	4.0	4.8

[a]Fed fry released into the zooplankton bloom in late April and early May.

[b]Fed fry released at the end of May and early June.

In each of 4 yr where coded-wire tag recoveries are available, the early release group exhibited consistently higher marine survivals than the later release group even though the longer term reared juveniles were larger at the time of release and entered warmer ocean conditions.

Discussion

Our evaluation of an admittedly small set of observations suggests that the duration of the annual emigration of pink salmon fry can be influenced by coastal ocean climate. Cloud cover appears to play a significant role in modifying spring air and water temperatures in Prince William Sound. Because most of the many thousands of pink salmon spawning streams around the northern Gulf of Alaska are short, 10 km or less, interactions between a cold or warm coastal ocean and the attendant meteorology can influence much of the pink salmon spawning habitat in the region. Although measurements of stream flow were not taken in this study, ADFG personnel manning the fry traps in 1990 and 1991 observed major differences in the spring breakup period. In the warmer year, highly variable stream flows occasionally flushed fry out in high numbers, whereas the slow breakup and melt of 1991 apparently contributed to the much extended emigration under lower stream flows.

Consequences of an extended emigration

The question of whether pink salmon fry are disadvantaged by a late entry during a cold year can be partially assessed by applying the results of the hatchery experimental release program. For whatever reason, late-released fry exhibit lower marine survivals than do fry released directly into high concentrations of zooplankton 2 or 3 wk earlier. We suggest that one contributing factor may be through a general decoupling of fry and their preferred pelagic forage resources under colder than average conditions. During colder coastal springs, a reduced percentage of the wild emigrants experience the highest early season zooplankton stocks.

The apparent disadvantage for fry entering late in the season or slowly over an extended period can also be explained in ways that do not necessarily involve a mismatch with early season food stocks. Mortensen and Wertheimer (1987) found that late-emerging fry in Auke Creek (southeastern Alaska) generally exhibited higher growth rates than those emerging under earlier, cooler conditions. However, even though growth was accelerated significantly by warmer conditions, in mixed schools of early and late emerging fry, the younger juveniles were always smaller than their early emerging and surviving counterparts. Under these conditions, the faster growing smaller fry (for a given age) will presumably be more susceptible to size-selective predation than will the larger, early emerging survivors (Hargreaves and LeBrasseur 1985).

Another possibility is raised by Peterman and Gatto (1978) who demonstrate that local predator fields can theoretically be swamped if the fry emigration is unusually large. While this phenomenon may be more the exception than the rule, an early spring breakup responding to warmer than average conditions could conceivably wash fry out in sufficient numbers to overwhelm the capacity of local predators. Under these conditions, a smaller proportion of the overall emigration would be lost to predation. Conversely, an extended breakup delivering fry in relatively small numbers over a long period of time might favor higher overall losses to local predators (fishes and birds). Because most pink salmon spawning streams in the northern Gulf of Alaska are relatively short with low volume flows, they probably support only modest levels of fish and bird predators.

Direct effects of temperature

It also seems reasonable to assume that colder than average coastal climate might directly impose elevated mortality rates on incubating eggs and alevins due to freezing and lowered stream flows. Pink salmon alevins have been censused by ADFG in Prince William Sound since 1960 (24 streams) and a pre-emergent index is used to predict the adult return. An evaluation of a possible relationship between ocean temperature (considered to be a proxy for coastal meteorological conditions) and alevin abundance for the months January, February, and March (1965–90) produces only one statistically significant regression (Table 3). January temperatures explain about 20% of the variability in alevin abundance in this time series; lower alevin densities are generally associated with colder temperatures. Weak correlations for the other two months occur in regressions that are not significant.

Table 3. Results of a linear regression of pink salmon alevin abundance on winter and spring coastal ocean temperatures, 1965–1990.

Month	Test of slope >0 (*P*-values)	R^2
January	<0.05	0.20
February	>0.05	0.13
March	>0.05	0.11

Odd–even year differences

We are unable to ascertain whether or not the observed differences in the emigration histories observed in 1990 and 1991 were merely variations associated with the reproductive strategies of the odd and even year brood lines of pink salmon in the region. Forty-three to 65% of the odd brood line fish spawn in freshwater portions of natal streams, whereas about 70% of the even brood line spawns in the intertidal reaches of these same streams (Helle 1970). If stream flow influences the emigration process as we suggest, it seems more likely that an accelerated spring season would have a greater effect on the upstream odd-year brood line emigrants because a higher percentage of these juveniles originate from the strictly freshwater portions of the spawning habitat. Emigration patterns for pink salmon in two other streams in Prince William Sound were reported by Kirkwood (1962). Here again, fry entering the coastal margin in the colder of the two years exhibited an extended emigration. However, these fry were also from an even year brood line so the question of an odd–even year effect remains unresolved.

A conceptual model

Our analysis of the emigration timing and duration for a small number of pink salmon natal streams in Prince William Sound prompts us to suggest that some portion of the decline in wild stock production occurring during periods of below-average coastal ocean temperatures in the northern Gulf of Alaska may be associated with temperature modification of freshwater survivals and the duration of the annual fry emigration process. We present evidence from an experimental hatchery release program that late emigrating pink salmon fry are disadvantaged in terms of their marine survival relative to fry released 2–3 wk earlier. This may be because optimal forage resources are in general decline by early summer, or because later released fry are unable to catch up in size with their earlier surviving siblings and therefore incur greater size-dependent mortality, or both. It is also suggested that wild fry entering the coastal ocean in high abundance as the result of an early and short break up may enjoy elevated survival rates because of the phenomenon of predator swamping.

Future studies

We urge continuing studies of factors influencing the survival of pink salmon in the Gulf of Alaska, taking advantage of the interannual and longer period cyclic character of the "noisy" growth environment for juveniles in this region. The latest measurements of subsurface ocean temperature near Seward, Alaska, indicate that a predicted return to colder than average conditions began in 1990 (T.C. Royer, personal communication). Although temperatures were apparently modified in the surface waters by the 1991–92 El Niño, cooler upper-layer temperatures and reduced wild pink salmon production can be expected in the northern Gulf during some years of the remaining decade.

Acknowledgments

Managers and staff at pink salmon hatcheries in Prince William Sound are thanked for their dedication to meteorological, oceanographic, and plankton monitoring programs supplying coastal ocean climate data for this manuscript. This research was funded by the Alaska Sea Grant College Program (grant number NA90AA-D-SGO66; project R/07-13), the Institute of Marine Science, the Alaska Department of Fish and Game, and the Prince William Sound Aquaculture Corporation.

References

Bailey, J.E. 1969. Alaska's fishery resources—the pink salmon. Bur. Commer. Fish. Leafl. 169. U.S. Fish and Wildlife Service, Washington, D.C. 8 p.

Bax, N.J. 1983. Early marine mortality of marked juvenile chum salmon (*Oncorhynchus keta*) released into Hood Canal, Puget Sound, Washington, in 1980. Can. J. Fish. Aquat. Sci. 40: 426–435.

Beacham, T.D., and C.B. Murray. 1987. Adaptive variations in body size, age, morphology, egg size, and developmental biology of chum salmon (*Oncorhynchus keta*) in British Columbia. Can. J. Fish. Aquat. Sci. 44: 244–261.

Brodeur, R.D., and D.M. Ware. 1992. Long-term variability in zooplankton biomass in the subarctic Pacific Ocean. Fish. Oceanogr. 1(1): 32–38.

Cooney, R.T., and T.M. Willette. 1991. Regional-level investigations of pink salmon production responses to interannual variations in ocean temperatures: Cooperative Fisheries and Oceanographic Studies (CFOS), p. 50–62. *In* B. White and I. Guthrie [eds.] Proceedings, 15th Pink and Chum Salmon Workshop (27 Feb. – 1 Mar. 91, Parkville, B.C.). Pacific Salmon Commission, Vancouver, B.C.

Francis, R.C., and T.H. Sibley. 1991. Climate change and fisheries: what are the real issues? Northwest Environ. J. 7(2): 295–307.

Hargreaves, N.B., and R.J. LeBrasseur. 1985. Species selective predation on juvenile pink (*Oncorhynchus gorbuscha*) and chum salmon (*O. keta*) by coho salmon (*O. kisutch*). Can. J. Fish. Aquat. Sci. 42: 659–668.

Hartt, A.C. 1980. Juvenile salmonids in the oceanic ecosystem—the critical first summer, p. 25–57. *In* W.J. McNeil and D.C. Himsworth [ed.] Salmonid ecosystems of the North Pacific. Oregon State University Press, Corvallis, OR.

Healey, M.C. 1980. The ecology of juvenile salmon in Georgia Strait, British Columbia, p. 203–230. *In* W.J. McNeil and D.C. Himsworth [ed.] Salmon ecosystems of the North Pacific. Oregon State University Press, Corvallis, OR.

Healey, M.C. 1982. Timing and relative intensity of size-selective mortality of juvenile chum salmon (*Oncorhynchus keta*) during early sea life. Can. J. Fish. Aquat. Sci. 39(7): 952–957.

Healey, M.C. 1991. Diets and feeding rates of juvenile pink, chum and sockeye salmon in Hecate Strait, British Columbia. Trans. Am. Fish. Soc. 120(3): 303–318.

Helle, J.H. 1970. Biological characteristics of intertidal and freshwater spawning pink salmon at Olsen Creek, Prince William Sound, Alaska, 1962–1963. U.S. Fish Wildl. Serv. Spec. Sci. Rep. Fish. 602: 19 p.

Kirkwood, J.B. 1962. Inshore-marine and freshwater life history phases of the pink salmon (*Oncorhynchus gorbuscha*) and the chum salmon (*Oncorhynchus keta*) in Prince William Sound, Alaska. Ph.D. thesis, University of Louisville, Louisville, KY. 300 p.

Mortensen, D.G. 1983. Laboratory studies on factors influencing the first feeding of newly emerged pink salmon (*Oncorhynchus gorbuscha*) fry. M.S. thesis, University of Alaska, Juneau, AK. 107 p.

Mortensen, D.G., and A.C. Wertheimer. 1987. Residency and growth of juvenile pink salmon in (*Oncorhynchus gorbuscha*) in Auke Bay. APPRISE Annual Report 1987. School of Fisheries and Ocean Sciences, University of Alaska, Fairbanks, AK. Tech. Rep. 1: 502–544.

Olson, J.M., and W.J. McNeil. 1967. Research on pink salmon at Little Port Walter, Alaska. U.S. Fish Wildl. Serv. Data Rep. 17: 301 p.

Parker, R.R. 1964. Estimation of the sea mortality rates for the 1960 brood-year pink salmon of Hook Nose Creek, British Columbia. J. Fish. Res. Board Can. 21(5): 1019–1034.

Parker, R.R. 1965. Estimation of sea mortality rates for the 1961 brood-year pink salmon of the Bella Coola Area, British Columbia. J. Fish. Res. Board Can. 22(6): 1523–1554.

Parker, R.R. 1968. Marine mortality schedules of pink salmon of the Bella Coola River, central British Columbia. J. Fish. Res. Board Can. 25(4): 757–794.

Parker, R.R. 1971. Size selective predation among juvenile salmonid fishes in a British Columbia inlet. J. Fish. Res. Board Can. 28: 1503–1510.

Peterman, R.M., and M. Gatto. 1978. Estimation of functional responses of predators on juvenile salmon. J. Fish. Res. Board Can. 35: 797–808.

Royer, T.C. 1989. Upper ocean temperature variability in the northeast Pacific Ocean: is it and indicator of global warming? J. Geophys. Res. 94(c12): 18 175 – 18 183.

Salmon, D.K. 1992. On interannual variability and climate change in the North Pacific. Ph.D thesis, University of Alaska Fairbanks, AK. 219 p.

Sharp, D. 1991. Methodology for assessing injury to wild pink salmon using coded-wire tagging, p. 168–172. *In* B. White and I. Guthrie [eds.] Proceedings 15th Northeast Pink and Chum Salmon Workshop (27 Feb. – 1 Mar. 91, Parkville, B.C.). Pacific Salmon Commission, Vancouver, B.C.

Taylor, S.G. 1988. Inter- and interannual survival of pink salmon (*Oncorhynchus gorbuscha*) returning to Auke Creek, Alaska in 1986 and 1987. APPRISE Annual Report 1987. School of Fisheries and Ocean Sciences, University of Alaska, Fairbanks, AK. Tech. Rep. 1: 547–571.

Urquhart, D.L. 1979. The feeding, movement and growth of pink salmon, *Oncorhynchus gorbuscha*, fry released from a hatchery in Prince William Sound, Alaska. M.S. thesis, University of Alaska, Fairbanks, AK. 111 p.

Walters, C.J., J. Holborn, R. M. Peterman, and M.J. Staley. 1978. Model for examining early ocean limitation on Pacific salmon production. J. Fish. Res. Board Can. 35: 1303–1315.

APPRISE: A multi-year investigation of environmental variation and its effects on larval recruitment

Paul K. Bienfang and David A. Ziemann

Bienfang, P.K., and D.A. Ziemann. 1995. APPRISE: A multi-year investigation of environmental variation and its effects on larval recruitment, p. 483–487. *In* R.J. Beamish [ed.] Climate change and northern fish populations. Can. Spec. Publ. Fish. Aquat. Sci. 121.

Abstract: From 1985–89, the APPRISE program conducted comprehensive environmental, water quality, and plankton sampling in Auke Bay in southeastern Alaska. The effects of environmental variability on the development patterns of the spring bloom and the resultant influences on the larvae of selected fish and shellfish were studied. Variability in the early spring wind and light patterns influenced characteristics of the spring phytoplankton bloom in ways that were significant to the growth and survival of larvae at higher trophic levels. Differences in the growth rates and survival of key herbivores (e.g., king crab larvae, *Paralithodes camtschatica*, and copepods, *Pseudocalanus* spp.) were associated with interannual variations in total phytoplankton production, species composition, and biomass. Early larval growth rates of walleye pollock, *Theragra chalcogramma*, were strongly correlated with the interannual differences in abundance of the principal prey (i.e., *Pseudocalanus* spp. nauplii). During the spring, this subarctic ecosystem was uncoupled between the first and second trophic levels, but tightly coupled between the herbivore and carnivore levels.

Résumé : De 1985 à 1989, le programme APPRISE a permis de procéder à un échantillonnage global de l'environnement, de la qualité de l'eau et du plancton dans la baie Auke au sud-est de l'Alaska. On a étudié les effets de la variabilité environnementale sur le développement de la prolifération printanière du phytoplancton et les conséquences qui en découlent pour les larves de poissons et de crustacés et mollusques choisis. La variabilité des conditions relatives au vent et à la lumière au début du printemps ont influé sur les caractéristiques de la prolifération printanière de phytoplancton de façon importante pour la croissance et pour la survie des larves à des niveaux trophiques supérieurs. Les différences dans les taux de croissance et dans la survie des herbivores essentiels (par exemple les larves de crabe royal, *Paralithodes camtschatica*, et les copépodes, espèce *Pseudocalanus*) ont été associées aux variations interannuelles de la production totale de phytoplancton, de la composition des espèces et de la biomasse. Les taux de croissance des premiers stades larvaires de morue du Pacifique occidental, *Theragra chalcogramma*, étaient en forte corrélation avec les différences interannuelles de l'abondance de la proie principale (c'est-à-dire l'espèce *Pseudocalanus* au stade de nauplius). Au printemps, cet écosystème subarctique était dissocié entre les premier et second niveaux trophiques, mais fermement associé entre les niveaux herbivore et carnivore.

Introduction

This paper highlights major technical findings from an interdisciplinary fisheries program. The purpose is to exemplify the value of fisheries recruitment programs that are comprehensive, institutionally cooperative, long term, and goal oriented by management design. We explain how one such effort, the Association of Primary Production and Recruitment in a Subarctic Ecosystem (APPRISE) Program, was organized and implemented. The remainder of the paper discusses key findings concerning elements of the food web that support commercially important stocks.

APPRISE Program Configuration

APPRISE was organized as a multiyear, multi-institutional, multidisciplinary program. Its goal was to improve the capability of predicting larval recruitment through the identification of relationships between environmental factors, primary and secondary production, and the recruitment success of selected economically important fish and shellfish. The program hypothesis was that the interannual variability in the recruitment potential of these selected larvae is related to the

P.K. Bienfang and **D.A. Ziemann.** The Oceanic Institute, Makapuu Point, P.O. Box 25280, Honolulu, Hawaii 96825, USA.

environmental variability during spring that affects food web dynamics. From this were derived a series of nine working hypotheses from which technical objectives were specified. Thus, the "recruitment" purview of this program was limited exclusively to larval dynamics occurring during the spring. In this, the focus was directed toward identification of circumstances associated with exceptional organism survival, rather than mortality.

The program used a high-frequency (three times per week) sampling regime to provide necessary resolution and used consistent methodology year to year (Ziemann and Fulton-Bennett 1990). Data gathering consisted of both continuous automated monitoring, and regular discrete samplings, and embraced a comprehensive suite of environmental and planktonic parameters. Specific sampling strategies and units used for data reporting were selected to assure maximum applicability among various components. During the sampling season, there was expeditious exchange of all forms of data. Annually, participating investigators within each component were required to produce yearly summaries of data, technical findings, and cumulative interpretations. The program overview and the annual technical summaries were regularly reviewed by both the executive committee and an external evaluation panel having expertise in fisheries recruitment programs.

The APPRISE investigations in Auke Bay were designed in part to be a pilot experiment from which results and strategies could be applied within more extensive, remote systems with significant fisheries populations. Details of the program and the technical results are presented in Ziemann and Fulton-Bennett (1990). Some of the technical findings are summarized below for the following key food web components: walleye pollock (*Theragra chalcogramma*) larvae, red king crab *(Paralithodes camtschatica)* larvae, calanoid copepods, and phytoplankton.

Results

Walleye pollock (*Theragra chalcogramma*) larvae

Haldorson et al. (1990) showed a tight coupling between the growth rate of early larval pollock and their principal prey, the nauplii of small copepods, primarily *Pseudocalanus* spp. Early larval growth rates are important in determining larval survival and thus recruitment potential because the smallest and slowest growing larvae suffer the highest mortality from predation. Correlations of spring temperature with growth rate refuted the hypothesis that temperature fluxes were significantly correlated with observed growth-rate variability, either intra- or inter-annually. In Auke Bay, prey availability was the primary cause of variations in growth rate.

L. Haldorson (University of Alaska, Fairbanks, AK, unpublished data) has shown a positive correlation between larval walleye pollock growth rate and *Pseudocalanus* spp. nauplii abundance for all cohorts of walleye pollock for all years. By comparing the temporal distributions of walleye pollock larvae and *Pseudocalanus* spp. nauplii throughout the spring seasons, Haldorson et al. (1990) showed that some walleye pollock cohorts coincided with the highest prevailing nauplii abundances. These synchronous cohorts had the highest growth rates within any given year. A plot of the growth rates of these synchronous cohorts versus prey density (Fig. 1) shows the close correlation ($r^2 = 0.98$) between growth rate and prey abundance. This indicates that even the maximum observed larval pollock growth rates were controlled by prey abundance. The absence of any plateau in Fig. 1 also indicates that food was not saturating even at the highest prey concentrations observed (i.e., <30 nauplii/L).

Thus, variations in the larval recruitment potential of walleye pollock were shown to be dependent on natural variations in prey abundance, and in turn were influenced by the forces affecting prey abundance.

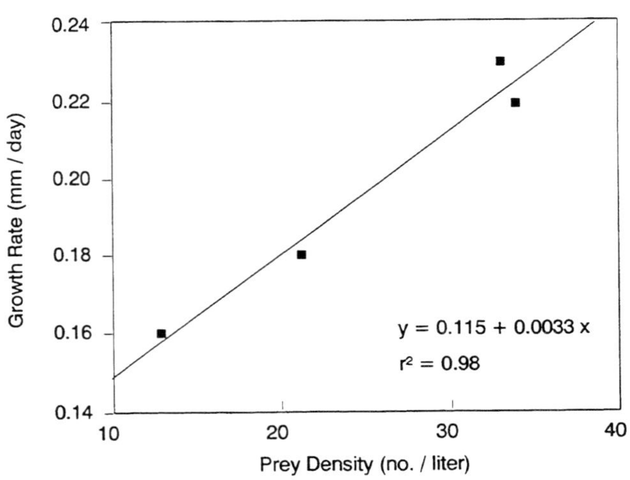

Fig. 1. Relationship between growth rates of larval walleye pollock cohorts that were synchronous with peaks in copepod nauplii abundance, and maximum copepod nauplii abundance during such peaks.

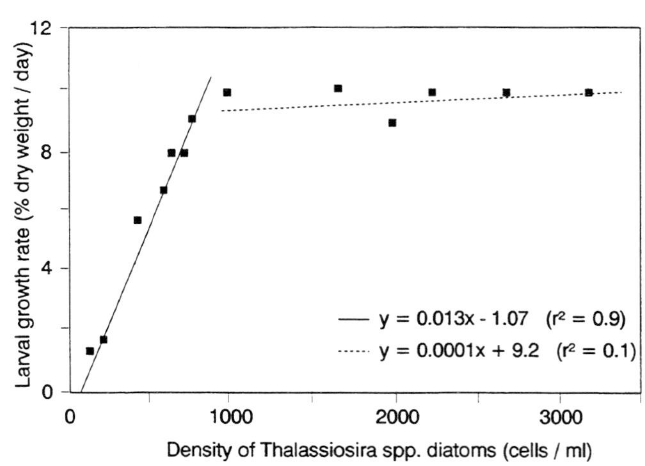

Fig. 2. Relationship between larval king crab growth rates and concentrations of *Thalassiosira* spp.

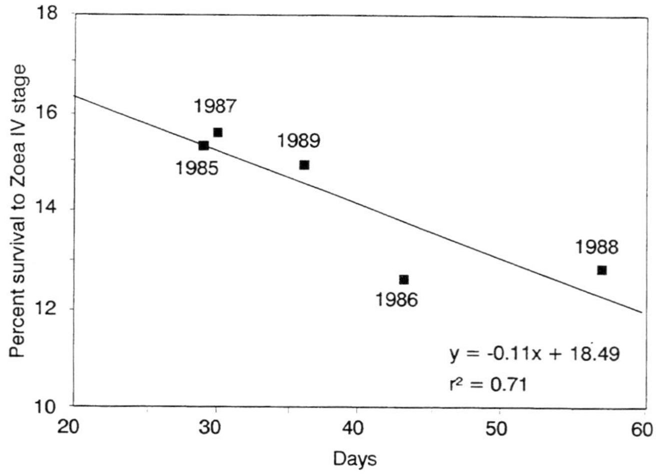

Fig. 3. Relationship between survival of king crab larvae to zoea IV stage and the duration of the larval period. The larval period was defined as the number of days that 90% of the larvae were observed in the water column. Values for 1985 were extrapolated from a regression for that year.

King crab *Paralithodes camtschatica* larvae

The investigations of Shirley and Shirley (1990), Paul (1990), and Paul et al. (1990) have focused on factors influencing larval crab survival during the spring. Studies of the herbivorous zoea I stage showed that larval growth rates were limited by phytoplankton concentrations up to 1000 *Thalassiosira* spp. cells/mL; above that, feeding was saturated (Fig. 2). Paul (1990) also showed that cells of *Thalassiosira* spp. provided superior nutrition to those of other genera. Survival of king crab larvae to the subsequent (carnivorous) zoea III stage was positively related to the density of copepod nauplii prey (Shirley and Shirley 1990).

Over the 5 yr study period, highly significant relationships were found between the duration of the larval period and the chlorophyll concentration (Shirley and Shirley 1990) and between the survival to zoea IV stage and the length of the larval period (Fig. 3). Other analyses showed a lack of correlation between spring temperature regimes and larval king crab survival. This is due in part to the small temperature variations during the period. Such findings support the hypothesis that the availability of herbivorous and carnivorous prey, rather than temperature variability, primarily influence the duration of the larval period, and thus the survival of king crab larvae during the spring.

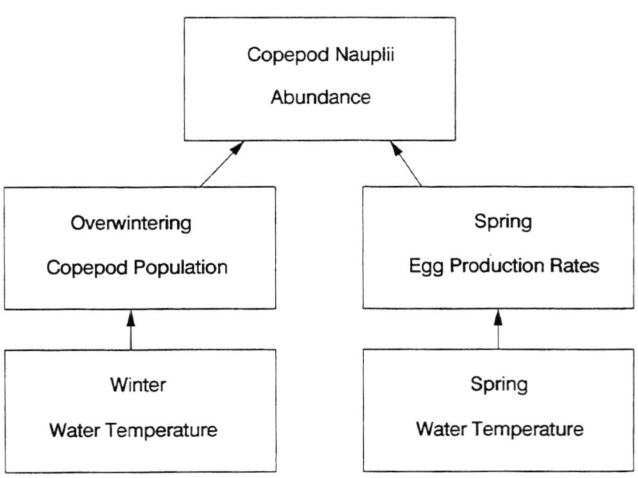

Fig. 4. Schematic of factors controlling *Pseudocalanus* nauplii abundance.

Copepod populations

In the Auke Bay ecosystem, the abundance of *Pseudocalanus* spp. nauplii (the principal prey items referred to above) was not appreciably influenced by variations in their prey, i.e., phytoplankton concentrations. Auke Bay was shown to be an uncoupled system during the spring; that is, the phytoplankton concentrations were nearly always far in excess of levels that would saturate copepod grazing (Paul 1990).

The abundance of copepod nauplii was primarily determined by (1) the numerical abundance of the initial (i.e., overwintering) population of *Pseudocalanus* spp. females, and (2) the rates of growth and egg production by females during the spring (Fig. 4). These two factors in turn were influenced primarily by physical factors in the environment. The size of the overwintering copepod population was affected by the water temperatures they experienced during the winter. The warmer the water in Auke Bay remained over the winter, the larger the initial population of females. Similarly, under the saturated feeding conditions prevailing, the rates of copepod growth and egg production were directly related to water temperatures during the spring.

Phytoplankton populations

From a fisheries recruitment perspective, the relevant aspects of the spring phytoplankton bloom are: (1) initiation date, (2) bloom duration, (3) maximum concentration, (4) species composition, and (5) fate of the phytoplankton biomass produced. The predominant environmental factors that exert control over the dynamics of the spring phytoplanton bloom in Auke Bay are summarized in Table 1. The timing of bloom-related events is summarized in Fig. 5. Variation in the initiation date of the spring bloom could affect the survival of herbivorous larvae, depending on the synchronicity of hatching with the bloom. However, Ziemann et al. (1991) showed that in the Auke Bay system the date of bloom initiation was strikingly consistent among years (coefficient of variation = 3%). This reflects the fact that available light depends on sun angle and day length, which are nearly identical from year to year.

Table 1. Controlling factors of spring phytoplankton variability.

Element	Variability	Control
Initiation Date	Low	Sun Angle Day Length
Duration	Medium	Wind Cloud Cover
Species Composition	Medium	Wind Cloud Cover
Bloom Peak Characteristics	High	Wind Cloud Cover
Fate of Algal Biomass	High	Wind Cloud Cover

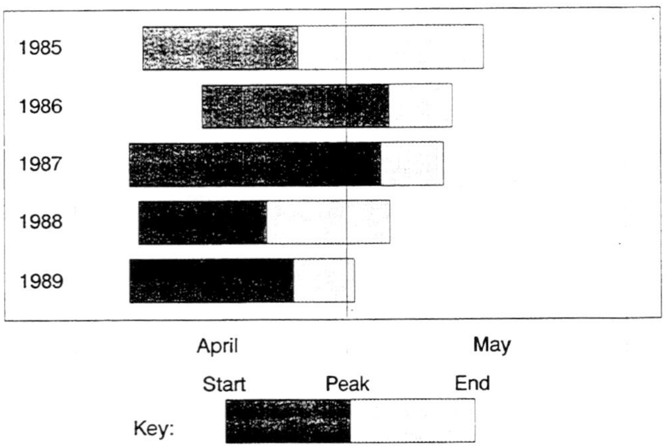

Fig. 5. Schematic of the phytoplankton spring bloom events in Auke Bay, Alaska.

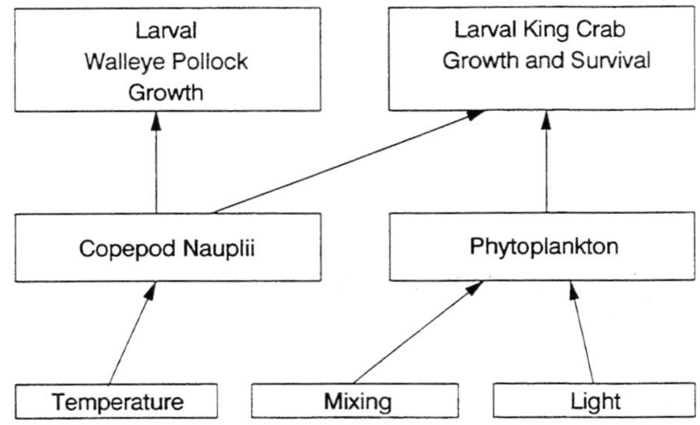

Fig. 6. Schematic of environmental and biotic factors controlling recruitment potential of selected species in Auke Bay.

The duration of the bloom is a function of short-term weather conditions, i.e., wind and cloud cover. Strong winds induce vertical mixing and can extend the length of the bloom by transporting nutrient-rich aphotic water to the photic zone. Extensive cloud cover reduces the light available to saturate the photosynthetic capability and thus the time required to exhaust available nutrients. Over the 5 yr, the bloom duration ranged from 25 to 38 days (Figure 6) and displayed more than five times the degree of interannual variation as the initiation date (Ziemann et al. 1991).

The time at which the peak phytoplankton concentration occurred was approximately constant among years and displayed a degree of interannual variability (coefficient of variation = 5%) similar to that for bloom initiation (Ziemann et al. 1991). Peak biomass levels showed a coefficient of variation of about 20%, but because the peak concentration levels exceed those that saturate grazing by the principal herbivores, this variability is not likely to influence larval recruitment within this environment.

The species composition of the phytoplankton bloom also appeared to be influenced by short-term weather conditions and vertical mixing. In years when wind mixing was low and stable water columns were the norm, a single genera (*Thalassiosira* spp.) vastly predominated the phytoplankton assemblage in Auke Bay. During years with greater vertical mixing, a more diverse phytoplankton assemblage was observed. As mentioned above, *Thalassiosira* spp. provide better nutrition to the copepods and crab larvae than other phytoplankton taxa.

About 65–80% of the algal biomass of the spring bloom was consumed by grazers or sank to the bottom uneaten; approximately 20–35% was transported from the bay to surrounding waters (Ziemann et al. 1993). The magnitude of the phytoplankton flux to either the pelagic food web or to the bethos via sedimentation showed considerable interannual variability. The amount of sedimentary flux varied directly with the amount of total production and with bloom duration (Ziemann et al. 1993). Conditions of high light and low wind were associated with short blooms, when a relatively low proportion of carbon was utilized within the pelagic food web. Extended periods of high sedimentation were associated with periods of high cloud cover and frequent wind-mixing events. Within the timeframe of APPRISE, investigations examined, but were unable to identify, any direct response of the benthic organisms to the arrival of the copious amounts of sedimented phytoplankton at the bottom (Fleeger et al. 1989).

The relationships between environmental factors, biological components of the ecosystem, and factors affecting larval recruitment potential are summarized in Fig. 6.

Summary

Recruitment challenges

Fisheries recruitment research faces a variety of challenges, ranging from technical to logistical and institutional. Recruitment is affected by both biotic and abiotic factors, and includes chemical, physical, biological, meteorological, and economic forces. The relative success of adding individuals to populations involves systematic but very complex interactions that occur over extremely wide temporal and spatial scales. The foraging milieu is typified by highly variable biological populations, which serve as prey, competition, or predators. The environments of interest are subject to a multitude of environmental variability, such as annual patterns of seasonal change, short-term weather patterns, and variable physical oceanographic states. Many of the prime fisheries areas are difficult to access, requiring research vessels of significant size to weather the difficult conditions. Because of the physical and biological complexity, adequate descriptions of the dynamics require frequent detailed sampling of long duration. The sampling techniques required are at or beyond the limits of the current technical capabilities. Because appropriate remote or automated sampling capability exists for only a few of the parameters, the physical presence of investigators is normally required throughout the period of interest. Finally, fisheries recruitment research must compete for funds and logistical support with other very visible and high priority research initiatives.

Recruitment program requirements

To address such challenges, ideal fisheries recruitment programs entail a minimum set of requirements. One necessity is a unifying program hypothesis that focuses the limited time and resources on critical points of the recruitment process. Without such focus, the predilections of individual researchers may prioritize related, but tangential, issues to those central to the program hypothesis. The diversity of capabilities needed to embrace the biological and physical complexity implies the requirement for multidisciplinary and, in fact, multi-institutional teams. Recruitment programs must be of multi-year duration. Interannual variations in the basic forcing functions and in the biological responses to those functions simply cannot be isolated and understood in a single year or in two or three. Programs must embrace appropriate temporal and spatial envelopes to include the relevant forcing functions and responsive processes. These processes may occur over time scales of days to months and spatial magnitudes encompassing tens to hundreds of kilometers. Sampling strategies that provide the highest resolution at the least cost of resources are critical to project success. Finally, program management, as distinct from the purely technical achievements, must be committed to assuring the requisite logistical, institutional, and financial support.

Scientists have a tendency to focus primarily on the technical aspects of recruitment problems. This unfortunate simplification minimizes the formidable managerial and institutional challenges presented by fisheries recruitment issues. Recruitment programs must be truly multidisciplinary and operated with continual retooling of technical priorities to match the changing understanding of the big picture. The amount and duration of required financial, labor, and logistical resources necessitates cooperative, multi-institutional efforts among academic, governmental, and private participants. A broad scope is crucial to drawing the defensible conclusions needed to further effective fisheries management.

Acknowledgments

The APPRISE program was a collaborative effort between the University of Alaska's School of Fisheries and Ocean Science, the Institute of Marine Science, Alaska, and The Oceanic Institute, Hawaii. Thanks are due to Executive Committee members Dr. V. Alexander, Dr. J. Goering, Dr. O. Mathison, and Dr. G. Snyder. The program was supported by contract NA-85-ABH-00022 from the U.S. Department of Commerce, National Oceanographic and Atmospheric Administration (NOAA), and benefitted from continuous logistical support from University of Alaska's Juneau School of Fisheries and Ocean Science, and the Auke Bay Laboratory, National Marine Fisheries Services (NMFS).

References

Fleeger, J.W., T. Shirley, and D. Ziemann. 1989. Meiofaunal responses to sedimentation from an Alaskan spring bloom. I. Major taxa. Mar. Ecol. Prog. Ser. 57: 137–145.

Haldorson, L., M. Pritchett, D. Sterritt, and J. Watts. 1990. Interannual variability in the recruitment potential of larval fishes in Auke Bay, Alaska, p. 319–355. *In* D.A. Ziemann and K.W. Fulton-Bennett [ed.] APPRISE — interannual variability and fisheries recruitment. The Oceanic Institute, Honolulu, HI.

Paul, A.J. 1990. Copepod nauplii as prey for larval pollock, p. 243–262. *In* D.A. Ziemann and KW. Fulton-Bennett [ed.] APPRISE — interannual variability and fisheries recruitment. The Oceanic Institute, Honolulu, HI.

Paul, A.J., K. Coyle, and D. Ziemann. 1990. Variations in egg production rates by *Pseudocalanus spp.* in a subarctic Alaskan bay during the onset of feeding by larval fish. J. of Crustacean Biol. 10(4): 648–658.

Shirley, T.C., and S. Shirley. 1990. Planktonic survival of red king crab larvae in a subarctic ecosystem, 1985–1989, p. 263285. *In* D.A. Ziemann and K.W. Fulton-Bennett [ed.] APPRISE — interannual variability and fisheries recruitment. The Oceanic Institute, Honolulu, HI.

Ziemann, D.A., P. Bienfang, K. Fulton-Bennett, W. Kottenmeier, and L. Conquest. 1993. Carbon budget for the spring bloom in Auke Bay, Alaska. Mar. Biol. 115: 509–521.

Ziemann, D.A., L. Conquest, M. Olaizola, and P.K. Bienfang. 1991. Interannual variability in the spring phytoplankton bloom in Auke Bay, Alaska. Mar. Biol. 109: 321–334.

Ziemann, D.A., and K.W. Fulton-Bennett [ed.]. 1990. APPRISE — interannual variability and fisheries recruitment. The Oceanic Institute, Honolulu, HI. 400 p.

A paleoscience approach to estimating the effects of climatic warming on salmonid fisheries of the Columbia River basin

James C. Chatters[1] and Virginia L. Butler[2], Michael J. Scott, David M. Anderson, and Duane A. Neitzel

Chatters, J.C., V.L. Butler, M.J. Scott, D.M. Anderson, and D.A. Neitzel. 1995. A plaeoscience approach to estimating the effects of climatic sarming on salmonid fisheries of the Columbia River basin, p. 489–496. *In* R.J. Beamish [ed.] Climate change and northern fish populations. Can. Spec. Publ. Fish. Aquat. Sci. 121.

Abstract: Efforts to estimate the effect of climate change on fisheries are hampered by the lack of models that project realistic aquatic habitat conditions at the regional scale. Data from the paleosciences are a suitable alternative both for environmental scenario development and model validation. We are using a paleoscience approach to calculate the potential effect of global warming on anadromous salmonid stocks of the Columbia River basin, western North America. First, archaeologically dated fluvial sediments and bivalves were used with terrestrial paleoecological data to reconstruct the flow, flow patterns, temperatures, and bed conditions 6000–7000 yr ago, when paleoclimatic indicators and atmospheric models suggest regional temperatures were up to 2°C warmer. Next, these conditions were imposed on Columbia system subbasins and their effects on salmon stocks were modeled. Results thus far indicate a 30–60% decline in salmon stocks relative to current conditions. Finally, fish remains from archaeological sites were analyzed for evidence of actual salmon production under the reconstructed stream conditions to assess the validity of model projections. Preliminary findings are comparable to model predictions.

Résumé : Les efforts visant à estimer les effets du changement climatique sur les pêches sont entravés par le manque de modèles exprimant des conditions réalistes pour les habitats aquatiques à l'échelle régionale. Les données des paléosciences constituent une option valable autant pour l'élaboration de scénarios environnementaux que pour la validation des modèles. Nous utilisons la démarche paléoscientifique pour calculer l'effet éventuel du réchauffement global sur les stocks de salmonidés anadromes du bassin du fleuve Columbia, dans l'ouest de l'Amérique du Nord. En premier lieu, nous nous sommes servis de sédiments fluviaux archéologiquement datés et de bivalves en plus des données paléoécologiques terrestres, pour reconstituer le débit, les configurations d'écoulement, les températures et les conditions du lit du fleuve il y a 6 ou 7 000 ans, époque où les indicateurs paléoclimatiques et les modèles atmosphériques permettent de croire que les températures régionales étaient jusqu'à 2° C plus chaudes. Par la suite, nous avons appliqué ces conditions sur les sous-bassins du bassin du Columbia et avons modélisé leurs effets sur les stocks de saumons. Les résultats jusqu'à maintenant font état d'un déclin de 30 à 60 % des stocks de saumons par rapport aux conditions actuelles. Enfin, nous avons analysé les fossiles de poissons de sites archéologiques pour trouver des preuves de la production réelle de saumons selon les conditions reconstituées du fleuve en vue d'évaluer la validité des projections faites avec le modèle. Les constatations préliminaires sont comparables à celles qui résultent du modèle.

J.C. Chatters and **V.L. Butler.** Environmental Sciences Department, Pacific Northwest Laboratory, P.O. Box 999, Richland, WA 99352, USA.

M.J. Scott and **D.M. Anderson.** Technology Policy Analysis Department, Pacific Northwest Laboratory, P.O. Box 999, Richland WA, 99352, USA.

D.A. Neitzel. Environmental Sciences Department, Pacific Northwest Laboratory, P.O. Box 999, Richland, WA 99352, USA.

[1] Present address: North American PaleoScience, 648 Saint Street, Richland, WA 99352, USA.

[2] Present address: University Museum, University of Colorado, Boulder, CO 80309, USA.

Introduction

The standard approach to predicting the effect of climatic changes on fish populations is a three-part process. The researcher begins either by conducting a careful analysis of the behaviour, habitat needs, and reproductive biology of the target organism and the influence of temperature on these variables, or by making a series of observations on population response to small-scale climatic variations occurring within the present-day equilibrium climate. Observations are then analyzed to generate an empirical model of a species' response to environmental conditions. Finally, the potential effects of future climates are inferred by driving the empirical model with scenarios generated by general circulation models (GCMs) and extrapolating from the effects of observed climatic extremes (see, for example, DeAngeles and Cushman 1990).

This approach produces thought-provoking results, as many of the papers in this symposium demonstrate. Certainly, the careful collection of empirical data and the derivation of empirical models therefrom is indispensable to our understanding of climatic effects on an important resource base. However, the plausibility of predictions based on GCMs and extrapolation from recent climatic variation is uncertain. Currently available GCM-generated climatic scenarios address the globe at grid scales that often mask regional-scale variability; regional-scale models derived from GCMs are still in the development stage. Therefore, to predict the effects of large-scale climate changes on fisheries, it is necessary to assume that great rises in air and sea-surface temperatures, for example, will produce regional climate changes of the same type, but simply of greater magnitude than those observed historically. This assumption may be invalid; historical records contain evidence of climatic fluctuations within one equilibrium state, but air temperature increases of 2–5°C may generate new equilibria, with regional effects that are not mere extrapolations of what we currently observe. In their present forms, no two GCMs agree on the regional-scale characteristics of future climates (Cushman and Spring 1989; Kalkstein 1991).

At Pacific Northwest Laboratory, we have been exploring the use of data from the paleosciences as a solution to this dilemma. Paleosciences are the disciplines, including geology, paleoecology, paleontology, paleohydrology, paleooceanography, and archaeology, that use the physical traces and remnants of processes and organisms to reconstruct and under-

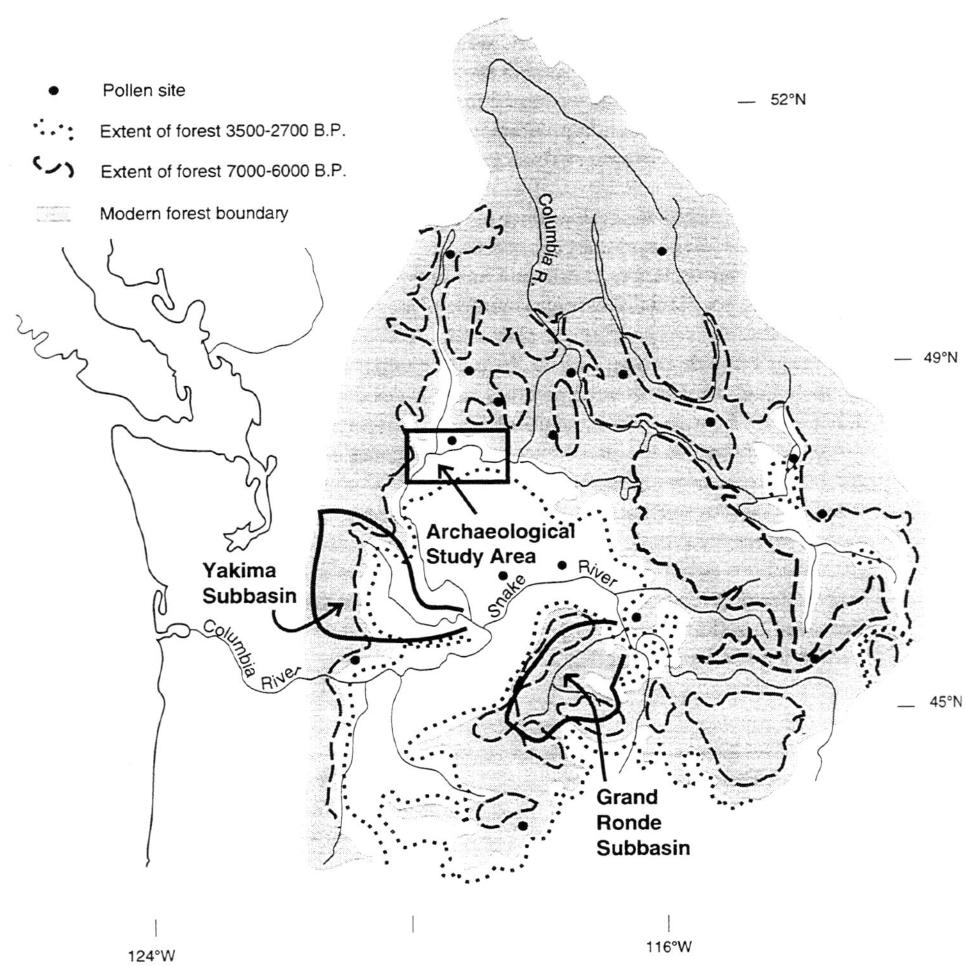

Fig. 1. The northern portion of the Columbia River Basin, showing the source of archaeological data used in this study, the Yakima and Grand Ronde subbasins, and generalized vegetation cover. Approximate forest distributions at 6000–7000 yr B.P. and 3500–2700 yr B.P. are reconstructed from pollen data.

stand ecosystems and cultures of the past. Such data can be used to generate plausibl, even probable, environmental analogs for possible future climate states and to validate dynamic environmental models (e.g., COHMAP members 1988).

We are applying the paleoscience approach to estimate the potential effect of climates 2°C warmer than today on the anadromous salmonids of the Columbia River basin, western North America (Chatters et al. 1991; Neitzel et al. 1991) and to validate the model used in that analysis. The results are intended to assist the Northwest Power Planning Council (NPPC), Bonneville Power Administration, and other resource-management agencies that are developing initiatives to enhance salmonid productivity in this basin. The study is also designed to help the U. S. Department of Energy and the U. S. Environmental Protection Agency identify the consequences global warming might have for water resources and national energy policy. This work is still in progress, so the results reported here are preliminary.

The paleoscience approach

To understand how climatic warming might affect salmonid populations, we are following a four-step procedure, incorporating paleoenvironmental data at the beginning and ending points. First, we used geomorphic, paleobotanical, and paleomalacological data to reconstruct stream conditions during the last 8000 yr. Second, we estimated the effect on salmon (*Oncorhynchus*) of conditions extant approximately 6000–7000 radiocarbon years before present (B.P.), when temperatures were as much as 2°C warmer than at present (Kutzbach 1983). This became an analog of future warmer climate and its effects on spawning, incubation, and rearing parameters of the NPPC's Tributary Parameter Model (TPM) for estimating salmonid production. Third, we ran the TPM in conjunction with the NPPC System Planning Model (SPM) to calculate the effect of these analog conditions on the population of returning adult fish in selected stream systems. Ultimately, we will run the models for all salmon-accessible subbasins of the Columbia River system. Finally, we are identifying fish remains obtained from archaeological sites along the Columbia River to compare variations in the taxonomic composition of ancient fish assemblages with model predictions.

Reconstructing ancient stream environments

Estimates of stream flows, annual flow patterns, bed characteristics, and temperatures have been obtained by analyzing past vegetation patterns, geologic stratigraphy of archaeological sites, and archaeological collections of freshwater mussels (Chatters 1989; Chatters and Hoover 1992). Data for this reconstruction come primarily from the Wells Reservoir Archaeological Project in Okanogan and Douglas Counties, Washington (Chatters 1986). Associated climate scenarios are derived from a synthesis of paleoclimatic indicators, including pollen sequences, cave and lake sediments, subfossil wood, and geomorphic evidence of glacial activity (see Chatters and Hoover 1992).

The distribution of vegetation, as indicated by studies of fossil pollen sequences throughout the river basin (Fig. 1; see Mehringer 1985), was used to estimate mean annual flows; the amount of water available for runoff is assumed to be proportional to that available for plant growth. Estimated flows ranged from 30 to 40% below modern levels under warm, arid climates prior to 6000 yr B.P. to as much as 30% above modern levels in the coolest, wettest times (Chatters and Hoover 1992).

Composition of the bed and suspended sediment loads of ancient streams were inferred partially from the texture, bedding structure, and rate of build-up of fluvial sediments, but primarily from the taxonomic composition of mussel assemblages (Chatters et al. 1991). There are two species of large mussels occurring in the flowing water of the U. S. Pacific northwest. *Margaritifera falcata* prefers swift-flowing, clear, gravel-bedded, stable streams, while *Gonidea angulata* favours slower water, higher suspended sediment loads, sand to silt bottoms, and unstable beds (Lyman 1980; Vannote and Minshall 1982). The percent *Gonidea* in assemblages from a time interval was taken as a crude indicator of the percent sand and silt in the river's bed. *Gonidea* ranged from zero between 300 and 700 yr B.P. to nearly 40% around 6700 yr B.P. (Fig. 2; Chatters 1986).

The annual distribution of flow, or hydrograph, was partially reconstructed from the season at death of river mussels from archaeological middens. Assuming, as one Native American elder stated, that people consumed mussels at any time of year, as long as they were available, and knowing that high stream flows make mussels unavailable as a result of water velocity and turbidity, the period when people were not collecting mussels indicates the period of peak flow. The pre-dam Columbia River, which was snowmelt fed, rose in May, peaked in mid-June during the freshet, and fell to near base flow by mid-July. Season at death for mussels was determined by measuring annual growth increments along the resilial tuberosity, where the hinge ligament attaches to the shell (Chatters 1987). A growth index was calculated by dividing the final growth increment by the expected growth for the death year. The growth indices were compared to the annual growth curve for modern *M. falcata* to estimate the time of death for each sample of mussel shells, or collection event. Finally, collection events were graphed by time period to determine the timing of the freshet. Results showed that freshets typically ended by late June between 7900 and 5500 yr B.P., and ended as late as early August between 3300 and 2200 yr B.P. (Fig. 2; Chatters 1986).

Stream temperatures are also derived from increments of mussel shell growth. Unpublished research by Chatters on modern *M. falcata* has shown that growth rate is closely correlated with water temperatures. Mean growth in the ninth and tenth years has a linear relationship with the number of days above 6, 8, and 10°C, with the strongest correlation ($R^2 = 0.89$) at 10°C. Growth rates in mussels from the Wells Reservoir area of the Columbia River indicate remarkable changes in water temperature over the past 8000 yr. Prior to 3900 yr B.P., the water was above 10°C for approximately 200 d, but since that time this has dropped to an average of less than 130 d (Fig. 2).

As an analog of possible future climate, we chose the period between 6000 and 7000 yr B.P., a time during which both paleoscientific indicators (Chatters 1989) and climate models (Kutzbach 1983) indicate temperatures in western North America were approximately 2°C warmer than at present.

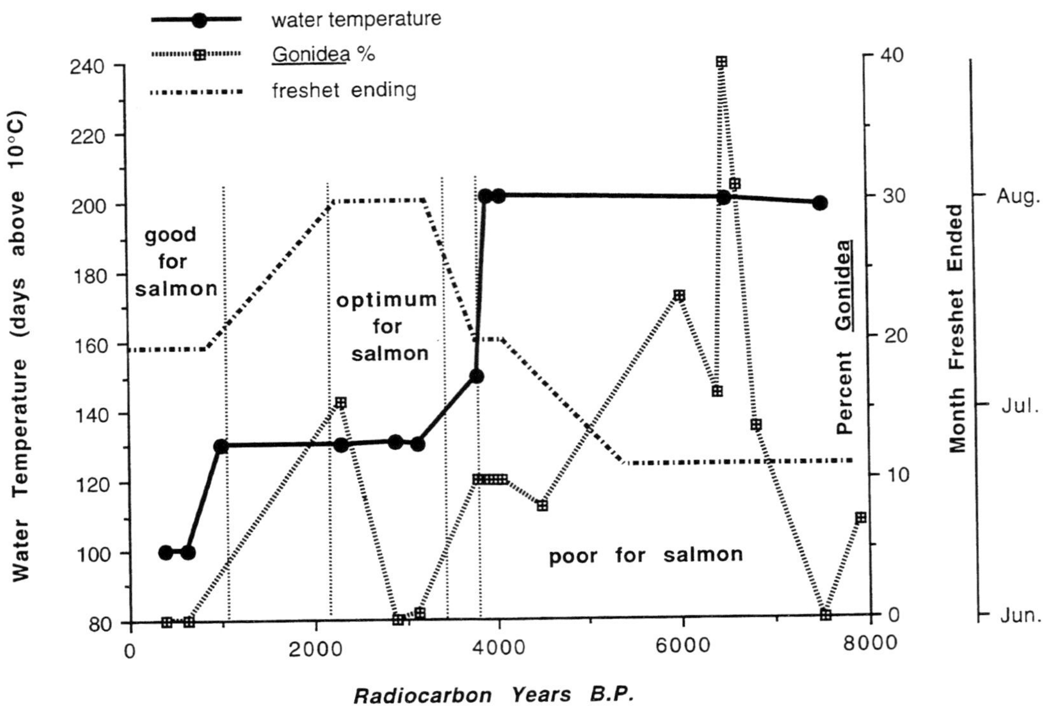

Fig. 2. A summary of evidence for stream conditions in the Wells Reach of the Columbia River since 8000 yr B.P.and their inferred effects on salmon production if they were manifested basinwide.

Based on the four lines of evidence described above, the Columbia River during this period had 30% lower flow, freshets ending nearly a month earlier, temperature above 10°C for nearly twice as long, and a bed with nearly 20% more sand and silt than found under predam conditions.

Altering parameters of the TPM

The NPPC system for simulating salmon production uses both the TPM, which calculates the effects of environmental changes on life stages critical to salmon survival in subbasins, and the SPM, which calculates cumulative salmon survival over multiple generations from TPM outputs, plus data on ocean and main-stem river survival (NPPC 1989, 1992). Although neither is biologically elegant, both models are readily understandable and, more importantly, guide the NPPC's salmon enhancement program.

The TPM contains four key parameters: adult prespawning survival, egg-to-smolt survival, smolt-to-smolt survival, and smolt capacity, which are computed for each reach of a tributary basin. We used reconstructions of 6000–7000 B.P. conditions in the Wells Reach of the Columbia River to modify each of these parameters (Chatters et al. 1991). Water temperature was used to estimate changes in adult prespawning survival because elevated temperatures increase the incidence and severity of disease at this life stage (Becker and Fujihara 1987). Texture of the stream bed was used to alter egg-to-smolt survival because increased sedimentation can suffocate eggs and displace fry (McNeil and Anhel 1964; Hall and Lantz 1969). Freshet timing affects smolt-to-smolt survival by accelerating or retarding outmigration to salt water. If the freshet is too early, smolt may reach salt water before they are physiologically prepared, or if left behind by the freshet, they may experience increased predation during a prolonged migration. Differences between present and past flow rates were used to directly modify smolt capacity, which was assumed to be proportional to stream volume (VanHyning 1973; Smith 1985). For the details of parameter value changes made in one simulation, see Chatters et al. (1991).

Simulating the effects of 6000–7000 yr B.P. stream conditions

Thus far, we have conducted simulations for two tributaries of the Columbia River system: the Yakima subbasin, on the east flank of the Cascade Mountains of Washington, and the Grand Ronde subbasin in mountainous northeastern Oregon (Fig. 1). Simulations estimated the productivity of the spring chinook salmon (*O. tshawytscha*) under a continuation of current environmental conditions and with the 2°C climate warming scenario represented by the 6000–7000 yr B.P. analog. Changes to the four TPM parameters differed slightly between basins largely because of differences in stream gradient and expected residence time of smolts. Relative to current conditions, productivity estimates under climate change were 60% lower for the Yakima subbasin (Chatters et al. 1991) and 30% lower for the Grand Ronde subbasin (Fig. 3). Although we cannot assert that these two subbasins sample the full range of variability in the Columbia River system, they indicate an average decline in productivity of 45%.

Evaluating the model with fish remains from archaeological sites

Human inhabitants of the Columbia River system have been exploiting the fishery for at least 10 000 yr (Cressman 1960; Butler 1993). Presumably, within the limits of available

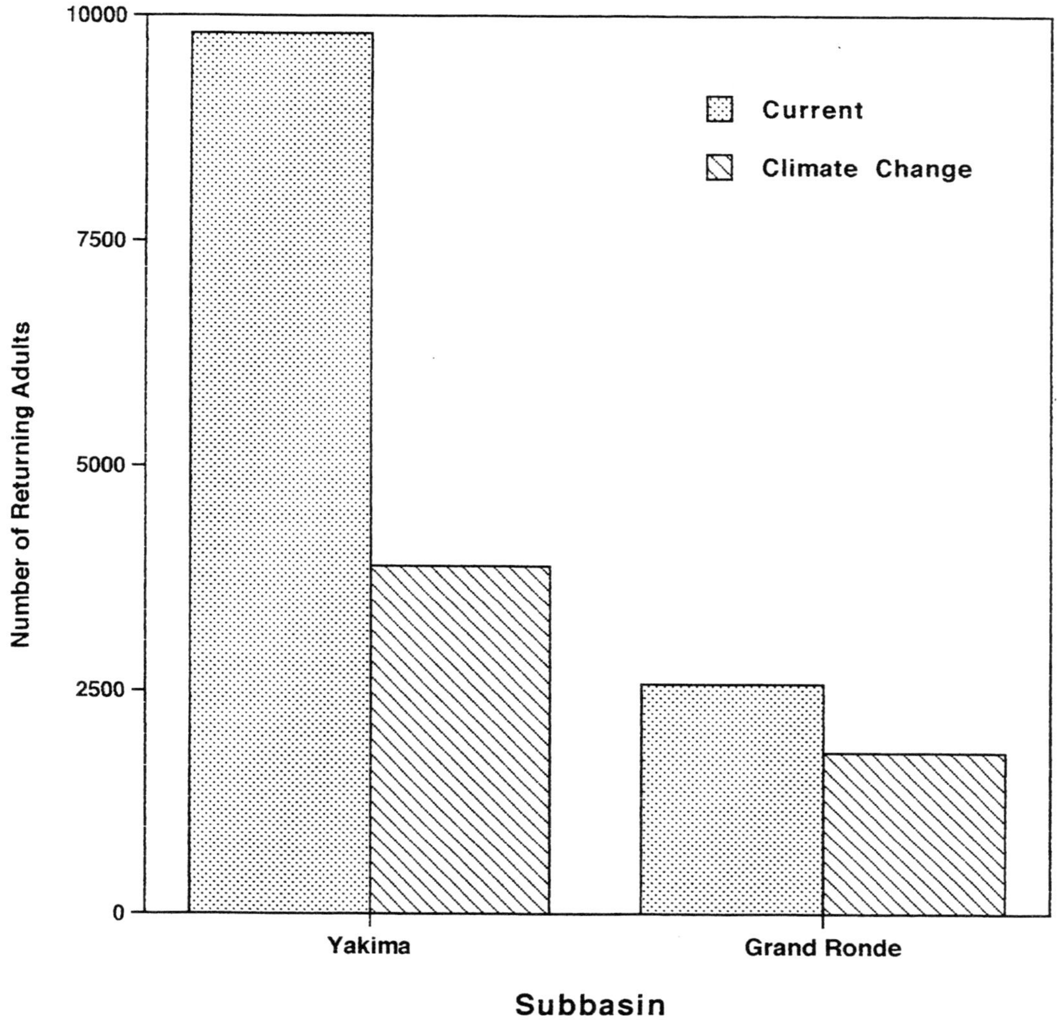

Fig. 3. Results of model runs under current and climate change analog stream conditions for two Columbia River system subbasins.

technology, fish exploitation patterns were consistent with the expectations of optimal foraging theory (MacArthur and Pianka 1966; Winterhalder and Smith 1981). Fish would have been taken in numbers proportional to their size and relative ease of capture. Salmon, among the largest and at times most numerous fishes in the ecosystem, would have been among the first exploited, followed by smaller and/or less numerous or accessible species. Assuming discard of fish remains in habitation sites or kitchen middens and equivalence in the durability of the remains of various species, taxonomic abundances among fish remains in archaeological bone assemblages (archaeofaunas) should reflect availability of various species, particularly the more desirable salmonids. Therefore, archaeofaunas offer data with which to evaluate models of the climate's effect on salmon productivity.

Archaeological research has been conducted in the Columbia River Basin for four decades, stimulated largely by hydroelectric projects along the Columbia and Snake rivers. Archaeofaunas have been collected systematically since the 1960s, providing an extensive data base for model evaluation. We are assembling and systematically identifying these fish archaeofaunas, in an attempt to generate a history of the fishery that is chronologically comparable with the reconstructions of stream conditions.

This part of our research program is the least developed. Thus far, we have analyzed samples from the Chief Joseph Dam Archaeological Project (CJDAP) (Campbell 1985) and Wells Reservoir Archaeological Project (WRAP) (Chatters 1986), two of the largest and most detailed archaeological data recovery efforts yet conducted in the region. Chatters (1986) has identified the fish remains from the WRAP; Butler has re-analyzed selected archaeofaunas from the CJDAP and, to make the data as comparable as possible, is reanalyzing WRAP assemblages. Comparability is somewhat affected by the collection procedures used on the two projects. The CJDAP used 3.2-mm mesh screen for data recovery and sifted the matrix dry; the WRAP used 1.0-mm mesh for 2.5% of the matrix and 6.4-mm mesh for the remainder, all screened using water. Quantification of the CJDAP data is preliminary; each site contained material from multiple habitation events, often spanning several thousand years. We have not yet separated the archaeofaunas by time period, but in each case, over 75%

of th fish remains come from one time interval. A general picture of fish use through time can be obtained by assuming that all fish in a site come from the dominant time interval.

To date, we have results on 13 archaeofaunas from 10 sites, ranging in age from 300 to nearly 8000 yr B.P. Although the oldest archaeofaunas date within the time period of our analog stream environment, we are uncertain how to use them. Before 4500 yr B.P., occupants of the region moved frequently throughout the year, exploiting foods that were in season at each locality (Campbell 1985; Chatters 1986). For this reason, the three archaeofaunas we have from this period may not fully represent the exploitation of Columbia River fish. Analyzed sites that postdate 4500 yr B.P., however, consist almost entirely of pithouse-containing base camps, which were either occupied year round or were the winter setting for consumption of foods gathered during all times of year (Lohse 1984a, b; Chatters 1986). The remainder appear to have been fishing camps (Miss 1984a, b).

As a preliminary evaluation of the model, we can compare the fish archaeofaunas from a period of time with stream conditions similar to those of 6000–7000 yr B.P. with those times when stream conditions existed under an essentially modern climate. Stream conditions comparable to those existing under the warmer climate persisted until approximately 3900 yr B.P. (Fig. 2). Although the freshet ended at approximately the same time as it does now, water temperatures continued to be warm and the stream bed contained a significant amount of sediment. Conditions for salmonid production before 3900 yr B.P. can be characterized as poor, equivalent to the results of climate change simulations for the Yakima and Grand Ronde subbasins. The period after 1200 yr B.P., with modern freshet timing, much cooler waters, and variable amounts of bed sedimentation, is equivalent to the current climate scenario. We characterize these conditions as good. Between 3400 and 2300 yr B.P., conditions were optimum for salmon production, with cool waters, a belated freshet, and virtually no bed sedimentation. Data are largely lacking for the periods between 1200 and yr 2300 yr B.P. and between 3400 and 3900 yr B.P., and we make no suggestions about their suitability for salmon.

Representatives of six genera have been identified from the fish archaeofaunas: *Oncorhynchus* (steelhead, trout, and salmon), *Prosopium* (whitefish), *Catostomus* (suckers), *Ptychocheilus* (squawfish), *Mylocheilus* (peamouth), and *Acipenser* (sturgeon). For this analysis, we present only the percentage of *Oncorhynchus* in each assemblage (Table 1).

Fish archaeofaunas dating between 300 and 4500 yr B.P. closely follow the expected pattern (Fig. 4). Results from the CJDAP and WRAP are remarkably similar, despite the differences in analysts and data-recovery techniques. The percentage of *Oncorhynchus* is lowest before 3700 yr B.P. (mean = 23.2%), highest from 3300 to 2200 yr B.P. (mean = 95.5%) and moderate thereafter (mean = 69.8%). The difference in *Oncorhynchus* frequencies between our analog for current conditions and the warmer climate scenario is 66.8%, which is remarkably close to the 60% difference between the current and climate change simulations for the Yakima subbasin.

Although this analysis is preliminary, it does indicate that our parameter manipulations and the TPM may accurately simulate effects of climate change. If anything, our results are conservative estimates of climate's effects. However, several nonenvironmental factors could also affect patterns of *Oncorhynchus* representation. First, differences in bone density between *Oncorhynchus* and other genera may cause relatively accelerated deterioration of salmon bone. Two of us (V.L. Butler and J.C. Chatters) are investigating this issue and its relationship to taxonomic composition of archaeofaunas. Second, human beings interact selectively with their environment, and may not assemble a representative sample of the available biota (Grayson 1981). Optimal-foraging theory and the general edibility of salmon argue against the importance of selectivity in this instance, however. Third, technological advances can improve the accessibility of fish species. For

Table 1. Data on fish archaeofaunas from Columbia River sites after 4500 yr B.P.

Site	Assemblage	Age range in radiocarbon yr B.P.	Screen size (mm)	Analyst	Percent *Oncorhynchus*	Number of fish bones	Reference for radiocarbon ages
45-DO-285	Zone 1 [a,b]	290±90 – 340±80	3.2	V.L. Butler	83.6	116 [d]	Miss 1984a:112
45-DO-372	Occupation II	630±70	6.4, 1	J.C. Chatters	61.0	2510	Chatters 1986:89
45-DO-214	Zones 2, 3 [a,b]	1035±65 – 1205±160	3.2	V.L. Butler	64.7	173 [d]	Miss 1984b:172
45-DO-372	Occupation III	2220±60 – 2400±80	6.4, 1	J.C. Chatters	97.1	1413	Chatters 1986: 89
45-DO-372	Occupation IV	2760±70 – 3000±90	6.4, 1	J.C. Chatters	95.0	1227	Chatters 1986:89
45-DO-211	Zone 4	2712±80 – 3117±119	3.2	V.L. Butler	93.4	394 [d]	Lohse 1984a:26
45-DO-372	Occupation V	2960±170 – 3340±80	6.4, 1	J.C. Chatters	96.4	56	Chatters 1986:89
45-OK-383	Occupation II	3770±80 – 4010±90	6.4, 1	J.C. Chatters	37.0 [c]	79	Chatters 1986:89
45-OK-11	Kartar	3720±80 – 4490±110	3.2	V.L. Butler	6.5	1537 [d]	Lohse 1984b:315
45-OK-382	Occupation II	4040±110	6.4, 1	J.C. Chatters	26.0 [c]	41	Chatters 1986:89

[a] These sites are not pithouse occupations, so depositional conditions and preservation maydiffer from the rest of the samples.
[b] Counts are from the site as a whole, which contains material from other time periods, but more than 72% of fish elements are from this zone.
[c] Butler has also identified these collections, obtaining *Oncorhynchus* percentages of 28.0% for 45-OK-383 and 14.3% for 45-OK-382.
[d] This number excludes vertebrae fragments, spines, ribs, and gill rakers.

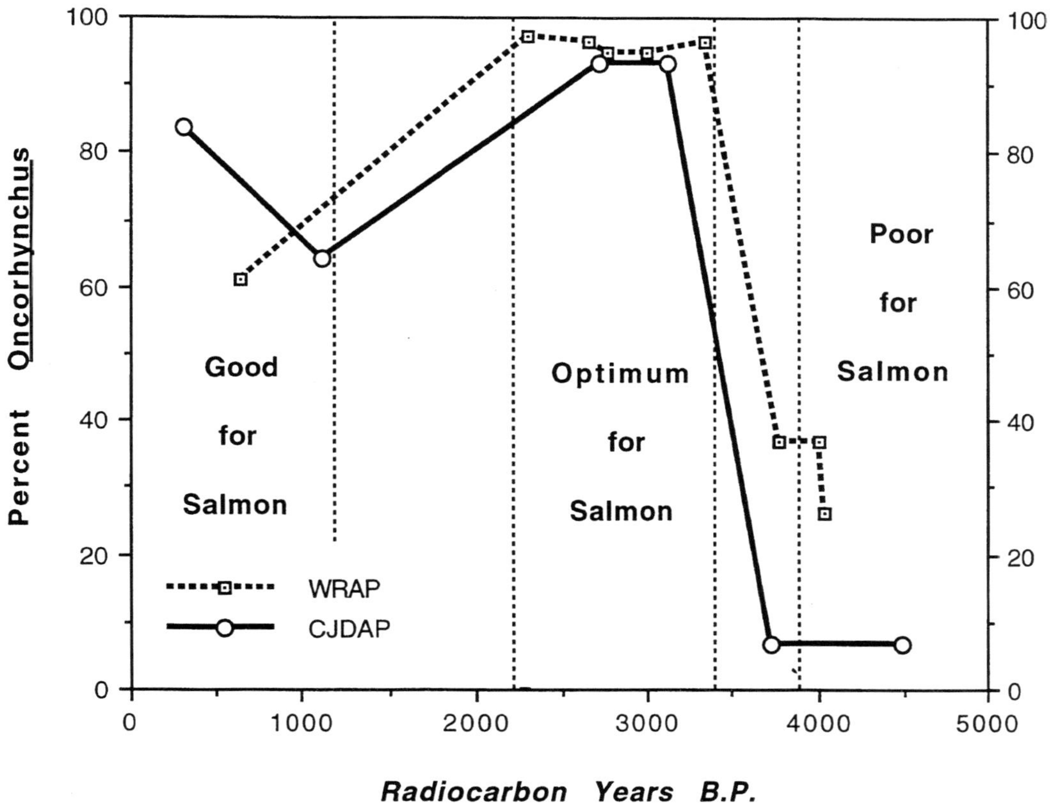

Fig. 4. The percentage of *Oncorhynchus* in archaeological collections after 4500 B.P. from the Wells Reservoir Archaeological Project (WRAP) and Chief Joseph Dam Archaeological Project (CJDAP) on the Columbia River, compared with expectations of salmon production based on inferred stream conditions. Where the radiocarbon age of a collection has a range of 200 yr or less, only the median of dates is shown; otherwise earliest and latest dates are both included.

example, salmon migrate in deep, swift water, whereas suckers can be found in shallower areas, making more complex tools necessary for salmon exploitation in some parts of the river system. Finally, human population density may be a factor. As the number of consumers increased relative to a constant prey population, people would have turned increasingly to less desirable prey species, lowering the relative contribution of salmon in archaeofaunal assemblages. We intend to address these last two issues by analyzing existing archaeological data from the Columbia River Basin, comparing the chronological pattern of fish use with evidence for technology and human population density. Intuitively, however, neither technology nor population pressure alone could generate the observed pattern; technological advances should produce a rising emphasis on salmon, while predation pressure from increasing population would produce a decline.

Summary

The paleosciences offer a viable alternative to GCMs as a source of information that can be used to model the effects of climate on northern fish populations. They also can provide direct evidence of fish communities that can be used for model validation or as a direct measure of climate's effects. Our research team is using information from the paleosciences to estimate anadromous salmonid production in the Columbia River Basin under warmer climatic conditions by (1) reconstructing stream conditions for a warm period of the past, (2) altering parameters of the NPPC's Tributary Parameter Model based on this analog environment, (3) running the TPM in conjunction with the SPM to simulate effects of current and warmer climate scenarios, and (4) comparing simulated salmon production with past salmon productivity indicated by the taxonomic composition of fish archaeofaunas. The results thus far are encouraging; simulated fish production under the analog climate compares closely with the archaeological evidence. If anything, the model underestimates the severity of climate's impact.

Analogs drawn from the paleosciences, like GCM-generated scenarios, have their limitations. Future warming, forced by greenhouse gas build-up rather than orbital variations, may have regional effects that differ from those of the past. Also, reconstructions are based on proxy indicators of environmental conditions, not direct measures, and may incorrectly represent past realities. Because they are based on empirical data, however, analogs from the past at the very least present us with plausible, readily understandable images of how important resources can be influenced by climate changes.

Acknowledgements

We thank Adeline Fredin of the Colville Confederated Tribes, Charles Robins of the P.U.D. No. 1 of Douglas County, Washington, and Lawr Salo of the Seattle District, Corps of Engineers, for the opportunity to analyze fish and mollusk remains from archaeological collections in their care. This work was supported by the U.S. Department of Energy (Contract DE-AC06-76RLO 1830) and the Northwest College and University Association for Science (Washington State University) under contract FG06-89ER-75522 with the U.S. Department of Energy.

References

Becker, C.D., and M.P. Fujihara. 1987. The bacterial pathogen *Flexibacter columnaris* and its epizoology among Columbia River fish. Am. Fish. Soc. Monogr. 2. American Fisheries Society, Bethesda, MD.

Butler, V.L. 1993. Natural versus cultural salmonid remains: origin of the Dalles Roadcut Bones, Columbia River, Oregon, U.S.A. J. Archaeol. Sci. (In press)

Campbell, S.K. 1985. Summary of results, Chief Joseph Dam Archaeological Project, Washington. Office of Public Archaeology, University of Washington, Seattle, WA.

Chatters, J.C. 1986. The Wells Reservoir Archaeological Project. Vol. 1. summary of findings. Cent. Wash. Archaeol. Surv. Archaeol. Rep. 86–6. Central Washington University, Ellensburg, WA.

Chatters J.C. 1987. Shell of *Margaritifera margaritifera falcata* as a source of paleoenvironmental and cultural data. Appendix F. *In* M.S. Kelly, E. Nilsson, and J. Cleland [ed.] Archaeological Investigations at Lake Britton, California. Pacific Gas and Electric Company, San Francisco, CA.

Chatters, J.C. 1989. Hypsithermal paleohydrology as an analog of greenhouse effects on the Columbia River Basin, Washington. Geol. Soc. Am. Abstr. Prog. 21(6).

Chatters, J.C., and K. A. Hoover. 1992. Response of the Columbia River fluvial system to Holocene climatic change. Quat. Res. 37: 42–49.

Chatters, J.C., D.A. Neitzel, M.J. Scott, and S.A. Schankle. 1991. The effect of climate change on stream environments: the salmonid resource of the Columbia River Basin. Northwest Environ. J. 7: 271–293.

COHMAP Members. 1988. Climatic changes of the last 18,000 years: observations and model simulations. Science (Wash., D.C.) 241: 1043–1052.

Cressman, L.S. 1960. Cultural sequences at The Dalles, Oregon. Trans. Am. Philos. Soc. 50(10).

Cushman, R.M., and P.N. Spring. 1989. Regional differences among GCM climate projections for resource analysis. Environ. Manage. 13: 739–795.

DeAngeles, D.L., and R.M. Cushman. Potential application of models in forecasting the effects of climate change on fisheries. Trans. Am. Fish. Soc. 119: 224–239.

Grayson, D.K. 1981. A critical review of the use of archaeological vertebrates in paleoenvironmental reconstruction. J. Ethnobiol. 1: 28–38.

Hall, J.C., and R.L. Lantz. 1969. Effects of logging on the habitat of coho salmon and cutthroat trout in coastal streams. H. R. MacMillan Lectures in Fisheries, University of British Columbia, Vancouver, B.C.

Kalkstein, L.S. [ed.]. 1991. Global comparisons of selected GCM control runs and observed climate data. Publ. 21P-2202 U. S. Environmental Protection Agency, Washington, DC.

Kutzbach, J.E. 1983. Modeling of Holocene climates, p. 271–277. *In* H.E. Wright, Jr. [ed.] Late Quaternary environments of the U.S. Vol. 2. The Holocene. University of Minnesota Press, Minneapolis, MN.

Lohse, E.S. 1984a. Archaeological investigations at site 45-DO-211, Chief Joseph Dam Project, Washington. Office of Public Archaeology, University of Washington, Seattle, WA.

Lohse, E.S. 1984b. Archaeological investigations at site 45-OK-11, Chief Joseph Dam Project, Washington. Office of Public Archaeology, University of Washington, Seattle, WA.

Lyman, R.L. 1980. Freshwater bivalve mollusks in southern Plateau prehistory. Northwest Sci. 54: 121–135.

MacArthur, R.H., and E.H. Pianka. 1966. On optimal use of a patchy environment. Am. Nat. 100: 603–609.

McNeil, W.H., and W.H. Anhel. 1964. Success of pink salmon spawning relative to size of spawning bed materials. Spec. Sci. Rep. Fish. 469. U.S. Fish and Wildlife Service, Corvallis, OR.

Mehringer, P.J. 1985. Late Quaternary Pollen Records from the Interior Pacific Northwest and Northern Great Basin of the United States, p. 167–189. *In* V. Bryant and R. Holloway [ed.] Pollen records of late Quaternary North American sediments. American Association of Stratigraphic Palynologists, Dallas, TX.

Miss, C.J. 1984a. Archaeological investigations at site 45-DO-285, Chief Joseph Dam Project, Washington. Office of Public Archaeology, University of Washington, Seattle, WA.

Miss, C.J. 1984b. Archaeological investigations at site 45-DO-214, Chief Joseph Dam Project, Washington. Office of Public Archaeology, University of Washington, Seattle, WA.

Neitzel, D.A., M.J. Scott, S.A. Schankle, and J.C. Chatters. 1991. The effect of climate change on stream environments: the salmonid resource of the Columbia River Basin. Northwest Environ. J. 7: 271–293.

Northwest Power Planning Council (NPPC). 1989. Salmon and steelhead system planning documentation, user information for the System Planning Model. Monitoring and Evaluation Group, Portland, OR.

Northwest Power Planning Council (NPPC). 1992. Salmon and steelhead system planning documentation, 1992 update of user information for the System Planning Model. Monitoring and Evaluation Group, Portland, OR.

Smith, R.J.F. 1985. The control of fish migration. Springer-Verlag, New York, NY.

VanHyning, J.M. 1973. Factors affecting the abundance of fall chinook salmon in the Columbia River. Fish Commission of Oregon, Portland, OR.

Vannote, R.L., and G.W. Minshall. 1982. Fluvial processes and local lithology controlling abundance, structure, and composition of mussel beds. Proc. Natl. Acad. Sci. U.S.A. 79: 4103–4107.

Winterhalder, B., and E.A. Smith. 1981. Hunter–gatherer foraging strategies. University of Chicago Press, Chicago, IL.

Relation of eastern Bering Sea walleye pollock (*Theragra chalcogramma*) recruitment to environmental and oceanographic variables

Terrance J. Quinn II and Henry J. Niebauer

Quinn, T.J., II, and H.J. Niebauer. 1995. Relation of eastern Bering Sea walleye pollock (*Theragra chalcogramma*) recruitment to environmental and oceanographic variables, p. 497–507. *In* R.J. Beamish [ed.] Climate change and northern fish populations. Can. Spec. Publ. Fish. Aquat. Sci. 121.

Abstract: We relate eastern Bering Sea walleye pollock (*Theragra chalcogramma*) recruitment (2 yr olds) from 1964–90 with coincident time series of anomalies of Pribilof Islands air temperature, eastern Bering Sea ice cover, sea-surface temperature near the Pribilof Islands, and bottom temperature in an area of the eastern Bering Sea. Significant cross correlations occur between the pollock recruitment series and the annual air temperature, ice cover, and bottom temperature anomaly series lagged by 1 yr, indicating that environmental variables are related to pollock during its early life history before recruitment at age 2. High pollock recruitment coincides with above-normal air and bottom temperatures and reduced ice cover. No lag is significant between pollock recruitment and sea-surface temperature. Using the logarithm of recruitment or residuals from a spawner–recruit relationship produced better relationships. Quarterly and monthly averages of the same environmental data also show persistent relationships with pollock recruitment and residuals. The highest correlations occur with lags in the middle of the early life history.

Résumé : Nous lions le recrutement des morues du Pacifique occidental (*Theragra chalcogramma*) d'âge 2 de l'est de la mer de Béring de 1964 à 1990 à une série chronologique coïncidente d'anomalies ayant caractérisé la température de l'air des îles Pribilof, la couverture de glace de l'est de la mer de Béring, la température de la surface de la mer près des îles Pribilof et la température du fond de la mer dans un secteur de l'est de la mer de Béring. Des corrélations significatives apparaissent entre la série sur le recrutement de la morue et celles sur les anomalies annuelles de la température de l'air, de la couverture de glace et de la température du fond de la mer, décalées d'un an, ce qui indique que les morues subissent l'influence des variables environnementales durant leurs premiers stades de vie, avant le recrutement à l'âge 2. Le recrutement élevé des morues coïncide avec des températures de l'air et du fond de la mer au-dessus des normales et avec une couverture de glace réduite. Il n'existe aucune décalage significatif entre le recrutement de la morue et la température à la surface de la mer. L'utilisation du logarithme de recrutement ou des résidus d'une relation géniteurs-recrues a produit de meilleures relations. En outre, les moyennes trimestrielles et mensuelles des mêmes données environnementales montrent une relation persistante avec le recrutement des morues et avec les résidus. Les corrélations les plus fortes ont lieu avec des décalages au milieu des premiers stades de vie.

Introduction

Fluctuations in fish recruitment may be caused by mechanisms resulting from interannual oceanographic and atmospheric changes and biological responses to population biomass. Co-analyses of physical and fisheries time series can provide a broad overview over a long time period of factors explaining variability in fish populations.

T.J. Quinn II. Juneau Center, School of Fisheries and Ocean Sciences, University of Alaska Fairbanks, Juneau AK 99801-8677

H.J. Niebauer. Institute for Marine Science, School of Fisheries and Ocean Sciences, University of Alaska Fairbanks, Fairbanks AK 99775

We brought together newly developed fisheries and oceanographic time series from the eastern Bering Sea to understand changes in the abundance of the commercially important walleye pollock (*Theragra chalcogramma*) population. We applied cross-correlation techniques to determine whether significant relationships exist between pollock recruitment and ice, air, and sea variables. Our work is complementary to other studies of the causes of recruitment fluctuations in Bering Sea pollock through field studies (Bering Sea Fisheries Oceanography Coordinated Investigations). Our results suggest appropriate mechanisms for further study, although they cannot establish cause and effect relationships. Our work provides a baseline model against which others can be evaluated. In this sense, basic null hypotheses are suggested that can be tested in a field setting by examining actual mechanisms.

Interest in factors affecting year-class strength in fish dates back at least to Hjort's published work in 1914 (see Sinclair et al. 1985). The assumption of a constant environment is a common feature in many population dynamics models, despite strong evidence of the influences of climate on fisheries (Cushing 1982). Recent studies with long-term data series have uncovered correlations of marine fish recruitment with environmental data for several species (Mysak et al. 1982; Sinclair et al. 1985; Norton 1987; Cabilio et al. 1987; Mendelssohn and Cury 1987; Stocker and Noakes 1988; Parker 1989; Hagen and Quinn 1991; Wespestad 1991). For anadromous salmonid populations, environmental correlates have been found for several populations (e.g., Barber and Walker 1980; Mysak et al. 1982; Willette 1985; Quinn and Marshall 1989; Geiger and Koenings 1991). Environmental phenomena thought to be important components of salmon catch and dynamics are given in several papers in Pearcy (1984).

In addition, several studies have noted synchrony in population response. In the Northeast Pacific Ocean, year-class data from 59 fish stocks showed similar, contemporaneous patterns in recruitment (Hollowed et al. 1987). Such a correspondence was attributed to environmental factors affecting recruitment. In a study of the five Pacific salmon species and steelhead trout (*Oncorhynchus* spp.) across several areas of the Pacific Ocean, trend analysis showed similarities between certain species and areas (Konkel and McIntyre 1987). Comparison of Pacific salmon catches in Alaska over a long time period showed synchrony among species within an area and among areas for some species (Quinn and Marshall 1989).

Bering Sea environment

The Bering Sea system is a complex environment governed by interactions among air, ice, and ocean. It has been shown (Niebauer 1988; Niebauer and Day 1989) that large-scale weather patterns in both the tropical South Pacific (El Niño – Southern Oscillation (ENSO) events) and the North Pacific (Pacific – North American (PNA) patterns) have strong teleconnection effects on the air, ice, and ocean environments of the Bering Sea. These signals come mainly via the atmosphere and not the ocean.

The connection between variability of the Bering Sea and ENSO and PNA appears to be the winter position of the Aleutian Low (Niebauer 1988; Niebauer and Day 1989), one of the two main low pressure systems or storm centers in the high latitude northern hemisphere (the other is the Icelandic Low over the North Atlantic). The Aleutian Low is located between the Asian and North American continents along the Aleutian Island chain in the North Pacific and is weak in summer but grows in strength in winter. It is a statistical low in the sense that it is a locus of the cyclonic storm centers passing from west to east over the North Pacific and in this sense its location is variable from year to year.

Interannual variabilities in air temperature, ice cover and surface wind in the Bering Sea generally are in phase with each other, whereas sea-surface temperature tends to lag these variables by 1–3 mo (Niebauer 1988; Niebauer and Day 1989). These time series are significantly correlated with the Southern Oscillation Index (SOI) time series (an indicator of ENSO events) when the Bering Sea data are lagged behind the SOI for up to 18 mo. These correlations suggest that warming in the Bering Sea follows negative anomalies in the SOI (i.e., El Niño events). Cooling in the Bering Sea tends to follow positive anomalies (i.e., precursors of El Niño's) in the SOI. Maximal correlations for the PNA also lag the SOI by 1 or 2 mo. Analyses of variance suggest that SOI can explain 30–40% of the variability in the Bering Sea time series.

The relationship between these interactions and pollock recruitment in the Bering Sea is not well understood. Bulatov (1989) suggested relationships between pollock and the environment but the bottom-trawl survey data used are probably not the best estimates of recruitment. Current practice is to determine pollock recruitment by using cohort analysis or catch-age analysis with auxiliary information from trawl and hydroacoustic surveys (Quinn and Collie 1990; Wespestad 1989). In addition, pollock recruitment has been modeled as a function of pollock spawning stock with the standard Ricker equation using stochastic log-normal variation (Wespestad 1989; Quinn et al. 1990; Quinn and Collie 1990). The estimated coefficient of variation in the relationship is about 50% (Quinn et al. 1990), suggesting that 50% of the variance in pollock recruitment remains unexplained after accounting for the biological factors of spawning and egg deposition. Such variation may be driven by environmental factors or other biological factors.

Methods

Analysis of catch-age data with auxiliary information from fishing effort, the nature of the spawner–recruit relationship, migration rates, and/or survey information provides reliable estimates of year-class strength (Deriso et al. 1985, 1989; Quinn et al. 1985). Catch-age analysis of walleye pollock data in the Bering Sea from 1964–90 was used to produce estimates of recruitment, abundance, and related parameters. We followed the approach described in Quinn and Collie (1990), updated to include 1989 and 1990 data. As in past studies, we defined recruitment to be the number of age 2 pollock, which is also the first age used in catch–age analysis. To account for biological factors affecting recruitment, we fitted a Ricker spawner–recruit model to the recruitment data and used residuals from the curve as adjusted data (Fig. 1).

It is important to use physical data sets that are long enough (>20 yr) to contain both interannual and interdecadal variability (Fritz et al. 1990). We selected variables with available time series at least as long as the pollock recruitment series, which represent air, ice, and sea effects in the Bering Sea. To this end, we used time series of Pribilof Islands air temperature, Bering Sea ice cover, and sea-surface temperature near the Pribilof Islands (Niebauer and Day 1989), which extend back to at least 1950. We constructed anomalies of the air temperature and ice cover by removing the seasonal means. Annual, quarterly, and monthly averages of physical variables were used in the analysis.

At the symposium from which this proceedings originated, we learned of another available oceanographic time series on bottom temperature in the eastern Bering Sea (Ohtani and Azumaya 1995). Summer observations of bottom temperature

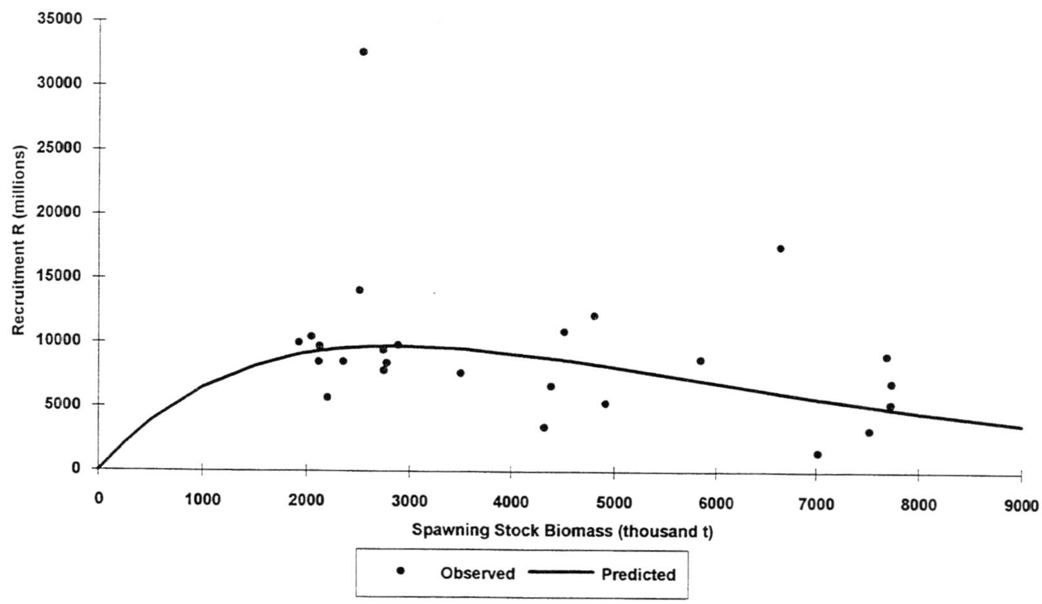

Fig. 1. Relationship between eastern Bering Sea pollock recruitment and spawning stock biomass and a fit of the Ricker spawner–recruit curve ($R = \alpha S e^{-\beta S}$, $\alpha = 9142$, $\beta = 3.45\times10^{-7}$).

were made from 1963 to 1990 from the survey vessel Oshoro Maru; Ohtani and Azumaya summarized an annual index using these data from a small rectangular area of the southern Bering Sea. We obtained their time series and compared our data with their's.

Cross-correlation analyses were performed to find which time series were related. As pollock spawn primarily in the period from January to June, the recruitment measured at age 2 could be affected by the environment during the life history from age 0 to age 2. Only correlations in the first two quarters of the year of recruitment and all quarters in the prior 2 yr were examined. For annual averages of environmental data, lags of 0, 1, and 2 yr would be of primary interest, although we report an interesting lag at 4 yr below. Recruitment, the logarithm of recruitment, and recruitment residuals were all used in the analyses to explore different aspects of the recruitment process. Some authors (e.g., Box and Jenkins 1976) do not recommend doing cross-correlation analyses unless at least 50 observations are available. Although our analyses are based on only 25–27 observations, we believe that cross-correlation analysis is still a viable tool. We are interested primarily in low-lag effects and the recruitment data represent at least two cycles of progress through low and high cycles.

In addition, we conducted analyses to explore the sensitivity of our findings to autocorrelation in the time series. Spurious cross correlations can appear if significant autocorrelation is present in either series (Chatfield 1991). The method we used to deal with this problem was to fit a time series model to the data and use the residuals from the fit as an alternative time series.

Results

Cross correlations between pollock recruitment and average annual anomalies of air, ice, and sea variables and their standard errors are shown in Table 1. We obtained equally strong and opposite results with both air and ice variables between recruitment and the environment. Significant correlations (greater than 1.96 standard errors) occurred when the air and ice temperature series are lagged 1 yr behind the pollock recruitment series, indicating that environmental variables are related to pollock during its early life history before recruitment at age 2. The signs of the correlation coefficients show that high pollock recruitment coincided with above normal air temperatures and reduced ice cover. A plot of recruitment and air temperature lagged 1 yr shows that the general pattern of both variables is the same (Fig. 2). In particular, the highest recruitment (1978 year-class) coincided with the highest temperature anomaly, and the low recruitment for the 1970 year-class coincided with the second lowest temperature anomaly. A second significant correlation occurred when pollock recruitment was lagged 4 yr behind air temperature and ice cover, but it may be spurious. The strong 1978 year-class (which recruited in 1980) has substantial influence on cross correlations because of its magnitude. As air temperature 4 yr earlier (in 1976) was low, this may have resulted in the significant lag-4 correlation.

In general, both logarithmic transformation of recruitment and adjustment of recruitment for biological factors by using spawner–recruit residuals decreased the influence of the 1978 year-class and thus result in better graphical relationships (cf. Figs. 2 and 3). The magnitudes of the lag-1 and lag-4

Table 1. Cross correlations and standard errors of age-2 pollock abundance (*R*), logarithm of abundance (ln(*R*)), and spawner–recruit residuals with **annual** averages of air temperature, ice cover, and sea-surface temperature anomalies. * Statistical significance at $\alpha = 0.05$.

Lag	*R*	SE	ln(R)	SE	Residual	SE
			Air temperature			
0	0.191	0.192	0.140	0.192	0.089	0.200
−1	0.527 *	0.196	0.404*	0.196	0.462*	0.204
−2	0.265	0.200	0.141	0.200	0.186	0.209
−3	−0.053	0.204	−0.124	0.204	−0.092	0.213
−4	−0.436*	0.209	−0.295	0.209	−0.245	0.218
−5	−0.216	0.213	−0.188	0.213	−0.024	0.224
−6	−0.207	0.218	−0.240	0.218	−0.014	0.229
−7	−0.096	0.224	−0.097	0.224	0.134	0.236
			Ice cover			
0	−0.077	0.192	−0.161	0.192	−0.025	0.200
−1	−0.445*	0.196	−0.359	0.196	−0.383	0.204
−2	−0.254	0.200	−0.139	0.200	−0.159	0.209
−3	0.126	0.204	0.192	0.204	0.179	0.213
−4	0.538*	0.209	0.409*	0.209	0.441*	0.218
−5	0.239	0.213	0.171	0.213	0.087	0.224
−6	0.045	0.218	0.055	0.218	−0.104	0.229
−7	0.001	0.224	0.007	0.224	−0.126	0.236
			Sea-surface temperature			
0	0.318	0.192	0.347	0.192	0.468*	0.200
−1	0.239	0.196	0.183	0.196	0.272	0.204
−2	−0.092	0.200	−0.088	0.200	−0.081	0.209
−3	0.194	0.204	0.113	0.204	0.229	0.213
−4	−0.279	0.209	−0.173	0.209	−0.064	0.218
−5	−0.206	0.213	−0.241	0.213	−0.088	0.224
−6	−0.269	0.218	−0.244	0.218	−0.084	0.229
−7	−0.283	0.224	−0.235	0.224	0.001	0.236

correlations were lower than when using untransformed data, but the same trends in the correlations occurred (Table 1). Nevertheless, relationships using the residuals adjustment showed the best correspondence in terms of patterns, as described below.

For air temperature, the lag-1 correlations were still significant, but the lag-4 correlations were not. A plot of residuals and air temperature lagged 1 yr (Fig. 3) reveals a very strong relationship, and the correspondence is better overall than for untransformed recruitment. The strong 1978 and weak 1970 year-classes have better correspondence with their environmental counterparts. However, for some ranges of years, the magnitude of the recruitment response does not match the change in the environment. For the 1970 to 1975 year-classes, the changes in the pollock residuals are paralleled by changes in the air temperatures; however, the magnitudes of recruitment would be expected to be much lower, given the low temperatures that occurred during this period. This suggests that the relationship between pollock and the environment may be nonlinear and that other factors may act to mediate the response.

For ice cover, the lag-1 correlations using the two transformations are no longer quite significant but the lag-4 correlations are. A plot of residuals and ice cover lagged 1 yr (Fig. 4) reveals a strong relationship opposite of air temperature, as would be expected. The strong 1978 year-class coincides with a large negative ice cover anomaly, and the weak 1964, 1970, and 1987 year-classes coincide with positive ice cover anomalies. However, large positive ice cover anomalies did not result in particularly weak 1974 and 1975 year-classes.

For sea-surface temperature, the only significant relationship is unlagged temperature with recruitment residuals, although the trends are similar to air temperature and negatively related to ice cover. That lag 0 is more significant than lag 1 may reflect the time it takes for air temperature to affect sea-surface temperature. Niebauer (1988) found a 1–3 mo lag of sea-surface temperature to air temperature. A plot of residuals and sea-surface temperature lagged 0 yr (Fig. 5) shows that strong pollock recruitment coincides with high relative sea-surface temperature in some years but not in others. In particular, the weak 1964, 1970, and 1987 year-classes and the strong 1982 year-class correspond well with sea-surface

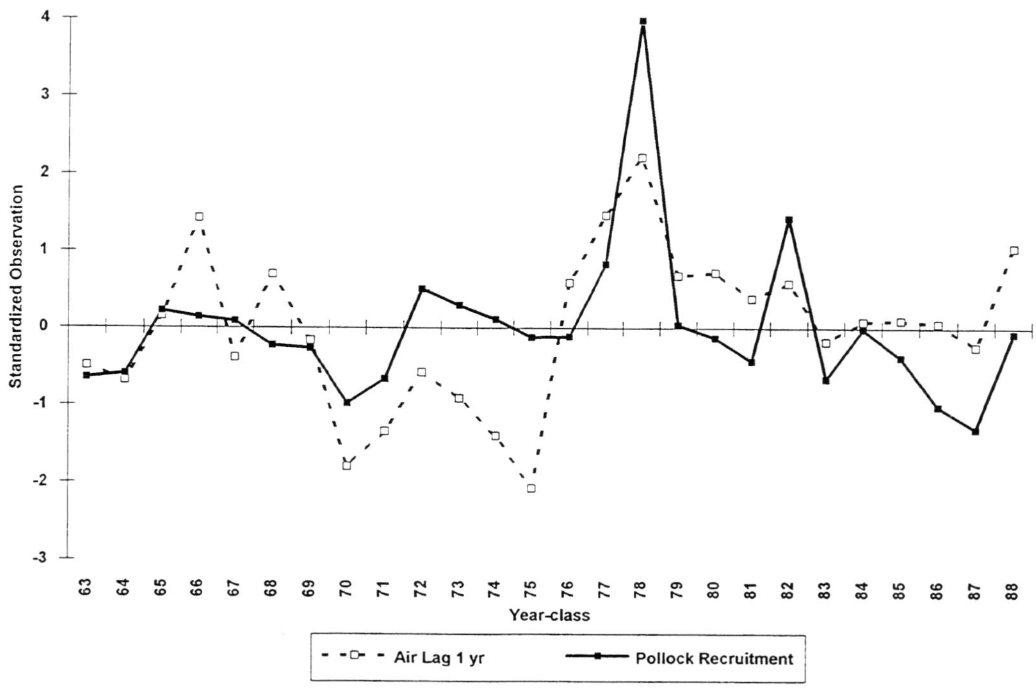

Fig. 2. Eastern Bering Sea pollock recruitment (age 2) in year $t + 2$ and Pribilof Islands air temperature anomalies lagged 1 yr in year $t + 1$ for each year-class t. For each series, observations are standardized by subtracting the mean, then dividing by the standard deviation.

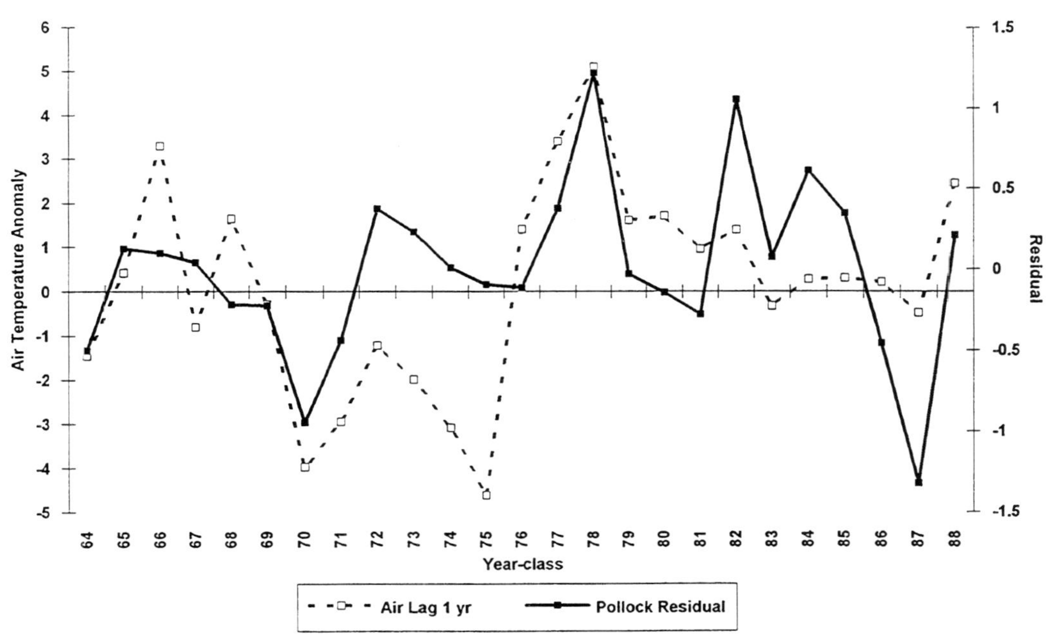

Fig. 3. Eastern Bering Sea pollock residuals from a Ricker spawner–recruit fit in year $t + 2$ and Pribilof Islands air temperature anomalies lagged 1 yr in year $t + 1$ for each year-class t.

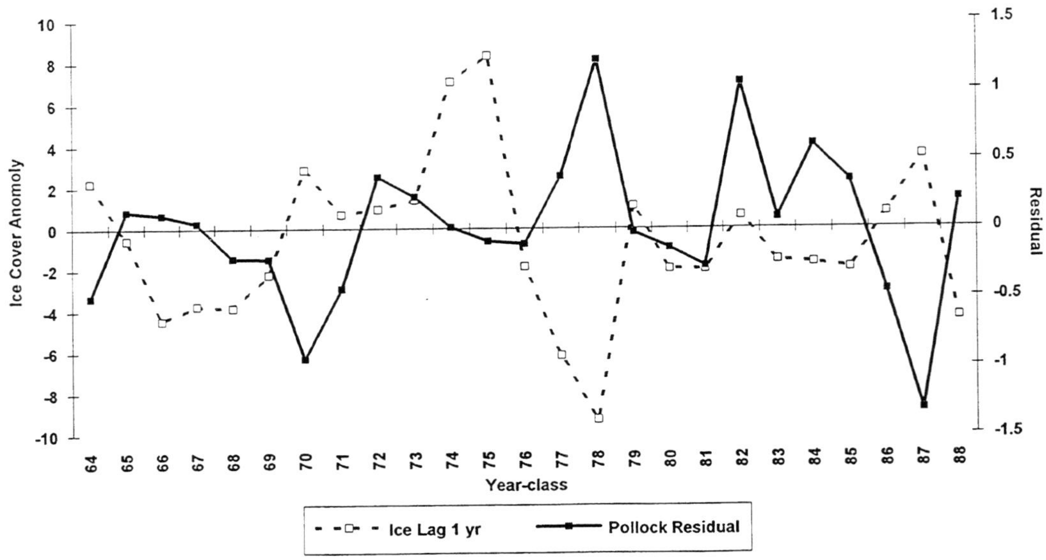

Fig. 4. Eastern Bering Sea pollock residuals from a Ricker spawner–recruit fit in year $t + 2$ and Bering and Chukchi Seas ice cover anomalies lagged 1 yr in year $t + 1$ for each year-class t.

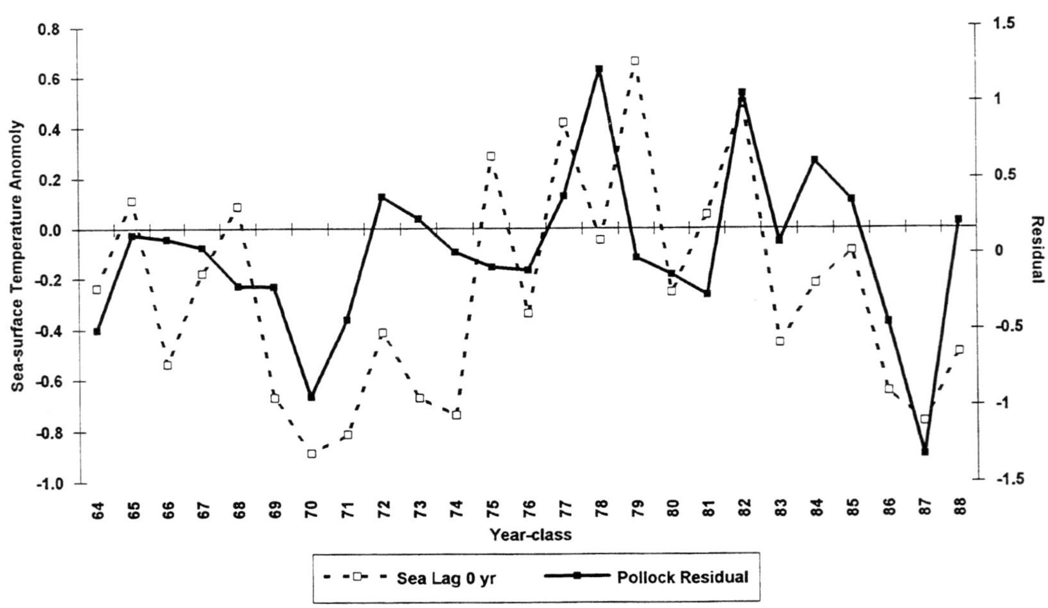

Fig. 5. Eastern Bering Sea pollock residuals from a Ricker spawner–recruit fit in year $t + 2$ and sea-surface temperature anomalies near the Pribilof Islands lagged 0 yr in year $t + 2$ for each year-class t.

Table 2. Cross correlations and standard errors of age-2 pollock abundance (R), and spawner–recruit residuals with **quarterly** averages of air temperature, ice cover, and sea-surface temperature anomalies. * Statistical significance at $\alpha = 0.05$.

Quarter	Lag	R	SE	Residual	SE
		Air temperature			
2	0	0.235	0.192	0.143	0.200
1	0	0.130	0.192	0.002	0.200
4	−1	0.433*	0.196	0.431*	0.204
3	−1	0.398*	0.196	0.329	0.204
2	−1	0.478*	0.196	0.397	0.204
1	−1	0.440*	0.196	0.387	0.204
4	−2	0.280	0.200	0.349	0.209
3	−2	0.197	0.200	0.159	0.209
2	−2	0.222	0.200	0.101	0.209
1	−2	0.210	0.200	0.116	0.209
		Ice cover			
2	0	0.031	0.192	0.045	0.200
1	0	0.039	0.192	0.005	0.200
4	−1	−0.348	0.196	−0.405*	0.204
3	−1	−0.188	0.196	−0.083	0.204
2	−1	−0.377	0.196	−0.307	0.204
1	−1	−0.331	0.196	−0.254	0.204
4	−2	−0.308	0.200	−0.395	0.209
3	−2	−0.271	0.200	−0.115	0.209
2	−2	−0.036	0.200	0.142	0.209
1	−2	−0.124	0.200	−0.079	0.209
		Sea-surface temperature			
2	0	0.217	0.192	0.285	0.200
1	0	0.322	0.192	0.498*	0.200
4	−1	0.257	0.196	0.391	0.204
3	−1	0.245	0.196	0.235	0.204
2	−1	0.223	0.196	0.133	0.204
1	−1	0.026	0.196	0.116	0.204
4	−2	−0.069	0.200	−0.028	0.209
3	−2	−0.033	0.200	0.025	0.209
2	−2	0.026	0.200	−0.102	0.209
1	−2	−0.231	0.200	−0.171	0.209

temperature anomalies, while the weak 1981 and strong 1972 and 1978 year-classes do not. Lagging sea-surface temperature by 1 yr gives a better relationship for the 1978 year-class, but disrupts many of the others. Thus, sea-surface temperature does not seem to be a sensitive indicator of pollock recruitment. This is perhaps due to the opportunistic nature of the data, which are obtained from reports of ships passing through a 300 km^2 area centered about the Pribilof Islands (Niebauer 1988).

Quarterly averages of air temperature and ice cover were also correlated with pollock recruitment (Table 2). Pollock recruitment was positively correlated with air temperature in 10 of 10 quarters and negatively correlated with ice cover in eight of 10 quarters between spawning and recruitment. The individual correlations were often statistically insignificant, but taken together, could not likely be produced by chance. The correlations with pollock residuals showed the same pattern (Table 2). For both pollock recruitment and residuals, the strongest correlations occurred in the middle of early life history. The relationship with sea surface temperature was not as clear cut, but positive correlations did occur at lags 0 and 1.

Monthly averages were also correlated with pollock recruitment (Table 3). Results are only shown for air temperature but are analogous to quarterly breakdowns for ice and sea variables. Positive correlations with air temperature occurred in 27 of the 30 mo examined corresponding to the early life history period. The highest correlations occurred with lags in the middle of early life history.

The summer bottom temperature time series of Ohtani and Azumaya (1995) produced similar results. This series was mostly highly correlated with air temperature in the same year (r = 0.797 ± 0.192 (±1 SE)), then sea surface temperature

Table 3. Cross correlations and standard errors of age-2 pollock abundance (*R*), and spawner–recruit residuals with **monthly** averages of air temperature anomalies. * Statistical significance at $\alpha = 0.05$.

Month	Lag	R	SE	Residual	SE
		Air temperature			
6	0	0.355	0.192	0.305	0.200
5	0	0.250	0.192	0.168	0.200
4	0	0.125	0.192	0.019	0.200
3	0	0.351	0.192	0.261	0.200
2	0	−0.126	0.192	−0.319	0.200
1	0	0.111	0.192	0.140	0.200
12	−1	0.247	0.196	0.352	0.204
11	−1	0.345	0.196	0.390	0.204
10	−1	0.420*	0.196	0.149	0.204
9	−1	0.423*	0.196	0.356	0.204
8	−1	0.327	0.196	0.258	0.204
7	−1	0.339	0.196	0.285	0.204
6	−1	0.464*	0.196	0.377	0.204
5	−1	0.444*	0.196	0.383	0.204
4	−1	0.433*	0.196	0.355	0.204
3	−1	0.343	0.196	0.390	0.204
2	−1	0.360	0.196	0.282	0.204
1	−1	0.319	0.196	0.212	0.204
12	−2	0.139	0.200	0.323	0.209
11	−2	0.356	0.200	0.343	0.209
10	−2	0.074	0.200	0.005	0.209
9	−2	0.344	0.200	0.255	0.209
8	−2	0.226	0.200	0.245	0.209
7	−2	−0.006	0.200	−0.041	0.209
6	−2	0.173	0.200	0.119	0.209
5	−2	0.187	0.200	0.089	0.209
4	−2	0.235	0.200	0.083	0.209
3	−2	0.041	0.200	−0.022	0.209
2	−2	0.143	0.200	0.023	0.209
1	−2	0.357	0.200	0.340	0.209

($r = 0.582 \pm 0.192$), and then ice cover ($r = -0.573 \pm 0.192$). Lag-1 and lead-1 correlations between air and bottom temperature were also significant; no other correlations between variables were significant. In a cross-correlation analysis with pollock recruitment, the lag-1 correlation was significant ($r = 0.452 \pm 0.196$) and the lag-4 correlation was large in magnitude although not significant ($r = -0.356 \pm 0.209$). For the log(recruitment) and recruitment residuals series, no correlations with bottom temperature were significant, but the lag-1 correlations ($r = 0.250 \pm 0.196$, 0.300 ± 0.204, respectively) were the highest of all the lags examined and were of intermediate magnitude compared to the air, ice, and sea-surface temperature series.

To explore the sensitivity of the results to autocorrelation, we computed autocorrelation functions for the series. The only annual environmental or biological series that had a significant autocorrelation at any lag was the air temperature series, which had a significant autocorrelation at lag 1. We fitted an autoregressive time series model of order 1 to the air temperature data (see Quinn and Marshall (1988) for a discussion of time series models) and used the residuals from the model fit in a new cross correlation analysis with pollock recruitment residuals. Results similar to Table 1 were obtained, except that the cross correlations were slightly lower. The lag-1 cross correlation was still significant ($r = 0.430 \pm 0.204$). We also examined the monthly air temperature series and found several significant autocorrelations. By fitting an autoregressive time-series model of order 4, we were able to remove most of the autocorrelations from the residual series. We repeated the cross-correlation analysis with pollock recruitment residuals shown in Table 3, replacing air temperature with residuals from the time series model fit. The cross correlations were lower than in the previous analysis, but 20 (rather than 27) of 30 cross correlations were still positive. This result still cannot by explained by chance (sign test, $P = 0.042$).

Discussion

Consistent relationships between environmental variables and pollock recruitment at age 2 occur regardless of whether the environmental breakdown is monthly, quarterly, or annual, and no matter which recruitment variable is used. The most consistent relationships occur at lag 1 yr with annual averages of environmental data, suggesting that an integration of the effects of environmental variables occurs in determining pollock recruitment. Our results are consistent with Ohtani and Azumaya (1995), who used a different recruitment series and a regression method. They also found a significant lag-1 term between pollock recruitment and bottom temperature.

The results showed that high pollock recruitment at age 2 coincided with above-normal air and sea temperatures and reduced ice cover 6–18 mo earlier. This suggests that warmer temperatures in the Bering Sea during this early life history period are conducive to strong recruitment. The lag-4 correlations, if not spurious, would correspond to environmental conditions 2 yr prior to the formation of the year-class at age 0, suggesting an opposite effect related to the parents of the year-class. As pollock are known to be cannibalistic, some type of density-dependent response may be measured by this significant correlation. However, the lag-4 correlation is no longer significant for air temperature when pollock residuals are used to account for this density-dependent response. The significant lag-4 correlation of pollock residuals with ice cover, however, suggests that further investigation is needed.

That environmental effects are strongest in the middle of the early life history of pollock before recruitment at age 2 rather than at the beginning (the egg and larval stages) is rather surprising, given that many researchers believe that recruitment is determined during the beginning stages. If interannual differences in survival at any early life history stage are a function of environmental variables, then recruitment should be correlated with those variables during those stages. If recruitment were determined at the beginning stages for pollock, then the strongest relationships between pollock recruitment and environmental variables should have occurred at a lag of 2 yr rather than 1 yr. Thus, research efforts directed at the middle of the early life history should be contemplated.

The major factors that affect recruitment to a fish population can be categorized as egg production effects, biological effects on early life survival to recruitment age, and environmental–oceanographic effects on early life survival to recruitment age. The first factor relates to maturity, fecundity, and spawning characteristics. In this study, spawning stock biomass is a proxy for egg production and should be reasonably accurate as long as fecundity is proportional to fish weight.

Biological effects on early life history are generally referred to as density-dependent effects and include such mechanisms are cannibalism and competition for food and space. Such effects are usually accounted for by fitting spawner–recruit relationships. For pollock, the significant Ricker fit to spawner–recruit data (Fig. 1) suggests that density dependence is present, and the most likely mechanism to explain this is cannibalism (Livingston 1993).

Environmental and oceanographic effects on recruitment are also well known, as mentioned in the introduction. These effects presumably filter through the food chain to affect fishes by providing more or better quality prey and through metabolic considerations in relation to temperature. Better environmental conditions may thus be conducive to better development and growth, and perhaps greater ability to avoid predation. One means of testing whether year effects (such as interannual variability in temperature) predominate over year-class effects (e.g., density-dependent growth) in affecting growth is to measure annual or daily increment widths on otoliths over several years and then perform permutation tests (Hagen and Quinn 1991). This has not yet been done for walleye pollock, although some short time series have been assembled (e.g., Yoklavich and Bailey 1989).

A physical mechanism that might lead to enhanced fish survival and/or growth is Ekman pumping (or suction) (Pond and Picard 1991), where cyclonic winds from the Aleutian Low cause Ekman transport away from a center of a storm, which results in upwelling in the upper ocean levels under the storm. This mechanism may affect pollock in that the upwelling under the storms can suck nutrients into the photic zone causing primary production, which may be beneficial in further trophic stages to pollock.

That the pollock residuals, which have density-dependent effects removed, showed a better pattern with air temperature (Fig. 3) than pollock recruitment itself (Fig. 2) suggests that both biological and environmental–oceanographic factors are important for walleye pollock and that these factors may be somewhat independent. Slight interannual differences in trend between the residuals and the air temperature series suggest that there may be some nonlinearities in the biological and environmental effects not accounted for by these essentially linear analyses.

Our results are also in accord with the current understanding of the physical processes of the eastern Bering Sea (e.g., Niebauer and Day 1989). As these processes are atmospherically driven, air temperature is likely to an omnibus variable to represent those physical processes, which follows empirically by noting that high correlations occur among these environmental variables. The higher correlation of bottom temperature to air temperature than to sea-surface temperature may reflect this point as well, or else, the quality of the bottom and air temperature data may be higher than sea-surface temperature, which is obtained opportunistically. At any rate, the most consistent results with pollock recruitment occurred by using air temperature.

Further work is needed to elucidate these relationships and to study mechanisms of pollock recruitment in relation to growth and population dynamics. The next step we intend to take is to develop models and estimation procedures that include biological processes, oceanographic relationships, and unexplained variability. Recent developments in time series analysis may be useful in developing the estimation framework for the models, such as nonlinear and robust time series methods (Franke et al. 1984). In addition, there are structural time series models that blend the nonlinear spawner–recruit relationships with time series errors (Criddle and Havenner 1989). Further refinement of mechanisms for pollock recruitment may require study at the juvenile stage and better understanding of the relationship of the physical environment and pollock prey.

This study concentrated on environmental data that go back to about 1950 and fish recruitment data that go back to 1964; further work is needed to extend both types of data backward in time. Research is also needed to determine which combinations of the physical data have the most utility in explaining fisheries data. The environmental data are available on a monthly or finer time scale, whereas fisheries data are available only on an annual time scale. Averages, medians, running averages, extremes, and robust smoothers could all be useful for data compression for comparison to fisheries data.

Acknowledgments

Mary Milkovich assisted with the processing of the oceanographic data and Lowell Fair assisted with the analyses. Vidar Wespestad provided the pollock data and helpful discussions about the results. Kiyotaka Ohtani kindly provided his bottom temperature data and analysis of pollock data. Warren Wooster suggested that Ekman pumping may be an important mechanism. We thank two anonymous reviewers and editor Richard Beamish for helpful comments. This publication is the result of research sponsored by Alaska Sea Grant with funds from the National Oceanic and Atmospheric Administration, Office of Sea Grant, Department of Commerce, under grant no. NA90AA-D-SG066, project number R/07-16, from the University of Alaska with funds appropriated by the state, and from the REFM Division, Alaska Fisheries Science Center. The U.S. Government is authorized to produce and distribute reprints for government purposes, notwithstanding any copyright notation that may appear hereon.

References

Barber, W.E., and R.J. Walker. 1980. Bristol Bay sockeye salmon — an exploration in factors influencing saltwater growth, p. 261–273. *In* Proceedings of the North Pacific Aquaculture Symposium, Aug. 1980, Anchorage, AK. Alaska Sea Grant College Program, University of Alaska, Fairbanks, AK.

Box, G.E.P., and G.M. Jenkins. 1976. Time series analysis: forecasting and control, revised edition. Holden-Day, Oakland, CA. 575 p.

Bulatov, O.A. 1989. The role of environmental factors in fluctuations of stocks of walleye pollock in the eastern Bering Sea, p. 353–357. *In* R.J. Beamish and G.A. McFarlane [ed.] Effects of ocean variability on recruitment and an evaluation of parameters used in stock assessment models. Can. Spec. Publ. Fish. Aquat. Sci. 108.

Cabilio, P., D.L. DeWolfe, and G.R. Daborn. 1987. Fish catches and long-term tidal cycles in Northwest Atlantic fisheries: a nonlinear regression approach. Can. J. Fish. Aquat. Sci. 44: 1890–1897.

Chatfield, C. 1991. The analysis of time series: an introduction. 4th ed. Chapman & Hall, London. 241 p.

Criddle, K.R., and A. M. Havenner. 1989. Forecasting halibut biomass using system theoretic time-series methods. Am. J. Agric. Econ. 71: 422–431.

Cushing, D. 1982. Climate and fisheries. Academic Press, New York, NY. 373 p.

Deriso, R.B., P.R. Neal, and T.J. Quinn, II. 1989. Further aspects of catch-age analysis with auxiliary information, p. 127–135. *In* R.J. Beamish and G.A. McFarlane [ed.] Effects of ocean variability on recruitment and an evaluation of parameters used in stock assessment models. Can. Spec. Publ. Fish. Aquat. Sci. 108.

Deriso, R.B., T.J. Quinn, II, and P.R. Neal. 1985. Catch-age analysis with auxiliary information. Can. J. Fish. Aquat. Sci. 42: 815–824.

Franke, J., W. Hardle, and D. Martin. 1984. Robust and nonlinear time series analysis. Lecture notes in Statistics. Vol. 26. Springer-Verlag, New York, NY. 286 p.

Fritz, E.S., L.B. Crowder, and R.C. Francis. 1990. The National Oceanic and Atmospheric Administration plan for recruitment fisheries oceanography research. Fisheries 15: 25–31.

Geiger, H.J., and J.P. Koenings. 1991. Escapement goals for sockeye salmon with informative prior probabilities based on habitat considerations. Fish. Res. 11: 239–256.

Hagen, P.T., and T.J. Quinn II. 1991. Long-term growth dynamics of young Pacific halibut: evidence of temperature-induced variation. Fish. Res. 11: 283–306.

Hollowed, A.B., K.M. Bailey, and W.S. Wooster. 1987. Patterns in recruitment of marine fishes in the Northeast Pacific Ocean. Biol. Oceanogr. 5: 99–131.

Konkel, G.W., and J.D. McIntyre. 1987. Trends in spawning populations of Pacific anadromous salmonids. Tech. Rep. 9. U.S. Fish Wildl. Service, Washington, D.C. 25 p.

Livingston, P.A. 1993. The importance of predation by groundfish, marine mammals, and birds on walleye pollock, *Theragra chalcogramma*, and Pacific herring, *Clupea pallasi*, in the eastern Bering Sea. Mar. Ecol. Progr. Ser. 102: 205–215.

Mendelssohn, R., and P. Cury. 1987. Fluctuations of a fortnightly abundance index of the Ivoirian coastal pelagic species and associated environmental conditions. Can. J. Fish. Aquat. Sci. 44: 408–421.

Mysak, L.A., W.W. Hsieh, and T.R. Parsons. 1982. On the relationship between interannual baroclinic waves and fish populations in the Northeast Pacific. Biol. Oceanogr. 2: 63–103.

Niebauer, H.J. 1988. Effects of El Niño – Southern Oscillation and North Pacific weather patterns on interannual variability in the subarctic Bering Sea. J. Geophys. Res. 93(C5): 5051–5068.

Niebauer, H.J., and R.H. Day. 1989. Causes of interannual variability in the sea ice cover of the eastern Bering Sea. GeoJournal 18: 45–59.

Norton, J. 1987. Ocean climate influences on groundfish recruitment in the California Current, p. 73–98. In Proceedings of the International Rockfish Symposium, Oct. 1986, Anchorage AK. Alaska Sea Grant College Program, University of Alaska, Fairbanks, AK.

Ohtani, K., and T. Azumaya. 1995. Influence of interannual changes in ocean conditions on the abundance of walleye pollock (*Theragra chalcogramma*) in the eastern Bering Sea, p. 87–95. In R.J. Beamish [ed.] Climate change and northern fish populations. Can. Spec. Publ. Fish Aquat. Sci. 121.

Parker, K.S. 1989. Influence of oceanographic and meteorological processes on the recruitment of Pacific halibut in the Gulf of Alaska., p. 221–237. *In* R.J. Beamish and G.A. McFarlane [ed.] Effects of ocean variability on recruitment and an evaluation of parameters used in stock assessment models. Can. Spec. Publ. Fish. Aquat. Sci. 108.

Pearcy, W. [ed.]. 1984. The influence of ocean conditions on the production of salmonids in the North Pacific. Sea Grant College Program, Oregon State University, Corvallis, OR. ORESU-W-83-001: 327 p.

Pond, S., and G.L. Picard. 1991. Introductory dynamical oceanography. Pergamon Press, New York, NY. 329 p.

Quinn, T.J., II, and J.S. Collie. 1990. Alternative estimators of walleye pollock in the eastern Bering Sea. Int. North Pac. Fish. Comm. Bull. 50: 243–257.

Quinn, T.J., II, R.B. Deriso, and S.H. Hoag. 1985. Methods of population assessment of Pacific halibut. Int. Pac. Halibut Comm. Sci. Rep. 72: 52 p.

Quinn, T.J., II, R. Fagen, and J. Zheng. 1990. Threshold management policies for exploited populations. Can. J. Fish. Aquat. Sci. 47: 2016–2029.

Quinn, T.J., II, and R.P. Marshall. 1989. Time series analysis: quantifying variability and correlation in Alaska salmon catches and environmental data, p. 67–80. *In* R.J. Beamish and G.A. McFarlane [ed.] Effects of ocean variability on recruitment and an evaluation of parameters used in stock assessment models. Can. Spec. Publ. Fish. Aquat. Sci. 108.

Sinclair, M., M.J. Tremblay, and P. Bernal. 1985. El Niño events and variability in a Pacific mackerel survival index: support for Hjort's second hypothesis. Can. J. Fish. Aquat. Sci. 42: 602–608.

Stocker, M., and D.J. Noakes. 1988. Evaluating forecasting procedures for predicting Pacific herring recruitment in British Columbia. Can. J. Fish. Aquat. Sci. 45: 928–935.

Wespestad, V. 1989. Walleye pollock, p. 24–45. *In* Stock assessment and fishery evaluation document for groundfish resources in the Bering Sea/Aleutian Islands region as projected for 1990. North Pacific Fishery Management Council, Anchorage, AK.

Wespestad, V. 1991. Pacific herring population dynamics, early life history, and recruitment variation relative to eastern Bering Sea oceanographic factors. Ph.D. dissertation, University of Washington, Seattle WA. 237 p.

Willette, T.M. 1985. The effects of ocean temperatures on the survival of the odd- and even-year pink salmon (*Oncorhynchus gorbuscha*) populations originating from Prince William Sound, Alaska. M.S. thesis, University of Alaska, Fairbanks, AK. 115 p.

Yoklavich, M.M., and K. Bailey. 1989. Growth of larval and juvenile walleye pollock from Shelikof Strait, Gulf of Alaska, as determined from daily increments in otoliths, p. 241–251. *In* Proceedings of the International Symposium on the Biological Management of Walleye Pollock. Alaska Sea Grant College Program, University of Alaska, Fairbanks, AK.

Climate-induced changes in Pacific hake (*Merluccius productus*) abundance and pelagic community interactions in the Vancouver Island upwelling system

Daniel M. Ware and Gordon A. McFarlane

Ware, D.M., and G.A, McFarlane. 1995. Climate-induced changes in Pacific hake (*Merluccius productus*) abundance and pelagic community interactions in the Vancouver Island Upwelling System. p. 509–521. *In* R.J. Beamish [ed.] Climate change and northern fish populations. Can. Spec. Publ. Fish. Aquat. Sci. 121.

Abstract: The west coast of Vancouver Island is one of the most productive fishing grounds off western North America. Pacific hake (*Merluccius productus*) migrate to this region to feed in summer, and dominate the trophic structure of the pelagic community. Hake feed primarily on the rich euphausiid and herring (*Clupea harengus*) stocks. An analysis of 13 summer, midwater trawl surveys indicates that significantly more hake migrate into the region in warm summers: with a 1°C rise in temperature above the long-term mean, the hake biomass doubles from 179 000 t to 353 500 t. We estimate that hake eat about 10 000 t of herring during August and September in cool summers, and 15 400 t in warm summers. With above-average summer temperatures since 1977, increased predation by hake appears to have had a major impact on the herring stock. Only four of the last 15 year-classes have been above the long-term average, and the herring stock has declined sharply. Our results demonstrate that hake predation is depensatory, and that small changes in ocean climate can have a large impact on the population dynamics of a key commercial and forage species like herring.

Résumé : La région de côte ouest de l'île de Vancouver représente une des pêcheries les plus productives de l'ouest de l'Amérique du Nord. Le merlu du Pacifique (*Merluccius productus*) migre dans cette région pour s'y alimenter pendant l'été et domine le réseau trophique de la communauté pélagique. Le merlu se nourrit principalement des riches stocks d'euphausiacés et de harengs (*Clupea harengus*). Une analyse de 13 relevés estivaux par chalutage pélagique révèle qu'un nombre considérablement plus élevé de merlus migrent dans cette région pendant les été chauds : avec une hausse de 10° C de la température au-dessus de la moyenne à long terme, la biomasse de merlu double et passe de 179 000 t à 353 500 t. Nous estimons que le merlu consomme environ 10 000 t de hareng en août et en septembre lorsque l'été est froid, et 15 400 t lorsque l'été est chaud. En raison de températures estivales plus élevées que la moyenne depuis 1977, une prédation accrue exercée par le merlu semble avoir eu une incidence majeure sur le stock de harengs. Seulement 4 des 15 dernières classes annuelles ont été au-dessus de la moyenne à long terme, et la ressource hareng a décliné brusquement. Nos résultats montrent que la prédation par le merlu est anticompensatoire et que de petits changements dans un climat océanique peuvent avoir d'importantes répercussions sur la dynamique d'une population d'une espèce commerciale et alimentaire essentielle comme le hareng.

Introduction

The La Perouse Bank region off the southwest coast of Vancouver Island (Fig. 1) lies near the northern terminus of the Coastal Upwelling Domain, which extends along the west coast of North America from Baja California to roughly the northern tip of Vancouver Island (Ware and McFarlane 1989). The La Perouse Bank region supports a productive multispecies fishery. During the 1980s the commercial fish catch in this region averaged 6.6 t · km^{-2} · yr^{-1}, which is about five times the average yield per unit area along the eastern boundary of the Gulf of Alaska, and two to three times the average (potential) yield in the Coastal Upwelling Domain (Ware and McFarlane 1989; Ware, 1992). Pacific hake (*Merluccius productus*) currently supports the largest single species fishery (averaging 276 000 t annually between 1987 and 1991) in the Coastal Upwelling Domain in general, and off British Columbia in particular (89 000 t).

The La Perouse Bank ecosystem is characterized by a relatively narrow continental shelf (average of 65 km), relatively intense wind-induced upwelling in the summer, and correspondingly high phytoplankton and euphausiid (principally *Thysanoessa spiniferia* and *Euphausia pacifica*) biomass (Mackas 1992). Winds in this region blow predominantly from the north in summer and from the south in winter.

D.M. Ware and **G.A. McFarlane.** Department of Fisheries and Oceans, Pacific Biological Station, Nanaimo, BC V9R 5K6, Canada.

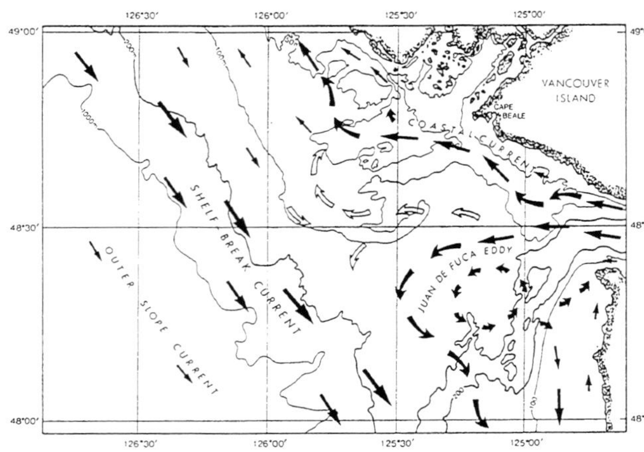

Fig. 1. General summer circulation pattern in the La Perouse Bank survey region (southwest coast of Vancouver Island). Stippled zone denotes an area of confused flow (after Thomson et al. 1989).

This change in wind direction from the typical summer to winter pattern occurs during a fairly abrupt (1–2 wk) transition period in the fall (October–November), and again in the spring (February–April; seeThomson and Ware 1988). Robinson et al. (1992) estimate from a simulation model that the primary production rate in the Juan de Fuca Eddyregion (Fig. 1) is of the order of 350 $gC \cdot m^{-2} \cdot yr^{-1}$. This relatively high production is supported by three sources of nutrient supply: (1) wind-induced upwelling along the outer shelf and subsequent cross-shelf mixing into the nearshore surface layer; (2) transport of nutrient-rich water seaward from Juan de Fuca Strait; and (3) intermittent, intense wind-induced upwelling at the northern tip of Vancouver Island and subsequent alongshore transport of nutrients and plankton by the equatorward flowing shelf-break current (Thomson and Ware 1990). Zooplankton biomass tends to be highest in the Juan de Fuca Eddy and along the shelf break and slope, and is dominated by calanoid and cyclopoid copepods and euphausiids (Mackas 1992). This rich plankton community supports year-round resident pelagic and demersal fish stocks, as well as large, migratory stocks of Pacific hake and sardine (*Sardinops sagax*) (historically) that travel from California to La Perouse Bank to feed from June to October. The biomass of pelagic fish in the region averages 65 $t \cdot km^{-2}$ in summer and 25 $t \cdot km^{-2}$ in winter, after the departure of the migratory stocks.

In 1985, a multidisciplinary investigation was begun to provide physical and biological data that could be used to improve the management of the fish stocks off the west coast of Vancouver Island (Thomson and Ware 1990). The La Perouse project has identified the dominant factors affecting the general circulation and water property structure, has quantified the seasonal and interannual variability in plankton biomass, and has made significant progress in determining the community interactions between the dominant plankton and fish species in the region. The purpose of this paper is to summarize the first 8 yr of data on the fisheries component of the program. Our results enable us to quantify interannual variability in the distribution and abundanceof the dominant fish species (hake, herring, and dogfish (*Squalus acanthias*), and predator–prey interactions in the fish community. We also identify the role of temperature and euphausiid biomass in regulating interannual changes in hake migration rates, distribution on the feeding grounds, and predation impact on Pacific herring, which is the key forage fish in the area.

Table 1. Average summer biomass of dominant pelagic and demersal fish species in the La Perouse Bank area during the 1980s.

Species	Biomass (10^3)	Source
Pacific hake	236 [a]	This paper
Pacific herring	55 [b]	Schweigert et al. 1992
Dogfish	38	This paper
Coho salmon1	5	M.C. Healey (personal communication, Univ. of B.C., Vancouver)
Pacific cod	10 [c]	Leaman 1992
Sablefish	8 [c]	Leaman 1992
Chinook salmon	11	Healey (personal communication)
Lingcod	8 [c]	Leaman 1992
Rockfish spp.	6 [d]	Leaman et al. 1988
Halibut	1	B.M. Leaman (personal communication, DFO, Pacific Biological Station, Nanaimo)
Total	388	

[a] Average hake biomass from stock assessments is 271 000 t.
[b] Biomass of age-3 and older fish.
[c] Estimated from data.
[d] Estimates of yellowtail rockfish biomass from R. Stanley (personal communication, DFO, Pacific Biological, Nanaimo)

Methods

Catch information was compiled by B.J. Waddell and D.M. Ware (unpublished information) from Canadian and U.S. sources. Estimates of the average biomass of the dominant fish species (Table 1) come from stock assesment analyses made by the Department of Fisheries and Oceans, except for Pacific hake and spiny dogfish biomass, which was estimated from the La Perouse midwater trawl surveys. These surveys have occurred annually since August 1985. In addition, we extended the time series by recovering an additional 6 yr of survey data back to 1968 (for approximately the same time of year: July–September) from other published sources. The procedure for calculating a swept-volume biomass estimate from the trawl data is outlined in Appendix 1. We believe this method is superior to other swept volume estimates, because we use an objective mapping program to interpolate the abundance over the entire survey area and then sum it to calculate the total biomass. The interpolated pattern closely matches the relative biomass pattern reconstructed from echograms collected during the surveys.

Historical Fisheries

Fisheries conducted off the west coast of Vancouver Island over the last 70 yr can be divided into pelagic fisheries: for sardine, Pacific herring, hake and salmon (*Oncorhynchus* spp.), and groundfish fisheries: for flatfish, rockfish, Pacific cod (*Gadus macrocephalus*), sablefish (*Anoplopoma fimbria*), lingcod (*Ophiodon elongatus*), and some minor species. Initially, Canadian and U.S. fisheries in the La Perouse region targeted sardine and herring (Fig. 2). Catches averaged 30 508 t annually from 1920 to 1966. By the mid-1960s, other nations were fishing in these waters (through special aggreements with Canada after 1977) so the species composition of the catch changed markedly (Fig. 2). Catches have averaged 63 106 t annually since 1966, and have increased substantially since the mid-1980s.

The catch during the first 50 yr was dominated by clupeoids. The Pacific sardine (or pilchard) fishery developed in the early 1920s and was the most important fishery in the region during the 1930s and early 1940s (Fig. 2). During the peak years (1926–44), Canadian sardine catches averaged 22 542 t annu-

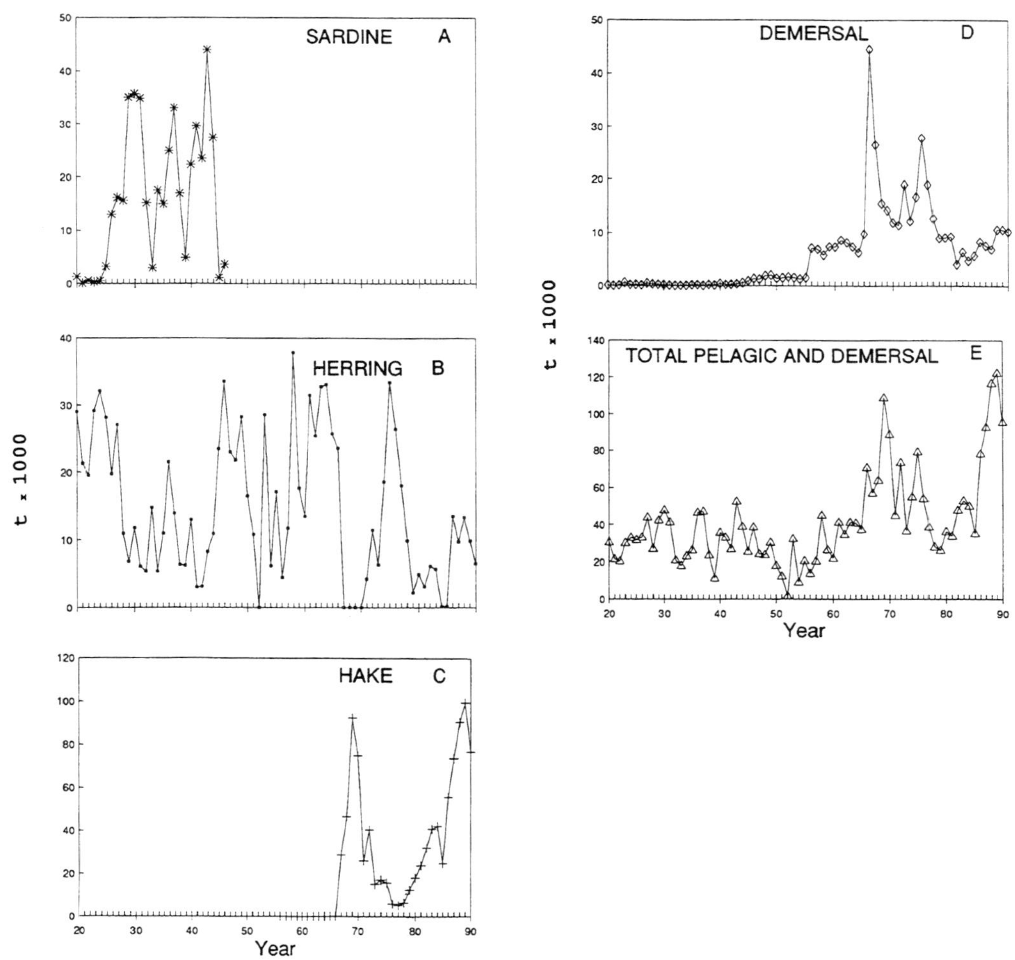

Fig. 2. Total landings of pelagic and demersal fish from the La Perouse Bank area 1920–90. Total pelagic landings include sardine, herring, hake, chinook and coho salmon, and pollock (B.J. Wadell and D.M. Ware, unpublished data).

ally. The northern Pacific sardine stock, which was harvested from California to British Columbia, collapsed after 1946 owing to a combination of overfishing and poor environmental conditions (MacCall 1979; Radovich 1982; Ware and Thomson 1991). Since the demise of this stock, Pacific herring has become the most economically important pelagic species in the La Perouse region. Catches have fluctuated between <1000 – 38 000 t annually, and averaged 15 000 t over the last 70 yr (Fig. 2). Since 1978 there has been a trend towards decreased catches (about 6600 t annually), primarily due to poor recruitment (Schweigert and Fort 1994), caused by unfavourable environmental conditions and probably increased predation (Ware 1991).

The most dramatic change in the pelagic fishery occurred when a large stock of Pacific hake was discovered offshore in the early 1960s. Since 1968, the largest catches have usually been hake (Fig. 2), averaging 42 000 t annually. After the Canadian extended fishing zone was created in 1977, hake catches increased steadily to 100 000 t in 1989.

The La Perouse Bank groundfish fishery (mainly hook and line) remainded small and stable from 1920 to 1944, averaging 236 t annually (Fig.2). Catches increased substantially in the late 1940s, after the introduction of bottom trawls, and averaged 1500 tonnes annually until 1955. The fishery peaked in 1966 when foreign vessels caught 45 000 t (mainly rockfish). During the late 1960s and 1970s catches remained high,

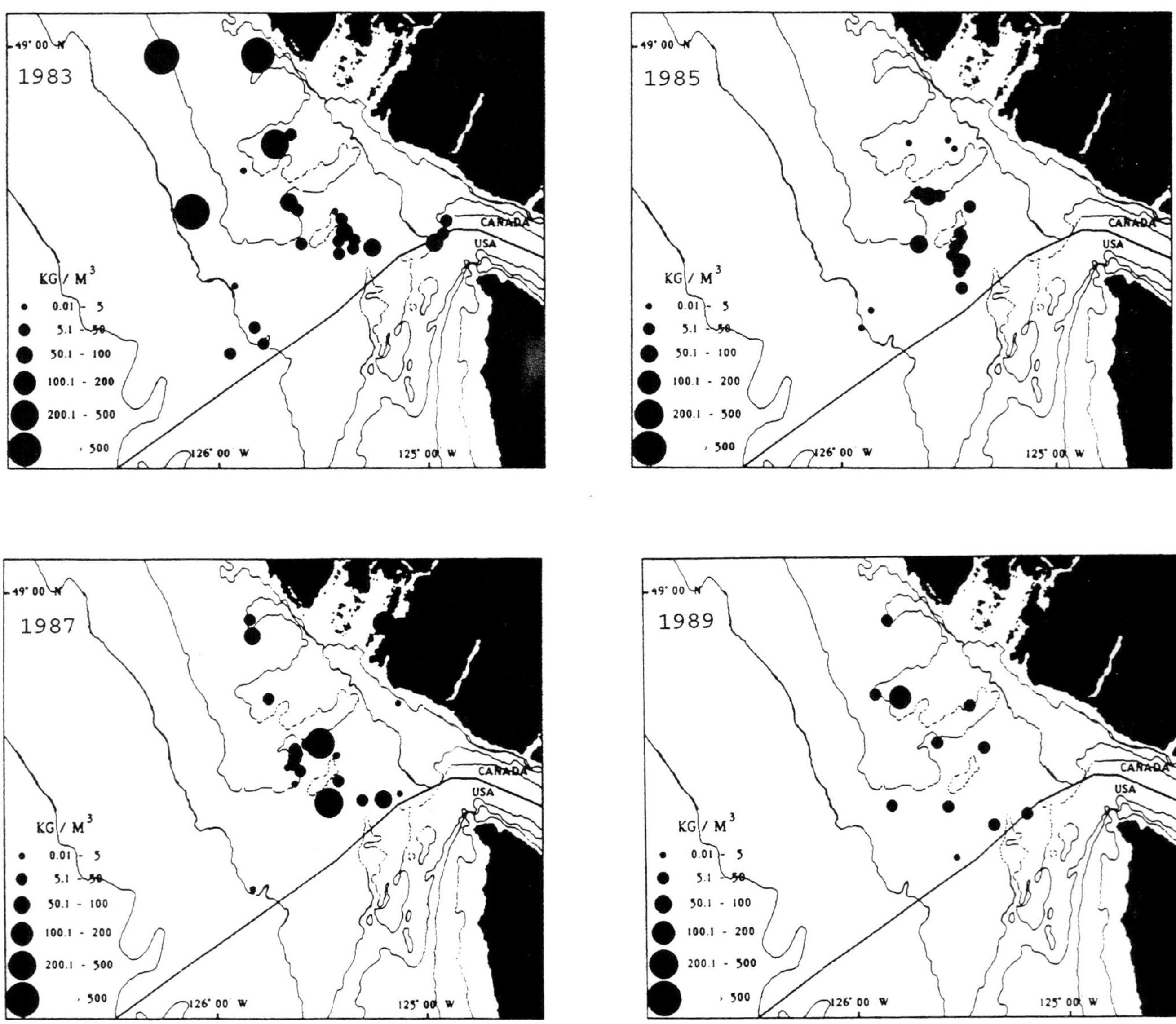

Fig. 3. August distribution (research trawl CPUE kg·m^{-3}) of Pacific hake in 1983 (warm summer) and 1985, 1987, and 1989 (average summers).

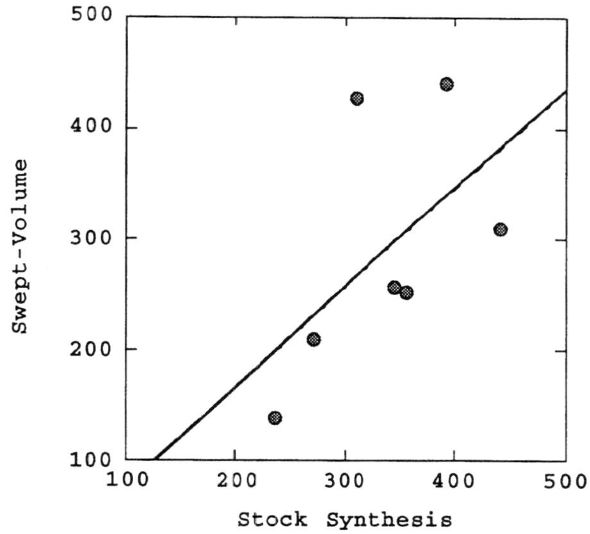

Fig. 4. Research trawl hake biomass estimate (swept volume) and adjusted stock synthesis estimate for 1983 and 1985–91 (p = 0.098).

averaging 16 000 t . Catches declined to approximately 8000 t during the 1980s, due to more stringent quotas resulting from the reduced biomass.

Results

Current dominant species and abundance

Pacific hake

Two types of hake occur around La Perouse Bank: residents and migrants. Resident hake remain in the region year round and spend the summer in the large sounds along the west coast of Vancouver Island. These stocks are small relative to the biomass of the migratory stock (Beamish and McFarlane 1985). By contrast, migratory hake currently dominate the pelagic fish biomass around La Perouse Bank in summer, numbering some 236 000 t, or 61% of the pelagic fish biomass (Table 1). Migrant hake spend the summer feeding in the deep basins on the continental shelf and along the shelf break. Around the time of the fall transition theybegin migrating southward to California, where they spawn in winter (January–March). After spawning, the older members of the stock begin their annual northward migration to feeding grounds off Washington and southern British Columbia. The summer distribution pattern in the La Perouse Bank region is shown in Fig. 3 for a few selected years.

Migratory hake biomass in the La Perouse area was estimated from the trawl surveys (Appendix 1). Our results indicate that the biomass has fluctuated significantly over the years reaching a high of 439 000 t during the summer of 1983, and averaging 211 000 t (CV = 0.60). The biomass of hake has been higher since 1983, averaging 262 000 t. Year-to-year changes in hake biomass according to our estimates agree fairly well (Fig. 4) with the adjusted biomass expected to be in the La Perouse Bank region according to the stock synthesis model (Dorn and Methot 1991; Saunders and McFarlane 1992). However, the swept-volume surveys appear to produce a lower estimate than the stock-synthesis model, particularly at low stock biomasses. A more intensive trawl survey might yield better results.

Pacific herring

Only one stock of herring is currently recognized for management purposes along the west coast of Vancouver Island. Herring is a resident species. It spawns nearshore in Barkley and Clayquot Sounds in late February – early March. After spawning, the adults migrate to the offshore banks on the continental shelf to feed. In summer, they typically occur near the eastern margins of the banks at depths of 100 m. Young-of-the-year herring remain inshore until mid-August, when they begin moving to the offshore feeding grounds. Juvenile herring (age 0+,1+, and 2+) remain offshore year round until they approach sexual maturity at the end of their third year of life. These maturing age-2+ juveniles join the schools of adult fish and migrate shoreward with them in the late winter where they form pre-spawning aggregations near the spawning grounds (Taylor 1964). During the 1980s, the herring stock on the west coast of Vancouver Island averaged 37 000 t, which is below the historical average stock size of 45 600 t. The reasons for this are discussed below. Probably about 67% of the herring on the west coast of Vancouver Island (25 000 t), and 50% of those in the Strait of Georgia (29 500 t) spend the summer feeding around La Perouse Bank. Thus in the 1980s, about 54 500 t of herring (age 2+ and older) summered on La Perouse Bank (Table 1). As a first approximation, there is probably an equivalent biomass of juveniles (age 0+ and 1+) also scattered over the offshore banks and basins.

Spiny dogfish

The southwest coast of Vancouver Island supports a significant biomass of spiny dogfish (Ketchen 1986). Juveniles and adults are distributed throughout the water column, and feed mainly on midwater species like euphausiids, herring, and hake (Tanasichuk et al. 1991). Large adults (>70 cm) are frequently caught on, or near, bottom. Although the biomass of adult spiny dogfish declined during the intense fisheries of the1930s and 1940s, the stock recovered quickly when the unexploited midwater biomass recruited to the population (Wood et al. 1979). Saunders (1989) estimated a total abundance of approximately 150 000 – 200 000 t in Canadian waters. About 30% of this biomass is off Vancouver Island (Ketchen 1986). Swept-volume estimates (Appendix 2) from the trawl surveys suggest an average midwater biomass of 38 100 t in the La Perouse region. The typical summer distribution pattern of spiny dogfish is shown in Fig. 5.

Other species

Coho (*O. kisutch*) and chinook (*O. tshawytscha*) salmon make up the fourth largest biomass of pelagic fish. These two species account for about 26 000 t in the1980s (Table 1). In summer, chinook tend to concentrate on the banks in close association with the main schools of herring. The demersal fish community is dominated by Pacific cod, sablefish, a mixture of rockfishes, lingcod, and Pacific halibut, which collectively amount to about 33 000 t.

Typical summer distribution pattern

The summer distribution of the major pelagic fishes (hake and herring) and their dominant food (euphausiids) was mapped using Compugrid Software, a geographic information system (Fig. 6). The frequency of commercial trawling locations, and research vessel hydroacoustic surface densities confirm that

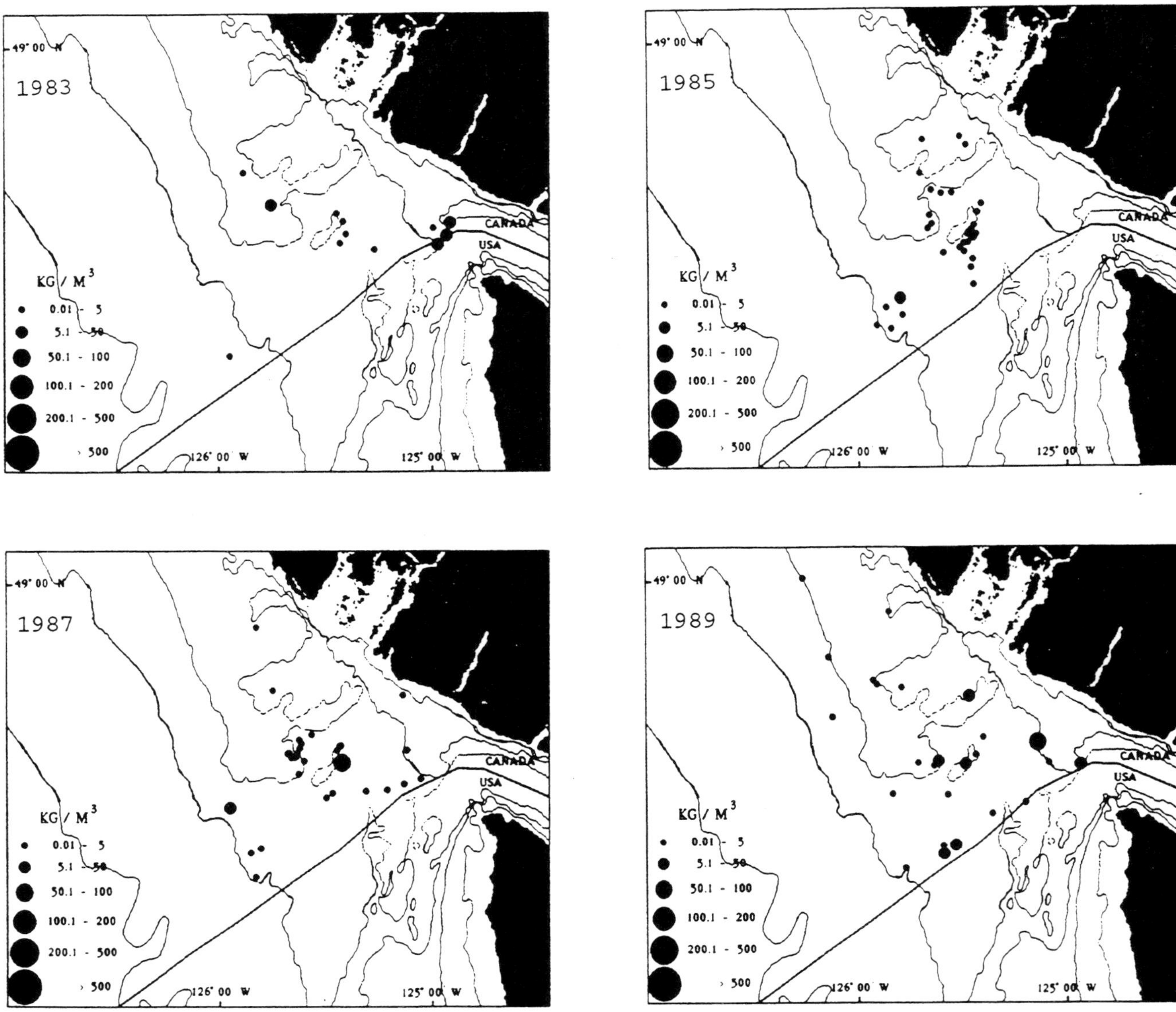

Fig. 5. August distribution (research trawl CPUE kg·m^{-3}) of spiny dogfish in 1983 (warm summer) and 1985, 1987, and 1989 (average summers).

the heaviest concentrations of hake in the La Perouse area occur along the shelf break, and in the basins, at depths greater than 100 m. There is a striking progression in the geographic distribution of hake over the summer (Fig. 7: in June, hake are caught predominately off Barkley Sound; in July and early August, the main concentration is further south near the Juan de Fuca Eddy; by August, hake begin moving to the shelf break). This offshore movement continues during September so, by October, the main concentrations occur in the eddy region and along the shelf break. The distribution of hake shows an almost perfect overlap with the distribution of euphausiids found by Simard and Mackas (1989). In contrast, herring occur predominantly on the majorbanks (La Perouse and Swiftsure) in August often associated with, or near, the 100 m isobath (Fig. 6). These patterns were consistent between 1985 and 1991.

Climate related changes in hake abundance and distribution

During the 13 yr of survey data collected between 1968 and 1991 there were large changes in oceanic conditions as indexed by summer water temperatures at Amphitrite Point, a coastal station adjacent to La Perouse Bank. June and July water temperatures ranged from a low of 11.4°C in 1975 to a high of 13.3°C in 1983 (Fig. 8). The lowest biomass of hake in the area occurred during the period of generally below-average water temperature in the early 1970s. Because hake is a migratory species, we hypothesized that the number moving into the La Perouse region may increase at higher water temperatures. We tested several possibilities and found that the hake biomass in August was most highly correlated with average temperature in the region 2 mo prior to the survey. The correlation was highly significant (p = 0.003; Fig. 9),

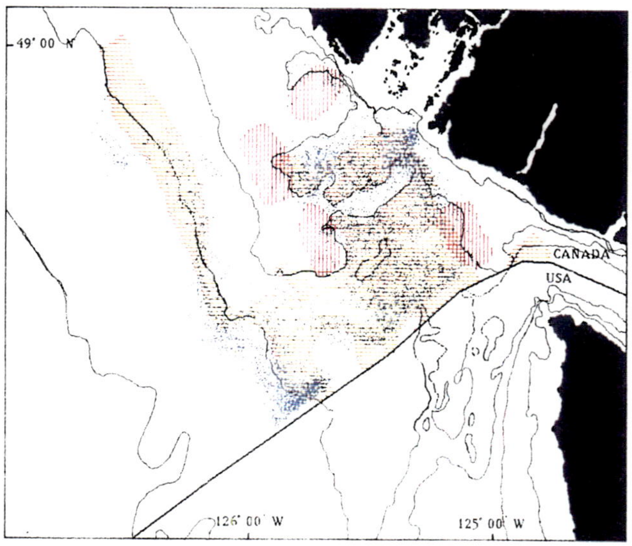

Fig. 6. Normal midsummer distribution (1985–91) of Pacific hake (dots), herring (red aggregations) and euphausiids (creme colored aggregations).

indicating that considerably more hake migrate into the La Perouse Bank area in warm summers. Because hake first appear in the region in June, this correlation suggests that the quantities of food (primarily euphausiids) they encounter in early June and July determine how many hake remain in the area by the time of the August trawl survey. The average surface water temperature in June and July around La Perouse Bank is 12.2°C (Fig. 8). At this temperature, 179 000 t of hake normally migrate into the area.

There were marked differences in the mid-summer distribution pattern of hake between the warm El Niño summer of 1983 (13.3°C) and years of average water temperatures such as 1985 (12.3°C), 1987 (12.5°C) and 1989 (12.8°C). In 1983, hake were more abundant near the outer edge of the continental margin and in the northernpart of the survey area (Fig. 3). In contrast, in 1985,1987, and 1989, they were concentrated midshelf in the southern part of the survey area, near the Juan de Fuca Eddy. We now recognize this as the typical midsummer distribution pattern under current climatic conditions. We tested for an association between water temperatures during the trawl surveys and the hake distribution pattern, by comparing the interpolated distribution pattern in the most extreme year (1983) with the pattern observed in the other 12 survey years. Conducting pairwise comparisons, we determined the number of co-occurrences (number of grid cells in the survey region in which hake appeared in both 1983 and in each of the other years). The frequency of co-occurences and water temperature were positively correlated (p = 0.07; Fig.10). Thus in cool years, hake tend to occur more in the southern, midshelf region of the survey area by midsummer (and therefore show a low amount of overlap with the 1983 distribution); whereas, in warm years, they tend to be found more frequently along the outer shelf and in the northern part of the survey area (thus showing an increasing amount of similarity to the 1983 pattern). Hydroacoustic surveys conducted in 1990, 1991, and 1992 support this conclusion (R. Kieser, personal communication, DFO, Pacific Biological Station, Nanaimo, B.C.). In addition, in August 1990 and 1992 (both warm years), the hydroacoustic surveys found that hake were abundant along the outer edge of the continental margin of the La Perouse area as far north as Queen Charlotte Sound.

Trophic interrelationships

The summer diet of the dominant fish species was measured in eight of the survey years (see Tanasichuk et al. 1991 for methods). Euphausiids (mainly *Thysanoessa spiniferia* and *Euphausia pacifica*) and herring are the main prey (Tables 2 and 3). On average, these two food items make up 92% of the diet of hake (by weight), 70% of the diet of dogfish, and 89% of the diet of chinook salmon. Euphausiids were the dominant food of most of the predator species sampled, except for chinook salmon and lingcod, where herring predominated.

Herring mortality caused by hake

Our surveys indicate that hake abundance in the La Perouse region is strongly modulated at interannual and decadal scales, by changes in water temperature, and that hake is a major herring predator because of its large biomass. What is the impact of this predation on the herring stock?

Stomach content information collected during the surveys indicates that, on average, hake eat 1.6% of their body weight in the late summer; 0.258% of the body weight of herring is eaten daily. Our data indicate that the fraction of herring in the hake diet is a dome-shaped function of the local temperature (Fig. 11). Because euphausiid productivity and biomass tend to decrease with increasing summer temperatures (C.L.K. Robinson and D.M. Ware, unpublished data, DFO, Pacific Biological Station, Nanaimo, B.C.), Fig. 11 implies that up to a temperature of about 14.1°C, hake consumption of their secondary prey (herring) increases, as the availability of their primary prey (euphausiids) decreases. During years with the highest August temperatures, hake are distributed more northward, and along the outer shelf (Fig. 3) where herring are less available (Fig. 6); consequently, the fraction of herring in the August diet diminishes.

The amount of herring eaten by hake can be estimated from the Elliott–Persson equation (Tanasichuk et al. 1991):

$$C_d = 24\mathrm{MSF}I(R)B_h$$

where C_d is the daily consumption of herring (tonnes); R is the average proportion of the body weight of herring digested per hour; B_h is the August hake biomass (tonnes, Table 4); and MSFI is the weighted mean stomach fullness index (expressed as proportion of body weight), estimated by weighting each diet sample by the relative abundance of hake in the region where the sample was taken.

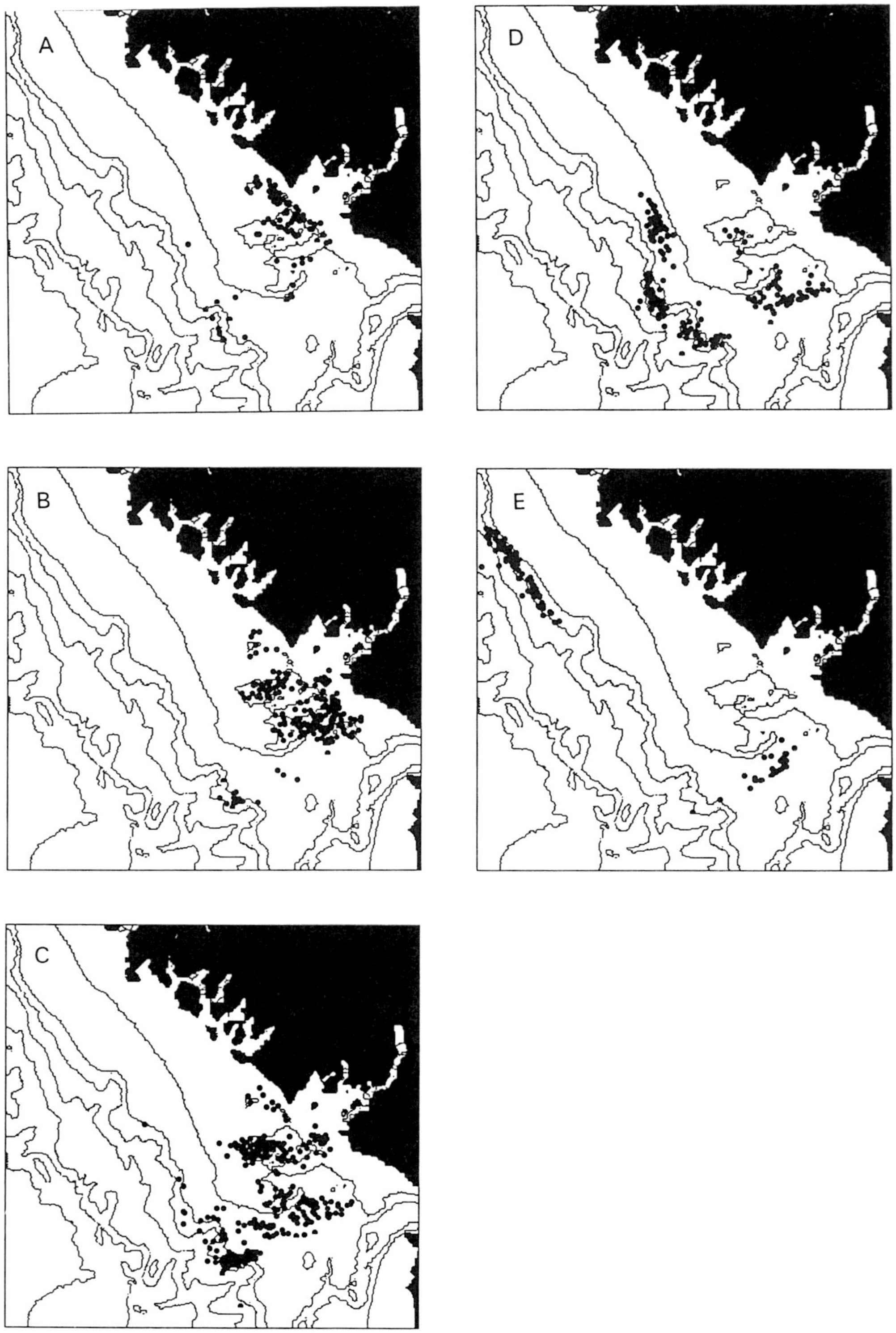

Fig. 7. Hake fleet distribution off southwest coast of Vancouver Island, 1988: (A) mid-June; (B) mid-July; (C) mid-August; (D) mid-September; and (E) mid-October.

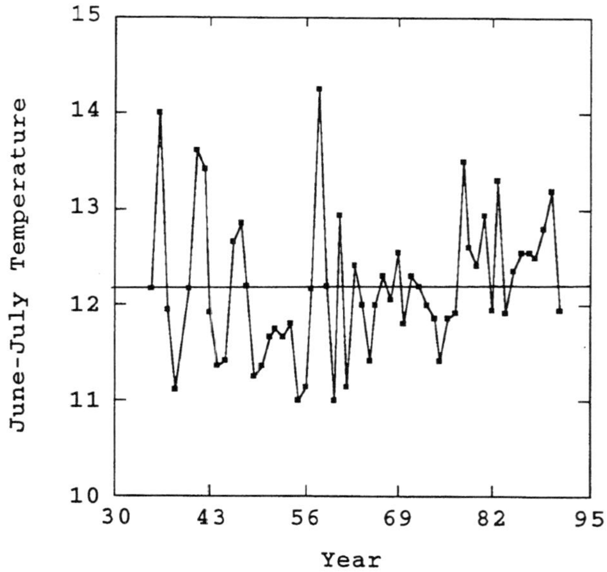

Fig. 8. Average June–July sea surface temperature at Amphitrite point (1935–91). Horizontal line represents the long term mean (12.2°C).

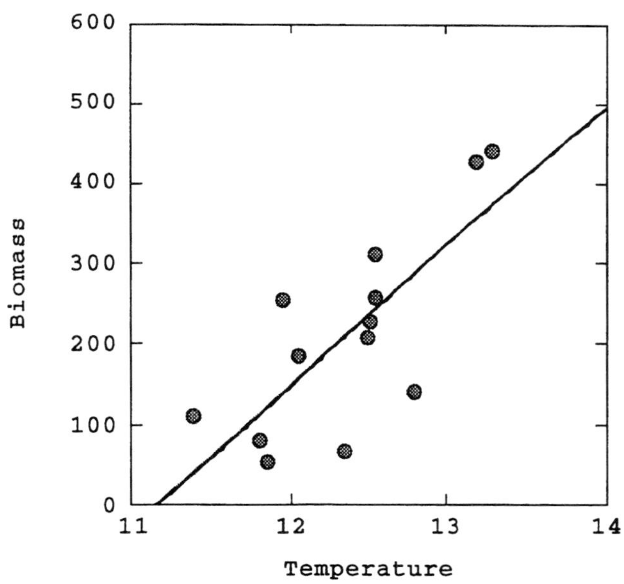

Fig. 9. Relationship between average June–July sea surface temperature and swept-volume hake biomass estimate for 13 survey years between 1968–91 ($p = 0.003$; Appendix 1).

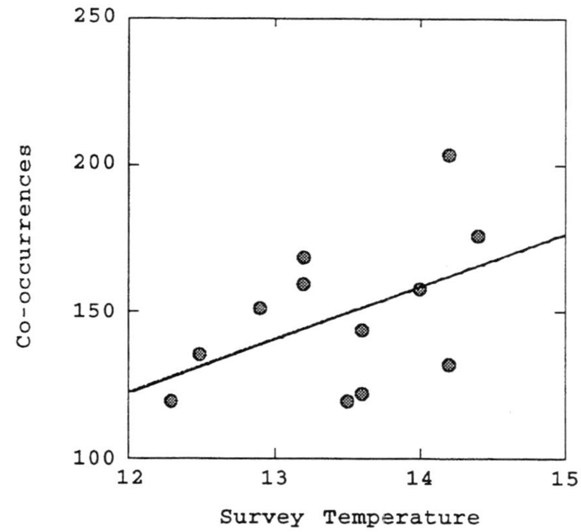

Fig. 10. Relationship between hake August distribution pattern and surface temperature at Amphitrite Point ($p = 0.12$). A large number of co-occurences indicates that the distribution in the survey year is similar to the most extreme distribution observed in the warm (El Niño) summer of 1983.

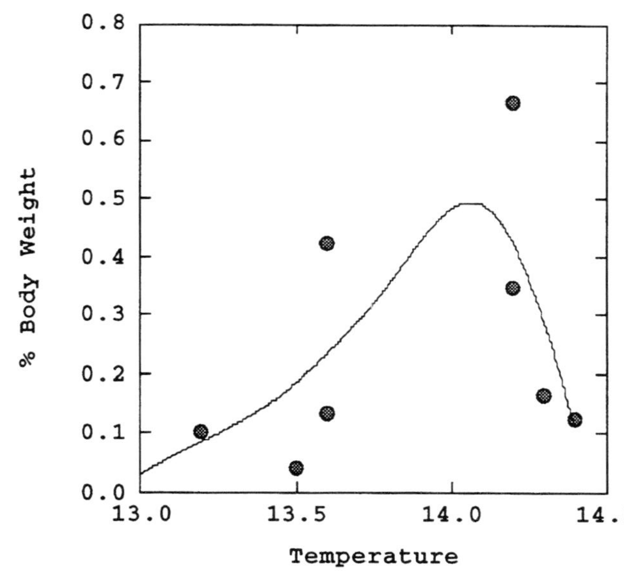

Fig. 11. Weighted amount of herring in hake diet in August as a percentage of the body weight, with respect to the sea surface temperature during the 1983, and 1985–91 surveys. Sample values were weighted by the relative abundance (catch rate) of hake in the area of each sample.

In an average summer, migratory hake eat about 208 t of herring daily, or about 12 700 t during the months of August and September. However, there are large differences in late summer herring mortality (as low as 3400 t in 1986, to as high as 28 200 t in 1987; Fig. 12). This variation arises from fluctuations in hake biomass, largely in response to inter-annual changes in water temperature, and from changes in the summer distribution pattern of hake, largely in response to changes in euphausiid availability. It is striking how much herring mortaltiy increases with a small rise in summer temperature. For example, during the surveys conducted when August was warm, 252 000 t of hake were in the survey area and ate an average of 15 400 t of herring in August and September. By comparison, during the surveys when August was cool, 197 000 t of hake were in the survey area and ate 9900 t of herring. Thus roughly a 1°C change in temperature results in a 155% increase in late summer herring mortality.

Table 2. Proportion of euphausids in diet (by weight) observed during August trawl surveys on La Perouse Bank.

	Year									
Species	1983	1985	1986	1987	1988	1989	1990	1991	Mean	CV (%)
Pollock	—	—	0.05	0.79	0.60	0.79	0.92	—	0.63	54
Hake	0.37	0.52	0.76	0.65	0.40	0.86	0.63	0.63	0.60	28
Coho	—	0.38	—	—	0.99	0.38	—	—	0.58	61
Dogfish	—	0.86	0.27	0.59	0.44	0.64	0.61	0.67	0.58	32
Sablefish	—	0.49	0.26	0.20	0.57	0.64	0.12	0.37	0.38	56
Flounder	—	—	—	—	—	0.33	—	—	0.33	—
Chinook	—	0.41	0.00	0.11	0.13	0.41	0.88	0.16	0.32	100
Pacific cod	—	0.02	0.61	—	0.03	0.15	0.24	—	0.21	115
Lingcod	—	0.00	0.00	0.00	0.00	0.00	0.00	—	0	0
Halibut	—	—	0.00	0.00	0.00	0.00	0.00	—	0	0

Table 3. Proportion of herring in diet (by weight) observed during August trawl surveys on La Perouse Bank. SS = small sample.

	Year									
Species	1983	1985	1986	1987	1988	1989	1990	1991	Mean	CV (%)
Lingcod	—	0.87	—	0.77	0.25	0.85	0.82	—	0.71	37
Chinook	—	—	—	0.85	0.78	0.54	0.11	0.82	0.62	50
Coho	—	—	—	—	(SS)	0.58	—	—	0.58	—
Halibut	—	—	—	0.31	0.88	0.78	0.15	—	0.53	67
Pacific cod	—	(SS)	(SS)	—	0.34	0.63	0.28	—	0.42	44
Hake	0.57	0.41	0.09	0.30	0.55	0.12	0.30	0.26	0.32	55
Sablefish	—	0.13	(SS)	0.14	0.22	0.25	0.16	(SS)	0.18	29
Dogfish	—	0.03	0.14	0.13	0.13	0.18	0.09	0.12	0.12	39

Table 4. Trawl survey estimates of hake biomass (t) and annual catch (t) in the La Perouse Bank area. See Appendix 1 for an explanation of the variables and calculations. The figures in parentheses in the catch column indicate the proportion of the total catch removed before the survey.

Year	*H/T*	TD	Weight	Biomass	Catch	Total
1968	1.75	40.2	827	156 300	61 361 (0.44)	183 300
1969	1.75	51.8	930	226 420	93 851 (0)	226 420
1970	1.25	15.5	827	60 250	75 009 (0.25)	79 000
1974	1.25	13.7	827	53 250	17 150 (0)	53 250
1975	1.25	27.1	827	105 340	15 704 (0.22)	108 800
1983	1.25	127.2	714	426 860	40 774 (0.29)	438 700
1985	2.37	11.6	1067	58 180	24 962 (0.29)	65 400
1986	2.37	84.4	762	302 270	24 962 (0.29)	309 500
1987	2.37	54.5	896	229 510	73 699 (0.36)	256 000
1988	3.12	50.8	658	157 100	90 491 (0.56)	207 800
1989	3.00	17.4	830	67 900	97 845 (0.72)	138 300
1990	3.25	109.5	713	366 900	73 269 (0.80)	425 500
1991	3.12	51.1	779	187 100	98 000 (0.66)	251 800
Mean	2.16	50.4	820	184 400	60 500	211 000
CV	0.37	0.73	0.13	0.66	0.53	0.60

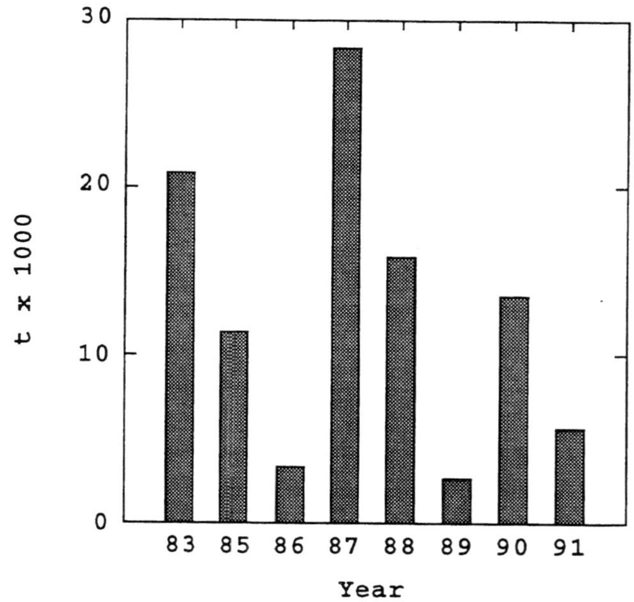

Fig. 12. Estimated amount of herring eaten by hake during August and September by year.

Discussion

The productive La Perouse upwelling region supports about 388 000 t (an average of 65 t · km–^{2}) of fish in the summer. Migratory hake account for 61% of this biomass and clearly dominate the system during their feeding season, which lasts about 120 days from mid-June to mid-October. During this period hake consume a daily ration of 1.6% of their body weight (Tanasichuk et al. 1991) which at the current stock size is about 3800 t·d^{-1}, 114 000 t·m^{-1}, and 456 000 t · season^{-1}. This large predation rate represents most of the local and imported euphausiid production (Robinson et al. 1992), and a significant quantity of herring. Thus, fluctuations in hake biomass in the La Perouse region arising from changes in ocean climate (and other factors) have a major impact on the resident euphausiid and herring stocks.

Perhaps the most important reason for the concentration of such a large biomass of fish around La Perouse Bank is that the prevailing surface and upwelling currents transport euphausiids along the shelf break, and into the Juan de Fuca Eddy (Robinson et al. 1992). This provides an abundant food supply for planktivorous fish like hake, and their typical summer distribution pattern shows a strong overlap with the distribution of euphausiids (Fig. 6). Euphausiids are also the principal prey of herring. However, the typical midsummer distribution pattern highlights the fact that major herring concentrations occur in specific locations on the banks. In these areas, herring are at the margins of the euphausiid and hake distribution (Fig. 6). Hence, the summer distribution of herring suggests a food–predation risk trade-off: herring concentrate in areas with lower food availability in return for a reduced risk of hake predation. This interpretation is also consistent with the observed seasonal movements of hake and herring. In late summer, when hake begin migrating offshore (Fig. 7), herring are found more frequently over the inner basins on the shelf, which were previouslyoccupied by hake (Thomson and Ware 1988). Thus, herring respond to the late summer departure of hake by moving into areas where the food supply is two to three times higher than it is on the banks (Mackas 1992), once the predation risk has been sharply lowered (but is still >0). Preliminary modelling of the predator–prey interaction between hake and herring and their seasonal movements around La Perouse Bank (C.L.K. Robinson and D.M. Ware, unpublished data, DFO, Pacific Biological Station, Nanaimo, B.C.) suggests that half the annual herring mortality occurs in late summer. However, the observed summer distribution of herrring probably results in a total mortality that is far less than it would have been if herring moved into the inner basins in the summer to forage with hake.

With respect to climate variability, our results highlight that two important changes occur in warm summers. First, more hake migrate into the region; on average, a 1°C rise in temperature results in a 174 000 t increase in hake biomass (Fig. 9). Second, modelling studies suggests that this increased abundance of hake quickly grazes down the euphausiid stocks on the shelf and in the Juan de Fuca Eddy. In response to this depletion of the local food supply, hake begin moving to comparatively better feeding grounds along the shelf break earlier than usual. Consequently in warm summers, hake tend to be more concentrated along the shelf break and in the northern part of the survey area a month earlier than they are in cooler summers (Figs. 3 and 7).

Migratory hake eat both juvenile and adult herring (Tanasichuk et al. 1991). The strongly skewed, dome-shaped curve (Fig. 11) indicates that more herring occur in hake stomachs in warm summers (except for the warmest years). Thus, the strong temperature dependent migration of hake into the La Perouse Bank area implies that the natural mortality rate of all age classes of herring are likely to be higher in warm summers. In general, average June–July temperatures have been warmer than normal in the La Perouse region since 1977 (Fig. 8). Concurrently, there has been a run of below-average

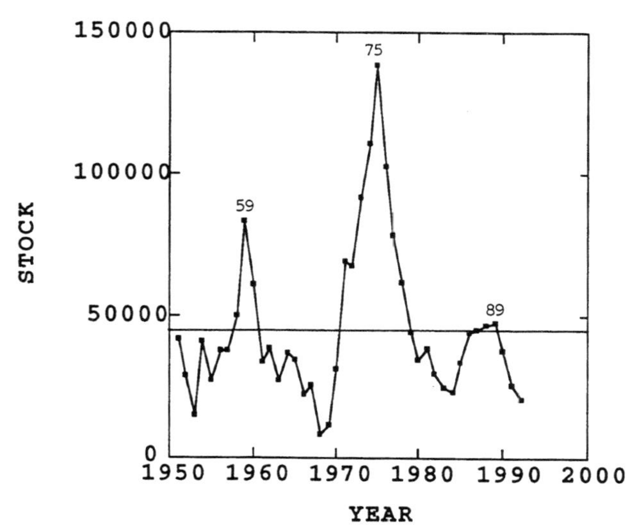

Fig. 13. Prefishery biomass (t) of west coast of Vancouver Island herring stock: 1951–92. Horizontal line indicates long-term mean bomass (45 600 t).

herring recruitment from 1977 to present; only 4 of the last 15 recruiting year-classes have been above the long-term (1951–92) average (Schweigert and Fort 1992). Owing to the recent run of below-average recruitment, and presumably higher hake predation on older herring age-classes, the herring stock declined from a peak of 139 000 t in 1975 to 20 500 t in 1992 (Fig. 13; Schweigert and Fort 1992). This negative correlation between year-class success in the west coast of Vancouver Island herring stock and water temperature has been a consistent feature of the 47-yr herring recruitment time series (Ware 1991).

After a detailed analysis of the west coast of Vancouver Island herring by a separable, sequential stock assessment model, Haist et al. (1993) concluded that a depensatory natural mortality rate (where a higher proportion of the herring stock is removed at low stock sizes) fit the stock age strucure and catch data significantly better than a model assuming constant natural mortality. Because the size of the Vancouver Island herring stock tends to fall during warm periods and rises during cool ones, the increased abundance of hake in warm summers suggests that their predation is depensatory, and therefore could be responsible, at least in part, for the process that Haist et al. (1993) discovered.

Acknowledgments

We thank Ron Tanasichuk, Bill Shaw, Mike Smith, Bill Andrews, and Cliff Robinson for their unflagging participation in the trawl surveys.

References

Beamish, R.J., and G.A. McFarlane. 1985. Pacific whiting stocks off the west coast of Vancouver Island. Mar. Fish. Rev. 47: 23–34.

Dorn M.W., and R.D. Methot. 1991. Status of the coastal Pacific whiting resource in 1991, p. A1–A77. *In* Pacific Fishery Management Council: status of the Pacific coast groundfish fishery through 1991 and recommended acceptable biological catches in 1992. Pacific Fishery Management Council, Metro Center, Suite 420, 2000 SW First Avenue, Portland, OR 97201.

Haist, V., D.A. Fournier, and J.F. Schweigert. 1993. Estimation of density-dependent natural mortality in British Columbia herring stocks through SSPA and its impact on sustainable harvesting strategies, p. 269–282. *In* S.J. Smith [ed.] Risk evaluation and biological reference points for fisheries management. Can. Spec. Publ. Fish. Aquat. Sci. 120.

Ketchen, K.S. 1986. The spiny dogfish (Squalus acanthias) in the Northeast Pacific and a history of its utilization. Can. Spec. Publ. Fish. Aquat. Sci. 88: 78 p.

Leaman, B.M. [ed.]. 1992. Groundfish stock assessments for the west coast of Canada in 1991 and recommended yield options for 1992. Can. Tech. Rep. Fish. Aquat. Sci. 1866: 304 p..

Leaman, B.M., G.E. Gillespie, D.A. Nagtegaal, and R.D. Stanley. 1988. Biomass survey of rockfish stocks off the Southwest coast of Vancouver Island, September 8–22, 1985 (M/V Howe Bay). Can. Tech. Rep. Fish. Aquat. Sci. 1611: 97 p.

MacCall, A.D. 1979. Population estimates for the waning years of the Pacific sardine fishery. Calif. Coop. Oceanic Fish. Invest. Rep. 20: 72–82.

Mackas, D.L. 1992. Seasonal cycle of zooplankton off southwestern British Columbia: 1979–89. Can. J. Fish. Aquat. Sci. 49: 903–921.

Radovich, J. 1982. The collapse of the California sardine fishery what have we learned? Calif. Coop. Oceanic Fish. Invest. Rep. 23: 56–78.

Robinson, C.L.K., D.M. Ware, and T.R. Parsons. 1992. Simulated annual plankton production in the northeast Pacific Coastal Upwelling Domain. J. Plankton. Res. 15: 161–183.

Sampson, R.J. 1984. Surface II Graphics System. Series on spatial analysis. Kansas Geological Survey, Lawrence, KS. 240 p.

Saunders, M.W. 1989. Dogfish. *In* J. Fargo and A.V.Tyler [ed.] Groundfish stock assessments for the west coast of Canada in 1988 and recommended yield options for 1989. Can. Tech. Rep. Fish. Aquat. Sci. 1646.

Saunders, M.W., and G.A. McFarlane. 1992. Pacific hake. *In* B.M. Leaman [ed.] Groundfish stock assessments for the west coast of Canada in 1991 and recommended yield options for 1992. Can. Tech. Rep. Fish. Aquat. Sci. 1866: 304 p.

Schweigert, J.F., and C. Fort. 1994. Stock assessment for British Columbia herring in 1993 and forecasts of the potential catch in 1994. Can. Tech. Rep. Fish. Aquat. Sci. 1971: 67 p.

Simard, Y., and D.L. Mackas. 1989. Mesoscale aggregations of euphausiid sound scattering layers on the continental shelf of Vancouver Island. Can. J. Fish. Aquat. Sci. 46: 1238–1249.

Tanasichuk, R.W., D.M. Ware, W. Shaw, and G.A. McFarlane. 1991. Variations in diet, daily rations, and feeding periodicity of Pacific hake(*Merluccius productus*) and spiny dogfish (*Squalus acanthias*) off the lower west coast of Vancouver Island. Can. J. Fish. Aquat. Sci. 48: 2118–2128.

Taylor, F.H.C. 1964. Life history and present status of British Columbia herring stocks. Bull. Fish. Res. Board Can. 143: 81 pp.

Thomson, R.E., B.M. Hickey, and P.H. LeBlond. 1989. The Vancouver Island coastal current: fisheries barrier and conduit, p. 265–296. *In* R.J. Beamish and G.A. McFarlane [ed.] Effects of ocean variability on recruitment and an evaluation of parameters used in stock assessment models. Can. Spec. Publ. Fish. Aquat. Sci. 108.

Thomson, R.E., and D.M. Ware. 1988. Oceanic factors affecting the distribution and recruitment of west coast fisheries. p. 31–65. *In* DFO National Workshop on Recruitment. Can. Tech. Rep. Fish. Aquat. Sci. 1626.

Thomson, R.E., and D.M. Ware. 1990. La Perouse Project: sixth annual progress report. Pacific Biological Station, Nanaimo. 67 p.

Ware, D.M. 1991. Climate, predators and prey: behaviour of a linked oscillating system, p. 279–291. *In* T. Kawaski [ed.] Long-term variability of pelagic fish populations and their environment. Pergamon Press, Oxford.

Ware, D.M. 1992. Production characteristics of upwelling systems and the trophodynamic role of hake, p. 501–513. *In* A.I.L. Payne, K.H. Brink, K.H. Mann, and R. Hilborn [ed.] Benguela trophic functioning. S. Afr. J. Mar. Sci. 12.

Ware, D.M., and G.A. McFarlane. 1989. Fisheries production domains in the northeast Pacific Ocean, p. 359–379. *In* R.J. Beamish and G.A. McFarlane [ed.] Effects of ocean variability and an evaluation of parameters used in stock assessment models. Can. Spec. Publ. Fish. Aquat. Sci. 108.

Ware, D.M., and R.E. Thomson. 1991. Link between long-term variability in upwelling and fish production in the northeast Pacific Ocean. Can. J. Fish. Aquat. Sci. 48: 2296–2306.

Wood, C.C., K.S. Ketchen, and R.J. Beamish. 1979. Population dynamics of spiny dogfish (*Squalus acanthias*) in British Columbia waters. J. Fish. Res. Board Can. 36: 647–656

Appendix 1

Trawl survey estimates of hake biomass

Hake distribution maps were generated by Surface II Graphics System (Sampson 1984). The interpolated maps were used to estimate the biomass of hake in the grid area (Fig. 1) during the trawl surveys. The total surface area of the Canadian portion of the grid (Fig. 1) out to the 200 m isobath is: 5932 km^2, or 82% of the total area. The Canadian portion of the grid was divided into 313 cells for the Surface II mapping interpolation. Hence, the surface area of each cell is: 18.9 km^2 (18.9' 10^6 m^2). The average density of hake (TD) in the layer they occupy during daylight is given by

$$\mathrm{TE} = \frac{HSE}{T}$$

where H is the height of mouth opening of the net; SE is the average density in the hake layer determined from trawl samples, and T is the average thickness of the hake layer. Echograms taken during the surveys indicate that hake occupy a distinct layer that is about 8 m thick during daylight. The thickness of this layer was similar between years.

Over the 13-yr time series, the average mouth height of the trawl was 16.5 m. Hence, TD in the hake layer is: 0.005 04 fish · m^{-3}. The volume of water potentially occupied by hake in the Canadian portion of the grid is 4.7×10^{10} m^3 ($313 \times 8 \times 18.9 \times 10^6$). Hence, the total number of hake in the La Perouse Bank survey area is 2.37×10^8 fish.

The average weight of an individual hake is 0.82 kg, so the average biomass during the surveys was about: $0.82 \times 2.37 \times 10^8$ kg = 194 400 tonnes. The total biomass was obtained by adding the biomass estimated during the survey to the amount of catch removed by the fishery before the survey (the fraction of the total annual catch removed prior to the survey is given in parentheses in Table 4).

Appendix 2

Trawl survey estimates of dogfish biomass

Dogfish distributions maps were also generated by the mapping program Surface II and the biomass of pelagic dogfish (primarily juveniles) in the La Perouse grid area during the 1985–91 trawl surveys was estimated from the interpolated maps. The same procedure as used for hake was followed.

Dogfish tend to be dispersed fairly evenly in a 20 m thick layer during daylight. Analysis of the Surface II distribution maps between 1985 and 1991 indicated that the average density of dogfish (TD) in this layer was 0.000 241 5 fish·m^{-3}. The volume of water potentially occupied was $313 \times 20 \times 18.9 \times 10^6$ = 11.8×10^{10} m^3. Hence the estimated total number of pelagic dogfish in the La Perouse Bank survey area is 2.85×10^7 fish. The average length and weight of this component of the stock was 71.83 cm and 1337 g, respectively. Hence the estimated biomass was 38 100 t. There is no significant commercial fishery for dogfish, so the biomass estimated during the survey represents the total biomass.

The impacts of climate change on Japanese fisheries

Tsuyoshi Kawasaki and Michio Omori

Kawasaki, T. and M. Omori. 1995. The impacts of climate change on Japanese fisheries, p. 523–528. *In* R.J. Beamish [ed.] Climate change and northern fish populations. Can. Spec. Publ. Fish. Aquat. Sci. 121.

Abstract: A history of Japanese fish production shows that an increase in total production was brought about primarily by increases in large-quantity species such as walleye pollock (*Theragra chalcogrammus*) and sardine (*Sardinops melanostictus*). After a decrease in production of walleye pollock because of political causes, sardine began to increase because of climate change, and Japanese fisheries became more dependent on large-quantity species. Now that the sardine stock is beginning to collapse, Japanese fisheries are faced by a serious situation. The impacts of the sardine collapse on the purse seine fishery, the Choshi fish market, and hamachi culture are discussed as case studies.

Résumé : Les antécédents de la production de poisson au Japon révèlent que l'augmentation de la production totale a été causée principalement par les augmentations de population des espèces qui s'y trouvent en grande quantité comme la morue du Pacifique occidental (*Theragra chalcogrammus*) et la sardine (*Sardinops melanostictus*). Après une diminution de la production de morue pour des raisons politiques, la population de sardines a commencé à augmenter par suite du changement climatique, et les pêcheries japonaises sont devenues plus tributaires des espèces de grande quantité. Maintenant que le stock de sardines commence à chuter, les pêcheries japonaises font face à une situation grave. Les effets de l'écroulement des prises de sardines sur la pêche à la senne coulissante, sur le marché du poisson de Choshi et sur la culture du hamachi sont analysés dans des études de cas.

Introduction

Wide fluctuations in fish catches have had an enormous impact not only on Japanese fisheries but also Japanese society. In the Tokugawa era (1600–1867), changes in the availability and abundance of pelagic fish stocks caused the development, as well as the collapse, of coastal fishery villages.

Japanese fisheries remain subject to fluctuations in stock abundance because they are largely dependent on widely fluctuating pelagic stocks distributed in the highly productive environments around Japan.

Kawasaki (1992*a*) reported on the fluctuations in the Far Eastern sardine (*Sardinops melanostictus*) caused by climate change and its effects on fisheries and society. In that report, a historical overview of the Far Eastern sardine is presented, and the relationship between the development of the purse seine fishery and the build-up of sardine stock is examined. Two other aspects are also discussed: the impact on Kushiro City of fluctuations in the sardine stock, and dependence of hamachi (yellowtail *Seriola quinqueradiata*) culture on sardine production.

T. Kawasaki. Tohoku University, Kugenuma-Fujigaya 1-10-6, Fujisawa 251, Japan.

M. Omori. Faculty of Agriculture, Tohoku University, Sendai 981, Japan.

In the present paper, the historical trend in Japanese fisheries after World War II and the impact on fisheries and society of climate change are described.

Data used in this work are obtained from Statistical Yearbook of Fisheries and Aquaculture Production, Statistical Yearbook of Fishery Products Circulation and Fisheries and Aquaculture Production (outline), issued by the Japanese Government.

Historical Overview

Figure 1 shows the total production of all types of marine fish and the production of large-quantity species such as walleye pollock (*Theragra chalcogrammus*), sardine, and chub mackerel (*Scomber japonicus*), and of other species. Here, "production" refers to marine catch, not including marine aquaculture and freshwater production.

Between 1951 and 1988, Japanese catches increased threefold from 38 to 113 million t. The increase was achieved primarily through the increase in large-quantity species, especially walleye pollock and sardine. Production of large-quantity species as a percentage of total production rose from 29 to 57% between 1951 and 1988. The catch of fish other than the large-quantity species has been almost stable since 1964. This shows that Japanese fisheries became more dependent on large-quantity species, and therefore, vulnerable to societal and environmental changes.

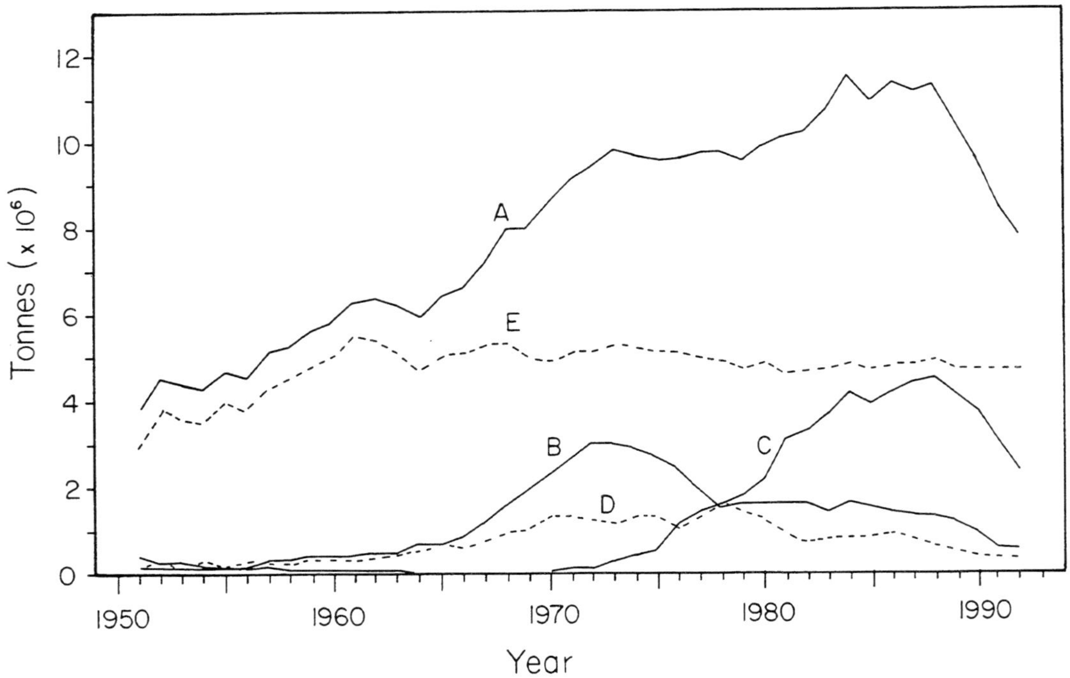

Fig. 1. Japanese fisheries 1951–92. A = total catch of marine fish; B = catch of walleye pollock; C = catch of sardine; D = catch of chub mackerel; E = catch of other species.

The catch of walleye pollock began to increase in the early 1960s and reached a maximum in 1972 through expansion of Japanese trawlers into the Okhotsk and Bering Seas. The trend in the walleye pollock catch, however, began decreasing in 1974 because of strengthening national jurisdiction of coastal countries (Japanese Government 1993), and the total production of all species levelled off. This occurred because, at the very beginning of the decrease in walleye pollock, the Far Eastern sardine began to recover. Japanese fisheries people were delighted by this coincidence. As the sardine stock continued to expand, total production began increasing in 1979 and peaked in 1984, at its historical high of 11.5 million t. However, the sardine stock began to collapse in the mid-1980s. Total production of all species in 1992 dropped to its 1969 value. History shows that total production by Japanese fisheries has been governed primarily by walleye pollock and sardine. Roughly speaking, throughout this period, production of fish other than walleye pollock and sardine remained about unchanged. Production of walleye pollock changed because of international politics, but that of sardine changed because of natural causes. The ratio of all marine fish to large-quantity species production had increased sharply since 1961, which shows that Japanese fisheries became more dependent on large-quantity species.

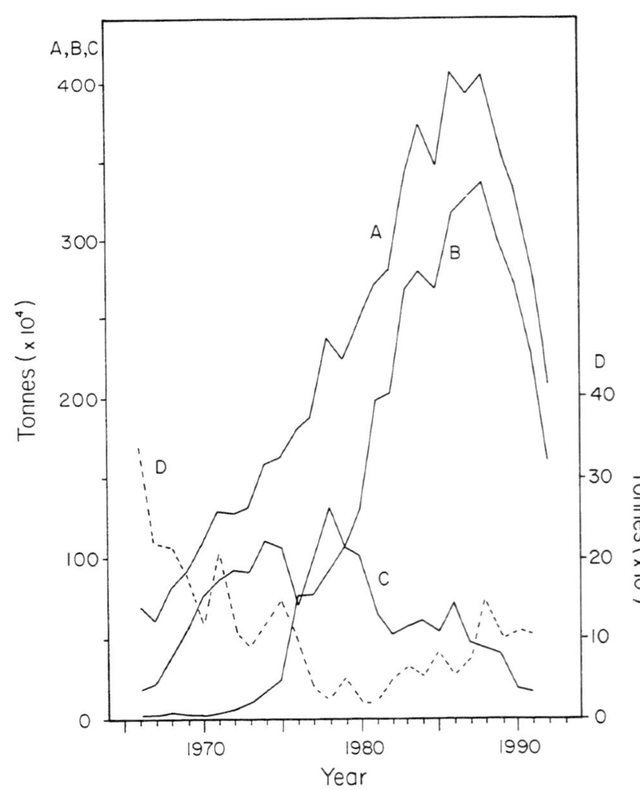

Fig. 2. Catch by large and medium size purse seiners by species, 1966–92. A = total catch of all species; B = catch of sardine; C = total catch of chub mackerel; D = total catch of jack mackerel.

Impacts on the purse seine fishery

The fishery for pelagic fish has been inseparably linked to the development of purse seiners. The annual catch of sardine by large and medium size one boat purse seiners in Japan, which have operated in offshore waters mostly within 100 nautical miles off the coasts, was about 0.7 million ts in 1966, but increased to over 4 million t in 1986 and 1988. This increase was made possible by targeting pelagic fish (Fig. 2). Before 1970, the major targets for purse seiners were chub mackerel and jack mackerel (*Trachurus japonicus*). In 1966, production of jack mackerel was as high as 337 000 t but it decreased until about 1980. Inversely, catch of chub mackerel increased until it reached 1.326 million t in 1978. After 1970, the sardine catch increased until 1988, when it reached 3.356 million t. Catch of chub mackerel began going down in 1979 and reached a low of 160 000 t in 1991. Sardine followed chub mackerel in 1989 and its catch dropped steeply to 1.660 million t in 1992 (half of its peak). On the other hand, catch of jack mackerel started increasing in 1982 and reached over 100 000 t in 1988–91.

Thus we can see an alternation between species in production by purse seiners during the period between 1961 and 1991 in the following order: jack mackerel, chub mackerel, sardine, and jack mackerel.

This alternation among species reflects that among populations. Mechanisms underlying the alternation were discussed at a symposium held on April 6, 1992, in Tokyo hosted by the Japanese Society of Fisheries Oceanography and a number of papers including Kawasaki's, one of the present authors, were presented.

Proceedings of the symposium were published in the *Bulletin of the Japanese Society of Fisheries Oceanography* in 1992. The crucial point of my hypothesis is shown below (Kawasaki 1991, 1992*a*, *b*, 1993). The sardine has undergone fluctuations of very large amplitude over long periods in a cyclic manner. The increase in sardine stock is directly affected by the global climate change in a density-independent manner, while its decrease is caused by a density-dependent mechanism.

Global atmospheric warming intensifies wind stress over the oceans, which leads to an active coastal upwelling or oceanic turbulence. These changes in oceanic circulation will result in a rise in marine productivity. The sardine, which is a rare herbivore in the marine environment, reacts sensitively to higher primary productivity. Its biomass increases rapidly and its range is extended widely, and it is transformed into a highly-migratory species. Larger-scale migration of the sardine brings about longer distance between the reproductive area and the feeding area. Thus, the population structure of sardine begins to change in a density-dependent manner. The range of sardine is reduced and it becomes less migratory. This transformation is called phase variation.

D94-043 Kawasaki Fig. 3 Reduce to 19.5 picas

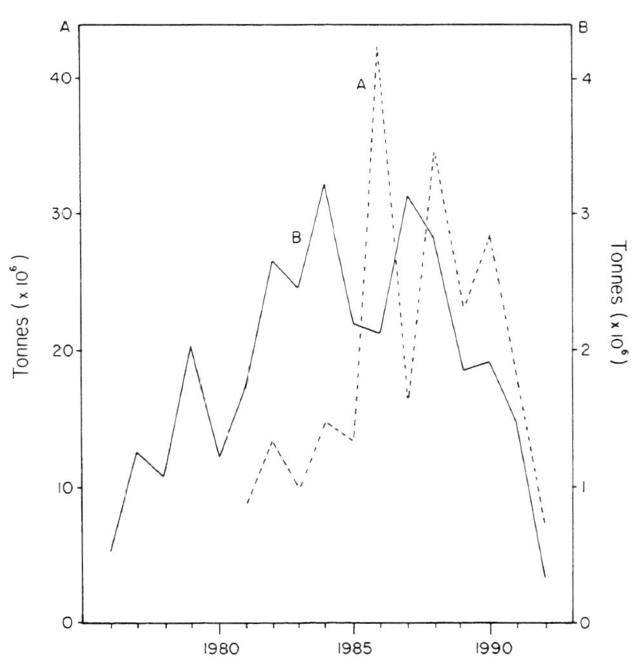

Fig. 3. Year-to-year change in sardine biomass (age 1 and older) in waters on the Pacific side of Japan, 1976–92. A = whole population; B = population southeast of Hokkaido. (Source: National Research Institute of Fisheries Science; Hokkaido National Fisheries Research Institute).

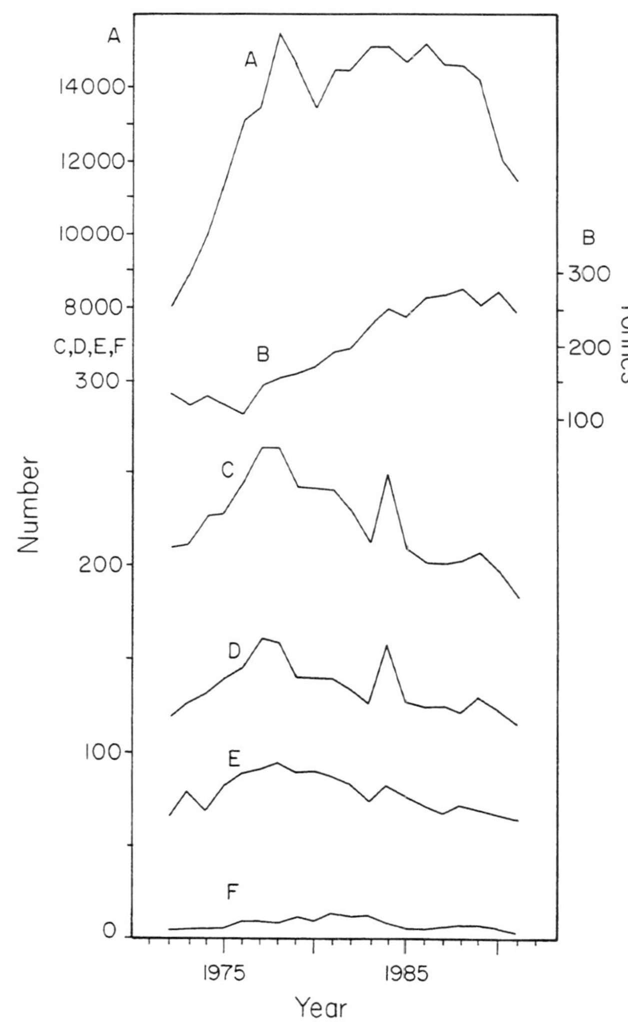

Fig. 4. Year-to-year change in fishing effort and catch per unit effort for large and medium size one-boat purse seiners, 1972–91. A = total number of fishing trips; B = catch per trip; C = total number of purse seiners; D = number of purse seiners over 100 tons; E = number of purse seiners of 50–100 tons; F = number of purse seiners under 50 tons.

Fluctuations in abundance of the sardine causes fluctuations in the other pelagic fishes such as chub mackerel and jack mackerel that constitute a pelagic fish community with the sardine around Japan, because the ranges and the feeding habits are completely or partly overlapped among these species and there must be competition for space and food among them. The sardine is the keystone species in the warm-water, pelagic fish community because of its phase-variational property and rare, herbivorous feeding habit. As the biomass and range of sardine varies, the community structure will vary, leading to the alternation between species.

Change in catches by the purse seiners has been caused mostly by variation in abundance of pelagic fish. In Fig. 3, year-to-year change in sardine biomass (age 1 and older), for both the whole range and southeast of Hokkaido, is shown. The change in catch of sardine by purse seiners in Fig. 2 well reflects that in biomass. Figure 4 shows an interannual change in the number of purse seiners by different tonnage categories (C–F). The purse seiners increased until 1978. It turned to a decreasing trend thereafter while the sardine biomass was increasing (Fig. 3). On the other hand, the number of fishing trips of the purse seiners had increased before 1978 and levelled off until 1986 (A in Fig. 4). Catch per trip continued to rise during the above period (B in Fig. 4), which shows that the increase in catch was brought about by natural causes at least after 1977.

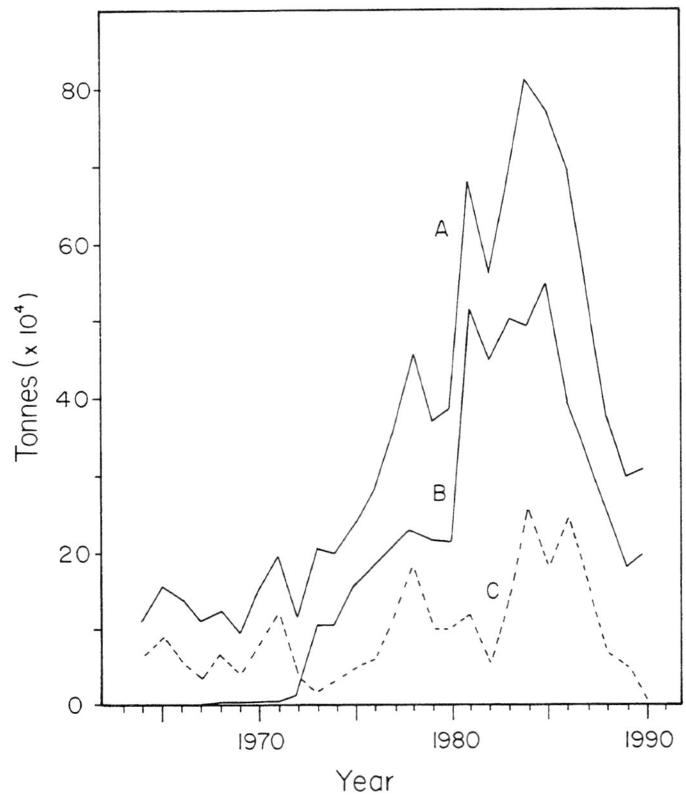

Fig. 6. Year-to-year changes in landings at the Choshi fish market by species in terms of weight, 1967–90. A = total species; B = sardine; C = chub mackerel.

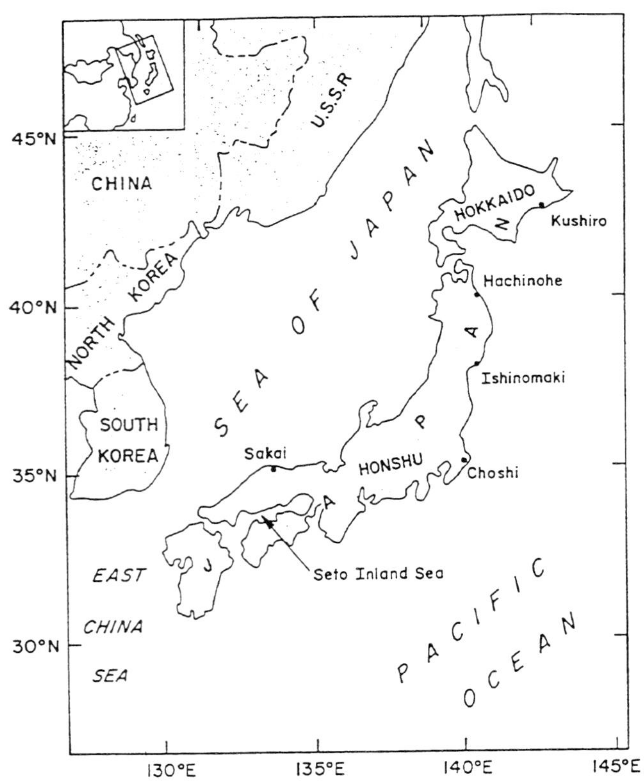

Fig. 5. Locations of ports where large quantities of sardine are landed.

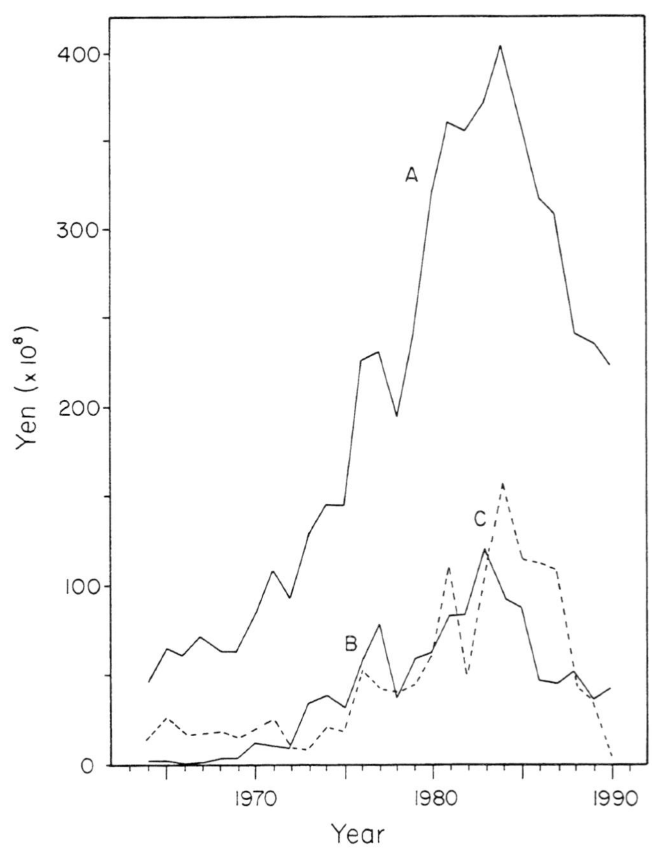

Fig. 7. Year-to-year changes in landings at the Choshi fish market by species in terms of value, 1967–90. A = total species; B = sardine; C = chub mackerel.

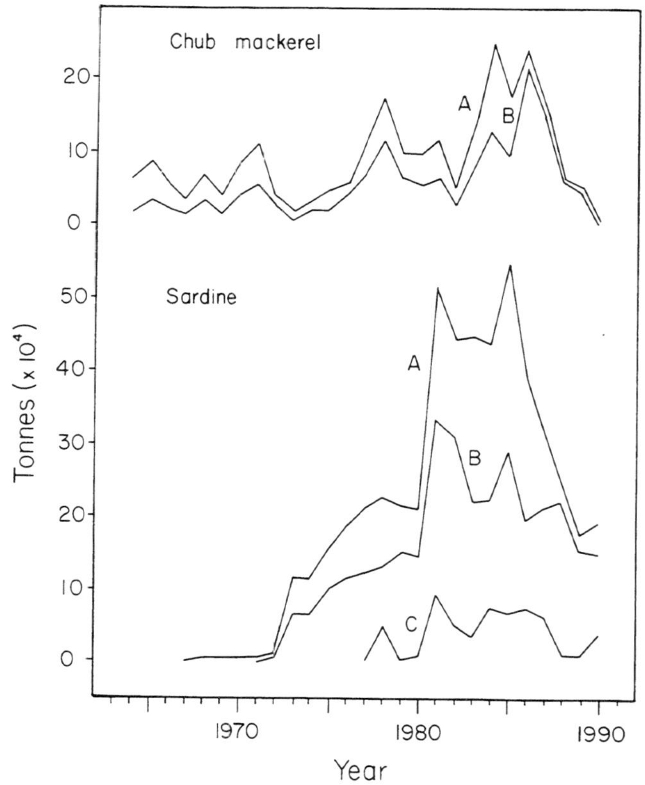

Fig. 8. Breakdown of landings at the Choshi fish market according to different utilization categories for each species, 1967–90. A = total; B = frozen; C = fish oil and fish meal.

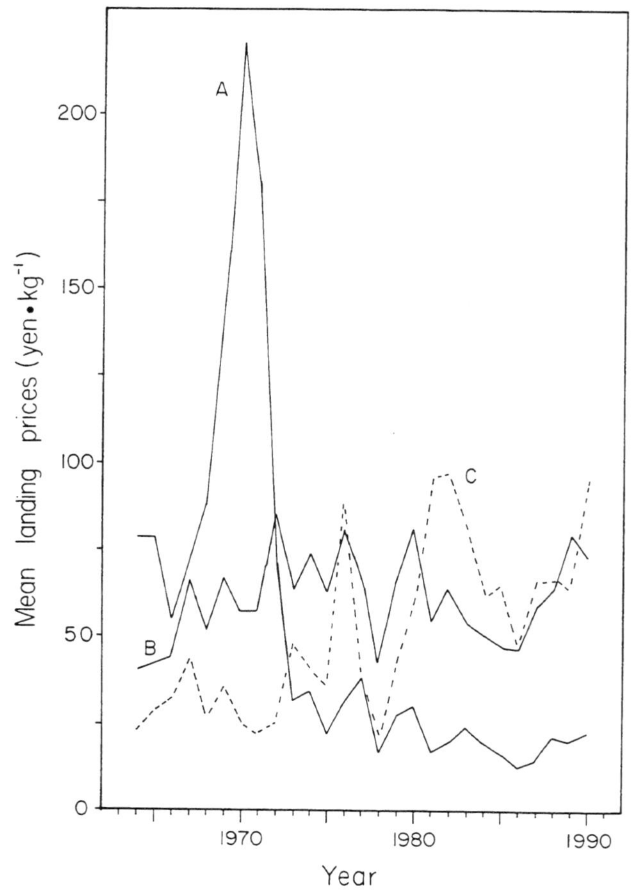

Fig. 9. Mean landing price at the Choshi fish market, 1967–90. A = all species; B = sardine; C = chub mackerel.

The decline in catch of the sardine by purse seiners after 1989 (Fig. 2) was caused by a recruitment failure. In spite of the rising trend in egg production, recruitment of new year classes has been nullified since 1988 (Kawasaki 1993). The situations are almost true of chum and jack mackerel.

A Case Study: Choshi

A great increase in the sardine stock resulted from a major expansion of the sardine's range: from south to north, from west to east and from inshore to offshore (Kawasaki 1992). The impacts on society of fluctuations in sardine abundance are found most markedly on the fringes of the sardine stock's distribution, east of 140°E. Choshi, a city located on the eastern side of Honshu and a base for the fishery operations, is used as a case study (Fig. 5).

Figure 6 shows how the total landings in terms of weight have risen and fallen at Choshi, influenced by the chub mackerel and especially sardine. From a total of 109 000 t, landings skyrocketed to 809 000 t in 1984, but dropped thereafter to 306 000 t in 1990. Landings in terms of value totalled 7.21 billion yen in 1967 but rose to 40.44 billion yen in 1984 (Fig.7). They declined to 22.33 billion yen in 1990. Choshi has become a city that depends on large-quantity, low-value species, especially sardine, for its economic well being.

Figure 8 shows the different utilization categories for sardine and mackerel at the Choshi fish market. Mackerel had been used for the most part for direct human consumption but landings at the Choshi fish market were almost zero in 1990. The landing price of mackerel at the Choshi fish market, with wide variations, rose from 22 yen·kg^{-1} in 1978 to 95 yen·kg^{-1} in 1990 (Fig. 9). Most sardine has been frozen as feed for hamachi (young yellowtail) culture or reduced to oil and meal, but sardine landings decreased from 0.55 million t in 1985 to 0.19 million t in 1990. The landing price of sardine, which declined drastically from 219 yen·kg^{-1} in 1970 to as low as 12 yen·kg^{-1} in 1986, rose to 22 yen·kg^{-1} in 1990.

There are many fish processing and reduction plants in Choshi, and many people work in these plants. Not only are facilities directly related to fisheries dependent on large-quantity fish landings but so are many ancillary industries such as transportation.

Hamachi Culture

Hamachi culture began in 1962. Production rose steeply until it reached about 160 000 t in 1979, after which it remained almost unchanged (Fig. 10). Figure 10 also shows trends in the amount of fresh feed for hamachi culture and trends in the production of frozen sardines. In 1973, frozen sardines made

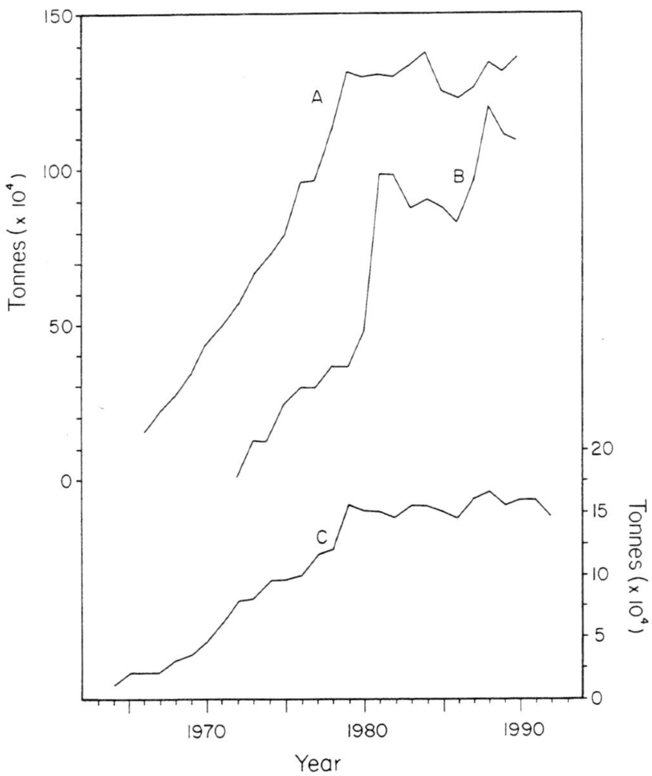

Fig. 10. Hamachi culture in Japan. A = amount of fresh feed for hamachi culture, 1966–90; B = production of frozen sardines, 1966–90; C = production of cultured hamachi, 1961–92.

up 20% of fresh feed. This proportion rose to 89% by 1988 and then dropped to 79% in 1990 because of decreased sardine landings in Japan (Fig. 1).

Hamachi culture in Japan has developed because of large amount of low-price, fresh feed was constantly available. Now that this condition is disappearing, hamachi culture will be faced with a serious situation. Furthermore, sardine meal has been widely used as trout feed and eel culture, and also livestock feed. There is no substitute for the sardine, because it is unlikely that other pelagic stocks reach such a high abundance as the sardine since they feed on zooplankton, not on phytoplankton. Japanese fisheries and ancillary industries are now at a turning point. They are suffering from imports of a huge amount of fisheries commodities, from the phasing out from Exclusive Economic Zones of foreign countries and from the high seas and from an aging and decreasing of work force. The collapse of the sardine stock would be another blow to the Japanese fisheries.

References

Japanese Government. 1993. 1992 white paper on the Japanese fisheries. Tokyo. 246 p.

Kawasaki, T. 1991. Effects of global climatic change on marine ecosystem and fisheries, p. 291–299. *In* J. Jäger and H.L. Furguson [ed.] Climate change: science, impacts and policy. Cambridge University Press, Cambridge, UK.

Kawasaki, T. 1992*a*. Climate-dependent fluctuations in the Far Eastern sardine population and their impacts on fisheries and society, p. 325–354. *In* M.H. Glantz [ed.] Climate variability, climate change and fisheries. Cambridge University Press, Cambridge, UK.

Kawasaki, T. 1992*b*. Mechanisms governing fluctuations in pelagic fish populations. S. Afr. J. mar. Sci. 12: 873–879.

Kawasaki, T. 1993. Recovery and collapse of the Far Eastern sardine. Fish. Oceanogr. 2: 244–253.

The extent of the effect of climate upon fisheries of Russia

S.A. Studenetsky

Studenenetsky, S.A. 1995. The extent of the effect of climate upon fisheries of Russia, p. 529–530. *In* R.J. Beamish [ed.] Climate change and northern fish populations. Can. Spec. Publ. Fish. Aquat. Sci. 121.

Abstract: The history, current issues and future of Russian fisheries is discussed. The fisheries expanded rapidly from the 1950s to the 1980s, but has suffered a decline in the early 1990's. The effects of the movement from a controlled economy to one based on market relationships has increased costs in the industry and will make the near future a difficult and dramatic period. The effect of climate change in addition to the economic changes is difficult to predict.

Résumé : L'histoire, les enjeux actuels et l'avenir des pêches en Russie sont analysés. Les pêches russes ont connu une expansion rapide à partir des années 1950 jusqu'aux années 1980, mais ont subi un déclin au début des années 1990. Les effets du passage d'une économie dirigée à une économie de marché ont fait grimper les coûts pour l'industrie et accentueront de façon spectaculaire les difficultés pour le proche avenir. L'effet du changement climatique conjugué aux changements économiques est difficile à prévoir.

Introduction

Fisheries have always been of particular importance among the numerous ways of supplying food to the population of the former USSR. Moreover, fisheries had special characteristics in a country with a planned and regulated economy. We have in mind the control of fisheries by quantitative indices, and the relation of this to the composition of our people's food ration.

As compared to western nations, the human diet in the former USSR was marked by shortage of products containing animal protein, and also of certain other products.

Since fishery products are important as a source of protein nutrition, a program of development of fisheries in inshore areas and on the high seas was worked out and put in force after the Second World War.

The rate of development of Soviet fisheries was rapid. In 1950, the total catch was 1.6 million tons; in 1989 it was 11.4 million tons.

It is difficult to determine the influence of climatic variations upon the Soviet (Russian) fisheries because there are several aspects of the development of Russian fisheries that make them different from those of the other nations that have a developed fishing industry.

In a number of cases these factors make it impossible to evaluate the effect of climatic changes on the economics of Russian fisheries. They have to be taken into account many analysis of those fisheries.

S.A. Studenetsky. Federal Research Institute of Fisheries and Oceanography (VNIRO), 17, Verkhnaya Krasnoselskaya, 107140 Moscow, Russia. Present address: AU-Union Academy of Agricultural Sciences (VASKHNIL), 17, Verkhnaya. Krasnoselskaya, Moscow, B-140, 107140, Russia.

In controlling the development of the USSR/Russian fisheries the state planning agencies had as a basic objective the maximum possible production of food fish products each year. The species composition of the catches was not adequately taken into account. By the 1980s, there appeared a distinct downward trend in our fisheries in respect to the number of species in the catches, which had an immediate impact on the variety of fish products in the domestic market.

By the late 1980s over half of the USSR catch consisted of five species: walleye pollock (*Theragra chalcogramma*), horse mackerel (*Trachurus symmetricus*), Pacific sardine (*Sardinops sagax*), Atlantic herring (*Clupea harengus*), and cod (*Arcotgadus glacialis*). An attempt was made by government authorities to exert control over the fishermen by rigid planning and inspection, to broaden the species composition of the catches, but it was not successful. A state planning of fisheries that was oriented toward a large quantity of food products was also the determining factor in developing the fishing fleet.

The core of the Soviet fishing fleet consisted of large trawlers and seiners that were not designed specifically for fishing in any particular area, or for any particular kind of fish. By the late 1980s, 75% of the national catch was taken with trawls and 15% with purse seines.

Large-tonnage trawlers and seiners were usually employed to take the limited number of very abundant species that occur in dense concentration. However, many species that are valued commercially, or as high-quality foods, cannot provide the high level of catch obtained from abundant species.

It was only during the late 1980s that vessels started to be constructed specially for catching shrimp, squid, and other species. However, their construction was delayed by the transition period in the countries' economics, and lack of investments.

The quantitative approach to the harvesting of aquatic products gave rise to a pattern of fishing such that large vessels used to be shifted by administrative order from one fishing area to another, and were continously deployed overthe densest concentrations so as to obtain the largest possible catches.

That is why variations in abundance, that were a result of climatic changes, did not directly affect the performance of the Soviet fish-catching fleet in terms of the total tonnage landed.

Even fishing in the Russian economic zone, i.e., on that is most subject to fluctuations caused by stock variations, was conducted in many cases by vessels large enough to fish in remote areas.

As the need might be, Soviet fishing vessels used to change not only their reas of fishing within the same geographic zone but also the zones themselves. For example, in 1960 some 87% of the Soviet catch was teken in the northern geographic zone, 2% in the Equatorial zone, and 11% in the southern zone; but only 10 yr later these figures were 71, 15, and 14%, respectively.

Given the planned and controlled economy of the country, directed management of the fisheries mad it possible to continually increase the total annual catch, and to reach a per capita consumption of fish and fishery products of 28.6 (average for 1986–88) (FAO 1991). The Soviet annual catch by the mid-1980s was distributed as follows: USSR 200-mile zone, 40–44%; 200-mile zones of other states, about 30%; areas beyone economic zones, 15–17%; inland waters, 10–12%.

Under the planned and controlled economy fishery products in the former USSR were in many cases supplied to the domestic market at prices much lower than the production cost. For instance, in the 1980s the consumer would purchase 1 kg of cod at prices five times lower than that of beef, or eight times lower than that of chicken. The cost of production naturally had to be reimbursed from the state budget, and this was actually done. The producers of fish products would get up to 3 billion roubles a year as subsidies.

What then were the effects of climatic changes upon Russian fisheries? Definite observations were available for the Northeast Atlantic and Northwest Pacific, i.e., for fisheries within the Russian 200-mile zone. Long-term fluctuations in the abundance of fish populations in these areas are usually expressed in terms of time periods which evidently depend on long-term climatic variations in production within the region.

Many species of the Norwegian–Barents Sea complex are marked by simultaneous emergence of year-classes of the same strength, which indicates a common cause of such fluctuations. Dragesund (1971) and Saetersdal (1984) noted that a comparative analysis of fecundity of cod, haddock, and herring in the 1950s to 1970s reveals a clear trend toward simultaneous appearance of strong year-classes in these species, and also in redfishes and halibut.

Recently the years 1982 and 1983 were favourable in producing strong year-classes of cod, haddock, plaice, redfishes, herring, capelin, and blue whiting, and this coincided with a temperature change from the preceding cold period.

Abundance fluctuations in cod, capelin, and herring in the Northeast Atlantic, and the ensuing rises and falls in catches, directly affected the employment rate among the coastal communities and the volume of production at the shore-based processing plants. The same is true for the decrease in the salmon fishery of the fisheries off Kamchatka and Sakhalin, as well as for the Okhotsk Sea herring.

At present, the Russian fisheries are in the throes of complex processes related to the change from a planned and controlled economy to one based on market relationships. The 1991 catches by the Russian Federation, taken alone, were lower than 1990 by 890 000 tons. Productivity dropped accordingly. The immediate reason was the drastic rise in the cost of engine fuel, fishing vessels, gear, etc., but a more basic reason was the nearly complete absence of investment in new or modernized fishing vessels.

The near future of Russian fisheries is expected to be a very difficult and dramatic period. One thing is clear: the changes to a market economy will greatly decrease the viability of Russian fisheries in distant waters. This in turn will decrease the possibility of mitigating the adverse economic effects of any climate-induced decrease in fish population abundance within the Russian 200-mile zone.

Reference

Food and Agriculture Organization of the United Nations (FAO). 1991. Fishery statistics: commodities, 1989. FAO Yearbook. Fish. Stat. 69: 397 p.

Influence of Climate Change on Living Resources in the Offshore Waters of China

J. Yang, L. Li, and S. Xia

Yang, J., L. Li, and S. Xia. 1995. Influence of climate change on living resources in the offshore waters of China, p. 531–535. *In* R.J. Beamish [ed.] Climate change and northern fish populations. Can. Spec. Publ. Fish. Aquat. Sci. 121.

Abstract: Data are provided showing long-term abundance changes in marine fisheries off China. For some species there is a close association between abundance changes and trends in precipitation.

Résumé : Des données sont fournies qui font état des changements dans l'abondance à long terme des pêcheries marines au large de la Chine. Pour certaines espèces, on note une association étroite entre les changements dans l'abondance et les tendances des précipitations.

Introduction

Scientific research shows that historical climate changes in China were synchronous with other changes around the globe. The climate in China has become warmer in this century (Xia 1991). For example, air and water temperatures in Qingdao were 0.2°C and 0.3°C higher, respectively, in the 1980s and 1960s. With the warming climate in the last 100 yr the world sea level has risen 12 cm on average, while the sea level in China has risen up to 14 cm (Lu 1992). The offshore Qingdao sea level has remained stable since the 1950s (Zhao et al. 1991).

The impact of climate change on the living marine resources of China must be examined using data series that extend over long periods. Therefore, data on climate, fisheries, and marine organisms have been collected; in addition, experienced fishermen were interviewed for related information. However, the data obtained are limited. In this report, we provide a preliminary report of the available data relating to fish abundance changes.

Abundance changes in China's offshore waters during the past five decades

Some species have disappeared from the offshore waters (Table 1). They include about 20 species, such as *Pristiophorus japonicus*, *Eleutheronema tetradactylum*, and *Rhabdosargus sarba* in the Bohai Sea; 10 species, such as *Heterodontus japonicus*, *Monocentrus japonicus*, and *Labotes surinamensis* in the Yellow Sea; 10 species, such as *Psephurus gladius*, *Bahaba flavolabiata*, and *Nibia japonica* in the East China Sea. Twenty-one species of the traditionally exploited fisheries resources have decreased in abundance or are almost depleted (Table 2). In contrast, the *Engraulis japonicus* population is apparently abundant (up to 3 million t). *Sardinops melanosticta* expanded its distribution westward to Haizhou Bay and adjacent waters in the Yellow Sea in recent years, and formed a new spawning ground there (Deng et al. 1991; Yang et al. 1980; Zhao 1985). These phenomena occurred simultaneously with climate changes, fishing activities, pollution, etc. However, the relationship of these changes and the fish abundance changes is unknown at this time. It is known from investigation in the fishing grounds (mainly spawning grounds of 14 traditional commercial species) that slight changes in temperature did not significantly affect these species (Table 3) (Zhao et al. 1990; Yang et al. 1990).

The Huanghe River (Yellow River) estuary is a spawning ground for *Penaeus orientalis* and a nursery ground for its juveniles. It spawns here during April–June and grows to market size by autumn (September–November). The higher discharge (1×10^8 m^3) in the river in April–June is associated with better catch of *Penaeus orientalis* in autumn, $r = 0.89$ (Table 4).

The habitat for *Palaemon* (*Exopalaemon*) *annandalei* in Hangzhou Bay is fresh or brackish water. The catch of this species in winter is positively correlated with the amount of annual precipitation, $r = 0.89$ (Fig. 1, X. Li, Zhejiang College of Fisheries, unpublished data).

June–September precipitation along the coast of Liaodong Bay is positively correlated with the production of *Acetes chinensis* the following year, $r = 0.93$ (Table 5). *Rhopilema esculanta* in Liaodong Bay is mainly distributed in the river mouth area. Its larva die when the salinity is less than 12‰, as does the scyphistoma larva when the salinity is less than 10‰. Therefore, excessive inflow of fresh water jeopardizes these larvae. Based on the analysis of runoff and *Rhopilema esculanta* catch in past years, we found that a particularly large runoff in August is usually associated with a significant decrease of *Rhopilema esculanta* catch the next year. For instance, the runoff in August 1975 was 2.8 times

J. Yang and **L. Li.** Institute of Oceanology, Academia Sinica, Qingdao 266071, People's Republic of China.

S. Xia. Yellow Sea Fisheries Research Institute, Chinese Academy of Fishery Sciences, Qingdao 266003, People's Republic of China.

Table 1. Fish species that have disappeared from three areas in China's offshore waters during the past five decades.

Species	Disappeared from		
	Bohai Sea	Yellow Sea	East China Sea
Notorhynchus platycephalus	-X-		
Heterodontus japonicus		-X-	
*Cephaloscyllium umbratile**		-X-	
Squalus acanthias	-X-		
Squalus fernadinus	-X-		
Pristiophorus japonicus	-X-		
*Rhinobatos schlegelii**	-X-		
Gymnura japonica	-X-		
*Myliobatis tobejei**	-X-		
*Chimaera phantasma**		-X-	
Psephurus gladius			-X-
Gadus macrocephasus	-X-		
Arius sinensis		-X-	
Hippocampus kelloggi			-X-
*Regalecus russellii**			-X-
Atherina bleekeri	-X-		
*Monocentrus japonicus**		-X-	-X-
*Zeus japonicus**		-X-	
Eleutheronema tetradactylum	-X-	-X-	
Lobotes surinamensis		-X-	
Bahaba flavolabiata			-X-
Nibea japonica			-X-
Hapalogenys nitens	-X-		
Rhabdosargus sarba	-X-		
Ditrema temmincki	-X-		
Draculo mirabilis	-X-		
Callionymus olidus	-X-		
Luciogobius guttatus	-X-		
Trachidermus fasciatus	-X-		
Lethotremus awae		-X-	-X-
Remora remora	-X-	-X-	
*Pseudorhombus cinnamomeus**	-X-		-X-
Ostracion tuberculatus			-X-
Lactoria cornuta			-X-
*Kentrocapros aculeatus**			
Diodon holacanthus	-X-		

*Collected by deep trawl from bottom area (120 to >1000 m in depth) around the continental slope of the East China Sea in 1980–81 (Xu et al. 1984).

that of the previous year and the *Rhopilema esculanta* catch dropped by 97.2% in the previous year. The runoff in August 1977 was 1.4 times that at the same period of the previous year and the *Rhopilema esculanta* catch decreased by 95.8% (N. Lu and J. Chen, Liaoning Province Marine Fisheries Research Institute, Chinese Academy of Fishery Sciences, unpublished data). Also, in June, the distance from the coast of the bottom trawl fishing ground of *Trichiurus haumela* is positively correlated with the runoff from the Changjiang River. The correlation coefficient $r = 0.61$ (Table 6) (Liu 1987).

Based on large quantities of unearthed fish bones and fish scales, it is estimated that Jiaozhou Bay and its vicinity in the Yellow Sea were rich in *Mugil so-iuy*, *Ilisha elongata*, *Sparus macrocephalus*, and *Scomberomorus niphonius* 5000 years ago (Cheng 1985), when air and water temperatures were 2–3°C higher and sea level was 2–4 m higher than at present (Geng 1981; Wang et al. 1980). Of the four species, the first three are extremely low in abundance or almost depleted in this area today.

From observations, we conclude that a long period of increasingly cold climate would have an unfavourable effect

Table 2. Change in catch (1000 t) of traditional fish species in China's offshore waters during the past five decades.

Species	Historical maximum catch (year)	Catch in 1990
Pseudosciaena crocea	222 (1934)	25
Pseudosciaena polyactis	270 (1934)	23
Gadus macrocephalus	40 (1934)	<1
Miichthys miiuy	10 (1934)	<1
Pagrosomus major	~10 (1934)	<1
Macrura reevesii	2 (1934)	<1
Cleisthenes herzensteni	>10 (1940s)	2
Lepidotrigla microptera	>3 (1950s)	<1
Lutianus erythopterus	4 (1957)	<1
Lateolabrax japonicus	2 (1960s)	<1
Upeneus moluccensis	14 (1966)	<5
Nibea albiflora	6 (1974)	<1
Clupea pallasi	182 (1972)	4
Pnematophorus japonicus	282 (1978)	197
Ilisha elongata	28 (1965)	23
Trichiurus haumela	577 (1974)	498
Fugu (*F. vermicularis*, *F. porphyreus*, *F. rubripes*, *F. pseudommus*, and *F. zanthopterus*)	>15 (1950s)	~2

Table 3. Locations of main fishing grounds of traditional species in China's offshore waters during the past five decades.

Species	1940s	Beginning of 1990s	Status
Triuchiurus haumela	Around Shengshan Island waters	Around Shengshan Island waters	Stable
Pseudosciaena crocea	Zhejiang offshore water	Zhejiang offshore waters, decreased to some extent	Almost stable
Pseudosciaena polyactis	Lusi-Dasha waters	Lusia-Dasha waters, decreased to some extent	Almost stable
Scomberomorus niphonius	Offshore waters, north of Zhejiang Province, China	Offshore waters, north of Zhejiang Province, China	Almost stable
Mugil so-iuy	Offshore waters of northern China	Offshore waters of northern China	Stable
Ilisha elongata	Offshore waters of China	Offshore waters of China	Stable
Lateolabrax japonicus	Offshore waters of northern China	Offshore waters of northern China	Stable
Stromateides argenteus	Offshore waters of China	Offshore waters of China	Stable
Coilia mystus	Big estuaries along China coast	Big estuaries along China coast	Stable
Gadus macrocephalus	Middle and northern Yellow Sea	Middle and northern Yellow Sea, decreased to some extent	Stable
Sepiella maindroni	Zhejiang–Mindong offshore waters	Zhejiang–Mindong offshore waters	Stable
Penaeus orientalis	Bohai and Yellow Sea	Bohai and Yellow Sea	Stable
Portunus trituberculatus	Offshore waters of China	Offshore waters of China	Stable
Rhopilema esculenta	Offshore waters of China	Offshore waters of China	Almost stable

Table 4. Runoff of Huanghe River (Yellow River) and catch of *Penaeus orientalis* in the estuary.

	Year			
	1956	1957	1958	1959
Runoff in April–June (1×10^8 m^3)	93	73	36	44
Catch in autumn (t)	6548	3112	2015	2876

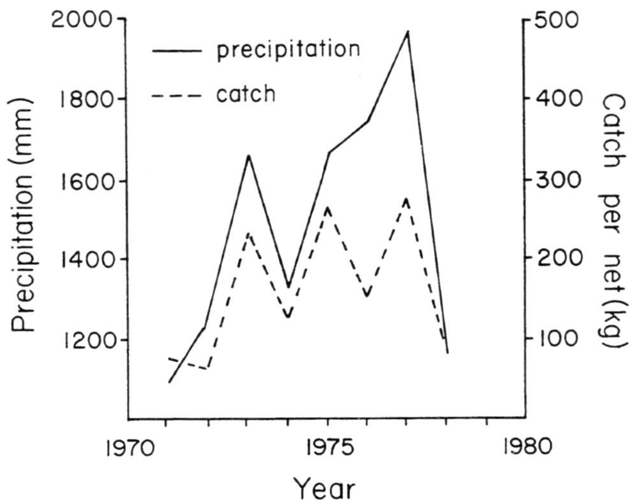

Fig. 1. Relationship between catch of *Palaemon* (*Exopalaemon*) *annandalei* and precipitation in Hangzhou Bay (after S. Li, Zhejiang Fisheries College, unpublished data).

on the abundance of *Mugil so-iuy*, *Ilisha elongata*, and *Sparus macrocephalus* and a warm climate would have a favourable effect. The long-distance migratory fish *Scomberomorus niphonius* would not be affected as strongly by climate change.

Acknowledgements

The authors are indebted to Dr. R. J. Beamish who, on behalf of the organizing committee for the Symposium on Climate Change and Northern Fish Populations, provided the financial support to Prof. Yang Jiming for his presentation in Victoria. Special thanks to Prof. Wang Cunxin and Prof. Tian Mingcheng who have assisted the authors with the information on the fish that have disappeared from China's seas.

References

Cheng, Q. 1985. Analysis of unearthed marine fish bones in China 5000 years ago. Sel. Works fish. hist. Ser. 1: 141–144.

Deng, J., and C. Zhao. 1991. Marine fishery biology. Agriculture Press, Beijing. 686 p.

Geng, X. 1981. Marine transgressions and regressions in east China Sea since the late Pleistocene Epoch. Acta Oceanol. Sin. 3: 114–130.

Liu, R., B. Luo, Y. Cui, and Y. Guo. 1987. Effects of the Three Gorge Project on estuarine biology and fishery resources, p. 403–446. *In* Collection of the studies of impacts of Changjiang River Three Gorge Project on the ecological environment and possible countermeasures. Science Press, Beijing.

Lu, B. 1992. "Greenhouse effect" on the future of the coastal zone of Hainan Province. Ocean Coastal Dev. 9: 62–64.

Table 5. Precipitation and catch of *Acetes chinensis* in Liaodong Bay from 1950 to 1957.

	1950	1951	1952	1953	1954	1955	1956	1957
Precipitation (mm) in June–September	509	570	308	734	499	495	534	
Catch in spring × 1000 t		20.4	35.8	7.2	42.8	24.6	20.7	35.0

Table 6. Distance from shore of *Trichiurus haumela* fishing ground and runoff from the Changjiang River in June.

	1972	1973	1974	1975	1976	1977	1978	1979	1980	1981
Distance from shore (km)	106	132	104	143	134	148	134	128	140	143
Runoff from Changjiang River ($m^3 \cdot s^{-1}$)	32 700	54 600	29 000	50 200	42 300	52 400	41 800	29 200	38 400	32 300

Wang, J., and P. Wang. 1980. Relationships between sea-level changes and climatic fluctuations in eastern China since the late Pleistocene. Acta Geogr. Sin. 35: 299–312.

Xia, G. 1991. The standpoints and countermeasures towards global climate change of China. China Environ. Sci. 11: 457–459.

Xu, C., S. Deng, H. Zhan, and K. Huang. 1984. Studies on the fish fauna of the outer region of the continental shelf and the continental slope in the East China Sea, p. 137–186. *In* Comprehensive investigations of the deep sea fishing ground at the continental slope and the edge of the continental shelf in the East China Sea.

Yang, J., B. Luo, G. Wu, and G. Cheng. 1980. The marine fishes of China. Science Press, Beijing. 94 p.

Yang, J., W. Yang, X. Wang, and G. Cheng. 1990. An estimate of fish biomass at the bottom layer of the Bohai Sea. Acta Oceanol. Sin. 12: 359–365.

Zhao, C., X. Liu, B. Zeng, and Y. Yu. 1990. Marine fishery resources of China. Zhejiang Science and Technology Press, Hanshou. 178 p.

Zhao, S., et al. 1991. Sea-level changes of recent 100 years. Mar. Sci. 4: 19–22.

Zhao, Y., H. Yu, and L. Yan. 1985. Preliminary study on the fishery development of Zhoushan Is. before the founding of P.R. China. Sel. Works Fish. Hist. Ser. 1: 27–41.

Warmwater and cool-water stocks of Pacific cod (*Gadus macrocephalus*): a comparative study of reproductive biology and stock dynamics

A. V. Tyler[1]

Tyler, A.V. 1995. Warmwater and cool-water stocks of Pacific cod (*Gadus macrocephalus*): a comparative study of reproductive biology and stock dynamics, p. 537–545. *In* R.J. Beamish [ed.] Climate change and northern fish populations. Can. Spec. Publ. Fish. Aquat. Sci. 121.

Abstract: The La Perouse Bank stock off southwest Vancouver Island is the most southerly stock of Pacific cod (*Gadus macrocephalus*) abundant enough to be a fishery target. It occupies water that is on average 2°C warmer in winter than that in Hecate Strait, the location of a comparable stock just to the north. Such a temperature difference is equivalent to a shift that could hypothetically be caused by global warming. Several biological responses were compared for the La Perouse and Hecate Strait stocks. Hecate Strait fish had greater fecundity-at-size than did those from La Perouse. Histological examination of ovary sections from fish taken just before spawning indicated that some of the mature individuals in the La Perouse sample actually would not spawn that year. This resting gonad state was observed much less frequently among fish from Hecate Strait. The higher temperatures of the La Perouse area may cause the cod to put more of their energy budget into maintenance and less into reproductive output. The species seems limited at the southern edge of its range by temperature stress acting on its reproductive capacity. Warming trends would increase the likelihood of overfishing, lead to the demise of the La Perouse stock for commercial use, and reduce the productivity of the more northerly Hecate Strait stock.

Résumé : Le stock de morues du Pacifique du banc La Perouse au large du sud-ouest de l'île de Vancouver est le stock le plus méridional dont l'effectif est suffisamment abondant pour faire l'objet d'une pêche dirigée. Il occupe un secteur dont l'eau est en moyenne de 2° C plus chaude en hiver que dans le détroit d'Hécate, où se trouve un stock comparable juste au nord. Une telle différence de température équivaut à la modification qui serait hypothétiquement causée par le réchauffement global. Plusieurs réactions biologiques ont été comparées entre les stocks de La Perouse et du détroit d'Hécate. Celui du détroit d'Hécate affiche une plus grande fécondité à maturité que celui de La Perouse. L'examen histologique de coupes d'ovaires de poissons femelles capturés juste avant le frai montre que certains individus matures de l'échantillon de La Perouse n'auraient pas frayé cette année-là. Cette inactivité gonadique a été observée beaucoup moins fréquemment chez le poisson du détroit d'Hécate. Les températures plus élevées du banc La Perouse pourraient forcer la morue à consacrer une plus grande partie de son budget énergétique à sa survie et moins à sa reproduction. À la limite sud de son aire de répartition, l'espèce semble limitée dans sa capacité reproductive par le stress dû à la température. Les tendances au réchauffement sont susceptibles d'augmenter la probabilité d'une surpêche, ce qui aboutirait à l'élimination du stock du banc La Perouse pour la pêche commerciale et réduirait la productivité de celui du détroit d'Hécate, plus au nord.

Introduction

The La Perouse Bank Pacific cod (*Gadus macrocephalus*) stock, sometimes referred to as the Big Bank or Amphitrite Bank stock, inhabits a relatively small but highly productive continental shelf area off southwest Vancouver Island. The Hecate Strait stock, the larger of the two stocks, occupies a much larger shelf area stretching from the most southerly to the most northerly ends of Hecate Strait (Fig. 1). The current systems of the two areas are distinct in summer (Ware and McFarlane 1989). La Perouse Bank has an offshore geostrophic current moving to the south that is enhanced in the summer by the winds from the north. In winter the Vancouver Island coastal current flows to the north driven by river discharges, and is enhanced by prevailing winds from the south (Fig. 1). The warm water in the La Perouse Bank region

A.V. Tyler. Pacific Biological Station, Department of Fisheries and Oceans, Nanaimo, BC V9R 5K6, Canada.

[1]Present Address: School of Fisheries and Ocean Sciences, University of Alaska Fairbanks, Fairbanks, AK 99775-7220, USA.

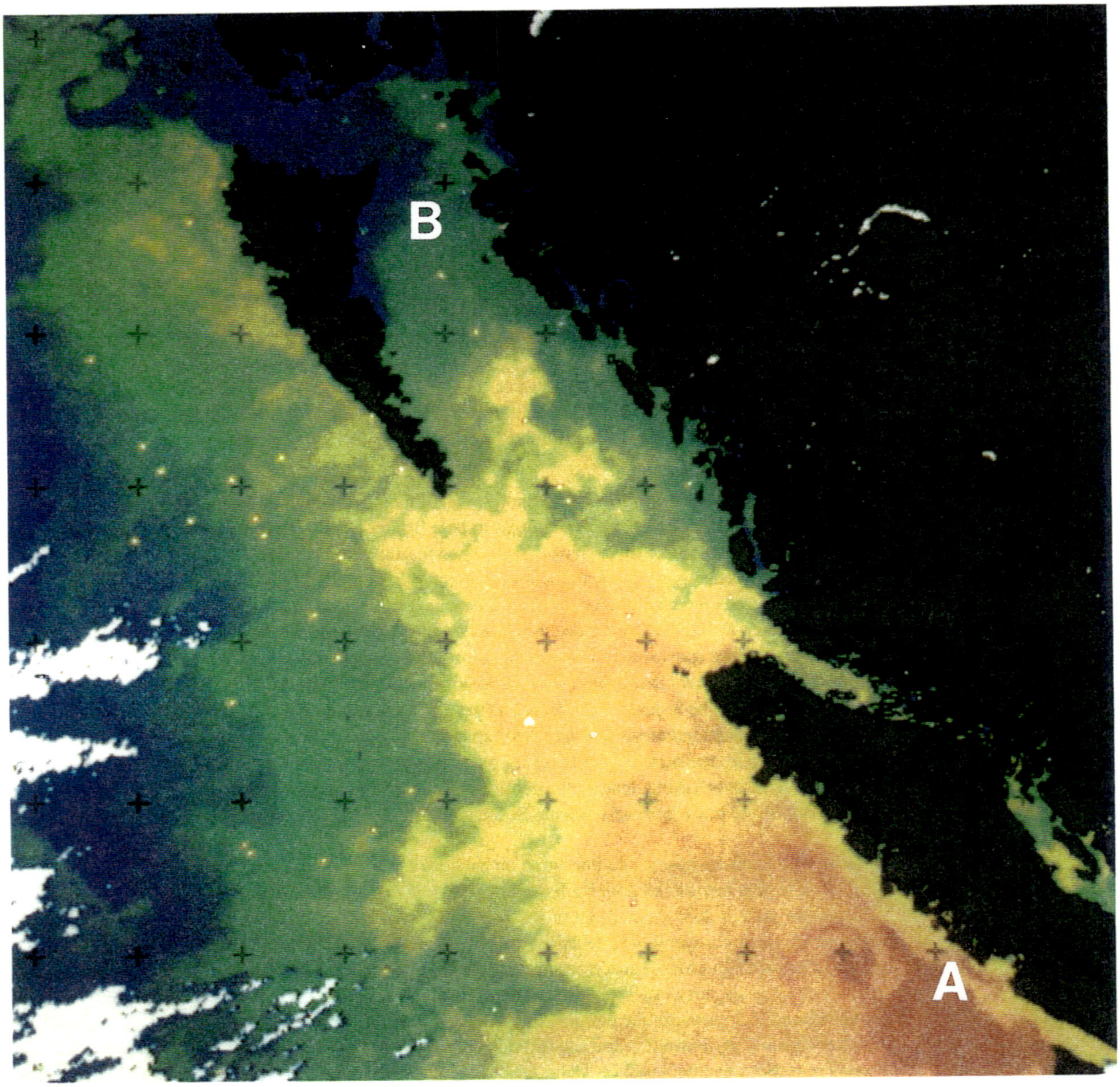

Fig. 1. Satellite image by Borstad of the west coast of Canada on February 13, 1986. Red in the south (A) corresponds to sea surface temperatures (SSTs) of about 9°C and this colour overlaps the La Perouse Bank area. Green corresponds to SST of about 7°C. Hecate Strait can be seen to the north (B), mostly green, and bounded to the west by the Queen Charlotte Islands.

stands out in the February satellite image (Fig. 1A), as does the northward moving coastal current that bifurcates and continues through Hecate Strait Fig. 1B)and simultaneously through the open sea. The image shows the water cooling as it moves northward, forming a temperature gradient. The La Perouse stock occupies a water mass that is on average 2–3°C warmer in any given month than is the water in Hecate Strait. A comparison of the two stocks is particularly significant because the indicated temperature gradient is equivalent to a shift that could hypothetically be caused by global warming. The gradient steepness might be maintained but the whole range could possibly become 2°C higher, so that the present southern end temperatures would become the temperatures of the north.

Geography suggests that the smaller production area of the La Perouse stock may be the simple cause of the difference in stock biomass and commercial production. The La Perouse stock is the most southerly stock of cod in the Pacific that is abundant enough to be a specific fishery target (INPFC 1985). To the north, the Hecate Strait stock is between four and five times larger in biomass and occupies an area about three times larger (Tyler and Foucher 1990). While the size of the production areas may be relevant, there is evidence that additional factors influence productivity of the stocks. I have been comparing of the biological responses of the two stocks as a potentially strong means of interpreting of the observations in relation to the contrasting physical conditions. Some differences in productivity have been noted previously. Hecate

Strait fish have lower growth rates and about the same mortality rates as do the La Perouse fish (Foucher and Tyler 1988). The new evidence presented in this paper considers differences in responses at the level of individual fish and their reproductive biology. I propose both trophic and ambient temperature related hypotheses to account for the observations, although at this point the temperature hypothesis seems to explain more.

Methods

Sample 1 came from a research cruise during January–February 1989 (Foucher et al. 1989) in Hecate Strait, with the majority of the fish coming from White Rocks Grounds. Sample 2, caught on Amphitrite Bank (part of the La Perouse stock), was from a charter landing to a fish plant in Ucluelet, B.C. on January 29, 1990. Sample 3 was from Amphitrite Bank off the southwest coast of Vancouver Island. It was sampled in a Nanaimo fish plant after the fish were landed and trucked from Ucluelet (Foucher and Tyler 1990).

A length-stratified sampling procedure was used after the first sample to ensure that selected fish were spread throughout the anticipated length range. The target sampling level was approximately 10 fish per 5-cm interval. All fish were measured for fork length to the nearest centimetre. Whole fish weights (less stomach contents) and ovary weights were taken to the nearest gram.

The method of fixation varied somewhat among samples. For a preliminary sample, Gilson's solution (equal parts chloroform, ethanol, and glacial acetic acid) was used. This solution hardened the oocytes very well. The ovaries were transferred to 30% ethanol. However, after 15 mo the oocytes remained difficult to separate because of the Gilson's fluid. The next sample was fixed in 30% ethanol. The ethanol fixed the samples adequately if sufficient volume was used. It also promoted separation of the oocytes from each other and from other ovarian tissue. For the last two samples, the fresh ovaries were sliced up extensively and immediately shaken vigorously in 70% ethanol. This seemed to liberate the oocytes very well from the ovarian tissue.

The procedure for determining fecundity involved separating the developed oocytes from any ovarian tissue and small primary oocytes by washing through sieves, with manual removal of larger pieces. Oocytes from individual fish were placed on a stacked sieve set consisting of either a 2.0- or 4.0-mm upper sieve, and a 0.850- or 1.0-mm lower sieve. A bucket under the sieve set collected the preservative which was then returned to the original sample jar.

Oocytes were removed from ovary wall tissue by gentle scraping with a sharp-edged spoon and washing with a water spray from a squeeze bottle. Larger clumps of oocytes were broken down by lightly running the flat surface of a spoon over them. Oocytes were separated from connective tissue with a fine spray or by placing the sieve set in a water bath and gently agitating the whole set at about 1 Hz to rapidly pass single oocytes through the larger sieve.

Samples with large amounts of fine connective tissue were decanted. For this process, oocytes were placed in a large bucket in 10 – 13 L of water and allowed to settle. Care was taken to ensure that no oocytes remained floating, trapped in the surface tension. After about 1 min, when all oocytes settled out, the water and much of the fine connective tissue were decanted. No oocyte loss was noted in this process.

Sieved samples were transferred to a 0.100-mm Nitex filter in a Buchner funnel. Two Buchner funnels were used, one with a 10-cm diameter, the other with a 17-cm diameter for larger samples. Samples were filtered at approximately 700 mm of mercury vacuum for at least 10 min after water had stopped dripping from the funnel. Sufficient dryness was determined by placing the sample, in the filter, on the clean dry surface of a table for 10–15 s. If water drops were left on the table top, the sample was not considered dry. When sufficiently dry, samples were transferred to a tared weighing boat, mixed thoroughly to ensure even distribution of connective tissue and consistent egg dryness (i.e., some oocytes, particularly those at the surface of the Buchner funnel dried more rapidly) and weighed to the nearest 0.01 g. Sufficiently dry samples did not cling to the filter and the oocytes could be transferred with no loss.

Four subsamples of approximately 500 oocytes each were removed from the sample, placed in tared weighing boats and weighed to the nearest 0.0001 g. Each of these subsamples was then placed in a gridded counting chamber and counted with the use of a tally counter. At random, one of the subsamples was recounted to ensure precision. If the second count differed by 2% or more the subsample was recounted. Resolved counts were taken as the mean of replicate counts.

Fecundity was determined by proportion using the four subsample counts and weights combined and the total sample weight:

$$N = \frac{W(\sum_{1}^{4} n_i)}{(\sum_{1}^{4} w_i)}$$

where N = estimated fecundity; W = weight of total sample; w = weight of subsample i; n = number of oocytes in subsample i; i = subsample number.

Additional samples for histological work were caught by otter trawl in Hecate Strait, on the northern coast of British Columbia during 1987–88 and off Ucluelet, on the southwest coast of Vancouver Island in 1988. Gonad maturity was determined macroscopically according to a maturity index (Table 1). Ovary samples were dissected and fixed in buffered formalin saline solution (130 g Na_2HPO_4, 80 g NaH_2PO_4, 180 g NaCl, 2 L formalin, and 18 L water).

Tissue from the middle section of the ovary was embedded, sectioned at 5µm, processed using routine histological techniques and stained with Harris' haematoxylin–eosin (Yasutake and Wales 1983). Prepared sections were examined to determine features on the microscopic level that are characteristic of maturity stages assigned at the macroscopic level, and then photographed. Oocyte size distributions were determined by direct measurement of an image of the section projected by a scale projector. Only those oocytes whose nuclei were included in the sections were measured across the longest diameter of the cell.

Table 1. Description of the maturity stages for female Pacific cod.

Stage	Maturity	Description[a]
1	Immature	Ovaries small, white, somewhat translucent, little more than smooth, clear strings from anterior origin to oviduct, sometimes thickening.
2	Immature (Maturing)	Ovaries small, but with definite elongated shape. May be becoming pinkish. No visible oocytes or prominent blood vessels. Diameter usually less than 2 cm. Still occur in fish in 40- to 50-cm range and occasionally larger.
3	Early ripe (Mature)	Ovaries large, opaque, creamy orange. Blood vessels prominent. Ovary size depends on size of fish but likely to be at least 1 cm in diameter by 5 cm in length, larger in older fish. Oocytes may be visible, at least if ovary is cut. May have some black or silver colouring to ovary wall if fish has spawned previously. Unlikely to occur in fish below 40 cm.
4	Late ripe (Mature)	Ovaries large, oocytes visible, creamy yellow. Ovary size could be 3 or 4 cm or larger in diameter and 10 cm or more in length. Individual oocytes visible through any translucent parts of the ovary wall. May contain a few translucent eggs.
5	Running ripe (Mature)	Ovaries very large, each ovary from a large fish may be 1 L in volume. Externally, the fish will look obvioulsy gravid. Ovary wall with thin, stretched appearance. Most eggs translucent. Eggs may flow from vent if fish is squeezed.
6	Spent (Mature)	Ovaries medium in outline but flaccid. Flaccid condition probably does not last long. Ovaries dark and bloodshot with reddish-brown interior. May be a few translucent eggs remaining.
7	Resting (Mature)	Ovaries small, firm, may have some black or silver color.

[a]Adapted from Westrheim (1977)

Results

Difference in Fecundity by Stock

The relationship between fish weight (without gonad) and the fecundity of the fish shows that, when cod first spawn, their weight is between 1 and 2 kg, and that between 0.6 and 1.2 million eggs are liberated. At the largest sizes, 6–7 k, spawning liberates between 4 and 5 million eggs (Fig 2). Linear regression parameters fitted to the two samples gave the following equations: for Hecate Strait (sample 1), $N = -0.416 + 0.741B$, $r = 0.90$; for La Perouse Bank (sample 2), $N = -0.071 + 0.500B$, $r = 0.77$, where N is the number of eggs per fish (fecundity) and B is body weight (kilograms) without ovary weight. The F-test for the comparison of the slopes shows a significant difference ($F = 3.47$, df) = 1, 105, $p = 0.10$). For ease of comparison, the fitted linear regression of the Hecate Strait fit in Fig. 2B was redrawn in Fig. 2A along with the fit for the La Perouse Bank. Another sample was taken from La Perouse Bank (sample 3) that also had a significantly different slope from the Hecate Strait Bank sample: $N = -0.071 + 0.500B$, $r = 0.92$; $F = 3.55$, df = 1, 102, $p = 0.10$. The two La Perouse samples were not significantly different; $F = 0.01$, df = 1, 79, $p = 0.10$. The variance of the scatter in Fig. 2A was noticeably greater for larger fish. The uneven variance was also found for the other La Perouse sample. This made the interpretation of the significant F-test questionable. Based on the point scatter I interpreted the significant difference in the slopes to mean that a portion of the La Perouse fish were less fecund than were the Hecate fish.

Because the populations could be tracked over time with length-frequency data from port sampling of the commercial fishery, I have carried out some analysis by length frequency rather than by weight frequency. The comparison of the fitted regressions of sample 1 and sample 2 fecundity (N) relationships show the diverging fecundities-at-length for the Hecate Strait and La Perouse stocks just as the fecundity-at-weight data did (Fig. 3A). The regression equations are: for Hecate Strait (sample 1), $N = -6.280 + 3.593L$, $r = 0.94$; for La Perouse Bank (sample 2), $N = -5.029 + 2.904L$, $r = 0.81$, where L is length in centimetres. The F-test for the comparison of the slopes shows a significant difference ($F = 3.66$; df = 1, 105; $p = 0.10$). The percentage difference between the two stocks gradually diverges from zero to 20% at the largest sizes (Fig. 3B). If two stocks had a predominance of old, large fish there would be more difference between their stock fecundities than if the stocks were dominated by small fish.

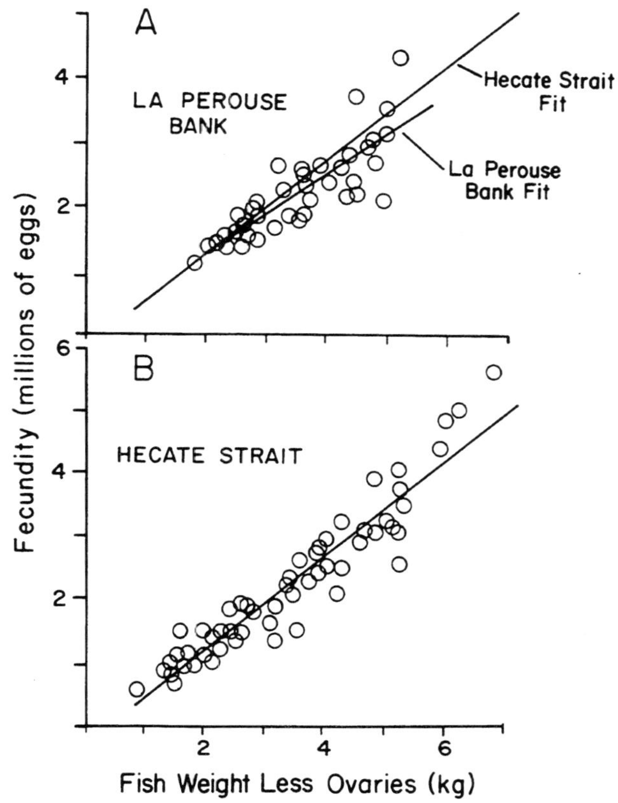

Fig. 2. Relationship between fecundity (millions of developed eggs per fish) and the weight of the fish (kg). The fitted linear regressions are given, the Hecate Strait (sample 1) fit in B being redrawn for comparison in A along with the fit for the La Perouse Bank (sample 2).

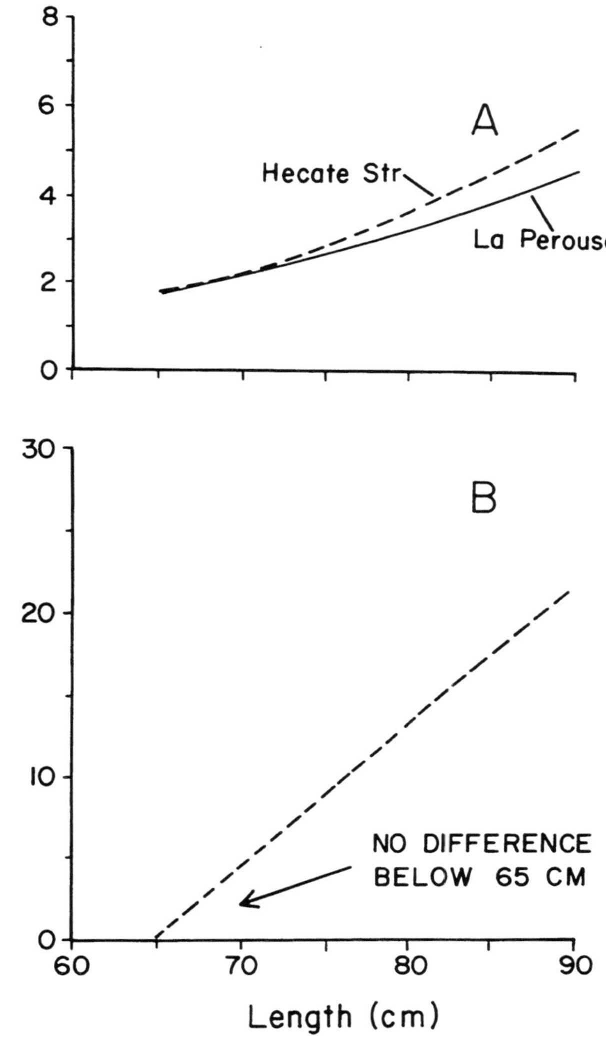

Fig. 3. (A) Fitted regression lines for fecundity (millions of developed eggs per fish) versus fish length. (B) Percentage difference in individual fecundity at length between the La Perouse stock (sample 2) and the Hecate Strait stock (sample 1), showing the increasing percentages by size of fish.

To show how the stock fecundity would be calculated, I have selected a length frequency from Hecate Strait taken in late winter of 1990, during the spawning period. Most of the fish were in the 58–75 cm range; the stock was dominated by medium-size fish. The length–fecundity relationship for Hecate Strait was superimposed as an increasing curve in Fig. 4. To calculate the stock fecundity, the number-at-length per 1000 fish was multiplied by the fecundity-at-length, giving the resulting stock fecundity-at-length (Fig. 5). The individual fecundity-at-length relationship weighted the lengthfrequency to the right, causing larger fish to be relatively more important to the reproductive potential of the stock than the length frequency indicated by itself.

I then calculated stock fecundities from a "what if" situation. Given that two stocks had approximately the same length frequencies where small fish predominate (Fig. 6A), and given that stock fecundities were as in Fig. 5, then by applying the two size–fecundity regressions, it was found that the two stocks would have had a 10% difference in stock fecundity. In the second case, given that two stocks had approximately the same length frequency as in Fig. 6B where large fish predominated, it was calculated that the two stocks had a 17% difference in stock fecundity.

It is clear that a shift in size frequencies as with temporary, high recruitment levels, would result in a dominance of young fish, and then the difference in stock fecundity would minimized for the two stocks. If there were a series of poor recruitmentyears, a size distribution as in Fig. 6B would result, and the difference in stock fecundity would maximized for the two stocks. If a stock actually had 17% less reproductive potential than was accounted for in a stock assessment, the chances of overfishing would be increased. That small differences of this sort can matter will be taken up in the Discussion section.

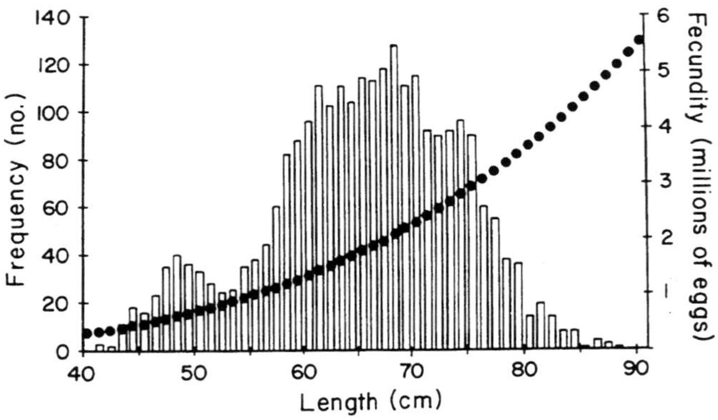

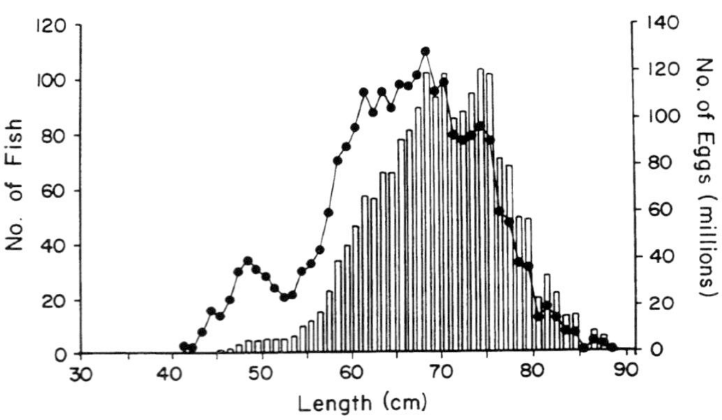

Fig. 5. Length-frequency sample (same as Fig. 5) rescaled as number of fish at length in a sample of 1000 fish (left axis), and number of eggs at length per sample of 1000 fish (right axis).

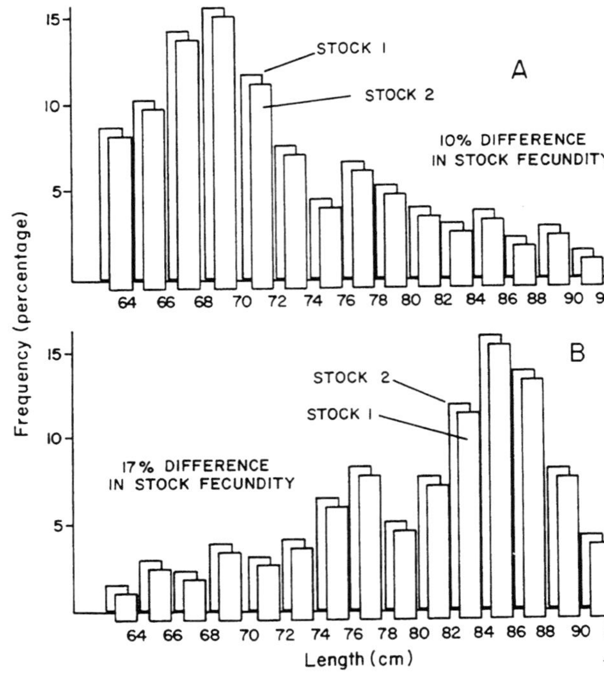

Fig. 6. (A) If two stocks had approximately the same length frequencies as in case A where small fish predominate, they would have a 10% difference in stock fecundity, given that the fish had a difference in fecundity at length as indicated in Fig. 3. Stock fecundities were calculated as in Fig. 5, by applying the separate size–fecundity regressions. (B) If two stocks had approximately the same length frequency as in case B where large fish predominated, the two stocks would have a 17% difference in stock fecundity (calculated as above).

Resting stage ovaries

Another aspect of reproductive biology that influences the two stocks to a different degree is the proportion of females that do not spawn in some years. I have often carried out field examinations of the stages of "ripeness" of ovaries in samples, and have developed a description of maturation stages (Table 1) that was partially based on earlier work (Westrheim 1977). Occasionally for Hecate Strait samples I found anindividual of mature body length that had a small, undeveloped ovary, with a dark, ovary wall (Fig. 7) that indicated prior spawning (Bowden et al. 1990). In one sample of 500 female Pacific cod from Hecate Strait taken in February 1989, we found two resting stage ovaries (0.2%). I have noticed these "resting" stage fish in other species as well, but have not found occasion for bringing the phenomenon into the context of a study. When I examined a sample of 50 mature female cod from the La Perouse Stock in January 1990, I found seven resting-stage, mature females, the equivalent of 14%. That these fish were all large for mature individuals (61, 68, 70, 71, 72, 77, 77 cm) suggested that it was mainly the older, larger fish that went into a resting stage. Small, mature, resting-stage fish have not been found so far in either location. The reduced productivity of some of the older fish owing to the resting stage, would be additive to the reduced fecundity of the older fish and cause a stock dominated by older fish to be even less productive. Unfortunately I was not able to take additional samples to carry out statistical hypothesis testing of these frequency differences between the two locations or obtain better estimates of the percentage levels. Though an estimate of percentage difference cannot be made at this time, it seems likely that the difference exists, and that equal biomasses in the two stocks do not have equal reproductive potential because of greater numbers of resting-stage fish in the south.

Histological preparations of resting ovaries were examined from both locations, andoocyte diameters were measured. The size frequencies indicated that the ovaries of the resting fish contained only primary oocytes (80–220 μm) as did the immature fish(Figs. 8A and 8C). By comparison the late-ripe mature fish (Table 1) contained both primary and yolk-filled oocytes (300–740 μm) (Fig. 8B). There were no old large oocytes or atretic oocytes being re-absorbed in the resting stage. The primary oocytes were the size that could conceivably develop and undergo yolk formation at some future time, perhaps the following year.

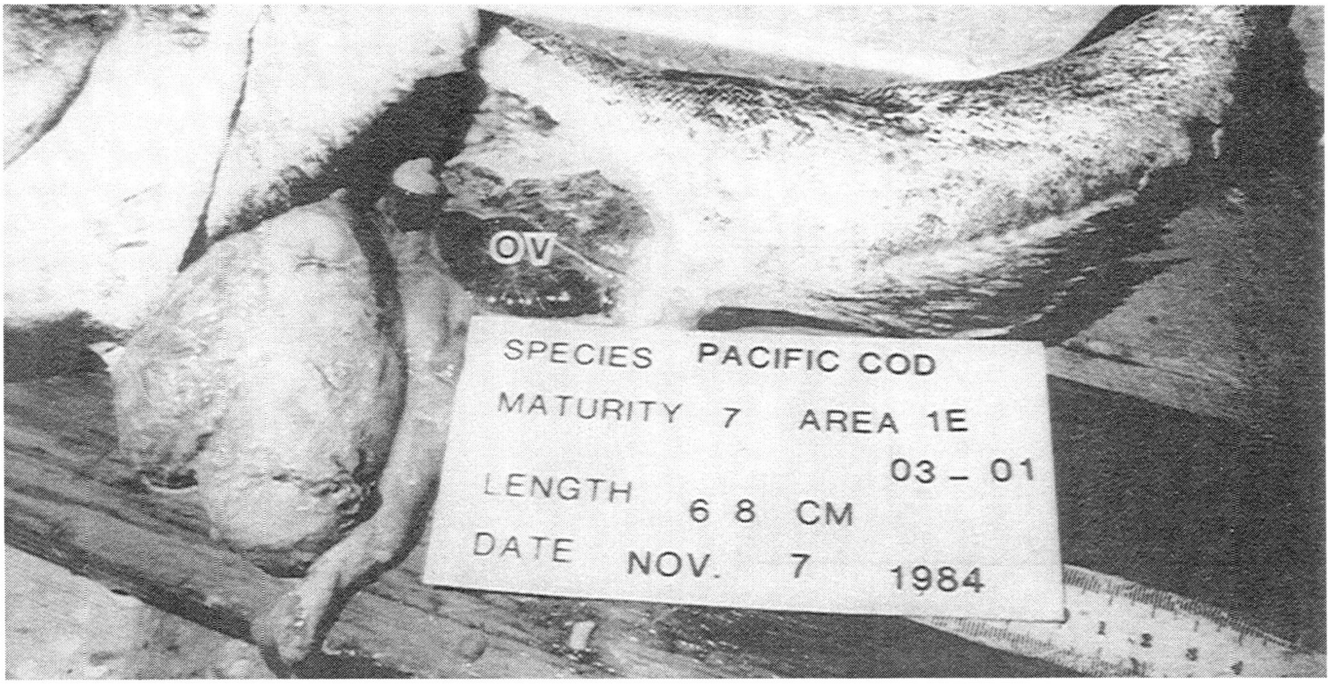

Fig. 7. Resting or nonspawning mature ovary of Pacific cod from Hecate Strait. Two of these were found in a sample of 1000 females in Hecate Strait. Resting-stage ovaries were found at a frequency of 7 in 50, equivalent to 140 per 1000 females, in a La Perouse sample of spawning cod.

An examination of histological preparations of the ovarian tissue and the cellular structure indicated that the resting ovaries were healthy. Cellular cytoplasm and nuclei of primary oocytes appeared to have the same consistency and staining properties in both the resting and immature stages (Figs. 9A and 9B). Though not pictured, the mature ovaries were also examined with a similar result. The smooth muscle cells of the ovary walls were the same in all preparations, as were the tissues between the primary oocytes. The only noticeable difference between immature and resting-stage ovaries was that the immature organs had a higher density of oocytes, which would be expected when comparing a fish that had never spawned with one that spawned more than once and had released millions of eggs. I concluded that the resting-stage individuals could spawn again were external conditions favourable, and that these individuals should not be considered senescent.

Discussion

The differences in fecundity and proportion of fish not spawning each year are additive and would likely make a difference to the reproductive potential of the two stocks. If there is no noticeable influence of stock size or its reproductive potential on resulting recruitment, the differences would be an interesting curiosity to a biologist, but would have no further impact. However, there is evidence now for a curvilinear, stock–recruitment relationship in Hecate Strait (Tyler and Crawford 1991). Our unpublished preliminary analysis indicates that a similar relationship is possible for the La Perouse stock. The curve (Fig. 10) shows that when stock is intermediate in size, 20% changes in stock size do not produce large effects on resulting recruitment. If a decrease in reproductive potential occurred from sample 1 to sample 2, the very small change in recruitment (6% in the graph) would be undetectable, yet the stock-size decrease represents a 20% change in a stock size of 500 units. On the other hand, if the same decrease went from sample 3 to sample 4, the response would be a decrease in recruitment of nearly 30%. The curvilinear relationship can intermittently translate small changes into large effects.

There are implications for fishery management from the above model of reproductive potential. The southern stock would have a different stock–recruitment relationship if it were dominated by old fish than if it were dominated by young mature fish, because the spawning biomass would be less productive for any given stock size. Consequently, if the spawning biomass were low either because of a series of poor recruitments or heavy fishing, the stock would rebuild more slowly than would a similar Hecate Strait stock. The fishery could then more easily cause an erosion of the stock biomass.

Hypotheses can be formulated that incorporate the findings on fecundity differences, the differences in the proportion of fish that are in reproductive resting stages, and the information that the La Perouse area is warmer than the Hecate Strait area.

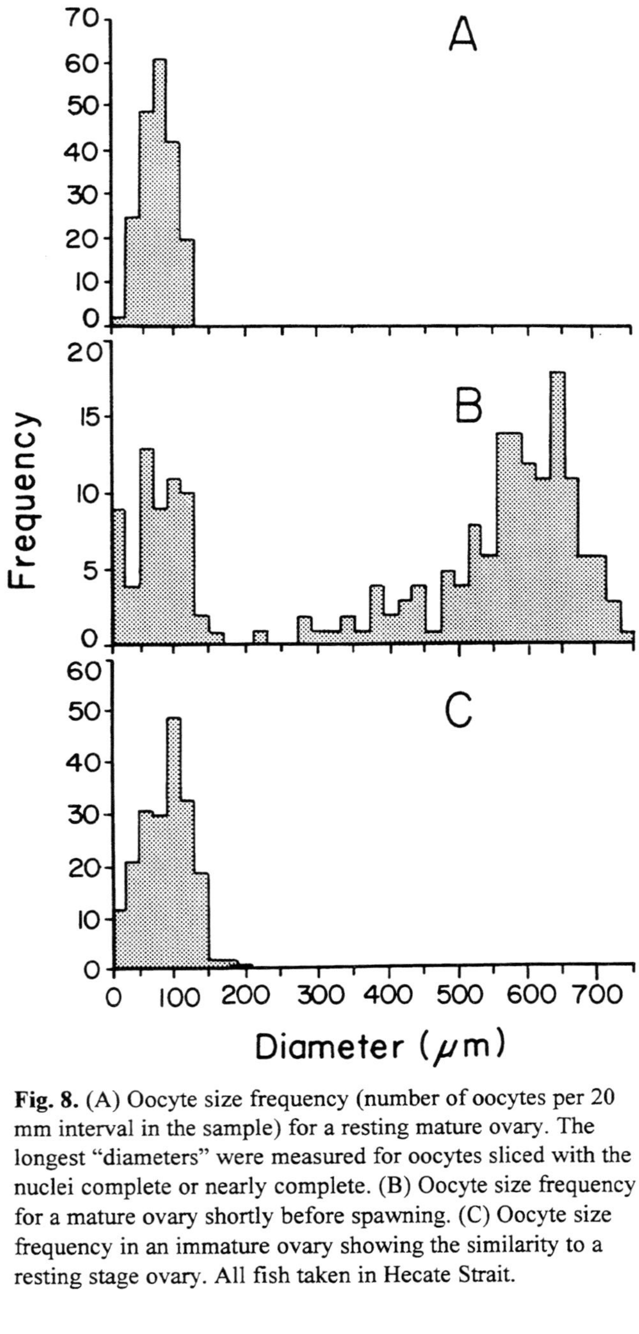

Fig. 8. (A) Oocyte size frequency (number of oocytes per 20 mm interval in the sample) for a resting mature ovary. The longest "diameters" were measured for oocytes sliced with the nuclei complete or nearly complete. (B) Oocyte size frequency for a mature ovary shortly before spawning. (C) Oocyte size frequency in an immature ovary showing the similarity to a resting stage ovary. All fish taken in Hecate Strait.

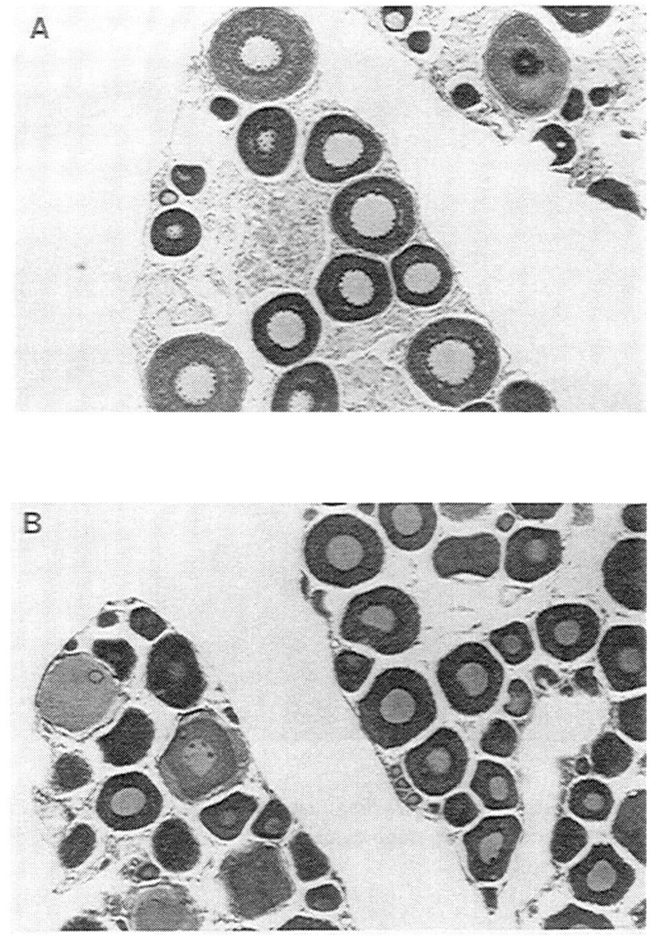

Fig. 9. (A) Photomicrograph of a resting mature ovary of a Pacific cod from Hecate Strait showing healthy ovarian tissue and a density of primary oocytes that is reduced compared to the oocyte density a fish that has not spawned (immature). (B) Photomicrograph of an immature ovary of a Pacific cod from Hecate Strait showing a higher density of primary oocytes than a resting mature ovary has.

Hypothesis 1 is that these fecundity differences are brought about by differences in feeding opportunities in the two areas. Adult Pacific cod eat fish, in particular Pacific herring (*Clupea pallasi*) and Pacific sand lance (*Ammodytes hexapterus*) (Tyler and Crawford 1991). It might be postulated that if there were less forage in the summer in some years, adults undergoing their maturation process would have more limited reproductive output. Walters et al. (1986) and Tyler and Crawford (1991) postulated that the apparent predator–prey cycle between cod and herring in Hecate Strait is due to juvenile herring being a limiting factor on cod production in some years. However, a correlation between cod and herring has not yet been found for the La Perouse stock (Ware and McFarlane 1986).

Hypothesis 2 would be that it is the warmer temperature in the La Perouse area that interferes with the reproductive response. That there are no substantial fisheries in the warmer areas to the south, only low volume landings for Pacific cod, fits the temperature hypothesis. The mechanism might be that the higher temperatures of the La Perouse area cause cod to put more of their energy budget into maintenance and less into reproductive output. The prognosis for the temperature hypothesis would be that warming trends increase the likelihood of overfishing. Decreased recruitment because of temperature increase would cause a change in size composition that would in turn bring a further decrease in reproductive potential: a positive feedback process that would lead to the elimination of the La Perouse stock for commercial use. The same warming trend of 2–3°C would reduce the productivity of the Hecate Strait stock as well and cause it to take on the present characteristics of the La Perouse stock.

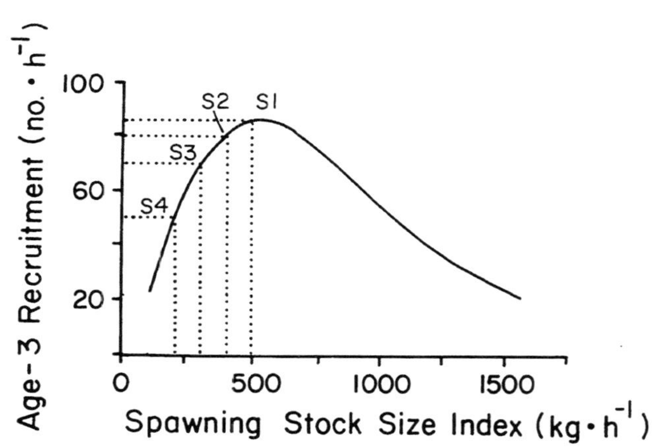

Fig. 10. The role of density dependence in translating small differences into large effects. The stock–recruitment curve for Pacific cod is taken from Tyler and Crawford 1991. If the fishery were to reduce the stock reproduction potential from sample 1 to sample 2, there would be no measurable effect on recruitment. But if the reduction in reproduction potential went from sample 3 to sample 4, there would be a disproportionately large reduction of recruits. The curvilinear relationship can translate small decrements of reproductive potential into large effects on recruitment.

Acknowledgements

Sampling and sample processing for this study were carried out while the author was a research scientist with the Pacific Region Science Directorate, Canada Department of Fisheries and Oceans. The research was generated as a part of the La Perouse Project of the Pacific Region's Science Directorate. I thank Ray Foucher for help in sampling fish and measuring oocytes. Photomicroscopy, analysis and write-up were carried out with the support of the School of Fisheries and Ocean Sciences, University of Alaska Fairbanks. Thanks are extended to Dr. Scott Smiley for his helpful expertise, and for making photos through the microscope. Aquametrix Research, Ltd, and Archipeligo Marine Research Ltd.Ö carried out the fecundity determinations.

References

Bowden, D.G., R.P. Foucher, and A.V. Tyler. 1990. A guide to ovarian histology of Pacific cod. Can. Tech. Rep. Fish. Aquat. Sci. 1723: 44 p.

Foucher, R.P. and A.V. Tyler. 1988. Pacific Cod. *In* J. Fargo, M.W. Saunders and A.V. Tyler [ed.] Groundfish stock assessments for the west coast of Canada in 1987 and recommended yield options for 1988. Can. Tech. Rep. Fish. Aquat. Sci. 1617: 304 p.

Foucher, R.P., and A.V. Tyler. 1990. Estimation of the fecundity of Pacific cod. Can. MS Rep. Fish Aquatic Sci. 2088: 49 p.

Foucher, R.P., A.V. Tyler, J. Fargo, and G.E. Gillespie. 1989. Reproductive biology of Pacific cod and English sole from the cruise of the FV BLUE WATERS to Hecate Strait, January 30 to February 11, 1989. Can. MS Rep. Fish. Aquat. Sci. 2026: 85 p.

International North Pacific Fisheries Commission (INPFC). 1985. Statistical yearbook. Vancouver, B.C. 123p.

Tyler, A.V., and W.R. Crawford. 1991. Modeling recruitment patterns for Pacific cod in Hecate Strait, British Columbia. Can. J. Fish. Aquat. Sci. 48: 2240–2249.

Tyler, A.V., and R.P. Foucher. 1990. Pacific cod. *In* A.V. Tyler, and J. Fargo [ed.] Groundfish stock assessments for the west coast of Canada in 1989 and recommended yield options for 1990. Can. Tech. Rep. Fish. Aquat. Sci. 1732: 343 p.

Walters, C.J., M. Stocker, A.V. Tyler, and S.J. Westrheim. 1986. Interaction between Pacific cod (*Gadus macrocephalus*) and herring (*Clupea harengus pallasi*) in the Hecate Strait, British Columbia. Can. J. Fish. Aquat. Sci. 43: 830–837.

Ware, D.M., and G.A. McFarlane. 1986. Relative impact of Pacific hake, sablefish and Pacific cod on west coast of Vancouver Island herring stocks. Int. North Pac. Fish. Comm. Bull. 47: 67–78.

Ware, D.M., and G.A. McFarlane. 1989. Fisheries production domains in the Northeast Pacific Ocean, p. 359–379. *In* R.J. Beamish and G.A. McFarlane [ed.] Effects of ocean variability on recruitment and an evaluation of parameters used in stock assessment models. Can. Spec. Publ. Fish. Aquat. Sci. 108.

Westrheim, S.J. 1977. Length–weight and length–girth relationships, maturity, spawning season and diet of Pacific cod (*Gadus macrocephalus*) collected in British Columbia waters during Apr. 1975 – Feb. 1976. MS Rep. Fish. Res. Board Can. 1420: 68 p.

Yasutake, W.T., and J.H. Wales. 1983. Microscopic anatomy of salmonids: an atlas. U.S. Fish Wild. Serv. Res. Publ. 150: 189 p.

Influence of wind-induced currents on the recruitment of Japanese sardine (*Sardinops melanostictus*) eggs and larvae

Akihide Kasai, Michio J. Kishi, and Takashige Sugimoto

Kasai, A., M.J. Kishi, and T. Sugimoto. 1995. Influence of wind-induced currents on the recruitment of Japanese sardine (*Sardinops melanostictus*) eggs and larvae, p. 547–552. *In* R.J. Beamish [ed.] Climate change and northern fish populations. Can. Spec. Publ. Fish. Aquat. Sci. 121.

Abstract: We propose an one-dimensional model for examining the advection and diffusion of Japanese sardine (*Sardinops melanostictus*) eggs and larvae from the spawning grounds in the south of Japan. Since the sardine's main spawning period is winter, the model mainly investigates the wind-induced current caused by the winter monsoon. The pattern of the path of the Kuroshio is also included in the model as a parameter, together with the location of the spawning ground. The results of the model suggest that the offshore transport of eggs and larvae caused by the winter monsoon and the distance of the spawning ground from the Kuroshio both have significant effects on the survival of eggs and larvae. The year-to-year variation in larval survival rates calculated by the model is consistent with that estimated previously by using field data of egg and larval abundance during 1978–88. This result seems to support our idea that the drift currents caused by the winter monsoon and the location of spawning grounds are of critical importance in determining recruitment success or failure.

Résumé : Nous proposons un modèle unidimensionnel pour l'examen de l'advection et de la diffusion des oeufs et des larves de sardines (*Sardinops melanostictus*) à partir des aires de frai du sud du Japon. Étant donné que l'hiver est la principale période de frai de la sardine, le modèle est principalement axé sur l'étude du courant dû au vent au cours de la mousson d'hiver. La configuration de la trajectoire du Kuroshio est également incluse dans le modèle en tant que paramètre, de même que l'emplacement de l'aire de frai. Les résultats de la modélisation donnent à croire que le déplacement au large des oeufs et des larves causé par la mousson d'hiver et la distance entre la frayère et le Kuroshio ont tous les deux des effets importants sur la survie des oeufs et des larves. La variation annuelle des taux de survie des larves calculée par le modèle est compatible avec l'estimation faite précédemment au moyen des données recueillies sur le terrain concernant l'abondance des oeufs et des larves durant la période 1978–1988. Ce résultat semble confirmer notre idée selon laquelle les courants de dérive causés par la mousson d'hiver et l'emplacement des frayères présentent une importance déterminante pour le succès ou l'échec du recrutement.

Introduction

Japanese sardine (*Sardinops melanostictus*) stock exhibits a very wide variation over a period of several decades (Fig. 1). The landings of sardine around Japan have exponentially increased since 1973 and reached a high level of more than 4 million tons in the early 1980s. However, since 1989 the landings have started to decline as seen in the figure.

This variation in the sardine stock is detrmined by the size of the parent stock and the amount of yearly recruitment. Yearly recruitment is mainly controlled by mortality from starvation and predation during egg and larval period. Among various physical processes affecting recruitment, transport of eggs and larvae from the spawning grounds to nurseries is of great importance (Hjort 1914).

A. Kasai, M.J. Kishi, and **T. Sugimoto.** Ocean Research Institute, University of Tokyo, 1-15-1 Minamidai, Nakanoku, Tokyo, 164, Japan.

In recent years, increased attention has been focused on numerical modeling of the egg and larval transport processes to understand the mechanism behind the variability of pelagic fish stocks. Bartsch et al. (1989) modeled the larval transport of Atlantic herring (*Clupea harengus*) in the North Sea, looking at the possibility of the larvae arriving at nursery grounds favorable to their survival. They suggested that high variability of the larval drift route is caused by meteorological force i.e., wind-driven drift. Power (1986), on the other hand, simulated larval drift of northern anchovy (*Engraulis mordax*) in the California Current and showed that the offshore Ekman transport did not significantly affect the drift of the larvae. In regard to the Japanese sardine, Kobayashi and Kuroda (1991) numerically examined the transport of sardine eggs and larvae by the Kuroshio Current, based on statistically averaged surface currents observed by geomagnetic electrokinetograph (GEK). They showed that the modification of the larval transport route corresponding to the variation in the path of the

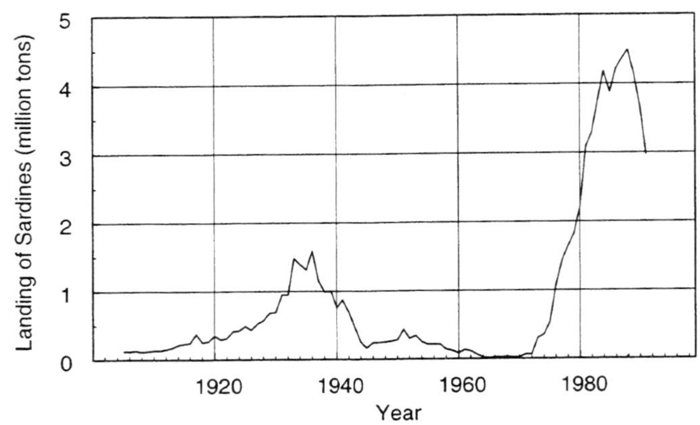

Fig. 1. Time sequence of the landing of Japanese sardine.

Kuroshio do not affect the survival of sardine larvae. In their model, however, the effect of the wind-driven offshore transport cannot be discriminated; therefore, they could not determine the influence of the wind on the survival of sardine larvae.

In this paper we develop a one-dimensional numerical model to examine the influence of offshore drift induced by the winter monsoon on the larval survival. Because the main spawning grounds of the Japanese sardine are located in and around the Kuroshio and has shifted every several years (Nakai 1962; Kuroda 1989), we also examined the combined effects of the variation in the pattern of the Kuroshio path and the location of the main spawning ground in this model. Year-to-year variation in the larval survival rate during 1978–88 simulated by the model was compared with previous estimates based on field data of egg and larval abundance.

Factors affecting the transport and survival of the sardine eggs and larvae

Variations in the Kuroshio path

The Kuroshio flowing along the southern coast of Japan shows various patterns in its path (Fig. 2). Because the majority of the sardine eggs and larvae spawned in the south of Japan are carried by the Kuroshio and associated currents, differences in the path of the Kuroshio appears to have an important effect on the drift of larvae (Watanabe 1982).

Shimizu et al. (1989) compared RNA/DNA ratios and protein contents of Japanese sardine larvae collected from coastal and offshore sides of the Kuroshio, and demonstrated that the nutritional condition of the larvae is lower in the offshore side than that in the coastal side. Moreover, predators on sardine eggs and larvae, such as *Portuguese* spp. are more abundant in the offshore side than in the coastal side of the Kuroshio (Matsukawa et al. 1991). Because these studies show that the coastal region seems most suitable for egg and larval survival, the mortality of Japanese sardine largely varies along the cross section of the Kuroshio compared with that in the direction of the Kuroshio flow. This suggests that changes in the pattern of the Kuroshio path may not seriously affect the survival of eggs and larvae. On the other hand, small-scale Kuroshio frontal disturbances and small eddies derived from the Kuroshio could have an effect on the transport of eggs and

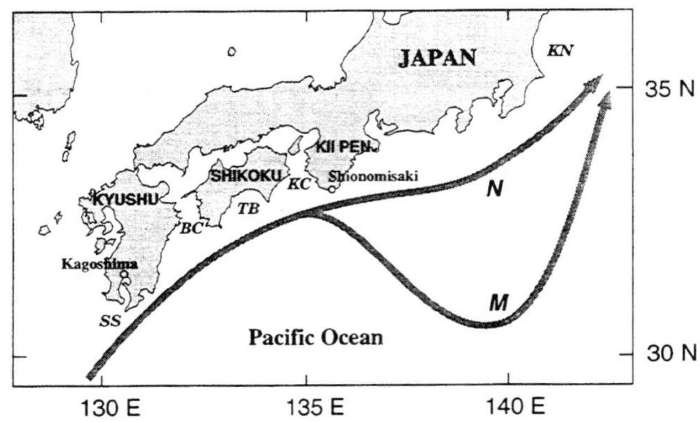

Fig. 2. Typical paths of the Kuroshio as it flows the southern coast of Japan. Arrows N and M denotermthe non-meander and meander type path, respectively. KN = Kashimanada, KC = Kii Channel, TB = Tosa Bay, BC = Bungo Channel, and SS = Satsunan Sea.

larvae. A large meander of the Kuroshio path causes those frontal disturbances and eddies more frequently than a straight path. Because the effects of such small eddies can be represented in terms of diffusivity, we looked at the effects of the variation in the pattern of the Kuroshio path by changing the diffusivity coefficient in this study.

Effects of the wind-induced drift current

The main spawning period of the Japanese sardine in the south of Japan (off Shikoku and Kyushu) is from December to March. During this period, strong westerly or northwesterly winds prevail, which may cause offshore (southward or south-westward) transport of the eggs and larvae (Sugimoto et al. 1991). Figure 3 shows the hodographs of the monthly wind velocity at the 850-mbar level (averaged over 1956–80) at

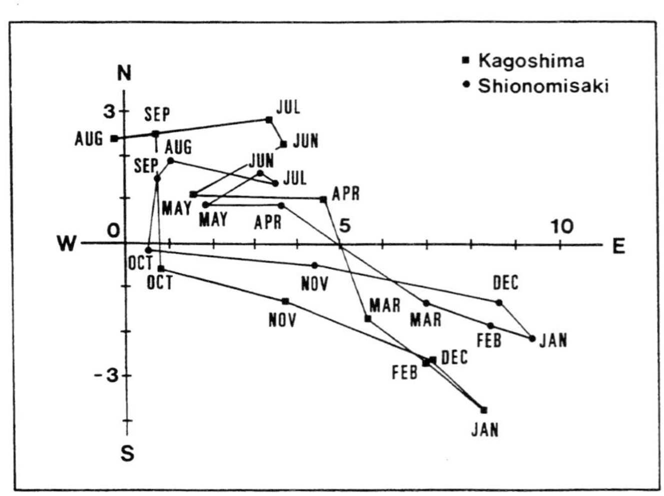

Fig. 3. Hodographs of the monthly mean wind velocity (m · s^{-1}) at 850 mbar. The data were collected at Shionomisaki and Kagoshima during 1956–80.

Cape Shionomisaki and Kagoshima. This indicates that the prevalent winds off Japan during the winter are almost always toward the east or southeast, which induce an offshore drift (Ekman drift) current of the order of several centimeters per second. In this study, the Ekman drift current with several centimeters per second was incorporated into the model experiment to examine the effect of the winter wind.

Changes in the main spawning grounds

Monitoring of the distribution of sardine eggs and larvae around Japan has been carried out every year since 1978 by the Japan Fisheries Agency. From these surveys, it has been shown that the main spawning grounds of the Japanese sardine extend widely from the Satsunan Sea to Kashimanada along the Pacific coast of southern Japan. In addition, it should be noted that the location of the main spawning grounds has shifted every several years (Nakai 1962; Kuroda 1989). The main spawning grounds were located around the Satsunan Sea and Tosa Bay after 1980. Since then they have tended to extend from northern coastal areas to the center of the Kuroshio, corresponding to the increase of sardine stock. The shift of the spawning grounds could seriously affect the transport of the eggs and larvae by the Kuroshio. Hence, two regions (i.e., the coastal and the Kuroshio frontal regions) were chosen for the initial location of the eggs to investigate the effect of the location of the spawning ground on the survival of Japanese sardine.

Model

An one-dimensional model for egg and larval advection and diffusion was used to investigate the effects of the three factors mentioned in the previous section and to calculate the survival rate of sardine eggs and larvae.

Eggs and larvae of the Japanese sardine are distributed near the surface, in depths shallower than 50 m (Konishi 1980). They are therefore carried and diffused by the surface current around the spawning area. In this study, eggs and larvae were considered to be passive and conservative drifters during the first several weeks. Hence, the advection-diffusion equation of the eggs and larvae is expressed by

$$(1)\quad \frac{\partial \hat{C}}{\partial t} + \frac{\partial}{\partial x}\left(u\hat{C} - \kappa\frac{\partial \hat{C}}{\partial x}\right) + \frac{\partial}{\partial y}\left(v\hat{C} - \kappa\frac{\partial \hat{C}}{\partial y}\right) = F(\hat{C}, x, y, t)$$

where C is the concentration of eggs and larvae, u and v are the velocities of x and y directions respectively, κ is the horizontal eddy diffusivity, and F designates the mortality of eggs and larvae.

Because the mortality rate is expected to depend on the location relative to the Kuroshio path, we assumed in this study that the Kuroshio flows straight and parallel to the x-axis, and that there is no other current except for the wind-induced drift current (Fig. 4). Because the survival conditions of the eggs and larvae are independent of x, the concentration C averaged along the x-axis was used instead of $\hat{C}$, then Equation 1 becomes

$$(2)\quad \frac{\partial C}{\partial t} - V_0(t)\frac{\partial C}{\partial y} - \kappa\frac{\partial^2 C}{\partial y^2} = -\alpha(y)C$$

where V_0 is the velocity of the drift current, and α the mortality rate of the eggs and larvae. Simulations of the egg and larval drift were carried out based on Equation 2, and survival rate (S_m) was calculated as

$$(3)\quad S_m(t) = \frac{\int_{-\infty}^{\infty} C(y, t)dy}{\int_{-\infty}^{\infty} C(y, 0)dy}$$

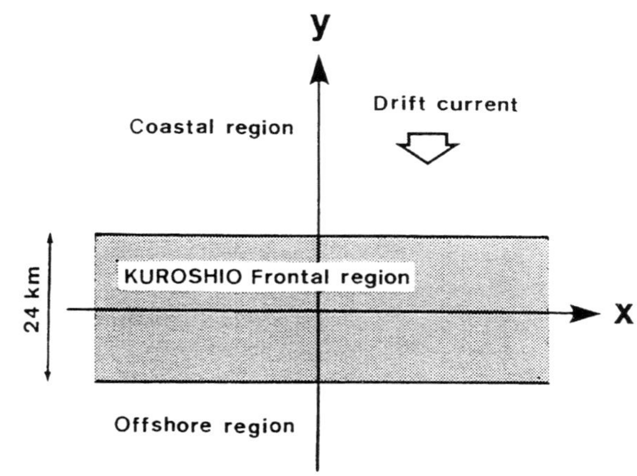

Fig. 4. Schematic view of the one-dimensional model of the current system in and around the Kuroshio Current.

Japanese sardines less than 10 mm in length are considerably affected by environmental conditions: they cannot swim well enough to feed and avoid predators until 20–30 d after hatching (Kuroda 1989). In this study, the distributions of eggs and larvae were calculated for 20 d after they were spawned. The initial concentration of eggs $C(y,0)$ was defined by Gaussian distribution with a variance of $2.5 \times 10^{-7} \cdot \text{cm}^{-1}$. The center points of the spawning grounds are located at 0 km and 50 km shoreward from the Kuroshio front. The grid interval, Δy, was taken to be 3 km and the time step, Δt, of 1 h. The velocities of the wind-induced drift current in y direction were taken to be 0, –5, and –10 cm · s^{-1}. The drift current velocity 5 cm · s^{-1} corresponds to the wind speed of about 10 m · s^{-1}. Taking the grid size into consideration, κ was estimated to be 1×10^5 cm^2 · s^{-1} in both coastal and offshore regions of the Kuroshio. The values of 1×10^6 and 5×10^6 cm^2 · s^{-1} were used in the Kuroshio frontal region 24 km in width during the periods of nonmeander and meander paths of the Kuroshio, respectively. The mortality rates were taken to be 0.01 · d^{-1} for the coastal and Kuroshio frontal regions, and 0.1 · d^{-1} for the offshore region.

The egg and larval survival rates were calculated for five cases, where different values of κ, the velocity of the wind-induced drift current, and the position of the spawning ground were used as shown in Table 1.

Results

Figure 5 shows the time sequence of the egg and larval survival rate calculated for five cases in Table 1. This figure shows that the horizontal eddy diffusivity does not affect the survival rate.

Table 1. The cases for the one-dimensional model. In the model, the retention rate is calculated for each case. Type = type of the Kuroshio path; κ = horizontal eddy diffusivity in the Kuroshio frontal region; V_0 = speed of the offshore drift current caused by the wind; y_0 = distance between the center of the spawning ground and the Kuroshio front.

Case	Type	κ ($cm^2 \cdot s^{-1}$)	V_0 ($cm \cdot s^{-1}$)	y_0 (km)
1	Nonmeander	1×10^6	0	50
2	Meander	5×10^6	0	50
3	Nonmeander	1×10^6	5	50
4	Nonmeander	1×10^6	10	50
5	Nonmeander	1×10^6	0	0

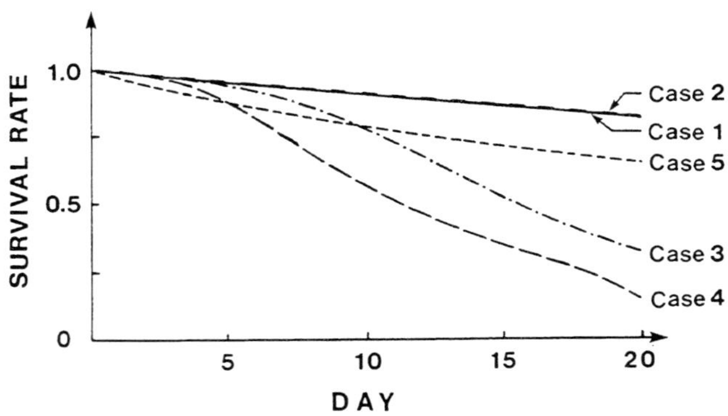

Fig. 5. Time sequences of the survival rate calculated by the model for five cases (see Table 1).

The survival of the eggs and larvae is independent of large changes in the Kuroshio path (compare cases 1 and 2). On the other hand, the intensity of the wind-induced current (compare cases 1, 3, and 4) and the location of the spawning ground (compare cases 1 and 5) do affect the survival. In particular, the offshoreward wind-induced current significantly reduces the sardine survival rate.

To clarify the effect of each factors mentioned above on the survival rate, an analytical model was developed, assuming that the eggs and larvae continue to survive for 20 d but that the larvae in the offshore side of the Kuroshio perish at the 20th day. The retention within the coastal side of the Kuroshio 20 d after the birth date is an important factor under the assumption. The retention rate, R, can be obtained from Equation 2 by the following error function (Erfc):

$$(4)\ R = \mathrm{Erfc}\left\{-\frac{\sigma}{\sqrt{\sigma^2 + 1/(4\kappa T)}}\ \frac{-V_0 T + y_0}{2\sqrt{\kappa T}}\right\}$$

where σ^2 is the variance of initial distribution of eggs, $-V_0$ is the time independent offshore velocity of the drift current, T is the time after the eggs were spawned, and y_0 is the distance of the center of the spawning ground from the Kuroshio axis.

From the time sequences of the retention rate calculated by Equation 4, the effects of the three factors on the retention rate are evident (Fig. 6). The velocity of the drift current and the distance of the spawning ground from the Kuroshio axis both have significant effects on the retention rate (Figs. 6A and 6B). Supposing $y_0 = 1.0 \times 10^6 \sim 5.0 \times 10^6$ cm, $-V_0 = 1 \sim 10\ cm \cdot s^{-1}$, and $T = 20$ d; then, the values of $-V_0 T (= 8.6 \times 10^5 \sim 8.6 \times 10^6)$ at $T = 20$ d and y_0 become of the same order.

On the other hand, κ has only a minor effect on the retention rate (Fig. 6C). As the coefficient value increases, the retention rate becomes closer to 0.5 because

$$(5)\ \frac{\sigma}{\sqrt{\sigma^2 + 1/(4\kappa T)}}\ \frac{-V_0 T + y_0}{2\sqrt{\kappa T}} \rightarrow 0 \qquad \text{as } \kappa \rightarrow \infty$$

This indicates that the variation of the Kuroshio path does not affect the survival of sardine eggs and larvae.

Discussion

The survival rate of eggs and larvae at 20 d after spawning S_m (20 d) was calculated using Equations 2 and 3 every month from 1978 to 1988. Since 1978, the Japan Fisheries Agency in cooperation with Prefectural Fisheries Experimental Stations has carried out an intensive sampling of sardine eggs and larvae around the spawning grounds. In the present analysis, data on the abundance of sardine eggs and larvae from 4000 to 5000 stations were used for determining the initial distribution of eggs which was given by the integrated egg density along the Kuroshio path. The *Kaiyou sokuhou*, a hydrographic data report published biweekly by the Japan Maritime Safety Agency, was used to determine the pattern of the Kuroshio path. In this study, the Kuroshio path was defined as contours of 15 and 16.5°C at a depth of 200 m in the eastern and western region of Kii Peninsula, respectively (Kawai 1969). Then, the Kuroshio paths were categorized into two types: the large meander type and the non-large meander type. Data were used from the regional meteorological stations at Kagoshima and Shionomisaki to know wind conditions. The offshore velocity of the wind-induced current was calculated by substituting the monthly average wind velocity into the bulk formula. After the monthly value of the survival rate was calculated, yearly survival rate S_n was also obtained by averaging four values in the winter season (i.e., from December to March).

To compare the numerical results with the field data, the yearly survival rate based on the field data S_f was estimated as

$$(6)\ S_f = L_a / (E_s + E_t)$$

where L_a is the amount of sardine larvae smaller than 10 mm collected in the southern area of Japan. These larvae were regarded as those having been spawned 20–30 d earlier. E_s and E_t were the amount of sardine eggs collected in the Satsunan Sea and around Tosa Bay, respectively. In recent years, nearly 90% of sardine larvae have been spawned in the Satsunan Sea and around Tosa Bay (Kuroda 1989).

Figure 7 shows the survival rates estimated from the numerical experiment (S_n) and the field data (S_f). The significant correlation ($r = 0.68, p = 0.02$) between them supports our idea that the location of the spawning ground and the offshore drift induced by the wind have serious influences on the recruitment of sardines. The year-to-year variation of the y-components of the drift current together with the location of the spawning

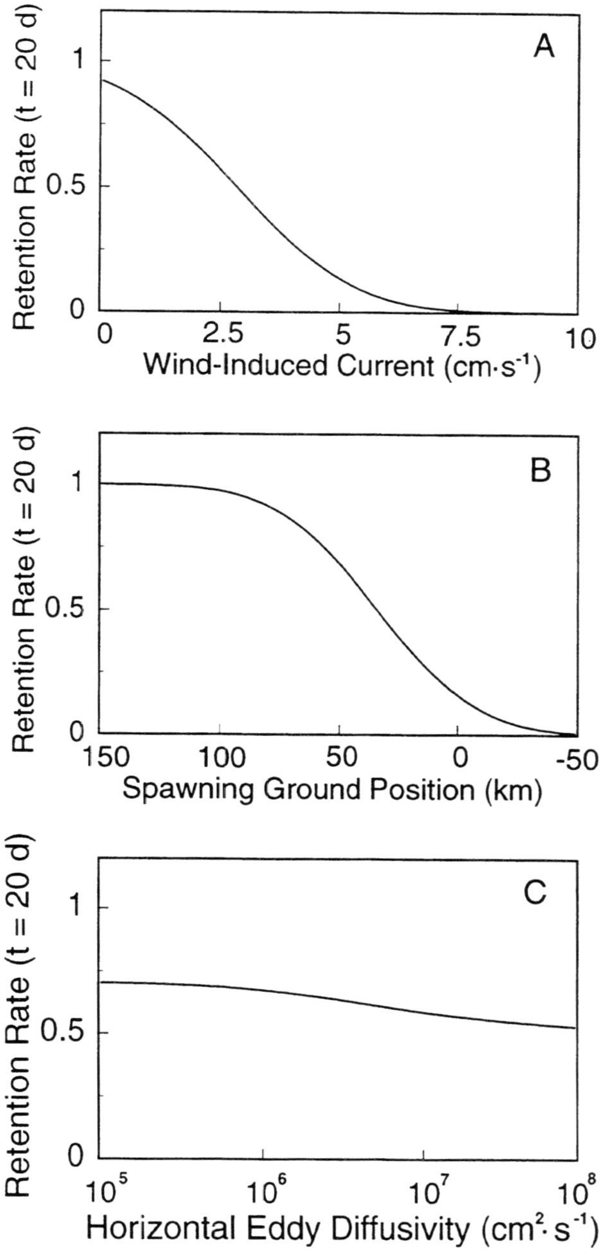

Fig. 6. Retention rate at $t = 20$ d calculated by the analytical model as a function of (A) wind-induced current, (B) shoreward distance from the spawning position to the Kuroshio front, and (C) horizontal eddy diffusivity. The standard parameter values are $\sigma = 2.5 \times 10^{-7} \cdot \text{cm}^{-1}$, $k = 1 \times 10^{6}\,\text{cm}^2 \cdot \text{s}^{-1}$, $y_0 = 50$ km, and $V_0 = 2\ \text{cm} \cdot \text{s}^{-1}$.

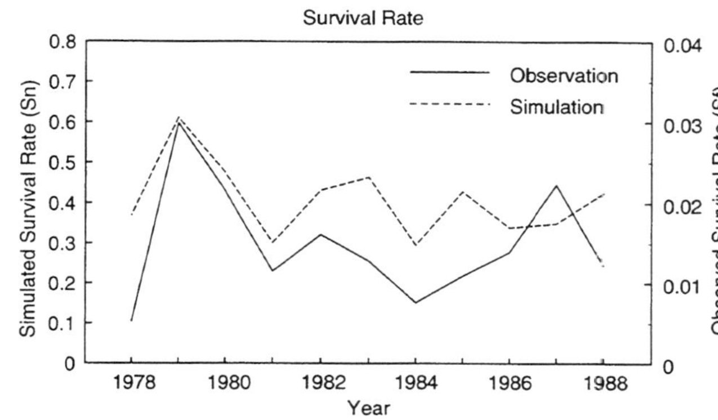

Fig. 7. Yearly variations in the survival rates S_n calculated in the one-dimensional model (broken line) and in the field survival rates S_f estimated from the field data (solid line).

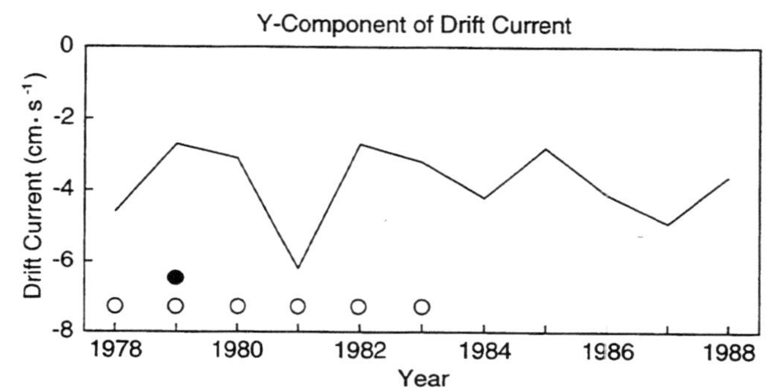

Fig. 8. Yearly variation in the average speed of the *y*-component of the wind-induced current in the south of Japan. The open circles indicate the years that the main spawning grounds around Tosa Bay were located in the coastal region, while the solid circle indicates the same for the Satsunan Sea.

grounds in the Satsunan Sea and around Tosa Bay are shown in Fig. 8. This figure shows that when the drift current was strong, recruitment failed. For recruitment to succeed, the wind must be reduced during the main spawning period, and adult sardines must deposit their eggs in the coastal region.

We now consider the reason for the rapid increase in the Japanese sardine stock in the 1970s. Kuroda (1989) reported that sardine spawned in the coastal region both before and after 1973 when the sardine stocks suddenly increased. According to our results, only the intensity of the wind-induced current might affect the survival of sardine eggs and larvae. Figure 9A shows the year-to-year variation in the east–west component of the wind speeds averaged from December to March at Cape Shionomisaki and Kagoshima. There is no significant difference between the wind intensity during the 1960s when the sardine stock was at low level and that during the 1970s when it started to increase exponentially. However, it should be noted that the winter monsoon was noticeably weak in 1972. After 1973, the landing of sardine suddenly increased by five times (Figs. 1 and 9B). This suggests that the 1972 year class of sardines might have been a strong one, and that caused the exponential increase in the sardine stock. On the other hand, the Japanese sardine has started to spawn in the offshore area in recent years, which ceases the recruitment, that may be the cause why the stock has started to decline.

It is necessary to pay more attention to the effect of the wind on the survival of sardine larvae. In this study, uniform time independent wind stress was used over the whole area.

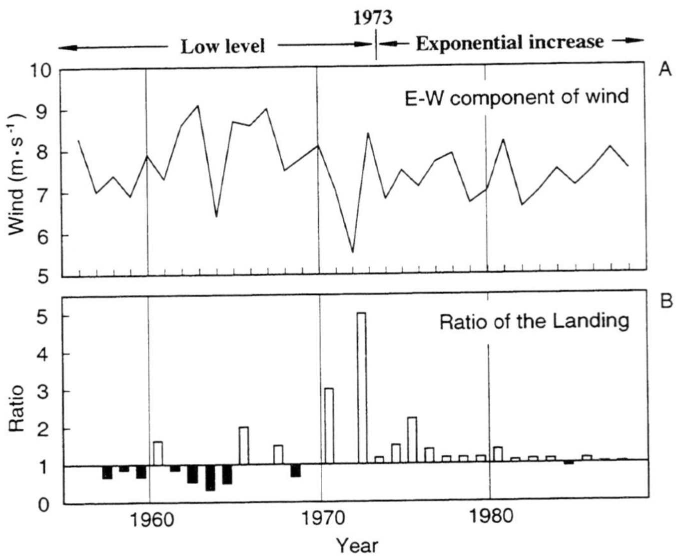

Fig. 9. (A) Interannual variations of the east-west components of the wind observed at the level of 850 mbar in winter during 1956–88. (B) Ratio of the landing of sardines in the year to that in the previous year.

However, consideration of spatial changes in the wind stress especially along the Kuroshio, is necessary, because there are some areas, namely Bungo and Kii Channels, where the winds blow strongly. The wind may have an effect on the vertical mixing and upwelling in addition to the effect on egg and larval drift. Since the wind-induced upwelling may supply nutrients from the bottom water and contribute to secondary and tertiary production, it is worth investigating these processes in the future by using a three-dimensional model.

Acknowledgments

We would like to thank Dr. H. Nakata and Dr. A. Masuda for helpful discussions and comments.

References

Bartsch, J., K. Brander, M. Heath, P. Munk, K. Richardson, and E. Svendsen. 1989. Modeling the advection of herring larvae in the North Sea. Nature (Lond.) 340(6235): 632–636.

Hjort, J. 1914. Fluctuations in the great fisheries of northern Europe viewed in the light of biological research. Rapp. P.-v. Réun. Cons. int. Explor. Mer. 20: 1–228.

Kawai, H. 1969. Statistical estimation of isotherms indicative of the Kuroshio axis. Deep-Sea Res. 16: 109–115.

Kobayashi, M., and K. Kuroda. 1991. Estimation of main spawning grounds of the Japanese sardine from a viewpoint of transport condition of its eggs and larvae, p. 109–116. *In* T. Kawasaki, S. Tanaka, Y. Toba, and A. Taniguchi [ed.] Proceedings of the International Symposium on the Long-Term Variability of Pelagic Fish Populations and Their Environment. Pergamon Press, London, UK.

Konishi, Y. 1980. Vertical distribution of eggs and larvae of sardine, *Sardinops melanosticta* (T. et S.), and round herring, *Etrumenus micropus* (T. et S.). Bull. Nansei Reg. Fish. Res. Lab. 12: 93–103. (in Japanese)

Kuroda, K. 1989. Studies on the reproduction process focusing on the early life history of Japanese sardine, *Sardinops melanostictus*. Ph.D. thesis, University of Tokyo, Tokyo, Japan. 293 p. (in Japanese)

Matsukawa, Y., K. Nakata, K. Kuroda, and T. Hada. 1991. Plankton composition and its yearly variation in and around the Kuroshio current, p. 222–223. *In* T. Asai [ed.] Proceedings of the 50th Anniversary Meeting of the Oceanographic Society of Japan. (in Japanese)

Nakai, Z. 1962. Studies relevant to mechanisms underlying the fluctuation in the catch of the Japanese sardine, *Sardinops melanosticta*. Jpn. J. Ichthyol. 9: 1–115.

Power, J.H. 1986. A model of the drift of northern anchovy, *Engraulis mordax*, larvae in the California current. Fish. Bull. U.S. 84: 585–603.

Shimizu, H., K. Nakata, and H. Nakano. 1989. Comparison of nutritional condition of sardine larvae, *Sardinops melanostictus* (T. & S.) taken from the coastal and offshore regions of the Kuroshio Current. Nippon Suisan Gakkaishi 55: 179.

Sugimoto, T., M. Kobayashi, K. Matsushita, S. Kimura, and H.S. Choo. 1991. Circulation and transport environment for sardine eggs and larvae in Tosa Bay, p. 91–99. *In* T. Kawasaki, S. Tanaka, Y. Toba, and A. Taniguchi [ed.] Proceedings of the International Symposium on the Long-Term Variability of Pelagic Fish Populations and Their Environment. Pergamon Press, London, UK.

Watanabe, T. 1982. Distribution of eggs and larvae of neurotic migratory fish in relation to the Kuroshio. Bull. Coastal Oceanogr. 19: 149–162. (in Japanese)

Variations in abundance of pelagic fishes in the Kuroshio Zone as related to climatic changes

V.A. Belyaev and T.A. Shatilina

Belaev, V.A., and T.A. Shatilina. 1995. Variations in abundance of pelagic fishes in the Kuroshio Zone as related to climatic changes, p. 553–559. *In* R.J. Beamish [ed.] Climate change and northern fish populations. Can. Spec. Publ. Fish. Aquat. Sci. 121.

Abstract: The complex and dynamic pelagic fish community in the Kuroshio zone, off Japan and the Kurile Islands, comprises some 637 species, of which the principal commercial species are anchovy (*Engraulis japonicus*), jack mackerel (*Trachurus japonicus*), mackeral (*Scomber japonicus*), sardine (*Sardinops sagax malanosticta*), and saury (*Cololadis saira*). Relative abundance was determined by commercial and ichthyoplankton catches. Long-term variations in dominance of individual species were related to geopotential pressure anomalies (GPAs) on the 500-mbar surface and related reiteration anomalies of various forms of atmospheric circulation (FACs) during winter–spring. For example, in 1972, when sardine abundance began to increase, large GPAs were observed as well as high reiterations of the eastern-type FAC. Presently, the change in FAC coincides with the decline in abundance of sardines and the rise of anchovies. Reiteration anomalies of FACs were reflected in water temperature patterns of the Northwest Pacific. Forecasting FACs could lead to predicting changes in prevalent species.

Résumé : L'ichtyofaune pélagique complexe et dynamique de la zone du Kuroshio, au large du Japon et des îles Kouriles, comprend quelque 637 espèces, dont les principales espèces commerciales sont l'anchois du Japon (*Engraulis japonicus*), le maquereau du Japon (*Trachurus japonicus*), le maquereau blanc (*Scomber japonicus*), la sardine (*Sardinops sagax melanosticta*) et le balaou (*Cololabis saira*). L'abondance relative a été déterminée par les prises commerciales et les prises d'ichtyoplancton. Les variations à long terme de la dominance d'espèces individuelles ont été rattachées aux anomalies de la pression géopotentielle (AGP) sur la surface de 500 mbar (1 mbar = 100 Pa) et aux anomalies connexes de réitération des diverses formes de circulation atmosphérique (FCA) durant la période hiver-printemps. Par exemple, en 1972, lorsque l'abondance de la sardine a commencé à augmenter, on a observé d'importantes AGP ainsi qu'un taux élevé de réitération de FCA du type de l'est. À l'heure actuelle, le changement de FCA coïncide avec le déclin de l'abondance des sardines et avec la hausse de celle des anchois. Les anomalies de réitération des FCA se répercutaient dans les tendances de la température de l'eau du nord-ouest du Pacifique. Les prévisions relatives aux FCA pourraient déboucher sur la prévision des changements dans les espèces dominantes.

Introduction

The area of the Northwest Pacific adjacent to Japan and the south Kurile Islands is characterized by dynamic oceanographic conditions and high productivity. The fish community in the upper pelagic layer, immediately related to the subarctic front, is a natural pelagic group. More than 637 species, in varous stages of development, have been reported, and they represent 189 families (Novikov 1980). High-abundance species comprise less than 3% of total known species, but are the so-called nucleus of fish community (Belyaev and Ivanov 1988). They are commercial, and mainly gregarious, plankton feeders. Principal species are anchovy (*Engraulis japonicus*), jack mackerel (*Trachurus japonicus*), mackerel (*Scomber japonicus*), sardine (*Sardinops sagax melanosticta*), and saury (*Cololadis saira*). These are characterized by interannual and long-range variations in abundance on a background of variations in climatic patterns in the area. Fluctuations in abundance are reflected in variations in interspecies dominance.

The purpose of this report is to investigate the dynamic processes involved in the fluctuations in abundance of the principal species in the upper pelagic community of the Kuroshio zone.

V.A. Belyaev and T.A. Shatilina. Pacific Research Institute of Fisheries and Oceanography (TINRO), 4 Shevchenko Allley, 690600 Vladivostok, Russia

Materials and Methods

Data sources used to examine the dynamic processes were commercial and ichtyoplankton catches of fish species, atmospheric records, and sea-surface temperatures (SSTs). The upper-atmospher factor was the geopotential anomalies (GPAs) of the 500-mbar surface. Forms of atmospheric circulation (FACs) were taken from Iliinskiy (1965).

Iliinskiy (1965) marks the main features of thermobaric field structure of troposphere.

1) Zonal, which is characterized by the latitudinal air mass transfer over the temperate zone of Asia and Far Eastern seas. In this form of circulation in space of the second natural synoptic region there are no macroscaled hollows and ridges in the structure of high field.

2) Western, which is characterized by a macroscaled high ridge over Siberia and a high hollow over the Far East (from 120 to 160°E).

3) Mixed, which is characterized by a macroscaled high baric hollow with ridges of high pressure to the west and east over the Far East.

4) Eastern, which is characterized by the latitudinal transfer of air masses over the temperate zone of Asia and strong ridge of high pressure over the Northeast Pacific and Bering Sea.

5) Central, which is characterized by high macroscaled ridge over the Far East (120–160°E).

This report deals with the basic traits of the temporal variations of atmospheric circulation over the North Pacific between 120 and 160°E. Analysis of long-term variations of the main FACs was limited to winter–spring, the period of reproduction and early development of eggs and larvae.

Results

Abundance trends

Periods of high and low catches of plankton-eating fish, during 1880–1992, are listed in Table 1. Chikuni (1985) reported that the species' proportions changed after the mid-1940s. Anchovy, jackmackerel, and saury were dominant species in the 1950s. Suary abundance decreased in the early 1960s, and that of jack mackerel somewhat later. At the same time, mackerel catches increased, to as much as 40–60% of the all-species catch in the early 1970s. While mackerel predominated, sardine was still important, and became the main species towards the early 1980s. Recently, (1989–92) mackerel catch declined, and that of anchovy increased (Table 1).

Periods of high catches in the ichthyoplankton are listed in Fig. 1. Species proportions in the ichthyoplankton during 1954–59 (Hattori 1964) and 1966–91 (this study) suggests a somewhat similar pattern to that of the commercial catches (Table 1). Predominant, or at least important, commercial species, by time period were: anchovy in 1954–59, 1966–70, and 1989–91; sardine in 1980–91; and saury in 1971–88. Since 1971, saury has comprised a relatively substantial proportion of catch, which suggest that its abundance has not varied as much as the other species. Thus, this species should be analyzed separately for species ration. Jack mackerel and mackerel did not comprise substantial proportions of the ichthyoplankton catches. Interestingly, myctophids maintained a relatively important proportion in the catches throughout the study period.

Trends in commercial catch records and qualitative ichthyoplankton data generally coincided for anchovy, sardine, and saury. The proportions of jack mackerel and mackerel in the ichthyoplankton were too small to detect trends. The ichthyoplankton data probably reflect more exactly the quantitative ration of species in the community, since variable market demand probably introduced bias in the commercial catch records.

Climate

Atmospheric circulation

This is the basic climate-forming factor that determines weather regime in the northwestern Pacific Ocean. During January–February and March–April 1954–85, reiteration anomalies of zonal (latitudinal) and mixed meridional (longitudinal) FACs were reciprocals of one another (Fig. 2). For January–February, zonal (mixed meridional) anomalies were negative (positive) during 1954–63, and generally positive (negative) during 1964–85. Maximum anomalies in

Table 1. Periods of high and low commercial, and high plankton catches of important plankton-eating fishes in the Kuroshio, 1880–91.

Species	Commercial High catches	Commercial Low catches	Plankton[a] High catches
Anchovy	1955–72	1973–79	1954–59
	1989–?		1966–70
			1989–91
Jack mackerel	1960–65	1952–56	—
		1972–73	
Mackerel	1970–79	1952–58	—
		1986–92	
Sardine	1932–41	1880–90	1976–80
	1976–90	1964–70	1980–91
Saury	1955–63	1964–72	1971–91
	1985–92	1979–83	

[a] See Fig. 1.

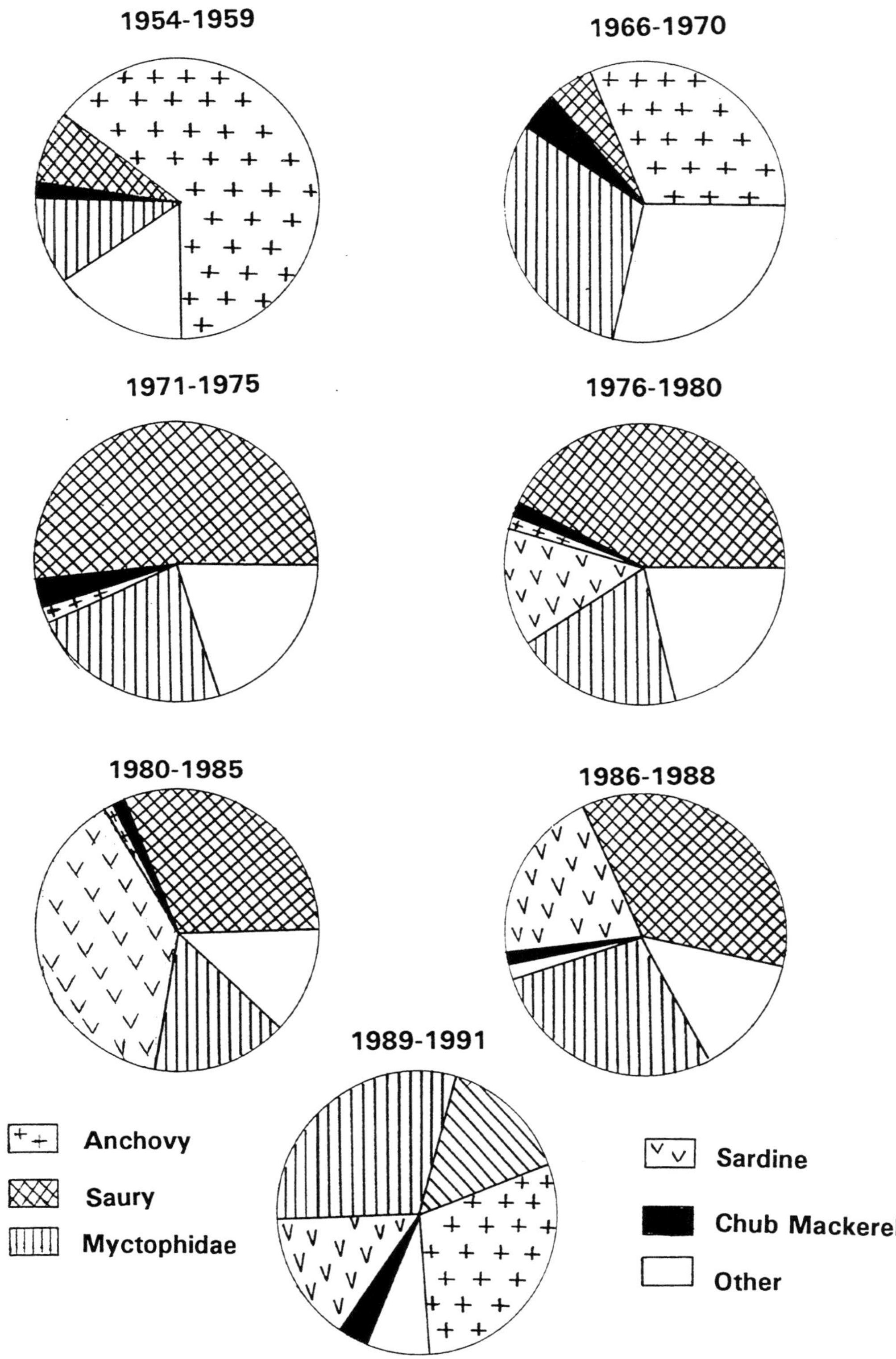

Fig. 1. Quantitative rations of dominating species of ichthyoplankton in the Kuroshio region, 1954–91.

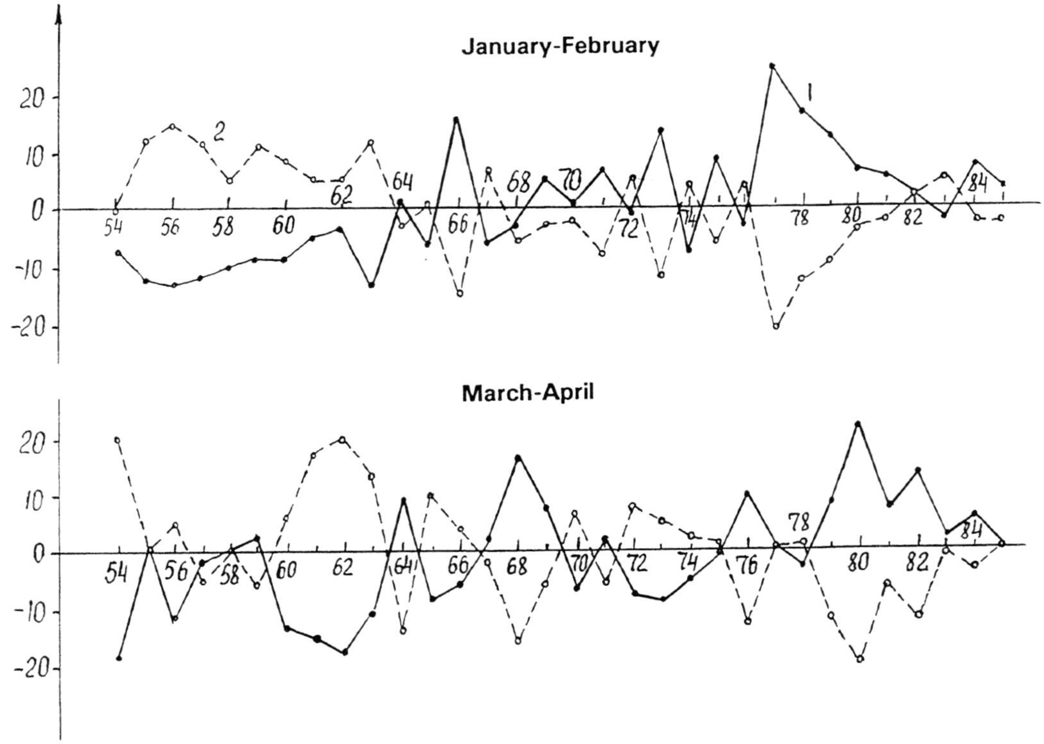

Fig. 2. Anomalies in the repeating forms of atmospheric circulation in the Northwest Pacific, 1954–85. 1 = zonal; 2 = meridional.

reiteration of meridional FACs in this period have been observed in 1955, 1956, 1957, and 1963; the duration of these anomalies was 12–15 and the most of it (25d) in reiteration FACs were observed in 1977. For March–April, the longest duration of meridonal FACs was seen in 1954 (anomaly in its reiteration reached 20 d), in 1961 it reached 17 d, in 1962 it reached 20 d, and in 1963 it reached 13 d, and the longest duration of zonal FACs reached in 1968 (17 d), 1980 (22 d), and 1982 (14 d). For March–April, neither FAC was maintained for a prolonged period. During 32 yr or observation, maximum zonal (type III) reiteration was observed in 1977 (58 d), and the minimum in 1963 (6 d).

Anomalies in reiteration of mixed, western, and eastern FACs are shown on Fig. 3. During January–February, the maximum reiteration of western form of atmospheric circulation was observed in 1964 (anomalies came to 19 d), 1981 (23 d), and 1983 (14 d), and low reiterations of this form of circulation in 1962 (−12 d), 1972 (−14 d), 1974 (−13 d), and 1982 (−9 d).

In January–February, the maximum reiteration of mixed forms of circulation may be divided to two different year periods: 1955–63 (high reiteration) and 1964–85 (low reiteration). The largest anomalies in reiteration were seen in 1962 and 1963 (26 d). In reiteration of the eastern form of circulation, the period of positive and negative anomalies has been observed (Fig. 3*a*). The maximum positive anomaly of circulation of the eastern form occurred in 1972 (20 d). In March–April, the largest reiteration of circulation of the western form occurred in 1961 (12 d), and in 1965 (13 d). In reiteration of mixed forms of circulation, the same periods of years with positive and negative anomalies as in January–March was observed.

High values of anomalies in reiteration of eastern form of circulation have been observed in 1954 (14 d), and in 1966 (15 d), and the lowest values of anomalies have been observed in 1976–84 (Fig. 3*b*). Another meridional FAC, eastern (type B), also exhibited periods of positive and negative anomalies. The maximum positive anomaly occurred in 1972.

Atmospheric topography

Variation in FAC was related to variation in the position of troughs and ridges on the 500-mbar surface. GPAs over this surface were calculated for January, 1950–90, for the eastern Okhotsk Sea and Japan (the area of reproduction). These data are presented in Table 2. Years of large positive anomalies were in 1956, 1963, 1970, and 1980, during extremely high pressure over eastern Okhotsk Sea. Significant negative anomalies occurred in 1967 and 1988 (−8). Above the southern part of Japan the low GPA values have been observed in 1950 (−9 GPA), and 1963 (−16 GPA); high values have been observed in 1964 (8 GPA), 1972 (11 GPA), and 1989 (9 GPA).

In February, two maxima were observed over the reproductive area: 1959 (13 GPA) and 1990 (13 GPA). Thus, since 1987, upper troposphere pressure increased over the reproduction area to a maximum in February 1990. The pressure increase suggested that high-altitude troughs did not move to

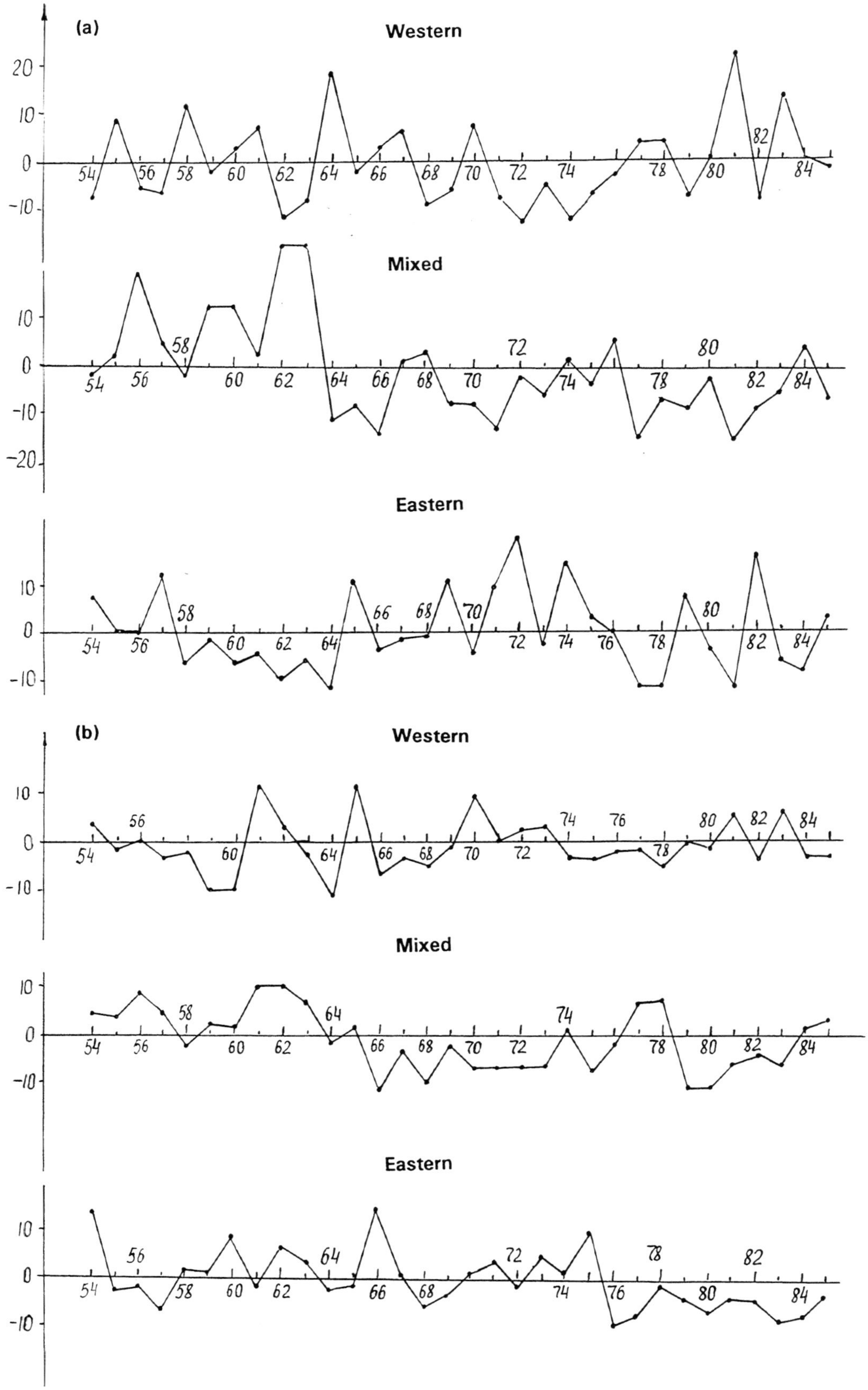

Fig. 3. Anomalies in the repeating forms of atmospheric circulation in the Northwest Pacific: (*a*) in January–February 1954–85; (*b*) in March–April 1954–85.

Table 2. The average anomaly of the 500-mbar surface geopotential in January–April 1950–90.

	East of Okhotsk				Sea South of Japan			
Year	Jan.	Feb.	March	April	Jan.	Feb.	March	April
1950	−1	−6	−1	5	−9	−8	−1	0
1951	−1	−3	5	4	1	3	1	−2
1952	−6	−6	8	1	1	0	0	0
1953	1	−7	5	−2	2	3	3	−7
1954	−5	−4	−3	8	4	5	3	1
1955	−5	2	3	−1	1	0	8	1
1956	9	8	2	−4	−2	−5	8	−1
1957	−2	10	0	−5	2	0	−3	1
1958	−6	2	−5	−4	5	−2	1	1
1959	1	−4	8	4	−1	13	6	−1
1960	−1	−6	7	1	2	7	4	−3
1961	−2	4	3	−5	−1	−1	4	−1
1962	0	−5	−6	14	−5	−1	−2	−1
1963	11	10	4	−1	−16	−2	1	5
1964	−4	−2	2	−5	8	3	2	9
1965	3	−3	−1	0	−1	−1	−6	−3
1966	−5	−12	8	−2	1	4	−2	0
1967	−11	−6	−6	0	4	1	4	3
1968	6	9	2	2	−2	−6	1	0
1969	−3	−6	−3	−6	2	4	2	0
1970	8	−4	−4	2	−2	2	−7	0
1971	—	4	0	−1	—	4	0	1
1972	0	9	7	7	11	2	0	0
1973	−5	−8	−1	−1	5	4	−2	3
1974	16	2	2	1	−1	4	−2	0
1975	3	0	5	2	1	−5	−2	−2
1976	−2	3	−4	−4	−2	6	1	0
1977	−1	−9	−4	2	1	−1	5	0
1978	−5	3	8	−1	−3	−5	2	−1
1979	−7	8	−2	−5	7	3	−1	−1
1980	9	−4	−6	−7	2	−3	3	−1
1981	−5	1	−6	−1	−3	0	2	−3
1982	1	6	−3	7	0	0	3	−3
1983	2	−3	2	−2	6	−1	−1	3
1984	7	−2	0	5	−5	−4	−9	−3
1985	−1	5	−6	6	−1	1	7	−2
1986	−1	6	3	4	−5	−6	−2	−1
1987	−6	−2	1	−6	4	0	1	−1
1988	−8	−2	2	1	5	0	−1	−7
1989	−5	−10	14	0	9	6	−1	−2
1990	−4	4	—	—	2	13	—	—

the southern Sea of Japan or Japan. Data on FACs proved it. For example, the high pressure in 1972 corresponded to high reiteration of the eastern form. In 1963, the trough (low pressure) expanded to the southern Sea of Japan and to Japan (high reiteration of mixed type of circulation). The same processes were characteristic of March 1984.

In April, over the eastern Okhotsk Sea, large positive GPAs occurred in 1954 (8 GPA) and in 1962 (14). These were years of numerous reiteration of the eastern form of circulation and mixed FACs, which suggested that in April the ridge of high pressure extended over the eastern Okhotsk Sea in those years. Negative GPAs were observed in 1956–58, 1969, 1979, 1980, and 1987.

Oceanic circulation

The abovementioned GPAs and FAC reiteration anomalies in January–March were reflected in the ocean-current pattern in the Northwest Pacific. SST anomalies in the Oyashio and Kuroshio zones exhibited groups of cold and mild years during January–March 1964–90, with some variation among months

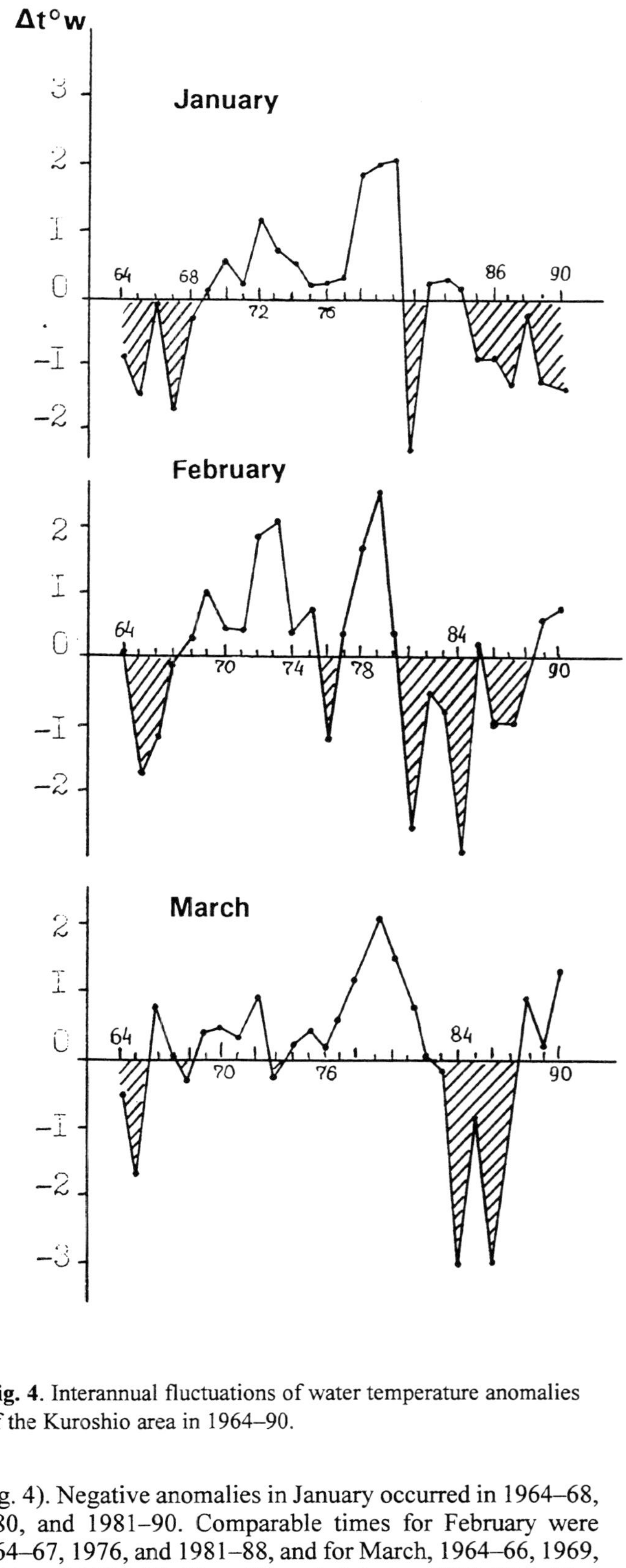

Fig. 4. Interannual fluctuations of water temperature anomalies of the Kuroshio area in 1964–90.

(Fig. 4). Negative anomalies in January occurred in 1964–68, 1980, and 1981–90. Comparable times for February were 1964–67, 1976, and 1981–88, and for March, 1964–66, 1969, 1974, and 1983–86. The mild period during 1967–75 (with a peak in 1972) marked the beginning of the increase in abundance of sardines. The large positive anomalies in 1972 and 1980 coincided with the production of strong year-classes of sardines, and large positive anomalies of zonal (1972) and meridional (1980) reiterations of atmospheric circulation during March–April (Fig. 2). Large positive anomalies of type C (mixed) circulation occurred in 1984, and in the low atmospheric strata.

Weather conditions over the Kuroshio zone were very mild during 1978–80, but very cold during 1984–86. Interestingly, the Kuroshio axis south of Honshu occupied its southernmost position during period 1988–89, while the meander occupied its southern position during abnormally cold years in the Oyashio–Kuroshio zone. Russian hydrographic surveys south of Honshu during February–March 1988 indicated milder water temperatures in the 0–100 m layer, compared with 1987. All layers as deep as 200 m were warmer in February–March 1989, compared with 1987–88. The process was accompanied by an increase in pressure at high altitudes.

Discussion

The data presented here suggest that large GPAs occurring over the Northwest Pacific, and over the entire Northern Hemisphere, correspond to margins of periods of change of dominating species (1955–56; 1963–66; 1972–78; 1988–91). Thus, in 1956, 1963, 1974, and 1988, GPAs were as large as 10–17 units. In 1972, when sardine abundance began to increase, large anomalies were observed as a result of high reiteration of the eastern FAC. Presently, the change in FAC coincides with the decline in abundance of sardine and the rise of anchovies.

Reiteration anomalies of FAC were reflected in water temperature patterns of the Northwest Pacific. In the Kuroshio zone, winter temperatures were mild during 197880, and extremely cold during 1984–86. In the Oyashio zone, mild years occurred during 1967–75 (peak in 1972); cold years after 1980 (minimum in 1984–86); and a warming trend after 1989.

Thus, changes in patterns of pelagic fish communities in the Kuroshio zone were related to weather patterns. The possibility of forecasting FACs may allow us to predict the change in abundance of important species.

References

Belyaev, V.A., and A.N. Ivanov. 1988. Fish resources of epipelagic zone of the subarctic front, p. 84–92. *In* N.S. Fadeev [ed.] Variability of ichthyophauna composition, yields of year-classes, and forecasting techniques of fishes in the western Pacific. TINRO, Vladivostok, Russia. (In Russian)

Chikuni, S. 1985. The fish resources of the northwest Pacific. FAO Fish. Tech. Pap. 266: 190 p.

Hattori, S. 1964. Studies on fish larvae in the Kuroshio and adjacent waters. Bull. Tokai Reg. Fish. Res. Lab. 40: 158 p. (In Japanese with English summary)

Iliinskiy, O.K. 1065. The practice of the main form definition of atmosphere circulation over the Far East. Proc. Tr. Dal'nevost. Nauchno-issled. Gidrometeorol. Inst. 20: 26–45.

Novikov, Yu.V. 1980. Composition, distribution, and biological characteristics of the neritic epipelagial community of fish and subarctic front of the northwestern Pacific. Izv. Tikhookean Nauchno-issled. Inst. Rybn. Khoz. Okeanogr. 104: 3–22. (In Russian)

Food web theory, marine food webs, and what climate change may do to northern marine fish populations

J. Rice

Rice, J. 1995. Food web theory, marine food webs, and what climate change may do to northern marine fish populations, p. 561–568. *In* R.J. Beamish [ed.] Climate change and northern fish populations. Can. Spec. Publ. Fish. Aquat. Sci. 121.

Abstract: Many marine fish and invertebrates feed at progressively higher trophic positions as they grow larger. Many marine food webs, particularly in boreal and subboreal seas, have "waists"; that is, a single taxon in a middle trophic position passes most of the food or energy from lower trophic levels to all higher predators. Life-history omnivory and waists have profound effects on system dynamics These effects seem not to depend on the details of the dynamic equations linking predators and prey. If the features are present in the real food web, any plausible representations of them in models are likely to dominate model dynamics. The abundances of many key species in marine food webs are conjectured to increase, decrease, or become more variable under climate change scenarios. Food web models are used to explore responses in other parts of the marine ecosystem to such changes in abundance of key taxa. Unless the dynamics of life-history omnivory and presence of waists are considered explicitly, models will fail to represent dominant ecosystem responses correctly. Unfortunately, if they are included it becomes difficult to make usefully specific predictions about trophically mediated responses to climate change.

Résumé : De nombreux poissons et invertébrés marins s'alimentent à des niveaux trophiques progressivement plus élevés à mesure qu'ils croissent. De nombreux réseaux alimentaires marins, particulièrement des mers boréales et subboréales, ont une «taille», c'est-à-dire un taxon simple de niveau trophique central qui transmet la plus grande partie des aliments ou de l'énergie de niveaux trophiques inférieurs à tous les prédateurs de niveau supérieur. Les cycles de vie des espèces omnivores et la présence de «tailles» ont de profondes répercussions sur la dynamique des systèmes. Ces répercussions semblent ne pas dépendre des détails des équations dynamiques reliant les prédateurs et leurs proies. Si les caractéristiques citées sont présentes dans le véritable réseau alimentaire, toute représentation plausible de ces dernières dans des modèles devrait vraisemblablement dominer la dynamique de ceux-ci. Les abondances de plusieurs espèces essentielles des réseaux alimentaires marins devraient censément augmenter, diminuer ou devenir plus variables selon les scénarios de changement climatique. Les modèles de réseaux alimentaires servent à analyser les réactions dans d'autres parties de l'écosystème marin à des changements apportés dans l'abondance des taxons clés. À moins que la dynamique du cycle de vie des omnivores et la présence de «tailles» ne soient prises en considération de façon explicite, les modèles ne permettront pas de représenter correctement les réactions des écosystèmes dominants. Malheureusement, lorsqu'on les inclut, il devient difficile de faire des prévisions spécifiques utiles quant aux réactions à médiation trophique provoquées par les changements climatiques.

Introduction

As the world's climate alters, the changes will have impacts on the physical properties of oceans, freshwater systems, and lands. Organisms occupying all of these systems will be affected by these impacts, as well as by how organisms with which they interact are affected by the physical changes. Planners and resource managers require advice from scientists on the nature and magnitude of the changes expected in populations, communities, and ecosystems. To do their jobs well the advice they receive must be specific and quantitative.

J. Rice. Department of Fisheries and Oceans, Biological Sciences Branch, Pacific Biological Station, Nanaimo, BC V9R 5K6, Canada.

Are the climate change studies we are conducting going to prepare us to provide the advice needed by managers, or only advice that is general and qualitative? Some studies look directly at how climate change affects populations of concern to managers. Models may forecast how the geographic distributions of species or stocks could change as physical properties of habitats change, or the population-level consequences of behavioural and physiological responses of individuals to their physical environment. Such forecasts can be specific and quantitative, with sound empirical bases, but reflect only single-species advice.

To many scientists, advice on impacts of climate change should include consideration of how the responses of groups of populations interact. Forecasting multispecies or ecosystem-level impacts requires building multispecies or ecosystem models; models that use equations of predator–prey or competitive interactions to link the dynamics of individual species or groups of species. These equations were developed for use in idealized, theoretical studies (cf. May 1974; Pimm 1991). If we are to use such equations and models as the foundation of applied advice, we need to consider their properties in applied contexts.

This paper will review some of the current problems in food web theory. These problems include choice of equations for modelling interactions, aggregation of populations in models, role of size and life history in feeding, and diffuse competition. The review will comment from two perspectives. The first is the implications of the problems for use of trophic models to evaluate and forecast biological consequences of climate change. The second is the types of advice on consequences of climate change that can be provided from trophic models, and how such advice differs from the types of advice resources managers are used to receiving from scientific advisors.

Equations for linking populations

Lotka–Volterra equations are used as the basis of many models of predator–prey interactions. They have been criticized frequently as too simplistic (Pimm 1982), and more complex alternatives have been proposed (Schoener 1976; Polis et al. 1989). However, in general the more complex equations of species interactions also have more complex solutions, if they can be solved at all (Pimm and Rice 1987; Pimm 1991). For the same degree of uncertainty in input parameters, such models are likely to produce more outcomes, and the possible outcomes will be more different. If we find that realistic levels of parameter uncertainty allow too much uncertainty in the outputs of models based on Lotka–Volterra equations, we have no reason to be optimistic that more complex models of these processes will produce fewer, more simple forecasts (Polis et al. 1989).

Unequal resolution of different trophic levels

Frequently, higher levels of food webs are disaggregated to species (cod (*Gadus morhua*), capelin (*Mallotus villosus*), puffin (*Fratercula arctica*), etc.), whereas intermediate levels are coarser groupings (copepods, jellyfish, molluscs, etc.). Primary producers, decomposers, and detritus often are grouped more coarsely yet. The practice of differential aggregation at different trophic levels persists even in recent studies of aquatic systems (McLusky and Sargent 1990; Vadas 1990; Moloney et al. 1991), but its consequences remain controversial (Sugihara et al. 1989; Hall and Raffaelli 1991; Martinez 1991; Polis 1991; Schoenly and Sugihara 1995). In general, some properties are affected substantially by aggregations, whereas others are not.

Many studies suggest that as long as species are trophically equivalent (i.e., they have identical predators and identical prey (Cohen 1978)), they can be aggregated without affecting properties of the web (Sugihara et al. 1989; Vadas 1990; Martinez 1991; Pimm et al. 1991). Unfortunately, species are shown to be trophically equivalent most easily if the web below them has already been aggregated severely. If effort is spent studying the trophic relations of lower trophic levels, few species remain trophic equivalents (Martinez 1991; Polis 1991). Moreover, the more disaggregated a web was to begin with, the greater the effects of aggregation on properties of the web. Conversely, the properties of food webs most resistant to aggregation are ones of coarsest information content, such as simply the mean food chain length.

The ultimate message on appropriate level of grouping of species is that food webs can be analyzed and modelled at any level of aggregation useful to the researcher. If patterns emerge from the analyses, they might be real. If two webs of similar aggregation differ, the differences probably are real. However, if lower trophic levels or taxa are aggregated, many other interactions will be occurring in the real system that the aggregated model will not provide any information about. If those interactions feed back on interactions that are represented in the web being analyzed, the results are likely to be misleading, even for the parts of the web represented. Because some of the major direct effects of climate change are expected to be on lower trophic levels (GLOBEC 1991), extreme caution is needed when constructing models to explore the trophic consequences of such direct effects. The models may forecast some consequences accurately, but it is almost certain that there will be important consequences which will not be forecast, as well.

Management consequences of aggregation

For the present, marine webs of northern seas are known about as well as most of the webs in the major catalogues (Cohen 1989; Cohen et al. 1990). A substantial amount is known about the feeding of adult fish, a few major invertebrates, and seabirds and marine mammals in some places and seasons. Little is known about the feeding of most pelagic and benthic invertebrates, and young (or small) fish. This produces a typical pattern of differential trophic aggregation, where at the top most species are represented individually, at intermediate levels a couple of key species are represented along with several aggregated groups, and at low levels each node is a very coarse grouping of taxa (Fig. 1).

AGGREGATION TYPICAL OF NORTHERN MARINE WEB

HIGHER PREDATORS	Pacific Cod	Pacific Hake	Rockfish	Spiny Dogfish	Harbour Seals	Sea Lions
INTERMEDIATE TAXA	Herring	Copepods	Epibenthos	Molluscs	Crabs	
Basal Groups	Phytoplankton	Detritus				

POSSIBLE STABLE PROPERTIES AT THAT LEVEL OF AGGREGATION

FOOD CHAIN LENGTH
PROPORTION OF SPECIES AT TOP, MIDDLE, BOTTOM
PREDATOR TO PREY RATIO
RIGID CIRCUIT FEATURES (IF ANY)

Fig. 1. Schematic illustration of level of aggregation typical of reports of marine food webs. This schemetic does not represent any specific marine system.

We can expect the properties of webs represented consistently in such cases are lengths of the food chains; proportions of species at top, intermediate, and bottom trophic levels; and ratios of predators to prey (Sugihara et al. 1989; Pimm et al. 1991). These properties give a very weak basis for forecasting the responses of the systems to climate change, in forms that managers can use readily. Moreover, if the literature on terrestrial webs is a guide, when we know more about the feeding relations of organisms in the 20 mm to 20 cm range, even these coarse features will prove variable.

Scientists will have to do more than study feeding relationships. They will have to address the question of what types of management advice can be constructed from information on lengths of food chains and ratios of top, intermediate, and basal species. Possibly some overall multispecies management strategies could be evaluated against such rough benchmarks. However, single species management actions, such as advised annual harvest levels, could not be refined quantitatively. The robust properties of food webs differentially aggregated at different levels would probably not contain enough information even to guide advice on how harvests of taxa at one trophic level should be balanced with harvests at another trophic level.

If aggregation were the only problem, one could proceed to use food webs cautiously to develop advice to managers on the consequences of climate change. Scientists would also hire hordes of experts in identification of marine invertebrates and plants, to disaggregate nodes in the food web. There are other problems, however.

FOR AGE STRUCTURE IN LOTKA - VOLTERRA MODELS (PIMM & RICE 1987)

LET:

X_1 = ABUNDANCE OF PRIMARY PRODUCER

X_2 = ABUNDANCE OF GRAZER

X_3 = ABUNDANCE OF YOUNG AGE GROUP OF PREDATOR

X_4 = ABUNDANCE OF ADULT AGE GROUP OF PREDATOR

$a_{i,j}$ and b_i = the standard Lotka - Volterra coefficients

and k, c_1, and c_2 = constants relating allocation of food to survivorship and maturation. Then:

$$\frac{dX_3}{dt} = + f(X_4) \quad \text{[Stock - recruit function]}$$

$$- (c_1 - [1-k]\, a_{31} X_1)\, X_3 \quad \text{[number that starve]}$$

$$- X_3 (c_2 + k a_{31} X_1) \quad \text{[maturation out of stage]}$$

AND

$$\frac{dX_4}{dt} = X_4(-b_4 + a_{42} X_2) \quad \text{[Adult feeding success]}$$

$$+ c_2 X_3 \quad \text{["Intrinsic" maturation]}$$

$$+ k\, a_{31} X_1 X_3 \quad \text{[Effect of young food on RATE of maturation]}$$

Fig 2. Summary of types of terms necessary in simplest predator-prey equations explicitly representing life-history omnivory. Details developed in Pimm and Rice (1987).

Life-history omnivory

Most theoretical modelling approaches to food webs indicate that omnivory is destabilizing to systems, and should be rare (Cohen 1978; Pimm 1982; Cohen et al. 1990). This result is surprisingly robust to alternative formulations of predator–prey dynamics (May 1974; Pimm 1982; Hastings 1983; Yodzis 1988). Despite the predictions of theory, omnivory is common in aquatic food webs. Moreover, individual species often display the omnivory ontogenetically. Different life-history stages have different diets (Steele 1974; Dayton 1984; Sprules and Bowerman 1988; Osman et al. 1990; Vadas 1990; Osenberg et al. 1992).

Both Polis (1984) and Werner and Gilliam (1984) describe a range of ways that feeding can change ontogenetically. At one extreme different life-history stages may be completely trophically distinct. At the other extreme feeding may be size dependent. At each life-history stage a fish adds different types, or just larger sizes, of prey to its suite of acceptable foods. It may drop some types or small sizes of prey from the list of prey consumed as well, but types and sizes are added faster than they are dropped. As a result the diets of large fish often overlap the diets of smaller fish of the same species, but include more items. Hardy (1924) reported such a pattern for North Sea herring some decades ago, and the pattern is apparent in the data on diets of most North Sea predatory fish (Anon. 1984).

Representing life-history omnivory requires adding significant complexity to simple predation models. Appropriate models require a term for the intrinsic rate of transition from each stage to the next, a term for the way this intrinsic rate is affected by food supply and its own density, and terms to make adults into new numbers of the youngest stage (Fig. 2).

Once added to a set of predator–prey equations, life-history omnivory has many consequences aside from contradicting stability analyses. If a food web, or a model of a food web, contains a species that changes its trophic role ontogenetically, that species dominates the mathematical properties of the web. Pimm and Rice (1987) examined diverse model webs containing one or two predators with 2, 3, or 4 life-history stages. Their analyses show all model dynamics become sensitive to crucial assumptions made about the way that food affects survivorship, life history stage duration, and the size at which transition between stages occurs (Fig. 3). When there is surplus food for any subadult stage, if the individuals simply grow larger before changing diet, then life-history omnivory is globally stabilizing. Each stage can serve as a reservoir for the population. Collapse is almost impossible unless food fails for all stages at once. However, if survivorship increases when the food available to sub-adult stages increases, then cohort strengths become highly variable, and time lags between changes in prey and changes in predators are common. Time delays are bad for model stability and for predictability of the populations (May 1974; Pimm 1982). Predictability becomes much worse if stage duration is sensitive to food supply (i.e., when food increases for a stage individuals grow faster and move to the next feeding stage sooner). The entire system is necessarily unstable. A system with such a species has little chance of ever reaching an equilibrium, and when disturbed even slightly, fluctuations amplify and spread.

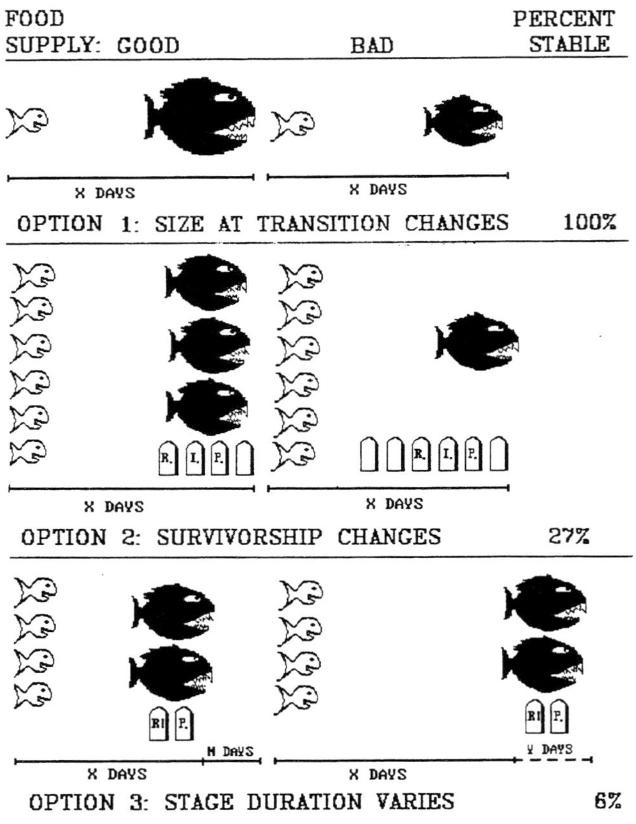

Fig 3. Schematic diagram of the three ways in which either good (left side) or poor (right side) feeding conditions for a sub-adult life history stage can affect the subsequent stage.

The models explored by Pimm and Rice (1987) used classic Lotka–Volterra predator–prey dynamics, but quite different formulations of predator–prey interactions are similarly unstable if a life-history omnivore is present (Hastings and Wollkind 1982; Wollkind et al. 1982; Polis 1984; Werner and Gilliam 1984; Polis et al. 1989). Model outcomes are parameter specific, and sensitive to starting conditions (Tschumy 1982; Hastings and Wollkind 1982; Wollkind et al. 1982; Hastings 1983; Cushing and Saleem 1983; Polis 1984; Werner and Gilliam 1984). That is, it is not possible to predict how a change in food to any life-history stage will, in the long run, alter the abundances of the predator with life-history structure, any of its prey, predators on any life-history stage, or any of its competitors unless every linkage and the starting populations of every life-history stage of every species has been determined.

Size dependent omnivory

It is not necessary to treat omnivory of fish in a rigid life history stage framework. The changes can be gradual and size dependent. Unfortunately, this framework does not make the problem simpler. Instead of representing a species or even a life-history stage as a node, models must represent frequency distributions of sizes of each species or trophic group (Werner and Gilliam 1984). When this is tried the theoretical models quickly become mathematically intractable (Hastings and Wollkind 1982; Wollkind et al. 1982; Hastings 1983; Werner and Gilliam 1984; Polis et al. 1989). It is possible to build models that at least track size categories of fish (Cousins 1985; Moloney and Field 1991; Moloney et al. 1991). When analyzed, these models also lack general predictive power. Results are completely dependent on initial conditions and parameters linking each size category of a predator with each of its prey.

Management implications of life-history – size-based omnivory

With huge commitments of resources, it is possible to quantify the initial conditions and feeding relationships for modest sets of species (Anon. 1989; Rice et al. 1991). Once initial parameters are estimated, age- or size-structured trophic models can forecast into the future. They have been used to evaluate consequences of management strategies for the North Sea (Gisleson 1991), to try to identify multispecies biological reference points (Anon. 1989), and to produce estimates of natural mortality for use in traditional models (Anon. 1989). It is noteworthy that quantitative management advice is still based on models without trophic linkages, however.

Trophic models may be useful for investigating consequences of climate change. Advice is likely to be restricted to evaluating large-scale management strategies. The limit occurs because indirect interactions can make projections unreliable at time scales greater than about twice the generation time of the major predators (Yodzis 1988). Hence, it would be necessary to reparameterize the models about twice a decade, requantifying the abundance of all taxa in the web, and all the feeding relationships.

Indirect or intraguild predation

Life-history and size-dependent omnivory are just two ways that marine food webs show indirect or intraguild predation. These terms apply to the large class of interactions where a species eats members of a species with which it also competes. Polis et al. (1989) list several ways such relationships can occur, including (Fig. 4):

1) Asymmetrical, not age structured: large predators (A) eat both somewhat smaller prey (B) and the even smaller taxa (C) which are prey of (B).
2) Asymmetrical, age structured: a young predator (A) eats prey (B) that are food of the prey eaten by the same predator as an adult (A).
3) Symmetrical - large predators (A) eat prey (B) that feed on smaller individuals of the predatory species (A).

Many combinations of those basic forms can and do occur (Werner and Gilliam 1984; Dayton 1984; Polis et al. 1989; Osman et al. 1990). Examples are particularly common in marine and freshwater invertebrates and benthos (Lane 1978; Conley and Turner 1985; Menge and Sutherland 1987; Sprules and Bowerman 1988).

Food webs that contain such interactions are extremely difficult to model. Even a simple two-stage, two-species model with a single predator on one life-history stage can show that an increase or a decrease in the juvenile stage can cause the adult stage to increase or decrease in abundance, and could stabilize or destabilize the two species system (Hastings 1983).

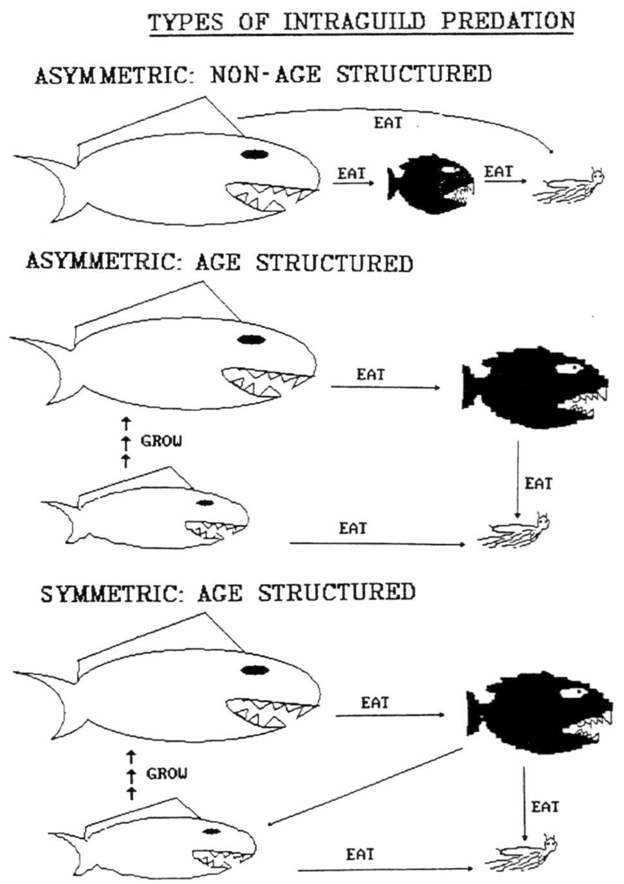

Fig 4. Schematic diagram of the forms of intraguid predation which must be considered in marine food web modelling.

The complexity seems to arise at least in part from resonances that can be set up or disrupted between the dynamics of stage duration and the oscillations of the predator–prey interactions (Hastings and Wollkind 1982; Wollkind et al. 1982).

As with life-history omnivory, when intraguild predation occurs, size-structured models are as complex as age-structured ones. Even in simple single-species models, if growth, mortality, or fecundity is density dependent, an increase in predation on small sizes can lead to either increases or decreases in abundances of larger sizes. Increases in predation on larger sizes will affect the smaller sizes in whatever way is assumed by the stock–recruit function used. Moving from linear to Michaelis–Menten or nonlinear predator–prey models, or to more complex formulations of the functional response of predators to prey does not simplify results (Wollkind et al. 1982; Polis et al. 1989; Moloney and Field 1991). More, rather than fewer, outcomes are possible with these more complex models (cf. Schoener 1976; Pimm and Rice 1987).

The trajectories followed by all the species in such models are sensitive to starting populations sizes, all the feeding linkages, and the all the ways that each stage or size group responds to changes in prey and competitor populations. For example, predation on any one node in the system can keep several other nodes below carrying capacity, allowing donor-control dynamics to dominate populations. If that predator is decreased in abundance, density dependence may appear in many places in the web. Intraguild predation can lead to coexistence of species that should not be able to coexist on the basis of their competitive interactions (Polis et al. 1989; Polis 1991) or to the extinction of either species despite apparently modest levels of competition and predation between them. In many types of interactions priority effects become important, with outcomes dependent on which species is initially more abundant (Hastings and Wollkind 1982; Werner and Gilliam 1984; Polis et al. 1989).

Food webs with waists

In typical food webs different predators feed on differing suites of prey. No one prey type dominates (Anon. 1989). In boreal seas, this is frequently not the case (Rice 1992). Capelin (*Mallotus villosus*) dominates as food in boreal seas of the North Atlantic (Lilly 1991; Mehl and Sunnana 1991; Magnusson and Pallson 1991). Pollock (*Theragra chalcogramma*) dominates in the Bering Sea (Springer et al. 1984, 1987). Even where there are several species of small fish that serve as prey for higher predators, often a single species of invertebrate dominates the next level down in the food web; often a copepod (Runge 1988; Ware and Thompson 1992).

Until recently, models and analyses of food webs and community dynamics have not considered cases where one or a few species at an intermediate trophic level filter production from all lower levels to predators above them. Pimm (1991) addresses the matter in a general way, and Runge (1988) presents an analysis of one marine fish – invertebrate – primary production system. Both show that the presence of one species as the primary channel for energy flow from lower to higher trophic levels makes it impossible in theory and in practice, to relate dynamics of any single upper level predator to any lower prey. It is even impossible to relate the dynamics of any single upper level predator to the species at the waist of the food web, except to describe how the predators will respond in the short term to large and abrupt changes in the species at the waist.

Models and analyses which will clarify the dynamics of food webs with waists (or "rose-shaped webs" (*sensu* Pimm 1991)) are still in their infancy. Much remains to be learned about even the simple, idealized cases used by theoreticians. It is clear already, however, that such webs do not behave like webs with networks of connections and more species at intermediate levels than at top or bottom positions. Moreover, the presence of one or a few species at the waist of a food web will determine the dynamic properties of the entire web. Structure elsewhere will have little influence on dynamics (Runge 1988; Pimm 1991).

Management implications of waists in food webs

Because many marine systems seem to have the feature of a waist, fisheries scientists should await a better understanding of its effects on web dynamics before food-web models are used to develop advice on dynamics of fish from information on lower trophic levels. They could prove either a blessing or a curse. They could filter system dynamics so selectively that depending on the state of the taxon at the waist, a system could appear resistant to a climate input at one time, yet respond greatly to another occurrence of a similar event. Such dynamics make advice and management difficult. On the other hand, taxa at the waist of a system may themselves be sensitive to climate events. For example, variation in year-class strength of capelin has a strong environmental component (Frank and Carscadden 1987). This may provide a window to measure and forecast important aspects of how climate events will affect all the species above it in the food web.

Prospects

The modelling results appear complex, possibly unnecessarily so. However, such model results are consistent with complex patterns of species replacements studied in intertidal systems (Paine 198; Roughgarden 1986), kelp beds (Elner and Vadas 1991), benthos (Wilson 1990), and lakes (Osenberg et al. 1992). They are consistent with the sensitivity of marine systems to small changes in the timing of plankton blooms and the responses of grazers (Townsend and Cammen 1988). They are consistent with many of the problems encountered in the attempts to link variation in recruitment to adult biomass and to populations of predators and prey (Bailey and Houde 1989). They are consistent with the observations that repeated removals at consistent levels do not have consistent impacts on the target population, or on related populations (Pimm 1994). They are consistent with the observations that species abundances, and mixes of species, show major changes over time with or without fishing, such as the "Russell Cycle," and the rises and falls of small pelagic organisms in boundary current systems (Souter and Isaacs 1974; Southward 1980; Sherman et al. 1981; Southward et al. 1988; Rice 1992). Although we may wish for systems that are more tractable, it may be necessary to accept the limits on predictability of marine systems.

What do these limits imply for studies of climate change in marine ecosystems, particularly in the context of fisheries science and fisheries management? Climate change will alter the physical environments of all species in the northern seas. Even direct abundance–environment relationships are rarely simple and linear (Rice 1993). When we try to go further, and predict trophic consequences of the environmentally driven changes in abundances, science quickly becomes fiction. We must deal with the intraguild and life-history patterns of predation and competition of fish stocks. These patterns make it intrinsically impossible to forecast population-specific longer term consequences of climate change. Therefore, we will not be able to advise managers on what specific changes in practice will mitigate undesirable effects of climate change, and enhance desirable ones. After an unexpected event has occurred, it may be possible to trace the proximate causes of the event. It will not be possible to identify which past event was the ultimate cause of the anomaly or whether the anomaly was caused by a fishery, an environmental event, or an interaction of events. Therefore, our studies may not even be able to give managers reliable feedback on the consequences of past management actions.

Pimm (1994) presents a clear set of arguments for why we frequently ignore these problems, and act as if we believe systems are more lawful than they actually are. Our expectations, and the tendency to record and report only the most marked events, gives a perception which is already severely filtered, and filtered selectively in favour of the rules we wish to apply (Pimm 1994; Finlayson 1994). We are quick enough to take credit when our models happen to approximate real systems. When they do not, we are equally quick to add whatever marked event was observed to coincide with the failure of our model, and treat the more complex model as "good enough," until the next surprise occurs.

However pessimistic science should be about modelling the trophic consequences of climate change, such work should continue. The problems posed by climate change are too important to ignore. We must ensure we take seriously not just the analysis and modelling tasks, however, but also the tasks of communcating with those who depend on our scientific advice.

We should take care to delude neither ourselves nor our clients about the realism of our results. Theory shows occasional surprises are inevitable given the interactions among the species and will occur whether the climate is changing or not. We have to ensure our clients understand the inevitability of occasional surprises, and ensure we don't hide actual model failures behind a cloak of "inevitable surprises." We also have to ensure our clients understand the limits on forecastability of our systems are to some extent intrinsic. At our best, the type of advice we will be able to provide on effects of climate change on fish populations is quite different than the advice clients are used to receiving from us. Many important clients of fisheries science are used to receiving estimates of biomass and advised harvest levels, with associated uncertainties, for each important fish stock. They are unlikely to be prepared to use advice on how changes in biomass arising from a climate event may be distributed across all species at a trophic level, with no ability to attribute a particular level of change to a particular species.

The inherent limits of our trophic models often will mean the best science will be able to do is advise on assemblage level consequences of climate change. Therefore, we must look at the usefulness of more assemblage level attributes as tools for ecosystem management. Many lines of study indicate that there are predictable properties of groups of species, even if the properties of individual species are not predictable (Yodzis 1988; Polis et al. 1989; Pimm 1991, 1994). Such assemblage-level properties are unlikely to be useful in the management of individual fisheries, but they may provide insights into the sustainability of the overall mix of fisheries, or the likelihood that climate-induced changes in a single population will have large effects on other stocks. Finally, it is only part of the job to give attention to the types of management decisions that can be based on the types of advice we will be able to provide. We must also give attention to the ways to market such management decisions to decision makers and the public, in cases when the decisions have user costs yet are justified in ways which will appear weaker than justifications of current decisions.

The results also indicate marine scientists should resist the temptation to use simple (or complex) food web models with many linkages known to, at best, a few orders of magnitude, and lots of aggregated groups of organisms at lower trophic levels, to estimate the effects of climate change on fish populations. These models can be built, and once built, will produce results. They can even be tinkered with until they produce some outputs that are intuitively satisfying and correspond to some field observations. The reality is that the models could produce dramatically different outputs (in terms of both magnitudes and directions of impacts), just by changing a few decisions on which linkages to include or exclude. The models will be particularly unreliable if they do not represent accurately the life-history or size-selective patterns of predation, and intraguild predation. If they do represent those processes accurately, however, the models will prove intractable. There is going to be no easy way for fisheries scientists to fufill their advisory role in dealing with climate change, even if they succeed in their research role.

References

Anon. 1984. Report of the coordinators of the stomach sampling project 1981. ICES C.M. 1984: G37. 22p.

Anon. 1989. Report of the Multispecies Assessment Working Group. ICES C.M. 1989/Assess: 20. 244p.

Bailey, K.M., and E.D. Houde. 1989. Predation on eggs and larvae of marine fishes and the recruitment problem. Adv. Mar. Biol. 25: 1–85.

Cohen, J.E. 1978. Food webs and niche space. Princeton University Press, Princeton, NJ. 109p.

Cohen, J.E. 1989. Ecologist Co-operative Web Bank (ECOWeB), version 1.0. Machine readable data base of food webs. Rockefeller University, New York, NY.

Cohen, J.E., F. Briand, and C.M. Newman. 1990. Community food webs: data and theory. Springer-Verlag, Berlin, Germany. 308p.

Conley, W., and J. Turner. 1985. Omnivory by the coastal marine copepods *Centrophages hamatus* and *Labidocera aestiva* Mar. Ecol. Prog. Ser. 21: 113–120.

Cousins, S.H. 1985. The trophic continuum in marine ecosystems: structure and equations for a predictive model, p. 76–93. *In* R.E. Ulanowicz and T. Platt [ed.] Ecosystem theory for biological oceanography. Can. Bull. Fish. Aquat. Sci. 213.

Cushing, J.G., and A. Saleem. 1983. A predator prey model with age structure, for fish populations. J. Math. Biol. 14: 231–250.

Dayton, P. 1984. Processes structuring some marine communities: are they general?, p. 181–197. *In* D. Strong, D. Simberloff, L.G. Abele, and A.B. Thistle [eds.] Ecological communities: conceptual issues and the evidence. Princeton University Press, Princeton, NJ.

Elner, R.W., and R.L. Vadas, Sr. 1990. Inference in ecology: the sea urchin phenomenon in the northwest Atlantic. Am. Nat. 136: 108–125.

Frank, K.T., and J.E. Carscadden. 1987. Factors affecting recruitment variability of capelin (*Mallotus villosus*) in the northwest Atlantic. J. Cons. Cons. int. Explor. Mer 45: 146–164.

Finlayson, A.C. 1994. A sociological analysis of a scientific crisis: fisheries science and advice for northern cod. Memorial University Press, St. John's, NF. 126 p.

Gisleson, H.A. 1991. The effect of recruitment variation on multispecies fisheries dynamics in the North Sea, p. 50–59. *In* N. Daan and M.P. Sissenwine [ed.] Multispecies models relevant to management of living resources. ICES MSS 193.

GLOBEC. 1991. Global ocean ecosystem dynamics, initial science plan, February 1991. Rep. 1. Joint Oceanography Institutes Incorporated, Washington, D.C. 93 p.

Hall, S.J., and D. Raffaelli. 1991. Food-web patterns: lessons from a species-rich web. J. Anim. Ecol. 60: 823–842.

Hardy, A.C. 1924. The herring in relation of its animate environment. Part I. The feed and feeding habits of the herring with special reference to the East Coast of England. Fish. Invest. Ser. II 7: 1–45.

Hastings, A. 1983. Age-dependent predation is not a simple process. II. Continuous time models. Theor. Popul. Biol. 23: 347–362.

Hastings, A., and D. Wollkind. 1982. Age structure in predator–prey systems. I. A general model and a specific example. Theor. Popul. Biol. 21: 44–56.

Lane, P. 1978. The role of invertebrate predation in structuring zooplankton communities. Verh. Int. Ver. Theor. Angew. Limnol. 20: 480–485.

Lilly, G.R. 1991. Interannual variability in predation by cod (*Gadus morhua*) on capelin (*Mallotus villosus*) and other prey off southern Labrador and northeastern Newfoundland, p. 133–146. *In* N. Daan and M.P. Sissenwine [ed.] Multispecies models relevant to management of living resources. ICES MSS 193.

Magnusson, K.G., and O.K. Palsson. 1991. Predator–prey interactions of cod and capelin in Icelandic waters, p. 153–171. *In* N. Daan and M.P. Sissenwine [ed.] Multispecies models relevant to management of living resources. ICES MSS 193.

Martinez, N.D. 1991. Artifacts or attributes? Effects of resolution on the Little Rock Lake food web. Ecol. Monogr. 61: 367–392.

May, R.M. 1974. Stability and complexity in model ecosystems. Princeton University Press, Princeton, NJ. 265 p.

McLusky, D.S., and J.R. Sargent. 1990. Trophic relationships in the marine environment - summary and way forward, p. 597–599. *In* M. Barnes and R.N. Gibson [ed.] Trophic relationships in the marine environment. Aberdeen University Press, Aberdeen, Scotland.

Mehl, S., and K. Sunnana. 1991. Changes in growth of Northeast Arctic cod in relation to food consumption in 1984–1988, p. 109–112. *In* N. Daan and M.P. Sissenwine [ed.] Multispecies models relevant to management of living resources. ICES MSS 193.

Menge, B., and J. Sutherland. 1987. Community regulation: variation in disturbance, competition and predation in relation to environmental stress and recruitment. Am. Nat. 130: 730–757.

Moloney, C.L., and J.G. Field. 1991. The size-based dynamics of plankton food webs. I. A simulation model of carbon and nitrogen flows. J. Plankton Res. 13: 1003–1038.

Moloney, C.L., J.G. Field, and M.I. Lucas. 1991. The size-based dynamics of food webs. II. Simulations of three contrasting southern Benguela food webs. J. Plankton Res. 13: 1039–1092.

Osenberg, C.W., G.G. Mittlebach, and P.C. Wainwright. 1992. Two-stage life histories in fish: the interaction between juvenile competition and adult performance. Ecology 73: 255–267.

Osman, R.W., R.B. Whitlatch, R.J. Malatesta, and R.N. Zajac. 1990. Ontogenetic changes in trophic relationships and their effects on recruitment, p. 117–129. *In* M. Barnes and R.N. Gibson [ed.] Trophic Relationships in the Marine Environment. Proceedings of the 24th European Marine Biology Symposium, Oban, Scotland, Oct. 1989. Aberdeen University Press, Aberdeen, Scotland.

Paine, R.T. 1984. Ecological determinism in the competition for space. Ecology 65: 1339–1357.

Pimm, S.L. 1982. Food webs. Chapman & Hall, London, UK.

Pimm, S.L. 1991. The balance of nature? The University of Chicago Press, Chicago, IL. 464 p.

Pimm. S.L. 1994. Understanding indirect effects: is it possible? *In* Y. Itoh [ed.] Proceedings of symposium on marine production and fisheries? Japan.

Pimm, S.L., J.H. Lawton, and J.E. Cohen. 1991. Food web patterns and their consequences. Nature (Lond.) 350: 669–674.

Pimm, S.L., and J.C. Rice. 1987. The dynamics of multispecies, multi-life-stage models of aquatic food webs. Theor. Popul. Biol. 32: 303–325.

Polis, G.A. 1984. Age structure component of niche width and intraspecific resource partitioning: can age groups function as ecological species. Am. Nat. 123: 541–564.

Polis, G.A. 1991. Complex trophic interactions in deserts: an empirical critique of food-web theory. Am. Nat. 138: 123–155.

Polis, G.A., C.A. Myers, and R.D. Holt. 1989. The ecology and evolution of intraguild predation: potential competitors that eat each other. Annu. Rev. Ecol. Syst. 20: 297–330.

Rice, J.C. 1992. Multispecies interactions in marine ecosystems: current approaches and implications for study of seabird populations, p. 586–601. *In* D.L. Nettleship and D.R. MacCullogh [ed.] Wildlife 2000: populations. Springer-Verlag, Berlin, Germany.

Rice, J.C. 1994. Forecasting abundance from habitat measures using nonparametric density estimation methods. Can. J. Fish. Aquat. Sci. 50: 1690–1698.

Rice, J.C., N. Daan, J.G. Pope, and H. Gislason. 1991. The stability of estimates of suitabilities in MSVPA over 4 years of data from predator stomachs, p. 34–45. *In* N. Daan and M.P. Sissenwine [ed.] Multispecies models relevant to management of living resources. ICES MSS 193.

Roughgarden, J. 1986. A comparison of food-limited and space-limited animal competition communities, p. 492–516. *In* J. Diamond and T.J. Case [ed.] Community ecology. Harper & Row, New York, NY.

Runge, J.A. 1988. Should we expect a relationship between primary production and fisheries? The role of copepod dynamics as a filter for trophic variability. Hydrobiologia 167/168: 61–71

Schoener, T.J. 1976. Alternatives to Lotka–Volterra competition models of intermediate complexity. Theor. Popul. Biol. 10: 309–333.

Schoenly, K., and G. Sugihara. 1995. The effects of aggregating food webs by taxonomic rather than trophic similarity. Ecology. 34 p. (In press)

Sherman, K., C. Jones, L. Sullivan, W. Smith, P. Berrlen, and L. Ejsymont. 1981. Congruent shifts in sand eel abundance in western and eastern North Atlantic ecosystems. Nature (Lond.) 291: 495–498.

Soutar, A., and J.D. Isaacs. 1974. Abundance of pelagic fish during the 19th and 20th centuries as recorded in anaerobic sediment off the Californias. Fish. Bull. U.S. 72: 257–273.

Southward, A.J. 1980. The Western English Channel—an inconstant ecosystem? Nature (Lond.) 285: 361–366.

Southward, A.J., G.T. Boalch, and L. Maddock. 1988. Fluctuations in the herring and pilchard fisheries of Devon and Cornwall linked to changes in climate since the 16th century. J. Mar. Biol. Assoc. UK. 68: 425–445.

Springer, A.M., E.C. Murphy, D.G. Roseneau, C.P. McRoy, and B.A. Cooper. 1987. The paradox of the pelagic food webs in the northern Bering Sea—seabird food habits. Cont. Shelf Res. 7: 895–911.

Springer, A.M., D.G. Roseneau, E.C. Murphy, and M.I. Springer. 1984. Environmental controls of marine food webs: food habits of seabirds in eastern Chukchi Sea. Can. J. Fish. Aquat. Sci. 41: 1202–1215.

Sprules, W.G., and J.E. Bowerman. 1988. Omnivory and food chain length in zooplankton food webs. Ecology 69: 418–426.

Steele, J.H. 1974. The Structure of marine ecosystems. Harvard University Press, Cambridge, MA. 128 p.

Sugihara, G., K. Schoenly, and A. Trombla. 1989. Scale invarance in food web properties. Science (Wash., D.C.) 245: 48–52.

Townsend, D.W., and L.M. Cammen. 1988. Potential importance of the timing of spring plankton blooms to benthic–pelagic coupling and recruitment of juvenile demersal fishes. Biol. Oceanogr. 5: 215–229.

Tschumy, W.O. 1982. Competition between juveniles and adults in age structured populations. Theor. Popul. Biol. 21: 255–268.

Vadas, R.L., Jr. 1990. The importance of omnivory and predator regulation of prey in freshwater fish assemblages of North America. Environ. Biol. Fish. 27: 285–302.

Ware, D.W., and R.E. Thomson. 1992. Link between long-term variability in upwelling and fish production in the northeast Pacific Ocean. Can. J. Fish. Aquat. Sci. 8: 2296–2306.

Werner, E.E., and J.F. Gilliam. 1984. The ontogenetic niche and species interactions in size-structured populations. Annu. Rev. Ecol. Syst. 15: 393–426.

Wilson, W.H. 1990. Competition and predation in marine soft-sediment communities. Annu. Rev. Ecol. Syst. 21: 221–241.

Wollkind, D., A. Hastings, and J. Logan. 1982. Age structure in predator–prey systems. II. Functional response and stability, and the paradox of enrichment. Theor. Popul. Biol. 21: 57–68.

Yodzis, P. 1988. The indeterminacy of ecological interactions as perceived through perturbation experiments. Ecology 69: 508–515.

Environmental effects on long-term population dynamics and recruitment to Pacific herring (*Clupea pallasi*) populations in southern British Columbia

Jacob F. Schweigert

Schweigert, J.F. 1995. Environmental effects on long-term population dynamics and recruitment to Pacific herring (*Clupea pallasi*) populations in southern British Columbia, p. 569–583. *In* R.J. Beamish [ed.] Climate change and northern fish populations. Can. Spec. Publ. Fish. Aquat. Sci. 121.

Abstract: Pacific herring (*Clupea pallasi*) stocks in southern British Columbia are near the southern limits of their natural range. Increases in temperature may be expected to result in a reduction of range and possibly abundance. Inverse relationships between sea-surface temperature for herring stocks in the Strait of Georgia and off the west coast Vancouver Island were confirmed. Unusually strong and weak year-classes appear to be linked with El Niño – Southern Oscillation events, and these may have an ameliorating effect on the anticipated negative effects of rising temperatures on stock productivity. The mechanism influencing recruitment may be linked to temperature influences on migration and predation by Pacific hake (*Merluccius productus*) and on the disappearance of Pacific sardines (*Sardinops sagax*) from the offshore summer ecosystem where juvenile and adult herring spend much of their lives. A clearer understanding of the relative impacts of biotic and abiotic factors on herring recruitment is critical to predicting the effect of long-term climatic changes on herring production.

Résumé : Les stocks de harengs du Pacifique (*Clupea pallasi*) dans le sud de la Colombie-Britannique sont rapprochés de la limite sud de leur aire de répartition naturelle. On peut s'attendre à ce que les hausses de température se traduisent par une réduction de l'aire de répartition et, éventuellement, de l'abondance de ce poisson. Des relations inverses entre la température de la surface de la mer dans le détroit de Géorgie et celle du large de la côte ouest de l'île de Vancouver ont été confirmées quant aux stocks de harengs. Les classes annuelles inhabituellement fortes et faibles semblent liées aux événements El Niño – Oscillation australe, qui peuvent minimiser les conséquences censément négatives de la hausse des températures sur la productivité des stocks. Le mécanisme qui influe sur le recrutement peut être rattaché à l'influence de la température sur la migration et sur les activités de prédation du merlu du Pacifique (*Merluccius productus*) ainsi que sur la disparition de la sardine du Pacifique (*Sardinops sagax*) de l'écosystème d'été au large, où les harengs juvéniles et adultes passent une grande partie de leur vie. Une compréhension plus précise des effets relatifs des facteurs biotiques et abiotiques sur le recrutement des harengs est essentielle à la prévision des effets des changements climatiques à long terme sur la production du hareng.

Introduction

The Pacific herring (*Clupea pallasi*) is one of the predominant north temperate pelagic species in the Pacific Ocean, ranging from the Arctic Ocean as far south as approximately 40°N. The southern extent of the species range in the eastern Pacific is southern British Columbia, although isolated populations are found as far south as San Diego, Calif. (Haegele and Schweigert 1985). As a result, any marked changes in environmental conditions particularly long-term increases in temperature may be expected to result in a reduction in the total range of the species and presumably the abundance and extent of Pacific herring stocks within southern British Columbia.

Pacific herring have supported a commercial fishery in British Columbia since at least 1877 (Taylor 1964). Initially, herring was taken for food, both local and export, with catches of up to 80 000 t (1904–34). In the early 1930s, demand for oil and meal from reduction fisheries increased the exploitation of British Columbia herring stocks and very large catches (200 000 t annually) were taken in the early 1960s, which combined with a series of poor recruitments and resulted in the collapse of the reduction fishery and its closure in 1968. A roe fishery began in 1971 as stocks recovered and has operated annually since that time. The current value of the fishery to the

J.F. Schweigert. Biological Sciences Branch, Pacific Biological Station, Department of Fisheries and Oceans, Nanaimo, BC V9R 5K6, Canada.

British Columbia economy is approximately \$100 million annually, and any long-term decline in stock abundance will have serious economic and social implications to coastal communities. Pacific herring also represent a significant food source for salmonids and various other commercially important fish stocks, which will be adversely affected by any long-term decline in food supplies.

The abundance of each Pacific herring stock at any given time is a function of the current and recent additions of first-spawning or recruiting cohorts (3 yr olds). Because herring are relatively short lived (generally less than 10 years old), the populations are very dynamic, fluctuating in abundance directly as a function of the number of recently added new recruits. Consequently, it is critical to understand the mechanisms that determine the success or failure of year-classes and how these may affect stock levels relative to long-term changes in environmental conditions in the North Pacific. The early work of Tester (1948), and more recently Ware (1991) suggests that herring year-class abundance for the west coast of Vancouver Island, the most southerly British Columbia herring stock, is inversely related to air and sea-surface temperatures.

In this paper, this result is investigated for a longer time series of data, as well as for the other major southern herring stock in the Strait of Georgia, British Columbia. In addition, survival, instantaneous population growth rate, and total production are examined in relation to long-term variations in both biotic and abiotic factors.

Materials and methods

The fishery for Pacific herring in British Columbia dates back to the late 1800s, and information on the biological characteristics of many of the stocks is available from near the turn of the century. However, complete and reliable information on total catch, spawn deposition, age structure, and length and weight-at-age are available only since 1950 in computer files at the Pacific Biological Station (PBS), in Nanaimo, B.C. In this paper, an attempt is made to reconstruct the stock abundances for the two major herring stocks in southern British Columbia (i.e., the Strait of Georgia and the west coast of Vancouver Island) consistently back to 1937. The basic data on the length, width, and intensity of egg deposition on spawning grounds are available on computer files and in hard copy at PBS from the late 1920s to the present (Hay and Kronlund 1987). Data on total catch prior to the 1950–51 season are taken from Taylor (1964) and data on age composition are those presented in Tester (1955). Data on weight-at-age are not available in published form prior to 1950–51. Instead, unpublished data summaries were used in the current study.

The methods used to reconstruct the stock abundances over the period from 1937 to 1992 are described in Schweigert and Stocker (1988). Estimates of catch and escapement, back calculated from total egg deposition, are combined with information on age structure and weights-at-age to determine total biomass-at-age using the following equation:

$$B_{at} = C_{at} + E_{at}$$

where B_{at} = total biomass at age a in year t; C_{at} = total catch at age a in year t; and E_{at} = total escapement at age a in year t.

Estimates of the total escapement are derived by determining the total area of egg deposition, estimating the average egg density, and assuming a constant population egg production rate of 100 eggs·g^{-1} (Schweigert 1993a). For some of the spawning ground surveys in the first years of the time series, no width data exist. For these data, the long-term mean was assumed (Hay and Kronlund 1987). Estimates of total abundance-at-age are estimated throughout the time series from estimates of age structure and average weights-at-age.

Herring in southern British Columbia generally spawn for the first time at age 3 (Taylor 1964), so the estimated number of herring at age 3 in each year may be used as an index of recruitment. However, uncertainty exists about how well the age structure in the catch reflects the spawning population during the pre-1950 period. There is also concern about the relative accuracy of the estimated proportion of fish at age 3 because of uncertainty about the extent and consistency of escapement data during this earlier period. Thus, an alternative recruitment index was developed. This index is based on the concept of the virtual population suggested by Fry (1949) except that the estimate of numbers-at-age rather than catch-at-age is used in the following equation:

$$\mathrm{RI}_{t-2} = \sum_{k=2,t}^{k=10,t+8} N_{kt}$$

where RI_{t-2} = estimated numbers of recruits in year-class $t-2$; N_{kt} = estimated numbers of fish in year-class k in year t. By amalgamating estimates of abundance throughout the course of a year-class's presence in the population, the potential effects of year-to-year inconsistancy in data quality during the early portion of the time series should be minimized.

In addition to recruitment variation, long-term environmental variation would be expected to affect total stock production, growth, and survival. Total production was estimated using the standard approach outlined by Chapman (1968) including an estimate for gonad production as described in Schweigert (1993b):

$$P_{at} = g_{at}\bar{B}_{at} + R_{at}$$

$$\bar{B}_{at} = \frac{B_{at} + B_{a+1,t+1}}{2}$$

where P_{at} = production of new biomass from age a to $a+1$ during the time interval t to $t+1$; g_{at} = instantaneous growth rate in weight from age a to $a+1$ during time interval t to $t+1$; $\bar{B}_{at}$ = average biomass at age a to $a+1$ during time interval t to $t+1$; and R_{at} = biomass of gonad produced from age a to $a+1$ during time interval t to $t+1$. In this study, average biomass and total production were calculated directly from estimated numbers of fish at age in the population, and their associated average weights at age. Instantaneous growth rate was calculated as the ratio of $P/\bar{B}$.

Estimates of survival were obtained by comparing the escapement of fish of age *a* in year *t* with the estimate of the number observed in the stock the following year:

$$SI_t = \frac{\sum (ESC_{a+1,t+1} + C_{a+1,t+1})}{\sum ESC_{at}}$$

where SI_t = survival index of fish of ages 3–9 in year *t* to *t* + 1; ESC_{at} = estimated number of spawning fish at age *a* in year *t*; $C_{a+1,t+1}$ = estimated number of age *a* + 1 fish in the catch in year *t* + 1. For these data, the survival index was estimated both for ages 3–9 and for age 4–9 fish. The latter estimate should be more accurate because the survival rate determined for age-classes 3 to 9 will implicitly include any partial recruitment at age older than 3 often resulting in survival index estimates exceeding one.

The environmental data series that are available during this time interval are limited. Data sets that were examined included average annual sea-surface temperatures and salinities from lighthouses at Amphitrite Point on the west coast of Vancouver Island (Fig. 1) and Entrance Island in the Strait of Georgia (1935–91), sea level in metres at Tofino on the west coast of Vancouver Island (1940–91) and Victoria in the Strait of Georgia (1935–91), Fraser River total annual discharge at Hope in millions of cubic decametres (1935–90), and Ekman transport at 48°N, 125°W (1946–89). The other two time series that were examined were the estimated relative biomass of Pacific hake (*Merluccius productus*) in Canadian waters (D. Ware, Pacific Biological Station, Nanaimo, B.C., personal communication) and the total catch of Pacific sardines (*Sardinops sagax*) in British Columbia in thousands of tonnes (Schweigert 1988).

A combination of bivariate plots and linear models for various combinations of environmental variables and recruitment, growth, survival, and production were examined for both herring stocks. Subsequently, significant relationships between combinations of these variables were further examined using a backward elimination multiple regression procedure available in the Statistical Analysis System (SAS). Finally, to further investigate the importance of the observed temperature link to Pacific herring recruitment, a general linear model was postulated linking recruitment to sea-surface temperature and El Niño – Southern Oscillation (ENSO) events. A linear model relating recruitment and sea-surface temperature was assumed and El Niño effects were examined as a treatment effect relative to "normal" or non-ENSO years, whereby moderate and strong or very strong ENSO events (see Quinn et. al 1987, Tables 1 and 2) were modeled as separate treatments.

Results

The history of the herring fishery in British Columbia plays an important role in understanding the dynamics of the stocks and the relationships presented here. The herring reduction fishery which was in operation from the early 1930s until the closure in 1968, focussed on all available age-classes during the fall inshore migration of spawning stocks. The fishery was essentially unlimited and resulted in a high exploitation rate of the stocks and a concomitant reduction in the number of age-classes supporting the population. With the beginning of

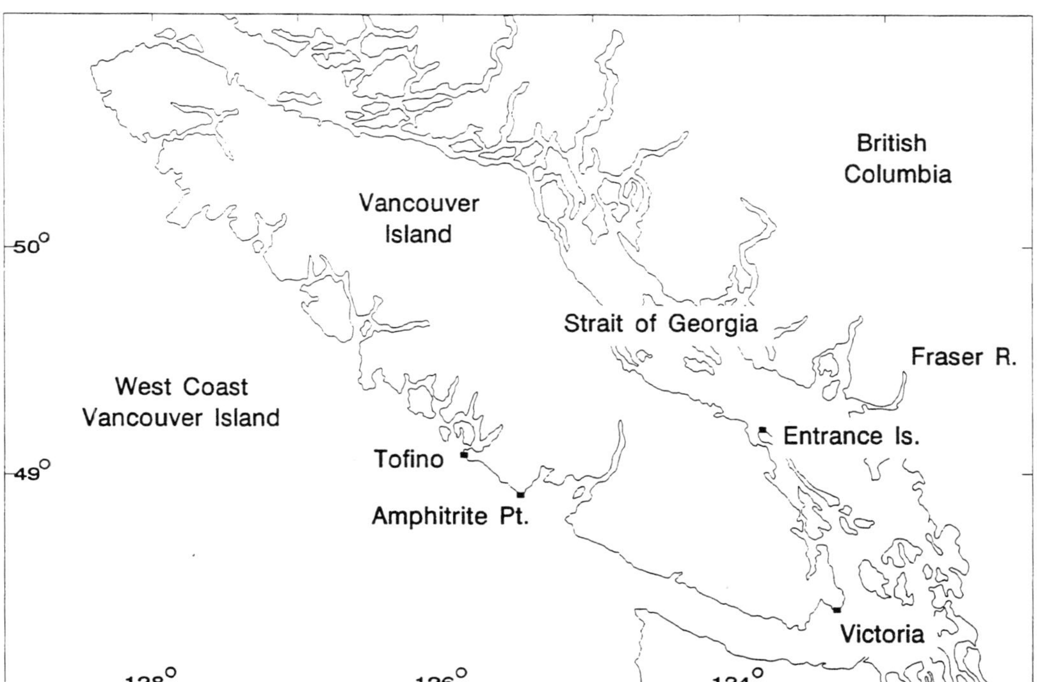

Fig. 1. Chart depicting the Strait of Georgia and west coast of Vancouver Island herring stocks and sites for environmental data series collections.

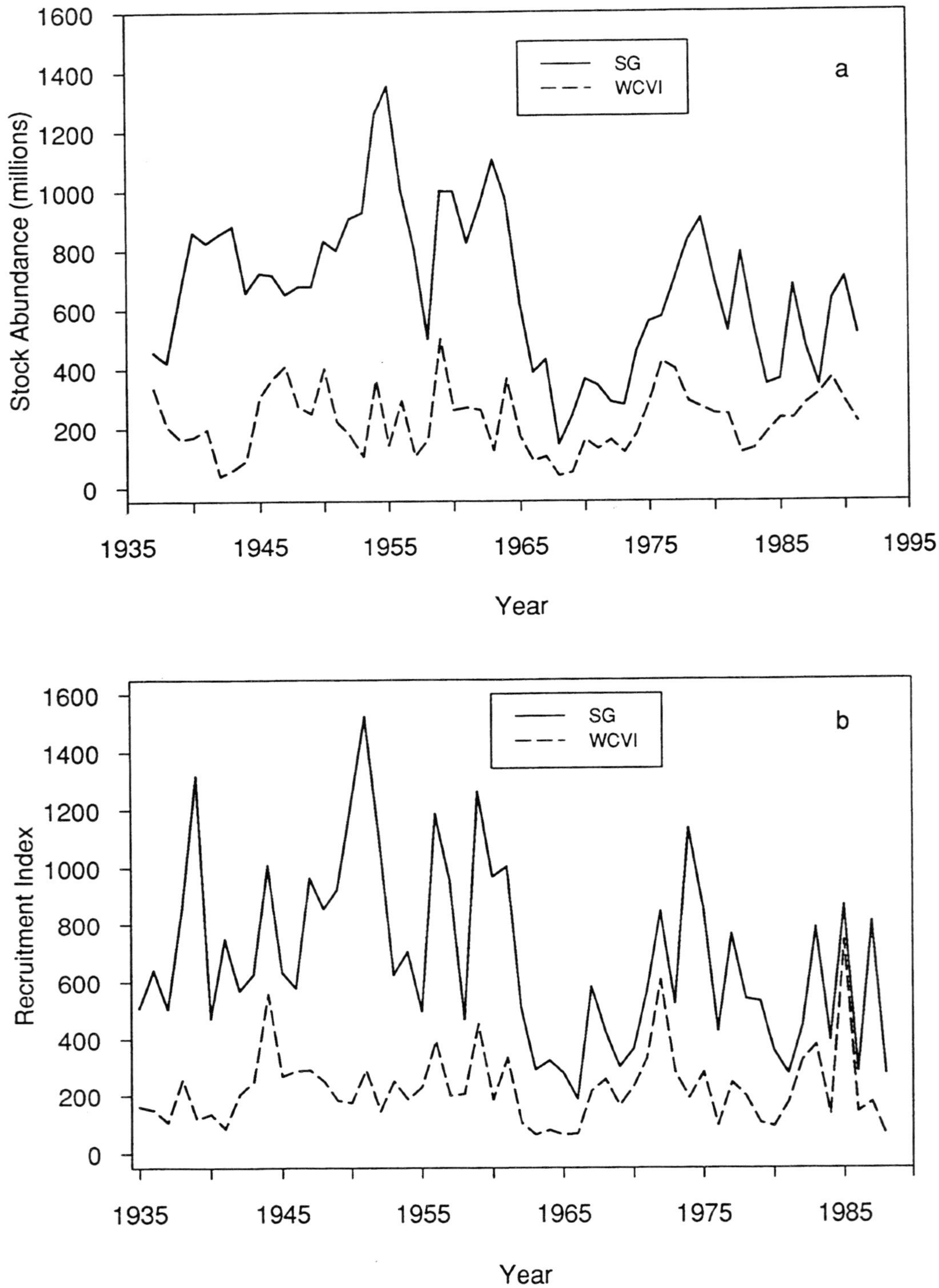

Fig. 2. (a) Estimated total stock biomass for herring stocks on the southern west coast of Vancouver Island and the Strait of Georgia from 1937–92. (b) Estimated recruitment index for herring stocks on the southern west coast of Vancouver Island and the Strait of Georgia for the 1935–88 year-classes.

the roe fishery in 1971 and eventual quota regulation in 1983, the harvest rate was much reduced and quickly led to the recovery of an expanded age structure. The reconstruction of total stock biomass estimates for the Strait of Georgia (SG) and for the southern west coast of Vancouver Island (WCVI) is shown in Fig. 2a. The two stocks demonstrate some similarities in biomass trends with high levels in the late 1940s, early 1950s, and late 1970s, and lows during the early 1940s, late 1960s, and early 1980s. The WCVI stock has averaged about 20 000 t of biomass annually, while the SG stock has averaged 50 000 – 60 000 t. Stock levels have remained stable or increased slightly during 1937–92 on the WCVI, while they have apparently decreased slightly in the SG. However, total biomass of the SG stock is currently comparable with or above historical levels whereas the WCVI stock is near the long-term average level.

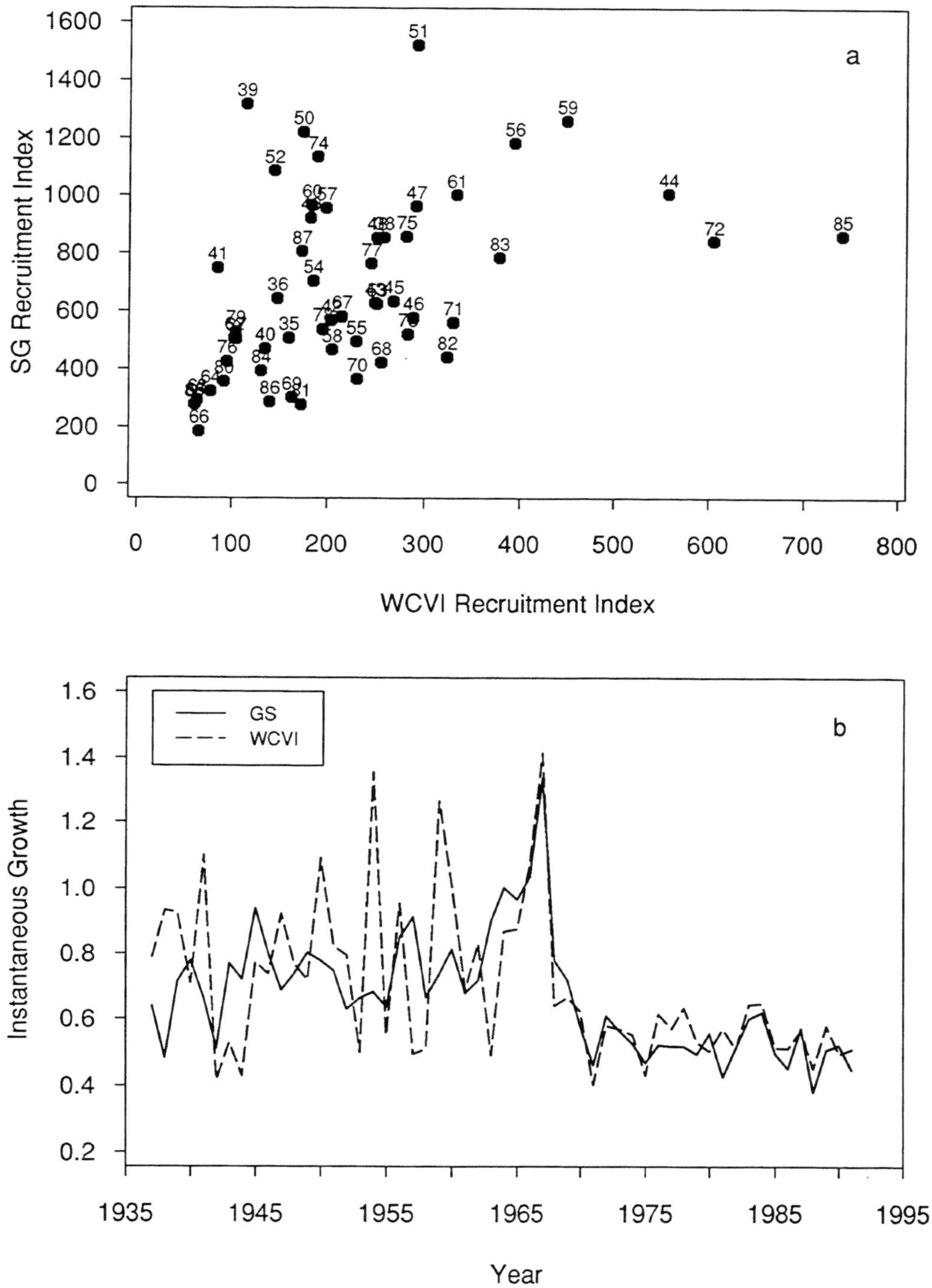

Fig. 3. (a) Estimated recruitment indices for herring stocks in the Strait of Georgia (SG) and on the west coast of Vancouver Island (WCVI) for the 1935–88 year-classes. (b) Estimated instantaneous population growth rate for herring stocks in the Strait of Georgia and on the west coast Vancouver Island for 1937–91.

The interpretation of these stock trends and all that follows depends on the acceptance of existing information on spawn-deposition data (see Hay and Kronlund (1987) for further discussion of the extent of this problem). If spawn-survey information prior to the 1950s is particularly incomplete, then estimates of total stock abundance, production, growth rates, and survival rates may be underestimated and any apparent stock declines in recent times are more extensive than the current interpretation of the data indicates. Alternatively, because much of the pre-1965 estimate of abundance is based on total catch data, which is likely to be more accurate than spawning deposition estimates, recent abundance estimates may be biased low due to incomplete or inadequate spawn assessments incorrectly suggesting a long-term decline in abundance. Unfortunately, these data do not permit the resolution of which interpretation of the information is correct.

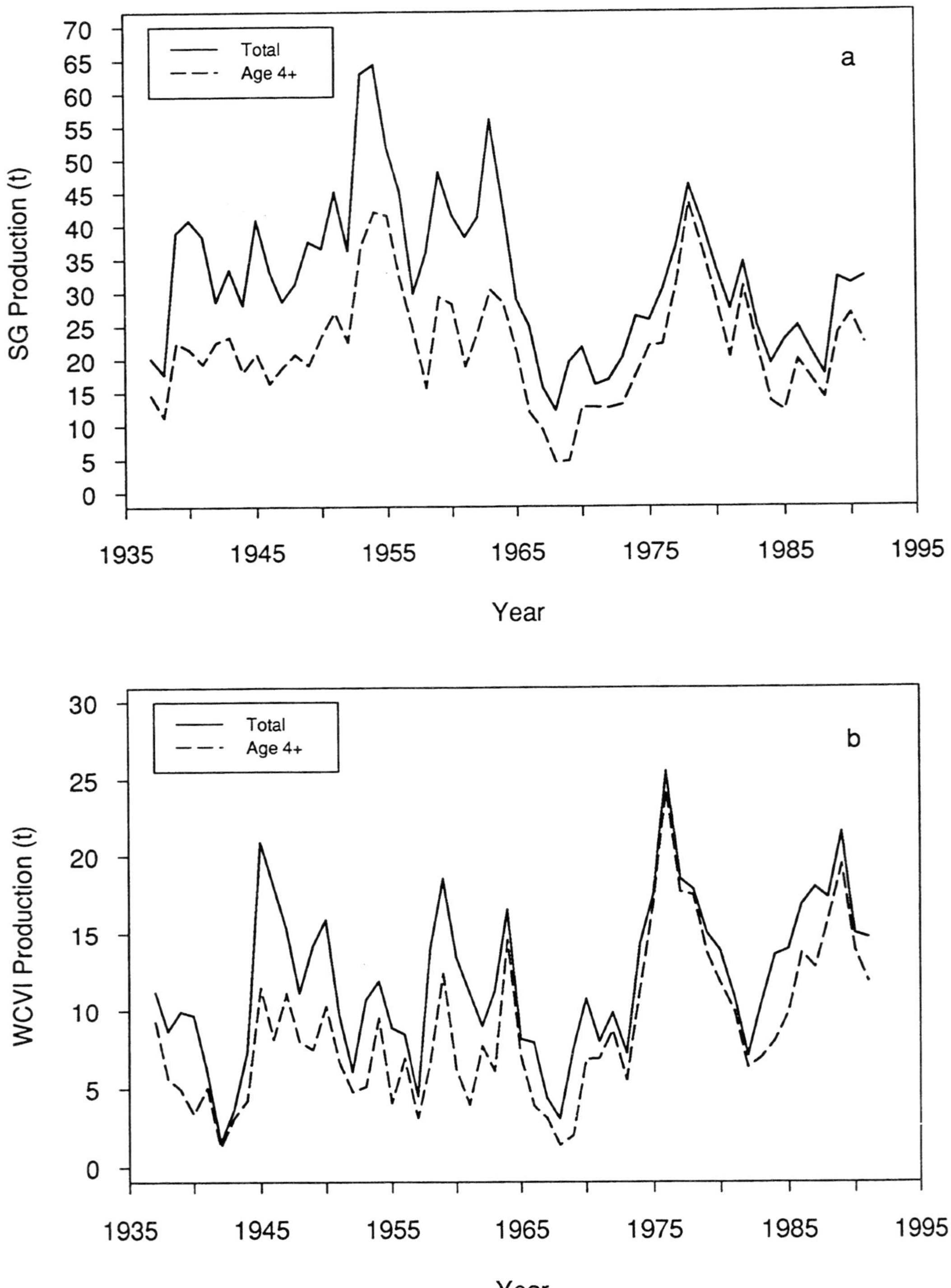

Fig. 4. (a) Estimated production (thousands of tonnes) for the total stock and for the age-4 and older stock in the Strait of Georgia from 1935–92. (b) Estimated production (thousands of tonnes) for the total stock and for the age-4 and older stock on the west coast of Vancouver Island from 1935–92.

The recruitment index and biomass estimates demonstrate similar trends (Figs. 2a and 2b) which is not surprising as recruitment is the dominant factor determining fluctuations in herring stock abundance. The recruitment indices were compared for the 1935 through 1988 year-classes, and it is evident that a strong positive relationship exists in the recruitment rates in the two stocks (Fig. 3a), except for some important outliers discussed in relation to ENSO events below. The importance of recruitment to the instantaneous population growth rate is shown in Fig. 3b and to total production in Figs. 4a and 4b. In both stocks, instantaneous growth rate increased during the 1950s and 1960s as the average age declined, peaking just prior to the stock collapse and fishery closure in 1968. As the stocks rebuilt, the age structure

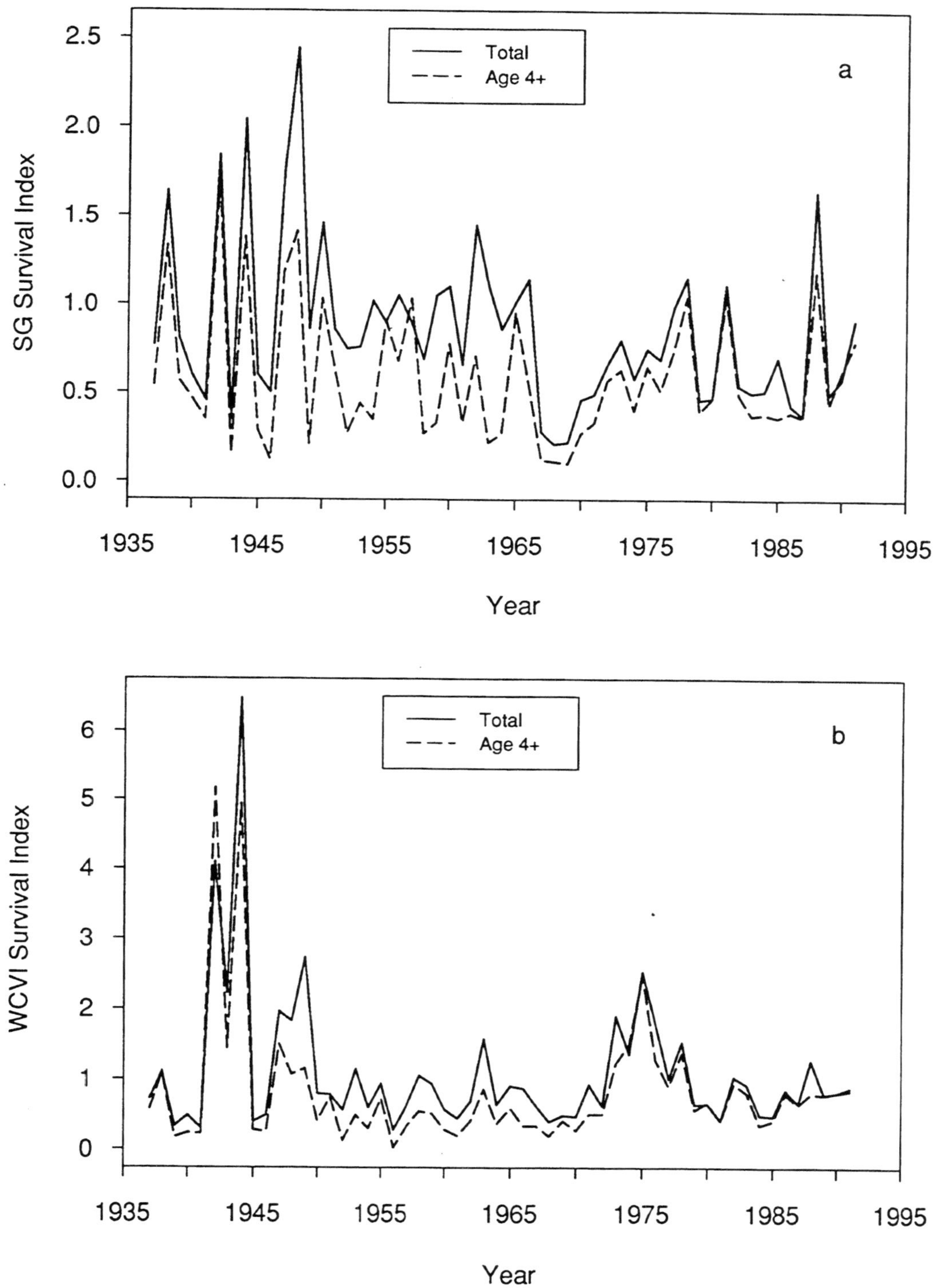

Fig. 5. (a) Estimated survival indices for the adult stock (broken line) and for the total stock (solid line) in the Strait of Georgia from 1937–91. (b) Estimated survival index for the adult stock and for the total stock on the west coast of Vancouver Island from 1937–91.

expanded and growth rate declined remaining stable at about 0.5 since 1970. The age-3 recruits also contributed significantly to the total production during the reduction fishery period ending in 1968 in both stocks (Figs. 4a and 4b). Subsequently, relatively little of the annual production was attributable to the recruiting age-class in either population. The production trends in the two stocks are broadly similar, but WCVI production has been more variable than in the Strait of Georgia, which is largely a reflection of the more variable recruitment rate and protracted age structure in the WCVI stock.

The survival indices for the two stocks show generally consistent trends over the time period from 1937 through 1991 (Figs. 5a and 5b). The survival-index estimates are exceptionally high and variable during the early 1940s,

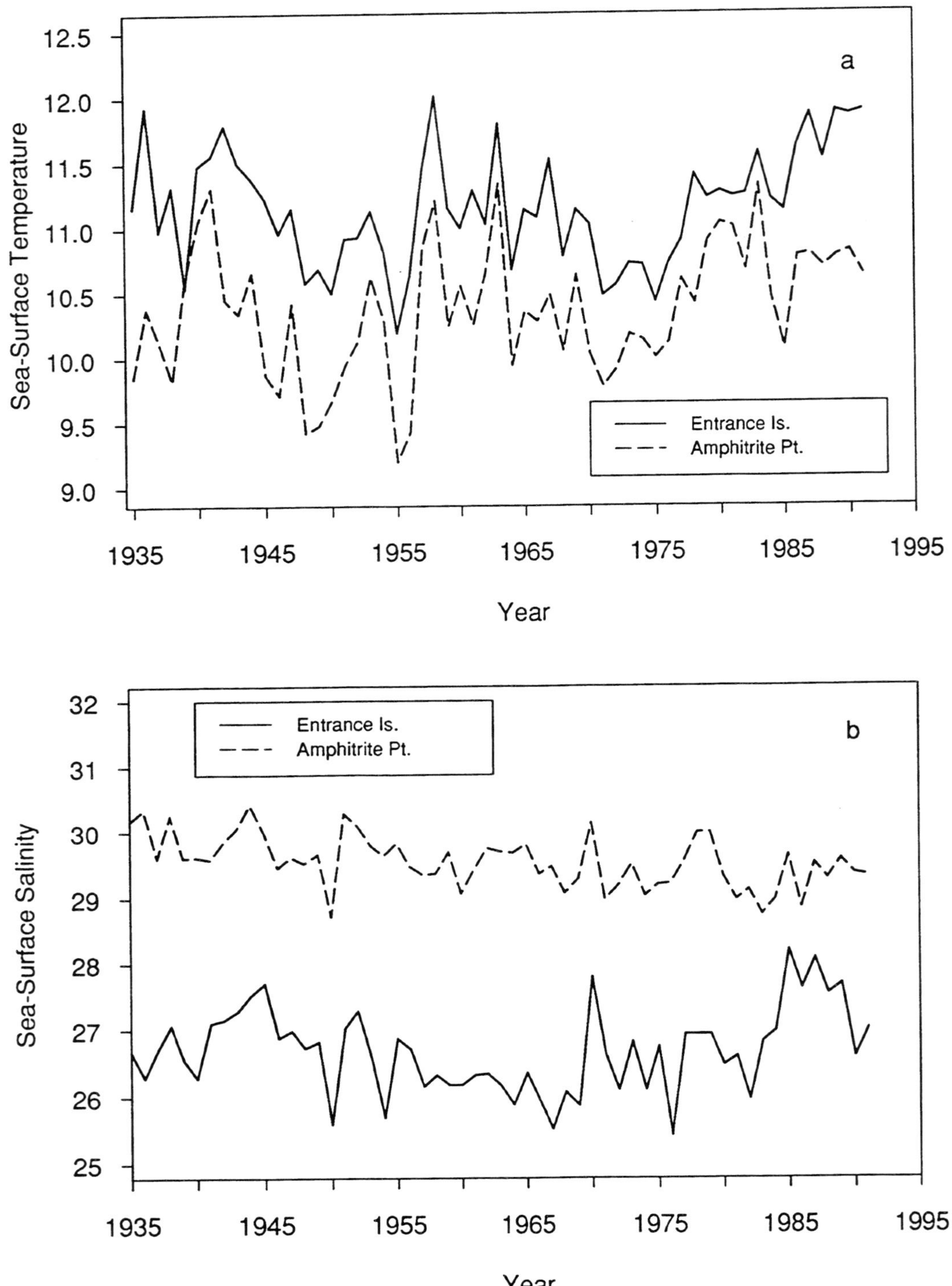

Fig. 6. (a) Long-term trend in annual sea-surface temperatures at Entrance Island and Amphitrite Point from 1935–91. (b) Long-term trend in annual sea-surface salinity at Entrance Island and Amphitrite Point from 1935–91.

probably reflecting poor data quality during this period. The survival-index estimates for both the adult age-classes and the entire stock are sensitive to the impact of strong year-classes. The survival index can be biased by the recruitment of year-classes at ages older than 3, resulting in some of the observed values exceeding one. Such high values can also be attributed to poor or incomplete age-structure or spawn survey data.

The environmental time series are presented in Figs. 6a–6d. Sea-surface temperatures for Entrance Island and Amphitrite Point indicate similar interannual variability and long-term trend (Fig. 6a). Figure 6b indicates a long-term decline in sea-surface salinity at Amphitrite Point and a decline and recovery at Entrance Island, with a low period during the mid-1960s. Similarly, sea level measured at Tofino has shown a long-term

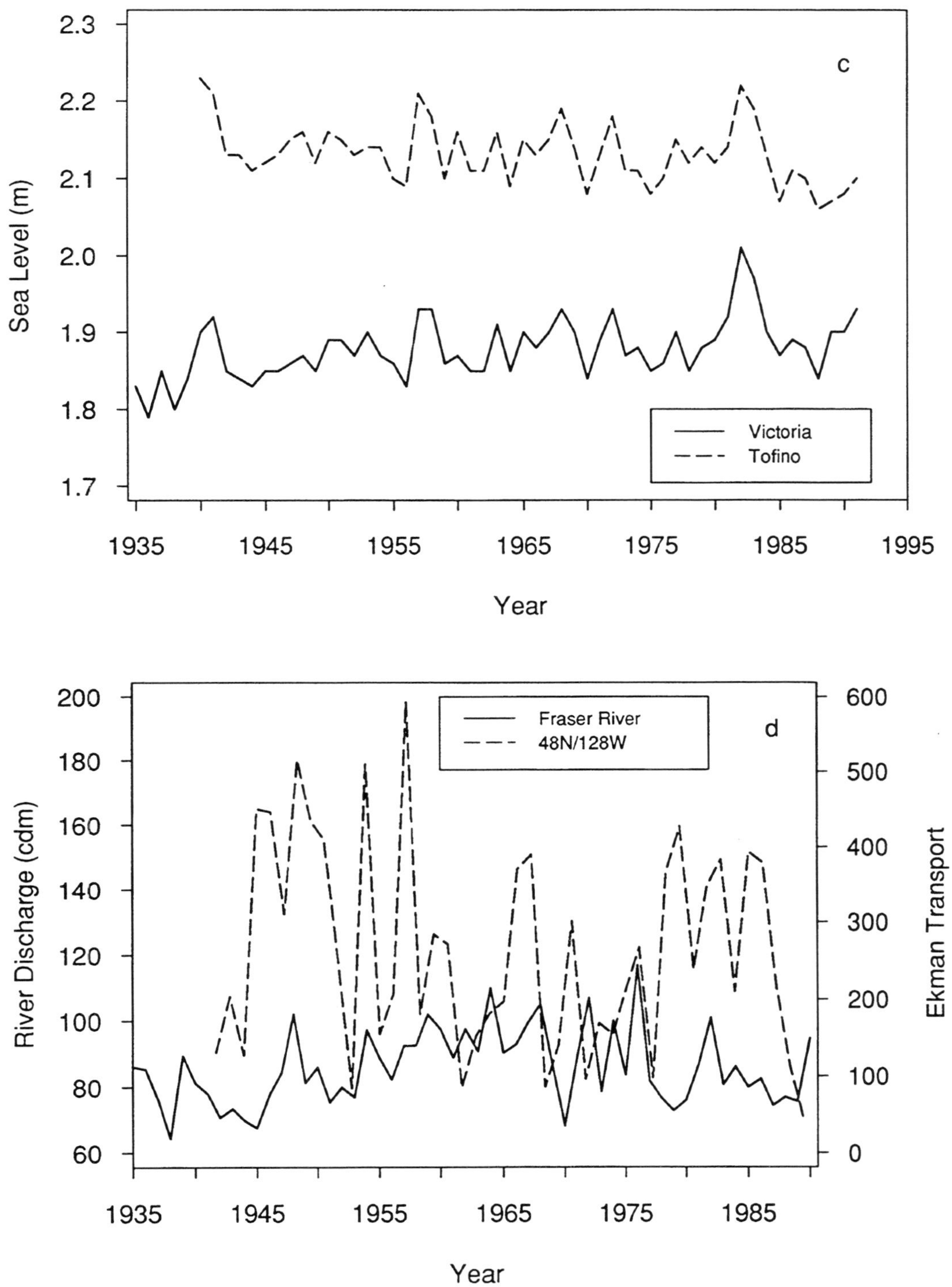

Fig. 6. (c) Long-term trend in sea level at Tofino and Victoria from 1935–91. (d) Long-term trend in Fraser River discharge and Ekman transport from 1935–90.

decline whereas the level at Victoria suggests an increasing trend throughout the 1940s and 1950s, remaining stable thereafter (Fig. 6c). Finally, Fraser River discharge increased from the 1930s to the mid-1970s and has decreased since, while Ekman transport has been above average in the 1950s and 1980s (Fig. 6d).

The relationship between the recruitment indices and various environmental variables was examined in detail but only a few are important. In both stocks, the strongest relationship is the inverse correlation between Amphitrite Point sea-surface temperature and the recruitment index (Figs. 7a and 7b; Table 1). Strait of Georgia recruitment is significantly

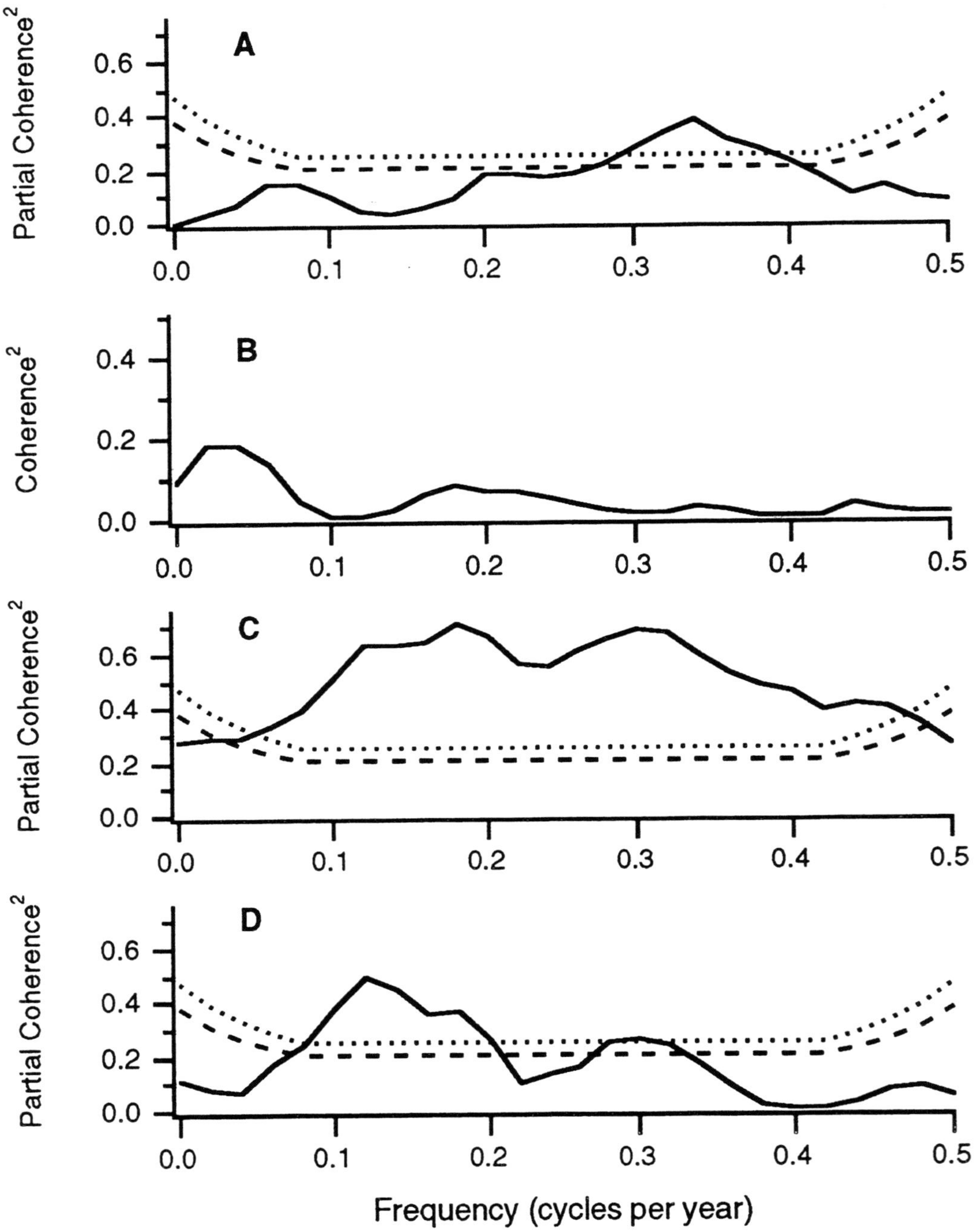

Fig. 7. (a) Relationship between recruitment index and Amphitrite Point sea-surface temperature for the 1935–88 year-classes of the herring stock in the Strait of Georgia. (b) Relationship between recruitment index and Amphitrite Point sea-surface temperature for the 1935–88 year-classes in the west coast of Vancouver Island stock.

inversely related only to sea-surface temperature at Amphitrite Point while WCVI recruitment is inversely related to Amphitrite sea-surface temperature and Pacific hake biomass (Table 1).

Statistically significant relationships were also observed among environmental variables and production, instantaneous growth rate, and the survival index for one or the other herring stock. Instantaneous population growth rate was significantly related to Amphitrite Point salinity, Fraser River discharge, and relative Pacific hake biomass in the SG stock and to Fraser River discharge and relative Pacific hake biomass in WCVI herring. The positive relationship with Fraser River discharge and relative hake biomass is shown in Figs. 8a and 8b. The negative relationship between Victoria sea level and the

Table 1. Models describing functional relationships among recruitment, growth rate, survival, and production for herring stocks on the west coast Vancouver Island (WCVI) and in the Strait of Georgia (SG).

Dependent variable	Model	*P*
SG recruitment	2462 – 173(Amphitrite temperature)	0.040
WCVI recruitment	1301 – 95(Amphitrite temperature) – 221(relative hake biomass)	0.016
SG growth rate	–4.08 + 0.46(relative hake biomass) + 0.01(Fraser discharge) + 0.14(Amphitrite salinity)	0.001
WCVI growth rate	0.12 + 0.50(relative hake biomass) + 0.004(Fraser discharge)	0.003
SG survival index	9.95 – 4.83(Victoria sea level)	0.006
WCVI survival index	1.78 if sardine present 1.33 if Sardine absent	0.006
SG adult production	–412 – 23(relative hake biomass) + 0.27 Fraser Discharge + 15 Amphitrite Salinity	0.006

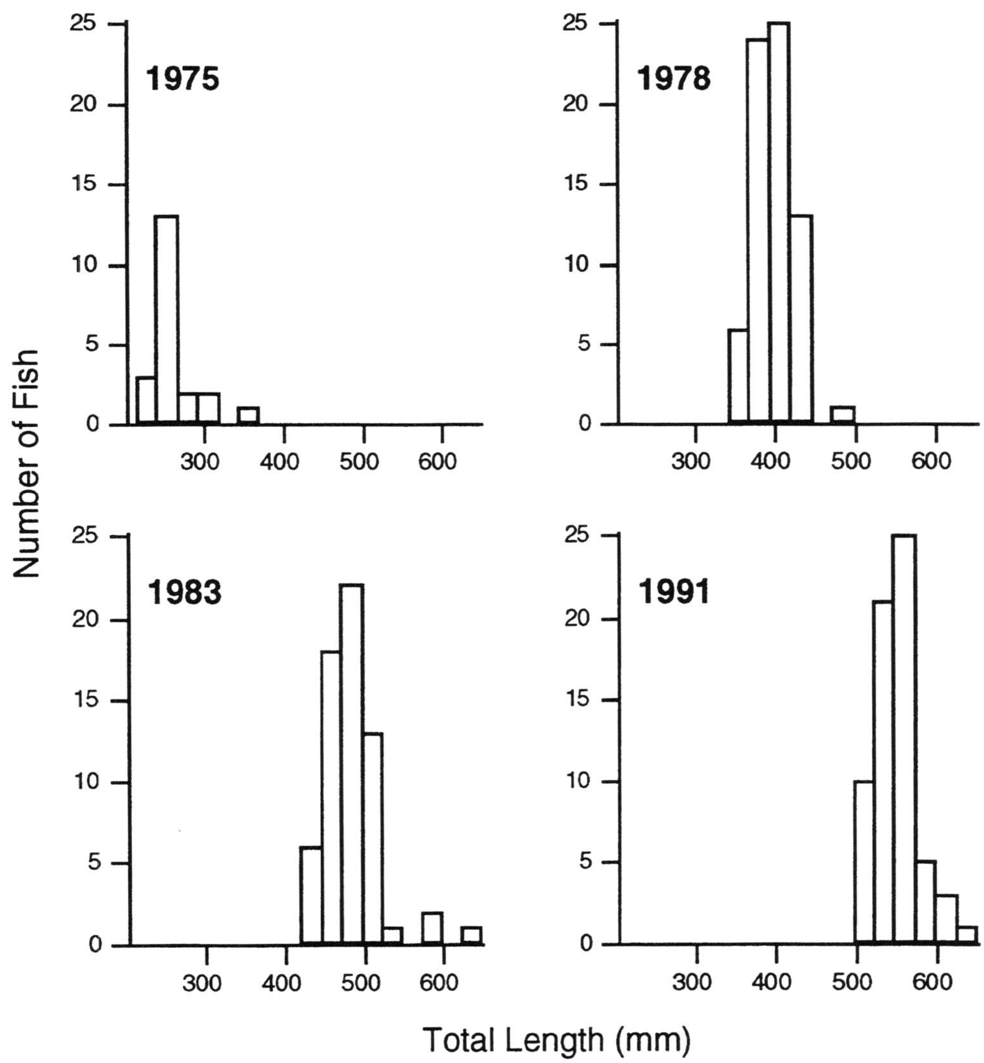

Fig. 8. (a) Relationship between instantaneous growth rate and Fraser River discharge for the 1935–88 year-classes in the Strait of Georgia stock. (b) Relationship between instantaneous growth rate and hake biomass for the 1935–88 year-classes in the west coast of Vancouver Island stock.

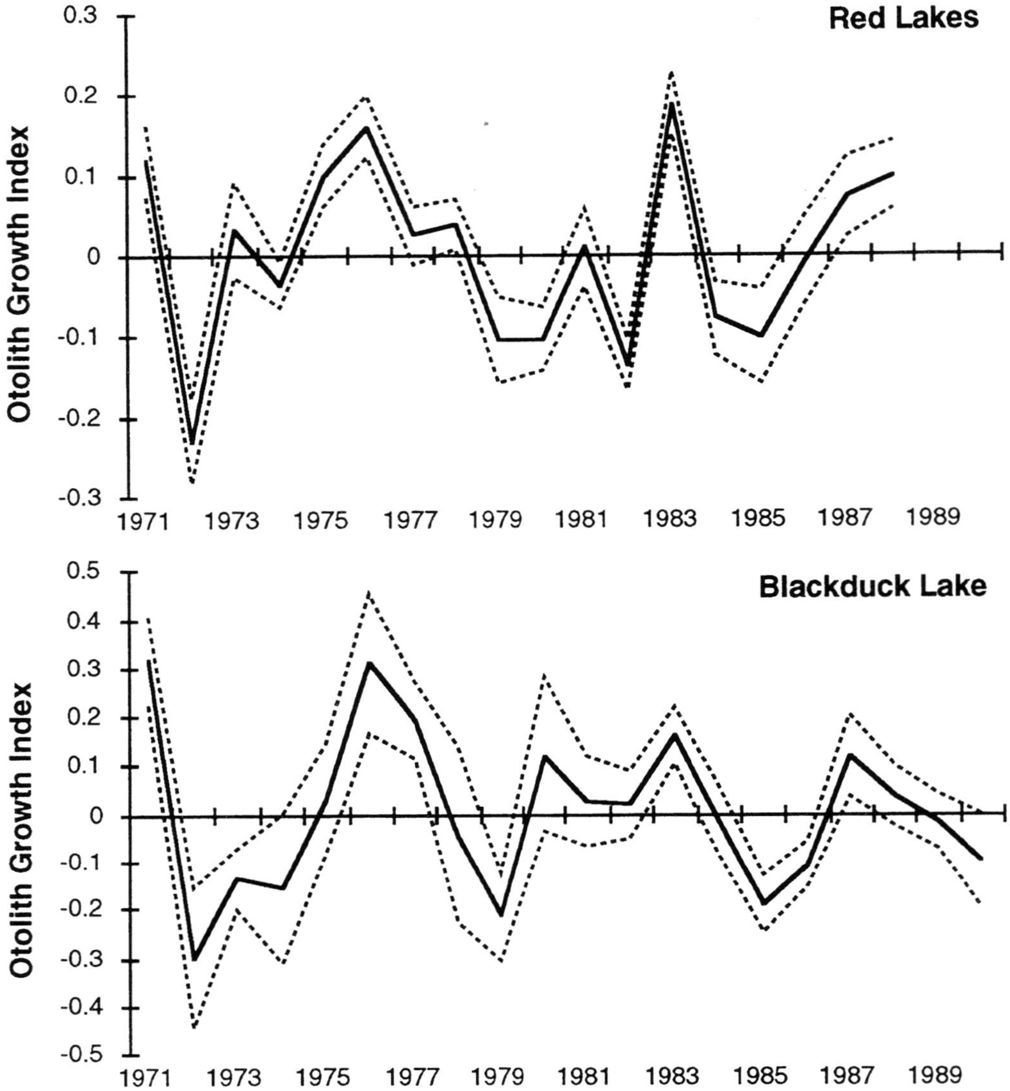

Fig. 9. (a) Relationship between survival index and Victoria sea level for the 1935–88 year-classes in the Strait of Georgia stock. (b) Relationship between total production and Amphitrite Point sea-surface salinity for the 1935–88 year-classes in the Strait of Georgia stock.

Table 2. Linear model relating herring recruitment in the Strait of Georgia (SG) and on the west coast of Vancouver Island (WCVI), Amphitrite Point sea-surface temperature (AMTMP) and El Niño – Southern Oscillation (ENSO) events for the 1935 to 1988 year-classes.

Dependent variable	Model parameters	$P>F$
SG recruitment	2754 – 204(AMTMP) + 0 (for non-ENSO years) + 110 (for strong ENSO) + 193 (for moderate ENSO)	0.085
WCVI Recruitment	1272 – 102(AMTMP) + 0 (for non-ENSO years) + 112 (for strong ENSO) – 47 (for moderate ENSO)	0.026

SG survival index is shown in Fig. 9a while the relationship of SG production and Amphitrite Point salinity is shown in Fig. 9b. The survival index for the WCVI was significantly related to whether or not sardines were present off southwest Vancouver Island (Table 1). No significant relationships were found to describe production for WCVI stocks with.

The impact of the ENSO events on recruitment in these stocks is statistically significant on the west coast of Vancouver Island but not in the Strait of Georgia. However, an examination of the parameter estimates suggests similar effects in both areas (Table 2). In both stocks, strong or very strong ENSO events have a positive effect on recruitment whereas moderate events appear to result in negative impacts on recruitment, and these effects are relatively more significant in the WCVI stock.

Discussion

The assertion that environmental factors are important in the determination of population fluctuations in marine species has been debated at various times (Ketchen 1956; Bell and Pruter 1958; Skud 1975). The understanding of the factors that determine fluctuations in herring abundance or recruitment has been the focus of research efforts on the Pacific coast of North America since the early 1930s (Tester 1948). These early studies demonstrated that year-class strength in Prince William Sound, Alaska, was positively related to air temperature in March to June (Rounsefell 1930, cited in Tester 1948), while it was inversely related to the July air temperature for WCVI herring stocks. This finding of conflicting evidence for an environmental relationship with year-class strength was not readily reconciled. Several subsequent studies of recruitment have also found conflicting relationships. Stocker et al. (1985) reported significant correlations between the residuals from a stock–recruitment curve and sea-surface temperature, Fraser River discharge, sea level at Point Atkinson, and total hours of sunlight in Nanaimo for SG herring stocks from 1951–80. Stocker and Noakes (1988) reported similar results for the WCVI stock, with sea-surface temperature, salinity, and sea level providing significant correlations with recruitment estimates for 1951–82. Schweigert and Noakes (1991) found Ekman transport, Tofino sea level, and Entrance Island salinity to be important predictors of recruitment for SG stocks and Amphitrite Point sea-surface temperature and Tofino sea level to be important for WCVI stocks.

Two interesting questions are why recruitment in these two stocks should be related to such an array of environmental variables and how they describe the underlying mechanism that determines recruitment success. Biologically, the recruitment process in these two stocks is somewhat different. On the west coast of Vancouver Island, most of the egg deposition occurs within Barkley Sound. The resulting larvae and juveniles spend their first summer inshore and apparently begin to migrate in their first fall or winter to offshore feeding grounds (Hourston and Haegele 1980). Stocks in Georgia Strait follow a similar pattern, but many juveniles apparently do not migrate out to the west coast of Vancouver Island until some time in their second summer. Thereafter, both stocks appear to intermingle on the offshore feeding grounds during the summer and subsequently return inshore on their fall and winter spawning migrations. Hence, unless recruitment success is determined exclusively during the first year of life, there should be some correlation between recruitment patterns in the two stocks, which has been shown by Ware and McFarlane (1989). In this study, the strong correlation in recruitment between the two stocks under average conditions is confirmed, but apparently it does not hold true for the strongest year-classes (Fig. 3a).

It is evident that the year-classes that are not well correlated between the stocks and are outliers in the relationship with sea-surface temperature are all year-classes associated with unusual oceanographic conditions (Figs. 3a, 7a and 7b). The strongest year-classes in the SG stock are the 1939, 1950, 1951, 1956, 1959 and 1974 year-classes, while the weakest are the 1963, 1965, 1966, 1981, 1986, and 1988 year-classes. Similarly, for the west coast of Vancouver Island the strongest year-classes are the 1944, 1956, 1959, 1972, 1983, and 1985 ones while the weakest are the 1941, 1963–66, 1976, 1980, and 1988 ones. The majority of these year-classes are produced in years of, or adjacent to, those characterized as moderate (1939, 1943, 1951, 1965, 1976, and 1987) to strong (1940–41, 1957–58, 1972–73, and 1982–83) ENSO events (Quinn et al. 1987). At present, there is no clear understanding of the mechanism that could lead to this relationship. Mysak (1986) speculates that moderate to strong El Niño years correspond to stormy conditions and high winter salinities, resulting in accentuation of the spring plankton bloom leading to enhanced survival of young herring.

Ware (1991) finds that recruitment is significantly related to sea-surface temperature and Pacific hake abundance and argues that predation pressure is directly related to the extent of northward migration of the hake stocks, which is a function of temperature. Presumably, their consumption rate would also be directly related to water temperature. Interestingly, this latter relationship is also evident in the longer time series (Table 1). However, hake abundance is not significantly related to recruitment for SG stocks and one would expect that given the strong correlation in recruitment between these two stocks, predation pressure should also be a prime determinant of year-class strength for SG stocks, unless recruitment is largely determined by the time juvenile herring migrate offshore rather than being modified by predation continuously throughout the prerecruitment stage. In this scenario, very strong year-classes must effectively swamp any depensatory predation effects.

Haist and Stocker (1985) investigated the relationship between juvenile herring growth rate and environmental variables for SG stocks. They identified an optimal summer temperature of 12.2°C at Amphitrite Point as the prime correlate of growth rate. Positive relationships of temperature and growth rate have been identified for many species particularly in fresh water (Le Cren 1958; Brett 1979). However, the mechanism involved is not well understood. In this study, the major factors determining instantaneous growth rates in SG and WCVI stocks were the Fraser River discharge and the relative hake biomass. There is also an inverse relationship between growth rate and Amphitrite Point salinity for the SG herring stock. Conversely, Ware (1991) reports a negative relationship between herring weight at age 4 and sea-surface temperature and summer salinity, which is suggested to be the result of higher plankton biomass associated with colder water temperatures. No explanation for these conflicting results is apparent although the instantaneous growth rate in this study is a measure of both survival and population growth rather than individual growth.

The estimated total production rates for the Strait of Georgia are about 20 000 – 30 000 t annually while for the WCVI they average about 10 000 – 20 000 t annually. These estimates are lower than the catches sustained by these stocks over the period from 1930–65. In fact, they are below the levels currently sustained by these stocks. No obvious explanation presents itself unless we accept that estimates of total stock biomass prior to 1970 were significantly lower than actual stock levels. Because much of the total production prior to 1970 was attributable to the influx of new recruits, it appears likely that estimates of the biomass of age-3 fish are too low. One potential reason for this could be the fact that reduction fisheries concentrated on the fall migration routes and likely would have targetted more on the adult components of the stocks. It is known that recruit spawners move to the spawning grounds some time after repeat spawners. If this pattern also applies during the inshore migration then age composition estimates from the fall and winter reduction fishery could significantly underestimate the relative abundance of recruits and also their total production.

The current study also identified a statistically significant difference in survival rate prior to and after 1950 when sardine had disappeared from British Columbia waters for the WCVI stock. However, this effect does not hold for SG fish whose survival appears to be inversely related to Victoria sea level, which is an indicator of the intensity of upwelling. As such, it is difficult to understand what the underlying mechanism might be. The large biomasses of sardines that annually migrated into British Columbia waters must have had some impacts on the zooplankton community to the detriment of other pelagic species in the area. On the positive side, they would also have acted as a buffer to hake predation, which is thought to have a significant impact on herring survival on the west coast of Vancouver Island (Ware 1991). In fact, it is tempting to speculate that the large recruitments that occurred in the WCVI stock the late 1930s and early 1940s were a direct result of the reduced competitive pressures of dwindling sardine stocks. Also, more intense upwelling could result in greater plankton production and enhance feeding success and survival for SG herring.

The relative importance of climatic variation and fishery induced or ecosystem driven changes in the abundance of various species has been debated for many years (Skud 1975). In the current study, herring population fluctuations are shown to be affected largely by variations in recruitment. Recruitment variation is inversely related to sea-surface temperature under average conditions. However, major oceanographic changes that occur either prior to or just after ENSO events are apparently either extremely favourable or detrimental to herring survival since both unusually strong and weak herring year-classes are coincident with these events. The mechanism underlying this result is not currently understood although more sophisticated statistical procedures may clarify some of these effects. Long-term increases in global temperatures appear to imply adverse affects on herring stock abundance in southern British Columbia, which may be ameliorated intermittantly through the impacts of moderate to strong ENSO events. There is also evidence that biotic factors such as hake or sardine abundance may have important effects on the community structure and population production, which may be related to the associated changes in oceanographic conditions. Skud (1982) has argued that populations may respond to environmental variation differently depending on the relative abundance of competitors. Given the current focus on the relative importance of environmental forces on stock productivity a greater understanding of the role of interspecific relationships in determining population abundance as a function of changing environmental conditions is critical to predicting the effects of long-term climatic changes on herring production in southern British Columbia. These questions can only be addressed by revisiting our historical data sets to reconstruct herring abundances to the turn of the century when Pacific sardines were more abundant in British Columbia waters and by monitoring the impacts of the currently rebuilding stocks of sardines off the west coast of North America on the two major herring stocks in southern British Columbia.

Acknowledgements

I am grateful to E. Taylor from the Co-operative Education Program at Simon Fraser University for assistance in collating the various data sets used in this study. My thanks to D. Ware for providing the hake biomass time series. The reviewer's comments and suggestions are greatly appreciated.

References

Bell, F.H., and A.T. Pruter. 1958. Climatic temperature changes and commercial yields of some marine fisheries. J. Fish. Res. Board Can. 15: 625–683.

Brett, J.R. 1979. Environmental factors and growth, p. 599–675. *In* W.S. Hoar, D.J. Randall, and J.R. Brett [ed.] Fish Physiology. Vol. 8. Academic Press, New York, NY.

Chapman, D.W. 1968. Production, p. 182–196. *In* W.E. Ricker [ed.] Methods for the assessement of fish production in fresh waters. Blackwell Scientific Publications, Oxford and Edinburgh, UK. IBP Handbo. 3.

Fry, F.E.J. 1949. Statistics of a lake trout fishery. Biometrics 5: 26–67.

Haegele, C.W., and J.F. Schweigert. 1985. Distribution and characteristics of herring spawning grounds and description of spawning behaviour. Can. J. Fish. Aquat. Sci. 42(Suppl. 1): 39–55.

Haist, V., and M. Stocker. 1985. Growth and maturation of Pacific herring (*Clupea harengus pallasi*) in the Strait of Georgia. Can. J. Fish. Aquat. Sci. 42(Suppl. 1): 138–146.

Hay, D.E., and A.R. Kronlund. 1987. Factors affecting the abundance and measurement of Pacific herring spawn. Can. J. Fish. Aquat. Sci. 44: 1181–1194.

Hourston, A.S., and C.W. Haegele. 1980. Herring on Canada's Pacific coast. Can. Spec. Publ. Fish. Aquat. Sci. 48: 23 p.

Ketchen, K.S. 1956. Climatic trends and fluctuations in yield of marine fisheries of the northeast Pacific. J. Fish. Res. Board Can. 13: 357–374.

Le Cren, E.D. 1958. Observations on the growth of perch (*Perca fluviatilis* L.) over twenty-two years with special reference to the effects of temperature and changes in population density. J. Anim. Ecol. 27: 287–334.

Mysak, L.A. 1986. El Niño, interannual variability and fisheries in the northeast Pacific Ocean. Can. J. Fish. Aquat. Sci. 43: 464–497.

Quinn, W.H., V.T. Neal, and S.E. Antunez de Mayola. 1987. El Niño occurences over the past four and a half centuries. J. Geophys. Res. 92: 14 449 – 14 461.

Schweigert, J.F. 1988. Status of the Pacific sardine, *Sardinops sagax*, in Canada. Can. Field-Nat. 102(2): 296–303.

Schweigert, J.F. 1993a. A review and evaluation of methodology for estimating Pacific herring egg deposition. Bull. Mar. Sci. 53: 880–923.

Schweigert, J.F. 1993b. Evaluation of harvesting policies for the management of Pacific herring (*Clupea pallasi*) stocks in British Columbia. *In* G. Kruse, D.M. Eggers, R.J. Marasco, C. Pautzke, and T.J. Quinn II [eds.] Proceedings of the International Symposium on Management Strategies for Exploited Fish Populations. Oct. 21–24, 1992, Anchorage, AK. Report #93-02. p. 167–190.

Schweigert, J.F., and D.J. Noakes. 1991. Forecasting Pacific herring (*Clupea harengus pallasi*) recruitment from spawner abundance and environmental information, p. 373–387. Proceedings of the International Herring Symposium. Oct. 23–25, 1990. Anchorage, Alaska. Report #91-01.

Schweigert, J.F., and M. Stocker. 1988. Escapement model for estimating Pacific herring stock size from spawn survey data and its management implications. N. Am. J. Fish. Manage. 8: 63–74.

Skud, B.E. 1975. Revised estimates of halibut abundance and the Thompson–Burkenroad debate. Int. Pac. Halibut Comm. Sci. Rep. 56: 36 p.

Skud, B.E. 1982. Dominance in fishes: the relation between environment and abundance. Science (Wash., D.C.) 216: 144–149.

Stocker, M., V. Haist, and D. Fournier. 1985. Environmental variation and recruitment of Pacific herring (*Clupea harengus pallasi*) in the Strait of Georgia. Can. J. Fish. Aquat. Sci. 42(Suppl. 1): 174–180.

Stocker, M., and D.J. Noakes. 1988. Evaluating forecasting procedures for predicting Pacific herring (*Clupea harengus pallasi*) recruitment in British Columbia. Can. J. Fish. Aquat. Sci. 45: 928–935.

Taylor, F.H.C. 1964. Life history and present status of British Columbia herring stocks. Bull. Fish. Res. Board Can. 143: 81 p.

Tester, A.L. 1948. The efficacy of catch limitations in regulating the British Columbia herring fishery. Trans. R. Soc. Can. 42: 135–163.

Tester, A.L. 1955. Estimation of recruitment and natural mortality rate from age-composition and catch data in British Columbia herring populations. J. Fish. Res. Board Can. 12(5): 649–681.

Ware, D.W. 1991. Climate, predators and prey: behaviour of a linked oscillating system, p. 279–291. *In* T. Kawasaki et al. [ed.] Long-term variability of pelagic fish populations and their environment. Pergamon Press, Tokyo, Japan.

Ware, D.W., and G.A. McFarlane. 1989. Fisheries production domains in the northeast Pacific Ocean, p. 359–379. *In* R.J. Beamish and G.A. McFarlane [ed.] Effects of ocean variability on recruitment and an evaluation of parameters used in stock assessment models. Can. Spec. Publ. Fish. Aquat. Sci. 108.

Marine fish production trends off the Pacific coast of Canada and the United States

R. J. Beamish and D.R. Bouillon[1]

Beamish, R.J., and D.R. Bouillon. 1995. Marine fish production trends off the Pacific coast of Canada and the United States, p. 585–591. *In* R.J. Beamish [ed.] Climate change and northern fish populations. Can. Spec. Publ. Fish. Aquat. Sci. 121.

Abstract: Trends in production were reviewed for salmon and nonsalmon species in the principal commercial fisheries off the Pacific coast of Canada and the United States. Salmon species included pink (*Oncorhynchus gorbuscha*), chum (*O. keta*), sockeye (*O. nerka*), and nonsalmon species included demersal fishes and small pelagic species. The catches of the three species of salmon from all the Pacific salmon producing countries showed parallel trends, which varied almost three times from the period of high catches to the period of low catches. The average nonsalmon production as indicated by annual recruitment patterns also showed a trend, despite some variation in the individual species response. The combined salmon and nonsalmon production trend follows a pattern similar to the Aleutian Low Pressure Index, an index of winter weather over the northern North Pacific Ocean; this indicates a linkage between long-term abundance trends for most of the major fisheries off the west coast of Canada and the United States and climate. Because the index of climate trends shows a fluctuating trend, it appears that in general, carrying capacities for the marine fishes in this study, also fluctuate.

Résumé : Les tendances dans la production du saumon et des non-salmonidés des principales pêcheries commerciales au large des côtes canadienne et américaine du Pacifique ont été examinées. Les espèces de saumons étudiées comprenaient le saumon rose (*Oncorhynchus gorbuscha*), le kéta (*O. keta*) et le sockeye (*O. nerka*), tandis que les non-salmonidés comprenaient des poissons démersaux et de petites espèces pélagiques. Les prises des trois espèces de saumons de tous les pays producteurs de saumon du Pacifique montraient des tendances parallèles qui variaient presque trois fois entre la période des fortes prises et la période des faibles prises. La production moyenne de non-salmonidés, exprimée par les tendances annuelles du recrutement, faisait également état d'une tendance, malgré certaines variations dans la réaction de chaque espèce. La tendance de la production combinée de saumons et de non-salmonidés a un comportement similaire à l'indice de la dépression des Aléoutiennes, un indice de température hivernale dans le nord du Pacifique Nord, ce qui indique la présence d'un lien entre les tendances de l'abondance à long terme pour la plupart des pêcheries majeures au large de la côte ouest du Canada et des États-Unis et le climat. Étant donné que l'indice des tendances du climat accuse des fluctuations, il semble qu'en général, les capacités d'accueil pour les poissons marins visée par l'étude fluctuent également.

Introduction

In this report, we review the results of two previous studies relating climate to fish production in the northern North Pacific Ocean and the Bering Sea. In one study (Beamish and Bouillon 1993), we showed that the trend of production of three species of Pacific salmon from 1925 to 1989 was approximately parallel to a long-term trend in an index of the Aleutian Low pressure system. In the other study, exceptional climate conditions from 1976 to 1978 were observed to be closely associated with synchronous production of strong year-classes of nonsalmon species (Beamish 1994). After reviewing the results of these two studies, we extend the data series for nonsalmon species to show that there is a similarity in the response of salmon and nonsalmon species that indicates that the dynamics of the major fish populations off the west coast of Canada and the United States can be closely associated with climate and the ocean environment.

Pacific salmon production trends

Beamish and Bouillon (1993) examined the trends of the all-nation catch of pink (*Oncorhynchus gorbuscha*), chum (*O. keta*), and sockeye salmon (*O. nerka*). In this earlier study, we excluded coho (*O. kisutch*) and chinook salmon (*O. tschawytscha*) from our analysis because they make up only about 10% of the catch, and in recent years there has been an extensive effort to rear these species artificially, making it

R.J. Beamish and **D.R. Bouillon.** Department of Fisheries and Oceans, Biological Sciences Branch, Pacific Biological Station, Nanaimo, BC V9R 5K6, Canada.

[1]Present address: ESSA (Environmental and Social Systems Analysts), 3rd Floor, 1765 West 8th Avenue, Vancouver, BC V6J 5C6, Canada.

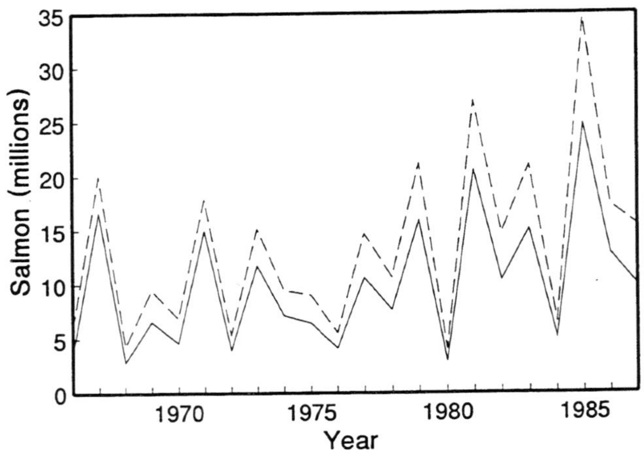

Fig. 1. Catch (solid line) and catch plus escapement (broken line) of pink, chum, and sockeye salmon produced in the Fraser River, British Columbia.

difficult to interpret abundance trends. Estimates of total returns of all stocks of pink, chum, and sockeye salmon to all rivers were not available because the fish that escape the coastal fisheries and reach their spawning ground are typically only enumerated in the largest or most important river systems. It is generally accepted, however, that 65–95% of all returning salmon are caught in the coastal fisheries. Commercial catch was therefore used as an index of total returns.

We tested this assumption using catch and escapement data for the Fraser River of British Columbia, one of the largest Pacific salmon producing rivers in the world. A reliable time series of catches exists because catch and escapement estimates for pink and sockeye salmon have been used by an international commission to manage Canadian and United States fisheries for these species. Estimates of catch of chum salmon are considered reliable because stocks are close to the southern limit of their distribution and catches of stocks from other drainages would be minimal in the coastal fisheries.

Sockeye salmon data from 1965 to 1989 were obtained from the International Pacific Salmon Fisheries Commission annual reports (IPSFC 1965–1985) and from the reports on the Fraser River sockeye salmon fishing season by the Fraser River Panel of the Pacific Salmon Commission (PSC 1986–1989). Pink salmon data were from the report on the 1989 Fraser River sockeye and pink salmon fishing season by the Fraser River Panel to the Pacific Salmon Commission (PSC 1990). Chum salmon data from 1966 to 1979 were from Beacham and Starr (1982). Chum salmon catches and escapements from 1980 to 1989 were from unpublished records of the Department of Fisheries and Oceans, Pacific Region, New Westminster, B.C.

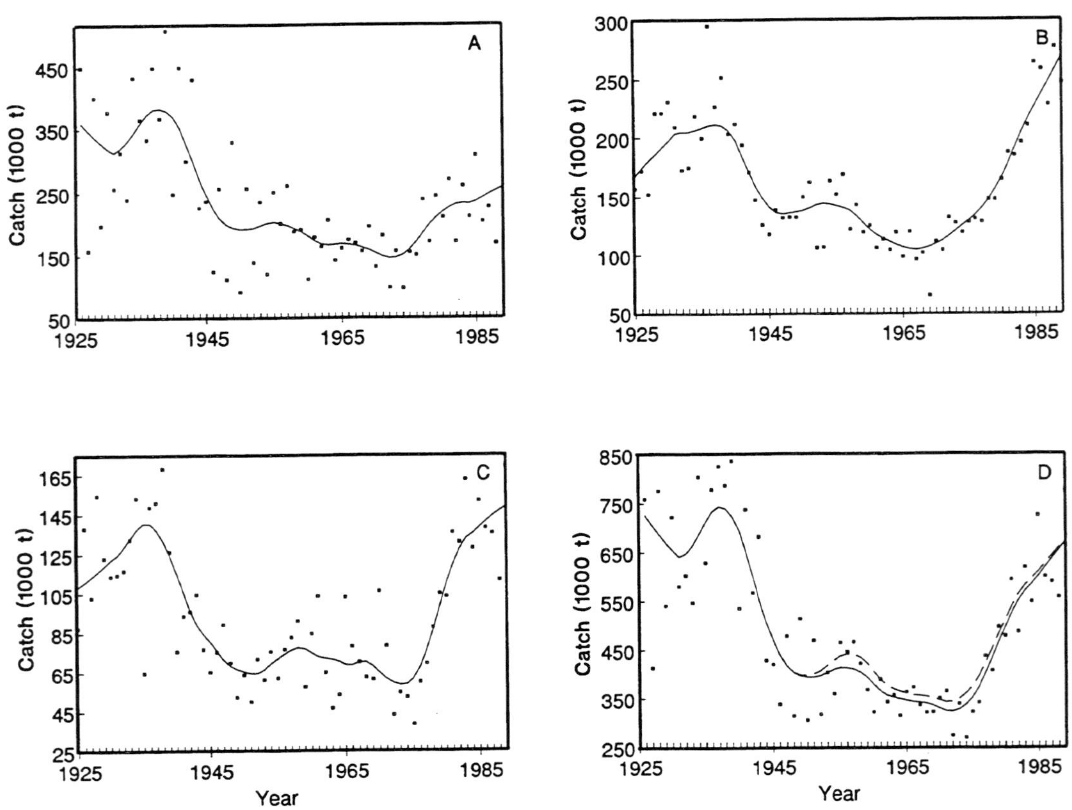

Fig. 2. Trends in catches of combined all-nation catches of (A) pink salmon, (B) chum salmon, (C) sockeye salmon, and (D) pink, chum, and sockeye salmon combined. Data points are actual observed coastal catch values. The broken line in (D) represents the expected increase in coastal catch if there was no high seas fishery. The solid line is the LOWESS smoother fit, using a band width (*f*) of 0.20.

(M. Joyce, personal communication). From the data, it was apparent that there is a close relationship between catch and catch plus escapement (Fig. 1). Thus we believe that, for Pacific salmon, catch is a reasonable indictor of abundance.

Catch data were collected for all Pacific salmon stocks from 1925 to 1989 for each of pink, chum, and sockeye salmon (see Beamish and Bouillon 1993 for data sources). The annual catch data were then smoothed using a LOWESS smoother (Fig. 2). The trends in pink (Fig. 2a), chum (Fig. 2b), and sockeye (Fig. 2c) abundance from 1925 to 1989 were very similar and, therefore, were combined to produce one trend of Pacific salmon abundance (Fig. 2d). Before production trends could be determined the total catch had to be corrected for lost coastal catches caused by Japanese high-seas driftnet fishing, which caught Pacific salmon originating from Russia, the United States, and Canada (Beamish and Bouillon 1993).

The smoothed trend (Fig. 2d) shows that abundance was high in the 1930s, declined until the mid-1950s when there was a small increase, and then declined to a historic low in the early 1970s. Beginning about 1977, the abundance increased, reaching levels almost three times the early 1972 level by the late 1980s. In the early 1990s abundance has declined slightly (R. Beamish, unpublished data), but it is too early to determine if the trend has changed. Once the trend changes, we would expect the decline to continue for a number of years as in previous trends.

Beamish and Bouillon (1993) developed an index of climate over the North Pacific Ocean. The climate index was based on the winter low pressure system referred to as the Aleutian Low Pressure System. This system develops in the fall of each year and breaks down about spring of the following year, when high pressure systems become established. The intensity of the system varies from year to year. We hypothesized that years of intense lows (a very large area of low pressure) would be years of above-average production, while years of less intense lows (a smaller area of low pressure or higher winter pressures) would be years of lower productivity. We proposed that intense lows were linked with increased salmon production through an increase in plankton abundance, in general, and copepod abundance, in particular.

The smoothed trend in all-nation salmon catch of all three species (Fig. 2d) showed a pattern very similar to the smoothed trend in the climate index (Figs. 3a and 3b). Periods of increasing low pressures were related to periods of increasing salmon abundance. Periods of declining low pressures (or increasing high pressures) were synchronous with periods of declining abundance (Fig. 3b). Beamish and Bouillon (1993) did not find that the relationship was significant ($P \leq 0.05$) because of the amount of variation associated with the data. However, the almost synchronous response in the salmon abundance trends and the smoothed climate index is strong evidence that a relationship exists between these two indices.

Reid (1962) proposed that horizontal divergence in the upper layer in the Alaskan Gyre brings nutrients into the mixed layer, and the horizontal divergence transports plankton and nutrients towards the edge of the gyre. An increase in the intensity of the Aleutian Low would intensify the winter winds, resulting in increased upwelling in the centre of the gyre (Thomson 1981) bringing nutrients into the mixed layer. Recently, it was shown that nitrate concentrations increase in the surface waters in the Northeast Pacific Ocean during intense Aleutian Lows (C.S. Wong, Institute of Ocean Sciences, Sidney, B.C., personal communication). Because nitrate in the surface water can be an indication of new production, an intensification of the Aleutian Low could increase primary production.

Copepods dominate the zooplankton biomass of the northeast Pacific Ocean and are a major item in the diet of marine fishes, particularly young fishes, including salmon. In one study, copepods accounted for 88–98% of the plankton in the upper 300 m (McFarlane and Beamish 1992). At other depths, copepods accounted for 63–84% of the plankton. Beamish and Bouillon (1993) identified a significant relationship between copepod abundance and the Aleutian Low Pressure Index (Fig. 4). Along the coast of North America, there was an increase in the abundance of plankton beginning in the late 1970s (Brodeur and Ware 1992) at about the same time as the intensification of the Aleutian Low indicating that the plankton abundance increases occurred over a large area. A relationship between the Aleutian Low and salmon production

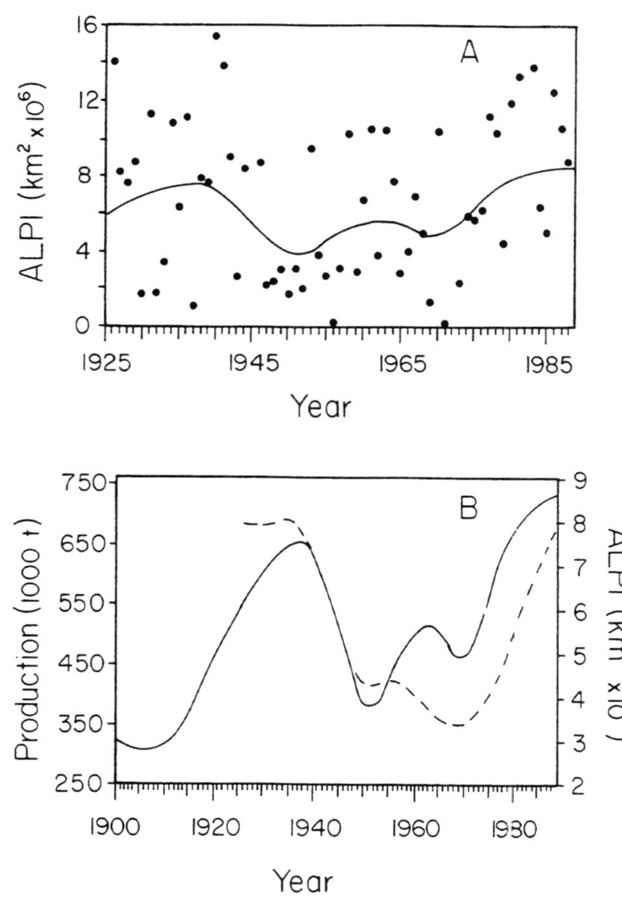

Fig. 3. (A) The smoothed trend of the Aleutian Low Pressure Index from Beamish and Bouillon (1993). The data points are the actual index values for the 6-mo period December to May, and the solid line is the LOWESS smoother fit, using a band width (*f*) of 0.20. (B) The relationship between the smoothed trend (*f* of 0.25) in total all-nation Pacific salmon abundance (broken line) and the Aleutian Low Pressure Index (solid line).

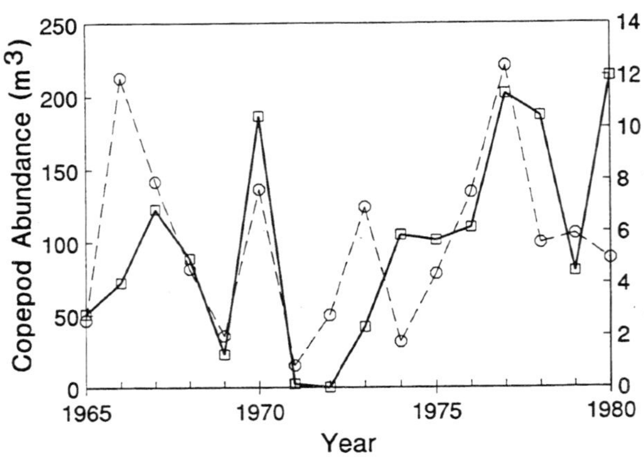

Fig. 4. The relationship (significant, $r = 0.50$; $p < 0.05$) between mean monthly copepod abundance (March–May) at Ocean Station P (broken line) in numbers per m^3 and the Aleutian Low Pressure Index (solid line) for December–May combined (from Beamish and Bouillon 1993).

is plausible because the Aleutian Low affects such a large area of the northern North Pacific Ocean, and there is a relationship between changes in the intensity of the low and the production of copepods, which are a main source of food for many fishes. The large variation in catch (Fig. 2d) over the time series, which was synchronous among species, would not be expected to result from synchronous management actions in all salmon-producing countries. A more likely explanation is that some other factor produced the synchronous abundance changes. We believe that the increase in plankton abundance and the similarity in the pattern of the salmon production and climate trends indicates that climate has a profound effect on the abundance of Pacific salmon.

Nonsalmon production trends

We use the term nonsalmon species to refer to groundfish species such as Pacific halibut (*Hippoglossus stenolepsis*) and Pacific cod (*Gadus macrocephalus*), as well as small pelagic species such as Pacific herring (*Clupea harengus*) and Pacific hake (*Merluccius productus*) that are typically found near the surface or in the midwater (Table 1). Studying the influence of climate on nonsalmon species is more difficult than studying the effects on salmon because the data bases for nonsalmon species tend to be short and, in some years, the data quality is poor. Some fisheries, such as Pacific halibut, have well-maintained data bases back to the 1920s, but most other data bases start in the 1960s and 1970s when major domestic fisheries began. Unfortunately, age estimates for fish during the early days of some fisheries can be unreliable (Chilton and Beamish 1982), making it difficult to use age structured models to estimate biomass. As more accurate age determination estimates became available in the late 1970s and early 1980s, more accurate estimates of age-specific biomass became available but these biomass data series are short.

Another difficulty with data series of some nonsalmon species is the tendency of some species to be long lived and to be recruited into a fishery at relatively old ages. For example, Pacific halibut are first recruited into the fishery at age 8 and fully recruited by age 11. Recruitment indices for species that are long lived typically end in the early to middle 1980s, and biomass information, calculated using age-structured models, is incomplete until most of the age groups have passed through the fishery. Trends in production of most nonsalmon species, therefore, cannot be examined using measurements of total biomass, instead we used indices of year-class strength similar to the technique used by Ebbesmeyer et al. (1991).

In this study, we extended the data series in Beamish (1994) and examined the recruitment pattern of 12 species from 13 data series (Table 1). All the species used are important in the fisheries off the Pacific coast of Canada and the United States.

Table 1. Data series for nonsalmon fishes from Beamish (1993), but extended for Pacific halibut, Pacific ocean perch, Pacific mackerel, and Pacific herring from British Columbia, with a new series for Pacific herring from Alaska (from Funk and Savikko (1990)). Data are available from R.J. Beamish.

Species	Data series
Pacific halibut (*Hippoglossus stenolepis*)	1927–80
Pacific ocean perch (*Sebastes alutus*)	1927–81
Pacific mackerel (*Scomber japonicas*)	1927–68, 1975–82
Pacific herring, (*Clupea pallasi*)	
British Columbia	1948–86
Alaska	1975–84
Pacific hake (*Merluccius productus*)	1956–87
Sablefish (*Anaplopoma fimbria*)	1960–82
Walleye pollock (*Theragra chalcogramma*)	1962–83
Northern anchovy (*Engraulis mordax*)	1964–87
Widow rockfish (*Sebastes entomelas*)	1966–80
Atka mackerel (*Pleurogrammus monopterygius*)	1972–87
Yellowtail rockfish (*Sebastes flavidus*)	1974–82
Pacific cod (*Gadus macrocephalus*)	1975–85

The lengths of the data series vary, depending on the quality of the data, the length of the fishery, and the age that the species is fully recruited into a fishery. The data series were longest for Pacific halibut, Pacific ocean perch (*Sebastes alutus*), Pacific herring, and Pacific mackerel (*Scomber japonicus*).

Relative year-class strength was estimated for the combined data series using the method described by Ebbesmeyer et al. (1991). For each stock the annual year-class strength value was converted to a "nondimensional" anomaly (no units) using the relationship $[\ln(R) - (\Sigma\ln(R/n)]\ \mathrm{SD}(\ln(R))$, where R is the index of relative year-class strength. To produce the combined nondimensional anomaly index of recruitment strength, we added these individual anomalies together for each year, for all species, and divided by the sample size. The recruitment anomaly produced by this method is an index of average year-class production for the combined species that is based on the recruitment anomalies for each species for any particular year. The year always represents the year of hatching for marine fish.

The recruitment anomalies follow a fluctuating trend of above-average recruitment from the late 1920s to about 1944, a below-average recruitment from 1945 to 1950, a return to above-average recruitment from 1951 to 1961, and a long period of low recruitment from 1962 to 1975 (Fig. 5). The period 1976–78 was exceptionally good for recruitment, and the period in the early 1980s appears to be below average. Within the various trends, there are a few anomalous years, such as 1933 and 1956, when below-average recruitment occurs during a trend towards above-average production. Years of above-average recruitment during trends of below-average recruitment occurred in 1970 and 1984. The recruitment index is an average of good and poor recruitment, and the significance of the anomalous years within the trends is difficult to determine without a more comprehensive review of the data for each species.

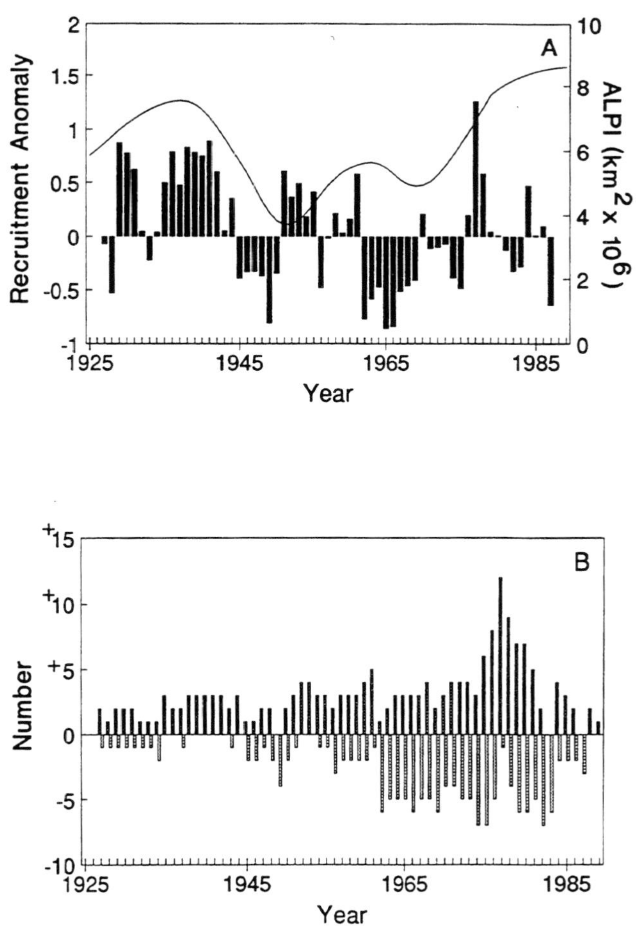

Fig. 5. (A) Recruitment anomalies (solid bars) for nonsalmon fishes and the Aleutian Low Pressure Index (solid line). (B) Number of above-average (+) and below-average (–) year-class survivals from 1927 to 1987.

The species used in the anomaly index have very different life histories, and are found in different oceanographic areas or regimes (Ware and McFarlane 1989). Despite these differences, there appears to be a general trend to the combined pattern of recruitment that indicates that there are periods of good and poor recruitment in nonsalmon species. The 1976–78 period is the most obvious example of a period of good recruitment. Beamish (1994) reported that in 1977, 10 of the species in the data series had positive anomalies and two had negative anomalies. In our study, the longer data series and the inclusion of Pacific herring from Alaska (the two herring fisheries were not combined because the populations have a very different age structure) resulted in only one (northern anchovy) of the 13 data series having a negative anomaly in 1977 (Fig. 5b). It is known that the late 1970s was a period of exceptional changes in the climate and ocean in the North Pacific (Royer 1989; Bakun 1990; Trenberth 1990; Ebbesmeyer et al. 1991).

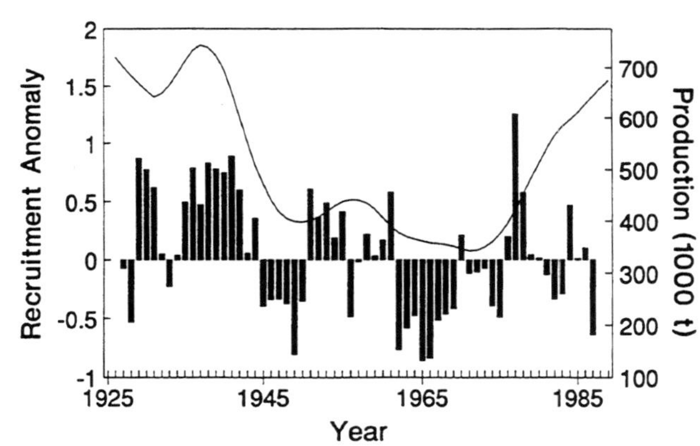

Fig. 6. Relationship between the smoothed Pacific salmon production trend index (solid line, *f* of 0.2) and the index of average year-class strength (recruitment anomalies) (bars) from 1925 to 1987.

There is a similarity between the trends in the recruitment anomaly and the pattern of the Aleutian Low Pressure Index (Fig. 5a). When the index is increasing (increasing low pressures), there are more species with above-average year-class survivals than below-average survival (Figs. 5a and 5b). When the index is decreasing (increasing high pressures), there is a tendency for survival to be below average (Figs. 5a and 5b). The synchrony of the changes is not exact, particularly in the early 1970s, but it is very similar. One problem with the recruitment anomaly index is that there are no data from the long-lived species in the mid-1980s, because they are just being recruited to fisheries, and the most recent data are seldom published. Thus, it is too early to conclude that the 1980s was a period of reduced recruitment. It is possible that direct measurements of biomass of nonsalmon species, which included commercial catch, would show an increasing trend in the 1980s. Walleye pollock catches, for example, increased dramatically in the 1980s to about 7 million or 8 million t and became the largest fishery in the world (Beamish and McFarlane 1994). The very large biomasses in the 1980s may have had a density-dependent influence on the strengths of subsequent year-classes resulting in fewer strong year classes, but it is difficult to know until the appropriate data are published.

Hollowed and Wooster (1992) identified alternating periods of strong and weak recruitment that lasted about 6–12 yr in some groundfish species. They showed that strong year-classes tended to occur at the onset of a warm period, and the northern stocks tend to be less productive during the cool periods. The periods of cool years (Wooster and Hollowed 1995) between 1946 and 1984 were 1946–52 and 1971–76, and years of synchronous strong year-classes were 1961, 1970, 1977, and 1984. We found, as did Wooster and Hollowed (1995) that 1945–50 and 1971–76 were periods of below-average production. However, we identified 1951 and 1952 as years of good production. We identified 1961 as a year that produced a large number of strong year-classes, but 1977 was the most outstanding year in the entire time series. Our study and the Wooster and Hollowed (1995) study both identified 1984 as a year of above average production for groundfish species.

Marine fish production trends

The salmon and nonsalmon species considered in this report represent most of the major marine fisheries off the Pacific coast of Canada and the United States. For both salmon and nonsalmon species, there was evidence of a long-term trend in production. When the two trends were compared (Fig. 6) there was an obvious similarity in response. Both salmon and nonsalmon species had above-average levels of survival from the late 1920s to the early 1940s, in the 1950s, and beginning in the late 1970s. The period from the mid-1940s to the early 1950s was a time of poor survival for both groups, as was the period from the early 1960s to the mid-1970s. In the 1980s, salmon biomass continued to increase while the number of nonsalmon species with strong year-classes may have been about average possibly for the reasons discussed earlier.

The Pacific salmon and nonsalmon data sets assembled here represent a large amount of data. The similarity in the production trends indicates that there is a pattern among the production trends of three Pacific salmon species, the year-class strength of 13 nonsalmon fish stocks, and the smoothed index of the Aleutian Low Pressure System over the North Pacific Ocean. The synchrony is not exact, but the relationship is not random. This indicates that there is an effect of the dominant winter atmospheric pressure system over the North Pacific Ocean, on the abundance of fish inhabiting the region.

The observation that climatic events are related to fish stock abundance suggests, at least on a gross scale, that there is a carrying capacity for fish in the northern North Pacific Ocean and the Bering Sea that fluctuates in response to long-term trends in climate. One obvious management implication of a fluctuating carrying capacity is that sustainable fishing levels must fluctuate. The concept of "rebuilding" stocks (or populations) to some historic level may be possible only during appropriate environmental conditions. We think it would be an oversimplification to assume that the relationship of recruitment with climate indicates that fishing is relatively unimportant, just as it is an oversimplification to assume that the abundance of stocks is relatively unrelated to climate trends. Understanding the relationships between trends in the environment and trends in marine fish production give us new insights into the mechanisms regulating fish populations that will enable us to better understand the population dynamics process and ultimately allows us to manage fisheries in the changing environment of the future.

References

Bakun, A. 1990. Global climate change and intensification of coastal ocean upwelling. Science (Wash., D.C.) 247: 198–201.

Beacham, T.D., and P. Starr. 1982. Population biology of chum salmon (*Oncorhynchus keta*) from the Fraser River, British Columbia. Fish. Bull. U.S. 80(4): 813–825.

Beamish, R.J. 1993. Climate change and exceptional fish production off the west coast of North America. Can. J. Fish. Aquat. Sci. 50: 2270–2291.

Beamish, R.J., and D.R. Bouillon. 1993. Pacific salmon production trends in relation to climate. Can. J. Fish. Aquat. Sci. 50: 1002–1016.

Beamish, R. J., and G.F. McFarlane. 1994. A discussion of the importance of aging errors, and an application to walleye pollock: the world's largest fishery. *In* Proceedings of the International Symposium on Fish Otolith Research and Application. (In press).

Brodeur, R.D., and D.M. Ware. 1992. Long-term variability in zooplankton biomass in the subarctic Pacific ocean. Fish. Oceanogr. 1: 32–38.

Chilton, D.E., and R.J. Beamish. 1982. Age determination methods for fishes studies by the groundfish program at the Pacific Biological Station. Can. Spec. Publ. Fish. Aquat. Sci. 60: 102 p.

Ebbesmeyer, C.C., D.R. Cayan, D.R. McLain, F.H. Nichols, D.H. Peterson, and K.T. Redmond. 1991. 1976 step in the Pacific climate: forty environmental changes between 1968–1975 and 1977–1984, p. 115–126. *In* J.L. Betancourt and V.L. Tharp [ed.] Proceedings of the Seventh Annual Pacific Climate (PACLIM) Workshop, April, 1990. California Department of Water Resources, Interagency Ecological Studies Program Tech. Rep. 26.

Funk, F., and H. Savikko. 1990. Preliminary forecasts of catch and stock abundance for 1990 Alaska herring fisheries. Alaska Dep. Fish Game Reg. Inf. Rep. 5J90-02: 92 p.

Hollowed, A.B., and W.S. Wooster. 1992. Variability of winter ocean conditions and strong year classes of Northeast Pacific groundfish. ICES Mar. Sci. Symp. 195: 433–444.

International Pacific Salmon Fisheries Commission (IPSFC). 1965–1985. Annual Reports. Vancouver, B.C.

McFarlane, G., and R.J. Beamish. 1992. Climatic influences linking copepod production with strong year-classes in sablefish, (*Anaplopoma fimbria*). Can. J. Fish. Aquat. Sci. 49: 743–753.

Pacific Salmon Commission (PSC). 1986–1989. Reports of the Fraser River Panels on the Fraser River Sockeye Salmon Fishing Season 1986–1989. Vancouver, B.C.

Pacific Salmon Commission (PSC). 1990. Report of the Fraser River panel to the Pacific Salmon Commission on the 1989 Fraser River sockeye and pink salmon fishing season. Vancouver, B.C.

Reid, J.L., Jr. 1962. On circulation, phosphate-phosphorus content, and zooplankton volumes in the upper part of the Pacific Ocean. Limnol. Oceanogr. 7: 287–306.

Royer, T.C. 1989. Upper ocean temperature variability in the northeast Pacific ocean: is it an indicator of global warming? J. Geophys. Res. 94: 18 175 – 18 183.

Thomson, R.E. 1981. Oceanography of the British Columbia coast. Can. Spec. Publ. Fish. Aquat. Sci. 56: 291 p.

Trenberth, K.E. 1990. Recent observed interdecadal climate changes, p. 91–95. *In* Preprints, Symposium on Global Change Systems. Special Sessions on Climate Variations and Hydrology, February 5–9, 1990, Anaheim, Calif. American Meteorological Society, Boston, MA.

Ware, D.M., and G.A. McFarlane. 1989. Fisheries production domains in the northeast Pacific Ocean, p. 359–379. *In* R.J. Beamish and G.A. McFarlane [ed.] Effects of ocean variability on recruitment and an evaluation of parameters used in stock assessment models. Can. Spec. Publ. Fish. Aquat. Sci. 108.

Wooster, W.S., and A.B. Hollowed. 199. Decadal scale variations in the eastern Subarctic Pacific: I. Winter ocean conditions, p. 81–85. *In* R.J. Beamish [ed.] Climate change and northern fish populations. Can. Spec. Publ. Fish. Aquat. Sci. 121.

Trends in the average size of Pacific salmon in Canadian catches

William E. Ricker

Ricker, W.E. 1995. Trends in the average size of Pacific salmon in Canadian catches, p. 593–602. *In* R.J. Beamish [ed.] Climate change and northern fish populations. Can. Spec. Publ. Fish. Aquat. Sci. 121.

Abstract: From 1951 through 1975 the pink (*Oncorhynchus gorbuscha*), coho (*O. kisutch*), and chinook salmon (*O. tshawytscha*) caught in British Columbia waters decreased in average weight in almost all sections of the province by 5–25%. During 1976–91 this trend slowed down, was arrested, or was reversed in most areas of the northern and central regions, and in the case of chinook salmon a loss of 3 kg or 25% has been fully restored in the far north, and less completely elsewhere. In the southern region, however, decrease in size has continued for coho and pink salmon, these species being now 20–40% lighter than in the 1950s. Sockeye (*O. nerka*) and chum salmon (*O. keta*) have had no sustained trends in size since 1951. As yet none of the observed changes in size have been shown to be correlated with changes in environmental and other factors over the long term, but much additional study is needed. A poor growth rate can be related to unusual abundance of pink salmon in local regions. Effects of selective fishing on heritable aspects of growth rate and age of maturity may play a role in some cases, although the evidence for this is circumstantial. However, it provides a possible explanation for the absence of sustained trends among sockeye and chum salmon, because the 2 or 3 ages present in their stocks make the effect of selection by size ambiguous and unpredictable: maturing fish that have spent any given number of years in the ocean are smaller, but have grown faster, than those that have spent any greater number of years there.

Résumé : De 1951 à 1975, les prises de saumon rose (*Oncorhynchus gorbuscha*), de coho (*O. kisutch*) et de quinnat (*O. tshawytscha*) dans les eaux de la Colombie-Britannique ont accusé une diminution moyenne de poids de 5–25 % dans presque toutes les parties de la province. Durant la période 1976–1991, cette tendance s'est ralentie, a stoppé ou a été inversée dans la plupart des régions du nord et du centre et, dans le cas du saumon quinnat, une perte de 3 kg, soit 25 %, a été complètement éliminée dans le grand nord, et pratiquement éliminée ailleurs. Toutefois, dans la région du sud, la réduction de taille s'est poursuivie pour le coho et pour le rose, ces espèces étant aujourd'hui de 20–40 % plus légères que dans les années 1950. Le sockeye (*O. nerka*) et le kéta (*O. keta*) n'accusent aucune tendance soutenue dans leur taille depuis 1951. Jusqu'ici, aucun des changements observés dans la taille n'a été corrélé à des changements à long termes touchant des facteurs environnementaux ou autres, mais la question devra être étudiée beaucoup plus à fond. Un faible taux de croissance peut être lié à l'abondance inhabituelle du saumon rose dans certaines régions. Les effets de la pêche sélective sur les aspects héréditaires du taux de croissance et de l'âge à maturité peuvent jouer un rôle dans certains cas, bien que les preuves à cet égard soient circonstancielles. Toutefois, ces effets fournissent une explication possible à l'absence de tendance soutenue chez le sockeye et le kéta, parce que le nombre d'individus d'âge 2 ou 3 présents dans les stocks rend ambigu et imprévisible l'effet de la sélection par la taille : les poissons en maturation qui ont passé un nombre donné d'années dans l'océan sont plus petits, mais ont grandi plus rapidement, que ceux qui y ont passé un nombre d'années plus élevé.

Introduction and methods

About 20 years ago I became interested, along with several colleagues, in the changes that had occurred in the size of our salmon. We prepared a series of reports, mainly based on information obtained from the sales slips of the commercial fishery, which is published annually by the Canadian Department of Fisheries and Oceans, starting with 1951. These include both the weight and the number of salmon landed, from which average weights were calculated for the catches of each kind of gear separately: gillnets, seines, and trolls. Troll-caught weights, reported as dressed weight head-on, were converted to round weight by dividing by 0.85 (Ricker 1980a, b, 1981, 1982, 1989; Ricker et al. 1978; Ricker and Wickett 1980). It turned out that while two species, chum (*Oncorhynchus keta*), and sockeye salmon (*O. nerka*), had not changed much in average weight up to 1975, the pink (*O. gorbuscha*), coho (*O. kisutch*), and chinook salmon (*O. tshawytscha*) had decreased significantly, some might say alarmingly, in most regions of British Columbia.

W.E. Ricker. Department of Fisheries and Oceans, Biological Sciences Branch, Pacific Biological Station, Nanaimo, BC V9R 5K6, Canada.

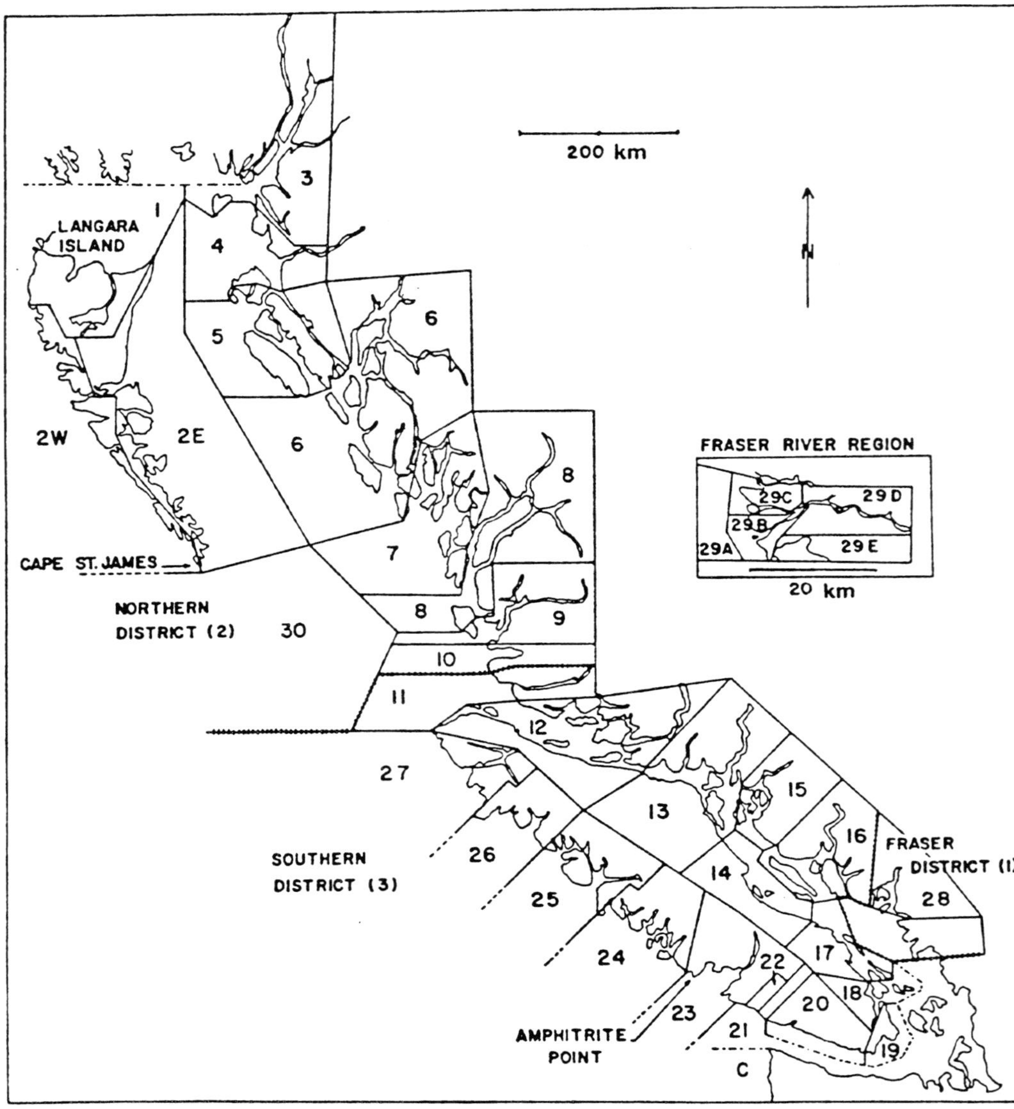

Fig. 1. Statistical areas and districts of British Columbia.

Interested in seeing what has happened since then, I have prepared graphs of trends in most of the Statistical Areas of the coast, based on the same kind of data, of which a selection is shown in Figs. 2–26. For pink and chinook salmon the data from two adjacent regions, southeastern Alaska and Puget Sound, are also included, obtained from the Statistical Yearbooks of the International North Pacific Fisheries Commission.

In the Canadian statistics, commercial catches are listed for 35 different areas and subareas, which are grouped into three districts (Fig. 1). Somewhat smaller groupings of areas are also used in this report, as follows:

Districts	Regional groups	Areas
2	Northern	1, 2W, 2E, 3–5
	Central	6–10, 30
3	Georgia and Johnstone straits	11–18
	Strait of Juan de Fuca	19, 20
	West coast of Vancouver Island	21–27
1	Fraser R. and adjacent salt water	28, 29A–29E

Of these, areas 18, 19, and 22 have not been open to commercial salmon fishing in recent years, and several others have produced only small catches.

From the approximately 120 available figures showing trends in average size of salmon caught, those presented in Figs. 2–26 have been selected because they represent areas and types of fishing gear in which large catches are usually made, or because they are representative of the region in question, or because they illustrate a special point mentioned in the text. Within this framework, preference has been given to seine catches, which are presumed to be less selective than those made by trolls or gillnets.

Usually salmon caught by troll or gillnet are larger than those caught by seine in the same area. Chum salmon are an exception in which the gillnetted fish are usually smaller than those seined, but for this there is a plausible explanation: a large part of the chum catch is made in gillnets whose mesh size is most efficient for catching sockeye and pinks, which are much smaller fish.

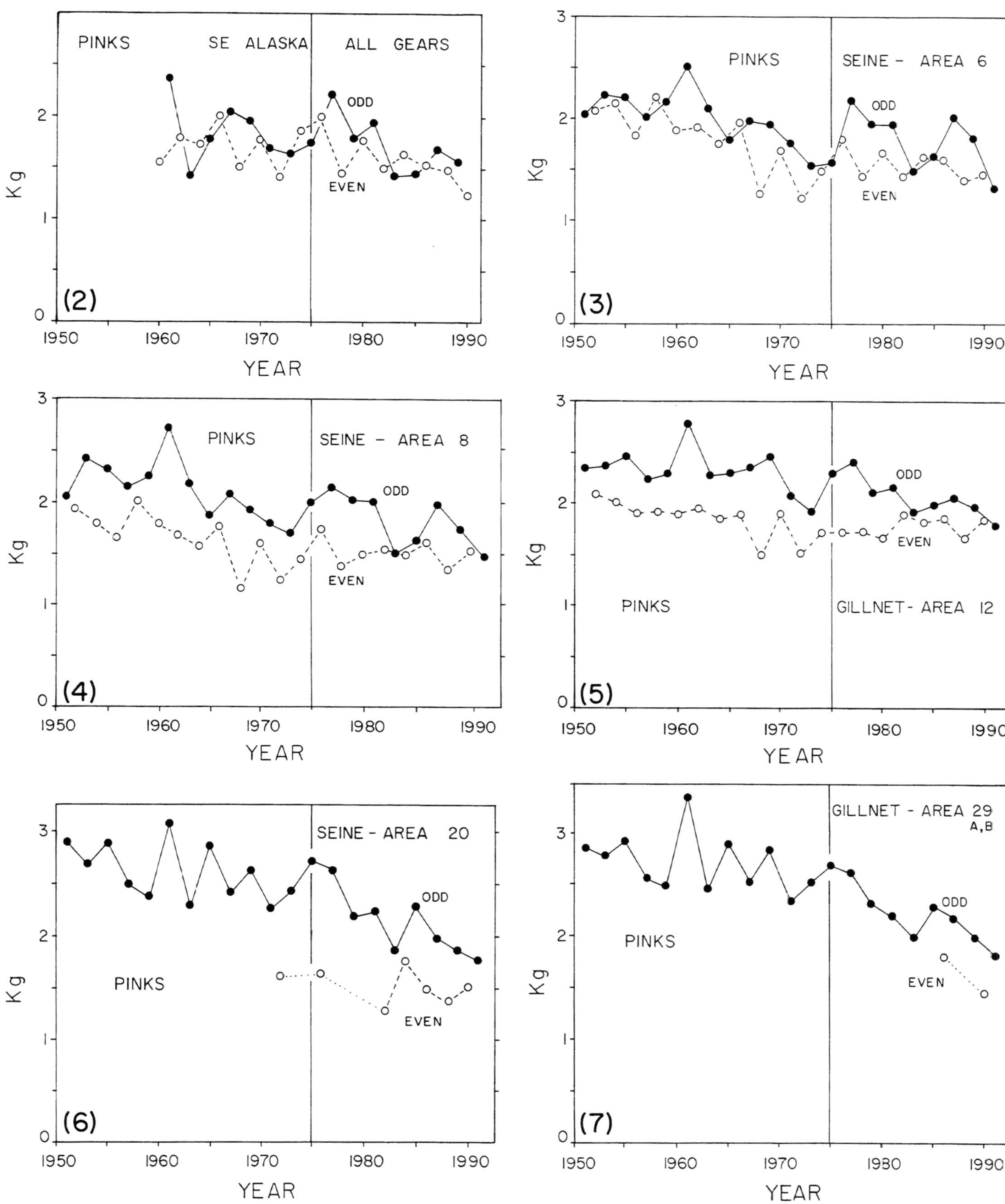

Figs. 2–26. Examples of the average weights of salmon caught in British Columbia and adjacent regions. Except in Alaska and Puget Sound, the weights of the fish taken by different years are shown separately.

Results

Pink salmon (*Oncorhynchus gorbuscha*)

Because of their 2-year life cycle, there is only one age-group of pink salmon in the catch in any particular year. Moreover, they feed mainly well out to sea, and do not enter shore-based catches until nearly all of their growth has been completed. The odd-year and even-year lines are plotted separately in Figs. 3-7 because odd years tend to have larger pinks almost everywhere along the British Columbia coast and southward; however, in southeastern Alaska there is little average size difference between alternate years (Fig. 2).

In the northern and central regions the 1951–75 decrease in size of pink salmon has been slowed in the odd years and halted in even years (Figs. 3–5). In the south, however, the decrease has accelerated since 1975 (Figs. 6 and 7), and the odd-year pinks are now about 1 kg lighter than in the 1950s, a decrease of 40%. (Even-year pinks are very scarce south of Area 12).

Coho salmon (*Oncorhynchus kisutch*)

A large majority of the cohos caught in British Columbia are in their 3rd year of life (2nd year in the ocean), and during that year they are growing rapidly. During the period 1951–75 the opening of the commercial trolling season for coho was twice put forward, which would tend to make average size in the catch larger. In fact, however, average size did not change much during that period; indeed, it had a small downward trend in most areas (Figs. 8–10), or a *large* trend downward in Area 12 at the northern tip of Vancouver Island (Fig. 11). Since then there has been little change in the north (Fig. 8), and in Area 5 even an increase in size in the troll catch (Fig. 9), although the opposite trend appears in the gillnet catch (Fig. 10). In Area 12 the earlier decrease was halted (Fig. 11), but in the south the average weight of seined cohos has decreased by about 1 kg since 1975 (Fig. 12). In the Fraser gillnet catch, mean size during 1976–90 had stabilized at a level about 0.5 kg lower than in 1951–75 (Fig. 13).

Chinook salmon (*Oncorhynchus tshawytscha*)

It is the chinooks that have provided the greatest surprises. They are big fish, of course, especially in northern British Columbia, but during 1951–75 they decreased by almost 3 kg in the troll catch (Figs. 15 and 16), and the outlook then seemed far from bright. But immediately afterward their sizes started to increase, and by 1990 they were back where they started. Similar sequences occurred for troll-caught fish from throughout the province, though in some cases the recovery has been less complete (Fig. 17). The chinooks caught by gillnet have changed less in size, but have been large in recent years (Fig. 18), Area 3 being the only exception.

In southeastern Alaska the picture is similar to that in our northern and central Areas (Fig. 14). In Puget Sound, however, a decrease of 1.5 kg occurred during 1975–80, with little recovery up to 1990, although they were larger during 1985–87 (Fig. 19).

Chum salmon (*Oncorhynchus keta*)

Chums have varied moderately in size, but differently in different areas (Figs. 20–24). In Areas 6 and 7, for example, they were doing well during 1976–82, then took a tumble; those of Area 6 have fully recovered from this, but not those of Area 7. Chums occur in our catches when in their 3rd or 4th year of life, with a very few in their 5th year. So if a year-class considerably larger than usual comes along, it means that average size in the catch will be small when they are maturing at age 3, and large a year later when they are nearing age 4. This might account for the sudden ups and downs of sizes in Areas 12 and 29AB in recent years (Figs. 23 and 24).

Sockeye salmon (*Oncorhynchus nerka*)

Sockeye catches may include fish of several life-history types. In British Columbia the commoner ones are 1/2, 1/3, 1/4, 2/3, and 2/4, where the first figure refers to the number of growing seasons spent in fresh water, and the second is the number of complete or partial growing seasons in the sea. Figure 29 shows the distribution of lengths of male sockeye of the different types in a typical region, females being usually about 2 cm shorter.

In northern and central British Columbia, south to Area 10, ages 1/3, 2/3, and 1/4 can all be important, although variable, components of the sockeye catch, while on the Fraser a great majority are 1/3. However, neither in the north nor in the south has there been any sustained trend in the average size of sockeye caught (Figs. 25 and 26). For the Fraser there are data back as far as 1912. The fish have been rather small during the last few years, but no worse than during the late 1950s or early 1980s. Since 1926 the 1902 line has been dominated by the Adams River run at Shuswap Lake, which usually had larger fish than the other three lines, especially since 1944, hence, the peaks at 4-year intervals in Fig. 26.

Possible causes of change in size

Oceanic conditions

Faced with this great diversity of direction and degree of change in size, we think first of changes in the ocean climate of the salmon, especially temperature and production of plankton organisms. It is tempting to associate the increase in size of the northern chinooks since 1975 with the rise in ocean temperature shown in Figs. 27 and 28. However, the size increase continued to 1990, whereas temperatures fell after 1983. Also, 1951–63 was a period of rising temperature but of decreasing chinook sizes. Of course the coastal surface temperatures of Figs. 27 and 28 are probably not the best indicators of temperature where the salmon live. Also, year-long average temperatures may not be the best indicators of how temperature affects the growth of the fish. So there is room for much additional study of the possible influence of temperature and other environmental conditions, whose effects need not be the same for all species, or for all regions of the coast.

Competition for food

Everywhere in British Columbia pink salmon tend to be larger in odd-numbered years than in even years. It is well known that pinks of the dominant lines in Kamchatka have always been smaller than those of the alternate poor years, even when dominance shifted from the odd-year to the even-year line (Ricker 1962, Fig. 21). However, in southern British Columbia pinks are very much more abundant in odd-numbered years, yet are *larger* then, which is the reverse of what food competition would accomplish.

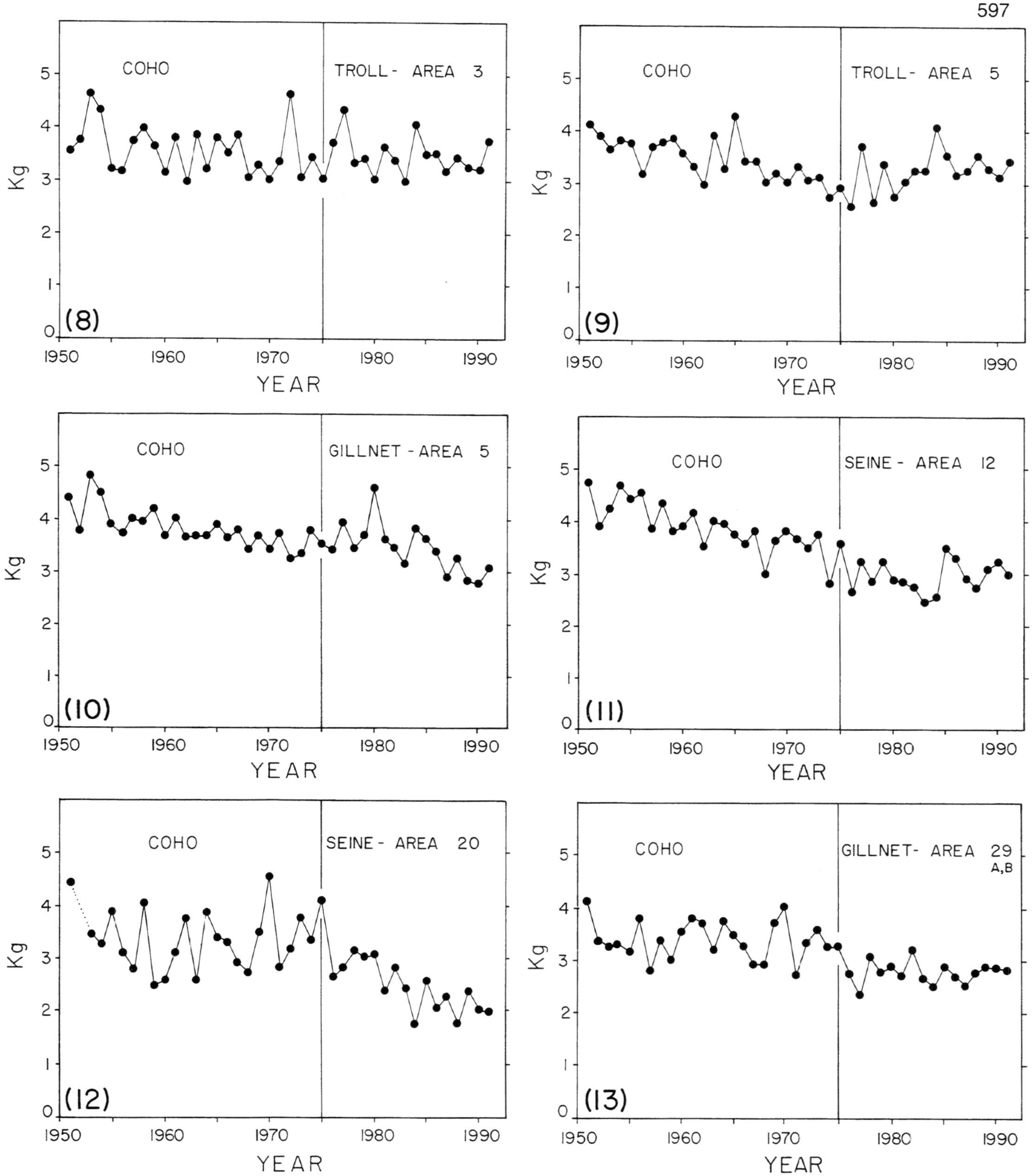

Figs. 2–26. (*Continued.*)

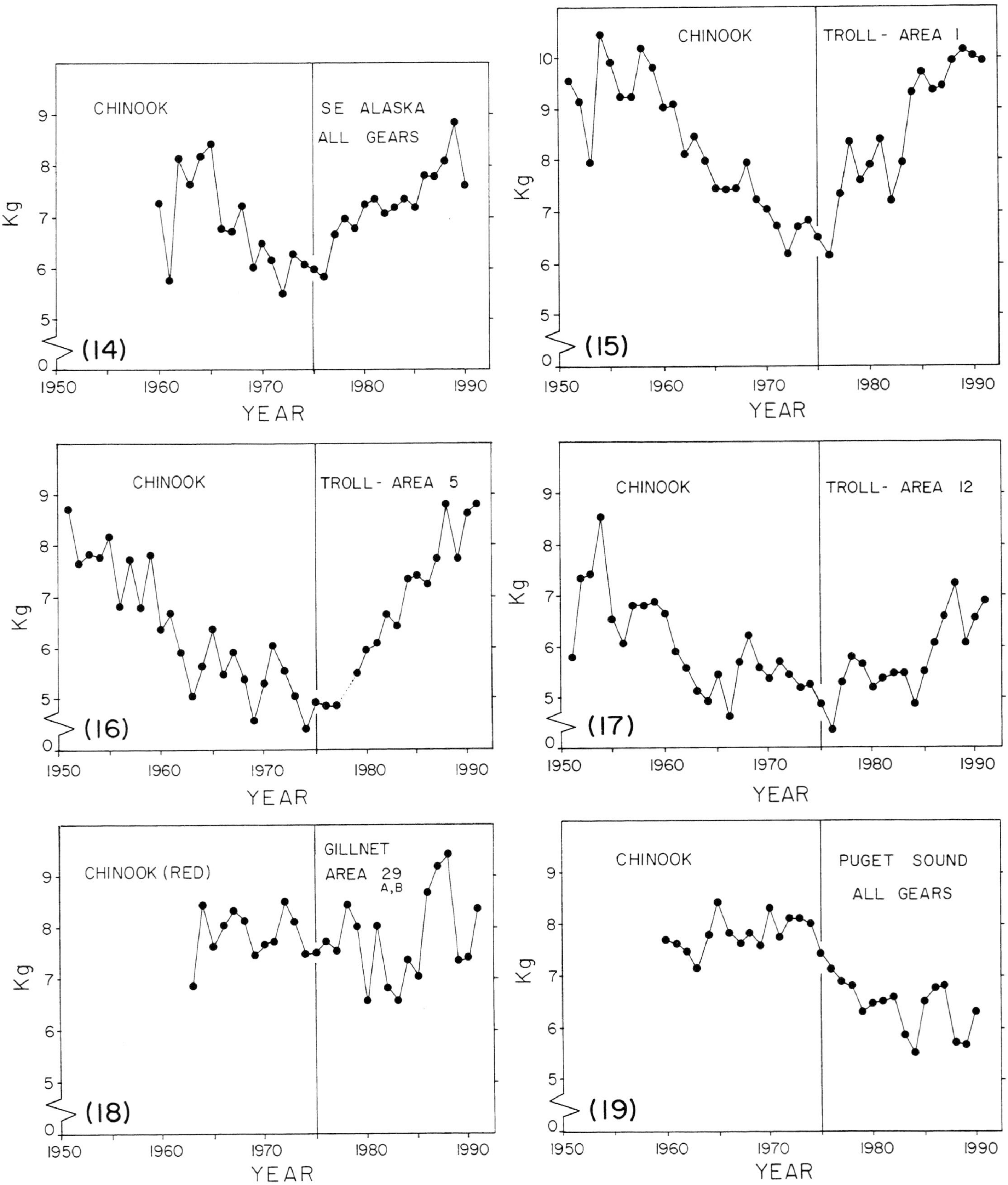

Figs. 2–26. (*Continued.*)

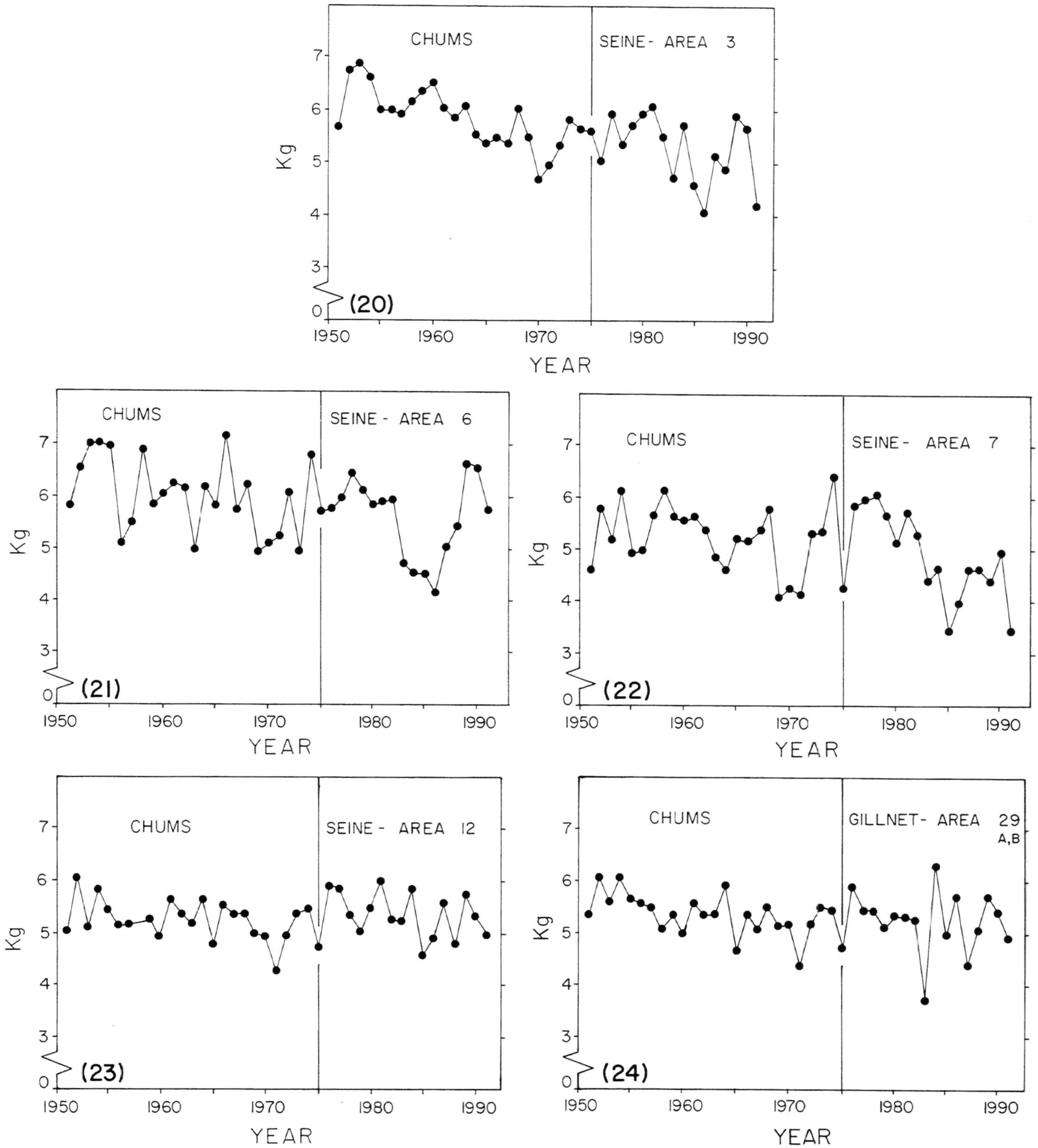

Figs. 2–26. (*Continued.*)

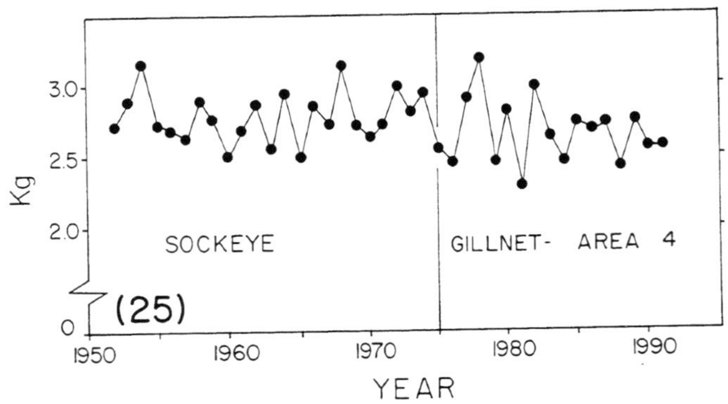

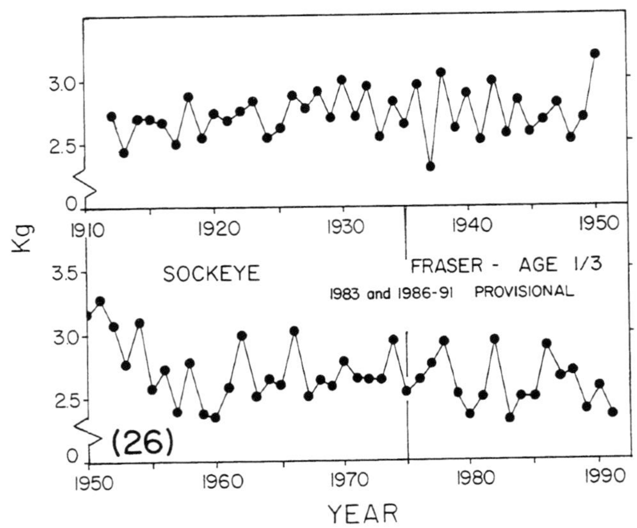

Figs. 2–26. (*Concluded.*)

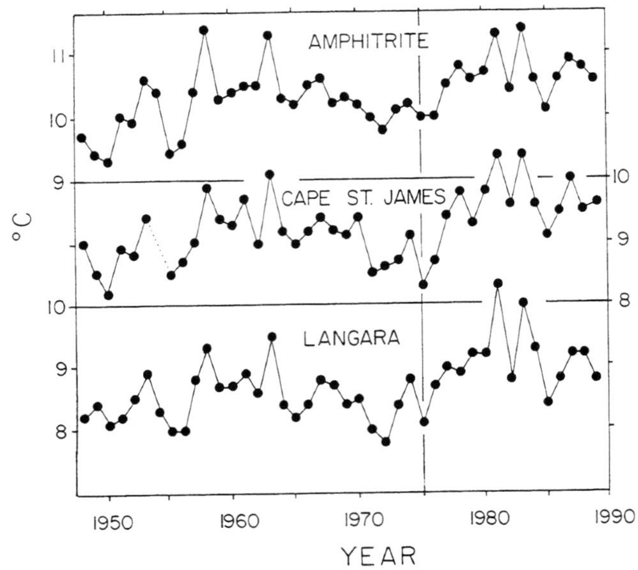

Fig. 27. Averages of monthly mean temperatures of near-surface sea water at three lighthouses, whose locations are shown in Fig. 1.

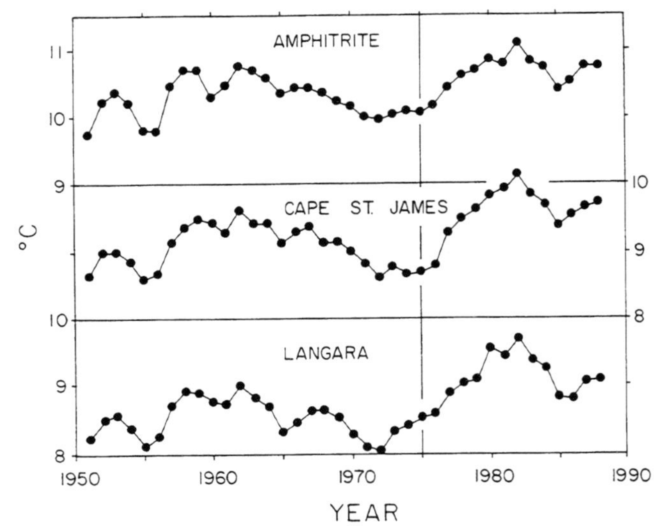

Fig. 28. The temperatures of Fig. 27 smoothed by a running average of three.

Nevertheless, there are other indications of an effect of competition. For one thing, Area 6 had a very large pink population in 1972, as did Area 8 in 1968, and these were also years of unusually small size (Figs. 2–4). The unusually large size of Fraser pinks in 1961 and 1965 may have been partly due to their unusual scarcity in those years, although a rather small run in 1987 consisted of small fish (Fig. 26). It would seem that any such effects must come from near-shore competition, because out on the high seas each individual run gets mixed up with far greater numbers of salmon from all along the coast.

But could changes in ocean food supply account for the steep decline and equally steep increase in the size of our northern chinooks? Many or most of these fish forage far out into the Aleutian region, where stocks of sockeye and other species of Alaska salmon have increased greatly in abundance during the last 15–20 years. Could the warming of the ocean since 1975 have increased basic food production in this region? Or could the increased supply of young sockeye, etc., be providing the chinooks with better feeding? Or could the removal of large quantities of pollock, rockfish, and Atka mackerel (*Pleurogrammus*) have reduced food competition for themselves and for salmon as well? Maybe so, but none of these possibilities seems to account for the *decrease* in size during 1951–75.

Selective fishing

From the data available up to 1975, I suggested that a change in the genetic constitution of salmon stocks was mainly responsible for the observed decreases in size. After all, if you are raising beef cattle, for example, you select breeding stock with a proven history of fast growth. Our fisheries have been doing exactly the opposite. Our breeding stocks, on the whole, contain more of the smaller individuals in any population. Experiments have shown that both rate of growth and age of maturity are determined, in part, by hereditary factors and, to some extent at least, independently so. Also, both average rate of growth and average age of maturity in a population can be

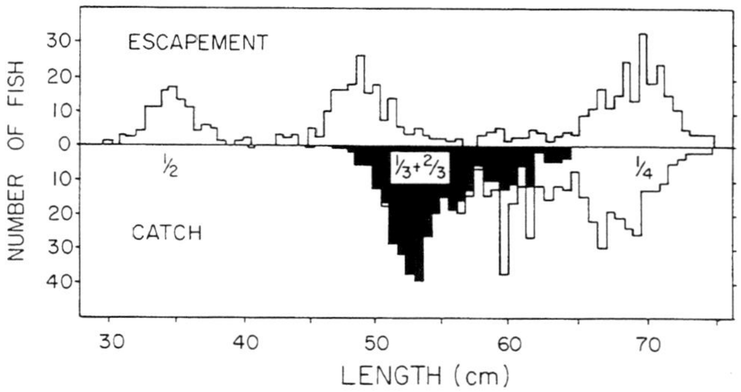

Fig.29. Mean fork lengths of samples of male Owikeno (Rivers Inlet) sockeye in 1952. The escapement lengths are reduced by dividing by 1.04 to adjust for the lengthening of the nose at maturity. The numbers of fish measured in the two samples do not indicate the relative size of catch and escapement. Except for age 1/2, it is likely that the fish caught, of any age, were at least twice as numerous as those that escaped. (After Foskett 1952, Fig. 7).

modified by selection (Ricker 1972). Moreover, estimates of the magnitude of the heritability of growth rate in salmon and other fishes suggested that existing selection differentials, among pink salmon at least, had been sufficient to cause changes of the magnitude observed (Ricker et al. 1978).

That gillnets can be very selective hardly needs demonstration, but a striking illustration was obtained many years ago by Foskett (1952). Figure 29 shows the size and age composition of male Owikeno or Rivers Inlet (Area 9) sockeye in the catch and on the spawning grounds. This gillnet fishery took almost no age 1/2 sockeye ("jacks"); it bore heavily on the middle part of the size range of the older fish, so that spawners consisted mainly of the smaller of those having 3 yr of ocean growth (1/3 and 2/3), and a somewhat larger than average selection of those with 4 years of ocean growth (mostly 1/4). Godfrey (1958) found that there had been an increase in the difference in size between sockeye of ocean age 3 and those of ocean age 4 in the Owikeno and Skeena runs, presumably a result of this selection.

In British Columbia up to 1950, all salmon except chinooks were bought by the piece from fishermen: during the 1930s it was 5 cents each for pinks and chums, 10 cents for cohos, and 25 cents for sockeye. Thus selection of the larger individuals of these four species would not increase anyone's income, so selection probably did not occur at that time. Since 1950 all salmon have been bought by weight, so it now makes sense for a fisherman to go after the larger ones. In a gillnet fishery this can be done by adjusting the size of mesh in the net used.

The first result of such a change would be an increase in the size of the fish in the gillnet catches for a few years. In the long series of Fraser sockeye weights there is an increase in 1950–54, but this quickly subsided and there was little change since (Fig. 26). For pink and coho salmon, which have only one important age in the catch, the initial increase would be followed by a long slow decline as genes favoring fast growth were gradually weeded out.

However, for species that mature at two or more ages, the effects of selective fishing on average size are complex. Pacific salmon, of any species, that mature at a particular age invariably have a faster average growth rate than those of the same sex, in the same population, and of the same year-class, that mature at any older age. In the case of sockeye, the fast-growing jacks of age 1/2, which are almost all males, have traditionally been fished very lightly, and in some stocks the proportion of jacks has increased greatly (Ricker 1972). Insofar as these small fish are effective in spawning, this should increase the representation of genes favoring fast growth in the population as a whole. Within the commercially desirable older ages, sockeye having 3 yr of ocean growth average smaller than those with 4 ocean yr, but they have grown faster, so that excess removal of 4-yr ocean fish will tend to increase growth rate in the population while it tends to decrease average age. Hence, the direction of change of average size in the catch as a whole becomes unpredictable until these opposed effects can be quantified.

Chum salmon are exceptional in that the gillnetted fish have been *smaller* than those caught by seine. Thus my 1980 prediction was that average size would *increase* after 1975. This did happen in most areas during 1976–82. Since then average sizes have held up in some areas and faltered in others (Figs. 20–24). For this species an excess of older fish on the spawning grounds will tend to increase the representation of their ages in subsequent generations. But since they are slow growing, the overall rate of growth in the population will be reduced. The net effect is again to make it impossible to predict the direction of change in average size in the population as a whole and in the chum catches, as in the sockeye, there have been no sustained trends in average size.

Summarizing, among pinks and cohos, the two species in which selection operates unambiguously on rate of growth, average sizes today are substantially less than in the 1950s, and selection is the only factor so far suggested that has been unidirectional during the whole of that time. The net effect of selection on average size in the other species cannot be confidently assessed or predicted.

Other factors

Other possible factors that could cause changes in average size and average age in salmon catches include changes in the relative abundance of the different stocks of a species that are present in a region, selection of breeding stock in hatcheries, and changes (even quite small ones) in the fishing gear used and the method of its deployment. Changes in fishing seasons and in minimum size limits can be of special importance for the troll fishery for chinooks and cohos, because in some places these species are caught while still making rapid growth, and they are subject to a growing sport fishery as well as the commercial effort. Such possibilities need to be assessed in detail for each region or each population, as far as possible. For example, the factor or factors responsible for the reversal of the size trend among chinooks might be apparent if a series of age composition data were available.

References

Canadian Department of Fisheries and Oceans, Pacific Region. 1951–1991. British Columbia catch statistics. Vancouver, B.C.

Foskett, D.R. 1958. The Rivers Inlet sockeye salmon. J. Fish. Res. Board Can. 15: 867–889.

Godfrey, H. 1958. A comparison of catches at Rivers Inlet and Skeena River, B.C., with particular reference to age at maturity. J. Fish. Res. Board Can. 15: 331–354.

International North Pacific Fisheries Commission. Statistical Yearbook 1960–1989. Vancouver, B.C.

International Pacific Salmon Fisheries Commission. Annual Reports 1937–1991. New Westminster, B.C.

Ricker, W.E. 1962. Regulation of the abundance of pink salmon populations, p. 155–201. *In* Symposium on Pink Salmon, 1960. H. R. MacMillan Lectures in Fisheries, University of British Columbia, Vancouver, B.C.

Ricker, W.E. 1972. Heriditary and environmental factors affecting certain salmonid populations, p. 27–160. *In* The stock concept in Pacific salmon. H.R. MacMillan Lectures in Fisheries, University of British Columbia, Vancouver, B.C.

Ricker, W.E. 1980a. Changes in the age and size of chum salmon (*Oncorhynchus keta*). Can. Tech. Rep. Fish. Aquat. Sci. 930: 99 p.

Ricker, W.E. 1980b. Causes of the decrease in age and size of chinook salmon (*Oncorhynchus tshawytscha*). Can. Tech. Rep. Fish. Aquat. Sci. 944: 25 p.

Ricker, W.E. 1981. Changes in the average size and average age of Pacific salmon. Can. J. Fish. Aquat. Sci. 58: 1636–1656.

Ricker, W.E. 1982. Size and age of sockeye salmon (*Oncorhynchus nerka*) in relation to environmental factors and the fishery. Can. Tech. Rep. Fish. Aquat. Sci. 1115: 117 p.

Ricker, W.E. 1989. History and present state of the odd-year pink salmon runs of the Fraser River region. Can. Tech. Rep. Fish. Aquat. Sci. 1702: 37 p.

Ricker, W.E., H.T. Bilton, and K.V. Aro. 1978. Causes of the decrease in size of pink salmon (*Oncorhynchus gorbuscha*). Can. Tech. Rep. Fish. Mar. Serv. 820: 93 p.

Ricker, W.E., and W.P. Wickett. 1980. Causes of the decrease in size of coho salmon (*Oncorhynchus kisutch*). Can. Tech. Rep. Fish. Aquat. Sci. 971: 63 p.

Interannual Variability of the zooplankton community off southern Vancouver Island

David L. Mackas

Mackas, D.L. 1995. Interannual variability of the zooplankton community off southern Vancouver Island, p. 603–615. *In* R.J. Beamish [ed.] Climate change and northern fish populations. Can. Spec. Publ. Fish. Aquat. Sci. 121.

Abstract: Time series of zooplankton anomalies (deviations from the long-term average seasonal cycle) are calculated from 1985–92 records of log-transformed biomass for dominant zooplankton taxa off the southwest coast of British Columbia. Statistically significant anomalies occur in all of the major taxa. Average persistence of the anomalies ranges from less than a year for gelantinous zooplankton, to 1–2 yr for the common copepods, to 3 yr or more for chaetognaths and euphausiids. Significant zooplankton anomalies occur throughout the time series; they are not confined to transient episodes such as the 1987 and 1992 El Niño events. Because of the long persistence of most of the anomalies, the length of the time series is still too short to permit stand-alone statistical proof of a linkage between environmental and zooplankton fluctuations. However, coincidence of timing suggests an association of local zooplankton and environmental anomalies with the 1988 change in winter-season atmospheric pressure patterns in the North Pacific Ocean.

Résumé : Les séries chronologiques sur les anomalies (écarts par rapport au cycle saisonnier moyen à long terme) du zooplancton du large de la côte sud-ouest de la Colombie-Britannique sont calculées à partir des données enregistrées de 1985 à 1992 sur de la biomasse des taxons dominants (après transformation logarithmique). Des anomalies statistiquement significatives sont présentes dans tous les principaux taxons. La persistance moyenne des anomalies varie de moins d'un an pour le zooplancton gélatineux à 1–2 ans pour les copépodes communs, et à 3 ans ou plus pour les chétognathes et les euphausiacés. Des anomalies importantes sont présentes tout au long de la série chronologique chez le zooplancton; ces anomalies ne sont pas confinées à des épisodes transitoires comme les événements El Niño de 1987 et de 1992. Étant donné la longue persistance de la plupart des anomalies, la durée des séries chronologiques est encore trop courte pour permettre à elle seule de faire la preuve statistique d'un lien entre les fluctuations environnementales et celles du zooplancton. Toutefois, la coïncidence chronologique porte à croire qu'il existe un lien entre les anomalies du zooplancton local et de l'environnement et le changement survenu à l'hiver de 1988 dans les tendances de la pression atmosphérique du Pacifique Nord.

Introduction

Most of the total harvest from the ocean comes from a moderate number of highly productive regional ecosystems (FAO 1981; Sherman and Alexander 1986; Couper 1989). What biotic and abiotic traits of the local ocean environment allow and limit the yield of these regions? How will these traits change under changing climate conditions? And how will the changes be linked as both cause and effect to larger scale variability elsewhere in the ocean?

These issues are an important practical justification for studies of ocean climate. One tool for studying them is the description and comparison of multiyear fishery and environmental time series. Much of the interannual and interdecadal variability of harvested fish stocks can be reconstructed from time-series information provided by commercial fishery statistics. But time-series data for causally related oceanographic environmental variables usually must come from relatively specialized and dedicated field sampling programs. Because of this, good environmental time series are fewer and shorter than the fishery time series they are used to interpret.

Where available, information about year-to-year and decade-to-decade differences in zooplankton biomass and community composition is especially useful. This is true for several reasons. Variations in zooplankton distributions and abundance reflect the effects of transport by changing current patterns, and can be used as models for direct physical effects on early life stages of fish and benthos. Zooplankton about 0.1–2.0 cm in body size also occupy a key intermediate position in the pelagic food web. They are the main consumers of phytoplankton cells larger than a few micrometres in diameter and of the microzooplankton that graze smaller phytoplankton. They are also the principal food source for many pelagic and some demersal fish stocks. Because of the short trophic linkage, zooplankton are very direct indices of food-web structure and carrying capacity for these higher trophic level predator species.

D.L. Mackas. Institute of Ocean Sciences, Biological Sciences Branch, Department of Fisheries and Oceans, Sidney, BC V8L 4B2, Canada.

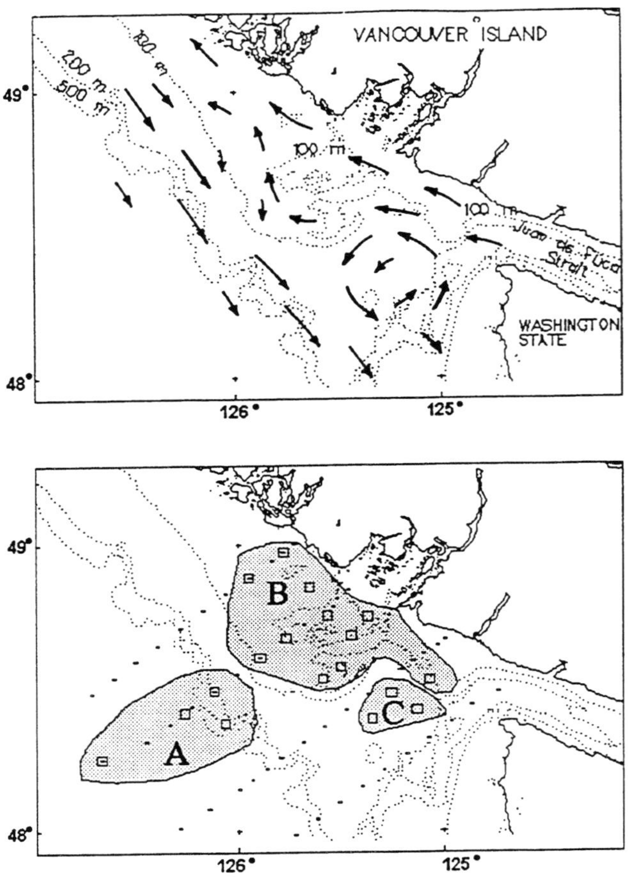

Fig. 1. Maps of the La Perouse Project study area off the southwest coast of Vancouver Island, British Columbia. Upper panel shows bathymetry and typical summer surface current pattern. Lower panel shows standard sites for conductivity–temperature–depth (CTD) profiles (solid circles); zooplankton new tows (open squares); and grouping of net tows into three statistical averaging regions: "offshore" sites over the continental shelf break and slope (A); "gyre" sites near the Juan de Fuca submarine canyon (C); and "inner shelf" sites (B) in areas strongly affected by the Vancouver Island Coastal Current.

Examples of significant and heavily used zooplankton time series include the Continuous Plankton Recorder surveys of the North Atlantic (e.g., Colebrook 1982; Colebrook et al. 1984), the California Cooperative Oceanic Fisheries Investigations (CalCOFI) surveys of the California Current (Colebrook 1977; Chelton et al. 1982; Roesler and Chelton 1987), and the Station P time series in the open subarctic Pacific (Frost 1983; Brodeur and Ware 1992). All of these show low-frequency zooplankton variability large enough to be both statistically and ecologically significant. Most also show evidence of strong coupling to large-scale variability in the climate of the surrounding ocean.

The continental shelf and adjoining deep-ocean waters off southern British Columbia (Fig. 1) are an important example of a highly productive regional ecosystem. Since 1985, Department of Fisheries and Oceans scientists participating in the La Perouse Project (Thomson and Ware 1988) have combined localized process studies with systematic time-series observations of this region. Some additional detailed but less-standardized sampling extends back to 1979. The La Perouse time series is now becoming long enough to allow some useful description of interannual variability. In this paper, I quantify for the 1985–92 period of standardized sampling the year-to-year deviations of the zooplankton time series from the long-term average seasonal cycles described by Mackas (1992). My emphasis here will be on magnitude, phase, and persistence of the zooplankton anomalies. A subsequent paper (D.L. Mackas and R.E. Thomson, unpublished data) will analyze correlations with local physical oceanographic fields.

Regional oceanographic setting

The continental margin off southern British Columbia lies at the northern and upstream end of the North American eastern boundary current system (the California Current). Although the region shares many species and oceanographic processes with more southerly parts of this current system (studied by the CalCOFI time series and a number of major programs of shorter duration), there are large and ecologically important latitudinal gradients within the California Current (GLOBEC 1992 and references cited therein). The seasonality of physical forcing, the peak rates of nutrient input and lower trophic level productivity, and the dynamic importance of coastal fresh water input are all higher off British Columbia than off California. Many species inhabit the entire system, but there are also indications that life history strategies differ systematically between latitudinal zones (Parrish et al. 1981).

The physical oceanography of the B.C. continental margin has been reviewed by Freeland et al. (1984), Thomson et al. (1989), and Crawford and Thomson (1991). Large-scale winds, and the surface-layer currents over the shelf break and outer shelf (roughly between the 100-m and 500-m isobaths) both reverse seasonally. They are downwelling favourable and toward the northwest in winter, upwelling favourable and toward the southeast in summer. Over the inner part of the continental shelf (roughly within 20–30 km from the coast), the Vancouver Island Coastal Current is driven by coastal freshwater input (Thomson et al. 1989) and maintains a northward flow year round.

Seasonal cycles of plankton and nutrients have been analyzed by Mackas (1992). The upper-layer concentration and input rate of "new" dissolved nutrients and the resulting phytoplankton production and biomass are strikingly high throughout a prolonged upwelling season lasting roughly from April through October. A variety of physical processes drive nutrient input. These include estuarine discharge from the Fraser River – Georgia Strait – Juan de Fuca system (Mackas et al. 1978; Crawford and Dewey 1989), interaction of currents with deep-shelf bathymetry (Freeland and Denman 1982; Freeland and McIntosh 1989), shelf-edge upwelling (Denman et al. 1981), and alongshore advection (Denman et al. 1989) in addition to conventional wind-driven coastal upwelling. In summer, average nutrient and phytoplankton pigment concentrations are highest (5–9 μM NO_3, 5.5–8.5 $mg \cdot m^{-3}$ chlorophyll*a*) over the inner part of the continental shelf. Seaward of the shelf break, average concentrations are lower but remain substantial for marine waters

(1–2.5 μM NO_3, 0.8–3 mg·m^{-3} chlorophyll*a*). There is much short-term variability about these averages caused by a sequence of several dense individual bloom "events" within eacy year's upwelling season (Denman et al. 1981; Forbes et al. 1987; Denman et al. 1989; Mackas 1992).

The zooplankton community off British Columbia integrates and smooths the rapid week-to-week cycling of phytoplankton biomass but retains substantial annual cycles of both total biomass and species composition (Mackas 1992). Amplitude and timing of the average seasonal cycle differ between shelf and offshore locations. Minimum biomass is in winter (about 1–2 g·m^{-2} dry weight) in all subregions. Maximum is in late spring on the continental shelf (7–8 g·m^{-2}) and in middle to late summer seaward of the shelf break (6–7 g·m^{-2}). Additional and updated information about the average zooplankton seasonal cycle will be provided in the Results section of this paper.

Annual fish catch per unit area from the La Perouse region is very high, averaging about 5–7 t·km^{-2} over the past several decades (D. Ware, Department of Fisheries and Oceans, Pacific Biological Station, Nanaimo, B.C., personal communication). As with the plankton, there is a very large seasonal cycle of total fish biomass. But in contrast to the plankton, this cycle is mostly caused by active migration into and out of the region. Major contributors to total fish biomass include Pacific hake (*Merluccius productus*), Pacific herring (*Clupea harrengus*), spiny dogfish (*Squalus acanthias*), and Pacific salmon (*Oncorhynchus* spp.).

Methods

Processes operating at several distinct space and time scales add and interact to produce the overall measured variability of zooplankton biomass and composition. The sampling and

Table 1. Components of overall zooplankton variability.

Source and/or scale of variability	How dealt with
Unresolved small-scale patchiness plus sampling error	Minimized by averaging of "replicates" at all levels
Persistent mesocale spatial structure	Regional stratification of samples into spatial averaging units
Annual seasonal cycle	Averaging within seasons and regions across years (1979–89 in Mackas (1992), updated here to 1979–91)
Interannual variability	Averaging within years and regions across seasons of deviations of cruise mean from seasonal mean (1985–92; this paper)

Table 2. Definitions of symbols used in mathematical equations.

Symbol	Definition
$N_{s(i),x(j),t}$	Raw abundance data; no·m^{-2} of stage *i* within species *s* at site *j* within mesoscale region *x* during sampling period *t*
$w_{s(i)}$	Dry weight per individual of stage *i* of species *s*
$D_{s,x(j),t}$	Raw biomass data: Total dryweight (g·m^{-2}) of taxon *s* at site *x*(*j*) during sampling period *t*
$\langle\langle D(\mathrm{reg})\rangle\rangle_{s,t}$	Within-region, within-cruise biomass of taxon *s* (averaged over all samples from region *x* during cruise *t*)
$\langle\langle D(\mathrm{reg,seas})\rangle\rangle_s$	Within-season, within-region biomass of taxon *s* (averaged across years within season)
$B_{s,x(i),t}$	Log_{10} transformation of $D_{s,x(i),t}$
$\langle\langle B(\mathrm{reg})\rangle\rangle_{s,t}$	Within-region, within-cruise average of log-transformed biomass
$\langle\langle B(\mathrm{reg,seas})\rangle\rangle_s$	Within-season, within-region average of log-transformed biomass
$b'(\mathrm{reg})_{s,t}$	Within-cruise, within-region anomaly from multi-year seasonal average of log-transformed biomass (used for calculation of autocovariance)
$\langle\langle b'(\mathrm{reg,year})\rangle\rangle_s$	Annual average anomaly for taxon *s* (used in plots of anomaly time-series)
$C_{s,\tau}$	Within-region autocovariance at lag τ of anomaly time-series of taxon *s*

Table 3. List of taxa included in this analysis of interannual zooplankton anomalies off southwestern British Columbia. Table columns give major group and species identification, dry weight coefficients used to convert from abundance to estimated biomass by taxa, and summary labels used in plots of regional seasonal cycles and anomaly time-series. Body-weight ranges are over the range of developmental stages (or body lengths) identified within individual taxa. A more complete list of taxa included in the analysis of multi-year average seasonal cycles is in Mackas (1992).

Taxon	Body weight (mg DW)	Group labelling for plots	
		Seasonal cycle	Anomalies
Euphausiids			
Euphausia pacifica	0.65 – 8.5	Euphausiids	EUPHpac
Thysanoessa inspinata	0.9 – 10	Euphausiids	THYSinsp
Thysanoessa spinifera	1.25 – 31	Euphausiids	THYSspin
Amphipods			
Parathemisto pacifica	0.75 – 3.7	Amphipods	PARATHEM
Primno spp.	0.32 – 40	Amphipods	PRIMNO
Cyphocaris challengeri	1 – 15	Amphipods	CYPHOCAR
Calanoid copepods			
Acartia longiremis	0.002 – 0.007	"Shelf" copepods	ACARlong
Calanus marshallae	0.008 – 0.23	"Shelf" copepods	CALmars
Pseudocalanus mimus	0.0018 – 0.012	"Shelf" copepods	PSEUDOCAL
Neocalanus cristatus	0.04 – 2.6	"Subarctic" copepods	NEOCALcr
Neocalanus plumchrus	0.013 – 0.55	"Subarctic" copepods	NEOCALpl
Mesocalanus tenuicornis	0.007 – 0.04	"Southern" copepods	MESOCALte
Clausocalanus spp.	0.0025 – 0.008	"Southern" copepods	CLAUSOCAL
Ctenocalanus vanus	0.004 – 0.012	"Southern" copepods	CTENOCAL
Chaetognaths			
Sagitta elegans	0.1 – 1.0	Chaetognaths	SAGITelde
Sagitta scrippsae	0.18 – 6.5	Chaetognaths	SAGITscr
Eukrohnia hamata	0.18 – 1.4	Chaetognaths	EUKROHNIZ
Gelatinous zooplankton			
Salpa spp.	7 – 130	Salps	SALPS
Hydromedusae	0.5 – 60.0	Hydromedusae	MEDUSAE
Pleurobrachia	0.9 – 6.9	Ctenophores	CYDIPPIDS

statistical methods used here and by Mackas (1992) were chosen to allow separate analysis and/or averaging out of four prominent components of variability: (1) sampling error combined with unresolved small scale patchiness; (2) persistent mesoscale spatial structure; (3) the annual seasonal cycle; and (4) interannual deviations from the long-term seasonal average (Table 1). This separation is important because the magnitude and form of biological response to environmental perturbation vary with the duration and predictability of exposure. In particular, small amplitude but unpredictable interannual climate differences may have impacts on marine organisms more extreme than those of larger but predictable environmental cycles associated with the annual change of seasons (Denman et al. 1989).

Sampling

Since 1985, zooplankton were collected with vertically integrated bongo net tows at all or most of the standard locations shown in Fig. 1. Additional vertically integrated tows collected at nearby locations from 1979–84 were sorted and averaged within surrounding statistical areas (see Mackas 1992). From 1985 through mid-1992, an average of about 11 samples per cruise were collected four to six times per year (a total of 38 research cruises). Zooplankton in quantitative subsamples were counted to species and developmental stage. Subsample size varied with taxa to give approximately uniform expected abundance estimation error (about 25–40% within a single sample) for all of the dominant groups. Symbols used in subsequent data reduction and statistical equations are summarized in Table 2. The most important taxonomic categories and their ranges of individual body size are listed in Table 3.

Data analysis

Abundance data $N_{s(i),x(j),t}$ were converted to "dry weight biomass" estimates $D_{s(i),x(j),t}$ using stage-and species-specific multiplicative size coefficients, and then summed within species or (for some taxa) broader taxonomic categories according to the equation:

$$(1)\ D_{s,x(j),t} = \sum w_{s(i)} N_{s(i),x(j),t}.$$

Within-cruise grouping of samples based on similarity of their zooplankton community composition had previously shown mesoscale spatial zonation that matched bathymetric and current patterns (Mackas 1992; Mackas and Sefton 1982; Mackas 1984; Mackas et al. 1985; Thomas and Emery 1986). For subsequent averaging, the samples were therefore classified into the three spatial subregions shown in Fig. 1. Subregion seasonal cycles for 1979–89 were estimated (both in Mackas 1992, and for the updated seasonal cycles described here) by spatial averaging of the biomass estimates within region and within cruise:

$$(2)\ <<D(\mathrm{reg})>>_{s,t} = (1/n_{x,t}) \sum D_{s,x(i),t}.$$

This was followed by weighted time averaging of these cruise means across years within roughly bimonthly seasonal blocks:

$$(3)\ <<D(\mathrm{reg, seas})>>_{s} = (\Sigma(\sqrt{n_{x,t}})\cdot\Sigma(\sqrt{n_{x,t}}\cdot<<D(x)>>_{s,t}).$$

Weighting of within-cruise means was proportional to the square root of the number of within-region samples per cruise (see Mackas (1992) for justification and comparison with alternative weighting schemes). In this paper, the seasonal cycles have been updated to include cruises in 1990–91. Arithmetic averaging was used to produce an unbiased estimate of long-term mean biomass.

For analysis of interannual zooplankton anomalies and their statistical significance, I preferred averages of $\log_{10}$-transformed data (geometric mean) to arithmetic averages for three reasons. First, I wanted to normalize the highly skewed raw statistical distribution of biomass frequency and thereby obtain symmetric and relatively uniform error bars for the cruise-by-cruise elements of the zooplankton anomaly time series. Second, I wanted to emphasize low-frequency variation (time scales longer than a year) by averaging sampling periods within years. In the presence of a strong cycle of absolute abundance, it is more reasonable to average percentage deviations from the seasonal norm than to average raw biomass anomalies. Third, I wanted to use this additional averaging step to further reduce the error component caused by incomplete filtering out of small-scale (within-cruise and within-region) patchiness. Within-cruise and multiyear seasonal averages were therefore recomputed as above but from $\log_{10}$-transformed data, i.e.,

$$(4)\ B_{s,x(j),t} = \log_{10}(D_{s,x(j),t})$$

$$(5)\ <<B(\mathrm{reg})>>_{s,t} = (1/n_{x,t}) \sum B_{s,x(i),t}$$

$$(6)\ <<B(\mathrm{reg, seas})>>_{s} = (\Sigma(\sqrt{n_{x,t}})\cdot\Sigma(\sqrt{n_{x,t}}\cdot<<B(\mathrm{reg})>>_{s,t})$$

Cruise-by-cruise anomaly time series were produced by subtracting the appropriate seasonal average from each cruise average:

$$(7)\ b'(\mathrm{reg})_{s,t} = <<B(\mathrm{reg})>>_{s,t} - <<B(\mathrm{reg, seas})>>_{s}$$

Note that anomalies derived from log-transformed data measure multiplicative deviation from the long-term within-season average; an anomaly of about +0.5 means that the taxon is about three times as abundant as the average for that particular season and region while an anomaly of -0.5 means an abundance about one third the long-term average; see Colebrook (1977), Roesler and Chelton (1987), and Mackas (1992) for additional discussion of arithmetic averaging versus averaging of log-transformed data.

The full cruise-by-cruise anomaly time series was used to estimate persistence of the zooplankton anomalies from time-lagged autocovariance of the anomaly time series:

$$(8)\ C_{s,\tau} = <<b'(\mathrm{reg})>>_{s,t}\cdot<<b'(\mathrm{reg})>>_{s,t+\tau}.$$

For each of the major taxa, individual autocovariance estimates were sorted and averaged within increasing intervals for time lag τ (0 lag, then 3 mo spacing to 1 yr, then 6 mo spacing; for clarity, plots were truncated at a lag of 3 yr). For a given species and region, the total variance of the cruise-by-cruise anomaly time series is measured by the autocovariance at zero time lag $C_{s,\tau} = 0$. The typical "duration" of an anomaly is approximately twice the time lag to the first zero crossing of the autocovariance function.

The cruise-by-cruise anomalies were further averaged within calendar years to produce time series of annual anomalies $<<b'(\mathrm{reg, year})_s>>$. Confidence intervals for the year-to-year anomalies were estimated assuming the four components of variability listed in Table 1, and assuming that both persistent spatial structure and within-year seasonal variability are removed effectively by subtraction of within-season, within-region averages. The anomaly time-series should therefore consist of any real interannual differences plus some added amount caused by incomplete averaging out of subsampling error and small-scale patchiness.

Estimated anomalies are significant if they exceed the expected absolute magnitude of this residual error, which can be estimated as follows. First, the residual variability of single samples is given by the standard deviation s_B of the individual log transformed biomass measurements $B_{s,x,t}$ about their "within treatment" (within-cruise, within-region) means $<<B(\mathrm{reg})_{s,t}>>$. Second, the reduced amount (standard error of a sample mean) leaking through to the final anomaly time series is approximately the within-treatment residual (sample standard deviation) adjusted for the number of samples used to form the annual mean anomaly:

$$(9)\ s<<b'(\mathrm{reg, year})>> \approx s_{B}(\Sigma\Sigma n_{x,t})^{-0.5}$$

where summations are of samples per cruise within region and of cruises within year.

To compare timing and persistence of the zooplankton anomalies with those of local physical oceanographic processes, I obtained "environmental" anomaly time-series data from H.J. Freeland (coastal surface temperature and salinity) and S. Tabata (Bakun upwelling index), both at the Institute of Ocean Sciences, Sidney, B.C. All three variables have been normalized to their 1979–91 mean value (as for the zooplankton time series). Temperature and salinity anomalies have been low-pass filtered with 7-month running means; this removes some high-frequency jitter but retains most of the variance in the time series (H.J. Freeland, personal communication). Upwelling-index anomalies have been seasonally averaged within October–March and April–September intervals for each year to emphasize the contrast between "winter" and "summer" components of interannual variability.

Results and discussion

Updated Average Seasonal Cycles

Average seasonal cycles updated to include 1990 and 1991 sampling periods are shown in Fig. 2. Compared with the averages reported by Mackas (1992), there are some differences in absolute and relative abundance of taxa (most notably higher biomass of euphausiids). This and other differences will be more obvious in the anomaly time series. But the major features of the seasonal and spatial patterns remain unchanged: "inner shelf," "gyre," and "offshore" regions, as derived from bathymetric and current patterns, can be distinguished during most sampling periods based on differing ratios of abundance of the major zooplankton species. The average annual maximum of zooplankton biomass is in late spring – early summer on the continental shelf, and in middle to late summer over the continental shelf break and slope. Mackas (1992) noted that the summer–autumn decline of herbivorous copepod biomass on the inner and middle parts of the shelf occurs during a period of sustained high food supply (3–8.5 mg·m^{-3} chlorophyll*a*), and interpreted this as evidence for rapid advective export of surface-layer zooplankton from the continental shelf during the summer upwelling season. Dominant shelf taxa include the euphausiid *Thysanoessa spinifera*; the copepods *Calanus marshallae*, *Pseudocalanus mimus*, and *Acartia longiremis*; the chaetognath *Sagitta elegans*; the ctenophore *Pleurobrachia*, and various hydromedusae. Along and seaward of the shelf break, the zooplankton community shows more oceanic influence. The copepods *C. marshallae* and *P. mimus* remain abundant but are joined by the subarctic oceanic copepods *Metridia pacifica* and (seasonally) *Neocalanus plumchrus* and *N. cristatus*; by the euphausiid *Euphausia pacifica*, the chaetognath *Eukrohnia hamata*; and intermittently by dense populations of salps (*Salpa fusiformis* and *S. aspera*) or of hydromedusae.

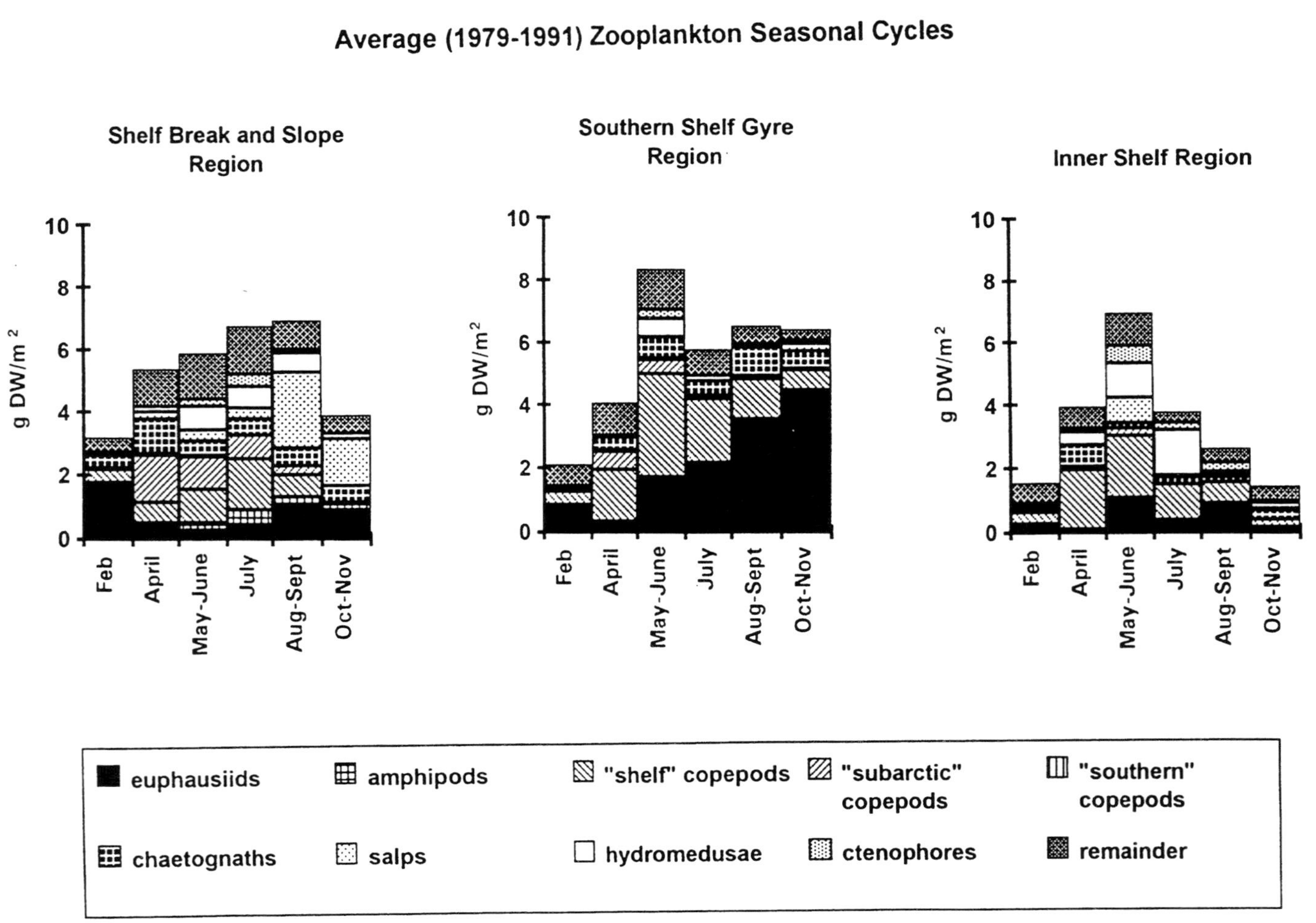

Fig. 2. Thirteen-year average (1979–91) seasonal cycle of zooplankton biomass and community composition in each of the three statistical regions. Most of the total zooplankton biomass is accounted for by about 20 species (see Table 1 for individual species forming each category; "Remainder" includes the total biomass from all other zooplankton taxa). There is a strong seasonal cycle in all three regions, but the regions differ in species mix and in the timing and duration of the biomass peak. This figure is a summary update of more detailed plots presented in Mackas (1992).

Confidence interval, magnitude, and duration of zooplankton anomalies

The 95% confidence level for the zooplankton anomalies varies among species. Significance thresholds for the annual mean anomalies ($2s_{<<b>>}$) are lowest for small, very abundant copepods (*Acartia* and *Pseudocalanus* both about ±0.2); slightly higher for large copepods and chaetognaths (*Calanus* about ±0.3, *Neocalanus* spp. and *Sagitta* spp. about ±0.35); and highest for large, numerically less abundant, and more patchy taxa such as medusae, salps, and euphausiids (about ±0.4 to ±0.55). Within taxa, the differences in significance threshold between regions are small and result mostly from differing numbers of sample sites in each region.

It is worth noting that in all cases fairly substantial numbers of samples need to be averaged (in the order of 16–25) to detect without statistical ambiguity two- to three-fold interannual variations in abundance. Without this averaging, added noise from aliased small scale patchiness will mask the pattern of year-to-year differences.

Persistence and mean-squared magnitude of the zooplankton anomalies are shown by time-lagged autocovariance plots for some of the dominant taxa (Fig. 3). These plots describe the average decay of within-region statistical resemblance between mean anomalies from individual sampling periods as a function of their separation in time. Note that these are autocovariances of the **anomaly** time-series $<<b'(\text{reg})>>_{s,t'}$, not of the biomass time series $<<B(\text{reg})>>_{s,t}$ or $<<D(\text{reg})>>_{s,t}$. Because time-correlated variability caused by the average annual cycle has already been removed, there is no a priori expection of nonzero autocovariance except at zero time lag. Any sequences of adjoining autocovariance estimates that are systematically greater than or less than zero provide evidence that the estimated anomalies are real (not aliased individual sample variability) and that they last longer than the average spacing between cruises.

Autocovariance curves vary both among regions and among taxa, but confirm the existance of persistent anomalies for nearly all of the dominant taxa. Within taxa, the anomalies tend to be stronger (larger autocovariance at zero lag) and/or more persistent (longer lag to zero autocovariance) seaward of the shelf break. There are also large differences in persistence of anomalies between taxa. Durations (twice the time lag to zero crossing of the autocovariance curve) are briefest for gelatinous zooplankton: essentially zero for medusae on the continental shelf, less than 1 yr for salps (shown for the

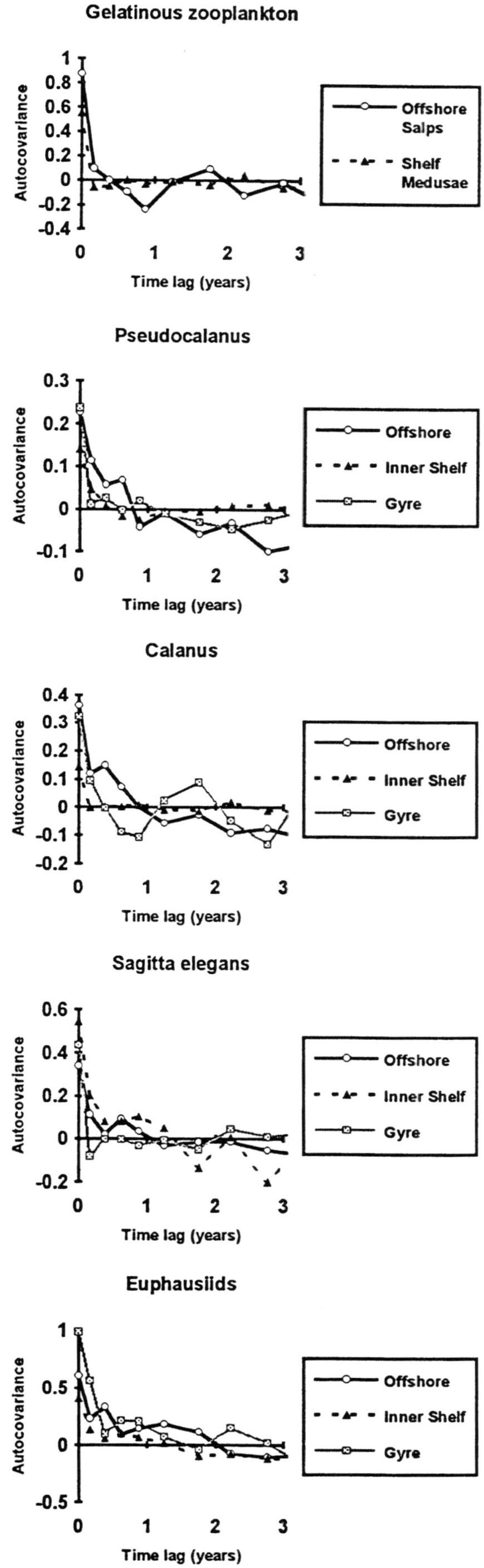

Fig. 3. Autocovariance of cruise-by-cruise anomaly time series of log-transformed zooplankton biomass for selected major taxa within each statistical region. The value at zero time lag is the mean square magnitude of anomalies within individual sampling periods. The remainder of the curve describes anomaly persistence by measuring the decrease of anomaly covariance as a function of increasing time separation between sampling periods. The expected duration of a positive or negative anomaly is twice the time lag to the first zero-crossing of the autocovariance curve. Gelatinous zooplankton have strong but brief episodes of anomalous abundance. Copepods (eg. *Calanus* and *Pseudocalanus*) have anomalies lasting about 1–2 yr. Chaetognath and euphausiid anomalies persist about 3–5 yr.

offshore region where they are most abundant) and ctenophores (not shown). Copepod anomalies persist about 1–2 yr for the dominant copepod species (*Calanus*, *Pseudocalanus*, *Neocalanus*, and *Acartia*). Anomaly duration increases to about 3–4 yr for euphausiids and chaetognaths. These estimates agree well with the 1 – 4 yr range of persistence times reported for anomalies of zooplankton biomass off central and southern California (Chelton et al. 1982; Roesler and Chelton 1987). They also provide some confidence that within-year averaging of anomaly time series is appropriate for many of the dominant taxa. However, annually averaged anomalies definitely underemphasize the intense but brief outbreaks of gelatinous zooplankton such as salps, medusae, and ctenophores, and may also do so for shorter lived copepods such as *Acartia*.

Zooplankton anomaly time series

Time-series of annual zooplankton anomalies are shown in Figs. 4–7. To save space and show shared trends, I have grouped as single graphs closely related taxa and/or those with similar ecological niches and spatial distribution patterns. As a rough guide to statistical significance of individual anomalies, tick spacing on the vertical axis for each graph was set to approximate or exceed the 95% significance threshold for taxa shown on that graph; column segments taller than the tick spacing represent anomalies larger than expected by change. The offshore (shelf-break and slope) region had the most diverse list of dominant taxa, and in general had the strongest year-to-year anomalies. Figures 4 and 5 show anomaly time series for the dominant offshore crustacean and noncrustacean taxa, respectively. Gyre region taxa are shown in Fig. 6 and inner-shelf region taxa in Fig. 7. The progressive offshore-to-coast reduction in mean-square anomaly amplitude mentioned in the previous section is apparent from comparison of euphausiid and copepod time series in Figs. 4, 6, and 7. Three features of anomaly timing and sequence are also noteworthy.

First, significant and substantial zooplankton anomalies (3 to 10-fold variation) occur throughout the time series, not just in 1 or 2 yr. For the abundant and well-censused taxa, the absolute magnitude of the annual anomalies exceeded their 95% confidence level threshold in about one third to one half the years. Conversely, a substantial fraction of the taxa (minimum about one third) had statistically significant anomalies in any single year. There was also no single time scale or mode of variation shared in a simple way by all taxa.

This result was somewhat surprising. I began this study expecting that the null hypothesis of "no significant interannual variation of La Perouse region zooplankton" would be rejected. But very strong interannual variation of zooplankton biomass and distribution coincident with El Niño – Southern Oscillation (ENSO) events had been observed in other studies of the west coast of North America (Chelton et al. 1982; Fulton and LeBrasseur 1985; McGowan 1984; Pearcy and Schoener 1987). A plausible alternative hypothesis for this study was therefore that La Perouse region zooplankton anomalies would be strongly significant and characteristic in El Niño years, but weak or absent in the intervening "normal" years. At least the second part of this alternate hypothesis is clearly not true.

To date, two El Niño events have reached the B.C. coast and been sampled by the La Perouse time series: 1986–87 (weak) and 1991–92 (moderate). The extent (0.8–1.5°C positive

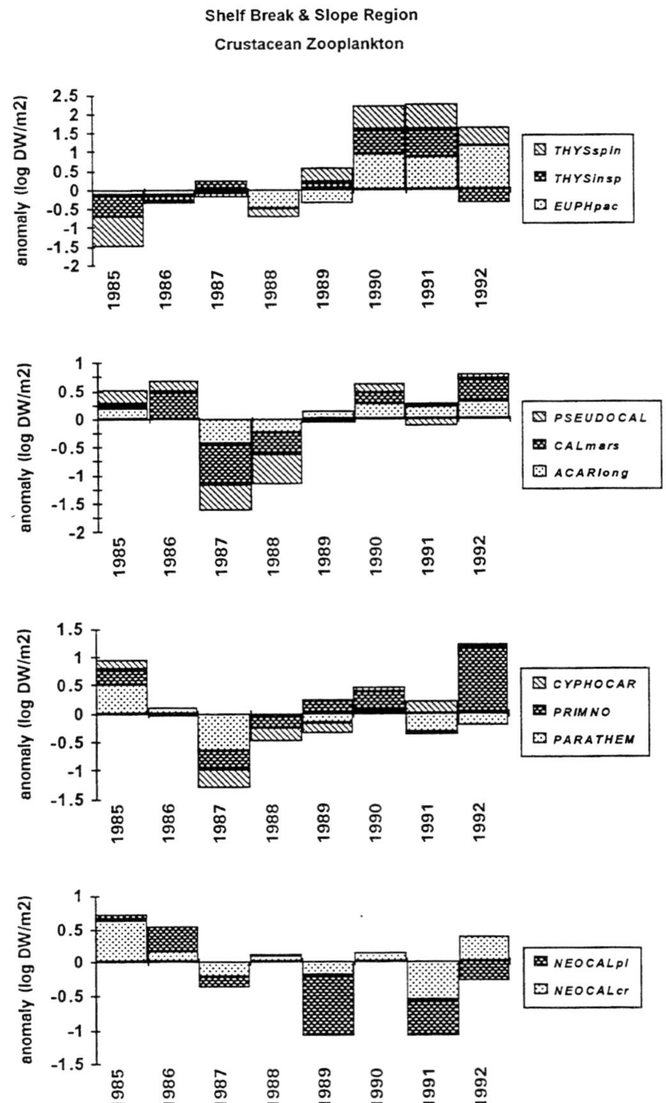

Fig. 4. Time-series of annual anomalies (average of all sampling periods within a given year) for major crustacean zooplankton taxa in the offshore region. To save space and show shared patterns, each graph groups individual time-series of two or three taxonomically and/or ecologically similar taxa. For most taxa, anomalies were largest in the offshore region. Euphausiids show a relatively steady biomass increase from 1985 through 1990. The "shelf copepods" *Calanus*, *Pseudocalanus*, and *Acartia*, and the amphipods share a 2-yr minimum (1987–88) followed by a gradual recovery. Offshore "subarctic" copepods (*Neocalanus* spp.) have a more prolonged and erratic minimum during 1987–91. See text for discussion of statistical significance levels for anomalies of various taxa. As a rough guide, anomalies are significant if their absolute magnitude (thickness of individual segments in the stacked bar graphs) exceeds the tick spacing on the vertical axis.

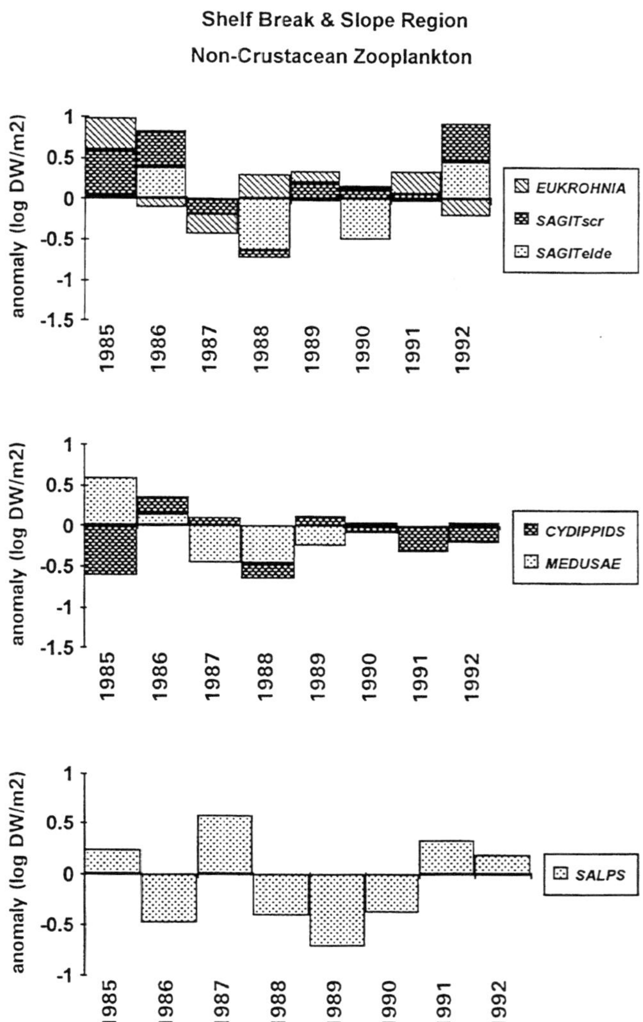

Fig. 5. Time series of annual anomalies for major noncrustacean zooplankton taxa in the offshore region. All three chaetognath species are on average relatively common in this region (Mackas 1992). Anomalies of chaetognath biomass were positive at the start and end of the series. *Sagitta elegans* (the species common on the continental shelf) had strongly negative anomalies in 1988 and 1990. Anomalies of ctenophores and medusae are mostly nonsignificant. Salps were extremely abundant in summer and autumn of 1987 and were unusually rare during 1988–90.

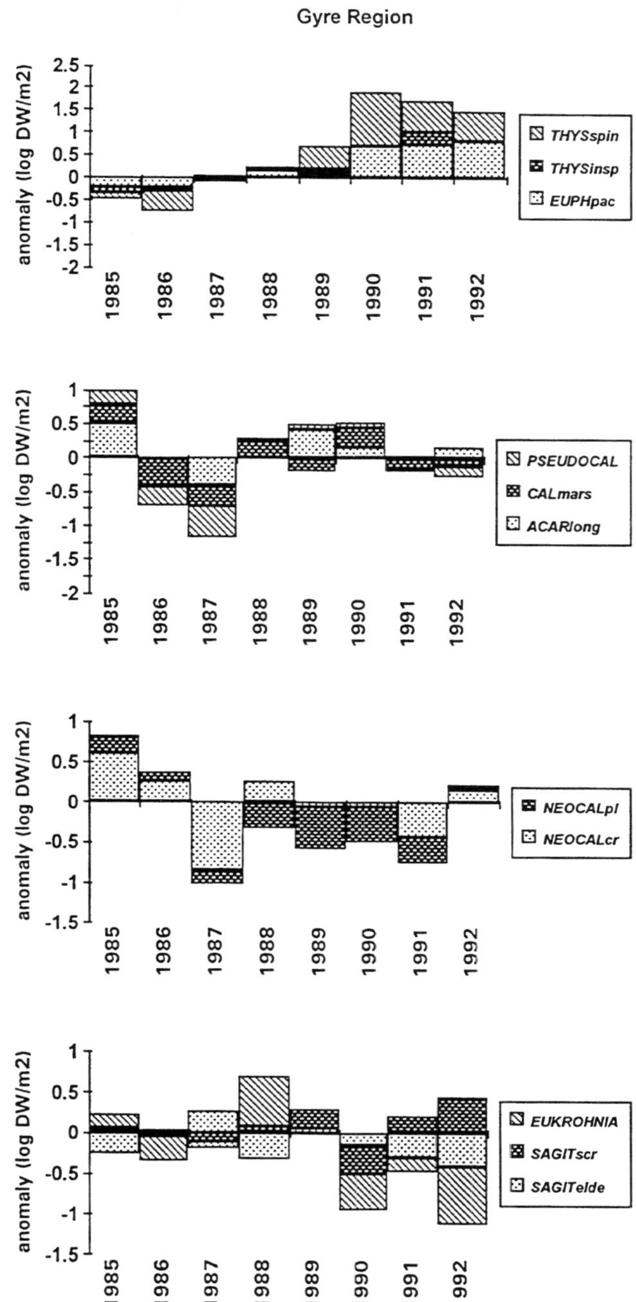

Fig. 6. Time series of annual anomalies for major zooplankton taxa in the southern shelf "gyre" region. *Thysanoessa spinifera* is the dominant euphausiid species. Euphausiid anomalies again show a trend of increasing biomass: negative in 1985–86 and strongly positive 1989–90. The "shelf" copepods have large negative anomalies in 1986–87. *Neocalanus* spp. have negative anomalies from 1987–91. Chaetognath anomalies are mostly small for the dominant species *Sagitta elegans* but are frequently significant for the "oceanic" species *Eukrohnia hamata* and *S. scrippsae*.

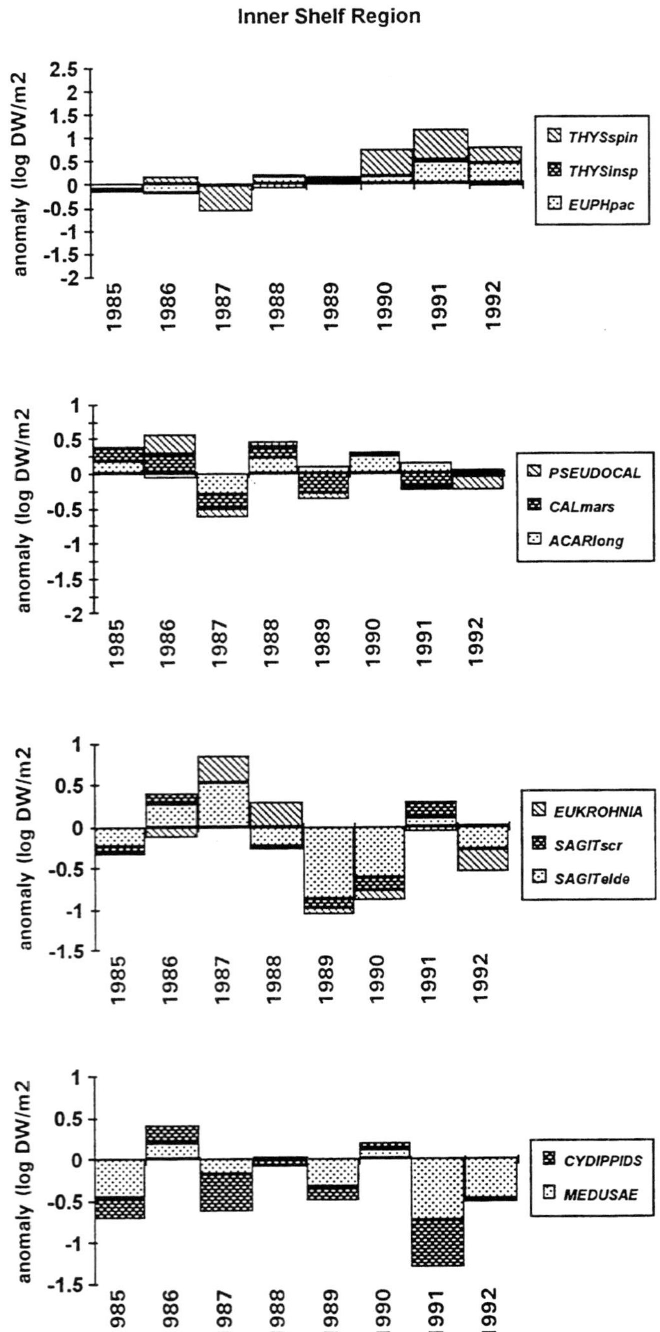

Fig. 7. Time series of annual anomalies for major zooplankton taxa in the inner shelf region. Despite the similarity of anomaly sequence between regions, the magnitude of the anomalies is typically smallest on the inner shelf and largest in the offshore region (compare Figs. 4, 6, and 7). The euphausiid *T. spinifera* again shows a trend of increasing biomass, and "shelf" copepods again have reduced abundance in 1987. The "shelf" chaetognath *S. elegans* has a strong positive anomaly in 1987 and negative anomalies during 1988–90. Annual anomalies of the gelatinous predators (ctenophores and medusae) are rarely significant, but do appear to covary with each other and with the biomass of the shelf copepod *Calanus*.

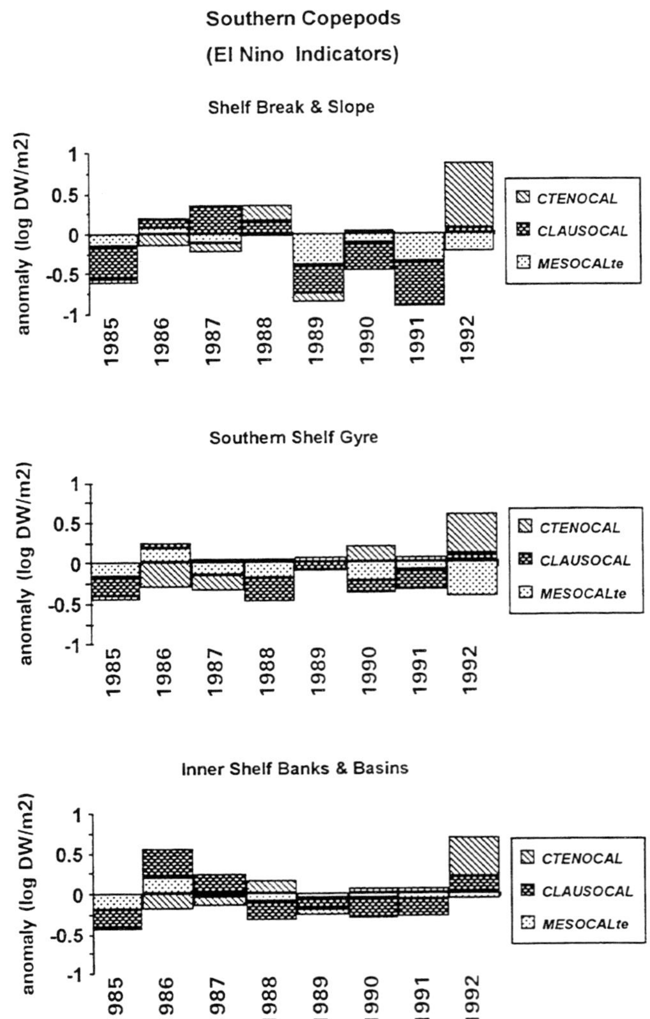

Fig. 8. Time series of annual anomalies for "southern" copepods endemic to the Oregon and California coast. High abundance of these species is indicative of increased transport of water and biota from the south. Their average abundance is high enough for them to be well censused, but they rarely contribute more than a few percent of total zooplankton biomass. Anomalies are positive in 1986–87 and in 1992, coincident with the two most recent El Niño events.

temperature anomalies) and duration (about 1 yr) of their effects on B.C. coastal water properties are illustrated in Fig. 9. Although some of the dominant zooplankton taxa had significant anomalies during the El Niño events (most notably high abundance of salps and low abundance of the major calanoid copepods in middle to late 1987), the early 1992 anomalies bear little qualitative resemblance to those in 1987, and anomalies of equally significant magnitude were present during the intervenine normal years. It is possible that the 1987 and 1992 events were too weak to produce major effects on the dominant zooplankton species and that the next very strong El Niño (i.e., comparable with the 1982–83 event) will have much larger within-year impact. But for the 1985 to mid-1992 period, the clearest evidence of immediate ENSO effects on zooplankton off the B.C. coast is in the abundance variations of rare rather than dominant species. Sampling coverage off British Columbia was poor in 1983; avoiding this situation in the future was part of the initial rationale for the La Perouse program. But we do know that several copepods endemic to the California and Oregon coasts (*Clausocalanus* spp., *Mesocalanus tenuicornis*, and *Ctenocalanus vanus*) were much more common than normal in 1983 (D.L. Mackas and H.A. Sefton, unpublished data). Between 1985 and 1992, the anomalies of these species were positive in 1986–87 and/or early 1992 and negative in other years (Fig. 8). Although larger than expected by chance, even these anomalies were not particularly large. I therefore conclude that interannual variability of most zooplankton off the B.C. coast is coupled only indirectly to ENSO forcing.

Second, euphausiid abundance in all three L Perouse Project regions increased strongly throughout most of the 1985–92 time span. The change from negative to positive anomalies (relative to the 1979–91 long-term average) was between 1987 and 1989. The upward trend may have peaked in 1990–91; concentrations in 1991–92 showed the first drop in several years. The total range of observed anomalies is equivalent to a 10- to 30-fold variation in abundance for each of the three local dominant euphausiid species (*Euphausia pacifica*, *Thysanoessa spinifera*, and *T. inspinata*). Although the bongo net tows used as the standard La Perouse sampling method probably underestimate euphausiid abundance (Fulton and LeBrasseur 1984; Robinson et al. 1993), there was no change in sampling methodology from 1985–92 and therefore no reason to attribute the observed trend to changes in capture efficiency.

Third, a number of taxa (including essentially all of the dominant crustacean species other than euphausiids) had their minimum abundance during the 1987–91 transition from low to high euphausiid abundance. Although most of the changes in abundance appear to begin in 1986 to 1987 (more or less coincident with the weak ENSO event), subsequent behaviour is extremely diverse. The three most abundant shelf copepods (*Calanus marshallae*, *Pseudocalanus*, and *Acartia longiremis*) all had strong negative anomalies of 1–2 yr duration in 1986–87 or 1987–88 (Figs. 4, 6, and 7). The gyre-region anomalies for these copepods (Fig. 6) were offset about 1 yr ahead of those in the offshore region (Fig. 4). Swings in abundance of several other taxa were more complex and prolonged (3–4 yr). Negative anomalies of the oceanic copepods *Neocalanus plumchrus* and *N. cristatus* were large in 1987 and 1989–91. Noncrustacean taxa with strong anomalies during the 1987–91 interval include salps (extremely abundant along and seaward of the shelf break in summer and autumn of 1987 but unusually rare by 1989) and chaetognaths.

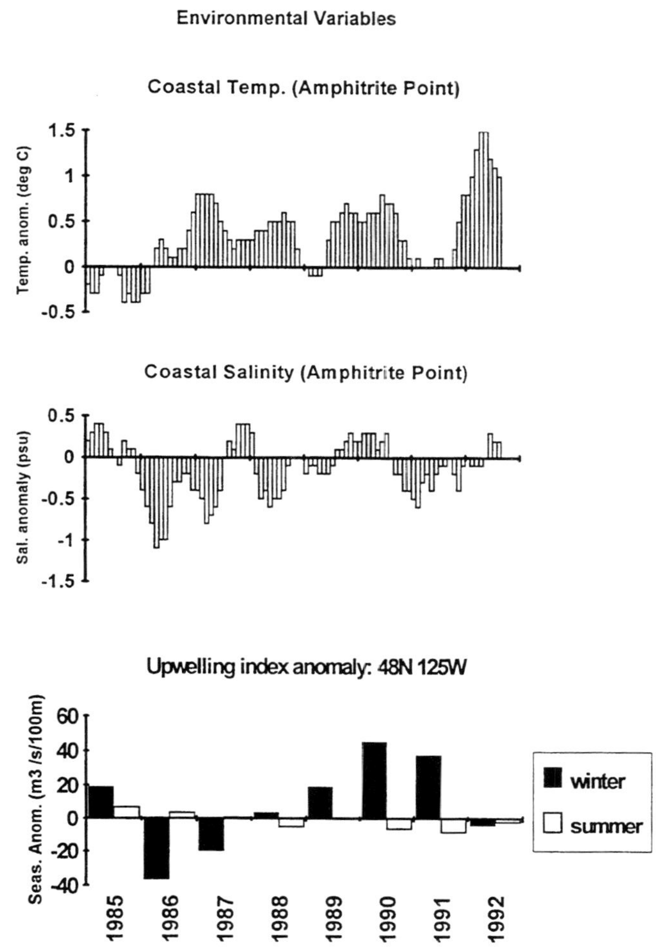

Fig. 9. Anomaly time-series for coastal surface temperature and salinity (measured at Amphitrite Point, 48° 55'N, 125° 32'W; data courtesy H. Freeland) and alongshore winds (Bakun upwelling index at 48°N, 125°W; data courtesy S. Tabata). All three series have been normalized to 1979–91 averages to match the zooplankton time series. Monthly upwelling index anomalies have been seasonally averaged within winter (October–March) and summer (April–September) intervals to emphasize the differing trends of the downwelling vs. upwelling seasons. The temperature time-series shows the 1–1.5°C positive anomalies associated with the 1986–87 and 1991–92 El Niño events. The upwelling index anomalies change sign in 1988. This is coincident with and probably part of the large-scale change in North Pacific atmospheric pressure patterns noted by Trenberth (1990). Note also the apparent 1988 sign change of the correlation between temperature and salinity anomalies.

Environmental Forcing

Anomaly time series for coastal surface temperature and salinity and Bakun upwelling index are shown in Fig. 9. As with the zooplankton anomalies, the three variables differ greatly in their sequence and in the persistence of high and low values. Both temperature and salinity anomalies contain considerable variability at time scales between 6 mo and 1 yr. Large positive anomalies in the temperature series include the winter–spring maxima associated with the 1987 and 1992 El Niños, but also a prolonged peak in 1989–90. Over the full 1985–92 period, correlation between the temperature and salinity anomalies is very weak ($r = 0.04$). But if the total series is divided into two roughly equal segments, the within-segment coherence between temperature and salinity is realtively strong and changes phase at about the end of 1988 ($r = -0.40$, 1985–88; $r = 0.51$, 1989–92). D.L. Mackas and S. Tabata are now examining this relationship for a longer time series.

Anomalies of the seasonally averaged Bakun upwelling index (bottom panel of Fig. 9) vary much more gradually than the coastal water properties, but again show indications of a change in 1988. Summer anomalies switch from weakly possitive to weakly negative (i.e., somewhat reduced summer upwelling in the most recent years). In winter, the normal wind pattern is from the south and downwelling favourable. In 1988, the winter anomalies switch from negative (i.e., stronger than average downwelling favourable winds from the south) to strongly positive (i.e., weaker than average winds from the south). Except for the similar low-frequency trends of upwelling index and euphausiid biomass, overall correlation of zooplankton anomalies with these environmental anomalies is relatively weak. The more striking association among the various time series is the common occurrence of a shift in anomaly pattern in or near 1988.

What can be concluded about linkage between low-frequence environmental forcing and zooplankton fluctuations off the B.C. coast? With statistical certainty, relatively little can be concluded from a time series of this length (5–10 additional yr of data would allow much more robust interpretation). I can state with considerable confidence that zooplankton-anomaly magnitudes, both as individual cases and root mean square, are larger than expected by chance. I can also conclude that, for most of the zooplankton taxa, individual abundance fluctuations (anomaly "events") persist for one to several years. This is equivalent to several generation times: life-cycle durations of the dominant taxa range from weeks to a few months for salps and small copepods, up to about a year for large copepods and euphausiids. The fact that two or more generations are included in any given anomaly "event" is evidence that the environmental causation is also persistent. But this very persistence makes statistical interpretation of the fluctuations more difficult. Long autocovariance time scales reduce the number of degrees of freedom because anomalies observed in adjacent years are not independent. The statistical power of an 8-yr time series is therefore also reduced. This is not because the signals are weak but because too few complete cycles of variation have been observed. Coherences between the multi-year anomaly sequences are provocative but must be interpreted as based on one or two rather than many observations.

The range of time scales of the zooplankton anomalies allow discounting of some possibilities. I have already discussed the evidence against direct ENSO forcing: mismatch of excitation and response anomaly time scales and relative dissimilarity of response between the two El Niño years. More generally, a primarily low-frequency (multiyear) response by species with short generation times to transient (<1 yr) environmental disturbance seems less likely than strong response at the environmental forcing frequency. Prey–predator oscillations are also implausible as a dominant source of interannual variability because of the lack of appropriate frequency and phase matching between known predator and prey pairs such as chaetognaths and copepods or fish and euphausiids.

A plausible cause of environmental variability matching both the timing and time scale of the zooplankton and local environmental anomalies is the interaction of the ocean with the intensity and location of North Pacific atmospheric pressure anomalies. The winter season averaged intensity of the low-pressure zone in the Northeast Pacific (the Aleutian Low) fluctuates strongly at decadal time scales, and this fluctuation in turn affects heat exchange, water properties, and currents in the Northeast Pacific (Trenberth 1990; Higuchi et al. 1991; Lin et al. 1992). Trenberth (1990) further notes that the Aleutian low was anomalously deep for a prolonged period from 1976–88 and argues that this produced a sustained and recognizable "regime" of weather conditions in the northeast Pacific. Wooster and Hollowed (1995) discuss in greater detail the physical oceanographic correlates of this regime. Various authors (Hollowed and Wooster 1995; Trenberth and Hurrell 1995) have also noted a range of biological responses to the start of this episode in 1976. The data reported in this paper may reflect local ecosystem responses to the end of the episode in 1988.

Acknowledgements

A number of people have contributed to the data presented here. I thank participants in the La Perouse Project for help with zooplankton sampling; M. Galbraith, H. Sefton, E. Anderson, and B. Boettger for taxonomic identifications; and R. Thomson, K. Denman, K. Burns, I. Perry, and an anonymous reviewer for valuable discussions and comments.

References

Brodeur, R.D., and D.M. Ware. 1992. Long-term variability in zooplankton biomass in the subarctic Pacific Ocean. Fish. Oceanogr. 1: 32–38.

Chelton, D.B., P.A. Bernal, and J.A. McGowan. 1982. Large scale interannual physical and biological interaction in the California Current. J. Mar. Res. 40: 1095–1125.

Colebrook, J.M. 1977. Annual fluctuations in biomass of taxonomic groups of zooplankton in the California Current, 1955–59. Fish. Bull. U.S. 75: 357–368.

Colebrook, J.M. 1982. Continuous plankton records: phytoplankton, zooplankton and environment, north-east Atlantic and North Sea, 1958–1980. Oceanol. Acta 5: 473–480.

Colebrook, J.M., G.A. Robinson, H.G. Hunt, J. Roskell, A.W.G. John, H.H. Bottrell, J.A. Lindley, N.R. Collins, and N.C. Halliday. 1984. Continuous plankton records: a possible reversal in the downward trend in the abundance of the plankton of the North Sea and the Northeast Atlantic. J. Cons. int. Explor. Mer 41: 304–306.

Couper, A. [ed.]. 1989. The Times atlas and encyclopedia of the sea. Times Books Ltd., London, UK. 272 p.

Crawford, W.R., and R.K. Dewey. 1989. Turbulence and mixing: Sources of nutrients on the Vancouver Island continental shelf. Atmos.-Ocean 27: 428–442.

Crawford, W.R., and R.E. Thomson. 1991. Physical oceanography of the western Canadian continental shelf. Cont. Shelf Res. 11: 669–683.

Denman, K.L., H.J. Freeland, and D.L. Mackas. 1989. Comparisons of time scales for biomass transfer up the marine food web and coastal transport processes, p. 255–264. *In* R.J. Beamish and G.A. McFarland [ed.] Effects of ocean variability on recruitment and an evaluation of parameters used in stock assessment models. Can. Spec. Publ. Fish. Aquat. Sci. 108.

Denman, K.L., D.L. Mackas, H.J. Freeland, M.J. Austin, and S.H. Hill. 1981. Persistent upwelling and mesoscale zones of high productivity off the west coast of Vancouver Island, Canada, p. 514–521. *In* F.A. Richards [ed.] Coastal upwelling. American Geophysical Union, Washington, D.C.

FAO. 1981. Atlas of the living resources of the sea. Rome.

Forbes, J.R., S.R. Buckingham, and A.T. Earme. 1987. Phytoplankton productivity experiments in British Columbia coastal waters, 1986. Can. Data Rep. Hydrogr. Ocean Sci. 56: 169 p.

Freeland, H.J., W.R. Crawford, and R.E. Thomson. 1984. Currents along the Pacific coast of Canada. Atmos.-Ocean 22: 151–172.

Freeland, H.J., and K.L. Denman. 1982. A topographically controlled upwelling center off Vancouver Island. J. Mar. Res. 40: 1069–1093.

Freeland, H.J., and P. McIntosh. 1989. The vorticity balance on the southern British Columbia continental shelf. Atmos.-Ocean 27: 643–657.

Frost, B.W. 1983. Interannual variation of zooplankton standing stock in the open Gulf of Alaska, p. 146–157. *In* W. Wooster [ed.] From year to year: interannual variability of the environment and fisheries of the Gulf of Alaska and eastern Bering Sea. University of Washington Press, Seattle, WA.

Fulton, J., and R. LeBrasseur. 1984. Euphausiids of the continental shelf and slope of the Pacific coast of Canada. La Mer 22: 268–276.

Fulton, J.D., and R.J. LeBrasseur. 1985. Interannual shifting of the subarctic boundary and some of the biotic effects on juvenile salmonids, p. 237–252. *In* W. Wooster and D. Fluharty [ed.] El Niño north: Niño effects in the eastern subarctic Pacific Ocean. Washington Sea Grant Program, Seattle, WA.

GLOBEC. 1992. Eastern boundary current program: report on climate change and the California Current ecosystem. University of California Press, Davis, CA. 99 p.

Higuchi, K., C.A. Lin, A. Shabbar, and J.L. Knox. 1991. Interannual variability of the January tropospheric meridional eddy sensible heat transport in the northern latitudes. J. Meteorol. Soc. Jpn. 69: 459–472.

Hollowed, A.B., and W.S. Wooster. 1995. Decadal scale variations in the subarctic Pacific. II. Response of northeast Pacific fish stocks, p. 373–385. *In* R.J. Beamish [ed.] Climate change and northern fish populations. Can. Spec. Publ. Fish. Aquat. Sci. 121.

Lin, C.A., K. Higuchi, and R. Michaud. 1992. Interannual variability of the surface heat budget over the northern North Atlantic and North Pacific oceans. Theor. Appl. Climatol. 45: 161–165.

Mackas, D.L. 1984. Spatial autocorrelation of plankton community composition in a continental shelf ecosystem. Limnol. Oceanogr. 29: 451–471.

Mackas, D.L. 1992. Seasonal cycle of zooplankton off southwestern British Columbia: 1979–89. Can. J. Fish. Aquat. Sci. 49: 903–921.

Mackas, D.L., K.L. Denman, and M.R. Abbott. 1985. Plankton patchiness: biology in the physical vernacular. Bull. Mar. Sci. 37: 652–674.

Mackas, D.L., G.C. Louttit, and M.J. Austin. 1978. Spatial distribution of phytoplankton and zooplankton in British Columbia coastal waters. Can. J. Fish. Aquat. Sci. 37: 1476–1487.

Mackas, D.L., and H.A. Sefton. 1982. Plankton species assemblages off southern Vancouver Island: geographic pattern and temporal variability. J. Mar. Res. 40: 1173–1200.

McGowan, J.A. 1984. The California El Niño, 1983. Oceanus 27: 48–51.

Parrish, R.H., C.S. Nelson, and A. Bakun. 1981. Transport mechanisms and reproductive success of fishes in the California Current. Biol. Oceanogr. 1: 175–203.

Pearcy, W.J., and A. Schoener. 1987. Changes in the marine biota coincident with the 1982–1983 El Niño in the northeastern subarctic Pacific Ocean. J. Geophys. Res. 92: 14 417 – 14 428.

Robinson, C.L.K., D.M. Ware, and T.R. Parsons. 1983. Simulated annual plankton production in the northeastern Pacific Coastal Upwelling Domain. J. Plankton Res. 15: 161–183.

Roesler, C.S., and D.B. Chelton. 1987. Zooplankton variability in the California Current, 1951–1982. Calif. Coop. Oceanic Fish. Invest. Rep. 28: 59–96.

Sherman, K., and L.M. Alexander [ed.]. 1986. Variability and management of large marine ecosystems. AAAS Sel. Symp. 99: 319 p.

Thomas, A.C., and W.J. Emery. 1986. Winter hydrography and plankton distribution on the southern British Columbia continental shelf. Can. J. Fish. Aquat. Sci. 43: 1249–1258.

Thomson, R.E., B.M. Hickey, and P.H. LeBlond. 1989. The Vancouver Island coastal current: fisheries barrier and conduit, p. 265–296. *In* R.J. Beamish and G.A. McFarlane [ed.] Effects of ocean variability on recruitment and an evaluation of parameters used in stock assessment models. Can. Spec. Publ. Fish. Aquat. Sci. 108.

Thomson, R.E., and D.M. Ware. 1988. Oceanic factors affecting the distribution and recruitment of west coast fisheries, p. 31–65. *In* M. Sinclair, J.T. Anderson, M. Chadwick, J. Gagne, W.D. McKone, J.C. Rice, and D. Ware [ed.] Report from the National Workshop on Recruitment. Can. Tech. Rep. Fish. Aquat. Sci. 1626: 261p.

Trenberth, K.E. 1990. Recent observed interdecadal climate changes in the northern hemisphere. Bull. Amer. Meterol. Soc. 71: 988–993.

Trenberth, K.E., and J.W. Hurrell. 1995. Decadal coupled atmosphere–ocean variations in the North Pacific Ocean, p. 15–24. *In* R.J. Beamish [ed.] Climate change and northern fish populations. Can. Spec. Publ. Fish. Aquat. Sci. 121.

Wooster, W.S., and A.B. Hollowed. 1995. Decadal scale variations in the subarctic Pacific: I. Winter ocean conditions, p. 81–85. *In* R.J. Beamish [ed.] Climate change and northern fish populations. Can. Spec. Publ. Fish. Aquat. Sci. 121.

Growth and recruitment of freshwater drum (*Aplodinotus grunniens*) as related to long-term temperature patterns[1]

Donald L. Pereira[2], Christopher Bingham, George R. Spangler, Yosef Cohen, David J. Conner, and Paul K. Cunningham

Pereira, D.L., C. Bingham, G.R. Spangler, Y. Cohen, D.J. Conner, and P.K. Cunningham. 1995. Growth and recruitment of freshwater drum (*Aplodinotus grunniens*) as related to long-term temperature patterns, p. 617–629. *In* R.J. Beamish [ed.] Climate change and northern fish populations. Can. Spec. Publ. Fish. Aquat. Sci. 121.

Abstract: Freshwater drum (*Aplodinotus grunniens*) in the Red Lakes, Minnesota, exhibit slow growth and episodic recruitment. Periods with no apparent recruitment are as long as 10 yr. Using sagittal otoliths, we constructed a growth history dating from 1878 to the present. The prominent, extant year-classes include 1955, 1970, and 1983. Spectral analysis indicates a dominant periodicity in otolith growth of 7 yr. Otolith growth is more closely correlated with summer than winter temperatures, and year-classes are formed in years with warm summers. Otolith growth was not related to the Southern Oscillation, though superposed epoch analysis indicates a possible linkage between these two variables. Dynamics (i.e., recruitment and otolith growth) of freshwater drum in Blackduck Lake, Minnesota, were quite similar to dynamics in the Red Lakes. In Blackduck Lake, the 1970 cohort was the only year-class detected from 1975 through 1991; growth of this year-class was correlated with growth of the same year-class from the Red Lakes. These results are consistent with the hypothesis that fish species located close to the northern limit of their range and with a wide latitudinal distribution may be candidates for indicators of climate change.

Résumé : Le malachigan (*Aplodinotus grunniens*) des lacs Red, au Minnesota, montre une croissance lente et un recrutement épisodique. Les périodes dépourvues de recrutement apparent peuvent atteindre 10 ans. Grâce aux otolites sagittaux, nous avons établi les antécédents de la croissance de ce poisson de 1878 à aujourd'hui. Les classes annuelles dominantes et présentes comprennent celles de 1955, de 1970 et de 1983. L'analyse spectrale révèle une périodicité dominante de sept ans dans la croissance des otolites. La croissance des otolites est en corrélation plus étroite avec les températures d'été qu'avec celles d'hiver, et les classes annuelles se sont formées au cours des années caractérisées par un été chaud. La croissance des otolites n'était pas liée à la présence de l'Oscillation australe, bien que l'analyse d'époques superposées révèle la présence d'un lien possible entre les deux variables. La dynamique (c'est-à-dire le recrutement et la croissance des otolites) du malachigan du lac Blackduck, au Minnesota, était très semblable à celle des lacs Red. Dans le premier, la cohorte de 1970 était la seule classe annuelle décelée entre 1975 et 1991; la croissance de cette classe annuelle était en corrélation avec celle de la même classe annuelle des lacs Red. Ces résultats sont compatibles avec l'hypothèse voulant que les espèces de poissons situées près de la limite nord de leur aire de répartition et qui sont distribuées sur plusieurs latitudes peuvent être candidates pour servir d'indicateurs du changement climatique.

D.L. Pereira, G.R. Spangler and **Y. Cohen.** Department of Fisheries and Wildlife, University of Minnesota, 200 Hodson Hall, 1980 Folwell Avenue, St. Paul, MN 55108, USA.

C. Bingham. School of Statistics, University of Minnesota, 352 Classroom-Office Building, 1994 Buford Avenue, St. Paul, MN 55108, USA.

D.J. Conner. Tribal Fisheries Assessment Unit, Red Lake Band of Chippewa Indians, P.O. Box 279, Red Lake, MN 56671, USA.

P.K. Cunningham. Bureau of Fish Management, Wisconsin Department of Natural Resources, 101 South Webster Street, P.O. Box 7921, Madison, WI 53707-7921, USA.

[1] Publication No. 20 436 in the Scientific Journal Series, University of Minnesota Agricultural Experiment Station.

[2] Present address: Minnesota Department of Natural Resources, 1200 Warner Road, St. Paul, MN 55106, USA.

Introduction

The Red Lakes are located in northwestern Minnesota at a latitude of approximately 48°N. They consist of two connected basins with a surface area of 116 876 ha (Fig. 1). Maximum and mean depths are 10.7 and 5.5 m, respectively. The lakes are subject to wind mixing and seldom stratify thermally during the ice-free period. The Red Lake Band of Chippewa Indians has operated a commercial fishery on the Red Lakes since the turn of the century. Since fisheries assessments by Smith (1977), changes have occurred in the fish community, the most notable of which is the increased commercial harvest of freshwater drum (*Aplodinotus grunniens*). Other, recent developments in this fishery include increased interannual variation in the harvests of walleye (*Stizostedion vitreum vitreum*) and yellow perch (*Perca flavescens*) (Pereira et al. 1992).

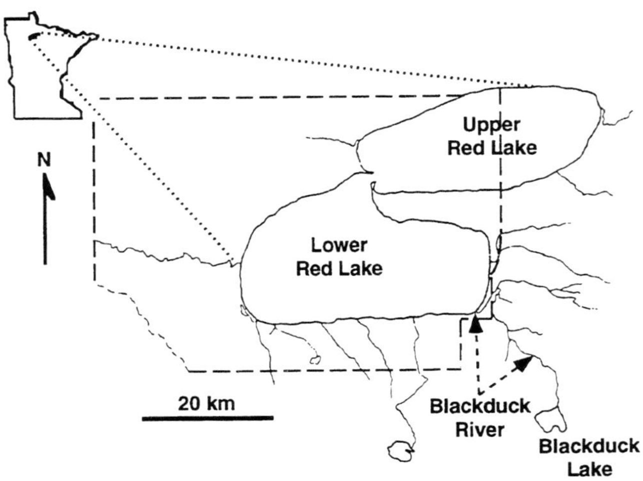

Fig. 1. Map of the Red Lakes and Blackduck Lake, Minnesota. Dashed lines indicate boundaries of the Red Lake Chippewa Indian Reservation. The Blackduck River flows north out of Blackduck Lake and is one of the primary tributaries to Lower Red Lake. Adapted from Smith (1977).

During earlier fisheries assessments (Smith 1977, and references therein), freshwater drum were ignored, and dynamics of this species in the Red Lakes were unknown. Red Lakes freshwater drum has low market value, and is considered a nuisance in years following its increase. Detailed stock assessment of this species began in 1988. At this time, we discovered that freshwater drum in the Red Lakes are longlived and grow slowly with episodic recruitment.

Blackduck Lake is located approximately 20 km southeast of Lower Red Lake (Fig. 1). It has a surface area of 1116 ha, maximum depth of 8.5 m, and is isothermal during the ice-free period. It is drained by the Blackduck River that is one of the primary tributaries of Lower Red Lake. Fish movement between these two systems is probably negligible due to numerous obstructions throughout most of the Blackduck River. In 1991, freshwater drum were sampled during standard fish community assessment conducted by the Minnesota Department of Natural Resources. Dynamics of freshwater drum in Blackduck Lake are similar to those in the Red Lakes.

In this paper we describe growth and recruitment of freshwater drum in the Red Lakes, and potential links to climate. The dynamics of drum may be affected by interactions with other species (Pereira et al. 1992) and by abiotic factors such as climate. We hypothesize that temperature is a strong determinant of drum dynamics, given that the species is close to the northern limit of its range in the Red Lakes, and its southern limit extends to Guatemala (Scott and Crossman 1973). Therefore, our primary objective is to ascertain the influence of climate on the dynamics of freshwater drum in the Red Lakes. We have included analysis of freshwater drum in Blackduck Lake as a comparison to the Red Lakes in an attempt to separate effects due to other species and climate.

Methods

We sampled freshwater drum from a variety of sources in the Red Lakes. The primary source of fish was the commercial gillnet fishery. Operation of this fishery is described in Pereira et al. (1992). As part of regular assessment, all fish species are randomly sampled in the fish processing facility. These sample data provide information on size and age distributions of all harvested species. The scarcity of large drum (i.e., greater than 360 mm) in the commercial fishery is partly due to low vulnerability of larger individuals to the 8.9 cm (stretched mesh) gill nets used in this fishery. Larger drum are occasionally sampled with other assessment gear types. For example, assessment biologists collected one drum with a beach seine in June 1989. This fish was 35 yr old (1955 year-class) and approximately 570 mm in total length. Because we rarely encountered fish of this size when sampling the commercial fishery, we started to purchase large drum directly from commercial fishers. Most of the fish reported here were sampled in 1990, with additional fish sampled in 1991. In addition to the commercial fishery gill nets, gear types used to sample drum in the Red Lakes included beach seines (6 mm mesh) and standard experimental gill nets consisting of five panels with graded mesh sizes. In Blackduck Lake, freshwater drum were sampled in 1991 with trap nets (19 mm mesh) and standard experimental gill nets as used in the Red Lakes.

We used sagittal otoliths for aging drum and for constructing a growth history or biochronology based on annular otolith growth. Casselman (1990) showed that otoliths continue to record seasonal growth in slow-growing fishes. We therefore assume that incremental otolith growth in old, slow-growing drum from the Red Lakes will serve as a general growth index. This assumption is supported by the finding that the relationship between otolith and body growth in freshwater drum is highly linear (Pereira et al. 1994). We fitted an exponential model to otolith increments by age. Residuals from this model provide an index of annual growth with the effects of age removed. Age validation, measurement accuracy, and methods used for estimating otolith growth histories are contained in Pereira (1992) and also in Pereira et al. (1994).

We applied time-series analysis in the frequency domain to test for relationships between otolith growth and a variety of climate indices. We employed spectral analysis using methods identical to those reported in Pereira et al. (1992). When necessary, we applied a simple least squares model to detrend time series with apparent trends. We did not find it necessary to transform any of the series before spectral estimation. We applied a triangular spectral window that spanned nine frequency ordinates.

Growth of freshwater drum in the Red Lakes may be affected by interannual climate. We analyzed two primary climate variables: air temperature and the Southern Oscillation (SO). The SO refers to a seesaw pattern in atmospheric pressure in the tropical Pacific. It is usually measured by taking the difference between atmospheric pressures at Tahiti and Darwin, Australia. The SO is strongly associated with the El Niño, the warming of surface waters in the Pacific Ocean. This process has been strongly implicated in marine biological perturbations throughout the western Pacific basin (Wooster and Fluharty 1985; Philander 1989; Demartini 1991), and teleconnections during strong El Niño events may extend to the interior of the North American continent (Ropelewski and Halpert 1986; Robertson 1989). We used the Tahiti–Darwin difference to index the SO (SOI) as described by Ropelewski and Jones (1987). These data are incomplete before 1933. Therefore, we established the 50-yr period from 1933 through 1982 for time-series analysis between growth and climate.

We used air temperature data from the Crookston, Minn. weather station as a surrogate for water temperature data because the latter were generally not available for the Red Lakes. This weather site is located approximately 105 km due west of the western limit of Lower Red Lake. Though there are closer stations (e.g., Red Lake, on the shores of Lower Red Lake), only the Crookston station provided continuous data suitable for time-series analysis that extended back to 1890. We calculated cumulative degree-days above and below a number of base temperatures and for different windows of time during the calendar year. The method for determining cumulative degree-days below a base temperature is

$$\text{If MDT} \langle \text{ BASE}, \quad \text{then CDD} = \text{CDD} + \text{BASE} - \text{MDT}$$
$$\text{else CDD} = \text{CDD}$$

where MDT is mean daily temperature (i.e., mean of minimum and maximum), BASE is a base temperature (e.g., 0°C is the base used for freezing degree-days, or growing degree-days), and CDD is cumulative degree-days. For CDD above a base, MDT less BASE is summed for days with MDT > BASE. We chose windows of time during both winter and summer seasons. Robertson (1989) has shown that SO teleconnections associated with El Niño events are primarily manifested in the winter climate in Madison, Wis. Wallace (1985) supports this finding.

Growth of drum can be influenced by either intensity of the winter season, which will partly determine the extent of ice cover, or by the warmth of the growing season. Growth of drum may therefore be linked to either winter climate (e.g., the period of ice cover may be shorter during an El Niño year, as shown by Robertson (1989)) or summer climate (e.g., drum may grow well during warm summers). The winter temperature index (WIN) that we report here is CDD below a base of 0°C from September through February. The summer index (SUM) is mean June through September temperature. We chose to report results from these temperature indices because they showed highest coherence with both otolith growth and with the SOI. Pereira (1992) provides analysis of additional climate indices.

The application of superposed epoch analysis (SEA) is described in Pereira et al. (1995). We used SEA (Haurwitz and Brier 1981) to test for significant relationships between climatic key events and otolith growth from Red Lakes drum, and between recruitment events of Red Lakes drum and climate and otolith growth. We used the *W* statistic as defined by Prager and Hoenig (1989), which compares the key event year with its own background years before and after the event. We conducted all SEA with the FORTRAN program EPOCH by Hoenig et al. (1989). Growth data derived from otoliths and temperature indices are listed in Table 1.

Results

Size and Age Distributions of Freshwater Drum from the Red Lakes

Most drum randomly sampled in the 1989 fishery were between 260 and 360 mm (Fig. 2A). The 1970 and 1983 year-classes accounted for approximately 94% of all drum harvested (Fig. 2B). A sample of 60 large drum purchased directly from fishers ranged from 322 to 657 mm (Fig. 3A). They ranged in age from 20 to 57 yr, representing eight year-classes in addition to the obvious 1970 year-class (Fig. 3B). Also, previous researchers archived four drum otoliths sampled by

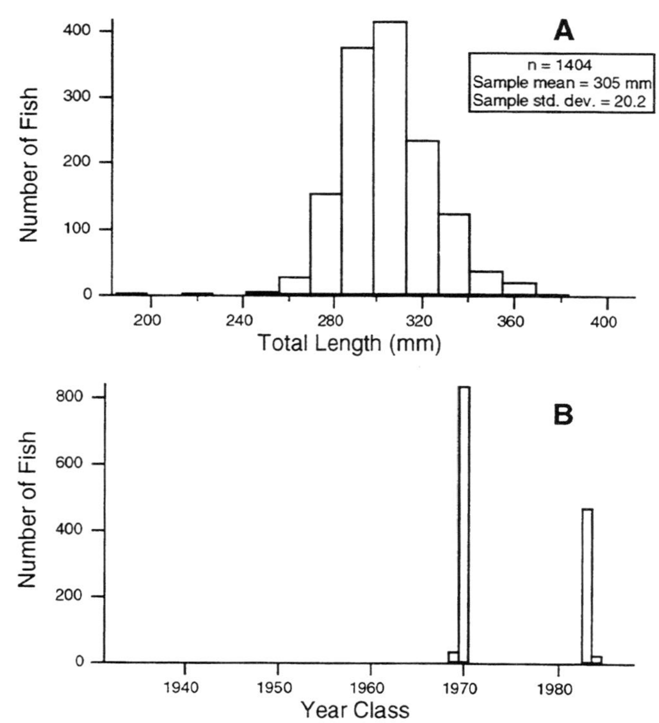

Fig. 2. (A) Length frequency of freshwater drum randomly sampled from the Red Lakes commercial fishery, 1989. This sample also includes two fish from the 1988 year-class. (B) Year-class distribution of freshwater drum from the Red Lakes commercial fishery, 1989.

Table 1. Freshwater drum otolith growth index from the Red Lakes and temperature indices (°C). See text for description of temperature indices.

Year	Growth index	Temperature indices Summer	Temperature indices Winter	Year	Growth index	Temperature indices Summer	Temperature indices Winter
1890	0.042	18.4	1008.3	1940	0.057	18.6	1228.1
1891	−0.079	18.2	1397.5	1941	0.093	18.7	921.1
1892	0.074	19.1	1924.2	1942	−0.049	16.5	1491.7
1893	−0.069	18.2	1798.1	1943	0.076	17.8	856.9
1894	0.155	19.5	1501.1	1944	−0.020	17.5	1024.7
1895	−0.078	16.9	1368.9	1945	−0.127	16.5	1422.5
1896	−0.286	17.3	1685.0	1946	0.071	17.6	1244.7
1897	0.096	18.2	1206.7	1947	0.026	17.8	1580.8
1898	−0.084	17.1	1541.9	1948	0.175	18.7	1453.3
1899	−0.178	17.1	1114.4	1949	0.064	18.2	1592.8
1900	0.057	18.6	1367.2	1950	−0.255	17.0	1468.9
1901	0.011	18.0	1125.6	1951	−0.126	16.5	1441.7
1902	−0.089	16.6	1346.9	1952	−0.033	18.2	948.1
1903	−0.073	16.1	1827.8	1953	0.062	18.0	1144.4
1904	−0.113	16.1	1391.4	1954	0.033	17.6	1158.3
1905	−0.007	17.7	1257.8	1955	0.163	18.8	1638.3
1906	−0.048	18.5	1563.6	1956	0.061	18.0	1276.7
1907	−0.035	16.6	1003.1	1957	0.047	18.3	918.6
1908	−0.033	18.2	1288.1	1958	−0.134	17.6	1510.8
1909	0.044	18.6	1433.1	1959	0.004	19.1	1131.9
1910	0.088	18.5	1576.4	1960	0.100	19.0	1180.3
1911	−0.066	17.4	1731.1	1961	0.074	19.0	1428.6
1912	−0.053	16.9	1321.7	1962	−0.002	18.2	1208.6
1913	0.093	18.4	1177.2	1963	0.164	20.2	1104.7
1914	−0.113	18.5	1334.7	1964	−0.008	17.7	1610.8
1915	−0.066	16.2	1557.2	1965	−0.149	16.7	1530.6
1916	0.025	18.1	1827.5	1966	0.034	19.0	1476.9
1917	−0.219	17.4	1646.7	1967	−0.074	18.4	1251.7
1918	−0.052	16.7	1002.5	1968	−0.073	18.2	1330.0
1919	0.151	19.1	1728.1	1969	−0.022	18.5	1300.0
1920	0.074	19.1	841.9	1970	0.110	20.1	1463.1
1921	0.131	20.0	1316.4	1971	0.086	18.7	1469.4
1922	−0.025	19.2	1291.7	1972	−0.141	18. 5	1209.4
1923	−0.009	19.1	1071.7	1973	−0.001	19.1	1427.5
1924	−0.208	16.8	1400.0	1974	−0.104	18.4	1044.2
1925	0.025	18.4	1000.0	1975	0.007	18.5	1082.2
1926	−0.116	17.1	1411.4	1976	0.099	20.0	1516.1
1927	0.027	17.4	1316.7	1977	0.073	17.8	1684.7
1928	−0.141	17.0	1412.5	1978	0.024	19.2	1912.8
1929	0.026	17.9	1386.9	1979	0.025	18.5	1156.7
1930	0.061	19.3	753.1	1980	−0.040	18.6	928.9
1931	0.089	19.7	1021.9	1981	−0.034	18.5	1454.7
1932	0.148	19.6	1373.6	1982	−0.124	16.8	969.2
1933	0.078	20.4	1343.6	1983	0.097	19.3	1279.2
1934	−0.089	17.6	1180.6	1984	−0.023	17.9	1464.2
1935	−0.130	18.1	2114.7	1985	−0.207	15.6	1659.2
1936	0.025	20.2	1642.2	1986	−0.113	17.6	898.6
1937	0.045	19.1	1407.2	1987	0.119	18.7	1228.1
1938	0.027	18.9	1463.1	1988	0.131	19.9	1460.0
1939	0.004	19.4	973.6	1989	0.050	20.1	1257.5

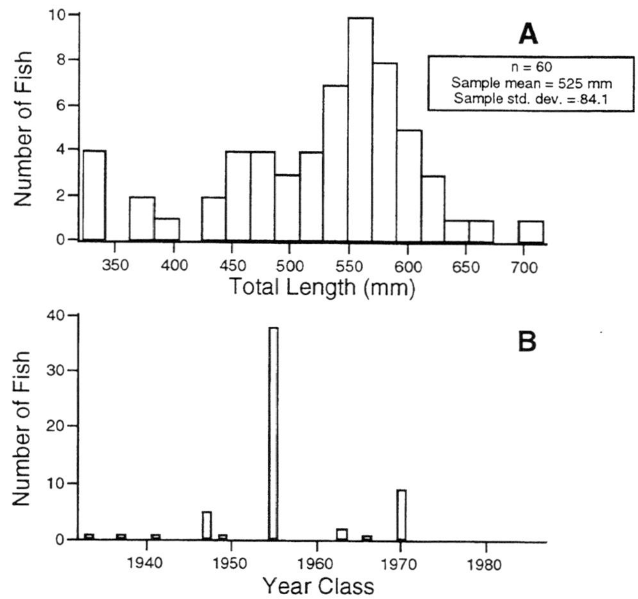

Fig. 3. (A) Length-frequency distribution of freshwater drum purchased directly from fishers, 1990. (B) Year-class distribution of freshwater drum purchased directly from fishers, 1990.

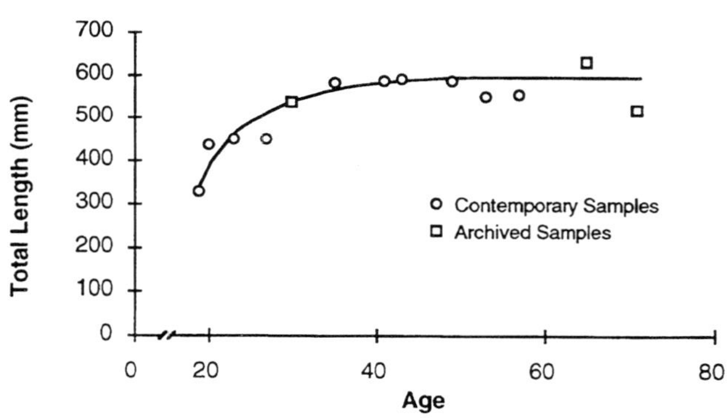

Fig. 4. Mean size at age of capture for 39 drum used for age and growth analysis. Archived samples included a total of four fish, with two being 71 yr old. All the contemporary samples were purchased from fishers in 1990 with the exception of eleven 19-yr old fish selected from the commercial fishery in 1989. The curve was drawn by eye.

a beach seine in 1949 (three fish) and 1951 (one fish). Two were 71 yr old at the time of capture (1878 year-class), one was 65 yr old (1886 year-class), and the fourth fish was 30 yr old (1919 year-class). Fish from three unique year-classes (1930, 1932, and 1940) were sampled from the commercial fishery in 1991. On the basis of age and length at capture, freshwater drum growth is slow in the Red Lakes (Fig. 4).

Spectral analysis of otolith growth

We subjected the index of otolith growth to spectral analysis to quantify periodicity in growth and to further test for potential correlates with climate. Table 2 lists year-class distribution

Table 2. Distribution of freshwater drum in the analysis, by year-class.

Year-class	No. of Fish
1878	2
1886	1
1919	1
1930	1
1932	1
1933	1
1937	1
1940	1
1941	1
1947	5
1949	1
1955	11
1963	2
1966	1
1970	12
Total	42

of freshwater drum in the growth analysis. The otolith-growth time series derived from these fish is illustrated in Fig. 5. A variety of biotic and abiotic factors affects body growth of fish. It is, therefore, likely that the forces influencing otolith growth may have changed during the 100-yr period analyzed here. We, therefore, decomposed the 100-yr series into the following overlapping 50-yr periods for estimating the autospectrum: 1890–1939, 1900–49, 1910–59, 1920–69, 1930–79, and 1940–89 (Fig. 5). For the 50-yr period beginning in 1930, the autospectrum indicates a significant amplitude at 0.14 (7.1 yr). Most of the variation at the frequency of 0.3 appears to occur during the period from 1890–1939, with amplitude equal to that at the frequency of 0.12. The 50-yr periods from 1900–49 and 1920–69 appear to have relatively more noise.

Spectral analysis of temperature and the Southern Oscillation

The two climate variables that we analyzed are air temperature and the SO. SO data were available starting in 1933; therefore, we chose the 50-yr period from 1933 through 1982 for time-series analysis between growth and climate. This period encompasses most of the 1930–79 period when drum otolith growth is relatively noise free (see Fig. 5). The otolith growth autospectrum for 1930–79 is similar to that for 1933–82, with one dominant, significant amplitude at a frequency of 0.14 (7.1 yr).

We selected the mean of the SOI from September through February for analysis. Table 3 lists mean SOIs for several 6-mo periods during 6 yr when strong El Niños occurred (1940, 1941, 1957, 1972, 1982, and 1986). We, therefore, chose the September–February period simply because it shows the greatest decrease in the SO during strong El Niños. Lough and Fritts (1985) reported that teleconnection patterns can be detected with several different variables associated with the SO. This suggests that our analyses with the SOI may not be too sensitive to the monthly time periods used. The autospectrum for the September–February SOI time series (Fig. 6)

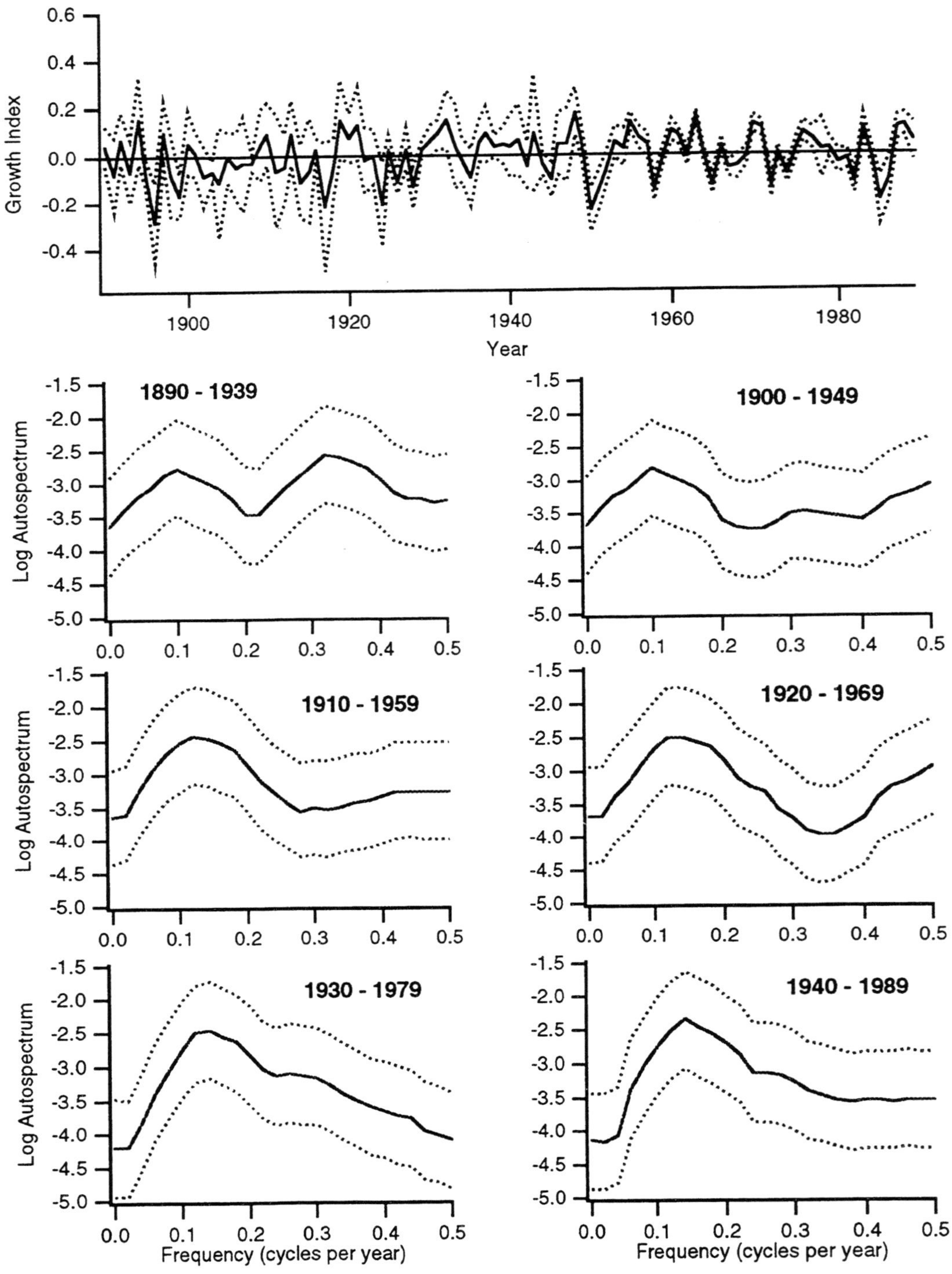

Fig. 5. Otolith growth index (top panel) and the log-transformed autospectrum of this time series estimated for overlapping 50-yr periods from 1890 through 1989. Dotted lines represent approximate 95% confidence intervals.

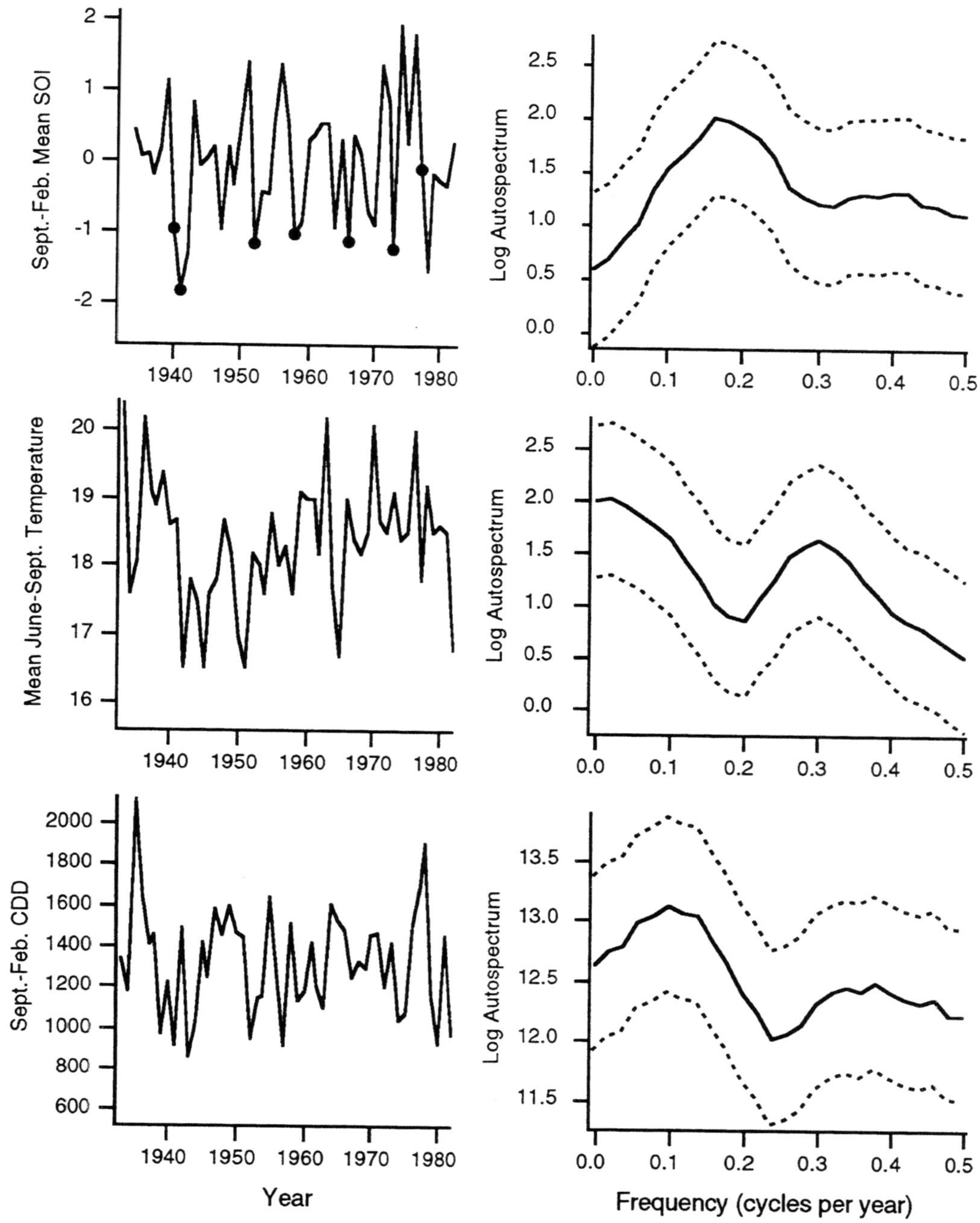

Fig. 6. Three climate series used for time-series analysis (1933–82) and their autospectra (dotted lines represent approximate 95% confidence intervals). For the Southern Oscillation Index (SOI Tahiti minus Darwin, mean of September through February, top graph), years correspond to the onset of established El Niño events. Solid dots represent years when a strong El Niño occurred. The middle graph is the summer temperature index (SUM, mean air temperature, June through September) and the bottom graph is the winter temperature index (WIN, freezing degree-days below a base of 0°C, September through February).

Table 3. Mean SOIs for 6-mo periods when strong El Niños occurred.

6-mo Period	Mean SOI
July–December	−1.41
August–January	−1.52
September–February	**−1.61**
October–March	−1.59
November–April	−1.52
Grand Mean	−1.53

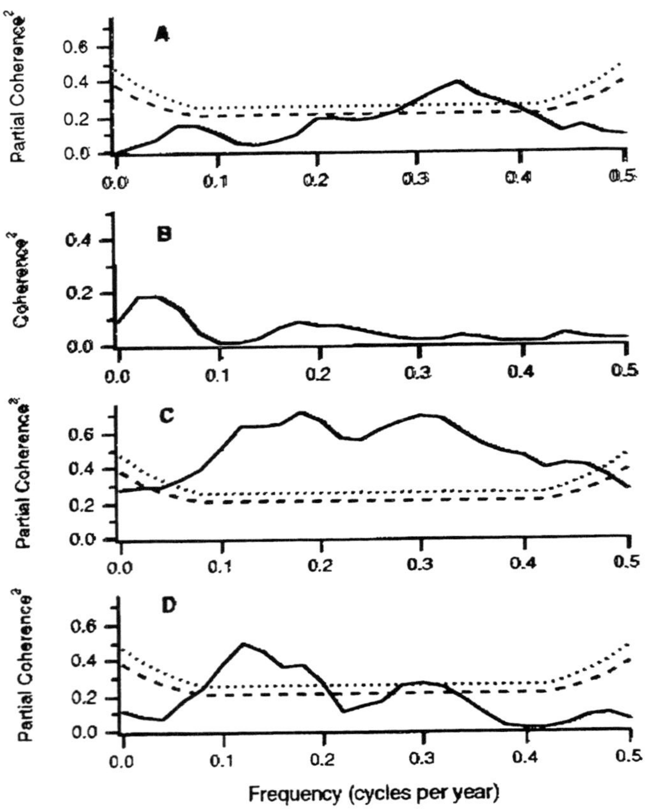

Fig. 7. Coherence between otolith growth and climate. Coherence is a measure of the linear correlation between two bivariate processes at a particular frequency (ω). For example, (A) is the partial squared coherence between SOI and WIN adjusted for SUM. This graph indicates significant coherence at a frequency around 0.3 (i.e., a 3-yr cycle). This indicates that cycles in these two time series at this particular frequency are correlated. Note, however, that the autospectra of these series in Fig. 6 indicate relatively low power at higher frequencies, suggesting that this coherence at frequencies around 0.3 translates to only a small amount of covariation between these variables. Partial coherence is directly analogous to partial correlation, where mutual correlation with other variables is removed for the two variables being considered. (B) Squared coherence between SOI and otolith growth index. (C) Partial squared coherence between otolith growth index and SUM, adjusted for WIN. (D) Partial squared coherence between otolith growth index and WIN, adjusted for SUM. Dotted lines represents 5% significance; broken lines represents 10% significance. See text for explanation of indices.

is similar to that for growth, with one dominant frequency at 0.16 (6.3 yr). The SO is considered to be quasiperiodic (Mysak 1986; Michaelsen 1989; Robertson 1989) with predominant frequencies changing between 0.1 and 0.35. The dominant frequency reported here is, therefore, within this range.

Autospectra for the temperature series are also illustrated in Fig. 6. WIN has a dominant frequency of 0.1 (10-yr periodicity). SUM has a dominant frequency of 0.3 (3-yr periodicity). Higher power at frequencies approaching zero in SUM was not due to nonstationarity because this series was detrended with a simple least squares model prior to estimating the autospectrum (Pereira 1992).

Coherence between growth and climate variables

We computed squared coherence between growth index and each of the climate indices, and between WIN and SOI (Fig. 7). SO teleconnections generally occur in North America during the winter months, so the SOI could potentially be coherent with winter temperature in Minnesota. Partial squared coherence between WIN and SOI adjusted for SUM (Fig. 7A) indicated significance only at a frequency of 0.35 (3-yr periodicity). This coherence does not correspond to the dominant frequencies in the autospectra of these two climate indices (see Fig. 6, top and bottom graphs). Squared coherence between SOI and otolith growth was also low at all frequencies (Fig. 7B). Therefore, there is no evidence that otolith growth is linked to teleconnections related to the SO. Partial squared coherence between SUM and otolith growth adjusted for WIN indicated a high band of coherence from frequencies around 0.1 to 0.3 (Fig. 7C). Partial squared coherence with WIN was highly significant at frequencies from 0.1 to 0.2 after adjusting for SUM (Fig. 7D). Therefore there is significant coherence between otolith growth and the winter index (CDD below 0°C September through February) and the summer index (mean temperature June through September). However, otolith growth is more coherent with summer than winter temperature, and the broad band of high coherence with summer temperature indicates high contemporaneous correlation.

Superposed epoch analysis

SEA is suited for testing possible associations between variables that are episodic, as opposed to periodic. SEA may, therefore, be appropriate to test for association between freshwater drum growth dynamics and El Niño events. We conducted all SEA with epochs of width seven. Thus, we compared the key event year to 6 background years, with 3 yr before the event and 3 after. We first subjected SEA to the otolith growth index (Table 4). We did not detect any association between the otolith growth index and the occurrence of El Niño events. However, when recalculated with epochs of width five, the test including moderate and strong El Niño events for the 100-yr period 1890-1989 was marginally significant (P = 0.0677). SEA indicated strong association with summer temperature, but no association with winter temperature. This result is consistent with coherency analysis, where otolith growth was more coherent with summer than winter temperature indices. Also, the test for low summer temperature was stronger than that for high summer temperature, suggesting the possibility that cool summers restrict growth more than warm summers enhance growth. This interpretation is confounded by the number of key events, where low summer

Table 4. Results of superposed epoch analysis of the otolith growth index. The width of the epoch for all tests was 7 yr, with 3 yr before and after the key event as background years. The number of key events in each test are indicated by (n).

Time period	Key event (n)	Hypothesis tested: H_0	Hypothesis tested: H_A	P
1890–1989	Moderate and strong El Niño (19)	Otolith growth is unrelated to El Niño	Otolith growth is high during El Niño	0.1209
	Strong El Niño only (9)		Otolith growth is high during strong El Niño	0.3744
1940–1989	Moderate and strong El Niño (9)		Otolith growth is high during El Niño	0.3246
	Strong El Niño only (4)		Otolith growth is high during strong El Niño	0.6292
1890–1989	Low summer temperature (7)[a]	Otolith growth is unrelated to low SUM	Otolith growth is low during low SUM	0.0018
	High summer temperature (5)	Otolith growth is unrelated to high SUM	Otolith growth is high during high SUM	0.0301
	Low winter temperature, time t (5)[b]	Otolith growth is unrelated to low WIN	Otolith growth is low following low WIN	0.1317
	Low winter temperature, time t - 1 (5)		Otolith growth is low preceding low WIN	0.3545
	High winter temperature, time t (5)	Otolith growth is unrelated to high WIN	Otolith growth is high following high WIN	0.4352
	High winter temperature, time t - 1 (5)		Otolith growth is high preceding high WIN	0.3764
1920–1989	Recruitment (12)[c]	Otolith growth is unrelated to recruitment	Otolith growth is high during recruitment years	0.0001

[a] Summer and winter temperature indices were the same as those used for spectral analysis SUM and WIN, respectively. We labeled low key events as years that were less than or equal to the 0.05 quantile of the distributions of the temperature indices, and high key events were years greater than or equal to the 0.95 quantile.

[b] Because the winter index (WIN) was CDD below a base of 0°C from September through February, we tested key events against growth during the year at the start of the index (September), denoted as t - 1, and against growth during the year following the end of the index (February), denoted as t. Tests at time t - 1 would therefore test for effects late in the growing season, while time t would test effects that may occur in the spring.

[c] Recruitment key events are simply years when the presence of a cohort has been determined through aging. This test is, therefore, weakened by the possibility that other year-classes either were at one time present and have been extirpated, or are still present and have not been detected.

temperature events numbered seven, while high summer temperature events numbered only five. The test for low summer temperature association, therefore, had higher power than that for high summer temperature. The test for association between otolith growth and recruitment had the smallest P value of all tests (0.0001), indicating that otolith growth was very good in years when documented year-classes were established.

We also applied SEA to recruitment of freshwater drum in the Red Lakes (Table 5). If the presence of a cohort was detected through aging, then we labeled the birth year of this cohort as a key event. This test is, therefore, weakened and perhaps biased over time by the possibility that some year-classes were extirpated before this study or that we did not detect all the extant year-classes. The tests for association between recruitment and the SO yield a P-value of 0.0427. This result is inconsistent with low coherence between otolith growth and the SO, and also with SEA of El Niño and otolith growth. SEA also indicated that recruitment is strongly associated with warm summer temperatures, but not associated with winter temperature. When recalculated with epochs of width five, the test for winter temperature following years with recruitment was nearly significant (P-value = 0.0544). These results for temperature are therefore consistent with results reported above and suggest that summer climate may be a stronger determinant of reproductive success than winter climate.

Growth and dynamics of freshwater drum in Blackduck Lake

Dynamics of drum in Blackduck Lake were quite similar to the Red Lakes. A random sample of fish from a population assessment conducted in 1991 indicated that a large majority and perhaps all fish sampled are from the 1970 year-class, which is also the first large cohort on record for the Red Lakes. This finding is supported by length frequencies from four surveys conducted in 1975, 1978, 1983, and 1991 (Fig. 8). Only one mode is present each year and increases monotonically through time. Annual increments from otoliths of nine Blackduck Lake drum (1970 year-class) were processed using the same methods as applied to Red Lakes freshwater drum (Pereira et al. 1994). The otolith growth indices for the 1970 year-classes of freshwater drum from Blackduck Lake and

Table 5. Results of superposed epoch analysis on climate data with recruitment as key events. Variables tested and key events are as described in Table 4. Widths of epochs were 7 yr, with 3 yr before and after the key event as background years.

Variable tested	Time period	Number of key events	Hypothesis tested: H_0	Hypothesis tested: H_A	P
Southern oscillation[a]	1934–90	12	Southern oscillation is unrelated to recruitment	Southern oscillation is lower during years with detected recruitment	0.0427
Summer temperature[b]	1927–89	15	Summer temperature is unrelated to recruitment	Summer temperature is higher during years with detected recruitment	0.0002
Winter temperature (t)[c]	1927–89	14	Winter temperature is unrelated to recruitment	Winter temperature is higher following years with detected recruitment	0.1251
Winter temperature (t – 1)[c]	1927–89	15	Winter temperature is unrelated to recruitment	Winter temperature is higher prior to years with detected recruitment	0.9962

[a] We used the same southern oscillation index (SOI) as used for spectral analysis.
[b] Summer temperature and winter temperature are the same indices (SUM and WIN respectively) as those used in spectral analysis.
[c] This notation is the same as that in Table 4.

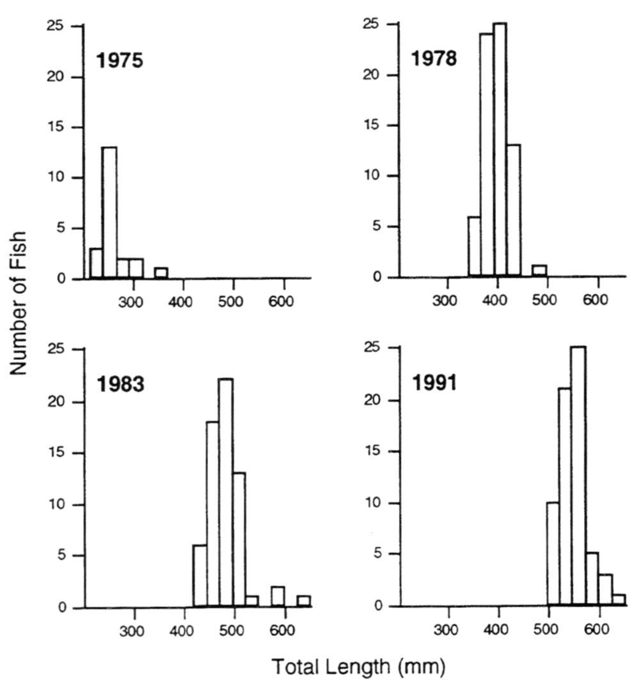

Fig. 8. Blackduck Lake drum length distribution, assessment gillnets. Length frequencies represent all the fish caught in standard assessment gillnets in 1975 and 1991; the distributions for 1978 and 1983 represent composites of trapnet and gillnet catches.

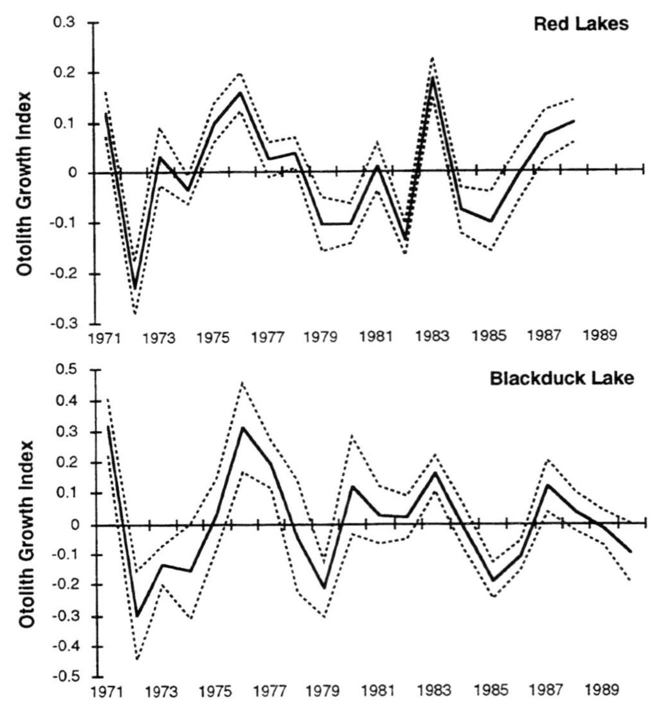

Fig. 9. Otolith growth index for the 1970 year-class of drum from Blackduck Lake and the Red Lakes. Broken lines indicate approximate 95% confidence intervals. The Blackduck Lake sample contained nine fish and the Red Lakes sample contained 12.

Red Lakes were similar in some features (Fig. 9), and had a correlation of 0.7. Similar fluctuations in otolith growth during early ages (2 and 3 yr) may be due to lack of fit of the exponential model used for removing age effects (see Pereira 1992; Pereira et al. 1994). However, drum of the 1970 year-class in both systems had positive otolith growth around 1976 and 1983, and relatively poor growth in 1979 and 1985. Positive growth in 1983 corresponds to a strong recruitment event in the Red Lakes. However, we did not detect this year-class in Blackduck Lake, which suggests that other factors (e.g., predation) are not consistent between the two study sites. This further suggests that some variation in Red Lakes drum dynamics may be attributed to perturbation of the fish community arising from the intensive commercial fishery (Pereira et al. 1992). Poor growth in 1985 is also consistent with growth of smallmouth bass in six northeastern Minnesota lakes (Pereira et al. 1995) and may indicate climate events occurring on spatial scales spanning major watersheds.

Discussion

Freshwater drum in the Red Lakes, Minnesota, are characterized by slow growth and infrequent, periodic recruitment. We provide evidence that growth and recruitment of drum in this system are positively related to summer air temperature. Otolith growth from the period 1930–79 showed a periodicity of about 7 yr. Recruitment appears to occur infrequently, but it is not clear whether it is periodic. One must keep in mind that this method of sampling is biased by time in that weak year-classes formed well in the past had a greater probability of being completely extinguished by the time contemporary sampling was underway. This further suggests that effects of summer temperature may be greater than indicated from this analysis.

Recruitment of this species appears to be a sensitive indicator of climate events. There are several mechanisms that may explain observed recruitment patterns. Because this species has a very wide latitudinal range, it is possible that reproduction is limited by the relatively short growing season in the Red Lakes. Therefore, the species may forgo reproduction for several years until sufficient energy accrues to permit production of gametes. This scenario becomes more complex when we consider the requirement of synchrony among spawners. An alternative hypothesis is that young-of-the-year (YOY) fish must attain a critical size to prevent overwinter mortality or starvation. This mechanism has been extensively developed in other fish species close to the northern limits of their range (Shuter et al. 1980). We, therefore, hypothesize that drum must attain a minimum size to permit survival until the onset of growth in the following growing season. This hypothesis is supported by observations on drum recruitment in the upper Missouri River. Swedburg and Walburg (1970) reported that strong year-classes were associated with warm water temperatures from June through August. This finding is, therefore, consistent with the strong association reported here between presence–absence of year-classes in the Red Lakes and summer air temperature. This hypothesis should be relatively easy to test by sampling drum throughout their first two growing seasons. This would require simply determining relative abundance and size of drum at several times during their first summer and then again during their second growing season. If drum YOY are present in late summer each year, then it is unlikely that the observed recruitment patterns arise from actual periodicity in spawning attempts. Mills and Mann (1985) discussed similar mechanisms that affect recruitment in cyprinid fishes. Our argument that recruitment (i.e., YOY fishes) may serve as sensitive indicators of climate is supported by Pereira et al. (1995).

Population dynamics are affected by factors internal to a lake basin, such as species interactions, and also by abiotic forces, such as climate acting on larger spatial scales. The fish community of the Red Lakes may be considered perturbed in part by the intensity of commercial fishing. Variation in the predator complex for drum larvae may also account for some of the variation in drum recruitment (Pereira et al. 1992). However, similar dynamics for this species in Blackduck Lake indicate that factors acting on a larger spatial scale are in part influencing species succession. It does not appear that the biomass of freshwater drum in Blackduck lake has increased significantly, as it has in the Red Lakes. Therefore, we suggest that perturbation to the fish community in the Red Lakes may act to amplify effects of other forces such as climate.

We hypothesized that recruitment of drum may be influenced by SO teleconnections. We note that the very large 1983 year-class followed the onset of one of the strongest El Niños on record (Philander 1989). However, the time-series methods employed may be insufficient for exploration of connections between drum dynamics and quasi-periodic climate processes such as El Niño or the SO. We simply have shown that drum otolith growth and teleconnections do not display correlated periodicities. Robertson (1989) used the SEA method to determine the influence of El Niño events on ice-cover duration. This methodology has proven useful for other studies of fish recruitment (Prager and Hoenig 1989). However, this analysis applied to the data reported here did not reveal any association between El Niño and drum otolith growth. On the other hand, SEA did indicate a possible association between the SO and the occurrence of year-classes of freshwater drum in the Red Lakes. This result is difficult to interpret and may be spurious considering the finding that SO teleconnections only affect the North American continent during El Niño events (Barnett 1981; Rasmusson and Carpenter 1990).

These results suggest that freshwater drum in the Red Lakes, Minnesota, may serve as sensitive indicators of climate events. The effect of northerly climates on life history has been investigated in other fish that display wide latitudinal ranges (Mills 1988; Johnson and Hatch 1991). While much climate change research focuses on community- and ecosystem-level processes, it appears obvious that species at the northern limits of their range may prove to be excellent research subjects for this problem. This research may involve quantifying changes in population rate processes within a given body of water, or determining factors affecting species range expansion (Shuter and Post 1990). Furthermore, freshwater drum in the Red Lakes may also provide a suitable species for climate change study based on the fact that the lakes are largely isothermal. This physical characteristic removes the opportunity for behavioural thermoregulation by the animal.

The time series of otolith growth presented here represents a biochronology, a record through time of natural events experienced by freshwater drum in the Red Lakes. Casselman (1990) has shown that otoliths continue to record seasonal growth in slow-growing fishes. This methodology is similar to dendrochronology, which has found extensive application in paleoecology and climate studies (Clark 1988; Briffa et al. 1990; Van Deusen 1990). It has long been recognized that fish record growth and environmental influences in the annual increments of their calcified structures (Casselman 1983). Recent analytical advances provide efficient methods for dividing growth of calcified structures into that owing to age allometry (i.e., fish growth slows with size and therefore age) and that owing to interannual variation in the environment (Weisberg and Frie 1987). Shroyer (1991) has also successfully applied this methodology to walleye in the Red Lakes, Minnesota. Development of biochronologies of fish will, therefore, allow us to test a broad suite of hypotheses pertaining to the ecology of fish and their surrounding environment. Examples include physical effects, such as climate change, and biological effects, such as those due to stock dynamics, interspecies relationships, and management activities.

Acknowledgements

This study was funded in part by the Red Lake Band of Chippewa Indians. Greg Spoden (State of Minnesota Climatology Office, St. Paul, Minn.) provided temperature data. C. Ropelewski (Climate Analysis Center, National Oceanic and Atmospheric Administration, Washington, D.C.) provided Southern Oscillation data. Personnel of the Minnesota Department of Natural Resources, Bemidji, Minn., Area Fisheries Station, provided freshwater drum otoliths and data from Blackduck Lake, Minn. An anonymous reviewer provided constructive suggestions that improved this manuscript.

References

Barnett, T.P. 1981. Statistical prediction of North American air temperatures from Pacific predictors. Mon. Weather Rev. 109: 1021–1041.

Briffa, K.R., T.S. Bartholin, D. Eckstein, P.D. Jones, W. Karlen, F.H. Schweingruber, and P. Zetterberg. 1990. A 1,400-year tree-ring record of summer temperatures in Fennoscandia. Nature (Lond.) 346: 434–439.

Casselman, J.M. 1983. Age and growth assessment of fish from their calcified structures: techniques and tools, p. 1–17. *In* E.D. Prince and L.M. Pulos [ed.] Proceedings of the International Workshop on Age Determination of Oceanic Pelagic Fishes: Tunas, Billfishes, and Sharks. NOAA Tech. Rep. NMFS 8.

Casselman, J.M. 1990. Growth and relative size of calcified structures of fish. Trans. Am. Fish. Soc. 119: 673–688.

Clark, J.S. 1988. Effect of climate change on fir regimes in northwestern Minnesota. Nature (Lond.) 334: 233–235.

Demartini, E.E. 1991. Annual variations in fecundity, egg size, and the gonadal and somatic conditions of queenfish *Seriphus politus* (Sciaenidae). Fish. Bull. U.S. 89: 9–18.

Haurwitz, M.W., and G.W. Brier. 1981. A critique of the superposed epoch analysis method: its application to solar–weather relations. Mon. Weather Rev. 109: 2074–2079.

Hoenig, J.M., M.H. Prager, and N.B. Payton. 1989. Computer programs for investigating the effects of environmental events on a time series of recruitment. Can. Tech. Rep. Fish. Aquat. Sci. 1713: 19 p.

Johnson, J.D., and J.T. Hatch. 1991. Life history of the least darter *Etheostoma microperca* at the northwestern limits of its range. Am. Middle Nat. 125: 87–103.

Lough, J.M., and H.C. Fritts. 1985. The southern oscillation and tree rings: 1600–1961. J. Clim. Appl. Meteorol. 24: 952–966.

Michaelsen, J. 1989. Long-period fluctuations in El Niño amplitude and frequency reconstructed from tree-rings. Geophys. Monogr. 55: 69–74.

Mills, C.A. 1988. The effect of extreme northerly climatic conditions on the life history of the minnow, *Phoxinus phoxinus* (L.). J. Fish. Biol. 33: 545–561.

Mills, C.A., and R.H.K. Mann. 1985. Environmentally-induced fluctuations in year-class strength and their implications for management. J. Fish Biol. 27(Suppl. A): 209–226.

Mysak, L.A. 1986. El Niño interannual variability and fisheries in the Northeast Pacific Ocean. Can. J. Fish. Aquat. Sci. 43: 464–497.

Pereira, D.L. 1992. Dynamics and biochronology of freshwater drum (*Aplodinotus grunniens*) in the Red Lakes, Minnesota. Ph.D. thesis, University of Minnesota, Saint Paul, MN. 115 p.

Pereira, D.L., C.S. Anderson, P. Radomski, T. Sak, and A. Stevens. 1995. Potential to index climate with growth and recruitment of temperate fish, p. 709–717. *In* R.J. Beamish [ed.] Climate change and northern fish populations. Can. Spec. Publ. Fish. Aquat. Sci. 121.

Pereira, D.L., C. Bingham, G.R. Spangler, D.J. Conner, and P.K. Cunningham. 1994. Construction of a 110-year biochronology from sagittae of freshwater drum (*Aplodinotus grunniens*). *In* D.H. Secor, J.M. Dean, and S.E. Campana [ed.] New developments in fish otolith research. University of South Carolina Press, Columbia, SC.

Pereira, D.L., Y. Cohen, and G.R. Spangler. 1992. Dynamics and species interactions in the commercial fishery of the Red Lakes, Minnesota. Can. J. Fish. Aquat. Sci. 49: 293–302.

Philander, G. 1989. El Niño and La Niña. Am. Sci. 77: 451–459.

Prager, M.H., and J.M. Hoenig. 1989. Superposed epoch analysis: a randomization test of environmental effects on recruitment with application to chub mackerel. Trans. Am. Fish. Soc. 118: 608–618.

Rasmusson, E.M., and T.M. Carpenter. 1982. Variations in tropical sea surface temperature and surface wind fields associated with the Southern Oscillation/El Niño. Mon. Weather Rev. 110: 354–384.

Robertson, D.M. 1989. The use of lake water temperature and ice cover as climatic indicators. Ph.D. thesis, University of Wisconsin, Madison, WI. 294 p.

Ropelewski, C.F., and M.S. Halpert. 1986. North American precipitation and temperature patterns associated with the El Niño/Southern Oscillation (ENSO). Mon. Weather Rev. 114: 2352–2362.

Ropelewski, C.F., and P.D. Jones. 1987. An extension of the Tahiti–Darwin southern oscillation index. Monthly Weather Rev. 115: 2161–2165.

Scott, W.B., and E.J. Crossman. 1973. Freshwater fishes of Canada. Fish. Res. Board Can. Bull. 184: 966 p.

Shuter, B.J., J.A. MacLean, F.E.J. Fry, and H.A. Regier. 1980. Stochastic simulation of temperature effects on first-year survival of smallmouth bass. Trans. Am. Fish. Soc. 109: 1–34.

Shuter, B.J., and J.R. Post. 1990. Climate, population viability, and the zoogeography of temperate fishes. Trans. Am. Fish. Soc. 119: 314–336.

Shroyer, S.M. 1991. Growth of Red Lakes walleyes from the 1940s to the 1980s. M.Sc. thesis, University of Minnesota, Saint Paul, MN. 104 p.

Smith, L.L., Jr. 1977. Walleye (*Stizostedion vitreum vitreum*) and yellow perch (*Perca flavescens*) populations and fisheries of the Red Lakes, Minnesota, 1930–75. J. Fish. Res. Board Can. 34: 1774–1783.

Swedburg, D.V., and C.H. Walburg. 1970. Spawning and early life history of the freshwater drum in Lewis and Clark Lake, Missouri River. Trans. Am. Fish. Soc. 99: 560–570.

Van Deusen, P.C. 1990. Evaluating time-dependent tree ring and climate relationships. J. Environ. Qual. 19: 481–488.

Wallace, J.M. 1985. Atmospheric response to equatorial sea-surface temperature anomalies, p. 9–21. *In* W.S. Wooster and D.L. Fluharty [ed.] El Niño north: Niño effects in the eastern subarctic Pacific ocean. Washington Sea Grant, University of Washington, Seattle, WA.

Weisberg, S., and R.V. Frie. 1987. Linear models for the growth of fish, p. 127–143. *In* R.C. Summerfelt and G.E. Hall [ed.] The age and growth of fish. Iowa State University Press, Ames, IO.

Wooster, W.S., and D.L. Fluharty [ed.] 1985. El Niño North: Niño effects in the eastern subarctic Pacific ocean. Washington Sea Grant, University of Washington, Seattle, WA. 312 p.

Biomass variations of walleye pollock of the Bering Sea in relation to oceanological conditions

O.A. Bulatov

Bulatov, O.A. 1995. Biomass variations of walleye pollock of the Bering Sea in relation to oceanological conditions, p. 631–640. *In* R.J. Beamish [ed.] Climate change and northern fish populations. Can. Spec. Publ. Fish. Aquat. Sci. 121.

Abstract: A review was undertaken of walleye pollock (*Theragra chalcogramma*) in the Bering Sea whose purpose was to develop a working hypothesis for explaining the fluctuations in abundance. The review involved the history of the fishery, spawning, population structure, size and age composition, and stock conditions. During 1970–91, total catch ranged from 1.251 million t (1977) to 4.014 million t (1989), and was 2.107 million t in 1991. Major fisheries developed in the eastern (1970–91; 0.9–1.9 million t/yr), western (1976–91; 0.3–1.3 million t/yr), and central (1984–91; 0.2–1.4 million t/yr) Bering Sea, and a minor fishery in the Aleutian Islands (1980–88; 29 000 – 82 000 t/yr). After 1989, substantial declines in catch occurred in both the central and western Bering Sea. Spawning occurred on the continental slope in winter, and on the shelf in spring. Major centers were located in both the eastern and western Bering Sea, but no spawning was detected in the central sector. Population structure, based on morphometrics, meristics, genetics, growth, length–weight, and spawning locations, indicated a complex mix throughout the Bering Sea. Size range in catches was about 30–60 cm, and age range, 2–10+ yr. Notable was the lack of young fish in the central area. Strong year-classes of pollock occurred nearly synchronously throughout the Bering Sea. Recent examples were 1978, 1984, and possibly 1989. Strong year-classes were forming during the first warm years on an ascending curve of solar activity, but other factors were probably important as well. Based on trawl-survey data, abundance was declining (Russian interpretation) or increasing (U.S. interpretation) in the eastern Bering Sea; and low (Russian interpretation) in the central and western sectors.

Résumé : On a procédé à un examen de la morue du Pacifique occidental (*Theragra chalcogramma*) dans la mer de Béring, examen dont l'objet était d'élaborer une hypothèse de travail pour expliquer les fluctuations dans l'abondance de ce poisson. L'examen a porté sur l'histoire de la pêcherie, du frai, de la structure des populations, de la composition selon la taille et l'âge et de l'état des stocks. Durant la période 1970–1991, le nombre total de prises est passé de 1,251 million de t (1977) à 4,014 millions de t (1989), puis est retombé à 2,107 millions de t en 1991. Des pêcheries majeures ont été ouvertes dans l'est (1970–1991; 0,9–1,9 million de t/an), dans l'ouest (1976–1991; 0,3–1,3 million de t/an) et dans le centre (1984–1991; 0,2–1,4 million de t/an) de la mer de Béring, et une mineure l'a été dans les îles Aléoutiennes (1980–1988; 29 000 – 82 000 t/an). Après 1989, des déclins substantiels des prises ont été enregistrés dans le centre et dans l'ouest de la mer de Béring. Le frai a eu lieu sur la pente continentale au cours de l'hiver et sur le plateau au printemps. Les centres majeurs étaient situés dans l'est et dans l'ouest de la mer de Béring, mais aucun frai n'a été décelé dans le secteur central. D'après les données sur la morphométrie, la méristique, la génétique, la croissance, la relation longueur-poids, et les lieux de frai, la structure des populations trahissait l'existence d'un mélange complexe dans toute la mer de Béring. La taille des poissons capturés variait de 30 à 60 cm, et les âges de 2 à 10+. L'absence de jeunes poissons dans la zone centrale était notable. Il y a eu des classes annuelles fortes de morues de façon presque synchrone dans toute la mer de Béring. Les exemples récents étaient 1978, 1984 et peut-être 1989. Les classes annuelles fortes se sont formées durant les premières années chaudes du cycle de l'activité solaire (courbe ascendante), mais d'autres facteurs ont probablement joué un rôle important. D'après des données recueillies par des chalutiers, l'abondance était en déclin (interprétation russe) ou en croissance (interprétation américaine) dans l'est de la mer de Béring, et faible (interprétation russe) dans les secteurs du centre et de l'ouest.

O.A. Bulatov. Pacific Research Institute of Fisheries and Oceanography (TINRO), 4 Shevchenko Alley, 690600 Vladivostok, Russia.

Introduction

Walleye pollock, *Theragra chalcogramma*, (hereafter referred to as pollock) is the most abundant gadid in the North Pacific Ocean. Since the 1980s, pollock has taken a leading place in world fisheries. Main fishing areas are associated with boreal basins: the Bering and Okhotsk seas. Comparative analysis of spawning activity has shown that pollock of the eastern Bering Sea is one of the most abundant populations in the North Pacific (Bulatov 1987). However, interannual egg production varies substantially, and reflects the equally variable abundance of spawners. Thus, investigations of the dynamics of pollock stocks are very important.

Numerous TINRO (Pacific Research Institute of Fisheries and Oceanography, Vladisvostok, Russia) studies have shown that strong year-classes are formed predominantly in warm years (Fadeev 1980; Kachina and Sergeeva 1981; Gavrilov and Bezlyudny 1983, 1986; Vasil'kov and Glebova 1984; Bulatov and Khen 1984; Sokolovsky and Glebova 1986; Balykin 1989; Bulatov 1988, 1989*a*; and Naumenko et al. 1990). Despite this voluminous information, at present there is no admissible hypotheses explaining the fluctuations of pollock abundance in the North Pacific Ocean, or their relationship to atmospheric–oceanological conditions.

The purpose of this report is to develop the main principles of a working hypothesis from a review and analysis of the available information on pollock in the Bering Sea.

Materials and methods

Information to be reviewed and analyzed involved the history of the fishery, spawning, population structure, size and age composition, and condition of stocks. A literature review provided information on the fishery history, population structure, and condition of stocks in the eastern, central, and western Bering Sea. (The central Bering Sea is defined as that portion of the sea lying seaward of the continental slope, some portions of which lie within the Extended Economic Zones (EEZs) of Russia and the United States. The so-called "Donut Hole", or international area, is that portion which lies outside the aforementioned EEZs).

Eastern Bering Sea

Demersal-trawl surveys with Russian research vessels provided information on spawning, size–age composition, and biomass estimates. Surveys were conducted during summer–autumn in 1980–81, 1983, 1986, and 1990–91 (Table 1). Biomass estimates were based on trawls conducted on a station grid used by U.S. scientists from the Northwest Alaska Fishery Center, Seattle, Wash. (Pereyra et al. 1976). The eastern Bering Sea was divided into seven statistical subregions, accepted by U.S. scientists. In each subregion, biomass was defined as:

$$P_i = \frac{Q_i x_i}{q_i k}$$

where P_i is the biomass (tonnes) in statistical subregion *i*; Q_i is the area (square kilometres) of statistical subregion *i*; x_i is the arithmetic mean catch per unit effort (CPUE) (tonnes per hour); q_i is the mean area (square kilometers) of ground trawled per hour; and k is the catchability coefficient (accepted as 1.0).

Total biomass (tonnes) of pollock was obtained by summing the biomass estimates in each statistical subregion. CPUE was calculated as tonnes per square kilometer.

Central Bering Sea

Estimates of abundance and biomass were obtained from Russian pelagic-trawl surveys conducted, during 1986–90, with large nets (108 × 528 m and 118 × 620 m) of small mesh (10 mm). In 1987–90, a preplanned station grid was occupied. Biomass was calculated as before, but with $k = 0.4$. In 1986, trawling was conducted on encountered concentrations of pollock, as well as on grid stations. Therefore, mean catches were calculated for each square of 1° latitude × 1° longitude.

Western Bering Sea

Estimates of abundance and biomass were obtained from the literature. See Balykin (1989) for details.

Table 1. Demersal-trawl surveys of walleye pollock by Russian research vessels in the eastern Bering Sea, by time, numbers of hauls, type of trawl, average trawling speed (km/h), and horizontal net opening (m), 1980–91.

Year	Months	No. of hauls	Trawl type	Mean speed (km/h)	Horizontal opening (m)
1980	Sept.–Nov.	299	31.0	5.6	20.0
1981	Sept.–Nov.	316	28.1	6.5	20.0
1983	July–Aug.	346	43.0	6.3	29.5
1985	July–Aug.	363	31.0	6.7	20.0
1986	June–July	361	31.0/43.0	7.4	18.0/22.0
1989	May–June	167	35/41.3	5.9	19.2
1990	May–July	192	35/41.3	5.7	19.2
1991	May–June	163	69/48	7.9	27.0

Results

History of the Fishery

The pollock fishery in the Bering Sea began in 1954, but was at a low level up to 1963, then sharply increased to 1.3 million in 1970. During 1970–91, catch rose to 2.015 million t in 1972; declined to 1.251 million t in 1977; rose irregularly to 4.014 million t in 1989; then declined to 2.107 million t in 1991 (Table 2). During 1970–75, Japanese, Russian, and other national fleets fished predominantly on the continental shelf and slope of the eastern Bering Sea. After promulgation of the U.S. EEZ, large-scale fisheries developed in the western Bering Sea in 1976, and the Donut Hole, of the central Bering Sea in 1986. A small-scale fishery, within the U.S. EEZ, developed in the Aleutian Islands in 1980.

Eastern Bering Sea

During 1970–75, the international fishery caught 1.2–1.9 million t/yr 88–97% of the total Bering Sea pollock catch (Table 2). Subsequently, catch declined to 914 000 t in 1979; rose to 1.386 million t in 1989; and was 1.3 million t in 1991. Catch quotas for foreign vessels were decreased after 1977, and all foreign fishing was eliminated in the mid-1980s.

Western Bering Sea

The Russian fishery rapidly expanded as a result of the promulgation of the U.S. EEZ in 1977, and the subsequent sharp reduction in the Russian pollock quota therein. During 1970–75, annual catches ranged from 40 000 t (1970) to 188 000 t (1975), and comprised 3.1–12.2% of the total Bering Sea catch. During 1976–91, catches ranged from 265 000 t (1977) to 1.253 million t (1988), and comprised 21.2% (1977) to 49.0% (1982) of the total catch. After 1988, catches dropped sharply to 514 000 t in 1991 (24.4%).

Donut Hole (Central Bering Sea)

The international fishery became important after all foreign fishing was eliminated from the U.S. EEZ in the mid-1980s. During 1980–91, catch was 15 000 t in 1980; rose to 1.448 million t in 1989; then declined sharply to 293 000 5 in 1991 (Table 2). The fishery's best contribution to the total catch was 29–37% during 1985–90.

Aleutian Islands

Records were available for 1977–88 (Table 2). Catch rose from 8 000 t in 1977 to a peak of 82 000 t in 1984; declined to 29 000 t in 1987; and was 43 000 t in 1988. The fishery's best contribution to the total catch was 2.6–3.9% during 1980–85. Records were not available for later years.

Spawning

Conditions of pollock spawning have been studied rather thoroughly (Musienko 1963; Serobaba 1968, 1971, 1974; Fadeev 1981; Waldron 1981; Bulatov 1986, 1987, 1989*a*, *b*; Hinkley 1987). Pollock spawn in winter (late February – early

Table 2. All-nation catches (10^3 t), and percentages of total annual catches, by region, of walleye pollock in the Bering Sea, 1970–91.

	Eastern[a]		Aleutian Islands[a]		Donut hole[b]		Western[c]		Total
Year	(10^3 t)	%	(10^3 t)	%	(10^3 t)	%	(10^3 t)	%	(10^3 t)
1970	1257	96.9	—	—	—	—	40	3.1	1297
1971	1745	95.1	—	—	—	—	89	4.9	1834
1972	1874	93.0	—	—	—	—	141	7.0	2015
1973	1759	95.8	—	—	—	—	77	4.2	1836
1974	1588	93.3	—	—	—	—	114	6.7	1702
1975	1357	87.8	—	—	—	—	188	12.2	1545
1976	1178	67.0	—	—	—	—	579	33.0	1757
1977	978	78.2	8	0.6	—	—	265	21.2	1251
1978	979	69.8	6	0.4	—	—	417	29.8	1402
1979	914	64.9	10	0.7	—	—	485	34.4	1409
1980	958	53.2	58	3.2	15	0.8	770	42.8	1801
1981	974	48.5	56	2.8	—	—	980	48.7	2010
1982	956	47.9	58	2.9	4	0.2	976	49.0	1994
1983	982	46.4	59	2.8	71	3.4	1006	47.4	2118
1984	1099	51.9	82	3.9	181	8.5	755	35.7	2117
1985	1180	52.7	59	2.6	336	15.0	662	29.7	2237
1986	1188	38.4	47	1.5	1032	33.3	828	26.8	3095
1987	1238	35.0	29	0.8	1326	37.4	947	26.8	3540
1988	1228	31.3	43	1.1	1397	35.6	1253	32.0	3921
1989	1386 [b]	34.5	NA [d]	—	1448	36.1	1180	29.4	4014
1990	1385 [b]	44.3	NA [d]	—	918	29.4	820	26.3	3123
1991	1300 [b]	61.7	NA [d]	—	293	13.9	514	24.4	2107

[a] U.S. EEZ. Source: North Pacific Fisheries Management Council (1989).
[b] Source: NOAA (1991, 1992).
[c] Source: GPO "Dalryba".
[d] NA = not available.

March) and spring (April–May). Ichthyoplankton investigations conducted by the author defined two nuclei of winter spawning, which were situated in the southeastern part of the Aleutian Basin, at >200 m. Eggs and larvae rose to the surface and moved shoreward to shallower water. Some spawning was also found on the continental shelf off Unimak Island (Bulatov 1988).

Spring spawning was located exclusively on the continental shelf. The main centers were in the eastern Bering Sea were Unimak Island and the Pribilov Islands; and in the western Bering Sea, the Olyutorsky region (Bulatov 1986, 1987, 1988). Localized spawning occurred in other areas. In each center, two spawning nuclei were found where the interannual quantity of eggs varied substantially. Up to the present, there is no information indicating that spawning occurs in the Donut Hole. Ichthyoplankton surveys yielded a single catch of eggs and larvae, which probably had been transported there by currents. Special investigations have not been conducted in the Aleutian Islands region; however, one may propose that there exists only local spawning.

Population structure

Studies, generally inadequate, have involved morphometrics and meristics, genetics, growth, length–weight, and spawning locations. Morphometric and meristic investigations by Serobaba (1977) indicated heterogeneity of pollock, and delineated eastern, western, northern, and Aleutian (Islands) groupings. Genetic investigations broadened concepts of the population structure in the Bering Sea, and identified six populations: Karaginskay–Olyutorskaya, Navarin, St. Matthew, Pribilov, Unimak, and Bristol (Flusova and Bogdanov 1986). The authors accepted Serobaba's Aleutian group as a valid population. Dawson (1990) identified pollock in the central Bering Sea, including off Bogoslov Island, as a single population. Growth rates were significantly different between pollock in the central area and the eastern shelf (Lunde et al. 1986), and between pollock inhabiting the shelf off Unimak Island and St. Matthew Island (Moiseev 1983). Length–weight relationships were significantly different between pollock from eastern and central Bering Sea (Hinkley 1987). Spatial-temporal isolation of 12 spawning sites (Bulatov 1987, 1988, 1989*b*), exclusive of the Aleutian Islands (no data), provided additional evidence of the high rate of differentiation of Bering Sea pollock. Thus, population structure is complex. Evidently, several populations exist within each of the eastern, central, and western regions of the Bering Sea.

Size–age structure

Eastern Bering Sea

Mean lengths of pollock, from the fishery, ranged from 31.4 cm (1975) to 42.6 cm (1985). A steady increase in mean length began in 1979 (34.0 cm) and ended in 1985. Length frequencies were unimodal during the late 1970s and early 1980s, and modal length steadily increased (Bakkala 1989). This indicated the presence of one or more strong year-classes passing through the fishery. The 1978 year-class was evidently strong.

Age range reported in research-vessel catches was 1–10 yr by Russian (Moiseev 1983) and U.S. (Sample and Bakkala 1989) scientists. Principal age-classes were 3–6 yr, but in some years the proportion of yearlings sharply increased. For 1964–88, U.S. scientists reported the strong year-classes were 1965–69, 1972, 1974, 1978, 1982, and 1984 (Table 3). The same reports indicated that the weakest, based on cohort analysis, were 1962–64, 1970, 1981, and 1983. For 1971–91, Russian scientists reported the strong year-classes were 1966–67, 1977–80, 1982, 1984, and 1989. Bulatov (1989*b*) listed 1971–75 and 1981 as weak year-classes.

Central Bering Sea (Aleutian Basin)

Size–age structure was unique in the catches of the large-scale fishery, which began in 1985. During 1986–90, overall size range was 36–60 cm (Bulatov and Sobolevsky 1989). Length frequencies were unimodal, and modal sizes increased with time in both the southwest sector and in the Donut Hole (Fig. 1). In the southwest sector, mean lengths increased from 45.10 cm in October 1986 to 50.75 cm in October–November 1990. Comparable values for the Donut Hole were 46.57 cm in August and October 1987 and 49.11 cm in October–November 1990. Similar results were reported by Japanese (Yoshimura and Sasaki 1990), Korean (G. Yeong, personal communication), and Polish (Jackowski and Trochinski 1989) scientists, and are compared with Russian data in Fig. 2. The relatively large modal sizes, and their steady increase with time suggested the absence of recent strong year-classes, and a population of predominantly old pollock.

Age composition reported for 1983 by U.S. scientists consisted primarily of ages 5 and 9–10: 1973–74 and 1978 year-classes (Dawson 1990). During 1986–88, the 1978 year-class was still predominant. Polish scientists reported that the age range in commercial catches was 5–23 yr, and the 1977 and 1978 year-classes comprised 21.8 and 16.4%, respectively (Kowalewska-Pahlke 1990). There were no pollock younger than 5 yr.

Table 3. Strong year-classes in the Bering Sea, by region, sample time, and source, 1964–91.

Region	Sample time	Year-classes	Source
Eastern	1964–88	1965–69, 1972, 1974, 1978, 1982, 1984	Wespestad & Traynor (1988); Bakkala (1989)
	1971–91	1966, 1967, 1977–80, 1982, 1984, 1989	Te.I. Moiseev (personal communication) Bulatov (1989*b*)
Donut Hole (Central)	1983	1973, 1974, 1978	Dawson (1990)
	NR[a]	1977, 1978	Kowalewska-Pahlke (1990)
Western	1978–87	1978, 1979, 1984	Balykin (1989)
	NR[a]	1989(?)	Pashchenko (personal communication)

[a] NR = not reported.

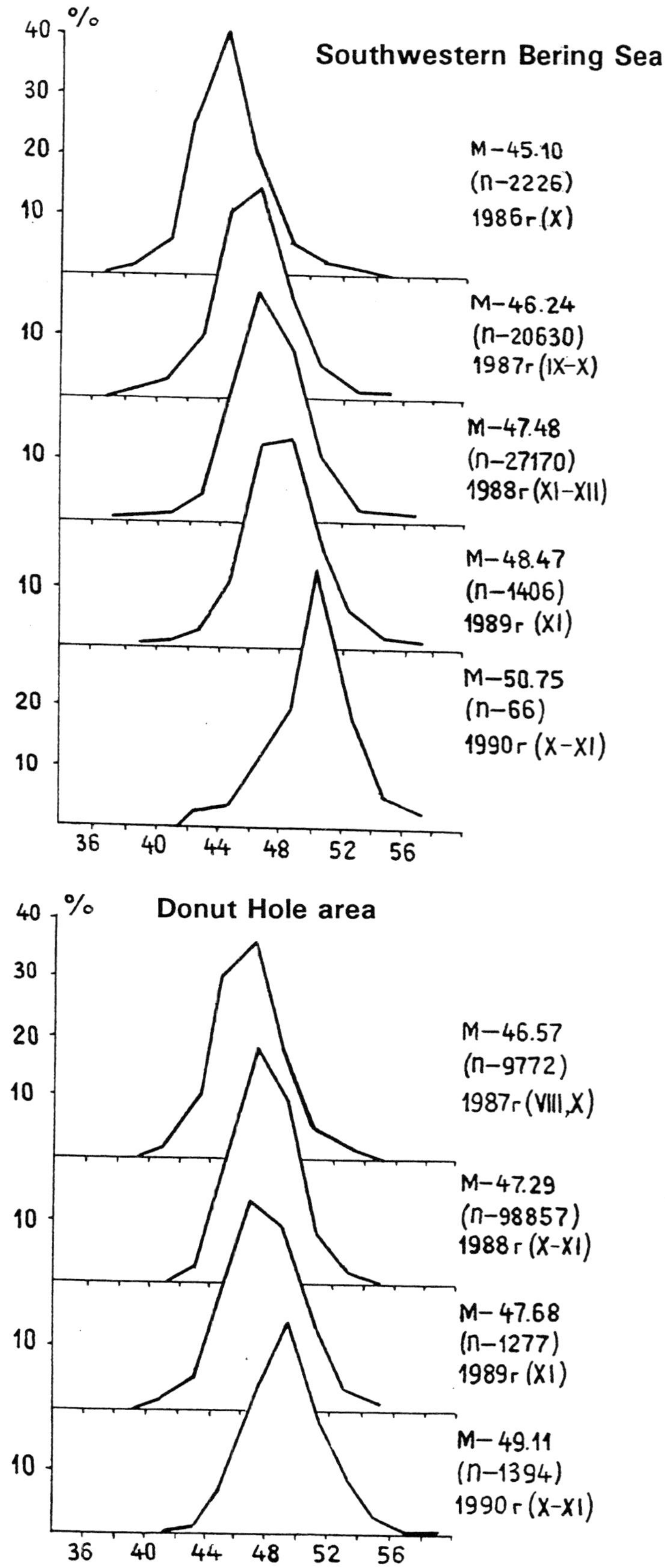

Fig. 1. Size composition of walleye pollock in research-vessel catches from the southwest sector and the Donut Hole area of the central Bering Sea, August–December 1986–90.

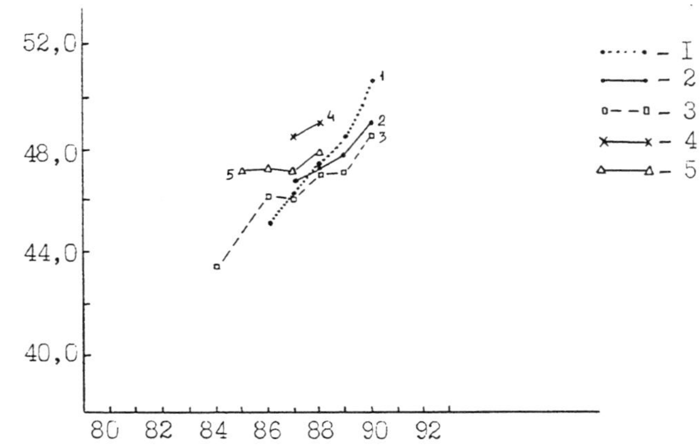

Fig. 2. Mean lengths (cm) of walleye pollock reported, by nation, for southwest Bering Sea (1) and the Donut Hole (2–5), 1986–90. 1 and 2 = Russia; 3 = Korea; 4 = Japan; 5 = Poland.

Western Bering Sea

Mean lengths in catches varied from 31.6 cm (1980) to 41.1 cm (1984) (Balykin 1989). The former value reflected the entry of the strong 1978 and 1979 year-classes, and the latter to their predominance during 1983–85. During 1978–87, the strong year-classes were 1978 and 1979 (Table 3). Recently, the 1984, and possibly the 1989, year-classes were reported to be strong. At age 1, the 1978 year-class comprised 12.8% of the catch, and the 1979 year-class, 16.4% (Balykin 1989). During 1982–87, the weakest year-class was 1981, and it remained so for 10 yr. Preliminary results of current Russian studies suggested that the 1989 year-class was strong (Pashchenko, personal communication).

Condition of stocks

Eastern Bering Sea

During 1964–86, two peaks of abundance were defined: 1966–69 and 1981–85 (Bulatov 1989*b*). Demersal-trawl surveys were initiated in 1975 by U.S. scientists, and in 1980 by Russian scientists. Despite the standardized procedures, results of the two programs differed for 1980–91. Annual estimates of biomass by Russian scientists ranged from 2.9 to

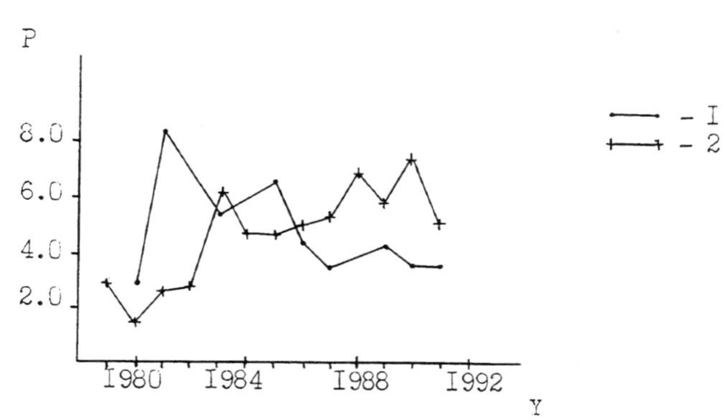

Fig. 3. Annual biomass estimates (′10^6 t) of walleye pollock in the eastern Bering Sea, based on demersal-trawl surveys by Russian (1; Bulatov 1989*b*) and U.S. (2; NOAA 1991) research vessels, 1980–91.

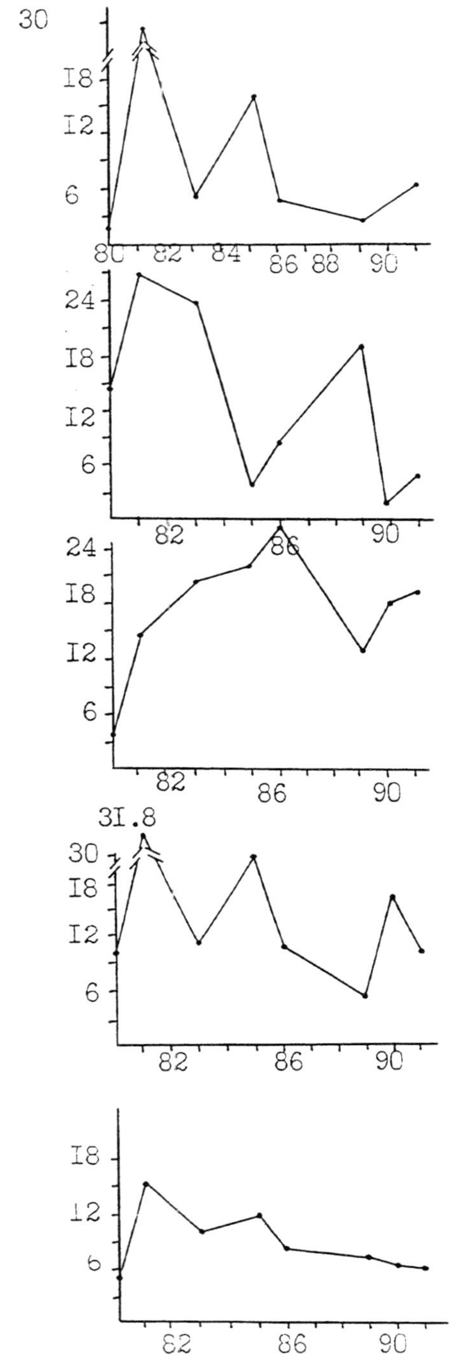

Fig. 4. CPUEs (t/km^2), by subarea, of walleye pollock from demersal-trawl surveys in the eastern Bering Sea, 1980–91.

7.4 million t, declining during 1981–85 and remaining stable after 1985 (Fig. 3). In contrast, during 1980–91, U.S. biomass estimates exhibited a general increase.

Because the dimensions of Russian trawl nets differed among years (Table 1), CPUEs were standardized to tonnes per square kilometer, and calculated for subareas 1, 2, 3S, and 3N (Unimak, Briston Bay, Pribiloff, and St. Matthew regions, respectively) (Fig. 4). Trends in CPUE were different in each

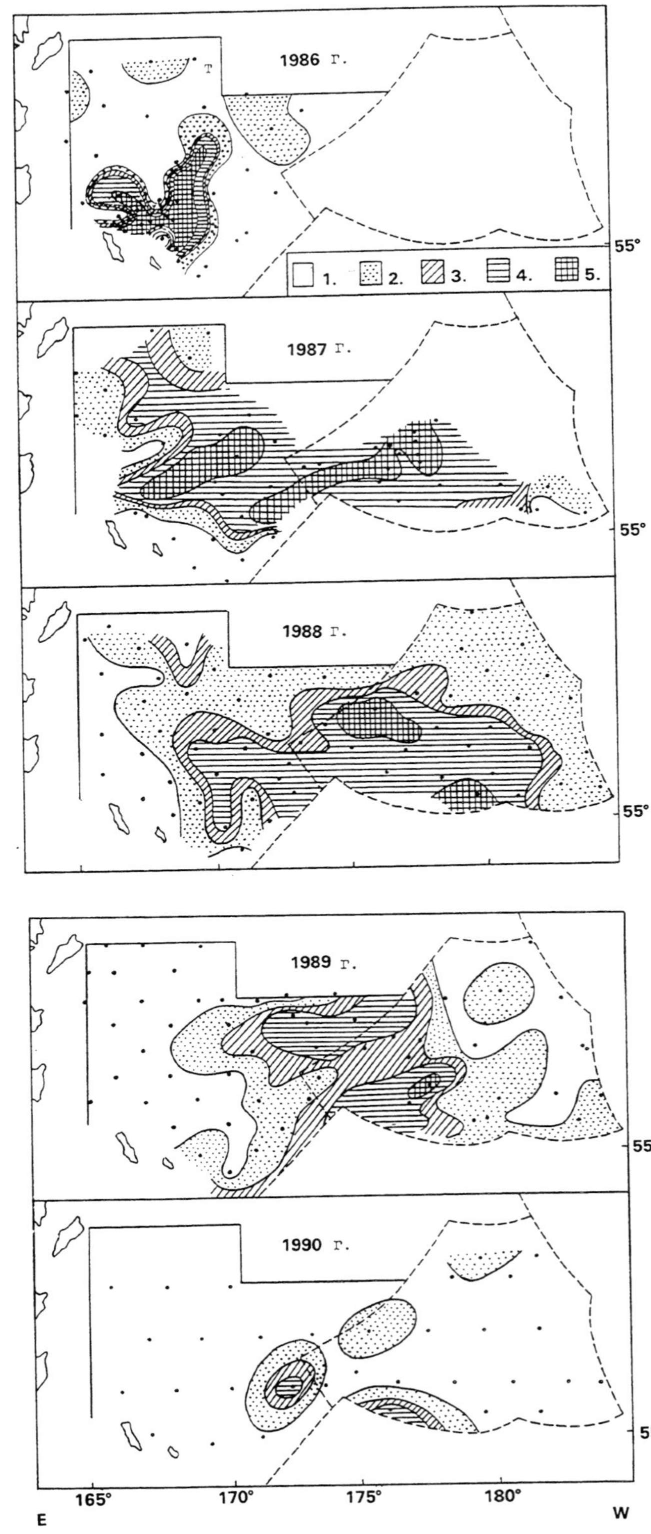

Fig. 5. Spatian distribution of annual CPUEs (t/h) for walleye pollock from pelagic-trawl surveys in the central Bering Sea, autumn 1986–90. (1 = <0.1; 2 = 0.1–0.5; 3 = 0.5–1.0; 4 = 1.0–5.0; >5.0. Dots = stations occupied.)

Table 4. CPUE (t/h) and biomass (10^3 t) of walleye pollock in the Donut Hole area and southwest sector of the central Bering Sea, based on pelagic-trawl surveys by Russian research vessels, 1986–90. (Source: Bulatov 1990)

Year	Month(s)	Donut Hole CPUE (t/h)	Donut Hole Biomass (10^3 t)	Southwest sector CPUE (t/h)	Southwest sector Biomass (10^3 t)
1986	October	—	—	2.1	1760
1987	October	4.6	1132	2.7	1960
1988	Oct.–Dec.	2.8	2802	0.6	371
1989	November	0.7	668	0.5	296
1990	Oct.–Nov.	0.3	194	0.2	185

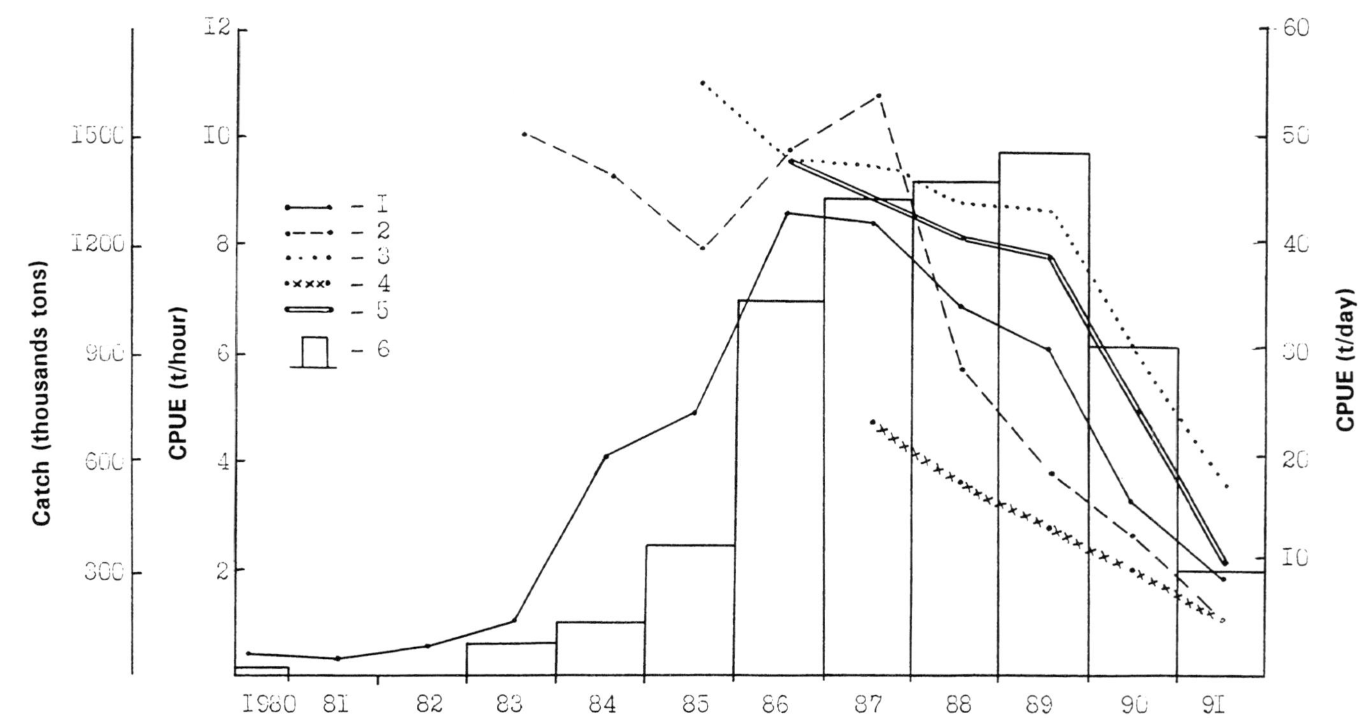

Fig. 6. Annual all-nation catch (t) and CPUE (t/h or t/d) by nation of walleye pollock in the Donut Hole area of the central Bering Sea, 1980–91. (Histogram = total catch; 1 = Japan (t/h); 2 = Korea (t/h); 3 = Poland (t/d); 4 = People's Republic of China (t/h); 5 = Russia (t/h).)

case, but for the total, a steady decline was evident during 1981–91. The relatively high CPUE in 1981 was attributed to the strong 1978 year-class, which comprised 60% of the catch in that year. The continued decline of the CPUE was attributed to normal attrition of the 1978 year-class, and the subsequent succession of relatively weak year-classes, such as that of 1981. The stabilization of abundance during 1989–91 was attributed to strong 1982 and 1984 year-classes. Nevertheless, the Russian scientists consider that pollock abundance in recent years is low.

Central Bering Sea

Annual distributions of pollock in the Donut Hole and adjacent southwestern sector from Russian pelagic-trawl surveys in autumn 1986–90 are shown in Fig. 5. In the Donut Hole during 1987–90, CPUE declined from 4.6 t/h in 1987 to 0.3 t/h in 1990 (Table 4). Biomass was 1.132 million t in 1987; rose to 2.802 million t in 1988; then declined to 194 000 t in 1990. For the southwest sector during 1986–90, CPUE was 2.1 t/h in 1986; rose to 2.7 t/h in 1987; then declined to 0.2 t/h in 1990. Comparable values for biomass were

Table 5. Abundance (10^6 fish) and biomass (10^3 t) of walleye pollock in the western Bering Sea, 1978–87. (Source: Balykin 1989)

Year	Abundance (10^6 fish)	Biomass (10^3 t)
1978	2715	1205
1979	4154	1483
1980	2586	1064
1981	3257	1421
1982	5443	2224
1983	3333	1688
1984	2939	1729
1985	1985	1105
1986	1422	856
1987	3027	1409

1.76 million, 1.96 million, and 185 000 t. During 1987–91, CPUEs for the commercial fleets of China, Japan, Korea, Poland, and Russia exhibited the same trend (Fig. 6).

Decline in CPUE and biomass was attributed to normal attrition of the strong 1978 year-class, unregulated fishing, and the absence of any new strong year-classes. The situation had been forecast in 1988 by Bulatov and Sobolevsky (1989).

Western Bering Sea

Abundance and biomass of pollock varied substantially during 1978–87 (Table 5). Fluctuations in the annual catches during 1978–91 (417 000 – 253 million t) reflected this (Table 2). Abundance and biomass peaked in 1982 (5.443 billion fish; 2.224 million t) due to the strong 1978–79 year-classes, which comprised 80% of the catch. Thereafter, the abundance declined through 1986 (1.422 billion fish; 856 000 t); then rose in 1987 (3.027 billion fish; 1.409 million t), when the relatively strong 1984 year-class entered the fishery.

Discussion

Strong year-classes of pollock occurred near synchronously throughout the Bering Sea. The absence of strong year-classes in the western Bering Sea during the 1960s was due in part to the lack of age composition data. Similarly, for the Donut Hole, early records wee fragmentary. For example, TINRO archives yielded records of successful pollock fishing during the winter of 1977, in the region of 57°N, 180°, corresponding to the Donut Hole. Predominant sizes in the catches were 44–46 cm, possibly the 1969 year-class. Recent data suggests that the 1989 year-class may be strong.

Fluctuations in pollock abundance were attributed to climatic conditions by Naumenko et al. (1990). Warm years favored pollock production, and cold years favored herring production.

The age composition in the eastern (Bulatov 1989*b*) and western (Balykin 1989; Naumenko et al. 1990) Bering Sea and positive anomalies of water temperature in spring–summer were investigated. It was found that strong year-classes were forming during the first warm years on an ascending curve of solar activity. Such a year-class (1978) is found in all three regions of the Bering Sea. Undoubtedly, strong year-classes do not depend on solar activity itself, but on interannual variability of biogenic concentration, presence of positive anomalies of water temperature in the post-spawning period, initial production, zooplankton availability, pressure of predators, conditions of first wintering, and other reasons.

References

Bakkala, R. 1989. Variability in the size and age composition of eastern Bering Sea walleye pollock, p. 307–322. *In* Proceedings of the International Symposium on Biology and Management of Walleye Pollock, Nov. 14–16, 1988, Anchorage, Alaska. Sea Grant Rep. University of Alaska, Fairbanks, Alaska. 89-1.

Balykin, P.A. 1989. Western Bering Sea walleye pollock population dynamics and stock condition, p. 559–568. *In* Proceedings of the International Symposium on Biology and Management of Walleye Pollock, Nov. 14–16, 1988, Anchorage Alaska. University of Alaska, Fairbanks, Alaska. Sea Grant Rep. 89-1.

Bulatov, O.A. 1986. Distribution of gadid eggs and larvae (subfamily Gadidae) in the Pacific Kamchatka and the western Bering Sea, p. 89–101. *In* N.S. Fadeev [ed.]. Gadidae of the Far Eastern seas. TINRO, Vladivostok. (In Russian)

1987. Distribution of pollock eggs and larvae in the eastern Bering Sea, p. 100–114. *In* N.S. Fadeev [ed.]. Population composition, abundance dynamics, and pollock ecology. TINRO, Vladivostok. (In Russian)

1988. Interannual variability of total stocks of pollock in the eastern Bering Sea, p. 4–16. *In* N.S. Fadeev [ed.]. Variability of ichthyofauna composition, yields of year-classes, and forecasting techniques of fishes in the western Pacific. TINRO, Vladivostok. (In Russian)

1989*a*. Reproduction and abundance of spawning pollock in the Bering Sea, p. 199–206. *In* Proceedings of the International Symposium on Biology and Management of Walleye Pollock. Nov. 14–16, 1988, Anchorage, Alaska. University of Alaska, Fairbanks, Alaska. Sea Grant Rep. 89-1.

1989*b*. The role of environmental factors in fluctuations of stocks of walleye pollock (*Theragra chalcogramma*) in the eastern Bering Sea, p. 353–357. *In* R.J. Beamish and G.A. McFarlane [ed.] Effects of ocean variability on recruitment and an evaluation of parameters used in stock assessment models. Can. Spec. Publ. Fish. Aquat. Sci. 108.

1990. The fishery and state of stocks of the most important species of fish in the Bering Sea, p. 143–147. *In* International Symposium on Bering Sea Fisheries, Apr. 2–8, 1990, Khabarovsk, Russia. Preliminary printing for in-house use by the U.S. Natl. Mar. Fish. Serv., NOAA, Seattle, Wash.

Bulatov, O.A., and G.V. Khen. 1984. Factors determining year-class strength of walleye pollock and yellowfin sole in the eastern Bering Sea, p. 58–69. *In* O.A. Skariato [ed.] Nature and biological resources of seas and oceans. Nauka, Leningrad. (In Russian)

Bulatov, O.A., and Ye.I. Sobolevsky. 1989. Distribution, condition of stocks, and outlook of the walleye pollock fishery in the high Bering Sea, p. 591–604. *In* Proceedings of the International Symposium on Biology and Management of Walleye Pollock, Nov. 14–16, 1988, Anchorage, Alaska. University of Alaska, Fairbanks, Alaska. Sea Grant Rep. 89-1.

Dawson, P. 1990. Information on the stock structure of Bering Sea pollock, p. 65–73. *In* International Symposium on Bering Sea Fisheries, Apr. 2–8, 1990. Khabarovsk, Russia. Preliminary printing for in-house use by the U.S. Natl. Mar. Fish. Serv., NOAA, Seattle, Wash.

Fadeev, N.S. 1980. Was there a 'flash' of the pollock in the North Pacific? Biol. (Vladivost.) 5: 66–71. (In Russian)

1981. Period of reproduction and spawning runs of Alaska pollock, p. 3–18. *In* V.P. Shuntov [ed.] Ecology of stocks and fishery of Alaska pollock. TINRO, Vladivostok, Russia. (In Russian)

Flusova, G.D., and L.V. Bogdanov. 1986. On the population structure according to genetic works, p. 79–88. *In* N.S. Fadeev [ed.] Gadidae of the Far Eastern seas. TINRO, Vladivostok, Russia. (In Russian)

Gavrilov, G.M., and A.M. Bezlyudny. 1983. Dynamics of catches of walleye pollock in the southeastern Sea of Japan. Rybn. Khoz. 7: 38. (In Russian)

Gavrilov, G.M., and A.M. Bezlyudny. 1986. Dynamics of abundance of pollock in the southwestern Sea of Japan, p. 5–38. *In* N.S. Fadeev [ed.] Gadidae of the Far Eastern seas. TINRO, Vladisvostok, Russia. (In Russian)

Hinkley, S. 1987. The reproduction biology of walleye pollock, *Theragra chalcogramma*, in the Bering Sea, with reference to spawning stock structure. Fish. Bull. U.S. 85: 481–498.

Jackowski, E., and B. Trochinski. 1989. Features of fishery and biology of Alaska pollock taken in open waters of the Bering Sea on the basis of Polish catches in 1985–1988, p. 280–305. *In* W. Aron [ed.] Proceedings of the International Scientific Symposium on Bering Sea Fisheries, July 19–21, 1988, Sitka, Alaska. U.S. Department of Commerce, Washington, D.C., NOAA Tech. Memo. F/NWC-163.

Kachina, T.F., and N.P. Sergeeva. 1981. Dynamics of abundance of the eastern Okhotsk walleye pollock, p. 19–27. *In* N.S. Fadeev [ed.]. Ecology of stocks, and fishery of Alaska pollock. TINRO, Vladivostok, Russia. (In Russian)

Kowalewska-Pahlke, M. 1990. Biological information on walleye pollock based on Polish catches in the international waters on the Bering Sea in 1988, p. 131–140. *In* Proceedings of the International Symposium on Bering Sea Fisheries, Apr. 2–8, 1990. Khabarovsk, Russia. Preliminary printing for in-house use by the U.S. Natl. Mar. Fish. Serv., NOAA, Seattle, Wash.

Lynde, C.M., M.V.H. Lynde, and R.C. Francis. 1986. Regional and temporal differences in growth of walleye pollock (*Theragra chalcogramma*) in the eastern Bering Sea and Aleutian Basin, with implications for management. Unpublished MS. U.S. Department of Commerce, Northwest and Alaska Fisheries Center, NMFS, NOAA, 7600 Sand Point Way NE, Seattle, Wash., 98115. 46 p.

Moiseev, Te.I. 1983. Age composition and growth rate of the eastern Bering Sea walleye pollock (*Theragra chalcogramma* Pallas). Izv. TINRO 107: 94–101. (In Russian)

Musienko, L.N. 1963. Ichthyoplankton of the Bering Sea (data of the Bering Sea Expedition of 1958–59, p. 251 – 285. *In* P.A. Moiseev [ed.] Soviet fisheries investigations in the Northeast Pacific. Part I. Tr. VINRO 48. Izv. TINRO 50. (Transl. from Russian by Israel Program for Sci. Transl., Jerusalem 1968).

Naumenko, N.I., P.A. Balykin, E.A. Naumenko, and E.R. Shaginyan. 1990. Year-to-year variation in stocks and community structure of western Bering Sea pelagic fishes, p. 223–224. *In* Proceedings of the International Symposium on Bering Sea Fisheries, Apr. 2–8, 1990, Khabarovsk, Russia. Preliminary printing for in-house use by the U.S. Nat. Mar. Fish. Serv., NOAA, Seattle, Wash.

NOAA 1991. Quarterly Report, July–August–September. 1991. U.S. Department of Commerce, Northwest and Alaska Fisheries Center, NMFS, NOAA, Seattle, Wash. 33p.

North Pacific Fisheries Management Council. 1989. Stock assessment and fishery evaluation document for groundfish resources in the Bering Sea/Aleutian Islands Region as projected for 1990. North Pac. Fish. Manage. Counc., P.O. Box 103136, Anchorage, Alaska, 99510. 213 p.

Pereyra, W.T., J.E. Reeves, and R.G. Bakkala. 1976. Demersal fish and shellfish resources of the eastern Bering Sea in the baseline. 1975. Progress Rep. Northwest and Alaska Fisheries Center, NMFS, NOAA, Seattle, Wash. 619 p.

Sample, T.M., and R.G. Bakkala. 1989. Assessment of walleye pollock of the eastern Bering Sea based on bottom trawl surveys, p. 457–470. *In* Proceedings of the International Symposium on Biology and Management of Walleye Pollock, Nov. 14–16, 1988, Anchorage, Alaska. University of Alaska, Fairbanks, Alaska. Sea Grant Rep. 89-1.

Serobaba, I.I. 1968. Pollock, *Theragra chalcogramma* Pallas, spawning in the northeastern Bering Sea. Vopr. Ichthyol. 8(6): 992–1003.

Serobaba, I.I. 1971. Pollock reproduction (*Theragra chalcogramma* Pallas) in the eastern Bering Sea. Izv. TINRO 75: 433–442. (In Russian)

Serobaba, I.I. 1974. Ecology of pollock (*Theragra chalcogramma* Pallas) spawning in the Bering Sea. Vopr. Ichthyol. 14(35): 544–552. (In Russian)

Serobaba, I.I. 1977. Fecundity of walleye pollock, *Theragra chalcogramma* Pallas, in the Bering Sea. Vopr. Ichthyol. 17(2): 219–231. (In Russian)

Sokolovsky, A.S., and S.E. Glebova. 1986. Long-term fluctuations of walleye pollock abundance in the Bering Sea. Izv. TINRO 110: 38–42. (In Russian)

Vasil'kov, V.V., and S.E. Glebova. 1984. Factors determining year-class strength of the walleye pollock, *Theragra chalcogramma* (Gadidae), of the western Kamchatka. Vopr. Ichthyol. 24(4): 561–570. (In Russian)

Waldron, K.D. 1981. Ichthyoplankton, p. 471–494. *In* D.W. Hood and J.A. Calder [ed.] The eastern Bering Sea Shelf: oceanography and resources. Vol. 1. U.S. Department of Commerce, NOAA Office of Marine Pollution, Washington, D.C.

Wespestad, V.G., and J.J. Traynor. 1988. Walleye pollock, p. 11–32. *In* R.G. Bakkala [ed.] Condition of groundfish resources of the eastern Bering Sea and Aleutian Islands region in 1987. U.S. Department of Commerce, Washington, D.C. NOAA Tech. Memo. NMFS F/NWC-139.

Yoshimura, T., and T. Sasaki. MS 1990. Outline of Japanese trawl fishery in the international waters of Bering Sea (1986–1989), p. 124–130. *In* Proceedings of the International Symposium on Bering Sea Fisheries, Apr. 28, 1990, Khabarovsk, Russia. Preliminary printing for in-house use by the U.S. Natl. Mar. Fish. Serv., NOAA, Seattle, Wash.

Influence of climate on recruitment and migration of fish stocks in the North Sea

E. Svendsen, A. Aglen, S.A. Iversen, D.W. Skagen, and O. Smestad

Svendsen, E., A. Aglen, S.A. Iversen, D.W. Skagen, and O. Smestad. 1995. Influence of climate on recruitment and migration of fish stocks in the North Sea, p. 641–653. *In* R.J. Beamish [ed.] Climate change and northern fish populations. Can. Spec. Publ. Fish. Aquat. Sci. 121.

Abstract: For several decades, the main aims of fishery science have been to understand and explain the great variability in survival from egg to mature fish. One of the reasons that efforts have not led to any clear and quantifiable conclusions is probably that the influence of the physical climate has so far been underestimated. Hydrographic data taken during 22 summers over most of the northern and central North Sea, and meteorological data from a station outside western Norway have been used to derive climatic parameters of prime importance for the biological productivity of the North Sea. These time series, together with recruitment and spawning-stock size data have been used to construct empirical models demonstrating, for example, that more than 70% of the year-to-year recruitment variability in several fish stocks may be explained by the climate or weather before and during larval stages. In addition to demonstrating which climatic factors are most important for biological processes, the models indicate that realistic forecasts of recruitment can be made within the summer of the spawning year. This possibility is of great importance to the better management of fish resources.

Résumé : Depuis plusieurs décennies, l'objet principal de la science de la pêche est de comprendre et d'expliquer la grande variabilité dans la survie des poissons, depuis le stade de l'oeuf jusqu'à la maturité. Une des raisons pour lesquelles ces efforts n'ont encore débouché sur aucune conclusion claire et quantifiable est probablement que l'influence du climat physique a été sous-estimée jusqu'ici. Les données hydrographiques recueillies durant 22 étés sur la plus grande partie du nord et du centre de la mer du Nord, et les données météorologiques provenant d'une station située au large de l'ouest de la Norvège ont servi à calculer des paramètres climatiques de première importance pour la productivité biologique de la mer du Nord. Ces séries chronologiques, de concert avec les données sur le recrutement et sur la taille des stocks en âge de frayer, ont servi à établir des modèles empiriques démontrant, par exemple, que plus de 70 % de la variabilité du recrutement d'une année à l'autre chez plusieurs espèces de poissons pouvait s'expliquer par le climat ou par les conditions météorologiques avant et durant le stade larvaire. Outre qu'ils mettent clairement en évidence les facteurs climatiques les plus importants pour les processus biologiques, les modèles indiquent que l'on peut faire des prévisions réalistes en matière de recrutement dans l'été de l'année de frai. Cette possibilité présente une grande importance pour l'amélioration de la gestion des ressources halieutiques.

Introduction

In studies of the ecology of the ocean, and especially in fishery biology, a common assumption is that the physical climate has a strong influence on population dynamics (e.g., recruitment success and migration pattern). It has, however, often been difficult to demonstrate such relations directly. This may be because many such investigations consider only one physical parameter at a time (often temperature), which is not necessarily a prime ecological steering factor. Corten (1990) suggested that most of the observed changes in the pelagic fish stocks in the North Sea (NS) could be explained by a theory assuming a long-term reduction in the inflow of water from the Atlantic Ocean to the NS during the period 1960–80, and an increase of this inflow in later years. However, he also stated that there is, as yet, no physical evidence that such a change occurred. The possible importance of Atlantic Water (AW) to the NS ecology is seen from the suggested general circulation pattern in Fig. 1. This schematic (more typical of summer than winter) is derived from a combination of earlier suggested circulation patterns, new information from several recent national and international investigations, and some numerical modeling experiments.

During the 1970s an ocean climate event occurred, known as the great salinity anomaly (Dickson et al. 1988). This event strongly affected most of the fish stocks in the North Atlantic. The anomaly manifested itself by both reduced salinity and temperature in the Northeast Atlantic, including the Norwegian Sea, the Barents Sea, and the NS, and was partly a direct effect of changes in the transport and mixed composition

E. Svendsen, A. Aglen, S.A. Iversen, D.W. Skagen, and **O. Smestad,** Institute of Marine Research, P.O.Box 1870 Nordnes, N-5024 Bergen, Norway.

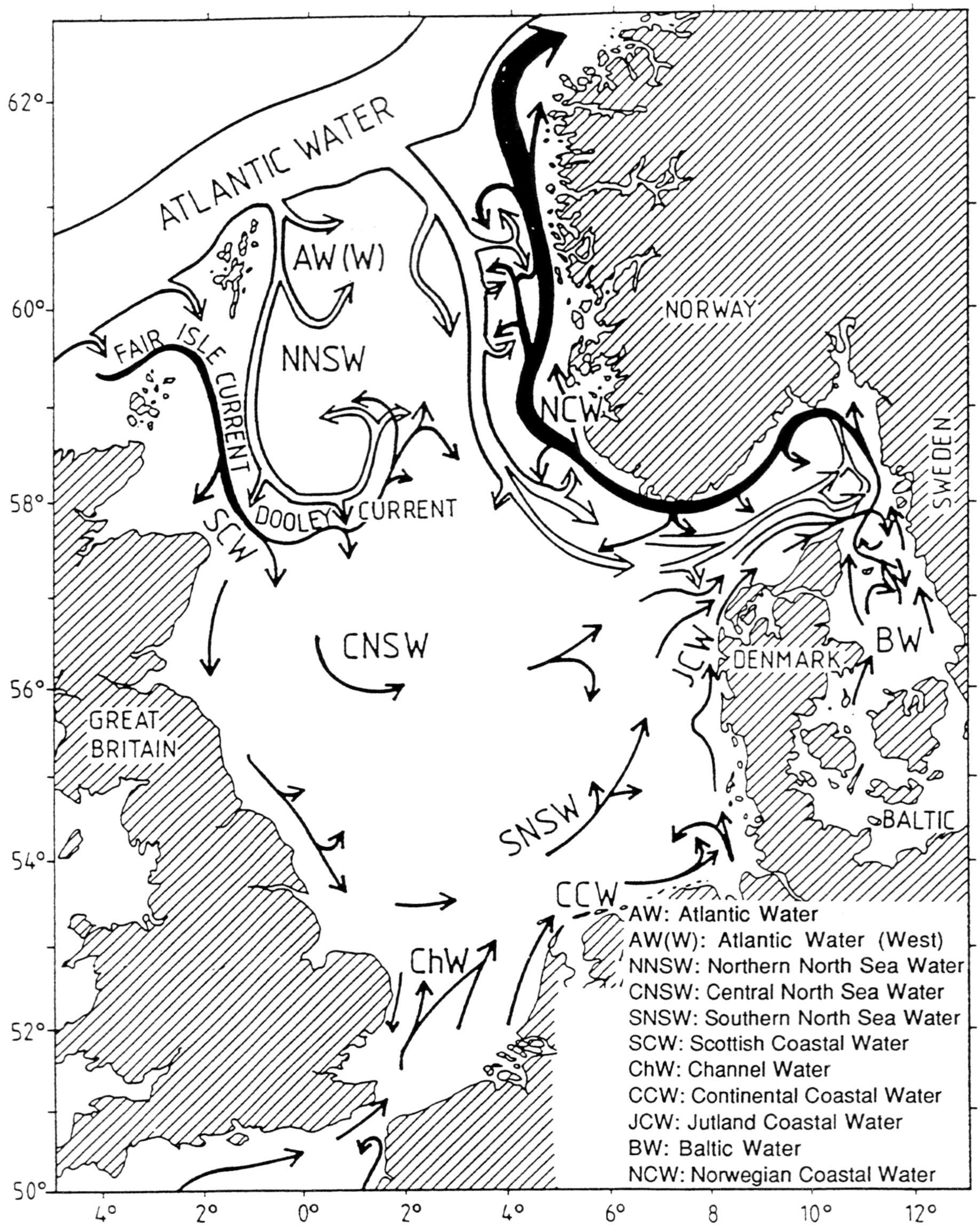

Fig. 1. Schematic general circulation in the North Sea. The width of the arrows indicates the magnitude of transport.

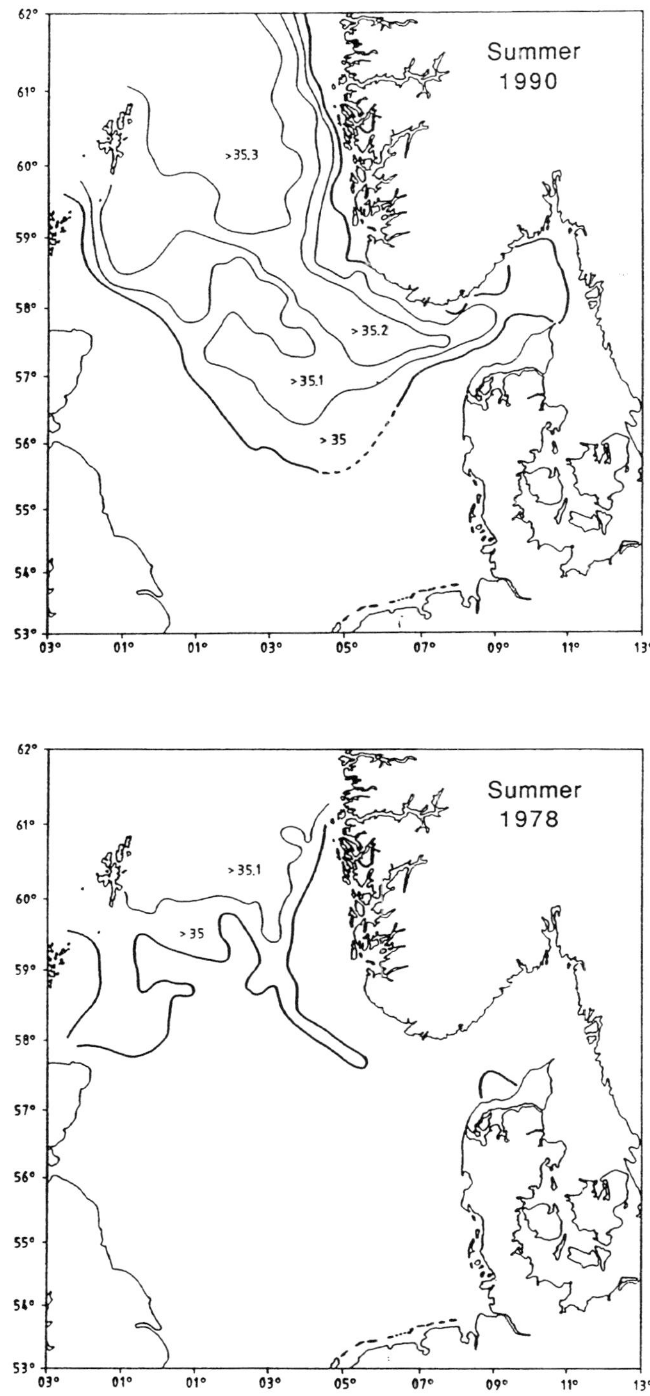

Fig. 2. Depth mean (50–200 m or bottom) salinity distribution, for the summers of 1990 (top) and 1978.

of AW and water from the Arctic Ocean. It was also coupled to an anomaly in the large-scale atmospheric climate, causing local anomalies (e.g., over the NS), of climate parameters such as air–sea heat exchange, wind speed, and wind direction.

There is no doubt that correlations between biological and physical variability related to such drastic climatic variations exist on a time scale of 5–10 yr. However, drastic biological variabilities are often observed from one year to the next. To understand which (of several partly correlated) climatic factors are of vital importance for biology, one must be able to quantify the possible connections from year to year.

The Institute of Marine Research (IMR) in Bergen, Norway, has yearly surveys around June–July in the northern and central NS. About 300–500 hydrographic (CTD (conductivity and temperature vs. depth)) stations are sampled in this period, giving generally good areal coverage, at least north of 57°N. These salinity and temperature data for the years 1968–90 are the physical basis for this study, together with wind and cloud observations (obtained from the Norwegian Meteorological Institute, DNMI) at Utsira, an island off western Norway. However, we do not believe that the actual salinity and temperature are necessarily prime ecological steering factors. Therefore, these data have been combined to give integrated measures of the physical conditions, such as the areal coverage of AW in the NS, believed to represent an index of the year-to-year variability of the inflow of AW.

Also one of our parameters is associated with heat exchange, which Colebrook and Taylor (1984) indicated was related to a 3 to 4-yr periodic fluctuation in plankton abundance. Colebrook (1985) also indicated that the influence of wind and temperature on zooplankton abundance might be primarily through processes involved in vertical mixing and stability, both represented by our derived parameters. A more detailed description of the derived physical variables used in this work is given by Svendsen and Magnusson (1992).

Physical variables

Because no time series of the above-mentioned physical processes are available, it has been a major task to derive parameters (from the data we have) that indirectly describe the physical variabilities believed to be of main importance for the biological variability. These are (Svendsen and Magnusson 1992):

1) Subsurface inflow of AW, represented by the area covered by water with high salinity (averaged from 50–200 m depth) greater than specified values (AW_{deep}).
2) Heat content, represented by the area covered by water with temperatures (averaged from 50–200 m depth) greater than specified values (AW_{heat}).
3) Surface inflow of AW, represented by the area covered by water with high salinity (averaged from 0–30 m depth) greater than 35.0 (AW_{surf}).
4) Stable surface layer, represented by the area covered by water with salinity (averaged from 0–30 m depth) less than 34.0 (NCW_{34}).

Examples of some relatively extreme years of these distributions are presented in Figs. 2–5. Clearly the subsurface inflow of AW (Fig. 2) was weak in 1978, being connected to the "late 1970s anomaly" (Dickson et al. 1988). In 1990 the inflow was quite strong (although not extreme), covering major parts of the northern NS and parts of Skagerrak. Although the measurements are taken during a relatively short period each year, it is believed that the distribution represents the integrated effect of the inflow that took place up to half a year prior to the observations.

Figure 3 shows the drastic differences in the heat content, from being low (cold) in 1987, to the extreme warm situation in 1990, probably the warmest during the past 50 yr. The

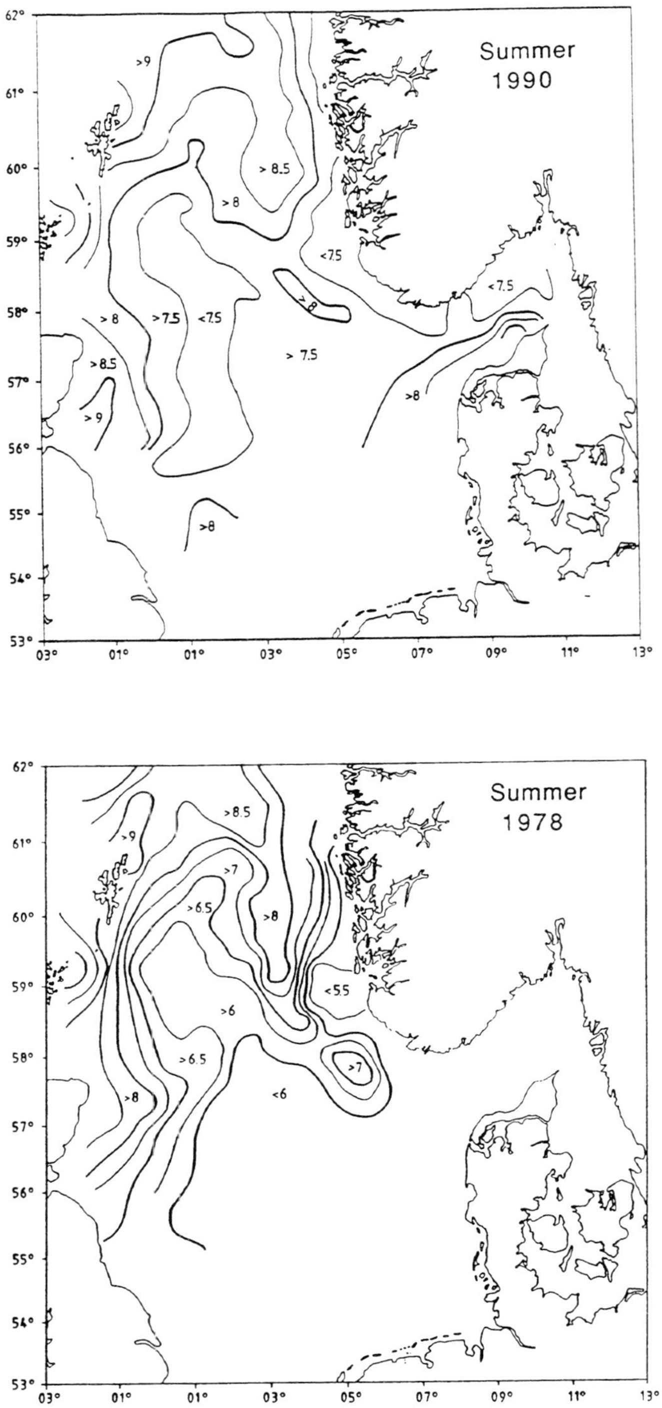

Fig. 3. Depth mean (50–200 m or bottom) temperature (°C) distribution, for the summers of 1990 (top) and 1978.

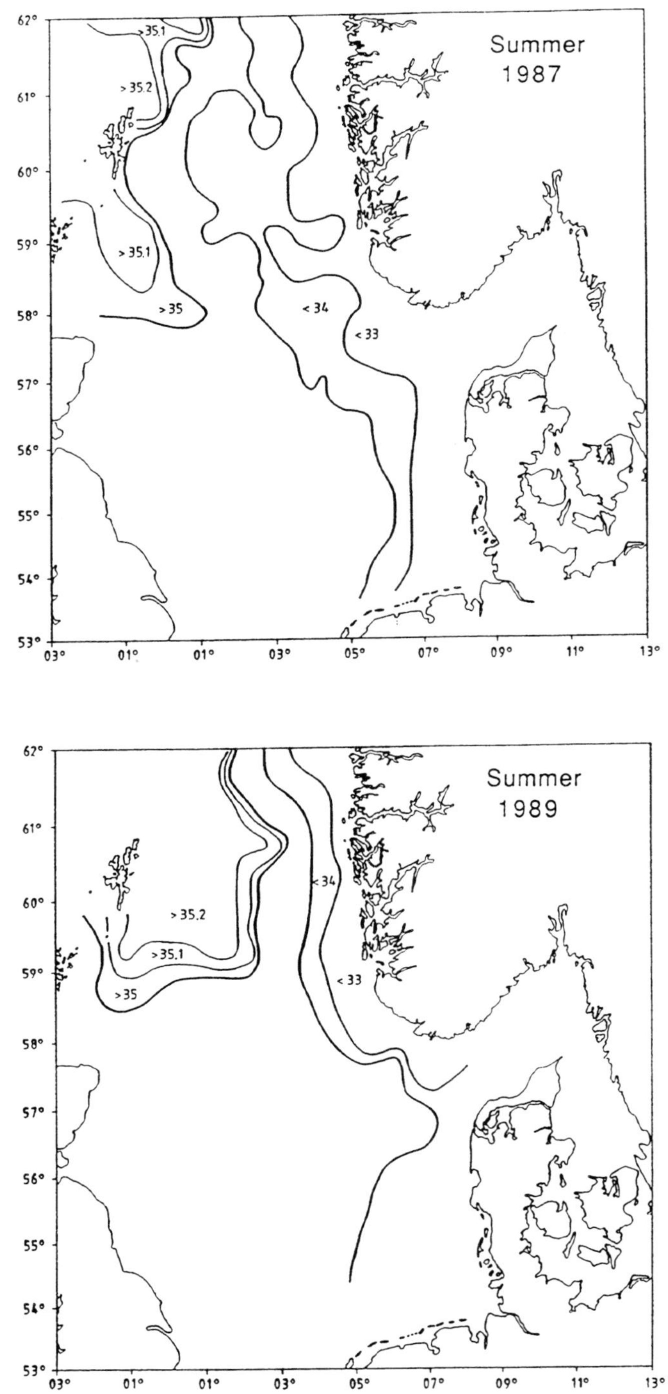

Fig. 4. Depth mean surface (0–30 m) salinity distribution, for the summers of 1987 (top) and 1989.

subsurface heat content is mainly a combined effect of the inflow of AW and surface cooling during the previous winter (Svendsen and Magnusson 1992).

The spread of relatively low-saline Norwegian coastal water (NCW) and/or continental water (Fig. 4) was at its maximum in 1987, covering more than 50% of the area north of 57°N and coinciding with an extreme maximum in the abundance of the phytoplankton *Ceratium* spp. (Dickson et al. 1992).

The surface inflow of AW east of Shetland might be of special interest in this context, since the nutrients in this water are directly available for primary production. Large

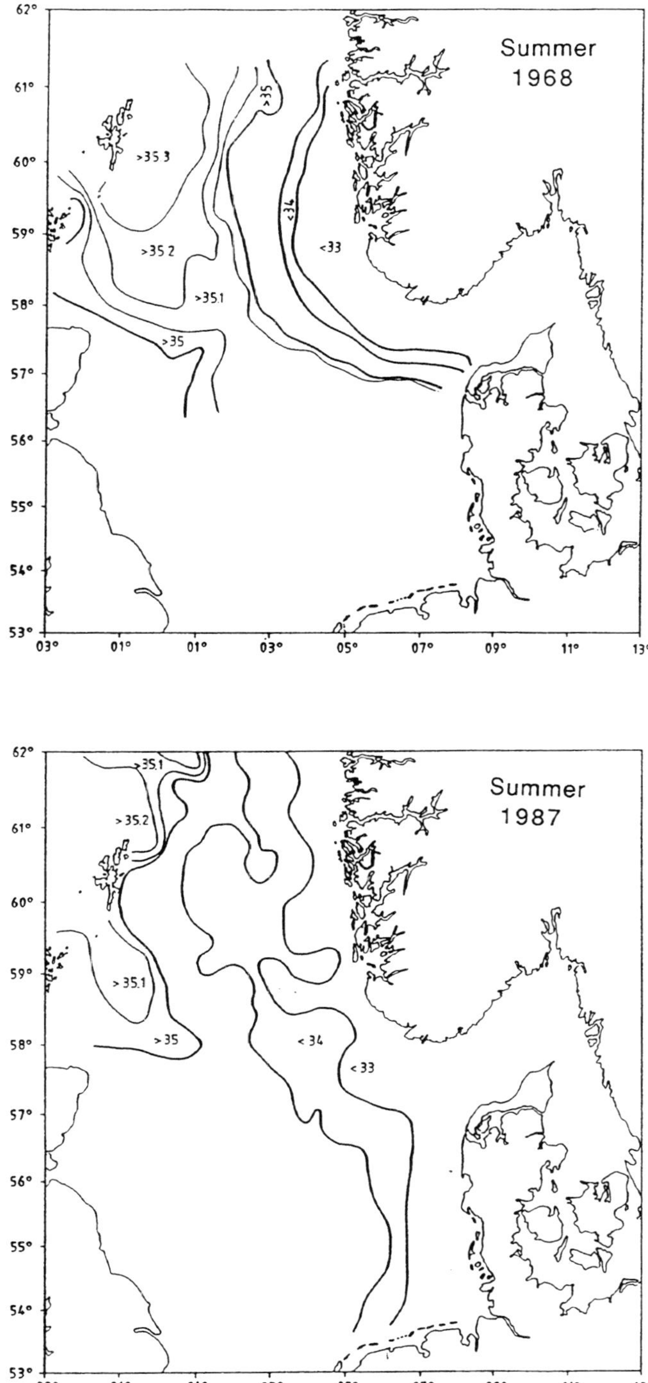

Fig. 5. Depth mean surface (0–30 m) salinity distribution, for the summers of 1968 (top) and 1987.

differences from year to year were observed (Fig. 5); however, it is uncertain if this parameter is representative of the inflow. In some years, a relative thin layer of NCW might mask the AW signature, which was probably the case in 1987.

The 22-yr time series of available ocean parameters are shown in Fig. 6 (from Svendsen and Magnusson 1992). The curves denoted, respectively, as S > 35.1, T > 6.5, S < 34, and S > 35.0 are taken as the main hydrographic climate variables, and it is clear that they are not highly correlated.

The meteorological variables assumed to influence biological parameters significantly are:

5) Wind stress (Newtons per square metre), represented by the monthly or seasonal mean stress components (proportional to squared wind velocity) toward the north (STRESSN) and toward the east (STRESSE).
6) Wind effect (EFF, Watts per square metre), represented by the monthly or seasonal means of nondirectional wind (proportional to the cubed wind speed).
7) Cloudiness (CLM), represented by the monthly or seasonal means of low and medium clouds.

Wind stress is chosen because it is the parameter driving major parts of the current system. The wind effect is more proportional to the turbulent energy input to the ocean, creating entrainment of nutrients into the euphotic zone. Turbulence also has a direct effect on contact rates between prey and predators (Rothschild and Osborn 1988; Sundby and Fossum 1990).

The cloudiness parameter is incorporated because it affects light conditions, which are particularly important during the spring bloom (Sakshaug and Skjoldal 1989), and are also important for prey–predator interactions depending on visibility (Aksnes and Giske 1990). Light conditions are especially sensitive to the concentration of low and medium clouds, which to a large extent regulates both cooling during winter and heating during summer.

Biological effects of climate

The different working groups of the International Council for the Exploration of the Seas (ICES) that assess fish stocks produce yearly updated time series of recruitment (for individual age groups) and spawning stock biomass or numbers for individual fish stocks. These data (and their quality) are essential for our task to resolve the effect climate has on year-to-year variability in fish stocks.

The most critical phase for fish is during the very early stages, within a few months after spawning. It seems that the so-called match–mismatch theory is particularly relevant. Here it is critical that a high abundance of food (zooplankton) is available where and when the larvae finish their yolk sac stage (Hjort 1914; Ellertsen et al. 1989). Recruitment success probably also depends on predation pressure from other fish stocks or by relatives (cannibalism) (Øyestad 1985). Clearly the climate affects processes that are relevant to both these theories. Later research comparing potential with actual fecundity of cod deprived of food prior to spawning showed that only 20–80% of potential fecundity was realized, depending on the nutritional status of the fish (Kjesbu et al. 1991). This indicates that the climate prior to spawning also might influence recruitment through its impact on the feeding condition of the spawners.

In this paper, these aspects are not deeply considered; rather, we discuss what climatic parameters provide good or bad chances for survival based on statistics and general knowledge. Because the oceanographical parameters used in this study have a time resolution of only 1 yr, they must be looked on as yearly measures of how the environment may be potentially good or bad for the total biological process leading to recruitment success or failure for each individual fish stock.

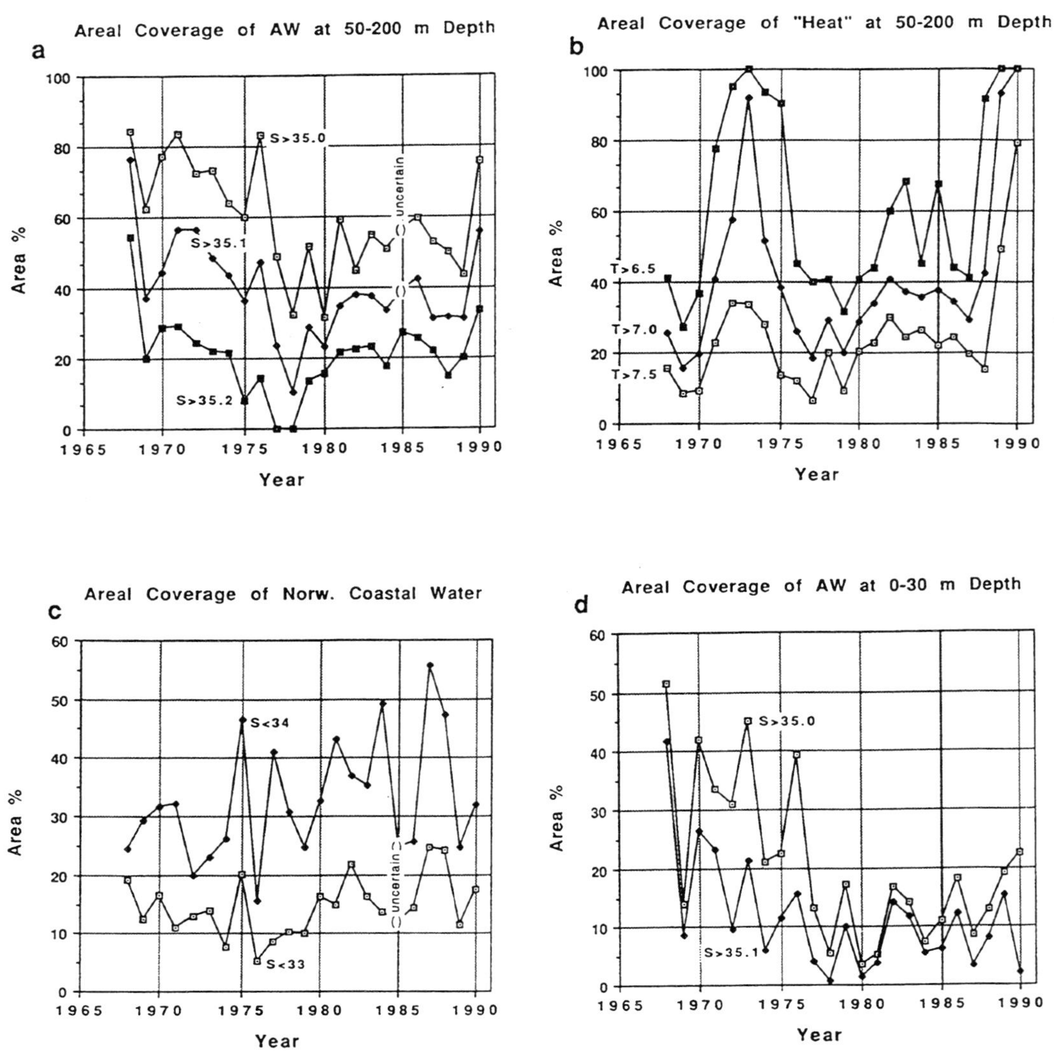

Fig. 6. Time series of the fraction of the area north of 57°N covered with (a) mean salinity >35.0, 35.1, and 35.2 (AW_{deep}, ref. Fig. 2); (b) mean temperature >6.5, 7.0, and 7.5°C (AW_{heat}, ref. Fig. 3); (c) mean salinity <34.0 and 33.0 ($NCW_{34,33}$, ref. Fig. 4); (d) mean salinity >35.1 and 35.0 (AW_{surf}, ref. Fig. 5).

The meteorological parameters used, having a time resolution of 1 mo, might pick up events each year that are critical for the survival of individual larvae populations.

General model

It seems that several explanatory climatic variables might be important from a biological point of view. To find a combination of a few (from many) parameters that are of main importance, multiple linear regression analysis (Wilkinson 1989) was chosen. This gives the coefficients *a*, *b*, *c*, and so on, based on the measured or estimated time series in the equation:

$$\text{(1) Biological Variable} = \text{Constant} + a(AW_{deep}) + b(AW_{heat}) + c(NCW) + d(AW_{surf}) + e_1(STRESSN_1) + e_2(STRESSN_2) + \ldots + f_1(STRESSE_1) + f_2(STRESSE_2) + \ldots + g_1(EFF_1) + g_2(EFF_2) + \ldots + h_1(CLM_1) + h_2(CLM_2) + \ldots + k_1 BIO_1 + k_2 BIO_2 + \ldots$$

where the indices (1, 2, ...) indicate specific months or "seasons." It should be noted that the statistics do not require that these variables be independent. The oceanic variables AW_{deep}, AW_{heat}, NCW_{34} and AW_{surf} might be any of the curves in Fig. 6, and the BIO variables can, for example, be a function of the spawning-stock biomass and the abundance of zooplankton.

The requirements set for selecting the most important explanatory variables are that:

1) Each selected coefficient (*a*, *b*, ...) is significantly different from zero, with a 95% confidence level (preferably better).
2) The number of variable is as few as possible (most of the coefficients in Equation 1 equal zero) to reach (3).
3) The total squared multiple correlation coefficient is preferably above 0.7.
4) The results are reasonable according to common oceanographical and biological knowledge.

The use of this type of correlation analysis could be discussed. One requirement is that each time series should not be autocorrelated. While autocorrelation seems to be weak for most of the time series, this is in general not the case for spawning stocks containing several year-classes. However, knowing that the variability of certain environmental parameters are in general a combined function of several variables, some kind of multivariate analysis is required to estimate the significance of each variable. (Because the spawning stock for herring is used here, it means that the numbers obtained for significance levels (*p* values) are somewhat too high). It is not claimed that the applied statistical method necessarily is the best for this purpose. The reasons for not examining several (all) stocks at once using principal components analysis, are because the results often are difficult to relate to the original environmental parameters and because of the authors lack of experience with such analysis.

Results

Because it is assumed that climate has a large impact on primary and secondary production, some time series on phytoplankton and zooplankton concentrations were obtained from the British Continuous Plankton Recorder (CPR) program (CPR Survey Team 1992). However, since the main focus in this paper is on spring-spawning fish and autumn-spawning herring *(Clupea harengus)*, only meteorological data from October to June–July have been used. Therefore, strong links between these "annual mean" plankton data and our selected climate variables were not expected. Although good correlations were obtained using six variables, not all the four requirements above were fulfilled. However, yearly phytoplankton abundance explains about 60% of the year-to-year zooplankton variability, and cloudy weather during the first quarter of the year (explaining an additional 10%) also contributes to a good zooplankton year. One explanation for this might be that cloudy weather extends the spring bloom, giving a prolonged time for the zooplankton to graze before the "phytoplankton" dies, or in general, there might be a better match if the spring bloom is delayed somewhat because of heavy clouds during the first months of the year.

Mackerel migration

During the initial phase of this work, the question was raised whether the climate had any influence on why variable fractions of the western mackerel *(Scomber scombus)* stock (WMACK), spawning west and southwest of Ireland, migrate into the NS to feed during summer and fall (Iversen and Skagen 1989). The spawning stock size of WMACK is in the order of 2 million t (ICES 1990a). For the time period considered, the main biomass of mackerel in the NS consists of WMACK. The local NS mackerel stock disappeared to a large extent during the 1970s, and is still on a level below 100 000 t (Iversen et al. 1991).

WMACK probably enters the NS along the western slope of the Norwegian Trench, which is associated with the area of the main influx of AW (Fig. 1). Walsh and Martin (1986) found that the main transport of AW (into Norwegian waters) is important for the migration of WMACK. They suggested that, in periods of weak flow, most of the AW and the mackerel population associated with it find their way into the NS, while in periods of strong flow more water and mackerel are likely to find their way into the Norwegian Sea. Although it is unlikely that most of the AW transport under any circumstances enters the NS, Walsh and Martin might be correct about the way that a larger fraction of AW enters the NS during periods of weak flow.

Our hypothesis is that the main mechanism for the WMACK to "choose" to enter the NS must be food availability (zooplankton), and also probably temperature since mackerel prefers warm water. This hypothesis, therefore, depends on good food production, which is assumed to be determined by: a previous large influx (spread) of AW carrying nutrients, zooplankton, and heat into the NS; a large spread of a stable surface layer providing large areas for possible good primary production; and relatively strong winds (wind effect) for vertical entrainment of nutrients and, therefore, continuation of good primary production necessary for good growth of the zooplankton. Experience shows that mackerel prefers water warmer than 9–10°C in the NS. In general, the subsurface core of the AW is colder than this, so migration into the NS probably takes place mainly in or above the pycnocline after the temperature in the surface layer has reached this temperature limit. A representative measure of the surface temperature is lacking because it increases very fast at this time of the year. However, the spread of heat in the lower layer might be an important preconditioning factor for the heating of the surface layer.

Figure 7 shows yearly measurements or estimates of WMACK compared with our empirical regression model. The model equation is

$$(2)\ \mathrm{MODWMACK} = -31.27 + 0.44(\mathrm{AW}_{heat}) + 37.26(\mathrm{EFF}_{2Q})$$

showing that large fractions occur in years with large areal coverage of "deep" water warmer than 6.5°C and high mean wind effect during the second quarter (EFF_{2Q}; April–June), and vice versa. This means that mackerel prefers entering the NS after a mild winter (probably connected to a relatively strong inflow of AW) and relatively strong turbulent mixing during the spring, the assumption being that this favours production of phytoplankton and thereby zooplankton, which is the main food for mackerel. Unfortunately, only 9 yr (1973–81) of reasonable migration estimates are available, but it seems that the "estimated" values (in parentheses in Fig. 7) for 1982–88 (Iversen and Skagen 1989) are not too far off, if the model is correct. The reason that the abundance of zooplankton from the CPR measurements does not come out as an explanatory variable might be that the yearly average is not representative for the situation around June–July when the WMACK normally enters the NS.

A closer look at the time series of the above explanatory variables, shows that the mean wind effect in April–June is the main cause of rapid and large year-to-year changes during the period 1973–76. The variable spread of heat (represented by AW_{heat}) seems to be the important factor for the somewhat slower changes in the mackerel migration taking place over a few years, including the effect of the late 1970s anomaly.

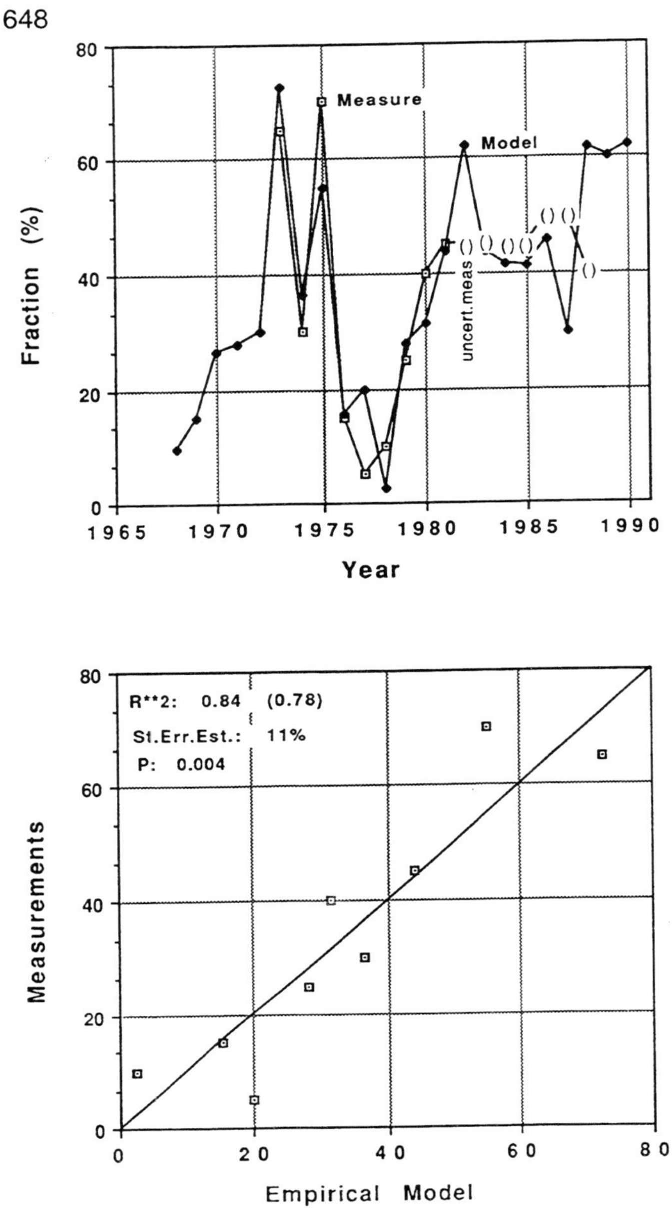

Fig. 7. Comparisons (annual time series and scatter plot) between measurements and the empirical model for the fraction of the western mackerel stock migrating into the North Sea.

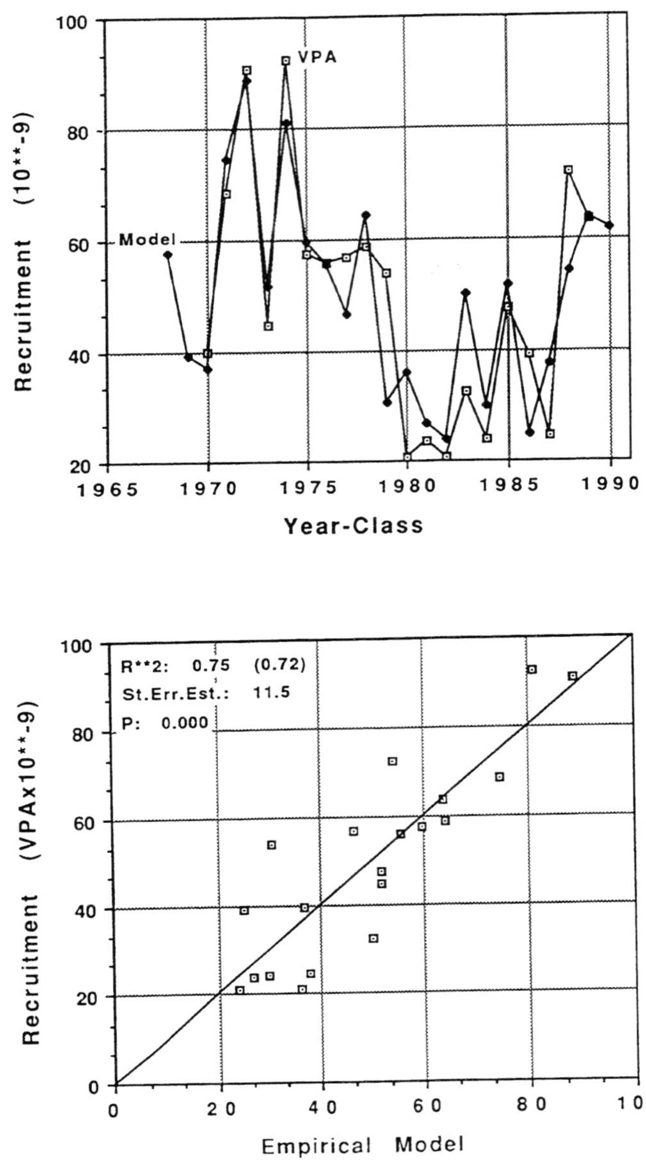

Fig. 8. Comparisons (annual time series and scatter plot) between measurements (VPA) and the empirical model for the recruitment of 0-group whiting.

Recruitment

The following fish stocks studied by this method are all assumed to be "pure" NS stocks. As mentioned earlier, the only hypothesis used is that the climate prior to and during the first months after spawning is most critical for the success of survival.

Whiting

The causes for good versus bad recruitment of 0-group whiting *(Merlangus merlangus)* (ICES 1991a) were obtained with the same two parameters as were used for mackerel migration. However, whiting prefers a weak mean wind effect during the second quarter, as seen from the model equation:

$$(3)\ \mathrm{MODWHIT}(0) = 57.72 + 0.52(\mathrm{AW}_{heat}) - 36.69(\mathrm{EFF}_{2Q})$$

It seems that, with just these two variables, we are able to explain more than 70% of the variability which has occurred during the years 1970–88 (Fig. 8). The inverse correlation with the wind effect is most obvious in the period 1970–75. At present, we cannot explain why generally weak winds during the egg and larvae stages of whiting are positive for producing a good year-class, but one answer could be that the western mackerel prey significantly on 0-group whiting. However, at present there are no biological observations to support this suggestion.

Herring

During the second half of the 1970s the NS herring stock (ICES 1991b) was very low, and under this situation, one should expect to find that recruitment (ICES 1991b) would be

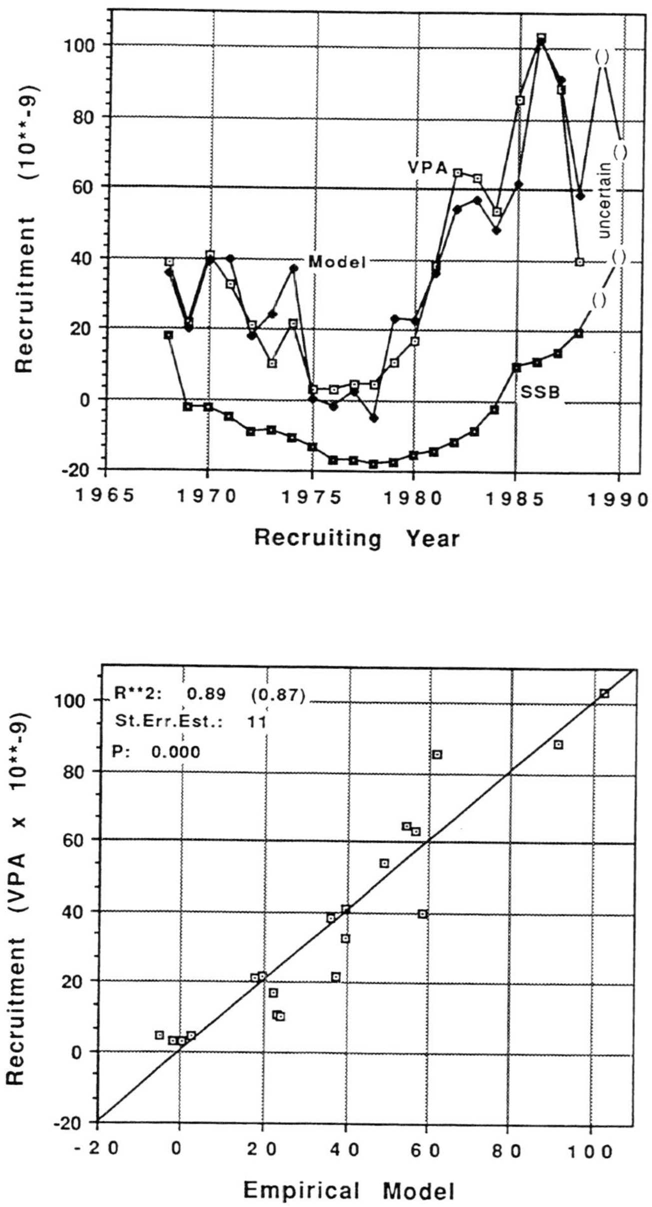

Fig. 9. Comparisons (annual time series and scatter plot) between measurements (VPA) and the empirical model for the recruitment of 0-ringer North Sea herring. Also shown are the parent-stock size (SSB). Note that the recruiting year is *n*, which means that the year-class and SSB refer to year *n* – 1.

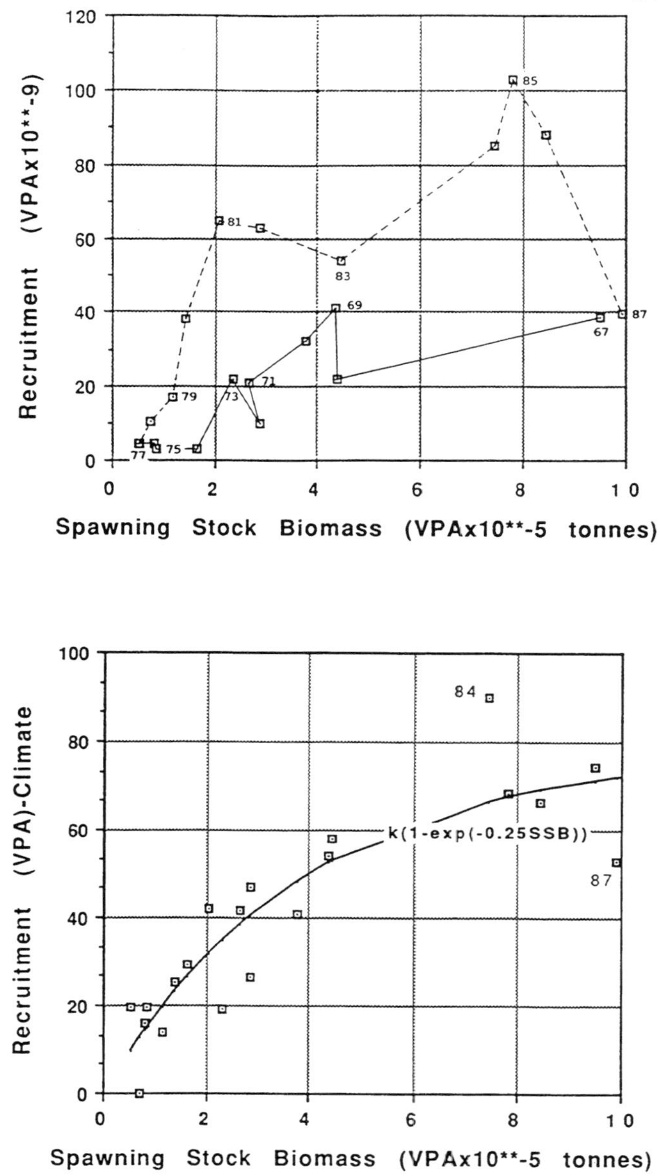

Fig. 10. Scatter plots of VPA estimates of parent stock versus recruitment (0-ringers) of North Sea herring (labels refer to year-class). In the lower plot, the recruitment is adjusted for the climatic (wind) terms in equation 4. The exponential curve relating to the SSB is shown for comparison.

affected by the spawning-stock size (SSB). This was the case according to the model equation:

$$(4)\ \mathrm{MODHERR}(0) = -57.76 + 13.34(\mathrm{EFF}_{4Q}) + 0.58(\mathrm{STRESSN}_{Mar}) + 78.69(1 - e^{-0.25(SSB)})$$

As Fig. 9 shows, the model picks up most of the variability in the virtual population analysis (VPA) recruitment estimates for the period 1968–88. Since the NS herring larvae are hatched in late autumn to early winter and stay as larvae through the winter and spring, we have chosen to relate recruitment to hydrographic conditions during the first half of the following year and to wind data for the fourth quarter (EFF_{4Q}; October–December) of the present year and the first half of the following year. It is interesting to note once more that the mean wind effect just after hatching (October– December) seems to be important for recruitment. In addition to strong winds during this period, good recruitment also seems to depend on strong or persistent southerly winds around March. This may imply that at the end of the larval stage, herring needs a final wind "push" to be transported to and/or kept at favourable feeding areas such as Skagerrak–Kattegat.

Studies of the individual time series show that the gross features of 0-group variability are mainly a result of spawning-stock size and the autumn wind effect. The northward wind stress in March seems to be important for explaining the more rapid and larger variability during the 1980s.

The stock–recruitment scatter plot (Fig. 10 top) clearly shows a different relationship for the year-classes 1967–76, compared with the year-classes 1978–86. When adjusting the VPA recruitment estimates by the climatic terms in Equation 4, the difference between these periods disappears, and a fairly good stock–recruitment relationship is obtained (Fig. 10 bottom). These findings support the hypothesis raised by Corten (1986, 1990) that the recruitment failure of NS herring in the 1970s was not only caused by a low spawning stock, but also to a large extent by climatic conditions.

Haddock

The next fish stock considered was haddock *(Melanugrammus aeglefinus)* (0-group; ICES 1990a). However, with the quality requirements set initially, it was impossible to explain the recruitment variabilities with the present climatic variables. Therefore, it seems that haddock behaves very differently from whiting.

Cod

In the ICES roundfish assessment working group report (ICES 1991a), the recruitment numbers for cod *(Gadus morhua)* and saithe *(Pollachius virens)* are given for the first time at the 1-group stage, so these numbers are used in the regression analysis with the climatic variables 1 yr earlier (during their larval stages). The results for cod are shown in Fig. 11, and the model equation is

$$(5)\ \mathrm{MODCOD}(1) = 901.05 - 17.78(\mathrm{AW}_{heat})^{>7.5} - 15.59(\mathrm{NCW}_{34}) + 169.70(\mathrm{EFF}_{1Q})$$

In this case, the best fit was obtained with AW_{heat} represented by the T > 7.5 curve in Fig. 6. This indicates that cod prefers situations with large areas of cold (at least colder than 7.5°C) subsurface water (and possibly reduced inflow of AW), reduced spread of the NCW, and strong winds during the first quarter of the year (EFF_{1Q}). This means that the NS cod stock may be severely reduced if the predictions about the greenhouse effect are correct. As Fig. 11 shows, the model produces unrealistically low estimates for the 1990 year-class (or actually the 1-group in 1991). These are caused by the heat content or temperatures in 1990 which were probably the highest (warmest) of the past 50 yr. The low estimate indicates that. under such extreme conditions, recruitment probably does not respond linearly to the AW_{heat}. Again it is evident that turbulent wind energy during the early stages is important for the recruitment success.

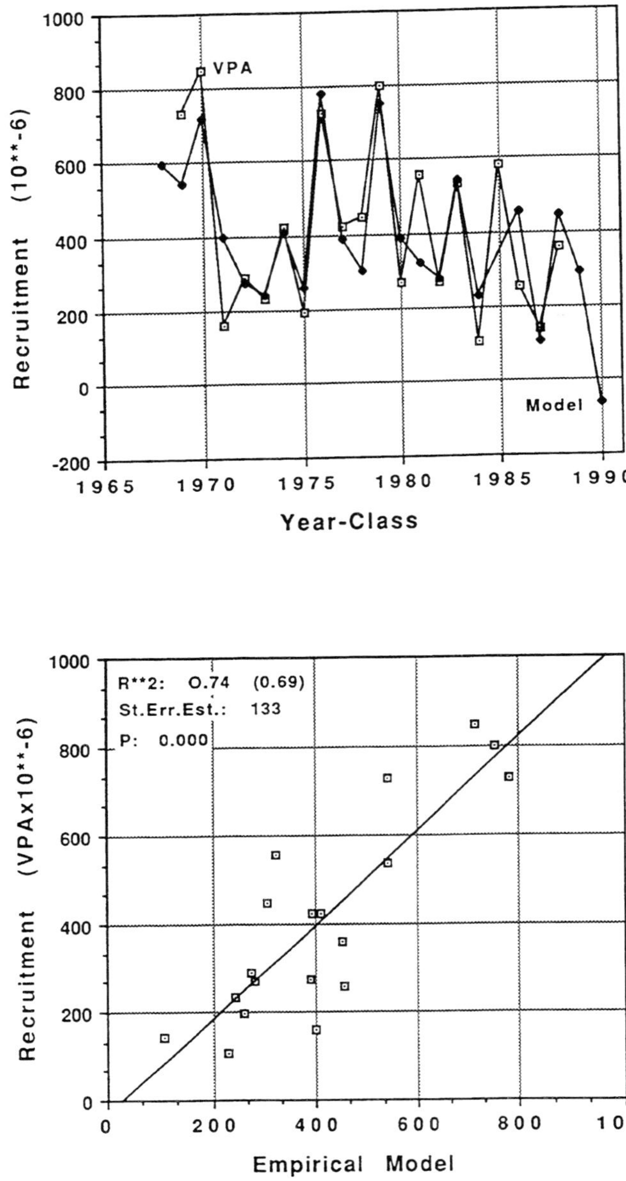

Fig. 11. Comparisons (annual time series and scatter plot) between measurements (VPA) and the empirical model for the recruitment of 1-group cod.

Saithe

For saithe, the response to heat is opposite to that for cod. The model equation is

$$(6)\ \mathrm{MODSAITHE}(1) = -125.09 + 4.95(\mathrm{AW}_{heat})^{>7.0} + 61.10(\mathrm{EFF}_{Mar}) + 67.23(\mathrm{EFF}_{Apr})$$

giving good recruitment when there are large areas of subsurface water warmer than 7°C, and strong wind mixing in March and April. This response to heat shows that the 1990 year-class (or actually the abundance of 1-group in 1991) of saithe was probably the best in the 20 yr considered (Fig. 12). In relation to cod and whiting, it seems that the wind effect for the critical stages of saithe comes somewhat later than for cod and somewhat earlier than for whiting. This might reflect differences in the typical peak spawning times of these fish stocks.

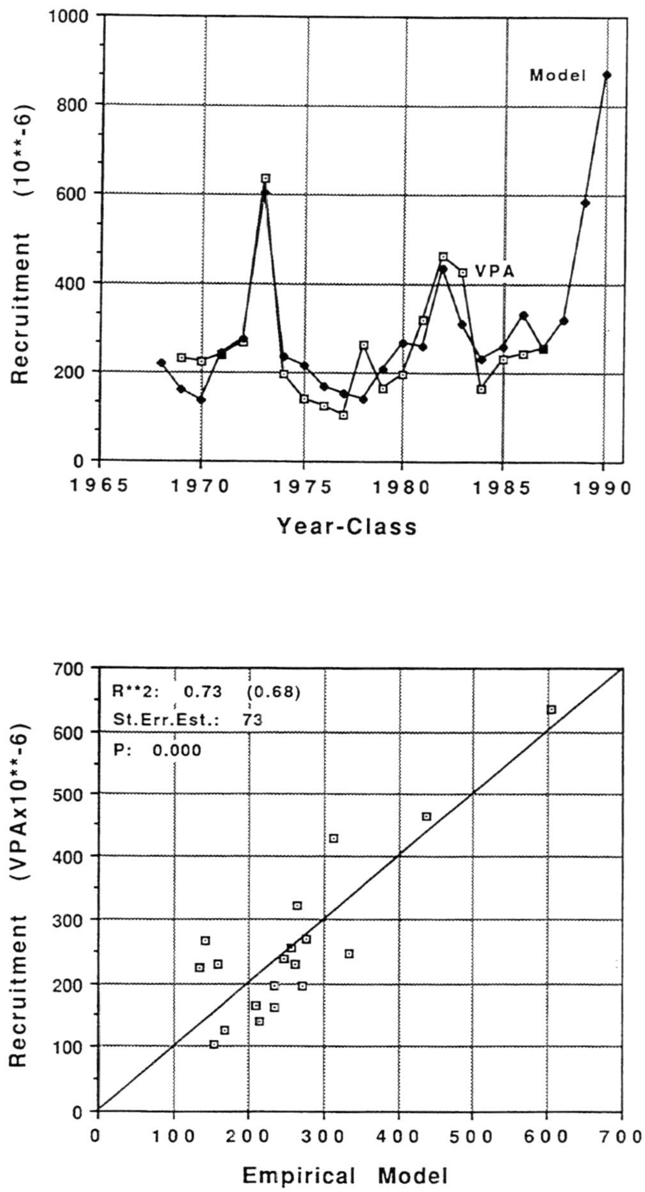

Fig. 12. Comparisons (annual time-series and scatter plot) between measurements (VPA) and the empirical model for the recruitment of 1-group saithe.

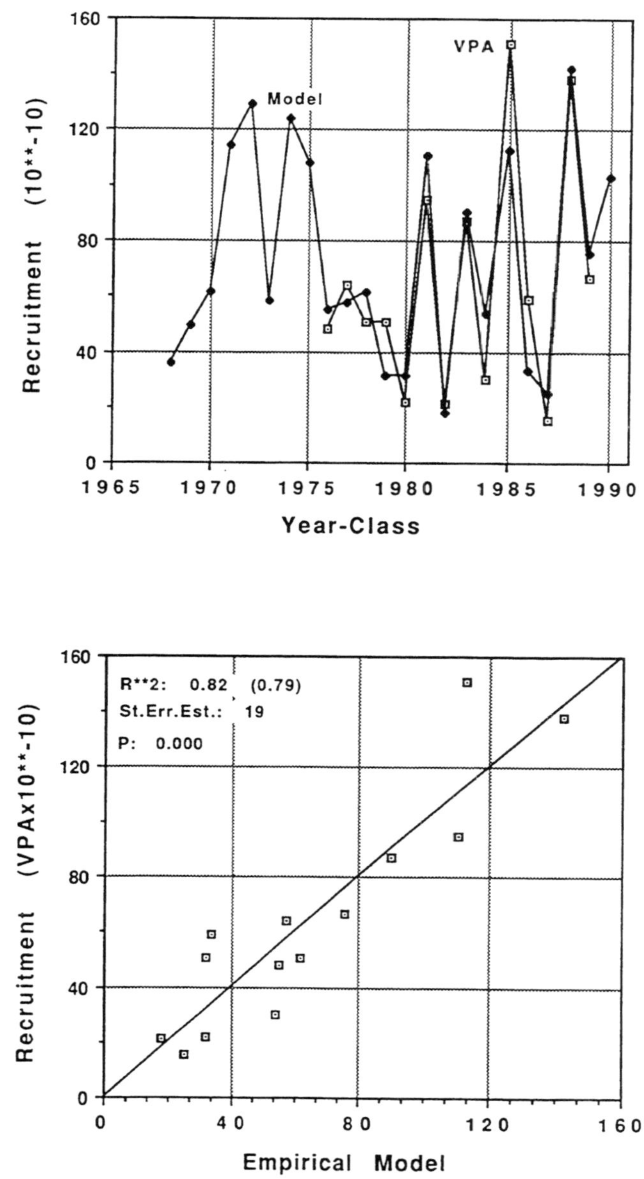

Fig. 13. Comparisons (annual time-series and scatter plot) between measurements (VPA) and the empirical model for the recruitment of 0-group sandeel.

Sandeel

The last stock considered is the NS sandeel *(Ammodytes marinus)* (0-group; ICES 1990b). Although smaller parts of the time series indicate correlation between SSB and recruitment, such a relation does not seem to be general. The model equation:

$$(7)\ \mathrm{MODSANDEEL}(0) = 2.36 - 4.89(\mathrm{STRESSE}_{May}) + 1.21(\mathrm{AW}_{heat})$$

indicates that reduced westerly (or increased easterly) wind stress in May and a large area of water warmer than 6.5°C (possibly in connection with increased inflow of AW) are positive for good recruitment (and vice versa). About 80% of the variability is explained by these two parameters. As Fig. 13 shows, only VPA estimates from 1976 to 1989 were available. Very large year-to-year fluctuations have taken place. The sandeel larvae are known to drift for only a few months after spawning and then settle to the bottom in July–August in specific areas. Based on the results from the model, we assumed that wind stress in about May is critical for this larval drift. This might be coupled to the wind-driven coastal–continental water sometimes flushing the larvae away from their preferred settling areas.

Evaluation of the models' predictive ability

The empirical models' results shown appear promising with respect to estimating the recruitment success of 0-group whiting and herring already within the summer of the spawning year (nowcasting), and predicting the recruitment of 1-group

Table 1. Comparison of prediction–nowcasting capability for the years 1986, 1987, and 1988 based on biological data up to 1985. * Those years where the prediction lies within one standard error of estimate obtained from the initial regression analysis.

	1986			1987			1988			
	PRED	MEAS	DIFF	PRED	MEAS	DIFF	PRED	MEAS	DIFF	SE
WHIT(0)	23	39	−16	36 *	24 *	12 *	49	72	23	12
HERR(0)	130	103	27	107	89	18	70	40	30	11
COD(1)	456	257	199	108 *	142 *	−34 *	459	360	99	73
SAITHE(1)	372 *	245 *	127 *	283 *	255 *	28 *	333	??	??	133

cod and saithe for the following year (forecasting). However, the actual forecasting or nowcasting capability has not been shown. To do so, the last 3 yr (near 20%) of VPA recruitment data (1986–88) were hidden and the regression analysis run again with the already selected parameters. Comparing these results with the actual data would be similar to going back to 1985, knowing which climatic factors were important, forecasting the recruitment for 1986, 1987, and 1988 as soon as the climatic data became available, and having no extra information on the size of the fish stocks except for the herring SSB. This should indicate the forecasting capability of models for the years to come, although such empirical models in principle are improving as more data become available.

The results are shown in Table 1, (mackerel and sandeel stocks are not included because the available time series are so short). This demonstrates that the results are not as promising as were indicated by the comparisons between the empirical models based on all the data and the data themselves. This means that the 3 yr being tested contain combinations of factors related to climate that were not experienced earlier in the time period considered. While running this predictability test, we observed that some of the initially selected variables did not have a statistically significant weight. This means that further work must be done to combine these methods to select the best combination of climate variables for each biological variable. However, most of the predictions are within two standard errors of the estimate (obtained from the initial regression analysis), and we believe that, at least for some of the species, we are on the right track. In this test, the herring came out clearly as the worst case, but this is caused by the very high correlation ($R^2 = 0.87$) obtained initially.

Discussion

From routine hydrographic surveys in the NS during the last 22 yr, integrated parameters were derived that we believe represent major features of the year-to-year climatic variability in the NS. The time series of these physical oceanic parameters, combined with some derived atmospheric variables, are used to explain significant parts of the variability in the fraction of western mackerel migrating into the NS to feed during the summer and autumn, and in the recruitment of whiting, cod, saithe, sandeel, and herring. Some of these parameters seem to be of prime importance for directly or indirectly regulating the year to year biological variability. The climate parameters are assumed to describe the variability in the: (1) subsurface and near-surface inflow of AW, (2) combined effect of this inflow and winter cooling on the subsurface heat or temperature distribution, (3) areal coverage of a stable surface layer, (4) directional wind stress, (5) nondirectional wind input of turbulent energy, (6) atmospheric light conditions.

Exploratory use of multiple linear regression statistics, gave the following quantified connections between these physical parameters and the different fish stocks.

$$\text{MODWMACK} = -31.27 + 0.44(\text{AW}_{\text{heat}}) + 37.26(\text{EFF}_{2Q})$$

$$\text{MODWHIT(0)} = 57.72 + 0.52(\text{AW}_{\text{heat}}) - 36.69(\text{EFF}_{2Q})$$

$$\text{MODHERR(0)} = -57.76 + 13.34(\text{EFF}_{4Q}) + 0.58(\text{STRESSN}_{\text{Mar}}) + 78.69(1-e^{-0.25(\text{SSB})})$$

$$\text{MODCOD(1)} = 901.05 - 17.78(\text{AW}_{\text{heat}})^{>7.5} + 169.70(\text{EFF}_{1Q}) - 15.59(\text{NCW}_{34})$$

$$\text{MODSAITHE(1)} = -125.09 + 4.95(\text{AW}_{\text{heat}})^{>7.0} + 61.10(\text{EFF}_{\text{Mar}}) + 67.23(\text{EFF}_{\text{Apr}})$$

$$\text{MODSANDEEL(0)} = 2.36 + 1.21(\text{AW}_{\text{heat}}) - 4.89(\text{STRESSE}_{\text{May}})$$

Are these extracted parameters, of which the heat and turbulence (EFF) parameters seems to be the most important, the best choices? This question could be discussed, but we have made a first try (in cooperation with Svendsen and Magnusson (1992)) to use an extensive amount of data to quantify the integrated effect of several (hopefully the most) important climatic factors affecting marine life.

Even if we cannot explain, through chemistry, primary and secondary production, fish spawning, areal distribution, and predation, the total connection between the physics and, for example, recruitment success, the results indicate that the climate has an effect on most (and maybe all) stages. They also indicate that using just one single climatic parameter to explain biological variability will, in general, be misleading. The result that the two most important parameters come out as explanatory factors for most of the stocks, strongly indicates that the physics these parameters represent are especially important for the ecology of the NS.

Today, realistic recruitment numbers are estimated (by back calculation through VPA) several years after the year of spawning. From our results, it seems that the climate during the early stages of some of the fish stocks is so important that recruitment numbers can be estimated as early as the summer of the spawning year. This provides an interesting outlook for better management of fishery resources.

References

Aksnes, D., and J. Giske. 1990. Habitat profitability in pelagic environments. Mar. Ecol. Prog. Ser. 64: 209–215.

Colebrook, J.M. 1985. Sea surface temperature and zooplankton, North Sea, 1948 to 1983. J. Cons. int. Explor. Mer 42: 179–185.

Colebrook, J.M., and A.H. Taylor. 1984. Significant time scales of long-term variability in the plankton and the environment. Rapp. P.-v. Réun. Cons. int. Explor. Mer 183: 20–26.

Corten, A. 1986. On the causes of the recruitment failure of herring in the central and northern North Sea in the years 1977–1978. J. Cons. int. Explor. Mer, 42: 281–294.

Corten, A. 1990. Long-term trends in pelagic fish stocks of the North Sea and adjacent waters and their possible connection to hydrographic changes. Netherlands J. Sea Res. 25(1/2): 227–235.

CPR Survey Team. 1992. Continuous plankton records: the North Sea in the 1980s. ICES Mar. Sci. Symp. 195: 243–248.

Dickson, R.R, J.M. Colebrook, and E. Svendsen. 1992. Recent changes in the summer plankton of the North Sea. ICES Mar. Sci. Symp. 195: 232–242.

Dickson, R.R., J. Meincke, S.A. Malmberg, and A. Lee. 1988. The "Great Salinity Anomaly" in the northern North Atlantic 1968–1982. Prog. Oceanogr. 20: 103–151.

Ellertsen, B., P. Fossum, P. Solemdal, and S. Sundby. 1989. Relation between temperature and survival of eggs and first-feeding larvae of northeast arctic cod. Rapp. P.-v. Réun. Cons. int. Explor. Mer 191: 209–219.

Hjort, J. 1914. Fluctuations in the great fisheries of northern Europe viewed in the light of biological research. Rapp. P.-v. Réun. Cons. int. Explor. Mer 20: 1–228.

ICES. 1990a. Report of the mackerel/horsemackerel egg production workshop. ICES Doc. C.M. 1990/H: 2, 89 p.

ICES. 1990b. Report of the Industrial Fisheries Working Group. ICES Doc. C.M. 1990/Assess:13.

ICES. 1991a. Report of the Roundfish Working Group. ICES Doc. C.M. 1991/ Assess: 4.

ICES. 1991b. Report of the Herring Assessment Working Group for the area south of 62°N. ICES Doc. C.M. 1991/ Assess: 15.

Iversen, S.A., A. Eltink, E. Kirkegaard, and D.W. Skagen. 1991. The egg production and spawning stock size of the North Sea mackerel stock in 1990. ICES Doc. C.M. 1991/ H: 11.

Iversen, S.A., and D. Skagen. 1989. Migration of western mackerel to the North Sea. ICES C.M. 1989, H: 20.

Kjesbu, O.S., J. Klungsøte, H. Kryvi, P.R. Witthames, and M. Greer Walker. 1991. Fecundity, atresia and egg size of captive Atlantic cod *(Gadus morhua)* in relation to proximate body composition. Can. J. Fish. Aquat. Sci. 48: 2333–2343.

Øyestad, V. 1985. Predation on fish larvae as regulatory force, illustrated in mesocosm studies with large groups of larvae. NAFO Sci. Counc. Stud. 8: 25–32.

Rothschild, B.J., and T.R. Osborn. 1988. Small scale turbulence and plankton contact rates. J. Plankton Res. 10: 465–474.

Sakshaug, E., and H.R. Skjoldal. 1989. Life at the ice edge. AMBIO 18(1): 60–67.

Sundby, S., and P. Fossum. 1990. Feeding condition of Arcto-norwegian larvae compared with the Rothschild–Osborn theory on small-scale turbulence and plankton contact rates. J. Plankton Res. 12(6): 1153–1162.

Svendsen, E., and A.K. Magnusson. 1992. Climatic variability in the North Sea. ICES Mar. Sci. Symp. 195: 144–158.

Walsh, M., and J.H. Martin. 1986. Recent changes in the distribution and migrations of the western mackerel stock in relation to hydrographical changes. ICES, Doc. C.M. 1986/H 17: 19 p.

Wilkinson, L. 1989. SYSTAT: the system for statistics. SYSTAT, Inc., Evanston, IL.

Effect of water temperature on emigration timing and size of Fraser River pink salmon (*Oncorhynchus gorbuscha*) fry: implications for marine survival

M.A. Henderson, R.E. Diewert, J.G. Stockner, and D.A. Levy

Henderson, M.A., R.E. Diewert, J.G. Stockner, and D.A. Levy. 1995. Effect of water temperature on emigration timing and size of Fraser River pink salmon (*Oncorhynchus gorbuscha*) fry: implications for marine survival, p. 655–664. *In* R.J. Beamish [ed.] Climate change and northern fish populations. Can. Spec. Publ. Fish. Aquat. Sci. 121.

Abstract: The effect of water temperature was evaluated on the date and dispersion of the emigration of Fraser River pink salmon (*Oncorhynchus gorbuscha*) fry and their size at the time of emigration. Increases in the number of degree-days over the first few weeks of embryo development resulted in an earlier date of emigration and smaller size of fry. However, the number of degree days over the entire period of embryo development had no effect on the date of emigration or fry size. Dispersion of the emigration was not affected by the number of degree-days over the early, late, or entire period of embryo development. Marine survival of pink salmon fry increased as the date of emigration occurred earlier in the year. Fry size at the time of emigration had no effect on marine survival. Warming of the waters of the Fraser River as the result of climate change should result in an earlier date of emigration of pink salmon fry, and as a result, possibly higher marine survival.

Résumé : On a évalué l'effet de la température de l'eau sur la date d'émigration, sur la répartition et sur la taille au moment de l'émigration de l'alevin du saumon rose (*Oncorhynchus gorbuscha*) du fleuve Fraser. Les hausses du nombre de degrés-jours au cours des deux premières semaines du développement des embryons se sont traduites par une date d'émigration hâtive et par une diminution de la taille des alevins. Toutefois, le nombre de degrés-jours n'a eu aucun effet sur la date de l'émigration ou sur la taille des alevins pendant toute la période de développement des embryons. La répartition des alevins émigrants n'a pas été touchée par le nombre de degrés-jours au début, à la fin ou sur toute la période de développement des embryons. Le taux de survie en mer des saumons roses s'est amélioré parce que la date d'émigration survenait plus tôt dans l'année. La grosseur des alevins au moment de l'émigration n'a eu aucun effet sur la survie en mer. Le réchauffement des eaux du fleuve Fraser par suite du changement climatique devrait se traduire par une date d'émigration hâtive des alevins de saumon rose et, en conséquence, par un taux de survie en mer plus élevé.

Introduction

Most marine mortality in Pacific salmon occurs during the early period of marine life (Parker 1968; Bax 1983). The absolute level of mortality is determined by various characteristics during this period, including the size of juvenile salmon when they make the transition from the freshwater to the marine environment (Henderson and Cass 1991; Ward et al. 1989; Healey 1982; Bilton et al. 1982) and the timing of ocean entry (Bilton et al. 1982; Taylor 1980). Because pink salmon (*Oncorhynchus gorbuscha*) migrate to the ocean immediately following emergence from the gravel, freshwater temperature conditions during egg development may exert a strong influence on both body size and the timing of ocean entry.

The size of juvenile salmon and the timing of their ocean entry are integral measures of the entire freshwater developmental process. Experimental studies have shown that a significant temperature effect can be demonstrated on the rate of development during various freshwater stages when other variables are controlled, and when the range in experimental temperature is large (see review by Beacham and Murray 1990). However, it is unclear how significant the temperature

M.A. Henderson and **R.E. Diewert.** Department of Fisheries and Oceans, Biological Sciences Branch, 555 West Hastings Street, Vancouver, BC V6B 5G3, Canada.

J.G. Stockner. Department of Fisheries and Oceans, Biological Sciences Branch, 4160 Marine Drive, West Vancouver, BC V7V 1N6, Canada.

D.A. Levy. Levy Research Services Limited, 102-2221 Folkstone Way, West Vancouver, BC V7S 2Y6, Canada.

effect is when measured across all freshwater developmental stages, and when measured for a wild population where numerous other variables affecting marine survival may have interactive and possibly confounding effects.

Most global climate models predict an average global warming of between 2 and 5°C by the middle of the next century when the CO_2 concentration in the earth's atmosphere will double pre-industrial levels. Anticipated changes in temperate freshwater environments include higher temperatures in response to atmospheric heat exchange, as well as altered stream flows caused by increased precipitation and a higher rain to snow ratio (Levy 1992). For salmon species that rear in fresh water as juveniles, global warming impacts may result in reduced freshwater growth and survival (Henderson et al. 1992). For pink salmon, global warming impacts will be expressed primarily through influences on egg development and survival rates.

The objective of this study is to test the hypothesis that ambient river temperature during the period of freshwater development has no effect on the time when Fraser River pink salmon fry make the transition from the freshwater to the marine environment or on their size at this time of transition. We further explore the effect on marine survival of fry size and timing at the point of ocean entry in those situations where the hypothesis of no freshwater temperature effect has been rejected. Finally, using historical variation in Fraser River water temperature as an analogue of climate change, we evaluate the potential implication of climate change on the size of pink salmon fry, their timing of ocean entry, and ultimately on their marine survival.

Methods

The Fraser River is located in British Columbia (Fig. 1) near the southern extreme of the distribution of pink salmon in North America (Heard 1991). The river supports the largest pink salmon stock complex in British Columbia (Henderson 1991) and among the largest in the world. Fraser River pink salmon spawn in October (Beacham and Murray 1988), primarily in gravel beds in the lower reaches of the river below Hell's Gate (Ricker 1989). Eggs incubate over the fall and winter and hatch in the spring, producing alevins. The alevins emerge from the gravel and immediately migrate down the river and into the ocean as fry. The peak in outmigration of pink salmon near the mouth of the Fraser River generally occurs in middle to late April. Fraser River pink salmon are

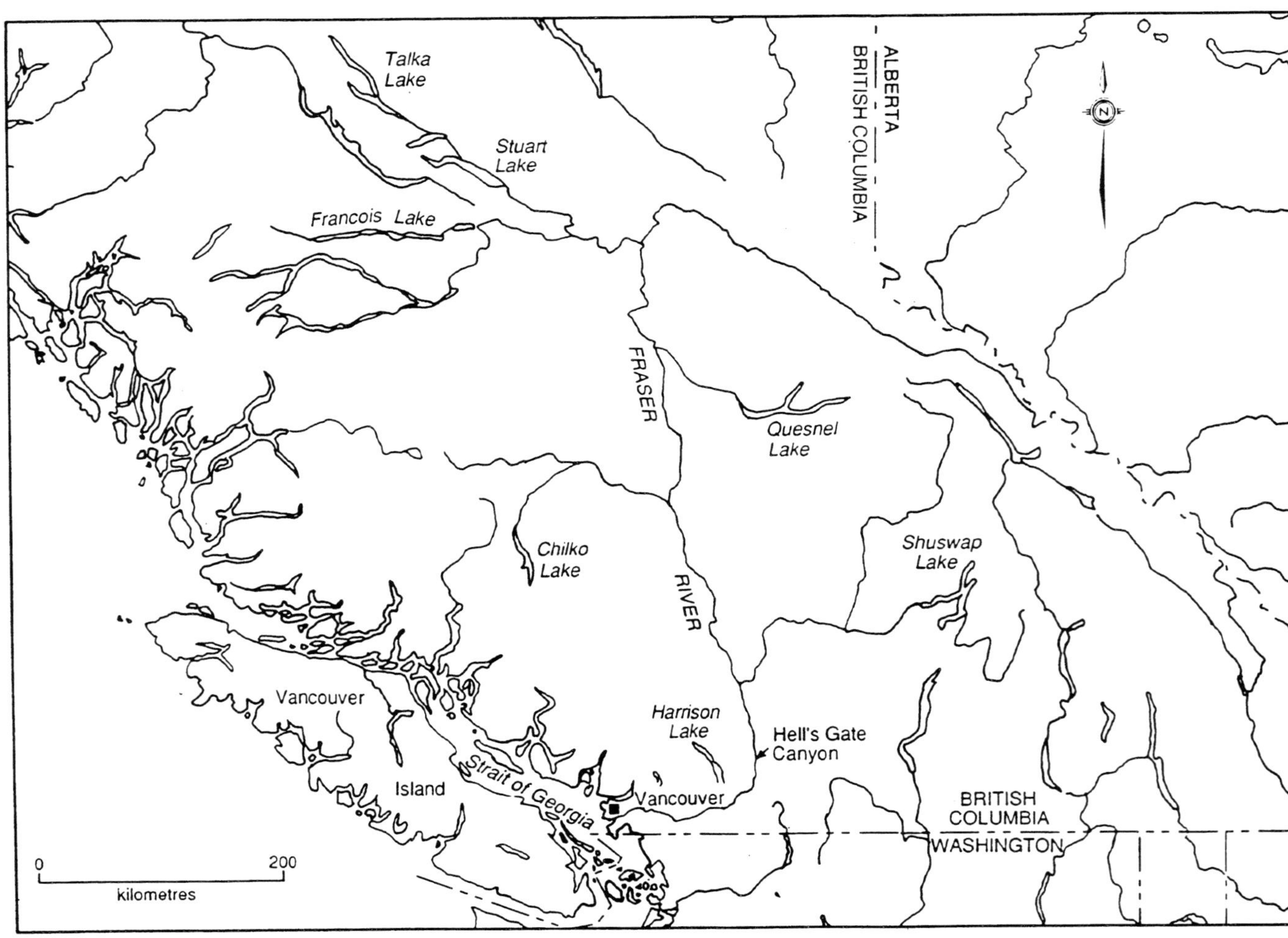

Fig. 1. The Fraser River.

somewhat unusual in that they return as adults almost exclusively in odd-numbered years at 2 yr of age (Vernon 1962). The total return of pink salmon to the Fraser River since 1983 (i.e., 5 return yr) has ranged from 7.1 to 18.9 million and averaged 14.7 million fish.

Estimates of the date of peak spawning for Fraser River pink salmon stocks were made in each odd-numbered year from 1959 to 1987 based on spawning-ground sampling programs (International Pacific Salmon Fisheries Commission 1962–1984; Pacific Salmon Commission 1986–1992). The largest Fraser River pink salmon stock spawns in the main stem of the Fraser River below Hell's Gate (Fig. 1). The other major stocks spawn in the Harrison, Chilliwack–Vedder, and Thompson river systems. The annual weighted mean peak day of spawning for the total Fraser River pink salmon stock complex was calculated by first determining the day of peak spawning for each stock. The peak day for each stock was then weighted relative to the proportion of the total spawning escapement contributed by that stock. Finally, the weighted peak days were summed across all stocks, producing an annual weighted mean peak day of spawning.

A downstream enumeration program for Fraser River pink salmon fry has been carried out on the lower Fraser River at Mission, B.C., on even-numbered years since 1962 (Vernon 1966). The sampling program provides a daily estimate of pink salmon fry moving out of the Fraser River system and an estimate of total fry production for each brood year. Pink salmon fry are also sampled for total length. The date on which 50% of the total pink salmon fry production had migrated past Mission was determined for each year. All dates used in our analyses are expressed numerically beginning with January 1 as day one.

Estimates of the spawning escapement of pink salmon to the Fraser River on odd-numbered years from 1959 to 1987 were made based on mark–recapture methodology (Roos 1991). The number of spawning fish was combined with the number of fish in the catch to generate estimates of the total return of pink salmon to the Fraser River (Roos 1991). The marine survival for Fraser River pink salmon was estimated as the ratio of total adult return to the total number of pink salmon fry that migrated past Mission for each brood year.

Daily estimates of Fraser River water temperature were made at Hell's Gate for the period from 1954 to 1985. Estimates were usually made from the output of a continuously recording thermograph, although daily temperatures determined from a thermometer were used in some cases when the thermograph was not in operation. Daily mean water temperatures were determined from thermograph recordings by calculating the midpoint between the daily maximum and minimum values. There were a few short periods within some years when there were no temperature data collected at Hell's Gate. In these situations, daily estimates of temperature were made by extrapolation. Daily temperature recordings were not made at Hell's Gate for the period from 1986 to 1990. Daily temperatures for this period were estimated based on a regression model between daily water temperatures at Hell's Gate and the Patullo Bridge (near the mouth of the Fraser River) for the period from 1954 to 1985 ($r^2 = 0.98$, $P < 0.001$) and daily temperatures at the Patullo Bridge for the period from 1986 to 1990.

Results

The total return (i.e., catch plus spawning escapement) of pink salmon to the Fraser River has increased discontinuously since the 1961 brood year with a particularly abrupt transition between the 1975 and 1977 brood years (Fig. 2). Total returns for the eight brood years prior to 1977 averaged 6.8 million fish, while returns for the most recent seven brood years (1977–89) averaged 15.2 million fish. Total pink salmon fry production exhibited a similar pattern, increasing from an average of approximately 244 million fry for the 1961 to 1975 brood years to 426 million fry for the brood years from 1977 to 1989 (Fig. 2).

The date of peak spawning for Fraser River pink salmon has ranged from October 6 to 16 with a median date of October 11 over the odd-numbered years from 1961 to 1989 (Fig. 3). As a result of the uncertainty in the actual peak spawning date for any year and the limited range in peak spawning dates over all years, the median was used to identify the peak spawning date for all years in the analyses described below. The range of dates on which 50% of the fry emigration was attained was more variable than peak spawning dates and ranged from April 10 to May 7 with a median date of April 24 (Fig. 3).

The average annual water temperature at Hell's Gate varied by more than 2°C from 1954 to 1990 (Fig. 4). Linear regression analysis indicated a significant increase in average water temperature over this period ($r^2 = 0.16$, $P = 0.0133$). However,

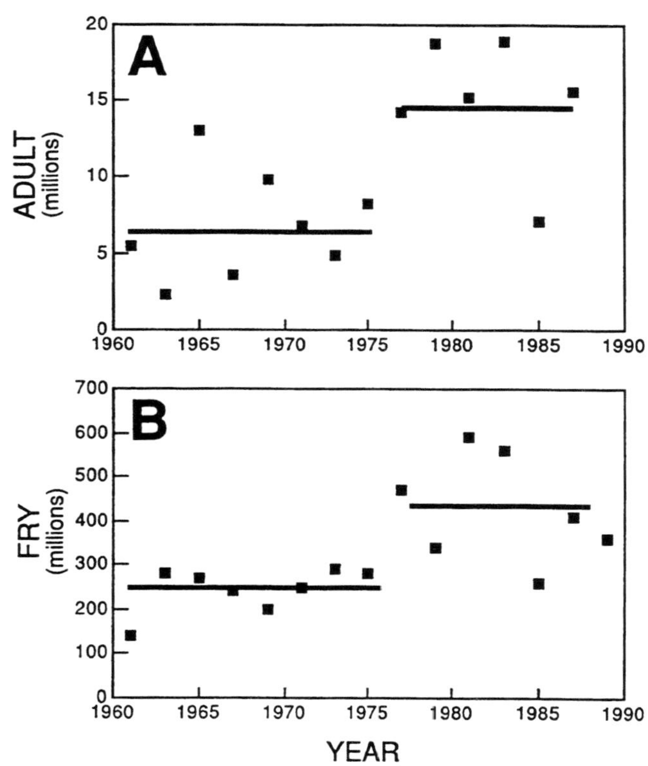

Fig. 2. Total adult (catch plus spawning escapement) (A), and fry production (B) by brood year for Fraser River pink salmon. Horizontal lines show the average adult and fry production for the periods 1961–75 and 1977–87.

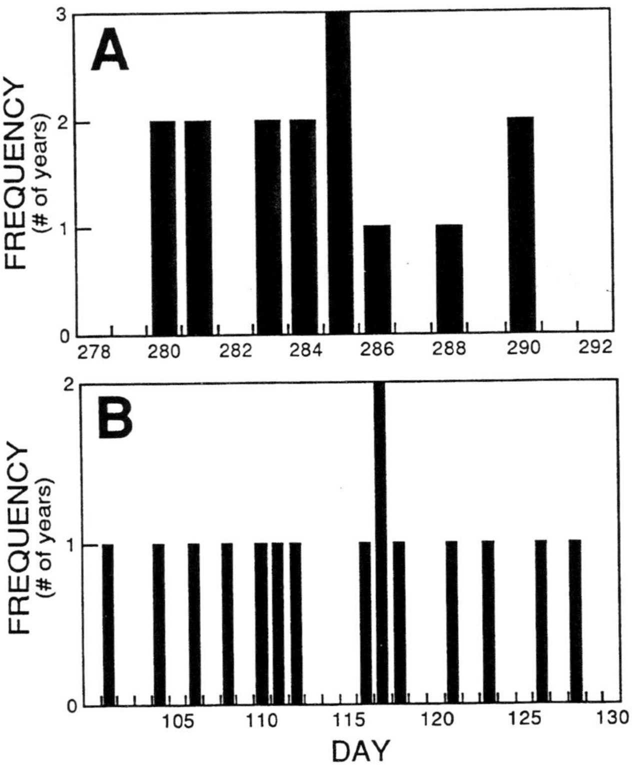

Fig.3. Frequency distribution of day of peak spawning (A), and of 50% fry emigration (B) for Fraser River pink salmon for brood years 1961–89 (January 1 = day 1).

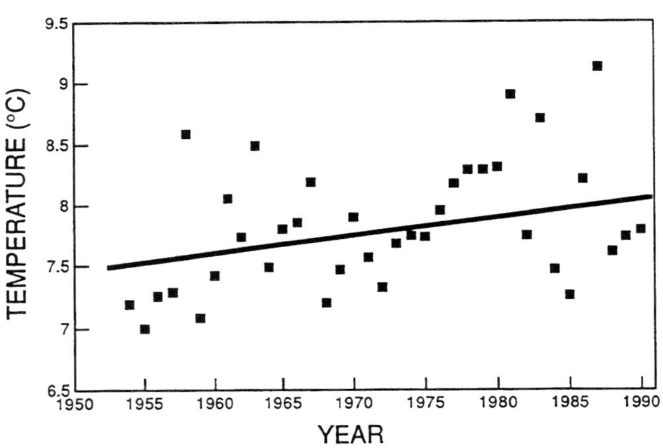

Fig. 4. Average annual water temperature (°C) at Hell's Gate for the period 1954–90 and the resulting regression line.

the absolute increase in water temperature varied by month over the period from egg deposition in October to hatching and emergence in March (Table 1). Average water temperature increased between the two periods 1954–59 and 1980–89 by 1.22, 1.15, and 1.32°C for the months of October, November, and March, respectively. Absolute increases in winter water temperatures between the same two periods were less and ranged from 0.49°C in December to 0.77°C in February.

The cumulative degree-days, as measured at Hell's Gate, from the peak of spawning to the point of 50% pink fry emigration ranged from 460 to 770 over the 15 brood yr considered in this study. However, despite the large interannual variation in cumulative degree-days, the slope of the regression between the number of days to 50% fry emigration and cumulative degree-days from the peak of spawning to 50% fry emigration was not significantly different from zero ($P = 0.9520$).

We explored further the possible effect of cumulative degree-days on the date of 50% fry emigration by restricting the number of days over which degree-days were accumulated to the early period of egg development. We did this because of two observations. First, a disproportionate amount of the increase in the total heat content of the Fraser River over the last four decades has occurred during the early period of egg development (Table 1). Also, others (Murray and Beacham 1986) have suggested initial incubation temperatures have a greater effect on the rate of development to hatching and emergence than do later incubation temperatures. As a consequence, cumulative degree-days were determined for four arbitrary periods: 10, 20, 30, and 45 d following the peak in spawning. The slope of the regression of the number of days from peak spawning to 50% fry outmigration on cumulative degree days over the first 20 d following peak spawning was significantly different from zero ($r^2 = 0.31$, $P = 0.0329$) (Fig. 5). This relationship indicated that an increase in cumulative degree-days from 142 to 213 resulted in a date for 50% fry emigration that was approximately 2 wk earlier.

Cumulative degree-days over the 10-d period following peak spawning had a marginally significant effect ($r^2 = 0.26$, $P = 0.0530$) on the number of days from peak spawning to 50% fry outmigration. Cumulative degree days over the first 30 and 45 d following peak spawning had no significant effect ($P > 0.05$) on the number of days between peak spawning and 50% fry emigration. Further, there was no significant ($P > 0.05$) relationship between cumulative degree-days for any of the four time periods described above and the date of peak spawning over the 15 brood yr considered in our study. This indicates that the earlier outmigration associated with a greater number of degree-days in the first few weeks following

Table 1. Average daily water temperature (°C) at Hell's Gate by month and decade for the period from October through March.

Years	October	November	December	January	February	March
1954–59	8.42	3.71	1.34	0.66	0.80	1.88
1960–69	9.14	4.51	2.02	0.91	1.40	2.49
1970–79	9.26	4.30	1.90	1.01	1.36	2.62
1980–89	9.64	4.87	1.83	1.20	1.57	3.20

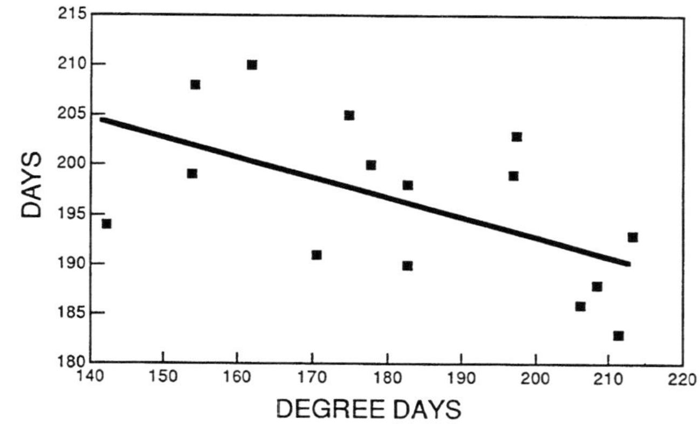

Fig. 5. Number of days from the mean peak spawning date (October 11) to day of 50% emigration of Fraser River pink salmon fry at different levels of cumulative degree days as determined from the first 20 d following the mean peak spawning date. The linear regression line between the two variables is shown.

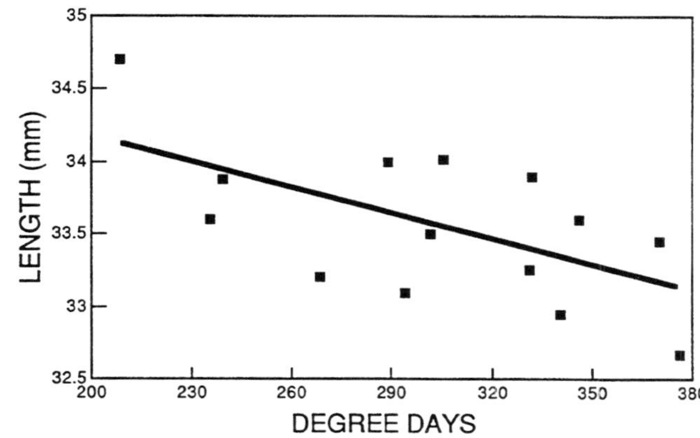

Fig. 6. The fork length (mm) of Fraser River pink salmon fry at different levels of cumulative degree days for the first 45 d following the mean peak spawning date (October 11). The linear regression line between the two variables is shown.

fertilization of the eggs is actually the result of an effect on embryo development and cannot be explained simply by changes in spawning time.

One standard deviation around the day of 50% fry outmigration of Fraser River pink salmon fry was used as the annual index of the dispersion of the run. Over the period of brood years from 1961 to 1989, the index of dispersion ranged from 9.2 to 18.3. The effect of Fraser River water temperature at Hell's Gate, as measured by cumulative degree-days, on the dispersion of each outmigration was evaluated using linear regression for three time periods: over the entire period from peak spawning to the date of 50% fry emigration; from January 1 to the date of 50% emigration; and for the last 30 d preceding the date of 50% emigration. For each of the three time periods, the slope of the regression between dispersion and cumulative degree-days was not significantly different ($P > 0.05$) from zero.

The average size of Fraser River pink salmon fry sampled at Mission varied from 32.7 to 34.7 mm over the period of brood years from 1961 to 1989. The effect of the temperature of the Fraser River, as measured by cumulative degree-days, on fry size was evaluated using linear regression analysis for five time periods: from the date of peak spawning to the date of 50% fry outmigration, and for 10, 20, 30, and 45 d following the date of peak spawning. The slope of the regression between fry length and cumulative degree-days over the period from the date of peak spawning to the date of 50% fry outmigration was negative, but only marginally different from zero ($P = 0.0544$). However, the negative slope of the regression was significant when based on the cumulative degree-days over the first 30 ($r^2 = 0.31$, $P = 0.0401$) and 45 d ($r^2 = 0.37$, $P = 0.0201$) (Fig. 6) following the date of peak spawning. The results indicated that the cumulative number of degree-days over the first several weeks following the date of peak spawning was an important determinant of the size of Fraser River pink salmon fry. The slope of the regressions between fry size and cumulative degree-days over the first 10 and 20 d following the date of peak spawning were not significantly different from zero ($P > 0.05$).

The foregoing analyses indicate that the time of 50% fry emigration and fry size is influenced by temperature during the initial weeks following egg fertilization. As a consequence, it is of interest to know what temperatures may be in the future during this critical incubation period. There has been a discontinuous but significant increase in average water temperature at Hell's Gate for the 20 d ($r^2 = 0.14$, $P = 0.0225$) and 45 d ($r^2 = 0.13$, $P = 0.0289$) periods following peak spawning over the period from 1954 to 1990 (Fig. 7). Although future temperatures cannot be predicted with certainty, a continuation of the trend should result in warmer waters during the period of early embryo development, and as a consequence, earlier emigration and smaller size of Fraser River pink salmon fry.

The effect of the date of 50% fry emigration on marine survival of Fraser River pink salmon was equivocal using data for all brood years from 1961 to 1987 (Fig. 8). The slope of the linear regression between the two variables was not significantly different from zero ($P = 0.5843$).

The marine survival of Fraser River pink salmon fry, as measured in this study, is dependent on many factors in addition to the date of 50% emigration. Further, these other factors may vary independently of the date of 50% fry emigration and thereby introduce a large amount of unexplained variability in the relationship between marine survival and date of 50% emigration. In an attempt to eliminate some of this variability, we arbitrarily eliminated the 3 yr of highest marine survival (1966, 1970, and 1980) and the 3 yr of lowest marine survival (1964, 1968, and 1974), and repeated the linear regression analysis between marine survival and the date of 50% fry emigration (Fig. 8). The slope of the regression for the abbreviated time series was significantly different from zero ($r^2 = 0.64$, $P = 0.0164$) and indicated a rapid decrease in marine survival from approximately 3.8 to 2.7% when the date of 50% emigration increased from day 104 to 128 of the year.

The effect of the size of pink salmon fry on marine survival using data from all brood years from 1961 to 1987 was not significantly different from zero ($P = 0.5698$). The regression analysis was repeated after eliminating the years of highest and

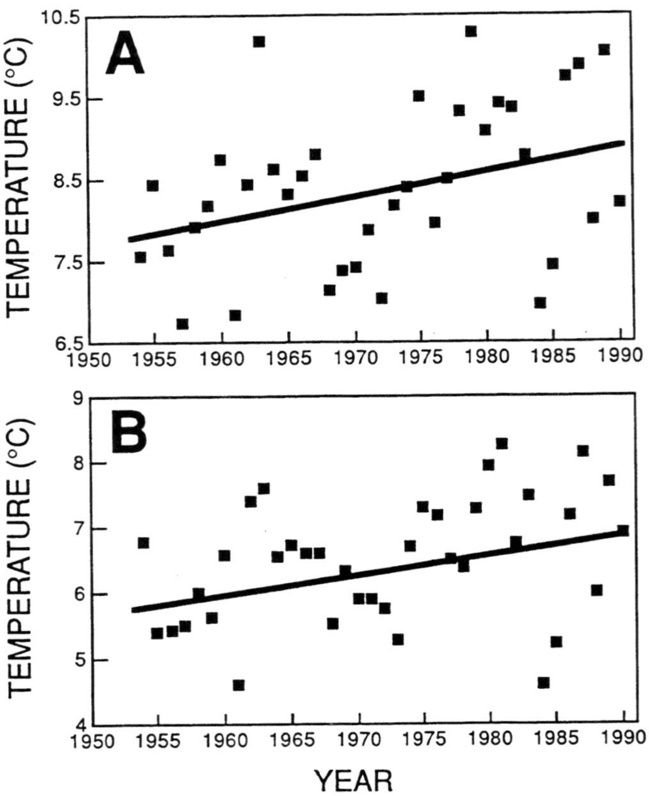

Fig. 7. Average water temperature (°C) at Hell's Gate for the first 20 (A) and 45 (B) d following the date of peak spawning of Fraser River pink salmon (October 11) for the period 1954–90. The corresponding regression lines are shown.

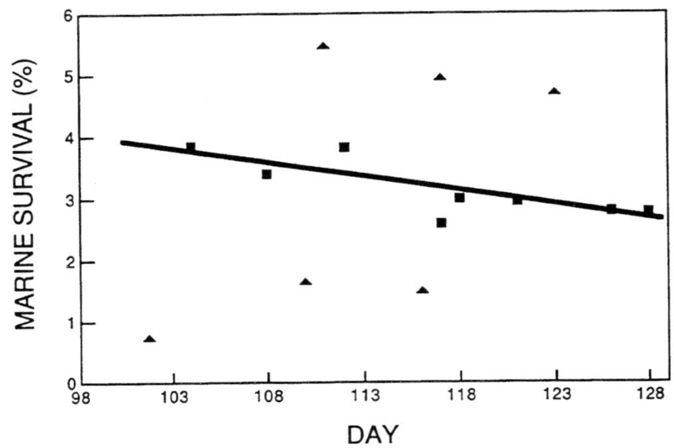

Fig. 8. Marine survival (%) of Fraser River pink salmon as a function of the day of 50% fry emigration. The regression line excludes the 3 yr with the highest and the 3 yr with the lowest estimates of marine survival.

lowest marine survival as described above; however, the slope of the regression was again not significantly different from zero ($P = 0.1704$).

Discussion

The results of this study provide evidence that the date of 50% fry emigration occurs earlier in the year (Fig. 5) and fry size decreases (Fig. 6) as the cumulative degree-days experienced by Fraser River pink salmon embryos during their first few weeks of development increases. Further, in most years, a later date of 50% emigration appears to contribute to a decline in marine survival (Fig. 8). Fry size at the time of emigration had no effect on marine survival. Further, the temperature of the Fraser River had no apparent effect on the dispersion of the emigration of pink salmon fry.

The date of 50% fry emigration reflects the integration of the time required for all earlier developmental stages from egg fertilization through the period of downriver migration of the fry. The effect of water temperature on several of the stages has been examined in the laboratory; however, the results are not entirely consistent with the findings from our study. Beacham and Murray (1986, 1988) have shown that, under constant temperatures, the time from egg fertilization to hatching decreases as the temperature increases from 4 to 12°C for several pink salmon stocks. The time from hatching to emergence of alevins also appears to decrease with increasing average temperatures from 4 to 12°C (Beacham and Murray 1986, 1988) and from 3 to 15°C (Godin 1980). Consequently, one would predict a negative relationship between the date of 50% emigration and cumulative degree-days over the entire developmental period, which is a correlate of average temperature over the same period.

Two conditions probably account for the lack of such a relationship in our study and illustrate the limitations when one attempts to make inferences about the behaviour of salmon in the natural environment based on reductionist-type experiments. Temperature is only one of several variables affecting the timing of hatching and emergence in salmon, and ultimately, the date of 50% emigration. Light intensity (Kwain 1975), oxygen concentrations (Silver et al. 1963; Garside 1966), alevin density in the gravel, river flow patterns and velocity (Heard 1991), additive genetic variation (McIntyre and Blanc 1973; Sato 1980; Beacham 1988), gravel characteristics (Bailey et al. 1980), and variation in water temperature (Alderdice and Velsen 1978; Heggberget and Wallace 1984; Beacham and Murray 1986) are all known to affect the timing of hatching and emergence of pink salmon. Interannual variation in these variables in the natural environment almost certainly adds an additional component of variability to the annual outmigration data that may preclude demonstrating a temperature effect. Also, the laboratory studies used a series of temperatures with a range of 8°C (Beacham and Murray 1985, 1986) to 12°C (Godin 1980) to demonstrate the effect of temperature on the early developmental stages. However, the average annual temperature at Hell's Gate varied by less than 4°C over the period from 1954 to 1990 (Fig. 4). This small range in temperature probably makes it very difficult to establish a temperature effect, particularly when other variables affecting the timing of the early developmental stages are also changing between years independently of temperature.

The significant ($P = 0.0329$) negative relationship between cumulative degree-days over the first 20 d following the date of peak spawning and the date of 50% emigration (Fig. 5) confirms earlier speculation by Beacham and Murray (1986) that initial incubation temperatures have a greater effect on the rate of development to hatching and emergence than do later incubation temperatures. The fact that it was possible to demonstrate this relationship despite the effect of the confounding variables described above indicates that temperature during the initial few weeks following spawning has a major effect on the rate of development of pink salmon over their entire period of freshwater residence. This is a particularly important observation for Fraser River pink salmon as the increases in average river temperature witnessed over the last four decades (Fig. 4) appear to be particularly high for the months of October and November, the time period corresponding to early egg incubation for pink salmon (Table 1). A continuation of this trend under the scenario of global warming should result in a progressively earlier date of 50% fry outmigration.

Godin (1980) has shown experimentally that the rate of emergence of pink salmon fry from a redd decreases as temperature increases from 3.4 to 15.0°C. Consequently, we had anticipated a significant positive effect ($P > 0.05$) of cumulative degree-days, particularly over the month preceding the date of 50% outmigration, on the dispersion of the outmigration of pink salmon fry from the Fraser River. However, the lack of such a relationship in our study may simply reflect the limited power of the test used to discern a relationship between temperature and dispersion of outmigration. Numerous variables affect the behaviour of pink salmon fry during emergence and downriver migration (Heard 1991), and probably confound attempts to describe a temperature effect from field data. Also, the production of pink salmon fry in the Fraser River originates from numerous stocks (Roos 1991) with genetically based differences in their response to changing temperatures (Beacham and Murray 1988). In addition, the relative contribution of each of these stocks to total Fraser River fry production varies among years. As a result, the emigration pattern of fry based on the composite of all stocks as measured in this study may bear little relationship to temperature.

The length of pink salmon fry reared from embryos in the laboratory has been shown to vary with the average temperature over the entire period of incubation (Beacham and Murray 1985, 1990). However, the relative pattern of variation with changes in temperature appears to be stock specific (Beacham and Murray 1986). The final size of pink salmon alevins and fry appears to be particularly sensitive to temperature during the early period of embryo development (Kazakov 1981; Murray and Beacham 1986). Although we found only a marginally significant ($P = 0.0544$) relationship between cumulative degree-days from the peak of spawning to 50% outmigration and fry size, increasing heat accumulation over the first 30 d ($P = 0.0401$) and 45 d ($P = 0.0201$) of embryo incubation had a very significant negative effect on fry size (Fig. 6). Fry size is also dependent on egg size (Beacham and Murray 1990) although the effect of this variable and any potential interaction with temperature was not evaluted in our study.

It is possible that the decrease in fry size with increases in cumulative degree-days over the early period of embryo development may reflect increased metabolic costs associated with higher temperatures (Beacham and Murray 1985). However, this explanation for the relationship between cumulative degree-days and fry size is not totally satisfactory. It does not account for the increasingly significant relationship between the two variables when the cumulative number of degree-days is based on the heat accumulation over only the early period of incubation. A more likely explanation is based on temperature effects during the early period of embryo development.

Ali and Lindsay (1974) and Lindsay and Arnason (1981) have shown that the number of vertebrae in fish is determined during the early period of embryo development, and that vertebrae number increases with decreases in incubation temperature. Hubbs (1926) suggested a mechanism for this response by postulating that temperature controls the number of vertebrae produced by embryos by differentially affecting the rates of growth and of differentiation. At low temperatures, growth is inhibited less than differentiation; the embryo becomes longer before the tissue forming the vertebrae segments, and as a consequence, the fish has a greater number of vertebrae. Finally, Lindsay (1975) has shown, based on an examination of over 3000 fish species, that maximum body length in fish increases with increases in vertebral number.

The above description of early growth and differentiation conforms to the observations in our study: larger fry are produced when the cumulative degree-days during the early period of embryo development is low. Further, it provides a possible explanation for the difference in the effect of temperature, when measured over the entire period of embryo development, on fry size between our study and laboratory studies. In the laboratory, temperature is usually held constant over the period of embryo development. As a result, there is a one-to-one correspondence between average temperature over the early period of embryo development and the entire period of embryo development. Therefore, the response of alevin or fry size to different temperatures would be the same regardless of the time period over which temperature was measured. The results of our study show that it is the number of cumulative degree-days over the first 30–45 d that determines fry size.

Estimates of the marine survival of Fraser River pink salmon ranged from 0.9–5.6% and averaged 3.2% over the 14 brood yr considered in our study (Fig. 8). The degree of interannual variation in marine survival and average marine survival are consistent with estimates for other pink salmon stocks (Heard 1991).

There is evidence that the time of ocean entry affects the marine survival of juvenile Pacific salmon (Taylor 1980; Bilton et al. 1982). Results of our study show the importance of the date of 50% fry emigration on subsequent marine survival to adulthood for Fraser River pink salmon (Fig. 8). However, we were only able to demonstrate a significant ($P = 0.0164$) relationship between date of 50% emigration and marine survival by eliminating the 3 yr with the highest and the 3 yr with the lowest estimates of marine survival. We believe this approach is necessary to eliminate those years when marine survival was largely controlled by other variables and independent of date of ocean entry. For example, the relative abundance of juvenile chum salmon in the marine environment has been shown to have a much greater effect on the loss of juvenile pink salmon to predation than has the size of the juvenile pink salmon (Hargreaves and LeBrasseur 1985). Consequently, the brood years with very high or low pink salmon survival could be associated with years of unusually high or low abundance of juvenile chum salmon. Also, salinity of coastal waters is thought to directly or indirectly control the marine survival of juvenile pink salmon (Vernon 1958; Wickett 1958). These variables and others that are independent of the date of ocean entry of juvenile pink salmon could obscure any effect of date of ocean entry on marine survival.

There is considerable evidence that marine survival increases with increases in the size of juvenile Pacific salmon at their point of ocean entry (Henderson and Cass 1991; Parker 1971; Healey 1982; Ward et al. 1989; Macdonald et al. 1987; Foerster 1954; Ricker 1962). Given the amount of support for a positive relationship between size at the time of ocean entry and marine survival, the lack of such a relationship for Fraser River pink salmon was surprising. Henderson and Cass (1991) made a similar observation, finding no significant relationship between the average annual size of sockeye salmon smolts produced from Chilko Lake and subsequent marine survival across a period of 34 brood yr. However, they were able to show a very strong effect of smolt size on marine survival when comparisons were made within brood years. The lack of a relationship using the averaged annual data over the 34 brood yr was attributable to the limited range in average smolt size and to numerous other factors affecting marine survival that vary independently of smolt size. We suspect a similar explanation accounts for the lack of a relationship between fry size and marine survival in Fraser River pink salmon. For example, the extremes in average annual fry size over the 15 brood yr considered in our study was only 2 mm (Fig. 6), or approximately 5% of the mean fry size over the 15-yr period.

The ultimate effect of climate change on the production characteristics of Fraser River pink salmon will depend on many factors. These include the actual extent of warming and other hydrological changes that may take place in the Fraser River basin and associated marine environment, and the ability of the pink salmon to develop and survive under these changing conditions.

There is a general consensus that doubling of atmospheric CO_2 levels will result in an increase in the average global temperature of 2–5°C. However there is a great deal of uncertainty about how the temperature change will be expressed at a regional level (Smith 1990), particularly in mountainous areas like the Fraser River basin where climatic gradients are steep (Slaymaker 1990; Moore 1991). It is also anticipated that warming of the climate will cause changes in numerous hydrological characteristics in the basin including annual flow patterns (Moore 1991) and rates of sedimentation (Slaymaker 1990), that will also affect the freshwater productivity of pink salmon. There is even greater uncertainty about the effects of climate change on the boundaries of the major zones in the North Pacific Ocean that are used by pink salmon, the productivity of these zones (Francis 1990), and the distribution of predators of pink salmon.

The high and low temperatures for the normal development and survival of pink salmon eggs and alevins are 15 and 4.5°C, respectively (Combs and Burrows 1957). Consequently, an increase in the average temperature of the Fraser River by several degrees because of global warming implies that temperature conditions in pink salmon spawning grounds will be adequate for egg development (Table 1). Moreover, the present results suggest that higher temperatures during egg incubation, particularly during the early developmental stages, may in fact reduce development time (Fig. 5) resulting in higher marine survival (Fig. 8). However, if the magnitude of warming is large (more then several degrees) because of simultaneous reductions in summer–fall stream flows for example, there may be a reduction in Fraser River pink salmon fry production.

References

Alderdice, D.F., and F.P.J. Velsen. 1978. Relation between temperature and incubation time for eggs of chinook salmon (*Oncorhynchus tshawytscha*). J. Fish. Res. Board Can. 35: 69–75.

Ali, M.Y., and C.C. Lindsay. 1974. Heritable and temperature-induced meristic variation in medaka, *Oryzias latipes*. Can. J. Zool. 52: 959–976.

Andrew, F.J., and G.H. Geen. 1960. Sockeye and pink salmon production in relation to proposed dams in the Fraser River system. Int. Pac. Salmon Fish. Comm. Bull. 11: 259 p.

Bailey, J.E., S.D. Rice, J.J. Pella, and S.G. Taylor. 1980. Effects of seeding density of pink salmon, *Oncorhynchus gorbuscha*, eggs on water chemistry, fry characteristics, and fry survival in gravel incubators. Fish. Bull. U.S. 78: 649–658.

Bax, N.J. 1983. Early marine mortality of marked juvenile chum salmon (*Oncorhynchus keta*) released into Hood Canal, Puget Sound, Washington, in 1980. Can. J. Fish. Aquat. Sci. 40: 426–435.

Beacham, T.D. 1988. A genetic analysis of early development in pink (*Oncorhynchus gorbuscha*) and chum salmon (*O. keta*) at three different temperatures. Genome 30: 89–96.

Beacham, T.D., and C.B. Murray. 1985. Effect of female size, egg size, and water temperature on developmental biology of chum salmon (*Oncorhynchus keta*) from the Nitinat River, British Columbia. Can. J. Fish. Aquat. Sci. 42: 1755–1765.

Beacham, T.D., and C.B. Murray. 1986. Comparative developmental biology of pink salmon, *Oncorhynchus gorbuscha*, in southern British Columbia. J. Fish. Biol. 28: 233–246.

Beacham, T.D., and C.B. Murray. 1988. Variation in the developmental biology of pink salmon (*Oncorhynchus gorbuscha*) in British Columbia. Can. J. Zool. 66: 2634–2648.

Beacham, T.D., and C.B. Murray. 1990. Temperature, egg size, and development of embryos and alevins of five species of Pacific salmon: a comparative analysis. Trans. Am. Fish. Soc. 119: 927–945.

Bilton, H.T., D.F. Alderdice, and J.T. Schnute. 1982. Influence of time and size at release of juvenile coho salmon (*Oncorhynchus kisutch*) on returns at maturity. Can. J. Fish. Aquat. Sci. 39: 426–447.

Combs, B.D., and R.E. Burrows. 1957. Threshold temperatures for the normal development of chinook salmon eggs. Prog. Fish. Cult. 19: 3–6.

Foerster, R.E. 1954. The sockeye salmon (*Oncorhynchus nerka*). Fish. Res. Board Can. Bull. 162: 422 p.

Francis, R.C. 1990. Climate change and marine fisheries. Bull. Am. Fish. Soc. 15(6): 7–9.

Variability in frontal boundaries, temperatures, and the geographic ranges of species and pelagic marine communities along 175°30'E, 1978–91

S.E. Ignell, S.R. Carlson, and R.A. Rumbaugh

Ignell, S.E., S.R. Carlson, and R.A. Rumbaugh. 1995. Variability in frontal boundaries, temperatures, and the geographic ranges of species and pelagic marine communities along 175°30'E, 1978–91, p. 667–674. *In* R.J. Beamish [ed.] Climate change and northern fish populations. Can. Spec. Publ. Fish. Aquat. Sci. 121.

Abstract: Data from cruises of the Japanese research vessel Hokusei maru were used to compare the relative importance of sea-surface temperature (SST) and frontal boundaries with annual changes in pelagic community boundaries and species ranges in the central North Pacific Ocean. The same set of stations was sampled approximately the same time each year from 1978 to 1991. Both oceanographic and gillnet stations were classified into three similar groupings; the oceanographic groupings correspond to subarctic domain, subarctic frontal zone (SFZ), and transition region waters. The northern boundary of the SFZ was stationary from 1978 to 1985, whereas the southern boundary of the SFZ, SST, and the location of the 15°C isotherm varied throughout the study period. Salmon range and pelagic marine community structure were strongly associated with oceanographic conditions. SST was a more effective environmental predictor of species ranges or communities than oceanographic zones. We conclude that the relationships between animals in the SFZ are better viewed from the continuum rather than the community paradigm.

Résumé : On a utilisé les données recueillies par le navire de recherche japonais Hokusei Maru pour comparer l'importance relative de la température de la surface de la mer (TSM) et des limites frontales par rapport aux changements annuels des limites des populations pélagiques et des aires de répartition des différentes espèces dans le centre du Pacifique Nord. La même série de stations a été échantillonnée à peu près la même période de chaque année de 1978 à 1991. Les stations océanographiques et les stations de filets maillants ont été classées en trois groupes similaires; ces groupements océanographiques correspondent au domaine subarctique, à la zone subarctique frontale (ZSF) et aux eaux de la région de transition. La limite nord de la ZSF est restée stationnaire de 1978 à 1985, tandis que la limite sud de la ZSF, la TSM et l'emplacement de l'isotherme de 15° C ont varié tout au long de la période d'étude. L'aire de répartition du saumon et la structure des populations marines pélagiques étaient fortement associées aux conditions océanographiques. La TSM s'est révélée un mécanisme de prévision environnementale des aires de répartition des espèces ou des populations bien plus efficace que les zones océanographiques. Nous en concluons que les relations entre les animaux dans les ZSF sont mieux perçues du point de vue du continuum plutôt que de celui du paradigme des populations.

Introduction

Directional changes of climate profoundly affect biological communities (Davis 1986). The effects are directly reflected in changes in species abundances and geographic ranges and are mediated through changes in physical habitat (Brown 1984; Gaston 1990). Changes in abundances and ranges may occur at an identifiable feature, such as an oceanic front, or be related to an oceanographic parameter, such as temperature.

S.E. Ignell, S.R. Carlson, and **R.A. Rumbaugh.** Auke Bay Laboratory, Alaska Fisheries Science Center, National Marine Fisheries Service, National Oceanic and Atmospheric Administration, 11305 Glacier Highway, Juneau, Alaska 99801-8626, USA.

The high intercorrelation between oceanographic parameters or features often makes it difficult to isolate the primary factor causing change in the geographic ranges of marine species. For example, the primary oceanographic factor limiting the southward range of salmonids (*Oncorhynchus* spp.) in the North Pacific Ocean was identified as sea-surface temperatures by Ogura and Takagi (1987), whereas Fujii et al. (1965) identified subsurface temperatures, and Mishima (1974), Favorite and Hanavan (1963), and Tabata (1983) identified oceanic fronts.

Such distinctions have practical implications for resource managers who depend on a clear understanding of species ranges to make responsible management decisions. This has been particularly true for the highly productive region in the

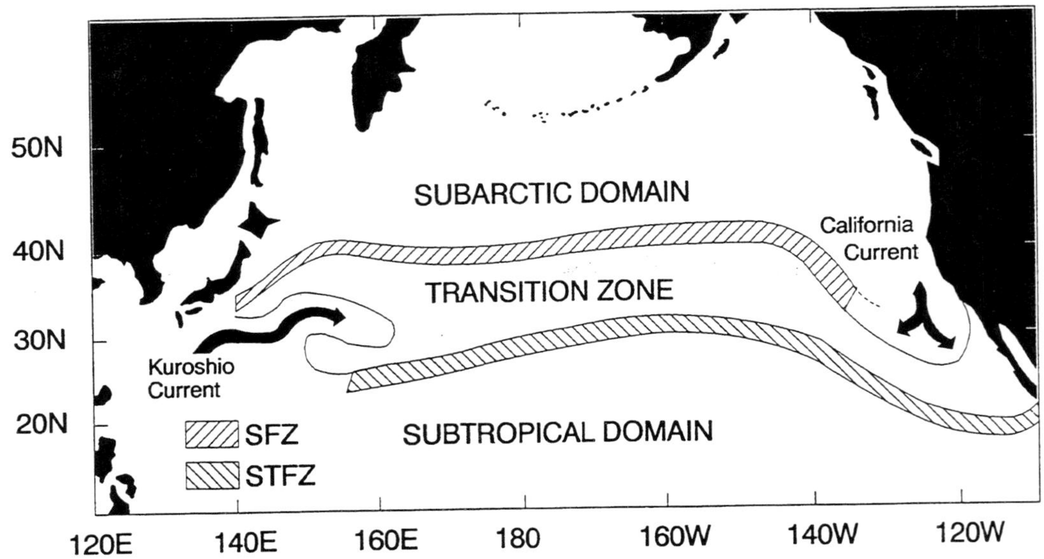

Fig. 1. Schematic diagram showing the locations of the subarctic and subtropical domains, the North Pacific transition zone, and the associated subarctic frontal zone (SFZ) and subtropical frontal zone (STFZ). Black arrows indicate intrusions of boundary currents. (Adapted from Roden 1991).

North Pacific Ocean called the subarctic frontal zone (SFZ). The SFZ has been the target of intense international and scientific interest since the early 1980s due to the controversial drift-net fishery there for neon flying squid (*Ommastrephes bartrami*; hereafter called flying squid). Although a United Nations sponsored moratorium on large-scale driftnetting has been observed since the start of 1993, abundant, valuable, yet underutilized, fisheries resources still inhabit SFZ waters. As a result, scientific interest in this region remains, ranging from the start-up of new, alternative-gear fisheries to ecosystem dynamics in the face of climatic change.

The SFZ is a water-mass boundary at the northernmost portion of the North Pacific transition zone (Fig. 1). This zone occurs across the North Pacific Ocean as a belt from 40 to 44°N that exhibits large and rapid changes in temperatures and salinities (Roden and Robinson 1988) (Fig. 1). In Baranov's terminology, the SFZ is a wide transitional and relatively time- and space-stationary zone between water masses (Fedorov 1986). At any given time, the SFZ may contain individual fronts (surfaces within the frontal zone corresponding to the maximum gradient of one or more oceanographic parameters, e.g., temperature).

To the north, the SFZ is bounded by the subarctic domain (SAD), a large region characterized by cool, low-salinity waters, a permanent halocline between 100 and 200 m, and large seasonal temperature variation. To the south is a transitional region (32–40°N) followed by the subtropical domain (south of 32°N), which is a large region characterized by stable, warm, high-salinity waters with a halocline opposite to that in the SAD and a permanent thermocline (Roden and Robinson 1988).

The north–south differences in the physical and chemical characteristics of the SFZ and adjoining regions often correspond to north–south differences in the distribution of the region's species. The SFZ delimits the southern or northern geographic ranges of many nektonic species found primarily in subarctic or subtropical waters (Shimazaki 1986), sharply separates different avifauna (Wahl et al. 1989), and is related to meridional differences in sexual composition and sizes of migratory species, such as flying squid and pomfret (*Brama japonica*) (Murakami et al. 1981; Pearcy 1991).

In this paper, we examine annual changes in the spatial distribution of species and species assemblages across the SFZ in the central North Pacific Ocean. Particular focus is placed on comparing the strength of association between either ocean sea-surface temperatures (SSTs) or frontal boundaries and changes in species ranges or assemblages. We have selected a time series from a standardized gillnet and oceanographic survey of the central North Pacific Ocean for these comparisons. The surveys, conducted from the Japanese research vessel Hokusei maru, have consisted of the same set of sampling stations approximately the same period (last week of July) each year since 1978 (Hokkaido University 1978–91). Our analyses begin by identifying the SFZ boundaries using both graphical and statistical methods. Next, we identify areas of similar species composition (hereafter called a "community," which, in this paper, is the part of an ecosystem composed of large-sized (e.g., nekton) pelagic animals), and we relate changes in species composition and ranges to physical oceanographic features. Spatial and temporal patterns of community and species ranges, oceanographic boundaries, and SST are also presented and compared.

Methods

Data collection

Thirteen oceanographic and seven gill-net stations were sampled along the 175°30'E meridian during late July each year from 1978 to 1991. Station locations ranged from 38°30'N to 47°30'N. Conductivity–temperature–depth (CTD) profiles were collected at approximately every 45' of latitude. Approximately 90–115 standard tans (50 m long and 6 m deep) of

Table 1. Common names and scientific names of taxa from Hokusei maru gillnet stations, 175°30'E transect.

Common name	Scientific name
Boreal clubhook squid	*Onychoteuthis borealijaponica*
Eight-armed squid	*Gonatopsis borealis*
Flying squid	*Ommastrephes bartrami*
Luminous flying squid	*Symplectoteuthis luminosa*
Gonate squids	Family Gonatidae
Salmon shark	*Lamna ditropis*
Blue shark	*Prionace glauca*
Spiny dogfish	*Squalus acanthias*
Salmon	*Oncorhynchus* spp.
Pacific sardine	*Sardinops sagax*
Longnose lancetfish	*Alepisaurus ferox*
Daggertooth	*Anotopterus pharao*
Lanternfishes	Family Myctophidae
Flying fishes	Family Exocoetidae
Pacific saury	*Cololabis saira*
Pilotfish	*Naucrates ductor*
Yellowtail	*Seriola lalandi*
Dolphin fish	*Coryphaena hippurus*
Pacific pomfret	*Brama japonica*
Skipjack tuna	*Katsuwonus pelamis*
Chub mackerel	*Scomber japonicus*
Albacore	*Thunnus alalunga*
Swordfish	*Xiphias gladius*
Striped marlin	*Tetrapturus audax*
Smalleye squaretail	*Tetragonurus cuvieri*
Ocean sunfish	*Mola mola*
Shearwaters	*Puffinus* spp.
Puffins	*Fratercula* spp.
Laysan albatross	*Diomedea immutabilis*
Dolphins and porpoises	Family Delphinidae
Northern fur seal	*Callorhinus ursinus*

variable-mesh gill nets were set at every 1°30' of latitude and soaked overnight for 10–12 h. Twenty-two different mesh sizes (19 to 233-mm stretched measure) were used in each gill net; about 50% of the tans were commercial fishing gear with mesh sizes 112–121 mm.

We identified 31 taxonomic groups from the gillnet catches (Table 1), after deleting extremely rare species and grouping a few species to higher taxonomic levels to increase occurrence rates (e.g., Clarke and Green 1988). Count data were converted to catch per unit effort (CPUE; one unit of effort = 100 tans of net) to account for differences in sampling effort among stations. The count data were then log-transformed to ln(CPUE + 1) to lessen the effect of extreme values and help control skewness (Clarke and Green 1988). Temperature and salinity data computed for standard oceanographic depth intervals (0–400 m) were extracted from the CTD profile data and used in our analyses.

SFZ boundary identification

We used two methods, one statistical and one graphical, to identify physical oceanographic gradients and boundaries. Oceanographic stations with similar temperature and salinity characteristics were grouped into distinct oceanographic regions using average linkage cluster analysis (procedure CLUS; SAS Institute 1990) on a Euclidean distance matrix (e.g., Ludwig and Reynolds 1988). Concurrent results of three statistics (a local peak in pseudo-F, high t^2 in the next cluster grouping, and a relatively high (or sharp) increase in R^2) were used to determine number of clusters in each subset of data (see Milligan and Cooper 1985). The resulting classification was then compared with temperature–salinity (T–S) diagrams (graphical method; see Favorite et al. 1976) before the geographic range of each oceanographic region was established.

Species community analysis

Gill-net stations with similar species composition were grouped into communities using average linkage cluster analysis on a Bray–Curtis distance matrix. The Bray–Curtis index avoids problems created in sparse matrices like ours when many of the species are absent from any two samples being compared (Clarke and Green 1988). Communities (clusters) were then cross-tabulated with the predefined oceanographic zones, and a measure of agreement K between community classification and oceanographic zone was computed (Bishop et al. 1975).

Discriminant analysis (procedure DISCRIM; SAS Institute 1990) was used to determine the separability of community clusters using several sets of environmental variables. Results were evaluated based on the success rate of reclassification using a quadratic discriminant function with a jackknife (cross-validation) procedure.

Correspondence analysis (e.g., Ludwig and Reynolds 1988) was used to ordinate stations and taxa simultaneously. The relationship between community indices (ordinates from correspondence analysis) and temperature and salinity data (to a depth of 400 m) was investigated using canonical-correlation analysis (procedure CANCORR; SAS Institute 1990). To interpret spatial patterns, we calculated pairwise correlation coefficients between species' geographic limits, community boundaries, and oceanographic variables.

Results

Ocean boundaries and temperatures

The oceanographic stations usually fell into three clusters. In some years, the groupings were distinct; the three groups in 1981 consisted of stations between 38°30'N and 39°15'N in one cluster, 40°N and 44°30'N in another cluster, and 45°15'N and 47°30'N in a third cluster (Fig. 2, upper graph). Based on comparisons with T–S plots for each transect and on previous work (e.g., Roden 1991), we designated the northernmost group as the subarctic domain, the intermediate group as the subarctic frontal zone, and the southernmost group as the transition region (TR). Only two clusters, however, were identified in 1980, 1986, and 1988. Based on analyses of the T–S plots, we concluded that the SFZ's southern boundary occurred south of the range of sampling for those years (40°N in 1980 and 38°30'N in 1986 and 1988).

The SFZ's northern boundary was more distinct (exhibiting a sharper gradient in temperature and salinity properties than the southern boundary) in 6 of the 11 yr in which both boundaries were located in the survey area. A runs test showed no significant order to the pattern of differences in boundary strength ($p = 0.79$).

Location of the SFZ's northern boundary appeared stationary from 1978 to 1985 (Fig. 3). Starting in 1986, this boundary not only shifted southward but varied in position from year to year. Average position during the study period was 44.1 ± 1.0°N

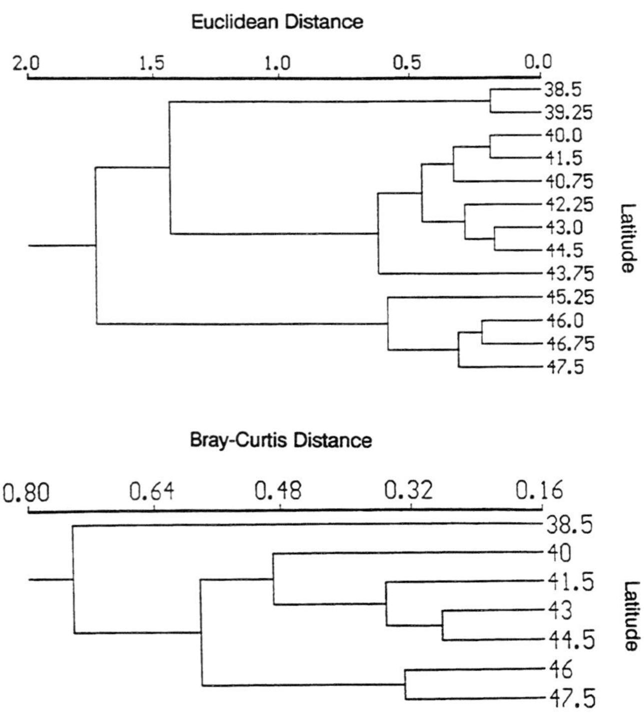

Fig. 2. Results of the cluster analysis revealing three main groupings of oceanographic (upper graph) and gillnet stations (lower graph) in 1981. The horizontal axis measure is Euclidean distance for the oceanographic stations, and Bray–Curtis distance for the gillnet stations.

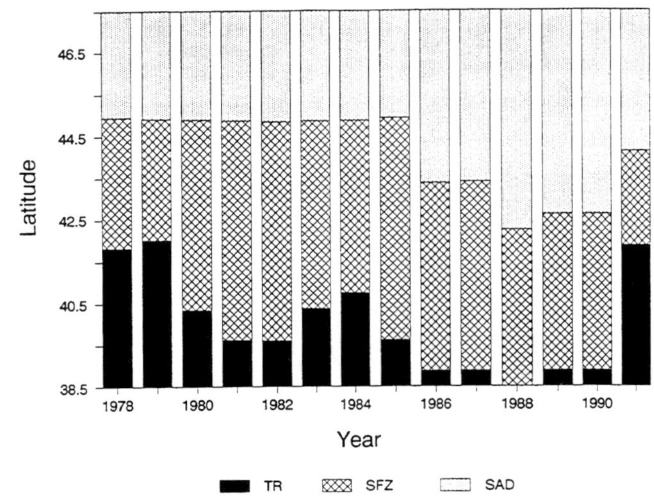

Fig. 3. Oceanographic zone positions for the 175°30'E transect, 1978–91, based on results of the cluster analyses and the interpretation of T–S plots; the groupings correspond to the subarctic domain (SAD), subarctic frontal zone (SFZ), and the transitional region (TR).

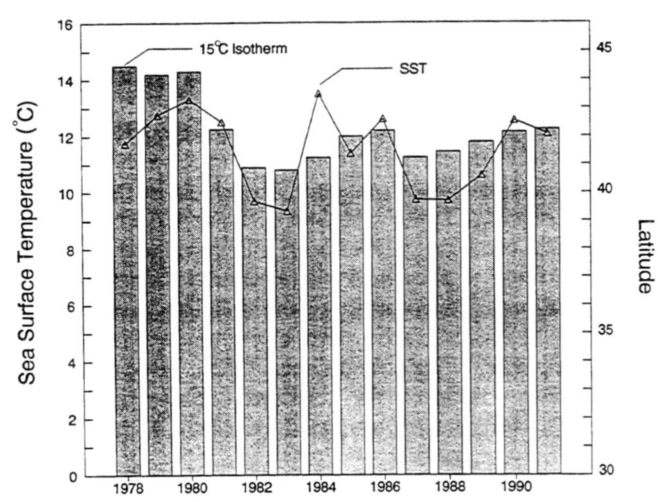

Fig. 4. Sea-surface temperature (SST) averaged over the latitude 44°30'N and 46°N oceanographic stations, and location of the 15°C isotherm along the 175°30'E transect of the Japanese research vessel Hokusei maru, 1978–91.

(mean ± SD). Location of the SFZ's southern boundary varied throughout the study period (Fig. 3). Average position was 40.1 ± 1.1°N; note that mean location is biased slightly upwards and standard deviation downward because the SFZ's southern boundary occurred south of the study area in 1988.

SST, on the other hand, was somewhat cyclic during the study period. SSTs along the SFZ's northern boundary (average of 44°30'N and 46°N stations) fluctuated with an approximate 5-yr period (Fig. 4). The location of the 15°C isotherm also appeared cyclic, but with a damped amplitude (Fig. 4).

Geographic ranges

The gillnet stations were also generally grouped in three clusters (for illustration purposes, see 1981 data in Fig. 2, lower graph), although only the two northernmost groupings were observed in 1983. Based on the frequency of occurrence and CPUE of each species caught, we classified the communities as (1) salmon – saury (*Cololabis saira*), (2) sardine (*Sardinops sagax*) – pomfret, or (3) skipjack tuna (*Katsuwonus pelamis*) – flying squid. The most distinct boundary occurred between communities 2 and 3, hereafter called the southern community boundary. The cluster analysis gave a relative distance of 1.3 for the southern community boundary versus 1.0 for the northern community boundary.

There was no noticeable trend in the location (latitude) of community boundaries (Fig. 5) or species ranges (Fig. 6) during the study period. Average position based on species composition data was 44.4 ± 0.7°N) for the northern community boundary and 40.2 ± 0.8°N) for the southern community boundary.

The boundaries between biological communities were identical to those between oceanographic zones in 3 of 14 yr. More generally, a cross tabulation of the three epipelagic communities with the three oceanographic zones showed that the communities were significantly associated with oceanographic zones ($\chi^2 = 86.6, p < 0.001$). Differences in classification (21 of 91 observations were misclassified) always occurred in spatially adjacent zones (Table 2). The central pelagic

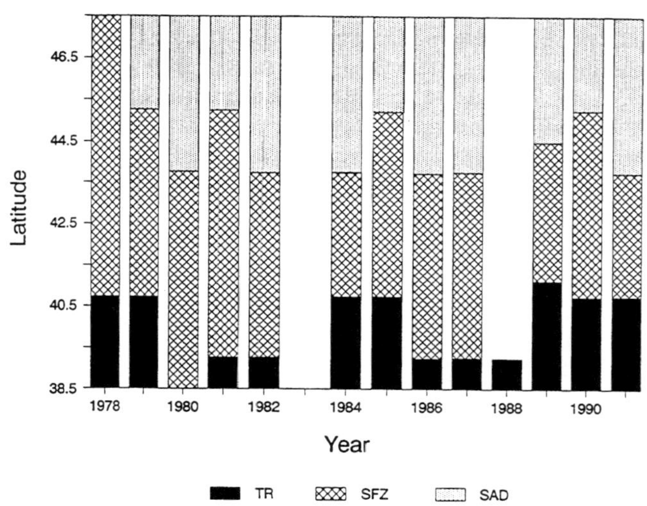

Fig. 5. Communities identified by average linkage cluster analysis (CA) plotted over the 14 yr of the study. Characteristic species for the northern, central, and southern communities are salmon and saury, sardine and pomfret, and skipjack and flying squid.

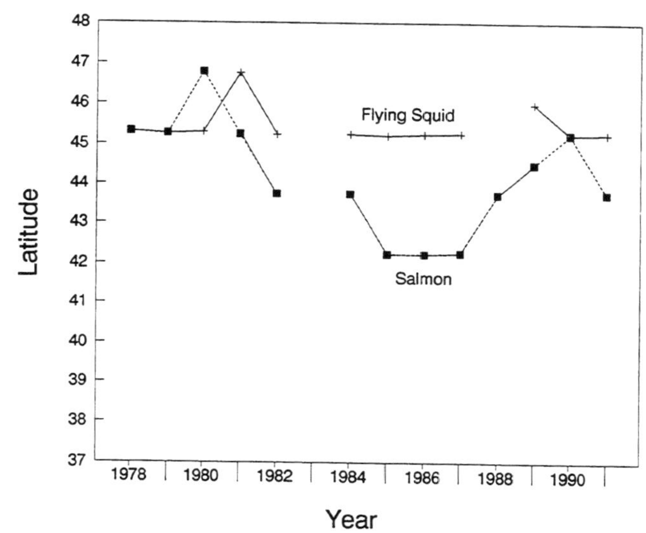

Fig. 6. The southernmost latitude in which salmonids were encountered and the northernmost latitude in which neon flying squid were encountered along the 175°30'E meridian, 1978–91. Data from the Japanese research vessel Hokusei maru.

Table 2. Cross tabulation of epipelagic communities (1978–91 data combined) and three oceanographic zones (subarctic domain (SAD), subarctic frontal zone (SFZ) and transitional region (TR)) along the 175°30'E transect ($\chi^2 = 86.6, p < 0.001$). Expected values are given in parentheses. Differences in classification always occurred in spatially adjacent areas; see text for discussion.

	Oceanographic zone		
Community type	SAD	SFZ	TR
Salmon–saury	30 (13.1)	4 (13.8)	0 (7.1)
Sardine–pomfret	5 (15.4)	29 (16.3)	6 (8.4)
Skipjack – flying squid	0 (6.5)	4 (6.9)	13 (3.5)

community, which was associated with the SFZ, had the highest misclassification rate (11 of 40 stations misclassified). Overall, the agreement K between estimated communities and oceanographic zones was 0.67 (95% confidence interval of 0.60 and 0.75), which is greater than that expected by chance ($K = 0.00$) but less than perfect agreement ($K = 1$).

Canonical correlation analysis revealed a strong relationship between community ordinates and oceanographic (temperature and salinity) variables ($R_1 = 0.97, p < 0.001; R_2 = 0.79, p < 0.001$). The first pair of canonical variables primarily represents the correlation between the first community index, which accounts for as much community variation as possible, and an oceanographic index nearly equally weighted by all temperature and salinity variables. The two variables with the highest weights were SST and temperature at 10 m; the correlations between these variables and the first canonical variate were 0.967 and 0.974, respectively. The second pair of canonical variables represents the relationship between the second community index and the deep-layer salinity (300–400 m).

A discriminant model based on the first three oceanographic indices from the canonical correlation analysis correctly classified 85.7% of the observations. When just the first index was used in the model, the classification rate dropped to 81.3%. A discriminate model based on SST correctly classified 83.5% of the observations.

When the community boundaries were examined individually, however, the northern community boundary was not associated with any environmental variable whereas the southern community boundary was possibly related to oceanographic zone ($r = 0.435, p < 0.2$) or SST ($r = 0.423, p < 0.20$). The southern range of salmon was significantly related to the location of the 15°C isotherm ($r = 0.668$, $p < 0.05$), whereas the northern range of flying squid was uncorrelated with any other variable.

Discussion

The location of the SFZ's northern boundary did not change for an 8-yr period (1978–85). We hypothesize that this lack of change is related to Roden's (1991) observation that the SFZ is topographically trapped by the Emperor Seamounts near 43°N and 170°E. SST, however, varied continuously from 1978 to 1991; some of the variability appeared to be contained in a moderately high-frequency (approximately 5 yr) fluctuation. This periodicity in SST is probably a response to El Niño – Southern Oscillation (ENSO) forcing. Years with cool surface temperatures (1982–83 and 1987–88) correspond to El Niño events, and there is evidence that variations in the upper mixed layer in the central North Pacific Ocean are related to changes in large-scale atmospheric systems, such as those associated with ENSO events (Ignell 1990).

Community structure across the SFZ appears to be highly related to oceanographic conditions. We found a near-perfect correlation ($R_1 = 0.97$) between two indices, one accounting for as much community variation as possible and the other for as much temperature–salinity variation from 0–400 m depth as possible. We also found that if we stratified the survey area into either three regions with similar oceanographic characteristics or three regions with similar species compositions, the resulting two sets of regions were similar in location; average width of the middle region, or the SFZ, is 4.0 (using oceanographic data) or 4.2° latitude (using biological data), which is similar to the 3 or 4° latitude width reported in Roden and Robinson (1988).

Our results also suggest that SST is a more effective environmental predictor of species range boundaries or communities than oceanographic zones. In our correlation analyses, oceanographic zone was not significantly related to any biological index whereas SST was significantly related to the southern range limit of salmon. A model based on SST better classified (84% accuracy rate) the three communities than a model based on oceanographic zones (77% accuracy rate). These results along with the high canonical correlations imply that in the SFZ, where our sampling occurred, the relationships between animals are best viewed from the continuum rather than the community paradigm (according to the continuum paradigm, groups of organisms change continuously and cannot be usefully differentiated into classes of well-defined integrated units (Pielou 1977)) and are particularly amenable to ordination-based statistical methods.

Although the high canonical correlations observed in our data are indicative of the overall influence of the ocean on species relationships throughout the area of sampling (38.5–47.5°N), correlations between specific locations (e.g., range limits or community boundaries) and our environmental variables generally lacked significance. This lack of association likely stemmed from two causes, one sampling and the other ecological related.

First, many of our biological or oceanographic measures were embedded with large uncertainties, raising the level of noise in our data. A portion of these uncertainties were based on the sampling resolution of the surveys, which typically exceeded 160 km for the gillnet stations and 80 km for the oceanographic stations, and the limited amount of gear (about 5 km per station) used for the survey. Other uncertainties arose through the choice of variable used in our analyses or through the measurement of those variables. Using presence–absence data, a species range limit is easy to identify, but highly dependent upon sampling effort and subject to large local variability. For example, in 1988 CPUE dropped from 113.9 salmon per 100 tans to 0.0 within one sampling station (1.5° of latitude) while it took a distance three times greater (4.5° of latitude) in 1987 to accomplish a similar drop in CPUE (Ignell 1989). The identification of oceanographic zone based on T–S diagrams, contour plots of temperature and salinity, or some other method is often subjective and can change according to the method used. We attempted to circumvent this subjectivity by the use of objective multivariate methods, but found that even those results are occasionally ambiguous. As a result, the oceanographic zone of some sampling stations may be misclassified, thereby adding an unknown amount of additional noise to our data.

Second, the SFZ, especially its northern boundary, is difficult to identify in surface waters during summer, because temperature and salinity gradients in the upper mixed layer are diminished by radiative heating and cooling (Roden and Robinson 1988). Below the mixed layer, the SFZ is detectable but is also beyond the depth range of species such as chum salmon (*O. keta*) and coho salmon (*O. kisutch*), which primarily inhabit subarctic waters (Ogura 1990; Ogura and Ishida 1992). The southern boundary of the SFZ, which is usually characterized by salinity, is more easily identifiable in both surface and subsurface waters (the upper 400 m) throughout the year, and therefore encountered by both surface-dwelling and vertically migrating nektonic species inhabiting that region.

Our results are consistent with this description of the physical environment. The southern range of salmon, a primarily surface-dwelling species, was significantly related to SST but not oceanographic zone. The southern community boundary was more distinct than the northern boundary. The northern community boundary was unrelated to any of our environmental variables, whereas the southern community boundary was weakly associated with oceanographic zone and SST.

Conclusion

The multivariate, multispecies-based methods presented in this paper provide a means to assess community-level responses to changes in a highly dimensional environment. The type of response we examined, the geographic range limit of a species or a group of species (community), is relevant to a variety of fishery management and scientific problems, including the recent concern over a changing climate and its impact on fishery resources. For example, two likely consequences of climate change are changes in community composition and shifts in the geographic distribution of species. The first consequence is based on the observation that biological communities do not typically respond to a changing climate as a singular unit; instead, variations in the rate of change among component species change the pattern of species abundances (Davis 1986). The second consequence stems from the observation that climatic factors are often more influential at the edges of species ranges where abundances are reduced (Brown 1984).

Our results indicate that changes in habitat will shift the geographic ranges of pelagic species in the central North Pacific Ocean, altering the species composition for a particular area. In our analyses we found that species ranges were more influenced by SST than oceanographic zone along the northern edge of the SFZ during the summer. We note, however, that the particular environmental factor(s) most affecting species ranges may vary from area to area and perhaps seasonally in accordance with the life history of a species. Hence, any prediction of change in the geographic range of a species needs, as its foundation, an in-depth understanding of the habitat of that species, including spatial and temporal variability in habitat characteristics.

References

Bishop, Y.M.M., S.E. Fienberg, and P.W. Holland. 1975. Discrete multivariate analysis: theory and practice. MIT Press, Cambridge, MA. 557 p.

Brown, J.H. 1984. On the relationship between abundance and distribution of species. Am. Nat. 124: 255–279.

Clarke, K.R., and R.H. Green. 1988. Statistical design and analysis for a ‘biological effects’ study. Mar. Ecol. Prog. Ser. 46: 213–226.

Davis, M.B. 1986. Climatic instability, time lags, and community disequilibrium, p. 269–284. *In* J. Diamond and T.J. Case [ed.] Community ecology. Harper & Row, New York, NY.

Favorite, F., A.J. Dodimead, and K. Nasu. 1976. Oceanography of the subarctic Pacific Region, 1960–1971. Int. North Pac. Fish. Comm. Bull. 33: 187 p.

Favorite, F., and M.G. Hanavan. 1963. Oceanographic conditions and salmon distribution south of the Alaska Peninsula and Aleutian Islands. Int. North Pac. Fish. Comm. Bull. 11: 57–72.

Fedorov, K.N. 1986. The physical nature and structure of oceanic fronts. Lect. Notes Coastal Estuarine Stud. 19. 333 p.

Fujii, T., K. Masuda, and T. Nishiyama. 1965. Salmon population and hydrography on the high sea fishing grounds developed in the southwest on the Attu Islands in the North Pacific, during the early period of the fishing season. Bull. Fac. Fish. Hokkaido Univ. 16(3): 154–163.

Gaston, K.J. 1990. Patterns in the geographical ranges of species. Biol. Rev. 65: 105–129.

Hokkaido University. 1978–91. Data record of oceanographic observations and exploratory fishing. Publ. No. 21–34. Faculty of Fisheries, Hokkaido University, Hakodate, Hokkaido, Japan.

Ignell, S.E. 1989. Annual variability in salmonid incidence and near-surface ocean temperatures in the central North Pacific Ocean, 1978–1988. INPFC Document. Auke Bay Laboratory, Alaska Fisheries Science Center, NMFS, NOAA, Juneau, AK 99801-8626. 23 p.

Ignell, S.E. 1990. Diagnostic indices of oceanographic and atmospheric conditions in the central North Pacific Ocean. INPFC Document. Auke Bay Laboratory, Alaska Fisheries Science Center, NMFS, NOAA, Juneau, AK 99801-8626. 14 p.

Ludwig, J.A., and J.F. Reynolds. 1988. Statistical ecology: a primer on methods and computing. John Wiley & Sons, New York, NY. 337 p.

Milligan, G.W., and M.C. Cooper. 1985. An examination of procedures for determining the number of clusters in a data set. Psychometrika 50: 159–179.

Mishima, S. 1974. Latitudinal distribution of salmon at longitude 167°30'E in the early spring in the northwestern North Pacific, 1971–1973. Bull. Fac. Fish. Hokkaido Univ. 25(1): 36–46.

Murakami, K., Y. Watanabe, and J. Nakata. 1981. Growth, distribution and migration of flying squid (*Ommastrephes bartrami*) in the North Pacific, p. 161–179. In S. Mishima [ed.] Pelagic animals and environments around the subarctic boundary in North Pacific. Spec. Vol. Res. Inst. North Pac. Fish. Hokkaido Univ.

Ogura, M. 1990. Outline of biotelemetry study of salmon by the *Shin riasu maru*, 1990. Document submitted to the Annual Meeting of the International North Pacific Fisheries Commission, Vancouver, B.C. October 1990. Fisheries Agency of Japan, National Research Institute of the Far Seas Fishery, 7–1 Orido 5 chome, Shimizu, Shizuoka, Japan 424. 20 p.

Ogura, M., and Y. Ishida. 1992. Swimming behavior of coho salmon, *Oncorhynchus kisutch*, in the open sea as determined by ultrasonic telemetry. Can. J. Fish. Aquat. Sci. 49: 453–457.

Ogura, M., and K. Takagi. 1987. Difference in area of distribution between salmon and flying squid based on data collected by Japanese salmon research vessels, 1983–1986. Document submitted to the Annual Meeting of the International North Pacific Fisheries Commission, Vancouver, B.C. October 1987. Fisheries Agency of Japan, Tokyo, Japan 100. 7 p.

Pearcy, W.G. 1991. Biology of the transition region. NOAA Tech. Rep. NMFS 105: 39–55.

Pielou, E.C. 1977. Mathematical ecology. John Wiley & Sons, New York, NY. 385 p.

Roden, G.I. 1991. Subarctic–subtropical transition zone of the North Pacific: large-scale aspects and mesoscale structure. NOAA Tech. Rep. NMFS 105: 1–38.

Roden, G.I., and A.R. Robinson. 1988. Subarctic frontal zone in the north-eastern Pacific: mesoscale structure and synoptic description. Harv. Ocean Model Rep. Rep. Meteorol. Oceanogr. 31. Harvard University, Cambridge, MA. 71 p.

SAS Institute Inc. 1990. SAS/STAT User's Guide, version 6. 4 ed. Vol. 1. SAS Institute Inc., Cary, NC. 943 p.

Shimazaki, K. 1986. Distribution of the pelagic fish community around the subarctic boundary in the North Pacific Ocean. Int. North Pac. Fish. Comm. Bull. 47: 247–264.

Tabata, S. 1983. Oceanographic factors influencing the distribution, migration, and survival of salmonids in the northeast Pacific Ocean—a review, p. 128–160. *In* W.G. Pearcy [ed.] The influence of ocean conditions on the production of salmonids in the North Pacific, a workshop. Sea Grant Coll. Program Rep. ORESU-W-83-001. Oregon State University, Newport, OR.

Wahl, T.R., D.G. Ainley, A.H. Benedict, and A.R. DeGange. 1989. Associations between seabirds and water-masses in the northern Pacific Ocean in summer. Mar. Biol. 103: 1–11.

Change in distribution of favourable water temperatures and migration of Japanese sardine (*Sardinops melanostictus*)

Makoto Kashiwai

Kashiwai, M. 1995. Change in the distribution of favourable water temperatures and migration of Japanese sardine (*Sardinops melanostictus*), p. 675–686. *In* R.J. Beamish [ed.] Climate change and northern fish populations. Can. Spec. Publ. Fish. Aquat. Sci. 121.

Abstract: This paper seeks to reestablish favourable temperature as a useful concept in analyzing the effects of environmental changes on pelagic migrant fish. Introducing the concept of fuzzy-set theory, this paper examines the effectiveness of using favourable temperature as an indicator of suitable habitat and proposes a method to determine favourable temperature using catch data from the Japanese sardine (*Sardinops melanostictus*) purse-seine fishery and hydrographic data. Simple simulation of temperature-driven migrants reproduced the movement of Japanese sardine and supported the hypothesis that Japanese sardine is a temperature-driven migrant. With the help of the fuzzy set operation rule, a method to determine favourable temperature from purse-seine catch and hydrographic data was developed. Estimation of the favourable temperature for Japanese sardine by this method coincides with the results of past research. Analysis of the volume of water of a favourable temperature showed that sea-surface temperature does not always indicate the habitat temperature of Japanese sardine and that the volume of water of favourable temperature range can be considered an index of habitat size.

Résumé : Le présent document vise à rétablir la notion de température favorable come concept utile pour l'analyse des effets des changements environnementaux sur les poissons pélagiques migrants. Introduisant le concept de la théorie des ensembles flous, nous examinons l'efficacité de l'utilisation de la température favorable comme indicateur d'un habitat adéquat et proposons une méthode permettant de déterminer cette température, à l'aide des données sur les prises de la pêche à la senne coulissante de la sardine du Japon (*Sardinops melanostictus*) et des données hydrographiques. Une simulation simple de migration causée par la température a permis de reproduire les déplacements de sardines du Japon et a corroboré l'hypothèse selon laquelle la sardine du Japon est un migrant influencé par la température. À l'aide de la règle régissant les opérations de la théorie des ensembles flous, nous avons élaboré une méthode permettant de déterminer la température favorable à partir des prises à la senne coulissante et des données hydrographiques. L'estimation de la température favorable pour la sardine du Japon selon cette méthode coïncide avec les résultats des recherches antérieures. L'analyse du volume de l'eau d'une température favorable montrait que la température de la surface de la mer n'indique pas toujours la température de l'habitat de la sardine du Japon et que le volume d'eau de la zone de température favorable peut être considéré comme un indice des dimensions de l'habitat.

Introduction

Discussions on the mechanisms through which the climatic changes of marine environment affect on the long-term variability of migrant pelagic fish populations are almost limited to the changes in food availabillity. However, it is well known that almost all fish show sharp temperature sensitivity and specific temperature preference, the so-called favourable temperature. For migrant fish to complete a life cycle, it is important that favourable water form possible habitat and migration paths at appropriate times and places.

M. Kashiwai. Hokkaido National Fisheries Research Institute, Fisheries Agency, Government of Japan, 116 Katsurakoi, Kushiro, Hokkaido 085, Japan.

These aspects must be taken into consideration when investigating potential mechanisms affecting long-term variations of pelagic fish populations. However, few scientists have used the concept of favourable temperature as a tool for elucidating the abundance and distribution dynamics of fish populations. This may result from the following reasons: (1) the methods of determining the favourable temperature for particular species are not established; and (2) the quantitative expressions of favourable temperature is not well enough developed to introduce in a model of population dynamics, either as lumped-parameter system or as a distributed-parameter system.

If favourable temperature, i.e., species-specific temperature range, exists as a major factor for habitat preference, the following phenomena must be observed: (1) the water of a specific temperature range prevails over distribution area of the fish; (2) the major fishing season occurs when a large water mass of the specific temperature range is present in the fishing ground; and (3) the migration of the fish appears to follow the geographical movement of specific temperature zone with season.

The existence of these phenomena can be the basis of an inductive demonstration of the existence of favourable temperature. The scientific proof for the existence of the favourable temperature is that the specific temperature range having above mentioned characteristics can be determined from available data.

Favourable temperature can be considered a fuzzy quantity because the degree of membership for any temperature belonging to the favourable temperature cannot take a value of 0 or 1 but takes a continuous value between 0 and 1. This paper describes the results of analysis, using the concept of fuzzy-set theory, that was designed to examine the effectiveness of favourable temperature and the method to determine the favourable temperature of the Japanese sardine (*Sardinops melanostictus*) using catch data from purse-seine fishery and hydrographic data.

The following parts of this paper are (1) the sources of data; (2) the evidence of the migration of Japanese sardine; (3) the formation of major fishing ground and seasonal change of sea-

Table 1. Special characteristics of catch reports.

Case	Search	Existence	Detect	Cast Net	Catch	Report
1	Made	Made	Made	Made	Made	Made
2	Made	Made	Made	Made	Made	Not done
3	Made	Made	Made	Made	Not done	
4	Made	Made	Made	Not done		
5	Made	Made	Not done			
6	Made	Not done				
7	Not done					

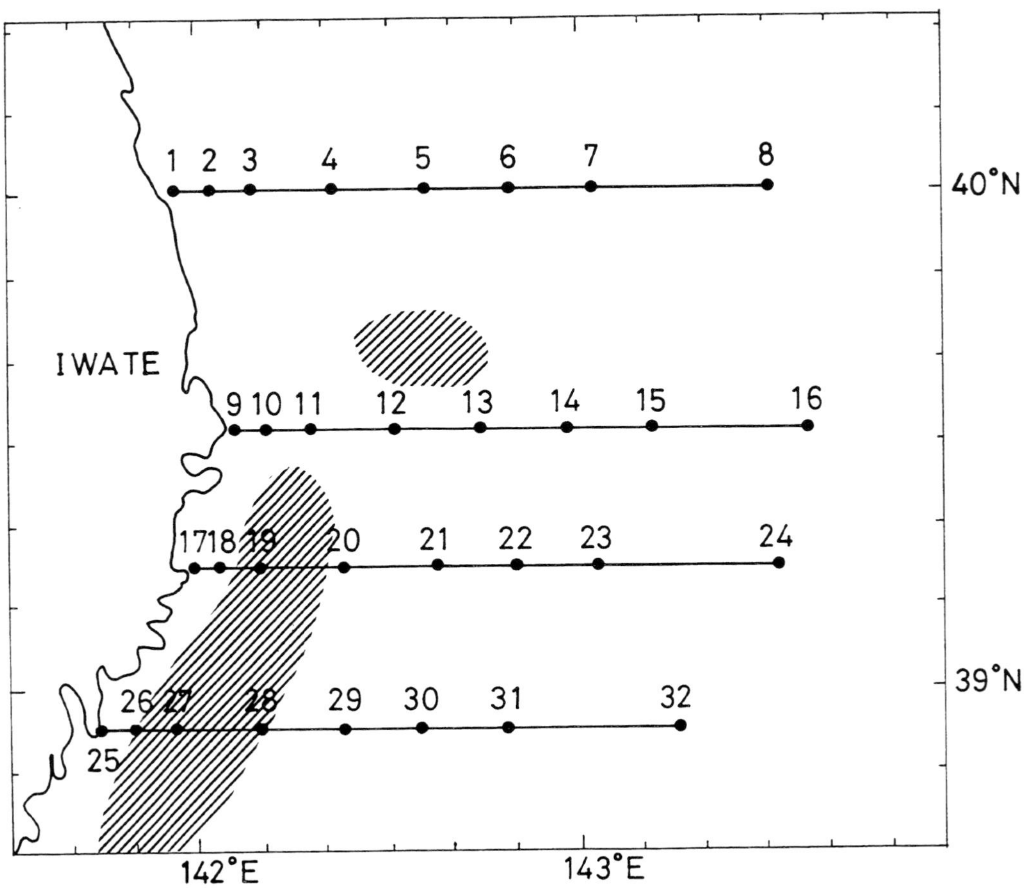

Fig. 1. Hydrographic observation stations in the water off Iwate and the areas where catches were reported during the observation period, July 1–5, 1988 (shown by hatching).

surface temperature (SST) at major fishing ground of Japanese sardine in the Pacific side of Japanese Islands; (4) the simulations of the migration of Japanese sardine by temperature-driven migrant in the varying sea-surface temperature (SST) field; (5) the method to determine the favourable temperature; and (6) the effectiveness of favourable temperature.

Sources of data

Around Japan, reports of the 1988 landings of Japanese sardine at ports around the Japanese Islands were used.

Analysis of favourable water volume and determination of favourable temperature were done for the Hyuga-nada area in the Nansei region and the coastal area off Iwate prefecture in the Sanriku region. There were SST data and fishing locations reported in the *Quick report of fisheries information* (Japan Fisheries Information Center 1988), and monthly hydrographic data (Iwate Prefecture Fisheries Experimental Station 1990; Miyazaki Prefecture Fisheries Experimental Station 1990).

These sites were selected because the purse-seine fisheries operate for most of the year, monthly hydrographic data are available for both regions, and these regions are the representative fishing grounds of the coastal water of the Kuroshio Current and Oyashio Current, respectively.

Catch information from the purse-seine fishery

From the information in the catch reports, the cases shown in Table 1 are possible. In general, the catch reports do not contain the time and place of unsuccessful catches. The existence of a catch indicates the existence of the fish, but the absence of a catch does not always indicate the absence of fish.

Cases 1 and 6 are desirable cases for analysis. To avoid other cases (especially case 7), the data must be collected from the potential fishing area where the search effort is frequent. The points of catch will show seasonal and/or interannual shift within the potential fishing area. This shift indicates the shift of the distribution of the fish, because the searching efforts of fishing boats are supposed to cover the potential fishing grounds.

If the abundance of the fish is not too low compared with the size of the favourable water mass, the fish will distribute over the favourable water and the favourable temperature will be the major factor controlling the distribution of fish. Therefore, the catch data of 1988, when Japanese sardine was in its high abundance stage, was analyzed.

Purse-seine catch reports contain no information on the depths at which the fish occur. This situation is the same as the plankton sampling by vertical or oblique tow of plankton net. It must be understood that a favourable water layer may exist at any depth within the water column at the place where a catch is reported.

Hydrographic data

To calculate temperature composition of the water, hydrographic data from the stations shown in Figs. 1 and 2 were used.

SST data

SST data were used in simulating temperature-driven migrants. The SST data set was taken on the grid point of 30' latitude and 30' longitude at 10-d intervals (Japan Fisheries Information Service Center 1988).

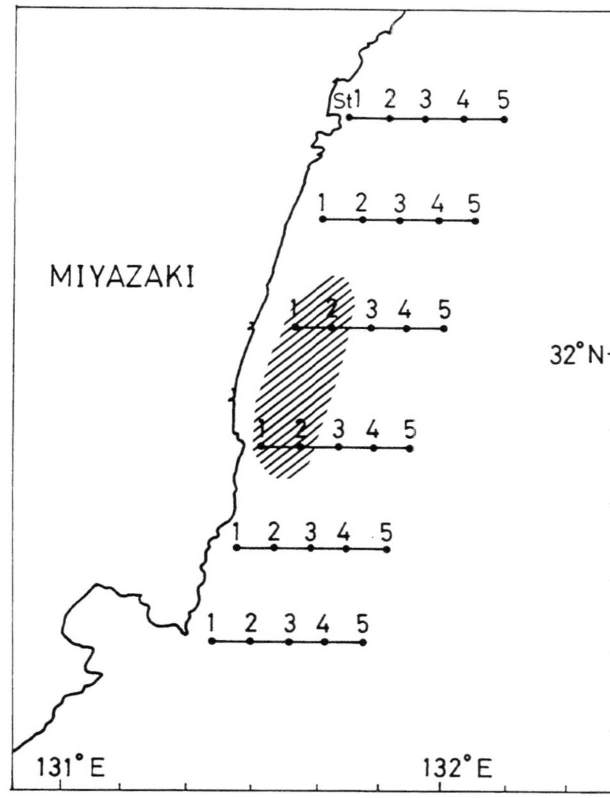

Fig. 2. Hydrographic observation stations in the water off Miyazaki and the areas where catches were reported during the observation period, February 5–13, 1988 (shown by hatching).

Results

Evidence of migration of Japanese sardine

Seasonal change in landings of Japanese sardine

To illustrate the migration of Japanese sardine, seasonal changes in landings of Japanese sardine are shown in Fig. 3. Monthly landings are shown by radar diagrams. For example, Japanese sardine migrated to the Okhotsk Sea coast of Hokkaido from June to August 1988. Total landings for each region are shown by the numbers besides the radar diagrams.

Migration of Japanese sardine is not a simple round-trip between the spawning ground and feeding ground. Judging from dominant landings in winter (Fig. 3), separate wintering regions exist: the Johban region and the region from the western Japan Sea through north Kyushu to west Kyushu.

This leads us to question what factors determine the southern or northern terminus of the annual migration: climatic boundaries in the ocean, a photoperiodic condition, or some internal clock independent from the environment.

Occurrence of catch in SST field

From the *Quick report of fisheries information* (Japan Fisheries Information Center 1988), the occurrence of catches of Japanese sardine by purse-seine in the SST field is shown in Fig. 4.

From the occurrence of catch in SST field (Fig. 4), the following can be pointed out:

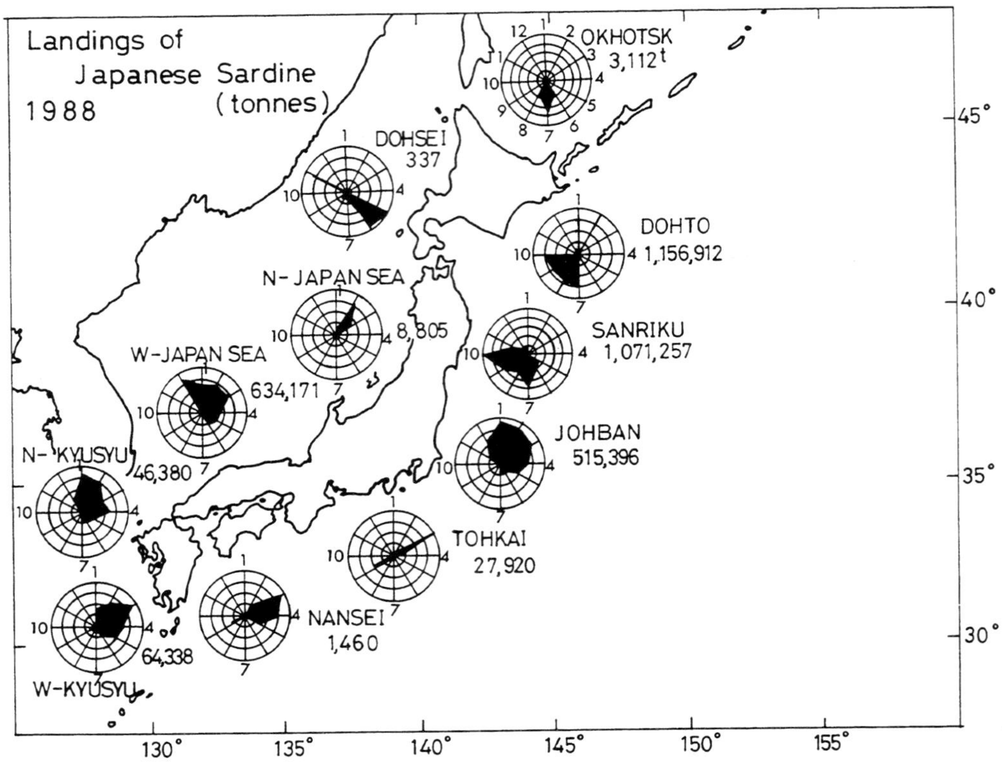

Fig. 3. Seasonal changes in landings of Japanese sardine in 1988.

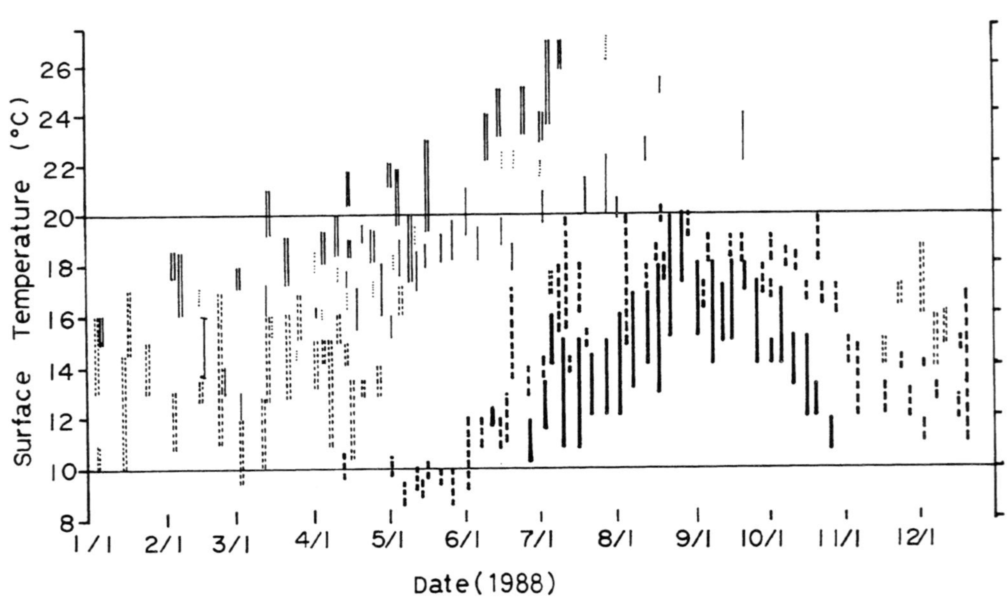

Fig. 4. Occurrence of Japanese sardine catch by purse seine in SST fields in the Pacific Ocean off Japan.

1) Two areas can be recognized. The warmer area belongs to the transition waters or the subtropic coastal waters, and the cooler area belongs to the subpolar waters.
2) SSTs at the fishing ground show seasonal changes.
3) The SST range at the fishing ground is very wide, ranging from 8 to 27°C.

These facts suggest the following possibilities:

1) Japanese sardine has a very wide temperature preference range or no temperature preference, and is not a temperature-driven migrant.
2) SST is not a direct index of habitat temperature for Japanese sardine because of the vertical temperature structure in the water column.
3) The temperature preference of Japanese sardine differs among subpopulations.

Simulation of temperature-driven migrant

Two-dimensional simulation

To test the hypothesis that Japanese sardine is a temperature-driven migrant, a simple simulation of temperature-driven migration was performed on a synoptic SST field.

Migrants can move toward neighboring grid points with separation of 30′ latitude and 30′ longitude, in eight directions, within the favourable temperature range (e.g. 10–18°C or 12–16°C). As for the migrating speed of the fish, three cases were adopted: the fish can move one, two, or three grid intervals in 10 d, which are nominally equal to speeds of 30, 60, or 90 mi (1 mi = 1.852 km) in 10 d. Migrants were started from the water off the Inubo-zaki, the southern end of Johban region, when the temperature was favourable.

Figs. 5a–5d show the distribution 1, 2, 3, and 4 mo after departure. In early July, the simulated migrants arrive at the Dohto fishing ground, as an actual Japanese sardine would. Some migrants in the water off Sanriku cannot find any suitable destination within reach of their travel range. These are considered to be in a state of dormancy, as stated by Hiramoto (1991),

From the results of all the other simulation runs, the following facts are evident:

1) Migrants with a speed of 30 mi in 10 days cannot follow the movement of isotherms.
2) Even for migrants with speed of 60 mi in 10 days, some lose the favourable water because of the rapid rise of temperature in the water off the Sanriku region in the latter half of July.
3) The main feature of the north–south movement of Japanese sardine in these waters is reproduced by simulating temperature-driven migrants.
4) These temperature-driven migrants spread east–west along the belt of favourable temperature zone. Without any additional mechanisms, most migrants cannot successfully return to their spawning or wintering ground.

One-dimensional simulation

To evaluate the effect of interannual temperature variations, a one-dimensional simulation for a coastal migration route was performed. From the results of the two-dimensional simulation of temperature-driven migration, we can conclude that the one-dimensional simulation provides enough answers about the north–south movement of Japanese sardine.

The coastal migration route was set from the Inubo-zaki through the waters of Sanriku region and the watter off the southeast coast of Hokkaido to 48°N. The migrants were designed to follow bidirectional dispersion, with a speed of 60 mi in 10 days, within the favourable temperature of 10–20°C. They started from the water off the Inubo-zaki when the temperature was favourable. SST data were reduced from the 30′ mesh data set for 1983.

Fig. 6 shows the results of the one-dimensional simulation. At the first rapid warming in the last 10 d of April 1983, migrants were not able to follow the enlargement of the favourable region. At the next rapid warming event in the first 10 d of August, the possible path to the north is very limited and a considerable number of migrants are left behind as the favourable water moves north. The latitudes of 42°N30′N–43°N, corresponding to the Dohto fishing ground, were reached by migrants in the last 10 d of July.

These results illustrate that temperture-driven migrants require the formation of migration path of favourable water at the appropriate time and place.

Method to determine favourable temperature

Algorithm

The favourable temperature (FVT) must be dominant at the time and place that catch is reported and not dominant when no catch is reported at the same point on the potential fishing ground. So, we can write

$$(1)\ \mathrm{FVT} = \mathrm{PVT}\ \text{and}\ (\text{not}\ \mathrm{AVT})$$

where PVT is the temperature composition of the water column where catch has been reported, and AVT is the temperature composition of the water column where no catch has been reported.

Temperature composition of the water mass was calculated from hydrographic data by counting the relative volume at each temperature step of 0.5°C and depth step of 25 m.

Every temperature on the FVT band need not appear in every observation; we must use <OR> PVT for observation. Furthermore, as it is enough for the existence of the fish that any part of the FVT range is dominant in any depth layer of the site, we take <OR> for depth steps. Therefore

$$(2)\ \mathrm{FVT} = \underset{d=1}{\overset{Nd}{\mathrm{OR}}} \{\underset{o=1}{\overset{No}{\mathrm{OR}}}\ \mathrm{PVT}\ \text{and}\ (\text{not}\ \underset{o=1}{\overset{No}{\mathrm{OR}}}\ \mathrm{AVT})\}$$

where d is the depth step, N_d is the total number of depth steps, o is the observation, and N_o the total number of observations.

As the association between PVT (or AVT) itself is the maximal positive association, we must use the Zadeh operator for the <OR> operation for observations and depth steps. On the other hand, relation between PVT and AVT is in maximal negative association, and we must use the Lukasievicz operator for the <AND> opertion in Equation 1 (Siler 1990).

This gives

$$(3)\ \mathrm{FVT} = \{t_r, \bigvee_{d=1}^{Nd} (0 \vee (\bigvee_{o=1}^{No} \mathrm{PVT}\,(t_r) - \bigvee_{o=1}^{No} \mathrm{AVT}\,(t_r))) \mid \forall\ t_r \in \mathrm{TR}\}$$

This is the fundamental algorithm to determine FVT from PVT and AVT.

A

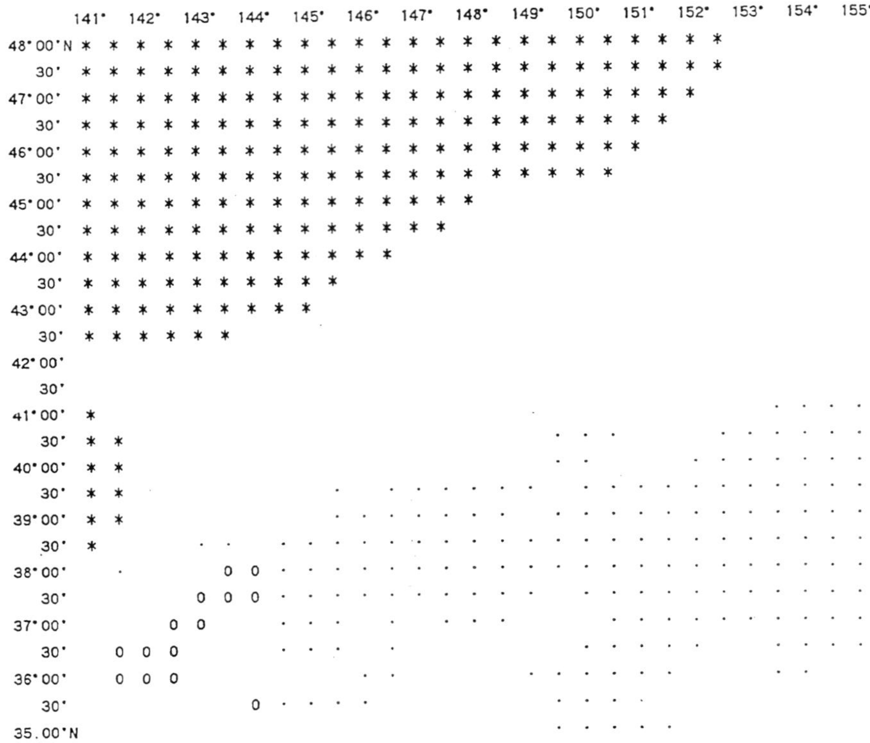

B

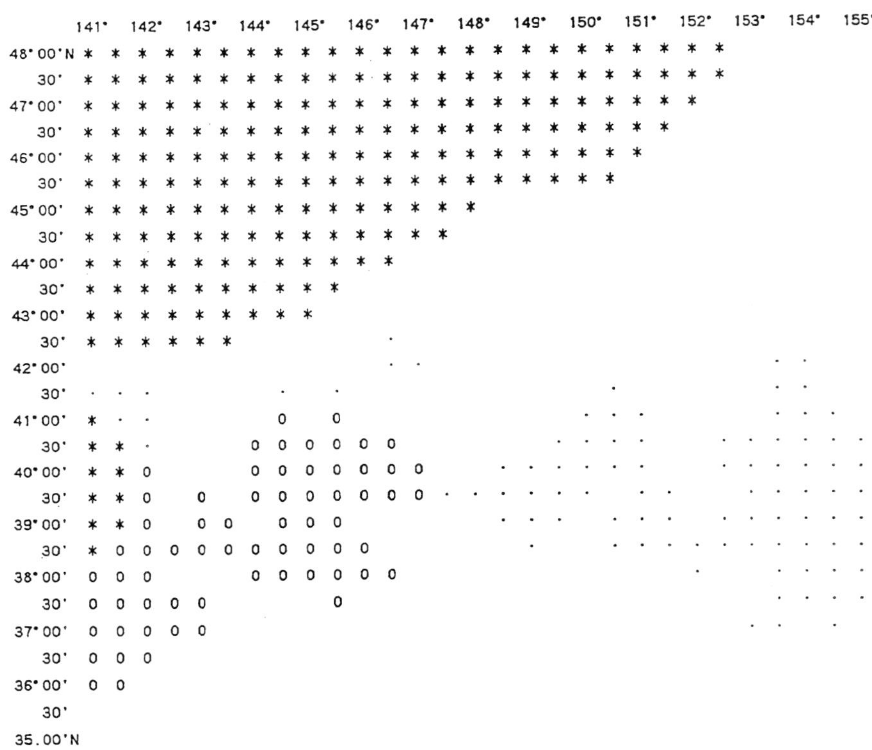

Fig. 5. Results of simulation of temperature-driven migrants in a synoptic SST field. Migrant of temperature preference is 10–16°C and migration speed is 60 mi. in 10 d. (A) 1 mo, (B) 2 mo, (C) 3 mo, and (D) 4 mo after leaving the Inubo-zaki (35°N) on April 1, 1986. o = migrant present; · = condition suitable; x = migrant dormant (no suitable zone in travel range); * = land.

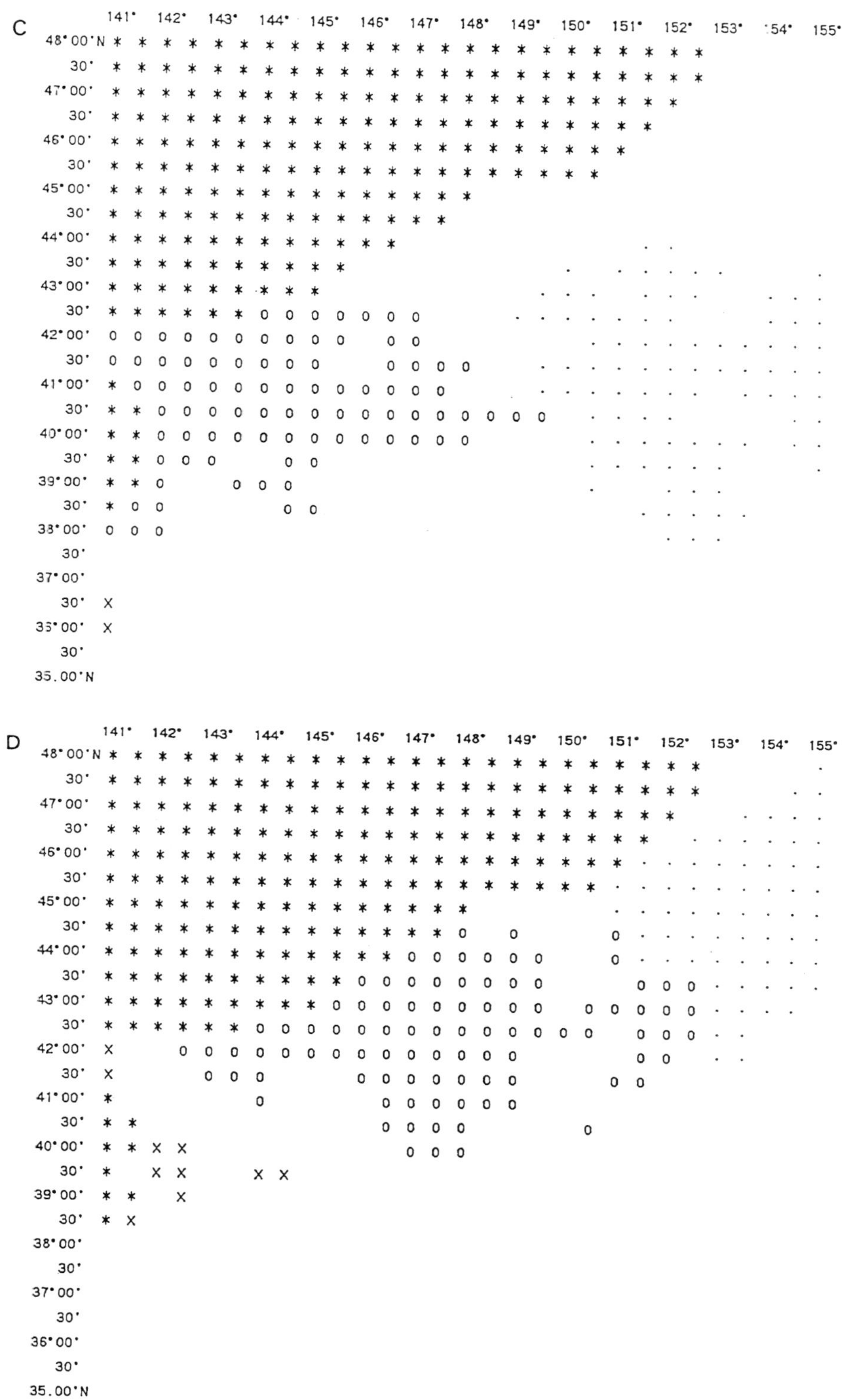

Fig. 5. (*Concluded.*)

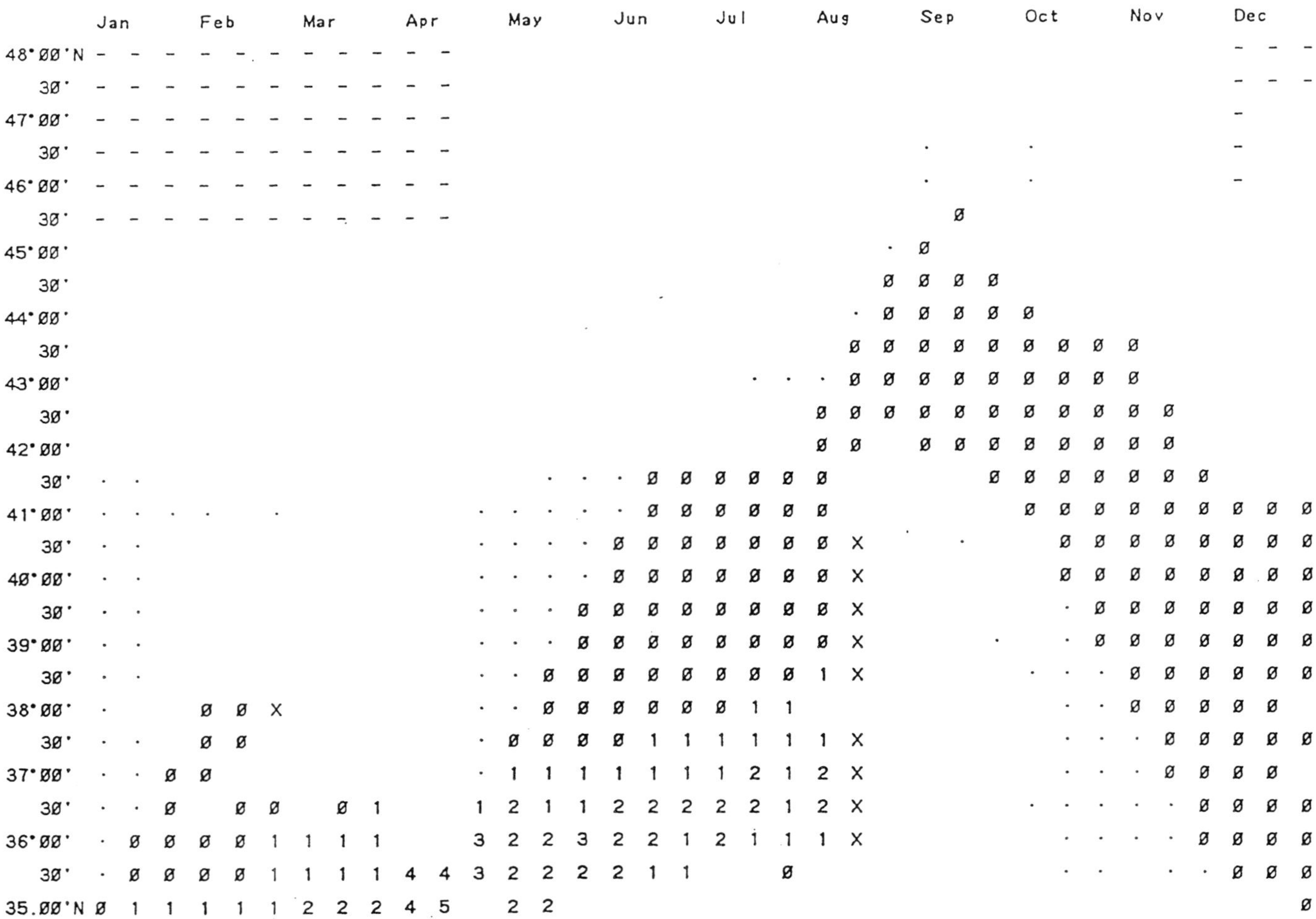

Fig. 6. Results of a one-dimensional simulation of temperature-driven migrants along the coastal route from the Inubo-zaki (35°N) through Sanriku region and the water off the southeast coast of Hokkaido to 48°N. Migrating speed is 60 mi. in 10 d and temperature preference is 10–20°C. Numerals indicate the relative number of migrants expressed in ranks between 0 and 5. · = condition suitable; x = migrant dormant (no suitable zone in travel range); - = no data. SST data are for 1983.

Determination of favourable temperature

FVT estimates using Equation 3, for the water off Iwate and Miyazaki are shown in Figs. 7 and 8.

In the Iwate water (Fig. 7), the FVT temperature range is 11–20°C. In the Miyazaki water (Fig. 8), the uppermost FVT is 20°C, while the upper limit of the temperature composition of the water (WVT) was 30°C. The lowest FVT in the Miyazaki water is limited by the lowest limit of the temperature composition of the water, and the lower part of the FVT cannot be determined from the data for this region.

These results show that the favourable temperature of Japanese sardine does not differ among subpopulations, but is common to the species. Uda and Okamoto (1936) analyzed the data from drift-net fisheries catching Japanese sardine around the Japan sea and reported that the favourable temperature for catching Japanese sardine ranges from 8 to 19°C, with the preferred range being 12–16°C and the optimum temperature being 14°C. Takase (1985) examined the distribution of Japanese sardine and vertical water temperature in the waters off the Boso to Joban regions and reported the results shown in Table 2. The estimate by Equation 3 coincides with the results of Uda and Okamoto (1936) and Takase (1985).

This method to determine the favourable temperature succeeded in indentifying the favourable temperature range, but did not determine the membership function of the favourable temperature. The estimates of favourable temperature (Figs. 7 and 8) show very sharp peaks and depressions. This ambiguity may be caused by incomplete catch data, bias from fishing site preferences of fishing boats, effect of other factors influencing the distribution of Japanese sardine, or insufficient number of samples.

The specific temperature range, peculiar to the water column at the point where Japanese sardine exist, can be identified from the data from purse-seine fishery. This is scientific proof of the existence of favourable temperature of Japanese sardine. The favourable temperature can be a powerful tool when incorporated in models of the distribution dynamics of pelagic fish.

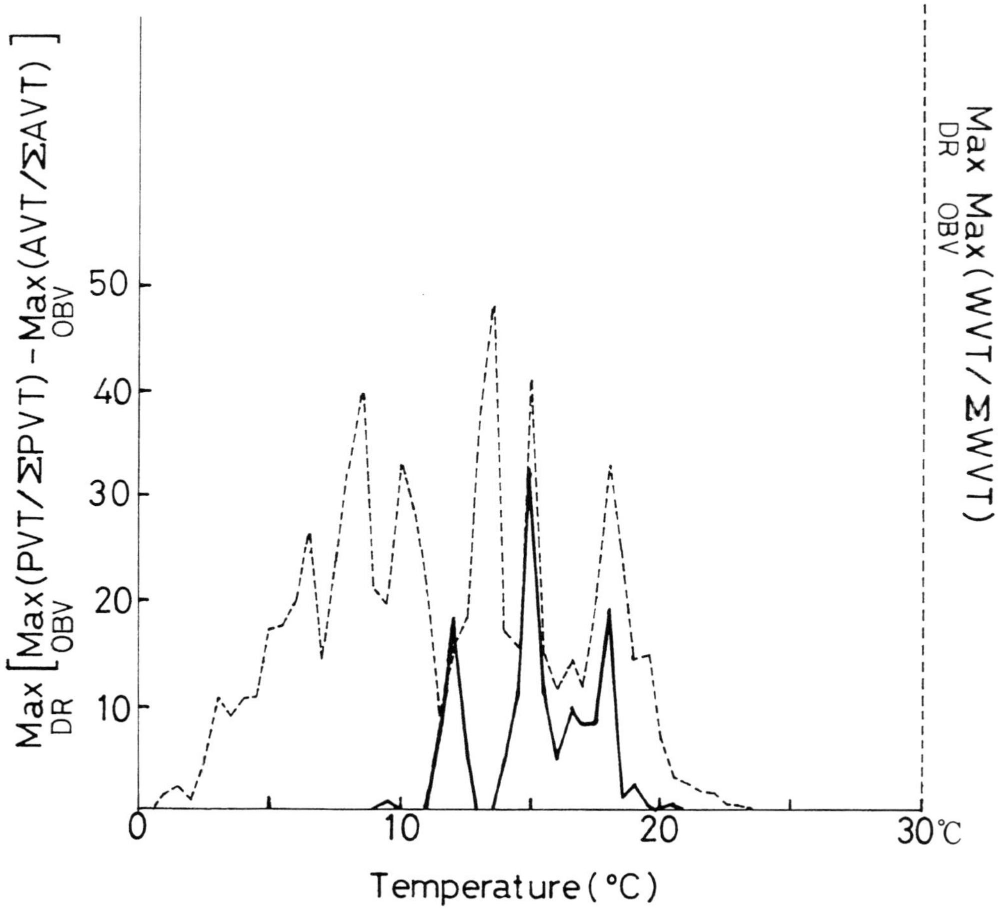

Fig. 7. Estimation from Equation 3 of the favourable temperature for Japanese sardine based on 1988 purse-seine catch and hydrographic data from the water off Iwate. Favourable temperature is shown by solid line, and the temperature composition of the water is shown by broken line.

Table 2. The temperature and depth of the distribution of Japanese sardine in the Bose–Joban retion (revised from Takase 1985). Dates are in 1976.

Date	Phase of life-cycle	Swimming layer (m)	Depth of sea (m)	Water Temperature (°C) Area	Center
Jan. 13	Adult prespawning	0–80	110–300	12.5–14.0	12.5–13.5
Feb. 2–5	Immature wintering	5–70	100–200	9.5–12.5	9.5–10.0
May 10–13	Adult northward migration	5–40	20–130	13.5–17.0	14.5–15.5
Nov. 8–11	Adult southward migration	0–55	60–110	19.5–21.0	20.0–20.5
Dec. 7–10	Immature southward migration	5–35	60–110	14.5–20.0	12.0–17.0

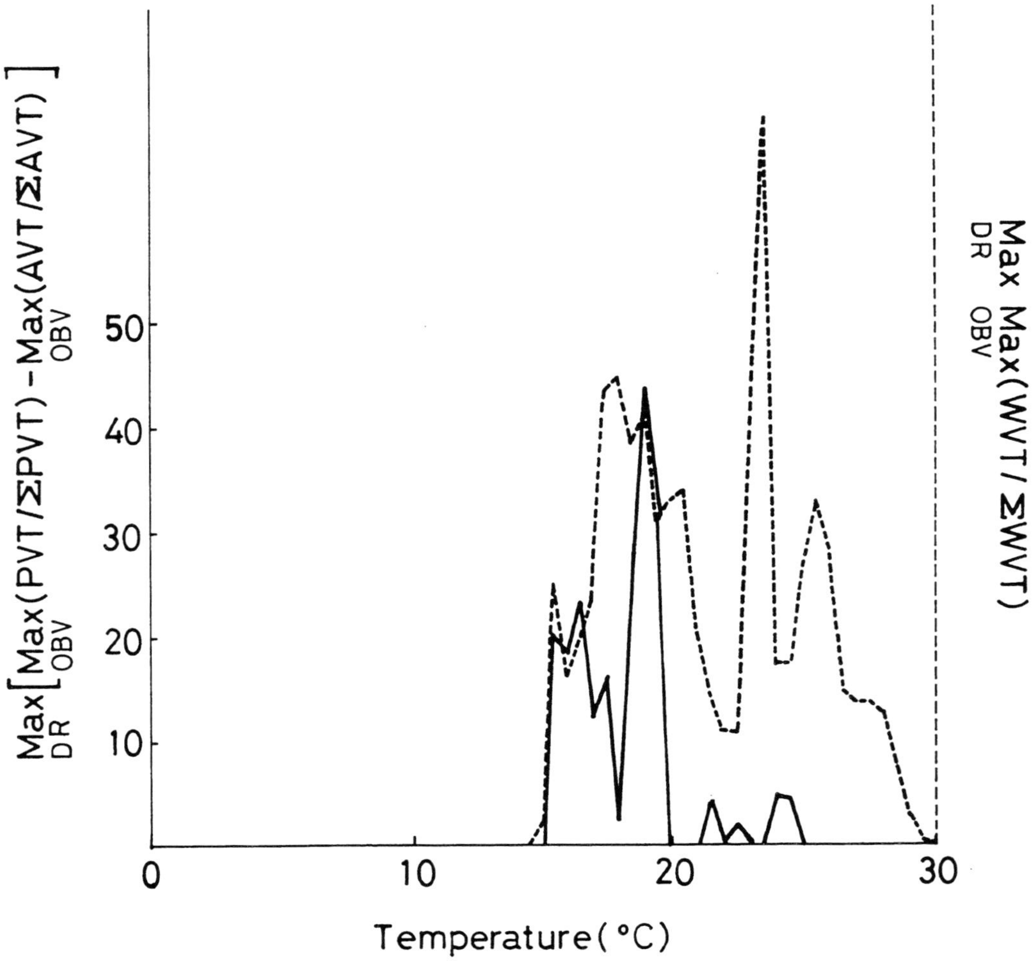

Fig. 8. Estimation from Equation 3 of the favourable temperature for Japanese sardine based on 1988 purse-seine catch and hydrographic data from the water off Miyazaki. Favourable temperature is shown by solid line, and the temperature composition of the water is shown by broken line.

Volume of favourable temperature water

Assuming that favourable temperature is an index of the favourability of the environment, then the volume of water at a favourable temperature can be an index of potential habitat.

In this case, we can expect some correlation between the abundance of fish and the volume of favourable water (FVV) in the potential fishing ground. FVV is given by

$$(4)\ \mathrm{FVV} = \sum_{tr=t1}^{t2} (\mathrm{FVT}\ (t_r) \times \mathrm{WVT}\ (t_r))$$

where FVT (t_r) is the degree of favourability of temperature t_r, WVT (t_r) is the temperature composition of potential fishing ground, and t_1 and t_2 are the lower and upper limit of temperature range.

Membership function of favourable temperature

Uda (1960) proposed a functional expression for the relative favourability of water temperature in an error function. This is the membership function of favourable temperature:

$$(5)\ \mathrm{FVT}\ (t_r) = \exp\left(\frac{-(t_r - t_\theta)^2}{2\sigma^2}\right)$$

According to the results of Uda and Okamoto (1936) $t_\theta = 14.0°\mathrm{C}$ and $\sigma = 0.5°\mathrm{C}$.

Correlation of favourable water with landings of Japanese sardine

Results of FVV calculations for the water off Iwate (Table 3) and Miyazaki (Table 4) show a clear difference between the two sites.

In the water off Iwate, most of the favourable water for Japanese sardine exists in the upper 100 m, while in the water off Miyazaki, the favourable water exists below 100 m. This difference can explain the difference in SST between fishing sites in subarctic waters and those in subtropical coastal waters as shown in Fig. 4, and suggest that SST does not directly indicate the existence of favourable water.

In the water off Iwate, the relative volume of favourable water for Japanese sardine increases in June through December and January, while in the water off Miyazaki, it is highest in the early half of the year. This annual pattern in each region coincides with the annual cycle of Japanese sardine landings from each region, suggesting a relationship between FVV and sardine landings.

Table 3. Volume of favourable water in the water off Iwate for 1988. (in arbitral unit)

Depth (m)	Jan.	Feb.	Mar.	Apr.	May	June	July	Aug.	Sept.	Oct.	Nov.	Dec.
0–25	0.19	0.05	0.00	0.01	0.12	0.54	0.48	0.42	0.22	0.16	0.89	0.88
0–50	0.38	0.09	0.00	0.01	0.19	0.72	0.71	1.02	0.52	0.54	1.74	1.73
0–75	0.54	0.12	0.01	0.01	0.26	0.83	0.90	1.46	0.76	0.71	2.46	2.56
0–100	0.69	0.13	0.01	0.01	0.32	0.93	1.07	1.79	0.92	0.85	3.03	3.24
0–125	0.79	0.15	0.01	0.01	0.34	1.01	1.19	2.06	0.99	0.93	3.43	3.71
0–150	0.86	0.15	0.01	0.01	0.34	1.08	1.28	2.28	1.03	0.95	3.72	4.04
0–175	0.89	0.16	0.01	0.01	0.34	1.12	1.36	2.48	1.03	0.95	3.87	4.20
0–200	0.90	0.16	0.01	0.01	0.34	1.14	1.44	2.63	1.04	0.95	3.98	4.29

Table 4. Volume of favourable water in the water off Miyazaki for 1988. (in arbitral unit)

Depth (m)	Jan.	Feb.	Mar.	Apr.	May	June	July	Aug.	Sept.	Oct.	Nov.	Dec.
0–25	0.03	0.11	0.09	0.04	0.01	0.01	0.00	0.00	0.01	0.00	0.00	0.00
0–50	0.00	0.22	0.20	0.12	0.05	0.02	0.01	0.00	0.01	0.00	0.00	0.01
0–75	0.14	0.37	0.34	0.27	0.18	0.06	0.06	0.04	0.10	0.00	0.00	0.01
0–100	0.25	0.64	0.56	0.52	0.40	0.18	0.23	0.19	0.39	0.00	0.00	0.03
0–125	0.47	1.08	0.87	0.89	0.92	0.37	0.51	0.48	0.80	0.01	0.02	0.10
0–150	0.82	1.79	1.30	1.48	1.59	0.72	0.94	0.92	1.33	0.08	0.12	0.30
0–175	1.37	2.67	1.84	2.32	2.42	1.24	1.58	1.52	1.88	0.31	0.44	0.74
0–200	2.19	3.54	2.48	3.27	3.36	1.93	2.32	2.17	2.52	0.83	1.09	1.46

Table 5. Correlation between volume of favourable water at different depths and landings of Japanese sardine in the waters off Iwate (IWT) and Miyazaki (MYZ) for 1988. ** = Correlation significant at 1% level; * = Correlation significant at 5% level.

Site	Depth (m) 0–25	0–50	0–75	0–100	0–125	0–150	0–175	0–200
IWT	0.26	0.11	0.02	−0.03	−0.07	−0.09	−0.09	−0.09
MYZ	0.54	0.62*	0.77**	0.74**	0.60*	0.53	0.50	0.50

The volume of water at favourable temperatures is much larger off Iwate than off Miyazaki, as is the annual total landing.

The results of correlation analysis between FVV and sardine landings for the water off Iwate and Miyazaki are shown in Table 5. The correlations for the water off Iwate are not significant and indicate a weak relationship only in the surface layer. This may be related to the fact that the water off Iwate is part of a continuous good fishing ground extending from the Inubozaki (35°N) to the northern end of Honshu Island (41°30'N). The results for the water off Miyazaki show significant high correlations at the 0–75 m and 0–100 m depth ranges, which suggest that the sardine sink to at least 75 m when the surface layer consists of unfavourable warm water.

Further use of favourable temperature as an index of habitat suitability must be in the model of migration dynamics in varying temperature fields.

Conclusions

1) The favourable temperature can be estimated from purse-seine catch data and hydrographic data.
2) The volume of favourable water can be a useful index of habitat size.

Acknowledgements

The author would like to thank Dr. D.M. Ware and Dr. T. Wada for their useful discussion and encouragement, and the staff of Japan Fisheries Information Service Center for providing the SST data set.

References

Hiramoto, K. 1991. The sardine fishery and ecology in the Joban and Boso waters of central Japan, p. 117–128. *In* T. Kawasaki, S. Tanaka, Y. Taba, and A. Taniguchi. [ed.] Long-term variability of Pelagic Fish Populations and their Environment. Proceedings of the International Symposium, Sendai, Japan, Nov. 14–18, 1989 Pergamon Press, Oxford, UK.

Iwate Prefecture Fishereries Experimental Station. 1990. Fishing and oceanographic information 1988, p. 30–41. (In Japanese)

Japan Fisheries Information Center. 1988. Annual report of the quick reports of fisheries information. No. 976-1037. (In Japanese)

Miyazaki Prefecture Fisheries Experimental Station. 1990. Results on the forecast of fishing and oceanographic condition in 1988, p. 39–86. (In Japanese)

Siler, W.M. 1990. Theory and practice of fuzzy expert system. (Trans. by L. Cui and K. Hirota, Denki-Shoin, Tokyo, Japan.) 198 p. (In Japanese)

Takase, H. 1985. Study on the distribution, especially schooling behavior of Japanese sardine. III. Bull. Fish. Exp. Stn. Ibaraki-ken 25: 135–145. (In Japanese)

Uda, M., and G. Okamoto. 1936. Relation between catch of Japanese sardine and oceanographic condition based on historical data (Survey of Communication Test). Rep. Fish. Exp. Stn. 7: 19–46. (In Japanese)

Uda, M. 1960. Fishings ground of the ocean. Fish. Sci. Ser. 16. Kohseisha-Kohseikaku, Tokyo, Japan. 36 p. (In Japanese)

Climate-dependent salmon and sardine stock fluctuations in the North Pacific

L. Klyashtorin and B. Smirnov

Klyashtorin, L., and B. Smirnov. 1995. Climate-dependent salmon and sardine stock fluctuations in the North Pacific, p. 687–689. *In* R.J. Beamish [ed.] Climate change and northern fish populations. Can. Spec. Publ. Fish. Aquat. Sci. 121.

Abstract: Analysis of all available statistical data on Pacific salmon commercial fisheries shows that the historic maximum harvest was reached in the late 1930s (about 1 million tons). There are three major periods in the commercial salmon catches: the maximum occurred in the 1930s, a minimum occurred in the 1960s and 1970s, and a recent period of salmon stock increase (middle 1970s to 1980s). Long-term changes in salmon catches corresponded well with those of mean surface air temperature in the northern hemisphere. The correlation coefficient between these two series is about 0.76. Far Eastern and Californian sardine catch fluctuations were in phase with both Pacific salmon catches and surface air temperature variations. The peaks in Far Eastern sardine populations practically coincide with those of salmon, but essentially were more pronounced. Apparently, the recent period of salmon and sardine population abundance is near its maximum. The unfavourable "cold" climatic period is now under way and there is evidence of an abrupt drop in sardine population abundance. The salmon stock changes are slower. During the next 5 yr, salmon catches will be kept on a rather stable level but in the first decade of the next century we can expect a decrease in salmon catches by 10–20% of recent levels.

Résumé : L'analyse de toutes les données statistiques disponibles sur les pêches commerciales de saumon du Pacifique montre que la récolte maximale historique a été atteinte à la fin des années 1930 (environ 1 million de t). On distingue trois grandes périodes de prises dans la pêche commerciale au saumon : la plus importante est survenue dans les années 1930, la moins importante dans les années 1960 et 1970, et on a assisté récemment à une période d'augmentation des stocks de saumons (du milieu des années 1970 à celui des années 1980). Les changements à long terme des prises de saumon correspondaient bien à ceux de la température moyenne de l'air à la surface dans l'hémisphère nord. Le coefficient de corrélation entre les deux séries est d'environ 0,76. Les fluctuations dans les prises de sardines de l'Extrême-Orient et de la Californie étaient en phase avec les prises de saumon du Pacifique et avec les variations de la température de l'air à la surface. Les pointes dans les populations de sardines de l'Extrême-Orient coïncident pratiquement avec celles du saumon, mais sont essentiellement plus prononcées. Il semble que la période récente d'abondance du saumon et de la sardine se rapproche de son niveau maximum. La période climatique «froide» défavorable est actuellement en cours et on note la présence de signes d'une baisse radicale de l'abondance de la population de sardines. Les changements dans les stocks de saumons sont plus lents. Au cours des cinq prochaines années, les prises de saumons resteront à un niveau plutôt stable mais au cours de la première décennie du siècle prochain, nous pouvons nous attendre à une diminution de 10 à 20 % des prises de saumons par rapport aux niveaux récents.

Introduction and methods

Very few of us need to be told that the salmon populations in the North Pacific Ocean are important natural resources. In this paper we attempt to estimate the climate impact on the abundance of Pacific salmon stocks. Unfortunately, the available data on commercial salmon catches contain missing values and in some cases data are unreliable. We carried out the analysis of all available statistical data since 1920 (Kasarnovsky 1987, 1989*a*, *b*, *c*; Janovskaja 1989).

L. Klyashtorin and **B. Smirnov.** Russian Institute for Fisheries and Oceanography (VNIRO), 107140 Moscow, Russia.

As characteristics of climatic changes during the period of 1890–1990, we used annual surface air temperature anomalies averaged over the northern hemisphere, which were smoothed by 5-yr running averaging from Krovnin and Rodionov (1992). These data are presented in Fig. 1, together with 5-yr running means of total North Pacific salmon catches, for the period of 1920–90. Figure 1 also shows data on the Far Eastern and Californian sardine catches (Kawasaki 1992).

Results and discussion

These data show that there are three rather large periods of changes of North Pacific salmon catches. The first period is characterized by the rise of catches in the late 1930s, when

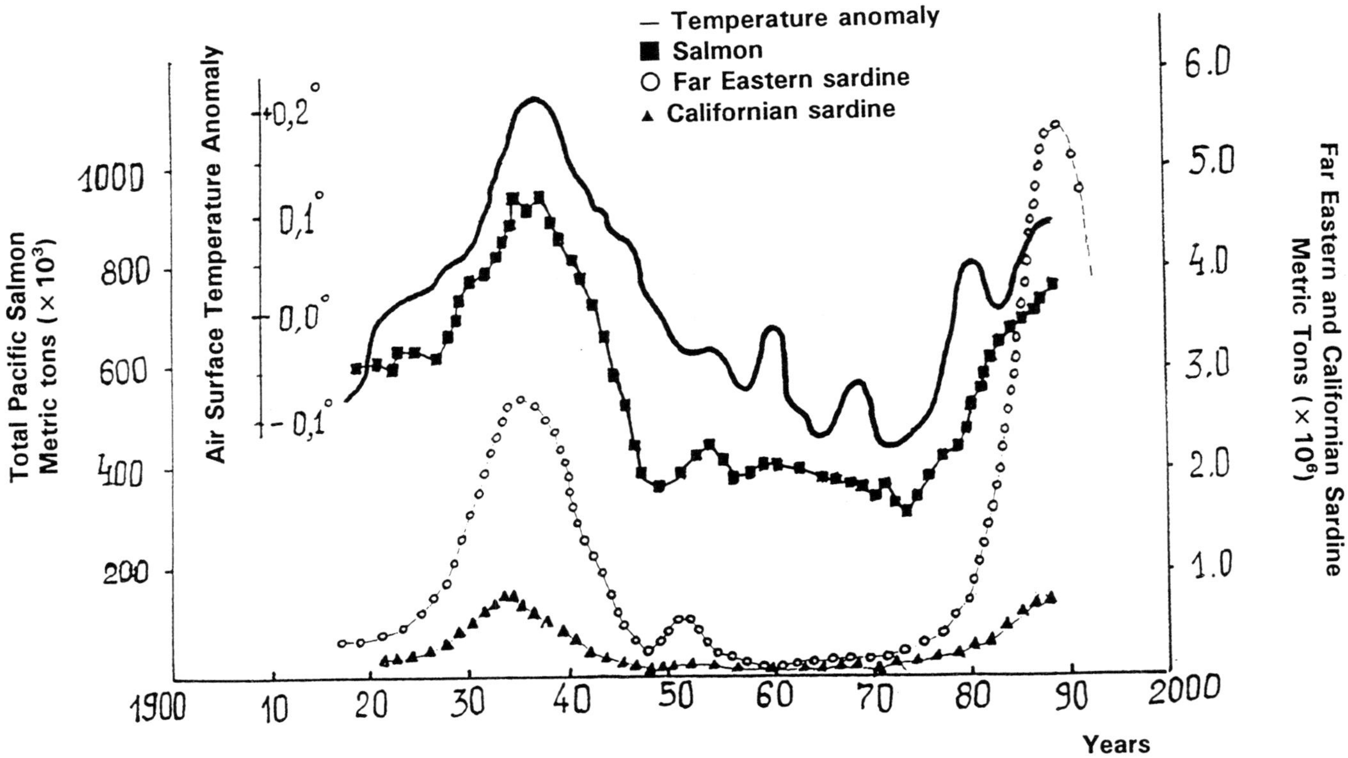

Fig. 1. Trends in catch of Pacific salmon, Far Eastern sardine, and California sardine and anomaly of the annual surface air temperature for the Northern Hemisphere.

commercial salmon harvests were at the maximum historic level of about 1 million tons. A period of salmon stock depression was observed in the 1960s and 1970s, when total Pacific salmon harvest dropped to approximately 0.4 million tons. The recent period of Pacific salmon population rise started in the middle of the 1970s. Especially fast increases in catches were observed in the middle 1980s. Recently, the rate of catch increase became essentially slower and has a tendency to stabilize at the mean level of about 0.75 million tons (maximum commercial harvest of 0.95 million tons was reached in 1991).

Long-term changes in surface air temperatures corresponded well to changes in salmon catches. The correlation coefficient between these two series is about 0.76. It should be noted also that fluctuations in Far Eastern and Californian sardine, during the period under consideration, were in phase with variation in Pacific salmon catches and surface air temperatures in the Northern Hemisphere. Both sardine and salmon catches started to rise in the 1920s, reached their maximum in the late 1930s, then dropped to the low level in the 1960s and 1970s, and, since the middle of the 1970s, started to rise again. The range of sardine catch variations drastically exceeds that of salmon. For example, the maximum total Pacific salmon commercial harvest exceeds the minimum harvest by a factor of 2–3 times (for 1.2 and 0.4 million tons, respectively), while for the Far Eastern sardine, this ratio is 50–100 (3–5 and 0.05 million tons, respectively). It can be assumed that apparent simultaneous changes in the abundance of sardines and Pacific salmon were induced by the identical natural mechanisms associated with long-term climate cycles. Unfortunately, these mechanisms are unknown. It is possible that the climate warming favours the extension of feeding area, thus resulting in the expansion of populations outside their usual area. For example, when the Far Eastern sardine stock was at a low level, sardine were confined to a small coastal area along southern Japan. When the sardine stock was abundant, their feeding area was broadened by a factor of 100–200 from the Sea of Japan to the east into the Pacific as far as 173°W.

Is it possible to forecast the salmon and sardine stock abundance? Kaeriyama (1989) presented the Japanese data on the chum salmon catches in Hokkaido since the 1870s. The catches increased until the middle of the 1880s, when they reached about 30 000 tons. Since that time, chum catch steadily decreased to a low level in the early 1900s. These data can be considered as some evidence for existence of periodicity in salmon stock fluctuations. It can be assumed that the Pacific salmon abundance maxima could be repeated every 50–60 years and observed at the end of the 1880s, 1930s, and 1980s (Klyashtorin and Smirnov 1992). Apparently, the recent period of salmon population abundance is near its maximum now. The peaks in Pacific sardine populations practically coincide with those in salmon populations but are more pronounced. According to recently published data, the total Far Eastern sardine harvest was 5.3 million tons in 1987, 5.4 million tons in 1988, 5.1 million tons in 1989, and 4.7 million tons in 1990 (FAO 1991). The increase in abundance is near its completion and the tendency toward an abrupt drop of the catch may have started. This is confirmed by sardine fishery experts. The leading Russian sardine fishery specialists (Belyaev et al. 1991) indicate the possibility of a drastic drop in Far Eastern sardine commercial stock in the near future. Recent research, carried out in the Far Eastern sardine fishing area, revealed the evidences of the beginning of a decrease of water temperature in this region (Rogachev et al. 1991). As salmon and sardine populations show the strong linear relationship with temperature anomalies, this may be considered as evidence for the possibility that an unfavourable climatic period is under way, and as a consequence, an abrupt drop in sardine population abundance may be observed (Kawasaki 1992). Unlike the Pacific sardine, Pacific salmon stocks change more slowly. It is possible that during the next 5–7 years salmon commercial harvest will remain on a rather stable level, but in the first decade of the next century we can expect (by analogy with the preceding maximum of the 1930s) a gradual decrease of salmon catch by 10–20% from recent levels.

References

Belyaev, V., Yu. Novikov, and V. Svirsky. 1991. Far Eastern sardine stocks and changes in ichthyocenosis of Northwest Pacific fishery economy. Rybn. Khoz. 9: 24–28. (In Russian)

FAO. 1991. Fishery statistics: commodities, 1989. FAO Yearbook Fish. Stat. 69: 397 p.

Janovskaya, N. [ed.] 1989. Commercial catches of Pacific salmon, 1900–1986, VNIRO, Moscow. 210 p. (In Russian)

Kasarnovsky, M. 1987. Japanese salmon catches in Northwest Pacific. VNIRO, Moscow. 72 p. (In Russian)

Kasarnovsky, M. 1989*a*. Soviet salmon catches in Northwest Pacific. VNIRO, Moscow. 44 p. (In Russian)

Kasarnovsky, M. 1989*b*. USA salmon catches in Northeast Pacific (1922–1984). VNIRO, Moscow. 63 p. (In Russian)

Kasarnovsky, M. 1989c. Canadian salmon catches in North Pacific (1951–1984). VNIRO, Moscow. 21 p. (In Russian)

Kaeriyama, M. 1989. Aspects of salmon ranching in Japan. Physiol. Ecol. Jpn. 1: 625–638.

Kawasaki, T. 1992. Climate-dependent fluctuations in the Far Eastern sardine population and their impacts on fisheries and society, p. 325–355. *In* M.H. Glantz [ed.] Climate variability, climate change, and fisheries. Cambridge University Press, Cambridge, UK.

Klyashtorin, L., and B. Smirnov. 1992. North Pacific salmon stocks dynamics and some aspects of salmon ranching. EQINAS, Russian Federation Committee of Fisheries, Moscow 2: Rev. Aquacult. ser. 36 p. (In Russian)

Krovnin, A., and S. Rodionov. 1992. Atlanto-Scandian herring: a case study, p. 232–260. *In* M.H. Glantz [ed.] Climate variability, climate change, and fisheries. Cambridge University Press, Cambridge, UK.

Rogachev, K.A. 1991. Climatic salinity changes in the Oyashio region. Meteorologiya i Gidrologiya, Moscow, Russia. (Translated from Russian by Meteorology and Hydrology, Washington, D.C. 1991(6)).

Larval fish growth in the Barents Sea

Harald Loeng, Herman Bjørke, and Geir Ottersen

Loeng, H., H. Bjørke, and G. Ottersen. 1995. Larval fish growth in the Barents Sea, p. 691–698. *In* R.J. Beamish [ed.] Climate change and northern fish populations. Can. Spec. Publ. Fish. Aquat. Sci. 121.

Abstract: Since 1965, the International 0-group fish survey has been carried out in the Barents Sea during late August and early September. Since 1977, an additional survey on early juveniles off northern Norway has been carried out, mainly in July. Based on data from these two yearly surveys, the growth of fish larvae of commercially important fish is estimated. The results showed a positive covariation in the annual length of larvae of cod, haddock, and herring. This is not surprising as these fish stocks reproduce in the same type of water masses with the same temperature variability and similar food conditions. There is a high correlation between larval length measured in August and the average daily increment during the preceding 1–2 mos. The results also unveiled a relatively close relation between temperature conditions and the length of larvae of cod, haddock, herring, and capelin.

Résumé : Depuis 1965, à la fin d'août et au début de septembre, se poursuit dans la mer de Barents l'étude sur le poisson du «International O-Group». Depuis 1977, une autre étude est en cours, principalement en juillet, sur les juvéniles précoces au large du nord de la Norvège. On se base sur les données de ces deux études annuelles pour estimer la croissance des larves d'espèces de poissons commercialement importantes. Les résultats ont montré une covariation positive dans la longueur annuelle des larves de morue, d'aiglefin et de hareng. Cela n'est pas surprenant puisque ces stocks se reproduisent dans le même type de masses d'eau, avec la même variabilité de température élevée et dans des conditions alimentaires similaires. On note la présence d'une corrélation élevée entre la longueur des larves mesurée en août et l'augmentation quotidienne moyenne durant le premier ou les deux mois précédents. Les résultats montraient également une relation relativement étroite entre les conditions de température et la longueur des larves de morue, d'aiglefin, de hareng et de capelan.

Introduction

During the late 1970s, temperature decreased in the Barents Sea. This led to investigations on the influence of temperature on fish population parameters such as recruitment, distribution, migration, and growth. The investigations focused especially on the temperature effect on the commercial important species: Arcto-Norwegian cod (*Gadus morhua*), haddock (*Melanogrammus aeglefinus*), herring (*Clupea harengus*), and capelin (*Mallotus villosus*). Some of the results were reviewed by Dragesund and Gjøsæter (1988) and Loeng (1989).

The present paper focuses on growth of fish larvae in the Barents Sea during the first half year of their life. Based on some preliminary results on growth of 0-group fish in relation to temperature (Loeng and Gjøsæter 1990) and growth of adult cod and capelin (Nakken and Raknes 1987; Gjøsæter and Loeng 1987), we would expect a faster larval growth in warm years than in cold ones. Consequently, the largest larvae should occur in warm years. If so, the results should give a close temperature–growth relationship.

Because the climatic variations are large scale, one would expect that fish stocks that reproduce in the same ecosystem would be similarly affected. Dragesund (1971) and Sætersdal and Loeng (1987) found several years of common high recruitment in cod, haddock, and herring, which gave a convincing demonstration of a close temporal relationship for larval survival. The favourable physical conditions seem to be related to increased heat transport to the Barents Sea by the warm Atlantic current (Mukhina et al. 1987; Sætersdal and Loeng 1987). We may therefore assume covarition in the interannual length variability of fish larvae.

Material and methods

Each year since 1965 the international 0-group survey has been carried out in the Barents Sea and adjacent waters. These joint Russian–Norwegian (up to 1976 also British) investigations have taken place in late August and early September, with only a few days variation from year to year (Table 1). The geographical distribution of 0-group fish is found by fishing regularly with a small-mesh midwater trawl. Until 1980, the fishing depth of the trawl was selected based on acoustic observations. From 1980, a standard trawling procedure recommended by ICES has been used (Anonymous 1983). The catches are sorted, the number of each species are counted and the length of 100 specimens are measured. The mean length is calculated as the arithmetic mean. All data from the 0-group surveys used in this paper are taken from the annual survey reports (e.g., Anonymous 1991). As seen from Fig. 2, the length of the different time series vary.

H. Loeng, H. Bjørke, and **G. Ottersen.** Institute of Marine Research, P.O. Box 1870 Nordnes, N-5024 Bergen, Norway.

Table 1. Middle date of 0-group survey, number of days between the juvenile and 0-group survey, daily increment of cod, haddock, and herring, and length of larvae normalized to 31 August each year.

Year	Middle day 0-group survey	Days between surveys	Daily growth increment ($mm \cdot d^{-1}$)			Length of larvae (mm) normalized to 31 August		
			Cod	Haddock	Herring	Cod	Haddock	Herring
1977	01.09	26	0.58	1.24		60.82	75.26	
1978	01.09	53	0.58	0.72	0.33	61.22	61.38	47.57
1979	02.09	64	0.66	1.05	0.41	63.18	82.70	54.68
1980	27.08	56	0.67	0.83	0.42	60.88	72.32	55.08
1981	27.08	44	0.81	0.91	0.62	63.84	69.94	61.52
1982	29.08	42	0.81	1.18	0.71	66.62	77.66	61.02
1983	29.08	58	0.74	1.12	0.73	79.48	90.94	84.06
1984	26.08	50	0.80	0.90	0.83	78.60	90.30	87.75
1985	27.08	54	0.95	1.17	0.97	77.20	90.28	93.58
1986	26.08	50	0.73	0.88	0.38	64.95	78.00	57.50
1987	25.08	44	0.66	0.78	0.31	59.16	63.18	48.46
1988	29.08	50	0.64	0.68	0.69	66.08	77.76	68.08
1989	31.08	44	1.01	1.28	0.72	78.00	87.00	79.00
1990	27.08	35	1.08	1.74	0.95	88.32	105.96	92.08
1991	27.08	49	0.82	1.10	0.70	83.28	88.40	81.80

Early juveniles (in this paper defined as fish 2–3 mo after hatching) have been surveyed each year during the period 1977–91. The surveys have been carried out in June and July by research vessels from the Institute of Marine Research, and the sampling period has been more variable than for the 0-group survey (Table 1). The trawl used for catching larvae is identical to the 0-group trawl. All the fish larvae are identified, and a subsample of 50 larvae is taken to measure the length. The methods used during the surveys are summarized by Bjørke and Sundby (1987) and Sundby et al. (1989). In this paper, only larvae sampled north of 68°N are considered.

To estimate growth of larvae, we calculated the difference in mean length measured during the juvenile fish survey and at the 0-group survey. We divided the difference in length by the number of days between the two surveys, for an estimate of the average daily growth rate. The juvenile fish are measured in standard length and the 0-group fish in total length. The number of days between cruises is defined as the difference between the middle day of the early juvenile and of the 0-group surveys. Assuming a linear growth, we have also normalized the length of the 0-group larvae to one certain day. For that purpose, we have choosen the 31 August. Owing to a poor coverage of the capelin larvae distribution area during the juvenile survey, the growth values of capelin larvae are considered to be uncertain.

During both surveys hydrographical observations are made along the survey tracks, either by CTD or Nansen bottles. To address the relation between length and temperature, it is necessary to determine the temperature corresponding to ambient conditions for the first half year of larval development. As the larvae inhabit the upper 60 m of the water column, we have calculated the mean temperature between 0 and 60 m, using temperature observations from late June and from the 0-group survey. In addition, mean temperatures for the depth layer 0–200 m from the fixed sections Fugløya – Bear Island, North Cape – Bear Island and the Kola–section (along 33°30′E) have been used (Fig. 1). The temperature values from the two last sections, are taken from the annual 0-group reports, e.g., Anonymous (1991).

The correlation between different parameters is estimated by calculating the sample Pearson product–moment correlation (r). We also worked out the significance probability (p) of the correlation under the null hypothesis that the statistic is zero. We have not taken possible autocorrelation or nonstationarity of the series into consideration during the calculations, but are aware of the bias this may cause. The calculations are done by means of the SAS package (SAS Institute 1988).

Results

Figure 1 shows the spawning areas for all species. Table 2 lists the main spawning periods. Cod and haddock are pelagic spawners, while herring and capelin have demersal eggs. The main spawning ground for cod is in the Lofoten area. Haddock spawns at different localities along the Norwegian continental shelf. The most important spawning area is found between 71 and 73°N, but there also seems to be spawning localities between 65 and 68°N. Norwegian spring-spawning herring spawn over a large area along the coast of western Norway. The capelin spawn along the coast of northern Norway and the western part of the Russian coast. The larvae of all four species drift with the Norwegian Coastal Current and the Atlantic Current to their nursery areas in the Barents Sea (Fig. 1). Capelin spawn in colder water than the other species and may, therefore, experience different environmental conditions.

The mean length distribution of fish larvae observed during the 0-group survey for the period 1965–92 is shown in Fig. 2. There is positive covariation of length among cod, haddock,

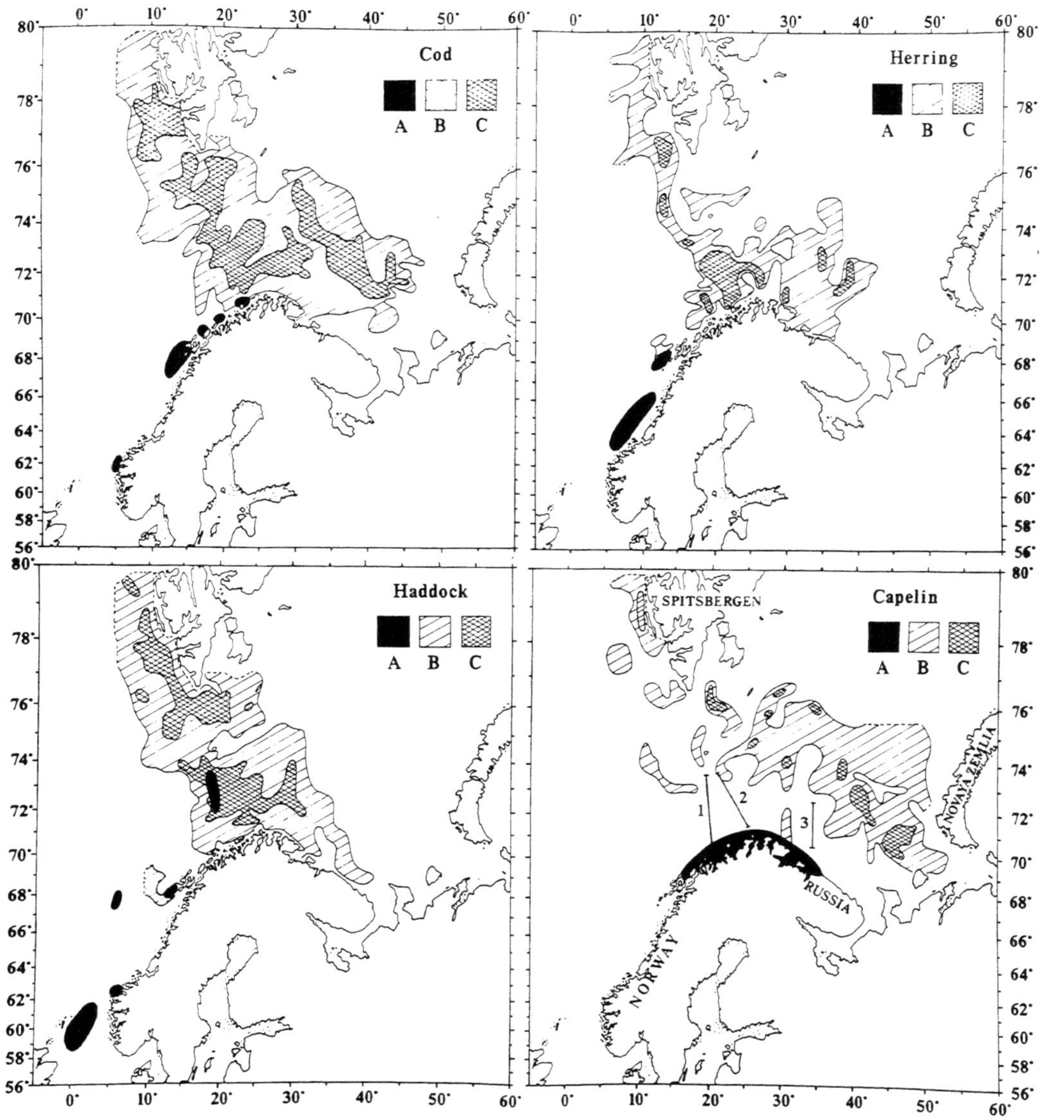

Fig. 1. Spawning grounds of cod, haddock, herring, and capelin (A). To illustrate a typical distribution of larvae during the 0-group survey, results from 1991 are presented (Anonymous 1991). The hatched areas (B), show low concentrations of larvae, while dense concentrations are double hatched (C). The positions of the hydrographic sections Fugløya – Bear Island (1), North Cape – Bear Island (2) and the Kola–section (3) are indicated in the capelin map.

Table 2. Approximate spawning time for the different species.

Species	Spawning period	Peak spawning
Cod	March–April	01 April – 10 April
Haddock	April–May	20 April – 10 May
Herring	February–March	01–15 March
Capelin	February–April	15 March – 15 April

and herring during most of the period. The capelin line, however, detracts from the trend of the other three species. The mean length of larvae was at a maximum in the early 1970s, after which it decreased to a minimum in the late 1970s. A new maximum was exhibited by all species during the period 1983–85; all species had a new minimum in 1987. From 1970 to 1976, the Barents Sea was warmer than average, while the second half of the decade was characterized by low temperatures. Throughout 1982 there was a strong temperature increase, which was followed by a couple of warm years. From early spring 1985 to the beginning of 1989 the temperature was below average. Since then the temperature has been higher than the long-term average (Loeng 1991). To look closer at temporal covariation of mean length, we have plotted the mean length of cod versus haddock, herring, and capelin (Fig. 3). Table 3 gives the correlation matrix between lengths of the different species at the 0-group stage. These results support the view that capelin growth is not quite synchronic with the other species among years, even though the relation between herring and capelin is rather good.

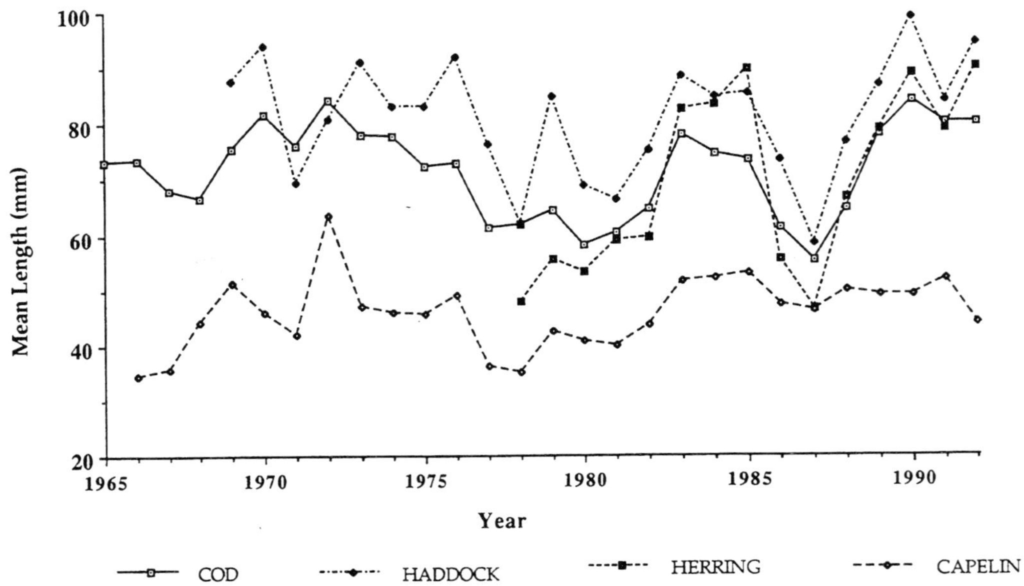

Fig. 2. Length distribution of 0-group fish in August–September during the period 1965–92.

The daily growth rate of larvae of cod, haddock, and herring are given in Table 1. The table shows a difference of 0.5 mm between minimum and maximum daily growth rate of cod. For haddock the difference is more than 1 mm, while herring has a difference of approximately 0.7 mm between minimum and maximum average daily increment. The interannual variability is quite similar. All species experienced low growth in 1978 and 1987, while growth was high in 1985 and 1990. If we compare growth rate of cod with that of haddock and herring, we find r = 0.73 and 0.78 (p = 0.001, n = 15), respectively. Assuming a linear growth rate, the daily increment is used to calculate the length of larvae at the last day of August. Table 1 suggests a relation between length observed in August and the average growth rate of the different species between the two surveys. The calculated correlation coeffecients given in Table 4 indicate a rather close relation between length measured during the 0-group survey and the daily growth during the last 1–2 mo, especially for herring.

Figure 4 shows the length and growth rate of 0-group fish plotted against the mean temperature from 0 to 200 m in the section North Cape – Bear Island. For all species, both length and growth rate of larvae increase with increasing temperature. Table 5 shows a significant relation between temperature and length of all species, while the relation between temperature and growth rate is significant only for cod.

Discussion

Larvae of cod, herring, and haddock experience similar environmental conditions during the first 6 mo of their life (Fig. 1). The relatively high covariation between the length of cod, haddock, and herring (Figs, 2 and 3, Table 3) indicate that these species are similary affected by the environment, as also shown by Dragesund (1971) and Sætersdal and Loeng (1987). The weaker relation between capelin and the other species is probably due to the fact that the capelin larvae partly grow up

Table 3. Correlation between length of different species of 0-group fish. r = correlation coeffecient; p = significance probability; n = number of years.

		Haddock	Herring	Capelin
Cod	r	0.80	0.93	0.53
	p	0.000	0.000	0.004
	n	24	15	27
Haddock	r		0.88	0.45
	p		0.000	0.029
	n		15	24
Herring	r			0.70
	p			0.004
	n			15

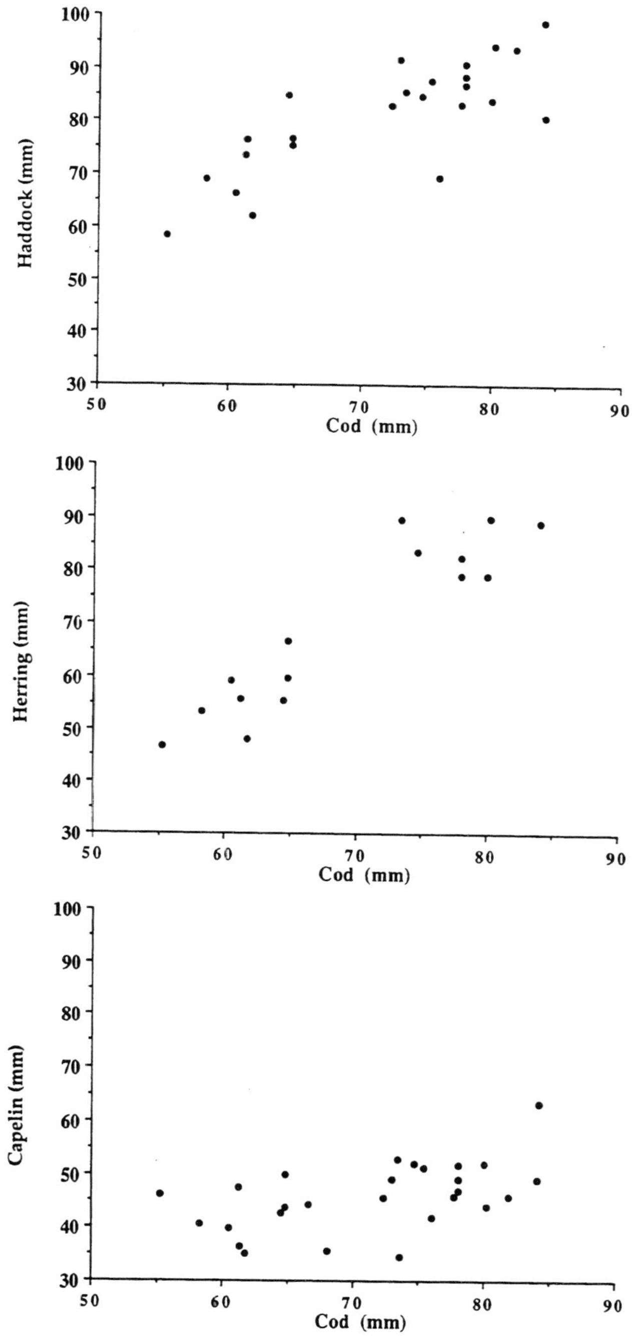

Fig. 3. Mean length of cod versus mean length of haddock, herring, and capelin. The years included may be seen from Figure 2.

Table 4. Relation between length of 0-group larvae in August and the growth of larvae the precceding 1–2 mo (see Table 3). r = correlation coeffecient; *p* = significance probability; *n* = number of years.

	Cod	Haddock	Herring
r	0.80	0.76	0.93
p	0.000	0.001	0.000
n	15	15	14

in a different area where the water masses have different characteristics (Loeng 1991). Another factor is that capelin has a much more variable spawning time than, for instance, cod (Sætre and Gjøsæter 1975). In some years, parts of the capelin stock also spawn during summer. In these years, the earliest hatched larvae from the summer spawning may be caught during the 0-group survey and included in the calculated mean length, which then will decrease.

There is a tendency for all species that both growth and length of larvae increase with increasing temperature (Fig. 4). The rather low correlation coeffecients (Table 5) may be explained in different ways. To address the relation between length and temperature, it is necessary to determine the temperature corresponding to ambient conditons for the first half year of larval development. As the larvae inhabit the upper 60 m of the water column, we first compared the growth with the mean temperature between 0 and 60 m, using temperature observations from late June and the 0-group survey. The results became approximately the same as using the 0 – 200 m mean for August from the North Cape – Bear Island section for the period 1977–91. We, therefore, chose the latter temperature values, because we then got longer time series for comparing length and temperature. Even the temperature from the North Cape section is not the most repesentative of what the larvae has experienced, the temperature shows the year-to-year variability in the environment very well.

One specific year contributes significantly to the low relationship observed between temperature and fish growth. In the cold year 1985, all species showed high average growth rate prior to the 0-group survey (Table 1). If the year 1985 is deleted, *r* is increased by more than 0.2 for haddock and herring and by 0.1 for cod compared with the values in Table 5. The high growth rate in the cold year 1985 is difficult to explain. At the spawning area, the 1985 year-class had lower

Table 5. Relation between temperature in the North Cape — Bear Island section and length and growth of larvae. *r* = correlation coeffecient; *p* = significance probability; *n* = number of years.

	Length				Daily growth increment		
	Cod	Haddock	Herring	Capelin	Cod	Haddock	Herring
r	0.70	0.59	0.60	0.56	0.66	0.45	0.45
p	0.000	0.002	0.019	0.003	0.007	0.091	0.107
n	28	24	15	27	15	15	14

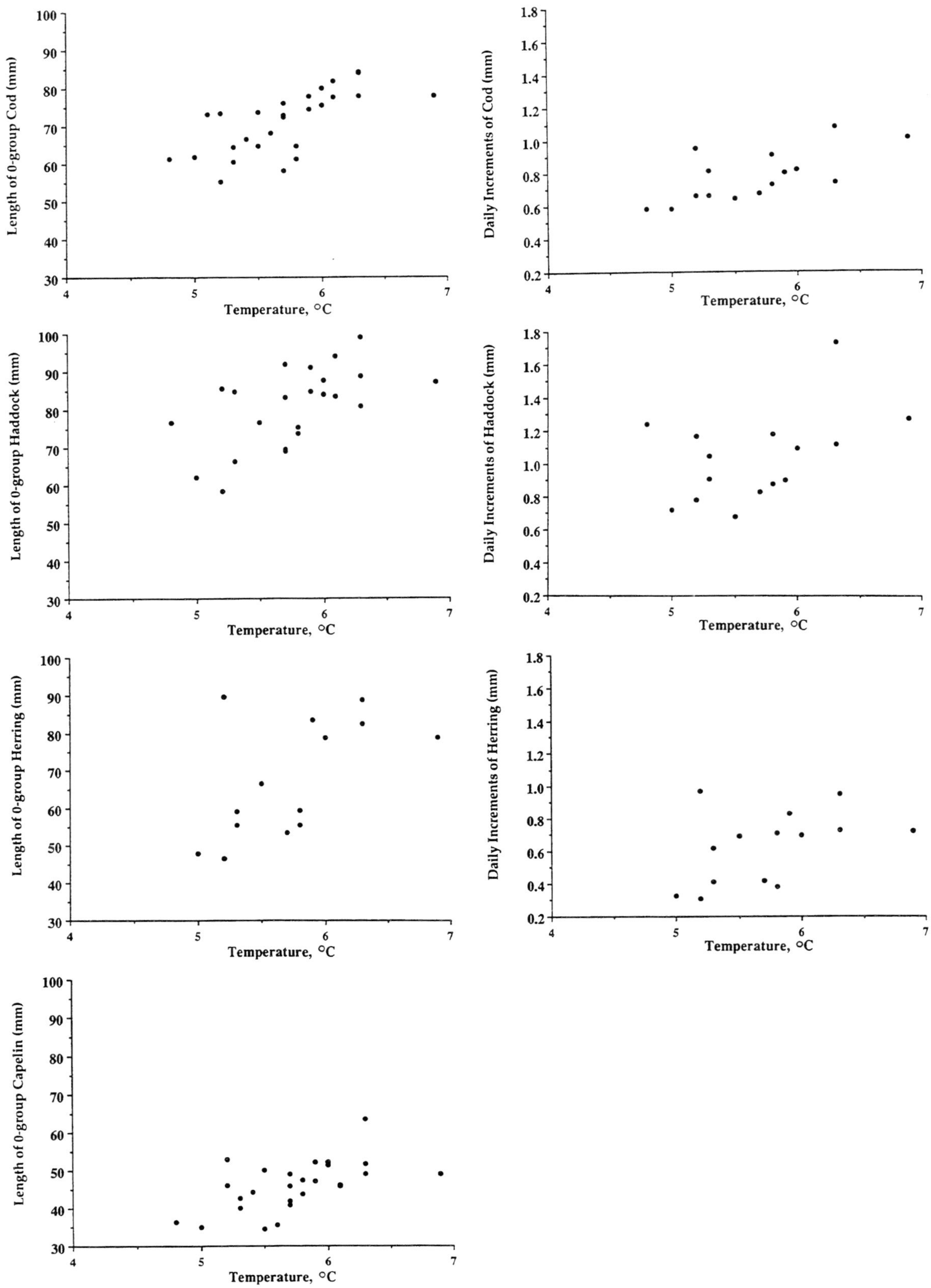

Fig. 4. Relation between temperature and length of 0-group fish (left side), and temperaure and daily increment (right side).

growth than the year-classes from 1982 to 1984 (Ellertsen et al. 1987). The Atlantic inflow to the Barents Sea may partly explain the high growth rate. According to Skjoldal and Rey (1989), the timing of major inflow events in relation to the seasonal vertical migration of *Calanus finmarchicus* in the Norwegian Sea would determine the amount of zooplankton found in the southwestern Barents Sea. The variability in the Atlantic inflow is therefore important in explainining the large interannual variation in zooplankton biomass in the southern Barents Sea. The high Atlantic inflow to the Barents Sea prior to April 1985 (Loeng 1991) may explain the high concentration of zooplankton reported by Nesterova (1990) during April–May that year. The zooplankton observations in June–July 1985 showed lower biomass values than the years before, indicating heavy feeding on zooplankton. This might have favoured the high growth rate in 1985.

In growth studies of adult cod, Nakken and Raknes (1987) concluded that the coupling between environmental changes and population parameters was far more complex than a simple length–temperature relationship. Gjøsæter and Loeng (1987) came to the same conclusion when they studied the growth of adult capelin. Both papers, however, indicated best growth at high temperatures. It is apparent that temperature effects linked to the availability of food may be as important as direct effects. The feeding conditions for fish larvae influence the year-class strength. The temperature-dependent spawning of the copepod *Calanus finmarchicus* at the spawning grounds of cod may be the most important process that causes variability in cod larval survival (Ellertsen et al. 1989). It is, therefore, reasonable to assume that food conditions also are important for the growth of larvae. Consequently, no simple relationship between temperature and growth should be expected.

A last factor that may influence the temperature–length relationship is the age of larvae. The spawning time may vary from year to year (Table 2), and this will, of course, influence the mean length observed during the 0-group survey. The spawning time, however, has no influence on the the temperature–growth relationship dicussed earlier.

The computation of daily growth increment of larvae (Table 1) is based on the assumption of a linear growth. The close relationship between length and growth (Table 4) supports this assumption. The variabilility in the number of days between the two surveys does not seem to influence the calculated daily increment, which also supports a linear growth rate. The interannual variability of more than 100% in the larval growth rate and consequently their size at the end of August (Table 1), will probably influence on their chance of survival. One direction for future work should, therefore, be to examine any relation between high growth rate and year-class strength.

Acknowledgements

This work has received financial support from the Norwegian Fisheries Research Council (NFFR).

References

Anonymous. 1983. Preliminary report on the International 0-group fish survey in the Barents Sea and adjacent waters in August–September 1980. Ann. Biol. 37: 259–266.

Anonymous. 1991. Preliminary report on the International 0-group fish survey in the Barents Sea and adjacent waters in August–September 1991. ICES CM 1991 (G:50): 33 p.

Bjørke, H., and S. Sundby. 1987. Distribution and abundance indices of postlarval and 0-group cod, p. 127–144. *In* H. Loeng [ed.] The Effect of Oceanographic Conditions on Distribution and Population Dynamics of Commercial Fish Stocks in the Barents Sea. Proceedings of the Third Soviet–Norwegian Symposium, Murmansk, Russia, May 26–28, 1986. Institute of Marine Research, Bergen, Norway.

Dragesund, O. 1971. Comparative analysis of year-class strength among fish stocks in the North Atlantic. Fiskeridir. Skr. Havunders. 16: 49–64.

Dragesund, O., and J. Gjøsæter. 1988. The Barents Sea, p. 339–361. *In* H. Postma and J.J. Zijlstra [ed.] Ecosystems of the world 27: continental shelves. Elsevier, Netherlands.

Ellertsen, B., P. Fossum, P. Solemdal, and S. Sundby. 1989. Relation between temperature and survival of eggs and first-feeding larvae of Northeast Arctic cod (*Gadus morhua* l.). ICES Rapp. P.-v. 191: 209–219.

Ellertsen, B., P. Fossum, P. Solemdal, S. Sundby, and S. Tilseth. 1987. The effect of biological and physical factors on the survival of Arcto-Norwegian cod and the influence on recruitment variability, p. 101–126. *In* H. Loeng [ed.] The Effect of Oceanographic Conditions on Distribution and Population Dynamics of Commercial Fish Stocks in the Barents Sea. Proceedings of the Third Soviet–Norwegian Symposium, Murmansk, Russia, May 26–28, 1986. Institute of Marine Research, Bergen, Norway.

Gjøsæter, H., and H. Loeng. 1987. Growth of the Barents Sea capelin, *Mallotus villosus,* in relation to climate. Environ. Biol. Fish. 20(4): 293–300.

Loeng, H. 1989. The influence of temperature on some fish population parametres in the Barents Sea. J. Northwest Atl. Fish. Sci. 9(2): 103–113.

Loeng, H. 1991. Features of the physical oceanographic conditions of the Barents Sea. Polar Res. 10(1): 5–18.

Loeng, H., and H. Gjøsæter. 1990. Growth of 0-group fish in relation to temperature conditions in the Barents Sea during the period 1965–1989. ICES CM 1990 (G:49): 9 p.

Mukhina, N.V., A.I. Mukhin, and E.A. Dvinina. 1987. Oceanographic conditions and reproduction of Arcto-Norwegian cod of the Barents Sea in 1980–1985, p. 145–158. *In* H. Loeng [ed.] The Effect of Oceanographic Conditions on Distribution and Population Dynamics of Commercial Fish Stocks in the Barents Sea. Proceedings of the Third Soviet–Norwegian Symposium, Murmansk, Russia, May 26–28, 1986. Institute of Marine Research, Bergen, Norway.

Nakken, O., and A. Raknes. 1987. The distribution and growth of Northeast Arctic cod in relation to bottom temperature in the Barents Sea, 1978–1984. Fish. Res. 5: 243–252.

Nestrova, V.N. 1990. Plankton biomass along the drift route of cod larvae. Report, PINRO, Murmansk, Russia. 64 p. (in Russian)

SAS Institute Inc. 1988. SAS procedures guide, release 6.03 edition. Cary, NC. 441 p.

Skjoldal, H.R., and F. Rey. 1989. Pelagic production and variability of the Barents Sea ecosystem, p. 241–286. *In* K. Sherman and V. Alexander [ed.] Biomass yields and geography of large ecosystems. American Association for the Advancement of Science, Washington, D.C.

Sundby, S., H. Bjørke, A. Soldal, and S. Olsen. 1989. Mortality rates during the early life stage and year-class strength of north-east Arctic cod (*Gadus morhua* L.). Rapp. P.-v. Réun. Cons. int. Explor. Mer 191: 351–358.

Sætersdal, G., and H. Loeng. 1987. Ecological adaption of reproduction in Northeast Arctic cod. Fish. Res. 5: 253–270.

Sætre, R., and J. Gjøsæter. 1975. Ecological investigations on the spawning grounds of the Barents Sea capelin. Fiskeridir. Skr. Havunders. 16: 203–227.

Potential impact of climate change on the habitat and population dynamics of juvenile Atlantic salmon (*Salmo salar*) in eastern Canada

Charles K. Minns, Robert G. Randall, E. Michael P. Chadwick, James E. Moore, and Robin Green[1]

Minns, C.K., R.G. Randall, E.M.P. Chadwick, J.E. Moore, and R. Green. 1995. Potential impact of climate change on the habitat and population dynamics of juvenile Atlantic salmon (*Salmo salar*) in Eastern Canada, p. 699–708. *In* R.J. Beamish [ed.] Climate change and northern fish populations. Can. Spec. Publ. Fish. Aquat. Sci. 121.

Abstract: The hypothesis that predicted climate changes in the next century will affect the population dynamics of juvenile Atlantic salmon (*Salmo salar)* in eastern Canada was addressed. Two issues were investigated. (1) With climate warming, runoff volume and patterns are predicted to change in eastern Canada. Known relationships between discharge and river width were used to estimate changes in the area of fluvial habitat available to juvenile salmon. By watershed, predicted changes ranged from –40 to 16%, with a net loss of habitat area of about 4%. (2) A published model of mean age of smoltification based on water temperature and latitude was used to predict changes in growth conditions for salmon in fresh water. The model predicted that growth rates would increase in all areas. One possible consequence is that mean age at smoltification may decrease from 8 to 29%, depending on the area. Other possible changes in the population dynamics of juvenile salmon and potential implications on freshwater survival are discussed. Our analyses indicate that climate warming could affect both the quantity and quality of salmon habitat in eastern Canada, presenting important challenges for fishery managers.

Résumé : On s'est penché sur l'hypothèse voulant que les changements climatiques prévus au cours du prochain siècle auraient une influence sur la dynamique des populations de saumons de l'Atlantique (*Salmo salar*) juvéniles de l'est du Canada. Deux aspects ont été examinés. (1) Avec le réchauffement du climat, le volume et la configuration de l'écoulement devraient changer dans l'est du Canada. On a utilisé les relations connues entre les décharges et les largeurs des rivières pour estimer les changements susceptibles de toucher l'habitat fluvial disponible pour le saumon juvénile. Selon le bassin hydrographique, les changements prévus variaient de –40 à 16 %, avec une perte nette d'habitat d'environ 4 %. (2) On a utilisé un modèle publié de l'âge moyen à la smoltification d'après la température de l'eau et la latitude pour prédire des changements dans les conditions de croissance du saumon dans l'eau douce. Le modèle a permis de prédire que les taux de croissance augmenteraient dans toutes les régions. Une conséquence possible de ce phénomène est que l'âge moyen à la smoltification pourrait diminuer de 8 à 29 %, selon la région envisagée. D'autres changements possibles dans la dynamique de la population du saumon juvénile et dans les conséquences éventuelles pour sa survie en eau douce sont analysés. Nos analyses indiquent que le réchauffement du climat pourrait avoir une influence autant sur la quantité que sur la qualité de l'habitat du saumon dans l'est du Canada, ce qui pose des défis importants pour les gestionnaires des pêcheries.

C.K. Minns, R.G. Randall, J.E. Moore, and **R. Green.** Great Lakes Laboratory for Fisheries and Aquatic Sciences, Department of Fisheries and Oceans, Bayfield Institute, 867 Lakeshore Road, P.O. Box 5050, Burlington, ON L7R 4A6, Canada.

E.M.P. Chadwick. Department of Fisheries and Oceans, Gulf Fisheries Centre, P.O. Box 5030, Moncton, NB E1C 9B6, Canada.

[1] Present address: 433A Midwood Crescent, Waterloo, ON N2L 3G1, Canada.

Introduction

Increasing concentrations of greenhouse gases are changing the earth's climate. By the year 2050, mean annual temperatures are predicted to increase by 1.5–4.5°C in Canada. The potential impacts of climate changes on fisheries and other aquatic resources are just beginning to be investigated (Regier et al. 1990), but they may be significant. Increases in air temperature associated with global warming will change the thermal habitats of fish, possibly driving populations further north or altering life-history traits such as growth rates and maturation. In addition, riverine species of fish may be affected by changes in runoff patterns associated with climate change. The vulnerability of fish stocks to changes in climate is of paramount importance to researchers and the fishing industry alike.

Atlantic salmon (*Salmo salar*) are native to the North Atlantic and, in Canada, self-sustaining populations inhabit rivers from Nova Scotia north to Labrador and the Ungava region of Quebec. Atlantic salmon are anadromous. Incubation and early life are spent in the cool, clean fluvial habitats of rivers. After 2–6 yr juveniles leave the rivers as smolts to undergo extensive feeding migrations in the North Atlantic before maturing and returning to their natal rivers to spawn. Each river and often the tributaries within larger rivers have their own unique genetic stock of Atlantic salmon, as the populations have adapted to the specific environmental conditions within their natal rivers (Power 1981; Saunders 1981). Thus there are a few hundred different stocks of Atlantic salmon in eastern Canada. Collectively, these stocks are an extremely valuable resource. Between 1910 and 1985, landings by commercial fishermen ranged from a high of 6101 t in 1930 to a low of 814 t in 1984 (Marshall 1988). Commercial fishing in coastal waters has now been discontinued in most areas, but salmon still support a lucrative angling fishery, and some fish are harvested by Native Canadians. Over past years, major stresses on the salmon populations included overfishing (Chadwick 1985a) and habitat degradation (Watt 1989). Global climate changes will place additional stresses on Atlantic salmon populations.

The potential effect of climate change on salmonines in Canada has not been well studied. Meisner et al. (1988) and Meisner (1990) predicted that elevated groundwater and summer air temperatures would reduce the summer thermal habitat available to brook trout (*Salvelinus fontinalis*), and would possibly eliminate trout from certain areas at the southern part of its native range. Power (1990) argued that climate warming in Quebec and Labrador may shift the distribution of salmonines northwards, and possibly make the rivers more susceptible to invasions of exotic salmonines such as *Oncorhynchus* spp. and brown trout (*Salmo trutta*). Potential shifts in distribution and species composition are of obvious interest to fishery managers.

Rather than consider possible changes in the distribution of Atlantic salmon as a result of climate warming, our objective was to investigate potential changes in the quantity and quality of habitat within the present geographic range of this species. Specifically, two issues are addressed. First, because climate change is expected to alter water-discharge patterns, present and projected future runoff values are used to estimate changes in the area of freshwater habitat available to juvenile salmon. Second, the effects of changes in water temperatures on the growth environment of salmon and on age at smoltification are investigated. Our results indicate that the potential impacts of climate change on Atlantic salmon populations are significant, and lead us to emphasize the need to initiate long-term monitoring programs to detect future changes in salmon habitat.

Materials and methods

Predictions of changes in mean age of smoltification (MAS) and summer-rearing habitat supply were developed in a series of discreet steps for the region enclosing the Canadian distribution of Atlantic salmon. Tertiary watersheds designated by the Water Survey of Canada (1977) were the basic unit for cross-linking climatic, hydrological, and fish information. The steps were integrated and linked using a geographic information system (Intera-Tydac's SPANS GIS). The steps were as follows.

1) Data sets describing mean annual and monthly air temperatures and predicted changes under climate change were assembled.
2) The temperature data sets were then used to predict the "growth opportunity index" (GOI) for both the current climate ($1 \times CO_2$) and that projected with a doubling of carbon dioxide levels in the atmosphere ($2 \times CO_2$) at all weather stations in the region. The results were then contoured to provide map coverage of all areas.
3) Area-mean GOI values were computed using the overlay features of the GIS, and the results were used with Metcalfe and Thorpe's (1990) regression equation predicting MAS from GOI for North American Atlantic salmon stocks.
4) Data sets describing long-term annual and monthly runoff at gauging stations throughout the region and predicted changes in precipitation and evapotranspiration were assembled.
5) The long-term hydrology data were used to construct a regional map of mean summer (July–September) unit runoff (metres per quarter) and the climate change data were used to construct a map of predicted change in summer runoff (metres per quarter).
6) Inventories of Atlantic salmon parr summer rearing habitat by river were assembled by tertiary watershed unit. The inventories provided the basis for the map of the distribution of salmon in Atlantic Canada.
7) Stream width (and hence habitat rearing area) was assumed to increase or decrease as the square root of the mean summer discharge. On this basis, changes in summer habitat area were calculated from projected changes in runoff by tertiary watershed.

Climate data

Long-term mean annual and monthly air temperatures (1951–80) at 147 weather stations in eastern Canada were obtained from the Atmospheric Environment Service (AES) (Climate Data Services Branch, AES, Environment Canada, Downsview, Ont.). The period 1951–80 is that used as a reference equivalent for $1 \times CO_2$ scenarios in global circulation models (GCM).

In this study, we used averaged output for $1 \times CO_2$ and $2 \times CO_2$ scenarios obtained with the Canadian Climate Centre Global Circulation Model (CCC-GCM), provided by AES (CCC-AES 1990). According to the Intergovernmental Panel on Climate Change (Houghton et al. 1990), the CCC-GCM is a second-generation model with a higher spatial resolution (3.75° latitude by 3.75° longitude), improved physics, and a more realistic 1 × CO_2 reference climate than most first-generation GCMs. A data set, drawn from that provided, consisted of 45 grid trapezoids covering eastern Canada. The output data represent areal averages for each grid trapezoid. To compute predicted changes in mean annual and monthly temperatures, we took the differences between the 1 × CO_2 and 2 × CO_2 scenarios. We then matched the positions of the 147 weather stations to the grid trapezoids in the CCC-GCM and added the temperature changes to the observed 1951–80 values to obtain a 2 × CO_2 temperature scenario taking into account finer scale climatic variation in the Atlantic region.

GOI and MAS

Metcalfe and Thorpe (1990) analyzed regional patterns of MAS for Atlantic salmon stocks in Europe and North America. They developed a series of regression equations relating MAS to GOI (degrees Celsius times hours), which incorporates both the seasonal air temperature patterns and seasonal–latitudinal variations in day length:

(1) $\mathrm{GOI} = \sum (T_i D_i)$

where T_i = mean monthly air temperature above 5.5°C and D_i = midmonth day length (hours) times the number of days in the month; i = 1–12 (months in the year). The value of 5.5°C was assumed to be the lower temperature limit for growth (Power 1981). The equation for North American stocks was

(2) $\mathrm{Log_e}\,(\mathrm{MAS}) = 1.548 - 0.0291\mathrm{GOI}$

We applied Equation 1 at each of the 147 weather stations in the AES temperature data both for the reference period 1951–80 and after the projected 2 × CO_2 increases. We then used the GIS to create contour maps of GOI using a 1000°C·h interval. We used equations described by Fee (1990) to calculate the midmonth day lengths (hours) based on the station latitudes.

Next we overlaid the GOI maps with the tertiary watershed map and computed area-mean GOI values. The calculations are based on integer contour interval class values. The means are then converted to GOIs by assigning the interval midpoints to integer values. Thus, in the worst case, the uncertainty surrounding a watershed mean GOI has a rectangular random distribution of ±500°C·h. Finally, we used the watershed GOI values with Equation 2 to predict MAS by tertiary watershed.

Surface runoff

Hydrological data on mean monthly and annual discharges by station and year were obtained from the Water Survey of Canada. Despite the spatial and temporal extent of the data set, there were insufficient data covering the period 1951–80 to produce a hydrological summary comparable to the climate reference data set. Instead, we used long-term data from 540 hydrology gauging stations in eastern Canada, with time series records of 15 yr or longer, to estimate the current reference hydrological runoff. Long-term means were calculated for each gauging station and flows were converted from cubic metres per second to metres per month or metres per yr.

In addition to temperature predictions, the CCC-GCM predicts monthly mean precipitation and evapotranspiration rates. Because, in simple terms, precipitation (P) equals evapotranspiration (AE) plus runoff (Q) in the long term (Dooge 1992), we used the difference (P – AE) as an estimate of Q in the 1 × CO_2 and 2 × CO_2 scenarios of the CCC-GCM output. We then took the difference between the two computed Q values to estimate the change in runoff, by month and in each GCM grid trapezoid. As with the temperature modelling, we matched hydrological stations with GCM grids and added the change in Q to that observed in the long term and summed over the 3 mo of summer (July–September). We then contoured the two sets of runoff values using a 0.02 m·quarter and did an overlay to determine area-mean runoff values by tertiary watershed for the current and 2 × CO_2 hydrological regimes.

Distribution and rearing habitat

For the Maritime provinces and insular Newfoundland, rivers with Atlantic salmon (Atlantic Salmon Foundation 1980) and estimates of rearing area were obtained from Atlantic Salmon Review Task Force (1978). Habitat area for juvenile salmon, depending on the river, was obtained from field surveys or from calculations based on watershed area. Rearing areas for four rivers in the Maritimes have been updated in recent years, and thus the updated estimates were used (Claytor and Leger 1986; Cutting and Jefferson 1986; Marshall and Penney 1983; Randall 1985). Data for rivers in Labrador were from Anderson (1985).

We then grouped the rearing habitats by tertiary watersheds. In some cases, only aggregate values were available for groups of tertiary watersheds, and those rearing areas were assigned to watersheds in proportion to the total areas of the watersheds. The watersheds with estimates of rearing habitat were assumed to represent the extant distribution of Atlantic salmon in eastern Canada. Stocks are known outside these watersheds but are considered to be insignificant contributors to overall habitat supply and salmon production.

To model climate change-induced changes in rearing habitat, we assumed that rearing habitat area increased or decreased as the square root of mean summer unit runoff by tertiary watershed. Average stream wet width tends to increase as the square root of the mean discharge (Hynes 1972) and hence habitat area should also be related to discharge in a similar manner.

The data set assembled in conjunction with these steps is listed by tertiary watershed unit in the Appendix and includes watershed code, latitude and longitude centroids, watershed area, rearing area, mean annual air temperature (1951–80), mean annual air temperature after 2 × CO_2, GOIs and MAS with and without climate change, and predicted rearing area after 2 × CO_2.

Results

Temperature effects

Tertiary watersheds in eastern Canada for which estimates of the rearing area for Atlantic salmon were available are illustrated in Fig. 1. Over this geographic area, the annual mean temperatures for 1951–80 ranged from –6.0 to 7.6°C north to south. After temperature changes of 3.5–5.4°C projected by the CCC-GCM, the range would be –1.5 to 11.1°C.

Fig. 1. Map showing the estimated distribution of Atlantic salmon by tertiary watershed (dark areas) used in this study. Watersheds for which estimates of rearing area were not available were excluded.

The greatest increases are predicted for the western New Brunswick and Quebec portions of the distribution, and the lowest for Labrador. Those annual temperature changes were reflected in changes in the GOI values. GOI values increased by a mean of 9700 with a range of 2700–11 500. This increased the mean GOI from 16 800 to 26 500.

Mean smolt ages in eastern Canada under current climate conditions, as calculated using Equation 2, ranged from 2.39 to 4.03 yr. These predicted smolt ages were compared with observed ages at eight rivers which were selected from the literature to cover a range of latitudes. Although Equation 2 underestimated smolt ages in the northern latitudes, particularly the salmon populations of Ungava Bay, the predicted ages were generally reasonable in other locations (Table 1). The distribution of smolt ages tended to follow the lines of latitude: the youngest ages occurred in southern Nova Scotia, while the oldest ages occurred in Labrador and Ungava bay. For Newfoundland and Labrador, the contours of smolt ages were

Table 1. Comparison of the mean smolt age of Atlantic salmon in selected rivers of eastern Canada. Observed ages are from field data (see references) and predicted ages were calculated from Equation 2. Watershed codes are for tertiary watersheds as designated by the Water Survey of Canada.

River	Latitude	Watershed code	Mean age at smoltification		Reference
			Observed	Predicted	
LaHave, Nova Scotia	44.6	1EF	2.1	2.4	Gray 1986
Big Salmon, New Brunswick	45.6	1BV	2.6	2.5	Jessop 1986
Morell, Prince Edward Island	46.0	1CD	3.0	2.6	Ducharme 1977
Miramichi, New Brunswick	46.5–47.2	1BM,1BN,1BO,1BP,1BQ	2.3–3.0	2.4–2.5	Randall et al. 1986
Western Arm, Newfoundland	51.3	2YA	3.9	3.5	Chadwick 1987
Bouleau, Quebec	51.6	2UC	3.1	3.3	Randall and Power 1979
Sand Hill, Labrador	53.0	3QC	4.1–4.6	3.5	Anderson 1985
Koksoak, Quebec	56.4–57.3	3KB,3KC	4.3–5.1	3.9	Robitaille et al. 1986

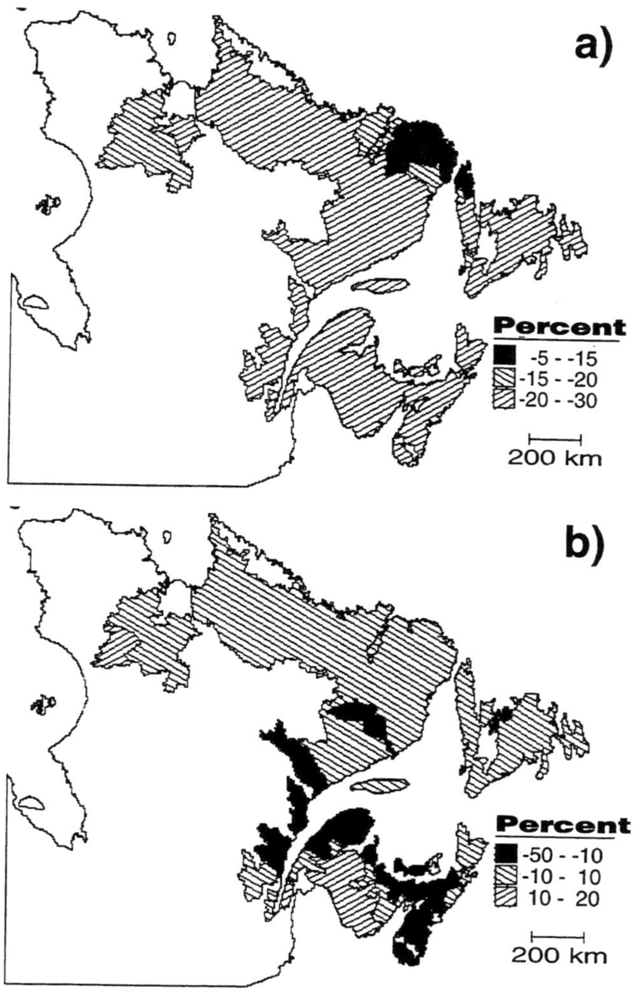

Fig. 2. Maps showing (a) the percent change in the mean age of Atlantic salmon smolts after climate warming, and (b) the percent change in area of fluvial habitat that may occur because of the changes in runoff associated with global warming in regions of eastern Canada.

generally similar to values published by Power (1981), with the exception of some local minor anomalies. Discrepancies probably resulted from differences in interpolation procedures of SPANS or from the use of different temperature and day-length data.

In a 2 × CO_2 environment, smolt ages are predicted to decrease between 8 and 29% (24 ± 3.6%; mean ± SD). Expressed in terms of years, these changes indicate a mean change of 0.7 yr (0.3–0.95). Changes in smolt ages varied from region to region. Maximum decreases of 20–30% were predicted to occur in Nova Scotia, New Brunswick, Quebec, and Prince Edward Island (Fig. 2a). Decreases in smolt age were predicted to be least (5–15%) in southeast Labrador and the Avalon Peninsula of Newfoundland.

Hydrologic effects

Overall, the CCC-GCM predicted a net decline in runoff in eastern Canada, with a maximum loss of 0.07 m·quarter and a mean loss of 0.02 m·quarter. The Gaspe coast and the lower Quebec north shore had the highest loss of runoff, while areas in southeast Labrador and western New Brunswick actually showed a gain of up to 0.07 m·quarter in runoff. Most areas of Nova Scotia, Newfoundland, and the upper Quebec north shore were predicted to have slight reductions in runoff (0 to −0.05 m·quarter).

Potential changes in available salmon habitat, as calculated from the runoff data, varied among regions (Fig. 2b). Predicted changes varied from reductions of up to 42% in certain areas of Quebec, to increases of 16% in western New Brunswick. Among watersheds, the predicted average change in rearing area was −7.0 (± 12.4%). Summed over the geographic range of salmon, the total area of habitat was predicted to decrease from 70.8 × 10^7 to 68.3 × 10^7 m^2, which represents a net reduction of about 4%.

Discussion

Predicted reductions in runoff and the area of fluvial habitat in certain rivers of eastern Canada could directly affect the production of Atlantic salmon. Chadwick (1985b) found that 86% of the variability in angling catches of salmon among different rivers in the eastern Canada could be explained by differences in rearing habitat. The amount of variability explained in Newfoundland rivers was less, but was still significant (62%). Thus, the production of adult salmon and potential yields to fisheries would be expected to decrease or increase in direct proportion to changes in discharge and habitat area. Although the overall reduction in habitat area in eastern Canada is predicted to be small (4%), changes within specific regions will be significant. Reductions in fluvial habitat areas could exceed 40% in some watersheds. Regional changes in fish production and shifts in the distribution of fishing effort are of obvious interest to fishery managers. Aside from the obvious impact of habitat area on production, changes in stream discharge can affect salmon in other ways. For example, the timing of adult salmon ascent and the distances travelled during upstream migrations could be affected (Power 1981). Discharge can also affect the catchability of salmon by anglers after the salmon have entered the rivers (Chadwick 1985b). Changes in runoff associated with climate change could thus affect salmon production both directly by changing the amount of habitat and indirectly by influencing behaviour.

Impacts of climate warming on the thermal habitat and the growth characteristics of salmon are more complex. Average increases in mean annual air temperatures of 4°C would affect the thermal habitat of salmon throughout the year, potentially affecting all life history stages, from egg incubation to adult spawning. We chose to focus on the summer growth period and the potential effects on smolting in order to illustrate one possible consequence of climate change. Increases in Metcalfe and Thorpe's (1990) GOI showed clearly that the conditions for growth would change significantly in all areas. Assuming that smoltification is affected by growth conditions, smolt ages could change appreciably. At present, the youngest age at smoltification in eastern Canada is age 2. Predicted mean smolt ages of less than 2 yr after climate warming implies that growth rates would be enhanced to the point that some salmon would smoltify after only 1 yr in fresh water. Age-1 smolts are observed in the warmer climates of Europe (Cuinat 1988), but whether or not Canadian rivers are capable of producing age-1 smolts is yet to be determined. Achievement of a younger age at smoltification may increase freshwater survival from the egg to smolt stages because of the reduced time spent in fresh water (Randall et al. 1987).

Whether or not smolt ages within populations would change significantly, however, is subject to conjecture. Smolt age is to some extent under genetic control (Refstie et al 1977; Saunders 1981), and thus it may not fluctuate beyond certain limits. Despite annual variability in growth conditions at Western Arm Brook, Newfoundland, Chadwick (1987) found that age at smoltification in this population was relatively fixed. Size at smoltification varied annually more so than age at smoltification. Although smolt ages have been shown to fluctuate within populations over time, this may have resulted from different exploitation rates on the different age groups or stocks, rather than from environmental influences (Chadwick 1987). Changes in the size of smolts rather than changes in age would also affect survival. In this case, freshwater survival to the smolt stage would be reduced as density is negatively correlated to parr size, at least in riverine habitats (Grant and Kramer 1990).

Another potential consequence of increases in freshwater growth could be changes in maturation. Juvenile salmon, particularly males, are capable of becoming sexually mature in fresh water. Under optimum and enhanced growth conditions, the incidence of freshwater maturation could increase (Thorpe 1987). Freshwater maturation would have a negative affect on the production of sea-run adults. Thus, regardless of whether growth increases would result in changes in smolt age or size, or freshwater maturation, adult salmon production would be affected in some way. Further research is needed to judge whether the net affect would be negative or positive.

Metcalfe and Thorpe (1990) emphasize that the GOI is a rather simplistic indicator of growth conditions, because it assumes growth rates increase positively with temperature. In actuality, growth rates would initially increase but then decrease when the optimum limit to growth is exceeded. With climate warming, it is safe to assume that certain populations, particularly those in the southern latitudes of Canada, would experience summer temperatures above the maximum for optimum growth. Because of this confounding factor, and because of the uncertainties of the actual response of juvenile salmon populations to increased growth as outlined above, our conclusions must remain tentative. Nevertheless, our results are important in describing the magnitude of the potential change in thermal habitats with climate warming, and hint at some of the possible consequences of these changes to the dynamics of the juvenile salmon populations.

Global climate models are still relatively new, and the data produced through the Canadian Climate Centre's GCM must be viewed with some caution. The Canadian model, like most other global climate models used in North America and elsewhere, is less accurate when predicting regional rather than larger scale consequences of climatic warming (Cushman and Spring 1989). To predict the full extent of climate change in localized watersheds, more refined climate models are needed. Limitations of climate models in general are discussed in detail by Houghton et al. (1990).

Despite the above limitations, our analyses show that the effects of climate change on runoff patterns and temperature will likely vary significantly from region to region in eastern Canada. Consequences to the dynamics and potential yield of salmon emphasize the need for biologists and managers to investigate the implications of climate change in more detail. Long-term monitoring programs designed to detect changes in runoff and thermal habitats should be initiated in areas judged to be most sensitive to climate change. To detect changes in runoff, monitoring sites in northeast New Brunswick and on the Gaspe coast would be useful because changes in runoff are expected to be the most extreme in these areas. Monitoring sites anywhere in New Brunswick or Nova Scotia would be useful for detecting changes in thermal habitat during summer. Index study rivers already established in eastern Canada (Chadwick 1985b) may be useful for this purpose.

References

Anderson, T.C. 1985. The rivers of Labrador. Can. Spec. Publ. Fish. Aquat. Sci. 81: 389 p.

Atlantic Salmon Foundation. 1980. Atlantic salmon rivers of North America (Map). Montreal, Que.

Atlantic Salmon Review Task Force. 1978. Biological Conservation Subcommittee report. Mimeographed report. Fisheries and Marine Service, Halifax, N.S. 203 p.

CCC-AES. 1990. Application of the Canadian Climate Centre general circulation model output for regional climate impact assessment. Canadian Climate Centre, Atmospheric Environment Service, Environment Canada, Downsview, Ont. Version 1, April 1990. 18 p.

Chadwick, E.M.P. 1985a. The influence of spawning stock on production and yield of Atlantic salmon (*Salmo salar* L.) in Canadian rivers. Aquacult. Fish. Manage. 1: 111–119.

Chadwick, E.M.P. 1985b. Fundamental research problems in the management of Atlantic salmon, *Salmo salar* L., in Atlantic Canada. J. Fish. Biol. 27(Suppl. A): 9–25.

Chadwick, E.M.P. 1987. Causes of variable recruitment in a small Atlantic salmon stock. Am. Fish. Soc. Symp. 1: 390–401.

Claytor, R.R., and Léger, C. 1986. Assessment of Atlantic salmon *Salmo salar* in the Margaree river, Nova Scotia, 1986. CAFSAC Res. Doc. 86/93

Cuinat, R. 1988. Atlantic salmon in an extensive French river system: the Loire–Allier, p. 389–399. *In* D. Mills and D. Piggins [ed.] Atlantic salmon: planning for the future. Timber Press, Portland, OR.

Cushman, M.R., and Spring, N.P. 1989. Differences among model simulations of climate change on the scale of resource regions. Environ. Manage. 13: 789–795

Cutting, R.E., and Jefferson, E.M. 1986. Status of Atlantic salmon the LaHave River, Nova Scotia, in 1985 and forecast of returns in 1986. CAFSAC Res. Doc. 86/17

Dooge, J.C. I. 1992. Hydrologic models and climate change. J. Geophys. Res. 97(D3): 2677–2686.

Fee, E.J. 1990. Computer programs for calculating *in situ* phytoplankton photosynthesis. Can. Tech. Rep. Fish. Aquat. Sci. 1740: 7 p.

Grant, J.W.A., and D.L. Kramer. 1990. Territory size as a predictor of the upper limit to population density of juvenile salmonids in streams. Can. J. Fish. Aquat. Sci. 47: 1724–1737.

Gray, R.W. 1986. Biological characteristics of Atlantic salmon (*Salmo salar* L) in the upper LaHave River basin. Can. Tech. Rep. Fish. Aquat. Sci. 1437: 45 p.

Ducharme, L.J.A. 1977. Atlantic salmon enhancement in the Morell River, Prince Edward Island. Tech. Rep. Ser. Mar/T 77-2: 19 p.

Houghton, J.T., G.T. Jenkins, and J.J. Ephrauems [ed.]. 1990. Climate change: the IPCC scientific assessment. Intergovernmental Panel on Climate Change. Cambridge University Press, Cambridge, UK. 365 p.

Hynes, H.B.N. 1972. The ecology of running waters. Liverpool University Press, Liverpool, U.K. 555 p.

Jessop, B.M. 1986. Atlantic salmon (*Salmo salar*) of the Big Salmon River, New Brunswick. Can. Tech. Rep. Fish. Aquat. Sci. 1415: 50 p.

Marshall, T.L. 1988. Harvest and recent management of Atlantic salmon in Canada, p. 117–142. *In* D. Mills and D. Piggins [ed.] Atlantic salmon: planning for the future. Timber Press, Portland, OR.

Marshall, T.L., and Penney, G.H. 1983. Spawning and river escapement requirements for Atlantic salmon of the Saint John River, New Brunswick. CAFSAC Res. Doc. 85/66

Meisner, D.J. 1990. Effect of climatic warming on the southern margins of the native range of brook trout, *Salvelinus fontinalis*. Can. J. Fish. Aquat. Sci. 47: 1065–1070.

Meisner, D.J., J.S. Rosenfeld, and H.A. Regier. 1988. The role of groundwater in the impact of climate warming on stream salmonids. Fisheries 13: 2–8.

Metcalfe, N.B., and Thorpe, J.E. 1990. Determinants of geographical variation in the age of seaward-migrating salmon, *Salmo salar*. J. Anim. Ecol. 59: 135–145.

Power, G. 1981. Stock characteristics and catches of Atlantic salmon (*Salmo salar*) in Quebec, and Newfoundland and Labrador in relation to environmental variables. Can. J. Fish. Aquat. Sci. 38: 1601–1611.

Power, G. 1990. Salmonid communities in Quebec and Labrador: temperature relations and climate change. Pol. Arch. Hydrobiol. 37: 13–28.

Randall, R.G. 1985. Spawning potential and spawning requirements of Atlantic salmon in the Miramichi River, New Brunswick. CAFSAC Res. Doc. 85/68.

Randall, R.G., M.C. Healey, and J.B. Dempson. 1987. Variability in length of freshwater residence of salmon, trout, and char. Am. Fish. Soc. Symp. 1: 27–41.

Randall, R.G., and G. Power. 1979. Atlantic salmon (*Salmo salar*) of the Pigou and Bouleau Rivers, Quebec. Environ. Biol. Fish. 4: 179–184.

Randall, R.G., J.E. Thorpe, R.J. Gibson, and D.G. Reddin. 1986. Biological factors affecting age at maturity in Atlantic salmon (*Salmo salar*), p. 90–96. *In* D.J. Meerburg [ed.] Salmonid age at maturity. Can. Spec. Publ. Fish. Aquat. Sci. 89.

Regier, H.A., J.J. Magnuson, and C.C. Coutant. 1990. Introduction to proceedings: symposium on effects of climate change on fish. Trans. Am. Fish. Soc. 119: 173–175.

Refstie, T., T.A. Steine, and T. Gjedrem. 1977. Selection experiments with salmon. II. Proportion of Atlantic salmon smoltifying at 1 year of age. Aquaculture 10: 231–242.

Robitaille, J.A., Y. Côté, G. Shooner, and G. Hayeur. 1986. Growth and maturation patterns of Atlantic salmon, *Salmo salar*, in the Koksoak River, Ungava, Quebec, p. 62–69. *In* D.J. Meerburg [ed.] Salmonid age at maturity. Can. Spec. Publ. Fish. Aquat. Sci. 89.

Saunders, R.L. 1981. Atlantic salmon (*Salmo salar*) stocks and management implications in the Canadian Atlantic provinces and New England, USA. Can. J. Fish. Aquat. Sci. 38: 1612–1625.

Thorpe, J.E. 1987. Smolting versus residency: developmental conflict in salmonids. Am. Fish. Soc. Symp. 1: 244–252.

Water Survey of Canada. 1977. Reference index, hydrometric map supplement, Canada 1977. Inland Waters Directorate, Water Survey of Canada, Ottawa, Ont.

Watt, W.D. 1989. The impact of habitat damage on Atlantic salmon (*Salmo salar*) catches, p. 154–163. *In* C.D. Levings, L.B. Holtby, and M.A. Henderson. [ed.] Proceedings of the National Workshop on Effects of Habitat Alteration on Salmonid Stocks. Can. Spec. Publ. Fish. Aquat. Sci. 105.

Appendix.

Tertiary watershed data and predictions for Atlantic salmon distribution

CODE = tertiary watershed code; LAT = latitude, geometric centroid of watershed; LONG = longitude, geometric centroid of watershed; AKM2 = watershed area (km^2); REARN = Atlantic salmon rearing area (10^3 m^2); TANN = mean annual air temperature (°C), 1951–80; TANF = predicted mean annual air temperature after $2 \times CO_2$; SUMRN = mean summer runoff (>15 yr records) (m); SUMRD = change in summer runoff ($1 \times CO_2 - 2 \times CO_2$) (m); GOIN = growth opportunity index with TANN temperatures; GOIF = growth opportunity index with TANF temperatures; AGEN = predicted mean smolt age using GOIN (years); AGEF = predicted mean smolt age using GOIF (years); REARF = predicted rearing area after $2 \times CO_2$ (10^3 m^2).

CODE	LAT	LONG	AKM2	REARN	TANN	TANF	SUMRN	SUMRD	GOIN	GOIF	AGEN	AGEF	REARF
01AF	47.3977	−67.9314	3762	3704.3	3.2	7.7	0.0875	0.0250	19600	30030	2.66	1.96	4200.3
01AH	47.1279	−67.1683	4337	4270.5	3.8	8.3	0.0765	0.0250	20630	31140	2.58	1.90	4919.0
01AJ	46.4465	−67.5516	2338	2302.1	4.5	9.0	0.0750	0.0250	21210	31670	2.54	1.87	2658.2
01AK	46.0295	−67.2061	3686	3629.5	4.7	8.8	0.0735	0.0250	21220	32110	2.54	1.85	4201.7
01AL	46.2928	−66.7754	1787	1759.6	4.9	9.4	0.0750	0.0250	22260	33030	2.46	1.80	2031.8
01AM	45.6893	−66.6516	2233	2198.8	5.5	9.0	0.0750	0.0250	22020	32980	2.48	1.80	2539.0
01AN	46.3551	−65.6785	1893	1864.0	5.5	10.0	0.0750	0.0015	22760	33450	2.42	1.78	1882.5
01AO	46.0443	−66.1382	2538	2499.1	5.5	9.7	0.0750	0.0250	23210	34040	2.39	1.75	2885.7
01AP	45.7869	−65.7041	6054	5961.2	5.5	9.3	0.0750	0.0035	22020	32900	2.48	1.81	6098.7
01AQ	45.3954	−66.7542	3417	695.2	5.4	8.9	0.0810	0.0250	20440	31440	2.59	1.88	795.3
01AR	45.4804	−67.3385	2655	524.8	5.3	8.8	0.0960	0.0250	20080	31080	2.62	1.90	589.2
01BA	47.5655	−67.6856	1688	6423.2	3.2	7.7	0.0820	0.0250	19290	29430	2.68	2.00	7337.3
01BB	47.9024	−67.8689	1535	5841.0	2.5	7.0	0.1220	0.0250	18250	28440	2.76	2.06	6411.6
01BC	47.954	−67.4469	2186	10318.1	2.7	7.2	0.1000	0.0250	18500	28640	2.74	2.04	11536.0
01BD	48.3393	−67.2509	3698	6010.0	2.5	7.0	0.1165	−0.0435	17760	2776	2.80	2.10	4757.4
01BE	47.6855	−66.8026	2414	9185.7	3.5	8.0	0.0750	0.0250	19550	29730	2.66	1.98	10606.7
01BF	48.2442	−66.5972	2215	550.0	2.6	7.1	0.1115	−0.0160	17950	27910	2.79	2.09	509.0
01BG	48.4975	−65.8887	7941	7396.0	2.9	7.4	0.1230	−0.0480	17200	27080	2.85	2.14	5775.3
01BH	48.6726	−64.8727	6915	10346.0	2.9	7.4	0.0975	−0.0250	16280	26010	2.93	2.21	8921.5
01BJ	47.7904	−66.1721	2971	2392.0	3.5	8.0	0.0750	0.0250	19310	29440	2.68	2.00	2762.0
01BK	47.4205	−66.2424	2368	5350.2	3.8	8.3	0.0750	0.0250	20500	30690	2.59	1.92	6177.9
01BL	47.5341	−65.1695	3475	2484.0	4.2	8.7	0.0740	−0.0240	19570	29690	2.66	1.98	2041.8
01BM	46.6306	−66.8766	2930	10920.3	4.5	9.0	0.0750	0.0250	21650	32460	2.50	1.83	12609.7
01BN	46.5054	−66.0407	2063	7688.9	5.2	9.7	0.0750	0.0200	22500	33100	2.44	1.79	8653.6
01BO	46.8955	−65.8861	4061	15135.5	4.5	9.0	0.0735	0.0095	21860	32350	2.49	1.83	16083.9
01BP	47.0231	−66.4622	1465	5460.1	4.5	9.0	0.0750	0.0250	21250	31750	2.53	1.87	6304.8
01BQ	47.1749	−66.0399	2110	7864.1	4.4	8.9	0.0750	0.0250	21180	31590	2.54	1.88	9080.7
01BR	46.8436	−65.1489	2022	8807.0	4.6	9.1	0.0750	−0.0250	21500	31780	2.52	1.86	7190.9
01BS	46.4505	−64.9346	2520	2527.3	5.4	9.9	0.0750	−0.0250	21580	32020	2.51	1.85	2063.5
01BT	46.0671	−64.2461	1471	336.7	5.5	9.8	0.0780	−0.0250	21440	31610	2.52	1.87	277.5
01BU	45.9842	−64.8706	2567	92.7	5.5	9.8	0.0770	−0.0250	21500	32170	2.52	1.84	76.2
01BV	45.5584	−65.2627	2280	547.3	5.5	9.0	0.1000	−0.0125	21390	32180	2.52	1.84	512.0
01CA	46.7377	−64.1231	1190	605.8	5.4	9.9	0.0750	−0.0250	21360	31500	2.53	1.88	494.6
01CB	46.39	−63.664	1289	656.2	5.5	10.0	0.0750	−0.0250	21670	32240	2.50	1.84	535.8
01CC	46.3175	−63.1258	1178	599.7	5.5	9.9	0.0685	−0.0250	21370	31500	2.52	1.88	477.9
01CD	46.3828	−62.4398	967	690.3	5.5	9.2	0.1050	−0.0250	20500	30700	2.59	1.92	602.5
01DA	44.2031	−66.0466	674	165.8	6.5	10.0	0.0900	−0.0250	19500	30600	2.67	1.93	140.9
01DC	44.8694	−65.1477	2022	1291.6	6.5	10.0	0.0750	−0.0230	22250	33250	2.46	1.79	1075.5
01DD	45.0825	−64.5234	1207	135.1	6.3	9.8	0.0720	−0.0250	22320	33260	2.46	1.79	109.2
01DE	44.9451	−64.1077	1313	345.8	6.2	9.7	0.0750	−0.0250	21950	32770	2.48	1.81	282.3
01DF	45.1839	−63.8555	1055	311.5	5.5	9.0	0.0750	−0.0250	20330	30750	2.60	1.92	254.3
01DG	45.1513	−63.3148	2649	2959.3	5.5	9.0	0.0750	−0.0250	20670	31370	2.58	1.89	2416.3
01DH	45.4784	−63.2686	1360	2165.5	5.5	9.0	0.0750	−0.0250	21360	31500	2.53	1.88	1768.1
01DJ	45.4834	−63.8743	1002	249.4	5.5	9.0	0.0750	−0.0250	20500	31170	2.59	1.90	203.6
01DK	45.4520	−64.5830	961	239.2	5.6	9.1	0.1185	−0.0250	20690	31370	2.58	1.89	212.5
01DL	45.6760	−64.2498	1313	694.8	5.5	9.0	0.0915	−0.0250	20670	31250	2.58	1.89	592.3
01DM	45.9139	−64.0036	387	51.7	5.5	10.0	0.0750	−0.0250	21500	31500	2.52	1.88	42.2
01DN	45.7299	−63.7269	1489	1750.6	5.5	9.0	0.0750	−0.0250	21290	31500	2.53	1.88	1429.4

Appendix (*continued*)

CODE	LAT	LONG	AKM2	REARN	TANN	TANF	SUMRN	SUMRD	GOIN	GOIF	AGEN	AGEF	REARF
01DO	45.7004	−63.1354	979	365.7	5.5	9.0	0.0750	−0.0250	21500	31500	2.52	1.88	298.6
01DP	45.5321	−62.6952	1154	583.1	5.5	9.0	0.0750	−0.0250	21300	31500	2.53	1.88	476.1
01DQ	45.6406	−62.3073	703	93.9	5.5	9.0	0.075	−0.0250	20720	31500	2.57	1.88	76.7
01DR	45.6112	−61.9969	967	411.1	5.5	9.0	0.075	−0.0250	20500	30860	2.59	1.92	335.7
01EA	44.0463	−65.8496	2233	1211.6	6.7	10.2	0.0835	−0.0250	19960	31210	2.63	1.90	1014.1
01EB	43.7710	−65.5910	1623	560.2	7.5	11.0	0.0515	−0.0250	20970	32500	2.55	1.83	401.8
01ED	44.1862	−65.1126	3247	326.6	7.6	11.1	0.062	−0.0250	22500	33730	2.44	1.76	252.3
01EE	44.3740	−64.8258	2151	1567.3	7.5	11.0	0.0605	−0.0250	23040	33790	2.40	1.76	1200.6
01EF	44.6350	−64.6934	1993	2246.5	6.8	10.3	0.06	−0.0250	22850	33650	2.42	1.77	1715.8
01EG	44.6456	−64.4133	1072	1170.0	6.9	10.4	0.0375	−0.0250	22330	33500	2.46	1.77	675.5
01EH	44.7344	−64.0111	1078	521.1	6.4	9.9	0.0625	−0.0250	20920	31920	2.56	1.86	403.6
01EJ	44.6376	−63.6883	1090	379.2	6.8	10.3	0.0095	−0.0250	21310	32560	2.53	1.82	0.00
01EK	44.9167	−63.2110	1600	1564.7	6.3	9.8	0.094	−0.0250	20440	31500	2.59	1.88	1340.6
01EL	44.8903	−62.8706	1207	659.6	6.1	9.6	0.143	−0.0250	19210	30000	2.69	1.96	599.2
01EM	45.0937	−62.6220	961	1313.2	5.5	9.0	0.125	−0.0250	19060	29720	2.70	1.98	1174.6
01EN	45.0902	−62.2597	1289	1639.4	5.5	9.0	0.1315	−0.0250	17900	28300	2.79	2.06	1475.4
01EO	45.3274	−62.2474	1635	3970.5	5.5	9.0	0.0845	−0.0250	19810	30120	2.64	1.96	3331.8
01EP	45.2392	−61.7770	727	397.3	5.6	9.1	0.1465	−0.0250	19640	30070	2.66	1.96	361.8
01EQ	45.3188	−61.4501	1119	611.5	5.9	9.4	0.1565	−0.0250	20030	30500	2.63	1.94	560.5
01ER	45.4978	−61.4736	592	323.5	5.5	9.0	0.125	−0.0250	20370	30500	2.60	1.94	289.3
01FA	45.8742	−61.3400	1289	514.3	5.5	9.0	0.125	−0.0250	20370	30830	2.60	1.92	460.0
01FB	46.3345	−61.0446	1295	3313.7	5.7	9.2	0.1515	−0.0250	20560	31500	2.58	1.88	3028.0
01FC	46.7379	−60.7904	803	642.4	5.7	9.2	0.2125	−0.0250	20080	30580	2.62	1.93	603.4
01FD	46.7742	−60.4938	1072	427.7	6.0	9.5	0.2040	−0.0250	20330	30670	2.60	1.93	400.6
01FE	46.4428	−60.6133	856	341.5	6.1	9.6	0.2125	−0.0250	20500	31250	2.59	1.89	320.8
01FF	46.2228	−60.8211	838	2684.3	5.5	9.0	0.1680	−0.0250	20500	30790	2.59	1.92	2476.5
01FG	45.9459	−61.0795	785	313.2	5.5	9.0	0.1250	−0.0250	20250	30500	2.61	1.94	280.1
01FH	45.7033	−60.7901	873	197.2	5.5	9.0	0.1350	−0.0250	19500	30500	2.67	1.94	178.0
01FJ	46.0027	−60.2782	2983	1137.8	5.5	9.0	0.1715	−0.0250	19630	30030	2.66	1.96	1051.6
02PE	47.3413	−70.7030	3182	912.0	3.7	8.2	0.1530	0.0250	21310	31780	2.53	1.86	983.7
02PG	47.4218	−69.7801	3047	1000.0	3.6	8.1	0.0835	0.0250	20330	31000	2.60	1.91	1139.9
02QA	48.2005	−68.5115	5702	1498.0	2.5	7.0	0.0935	−0.0200	17930	28080	2.79	2.08	1328.2
02QB	48.7753	−67.1628	3721	4710.0	2.5	7.0	0.1180	−0.0750	17150	26870	2.85	2.15	2843.2
02QC	49.0619	−65.6617	3950	3167.0	2.4	6.9	0.1215	−0.0570	15610	25320	2.99	2.25	2307.5
02RH	48.3700	−70.8870	14897	4240.0	2.4	6.9	0.1590	−0.0345	19320	29400	2.68	2.00	3751.9
02SB	49.4173	−69.2685	6388	6800.0	1.7	6.2	0.1260	−0.0750	16710	26410	2.89	2.18	4326.2
02SC	48.8190	−69.6666	7777	646.0	2.3	6.8	0.1295	−0.0710	18080	28040	2.78	2.08	434.2
02UA	50.0665	−67.4387	11428	4052.0	1.2	5.7	0.1515	−0.0750	14360	23890	3.10	2.35	2879.3
02UC	51.5809	−66.3605	21109	26255.0	−0.6	3.9	0.1920	−0.0580	11810	20820	3.33	2.57	21933.8
02VA	50.7571	−65.1618	6083	460.0	1.0	5.5	0.1750	−0.0250	13300	22500	3.19	2.44	425.9
02VB	51.1518	−64.4948	15893	7943.0	0.7	5.2	0.1750	−0.0265	13120	22550	3.21	2.44	7316.9
02VC	51.6415	−63.5929	14516	17880.0	0.5	5.0	0.1750	−0.0415	13110	22890	3.21	2.42	15616.7
02WA	50.9525	−62.5295	13162	4008.0	1.4	5.9	0.1750	−0.0250	13290	22830	3.19	2.42	3710.7
02WB	51.7463	−62.1401	16069	18600.0	0.9	5.4	0.1750	−0.0445	13330	23280	3.19	2.39	16062.0
02WC	50.8524	−60.7366	15718	12963.0	2.1	6.6	0.1915	−0.0250	12810	22640	3.24	2.43	12087.3
02WD	49.4938	−63.0174	7888	7212.0	2.7	7.2	0.1665	−0.0250	14510	23230	3.08	2.39	6648.5
02XA	51.8905	−60.5990	22527	390.0	1.1	5.4	0.1850	−0.0330	13140	23210	3.21	2.39	353.5
02XB	51.8375	−59.0144	12178	14635.0	1.6	5.4	0.1785	−0.0250	12720	22610	3.25	2.44	13571.5
02XC	52.0103	−57.9105	9564	9746.0	1.5	5.0	0.1535	−0.0105	11560	18820	3.36	2.72	9406.8
02XD	51.8667	−56.6916	4788	4955.3	1.6	5.1	0.1220	−0.0085	10270	14210	3.49	3.11	4779.6
02YA	51.3355	−56.3226	1752	839.6	2.4	5.9	0.1555	−0.0250	10200	13630	3.49	3.16	769.2
02YB	51.1948	−56.0644	2262	550.2	2.5	6.0	0.1395	−0.0250	10440	13120	3.47	3.21	498.5
02YC	50.7660	−56.9250	1776	2292.8	2.5	6.0	0.233	−0.0250	11500	17760	3.36	2.80	2166.3
02YD	50.7934	−56.3490	2080	871.9	2.5	6.1	0.168	−0.0250	11170	16170	3.40	2.94	804.4
02YE	50.1922	−57.3884	3047	1418.2	3.4	7.7	0.2345	−0.0250	12890	21400	3.23	2.52	1340.5
02YF	50.2494	−56.7916	2543	618.5	3.4	7.8	0.1945	−0.0250	12500	20210	3.27	2.61	577.4

Appendix *(concluded)*

CODE	LAT	LONG	AKM2	REARN	TANN	TANF	SUMRN	SUMRD	GOIN	GOIF	AGEN	AGEF	REARF
02YG	49.7942	−56.8507	2362	2388.5	3.5	8.0	0.1605	−0.0250	13630	22760	3.16	2.42	2194.6
02YH	49.3782	−57.8457	2508	12990.3	4.3	8.8	0.225	−0.0250	15260	24100	3.02	2.33	12247.4
02YJ	48.7552	−58.4442	2620	5467.0	4.5	8.3	0.1945	−0.0250	16340	23820	2.92	2.35	5103.6
02YL	49.3590	−57.4739	3569	774.0	4.4	8.9	0.1905	−0.0250	15680	24880	2.98	2.28	721.4
02YM	49.6947	−56.2043	3042	2827.5	3.6	8.1	0.107	−0.0250	13940	22360	3.13	2.45	2475.2
02YO	48.8941	−55.8234	7495	7591.4	4.3	8.7	0.119	−0.0250	15700	24860	2.98	2.28	6747.0
02YP	49.2884	−55.9614	2743	946.1	4.0	8.5	0.0915	−0.0250	14970	24100	3.04	2.33	806.6
02YQ	48.7944	−55.0496	5796	16507.4	4.3	8.7	0.1275	−0.0235	15280	24570	3.01	2.30	14908.7
02YR	49.0101	−54.1876	4694	5113.4	3.9	7.6	0.0675	0.0140	14450	22170	3.09	2.47	5618.7
02YS	48.4304	−54.4283	4266	3282.1	4.5	8.7	0.1145	0.0005	15220	24660	3.02	2.29	3289.3
02ZA	48.1995	−58.4858	5462	13766.2	4.7	8.2	0.2455	−0.0250	16830	24630	2.88	2.30	13046.5
02ZB	47.8032	−58.5238	2983	2036.3	5.3	8.8	0.2925	−0.0250	17060	25380	2.86	2.25	1947.3
02ZC	47.9462	−57.4367	4219	4220.9	4.9	8.4	0.2215	−0.0250	16500	24770	2.91	2.29	3975.6
02ZD	47.9544	−56.6795	4647	1992.5	4.9	8.4	0.1925	−0.0250	16080	24730	2.94	2.29	1858.6
02ZE	48.1271	−55.9681	4753	2149.4	4.6	8.2	0.1680	−0.0250	15540	25220	2.99	2.26	1983.0
02ZF	47.8653	−55.2270	4817	4017.5	4.9	8.6	0.1505	−0.0250	15200	25080	3.02	2.27	3668.7
02ZG	47.2210	−55.2417	3170	2261.6	5.5	9.0	0.2130	−0.0250	14970	25500	3.04	2.24	2124.7
02ZH	47.8918	−54.4050	2989	999.9	4.9	8.8	0.1440	−0.0250	14550	25060	3.08	2.27	909.0
02ZJ	48.2951	−53.6468	3053	571.2	4.5	8.1	0.0975	0.0070	14340	23550	3.10	2.37	591.3
02ZK	47.2981	−53.7828	3206	3067.7	5.1	8.6	0.2040	−0.0250	14320	24650	3.10	2.29	2873.6
02ZL	47.7192	−53.2647	1864	709.8	4.5	8.0	0.1460	−0.0250	14460	24270	3.09	2.32	646.2
02ZM	47.1927	−52.9882	2907	1924.2	5.2	8.7	0.2335	−0.0250	14070	24640	3.12	2.30	1818.3
02ZN	46.9790	−53.4092	1799	1822.1	5.5	9.0	0.2795	−0.0250	13540	24500	3.17	2.30	1738.7
03JA	57.4318	−73.9728	17499	12884.4	−5.7	−0.3	0.2425	0.0690	5470	11430	4.01	3.37	14602.8
03JB	58.3419	−71.9571	21203	15611.6	−6.0	−1.5	0.1950	0.0310	5280	12310	4.03	3.29	16806.8
03KA	57.1850	−72.2122	8199	7813.0	−5.5	−0.9	0.1835	0.0310	6180	13080	3.93	3.21	8447.2
03KB	56.3596	−71.3328	20605	19635.0	−5.4	−0.9	0.1635	0.0190	6880	14450	3.85	3.09	20744.5
03KC	57.3287	−70.2738	7138	6802.0	−5.5	−1.0	0.1490	0.0250	6410	14500	3.90	3.08	7350.5
03LF	57.2335	−69.2610	5093	8320.0	−5.5	−1.0	0.1750	0.0250	6500	15360	3.89	3.01	8894.5
03MA	55.8728	−64.6064	27368	21778.1	−3.9	0.6	0.2400	−0.0445	6730	15520	3.87	2.99	19655.7
03MB	56.2535	−66.3317	24678	20498.5	−4.6	−0.1	0.1915	0.0085	6810	15500	3.86	3.00	20948.5
03MC	57.5882	−67.7775	9031	7501.5	−5.5	−1.0	0.1810	0.0250	6390	15390	3.90	3.00	8002.8
03MD	58.1849	−65.5002	39487	31421.9	−5.1	−0.6	0.4000	−0.0040	5470	14330	4.01	3.10	31264.4
03ND	56.9081	−62.9504	7988	6617.4	−3.5	1.0	0.4220	−0.0250	5680	15010	3.99	3.04	6418.4
03NE	56.0671	−62.7646	15958	10028.2	−3.2	1.3	0.3445	−0.0300	6330	15490	3.91	3.00	9581.6
03NF	55.2807	−61.7594	15202	18374.1	−2.6	1.9	0.3010	−0.0445	6900	16130	3.85	2.94	16961.6
03NG	54.7342	−61.7602	18202	5335.0	−2.1	2.3	0.2395	−0.0475	8280	17500	3.70	2.83	4776.7
03OE	53.0828	−61.3990	16069	5330.9	−0.2	4.3	0.1780	−0.0395	13000	23680	3.22	2.36	4702.4
03PB	54.0100	−61.7070	16737	11917.5	−1.3	3.2	0.2035	−0.0450	11030	21240	3.41	2.53	10517.6
03PC	54.3683	−59.1684	13860	11602.7	−0.8	3.4	0.2295	−0.0180	8800	16010	3.64	2.95	11138.4
03QA	53.0274	−60.0384	5591	1157.0	0.3	4.8	0.1740	−0.0250	12960	23090	3.22	2.40	1070.7
03QB	53.9096	−58.3896	3628	66.2	−0.5	3.5	0.1800	−0.0085	9550	15290	3.56	3.01	64.6
03QC	53.0207	−57.5672	35760	32131.7	0.5	4.1	0.1235	0.0115	10160	14400	3.50	3.09	33594.4

Potential to index climate with growth and recruitment of temperate fish

Donald L. Pereira, Charles S. Anderson, Paul J. Radomski, Thomas Sak, and Allen Stevens

Pereira, D.L., C.S. Anderson, P. Radomski, T. Sak, and A. Stevens. 1995. Potential to index climate with growth and recruitment of temperate fish, p. 709–717. *In* R.J. Beamish [ed.] Climate change and northern fish populations. Can. Spec. Publ. Fish. Aquat. Sci. 121.

Abstract: Those fish and communities that are measurably sensitive to episodic climate events on time scales of months to seasons may be the best candidates for examining climate change effects on longer time scales. Application of linear models to routinely collected data on age structure and scale growth allows calculation of indices of yearly recruitment and growth. Walleye (*Stizostedion vitreum vitreum*) growth indices were associated with several measures of extreme temperatures, but first-year growth was most sensitive. Changes in smallmouth bass (*Micropterus dolomieu*) growth were synchronous among six lakes near the northern limit of their range and appeared to be related to summer temperatures. Relatively short time series (<10 yr) of growth data may suffice to examine spatial scales of responses to climate events, while longer series are necessary for superposed epoch analysis of growth or recruitment.

Résumé : Les poissons et communautés dont la sensibilité aux événements climatiques épisodiques peut se mesurer sur des échelles de temps assez courtes (mois à saisons) sont peut-être les meilleurs candidats pour l'examen des effets du changement climatique sur des échelles de temps plus longues. L'application de modèles linéaires à des données recueillies systématiquement sur la structure par âge et sur la croissance des écailles permet de calculer des indices annuels de recrutement et de croissance. Les indices de croissance du doré jaune (*Stizostedion vitreum vitreum*) ont été reliés à plusieurs mesures de températures extrêmes, mais la croissance de la première année a été des plus sensible. Les changements dans la croissance de l'achigan à petite bouche (*Micropterus dolomieu*) étaient synchrones dans six lacs situés près de la limite nord de l'aire de répartition de cette espèce et semblaient liés aux températures estivales. Une série chronologique relativement courte (< 10 ans) de données sur la croissance pourrait suffire à l'examen des échelles spatiales de réactions aux événements climatiques, tandis que des séries plus longues sont nécessaires pour l'analyse par époques superposées de la croissance ou du recrutement.

Introduction

Standardized assessment of fish populations has been conducted in Minnesota lakes for approximately 40 yr (Scidmore 1970), with some lakes in the program being surveyed on a 5-yr rotation. More recently, an annual assessment program was established for Minnesota's large lakes (i.e., over 6100 ha; Wingate and Schupp 1985). Together, these survey programs provide a growing data base including samples of calcified structures and relative abundance data from key species. This sampling regime, therefore, is of potential value for establishing baseline information on growth and recruitment needed to detect significant trends arising from changes in global climate.

Temperature is one facet of climate that has important effects on the dynamics of fish populations. Regier et al. (1990) noted that many studies of the relationship of fish to climate examine indirect causal pathways involving links between climatological, hydrological, and biotic subsystems. In this paper, we do not examine causal pathways directly, but rather examine whether the effects of episodic climatological events can be detected in fisheries data providing time series of varying lengths. Those fish and communities that are measurably sensitive to episodic events occurring on a time scale of months to seasons may be the best candidates for examining climate change effects on somewhat longer time scales. Such episodic events may include years with abnormal temperatures or years during the mature phase of an El Niño event. Teleconnections from the southern hemisphere, including El Niño, influence fish stocks in the northern hemisphere (Wooster and Fluharty 1985). Teleconnections during strong El Niño events may extend to the interior of the North American continent (Ropelewski and Halpert 1986; Robertson 1989).

The best candidate lakes may be isothermal, offering little opportunity for behavioural thermoregulation. Several of Minnesota's large walleye (*Stizostedion vitreum vitreum*) lakes do not normally undergo temperature stratification during the

D.L. Pereira, C.S. Anderson, P. Radomski, T. Sak, and **A. Stevens.** Minnesota Department of Natural Resources, Section of Fisheries, 1200 Warner Rd., St. Paul, MN 55106, USA.

summer, although walleye growth may vary between basins within these lakes (Schupp 1972). We examined walleye growth in two of these lakes (Lake Pepin and the Red Lakes) and in a large walleye lake that stratifies (Rainy Lake). Alternatively, climate change effects may be most evident for species near the limits of their distribution. Smallmouth bass (*Micropterus dolomieu*) in northern Minnesota are near the limit of their native range and may be in thermally marginal habitat (Shuter and Post 1990), especially in deep lakes. This study examines the potential of these candidate lake types and fish species for future measurement of climate change effects in Minnesota. We formally tested whether changes in recruitment or growth (at different ages) of walleye or sauger (*S. canadense*) coincide with seasonal and annual temperature changes, and we present analytical methods useful for detecting climate-induced impacts in fisheries data. Analysis of smallmouth bass growth illustrates a possible way to examine spatial scales of climatic events with relatively rapidly obtainable data.

Study Sites

Lake Pepin is part of navigational pool 4 of the Mississippi River. The lake, however, was a natural reservoir created by a delta at the confluence of the Chippewa and Mississippi Rivers. The lake is long and narrow with a fairly regular shoreline and few backwater bays. Shoal water substrates are mostly sand and gravel. Submerged aquatic vegetation is not abundant. The lake has a diverse fish community with at least 83 species recorded.

The Red Lakes (Upper and Lower) are two large, shallow oval basins in northwestern Minnesota. Shoal water substrates are mostly sand and submerged aquatic vegetation is sparse. The lakes do not stratify in the summer and are mesotrophic. The fish community is similar to that of Rainy Lake (described below). The Red Lake Band of Chippewa Indians has operated a commercial fishery since 1917. Pereira et al. (1992) explored the dynamics of this fishery.

Rainy Lake is located on the border between Minnesota and Ontario with 75% of the surface area in Ontario. Water levels in the lake have been controlled since 1909 by a dam on the outlet. The lake has a rocky, irregular shoreline and three distinct basins: the North Arm, Redgut Bay, and the South Arm. The lake is dimictic, and the South Arm is generally thermally stratified from June to September. Rainy Lake is typically ice covered from 5 to 6 mo of the year. Approximately 50 fish species occur in Rainy Lake. The most abundant species in experimental gillnet catches have been northern pike (*Esox lucius*), yellow perch (*Perca flavescens*), sauger, cisco (*Coregonus artedii*), walleye, and white sucker (*Catostomous commersoni*). Cohen et al. (1993), in analyzing patterns in the fish communities in Rainy Lake over time, found that walleye gillnet catch per unit effort (CPUE) fluctuations were more synchronized with the fish assemblage in the South Arm than in the other basins.

We also analyzed growth data for smallmouth bass from six oligotrophic lakes in northeastern Minnesota. These lakes were part of a study to evaluate interactions between lake trout (*Salvelinus namaycush*) and smallmouth bass (Eiler and Sak 1993). Physical and chemical data for all lakes included in the analysis are listed in Table 1.

Methods

All growth and CPUE data collected on walleye from Rainy Lake and walleye and sauger from Lake Pepin were derived from regular stock assessment with five-panel, variable mesh gillnets (Wingate and Schupp 1985). Most sampling was conducted during late summer. We assumed that gillnet CPUE is an index of relative abundance. Growth data for the Red

Table 1. Physical and chemical characteristics of study lakes.

Lake	Latitude	Longitude	Surface area (km^2)	Mean depth (m)	Maximum depth (m)	Total Dissolved solids (mg/L)	phosphorous (mg/L)	chlorophyll *a* (mg/L)	Alkalinity (mg/L)	pH
Walleye lakes										
Peppin	44°22'N	92°16'W	158.9	6.40	18.29	312.1	0.210	18.7	150.0	8.4
Rainy[a]	48°37'N	93°10'W	489.3	11.46	49.07	51.9	0.018	5.0	18.2	6.8
Red Lakes[b]	48°08'N	94°47'W	1168.8	4.36	10.70	190.5	0.057	16.2	112.5	7.5
Smallmouth bass lakes[c]										
Caribou	47°43'N	90°40'W	2.95	na[d]	8.2	36.5	na	na	30.5	7.9
Flour	48°03'N	90°21'W	1.35	8.4	22.9	30.7	na	na	18.3	7.6
Greenwood	48°00'N	90°07'W	8.18	9.9	30.8	14.1	na	na	6.0	7.1
Loon	48°05'N	90°38'W	3.15	20.6	61.6	22.4	na	na	12.5	7.5
Two Island	47°52'N	90°28'W	2.96	2.6	8.2	17.8	na	na	10.7	7.1
West Bearskin	48°04'N	90°24'W	2.00	10.2	23.8	26.5	na	na	17.3	7.7

[a] Data for Rainy Lake are from the South Arm.
[b] Water quality data and latitude–longitude pertains to Upper Red Lake.
[c] Total dissolved solids for smallmouth bass lakes were estimated by multiplying conductivity (micromhos·cm^{-1}) by 0.65.
[d] Not available.

Lakes were obtained from Shroyer (1991). Shroyer randomly sampled the commercial fishery in 1987 and 1988. He also used random subsamples of scales archived at the University of Minnesota that were originally collected in 1949, 1950, 1968, 1969, 1972, and 1973. Smallmouth bass were sampled by electrofishing in May and June 1988–90, using a spherical anode with the boat as a cathode. All walleye, sauger, and smallmouth bass were aged using scales.

To index recruitment, we fitted the log-linear model:

$$(1)\ \mathbf{Log_e}\,(\mathbf{Y}_{ij}) = \mu + \alpha_i + \beta_j + \varepsilon_{ij}, \qquad \varepsilon_{ij} \sim N(0,\sigma^2)$$

where Y_{ij} is CPUE (number of fish per gillnet) of year-class j caught at age i, μ is the main effect, α_i is the age effect, β_j is the year-class effect we consider to be an index of recruitment, and ε_{ij} is the error. We fit this model to field data with JMP software (SAS Institute Inc. 1989). The model is statistically similar to that presented by Kimura (1988).

Scale increments were digitized to construct growth records for walleye and smallmouth bass. We applied the linear modeling system of Weisberg (1993), because it appropriately partitions variation in scale data between age and year effects. We estimated growth series with the linear model:

$$(2)\ \mathbf{Log_e}\,(S_{ijk}) = \mu + \alpha_i + \beta_j + \gamma_{ij} + \varepsilon_{ijk}; \qquad \varepsilon_{ijk} \sim N(0, \sigma^2)$$

where S_{ijk} is a scale increment for fish k at age i in year j, μ is the main effect, α_i is the age effect; β_j is the year effect, γ_{ij} is the age-by-year interaction effect, and ε_{ijk} is the error. The vector of year effects (i.e., the β_js) represents the growth index that we subjected to further analyses with climate data. Walleye estimated to be older than 8 yr were excluded from the analyses for Rainy and Red lakes, and those greater than 500 mm were excluded for Lake Pepin because of uncertainty in ageing. For walleye, we fit the model after excluding the first increment because there was no significant correlation between growth of young-of-the-year and older fish. Subsequent analyses of walleye growth were done separately on the mean first-year log scale increments (age-0 growth) and on the year coefficients (adult growth).

To test whether key years of extreme environmental conditions had unusual fish growth or recruitment, we applied superposed epoch analysis as described by Prager and Hoenig (1989), using software by Hoenig et al. (1989). This technique compares the average response for the years just prior to, during, and after a key event and determines whether the observed differences are statistically significant using Monte Carlo randomization to formulate the null distribution. We used the *W*-statistic of Prager and Hoenig (1989) where each key-event year is compared with its own background years; this is analogous to the Student's *t*-statistic computed for paired data. Prager and Hoenig (1992) estimated the statistical power of this technique from 0.35 to 0.50 for simulated data sets with a 35-yr time series, three or five key-events, and an epoch width of 5 yr.

We explored several climate variables with air temperature data provided by the State of Minnesota Climatology Office. For all study sites with the exception of the Red Lakes, air temperatures were recorded at stations with in approximately 55 km of the lake basins. For analysis of Red Lakes data, we obtained air temperature data from the Crookston weather station, which is located approximately 105 km due west of lower Red Lake. This station is situated in a geographic area that is primarily agricultural and contains less forest cover than the area to the east around the Red Lakes. This station, therefore, has maximum and minimum daily air temperatures that are 2.5 and 1.0°C higher, respectively, than those recorded at the town of Red Lake, Minn. There are no trends through time in these differences. We used data from the Crookston station because it provided the most complete temperature data near the Red Lakes. We used cumulative degree-days above 20°C (CDD) for the entire year, mean annual air temperature, and mean July air temperature for analysis of walleye and sauger recruitment and growth. For smallmouth bass growth, we used CDD and mean temperature from June through September. The method for determining cumulative degree-days above a base temperature is

$$(3)\ \text{If MDT} > \text{BASE, then CDD} = \text{CDD} + \text{MDT} - \text{BASE} \quad \text{else CDD} = \text{CDD}$$

where MDT is mean daily temperature (i.e., mean of minimum and maximum) and BASE is the base temperature of 20°C. Our use of CDD is, therefore, similar to that used by Colby and Nepszy (1981) and McCauley and Murdoch (1987). Key-event years for temperature were those with temperatures more than one standard deviation (SD) above or below the mean. For key events related to the El Niño – Southern Oscillation (ENSO), we used years during the mature phase of strong and moderate intensity El Niños as identified by Quinn et al. (1978) and Robertson (1989). Moderate ENSO events occurred in 1966, 1977, and 1987; strong ENSO events occurred in 1973 and 1983. We defined background years as 1 yr before and 1 yr after a key event for mean temperature and CDD (epoch width = 3 yr), and 2 yr before and after an ENSO year (epoch width = 5 yr).

Results

Lake Pepin walleye and sauger

Climate and biological series analyzed from Lake Pepin are illustrated in Fig. 1. Both temperature indices indicate the warmer period beginning in the mid-1980s. Sauger and walleye recruitment appear to be closely correlated, with both species showing downward trends in recruitment. Growth of adult walleye shows a positive trend beginning in the early 1970s, which follows a possible period of relatively fast growth in the mid-1960s. Age-0 growth of walleye calculated from fish of all capture ages was variable and does not show a consistent trend, although growth appears to increase starting around 1985. There was no recent trend apparent in first-year growth when calculated only from age 1+ fish.

Superposed epoch analysis for Lake Pepin showed that walleye recruitment was positively related to high CDD events. However, sauger recruitment showed no relationship to any climate variable (Table 2). Walleye adult growth appears to be slower during cool years, as indicated by significant tests for mean annual and July temperatures. Only mean annual temperature was related to first-year growth, with growth being faster and slower during warmer and cooler years

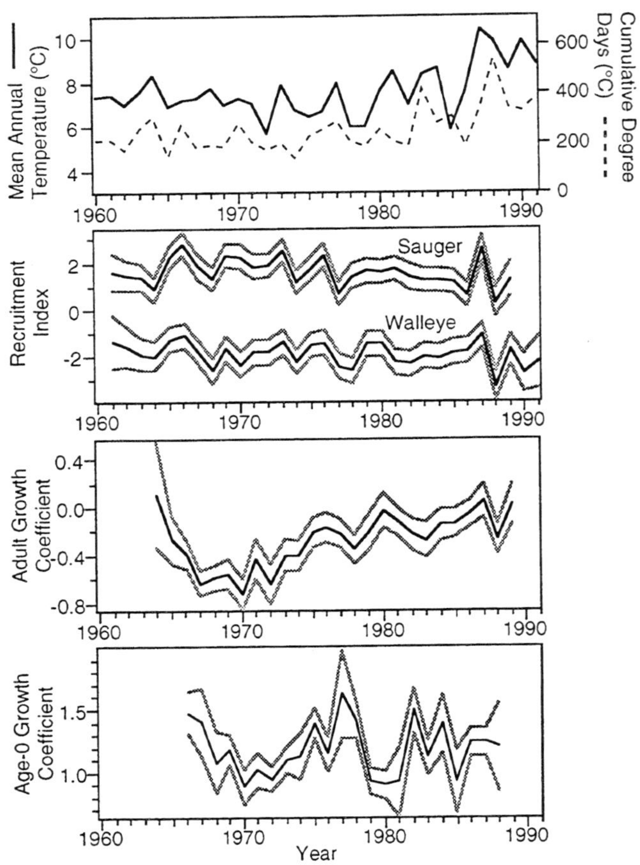

Fig. 1. Climate, recruitment, and growth series used for analysis of Lake Pepin walleye and sauger. Only recruitment data are available for sauger. Temperature data are from Winona, Minnesota. Shaded lines above and below the solid lines for recruitment and growth represent approximate 95% confidence intervals. Separate analyses were made for age-0 growth and subsequent growth of walleye. Age-0 growth was analyzed separately for fish captured at age 1+ and for fish captured at all ages. Only the data for age-0 growth estimated from fish captured at age 1+ is illustrated here. CDD is cumulative degree-days above a base temperature of 20°C.

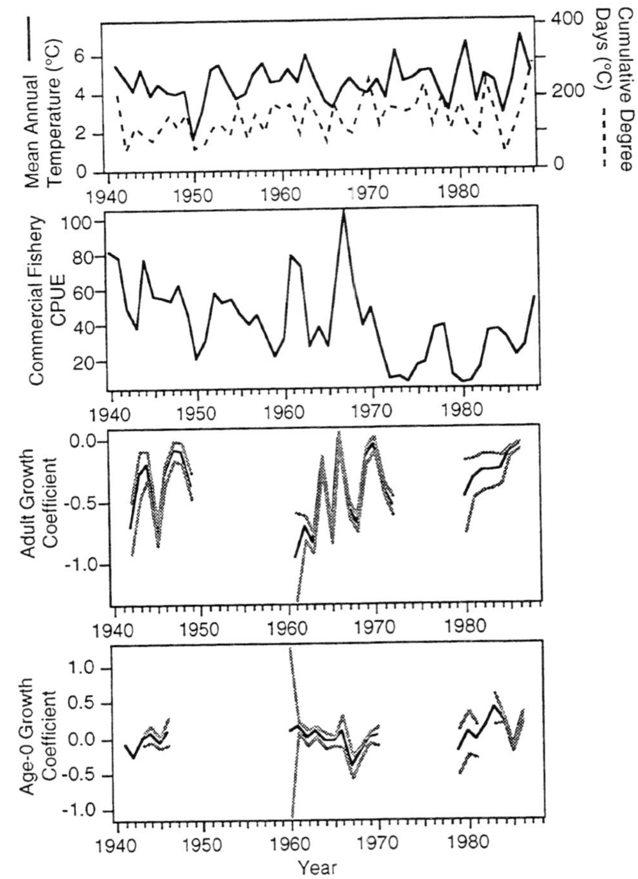

Fig. 2. Climate and growth series used for analysis of Red Lake walleye. Temperature data are from Crookston, Minnesota. CPUE data are for walleye from the Red Lakes commercial fishery, as described in Pereira et al. (1992). Growth data are from Shroyer (1991), with first-year growth coefficients estimated from fish captured at ages 2–8. Separate analyses were made for age-0 growth and for subsequent growth. CDD and approximate confidence intervals are as described in Fig. 1.

respectively. Estimates of first-year growth with fish of all capture ages showed no effect with high mean annual temperature ($P = 0.3371$), while the effect during years with low mean annual temperature was slightly significant ($P = 0.0454$). Therefore, it appears that first-year growth indexed from the scale increment of yearling fish (i.e., age 1+) is more sensitive to temperature than when estimated from fish of older ages. At the mature phase of El Niño, strong and moderate events were not related to adult growth and were only marginally related to recruitment ($P = 0.1303$) and first-year growth ($P = 0.1018$).

Red Lakes walleye

From the Red Lakes, we analyzed growth of adult and age-0 walleye (Fig. 2). The Red Lakes have supported an intensive, commercial fishery for walleye since the turn of this century (Smith 1977). Walleye dynamics are partially affected by this fishery; recruitment is currently episodic, with large year-classes following periods of 4–5 yr with little or no recruitment (Pereira et al. 1992). This pattern is illustrated in the time series of commercial CPUE (Fig. 2).

Superposed epoch analysis of Red Lakes walleye showed no significant results between adult growth and any of the climate variables (Table 3). Age-0 growth, however, was related to two temperature indices: low CDD years had slower growth, and years of high mean July temperature had higher growth.

Rainy Lake walleye

Climate, recruitment, and growth series used for analysis of Rainy Lake walleye are illustrated in Fig. 3. There were no apparent trends in recruitment, though recruitment appears to be relatively variable compared with Lake Pepin. Both growth series are relatively short, with years of good growth occurring in 1983 and 1988 for both adults and age-0 fish. Age-0 growth

Table 2. Results from epoch analysis of Lake Pepin walleye and sauger.

Population attribute	Key event (n)	*P* value	
		Walleye	Sauger
Recruitment[a]	CDD[b] ≥1 SD[c] above mean (5)	0.0139*	0.2816[d]
	CDD ≥1 SD below mean (2)	0.3075	0.3846
	Annual temp.[e] ≥1 SD above mean (4)	0.7333	0.4469[f]
	Annual temp. ≥1 SD below mean (4)	0.2752	0.6575
	July temp.[g] ≥1 SD above mean (6)	0.4380	0.7009
	July temp. ≥1 SD below mean (6)	0.0643	0.4785
	Mature phase El Niño (5)	0.1303	0.1294
Growth[h]	CDD ≥1 SD above mean (4)	0.5622	
	CDD ≥1 SD below mean (2)	0.2274	
	Annual temp. ≥1 SD above mean (3)	0.6780	
	Annual temp. ≥1 SD below mean (4)	0.0102*	
	July temp. ≥1 SD above mean (6)	0.2113	
	July temp. ≥1 SD below mean (5)	0.0192*	
	Mature phase El Niño (5)	0.4268	
Age-0 Growth[i]	CDD ≥1 SD above mean (5)	0.7646	
	CDD ≥1 SD below mean (1)	0.3922	
	Annual temp. ≥1 SD above mean (3)	0.0001*	
	Annual temp. ≥1 SD below mean (4)	0.0558*	
	July temp. ≥1 SD above mean (5)	0.7659	
	July temp. ≥1 SD below mean (4)	0.2121	
	Mature phase El Niño (5)	0.1018	

[a] Year-class main effect coefficients in Equation 1 are used for the recruitment index.
[b] CDD are cumulative degree-days throughout the year, above a base of 20°C.
[c] SD is standard deviation.
[d] This test had only four key events.
[e] Annual mean temperature (°C).
[f] This test had only two key events.
[g] July mean temperature (°C).
[h] Growth beginning in the second growing season was indexed with the year coefficients from the linear growth model using scale increments.
[i] First-year growth was indexed by the scale increment from the first growing season. Results reported in this table are from tests that included only fish captured at age 1+ (i.e., in their second growing season). Additional tests were done with this index derived from fish of all capture ages; these results are discussed further in the text.
* Significant *P* values.

was also fast in 1977. Recruitment and first-year growth had significant superposed epoch tests (Table 4). Recruitment was related to both high and low mean July temperatures. Age-0 growth was faster in years with high CDD and slower in years with low mean annual temperatures.

Walleye in Rainy Lake were the only species to have significant superposed epoch tests for the mature phase of El Niño (Table 4). The test for age-0 growth was highly significant ($P = 0.0115$). A nearly significant test for adult growth ($P = 0.0785$) suggests that the result for age-0 is not spurious.

Smallmouth bass in six northeastern lakes

Growth and climate series for smallmouth bass in six northeastern lakes are depicted in Fig. 4. Electrofishing yielded samples consisting almost entirely of fish less than 300 mm in total length. However, because growth was extremely slow, ages 1+ to 7+ were well represented. Because each of these series was quite short, we did not subject them to superposed epoch analysis. However, it is apparent that 1985 was a poor year for smallmouth bass growth in all six of these lakes. Patterns in growth also follow temperature, with 1985 and 1986 having the coolest summers for the 6 yr included in this analysis. Age distributions also indicated that there was poor recruitment of smallmouth bass in 1985, as fewer fish of the 1985 year-class were captured than of any other year-class.

Discussion

Results from superposed epoch analysis for walleye in Pepin, Red, and Rainy lakes suggest that growth of young-of-the-year may serve as a more sensitive indicator of climate events and change than growth of adult fish. Greater sensitivity of young fish is supported by evidence that the degree to which

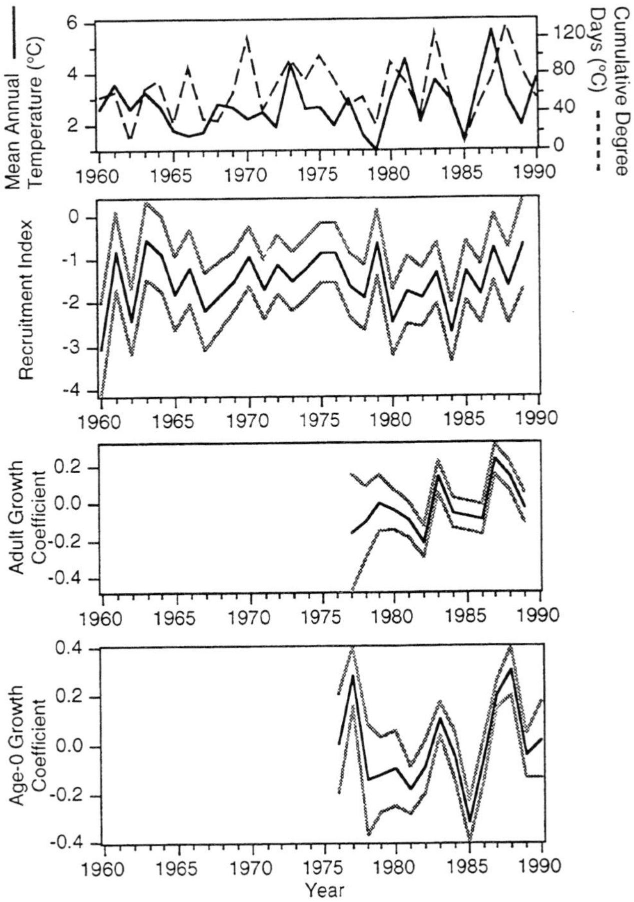

Fig. 3. Climate, recruitment, and growth series used for analysis of Rainy Lake walleye. Temperature data are from International Falls, Minnesota. Age-0 growth coefficients were estimated from fish captured at age 1+ only. CDD and approximate confidence intervals are as described in Fig. 1.

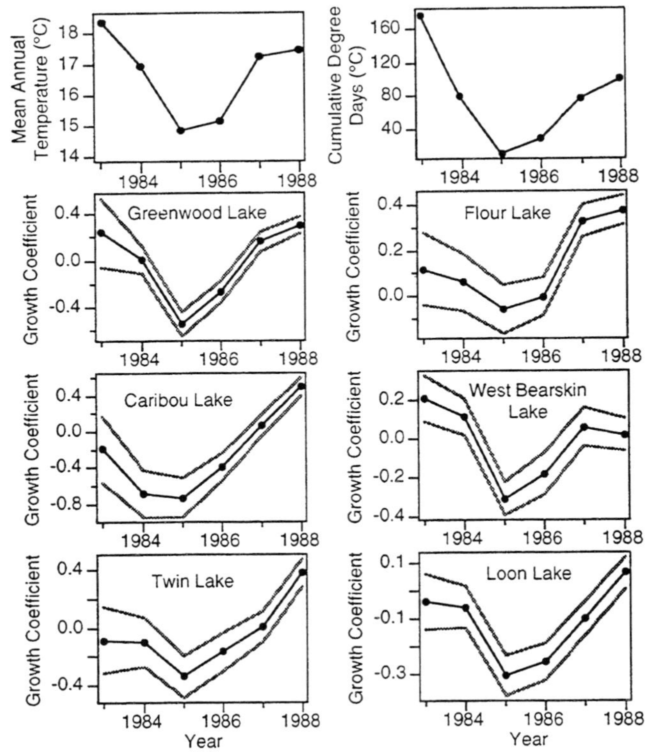

Fig. 4. Temperature and growth indices for smallmouth bass from six lakes in northeastern Minnesota. Temperature data are from the Gunflint weather station in northeastern Minnesota (latitude 48°10'N, longitude 90°53'W). CDD and approximate confidence intervals are as described in Fig. 1.

temperature affects metabolism is dependent on the size of the fish (Shuter and Post 1990). In this study, superposed epoch analysis yielded seven significant tests (of 21) for age-0 growth, while analysis of adult growth only resulted in two significant tests (both on Lake Pepin). A climate–growth link may indirectly influence the climate–recruitment relationship, as it is a common finding that walleye recruitment is positively associated with first-year growth (Forney 1976, 1980; Madenjian 1991; D. Pereira, unpublished data). Results from Pepin and Rainy lakes showed significant temperature–recruitment tests. Factors involved in the link may include prey size suitability, lipid content, or vulnerability to predation. Chevalier (1973) indicated that growth rates may influence mortality from cannibalism in Oneida Lake; that is, faster growing young-of-the-year walleye would be less vulnerable to adult predation.

Testing of temperature extremes with superposed epoch analysis may not show the form of nonlinear temperature associations with growth or recruitment. If mortality is moderated by growth rates, then year-class strength may often have a nonlinear relationship to temperature. Species have thermal optima for growth rate (Christie and Regier 1988), but additional nonlinearity may be introduced by growth-dependent predator–prey interactions. However, nonparametric superposed epoch analysis has a moderate level of statistical power to reject the null hypothesis where there are associations between extremes in environmental events and recruitment (Prager and Hoenig 1992). In addition, the selection of a temperature index may obscure associations. Colby et al. (1979) cited investigations of changes in walleye growth rates within the year as being dependent on suitable thermal conditions.

The smallmouth bass analyzed were all small and immature, and their growth fluctuations appeared synchronous in lakes in proximity. Because the fluctuations also appeared related to a climate variable, young smallmouth bass would be a good model with which to examine the spatial scale of climatic events as they drive fish production or the relative sensitivity of populations in different types of lakes.

El Niño key events had a significant relationship only with young-of-the-year growth in Rainy Lake. Ropelewski and Halpert (1986) state that teleconnections during El Niño do occur in northwestern North America as positive temperature anomalies, and Robertson (1989) reported shorter periods of ice cover for Lake Mendota in southern Wisconsin during El Niño events. Robertson also reported that effects of El Niño appear stronger in late winter, with ice breakup dates that

Table 3. Results from epoch analysis of Red Lakes walleye.

Population attribute	Key event (n)	*P* value
Growth[a]	CDD[b] ≥1 SD[c] above mean (5)	0.6162
	CDD ≥1 SD below mean (8)	0.2156
	Annual temp.[d] ≥1 SD above mean (5)	0.7757
	Annual temp. ≥1 SD below mean (6)	0.1234
	July temp.[e] ≥1 SD above mean (10)	0.2066
	July temp. ≥1 SD below mean (9)	0.2198
	Mature phase El Niño (8)	0.8169
Age-0 Growth[f]	CDD ≥1 SD above mean (6)	0.7676
	CDD ≥1 SD below mean (8)	0.0246*
	Annual temp. ≥1 SD above mean (4)	0.5604
	Annual temp. ≥1 SD below mean (6)	0.3481
	July temp. ≥1 SD above mean (10)	0.0075*
	July temp. ≥1 SD below mean (9)	0.2398
	Mature phase El Niño (7)	0.8013

[a] Growth beginning in the second growing season was indexed with the year coefficients from the linear growth model using scale increments.
[b] CDD are cumulative degree-days throughout the year, above a base of 20°C.
[c] SD is standard deviation.
[d] Annual mean temperature (°C).
[e] July mean temperature (°C).
[f] First-year growth was estimated with scale increments from fish of capture ages 2–8 (see Shroyer 1991).
* Significant *P* values.

Table 4. Results of epoch analysis of Rainy Lake walleye.

Population attribute	Key event (n)	*P* value
Recruitment[a]	CDD[b] ≥1 SD[c] above mean (5)	0.3781
	CDD ≥1 SD below mean (6)	0.4092
	Annual temp.[d] ≥1 SD above mean (3)	0.1997
	Annual temp. ≥1 SD below mean (5)	0.8711
	July temp.[e] ≥1 SD above mean (6)	0.0041*
	July temp. ≥1 SD below mean (5)	0.0287*
	Mature phase El Niño (5)	0.2546
Growth[f]	CDD ≥1 SD above mean (2)	0.1068
	CDD ≥1 SD below mean (2)	0.7764
	Annual temp. ≥1 SD above mean (3)	0.1247
	Annual temp. ≥1 SD below mean (3)	0.8287
	July temp. ≥1 SD above mean (2)	0.2781
	July temp. ≥1 SD below mean (1)	0.3028
	Mature phase El Niño (3)	0.0785
Age-0 Growth[g]	CDD ≥1 SD above mean (2)	0.0562*
	CDD ≥1 SD below mean (2)	0.1285
	Annual temp. ≥1 SD above mean (3)	0.4683
	Annual temp. ≥1 SD below mean (3)	0.0346*
	July temp. ≥1 SD above mean (2)	0.5355
	July temp. ≥1 SD below mean (1)	0.2075
	Mature phase El Niño (3)	0.0115*

[a] Year-class main effect coefficients in Equation 1 are used for the recruitment index.
[b] CDD are cumulative degree-days throughout the year, above a base of 20°C.
[c] SD is standard deviation.
[d] Annual mean temperature (°C).
[e] July mean temperature (°C).
[f] Growth beginning in the second growing season was indexed with the year coefficients from the linear growth model using scale increments.
[g] First-year growth was indexed with scale increments from fish capture at age 1+ only.
* Significant *P* values.

were 15 d earlier during El Niño events. On the basis of findings by Ropelewski and Halpert (1986), El Niño effects may be stronger when moving northwesterly from southern Wisconsin through northern Minnesota. A possible mechanism underlying the effect of El Niño on age-0 walleye growth in Rainy Lake could be a lengthening of the growing season. Madenjian (1991) showed the length of growing season and prey encounter rates determined the potential for young-of-the-year growth. Early ice breakup during El Niño events may significantly increase the length of the growing season, and therefore account for larger first-year increments.

Tests for the significance of El Niño events with Red Lakes walleye growth were not significant, which is consistent with results for freshwater drum (*Aplodinotus grunniens*) from the Red Lakes (Pereira et al. 1994). Because the Red Lakes are also located in northern Minnesota approximately 80 km south of Rainy Lake, we may expect to see results similar to those for Rainy Lake. However, the Red Lakes support an intensive fishery for walleye that is believed to induce considerable variation in recruitment and first-year dynamics of walleye (Pereira et al. 1992). Effects of other processes, such as exploitation, would hinder our ability to delineate climate signals in growth and recruitment, so analyses such as ours should be considered in light of results from other methods of stock assessment. This argument is also of concern for the analysis of Lake Pepin walleye and sauger. In that case, both species appear to have slight, downward trends in recruitment, while growth appears to increase. We suggest that trends should be explained before further analysis of climate associations is undertaken. Finally, application of these analyses to other large lakes in northern Minnesota, such as Lake of the Woods, may provide additional insight into potential teleconnections related to El Niño.

Possible synchrony in walleye year-class across wide geographical areas has been suggested (Colby et al. 1979; Serns 1982); however, year-class strengths of Rainy Lake walleye and Lake Pepin walleye were not correlated ($P > 0.05$). Böhling et al. (1991) found that patterns in the variation of year-class strength of perch were similar over large geographic areas, and attributed this to large-scale weather variations influencing water temperature. They did, however, find deviating patterns in areas exposed to environmental disturbances, such as water-level fluctuations and pollution. Water-level fluctuations in both Rainy and Pepin Lakes have been shown to influence walleye populations (Thorn 1984; Cohen and Radomski 1993). Cohen et al. (1991) found that dominant factors influencing Atlantic cod recruitment were operating on more local scales. They stated that "since recruitment is a multivariate process with different processes operating to varying degrees in different years, it is not surprising that the local scale of events is more important than the large scale." The primary factors determining year-to-year variation in fish growth include changes in stock abundance, fluctuations in the amount of available food, and climate factors such as temperature and length of growing season. With short time series such as ours, coherent variation over wide areas may be hidden by local disturbances, such as water levels and exploitation.

The methods we have applied here require the assumptions that fluctuations in predator–prey abundance are simply a source of random background variation, and that with enough data one can detect effects of climate. It appears that both walleye and smallmouth bass may serve as suitable species, with young-of-the-year providing the most sensitive life stage, and that the relative sensitivity of populations in different types of lakes may be measurable. Bioenergetic analysis (Hewitt and Johnson 1987) provides a method for separating variation in growth caused by temperature, length of growing season, and food availability. Madenjian (1991) has successfully applied similar methods to explain variation in young-of-the-year walleye growth in lakes Erie and Oneida. We propose the application of similar methods to suitable large walleye lakes in Minnesota. A review of scale increments suggests that growth of Lake Pepin age 1+ sauger during 1988 (i.e., the second growing season for the 1987 year-class) was curtailed by water temperatures exceeding 28–29°C. Bioenergetic analysis over the range of thermal conditions provided by Minnesota lakes may more clearly define temperature constraints and climate impacts on growth.

Acknowledgements

This work was supported in part by the Federal Aid in Sport Fish Restoration (Dingel-Johnson) Program. Greg Spoden of the Minnesota Climatology Office provided air temperature data. Two anonymous reviewers provided constructive suggestions that improved this manuscript.

References

Böhling, P., R. Hudd, H. Lehtonen, P. Karas, E. Neuman, and G. Thoresson. 1991. Variations in year-class strength of different perch (*Perca fluviatilis*) populations in the Baltic Sea with special reference to temperature and pollution. Can. J. Fish. Aquat. Sci. 48: 1181–1187.

Chevalier, J.R. 1973. Cannibalism as a factor in first year survival of walleye in Oneida Lake. Trans. Am. Fish. Soc. 102: 739–744.

Christie, G.C., and H.A. Regier. 1988. Measures of optimal thermal habitat and their relationships to yields of four commercial fish species. Can. J. Fish. Aquat. Sci. 45: 301–314.

Cohen, E.B., D.G. Mountain, and R. Oboyle. 1991. Local-scale versus large-scale factors affecting recruitment. Can. J. Fish. Aquat. Sci. 48: 1003–1006.

Cohen, Y., and P. Radomski. 1993. Water level regulations and fisheries in Rainy Lake and the Namakan Reservoir. Can. J. Fish. Aquat. Sci. 50: 1934–1945.

Cohen, Y., P. Radomski, and R. Moen. 1993. Assessing the interdependence of assemblages from Rainy Lake fisheries data. Can. J. Fish. Aquat. Sci. 50: 402–409.

Colby, P.J., R.E. McNicol, and R.A. Ryder. 1979. Synopsis of biological data on the walleye *Stizostedion vitreum vitreum* (Mitchill 1818). FAO Fish. Synop. 119: 139 p.

Colby, P.J., and S.J. Nepszy. 1981. Variation among stocks of walleye (*Stizostedion vitreum vitreum*): management implications. J. Fish. Res. Board Can. 38: 1814–1831.

Eiler, P.D., and T.J. Sak. 1993. Potential interactions between lake trout and smallmouth bass in lakes of northeastern Minnesota. Minn. Dep. Nat. Res. Sect. Fish. Invest. Rep. 426: 25 p.

Forney, J.L. 1976. Year-class formation in the walleye (*Stizostedion vitreum vitreum*) population of Oneida Lake, New York, 1966–73. J. Fish. Res. Board Can. 33: 783–792.

Forney, J.L. 1980. Evolution of a management strategy for the walleye in Oneida Lake, New York. N.Y. Fish Game J. 27: 105–141.

Hewitt, S.W., and B.L. Johnson. 1987. A generalized bioenergetics model of fish growth for microcomputers. Univ. Wis. Sea Grant Inst. Tech. Rep. WIS-SG-87-245: 69 p.

Hoenig, J.M., M.H. Prager, and N.B. Payton. 1989. Computer programs for investigating the effects of environmental events on a time series of recruitment. Can. Tech. Rep. Fish. Aquat. Sci. 1713: 19 p.

Kimura, D.K. 1988. Analyzing relative abundance indices with log-linear models. N. Am. J. Fish. Manage. 8: 175–180.

Madenjian, C.P. 1991. Limits to growth of young-of-the-year walleye (*Stizostedion vitreum vitreum*): an individual-based model perspective. Can. J. Fish. Aquat. Sci. 48: 1492–1499.

McCauley, E., and W.W. Murdoch. 1987. Cyclic and stable populations: plankton as paradigm. Am. Nat. 129: 97–121.

Pereira, D.L., C. Bingham, G.R. Spangler, Y. Cohen, D.J. Conner, and P.K. Cunningham. 1995. Growth and recruitment of freshwater drum (*Aplodinotus grunniens*) as related to long-term temperature patterns, p. 617–629. *In* R.J. Beamish [ed.] Climate change and northern fish populations. Can. Spec. Publ. Fish. Aquat. Sci. 121.

Pereira, D.L., Y. Cohen, and G.R. Spangler. 1992. Dynamics and species interactions in the commercial fishery of the Red Lakes, Minnesota. Can. J. Fish. Aquat. Sci. 49: 293–302.

Prager, M.H., and J.M. Hoenig. 1989. Superposed epoch analysis: a randomization test of environmental effects on recruitment with application to chub mackerel. Trans. Am. Fish. Soc. 118: 608–618.

Prager, M.H., and J.M. Hoenig. 1992. Can we determine the significance of key-event effects on a recruitment time series?—a power study of superposed epoch analysis. Trans. Am. Fish. Soc. 121: 123–131.

Quinn, W.H., D.O. Zopf, K.S. Short, and R.T.W. Kuo Yang. 1978. Historical trends and statistics of the southern oscillation, El Niño, and Indonesian droughts. Fish. Bull. U.S. 76: 663–678.

Regier, H.A., J.J. Magnuson, and C.C. Coutant. 1990. Introduction to proceedings: symposium on effects of climate change on fish. Trans. Am. Fish. Soc. 119: 173–175.

Robertson, D.M. 1989. The use of lake water temperature and ice cover as climatic indicators. Ph.D. thesis, University of Wisconsin, Madison, WI. 294 p.

Ropelewski, C.F., and M.S. Halpert. 1986. North American precipitation and temperature patterns associated with the El Niño – Southern Oscillation (ENSO). Mon. Weather Rev. 114: 2352–2362.

SAS Institute Inc. 1989. JMP® user's guide, version 2 edition. Cary, NC. 584 p.

Schupp, D.H. 1972. The walleye fishery of Leech Lake, Minnesota. Minn. Dep. Nat. Res. Sect. Fish. Invest. Rep. 317: 11 p.

Scidmore, W. J. 1970. Manual of instructions for lake survey. Minn. Dept. Conserv. Spec. Pub. 1 (Revised 1970): 47 p.

Serns, S.L. 1982. Influence of various factors on density and growth of age-0 walleyes in Escanaba Lake, Wisconsin, 1958–1980. Trans. Am. Fish. Soc. 111: 299–306.

Shroyer, S.M. 1991. Growth of Red Lakes walleyes from the 1940s to the 1980s. M.S. thesis, University of Minnesota, St. Paul, MN. 104 p.

Shuter, B.J., and J.R. Post. 1990. Climate, population viability, and the zoogeography of temperate fishes. Trans. Am. Fish. Soc. 119: 314–336.

Smith, L.L., Jr. 1977. Walleye (*Stizostedion vitreum vitreum*) and yellow perch (*Perca flavescens*) populations and fisheries of the Red Lakes, Minnesota, 1930–75. J. Fish. Res. Board Can. 34: 1774–1783.

Thorn, W.C. 1984. Effects of continuous fishing on the walleye and sauger population in Pool 4, Mississippi River. Minn. Dep. Nat. Resour. Sect. Fish. Invest. Rep. 378: 52 p.

Weisberg, S. 1993. Using hard-part increment data to estimate age and environmental effects. Can. J. Fish. Aquat. Sci. 50: 1229–1237.

Wingate, P.J., and D.H. Schupp. 1985. Large lake sampling guide. Minn. Dep. Nat. Resour. Sect. Fish. Spec. Pub.. 140: 27 p.

Wooster, W.S., and D.L. Fluharty [ed.]. 1985. El Niño North: Niño effects in the eastern subarctic Pacific ocean. Washington Sea Grant, University of Washington. Seattle, WA. 312 p.

Modeling the impact of climate variability on the advection of larval walleye pollock (*Theragra chalcogramma*) in the Gulf of Alaska

P.J. Stabeno, A.J. Hermann, N.A. Bond, and S.J. Bograd

Stabeno, P.J., A.J. Hermann, N.A. Bond, and S.J. Bograd. 1995. Modeling the impact of climate variability on the advection of larval walleye pollock (*Theragra chalcogramma*) in the Gulf of Alaska, p. 719–727. *In* R.J. Beamish [ed.] Climate change and northern fish populations. Can. Spec. Publ. Fish. Aquat. Sci. 121.

Abstract: An atmospherically driven primitive equation ocean model was integrated for two 90-d periods: the first, the spring of 1978, was a year of excellent recruitment for Gulf of Alaska walleye pollock (*Theragra chalcogramma*); the second, the spring of 1990, was a year of poor recruitment. There were distinct differences in the atmospheric forcing between the 2 yr, which produced accompanying differences in the currents. During 1978, larvae were more likely to be transported onto the shallow shelves adjacent to the sea valley, while during 1990 they were more likely to remain in the sea valley. The relevance of these results to the effects of future climate trends on the Alaska Coastal Current and larval survival are discussed.

Résumé : Un modèle océanique à équations primitives sur les effets de l'atmosphère a été intégré pour deux périodes de 90 jours : la première période, le printemps de 1978, se trouvait dans une année où le recrutement a été excellent pour le goberge du golfe d'Alaska; la seconde, le printemps de 1990, se trouvait dans une année où le recrutement était médiocre. On a noté des différences distinctes dans le forçage atmosphérique entre les deux années, différences qui ont produit les écarts résultants dans les courants. En 1978, les larves étaient plus susceptibles d'être transportées sur les plateaux peu profonds adjacents à la vallée sous-marine, tandis qu'en 1990, elles étaient plus portées à demeurer dans la vallée. La pertinence de ces résultats quant aux effets des futures tendances climatiques sur le courant côtier de l'Alaska et sur la survie des larves fait l'objet d'une analyse.

Introduction

Fisheries Oceanography Coordinated Investigations (FOCI) conducts research on the influence of physical and biological factors on the early life history of walleye pollock (*Theragra chalcogramma*) in the northern Gulf of Alaska. Large concentrations of adult pollock congregate each March in Shelikof Strait (Fig. 1). They spawn in early April, and by late April patches of larvae can usually be found southwest of the spawning area. Because spawning occurs within such small temporal and spatial windows, population and dynamic studies are tractable in this region.

P.J. Stabeno. National Oceanic and Atmospheric Administration, Pacific Marine Environmental Laboratory, NOAA Building No. 3, 7600 Sand Point Way N.E., Seattle, WA 98115, USA.

A.J. Hermann, N.A. Bond and **S.J. Bograd.** Joint Institute for the Study of the Atmosphere and Ocean, University of Washington, Seattle, WA 98195, USA.

Contribution No. 1432 from NOAA/Pacific Marine Environmental Laboratory

One FOCI hypothesis is that the survival rate is greatest for those pollock larvae that remain in shallow shelf waters along the Alaska Peninsula. Because eggs and larvae are mainly planktonic, transport to this location is largely determined by advection and diffusion. Transport on the shelf is dominated by the Alaska Coastal Current (ACC). Mean transport in the ACC is 0.85×10^6 m^3·s^{-1}, but can exceed 3×10^6 m^3·s^{-1}. There is also evidence that larval patches are associated with eddies (Schumacher and Kendall 1991; Schumacher et al. 1993). If systematic changes in the mean currents and eddy activity occur in this region, they will likely have important ramifications for pollock recruitment.

To directly assess the sensitivity of the ACC to changes in atmospheric forcing, a three-dimensional, primitive equation hydrodynamic model driven by winds and freshwater discharge has been tuned to this region (Hermann and Stabeno 1995). The results of this model were analyzed with a focus on the currents and eddies that are likely to have a direct bearing on the survivability of pollock larvae.

A change in the regional climate of the North Pacific during the middle to late 1970s presents a prime opportunity for this kind of sensitivity test. Niebauer (1988) documented mean increases of 3°C in sea-surface temperature and 10°C in air temperature, and a 20% decrease in ice cover for the southeastern Bering Sea from 1975 to 1979. During the same period, Trenberth (1990) found that the mean sea-level pressure for a region encompassing virtually the entire North Pacific decreased by about 2 mbar (1 mbar = 100 Pa). Associated with these changes were a strengthening and eastward displacement of the Aleutian Low. Of special relevance to FOCI, the period from 1975 to 1979 produced the best pollock year-classes observed in the northwestern Gulf of Alaska (Megrey et al. 1994). Since this period, the pollock recruitment has ranged from poor to mediocre and, in general, has been relatively weak. It is important to gauge how much of these variations in recruitment can be attributed to variations in the physical environment.

This paper investigates the possible impact of climate change by examining two different years. As a first step toward studying the details of the ocean's response, the numerical ocean model was driven with the two different periods of atmospheric forcing. These periods were the springs of 1978 and 1990, years in which there were distinct differences between the atmospheric forcing. They are of special interest because recruitment during 1978 was extremely strong while during 1990 it was poor.

Physical setting

Shelikof Strait and its associated sea valley are less than 60 km wide and extend for about 450 km between the Alaska Peninsula and the Kodiak Island plateau (Fig. 1). The sea valley forms a natural guide for circulation, connecting the inner shelf to the continental slope. The ACC is a buoyancy-driven current that is strongly modified by the winds (Schumacher et al. 1990). The high mountains along the Alaska Peninsula and on Kodiak Island interact with the strong atmospheric pressure gradients to produce ageostrophic winds, which result in periods of strong convergence (or divergence) at the exit to Shelikof Strait. The onset of the most common meteorological pattern (high pressure over Alaska and low pressure southwest

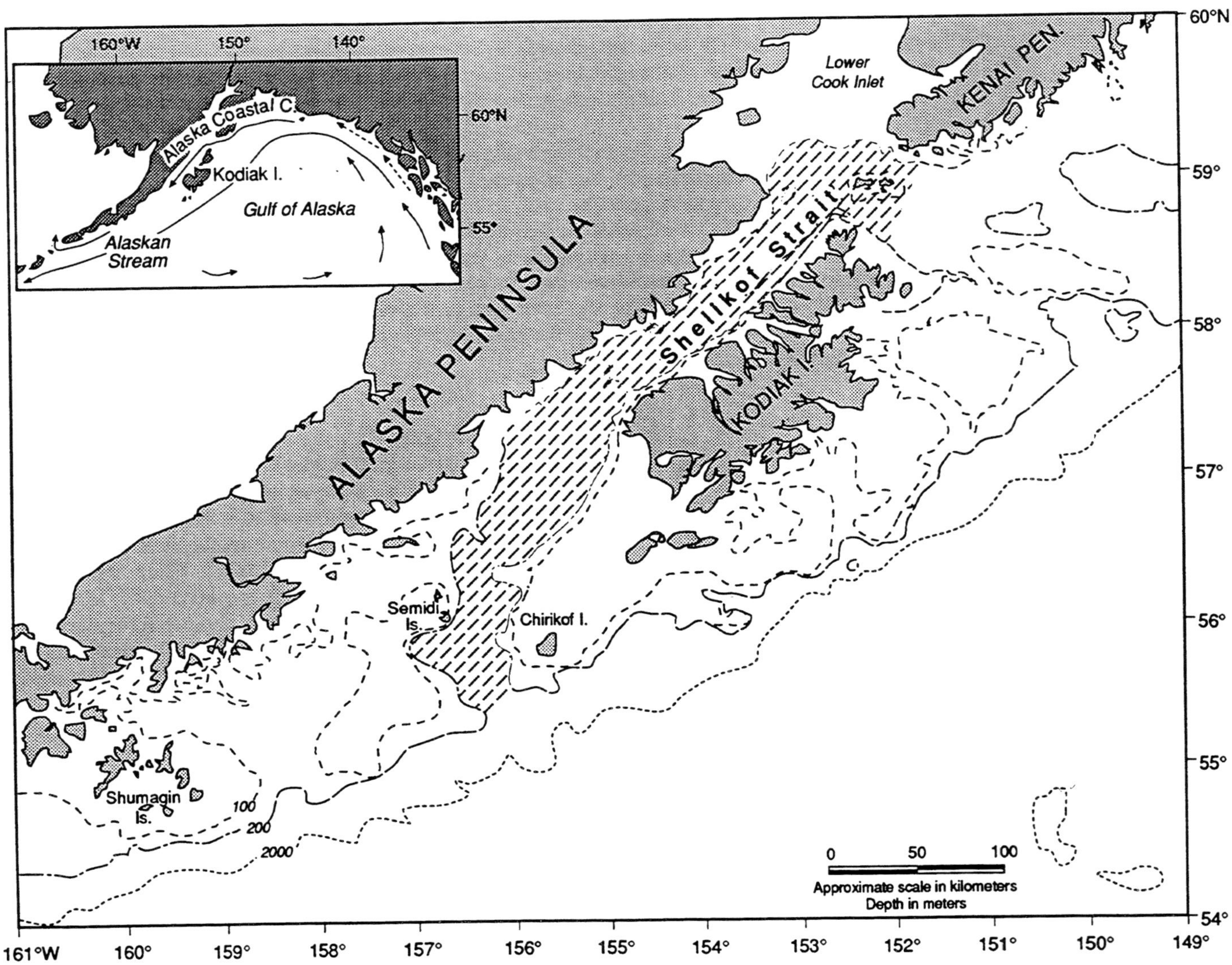

Fig. 1. The study area in the western Gulf of Alaska, with the Shelikof sea valley indicated by hatched lines. The insert shows the location of the Alaska Coastal Current. The depth contours are in metres.

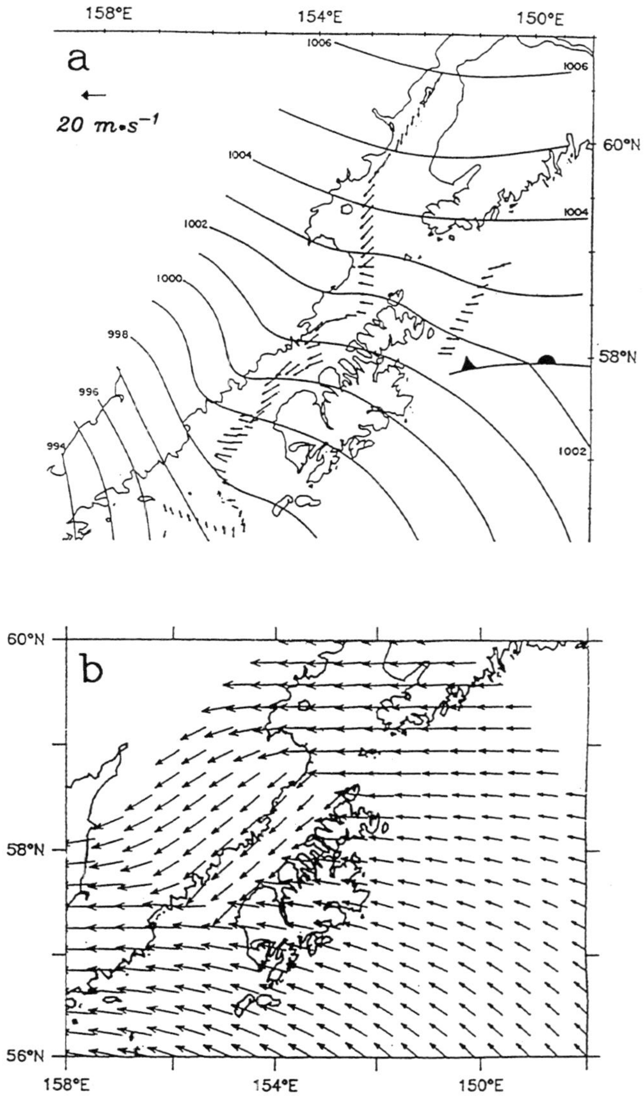

Fig. 2. (a) Winds observed by aircraft at 90 m, and (b) surface stresses (adjusted as in Table 1) based on large-scale sea-level pressure (db) distributions for 0000 UTC 4 March 1983. Note the flow parallel to the terrain within Shelikof Strait and the abrupt wind shift at its southwest exit.

of Kodiak Island (Overland and Hiester 1980)), can force rapid fluctuations of over 1×10^6 $m^3 \cdot s^{-1}$ in the transport of the ACC, doubling transport over the course of a day (Schumacher et al. 1990).

Wind forcing

Although direct measurements of winds in the northern Gulf of Alaska are rare, geotriptic winds calculated from sea-level pressure provide fair representation of winds in the region west of Kodiak Island (Macklin et al. 1993). Surface winds for March–May of 1978 and 1990 were computed from 6-hr atmospheric surface pressure supplied by the Fleet Numerical Oceanographic Center. The geostrophic winds were rotated 15° counterclockwise, speeds were reduced by 30% from the geostrophic value, and the results were interpolated to the hydrodynamic model grid.

The geotriptic winds in Shelikof Strait proper must be modified because ageostrophic (down gradient) winds are common here. The terrain-induced wind variations in Shelikof Strait have been documented with research aircraft (Lackmann and Overland 1989) and a special observing network (Macklin et al. 1993). Because these variations are large and persistent, they can have a significant local impact on the upper ocean. A simple algorithm has been developed from these observations to approximate the winds in Shelikof Strait from the large-scale sea level pressure field. For a prescribed range of geostrophic wind directions, the surface winds are assumed to be channeled and enhanced within the strait. The exact modifications to the geostrophic winds are given in Table 1. An example of ageostrophic winds in Shelikof Strait is shown in Fig. 2a. There is strong wind shear at the mouth of the sea valley, where the winds change abruptly from ageostrophic to geostrophic. This behaviour has been observed in aircraft observations (Fig. 2b).

Because the oceanic model is sensitive to the winds in the strait, we compared time series at 57.8°N, 155.2°W during 1978 and 1990 (Fig. 3). Throughout the 1978 time series, the winds were variable with downstrait winds occurring from the middle of April through the middle of May. During the first 45 d of the 1990 time series, the winds were generally downstrait, while from mid-April through May, the winds were relatively weak.

Table 1. The rotation and change in magnitude of the winds in the northeastern and southwestern portions of Shelikof Strait (Fig. 1) necessary to obtain appropriate ageostrophic winds. Winds are in the oceanographic convention.

Geostrophic winds direction (*T*)	Northeastern Strait		Southwestern Strait	
	Direction	Magnitude	Direction	Magnitude
150–210°	No change		No change	
210–270°	225°	1.1	225°	1.4
270–360°	T - 45°	0.8	T - 45°	0.9
0–30°	No change		No change	
30–90°	45°	1.3	45°	1.1
90–150°	T - 45°	0.9	T - 45°	0.8

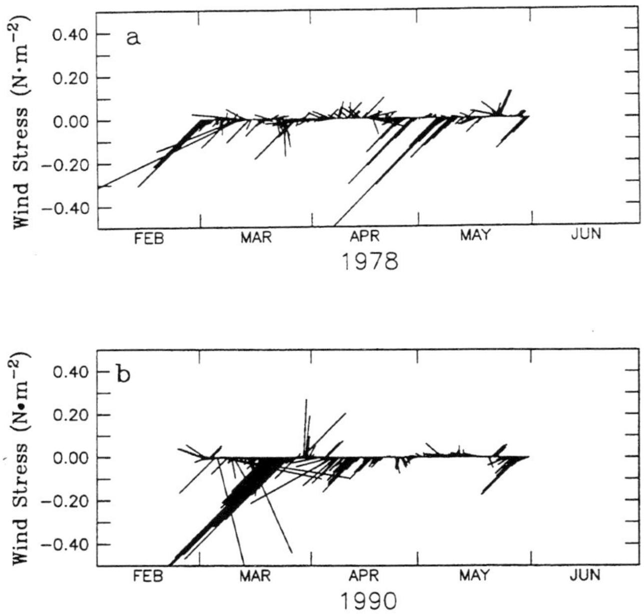

Fig. 3. Wind-stress times series at the exit to Shelikof Strait (57.8°N, 155.2°W) for (a) 1978 and (b) 1990.

Buoyancy forcing

High precipitation rates along the coast of the northeast Pacific Ocean (from 54°N to 150°W) produce large freshwater discharge, which enters the shelf waters through many streams and rivers. This line source of buoyancy at the coast, along with the winds, drives the ACC (Royer 1982). The March–May mean discharge for 1931–79 was 11×10^3 $m^3 \cdot s^{-1}$. During these 3 mo in 1978 the mean discharge was 13×10^3 $m^3 \cdot s^{-1}$ and in 1990 it was 9.3×10^3 $m^3 \cdot s^{-1}$ (T.C. Royer, School of Fisheries and Ocean Sciences, University of Alaska, Fairbanks, AK, personal communication). Thus runoff during spring 1978 was slightly greater than normal and during spring 1990 it was less than normal. The model was initialized and forced with salinity profiles normalized by the spring runoff values for each year. The average spring salinity field was obtained from conductivity–temperature–depth profiler (CTD) surveys (Reed 1984; Reed et al. 1987; Johnson et al. 1988). The initial conditions are shown in Fig. 4.

The Model

The eddy-resolving circulation model that we applied to the Shelikof Strait region is the rigid lid semispectral primitive equation model (SPEM) of Haidvogel et al. (1991a), which employs a topography-following (σ) vertical coordinate system. Details of the model's implementation in this region may be found in Hermann and Stabeno (1995). The horizontal grid was formulated in curvilinear-orthogonal coordinates; 257×97 horizontal grid points and nine vertical levels were employed. The full model domain was approximately 1500×500 km, with grid spacing telescoped (gradually increased) near the southern, eastern, and western boundaries. In the central near-coastal region of interest, the mean grid spacing was ~4 km. The baroclinic and barotropic modes were treated separately, the latter through the use of a barotropic streamfunction. A free-slip condition was applied at all boundaries, and bottom stress is proportional to velocity at the lowest grid point (Haidvogel et al. 1991b). Only Kodiak Island was treated explicitly as an island; the barotropic streamfunction value at this island was updated at each time step (J.I. Wilkin, K. Hedström, and J. Mansbridge, unpublished data). All other islands in the model domain were treated as subsurface topographic features (i.e., seamounts).

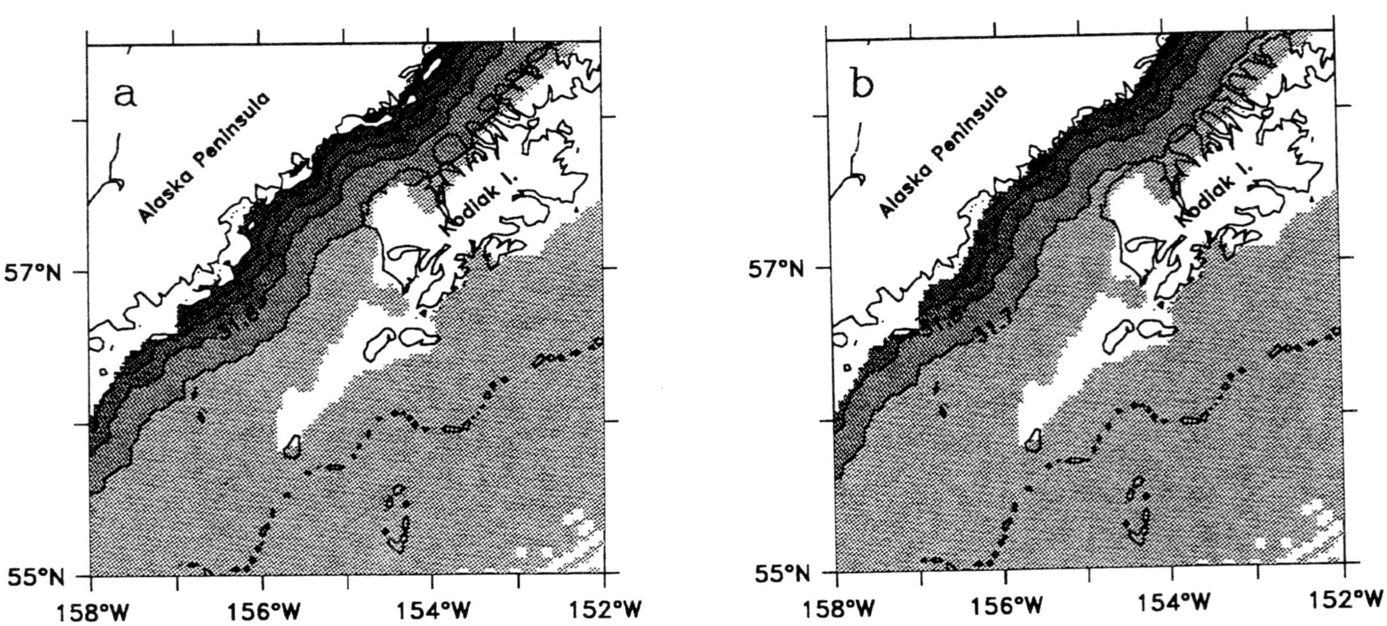

Fig. 4. The salinity (practical salinity units (psu)) distributions for (a) 1978 and (b) 1990, which served to initiate the two model runs. Darker shades represent lower salinity. Contour intervals are 0.1 psu.

The steeply sloping bathymetry of the region increased the difficulty of obtaining accurate horizontal pressure gradients. This problem, common with σ coordinate models, was partially solved by use of a high-order interpolating scheme for hydrostatic pressure between vertical grid points (Hedström 1990). This method enhanced the accuracy of the horizontal gradients computed at any vertical level.

Horizontal mixing of momentum and salinity was effected with a biharmonic operator scaled as grid spacing cubed (mean value 2×10^9 $m^3 \cdot s^{-1}$). Vertical viscosity and diffusivity were set constant at 15×10^{-4} $m^2 \cdot s^{-1}$ and 5×10^{-4} $m^2 \cdot s^{-1}$, respectively. Convective adjustment of statically unstable density profiles was performed every fifth time step ($\Delta t = 135$ sec). At the temperatures typical of this region in the spring, the density is largely determined by salinity; hence, an equation of state was employed that is a linear function of salinity only.

Wind forcing was achieved by distributing the calculated wind stress evenly in the top 20 m of the water column. The strength of the Alaskan Stream was fixed at 1×10^7 $m^3 \cdot s^{-1}$, which agrees with observed transport (Reed 1984). Buoyancy forcing was achieved by damping the salinity field back to the climatological mean (appropriate for that year) near the eastern and western boundaries with a damping time of 1 d.

Results

We examined the flow at 40 m because most larvae are found at this depth (Schumacher and Kendall 1991). Both years show mesoscale meanders in the flow with similar horizontal scales (approximately 40 km), which are also observed in satellite-tracked buoys and satellite images (Schumacher and Kendall 1991; Schumacher et al. 1991; Vastano et al. 1992). The mean flow generally followed contours of constant depth.

Three snapshots of salinity at 40 m, currents at 40 m and the barotropic stream function for 1978 (Fig. 5) and 1990 (Fig. 6) are shown. While major characteristics of flow and density are evident in all figures, there were also significant differences among snapshots.

Although the freshest water was confined along the peninsula in both years, the lowest salinities (less than 31.5 practical salinity units) were observed in 1978. This was due to higher runoff during 1978 compared to 1990. The horizontal density gradient (which is dominated by salinity) was greatest in 1978.

Currents at 40 m were generally stronger in 1978, exceeding 20 $cm \cdot s^{-1}$ at times. Mesoscale features (eddies) were common in both years, but were more numerous in 1978. Two eddies were evident in 1990, one near the exit to Shelikof Strait and the other approximately 200 km further southwest in the sea valley. These features were quasi stationary, appearing at similar locations in all snapshots for 1990. In contrast the eddies in 1978 generally occurred as dipole pairs and translated down the sea valley. One marked difference in flow patterns between 1978 and 1990, was the occurrence of flow onto the shallow (less than 120 m) shelves to the northwest and east of the sea valley in 1978. There was also stronger return flow at the shelf break onto the shelf in 1978.

The barotropic stream function illustrates the total, depth-integrated transport rather than currents at 40 m. The two stationary eddies of 1990 are clearly evident in the stream function maps of that year. Depth-integrated transport in Shelikof Strait was consistently stronger in 1978 than in 1990.

The model was seeded with neutrally buoyant floats near the exit to Shelikof Strait (Fig. 7a) on April 15 in both years. This is the region where larvae hatch each spring (Schumacher and Kendall 1991). Larvae at this age are mainly planktonic and the float tracks demonstrate the integrated effect of currents on larval transport. Generally, the floats in 1978 tended to progress further onto the shallow shelves to the northwest and southeast of the sea valley (Fig. 7b). There were also entrapment by mesoscale eddies near the exit of the strait. Conversely, floats in 1990 (Fig. 7c) remained near the center of the sea valley, with entrapment in the quasi-stationary eddy at the exit of the strait. Floats deployed on April 15, 1990, were not transported far down the sea valley due to the weak winds and baroclinic structure in April and May. Some floats deployed in late March, however, were transported to the exit of the sea valley during April (not shown).

Discussion

The Shelikof region with its predictable spawning provides a natural laboratory for exploring recruitment questions. This region has been studied over the last 7 yr, and a connection between transport, eddies, winds, and recruitment has been noted. Strong currents that flow down the sea valley joining the Alaskan Stream appear to produce poor year-classes (Schumacher and Kendall 1991); the occurrence of eddies is often associated with patches of larvae (Vastano et al 1992; Schumacher et al. 1993), and strong winds during the first days after hatching appear detrimental to the survival of larvae (Bailey and Macklin 1994).

During April and May, larval abundance is at its peak in the sea valley. There is some evidence that larval survival is enhanced within eddies (Schumacher et al. 1993); if so, the numerous eddies hindcasted by the model could have increased larval survival during 1978. Such eddies help retain larvae on the shelf. The tendency for flow onto the shallow shelves adjacent to the sea valley in 1978 (as seen in the float tracks) would have transported the larvae to appropriate nursery grounds, thus enhancing their survival. By contrast, the model results suggest that a greater proportion of larvae in 1990 were ultimately transported down the sea valley into the Alaskan Stream, where their fate is uncertain.

To anticipate the influence of future climate change on the pollock fishery, we need to be able to predict the changes that will occur in the physical environment and understand how these changes will affect recruitment. Regarding the first requirement, even the large-scale changes in the climate that are likely to occur are uncertain, but some consensus has been reached (Houghton et al. 1990). With increases in the concentration of greenhouse gases, most recent general circulation models (GCM) predict a mean surface-temperature increase, especially at higher latitudes such as the Gulf of Alaska. These models also predict increased winter precipitation for the gulf and the coastal regions along its rim (McBean et al. 1991). Because the resolution of the GCMs are coarse, regional interpretation of their results is limited. The changes in the local wind forcing near Shelikof Strait are uncertain, because

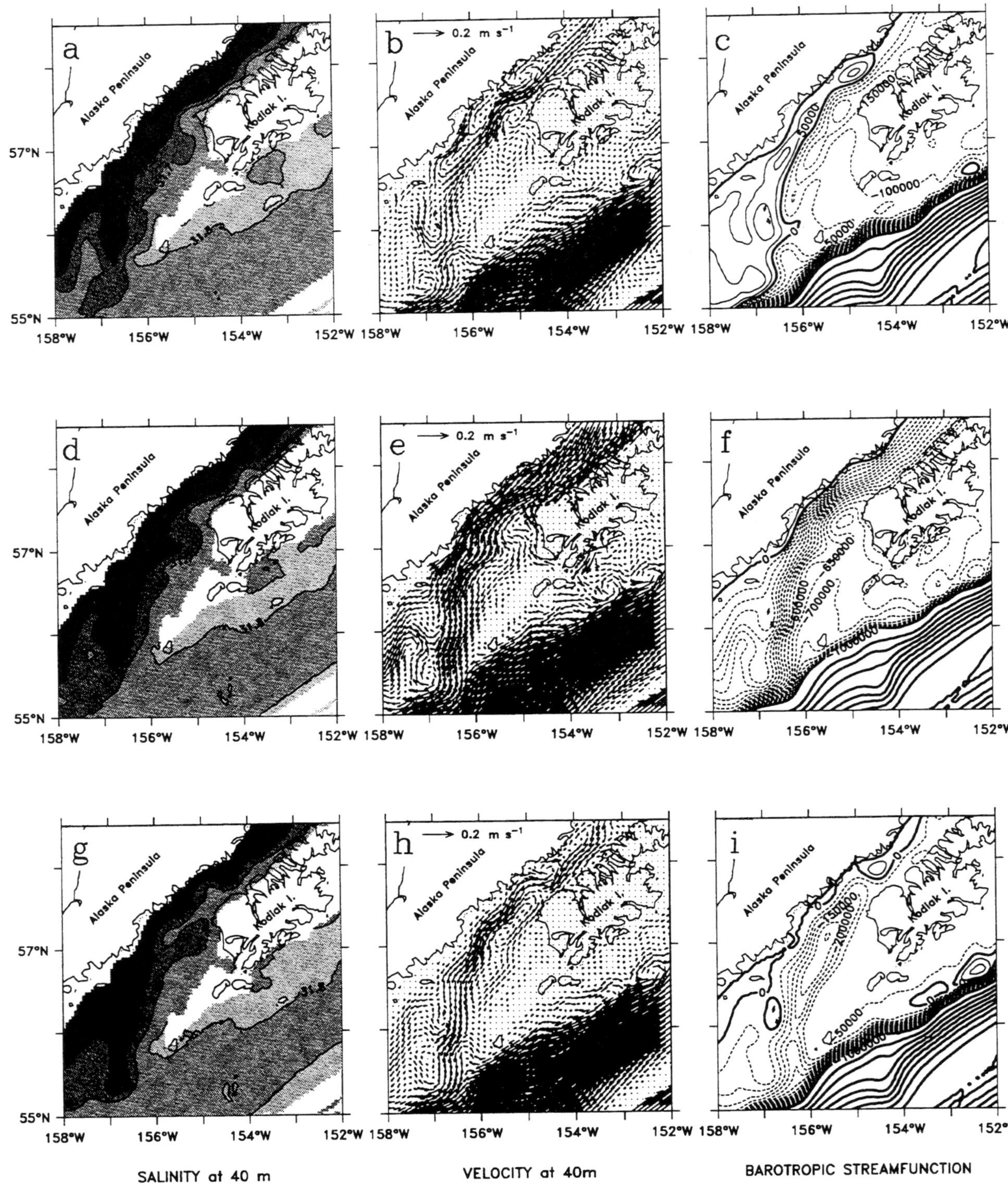

Fig. 5. Salinity (psu) at 40 m, currents at 40 m, and the barotropic stream function (a–c) April 15, 1978; (d–f) May 1, 1978 and (g–i) May 15, 1978.

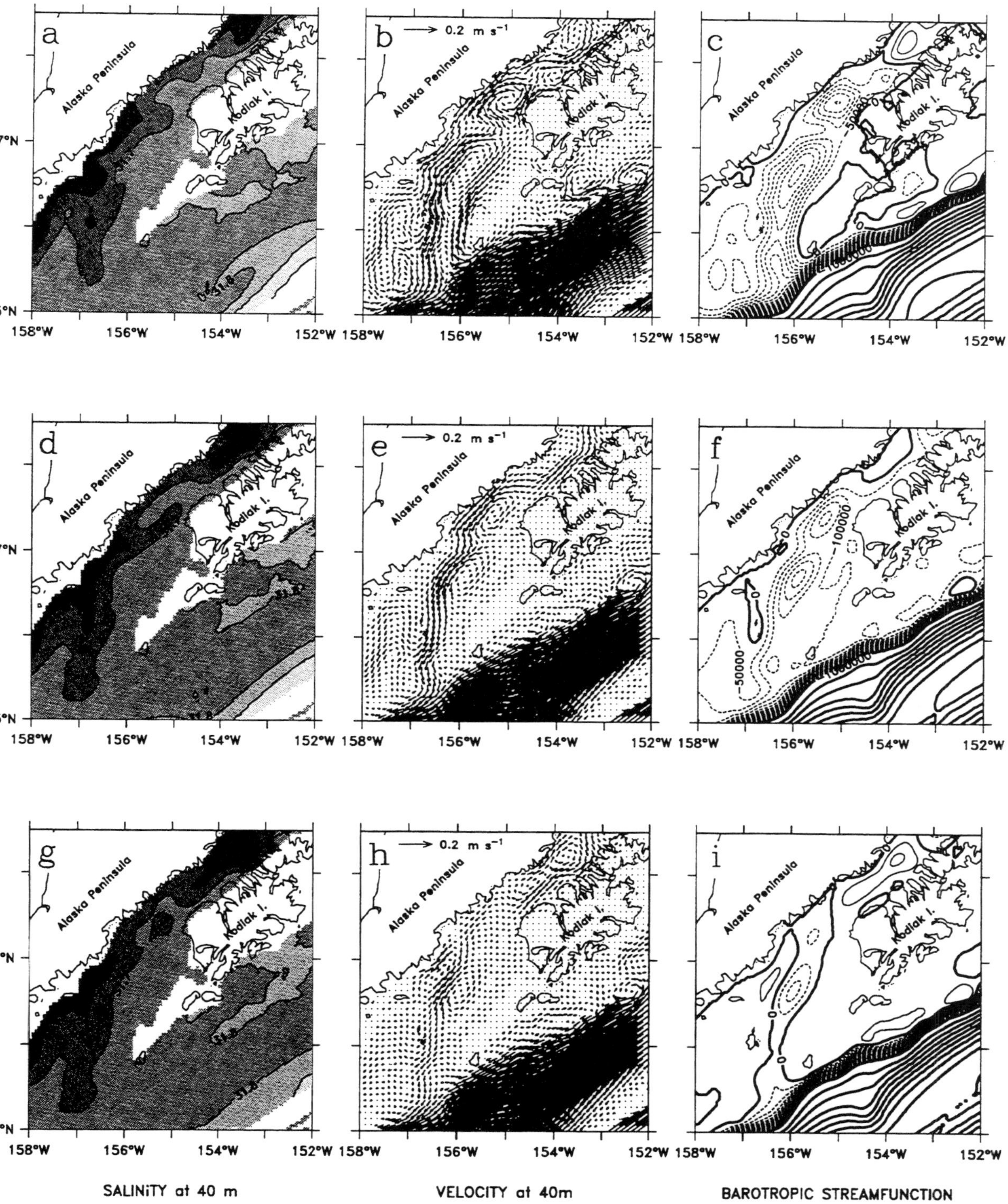

Fig. 6. Salinity (psu) at 40 m, currents at 40 m, and the barotropic stream function (a–c) April 15, 1990, (d–f) May 1, 1990, and (g–i) May 15, 1990.

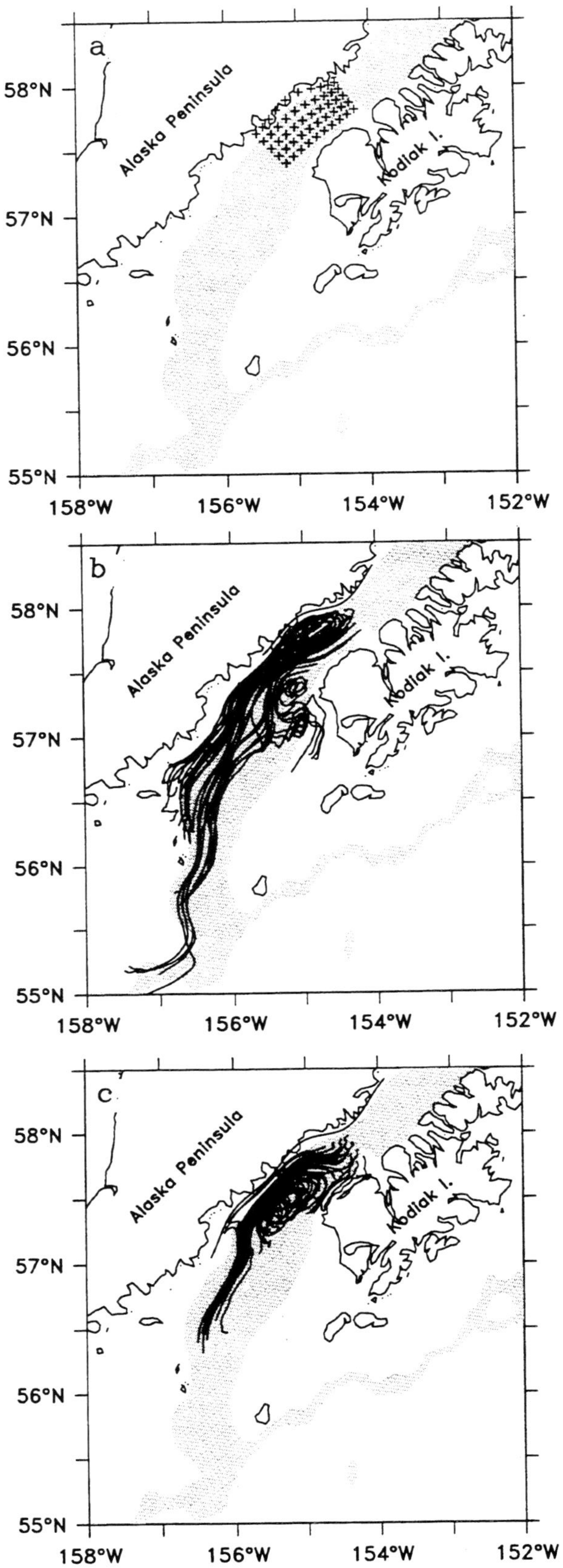

Fig. 7. (a) The starting positions of 50 floats that were seeded in the model on April 15 of each year. The paths of floats for (b) April 15–June 1, 1978, (c) April 15 – June 1, 1990. All buoys were drogued at 40 m. Depth range 150–130 m is shaded.

the winds there are strongly affected by the topography and are sensitive to small changes in the storm track. This is of particular relevance to the present study.

Nevertheless, some plausible speculations are possible. Warmer temperatures and increased winter precipitation would tend to increase the freshwater discharge into the ACC during the spring, hence increasing its buoyancy forcing. The limited results from the present study suggest that when the buoyancy forcing is stronger, such as in 1978, there may be a tendency for flows onto the shallow shelves adjacent to the sea valley, which is believed to be nursery grounds for pollock larvae. This study also shows that the along-strait and cross-shelf flow is strongly modulated by eddies. It is unknown what kinds of systematic changes in the eddies would occur with a warmer and wetter climate, because these eddies are sensitive to oceanic processes and atmospheric forcing on relatively small scales. We might speculate, however, that the enhanced salinity gradient from a wetter climate (as in 1978) would generally enhance eddy formation, since more potential energy would be available in the density field.

Continuation and expansion of this work is planned. A recruitment time series has been constructed from 1962 to 1989 (Megrey et al. 1995). This, coupled with modeled current patterns over these years, will permit us to investigate effects of the physical environment on larval survival over several decades.

Acknowledgements

This research is contribution FOCI-0178 to NOAA's Fisheries Oceanography Coordinated Investigations, contribution 1432 to NOAA's Pacific Marine Environmental Laboratory, and contribution 253 from the Joint Institute for the Study of the Atmosphere and Ocean (JISAO) of the University of Washington. Computation support was provided by the Arctic Region Supercomputing Center of the University of Alaska.

References

Bailey, K.M., and S.A. Macklin. 1994. Relationships between survival of larval walleye larval pollock (*Theragra chalcogramma*) and wind mixing in Shelikof Strait, Gulf of Alaska. Mar. Ecol. Prog. Ser. 113: 1–12.

Haidvogel, D.B., A. Beckman, and K.S. Hedström. 1991b. Dynamical simulations of filament formation and evolution in the coastal transition zone. J. Geophys. Res. 96: 15 017 – 15 040.

Haidvogel, D.B., J.L. Wilkin, and R. Young. 1991a. A semi-spectral primitive equation ocean circulation model using vertical sigma and orthogonal curvilinear horizontal coordinates. J. Comput. Phys. 94: 151–185.

Hedström, K.S. 1990. Users manual for a semi-spectral primitive equation ocean circulation model. Institute for Naval Oceanography, Stennis Space Center, MS.

Hermann, A.J., and P.J. Stabeno. 1995. An eddy-resolving, circulation model of the western Gulf of Alaska Shelf, I: Model development and sensitivity. J. Geophys. Res. (submitted).

Houghton, J.T., M. Seck, and A.D. Moura [ed.]. 1990. Scientific assessment of climate change. Intergovernmental Panel on Climate Change Working Group I. Cambridge University Press, New York, NY.

Johnson, W.R., T.C. Royer, and J.L. Luick. 1988. On the seasonal variability of the Alaska Coastal Current. J. Geophys. Res. 93: 12 423 – 12 437.

Lackmann, G.M., and J.E. Overland. 1989. Atmospheric structure and momentum balance during a gap-wind event in Shelikof Strait, Alaska. Mon. Weather Rev. 117: 1817–1833.

Macklin, S.A., P.J. Stabeno, and J.D. Schumacher. 1993. A comparison of gradient and observed over-the-water winds along a mountainous coast. J. Geophys. Res. 98: 16 555 – 16 569.

McBean, G.A., O. Slaymaker, T. Northcote, P. Leblond, and T.S. Parsons. 1991. Review of models for climate change and impacts on hydrology, coastal currents and fisheries in British Columbia. Unpublished manuscript. Rep. No. 91–11. Canadian Climate Centre, Downsview, Ont.

Megrey, B.A., S.J. Bograd, W.C. Rugen, A.B. Hollowed, P.J. Stabeno, S.A. Macklin, J.D. Schumacher, and W.J. Ingraham, Jr. 1995. An exploratory analysis of associations between biotic and abiotic factors and year class strength of Gulf of Alaska walleye pollock, p. 227–243. *In* R.J. Beamish [ed.] Climate change and northern fish populations. Can. Spec. Publ. Fish. Aquat. Sci. 121.

Niebauer, H.J. 1988. Effects of El Niño – Southern Oscillation and North Pacific weather patterns on interannual variability in the subarctic Bering Sea. J. Geophys. Res. 93: 5051–5068.

Overland, J.E., and T.R. Hiester. 1980. Development of a synoptic climatology for the northeast Gulf of Alaska. J. Appl. Meteorol. 19: 1–14.

Reed, R.K. 1984. Flow of the Alaskan Stream and its variations. Deep-Sea Res. 31: 369–386.

Reed, R.K., J.D. Schumacher, and L.S. Incze. 1987. Circulation in Shelikof Strait, Alaska. J. Phys. Oceanogr. 17: 1546–1556.

Royer, T.C. 1982. Coastal freshwater discharge in the northeast Pacific. J. Geophys. Res. 87: 2017–2021.

Schumacher, J.D., W.E. Barber, B. Holt and A.K. Kiu. 1991. Satellite observations of mesoscale features in lower Cook Inlet and Shelikof Strait, Gulf of Alaska. NOAA Tech. Rep. ERL 445-PMEL: 40 p.

Schumacher, J.D., and A.W. Kendall, Jr. 1991. Some interactions between young walleye pollock and their physical environment in the western Gulf of Alaska. Calif. Coop. Oceanic Fish Invest. Rep. 32: 22–40.

Schumacher, J.D., P.J. Stabeno, and S.J. Bograd. 1993. Characteristics of an eddy over a continental shelf: Shelikof Strait, Alaska. J. Geophys. Res. 98: 8395–8404.

Schumacher, J.D., P.J. Stabeno, and A.T. Roach. 1990. Volume transport in the Alaska Coastal Current. Cont. Shelf Res. 95: 1071–1083.

Trenberth, K.E. 1990. Recent observed interdecadal climate changes in the Northern Hemisphere. Bull. Am. Meteorol. Soc. 71: 988–993.

Vastano, A.C., L.S. Incze, and J.D. Schumacher. 1992. Observations and analysis of fishery processes: larval pollock at Shelikof Strait, Alaska. Fish. Oceanogr. 1: 20–31.

TOPIC 3.

Preparing for climate change: workshops

Workshop No. 1

Workshop No. 2

Workshop No. 3

Workshop No. 1

Preparing for climate change
Fisheries science and climate change

Coordinator: B.J. Rothschild
Rapporteur: D.R. Bouillon

At the outset of the workshop, we discussed the agenda of issues that the working group believed were relevant for discussion. Below is a list of the agenda topics followed by consensus viewpoints and some individual opinions from the working group.

1) Concepts and or processes

The participants emphasized (a) environment versus fisheries; (b) demonstration of climate change and temporal and spatial time scales; and (c) correlation analysis.

What does "climate change" mean and how should we use this term? Climatic change is a relative term and must be used within the context of the problem. The time scale of importance will be different depending upon whether we are studying copepods, fish, or whales. We still don't know if the low-frequency changes that we have been observing in such variables as mean global temperature represent true long-term changes in the trend caused by increasing CO_2 emissions or if they are part of a low-frequency natural cyclic pattern. Evidence since the last ice age does suggest that in previous times global temperatures were higher than they are now. What do these changes mean to us over the time scale that we are interested in (one or two human generations)? There is too much uncertainty to really know. Climate change may not be the real issue. The real issue may be climate variability. Many of the presentations at the symposium really dealt with climate variability and tried to address indirectly how we could use this variability to explain, and perhaps predict, patterns in fish population response. Evidence presented at the symposium suggested that phytoplankton and many fish stocks appear to exhibit decadal and multidecadal cyclic changes.

There appeared to be a general consensus that while exploratory correlation analysis will continue to be done it is important to keep in mind that enough exploratory correlation analysis will inevitably lead to a significant coefficient of determination. It is perhaps not good enough to simply report on these correlations without specifically addressing the limitations of the data and explicitly attempting to describe the mechanism(s). Many investigators neglect this important part of the exercise probably because there is no apparent logic or theory to explain the relationship. For example, what is the mechanistic link between sea-surface salinity and the year-class strength of some fish stock? Once there has been an attempt to explain the relationship (the theory), others can then test the theory. After exploratory correlation analysis we must speculate on theories and then formulate testable hypotheses to come to a conclusion about the legitimacy of the original correlations. Such studies will ultimately lead to more refined hypotheses and lead us down the proper path of research. If the causal mechanism is robust then one would expect that significant coefficients of determination will appear for other similar species in other regions of the world and from one fisheries system to the next. On this same topic one should look for similar relationships in various regions of the world, i.e., take a more global approach to their theories.

Interactions between the life history of creatures in the sea and the physical world must be complex. Perhaps one should not look only at single variables but should take a multivariate approach to investigating how changes in climate affects fish populations. One should also keep in mind that climate alone does not regulate fish populations and that the influence of human fishing activity likely plays an important role in the patterns of fish population response. Perhaps our goal is to use any new found knowledge of robust relationships between climate and fisheries to better manage populations under exploitation. Perhaps we want to design our fishing fleets so that they are more adaptable to the changing nature of fish populations! In general, there was the impression that we have much to learn about air–sea interactions.

2) Interactions with other disciplines

There was a strong feeling that interdisciplinary research must continue to be promoted and that this should take second priority only to concepts and processes. The complexity of the question of how fish populations respond to regional and global atmosphere–ocean processes requires the expert input of researchers from many different disciplines.

3) Data collection

The question was asked whether we are collecting enough and the right kind of data. The response was that there was always much room for improvement but that data collection was governed by budgetary limits and that it might simply be most effective to fine-tune data collection to specific needs. To do this we have to be more explicit about the questions we are asking and the theories that we want to test. The suggestion was made that we should make all attempts we can to extend times series backward using archaeological information. The anaerobic ocean sediment study of Soutar and Isaacs was cited.

4) Statistical techniques

The topic of statistical techniques did not receive a lot of comment. Bill Ricker suggested that statistical methods should be kept simple so that other researchers can understand the study. Another comment made was that we should use the most appropriate statistical technique and build on our statistical knowledge. In addition there were the comments that we should promote a greater awareness of the effects of autocorrelation on time series and make greater use of smoothing techniques such as LOWESS analysis to examine patterns and less to investigate correlations with linear regression.

5) Numerical modelling

Yes, it is time to get on with the numerical modelling process, because it is likely a profitable area of research. These models will be complex which brings into question their feasibility. Before they can be used we must resolve some of the spatial issues and data shortages. We must start to collect the empirical data required for such models if we ever hope to truly understand the complex interrelationships between climate variability and fish populations.

Workshop No. 2

Fisheries management and climate change

Coordinator: W. Aron
Rapporteur: S. McKinnell

The discussion of the working group focussed on three major points.

1) *What is the state of our knowledge?*

Exploited fish populations can vary in abundance by orders of magnitude. For the routine interannual management of fisheries, it is unlikely that changes in population size can be attributed to changes in climate change unless the change is extreme. Other factors such as fishing, stock size, or habitat alteration are varying coincident with climate and it will be difficult to separate the effects of climate from the effects of other factors. Forecasting abilities are generally limited at present with annual forecasts of abundance varying in quality by species group.

2) *What kinds of advice should we give to fisheries managers in light of what we know vis-a-vis climate change?*

At this time, the working group suggested that fisheries managers not modify existing practices in consideration of the current state of our knowledge on the effects of climate change provided at the Symposium. However, there is a need for managers to be aware of long-term species-specific cycles in abundance that render the concept of long-term sustainable yields inappropriate for some species. Harvesters need to adapt to the time scale of change in abundance of target species.

3) *What scientific research should be undertaken to provide advice to managers?*

The participants offered a variety of opinions concerning activities that should be undertaken in future. The most common theme was the need to establish routine monitoring programs at varying time or space scales to track changes in physical and biological factors. There was general agreement that long-term monitoring and regular surveys are important in the study of climate-related changes in the abundance of fish populations. There is a need to establish scientific working groups to determine the data required to monitor changes on various time and space scales and how these data may be collected, coordinated, and shared among researchers. Benefits can be derived from research on fish populations, where physical oceanographers, physical limnologists, climatologists, meteorologists, and fisheries scientists work cooperatively to improve our knowledge of the mechanisms underlying variation in fish population abundance. Owing to the cost, particularly for basin scale projects, it will be important to undertake cooperative and/or coordinated studies between interested researchers. Long-term forecasts of abundance, changes in species dominance in ecosystems, and awareness of trends will be important in planning long-term (>20 to 30-year) fisheries policies.

Workshop No. 3

Large-scale atmospheric anomalies and climate change

Coordinator: W.S. Wooster
Rapporteur: J. Boutillier

Exploration of the topic by workshop participants led to several preliminary conclusions.

1) Although the assigned topic was "large-scale atmospheric anomalies," the discussion should also cover large-scale oceanic anomalies.
2) Although many papers of the Symposium were concerned with the biological effects of climate change, this workshop would focus on changes in the physical system.
3) Although in the subarctic Pacific, the behaviour of the Aleutian Low and its relation to the Southern Oscillation were of great significance, it would be a mistake to consider that all large-scale anomalies in the region were driven by tropical events.

It was generally agreed that large-scale anomalies such as those mentioned above are of great physical and biological importance and interest, and that the most appropriate task for the workshop was to consider how to accelerate gaining understanding of these matters. The following categories of effort were considered:

Observations

There is a need for long time series at annual, and preferably monthly, resolution. This can be achieved by addressing three aspects.

Historical data: The only quick way to get long time series is retrospectively. However, the data are usually proxies and difficult to interpret unequivocally. Marine sources include paleoclimatological data from laminar flow coring samplers. There is a draft sampling protocol ("climate barometer") that recommends coring anoxic basins along a linear transect extending from Antarctica to the Bering Sea; such studies between Baja, California, and Vancouver Island are envisioned in the GLOBEC program in the California Current system. Important clues are also available from historical atmospheric data records, e.g., timing of first snow falls in Edo, and from historical accounts.

Contemporary data: The system for monitoring atmospheric and ocean variability at climate scales needs to be improved. In the ocean, this is the purpose of the Global Ocean Observing System being developed by IOC. The use of platforms of opportunity should increase and satellite data will be important, despite limitations to their usefulness in regions of frequent cloud cover such as the subarctic Pacific. Moorings and drifters will also play a role.

Future data: Other kinds of observations not now routine, such as measures of light penetration, should be incorporated in observing systems. Technological advances are likely to increase effectiveness and perhaps reduce costs of the system.

Theory and Modelling

The supply of theoreticians and modelers is finite, and many are working on equatorial problems, so to focus their attention on high-latitude anomalies on appropriate time and space scales will require some change in priorities. Plans to establish on the Pacific coast an ocean–atmosphere modelling node of the Canadian climate modelling network were applauded as this would permit enhanced interaction with fishery scientists in the Pacific. It was noted that modelling at this time is, to some extent, technology limited, but the computer field is advancing so rapidly that this may not be a problem for long.

Priorities

Senior government officials, as well as the public, need to understand the importance of research on large-scale atmosphere and ocean anomalies. There has been a change in the thinking over the last 10 yr of fisheries scientists who are beginning to recognize the importance of environmental factors in trying to forecast fishery production. The strong evidence now available on fisheries impact should be publicized in the attempt to gain support for the further research needed to understand causes and effects.

Linkages

There are a number of areas in which linkages may provide opportunities for enhanced understanding, including comparative studies of analogous systems. Such studies are ultimately interdisciplinary in nature and require interinstitutional as well as international cooperation. It was also suggested that ocean users should be more closely integrated with science, e.g., ships of opportunity not only provide economical platforms but also provide important linkages in the education of investigators and users.

Information and data exchange

Information exchange is an essential element in strengthening the scientific knowledge base. Scientific meetings, such as the present symposium, promote the exchange of both information and ideas. PICES committee sessions and annual meetings will foster such exchanges in a continuing way. Data sharing, especially in near-real time, as is common in atmospheric monitoring, is developing in ocean studies such as TOGA (e.g., system for collection of data across the Pacific using drifters and moorings along specified grids). There is continuing use for historical data which should also be exchanged. There is also a role for user education on what data are available and how they can be obtained and used; PICES should undertake activities of this sort. Workshops on the mathematics and statistical procedures and collection methodology should be organized.

Coordination

Some concern was expressed over what might be called "ceremonial coordination," where a ponderous bureaucracy is imposed on the selection of problems and approaches by scientists. However, there was general agreement on the need for coordination in the positive sense where clear objectives, roles, and responsibilities are developed and where the best use is made of available resources: funds, people, ideas, equipment, and facilities. An essential component of such coordination is the timely exchange of information on what all players in the joint effort are doing. The hope was expressed that PICES might become an effective mechanism for coordination of international research activities in the northern North Pacific.

Symposium summary

Symposium summary

Michael Sinclair and Kenneth T. Frank

Physical oceanographic variability

An overview of the variability in the oceanographic conditions of the North Pacific was provided in three papers (Hanawa, Trenberth and Hurrell, Wooster and Hollowed). There is considerable evidence from salinity, temperature, and wind observations that there are two patterns or modes in circulation and mixing in the North Pacific, the modes lasting about a decade. The alternating oceanographic conditions are driven in part by shifts in the atmospheric circulation, specifically the changes in the location and level of the so-called Aleutian Low pressure system. During the time period 1976–88, for example, the system was deeper and more eastward than during the previous decade. Under these atmospheric conditions, the North Pacific Gyre is anomalously cold, with a stronger circulation and upward doming of the pycnocline. In contrast, the Alaskan Gyre is warmer with less coastal upwelling. The Bering Sea is somewhat warmer, with stronger inflow from the Alaskan Gyre. It is also concluded that there has been increased high frequency variability in oceanographic conditions during the 1976–88 time period.

The shifts in atmospheric circulation on decadal time scales are hypothesized to be caused by El Niño – Southern Oscillation (ENSO) dynamics in the tropical Pacific and Indian Oceans, which occur on interannual time scales (i.e., every few years there is an El Niño). It is not yet understood how the decadal time-scale shifts in the North Pacific atmospheric circulation are coupled to the interannual shifts in ENSO phenomena. Also, at this stage the coupled ocean–atmosphere circulation models do not reproduce the observed El Niño patterns. Thus, we cannot realistically use existing models to predict the response of the North Pacific to a doubling of atmospheric CO_2 concentrations.

In sum, the physical oceanographic presentations indicated large-scale shifts in circulation and mixing in the North Pacific in response to changes in the atmospheric circulation, with the patterns persisting for about a decade. At this stage, no predictions are available on whether one mode or the other would be favoured under conditions of global warming. The published study by A. Bakun (which was discussed at the symposium, but not presented) predicts increased upwelling in eastern boundary currents, with increases in atmospheric CO_2 concentration.

The above conclusions from the physical oceanographic presentations constrain the discussion on the impacts of climate change (in contrast to climate variability) on northern fish populations. The El Niño events and decadal scale shifts in circulation can be used to evaluate the response of fish populations to different oceanographic conditions, but at this stage we cannot infer which conditions might be favoured under "climate change." Furthermore, even though the large-scale description of shifts in atmospheric conditions and in sea-surface temperature (SST) are robust, the more detailed descriptions of circulation and mixing on time and space scales of relevance to specific fish populations are lacking. The lack of detail makes it difficult to consider the underlying processes by which the physical environment influences population responses.

Krovnin compared sea-surface conditions in the North Pacific and North Atlantic. Ward et al. showed that the location of the subarctic front in the North Pacific has been relatively fixed in spite of the decadal-scale shifts in circulation and mixing. Bernard et al. and Verkhunov described recent changes in the circulation of the Bering Sea. The physical oceanographic papers and their subject areas are listed in Table 1.

Table 1. Papers on physical oceanography.

North Pacific decadal-scale variability	K.E. Trenberth and J.W. Hurrell K. Hanawa W.S. Wooster and A.B. Hollowed
Subarctic front	M.B. Ward, S. Riser, and R.C. Francis
Bering Sea	E.N. Bernard, P.J. Stabeno, R.K. Read, A.V. Verkhunov, and J.E. Overland
Comparison of North Pacific and North Atlantic	A.S. Krovnin

Biological responses to variability in the environment

There were a large number of papers that addressed biological responses to physical oceanographic variability. These papers have been grouped in several categories (Table 2). The first group describes changes in biological productivity. Wong et al. concluded from an analysis of all available data in the eastern North Pacific that phytoplankton productivity has not changed in response to the decadal time-scale shifts in oceanographic conditions. However, the El Niño events that propogate up the eastern margin of the North Pacific generate a reduction in primary production by about 50%. In sum, an interannual response of phytoplankton to oceanographic conditions has been observed, but not a decadal-time scale response. In contrast, there is strong published evidence that primary production in the North Pacific Gyre itself has responded to the shift in modes (with higher production during the cold periods for the gyre). Thus, it appears that the plankton response in the North Pacific Gyre is different from that in the contiguous water masses. Mackas described the higher

M. Sinclair and **K.T. Frank.** Department of Fisheries and Oceans, Bedford Institute of Oceanography, P.O. Box 1006, Dartmouth, NS B2Y 4A2, Canada.

frequency seasonal variability in plankton, as well as the interannual variability south of Vancouver Island. He noted reduced copepod biomass seaward of the shelf break during the middle to late 1980s, with an increasing trend in euphausiid biomass shoreward of the shelf break. The duration of the time series is not adequate to evaluate the response of the zooplankton to the decadal-scale variability in the physics. Species-specific responses were observed to the higher frequency seasonal and interannual variability, with some population responses lasting longer than the El Niño events. Brodeur and Ware compared nekton distribution and abundance in the Gulf of Alaska during a cold mode (late 1950s) and a warmer mode (late 1980s). There was a doubling of the total abundance and biomass of 14 species of salmon, nonsalmonid fishes and cephalopods during the more recent time period. Almost all species (with the exception of chub mackerel (*Scomber japonicus*)) responded in the same manner and a ring structure (i.e., with higher abundance levels at the perimeter of the Alaskan Gyre) was a persistent feature. Finally, Ishida et al. presented somewhat anomalous results on chum salmon (*Oncorhynchus keta*) in the central North Pacific. Growth rates were lower during the cold mode (which is characterized by higher plankton productivity) and return rates were inversely related to SST in the gyre. These results suggest that the cool, high-production conditions are not particularly beneficial to chum salmon. The observations, however, may be confounded by the temporal trend in the numbers of released juvenile salmon and their higher return rate. Ricker described changes in the mean weight of salmon in the catch for different rivers in the North Pacific from 1950 to the present. Even though there was a decline until 1975, followed by an upwards trend to present, there has been considerable variability in the patterns between species and within species among rivers. Thus, the salmon responses in their average weight-at-age do not appear to respond to the shift in marine environmental conditions on decadal time scales.

A second set of papers addressed changes in fish migration in response to variability in oceanographic conditions. Narayanan et al. described shifts in the geographic patterns of migration of Atlantic salmon (*Salmo salar*) and capelin (*Mallotus villosus*) in response to colder conditions off the coast of Labrador and Newfoundland (the ice conditions have been more extreme in the recent cold period and the volume of the cold intermediate layer greater than normal). Ware and McFarlane showed that Pacific hake (*Merluccius productus*) availability increased around La Perouse Bank during warmer periods and inferred that this extended distribution generated a higher natural mortality for Pacific herring (*Clupea pallasi*). Svendsen et al. summarized the distribution of mackerel, whiting, and Atlantic herring (*Clupea harengus*) in the Northeast Atlantic from 1968 to 1990. They concluded that migration into the North Sea for these species was favoured during time periods when the percentage of Atlantic water in the North Sea was higher. D'Amours, from an analysis of Atlantic cod (*Gadus morhua*) distributions, oxygen concentrations and temperature in the Gulf of St. Lawrence, demonstrated that cod are not observed in water with less than 3.4 $mg \cdot L^{-1}$ oxygen concentration and that they follow a preferred temperature.

Four papers addressed shifts in community structure as a function of environmental variability. For example, Godo and Skjoldal, and Skjoldal, described alternating relatively cold and warm periods in the Barents Sea. The changes in environmental conditions are associated with substantive changes in community structure and function (abundance changes in cod and capelin, shifts in sea migrations, diet changes, growth variation, life history parameter changes, etc.). These decadal time-scale shifts in the Barents Sea, of both the environment and the biological communities, have been described since the late 1800s.

Two papers described changes in the timing of life-history events in response to environmental variability. Henderson et al. demonstrated that the length of salmon fry at the time of outmigration from the rivers to the sea is an inverse function of "degree-days." Also, the date of 50% outmigration is an inverse function of accumulated degree-days, with earlier migration occurring when the rivers have been warmer. However, the marine survival index does not appear to be influenced by the above indices. Cooney et al. took a similar approach. They hypothesize that interannual variability caused by temperature differences in outmigration of pink salmon (*Oncorhynchus gorbuscha*) into coastal nursery areas may result in a match or mismatch in availability of food.

Two papers addressed changes in time and/or location of spawning and subsequent larval distributions. Wyllie Echeverria showed that seasonal extent of ice cover varies substantially and that this in turn influences the distribution of spawning of a number of species. Alaskan walleye pollock (*Theragra chalcogramma*) were shown to shift their distribution to the open area during years of reduced ice cover. It was inferred that global warming, if it changed the distribution of ice cover, would substantially modify the spawning distribution of certain species. The observations suggest a rapid redistribution of fish populations to short-term environmental variability (and thus expansion of ranges of distribution are to be expected very rapidly in such extreme environments). The second paper showed the changes in fish egg and larval distributions off Washington, Oregon, and Northern California from 1980 to 1987, with particular emphasis on the impacts of El Niños. There were large shifts in the distribution of eggs and larvae during El Niño years, but there is no information on whether these changes in location of spawning and larval distributions resulted in fluctuations in population abundances, as most of the species considered were noncommercial.

Tyler showed that fecundity differed between two populations of Pacific cod (*Gadus macrocephalus*), with lower fecundity observed in the higher temperature area. He inferred that increased temperature under climate change may influence fish population dynamics through changes in reproductive physiology.

Finally, two papers demonstrated that analyses of marine bird and mammal distributions and abundance may be useful in the analysis of responses of pelagic communities to environmental variability. For example, Decker et al. described greater reproductive success for surface-feeding birds during the cold years in the Bering Sea. This suggests that the warmer periods for the Bering Sea (as observed during the 1976–88 mode) may generate lower biological productivity. Also, some diet changes for certain bird species suggest shifts in species composition from the cold to warm oceanographic modes.

In sum, a wide range of examples were presented at the symposium in which a range of biological processes (productivity, distributions, life-history events, community structure and function) respond to variability in the oceanographic and freshwater environment. The biological responses can be substantive and the time lags short, to relatively modest changes in the environment. Thus, it can be inferred that equally significant biological responses would occur following climate change.

Table 2. Papers on biological responses to environmental variability.

Production processes	C.S. Wong, F.A. Whitney, K. Iseki, J.S. Page, and J. Zeng
	D.L. Mackas
	R.D. Brodeur and D.M. Ware
	Y. Isheda, D.W. Welch, and M. Ogura
	W. Ricker
	J.H. Helle and M.S. Hoffman
Migrations	S. Narayanan, J.B. Dempson, M.F. O'Connell, S. Prinsenberg, D.G. Reddin, and N. Shackell
	D.M. Ware and G.A. McFarlane
	E. Svendsen, A. Aglen, S.A. Iverson, D.W. Skagen, D. D'Amours, and O. Smestad
Community structure changes	O.R. Godo and H.R. Skjoldal
	H.R. Skjoldal
	V.V. Kuznetsov
	J. Yang, L. Li, and S. Xia
Timing of life-history events	M.A. Henderson, R.E. Diewert, J.C. Stockner, and D.A. Levy
	R.T. Cooney, T.M. Willette, S. Sharr, D. Sharp, and J. Olsen
Abundance changes	V.A. Belyaev and T.A. Shatilina
	O.A. Bulatov
Spawning and larval distributional changes	T. Wyllie Echeverria
	M.J. Doyle
Fecundity differences	A.V. Tyler
Marine birds and mammals as indicators of changes in production	M.B. Decker, A.E. York, G.L. Hunt, Jr., and G.V. Byrd, Jr.

Correlations between fish abundance and oceanographic parameters

The papers that have correlated temporal data series in fish abundance and oceanographic parameters are listed in Table 3. Three papers analyzed fish population responses in the northeastern Pacific to the decadal-scale shifts in oceanographic conditions. The results and general conclusions were consistent among studies, even though different approaches and data sets were used. This is reassuring, and suggests that the conclusions are not artifacts of the particular data sets or statistical techniques chosen for the analysis. The fish populations in the Alaskan Gyre and associated coastal zone do well during the warm periods (i.e., cold periods for North Pacific Gyre and eastward displacement of the Aleutian Low). The favoured interpretation for the mechanism underlying the correlations between the fish and physics is the classical food-chain hypothesis. Under this hypothesis the Aleutian Low generates increased circulation of the North Pacific Gyre (which is also relatively colder during this mode) and increased vertical fluxes of nutrients (N, P, Si) important for phytoplankton production. This higher production at the base of the food chain in the central gyre (which is well documented by Venrick) is then inferred to generate increased zooplankton production. The circulation features are then interpreted to transport this enhanced plankton productivity towards the Alaskan Gyre and associated coastal zone.

The evidence for the hypothesis is somewhat contradictory. Under the warmer conditions in the Gulf of Alaska, one might predict reduced upwelling and no increase in primary production (as shown by Wong et al. for phytoplankton). Brodeur and Ware, however, have shown increased zooplankton and nekton in the Alaskan Gyre during the warm period. It seems improbable that this response is generated by the North Pacific Gyre to the west. It may be that the food chain is somewhat uncoupled, and the zooplankton and nekton populations respond directly to an enhanced persistence of the circulation features during the warm periods of the Alaskan Gyre. At this stage, given the limited observations on the circulation and food-chain dynamics in this large area, it would seem appropriate to keep an open mind with respect to causal mechanisms. At the interannual time scale, there was the intriguing observation that year-classes born the year after an El Niño are frequently abundant. There was no interpretation of this particular observation.

Two papers correlated environmental data and fish abundance trends with the 18.6-year nodal cycle (low-frequency tidal cycle), both in the northeastern Atlantic and Pacific. Surprisingly, the tidal signal was even correlated with air temperature. Previous studies have inferred that changes in mixing at this frequency in the Bay of Fundy also influence the intensity of toxic algal blooms, as well as shad and scallop abundances. The surprise is that such a small difference in tidal mixing (a few percent), can cause (if this is indeed the case) such marked responses in the biological processes. It is to be noted that the above studies are not part of climate change per se, but may well be background variability that needs to be taken into consideration when discussing the impacts of global warming.

There were 10 papers that used correlation analyses to explore biological processes. For example, Zebdi and Collie analyzed the geographic scales of coherence of Pacific herring recruitment. Stocks above and below 54°N tended to cluster together and, interestingly, this latitude coincides with the bifurcation of the North Pacific Current. Schweigert, for the same species, demonstrated that there were no population responses to the two large scale oceanographic modes.

Table 3. Papers involving correlations between population abundances and environmental variability.

Northeast Pacific "regime shifts"	R.J. Beamish and D.R. Bouillon
	A.B. Hollowed and W.S. Wooster
	S.R. Hare and R.C. Francis
18.6 year nodal cycle in tides	K.S. Parker, T.C. Royer, and R.B. Deriso
	T. Wyatt and R.G. Currie
Exploratory analyses	J. Schweigert
	T.J. Quinn II and H.J. Niebauer
	B.A. Megrey, S.J. Bograd, W.C. Rugen, A.B. Hollowed, P.J. Stabeno, S.A. Macklin, and J.D. Schumacher
	W.J. Ingraham, Jr.
	K. Ohtani and T. Azumaya
	J. Wang
	Q. Tang
	C. Lin, B. Xu, and S. Huang
	S.N. Rodionov
	F.J. Müter, B.L. Norcross, and T.C. Royer
	A. Zebdi and J.S. Collie

Process-oriented studies

Several process-oriented studies evaluated the mechanisms by which environmental variability influences year-class strengths and availability (Table 4). Bienfang and Ziemann conclude that classical food-chain processes in the plankton (driven by wind mixing and light availability) generate the observed interannual fluctuations in recruitment. In contrast, McConnaughey and Anderson, Kasai et al., and Wada et al. infer that wind-forced (or buoyancy driven) variability in the circulation patterns generates interannual variability in the distribution of eggs and larvae from appropriate geographic areas, which may be necessary for the persistence of the population (along the lines argued under the member–vagrant hypothesis). Kasai et al. showed that during 1972 the monsoon off Japan was very weak, at which time an exceptionally large year-class occurred. Also, larval retention was argued to be influenced by frontal processes. It is inferred that exceptional events may be required to "kick-start" a population recovery and broad-scale spawning. Thomson et al. used a modelling approach to simulate changes in migration patterns of sockeye salmon in the Alaskan Gyre under varying circulation conditions. The model results indicate that variability in circulation generates substantive differences in migration routes.

A general point of significance to the theme of the symposium is the close coupling between physical processes and population abundance. If this occurs in a routine manner, it would not always be necessary to understand the complex biological interactions in food chains to provide predictions on the direction of change of populations following changes in the circulation and mixing patterns of the ocean.

Table 4. Papers dealing with process-oriented studies.

Process-oriented studies	P.K. Bienfang and D.A. Ziemann
	R.A. McConnaughey, and D.A. Armstrong
	A. Kasai, M.J. Kishi, and T. Sugïmoto
	K.A. Thomson, W.J. Ingraham, Jr., P.H. LeBlond, M.C. Healey, C. Groot, and C.G. Healey,
	T. Wada, T. Matsubara, N. Koigumi, and Y. Matsumiya.

Freshwater Studies

Three papers dealt with biological responses to variability in freshwater temperature (Table 5). Given that some scenarios of climate change predict increases in freshwater temperatures in northern areas by 4–5°C, in contrast to 1–2°C for marine environments, the physiological constraints might be expected to be greater for freshwater systems. However, there are threshold levels for both environments for which small changes in temperature can have a very large impact (such as lower lethal temperatures). Thus, the generalization on the potential for greater physiological impacts in freshwater systems needs to be tempered.

Lehtonen and Lappalainen showed coherence in recruitment time series for four freshwater species in Finland. Summer air temperature and ice cover were significant variables. Species at the northern edge of their distribution respond positively to increased temperatures.

Table 5. Papers dealing with freshwater studies.

Physiological constraints	P. Lin and H. Regier
Biological changes to environmental variability	H. Lehtonen and J. Lappaleinen
	J. Hartmann

Methods and Theory

Several papers dealt with methods and theory for the investigation of climate change impacts on northern fish populations (Table 6). For example, Cury et al. argued that new statistical approaches would be required to efficiently analyze nonlinear responses. A second example of methodology was the use of archaeologically dated sequences in fluvial sediments (freshwater clams) to reconstruct historical impacts of climate change on freshwater populations (Chatters et al.). As an example of the theory papers, Sundby extended the conclusions of earlier work on the role of turbulence on contact rates between fish

Table 6. Papers dealing with methods and theory.

Methods	J.C. Chatters, V.L. Butler, M.J. Scott, D.M. Anderson, and D.H. Neitzel
	D.L. Pereira, C. Bingham, G.R. Spangler, Y. Cohen, D.J. Conner, and P.K. Cunningham
	P. Cury, C. Roy, R. Mendelssohn, A. Bakun, D.M. Husby, and R.H. Parrish
Theory	S. Sundby
	J. Rice
	B.J. Rothschild
	J.H. Steele
	Y.I. Zuenko

larvae and their food (the Rothschild–Osborn hypothesis of recruitment variability). He showed that variability in mixing should be important for the first 60 d of the larval stage.

Climate Impact Models

The final category of papers addressed the key theme of the symposium, the impact of climate change itself rather than environmental variability (Table 7). Two of the modelling papers are for fresh water. Stefan et al. conclude that for unshaded streams in Minnesota, cold- and cool-water fish species will be eliminated under a doubling of atmospheric CO_2 scenario, whereas warmwater species will broaden their distributional range to include Minnesota. For shaded streams, some cold-water species will remain. In summary, the habitat for warmwater species will increase by 53%, whereas there will be a 40% reduction for cold-water species. Minns and Moore indicate that 33 of 61 species in Ontario lakes show a significant response to temperature. In their model, most of the changes in distribution occur at mean annual air temperatures between 1 and 6°C. They conclude that within a band across northern Ontario there will be substantive changes in species composition, with increases in some undesirable species from an angling perspective (i.e., carp).

Holtby and Scrivener used a complex life-history model of chum salmon on West Vancouver Island to predict population impacts. Six critical events on the life history are identified along with three related physical processes (late winter temperatures in streams, peak winter discharge, and surface salinity in the coastal zone). Using projections into the future on how these physical properties could change the model predicts decreasing abundance of chum salmon on the Island, as well as some local extinctions. The strengths of the model are the biological components, whereas weaknesses exist in the provision of realistic scenarios on changes in the physical environment.

Hinch et al. presented the only marine impact model for climate change. The species chosen is sockeye salmon, the model being based on bio-energetic responses to temperature. A range of ±5°C is used to generate model results. One model output was the size of salmon at the time of return migration to the river. A 2°C increase in marine surface temperature generates a reduction in size of returning salmon that is outside of the range of empirical results. The authors infer that size changes could have subtle effects (i.e., reduction in energy storage for upstream migration, extra year at sea).

Table 7. Papers dealing with climate impact models.

Freshwater resources	H.G. Stefan, J.G. Eaton, M. Hondzo, and J.H. McCormick
	C.K. Minns and J.E. Moore
Anadromous resources	L.B. Holtby and J.C. Scrivener
	S. Hinch, M. Healy, R.E. Diewert, and M. Henderson

Concluding Remarks

There has been a rapid increase in understanding of the role of decadal-scale oceanographic variability in generating changes in both abundance and distribution of fish in the North Pacific. During a Department of Fisheries and Oceans workshop on this topic in 1990, the "picture" was very vague indeed. The analyses carried out in the subsequent 2 yr, many of which were presented at this symposium, provide coherent descriptions of low-frequency variability on ocean-basin scales. The physics and the fish are marching, to a certain degree, to the same tune. There are, however, some major gaps in understanding of both the physical and biological processes (let alone the key interactions).

There has, in contrast to the above progress in descriptions of decadal time-scale variability, been very little increase in understanding of interannual variability in recruitment. Given the limitations in the state of the art in fisheries oceanography, it is perhaps not appropriate to select a favoured hypothesis on the mechanism by which physical variability causes fluctuations in population abundance. The impact models for freshwater habitats are, in comparison with marine environments, at an advanced stage of development. However, considerable uncertainty exists at local scales in terms of climate change impacts.

In closing, one must end this symposium on a very positive note. The studies on fisheries oceanography, particularly in the North Pacific, indicate that large scale oceanographic processes have a major impact on fish population fluctuations. Given the rapid increases in understanding of the ocean–atmosphere coupling, it is to be expected that we will soon have meaningful predictions on the response of the circulation and mixing of the North Pacific to climate change. Even without a full understanding of biological interactions, some predictions on biological trends should then be forthcoming.